DATE DUE

			PRINTED IN U.S.A.

Statistical Abstract of the World

Statistical Abstract of the World

Third Edition

Annmarie Muth, Editor

GALE

DETROIT · NEW YORK · TORONTO · LONDON

Annemarie S. Muth, *Editor*

Editorial Code and Data, Inc. Staff

Kenneth J. Muth, *Associate Editor/Programmer/Analyst*
David Smith, *Assistant Editor*
Kenneth J. Muth, *Cartography*
Sherae R. Carroll, *Data Entry Associate*

Gale Research Staff

Camille A. Killens, *Associate Editor*
Lawrence W. Baker, *Managing Editor*

Mary Beth Trimper, *Production Director*
Evi Seoud, *Assistant Production Manager*
Deborah Milliken, *Production Assistant*
Cynthia Baldwin, *Production Design Manager*
Barbara J. Yarrow, *Graphic Services Supervisor*
C. J. Jonik, *Desktop Publisher*

Copyright © 1997
Gale Research
835 Penobscot Building
Detroit, MI 48226-4094
All rights reserved including the right of reproduction in whole or in part in any form.

ISBN 0-8103-6434-4
ISSN 1077-1360
Printed in the United States of America

TABLE OF CONTENTS

INTRODUCTION

The third edition of the *Statistical Abstract of the World (SAW)*, profiles 185 UN member nations and three equally important nonmember nations: Hong Kong, Switzerland, and Taiwan. As in previous editions, the purpose of this year's *SAW* is to serve as a comprehensive, easy-to-access guide to a world of information, covering such topics as the geography, demographics, healthcare, organization of government, science and technology, and trade of the country. A further objective is to update this edition to better reflect the current events touching on those topic areas.

Thanks to the contributions of eight additional sources, many of the 42 topic areas have been updated and enhanced in this manner without sacrificing the integrity of content or uniformity of format. Several of these enhancements deserve mention.

New this year, for example, *SAW* takes a more in-depth look at world health. As the population increases and natural resources are pushed to the limit, the reader is referred to our new Health Indicators section to examine timely UNICEF data concerning contraceptive prevalence, access to safe drinking water, and the use rate of oral rehydration therapy (ORT) (crucial in combating life- threatening dehydration in children). The reader may also browse our Burden of Disease section to compare the incidence of AIDS and malaria in developing countries. Or he can turn to the industrialized countries and analyze similar findings on AIDS and heart disease. The Women and Children section, also new to this edition, offers recent statistics on the immuniza-tion of pregnant women as well as maternal and child mortality.

For the first time, the Demographics section tracks the net migration rate, recording recent mass movements of peoples fleeing civil war in Rwanda, for example, and gives projections of these movements into the year 2030. Turning to the Arts, the Cinema section has been renamed Culture, thanks to UNESCO's museum statistics supplementing that section. Finally, the United Nations Statistical Division has contributed new Manufacturing statistics, and the United Nations Criminal Justice Network (UNCJIN) has added to our Crime statistics.

Besides new data and sources, there are also a few new countries (and some reorganized old ones with new names). Palau is featured in this edition of *SAW* for the first time, having joined the UN in 1994 shortly after winning its independence; Andorra, another new entry, actually joined the UN in 1993. Also in this edition, Serbia and Montenegro now make up the UN-recognized Federal Republic of Yugoslavia; and Burma is now officially Myanmar. The reader is cautioned to note these changes when conducting research.

Lastly, it is worth mentioning the location of some of this edition's sources. Because many of these reside on the World Wide Web, they are updated more frequently than their hardcover companions. Consequently, the reader may rely on the fact that he is examining the most up-to-date statistics available.

General Design

To provide uniform organization, countries are arranged in alphabetical order. Data for each country cover five or six pages, beginning with a regional and country map and followed by 42 panels. The panels are grouped by topic, numbered, and titled (even if empty), and feature either tabular data or blocks of text. In the preparation of this edition of *SAW* as in previous editions, the editor intends to present statistical information on the countries of the world in a standard format—despite the fact that using such a format may highlight the absence of data. For when data are not available in a particular topic area, that panel is always shaded grey. Although coverage has improved as the editor discovers new sources, smaller countries, and those whose data gathering infrastructure has been destroyed by political upheaval or natural disasters, inevitably receive less coverage.

Variant Pages

Often tabular data are successfully replenished with textual data from an alternate source (usually the CIA). This presents a different appearing format and page count for that particular country. Rather than presenting large blocks of empty space, the fourth page of such a presentation is redesigned to hold an Energy Resource Summary if detailed energy consumption and production data are unavailable. Similarly, for those countries with limited manufacturing data, the fifth and sixth pages of the presentation are combined. In these cases, the manufacturing page is replaced by a single panel, Industrial Summary. (Thus the reason for the country presentation of either five or six pages.) Panel numbers, however, remain the same; the general location of the panels follows the same pattern whether variant pages are used or not.

Sources, Notes, and Footnotes

Sources, notes, and footnote texts reside in a separate Annotated Source Appendix accessible by panel number and the Table of Contents provided at the beginning of the appendix.

Data Characteristics

Variability of Statistical Information

Another possible variation concerns content and results from a variety of circumstances. Content variation may arise from supplementing tabular data from the original source with comparable tabular data from a second source. Since data from each source are calculated based on a particluar statistical methodology (for easier comparison from country to country) this is not an ideal solution to the problem of a data shortage. Generally, however, statistics for some countries in certain topic areas will always be lacking. The editor believes it is more desirable to provide comparable data for these topic areas rather than leaving them blank.

Many other factors can affect the quality of available data as well. Often demographic data are suspect because they are based on very rough estimates or they may have been collected over a wide range of time. Comparability of economic data for manufacturing, the production sectors, international trade, military expenditures, etc., may be complicated by the great diversity of the world's economies. Difficulties may arise in translating local values to a global standard or in comparing the "cost of living" between two or more countries with very different standards of living. The reader is referred to the Annotated Source Index for explanations, analogies, and pertinent definitions of terms to help interpret the data as accurately as possible.

Dated Information

The continual frustration of statistical reference book editors is that the "most current" statistics, and especially international statistics are, by their very nature, retrospective. This is the case because statistics can only measure moments in time. While those moments are being gathered and documented, people and events change the world, obsoleting all that went before. Thus information regarding say, Zaire, and other politically evolving locales may be of historical value only by the time it reaches the reader.

Content

The contents of *SAW* reflect the availability of statistical information. The main topics covered under 8 major headings are shown in the inset.

In general, coverage is quite comprehensive. One area of difficulty is the Arts. Information on Libraries, Newspapers, and Culture is scarce. These topics are grouped under the heading Education.

Sources

A number of international and U.S. federal agencies provided data for this edition of *SAW*. Because the goal of this book is to provide a uniform format for as many of the 188 countries as possible, the editor has chosen sources that can provide meaningful and consistent data for as many countries as possible.

The major entities that collect statistical data on countries worldwide are the United Nations, the International Bank for Reconstruction and Development/World Bank, the International Monetary Fund (IMF), the U.S. Central Intelligence Agency (CIA), and the U.S. State Department.

Geography

Human Factors
- Demographics
- Health Indicators
- Health Expenditures
- Women and Children
- Burden of Disease
- Ethnic Division
- Religion
- Major Languages

Education
- Public Education Expenditures
- Educational Attainment
- Literacy Rate
- Libraries
- Daily Newspapers
- Culture

Science and Technology
- Scientific/Technical Forces
- R&D Expenditures
- U.S. Patents Issued

Government and Law
- Organization of Government
- Elections
- Government Budget
- Military Affairs

- Crime
- Human Rights

Labor Force
- Total Labor Force
- Labor Force by Occupation
- Unemployment Rate

Production Sectors
- Energy Production
- Energy Consumption
- Telecommunications
- Transportation
- Top Agricultural Products
- Top Mining Products
- Tourism

Manufacturing Sector

Finance, Economics, Trade
- Economic Indicators
- Balance of Payments
- Exchange Rates
- Top Import Origins
- Top Export Destinations
- Foreign Aid
- Import/Export Commodities

While these agencies provide a broad range of statistical data, various branches of the United Nations, such as regional and specialized agencies, and executive branches of the U.S. government also compile statistics relating to subject matter within their jurisdiction. International agencies include organizations such as the Organization for Economic Cooperation and Development (OECD), International Civil Aviation Organization (ICAO), Organization of American States (OAS), Asian Development Bank, and specialized agencies of the United Nations, including the International Labor Organization (ILO), the World Health Organization (WHO), the United Nations Educational, Scientific, and Cultural Organization (UNESCO), United Nations Children's Fund (UNICEF), United Nations Criminal Justice Information Network (UNCJIN), the United Nations Development Programme (UNDP), the United Nations Conference for Trade and Development (UNCTAD), and the Economic Commission for Latin America and the Caribbean (ECLAC).

Federal agencies of the U.S. government that publish international statistics include the U.S. Department of Commerce (Bureau of the Census, International Trade Administration, Office of Patents and Trademarks), the U.S. Department of Labor (Bureau of Labor Statistics), the U.S. Department of Agriculture, the U.S. Department of Defense, the U.S. Department of Energy, the U.S. Department of Education, the U.S. Department of State, the U.S. Department of the Interior (Bureau of Mines and U.S. Geological Survey), and the U.S. National Science Foundation. Agencies not listed here also publish some international statistics from time to time in their areas of expertise.

Specialized and regional international agencies offer excellent resources for information in their own right. The OECD, for instance, offers a wide range of reports on scientific and economic subjects. These publications are timely and informative. ECLAC also reports on a variety of important issues. However, regional reports tend to reflect the issues that are of direct concern to a particular geographic or economic area. Also, methods of statistical collection may vary among agencies, depending on the purpose of the report.

With the exception of tourism information, no data from associations or private organizations were used.

The most comprehensive source for *SAW* data in terms of geographical range is the CIA *World Factbook*, which provides largely textual (nontabular) information for all countries included in this edition. This information, with a great deal of editorial and computer processing, has been fortuitous in the production of this book.

Arrangement of the Book

Access to *SAW* is provided in the **Table of Contents**. It lists the countries featured in alphabetical order by the page on which the country's presentation starts.

The **Introduction** follows the Table of Contents.

Two pages follow showing the **Regions of the World**. These maps locate every country profiled in *SAW*.

A brief section showing **Abbreviations and Acronyms** follows. Additional abbreviations and their definitions also appear in the Annotated Source Appendix (see below).

The body of the book begins immediately after the Abbreviations and Acronyms. Countries are arranged in alphabetical order. Each country is shown in five or six standard pages. Information is presented in the order depicted in the inset on the previous page.

Annotated Source Appendix

The **Annotated Source Appendix** follows the last country. This section documents the source or sources used, presents notes and definitions of terms and acronyms pertinent to the data, and presents the texts of footnotes sequentially.

The Source Appendix is largely drawn, with minor editorial changes, from the primary sources used in the compilation of this book.

Keyword Index

The last item in *SAW* is the **Keyword Index**. The index holds more than 3,000 subject references and the names of languages, religions, ethnicities, and political parties. The names of political parties are sometimes difficult to identify; for this reason, each entry is identified with the tag, *pol*. In all cases, page references are provided. In most cases, the name of the country is also provided with the page reference.

Acknowledgments

Many thanks to all (too numerous to mention here) who helped in the compilation of this edition of *SAW* by providing data, advice, many hours of labor, and moral support.

Comments and Suggestions

Comments on *SAW* or suggestions for improvement concerning its format and coverage are welcome. Although every effort has been made to maintain accuracy, errors may have occurred; the editor would be grateful if these are called to her attention. Please contact:

Editor, *Statistical Abstract of the World*
Gale Research
835 Penobscot Building
Detroit, MI 48226-4094
Phone: (313) 961-2242 or (800) 347-GALE
Fax: (313) 961-6815

Regions of the World

North America

Europe

South America

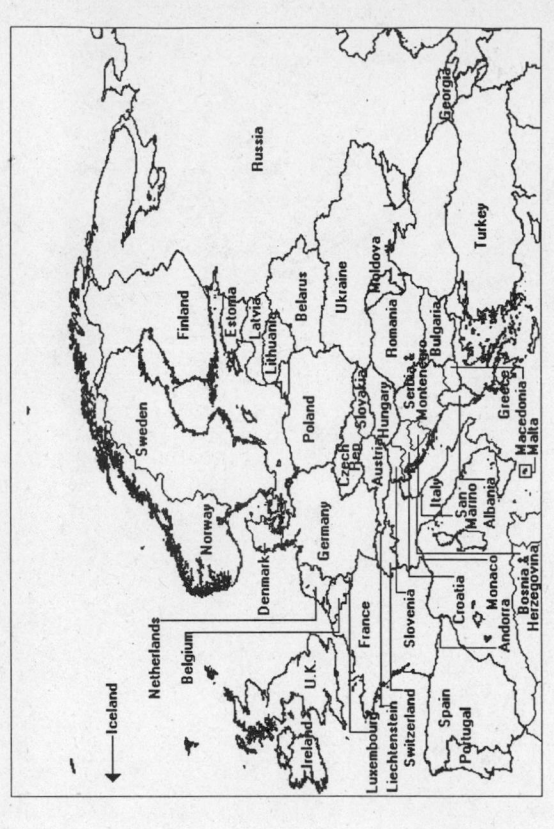

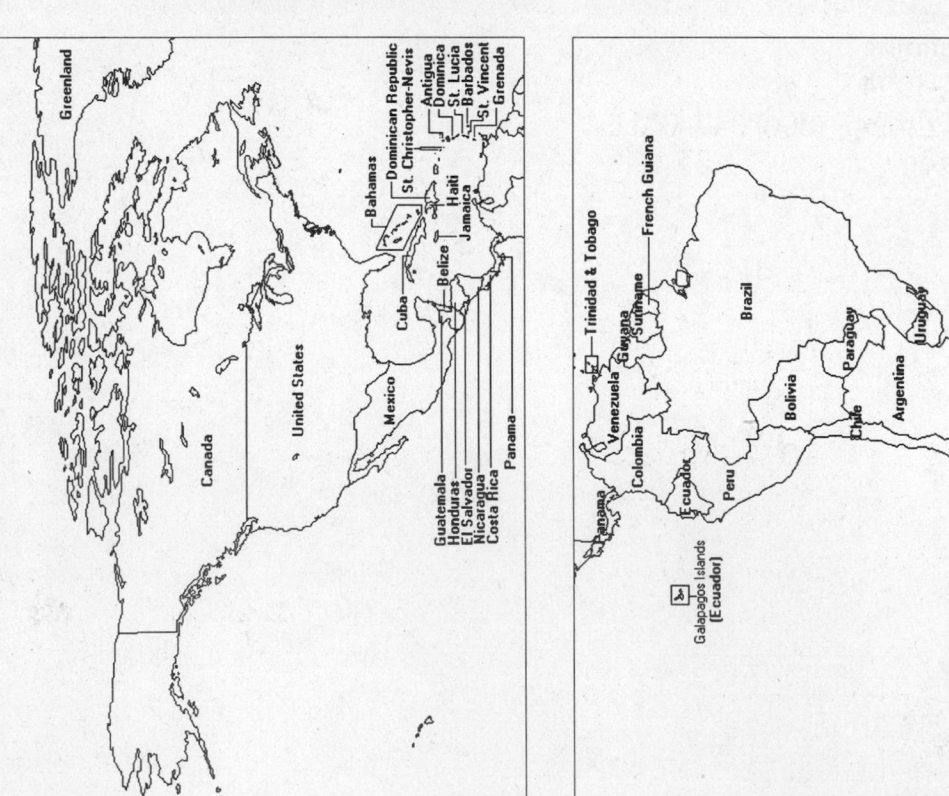

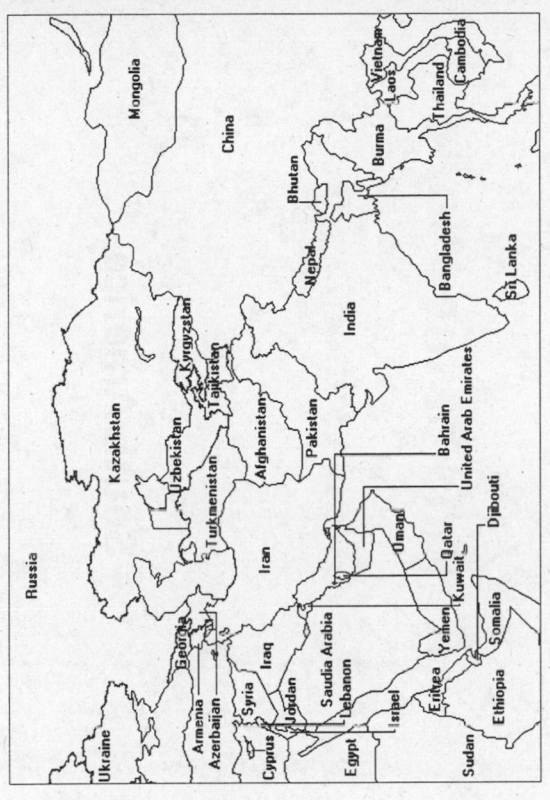

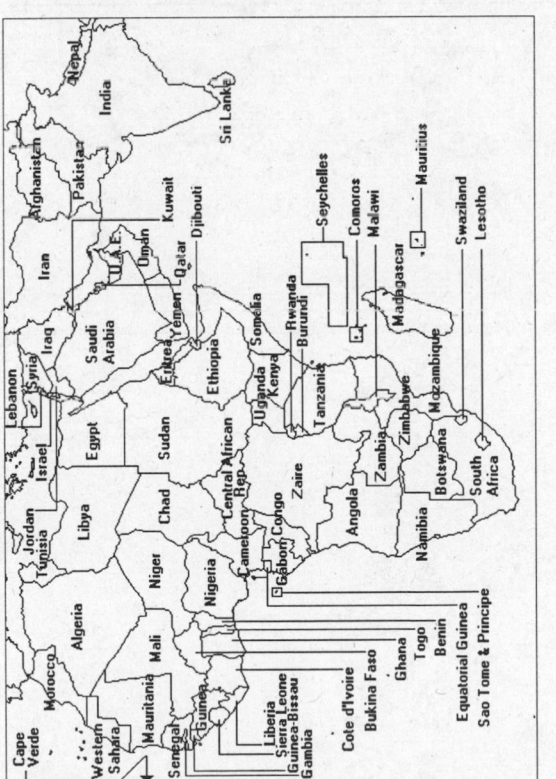

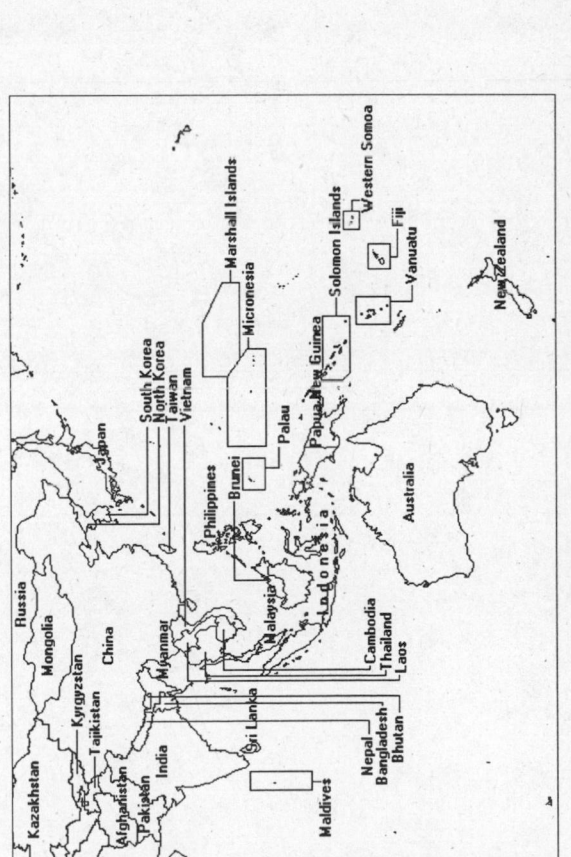

Africa

Southwest Asia

Southeast Asia

ABBREVIATIONS AND ACRONYMS

Additional explanations of some of the terms defined here may be found in the *Annotated Source Appendix*, page 1061.

ACHR	American Convention on Human Rights
Admin.	Administrative
AFL	Convention Concerning the Abolition of Forced Labor
AG	Silver
AIDS	Acquired Immune Deficiency Syndrome
AM	Amplitude modulation
APROBC	Convention Concerning the Application of the Principles of the Right to Organize and Bargain Collectively
ARABSAT	Arab Satellite Communications Organization
ASEAN	Association of Southeast Asian Nations
ASST	Supplementary Convention on the Abolition of Slavery, the Slave Trade, and Institutions and Practices Similar to Slavery
AU	Gold
avg.	Average
bd. ft.	Board foot
bil.	Billion
BLEU	Belgium-Luxembourg Economic Union
BMR	Black market rate
BTU	British thermal unit
C	Canadian
CARICOM	Caribbean Community and Common Market
CEMA	Council for Mutual Economic Assistance
C.F.A.	Communaute Financiere Africaine
CGE	Central government expenditure
CIA	Central Intelligence Agency
c.i.f.	Cost, insurance, and freight
Circ.	Circulation
CIS	Commonwealth of Independent States

CO	Cobalt
const.	Constant
CPI	Consumer price index
CPR	International Covenant on Civil and Political Rights
CU	Copper
curr.	Current
Dec	December
DOC	Department of Commerce
DOD	Department of Defense
DOE	Department of Energy
DOI	Department of the Interior
DOS	Department of State
DM	Deutsche Marks
dom.	Domestic
DPT	Diphtheria, pertussis, and tetanus
DWT	Deadweight ton
EAFDAW	Convention on the Elimination of All Forms of Discrimination Against Women
EAFRD	International Convention on the Elimination of All Forms of Racial Discrimination
EC	European Community
ECOWAS	Economic Community of West African States
ed.	Education
EFTA	European Free Trade Association
equiv.	Equivalent
ESCR	International Covenant on Economic, Social, and Cultural Rights
est.	Estimate
EU	European Union
EUTELSAT	European Telecommunications Satellite Organization
Ex-Im	Export-Import Bank of the United States
Expend.	Expenditure
FAPRO	Convention Concerning Freedom of Association and Protection of the Right to Organize
FL	Convention Concerning Forced Labor
f.o.b.	Free on board
FM	Frequency modulation
FRG	Federal Republic of Germany

FSU	Former Soviet Union
FY	Fiscal year
gal.	Gallon
GDP	Gross domestic product
GDR	[former] German Democratic Republic (East Germany)
GFCF	Gross fixed capital formation
GNP	Gross national product
Govt.	Government
GRT	Gross register ton
GSP	Generalized system of preferences
hl	Hectoliter
ICJ	International Court of Justice
IEA	International Energy Administration
ILO	International Labor Organization
IMF	International Monetary Fund
incl.	Including
INMARSAT	International Mobile Satellite Organization
INTELSAT	International Telecommunications Satellite Organization
INTERSPUTNIK	International Organization of Space Communications
ISCED	International Standard Classification of Education
Jan	January
kg	Kilogram
km	Kilometer
km^2	Square kilometer
kW	Kilowatt
kWh	Kilowatt hour
lb.	Pound
LDCs	Less-developed countries
LPG	Liquid petroleum gas
m	Meter
MAAE	Convention Concerning Minimum Age for Admission to Employment
MARECS	Maritime European Communications Satellite
Medarabtel	Middle East Telecommunications Project of the International Telecommunications Union
mfg.	Manufacturing
Mg	Magnesia

mkt	Market
max.	Maximum
mi.	Mile
mil.	Million
Mn	Manganese
Mo	Molybdenum
Mt.	Metric tons
N	Nitrogen
NA	Not available
NATO	North Atlantic Treaty Organization
NEGL	Negligible
NG	Natural gas
NGO	Nongovernmental organization
Ni	Nickel
nm	Nautical mile
NMT	Nordic Mobile Telecommunications
nom.	Nominal
NZ	New Zealand
ODA	Official development assistance
OECD	Organization for Economic Cooperation and Development
OECS	Organization of Eastern Caribbean States
OOF	Other official flows
OPEC	Organization of Petroleum Exporting Countries
ORT	Oral Rehydration Therapy
Pb	Lead
pct	Percent
PCPTW	Geneva Convention Relative to the Protection of Civilian Persons in Time of War
PHRFF	European Convention for the Protection of Human Rights and Fundamental Freedoms
Pop.	Population
PPCG	Convention on the Prevention and Punishment of the Crime of Genocide
PPP	Purchasing power parity
prod.	Production
PRW	Convention on the Political Rights of Women

PVIAC	Protocol Additional to the Geneva Conventions and Relating to the Protection of Victims of International Armed Conflicts
PVNAC	Protocol Additional to the Geneva Conventions and Relating to the Protection of Victims of Non-International Armed Conflicts
RC	Convention on the Rights of the Child
Reg.	Registered
RMB	Renminbi
SAAR	Seasonally adjusted annual rate
SACU	South African Customs Union
Sb	Antimony
SHF	Super High Frequency
Sn	Tin
sq.	Square
SR	Protocol Relating to the Status of Refugees
SSTS	Convention to Suppress the Slave Trade and Slavery
STPEP	Convention for the Suppression of the Traffic in Persons and of the Exploitation of the Prostitution of Others
Svc. Pts.	Service points
svgs.	Savings
TAT	Trans-Atlantic Telephone
TCIDTP	Convention Against Torture and Other Cruel, Inhuman or Degrading Treatment or Punishment
TPW	Geneva Convention Relative to the Treatment of Prisoners of War
T	Trillion
TT	Trinidad and Tobago
TV	Television
UHF	Ultrahigh frequency
VHF	Very High Frequency
Vols.	Volumes
UAE	United Arab Emirates
UK	United Kingdom
UN	United Nations
UNDP	United Nations Development Programme
UNESCO	United Nations Educational, Scientific, and Cultural Organization
UNICEF	United Nations Children's Fund
UNCTAD	United Nations Conference on Trade and Development
US	United States

USSR	[former] Union of Soviet Socialist Republics (Soviet Union)
W	Tungsten
WHO	World Health Organization
WPI	Wholesale price index
Zn	Zinc

Statistical Abstract of the World

Afghanistan

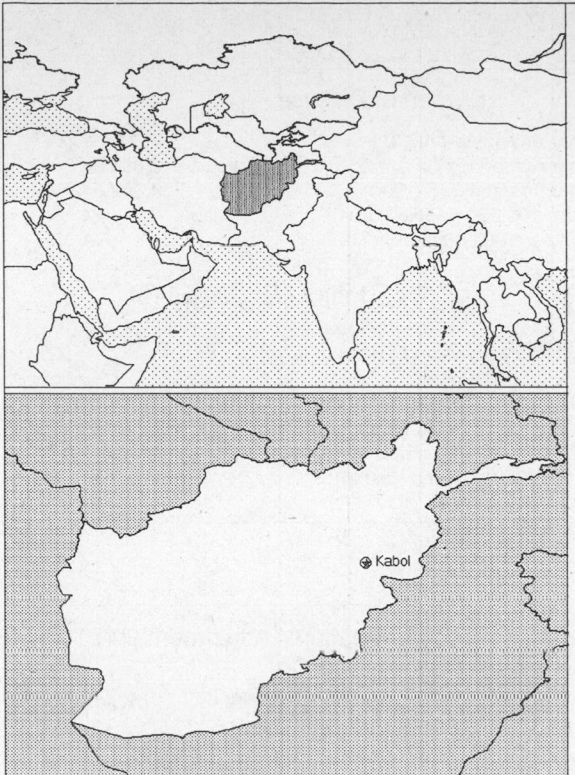

Geography [1]

Total area:
647,500 sq km   250,001 sq mi
Land area:
647,500 sq km   250,001 sq mi
Comparative area:
Slightly smaller than Texas
Land boundaries:
Total 5,529 km, China 76 km, Iran 936 km, Pakistan 2,430 km, Tajikistan 1,206 km, Turkmenistan 744 km, Uzbekistan 137 km
Coastline:
0 km (landlocked)
Climate:
Arid to semiarid; cold winters and hot summers
Terrain:
Mostly rugged mountains; plains in North and Southwest
Natural resources:
Natural gas, petroleum, coal, copper, talc, barites, sulfur, lead, zinc, iron ore, salt, precious and semiprecious stones
Land use:
Arable land: 12%
Permanent crops: 0%
Meadows and pastures: 46%
Forest and woodland: 3%
Other: 39%

Demographics [2]

	1970	1980	1990	1995[1]	1996	2000	2010	2020	2030
Population	12,431	14,985	14,767	21,571	22,664	26,668	34,098	43,050	53,334
Population density (persons per sq. mi.)	50	60	59	86	91	107	136	172	213
(persons per sq. km.)	19	23	23	33	35	41	53	66	82
Net migration rate (per 1,000 population)	NA	-86.3	-31.7	26.3	22.9	0.0	0.0	0.0	0.0
Births	NA	NA	NA	NA	975	NA	NA	NA	NA
Deaths	NA	NA	NA	NA	412	NA	NA	NA	NA
Life expectancy - males	NA	41.6	43.8	46.0	46.4	48.3	53.1	57.9	62.4
Life expectancy - females	NA	39.9	42.2	44.7	45.2	47.4	53.1	59.0	64.7
Birth rate (per 1,000)	NA	47.7	44.7	43.3	43.0	41.6	37.1	32.5	28.3
Death rate (per 1,000)	NA	22.7	20.4	18.5	18.2	16.6	13.0	10.1	8.1
Women of reproductive age (15-49 yrs.)	NA	3,361	3,323	4,900	5,156	6,102	7,969	10,383	13,278
of which are currently married[2]	NA	NA	2,674	NA	4,161	4,926	6,448	NA	NA
Fertility rate	NA	7.0	6.5	6.2	6.1	5.9	5.1	4.3	3.6

Except as noted, values for vital statistics are in thousands; life expectancy is in years.

Health

Health Indicators [3]

% of population with access to	
safe water (1990-95)	12
adequate sanitation (1990-95)	NA
health services (1985-95)	29
% of 1-year-olds immunized (1990-94) against	
TB (tuberculosis)	44
DPT (diphtheria, pertussis, tetanus)	18
polio	18
measles	40
% of contraceptive prevalence (1980-94)[1]	2
ORT use rate (1990-94)	26

Health Expenditures [4]

Total health expenditure, 1990 (official exchange rate)	
Millions of dollars	NA
Dollars per capita	NA
Health expenditures as a percentage of GDP	
Total	NA
Public sector	NA
Private sector	NA
Development assistance for health	
Total aid flows (millions of dollars)[1]	53
Aid flows per capita (dollars)	2.6
Aid flows as a percentage of total health expenditure	NA

For sources, notes, and explanations, see Annotated Source Appendix, page 1061.

Human Factors

Women and Children [5]

% of pregnant women immunized (tetanus 1990-94)	6
% of births attended by trained health personnel (1983-94)	9
Maternal mortality rate (1980-92)	640
Under-5 mortality rate (1994)	257
% under-5 moderately/severely underweight (1980-1994)	NA

Burden of Disease [6]

Population per physician (1993)	7,000.63
Population per nurse (1987)	8,899.30
Population per hospital bed (1990)	4,002.54
AIDS cases per 100,000 people (1994)	*
Malaria cases per 100,000 people (1992)	NA

Ethnic Division [7]

Minor ethnic groups (Chahar Aimaks, Turkmen, Baloch, and others).	
Pashtun	38%
Tajik	25%
Uzbek	6%
Hazara	19%

Religion [8]

Sunni Muslim	84%
Shi'a Muslim	15%
Other	1%

Major Languages [9]

Much bilingualism.	
Afghan Persian (Dari)	50%
Pashtu	35%
Turkic languages (primarily Uzbek and Turkmen)	11%
30 minor languages (primarily Balochi and Pashai)	4%

Education

Public Education Expenditures [10]

Million (Afghani)	1980	1989	1990	1991	1992	1994
Total education expenditure	3,205	NA	5,667	NA	NA	NA
as percent of GNP	2.0	NA	NA	NA	NA	NA
as percent of total govt. expend.	12.7	NA	NA	NA	NA	NA
Current education expenditure	2,886	NA	5,282	NA	NA	NA
as percent of GNP	1.8	NA	NA	NA	NA	NA
as percent of current govt. expend.	14.4	NA	NA	NA	NA	NA
Capital expenditure	319	NA	385	NA	NA	NA

Educational Attainment [11]

Age group (1979)	25+
Total population	4,891,473
Highest level attained (%)	
No schooling	89.0
First level	
Not completed	6.5
Completed	0.3
Entered second level	
S-1	1.1
S-2	NA
Postsecondary	3.0

Illiteracy [12]

In thousands and percent[1]	1990	1995	2000
Illiterate population (15+ yrs.)	6,173	8,169	10,191
Illiteracy rate - total pop. (%)	73.5	66.6	66.7
Illiteracy rate - males (%)	57.5	50.6	49.3
Illiteracy rate - females (%)	91.4	83.8	85.4

Libraries [13]

Daily Newspapers [14]

	1980	1985	1990	1994
Number of papers	13	13	14	15
Circ. (000)	90[e]	110[e]	180[e]	216

Culture [15]

Science and Technology

Scientific/Technical Forces [16]

R&D Expenditures [17]

U.S. Patents Issued [18]

                                         For sources, notes, and explanations, see Annotated Source Appendix, page 1061.

Government and Law

Organization of Government [19]

Long-form name:
Islamic State of Afghanistan
Type:
Transitional government
Independence:
19 August 1919 (from UK)
National holiday:
Victory of the Muslim Nation, 28 April;
Remembrance Day for Martyrs and
Disabled, 4 May; Independence Day, 19
August
Constitution:
None
Legal system:
A new legal system has not been adopted
but the transitional government has
declared it will follow Islamic law (Shari'a)
Executive branch:
President; Vice President; Prime Minister;
First Deputy Prime Minister; Deputy Prime
Minister; Council of Ministers
Legislative branch:
A unicameral parliament of 205 members
chosen by a national shura in January
1993; non-functioning as of June 1993
Judicial branch:
Interim Chief Justice of the Supreme
Court appointed; court system not yet
organized

Elections [20]

A unicameral parliament consisting of
205 members was chosen by the shura
in January 1993; non-functioning as of
June 1993.

Government Budget [21]

Crime [23]

Military Expenditures and Arms Transfers [22]

	1990	1991	1992	1993	1994
Military expenditures					
Current dollars (mil.)	400[e]	NA	NA	NA	NA
1994 constant dollars (mil.)	445[e]	NA	NA	NA	NA
Armed forces (000)	58	45	45	45	45
Gross national product (GNP)					
Current dollars (mil.)	NA	NA	NA	NA	NA
1994 constant dollars (mil.)	NA	NA	NA	NA	NA
Central government expenditures (CGE)					
1994 constant dollars (mil.)	NA	NA	NA	NA	NA
People (mil.)	15.6	15.9	16.6	17.9	19.3
Military expenditure as % of GNP	NA	NA	NA	NA	NA
Military expenditure as % of CGE	NA	NA	NA	NA	NA
Military expenditure per capita (1994 $)	NA	NA	NA	NA	NA
Armed forces per 1,000 people (soldiers)	3.7	2.8	2.7	2.5	2.3
GNP per capita (1994 $)	NA	NA	NA	NA	NA
Arms imports[6]					
Current dollars (mil.)	3,500	1,900	0	0	20
1994 constant dollars (mil.)	3,895	2,037	0	0	20
Arms exports[6]					
Current dollars (mil.)	0	0	0	0	20
1994 constant dollars (mil.)	0	0	0	0	20
Total imports[7]					
Current dollars (mil.)	936	616	NA	NA	NA
1994 constant dollars (mil.)	1,042	660	NA	NA	NA
Total exports[7]					
Current dollars (mil.)	235	188	NA	NA	NA
1994 constant dollars (mil.)	262	202	NA	NA	NA
Arms as percent of total imports[8]	373.8	308.4	0	0	NA
Arms as percent of total exports[8]	0	0	0	0	NA

Human Rights [24]

	SSTS	FL	FAPRO	PPCG	APROBC	TPW	PCPTW	STPEP	PHRFF	PRW	ASST	AFL
Observes	P			P		P	P	P		P	P	P
		EAFRD	CPR	ESCR	SR	ACHR	MAAE	PVIAC	PVNAC	EAFDAW	TCIDTP	RC
Observes		P	P	P						S	P	P

P = Party; S = Signatory; see Appendix for meaning of abbreviations.

Labor Force

Total Labor Force [25]

4.98 million

Labor Force by Occupation [26]

Agriculture and animal husbandry	67.8%
Industry	10.2
Construction	6.3
Commerce	5.0
Services and other	10.7

Date of data: 1980 est.

Unemployment Rate [27]

For sources, notes, and explanations, see Annotated Source Appendix, page 1061.

3

Production Sectors

Commercial Energy Production and Consumption

Data are shown in quadrillion (10^{15}) BTUs and percent for 1995
Values for hydroelectric, nuclear, geothermal, solar, and wind power refer to electrical generation.

Production [28]

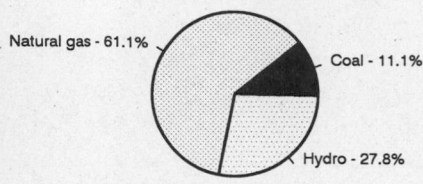

Natural gas - 61.1%
Coal - 11.1%
Hydro - 27.8%

Consumption [29]

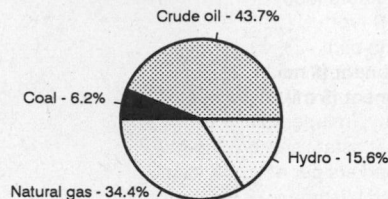

Crude oil - 43.7%
Coal - 6.2%
Hydro - 15.6%
Natural gas - 34.4%

Dry natural gas	0.011
Coal	0.002
Net hydroelectric power	0.005
Total	0.018

Crude oil	0.014
Dry natural gas	0.011
Coal	0.002
Net hydroelectric power	0.005
Total	0.032

Telecommunications [30]

- 31,200 (1983 est.) telephones
- Domestic: very limited telephone and telegraph service; 1 public telephone in Kabul
- International: satellite earth stations - 1 Intelsat (Indian Ocean) linked only to Iran and 1 Intersputnik (Atlantic Ocean Region)
- Radio: Broadcast stations: AM 5, FM 0, shortwave 2
- Television: Televisions: 100,000 (1993 est.)

Transportation [31]

Railways: total: 24.6 km; broad gauge: 9.6 km 1.524-m gauge from Gushgy (Turkmenistan) to Towraghondi; 15 km 1,524-m gauge from Termiz (Uzbekistan) to Kheyrabad transshipment point on south bank of Amu Darya

Highways: total: 21,000 km; paved: 2,800 km; unpaved: 18,200 km (1984 est.)

Airports

Total:	35
With paved runways over 3,047 m:	3
With paved runways 2,438 to 3,047 m:	4
With paved runways 1,524 to 2,437 m:	2
With paved runways under 914 m:	7
With unpaved runways 2,438 to 3,047 m:	3

Top Agricultural Products [32]

Agriculture accounts for 65% of the GDP; produces wheat, fruits, nuts, karakul pelts; wool, mutton.

Top Mining Products [33]

Metric tons except as noted	8/31/95[*]
Barite	2,000
Cement, hydraulic	115,000
Coal, bituminous	180,000
Copper, mine output, Cu content	5,000
Gas, natural, gross (mil. cu. meters)	2,700
Gypsum	3,000
Natural gas liquids (000 42-gal. bls.)	40
Nitrogen, N content of ammonia	30,000
Salt, rock	13,000

Tourism [34]

For sources, notes, and explanations, see Annotated Source Appendix, page 1061.

Manufacturing Sector

Manufacturing Summary [35]

	1987		1988		1989		1990		1991	
	$ bil.	%	$ bil.	%	$ bil.	%	$ bil.	%	$ bil.	%
Establishments or enterprises (number)	390	0.017	425	0.020	-	-	-	-	-	-
Total employment (000)	40	0.029	38	0.028	-	-	-	-	-	-
Production workers (000)	-	-	-	-	-	-	-	-	-	-
Output ($ bil.)	0.479	0.005	0.460	0.004	-	-	-	-	-	-
Value added ($ bil.)	-	-	-	-	-	-	-	-	-	-
Capital investment ($ mil.)	-	-	-	-	-	-	-	-	-	-
M & E investment ($ mil.)	-	-	-	-	-	-	-	-	-	-
Employees per establishment (number)	102	176.104	89	136.148	-	-	-	-	-	-
Production workers per establishment	-	-	-	-	-	-	-	-	-	-
Output per establishment ($ mil.)	1	28.322	1	19.499	-	-	-	-	-	-
Capital investment per estab. ($ mil.)	-	-	-	-	-	-	-	-	-	-
M & E per establishment ($ mil)	-	-	-	-	-	-	-	-	-	-
Payroll per employee ($)	1,564	17.441	1,917	19.242	-	-	-	-	-	-
Wages per production worker ($)	-	-	-	-	-	-	-	-	-	-
Hours per production worker (hours)	-	-	-	-	-	-	-	-	-	-
Output per employee ($)	12,045	16.083	12,199	14.322	-	-	-	-	-	-
Capital investment per employee ($)	-	-	-	-	-	-	-	-	-	-
M & E per employee ($)	-	-	-	-	-	-	-	-	-	-

Note: Columns headed % show percent of world total or ratio. Ratios closest to 100 are closest to world average. M & E stands for machinery & equipment.

Output in Manufacturing

	1987		1988		1989		1990		1991	
	$ bil.	%	$ bil.	%	$ bil.	%	$ bil.	%	$ bil.	%
3110 - Food products	0.101	21.09	0.079	17.17	-	-	-	-	-	-
3130 - Beverages	0.010	2.09	0.008	1.74	-	-	-	-	-	-
3210 - Textiles	0.050	10.44	0.035	7.61	-	-	-	-	-	-
3220 - Wearing apparel	0.000	0.00	0.002	0.43	-	-	-	-	-	-
3230 - Leather and products	0.016	3.34	0.053	11.52	-	-	-	-	-	-
3240 - Footwear	0.014	2.92	0.020	4.35	-	-	-	-	-	-
3410 - Paper and products	0.002	0.42	0.004	0.87	-	-	-	-	-	-
3420 - Printing, publishing	0.017	3.55	0.021	4.57	-	-	-	-	-	-
3510 - Industrial chemicals	0.017	3.55	0.021	4.57	-	-	-	-	-	-
3511 - Basic chemicals, excl fertilizers	0.017	3.55	0.021	4.57	-	-	-	-	-	-
3520 - Chemical products nec	0.021	4.38	0.012	2.61	-	-	-	-	-	-
3522 - Drugs and medicines	0.020	4.18	0.012	2.61	-	-	-	-	-	-
3560 - Plastic products nec	0.009	1.88	0.009	1.96	-	-	-	-	-	-
3710 - Iron and steel	0.001	0.21	0.002	0.43	-	-	-	-	-	-
3840 - Transportation equipment	-	-	0.000	0.00	-	-	-	-	-	-
3843 - Motor vehicles	-	-	0.000	0.00	-	-	-	-	-	-
3900 - Industries nec	0.184	38.41	0.161	35.00	-	-	-	-	-	-

Note: Codes are International Standard Industry codes (ISIC). Percentages are % of total Output. [f]: Factor Prices; [p]: Producer Prices.

Finance, Economics, and Trade

Economic Indicators [36]

- **National product**: GDP—purchasing power parity—$12.8 billion (1995 est.)
- **National product real growth rate**: NA%
- **National product per capita**: $600 (1995 est.)
- **Inflation rate (consumer prices)**: NA%
- **External debt**: $2.3 billion (March 1991 est.)

Balance of Payments Summary [37]

Values in millions of dollars.

	1985	1986	1987	1988	1989
Exports of goods (f.o.b.)	628.2	497.0	538.7	453.8	252.3
Imports of goods (f.o.b.)	-921.6	-1,138.8	-904.5	-731.8	-623.5
Trade balance	-293.4	-641.8	-365.8	-278.0	-371.2
Services - debits	-162.7	-215.9	-167.6	-131.5	-111.3
Services - credits	69.2	53.1	54.8	92.9	28.3
Private transfers (net)	NA	NA	NA	NA	-1.2
Government transfers (net)	143.7	267.4	311.7	342.8	312.1
Long-term capital (net)	77.6	217.6	113.6	22.4	-186.4
Short-term capital (net)	23.2	84.5	-147.5	-26.5	126.8
Errors and omissions	168.4	216.4	211.6	-47.9	182.8
Overall balance	26.0	-18.7	10.8	-25.8	-20.1

Exchange Rates [38]

Currency: **afghani.**
Symbol: **AF.**

Data are currency units per $1. These rates reflect the free market exchange rates rather than the official exchange rate, which is a fixed rate of 50.600 afghanis to the dollar.

January 1995	7,000
January 1994	1,900
March 1993	1,019
1991	850

Imports and Exports

Top Import Origins [39]

$616.4 million (c.i.f., 1991).

Origins	%
FSU countries	NA
Pakistan	NA
Iran	NA
Japan	NA
Singapore	NA
India	NA
South Korea	NA
Germany	NA

Top Export Destinations [40]

$188.2 million (f.o.b., 1991).

Destinations	%
FSU countries	NA
Pakistan	NA
Iran	NA
Germany	NA
India	NA
UK	NA
Belgium	NA
Luxembourg	NA
Czechoslovakia	NA

Foreign Aid [41]

UN provides assistance in the form of food aid, immunization, land mine removal, and a wide range of aid to refugees and displaced persons.

	U.S. $	
ODA	NA	
US assistance (1985-93)	450	million

Import and Export Commodities [42]

Import Commodities

Food and petroleum products
Most consumer goods

Export Commodities

Fruits and nuts
Handwoven carpets
Wool
Cotton
Hides and pelts
Precious and semiprecious gems

For sources, notes, and explanations, see Annotated Source Appendix, page 1061.

Albania

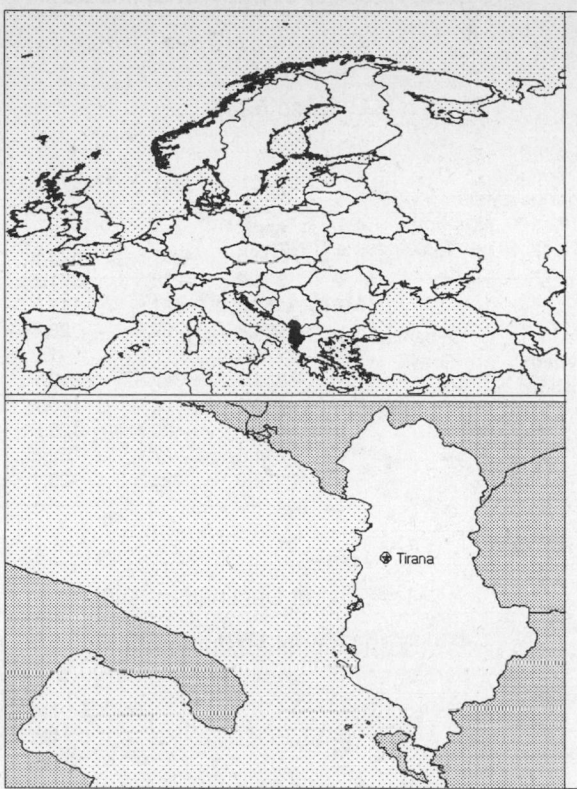

Geography [1]

Total area:
28,750 sq km   11,100 sq mi
Land area:
27,400 sq km   10,579 sq mi
Comparative area:
Slightly larger than Maryland
Land boundaries:
Total 720 km, Greece 282 km, Macedonia 151 km, Yugoslavia 287 km
(114 km with Serbia, 173 km with Montenegro)
Coastline:
362 km
Climate:
Mild temperate; cool, cloudy, wet winters; hot, clear, dry summers;
interior is cooler and wetter
Terrain:
Mostly mountains and hills; small plains along coast
Natural resources:
Petroleum, natural gas, coal, chromium, copper, timber, nickel
Land use:
Arable land: 21%
Permanent crops: 4%
Meadows and pastures: 15%
Forest and woodland: 38%
Other; 22%

Demographics [2]

	1970	1980	1990	1995[1]	1996	2000	2010	2020	2030
Population	2,157	2,699	3,273	3,207	3,249	3,427	3,858	4,257	4,563
Population density (persons per sq. mi.)	204	255	309	303	307	324	365	402	431
(persons per sq. km.)	79	99	119	117	119	125	141	155	167
Net migration rate (per 1,000 population)	0.0	0.0	-8.8	-2.4	-1.2	-0.1	0.0	0.0	0.0
Births	70	71	NA	NA	72	NA	NA	NA	NA
Deaths	20	17	NA	NA	25	NA	NA	NA	NA
Life expectancy - males	NA	67.0	66.0	64.6	64.9	66.3	69.2	72.9	75.6
Life expectancy - females	NA	72.0	72.6	70.8	71.2	72.7	75.6	79.1	81.8
Birth rate (per 1,000)	32.6	26.5	25.6	22.8	22.2	20.2	17.6	14.8	12.4
Death rate (per 1,000)	9.3	6.4	7.4	7.7	7.6	7.2	6.7	6.4	6.8
Women of reproductive age (15-49 yrs.)	NA	NA	828	855	871	947	1,083	1,129	1,148
of which are currently married	NA	NA	563	NA	602	658	765	NA	NA
Fertility rate	NA	3.6	3.0	2.7	2.7	2.4	2.1	1.9	1.8

Except as noted, values for vital statistics are in thousands; life expectancy is in years.

Health

Health Indicators [3]

% of population with access to	
safe water (1990-95)	NA
adequate sanitation (1990-95)	NA
health services (1985-95)	NA
% of 1-year-olds immunized (1990-94) against	
TB (tuberculosis)	81
DPT (diphtheria, pertussis, tetanus)	96
polio	97
measles	81
% of contraceptive prevalence (1980-94)	NA
ORT use rate (1990-94)	NA

Health Expenditures [4]

Total health expenditure, 1990 (official exchange rate)	
Millions of dollars	84
Dollars per capita	26
Health expenditures as a percentage of GDP	
Total	4.0
Public sector	3.4
Private sector	0.6
Development assistance for health	
Total aid flows (millions of dollars)[1]	NA
Aid flows per capita (dollars)	NA
Aid flows as a percentage of total health expenditure	NA

For sources, notes, and explanations, see Annotated Source Appendix, page 1061.

7

Human Factors

Women and Children [5]

% of pregnant women immunized (tetanus 1990-94)	92
% of births attended by trained health personnel (1983-94)	99
Maternal mortality rate (1980-92)	NA
Under-5 mortality rate (1994)	41
% under-5 moderately/severely underweight (1980-1994)	NA

Burden of Disease [6]

Population per physician	NA
Population per nurse	NA
Population per hospital bed (1990)	248.11
AIDS cases per 100,000 people (1994)	0.1
Heart disease cases per 1,000 people (1990-93)	305.5

Ethnic Division [7]

Other (Vlachs, Gypsies, Serbs, and Bulgarians).

Albanian	95%
Greeks	3%
Other	2%

(1989 est.)

Religion [8]

All mosques and churches were closed in 1967 and religious observances prohibited; in November 1990, Albania began allowing private religious practice.

Muslim	70%
Albanian Orthodox	20%
Roman Catholic	10%

Major Languages [9]

Albanian (Tosk is the official dialect) and Greek.

Education

Public Education Expenditures [10]

	1980	1988	1989	1990	1991	1994
Million (Lek)						
Total education expenditure	767	952	937	984	NA	5,893
as percent of GNP	NA	NA	NA	5.8	NA	3.0
as percent of total govt. expend.	10.3	11.2	NA	NA	NA	NA
Current education expenditure	NA	NA	NA	NA	NA	5,353
as percent of GNP	NA	NA	NA	NA	NA	2.7
as percent of current govt. expend.	NA	NA	NA	NA	NA	NA
Capital expenditure	NA	NA	NA	NA	NA	540

Educational Attainment [11]

Illiteracy [12]

Libraries [13]

	Admin. Units	Svc. Pts.	Vols. (000)	Shelving (meters)	Vols. Added	Reg. Users
National (1986)	1	NA	883	17,049	19,701	8,710
Nonspecialized (1986)	40	NA	2,344	NA	NA	116,755
Public (1988)	45	NA	4,072	NA	NA	228,786
Higher ed.	NA	NA	NA	NA	NA	NA
School	NA	NA	NA	NA	NA	NA

Daily Newspapers [14]

	1980	1985	1990	1994
Number of papers	2	2	2	3
Circ. (000)	145	135	135	185

Culture [15]

Cinema (seats per 1,000)	8.9
Annual attendance per person	2.2
Gross box office receipts (mil. Lek)	5.5
Museums (reporting)	NA
Visitors (000)	NA
Annual receipts (000 Lek)	NA

Science and Technology

Scientific/Technical Forces [16]

R&D Expenditures [17]

U.S. Patents Issued [18]

For sources, notes, and explanations, see Annotated Source Appendix, page 1061.

Government and Law

Organization of Government [19]

Long-form name:
Republic of Albania
Type:
Emerging democracy
Independence:
28 November 1912 (from Ottoman Empire)
National holiday:
Independence Day, 28 November (1912)
Constitution:
An interim basic law was approved by the People's Assembly on 29 April 1991; a draft constitution was rejected by popular referendum in the fall of 1994 and a new draft is pending
Legal system:
Has not accepted compulsory ICJ jurisdiction
Executive branch:
President of the Republic; Prime Minister of the Council of Ministers ; Council of Ministers
Legislative branch:
Unicameral: People's Assembly (Kuvendi Popullor):
Judicial branch:
Supreme Court

Crime [23]

Elections [20]

People's Assembly	% of votes
Democratic Party (DP)	62.3
Albanian Socialist Party (ASP)	25.6
Social Democratic Party (SDP)	4.3
Albanian Republican Party (RP)	3.2
Unity for Human Rights (UHP)	2.9
Others	1.7

Government Expenditures [21]

Housing - 3.5%
Other - 33.2%
Edu./Health - 29.6%
Industry - 12.8%
Defense - 7.1%
Gen. Services - 13.8%

(% distribution). Expend. for CY95: 69,687 (Lek mil.)

Military Expenditures and Arms Transfers [22]

	1990	1991	1992	1993	1994
Military expenditures					
Current dollars (mil.)	NA	NA	NA	NA	NA
1994 constant dollars (mil.)	NA	NA	NA	NA	NA
Armed forces (000)	NA	NA	65[e]	65[e]	75
Gross national product (GNP)					
Current dollars (mil.)[e]	4,100	2,700	2,500	3,300	3,800
1994 constant dollars (mil.)[e]	4,563	2,894	2,607	3,368	3,800
Central government expenditures (CGE)					
1994 constant dollars (mil.)[e]	NA	1,501	NA	NA	1,400
People (mil.)	3.2	3.3	3.3	3.3	3.4
Military expenditure as % of GNP	NA	NA	NA	NA	NA
Military expenditure as % of CGE	NA	NA	NA	NA	NA
Military expenditure per capita (1994 $)	NA	NA	NA	NA	NA
Armed forces per 1,000 people (soldiers)	NA	NA	19.7	19.5	22.2
GNP per capita (1994 $)	1,404	884	792	1,010	1,126
Arms imports[6]					
Current dollars (mil.)	0	0	0	0	0
1994 constant dollars (mil.)	0	0	0	0	0
Arms exports[6]					
Current dollars (mil.)	0	0	0	0	0
1994 constant dollars (mil.)	0	0	0	0	0
Total imports[7]					
Current dollars (mil.)[e]	444	467	655	629	NA
1994 constant dollars (mil.)[e]	494	501	683	642	NA
Total exports[7]					
Current dollars (mil.)[e]	222	168	174	117	NA
1994 constant dollars (mil.)[e]	247	180	181	119	NA
Arms as percent of total imports[8]	0	0	0	0	0
Arms as percent of total exports[8]	0	0	0	0	0

Human Rights [24]

	SSTS	FL	FAPRO	PPCG	APROBC	TPW	PCPTW	STPEP	PHRFF	PRW	ASST	AFL
Observes	P	P	P	P	P	P	P	P	P	P	P	
	EAFRD	CPR	ESCR	SR	ACHR	MAAE	PVIAC	PVNAC	EAFDAW	TCIDTP	RC	
Observes		P	P	P	P		P	P	P	P	P	

P=Party; S=Signatory; see Appendix for meaning of abbreviations.

Labor Force

Total Labor Force [25]

1.692 million (1994 est.) (including 352,000 emigrant workers and 261,000 domestically unemployed)

Labor Force by Occupation [26]

Agriculture (nearly all private)	49.5%
Private sector	22.2
State (nonfarm) sector	28.3
including state-owned industry	7.8

Unemployment Rate [27]

19% (1994 est.)

For sources, notes, and explanations, see Annotated Source Appendix, page 1061.

9

Production Sectors

Commercial Energy Production and Consumption

Data are shown in quadrillion (10^{15}) BTUs and percent for 1995
Values for hydroelectric, nuclear, geothermal, solar, and wind power refer to electrical generation.

Production [28]

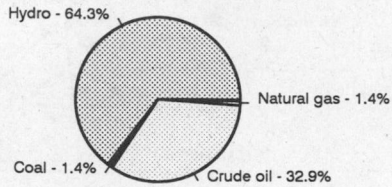

Hydro - 64.3%
Natural gas - 1.4%
Coal - 1.4%
Crude oil - 32.9%

Consumption [29]

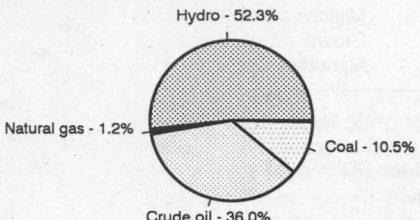

Hydro - 52.3%
Natural gas - 1.2%
Coal - 10.5%
Crude oil - 36.0%

Crude oil	0.023
Dry natural gas	0.001
Coal	0.001
Net hydroelectric power	0.045
Total	0.070

Crude oil	0.031
Dry natural gas	0.001
Coal	0.009
Net hydroelectric power	0.045
Total	0.086

Telecommunications [30]

- 55,000 telephones
- Domestic: obsolete wire system; no longer provides a telephone for every village; in 1992, following the fall of the communist government, peasants cut the wire to about 1,000 villages and used it to build fences
- International: inadequate; international traffic carried by microwave radio relay from the Tirane exchange to Italy and Greece
- Radio: Broadcast stations: AM 17, FM 1, shortwave 0 Radios: 577,000 (1991 est.)
- Television: Broadcast stations: 9 Televisions: 300,000 (1993 est.)

Transportation [31]

Railways: total: 670 km; standard gauge: 670 km 1.435-m gauge (1995)

Highways: total: 18,450 km; paved: 17,450 km; unpaved: 1,000 km (1991 est.)

Merchant marine: total: 11 cargo ships (1,000 GRT or over) totaling 52,967 GRT/76,887 DWT (1995 est.)

Airports

Total:	11
With paved runways 2,438 to 3,047 m:	3
With paved runways 914 to 1,523 m:	2
With unpaved runways over 3,047 m:	2
With unpaved runways 2,438 to 3,047 m:	1
With unpaved runways 1,524 to 2,437 m:	1

Top Agricultural Products [32]

Agriculture accounts for 55% of the GDP; produces wide range of temperate-zone crops and livestock.

Top Mining Products [33]

Metric tons except as noted[e]	5/95[*]
Copper concentrate	3,500
Pig iron	10,000
Steel, crude	5,000
Dolomite	50,000
Fertilizer, manufactured	14,000
Nitrogen, N content of ammonia	15,000
Pyrite, unroasted	7,000
Salt	10,000
Petroleum, crude, converted (000 42-gal. bls.)	3,300

Tourism [34]

	1990	1991	1992	1993	1994
Visitors[1]	NA	80	148	218	226
Tourists[2]	30	13	28	45	28
Excursionists	11	7	4	10	31
Tourism receipts[3]	7	5	9	8	5
Tourism expenditures[3]	4	3	5	6	4
Fare receipts	NA	NA	1	1	2
Fare expenditures	NA	NA	4	8	8

Travelers are in thousands, money in million U.S. dollars.

For sources, notes, and explanations, see Annotated Source Appendix, page 1061.

Finance, Economics, and Trade

GDP and Manufacturing Summary [35]

	1980	1985	1990	1991	1992
Gross Domestic Product					
Millions of 1980 dollars	2,373	2,711	3,160	2,718	2,283[e]
Growth rate in percent	6.29	1.48	2.92	-14.00	-16.00[e]
Manufacturing Value Added					
Millions of 1980 dollars	912	1,111	1,350	776	621[e]
Growth rate in percent	6.08	1.57	3.50	-42.50	-20.00[e]
Manufacturing share in percent of current prices	36.1	34.8	33.8	NA	NA

Economic Indicators [36]

- **National product**: GDP—purchasing power parity—$4.1 billion (1995 est.)

- **National product real growth rate**: 6% (1995 est.)

- **National product per capita**: $1,210 (1995 est.)

- **Inflation rate (consumer prices)**: 16% (1994 est.)

- **External debt**: $977 million (1994 est.)

Balance of Payments Summary [37]

Values in millions of dollars.

	1989	1990	1991	1992	1993
Exports of goods (f.o.b.)	394	322	73	70	112
Imports of goods (f.o.b.)	-456	-456	-281	-540	-601
Trade balance	-62	-134	-208	-470	-490
Services - debits	-28	-31	-59	-127	-193
Services - credits	41	32	10	23	143
Private transfers (net)	9	15	8	150	273
Government transfers (net)	2	NA	81	374	282
Long term capital (net)	-2	27	21	41	107
Short-term capital (net)	361	61	-7	-24	13
Errors and omissions	5	-2	125	47	-10
Overall balance	325	-32	-28	13	98

Exchange Rates [38]

Currency: **lek.**
Symbol: **L.**

Data are currency units per $1.

January 1996	95.65
January 1995	100.00
January 1994	99.00
January 1993	97.00
January 1992	50.00
September 1991	25.00

Imports and Exports

Top Import Origins [39]

$601 million (f.o.b., 1993 est.).

Origins	%
Italy	NA
Greece	NA
Bulgaria	NA
Turkey	NA
Macedonia	NA

Top Export Destinations [40]

$141 million (f.o.b., 1994 est.).

Destinations	%
Italy	NA
US	NA
Greece	NA
Macedonia	NA

Foreign Aid [41]

Recipient: ODA, $NA.

Import and Export Commodities [42]

Import Commodities	Export Commodities
Machinery	Asphalt
Consumer goods	Metals and metallic ores
Grains	Electricity
	Crude oil
	Vegetables
	Fruits
	Tobacco

Algeria

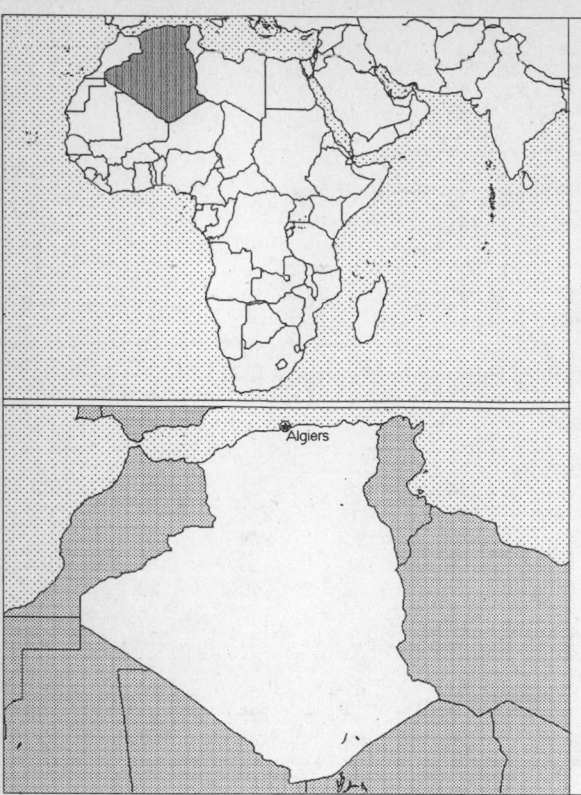

Geography [1]

Total area:
2,381,740 sq km   919,595 sq mi
Land area:
2,381,740 sq km   919,595 sq mi
Comparative area:
Slightly less than 3.5 times the size of Texas
Land boundaries:
Total 6,343 km, Libya 982 km, Mali 1,376 km, Mauritania 463 km, Morocco 1,559 km, Niger 956 km, Tunisia 965 km, western Sahara 42 km
Coastline:
998 km
Climate:
Arid to semiarid; mild, wet winters with hot, dry summers along coast; drier with cold winters and hot summers on high plateau; sirocco is a hot, dust/sand-laden wind especially common in summer
Terrain:
Mostly high plateau and desert; some mountains; narrow, discontinuous coastal plain
Natural resources:
Petroleum, natural gas, iron ore, phosphates, uranium, lead, zinc
Land use:
Arable land: 3%
Permanent crops: 0%
Meadows and pastures: 13%
Forest and woodland: 2%
Other: 82%

Demographics [2]

	1970	1980	1990	1995[1]	1996	2000	2010	2020	2030
Population	13,932	18,862	25,352	28,539	29,183	31,788	38,479	44,783	50,409
Population density (persons per sq. mi.)	15	21	28	31	32	35	42	49	55
(persons per sq. km.)	6	8	11	12	12	13	16	19	21
Net migration rate (per 1,000 population)	-4.0	0.0	-0.4	-0.5	-0.5	-0.5	-0.3	-0.2	-0.2
Births[3]	689	NA	NA	NA	832	NA	NA	NA	NA
Deaths[4]	226	NA	NA	NA	172	NA	NA	NA	NA
Life expectancy - males	52.8	55.9	65.4	66.9	67.2	68.4	70.9	73.0	74.7
Life expectancy - females	53.6	58.8	67.4	69.1	69.5	70.8	73.8	76.4	78.6
Birth rate (per 1,000)	50.0	43.4	32.2	29.0	28.5	26.5	22.3	18.2	16.1
Death rate (per 1,000)	16.4	12.1	7.0	6.1	5.9	5.4	4.8	4.7	5.4
Women of reproductive age (15-49 yrs.)	NA	NA	5,740	6,900	7,158	8,268	10,481	12,211	13,322
of which are currently married	NA	NA	2,910	NA	3,701	4,294	5,952	NA	NA
Fertility rate	7.9	NA	4.4	3.7	3.6	3.2	2.5	2.2	2.1

Except as noted, values for vital statistics are in thousands; life expectancy is in years.

Health

Health Indicators [3]

% of population with access to	
safe water (1990-95)	79
adequate sanitation (1990-95)	77
health services (1985-95)	98
% of 1-year-olds immunized (1990-94) against	
TB (tuberculosis)	92
DPT (diphtheria, pertussis, tetanus)	72
polio	72
measles	65
% of contraceptive prevalence (1980-94)	51
ORT use rate (1990-94)	27

Health Expenditures [4]

Total health expenditure, 1990 (official exchange rate)	
Millions of dollars	4,159
Dollars per capita	166
Health expenditures as a percentage of GDP	
Total	7.0
Public sector	5.4
Private sector	1.6
Development assistance for health	
Total aid flows (millions of dollars)[1]	2
Aid flows per capita (dollars)	0.1
Aid flows as a percentage of total health expenditure	0.1

For sources, notes, and explanations, see Annotated Source Appendix, page 1061.

Human Factors

Women and Children [5]

% of pregnant women immunized (tetanus 1990-94)	NA
% of births attended by trained health personnel (1983-94)	15
Maternal mortality rate (1980-92)[1]	140
Under-5 mortality rate (1994)	65
% under-5 moderately/severely underweight (1980-1994)	9

Burden of Disease [6]

Population per physician (1988)	2,332.35
Population per nurse (1988)	330.36
Population per hospital bed (1990)	400.11
AIDS cases per 100,000 people (1994)	0.3
Malaria cases per 100,000 people (1992)	NA

Ethnic Division [7]

Arab-Berber	99%
European	< 1%

Religion [8]

Sunni Muslim (state religion)	99%
Christian and Jewish	1%

Major Languages [9]

Arabic (official), French, Berber dialects.

Education

Public Education Expenditures [10]

	1980	1985	1990	1992	1993	1994
Million (Dinar)[1]						
Total education expenditure	12,355	24,248	29,504	53,516	80,841	79,889
as percent of GNP	7.8	8.5	5.7	5.4	7.2	5.6
as percent of total govt. expend.	24.3	20.7	21.1	16.3	19.6	17.6
Current education expenditure	8,259	16,814	24,953	NA	70,134	69,689
as percent of GNP	5.2	5.9	4.8	NA	6.3	4.9
as percent of current govt. expend.	29.8	26.2	29.7	NA	23.1	21.6
Capital expenditure	4,095	7,434	4,551	NA	10,707	10,200

Educational Attainment [11]

Illiteracy [12]

In thousands and percent[1]	1990	1995	2000
Illiterate population (15+ yrs.)	6,570	6,582	6,484
Illiteracy rate - total pop. (%)	45.9	38.8	32.1
Illiteracy rate - males (%)	32.1	26.4	21.2
Illiteracy rate - females (%)	59.7	51.3	43.1

Libraries [13]

	Admin. Units	Svc. Pts.	Vols. (000)	Shelving (meters)	Vols. Added	Reg. Users
National (1987)	1	1	1,020	NA	7,307	13,130
Nonspecialized	NA	NA	NA	NA	NA	NA
Public	NA	NA	NA	NA	NA	NA
Higher ed.	NA	NA	NA	NA	NA	NA
School	NA	NA	NA	NA	NA	NA

Daily Newspapers [14]

	1980	1985	1990	1994
Number of papers	4	5	10	6
Circ. (000)	448	570	1,274	1,250

Culture [15]

Cinema (seats per 1,000)	NA
Annual attendance per person	0.9
Gross box office receipts (mil. Dinar)	181
Museums (reporting)	NA
Visitors (000)	NA
Annual receipts (000 Dinar)	NA

Science and Technology

Scientific/Technical Forces [16]

R&D Expenditures [17]

U.S. Patents Issued [18]

For sources, notes, and explanations, see Annotated Source Appendix, page 1061.

Government and Law

Organization of Government [19]

Long-form name:
Democratic and Popular Republic of Algeria
Type:
Republic
Independence:
5 July 1962 (from France)
National holiday:
Anniversary of the Revolution, 1 November (1954)
Constitution:
19 November 1976, effective 22 November 1976; revised 3 November 1988 and 23 February 1989
Legal system:
Socialist, based on French and Islamic law; judicial review of legislative acts in ad hoc Constitutional Council composed of various public officials, including several Supreme Court justices
Executive branch:
President; Prime Minister; Council of Ministers
Legislative branch:
Unicameral: National People's Assembly (Al-Majlis Ech-Chaabi Al-Watani) Note: suspended since 1992
Judicial branch:
Supreme Court (Cour Supreme)

Crime [23]

Elections [20]

National People's Assembly. The government established a multiparty system in September 1989 and, as of 31 December 1990, over 50 legal parties existed. First round held on 26 December 1991 (second round canceled by the military after President Bendjedid resigned 11 January 1992); results - seats (281 total); the fundamentalist Islamic Salvation Front won 188 of the 231 seats contested in the first round.

Government Budget [21]

For 1995 est.

	$ bil.
Revenues	14.3
Expenditures	17.9
Capital expenditures	NA

Military Expenditures and Arms Transfers [22]

	1990	1991	1992	1993	1994
Military expenditures					
Current dollars (mil.)[1]	767[e]	636[e]	754[e]	1,135[e]	1,335
1994 constant dollars (mil.)[1]	853[e]	682[e]	786[e]	1,159[e]	1,335
Armed forces (000)	126	126	139	139	126
Gross national product (GNP)					
Current dollars (mil.)	37,450	37,840	39,700	40,080	40,510
1994 constant dollars (mil.)	41,680	40,560	41,400	40,910	40,510
Central government expenditures (CGE)					
1994 constant dollars (mil.)	11,100	NA	13,220[e]	NA	17,900[e]
People (mil.)	25.4	26.0	26.6	27.3	27.9
Military expenditure as % of GNP	2.0	1.7	1.9	2.8	3.3
Military expenditure as % of CGE	7.7	NA	5.9	NA	7.5
Military expenditure per capita (1994 $)	34	26	30	43	48
Armed forces per 1,000 people (soldiers)	5.0	4.8	5.2	5.1	4.5
GNP per capita (1994 $)	1,644	1,561	1,555	1,501	1,452
Arms imports[6]					
Current dollars (mil.)	310	130	5	20	140
1994 constant dollars (mil.)	345	139	5	20	140
Arms exports[6]					
Current dollars (mil.)	0	0	0	0	0
1994 constant dollars (mil.)	0	0	0	0	0
Total imports[7]					
Current dollars (mil.)	9,715	7,538	8,573	7,770	9,230[e]
1994 constant dollars (mil.)	10,810	8,080	8,940	7,930	9,230[e]
Total exports[7]					
Current dollars (mil.)	12,930	12,570	11,130	10,230	9,100
1994 constant dollars (mil.)	14,390	13,470	11,610	10,440	9,100
Arms as percent of total imports[8]	3.2	1.7	0.1	0.3	1.5
Arms as percent of total exports[8]	0	0	0	0	0

Human Rights [24]

	SSTS	FL	FAPRO	PPCG	APROBC	TPW	PCPTW	STPEP	PHRFF	PRW	ASST	AFL
Observes	P	P	P	P	P	P	P	P			P	P
	EAFRD	CPR	ESCR	SR	ACHR	MAAE	PVIAC	PVNAC	EAFDAW	TCIDTP	RC	
Observes		P	P	P	P		P	P	P	P	P	

P = Party; S = Signatory; see Appendix for meaning of abbreviations.

Labor Force

Total Labor Force [25]

6.2 million (1992 est.)

Labor Force by Occupation [26]

Government	29.5%
Agriculture	22
Construction and public works	16.2
Industry	13.6
Commerce and services	13.5
Transportation and communication	5.2

Date of data: 1989

Unemployment Rate [27]

25% (1995 est.)

For sources, notes, and explanations, see Annotated Source Appendix, page 1061.

Production Sectors

Commercial Energy Production and Consumption

Data are shown in quadrillion (10^{15}) BTUs and percent for 1995
Values for hydroelectric, nuclear, geothermal, solar, and wind power refer to electrical generation.

Production [28]

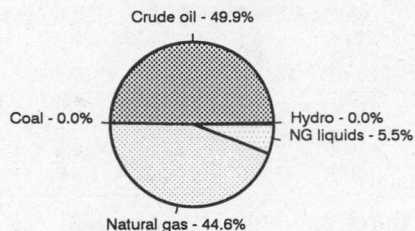

Crude oil - 49.9%
Coal - 0.0%
Hydro - 0.0%
NG liquids - 5.5%
Natural gas - 44.6%

Consumption [29]

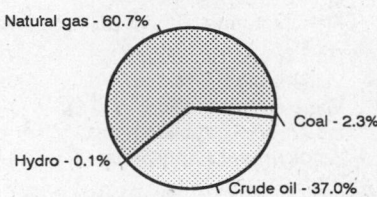

Natural gas - 60.7%
Coal - 2.3%
Hydro - 0.1%
Crude oil - 37.0%

Crude oil	2.437
Natural gas liquids	0.269
Dry natural gas	2.179
Coal	0.001
Net hydroelectric power	0.001
Total	4.887

Crude oil	0.471
Dry natural gas	0.773
Coal	0.029
Net hydroelectric power	0.001
Total	1.274

Telecommunications [30]

- 862,000 (1991 est.) telephones
- Domestic: excellent service in North but sparse in South; domestic satellite system with 12 earth stations (20 additional domestic earth stations are planned)
- International: 5 submarine cables; microwave radio relay to Italy, France, Spain, Morocco, and Tunisia; coaxial cable to Morocco and Tunisia; participant in Medarabtel; satellite earth stations - 2 Intelsat (1 Atlantic Ocean and 1 Indian Ocean), 1 Intersputnik, and 1 Arabsat
- Radio: Broadcast stations: AM 26, FM 0, shortwave 0 Radios: 6 million (1991 est.)
- Television: Broadcast stations: 18 Televisions: 2 million (1993 est.)

Transportation [31]

Railways: total: 4,772 km; standard gauge: 3,616 km 1.435-m gauge (301 km electrified; 215 km double track); narrow gauge: 1,156 km 1.055-m gauge

Highways: total: 95,576 km; paved: 63,080 km (including 400 km of expressways); unpaved: 32,496 km (1992 est.)

Merchant marine: total: 77 ships (1,000 GRT or over) totaling 916,701 GRT/1,086,324 DWT; ships by type: bulk 9, cargo 27, chemical tanker 7, liquefied gas tanker 10, oil tanker 5, roll-on/roll-off cargo 13, short-sea passenger 5, specialized tanker 1 (1995 est.)

Airports

Total:	119
With paved runways over 3,047 m:	8
With paved runways 2,438 to 3,047 m:	24
With paved runways 1,524 to 2,437 m:	13
With paved runways 914 to 1,523 m:	4
With paved runways under 914 m:	17

Top Agricultural Products [32]

Agriculture accounts for 12% of the GDP; produces wheat, barley, oats, grapes, olives, citrus, fruits; sheep, cattle.

Top Mining Products [33]

Metric tons except as noted[e]	5/1/95*
Lime, hydraulic	62,000
Nitrogen, N content of ammonia	400,000
Gas, natural (mil. cu. meters)	
gross	127,000
dry	56,000[1]
Natural gas plant liquids (000 42-gal. bls.)	53,000
Petroleum (000 42-gal. bls.)	
crude	275,000
condensate	155,000
refinery products	164,000

Tourism [34]

	1990	1991	1992	1993	1994
Visitors[4]	1,137	1,193	1,120	1,128	805
Tourism receipts	64	84	75	55	49
Tourism expenditures	149	140	163	NA	135
Fare receipts	86	86	NA	NA	NA
Fare expenditures	124	89	NA	NA	NA

Travelers are in thousands, money in million U.S. dollars.

Manufacturing Sector

GDP and Manufacturing Summary [35]

	1980	1985	1989	1990	% change 1980-1990	% change 1989-1990
GDP (million 1980 $)	42,342	53,959	56,895	56,964	34.5	0.1
GDP per capita (1980 $)	2,259	2,477	2,342	2,282	1.0	-2.6
Manufacturing as % of GDP (current prices)	10.9	11.6	11.5[e]	9.1	-16.5	-20.9
Gross output (million $)	9,122	13,978[e]	13,765[e]	13,238[e]	45.1	-3.8
Value added (million $)	3,644	6,157	5,997[e]	5,739[e]	57.5	-4.3
Value added (million 1980 $)	3,286	5,029	6,090[e]	5,326	62.1	-12.5
Industrial production index	100	154	161[e]	157	57.0	-2.5
Employment (thousands)	312	413[e]	467[e]	470[e]	50.6	0.6

Note: GDP stands for Gross Domestic Product. 'e' stands for estimated value.

Profitability and Productivity

	1980	1985	1989	1990	% change 1980-1990	% change 1989-1990
Intermediate input (%)	60	56[e]	56[e]	57[e]	-5.0	1.8
Wages, salaries, and supplements (%)	22	25[e]	28[e]	26[e]	18.2	-7.1
Gross operating surplus (%)	18	19[e]	16[e]	17[e]	-5.6	6.3
Gross output per worker ($)	29,246	33,067[e]	29,445[e]	2,763[e]	-90.6	-90.6
Value added per worker ($)	11,682	14,740[e]	12,828[e]	12,019[e]	2.9	-6.3
Average wage (incl. benefits) ($)	6,523	8,303[e]	8,199[e]	7,377[e]	13.1	-10.0

Profitability is in percent of gross output. Productivity is in U.S. $. 'e' stands for estimated value.

Profitability - 1990

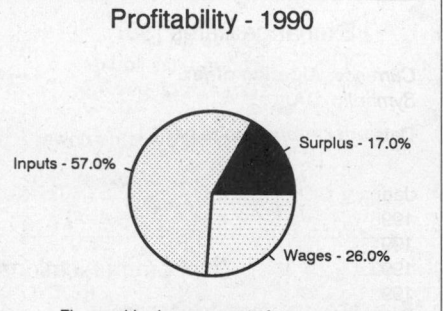

Surplus - 17.0%
Inputs - 57.0%
Wages - 26.0%

The graphic shows percent of gross output.

Value Added in Manufacturing

	1980 $ mil.	1980 %	1985 $ mil.	1985 %	1989 $ mil.	1989 %	1990 $ mil.	1990 %	% change 1980-1990	% change 1989-1990
311 Food products	655	18.0	852	13.8	847[e]	14.1	815[e]	14.2	24.4	-3.8
313 Beverages	135	3.7	176	2.9	185[e]	3.1	172[e]	3.0	27.4	-7.0
314 Tobacco products	176	4.8	229	3.7	211[e]	3.5	216[e]	3.8	22.7	2.4
321 Textiles	291	8.0	450	7.3	399[e]	6.7	421[e]	7.3	44.7	5.5
322 Wearing apparel	234	6.4	362	5.9	319[e]	5.3	370[e]	6.4	58.1	16.0
323 Leather and fur products	52	1.4	80	1.3	84[e]	1.4	73[e]	1.3	40.4	-13.1
324 Footwear	90	2.5	140	2.3	146[e]	2.4	127[e]	2.2	41.1	-13.0
331 Wood and wood products	120	3.3	205	3.3	216[e]	3.6	189[e]	3.3	57.5	-12.5
332 Furniture and fixtures	57	1.6	97	1.6	101[e]	1.7	89[e]	1.6	56.1	-11.9
341 Paper and paper products	143	3.9	242	3.9	266[e]	4.4	223[e]	3.9	55.9	-16.2
342 Printing and publishing	16	0.4	27	0.4	30[e]	0.5	25[e]	0.4	56.3	-16.7
351 Industrial chemicals	14	0.4	25	0.4	26[e]	0.4	22[e]	0.4	57.1	-15.4
352 Other chemical products	93	2.6	167	2.7	172[e]	2.9	170[e]	3.0	82.8	-1.2
353 Petroleum refineries	83	2.3	150	2.4	149[e]	2.5	162[e]	2.8	95.2	8.7
354 Miscellaneous petroleum and coal products	4	0.1	7	0.1	8[e]	0.1	7[e]	0.1	75.0	-12.5
355 Rubber products	17	0.5	30	0.5	30[e]	0.5	25[e]	0.4	47.1	-16.7
356 Plastic products	34	0.9	61	1.0	66[e]	1.1	58[e]	1.0	70.6	-12.1
361 Pottery, china, and earthenware	10	0.3	14	0.2	15[e]	0.3	14[e]	0.2	40.0	-6.7
362 Glass and glass products	36	1.0	51	0.8	55[e]	0.9	52[e]	0.9	44.4	-5.5
369 Other nonmetal mineral products	355	9.7	497	8.1	543[e]	9.1	510[e]	8.9	43.7	-6.1
371 Iron and steel	323	8.9	727	11.8	795[e]	13.3	574[e]	10.0	77.7	-27.8
372 Nonferrous metals	19	0.5	42	0.7	47[e]	0.8	37[e]	0.6	94.7	-21.3
381 Metal products	265	7.3	598	9.7	496[e]	8.3	576[e]	10.0	117.4	16.1
382 Nonelectrical machinery	46	1.3	105	1.7	87[e]	1.5	89[e]	1.6	93.5	2.3
383 Electrical machinery	123	3.4	278	4.5	231[e]	3.9	265[e]	4.6	115.4	14.7
384 Transport equipment	181	5.0	407	6.6	338[e]	5.6	317[e]	5.5	75.1	-6.2
385 Professional and scientific equipment	30	0.8	67	1.1	56[e]	0.9	69[e]	1.2	130.0	23.2
390 Other manufacturing industries	42	1.2	72	1.2	79[e]	1.3	72[e]	1.3	71.4	-8.9

Note: The industry codes shown are International Standard Industry codes (ISIC). Percentages are percent of total Value Added. 'e' stands for estimated value

For sources, notes, and explanations, see Annotated Source Appendix, page 1061.

Finance, Economics, and Trade

Economic Indicators [36]

- **National product**: GDP—purchasing power parity—$108.7 billion (1995 est.)

- **National product real growth rate**: 3.5% (1995 est.)

- **National product per capita**: $3,800 (1995 est.)

- **Inflation rate (consumer prices)**: 28% (1995 est.)

- **External debt**: $26 billion (1994)

Balance of Payments Summary [37]

Values in millions of dollars.

	1987	1988	1989	1990	1991
Exports of goods (f.o.b.)	9,029	7,620	9,534	12,964	12,330
Imports of goods (f.o.b.)	-6,616	-6,675	-8,372	-8,777	-6,852
Trade balance	2,413	945	1,162	4,187	5,478
Services - debits	-3,464	-3,917	-3,390	-3,671	-3,790
Services - credits	675	542	607	571	463
Private transfers (net)	522	385	535	332	239
Government transfers (net)	-5	5	6	1	-23
Long-term capital (net)	21	767	715	-872	-999
Short-term capital (net)	289	-23	40	-74	-21
Errors and omissions	-802	337	-448	-336	-299
Overall balance	-351	-959	-773	138	1,048

Exchange Rates [38]

Currency: **Algerian dinar.**
Symbol: **DA.**

Data are currency units per $1.

January 1996	53.003
1995	47.663
1994	35.059
1993	23.345
1992	21.836
1991	18.473

Imports and Exports

Top Import Origins [39]

$10.6 billion (f.o.b., 1995 est.).

Origins	%
France	29
Italy	14
Spain	9
US	9
Germany	7

Top Export Destinations [40]

$9.5 billion (f.o.b., 1995 est.).

Destinations	%
Italy	21
France	16
US	14
Germany	13
Spain	9

Foreign Aid [41]

Recipient: ODA, $316 million (1993).

Import and Export Commodities [42]

Import Commodities

Capital goods 39.7%
Food and beverages 21.7%
Consumer goods 11.8%

Export Commodities

Petroleum and
 natural gas 97%

Andorra

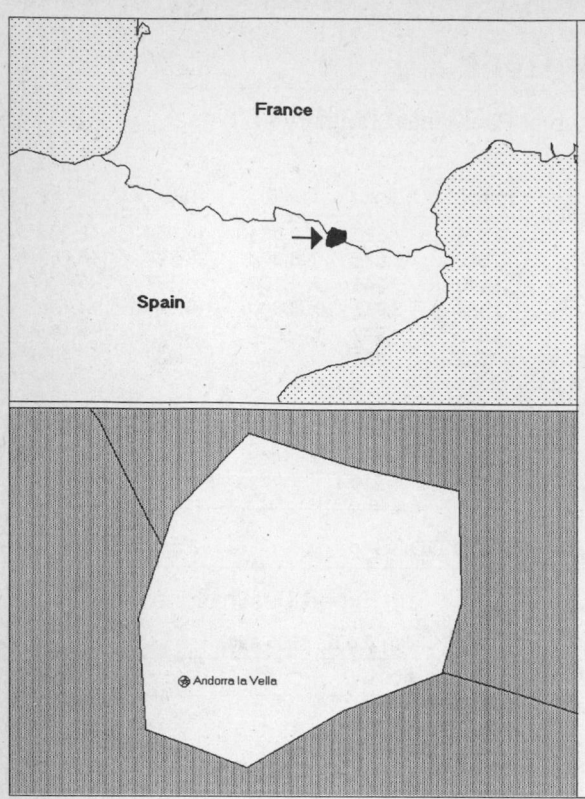

France

Spain

Andorra la Vella

Geography [1]

Total area:
  450 sq km   174 sq mi
Land area:
  450 sq km   174 sq mi
Comparative area:
  2.5 times the size of Washington, DC
Land boundaries:
  Total 125 km, France 60 km, Spain 65 km
Coastline:
  0 km (landlocked)
Climate:
  Temperate; snowy, cold winters and warm, dry summers
Terrain:
  Rugged mountains dissected by narrow valleys
Natural resources:
  Hydropower, mineral water, timber, iron ore, lead
Land use:
  Arable land: 2%
  Permanent crops: 0%
  Meadows and pastures: 56%
  Forest and woodland: 22%
  Other: 20%

Demographics [2]

	1970	1980	1990	1995[1]	1996	2000	2010	2020	2030
Population	20	34	53	71	73	80	92	97	96
Population density (persons per sq. mi.)	112	193	303	405	418	460	527	557	549
(persons per sq. km.)	43	74	117	156	161	178	204	215	212
Net migration rate (per 1,000 population)	NA	NA	60.6	25.2	22.3	12.5	6.5	2.1	0.0
Births	0	0	NA	NA	1	NA	NA	NA	NA
Deaths	0	0	NA	NA	Z	NA	NA	NA	NA
Life expectancy - males	NA	NA	74.8	86.6	86.5	85.9	84.7	83.7	82.7
Life expectancy - females	NA	NA	80.8	95.4	95.2	94.3	92.3	90.7	89.3
Birth rate (per 1,000)	NA	NA	14.6	10.3	10.2	9.6	7.8	7.3	6.8
Death rate (per 1,000)	NA	NA	6.4	2.8	2.9	3.5	5.2	7.5	10.8
Women of reproductive age (15-49 yrs.)	NA	NA	14	20	20	22	23	20	17
of which are currently married	NA	NA	9	NA	12	12	11	NA	NA
Fertility rate	NA	NA	1.7	1.1	1.1	1.2	1.2	1.3	1.4

Except as noted, values for vital statistics are in thousands; life expectancy is in years.

Health

Health Indicators [3]

Health Expenditures [4]

For sources, notes, and explanations, see Annotated Source Appendix, page 1061.

Human Factors

Women and Children [5]	Burden of Disease [6]

Ethnic Division [7]
Spanish	61%
Andorran	30%
French	6%
Other	3%

Religion [8]
Roman Catholic (predominant).

Major Languages [9]
Catalan (official), French, Castilian.

Education

Public Education Expenditures [10]	Educational Attainment [11]

Illiteracy [12]	Libraries [13]

Daily Newspapers [14]
	1980	1985	1990	1994
Number of papers	-	-	-	3
Circ. (000)	-	-	-	4[e]

Culture [15]
Cinema (seats per 1,000)	NA
Annual attendance per person	NA
Gross box office receipts (mil. Franc)	NA
Museums (reporting)	3
Visitors (000)	28
Annual receipts (000 Franc)	3,425

Science and Technology

Scientific/Technical Forces [16]	R&D Expenditures [17]	U.S. Patents Issued [18]

U.S. Patents Issued [18]
Values show patents issued to citizens of the country by the U.S. Patents Office.

	1993	1994	1995
Number of patents	1	0	1

For sources, notes, and explanations, see Annotated Source Appendix, page 1061.

19

Government and Law

Organization of Government [19]

Long-form name:
  Principality of Andorra
Type:
  Parliamentary democracy (March 1993)
  with a coprincipality (president of France
  and Spanish bishop of Seo de Urgel)
Independence:
  1278
National holiday:
  Mare de Deu de Meritxell, 8 September
Constitution:
  Andorra's first written constitution was
  drafted in 1991; adopted 14 March 1993
Legal system:
  Based on French and Spanish civil codes;
  no judicial review of legislative acts
Executive branch:
  Spanish Episcopal and French Coprinces
  (represented by veugers); Executive
  Council President; Executive Council
Legislative branch:
  Unicameral: General Council of the
  Valleys (Consell General de las Valls)
Judicial branch:
  Supreme Court of Andorra at Perpignan
  (France) and Ecclesiastical Court of the
  Bishop of Seo de Urgel (Spain) for civil
  cases; Tribunal of the Courts (Tribunal
  des Cortes) for criminal cases

Elections [20]

General Council of the Valleys	% of seats
National Democratic Group (AND)	28.6
Liberal Union (UL)	17.9
New Democracy (ND)	17.9
Andorran National Coalition (CNA)	7.1
National Democratic Initiative (IDN)	7.1
Others	21.4

Government Budget [21]

For 1993.

	$ mil.
Revenues	138
Expenditures	177
Capital expenditures	NA

Defense Summary [22]

Note: Defense is the responsibility of France and Spain

Crime [23]

	1994
Crime volume	
Cases known to police	1,761
Attempts (percent)	NA
Percent cases solved	81.40
Crimes per 100,000 persons	2,795.24
Persons responsible for offenses	
Total number offenders	1,052
Percent female	14.70
Percent juvenile (16-20 yrs.)	18.20
Percent foreigners	89.00

Human Rights [24]

	SSTS	FL	FAPRO	PPCG	APROBC	TPW	PCPTW	STPEP	PHRFF	PRW	ASST	AFL
Observes						P	P		P			
	EAFRD	CPR	ESCR	SR	ACHR	MAAE	PVIAC	PVNAC	EAFDAW	TCIDTP	RC
Observes											P

P=Party; S=Signatory; see Appendix for meaning of abbreviations.

Labor Force

Total Labor Force [25]

Labor Force by Occupation [26]

Unemployment Rate [27]

0%

For sources, notes, and explanations, see Annotated Source Appendix, page 1061.

Production Sectors

Energy Resource Summary [28]

Energy resources: Hydropower. **Electricity**: Capacity: 35,000 kW. Production: 140 million kWh. Consumption per capita: 2,570 kWh (1992).

Telecommunications [30]

- 21,258 (1983 est.) telephones
- Domestic: modern system with microwave radio relay connections between exchanges
- International: landline circuits to France and Spain
- Radio: Broadcast stations: AM 1, FM 0, shortwave 0 Radios: 10,000 (1993 est.)
- Television: Broadcast stations: 0 Televisions: 7,000 (1991 est.)

Transportation [31]

Railways: 0 km

Highways: total: 269 km; paved: 198 km; unpaved: 71 km (1991 est.)

Airports

Total: None

Top Agricultural Products [32]

Produces small quantities of tobacco, rye, wheat, barley, oats, vegetables; sheep raising.

Top Mining Products [33]

Detailed information is not available. A summary of natural resources follows. **Mineral Resources**: Iron ore and lead.

Tourism [34]

Finance, Economics, and Trade

Industrial Summary [35]

Industrial Production: Growth rate not available. **Industries**: Tourism (particularly skiing), sheep, timber, tobacco, banking.

Economic Indicators [36]

- **National product**: GDP—purchasing power parity—$1 billion (1993 est.)
- **National product real growth rate**: NA%
- **National product per capita**: $16,200 (1993 est.)
- **Inflation rate (consumer prices)**: NA%
- **External debt**: $NA

Balance of Payments Summary [37]

Exchange Rates [38]

Currency: **French franc and Spanish peseta**
Symbol: **F and Pta.**

Data are currency units per $1. Both French and Spanish currencies are used.

French francs	
January 1996	5.0056
1995	4.9915
1994	5.5520
1993	5.6632
1992	5.2938
Spanish pesetas	
January 1996	123.19
1995	124.69
1994	133.96
1993	127.26
1992	102.38

Imports and Exports

Top Import Origins [39]

$920.2 million (1993) Data are for 1992.

Origins	%
US	2.6
France	NA
Spain	NA

Top Export Destinations [40]

$46.2 million (f.o.b.; 1993).

Destinations	%
Spain	59
France	35

Foreign Aid [41]

Recipient: None; Donor: None.

Import and Export Commodities [42]

Import Commodities	Export Commodities
Consumer goods	Electricity
Food	Tobacco products
	Furniture

Angola

Geography [1]

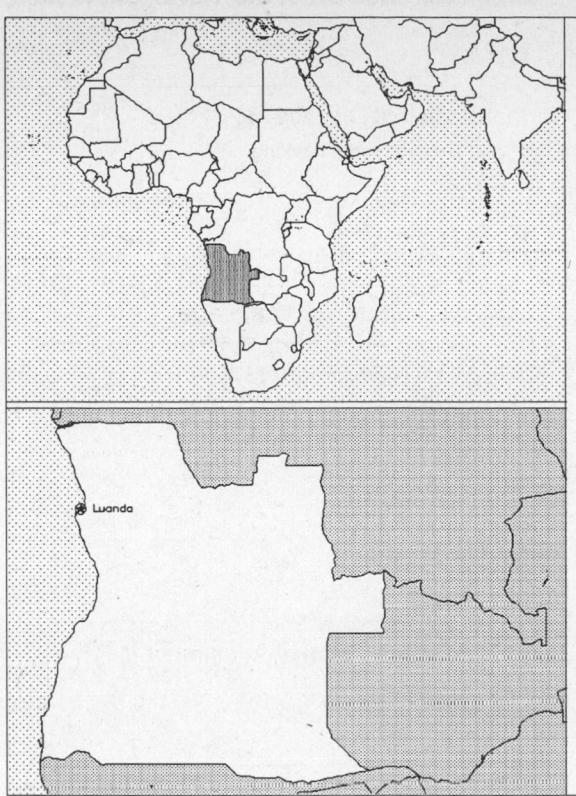

Total area:
   1,246,700 sq km   481,354 sq mi
Land area:
   1,246,700 sq km   481,354 sq mi
Comparative area:
   Slightly less than twice the size of Texas
Land boundaries:
   Total 5,198 km, Congo 201 km, Namibia 1,376 km, Zaire 2,511 km, Zambia 1,110 km
Coastline:
   1,600 km
Climate:
   Semiarid in South and along coast to Luanda; North has cool, dry season
   (May to October) and hot, rainy season (November to April)
Terrain:
   Narrow coastal plain rises abruptly to vast interior plateau
Natural resources:
   Petroleum, diamonds, iron ore, phosphates, copper, feldspar, gold,
   bauxite, uranium
Land use:
   Arable land: 2%
   Permanent crops: 0%
   Meadows and pastures: 23%
   Forest and woodland: 43%
   Other: 32%

Demographics [2]

	1970	1980	1990	1995[1]	1996	2000	2010	2020	2030
Population	5,606	6,794	8,430	10,070	10,343	11,513	14,982	19,272	24,174
Population density (persons per sq. mi.)	12	14	18	21	21	24	31	40	50
(persons per sq. km.)	4	5	7	8	8	9	12	15	19
Net migration rate (per 1,000 population)	5.6	-0.2	-0.1	-0.2	-0.1	0.0	0.0	0.0	0.0
Births	NA	NA	NA	NA	461	NA	NA	NA	NA
Deaths	NA	NA	NA	NA	183	NA	NA	NA	NA
Life expectancy - males	34.6	37.9	41.9	44.2	44.7	46.6	51.6	56.8	61.6
Life expectancy - females	36.5	40.7	45.7	48.5	49.1	51.4	57.5	63.5	68.9
Birth rate (per 1,000)	48.6	45.1	46.9	45.1	44.6	42.6	38.0	33.2	27.9
Death rate (per 1,000)	28.1	23.5	20.3	18.1	17.7	15.9	12.1	9.0	6.9
Women of reproductive age (15-49 yrs.)	1,290	1,553	1,895	2,242	2,302	2,578	3,559	4,757	6,250
of which are currently married[5]	905	NA	1,396	NA	1,701	1,903	2,617	NA	NA
Fertility rate	6.7	6.7	6.7	6.4	6.3	6.1	5.2	4.2	3.4

Except as noted, values for vital statistics are in thousands; life expectancy is in years.

Health

Health Indicators [3]

% of population with access to	
safe water (1990-95)	32
adequate sanitation (1990-95)	16
health services (1985-95)[1]	30
% of 1-year-olds immunized (1990-94) against	
TB (tuberculosis)	48
DPT (diphtheria, pertussis, tetanus)	27
polio	28
measles	44
% of contraceptive prevalence (1980-94)[1]	1
ORT use rate (1990-94)	48

Health Expenditures [4]

Total health expenditure, 1990 (official exchange rate)	
Millions of dollars	NA
Dollars per capita	NA
Health expenditures as a percentage of GDP	
Total	NA
Public sector	NA
Private sector	NA
Development assistance for health	
Total aid flows (millions of dollars)[1]	28
Aid flows per capita (dollars)	2.8
Aid flows as a percentage of total health expenditure	NA

For sources, notes, and explanations, see Annotated Source Appendix, page 1061.

Human Factors

Women and Children [5]

% of pregnant women immunized (tetanus 1990-94)	18
% of births attended by trained health personnel (1983-94)	15
Maternal mortality rate (1980-92)	NA
Under-5 mortality rate (1994)	292
% under-5 moderately/severely underweight (1980-1994)	NA

Burden of Disease [6]

Population per physician (1984)	16,143.45
Population per nurse (1984)	921.33
Population per hospital bed (1990)	771.54
AIDS cases per 100,000 people (1994)	1.4
Malaria cases per 100,000 people (1992)	NA

Ethnic Division [7]

Ovimbundu	37%
Kimbundu	25%
Bakongo	13%
Mestico	2%
European	1%
Other	22%

Religion [8]

Indigenous beliefs	47%
Roman Catholic	38%
Protestant	15%
(est.)	

Major Languages [9]

Portuguese (official), Bantu and other African languages.

Education

Public Education Expenditures [10]

Million (Kwansa)[2]	1980	1985	1987	1990	1991	1994
Total education expenditure	NA	9,643	12,854	12,076	NA	NA
as percent of GNP	NA	NA	NA	NA	NA	NA
as percent of total govt. expend.	NA	10.8	13.8	10.7	NA	NA
Current education expenditure	NA	9,419	11,567	10,856	NA	NA
as percent of GNP	NA	NA	NA	NA	NA	NA
as percent of current govt. expend.	NA	14.0	15.5	NA	NA	NA
Capital expenditure	NA	224	1,287	1,220	NA	NA

Educational Attainment [11]

Illiteracy [12]

In thousands and percent[3]	1985	1991	2000
Illiterate population (15+ yrs.)	3,117	3,221	3,395
Illiteracy rate - total pop. (%)	64.3	58.3	46.6
Illiteracy rate - males (%)	50.4	44.4	33.6
Illiteracy rate - females (%)	77.4	71.5	59.1

Libraries [13]

Daily Newspapers [14]

	1980	1985	1990	1994
Number of papers	4	4	4	4
Circ. (000)	143	103	115[e]	117[e]

Culture [15]

Science and Technology

Scientific/Technical Forces [16]

R&D Expenditures [17]

U.S. Patents Issued [18]

Government and Law

Organization of Government [19]

Long-form name:
Republic of Angola
Type:
Transitional government nominally a
multiparty democracy with a strong
presidential system
Independence:
11 November 1975 (from Portugal)
National holiday:
Independence Day, 11 November (1975)
Constitution:
11 November 1975; revised 7 January
1978, 11 August 1980, 6 March 1991, and
26 August 1992
Legal system:
Based on Portuguese civil law system and
customary law; recently modified to
accommodate political pluralism and
increased use of free markets
Executive branch:
President; Prime Minister; Council of
Ministers
Legislative branch:
Unicameral: National Assembly
(Assembleia Nacional):
Judicial branch:
Supreme Court (Tribunal da Relacao),
judges of the Supreme Court are
appointed by the president

Crime [23]

	1994
Crime volume	
Cases known to police	3,292
Attempts (percent)	NA
Percent cases solved	53.50
Crimes per 100,000 persons	30.77
Persons responsible for offenses	
Total number offenders	2,575
Percent female	13.70
Percent juvenile (0-17 yrs.)	21.50
Percent foreigners	0.30

Elections [20]

Popular Movement for the Liberation of
Angola (MPLA) is the ruling party and
has been in power since 1975; National
Union for the Total Independence of
Angola (UNITA) remains a legal party
despite its return to armed resistance to
the government; five minor parties have
small numbers of seats in the National
Assembly. First nationwide, multiparty
elections were held in late September
1992 with disputed results; further
elections are being discussed.

Government Budget [21]

For 1992 est.

	$ bil.
Revenues	0.928
Expenditures	2.5
Capital expenditures	0.963

Military Expenditures and Arms Transfers [22]

	1990	1991	1992	1993	1994
Military expenditures					
Current dollars (mil.)	NA	NA	NA	NA	NA
1994 constant dollars (mil.)	NA	NA	NA	NA	NA
Armed forces (000)	115	150	128	128	120
Gross national product (GNP)					
Current dollars (mil.)	NA	NA	NA	NA	NA
1994 constant dollars (mil.)	NA	NA	NA	NA	NA
Central government expenditures (CGE)					
1994 constant dollars (mil.)	3,654[e]	NA	4,267[e]	NA	NA
People (mil.)	8.4	8.7	9.1	9.5	9.8
Military expenditure as % of GNP	NA	NA	NA	NA	NA
Military expenditure as % of CGE	NA	NA	NA	NA	NA
Military expenditure per capita (1994 $)	NA	NA	NA	NA	NA
Armed forces per 1,000 people (soldiers)	13.6	17.3	14.1	13.4	12.2
GNP per capita (1994 $)	NA	NA	NA	NA	NA
Arms imports[6]					
Current dollars (mil.)	525	50	40	240	600
1994 constant dollars (mil.)	584	54	42	245	600
Arms exports[6]					
Current dollars (mil.)	0	0	0	0	0
1994 constant dollars (mil.)	0	0	0	0	0
Total imports[7]					
Current dollars (mil.)	1,577	1,909[e]	1,600[e]	2,046[e]	NA
1994 constant dollars (mil.)	1,755	2,046[e]	1,668[e]	2,088[e]	NA
Total exports[7]					
Current dollars (mil.)	3,944	3,449	3,788	3,182[e]	NA
1994 constant dollars (mil.)	4,389	3,697	3,950	3,284[e]	NA
Arms as percent of total imports[8]	33.3	2.6	2.5	11.7	30.0
Arms as percent of total exports[8]	0	0	0	0	0

Human Rights [24]

	SSTS	FL	FAPRO	PPCG	APROBC	TPW	PCPTW	STPEP	PHRFF	PRW	ASST	AFL
Observes	P			P	P	P				P		P
	EAFRD	CPR	ESCR	SR	ACHR	MAAE	PVIAC	PVNAC	EAFDAW	TCIDTP	RC	
Observes		P	P	P			P		P		P	

P = Party; S = Signatory; see Appendix for meaning of abbreviations.

Labor Force

Total Labor Force [25]

2.783 million economically active

Labor Force by Occupation [26]

Agriculture	85%
Industry	15

Date of data: 1985 est.

Unemployment Rate [27]

24% with extensive underemployment (1993
est.)

For sources, notes, and explanations, see Annotated Source Appendix, page 1061.

25

Production Sectors

Commercial Energy Production and Consumption

Data are shown in quadrillion (10^{15}) BTUs and percent for 1995
Values for hydroelectric, nuclear, geothermal, solar, and wind power refer to electrical generation.

Production [28]

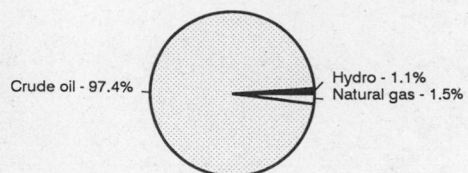

Crude oil - 97.4%
Hydro - 1.1%
Natural gas - 1.5%

Consumption [29]

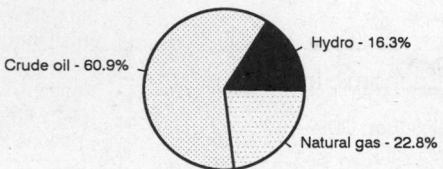

Hydro - 16.3%
Crude oil - 60.9%
Natural gas - 22.8%

Crude oil	1.374
Dry natural gas	0.021
Net hydroelectric power	0.015
Total	1.410

Crude oil	0.056
Dry natural gas	0.021
Net hydroelectric power	0.015
Total	0.092

Telecommunications [30]

- 78,000 (1991 est.) telephones; telephone service limited mostly to government and business use; HF radiotelephone used extensively for military links
- Domestic: limited system of wire, microwave radio relay, and tropospheric scatter
- International: satellite earth stations - 2 Intelsat (Atlantic Ocean)
- Radio: Broadcast stations: AM 17, FM 13, shortwave 0
- Television: Broadcast stations: 6 Televisions: 50,000 (1993 est.)

Top Agricultural Products [32]

Agriculture accounts for 12% of the GDP; produces bananas, sugarcane, coffee, sisal, corn, cotton, manioc (tapioca), tobacco, vegetables, plantains; livestock; forest products; fish.

Transportation [31]

Railways: total: 2,952 km (1995 est.); note - limited trackage in use because of landmines still in place from the civil war; narrow gauge: 2,798 km 1.067-m gauge; 154 km 0.600-m gauge

Highways: total: 72,626 km; paved: 18,157 km; unpaved: 54,469 km (1992 est.)

Merchant marine: total: 12 ships (1,000 GRT or over) totaling 63,776 GRT/99,863 DWT; ships by type: cargo 11, oil tanker 1 (1995 est.)

Airports

Total:	143
With paved runways over 3,047 m:	3
With paved runways 2,438 to 3,047 m:	8
With paved runways 1,524 to 2,437 m:	11
With paved runways 914 to 1,523 m:	4
With paved runways under 914 m:	40

Top Mining Products [33]

Metric tons except as noted[e]	8/1/95[*]
Diamond (000 carats)	300[2]
Gas, natural, gross (mil. cu. meters)	2,800[3]
Granite (000 cu. meters)	1,490
Steel, crude	9,000
Marble (000 cu. meters)	91,000
Natural gas plant liquids (000 42-gal. bls.)	2,000
Petroleum (000 42-gal. bls.)	
crude	199,000
refinery products	9,000[4]
Salt	30,000

Tourism [34]

	1990	1991	1992	1993	1994
Visitors[5]	NA	55	NA	21	11
Tourism receipts	13	NA	NA	20	13
Tourism expenditures	38	65	75	66	NA
Fare receipts	14	NA	NA	NA	NA
Fare expenditures	144	52	80	47	NA

Travelers are in thousands, money in million U.S. dollars.

For sources, notes, and explanations, see Annotated Source Appendix, page 1061.

Finance, Economics, and Trade

Industrial Summary [35]

Industrial Production: Growth rate not available; accounts for 56% of the GDP. **Industries**: Petroleum; diamonds, iron ore, phosphates, feldspar, bauxite, uranium, and gold; fish processing; food processing; brewing; tobacco; sugar; textiles; cement; basic metal products.

Economic Indicators [36]

- **National product**: GDP—purchasing power parity—$7.4 billion (1995 est.)
- **National product real growth rate**: 4% (1995 est.)
- **National product per capita**: $700 (1995 est.)
- **Inflation rate (consumer prices)**: 20% monthly average (1994 est.)
- **External debt**: $12 billion (1995 est.)

Balance of Payments Summary [37]

Values in millions of dollars.

	1989	1990	1991	1992	1993
Exports of goods (f.o.b.)	3,014	3,884	3,449	3,833	2,900
Imports of goods (f.o.b.)	-1,338	-1,578	-1,347	-1,988	-1,463
Trade balance	1,676	2,306	2,102	1,845	1,438
Services - debits	-1,954	-2,583	-2,896	2,840	-2,389
Services - credits	150	119	186	159	117
Private transfers (net)	-68	-140	-30	55	136
Government transfers (net)	64	63	58	47	30
Long-term capital (net)	1,699	61	-533	-339	-496
Short-term capital (net)	-895	189	1,134	1,258	1,349
Errors and omissions	-678	-19	27	43	-377
Overall balance	-6	-4	48	227	-193

Exchange Rates [38]

Currency: **new kwanza.**
Symbol: **NKz.**

Data are currency units per $1.

25 April 1995 (official rate)	900,000
6 April 1995 (BMR)	1,900,000
10 Jan 1995 (official rate)	600,000
1 June 1994 (official rate)	90,000
1 June 1994 (BMR)	180,000
16 Dec 1993 (official rate)	7,000
16 Dec 1993 (BMR)	50,000
July 1993	3,884
April 1992	550
November 1991	90
October 1990	60

Imports and Exports

Top Import Origins [39]

$1.6 billion (f.o.b., 1992 est.).

Origins	%
Portugal	NA
Brazil	NA
US	NA
France	NA
Spain	NA

Top Export Destinations [40]

$3 billion (f.o.b., 1993 est.).

Destinations	%
US	NA
France	NA
Germany	NA
Netherlands	NA
Brazil	NA

Foreign Aid [41]

Recipient: ODA, $189 million (1993).

Import and Export Commodities [42]

Import Commodities	Export Commodities
Capital equipment (machinery and electrical equipment)	Oil
Food	Diamonds
Vehicles and spare parts	Refined petroleum products
Textiles and clothing	Gas
Medicines	Coffee
Substantial military deliveries	Sisal
	Fish and fish products
	Timber
	Cotton

For sources, notes, and explanations, see Annotated Source Appendix, page 1061.

27

Antigua and Barbuda

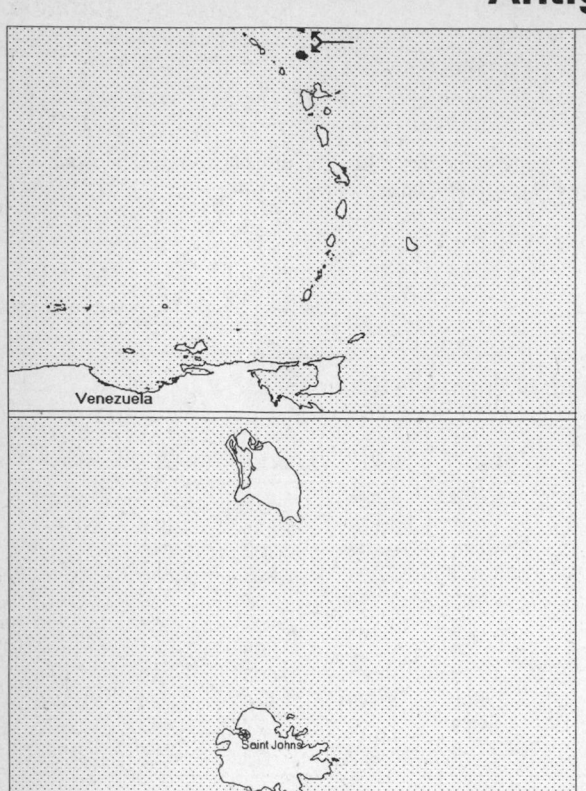

Geography [1]

Total area:
440 sq km   170 sq mi
Land area:
440 sq km   170 sq mi
Comparative area:
2.5 times the size of Washington, DC
Note: Includes Redonda
Land boundaries:
0 km
Coastline:
153 km
Climate:
Tropical marine; little seasonal temperature variation
Terrain:
Mostly low-lying limestone and coral islands with some higher volcanic areas
Natural resources:
Negligible; pleasant climate fosters tourism
Land use:
Arable land: 18%
Permanent crops: 0%
Meadows and pastures: 7%
Forest and woodland: 16%
Other: 59%

Demographics [2]

	1970	1980	1990	1995[1]	1996	2000	2010	2020	2030
Population	66	69	64	65	66	68	74	80	83
Population density (persons per sq. mi.)	386	403	375	384	386	401	438	468	486
(persons per sq. km.)	149	156	145	148	149	155	169	181	188
Net migration rate (per 1,000 population)	NA	NA	-10.1	-4.9	-3.9	0.0	0.0	0.0	0.0
Births	2	1	NA	NA	1	NA	NA	NA	NA
Deaths	0	0	NA	NA	Z	NA	NA	NA	NA
Life expectancy - males	NA	NA	70.1	71.3	71.6	72.5	74.4	76.0	77.2
Life expectancy - females	NA	NA	74.1	75.6	75.8	76.9	79.3	81.1	82.6
Birth rate (per 1,000)	NA	NA	17.9	17.1	16.8	15.7	12.7	11.2	10.1
Death rate (per 1,000)	NA	NA	5.7	5.4	5.3	5.2	5.1	5.7	8.3
Women of reproductive age (15-49 yrs.)	NA	NA	19	20	20	21	21	19	18
of which are currently married	NA	NA	9	NA	10	11	11	NA	NA
Fertility rate	NA	NA	1.7	1.7	1.7	1.7	1.7	1.7	1.7

Except as noted, values for vital statistics are in thousands; life expectancy is in years.

Health

Health Indicators [3]

Health Expenditures [4]

For sources, notes, and explanations, see Annotated Source Appendix, page 1061.

Human Factors

Women and Children [5]	Burden of Disease [6]	
	Population per physician (1993)	283.90
	Population per nurse	NA
	Population per hospital bed (1993)	169.62
	AIDS cases per 100,000 people (1994)	7.3
	Malaria cases per 100,000 people (1992)	NA

Ethnic Division [7]	Religion [8]	Major Languages [9]
Black, British, Portuguese, Lebanese, Syrian.	Anglican (predominant), other Protestant sects, some Roman Catholic.	English (official) and local dialects.

Education

Public Education Expenditures [10]

Million (E. Carib. Dollar)	1980	1984	1988	1990	1993	1994
Total education expenditure	9	12	NA	NA	NA	NA
as percent of GNP	3.0	2.7	NA	NA	NA	NA
as percent of total govt. expend.	NA	NA	NA	NA	NA	NA
Current education expenditure	9	11	31	NA	NA	NA
as percent of GNP	3.0	2.6	3.7	NA	NA	NA
as percent of current govt. expend.	NA	NA	NA	NA	NA	NA
Capital expenditure	0.0	1	NA	NA	NA	NA

Educational Attainment [11]

Illiteracy [12]

Libraries [13]

Daily Newspapers [14]

	1980	1985	1990	1994
Number of papers	1	1	1	-
Circ. (000)	6	6	6	-

Culture [15]

Science and Technology

Scientific/Technical Forces [16]

R&D Expenditures [17]

U.S. Patents Issued [18]

Values show patents issued to citizens of the country by the U.S. Patents Office.

	1993	1994	1995
Number of patents	0	1	2

For sources, notes, and explanations, see Annotated Source Appendix, page 1061.

29

Government and Law

Organization of Government [19]

Long-form name:
None
Type:
Parliamentary democracy
Independence:
1 November 1981 (from UK)
National holiday:
Independence Day, 1 November (1981)
Constitution:
1 November 1981
Legal system:
Based on English common law
Executive branch:
British Monarch (represented By Governor General); Prime Minister; Council of Ministers
Legislative branch:
Bicameral Parliament: Senate and House of Representatives:
Judicial branch:
Eastern Caribbean Supreme Court (based in Saint Lucia), one judge of the Supreme Court is a resident of the islands and presides over the Court of Summary Jurisdiction

Elections [20]

House of Representatives	% of seats
Antigua Labor Party (ALP)	64.7
United Progressive Party (UPP)	29.4
Independent	5.8

Government Budget [21]

For 1995.

	$ mil.
Revenues	134
Expenditures	135.4
Capital expenditures	NA

Defense Summary [22]

Branches: Royal Antigua and Barbuda Defense Force, Royal Antigua and Barbuda Police Force (includes the Coast Guard)

Manpower Availability: Males age 15-49 NA; males fit for military service NA

Defense Expenditures: $1.4 million, 1% of GDP (FY90/91)

Crime [23]

Human Rights [24]

	SSTS	FL	FAPRO	PPCG	APROBC	TPW	PCPTW	STPEP	PHRFF	PRW	ASST	AFL
Observes	P	P	P	P	P	P	P			P	P	P
	EAFRD	CPR	ESCR	SR	ACHR	MAAE	PVIAC	PVNAC	EAFDAW	TCIDTP	RC
Observes		P		P		P	P	P	P	P	P

P = Party; S = Signatory; see Appendix for meaning of abbreviations.

Labor Force

Total Labor Force [25]

30,000

Labor Force by Occupation [26]

Commerce and services	82%
Agriculture	11
Industry	7

Date of data: 1983

Unemployment Rate [27]

5%-10% (1995 est.)

Production Sectors

Energy Resource Summary [28]

Electricity: Capacity: 52,100 kW. Production: 95 million kWh. Consumption per capita: 1,242 kWh (1993).

Telecommunications [30]

- 6,700 telephones
- Domestic: good automatic telephone system
- International: 1 coaxial submarine cable; satellite earth station - 1 Intelsat (Atlantic Ocean); tropospheric scatter to Saba (Netherlands Antilles) and Guadeloupe
- Radio: Broadcast stations: AM 4, FM 2, shortwave 2
- Television: Broadcast stations: 2 Televisions: 28,000 (1993 est.)

Top Agricultural Products [32]

Agriculture accounts for 3.5% of the GDP; produces cotton, fruits, vegetables, bananas, coconuts, cucumbers, mangoes, sugarcane; livestock.

Top Mining Products [33]

Detailed information is not available. A summary of mineral resources follows. **Mineral Resources**: Antigua: sand, gravel, crushed stone, limestone. Barbuda: salt.

Transportation [31]

Railways: total: 77 km; narrow gauge: 64 km 0.760-m gauge; 13 km 0.610-m gauge (used almost exclusively for handling sugarcane)

Highways: total: 240 km; paved: NA km; unpaved: NA km

Merchant marine: total: 367 ships (1,000 GRT or over) totaling 1,573,063 GRT/2,147,243 DWT; ships by type: bulk 6, cargo 247, chemical tanker 6, combination bulk 1, container 72, liquefied gas tanker 2, oil tanker 3, refrigerated cargo 14, roll-on/roll-off cargo 16

Airports

Total: 3
With paved runways 2,438 to 3,047 m: 1
With paved runways under 914 m: 2 (1995 est.)

Tourism [34]

	1990	1991	1992	1993	1994
Visitors	458	487	493	515	529
Tourists[6]	197	197	210	240	255
Cruise passengers[7]	260	282	275	266	266
Tourism receipts	298	314	329	372	394
Tourism expenditures	18	20	23	23	24
Fare receipts	42	55	53	52	53
Fare expenditures	10	11	9	11	12

Travelers are in thousands, money in million U.S. dollars.

Finance, Economics, and Trade

Industrial Summary [35]

Industrial Production: Growth rate - 4.9% (1993 est.); accounts for 19.3% of the GDP. **Industries**: Tourism, construction, light manufacturing (clothing, alcohol, household appliances).

Economic Indicators [36]

- **National product**: GDP—purchasing power parity—$425 million (1994 est.)
- **National product real growth rate**: 4.2% (1994 est.)
- **National product per capita**: $6,600 (1994 est.)
- **Inflation rate (consumer prices)**: 3.5% (1994)
- **External debt**: $377 million (1995 est.)

Balance of Payments Summary [37]

Values in millions of dollars.

	1988	1989	1990	1991	1992
Exports of goods (f.o.b.)	17.0	15.7	19.0	35.4	54.7
Imports of goods (f.o.b.)	-201.1	-242.3	-230.7	-252.6	-260.9
Trade balance	-184.1	-226.5	-211.8	-217.1	-206.3
Services - debits	-108.4	-139.4	-155.7	-150.5	-148.6
Services - credits	237.2	274.5	318.8	324.2	339.9
Private transfers (net)	9.4	9.0	7.9	6.6	4.9
Government transfers (net)	2.8	3.3	2.5	2.7	1.4
Long-term capital (net)	28.7	28.0	15.8	29.0	-4.0
Short-term capital (net)	16.8	47.0	46.5	15.9	48.8
Errors and omissions	0.3	4.2	-23.6	-4.7	-14.4
Overall balance	2.7	NA	0.5	6.0	21.7

Exchange Rates [38]

Currency: **EC dollar.**
Symbol: **EC$.**

Data are currency units per $1.

Fixed rate since 1976   2.70

Imports and Exports

Top Import Origins [39]

$443.8 million (f.o.b., 1994 est.).

Origins	%
US	27
UK	16
Canada	4
OECS	3
Other	50

Top Export Destinations [40]

$40.9 million (f.o.b., 1994 est.).

Destinations	%
OECS	26
Barbados	15
Guyana	4
Trinidad and Tobago	2
US	0.3

Foreign Aid [41]

	U.S. $	
US commitments (1985-88)	10	million
Western (non-US) countries, ODA and OOF bilateral commitments (1970-89)	50	million

Import and Export Commodities [42]

Import Commodities
Food and live animals
Machinery and transport equipment
Manufactures
Chemicals
Oil

Export Commodities
Petroleum products 48%
Manufactures 23%
Food and live animals 4%
Machinery and transport equipment 17%

Argentina

Geography [1]

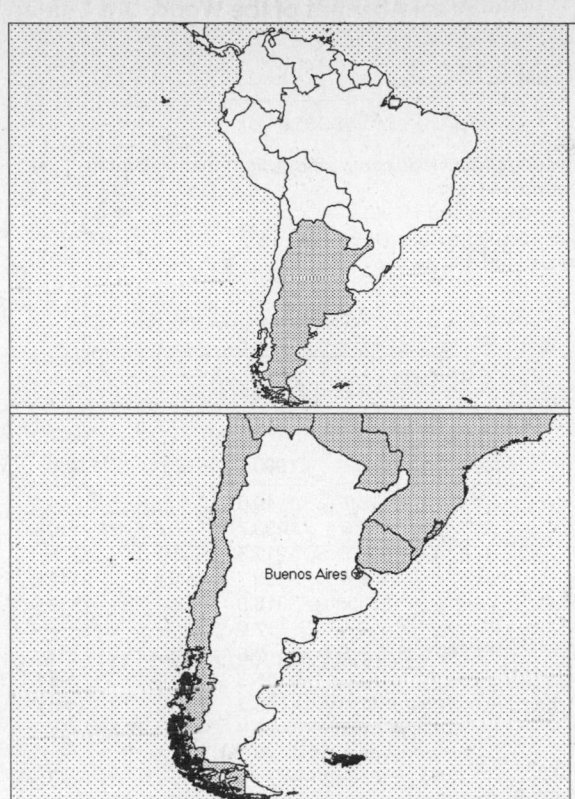

Total area:
2,766,890 sq km   1,068,302 sq mi
Land area:
2,736,690 sq km   1,056,642 sq mi
Comparative area:
Slightly less than three-tenths the size of the US
Land boundaries:
Total 9,665 km, Bolivia 832 km, Brazil 1,224 km, Chile 5,150 km, Paraguay 1,880 km, Uruguay 579 km
Coastline:
4,989 km
Climate:
Mostly temperate; arid in Southeast; subantarctic in Southwest
Terrain:
Rich plains of the Pampas in northern half, flat to rolling plateau of Patagonia in South, rugged Andes along western border
Natural resources:
Fertile plains of the pampas, lead, zinc, tin, copper, iron ore, manganese, petroleum, uranium
Land use:
Arable land: 9%
Permanent crops: 4%
Meadows and pastures: 52%
Forest and woodland: 22%
Other: 13%

Demographics [2]

	1970	1980	1990	1995[1]	1996	2000	2010	2020	2030
Population	23,962	28,237	32,386	34,293	34,673	36,202	39,947	43,190	45,873
Population density (persons per sq. mi.)	23	27	31	32	33	34	38	41	43
(persons per sq. km.)	9	10	12	13	13	13	15	16	17
Net migration rate (per 1,000 population)	NA	0.5	0.3	0.2	0.2	0.0	0.0	0.0	0.0
Births	NA	697	NA	NA	673	NA	NA	NA	NA
Deaths	NA	241	260	NA	299	NA	NA	NA	NA
Life expectancy - males	NA	65.8	67.4	68.2	68.4	69.0	70.4	71.7	72.8
Life expectancy - females	NA	72.5	74.2	75.0	75.1	75.7	77.1	78.4	79.5
Birth rate (per 1,000)	NA	24.1	20.3	19.5	19.4	19.1	17.4	15.4	14.2
Death rate (per 1,000)	NA	8.8	8.7	8.6	8.6	8.6	8.5	8.6	9.0
Women of reproductive age (15-49 yrs.)	NA	6,752	7,705	8,410	8,539	9,001	9,891	10,779	11,142
of which are currently married	3,164	3,552	4,793	NA	5,232	5,552	6,280	NA	NA
Fertility rate	NA	3.3	2.8	2.7	2.6	2.5	2.3	2.1	2.0

Except as noted, values for vital statistics are in thousands; life expectancy is in years.

Health

Health Indicators [3]

% of population with access to	
safe water (1990-95)	71
adequate sanitation (1990-95)	68
health services (1985-95)	71
% of 1-year-olds immunized (1990-94) against	
TB (tuberculosis)	100
DPT (diphtheria, pertussis, tetanus)	97
polio	84
measles	95
% of contraceptive prevalence (1980-94)	74
ORT use rate (1990-94)	80

Health Expenditures [4]

Total health expenditure, 1990 (official exchange rate)	
Millions of dollars	4,441
Dollars per capita	138
Health expenditures as a percentage of GDP	
Total	4.2
Public sector	2.5
Private sector	1.7
Development assistance for health	
Total aid flows (millions of dollars)[1]	11
Aid flows per capita (dollars)	0.3
Aid flows as a percentage of total health expenditure	0.2

For sources, notes, and explanations, see Annotated Source Appendix, page 1061.

Human Factors

Women and Children [5]

% of pregnant women immunized (tetanus 1990-94)	NA
% of births attended by trained health personnel (1983-94)[1]	87
Maternal mortality rate (1980-92)	140
Under-5 mortality rate (1994)	27
% under-5 moderately/severely underweight (1980-1994)	NA

Burden of Disease [6]

Population per physician (1986)	335.27
Population per nurse (1985)	979.81
Population per hospital bed (1990)	217.80
AIDS cases per 100,000 people (1994)	5.6
Malaria cases per 100,000 people (1992)	2

Ethnic Division [7]

White	85%
Mestizo, Indian, or other nonwhite groups	15%

Religion [8]

Nominally Roman Catholic	90%
Protestant	2%
Jewish	2%
Other	6%

Major Languages [9]

Spanish (official), English, Italian, German, French.

Education

Public Education Expenditures [10]

	1980	1985	1990	1992	1993	1994
Million or Trillion (T) (Peso)[3]						
Total education expenditure	1.018	740	7.36 T	6,962	8,310	10,471
as percent of GNP	2.7	1.5	1.1	3.1	3.3	3.8
as percent of total govt. expend.	15.1	NA	NA	15.7	12.4	14.0
Current education expenditure	0.860	667	7.06 T	NA	NA	NA
as percent of GNP	2.3	1.3	1.1	NA	NA	NA
as percent of current govt. expend.	18.8	NA	NA	NA	NA	NA
Capital expenditure	0.158	73	293,621	NA	NA	NA

Educational Attainment [11]

Age group (1991)	25+
Total population	17,340,713
Highest level attained (%)	
No schooling	5.7
First level	
Not completed	22.3
Completed	34.6
Entered second level	
S-1	25.3
S-2	NA
Postsecondary	12.0

Illiteracy [12]

In thousands and percent[1]	1990	1995	2000
Illiterate population (15+ yrs.)	983	935	891
Illiteracy rate - total pop. (%)	4.3	3.8	3.4
Illiteracy rate - males (%)	4.2	3.7	3.4
Illiteracy rate - females (%)	4.5	3.8	3.3

Libraries [13]

	Admin. Units	Svc. Pts.	Vols. (000)	Shelving (meters)	Vols. Added	Reg. Users
National (1989)	3	11	1,950	NA	115,419	NA
Nonspecialized	NA	NA	NA	NA	NA	NA
Public	NA	NA	NA	NA	NA	NA
Higher ed.	NA	NA	NA	NA	NA	NA
School	NA	NA	NA	NA	NA	NA

Daily Newspapers [14]

	1980	1985	1990	1994
Number of papers	220	218	159	187
Circ. (000)	4,000[e]	3,940[e]	4,000[e]	4,705[e]

Culture [15]

Cinema (seats per 1,000)	2.1
Annual attendance per person	0.5[e]
Gross box office receipts (mil. Austral)	NA
Museums (reporting)	NA
Visitors (000)	NA
Annual receipts (000 Austral)	NA

Science and Technology

Scientific/Technical Forces [16]

Scientists/engineers	11,088[e]
Number female	4,798
Technicians	6,241[e]
Number female	NA
Total	17,329[e]

R&D Expenditures [17]

	Austral (000) 1992
Total expenditure	664,700[e]
Capital expenditure	142,300[e]
Current expenditure	522,400[e]
Percent current	78.6[e]

U.S. Patents Issued [18]

Values show patents issued to citizens of the country by the U.S. Patents Office.

	1993	1994	1995
Number of patents	25	37	32

　　　　　　　　　For sources, notes, and explanations, see Annotated Source Appendix, page 1061.

Government and Law

Organization of Government [19]

Long-form name:
Argentine Republic
Type:
Republic
Independence:
9 July 1816 (from Spain)
National holiday:
Revolution Day, 25 May (1810)
Constitution:
1 May 1853; revised August 1994
Legal system:
Mixture of US and West European legal systems; has not accepted compulsory ICJ jurisdiction
Executive branch:
President; Vice President; Cabinet
Legislative branch:
Bicameral National Congress (Congreso Nacional): Senate and Chamber of Deputies
Judicial branch:
Supreme Court (Corte Suprema)

Elections [20]

Chamber of Deputies	% of seats
Justicialist Party (PJ)	51.3
Radical Civic Union (UCR)	26.5
Front for a Country in Solidarity (Frepaso)	10.1
Others	12.1

Government Expenditures [21]

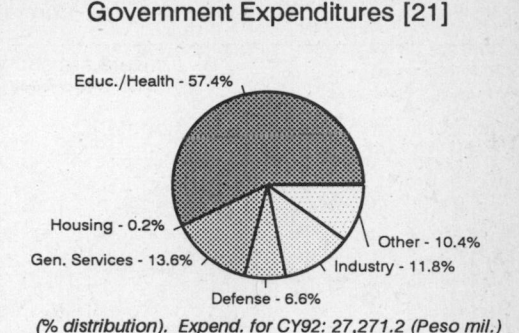

Educ./Health - 57.4%
Housing - 0.2%
Gen. Services - 13.6%
Defense - 6.6%
Industry - 11.8%
Other - 10.4%

(% distribution). Expend. for CY92: 27,271.2 (Peso mil.)

Military Expenditures and Arms Transfers [22]

	1990	1991	1992	1993	1994
Military expenditures					
Current dollars (mil.)	3,407	2,745	4,475	4,240	4,716
1994 constant dollars (mil.)	3,792	2,942	4,666	4,328	4,716
Armed forces (000)	85	70	65	65	69
Gross national product (GNP)					
Current dollars (mil.)	179,900	206,000	232,900	254,300	278,200
1994 constant dollars (mil.)	200,200	220,900	242,900	259,500	278,200
Central government expenditures (CGE)					
1994 constant dollars (mil.)	9,014	11,330	14,160	17,460	17,490
People (mil.)	32.4	32.8	33.2	33.5	33.9
Military expenditure as % of GNP	1.9	1.3	1.9	1.7	1.7
Military expenditure as % of CGE	42.1	26.0	33.0	24.8	27.0
Military expenditure per capita (1994 $)	117	90	141	129	139
Armed forces per 1,000 people (soldiers)	2.6	2.1	2.0	1.9	2.0
GNP per capita (1994 $)	6,181	6,739	7,327	7,739	8,205
Arms imports[6]					
Current dollars (mil.)	20	10	10	20	10
1994 constant dollars (mil.)	22	11	10	20	10
Arms exports[6]					
Current dollars (mil.)	20	5	5	10	0
1994 constant dollars (mil.)	22	5	5	10	0
Total imports[7]					
Current dollars (mil.)	4,076	8,275	14,870	16,780	21,530
1994 constant dollars (mil.)	4,536	8,870	15,510	17,130	21,530
Total exports[7]					
Current dollars (mil.)	12,350	11,980	12,230	13,120	15,660
1994 constant dollars (mil.)	13,750	12,840	12,760	13,390	15,660
Arms as percent of total imports[8]	0.5	0.1	0.1	0.1	0
Arms as percent of total exports[8]	0.2	0	0	0.1	0

Crime [23]

	1994
Crime volume	
Cases known to police	61,444
Attempts (percent)	0.04
Percent cases solved	12.21
Crimes per 100,000 persons	186.19
Persons responsible for offenses	
Total number offenders	7,506
Percent female	12.07
Percent juvenile (0-18 yrs.)	1.61
Percent foreigners	NA

Human Rights [24]

	SSTS	FL	FAPRO	PPCG	APROBC	TPW	PCPTW	STPEP	PHRFF	PRW	ASST	AFL
Observes		P	P	P	P	P	P	P		P	P	P
	EAFRD	CPR	ESCR	SR	ACHR	MAAE	PVIAC	PVNAC	EAFDAW	TCIDTP	RC
Observes	P	P	P	P	P		P	P	P	P	P

P = Party; S = Signatory; see Appendix for meaning of abbreviations.

Labor Force

Total Labor Force [25]

10.9 million

Labor Force by Occupation [26]

Agriculture	12%
Industry	31
Services	57

Date of data: 1985 est.

Unemployment Rate [27]

16% (1995 est.)

For sources, notes, and explanations, see Annotated Source Appendix, page 1061.

35

Production Sectors

Commercial Energy Production and Consumption

Data are shown in quadrillion (10^{15}) BTUs and percent for 1995
Values for hydroelectric, nuclear, geothermal, solar, and wind power refer to electrical generation.

Production [28]

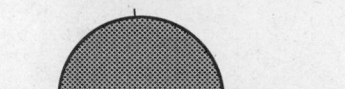

Crude oil - 52.4%
Coal - 0.1%
Nuclear - 2.8%
Hydro - 10.5%
NG liquids - 2.0%
Natural gas - 32.2%

Consumption [29]

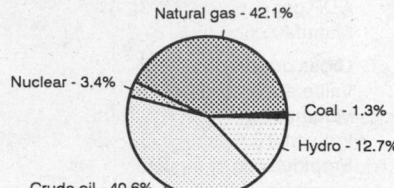

Natural gas - 42.1%
Nuclear - 3.4%
Coal - 1.3%
Hydro - 12.7%
Crude oil - 40.6%

Crude oil	1.564
Natural gas liquids	0.059
Dry natural gas	0.960
Coal	0.004
Net hydroelectric power	0.312
Net nuclear power	0.083
Total	2.982

Crude oil	0.998
Dry natural gas	1.035
Coal	0.032
Net hydroelectric power	0.312
Net nuclear power	0.083
Total	2.460

Telecommunications [30]

- 2.7 million (1983 est.) telephones; 12,000 public telephones; extensive modern system but many families do not have telephones; despite extensive use of microwave radio relay, the telephone system frequently grounds out during rainstorms, even in Buenos Aires
- Domestic: microwave radio relay and a domestic satellite system with 40 earth stations serve the trunk network
- International: satellite earth stations - 2 Intelsat (Atlantic Ocean)
- Radio: Broadcast stations: AM 171, FM 0, shortwave 13 Radios: 22.3 million (1991 est.)
- Television: Broadcast stations: 231 Televisions: 7.165 million (1991 est.)

Transportation [31]

Railways: total: 37,910 km; broad gauge: 24,124 km 1.676-m gauge (142 km electrified); standard gauge: 2,765 km 1.435-m gauge; narrow gauge: 11,021 km 1.000-m gauge (26 km electrified)

Highways: total: 215,578 km; paved: 61,440 km; unpaved: 154,138 km

Merchant marine: total: 37 ships (1,000 GRT or over) totaling 303,448 GRT/458,864 DWT; ships by type: bulk 1, cargo 11, chemical tanker 1, container 3, oil tanker 14, railcar carrier 1, refrigerated cargo 5, roll-on/roll-off cargo 1 (1995 est.)

Airports

Total:	1,253
With paved runways over 3,047 m:	5
With paved runways 2,438 to 3,047 m:	25
With paved runways 1,524 to 2,437 m:	54
With paved runways 914 to 1,523 m:	46
With paved runways under 914 m:	511

Top Agricultural Products [32]

Agriculture accounts for 6% of the GDP; produces wheat, corn, sorghum, soybeans, sugar beets; livestock.

Top Mining Products [33]

Metric tons except as noted[e]	7/95[*]
Aluminum, primary	165,000
Silver, metal, smelter (kg.)	108,000
Uranium, mine output, U308 content (kg.)	150,000
Boron materials, crude	140,000
Gypsum, crude	520,000
Dolomite	390,000
Water, mineral containing	130,000
Petroleum (000 42-gal. bls.)	
crude	237,000
refinery fuel and losses	133,000

Tourism [34]

	1990	1991	1992	1993	1994
Tourists[8]	2,728	2,870	3,031	3,532	3,866
Tourism receipts	1,976	2,336	3,090	3,614	3,970
Tourism expenditures	1,171	1,739	2,211	2,445	NA
Fare receipts[9]	389	375	373	398	430
Fare expenditures[9]	270	418	504	606	621

Travelers are in thousands, money in million U.S. dollars.

For sources, notes, and explanations, see Annotated Source Appendix, page 1061.

Manufacturing Sector

GDP and Manufacturing Summary [35]

	1980	1985	1989	1990	% change 1980-1990	% change 1989-1990
GDP (million 1980 $)	60,917	54,708	53,638	55,101	-9.5	2.7
GDP per capita (1980 $)	2,157	1,804	1,680	1,704	-21.0	1.4
Manufacturing as % of GDP (current prices)	25.0	27.3	19.2[e]	18.3[e]	-26.8	-4.7
Gross output (million $)	55,936[e]	48,084[e]	43,284[e]	79,001[e]	41.2	82.5
Value added (million $)	24,511	28,891	24,712[e]	31,156	27.1	26.1
Value added (million 1980 $)	15,224	12,506	12,136	11,586	-23.9	-4.5
Industrial production index	100	86	82	90	-10.0	9.8
Employment (thousands)	1,346[e]	1,127[e]	1,068[e]	942[e]	-30.0	-11.8

Note: GDP stands for Gross Domestic Product. 'e' stands for estimated value.

Profitability and Productivity

	1980	1985	1989	1990	% change 1980-1990	% change 1989-1990
Intermediate input (%)	56[e]	40[e]	43[e]	61[e]	8.9	41.9
Wages, salaries, and supplements (%)	10[e]	11[e]	9[e]	8[e]	-20.0	-11.1
Gross operating surplus (%)	33[e]	49[e]	48[e]	31[e]	-6.1	-35.4
Gross output per worker ($)	41,552[e]	42,656[e]	40,519[e]	83,878[e]	101.9	107.0
Value added per worker ($)	18,208[e]	25,630[e]	23,134[e]	33,080[e]	81.7	43.0
Average wage (incl. benefits) ($)	4,301[e]	4,596[e]	3,806[e]	6,767[e]	57.3	77.8

Profitability is in percent of gross output. Productivity is in U.S. $. 'e' stands for estimated value.

Profitability - 1990

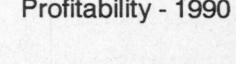
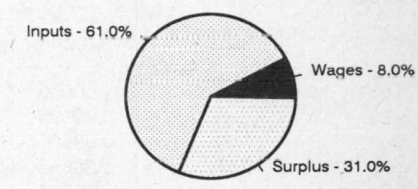

Inputs - 61.0%
Wages - 8.0%
Surplus - 31.0%

The graphic shows percent of gross output.

Value Added in Manufacturing

	1980 $ mil.	1980 %	1985 $ mil.	1985 %	1989 $ mil.	1989 %	1990 $ mil.	1990 %	% change 1980-1990	% change 1989-1990
311 Food products	3,544	14.5	4,912	17.0	4,256[e]	17.2	4,695	15.1	32.5	10.3
313 Beverages	703	2.9	942	3.3	1,033[e]	4.2	932	3.0	32.6	-9.8
314 Tobacco products	498	2.0	719	2.5	435[e]	1.8	480	1.5	-3.6	10.3
321 Textiles	1,703	6.9	1,832	6.3	1,960[e]	7.9	2,209	7.1	29.7	12.7
322 Wearing apparel	919	3.7	558	1.9	403[e]	1.6	492	1.6	-46.5	22.1
323 Leather and fur products	284	1.2	350	1.2	283[e]	1.1	336	1.1	18.3	18.7
324 Footwear	245	1.0	240	0.8	146[e]	0.6	190	0.6	-22.4	30.1
331 Wood and wood products	363	1.5	283	1.0	237[e]	1.0	255	0.8	-29.8	7.6
332 Furniture and fixtures	225	0.9	185	0.6	232[e]	0.9	246	0.8	9.3	6.0
341 Paper and paper products	554	2.3	763	2.6	636[e]	2.6	882	2.8	59.2	38.7
342 Printing and publishing	679	2.8	800	2.8	561[e]	2.3	695	2.2	2.4	23.9
351 Industrial chemicals	914	3.7	1,367	4.7	1,271[e]	5.1	1,844	5.9	101.8	45.1
352 Other chemical products	1,206	4.9	1,916	6.6	1,409[e]	5.7	1,791	5.7	48.5	27.1
353 Petroleum refineries	3,647	14.9	5,120	17.7	3,295[e]	13.3	6,069	19.5	66.4	84.2
354 Miscellaneous petroleum and coal products	86	0.4	121	0.4	153[e]	0.6	122	0.4	41.9	-20.3
355 Rubber products	331	1.4	327	1.1	272[e]	1.1	368	1.2	11.2	35.3
356 Plastic products	424	1.7	485	1.7	371[e]	1.5	436	1.4	2.8	17.5
361 Pottery, china, and earthenware	189	0.8	130	0.4	149[e]	0.6	156	0.5	-17.5	4.7
362 Glass and glass products	199	0.8	153	0.5	210[e]	0.8	249	0.8	25.1	18.6
369 Other nonmetal mineral products	659	2.7	587	2.0	740[e]	3.0	932	3.0	41.4	25.9
371 Iron and steel	900	3.7	1,239	4.3	1,523[e]	6.2	1,651	5.3	83.4	8.4
372 Nonferrous metals	235	1.0	257	0.9	229[e]	0.9	305	1.0	29.8	33.2
381 Metal products	1,272	5.2	1,499	5.2	1,593[e]	6.4	1,611	5.2	26.7	1.1
382 Nonelectrical machinery	1,358	5.5	930	3.2	640[e]	2.6	835	2.7	-38.5	30.5
383 Electrical machinery	902	3.7	936	3.2	914[e]	3.7	1,025	3.3	13.6	12.1
384 Transport equipment	2,289	9.3	2,054	7.1	1,586[e]	6.4	2,140	6.9	-6.5	34.9
385 Professional and scientific equipment	86	0.4	95	0.3	91[e]	0.4	112	0.4	30.2	23.1
390 Other manufacturing industries	96	0.4	92	0.3	84[e]	0.3	97	0.3	1.0	15.5

Note: The industry codes shown are International Standard Industry codes (ISIC). Percentages are percent of total Value Added. 'e' stands for estimated value

For sources, notes, and explanations, see Annotated Source Appendix, page 1061.

37

Finance, Economics, and Trade

Economic Indicators [36]

- **National product**: GDP—purchasing power parity— $278.5 billion (1995 est.)

- **National product real growth rate**: - 4.4%

- **National product per capita**: $8,100 (1995 est.)

- **Inflation rate (consumer prices)**: 1.7% (1995 est.)

- **External debt**: $90 billion (December 1995)

Balance of Payments Summary [37]

Values in millions of dollars.

	1989	1990	1991	1992	1993
Exports of goods (f.o.b.)	9,573	12,354	11,978	12,235	13,117
Imports of goods (f.o.b.)	-3,864	-3,726	-7,559	-13,685	-15,545
Trade balance	5,709	8,628	4,419	-1,450	-2,428
Services - debits	-9,491	-9,374	-10,013	-9,774	-9,628
Services - credits	2,469	4,300	4,154	3,929	4,158
Private transfers (net)	8	998	793	749	535
Government transfers (net)	NA	NA	NA	NA	-89
Long-term capital (net)	4,747	4,206	4,058	3,784	13,081
Short-term capital (net)	-4,466	-3,060	-462	7,196	-3,162
Errors and omissions	-249	715	-341	137	87
Overall balance	-1,273	3,413	2,608	4,571	2,554

Exchange Rates [38]

Currency: **nuevo peso.**
Symbol: **P.**

Data are currency units per $1.

January 1996	1.00000
1995	0.99975
1994	0.99901
1993	0.99895
1992	0.99064
1991	0.95355

Imports and Exports

Top Import Origins [39]

$19.5 billion (c.i.f., 1995).

Origins	%
US	21
Brazil	NA
Germany	NA
Bolivia	NA
Japan	NA
Italy	NA
Netherlands	NA

Top Export Destinations [40]

$20.7 billion (f.o.b., 1995).

Destinations	%
US	9
Brazil	NA
Italy	NA
Japan	NA
Netherlands	NA

Foreign Aid [41]

	U.S. $	
US commitments, including Ex-Im (FY70-89)	1	billion
Western (non-US) countries, ODA and OOF bilateral commitments (1970-89)	4.4	billion
Communist countries (1970-89)	718	million

Import and Export Commodities [42]

Import Commodities	**Export Commodities**
Machinery and equipment	Meat
Chemicals	Wheat
Metals	Corn
Fuels and lubricants	Oilseed
Agricultural products	Manufactures

Armenia

Geography [1]

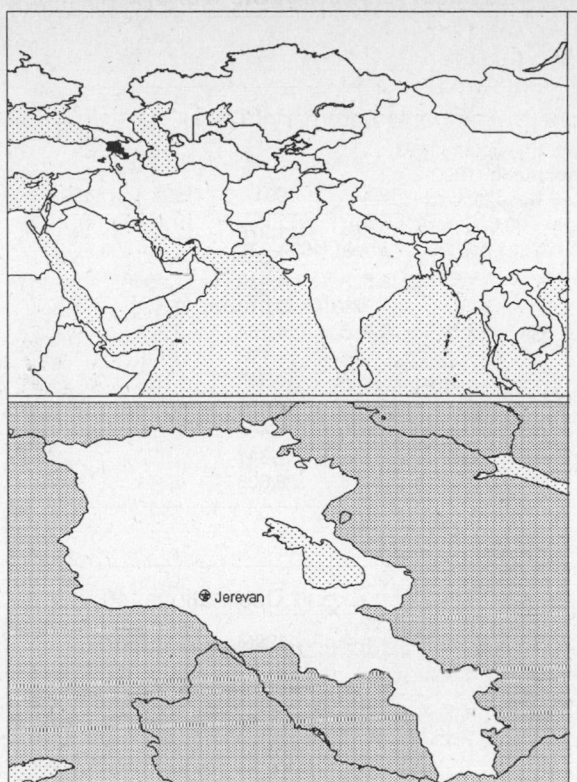

Total area:
  29,800 sq km   11,506 sq mi
Land area:
  28,400 sq km   10,965 sq mi
Comparative area:
  Slightly larger than Maryland
Land boundaries:
  Total 1,254 km, Azerbaijan-proper 566 km, Azerbaijan-Naxcivan exclave 221 km, Georgia 164 km, Iran 35 km, Turkey 268 km
Coastline:
  0 km (landlocked)
Climate:
  Highland continental, hot summers, cold winters
Terrain:
  High Armenian Plateau with mountains; little forest land; fast flowing rivers; good soil in Aras River valley
Natural resources:
  Small deposits of gold, copper, molybdenum, zinc, alumina
Land use:
  Arable land: 17%
  Permanent crops: 3%
  Meadows and pastures: 20%
  Forest and woodland: 0%
  Other: 60%

Demographics [2]

	1970	1980	1990	1995[1]	1996	2000	2010	2020	2030
Population	2,520	3,115	3,366	3,464	3,464	3,481	3,577	3,665	3,822
Population density (persons per sq. mi.)	219	271	293	301	301	303	311	319	332
(persons per sq. km.)	85	105	113	116	116	117	120	123	128
Net migration rate (per 1,000 population)	NA	NA	-0.6	-8.4	-8.3	-8.0	-4.7	-1.5	0.0
Births	NA	NA	NA	NA	56	NA	NA	NA	NA
Deaths	NA	NA	NA	NA	27	NA	NA	NA	NA
Life expectancy - males	NA	NA	67.2	64.3	64.4	65.0	66.5	70.2	73.2
Life expectancy - females	NA	NA	74.0	73.8	73.9	74.4	75.4	78.3	80.6
Birth rate (per 1,000)	NA	NA	22.4	15.6	16.3	18.3	16.6	13.4	12.9
Death rate (per 1,000)	NA	NA	6.9	7.6	7.7	8.2	9.3	8.7	8.6
Women of reproductive age (15-49 yrs.)	NA	NA	851	912	922	954	929	910	940
of which are currently married	NA	NA	579	NA	632	646	644	NA	NA
Fertility rate	NA	NA	2.6	2.0	2.1	2.3	2.0	1.9	1.9

Except as noted, values for vital statistics are in thousands; life expectancy is in years.

Health

Health Indicators [3]

% of population with access to	
safe water (1990-95)	NA
adequate sanitation (1990-95)	NA
health services (1985-95)	NA
% of 1-year-olds immunized (1990-94) against	
TB (tuberculosis)	83
DPT (diphtheria, pertussis, tetanus)	83
polio	92
measles	95
% of contraceptive prevalence (1980-94)	NA
ORT use rate (1990-94)	NA

Health Expenditures [4]

Total health expenditure, 1990 (official exchange rate)	
Millions of dollars	506
Dollars per capita	152
Health expenditures as a percentage of GDP	
Total	4.2
Public sector	2.5
Private sector	1.7
Development assistance for health	
Total aid flows (millions of dollars)[1]	NA
Aid flows per capita (dollars)	NA
Aid flows as a percentage of total health expenditure	NA

For sources, notes, and explanations, see Annotated Source Appendix, page 1061.

Human Factors

Women and Children [5]

% of pregnant women immunized (tetanus 1990-94)	NA
% of births attended by trained health personnel (1983-94)	NA
Maternal mortality rate (1980-92)	NA
Under-5 mortality rate (1994)	32
% under-5 moderately/severely underweight (1980-1994)	NA

Burden of Disease [6]

Population per physician (1993)	260.98
Population per nurse (1992)	100.74
Population per hospital bed (1993)	120.39
AIDS cases per 100,000 people (1994)	*
Heart disease cases per 1,000 people (1990-93)	499.5

Ethnic Division [7]

Armenian	93%
Azeri	3%
Russian	2%
Other (mostly Yezidi Kurds)	2%
(1989)	

Religion [8]

Armenian Orthodox	94%

Major Languages [9]

Armenian	96%
Russian	2%
Other	2%

Education

Public Education Expenditures [10]

	1980	1989	1990	1991	1993	1994
Million (Ruble)						
Total education expenditure	NA	618	4	NA	NA	NA
as percent of GNP	NA	NA	7.4	NA	NA	NA
as percent of total govt. expend.	NA	13.2	20.5	NA	NA	NA
Current education expenditure	NA	NA	NA	NA	136	NA
as percent of GNP	NA	NA	NA	NA	3.5	NA
as percent of current govt. expend.	NA	NA	NA	NA	NA	NA
Capital expenditure	NA	NA	NA	NA	NA	NA

Educational Attainment [11]

Illiteracy [12]

	1985	1989	1995
Illiterate population (15+ yrs.)[2]	NA	NA	9,000
Illiteracy rate - total pop. (%)	NA	1.2	0.4
Illiteracy rate - males (%)	NA	0.6	0.3
Illiteracy rate - females (%)	NA	1.9	0.5

Libraries [13]

	Admin. Units	Svc. Pts.	Vols. (000)	Shelving (meters)	Vols. Added	Reg. Users
National (1992)	2	NA	4,094	NA	24,727	33,940
Nonspecialized	NA	NA	NA	NA	NA	NA
Public (1992)	42	1,306	14,685	NA	151,487	1 mil
Higher ed.	NA	NA	NA	NA	NA	NA
School	NA	NA	NA	NA	NA	NA

Daily Newspapers [14]

	1980	1985	1990	1994
Number of papers	NA	NA	NA	7
Circ. (000)	NA	NA	NA	80[e]

Culture [15]

Cinema (seats per 1,000)	33.9
Annual attendance per person	NA
Gross box office receipts (mil. Ruble)	44
Museums (reporting)	54
Visitors (000)	300
Annual receipts (000 Ruble)	75,969

Science and Technology

Scientific/Technical Forces [16]

R&D Expenditures [17]

U.S. Patents Issued [18]

Values show patents issued to citizens of the country by the U.S. Patents Office.

	1993	1994	1995
Number of patents	0	0	1

For sources, notes, and explanations, see Annotated Source Appendix, page 1061.

Government and Law

Organization of Government [19]

Long-form name:
Republic of Armenia
Type:
Republic
Independence:
28 May 1918 (First Armenian Republic);
23 September 1991 (from Soviet Union)
National holiday:
Referendum Day, 21 September
Constitution:
Adopted by nationwide referendum 5 July
1995
Legal system:
Based on civil law system
Executive branch:
President; Prime Minister; First Deputy
Prime Minister; Council of Ministers
Legislative branch:
Unicameral: National Assembly
Judicial branch:
Supreme Court

Elections [20]

National Assembly	% of seats
Republican Block	83.7
Shamiram Women's Movement	4.2
Armenian Communist Party (ACP)	3.7
National Democratic Union (NDU)	2.6
Union of Self-Determination (NSDU)	1.6
Democratic Liberal Party (DLP)	0.5
Armenian Revolutionary Fed. (ARF)	0.5
Others	2.1
Vacant	1.1

Government Budget [21]

Military Expenditures and Arms Transfers [22]

	1990	1991	1992[14]	1993	1994
Military expenditures					
Current dollars (mil.)[e]	NA	NA	NA	NA	71
1994 constant dollars (mil.)[e]	NA	NA	NA	NA	71
Armed forces (000)	NA	NA	20	21	45
Gross national product (GNP)					
Current dollars (mil.)[e]	NA	NA	8,846	7,760	8,187
1994 constant dollars (mil.)[e]	NA	NA	9,225	7,920	8,187
Central government expenditures (CGE)					
1994 constant dollars (mil.)[e]	NA	NA	4,508	NA	NA
People (mil.)	NA	NA	3.4	3.5	3.5
Military expenditure as % of GNP	NA	NA	NA	NA	0.9
Military expenditure as % of CGE	NA	NA	NA	NA	NA
Military expenditure per capita (1994 $)	NA	NA	NA	NA	20
Armed forces per 1,000 people (soldiers)	NA	NA	5.8	6.0	12.8
GNP per capita (1994 $)	NA	NA	2,682	2,275	2,325
Arms imports[6]					
Current dollars (mil.)	NA	NA	0	0	10
1994 constant dollars (mil.)	NA	NA	0	0	10
Arms exports[6]					
Current dollars (mil.)	NA	NA	0	0	0
1994 constant dollars (mil.)	NA	NA	0	0	0
Total imports[7]					
Current dollars (mil.)	NA	NA	328	191	345
1994 constant dollars (mil.)	NA	NA	342	195	345
Total exports[7]					
Current dollars (mil.)	NA	NA	256	57	196
1994 constant dollars (mil.)	NA	NA	267	58	196
Arms as percent of total imports[8]	NA	NA	0	0	2.9
Arms as percent of total exports[8]	NA	NA	0	0	0

Crime [23]

	1994
Crime volume	
Cases known to police	6,022
Attempts (percent)	3.05
Percent cases solved	73.00
Crimes per 100,000 persons	160.44
Persons responsible for offenses	
Total number offenders	8,150
Percent female	7.01
Percent juvenile (14-18 yrs.)	5.74
Percent foreigners	0.15

Human Rights [24]

	SSTS	FL	FAPRO	PPCG	APROBC	TPW	PCPTW	STPEP	PHRFF	PRW	ASST	AFL
Observes				P								

	EAFRD	CPR	ESCR	SR	ACHR	MAAE	PVIAC	PVNAC	EAFDAW	TCIDTP	RC
Observes	P	P	P	P			P	P	P	P	P

P = Party; S = Signatory; see Appendix for meaning of abbreviations.

Labor Force

Total Labor Force [25]

1.012 million

Labor Force by Occupation [26]

Industry and construction	46%
Agriculture	2
Transportation and communication	7
Other	45

Date of data: 1992

Unemployment Rate [27]

8% officially registered unemployed, but large
numbers of underemployed (December 1995)

For sources, notes, and explanations, see Annotated Source Appendix, page 1061.

41

Production Sectors

Commercial Energy Production and Consumption

Data are shown in quadrillion (10^{15}) BTUs and percent for 1995
Values for hydroelectric, nuclear, geothermal, solar, and wind power refer to electrical generation.

Production [28]

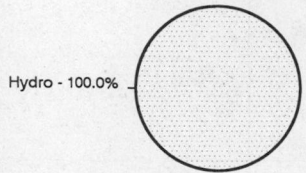

Hydro - 100.0%

Consumption [29]

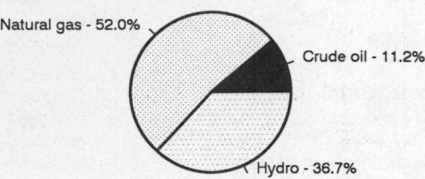

Natural gas - 52.0%
Crude oil - 11.2%
Hydro - 36.7%

Net hydroelectric power	0.036
Total	0.036

Crude oil	0.011
Dry natural gas	0.051
Net hydroelectric power	0.036
Total	0.098

Telecommunications [30]

- 650,000 telephones; joint venture agreement to install fiber-optic cable and construct facilities for cellular telephone service remains in the negotiation phase
- International: international connections to other former Soviet republics are by landline or microwave radio relay and to other countries by satellite and by leased connection through the Moscow international gateway switch; satellite earth station - 1 Intelsat
- Radio: Broadcast stations: AM 10, FM 3, shortwave NA (1991)
- Television: Broadcast stations: 1

Transportation [31]

Railways: total: 825 km in common carrier service; does not include industrial lines; broad gauge: 825 km 1.520-m gauge (1992)

Highways: total: 11,300 km; paved: 10,500 km (including graveled); unpaved: 800 km (1990 est.)

Airports

Total:	11
With paved runways over 3,047 m:	2
With paved runways 1,524 to 2,437 m:	1
With paved runways 914 to 1,523 m:	2
With unpaved runways 1,524 to 2,437 m:	2
With unpaved runways 914 to 1,523 m:	3

Top Agricultural Products [32]

Agriculture accounts for 57% of the GDP; produces fruit (especially grapes), vegetables; vineyards near Yerevan are famous for brandy and other liqueurs; minor livestock sector.

Top Mining Products [33]

Metric tons except as noted	6/23/95[*]
Bentonite	100,000
Cement	200,000
Copper ore, gross weight, 1% Cu	50,000
Gold (kg.)	500
Limestone	500,000
Molybdenum, mine output, Mo content	500
Perlite	10,000
Salt	50,000

Tourism [34]

For sources, notes, and explanations, see Annotated Source Appendix, page 1061.

Finance, Economics, and Trade

Industrial Summary [35]

Industrial Production: Growth rate 2.4% (1995 est.); accounts for 36% of the GDP. **Industries**: Much of industry is shut down; metal-cutting machine tools, forging-pressing machines, electric motors, tires, knitted wear, hosiery, shoes, silk fabric, washing machines, chemicals, trucks, watches, instruments, microelectronics.

Economic Indicators [36]

- **National product**: GDP—purchasing power parity—$9.1 billion (1995 estimate as extrapolated GDP—from World Bank estimate for 1994)

- **National product real growth rate**: 5.2% (1995 est.)

- **National product per capita**: $2,560 (1995 est.)

- **Inflation rate (consumer prices)**: 32.2% (1995 est.)

- **External debt**: $850 million (of which $75 million to Russia) (1995 est.)

Balance of Payments Summary [37]

Values in millions of dollars.

	1989	1990	1991	1992	1993[1]
Exports of goods (f.o.b.)	NA	NA	NA	NA	156
Imports of goods (f.o.b.)	NA	NA	NA	NA	-254
Trade balance	NA	NA	NA	NA	-98
Services - debits	NA	NA	NA	NA	-41
Services - credits	NA	NA	NA	NA	17
Private transfers (net)	NA	NA	NA	NA	36
Government transfers (net)	NA	NA	NA	NA	24
Long-term capital (net)	NA	NA	NA	NA	48
Short-term capital (net)	NA	NA	NA	NA	-28
Errors and omissions	NA	NA	NA	NA	59
Overall balance	NA	NA	NA	NA	17

Exchange Rates [38]

Currency: **dram.**

Data are currency units per $1. Introduced new currency in November 1993.

End December 1995	401.8
End December 1994	406

Imports and Exports

Top Import Origins [39]

$661 million (c.i.f., 1995).

Origins	%
Iran	NA
Russia	NA
Turkmenistan	NA
Georgia	NA
US	NA
EU	NA

Top Export Destinations [40]

$248 million (f.o.b., 1995).

Destinations	%
Iran	NA
Russia	NA
Turkmenistan	NA
Georgia	NA

Foreign Aid [41]

	U.S. $	
ODA (1993)	30	million
Commitments excluding Russia (1992-95)	1,385	million

Import and Export Commodities [42]

Import Commodities	Export Commodities
Grain	Gold and jewelry
Other foods	Aluminum
Fuel	Transport equipment
Other energy	Electrical equipment
	Scrap metal

For sources, notes, and explanations, see Annotated Source Appendix, page 1061.

43

Australia

Geography [1]

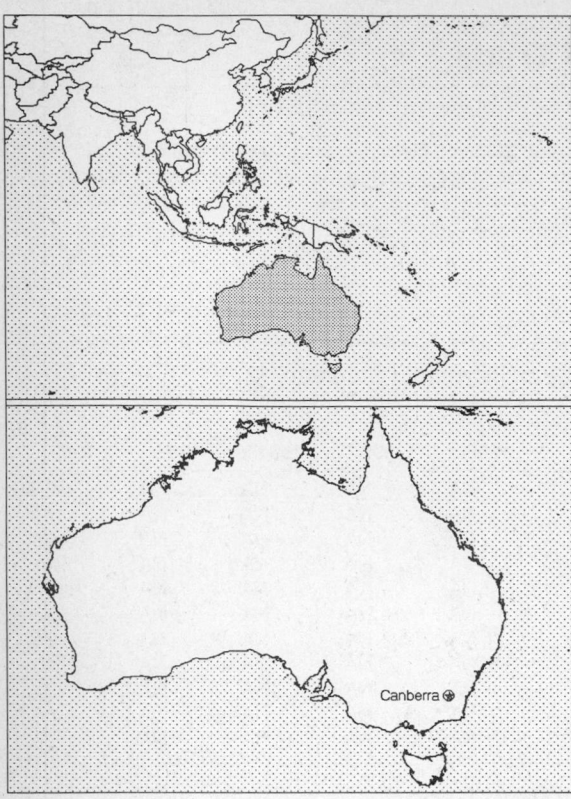

Total area:
7,686,850 sq km   2,967,909 sq mi
Land area:
7,617,930 sq km   2,941,299 sq mi
Comparative area:
Slightly smaller than the US
Note: Includes Macquarie Island
Land boundaries:
0 km
Coastline:
25,760 km
Climate:
Generally arid to semiarid; temperate in South and East; tropical in North
Terrain:
Mostly low plateau with deserts; fertile plain in Southeast
Natural resources:
Bauxite, coal, iron ore, copper, tin, silver, uranium, nickel, tungsten, mineral sands, lead, zinc, diamonds, natural gas, petroleum
Land use:
Arable land: 6%
Permanent crops: 0%
Meadows and pastures: 58%
Forest and woodland: 14%
Other: 22%

Demographics [2]

	1970	1980	1990	1995[1]	1996	2000	2010	2020	2030
Population	12,660	14,616	17,033	18,079	18,261	18,950	20,434	21,696	22,541
Population density (persons per sq. mi.)	4	5	6	6	6	6	7	7	8
(persons per sq. km.)	2	2	2	2	2	2	3	3	3
Net migration rate (per 1,000 population)	9.7	6.4	5.7	2.8	2.7	2.6	2.5	2.3	2.2
Births[6]	NA	NA	NA	NA	255	NA	NA	NA	NA
Deaths	NA	NA	NA	NA	126	NA	NA	NA	NA
Life expectancy - males	NA	71.0	74.2	76.2	76.4	77.5	79.2	80.1	80.5
Life expectancy - females	NA	78.1	80.8	82.3	82.5	83.5	85.1	86.0	86.5
Birth rate (per 1,000)	20.3	15.4	15.4	14.2	14.0	13.0	11.7	11.1	10.2
Death rate (per 1,000)	8.9	7.4	7.1	6.9	6.9	6.9	7.5	8.3	10.0
Women of reproductive age (15-49 yrs.)	NA	NA	4,474	4,702	4,724	4,774	4,853	4,797	4,757
of which are currently married	NA	NA	2,560	NA	2,808	2,860	2,875	NA	NA
Fertility rate	NA	1.9	1.9	1.9	1.8	1.8	1.8	1.8	1.7

Except as noted, values for vital statistics are in thousands; life expectancy is in years.

Health

Health Indicators [3]

% of population with access to	
safe water (1990-95)	NA
adequate sanitation (1990-95)	NA
health services (1985-95)	NA
% of 1-year-olds immunized (1990-94) against	
TB (tuberculosis)	NA
DPT (diphtheria, pertussis, tetanus)	95
polio	72
measles	86
% of contraceptive prevalence (1980-94)	76
ORT use rate (1990-94)	NA

Health Expenditures [4]

Total health expenditure, 1990 (official exchange rate)	
Millions of dollars	22,736
Dollars per capita	1,331
Health expenditures as a percentage of GDP	
Total	7.7
Public sector	5.4
Private sector	2.3
Development assistance for health	
Total aid flows (millions of dollars)[1]	NA
Aid flows per capita (dollars)	NA
Aid flows as a percentage of total health expenditure	NA

For sources, notes, and explanations, see Annotated Source Appendix, page 1061.

Human Factors

Women and Children [5]

% of pregnant women immunized (tetanus 1990-94)	NA
% of births attended by trained health personnel (1983-94)[1]	99
Maternal mortality rate (1980-92)	3
Under-5 mortality rate (1994)	8
% under-5 moderately/severely underweight (1980-1994)	NA

Burden of Disease [6]

Population per physician (1986)	437.53
Population per nurse (1986)	114.88
Population per hospital bed (1990)	183.28
AIDS cases per 100,000 people (1994)	4.6
Heart disease cases per 1,000 people (1990-93)	356

Ethnic Division [7]

Caucasian	95%
Asian	4%
Aboriginal and other	1%

Religion [8]

Anglican	26.1%
Roman Catholic	26%
Other Christian	24.3%
Other	23.6%

Major Languages [9]

English and native languages.

Education

Public Education Expenditures [10]

Million (Dollar)	1980	1985	1990	1991	1992	1994
Total education expenditure	7,592	12,925	19,364	20,417	23,304	NA
as percent of GNP	5.5	5.6	5.4	5.5	6.0	NA
as percent of total govt. expend.	14.8	12.8	14.8	14.1	14.2	NA
Current education expenditure	6,899	11,848	17,889	18,983	21,252	NA
as percent of GNP	5.0	5.1	4.9	5.1	5.4	NA
as percent of current govt. expend.	16.8	14.2	14.8	14.6	NA	NA
Capital expenditure	693	1,077	1,475	1,434	2,052	NA

Educational Attainment [11]

Illiteracy [12]

Libraries [13]

	Admin. Units	Svc. Pts.	Vols. (000)	Shelving (meters)	Vols. Added	Reg. Users
National (1991)[1]	1	NA	4,625	NA	NA	NA
Nonspecialized	NA	NA	NA	NA	NA	NA
Public (1987)	497	1,804	27,000	NA	118,800	NA
Higher ed. (1994)	43	231	33,000	NA	2 mil[e]	550,000[e]
School	NA	NA	NA	NA	NA	NA

Daily Newspapers [14]

	1980	1985	1990	1994
Number of papers	62	62	62	69
Circ. (000)	4,700[e]	4,300[e]	4,200[e]	4,600[e]

Culture [15]

Cinema (seats per 1,000)	16.9
Annual attendance per person	3.0
Gross box office receipts (mil. Dollar)	369
Museums (reporting)	NA
Visitors (000)	17,960
Annual receipts (000 Dollar)	NA

Science and Technology

Scientific/Technical Forces [16]

Scientists/engineers	41,837
Number female	NA
Technicians	15,922
Number female	NA
Total	57,759

R&D Expenditures [17]

	Dollar (000) 1990
Total expenditure	5,087,600
Capital expenditure	698,800
Current expenditure	4,388,800
Percent current	86.3

U.S. Patents Issued [18]

Values show patents issued to citizens of the country by the U.S. Patents Office.

	1993	1994	1995
Number of patents	470	564	548

For sources, notes, and explanations, see Annotated Source Appendix, page 1061.

Government and Law

Organization of Government [19]

Long-form name:
Commonwealth of Australia
Type:
Federal parliamentary state
Independence:
1 January 1901 (federation of UK colonies)
National holiday:
Australia Day, 26 January (1788)
Constitution:
9 July 1900, effective 1 January 1901
Legal system:
Based on English common law; accepts compulsory ICJ jurisdiction, with reservations
Executive branch:
British Monarch (represented by Governor General); Prime Minister; Deputy Prime Minister; Cabinet
Legislative branch:
Bicameral Federal Parliament: Senate and House of Representatives
Judicial branch:
High Court

Elections [20]

House of Representatives	% of seats
Liberal-National	63.5
Labor	33.1
Independent	3.4

Government Expenditures [21]

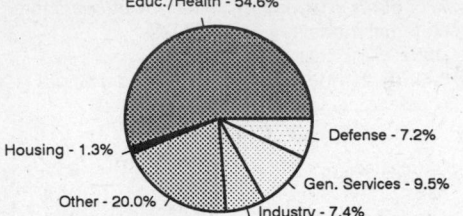

(% distribution). Expend. for FY95: 129.8 (Dollar bil.)

Crime [23]

	1990
Crime volume	
Cases known to police	1,149,254
Attempts (percent)	NA
Percent cases solved	NA
Crimes per 100,000 persons	6,747.1
Persons responsible for offenses	
Total number offenders	NA
Percent female	NA
Percent juvenile	NA
Percent foreigners	NA

Military Expenditures and Arms Transfers [22]

	1990	1991	1992	1993	1994
Military expenditures					
Current dollars (mil.)	5,845	6,687	7,034	7,926	8,270
1994 constant dollars (mil.)	6,505	7,167	7,335	8,090	8,270
Armed forces (000)	68	68	68	68	59
Gross national product (GNP)					
Current dollars (mil.)	249,500	261,700	279,400	297,800	320,400
1994 constant dollars (mil.)	277,700	280,500	291,400	304,000	320,400
Central government expenditures (CGE)					
1994 constant dollars (mil.)	70,750	76,040	80,700	85,050	88,020
People (mil.)	17.1	17.3	17.6	17.8	18.1
Military expenditure as % of GNP	2.3	2.6	2.5	2.7	2.6
Military expenditure as % of CGE	9.2	9.4	9.1	9.5	9.4
Military expenditure per capita (1994 $)	381	414	417	454	457
Armed forces per 1,000 people (soldiers)	4.0	3.9	3.9	3.8	3.3
GNP per capita (1994 $)	16,270	16,190	16,580	17,050	17,720
Arms imports[6]					
Current dollars (mil.)	600	290	230	330	430
1994 constant dollars (mil.)	668	311	240	337	430
Arms exports[6]					
Current dollars (mil.)	60	40	60	40	20
1994 constant dollars (mil.)	67	43	63	41	20
Total imports[7]					
Current dollars (mil.)	41,290	41,690	43,810	45,580	53,430
1994 constant dollars (mil.)	45,950	44,690	45,680	46,520	53,430
Total exports[7]					
Current dollars (mil.)	39,750	41,850	42,820	42,720	47,570
1994 constant dollars (mil.)	44,240	44,860	44,660	43,600	47,570
Arms as percent of total imports[8]	1.5	0.7	0.5	0.7	0.8
Arms as percent of total exports[8]	0.2	0.1	0.1	0.1	0

Human Rights [24]

	SSTS	FL	FAPRO	PPCG	APROBC	TPW	PCPTW	STPEP	PHRFF	PRW	ASST	AFL
Observes	P	P	P	P	P	P	P			P	P	P
	EAFRD	CPR	ESCR	SR	ACHR	MAAE	PVIAC	PVNAC	EAFDAW	TCIDTP	RC	
Observes	P	P	P	P			P	P	P	P	P	

P = Party; S = Signatory; see Appendix for meaning of abbreviations.

Labor Force

Total Labor Force [25]

8.63 million (September 1991)

Labor Force by Occupation [26]

Finance and services	33.8%
Public and community services	22.3
Wholesale and retail trade	20.1
Manufacturing and industry	16.2
Agriculture	6.1

Date of data: 1987

Unemployment Rate [27]

8.1% (December 1995)

For sources, notes, and explanations, see Annotated Source Appendix, page 1061.

Production Sectors

Commercial Energy Production and Consumption

Data are shown in quadrillion (10^{15}) BTUs and percent for 1995
Values for hydroelectric, nuclear, geothermal, solar, and wind power refer to electrical generation.

Production [28]

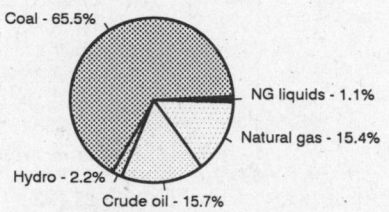

Coal - 65.5%
NG liquids - 1.1%
Natural gas - 15.4%
Hydro - 2.2%
Crude oil - 15.7%

Consumption [29]

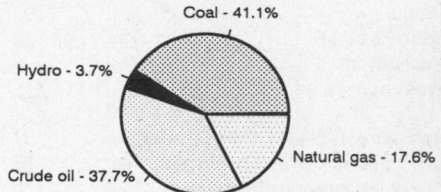

Coal - 41.1%
Hydro - 3.7%
Natural gas - 17.6%
Crude oil - 37.7%

Crude oil	1.144
Natural gas liquids	0.081
Dry natural gas	1.126
Coal	4.776
Net hydroelectric power	0.164
Total	7.291

Crude oil	1.669
Dry natural gas	0.778
Coal	1.821
Net hydroelectric power	0.164
Total	4.432

Telecommunications [30]

- 8.7 million (1987 est.) telephones; good domestic and international service
- Domestic: domestic satellite system
- International: submarine cables to New Zealand, Papua New Guinea, and Indonesia; satellite earth stations - 10 Intelsat (4 Indian Ocean and 6 Pacific Ocean), 2 Inmarsat (Indian and Pacific Ocean Regions)
- Radio: Broadcast stations: AM 258, FM 67, shortwave 0
- Television: Broadcast stations: 134 (1987 est.) Televisions: 9.2 million (1992 est.)

Transportation [31]

Railways: total: 38,563 km (2,914 km electrified; 172 km dual gauge); broad gauge: 6,083 km 1.600-m gauge; standard gauge: 16,752 km 1.435-m gauge; narrow gauge: 15,728 km 1.067-m gauge

Highways: total: 810,264 km; paved: 283,592 km (including 1,200 km of expressways); unpaved: 526,672 km (1989 est.)

Merchant marine: total: 76 ships (1,000 GRT or over) totaling 2,547,869 GRT/3,679,534 DWT; ships by type: bulk 30, cargo 4, chemical tanker 3, combination bulk 1, container 6, liquefied gas tanker 6, oil tanker 18, roll-on/roll-off cargo 7, short-sea passenger 1 (1995 est.)

Airports

Total:	442
With paved runways over 3,047 m:	9
With paved runways 2,438 to 3,047 m:	13
With paved runways 1,524 to 2,437 m:	106
With paved runways 914 to 1,523 m:	116
With paved runways under 914 m:	30

Top Agricultural Products [32]

Agriculture accounts for 3.1% of the GDP; produces wheat, barley, sugarcane, fruits; cattle, sheep, poultry.

Top Mining Products [33]

Metric tons except as noted[e]	8/4/95*
Gold (kg.)[r]	
mine output, Au content	256,000
refined, primary	303,000
Ferroalloys	200,000[5]
Rutile, gross weight	233,000[r]
Lime	1,500,000
Magnesite	286,000[r]
Nitrogen, N content of ammonia	400,000
Talc, chlorite, pyrophyllite, steatite	215,000
Petroleum refinery products (000 42-gal. bls.)	250,000

Tourism [34]

	1990	1991	1992	1993	1994
Visitors[10]	2,215	2,370	2,603	2,996	3,362
Tourism receipts	4,088	4,484	4,405	4,655	5,955
Tourism expenditures	4,535	4,247	4,301	4,100	4,339
Fare receipts	1,188	1,290	1,425	1,629	1,822
Fare expenditures	1,798	1,821	1,837	1,643	1,800

Travelers are in thousands, money in million U.S. dollars.

For sources, notes, and explanations, see Annotated Source Appendix, page 1061.

Manufacturing Sector

Manufacturing Summary [35]

	1987		1988		1989		1990		1991	
	$ bil.	%	$ bil.	%	$ bil.	%	$ bil.	%	$ bil.	%
Establishments or enterprises (number)	30,974	1.320	33,968	1.618	33,687	1.798	32,791	1.831	43,760	5.725
Total employment (000)	1,157	0.853	1,200	0.877	1,216	0.988	1,160	1.048	1,092	1.572
Production workers (000)	845	1.253	-	-	-	-	-	-	-	-
Output ($ bil.)	95	0.937	123	1.051	141	1.191	151	1.315	155	1.524
Value added ($ bil.)	37	0.860	-	-	-	-	63	1.264	-	-
Capital investment ($ mil.)	-	-	-	-	-	-	6,215	1.118	-	-
M & E investment ($ mil.)	-	-	-	-	-	-	4,673	1.108	-	-
Employees per establishment (number)	37	64.567	35	54.188	36	54.936	35	57.253	25	27.459
Production workers per establishment	27	94.886	-	-	-	-	-	-	-	-
Output per establishment ($ mil.)	3	70.999	4	64.948	4	66.249	5	71.844	4	26.616
Capital investment per estab. ($ mil.)	-	-	-	-	-	-	0.190	61.090	-	-
M & E per establishment ($ mil)	-	-	-	-	-	-	0.143	60.552	-	-
Payroll per employee ($)	15,849	176.749	19,058	191.294	20,338	182.044	22,332	160.991	24,421	167.259
Wages per production worker ($)	15,392	195.062	-	-	-	-	-	-	-	-
Hours per production worker (hours)	-	-	-	-	-	-	-	-	-	-
Output per employee ($)	82,356	109.962	102,088	119.856	115,575	120.593	130,045	125.485	142,302	96.931
Capital investment per employee ($)	-	-	-	-	-	-	5,358	106.702	-	-
M & E per employee ($)	-	-	-	-	-	-	4,028	105.762	-	-

Note: Columns headed % show percent of world total or ratio. Ratios closest to 100 are closest to world average. M & E stands for machinery & equipment.

Output in Manufacturing

	1987		1988		1989		1990		1991	
	$ bil.[f]	%	$ bil.[f]	%	$ bil.[f]	%	$ bil.[f]	%	$ bil.[f]	%
3110 - Food products	14.000	14.74	18.000	14.63	20.000	14.18	21.000	13.91	22.000	14.19
3130 - Beverages	2.487	2.62	3.160	2.57	3.723	2.64	3.741	2.48	4.310	2.78
3140 - Tobacco	0.459	0.48	0.532	0.43	0.563	0.40	0.630	0.42	0.634	0.41
3210 - Textiles	3.064	3.23	3.812	3.10	4.080	2.89	3.947	2.61	3.892	2.51
3211 - Spinning, weaving, etc.	1.448	1.52	1.847	1.50	1.896	1.34	1.760	1.17	1.707	1.10
3220 - Wearing apparel	1.790	1.88	2.315	1.88	2.648	1.88	2.609	1.73	2.605	1.68
3230 - Leather and products	0.406	0.43	0.518	0.42	0.473	0.34	0.429	0.28	0.425	0.27
3240 - Footwear	0.510	0.54	0.587	0.48	0.617	0.44	0.551	0.36	0.529	0.34
3310 - Wood products	2.694	2.84	3.480	2.83	4.095	2.90	3.907	2.59	3.795	2.45
3320 - Furniture, fixtures	1.412	1.49	2.014	1.64	2.164	1.53	2.317	1.53	2.304	1.49
3410 - Paper and products	2.518	2.65	3.127	2.54	3.574	2.53	3.268	2.16	3.811	2.46
3411 - Pulp, paper, etc.	1.138	1.20	1.389	1.13	1.717	1.22	1.525	1.01	1.736	1.12
3420 - Printing, publishing	4.772	5.02	6.298	5.12	7.019	4.98	7.281	4.82	7.714	4.98
3510 - Industrial chemicals	3.385	3.56	4.459	3.63	4.713	3.34	4.593	3.04	4.783	3.09
3511 - Basic chemicals, excl fertilizers	1.276	1.34	1.686	1.37	1.842	1.31	1.923	1.27	2.049	1.32
3513 - Synthetic resins, etc.	1.114	1.17	1.415	1.15	1.503	1.07	1.459	0.97	1.507	0.97
3520 - Chemical products nec	3.203	3.37	4.182	3.40	4.685	3.32	4.878	3.23	5.657	3.65
3522 - Drugs and medicines	0.923	0.97	1.259	1.02	1.417	1.00	1.438	0.95	1.794	1.16
3530 - Petroleum refineries	1.216	1.28	1.330	1.08	1.465	1.04	6.264	4.15	7.722	4.98
3540 - Petroleum, coal products	0.102	0.11	0.127	0.10	0.112	0.08	0.101	0.07	0.135	0.09
3550 - Rubber products	0.662	0.70	0.832	0.68	0.987	0.70	1.031	0.68	0.989	0.64
3560 - Plastic products nec	2.549	2.68	3.450	2.80	4.034	2.86	3.856	2.55	4.146	2.67
3610 - Pottery, china, etc.	0.090	0.09	0.099	0.08	0.115	0.08	0.118	0.08	0.134	0.09
3620 - Glass and products	0.555	0.58	0.692	0.56	0.835	0.59	0.869	0.58	0.802	0.52
3690 - Nonmetal products nec	3.145	3.31	3.930	3.20	4.810	3.41	5.125	3.39	4.919	3.17
3710 - Iron and steel	4.728	4.98	5.891	4.79	7.121	5.05	7.374	4.88	7.293	4.71
3720 - Nonferrous metals	5.121	5.39	6.937	5.64	8.260	5.86	9.463	6.27	9.416	6.07
3810 - Metal products	5.932	6.24	7.931	6.45	9.501	6.74	9.840	6.52	9.615	6.20
3820 - Machinery nec	4.290	4.52	5.367	4.36	6.426	4.56	6.632	4.39	6.989	4.51
3825 - Office, computing machinery	1.217	1.28	1.535	1.25	1.940	1.38	1.988	1.32	2.298	1.48
3830 - Electrical machinery	3.526	3.71	4.537	3.69	5.288	3.75	5.226	3.46	5.528	3.57
3832 - Radio, television, etc.	0.189	0.20	0.242	0.20	0.297	0.21	0.245	0.16	0.276	0.18
3840 - Transportation equipment	7.495	7.89	9.617	7.82	11.000	7.80	12.000	7.95	11.000	7.10
3841 - Shipbuilding, repair	0.553	0.58	0.862	0.70	1.052	0.75	1.735	1.15	1.275	0.82
3843 - Motor vehicles	5.567	5.86	7.331	5.96	8.530	6.05	9.171	6.07	8.847	5.71
3850 - Professional goods	0.663	0.70	0.840	0.68	0.910	0.65	0.980	0.65	0.954	0.62
3900 - Industries nec	0.606	0.64	0.794	0.65	0.885	0.63	0.924	0.61	1.001	0.65

Note: Codes are International Standard Industry codes (ISIC). Percentages are % of total Output. [f]: Factor Prices; [p]: Producer Prices.

For sources, notes, and explanations, see Annotated Source Appendix, page 1061.

Finance, Economics, and Trade

Economic Indicators [36]

- **National product**: GDP—purchasing power parity—$405.4 billion (1995 est.)
- **National product real growth rate**: 3.3% (1995 est.)
- **National product per capita**: $22,100 (1995 est.)
- **Inflation rate (consumer prices)**: 4.75% (1995)
- **External debt**: $147.2 billion (1994)

Balance of Payments Summary [37]

Values in millions of dollars.

	1989	1990	1991	1992	1993
Exports of goods (f.o.b.)	36,883	39,332	42,005	42,375	42,240
Imports of goods (f.o.b.)	-40,329	-38,964	-38,491	-40,820	-42,363
Trade balance	-3,446	368	3,514	1,555	-123
Services - debits	-28,403	-31,145	-29,737	-28,353	-26,392
Services - credits	12,222	14,121	14,583	15,089	15,822
Private transfers (net)	2,124	1,965	2,075	1,524	738
Government transfers (net)	-173	-158	-245	-361	-414
Long-term capital (net)	15,182	14,348	11,235	9,882	8,583
Short-term capital (net)	837	-526	1,366	149	308
Errors and omissions	2,276	2,754	-3,107	-4,223	1,423
Overall balance	629	1,727	-316	-4,738	-55

Exchange Rates [38]

Currency: **Australian dollar.**
Symbol: **$A.**

Data are currency units per $1.

January 1996	1.3477
1995	1.3486
1994	1.3668
1993	1.4704
1992	1.3600
1991	1.2835

Imports and Exports

Top Import Origins [39]

$57.41 billion (f.o.b., 1995) Data are for 1992.

Origins	%
US	23
Japan	18
UK	6
Germany	5.7
New Zealand	4

Top Export Destinations [40]

$51.57 billion (f.o.b., 1995) Data are for 1992.

Destinations	%
Japan	25
US	11
South Korea	6
New Zealand	5.7
UK	NA
Taiwan	NA
Singapore	NA
Hong Kong	NA

Foreign Aid [41]

Donor: ODA, $953 million (1993).

Import and Export Commodities [42]

Import Commodities	Export Commodities
Machinery and transport equipment	Coal
Computers and office machines	Gold
Crude oil and petroleum products	Meat
	Wool
	Alumina
	Wheat
	Machinery and transport equipment

Austria

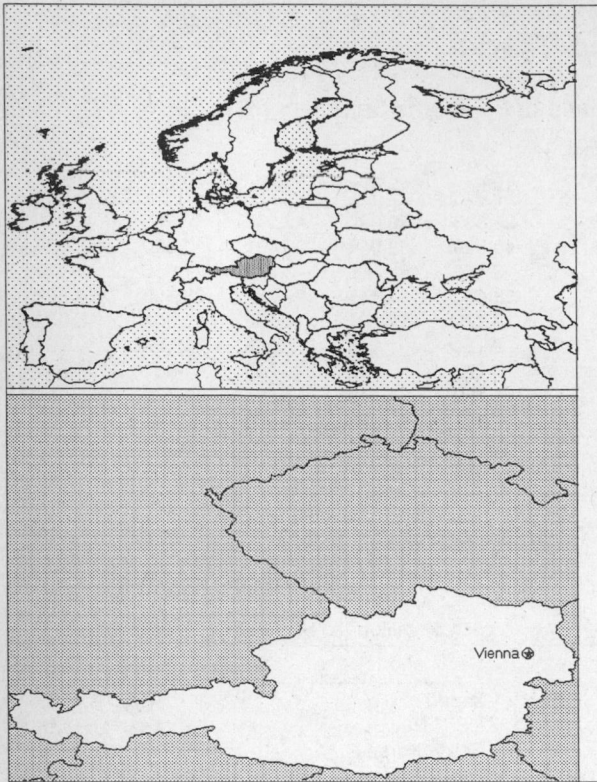

Geography [1]

Total area:
83,850 sq km   32,375 sq mi
Land area:
82,730 sq km   31,942 sq mi
Comparative area:
Slightly smaller than Maine
Land boundaries:
Total 2,558 km, Czech Republic 362 km, Germany 784 km, Hungary 366 km, Italy 430 km, Liechtenstein 37 km, Slovakia 91 km, Slovenia 324 km, Switzerland 164 km
Coastline:
0 km (landlocked)
Climate:
Temperate; continental, cloudy; cold winters with frequent rain in lowlands and snow in mountains; cool summers with occasional showers
Terrain:
In the West and South mostly mountains (Alps); along the eastern and northern margins mostly flat or gently sloping
Natural resources:
Iron ore, oil, timber, magnesite, lead, coal, lignite, copper, hydropower
Land use:
Arable land: 17%
Permanent crops: 1%
Meadows and pastures: 24%
Forest and woodland: 39%
Other: 19%

Demographics [2]

	1970	1980	1990	1995[1]	1996	2000	2010	2020	2030
Population	7,467	7,549	7,718	7,988	8,023	8,124	8,223	8,262	8,070
Population density (persons per sq. mi.)	234	236	242	250	251	254	257	259	253
(persons per sq. km.)	90	91	93	97	97	98	99	100	98
Net migration rate (per 1,000 population)	NA	NA	16.0	3.7	3.3	2.1	2.1	2.1	0.0
Births	112	91	NA	NA	88	NA	NA	NA	NA
Deaths	99	92	83	NA	82	NA	NA	NA	NA
Life expectancy - males	NA	NA	72.4	73.2	73.4	74.2	75.8	78.0	79.5
Life expectancy - females	NA	NA	79.1	79.7	79.8	80.5	81.9	84.0	85.5
Birth rate (per 1,000)	NA	NA	11.7	11.4	11.2	10.3	9.0	8.8	7.8
Death rate (per 1,000)	NA	NA	10.8	10.5	10.4	10.3	10.6	10.6	11.9
Women of reproductive age (15-49 yrs.)	NA	NA	1,966	1,980	1,990	1,985	1,926	1,703	1,567
of which are currently married	NA	1,058	1,220	NA	1,289	1,292	1,246	NA	NA
Fertility rate	NA	NA	1.5	1.5	1.5	1.5	1.5	1.5	1.5

Except as noted, values for vital statistics are in thousands; life expectancy is in years.

Health

Health Indicators [3]

% of population with access to	
safe water (1990-95)	NA
adequate sanitation (1990-95)	NA
health services (1985-95)	NA
% of 1-year-olds immunized (1990-94) against	
TB (tuberculosis)	NA
DPT (diphtheria, pertussis, tetanus)	90
polio	90
measles	60
% of contraceptive prevalence (1980-94)	71
ORT use rate (1990-94)	NA

Health Expenditures [4]

Total health expenditure, 1990 (official exchange rate)	
Millions of dollars	13,193
Dollars per capita	1,711
Health expenditures as a percentage of GDP	
Total	8.3
Public sector	5.5
Private sector	2.8
Development assistance for health	
Total aid flows (millions of dollars)[1]	NA
Aid flows per capita (dollars)	NA
Aid flows as a percentage of total health expenditure	NA

Human Factors

Women and Children [5]

% of pregnant women immunized (tetanus 1990-94)	NA
% of births attended by trained health personnel (1983-94)	NA
Maternal mortality rate (1980-92)	8
Under-5 mortality rate (1994)	7
% under-5 moderately/severely underweight (1980-1994)	NA

Burden of Disease [6]

Population per physician (1990)	231.03
Population per nurse (1985)	184.09
Population per hospital bed (1990)	94.54
AIDS cases per 100,000 people (1994)	2.1
Heart disease cases per 1,000 people (1990-93)	391

Ethnic Division [7]

German	99.4%
Croatian	0.3%
Slovene	0.2%
Other	0.1%

Religion [8]

Roman Catholic	85%
Protestant	6%
Other	9%

Major Languages [9]

German.

Education

Public Education Expenditures [10]

	1980	1990	1991	1992	1993	1994
Million (Schilling)						
Total education expenditure	55,016	97,301	107,329	117,519	115,780	NA
as percent of GNP	5.6	5.4	5.6	5.8	5.5	NA
as percent of total govt. expend.	8.0	7.6	7.6	7.7	NA	NA
Current education expenditure	46,955	89,858	98,054	103,693	102,207	NA
as percent of GNP	4.8	5.0	5.1	5.1	4.8	NA
as percent of current govt. expend.	8.4	8.8	8.6	8.6	NA	NA
Capital expenditure	8,061	7,443	9,275	13,826	13,573	NA

Educational Attainment [11]

Age group (1981)	25 +
Total population	4,558,681
Highest level attained (%)	
No schooling	NA
First level	
Not completed	49.3
Completed	NA
Entered second level	
S-1	NA
S-2	47.5
Postsecondary	3.3

Illiteracy [12]

Libraries [13]

	Admin. Units	Svc. Pts.	Vols. (000)	Shelving (meters)	Vols. Added	Reg. Users
National (1992)	1	1	3,069	NA	40,748	360,656
Nonspecialized (1992)	7	9	1,562	NA	37,335	84,351
Public (1992)	2,129	2,129	9,202	NA	NA	960,125
Higher ed. (1993)[2]	21	NA	16,953	NA	459,477	4 mil
School	NA	NA	NA	NA	NA	NA

Daily Newspapers [14]

	1980	1985	1990	1994
Number of papers	30	33	25	23
Circ. (000)	2,651	2,729	2,706	3,736

Culture [15]

Cinema (seats per 1,000)	9.0
Annual attendance per person	1.5
Gross box office receipts (mil. Schilling)	809
Museums (reporting)	685
Visitors (000)	18,277
Annual receipts (000 Schilling)	NA

Science and Technology

Scientific/Technical Forces [16]

Scientists/engineers	12,820
Number female	NA
Technicians	6,397
Number female	NA
Total	19,217

R&D Expenditures [17]

	Schilling (000) 1993
Total expenditure	30,692,586
Capital expenditure	4,524,986
Current expenditure	26,167,600
Percent current	85.3

U.S. Patents Issued [18]

Values show patents issued to citizens of the country by the U.S. Patents Office.

	1993	1994	1995
Number of patents	341	316	360

For sources, notes, and explanations, see Annotated Source Appendix, page 1061.

Government and Law

Organization of Government [19]

Long-form name:
Republic of Austria
Type:
Federal republic
Independence:
12 November 1918 (from Austro-Hungarian Empire)
National holiday:
National Day, 26 October (1955)
Constitution:
1920; revised 1929 (reinstated 1 May 1945)
Legal system:
Civil law system with Roman law origin; judicial review of legislative acts by the Constitutional Court
Executive branch:
President; Chancellor; Vice Chancellor; Council of Ministers
Legislative branch:
Bicameral Federal Assembly (Bundesversammlung): Federal Council (Bundesrat) and National Council (Nationalrat)
Judicial branch:
Supreme Judicial Court for civil and criminal cases; Administrative Court for bureaucratic cases; Constitutional Court for constitutional cases

Elections [20]

National Council	% of votes
Social Democratic Party (SPOE)	38.3
Austrian People's Party (OEVP)	28.3
Freedom Party of Austria (FPOE)	22.1
Liberal Forum (LF)	5.3
The Greens	4.6
Others	1.4

Government Expenditures [21]

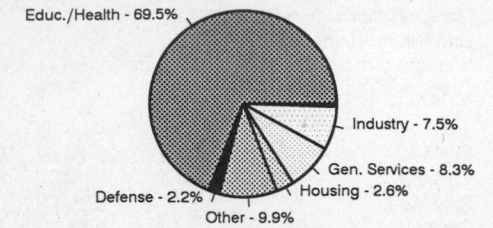

Educ./Health - 69.5%
Industry - 7.5%
Gen. Services - 8.3%
Housing - 2.6%
Defense - 2.2%
Other - 9.9%

(% distribution). Expend. for CY94 est.: 915.90 (Schilling bil.)

Crime [23]

	1994
Crime volume	
Cases known to police	504,568
Attempts (percent)	4.70
Percent cases solved	49.60
Crimes per 100,000 persons	6,313.82
Persons responsible for offenses	
Total number offenders	201,757
Percent female	19.10
Percent juvenile (14-19 yrs.)	12.40
Percent foreigners	20.80

Military Expenditures and Arms Transfers [22]

	1990	1991	1992	1993	1994
Military expenditures					
Current dollars (mil.)	1,750	1,771	1,747	1,782	1,863
1994 constant dollars (mil.)	1,948	1,898	1,822	1,818	1,863
Armed forces (000)	43	44	52	NA	45
Gross national product (GNP)					
Current dollars (mil.)	162,900	173,600	182,800	186,700	195,700
1994 constant dollars (mil.)	181,300	186,100	190,600	190,600	195,700
Central government expenditures (CGE)					
1994 constant dollars (mil.)	72,020	75,300	77,650	80,280	NA
People (mil.)	7.7	7.8	7.9	7.9	8.0
Military expenditure as % of GNP	1.1	1.0	1.0	1.0	1.0
Military expenditure as % of CGE	2.7	2.5	2.3	2.3	NA
Military expenditure per capita (1994 $)	252	243	232	230	234
Armed forces per 1,000 people (soldiers)	5.6	5.6	6.6	NA	5.7
GNP per capita (1994 $)	23,490	23,820	24,220	24,070	24,600
Arms imports[6]					
Current dollars (mil.)	40	50	50	5	10
1994 constant dollars (mil.)	45	54	52	5	10
Arms exports[6]					
Current dollars (mil.)	160	130	110	50	50
1994 constant dollars (mil.)	178	139	115	51	50
Total imports[7]					
Current dollars (mil.)	49,150	50,810	54,110	48,580	55,340
1994 constant dollars (mil.)	54,700	54,470	56,430	49,580	55,340
Total exports[7]					
Current dollars (mil.)	41,260	41,110	47,270	40,170	45,210
1994 constant dollars (mil.)	45,920	44,070	49,290	41,000	45,210
Arms as percent of total imports[8]	0.1	0.1	0.1	0	0
Arms as percent of total exports[8]	0.4	0.3	0.2	0.1	0.1

Human Rights [24]

	SSTS	FL	FAPRO	PPCG	APROBC	TPW	PCPTW	STPEP	PHRFF	PRW	ASST	AFL
Observes	P	P	P	P	P	P	P		P	P	P	P
		EAFRD	CPR	ESCR	SR	ACHR	MAAE	PVIAC	PVNAC	EAFDAW	TCIDTP	RC
Observes		P	P	P	P			P	P	P	P	P

P=Party; S=Signatory; see Appendix for meaning of abbreviations.

Labor Force

Total Labor Force [25]

3.47 million (1989)

Labor Force by Occupation [26]

Services
Industry and crafts
Agriculture and forestry
An estimated 200,000 Austrians are employed in other European countries.
Date of data: 1988

Unemployment Rate [27]

4.6% (1995 est.)

Production Sectors

Commercial Energy Production and Consumption

Data are shown in quadrillion (10^{15}) BTUs and percent for 1995
Values for hydroelectric, nuclear, geothermal, solar, and wind power refer to electrical generation.

Production [28]

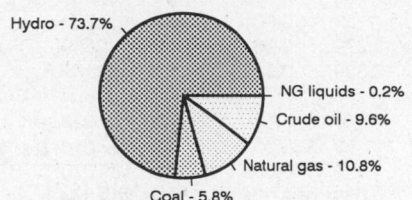

Hydro - 73.7%
NG liquids - 0.2%
Crude oil - 9.6%
Natural gas - 10.8%
Coal - 5.8%

Consumption [29]

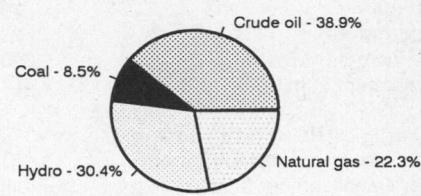

Crude oil - 38.9%
Coal - 8.5%
Natural gas - 22.3%
Hydro - 30.4%

Crude oil	0.050
Natural gas liquids	0.001
Dry natural gas	0.056
Coal	0.030
Net hydroelectric power	0.383
Total	0.520

Crude oil	0.490
Dry natural gas	0.281
Coal	0.107
Net hydroelectric power	0.383
Total	1.261

Telecommunications [30]

- 3.47 million (1986 est.) telephones
- Domestic: highly developed and efficient
- International: satellite earth stations - 2 Intelsat (1 Atlantic Ocean and 1 Indian Ocean) and 2 Eutelsat
- Radio: Broadcast stations: AM 6, FM 21 (repeaters 545), shortwave 0
- Television: Broadcast stations: 47 (repeaters 870) Televisions: 2,418,584 (1984 est.)

Top Agricultural Products [32]

Agriculture accounts for 2% of the GDP; produces grains, fruit, potatoes, sugar beets; cattle, pigs, poultry; sawn wood.

Top Mining Products [33]

Metric tons except as noted	5/95[*]
Aluminum, secondary	46,800
Copper	
smelter, secondary	51,600
refined	50,500
Talc and soapstone	131,000
Stone (000 tons)	61,800[6]
Sand and gravel (000 tons)	64,500
Manganese, Mn content of domestic ore	33,000[e]
Petroleum refinery products (000 42-gal. bls.)	65,600[e]
Lead, metal, smelter	17,200

Transportation [31]

Railways: total: 5,624 km; standard gauge: 5,269 km 1.435-m gauge (3,263 km electrified); narrow gauge: 355 km 1.000-m and 0.760-m gauge (86 km electrified) (1995)

Highways: total: 108,000 km; paved: 22,000 km (including 1,800 km of expressways); unpaved: 86,000 km (1992 est.)

Merchant marine: total: 29 ships (1,000 GRT or over) totaling 88,617 GRT/122,475 DWT; ships by type: bulk 1, cargo 23, combination bulk 2, container 1, refrigerated cargo 2 (1995 est.)

Airports

Total:	55
With paved runways over 3,047 m:	1
With paved runways 2,438 to 3,047 m:	5
With paved runways 1,524 to 2,437 m:	1
With paved runways 914 to 1,523 m:	3
With paved runways under 914 m:	41

Tourism [34]

	1990	1991	1992	1993	1994
Tourists[11]	19,011	19,092	19,098	18,257	17,894
Tourism receipts	13,410	13,800	14,526	13,566	13,160
Tourism expenditures	7,723	7,392	8,393	8,180	9,330

Travelers are in thousands, money in million U.S. dollars.

For sources, notes, and explanations, see Annotated Source Appendix, page 1061.

53

Manufacturing Sector

Manufacturing Summary [35]

	1987		1988		1989		1990		1991	
	$ bil.	%	$ bil.	%	$ bil.	%	$ bil.	%	$ bil.	%
Establishments or enterprises (number)	9,762	0.416	9,712	0.463	9,598	0.512	9,725	0.543	-	-
Total employment (000)	716	0.527	715	0.522	728	0.592	739	0.667	-	-
Production workers (000)	498	0.739	496	0.793	504	0.685	511	0.719	-	-
Output ($ bil.)	76	0.751	83	0.710	85	0.720	107	0.930	-	-
Value added ($ bil.)	27	0.618	29	0.609	29	0.601	37	0.732	-	-
Capital investment ($ mil.)	5,393	1.237	5,361	1.126	5,279	0.964	7,552	1.359	-	-
M & E investment ($ mil.)	3,462	1.111	3,614	1.044	3,589	0.919	5,011	1.189	-	-
Employees per establishment (number)	73	126.726	74	112.925	76	115.451	76	122.935	-	-
Production workers per establishment	51	177.540	51	171.458	52	133.651	53	132.514	-	-
Output per establishment ($ mil.)	8	180.370	9	153.495	9	140.609	11	171.220	-	-
Capital investment per estab. ($ mil.)	0.552	297.335	0.552	243.337	0.550	188.098	0.777	250.300	-	-
M & E per establishment ($ mil)	0.355	266.980	0.372	225.663	0.374	179.384	0.515	218.959	-	-
Payroll per employee ($)	21,171	236.096	22,125	222.085	21,282	190.493	26,584	191.639	-	-
Wages per production worker ($)	17,382	220.284	18,298	216.544	17,347	172.973	21,644	182.250	-	-
Hours per production worker (hours)	3,514	185.182	3,333	173.420	3,193	174.178	3,228	172.560	-	-
Output per employee ($)	106,599	142.331	115,775	135.926	116,723	121.790	144,338	139.277	-	-
Capital investment per employee ($)	7,535	234.628	7,498	215.485	7,251	162.924	10,224	203.604	-	-
M & E per employee ($)	4,837	210.676	5,054	199.834	4,929	155.377	6,784	178.110	-	-

Note: Columns headed % show percent of world total or ratio. Ratios closest to 100 are closest to world average. M & E stands for machinery & equipment.

Output in Manufacturing

	1987		1988		1989		1990		1991	
	$ bil.	%	$ bil.	%	$ bil.	%	$ bil.	%	$ bil.	%
3110 - Food products	8.301	10.92	8.652	10.42	8.303	9.77	10.000	9.35	-	-
3130 - Beverages	1.732	2.28	1.829	2.20	1.744	2.05	2.287	2.14	-	-
3140 - Tobacco	1.464	1.93	1.488	1.79	1.401	1.65	1.671	1.56	-	-
3210 - Textiles	2.813	3.70	2.959	3.57	2.890	3.40	3.597	3.36	-	-
3211 - Spinning, weaving, etc.	1.550	2.04	1.679	2.02	1.652	1.94	1.988	1.86	-	-
3220 - Wearing apparel	1.220	1.61	1.200	1.45	1.126	1.32	1.390	1.30	-	-
3230 - Leather and products	0.244	0.32	0.247	0.30	0.229	0.27	0.308	0.29	-	-
3240 - Footwear	0.656	0.86	0.621	0.75	0.578	0.68	0.704	0.66	-	-
3310 - Wood products	1.936	2.55	2.161	2.60	2.354	2.77	3.219	3.01	-	-
3320 - Furniture, fixtures	1.864	2.45	2.060	2.48	2.114	2.49	2.480	2.32	-	-
3410 - Paper and products	3.052	4.02	3.379	4.07	3.458	4.07	4.178	3.90	-	-
3411 - Pulp, paper, etc.	2.356	3.10	2.624	3.16	2.673	3.14	3.157	2.95	-	-
3420 - Printing, publishing	1.990	2.62	2.070	2.49	2.295	2.70	2.832	2.65	-	-
3510 - Industrial chemicals	3.354	4.41	4.112	4.95	4.014	4.72	4.617	4.31	-	-
3513 - Synthetic resins, etc.	1.496	1.97	2.218	2.67	2.166	2.55	2.612	2.44	-	-
3520 - Chemical products nec	2.537	3.34	2.800	3.37	2.801	3.30	3.386	3.16	-	-
3522 - Drugs and medicines	1.047	1.38	1.210	1.46	1.225	1.44	1.495	1.40	-	-
3530 - Petroleum refineries	2.644	3.48	2.178	2.62	2.425	2.85	3.175	2.97	-	-
3540 - Petroleum, coal products	0.141	0.19	0.181	0.22	0.169	0.20	0.193	0.18	-	-
3550 - Rubber products	0.784	1.03	0.877	1.06	0.835	0.98	0.941	0.88	-	-
3560 - Plastic products nec	1.120	1.47	1.156	1.39	1.149	1.35	1.495	1.40	-	-
3610 - Pottery, china, etc.	0.147	0.19	0.185	0.22	0.169	0.20	0.202	0.19	-	-
3620 - Glass and products	0.755	0.99	0.736	0.89	0.788	0.93	0.976	0.91	-	-
3690 - Nonmetal products nec	2.902	3.82	3.166	3.81	3.086	3.63	3.782	3.53	-	-
3710 - Iron and steel	3.797	5.00	4.134	4.98	4.387	5.16	4.837	4.52	-	-
3720 - Nonferrous metals	1.380	1.82	1.715	2.07	2.093	2.46	2.559	2.39	-	-
3810 - Metal products	3.960	5.21	4.593	5.53	4.672	5.50	5.919	5.53	-	-
3820 - Machinery nec	6.283	8.27	6.342	7.64	6.636	7.81	9.120	8.52	-	-
3830 - Electrical machinery	7.051	9.28	7.368	8.88	8.187	9.63	11.000	10.28	-	-
3832 - Radio, television, etc.	1.599	2.10	1.959	2.36	2.144	2.52	2.858	2.67	-	-
3840 - Transportation equipment	2.806	3.69	3.196	3.85	3.398	4.00	4.749	4.44	-	-
3841 - Shipbuilding, repair	0.114	0.15	0.115	0.14	0.079	0.09	0.132	0.12	-	-
3843 - Motor vehicles	2.217	2.92	2.562	3.09	2.780	3.27	3.914	3.66	-	-
3850 - Professional goods	0.414	0.54	0.431	0.52	0.390	0.46	0.457	0.43	-	-
3900 - Industries nec	0.568	0.75	0.578	0.70	0.574	0.68	0.616	0.58	-	-

Note: Codes are International Standard Industry codes (ISIC). Percentages are % of total Output. [f]: Factor Prices; [p]: Producer Prices.

Finance, Economics, and Trade

Economic Indicators [36]

- **National product**: GDP—purchasing power parity—$152 billion (1995 est.)

- **National product real growth rate**: 2.4% (1995 est.)

- **National product per capita**: $19,000 (1995 est.)

- **Inflation rate (consumer prices)**: 2.3% (1995 est.)

- **External debt**: $28.7 billion (1995 est.)

Balance of Payments Summary [37]

Values in millions of dollars.

	1989	1990	1991	1992	1993
Exports of goods (f.o.b.)	31,901	40,336	40,285	43,386	39,257
Imports of goods (f.o.b.)	-37,482	-47,348	-48,882	-52,228	-47,082
Trade balance	-5,581	-7,012	-8,597	-8,842	-7,825
Services - debits	-19,280	-24,319	-26,383	-29,582	-30,593
Services - credits	25,225	32,502	35,172	38,661	38,464
Private transfers (net)	-57	110	26	-742	-684
Government transfers (net)	-71	-108	-102	-198	-236
Long-term capital (net)	510	-967	-2,092	-549	6,691
Short-term capital (net)	857	949	2,080	1,592	-2,832
Errors and omissions	-613	-1,170	731	2,253	-782
Overall balance	990	-15	835	-2,593	2,203

Exchange Rates [38]

Currency: **Austrian schilling.**
Symbol: **S.**

Data are currency units per $1.

January 1996	10.314
1995	10.081
1994	11.422
1993	11.632
1992	10.989
1991	11.676

Imports and Exports

Top Import Origins [39]

$55.3 billion (1994) Data are for 1994.

Origins	%
EU	68.4
Germany	40
Italy	8.8
Eastern Europe	6.5
Japan	4.3
US	4.4

Top Export Destinations [40]

$45.2 billion (1994) Data are for 1994.

Destinations	%
EU	64.8
Germany	38.1
Italy	8.1
Eastern Europe	11.8
Japan	1.6
US	3.5

Foreign Aid [41]

Donor: ODA, $544 million (1993).

Import and Export Commodities [42]

Import Commodities	**Export Commodities**
Petroleum	Machinery and equipment
Foodstuffs	Iron and steel
Machinery and equipment	Lumber
Vehicles	Textiles
Chemicals	Paper products
Textiles and clothing	Chemicals
Pharmaceuticals	

Azerbaijan

Geography [1]

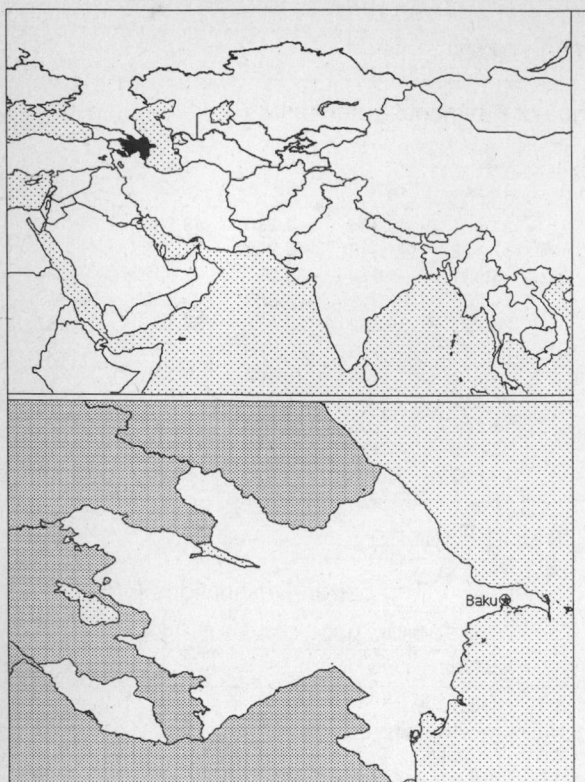

Total area:
  86,600 sq km   33,436 sq mi
Land area:
  86,100 sq km   33,243 sq mi
Comparative area:
  Slightly larger than Maine
  Note: Includes Naxcivan Autonomous Republic and Nagorno-Karabakh region
Land boundaries:
  Total 2,013 km, Armenia (Azerbaijan-proper) 566 km, Armenia (Azerbaijan-Naxcivan
  exclave) 221 km, Georgia 322 km, Iran (Azerbaijan-proper) 432 km, Iran
  (Azerbaijan-Naxcivan exclave) 179 km, Russia 284 km, Turkey 9 km
Coastline:
  0 km (landlocked)
  Note: Azerbaijan borders the Caspian Sea (800 km, est.)
Climate:
  Dry, semiarid steppe
Terrain:
  Large, flat Kur-Araz Lowland (much of it below sea level) with Great
  Caucasus Mountains to the North, Qarabag (Karabakh) Upland in West;
  Baku lies on Abseron (Apsheron) Peninsula that juts into Caspian Sea
Natural resources:
  Petroleum, natural gas, iron ore, nonferrous metals, alumina
Land use:
  Arable land: 18%
  Permanent crops: 4%
  Meadows and pastures: 25%
  Forest and woodland: 0%
  Other: 53%

Demographics [2]

	1970	1980	1990	1995[1]	1996	2000	2010	2020	2030
Population	5,169	6,173	7,200	7,616	7,677	7,902	8,410	9,007	9,610
Population density (persons per sq. mi.)	155	185	215	228	230	236	252	269	287
(persons per sq. km.)	60	71	83	88	89	91	97	104	111
Net migration rate (per 1,000 population)	NA	NA	-7.5	-5.8	-5.8	-5.8	-3.3	-1.0	0.0
Births	NA	NA	NA	NA	171	NA	NA	NA	NA
Deaths	NA	NA	NA	NA	67	NA	NA	NA	NA
Life expectancy - males	NA	NA	64.3	60.0	60.1	60.7	62.2	67.1	71.2
Life expectancy - females	NA	NA	72.8	69.7	69.8	70.2	71.3	75.4	78.7
Birth rate (per 1,000)	NA	NA	27.2	22.7	22.3	21.1	18.5	15.9	13.5
Death rate (per 1,000)	NA	NA	7.3	8.7	8.7	8.6	8.9	8.0	7.8
Women of reproductive age (15-49 yrs.)	NA	NA	1,824	1,987	2,023	2,145	2,334	2,347	2,453
of which are currently married	NA	NA	1,122	NA	1,294	1,369	1,479	NA	NA
Fertility rate	NA	NA	2.9	2.6	2.6	2.6	2.2	2.0	1.9

Except as noted, values for vital statistics are in thousands; life expectancy is in years.

Health

Health Indicators [3]

% of population with access to	
safe water (1990-95)	NA
adequate sanitation (1990-95)	NA
health services (1985-95)	NA
% of 1-year-olds immunized (1990-94) against	
TB (tuberculosis)	50
DPT (diphtheria, pertussis, tetanus)	90
polio	94
measles	91
% of contraceptive prevalence (1980-94)	NA
ORT use rate (1990-94)	NA

Health Expenditures [4]

Total health expenditure, 1990 (official exchange rate)	
Millions of dollars	785
Dollars per capita	98
Health expenditures as a percentage of GDP	
Total	4.3
Public sector	2.6
Private sector	1.7
Development assistance for health	
Total aid flows (millions of dollars)[1]	NA
Aid flows per capita (dollars)	NA
Aid flows as a percentage of total health expenditure	NA

For sources, notes, and explanations, see Annotated Source Appendix, page 1061.

Human Factors

Women and Children [5]

% of pregnant women immunized (tetanus 1990-94)	NA
% of births attended by trained health personnel (1983-94)	NA
Maternal mortality rate (1980-92)	NA
Under-5 mortality rate (1994)	51
% under-5 moderately/severely underweight (1980-1994)	NA

Burden of Disease [6]

Population per physician (1993)	257.08
Population per nurse (1993)	105.62
Population per hospital bed (1993)	96.28
AIDS cases per 100,000 people (1994)	*
Heart disease cases per 1,000 people (1990-93)	NA

Ethnic Division [7]

Azeri	90%
Dagestani peoples	3.2%
Russian	2.5%
Armenian	2.3%
Other	2%
(1995 est.)	

Religion [8]

Religious affiliation is still nominal in Azerbaijan; practicing adherents are few.

Muslim	93.4%
Russian Orthodox	2.5%
Armenian Orthodox	2.3%
Other	1.8%
(1995 est.)	

Major Languages [9]

Azeri	89%
Russian	3%
Armenian	2%
Other	6%
(1995 est.)	

Education

Public Education Expenditures [10]

	1980	1989	1990	1991	1992	1994
Million (Ruble)						
Total education expenditure	NA	NA	1,132	2,066	16,205	917,980
as percent of GNP	NA	NA	7.7	7.7	6.5	5.5
as percent of total govt. expend.	NA	NA	24.2	24.7	NA	13.7
Current education expenditure	NA	NA	NA	NA	15,704	NA
as percent of GNP	NA	NA	NA	NA	6.3	NA
as percent of current govt. expend.	NA	NA	NA	NA	NA	NA
Capital expenditure	NA	NA	NA	NA	501	NA

Educational Attainment [11]

Illiteracy [12]

	1985	1989	1995
Illiterate population (15+ yrs.)[2]	NA	NA	18,000
Illiteracy rate - total pop. (%)	NA	2.7	0.4
Illiteracy rate - males (%)	NA	1.1	0.3
Illiteracy rate - females (%)	NA	4.1	0.5

Libraries [13]

	Admin. Units	Svc. Pts.	Vols. (000)	Shelving (meters)	Vols. Added	Reg. Users
National (1992)	1	13	2,360	NA	14,508	25,028
Nonspecialized (1992)	1	14	1,725	NA	13,124	16,160
Public (1992)	4,650	NA	40,087	NA	1 mil	3 mil
Higher ed.	NA	NA	NA	NA	NA	NA
School	NA	NA	NA	NA	NA	NA

Daily Newspapers [14]

	1980	1985	1990	1994
Number of papers	NA	NA	NA	3
Circ. (000)	NA	NA	NA	210[e]

Culture [15]

Cinema (seats per 1,000)	47.1
Annual attendance per person	0.3
Gross box office receipts (mil. Rubles)	NA
Museums (reporting)	115
Visitors (000)	2,290
Annual receipts (000 Rubles)	57,447

Science and Technology

Scientific/Technical Forces [16]

R&D Expenditures [17]

U.S. Patents Issued [18]

For sources, notes, and explanations, see Annotated Source Appendix, page 1061.

57

Government and Law

Organization of Government [19]

Long-form name:
  Azerbaijani Republic
Type:
  Republic
Independence:
  30 August 1991 (from Soviet Union)
National holiday:
  Independence Day, 28 May
Constitution:
  Adopted 12 November 1995
Legal system:
  Based on civil law system
Executive branch:
  President; Prime Minister ; 4 First Deputy
  Prime Ministers; Council of Ministers
Legislative branch:
  Unicameral: National Assembly (Milli
  Mejlis)
Judicial branch:
  Supreme Court

Elections [20]

Elections for the unicameral National
Assembly consisting of 125 members
were held on the 12th and 26th of
November 1995. The results by party
either for percent of vote or seats are not
available.

Government Budget [21]

For 1995 est.

	$ mil.
Revenues	465
Expenditures	488
Capital expenditures	NA

Military Expenditures and Arms Transfers [22]

	1990	1991	1992[14]	1993	1994
Military expenditures					
Current dollars (mil.)[e]	NA	NA	208	540	132
1994 constant dollars (mil.)[e]	NA	NA	217	551	132
Armed forces (000)	NA	NA	43	63	50
Gross national product (GNP)					
Current dollars (mil.)[e]	NA	NA	18,380	16,580	12,850
1994 constant dollars (mil.)[e]	NA	NA	19,170	16,920	12,850
Central government expenditures (CGE)					
1994 constant dollars (mil.)[e]	NA	NA	5,770	NA	NA
People (mil.)	NA	NA	7.5	7.6	7.7
Military expenditure as % of GNP	NA	NA	1.1	3.3	1.0
Military expenditure as % of CGE	NA	NA	3.8	NA	NA
Military expenditure per capita (1994 $)	NA	NA	29	73	17
Armed forces per 1,000 people (soldiers)	NA	NA	5.7	8.3	6.5
GNP per capita (1994 $)	NA	NA	2,571	2,234	1,672
Arms imports[6]					
Current dollars (mil.)	NA	NA	0	10	60
1994 constant dollars (mil.)	NA	NA	0	10	60
Arms exports[6]					
Current dollars (mil.)	NA	NA	0	0	0
1994 constant dollars (mil.)	NA	NA	0	0	0
Total imports[7]					
Current dollars (mil.)	NA	NA	875	490	781
1994 constant dollars (mil.)	NA	NA	913	500	781
Total exports[7]					
Current dollars (mil.)	NA	NA	1,284	626	621
1994 constant dollars (mil.)	NA	NA	1,338	639	621
Arms as percent of total imports[8]	NA	NA	0	2.0	7.7
Arms as percent of total exports[8]	NA	NA	0	0	0

Crime [23]

	1994
Crime volume	
Cases known to police	18,553
Attempts (percent)	2
Percent cases solved	78.80
Crimes per 100,000 persons	247.42
Persons responsible for offenses	
Total number offenders	14,637
Percent female	6.20
Percent juvenile (14-18 yrs.)	5.60
Percent foreigners	0.30

Human Rights [24]

	SSTS	FL	FAPRO	PPCG	APROBC	TPW	PCPTW	STPEP	PHRFF	PRW	ASST	AFL
Observes		P	P	P	P							
	EAFRD	CPR	ESCR	SR	ACHR	MAAE	PVIAC	PVNAC	EAFDAW	TCIDTP	RC	
Observes		P	P	P		P				P		P

P = Party; S = Signatory; see Appendix for meaning of abbreviations.

Labor Force

Total Labor Force [25]

2.789 million

Labor Force by Occupation [26]

Agriculture and forestry	32%
Industry and construction	26
Other	42

Date of data: 1990

Unemployment Rate [27]

2.3% includes officially registered
unemployed; also large numbers of
unregistered unemployed and
underemployed workers (December 1995)

Production Sectors

Commercial Energy Production and Consumption

Data are shown in quadrillion (10^{15}) BTUs and percent for 1995
Values for hydroelectric, nuclear, geothermal, solar, and wind power refer to electrical generation.

Production [28]

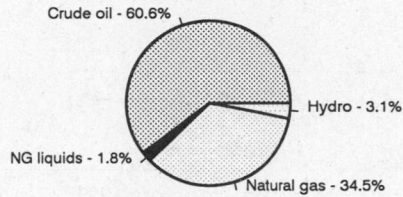

Crude oil - 60.6%
Hydro - 3.1%
NG liquids - 1.8%
Natural gas - 34.5%

Consumption [29]

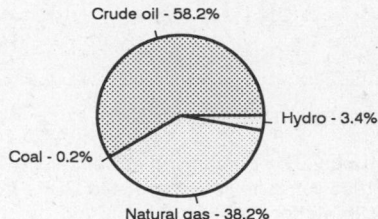

Crude oil - 58.2%
Hydro - 3.4%
Coal - 0.2%
Natural gas - 38.2%

Crude oil	0.376
Natural gas liquids	0.011
Dry natural gas	0.214
Net hydroelectric power	0.019
Total	0.620

Crude oil	0.326
Dry natural gas	0.214
Coal	0.001
Net hydroelectric power	0.019
Total	0.560

Telecommunications [30]

- 710,000 (1991 est.) telephones; 202,000 persons waiting for telephone installations (January 1991 est.)
- Domestic: telephone service is inadequate; a joint venture to establish a cellular telephone system in the Baku area was supposed to become operational in 1994
- International: cable and microwave radio relay connections to former Soviet republics; connection through Moscow international gateway switch to other countries; satellite earth stations - 1 Intelsat and 1 Intersputnik (Intelsat provides service to Turkey and to 200 other countries; Intersputnik provides direct service to New York)
- Radio: Broadcast stations: AM NA, FM NA, shortwave NA (1 state-owned radio broadcast station)
- Television: Broadcast stations: 2

Top Agricultural Products [32]

Produces cotton, grain, rice, grapes, fruit, vegetables, tea, tobacco; cattle, pigs, sheep, goats.

Transportation [31]

Railways: total: 2,125 km in common carrier service; does not include industrial lines; broad gauge: 2,125 km 1.520-m gauge (1,278 km electrified) (1993)

Highways: total: 36,700 km; paved: 31,800 km (includes graveled); unpaved: 4,900 km (1990 est.)

Airports

Total:	69
With paved runways over 3,047 m:	2
With paved runways 2,438 to 3,047 m:	6
With paved runways 1,524 to 2,437 m:	17
With paved runways 914 to 1,523 m:	3
With paved runways under 914 m:	1

Top Mining Products [33]

Metric tons except as noted	1994[*]
Alumina	150,000
Aluminum	15,000
Cement	300,000
Gypsum	60,000
Iron ore, marketable	100,000
Limestone	50,000
Gas, natural (mil. cu. meters)	6,380
Petroleum	9,560
Salt	80,000
Steel, crude	36,000

Tourism [34]

	1990	1991	1992	1993	1994
Visitors	1,198	1,469	NA	1,793	3,258
Tourists	57	123	NA	298	321
Tourism receipts	228	390	511	NA	2,042

Travelers are in thousands, money in million U.S. dollars.

For sources, notes, and explanations, see Annotated Source Appendix, page 1061.

Finance, Economics, and Trade

Industrial Summary [35]

Industrial Production: Growth rate - 21% (1995 est.). **Industries:** Petroleum and natural gas, petroleum products, oilfield equipment; steel, iron ore, cement; chemicals and petrochemicals; textiles.

Economic Indicators [36]

- **National product:** GDP—purchasing power parity— $11.5 billion (1995 estimate as extrapolated GDP—from World Bank estimate for 1994)

- **National product real growth rate:** - 17% (1995 est.)

- **National product per capita:** $1,480 (1995 est.)

- **Inflation rate (consumer prices):** 85% (1995 est.)

- **External debt:** $100 million (of which $75 million to Russia)

Balance of Payments Summary [37]

Exchange Rates [38]

Currency: **manat.**

Data are currency units per $1.

April 1996	4,375
April 1995	4,500
End of December 1994	4,168

Imports and Exports

Top Import Origins [39]

$681.5 million (c.i.f., 1995).

Origins	%
European countries	NA

Top Export Destinations [40]

$549.9 million (f.o.b., 1995).

Destinations	%
CIS	NA
European countries	NA

Foreign Aid [41]

	U.S. $	
ODA (1993)	14	million
Turkey (1992-95)	1,000	million

Import and Export Commodities [42]

Import Commodities	**Export Commodities**
Machinery and parts	Oil and gas
Consumer durables	Chemicals
Foodstuffs	Oilfield equipment
Textiles	Textiles
	Cotton

The Bahamas

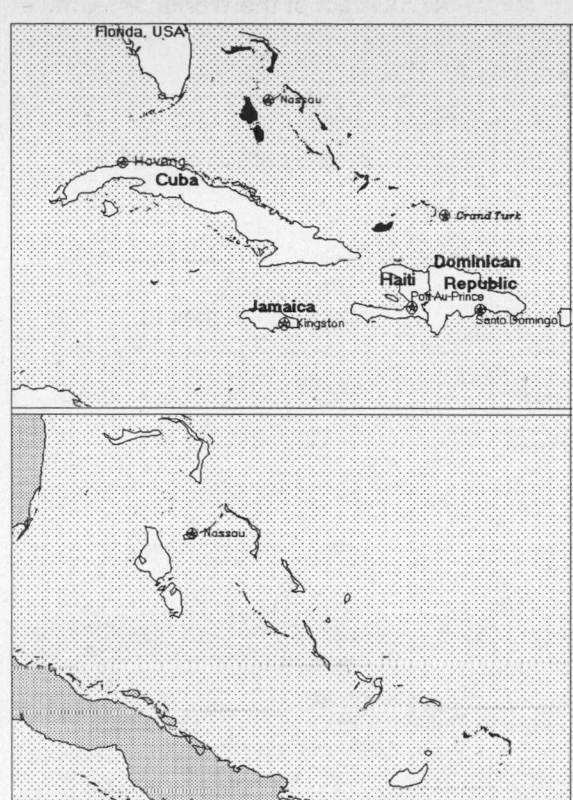

Geography [1]

Total area:
  13,940 sq km   5,382 sq mi
Land area:
  10,070 sq km   3,888 sq mi
Comparative area:
  Slightly larger than Connecticut
Land boundaries:
  0 km
Coastline:
  3,542 km
Climate:
  Tropical marine; moderated by warm waters of Gulf Stream
Terrain:
  Long, flat coral formations with some low rounded hills
Natural resources:
  Salt, aragonite, timber
Land use:
  Arable land: 1%
  Permanent crops: 0%
  Meadows and pastures: 0%
  Forest and woodland: 32%
  Other: 67%

Demographics [2]

	1970	1980	1990	1995[1]	1996	2000	2010	2020	2030
Population	170	210	241	257	259	269	293	314	327
Population density (persons per sq. mi.)	44	54	62	66	67	69	75	81	84
(persons per sq. km.)	17	21	24	25	26	27	29	31	32
Net migration rate (per 1,000 population)	NA	-2.8	-2.3	-2.6	-2.5	-2.4	0.0	0.0	0.0
Births	4	NA	NA	NA	5	NA	NA	NA	NA
Deaths	1	NA	NA	NA	1	NA	NA	NA	NA
Life expectancy - males	NA	62.7	67.5	67.4	68.0	70.6	72.7	74.5	75.9
Life expectancy - females	NA	71.1	75.3	77.0	77.2	77.9	79.6	81.1	82.2
Birth rate (per 1,000)	NA	24.3	20.8	19.2	18.7	16.5	14.2	12.6	11.3
Death rate (per 1,000)	NA	6.4	5.6	5.8	5.7	5.5	6.0	7.1	8.5
Women of reproductive age (15-49 yrs.)	NA	54	69	75	75	79	83	80	76
of which are currently married	19	18	30	NA	35	38	41	NA	NA
Fertility rate	NA	2.8	2.2	2.0	2.0	1.8	1.8	1.8	1.7

Except as noted, values for vital statistics are in thousands; life expectancy is in years.

Health

Health Indicators [3]

Health Expenditures [4]

For sources, notes, and explanations, see Annotated Source Appendix, page 1061.

61

Human Factors

Women and Children [5]

Burden of Disease [6]

Population per physician (1993)	691.71
Population per nurse (1993)	210.73
Population per hospital bed (1993)	257.47
AIDS cases per 100,000 people (1994)	131.4
Malaria cases per 100,000 people (1992)	NA

Ethnic Division [7]

Black	85%
White	15%

Religion [8]

Baptist	32%
Anglican	20%
Roman Catholic	19%
Methodist	6%
Church of God	6%
Other Protestant	12%
Other	5%

Major Languages [9]

English, Creole (among Haitian immigrants).

Education

Public Education Expenditures [10]

	1980	1985	1989	1990	1991	1994
Million (Dollar)						
Total education expenditure	NA	86	NA	125	114	NA
as percent of GNP	NA	4.0	NA	4.2	3.9	NA
as percent of total govt. expend.	NA	18.0	NA	17.8	16.3	NA
Current education expenditure	53	81	105	112	103	NA
as percent of GNP	4.4	3.8	3.8	3.7	3.5	NA
as percent of current govt. expend.	22.1	NA	18.4	NA	NA	NA
Capital expenditure	NA	5	NA	14	11	NA

Educational Attainment [11]

Age group (1990)	25+
Total population	104,472
Highest level attained (%)	
No schooling	3.5
First level	
Not completed	25.4
Completed	NA
Entered second level	
S-1	57.7
S-2	NA
Postsecondary	13.5

Illiteracy [12]

In thousands and percent[1]	1990	1995	2000
Illiterate population (15+ yrs.)	4	3	3
Illiteracy rate - total pop. (%)	2.4	1.6	1.5
Illiteracy rate - males (%)	1.3	1.1	1.0
Illiteracy rate - females (%)	2.3	2.1	1.9

Libraries [13]

	Admin. Units	Svc. Pts.	Vols. (000)	Shelving (meters)	Vols. Added	Reg. Users
National	NA	NA	NA	NA	NA	NA
Nonspecialized	NA	NA	NA	NA	NA	NA
Public	NA	NA	NA	NA	NA	NA
Higher ed. (1987)	2	2	68	NA	300	2,500
School	NA	NA	NA	NA	NA	NA

Daily Newspapers [14]

	1980	1985	1990	1994
Number of papers	3	3	3	3
Circ. (000)	33	39	35	35

Culture [15]

Cinema (seats per 1,000)	NA
Annual attendance per person	NA
Gross box office receipts (mil. Dollar)	NA
Museums (reporting)	2
Visitors (000)	11,303
Annual receipts (000 Dollar)	8,502

Science and Technology

Scientific/Technical Forces [16]

R&D Expenditures [17]

U.S. Patents Issued [18]

Values show patents issued to citizens of the country by the U.S. Patents Office.

	1993	1994	1995
Number of patents	3	2	4

For sources, notes, and explanations, see Annotated Source Appendix, page 1061.

Government and Law

Organization of Government [19]

Long-form name:
Commonwealth of The Bahamas
Type:
Commonwealth
Independence:
10 July 1973 (from UK)
National holiday:
National Day, 10 July (1973)
Constitution:
10 July 1973
Legal system:
Based on English common law
Executive branch:
Chief of state: British Monarch
(represented by Governor General); Prime
Minister; Deputy Prime Minister; Cabinet
Legislative branch:
Bicameral Parliament: Senate and House
of Assembly
Judicial branch:
Supreme Court

Elections [20]

House of Assembly	% of seats
Free National Movement (FNM)	65.3
Progressive Liberal Party (PLP)	34.7

Government Expenditures [21]

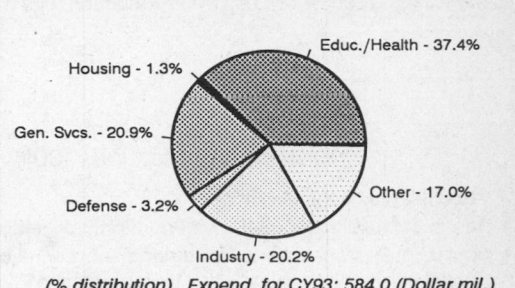

Housing - 1.3%
Educ./Health - 37.4%
Gen. Svcs. - 20.9%
Other - 17.0%
Defense - 3.2%
Industry - 20.2%

(% distribution). Expend. for CY93: 584.0 (Dollar mil.)

Defense Summary [22]

Branches: Royal Bahamas Defense Force (Coast Guard only), Royal Bahamas Police Force

Manpower Availability: Males age 15-49 NA; males fit for military service NA

Defense Expenditures: $20 million, 3.8% of GDP (FY95/96)

Crime [23]

	1990
Crime volume	
Cases known to police	17,409
Attempts (percent)	2.19
Percent cases solved	29.39
Crimes per 100,000 persons	6,835.50
Persons responsible for offenses	
Total number offenders	5,003
Percent female	10.13
Percent juvenile (7-17 yrs.)	10.11
Percent foreigners	NA

Human Rights [24]

	SSTS	FL	FAPRO	PPCG	APROBC	TPW	PCPTW	STPEP	PHRFF	PRW	ASST	AFL
Observes	P	P		P	P	P	P			P	P	P
	EAFRD	CPR		ESCR	SR	ACHR	MAAE	PVIAC	PVNAC	EAFDAW	TCIDTP	RC
Observes		P			P			P	P	P		P

P = Party; S = Signatory; see Appendix for meaning of abbreviations.

Labor Force

Total Labor Force [25]

136,900 (1993)

Labor Force by Occupation [26]

Government	30%
Tourism	40
Business services	10
Agriculture	5

Date of data: 1995 est.

Unemployment Rate [27]

15% (1995 est.)

For sources, notes, and explanations, see Annotated Source Appendix, page 1061.

63

Production Sectors

Energy Resource Summary [28]

Electricity: Capacity: 424,000 kW. Production: 929 million kWh. Consumption per capita: 3,200 kWh (1993).

Telecommunications [30]

- 119,000 (1987 est.) telephones
- Domestic: totally automatic system; highly developed
- International: tropospheric scatter and submarine cable to Florida; 3 coaxial submarine cables; satellite earth station - 1 Intelsat (Atlantic Ocean)
- Radio: Broadcast stations: AM 3, FM 2, shortwave 0 Radios: 200,000 (1993 est.)
- Television: Broadcast stations: 1 (1986 est.) Televisions: 60,000 (1993 est.)

Top Agricultural Products [32]

Agriculture accounts for 3% of the GDP; produces citrus, vegetables; poultry.

Top Mining Products [33]

Thousand metric tons[e]	3/31/95[*]
Salt	900
Stone: Aragonite	1,200

Transportation [31]

Railways: 0 km

Highways: total: 2,386 km; paved: 1,342 km; unpaved: 1,044 km (1986 est.)

Merchant marine: total: 956 ships (1,000 GRT or over) totaling 22,592,285 GRT/35,765,965 DWT; ships by type: bulk 176, cargo 182, chemical tanker 43, combination bulk 9, combination ore/oil 19, container 53, liquefied gas tanker 20, oil tanker 180, passenger 53, refrigerated cargo 147, roll-on/roll-off cargo 47, short-sea passenger 13, vehicle carrier 14

Airports

Total:	55
With paved runways over 3,047 m:	2
With paved runways 2,438 to 3,047 m:	1
With paved runways 1,524 to 2,437 m:	16
With paved runways 914 to 1,523 m:	11
With paved runways under 914 m:	17

Tourism [34]

	1990	1991	1992	1993	1994
Visitors	3,629	3,622	3,687	3,680	3,444
Tourists[12]	1,562	1,427	1,399	1,489	1,516
Excursionists	213	175	149	144	122
Cruise passengers	1,854	2,020	2,139	2,047	1,806
Tourism receipts	1,333	1,193	1,244	1,304	1,333
Tourism expenditures	196	200	187	171	192
Fare receipts	12	9	9	9	10
Fare expenditures	21	20	19	16	17

Travelers are in thousands, money in million U.S. dollars.

Manufacturing Sector

Manufacturing Summary [35]

	1987 $ bil.	1987 %	1988 $ bil.	1988 %	1989 $ bil.	1989 %	1990 $ bil.	1990 %	1991 $ bil.	1991 %
Establishments or enterprises (number)	-	-	-	-	115	0.006	102	0.006	105	0.014
Total employment (000)	-	-	-	-	-	-	-	-	-	-
Production workers (000)	-	-	-	-	-	-	-	-	-	-
Output ($ bil.)	0.320	0.003	-	-	0.144	0.001	0.158	0.001	0.177	0.002
Value added ($ bil.)	0.116	0.003	-	-	4	0.084	6	0.124	5	0.153
Capital investment ($ mil.)	27	0.006	-	-	2,678	0.489	2,597	0.467	3,785	0.994
M & E investment ($ mil.)	-	-	-	-	-	-	-	-	-	-
Employees per establishment (number)	-	-	-	-	-	-	-	-	-	-
Production workers per establishment	-	-	-	-	-	-	-	-	-	-
Output per establishment ($ mil.)	-	-	-	-	1	19.884	2	24.222	2	12.671
Capital investment per estab. ($ mil.)	-	-	-	-	23	7,963	25	8,207	36	7,236
M & E per establishment ($ mil)	-	-	-	-	-	-	-	-	-	-
Payroll per employee ($)	-	-	-	-	-	-	-	-	-	-
Wages per production worker ($)	-	-	-	-	-	-	-	-	-	-
Hours per production worker (hours)	-	-	-	-	-	-	-	-	-	-
Output per employee ($)	-	-	-	-	-	-	-	-	-	-
Capital investment per employee ($)	-	-	-	-	-	-	-	-	-	-
M & E per employee ($)	-	-	-	-	-	-	-	-	-	-

Note: Columns headed % show percent of world total or ratio. Ratios closest to 100 are closest to world average. M & E stands for machinery & equipment.

Output in Manufacturing

	1987 $ bil.	1987 %	1988 $ bil.	1988 %	1989 $ bil.	1989 %	1990 $ bil.	1990 %	1991 $ bil.	1991 %
3110 - Food products	0.047	14.69	-	-	0.011	7.64	0.013	8.23	0.013	7.34
3130 - Beverages	0.070	21.88	-	-	0.073	50.69	0.083	52.53	0.102	57.63
3210 - Textiles	0.000	0.00	-	-	0.001	0.69	0.002	1.27	0.000	0.00
3220 - Wearing apparel	0.002	0.62	-	-	0.001	0.69	0.001	0.63	0.002	1.13
3230 - Leather and products	-	-	-	-	-	-	-	-	-	-
3240 - Footwear	0.000	0.00	-	-	-	-	-	-	-	-
3320 - Furniture, fixtures	0.006	1.87	-	-	0.004	2.78	0.002	1.27	0.002	1.13
3410 - Paper and products	0.001	0.31	-	-	0.000	0.00	0.000	0.00	0.002	1.13
3411 - Pulp, paper, etc.	-	-	-	-	0.002	1.39	0.001	0.63	-	-
3420 - Printing, publishing	0.017	5.31	-	-	0.009	6.25	0.011	6.96	0.011	6.21
3520 - Chemical products nec	0.075	23.44	-	-	0.003	2.08	0.015	9.49	0.018	10.17
3522 - Drugs and medicines	0.073	22.81	-	-	0.013	9.03	-	-	-	-
3530 - Petroleum refineries	0.006	1.87	-	-	-	-	-	-	-	-
3560 - Plastic products nec	0.001	0.31	-	-	0.002	1.39	0.002	1.27	0.000	0.00
3610 - Pottery, china, etc.	0.001	0.31	-	-	0.000	0.00	-	-	-	-
3620 - Glass and products	-	-	-	-	0.008	5.56	0.005	3.16	0.006	3.39
3690 - Nonmetal products nec	0.013	4.06	-	-	0.016	11.11	0.020	12.66	0.017	9.60
3710 - Iron and steel	-	-	-	-	0.001	0.69	0.003	1.90	0.003	1.69
3830 - Electrical machinery	0.002	0.62	-	-	-	-	-	-	-	-
3832 - Radio, television, etc.	0.001	0.31	-	-	0.001	0.69	0.000	0.00	0.000	0.00
3841 - Shipbuilding, repair	-	-	-	-	0.000	0.00	-	-	-	-
3900 - Industries nec	0.006	1.87	-	-	0.000	0.00	0.000	0.00	0.001	0.56

Note: Codes are International Standard Industry codes (ISIC). Percentages are % of total Output. [f]: Factor Prices; [p]: Producer Prices.

For sources, notes, and explanations, see Annotated Source Appendix, page 1061.

65

Finance, Economics, and Trade

Economic Indicators [36]

- **National product**: GDP—purchasing power parity—$4.8 billion (1995 est.)
- **National product real growth rate**: 2% (1995 est.)
- **National product per capita**: $18,700 (1995 est.)
- **Inflation rate (consumer prices)**: 1.5% (1994)
- **External debt**: $407.8 million (December 1994)

Balance of Payments Summary [37]

Values in millions of dollars.

	1989	1990	1991	1992	1993
Exports of goods (f.o.b.)	259.2	307.6	319.8	310.2	256.8
Imports of goods (f.o.b.)	-1,203.5	-1,190.2	1,045.6	-1,069.2	-1,080.9
Trade balance	-944.3	-882.6	-725.8	-759.0	-824.1
Services - debits	-721.2	-773.6	-775.3	-708.8	-740.4
Services - credits	1,506.7	1,570.3	1,395.4	1,420.7	1,479.3
Private transfers (net)	-17.9	-10.6	-7.8	-12.8	-12.6
Government transfers (net)	18.9	21.2	27.4	26.2	25.3
Long-term capital (net)	46.5	24.2	165.0	50.8	-3.8
Short-term capital (net)	50.0	32.9	11.9	-37.9	1.1
Errors and omissions	48.1	27.5	-77.8	-7.9	94.2
Overall balance	-13.2	9.3	13.0	-28.7	19.0

Exchange Rates [38]

Currency: **Bahamian dollar.**
Symbol: **B$.**

Data are currency units per $1.

Fixed rate	1.00

Imports and Exports

Top Import Origins [39]

$1.08 billion (c.i.f., 1994).

Origins	%
US	55
Japan	17
Nigeria	12
Denmark	7
Norway	6

Top Export Destinations [40]

$224.257 million (f.o.b., 1994).

Destinations	%
US	51
UK	7
Norway	7
France	6
Italy	5

Foreign Aid [41]

	U.S. $	
US commitments, including Ex-Im (FY85-89)	1	million
Western (non-US) countries, ODA and OOF bilateral commitments (1970-89)	345	million

Import and Export Commodities [42]

Import Commodities	Export Commodities
Foodstuffs	Pharmaceuticals
Manufactured goods	Cement
Crude oil	Rum
Vehicles	Crawfish
Electronics	Refined petroleum products

Bahrain

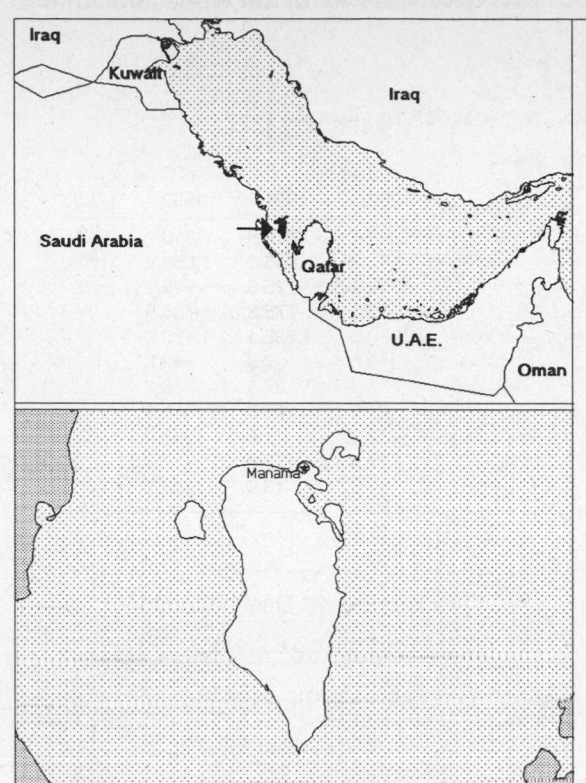

Geography [1]

Total area:
  620 sq km   239 sq mi
Land area:
  620 sq km   239 sq mi
Comparative area:
  3.5 times the size of Washington, DC
Land boundaries:
  0 km
Coastline:
  161 km
Climate:
  Arid; mild, pleasant winters; very hot, humid summers
Terrain:
  Mostly low desert plain rising gently to low central escarpment
Natural resources:
  Oil, associated and nonassociated natural gas, fish
Land use:
  Arable land: 2%
  Permanent crops: 2%
  Meadows and pastures: 6%
  Forest and woodland: 0%
  Other: 90%

Demographics [2]

	1970	1980	1990	1995[1]	1996	2000	2010	2020	2030
Population	220	348	502	576	590	642	759	870	970
Population density (persons per sq. mi.)	919	1,454	2,099	2,410	2,469	2,684	3,175	3,641	4,057
(persons per sq. km.)	355	561	811	930	953	1,036	1,226	1,406	1,566
Net migration rate (per 1,000 population)	NA	NA	8.5	5.0	2.4	1.1	0.0	0.0	0.0
Births	NA	NA	NA	NA	14	NA	NA	NA	NA
Deaths	NA	NA	NA	NA	2	NA	NA	NA	NA
Life expectancy - males	NA	NA	69.6	71.5	71.8	73.1	75.6	77.4	78.6
Life expectancy - females	NA	NA	74.4	76.5	76.8	78.3	81.2	83.2	84.6
Birth rate (per 1,000)	NA	NA	26.7	24.1	23.6	21.3	18.3	17.4	15.5
Death rate (per 1,000)	NA	NA	3.7	3.3	3.3	3.2	3.7	4.8	6.6
Women of reproductive age (15-49 yrs.)	NA	NA	117	137	141	155	182	199	221
of which are currently married	NA	NA	67	NA	85	95	108	NA	NA
Fertility rate	NA	NA	3.4	3.1	3.1	2.9	2.7	2.4	2.3

Except as noted, values for vital statistics are in thousands; life expectancy is in years.

Health

Health Indicators [3]

Health Expenditures [4]

For sources, notes, and explanations, see Annotated Source Appendix, page 1061.

67

Human Factors

Women and Children [5]	Burden of Disease [6]	
	Population per physician (1985)	820.46
	Population per nurse (1985)	370.21
	Population per hospital bed	NA
	AIDS cases per 100,000 people (1994)	0.9
	Malaria cases per 100,000 people (1992)	NA

Ethnic Division [7]

Bahraini	63%
Asian	13%
Other Arab	10%
Iranian	8%
Other	6%

Religion [8]

Shi'a Muslim	75%
Sunni Muslim	25%

Major Languages [9]

Arabic, English, Farsi, Urdu.

Education

Public Education Expenditures [10]

Million (Dinar)[4]	1980	1990	1991	1992	1993	1994
Total education expenditure	33	65	68	72	71	NA
as percent of GNP	2.9	5.0	5.1	5.0	4.7	NA
as percent of total govt. expend.	10.3	NA	12.8	12.3	NA	NA
Current education expenditure	28	61	65	68	69	NA
as percent of GNP	2.5	4.7	4.8	4.7	4.6	NA
as percent of current govt. expend.	14.7	NA	14.9	14.4	NA	NA
Capital expenditure	4	4	4	4	2	NA

Educational Attainment [11]

Age group (1991)	25+
Total population	263,720
Highest level attained (%)	
No schooling	38.4
First level	
Not completed	26.2
Completed	NA
Entered second level	
S-1	25.1
S-2	NA
Postsecondary	10.3

Illiteracy [12]

In thousands and percent[1]	1990	1995	2000
Illiterate population (15+ yrs.)	59	56	52
Illiteracy rate - total pop. (%)	17.1	14.0	11.6
Illiteracy rate - males (%)	12.9	10.5	8.7
Illiteracy rate - females (%)	24.3	19.5	15.8

Libraries [13]

	Admin. Units	Svc. Pts.	Vols. (000)	Shelving (meters)	Vols. Added	Reg. Users
National	NA	NA	NA	NA	NA	NA
Nonspecialized	NA	NA	NA	NA	NA	NA
Public (1989)	1	10	218	NA	NA	NA
Higher ed. (1990)[3]	1	1	35	790	1,267	1,800
School (1990)	184[e]	184[e]	166[e]	5,520[e]	100,000[e]	80,000[e]

Daily Newspapers [14]

	1980	1985	1990	1994
Number of papers	3	2	2	3
Circ. (000)	14[e]	19	29	70[e]

Culture [15]

Cinema (seats per 1,000)	NA
Annual attendance per person	NA
Gross box office receipts (mil. Dina)	NA
Museums (reporting)	2
Visitors (000)	110
Annual receipts (000 Dina)	NA

Science and Technology

Scientific/Technical Forces [16]

Scientists/engineers	10,747
Number female	3,184
Technicians	11,615
Number female	4,215
Total[36]	22,362

R&D Expenditures [17]

U.S. Patents Issued [18]

Government and Law

Organization of Government [19]

Long-form name:
State of Bahrain
Type:
Traditional monarchy
Independence:
15 August 1971 (from UK)
National holiday:
Independence Day, 16 December (1971)
Constitution:
26 May 1973, effective 6 December 1973
Legal system:
Based on Islamic law and English
common law
Executive branch:
Emir; Heir Apparent; Prime Minister;
Cabinet
Legislative branch:
Unicameral National Assembly was
dissolved 26 August 1975 and legislative
powers were assumed by the Cabinet;
appointed Advisory Council established
16 December 1992
Judicial branch:
High Civil Appeals Court

Elections [20]

Political parties prohibited; several
small, clandestine leftist and Islamic
fundamentalist groups are active. No
elections.

Government Expenditures [21]

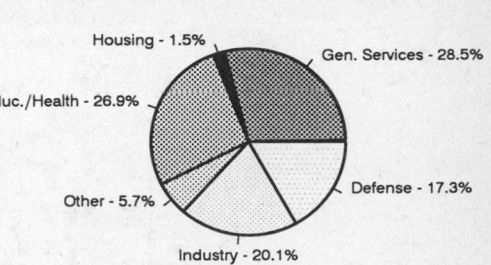

Housing - 1.5%
Gen. Services - 28.5%
Educ./Health - 26.9%
Defense - 17.3%
Other - 5.7%
Industry - 20.1%

(% distribution). Expend. for CY95: 594.1 (Dinar mil.)

Military Expenditures and Arms Transfers [22]

	1990	1991	1992	1993	1994
Military expenditures					
Current dollars (mil.)	216	237	252	251	256
1994 constant dollars (mil.)	240	254	262	256	256
Armed forces (000)	8	8	7	7	8
Gross national product (GNP)					
Current dollars (mil.)[e]	3,448	3,583	3,835	4,006	4,006
1994 constant dollars (mil.)[e]	3,837	3,841	3,999	4,088	4,006
Central government expenditures (CGE)					
1994 constant dollars (mil.)	1,771	1,643	1,711	1,535[e]	1,522
People (mil.)	0.5	0.5	0.5	0.5	0.6
Military expenditure as % of GNP	6.3	6.6	6.6	6.3	6.4
Military expenditure as % of CGE	13.6	15.5	15.3	16.7	16.8
Military expenditure per capita (1994 $)	479	492	494	469	456
Armed forces per 1,000 people (soldiers)	15.9	15.5	13.2	12.8	14.3
GNP per capita (1994 $)	7,649	7,432	7,528	7,486	7,140
Arms imports[6]					
Current dollars (mil.)	290	50	130	60	80
1994 constant dollars (mil.)	323	54	136	61	80
Arms exports[6]					
Current dollars (mil.)	0	0	0	0	0
1994 constant dollars (mil.)	0	0	0	0	0
Total imports[7]					
Current dollars (mil.)	3,711	4,115	4,263	3,858	3,737
1994 constant dollars (mil.)	4,130	4,411	4,445	3,938	3,737
Total exports[7]					
Current dollars (mil.)	3,761	3,513	3,464	3,710	2,656
1994 constant dollars (mil.)	4,186	3,766	3,612	3,786	2,656
Arms as percent of total imports[8]	7.8	1.2	3.0	1.6	2.1
Arms as percent of total exports[8]	0	0	0	0	0

Crime [23]

	1994
Crime volume	
Cases known to police	9,629
Attempts (percent)	NA
Percent cases solved	NA
Crimes per 100,000 persons	1,713.74
Persons responsible for offenses	
Total number offenders	NA
Percent female	NA
Percent juvenile (3-15 yrs.)	NA
Percent foreigners	NA

Human Rights [24]

	SSTS	FL	FAPRO	PPCG	APROBC	TPW	PCPTW	STPEP	PHRFF	PRW	ASST	AFL
Observes	P	P		P		P	P				P	
	EAFRD	CPR	ESCR	SR	ACHR	MAAE	PVIAC	PVNAC	EAFDAW	TCIDTP	RC
Observes		P					P	P			P

P = Party; S = Signatory; see Appendix for meaning of abbreviations.

Labor Force

Total Labor Force [25]

140,000

Labor Force by Occupation [26]

Industry and commerce	85%
Agriculture	5
Services	5
Government	3
(1982)	

Unemployment Rate [27]

25% (1994 est.)

For sources, notes, and explanations, see Annotated Source Appendix, page 1061.

69

Production Sectors

Commercial Energy Production and Consumption

Data are shown in quadrillion (10^{15}) BTUs and percent for 1995
Values for hydroelectric, nuclear, geothermal, solar, and wind power refer to electrical generation.

Production [28]

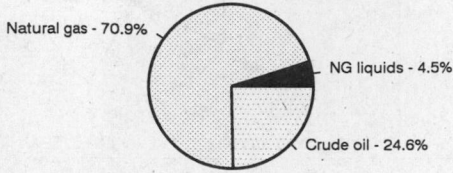

Natural gas - 70.9%

NG liquids - 4.5%

Crude oil - 24.6%

Consumption [29]

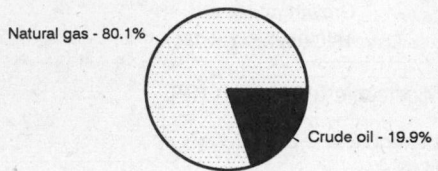

Natural gas - 80.1%

Crude oil - 19.9%

Crude oil	0.088
Natural gas liquids	0.016
Dry natural gas	0.253
Total	0.357

Crude oil	0.063
Dry natural gas	0.253
Total	0.316

Telecommunications [30]

- 73,552 (1987 est.) telephones; modern system; good domestic services and excellent international connections
- International: tropospheric scatter to Qatar and UAE; microwave radio relay to Saudi Arabia; submarine cable to Qatar, UAE, and Saudi Arabia; satellite earth stations - 2 Intelsat (1 Atlantic Ocean and 1 Indian Ocean) and 1 Arabsat
- Radio: Broadcast stations: AM 2, FM 3, shortwave 0 Radios: 320,000 (1993 est.)
- Television: Broadcast stations: 2 (1988 est.) Televisions: 270,000 (1993 est.)

Top Agricultural Products [32]

Produces fruit, vegetables; poultry, dairy products; shrimp, fish.

Top Mining Products [33]

Thousand metric tons except as noted[e]	6/1/95[*]
Aluminum, smelter output, primary metal	447,000
Gas, natural, gross (mil. cu. meters)	9,800
Petroleum (000 42-gal. bls.)	
crude	14,700
refinery products	88,000
Methanol	419,000
Nitrogen, N content of ammonia	338,000
Cement	225,000

Transportation [31]

Railways: 0 km

Highways: total: 2,671 km; paved: 2,011 km; unpaved: 660 km (1991 est.)

Merchant marine: total: 6 ships (1,000 GRT or over) totaling 117,060 GRT/194,061 DWT; ships by type: bulk 1, cargo 3, chemical tanker 1, oil tanker 1 (1995 est.)

Airports

Total: 3
With paved runways over 3,047 m: 2
With unpaved runways 1,524 to 2,437 m: 1 (1995 est.)

Tourism [34]

	1990	1991	1992	1993	1994
Visitors[1]	2,051	2,416	2,474	2,869	3,323
Tourists	1,376	1,674	1,419	1,761	2,270
Tourism receipts	135	162	177	222	302
Tourism expenditures	94	98	141	130	146

Travelers are in thousands, money in million U.S. dollars.

For sources, notes, and explanations, see Annotated Source Appendix, page 1061.

Finance, Economics, and Trade

GDP and Manufacturing Summary [35]

	1980	1985	1990	1991	1992
Gross Domestic Product					
Millions of 1980 dollars	3,072	2,902	3,334	3,423	3,469[e]
Growth rate in percent	0.24	-3.94	1.24	2.68	1.34[e]
Manufacturing Value Added					
Millions of 1980 dollars	498	457	577	591	600[e]
Growth rate in percent	-3.38	-16.13	5.23	2.35	1.64[e]
Manufacturing share in percent of current prices	14.8	8.5	15.9	NA	NA

Economic Indicators [36]

- **National product**: GDP—purchasing power parity—$7.3 billion (1995 est.)
- **National product real growth rate**: - 2% (1995 est.)
- **National product per capita**: $12,000 (1995 est.)
- **Inflation rate (consumer prices)**: 3% (1995 est.)
- **External debt**: $2.6 billion (1993)

Balance of Payments Summary [37]

Values in millions of dollars.

	1988	1989	1990	1991	1992
Exports of goods (f.o.b.)	2,411.4	2,831.1	3,761	3,513	3,417
Imports of goods (f.o.b.)	-2,334.0	-2,820.2	-3,340	-3,703	-3,730
Trade balance	77.4	10.9	421	-190	-313
Services - debits	-1,223.4	-1,294.4	-1,558	-1,620	-1,774
Services - credits	1,162.5	1,250.5	1,196	1,216	1,264
Private transfers (net)	-193.1	-198.9	-272	-303	-271
Government transfers (net)	366.5	102.1	459	102	100
Long-term capital (net)	205.1	93.6	-97	-58	-14
Short-term capital (net)	-419.4	-359.3	553	-289	380
Errors and omissions	117.0	207.0	-520	1,425	545
Overall balance	92.6	-188.5	182	282	-82

Exchange Rates [38]

Currency: **Bahraini dinar.**
Symbol: **BD.**

Data are currency units per $1.

Fixed rate	0.3760

Imports and Exports

Top Import Origins [39]

$3.29 billion (c.i.f., 1995 est.) Data are for 1994.

Origins	%
Saudi Arabia	37
US	12
UK	6
Japan	5
Germany	4

Top Export Destinations [40]

$3.2 billion (f.o.b., 1995 est.) Data are for 1994.

Destinations	%
India	20
Japan	14
Saudi Arabia	7
US	6
UAE	5

Foreign Aid [41]

	U.S. $	
US commitments, including Ex-Im (FY70-79)	24	million
Western (non-US) countries, ODA and OOF bilateral commitments (1970-89)	45	million
OPEC bilateral aid (1979-89)	9.8	billion

Import and Export Commodities [42]

Import Commodities

Nonoil 59%
Crude oil 41%

Export Commodities

Petroleum and petroleum products 80%
Aluminum 7%

For sources, notes, and explanations, see Annotated Source Appendix, page 1061.

71

Bangladesh

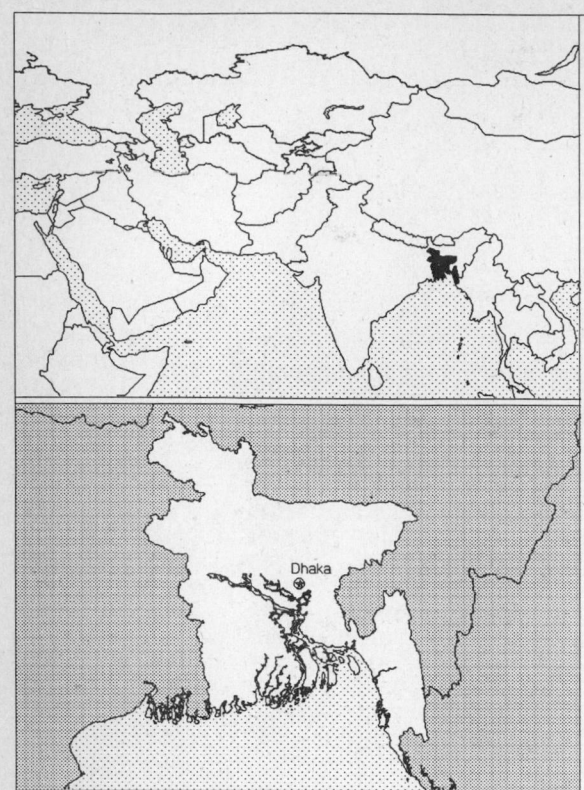

Geography [1]

Total area:
144,000 sq km 55,599 sq mi
Land area:
133,910 sq km 51,703 sq mi
Comparative area:
Slightly smaller than Wisconsin
Land boundaries:
Total 4,246 km, Myanmar 193 km, India 4,053 km
Coastline:
580 km
Climate:
Tropical; cool, dry winter (October to March); hot, humid summer (March to June); cool, rainy monsoon (June to October)
Terrain:
Mostly flat alluvial plain; hilly in Southeast
Natural resources:
Natural gas, arable land, timber
Land use:
Arable land: 67%
Permanent crops: 2%
Meadows and pastures: 4%
Forest and woodland: 16%
Other: 11%

Demographics [2]

	1970	1980	1990	1995[1]	1996	2000	2010	2020	2030
Population	67,403	88,077	110,118	120,788	123,063	132,081	153,195	172,041	188,318
Population density (persons per sq. mi.)	1,304	1,704	2,130	2,336	2,380	2,555	2,963	3,327	3,642
(persons per sq. km.)	503	658	822	902	919	986	1,144	1,285	1,406
Net migration rate (per 1,000 population)	NA	NA	-3.2	-0.8	-0.8	-0.6	-0.3	0.0	0.0
Births	NA	NA	NA	NA	3,753	NA	NA	NA	NA
Deaths	NA	NA	NA	NA	1,380	NA	NA	NA	NA
Life expectancy - males	NA	NA	54.0	55.7	56.0	57.4	60.6	63.7	66.4
Life expectancy - females	NA	NA	52.8	55.2	55.7	57.6	62.3	66.7	70.6
Birth rate (per 1,000)	NA	NA	35.8	31.2	30.5	27.4	21.6	18.2	15.9
Death rate (per 1,000)	NA	NA	13.3	11.5	11.2	10.1	8.4	7.8	8.0
Women of reproductive age (15-49 yrs.)	NA	NA	25,287	29,425	30,400	34,185	42,230	48,032	50,521
of which are currently married	NA	NA	19,847	NA	23,918	26,979	34,031	NA	NA
Fertility rate	NA	NA	4.5	3.7	3.6	3.1	2.4	2.1	2.0

Except as noted, values for vital statistics are in thousands; life expectancy is in years.

Health

Health Indicators [3]

% of population with access to	
safe water (1990-95)	97
adequate sanitation (1990-95)	34
health services (1985-95)	45
% of 1-year-olds immunized (1990-94) against	
TB (tuberculosis)	95
DPT (diphtheria, pertussis, tetanus)	94
polio	94
measles	95
% of contraceptive prevalence (1980-94)	45
ORT use rate (1990-94)	91

Health Expenditures [4]

Total health expenditure, 1990 (official exchange rate)	
Millions of dollars	715
Dollars per capita	7
Health expenditures as a percentage of GDP	
Total	3.2
Public sector	1.4
Private sector	1.8
Development assistance for health	
Total aid flows (millions of dollars)[1]	128
Aid flows per capita (dollars)	1.2
Aid flows as a percentage of total health expenditure	17.9

For sources, notes, and explanations, see Annotated Source Appendix, page 1061.

Human Factors	

Human Factors

Women and Children [5]

% of pregnant women immunized (tetanus 1990-94)	81
% of births attended by trained health personnel (1983-94)	10
Maternal mortality rate (1980-92)	600
Under-5 mortality rate (1994)	117
% under-5 moderately/severely underweight (1980-1994)	67

Burden of Disease [6]

Population per physician (1991)	5,308.69
Population per nurse (1991)	11,548.69
Population per hospital bed (1991)	3,245.62
AIDS cases per 100,000 people (1994)	*
Malaria cases per 100,000 people (1992)	103

Ethnic Division [7]

Bengali	98%
Tribals	< 1 mil
Biharis	250,000

Religion [8]

Muslim	83%
Hindu	16%
Buddhist, Christian, and other	1%

Major Languages [9]

Bangla (official), English.

Education

Public Education Expenditures [10]

Million (Taka)[5]	1980	1985	1990	1991	1992	1994
Total education expenditure	3,009	7,782	14,942	18,184	20,996	NA
as percent of GNP	1.5	1.9	2.0	2.2	2.3	NA
as percent of total govt. expend.	7.8	9.7	10.3	11.3	8.7	NA
Current education expenditure	2,010	6,005	11,820	13,816	16,744	NA
as percent of GNP	1.0	1.5	1.6	1.7	1.9	NA
as percent of current govt. expend.	13.6	15.3	14.4	15.9	10.4	NA
Capital expenditure	999	1,777	3,122	4,368	4,252	NA

Educational Attainment [11]

Age group (1981)	25+
Total population	31,593,122
Highest level attained (%)	
No schooling	70.4
First level	
Not completed	16.7
Completed	NA
Entered second level	
S-1	7.4
S-2	4.2
Postsecondary	1.3

Illiteracy [12]

In thousands and percent[1]	1990	1995	2000
Illiterate population (15+ yrs.)	40,704	45,082	49,983
Illiteracy rate - total pop. (%)	64.5	61.9	59.2
Illiteracy rate - males (%)	52.6	50.7	48.3
Illiteracy rate - females (%)	77.2	73.8	70.7

Libraries [13]

	Admin. Units	Svc. Pts.	Vols. (000)	Shelving (meters)	Vols. Added	Reg. Users
National (1989)	1	1	15	328	NA	NA
Nonspecialized	NA	NA	NA	NA	NA	NA
Public (1989)	57	61	521	NA	26,600	NA
Higher ed. (1987)	946	983	4,076	NA	19,848	680,639
School	NA	NA	NA	NA	NA	NA

Daily Newspapers [14]

	1980	1985	1990	1994
Number of papers	44	60	52	51
Circ. (000)	274	591	700[e]	710[e]

Culture [15]

Science and Technology

Scientific/Technical Forces [16]	R&D Expenditures [17]	U.S. Patents Issued [18]

For sources, notes, and explanations, see Annotated Source Appendix, page 1061.

73

Government and Law

Organization of Government [19]

Long-form name:
People's Republic of Bangladesh
Type:
Republic
Independence:
16 December 1971 (from Pakistan)
National holiday:
Independence Day, 26 March (1971)
Constitution:
4 November 1972, effective 16 December 1972, suspended following coup of 24 March 1982, restored 10 November 1986, amended many times
Legal system:
Based on English common law
Executive branch:
President; Caretaker Prime Minister; Advisory Council
Legislative branch:
Unicameral: National Parliament (Jatiya Sangsad): RAHMAN
Judicial branch:
Supreme Court

Elections [20]

Elections for the unicameral National Parliment consisting of 330 members were held on 15 February 1996. The elections were held despite the fact that all major opposition parties boycotted them. The President dissolved parliament and named a caretaker prime minister. A date for new elections is not known.

Government Budget [21]

For FY92/93.

	$ bil.
Revenues	2.8
Expenditures	4.1
Capital expenditures	1.8

Military Expenditures and Arms Transfers [22]

	1990	1991	1992	1993	1994
Military expenditures					
Current dollars (mil.)	301	288	327	410	448
1994 constant dollars (mil.)	336	309	341	418	448
Armed forces (000)	103	107	107	107	113
Gross national product (GNP)					
Current dollars (mil.)	19,540	20,990	22,430	24,050	25,730
1994 constant dollars (mil.)	21,750	22,500	23,390	24,540	25,730
Central government expenditures (CGE)					
1994 constant dollars (mil.)	NA	3,667	3,732	4,217	NA
People (mil.)	114.0	116.7	119.4	122.3	125.1
Military expenditure as % of GNP	1.5	1.4	1.5	1.7	1.7
Military expenditure as % of CGE	NA	8.4	9.1	9.9	NA
Military expenditure per capita (1994 $)	3	3	3	3	4
Armed forces per 1,000 people (soldiers)	0.9	0.9	0.9	0.9	0.9
GNP per capita (1994 $)	191	193	196	201	206
Arms imports[6]					
Current dollars (mil.)	30	80	40	30	10
1994 constant dollars (mil.)	33	86	42	31	10
Arms exports[6]					
Current dollars (mil.)	0	0	0	0	0
1994 constant dollars (mil.)	0	0	0	0	0
Total imports[7]					
Current dollars (mil.)	3,598	3,401	3,888	4,001	4,701
1994 constant dollars (mil.)	4,004	3,645	4,054	4,083	4,701
Total exports[7]					
Current dollars (mil.)	1,671	1,689	2,098	2,272	2,650
1994 constant dollars (mil.)	1,860	1,810	2,188	2,319	2,650
Arms as percent of total imports[8]	0.8	2.4	1.0	0.7	0.2
Arms as percent of total exports[8]	0	0	0	0	0

Crime [23]

	1994
Crime volume	
Cases known to police	75,309
Attempts (percent)	NA
Percent cases solved	93.45
Crimes per 100,000 persons	64.37
Persons responsible for offenses	
Total number offenders	210,063
Percent female	0.59
Percent juvenile (7-16 yrs.)	0.09
Percent foreigners	NA

Human Rights [24]

	SSTS	FL	FAPRO	PPCG	APROBC	TPW	PCPTW	STPEP	PHRFF	PRW	ASST	AFL
Observes	P	P	P		P	P	P	P			P	P
		EAFRD	CPR	ESCR	SR	ACHR	MAAE	PVIAC	PVNAC	EAFDAW	TCIDTP	RC
Observes		P						P	P	P		P

P = Party; S = Signatory; see Appendix for meaning of abbreviations.

Labor Force

Total Labor Force [25]

50.1 million

Labor Force by Occupation [26]

Agriculture	65%
Services	21
Industry and mining	14

Extensive export of labor to Saudi Arabia, UAE, and Oman.
Date of data: 1989

Unemployment Rate [27]

For sources, notes, and explanations, see Annotated Source Appendix, page 1061.

Production Sectors

Commercial Energy Production and Consumption

Data are shown in quadrillion (10^{15}) BTUs and percent for 1995
Values for hydroelectric, nuclear, geothermal, solar, and wind power refer to electrical generation.

Production [28]	Consumption [29]
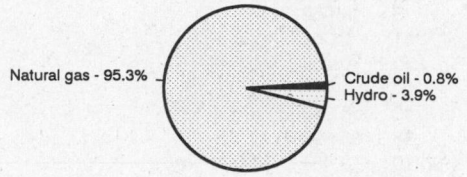	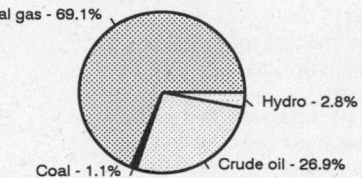
Natural gas - 95.3% Crude oil - 0.8% Hydro - 3.9%	Natural gas - 69.1% Hydro - 2.8% Crude oil - 26.9% Coal - 1.1%

Crude oil	0.002	Crude oil	0.095
Dry natural gas	0.244	Dry natural gas	0.244
Net hydroelectric power	0.010	Coal	0.004
Total	0.256	Net hydroelectric power	0.010
		Total	0.353

Telecommunications [30]

- 249,800 (1994 est.) telephones
- Domestic: poor domestic telephone service
- International: satellite earth stations - 2 Intelsat (Indian Ocean); international radiotelephone communications and landline service to neighboring countries
- Radio: Broadcast stations: AM 9, FM 6, shortwave 0
- Television: Broadcast stations: 11 Televisions: 350,000 (1993 est.)

Top Agricultural Products [32]

Produces jute, rice, wheat, tea, sugarcane, potatoes; beef, milk, poultry.

Top Mining Products [33]

Metric tons except as noted[e]	8/10/95[*]
Cement, hydraulic	280,000[7]
Clay, kaolin	7,500[7]
Gas, natural, marketed (mil. cu. meters)	5,970[7,8,r]
Steel products	87,000[7]
Nitrogen, N content of urea, ammonia, and ammonium sulfate	995,000
Petroleum refinery products (000 42-gal. bls.)	7,600
Salt, marine	350,000[7]
Limestone	52,000[7]

Transportation [31]

Railways: total: 2,892 km; broad gauge: 978 km 1.676-m gauge; narrow gauge: 1,914 km 1.000-m gauge (1992)

Highways: total: 13,627 km; paved: 8,546 km; unpaved: 5,081 km (1992)

Merchant marine: total: 37 ships (1,000 GRT or over) totaling 296,503 GRT/423,274 DWT; ships by type: bulk 3, cargo 29, oil tanker 2, refrigerated cargo 3 (1995 est.)

Airports

Total:	15
With paved runways over 3,047 m:	2
With paved runways 2,438 to 3,047 m:	2
With paved runways 1,524 to 2,437 m:	4
With paved runways 914 to 1,523 m:	1
With paved runways under 914 m:	6 (1995 est.)

Tourism [34]

	1990	1991	1992	1993	1994
Tourists[8]	115	113	110	127	140
Tourism receipts	11	9	8	15	19
Tourism expenditures	78	83	111	153	210

Travelers are in thousands, money in million U.S. dollars.

For sources, notes, and explanations, see Annotated Source Appendix, page 1061.

75

Manufacturing Sector

Manufacturing Summary [35]

	1987		1988		1989		1990		1991	
	$ bil.	%	$ bil.	%	$ bil.	%	$ bil.	%	$ bil.	%
Establishments or enterprises (number)	4,909	0.209	7,071	0.337	32,523	1.736	-	-	-	-
Total employment (000)	761	0.560	808	0.590	1,473	1.197	-	-	-	-
Production workers (000)	609	0.904	650	1.041	1,225	1.665	-	-	-	-
Output ($ bil.)	4	0.035	4	0.033	7	0.060	-	-	-	-
Value added ($ bil.)	1	0.033	1	0.030	3	0.052	-	-	-	-
Capital investment ($ mil.)	181	0.042	123	0.026	341	0.062	-	-	-	-
M & E investment ($ mil.)	128	0.041	70	0.020	220	0.056	-	-	-	-
Employees per establishment (number)	155	267.823	114	175.298	45	68.943	-	-	-	-
Production workers per establishment	124	431.843	92	309.102	38	95.897	-	-	-	-
Output per establishment ($ mil.)	0.726	16.761	0.539	9.699	0.219	3.481	-	-	-	-
Capital investment per estab. ($ mil.)	0.037	19.865	0.017	7.657	0.010	3.584	-	-	-	-
M & E per establishment ($ mil)	0.026	19.587	0.010	5.990	0.007	3.242	-	-	-	-
Payroll per employee ($)	841	9.378	912	9.157	865	7.746	-	-	-	-
Wages per production worker ($)	739	9.367	795	9.407	712	7.099	-	-	-	-
Hours per production worker (hours)	-	-	-	-	-	-	-	-	-	-
Output per employee ($)	4,687	6.258	4,713	5.533	4,839	5.050	-	-	-	-
Capital investment per employee ($)	238	7.417	152	4.368	231	5.198	-	-	-	-
M & E per employee ($)	168	7.313	86	3.417	149	4.702	-	-	-	-

Note: Columns headed % show percent of world total or ratio. Ratios closest to 100 are closest to world average. M & E stands for machinery & equipment.

Output in Manufacturing

	1987		1988		1989		1990		1991	
	$ bil.	%	$ bil.	%	$ bil.	%	$ bil.	%	$ bil.	%
3110 - Food products	0.370	9.25	0.408	10.20	1.201	17.16	-	-	-	-
3130 - Beverages	0.007	0.18	0.007	0.18	0.019	0.27	-	-	-	-
3140 - Tobacco	0.211	5.28	0.145	3.63	0.191	2.73	-	-	-	-
3210 - Textiles	0.719	17.98	0.771	19.27	1.279	18.27	-	-	-	-
3211 - Spinning, weaving, etc.	0.596	14.90	0.742	18.55	1.140	16.29	-	-	-	-
3220 - Wearing apparel	0.043	1.08	0.069	1.72	0.545	7.79	-	-	-	-
3230 - Leather and products	0.135	3.38	0.148	3.70	0.265	3.79	-	-	-	-
3240 - Footwear	0.023	0.58	0.028	0.70	0.053	0.76	-	-	-	-
3310 - Wood products	0.019	0.48	0.020	0.50	0.047	0.67	-	-	-	-
3320 - Furniture, fixtures	0.006	0.15	0.005	0.13	0.005	0.07	-	-	-	-
3410 - Paper and products	0.088	2.20	0.090	2.25	0.167	2.39	-	-	-	-
3411 - Pulp, paper, etc.	0.082	2.05	0.083	2.08	0.133	1.90	-	-	-	-
3420 - Printing, publishing	0.022	0.55	0.022	0.55	0.142	2.03	-	-	-	-
3510 - Industrial chemicals	0.167	4.17	0.287	7.17	0.525	7.50	-	-	-	-
3511 - Basic chemicals, excl fertilizers	0.001	0.02	0.113	2.83	0.222	3.17	-	-	-	-
3513 - Synthetic resins, etc.	-	-	-	-	0.000	0.00	-	-	-	-
3520 - Chemical products nec	0.206	5.15	0.111	2.78	0.182	2.60	-	-	-	-
3522 - Drugs and medicines	0.116	2.90	0.004	0.10	0.050	0.71	-	-	-	-
3530 - Petroleum refineries	0.258	6.45	0.261	6.53	0.015	0.21	-	-	-	-
3540 - Petroleum, coal products	0.010	0.25	0.009	0.23	0.003	0.04	-	-	-	-
3550 - Rubber products	0.006	0.15	0.006	0.15	0.021	0.30	-	-	-	-
3560 - Plastic products nec	0.014	0.35	0.015	0.38	0.028	0.40	-	-	-	-
3610 - Pottery, china, etc.	0.008	0.20	0.012	0.30	0.021	0.30	-	-	-	-
3620 - Glass and products	0.009	0.23	0.008	0.20	0.006	0.09	-	-	-	-
3690 - Nonmetal products nec	0.027	0.67	0.026	0.65	0.067	0.96	-	-	-	-
3710 - Iron and steel	0.127	3.17	0.110	2.75	0.306	4.37	-	-	-	-
3720 - Nonferrous metals	-	-	0.005	0.13	-	-	-	-	-	-
3810 - Metal products	0.045	1.13	0.047	1.18	0.111	1.59	-	-	-	-
3820 - Machinery nec	0.035	0.88	0.029	0.73	0.024	0.34	-	-	-	-
3830 - Electrical machinery	0.083	2.08	0.086	2.15	0.152	2.17	-	-	-	-
3832 - Radio, television, etc.	0.026	0.65	0.028	0.70	0.024	0.34	-	-	-	-
3840 - Transportation equipment	0.039	0.98	0.042	1.05	0.093	1.33	-	-	-	-
3841 - Shipbuilding, repair	0.006	0.15	0.006	0.15	0.016	0.23	-	-	-	-
3843 - Motor vehicles	0.017	0.43	0.021	0.53	0.043	0.61	-	-	-	-
3850 - Professional goods	0.000	0.00	0.000	0.00	0.002	0.03	-	-	-	-
3900 - Industries nec	0.045	1.13	0.040	1.00	0.034	0.49	-	-	-	-

Note: Codes are International Standard Industry codes (ISIC). Percentages are % of total Output. [f]: Factor Prices; [p]: Producer Prices.

                                    For sources, notes, and explanations, see Annotated Source Appendix, page 1061.

Finance, Economics, and Trade

Economic Indicators [36]

- **National product**: GDP—purchasing power parity—$144.5 billion (1995 est.)
- **National product real growth rate**: 4.6% (1995 est.)
- **National product per capita**: $1,130 (1995 est.)
- **Inflation rate (consumer prices)**: 4.5% (1995 est.)
- **External debt**: $15.7 billion (1995 est.)

Balance of Payments Summary [37]

Values in millions of dollars.

	1989	1990	1991	1992	1993
Exports of goods (f.o.b.)	1,304.8	1,672.4	1,688.7	2,097.9	2,277.9
Imports of goods (f.o.b.)	-3,300.1	-3,259.4	-3,074.5	-3,353.8	-3,560.9
Trade balance	-1,995.3	-1,587.0	-1,385.8	-1,255.9	-1,283.0
Services - debits	-923.4	-880.2	-862.2	-954.7	-1,095.8
Services - credits	423.1	455.8	501.0	583.4	630.7
Private transfers (net)	806.8	828.3	901.8	1,019.8	1,132.4
Government transfers (net)	589.1	785.2	909.8	788.2	812.9
Long-term capital (net)	889.8	849.6	551.3	731.1	452.7
Short-term capital (net)	-56.5	-151.8	-83.7	-192.7	38.1
Errors and omissions	-43.1	-75.7	-98.4	-84.0	12.0
Overall balance	-309.5	224.2	433.8	635.2	700.0

Exchange Rates [38]

Currency: **taka.**
Symbol: **Tk.**

Data are currency units per $1.

January 19965	40.933
1995	40.278
1994	40.212
1993	39.567
1992	38.951
1991	36.596

Imports and Exports

Top Import Origins [39]

$4.7 billion (1995 est.) Data are for FY91/92 est.

Origins	%
Hong Kong	7.5
Singapore	7.4
China	7.4
Japan	7.1

Top Export Destinations [40]

$2.7 billion (1995 est.) Data are for FY91/92 est.

Destinations	%
Western Europe	39
Germany	8.4
Italy	6
US	33

Foreign Aid [41]

Recipient: ODA, $1.099 billion (1993).

Import and Export Commodities [42]

Import Commodities	Export Commodities
Capital goods	Garments
Petroleum	Jute and jute goods
Food	Leather
Textiles	Shrimp

For sources, notes, and explanations, see Annotated Source Appendix, page 1061.

77

Barbados

Geography [1]

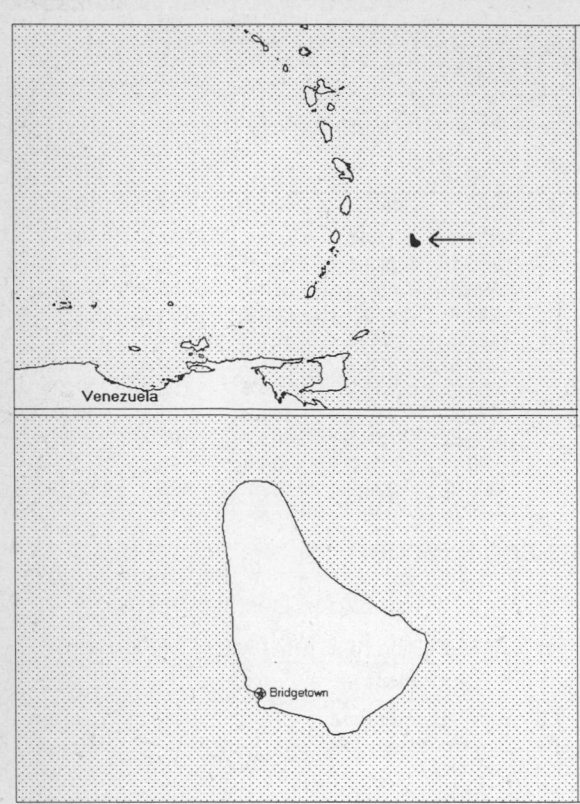

Total area:
  430 sq km   166 sq mi
Land area:
  430 sq km   166 sq mi
Comparative area:
  2.5 times the size of Washington, DC
Land boundaries:
  0 km
Coastline:
  97 km
Climate:
  Tropical; rainy season (June to October)
Terrain:
  Relatively flat; rises gently to central highland region
Natural resources:
  Petroleum, fish, natural gas
Land use:
  Arable land: 77%
  Permanent crops: 0%
  Meadows and pastures: 9%
  Forest and woodland: 0%
  Other: 14%

Demographics [2]

	1970	1980	1990	1995[1]	1996	2000	2010	2020	2030
Population	239	252	254	256	257	260	272	284	290
Population density (persons per sq. mi.)	1,438	1,518	1,532	1,544	1,548	1,567	1,639	1,712	1,746
(persons per sq. km.)	555	586	592	596	598	605	633	661	674
Net migration rate (per 1,000 population)	NA	-11.7	-6.5	-4.8	-4.5	-3.2	0.0	0.0	0.0
Births	5	NA	NA	NA	4	NA	NA	NA	NA
Deaths[7]	2	2	2	NA	2	NA	NA	NA	NA
Life expectancy - males	NA	70.2	69.7	71.5	71.7	72.4	74.1	75.4	76.6
Life expectancy - females	NA	73.9	75.6	77.1	77.3	78.0	79.7	81.0	82.2
Birth rate (per 1,000)	NA	18.0	16.0	15.5	15.3	14.7	12.7	11.4	10.3
Death rate (per 1,000)	NA	8.5	8.9	8.3	8.2	7.9	7.5	8.0	9.9
Women of reproductive age (15-49 yrs.)	NA	63	70	72	72	74	70	66	63
of which are currently married	16	16	33	NA	36	37	36	NA	NA
Fertility rate	NA	2.1	1.8	1.8	1.8	1.8	1.8	1.8	1.7

Except as noted, values for vital statistics are in thousands; life expectancy is in years.

Health

Health Indicators [3]

Health Expenditures [4]

For sources, notes, and explanations, see Annotated Source Appendix, page 1061.

Human Factors

Women and Children [5]	Burden of Disease [6]	
	Population per physician (1984)	1,124.44
	Population per nurse (1984)	223.10
	Population per hospital bed	NA
	AIDS cases per 100,000 people (1994)	44.1
	Malaria cases per 100,000 people (1992)	NA

Ethnic Division [7]		Religion [8]		Major Languages [9]
African	80%	Protestant	67%	English.
European	4%	Roman Catholic	4%	
Other	16%	None	17%	
		Unknown	3%	
		Other	9%	
		(1980)		

Education

Public Education Expenditures [10]

Million (Dollar)	1980	1990	1991	1992	1993	1994
Total education expenditure	109	269	248	215	238	NA
as percent of GNP	6.5	7.9	7.6	7.0	7.5	NA
as percent of total govt. expend.	20.5	22.2	18.4	16.9	18.6	NA
Current education expenditure	90	218	212	192	230	NA
as percent of GNP	5.4	6.4	6.5	6.3	7.3	NA
as percent of current govt. expend.	NA	22.1	18.7	17.5	20.0	NA
Capital expenditure	19	51	36	23	8	NA

Educational Attainment [11]

Age group (1980)	25+
Total population	116,874
Highest level attained (%)	
No schooling	0.8
First level	
Not completed	63.5
Completed	NA
Entered second level	
S-1	32.3
S-2	NA
Postsecondary	3.3

Illiteracy [12]

In thousands and percent[1]	1990	1995	2000
Illiterate population (15+ yrs.)	6	5	4
Illiteracy rate - total pop. (%)	3.2	2.6	2.0
Illiteracy rate - males (%)	2.2	2.2	2.1
Illiteracy rate - females (%)	4.0	2.9	2.9

Libraries [13]

	Admin. Units	Svc. Pts.	Vols. (000)	Shelving (meters)	Vols. Added	Reg. Users
National (1992)	1	NA	38	NA	1,011	NA
Nonspecialized	NA	NA	NA	NA	NA	NA
Public (1992)	1	9	175	NA	5,069	56,094
Higher ed. (1993)[4]	1	1	29	365	616	195
School	NA	NA	NA	NA	NA	NA

Daily Newspapers [14]

	1980	1985	1990	1994
Number of papers	2	2	2	2
Circ. (000)	39	40	30	41

Culture [15]

Cinema (seats per 1,000)	NA
Annual attendance per person	0.0
Gross box office receipts (mil. Dollar)	NA
Museums (reporting)	4
Visitors (000)	62
Annual receipts (000 Dollar)	587

Science and Technology

Scientific/Technical Forces [16]	R&D Expenditures [17]	U.S. Patents Issued [18]

For sources, notes, and explanations, see Annotated Source Appendix, page 1061.

79

Government and Law

Organization of Government [19]

Long-form name:
None
Type:
Parliamentary democracy
Independence:
30 November 1966 (from UK)
National holiday:
Independence Day, 30 November (1966)
Constitution:
30 November 1966
Legal system:
English common law; no judicial review of legislative acts
Executive branch:
British Monarch (represented by Acting Governor General); Prime Minister; Deputy Prime Minister; Cabinet
Legislative branch:
Bicameral Parliament: Senate and House of Assembly
Judicial branch:
Supreme Court of Judicature

Elections [20]

House of Assembly	% of seats
Barbados Labor Party (BLP)	67.9
Democratic Labor Party (DLP)	28.6
National Democratic Party (NDP)	3.6

Government Expenditures [21]

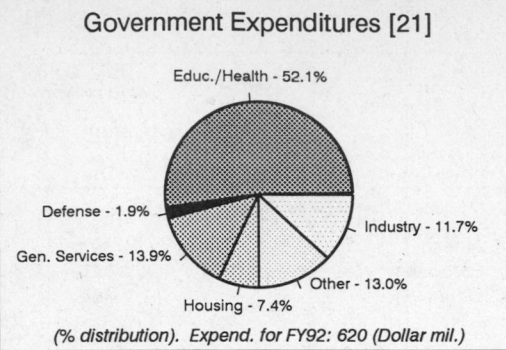

Educ./Health - 52.1%
Defense - 1.9%
Industry - 11.7%
Gen. Services - 13.9%
Other - 13.0%
Housing - 7.4%

(% distribution). Expend. for FY92: 620 (Dollar mil.)

Crime [23]

	1994
Crime volume	
Cases known to police	11,428
Attempts (percent)	NA
Percent cases solved	36.40
Crimes per 100,000 persons	4,337.00
Persons responsible for offenses	
Total number offenders	4,165
Percent female	9.50
Percent juvenile (7-16 yrs.)	2.50
Percent foreigners	NA

Military Expenditures and Arms Transfers [22]

	1990	1991	1992	1993	1994
Military expenditures					
Current dollars (mil.)	NA	NA	NA	12	13
1994 constant dollars (mil.)	NA	NA	NA	13	13
Armed forces (000)	0	0	0	0	0
Gross national product (GNP)					
Current dollars (mil.)	1,551	1,554	1,531	1,569	1,671
1994 constant dollars (mil.)	1,726	1,666	1,596	1,601	1,671
Central government expenditures (CGE)					
1994 constant dollars (mil.)[e]	594	552	524	553	546
People (mil.)	0.3	0.3	0.3	0.3	0.3
Military expenditure as % of GNP	NA	NA	NA	0.8	0.8
Military expenditure as % of CGE	NA	NA	NA	2.3	2.4
Military expenditure per capita (1994 $)	NA	NA	NA	49	51
Armed forces per 1,000 people (soldiers)	1.8	1.6	1.6	1.6	0
GNP per capita (1994 $)	6,784	6,543	6,261	6,271	6,530
Arms imports[6]					
Current dollars (mil.)	0	0	0	0	0
1994 constant dollars (mil.)	0	0	0	0	0
Arms exports[6]					
Current dollars (mil.)	0	0	0	0	0
1994 constant dollars (mil.)	0	0	0	0	0
Total imports[7]					
Current dollars (mil.)	700	694	521	574	608
1994 constant dollars (mil.)	779	744	543	586	608
Total exports[7]					
Current dollars (mil.)	209	205	191	179	181
1994 constant dollars (mil.)	233	220	199	183	181
Arms as percent of total imports[8]	0	0	0	0	0
Arms as percent of total exports[8]	0	0	0	0	0

Human Rights [24]

	SSTS	FL	FAPRO	PPCG	APROBC	TPW	PCPTW	STPEP	PHRFF	PRW	ASST	AFL
Observes	P	P	P	P	P	P	P			P	P	P
		EAFRD	CPR	ESCR	SR	ACHR	MAAE	PVIAC	PVNAC	EAFDAW	TCIDTP	RC
Observes		P	P	P		P		P	P	P		P

P = Party; S = Signatory; see Appendix for meaning of abbreviations.

Labor Force

Total Labor Force [25]

126,000 (1993)

Labor Force by Occupation [26]

Services and government	41%
Commerce	15
Manufacturing and construction	18
Transportation, storage, communications, and financial	8
Agriculture	6
Utilities	2

Date of data: 1992 est.

Unemployment Rate [27]

19.9% (September 1995)

For sources, notes, and explanations, see Annotated Source Appendix, page 1061.

Production Sectors

Commercial Energy Production and Consumption

Data are shown in quadrillion (10^{15}) BTUs and percent for 1995
Values for hydroelectric, nuclear, geothermal, solar, and wind power refer to electrical generation.

Production [28]	Consumption [29]
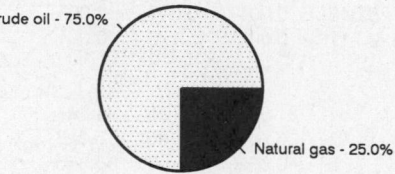	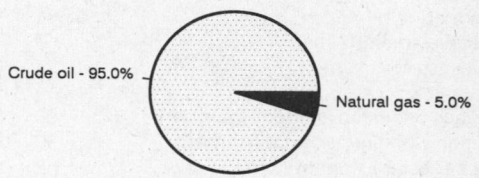

Production [28]: Crude oil - 75.0%; Natural gas - 25.0%

Consumption [29]: Crude oil - 95.0%; Natural gas - 5.0%

Crude oil	0.003
Dry natural gas	0.001
Total	0.004

Crude oil	0.019
Dry natural gas	0.001
Total	0.020

Telecommunications [30]

- 87,343 (1991 est.) telephones
- Domestic: island wide automatic telephone system
- International: satellite earth station - 1 Intelsat (Atlantic Ocean); tropospheric scatter to Trinidad and Saint Lucia
- Radio: Broadcast stations: AM 3, FM 2, shortwave 0
- Television: Broadcast stations: 2 (1 pay) Televisions: 69,350 (1993 est.)

Transportation [31]

Railways: 0 km

Highways: total: 1,550 km; paved: 1,550 km

Merchant marine: total: 34 ships (1,000 GRT or over) totaling 183,937 GRT/271,707 DWT; ships by type: bulk 6, cargo 21, combination bulk 3, oil tanker 3, roll-on/roll-off cargo 1 (1995 est.)

Airports
Total: 1
With paved runways over 3,047 m: 1 (1995 est.)

Top Agricultural Products [32]

Agriculture accounts for 6.4% of the GDP; produces sugarcane, vegetables, cotton.

Top Mining Products [33]

Thousand metric tons except as noted	3/31/95[*]
Cement, hydraulic	200
Gas, liquefied petroleum (42-gal. bls.)	20,000
Gas, natural, gross (mil. cu. meters)	35
Petroleum (000 42-gal. bls.)	
crude	475
refinery products	2,250

Tourism [34]

	1990	1991	1992	1993	1994
Visitors	795	766	785	825	885
Tourists[12]	432	394	385	396	426
Cruise passengers	363	372	400	429	459
Tourism receipts	494	460	463	528	598
Tourism expenditures	47	44	41	52	NA
Fare receipts	NA	2	3	3	NA
Fare expenditures	25	26	26	29	NA

Travelers are in thousands, money in million U.S. dollars.

For sources, notes, and explanations, see Annotated Source Appendix, page 1061.

81

Manufacturing Sector

Manufacturing Summary [35]

	1987		1988		1989		1990		1991	
	$ bil.	%	$ bil.	%	$ bil.	%	$ bil.	%	$ bil.	%
Establishments or enterprises (number)	101	0.004	97	0.005	73	0.004	57	0.003	-	-
Total employment (000)	5	0.004	5	0.004	4	0.003	3	0.003	-	-
Production workers (000)	4	0.006	4	0.006	3	0.004	3	0.004	-	-
Output ($ bil.)	0.181	0.002	0.199	0.002	0.184	0.002	0.184	0.002	-	-
Value added ($ bil.)	0.051	0.001	0.054	0.001	0.048	0.001	0.047	0.001	-	-
Capital investment ($ mil.)	7	0.002	14	0.003	11	0.002	6	0.001	-	-
M & E investment ($ mil.)	-	-	-	-	-	-	-	-	-	-
Employees per establishment (number)	50	86.443	50	77.327	54	81.933	58	94.494	-	-
Production workers per establishment	40	139.814	39	130.762	42	106.104	46	115.213	-	-
Output per establishment ($ mil.)	2	41.440	2	36.955	3	40.009	3	50.365	-	-
Capital investment per estab. ($ mil.)	0.071	38.409	0.140	61.643	0.148	50.560	0.100	32.342	-	-
M & E per establishment ($ mil)	-	-	-	-	-	-	-	-	-	-
Payroll per employee ($)	7,315	81.571	6,936	69.616	7,511	67.231	9,218	66.455	-	-
Wages per production worker ($)	-	-	-	-	-	-	-	-	-	-
Hours per production worker (hours)	-	-	-	-	-	-	-	-	-	-
Output per employee ($)	35,904	47.940	40,706	47.791	46,799	48.831	55,236	53.300	-	-
Capital investment per employee ($)	1,427	44.432	2,774	79.717	2,746	61.710	1,719	34.226	-	-
M & E per employee ($)	-	-	-	-	-	-	-	-	-	-

Note: Columns headed % show percent of world total or ratio. Ratios closest to 100 are closest to world average. M & E stands for machinery & equipment.

Output in Manufacturing

	1987		1988		1989		1990		1991	
	$ bil.	%	$ bil.	%	$ bil.	%	$ bil.	%	$ bil.	%
3110 - Food products	0.113	62.43	0.121	60.80	0.125	67.93	0.126	68.48	-	-
3220 - Wearing apparel	0.021	11.60	0.023	11.56	0.014	7.61	0.009	4.89	-	-
3520 - Chemical products nec	0.016	8.84	0.016	8.04	0.012	6.52	0.018	9.78	-	-
3810 - Metal products	0.029	16.02	0.038	19.10	0.031	16.85	0.030	16.30	-	-
3900 - Industries nec	0.002	1.10	0.002	1.01	0.002	1.09	0.001	0.54	-	-

Note: Codes are International Standard Industry codes (ISIC). Percentages are % of total Output. [f]: Factor Prices; [p]: Producer Prices.

Finance, Economics, and Trade

Economic Indicators [36]

- **National product**: GDP—purchasing power parity—$2.5 billion (1995 est.)
- **National product real growth rate**: 2% (1995 est.)
- **National product per capita**: $9,800 (1995 est.)
- **Inflation rate (consumer prices)**: 1.7% (1995 est.)
- **External debt**: $408 million (1995 est.)

Balance of Payments Summary [37]

Values in millions of dollars.

	1989	1990	1991	1992	1993
Exports of goods (f.o.b.)	146.9	151.0	143.6	158.0	152.4
Imports of goods (f.o.b.)	-599.3	-624.1	-617.4	-464.7	-511.3
Trade balance	-452.4	-473.1	-473.8	-306.7	-358.9
Services - debits	-331.0	-271.2	-272.8	-254.2	-327.3
Services - credits	774.8	685.3	688.4	664.3	729.7
Private transfers (net)	31.9	34.8	32.0	39.6	26.0
Government transfers (net)	-26.1	7.8	1.0	0.6	-5.2
Long-term capital (net)	1.1	31.5	-2.9	-19.0	-25.2
Short-term capital (net)	-1.2	16.6	20.7	-75.0	26.4
Errors and omissions	-39.5	-70.6	-32.6	-21.5	-44.7
Overall balance	-42.4	-38.9	-40.0	28.1	20.8

Exchange Rates [38]

Currency: **Barbadian dollar.**
Symbol: **Bds$.**

Data are currency units per $1.

Fixed rate	2.0113

Imports and Exports

Top Import Origins [39]

$693 million (c.i.f., 1995 est.).

Origins	%
US	36
UK	11
Trinidad and Tobago	11
Japan	3

Top Export Destinations [40]

$158.6 million (f.o.b., 1995 est.).

Destinations	%
US	13
UK	10
Trinidad and Tobago	9
Windward Islands	8

Foreign Aid [41]

	U.S. $	
US commitments, including Ex-Im (FY70-89)	15	million
Western (non-US) countries, ODA and OOF bilateral commitments (1970-89)	171	million

Import and Export Commodities [42]

Import Commodities
- Consumer goods
- Machinery
- Foodstuffs
- Construction materials
- Chemicals
- Fuel
- Electrical components

Export Commodities
- Sugar and molasses
- Rum
- Other foods and beverages
- Chemicals
- Electrical components
- Clothing

For sources, notes, and explanations, see Annotated Source Appendix, page 1061.

Belarus

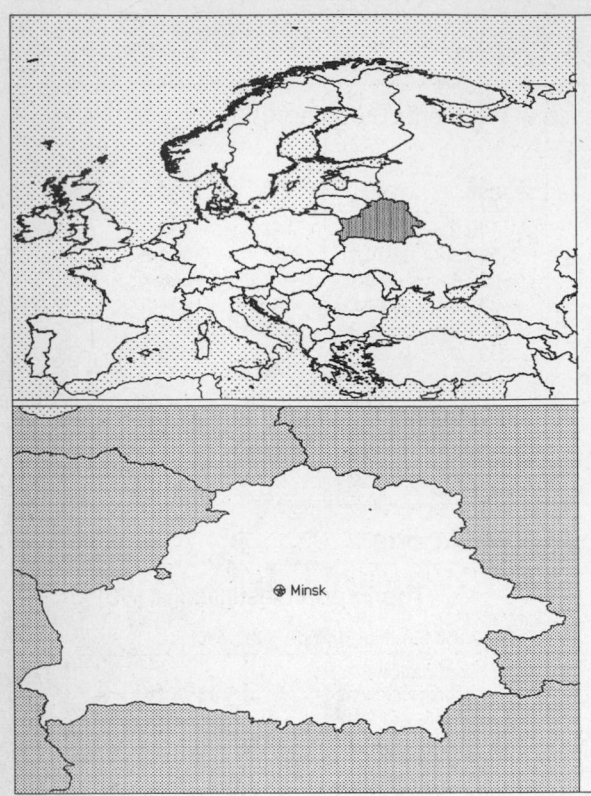

Geography [1]

Total area:
  207,600 sq km   80,155 sq mi
Land area:
  207,600 sq km   80,155 sq mi
Comparative area:
  Slightly smaller than Kansas
Land boundaries:
  Total 3,098 km, Latvia 141 km, Lithuania 502 km, Poland 605 km, Russia 959 km, Ukraine 891 km
Coastline:
  0 km (landlocked)
Climate:
  Cold winters, cool and moist summers; transitional between continental and maritime
Terrain:
  Generally flat and contains much marshland
Natural resources:
  Forests, peat deposits, small quantities of oil and natural gas
Land use:
  Arable land: 29%
  Permanent crops: 1%
  Meadows and pastures: 15%
  Forest and woodland: 0%
  Other: 55%

Demographics [2]

	1970	1980	1990	1995[1]	1996	2000	2010	2020	2030
Population	9,027	9,644	10,215	10,398	10,416	10,545	10,924	11,059	11,101
Population density (persons per sq. mi.)	113	120	127	130	130	132	136	138	138
(persons per sq. km.)	43	46	49	50	50	51	53	53	53
Net migration rate (per 1,000 population)	NA	NA	-1.9	3.6	3.5	3.0	1.7	0.6	0.0
Births	NA	NA	NA	NA	127	NA	NA	NA	NA
Deaths	NA	NA	NA	NA	142	NA	NA	NA	NA
Life expectancy - males	NA	NA	66.2	62.9	63.2	64.4	67.3	70.7	73.6
Life expectancy - females	NA	NA	75.8	74.1	74.2	74.7	75.9	78.6	80.8
Birth rate (per 1,000)	NA	NA	14.0	11.6	12.2	14.0	13.1	11.1	11.0
Death rate (per 1,000)	NA	NA	10.9	13.8	13.6	12.9	12.4	11.1	10.8
Women of reproductive age (15-49 yrs.)	NA	NA	2,462	2,624	2,662	2,748	2,710	2,650	2,554
of which are currently married	NA	NA	1,688	NA	1,828	1,870	1,895	NA	NA
Fertility rate	NA	NA	1.9	1.6	1.7	1.9	1.8	1.7	1.7

Except as noted, values for vital statistics are in thousands; life expectancy is in years.

Health

Health Indicators [3]

% of population with access to	
safe water (1990-95)	NA
adequate sanitation (1990-95)	NA
health services (1985-95)	NA
% of 1-year-olds immunized (1990-94) against	
TB (tuberculosis)	93
DPT (diphtheria, pertussis, tetanus)	92
polio	93
measles	97
% of contraceptive prevalence (1980-94)	NA
ORT use rate (1990-94)	NA

Health Expenditures [4]

Total health expenditure, 1990 (official exchange rate)	
Millions of dollars	1,613
Dollars per capita	157
Health expenditures as a percentage of GDP	
Total	3.2
Public sector	2.2
Private sector	1.0
Development assistance for health	
Total aid flows (millions of dollars)[1]	NA
Aid flows per capita (dollars)	NA
Aid flows as a percentage of total health expenditure	NA

For sources, notes, and explanations, see Annotated Source Appendix, page 1061.

Human Factors

Women and Children [5]

% of pregnant women immunized (tetanus 1990-94)	NA
% of births attended by trained health personnel (1983-94)	NA
Maternal mortality rate (1980-92)	NA
Under-5 mortality rate (1994)	21
% under-5 moderately/severely underweight (1980-1994)	NA

Burden of Disease [6]

Population per physician (1993)	235.92
Population per nurse (1993)	89.28
Population per hospital bed (1993)	80.29
AIDS cases per 100,000 people (1994)	*
Heart disease cases per 1,000 people (1990-93)	396.5

Ethnic Division [7]

Byelorussian	77.9%
Russian	13.2%
Polish	4.1%
Ukrainian	2.9%
Other	1.9%

Religion [8]

Eastern Orthodox	60%
Other (primarily Roman Catholic and Muslim)	40%
(early 1990s)	

Major Languages [9]

Byelorussian, Russian, other.

Education

Public Education Expenditures [10]

Million or Trillion (T) (Ruble)[6]	1980	1985	1990	1992	1993	1994
Total education expenditure	1,253	1,499	2,093	60,520	662,987	1.23 T
as percent of GNP	5.2	4.8	5.0	6.6	5.3	6.1
as percent of total govt. expend.	NA	NA	NA	19.3	15.9	17.3
Current education expenditure	1,051	1,264	1,758	49,528	527,327	1.02 T
as percent of GNP	4.3	4.0	4.2	5.4	4.2	5.0
as percent of current govt. expend.	NA	NA	NA	20.0	NA	NA
Capital expenditure	202	235	335	10,992	135,660	208,220

Educational Attainment [11]

Age group (1979)	10+
Total population	NA
Highest level attained (%)	
No schooling	40.6
First level	
Not completed	NA
Completed	NA
Entered second level	
S-1	41.5
S-2	10.2
Postsecondary	7.7

Illiteracy [12]

	1985	1989	1995
Illiterate population (15+ yrs.)[2]	NA	165,406	38,000
Illiteracy rate - total pop. (%)	NA	2.1	0.5
Illiteracy rate - males (%)	NA	0.6	0.3
Illiteracy rate - females (%)	NA	3.4	0.6

Libraries [13]

	Admin. Units	Svc. Pts.	Vols. (000)	Shelving (meters)	Vols. Added	Reg. Users
National (1992)	1	2	6,868	NA	109,900	37,000
Nonspecialized (1992)	7	16	4,511	NA	115,900	226,500
Public (1992)	5,743	11,329	77,142	NA	NA	4 mil
Higher ed. (1993)	38	NA	20,428	NA	636,429	129,683
School (1990)	5,198	NA	76,281	NA	NA	NA

Daily Newspapers [14]

	1980	1985	1990	1994
Number of papers	27	28	28	10
Circ. (000)	2,343	2,446	2,937	1,899[e]

Culture [15]

Cinema (seats per 1,000)	73.1
Annual attendance per person	2.9
Gross box office receipts (mil. Ruble)	2,581
Museums (reporting)	117
Visitors (000)	2,682
Annual receipts (000 Ruble)	4,343,672

Science and Technology

Scientific/Technical Forces [16]

Scientists/engineers	33,685
Number female	5,463
Technicians	5,254
Number female	NA
Total	38,939

R&D Expenditures [17]

	Ruble (000) 1992
Total expenditure	8,590,900
Capital expenditure	2,258,700
Current expenditure	6,332,200
Percent current	73.7

U.S. Patents Issued [18]

Values show patents issued to citizens of the country by the U.S. Patents Office.

	1993	1994	1995
Number of patents	0	2	3

For sources, notes, and explanations, see Annotated Source Appendix, page 1061.

Government and Law

Organization of Government [19]

Long-form name:
Republic of Belarus
Type:
Republic
Independence:
25 August 1991 (from Soviet Union); the Belarussian Supreme Soviet issued a proclamation of independence; on 17 July 1990 Belarus issued a declaration of sovereignty
National holiday:
Independence Day, 27 July (1990)
Constitution:
Adopted 15 March 1994; replaces constitution of April 1978
Legal system:
Based on civil law system
Executive branch:
President; Prime Minister; 5 Deputy Prime Ministers; Council of Ministers
Legislative branch:
Unicameral: Supreme Soviet
Judicial branch:
Supreme Court

Elections [20]

Supreme Soviet	% of votes
Independents	36.5
Belarusian Communist Party (KPB)	16.2
Agrarian Party	12.7
Civic Accord Bloc(CAB)	3.5
Party of People's Concord	3.1
All-Belar. Unity and Concord	0.8
Belarusian Social-Democrat	0.8
Other	2.7
Vacant	23.8

Government Expenditures [21]

Educ./Health - 56.6%
Gen. Services - 5.2%
Other - 9.5%
Defense - 4.1%
Industry - 23.4%
Housing - 1.2%

(% distribution). Expend. for CY92: 344,794 (Ruble mil.)

Military Expenditures and Arms Transfers [22]

	1990	1991	1992[14]	1993	1994
Military expenditures					
Current dollars (mil.)[e]	NA	NA	NA	200	173
1994 constant dollars (mil.)[e]	NA	NA	NA	204	173
Armed forces (000)	NA	NA	102[e]	115	90
Gross national product (GNP)					
Current dollars (mil.)[e]	NA	NA	70,010	65,630	50,920
1994 constant dollars (mil.)[e]	NA	NA	73,010	66,980	50,920
Central government expenditures (CGE)					
1994 constant dollars (mil.)[e]	NA	NA	36,140	NA	NA
People (mil.)	NA	NA	10.3	10.4	10.4
Military expenditure as % of GNP	NA	NA	NA	0.3	0.3
Military expenditure as % of CGE	NA	NA	NA	NA	NA
Military expenditure per capita (1994 $)	NA	NA	NA	20	17
Armed forces per 1,000 people (soldiers)	NA	NA	9.9	11.1	8.6
GNP per capita (1994 $)	NA	NA	7,065	6,459	4,894
Arms imports[6]					
Current dollars (mil.)	NA	NA	0	0	0
1994 constant dollars (mil.)	NA	NA	0	0	0
Arms exports[6]					
Current dollars (mil.)	NA	NA	0	0	0
1994 constant dollars (mil.)	NA	NA	0	0	0
Total imports[7]					
Current dollars (mil.)	NA	NA	2,929	4,654	4,296
1994 constant dollars (mil.)	NA	NA	3,054	4,750	4,296
Total exports[7]					
Current dollars (mil.)	NA	NA	3,064	3,659	2,413
1994 constant dollars (mil.)	NA	NA	3,195	3,734	2,413
Arms as percent of total imports[8]	NA	NA	0	0	0
Arms as percent of total exports[8]	NA	NA	0	0	0

Crime [23]

	1990
Crime volume	
Cases known to police	NA
Attempts (percent)	NA
Percent cases solved	NA
Crimes per 100,000 persons	NA
Persons responsible for offenses	
Total number offenders	29,840
Percent female	9.9
Percent juvenile	14.6
Percent foreigners	NA

Human Rights [24]

	SSTS	FL	FAPRO	PPCG	APROBC	TPW	PCPTW	STPEP	PHRFF	PRW	ASST	AFL
Observes	P	P	P	P	P	P	P	P		P	P	

	EAFRD	CPR	ESCR	SR	ACHR	MAAE	PVIAC	PVNAC	EAFDAW	TCIDTP	RC
Observes	P	P	P			P	P	P	P	P	P

P = Party; S = Signatory; see Appendix for meaning of abbreviations.

Labor Force

Total Labor Force [25]

4.259 million

Labor Force by Occupation [26]

Industry and construction	40%
Agriculture and forestry	21
Other	39

Date of data: 1992

Unemployment Rate [27]

2.6% officially registered unemployed (December 1994); large numbers of underemployed workers

For sources, notes, and explanations, see Annotated Source Appendix, page 1061.

Production Sectors

Commercial Energy Production and Consumption

Data are shown in quadrillion (10^{15}) BTUs and percent for 1995
Values for hydroelectric, nuclear, geothermal, solar, and wind power refer to electrical generation.

Production [28]	Consumption [29]
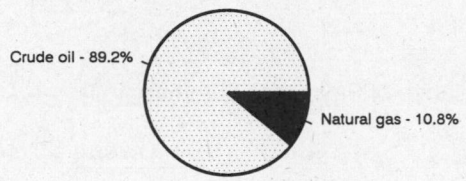	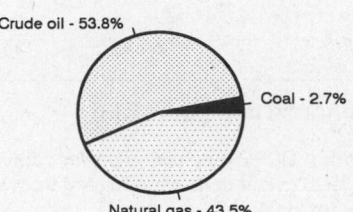

Crude oil	0.083	Crude oil	0.522
Dry natural gas	0.010	Dry natural gas	0.422
Total	0.093	Coal	0.026
		Total	0.970

Telecommunications [30]

- 1.849 million (1991 est.) telephones; telephone service inadequate for the purposes of either business or the population; about 70% of the telephones are in homes; over 750,000 applications from households for telephones remain unsatisfied (1992 est.); new investment centers on international connections and business needs
- Domestic: new NMT-450 analog cellular system now operating in Minsk
- International: international traffic is carried by the Moscow international gateway switch and also by satellite; satellite earth stations - 1 Intelsat (through Canada) and 1 Eutelsat (through the UK)
- Radio: Broadcast stations: AM 35, FM 18, shortwave 0 Radios: 3.17 million (1991 est.) (5,615,000 with multiple speaker systems for program diffusion)
- Television: Broadcast stations: 2 (one national and one private; the license of the private station was suspended during the parliamentary elections of 1994) Televisions: 3.5 million (1992 est.)

Top Agricultural Products [32]

Agriculture accounts for 21% of the GDP; produces grain, potatoes, vegetables; meat, milk.

Top Mining Products [33]

Thousand metric tons except as noted	3/31/95[*]
Cement	1,500
Nitrogen, N content of ammonia	400
Peat (fuel use)	4,000
Petroleum	
crude	2,000
refined	14,000
Potash, K2O content	2,500
Salt	300
Steel, crude	873[r]
Gas, natural (mil. cu. meters)	300

Transportation [31]

Railways: total: 5,488 km; broad gauge: 5,488 km 1.520-m gauge (873 km electrified) (1993)

Highways: total: 92,200 km; paved: 61,000 km (including graveled); unpaved: 31,200 km (1994 est.)

Merchant marine: Note: claims 5% of former Soviet fleet (1995 est.)

Airports

Total:	118
With paved runways over 3,047 m:	2
With paved runways 2,438 to 3,047 m:	18
With paved runways 1,524 to 2,437 m:	5
With paved runways under 914 m:	11
With unpaved runways over 3,047 m:	1

Tourism [34]

Finance, Economics, and Trade

Industrial Summary [35]

Industrial Production: Growth rate - 11% (1995 est.); accounts for 49% of the GDP. **Industries**: Tractors, metal-cutting machine tools, off-highway dump trucks up to 110-metric-ton load capacity, wheel-type earth movers for construction and mining, eight-wheel-drive, high-flotation trucks with cargo capacity of 25 metric tons for use in tundra and roadless areas, equipment for animal husbandry and livestock feeding, motorcycles, television sets, chemical fibers, fertilizer, linen fabric, wool fabric, radios, refrigerators, other consumer goods.

Economic Indicators [36]

- **National product**: GDP—purchasing power parity—$49.2 billion (1995 estimate as extrapolated from World Bank estimate for 1994)

- **National product real growth rate**: - 10% (1995 est.)

- **National product per capita**: $4,700 (1995 est.)

- **Inflation rate (consumer prices)**: 244% (1995 est.)

- **External debt**: $2 billion (September 1995 est.)

Balance of Payments Summary [37]

Exchange Rates [38]

Currency: **Belarusian ruble.**
Symbol: **BR.**

Data are currency units per $1.

Year-end 1995	11,500
Year-end 1994	10,600

Imports and Exports

Top Import Origins [39]

$4.6 billion (c.i.f., 1995).

Origins	%
Russia	NA
Ukraine	NA
Poland	NA
Germany	NA

Top Export Destinations [40]

$4.2 billion (f.o.b., 1995).

Destinations	%
Russia	NA
Ukraine	NA
Poland	NA
Germany	NA

Foreign Aid [41]

	U.S. $	
Commitments (1992-95)	3,930	million

Import and Export Commodities [42]

Import Commodities	Export Commodities
Fuel	Machinery and transport equipment
Natural gas	Chemicals
Industrial raw materials	Foodstuffs
Textiles	
Sugar	

Belgium

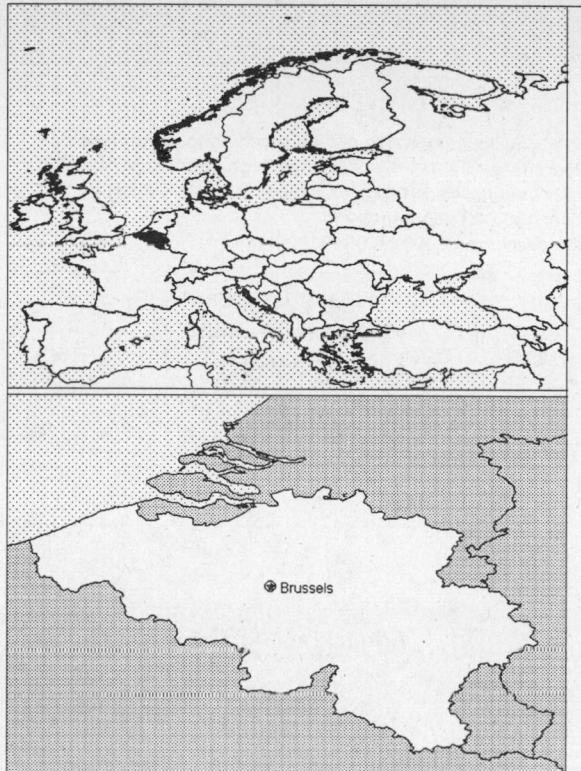

Geography [1]

Total area:
30,510 sq km   11,780 sq mi
Land area:
30,230 sq km   11,672 sq mi
Comparative area:
Slightly larger than Maryland
Land boundaries:
Total 1,385 km, France 620 km, Germany 167 km, Luxembourg 148 km,
Netherlands 450 km
Coastline:
64 km
Climate:
Temperate; mild winters, cool summers; rainy, humid, cloudy
Terrain:
Flat coastal plains in Northwest, central rolling hills, rugged mountains
of Ardennes Forest in Southeast
Natural resources:
Coal, natural gas
Land use:
Arable land: 24%
Permanent crops: 1%
Meadows and pastures: 20%
Forest and woodland: 21%
Other: 34%

Demographics [2]

	1970	1980	1990	1995[1]	1996	2000	2010	2020	2030
Population	9,638	9,847	9,962	10,136	10,170	10,286	10,358	10,271	10,046
Population density (persons per sq. mi.)	826	844	854	868	871	881	887	880	861
(persons per sq. km.)	319	326	330	335	336	340	343	340	332
Net migration rate (per 1,000 population)	NA	NA	1.0	1.7	1.6	1.3	0.7	0.3	0.0
Births	141	125	NA	NA	113	NA	NA	NA	NA
Deaths	119	114	60	NA	103	NA	NA	NA	NA
Life expectancy - males	NA	NA	72.7	73.7	73.9	74.5	75.9	78.0	79.5
Life expectancy - females	NA	NA	79.6	80.4	80.5	81.0	82.0	84.1	85.5
Birth rate (per 1,000)	NA	NA	12.4	12.0	12.0	11.2	9.8	9.5	8.6
Death rate (per 1,000)	NA	NA	10.5	10.3	10.3	10.4	11.1	11.0	12.0
Women of reproductive age (15-49 yrs.)	NA	NA	2,438	2,488	2,490	2,460	2,337	2,137	1,979
of which are currently married[8]	1,591	NA	1,712	NA	1,769	1,742	1,614	NA	NA
Fertility rate	NA	NA	1.6	1.7	1.7	1.7	1.6	1.6	1.6

Except as noted, values for vital statistics are in thousands; life expectancy is in years.

Health

Health Indicators [3]

% of population with access to	
safe water (1990-95)	NA
adequate sanitation (1990-95)	NA
health services (1985-95)	NA
% of 1-year-olds immunized (1990-94) against	
TB (tuberculosis)	NA
DPT (diphtheria, pertussis, tetanus)	85
polio	100
measles	67
% of contraceptive prevalence (1980-94)	79
ORT use rate (1990-94)	NA

Health Expenditures [4]

Total health expenditure, 1990 (official exchange rate)	
Millions of dollars	14,428
Dollars per capita	1,449
Health expenditures as a percentage of GDP	
Total	7.5
Public sector	6.2
Private sector	1.3
Development assistance for health	
Total aid flows (millions of dollars)[1]	NA
Aid flows per capita (dollars)	NA
Aid flows as a percentage of total health expenditure	NA

For sources, notes, and explanations, see Annotated Source Appendix, page 1061.

89

Human Factors

Women and Children [5]

% of pregnant women immunized (tetanus 1990-94)	NA
% of births attended by trained health personnel (1983-94)	100
Maternal mortality rate (1980-92)	3
Under-5 mortality rate (1994)	10
% under-5 moderately/severely underweight (1980-1994)	NA

Burden of Disease [6]

Population per physician (1987)	311.18
Population per nurse	NA
Population per hospital bed (1990)	120.83
AIDS cases per 100,000 people (1994)	2.3
Heart disease cases per 1,000 people (1990-93)	NA

Ethnic Division [7]

Fleming	55%
Walloon	33%
Mixed or other	12%

Religion [8]

Roman Catholic	75%
Protestant or other	25%

Major Languages [9]

Divided along ethnic lines.

Dutch	56%
French	32%
German	1%
Legally bilingual	11%

Education

Public Education Expenditures [10]

Million (Franc)[7]	1980	1990	1991	1992	1993	1994
Total education expenditure	208,469	325,282	342,383	361,584	409,254	NA
as percent of GNP	6.1	5.1	5.1	5.1	5.6	NA
as percent of total govt. expend.	NA	NA	NA	9.0	9.9	NA
Current education expenditure	206,227	321,427	338,707	358,648	403,466	NA
as percent of GNP	6.0	5.1	5.0	5.1	5.5	NA
as percent of current govt. expend.	NA	NA	NA	NA	NA	NA
Capital expenditure	2,243	3,855	3,676	2,936	5,788	NA

Educational Attainment [11]

Illiteracy [12]

Libraries [13]

	Admin. Units	Svc. Pts.	Vols. (000)	Shelving (meters)	Vols. Added	Reg. Users
National (1990)[5]	1	1	4,278[e]	57,591[e]	NA	104,544[e]
Nonspecialized (1990)	5[e]	6[e]	1,402[e]	24,139[e]	NA	55,262[e]
Public (1990)	38	1,151	29,678	823,936[e]	NA	2 mil[e]
Higher ed. (1990)	16	140	5,988[e]	147,413[e]	65,853	87,400
School	NA	NA	NA	NA	NA	NA

Daily Newspapers [14]

	1980	1985	1990	1994
Number of papers	26	24	33	32
Circ. (000)	2,289	2,171	3,000[e]	3,231

Culture [15]

Cinema (seats per 1,000)	NA
Annual attendance per person	1.9
Gross box office receipts (mil. Franc)	3,247
Museums (reporting)[1]	75
Visitors (000)	2,413
Annual receipts (000 Franc)	182,380

Science and Technology

Scientific/Technical Forces [16]

Scientists/engineers	18,105[e]
Number female	NA
Technicians[2]	21,958[e]
Number female	NA
Total[2]	40,063[e]

R&D Expenditures [17]

	Franc (000) 1991
Total expenditure	112,065,000[e]
Capital expenditure	NA
Current expenditure	NA
Percent current	NA

U.S. Patents Issued [18]

Values show patents issued to citizens of the country by the U.S. Patents Office.

	1993	1994	1995
Number of patents	377	391	419

For sources, notes, and explanations, see Annotated Source Appendix, page 1061.

Government and Law

Organization of Government [19]

Long-form name:
  Kingdom of Belgium
Type:
  Constitutional monarchy
Independence:
  4 October 1830 (from the Netherlands)
National holiday:
  National Day, 21 July (ascension of King
  Leopold to the throne in 1831)
Constitution:
  7 February 1831, last revised 14 July
  1993; parliament approved a
  constitutional package creating a federal
  state
Legal system:
  Civil law system influenced by English
  constitutional theory; judicial review of
  legislative acts; accepts compulsory ICJ
  jurisdiction, with reservations
Executive branch:
  King; Prime Minister; Cabinet
Legislative branch:
  Bicameral Parliament: Senate and
  Chamber of Deputies
Judicial branch:
  Supreme Court of Justice (Flemish - Hof
  van Cassatie, French - Cour de Cassation)

Elections [20]

Chamber of Deputies	% of votes
Flemish Social Christian (CVP)	17.2
Flemish Liberals and Dem.'s (VLD)	13.1
Flemish Socialist (SP)	12.6
Walloon Socialist (PS)	11.9
Walloon Liberal (PRL)	10.3
Vlaams Blok (VB)	7.8
Walloon Social Christian (PSC)	7.7
Volksunie (VU)	4.7
Flemish Greens (AGALEV)	4.4
Other	6.3

Government Budget [21]

Military Expenditures and Arms Transfers [22]

	1990	1991	1992	1993	1994
Military expenditures					
Current dollars (mil.)	4,729	4,865	4,068	3,888	3,944
1994 constant dollars (mil.)	5,263	5,215	4,242	3,969	3,944
Armed forces (000)	106	101	79	70	53
Gross national product (GNP)					
Current dollars (mil.)	193,700	206,900	216,200	219,600	229,400
1994 constant dollars (mil.)	215,500	221,800	225,500	224,100	229,400
Central government expenditures (CGE)					
1994 constant dollars (mil.)	108,000	111,700	115,000	113,700	NA
People (mil.)	10.0	10.0	10.0	10.0	10.1
Military expenditure as % of GNP	2.4	2.4	1.9	1.8	1.7
Military expenditure as % of CGE	4.9	4.7	3.7	3.5	NA
Military expenditure per capita (1994 $)	528	522	423	395	392
Armed forces per 1,000 people (soldiers)	10.6	10.1	7.9	7.0	5.3
GNP per capita (1994 $)	21,630	22,200	22,510	22,320	22,800
Arms imports[6]					
Current dollars (mil.)	280	300	100	320	40
1994 constant dollars (mil.)	312	322	104	327	40
Arms exports[6]					
Current dollars (mil.)	190	80	390	50	30
1994 constant dollars (mil.)	211	86	407	51	30
Total imports[7]					
Current dollars (mil.)	119,700	120,200	125,000	NA	NA
1994 constant dollars (mil.)	133,200	128,800	130,400	NA	NA
Total exports[7]					
Current dollars (mil.)	117,700	118,200	123,100	NA	NA
1994 constant dollars (mil.)	131,000	126,700	128,400	NA	NA
Arms as percent of total imports[8]	0.2	0.2	0.1	NA	NA
Arms as percent of total exports[8]	0.2	0.1	0.3	NA	NA

Crime [23]

	1994
Crime volume	
Cases known to police	583,309
Attempts (percent)	NA
Percent cases solved	NA
Crimes per 100,000 persons	58,330.90
Persons responsible for offenses	
Total number offenders	NA
Percent female	NA
Percent juvenile (0-18 yrs.)	NA
Percent foreigners	NA

Human Rights [24]

	SSTS	FL	FAPRO	PPCG	APROBC	TPW	PCPTW	STPEP	PHRFF	PRW	ASST	AFL
Observes	P	P	P	P	P	P	P	P	P	P	P	P
	EAFRD	CPR	ESCR	SR	ACHR	MAAE	PVIAC	PVNAC	EAFDAW	TCIDTP	RC	
Observes	P	P	P	P		P	P	P	P	S	P	

P = Party; S = Signatory; see Appendix for meaning of abbreviations.

Labor Force

Total Labor Force [25]

4.126 million

Labor Force by Occupation [26]

Services	63.6%
Industry	28
Construction	6.1
Agriculture	2.3
Date of data: 1988

Unemployment Rate [27]

14% (1995 est.)

For sources, notes, and explanations, see Annotated Source Appendix, page 1061.

91

Production Sectors

Commercial Energy Production and Consumption

Data are shown in quadrillion (10^{15}) BTUs and percent for 1995
Values for hydroelectric, nuclear, geothermal, solar, and wind power refer to electrical generation.

Production [28]

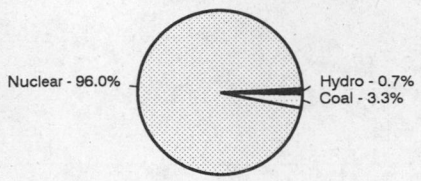

Nuclear - 96.0%
Hydro - 0.7%
Coal - 3.3%

Consumption [29]

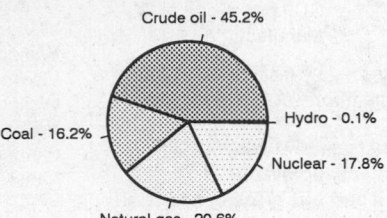

Crude oil - 45.2%
Hydro - 0.1%
Coal - 16.2%
Nuclear - 17.8%
Natural gas - 20.6%

Coal	0.014
Net hydroelectric power	0.003
Net nuclear power	0.408
Total	0.425

Crude oil	1.036
Dry natural gas	0.471
Coal	0.372
Net hydroelectric power	0.003
Net nuclear power	0.408
Total	2.290

Telecommunications [30]

- 5.691 million (1992 est.) telephones; highly developed, technologically advanced, and completely automated domestic and international telephone and telegraph facilities
- Domestic: nationwide cellular telephone system; extensive cable network; limited microwave radio relay network
- International: 5 submarine cables; satellite earth stations - 2 Intelsat (Atlantic Ocean) and 1 Eutelsat
- Radio: Broadcast stations: AM 3, FM 39, shortwave 0 Radios: 100,000 (1992 est.)
- Television: Broadcast stations: 32 (1987 est.) Televisions: 3,315,662 (1993 est.)

Top Agricultural Products [32]

Agriculture accounts for 2% of the GDP; produces sugar beets, fresh vegetables, fruits, grain, tobacco; beef, veal, pork, milk.

Transportation [31]

Railways: total: 3,396 km (2,363 km electrified; 2,563 km double track); standard gauge: 3,396 km 1.435-m gauge (1995)

Highways: total: 137,876 km; paved: 129,603 km (including 1,667 km of expressways); unpaved: 8,273 km (1992 est.)

Merchant marine: total: 23 ships (1,000 GRT or over) totaling 64,220 GRT/83,360 DWT; ships by type: bulk 1, cargo 8, chemical tanker 5, liquefied gas tanker 3, oil tanker 6 (1995 est.)

Airports

Total:	42
With paved runways over 3,047 m:	6
With paved runways 2,438 to 3,047 m:	9
With paved runways 1,524 to 2,437 m:	2
With paved runways 914 to 1,523 m:	1
With paved runways under 914 m:	21

Top Mining Products [33]

Thousand metric tons except as noted	6/96[*]
Pig iron	9,030
Steel	
crude	11,319[r]
hot-rolled products	11,000
Cement, hydraulic	8,000[e]
Dolomite	4,000[e]
Limestone	33,500[e]
Porphyry, all types	4,000[e]
Construction sand	9,200[e]
Gravel, dredged	5,000[e]

Tourism [34]

	1990	1991	1992	1993	1994
Tourism receipts	3,721	3,612	4,101	4,054	5,182
Tourism expenditures	5,477	5,543	6,714	6,338	7,782
Fare receipts[13]	1,077	1,156	1,076	1,094	1,402
Fare expenditures[13]	880	1,002	953	873	1,227

Travelers are in thousands, money in million U.S. dollars.

For sources, notes, and explanations, see Annotated Source Appendix, page 1061.

Manufacturing Sector

GDP and Manufacturing Summary [35]

	1980	1985	1989	1990	% change 1980-1990	% change 1989-1990
GDP (million 1980 $)	118,016	122,611	141,464	143,454	21.6	1.4
GDP per capita (1980 $)	11,979	12,436	14,366	14,573	21.7	1.4
Manufacturing as % of GDP (current prices)	25.5	24.6	23.3	23.7	-7.1	1.7
Gross output (million $)	84,723[e]	59,419	116,888	146,712	73.2	25.5
Value added (million $)	28,130	18,232	33,454	42,392	50.7	26.7
Value added (million 1980 $)	25,772	29,229	37,645	34,819	35.1	-7.5
Industrial production index	100	107	123	125	25.0	1.6
Employment (thousands)	872	755	718[e]	735	-15.7	2.4

Note: GDP stands for Gross Domestic Product. 'e' stands for estimated value.

Profitability and Productivity

	1980	1985	1989	1990	% change 1980-1990	% change 1989-1990
Intermediate input (%)	67[e]	69	71	71	6.0	0.0
Wages, salaries, and supplements (%)	17[e]	13[e]	12[e]	11[e]	-35.3	-8.3
Gross operating surplus (%)	17[e]	17[e]	16[e]	17[e]	0.0	6.3
Gross output per worker ($)	91,198[e]	78,700	162,831[e]	199,636	118.9	22.6
Value added per worker ($)	30,345	24,149	46,603[e]	57,684	90.1	23.8
Average wage (incl. benefits) ($)	16,066[e]	10,618[e]	19,824[e]	22,774[e]	41.8	14.9

Profitability is in percent of gross output. Productivity is in U.S. $. 'e' stands for estimated value.

Profitability - 1990

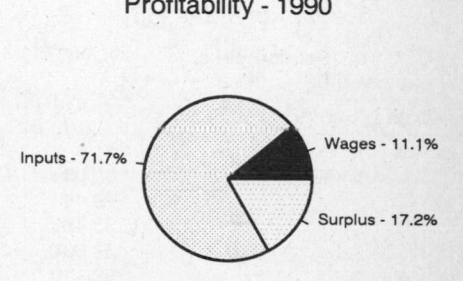

Inputs - 71.7%
Wages - 11.1%
Surplus - 17.2%

The graphic shows percent of gross output.

Value Added in Manufacturing

	1980 $ mil.	1980 %	1985 $ mil.	1985 %	1989 $ mil.	1989 %	1990 $ mil.	1990 %	% change 1980-1990	% change 1989-1990
311 Food products	3,991	14.2	2,863	15.7	5,588	16.7	6,015	14.2	50.7	7.6
313 Beverages	549	2.0	359	2.0	627	1.9	717	1.7	30.6	14.4
314 Tobacco products	199	0.7	123	0.7	188	0.6	257	0.6	29.1	36.7
321 Textiles	1,445	5.1	937	5.1	1,612	4.8	1,786	4.2	23.6	10.8
322 Wearing apparel	671	2.4	392	2.2	694	2.1	960	2.3	43.1	38.3
323 Leather and fur products	136	0.5	93	0.5	81	0.2	183	0.4	34.6	125.9
324 Footwear	67	0.2	35	0.2	41	0.1	61	0.1	-9.0	48.8
331 Wood and wood products	226	0.8	131	0.7	266[e]	0.8	458	1.1	102.7	72.2
332 Furniture and fixtures	1,123	4.0	614	3.4	1,257[e]	3.8	1,514	3.6	34.8	20.4
341 Paper and paper products	612	2.2	441	2.4	823[e]	2.5	1,095	2.6	78.9	33.0
342 Printing and publishing	926	3.3	602	3.3	1,111	3.3	1,677	4.0	81.1	50.9
351 Industrial chemicals	2,401	8.5	2,253	12.4	3,603[e]	10.8	4,483	10.6	86.7	24.4
352 Other chemical products	665	2.4	467	2.6	960[e]	2.9	1,186	2.8	78.3	23.5
353 Petroleum refineries	517	1.8	212	1.2	467	1.4	305	0.7	-41.0	-34.7
354 Miscellaneous petroleum and coal products	72	0.3	36	0.2	39	0.1	69	0.2	-4.2	76.9
355 Rubber products	193	0.7	130	0.7	308[e]	0.9	314	0.7	62.7	1.9
356 Plastic products	819	2.9	633	3.5	1,431[e]	4.3	2,197	5.2	168.3	53.5
361 Pottery, china, and earthenware	107	0.4	61	0.3	147	0.4	151	0.4	41.1	2.7
362 Glass and glass products	516	1.8	289	1.6	464	1.4	770	1.8	49.2	65.9
369 Other nonmetal mineral products	654	2.3	307	1.7	722	2.2	881	2.1	34.7	22.0
371 Iron and steel	2,294	8.2	985	5.4	1,548	4.6	2,510	5.9	9.4	62.1
372 Nonferrous metals	487	1.7	417	2.3	738	2.2	1,140	2.7	134.1	54.5
381 Metal products	2,071	7.4	1,228	6.7	2,374	7.1	2,835	6.7	36.9	19.4
382 Nonelectrical machinery	2,490	8.9	1,556	8.5	2,615	7.8	3,673	8.7	47.5	40.5
383 Electrical machinery	2,303	8.2	1,451	8.0	2,433	7.3	2,913	6.9	26.5	19.7
384 Transport equipment	1,892	6.7	1,217	6.7	2,443	7.3	3,196	7.5	68.9	30.8
385 Professional and scientific equipment	170	0.6	106	0.6	148	0.4	271	0.6	59.4	83.1
390 Other manufacturing industries	537	1.9	294	1.6	725[e]	2.2	772	1.8	43.8	6.5

Note: The industry codes shown are International Standard Industry codes (ISIC). Percentages are percent of total Value Added. 'e' stands for estimated value

For sources, notes, and explanations, see Annotated Source Appendix, page 1061.

93

Finance, Economics, and Trade

Economic Indicators [36]

- **National product**: GDP—purchasing power parity—$197 billion (1995 est.)
- **National product real growth rate**: 2.3% (1995 est.)
- **National product per capita**: $19,500 (1995 est.)
- **Inflation rate (consumer prices)**: 1.6% (1995 est.)
- **External debt**: $31.3 billion (1992 est.)

Balance of Payments Summary [37]

Values in millions of dollars.

	1989	1990	1991	1992	1993
Exports of goods (f.o.b.)	89,988	107,654	106,019	113,638	103,873
Imports of goods (f.o.b.)	-89,020	-107,064	-106,085	-112,307	-99,905
Trade balance	968	590	-66	1,331	3,932
Services - debits	-67,902	-90,646	-101,703	-117,520	-109,283
Services - credits	71,850	96,971	108,251	125,144	120,527
Private transfers (net)	47	-597	-280	-503	-602
Government transfers (net)	-1,765	-1,369	-1,470	-1,984	-1,986
Long-term capital (net)	-2,437	-4	4,049	-7,710	9,503
Short-term capital (net)	-2,767	-1,650	-7,824	60	-23,234
Errors and omissions	-86	-2,844	-992	1,847	-938
Overall balance	-2,092	451	505	665	-2,081

Exchange Rates [38]

Currency: **Belgian franc.**
Symbol: **BF.**

Data are currency units per $1.

January 1996	30.036
1995	29.480
1994	33.456
1993	34.597
1992	32.150
1991	34.148

Imports and Exports

Top Import Origins [39]

$140 billion (c.i.f., 1994) BLEU. Data are for 1994.

Origins	%
EU	68
Germany	22.1
US	8.8
Former Communist countries	0.8

Top Export Destinations [40]

$108 billion (f.o.b., 1994) BLEU. Data are for 1994.

Destinations	%
EU	67.2
Germany	19
US	5.8
Former Communist countries	1.4

Foreign Aid [41]

Donor: ODA, $808 million (1993).

Import and Export Commodities [42]

Import Commodities	Export Commodities
Fuels	Iron and steel
Grains	Transportation equipment
Chemicals	Tractors
Foodstuffs	Diamonds
	Petroleum products

Belize

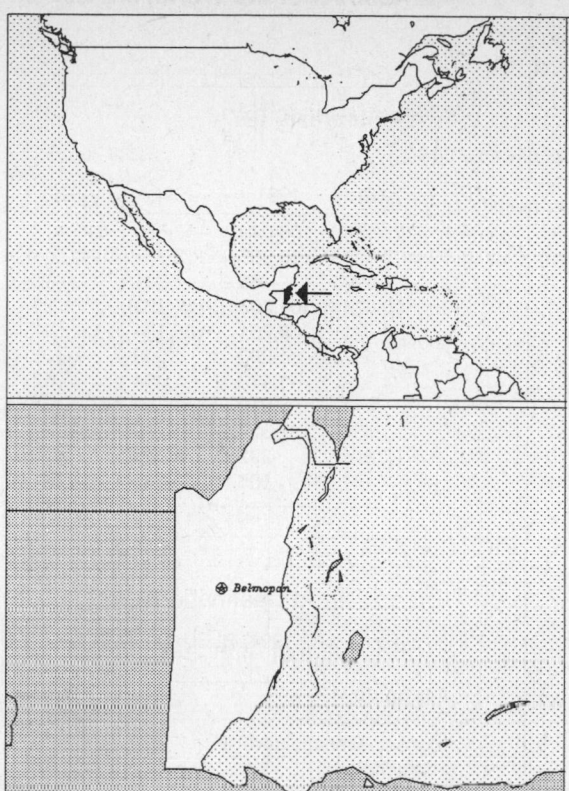

Geography [1]

Total area:
  22,960 sq km   8,865 sq mi
Land area:
  22,800 sq km   8,803 sq mi
Comparative area:
  Slightly larger than Massachusetts
Land boundaries:
  Total 516 km, Guatemala 266 km, Mexico 250 km
Coastline:
  386 km
Climate:
  Tropical; very hot and humid; rainy season (May to February)
Terrain:
  Flat, swampy coastal plain; low mountains in South
Natural resources:
  Arable land potential, timber, fish
Land use:
  Arable land: 2%
  Permanent crops: 0%
  Meadows and pastures: 2%
  Forest and woodland: 44%
  Other: 52%

Demographics [2]

	1970	1980	1990	1995[1]	1996	2000	2010	2020	2030
Population	122	144	190	214	219	242	299	356	409
Population density (persons per sq. mi.)	14	16	22	24	25	27	34	40	46
(persons per sq. km.)	5	6	8	9	10	11	13	16	18
Net migration rate (per 1,000 population)	NA	-1.0	-8.4	-3.7	-2.9	0.0	0.0	0.0	0.0
Births	NA	NA	NA	NA	7	NA	NA	NA	NA
Deaths	NA	NA	NA	NA	1	NA	NA	NA	NA
Life expectancy - males	NA	62.8	65.2	66.4	66.6	67.5	69.4	71.2	72.8
Life expectancy - females	NA	66.7	69.2	70.4	70.6	71.5	73.6	75.5	77.2
Birth rate (per 1,000)	38.5	42.7	38.7	33.7	32.8	29.4	23.9	20.2	17.1
Death rate (per 1,000)	NA	8.7	6.6	5.9	5.7	5.3	4.7	4.7	5.2
Women of reproductive age (15-49 yrs.)	NA	30	42	49	51	59	81	97	111
of which are currently married	10	12	18	NA	22	26	37	NA	NA
Fertility rate	6.6	6.2	5.0	4.3	4.1	3.6	2.7	2.3	2.1

Except as noted, values for vital statistics are in thousands; life expectancy is in years.

Health

Health Indicators [3]

Health Expenditures [4]

For sources, notes, and explanations, see Annotated Source Appendix, page 1061.

95

Human Factors

Women and Children [5]

Burden of Disease [6]

Population per physician (1993)	2,029.70
Population per nurse (1993)	490.43
Population per hospital bed (1993)	357.14
AIDS cases per 100,000 people (1994)	4.5
Malaria cases per 100,000 people (1992)	2684

Ethnic Division [7]

Mestizo	44%
Creole	30%
Maya	11%
Garifuna	7%
Other	8%

Religion [8]

Roman Catholic	62%
Protestant	30%
None	2%
Other	6%
(1980)	

Major Languages [9]

English (official), Spanish, Mayan, Garifuna (Carib).

Education

Public Education Expenditures [10]

Million (Dollar)	1980	1986	1990	1991	1993	1994
Total education expenditure	NA	NA	NA	48	NA	NA
as percent of GNP	NA	NA	NA	5.7	NA	NA
as percent of total govt. expend.	NA	NA	NA	15.5	NA	NA
Current education expenditure	NA	18	NA	38	NA	NA
as percent of GNP	NA	4.0	NA	4.5	NA	NA
as percent of current govt. expend.	NA	NA	NA	21.8	NA	NA
Capital expenditure	NA	NA	NA	10	NA	NA

Educational Attainment [11]

Age group (1991)	25+
Total population	66,520
Highest level attained (%)	
No schooling	13.0
First level	
Not completed	64.3
Completed	NA
Entered second level	
S-1	14.9
S-2	NA
Postsecondary	6.6

Illiteracy [12]

	1989	1991	1995
Illiterate population (14+ yrs.)[2]	NA	31,879	NA
Illiteracy rate - total pop. (%)	NA	29.7	NA
Illiteracy rate - males (%)	NA	29.7	NA
Illiteracy rate - females (%)	NA	29.7	NA

Libraries [13]

	Admin. Units	Svc. Pts.	Vols. (000)	Shelving (meters)	Vols. Added	Reg. Users
National (1992)	1	NA	150[e]	NA	1,011	25,096
Nonspecialized	NA	NA	NA	NA	NA	NA
Public (1992)[6]	1	26	130[e]	47,079	NA	14,999[e]
Higher ed. (1990)	1	1	6	NA	NA	312
School	NA	NA	NA	NA	NA	NA

Daily Newspapers [14]

	1980	1985	1990	1994
Number of papers	1	1	-	-
Circ. (000)	3	3	-	-

Culture [15]

Science and Technology

Scientific/Technical Forces [16]

R&D Expenditures [17]

U.S. Patents Issued [18]

For sources, notes, and explanations, see Annotated Source Appendix, page 1061.

Government and Law

Organization of Government [19]

Long-form name:
None
Type:
Parliamentary democracy
Independence:
21 September 1981 (from UK)
National holiday:
Independence Day, 21 September (1981)
Constitution:
21 September 1981
Legal system:
English law
Executive branch:
British Monarch (represented by Governor General); Prime Minister; Deputy Prime Minister; Cabinet
Legislative branch:
Bicameral National Assembly: Senate and National Assembly
Judicial branch:
Supreme Court

Elections [20]

National Assembly	% of seats
United Democratic Party (UDP)	53.5
People's United Party (PUP)	46.4

Government Expenditures [21]

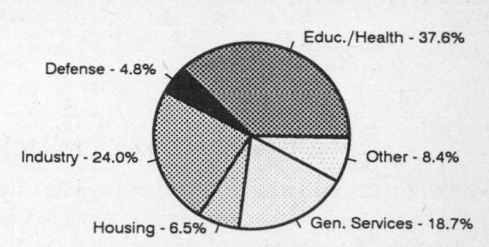

Educ./Health - 37.6%
Defense - 4.8%
Industry - 24.0%
Other - 8.4%
Housing - 6.5%
Gen. Services - 18.7%

(% distribution). Expend. for FY94: 348,173 (Dollar mil.)

Crime [23]

Military Expenditures and Arms Transfers [22]

	1990	1991	1992	1993	1994
Military expenditures					
Current dollars (mil.)	5	5	6	7	9
1994 constant dollars (mil.)	6	5	6	7	9
Armed forces (000)	1	1	1	1	1
Gross national product (GNP)					
Current dollars (mil.)	401	433	484	513	533
1994 constant dollars (mil.)	446	464	505	524	533
Central government expenditures (CGE)					
1994 constant dollars (mil.)	125	142	155	189	207
People (mil.)	0.2	0.2	0.2	0.2	0.2
Military expenditure as % of GNP	1.3	1.1	1.2	1.3	1.7
Military expenditure as % of CGE	4.5	3.6	3.8	3.7	4.4
Military expenditure per capita (1994 $)	29	27	30	34	44
Armed forces per 1,000 people (soldiers)	5.2	5.0	5.0	4.9	4.8
GNP per capita (1994 $)	2,350	2,385	2,535	2,567	2,551
Arms imports[6]					
Current dollars (mil.)	0	0	0	0	0
1994 constant dollars (mil.)	0	0	0	0	0
Arms exports[6]					
Current dollars (mil.)	0	0	0	0	0
1994 constant dollars (mil.)	0	0	0	0	0
Total imports[7]					
Current dollars (mil.)	211	251	273	281	260
1994 constant dollars (mil.)	235	269	285	287	260
Total exports[7]					
Current dollars (mil.)	105	102	116	115	119
1994 constant dollars (mil.)	117	109	121	117	119
Arms as percent of total imports[8]	0	0	0	0	0
Arms as percent of total exports[8]	0	0	0	0	0

Human Rights [24]

	SSTS	FL	FAPRO	PPCG	APROBC	TPW	PCPTW	STPEP	PHRFF	PRW	ASST	AFL
Observes	1	P	P		P	P	P				1	P
	EAFRD	CPR	ESCR	SR	ACHR	MAAE	PVIAC	PVNAC	EAFDAW	TCIDTP	RC
Observes	P	1	1	P			P	P	P	P	P

P = Party; S = Signatory; see Appendix for meaning of abbreviations.

Labor Force

Total Labor Force [25]

51,500

Labor Force by Occupation [26]

Agriculture	30%
Services	16
Government	15.4
Commerce	11.2
Manufacturing	10.3

Date of data: 1985

Unemployment Rate [27]

10% (1993 est.)

For sources, notes, and explanations, see Annotated Source Appendix, page 1061.

97

Production Sectors

Energy Resource Summary [28]
Electricity: Capacity: 34,532 kW. Production: 110 million kWh. Consumption per capita: 490 kWh (1993).

Telecommunications [30]
- 15,917 (1990 est.) telephones; above-average system
- Domestic: trunk network depends primarily on microwave radio relay
- International: satellite earth station - 1 Intelsat (Atlantic Ocean)
- Radio: Broadcast stations: AM 6, FM 5, shortwave 1
- Television: Broadcast stations: 1 Televisions: 27,048 (1993 est.)

Top Agricultural Products [32]
Agriculture accounts for 30% of the GDP; produces bananas, coca, citrus, sugarcane; lumber; fish, cultured shrimp.

Top Mining Products [33]

Metric tons except as noted	12/94[*]
Clays	2,100,000
Dolomite	30,000[e]
Gold (kg.)	5[e]
Lime	1,000[e]
Limestone	300,000
Marl (000 tons)	1,050
Sand and gravel	300,000

Transportation [31]
Railways: 0 km

Highways: total: 2,560 km; paved: 336 km; unpaved: 2,224 km (1987 est.)

Merchant marine: total: 89 ships (1,000 GRT or over) totaling 311,731 GRT/470,272 DWT; ships by type: bulk 9, cargo 60, container 6, liquefied gas tanker 1, oil tanker 3, refrigerated cargo 4, roll-on/roll-off cargo 4, specialized tanker 1, vehicle carrier 1 (1995 est.)

Airports
Total: 35
With paved runways 1,524 to 2,437 m: 1
With paved runways under 914 m: 25
With unpaved runways 2,438 to 3,047 m: 1
With unpaved runways 914 to 1,523 m: 8 (1995 est.)

Tourism [34]

	1990	1991	1992	1993	1994
Visitors[14]	NA	223	253	295	338
Tourists	NA	87	113	117	129
Excursionists[15]	NA	136	138	172	196
Cruise passengers	NA	1	1	6	13
Tourism receipts	51	53	60	70	74
Tourism expenditures	7	8	14	21	NA
Fare expenditures	8	9	10	9	NA

Travelers are in thousands, money in million U.S. dollars.

For sources, notes, and explanations, see Annotated Source Appendix, page 1061.

Manufacturing Sector

Manufacturing Summary [35]

	1987		1988		1989		1990		1991	
	$ bil.	%	$ bil.	%	$ bil.	%	$ bil.	%	$ bil.	%
Establishments or enterprises (number)	-	-	-	-	297	0.016	313	0.017	333	0.044
Total employment (000)	-	-	-	-	-	-	-	-	-	-
Production workers (000)	-	-	-	-	-	-	-	-	-	-
Output ($ bil.)	-	-	-	-	0.110	0.001	0.170	0.001	0.177	0.002
Value added ($ bil.)	-	-	-	-	0.032	0.001	0.043	0.001	0.051	0.001
Capital investment ($ mil.)	-	-	-	-	-	-	-	-	-	-
M & E investment ($ mil.)	-	-	-	-	-	-	-	-	-	-
Employees per establishment (number)	-	-	-	-	-	-	-	-	-	-
Production workers per establishment	-	-	-	-	-	-	-	-	-	-
Output per establishment ($ mil.)	-	-	-	-	0.372	5.906	0.543	8.472	0.532	3.987
Capital investment per estab. ($ mil.)	-	-	-	-	-	-	-	-	-	-
M & E per establishment ($ mil)	-	-	-	-	-	-	-	-	-	-
Payroll per employee ($)	-	-	-	-	-	-	-	-	-	-
Wages per production worker ($)	-	-	-	-	-	-	-	-	-	-
Hours per production worker (hours)	-	-	-	-	-	-	-	-	-	-
Output per employee ($)	-	-	-	-	-	-	-	-	-	-
Capital investment per employee ($)	-	-	-	-	-	-	-	-	-	-
M & E per employee ($)	-	-	-	-	-	-	-	-	-	-

Note: Columns headed % show percent of world total or ratio. Ratios closest to 100 are closest to world average. M & E stands for machinery & equipment.

Output in Manufacturing

	1987		1988		1989		1990		1991	
	$ bil.	%	$ bil.	%	$ bil.	%	$ bil.	%	$ bil.	%
3110 - Food products	-	-	-	-	0.071	64.55	0.107	62.94	0.108	61.02
3130 - Beverages	-	-	-	-	0.016	14.55	0.019	11.18	0.022	12.43
3140 - Tobacco	-	-	-	-	0.005	4.55	0.005	2.94	0.005	2.82
3310 - Wood products	-	-	-	-	0.002	1.82	0.007	4.12	0.007	3.95
3320 - Furniture, fixtures	-	-	-	-	0.000	0.00	0.006	3.53	0.008	4.52
3410 - Paper and products	-	-	-	-	0.002	1.82	0.002	1.18	0.001	0.56
3420 - Printing, publishing	-	-	-	-	0.000	0.00	0.002	1.18	0.002	1.13
3510 - Industrial chemicals	-	-	-	-	0.005	4.55	-	-	-	-
3520 - Chemical products nec	-	-	-	-	0.000	0.00	-	-	-	-
3690 - Nonmetal products nec	-	-	-	-	0.003	2.73	0.008	4.71	0.009	5.08
3810 - Metal products	-	-	-	-	0.001	0.91	0.004	2.35	0.004	2.26
3830 - Electrical machinery	-	-	-	-	-	-	0.001	0.59	0.001	0.56
3840 - Transportation equipment	-	-	-	-	0.005	4.55	0.009	5.29	0.009	5.08
3841 - Shipbuilding, repair	-	-	-	-	0.000	0.00	-	-	-	-
3843 - Motor vehicles	-	-	-	-	0.000	0.00	-	-	-	-
3900 - Industries nec	-	-	-	-	0.000	0.00	0.000	0.00	0.000	0.00

Note: Codes are International Standard Industry codes (ISIC). Percentages are % of total Output. [f]: Factor Prices; [p]: Producer Prices.

For sources, notes, and explanations, see Annotated Source Appendix, page 1061.

99

Finance, Economics, and Trade

Economic Indicators [36]

- **National product**: GDP—purchasing power parity— $575 million (1994 est.)
- **National product real growth rate**: 2% (1994 est.)
- **National product per capita**: $2,750 (1994 est.)
- **Inflation rate (consumer prices)**: 2.3% (1994 est.)
- **External debt**: $167.5 million (1992)

Balance of Payments Summary [37]

Values in millions of dollars.

	1989	1990	1991	1992	1993
Exports of goods (f.o.b.)	124.4	129.2	126.1	140.6	132.0
Imports of goods (f.o.b.)	-188.5	-188.4	-223.6	-244.5	-250.5
Trade balance	-64.1	-59.2	-97.5	103.9	-118.5
Services - debits	-81.6	-80.8	-87.3	-104.4	-115.9
Services - credits	95.5	125.9	131.0	149.3	156.4
Private transfers (net)	20.7	16.3	15.4	17.6	15.4
Government transfers (net)	10.4	13.0	12.6	12.8	14.1
Long-term capital (net)	29.9	29.3	25.5	24.5	35.0
Short-term capital (net)	-4.4	-4.2	-3.3	-2.1	-2.2
Errors and omissions	9.1	-25.0	-12.8	6.3	1.5
Overall balance	15.5	15.3	-16.4	0.1	-14.2

Exchange Rates [38]

Currency: **Belizean dollar.**
Symbol: **Bz$.**

Data are currency units per $1.

Fixed rate	2.00

Imports and Exports

Top Import Origins [39]

$281 million (c.i.f., 1993) Data are for 1994.

Origins	%
US	53
UK	NA
Other EU	NA
Mexico	NA

Top Export Destinations [40]

$115 million (f.o.b., 1993) Data are for 1994.

Destinations	%
US	38
UK	NA
Other EU	NA

Foreign Aid [41]

Recipient: ODA, $NA.

Import and Export Commodities [42]

Import Commodities	Export Commodities
Machinery and transportation equipment	Sugar
Food	Citrus fruits
Manufactured goods	Bananas
Fuels	Clothing
Chemicals	Fish products
Pharmaceuticals	Molasses
	Wood

For sources, notes, and explanations, see Annotated Source Appendix, page 1061.

Benin

Geography [1]

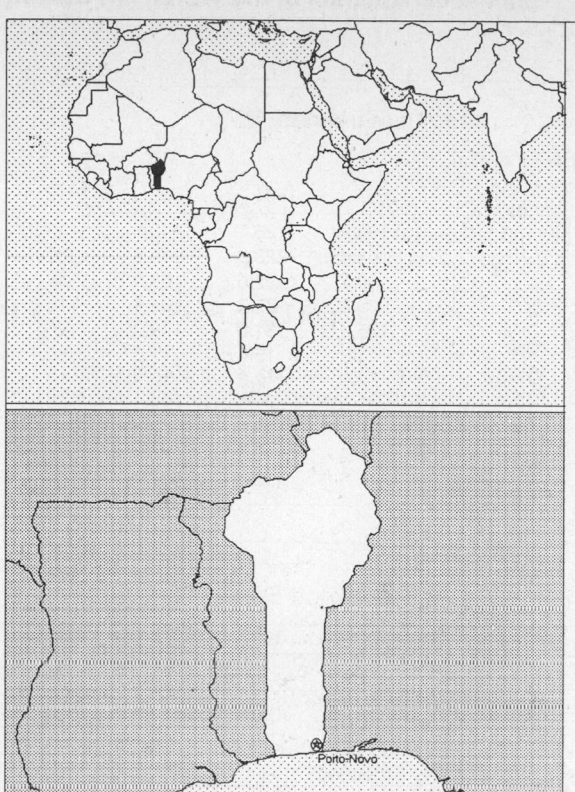

Total area:
112,620 sq km   43,483 sq mi
Land area:
110,620 sq km   42,711 sq mi
Comparative area:
Slightly smaller than Pennsylvania
Land boundaries:
Total 1,989 km, Burkina Faso 306 km, Niger 266 km, Nigeria 773 km, Togo 644 km
Coastline:
121 km
Climate:
Tropical; hot, humid in South; semiarid in North
Terrain:
Mostly flat to undulating plain; some hills and low mountains
Natural resources:
Small offshore oil deposits, limestone, marble, timber
Land use:
Arable land: 12%
Permanent crops: 4%
Meadows and pastures: 4%
Forest and woodland: 35%
Other: 45%

Demographics [2]

	1970	1980	1990	1995[1]	1996	2000	2010	2020	2030
Population	2,620	3,444	4,676	5,523	5,710	6,517	8,955	11,920	15,224
Population density (persons per sq. mi.)	61	81	109	129	134	153	210	279	356
(persons per sq. km.)	24	31	42	50	52	59	81	108	138
Net migration rate (per 1,000 population)	NA	0.0	0.0	0.0	0.0	0.0	0.0	0.0	0.0
Births	NA	NA	NA	NA	267	NA	NA	NA	NA
Deaths	NA	NA	NA	NA	77	NA	NA	NA	NA
Life expectancy - males	NA	44.4	48.3	50.3	50.7	52.4	56.5	60.4	63.9
Life expectancy - females	NA	46.7	51.6	54.2	54.7	56.8	61.8	66.6	70.7
Birth rate (per 1,000)	NA	50.4	49.5	47.3	46.8	44.9	39.5	33.6	28.2
Death rate (per 1,000)	NA	23.4	16.2	13.9	13.5	12.0	9.1	7.0	5.8
Women of reproductive age (15-49 yrs.)	NA	783	1,064	1,249	1,291	1,475	2,101	2,953	3,966
of which are currently married	NA	NA	758	NA	918	1,051	1,501	NA	NA
Fertility rate	NA	7.1	7.1	6.7	6.6	6.3	5.4	4.3	3.5

Except as noted, values for vital statistics are in thousands; life expectancy is in years.

Health

Health Indicators [3]

% of population with access to	
safe water (1990-95)	50
adequate sanitation (1990-95)	20
health services (1985-95)	18
% of 1-year-olds immunized (1990-94) against	
TB (tuberculosis)	90
DPT (diphtheria, pertussis, tetanus)	81
polio	81
measles	75
% of contraceptive prevalence (1980-94)	9
ORT use rate (1990-94)	77

Health Expenditures [4]

Total health expenditure, 1990 (official exchange rate)	
Millions of dollars	79
Dollars per capita	17
Health expenditures as a percentage of GDP	
Total	4.3
Public sector	2.8
Private sector	1.6
Development assistance for health	
Total aid flows (millions of dollars)[1]	33
Aid flows per capita (dollars)	7.0
Aid flows as a percentage of total health expenditure	41.8

For sources, notes, and explanations; see Annotated Source Appendix, page 1061.

Human Factors

Women and Children [5]

% of pregnant women immunized (tetanus 1990-94)	85
% of births attended by trained health personnel (1983-94)	45
Maternal mortality rate (1980-92)	160
Under-5 mortality rate (1994)	142
% under-5 moderately/severely underweight (1980-1994)	NA

Burden of Disease [6]

Population per physician (1994)	16,435.19
Population per nurse (1994)	4,179.75
Population per hospital bed (1994)	4,280.55
AIDS cases per 100,000 people (1994)	6.0
Malaria cases per 100,000 people (1992)	NA

Ethnic Division [7]

African includes 42 ethnic groups, the most important being Fon, Adja Yoruba, and Barbiba.

African	99%
Europeans	1%

Religion [8]

Indigenous beliefs	70%
Muslim	15%
Christian	15%

Major Languages [9]

French (official), Fon and Yoruba (most common vernaculars in the South), tribal languages (at least six major ones in the North).

Education

Public Education Expenditures [10]

	1980	1985	1990	1992	1993	1994
Million (Franc C.F.A.)						
Total education expenditure	NA	NA	NA	NA	NA	NA
as percent of GNP	NA	NA	NA	NA	NA	NA
as percent of total govt. expend.	NA	NA	NA	NA	NA	NA
Current education expenditure	12,426	NA	NA	NA	NA	NA
as percent of GNP	4.2	NA	NA	NA	NA	NA
as percent of current govt. expend.	36.8	NA	NA	NA	NA	NA
Capital expenditure	NA	NA	NA	NA	NA	NA

Educational Attainment [11]

Age group (1992)	25+
Total population	1,700,914
Highest level attained (%)	
No schooling	78.5
First level	
Not completed	10.8
Completed	NA
Entered second level	
S-1	8.2
S-2	NA
Postsecondary	1.3

Illiteracy [12]

In thousands and percent[1]	1990	1995	2000
Illiterate population (15+ yrs.)	1,722	1,792	1,839
Illiteracy rate - total pop. (%)	71.4	62.6	53.9
Illiteracy rate - males (%)	62.3	52.5	43.7
Illiteracy rate - females (%)	79.3	71.6	63.3

Libraries [13]

	Admin. Units	Svc. Pts.	Vols. (000)	Shelving (meters)	Vols. Added	Reg. Users
National (1989)	1	NA	6	177	1,102	158
Nonspecialized (1989)	4	4	40	1,100	NA	NA
Public (1989)	12	12	28	777	2,008	689
Higher ed. (1987)	10	10	95	2,611	NA	NA
School (1987)	5	5	16	797	NA	NA

Daily Newspapers [14]

	1980	1985	1990	1994
Number of papers	1	1	1	1
Circ. (000)	1	1	12	12

Culture [15]

Cinema (seats per 1,000)	1.8
Annual attendance per person	0.1
Gross box office receipts (mil. Franc C.F.A.)	116
Museums (reporting)	6
Visitors (000)	8
Annual receipts (000 Franc C.F.A.)	1,884

Science and Technology

Scientific/Technical Forces [16]

Scientists/engineers	794
Number female	100
Technicians	242
Number female	64
Total[3]	1,036

R&D Expenditures [17]

	(000) 1989
Total expenditure[1]	NA
Capital expenditure	NA
Current expenditure	3,347,695
Percent current	NA

U.S. Patents Issued [18]

For sources, notes, and explanations, see Annotated Source Appendix, page 1061.

Government and Law

Organization of Government [19]

Long-form name:
Republic of Benin
Type:
Republic under multiparty democratic rule dropped Marxism-Leninism December 1989; democratic reforms adopted February 1990; transition to multiparty system completed 4 April 1991
Independence:
1 August 1960 (from France)
National holiday:
National Day, 1 August (1990)
Constitution:
2 December 1990
Legal system:
Based on French civil law and customary law; has not accepted compulsory ICJ jurisdiction
Executive branch:
President; Executive Council
Legislative branch:
Unicameral: National Assembly (Assemblee Nationale)
Judicial branch:
Supreme Court (Cour Supreme)

Crime [23]

Elections [20]

National Assembly	% of seats
Renaissance Party and allies	24.1
Democratic Renewal Party (PRD)	22.9
Action for Renewal and Development (FARD-ALAFIA)	12.0
All. of the Soc. Dem. Party (PSD)	8.4
Our Common Cause (NCC)	3.6
RDL-VIVOTEN	3.6
Communist Party	2.4
Others	22.9

Government Budget [21]

For 1993 est.

	$ mil.
Revenues	272
Expenditures	375
Capital expenditures	84

Military Expenditures and Arms Transfers [22]

	1990	1991	1992	1993	1994
Military expenditures					
Current dollars (mil.)	23	NA	17[e]	22[e]	34
1994 constant dollars (mil.)	26	NA	18[e]	22[e]	34
Armed forces (000)	6	7	6	6	6
Gross national product (GNP)					
Current dollars (mil.)	1,141	1,246	1,320	1,406	1,481
1994 constant dollars (mil.)	1,270	1,336	1,376	1,435	1,481
Central government expenditures (CGE)					
1994 constant dollars (mil.)	NA	254[e]	NA	258[e]	NA
People (mil.)	4.7	4.8	5.0	5.2	5.3
Military expenditure as % of GNP	2.0	NA	1.3	1.5	2.3
Military expenditure as % of CGE	NA	NA	NA	8.6	NA
Military expenditure per capita (1994 $)	6	NA	4	4	6
Armed forces per 1,000 people (soldiers)	1.3	1.4	1.2	1.2	1.1
GNP per capita (1994 $)	272	276	275	278	277
Arms imports[6]					
Current dollars (mil.)	5	0	0	0	0
1994 constant dollars (mil.)	6	0	0	0	0
Arms exports[6]					
Current dollars (mil.)	0	0	0	0	0
1994 constant dollars (mil.)	0	0	0	0	0
Total imports[7]					
Current dollars (mil.)	265	241	440	344	519[e]
1994 constant dollars (mil.)	295	258	459	351	519[e]
Total exports[7]					
Current dollars (mil.)	122	21	63	122	194[e]
1994 constant dollars (mil.)	136	23	66	125	194[e]
Arms as percent of total imports[8]	1.9	0	0	0	0
Arms as percent of total exports[8]	0	0	0	0	0

Human Rights [24]

	SSTS	FL	FAPRO	PPCG	APROBC	TPW	PCPTW	STPEP	PHRFF	PRW	ASST	AFL
Observes	2	P	P		P	P	P					P
	EAFRD	CPR	ESCR	SR	ACHR	MAAE	PVIAC	PVNAC	EAFDAW	TCIDTP	RC	
Observes	S	P	P	P			P	P	P	P	P	

P=Party; S=Signatory; see Appendix for meaning of abbreviations.

Labor Force

Total Labor Force [25]

1.9 million (1987)

Labor Force by Occupation [26]

Agriculture	60%
Transport, commerce, and public services	38
Industry	<2

Unemployment Rate [27]

For sources, notes, and explanations, see Annotated Source Appendix, page 1061.

103

Production Sectors

Commercial Energy Production and Consumption

Data are shown in quadrillion (10^{15}) BTUs and percent for 1995
Values for hydroelectric, nuclear, geothermal, solar, and wind power refer to electrical generation.

Production [28]	Consumption [29]
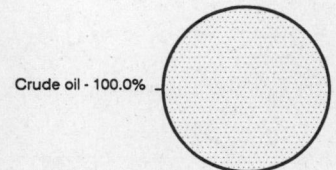	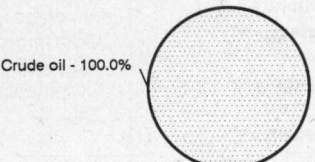
Crude oil - 100.0%	Crude oil - 100.0%

Crude oil	0.007	Crude oil	0.007
Total	0.007	Total	0.007

Telecommunications [30]

- 16,200 (1986 est.) telephones
- Domestic: fair system of open wire and microwave radio relay
- International: satellite earth station - 1 Intelsat (Atlantic Ocean); submarine cable
- Radio: Broadcast stations: AM 2, FM 2, shortwave 0
- Television: Broadcast stations: 2 Televisions: 20,000 (1993 est.)

Transportation [31]

Railways: total: 578 km (single track) (1995 est.); narrow gauge: 578 km 1.000-m gauge

Highways: total: 6,070 km; paved: 1,214 km; unpaved: 4,856 km (1992 est.)

Merchant marine: none

Airports
Total: 5
With paved runways 2,438 to 3,047 m: 2
With unpaved runways 1,524 to 2,437 m: 1
With unpaved runways 914 to 1,523 m: 2 (1995 est.)

Top Agricultural Products [32]

Agriculture accounts for 36.8% of the GDP; produces corn, sorghum, cassava (tapioca), yams, beans, rice, cotton, palm oil, peanuts; poultry, livestock.

Top Mining Products [33]

Metric tons except as noted[e]	2/10/95*
Cement, hydraulic	380,000
Petroleum, crude (000 42-gal. bls.)	900

Tourism [34]

	1990	1991	1992	1993	1994
Visitors	247	401	433	653	542
Tourists[16]	110	117	130	140	NA
Excursionists	137	284	303	513	NA
Tourism receipts	28	29	32	38	55
Tourism expenditures	12	10	12	13	19
Fare expenditures	11	11	11	11	NA

Travelers are in thousands, money in million U.S. dollars.

For sources, notes, and explanations, see Annotated Source Appendix, page 1061.

Finance, Economics, and Trade

GDP and Manufacturing Summary [35]

	1980	1985	1990	1991	1992
Gross Domestic Product					
Millions of 1980 dollars	1,163	1,225	1,268	1,306	1,343[e]
Growth rate in percent	10.16	-2.47	3.30	3.03	2.80[e]
Manufacturing Value Added					
Millions of 1980 dollars	78	129	91	94	98[e]
Growth rate in percent	-3.47	1.34	5.83	3.26	4.99[e]
Manufacturing share in percent of current prices	12.9	8.2	9.2	9.1	NA

Economic Indicators [36]

- **National product**: GDP—purchasing power parity—$7.6 billion (1995 est.)

- **National product real growth rate**: 6% (1995 est.)

- **National product per capita**: $1,380 (1995 est.)

- **Inflation rate (consumer prices)**: 55% (1994 est.)

- **External debt**: $1.5 billion (1993 est.)

Balance of Payments Summary [37]

Values in millions of dollars.

	1989	1990	1991	1992	1993
Exports of goods (f.o.b.)	178.4	287.2	329.0	369.1	332.7
Imports of goods (f.o.b.)	-316.6	-427.9	-482.4	-551.6	-571.4
Trade balance	-138.2	-140.7	-153.4	-182.5	-238.7
Services - debits	-149.8	-174.1	-169.4	-220.3	-195.3
Services - credits	87.8	114.6	122.6	141.7	134.6
Private transfers (net)	65.8	86.3	87.9	99.0	87.2
Government transfers (net)	121.3	112.4	101.7	133.0	160.0
Long-term capital (net)	321.9	87.8	174.0	122.4	83.0
Short-term capital (net)	-205.3	7.4	-79.0	-26.3	-4.6
Errors and omissions	-109.8	-5.0	106.2	126.6	-32.6
Overall balance	-6.3	88.7	190.6	193.6	-6.4

Exchange Rates [38]

Currency: **Communaute Financiere Africaine franc.**
Symbol: **CFAF.**

Data are currency units per $1.

January 1996	500.56
1995	499.15
1994	555.20
1993	283.16
1992	264.69
1991	282.11

Imports and Exports

Top Import Origins [39]

$439 million (c.i.f., 1994 est.).

Origins	%
France	24
Thailand	12
Netherlands	7
US	5
China	NA
Hong Kong	NA

Top Export Destinations [40]

$310 million (f.o.b., 1994 est.).

Destinations	%
Morocco	37
Portugal	14
France	NA
Spain	NA
Italy	NA
UK	NA
US	NA
Libya	NA

Foreign Aid [41]

Recipient: ODA, $NA.

Import and Export Commodities [42]

Import Commodities	Export Commodities
Foodstuffs	Cotton
Beverages	Crude oil
Tobacco	Palm products
Petroleum products	Cocoa
Intermediate goods	
Capital goods	
Light consumer goods	

For sources, notes, and explanations, see Annotated Source Appendix, page 1061.

105

Bhutan

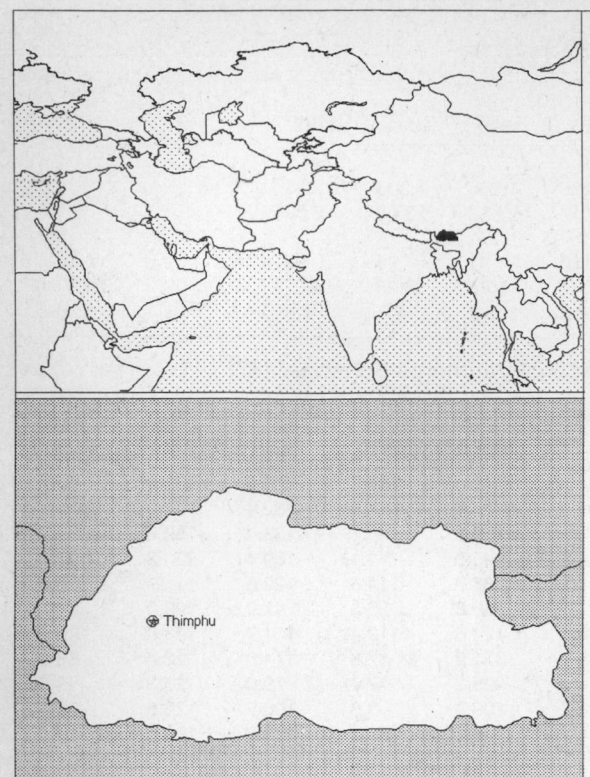

Geography [1]

Total area:
47,000 sq km   18,147 sq mi
Land area:
47,000 sq km   18,147 sq mi
Comparative area:
Slightly more than half the size of Indiana
Land boundaries:
Total 1,075 km, China 470 km, India 605 km
Coastline:
0 km (landlocked)
Climate:
Varies; tropical in southern plains; cool winters and hot summers in central valleys; severe winters and cool summers in Himalayas
Terrain:
Mostly mountainous with some fertile valleys and savanna
Natural resources:
Timber, hydropower, gypsum, calcium carbide
Land use:
Arable land: 2%
Permanent crops: 0%
Meadows and pastures: 5%
Forest and woodland: 70%
Other: 23%

Demographics [2]

	1970	1980	1990	1995[1]	1996	2000	2010	2020	2030
Population	1,045	1,281	1,585	1,781	1,823	1,996	2,474	3,035	3,655
Population density (persons per sq. mi.)	58	71	87	98	100	110	136	167	201
(persons per sq. km.)	22	27	34	38	39	42	53	65	78
Net migration rate (per 1,000 population)	NA	0.0	0.0	0.0	0.0	0.0	0.0	0.0	0.0
Births	NA	NA	NA	NA	70	NA	NA	NA	NA
Deaths	NA	NA	NA	NA	28	NA	NA	NA	NA
Life expectancy - males	NA	45.6	49.5	51.6	52.0	53.6	57.7	61.5	65.0
Life expectancy - females	NA	44.1	48.3	50.5	50.9	52.8	57.4	61.9	66.3
Birth rate (per 1,000)	NA	39.6	40.3	39.0	38.5	36.3	32.1	29.0	25.3
Death rate (per 1,000)	NA	20.2	17.2	15.6	15.3	13.9	11.2	9.3	7.9
Women of reproductive age (15-49 yrs.)	NA	293	371	412	422	462	584	732	900
of which are currently married	NA	NA	305	NA	350	382	480	NA	NA
Fertility rate	NA	5.6	5.5	5.4	5.3	5.1	4.5	3.8	3.3

Except as noted, values for vital statistics are in thousands; life expectancy is in years.

Health

Health Indicators [3]

% of population with access to	
safe water (1990-95)	NA
adequate sanitation (1990-95)	NA
health services (1985-95)	65
% of 1-year-olds immunized (1990-94) against	
TB (tuberculosis)	96
DPT (diphtheria, pertussis, tetanus)	86
polio	84
measles	81
% of contraceptive prevalence (1980-94)	2
ORT use rate (1990-94)	85

Health Expenditures [4]

For sources, notes, and explanations, see Annotated Source Appendix, page 1061.

Human Factors

Women and Children [5]

% of pregnant women immunized (tetanus 1990-94)	60
% of births attended by trained health personnel (1983-94)	7
Maternal mortality rate (1980-92)	620
Under-5 mortality rate (1994)	193
% under-5 moderately/severely underweight (1980-1994)	38

Burden of Disease [6]

Population per physician (1994)	6,250.00
Population per nurse	NA
Population per hospital bed (1994)	619.27
AIDS cases per 100,000 people (1994)	*
Malaria cases per 100,000 people (1992)	1,827

Ethnic Division [7]

Bhote	50%
Ethnic Nepalese	35%
Indigenous or migrant tribes	15%

Religion [8]

Lamaistic Buddhism	75%
Indian and Nepalese	
influenced Hinduism	25%

Major Languages [9]

Dzongkha (official), Bhotes speak various Tibetan dialects, Nepalese speak various Nepalese dialects.

Education

Public Education Expenditures [10]

	1980	1988	1990	1992	1993	1994
Million (Ngultrum)						
Total education expenditure	NA	125	NA	NA	NA	NA
as percent of GNP	NA	3.4	NA	NA	NA	NA
as percent of total govt. expend.	NA	NA	NA	NA	NA	NA
Current education expenditure	NA	NA	NA	187	192	NA
as percent of GNP	NA	NA	NA	2.9	2.7	NA
as percent of current govt. expend.	NA	NA	NA	NA	NA	NA
Capital expenditure	NA	NA	NA	NA	NA	NA

Educational Attainment [11]

Illiteracy [12]

In thousands and percent[1]	1990	1995	2000
Illiterate population (15+ yrs.)	575	558	574
Illiteracy rate - total pop. (%)	59.7	52.2	48.0
Illiteracy rate - males (%)	45.4	38.4	34.5
Illiteracy rate - females (%)	74.7	66.7	62.2

Libraries [13]

Daily Newspapers [14]

Culture [15]

Science and Technology

Scientific/Technical Forces [16]

R&D Expenditures [17]

U.S. Patents Issued [18]

For sources, notes, and explanations, see Annotated Source Appendix, page 1061.

107

Government and Law

Organization of Government [19]

Long-form name:
Kingdom of Bhutan
Type:
Monarchy; special treaty relationship with India
Independence:
8 August 1949 (from India)
National holiday:
National Day, 17 December (1907) (Ugyen Wangchuck became first hereditary king)
Constitution:
No written constitution or bill of rights
Note: Bhutan uses 1953 Royal decree for the Constitution of the National Assembly
Legal system:
Based on Indian law and English common law; has not accepted compulsory ICJ jurisdiction
Executive branch:
King; Royal Advisory Council (Lodoi Tsokde); Council of Ministers
Legislative branch:
Unicameral: National Assembly (Tshogdu):
Judicial branch:
The Supreme Court of Appeal is the king; High Court

Elections [20]

No legal parties. No national elections. Each family has one vote in village-level elections for the unicameral National Assembly which has 150 seats. 105 are from these village elections, 12 represent religious bodies, and 33 designated by the king to represent government and other secular interests.

Government Expenditures [21]

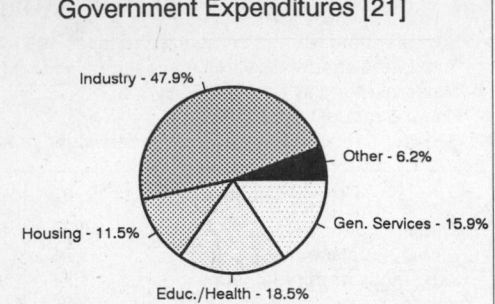

Industry - 47.9%
Other - 6.2%
Gen. Services - 15.9%
Housing - 11.5%
Educ./Health - 18.5%

(% distribution). Expend. for CY95: 4,716.7 (Ngultrum mil.)

Crime [23]

Military Expenditures and Arms Transfers [22]

	1990	1991	1992	1993	1994
Military expenditures					
Current dollars (mil.)	NA	NA	NA	NA	NA
1994 constant dollars (mil.)	NA	NA	NA	NA	NA
Armed forces (000)	NA	NA	5	5	5
Gross national product (GNP)					
Current dollars (mil.)	204	223	237	254	272
1994 constant dollars (mil.)	227	239	248	259	272
Central government expenditures (CGE)					
1994 constant dollars (mil.)	83	78	85	91	110
People (mil.)	1.6	1.6	1.7	1.7	1.7
Military expenditure as % of GNP	NA	NA	NA	NA	NA
Military expenditure as % of CGE	NA	NA	NA	NA	NA
Military expenditure per capita (1994 $)	NA	NA	NA	NA	NA
Armed forces per 1,000 people (soldiers)	NA	NA	3.0	2.9	2.9
GNP per capita (1994 $)	143	147	149	152	156
Arms imports[6]					
Current dollars (mil.)	0	0	0	0	0
1994 constant dollars (mil.)	0	0	0	0	0
Arms exports[6]					
Current dollars (mil.)	0	0	0	0	0
1994 constant dollars (mil.)	0	0	0	0	0
Total imports[7]					
Current dollars (mil.)	108	102	NA	125[e]	98[e]
1994 constant dollars (mil.)	120	109	NA	128[e]	98[e]
Total exports[7]					
Current dollars (mil.)	75	72	NA	67[e]	NA
1994 constant dollars (mil.)	83	77	NA	68[e]	NA
Arms as percent of total imports[8]	0	0	0	0	0
Arms as percent of total exports[8]	0	0	0	0	0

Human Rights [24]

	SSTS	FL	FAPRO	PPCG	APROBC	TPW	PCPTW	STPEP	PHRFF	PRW	ASST	AFL
Observes												

	EAFRD	CPR	ESCR	SR	ACHR	MAAE	PVIAC	PVNAC	EAFDAW	TCIDTP	RC
Observes	S							P			P

P = Party; S = Signatory; see Appendix for meaning of abbreviations.

Labor Force

Total Labor Force [25]

Labor Force by Occupation [26]

Agriculture	93%
Services	5
Industry and commerce	2

Massive lack of skilled labor.

Unemployment Rate [27]

For sources, notes, and explanations, see Annotated Source Appendix, page 1061.

Production Sectors

Commercial Energy Production and Consumption

Data are shown in quadrillion (10^{15}) BTUs and percent for 1995
Values for hydroelectric, nuclear, geothermal, solar, and wind power refer to electrical generation.

Production [28]

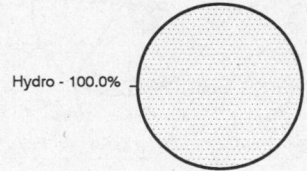

Hydro - 100.0%

Consumption [29]

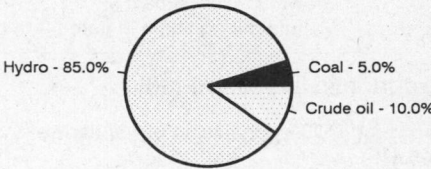

Hydro - 85.0% Coal - 5.0% Crude oil - 10.0%

Net hydroelectric power	0.017
Total	0.017

Crude oil	0.002
Coal	0.001
Net hydroelectric power	0.017
Total	0.020

Telecommunications [30]

- 4,620 (1991 est.) telephones
- Domestic: domestic telephone service is very poor with very few telephones in use
- International: international telephone and telegraph service is by landline through India; a satellite earth station was planned (1990)
- Radio: Broadcast stations: AM 1, FM 1, shortwave 0 (1990) Radios: 23,000 (1989 est.)
- Television: Broadcast stations: 0 (1990 est.) Televisions: 200 (1985 est.)

Transportation [31]

Railways: 0 km

Highways: total: 1,296 km; paved: 416 km; unpaved: 880 km (1988 est.)

Airports
Total: 2
With paved runways 1,524 to 2,437 m: 1
With unpaved runways 914 to 1,523 m: 1 (1995 est.)

Top Agricultural Products [32]

Produces rice, corn, root crops, citrus, foodgrains; dairy products, eggs.

Top Mining Products [33]

Metric tons except as noted[e]	3/17/95*
Cement	140,000
Dolomite	95,000
Gypsum	20,000
Limestone	200,000

Tourism [34]

	1990	1991	1992	1993	1994
Tourists	2	2	3	3	4
Tourism receipts	2	2	3	3	NA

Travelers are in thousands, money in million U.S. dollars.

For sources, notes, and explanations, see Annotated Source Appendix, page 1061.

109

Finance, Economics, and Trade

GDP and Manufacturing Summary [35]

	1980	1985	1990	1991	1992
Gross Domestic Product					
Millions of 1980 dollars	142	196	283	291	306
Growth rate in percent	17.63	3.69	4.56	2.96	5.00
Manufacturing Value Added					
Millions of 1980 dollars	5	10	20	21	22[e]
Growth rate in percent	-11.49	12.20	21.44	4.00	6.00[e]
Manufacturing share in percent of current prices	3.2	5.3	9.7	9.7	NA

Economic Indicators [36]

- **National product**: GDP—purchasing power parity—$1.3 billion (1995 est.)
- **National product real growth rate**: 6% (1995 est.)
- **National product per capita**: $730 (1995 est.)
- **Inflation rate (consumer prices)**: 8.6% (FY94/95 est.)
- **External debt**: $141 million (October 1994)

Balance of Payments Summary [37]

Exchange Rates [38]

Currency: **ngultrum.**
Symbol: **Nu.**

Data are currency units per $1. The Bhutanese ngultrum is at par with the Indian rupee.

January 1996	35.766
1995	32.427
1994	31.374
1993	30.493
1992	25.918
1991	22.742

Imports and Exports

Top Import Origins [39]

$113.6 million (c.i.f., FY94/95 est.).

Origins	%
India	77
Japan	NA
UK	NA
Germany	NA
US	NA

Top Export Destinations [40]

$70.9 million (f.o.b., FY94/95 est.).

Destinations	%
India	94
Bangladesh	NA

Foreign Aid [41]

	U.S. $	
Western (non-US) countries, ODA and OOF bilateral commitments (1970-89)	115	million
OPEC bilateral aid (1979-89)	11	million

Import and Export Commodities [42]

Import Commodities
Fuel and lubricants
Grain
Machinery and parts
Vehicles
Fabrics
Rice

Export Commodities
Cardamon
Gypsum
Timber
Handicrafts
Cement
Fruit
Electricity (to India)
Precious stones
Spices

For sources, notes, and explanations, see Annotated Source Appendix, page 1061.

Bolivia

Geography [1]

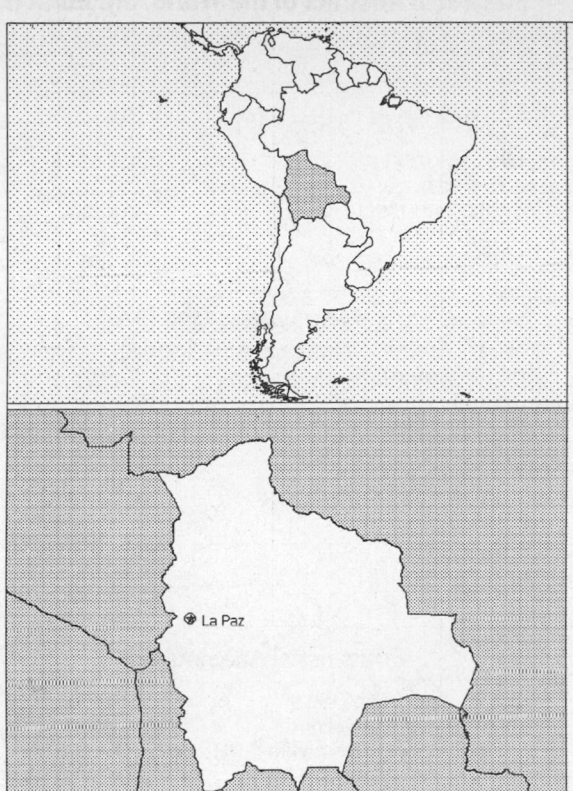

Total area:
1,098,580 sq km   424,164 sq mi
Land area:
1,084,390 sq km   418,685 sq mi
Comparative area:
Slightly less than three times the size of Montana
Land boundaries:
Total 6,743 km, Argentina 832 km, Brazil 3,400 km, Chile 861 km,
Paraguay 750 km, Peru 900 km
Coastline:
0 km (landlocked)
Climate:
Varies with altitude; humid and tropical to cold and semiarid
Terrain:
Rugged Andes Mountains with a highland plateau (Altiplano), hills,
lowland plains of the Amazon Basin
Natural resources:
Tin, natural gas, petroleum, zinc, tungsten, antimony, silver, iron,
lead, gold, timber
Land use:
Arable land: 3%
Permanent crops: 0%
Meadows and pastures: 25%
Forest and woodland: 52%
Other: 20%

Demographics [2]

	1970	1980	1990	1995[1]	1996	2000	2010	2020	2030
Population	4,270	5,296	6,388	7,035	7,165	7,680	8,941	10,246	11,574
Population density (persons per sq. mi.)	10	13	15	17	17	18	21	24	28
(persons per sq. km.)	4	5	6	6	7	7	8	9	11
Net migration rate (per 1,000 population)	-1.9	-4.6	-3.8	-3.5	-3.4	-3.2	-1.6	-0.5	0.0
Births	NA	NA	NA	NA	232	NA	NA	NA	NA
Deaths	NA	NA	NA	NA	77	NA	NA	NA	NA
Life expectancy - males	45.0	48.4	53.8	56.4	56.9	59.0	63.9	68.1	71.5
Life expectancy - females	48.1	53.4	59.4	62.3	62.8	65.1	70.2	74.5	77.9
Birth rate (per 1,000)	44.7	38.7	35.7	33.1	32.4	29.3	23.5	20.3	17.5
Death rate (per 1,000)	19.1	15.8	12.5	11.0	10.8	9.6	7.7	6.8	6.4
Women of reproductive age (15-49 yrs.)	980	1,209	1,516	1,697	1,732	1,879	2,335	2,778	3,087
of which are currently married	NA	NA	908	NA	1,049	1,152	1,455	NA	NA
Fertility rate	6.5	5.5	4.9	4.4	4.3	3.8	2.8	2.4	2.2

Except as noted, values for vital statistics are in thousands; life expectancy is in years.

Health

Health Indicators [3]

% of population with access to	
safe water (1990-95)	55
adequate sanitation (1990-95)	55
health services (1985-95)	67
% of 1-year-olds immunized (1990-94) against	
TB (tuberculosis)	91
DPT (diphtheria, pertussis, tetanus)	80
polio	86
measles	86
% of contraceptive prevalence (1980-94)	45
ORT use rate (1990-94)	63

Health Expenditures [4]

Total health expenditure, 1990 (official exchange rate)	
Millions of dollars	181
Dollars per capita	25
Health expenditures as a percentage of GDP	
Total	4.0
Public sector	2.4
Private sector	1.6
Development assistance for health	
Total aid flows (millions of dollars)[1]	37
Aid flows per capita (dollars)	5.1
Aid flows as a percentage of total health expenditure	20.3

For sources, notes, and explanations, see Annotated Source Appendix, page 1061.

111

Human Factors

Women and Children [5]

% of pregnant women immunized (tetanus 1990-94)	52
% of births attended by trained health personnel (1983-94)	47
Maternal mortality rate (1980-92)	390
Under-5 mortality rate (1994)	110
% under-5 moderately/severely underweight (1980-1994)	16

Burden of Disease [6]

Population per physician (1984)	1,433.53
Population per nurse (1985)	2,291.99
Population per hospital bed (1990)	758.83
AIDS cases per 100,000 people (1994)	0.2
Malaria cases per 100,000 people (1992)	355

Ethnic Division [7]

Quechua	30%
Aymara	25%
Mestizo	25%-30%
European	5%-15%

Religion [8]

Roman Catholic	95%
Protestant (Evangelical Methodist)	5%

Major Languages [9]

Spanish (official), Quechua (official), Aymara (official).

Education

Public Education Expenditures [10]

Million (Boliviano)[8]	1980	1985	1989	1990	1991	1994
Total education expenditure	0.0051	43	269	363	467	1,322
as percent of GNP	4.4	2.1	2.3	2.7	2.7	5.4
as percent of total govt. expend.	NA	NA	NA	NA	NA	11.2
Current education expenditure	0.0049	NA	268	NA	NA	1,293
as percent of GNP	4.2	NA	2.3	NA	NA	5.3
as percent of current govt. expend.	NA	NA	NA	NA	NA	15.1
Capital expenditure	0.0002	NA	1	NA	NA	28

Educational Attainment [11]

Age group (1992)[1]	25+
Total population	2,533,393
Highest level attained (%)	
No schooling	23.5
First level	
Not completed	20.4
Completed	6.6
Entered second level	
S-1	15.2
S-2	15.7
Postsecondary	9.9

Illiteracy [12]

In thousands and percent[1]	1990	1995	2000
Illiterate population (15+ yrs.)	809	745	676
Illiteracy rate - total pop. (%)	21.3	17.5	14.2
Illiteracy rate - males (%)	12.5	9.9	7.7
Illiteracy rate - females (%)	29.6	24.7	20.4

Libraries [13]

Daily Newspapers [14]

	1980	1985	1990	1994
Number of papers	14	14	17	11
Circ. (000)	226	290[e]	400[e]	500[e]

Culture [15]

Cinema (seats per 1,000)	NA
Annual attendance per person	0.3
Gross box office receipts (mil. Boliviano)	NA
Museums (reporting)	NA
Visitors (000)	NA
Annual receipts (000 Boliviano)	NA

Science and Technology

Scientific/Technical Forces [16]

Scientists/engineers	1,681
Number female	700
Technicians	1,039
Number female	500
Total[4]	2,720

R&D Expenditures [17]

	Boliviano (000) 1991
Total expenditure	282,899,000
Capital expenditure	200,536,000
Current expenditure	82,363,000
Percent current	29.1

U.S. Patents Issued [18]

For sources, notes, and explanations, see Annotated Source Appendix, page 1061.

Government and Law

Organization of Government [19]

Long-form name:
Republic of Bolivia

Type:
Republic

Independence:
6 August 1825 (from Spain)

National holiday:
Independence Day, 6 August (1825)

Constitution:
2 February 1967

Legal system:
Based on Spanish law and Napoleonic Code; has not accepted compulsory ICJ jurisdiction

Executive branch:
President; Vice President; Cabinet

Legislative branch:
Bicameral National Congress (Congreso Nacional): Chamber of Deputies (Camara de Diputados) and Chamber of Senators (Camara de Senadores)

Judicial branch:
Supreme Court (Corte Suprema)

Crime [23]

Elections [20]

Chamber of Deputies	% of seats
Nationalist Revolutionary Movement (MNR)	40.0
Nationalist Dem. Action (ADN)	13.1
Movement of the Revolutionary Left (MIR)	13.1
Conscience of the Fatherland (CONDEPA)	10.0
Free Bolivia Movement (MBL)	5.4
Others	2.4

Government Expenditures [21]

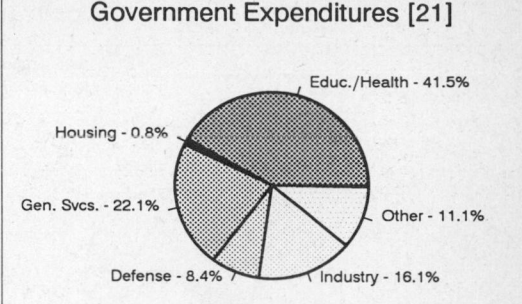

Educ./Health - 41.5%
Housing - 0.8%
Gen. Svcs. - 22.1%
Other - 11.1%
Defense - 8.4%
Industry - 16.1%

(% distribution). Expend. for CY95: 6,801.5 (Boliviano mil.)

Military Expenditures and Arms Transfers [22]

	1990	1991	1992	1993	1994
Military expenditures					
Current dollars (mil.)	142	114	102	121	130
1994 constant dollars (mil.)	158	122	106	124	130
Armed forces (000)	30	33	32	32	28
Gross national product (GNP)					
Current dollars (mil.)	4,021	4,386	4,688	4,981	5,315
1994 constant dollars (mil.)	4,475	4,702	4,888	5,083	5,315
Central government expenditures (CGE)					
1994 constant dollars (mil.)	843	872	1,130	1,314	1,347
People (mil.)	7.0	7.2	7.4	7.5	7.7
Military expenditure as % of GNP	3.5	2.6	2.2	2.4	2.4
Military expenditure as % of CGE	18.8	14.0	9.4	9.4	9.6
Military expenditure per capita (1994 $)	22	17	14	16	17
Armed forces per 1,000 people (soldiers)	4.3	4.6	4.3	4.2	3.6
GNP per capita (1994 $)	637	653	663	674	688
Arms imports[6]					
Current dollars (mil.)	10	20	10	10	0
1994 constant dollars (mil.)	11	21	10	10	0
Arms exports[6]					
Current dollars (mil.)	0	0	0	0	0
1994 constant dollars (mil.)	0	0	0	0	0
Total imports[7]					
Current dollars (mil.)	687	970	1,090	1,206	1,209
1994 constant dollars (mil.)	765	1,040	1,137	1,231	1,209
Total exports[7]					
Current dollars (mil.)	926	849	710	728	1,032
1994 constant dollars (mil.)	1,031	910	740	743	1,032
Arms as percent of total imports[8]	1.5	2.1	0.9	0.8	0
Arms as percent of total exports[8]	0	0	0	0	0

Human Rights [24]

	SSTS	FL	FAPRO	PPCG	APROBC	TPW	PCPTW	STPEP	PHRFF	PRW	ASST	AFL
Observes	P		P	S	P	P	P	P		P	P	P
	EAFRD	CPR	ESCR	SR	ACHR	MAAE	PVIAC	PVNAC	EAFDAW	TCIDTP	RC	
Observes		P	P	P	P	P		P	P	P	S	P

P = Party; S = Signatory; see Appendix for meaning of abbreviations.

Labor Force

Total Labor Force [25]

3.54 million

Labor Force by Occupation [26]

Services and utilities	20%
Manufacturing, mining, and construction	7
Agriculture	

Date of data: 1993

Unemployment Rate [27]

8% urban rate (1995 est.)

For sources, notes, and explanations, see Annotated Source Appendix, page 1061.

113

Production Sectors

Commercial Energy Production and Consumption

Data are shown in quadrillion (10^{15}) BTUs and percent for 1995
Values for hydroelectric, nuclear, geothermal, solar, and wind power refer to electrical generation.

Production [28]	Consumption [29]
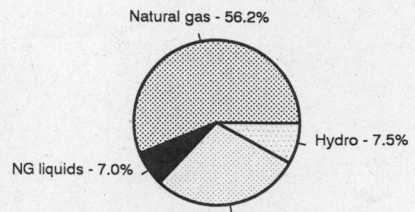	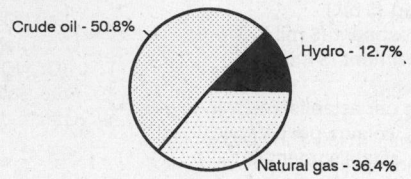
Natural gas - 56.2%	Crude oil - 50.8%
Hydro - 7.5%	Hydro - 12.7%
NG liquids - 7.0%	Natural gas - 36.4%
Crude oil - 29.4%	

Crude oil	0.059	Crude oil	0.060
Natural gas liquids	0.014	Dry natural gas	0.043
Dry natural gas	0.113	Net hydroelectric power	0.015
Net hydroelectric power	0.015	Total	0.118
Total	0.201		

Telecommunications [30]

- 144,300 (1987 est.) telephones; new subscribers face bureaucratic difficulties; most telephones are concentrated in La Paz and other cities
- Domestic: microwave radio relay system being expanded
- International: satellite earth station - 1 Intelsat (Atlantic Ocean)
- Radio: Broadcast stations: AM 129, FM 0, shortwave 68
- Television: Broadcast stations: 43 Televisions: 500,000 (1993 est.)

Transportation [31]

Railways: total: 3,691 km (single track); narrow gauge: 3,652 km 1.000-m gauge; 39 km 0.760-m gauge (13 km electrified) (1995)

Highways: total: 46,311 km; paved: 1,940 km (including 27 km of expressways); unpaved: 44,371 km (1991 est.)

Merchant marine: total: 1 cargo ship (1,000 GRT or over) totaling 4,214 GRT/6,390 DWT (1995 est.)

Airports

Total:	1,017
With paved runways over 3,047 m:	3
With paved runways 2,438 to 3,047 m:	4
With paved runways 1,524 to 2,437 m:	3
With paved runways under 914 m:	750
With unpaved runways 2,438 to 3,047 m:	2

Top Agricultural Products [32]

Produces coffee, coca, cotton, corn, sugarcane, rice, potatoes; timber.

Top Mining Products [33]

Metric tons except as noted[e]	5/95*
Gold, mine output, Au content (kg.)	12,800[r,9]
Lead, mine output, Pb content	19,700[r]
Silver, mine output, Ag content (kg.)	352,000[r,10]
Tin	
mine output, Sn content	16,200[r]
metal, smelter	15,300[r]
Zinc, mine output, Zn content	101,000[r]
Cement, hydraulic	500,000
Ulexite	10,400[r]
Petroleum refinery products (000 42-gal. bls.)	9,760

Tourism [34]

	1990	1991	1992	1993	1994
Tourists[16]	217	221	245	269	320
Tourism receipts	84	91	107	115	135
Tourism expenditures	130	129	141	137	140
Fare receipts	21	23	24	22	25
Fare expenditures	23	24	25	23	24

Travelers are in thousands, money in million U.S. dollars.

Manufacturing Sector

Manufacturing Summary [35]

	1987		1988		1989		1990		1991	
	$ bil.	%	$ bil.	%	$ bil.	%	$ bil.	%	$ bil.	%
Establishments or enterprises (number)	429	0.018	451	0.021	442	0.024	-	-	-	-
Total employment (000)	29	0.021	28	0.020	30	0.024	-	-	-	-
Production workers (000)	-	-	18	0.029	20	0.027	-	-	-	-
Output ($ bil.)	0.987	0.010	1	0.009	1	0.010	-	-	-	-
Value added ($ bil.)	0.460	0.011	0.643	0.013	0.565	0.012	-	-	-	-
Capital investment ($ mil.)	2,554	0.586	983	0.206	1,308	0.239	-	-	-	-
M & E investment ($ mil.)	-	-	-	-	-	-	-	-	-	-
Employees per establishment (number)	68	117.176	62	95.254	67	102.378	-	-	-	-
Production workers per establishment	-	-	41	137.155	46	116.400	-	-	-	-
Output per establishment ($ mil.)	2	53.105	2	42.428	3	43.326	-	-	-	-
Capital investment per estab. ($ mil.)	6	3,204	2	961	3	1,012	-	-	-	-
M & E per establishment ($ mil)	-	-	-	-	-	-	-	-	-	-
Payroll per employee ($)	2,441	27.221	2,898	29.088	2,496	22.340	-	-	-	-
Wages per production worker ($)	-	-	1,404	16.611	1,224	12.204	-	-	-	-
Hours per production worker (hours)	-	-	-	-	-	-	-	-	-	-
Output per employee ($)	33,943	45.320	37,939	44.542	40,559	42.319	-	-	-	-
Capital investment per employee ($)	87,817	2,735	35,110	1,009	43,995	989	-	-	-	-
M & E per employee ($)	-	-	-	-	-	-	-	-	-	-

Note: Columns headed % show percent of world total or ratio. Ratios closest to 100 are closest to world average. M & E stands for machinery & equipment.

Output in Manufacturing

	1987		1988		1989		1990		1991	
	$ bil.[p]	%	$ bil.[p]	%	$ bil.[p]	%	$ bil.[p]	%	$ bil.[p]	%
3110 - Food products	0.261	26.44	0.240	24.00	0.246	24.60	-	-	-	-
3130 - Beverages	0.121	12.26	0.123	12.30	0.142	14.20	-	-	-	-
3140 - Tobacco	0.014	1.42	0.011	1.10	0.012	1.20	-	-	-	-
3210 - Textiles	-	-	0.039	3.90	0.043	4.30	-	-	-	-
3211 - Spinning, weaving, etc.	0.026	2.63	0.028	2.80	0.029	2.90	-	-	-	-
3220 - Wearing apparel	0.005	0.51	0.003	0.30	0.004	0.40	-	-	-	-
3230 - Leather and products	0.012	1.22	0.013	1.30	0.017	1.70	-	-	-	-
3240 - Footwear	0.011	1.11	0.013	1.30	0.014	1.40	-	-	-	-
3310 - Wood products	0.023	2.33	0.022	2.20	0.039	3.90	-	-	-	-
3320 - Furniture, fixtures	0.002	0.20	0.002	0.20	0.001	0.10	-	-	-	-
3410 - Paper and products	0.009	0.91	0.004	0.40	0.005	0.50	-	-	-	-
3411 - Pulp, paper, etc.	0.003	0.30	0.001	0.10	0.002	0.20	-	-	-	-
3420 - Printing, publishing	0.019	1.93	0.028	2.80	0.032	3.20	-	-	-	-
3510 - Industrial chemicals	0.004	0.41	0.004	0.40	0.004	0.40	-	-	-	-
3511 - Basic chemicals, excl fertilizers	0.004	0.41	0.003	0.30	0.003	0.30	-	-	-	-
3513 - Synthetic resins, etc.	0.000	0.00	0.001	0.10	0.001	0.10	-	-	-	-
3520 - Chemical products nec	0.028	2.84	0.034	3.40	0.033	3.30	-	-	-	-
3522 - Drugs and medicines	0.015	1.52	0.019	1.90	0.017	1.70	-	-	-	-
3530 - Petroleum refineries	0.318	32.22	0.341	34.10	0.366	36.60	-	-	-	-
3550 - Rubber products	0.003	0.30	-	-	-	-	-	-	-	-
3560 - Plastic products nec	0.014	1.42	0.018	1.80	0.019	1.90	-	-	-	-
3610 - Pottery, china, etc.	-	-	0.000	0.00	0.012	1.20	-	-	-	-
3620 - Glass and products	0.009	0.91	0.006	0.60	0.006	0.60	-	-	-	-
3690 - Nonmetal products nec	0.038	3.85	0.040	4.00	0.046	4.60	-	-	-	-
3710 - Iron and steel	0.000	0.00	0.000	0.00	0.000	0.00	-	-	-	-
3720 - Nonferrous metals	0.022	2.23	0.039	3.90	0.083	8.30	-	-	-	-
3810 - Metal products	0.010	1.01	0.015	1.50	0.017	1.70	-	-	-	-
3820 - Machinery nec	0.001	0.10	0.001	0.10	0.001	0.10	-	-	-	-
3830 - Electrical machinery	0.003	0.30	0.004	0.40	0.003	0.30	-	-	-	-
3832 - Radio, television, etc.	0.001	0.10	0.001	0.10	0.000	0.00	-	-	-	-
3840 - Transportation equipment	0.004	0.41	0.003	0.30	0.003	0.30	-	-	-	-
3843 - Motor vehicles	0.004	0.41	0.003	0.30	0.003	0.30	-	-	-	-
3850 - Professional goods	0.001	0.10	0.001	0.10	0.001	0.10	-	-	-	-
3900 - Industries nec	0.001	0.10	0.001	0.10	-	-	-	-	-	-

Note: Codes are International Standard Industry codes (ISIC). Percentages are % of total Output. [f]: Factor Prices; [p]: Producer Prices.

For sources, notes, and explanations, see Annotated Source Appendix, page 1061.

115

Finance, Economics, and Trade

Economic Indicators [36]

- **National product**: GDP—purchasing power parity—$20 billion (1995 est.)

- **National product real growth rate**: 3.7% (1995 est.)

- **National product per capita**: $2,530 (1995 est.)

- **Inflation rate (consumer prices)**: 12% (1995 est.)

- **External debt**: $4.4 billion (November 1995)

Balance of Payments Summary [37]

Values in millions of dollars.

	1988	1989	1990	1991	1992
Exports of goods (f.o.b.)	542.5	723.5	830.8	760.3	608.4
Imports of goods (f.o.b.)	-590.9	-729.5	-775.6	-804.2	-1,040.8
Trade balance	-48.4	-6.0	55.2	-43.9	-432.4
Services - debits	-537.8	-581.2	-578.0	-582.8	-526.4
Services - credits	146.5	167.2	164.7	181.6	182.3
Private transfers (net)	12.7	20.6	21.6	23.0	22.7
Government transfers (net)	171.6	135.7	145.0	160.0	220.5
Long-term capital (net)	266.2	228.0	319.0	320.4	528.2
Short-term capital (net)	-128.9	-193.1	-62.0	-36.3	3.8
Errors and omissions	46.6	-32.1	-11.4	53.3	34.3
Overall balance	-71.5	-260.9	54.1	75.3	33.0

Exchange Rates [38]

Currency: **boliviano.**
Symbol: **$B.**

Data are currency units per $1.

December 1995	4.9137
1995	4.8003
1994	4.6205
1993	4.2651
1992	3.9005
1991	3.5806

Imports and Exports

Top Import Origins [39]

$1.21 billion (c.i.f., 1994 est.) Data are for 1993 est.

Origins	%
US	24
Argentina	13
Brazil	11
Japan	11

Top Export Destinations [40]

$1.1 billion (f.o.b., 1994 est.) Data are for 1993 est.

Destinations	%
US	26
Argentina	15

Foreign Aid [41]

Recipient: ODA, $362 million (1993).

Import and Export Commodities [42]

Import Commodities	Export Commodities
Capital goods 48%	Metals 39%
Chemicals 11%	Natural gas 9%
Petroleum 5%	Soybeans 11%
Food 5%	Jewelry 11%
	Wood 8%

Bosnia and Herzegovina

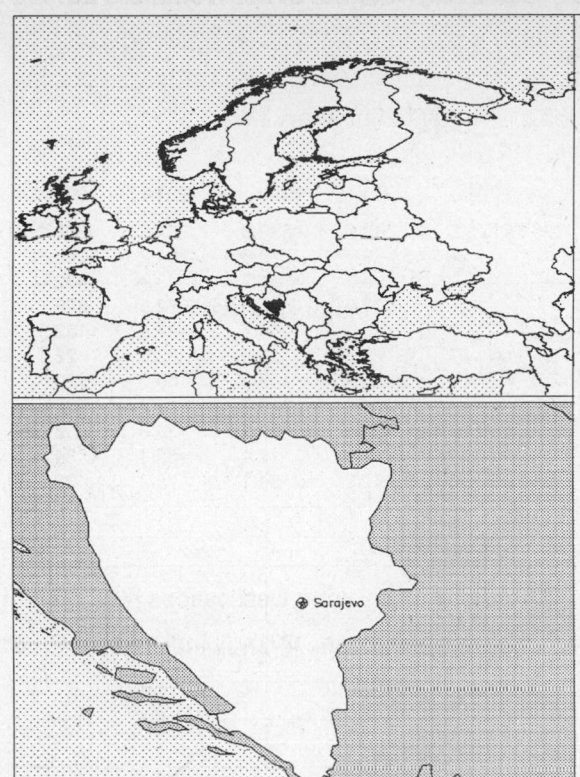

Geography [1]

Total area:
51,233 sq km   19,781 sq mi
Land area:
51,233 sq km   19,781 sq mi
Comparative area:
Slightly smaller than West Virginia
Land boundaries:
Total 1,459 km, Croatia 932 km, Yugoslavia 527 km (312 km with Serbia, 215 km with Montenegro)
Coastline:
20 km
Climate:
Hot summers and cold winters; areas of high elevation have short, cool summers and long, severe winters; mild, rainy winters along coast
Terrain:
Mountains and valleys
Natural resources:
Coal, iron, bauxite, manganese, forests, copper, chromium, lead, zinc
Land use:
Arable land: 20%
Permanent crops: 2%
Meadows and pastures: 25%
Forest and woodland: 36%
Other: 17%

Demographics [2]

	1970	1980	1990	1995[1]	1996	2000	2010	2020	2030
Population	3,703	4,092	4,360	2,782	2,656	2,618	2,892	2,966	2,898
Population density (persons per sq. mi.)	188	207	221	141	135	133	147	150	147
(persons per sq. km.)	72	80	85	54	52	51	57	58	57
Net migration rate (per 1,000 population)	NA	NA	NA	-53.9	-18.8	9.9	6.0	2.9	0.0
Births	NA	NA	NA	NA	17	NA	NA	NA	NA
Deaths	NA	NA	NA	NA	42	NA	NA	NA	NA
Life expectancy - males	NA	NA	NA	51.2	51.2	70.9	72.2	74.7	76.7
Life expectancy - females	NA	NA	NA	61.4	61.4	77.2	79.6	81.6	83.2
Birth rate (per 1,000)	NA	NA	NA	6.5	6.3	11.8	11.9	8.3	8.8
Death rate (per 1,000)	NA	NA	NA	15.5	15.9	10.6	10.7	11.3	11.9
Women of reproductive age (15-49 yrs.)	NA	NA	NA	688	663	671	684	671	637
of which are currently married	NA	NA	NA	NA	460	458	500	NA	NA
Fertility rate	NA	NA	NA	1.0	1.0	1.7	1.6	1.5	1.5

Except as noted, values for vital statistics are in thousands; life expectancy is in years.

Health

Health Indicators [3]

% of population with access to	
safe water (1990-95)	NA
adequate sanitation (1990-95)	NA
health services (1985-95)	NA
% of 1-year-olds immunized (1990-94) against	
TB (tuberculosis)	24
DPT (diphtheria, pertussis, tetanus)	38
polio	45
measles	48
% of contraceptive prevalence (1980-94)	NA
ORT use rate (1990-94)	NA

Health Expenditures [4]

Total health expenditure, 1990 (official exchange rate)[3]	
Millions of dollars	4,512
Dollars per capita	205
Health expenditures as a percentage of GDP	
Total	3.0
Public sector	4.0
Private sector	1.0
Development assistance for health	
Total aid flows (millions of dollars)[1]	NA
Aid flows per capita (dollars)	NA
Aid flows as a percentage of total health expenditure	NA

For sources, notes, and explanations, see Annotated Source Appendix, page 1061.

117

Human Factors

Women and Children [5]

% of pregnant women immunized (tetanus 1990-94)	NA
% of births attended by trained health personnel (1983-94)	NA
Maternal mortality rate (1980-92)	NA
Under-5 mortality rate (1994)	17
% under-5 moderately/severely underweight (1980-1994)	NA

Burden of Disease [6]

Ethnic Division [7]

Serb	40%
Muslim	38%
Croat (est.)	22%

Religion [8]

Muslim	40%
Orthodox	31%
Catholic	15%
Protestant	4%
Other	10%

Major Languages [9]

Serbo-Croatian	99%
Other	1%

Education

Public Education Expenditures [10]

Educational Attainment [11]

Illiteracy [12]

In thousands and percent[3]	1985	1991	2000
Illiterate population (15+ years)	1,614	1,342	942
Illiteracy rate - total pop. (%)	9.2	7.3	4.7
Illiteracy rate - males (%)	3.5	2.6	1.3
Illiteracy rate - females (%)	14.6	11.9	7.9

Libraries [13]

	Admin. Units	Svc. Pts.	Vols. (000)	Shelving (meters)	Vols. Added	Reg. Users
National (1989)	8	8	12,316	303,555	305,462	163,169
Nonspecialized (1989)	20	20	3,488	60,443	72,503	49,262
Public (1989)	808	1,937	30,238	552,866	1 mil	20 mil
Higher ed. (1989)	409	421	14,462	319,329	469,142	529,549
School (1989)	7,784	7,784	38,430	NA	1,680	NA

Daily Newspapers [14]

	1980	1985	1990[1]	1994
Number of papers	NA	NA	NA	2
Circ. (000)	NA	NA	NA	518

Culture [15]

Science and Technology

Scientific/Technical Forces [16]

R&D Expenditures [17]

	Dinar (000) 1992
Total expenditure	2,152,032
Capital expenditure	815,082
Current expenditure	1,336,950
Percent current	62.1

U.S. Patents Issued [18]

Government and Law

Organization of Government [19]

Long-form name:
Republic of Bosnia and Herzegovina
Type:
Emerging democracy
Independence:
NA April 1992 (from Yugoslavia)
National holiday:
NA
Constitution:
First promulgated in 1974 (under the Communists), amended 1989, 1990, and 1991; constitution of Muslim/Croat Federation of Bosnia and Herzegovina ratified April 1994; in Dayton Agreement of November 1995, the Federation and Serb republic accepted new basic constitutional principles
Legal system:
Based on civil law system
Executive branch:
President; Seven member collective presidency; Prime Minister; Cabinet; Executive body of ministers
Legislative branch:
Bicameral National Assembly: Chamber of Municipalities (Vijece Opeina) and Chamber of Citizens (Vijece Gradanstvo)
Judicial branch:
Supreme Court; Constitutional Court

Crime [23]

	1994
Crime volume	
Cases known to police	14,193
Attempts (percent)	NA
Percent cases solved	10,819
Crimes per 100,000 persons	NA
Persons responsible for offenses	
Total number offenders	12,422
Percent female	NA
Percent juvenile (14-18 yrs.)	1,434
Percent foreigners	NA

Elections [20]

Chamber of Citizens	% of seats
Party of Democratic Action (SDA)	33.1
Serbian Democ. Party (SDS BiH)	26.2
Croatian Democ. Union (HDZ BiH)	16.2
Party of Democratic Changes	11.5
All. of Reform Forces (SRSJ BiH)	9.2
Muslim-Bosnian Organization (MBO)	1.5
Democratic Party of Socialists (DSS)	0.8
Democratic League of Greens (DSZ)	0.8
Liberal Party (LS)	0.8

Government Budget [21]

Military Expenditures and Arms Transfers [22]

	1990	1991	1992[15]	1993	1994
Military expenditures					
Current dollars (mil.)	NA	NA	NA	NA	NA
1994 constant dollars (mil.)	NA	NA	NA	NA	NA
Armed forces (000)	NA	NA	60	NA	70
Gross national product (GNP)					
Current dollars (mil.)	NA	NA	NA	NA	NA
1994 constant dollars (mil.)	NA	NA	NA	NA	NA
Central government expenditures (CGE)					
1994 constant dollars (mil.)	NA	NA	NA	NA	NA
People (mil.)	NA	NA	4.0	3.4	3.2
Military expenditure as % of GNP	NA	NA	NA	NA	NA
Military expenditure as % of CGE	NA	NA	NA	NA	NA
Military expenditure per capita (1994 $)	NA	NA	NA	NA	NA
Armed forces per 1,000 people (soldiers)	NA	NA	15.1	NA	22.0
GNP per capita (1994 $)	NA	NA	NA	NA	NA
Arms imports[6]					
Current dollars (mil.)	NA	NA	0	0	100
1994 constant dollars (mil.)	NA	NA	0	0	100
Arms exports[6]					
Current dollars (mil.)	NA	NA	0	0	0
1994 constant dollars (mil.)	NA	NA	0	0	0
Total imports[7]					
Current dollars (mil.)	NA	NA	NA	155	200
1994 constant dollars (mil.)	NA	NA	NA	158	200
Total exports[7]					
Current dollars (mil.)	NA	NA	NA	41	21
1994 constant dollars (mil.)	NA	NA	NA	42	21
Arms as percent of total imports[8]	NA	NA	0	0	50.0
Arms as percent of total exports[8]	NA	NA	0	0	0

Human Rights [24]

	SSTS	FL	FAPRO	PPCG	APROBC	TPW	PCPTW	STPEP	PHRFF	PRW	ASST	AFL
Observes	P	P	P	P	P			P		P	P	
	EAFRD	CPR	ESCR	SR	ACHR	MAAE	PVIAC	PVNAC	EAFDAW	TCIDTP	RC	
Observes	P	P	P	P		P	P	P	P	P	P	

P=Party; S=Signatory; see Appendix for meaning of abbreviations.

Labor Force

Total Labor Force [25]

1.026 million

Labor Force by Occupation [26]

Unemployment Rate [27]

For sources, notes, and explanations, see Annotated Source Appendix, page 1061.

119

Production Sectors

Commercial Energy Production and Consumption

Data are shown in quadrillion (10^{15}) BTUs and percent for 1995
Values for hydroelectric, nuclear, geothermal, solar, and wind power refer to electrical generation.

Production [28]

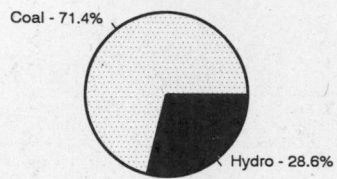

Coal - 71.4%
Hydro - 28.6%

Consumption [29]

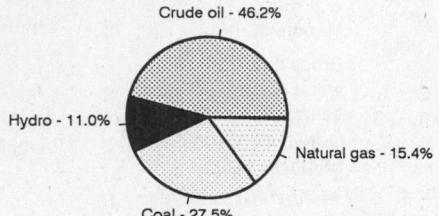

Crude oil - 46.2%
Hydro - 11.0%
Natural gas - 15.4%
Coal - 27.5%

Coal	0.025
Net hydroelectric power	0.010
Total	0.035

Crude oil	0.042
Dry natural gas	0.014
Coal	0.025
Net hydroelectric power	0.010
Total	0.091

Telecommunications [30]

- 727,000 telephones; telephone and telegraph network is in need of modernization and expansion; many urban areas are below average when compared with services in other former Yugoslav republics
- International: no satellite earth stations
- Radio: Broadcast stations: AM 9, FM 2, shortwave 0 Radios: 840,000
- Television: Broadcast stations: 6 Televisions: 1,012,094

Transportation [31]

Railways: total: 1,021 km (electrified 795 km); standard gauge: 1,021 km 1.435-m gauge (1991)

Highways: total: 21,168 km; paved: 11,436 km; unpaved: 9,732 km (1991 est.)

Merchant marine: none

Airports
Total:	24
With paved runways 2,438 to 3,047 m:	3
With paved runways 1,524 to 2,437 m:	3
With paved runways 914 to 1,523 m:	1
With paved runways under 914 m:	7
With unpaved runways 1,524 to 2,437 m:	1

Top Agricultural Products [32]

Produces wheat, corn, fruits, vegetables; livestock.

Top Mining Products [33]

Metric tons except as noted[e]	3/95[*]
Bauxite	75,000
Alumina	50,000
Iron ore	
gross weight	200,000
Fe content	70,000
Pig iron	100,000
Steel, crude	100,000
Semimanufactures	100,000
Glass sand	50,000
Salt, all sources	50,000

Tourism [34]

	1987	1988	1989	1990	1991
Visitors[111]	26,151	29,635	34,118	39,573	NA
Tourists[11,111]	8,907	9,018	8,644	7,880	1,459

Travelers are in thousands, money in million U.S. dollars.

Manufacturing Sector

GDP and Manufacturing Summary [35]

	1980	1985	1989	1990	% change 1980-1990	% change 1989-1990
GDP (million 1980 $)	69,958	71,058	72,234	66,371	-5.1	-8.1
GDP per capita (1980 $)	3,136	3,073	3,050	2,786	-11.2	-8.7
Manufacturing as % of GDP (current prices)	30.6	37.2	39.5	42.0	37.3	6.3
Gross output (million $)	72,629	57,020	65,078	62,136[e]	-14.4	-4.5
Value added (million $)	21,750	17,171	30,245	27,660[e]	27.2	-8.5
Value added (million 1980 $)	19,526	22,283	24,021	21,703	11.1	-9.6
Industrial production index	100	116	120	108	8.0	-10.0
Employment (thousands)	2,106	2,467	2,658	2,537[e]	20.5	-4.6

Note: GDP stands for Gross Domestic Product. 'e' stands for estimated value.

Profitability and Productivity

	1980	1985	1989	1990	% change 1980-1990	% change 1989-1990
Intermediate input (%)	70	70	54	55[e]	-21.4	1.9
Wages, salaries, and supplements (%)	14	12	12[e]	18[e]	28.6	50.0
Gross operating surplus (%)	15	18	34[e]	26[e]	73.3	-23.5
Gross output per worker ($)	34,487	23,113	24,484	24,248[e]	-29.7	-1.0
Value added per worker ($)	10,328	6,960	11,379	10,796[e]	4.5	-5.1
Average wage (incl. benefits) ($)	4,991	2,703	2,986[e]	4,488[e]	-10.1	50.3

Profitability is in percent of gross output. Productivity is in U.S. $. 'e' stands for estimated value.

Profitability - 1990

Inputs - 55.6%
Wages - 18.2%
Surplus - 26.3%

The graphic shows percent of gross output.

Value Added in Manufacturing

	1980 $ mil.	1980 %	1985 $ mil.	1985 %	1989 $ mil.	1989 %	1990 $ mil.	1990 %	% change 1980-1990	% change 1989-1990
311 Food products	1,897	8.7	1,458	8.5	3,916	12.9	3,484[e]	12.6	83.7	-11.0
313 Beverages	459	2.1	353	2.1	663	2.2	589[e]	2.1	28.3	-11.2
314 Tobacco products	184	0.8	221	1.3	344	1.1	308[e]	1.1	67.4	-10.5
321 Textiles	1,759	8.1	1,428	8.3	2,881	9.5	2,663[e]	9.6	51.4	-7.6
322 Wearing apparel	903	4.2	718	4.2	1,593	5.3	1,427[e]	5.2	58.0	-10.4
323 Leather and fur products	226	1.0	231	1.3	383	1.3	340[e]	1.2	50.4	-11.2
324 Footwear	482	2.2	503	2.9	1,022	3.4	899[e]	3.3	86.5	-12.0
331 Wood and wood products	977	4.5	530	3.1	794	2.6	706[e]	2.6	-27.7	-11.1
332 Furniture and fixtures	730	3.4	438	2.6	1,030	3.4	1,065[e]	3.9	45.9	3.4
341 Paper and paper products	529	2.4	394	2.3	759	2.5	674[e]	2.4	27.4	-11.2
342 Printing and publishing	876	4.0	462	2.7	761	2.5	678[e]	2.5	-22.6	-10.9
351 Industrial chemicals	694	3.2	631	3.7	1,107	3.7	992[e]	3.6	42.9	-10.4
352 Other chemical products	681	3.1	525	3.1	1,419	4.7	1,315[e]	4.8	93.1	-7.3
353 Petroleum refineries	454	2.1	415	2.4	260	0.9	233[e]	0.8	-48.7	-10.4
354 Miscellaneous petroleum and coal products	101	0.5	101	0.6	104	0.3	91[e]	0.3	-9.9	-12.5
355 Rubber products	276	1.3	269	1.6	479	1.6	456[e]	1.6	65.2	-4.8
356 Plastic products	413	1.9	258	1.5	397	1.3	350[e]	1.3	-15.3	-11.8
361 Pottery, china, and earthenware	128	0.6	72	0.4	162	0.5	144[e]	0.5	12.5	-11.1
362 Glass and glass products	163	0.7	113	0.7	224	0.7	204[e]	0.7	25.2	-8.9
369 Other nonmetal mineral products	906	4.2	513	3.0	683	2.3	604[e]	2.2	-33.3	-11.6
371 Iron and steel	1,221	5.6	1,000	5.8	1,343	4.4	1,171[e]	4.2	-4.1	-12.8
372 Nonferrous metals	480	2.2	509	3.0	944	3.1	927[e]	3.4	93.1	-1.8
381 Metal products	2,105	9.7	1,577	9.2	1,293	4.3	1,130[e]	4.1	-46.3	-12.6
382 Nonelectrical machinery	1,828	8.4	1,463	8.5	2,372	7.8	2,378[e]	8.6	30.1	0.3
383 Electrical machinery	1,600	7.4	1,544	9.0	2,640	8.7	2,334[e]	8.4	45.9	-11.6
384 Transport equipment	1,441	6.6	1,263	7.4	2,389	7.9	2,241[e]	8.1	55.5	-6.2
385 Professional and scientific equipment	101	0.5	93	0.5	154	0.5	146[e]	0.5	44.6	-5.2
390 Other manufacturing industries	134	0.6	88	0.5	128	0.4	114[e]	0.4	-14.9	-10.9

Note: The industry codes shown are International Standard Industry codes (ISIC). Percentages are percent of total Value Added. 'e' stands for estimated value

Finance, Economics, and Trade

Economic Indicators [36]

- **National product**: GDP—purchasing power parity—$1 billion (1995 est.)
- **National product real growth rate**: NA%
- **National product per capita**: $300 (1995 est.)
- **Inflation rate (consumer prices)**: NA%
- **External debt**: $NA

Balance of Payments Summary [37]

Exchange Rates [38]

Currency: **dinar.**
Symbol: **D.**

Croatian dinar used in Croat-held area, presumably to be replaced by new Croatian kuna; old and new Serbian dinars used in Serb-held area; hard currencies probably supplanting local currencies in areas held by Bosnian Government.

Imports and Exports

Top Import Origins [39]

Top Export Destinations [40]

Foreign Aid [41]

Recipient: ODA, $NA.

Import and Export Commodities [42]

Import Commodities

Fuels and lubricants 32%
Machinery, transp. equip. 23.3%
Other manufactures 21.3%
Chemicals 10%
Raw materials 6.7%
Food and live animals 5.5%
Beverages and tobacco 1.9%

Export Commodities

Manufactured goods 31%
Machinery, transp. equip. 20.8%
Raw materials 18%
Misc. articles 17.3%
Chemicals 9.4%
Fuel and lubricants 1.4%
Food and live animals 1.2%

Botswana

Geography [1]

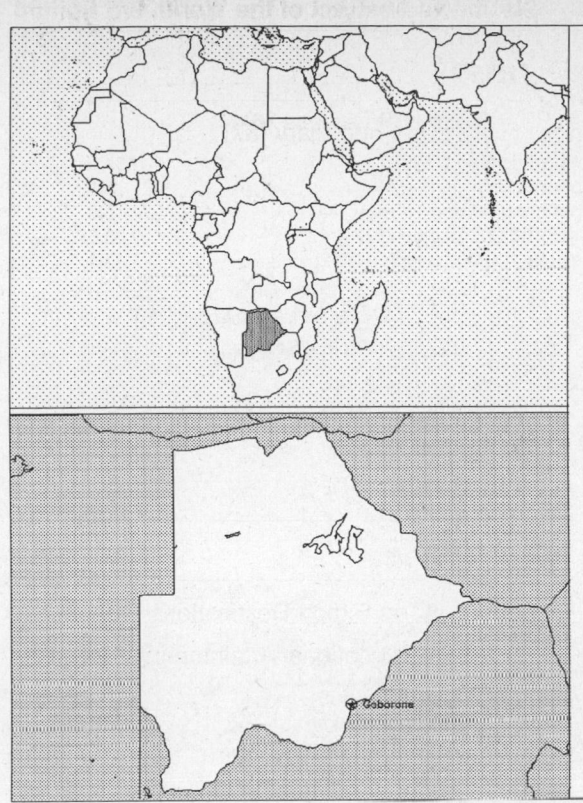

Total area:
600,370 sq km   231,804 sq mi
Land area:
585,370 sq km   226,013 sq mi
Comparative area:
Slightly smaller than Texas
Land boundaries:
Total 4,013 km, Namibia 1,360 krn, South Africa 1,840 km, Zimbabwe 813 km
Coastline:
0 km (landlocked)
Climate:
Semiarid; warm winters and hot summers
Terrain:
Predominately flat to gently rolling tableland; Kalahari Desert in Southwest
Natural resources:
Diamonds, copper, nickel, salt, soda ash, potash, coal, iron ore, silver
Land use:
Arable land: 2%
Permanent crops: 0%
Meadows and pastures: 75%
Forest and woodland: 2%
Other: 21%

Demographics [2]

	1970	1980	1990	1995[1]	1996	2000	2010	2020	2030
Population	584	903	1,304	1,453	1,478	1,557	1,598	1,553	1,618
Population density (persons per sq. mi.)	3	4	6	6	7	7	7	7	7
(persons per sq. km.)	1	2	2	2	3	3	3	3	3
Net migration rate (per 1,000 population)	NA	NA	0.0	0.0	0.0	0.0	0.0	0.0	0.0
Births	NA	NA	NA	NA	49	NA	NA	NA	NA
Deaths	NA	NA	NA	NA	25	NA	NA	NA	NA
Life expectancy - males	NA	NA	54.0	46.5	44.9	39.9	34.4	39.0	53.8
Life expectancy - females	NA	NA	61.9	48.8	47.1	41.5	32.3	37.6	56.8
Birth rate (per 1,000)	NA	NA	35.5	34.0	33.3	30.5	25.1	23.7	21.6
Death rate (per 1,000)	NA	NA	12.2	16.2	17.0	20.7	29.2	24.5	12.5
Women of reproductive age (15-49 yrs.)	NA	NA	310	364	372	397	409	416	453
of which are currently married	NA	NA	118	NA	143	152	151	NA	NA
Fertility rate	NA	NA	4.8	4.4	4.3	3.8	2.9	2.4	2.2

Except as noted, values for vital statistics are in thousands; life expectancy is in years.

Health

Health Indicators [3]

% of population with access to	
safe water (1990-95)[1]	93
adequate sanitation (1990-95)	55
health services (1985-95)[1]	89
% of 1-year-olds immunized (1990-94) against	
TB (tuberculosis)	92
DPT (diphtheria, pertussis, tetanus)	78
polio	78
measles	71
% of contraceptive prevalence (1980-94)	33
ORT use rate (1990-94)	64

Health Expenditures [4]

For sources, notes, and explanations, see Annotated Source Appendix, page 1061.

123

Human Factors

Women and Children [5]

% of pregnant women immunized (tetanus 1990-94)	97
% of births attended by trained health personnel (1983-94)	78
Maternal mortality rate (1980-92)	250
Under-5 mortality rate (1994)	54
% under-5 moderately/severely underweight (1980-1994)[1]	15

Burden of Disease [6]

Population per physician (1989)	5,145.83
Population per nurse	NA
Population per hospital bed	NA
AIDS cases per 100,000 people (1994)	65.6
Malaria cases per 100,000 people (1992)	NA

Ethnic Division [7]

Batswana	95%
Kalanga, Basarwa, and Kgalagadi	4%
White	1%

Religion [8]

Indigenous beliefs	50%
Christian	50%

Major Languages [9]

English (official), Setswana.

Education

Public Education Expenditures [10]

	1980	1985	1990	1992	1993	1994
Million (Pula)[9]						
Total education expenditure	NA	111	437	611	758	872
as percent of GNP	NA	6.8	7.6	7.6	9.0	8.5
as percent of total govt. expend.	NA	15.4	17.0	15.1	17.6	NA
Current education expenditure	NA	88	311	493	610	708
as percent of GNP	NA	5.4	5.4	6.2	7.3	6.9
as percent of current govt. expend.	NA	18.6	21.1	21.6	21.8	NA
Capital expenditure	NA	23	126	118	148	164

Educational Attainment [11]

Age group (1981)	25+
Total population	310,303
Highest level attained (%)	
No schooling	54.7
First level	
Not completed	31.1
Completed	9.4
Entered second level	
S-1	3.1
S-2	1.3
Postsecondary	0.5

Illiteracy [12]

In thousands and percent[1]	1990	1995	2000
Illiterate population (15+ yrs.)	242	255	265
Illiteracy rate - total pop. (%)	33.3	30.6	28.9
Illiteracy rate - males (%)	21.9	20.4	19.2
Illiteracy rate - females (%)	43.3	39.8	37.9

Libraries [13]

	Admin. Units	Svc. Pts.	Vols. (000)	Shelving (meters)	Vols. Added	Reg. Users
National (1988)	NA	NA	NA	NA	NA	NA
Nonspecialized	NA	NA	NA	NA	NA	NA
Public	NA	NA	NA	NA	NA	NA
Higher ed.	8	8	9	NA	2,834	680
School	NA	NA	NA	NA	NA	NA

Daily Newspapers [14]

	1980	1985	1990	1994
Number of papers	1	1	1	1
Circ. (000)	19	18	18	35

Culture [15]

Cinema (seats per 1,000)	NA
Annual attendance per person	NA
Gross box office receipts (mil. Pula)	NA
Museums (reporting)	1
Visitors (000)	40
Annual receipts (000 Pula)	40

Science and Technology

Scientific/Technical Forces [16]

R&D Expenditures [17]

U.S. Patents Issued [18]

For sources, notes, and explanations, see Annotated Source Appendix, page 1061.

Government and Law

Organization of Government [19]

Long-form name:
Republic of Botswana
Type:
Parliamentary republic
Independence:
30 September 1966 (from UK)
National holiday:
Independence Day, 30 September (1966)
Constitution:
March 1965, effective 30 September 1966
Legal system:
Based on Roman-Dutch law and local customary law; judicial review limited to matters of interpretation; has not accepted compulsory ICJ jurisdiction
Executive branch:
President; Vice President; Cabinet
Legislative branch:
Bicameral Parliament: House of Chiefs and National Assembly
Judicial branch:
High Court; Court of Appeal

Elections [20]

National Assembly	% of votes
Botswana Democratic Party (BDP)	67.5
Botswana National Front (BNF)	32.5

Government Expenditures [21]

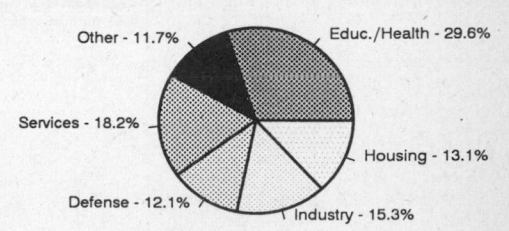

Other - 11.7%
Educ./Health - 29.6%
Services - 18.2%
Housing - 13.1%
Defense - 12.1%
Industry - 15.3%

(% distribution). Expend. for FY93: 3,924.5 (Pula mil.)

Crime [23]

	1994
Crime volume	
Cases known to police	124,738
Attempts (percent)	NA
Percent cases solved	NA
Crimes per 100,000 persons	8,281.46
Persons responsible for offenses	
Total number offenders	NA
Percent female	NA
Percent juvenile (8-18 yrs.)	NA
Percent foreigners	NA

Military Expenditures and Arms Transfers [22]

	1990	1991	1992	1993	1994
Military expenditures					
Current dollars (mil.)	148	167	174	228	229
1994 constant dollars (mil.)	165	179	181	233	229
Armed forces (000)	6	7	6	6	8
Gross national product (GNP)					
Current dollars (mil.)	2,706	3,134	3,636	3,583	3,806
1994 constant dollars (mil.)	3,012	3,360	3,792	3,657	3,806
Central government expenditures (CGE)					
1994 constant dollars (mil.)	1,526	1,633	1,756	1,913[e]	1,629[e]
People (mil.)	1.2	1.3	1.3	1.3	1.4
Military expenditure as % of GNP	5.5	5.3	4.8	6.4	6.0
Military expenditure as % of CGE	10.8	11.0	10.3	12.2	14.1
Military expenditure per capita (1994 $)	135	142	140	176	169
Armed forces per 1,000 people (soldiers)	4.9	5.2	4.6	4.5	5.9
GNP per capita (1994 $)	2,460	2,670	2,935	2,758	2,800
Arms imports[6]					
Current dollars (mil.)	20	20	20	20	20
1994 constant dollars (mil.)	22	21	21	20	20
Arms exports[6]					
Current dollars (mil.)	0	0	0	0	0
1994 constant dollars (mil.)	0	0	0	0	0
Total imports[7]					
Current dollars (mil.)	1,946	1,947	1,861	1,776	NA
1994 constant dollars (mil.)	2,166	2,087	1,941	1,813	NA
Total exports[7]					
Current dollars (mil.)	1,784	1,849	1,742	1,725	1,800
1994 constant dollars (mil.)	1,985	1,982	1,817	1,761	1,800
Arms as percent of total imports[8]	1.0	1.0	1.1	1.1	NA
Arms as percent of total exports[8]	0	0	0	0	0

Human Rights [24]

	SSTS	FL	FAPRO	PPCG	APROBC	TPW	PCPTW	STPEP	PHRFF	PRW	ASST	AFL
Observes	1					P	P				1	

	EAFRD	CPR	ESCR	SR	ACHR	MAAE	PVIAC	PVNAC	EAFDAW	TCIDTP	RC
Observes	P			P			P	P			P

P = Party; S = Signatory; see Appendix for meaning of abbreviations.

Labor Force

Total Labor Force [25]

428,000 (1992)

Labor Force by Occupation [26]

Various mining in South Africa
Cattle raising
Subsistence agriculture
Date of data: 1992 est.

Unemployment Rate [27]

21% (1995 est.)

For sources, notes, and explanations, see Annotated Source Appendix, page 1061.

125

Production Sectors

Commercial Energy Production and Consumption

Data are shown in quadrillion (10^{15}) BTUs and percent for 1995
Values for hydroelectric, nuclear, geothermal, solar, and wind power refer to electrical generation.

Production [28]

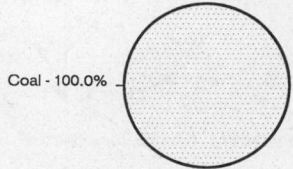

Coal - 100.0%

Coal	0.021
Total	0.021

Consumption [29]

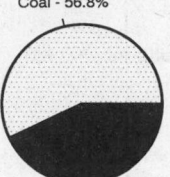

Coal - 56.8%
Crude oil - 43.2%

Crude oil	0.016
Coal	0.021
Total	0.037

Telecommunications [30]

- 19,109 (1985 est.) telephones; sparse system
- Domestic: small system of open-wire lines, microwave radio relay links, and a few radiotelephone communication stations
- International: microwave radio relay links to Zambia, Zimbabwe and South Africa; satellite earth station - 1 Intelsat (Indian Ocean)
- Radio: Broadcast stations: AM 7, FM 13, shortwave 0
- Television: Broadcast stations: 0 (1988 est.) Televisions: 13,800 (1993 est.)

Transportation [31]

Railways: total: 971 km; narrow gauge: 971 km 1.067-m gauge (1995)

Highways: total: 11,448 km; paved: 1,590 km; unpaved: 9,858 km (1988 est.)

Airports

Total:	81
With paved runways over 3,047 m:	1
With paved runways 2,438 to 3,047 m:	1
With paved runways 1,524 to 2,437 m:	9
With paved runways 914 to 1,523 m:	1
With paved runways under 914 m:	22

Top Agricultural Products [32]

Agriculture accounts for 5% of the GDP; produces sorghum, maize, millet, pulses, groundnuts (peanuts), beans, cowpeas, sunflower seed; livestock.

Top Mining Products [33]

Metric tons except as noted	10/1/95[*]
Coal, bituminous	900,000
Copper, mine output, Cu content/ore milled	25,300[11]
Gemstones, semiprecious (kg.)	67,000[12]
Nickel	
mine output, Ni content/ore milled	22,800
smelter output, matte, gross weight	51,500[13]
Salt	186,000[14]
Sand, construction (cu. meters)	140,000[15]
Soda ash, natural	174,000
Stone, crushed (cu. meters)	572,000

Tourism [34]

	1990	1991	1992	1993	1994
Visitors[17]	844	899	916	962	924
Tourists	543	592	590	607	NA
Excursionists	301	307	326	355	NA
Tourism receipts	19	13	24	31	35
Tourism expenditures	41	49	54	59	52
Fare receipts	12	12	11	1	1
Fare expenditures	20	19	17	2	13

Travelers are in thousands, money in million U.S. dollars.

For sources, notes, and explanations, see Annotated Source Appendix, page 1061.

Manufacturing Sector

Manufacturing Summary [35]

	1987		1988		1989		1990		1991	
	$ bil.	%	$ bil.	%	$ bil.	%	$ bil.	%	$ bil.	%
Establishments or enterprises (number)	471	0.020	131	0.006	102	0.005	114	0.006	132	0.017
Total employment (000)	7	0.005	8	0.006	11	0.009	11	0.010	12	0.017
Production workers (000)	-	-	-	-	-	-	-	-	-	-
Output ($ bil.)	0.210	0.002	0.231	0.002	-	-	-	-	-	-
Value added ($ bil.)	0.072	0.002	0.072	0.001	-	-	-	-	-	-
Capital investment ($ mil.)	-	-	-	-	-	-	-	-	-	-
M & E investment ($ mil.)	-	-	-	-	-	-	-	-	-	-
Employees per establishment (number)	15	25.689	58	88.989	105	159.651	97	157.585	89	98.367
Production workers per establishment	-	-	-	-	-	-	-	-	-	-
Output per establishment ($ mil.)	0.446	10.292	2	31.781	-	-	-	-	-	-
Capital investment per estab. ($ mil.)	-	-	-	-	-	-	-	-	-	-
M & E per establishment ($ mil)	-	-	-	-	-	-	-	-	-	-
Payroll per employee ($)	-	-	-	-	-	-	-	-	-	-
Wages per production worker ($)	-	-	-	-	-	-	-	-	-	-
Hours per production worker (hours)	-	-	-	-	-	-	-	-	-	-
Output per employee ($)	30,004	40.062	30,419	35.713	-	-	-	-	-	-
Capital investment per employee ($)	-	-	-	-	-	-	-	-	-	-
M & E per employee ($)	-	-	-	-	-	-	-	-	-	-

Note: Columns headed % show percent of world total or ratio. Ratios closest to 100 are closest to world average. M & E stands for machinery & equipment.

Output in Manufacturing

	1987		1988		1989		1990		1991	
	$ bil.[f]	%	$ bil.[f]	%	$ bil.[f]	%	$ bil.[f]	%	$ bil.[f]	%
3110 - Food products	0.111	52.86	0.109	47.19	-	-	-	-	-	-
3130 - Beverages	0.046	21.90	0.053	22.94	-	-	-	-	-	-
3900 - Industries nec	0.052	24.76	0.069	29.87	-	-	-	-	-	-

Note: Codes are International Standard Industry codes (ISIC). Percentages are % of total Output. [f]: Factor Prices; [p]: Producer Prices.

For sources, notes, and explanations, see Annotated Source Appendix, page 1061.

127

Finance, Economics, and Trade

Economic Indicators [36]

- **National product**: GDP—purchasing power parity—$4.5 billion (1995 est.)
- **National product real growth rate**: 1% (1995 est.)
- **National product per capita**: $3,200 (1995 est.)
- **Inflation rate (consumer prices)**: 10% (1994 est.)
- **External debt**: $691 million (1994)

Balance of Payments Summary [37]

Values in millions of dollars.

	1986	1987	1988	1989	1990
Exports of goods (f.o.b.)	852.5	1,586.6	1,468.9	1,819.7	1,753.2
Imports of goods (f.o.b.)	-608.4	-803.9	-986.9	-1,185.1	-1,606.2
Trade balance	244.1	782.7	482.0	634.6	147.0
Services - debits	-400.9	-589.6	-791.9	-711.5	-782.1
Services - credits	216.6	296.9	330.4	355.3	497.3
Private transfers (net)	-2.6	6.8	-17.5	-30.6	-40.8
Government transfers (net)	53.8	166.7	184.6	250.5	316.2
Long-term capital (net)	99.6	-84.3	39.9	162.4	252.7
Short-term capital (net)	6.0	-5.5	-65.2	-49.5	-61.3
Errors and omissions	90.5	-12.2	220.0	-34.8	-21.7
Overall balance	307.1	561.5	382.3	576.4	307.3

Exchange Rates [38]

Currency: **pula.**
Symbol: **P.**

Data are currency units per $1.

January 1996	2.8305
1995	2.7716
1994	2.6831
1993	2.4190
1992	2.1327
1991	2.0173

Imports and Exports

Top Import Origins [39]

$1.8 billion (c.i.f., 1992).

Origins	%
Switzerland	NA
Southern African Customs Union (SACU)	NA
UK	NA
US	NA

Top Export Destinations [40]

$1.8 billion (f.o.b. 1994).

Destinations	%
Switzerland	NA
UK	NA
Southern African Customs Union (SACU)	NA

Foreign Aid [41]

Recipient: ODA, $189 million (1993).

Import and Export Commodities [42]

Import Commodities	Export Commodities
Foodstuffs	Diamonds 78%
Vehicles and transport equipment	Copper and nickel 6%
Textiles	Meat 5%
Petroleum products	

Brazil

Geography [1]

Total area:
8,511,965 sq km   3,286,488 sq mi
Land area:
8,456,510 sq km   3,265,077 sq mi
Comparative area:
Slightly smaller than the US
Note: Includes Arquipelago de Fernando de Noronha, Atol das Rocas, Ilha da Trindade, Ilhas Martin Vaz, and Penedos de Sao Pedro e Sao Paulo
Land boundaries:
Total 14,691 km, Argentina 1,224 km, Bolivia 3,400 km, Colombia 1,643 km, French Guiana 673 km, Guyana 1,119 km, Paraguay 1,290 km, Peru 1,560 km, Suriname 597 km, Uruguay 985 km, Venezuela 2,200 km
Coastline:
7,491 km
Climate:
Mostly tropical, but temperate in South
Terrain:
Mostly flat to rolling lowlands in North; some plains, hills, mountains, and narrow coastal belt
Natural resources:
Bauxite, gold, iron ore, manganese, nickel, phosphates, platinum, tin, uranium, petroleum, hydropower, timber
Land use:
Arable land: 7%
Permanent crops: 1%
Meadows and pastures: 19%
Forest and woodland: 67%
Other: 6%

Demographics [2]

	1970	1980	1990	1995[1]	1996	2000	2010	2020	2030
Population	95,684	122,830	150,062	160,738	162,661	169,545	183,747	194,246	201,874
Population density (persons per sq. mi.)	29	38	46	49	50	52	56	59	62
(persons per sq. km.)	11	15	18	19	19	20	22	23	24
Net migration rate (per 1,000 population)	0.0	0.0	0.0	0.0	0.0	0.0	0.0	0.0	0.0
Births	NA	NA	NA	NA	3,383	NA	NA	NA	NA
Deaths	NA	NA	848	NA	1,495	NA	NA	NA	NA
Life expectancy - males	55.3	58.2	61.1	56.6	56.7	57.1	61.9	64.4	70.0
Life expectancy - females	59.9	64.6	67.4	67.3	66.8	64.8	68.5	70.9	76.1
Birth rate (per 1,000)	37.5	31.8	22.6	21.2	20.8	19.2	16.5	14.1	12.8
Death rate (per 1,000)	10.9	8.6	7.4	9.0	9.2	10.1	9.5	9.8	9.3
Women of reproductive age (15-49 yrs.)	22,357	30,158	39,466	44,132	45,027	47,806	51,275	51,714	50,426
of which are currently married	11,419	17,003	23,660	NA	27,467	29,531	32,808	NA	NA
Fertility rate	5.3	4.1	2.6	2.4	2.3	2.1	1.9	1.8	1.8

Except as noted, values for vital statistics are in thousands; life expectancy is in years.

Health

Health Indicators [3]

% of population with access to	
safe water (1990-95)	87
adequate sanitation (1990-95)	83
health services (1985-95)	NA
% of 1-year-olds immunized (1990-94) against	
TB (tuberculosis)	92
DPT (diphtheria, pertussis, tetanus)	73
polio	68
measles	76
% of contraceptive prevalence (1980-94)	66
ORT use rate (1990-94)	63

Health Expenditures [4]

Total health expenditure, 1990 (official exchange rate)	
Millions of dollars	19,871
Dollars per capita	132
Health expenditures as a percentage of GDP	
Total	4.2
Public sector	2.8
Private sector	1.4
Development assistance for health	
Total aid flows (millions of dollars)[1]	84
Aid flows per capita (dollars)	0.6
Aid flows as a percentage of total health expenditure	0.4

For sources, notes, and explanations, see Annotated Source Appendix, page 1061.

129

Human Factors

Women and Children [5]

% of pregnant women immunized (tetanus 1990-94)	NA
% of births attended by trained health personnel (1983-94)	95
Maternal mortality rate (1980-92)	200
Under-5 mortality rate (1994)	61
% under-5 moderately/severely underweight (1980-1994)	7

Burden of Disease [6]

Population per physician (1985)	680.90
Population per nurse (1984)	1,202.19
Population per hospital bed (1990)	299.88
AIDS cases per 100,000 people (1994)	7.3
Malaria cases per 100,000 people (1992)	396

Ethnic Division [7]

White (mostly Portuguese, German, Italian, Spanish, Polish)	55%
Mixed white and African	38%
African	6%
Other (mostly Japanese, Arab, Amerindian)	1%

Religion [8]

Nominal Roman Catholic	70%
Other	30%

Major Languages [9]

Portuguese (official), Spanish, English, French.

Education

Public Education Expenditures [10]

Million (Cruzeiro)[10]	1980	1985	1989	1990	1991	1994
Total education expenditure	0.431	50	56,101	NA	NA	5,761
as percent of GNP	3.6	3.8	4.6	NA	NA	1.6
as percent of total govt. expend.	NA	NA	NA	NA	NA	NA
Current education expenditure	NA	NA	NA	NA	NA	NA
as percent of GNP	NA	NA	NA	NA	NA	NA
as percent of current govt. expend.	NA	NA	NA	NA	NA	NA
Capital expenditure	NA	NA	NA	NA	NA	NA

Educational Attainment [11]

Age group (1989)[2]	10+
Total population	110,157,487
Highest level attained (%)	
No schooling	18.7
First level	
Not completed	57.0
Completed	6.9
Entered second level	
S-1	11.9
S-2	5.5
Postsecondary	NA

Illiteracy [12]

In thousands and percent[1]	1990	1995	2000
Illiterate population (15+ yrs.)	18,514	18,331	17,842
Illiteracy rate - total pop. (%)	18.8	16.6	14.7
Illiteracy rate - males (%)	18.5	16.7	15.3
Illiteracy rate - females (%)	19.1	16.4	14.2

Libraries [13]

	Admin. Units	Svc. Pts.	Vols. (000)	Shelving (meters)	Vols. Added	Reg. Users
National (1993)	1	2	5,280	38,548	67,175	465,506
Nonspecialized	NA	NA	NA	NA	NA	NA
Public (1994)	2,739	NA	NA	NA	NA	NA
Higher ed.	NA	NA	NA	NA	NA	NA
School	NA	NA	NA	NA	NA	NA

Daily Newspapers [14]

	1980	1985	1990	1994
Number of papers	343	322	356	317
Circ. (000)	5,482	6,534	8,100[e]	7,200[e]

Culture [15]

Science and Technology

Scientific/Technical Forces [16]

Scientists/engineers[5]	26,754
Number female	NA
Technicians	9,327
Number female	NA
Total	36,081

R&D Expenditures [17]

	Reais (000) 1994
Total expenditure[2]	1,491,165
Capital expenditure	NA
Current expenditure	NA
Percent current	NA

U.S. Patents Issued [18]

Values show patents issued to citizens of the country by the U.S. Patents Office.

	1993	1994	1995
Number of patents	59	61	70

For sources, notes, and explanations, see Annotated Source Appendix, page 1061.

Government and Law

Organization of Government [19]

Long-form name:
Federative Republic of Brazil
Type:
Federal republic
Independence:
7 September 1822 (from Portugal)
National holiday:
Independence Day, 7 September (1822)
Constitution:
5 October 1988
Legal system:
Based on Roman codes; has not
accepted compulsory ICJ jurisdiction
Executive branch:
President; Vice President; Cabinet
Legislative branch:
Bicameral National Congress (Congresso
Nacional): Federal Senate (Senado
Federal) and Chamber of Deputies
(Camara dos Deputados)
Judicial branch:
Supreme Federal Tribunal

Crime [23]

Elections [20]

Chamber of Deputies	% of seats
Democratic Movement (PMDB)	21.0
Liberal Front Party (PFL)	18.0
Brazilian Soc. Dem. Party (PSDB)	12.0
Progressive Renewal Party (PDR)	10.0
Workers' Party (PT)	10.0
Democratic Labor Party (PDT)	7.0
Brazilian Labor Party (PTB)	6.0
Others	16.0

Government Expenditures [21]

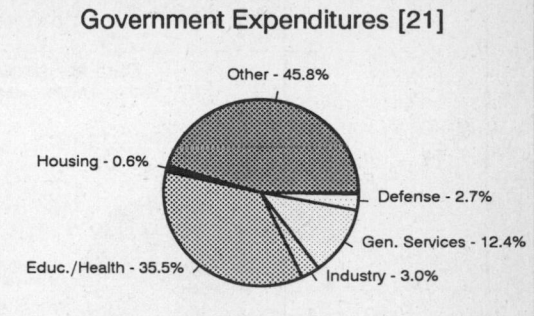

Other - 45.8%
Housing - 0.6%
Defense - 2.7%
Gen. Services - 12.4%
Educ./Health - 35.5%
Industry - 3.0%

(% distribution). Expend. for CY93: 5,250.4 (Real mil.)

Military Expenditures and Arms Transfers [22]

	1990	1991	1992	1993	1994
Military expenditures					
Current dollars (mil.)	8,157	6,712	5,482	6,466	6,427
1994 constant dollars (mil.)	9,078	7,194	5,717	6,599	6,427
Armed forces (000)	295	295	296	296	196
Gross national product (GNP)					
Current dollars (mil.)	478,200	501,100	511,800	547,200	580,200
1994 constant dollars (mil.)	532,200	537,100	533,700	558,500	580,200
Central government expenditures (CGE)					
1994 constant dollars (mil.)	197,700	148,300	161,100	NA	NA
People (mil.)	150.1	152.3	154.5	156.7	158.7
Military expenditure as % of GNP	1.7	1.3	1.1	1.2	1.1
Military expenditure as % of CGE	4.6	4.9	3.5	NA	NA
Military expenditure per capita (1994 $)	60	47	37	42	40
Armed forces per 1,000 people (soldiers)	2.0	1.9	1.9	1.9	1.2
GNP per capita (1994 $)	3,546	3,526	3,454	3,565	3,655
Arms imports[6]					
Current dollars (mil.)	140	160	90	60	90
1994 constant dollars (mil.)	156	172	94	61	90
Arms exports[6]					
Current dollars (mil.)	70	80	200	50	80
1994 constant dollars (mil.)	78	86	209	51	80
Total imports[7]					
Current dollars (mil.)	22,520	22,960	23,070	27,740	36,000
1994 constant dollars (mil.)	25,070	24,610	24,060	28,310	36,000
Total exports[7]					
Current dollars (mil.)	31,410	31,620	35,790	38,600	43,560
1994 constant dollars (mil.)	34,960	33,890	37,330	39,390	43,560
Arms as percent of total imports[8]	0.6	0.7	0.4	0.2	0.3
Arms as percent of total exports[8]	0.2	0.3	0.6	0.1	0.2

Human Rights [24]

	SSTS	FL	FAPRO	PPCG	APROBC	TPW	PCPTW	STPEP	PHRFF	PRW	ASST	AFL
Observes	P	P		P	P	P	P	P		P	P	P
	EAFRD	CPR	ESCR	SR	ACHR	MAAE	PVIAC	PVNAC	EAFDAW	TCIDTP	RC	
Observes	P	P	P	P	P	P	P	P	P	P	P	

P = Party; S = Signatory; see Appendix for meaning of abbreviations.

Labor Force

Total Labor Force [25]

57 million (1989 est.)

Labor Force by Occupation [26]

Services	42%
Agriculture	31
Industry	27

Unemployment Rate [27]

5% (1995 est.)

Production Sectors

Commercial Energy Production and Consumption

Data are shown in quadrillion (10^{15}) BTUs and percent for 1995
Values for hydroelectric, nuclear, geothermal, solar, and wind power refer to electrical generation.

Production [28]

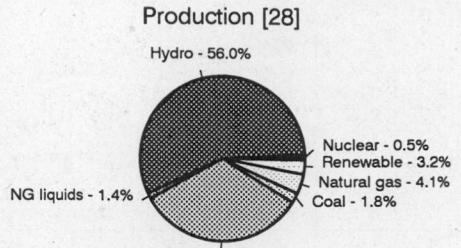

Hydro - 56.0%
Nuclear - 0.5%
Renewable - 3.2%
Natural gas - 4.1%
Coal - 1.8%
NG liquids - 1.4%
Crude oil - 32.9%

Consumption [29]

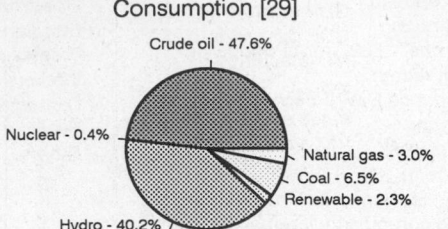

Crude oil - 47.6%
Nuclear - 0.4%
Natural gas - 3.0%
Coal - 6.5%
Renewable - 2.3%
Hydro - 40.2%

Crude oil	1.500
Natural gas liquids	0.062
Dry natural gas	0.187
Coal	0.084
Net hydroelectric power	2.548
Net nuclear power	0.025
Geothermal, solar, wind	0.147
Total	4.553

Crude oil	3.020
Dry natural gas	0.187
Coal	0.411
Net hydroelectric power	2.548
Net nuclear power	0.025
Geothermal, solar, wind	0.147
Total	6.338

Telecommunications [30]

- 14,426,673 (1992 est.) telephones; good working system
- Domestic: extensive microwave radio relay system and a domestic satellite system with 64 earth stations
- International: 3 coaxial submarine cables; satellite earth stations - 3 Intelsat (Atlantic Ocean), 1 Inmarsat (Atlantic Ocean Region East)
- Radio: Broadcast stations: AM 1,223, FM 0, shortwave 151 Radios: 60 million (1993 est.)
- Television: Broadcast stations: 112 Televisions: 30 million (1993 est.)

Transportation [31]

Railways: total: 27,418 km (1,750 km electrified); broad gauge: 5,730 km 1.600-m gauge; standard gauge: 194 km 1.440-m gauge; narrow gauge: 20,958 km 1.000-m gauge; 13 km 0.760-m gauge

Highways: total: 1,661,850 km; paved: 142,919 km; unpaved: 1,518,931 km (1992 est.)

Merchant marine: total: 207 ships (1,000 GRT or over) totaling 5,108,543 GRT/8,477,760 DWT; ships by type: bulk 48, cargo 29, chemical tanker 11, combination ore/oil 12, container 14, liquefied gas tanker 11, multifunction large-load carrier 1, oil tanker 64, passenger-cargo 5, refrigerated cargo 1, roll-on/roll-off cargo 11 (1995 est.)

Airports

Total:	2,950
With paved runways over 3,047 m:	5
With paved runways 2,438 to 3,047 m:	19
With paved runways 1,524 to 2,437 m:	122
With paved runways 914 to 1,523 m:	295
With paved runways under 914 m:	1,298

Top Agricultural Products [32]

Agriculture accounts for 16% of the GDP; produces coffee, soybeans, wheat, rice, corn, sugarcane, cocoa, citrus; beef.

Top Mining Products [33]

Metric tons except as noted[16]	8/31/95*
Bauxite, dry basis, gross weight	8,280,000[r]
Alumina	1,870,000[r]
Ferroalloys, electric furnace	1,050,000
Manganese ore/concentrate, gr. wt.	2,320,000[r]
Asbestos, crude ore	3,950,000[e]
Clays, crude	2,200,000
Magnesite, crude	1,500,000
Gneiss (cu. meters)	1,100,000
Basalt (cu. meters)	1,200,000
Sand, industrial	2,700,000

Tourism [34]

	1990	1991	1992	1993	1994
Tourists	1,091	1,228	1,688	1,572	1,700
Tourism receipts[18]	1,444	1,559	1,307	1,091	1,925
Tourism expenditures	1,559	1,224	1,221	1,892	2,931
Fare receipts	32	71	296	149	229
Fare expenditures	346	496	392	490	437

Travelers are in thousands, money in million U.S. dollars.

132

For sources, notes, and explanations, see Annotated Source Appendix, page 1061.

Manufacturing Sector

GDP and Manufacturing Summary [35]

	1980	1985	1989	1990	% change 1980-1990	% change 1989-1990
GDP (million 1980 $)	233,962	254,528	296,273	280,406	19.9	-5.4
GDP per capita (1980 $)	1,929	1,878	2,010	1,863	-3.4	-7.3
Manufacturing as % of GDP (current prices)	31.1	30.0	27.4[e]	23.3	-25.1	-15.0
Gross output (million $)	176,175	174,241	330,962[e]	291,993[e]	65.7	-11.8
Value added (million $)	71,690	77,082	148,882[e]	73,294	2.2	-50.8
Value added (million 1980 $)	70,679	68,069	73,225	69,529	-1.6	-5.0
Industrial production index	100	98	112	103	3.0	-8.0
Employment (thousands)	4,449	5,501	4,165[e]	5,213	17.2	25.2

Note: GDP stands for Gross Domestic Product. 'e' stands for estimated value.

Profitability and Productivity

	1980	1985	1989	1990	% change 1980-1990	% change 1989-1990
Intermediate input (%)	59	56	55[e]	75[e]	27.1	36.4
Wages, salaries, and supplements (%)	7	9[e]	9[e]	8[e]	14.3	-11.1
Gross operating surplus (%)	34	36[e]	36[e]	17[e]	50.0	-52.8
Gross output per worker ($)	39,599	31,674	79,462[e]	56,015[e]	41.5	-29.5
Value added per worker ($)	16,114	14,012	35,745[e]	14,061	-12.7	-60.7
Average wage (incl. benefits) ($)	2,773	2,753[e]	7,008[e]	4,334[e]	56.3	-38.2

Profitability is in percent of gross output. Productivity is in U.S. $. 'e' stands for estimated value.

Profitability - 1990

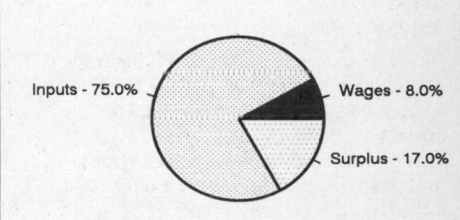

Inputs - 75.0%
Wages - 8.0%
Surplus - 17.0%

The graphic shows percent of gross output.

Value Added in Manufacturing

	1980 $ mil.	1980 %	1985 $ mil.	1985 %	1989 $ mil.	1989 %	1990 $ mil.	1990 %	% change 1980-1990	% change 1989-1990
311 Food products	7,996	11.2	9,259	12.0	14,703[e]	9.9	8,687	11.9	8.6	-10.9
313 Beverages	1,375	1.9	957	1.2	1,418[e]	1.0	1,388	1.9	0.9	-2.1
314 Tobacco products	495	0.7	587	0.8	962[e]	0.6	726	1.0	46.7	-24.5
321 Textiles	4,860	6.8	4,586	5.9	10,010[e]	6.7	3,862	5.3	-20.5	-61.4
322 Wearing apparel	2,307	3.2	2,639	3.4	4,198[e]	2.8	2,425[e]	3.3	5.1	-42.2
323 Leather and fur products	309	0.4	464	0.6	1,023[e]	0.7	371	0.5	20.1	-63.7
324 Footwear	985	1.4	1,353	1.8	3,572[e]	2.4	1,243[e]	1.7	26.2	-65.2
331 Wood and wood products	1,903	2.7	1,220	1.6	1,619[e]	1.1	951	1.3	-50.0	-41.3
332 Furniture and fixtures	1,087	1.5	949	1.2	1,206[e]	0.8	843	1.2	-22.4	-30.1
341 Paper and paper products	2,238	3.1	2,260	2.9	4,904[e]	3.3	2,556	3.5	14.2	-47.9
342 Printing and publishing	1,901	2.7	1,496	1.9	3,172[e]	2.1	2,305	3.1	21.3	-27.3
351 Industrial chemicals	3,428	4.8	5,933[e]	7.7	9,823[e]	6.6	3,930[e]	5.4	14.6	-60.0
352 Other chemical products	3,544	4.9	6,465[e]	8.4	9,773[e]	6.6	4,560[e]	6.2	28.7	-53.3
353 Petroleum refineries	3,075	4.3	1,956[e]	2.5	9,355[e]	6.3	1,343[e]	1.8	-56.3	-85.6
354 Miscellaneous petroleum and coal products	1,216	1.7	990[e]	1.3	1,295[e]	0.9	714[e]	1.0	-41.3	-44.9
355 Rubber products	941	1.3	1,420	1.8	3,154[e]	2.1	1,059	1.4	12.5	-66.4
356 Plastic products	1,994	2.8	1,742	2.3	3,911[e]	2.6	1,847	2.5	-7.4	-52.8
361 Pottery, china, and earthenware	190	0.3	844	1.1	323[e]	0.2	761[e]	1.0	300.5	135.6
362 Glass and glass products	558	0.8	525	0.7	1,022[e]	0.7	447[e]	0.6	-19.9	-56.3
369 Other nonmetal mineral products	3,447	4.8	1,941	2.5	4,956[e]	3.3	1,930[e]	2.6	-44.0	-61.1
371 Iron and steel	4,128	5.8	4,927	6.4	10,707[e]	7.2	5,811[e]	7.9	40.8	-45.7
372 Nonferrous metals	1,115	1.6	1,564	2.0	3,476[e]	2.3	2,172[e]	3.0	94.8	-37.5
381 Metal products	3,599	5.0	3,063	4.0	5,893[e]	4.0	3,714[e]	5.1	3.2	-37.0
382 Nonelectrical machinery	7,171	10.0	7,092	9.2	13,879[e]	9.3	6,282[e]	8.6	-12.4	-54.7
383 Electrical machinery	4,536	6.3	5,831	7.6	11,893[e]	8.0	6,341	8.7	39.8	-46.7
384 Transport equipment	3,625	5.1	4,954	6.4	9,578[e]	6.4	5,652	7.7	55.9	-41.0
385 Professional and scientific equipment	453	0.6	532[e]	0.7	1,422[e]	1.0	637[e]	0.9	40.6	-55.2
390 Other manufacturing industries	1,216	1.7	1,532[e]	2.0	1,628[e]	1.1	1,738[e]	2.4	42.9	6.8

Note: The industry codes shown are International Standard Industry codes (ISIC). Percentages are percent of total Value Added. 'e' stands for estimated value

Finance, Economics, and Trade

Economic Indicators [36]

- **National product**: GDP—purchasing power parity—$976.8 billion (1995 est.)
- **National product real growth rate**: 4.2% (1995)
- **National product per capita**: $6,100 (1995 est.)
- **Inflation rate (consumer prices)**: 23% (1995)
- **External debt**: $94 billion (1995 est.)

Balance of Payments Summary [37]

Values in millions of dollars.

	1989	1990	1991	1992	1993
Exports of goods (f.o.b.)	34,375	31,408	31,619	35,793	38,783
Imports of goods (f.o.b.)	-18,263	-20,661	-21,041	-20,554	-25,711
Trade balance	16,112	10,747	10,578	15,239	13,072
Services - debits	-19,773	-20,288	-17,765	-16,545	-20,525
Services - credits	4,442	4,919	4,223	5,206	5,163
Private transfers (net)	226	813	1,521	2,240	1,682
Government transfers (net)	18	21	35	3	-29
Long-term capital (net)	-3,025	-4,359	-1,343	22,546	10,111
Short-term capital (net)	3,421	9,563	2,835	-11,600	592
Errors and omissions	-819	-296	852	-1,393	-853
Overall balance	602	1,120	936	15,696	9,213

Exchange Rates [38]

Currency: **real.**
Symbol: **R$.**

Data are currency units per $1. January 1994 and prior are $CR per $1. On 1 August 1993 the cruzeiro real (CR$), equal to 1,000 cruzeiros, was introduced; another new currency, the real (R$) was introduced on 1 July 1994, equal to 2,750 cruzeiro reals.

January 1996	0.975
1995	0.918
1994	0.639
January 1994	390.845
1993	88.449
1992	4.513
1991	0.407

Imports and Exports

Top Import Origins [39]

$49.7 billion (f.o.b., 1995) Data are for 1993.

Origins	%
US	23.3
EU	22.5
Middle East	13.0
Latin America	11.8
Japan	6.5

Top Export Destinations [40]

$46.5 billion (f.o.b., 1995) Data are for 1993.

Destinations	%
EU	27.6
Latin America	21.8
US	17.4
Japan	6.3

Foreign Aid [41]

Recipient: ODA, $107 million (1993).

Import and Export Commodities [42]

Import Commodities	Export Commodities
Crude oil	Iron ore
Capital goods	Soybean bran
Chemical products	Orange juice
Foodstuffs	Footwear
Coal	Coffee
	Motor vehicle parts

For sources, notes, and explanations, see Annotated Source Appendix, page 1061.

Brunei

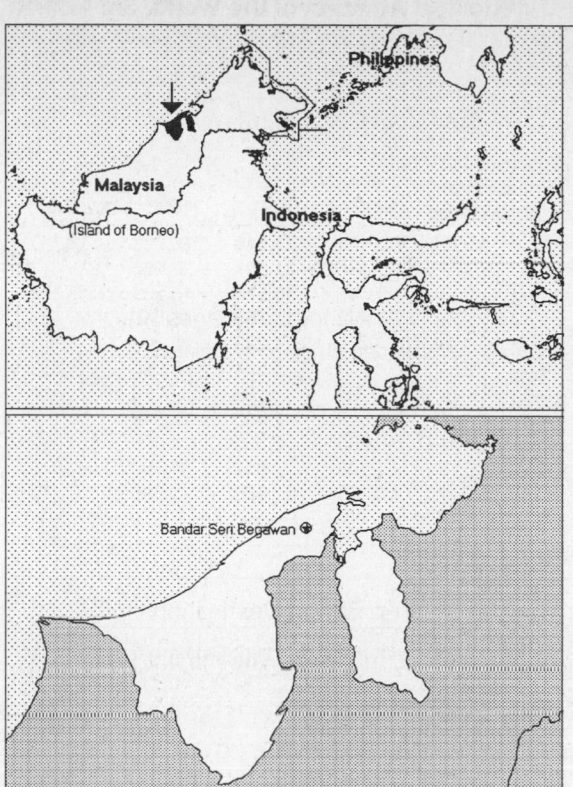

Geography [1]

Total area:
  5,770 sq km   2,228 sq mi
Land area:
  5,270 sq km   2,035 sq mi
Comparative area:
  Slightly larger than Delaware
Land boundaries:
  Total 381 km, Malaysia 381 km
Coastline:
  161 km
Climate:
  Tropical; hot, humid, rainy
Terrain:
  Flat coastal plain rises to mountains in East; hilly lowland in West
Natural resources:
  Petroleum, natural gas, timber
Land use:
  Arable land: 1%
  Permanent crops: 1%
  Meadows and pastures: 1%
  Forest and woodland: 79%
  Other: 18%

Demographics [2]

	1970	1980	1990	1995[1]	1996	2000	2010	2020	2030
Population	128	185	254	292	300	331	410	490	569
Population density (persons per sq. mi.)	63	91	125	144	147	162	201	241	279
(persons per sq. km.)	24	35	48	55	57	63	78	93	108
Net migration rate (per 1,000 population)	NA	11.4	7.3	5.5	5.2	4.1	2.7	1.7	1.0
Births	NA	NA	NA	NA	8	NA	NA	NA	NA
Deaths[9]	NA	1	NA	NA	2	NA	NA	NA	NA
Life expectancy - males	NA	NA	68.7	69.7	69.8	70.5	72.1	73.5	74.7
Life expectancy - females	NA	NA	72.3	72.9	73.0	73.6	74.8	75.9	76.9
Birth rate (per 1,000)	NA	31.4	27.7	25.8	25.5	24.5	22.9	21.2	19.9
Death rate (per 1,000)	NA	NA	5.0	5.1	5.1	5.3	5.9	6.5	7.4
Women of reproductive age (15-49 yrs.)	NA	NA	63	74	76	84	99	117	135
of which are currently married	NA	NA	40	NA	49	54	63	NA	NA
Fertility rate	NA	NA	3.5	3.4	3.4	3.3	3.1	3.0	2.8

Except as noted, values for vital statistics are in thousands; life expectancy is in years.

Health

Health Indicators [3]

Health Expenditures [4]

For sources, notes, and explanations, see Annotated Source Appendix, page 1061.

135

Human Factors

Women and Children [5]	Burden of Disease [6]	
	Population per physician (1984)	1,887.93
	Population per nurse (1984)	273.41
	Population per hospital bed	NA
	AIDS cases per 100,000 people (1994)	*
	Malaria cases per 100,000 people (1992)	NA

Ethnic Division [7]
Malay	64%
Chinese	20%
Other	16%

Religion [8]
Muslim (official)	63%
Buddhism	14%
Christian	8%
Indigenous beliefs and other (1981)	15%

Major Languages [9]
Malay (official), English, Chinese.

Education

Public Education Expenditures [10]

Million (Dollar)[11]	1980	1990	1991	1992	1993	1994
Total education expenditure	129	253	235	258	287	NA
as percent of GNP	1.2	3.9	3.6	NA	NA	NA
as percent of total govt. expend.	11.8	NA	NA	NA	NA	NA
Current education expenditure	115	229	217	239	273	NA
as percent of GNP	1.1	3.5	3.3	NA	NA	NA
as percent of current govt. expend.	12.5	NA	NA	NA	NA	NA
Capital expenditure	15	24	17	19	14	NA

Educational Attainment [11]
Age group (1981)	25+
Total population	75,283
Highest level attained (%)	
No schooling	32.1
First level	
Not completed	28.3
Completed	NA
Entered second level	
S-1	30.1
S-2	NA
Postsecondary	9.4

Illiteracy [12]
In thousands and percent[1]	1990	1995	2000
Illiterate population (15+ yrs.)	24	22	20
Illiteracy rate - total pop. (%)	14.7	11.4	9.0
Illiteracy rate - males (%)	9.2	6.9	5.1
Illiteracy rate - females (%)	21.1	16.5	12.3

Libraries [13]
	Admin. Units	Svc. Pts.	Vols. (000)	Shelving (meters)	Vols. Added	Reg. Users
National	NA	NA	NA	NA	NA	NA
Nonspecialized	NA	NA	NA	NA	NA	NA
Public (1992)	1	5	285	NA	23,426	41,001
Higher ed. (1990)	2	3	150	NA	10,854	1,504
School (1991)	23	NA	287	2,228	18,941	36,890

Daily Newspapers [14]
	1980	1985	1990	1994
Number of papers	-	-	1	1
Circ. (000)	-	-	10	20

Culture [15]
Cinema (seats per 1,000)	NA
Annual attendance per person	NA
Gross box office receipts (mil. Dollar)	NA
Museums (reporting)	5
Visitors (000)	142
Annual receipts (000 Dollar)	NA

Science and Technology

Scientific/Technical Forces [16]
Scientists/engineers	20
Number female	NA
Technicians	116
Number female	NA
Total[6]	136

R&D Expenditures [17]
	Dollar (000) 1984
Total expenditure[3]	10,880
Capital expenditure	2,660
Current expenditure	8,220
Percent current	75.6

U.S. Patents Issued [18]

For sources, notes, and explanations, see Annotated Source Appendix, page 1061.

Government and Law

Organization of Government [19]

Long-form name:
Negara Brunei Darussalam
Type:
Constitutional sultanate
Independence:
1 January 1984 (from UK)
National holiday:
National Day, 23 February (1984)
Constitution:
29 September 1959 (some provisions suspended under a State of Emergency since December 1962, others since independence on 1 January 1984)
Legal system:
Based on Islamic law
Executive branch:
Sultan and Prime Minister; Council of Cabinet Ministers; Religious Council; Privy Council; The Council of Succession
Legislative branch:
Unicameral: Legislative Council (Majlis Masyuarat Megeri)
Judicial branch:
Supreme Court

Elections [20]

Legislative Council. The Brunei National Democratic Party, the first legal political party, is now banned. Elections were last held in March 1962; in 1970 the Council was changed to an appointive body by decree of the sultan; an elective legislative Council is being considered as part of constitutional reform, but elections are unlikely for several years.

Government Budget [21]

For 1993.

	$ bil.
Revenues	2.1
Expenditures	2.1
Capital expenditures	0.427

Military Expenditures and Arms Transfers [22]

	1990	1991	1992	1993	1994
Military expenditures					
Current dollars (mil.)	305	NA	399	NA	309
1994 constant dollars (mil.)	339	NA	416	NA	309
Armed forces (000)	4	4	4	4	4
Gross national product (GNP)					
Current dollars (mil.)	3,649	3,927	3,995	3,917	3,917
1994 constant dollars (mil.)	4,062	4,209	4,166	3,997	3,917
Central government expenditures (CGE)					
1994 constant dollars (mil.)	1,685[e]	NA	NA	NA	NA
People (mil.)	0.3	0.3	0.3	0.3	0.3
Military expenditure as % of GNP	8.4	NA	10.0	NA	7.9
Military expenditure as % of CGE	20.1	NA	NA	NA	NA
Military expenditure per capita (1994 $)	1,336	NA	1,546	NA	1,087
Armed forces per 1,000 people (soldiers)	16.7	17.0	14.9	14.4	14.1
GNP per capita (1994 $)	15,990	16,090	15,470	14,430	13,760
Arms imports[6]					
Current dollars (mil.)	10	0	0	0	0
1994 constant dollars (mil.)	11	0	0	0	0
Arms exports[6]					
Current dollars (mil.)	0	0	0	0	0
1994 constant dollars (mil.)	0	0	0	0	0
Total imports[7]					
Current dollars (mil.)	1,019	1,084	NA	1,200[e]	NA
1994 constant dollars (mil.)	1,134	1,162	NA	1,225[e]	NA
Total exports[7]					
Current dollars (mil.)	2,226	2,480	2,496[e]	2,373[e]	NA
1994 constant dollars (mil.)	2,477	2,658	2,603[e]	2,422[e]	NA
Arms as percent of total imports[8]	1.0	0	0	0	0
Arms as percent of total exports[8]	0	0	0	0	0

Crime [23]

	1994
Crime volume	
Cases known to police	2,986
Attempts (percent)	NA
Percent cases solved	44.00
Crimes per 100,000 persons	1,148.46
Persons responsible for offenses	
Total number offenders	2,160
Percent female	10.80
Percent juvenile (7-16 yrs.)	NA
Percent foreigners	41.00

Human Rights [24]

	SSTS	FL	FAPRO	PPCG	APROBC	TPW	PCPTW	STPEP	PHRFF	PRW	ASST	AFL
Observes	1					P	P			1	1	
	EAFRD	CPR	ESCR	SR	ACHR	MAAE	PVIAC	PVNAC	EAFDAW	TCIDTP	RC	
Observes							P	P			P	

P = Party; S = Signatory; see Appendix for meaning of abbreviations.

Labor Force

Total Labor Force [25]

119,000 (1993 est.); note - includes members of the Army

Labor Force by Occupation [26]

Government	47.5%
Production of oil, natural gas, services and construction	41.9
Agriculture, forestry, and fishing	3.8

Date of data: 1986

Unemployment Rate [27]

4.8% (1994 est.)

For sources, notes, and explanations, see Annotated Source Appendix, page 1061.

137

Production Sectors

Commercial Energy Production and Consumption

Data are shown in quadrillion (10^{15}) BTUs and percent for 1995
Values for hydroelectric, nuclear, geothermal, solar, and wind power refer to electrical generation.

Production [28]

Natural gas - 51.6%

NG liquids - 2.2%

Crude oil - 46.3%

Consumption [29]

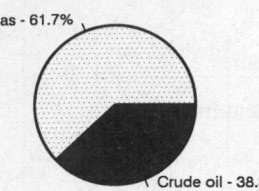

Natural gas - 61.7%

Crude oil - 38.3%

Crude oil	0.342
Natural gas liquids	0.016
Dry natural gas	0.381
Total	0.739

Crude oil	0.023
Dry natural gas	0.037
Total	0.060

Telecommunications [30]

- 76,900 (1993) telephones; service throughout country is adequate for present needs; international service good to adjacent Malaysia
- International: satellite earth stations - 2 Intelsat (1 Indian Ocean and 1 Pacific Ocean)
- Radio: Broadcast stations: AM 4, FM 4, shortwave 0 Radios: 115,000 (1993)
- Television: Broadcast stations: 1 (1984 est.) Televisions: 78,000 (1993 est.)

Transportation [31]

Railways: total: 13 km private line; narrow gauge: 13 km 0.610-m gauge

Highways: total: 2,443 km; paved: 1,296 km; unpaved: 1,147 km (1993)

Merchant marine: total: 7 liquefied gas tankers (1,000 GRT or over) totaling 348,476 GRT/340,635 DWT (1994 est.)

Airports

Total: 2
With paved runways over 3,047 m: 1
With unpaved runways 914 to 1,523 m: 1 (1995 est.)

Top Agricultural Products [32]

Agriculture accounts for 3% of the GDP; produces rice, cassava (tapioca), bananas; water buffalo, pigs.

Top Mining Products [33]

Values as noted[e]	5/12/95[*]
Gas, natural, gross (mil. cu. meters)	9,800
Natural gas liquids (000 42-gal. bls.)	5,080
Petroleum (000 42-gal. bls.)	
crude	59,200
refined products	1,390

Tourism [34]

	1990	1991	1992	1993	1994
Visitors	377	344	412	489	NA

Travelers are in thousands, money in million U.S. dollars.

Finance, Economics, and Trade

GDP and Manufacturing Summary [35]

	1980	1985	1990	1991	1992
Gross Domestic Product					
Millions of 1980 dollars	4,848	4,115	3,778	3,846	3,923
Growth rate in percent	-7.00	0.73	-1.15	3,846	3,923
Manufacturing Value Added					
Millions of 1980 dollars	573	339	367	376	385[e]
Growth rate in percent	-8.35	-5.42	0.09	2.33	2.44[e]
Manufacturing share in percent of current prices	11.7	10.0	NA	NA	NA

Economic Indicators [36]

- **National product**: GDP—purchasing power parity—$4.6 billion (1995 est.)
- **National product real growth rate**: 2% (1995 est.)
- **National product per capita**: $15,800 (1995 est.)
- **Inflation rate (consumer prices)**: 2.4% (1994 est.)
- **External debt**: 0

Balance of Payments Summary [37]

Exchange Rates [38]

Currency: **Bruneian dollar.**
Symbol: **B$.**

Data are currency units per $1. The Bruneian dollar is at par with the Singapore dollar.

January 1996	1.4214
1995	1.4174
1994	1.5274
1993	1.6158
1992	1.6290
1991	1.7276

Imports and Exports

Top Import Origins [39]

$1.8 billion (c.i.f., 1994 est.) Data are for 1994 est.

Origins	%
Singapore	29
UK	19
US	13
Malaysia	9
Japan	5

Top Export Destinations [40]

$2.4 billion (f.o.b., 1994 est.) Data are for 1994 est.

Destinations	%
Japan	50
UK	19
Thailand	10
Singapore	9

Foreign Aid [41]

	U.S. $	
US commitments, including Ex-Im (FY70-87)	20.6	million
Western (non-US) countries, ODA and OOF bilateral commitments (1970-89)	153	million

Import and Export Commodities [42]

Import Commodities
Machinery and transport equipment
Manufactured goods
Food
Chemicals

Export Commodities
Crude oil
Liquefied natural gas
Petroleum products

Bulgaria

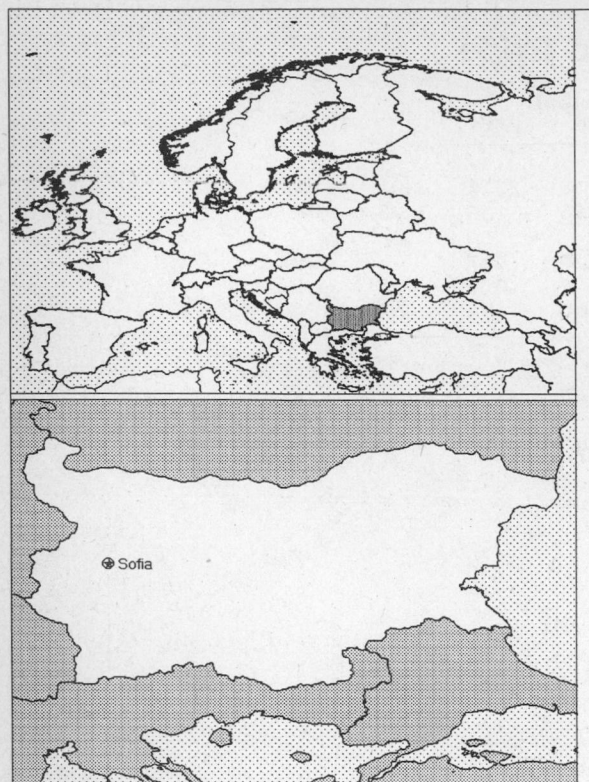

Geography [1]

Total area:
110,910 sq km   42,823 sq mi
Land area:
110,550 sq km   42,684 sq mi
Comparative area:
Slightly larger than Tennessee
Land boundaries:
Total 1,808 km, Greece 494 km, Macedonia 148 km, Romania 608 km, Yugoslavia 318 km (all with Serbia), Turkey 240 km
Coastline:
354 km
Climate:
Temperate; cold, damp winters; hot, dry summers
Terrain:
Mostly mountains with lowlands in North and Southeast
Natural resources:
Bauxite, copper, lead, zinc, coal, timber, arable land
Land use:
Arable land: 34%
Permanent crops: 3%
Meadows and pastures: 18%
Forest and woodland: 35%
Other: 10%

Demographics [2]

	1970	1980	1990	1995[1]	1996	2000	2010	2020	2030
Population	8,490	8,844	8,966	8,574	8,613	8,769	8,928	8,777	8,502
Population density (persons per sq. mi.)	199	207	210	201	202	205	209	206	199
(persons per sq. km.)	77	80	81	78	78	79	81	79	77
Net migration rate (per 1,000 population)	-1.3	0.0	NA	9.9	9.8	5.2	3.1	1.0	0.0
Births	139	128	NA	NA	72	NA	NA	NA	NA
Deaths	77	98	NA	NA	117	NA	NA	NA	NA
Life expectancy - males	NA	NA	NA	67.0	67.1	67.6	68.7	72.6	75.4
Life expectancy - females	NA	NA	NA	75.0	75.1	75.8	77.2	80.1	82.3
Birth rate (per 1,000)	16.3	14.5	NA	8.2	8.3	12.5	10.2	8.7	8.6
Death rate (per 1,000)	9.1	11.1	NA	13.6	13.6	13.6	13.8	12.3	12.5
Women of reproductive age (15-49 yrs.)	NA	NA	NA	2,101	2,120	2,155	2,119	2,024	1,781
of which are currently married	NA	NA	NA	NA	1,628	1,672	1,690	NA	NA
Fertility rate	2.2	2.1	NA	1.2	1.2	1.7	1.6	1.6	1.5

Except as noted, values for vital statistics are in thousands; life expectancy is in years.

Health

Health Indicators [3]

% of population with access to	
safe water (1990-95)	NA
adequate sanitation (1990-95)	NA
health services (1985-95)	NA
% of 1-year-olds immunized (1990-94) against	
TB (tuberculosis)	98
DPT (diphtheria, pertussis, tetanus)	98
polio	97
measles	87
% of contraceptive prevalence (1980-94)[1]	76
ORT use rate (1990-94)	NA

Health Expenditures [4]

Total health expenditure, 1990 (official exchange rate)	
Millions of dollars	1,154
Dollars per capita	131
Health expenditures as a percentage of GDP	
Total	5.4
Public sector	4.4
Private sector	1.0
Development assistance for health	
Total aid flows (millions of dollars)[1]	NA
Aid flows per capita (dollars)	NA
Aid flows as a percentage of total health expenditure	NA

For sources, notes, and explanations, see Annotated Source Appendix, page 1061.

Human Factors

Women and Children [5]

% of pregnant women immunized (tetanus 1990-94)	NA
% of births attended by trained health personnel (1983-94)	100
Maternal mortality rate (1980-92)	9
Under-5 mortality rate (1994)	19
% under-5 moderately/severely underweight (1980-1994)	NA

Burden of Disease [6]

Population per physician (1989)	314.59
Population per nurse (1984)	155.22
Population per hospital bed (1990)	100.93
AIDS cases per 100,000 people (1994)	0.1
Heart disease cases per 1,000 people (1990-93)	350

Ethnic Division [7]

Bulgarian	85.3%
Turk	8.5%
Gypsy	2.6%
Macedonian	2.5%
Armenian	0.3%
Russian	0.2%
Other	0.6%

Religion [8]

Bulgarian Orthodox	85%
Muslim	13%
Jewish	0.8%
Roman Catholic	0.5%
Uniate Catholic	0.2%
Protestant, Gregorian-Armenian, and other	0.5%

Major Languages [9]

Bulgarian, secondary languages closely correspond to ethnic breakdown.

Education

Public Education Expenditures [10]

	1980	1985	1990	1992	1993	1994
Million (Lev)						
Total education expenditure	1,145	1,784	2,357	11,729	16,307	23,999
as percent of GNP	4.5	5.5	5.6	5.9	5.5	4.5
as percent of total govt. expend.	NA	NA	NA	NA	NA	NA
Current education expenditure	1,098	1,598	2,183	11,211	15,210	22,620
as percent of GNP	4.3	4.9	5.2	5.6	5.1	4.3
as percent of current govt. expend.	NA	NA	NA	NA	NA	NA
Capital expenditure	47	186	174	518	1,097	1,379

Educational Attainment [11]

Age group (1992)[3]	25+
Total population	5,649,672
Highest level attained (%)	
No schooling	4.7
First level	
Not completed	12.5
Completed	31.9
Entered second level	
S-1	35.7
S-2	NA
Postsecondary	15.0

Illiteracy [12]

	1989	1992	1995
Illiterate population (15+ yrs.)[2]	NA	147,389	125,000
Illiteracy rate - total pop. (%)	NA	2.1	1.7
Illiteracy rate - males (%)	NA	1.3	1.1
Illiteracy rate - females (%)	NA	2.9	2.3

Libraries [13]

	Admin. Units	Svc. Pts.	Vols. (000)	Shelving (meters)	Vols. Added	Reg. Users
National (1993)	1	NA	2,266[e]	NA	NA	24,000[e]
Nonspecialized (1992)	27	NA	10,655	NA	301,859	232,546
Public (1992)	4,879	NA	57,092	NA	1 mil	1 mil
Higher ed. (1993)	81	NA	7,520	NA	154,190	149,946
School (1990)	3,208	NA	16,625	NA	520,286	761,532

Daily Newspapers [14]

	1980	1985	1990	1994
Number of papers	14	17	24	17
Circ. (000)	2,244	2,626	4,065	1,843

Culture [15]

Cinema (seats per 1,000)	13.8
Annual attendance per person	1.2
Gross box office receipts (mil. Lev)	103
Museums (reporting)	221
Visitors (000)	3,435
Annual receipts (000 Lev)	197,322

Science and Technology

Scientific/Technical Forces [16]

Scientists/engineers	37,825
Number female	17,362
Technicians	10,752
Number female	6,555
Total	48,577

R&D Expenditures [17]

	Lev (000) 1992
Total expenditure	3,103,800
Capital expenditure	256,100
Current expenditure	2,847,700
Percent current	91.7

U.S. Patents Issued [18]

Values show patents issued to citizens of the country by the U.S. Patents Office.

	1993	1994	1995
Number of patents	5	4	1

For sources, notes, and explanations, see Annotated Source Appendix, page 1061.

Government and Law

Organization of Government [19]

Long-form name:
Republic of Bulgaria
Type:
Emerging democracy
Independence:
22 September 1908 (from Ottoman Empire)
National holiday:
Independence Day, 3 March (1878)
Constitution:
Adopted 12 July 1991
Legal system:
Based on civil law system with Soviet law influence; accepts compulsory ICJ jurisdiction
Executive branch:
President; Vice President; Chairman of the Council of Ministers (Prime Minister); 4 Deputy Prime Ministers; Council of Ministers
Legislative branch:
Unicameral: National Assembly (Narodno Sobranie)
Judicial branch:
Supreme Court; Constitutional Court

Elections [20]

National Assembly	% of votes
Bulgarian Socialist Party (BSP)	43.5
Union of Democratic Forces (UDF)	24.2
People's Union (PU)	6.5
Movement for Rights and Freedoms (MRF)	5.4
Bulgarian Business Bloc (BBB)	4.7

Government Expenditures [21]

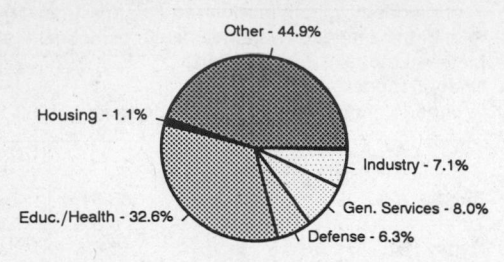

Other - 44.9%
Housing - 1.1%
Industry - 7.1%
Gen. Services - 8.0%
Defense - 6.3%
Educ./Health - 32.6%

(% distribution). Expend. for CY95: 360,607 (Lev mil.)

Crime [23]

	1994
Crime volume	
Cases known to police	213,390
Attempts (percent)	NA
Number cases solved	NA
Crimes per 100,000 persons	2,522.41
Persons responsible for offenses	
Total number offenders	NA
Percent female	NA
Percent juvenile (16-18 yrs.)	NA
Percent foreigners	NA

Military Expenditures and Arms Transfers [22]

	1990	1991	1992	1993	1994
Military expenditures					
Current dollars (mil.)	3,887[r]	1,464[r]	935[r]	1,010[r]	1,001[e]
1994 constant dollars (mil.)	4,326[r]	1,569[r]	975[r]	1,031[r]	1,001[e]
Armed forces (000)	129	107	99	52	80
Gross national product (GNP)					
Current dollars (mil.)[e]	45,290	36,900	32,160	31,550	37,300
1994 constant dollars (mil.)[e]	50,400	39,550	33,540	32,200	37,300
Central government expenditures (CGE)					
1994 constant dollars (mil.)[e]	NA	NA	13,010	NA	16,390
People (mil.)	9.0	8.9	8.9	8.8	8.8
Military expenditure as % of GNP	8.6	4.0	2.9	3.2	2.7
Military expenditure as % of CGE	NA	NA	7.5	NA	6.1
Military expenditure per capita (1994 $)	483	176	110	117	114
Armed forces per 1,000 people (soldiers)	14.4	12.0	11.2	5.9	9.1
GNP per capita (1994 $)	5,622	4,437	3,782	3,646	4,239
Arms imports[6]					
Current dollars (mil.)	675	0	0	5	0
1994 constant dollars (mil.)	751	0	0	5	0
Arms exports[6]					
Current dollars (mil.)	60	50	40	20	50
1994 constant dollars (mil.)	67	54	42	20	50
Total imports[7]					
Current dollars (mil.)	9,600	2,800[e]	4,346[e]	5,059	4,377
1994 constant dollars (mil.)	10,680	3,001[e]	4,532[e]	5,163	4,377
Total exports[7]					
Current dollars (mil.)	8,400	3,500[e]	3,600[e]	3,738	4,226
1994 constant dollars (mil.)	9,349	3,752[e]	3,754[e]	3,815	4,226
Arms as percent of total imports[8]	7.0	0	0	0.1	0
Arms as percent of total exports[8]	0.7	1.4	1.1	0.5	1.2

Human Rights [24]

	SSTS	FL	FAPRO	PPCG	APROBC	TPW	PCPTW	STPEP	PHRFF	PRW	ASST	AFL
Observes	2	P	P	P	P	P	P	P	P	P	P	

	EAFRD	CPR	ESCR	SR	ACHR	MAAE	PVIAC	PVNAC	EAFDAW	TCIDTP	RC
Observes	P	P	P	P		P	P	P	P	P	P

P = Party; S = Signatory; see Appendix for meaning of abbreviations.

Labor Force

Total Labor Force [25]

3.1 million

Labor Force by Occupation [26]

Industry	41%
Agriculture	18
Other	41

Date of data: 1992

Unemployment Rate [27]

11.9% (1995 est.)

For sources, notes, and explanations, see Annotated Source Appendix, page 1061.

Production Sectors

Commercial Energy Production and Consumption

Data are shown in quadrillion (10^{15}) BTUs and percent for 1995
Values for hydroelectric, nuclear, geothermal, solar, and wind power refer to electrical generation.

Production [28]

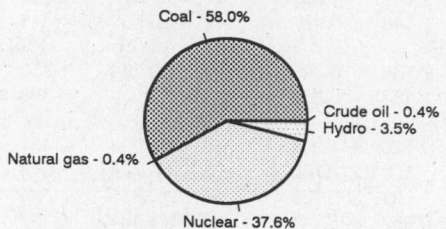

Coal - 58.0%
Crude oil - 0.4%
Hydro - 3.5%
Natural gas - 0.4%
Nuclear - 37.6%

Consumption [29]

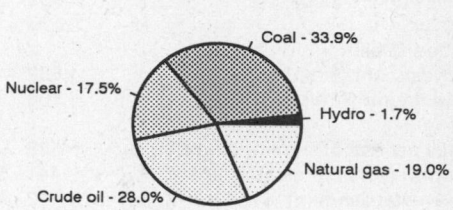

Coal - 33.9%
Nuclear - 17.5%
Hydro - 1.7%
Natural gas - 19.0%
Crude oil - 28.0%

Crude oil	0.002
Dry natural gas	0.002
Coal	0.278
Net hydroelectric power	0.017
Net nuclear power	0.100
Total	0.479

Crude oil	0.288
Dry natural gas	0.195
Coal	0.348
Net hydroelectric power	0.017
Net nuclear power	0.180
Total	1.028

Telecommunications [30]

- 2,773,293 (1993 est.) telephones; almost two-thirds of the lines are residential; 67% of Sofia households have telephones (November 1988 est.)
- Domestic: extensive but antiquated transmission system of coaxial cable and microwave radio relay; telephone service is available in most villages
- International: direct dialing to 36 countries; satellite earth stations - 1 Intersputnik (Atlantic Ocean Region); Intelsat available through a Greek earth station
- Radio: Broadcast stations: AM 20, FM 15, shortwave 0
- Television: Broadcast stations: 29 (Russian repeater in Sofia 1) Televisions: 2.1 million (May 1990 est.)

Top Agricultural Products [32]

Agriculture accounts for 12% of the GDP; produces grain, oilseed, vegetables, fruits, tobacco; livestock.

Transportation [31]

Railways: total: 4,292 km; standard gauge: 4,047 km 1.435-m gauge (2,650 km electrified; 917 double track)

Highways: total: 36,932 km; paved: 33,904 km (including 276 km of expressways); unpaved: 3,028 km (1992 est.)

Merchant marine: total: 103 ships (1,000 GRT or over) totaling 1,084,090 GRT/1,596,735 DWT; ships by type: bulk 45, cargo 27, chemical tanker 4, container 2, oil tanker 13, passenger-cargo 1, railcar carrier 2, roll-on/roll-off cargo 6, short-sea passenger 2, refrigerated cargo 1

Airports

Total:	355
With paved runways over 3,047 m:	1
With paved runways 2,438 to 3,047 m:	17
With paved runways 1,524 to 2,437 m:	10
With paved runways under 914 m:	88
With unpaved runways 2,438 to 3,047 m:	2

Top Mining Products [33]

Metric tons except as noted[e]	9/95[*]
Copper, metal, smelter	40,000
Lead	
mine output, Pb content	50,000
concentrate, gross weight	65,000
Zinc	
concentrate, gross weight	75,000
metal, smelter	64,000[r]
Sulphur	100,000
Coal, marketable (000 tons)	29,800
Petroleum refinery products (000 42-gal. bls.)	25,000

Tourism [34]

	1990	1991	1992	1993	1994
Visitors	10,330	6,818	6,124	8,302	10,068
Tourism receipts	320	44	215	307	358
Tourism expenditures	189	128	313	257	242

Travelers are in thousands, money in million U.S. dollars.

For sources, notes, and explanations, see Annotated Source Appendix, page 1061.

Manufacturing Sector

Manufacturing Summary [35]

	1987		1988		1989		1990		1991	
	$ bil.	%	$ bil.	%	$ bil.	%	$ bil.	%	$ bil.	%
Establishments or enterprises (number)	2,045	0.087	2,118	0.101	2,274	0.121	2,384	0.133	2,495	0.326
Total employment (000)	1,101	0.811	1,107	0.809	1,411	1.146	1,343	1.214	1,081	1.556
Production workers (000)	903	1.338	905	1.449	1,141	1.551	1,077	1.518	867	2.633
Output ($ bil.)	38	0.377	44	0.378	48	0.407	81	0.705	11	0.106
Value added ($ bil.)	-	-	-	-	-	-	-	-	-	-
Capital investment ($ mil.)	1,682	0.386	2,135	0.448	2,166	0.395	2,522	0.454	330	0.087
M & E investment ($ mil.)	1,108	0.356	1,673	0.483	1,483	0.380	-	-	256	0.091
Employees per establishment (number)	538	930.692	523	801.559	620	944.266	563	911.933	433	476.622
Production workers per establishment	441	1,535.306	427	1,435.997	502	1,277.700	452	1,140.724	348	806.775
Output per establishment ($ mil.)	19	432.028	21	374.365	21	335.136	34	529.855	4	32.575
Capital investment per estab. ($ mil.)	0.822	442.701	1	444.428	0.953	325.741	1	341.004	0.132	26.571
M & E per establishment ($ mil)	0.542	407.967	0.790	479.196	0.652	312.914	-	-	0.102	27.780
Payroll per employee ($)	3,267	36.429	3,882	38.963	4,110	36.786	5,610	40.441	729	4.994
Wages per production worker ($)	3,179	40.285	3,749	44.373	4,040	40.287	5,466	46.028	703	6.395
Hours per production worker (hours)	1,802	94.964	1,750	91.025	1,734	94.583	1,714	91.647	1,633	91.348
Output per employee ($)	34,766	46.420	39,781	46.705	34,015	35.492	60,214	58.102	10,034	6.835
Capital investment per employee ($)	1,528	47.567	1,929	55.445	1,535	34.497	1,878	37.394	306	5.575
M & E per employee ($)	1,006	43.835	1,512	59.783	1,051	33.138	-	-	237	5.829

Note: Columns headed % show percent of world total or ratio. Ratios closest to 100 are closest to world average. M & E stands for machinery & equipment.

Output in Manufacturing

	1987		1988		1989		1990		1991	
	$ bil.	%	$ bil.	%	$ bil.	%	$ bil.	%	$ bil.	%
3110 - Food products	13.000	34.21	14.000	31.82	14.000	29.17	13.000	16.05	1.448	13.16
3130 - Beverages	-	-	-	-	-	-	2.151	2.66	0.298	2.71
3140 - Tobacco	-	-	-	-	-	-	2.439	3.01	0.499	4.54
3210 - Textiles	2.841	7.48	3.259	7.41	3.310	6.90	4.481	5.53	0.414	3.76
3211 - Spinning, weaving, etc.	-	-	-	-	-	-	3.051	3.77	0.289	2.63
3220 - Wearing apparel	1.199	3.16	1.402	3.19	1.381	2.88	2.012	2.48	0.158	1.44
3230 - Leather and products	-	-	-	-	-	-	0.554	0.68	0.057	0.52
3240 - Footwear	-	-	-	-	-	-	0.585	0.72	0.066	0.60
3310 - Wood products	1.530	4.03	1.722	3.91	1.699	3.54	0.857	1.06	0.122	1.11
3320 - Furniture, fixtures	-	-	-	-	-	-	0.830	1.02	0.093	0.85
3410 - Paper and products	0.641	1.69	0.743	1.69	0.689	1.44	0.859	1.06	0.194	1.76
3411 - Pulp, paper, etc.	-	-	-	-	-	-	0.550	0.68	0.133	1.21
3420 - Printing, publishing	0.221	0.58	0.245	0.56	0.271	0.56	0.441	0.54	0.082	0.75
3510 - Industrial chemicals	-	-	-	-	-	-	1.919	2.37	0.562	5.11
3511 - Basic chemicals, excl fertilizers	-	-	-	-	-	-	0.365	0.45	0.151	1.37
3513 - Synthetic resins, etc.	-	-	-	-	-	-	0.299	0.37	0.104	0.95
3520 - Chemical products nec	-	-	-	-	-	-	2.116	2.61	0.520	4.73
3522 - Drugs and medicines	-	-	-	-	-	-	0.949	1.17	0.361	3.28
3540 - Petroleum, coal products	-	-	-	-	-	-	3.618	4.47	0.848	7.71
3550 - Rubber products	-	-	-	-	-	-	0.969	1.20	0.129	1.17
3560 - Plastic products nec	-	-	-	-	-	-	0.891	1.10	0.122	1.11
3610 - Pottery, china, etc.	-	-	-	-	-	-	0.195	0.24	0.029	0.26
3620 - Glass and products	-	-	-	-	-	-	0.561	0.69	0.080	0.73
3690 - Nonmetal products nec	2.026	5.33	2.213	5.03	2.031	4.23	2.104	2.60	0.238	2.16
3710 - Iron and steel	1.720	4.53	1.933	4.39	1.835	3.82	2.172	2.68	0.575	5.23
3720 - Nonferrous metals	-	-	-	-	1.411	2.94	1.136	1.40	0.241	2.19
3810 - Metal products	-	-	-	-	-	-	3.003	3.71	0.354	3.22
3820 - Machinery nec	8.434	22.19	9.396	21.35	11.000	22.92	8.222	10.15	0.645	5.86
3830 - Electrical machinery	7.038	18.52	9.009	20.48	10.000	20.83	5.486	6.77	0.691	6.28
3832 - Radio, television, etc.	-	-	-	-	-	-	2.470	3.05	0.213	1.94
3840 - Transportation equipment	-	-	-	-	-	-	4.045	4.99	0.386	3.51
3841 - Shipbuilding, repair	-	-	-	-	-	-	0.432	0.53	0.098	0.89
3843 - Motor vehicles	-	-	-	-	-	-	1.805	2.23	0.127	1.15
3900 - Industries nec	-	-	-	-	-	-	5.950	7.35	0.518	4.71

Note: Codes are International Standard Industry codes (ISIC). Percentages are % of total Output. [f]: Factor Prices; [p]: Producer Prices.

Finance, Economics, and Trade

Economic Indicators [36]

- **National product**: GDP—purchasing power parity—$43.2 billion (1995 est.)
- **National product real growth rate**: 2.4% (1995 est.)
- **National product per capita**: $4,920 (1995 est.)
- **Inflation rate (consumer prices)**: 35% (1995)
- **External debt**: $10.4 billion (1995)

Balance of Payments Summary [37]

Values in millions of dollars.

	1989	1990	1991	1992	1993
Exports of goods (f.o.b.)	8,268	6,113	3,737	3,956	3,971
Imports of goods (f.o.b.)	-8,960	-7,427	-3,769	-4,169	-4,301
Trade balance	-692	-1,314	-32	-213	-330
Services - debits	-1,504	-1,478	-569	-1,386	-1,507
Services - credits	1,350	957	455	1,195	1,276
Private transfers (net)	77	125	50	40	37
Government transfers (net)	NA	NA	19	3	NA
Long-term capital (net)	319	-3,535	496	563	388
Short-term capital (net)	-359	4,297	-401	324	-195
Errors and omissions	375	70	NA	-85	28
Overall balance	-434	-878	18	441	-303

Exchange Rates [38]

Currency: **lev.**
Symbol: **Lv.**

Data are currency units per $1. Floating exchange rate since February 1991.

December 1995	70.5
1994	54.2
1993	27.1
1992	23.3
1991	18.4

Imports and Exports

Top Import Origins [39]

$4 billion (c.i.f., 1994).

Origins	%
OECD	48.3
EU	34.1
Former CEMA countries	40.3
Arab countries	1.7
Other	9.7

Top Export Destinations [40]

$4.2 billion (f.o.b., 1994).

Destinations	%
OECD	46.6
EU	33.5
Former CEMA countries	35.7
Arab countries	5.1
Other	12.6

Foreign Aid [41]

	U.S. $	
ODA (1993)	39	million
Western nations, bal. of pay. (1994)	700	million

Import and Export Commodities [42]

Import Commodities

Fuels, minerals, and raw materials 30.1%
Machinery and equipment 23.6%
Textiles and apparel 11.6%
Agricultural products 10.8%
Metals and ores 6.8%
Chemicals 12.3%
Other 4.8%

Export Commodities

Machinery and equipment 12.8%
Agricultural products and food 21.9%
Textiles and apparel 14%
Metals and ores 19.7%
Chemicals 16.9%
Minerals and fuels 9.3%

For sources, notes, and explanations, see Annotated Source Appendix, page 1061.

145

Burkina Faso

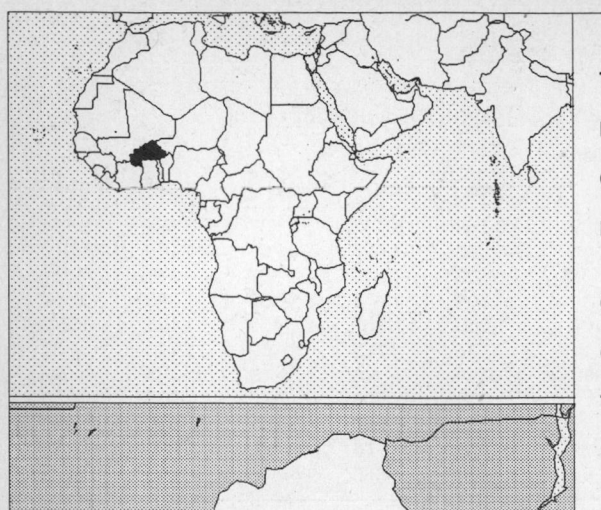

Geography [1]

Total area:
274,200 sq km   105,869 sq mi
Land area:
273,800 sq km   105,715 sq mi
Comparative area:
Slightly larger than Colorado
Land boundaries:
Total 3,192 km, Benin 306 km, Ghana 548 km, Cote d'Ivoire 584 km,
Mali 1,000 km, Niger 628 km, Togo 126 km
Coastline:
0 km (landlocked)
Climate:
Tropical; warm, dry winters; hot, wet summers
Terrain:
Mostly flat to dissected, undulating plains; hills in West and Southeast
Natural resources:
Manganese, limestone, marble; small deposits of gold, antimony, copper,
nickel, bauxite, lead, phosphates, zinc, silver
Land use:
Arable land: 10%
Permanent crops: 0%
Meadows and pastures: 37%
Forest and woodland: 26%
Other: 27%

Demographics [2]

	1970	1980	1990	1995[1]	1996	2000	2010	2020	2030
Population	5,626	6,939	9,033	10,354	10,623	11,684	14,150	16,569	19,805
Population density (persons per sq. mi.)	53	66	85	98	100	111	134	157	187
(persons per sq. km.)	21	25	33	38	39	43	52	61	72
Net migration rate (per 1,000 population)	NA	NA	-2.9	-1.9	-1.7	-1.1	0.0	0.0	0.0
Births	NA	NA	NA	NA	500	NA	NA	NA	NA
Deaths	NA	NA	NA	NA	212	NA	NA	NA	NA
Life expectancy - males	NA	NA	48.0	44.5	43.5	39.8	35.8	40.1	53.3
Life expectancy - females	NA	NA	48.7	43.8	43.0	39.9	34.6	39.4	55.9
Birth rate (per 1,000)	NA	NA	49.9	47.6	47.0	44.9	40.3	35.7	30.6
Death rate (per 1,000)	NA	NA	18.5	19.7	20.0	21.4	24.1	19.9	10.6
Women of reproductive age (15-49 yrs.)	NA	NA	2,031	2,304	2,357	2,573	3,182	3,973	5,177
of which are currently married	NA	NA	1,676	NA	1,935	2,099	2,555	NA	NA
Fertility rate	NA	NA	7.2	6.9	6.8	6.5	5.4	4.3	3.4

Except as noted, values for vital statistics are in thousands; life expectancy is in years.

Health

Health Indicators [3]

% of population with access to	
safe water (1990-95)	78
adequate sanitation (1990-95)	18
health services (1985-95)	90
% of 1-year-olds immunized (1990-94) against	
TB (tuberculosis)	63
DPT (diphtheria, pertussis, tetanus)	41
polio	NA
measles	45
% of contraceptive prevalence (1980-94)	8
ORT use rate (1990-94)	15

Health Expenditures [4]

Total health expenditure, 1990 (official exchange rate)	
Millions of dollars	219
Dollars per capita	24
Health expenditures as a percentage of GDP	
Total	8.5
Public sector	7.0
Private sector	1.5
Development assistance for health	
Total aid flows (millions of dollars)[1]	42
Aid flows per capita (dollars)	4.7
Aid flows as a percentage of total health expenditure	19.4

For sources, notes, and explanations, see Annotated Source Appendix, page 1061.

Human Factors

Women and Children [5]

% of pregnant women immunized (tetanus 1990-94)	41
% of births attended by trained health personnel (1983-94)	42
Maternal mortality rate (1980-92)	810
Under-5 mortality rate (1994)	169
% under-5 moderately/severely underweight (1980-1994)	30

Burden of Disease [6]

Population per physician (1989)	57,313.73
Population per nurse (1989)	1,681.82
Population per hospital bed (1990)	3,392.02
AIDS cases per 100,000 people (1994)	*
Malaria cases per 100,000 people (1992)	NA

Ethnic Division [7]

Mossi (about 24%), Gurumsi, Senufo, Lobi, Bobo, Mande, Fulani.

Religion [8]

Indigenous beliefs	40%
Muslim	50%
Christian (mainly Roman Catholic)	10%

Major Languages [9]

French (official), tribal languages belonging to Sudanic family spoken by 90% of the population.

Education

Public Education Expenditures [10]

	1980	1985	1989	1990	1992	1994
Million (Franc C.F.A.)						
Total education expenditure	7,994	12,901	18,780	NA	23,577	36,315
as percent of GNP	2.6	2.3	2.7	NA	3.1	3.6
as percent of total govt. expend.	19.8	21.0	17.5	NA	NA	11.1
Current education expenditure	7,436	12,292	18,727	NA	22,002	30,016
as percent of GNP	2.4	2.2	2.7	NA	2.9	3.0
as percent of current govt. expend.	21.1	22.4	21.9	NA	NA	17.4
Capital expenditure	558	609	53	NA	1,575	6,299

Educational Attainment [11]

Illiteracy [12]

In thousands and percent[1]	1990	1995	2000
Illiterate population (15+ yrs.)	4,207	4,597	4,993
Illiteracy rate - total pop. (%)	88.8	85.4	82.3
Illiteracy rate - males (%)	83.8	78.3	72.4
Illiteracy rate - females (%)	93.2	91.7	91.1

Libraries [13]

Daily Newspapers [14]

	1980	1985	1990	1994
Number of papers	1	2	1	1
Circ. (000)	2	4	3	3

Culture [15]

Cinema (seats per 1,000)	NA
Annual attendance per person	NA
Gross box office receipts (mil. Franc C.F.A.)	NA
Museums (reporting)	1
Visitors (000)	300
Annual receipts (000 Franc C.F.A.)	807

Science and Technology

Scientific/Technical Forces [16]

R&D Expenditures [17]

U.S. Patents Issued [18]

For sources, notes, and explanations, see Annotated Source Appendix, page 1061.

Government and Law

Organization of Government [19]

Long-form name:
None

Type:
Parliamentary

Independence:
5 August 1960 (from France)

National holiday:
Anniversary of the Revolution, 4 August (1983)

Constitution:
2 June 1991

Legal system:
Based on French civil law system and customary law

Executive branch:
President; Prime Minister; Council of Ministers

Legislative branch:
Unicameral: Assembly of People's Deputies Note:The current law also provides for a second consultative chamber, which has not been formally constituted

Judicial branch:
Appeals Court

Crime [23]

Elections [20]

Assembly of People's Deputies	% of seats
Organization People's Dem. (ODP-MT)	72.9
National Convention (CNPP-PSD)	11.2
Democratic Assembly (RDA)	5.6
Alliance for Democracy (ADF)	3.7
Other	6.5

Government Expenditures [21]

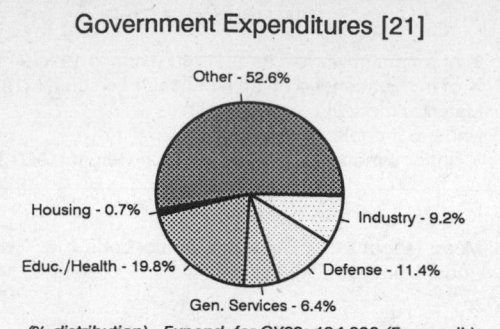

Other - 52.6%
Housing - 0.7%
Industry - 9.2%
Educ./Health - 19.8%
Defense - 11.4%
Gen. Services - 6.4%

(% distribution). Expend. for CY92: 134,828 (Franc mil.)

Military Expenditures and Arms Transfers [22]

	1990	1991	1992	1993	1994
Military expenditures					
Current dollars (mil.)	50	43	42	39	43
1994 constant dollars (mil.)	56	46	44	40	43
Armed forces (000)	10	10	9	9	9
Gross national product (GNP)					
Current dollars (mil.)	1,530	1,678	1,734	1,778	1,824
1994 constant dollars (mil.)	1,703	1,799	1,808	1,814	1,824
Central government expenditures (CGE)					
1994 constant dollars (mil.)	274	NA	316	334[e]	NA
People (mil.)	9.0	9.3	9.6	9.9	10.1
Military expenditure as % of GNP	3.3	2.5	2.4	2.2	2.3
Military expenditure as % of CGE	20.4	NA	14.0	12.0	NA
Military expenditure per capita (1994 $)	6	5	5	4	4
Armed forces per 1,000 people (soldiers)	1.1	1.1	0.9	0.9	0.9
GNP per capita (1994 $)	188	193	189	184	180
Arms imports[6]					
Current dollars (mil.)	20	0	5	10	5
1994 constant dollars (mil.)	22	0	5	10	5
Arms exports[6]					
Current dollars (mil.)	0	0	0	0	0
1994 constant dollars (mil.)	0	0	0	0	0
Total imports[7]					
Current dollars (mil.)	536	533	564[e]	573[e]	NA
1994 constant dollars (mil.)	597	571	588[e]	585[e]	NA
Total exports[7]					
Current dollars (mil.)	152	106	NA	273[e]	NA
1994 constant dollars (mil.)	169	114	NA	279[e]	NA
Arms as percent of total imports[8]	3.7	0	0.9	1.7	NA
Arms as percent of total exports[8]	0	0	0	0	0

Human Rights [24]

	SSTS	FL	FAPRO	PPCG	APROBC	TPW	PCPTW	STPEP	PHRFF	PRW	ASST	AFL
Observes	P	P	P	P	P	P	P					
	EAFRD	CPR	ESCR	SR	ACHR	MAAE	PVIAC	PVNAC	EAFDAW	TCIDTP	RC	
Observes	P		P				P	P	P		P	

P = Party; S = Signatory; see Appendix for meaning of abbreviations.

Labor Force

Total Labor Force [25]

Labor Force by Occupation [26]

Agriculture	80%
Industry	15
Commerce, services, and government	5
Migratory seasonal labor	

Date of data: 1984

Unemployment Rate [27]

148

For sources, notes, and explanations, see Annotated Source Appendix, page 1061.

Production Sectors

Commercial Energy Production and Consumption

Data are shown in quadrillion (10^{15}) BTUs and percent for 1995
Values for hydroelectric, nuclear, geothermal, solar, and wind power refer to electrical generation.

Production [28]

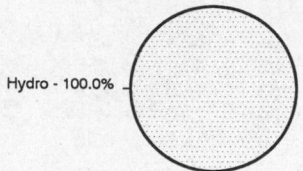

Hydro - 100.0%

Consumption [29]

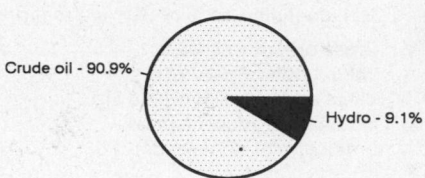

Crude oil - 90.9%

Hydro - 9.1%

Net hydroelectric power	0.001
Total	0.001

Crude oil	0.010
Net hydroelectric power	0.001
Total	0.011

Telecommunications [30]

- 21,000 (1993 est.) telephones; all services only fair
- Domestic: microwave radio relay, open wire, and radiotelephone communication stations
- International: satellite earth station - 1 Intelsat (Atlantic Ocean)
- Radio: Broadcast stations: AM 2, FM 1, shortwave 0
- Television: Broadcast stations: 2 (1987 est.) Televisions: 49,000 (1991 est.)

Top Agricultural Products [32]

Agriculture accounts for 32% of the GDP; produces peanuts, shea nuts, sesame, cotton, sorghum, millet, corn, rice; livestock.

Top Mining Products [33]

Metric tons except as noted[e]	4/18/95[*]
Gold (kg.)	6,000
Pumice and related volcanic materials	11,000
Salt	6,500
Marble	110,000
Manganese, Mn content of ore	30,000

Transportation [31]

Railways: total: 622 km (1995 est.); narrow gauge: 622 km 1.000-m gauge (517 km Ouagadougou to Cote d'Ivoire border and 105 km opened in 1993 from Ouagadougou to Kaya)

Highways: total: 16,400 km; paved: 1,280 km; unpaved: 15,120 km (1987 est.)

Airports

Total:	23
With paved runways over 3,047 m:	1
With paved runways 2,438 to 3,047 m:	1
With paved runways under 914 m:	8
With unpaved runways 1,524 to 2,437 m:	3
With unpaved runways 914 to 1,523 m:	10 (1995 est.)

Tourism [34]

	1990	1991	1992	1993	1994
Tourists[16]	74	80	92	111	133
Tourism receipts	8	8	9	8	22
Tourism expenditures	32	22	37	35	NA
Fare receipts	3	2	NA	NA	NA
Fare expenditures	13	17	31	30	NA

Travelers are in thousands, money in million U.S. dollars.

For sources, notes, and explanations, see Annotated Source Appendix, page 1061.

149

Manufacturing Sector

GDP and Manufacturing Summary [35]

	1980	1985	1989	1990	% change 1980-1990	% change 1989-1990
GDP (million 1980 $)	1,287	1,435	1,767	1,607	24.9	-9.1
GDP per capita (1980 $)	185	182	202	179	-3.2	-11.4
Manufacturing as % of GDP (current prices)	11.4	11.2	12.5^e	11.5	0.9	-8.0
Gross output (million $)	391	318^e	524^e	596^e	52.4	13.7
Value added (million $)	144	121^e	218^e	206^e	43.1	-5.5
Value added (million 1980 $)	141	144	183^e	163	15.6	-10.9
Industrial production index	100	110	140^e	129	29.0	-7.9
Employment (thousands)	8	9^e	10^e	9^e	12.5	-10.0

Note: GDP stands for Gross Domestic Product. 'e' stands for estimated value.

Profitability and Productivity

	1980	1985	1989	1990	% change 1980-1990	% change 1989-1990
Intermediate input (%)	63	62^e	58^e	65^e	3.2	12.1
Wages, salaries, and supplements (%)	8	7^e	7^e	8^e	0.0	14.3
Gross operating surplus (%)	28	31^e	35^e	27^e	-3.6	-22.9
Gross output per worker ($)	47,326	36,452^e	54,487^e	64,078^e	35.4	17.6
Value added per worker ($)	17,465	13,890^e	22,679^e	22,195^e	27.1	-2.1
Average wage (incl. benefits) ($)	4,021	2,712^e	3,669^e	5,088^e	26.5	38.7

Profitability is in percent of gross output. Productivity is in U.S. $. 'e' stands for estimated value.

Profitability - 1990

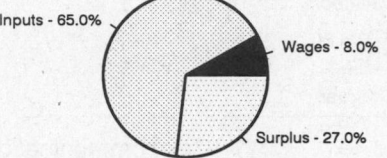

Inputs - 65.0%
Wages - 8.0%
Surplus - 27.0%

The graphic shows percent of gross output.

Value Added in Manufacturing

	1980 $ mil.	1980 %	1985 $ mil.	1985 %	1989 $ mil.	1989 %	1990 $ mil.	1990 %	% change 1980-1990	% change 1989-1990
311 Food products	55	38.2	55^e	45.5	108^e	49.5	98^e	47.6	78.2	-9.3
313 Beverages	29	20.1	21^e	17.4	33^e	15.1	36^e	17.5	24.1	9.1
314 Tobacco products	1	0.7	1^e	0.8	2^e	0.9	2^e	1.0	100.0	0.0
321 Textiles	20	13.9	18^e	14.9	34^e	15.6	28^e	13.6	40.0	-17.6
322 Wearing apparel	2	1.4	2^e	1.7	4^e	1.8	3^e	1.5	50.0	-25.0
323 Leather and fur products	2	1.4	1^e	0.8	2^e	0.9	3^e	1.5	50.0	50.0
324 Footwear	3	2.1	3^e	2.5	4^e	1.8	5^e	2.4	66.7	25.0
331 Wood and wood products	NA	0.0	NA	0.0	NA	0.0	NA	0.0	NA	NA
332 Furniture and fixtures	2	1.4	1^e	0.8	3^e	1.4	3^e	1.5	50.0	0.0
341 Paper and paper products	NA	0.0	NA	0.0	NA	0.0	NA	0.0	NA	NA
342 Printing and publishing	1	0.7	1^e	0.8	2^e	0.9	1^e	0.5	0.0	-50.0
351 Industrial chemicals	1	0.7	1^e	0.8	1^e	0.5	1^e	0.5	0.0	0.0
352 Other chemical products	NA	0.0	NA	0.0	NA	0.0	NA	0.0	NA	NA
353 Petroleum refineries	NA	0.0	NA	0.0	NA	0.0	NA	0.0	NA	NA
354 Miscellaneous petroleum and coal products	NA	0.0	NA	0.0	NA	0.0	NA	0.0	NA	NA
355 Rubber products	4	2.8	2^e	1.7	3^e	1.4	3^e	1.5	-25.0	0.0
356 Plastic products	2	1.4	1^e	0.8	2^e	0.9	1^e	0.5	-50.0	-50.0
361 Pottery, china, and earthenware	NA	0.0	NA	0.0	NA	0.0	NA	0.0	NA	NA
362 Glass and glass products	NA	0.0	NA	0.0	NA	0.0	NA	0.0	NA	NA
369 Other nonmetal mineral products	NA	0.0	NA	0.0	NA	0.0	NA	0.0	NA	NA
371 Iron and steel	1^e	0.7	1^e	0.8	2^e	0.9	1^e	0.5	0.0	-50.0
372 Nonferrous metals	1^e	0.7	NA	0.0	1^e	0.5	1^e	0.5	0.0	0.0
381 Metal products	1	0.7	NA	0.0	1^e	0.5	1^e	0.5	0.0	0.0
382 Nonelectrical machinery	1	0.7	NA	0.0	NA	0.0	NA	0.0	NA	NA
383 Electrical machinery	1	0.7	NA	0.0	1^e	0.5	1^e	0.5	0.0	0.0
384 Transport equipment	3	2.1	1^e	0.8	2^e	0.9	3^e	1.5	0.0	50.0
385 Professional and scientific equipment	NA	0.0	NA	0.0	NA	0.0	NA	0.0	NA	NA
390 Other manufacturing industries	12	8.3	9^e	7.4	12^e	5.5	15^e	7.3	25.0	25.0

Note: The industry codes shown are International Standard Industry codes (ISIC). Percentages are percent of total Value Added. 'e' stands for estimated value

Finance, Economics, and Trade

Economic Indicators [36]

- **National product**: GDP—purchasing power parity—$7.4 billion (1995 est.)
- **National product real growth rate**: 4% (1995 est.)
- **National product per capita**: $700 (1995 est.)
- **Inflation rate (consumer prices)**: 5% (1995 est.)
- **External debt**: $1 billion (December 1993 est.)

Balance of Payments Summary [37]

Values in millions of dollars.

	1989	1990	1991	1992	1993
Exports of goods (f.o.b.)	215.7	272.2	283.2	287.5	276.5
Imports of goods (f.o.b.)	-501.6	-593.2	-601.5	-642.3	-643.4
Trade balance	-285.9	-321.0	-318.3	-354.8	-366.9
Services - debits	-237.3	-272.9	-277.6	-291.7	-289.5
Services - credits	52.7	65.0	63.5	72.9	66.2
Private transfers (net)	97.5	113.9	106.4	106.9	96.8
Government transfers (net)	460.9	312.9	335.6	369.5	376.5
Long-term capital (net)	-206.3	75.8	276.5	170.8	127.1
Short-term capital (net)	52.4	33.5	-169.3	-48.7	20.3
Errors and omissions	5.1	-5.2	18.8	-9.0	21.9
Overall balance	-60.9	2.0	35.6	15.9	52.4

Exchange Rates [38]

Currency: **Communaute Financiere Africaine franc.**
Symbol: **CFAF.**

Data are currency units per $1.

January 1996	500.56
1995	499.15
1994	555.20
1993	283.16
1992	264.69
1991	282.11

Imports and Exports

Top Import Origins [39]

$636 million (f.o.b., 1993).

Origins	%
EU	NA
Africa	NA
Japan	NA

Top Export Destinations [40]

$273 million (f.o.b., 1993).

Destinations	%
EU	NA
Cote d'Ivoire	NA
Taiwan	NA
Thailand	NA

Foreign Aid [41]

Recipient: ODA, $NA.

Import and Export Commodities [42]

Import Commodities	Export Commodities
Machinery	Cotton
Food products	Gold
Petroleum	Animal products

For sources, notes, and explanations, see Annotated Source Appendix, page 1061.

151

Burundi

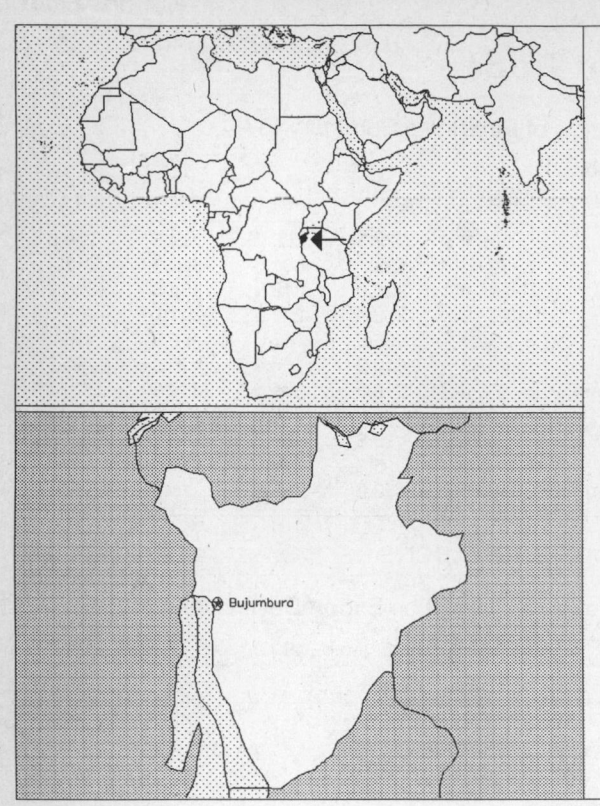

Geography [1]

Total area:
27,830 sq km   10,745 sq mi
Land area:
25,650 sq km   9,904 sq mi
Comparative area:
Slightly larger than Maryland
Land boundaries:
Total 974 km, Rwanda 290 km, Tanzania 451 km, Zaire 233 km
Coastline:
0 km (landlocked)
Climate:
Temperate; warm; occasional frost in uplands; dry season from June to September
Terrain:
Hilly and mountainous, dropping to a plateau in East, some plains
Natural resources:
Nickel, uranium, rare earth oxides, peat, cobalt, copper, platinum (not yet exploited), vanadium
Land use:
Arable land: 43%
Permanent crops: 8%
Meadows and pastures: 35%
Forest and woodland: 2%
Other: 12%

Demographics [2]

	1970	1980	1990	1995[1]	1996	2000	2010	2020	2030
Population	3,513	4,138	5,633	5,924	5,943	6,493	8,229	10,197	12,514
Population density (persons per sq. mi.)	355	418	569	598	600	656	831	1,030	1,264
(persons per sq. km.)	137	161	220	231	232	253	321	398	488
Net migration rate (per 1,000 population)	1.8	2.3	-0.7	-37.7	-12.5	0.0	0.0	0.0	0.0
Births	NA	NA	NA	NA	256	NA	NA	NA	NA
Deaths	NA	NA	NA	NA	90	NA	NA	NA	NA
Life expectancy - males	40.5	47.5	48.7	48.7	48.3	46.9	43.7	47.4	57.3
Life expectancy - females	43.1	50.8	52.6	50.7	50.4	49.3	46.0	50.1	62.1
Birth rate (per 1,000)	46.7	47.2	47.2	43.8	43.0	41.0	38.8	34.8	29.7
Death rate (per 1,000)	NA	18.1	15.6	15.2	15.2	15.3	16.8	14.4	8.8
Women of reproductive age (15-49 yrs.)	NA	975	1,291	1,335	1,342	1,481	1,925	2,451	3,237
of which are currently married[10]	544	NA	832	NA	856	923	1,172	NA	NA
Fertility rate	6.6	6.8	7.0	6.6	6.6	6.3	5.3	4.4	3.5

Except as noted, values for vital statistics are in thousands; life expectancy is in years.

Health

Health Indicators [3]

% of population with access to	
safe water (1990-95)[1]	70
adequate sanitation (1990-95)	51
health services (1985-95)	80
% of 1-year-olds immunized (1990-94) against	
TB (tuberculosis)	62
DPT (diphtheria, pertussis, tetanus)	48
polio	50
measles	43
% of contraceptive prevalence (1980-94)	9
ORT use rate (1990-94)	49

Health Expenditures [4]

Total health expenditure, 1990 (official exchange rate)	
Millions of dollars	36
Dollars per capita	7
Health expenditures as a percentage of GDP	
Total	3.3
Public sector	1.7
Private sector	1.6
Development assistance for health	
Total aid flows (millions of dollars)[1]	15
Aid flows per capita (dollars)	2.8
Aid flows as a percentage of total health expenditure	42.7

For sources, notes, and explanations, see Annotated Source Appendix, page 1061.

Human Factors

Women and Children [5]

% of pregnant women immunized (tetanus 1990-94)	19
% of births attended by trained health personnel (1983-94)	19
Maternal mortality rate (1980-92)	NA
Under-5 mortality rate (1994)	176
% under-5 moderately/severely underweight (1980-1994)[1]	38

Burden of Disease [6]

Population per physician (1991)	17,191.49
Population per nurse (1991)	4,789.16
Population per hospital bed (1991)	1,522.07
AIDS cases per 100,000 people (1994)	2.4
Malaria cases per 100,000 people (1992)	NA

Ethnic Division [7]

Africans:	
Hutu (Bantu)	85%
Tutsi (Hamitic)	14%
Twa (Pygmy)	1%
Non-Africans:	
Europeans	3,000
South Asians	2,000

Religion [8]

Christian	67%
Roman Catholic	62%
Protestant	5%
Indigenous beliefs	32%
Muslim	1%

Major Languages [9]

Kirundi (official), French (official), Swahili (along Lake Tanganyika and in the Bujumbura area).

Education

Public Education Expenditures [10]

Million (Franc)[12]	1980	1985	1990	1991	1992	1994
Total education expenditure	NA	3,467	6,570	7,403	8,586	NA
as percent of GNP	NA	2.5	3.4	3.5	3.8	NA
as percent of total govt. expend.	NA	15.5	16.7	17.7	12.2	NA
Current education expenditure	NA	3,212	6,370	7,208	8,023	NA
as percent of GNP	NA	2.4	3.3	3.4	3.6	NA
as percent of current govt. expend.	NA	17.1	19.4	20.2	24.2	NA
Capital expenditure	NA	254	200	195	563	NA

Educational Attainment [11]

Age group (1990)	25+
Total population	1,897,323
Highest level attained (%)	
No schooling	75.4
First level	
Not completed	19.9
Completed	NA
Entered second level	
S-1	2.5
S-2	NA
Postsecondary	0.6

Illiteracy [12]

In thousands and percent[1]	1990	1995	2000
Illiterate population (15+ yrs.)	2,062	2,221	2,399
Illiteracy rate - total pop. (%)	68.1	71.1	68.9
Illiteracy rate - males (%)	54.0	55.4	53.0
Illiteracy rate - females (%)	81.0	85.6	84.0

Libraries [13]

	Admin. Units	Svc. Pts.	Vols. (000)	Shelving (meters)	Vols. Added	Reg. Users
National	NA	NA	NA	NA	NA	NA
Nonspecialized (1989)	174	NA	174	NA	NA	500
Public (1989)	60	60	NA	NA	NA	20,000
Higher ed. (1987)	1	1	0.2	NA	224	NA
School	NA	NA	NA	NA	NA	NA

Daily Newspapers [14]

	1980	1985	1990	1994
Number of papers	1	1	1	1
Circ. (000)	1	2	20	20

Culture [15]

Cinema (seats per 1,000)	NA
Annual attendance per person	NA
Gross box office receipts (mil. Franc C.F.A.)	NA
Museums (reporting)	2
Visitors (000)	18
Annual receipts (000 Franc C.F.A.)	958

Science and Technology

Scientific/Technical Forces [16]

Scientists/engineers	170
Number female	17
Technicians	168
Number female	NA
Total[1,7]	338

R&D Expenditures [17]

	Franc (000) 1989
Total expenditure[4]	536,187
Capital expenditure	NA
Current expenditure	NA
Percent current	NA

U.S. Patents Issued [18]

For sources, notes, and explanations, see Annotated Source Appendix, page 1061.

153

Government and Law

Organization of Government [19]

Long-form name:
Republic of Burundi
Type:
Republic
Independence:
1 July 1962 (from UN trusteeship under Belgian administration)
National holiday:
Independence Day, 1 July (1962)
Constitution:
13 March 1992; provides for establishment of a plural political system
Legal system:
Based on German and Belgian civil codes and customary law; does not accept compulsory ICJ jurisdiction
Executive branch:
President; Prime Minister; Council of Ministers
Legislative branch:
Unicameral: National Assembly (Assemblee Nationale)
Judicial branch:
Supreme Court (Cour Supreme)

Elections [20]

National Assembly	% of votes
Burundi Democratic Front (FRODEBU)	71.0
Unity for National Progress (UPRONA)	21.4

Government Budget [21]

For 1991 est.

	$ mil.
Revenues	318
Expenditures	326
Capital expenditures	150

Crime [23]

	1990
Crime volume	
Cases known to police	4,916
Attempts (percent)	1.3
Percent cases solved	NA
Crimes per 100,000 persons	87.29
Persons responsible for offenses	
Total number offenders	NA
Percent female	NA
Percent juvenile (1-17 yrs.)	NA
Percent foreigners	NA

Military Expenditures and Arms Transfers [22]

	1990	1991	1992	1993	1994
Military expenditures					
Current dollars (mil.)	21	23	27[e]	27[e]	32[e]
1994 constant dollars (mil.)	24	25	28[e]	28[e]	32[e]
Armed forces (000)	12	12	7	7	17
Gross national product (GNP)					
Current dollars (mil.)	869	947	1,000	965	871
1994 constant dollars (mil.)	967	1,015	1,043	985	871
Central government expenditures (CGE)					
1994 constant dollars (mil.)	174	176	204	201	166
People (mil.)	5.6	5.7	5.8	6.0	6.1
Military expenditure as % of GNP	2.4	2.4	2.7	2.8	3.7
Military expenditure as % of CGE	13.6	13.9	13.7	13.8	19.2
Military expenditure per capita (1994 $)	4	4	5	5	5
Armed forces per 1,000 people (soldiers)	2.2	2.1	1.2	1.2	2.8
GNP per capita (1994 $)	174	178	178	164	142
Arms imports[6]					
Current dollars (mil.)	5	10	0	10	5
1994 constant dollars (mil.)	6	11	0	10	5
Arms exports[6]					
Current dollars (mil.)	0	0	0	0	0
1994 constant dollars (mil.)	0	0	0	0	0
Total imports[7]					
Current dollars (mil.)	231	248	221	204	224
1994 constant dollars (mil.)	257	266	230	208	224
Total exports[7]					
Current dollars (mil.)	75	90	72	68	106
1994 constant dollars (mil.)	83	96	75	69	106
Arms as percent of total imports[8]	2.2	4.0	0	4.9	2.2
Arms as percent of total exports[8]	0	0	0	0	0

Human Rights [24]

	SSTS	FL	FAPRO	PPCG	APROBC	TPW	PCPTW	STPEP	PHRFF	PRW	ASST	AFL
Observes		P	P			P	P			P		P
	EAFRD	CPR	ESCR	SR	ACHR	MAAE	PVIAC	PVNAC	EAFDAW	TCIDTP	RC	
Observes		P	P	P			P	P	P	P	P	

P = Party; S = Signatory; see Appendix for meaning of abbreviations.

Labor Force

Total Labor Force [25]

1.9 million (1983 est.)

Labor Force by Occupation [26]

Agriculture	93.0%
Government	4.0
Industry and commerce	1.5
Services	1.5

Unemployment Rate [27]

For sources, notes, and explanations, see Annotated Source Appendix, page 1061.

Production Sectors

Commercial Energy Production and Consumption

Data are shown in quadrillion (10^{15}) BTUs and percent for 1995
Values for hydroelectric, nuclear, geothermal, solar, and wind power refer to electrical generation.

Production [28]

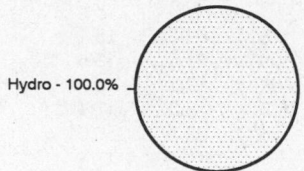

Hydro - 100.0%

Consumption [29]

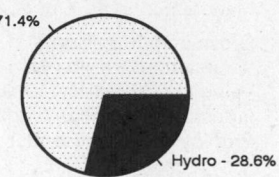

Crude oil - 71.4%

Hydro - 28.6%

Net hydroelectric power	0.002
Total	0.002

Crude oil	0.005
Net hydroelectric power	0.002
Total	0.007

Telecommunications [30]

- 7,200 (1987 est.) telephones; primitive system
- Domestic: sparse system of open wire, radiotelephone communications, and low-capacity microwave radio relay
- International: satellite earth station - 1 Intelsat (Indian Ocean)
- Radio: Broadcast stations: AM 2, FM 2, shortwave 0
- Television: Broadcast stations: 1 Televisions: 4,500 (1993 est.)

Transportation [31]

Railways: 0 km

Highways: total: 14,473 km; paved: 1,028 km; unpaved: 13,445 km (1992 est.)

Airports

Total: 3
With paved runways over 3,047 m: 1
With unpaved runways 914 to 1,523 m: 2 (1995 est.)

Top Agricultural Products [32]

Agriculture accounts for 54.1% of the GDP; produces coffee, cotton, tea, corn, sorghum, sweet potatoes, bananas, manioc; meat, milk, hides.

Top Mining Products [33]

Metric tons except as noted[e]	8/1/95*
Clay, kaolin	5,000
Gold (kg.)	20
Lime	150
Peat	10,000
Tin, mine output, ore, Sn content	50

Tourism [34]

	1990	1991	1992	1993	1994
Visitors	212	246	161	143	NA
Tourists[19]	109	125	86	75	29
Excursionists	103	121	75	NA	NA
Tourism receipts	3	4	4	3	3
Tourism expenditures	17	18	21	20	4
Fare receipts	2	1	1	1	NA

Travelers are in thousands, money in million U.S. dollars.

For sources, notes, and explanations, see Annotated Source Appendix, page 1061.

155

Manufacturing Sector

GDP and Manufacturing Summary [35]

	1980	1985	1989	1990	% change 1980-1990	% change 1989-1990
GDP (million 1980 $)	951	1,208	1,374	1,421	49.4	3.4
GDP per capita (1980 $)	230	255	259	260	13.0	0.4
Manufacturing as % of GDP (current prices)	9.0	13.6	10.0[e]	14.0	55.6	40.0
Gross output (million $)	95	137[e]	152[e]	175[e]	84.2	15.1
Value added (million $)	56	84[e]	96[e]	109[e]	94.6	13.5
Value added (million 1980 $)	77	80	101[e]	94	22.1	-6.9
Industrial production index	100	144[e]	152[e]	178[e]	78.0	17.1
Employment (thousands)	3	4[e]	5[e]	5[e]	66.7	0.0

Note: GDP stands for Gross Domestic Product. 'e' stands for estimated value.

Profitability and Productivity

	1980	1985	1989	1990	% change 1980-1990	% change 1989-1990
Intermediate input (%)	41	39[e]	37[e]	38[e]	-7.3	2.7
Wages, salaries, and supplements (%)	9[e]	10[e]	6[e]	9[e]	0.0	50.0
Gross operating surplus (%)	51[e]	51[e]	57[e]	54[e]	5.9	-5.3
Gross output per worker ($)	27,640	27,581[e]	29,956[e]	30,394[e]	10.0	1.5
Value added per worker ($)	16,370	17,238[e]	18,869[e]	19,680[e]	20.2	4.3
Average wage (incl. benefits) ($)	2,357[e]	3,378[e]	1,822[e]	3,008[e]	27.6	65.1

Profitability is in percent of gross output. Productivity is in U.S. $. 'e' stands for estimated value.

Profitability - 1990

Surplus - 53.5%
Wages - 8.9%
Inputs - 37.6%

The graphic shows percent of gross output.

Value Added in Manufacturing

	1980 $ mil.	1980 %	1985 $ mil.	1985 %	1989 $ mil.	1989 %	1990 $ mil.	1990 %	% change 1980-1990	% change 1989-1990
311 Food products	27[e]	48.2	40[e]	47.6	41[e]	42.7	56[e]	51.4	107.4	36.6
313 Beverages	13[e]	23.2	19[e]	22.6	26[e]	27.1	25[e]	22.9	92.3	-3.8
314 Tobacco products	4[e]	7.1	6[e]	7.1	13[e]	13.5	7[e]	6.4	75.0	-46.2
321 Textiles	2	3.6	3[e]	3.6	2[e]	2.1	3[e]	2.8	50.0	50.0
322 Wearing apparel	3	5.4	4[e]	4.8	3[e]	3.1	4[e]	3.7	33.3	33.3
323 Leather and fur products	1	1.8	1[e]	1.2	NA	0.0	NA	0.0	NA	NA
324 Footwear	NA	0.0	NA	0.0	NA	0.0	NA	0.0	NA	NA
331 Wood and wood products	NA	0.0	NA	0.0	NA	0.0	NA	0.0	NA	NA
332 Furniture and fixtures	NA	0.0	NA	0.0	NA	0.0	NA	0.0	NA	NA
341 Paper and paper products	NA	0.0	NA	0.0	NA	0.0	NA	0.0	NA	NA
342 Printing and publishing	1	1.8	1[e]	1.2	1[e]	1.0	1[e]	0.9	0.0	0.0
351 Industrial chemicals	1	1.8	3[e]	3.6	3[e]	3.1	3[e]	2.8	200.0	0.0
352 Other chemical products	NA	0.0	1[e]	1.2	1[e]	1.0	1[e]	0.9	NA	0.0
353 Petroleum refineries	NA	0.0	NA	0.0	NA	0.0	NA	0.0	NA	NA
354 Miscellaneous petroleum and coal products	NA	0.0	NA	0.0	NA	0.0	NA	0.0	NA	NA
355 Rubber products	NA	0.0	NA	0.0	NA	0.0	NA	0.0	NA	NA
356 Plastic products	NA	0.0	NA	0.0	NA	0.0	NA	0.0	NA	NA
361 Pottery, china, and earthenware	NA	0.0	NA	0.0	NA	0.0	NA	0.0	NA	NA
362 Glass and glass products	NA	0.0	NA	0.0	NA	0.0	NA	0.0	NA	NA
369 Other nonmetal mineral products	1	1.8	2[e]	2.4	1[e]	1.0	2[e]	1.8	100.0	100.0
371 Iron and steel	NA	0.0	NA	0.0	NA	0.0	NA	0.0	NA	NA
372 Nonferrous metals	NA	0.0	NA	0.0	NA	0.0	NA	0.0	NA	NA
381 Metal products	2	3.6	4[e]	4.8	4[e]	4.2	5[e]	4.6	150.0	25.0
382 Nonelectrical machinery	NA	0.0	NA	0.0	NA	0.0	NA	0.0	NA	NA
383 Electrical machinery	NA	0.0	NA	0.0	NA	0.0	NA	0.0	NA	NA
384 Transport equipment	NA	0.0	NA	0.0	NA	0.0	NA	0.0	NA	NA
385 Professional and scientific equipment	NA	0.0	NA	0.0	NA	0.0	NA	0.0	NA	NA
390 Other manufacturing industries	NA	0.0	NA	0.0	NA	0.0	NA	0.0	NA	NA

Note: The industry codes shown are International Standard Industry codes (ISIC). Percentages are percent of total Value Added. 'e' stands for estimated value

For sources, notes, and explanations, see Annotated Source Appendix, page 1061.

Finance, Economics, and Trade

Economic Indicators [36]

- **National product**: GDP—purchasing power parity—$4 billion (1995 est.)

- **National product real growth rate**: 2.7% (1995 est.)

- **National product per capita**: $600 (1995 est.)

- **Inflation rate (consumer prices)**: 10% (1993 est.)

- **External debt**: $1.05 billion (1994 est.)

Balance of Payments Summary [37]

Values in millions of dollars.

	1989	1990	1991	1992	1993
Exports of goods (f.o.b.)	93.2	72.9	90.7	79.3	75.0
Imports of goods (f.o.b.)	-151.4	-189.0	-195.9	-181.8	-172.8
Trade balance	-58.2	-116.1	-105.2	-102.5	-97.8
Services - debits	-119.1	-148.5	-157.7	-161.3	-134.6
Services - credits	24.2	24.8	35.1	31.3	25.3
Private transfers (net)	8.6	10.0	13.2	12.8	17.4
Government transfers (net)	132.4	163.6	182.5	164.9	163.4
Long-term capital (net)	64.0	62.3	56.8	90.6	46.8
Short-term capital (net)	0.4	15.7	13.7	8.2	6.0
Errors and omissions	-14.3	-15.1	-5.8	-18.7	-16.3
Overall balance	38.0	-3.3	-32.6	25.3	10.2

Exchange Rates [38]

Currency: **Burundi franc.**
Symbol: **FBu.**

Data are currency units per $1.

November 1995	268.13
1994	252.66
1993	242.78
1992	208.30
1991	181.51
1990	171.26

Imports and Exports

Top Import Origins [39]

$203 million (c.i.f., 1993).

Origins	%
EU	45
Asia	29
US	2

Top Export Destinations [40]

$68 million (f.o.b., 1993).

Destinations	%
EU	57
US	19
Asia	1

Foreign Aid [41]

Recipient: ODA, $NA.

Import and Export Commodities [42]

Import Commodities	Export Commodities
Capital goods 31%	Coffee 81%
Petroleum products 15%	Tea
Foodstuffs	Cotton
Consumer goods	Hides

Cambodia

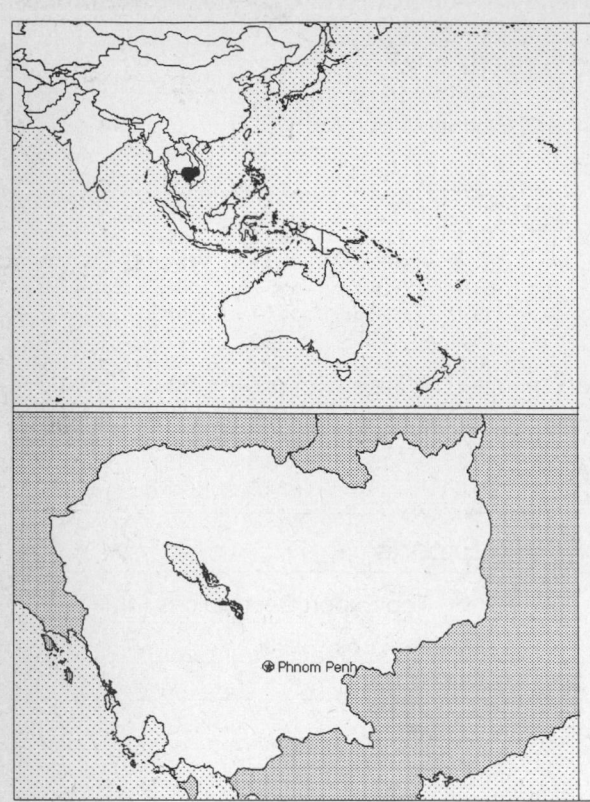

Geography [1]

Total area:
181,040 sq km   69,900 sq mi
Land area:
176,520 sq km   68,155 sq mi
Comparative area:
Slightly smaller than Oklahoma
Land boundaries:
Total 2,572 km, Laos 541 km, Thailand 803 km, Vietnam 1,228 km
Coastline:
443 km
Climate:
Tropical; rainy, monsoon season (May to November); dry season (December to April); little seasonal temperature variation
Terrain:
Mostly low, flat plains; mountains in Southwest and North
Natural resources:
Timber, gemstones, some iron ore, manganese, phosphates, hydropower potential
Land use:
Arable land: 16%
Permanent crops: 1%
Meadows and pastures: 3%
Forest and woodland: 76%
Other: 4%

Demographics [2]

	1970	1980	1990	1995[1]	1996	2000	2010	2020	2030
Population	6,996	6,499	8,731	10,561	10,861	12,098	15,679	20,208	25,574
Population density (persons per sq. mi.)	103	95	128	155	159	178	230	297	375
(persons per sq. km.)	40	37	49	60	62	69	89	114	145
Net migration rate (per 1,000 population)	-27.9	1.1	4.6	0.0	0.0	0.0	0.0	0.0	0.0
Births	NA	NA	NA	NA	472	NA	NA	NA	NA
Deaths	NA	NA	NA	NA	171	NA	NA	NA	NA
Life expectancy - males	33.6	34.1	47.0	48.0	48.4	50.0	55.0	59.9	64.3
Life expectancy - females	39.4	37.5	50.0	51.0	51.4	53.0	58.5	63.9	68.7
Birth rate (per 1,000)	47.0	49.2	46.8	44.4	43.5	41.0	36.5	33.3	28.8
Death rate (per 1,000)	26.6	26.3	17.2	16.2	15.8	14.3	10.9	8.5	6.9
Women of reproductive age (15-49 yrs.)	1,591	1,694	2,275	2,544	2,611	2,891	3,797	4,852	6,372
of which are currently married	NA	NA	1,379	NA	1,676	1,798	2,269	NA	NA
Fertility rate	7.1	6.3	5.8	5.8	5.8	5.8	5.2	4.6	3.9

Except as noted, values for vital statistics are in thousands; life expectancy is in years.

Health

Health Indicators [3]

% of population with access to	
safe water (1990-95)	36
adequate sanitation (1990-95)	14
health services (1985-95)	53
% of 1-year-olds immunized (1990-94) against	
TB (tuberculosis)	78
DPT (diphtheria, pertussis, tetanus)	53
polio	54
measles	53
% of contraceptive prevalence (1980-94)	NA
ORT use rate (1990-94)	6

Health Expenditures [4]

For sources, notes, and explanations, see Annotated Source Appendix, page 1061.

Human Factors

Women and Children [5]

% of pregnant women immunized (tetanus 1990-94)	28
% of births attended by trained health personnel (1983-94)	47
Maternal mortality rate (1980-92)	500
Under-5 mortality rate (1994)	177
% under-5 moderately/severely underweight (1980-1994)	40

Burden of Disease [6]

Population per physician (1994)	9,727.27
Population per nurse (1994)	669.56
Population per hospital bed (1990)	490.70
AIDS cases per 100,000 people (1994)	0.1
Malaria cases per 100,000 people (1992)	968

Ethnic Division [7]

Khmer	90%
Vietnamese	5%
Chinese	1%
Other	4%

Religion [8]

Theravada Buddhism	95%
Other	5%

Major Languages [9]

Khmer (official), French.

Education

Public Education Expenditures [10]

Educational Attainment [11]

Age group (1993)	5+
Total population	8,664,920
Highest level attained (%)	
No schooling	30.5
First level	
Not completed	47.0
Completed	NA
Entered second level	
S-1	16.2
S-2	4.1
Postsecondary	1.0

Illiteracy [12]

In thousands and percent[3,5]	1985	1991	2000
Illiterate population (15+ yrs.)	3,498	3,479	3,213
Illiteracy rate - total pop. (%)	71.2	64.8	52.0
Illiteracy rate - males (%)	58.7	51.8	38.9
Illiteracy rate - females (%)	83.4	77.6	64.9

Libraries [13]

Daily Newspapers [14]

Culture [15]

Science and Technology

Scientific/Technical Forces [16]

R&D Expenditures [17]

U.S. Patents Issued [18]

For sources, notes, and explanations, see Annotated Source Appendix, page 1061.

159

Government and Law

Organization of Government [19]

Long-form name:
Kingdom of Cambodia
Type:
Multiparty liberal democracy under a
constitutional monarchy established in
September 1993
Independence:
9 November 1949 (from France)
National holiday:
Independence Day, 9 November 1949
Constitution:
Promulgated 21 September 1993
Legal system:
Currently being defined
Executive branch:
King; First Prime Minister Prince; Second
Prime Minister; Council of Ministers
Legislative branch:
Unicameral: National Assembly Note: The
May 1993 elections were for the
Constituent Assembly which became the
National Assembly after the new
constitution was promulgated in
September 1993
Judicial branch:
Supreme Court provided for by the
constitution has not yet been established
and the future judicial system is yet to be
defined by law

Crime [23]

Elections [20]

National Assembly	% of seats
National Front for an Independent, Neutral, Peaceful, and Cooperative Cambodia	48.3
Cambodian Peoples Party	42.5
Buddhist Liberal Democrat	8.3
Molinaka	0.8

Government Budget [21]

For 1994 est.

	$ mil.
Revenues	210
Expenditures	346
Capital expenditures	NA

Military Expenditures and Arms Transfers [22]

	1990	1991	1992	1993	1994
Military expenditures					
Current dollars (mil.)[e]	NA	64	98	73	61
1994 constant dollars (mil.)[e]	NA	69	102	75	61
Armed forces (000)	112	112	135	102	70
Gross national product (GNP)					
Current dollars (mil.)[e]	1,637	1,829	2,012	2,140	2,271
1994 constant dollars (mil.)[e]	1,822	1,960	2,098	2,184	2,271
Central government expenditures (CGE)					
1994 constant dollars (mil.)	NA	NA	NA	NA	365[e]
People (mil.)	8.7	9.0	9.4	9.9	10.3
Military expenditure as % of GNP	NA	3.5	4.9	3.4	2.7
Military expenditure as % of CGE	NA	NA	NA	NA	16.7
Military expenditure per capita (1994 $)	NA	8	11	8	6
Armed forces per 1,000 people (soldiers)	12.8	12.4	14.3	10.3	6.8
GNP per capita (1994 $)	209	217	222	221	221
Arms imports[6]					
Current dollars (mil.)	230	40	0	10	10
1994 constant dollars (mil.)	256	43	0	10	10
Arms exports[6]					
Current dollars (mil.)	0	0	0	0	0
1994 constant dollars (mil.)	0	0	0	0	0
Total imports[7]					
Current dollars (mil.)	111	100	138	223	427
1994 constant dollars (mil.)	123	107	143	227	427
Total exports[7]					
Current dollars (mil.)	53	145	213	181	167
1994 constant dollars (mil.)	59	156	222	185	167
Arms as percent of total imports[8]	208.0	40.1	0	4.5	2.3
Arms as percent of total exports[8]	0	0	0	0	0

Human Rights [24]

	SSTS	FL	FAPRO	PPCG	APROBC	TPW	PCPTW	STPEP	PHRFF	PRW	ASST	AFL
Observes		P		P		P	P				P	
	EAFRD	CPR	ESCR	SR	ACHR	MAAE	PVIAC	PVNAC	EAFDAW	TCIDTP	RC	
Observes	P	P	P	P					P	P	P	

P = Party; S = Signatory; see Appendix for meaning of abbreviations.

Labor Force

Total Labor Force [25]

2.5 million to 3 million

Labor Force by Occupation [26]

Agriculture   80%
Date of data: 1988 est.

Unemployment Rate [27]

For sources, notes, and explanations, see Annotated Source Appendix, page 1061.

Production Sectors

Commercial Energy Production and Consumption

Data are shown in quadrillion (10^{15}) BTUs and percent for 1995
Values for hydroelectric, nuclear, geothermal, solar, and wind power refer to electrical generation.

Production [28]

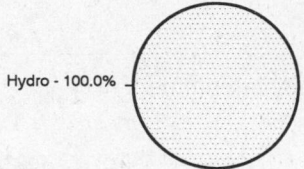

Hydro - 100.0%

Consumption [29]

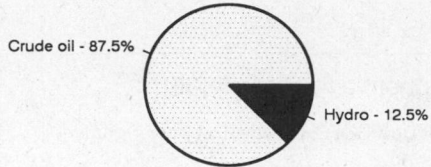

Crude oil - 87.5%

Hydro - 12.5%

Net hydroelectric power	0.001
Total	0.001

Crude oil	0.007
Net hydroelectric power	0.001
Total	0.008

Telecommunications [30]

- 7,000 (1981 est.) telephones; service barely adequate for government requirements and virtually nonexistent for general public
- International: landline international service limited to Vietnam and other adjacent countries; satellite earth station - 1 Intersputnik (Indian Ocean Region)
- Radio: Broadcast stations: AM 1, FM 0, shortwave 0
- Television: Broadcast stations: 1 (1986 est.) Televisions: 70,000 (1993 est.)

Top Agricultural Products [32]

Agriculture accounts for 52% of the GDP; produces rice, rubber, corn, vegetables.

Top Mining Products [33]

Metric tons[e]	1994[*]
Salt	40,000

Transportation [31]

Railways: total: 603 km; narrow gauge: 603 km 1.000-m gauge

Highways: total: 34,100 km; paved: 3,000 km; unpaved: 31,100 km (1994 est.)

Merchant marine: total: 5 cargo ships (1,000 GRT or over) totaling 17,451 GRT/18,280 DWT (1995 est.)

Airports

Total:	14
With paved runways 2,438 to 3,047 m:	2
With paved runways 1,524 to 2,437 m:	2
With paved runways 914 to 1,523 m:	2
With unpaved runways 1,524 to 2,437 m:	1
With unpaved runways 914 to 1,523 m:	7 (1995 est.)

Tourism [34]

	1990	1991	1992	1993	1994
Tourists[20]	17	25	88	118	177
Tourism receipts	NA	NA	50	48	70
Tourism expenditures	NA	NA	NA	4	7
Fare receipts	NA	NA	NA	3	4
Fare expenditures	NA	NA	16	11	14

Travelers are in thousands, money in million U.S. dollars.

For sources, notes, and explanations, see Annotated Source Appendix, page 1061.

161

Finance, Economics, and Trade

Industrial Summary [35]

Industrial Production: Growth rate 7.9% (1993 est.); accounts for 13.5% of the GDP. **Industries**: Rice milling, fishing, wood and wood products, rubber, cement, gem mining.

Economic Indicators [36]

- **National product**: GDP—purchasing power parity—$7 billion (1995 est.)
- **National product real growth rate**: 6.7% (1995 est.)
- **National product per capita**: $660 (1995 est.)
- **Inflation rate (consumer prices)**: 6% (1995 est.)
- **External debt**: $383 million to OECD members (1993)

Balance of Payments Summary [37]

Values in millions of dollars.

	1989	1990	1991	1992	1993
Exports of goods (f.o.b.)	NA	NA	NA	265	284
Imports of goods (f.o.b.)	NA	NA	NA	-443	-479
Trade balance	NA	NA	NA	-178	-195
Services - debits	NA	NA	NA	-110	-143
Services - credits	NA	NA	NA	50	64
Private transfers (net)	NA	NA	NA	9	3
Government transfers (net)	NA	NA	NA	262	326
Long-term capital (net)	NA	NA	NA	38	29
Short-term capital (net)	NA	NA	NA	-24	-7
Errors and omissions	NA	NA	NA	-39	-37
Overall balance	NA	NA	NA	8	40

Exchange Rates [38]

Currency: **new riel.**
Symbol: **CR.**

Data are currency units per $1.

December 1994	2,585
December 1993	2,470
September 1992	2,800
December 1991	500
1990	560

Imports and Exports

Top Import Origins [39]

$630.5 million (1995 est.).

Origins	%
Singapore	NA
Vietnam	NA
Japan	NA
Australia	NA
Hong Kong	NA
Indonesia	NA

Top Export Destinations [40]

$240.7 million (1995 est.).

Destinations	%
Singapore	NA
Japan	NA
Thailand	NA
Hong Kong	NA
Indonesia	NA
Malaysia	NA

Foreign Aid [41]

	U.S. $	
ODA	NA	
IMF aid pledge (1995-98)	120	million

Import and Export Commodities [42]

Import Commodities	Export Commodities
Cigarettes	Timber
Construction materials	Rubber
Petroleum products	Soybeans
Machinery	Sesame
Motor vehicles	

Cameroon

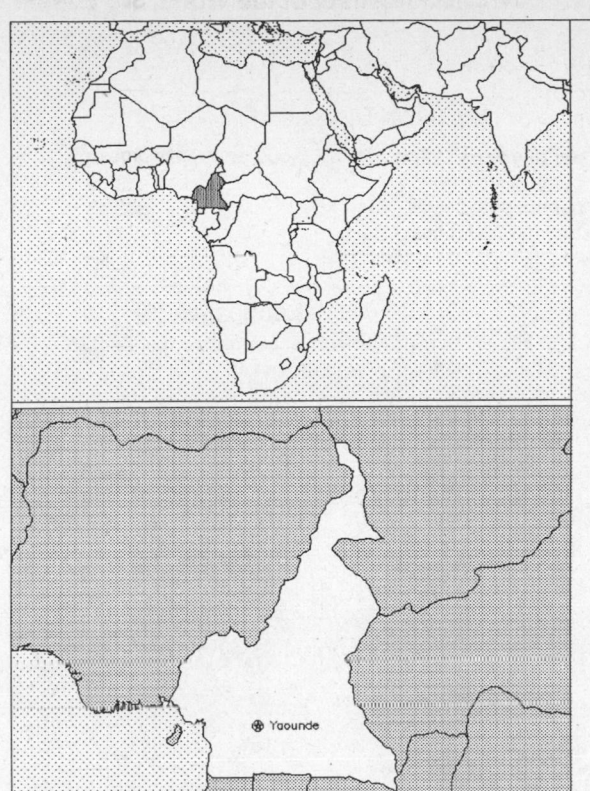

Geography [1]

Total area:
  475,440 sq km   183,568 sq mi
Land area:
  469,440 sq km   181,252 sq mi
Comparative area:
  Slightly larger than California
Land boundaries:
  Total 4,591 km, Central African Republic 797 km, Chad 1,094 km, Congo 523 km, Equatorial Guinea 189 km, Gabon 298 km, Nigeria 1,690 km
Coastline:
  402 km
Climate:
  Varies with terrain, from tropical along coast to semiarid and hot in North
Terrain:
  Diverse, with coastal plain in Southwest, dissected plateau in center, mountains in West, plains in North
Natural resources:
  Petroleum, bauxite, iron ore, timber, hydropower potential
Land use:
  Arable land: 13%
  Permanent crops: 2%
  Meadows and pastures: 18%
  Foroot and woodland: 54%
  Other: 13%

Demographics [2]

	1970	1980	1990	1995[1]	1996	2000	2010	2020	2030
Population	6,727	8,761	11,905	13,852	14,262	15,966	20,630	25,896	32,163
Population density (persons per sq. mi.)	37	48	66	76	79	88	114	143	177
(persons per sq. km.)	14	19	25	30	30	34	44	55	69
Net migration rate (per 1,000 population)	NA	10.3	0.0	0.0	0.0	0.0	0.0	0.0	0.0
Births	NA	NA	NA	NA	606	NA	NA	NA	NA
Deaths	NA	NA	NA	NA	193	NA	NA	NA	NA
Life expectancy - males	NA	49.7	52.1	51.9	51.6	50.3	47.6	50.9	59.6
Life expectancy - females	NA	53.2	55.9	54.0	53.7	52.4	50.2	53.8	63.8
Birth rate (per 1,000)	NA	46.2	44.3	42.9	42.5	41.4	38.5	34.5	30.0
Death rate (per 1,000)	NA	15.3	13.3	13.5	13.6	13.9	14.7	12.7	8.3
Women of reproductive age (15-49 yrs.)	NA	2,002	2,586	3,016	3,108	3,500	4,670	6,203	8,243
of which are currently married	NA	NA	1,940	NA	2,293	2,574	3,426	NA	NA
Fertility rate	NA	6.4	6.3	6.1	6.0	5.7	5.0	4.2	3.5

Except as noted, values for vital statistics are in thousands; life expectancy is in years.

Health

Health Indicators [3]

% of population with access to	
safe water (1990-95)	50
adequate sanitation (1990-95)	50
health services (1985-95)	70
% of 1-year-olds immunized (1990-94) against	
TB (tuberculosis)	46
DPT (diphtheria, pertussis, tetanus)	31
polio	31
measles	31
% of contraceptive prevalence (1980-94)	16
ORT use rate (1990-94)	84

Health Expenditures [4]

Total health expenditure, 1990 (official exchange rate)	
Millions of dollars	286
Dollars per capita	24
Health expenditures as a percentage of GDP	
Total	2.6
Public sector	1.0
Private sector	1.6
Development assistance for health	
Total aid flows (millions of dollars)[1]	38
Aid flows per capita (dollars)	3.3
Aid flows as a percentage of total health expenditure	13.4

For sources, notes, and explanations, see Annotated Source Appendix, page 1061.

163

Human Factors

Women and Children [5]

% of pregnant women immunized (tetanus 1990-94)	9
% of births attended by trained health personnel (1983-94)	64
Maternal mortality rate (1980-92)	430
Under-5 mortality rate (1994)	109
% under-5 moderately/severely underweight (1980-1994)	14

Burden of Disease [6]

Population per physician (1992)	12,060.19
Population per nurse (1992)	2,009.69
Population per hospital bed (1990)	393.43
AIDS cases per 100,000 people (1994)	11.3
Malaria cases per 100,000 people (1992)	NA

Ethnic Division [7]

Cameroon Highlanders	31%
Equatorial Bantu	19%
Kirdi	11%
Fulani	10%
Northwestern Bantu	8%
Eastern Nigritic	7%
Other African	13%
Non-African	< 1%

Religion [8]

Indigenous beliefs	51%
Christian	33%
Muslim	16%

Major Languages [9]

24 major African language groups, English (official), French (official).

Education

Public Education Expenditures [10]

	1980	1985	1990	1992	1993	1994
Million (Franc C.F.A.)[13]						
Total education expenditure	45,099	109,344	107,968	88,000	84,000	92,975
as percent of GNP	3.2	3.0	3.4	3.0	3.1	3.1
as percent of total govt. expend.	20.3	14.8	19.6	16.1	NA	NA
Current education expenditure	36,653	90,045	97,948	NA	NA	NA
as percent of GNP	2.6	2.5	3.1	NA	NA	NA
as percent of current govt. expend.	NA	20.9	26.9	NA	NA	NA
Capital expenditure	8,446	19,299	10,020	NA	NA	NA

Educational Attainment [11]

Illiteracy [12]

In thousands and percent[1]	1990	1995	2000
Illiterate population (15+ yrs.)	2,741	2,712	2,660
Illiteracy rate - total pop. (%)	42.7	36.3	30.7
Illiteracy rate - males (%)	29.4	24.5	20.5
Illiteracy rate - females (%)	55.8	47.9	40.8

Libraries [13]

	Admin. Units	Svc. Pts.	Vols. (000)	Shelving (meters)	Added	Reg. Users
National (1989)	1	7	40	1,000	25,000	5,000
Nonspecialized (1989)	1	14	50	NA	3,500	3,000
Public	NA	NA	NA	NA	NA	NA
Higher ed. (1988)	3	3	13	NA	2,939	617
School	NA	NA	NA	NA	NA	NA

Daily Newspapers [14]

	1980	1985	1990	1994
Number of papers	2	1	2	1
Circ. (000)	65	35	80[e]	50[e]

Culture [15]

Cinema (seats per 1,000)	3.4
Annual attendance per person	NA
Gross box office receipts (mil. Franc C.F.A.)	NA
Museums (reporting)	NA
Visitors (000)	NA
Annual receipts (000 Franc C.F.A.)	NA

Science and Technology

Scientific/Technical Forces [16]

R&D Expenditures [17]

U.S. Patents Issued [18]

For sources, notes, and explanations, see Annotated Source Appendix, page 1061.

Government and Law

Organization of Government [19]

Long-form name:
Republic of Cameroon
Type:
Unitary republic; multiparty presidential regime (opposition parties legalized 1990)
Independence:
1 January 1960 (from UN trusteeship under French administration)
National holiday:
National Day, 20 May (1972)
Constitution:
20 May 1972
Legal system:
Based on French civil law system, with common law influence; does not accept compulsory ICJ jurisdiction
Executive branch:
President; Prime Minister; Cabinet
Legislative branch:
Unicameral: National Assembly (Assemblee Nationale)
Judicial branch:
Supreme Court

Elections [20]

National Assembly	% of seats
People's Democratic Movement	48.9
National Union	37.8
Democratic Union	10.0
MDR	3.3

Government Expenditures [21]

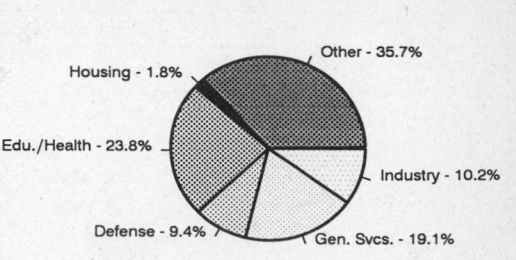

Other - 35.7%
Housing - 1.8%
Edu./Health - 23.8%
Industry - 10.2%
Defense - 9.4%
Gen. Svcs. - 19.1%

(% distribution). Expend. for FY93: 501.15 (Franc bil.)

Crime [23]

	1994
Crime volume	
Cases known to police	1,318
Attempts (percent)	1.37
Percent cases solved	92.56
Crimes per 100,000 persons	10.98
Persons responsible for offenses	
Total number offenders	1,478
Percent female	9.00
Percent juvenile (10-18 yrs.)	12.25
Percent foreigners	2.91

Military Expenditures and Arms Transfers [22]

	1990	1991	1992	1993	1994
Military expenditures					
Current dollars (mil.)	100	91	97	103	102
1994 constant dollars (mil.)	111	98	101	105	102
Armed forces (000)	23	24	12	12	12
Gross national product (GNP)					
Current dollars (mil.)	6,106	5,927	5,973	5,527	5,319
1994 constant dollars (mil.)	6,796	6,353	6,229	5,641	5,319
Central government expenditures (CGE)					
1994 constant dollars (mil.)	1,524	1,458	1,175	1,033	NA
People (mil.)	11.7	12.0	12.4	12.8	13.1
Military expenditure as % of GNP	1.6	1.5	1.6	1.9	1.9
Military expenditure as % of CGE	7.3	6.7	8.6	10.2	NA
Military expenditure per capita (1994 $)	10	8	8	8	8
Armed forces per 1,000 people (soldiers)	2.0	2.0	1.0	0.9	0.9
GNP per capita (1994 $)	581	528	503	442	405
Arms imports[6]					
Current dollars (mil.)	10	0	0	0	5
1994 constant dollars (mil.)	11	0	0	0	5
Arms exports[6]					
Current dollars (mil.)	0	0	0	0	0
1994 constant dollars (mil.)	0	0	0	0	0
Total imports[7]					
Current dollars (mil.)	1,400	1,173	1,163	1,102	NA
1994 constant dollars (mil.)	1,558	1,257	1,213	1,125	NA
Total exports[7]					
Current dollars (mil.)	2,002	1,834	1,840	1,883	NA
1994 constant dollars (mil.)	2,228	1,966	1,919	1,922	NA
Arms as percent of total imports[8]	0.7	0	0	0	NA
Arms as percent of total exports[8]	0	0	0	0	0

Human Rights [24]

	SSTS	FL	FAPRO	PPCG	APROBC	TPW	PCPTW	STPEP	PHRFF	PRW	ASST	AFL
Observes	P	P	P		P	P	P	P			P	P
	EAFRD	CPR	ESCR	SR	ACHR	MAAE	PVIAC	PVNAC	EAFDAW	TCIDTP	RC	
Observes	P	P	P	P			P	P	P	P	P	

P = Party; S = Signatory; see Appendix for meaning of abbreviations.

Labor Force

Total Labor Force [25]

Labor Force by Occupation [26]

Agriculture	74.4%
Industry and transport	11.4
Other services	14.2

Date of data: 1983

Unemployment Rate [27]

For sources, notes, and explanations, see Annotated Source Appendix, page 1061.

165

Production Sectors

Commercial Energy Production and Consumption

Data are shown in quadrillion (10^{15}) BTUs and percent for 1995
Values for hydroelectric, nuclear, geothermal, solar, and wind power refer to electrical generation.

Production [28]

Crude oil - 88.9%
Hydro - 11.1%

Consumption [29]

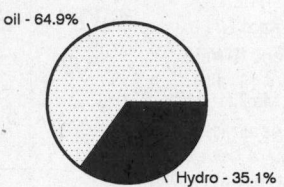

Crude oil - 64.9%
Hydro - 35.1%

Crude oil	0.217
Net hydroelectric power	0.027
Total	0.244

Crude oil	0.050
Net hydroelectric power	0.027
Total	0.077

Telecommunications [30]

- 36,737 (1991 est.) telephones; available only to business and government
- Domestic: cable, microwave radio relay, and tropospheric scatter
- International: satellite earth stations - 2 Intelsat (Atlantic Ocean)
- Radio: Broadcast stations: AM 11, FM 11, shortwave 0 Radios: 2 million (1993 est.)
- Television: Broadcast stations: 1 (1995)

Top Agricultural Products [32]

Agriculture accounts for 29% of the GDP; produces coffee, cocoa, cotton, rubber, bananas, oilseed, grains, root starches; livestock; timber.

Top Mining Products [33]

Metric tons except as noted[e]	8/95*
Aluminum metal, primary	78,000
Cement, hydraulic	620,000
Gold, mine output, Au content (kg.)	10
Petroleum, crude (000 42-gal. bls.)	34,600
Pozzolana	130,000
Limestone	57,000
Marble	200
Tin, ore and concentrate (kg.)	
gross weight	4,300
Sn content	3,050

Transportation [31]

Railways: total: 1,104 km (1995 est.); narrow gauge: 1,104 km 1.000-m gauge

Highways: total: 64,626 km; paved: 2,666 km; unpaved: 61,960 km (1987 est.)

Merchant marine: total: 2 cargo ships (1,000 GRT or over) totaling 24,122 GRT/33,509 DWT (1995 est.)

Airports

Total:	45
With paved runways over 3,047 m:	2
With paved runways 2,438 to 3,047 m:	4
With paved runways 1,524 to 2,437 m:	3
With paved runways 914 to 1,523 m:	1
With paved runways under 914 m:	13

Tourism [34]

	1990	1991	1992	1993	1994
Tourists[16]	89	84	62	81	NA
Tourism receipts	53	73	59	47	NA
Tourism expenditures	279	414	228	225	NA
Fare receipts	43	36	39	30	NA
Fare expenditures	118	54	49	26	NA

Travelers are in thousands, money in million U.S. dollars.

For sources, notes, and explanations, see Annotated Source Appendix, page 1061.

Manufacturing Sector

Manufacturing Summary [35]

	1987		1988		1989		1990		1991	
	$ bil.	%	$ bil.	%	$ bil.	%	$ bil.	%	$ bil.	%
Establishments or enterprises (number)	-	-	-	-	97	0.005	87	0.005	-	-
Total employment (000)	-	-	-	-	35	0.028	30	0.028	-	-
Production workers (000)	-	-	-	-	26	0.035	22	0.031	-	-
Output ($ bil.)	-	-	-	-	2	0.013	2	0.018	-	-
Value added ($ bil.)	-	-	-	-	0.423	0.009	0.768	0.015	-	-
Capital investment ($ mil.)	-	-	-	-	119	0.022	243	0.044	-	-
M & E investment ($ mil.)	-	-	-	-	-		-		-	-
Employees per establishment (number)	-	-	-	-	360	548.405	350	566.695	-	-
Production workers per establishment	-	-	-	-	263	670.023	252	635.903	-	-
Output per establishment ($ mil.)	-	-	-	-	16	256.152	23	363.499	-	-
Capital investment per estab. ($ mil.)	-	-	-	-	1	419.862	3	898.819	-	-
M & E per establishment ($ mil)	-	-	-	-	-		-		-	-
Payroll per employee ($)	-	-	-	-	7,295	65.300	8,641	62.290	-	-
Wages per production worker ($)	-	-	-	-	3,293	32.841	3,874	32.619	-	-
Hours per production worker (hours)	-	-	-	-	-		-		-	-
Output per employee ($)	-	-	-	-	44,765	46.709	66,475	64.144	-	-
Capital investment per employee ($)	-	-	-	-	3,407	76.561	7,964	158.607	-	-
M & E per employee ($)	-	-	-	-	-		-		-	-

Note: Columns headed % show percent of world total or ratio. Ratios closest to 100 are closest to world average. M & E stands for machinery & equipment.

Output in Manufacturing

	1987		1988		1989		1990		1991	
	$ bil.[p]	%	$ bil.[p]	%	$ bil.[p]	%	$ bil.[p]	%	$ bil.[p]	%
3110 - Food products	-	-	-	-	0.340	17.00	0.183	9.15	-	-
3130 - Beverages	-	-	-	-	0.503	25.15	0.523	26.15	-	-
3140 - Tobacco	-	-	-	-	0.068	3.40	0.074	3.70	-	-
3210 - Textiles	-	-	-	-	0.180	9.00	0.190	9.50	-	-
3220 - Wearing apparel	-	-	-	-	0.004	0.20	0.002	0.10	-	-
3520 - Chemical products nec	-	-	-	-	0.443	22.15	1.038	51.90	-	-
3840 - Transportation equipment	-	-	-	-	0.013	0.65	0.007	0.35	-	-
3841 - Shipbuilding, repair	-	-	-	-	0.013	0.65	0.007	0.35	-	-

Note: Codes are International Standard Industry codes (ISIC). Percentages are % of total Output. [f]: Factor Prices; [p]: Producer Prices.

Finance, Economics, and Trade

Economic Indicators [36]

- **National product**: GDP—purchasing power parity— $16.5 billion (1995 est.)

- **National product real growth rate**: 1.8% (1995 est.)

- **National product per capita**: $1,200 (1995 est.)

- **Inflation rate (consumer prices)**: 48% (1994)

- **External debt**: $6.6 billion (1993)

Balance of Payments Summary [37]

Values in millions of dollars.

	1988	1990	1991	1992	1993
Exports of goods (f.o.b.)	1,841.2	2,125.4	1,957.5	1,934.1	1,144.2
Imports of goods (f.o.b.)	-1,220.8	-1,347.2	-1,173.1	-983.3	-927.5
Trade balance	620.4	778.2	784.4	950.8	216.7
Services - debits	-1,416.4	-1,611.6	-1,572.9	-1,729.3	-1,321.7
Services - credits	473.0	389.4	424.3	432.7	367.8
Private transfers (net)	-134.9	-51.9	-65.8	-94.4	-79.8
Government transfers (net)	29.0	28.9	28.4	134.9	75.3
Long-term capital (net)	304.8	-45.4	129.9	-110.1	-156.7
Short-term capital (net)	-27.8	616.6	268.5	1,071.7	725.5
Errors and omissions	166.2	-165.4	34.9	-626.0	210.9
Overall balance	14.3	-61.2	31.7	30.3	38.0

Exchange Rates [38]

Currency: **Communaute Financiere Africaine franc.**
Symbol: **CFAF.**

Data are currency units per $1.

January 1996	500.56
1995	499.15
1994	555.20
1993	283.16
1992	264.69
1991	282.11

Imports and Exports

Top Import Origins [39]

$810 million (f.o.b., 1994).

Origins	%
EU	NA
France	38
Germany	NA
African countries	NA
Japan	5
US	5

Top Export Destinations [40]

$1.2 billion (f.o.b., 1994).

Destinations	%
EU (mainly France)	50
African countries	NA
US	NA

Foreign Aid [41]

Recipient: ODA, $449 million (1993).

Import and Export Commodities [42]

Import Commodities	Export Commodities
Machines and electrical equipment	Crude oil and petroleum products
Food	Lumber
Consumer goods	Aluminum
Transport equipment	Cocoa beans
Petroleum products	Coffee
	Cotton

Canada

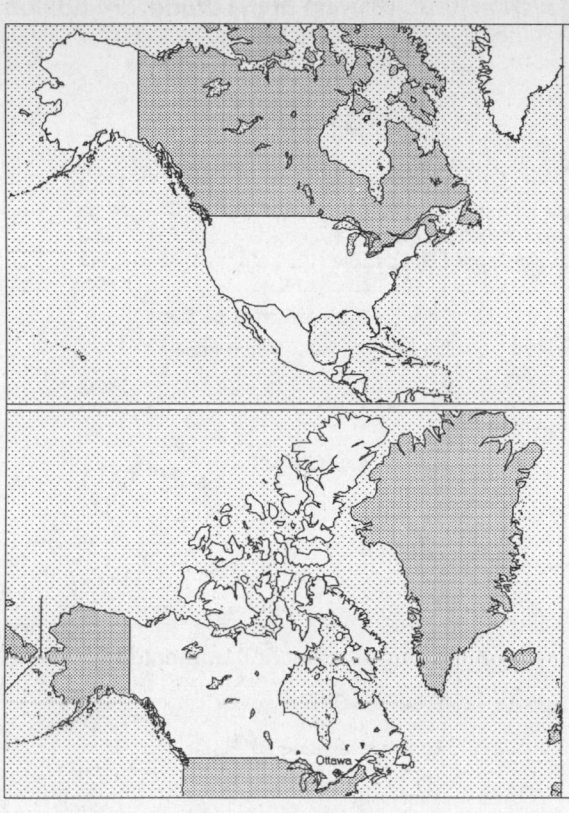

Geography [1]

Total area:
9,976,140 sq km   3,851,809 sq mi
Land area:
9,220,970 sq km   3,560,236 sq mi
Comparative area:
Slightly larger than US
Land boundaries:
Total 8,893 km, US 8,893 km (includes 2,477 km with Alaska)
Coastline:
243,791 km
Climate:
Varies from temperate in South to subarctic and arctic in North
Terrain:
Mostly plains with mountains in West and lowlands in Southeast
Natural resources:
Nickel, zinc, copper, gold, lead, molybdenum, potash, silver, fish, timber, wildlife, coal, petroleum, natural gas
Land use:
Arable land: 9%
Permanent crops: 0%
Meadows and pastures: 3%
Forest and woodland: 45%
Other: 43%

Demographics [2]

	1970	1980	1990	1995[1]	1996	2000	2010	2020	2030
Population	21,324	24,070	26,620	28,511	28,821	29,989	32,534	34,753	36,294
Population density (persons per sq. mi.)	6	7	7	8	8	8	9	10	10
(persons per sq. km.)	2	3	3	3	3	3	4	4	4
Net migration rate (per 1,000 population)	NA	3.8	6.6	4.5	4.5	4.2	3.7	3.3	3.1
Births	372	371	NA	NA	384	NA	NA	NA	NA
Deaths	156	171	192	NA	207	NA	NA	NA	NA
Life expectancy - males	NA	NA	74.1	75.4	75.7	76.7	78.5	79.5	80.2
Life expectancy - females	NA	NA	81.3	82.5	82.7	83.5	84.9	85.8	86.3
Birth rate (per 1,000)	17.4	15.4	15.0	13.6	13.3	12.3	11.4	11.0	10.1
Death rate (per 1,000)	7.3	7.1	7.2	7.2	7.2	7.2	7.8	8.6	10.3
Women of reproductive age (15-49 yrs.)	NA	NA	7,154	7,535	7,583	7,657	7,649	7,506	7,601
of which are currently married	NA	NA	4,326	NA	4,679	4,716	4,602	NA	NA
Fertility rate	NA	1.7	1.8	1.8	1.8	1.8	1.8	1.8	1.7

Except as noted, values for vital statistics are in thousands; life expectancy is in years.

Health

Health Indicators [3]

% of population with access to	
safe water (1990-95)	NA
adequate sanitation (1990-95)	NA
health services (1985-95)	NA
% of 1-year-olds immunized (1990-94) against	
TB (tuberculosis)	NA
DPT (diphtheria, pertussis, tetanus)	93
polio	89
measles	98
% of contraceptive prevalence (1980-94)	73
ORT use rate (1990-94)	NA

Health Expenditures [4]

Total health expenditure, 1990 (official exchange rate)	
Millions of dollars	51,594
Dollars per capita	1,945
Health expenditures as a percentage of GDP	
Total	9.1
Public sector	6.8
Private sector	2.4
Development assistance for health	
Total aid flows (millions of dollars)[1]	NA
Aid flows per capita (dollars)	NA
Aid flows as a percentage of total health expenditure	NA

For sources, notes, and explanations, see Annotated Source Appendix, page 1061.

169

Human Factors

Women and Children [5]

% of pregnant women immunized (tetanus 1990-94)	NA
% of births attended by trained health personnel (1983-94)	99
Maternal mortality rate (1980-92)	5
Under-5 mortality rate (1994)	8
% under-5 moderately/severely underweight (1980-1994)	NA

Burden of Disease [6]

Population per physician (1988)	469.30
Population per nurse	NA
Population per hospital bed (1990)	67.30
AIDS cases per 100,000 people (1994)	4.6
Heart disease cases per 1,000 people (1990-93)	NA

Ethnic Division [7]

British Isles origin	40%
French origin	27%
Other European	20%
Indigenous Indian and Eskimo	1.5%
Other (mostly Asian)	11.5%

Religion [8]

Roman Catholic	45%
United Church	12%
Anglican	8%
Other	35%
(1991)	

Major Languages [9]

English (official), French (official).

Education

Public Education Expenditures [10]

	1980	1985	1990	1991	1992	1994
Million (Dollar)						
Total education expenditure	20,833	30,287	43,487	47,764	49,955	NA
as percent of GNP	6.9	6.6	6.8	7.4	7.6	NA
as percent of total govt. expend.	16.3	11.9	14.2	NA	14.3	NA
Current education expenditure	19,295	28,202	40,288	44,356	46,515	NA
as percent of GNP	6.4	6.1	6.3	6.9	7.1	NA
as percent of current govt. expend.	NA	NA	NA	NA	NA	NA
Capital expenditure	1,538	2,085	3,199	3,408	3,440	NA

Educational Attainment [11]

Age group (1991)	25+
Total population	17,471,920
Highest level attained (%)	
No schooling	1.0
First level	
Not completed	4.0
Completed	11.7
Entered second level	
S-1	34.3
S-2	27.7
Postsecondary	21.4

Illiteracy [12]

	1986	1989	1995
Illiterate population (15+ yrs.)[2]	659,745	NA	NA
Illiteracy rate - total pop. (%)	3.4	NA	NA
Illiteracy rate - males (%)	NA	NA	NA
Illiteracy rate - females (%)	NA	NA	NA

Libraries [13]

	Admin. Units	Svc. Pts.	Vols. (000)	Shelving (meters)	Vols. Added	Reg. Users
National (1991)	1	NA	NA	NA	207,297	7,500[e]
Nonspecialized	NA	NA	NA	NA	NA	NA
Public (1990)[7]	1,027	3,301	60,955	NA	5 mil	NA
Higher ed.	NA	NA	NA	NA	NA	NA
School	NA	NA	NA	NA	NA	NA

Daily Newspapers [14]

	1980	1985	1990	1994
Number of papers	123	117	108	107
Circ. (000)	5,425	5,566	5,800[e]	5,500[e]

Culture [15]

Cinema (seats per 1,000)	26.0
Annual attendance per person	2.8
Gross box office receipts (mil. Dollar)[2]	439
Museums (reporting)[3]	1,327
Visitors (000)	27,302
Annual receipts (000 Dollar)	431,319

Science and Technology

Scientific/Technical Forces [16]

Scientists/engineers	65,350
Number female	NA
Technicians	27,520
Number female	NA
Total[8]	92,870

R&D Expenditures [17]

	Dollar (000) 1994
Total expenditure	11,649,000[e]
Capital expenditure	NA
Current expenditure	NA
Percent current	NA

U.S. Patents Issued [18]

Values show patents issued to citizens of the country by the U.S. Patents Office.

	1993	1994	1995
Number of patents	2,231	2,380	2,447

For sources, notes, and explanations, see Annotated Source Appendix, page 1061.

Government and Law

Organization of Government [19]

Long-form name:
None
Type:
Confederation with parliamentary
democracy
Independence:
1 July 1867 (from UK)
National holiday:
Canada Day, 1 July (1867)
Constitution:
Amended British North America Act 1867
patriated to Canada 17 April 1982; charter
of rights and unwritten customs
Legal system:
Based on English common law, except in
Quebec, where civil law system based on
French law prevails; accepts compulsory
ICJ jurisdiction, with reservations
Executive branch:
British Monarch (represented by Governor
General), Prime Minister, Deputy Prime
Minister; Federal Ministry
Legislative branch:
Bicameral Parliament (Parlement): Senate
(Senat) and House of Commons
(Chambre des Communes) Supreme
Court

Elections [20]

House of Commons	% of votes
Liberal Party	60.7
Bloc Quebecois	17.9
Reform Party	17.6
New Democratic Party	2.7
Progressive Conservative Party	0.7
Independents	0.3

Government Expenditures [21]

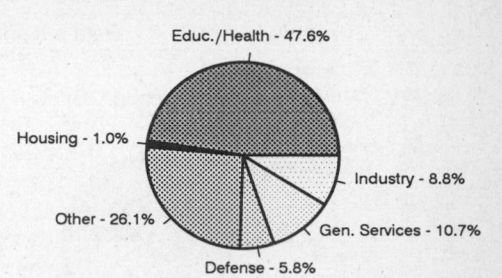

Educ./Health - 47.6%
Housing - 1.0%
Industry - 8.8%
Other - 26.1%
Gen. Services - 10.7%
Defense - 5.8%

(% distribution). Expend. for FY94: 181,947 (Dollar mil.)

Military Expenditures and Arms Transfers [22]

	1990	1991	1992	1993	1994
Military expenditures					
Current dollars (mil.)	9,368	9,027	9,350	9,576	9,525
1994 constant dollars (mil.)	10,430	9,676	9,750	9,773	9,525
Armed forces (000)	87	86	82	76	75
Gross national product (GNP)					
Current dollars (mil.)	444,200	454,200	468,600	490,300	523,000
1994 constant dollars (mil.)	494,400	486,800	488,600	500,400	523,000
Central government expenditures (CGE)					
1994 constant dollars (mil.)	123,600	126,000	129,500	121,200[e]	120,600[e]
People (mil.)	26.6	27.0	27.4	27.8	28.1
Military expenditure as % of GNP	2.1	2.0	2.0	2.0	1.8
Military expenditure as % of CGE	8.4	7.7	7.5	8.1	7.9
Military expenditure per capita (1994 $)	392	358	356	352	339
Armed forces per 1,000 people (soldiers)	3.3	3.2	3.0	2.7	2.7
GNP per capita (1994 $)	18,570	18,020	17,830	18,020	18,600
Arms imports[6]					
Current dollars (mil.)	190	280	410	190	170
1994 constant dollars (mil.)	211	300	428	194	170
Arms exports[6]					
Current dollars (mil.)	625	550	1,200	775	230
1994 constant dollars (mil.)	696	590	1,251	791	230
Total imports[7]					
Current dollars (mil.)	123,200	124,800	129,300	139,000	155,100
1994 constant dollars (mil.)	137,200	133,800	134,800	141,900	155,100
Total exports[7]					
Current dollars (mil.)	127,600	127,200	134,400	145,200	165,400
1994 constant dollars (mil.)	142,000	136,300	140,200	148,200	165,400
Arms as percent of total imports[8]	0.2	0.2	0.3	0.1	0.1
Arms as percent of total exports[8]	0.5	0.4	0.9	0.5	0.1

Crime [23]

	1994
Crime volume	
Cases known to police	3,027,636
Attempts (percent)	NA
Percent cases solved	41.60
Crimes per 100,000 persons	10,351.56
Persons responsible for offenses	
Total number offenders	754,371
Percent female	16.88
Percent juvenile (12-17 yrs.)	19.01
Percent foreigners	NA

Human Rights [24]

	SSTS	FL	FAPRO	PPCG	APROBC	TPW	PCPTW	STPEP	PHRFF	PRW	ASST	AFL
Observes	P		P	P		P	P			P	P	P
	EAFRD	CPR	ESCR	SR	ACHR	MAAE	PVIAC	PVNAC	EAFDAW	TCIDTP	RC	
Observes	P	P	P	P				P	P	P	P	P

P = Party; S = Signatory; see Appendix for meaning of abbreviations.

Labor Force

Total Labor Force [25]

13.38 million

Labor Force by Occupation [26]

Services	75%
Manufacturing	14
Agriculture	4
Construction	3
Other	4

Date of data: 1988

Unemployment Rate [27]

9.5% (1995)

For sources, notes, and explanations, see Annotated Source Appendix, page 1061.

171

Production Sectors

Commercial Energy Production and Consumption

Data are shown in quadrillion (10^{15}) BTUs and percent for 1995
Values for hydroelectric, nuclear, geothermal, solar, and wind power refer to electrical generation.

Production [28]

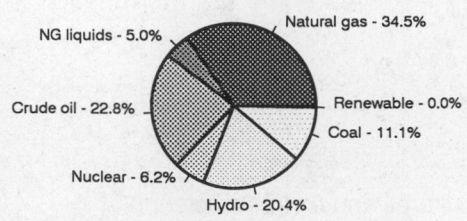

NG liquids - 5.0%
Natural gas - 34.5%
Crude oil - 22.8%
Renewable - 0.0%
Coal - 11.1%
Nuclear - 6.2%
Hydro - 20.4%

Consumption [29]

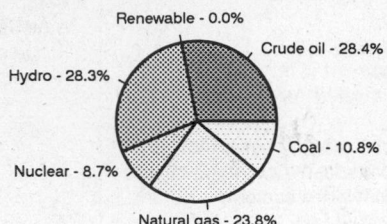

Renewable - 0.0%
Crude oil - 28.4%
Hydro - 28.3%
Coal - 10.8%
Nuclear - 8.7%
Natural gas - 23.8%

Production		Consumption	
Crude oil	3.828	Crude oil	3.435
Natural gas liquids	0.843	Dry natural gas	2.886
Dry natural gas	5.796	Coal	1.302
Coal	1.859	Net hydroelectric power	3.429
Net hydroelectric power	3.429	Net nuclear power	1.049
Net nuclear power	1.049	Geothermal, solar, wind	0.001
Geothermal, solar, wind	0.001	Total	12.102
Total	16.805		

Telecommunications [30]

- 15.3 million (1990) telephones; excellent service provided by modern technology
- Domestic: domestic satellite system with about 300 earth stations
- International: 5 coaxial submarine cables; satellite earth stations - 5 Intelsat (4 Atlantic Ocean and 1 Pacific Ocean) and 2 Intersputnik (Atlantic Ocean Region)
- Radio: Broadcast stations: AM 900, FM 29, shortwave 0
- Television: Broadcast stations: 70 (repeaters 1,400) (1991) Televisions: 11.53 million (1983 est.)

Transportation [31]

Railways: total: 70,176 km; note - there are two major transcontinental freight railway systems: Canadian National (privatized November 1995) and Canadian Pacific Railway; passenger service provided by government-operated firm VIA, which has no trackage of its own; standard gauge: 70,000 km 1.435-m gauge (63 km electrified); narrow gauge: 176 km 0.914-m gauge (1995)

Highways: total: 849,404 km; paved: 297,291 km (including 15,983 km of expressways); unpaved: 552,113 km (1991 est.)

Merchant marine: total: 62 ships (1,000 GRT or over) totaling 573,089 GRT/804,436 DWT; ships by type: bulk 17, cargo 9, chemical tanker 4, oil tanker 15, passenger 2, passenger-cargo 1, railcar carrier 2, roll-on/roll-off cargo 7, short-sea passenger 3, specialized tanker 2

Airports

Total:	1,138
With paved runways over 3,047 m:	17
With paved runways 2,438 to 3,047 m:	15
With paved runways 1,524 to 2,437 m:	136
With paved runways 914 to 1,523 m:	226
With paved runways under 914 m:	422

Top Agricultural Products [32]

Agriculture accounts for 2% of the GDP; produces wheat, barley, oilseed, tobacco, fruits, vegetables; dairy products; forest products; commercial fisheries provide annual catch of 1.5 million metric tons, of which 75% is exported.

Top Mining Products [33]

Metric tons except as noted	7/95[*]
Aluminum metal, primary	2,250,000
Copper, mine output, Cu content	626,000[17]
Petroleum, crude (000 42-gal. bls.)	636,000
Selenium, refined (kg.)	600,000[e,18]
Silver, refined (kg.)	1,000,000[e]
Titanium, Sorel slag	764,000[19]
Zinc, mine output, Zn content	984,000
Nepheline syenite	544,000
Nitrogen, N content of ammonia	3,470,000
Peat	1,020,000

Tourism [34]

	1990	1991	1992	1993	1994
Visitors[21]	37,990	36,818	35,731	36,100	38,651
Tourists[21,12]	15,209	14,912	14,741	15,105	15,971
Excursionists[21]	22,781	21,906	20,990	20,995	22,679
Tourism receipts[22]	5,612	5,886	5,712	5,897	6,309
Tourism expenditures[22]	10,401	11,367	11,289	10,629	11,676
Fare receipts	1,028	923	957	939	1,154
Fare expenditures	2,032	2,043	2,131	2,304	2,243

Travelers are in thousands, money in million U.S. dollars.

For sources, notes, and explanations, see Annotated Source Appendix, page 1061.

Manufacturing Sector

Manufacturing Summary [35]

	1987		1988		1989		1990		1991	
	$ bil.	%	$ bil.	%	$ bil.	%	$ bil.	%	$ bil.	%
Establishments or enterprises (number)	39,554	1.686	43,376	2.066	42,252	2.256	42,921	2.396	-	-
Total employment (000)	2,239	1.650	2,343	1.712	2,373	1.928	2,250	2.033	-	-
Production workers (000)	-	-	1,764	2.823	1,792	2.436	1,669	2.351	-	-
Output ($ bil.)	312	3.074	375	3.217	401	3.402	390	3.400	-	-
Value added ($ bil.)	114	2.638	139	2.898	148	3.031	143	2.866	-	-
Capital investment ($ mil.)	14,775	3.390	18,688	3.924	23,310	4.255	21,514	3.871	-	-
M & E investment ($ mil.)	11,936	3.830	15,030	4.343	18,477	4.733	16,706	3.962	-	-
Employees per establishment (number)	57	97.845	54	82.854	56	85.475	52	84.842	-	-
Production workers per establishment	-	-	41	136.642	42	107.991	39	98.142	-	-
Output per establishment ($ mil.)	8	182.284	9	155.689	9	150.814	9	141.924	-	-
Capital investment per estab. ($ mil.)	0.374	201.058	0.431	189.916	0.552	188.662	0.501	161.552	-	-
M & E per establishment ($ mil)	0.302	227.173	0.347	210.164	0.437	209.815	0.389	165.379	-	-
Payroll per employee ($)	22,375	249.525	24,954	250.474	27,681	247.769	29,348	211.566	-	-
Wages per production worker ($)	-	-	23,131	273.743	25,180	251.087	26,526	223.366	-	-
Hours per production worker (hours)	-	-	1,902	98.967	1,893	103.289	1,881	100.569	-	-
Output per employee ($)	139,529	186.300	160,050	187.907	169,100	176.442	173,359	167.281	-	-
Capital investment per employee ($)	6,599	205.188	7,976	220.217	9,823	220.722	9,562	100.416	-	-
M & E per employee ($)	5,331	232.177	6,415	253.655	7,786	245.469	7,425	194.927	-	-

Note: Columns headed % show percent of world total or ratio. Ratios closest to 100 are closest to world average. M & E stands for machinery & equipment.

Output in Manufacturing

	1987		1988		1989		1990		1991	
	$ bil.	%	$ bil.	%	$ bil.	%	$ bil.	%	$ bil.	%
3110 - Food products	32.000	10.26	36.000	9.60	38.000	9.48	39.000	10.00	-	-
3130 - Beverages	4.630	1.48	5.639	1.50	5.431	1.35	5.108	1.31	-	-
3140 - Tobacco	1.870	0.60	2.129	0.57	2.306	0.58	2.425	0.62	-	-
3210 - Textiles	6.297	2.02	7.029	1.87	7.399	1.85	7.182	1.84	-	-
3211 - Spinning, weaving, etc.	2.511	0.80	2.763	0.74	2.804	0.70	2.751	0.71	-	-
3220 - Wearing apparel	4.819	1.54	5.428	1.45	6.014	1.50	5.965	1.53	-	-
3230 - Leather and products	0.370	0.12	0.423	0.11	0.448	0.11	0.403	0.10	-	-
3240 - Footwear	0.762	0.24	0.764	0.20	0.760	0.19	0.720	0.18	-	-
3310 - Wood products	11.000	3.53	12.000	3.20	13.000	3.24	13.000	3.33	-	-
3320 - Furniture, fixtures	3.522	1.13	4.079	1.09	4.510	1.12	4.414	1.13	-	-
3410 - Paper and products	18.000	5.77	21.000	5.60	22.000	5.49	21.000	5.38	-	-
3411 - Pulp, paper, etc.	14.000	4.49	17.000	4.53	18.000	4.49	16.000	4.10	-	-
3420 - Printing, publishing	8.816	2.83	11.000	2.93	12.000	2.99	12.000	3.08	-	-
3510 - Industrial chemicals	9.532	3.06	12.000	3.20	13.000	3.24	13.000	3.33	-	-
3511 - Basic chemicals, excl fertilizers	5.875	1.88	7.784	2.08	8.176	2.04	8.236	2.11	-	-
3513 - Synthetic resins, etc.	2.232	0.72	2.925	0.78	3.150	0.79	3.094	0.79	-	-
3520 - Chemical products nec	9.012	2.89	10.000	2.67	11.000	2.74	12.000	3.08	-	-
3522 - Drugs and medicines	2.775	0.89	3.242	0.86	3.429	0.86	3.848	0.99	-	-
3530 - Petroleum refineries	13.000	4.17	12.000	3.20	13.000	3.24	17.000	4.36	-	-
3540 - Petroleum, coal products	0.679	0.22	0.642	0.17	0.794	0.20	0.720	0.18	-	-
3550 - Rubber products	2.534	0.81	2.917	0.78	3.193	0.80	3.257	0.84	-	-
3560 - Plastic products nec	5.377	1.72	6.663	1.78	7.120	1.78	6.616	1.70	-	-
3610 - Pottery, china, etc.	0.151	0.05	0.130	0.03	0.118	0.03	0.111	0.03	-	-
3620 - Glass and products	1.184	0.38	1.292	0.34	1.385	0.35	1.380	0.35	-	-
3690 - Nonmetal products nec	4.992	1.60	5.769	1.54	6.258	1.56	5.819	1.49	-	-
3710 - Iron and steel	7.881	2.53	9.856	2.63	11.000	2.74	8.570	2.20	-	-
3720 - Nonferrous metals	7.006	2.25	9.491	2.53	9.628	2.40	8.305	2.13	-	-
3810 - Metal products	12.000	3.85	14.000	3.73	16.000	3.99	15.000	3.85	-	-
3820 - Machinery nec	12.000	3.85	15.000	4.00	17.000	4.24	17.000	4.36	-	-
3825 - Office, computing machinery	1.998	0.64	2.876	0.77	3.083	0.77	3.197	0.82	-	-
3830 - Electrical machinery	11.000	3.53	14.000	3.73	15.000	3.74	15.000	3.85	-	-
3832 - Radio, television, etc.	5.694	1.83	6.655	1.77	7.601	1.90	7.551	1.94	-	-
3840 - Transportation equipment	44.000	14.10	56.000	14.93	60.000	14.96	57.000	14.62	-	-
3841 - Shipbuilding, repair	1.026	0.33	1.430	0.38	1.774	0.44	1.697	0.44	-	-
3843 - Motor vehicles	39.000	12.50	49.000	13.07	51.000	12.72	48.000	12.31	-	-
3850 - Professional goods	1.365	0.44	1.593	0.42	1.731	0.43	1.740	0.45	-	-
3900 - Industries nec	2.504	0.80	2.998	0.80	3.302	0.82	3.265	0.84	-	-

Note: Codes are International Standard Industry codes (ISIC). Percentages are % of total Output. [f]: Factor Prices; [p]: Producer Prices.

For sources, notes, and explanations, see Annotated Source Appendix, page 1061.

173

Finance, Economics, and Trade

Economic Indicators [36]

- **National product**: GDP—purchasing power parity—$694 billion (1995 est.)
- **National product real growth rate**: 2.1% (1995 est.)
- **National product per capita**: $24,400 (1995 est.)
- **Inflation rate (consumer prices)**: 2.4% (1995 est.)
- **External debt**: $233 billion (1994)

Balance of Payments Summary [37]

Values in millions of dollars.

	1989	1990	1991	1992	1993
Exports of goods (f.o.b.)	122,971	128,438	126,003	132,351	144,030
Imports of goods (f.o.b.)	-116,985	-120,108	-122,308	-126,370	-136,418
Trade balance	5,986	8,330	3,695	5,981	7,612
Services - debits	-50,158	-55,727	-53,950	-53,232	-56,323
Services - credits	24,196	25,901	26,187	25,095	24,661
Private transfers (net)	763	731	656	590	545
Government transfers (net)	-523	-783	-640	-494	-363
Long-term capital (net)	18,504	14,788	13,505	10,272	19,975
Short-term capital (net)	1,037	8,786	10,374	4,585	8,729
Errors and omissions	486	-1,401	-2,314	1,397	-5,327
Overall balance	291	625	-2,487	-5,806	-491

Exchange Rates [38]

Currency: **Canadian dollar.**
Symbol: **Can$.**

Data are currency units per $1.

January 1996	1.3666
1995	1.3724
1994	1.3656
1993	1.2901
1992	1.2087
1991	1.1457

Imports and Exports

Top Import Origins [39]

$166.7 billion (c.i.f., 1995 est.).

Origins	%
US	NA
Japan	NA
UK	NA
Germany	NA
France	NA
Mexico	NA
Taiwan	NA
South Korea	NA

Top Export Destinations [40]

$185 billion (f.o.b., 1995 est.).

Destinations	%
US	NA
Japan	NA
UK	NA
Germany	NA
South Korea	NA
Netherlands	NA
China	NA

Foreign Aid [41]

	U.S. $	
ODA (1993)	2.373	billion
ODA and OOF commitments (1986-91)	10.1	billion

Import and Export Commodities [42]

Import Commodities
- Crude oil
- Chemicals
- Motor vehicles and parts
- Durable consumer goods
- Electronic computers
- Telecommunications equipment and parts

Export Commodities
- Newsprint
- Wood pulp
- Timber
- Crude petroleum
- Machinery
- Natural gas
- Aluminum
- Motor vehicles and parts
- Telecommunications equipment

174

For sources, notes, and explanations, see Annotated Source Appendix, page 1061.

Cape Verde

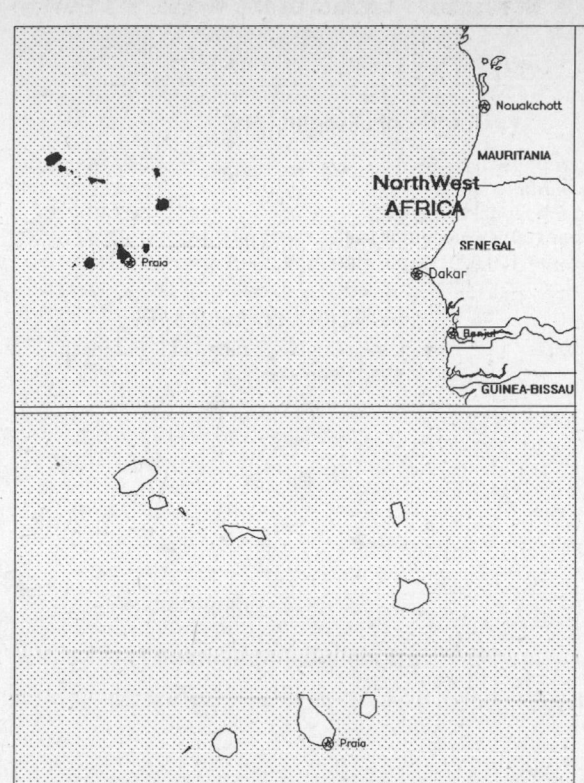

Geography [1]

Total area:
  4,030 sq km   1,556 sq mi
Land area:
  4,030 sq km   1,556 sq mi
Comparative area:
  Slightly larger than Rhode Island
Land boundaries:
  0 km
Coastline:
  965 km
Climate:
  Temperate; warm, dry summer; precipitation meager and very erratic
Terrain:
  Steep, rugged, rocky, volcanic
Natural resources:
  Salt, basalt rock, pozzolana, limestone, kaolin, fish
Land use:
  Arable land: 9%
  Permanent crops: 0%
  Meadows and pastures: 6%
  Forest and woodland: 0%
  Other: 85%

Demographics [2]

	1970	1980	1990	1995[1]	1996	2000	2010	2020	2030
Population	269	296	375	436	449	503	646	812	986
Population density (persons per sq. mi.)	173	190	241	280	289	323	415	522	634
(persons per sq. km.)	67	73	93	108	111	125	160	202	245
Net migration rate (per 1,000 population)	NA	-10.1	-8.0	-6.9	-6.7	-6.0	-4.6	-3.7	-3.0
Births	NA	NA	NA	NA	20	NA	NA	NA	NA
Deaths	NA	NA	NA	NA	4	NA	NA	NA	NA
Life expectancy - males	NA	55.1	59.1	61.1	61.5	63.0	66.5	69.5	72.0
Life expectancy - females	NA	58.5	62.9	65.0	65.4	67.0	70.8	74.1	76.8
Birth rate (per 1,000)	NA	40.0	48.7	45.3	44.3	40.3	33.3	29.3	23.5
Death rate (per 1,000)	NA	13.7	10.6	8.7	8.3	7.0	5.1	4.0	3.6
Women of reproductive age (15-49 yrs.)	NA	68	86	100	103	118	165	212	267
of which are currently married	NA	18	36	NA	45	52	70	NA	NA
Fertility rate	NA	6.7	6.7	6.2	6.1	5.7	4.5	3.5	2.8

Except as noted, values for vital statistics are in thousands; life expectancy is in years.

Health

Health Indicators [3]

Health Expenditures [4]

For sources, notes, and explanations, see Annotated Source Appendix, page 1061.

175

Human Factors

Women and Children [5]

Burden of Disease [6]

Population per physician (1992)	4,289.16
Population per nurse (1992)	1,640.55
Population per hospital bed (1992)	633.45
AIDS cases per 100,000 people (1994)	2.3
Malaria cases per 100,000 people (1992)	NA

Ethnic Division [7]

Creole (mulatto)	71%
African	28%
European	1%

Religion [8]

Roman Catholicism fused with indigenous beliefs.

Major Languages [9]

Portuguese, Crioulo, a blend of Portuguese and West African words.

Education

Public Education Expenditures [10]

	1980	1985	1987	1990	1991	1994
Million (Escudo)						
Total education expenditure	NA	341	493	NA	903	NA
as percent of GNP	NA	3.6	3.8	NA	4.4	NA
as percent of total govt. expend.	NA	NA	14.8	NA	19.9	NA
Current education expenditure	NA	325	472	NA	890	NA
as percent of GNP	NA	3.5	3.6	NA	4.3	NA
as percent of current govt. expend.	NA	15.2	15.3	NA	20.0	NA
Capital expenditure	NA	16	21	NA	13	NA

Educational Attainment [11]

Illiteracy [12]

In thousands and percent[1]	1990	1995	2000
Illiterate population (15+ yrs.)	70	64	61
Illiteracy rate - total pop. (%)	35.7	29.2	24.2
Illiteracy rate - males (%)	24.7	20.2	16.4
Illiteracy rate - females (%)	43.8	36.8	30.3

Libraries [13]

Daily Newspapers [14]

Culture [15]

Science and Technology

Scientific/Technical Forces [16]

R&D Expenditures [17]

U.S. Patents Issued [18]

For sources, notes, and explanations, see Annotated Source Appendix, page 1061.

Government and Law

Organization of Government [19]

Long-form name:
Republic of Cape Verde
Type:
Republic
Independence:
5 July 1975 (from Portugal)
National holiday:
Independence Day, 5 July (1975)
Constitution:
New constitution came into force 25
September 1992
Legal system:
NA
Executive branch:
President; Prime Minister; Council of
Ministers
Legislative branch:
Unicameral: People's National Assembly
(Assembleia Nacional Popular)
Judicial branch:
Supreme Tribunal of Justice (Supremo
Tribunal de Justia)

Elections [20]

People's National Assembly	% of votes
Movement for Democracy (MPD)	59.0
African Party for Independence (PAICV)	28.0
Party for Democratic Convergence (PCD)	6.0
Others	7.0

Government Budget [21]

For 1993 est.

	$ mil.
Revenues	174
Expenditures	235
Capital expenditures	165

Crime [23]

Military Expenditures and Arms Transfers [22]

	1990	1991	1992	1993	1994
Military expenditures					
Current dollars (mil.)	NA	3	3	3	3
1994 constant dollars (mil.)	NA	3	3	3	3
Armed forces (000)	1	1	1	1	1
Gross national product (GNP)					
Current dollars (mil.)	232	251	295	309	330
1994 constant dollars (mil.)	258	269	308	315	330
Central government expenditures (CGE)					
1994 constant dollars (mil.)	NA	124[e]	NA	NA	NA
People (mil.)	0.4	0.4	0.4	0.4	0.4
Military expenditure as % of GNP	NA	1.3	1.1	1.0	1.0
Military expenditure as % of CGE	NA	2.8	NA	NA	NA
Military expenditure per capita (1994 $)	NA	9	8	8	8
Armed forces per 1,000 people (soldiers)	2.7	2.6	2.5	2.4	2.4
GNP per capita (1994 $)	687	697	773	768	779
Arms imports[6]					
Current dollars (mil.)	5	0	0	0	0
1994 constant dollars (mil.)	6	0	0	0	0
Arms exports[6]					
Current dollars (mil.)	0	0	0	0	0
1994 constant dollars (mil.)	0	0	0	0	0
Total imports[7]					
Current dollars (mil.)	136	147	180	189[e]	NA
1994 constant dollars (mil.)	151	158	188	193[e]	NA
Total exports[7]					
Current dollars (mil.)	6	6	5	6[e]	NA
1994 constant dollars (mil.)	7	6	5	6[e]	NA
Arms as percent of total imports[8]	3.7	0	0	0	0
Arms as percent of total exports[8]	0	0	0	0	0

Human Rights [24]

	SSTS	FL	FAPRO	PPCG	APROBC	TPW	PCPTW	STPEP	PHRFF	PRW	ASST	AFL
Observes		P			P	P	P					P
	EAFRD	CPR	ESCR	SR	ACHR	MAAE	PVIAC	PVNAC	EAFDAW	TCIDTP	RC	
Observes	P	P	P	P			P	P	P	P	P	

P = Party; S = Signatory; see Appendix for meaning of abbreviations.

Labor Force

Total Labor Force [25]

102,000 (1985 est.)

Labor Force by Occupation [26]

Agriculture (mostly subsistence)	57%
Services	29
Industry	14

Date of data: 1981

Unemployment Rate [27]

35% (1994 est.)

For sources, notes, and explanations, see Annotated Source Appendix, page 1061.

Production Sectors

Energy Resource Summary [28]

Electricity: Capacity: 15,000 kW. Production: 40 million kWh. Consumption per capita: 73 kWh (1993).

Telecommunications [30]

- 1,740 (1987 est.) telephones
- Domestic: interisland microwave radio relay system
- International: 2 coaxial submarine cables; HF radiotelephone to Senegal and Guinea-Bissau; satellite earth station - 1 Intelsat (Atlantic Ocean)
- Radio: Broadcast stations: AM 1, FM 6, shortwave 0
- Television: Broadcast stations: 1 (1987 est.) Televisions: 7,000 (1991 est.)

Transportation [31]

Railways: 0 km

Highways: total: 1,100 km; paved: 680 km; unpaved: 420 km (1992 est.)

Merchant marine: cargo 3, chemical tanker 1 (1995 est.); total: 4 (1,000 GRT or over) totaling 5,632 GRT/8,872 DWT

Airports

Total: 6
With paved runways over 3,047 m: 1
With paved runways 914 to 1,523 m: 5 (1995 est.)

Top Agricultural Products [32]

Agriculture accounts for 13% of the GDP; produces bananas, corn, beans, sweet potatoes, sugarcane, coffee, peanuts; fish.

Top Mining Products [33]

Metric tons except as noted[e]	2/24/95[*]
Salt	4,000
Pozzolana	5,000

Tourism [34]

	1990	1991	1992	1993	1994
Tourists[23]	24	20	19	27	31
Tourism receipts	6	8	7	10	10
Tourism expenditures	5	3	6	8	9
Fare receipts	4	2	3	NA	NA
Fare expenditures	3	3	4	NA	NA

Travelers are in thousands, money in million U.S. dollars.

Manufacturing Sector

Manufacturing Summary [35]

	1987		1988		1989		1990		1991	
	$ bil.	%	$ bil.	%	$ bil.	%	$ bil.	%	$ bil.	%
Establishments or enterprises (number)	31	0.001	253	0.012	23	0.001	309	0.017	308	0.040
Total employment (000)	-	-	-	-	-	-	-	-	-	-
Production workers (000)	-	-	-	-	-	-	-	-	-	-
Output ($ bil.)	0.011	0.000	0.010	0.000	0.013	0.000	0.016	0.000	0.011	0.000
Value added ($ bil.)	-	-	-	-	-	-	-	-	-	-
Capital investment ($ mil.)	-	-	-	-	-	-	0.025	0.000	0.011	0.000
M & E investment ($ mil.)	-	-	-	-	-	-	-	-	-	-
Employees per establishment (number)	-	-	-	-	-	-	-	-	-	-
Production workers per establishment	-	-	-	-	-	-	-	-	-	-
Output per establishment ($ mil.)	0.368	8.488	0.039	0.709	0.570	9.055	0.052	0.806	0.037	0.280
Capital investment per estab. ($ mil.)	-	-	-	-	-	-	0.000	0.026	0.000	0.007
M & E per establishment ($ mil)	-	-	-	-	-	-	-	-	-	-
Payroll per employee ($)	-	-	-	-	-	-	-	-	-	-
Wages per production worker ($)	-	-	-	-	-	-	-	-	-	-
Hours per production worker (hours)	-	-	-	-	-	-	-	-	-	-
Output per employee ($)	-	-	-	-	-	-	-	-	-	-
Capital investment per employee ($)	-	-	-	-	-	-	-	-	-	-
M & E per employee ($)	-	-	-	-	-	-	-	-	-	-

Note: Columns headed % show percent of world total or ratio. Ratios closest to 100 are closest to world average. M & E stands for machinery & equipment.

Output in Manufacturing

	1987		1988		1989		1990		1991	
	$ bil.	%	$ bil.	%	$ bil.	%	$ bil.	%	$ bil.	%
3110 - Food products	-	-	-	-	-	-	-	-	-	-
3130 - Beverages	0.000	0.00	0.000	0.00	0.001	7.69	0.001	6.25	-	-
3140 - Tobacco	0.003	27.27	0.004	40.00	0.004	30.77	0.005	31.25	0.005	45.45
3220 - Wearing apparel	0.003	27.27	-	-	0.001	7.69	0.001	6.25	0.001	9.09
3240 - Footwear	0.000	0.00	0.000	0.00	0.000	0.00	0.001	6.25	-	-
3310 - Wood products	-	-	-	-	-	-	-	-	0.000	0.00
3320 - Furniture, fixtures	-	-	-	-	-	-	-	-	-	-
3420 - Printing, publishing	-	-	-	-	0.001	7.69	0.001	6.25	0.002	18.18
3522 - Drugs and medicines	0.000	0.00	0.000	0.00	0.001	7.69	0.001	6.25	0.000	0.00
3610 - Pottery, china, etc.	-	-	-	-	-	-	-	-	0.000	0.00
3810 - Metal products	0.000	0.00	0.000	0.00	0.000	0.00	0.001	6.25	-	-
3841 - Shipbuilding, repair	0.005	45.45	0.005	50.00	0.005	38.46	0.006	37.50	0.004	36.36
3900 - Industries nec	-	-	-	-	-	-	-	-	-	-

Note: Codes are International Standard Industry codes (ISIC). Percentages are % of total Output. [f]: Factor Prices; [p]: Producer Prices.

For sources, notes, and explanations, see Annotated Source Appendix, page 1061.

179

Finance, Economics, and Trade

Economic Indicators [36]

- **National product**: GDP—purchasing power parity—$440 million (1994 est.)
- **National product real growth rate**: 4.6% (1994 est.)
- **National product per capita**: $1,040 (1994 est.)
- **Inflation rate (consumer prices)**: 5% (1994 est.)
- **External debt**: $156 million (1991)

Balance of Payments Summary [37]

Values in millions of dollars.

	1988	1989	1990	1991	1992
Exports of goods (f.o.b.)	3.3	6.7	5.6	4.1	4.4
Imports of goods (f.o.b.)	-101.8	-106.9	-119.5	-132.2	-173.3
Trade balance	-98.5	-100.1	-113.8	-128.0	-168.9
Services - debits	-24.2	-32.5	-36.9	-23.2	-31.1
Services - credits	45.0	57.3	61.3	54.3	59.1
Private transfers (net)	39.5	43.2	52.1	57.8	69.9
Government transfers (net)	38.7	27.6	25.5	31.3	67.4
Long-term capital (net)	-1.9	1.1	5.3	1.4	5.4
Short-term capital (net)	2.7	-1.9	3.7	2.0	6.0
Errors and omissions	-0.7	-1.2	5.2	-7.3	5.5
Overall balance	0.6	-6.5	2.2	-11.9	13.3

Exchange Rates [38]

Currency: **Cape Verdean escudo.**
Symbol: **CVEsc.**

Data are currency units per $1.

December 1995	77.860
1995	76.853
1994	81.891
1993	80.427
1992	68.018
1991	71.408

Imports and Exports

Top Import Origins [39]

$173 million (f.o.b., 1992 est.).

Origins	%
Portugal	NA
Netherlands	NA
Germany	NA
Spain	NA
Brazil	NA
France	NA
Cote d'Ivoire	NA

Top Export Destinations [40]

$4.4 million (f.o.b., 1992 est.).

Destinations	%
Netherlands	NA
Portugal	NA
Angola	NA
Spain	NA

Foreign Aid [41]

Recipient: ODA, $NA.

Import and Export Commodities [42]

Import Commodities	Export Commodities
Foodstuffs	Fish
Consumer goods	Bananas
Industrial products	
Transport equipment	

Central African Republic

Geography [1]

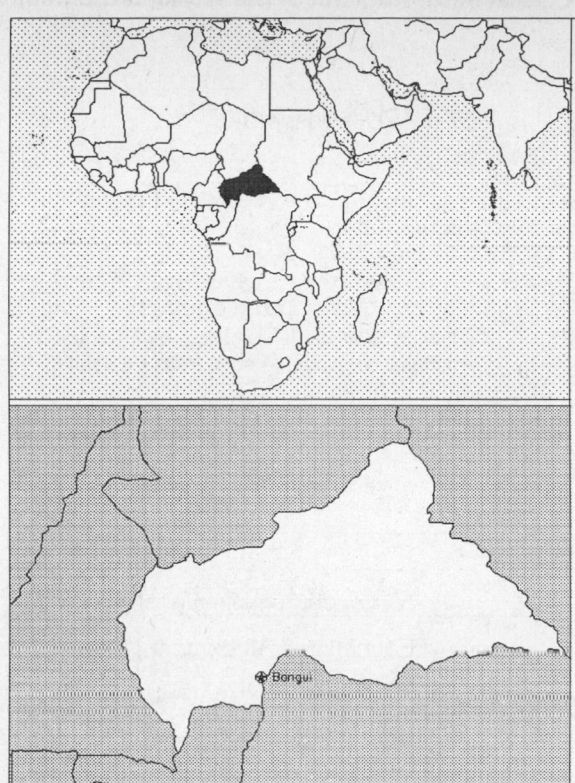

Total area:
622,980 sq km   240,534 sq mi
Land area:
622,980 sq km   240,534 sq mi
Comparative area:
Slightly smaller than Texas
Land boundaries:
Total 5,203 km, Cameroon 797 km, Chad 1,197 km, Congo 467 km, Sudan 1,165 km, Zaire 1,577 km
Coastline:
0 km (landlocked)
Climate:
Tropical; hot, dry winters; mild to hot, wet summers
Terrain:
Vast, flat to rolling, monotonous plateau; scattered hills in Northeast and Southwest
Natural resources:
Diamonds, uranium, timber, gold, oil
Land use:
Arable land: 3%
Permanent crops: 0%
Meadows and pastures: 5%
Forest and woodland: 64%
Other: 28%

Demographics [2]

	1970	1980	1990	1995[1]	1996	2000	2010	2020	2030
Population	1,827	2,244	2,806	3,213	3,274	3,539	4,177	4,780	5,562
Population density (persons per sq. mi.)	8	9	12	13	14	15	17	20	23
(persons per sq. km.)	3	4	5	5	5	6	7	8	9
Net migration rate (per 1,000 population)	NA	NA	0.6	-6.1	-1.5	-1.4	0.0	0.0	0.0
Births	NA	NA	NA	NA	131	NA	NA	NA	NA
Deaths	NA	NA	NA	NA	58	NA	NA	NA	NA
Life expectancy - males	NA	NA	47.7	45.7	45.0	42.6	39.7	43.8	55.9
Life expectancy - females	NA	NA	50.6	47.3	46.7	44.5	40.1	44.7	59.6
Birth rate (per 1,000)	NA	NA	43.1	40.5	40.0	38.4	35.2	31.5	27.8
Death rate (per 1,000)	NA	NA	16.3	17.3	17.6	18.8	21.2	18.1	10.7
Women of reproductive age (15-49 yrs.)	NA	NA	655	741	755	817	984	1,173	1,460
of which are currently married	NA	NA	491	NA	562	604	722	NA	NA
Fertility rate	NA	NA	5.8	5.5	5.4	5.2	4.4	3.7	3.1

Except as noted, values for vital statistics are in thousands; life expectancy is in years.

Health

Health Indicators [3]

% of population with access to	
safe water (1990-95)	18
adequate sanitation (1990-95)	45
health services (1985-95)	45
% of 1-year-olds immunized (1990-94) against	
TB (tuberculosis)	82
DPT (diphtheria, pertussis, tetanus)	31
polio	29
measles	44
% of contraceptive prevalence (1980-94)	15
ORT use rate (1990-94)	24

Health Expenditures [4]

Total health expenditure, 1990 (official exchange rate)	
Millions of dollars	55
Dollars per capita	18
Health expenditures as a percentage of GDP	
Total	4.2
Public sector	2.6
Private sector	1.6
Development assistance for health	
Total aid flows (millions of dollars)[1]	20
Aid flows per capita (dollars)	6.5
Aid flows as a percentage of total health expenditure	35.8

For sources, notes, and explanations, see Annotated Source Appendix, page 1061.

Human Factors

Women and Children [5]

% of pregnant women immunized (tetanus 1990-94)	41
% of births attended by trained health personnel (1983-94)	46
Maternal mortality rate (1980-92)	600
Under-5 mortality rate (1994)	175
% under-5 moderately/severely underweight (1980-1994)	NA

Burden of Disease [6]

Population per physician (1989)	25,265.49
Population per nurse	NA
Population per hospital bed (1990)	1,139.35
AIDS cases per 100,000 people (1994)	1.6
Malaria cases per 100,000 people (1992)	NA

Ethnic Division [7]

Baya	34%
Banda	27%
Sara	10%
Mandjia	21%
Mboum	4%
M'Baka	4%
Europeans	6,500

Religion [8]

Animistic beliefs and practices strongly influence the Christian majority.

Indigenous beliefs	24%
Protestant	25%
Roman Catholic	25%
Muslim	15%
Other	11%

Major Languages [9]

French (official), Sangho (lingua franca and national language), Arabic, Hunsa, Swahili.

Education

Public Education Expenditures [10]

	1980	1986	1987	1988	1990	1994
Million (Franc C.F.A.)						
Total education expenditure	NA	9,553	9,179	8,475	9,862	NA
as percent of GNP	NA	2.8	3.0	2.6	2.8	NA
as percent of total govt. expend.	NA	NA	16.8	NA	NA	NA
Current education expenditure	NA	9,313	9,002	8,227	9,622	NA
as percent of GNP	NA	2.8	2.9	2.6	2.8	NA
as percent of current govt. expend.	NA	25.6	24.7	21.7	NA	NA
Capital expenditure	NA	240	195	248	240	NA

Educational Attainment [11]

Age group (1988)	25+
Total population	920,929
Highest level attained (%)	
No schooling	70.7
First level	
Not completed	19.5
Completed	NA
Entered second level	
S-1	7.3
S-2	NA
Postsecondary	2.0

Illiteracy [12]

In thousands and percent[1]	1990	1995	2000
Illiterate population (15+ yrs.)	841	760	671
Illiteracy rate - total pop. (%)	53.9	42.3	33.6
Illiteracy rate - males (%)	42.5	32.6	25.3
Illiteracy rate - females (%)	64.5	51.6	41.7

Libraries [13]

Daily Newspapers [14]

	1980	1985	1990	1994
Number of papers	-	-	1	1
Circ. (000)	-	-	2[e]	2[e]

Culture [15]

Science and Technology

Scientific/Technical Forces [16]

Scientists/engineers	162
Number female	16
Technicians	92
Number female	5
Total[1]	254

R&D Expenditures [17]

	(000) 1984
Total expenditure[5]	680,791
Capital expenditure	NA
Current expenditure	NA
Percent current	NA

U.S. Patents Issued [18]

For sources, notes, and explanations, see Annotated Source Appendix, page 1061.

Government and Law

Organization of Government [19]

Long-form name:
Central African Republic
Type:
Republic;
Independence:
13 August 1960 (from France)
National holiday:
National Day, 1 December (1958)
(proclamation of the republic)
Constitution:
Passed by referendum 29 December
1994; adopted 7 January 1995
Legal system:
Based on French law
Executive branch:
President; Prime Minister; Council of
Ministers
Legislative branch:
Unicameral: National Assembly
(Assemblee Nationale) Note: The National
Assembly is advised by the Economic and
Regional Council (Conseil Economique et
Regional); when they sit together they are
called the Congress (Congres)
Judicial branch:
Supreme Court (Cour Supreme);
Constitutional Court

Elections [20]

National Assembly	% of seats
Movement for the Liberation of Central African People (MLPC)	38.8
Central African Democratic Assembly (RDC)	16.5
PLD	8.2
PSD	3.5
ADP	7.1
Other	25.9

Government Budget [21]

Crime [23]

	1990
Crime volume	
Cases known to police	4,076
Attempts (percent)	157.9
Percent cases solved	NA
Crimes per 100,000 persons	135.8
Persons responsible for offenses	
Total number offenders	4,685
Percent female	94.6
Percent juvenile (0-17 yrs.)	99.4
Percent foreigners	NA

Military Expenditures and Arms Transfers [22]

	1990	1991	1992	1993	1994
Military expenditures					
Current dollars (mil.)	14	NA	19[e]	22[e]	30[e]
1994 constant dollars (mil.)	16	NA	20[e]	23[e]	30[e]
Armed forces (000)	4	4	7	7	5
Gross national product (GNP)					
Current dollars (mil.)	879	896	899	886	950
1994 constant dollars (mil.)	978	960	938	904	950
Central government expenditures (CGE)					
1994 constant dollars (mil.)	NA	239[e]	NA	NA	NA
People (mil.)	2.9	2.9	3.0	3.1	3.1
Military expenditure as % of GNP	1.6	NA	2.1	2.5	3.2
Military expenditure as % of CGE	NA	NA	NA	NA	NA
Military expenditure per capita (1994 $)	6	NA	7	7	10
Armed forces per 1,000 people (soldiers)	1.4	1.4	2.3	2.3	1.6
GNP per capita (1994 $)	341	327	312	294	302
Arms imports[6]					
Current dollars (mil.)	0	0	0	0	0
1994 constant dollars (mil.)	0	0	0	0	0
Arms exports[6]					
Current dollars (mil.)	0	0	0	0	0
1994 constant dollars (mil.)	0	0	0	0	0
Total imports[7]					
Current dollars (mil.)	154	179	165	145	NA
1994 constant dollars (mil.)	171	192	172	148	NA
Total exports[7]					
Current dollars (mil.)	120	126	124	129	NA
1994 constant dollars (mil.)	134	135	129	132	NA
Arms as percent of total imports[8]	0	0	0	0	0
Arms as percent of total exports[8]	0	0	0	0	0

Human Rights [24]

	SSTS	FL	FAPRO	PPCG	APROBC	TPW	PCPTW	STPEP	PHRFF	PRW	ASST	AFL
Observes	2	P	P		P	P	P	P		P	P	P
	EAFRD	CPR	ESCR	SR	ACHR	MAAE	PVIAC	PVNAC	EAFDAW	TCIDTP		RC
Observes		P	P	P	P			P	P	P		P

P = Party; S = Signatory; see Appendix for meaning of abbreviations.

Labor Force

Total Labor Force [25]

775,413 (1986 est.)

Labor Force by Occupation [26]

Agriculture	85%
Commerce and services	9
Industry	3
Government	3

About 64,000 salaried workers (1985)

Unemployment Rate [27]

For sources, notes, and explanations, see Annotated Source Appendix, page 1061.

Production Sectors

Commercial Energy Production and Consumption

Data are shown in quadrillion (10^{15}) BTUs and percent for 1995
Values for hydroelectric, nuclear, geothermal, solar, and wind power refer to electrical generation.

Production [28]

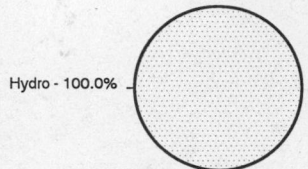

Hydro - 100.0%

Consumption [29]

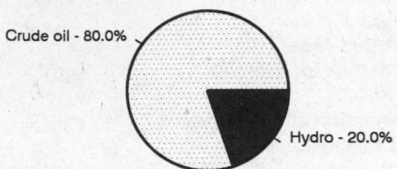

Crude oil - 80.0%

Hydro - 20.0%

Net hydroelectric power	0.001
Total	0.001

Crude oil	0.004
Net hydroelectric power	0.001
Total	0.005

Telecommunications [30]

- 16,867 (1992 est.) telephones; fair system
- Domestic: network consists principally of microwave radio relay and low-capacity, low-powered radiotelephone communication
- International: satellite earth station - 1 Intelsat (Atlantic Ocean)
- Radio: Broadcast stations: AM 1, FM 1, shortwave 0
- Television: Broadcast stations: 1 (1987 est.) Televisions: 7,500 (1993 est.)

Transportation [31]

Railways: 0 km

Highways: total: 23,738 km; paved: 427 km; unpaved: 23,311 km (1991 est.)

Airports

Total:	48
With paved runways 2,438 to 3,047 m:	1
With paved runways 1,524 to 2,437 m:	2
With paved runways under 914 m:	11
With unpaved runways 2,438 to 3,047 m:	1
With unpaved runways 1,524 to 2,437 m:	9

Top Agricultural Products [32]

Agriculture accounts for 50% of the GDP; produces cotton, coffee, tobacco, manioc (tapioca), yams, millet, corn, bananas; timber.

Top Mining Products [33]

Values as noted[e]	6/95[*]
Diamond (carats)	
gem	400,000
industrial	131,000
Gold (kg.)	87

Tourism [34]

For sources, notes, and explanations, see Annotated Source Appendix, page 1061.

Manufacturing Sector

Manufacturing Summary [35]

	1987 $ bil.	%	1988 $ bil.	%	1989 $ bil.	%	1990 $ bil.	%	1991 $ bil.	%
Establishments or enterprises (number)	7	0.000	7	0.000	5	0.000	5	0.000	6	0.001
Total employment (000)	0.211	0.000	0.216	0.000	0.181	0.000	0.139	0.000	-	-
Production workers (000)	-	-	-	-	-	-	-	-	-	-
Output ($ bil.)	0.003	0.000	0.003	0.000	0.003	0.000	0.003	0.000	-	-
Value added ($ bil.)	0.002	0.000	0.001	0.000	0.001	0.000	0.001	0.000	-	-
Capital investment ($ mil.)	-	-	-	-	-	-	-	-	-	-
M & E investment ($ mil.)	-	-	-	-	-	-	-	-	-	-
Employees per establishment (number)	30	52.102	31	47.331	36	55.093	28	44.993	-	-
Production workers per establishment	-	-	-	-	-	-	-	-	-	-
Output per establishment ($ mil.)	0.486	11.223	0.453	8.154	0.611	9.697	0.529	8.260	-	-
Capital investment per estab. ($ mil.)	-	-	-	-	-	-	-	-	-	-
M & E per establishment ($ mil)	-	-	-	-	-	-	-	-	-	-
Payroll per employee ($)	4,983	55.571	4,368	43.841	5,161	46.195	9,063	65.338	-	-
Wages per production worker ($)	-	-	-	-	-	-	-	-	-	-
Hours per production worker (hours)	-	-	-	-	-	-	-	-	-	-
Output per employee ($)	16,132	21.540	14,673	17.227	16,868	17.601	19,025	18.358	-	-
Capital investment per employee ($)	-	-	-	-	-	-	-	-	-	-
M & E per employee ($)	-	-	-	-	-	-	-	-	-	-

Note: Columns headed % show percent of world total or ratio. Ratios closest to 100 are closest to world average. M & E stands for machinery & equipment.

Output in Manufacturing

	1987 $ bil.[f]	%	1988 $ bil.[f]	%	1989 $ bil.[f]	%	1990 $ bil.[f]	%	1991 $ bil.[f]	%
3900 - Industries nec	0.003	100.00	0.003	100.00	0.003	100.00	0.003	100.00	-	-

Note: Codes are International Standard Industry codes (ISIC). Percentages are % of total Output. [f]: Factor Prices; [p]: Producer Prices.

Finance, Economics, and Trade

Economic Indicators [36]

- **National product**: GDP—purchasing power parity—$2.5 billion (1995 est.)
- **National product real growth rate**: 4.1% (1995 est.)
- **National product per capita**: $800 (1995 est.)
- **Inflation rate (consumer prices)**: 45% (1994 est.)
- **External debt**: $904.3 million (1993 est.)

Balance of Payments Summary [37]

Values in millions of dollars.

	1988	1989	1990	1991	1992
Exports of goods (f.o.b.)	133.7	148.1	150.5	125.6	123.5
Imports of goods (f.o.b.)	-178.2	-186.0	-241.6	-178.7	-165.1
Trade balance	-44.5	-37.9	-91.1	-53.1	-41.6
Services - debits	NA	-165.9	-190.9	-155.6	-169.3
Services - credits	NA	66.2	69.8	56.0	60.1
Private transfers (net)	NA	-24.9	-32.9	-29.8	-32.1
Government transfers (net)	NA	129.0	155.9	120.7	125.4
Long-term capital (net)	NA	47.9	90.5	43.7	52.9
Short-term capital (net)	-3.8	-0.3	-9.3	3.1	1.2
Errors and omissions	149.2	1.3	1.1	-1.9	0.7
Overall balance	153.8	15.4	-6.9	-16.9	-2.7

Exchange Rates [38]

Currency: **Communaute Financiere Africaine franc.**
Symbol: **CFAF.**

Data are currency units per $1.

January 1996	500.56
1995	499.15
1994	555.20
1993	283.16
1992	264.69
1991	282.11

Imports and Exports

Top Import Origins [39]

$215 million (f.o.b., 1994 est.).

Origins	%
France	NA
Other EU countries	NA
Japan	NA
Algeria	NA
Cameroon	NA
Namibia	NA

Top Export Destinations [40]

$154 million (f.o.b., 1994 est.).

Destinations	%
France	NA
Belgium	NA
Italy	NA
Japan	NA
US	NA
Spain	NA
Iran	NA

Foreign Aid [41]

Recipient: ODA, $NA.

Import and Export Commodities [42]

Import Commodities	Export Commodities
Food	Diamonds
Textiles	Timber
Petroleum products	Cotton
Machinery	Coffee
Electrical equipment	Tobacco
Motor vehicles	
Chemicals	
Pharmaceuticals	
Consumer goods	
Industrial products	

For sources, notes, and explanations, see Annotated Source Appendix, page 1061.

Chad

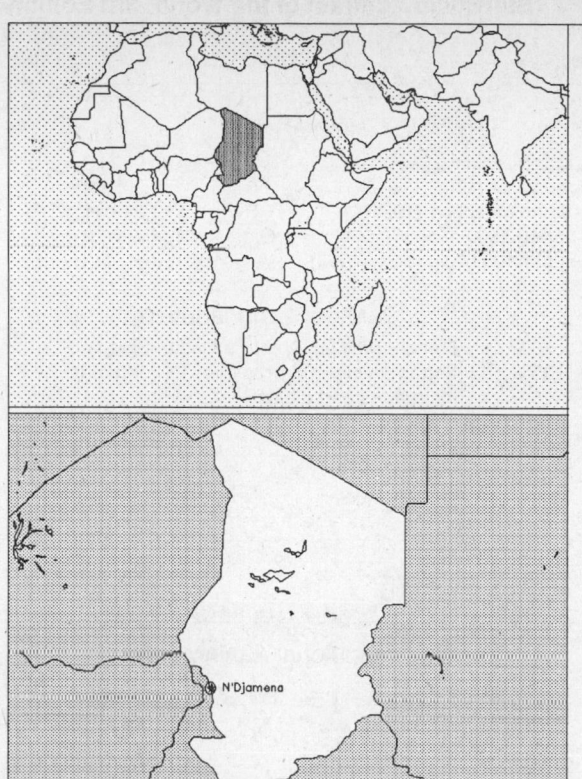

Geography [1]

Total area:
1.284 million sq km   0 sq mi
Land area:
1,259,200 sq km   486,180 sq mi
Comparative area:
Slightly more than three times the size of California
Land boundaries:
Total 5,968 km, Cameroon 1,094 km, Central African Republic 1,197 km, Libya 1,055 km, Niger 1,175 km, Nigeria 87 km, Sudan 1,360 km
Coastline:
0 km (landlocked)
Climate:
Tropical in South, desert in North
Terrain:
Broad, arid plains in center, desert in North, mountains in Northwest, lowlands in South
Natural resources:
Petroleum (unexploited but exploration under way), uranium, natron, kaolin, fish (Lake Chad)
Land use:
Arable land: 2%
Permanent crops: 0%
Meadows and pastures: 36%
Forest and woodland: 11%
Other: 51%

Demographics [2]

	1970	1980	1990	1995[1]	1996	2000	2010	2020	2030
Population	3,733	4,507	5,889	6,784	6,977	7,760	10,055	12,831	15,951
Population density (persons per sq. mi.)	8	9	12	14	14	16	21	26	33
(persons per sq. km.)	3	4	5	5	6	6	8	10	13
Net migration rate (per 1,000 population)	0.0	-21.8	0.5	2.2	0.0	0.0	0.0	0.0	0.0
Births	NA	NA	NA	NA	309	NA	NA	NA	NA
Deaths	NA	NA	NA	NA	122	NA	NA	NA	NA
Life expectancy - males	37.9	40.5	43.4	44.9	45.2	46.5	49.7	53.1	56.5
Life expectancy - females	41.5	44.5	47.9	49.7	50.0	51.5	55.2	58.9	62.5
Birth rate (per 1,000)	46.4	44.8	44.8	44.6	44.3	42.8	38.9	34.6	29.8
Death rate (per 1,000)	23.9	21.3	18.9	17.7	17.4	16.3	13.6	11.4	9.6
Women of reproductive age (15-49 yrs.)	882	1,052	1,371	1,570	1,613	1,795	2,385	3,167	4,107
of which are currently married	NA	NA	1,126	NA	1,325	1,473	1,958	NA	NA
Fertility rate	6.0	5.9	5.9	5.9	5.8	5.6	5.0	4.3	3.6

Except as noted, values for vital statistics are in thousands; life expectancy is in years.

Health

Health Indicators [3]

% of population with access to	
safe water (1990-95)	24
adequate sanitation (1990-95)	NA
health services (1985-95)	30
% of 1-year-olds immunized (1990-94) against	
TB (tuberculosis)	43
DPT (diphtheria, pertussis, tetanus)	18
polio	18
measles	23
% of contraceptive prevalence (1980-94)[1]	1
ORT use rate (1990-94)	15

Health Expenditures [4]

Total health expenditure, 1990 (official exchange rate)	
Millions of dollars	76
Dollars per capita	13
Health expenditures as a percentage of GDP	
Total	6.3
Public sector	4.7
Private sector	1.6
Development assistance for health	
Total aid flows (millions of dollars)[1]	33
Aid flows per capita (dollars)	5.8
Aid flows as a percentage of total health expenditure	43.0

For sources, notes, and explanations, see Annotated Source Appendix, page 1061.

Human Factors

Women and Children [5]

% of pregnant women immunized (tetanus 1990-94)	NA
% of births attended by trained health personnel (1983-94)	15
Maternal mortality rate (1980-92)	960
Under-5 mortality rate (1994)	202
% under-5 moderately/severely underweight (1980-1994)	NA

Burden of Disease [6]

Population per physician (1991)	30,030.93
Population per nurse (1986)	3,395.64
Population per hospital bed (1991)	1,373.41
AIDS cases per 100,000 people (1994)	20.2
Malaria cases per 100,000 people (1992)	NA

Ethnic Division [7]

North and Center: Muslims (Arabs, Toubou, Hadjerai, Fulbe, Kotoko, Kanembou, Baguirmi, Boulala, Zaghawa, and Maba). South: Non-Muslims (Sara, Ngambaye, Mbaye, Goulaye, Moundang, Moussei, Massa), 150,000 non-indigenous (1,000 French).

Religion [8]

Muslim	50%
Christian	25%
Indigenous beliefs (mostly animism)	25%

Major Languages [9]

French (official), Arabic (official), Sara and Sango (in south), more than 100 different languages and dialects.

Education

Public Education Expenditures [10]

	1980	1985	1990	1991	1993	1994
Million (Franc C.F.A.)						
Total education expenditure	NA	NA	NA	8,284	NA	10,796
as percent of GNP	NA	NA	NA	2.3	NA	2.2
as percent of total govt. expend.	NA	NA	NA	NA	NA	NA
Current education expenditure	NA	NA	NA	8,212	8,768	10,691
as percent of GNP	NA	NA	NA	2.3	2.6	2.1
as percent of current govt. expend.	NA	NA	NA	NA	NA	NA
Capital expenditure	NA	NA	NA	72	NA	105

Educational Attainment [11]

Illiteracy [12]

In thousands and percent[1]	1990	1995	2000
Illiterate population (15+ yrs.)	1,804	1,868	1,939
Illiteracy rate - total pop. (%)	54.2	49.2	44.8
Illiteracy rate - males (%)	40.5	36.0	32.1
Illiteracy rate - females (%)	67.4	61.8	56.8

Libraries [13]

	Admin. Units	Svc. Pts.	Vols. (000)	Shelving (meters)	Vols. Added	Reg. Users
National	NA	NA	NA	NA	NA	NA
Nonspecialized	NA	NA	NA	NA	NA	NA
Public	NA	NA	NA	NA	NA	NA
Higher ed. (1987)	1	1	10	128	324	350
School	NA	NA	NA	NA	NA	NA

Daily Newspapers [14]

	1980	1985	1990	1994
Number of papers	1	1	1	1
Circ. (000)	1	1	2	2

Culture [15]

Science and Technology

Scientific/Technical Forces [16]

R&D Expenditures [17]

U.S. Patents Issued [18]

For sources, notes, and explanations, see Annotated Source Appendix, page 1061.

Government and Law

Organization of Government [19]

Long-form name:
Republic of Chad
Type:
Republic
Independence:
11 August 1960 (from France)
National holiday:
Independence Day, 11 August (1960)
Constitution:
31 March 1995, passed by referendum
Legal system:
Based on French civil law system and
Chadian customary law; does not accept
compulsory ICJ jurisdiction
Executive branch:
President; Prime Minister; Council of
State
Legislative branch:
Unicameral: Higher Transitional Council
(Conseil Superieur de Transition)
Judicial branch:
Court of Appeal

Elections [20]

National Consultative Council. Last
held 8 July 1990; disbanded 3
December 1990 and replaced by the
Provisional Council of the Republic
having 30 members appointed by
President DEBY on 8 March 1991; this,
in turn, was replaced by a 57-member
Higher Transitional Council (Conseil
Superieur de Transition) elected by a
specially convened Sovereign National
Conference on 6 April 1993.

Government Budget [21]

For 1992 est.

	$ mil.
Revenues	120
Expenditures	363
Capital expenditures	104

Crime [23]

Military Expenditures and Arms Transfers [22]

	1990	1991	1992	1993	1994
Military expenditures					
Current dollars (mil.)	NA	45	23	25	24[e]
1994 constant dollars (mil.)	NA	48	24	26	24[e]
Armed forces (000)	50	50	30	30	30
Gross national product (GNP)					
Current dollars (mil.)	743	838	863	845	897
1994 constant dollars (mil.)	827	898	900	863	897
Central government expenditures (CGE)					
1994 constant dollars (mil.)	269	288	252[e]	NA	NA
People (mil.)	5.0	5.1	5.2	5.4	5.5
Military expenditure as % of GNP	NA	5.3	2.7	3.0	2.7
Military expenditure as % of CGE	NA	16.6	9.7	NA	NA
Military expenditure per capita (1994 $)	NA	9	5	5	4
Armed forces per 1,000 people (soldiers)	10.0	9.7	5.7	5.6	5.5
GNP per capita (1994 $)	165	175	172	161	164
Arms imports[6]					
Current dollars (mil.)	60	5	10	5	40
1994 constant dollars (mil.)	67	5	10	5	40
Arms exports[6]					
Current dollars (mil.)	0	0	0	0	0
1994 constant dollars (mil.)	0	0	0	0	0
Total imports[7]					
Current dollars (mil.)	286	297	243	201	NA
1994 constant dollars (mil.)	318	318	253	205	NA
Total exports[7]					
Current dollars (mil.)	188	194	182	132	NA
1994 constant dollars (mil.)	209	208	190	135	NA
Arms as percent of total imports[8]	21.0	1.7	4.1	2.5	NA
Arms as percent of total exports[8]	0	0	0	0	0

Human Rights [24]

	SSTS	FL	FAPRO	PPCG	APROBC	TPW	PCPTW	STPEP	PHRFF	PRW	ASST	AFL
Observes		P	P		P	P	P					P
	EAFRD	CPR	ESCR	SR	ACHR	MAAE	PVIAC	PVNAC	EAFDAW	TCIDTP	RC	
Observes	P	P	P	P					P	P	P	

P = Party; S = Signatory; see Appendix for meaning of abbreviations.

Labor Force

Total Labor Force [25]

Labor Force by Occupation [26]

Agriculture (subsistence farming, herding, and fishing)	85%

Unemployment Rate [27]

For sources, notes, and explanations, see Annotated Source Appendix, page 1061.

189

Production Sectors

Energy Resource Summary [28]

Energy resources: Petroleum (unexploited but exploration under way), uranium. **Electricity:** Capacity: 40,000 kW. Production: 80 million kWh. Consumption per capita: 13 kWh (1993).

Telecommunications [30]

- 5,000 (1987 est.) telephones; primitive system
- Domestic: fair system of radiotelephone communication stations
- International: satellite earth station - 1 Intelsat (Atlantic Ocean)
- Radio: Broadcast stations: AM 6, FM 1, shortwave 0
- Television: Broadcast stations: 1 (1987 est.) Televisions: 7,000 (1991 est.)

Transportation [31]

Railways: 0 km

Highways: total: 31,141 km; paved: 32 km; unpaved: 31,109 km (1987 est.)

Airports

Total:	47
With paved runways 2,438 to 3,047 m:	3
With paved runways 1,524 to 2,437 m:	1
With paved runways under 914 m:	11
With unpaved runways over 3,047 m:	1
With unpaved runways 1,524 to 2,437 m:	13

Top Agricultural Products [32]

Agriculture accounts for 49% of the GDP; produces cotton, sorghum, millet, peanuts, rice, potatoes, manioc (tapioca); cattle, sheep, goats, camels.

Top Mining Products [33]

Detailed information is not available. A summary of mineral resources follows. **Mineral Resources:** Unexploited crude oil, bauxite, natron, tin, tungsten, uranium, kaolin, diatomite, dolomite, granite, and marble.

Tourism [34]

	1990	1991	1992	1993	1994
Tourists	9	21	17	21	19
Tourism receipts	12	10	21	23	36
Tourism expenditures	36	32	30	12	NA
Fare receipts	4	4	2	2	NA
Fare expenditures	11	9	5	5	NA

Travelers are in thousands, money in million U.S. dollars.

Finance, Economics, and Trade

GDP and Manufacturing Summary [35]

	1980	1985	1990	1991	1992
Gross Domestic Product					
Millions of 1980 dollars	1,005	804	952	1,017	1,048[e]
Growth rate in percent	-7.40	6.86	-2.70	6.81	3.04[e]
Manufacturing Value Added					
Millions of 1980 dollars	92	69	83	68	70[e]
Growth rate in percent	-12.00	5.39	-1.90	-18.00	2.61[e]
Manufacturing share in percent of current prices	10.7[e]	11.1	15.4	11.1	NA

Economic Indicators [36]

- **National product**: GDP—purchasing power parity—$3.3 billion (1995 est.)
- **National product real growth rate**: 4% (1994 est.)
- **National product per capita**: $600 (1995 est.)
- **Inflation rate (consumer prices)**: 41% (1994 est.)
- **External debt**: $757 million (December 1993)

Balance of Payments Summary [37]

Values in millions of dollars.

	1989	1990	1991	1992	1993
Exports of goods (f.o.b.)	155.4	230.3	193.5	182.3	135.8
Imports of goods (f.o.b.)	-240.3	-259.5	-249.9	-243.0	-201.3
Trade balance	-84.9	-29.2	-56.4	-60.7	-65.5
Services - debits	-220.7	-252.0	-219.3	-239.1	-214.6
Services - credits	43.6	44.0	39.8	44.3	37.6
Private transfers (net)	-20.2	-12.9	-19.8	-34.5	-21.2
Government transfers (net)	230.9	204.5	195.5	204.3	180.0
Long-term capital (net)	81.0	190.0	86.8	74.5	55.3
Short-term capital (net)	-19.0	-119.3	-21.2	-24.6	12.0
Errors and omissions	23.7	-33.3	-13.0	9.2	-34.8
Overall balance	34.4	-8.2	-7.6	-26.6	-51.2

Exchange Rates [38]

Currency: **Communaute Financi-ere Africaine franc.**
Symbol: **CFAF.**

Data are currency units per $1.

January 1996	500.56
1995	499.15
1994	555.20
1993	283.16
1992	264.69
1991	282.11

Imports and Exports

Top Import Origins [39]

$201 million (f.o.b., 1993).

Origins	%
US	NA
France	NA
Nigeria	NA
Cameroon	NA
Italy	NA
Germany	NA

Top Export Destinations [40]

$132 million (f.o.b., 1993).

Destinations	%
France	NA
Nigeria	NA
Cameroon	NA
Zaire	NA
Sudan	NA
Central African Republic	NA

Foreign Aid [41]

Recipient: ODA, $NA.

Import and Export Commodities [42]

Import Commodities	Export Commodities
Machinery and transportation equipment 39%	Cotton
Industrial goods 20%	Cattle
Petroleum products 13%	Textiles
Foodstuffs 9%	Fish
Textiles	
Note - excludes military equipment	

For sources, notes, and explanations, see Annotated Source Appendix, page 1061.

191

Chile

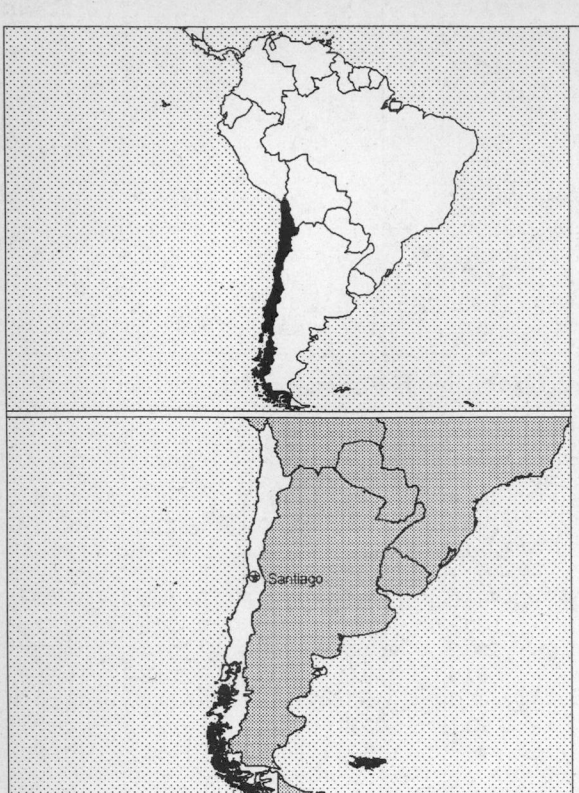

Geography [1]

Total area:
756,950 sq km 292,260 sq mi
Land area:
748,800 sq km 289,113 sq mi
Comparative area:
Slightly smaller than twice the size of Montana
Note: Includes Isla de Pascua (Easter Island) and Isla Sala y Gomez
Land boundaries:
Total 6,171 km, Argentina 5,150 km, Bolivia 861 km, Peru 160 km
Coastline:
6,435 km
Climate:
Temperate; desert in North; cool and damp in South
Terrain:
Low coastal mountains; fertile central valley; rugged Andes in East
Natural resources:
Copper, timber, iron ore, nitrates, precious metals, molybdenum
Land use:
Arable land: 7%
Permanent crops: 0%
Meadows and pastures: 16%
Forest and woodland: 21%
Other: 56%

Demographics [2]

	1970	1980	1990	1995[1]	1996	2000	2010	2020	2030
Population	9,369	11,094	13,121	14,152	14,333	14,996	16,382	17,535	18,256
Population density (persons per sq. mi.)	32	38	45	49	50	52	57	61	63
(persons per sq. km.)	13	15	18	19	19	20	22	23	24
Net migration rate (per 1,000 population)	1.6	0.5	0.0	0.0	0.0	0.0	0.0	0.0	0.0
Births	NA	NA	NA	NA	259	NA	NA	NA	NA
Deaths	NA	74	78	NA	81	NA	NA	NA	NA
Life expectancy - males	NA	NA	69.1	71.0	71.3	72.3	74.4	76.0	77.3
Life expectancy - females	NA	NA	75.9	77.5	77.7	78.7	80.6	82.1	83.3
Birth rate (per 1,000)	25.5	22.3	22.3	18.7	18.1	15.9	14.0	12.7	11.1
Death rate (per 1,000)	9.0	6.8	6.1	5.7	5.7	5.7	6.2	7.1	8.6
Women of reproductive age (15-49 yrs.)	NA	NA	3,485	3,734	3,786	3,985	4,370	4,389	4,346
of which are currently married	1,093	1,514	2,027	NA	2,227	2,345	2,575	NA	NA
Fertility rate	NA	NA	2.6	2.3	2.2	2.0	1.8	1.7	1.7

Except as noted, values for vital statistics are in thousands; life expectancy is in years.

Health

Health Indicators [3]

% of population with access to	
safe water (1990-95)	85
adequate sanitation (1990-95)	83
health services (1985-95)	97
% of 1-year-olds immunized (1990-94) against	
TB (tuberculosis)	96
DPT (diphtheria, pertussis, tetanus)	92
polio	92
measles	96
% of contraceptive prevalence (1980-94)[1]	43
ORT use rate (1990-94)	90

Health Expenditures [4]

Total health expenditure, 1990 (official exchange rate)	
Millions of dollars	1,315
Dollars per capita	100
Health expenditures as a percentage of GDP	
Total	4.7
Public sector	3.4
Private sector	1.4
Development assistance for health	
Total aid flows (millions of dollars)[1]	10
Aid flows per capita (dollars)	0.7
Aid flows as a percentage of total health expenditure	0.7

For sources, notes, and explanations, see Annotated Source Appendix, page 1061.

Human Factors

Women and Children [5]

% of pregnant women immunized (tetanus 1990-94)	NA
% of births attended by trained health personnel (1983-94)	98
Maternal mortality rate (1980-92)	35
Under-5 mortality rate (1994)	15
% under-5 moderately/severely underweight (1980-1994)[1]	3

Burden of Disease [6]

Population per physician (1990)	2,152.45
Population per nurse (1990)	335.40
Population per hospital bed (1992)	319.98
AIDS cases per 100,000 people (1994)	1.5
Malaria cases per 100,000 people (1992)	NA

Ethnic Division [7]

European and European-Indian	95%
Indian	3%
Other	2%

Religion [8]

A small Jewish population is present. Data are unavailable.

Roman Catholic	89%
Protestant	11%

Major Languages [9]

Spanish.

Education

Public Education Expenditures [10]

	1980	1985	1990	1992	1993	1994
Million (Peso)						
Total education expenditure	47,961	101,493	232,516	407,645	477,307	620,095
as percent of GNP	4.6	4.4	2.7	2.8	2.7	2.9
as percent of total govt. expend.	11.9	15.3	10.4	12.9	NA	13.4
Current education expenditure	45,504	NA	225,620	392,411	468,452	585,399
as percent of GNP	4.4	NA	2.6	2.7	2.6	2.8
as percent of current govt. expend.	13.4	NA	NA	15.0	NA	14.7
Capital expenditure	2,457	NA	6,896	15,234	8,855	34,696

Educational Attainment [11]

Age group (1992)	25 i
Total population	NA
Highest level attained (%)	
No schooling	5.8
First level	
Not completed	48.0
Completed	NA
Entered second level	
S-1	33.9
S-2	NA
Postsecondary	12.3

Illiteracy [12]

In thousands and percent[1]	1990	1995	2000
Illiterate population (15+ yrs.)	546	485	437
Illiteracy rate - total pop. (%)	6.0	4.8	4.0
Illiteracy rate - males (%)	5.6	4.6	3.8
Illiteracy rate - females (%)	6.3	5.1	4.2

Libraries [13]

	Admin. Units	Svc. Pts.	Vols. (000)	Shelving (meters)	Vols. Added	Reg. Users
National (1993)	1	3	3,554	NA	2,590	NA
Nonspecialized	NA	NA	NA	NA	NA	NA
Public (1992)	289	289	1,121	NA	NA	2 mil
Higher ed. (1989)	178	NA	5,669	NA	NA	NA
School (1989)	821	NA	3,820	NA	NA	NA

Daily Newspapers [14]

	1980	1985	1990	1994
Number of papers	34	38[e]	45[e]	32
Circ. (000)	NA	NA	1,923[e]	1,411

Culture [15]

Cinema (seats per 1,000)	5.5[e]
Annual attendance per person	0.6[e]
Gross box office receipts (mil. Peso)	NA
Museums (reporting)	25
Visitors (000)	648
Annual receipts (000 Peso)	133,799

Science and Technology

Scientific/Technical Forces [16]

Scientists/engineers	4,630
Number female	NA
Technicians	2,940
Number female	NA
Total[1,9]	7,570

R&D Expenditures [17]

	Peso (000) 1992
Total expenditure[6]	102,196,000[e]
Capital expenditure	NA
Current expenditure	NA
Percent current	NA

U.S. Patents Issued [18]

Values show patents issued to citizens of the country by the U.S. Patents Office.

	1993	1994	1995
Number of patents	10	8	7

For sources, notes, and explanations, see Annotated Source Appendix, page 1061.

Government and Law

Organization of Government [19]

Long-form name:
Republic of Chile
Type:
Republic
Independence:
18 September 1810 (from Spain)
National holiday:
Independence Day, 18 September (1810)
Constitution:
11 September 1980, effective 11 March 1981; amended 30 July 1989
Legal system:
Based on Code of 1857 derived from Spanish law and subsequent codes influenced by French and Austrian law; judicial review of legislative acts in the Supreme Court; does not accept compulsory ICJ jurisdiction
Executive branch:
President; Cabinet
Legislative branch:
Bicameral National Congress (Congreso Nacional): Senate (Senado) and Chamber of Deputies (Camara de Diputados)
Judicial branch:
Supreme Court (Corte Suprema)

Elections [20]

Chamber of Deputies	% of votes
Coalition of Parties for Democracy	54.0
PDC	26.2
PPD	11.8
PR	3.0
Union for the Progress of Chile	30.6
National Renovation (RN)	15.3
Independent Demo. Union (UDI)	12.1
Center Center Union	3.2

Government Expenditures [21]

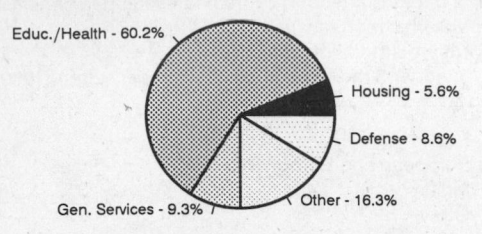

Educ./Health - 60.2%
Housing - 5.6%
Defense - 8.6%
Other - 16.3%
Gen. Services - 9.3%

(% distribution). Expend. for CY95: 5,137.05 (Peso bil.)

Military Expenditures and Arms Transfers [22]

	1990	1991	1992	1993	1994
Military expenditures					
Current dollars (mil.)[3]	1,072[e]	1,136[e]	1,043	1,042	966
1994 constant dollars (mil.)[3]	1,193[e]	1,218[e]	1,088	1,063	966
Armed forces (000)	95	90	92	92	102
Gross national product (GNP)					
Current dollars (mil.)	32,970	36,590	41,870	45,840	50,350
1994 constant dollars (mil.)	36,690	39,220	43,660	46,790	50,350
Central government expenditures (CGE)					
1994 constant dollars (mil.)	7,758	8,688	9,297	10,040	10,680
People (mil.)	13.1	13.3	13.5	13.7	14.0
Military expenditure as % of GNP	3.3	3.1	2.5	2.3	1.9
Military expenditure as % of CGE	15.4	14.0	11.7	10.6	9.1
Military expenditure per capita (1994 $)	91	91	80	77	69
Armed forces per 1,000 people (soldiers)	7.2	6.8	6.8	6.7	7.3
GNP per capita (1994 $)	2,799	2,945	3,227	3,405	3,609
Arms imports[6]					
Current dollars (mil.)	70	70	50	40	90
1994 constant dollars (mil.)	78	75	52	41	90
Arms exports[6]					
Current dollars (mil.)	10	0	5	10	0
1994 constant dollars (mil.)	11	0	5	10	0
Total imports[7]					
Current dollars (mil.)	7,678	8,094	10,130	11,120	11,820
1994 constant dollars (mil.)	8,545	8,676	10,560	11,350	11,820
Total exports[7]					
Current dollars (mil.)	8,310	8,929	9,989	9,202	11,570
1994 constant dollars (mil.)	9,248	9,571	10,420	9,392	11,570
Arms as percent of total imports[8]	0.9	0.9	0.5	0.4	0.8
Arms as percent of total exports[8]	0.1	0	0.1	0.1	0

Crime [23]

	1994
Crime volume	
Cases known to police	152,044
Attempts (percent)	NA
Percent cases solved	48.11
Crimes per 100,000 persons	1,086.47
Persons responsible for offenses	
Total number offenders	29,337
Percent female	14.07
Percent juvenile (0-17 yrs.)	6.59
Percent foreigners	0.51

Human Rights [24]

	SSTS	FL	FAPRO	PPCG	APROBC	TPW	PCPTW	STPEP	PHRFF	PRW	ASST	AFL
Observes	P	P		P		P	P			P	P	
	EAFRD	CPR	ESCR	SR	ACHR	MAAE	PVIAC	PVNAC	EAFDAW	TCIDTP	RC	
Observes	P	P	P	P	P		P	P	P	P	P	

P = Party; S = Signatory; see Appendix for meaning of abbreviations.

Labor Force

Total Labor Force [25]

4.728 million

Labor Force by Occupation [26]

Services	38.3%
includes government	12
Industry and commerce	33.8
Agriculture, forestry, and fishing	19.2
Mining	2.3
Construction	6.4

Date of data: 1990

Unemployment Rate [27]

5.4% (1995 est.)

For sources, notes, and explanations, see Annotated Source Appendix, page 1061.

Production Sectors

Commercial Energy Production and Consumption

Data are shown in quadrillion (10^{15}) BTUs and percent for 1995
Values for hydroelectric, nuclear, geothermal, solar, and wind power refer to electrical generation.

Production [28]

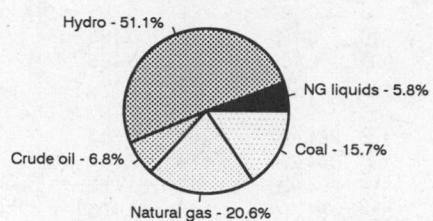

Hydro - 51.1%
NG liquids - 5.8%
Coal - 15.7%
Crude oil - 6.8%
Natural gas - 20.6%

Consumption [29]

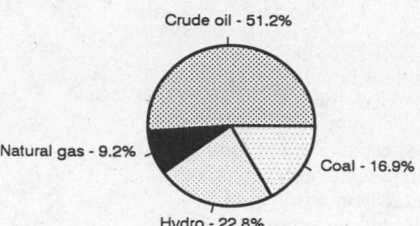

Crude oil - 51.2%
Natural gas - 9.2%
Coal - 16.9%
Hydro - 22.8%

Crude oil	0.022
Natural gas liquids	0.019
Dry natural gas	0.067
Coal	0.051
Net hydroelectric power	0.166
Total	0.325

Crude oil	0.373
Dry natural gas	0.067
Coal	0.123
Net hydroelectric power	0.166
Total	0.729

Telecommunications [30]

- 1.5 million (1994 est.) telephones; modern system based on extensive microwave radio relay facilities
- Domestic: extensive microwave radio relay links; domestic satellite system with 3 earth stations
- International: satellite earth stations - 2 Intelsat (Atlantic Ocean)
- Radio: Broadcast stations: AM 159, FM 0, shortwave 11
- Television: Broadcast stations: 131 Televisions: 2.85 million (1992 est.)

Transportation [31]

Railways: total: 6,782 km; broad gauge: 3,743 km 1.676-m gauge (1,653 km electrified); narrow gauge: 116 km 1.067-m gauge; 2,923 km 1.000-m gauge (40 km electrified) (1995)

Highways: total: 79,593 km; paved: 10,984 km; unpaved: 68,609 km (1991 est.)

Merchant marine: total: 37 ships (1,000 GRT or over) totaling 529,512 GRT/925,364 DWT; ships by type: bulk 11, cargo 8, chemical tanker 4, combination ore/oil 2, container 1, liquefied gas tanker 2, oil tanker 4, roll-on/roll-off cargo 3, vehicle carrier 2 (1995 est.)

Airports
Total:	344
With paved runways over 3,047 m:	5
With paved runways 2,438 to 3,047 m:	5
With paved runways 1,524 to 2,437 m:	17
With paved runways 914 to 1,523 m:	16
With paved runways under 914 m:	220

Top Agricultural Products [32]

Agriculture accounts for 7.4% of the GDP; produces wheat, corn, grapes, beans, sugar beets, potatoes, fruit; beef, poultry, wool; timber; 1991 fish catch of 6.6 million metric tons.

Top Mining Products [33]

Metric tons except as noted[e]	9/95[*]
Limestone (calcium carbonate) (000 tons)	6,300[r]
Sulfur	351,000
Talc	5,400[r]
Coal, bituminous and lignite (000 tons)	1,700[r]
Gas, natural, gross (mil. cu. meters)	4,150
Natrual gas liquids (000 42-gal. bls.)	2,650
Petroleum refinery products	48,500

Tourism [34]

	1990	1991	1992	1993	1994
Tourists	943	1,349	1,283	1,412	1,623
Tourism receipts	540	700	706	744	833
Tourism expenditures	426	409	535	560	639
Fare receipts	135	139	150	182	192
Fare expenditures	169	174	169	186	193

Travelers are in thousands, money in million U.S. dollars.

For sources, notes, and explanations, see Annotated Source Appendix, page 1061.

195

Manufacturing Sector

Manufacturing Summary [35]

	1987		1988		1989		1990		1991	
	$ bil.	%	$ bil.	%	$ bil.	%	$ bil.	%	$ bil.	%
Establishments or enterprises (number)	-	-	-	-	1,777	0.095	1,825	0.102	-	-
Total employment (000)	-	-	-	-	329	0.267	336	0.303	-	-
Production workers (000)	-	-	-	-	240	0.327	244	0.343	-	-
Output ($ bil.)	-	-	-	-	23	0.194	24	0.205	-	-
Value added ($ bil.)	-	-	-	-	10	0.198	10	0.198	-	-
Capital investment ($ mil.)	-	-	-	-	985	0.180	916	0.165	-	-
M & E investment ($ mil.)	-	-	-	-	760	0.195	715	0.170	-	-
Employees per establishment (number)	-	-	-	-	185	281.497	184	297.553	-	-
Production workers per establishment	-	-	-	-	135	344.234	134	337.011	-	-
Output per establishment ($ mil.)	-	-	-	-	13	204.920	13	201.293	-	-
Capital investment per estab. ($ mil.)	-	-	-	-	0.554	189.609	0.502	161.703	-	-
M & E per establishment ($ mil)	-	-	-	-	0.427	205.090	0.392	166.538	-	-
Payroll per employee ($)	-	-	-	-	5,362	47.995	5,922	42.689	-	-
Wages per production worker ($)	-	-	-	-	3,373	33.636	3,826	32.220	-	-
Hours per production worker (hours)	-	-	-	-	-	-	-	-	-	-
Output per employee ($)	-	-	-	-	69,768	72.796	70,108	67.649	-	-
Capital investment per employee ($)	-	-	-	-	2,998	67.357	2,729	54.344	-	-
M & E per employee ($)	-	-	-	-	2,311	72.857	2,132	55.969	-	-

Note: Columns headed % show percent of world total or ratio. Ratios closest to 100 are closest to world average. M & E stands for machinery & equipment.

Output in Manufacturing

	1987		1988		1989		1990		1991	
	$ bil.	%	$ bil.	%	$ bil.	%	$ bil.	%	$ bil.	%
3110 - Food products	-	-	-	-	3.922	17.05	4.072	16.97	-	-
3130 - Beverages	-	-	-	-	0.639	2.78	0.768	3.20	-	-
3140 - Tobacco	-	-	-	-	0.303	1.32	0.348	1.45	-	-
3210 - Textiles	-	-	-	-	0.679	2.95	0.685	2.85	-	-
3211 - Spinning, weaving, etc.	-	-	-	-	0.510	2.22	0.492	2.05	-	-
3220 - Wearing apparel	-	-	-	-	0.332	1.44	0.339	1.41	-	-
3230 - Leather and products	-	-	-	-	0.082	0.36	0.086	0.36	-	-
3240 - Footwear	-	-	-	-	0.244	1.06	0.253	1.05	-	-
3310 - Wood products	-	-	-	-	0.657	2.86	0.679	2.83	-	-
3320 - Furniture, fixtures	-	-	-	-	0.099	0.43	0.106	0.44	-	-
3410 - Paper and products	-	-	-	-	1.102	4.79	1.126	4.69	-	-
3411 - Pulp, paper, etc.	-	-	-	-	0.785	3.41	0.771	3.21	-	-
3420 - Printing, publishing	-	-	-	-	0.349	1.52	0.398	1.66	-	-
3510 - Industrial chemicals	-	-	-	-	0.459	2.00	0.463	1.93	-	-
3511 - Basic chemicals, excl fertilizers	-	-	-	-	0.216	0.94	0.198	0.83	-	-
3513 - Synthetic resins, etc.	-	-	-	-	0.094	0.41	0.100	0.42	-	-
3520 - Chemical products nec	-	-	-	-	1.032	4.49	1.151	4.80	-	-
3522 - Drugs and medicines	-	-	-	-	0.309	1.34	0.348	1.45	-	-
3530 - Petroleum refineries	-	-	-	-	1.549	6.73	1.861	7.75	-	-
3540 - Petroleum, coal products	-	-	-	-	0.198	0.86	0.151	0.63	-	-
3550 - Rubber products	-	-	-	-	0.144	0.63	0.147	0.61	-	-
3560 - Plastic products nec	-	-	-	-	0.308	1.34	0.419	1.75	-	-
3610 - Pottery, china, etc.	-	-	-	-	0.108	0.47	0.017	0.07	-	-
3620 - Glass and products	-	-	-	-	0.083	0.36	0.082	0.34	-	-
3690 - Nonmetal products nec	-	-	-	-	0.402	1.75	0.414	1.73	-	-
3710 - Iron and steel	-	-	-	-	0.697	3.03	0.637	2.65	-	-
3720 - Nonferrous metals	-	-	-	-	5.131	22.31	4.983	20.76	-	-
3810 - Metal products	-	-	-	-	0.758	3.30	0.864	3.60	-	-
3820 - Machinery nec	-	-	-	-	0.399	1.73	0.449	1.87	-	-
3830 - Electrical machinery	-	-	-	-	0.249	1.08	0.242	1.01	-	-
3832 - Radio, television, etc.	-	-	-	-	0.057	0.25	0.019	0.08	-	-
3840 - Transportation equipment	-	-	-	-	0.511	2.22	0.430	1.79	-	-
3841 - Shipbuilding, repair	-	-	-	-	0.054	0.23	0.077	0.32	-	-
3843 - Motor vehicles	-	-	-	-	0.432	1.88	0.304	1.27	-	-
3850 - Professional goods	-	-	-	-	0.015	0.07	0.018	0.08	-	-
3900 - Industries nec	-	-	-	-	0.024	0.10	0.026	0.11	-	-

Note: Codes are International Standard Industry codes (ISIC). Percentages are % of total Output. [f]: Factor Prices; [p]: Producer Prices.

For sources, notes, and explanations, see Annotated Source Appendix, page 1061.

Finance, Economics, and Trade

Economic Indicators [36]

- **National product**: GDP—purchasing power parity—$113.2 billion (1995 est.)
- **National product real growth rate**: 8.5% (1995 est.)
- **National product per capita**: $8,000 (1995 est.)
- **Inflation rate (consumer prices)**: 8.1% (1995 est.)
- **External debt**: $21.1 billion (1995 est.)

Balance of Payments Summary [37]

Values in millions of dollars.

	1989	1990	1991	1992	1993
Exports of goods (f.o.b.)	8,080	8,310	8,928	9,986	9,202
Imports of goods (f.o.b.)	-6,502	-7,037	-7,354	-9,236	-10,181
Trade balance	1,578	1,273	1,574	750	-979
Services - debits	-4,337	-4,381	-4,494	-4,797	-4,631
Services - credits	1,777	2,260	2,593	2,873	3,131
Private transfers (net)	58	54	40	74	61
Government transfers (net)	157	146	299	357	325
Long-term capital (net)	666	1,633	340	942	1,673
Short-term capital (net)	612	1,415	448	1,942	1,091
Errors and omissions	-71	-32	400	359	-95
Overall balance	440	2,368	1,240	2,500	576

Exchange Rates [38]

Currency: **Chilean peso.**
Symbol: **Ch$.**

Data are currency units per $1.

December 1995	408.64
1995	396.78
1994	420.08
1993	404.35
1992	362.59
1991	349.37

Imports and Exports

Top Import Origins [39]

$14.3 billion (f.o.b., 1995 est.) Data are for 1995 est.

Origins	%
Latin America	26
US	25
EU	18
Asia	16

Top Export Destinations [40]

$15.9 billion (f.o.b., 1995 est.) Data are for 1995 est.

Destinations	%
Asia	34
EU	25
Latin America	20
US	15

Foreign Aid [41]

Recipient: ODA, $62 million (1993).

Import and Export Commodities [42]

Import Commodities

Capital goods 25.2%
Spare parts 24.8%
Raw materials 15.4%
Petroleum 10%
Foodstuffs 5.7%

Export Commodities

Copper 41%
Other metals and minerals 8.7%
Wood products 7.1%
Fish and fishmeal 9.8%
Fruits 8.4%

China

Geography [1]

Total area:
9,596,960 sq km   3,705,407 sq mi
Land area:
9,326,410 sq km   3,600,947 sq mi
Comparative area:
Slightly larger than the US
Land boundaries:
Total 22,143.34 km, Afghanistan 76 km, Bhutan 470 km, Myanmar 2,185 km, Hong Kong 30 km, India 3,380 km, Kazakhstan 1,533 km, North Korea 1,416 km, Kyrgyzstan 858 km, Laos 423 km, Macau 0.34 km, Mongolia 4,673 km, Nepal 1,236 km, Pakistan 523 km, Russia (Northeast) 3,605 km, Russia (Northwest) 40 km, Tajikistan 414 km, Vietnam 1,281 km
Coastline:
14,500 km
Climate:
Extremely diverse; tropical in South to subarctic in North
Terrain:
Mostly mountains, high plateaus, deserts in West; plains, deltas, and hills in East
Natural resources:
Coal, iron ore, petroleum, mercury, tin, tungsten, antimony, manganese, molybdenum, vanadium, magnetite, aluminum, lead, zinc, uranium, hydropower potential (world's largest)
Land use:
Arable land: 10%
Permanent crops: 0%
Meadows and pastures: 31%
Forest and woodland: 14%
Other: 45%

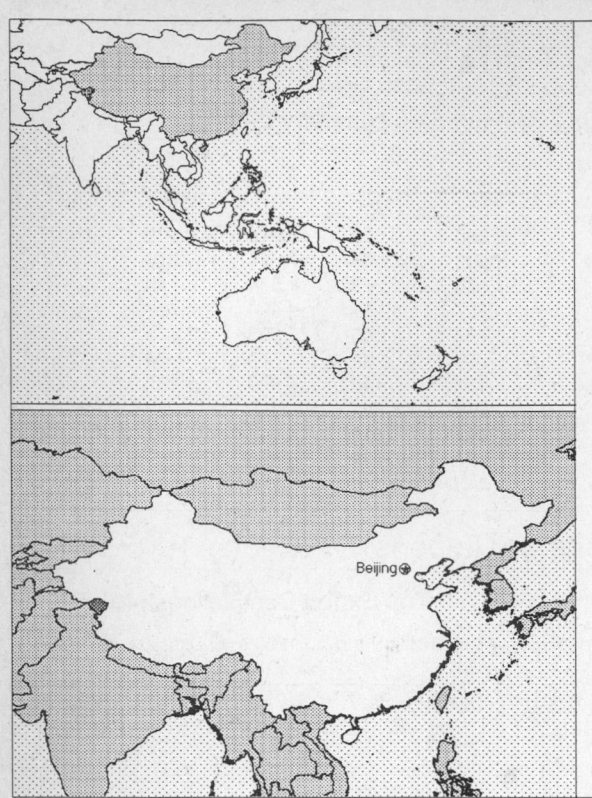

Beijing

Demographics [2]

	1970	1980	1990	1995[1]	1996	2000	2010	2020	2030
Population (mil.)	820	985	1,134	1,198	1,210	1,253	1,340	1,413	1,444
Population density (persons per sq. mi.)	228	273	315	333	336	348	372	392	401
(persons per sq. km.)	88	106	122	128	130	134	144	152	155
Net migration rate (per 1,000 population)	NA	NA	-0.4	-0.3	-0.3	-0.3	-0.2	-0.1	0.0
Births (mil.)	NA	NA	NA	NA	21	NA	NA	NA	NA
Deaths (mil.)	NA	NA	NA	NA	8	NA	NA	NA	NA
Life expectancy - males (yrs.)	NA	NA	66.5	68.0	68.3	69.5	72.0	74.1	75.7
Life expectancy - females (yrs.)	NA	NA	68.2	70.6	71.1	72.9	76.6	79.6	81.4
Birth rate (per 1,000)	NA	NA	21.1	17.4	17.0	15.0	13.6	11.6	10.1
Death rate (per 1,000)	NA	NA	7.5	7.0	6.9	6.8	7.0	7.7	9.3
Women of reproductive age (15-49 yrs.)	NA	NA	306	328	330	343	361	335	315
of which are currently married (mil.)	NA	NA	191	NA	231	249	260	NA	NA
Fertility rate	NA	NA	2.2	1.8	1.8	1.8	1.8	1.8	1.7

Except as noted, values for vital statistics are in thousands; life expectancy is in years.

Health

Health Indicators [3]

% of population with access to	
safe water (1990-95)	67
adequate sanitation (1990-95)	24
health services (1985-95)	92
% of 1-year-olds immunized (1990-94) against	
TB (tuberculosis)	94
DPT (diphtheria, pertussis, tetanus)	93
polio	94
measles	89
% of contraceptive prevalence (1980-94)	83
ORT use rate (1990-94)	84

Health Expenditures [4]

Total health expenditure, 1990 (official exchange rate)	
Millions of dollars	12,969
Dollars per capita	11
Health expenditures as a percentage of GDP	
Total	3.5
Public sector	2.1
Private sector	1.4
Development assistance for health	
Total aid flows (millions of dollars)[1]	77
Aid flows per capita (dollars)	0.1
Aid flows as a percentage of total health expenditure	0.6

For sources, notes, and explanations, see Annotated Source Appendix, page 1061.

Human Factors

Women and Children [5]

% of pregnant women immunized (tetanus 1990-94)	3
% of births attended by trained health personnel (1983-94)	94
Maternal mortality rate (1980-92)	95
Under-5 mortality rate (1994)	43
% under-5 moderately/severely underweight (1980-1994)	17

Burden of Disease [6]

Population per physician (1993)	1,063.00
Population per nurse (1993)	1,490.00
Population per hospital bed (1993)	612.00
AIDS cases per 100,000 people (1994)	*
Malaria cases per 100,000 people (1992)	6

Ethnic Division [7]

Other (Zhuang, Uygur, Hui, Yi, Tibetan, Miao, Manchu, Mongol, Buyi, Korean, and other nationalities).

Han Chinese	91.9%
Other	8.1%

Religion [8]

Officially atheist, but traditionally pragmatic and eclectic.

Daoism (Taoism), Buddhism, or Muslim	2%-3%
Christian	1%

(est.)

Major Languages [9]

Standard Chinese or Mandarin (Putonghua, based on the Beijing dialect), Yue (Cantonese), Wu (Shanghaiese), Minbei (Fuzhou), Minnan (Hokkien-Taiwanese), Xiang, Gan, Hakka dialects, minority languages.

Education

Public Education Expenditures [10]

	1980	1985	1990	1992	1993	1994
Million (Yuan)						
Total education expenditure	11,319	22,489	43,386	53,874	64,439	111,769
as percent of GNP	2.5	2.6	2.3	2.0	1.9	2.6
as percent of total govt. expend.	9.3	12.2	12.8	12.2	12.2	NA
Current education expenditure	10,263	19,770	40,423	48,976	59,104	100,128
as percent of GNP	2.3	2.3	2.2	1.8	1.7	2.3
as percent of current govt. expend.	NA	NA	NA	13.5	NA	NA
Capital expenditure	1,056	2,719	2,963	4,898	5,335	11,641

Educational Attainment [11]

Age group (1990)	25+
Total population	571,589,800
Highest level attained (%)	
No schooling	29.3
First level	
Not completed	34.3
Completed	NA
Entered second level	
S-1	34.4
S-2	NA
Postsecondary	2.0

Illiteracy [12]

In thousands and percent[1]	1990	1995	2000
Illiterate population (15+ yrs.)	181,609	166,173	143,458
Illiteracy rate - total pop. (%)	22.1	18.9	15.3
Illiteracy rate - males (%)	12.9	10.3	7.9
Illiteracy rate - females (%)	31.9	0.3	23.0

Libraries [13]

	Admin. Units	Svc. Pts.	Vols. (000)	Shelving (meters)	Vols. Added	Reg. Users
National (1989)	1	NA	13,768	290,000	470,790	166,861
Nonspecialized	NA	NA	NA	NA	NA	NA
Public (1993)	2,579	NA	314,100	8 mil	8 mil	7 mil
Higher ed. (1993)[8]	1,075	5,000[e]	406,471	NA	10 mil	4 mil
School	NA	NA	NA	NA	NA	NA

Daily Newspapers [14]

	1980	1985	1990	1994
Number of papers	50	70	44	38
Circ. (000)	34,375	39,000[e]	48,000[e]	27,790

Culture [15]

Cinema (seats per 1,000)	NA
Annual attendance per person	12.3
Gross box office receipts (mil. Yuan)	2,365
Museums (reporting)	NA
Visitors (000)	NA
Annual receipts (000 Yuan)	NA

Science and Technology

Scientific/Technical Forces [16]

Scientists/engineers	418,500
Number female	NA
Technicians[2]	224,000
Number female	NA
Total[2]	442,500

R&D Expenditures [17]

	Yuan (000) 1993
Total expenditure	19,600,000
Capital expenditure	NA
Current expenditure	NA
Percent current	NA

U.S. Patents Issued [18]

Values show patents issued to citizens of the country by the U.S. Patents Office.

	1993	1994	1995
Number of patents	53	48	63

For sources, notes, and explanations, see Annotated Source Appendix, page 1061.

Government and Law

Organization of Government [19]

Long-form name:
People's Republic of China
Type:
Communist state
Independence:
221 BC (unification under the Qin or Ch'in Dynasty 221 BC; Qing or Ch'ing Dynasty replaced by the Republic on 12 February 1912; People's Republic established 1 October 1949)
National holiday:
National Day, 1 October (1949)
Constitution:
Most recent promulgated 12/4/1982
Legal system:
A complex amalgam of custom and statute, largely criminal law; rudimentary civil code in effect since 1 January 1987; new legal codes in effect since 1 January 1980; continuing efforts to improve civil, administrative, criminal, and commercial law
Executive branch:
President; Vice President; Premier; 6 Vice Premiers; State Council
Legislative branch:
Unicameral: National People's Congress
Judicial branch:
Supreme People's Court

Elections [20]

National People's Congress. Last held March 1993 (next to be held March 1998); results - CCP is the only party but there are also independents; seats - (2,977 total) (elected at county or xian level).

Government Expenditures [21]

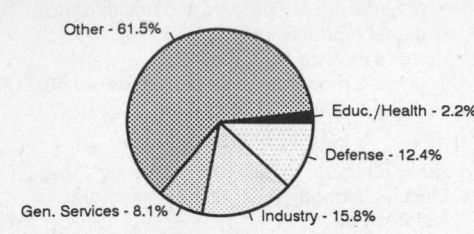

(% distribution). Expend. for CY94: 440.1 (Yuan bil.)

Crime [23]

	1994
Crime volume	
Cases known to police	1,660,734
Attempts (percent)	NA
Percent cases solved	78.20
Crimes per 100,000 persons	127.75
Persons responsible for offenses	
Total number offenders	NA
Percent female	3.50
Percent juvenile (14-18 yrs.)	12.70
Percent foreigners	0.02

Military Expenditures and Arms Transfers [22]

	1990	1991	1992	1993	1994
Military expenditures					
Current dollars (mil.)[r]	47,270	46,150	49,100	51,620	52,840
1994 constant dollars (mil.)[r]	52,610	49,470	51,210	52,680	52,840
Armed forces (000)	3,500	3,200	3,160	3,031	2,930
Gross national product (GNP)					
Current dollars (bil.)[e]	1,351	1,518	1,770	2,047	2,214
1994 constant dollars (bil.)[e]	1,504	1,628	1,846	2,089	2,214
Central government expenditures (CGE)					
1994 constant dollars (mil.)	279,400	285,900[e]	303,800[e]	323,900[e]	294,000
People (mil.)	1,136.6	1,151.4	1,164.8	1,177.6	1,190.4
Military expenditure as % of GNP	3.5	3.0	2.8	2.5	2.4
Military expenditure as % of CGE	18.8	17.3	16.9	16.3	18.0
Military expenditure per capita (1994 $)	46	43	44	45	44
Armed forces per 1,000 people (soldiers)	3.1	2.8	2.7	2.6	2.5
GNP per capita (1994 $)	1,323	1,413	1,585	1,774	1,860
Arms imports[6]					
Current dollars (mil.)	200	200	1,200	500	130
1994 constant dollars (mil.)	223	214	1,251	510	130
Arms exports[6]					
Current dollars (mil.)	1,500	1,400	925	1,100	800
1994 constant dollars (mil.)	1,669	1,501	965	1,123	800
Total imports[7]					
Current dollars (mil.)	53,340	63,790	80,580	103,100	114,600
1994 constant dollars (mil.)	59,370	68,380	84,030	105,200	114,600
Total exports[7]					
Current dollars (mil.)	61,270	70,450	80,520	90,970	119,800
1994 constant dollars (mil.)	68,190	75,510	83,960	92,840	119,800
Arms as percent of total imports[8]	0.4	0.3	1.5	0.5	0.1
Arms as percent of total exports[8]	2.4	2.0	1.1	1.2	0.7

Human Rights [24]

	SSTS	FL	FAPRO	PPCG	APROBC	TPW	PCPTW	STPEP	PHRFF	PRW	ASST	AFL
Observes				P		P	P			P	P	
	EAFRD	CPR	ESCR	SR	ACHR	MAAE	PVIAC	PVNAC	EAFDAW	TCIDTP	RC	
Observes	P			P			P	P	P	P	P	

P = Party; S = Signatory; see Appendix for meaning of abbreviations.

Labor Force

Total Labor Force [25]

583.6 million (1991)

Labor Force by Occupation [26]

Agriculture and forestry	60%
Industry and commerce	25
Construction and mining	5
Social services	5
Other	5

Date of data: 1990 est.

Unemployment Rate [27]

5.2% in urban areas (1995 est.); substantial underemployment

For sources, notes, and explanations, see Annotated Source Appendix, page 1061.

Production Sectors

Commercial Energy Production and Consumption

Data are shown in quadrillion (10^{15}) BTUs and percent for 1995
Values for hydroelectric, nuclear, geothermal, solar, and wind power refer to electrical generation.

Production [28]

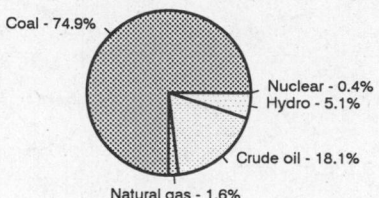

Coal - 74.9%
Nuclear - 0.4%
Hydro - 5.1%
Crude oil - 18.1%
Natural gas - 1.6%

Consumption [29]

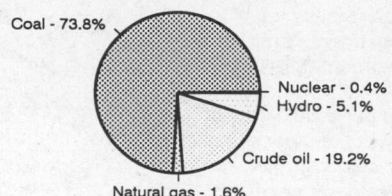

Coal - 73.8%
Nuclear - 0.4%
Hydro - 5.1%
Crude oil - 19.2%
Natural gas - 1.6%

Crude oil	6.416		Crude oil	6.892
Dry natural gas	0.560		Dry natural gas	0.560
Coal	26.57		Coal	26.423
Net hydroelectric power	1.820		Net hydroelectric power	1.820
Net nuclear power	0.126		Net nuclear power	0.126
Total	35.492		Total	35.821

Telecommunications [30]

- 20 million (1994 est.) telephones; domestic and international services are increasingly available for private use; unevenly distributed domestic system serves principal cities, industrial centers, and most townships
- Domestic: telephone lines are being expanded; interprovincial fiber-optic trunk lines and cellular telephone systems have been installed; a domestic satellite system with 55 earth stations is in place
- International: satellite earth stations - 5 Intelsat (4 Pacific Ocean and 1 Indian Ocean), 1 Intersputnik (Indian Ocean Region) and 1 Inmarsat (Pacific and Indian Ocean Regions); several international fiber-optic links to Japan, South Korea, and Hong Kong
- Radio: Broadcast stations: AM 274, FM NA, shortwave 0 Radios: 216.5 million (1992 est.)
- Television: Broadcast stations: 202 (repeaters 2,050) Televisions: 75 million

Transportation [31]

Railways: total: 58,399 km; standard gauge: 54,799 km 1.435-m gauge (7,174 km electrified; more than 11,000 km double track); narrow gauge: 3,600 km 0.762-m gauge local industrial lines (1995)

Highways: total: 1.029 million km; paved: 170,000 km; unpaved: 859,000 km (1990 est.)

Merchant marine: total: 1,700 ships (1,000 GRT or over) totaling 16,663,260 GRT/25,026,090 DWT; ships by type: barge carrier 2, bulk 316, cargo 876, chemical tanker 15, combination bulk 11, container 103, liquefied gas tanker 4, multifunction large-load carrier 3, oil tanker 227, passenger 24, passenger-cargo 28, refrigerated cargo 22, roll-on/roll-off cargo 24, short-sea passenger 45

Airports

Total:	204
With paved runways over 3,047 m:	17
With paved runways 2,438 to 3,047 m:	69
With paved runways 1,524 to 2,437 m:	89
With paved runways 914 to 1,523 m:	9
With paved runways under 914 m:	7

Top Agricultural Products [32]

Agriculture accounts for 19% of the GDP; produces rice, potatoes, sorghum, peanuts, tea, millet, barley, cotton, other fibers, oilseed; pork and other livestock products; fish.

Top Mining Products [33]

Metric tons except as noted	5/10/95[*]
Copper, mine output, Cu content	350,000
Iron ore, gross weight (000 tons)	240,000
Lead, mine output, Pb content	340,000
Zinc, mine output, Zn content	780,000
Asbestos	240,000
Boron, mine B2O3 equivalent	120,000
Cement, hydraulic (000 tons)	400,000
Graphite	360,000
Coal, all types (000 tons)	1,210,000
Coke, all types (000 tons)	90,000

Tourism [34]

	1990	1991	1992	1993	1994
Visitors[24]	27,462	33,350	38,115	41,527	43,684
Tourists[25]	1,747	2,710	4,006	4,656	5,182
Tourism receipts	2,218	2,845	3,948	4,683	7,323
Tourism expenditures	470	511	2,512	2,797	3,036
Fare receipts	480	494	417	294	254

Travelers are in thousands, money in million U.S. dollars.

For sources, notes, and explanations, see Annotated Source Appendix, page 1061.

201

Manufacturing Sector

Manufacturing Summary [35]

	1987 $ bil.	1987 %	1988 $ bil.	1988 %	1989 $ bil.	1989 %	1990 $ bil.	1990 %	1991 $ bil.	1991 %
Establishments or enterprises (number)	438,835	18.707	447,437	21.314	453,269	24.197	451,593	25.210	-	-
Total employment (000)	-	-	-	-	-	-	-	-	-	-
Production workers (000)	-	-	-	-	-	-	-	-	-	-
Output ($ bil.)	267	2.623	319	2.733	334	2.835			-	-
Value added ($ bil.)	90	2.081	-	-	-	-	-	-	-	-
Capital investment ($ mil.)	-	-	-	-	-	-	-	-	-	-
M & E investment ($ mil.)	-	-	-	-	-	-	-	-	-	-
Employees per establishment (number)	-	-	-	-	-	-	-	-	-	-
Production workers per establishment	-	-	-	-	-	-	-	-	-	-
Output per establishment ($ mil.)	0.608	14.024	0.712	12.824	0.738	11.715	-	-	-	-
Capital investment per estab. ($ mil.)	-	-	-	-	-	-	-	-	-	-
M & E per establishment ($ mil)	-	-	-	-	-	-	-	-	-	-
Payroll per employee ($)	-	-	-	-	-	-	-	-	-	-
Wages per production worker ($)	-	-	-	-	-	-	-	-	-	-
Hours per production worker (hours)	-	-	-	-	-	-	-	-	-	-
Output per employee ($)	-	-	-	-	-	-	-	-	-	-
Capital investment per employee ($)	-	-	-	-	-	-	-	-	-	-
M & E per employee ($)	-	-	-	-	-	-	-	-	-	-

Note: Columns headed % show percent of world total or ratio. Ratios closest to 100 are closest to world average. M & E stands for machinery & equipment.

Output in Manufacturing

	1987 $ bil.	1987 %	1988 $ bil.	1988 %	1989 $ bil.	1989 %	1990 $ bil.	1990 %	1991 $ bil.	1991 %
3110 - Food products	20.000	7.49	23.000	7.21	23.000	6.89	-	-	-	-
3130 - Beverages	4.859	1.82	5.913	1.85	5.650	1.69	-	-	-	-
3140 - Tobacco	5.558	2.08	6.588	2.07	7.082	2.12	-	-	-	-
3210 - Textiles	37.000	13.86	41.000	12.85	42.000	12.57	-	-	-	-
3230 - Leather and products	2.803	1.05	3.282	1.03	3.531	1.06	-	-	-	-
3310 - Wood products	1.787	0.67	2.075	0.65	2.166	0.65	-	-	-	-
3320 - Furniture, fixtures	1.718	0.64	2.077	0.65	2.110	0.63	-	-	-	-
3410 - Paper and products	5.265	1.97	6.258	1.96	6.751	2.02	-	-	-	-
3420 - Printing, publishing	2.979	1.12	3.522	1.10	3.760	1.13	-	-	-	-
3510 - Industrial chemicals	19.000	7.12	22.000	6.90	24.000	7.19	-	-	-	-
3511 - Basic chemicals, excl fertilizers	2.869	1.07	3.329	1.04	3.668	1.10	-	-	-	-
3520 - Chemical products nec	9.722	3.64	12.000	3.76	12.000	3.59	-	-	-	-
3522 - Drugs and medicines	5.748	2.15	7.414	2.32	7.585	2.27	-	-	-	-
3530 - Petroleum refineries	6.960	2.61	7.772	2.44	8.289	2.48	-	-	-	-
3540 - Petroleum, coal products	0.293	0.11	0.328	0.10	0.339	0.10	-	-	-	-
3550 - Rubber products	4.549	1.70	5.312	1.67	5.823	1.74	-	-	-	-
3560 - Plastic products nec	5.597	2.10	7.210	2.26	7.551	2.26	-	-	-	-
3610 - Pottery, china, etc.	1.112	0.42	1.296	0.41	1.444	0.43	-	-	-	-
3620 - Glass and products	1.926	0.72	2.255	0.71	2.321	0.69	-	-	-	-
3690 - Nonmetal products nec	9.847	3.69	11.000	3.45	12.000	3.59	-	-	-	-
3710 - Iron and steel	14.000	5.24	16.000	5.02	16.000	4.79	-	-	-	-
3720 - Nonferrous metals	6.032	2.26	6.421	2.01	6.932	2.08	-	-	-	-
3810 - Metal products	8.711	3.26	10.000	3.13	11.000	3.29	-	-	-	-
3820 - Machinery nec	31.000	11.61	37.000	11.60	38.000	11.38	-	-	-	-
3825 - Office, computing machinery	0.183	0.07	0.259	0.08	0.244	0.07	-	-	-	-
3830 - Electrical machinery	25.000	9.36	33.000	10.34	35.000	10.48	-	-	-	-
3832 - Radio, television, etc.	5.492	2.06	7.928	2.49	7.704	2.31	-	-	-	-
3840 - Transportation equipment	11.000	4.12	14.000	4.39	15.000	4.49	-	-	-	-
3841 - Shipbuilding, repair	1.085	0.41	1.148	0.36	1.271	0.38	-	-	-	-
3843 - Motor vehicles	5.649	2.12	7.529	2.36	7.863	2.35	-	-	-	-
3850 - Professional goods	2.299	0.86	2.766	0.87	2.869	0.86	-	-	-	-
3900 - Industries nec	7.444	2.79	9.524	2.99	11.000	3.29	-	-	-	-

Note: Codes are International Standard Industry codes (ISIC). Percentages are % of total Output. [f]: Factor Prices; [p]: Producer Prices.

Finance, Economics, and Trade

Economic Indicators [36]

- **National product**: GDP—purchasing power parity—$3.5 trillion (1995 estimate as extrapolated from World Bank estimate with use of official Chinese growth figures for 1993-95; the result may overstate China's GDP by as much as 25%)

- **National product real growth rate**: 10.3% (1995 est.)

- **National product per capita**: $2,900 (1995 est.)

- **Inflation rate (consumer prices)**: 10.1% (December 1995 over December 1994)

- **External debt**: $92 billion (1994 est.)

Balance of Payments Summary [37]

Values in millions of dollars.

	1989	1990	1991	1992	1993
Exports of goods (f.o.b.)	43,220	51,519	58,919	69,568	75,659
Imports of goods (f.o.b.)	-48,840	-42,354	-50,176	-64,385	-86,313
Trade balance	-5,620	9,165	8,743	5,183	-10,654
Services - debits	-5,575	-6,314	-7,000	-14,781	-17,710
Services - credits	6,497	8,872	10,698	14,884	-15,583
Private transfers (net)	238	222	444	804	883
Government transfers (net)	143	52	387	351	289
Long-term capital (net)	5,241	6,453	7,670	656	27,412
Short-term capital (net)	-1,518	-3,198	362	-906	-3,938
Errors and omissions	115	-3,205	-6,767	-8,211	-10,096
Overall balance	-479	12,047	14,537	-2,060	1,769

Exchange Rates [38]

Currency: **yuan.**
Symbol: **¥.**

Data are currency units per $1.

January 1996	8.3186
1995	8.3514
1994	8.6187
1993	5.7620
1992	5.5146
1991	5.3234

Imports and Exports

Top Import Origins [39]

$132.1 billion (c.i.f., 1995) Data are for 1994.

Origins	%
Japan	NA
Taiwan	NA
US	NA
Hong Kong	NA
South Korea	NA
Germany	NA

Top Export Destinations [40]

$148.8 billion (f.o.b., 1995) Data are for 1994.

Destinations	%
Hong Kong	NA
Japan	NA
US	NA
Germany	NA
South Korea	NA
Singapore	NA

Foreign Aid [41]

	U.S. $	
Donor:		
To less developed countries (1970-89)	NA	
Recipient:		
ODA (1993)	1.977	billion

Import and Export Commodities [42]

Import Commodities	Export Commodities
Industrial machinery	Garments
Textiles	Textiles
Plastics	Footwear
Telecommunications equipment	Toys
Steel bars	Machinery and equipment
Aircraft	

For sources, notes, and explanations, see Annotated Source Appendix, page 1061.

203

Colombia

Geography [1]

Total area:
1,138,910 sq km   439,736 sq mi
Land area:
1,038,700 sq km   401,044 sq mi
Comparative area:
Slightly less than three times the size of Montana
Note: Includes Isla de Malpelo, Roncador Cay, Serrana Bank, and
Serranilla Bank
Land boundaries:
Total 7,408 km, Brazil 1,643 km, Ecuador 590 km, Panama 225 km, Peru 2,900 km,
Venezuela 2,050 km
Coastline:
3,208 km (Caribbean Sea 1,760 km, North Pacific Ocean 1,448 km)
Climate:
Tropical along coast and eastern plains; cooler in highlands
Terrain:
Flat coastal lowlands, central highlands, high Andes Mountains, eastern
lowland plains
Natural resources:
Petroleum, natural gas, coal, iron ore, nickel, gold, copper, emeralds
Land use:
Arable land: 4%
Permanent crops: 2%
Meadows and pastures: 29%
Forest and woodland: 49%
Other: 16%

Demographics [2]

	1970	1980	1990	1995[1]	1996	2000	2010	2020	2030
Population	21,430	26,580	32,983	36,200	36,813	39,172	44,504	49,266	52,895
Population density (persons per sq. mi.)	53	66	82	90	92	98	111	123	132
(persons per sq. km.)	21	26	32	35	35	38	43	47	51
Net migration rate (per 1,000 population)	NA	-1.0	-0.4	-0.2	-0.1	0.0	0.0	0.0	0.0
Births	NA	NA	NA	NA	786	NA	NA	NA	NA
Deaths	NA	NA	NA	NA	614	NA	NA	NA	NA
Life expectancy - males	NA	62.4	68.0	69.7	70.0	71.2	73.7	75.7	77.2
Life expectancy - females	NA	66.6	73.3	75.4	75.7	77.2	80.1	82.3	83.8
Birth rate (per 1,000)	NA	28.8	25.5	21.9	21.3	19.0	16.2	14.2	12.3
Death rate (per 1,000)	NA	6.8	5.0	4.7	4.7	4.6	4.8	5.4	6.8
Women of reproductive age (15-49 yrs.)	NA	6,694	9,022	10,019	10,209	10,991	12,467	12,748	12,778
of which are currently married	NA	NA	4,827	NA	5,699	6,197	7,129	NA	NA
Fertility rate	NA	3.6	2.8	2.4	2.4	2.2	1.9	1.9	1.8

Except as noted, values for vital statistics are in thousands; life expectancy is in years.

Health

Health Indicators [3]

% of population with access to	
safe water (1990-95)	87
adequate sanitation (1990-95)	63
health services (1985-95)	60
% of 1-year-olds immunized (1990-94) against	
TB (tuberculosis)	99
DPT (diphtheria, pertussis, tetanus)	91
polio	95
measles	87
% of contraceptive prevalence (1980-94)	66
ORT use rate (1990-94)	40

Health Expenditures [4]

Total health expenditure, 1990 (official exchange rate)	
Millions of dollars	1,604
Dollars per capita	50
Health expenditures as a percentage of GDP	
Total	4.0
Public sector	1.8
Private sector	2.2
Development assistance for health	
Total aid flows (millions of dollars)[1]	26
Aid flows per capita (dollars)	0.8
Aid flows as a percentage of total health expenditure	1.6

For sources, notes, and explanations, see Annotated Source Appendix, page 1061.

Human Factors

Women and Children [5]

% of pregnant women immunized (tetanus 1990-94)	52
% of births attended by trained health personnel (1983-94)	81
Maternal mortality rate (1980-92)	200
Under-5 mortality rate (1994)	19
% under-5 moderately/severely underweight (1980-1994)	10

Burden of Disease [6]

Population per physician (1986)	1,198.58
Population per nurse (1984)	676.71
Population per hospital bed (1990)	733.43
AIDS cases per 100,000 people (1994)	3.3
Malaria cases per 100,000 people (1992)	551

Ethnic Division [7]

Mestizo	58%
White	20%
Mulatto	14%
Black	4%
Mixed black-Indian	3%
Indian	1%

Religion [8]

Roman Catholic	95%
Other	5%

Major Languages [9]

Spanish.

Education

Public Education Expenditures [10]

Million or Trillion (T) (Peso)[14]	1980	1985	1990	1992	1993	1994
Total education expenditure	29,240	136,570	526,686	996,567	1.40 T	1.93 T
as percent of GNP	1.9	2.9	2.8	3.1	3.5	3.7
as percent of total govt. expend.	14.3	NA	12.4	NA	12.3	12.9
Current education expenditure	27,286	127,908	NA	NA	1.23 T	1.58 T
as percent of GNP	1.7	2.7	NA	NA	3.0	3.0
as percent of current govt. expend.	19.9	NA	NA	NA	19.2	10.6
Capital expenditure	1,954	8,662	NA	NA	164,515	348,028

Educational Attainment [11]

Illiteracy [12]

In thousands and percent[1]	1990	1995	2000
Illiterate population (15+ yrs.)	2,164	2,046	1,944
Illiteracy rate - total pop. (%)	10.0	8.4	7.1
Illiteracy rate - males (%)	9.8	8.4	7.3
Illiteracy rate - females (%)	10.1	8.3	6.9

Libraries [13]

	Admin. Units	Svc. Pts.	Vols. (000)	Shelving (meters)	Vols. Added	Reg. Users
National (1993)	1	1	463	17,534	12,636	3,676
Nonspecialized	NA	NA	NA	NA	NA	NA
Public (1993)	1,378	NA	NA	NA	NA	NA
Higher ed.	NA	NA	NA	NA	NA	NA
School	NA	NA	NA	NA	NA	NA

Daily Newspapers [14]

	1980	1985	1990	1994
Number of papers	36	46[e]	45	46
Circ. (000)	1,400[e]	1,800[e]	2,000[e]	2,200[e]

Culture [15]

Cinema (seats per 1,000)	7.5
Annual attendance per person	1.3
Gross box office receipts (mil. Peso)	12,078
Museums (reporting)	NA
Visitors (000)	NA
Annual receipts (000 Peso)	NA

Science and Technology

Scientific/Technical Forces [16]

Scientists/engineers	1,083
Number female	NA
Technicians	1,024
Number female	NA
Total[3]	2,107

R&D Expenditures [17]

	Peso (000) 1982
Total expenditure[1]	2,754,273
Capital expenditure	NA
Current expenditure	NA
Percent current	NA

U.S. Patents Issued [18]

Values show patents issued to citizens of the country by the U.S. Patents Office.

	1993	1994	1995
Number of patents	6	5	5

For sources, notes, and explanations, see Annotated Source Appendix, page 1061.

205

Government and Law

Organization of Government [19]

Long-form name:
Republic of Colombia
Type:
Republic; executive branch dominates government structure
Independence:
20 July 1810 (from Spain)
National holiday:
Independence Day, 20 July (1810)
Constitution:
5 July 1991
Legal system:
Based on Spanish law; a new criminal code modeled after US procedures was enacted in 1992-93; judicial review of executive and legislative acts; accepts compulsory ICJ jurisdiction, with reservations
Executive branch:
President; Vice President; Cabinet
Legislative branch:
Bicameral Congress (Congreso): Senate (Senado) and House of Representatives (Camara de Representantes)
Judicial branch:
Supreme Court of Justice (Corte Suprema de Justical) - criminal law; Council of State - administrative law; Constitutional Court

Elections [20]

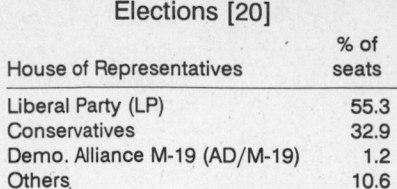

House of Representatives	% of seats
Liberal Party (LP)	55.3
Conservatives	32.9
Demo. Alliance M-19 (AD/M-19)	1.2
Others	10.6

Government Expenditures [21]

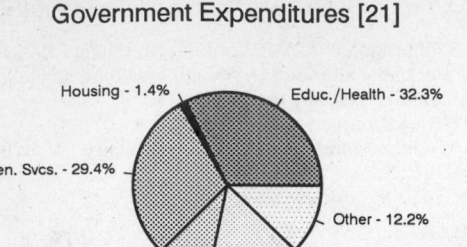

Housing - 1.4%
Educ./Health - 32.3%
Gen. Svcs. - 29.4%
Other - 12.2%
Defense - 8.7%
Industry - 16.0%

(% distribution). Expend. for CY93: 6,309.4 (Peso bil.)

Crime [23]

Military Expenditures and Arms Transfers [22]

	1990	1991	1992	1993	1994
Military expenditures					
Current dollars (mil.)	1,081[e]	1,294[e]	1,214[e]	1,484[e]	1,190
1994 constant dollars (mil.)	1,203[e]	1,387[e]	1,265[e]	1,515[e]	1,190
Armed forces (000)	110	110	139	139	146
Gross national product (GNP)					
Current dollars (mil.)	46,030	49,170	53,130	56,370	62,380
1994 constant dollars (mil.)	51,230	52,710	55,400	57,530	62,380
Central government expenditures (CGE)					
1994 constant dollars (mil.)	6,002[e]	6,371	9,072	9,590	NA
People (mil.)	33.0	33.6	34.3	34.9	35.6
Military expenditure as % of GNP	2.3	2.6	2.3	2.6	1.9
Military expenditure as % of CGE	20.0	21.8	13.9	15.8	NA
Military expenditure per capita (1994 $)	36	41	37	43	33
Armed forces per 1,000 people (soldiers)	3.3	3.3	4.1	4.0	4.1
GNP per capita (1994 $)	1,553	1,567	1,615	1,646	1,753
Arms imports[6]					
Current dollars (mil.)	90	220	120	40	30
1994 constant dollars (mil.)	100	236	125	41	30
Arms exports[6]					
Current dollars (mil.)	0	0	0	0	0
1994 constant dollars (mil.)	0	0	0	0	0
Total imports[7]					
Current dollars (mil.)	5,590	4,906	6,516	9,832	11,880
1994 constant dollars (mil.)	6,221	5,259	6,795	10,030	11,880
Total exports[7]					
Current dollars (mil.)	6,766	7,232	6,917	7,116	8,399
1994 constant dollars (mil.)	7,530	7,752	7,213	7,263	8,399
Arms as percent of total imports[8]	1.6	4.5	1.8	0.4	0.3
Arms as percent of total exports[8]	0	0	0	0	0

Human Rights [24]

	SSTS	FL	FAPRO	PPCG	APROBC	TPW	PCPTW	STPEP	PHRFF	PRW	ASST	AFL
Observes		P	P	P	P	P	P			P		P
	EAFRD	CPR	ESCR	SR	ACHR	MAAE	PVIAC	PVNAC	EAFDAW	TCIDTP	RC	
Observes		P	P	P	P	P		P	P	P	P	P

P = Party; S = Signatory; see Appendix for meaning of abbreviations.

Labor Force

Total Labor Force [25]

12 million (1990)

Labor Force by Occupation [26]

Services	46%
Agriculture	30
Industry	24

Date of data: 1990

Unemployment Rate [27]

9.5% (1995)

For sources, notes, and explanations, see Annotated Source Appendix, page 1061.

Production Sectors

Commercial Energy Production and Consumption

Data are shown in quadrillion (10^{15}) BTUs and percent for 1995
Values for hydroelectric, nuclear, geothermal, solar, and wind power refer to electrical generation.

Production [28]	Consumption [29]
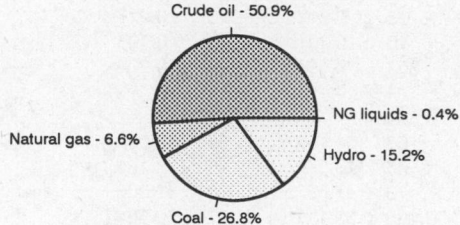	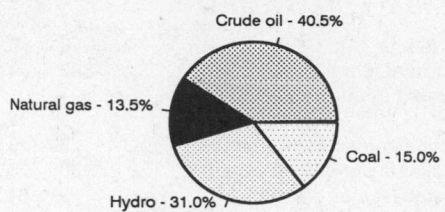

Production [28]
- Crude oil - 50.9%
- NG liquids - 0.4%
- Hydro - 15.2%
- Coal - 26.8%
- Natural gas - 6.6%

Consumption [29]
- Crude oil - 40.5%
- Natural gas - 13.5%
- Coal - 15.0%
- Hydro - 31.0%

Crude oil	1.286	Crude oil	0.503
Natural gas liquids	0.010	Dry natural gas	0.167
Dry natural gas	0.167	Coal	0.186
Coal	0.677	Net hydroelectric power	0.385
Net hydroelectric power	0.385	Total	1.241
Total	2.525		

Telecommunications [30]

- 1.89 million (1986 est.) telephones; modern system in many respects
- Domestic: nationwide microwave radio relay system; domestic satellite system with 11 earth stations
- International: satellite earth stations - 2 Intelsat (Atlantic Ocean)
- Radio: Broadcast stations: AM 413 (licensed), FM 217 (licensed), shortwave 28
- Television: Broadcast stations: 33 Televisions: 5.5 million (1993 est.)

Transportation [31]

Railways: total: 3,386 km; standard gauge: 150 km 1.435-m gauge (connects Cerrejon coal mines to maritime port at Bahia Portete); narrow gauge: 3,236 km 0.914-m gauge (1830 km in use) (1995)

Highways: total: 107,200 km; paved: 12,600 km; unpaved: 94,600 km

Merchant marine: total: 19 ships (1,000 GRT or over) totaling 97,037 GRT/129,404 DWT; ships by type: bulk 5, cargo 8, container 3, oil tanker 3 (1995 est.)

Airports
Total:	989
With paved runways over 3,047 m:	2
With paved runways 2,438 to 3,047 m:	9
With paved runways 1,524 to 2,437 m:	33
With paved runways 914 to 1,523 m:	35
With paved runways under 914 m:	557

Top Agricultural Products [32]

Agriculture accounts for 21.5% of the GDP; produces coffee, cut flowers, bananas, rice, tobacco, corn, sugarcane, cocoa beans, oilseed, vegetables; forest products; shrimp farming.

Top Mining Products [33]

Metric tons except as noted[e]	7/14/95[*]
Gold (kg.)	27,500[r]
Nickel, mine output, Ni content	23,500
Feldspar	70,000
Nitrogen, N content of ammonia	90,000
Sodium carbonate	121,000
Marble	35,000
Sand, excluding metal-bearing	850,000
Sulfur	65,600
Petroleum, crude (000 42-gal. bls.)	166,000

Tourism [34]

	1990	1991	1992	1993	1994
Tourists[4]	813	857	1,076	1,047	1,207
Tourism receipts	406	468	705	755	794
Tourism expenditures	454	509	641	644	756
Fare receipts	168	176	225	NA	NA
Fare expenditures	137	143	124	NA	NA

Travelers are in thousands, money in million U.S. dollars.

For sources, notes, and explanations, see Annotated Source Appendix, page 1061.

207

Manufacturing Sector

Manufacturing Summary [35]

	1987		1988		1989		1990		1991	
	$ bil.	%	$ bil.	%	$ bil.	%	$ bil.	%	$ bil.	%
Establishments or enterprises (number)	7,659	0.326	7,960	0.379	8,344	0.445	8,279	0.462	-	-
Total employment (000)	557	0.410	559	0.408	571	0.464	573	0.518	-	-
Production workers (000)	383	0.567	384	0.614	388	0.527	384	0.541	-	-
Output ($ bil.)	21	0.211	25	0.211	25	0.212	25	0.217	-	-
Value added ($ bil.)	8	0.188	10	0.199	10	0.201	10	0.193	-	-
Capital investment ($ mil.)	832	0.191	1,071	0.225	857	0.156	1,299	0.234	-	-
M & E investment ($ mil.)	-	-	-	-	-	-	-	-	-	-
Employees per establishment (number)	73	125.638	70	107.719	68	104.166	69	112.092	-	-
Production workers per establishment	50	173.747	48	161.962	46	118.278	46	117.125	-	-
Output per establishment ($ mil.)	3	64.773	3	55.696	3	47.697	3	47.006	-	-
Capital investment per estab. ($ mil.)	0.109	58.504	0.135	59.310	0.103	35.133	0.157	50.564	-	-
M & E per establishment ($ mil)	-	-	-	-	-	-	-	-	-	-
Payroll per employee ($)	2,605	29.055	2,699	27.087	2,696	24.136	2,637	19.010	-	-
Wages per production worker ($)	2,163	27.419	2,132	25.236	2,087	20.810	2,024	17.041	-	-
Hours per production worker (hours)	-	-	-	-	-	-	-	-	-	-
Output per employee ($)	38,612	51.555	44,040	51.705	43,884	45.789	43,459	41.935	-	-
Capital investment per employee ($)	1,495	46.565	1,916	55.060	1,501	33.728	2,265	45.109	-	-
M & E per employee ($)	-	-	-	-	-	-	-	-	-	-

Note: Columns headed % show percent of world total or ratio. Ratios closest to 100 are closest to world average. M & E stands for machinery & equipment.

Output in Manufacturing

	1987		1988		1989		1990		1991	
	$ bil.	%	$ bil.	%	$ bil.	%	$ bil.	%	$ bil.	%
3110 - Food products	4.056	19.31	4.352	17.41	4.632	18.53	4.882	19.53	-	-
3130 - Beverages	1.422	6.77	1.538	6.15	1.610	6.44	1.577	6.31	-	-
3140 - Tobacco	0.227	1.08	0.241	0.96	0.243	0.97	0.245	0.98	-	-
3210 - Textiles	1.340	6.38	1.544	6.18	1.600	6.40	1.629	6.52	-	-
3211 - Spinning, weaving, etc.	0.989	4.71	1.173	4.69	1.221	4.88	1.209	4.84	-	-
3220 - Wearing apparel	0.486	2.31	0.525	2.10	0.554	2.22	0.569	2.28	-	-
3230 - Leather and products	0.173	0.82	0.201	0.80	0.201	0.80	0.223	0.89	-	-
3240 - Footwear	0.157	0.75	0.247	0.99	0.248	0.99	0.263	1.05	-	-
3310 - Wood products	0.099	0.47	0.117	0.47	0.123	0.49	0.111	0.44	-	-
3320 - Furniture, fixtures	0.070	0.33	0.087	0.35	0.089	0.36	0.084	0.34	-	-
3410 - Paper and products	0.680	3.24	0.782	3.13	0.805	3.22	0.840	3.36	-	-
3411 - Pulp, paper, etc.	0.412	1.96	0.465	1.86	0.471	1.88	0.478	1.91	-	-
3420 - Printing, publishing	0.466	2.22	0.491	1.96	0.491	1.96	0.468	1.87	-	-
3510 - Industrial chemicals	1.274	6.07	1.484	5.94	1.555	6.22	1.591	6.36	-	-
3511 - Basic chemicals, excl fertilizers	0.256	1.22	0.318	1.27	0.345	1.38	0.331	1.32	-	-
3513 - Synthetic resins, etc.	0.507	2.41	0.658	2.63	0.672	2.69	0.603	2.41	-	-
3520 - Chemical products nec	1.195	5.69	1.284	5.14	1.320	5.28	1.388	5.55	-	-
3522 - Drugs and medicines	0.507	2.41	0.495	1.98	0.510	2.04	0.508	2.03	-	-
3530 - Petroleum refineries	0.816	3.89	0.819	3.28	0.800	3.20	0.812	3.25	-	-
3540 - Petroleum, coal products	0.095	0.45	0.104	0.42	0.115	0.46	0.100	0.40	-	-
3550 - Rubber products	0.272	1.30	0.308	1.23	0.306	1.22	0.307	1.23	-	-
3560 - Plastic products nec	0.614	2.92	0.645	2.58	0.635	2.54	0.621	2.48	-	-
3610 - Pottery, china, etc.	0.095	0.45	0.107	0.43	0.110	0.44	0.108	0.43	-	-
3620 - Glass and products	0.190	0.90	0.204	0.82	0.196	0.78	0.205	0.82	-	-
3690 - Nonmetal products nec	0.569	2.71	0.642	2.57	0.708	2.83	0.657	2.63	-	-
3710 - Iron and steel	0.610	2.90	0.789	3.16	0.774	3.10	0.637	2.55	-	-
3720 - Nonferrous metals	0.103	0.49	0.140	0.56	0.118	0.47	0.129	0.52	-	-
3810 - Metal products	0.556	2.65	0.679	2.72	0.685	2.74	0.693	2.77	-	-
3820 - Machinery nec	0.305	1.45	0.354	1.42	0.345	1.38	0.354	1.42	-	-
3825 - Office, computing machinery	0.004	0.02	0.003	0.01	0.005	0.02	0.004	0.02	-	-
3830 - Electrical machinery	0.577	2.75	0.669	2.68	0.672	2.69	0.643	2.57	-	-
3832 - Radio, television, etc.	0.107	0.51	0.134	0.54	0.139	0.56	0.127	0.51	-	-
3840 - Transportation equipment	1.076	5.12	1.434	5.74	1.286	5.14	1.165	4.66	-	-
3841 - Shipbuilding, repair	0.025	0.12	0.060	0.24	0.052	0.21	0.062	0.25	-	-
3843 - Motor vehicles	0.948	4.51	1.274	5.10	1.124	4.50	0.997	3.99	-	-
3850 - Professional goods	0.087	0.41	0.110	0.44	0.123	0.49	0.123	0.49	-	-
3900 - Industries nec	0.132	0.63	0.144	0.58	0.180	0.72	0.177	0.71	-	-

Note: Codes are International Standard Industry codes (ISIC). Percentages are % of total Output. [f]: Factor Prices; [p]: Producer Prices.

Finance, Economics, and Trade

Economic Indicators [36]

- **National product**: GDP—purchasing power parity— $192.5 billion (1995 est.)
- **National product real growth rate**: 5.3% (1995 est.)
- **National product per capita**: $5,300 (1995 est.)
- **Inflation rate (consumer prices)**: 19.5% (1995 est.)
- **External debt**: $14 billion (1995 est.)

Balance of Payments Summary [37]

Values in millions of dollars.

	1988	1989	1990	1991	1992
Exports of goods (f.o.b.)	5,343	6,031	7,263	7,507	7,263
Imports of goods (f.o.b.)	-4,516	-4,557	-6,030	-4,548	-6,030
Trade balance	827	1,474	1,233	2,959	1,233
Services - debits	-3,672	-4,151	-4,487	-4,292	-4,487
Services - credits	1,665	1,578	2,432	1,984	2,432
Private transfers (net)	975	912	1,747	1,712	1,747
Government transfers (net)	-11	-14	-13	-14	-13
Long-term capital (net)	834	653	196	143	337
Short-term capital (net)	106	-246	-170	-928	-54
Errors and omissions	-530	157	70	269	14
Overall balance	194	363	638	1,833	1,209

Exchange Rates [38]

Currency: **Colombian peso.**
Symbol: **Col$.**

Data are currency units per $1

January 1996	1,011.11
1995	912.83
1994	844.84
1993	863.06
1992	759.28
1991	633.05

Imports and Exports

Top Import Origins [39]

$13.5 billion (c.i.f., 1995 est.) Data are for 1992

Origins	%
US	36
EU	18
Venezuela	6.5
Japan	8.7
Brazil	4

Top Export Destinations [40]

$10.5 billion (f.o.b., 1995 est.) Data are for 1992

Destinations	%
US	39
EU	25.7
Venezuela	8.5
Japan	2.9

Foreign Aid [41]

Recipient: ODA, $30 million (1993).

Import and Export Commodities [42]

Import Commodities	**Export Commodities**
Industrial equipment	Petroleum
Transportation equipment	Coffee
Consumer goods	Coal
Chemicals	Bananas
Paper products	Fresh cut flowers

For sources, notes, and explanations, see Annotated Source Appendix, page 1061.

209

Comoros

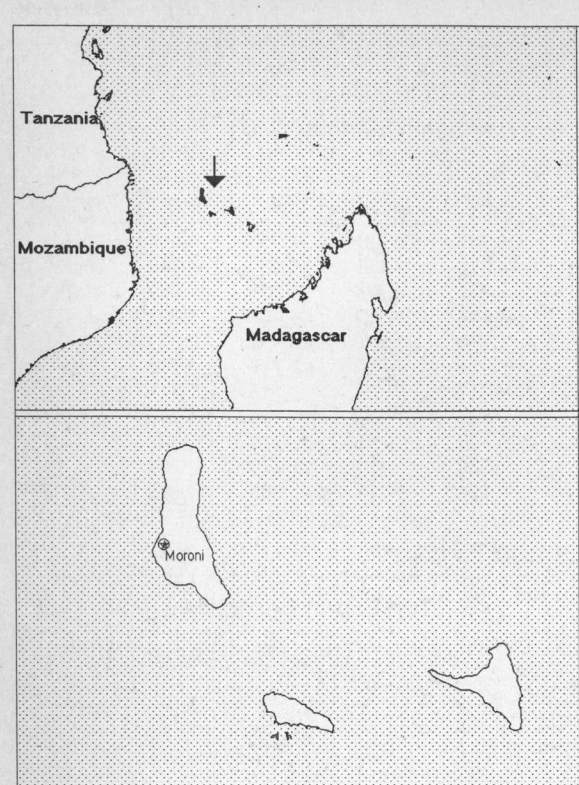

Geography [1]

Total area:
2,170 sq km 838 sq mi
Land area:
2,170 sq km 838 sq mi
Comparative area:
Slightly more than 12 times the size of Washington, DC
Land boundaries:
0 km
Coastline:
340 km
Climate:
Tropical marine; rainy season (November to May)
Terrain:
Volcanic islands, interiors vary from steep mountains to low hills
Natural resources:
Negligible
Land use:
Arable land: 35%
Permanent crops: 8%
Meadows and pastures: 7%
Forest and woodland: 16%
Other: 34%

Demographics [2]

	1970	1980	1990	1995[1]	1996	2000	2010	2020	2030
Population	236	334	460	549	569	656	919	1,249	1,624
Population density (persons per sq. mi.)	282	398	549	656	679	782	1,097	1,490	1,939
(persons per sq. km.)	109	154	212	253	262	302	424	575	748
Net migration rate (per 1,000 population)	NA	0.0	0.0	0.0	0.0	0.0	0.0	0.0	0.0
Births	NA	NA	NA	NA	26	NA	NA	NA	NA
Deaths	NA	NA	NA	NA	6	NA	NA	NA	NA
Life expectancy - males	NA	50.0	54.0	56.0	56.4	58.0	61.8	65.3	68.3
Life expectancy - females	NA	53.0	58.1	60.6	61.1	63.0	67.6	71.7	75.1
Birth rate (per 1,000)	NA	46.6	47.6	46.2	45.8	44.1	39.2	33.8	28.3
Death rate (per 1,000)	NA	17.8	12.4	10.6	10.3	9.0	6.7	5.2	4.4
Women of reproductive age (15-49 yrs.)	NA	71	101	120	125	144	211	304	416
of which are currently married	NA	47	67	NA	83	96	141	NA	NA
Fertility rate	NA	7.0	7.0	6.7	6.7	6.3	5.4	4.3	3.4

Except as noted, values for vital statistics are in thousands; life expectancy is in years.

Health

Health Indicators [3]

Health Expenditures [4]

For sources, notes, and explanations, see Annotated Source Appendix, page 1061.

Human Factors

Women and Children [5]

Burden of Disease [6]

Population per physician (1990)	8,836.73
Population per nurse (1990)	621.23
Population per hospital bed (1990)	362.95
AIDS cases per 100,000 people (1994)	0.3
Malaria cases per 100,000 people (1992)	NA

Ethnic Division [7]

Antalote, Cafre, Makoa, Oimatsaha, Sakalava.

Religion [8]

Sunni Muslim	86%
Roman Catholic	14%

Major Languages [9]

Arabic (official), French (official), Comoran (a blend of Swahili and Arabic).

Education

Public Education Expenditures [10]

Million (Franc C.F.A.)[15]	1980	1985	1990	1992	1993	1994
Total education expenditure	NA	NA	NA	NA	NA	NA
as percent of GNP	NA	NA	NA	NA	NA	NA
as percent of total govt. expend.	NA	NA	NA	NA	NA	NA
Current education expenditure	NA	2,105	2,666	2,829	2,938	3,285
as percent of GNP	NA	4.1	4.0	4.0	4.1	3.7
as percent of current govt. expend.	NA	23.1	24.3	22.0	20.5	21.6
Capital expenditure	NA	NA	NA	NA	NA	NA

Educational Attainment [11]

Illiteracy [12]

In thousands and percent[1]	1990	1995	2000
Illiterate population (15+ yrs.)	128	143	160
Illiteracy rate - total pop. (%)	53.1	50.0	46.9
Illiteracy rate - males (%)	45.4	42.6	40.5
Illiteracy rate - females (%)	60.7	57.2	53.2

Libraries [13]

Daily Newspapers [14]

Culture [15]

Science and Technology

Scientific/Technical Forces [16]

R&D Expenditures [17]

U.S. Patents Issued [18]

For sources, notes, and explanations, see Annotated Source Appendix, page 1061.

211

Government and Law

Organization of Government [19]

Long-form name:
Federal Islamic Republic of the Comoros
Type:
Independent republic
Independence:
6 July 1975 (from France)
National holiday:
Independence Day, 6 July (1975)
Constitution:
7 June 1992
Legal system:
French and Muslim law in a new consolidated code
Executive branch:
President; Prime Minister; Council of Ministers
Legislative branch:
Unicameral: Federal Assembly (Assemblee Federale)
Judicial branch:
Supreme Court (Cour Supreme)

Elections [20]

Federal Assembly	% of seats
RDR	35.7
UDZIMA	19.0
UNDC	11.9
DPA/MWANGAZA	4.8
Unfilled seats	4.8
Others	23.8

Government Budget [21]

For 1992.

	$ mil.
Revenues	83
Expenditures	92
Capital expenditures	32

Defense Summary [22]

Branches: Comoran Security Force

Manpower Availability: Males age 15-49 121,854; males fit for military service 72,873 (1996 est.)

Defense Expenditures: NA

Crime [23]

Human Rights [24]

	SSTS	FL	FAPRO	PPCG	APROBC	TPW	PCPTW	STPEP	PHRFF	PRW	ASST	AFL
Observes		P	P		P	P	P					P
	EAFRD	CPR	ESCR	SR	ACHR	MAAE	PVIAC	PVNAC	EAFDAW	TCIDTP	RC	
Observes							P	P	P		P	

P = Party; S = Signatory; see Appendix for meaning of abbreviations.

Labor Force

Total Labor Force [25]

140,000 (1982)

Labor Force by Occupation [26]

Agriculture	80%
Government	3

Unemployment Rate [27]

15.8% (1989)

For sources, notes, and explanations, see Annotated Source Appendix, page 1061.

Production Sectors

Energy Resource Summary [28]

Electricity: Capacity: 16,000 kW. Production: 17 million kWh. Consumption per capita: 27 kWh (1993).

Telecommunications [30]

- 3,770 (1991 est.) telephones; sparse system of microwave radio relay and HF radiotelephone communication stations
- Domestic: HF radiotelephone communications and microwave radio relay
- International: HF radiotelephone communications to Madagascar and Reunion
- Radio: Broadcast stations: AM 2, FM 1, shortwave 0
- Television: Broadcast stations: 0 Televisions: 200 (1991 est.)

Top Agricultural Products [32]

Produces vanilla, cloves, perfume essences, copra, coconuts, bananas, cassava (tapioca).

Top Mining Products [33]

Detailed information is not available. A summary of mineral resources follows. **Mineral Resources**: Clay, sand, gravel, and crushed stone for construction.

Transportation [31]

Railways: 0 km

Highways: total: 1,104 km; paved: 400 km; unpaved: 704 km (1988 est.)

Merchant marine: none

Airports

Total: 4
With paved runways 2,438 to 3,047 m: 1
With paved runways 914 to 1,523 m: 3 (1995 est.)

Tourism [34]

	1990	1991	1992	1993	1994
Tourists[20]	8	17	19	24	27
Tourism receipts	2	9	8	NA	NA
Tourism expenditures	6	7	6	NA	NA
Fare receipts	2	2	5	NA	NA
Fare expenditures	7	5	6	NA	NA

Travelers are in thousands, money in million U.S. dollars.

For sources, notes, and explanations, see Annotated Source Appendix, page 1061.

213

Finance, Economics, and Trade

Industrial Summary [35]

Industrial Production: Growth rate - 6.5% (1989 est.). **Industries**: Tourism, perfume distillation, textiles, furniture, jewelry, construction materials, soft drinks.

Economic Indicators [36]

- **National product**: GDP—purchasing power parity— $370 million (1994 est.)
- **National product real growth rate**: 0.9% (1994 est.)
- **National product per capita**: $700 (1994 est.)
- **Inflation rate (consumer prices)**: 15% (1993 est.)
- **External debt**: $160 million (1992 est.)

Balance of Payments Summary [37]

Values in millions of dollars.

	1987	1988	1989	1990	1991
Exports of goods (f.o.b.)	11.6	21.5	18.1	17.9	24.4
Imports of goods (f.o.b.)	-44.2	-44.3	-35.7	-45.2	-53.6
Trade balance	-32.5	-22.8	-17.6	-27.3	-29.2
Services - debits	-44.7	-46.2	-42.6	-46.9	-48.1
Services - credits	16.2	18.6	21.8	20.2	27.5
Private transfers (net)	0.9	3.1	2.7	5.5	3.7
Government transfers (net)	38.7	40.8	41.1	39.2	37.2
Long-term capital (net)	18.1	5.8	6.9	0.2	13.8
Short-term capital (net)	11.4	-1.6	0.7	13.5	-9.6
Errors and omissions	0.7	-1.4	-7.6	-9.2	1.7
Overall balance	8.8	-3.7	5.5	-4.9	-3.0

Exchange Rates [38]

Currency: **Comoran franc.**
Symbol: **CF.**

Data are currency units per $1.

January 1996	375.42
1995	374.36
1994	416.40
1993	283.16
1992	264.69
1991	282.11

Imports and Exports

Top Import Origins [39]

$40.9 million (f.o.b., 1993 est.) Data are for 1992.

Origins	%
France	34
South Africa	14
Kenya	8
Japan	4

Top Export Destinations [40]

$13.7 million (f.o.b., 1993 est.) Data are for 1992.

Destinations	%
US	44
France	40
Germany	6
Africa	5

Foreign Aid [41]

Recipient: ODA, $NA.

Import and Export Commodities [42]

Import Commodities	Export Commodities
Rice and other foodstuffs	Vanilla
Petroleum products	Ylang-ylang
Cement	Cloves
Consumer goods	Perfume oil
	Copra

For sources, notes, and explanations, see Annotated Source Appendix, page 1061.

Congo

Geography [1]

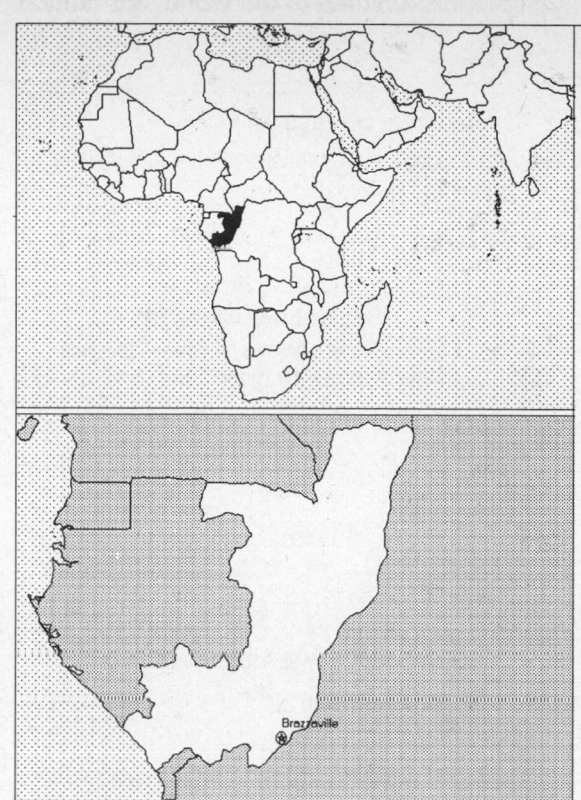

Total area:
  342,000 sq km   132,047 sq mi
Land area:
  341,500 sq km   131,854 sq mi
Comparative area:
  Slightly smaller than Montana
Land boundaries:
  Total 5,504 km, Angola 201 km, Cameroon 523 km, Central African Republic 467 km, Gabon 1,903 km, Zaire 2,410 km
Coastline:
  169 km
Climate:
  Tropical; rainy season (March to June); dry season (June to October); constantly high temperatures and humidity; particularly enervating climate astride the Equator
Terrain:
  Coastal plain, southern basin, central plateau, northern basin
Natural resources:
  Petroleum, timber, potash, lead, zinc, uranium, copper, phosphates, natural gas
Land use:
  Arable land: 2%
  Permanent crops: 0%
  Meadows and pastures: 29%
  Forest and woodland: 62%
  Other: 7%

Demographics [2]

	1970	1980	1990	1995[1]	1996	2000	2010	2020	2030
Population	1,183	1,620	2,204	2,473	2,528	2,750	3,298	3,817	4,400
Population density (persons per sq. mi.)	9	12	17	19	19	21	25	29	33
(persons per sq. km.)	3	5	6	7	7	8	10	11	13
Net migration rate (per 1,000 population)	NA	NA	0.6	0.0	0.0	0.0	0.0	0.0	0.0
Births	NA	NA	NA	NA	99	NA	NA	NA	NA
Deaths	NA	NA	NA	NA	44	NA	NA	NA	NA
Life expectancy - males	NA	NA	45.3	44.2	44.2	44.3	45.3	48.9	58.6
Life expectancy - females	NA	NA	47.0	47.5	47.4	47.0	48.3	52.2	63.6
Birth rate (per 1,000)	NA	NA	41.6	39.6	39.2	37.3	32.2	27.7	24.1
Death rate (per 1,000)	NA	NA	17.5	17.4	17.4	17.2	16.2	14.1	9.1
Women of reproductive age (15-49 yrs.)	NA	NA	513	582	599	666	824	999	1,200
of which are currently married	NA	NA	379	NA	445	493	615	NA	NA
Fertility rate	NA	NA	5.6	5.2	5.1	4.8	4.0	3.2	2.7

Except as noted, values for vital statistics are in thousands; life expectancy is in years.

Health

Health Indicators [3]

% of population with access to	
safe water (1990-95)[1]	38
adequate sanitation (1990-95)	NA
health services (1985-95)	83
% of 1-year-olds immunized (1990-94) against	
TB (tuberculosis)	94
DPT (diphtheria, pertussis, tetanus)	79
polio	79
measles	70
% of contraceptive prevalence (1980-94)	NA
ORT use rate (1990-94)	67

Health Expenditures [4]

For sources, notes, and explanations, see Annotated Source Appendix, page 1061.

215

Human Factors

Women and Children [5]

% of pregnant women immunized (tetanus 1990-94)	NA
% of births attended by trained health personnel (1983-94)	NA
Maternal mortality rate (1980-92)	900
Under-5 mortality rate (1994)	109
% under-5 moderately/severely underweight (1980-1994)	24

Burden of Disease [6]

Population per physician (1986)	4,000.00
Population per nurse	NA
Population per hospital bed	NA
AIDS cases per 100,000 people (1994)	58.4
Malaria cases per 100,000 people (1992)	NA

Ethnic Division [7]

Kongo	48%
Sangha	20%
Teke	17%
M'Bochi	12%
Europeans	3%

Religion [8]

Christian	50%
Animist	48%
Muslim	2%

Major Languages [9]

French (official), African languages (Lingala and Kikongo are the most widely used).

Education

Public Education Expenditures [10]

	1980	1984	1989	1990	1991	1994
Million (Franc C.F.A.)						
Total education expenditure	22,942	44,442	38,989	37,899	57,092	NA
as percent of GNP	7.0	5.1	5.9	5.7	8.3	NA
as percent of total govt. expend.	23.6	9.8	16.5	14.4	NA	NA
Current education expenditure	21,517	41,033	38,033	36,906	56,537	NA
as percent of GNP	6.6	4.7	5.7	5.6	8.3	NA
as percent of current govt. expend.	24.1	14.7	20.2	18.0	19.3	NA
Capital expenditure	1,424	3,409	956	993	555	NA

Educational Attainment [11]

Age group (1984)	25+
Total population	646,626
Highest level attained (%)	
No schooling	58.8
First level	
Not completed	13.0
Completed	8.5
Entered second level	
S-1	11.0
S-2	5.9
Postsecondary	3.0

Illiteracy [12]

In thousands and percent[1]	1990	1995	2000
Illiterate population (15+ yrs.)	393	354	306
Illiteracy rate - total pop. (%)	32.0	25.4	19.3
Illiteracy rate - males (%)	22.3	17.0	12.5
Illiteracy rate - females (%)	40.9	33.2	25.9

Libraries [13]

	Admin. Units	Svc. Pts.	Vols. (000)	Shelving (meters)	Vols. Added	Reg. Users
National (1989)	1	1	9	125	2,000	995
Nonspecialized	NA	NA	NA	NA	NA	NA
Public (1989)	1	4	15	360	1,200	22,365
Higher ed. (1987)	1	9	78	2,446	650	12,000
School (1990)	1	1	10	96	155	492

Daily Newspapers [14]

	1980	1985	1990	1994
Number of papers	1	1	5	6
Circ. (000)	3	8	17[e]	19[e]

Culture [15]

Science and Technology

Scientific/Technical Forces [16]

Scientists/engineers[10]	862
Number female	NA
Technicians	1,473
Number female	NA
Total[1]	2,335

R&D Expenditures [17]

	(000) 1984
Total expenditure[6]	25,530
Capital expenditure	14,263
Current expenditure	11,267
Percent current	44.1

U.S. Patents Issued [18]

For sources, notes, and explanations, see Annotated Source Appendix, page 1061.

Government and Law

Organization of Government [19]

Long-form name:
Republic of the Congo
Type:
Republic
Independence:
15 August 1960 (from France)
National holiday:
Congolese National Day, 15 August (1960)
Constitution:
New constitution approved by referendum March 1992
Legal system:
Based on French civil law system and customary law
Executive branch:
President; Prime Minister; Council of Ministers
Legislative branch:
Bicameral: National Assembly (Assemblee Nationale) and Senate
Judicial branch:
Supreme Court (Cour Supreme)

Elections [20]

National Assembly	% of seats
Pan-African Union (UPADS)	51.2
URD/PCT	46.4
Others	2.4

Government Budget [21]

For 1994 est.

	$ bil.
Revenues	2.18
Expenditures	NA
Capital expenditures	NA

Military Expenditures and Arms Transfers [22]

	1990	1991	1992	1993	1994
Military expenditures					
Current dollars (mil.)	52	86	80	71[e]	28[e]
1994 constant dollars (mil.)	58	92	83	72[e]	28[e]
Armed forces (000)	9	9	10	10	10
Gross national product (GNP)					
Current dollars (mil.)	1,204	1,314	1,421	1,376	1,162
1994 constant dollars (mil.)	1,340	1,408	1,481	1,404	1,162
Central government expenditures (CGE)					
1994 constant dollars (mil.)	523	NA	NA	NA	NA
People (mil.)	2.2	2.3	2.3	2.4	2.4
Military expenditure as % of GNP	4.3	6.5	5.6	5.1	2.4
Military expenditure as % of CGE	11.1	NA	NA	NA	NA
Military expenditure per capita (1994 $)	26	41	36	30	11
Armed forces per 1,000 people (soldiers)	4.1	4.0	4.3	4.2	4.1
GNP per capita (1994 $)	605	620	636	588	475
Arms imports[6]					
Current dollars (mil.)	5	0	0	5	20
1994 constant dollars (mil.)	6	0	0	5	20
Arms exports[6]					
Current dollars (mil.)	0	0	0	0	0
1994 constant dollars (mil.)	0	0	0	0	0
Total imports[7]					
Current dollars (mil.)	621	472	688	582	NA
1994 constant dollars (mil.)	691	506	717	594	NA
Total exports[7]					
Current dollars (mil.)	981	1,029	1,183	1,069	NA
1994 constant dollars (mil.)	1,092	1,103	1,234	1,091	NA
Arms as percent of total imports[8]	0.8	0	0	0.9	NA
Arms as percent of total exports[8]	0	0	0	0	0

Crime [23]

	1990
Crime volume	
Cases known to police	649
Attempts (percent)	21
Percent cases solved	4.28
Crimes per 100,000 persons	32.45
Persons responsible for offenses	
Total number offenders	404
Percent female	40
Percent juvenile (0-17 yrs.)	10
Percent foreigners	88

Human Rights [24]

	SSTS	FL	FAPRO	PPCG	APROBC	TPW	PCPTW	STPEP	PHRFF	PRW	ASST	AFL
Observes	2	P	P			P	P	P		P	P	
	EAFRD	CPR	ESCR	SR	ACHR	MAAE	PVIAC	PVNAC	EAFDAW	TCIDTP	RC	
Observes	P	P	P	P			P	P	P		P	

P = Party; S = Signatory; see Appendix for meaning of abbreviations.

Labor Force

Total Labor Force [25]

79,100 wage earners

Labor Force by Occupation [26]

Agriculture	75%
Commerce, industry, and government	25

Unemployment Rate [27]

For sources, notes, and explanations, see Annotated Source Appendix, page 1061.

217

Production Sectors

Commercial Energy Production and Consumption

Data are shown in quadrillion (10^{15}) BTUs and percent for 1995
Values for hydroelectric, nuclear, geothermal, solar, and wind power refer to electrical generation.

Production [28]

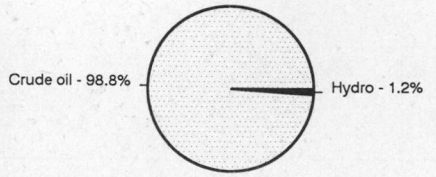

Crude oil - 98.8%   Hydro - 1.2%

Consumption [29]

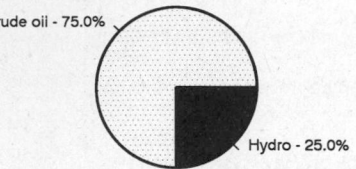

Crude oil - 75.0%   Hydro - 25.0%

Crude oil	0.397
Net hydroelectric power	0.005
Total	0.402

Crude oil	0.015
Net hydroelectric power	0.005
Total	0.020

Telecommunications [30]

- 18,000 (1983 est.) telephones; services adequate for government use; key exchanges are in Brazzaville, Pointe-Noire, and Loubomo
- Domestic: primary network consists of microwave radio relay and coaxial cable
- International: satellite earth station - 1 Intelsat (Atlantic Ocean)
- Radio: Broadcast stations: AM 4, FM 1, shortwave 0
- Television: Broadcast stations: 4 (1987 est.) Televisions: 8,500 (1993 est.)

Transportation [31]

Railways: total: 795 km (1995 est.); narrow gauge: 795 km 1.067-m gauge (includes 285 km that are privately owned)

Highways: total: 12,745 km; paved: 1,236 km; unpaved: 11,509 km (1992 est.)

Merchant marine: total: 1 cargo ship (1,000 GRT or over) totaling 2,218 GRT/4,100 DWT (1995 est.)

Airports
Total: 34
With paved runways over 3,047 m: 1
With paved runways 1,524 to 2,437 m: 3
With paved runways under 914 m: 9
With unpaved runways 1,524 to 2,437 m: 7
With unpaved runways 914 to 1,523 m: 14 (1995 est.)

Top Agricultural Products [32]

Agriculture accounts for 11.4% of the GDP; produces cassava (tapioca) accounts for 90% of food output, sugar, rice, corn, peanuts, vegetables, coffee, cocoa; forest products.

Top Mining Products [33]

Metric tons except as noted[e]	3/31/95*
Cement, hydraulic	114,000
Gas, natural, gross (mil. cu. meters)	360
Gold, mine output, Au content (kg.)	5
Lime	240
Petroleum, crude (000 42-gal. bls.)	67,500

Tourism [34]

	1990	1991	1992	1993	1994
Tourists[26]	33	33	36	34	30
Tourism receipts	8	8	7	6	3
Tourism expenditures	113	106	93	68	36
Fare receipts	NA	NA	31	14	7
Fare expenditures	33	47	32	31	16

Travelers are in thousands, money in million U.S. dollars.

For sources, notes, and explanations, see Annotated Source Appendix, page 1061.

Manufacturing Sector

GDP and Manufacturing Summary [35]

	1980	1985	1989	1990	% change 1980-1990	% change 1989-1990
GDP (million 1980 $)	1,706	2,860	2,455	2,784	63.2	13.4
GDP per capita (1980 $)	1,022	1,474	1,116	1,227	20.1	9.9
Manufacturing as % of GDP (current prices)	7.7	5.7	7.6^e	7.9	2.6	3.9
Gross output (million $)	193^e	154	281^e	276^e	43.0	-1.8
Value added (million $)	69^e	57	80^e	104^e	50.7	30.0
Value added (million 1980 $)	128	256	231^e	272	112.5	17.7
Industrial production index	100	183	121^e	168	68.0	38.8
Employment (thousands)	5^e	9	9^e	8^e	60.0	-11.1

Note: GDP stands for Gross Domestic Product. 'e' stands for estimated value.

Profitability and Productivity

	1980	1985	1989	1990	% change 1980-1990	% change 1989-1990
Intermediate input (%)	64^e	63	72^e	62^e	-3.1	-13.9
Wages, salaries, and supplements (%)	15^e	17^e	17^e	15^e	0.0	-11.8
Gross operating surplus (%)	20^e	20^e	11^e	22^e	10.0	100.0
Gross output per worker ($)	16,482^e	17,590	30,826^e	34,628^e	110.1	12.3
Value added per worker ($)	5,895^e	6,525	8,736^e	13,059^e	121.5	49.5
Average wage (incl. benefits) ($)	5,463^e	3,032^e	5,234^e	5,320^e	-2.6	1.6

Profitability is in percent of gross output. Productivity is in U.S. $. 'e' stands for estimated value.

Profitability - 1990

Wages - 15.2%
Inputs - 62.6%
Surplus - 22.2%

The graphic shows percent of gross output.

Value Added in Manufacturing

	1980 $ mil.	1980 %	1985 $ mil.	1985 %	1989 $ mil.	1989 %	1990 $ mil.	1990 %	% change 1980-1990	% change 1989-1990
311 Food products	11^e	15.9	10	17.5	11^e	13.8	23^e	22.1	109.1	109.1
313 Beverages	12^e	17.4	11	19.3	13^e	16.3	24^e	23.1	100.0	84.6
314 Tobacco products	3^e	4.3	3	5.3	5^e	6.3	9^e	8.7	200.0	80.0
321 Textiles	4^e	5.8	2^e	3.5	6^e	7.5	2^e	1.9	-50.0	-66.7
322 Wearing apparel	1^e	1.4	1^e	1.8	2^e	2.5	1^e	1.0	0.0	-50.0
323 Leather and fur products	NA	0.0	NA	0.0	NA	0.0	NA	0.0	NA	NA
324 Footwear	3^e	4.3	2	3.5	5^e	6.3	2^e	1.9	-33.3	-60.0
331 Wood and wood products	7^e	10.1	5^e	8.8	7^e	8.8	7^e	6.7	0.0	0.0
332 Furniture and fixtures	4^e	5.8	3^e	5.3	4^e	5.0	4^e	3.8	0.0	0.0
341 Paper and paper products	1^e	1.4	NA	0.0	1^e	1.3	1^e	1.0	0.0	0.0
342 Printing and publishing	1^e	1.4	NA	0.0	1^e	1.3	1^e	1.0	0.0	0.0
351 Industrial chemicals	6^e	8.7	4^e	7.0	1^e	1.3	8^e	7.7	33.3	700.0
352 Other chemical products	3^e	4.3	2^e	3.5	3^e	3.8	4^e	3.8	33.3	33.3
353 Petroleum refineries	1^c	1.4	1^e	1.8	1^e	1.3	2^e	1.9	100.0	100.0
354 Miscellaneous petroleum and coal products	NA	0.0	NA	0.0	NA	0.0	NA	0.0	NA	NA
355 Rubber products	1^e	1.4	1^e	1.8	1^e	1.3	1^e	1.0	0.0	0.0
356 Plastic products	NA	0.0	NA	0.0	NA	0.0	1^e	1.0	NA	NA
361 Pottery, china, and earthenware	1^e	1.4	2^e	3.5	NA	0.0	1^e	1.0	0.0	NA
362 Glass and glass products	NA	0.0	NA	0.0	NA	0.0	NA	0.0	NA	NA
369 Other nonmetal mineral products	NA	0.0	NA	0.0	2^e	2.5	NA	0.0	NA	NA
371 Iron and steel	NA	0.0	NA	0.0	NA	0.0	NA	0.0	NA	NA
372 Nonferrous metals	NA	0.0	NA	0.0	NA	0.0	NA	0.0	NA	NA
381 Metal products	4^e	5.8	4^e	7.0	7^e	8.8	5^e	4.8	25.0	-28.6
382 Nonelectrical machinery	1^e	1.4	1^e	1.8	3^e	3.8	2^e	1.9	100.0	-33.3
383 Electrical machinery	2^e	2.9	2^e	3.5	3^e	3.8	2^e	1.9	0.0	-33.3
384 Transport equipment	3^e	4.3	2	3.5	4^e	5.0	3^e	2.9	0.0	-25.0
385 Professional and scientific equipment	NA	0.0	NA	0.0	NA	0.0	NA	0.0	NA	NA
390 Other manufacturing industries	NA	0.0	NA	0.0	NA	0.0	NA	0.0	NA	NA

Note: The industry codes shown are International Standard Industry codes (ISIC). Percentages are percent of total Value Added. 'e' stands for estimated value

Finance, Economics, and Trade

Economic Indicators [36]

- **National product**: GDP—purchasing power parity—$7.7 billion (1995 est.)
- **National product real growth rate**: 3.3% (1995 est.)
- **National product per capita**: $3,100 (1995 est.)
- **Inflation rate (consumer prices)**: 61% (1994 est.)
- **External debt**: $5 billion (1993)

Balance of Payments Summary [37]

Values in millions of dollars.

	1989	1990	1991	1992	1993
Exports of goods (f.o.b.)	1,160.5	1,388.7	1,107.7	1,178.7	1,107.5
Imports of goods (f.o.b.)	-532.0	-512.7	-494.5	-438.2	-490.9
Trade balance	628.5	876.0	613.2	740.5	616.6
Services - debits	-857.3	-1,244.0	-1,188.2	-1,117.9	-1,130.5
Services - credits	97.5	113.9	118.0	78.6	65.0
Private transfers (net)	-46.7	-62.8	-59.9	-63.1	-83.3
Government transfers (net)	93.1	65.7	55.3	44.6	24.7
Long-term capital (net)	-279.9	862.8	-34.4	-250.9	87.6
Short-term capital (net)	361.1	456.5	475.0	553.9	457.7
Errors and omissions	8.5	-40.6	-6.3	41.2	-35.4
Overall balance	4.8	114.5	-27.3	26.9	2.4

Exchange Rates [38]

Currency: **Communaute Financiere Africaine franc.**
Symbol: **CFAF.**

Data are currency units per $1.

January 1996	500.56
1995	499.15
1994	555.20
1993	283.16
1992	264.69
1991	282.11

Imports and Exports

Top Import Origins [39]

$600 million (c.i.f., 1995).

Origins	%
France	NA
Italy	NA
Other EU countries	NA
US	NA
Japan	NA
Thailand	NA

Top Export Destinations [40]

$1 billion (f.o.b., 1995).

Destinations	%
Italy	NA
France	NA
Spain	NA
Other EU countries	NA
US	NA
Taiwan	NA

Foreign Aid [41]

Recipient: ODA, $NA.

Import and Export Commodities [42]

Import Commodities	Export Commodities
Intermediate manufactures	Crude oil 90%
Capital equipment	Lumber
Construction materials	Plywood
Foodstuffs	Sugar
Petroleum products	Cocoa
	Coffee
	Diamonds

Costa Rica

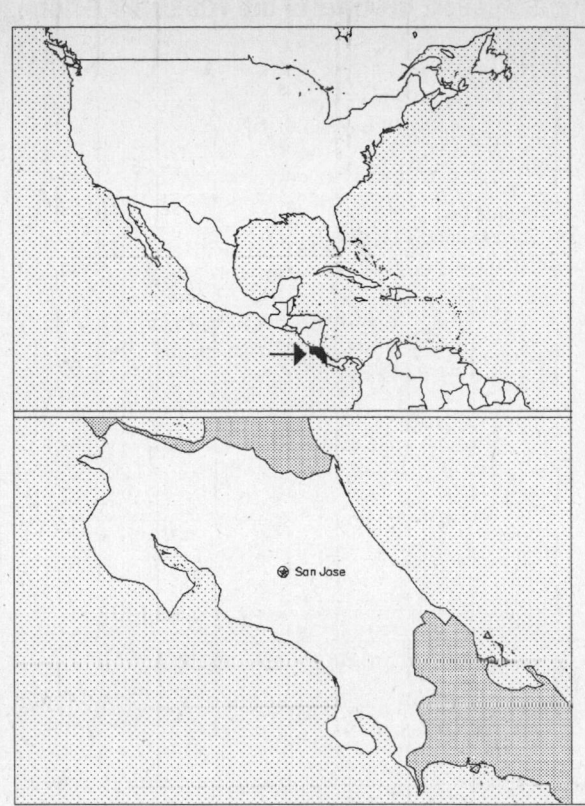

Geography [1]

Total area:
51,100 sq km 19,730 sq mi
Land area:
50,660 sq km 19,560 sq mi
Comparative area:
Slightly smaller than West Virginia
Note: Includes Isla del Coco
Land boundaries:
Total 639 km, Nicaragua 309 km, Panama 330 km
Coastline:
1,290 km
Climate:
Tropical; dry season (December to April); rainy season (May to November)
Terrain:
Coastal plains separated by rugged mountains
Natural resources:
Hydropower potential
Land use:
Arable land: 6%
Permanent crops: 7%
Meadows and pastures: 45%
Forest and woodland: 34%
Other: 8%

Demographics [2]

	1970	1980	1990	1995[1]	1996	2000	2010	2020	2030
Population	1,736	2,307	3,022	3,391	3,463	3,744	4,416	5,044	5,585
Population density (persons per sq. mi.)	89	118	155	173	177	191	226	258	286
(persons per sq. km.)	34	46	60	67	68	74	87	100	110
Net migration rate (per 1,000 population)	3.0	NA	1.7	1.0	0.9	0.5	0.0	0.0	0.0
Births	NA	NA	NA	NA	83	NA	NA	NA	NA
Deaths	NA	NA	11	NA	14	NA	NA	NA	NA
Life expectancy - males	NA	NA	73.3	73.2	73.3	73.7	74.6	75.3	76.0
Life expectancy - females	NA	NA	78.2	78.1	78.2	78.7	79.9	80.8	81.7
Birth rate (per 1,000)	33.0	NA	27.5	24.3	23.8	22.0	19.4	17.0	15.2
Death rate (per 1,000)	7.0	NA	4.1	4.2	4.1	4.2	4.6	5.3	6.5
Women of reproductive age (15-49 yrs.)	NA	NA	773	875	895	981	1,162	1,277	1,378
of which are currently married	NA	NA	466	NA	550	602	716	NA	NA
Fertility rate	NA	NA	3.2	2.9	2.9	2.7	2.4	2.2	2.1

Except as noted, values for vital statistics are in thousands; life expectancy is in years.

Health

Health Indicators [3]

% of population with access to	
safe water (1990-95)	92
adequate sanitation (1990-95)	97
health services (1985-95)[1]	80
% of 1-year-olds immunized (1990-94) against	
TB (tuberculosis)	97
DPT (diphtheria, pertussis, tetanus)	88
polio	88
measles	88
% of contraceptive prevalence (1980-94)	75
ORT use rate (1990-94)	78

Health Expenditures [4]

For sources, notes, and explanations, see Annotated Source Appendix, page 1061.

221

Human Factors

Women and Children [5]

% of pregnant women immunized (tetanus 1990-94)	NA
% of births attended by trained health personnel (1983-94)	93
Maternal mortality rate (1980-92)	36
Under-5 mortality rate (1994)	16
% under-5 moderately/severely underweight (1980-1994)	2

Burden of Disease [6]

Population per physician (1987)	1,031.72
Population per nurse (1984)	475.37
Population per hospital bed	NA
AIDS cases per 100,000 people (1994)	4.7
Malaria cases per 100,000 people (1992)	218

Ethnic Division [7]

White (including mestizo)	96%
Black	2%
Chinese	1%
Indian	1%

Religion [8]

Roman Catholic	95%
Other	5%

Major Languages [9]

Spanish (official), English spoken around Puerto Limon.

Education

Public Education Expenditures [10]

Million (Colon)[16]	1980	1985	1990	1992	1993	1994
Total education expenditure	3,069	8,181	22,907	38,552	47,656	58,699
as percent of GNP	7.8	4.5	4.6	4.4	4.6	4.7
as percent of total govt. expend.	22.2	22.7	20.8	21.4	20.2	19.2
Current education expenditure	2,802	7,787	22,188	36,636	45,717	56,323
as percent of GNP	7.1	4.2	4.5	4.2	4.4	4.5
as percent of current govt. expend.	26.7	26.2	26.3	25.4	25.8	22.9
Capital expenditure	267	394	719	170	902	1,256

Educational Attainment [11]

Illiteracy [12]

In thousands and percent[1]	1990	1995	2000
Illiterate population (15+ yrs.)	117	115	111
Illiteracy rate - total pop. (%)	6.1	5.2	4.4
Illiteracy rate - males (%)	6.1	5.3	4.6
Illiteracy rate - females (%)	6.0	5.1	4.3

Libraries [13]

	Admin. Units	Svc. Pts.	Vols. (000)	Shelving (meters)	Vols. Added	Reg. Users
National (1986)	1	1	7	92	11,982	190,419
Nonspecialized	NA	NA	NA	NA	NA	NA
Public (1986)	81	87	321	NA	16,588	293,615
Higher ed. (1990)	4	4	633	NA	25,145	NA
School	NA	NA	NA	NA	NA	NA

Daily Newspapers [14]

	1980	1985	1990	1994
Number of papers	4	6	5	5
Circ. (000)	251	280[e]	306[e]	333

Culture [15]

Cinema (seats per 1,000)	NA
Annual attendance per person	NA
Gross box office receipts (mil. Colon)	NA
Museums (reporting)	19
Visitors (000)	733
Annual receipts (000 Colon)	NA

Science and Technology

Scientific/Technical Forces [16]

Scientists/engineers	17,222
Number female	NA
Technicians	NA
Number female	NA
Total[1]	NA

R&D Expenditures [17]

	Colon (000) 1986
Total expenditure	612,000
Capital expenditure	NA
Current expenditure	NA
Percent current	NA

U.S. Patents Issued [18]

Values show patents issued to citizens of the country by the U.S. Patents Office.

	1993	1994	1995
Number of patents	5	7	10

For sources, notes, and explanations, see Annotated Source Appendix, page 1061.

Government and Law

Organization of Government [19]

Long-form name:
Republic of Costa Rica
Type:
Democratic republic
Independence:
15 September 1821 (from Spain)
National holiday:
Independence Day, 15 September (1821)
Constitution:
9 November 1949
Legal system:
Based on Spanish civil law system; judicial review of legislative acts in the Supreme Court; has not accepted compulsory ICJ jurisdiction
Executive branch:
President; First Vice President; Second Vice President; Cabinet
Legislative branch:
Unicameral: Legislative Assembly (Asamblea Legislativa)
Judicial branch:
Supreme Court (Corte Suprema)

Elections [20]

Legislative Assembly	% of seats
Social Christian Unity (PUSC)	47.5
National Liberation Party (PLN)	45.9
Minority parties	6.6

Government Expenditures [21]

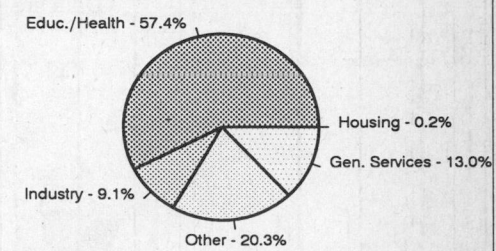

Educ./Health - 57.4%
Housing - 0.2%
Gen. Services - 13.0%
Industry - 9.1%
Other - 20.3%

(% distribution). Expend. for CY95: 472.25 (Colon bil.)

Crime [23]

	1990
Crime volume	
Cases known to police	32,144
Attempts (percent)	NA
Percent cases solved	19.9
Crimes per 100,000 persons	1,063.5
Persons responsible for offenses	
Total number offenders	6,383
Percent female	9.2
Percent juvenile	NA
Percent foreigners	NA

Military Expenditures and Arms Transfers [22]

	1990	1991	1992	1993	1994
Military expenditures					
Current dollars (mil.)	23	22	24	28	NA
1994 constant dollars (mil.)	26	24	25	29	NA
Armed forces (000)	8	8	8	8	8
Gross national product (GNP)					
Current dollars (mil.)	5,749	6,226	6,887	7,489	7,987
1994 constant dollars (mil.)	6,399	6,674	7,182	7,643	7,987
Central government expenditures (CGE)					
1994 constant dollars (mil.)	1,764	1,755	1,796	2,066	2,552
People (mil.)	3.0	3.1	3.2	3.3	3.3
Military expenditure as % of GNP	0.4	0.4	0.3	0.4	NA
Military expenditure as % of CGE	1.5	1.3	1.4	1.4	NA
Military expenditure per capita (1994 $)	8	8	8	9	NA
Armed forces per 1,000 people (soldiers)	2.6	2.6	2.5	2.5	2.4
GNP per capita (1994 $)	2,111	2,147	2,254	2,341	2,390
Arms imports[6]					
Current dollars (mil.)	0	0	0	0	0
1994 constant dollars (mil.)	0	0	0	0	0
Arms exports[6]					
Current dollars (mil.)	0	0	0	0	0
1994 constant dollars (mil.)	0	0	0	0	0
Total imports[7]					
Current dollars (mil.)	1,990	1,877	2,440	2,885	3,089
1994 constant dollars (mil.)	2,215	2,012	2,544	2,944	3,089
Total exports[7]					
Current dollars (mi)	1,448	1,598	1,829	2,049	2,233
1994 constant dollars (mil.)	1,612	1,713	1,907	2,091	2,233
Arms as percent of total imports[8]	0	0	0	0	0
Arms as percent of total exports[8]	0	0	0	0	0

Human Rights [24]

	SSTS	FL	FAPRO	PPCG	APROBC	TPW	PCPTW	STPEP	PHRFF	PRW	ASST	AFL
Observes		P	P	P	P	P	P			P		P
	EAFRD	CPR	ESCR	SR	ACHR	MAAE	PVIAC	PVNAC	EAFDAW	TCIDTP		RC
Observes	P	P	P	P	P	P	P	P	P	P		P

P = Party; S = Signatory; see Appendix for meaning of abbreviations.

Labor Force

Total Labor Force [25]

868,300

Labor Force by Occupation [26]

Industry and commerce	35.1%
Government and services	33
Agriculture	27
Other	4.9

Date of data: 1985 est.

Unemployment Rate [27]

5.2% (1995 est.); much underemployment

For sources, notes, and explanations, see Annotated Source Appendix, page 1061.

223

Production Sectors

Commercial Energy Production and Consumption

Data are shown in quadrillion (10^{15}) BTUs and percent for 1995
Values for hydroelectric, nuclear, geothermal, solar, and wind power refer to electrical generation.

Production [28]
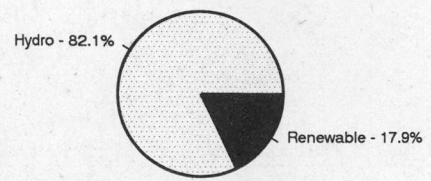
Hydro - 82.1%
Renewable - 17.9%

Consumption [29]
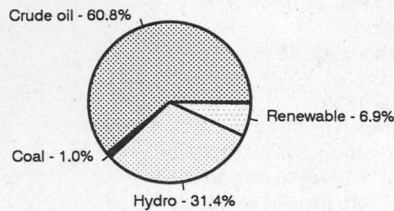
Crude oil - 60.8%
Renewable - 6.9%
Coal - 1.0%
Hydro - 31.4%

Net hydroelectric power	0.032
Geothermal, solar, wind	0.007
Total	0.039

Crude oil	0.062
Coal	0.001
Net hydroelectric power	0.032
Geothermal, solar, wind	0.007
Total	0.102

Telecommunications [30]

- 281,042 (1983 est.) telephones; very good domestic telephone service
- International: connected to Central American Microwave System; satellite earth station - 1 Intelsat (Atlantic Ocean)
- Radio: Broadcast stations: AM 71, FM 0, shortwave 13
- Television: Broadcast stations: 18 Televisions: 340,000 (1993 est.)

Transportation [31]

Railways: total: 950 km; narrow gauge: 950 km 1.067-m gauge (260 km electrified)

Highways: total: 35,560 km; paved: 5,608 km; unpaved: 29,952 km (1992 est.)

Merchant marine: none

Airports
Total: 145
With paved runways 2,438 to 3,047 m: 2
With paved runways 1,524 to 2,437 m: 1
With paved runways 914 to 1,523 m: 16
With paved runways under 914 m: 97
With unpaved runways 914 to 1,523 m: 29 (1995 est.)

Top Agricultural Products [32]

Produces coffee, bananas, sugar, corn, rice, beans, potatoes; beef; timber (depletion of forest resources has resulted in declining timber output).

Top Mining Products [33]

Metric tons except as noted[e]	3/31/95[*]
Cement	780,000
Clays, common	401,000
Diatomite	12,000
Iron and steel, semimanufatures	87,000
Lime	9,500
Petroleum refinery products (000 42-gal. bls.)	4,000
Salt, marine	45,000
Limestone and other calcareous materials (000 tons)	1,700
Sand and gravel (000 tons)	1,400

Tourism [34]

	1990	1991	1992	1993	1994
Tourists	435	505	610	684	761
Cruise passengers	55	68	71	112	156
Tourism receipts	275	331	431	577	626
Tourism expenditures	148	149	223	267	300
Fare receipts	45	45	65	74	81
Fare expenditures	11	11	16	11	12

Travelers are in thousands, money in million U.S. dollars.

For sources, notes, and explanations, see Annotated Source Appendix, page 1061.

Manufacturing Sector

Manufacturing Summary [35]

	1987		1988		1989		1990		1991	
	$ bil.	%	$ bil.	%	$ bil.	%	$ bil.	%	$ bil.	%
Establishments or enterprises (number)	4,956	0.211	4,414	0.210	-	-	8,801	0.491	-	-
Total employment (000)	124	0.091	127	0.093	-	-	125	0.113	-	-
Production workers (000)	-		-		-		-		-	
Output ($ bil.)	3	0.030	3	0.026	3	0.028	3	0.030	-	-
Value added ($ bil.)	0.900	0.021	0.869	0.018	0.923	0.019	0.982	0.020	-	-
Capital investment ($ mil.)	-		-		-		-		-	
M & E investment ($ mil.)	-		-		-		-		-	
Employees per establishment (number)	25	43.296	29	44.250	-	-	14	23.024	-	-
Production workers per establishment	-		-		-		-		-	
Output per establishment ($ mil.)	0.608	14.041	0.695	12.516	-	-	0.386	6.024	-	-
Capital investment per estab. ($ mil.)	-		-		-		-		-	
M & E per establishment ($ mil)	-		-		-		-		-	
Payroll per employee ($)	2,731	30.451	2,458	24.675	-	-	-	-	-	-
Wages per production worker ($)	-		-		-		-		-	
Hours per production worker (hours)	-		-		-		-		-	
Output per employee ($)	24,288	32.429	24,092	28.285	-	-	27,114	26.163	-	-
Capital investment per employee ($)	-		-		-		-		-	
M & E per employee ($)	-		-		-		-		-	

Note: Columns headed % show percent of world total or ratio. Ratios closest to 100 are closest to world average. M & E stands for machinery & equipment.

Output in Manufacturing

	1987		1988		1989		1990		1991	
	$ bil.	%	$ bil.	%	$ bil.	%	$ bil.	%	$ bil.	%
3110 - Food products	1.138	37.93	1.203	40.10	1.233	41.10	1.259	41.97	-	-
3130 - Beverages	0.192	6.40	0.188	6.27	0.210	7.00	0.237	7.90	-	-
3140 - Tobacco	0.040	1.33	0.042	1.40	0.047	1.57	0.050	1.67	-	-
3210 - Textiles	0.075	2.50	0.080	2.67	0.087	2.90	0.093	3.10	-	-
3211 - Spinning, weaving, etc.	0.035	1.17	0.038	1.27	0.040	1.33	0.043	1.43	-	-
3220 - Wearing apparel	0.059	1.97	0.065	2.17	0.078	2.60	0.084	2.80	-	-
3230 - Leather and products	0.012	0.40	0.012	0.40	0.014	0.47	0.013	0.43	-	-
3240 - Footwear	0.018	0.60	0.021	0.70	0.021	0.70	0.020	0.67	-	-
3310 - Wood products	0.055	1.83	0.054	1.80	0.047	1.57	0.052	1.73	-	-
3320 - Furniture, fixtures	0.058	1.93	0.061	2.03	0.065	2.17	0.068	2.27	-	-
3410 Paper and products	0.106	3.53	0.133	4.43	0.153	5.10	0.180	6.00	-	-
3420 - Printing, publishing	0.092	3.07	0.097	3.23	0.102	3.40	0.098	3.27	-	-
3510 - Industrial chemicals	0.093	3.10	0.104	3.47	0.121	4.03	0.136	4.53	-	-
3511 - Basic chemicals, excl fertilizers	0.000	0.00	0.001	0.03	0.001	0.03	0.001	0.03	-	-
3513 - Synthetic resins, etc.	0.007	0.23	0.007	0.23	0.007	0.23	0.007	0.23	-	-
3520 - Chemical products nec	0.150	5.00	0.145	4.83	0.162	5.40	0.168	5.60	-	-
3522 - Drugs and medicines	0.053	1.77	0.050	1.67	0.054	1.80	0.048	1.60	-	-
3530 - Petroleum refineries	0.241	8.03	0.207	6.90	0.182	6.07	0.132	4.40	-	-
3540 - Petroleum, coal products	0.000	0.00	0.000	0.00	0.000	0.00	0.000	0.00	-	-
3550 - Rubber products	0.047	1.57	0.050	1.67	0.053	1.77	0.050	1.67	-	-
3560 - Plastic products nec	0.087	2.90	0.095	3.17	0.103	3.43	0.103	3.43	-	-
3610 - Pottery, china, etc.	0.005	0.17	0.005	0.17	0.007	0.23	0.010	0.33	-	-
3620 - Glass and products	0.026	0.87	0.020	0.67	0.025	0.83	0.030	1.00	-	-
3690 - Nonmetal products nec	0.067	2.23	0.076	2.53	0.083	2.77	0.092	3.07	-	-
3720 - Nonferrous metals	-	-	0.002	0.07	0.004	0.13	0.005	0.17	-	-
3810 - Metal products	0.057	1.90	0.059	1.97	0.078	2.60	0.085	2.83	-	-
3820 - Machinery nec	0.047	1.57	0.041	1.37	0.037	1.23	0.039	1.30	-	-
3830 - Electrical machinery	0.118	3.93	0.100	3.33	0.111	3.70	0.131	4.37	-	-
3832 - Radio, television, etc.	0.085	2.83	0.067	2.23	0.080	2.67	0.099	3.30	-	-
3840 - Transportation equipment	0.028	0.93	0.025	0.83	0.030	1.00	0.033	1.10	-	-
3841 - Shipbuilding, repair	0.003	0.10	0.004	0.13	0.005	0.17	0.004	0.13	-	-
3843 - Motor vehicles	0.014	0.47	0.011	0.37	0.010	0.33	0.015	0.50	-	-
3900 - Industries nec	0.008	0.27	0.007	0.23	0.010	0.33	0.010	0.33	-	-

Note: Codes are International Standard Industry codes (ISIC). Percentages are % of total Output. [f]: Factor Prices; [p]: Producer Prices.

Finance, Economics, and Trade

Economic Indicators [36]

- **National product**: GDP—purchasing power parity—$18.4 billion (1995 est.)

- **National product real growth rate**: 2.5% (1995 est.)

- **National product per capita**: $5,400 (1995 est.)

- **Inflation rate (consumer prices)**: 22.5% (1995 est.)

- **External debt**: $4 billion (1995 est.)

Balance of Payments Summary [37]

Values in millions of dollars.

	1989	1990	1991	1992	1993
Exports of goods (f.o.b.)	1,333.4	1,354.2	1,498.1	1,739.1	1,944.6
Imports of goods (f.o.b.)	-1,572.0	-1,796.7	-1,697.6	-2,210.9	-2,610.4
Trade balance	-238.6	-442.5	-199.5	-471.8	-665.8
Services - debits	-985.5	-912.7	-820.1	-1,026.0	-1,094.2
Services - credits	617.8	739.3	802.8	954.1	1,137.1
Private transfers (net)	39.2	55.4	50.3	88.3	85.9
Government transfers (net)	152.2	136.5	91.3	85.0	67.0
Long-term capital (net)	59.7	548.0	463.8	221.1	403.7
Short-term capital (net)	296.1	-370.5	-147.2	129.1	16.0
Errors and omissions	208.9	56.4	99.9	201.9	19.7
Overall balance	149.8	-190.1	341.3	181.7	-30.6

Exchange Rates [38]

Currency: **Costa Rican colon.**
Symbol: **C.**

Data are currency units per $1.

December 1995	193.93
1995	179.73
1994	157.07
1993	142.17
1992	134.51
1991	122.43

Imports and Exports

Top Import Origins [39]

$3 billion (c.i.f., 1995 est.).

Origins	%
US	NA
Japan	NA
Mexico	NA
Guatemala	NA
Venezuela	NA
Germany	NA

Top Export Destinations [40]

$2.4 billion (f.o.b., 1995 est.).

Destinations	%
US	NA
Germany	NA
Italy	NA
Guatemala	NA
El Salvador	NA
Netherlands	NA
UK	NA
France	NA

Foreign Aid [41]

Recipient: ODA, $NA.

Import and Export Commodities [42]

Import Commodities	Export Commodities
Raw materials	Coffee
Consumer goods	Bananas
Capital equipment	Textiles
Petroleum	Sugar

Cote d'Ivoire

Geography [1]

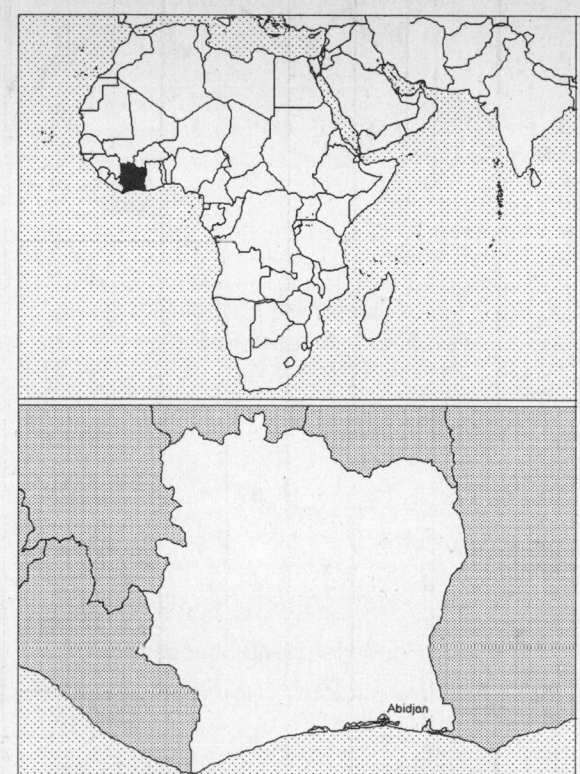

Total area:
  322,460 sq km   124,503 sq mi
Land area:
  318,000 sq km   122,780 sq mi
Comparative area:
  Slightly larger than New Mexico
Land boundaries:
  Total 3,110 km, Burkina Faso 584 km, Ghana 668 km, Guinea 610 km, Liberia 716 km,
  Mali 532 km
Coastline:
  515 km
Climate:
  Tropical along coast, semiarid in far North; three seasons - warm
  and dry (November to March), hot and dry (March to May), hot and wet
  (June to October)
Terrain:
  Mostly flat to undulating plains; mountains in Northwest
Natural resources:
  Petroleum, diamonds, manganese, iron ore, cobalt, bauxite, copper
Land use:
  Arable land: 9%
  Permanent crops: 4%
  Meadows and pastures: 9%
  Forest and woodland: 26%
  Other: 52%

Demographics [2]

	1970	1980	1990	1995[1]	1996	2000	2010	2020	2030
Population	5,427	8,276	11,926	14,283	14,762	16,172	20,261	24,634	29,810
Population density (persons per sq. mi.)	44	67	97	116	120	132	165	201	243
(persons per sq. km.)	17	26	38	45	46	51	64	77	94
Net migration rate (per 1,000 population.)	NA	NA	27.1	9.5	2.4	-3.7	0.0	0.0	0.0
Births	NA	NA	NA	NA	627	NA	NA	NA	NA
Deaths	NA	NA	NA	NA	232	NA	NA	NA	NA
Life expectancy - males	NA	NA	48.1	46.9	46.2	43.9	44.4	48.1	58.7
Life expectancy - females	NA	NA	48.7	47.5	47.3	46.3	45.3	49.5	62.3
Birth rate (per 1,000)	NA	NA	46.1	43.0	42.5	40.9	37.2	32.6	28.3
Death rate (per 1,000)	NA	NA	15.5	15.5	15.7	16.4	16.3	14.1	8.5
Women of reproductive age (15-44 yrs.)	NA	NA	2,411	2,884	2,989	3,335	4,398	5,682	7,278
of which are currently married	NA	NA	1,852	NA	2,282	2,523	3,299	NA	NA
Fertility rate	NA	6.9	6.7	6.2	6.1	5.8	4.9	4.0	3.3

Except as noted, values for vital statistics are in thousands; life expectancy is in years.

Health

Health Indicators [3]

% of population with access to	
safe water (1990-95)	72
adequate sanitation (1990-95)	54
health services (1985-95)[1]	30
% of 1-year-olds immunized (1990-94) against	
TB (tuberculosis)	49
DPT (diphtheria, pertussis, tetanus)	44
polio	44
measles	49
% of contraceptive prevalence (1980-94)	11
ORT use rate (1990-94)	15

Health Expenditures [4]

Total health expenditure, 1990 (official exchange rate)	
Millions of dollars	332
Dollars per capita	28
Health expenditures as a percentage of GDP	
Total	3.3
Public sector	1.7
Private sector	1.6
Development assistance for health	
Total aid flows (millions of dollars)[1]	11
Aid flows per capita (dollars)	0.9
Aid flows as a percentage of total health expenditure	3.4

For sources, notes, and explanations, see Annotated Source Appendix, page 1061.

227

Human Factors

Women and Children [5]

% of pregnant women immunized (tetanus 1990-94)	NA
% of births attended by trained health personnel (1983-94)	45
Maternal mortality rate (1980-92)	NA
Under-5 mortality rate (1994)	150
% under-5 moderately/severely underweight (1980-1994)	12

Burden of Disease [6]

Population per physician (1985)	14,251.08
Population per nurse	NA
Population per hospital bed (1990)	1,268.03
AIDS cases per 100,000 people (1994)	44.6
Malaria cases per 100,000 people (1992)	NA

Ethnic Division [7]

Present: Agni, about 3 million foreign Africans (mostly Burkinabe and Maliains), 130,000 to 330,000 non-Africans (mostly French and Lebanese).

Baoule	23%
Bete	18%
Senoufou	15%
Malinke	11%

Religion [8]

Indigenous	25%
Muslim	60%
Christian	12%

Major Languages [9]

French (official), 60 native dialects with Dioula the most widely spoken.

Education

Public Education Expenditures [10]

Million (Franc C.F.A.)[17]	1980	1985	1990	1992	1993	1994
Total education expenditure	147,478	NA	NA	NA	NA	NA
as percent of GNP	7.2	NA	NA	NA	NA	NA
as percent of total govt. expend.	22.6	NA	NA	NA	NA	NA
Current education expenditure	123,196	179,447	NA	153,004	NA	168,923
as percent of GNP	6.0	6.3	NA	6.7	NA	5.6
as percent of current govt. expend.	36.4	NA	NA	NA	NA	NA
Capital expenditure	24,282	NA	NA	NA	NA	NA

Educational Attainment [11]

Age group (1988)[4]	25+
Total population	739,179
Highest level attained (%)	
No schooling	NA
First level	
Not completed	NA
Completed	48.2
Entered second level	
S-1	43.1
S-2	NA
Postsecondary	8.7

Illiteracy [12]

In thousands and percent[1]	1990	1995	2000
Illiterate population (15+ yrs.)	4,077	4,339	4,532
Illiteracy rate - total pop. (%)	65.6	58.1	52.7
Illiteracy rate - males (%)	55.6	48.3	43.2
Illiteracy rate - females (%)	76.2	68.5	62.7

Libraries [13]

Daily Newspapers [14]

	1980	1985	1990	1994
Number of papers	2	1	1	1
Circ. (000)	81	90	90	90

Culture [15]

Cinema (seats per 1,000)	5.3
Annual attendance per person	0.5
Gross box office receipts (mil. Franc C.F.A.)	NA
Museums (reporting)	NA
Visitors (000)	NA
Annual receipts (000 Franc C.F.A.)	NA

Science and Technology

Scientific/Technical Forces [16]

R&D Expenditures [17]

U.S. Patents Issued [18]

For sources, notes, and explanations, see Annotated Source Appendix, page 1061.

Government and Law

Organization of Government [19]

Long-form name:
Republic of Cote d'Ivoire
Type:
Republic; multiparty presidential regime established 1960
Independence:
7 August 1960 (from France)
National holiday:
National Day, 7 August
Constitution:
3 November 1960; has been amended numerous times, last time November 1990
Legal system:
Based on French civil law system and customary law; judicial review in the Constitutional Chamber of the Supreme Court; has not accepted compulsory ICJ jurisdiction
Executive branch:
President; Prime Minister; Council of Ministers
Legislative branch:
Unicameral: National Assembly (Assemblee Nationale)
Judicial branch:
Supreme Court (Cour Supreme)

Elections [20]

National Assembly	% of seats
Democratic Party (PDCI)	84.0
Rally of the Republicans (RDR)	8.0
Ivorian Popular Front (FPI)	5.7
Unfilled	2.3

Government Budget [21]

For 1993.

	$ bil.
Revenues	1.9
Expenditures	3.4
Capital expenditures	0.408

Military Expenditures and Arms Transfers [22]

	1990	1991	1992	1993	1994
Military expenditures					
Current dollars (mil.)	85	86	94	88[e]	61[e]
1994 constant dollars (mil.)	95	93	98	90[e]	61[e]
Armed forces (000)	15	15	15	15	15
Gross national product (GNP)					
Current dollars (mil.)	5,211	5,347	5,492	5,515	5,445
1994 constant dollars (mil.)	5,799	5,732	5,727	5,628	5,445
Central government expenditures (CGE)					
1994 constant dollars (mil.)	NA	NA	NA	NA	NA
People (mil.)	12.4	12.9	13.3	13.8	14.3
Military expenditure as % of GNP	1.6	1.6	1.7	1.6	1.1
Military expenditure as % of CGE	NA	NA	NA	NA	NA
Military expenditure per capita (1994 $)	8	7	7	7	4
Armed forces per 1,000 people (soldiers)	1.2	1.2	1.1	1.1	1.0
GNP per capita (1994 $)	468	446	430	408	381
Arms imports[6]					
Current dollars (mil.)	0	0	0	0	0
1994 constant dollars (mil.)	0	0	0	0	0
Arms exports[6]					
Current dollars (mil.)	0	0	0	0	0
1994 constant dollars (mil.)	0	0	0	0	0
Total imports[7]					
Current dollars (mil.)	2,098	2,103	5,347	2,212[e]	NA
1994 constant dollars (mil.)	2,335	2,254	5,576	2,258[e]	NA
Total exports[7]					
Current dollars (mil.)	2,817[e]	2,777[e]	6,220	3,272[e]	NA
1994 constant dollars (mil.)	3,135[e]	2,977[e]	6,486	3,339[e]	NA
Arms as percent of total imports[8]	0	0	0	0	0
Arms as percent of total exports[8]	0	0	0	0	0

Crime [23]

	1990
Crime volume	
Cases known to police	15,502
Attempts (percent)	NA
Percent cases solved	50.58
Crimes per 100,000 persons	124.92
Persons responsible for offenses	
Total number offenders	11,120
Percent female	NA
Percent juvenile	NA
Percent foreigners	NA

Human Rights [24]

	SSTS	FL	FAPRO	PPCG	APROBC	TPW	PCPTW	STPEP	PHRFF	PRW	ASST	AFL
Observes	2	P	P	P	P	P	P			P	P	P
	EAFRD	CPR	ESCR	SR	ACHR	MAAE	PVIAC	PVNAC	EAFDAW	TCIDTP	RC
Observes	P	P	P	P			P	P	P	P	P

P = Party; S = Signatory; see Appendix for meaning of abbreviations.

Labor Force

Total Labor Force [25]

5.718 million

Labor Force by Occupation [26]

Agriculture
Forestry
Livestock raising
Government
Industry
Commerce
Professions

Unemployment Rate [27]

For sources, notes, and explanations, see Annotated Source Appendix, page 1061.

229

Production Sectors

Commercial Energy Production and Consumption

Data are shown in quadrillion (10^{15}) BTUs and percent for 1995
Values for hydroelectric, nuclear, geothermal, solar, and wind power refer to electrical generation.

Production [28]
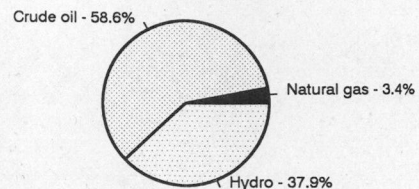
Crude oil - 58.6%
Natural gas - 3.4%
Hydro - 37.9%

Consumption [29]
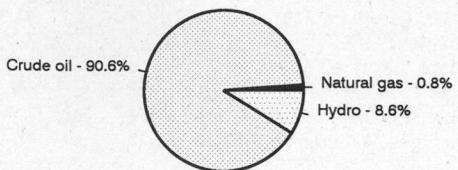
Crude oil - 90.6%
Natural gas - 0.8%
Hydro - 8.6%

Crude oil	0.017
Dry natural gas	0.001
Net hydroelectric power	0.011
Total	0.029

Crude oil	0.116
Dry natural gas	0.001
Net hydroelectric power	0.011
Total	0.128

Telecommunications [30]

- 87,700 (1987 est.) telephones; well-developed by African standards but operating well below capacity
- Domestic: open-wire lines and microwave radio relay
- International: satellite earth stations - 2 Intelsat (1 Atlantic Ocean and 1 Indian Ocean); 2 coaxial submarine cables
- Radio: Broadcast stations: AM 71, FM 0, shortwave 13
- Television: Broadcast stations: 18 Televisions: 810,000 (1993 est.)

Top Agricultural Products [32]

Agriculture accounts for 37% of the GDP; produces coffee, cocoa beans, bananas, palm kernels, corn, rice, manioc, sweet potatoes, sugar; cotton, rubber; timber.

Transportation [31]

Railways: total: 660 km (1995 est.); narrow gauge: 660 km 1.000-meter gauge; 25 km double track

Highways: total: 46,331 km; paved: 3,579 km; unpaved: 42,752 km (1984 est.)

Merchant marine: total: 3 ships (1,000 GRT or over) totaling 27,726 GRT/34,711 DWT; ships by type: container 2, oil tanker 1 (1995 est.)

Airports
Total: 35
With paved runways over 3,047 m: 1
With paved runways 2,438 to 3,047 m: 2
With paved runways 1,524 to 2,437 m: 4
With paved runways under 914 m: 10
With unpaved runways 1,524 to 2,437 m: 6

Top Mining Products [33]

Metric tons except as noted[e]	6/95*
Cement	500[20]
Diamond (carats)	15,000[21]
Gold (kg.)	1,500[21]
Petroleum refinery products (000 42-gal. bls.)	10,300

Tourism [34]

	1990	1991	1992	1993	1994
Tourists[27]	196	200	217	159	157
Tourism receipts	51	62	66	53	43
Tourism expenditures	169	163	168	160	118
Fare receipts	3	3	3	3	2
Fare expenditures	136	95	102	95	73

Travelers are in thousands, money in million U.S. dollars.

Manufacturing Sector

GDP and Manufacturing Summary [35]

	1980	1985	1989	1990	% change 1980-1990	% change 1989-1990
GDP (million 1980 $)	10,176	10,660	9,962	10,184	0.1	2.2
GDP per capita (1980 $)	1,242	1,073	862	849	-31.6	-1.5
Manufacturing as % of GDP (current prices)	11.7	13.8	13.9[e]	14.1[e]	20.5	1.4
Gross output (million $)	4,006	2,869[e]	6,031[e]	5,423[e]	35.4	-10.1
Value added (million $)	1,273	719[e]	1,334[e]	1,409[e]	10.7	5.6
Value added (million 1980 $)	1,141	1,226	970	1,178	3.2	21.4
Industrial production index	100	100	111	97	-3.0	-12.6
Employment (thousands)	67	55[e]	50[e]	51[e]	-23.9	2.0

Note: GDP stands for Gross Domestic Product. 'e' stands for estimated value.

Profitability and Productivity

	1980	1985	1989	1990	% change 1980-1990	% change 1989-1990
Intermediate input (%)	68	75[e]	78[e]	74[e]	8.8	-5.1
Wages, salaries, and supplements (%)	10[e]	9[e]	9[e]	7[e]	-30.0	-22.2
Gross operating surplus (%)	22[e]	16[e]	14[e]	19[e]	-13.6	35.7
Gross output per worker ($)	59,631	51,722[e]	124,054[e]	104,503[e]	75.2	-15.8
Value added per worker ($)	18,950	12,964[e]	27,435[e]	27,184[e]	43.5	-0.9
Average wage (incl. benefits) ($)	5,744[e]	4,926[e]	10,586[e]	7,859[e]	36.8	-25.8

Profitability is in percent of gross output. Productivity is in U.S. $. 'e' stands for estimated value.

Profitability - 1990

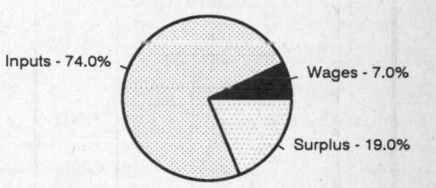

Inputs - 74.0%
Wages - 7.0%
Surplus - 19.0%

The graphic shows percent of gross output.

Value Added in Manufacturing

	1980 $ mil.	1980 %	1985 $ mil.	1985 %	1989 $ mil.	1989 %	1990 $ mil.	1990 %	% change 1980-1990	% change 1989-1990
311 Food products	303[e]	23.8	171[e]	23.8	278[e]	20.8	339[e]	24.1	11.9	21.9
313 Beverages	75	5.9	35[e]	4.9	85[e]	6.4	68[e]	4.8	-9.3	-20.0
314 Tobacco products	66[e]	5.2	32[e]	4.5	40[e]	3.0	59[e]	4.2	-10.6	47.5
321 Textiles	169[e]	13.3	97[e]	13.5	138[e]	10.3	162[e]	11.5	-4.1	17.4
322 Wearing apparel	8[e]	0.6	5[e]	0.7	7[e]	0.5	8[e]	0.6	0.0	14.3
323 Leather and fur products	5[e]	0.4	6[e]	0.8	10[e]	0.7	12[e]	0.9	140.0	20.0
324 Footwear	5[e]	0.4	6[e]	0.8	23[e]	1.7	7[e]	0.5	40.0	-69.6
331 Wood and wood products	67[e]	5.3	24[e]	3.3	59[e]	4.4	32[e]	2.3	-52.2	-45.8
332 Furniture and fixtures	21[e]	1.6	8[e]	1.1	18[e]	1.3	11[e]	0.8	-47.6	-38.9
341 Paper and paper products	14[e]	1.1	7[e]	1.0	NA	0.0	10[e]	0.7	-28.6	NA
342 Printing and publishing	22[e]	1.7	9[e]	1.3	NA	0.0	13[e]	0.9	-40.9	NA
351 Industrial chemicals	22[e]	1.7	10[e]	1.4	18[e]	1.3	18[e]	1.3	-18.2	0.0
352 Other chemical products	53[e]	4.2	29[e]	4.0	82[e]	6.1	80[e]	5.7	50.9	-2.4
353 Petroleum refineries	181[e]	14.2	119[e]	16.6	201[e]	15.1	233[e]	16.5	28.7	15.9
354 Miscellaneous petroleum and coal products	NA	0.0	NA	0.0	NA	0.0	NA	0.0	NA	NA
355 Rubber products	4	0.3	2[e]	0.3	2[e]	0.1	4[e]	0.3	0.0	100.0
356 Plastic products	1[e]	0.1	NA	0.0	NA	0.0	NA	0.0	NA	NA
361 Pottery, china, and earthenware	2[e]	0.2	1[e]	0.1	4[e]	0.3	2[e]	0.1	0.0	-50.0
362 Glass and glass products	NA	0.0	NA	0.0	NA	0.0	NA	0.0	NA	NA
369 Other nonmetal mineral products	27[e]	2.1	12[e]	1.7	24[e]	1.8	26[e]	1.8	-3.7	8.3
371 Iron and steel	5[e]	0.4	1[e]	0.1	2[e]	0.1	3[e]	0.2	-40.0	50.0
372 Nonferrous metals	3[e]	0.2	1[e]	0.1	2[e]	0.1	2[e]	0.1	-33.3	0.0
381 Metal products	70	5.5	33[e]	4.6	85[e]	6.4	56[e]	4.0	-20.0	-34.1
382 Nonelectrical machinery	3	0.2	1[e]	0.1	3[e]	0.2	2[e]	0.1	-33.3	-33.3
383 Electrical machinery	20	1.6	9[e]	1.3	22[e]	1.6	16[e]	1.1	-20.0	-27.3
384 Transport equipment	106	8.3	88[e]	12.2	209[e]	15.7	223[e]	15.8	110.4	6.7
385 Professional and scientific equipment	NA	0.0	NA	0.0	NA	0.0	NA	0.0	NA	NA
390 Other manufacturing industries	20	1.6	14[e]	1.9	21[e]	1.6	23[e]	1.6	15.0	9.5

Note: The industry codes shown are International Standard Industry codes (ISIC). Percentages are percent of total Value Added. 'e' stands for estimated value

For sources, notes, and explanations, see Annotated Source Appendix, page 1061.

231

Finance, Economics, and Trade

Economic Indicators [36]

- **National product**: GDP—purchasing power parity—$21.9 billion (1995 est.)
- **National product real growth rate**: 5% (1995 est.)
- **National product per capita**: $1,500 (1995 est.)
- **Inflation rate (consumer prices)**: 10% (1995 est.)
- **External debt**: $19 billion (1993)

Balance of Payments Summary [37]

Values in millions of dollars.

	1989	1990	1991	1992	1993
Exports of goods (f.o.b.)	2,696.8	3,027.9	2,686.2	2,880.0	2,734.1
Imports of goods (f.o.b.)	-1,777.1	-1,700.6	-1,706.8	-1,885.6	-1,662.3
Trade balance	919.7	1,327.3	979.4	994.4	1,071.8
Services - debits	-2,210.6	-2,792.5	-2,788.0	-2,764.0	-2,644.1
Services - credits	542.3	581.8	553.0	588.2	564.0
Private transfers (net)	-375.5	-553.5	-464.4	-492.3	-394.1
Government transfers (net)	201.9	227.4	230.1	245.2	173.4
Long-term capital (net)	267.4	407.7	744.0	448.4	-50.1
Short-term capital (net)	791.2	794.1	762.1	1,154.6	1,245.6
Errors and omissions	-22.9	-97.1	53.0	35.0	61.9
Overall balance	113.5	-104.8	69.2	209.5	28.4

Exchange Rates [38]

Currency: **Communaute Financiere Africaine franc.**
Symbol: **CFAF.**

Data are currency units per $1.

January 1996	500.56
1995	499.15
1994	555.20
1993	283.16
1992	264.69
1991	282.11

Imports and Exports

Top Import Origins [39]

$1.6 billion (f.o.b., 1994 est.).

Origins	%
France	NA
Nigeria	NA
Japan	NA
Netherlands	NA
US	NA
Italy	NA

Top Export Destinations [40]

$2.9 billion (f.o.b., 1994 est.).

Destinations	%
France	NA
Netherlands	NA
Germany	NA
Italy	NA
Burkina Faso	NA
US	NA
UK	NA

Foreign Aid [41]

Recipient: ODA, $552 million (1993).

Import and Export Commodities [42]

Import Commodities	Export Commodities
Food	Cocoa 55%
Capital goods	Coffee 12%
Consumer goods	Tropical woods 11%
Fuel	Petroleum
	Cotton
	Bananas
	Pineapples
	Palm oil
	Cotton
	Fish

For sources, notes, and explanations, see Annotated Source Appendix, page 1061.

Croatia

Geography [1]

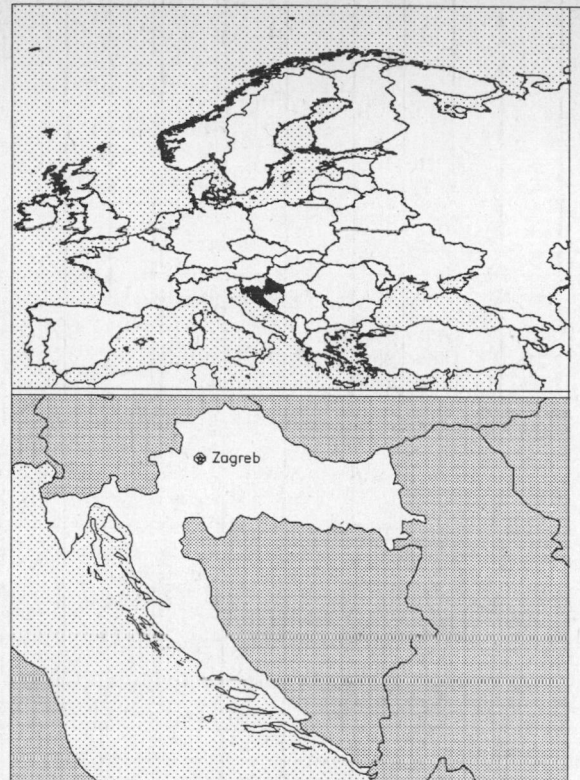

Total area:
56,538 sq km 21,829 sq mi
Land area:
56,410 sq km 21,780 sq mi
Comparative area:
Slightly smaller than West Virginia
Land boundaries:
Total 2,073 km, Bosnia and Herzegovina 932 km, Hungary 329 km, Yugoslavia 266 km (241 km with Serbia; 25 km with Montenego), Slovenia 546 km
Coastline:
5,790 km (mainland 1,778 km, islands 4,012 km)
Climate:
Mediterranean and continental; continental climate predominant with hot summers and cold winters; mild winters, dry summers along coast
Terrain:
Geographically diverse; flat plains along Hungarian border, low mountains and highlands near Adriatic coast, coastline, and islands
Natural resources:
Oil, some coal, bauxite, low-grade iron ore, calcium, natural asphalt, silica, mica, clays, salt
Land use:
Arable land: 32%
Permanent crops: 20%
Meadows and pastures: 18%
Forest and woodland: 15%
Other: 15%

Demographics [2]

	1970	1980	1990	1995[1]	1996	2000	2010	2020	2030
Population	4,411	4,593	4,754	4,969	5,004	5,044	4,986	4,821	4,633
Population density (persons per sq. mi.)	202	210	218	228	229	231	228	221	212
(persons per sq. km.)	78	81	84	88	89	89	88	85	82
Net migration rate (per 1,000 population)	NA	NA	NA	9.8	7.3	-0.1	0.1	0.0	0.0
Births	NA	NA	NA	NA	49	NA	NA	NA	NA
Deaths	NA	NA	NA	NA	57	NA	NA	NA	NA
Life expectancy - males	NA	NA	NA	69.0	69.1	69.6	70.6	73.7	76.1
Life expectancy - females	NA	NA	NA	76.6	76.7	77.3	78.5	80.9	82.8
Birth rate (per 1,000)	NA	NA	NA	9.8	9.8	11.8	10.1	8.6	8.5
Death rate (per 1,000)	NA	NA	NA	11.3	11.3	11.6	12.9	12.4	12.9
Women of reproductive age (15-49 yrs.)	NA	NA	NA	1,246	1,265	1,267	1,156	1,053	945
of which are currently married	NA	NA	NA	NA	896	902	840	NA	NA
Fertility rate	NA	NA	NA	1.4	1.4	1.7	1.6	1.5	1.5

Except as noted, values for vital statistics are in thousands; life expectancy is in years.

Health

Health Indicators [3]

% of population with access to
  safe water (1990-95) — NA
  adequate sanitation (1990-95) — NA
  health services (1985-95) — NA
% of 1-year-olds immunized (1990-94) against
  TB (tuberculosis) — 92
  DPT (diphtheria, pertussis, tetanus) — 85
  polio — 85
  measles — 90
% of contraceptive prevalence (1980-94) — NA
ORT use rate (1990-94) — NA

Health Expenditures [4]

For sources, notes, and explanations, see Annotated Source Appendix, page 1061.

Human Factors

Women and Children [5]

% of pregnant women immunized (tetanus 1990-94)	NA
% of births attended by trained health personnel (1983-94)	NA
Maternal mortality rate (1980-92)	NA
Under-5 mortality rate (1994)	14
% under-5 moderately/severely underweight (1980-1994)	NA

Burden of Disease [6]

Ethnic Division [7]

Croat	78%
Serb	12%
Muslim	0.9%
Hungarian	0.5%
Slovenian	0.5%
Others (1991)	8.1%

Religion [8]

Catholic	76.5%
Orthodox	11.1%
Slavic Muslim	1.2%
Protestant	0.4%
Others and unknown	10.8%

Major Languages [9]

Serbo-Croatian	96%
Other (primarily Italian, Hungarian, Czechoslovak and German)	4%

Education

Public Education Expenditures [10]

Educational Attainment [11]

Age group (1991)[5]	25+
Total population	2,969,584
Highest level attained (%)	
No schooling	10.2
First level	
Not completed	43.6
Completed	NA
Entered second level	
S-1	39.5
S-2	NA
Postsecondary	6.4

Illiteracy [12]

	1989	1991	1995
Illiterate population (15+ yrs.)[2]	NA	126,624	87,000
Illiteracy rate - total pop. (%)	NA	3.3	2.4
Illiteracy rate - males (%)	NA	1.2	1.8
Illiteracy rate - females (%)	NA	5.1	2.9

Libraries [13]

	Admin. Units	Svc. Pts.	Vols. (000)	Shelving (meters)	Vols. Added	Reg. Users
National (1992)	1	NA	2,333	61,725	19,785	223,239
Nonspecialized (1992)	5	NA	2,457	44,456	23,703	96,366
Public (1992)	250	NA	4,631	104,082	232,784	6 mil
Higher ed. (1992)	128	136	3,433	75,073	80,464	101,500
School (1989)	7,784	7,784	38,430	NA	1,680	NA

Daily Newspapers [14]

	1980	1985	1990	1994
Number of papers	NA	8	9	6
Circ. (000)	NA	NA	2,400[e]	2,600[e]

Culture [15]

Cinema (seats per 1,000)	11.5
Annual attendance per person	0.8
Gross box office receipts (mil. Dinar)	16,196
Museums (reporting)	90
Visitors (000)	580
Annual receipts (000 Dinar)	65,613

Science and Technology

Scientific/Technical Forces [16]

Scientists/engineers	8,928
Number female	3,339
Technicians	3,818
Number female	2,332
Total	12,746

R&D Expenditures [17]

	(000) 1992
Total expenditure	NA
Capital expenditure	NA
Current expenditure	21,874,940
Percent current	NA

U.S. Patents Issued [18]

Values show patents issued to citizens of the country by the U.S. Patents Office.

	1993	1994	1995
Number of patents	2	1	6

For sources, notes, and explanations, see Annotated Source Appendix, page 1061.

Government and Law

Organization of Government [19]

Long-form name:
  Republic of Croatia
Type:
  Parliamentary democracy
Independence:
  25 June 1991 (from Yugoslavia)
National holiday:
  Statehood Day, 30 May (1990)
Constitution:
  Adopted on 22 December 1990
Legal system:
  Based on civil law system
Executive branch:
  President; Prime Minister; 5 Deputy Prime
  Ministers; Council of Ministers
Legislative branch:
  Bicameral parliament Assembly (Sabor):
  House of Districts (Zupanije Dom) and
  House of Representatives (Zastupnicki
  Dom)
Judicial branch:
  Supreme Court; Constitutional Court

Elections [20]

House of Representatives	% of seats
Croatian Democratic Union (HDZ)	59.1
Croatian Soc. Liberal Party (HSLS)	9.4
Social Democratic Party (SDP)	7.9
Croatian Peasants' Party (HSS)	7.9
Istrian Dem. Assembly (IDS)	3.1
Croatian Party of Rights (HSP)	3.1
Independents	3.1
Others	6.3

Government Expenditures [21]

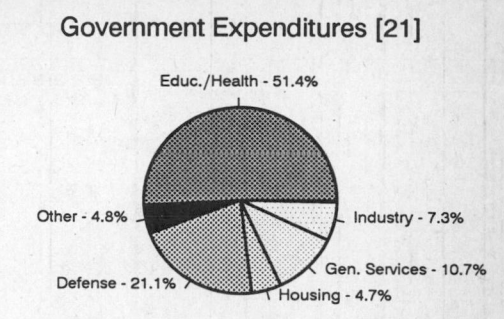

Educ./Health - 51.4%
Other - 4.8%
Industry - 7.3%
Gen. Services - 10.7%
Defense - 21.1%
Housing - 4.7%

(% distribution). Expend. for CY95: 43,946 (Kuna mil.)

Crime [23]

	1994
Crime volume	
Cases known to police	64,051
Attempts (percent)	3.40
Percent cases solved	69.10
Crimes per 100,000 persons	1,334.40
Persons responsible for offenses	
Total number offenders	37,710
Percent female	7.20
Percent juvenile (14-18 yrs.)	9.10
Percent foreigners	4.80

Military Expenditures and Arms Transfers [22]

	1990	1991	1992[15]	1993	1994
Military expenditures					
Current dollars (mil.)	NA	NA	NA	NA	NA
1994 constant dollars (mil.)	NA	NA	NA	NA	NA
Armed forces (000)	NA	NA	103[e]	NA	80
Gross national product (GNP)					
Current dollars (mil.)[e]	NA	NA	18,000	10,700	12,400
1994 constant dollars (mil.)[e]	NA	NA	18,770	10,920	12,400
Central government expenditures (CGE)					
1994 constant dollars (mil.)	NA	NA	NA	NA	NA
People (mil.)	NA	NA	4.5	4.6	4.7
Military expenditure as % of GNP	NA	NA	NA	NA	NA
Military expenditure as % of CGE	NA	NA	NA	NA	NA
Military expenditure per capita (1994 $)	NA	NA	NA	NA	NA
Armed forces per 1,000 people (soldiers)	NA	NA	23.2	NA	17.2
GNP per capita (1994 $)	NA	NA	4,217	2,375	2,660
Arms imports[6]					
Current dollars (mil.)	NA	NA	0	20	30
1994 constant dollars (mil.)	NA	NA	0	20	30
Arms exports[6]					
Current dollars (mil.)	NA	NA	0	0	0
1994 constant dollars (mil.)	NA	NA	0	0	0
Total imports[7]					
Current dollars (mil.)	NA	NA	4,500	4,666	5,231
1994 constant dollars (mil.)	NA	NA	4,693	4,762	5,231
Total exports[7]					
Current dollars (mil.)	NA	NA	4,597	3,913	4,259
1994 constant dollars (mil.)	NA	NA	4,794	3,994	4,259
Arms as percent of total imports[8]	NA	NA	0	0.4	0.6
Arms as percent of total exports[8]	NA	NA	0	0	0

Human Rights [24]

	SSTS	FL	FAPRO	PPCG	APROBC	TPW	PCPTW	STPEP	PHRFF	PRW	ASST	AFL
Observes	P	P	P	P	P	P	P	P	P	P	P	
	EAFRD	CPR	ESCR	SR	ACHR	MAAE	PVIAC	PVNAC	EAFDAW	TCIDTP	RC	
Observes	P	P	P	P		P	P	P	P	P	P	

P = Party; S = Signatory; see Appendix for meaning of abbreviations.

Labor Force

Total Labor Force [25]

1.444 million (1995)

Labor Force by Occupation [26]

Industry and mining	31.1%
Agriculture	4.3
Government (including education and health)	19.1
Other	45.5

Date of data: 1993

Unemployment Rate [27]

18.1% (January 1996)

For sources, notes, and explanations, see Annotated Source Appendix, page 1061.

235

Production Sectors

Commercial Energy Production and Consumption

Data are shown in quadrillion (10^{15}) BTUs and percent for 1995

Values for hydroelectric, nuclear, geothermal, solar, and wind power refer to electrical generation.

Production [28]

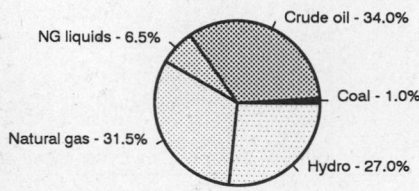

NG liquids - 6.5%
Crude oil - 34.0%
Coal - 1.0%
Natural gas - 31.5%
Hydro - 27.0%

Consumption [29]

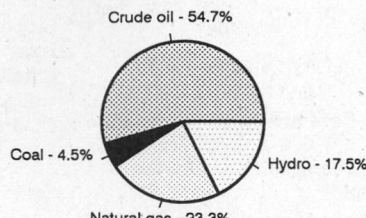

Crude oil - 54.7%
Coal - 4.5%
Hydro - 17.5%
Natural gas - 23.3%

Crude oil	0.068
Natural gas liquids	0.013
Dry natural gas	0.063
Coal	0.002
Net hydroelectric power	0.054
Total	0.200

Crude oil	0.169
Dry natural gas	0.072
Coal	0.014
Net hydroelectric power	0.054
Total	0.309

Telecommunications [30]

- 1.216 million (1993 est.) telephones
- International: no satellite earth stations
- Radio: Broadcast stations: AM 14, FM 8, shortwave 0 Radios: 1.1 million
- Television: Broadcast stations: 12 (repeaters 2) Televisions: 1.52 million (1992 est.)

Transportation [31]

Railways: total: 2,699 km; standard gauge: 2,699 km 1.435-m gauge (1213 km electrified)

Highways: total: 27,378 km; paved: 22,176 km (including 302 km of expressways); unpaved: 5,202 km (1991 est.)

Merchant marine: total: 39 ships (1,000 GRT or over) totaling 203,495 GRT/252,818 DWT; ships by type: bulk 2, cargo 23, chemical tanker 1, container 3, oil tanker 1, passenger 2, refrigerated cargo 1, roll-on/roll-off cargo 2, short-sea passenger 4

Airports

Total:	68
With paved runways over 3,047 m:	2
With paved runways 2,438 to 3,047 m:	6
With paved runways 1,524 to 2,437 m:	2
With paved runways 914 to 1,523 m:	3
With paved runways under 914 m:	47

Top Agricultural Products [32]

Agriculture accounts for 12.7% of the GDP; produces wheat, corn, sugar beets, sunflower seed, alfalfa, clover, olives, citrus, grapes, vegetables; livestock breeding, dairy farming.

Top Mining Products [33]

Metric tons except as noted[e]	3/95*
Aluminum ingots, primary/secondary	25,000
Ferrochromium	31,700
Ferrosilicomanganese	30,000
Steel, crude	73,000
Fire clay, crude	30,000
Gypsum, crude	50,000
Quartz, quartzite, glass sand	25,000
Salt, all sources	30,000
Ornamental dimension stone, crude (cu. m.)	1,100,000
Carbon black	15,000

Tourism [34]

	1990	1991	1992	1993	1994
Tourists[11]	7,049	1,346	1,271	1,521	2,293
Tourism receipts	1,704	300	543	832	1,427
Tourism expenditures	729	231	158	298	552

Travelers are in thousands, money in million U.S. dollars.

Manufacturing Sector

GDP and Manufacturing Summary [35]

	1980	1985	1989	1990	% change 1980-1990	% change 1989-1990
GDP (million 1980 $)	69,958	71,058	72,234	66,371	-5.1	-8.1
GDP per capita (1980 $)	3,136	3,073	3,050	2,786	-11.2	-8.7
Manufacturing as % of GDP (current prices)	30.6	37.2	39.5	42.0	37.3	6.3
Gross output (million $)	72,629	57,020	65,078	62,136[e]	-14.4	-4.5
Value added (million $)	21,750	17,171	30,245	27,660[e]	27.2	-8.5
Value added (million 1980 $)	19,526	22,283	24,021	21,703	11.1	-9.6
Industrial production index	100	116	120	108	8.0	-10.0
Employment (thousands)	2,106	2,467	2,658	2,537[e]	20.5	-4.6

Note: GDP stands for Gross Domestic Product. 'e' stands for estimated value.

Profitability and Productivity

	1980	1985	1989	1990	% change 1980-1990	% change 1989-1990
Intermediate input (%)	70	70	54	55[e]	-21.4	1.9
Wages, salaries, and supplements (%)	14	12	12[e]	18[e]	28.6	50.0
Gross operating surplus (%)	15	18	34[e]	26[e]	73.3	-23.5
Gross output per worker ($)	34,487	23,113	24,484	24,248[e]	-29.7	-1.0
Value added per worker ($)	10,328	6,960	11,379	10,796[e]	4.5	-5.1
Average wage (incl. benefits) ($)	4,991	2,703	2,986[e]	4,488[e]	-10.1	50.3

Profitability is in percent of gross output. Productivity is in U.S. $. 'e' stands for estimated value.

Profitability - 1990

Wages - 10.2%
Inputs - 55.6%
Surplus - 26.3%

The graphic shows percent of gross output.

Value Added in Manufacturing

	1980 $ mil.	1980 %	1985 $ mil.	1985 %	1989 $ mil.	1989 %	1990 $ mil.	1990 %	% change 1980-1990	% change 1989-1990
311 Food products	1,897	8.7	1,458	8.5	3,916	12.9	3,484[e]	12.6	83.7	-11.0
313 Beverages	459	2.1	353	2.1	663	2.2	589[e]	2.1	28.3	-11.2
314 Tobacco products	184	0.8	221	1.3	344	1.1	308[e]	1.1	67.4	-10.5
321 Textiles	1,759	8.1	1,428	8.3	2,881	9.5	2,663[e]	9.6	51.4	-7.6
322 Wearing apparel	903	4.2	718	4.2	1,593	5.3	1,427[e]	5.2	58.0	-10.4
323 Leather and fur products	226	1.0	231	1.3	383	1.3	340[e]	1.2	50.4	-11.2
324 Footwear	482	2.2	503	2.9	1,022	3.4	899[e]	3.3	86.5	-12.0
331 Wood and wood products	977	4.5	530	3.1	794	2.6	706[e]	2.6	-27.7	-11.1
332 Furniture and fixtures	730	3.4	438	2.6	1,030	3.4	1,065[e]	3.9	45.9	3.4
341 Paper and paper products	529	2.4	394	2.3	759	2.5	674[e]	2.4	27.4	-11.2
342 Printing and publishing	876	4.0	462	2.7	761	2.5	678[e]	2.5	-22.6	-10.9
351 Industrial chemicals	694	3.2	631	3.7	1,107	3.7	992[e]	3.6	42.9	-10.4
352 Other chemical products	681	3.1	525	3.1	1,419	4.7	1,315[e]	4.8	93.1	-7.3
353 Petroleum refineries	454	2.1	415	2.4	260	0.9	233[e]	0.8	-48.7	-10.4
354 Miscellaneous petroleum and coal products	101	0.5	101	0.6	104	0.3	91[e]	0.3	-9.9	-12.5
355 Rubber products	276	1.3	269	1.6	479	1.6	456[e]	1.6	65.2	-4.8
356 Plastic products	413	1.9	258	1.5	397	1.3	350[e]	1.3	-15.3	-11.8
361 Pottery, china, and earthenware	128	0.6	72	0.4	162	0.5	144[e]	0.5	12.5	-11.1
362 Glass and glass products	163	0.7	113	0.7	224	0.7	204[e]	0.7	25.2	-8.9
369 Other nonmetal mineral products	906	4.2	513	3.0	683	2.3	604[e]	2.2	-33.3	-11.6
371 Iron and steel	1,221	5.6	1,000	5.8	1,343	4.4	1,171[e]	4.2	-4.1	-12.8
372 Nonferrous metals	480	2.2	509	3.0	944	3.1	927[e]	3.4	93.1	-1.8
381 Metal products	2,105	9.7	1,577	9.2	1,293	4.3	1,130[e]	4.1	-46.3	-12.6
382 Nonelectrical machinery	1,828	8.4	1,463	8.5	2,372	7.8	2,378[e]	8.6	30.1	0.3
383 Electrical machinery	1,600	7.4	1,544	9.0	2,640	8.7	2,334[e]	8.4	45.9	-11.6
384 Transport equipment	1,441	6.6	1,263	7.4	2,389	7.9	2,241[e]	8.1	55.5	-6.2
385 Professional and scientific equipment	101	0.5	93	0.5	154	0.5	146[e]	0.5	44.6	-5.2
390 Other manufacturing industries	134	0.6	88	0.5	128	0.4	114[e]	0.4	-14.9	-10.9

Note: The industry codes shown are International Standard Industry codes (ISIC). Percentages are percent of total Value Added. 'e' stands for estimated value

For sources, notes, and explanations, see Annotated Source Appendix, page 1061.

Finance, Economics, and Trade

Economic Indicators [36]

- **National product**: GDP—purchasing power parity—$20.1 billion (1995 est.)

- **National product real growth rate**: 1.5% (1995 est.)

- **National product per capita**: $4,300 (1995 est.)

- **Inflation rate (consumer prices)**: 3.7% (1995)

- **External debt**: $3.15 billion (September 1995)

Balance of Payments Summary [37]

Values in millions of dollars.

	1989	1990	1991	1992	1993[3]
Exports of goods (f.o.b.)	NA	NA	NA	NA	43,711
Imports of goods (f.o.b.)	NA	NA	NA	NA	34,163
Trade balance	NA	NA	NA	NA	9,548
Services - debits	NA	NA	NA	NA	-14,580
Services - credits	NA	NA	NA	NA	8,939
Private transfers (net)	NA	NA	NA	NA	2,324
Government transfers (net)	NA	NA	NA	NA	NA
Long-term capital (net)	NA	NA	NA	NA	-883
Short-term capital (net)	NA	NA	NA	NA	-7,323
Errors and omissions	NA	NA	NA	NA	4,850
Overall balance	NA	NA	NA	NA	2,875

Exchange Rates [38]

Currency: **Croatian kuna.**
Symbol: **HRK.**

Data are currency units per $1.

January 1996	5.405
1995	5.230
1994	5.996
1993	3.577

Imports and Exports

Top Import Origins [39]

$5.2 billion (c.i.f., 1994).

Origins	%
Germany	NA
Italy	NA
Slovenia	NA
Iran	NA

Top Export Destinations [40]

$4.3 billion (f.o.b., 1994) Data are for 1993.

Destinations	%
Germany	22.9
Italy	21.2
Slovenia	18.3

Foreign Aid [41]

	U.S. $	
ODA	NA	
IMF	192	million
World Bank	100	million

Import and Export Commodities [42]

Import Commodities

Machinery and transport equipment 23.1%
Fuels and lubricants 8.8%
Food and live animals 9.0%
Chemicals 14.2%
Miscellaneous manufactured articles 16.0%
Raw materials 3.5%
Beverages and tobacco 1.4%

Export Commodities

Machinery and transport equipment 13.6%
Misc. manufactures 27.6%
Chemicals 14.2%
Food and live animals 12.2%
Raw materials 6.1%
Fuels and lubricants 9.4%
Beverages and tobacco 2.7%

Cuba

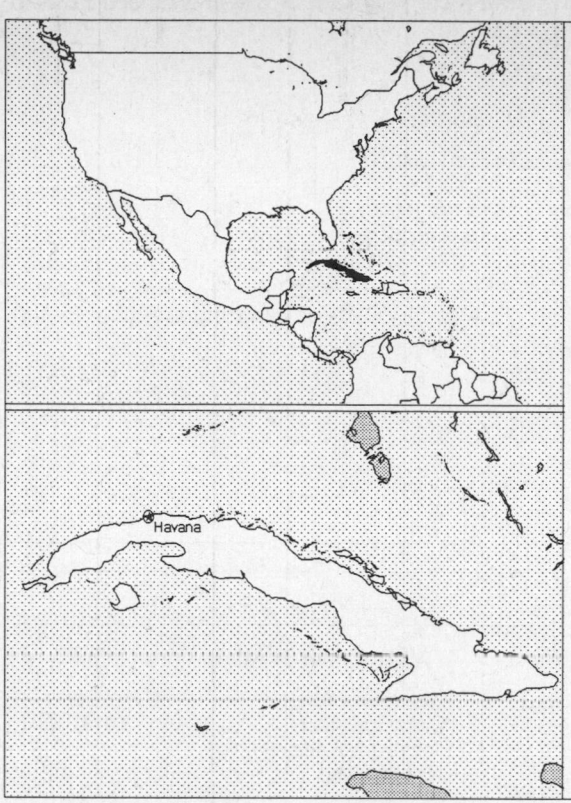

Geography [1]

Total area:
110,860 sq km 42,803 sq mi
Land area:
110,860 sq km 42,803 sq mi
Comparative area:
Slightly smaller than Pennsylvania
Land boundaries:
Total 29 km, US Naval Base at Guantanamo Bay 29 km
Note: Guantanamo Naval Base is leased by the US and thus remains part of Cuba
Coastline:
3,735 km
Climate:
Tropical; moderated by trade winds; dry season (November to April); rainy season (May to October)
Terrain:
Mostly flat to rolling plains with rugged hills and mountains in the Southeast
Natural resources:
Cobalt, nickel, iron ore, copper, manganese, salt, timber, silica, petroleum
Land use:
Arable land: 23%
Permanent crops: 6%
Meadows and pastures: 23%
Forest and woodland: 17%
Other: 31%

Demographics [2]

	1970	1980	1990	1995[1]	1996	2000	2010	2020	2030
Population	8,543	9,653	10,545	10,902	10,951	11,131	11,481	11,699	11,614
Population density (persons per sq. mi.)	200	226	246	255	256	260	268	273	271
(persons per sq. km.)	77	87	95	98	99	100	104	106	105
Net migration rate (per 1,000 population)	-6.4	-14.7	-1.0	-1.6	-1.5	-1.5	0.0	0.0	0.0
Births	237	137	NA	NA	158	NA	NA	NA	NA
Deaths	NA	NA	72	NA	72	NA	NA	NA	NA
Life expectancy - males	NA	NA	72.8	72.6	72.7	73.2	74.3	75.3	76.2
Life expectancy - females	NA	NA	77.6	77.4	77.5	78.2	79.7	81.0	82.1
Birth rate (per 1,000)	27.8	14.2	17.7	13.6	13.4	12.7	11.0	10.1	9.1
Death rate (per 1,000)	7.1	5.7	6.8	7.4	7.4	7.5	8.2	9.4	11.2
Women of reproductive age (15-49 yrs.)	NA	NA	2,978	3,022	3,010	3,021	3,072	2,660	2,404
of which are currently married	838	NA	1,941	NA	2,083	2,108	2,137	NA	NA
Fertility rate	3.8	1.7	1.8	1.5	1.5	1.6	1.6	1.6	1.7

Except as noted, values for vital statistics are in thousands; life expectancy is in years.

Health

Health Indicators [3]

% of population with access to	
safe water (1990-95)	93
adequate sanitation (1990-95)	66
health services (1985-95)	98
% of 1-year-olds immunized (1990-94) against	
TB (tuberculosis)	99
DPT (diphtheria, pertussis, tetanus)	100
polio	NA
measles	NA
% of contraceptive prevalence (1980-94)	70
ORT use rate (1990-94)	80

Health Expenditures [4]

Total health expenditure, 1990 (official exchange rate)	
Millions of dollars	NA
Dollars per capita	NA
Health expenditures as a percentage of GDP	
Total	NA
Public sector	NA
Private sector	NA
Development assistance for health	
Total aid flows (millions of dollars)[1]	3
Aid flows per capita (dollars)	0.3
Aid flows as a percentage of total health expenditure	NA

For sources, notes, and explanations, see Annotated Source Appendix, page 1061.

239

Human Factors

Women and Children [5]

% of pregnant women immunized (tetanus 1990-94)	61
% of births attended by trained health personnel (1983-94)	90
Maternal mortality rate (1980-92)	39
Under-5 mortality rate (1994)	10
% under-5 moderately/severely underweight (1980-1994)	NA

Burden of Disease [6]

Population per physician (1990)	274.62
Population per nurse (1990)	81.85
Population per hospital bed (1990)	206.55
AIDS cases per 100,000 people (1994)	0.8
Malaria cases per 100,000 people (1992)	NA

Ethnic Division [7]

Mulatto	51%
White	37%
Black	11%
Chinese	1%

Religion [8]

Data are prior to Castro assuming power.

Nominally Roman Catholic	85%
Other (Protestants, Jehovah's Witnesses, Jews, and Santeria)	15%

Major Languages [9]

Spanish.

Education

Public Education Expenditures [10]

Million (Peso)[18]	1980	1985	1989	1990	1993	1994
Total education expenditure	1,267	1,690	1,778	1,748	NA	NA
as percent of GNP	7.2	6.3	6.6	6.6	NA	NA
as percent of total govt. expend.	NA	NA	12.8	12.3	NA	NA
Current education expenditure	1,135	1,588	1,659	1,627	NA	NA
as percent of GNP	6.4	5.9	6.2	6.1	NA	NA
as percent of current govt. expend.	NA	NA	15.4	14.4	NA	NA
Capital expenditure	133	103	119	121	NA	NA

Educational Attainment [11]

Age group (1981)	25-49
Total population	3,013,315
Highest level attained (%)	
No schooling	3.7
First level	
Not completed	22.6
Completed	27.6
Entered second level	
S-1	40.2
S-2	NA
Postsecondary	5.9

Illiteracy [12]

In thousands and percent[1]	1990	1995	2000
Illiterate population (15+ yrs.)	468	364	279
Illiteracy rate - total pop. (%)	5.8	4.3	3.2
Illiteracy rate - males (%)	5.0	3.9	3.0
Illiteracy rate - females (%)	6.5	4.7	3.4

Libraries [13]

	Admin. Units	Svc. Pts.	Vols. (000)	Shelving (meters)	Vols. Added	Reg. Users
National (1992)	1	1	2,431	NA	55,900	13,299
Nonspecialized	NA	NA	NA	NA	NA	NA
Public (1992)	353	4,627	5,326	NA	337,500	207,411
Higher ed. (1993)[9]	86	86	2,525	NA	2,340	270,540
School (1990)	3,780	3,780	NA	NA	NA	NA

Daily Newspapers [14]

	1980	1985	1990	1994
Number of papers	17	17	19	17
Circ. (000)	1,050	1,207	1,824	1,315

Culture [15]

Cinema (seats per 1,000)	17.3
Annual attendance per person	2.2
Gross box office receipts (mil. Peso)	6.4
Museums (reporting)[4]	216
Visitors (000)	6,734
Annual receipts (000 Peso)	NA

Science and Technology

Scientific/Technical Forces [16]

Scientists/engineers	14,770
Number female	6,383
Technicians	9,465
Number female	5,251
Total[9]	24,235

R&D Expenditures [17]

	Peso (000) 1992
Total expenditure[7]	247,925
Capital expenditure	91,216
Current expenditure	156,709
Percent current	63.2

U.S. Patents Issued [18]

Values show patents issued to citizens of the country by the U.S. Patents Office.

	1993	1994	1995
Number of patents	2	4	0

For sources, notes, and explanations, see Annotated Source Appendix, page 1061.

Government and Law

Organization of Government [19]

Long-form name:
Republic of Cuba
Type:
Communist state
Independence:
20 May 1902 (from Spain 10 December 1898; administered by US 1898 to 1902)
National holiday:
Rebellion Day, 26 July (1953); Liberation Day, 1 January (1959)
Constitution:
24 February 1976
Legal system:
Based on Spanish and American law, with large elements of Communist legal theory; does not accept compulsory ICJ jurisdiction
Executive branch:
President of the Council of State; President of the Council of Ministers; First Vice President of the Council of State; First Vice President of the Council of Ministers; Council of Ministers; Council of State
Legislative branch:
Unicameral: National Assembly of People's Power
Judicial branch:
People's Supreme Court

Elections [20]

National Assembly of People's Power. Elections last held February 1993; seats - 589 total, indirectly elected from slates approved by special candicacy commissions. Only one party, Cuban Communist Party (PCC).

Government Budget [21]

Crime [23]

Military Expenditures and Arms Transfers [22]

	1990	1991	1992	1993	1994
Military expenditures					
Current dollars (mil.)[3]	1,400	1,160[e]	500[e]	426[e]	350[e]
1994 constant dollars (mil.)[3]	1,558	1,243[e]	521[e]	435[e]	350[e]
Armed forces (000)	297	297	175	175	140
Gross national product (GNP)					
Current dollars (mil.)[e]	33,690	26,950	23,850	21,460	22,000
1994 constant dollars (mil.)[e]	37,490	28,890	24,870	21,910	22,000
Central government expenditures (CGE)					
1994 constant dollars (mil.)	NA	NA	NA	NA	NA
People (mil.)	10.5	10.6	10.7	10.8	10.9
Military expenditure as % of GNP	4.2	4.3	2.1	2.0	1.6
Military expenditure as % of CGE	NA	NA	NA	NA	NA
Military expenditure per capita (1994 $)	148	117	49	40	32
Armed forces per 1,000 people (soldiers)	28.1	27.9	16.3	16.2	12.9
GNP per capita (1994 $)	3,556	2,715	2,319	2,030	2,025
Arms imports[6]					
Current dollars (mil.)	1,400	525	100	100	0
1994 constant dollars (mil.)	1,558	563	104	102	0
Arms exports[6]					
Current dollars (mil.)	0	0	0	0	0
1994 constant dollars (mil.)	0	0	0	0	0
Total imports[7]					
Current dollars (mil.)	6,745	3,690	2,245	1,855	1,855
1994 constant dollars (mil.)	7,507	3,955	2,341	1,893	1,855
Total exports[7]					
Current dollars (mil.)	4,910	3,550	2,030	1,310	1,360
1994 constant dollars (mil.)	5,464	3,805	2,117	1,337	1,360
Arms as percent of total imports[8]	20.8	14.2	4.5	5.4	0
Arms as percent of total exports[8]	0	0	0	0	0

Human Rights [24]

	SSTS	FL	FAPRO	PPCG	APROBC	TPW	PCPTW	STPEP	PHRFF	PRW	ASST	AFL
Observes	P	P	P	P	P	P	P	P		P	P	P

	EAFRD	CPR	ESCR	SR	ACHR	MAAE	PVIAC	PVNAC	EAFDAW	TCIDTP	RC
Observes		P					P	P		P	P	P

P=Party; S=Signatory; see Appendix for meaning of abbreviations.

Labor Force

Total Labor Force [25]

4.71 million economically active population (1989); 3.527 million employed in state civilian sector (1989)

Labor Force by Occupation [26]

Services and government	30%
Industry	22
Agriculture	20
Commerce	11
Construction	10
Transportation and communications	7

Date of data: June 1990

Unemployment Rate [27]

For sources, notes, and explanations, see Annotated Source Appendix, page 1061.

241

Production Sectors

Commercial Energy Production and Consumption

Data are shown in quadrillion (10^{15}) BTUs and percent for 1995
Values for hydroelectric, nuclear, geothermal, solar, and wind power refer to electrical generation.

Production [28]

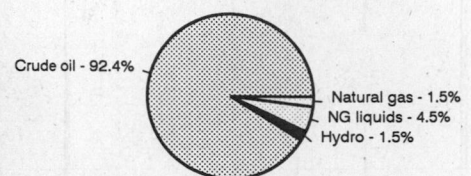

Crude oil - 92.4%
Natural gas - 1.5%
NG liquids - 4.5%
Hydro - 1.5%

Consumption [29]

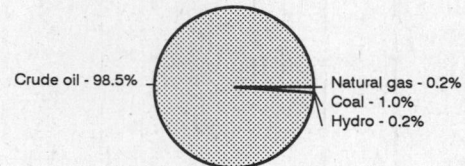

Crude oil - 98.5%
Natural gas - 0.2%
Coal - 1.0%
Hydro - 0.2%

Crude oil	0.061
Natural gas liquids	0.003
Dry natural gas	0.001
Net hydroelectric power	0.001
Total	0.066

Crude oil	0.401
Dry natural gas	0.001
Coal	0.004
Net hydroelectric power	0.001
Total	0.407

Telecommunications [30]

- 430,000 (1987 est.) telephones; among the world's least developed telephone systems
- International: satellite earth station - 1 Intersputnik (Atlantic Ocean Region)
- Radio: Broadcast stations: AM 150, FM 5, shortwave 0 Radios: 2.14 million (1993 est.)
- Television: Broadcast stations: 58 Televisions: 2.5 million (1993 est.)

Transportation [31]

Railways: total: 4,677 km; standard gauge: 4,677 km 1.435-m gauge (132 km electrified)

Highways: total: 26,500 km; paved: 14,575 km; unpaved: 11,925 km (1996 est.)

Merchant marine: total: 41 ships (1,000 GRT or over) totaling 220,870 GRT/310,169 DWT; ships by type: cargo 17, chemical tanker 1, liquefied gas tanker 4, oil tanker 9, passenger-cargo 1, refrigerated cargo 9

Airports

Total:	156
With paved runways over 3,047 m:	7
With paved runways 2,438 to 3,047 m:	7
With paved runways 1,524 to 2,437 m:	14
With paved runways 914 to 1,523 m:	9
With paved runways under 914 m:	87

Top Agricultural Products [32]

Agriculture accounts for 7% of the GDP; produces sugarcane, tobacco, citrus, coffee, rice, potatoes and other tubers, beans; livestock.

Top Mining Products [33]

Metric tons except as noted[e]	2/95[*]
Cement, hydraulic (000 tons)	1,000
Cobalt	1,000[22]
Copper, mine output, Cu content	1,400
Gas, natural (000 cu. meters)	
gross	37,000
marketed	4,000
Nickel, mine output, Ni-Co content of oxide and sulfide	26,000
Petroleum refinery products (000 42-gal. bls.)	55,500

Tourism [34]

	1990	1991	1992	1993	1994
Visitors[1]	340	424	461	546	619
Tourists[12]	327	418	455	544	617
Excursionists[28]	13	6	6	2	2
Cruise passengers	7	NA	NA	NA	NA
Tourism receipts	243	387	567	720	850

Travelers are in thousands, money in million U.S. dollars.

Manufacturing Sector

Manufacturing Summary [35]

	1987		1988		1989		1990		1991	
	$ bil.	%	$ bil.	%	$ bil.	%	$ bil.	%	$ bil.	%
Establishments or enterprises (number)	452	0.019	472	0.022	467	0.025	-	-	-	-
Total employment (000)	408	0.301	417	0.305	429	0.349	-	-	-	-
Production workers (000)	311	0.461	325	0.520	341	0.463	-	-	-	-
Output ($ bil.)	6	0.058	6	0.052	6	0.050	-	-	-	-
Value added ($ bil.)	1	0.027	1	0.029	1	0.025	-	-	-	-
Capital investment ($ mil.)	666	0.153	757	0.159	713	0.130	-	-	-	-
M & E investment ($ mil.)	282	0.091	292	0.084	-	-	-	-	-	-
Employees per establishment (number)	903	1,561	883	1,355	919	1,399	-	-	-	-
Production workers per establishment	688	2,394	688	2,313	729	1,857	-	-	-	-
Output per establishment ($ mil.)	13	299	13	233	13	201	-	-	-	-
Capital investment per estab. ($ mil.)	1	793	2	707	2	522	-	-	-	-
M & E per establishment ($ mil)	0.625	470	0.618	375	-	-	-	-	-	-
Payroll per employee ($)	2,201	24.549	2,237	22.449	2,265	20.270	-	-	-	-
Wages per production worker ($)	-	-	-	-	-	-	-	-	-	-
Hours per production worker (hours)	1,944	102.438	1,898	98.755	1,917	104.580	-	-	-	-
Output per employee ($)	14,364	19.179	14,628	17.174	13,755	14.352	-	-	-	-
Capital investment per employee ($)	1,631	50.800	1,816	52.191	1,660	37.298	-	-	-	-
M & E per employee ($)	692	30.123	700	27.659	-	-	-	-	-	-

Note: Columns headed % show percent of world total or ratio. Ratios closest to 100 are closest to world average. M & E stands for machinery & equipment.

Output in Manufacturing

	1987		1988		1989		1990		1991	
	$ bil.	%	$ bil.	%	$ bil.	%	$ bil.	%	$ bil.	%
3110 - Food products	3.811	63.52	4.018	66.97	3.894	64.90	-	-	-	-
3210 - Textiles	0.332	5.53	0.343	5.72	0.296	4.93	-	-	-	-
3220 - Wearing apparel	0.224	3.73	0.219	3.65	0.180	3.00	-	-	-	-
3410 - Paper and products	0.190	3.17	0.131	2.18	0.141	2.35	-	-	-	-
3420 - Printing, publishing	0.099	1.65	0.144	2.40	0.140	2.33	-	-	-	-
3710 - Iron and steel	0.190	3.17	0.172	2.87	0.181	3.02	-	-	-	-
3720 - Nonferrous metals	0.178	2.97	0.196	3.27	0.225	3.75	-	-	-	-
3810 - Metal products	0.192	3.20	0.195	3.25	0.170	2.83	-	-	-	-
3830 - Electrical machinery	0.139	2.32	0.169	2.82	0.137	2.28	-	-	-	-
3900 - Industries nec	0.510	8.50	0.513	8.55	0.541	9.02	-	-	-	-

Note. Codes are International Standard Industry codes (ISIC). Percentages are % of total Output. [f]: Factor Prices; [p]: Producer Prices

Finance, Economics, and Trade

Economic Indicators [36]

- **National product**: GDP—purchasing power parity— $14.7 billion (1995 est.)

- **National product real growth rate**: 2.5% (1995 est.)

- **National product per capita**: $1,300 (1995 est.)

- **Inflation rate (consumer prices)**: NA%

- **External debt**: $9.1 billion (convertible currency,1995); another $20 billion owed to Russia (1995)

Balance of Payments Summary [37]

Exchange Rates [38]

Currency: **Cuban peso.**
Symbol: **Cu$.**

Data are currency units per $1. The peso is non-convertible and officially linked to the US dollar.

Official rate   1.0000

Imports and Exports

Top Import Origins [39]

$2.4 billion (c.i.f., 1995 est.) Data are for 1995 est.

Origins	%
Spain	15
Mexico	15
Russia	10

Top Export Destinations [40]

$1.6 billion (f.o.b., 1995 est.) Data are for 1995 est.

Destinations	%
Canada	15
China	15
Russia	15

Foreign Aid [41]

Recipient: ODA, $NA.

Import and Export Commodities [42]

Import Commodities	Export Commodities
Petroleum	Sugar
Food	Nickel
Machinery	Shellfish
Chemicals	Tobacco
	Medical products
	Citrus
	Coffee

Cyprus

Geography [1]

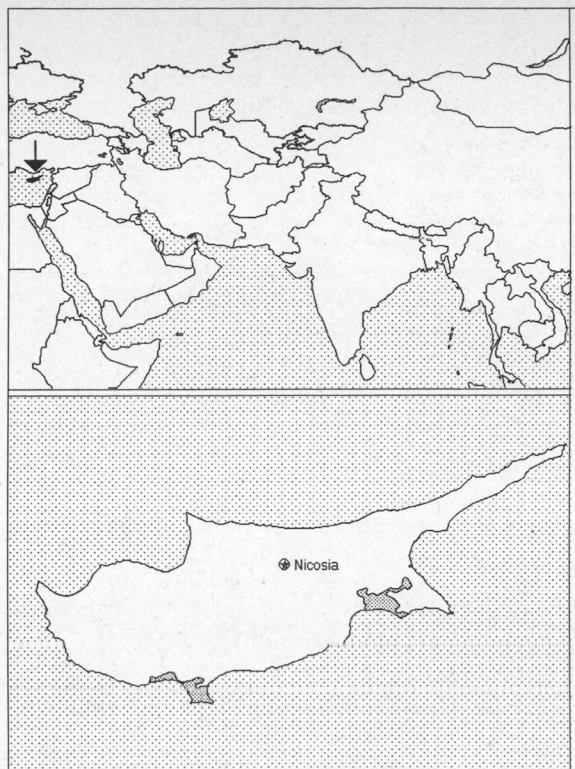

○ Nicosia

Total area:
   9,250 sq km   3,571 sq mi
   Note: 3,355 sq km (1295 sq mi) are in the Turkish area
Land area:
   9,240 sq km   3,568 sq mi
Comparative area:
   About 0.7 times the size of Connecticut
Land boundaries:
   0 km
Coastline:
   648 km
Climate:
   Temperate, Mediterranean with hot, dry summers and cool, wet winters
Terrain:
   Central plain with mountains to North and South; scattered but significant plains along southern coast
Natural resources:
   Copper, pyrites, asbestos, gypsum, timber, salt, marble, clay earth pigment
Land use:
   Arable land: 40%
   Permanent crops: 7%
   Meadows and pastures: 10%
   Forest and woodland: 18%
   Other: 25%

Demographics [2]

	1970	1980	1990	1995[1]	1996	2000	2010	2020	2030
Population	615	627	681	736	745	777	858	936	993
Population density (persons per sq. mi.)	172	176	191	206	209	218	240	262	278
(persons per sq. km.)	67	68	74	80	81	84	93	101	107
Net migration rate (per 1,000 population)	NA	0.5	7.7	3.4	3.4	3.2	2.9	2.7	2.5
Births	NA	13	NA	NA	11	NA	NA	NA	NA
Deaths	NA	NA	NA	NA	6	NA	NA	NA	NA
Life expectancy - males	NA	NA	72.4	73.8	74.1	75.2	77.3	78.7	79.5
Life expectancy - females	NA	NA	76.7	78.2	78.5	79.7	82.0	83.6	84.8
Birth rate (per 1,000)	NA	20.4	18.7	15.8	15.4	14.3	14.0	12.7	11.4
Death rate (per 1,000)	NA	9.3	8.4	7.7	7.7	7.3	7.3	8.0	9.2
Women of reproductive age (15-49 yrs.)	NA	NA	172	182	184	192	206	212	223
of which are currently married[11]	NA	NA	118	NA	127	130	139	NA	NA
Fertility rate	2.5	2.5	2.4	2.2	2.2	2.1	2.0	1.9	1.8

Except as noted, values for vital statistics are in thousands; life expectancy is in years.

Health

Health Indicators [3]

Health Expenditures [4]

For sources, notes, and explanations, see Annotated Source Appendix, page 1061.

245

Human Factors

Women and Children [5]

Burden of Disease [6]

Population per physician (1987)	724.48
Population per nurse (1986)	265.10
Population per hospital bed	NA
AIDS cases per 100,000 people (1994)	1.2
Malaria cases per 100,000 people (1992)	NA

Ethnic Division [7]

Greek	78%
Turkish	18%
Other	4%

Religion [8]

Greek Orthodox	78%
Muslim	18%
Maronite, Armenian Apostolic, and other	4%

Major Languages [9]

Greek, Turkish, English.

Education

Public Education Expenditures [10]

Million (Pound)[19]	1980	1985	1990	1992	1993	1994
Total education expenditure	27	55	89	120	139	156
as percent of GNP	3.5	3.7	3.5	3.9	4.3	NA
as percent of total govt. expend.	12.9	12.2	11.3	12.5	12.6	14.2
Current education expenditure	25	53	84	112	127	143
as percent of GNP	3.3	3.6	3.3	3.6	3.9	NA
as percent of current govt. expend.	16.2	13.4	12.3	13.1	12.9	14.8
Capital expenditure	2	3	5	9	13	14

Educational Attainment [11]

Age group (1992)	25+
Total population	363,573
Highest level attained (%)	
No schooling	5.1
First level	
Not completed	13.0
Completed	30.6
Entered second level	
S-1	34.2
S-2	NA
Postsecondary	17.0

Illiteracy [12]

	1989	1992	1995
Illiterate population (15+ yrs.)[2]	NA	25,216	NA
Illiteracy rate - total pop. (%)	NA	5.6	NA
Illiteracy rate - males (%)	NA	2.2	NA
Illiteracy rate - females (%)	NA	8.9	NA

Libraries [13]

	Admin. Units	Svc. Pts.	Vols. (000)	Shelving (meters)	Vols. Added	Reg. Users
National (1992)	1	2	54	NA	2,070	6,500
Nonspecialized	NA	NA	NA	NA	NA	NA
Public (1992)	20	NA	318	NA	20,800	43,050
Higher ed.	NA	NA	NA	NA	NA	NA
School	NA	NA	NA	NA	NA	NA

Daily Newspapers [14]

	1980	1985	1990	1994
Number of papers	12	10	11	15
Circ. (000)	80[e]	83	78	81

Culture [15]

Cinema (seats per 1,000)	12.0
Annual attendance per person	NA
Gross box office receipts (mil. Pound)	NA
Museums (reporting)	19
Visitors (000)	515
Annual receipts (000 Pound)	224

Science and Technology

Scientific/Technical Forces [16]

Scientists/engineers	147
Number female	NA
Technicians	165
Number female	NA
Total	312

R&D Expenditures [17]

	Pound (000) 1992
Total expenditure	5,578
Capital expenditure	772
Current expenditure	4,806
Percent current	86.2

U.S. Patents Issued [18]

Values show patents issued to citizens of the country by the U.S. Patents Office.

	1993	1994	1995
Number of patents	1	3	3

For sources, notes, and explanations, see Annotated Source Appendix, page 1061.

Government and Law

Organization of Government [19]

Long-form name:
Republic of Cyprus

Type:
Republic, Note: Ethnic disputes resulted in separate Turkish and Greek areas. Greek area is only internationally recognized government.

Independence:
16 August 1960 (from UK) Note: Turkish area proclaimed self-rule in February 1975

National holiday:
Independence Day, 1 October Note: Turkish area celebrates 15 November

Constitution:
16 August 1960; negotiations for an new constitution to govern both Greek and Turkish areas continue intermittently

Legal system:
Based on common law, with civil law modifications

Executive branch:
President; Council of Ministers

Legislative branch:
Unicameral: Greek area: House of Representatives; Turkish area: Assembly of the Republic

Judicial branch:
Supreme Court; Note: There is also a Supreme Court in the Turkish area

Elections [20]

House of Representatives: *Greek Area*: Last held 19 May 1991; results - DISY 35.8%, AKEL (Communist) 30.6%, DIKO 19.5%, EDEK 10.9%, others, 3.2%;
Assembly of the Republic: *Turkish Area*: Last held 12 September 1993 (next to be held NA); results - UBP 29.9%, DP 29.2%, CTP 24.2%, TKP 13.3%, others 3.4%

Government Expenditures [21]

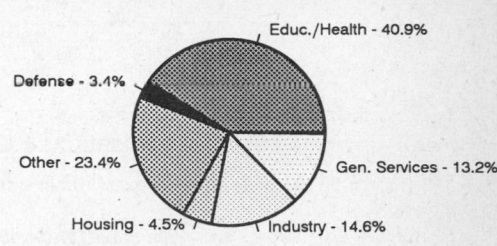

Educ./Health - 40.9%
Defense - 3.4%
Other - 23.4%
Gen. Services - 13.2%
Housing - 4.5%
Industry - 14.6%

(% distribution). Expend. for CY94: 1,192 (Pound mil.)

Military Expenditures and Arms Transfers [22]

	1990	1991	1992	1993	1994
Military expenditures					
Current dollars (mil.)	202	400	NA	240	338
1994 constant dollars (mil.)	224	429	NA	245	338
Armed forces (000)	10	10	10	NA	NA
Gross national product (GNP)					
Current dollars (mil.)	5,398	5,649	6,393	6,641	NA
1994 constant dollars (mil.)	6,008	6,055	6,667	6,778	NA
Central government expenditures (CGE)					
1994 constant dollars (mil.)	1,804	1,903	2,127	2,243	NA
People (mil.)	0.7	0.7	0.7	0.7	0.7
Military expenditure as % of GNP	3.7	7.1	NA	3.6	NA
Military expenditure as % of CGE	12.4	22.5	NA	10.9	NA
Military expenditure per capita (1994 $)	320	605	NA	338	463
Armed forces per 1,000 people (soldiers)	14.2	14.1	14.0	NA	NA
GNP per capita (1994 $)	8,555	8,534	9,305	9,369	NA
Arms imports[6]					
Current dollars (mil.)	50	40	150	30	60
1994 constant dollars (mil.)	56	43	156	31	60
Arms exports[6]					
Current dollars (mil.)	0	0	0	0	0
1994 constant dollars (mil.)	0	0	0	0	0
Total imports[7]					
Current dollars (mil.)	2,568	2,659	3,313	2,590	3,018
1994 constant dollars (mil.)	2,858	2,850	3,455	2,643	3,018
Total exports[7]					
Current dollars (mil.)	957	980	987	867	968
1994 constant dollars (mil.)	1,065	1,050	1,029	885	968
Arms as percent of total imports[8]	1.9	1.5	4.5	1.2	2.0
Arms as percent of total exports[8]	0	0	0	0	0

Crime [23]

	1994
Crime volume	
Cases known to police	4,369
Attempts (percent)	NA
Percent cases solved	110.10
Crimes per 100,000 persons	689.23
Persons responsible for offenses	
Total number offenders	1,173
Percent female	5.90
Percent juvenile (7-15 yrs.)	12.00
Percent foreigners	18.30

Human Rights [24]

	SSTS	FL	FAPRO	PPCG	APROBC	TPW	PCPTW	STPEP	PHRFF	PRW	ASST	AFL
Observes	P	P	P	P	P	P	P	P	P	P	P	P
	EAFRD	CPR	ESCR	SR	ACHR	MAAE	PVIAC	PVNAC	EAFDAW	TCIDTP	RC	
Observes	P	P	P	P			P	P	P	P	P	

P = Party; S = Signatory; see Appendix for meaning of abbreviations.

Labor Force

Total Labor Force [25]	Labor Force by Occupation [26]	Unemployment Rate [27]

For sources, notes, and explanations, see Annotated Source Appendix, page 1061.

247

Production Sectors

Energy Resource Summary [28]

Electricity: Capacity: 550,000 kW. Production: 2.3 billion kWh. Consumption per capita: 2,903 kWh (1993).

Telecommunications [30]

- 331,000 (1995 est.) telephones; excellent in both Greek and Turkish areas
- Domestic: open wire, fiber-optic cable, and microwave radio relay
- International: trophospheric scatter; 3 coaxial and 5 fiber-optic submarine cables; satellite earth stations - 3 Intelsat (1 Atlantic Ocean, 2 Indian Ocean), 2 Eutelsat, 2 Intersputnik, and 1 Arabsat
- Radio: Radio broadcast stations: Greek area: AM 11, FM 8; Turkish area: AM 2, FM 6 Radios: Greek area: 270,000 (1993 est.); Turkish area: 42,170 (1985 est.)
- Television: Television broadcast stations: Greek area: 1 (repeaters 34); Turkish area: 1 Televisions: Greek area: 107,000 (1992 est.); Turkish area: 75,000 (1993 est.)

Top Agricultural Products [32]

Agriculture accounts for 5.6% of the GDP in the Greek area and 11.4% of the GDP in the Turkish area; produces potatoes, vegetables, barley, grapes, olives, citrus, vegetables.

Top Mining Products [33]

Thousand metric tons except as noted[23]	5/8/95*
Cement, hydraulic	1,040
Clay, bentonite, crude	50
Gypsum, crude	180
Lime, hydrated	6,000
Limestone, crushed (Havara)	3,000
Marl, for cement production	1,600
Sand and gravel	6,000[24]
Umber	9
Petroleum refinery products (000 42-gal. bls.)	5,650[e]

Transportation [31]

Railways: 0 km

Highways: Greek area - total: 10,448 km; Greek area - paved: 5,694 km; Greek area - unpaved: 4,754 km; Turkish area - total: 6,116 km; Turkish area - paved: 5,278 km; Turkish area - unpaved: 838 km

Merchant marine: total: 1,524 ships (1,000 GRT or over) totaling 23,949,242 GRT/40,236,638 DWT; ships by type: bulk 490, cargo 562, chemical tanker 27, combination bulk 53, combination ore/oil 22, container 115, liquefied gas tanker 3, multifunction large-load carrier 4, oil tanker 129, passenger 6, passenger-cargo 1, refrigerated cargo 62, roll-on/roll-off cargo 28, short-sea passenger 17, specialized tanker 3, vehicle carrier 2

Airports

Total:	15
With paved runways 2,438 to 3,047 m:	8
With paved runways 914 to 1,523 m:	3
With paved runways under 914 m:	3
With unpaved runways 914 to 1,523 m:	1 (1995 est.)

Tourism [34]

	1990	1991	1992	1993	1994
Visitors[1]	1,676	1,473	2,117	1,984	2,216
Tourists[12]	1,561	1,385	1,991	1,841	2,069
Excursionists[28]	114	88	126	143	147
Cruise passengers	31	12	26	40	47
Tourism receipts	1,258	1,026	1,539	1,396	1,700
Tourism expenditures	111	113	132	133	176
Fare receipts	163	200	229	197	187
Fare expenditures	64	82	96	97	101

Travelers are in thousands, money in million U.S. dollars.

For sources, notes, and explanations, see Annotated Source Appendix, page 1061.

Manufacturing Sector

Manufacturing Summary [35]

	1987		1988		1989		1990		1991	
	$ bil.	%	$ bil.	%	$ bil.	%	$ bil.	%	$ bil.	%
Establishments or enterprises (number)	6,822	0.291	7,175	0.342	7,198	0.384	7,337	0.410	7,300	0.955
Total employment (000)	40	0.030	42	0.031	44	0.035	44	0.040	44	0.063
Production workers (000)	35	0.052	36	0.057	37	0.050	37	0.052	37	0.111
Output ($ bil.)	2	0.016	2	0.016	2	0.016	2	0.020	2	0.023
Value added ($ bil.)	0.581	0.013	0.678	0.014	0.690	0.014	0.814	0.016	0.848	0.025
Capital investment ($ mil.)	79	0.018	95	0.020	108	0.020	115	0.021	116	0.031
M & E investment ($ mil.)	53	0.017	66	0.019	73	0.019	77	0.018	80	0.028
Employees per establishment (number)	6	10.212	6	9.067	6	9.214	6	9.787	6	6.617
Production workers per establishment	5	17.797	5	16.645	5	12.970	5	12.742	5	11.653
Output per establishment ($ mil.)	0.234	5.408	0.257	4.622	0.264	4.197	0.307	4.799	0.323	2.423
Capital investment per estab. ($ mil.)	0.012	6.269	0.013	5.808	0.015	5.137	0.016	5.042	0.016	3.195
M & E per establishment ($ mil)	0.008	5.881	0.009	5.554	0.010	4.838	0.010	4.432	0.011	2.971
Payroll per employee ($)	7,485	83.477	8,427	84.582	8,427	75.430	10,028	72.293	10,722	73.436
Wages per production worker ($)	7,049	89.334	7,933	93.889	7,979	79.563	9,645	81.218	10,217	92.890
Hours per production worker (hours)	-	-	-	-	-	-	-	-	-	-
Output per employee ($)	39,658	52.952	43,414	50.970	43,660	45.556	50,813	49.031	53,770	36.626
Capital investment per employee ($)	1,971	61.389	2,229	64.056	2,481	55.755	2,587	51.512	2,647	48.280
M & E per employee ($)	1,322	57.584	1,549	61.254	1,666	52.506	1,725	45.280	1,823	44.907

Note: Columns headed % show percent of world total or ratio. Ratios closest to 100 are closest to world average. M & E stands for machinery & equipment.

Output in Manufacturing

	1987		1988		1989		1990		1991	
	$ bil.	%	$ bil.	%	$ bil.	%	$ bil.	%	$ bil.	%
3110 - Food products	0.269	13.45	0.295	14.75	0.291	14.55	0.338	16.90	0.358	17.90
3130 - Beverages	0.097	4.85	0.111	5.55	0.130	6.50	0.156	7.80	0.153	7.65
3140 - Tobacco	0.063	3.15	0.071	3.55	0.065	3.25	0.075	3.75	0.078	3.90
3210 - Textiles	0.052	2.60	0.063	3.15	0.068	3.40	0.079	3.95	0.085	4.25
3211 - Spinning, weaving, etc.	0.010	0.50	0.013	0.65	0.016	0.80	0.020	1.00	0.018	0.90
3220 - Wearing apparel	0.218	10.90	0.259	12.95	0.254	12.70	0.311	15.55	0.327	16.35
3230 - Leather and products	0.026	1.30	0.030	1.50	0.026	1.30	0.030	1.50	0.026	1.30
3240 - Footwear	0.068	3.40	0.077	3.85	0.068	3.40	0.078	3.90	0.079	3.95
3310 - Wood products	0.067	3.35	0.072	3.60	0.079	3.95	0.092	4.60	0.101	5.05
3320 - Furniture, fixtures	0.065	3.25	0.072	3.60	0.064	3.20	0.079	3.95	0.081	4.05
3410 - Paper and products	0.040	2.00	0.045	2.25	0.043	2.15	0.064	3.20	0.054	2.70
3420 - Printing, publishing	0.047	2.35	0.063	3.15	0.068	3.40	0.087	4.35	0.086	4.30
3510 - Industrial chemicals	0.014	0.70	0.033	1.65	0.017	0.85	0.009	0.45	0.007	0.35
3511 - Basic chemicals, excl fertilizers	0.001	0.05	0.001	0.05	0.001	0.05	0.002	0.10	0.002	0.10
3520 - Chemical products nec	0.059	2.95	0.069	3.45	0.084	4.20	0.101	5.05	0.120	6.00
3522 - Drugs and medicines	0.010	0.50	0.014	0.70	0.013	0.65	0.018	0.90	0.027	1.35
3530 - Petroleum refineries	0.096	4.80	0.092	4.60	0.101	5.05	0.117	5.85	0.129	6.45
3550 - Rubber products	0.007	0.35	0.008	0.40	0.007	0.35	0.007	0.35	0.006	0.30
3560 - Plastic products nec	0.044	2.20	0.053	2.65	0.054	2.70	0.065	3.25	0.072	3.60
3610 - Pottery, china, etc.	0.001	0.05	0.002	0.10	0.003	0.15	0.003	0.15	0.003	0.15
3620 - Glass and products	0.000	0.00	0.000	0.00	0.003	0.15	0.004	0.20	0.006	0.30
3690 - Nonmetal products nec	0.108	5.40	0.125	6.25	0.141	7.05	0.165	8.25	0.168	8.40
3810 - Metal products	0.109	5.45	0.124	6.20	0.142	7.10	0.159	7.95	0.172	8.60
3820 - Machinery nec	0.034	1.70	0.044	2.20	0.048	2.40	0.058	2.90	0.061	3.05
3825 - Office, computing machinery	0.000	0.00	0.000	0.00	-	-	-	-	-	-
3830 - Electrical machinery	0.024	1.20	0.030	1.50	0.033	1.65	0.033	1.65	0.031	1.55
3832 - Radio, television, etc.	0.001	0.05	0.000	0.00	0.001	0.05	0.000	0.00	0.001	0.05
3840 - Transportation equipment	0.015	0.75	0.017	0.85	0.017	0.85	0.019	0.95	0.019	0.95
3841 - Shipbuilding, repair	0.004	0.20	0.004	0.20	0.004	0.20	0.005	0.25	0.006	0.30
3843 - Motor vehicles	0.011	0.55	0.012	0.60	0.012	0.60	0.014	0.70	0.013	0.65
3850 - Professional goods	-	-	-	-	0.000	0.00	0.000	0.00	0.001	0.05
3900 - Industries nec	0.038	1.90	0.043	2.15	0.049	2.45	0.065	3.25	0.071	3.55

Note: Codes are International Standard Industry codes (ISIC). Percentages are % of total Output. [f]: Factor Prices; [p]: Producer Prices.

For sources, notes, and explanations, see Annotated Source Appendix, page 1061.

249

Finance, Economics, and Trade

Economic Indicators [36]

- **National product**: GDP—purchasing power parity—Greek area $7.8 billion (1995 est.); Turkish area $520 million (1995 est.)

- **National product real growth rate**: Greek area 5% (1995 est.); Turkish area 0.5% (1995 est.)

- **National product per capita**: Greek area $13,000 (1995 est.); Turkish area $3,900 (1995 est.)

- **Inflation rate (consumer prices)**: Greek area 3% (1995 est.); Turkish area 215% (1994)

- **External debt**: Greek area $1.4 billion (1994)

Balance of Payments Summary [37]

Values in millions of dollars.

	1988	1989	1990	1991	1992
Exports of goods (f.o.b.)	645.5	717.3	846.7	875.0	903.2
Imports of goods (f.o.b.)	-1,666.8	-2,072.1	-2,308.9	-2,363.1	-2,989.9
Trade balance	-1,021.3	-1,354.8	-1,462.2	-1,488.1	-2,086.7
Services - debits	-650.6	-696.6	-859.4	-891.8	-1,039.9
Services - credits	1,628.3	1,878.6	2,345.1	2,156.6	2,854.0
Private transfers (net)	24.6	22.9	21.9	20.6	22.0
Government transfers (net)	27.4	17.0	20.8	16.2	9.0
Long-term capital (net)	51.2	376.7	108.1	191.1	215.5
Short-term capital (net)	97.7	33.6	353.1	54.5	111.1
Errors and omissions	-86.6	-48.9	-230.1	-148.9	-299.0
Overall balance	70.7	228.5	297.3	-89.8	-217.0

Exchange Rates [38]

Currency: **Cypriot pound and Turkish lira.**
Symbol: **#C and TL.**

Data are currency units per $1.

Cypriot pound	
January 1996	0.4628
1995	0.4522
1994	0.4915
1993	0.4970
1992	0.4502
1991	0.4633
Turkish lira	
January 1996	60,502.1
1995	45,845.1
1994	29,608.7
1993	10,984.6
1992	6,872.4
1991	4,171.8

Imports and Exports

Top Import Origins [39]

Greek Area: $2.7 billion (f.o.b., 1994); Turkish Area: $330 million (f.o.b., 1994).

Origins	%
Greek Area	
UK	12
Italy	10
Japan	9
Germany	9
US	8
Turkish Area	
Turkey	48
UK	19

Top Export Destinations [40]

Greek Area: $968 million (f.o.b., 1994); Turkish Area: $59 million (f.o.b., 1994).

Destinations	%
Greek Area	
UK	16
Russia	12
Lebanon	9
Greece	8
Turkish Area	
UK	48
Turkey	22

Foreign Aid [41]

	U.S. $	
US commitments, including Ex-Im (FY70-89)	292	million
Western (non-US) countries, ODA and OOF bilateral commitments (1970-89)	250	million
OPEC bilateral aid (1979-89)	62	million
Communist countries (1970-89)	24	million

Import and Export Commodities [42]

Import Commodities	Export Commodities
Consumer goods	Citrus
Petroleum and lubricants	Potatoes
Food and feed grains	Grapes
Machinery	Wine
Food	Cement
Minerals	Clothing and shoes
Chemicals	Potatoes
Machinery	Textiles

Czech Republic

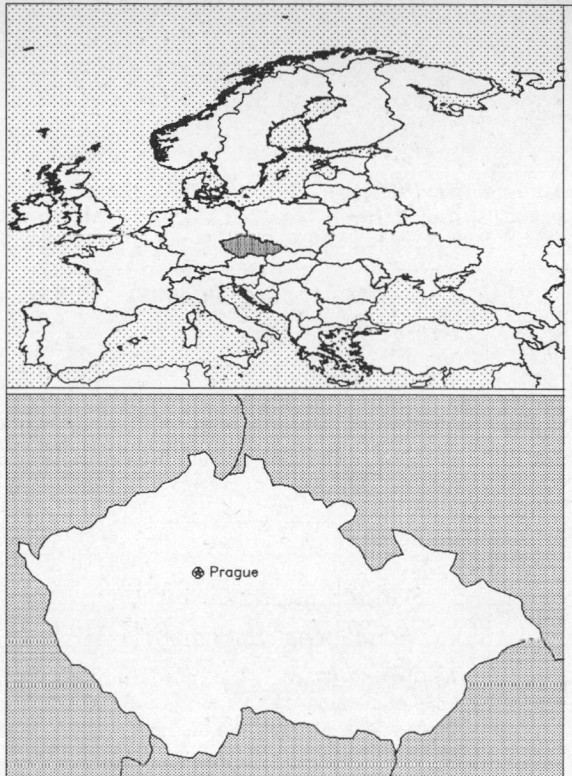

Geography [1]

Total area:
78,703 sq km   30,387 sq mi
Land area:
78,645 sq km   30,365 sq mi
Comparative area:
Slightly smaller than South Carolina
Land boundaries:
Total 1,880 km, Austria 362 km, Germany 646 km, Poland 658 km, Slovakia 214 km
Coastline:
0 km (landlocked)
Climate:
Temperate; cool summers; cold, cloudy, humid winters
Terrain:
Bohemia in the West consists of rolling plains, hills, and plateaus surrounded by low mountains; Moravia in the East consists of very hilly country
Natural resources:
Hard coal, soft coal, kaolin, clay, graphite
Land use:
Arable land: NA%
Permanent crops: NA%
Meadows and pastures: NA%
Forest and woodland: NA%
Other: NA%

Demographics [2]

	1970	1980	1990	1995[1]	1996	2000	2010	2020	2030
Population	9,795	10,289	10,310	10,325	10,321	10,358	10,445	10,271	10,009
Population density (persons per sq. mi.)	322	338	339	339	339	340	343	337	329
(persons per sq. km.)	124	130	131	131	131	131	132	130	127
Net migration rate (per 1,000 population)	NA	NA	NA	0.2	0.2	0.2	0.1	0.1	0.0
Births	NA	NA	NA	NA	107	NA	NA	NA	NA
Deaths	NA	NA	NA	NA	112	NA	NA	NA	NA
Life expectancy - males	NA	NA	NA	70.0	70.1	70.5	71.4	74.2	76.4
Life expectancy - females	NA	NA	NA	77.5	77.7	78.2	79.3	81.4	83.1
Birth rate (per 1,000)	NA	NA	NA	10.2	10.4	13.3	10.5	9.2	9.0
Death rate (per 1,000)	NA	NA	NA	10.8	10.9	11.0	11.6	11.3	12.4
Women of reproductive age (15-49 yrs.)	NA	NA	NA	2,660	2,652	2,580	2,432	2,364	2,062
of which are currently married	NA	NA	NA	NA	1,770	1,764	1,709	NA	NA
Fertility rate	NA	NA	NA	1.4	1.4	1.7	1.7	1.6	1.6

Except as noted, values for vital statistics are in thousands; life expectancy is in years.

Health

Health Indicators [3]

% of population with access to	
safe water (1990-95)	NA
adequate sanitation (1990-95)	NA
health services (1985-95)	NA
% of 1-year-olds immunized (1990-94) against	
TB (tuberculosis)	98
DPT (diphtheria, pertussis, tetanus)	98
polio	98
measles	97
% of contraceptive prevalence (1980-94)	69
ORT use rate (1990-94)	NA

Health Expenditures [4]

Total health expenditure, 1990 (official exchange rate)[2]	
Millions of dollars	2,711
Dollars per capita	173
Health expenditures as a percentage of GDP	
Total	5.9
Public sector	5.0
Private sector	0.9
Development assistance for health	
Total aid flows (millions of dollars)[1]	NA
Aid flows per capita (dollars)	NA
Aid flows as a percentage of total health expenditure	NA

For sources, notes, and explanations, see Annotated Source Appendix, page 1061.

251

Human Factors

Women and Children [5]

% of pregnant women immunized (tetanus 1990-94)	NA
% of births attended by trained health personnel (1983-94)	NA
Maternal mortality rate (1980-92)	NA
Under-5 mortality rate (1994)	10
% under-5 moderately/severely underweight (1980-1994)	NA

Burden of Disease [6]

Population per physician (1993)	272.90
Population per nurse	NA
Population per hospital bed (1993)	122.40
AIDS cases per 100,000 people (1994)	0.1
Heart disease cases per 1,000 people (1990-93)	322

Ethnic Division [7]

Czech	94.4%
Slovak	3%
Polish	0.6%
German	0.5%
Gypsy	0.3%
Hungarian	0.2%
Other	1%

Religion [8]

Atheist	39.8%
Roman Catholic	39.2%
Protestant	4.6%
Orthodox	3%
Other	13.4%

Major Languages [9]

Czech, Slovak.

Education

Public Education Expenditures [10]

	1980	1990	1991	1992	1993	1994
Million (Koruna)[20]						
Total education expenditure	30,549	37,323	33,546	36,915	53,393	NA
as percent of GNP	4.3	4.6	3.5	4.7	5.9	NA
as percent of total govt. expend.	8.0	8.2	6.6	NA	12.7	NA
Current education expenditure	29,247	35,482	30,816	33,631	49,198	NA
as percent of GNP	4.1	4.4	3.2	4.3	5.4	NA
as percent of current govt. expend.	8.7	8.7	6.8	NA	NA	NA
Capital expenditure	1,302	1,841	2,730	3,284	4,195	NA

Educational Attainment [11]

Age group (1991)	25+
Total population	6,580,525
Highest level attained (%)	
No schooling	0.3
First level	
Not completed	31.4
Completed	NA
Entered second level	
S-1	58.6
S-2	NA
Postsecondary	8.5

Illiteracy [12]

Libraries [13]

	Admin. Units	Svc. Pts.	Vols. (000)	Shelving (meters)	Vols. Added	Reg. Users
National (1993)	11	NA	16,184	NA	NA	279,500
Nonspecialized	NA	NA	NA	NA	NA	NA
Public (1993)	6,227	7,848	37,479	NA	NA	1 mil
Higher ed. (1993)[10]	58	1,149	8,551	NA	208,862	183,834
School	NA	NA	NA	NA	NA	NA

Daily Newspapers [14]

	1980	1985	1990	1994
Number of papers	NA	NA	NA	23
Circ. (000)	NA	NA	NA	2,259

Culture [15]

Cinema (seats per 1,000)	35.0
Annual attendance per person	2.1
Gross box office receipts (mil. Koruna)	433
Museums (reporting)	254
Visitors (000)	9,029
Annual receipts (000 Koruna)	1,647

Science and Technology

Scientific/Technical Forces [16]

Scientists/engineers	13,225
Number female	NA
Technicians	9,771
Number female	NA
Total[9]	23,096

R&D Expenditures [17]

	Koruna (000) 1994
Total expenditure[8]	12,983,000
Capital expenditure	1,768,000
Current expenditure	11,215,000
Percent current	86.4

U.S. Patents Issued [18]

Values show patents issued to citizens of the country by the U.S. Patents Office.

	1993	1994	1995
Number of patents	0	1	1

For sources, notes, and explanations, see Annotated Source Appendix, page 1061.

Government and Law

Organization of Government [19]

Long-form name:
Czech Republic
Type:
Parliamentary democracy
Independence:
1 January 1993 (from Czechoslovakia)
National holiday:
National Liberation Day, 8 May; Founding of the Republic, 28 October
Constitution:
Ratified 16 December 1992; effective 1 January 1993
Legal system:
Civil law system based on Austro-Hungarian codes; has not accepted compulsory ICJ jurisdiction; legal code modified to bring it in line with Organization on Security and Cooperation in Europe (OSCE) obligations and to expunge Marxist-Leninist legal theory
Executive branch:
President; Prime Minister; 3 Deputy Prime Ministers; Cabinet
Legislative branch:
Bicameral Parliament (Parlament): Senate (Senate) and Chamber of Deputies (Snemovna Poslancu):
Judicial branch:
Supreme Court; Constitutional Court

Elections [20]

Chamber of Deputies	% of seats
Civic Democratic Party (ODS)	32.5
Left Bloc (LB)	12.5
Christian Soc. Democrats	9.0
Civic Dem. Alliance (ODA)	8.0
Christian Democratic Union/ Czech People's (KDU/CSL)	7.5
Christian Dem. (KDS)	5.0
Communist Party (KSDM)	5.0
Bohemian-Moravian (CMSS)	4.5
Others	22.0

Government Expenditures [21]

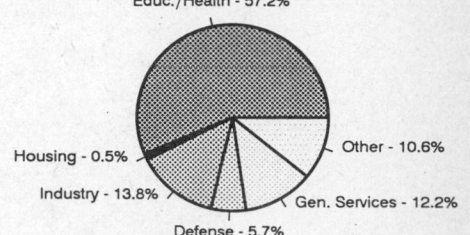

Educ./Health - 57.2%
Other - 10.6%
Housing - 0.5%
Gen. Services - 12.2%
Industry - 13.8%
Defense - 5.7%

(% distribution). Expend. for CY95: 499,358 (Koruna mil.)

Military Expenditures and Arms Transfers [22]

	1990	1991	1992	1993[16]	1994
Military expenditures					
Current dollars (mil.)	NA	NA	NA	2,040[r]	2,165
1994 constant dollars (mil.)	NA	NA	NA	2,082[r]	2,165
Armed forces (000)	NA	NA	NA	107	90
Gross national product (GNP)					
Current dollars (mil.)[e]	NA	NA	NA	79,490	81,430
1994 constant dollars (mil.)[e]	NA	NA	NA	81,120	81,430
Central government expenditures (CGE)					
1994 constant dollars (mil.)	NA	NA	NA	30,700	29,190
People (mil.)	NA	NA	NA	10.4	10.4
Military expenditure as % of GNP	NA	NA	NA	2.6	2.7
Military expenditure as % of CGE	NA	NA	NA	6.8	7.4
Military expenditure per capita (1994 $)	NA	NA	NA	200	208
Armed forces per 1,000 people (soldiers)	NA	NA	NA	10.3	8.6
GNP per capita (1994 $)	NA	NA	NA	7,809	7,824
Arms imports[6]					
Current dollars (mil.)	NA	NA	NA	5	90
1994 constant dollars (mil.)	NA	NA	NA	5	90
Arms exports[6]					
Current dollars (mil.)	NA	NA	NA	210	300
1994 constant dollars (mil.)	NA	NA	NA	214	300
Total imports[7]					
Current dollars (mil.)	NA	NA	NA	13,340	15,460
1994 constant dollars (mil.)	NA	NA	NA	13,620	15,460
Total exports[7]					
Current dollars (mil.)	NA	NA	NA	12,800	14,290
1994 constant dollars (mil.)	NA	NA	NA	13,070	14,290
Arms as percent of total imports[8]	NA	NA	NA	0	0.6
Arms as percent of total exports[8]	NA	NA	NA	1.6	2.1

Crime [23]

	1990[1]
Crime volume	
Cases known to police	170,257
Attempts (percent)	NA
Percent cases solved	9.7
Crimes per 100,000 persons	1,651.5
Persons responsible for offenses	
Total number offenders	16,521
Percent female	8.9
Percent juvenile	12.0
Percent foreigners	NA

Human Rights [24]

	SSTS	FL	FAPRO	PPCG	APROBC	TPW	PCPTW	STPEP	PHRFF	PRW	ASST	AFL
Observes	2	P	P	P	P	P	P	P	P	P	P	

	EAFRD	CPR	ESCR	SR	ACHR	MAAE	PVIAC	PVNAC	EAFDAW	TCIDTP	RC
Observes	P	P	P	P			P	P	P	P	P

P = Party; S = Signatory; see Appendix for meaning of abbreviations.

Labor Force

Total Labor Force [25]

5.389 million

Labor Force by Occupation [26]

Industry	37.9%
Agriculture	8.1
Construction	8.8
Communications and other	45.2

Date of data: 1990

Unemployment Rate [27]

2.9% (1995 est.)

For sources, notes, and explanations, see Annotated Source Appendix, page 1061.

Statistical Abstract of the World, 3rd Edition — Czech Republic — 253

Production Sectors

Commercial Energy Production and Consumption

Data are shown in quadrillion (10^{15}) BTUs and percent for 1995
Values for hydroelectric, nuclear, geothermal, solar, and wind power refer to electrical generation.

Production [28]

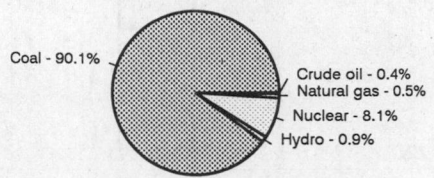

- Coal - 90.1%
- Crude oil - 0.4%
- Natural gas - 0.5%
- Nuclear - 8.1%
- Hydro - 0.9%

Consumption [29]

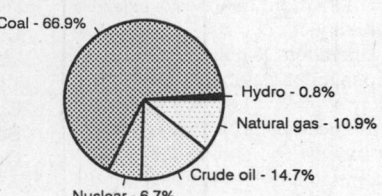

- Coal - 66.9%
- Hydro - 0.8%
- Natural gas - 10.9%
- Crude oil - 14.7%
- Nuclear - 6.7%

Crude oil	0.007		Crude oil	0.325
Dry natural gas	0.009		Dry natural gas	0.241
Coal	1.645		Coal	1.475
Net hydroelectric power	0.017		Net hydroelectric power	0.017
Net nuclear power	0.147		Net nuclear power	0.147
Total	1.825		Total	2.205

Telecommunications [30]

- 3,349,539 (1993 est.) telephones
- International: satellite earth stations - 2 Intersputnik (Atlantic and Indian Ocean Regions)
- Radio: Broadcast stations: AM NA, FM NA, shortwave NA

Transportation [31]

Railways: total: 9,413 km; standard gauge: 9,316 km 1.435-m standard gauge (2640 km electrified); narrow gauge: 97 km several narrow gauges (1995)

Highways: total: 55,557 km (1994 est.); paved: NA km; unpaved: NA km

Merchant marine: total: 10 ships (1,000 GRT or over) totaling 155,946 GRT/251,624 DWT; ships by type: bulk 5, cargo 5 (1995 est.)

Airports

Total:	116
With paved runways over 3,047 m:	2
With paved runways 2,438 to 3,047 m:	9
With paved runways 1,524 to 2,437 m:	13
With paved runways under 914 m:	5
With unpaved runways over 3,047 m:	1

Top Agricultural Products [32]

Agriculture accounts for 5.8% of the GDP; produces grains, potatoes, sugar beets, hops, fruit; pigs, cattle, poultry; forest products.

Top Mining Products [33]

Metric tons except as noted[e]	6/95[*]
Sulfur, byproducts, all sources	20,000
Sulfuric acid	350,000[r]
Coal (000 tons)	
bituminous	20,900[r]
brown and lignite	60,700[r]
Coke, metallurgical (000 tons)	5,150[r]
Gas, manufactured, all types (mil. cu. meters)	5,000
Petroleum refinery products (000 42-gal. bls.)	95,000

Tourism [34]

	1990	1991	1992	1993	1994
Visitors[29]	36,571	50,863	69,412	71,736	101,140
Tourists	NA	NA	10,900	11,500	17,000
Tourism receipts	419	714	1,126	1,558	1,966
Tourism expenditures	455	274	467	525	832

Travelers are in thousands, money in million U.S. dollars.

For sources, notes, and explanations, see Annotated Source Appendix, page 1061.

Manufacturing Sector

Manufacturing Summary [35]

	1987		1988		1989		1990		1991	
	$ bil.	%	$ bil.	%	$ bil.	%	$ bil.	%	$ bil.	%
Establishments or enterprises (number)	853	0.036	944	0.045	990	0.053	1,399	0.078	1,743	0.228
Total employment (000)	2,763	2.036	2,755	2.013	2,736	2.223	2,603	2.352	2,081	2.996
Production workers (000)	1,983	2.940	1,969	3.152	1,954	2.656	1,863	2.625	1,465	4.449
Output ($ bil.)	62	0.612	60	0.518	56	0.475	47	0.411	21	0.203
Value added ($ bil.)	18	0.421	18	0.377	16	0.336	13	0.255	-	-
Capital investment ($ mil.)	3,500	0.803	3,673	0.771	3,443	0.629	3,538	0.637	-	-
M & E investment ($ mil.)	2,569	0.824	2,731	0.789	2,547	0.652	2,665	0.632	-	-
Employees per establishment (number)	3,239	5,599	2,918	4,477	2,764	4,206	1,861	3,011	1,194	1,314
Production workers per establishment	2,325	8,086	2,086	7,008	1,974	5,026	1,332	3,361	841	1,951
Output per establishment ($ mil.)	73	1,684	64	1,151	57	900	34	526	12	89
Capital investment per estab. ($ mil.)	4	2,208	4	1,715	3	1,189	3	815	-	-
M & E per establishment ($ mil)	3	2,267	3	1,755	3	1,234	2	809	-	-
Payroll per employee ($)	2,584	28.815	2,536	25.455	2,490	22.285	2,164	15.600	1,587	10.867
Wages per production worker ($)	2,516	31.883	2,458	29.094	2,412	24.053	2,101	17.692	1,508	13.707
Hours per production worker (hours)	1,890	99.616	1,874	97.494	1,843	100.552	1,810	96.761	1,668	93.318
Output per employee ($)	22,525	30.076	21,897	25.708	20,497	21.387	18,093	17.459	9,964	6.787
Capital investment per employee ($)	1,267	39.441	1,333	38.311	1,258	28.278	1,359	27.069	-	-
M & E per employee ($)	930	40.496	991	39.199	931	29.346	1,024	26.875	-	-

Note: Columns headed % show percent of world total or ratio. Ratios closest to 100 are closest to world average. M & E stands for machinery & equipment.

Output in Manufacturing

	1987		1988		1989		1990		1991	
	$ bil.	%	$ bil.	%	$ bil.	%	$ bil.	%	$ bil.	%
3110 - Food products	9.375	15.12	8.918	14.86	10.000	17.86	6.633	14.11	3.818	18.18
3130 - Beverages	0.955	1.54	0.915	1.52	1.017	1.82	0.783	1.67	0.628	2.99
3140 - Tobacco	0.155	0.25	0.150	0.25	0.160	0.29	0.129	0.27	0.088	0.42
3210 - Textiles	2.923	4.71	2.843	4.74	2.740	4.89	2.283	4.86	0.967	4.60
3211 - Spinning, weaving, etc.	2.317	3.74	2.245	3.74	2.178	3.89	1.802	3.83	0.773	3.68
3220 - Wearing apparel	0.795	1.28	0.719	1.20	0.698	1.25	0.575	1.22	0.199	0.95
3230 - Leather and products	0.390	0.63	0.378	0.63	0.403	0.72	0.320	0.68	0.162	0.77
3240 - Footwear	0.841	1.36	0.822	1.37	0.876	1.56	0.640	1.36	0.249	1.19
3310 - Wood products	1.087	1.75	1.058	1.76	1.060	1.89	0.697	1.48	0.381	1.81
3320 - Furniture, fixtures	0.576	0.93	0.566	0.94	0.581	1.04	0.413	0.88	0.222	1.06
3410 - Paper and products	1.329	2.14	1.308	2.18	1.336	2.39	1.040	2.21	0.555	2.64
3420 - Printing, publishing	0.325	0.52	0.335	0.56	0.330	0.59	0.339	0.72	0.166	0.79
3510 - Industrial chemicals	3.687	5.95	3.619	6.03	3.153	5.63	3.090	6.57	0.941	4.48
3520 - Chemical products nec	0.659	1.06	0.646	1.08	0.658	1.18	0.693	1.47	0.255	1.21
3522 - Drugs and medicines	0.456	0.74	0.446	0.74	0.423	0.76	0.379	0.81	0.150	0.71
3530 - Petroleum refineries	4.360	7.03	4.199	7.00	2.829	5.05	2.320	4.94	1.235	5.88
3540 - Petroleum, coal products	0.420	0.68	0.390	0.65	0.407	0.73	0.365	0.78	0.168	0.80
3550 - Rubber products	0.750	1.21	0.724	1.21	0.668	1.19	0.597	1.27	0.215	1.02
3560 - Plastic products nec	0.131	0.21	0.130	0.22	0.109	0.19	0.124	0.26	0.054	0.26
3610 - Pottery, china, etc.	0.079	0.13	0.075	0.13	0.076	0.14	0.074	0.16	0.027	0.13
3620 - Glass and products	0.766	1.24	0.780	1.30	0.817	1.46	0.670	1.43	0.318	1.51
3690 - Nonmetal products nec	1.666	2.69	1.568	2.61	1.443	2.58	1.180	2.51	0.502	2.39
3710 - Iron and steel	6.430	10.37	6.184	10.31	5.385	9.62	5.020	10.68	2.149	10.23
3720 - Nonferrous metals	1.447	2.33	1.405	2.34	1.233	2.20	1.047	2.23	0.384	1.83
3810 - Metal products	2.191	3.53	2.194	3.66	2.055	3.67	1.824	3.88	0.709	3.38
3820 - Machinery nec	8.651	13.95	8.389	13.98	7.073	12.63	6.312	13.43	2.531	12.05
3830 - Electrical machinery	3.115	5.02	3.111	5.18	2.695	4.81	2.971	6.32	0.920	4.38
3840 - Transportation equipment	5.359	8.64	5.234	8.72	4.665	8.33	3.984	8.48	1.595	7.60
3850 - Professional goods	0.262	0.42	0.255	0.43	0.230	0.41	0.192	0.41	0.077	0.37
3900 - Industries nec	0.739	1.19	0.719	1.20	0.744	1.33	0.601	1.28	0.299	1.42

Note: Codes are International Standard Industry codes (ISIC). Percentages are % of total Output. [f]: Factor Prices; [p]: Producer Prices.

For sources, notes, and explanations, see Annotated Source Appendix, page 1061.

255

Finance, Economics, and Trade

Economic Indicators [36]

- **National product**: GDP—purchasing power parity—$106.2 billion (1995 est.)
- **National product real growth rate**: 5% (1995 est.)
- **National product per capita**: $10,200 (1995 est.)
- **Inflation rate (consumer prices)**: 9.1% (1995 est.)
- **External debt**: $14.9 billion (June 1995)

Balance of Payments Summary [37]

Values in millions of dollars.

	1989	1990	1991	1992	1993[2]
Exports of goods (f.o.b.)	NA	NA	NA	NA	43,711
Imports of goods (f.o.b.)	NA	NA	NA	NA	34,163
Trade balance	NA	NA	NA	NA	9,548
Services - debits	NA	NA	NA	NA	-14,580
Services - credits	NA	NA	NA	NA	8,939
Private transfers (net)	NA	NA	NA	NA	2,324
Government transfers (net)	NA	NA	NA	NA	NA
Long-term capital (net)	NA	NA	NA	NA	-883
Short-term capital (net)	NA	NA	NA	NA	-7,323
Errors and omissions	NA	NA	NA	NA	4,850
Overall balance	NA	NA	NA	NA	2,875

Exchange Rates [38]

Currency: **koruna.**
Symbol: **Kc.**

Data are currency units per $1.

January 1996	26.967
1995	26.541
1994	28.785
1993	29.153
1992	28.26
1991	29.53
1990	17.95

Imports and Exports

Top Import Origins [39]

$21.3 billion (f.o.b., 1995 est.) Data are for January-September 1995.

Origins	%
Germany	26
Slovakia	13.2
Russia	9.2
Austria	7
Italy	5.6
France	4.1
US	3.8
Poland	3.1
Netherlands	2.9
UK	2.9
Switzerland	2.1
Belgium	2.0

Top Export Destinations [40]

$17.4 billion (f.o.b., 1995 est.) Data are for January-September 1995.

Destinations	%
Germany	32.4
Slovakia	16.1
Austria	6.7
Poland	5.3
Italy	4
Russia	3.3
Netherlands	2.8
France	2.6
UK	2.2
Hungary	2.1
US	1.8
Belgium	1.5

Foreign Aid [41]

Recipient: ODA, $27 million (1993).

Import and Export Commodities [42]

Import Commodities

Machinery and transport equipment
Manufactured goods
Chemicals
Fuels and lubricants
Raw materials
Agricultural products

Export Commodities

Manufactured goods
Machinery and transport equipment
Chemicals
Fuels
Minerals
Metals
Agricultural products

Denmark

Geography [1]

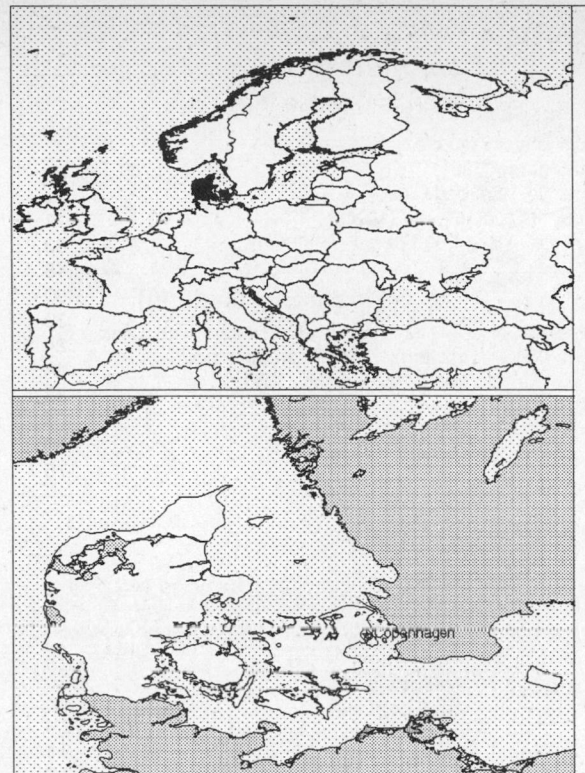

Total area:
43,070 sq km   16,629 sq mi
Land area:
42,370 sq km   16,359 sq mi
Comparative area:
Slightly more than twice the size of Massachusetts
Note: Includes the island of Bornholm in the Baltic Sea and the rest of
metropolitan Denmark, but excludes the Faroe Islands and Greenland
Land boundaries:
Total 68 km, Germany 68 km
Coastline:
3,379 km
Climate:
Temperate; humid and overcast; mild, windy winters and cool summers
Terrain:
Low and flat to gently rolling plains
Natural resources:
Petroleum, natural gas, fish, salt, limestone
Land use:
Arable land: 61%
Permanent crops: 0%
Meadows and pastures: 6%
Forest and woodland: 12%
Other: 21%

Demographics [2]

	1970	1980	1990	1995[1]	1996	2000	2010	2020	2030
Population	4,929	5,123	5,141	5,229	5,250	5,320	5,417	5,458	5,383
Population density (persons per sq. mi.)	301	313	314	320	321	325	331	334	329
(persons per sq. km.)	116	121	121	123	124	126	128	129	127
Net migration rate (per 1,000 population)	NA	NA	1.7	2.0	2.0	2.1	2.0	2.0	0.0
Births	71	57	NA	NA	64	NA	NA	NA	NA
Deaths	48	56	36	NA	58	NA	NA	NA	NA
Life expectancy - males	NA	NA	71.8	73.6	73.8	74.4	75.8	78.0	79.5
Life expectancy - females	NA	NA	77.7	80.9	81.0	81.4	82.4	84.3	85.6
Birth rate (per 1,000)	NA	NA	12.4	12.6	12.2	11.3	9.2	9.6	8.9
Death rate (per 1,000)	NA	NA	11.9	10.4	10.4	10.4	10.6	10.7	12.1
Women of reproductive age (15-49 yrs.)	NA	NA	1,310	1,296	1,284	1,244	1,224	1,159	1,063
of which are currently married[12]	750	705	646	NA	652	644	617	NA	NA
Fertility rate	NA	NA	1.7	1.7	1.7	1.6	1.6	1.6	1.5

Except as noted, values for vital statistics are in thousands; life expectancy is in years.

Health

Health Indicators [3]

% of population with access to	
safe water (1990-95)	NA
adequate sanitation (1990-95)	NA
health services (1985-95)	NA
% of 1-year-olds immunized (1990-94) against	
TB (tuberculosis)	NA
DPT (diphtheria, pertussis, tetanus)	88
polio	95
measles	81
% of contraceptive prevalence (1980-94)	78
ORT use rate (1990-94)	NA

Health Expenditures [4]

Total health expenditure, 1990 (official exchange rate)	
Millions of dollars	8,160
Dollars per capita	1,588
Health expenditures as a percentage of GDP	
Total	6.3
Public sector	5.3
Private sector	1.0
Development assistance for health	
Total aid flows (millions of dollars)[1]	NA
Aid flows per capita (dollars)	NA
Aid flows as a percentage of total health expenditure	NA

For sources, notes, and explanations, see Annotated Source Appendix, page 1061.

257

Human Factors

Women and Children [5]

% of pregnant women immunized (tetanus 1990-94)	NA
% of births attended by trained health personnel (1983-94)[1]	100
Maternal mortality rate (1980-92)	3
Under-5 mortality rate (1994)	7
% under-5 moderately/severely underweight (1980-1994)	NA

Burden of Disease [6]

Population per physician (1987)	390.06
Population per nurse (1984)	60.86
Population per hospital bed (1990)	177.02
AIDS cases per 100,000 people (1994)	4.5
Heart disease cases per 1,000 people (1990-93)	312.5

Ethnic Division [7]

Scandinavian, Eskimo, Faroese, German.

Religion [8]

Evangelical Lutheran	91%
Other Protestant and Roman Catholic	2%
Other	7%
(1988)	

Major Languages [9]

Danish, Faroese, Greenlandic (an Eskimo dialect), German (small minority).

Education

Public Education Expenditures [10]

Million (Krone)	1980	1989	1990	1991	1993	1994
Total education expenditure	25,020	55,448	NA	58,960	72,168	NA
as percent of GNP	6.9	7.5	NA	7.4	8.5	NA
as percent of total govt. expend.	9.5	13.0	NA	11.8	13.0	NA
Current education expenditure	22,188	52,270	NA	54,896	67,809	NA
as percent of GNP	6.1	7.1	NA	6.9	8.0	NA
as percent of current govt. expend.	9.0	12.9	NA	11.5	NA	NA
Capital expenditure	2,832	3,178	NA	4,064	4,359	NA

Educational Attainment [11]

Age group (1991)[6]	25+
Total population	2,742,734
Highest level attained (%)	
No schooling	
First level	
Not completed	38.7
Completed	NA
Entered second level	
S-1	3.4
S-2	38.2
Postsecondary	19.6

Illiteracy [12]

Libraries [13]

	Admin. Units	Svc. Pts.	Vols. (000)	Shelving (meters)	Vols. Added	Reg. Users
National (1992)	1	4	4,388	103,745	72,414	NA
Nonspecialized (1992)	5	13	5,252[e]	131,314	136,702	NA
Public (1992)	251	904	32,479	NA	2,059	NA
Higher ed. (1993)[11]	20	52	12,584	290,172	275,499	116,932
School (1990)	275	1,773	32,235	NA	1 mil	NA

Daily Newspapers [14]

	1980	1985	1990	1994
Number of papers	48	47	47	51
Circ. (000)	1,874	1,855	1,810	1,886

Culture [15]

Cinema (seats per 1,000)	10.1
Annual attendance per person	2.0
Gross box office receipts (mil. Krone)	312
Museums (reporting)	288
Visitors (000)	11,197
Annual receipts (000 Krone)	NA

Science and Technology

Scientific/Technical Forces [16]

Scientists/engineers	13,673
Number female	NA
Technicians[2]	13,717
Number female	NA
Total[2]	27,390

R&D Expenditures [17]

	Krone (000) 1993
Total expenditure	15,695,000
Capital expenditure	1,760,000
Current expenditure[9]	13,935,000
Percent current	88.8

U.S. Patents Issued [18]

Values show patents issued to citizens of the country by the U.S. Patents Office.

	1993	1994	1995
Number of patents	261	313	315

Government and Law

Organization of Government [19]

Long-form name:
Kingdom of Denmark
Type:
Constitutional monarchy
Independence:
10th century first organized as a unified state; in 1849 became a constitutional monarchy
National holiday:
Birthday of the Queen, 16 April (1940)
Constitution:
1849 was the original constitution; there was a major overhaul 5 June 1953, allowing for a unicameral legislature and a female chief of state
Legal system:
Civil law system; judicial review of legislative acts; accepts compulsory ICJ jurisdiction, with reservations
Executive branch:
Queen; Heir Apparent Crown Prince; Prime Minister; Cabinet
Legislative branch:
Unicameral: Parliament (Folketing)
Judicial branch:
Supreme Court

Elections [20]

Parliament	% of votes
Social Democratic Party	34.6
Liberal	23.3
Conservative Party	15.0
Socialist People's Party	7.3
Progress Party	6.4
Radical Liberal Party	4.6
Center Democratic Party	2.8
Christian People's Party	1.8

Government Expenditures [21]

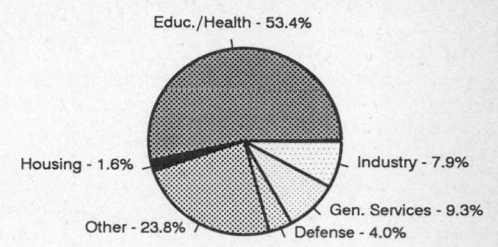

Educ./Health - 53.4%
Housing - 1.6%
Industry - 7.9%
Gen. Services - 9.3%
Defense - 4.0%
Other - 23.8%

(% distribution). Expend. for CY95: 419,725 (Krone mil.) |p|

Military Expenditures and Arms Transfers [22]

	1990	1991	1992	1993	1994
Military expenditures					
Current dollars (mil.)	2,484	2,631	2,658	2,728	2,719
1994 constant dollars (mil.)	2,765	2,820	2,772	2,784	2,719
Armed forces (000)	31	30	28	27	28
Gross national product (GNP)					
Current dollars (mil.)	115,900	122,200	127,100	132,500	141,200
1994 constant dollars (mil.)	129,000	131,000	132,600	135,200	141,200
Central government expenditures (CGE)					
1994 constant dollars (mil.)	54,620	56,380	58,220	62,050	64,940
People (mil.)	5.1	5.2	5.2	5.2	5.2
Military expenditure as % of GNP	2.1	2.2	2.1	2.1	1.9
Military expenditure as % of CGE	5.1	5.0	4.8	4.5	4.2
Military expenditure per capita (1994 $)	538	547	537	538	524
Armed forces per 1,000 people (soldiers)	6.0	5.8	5.4	5.2	5.4
GNP per capita (1994 $)	25,080	25,420	25,670	26,120	27,210
Arms imports[6]					
Current dollars (mil.)	120	50	80	90	40
1994 constant dollars (mil.)	134	54	83	92	40
Arms exports[6]					
Current dollars (mil.)	80	20	10	10	10
1994 constant dollars (mil.)	89	21	10	10	10
Total imports[7]					
Current dollars (mil.)	32,230	32,400	35,170	30,540	34,880
1994 constant dollars (mil.)	35,870	34,730	36,680	31,170	34,880
Total exports[7]					
Current dollars (mil.)	35,130	36,000	41,050	37,170	41,420
1994 constant dollars (mil.)	39,100	38,590	42,810	37,930	41,420
Arms as percent of total imports[8]	0.4	0.2	0.2	0.3	0.1
Arms as percent of total exports[8]	0.2	0.1	0	0	0

Crime [23]

	1994
Crime volume	
Cases known to police	546,928
Attempts (percent)	NA
Percent cases solved	20.70
Crimes per 100,000 persons	10,524.64
Persons responsible for offenses	
Total number offenders	70,020
Percent female	18.60
Percent juvenile (15-19 yrs.)	15.80
Percent foreigners	NA

Human Rights [24]

	SSTS	FL	FAPRO	PPCG	APROBC	TPW	PCPTW	STPEP	PHRFF	PRW	ASST	AFL
Observes	P	P	P	P	P	P	P	S	P	P	P	P
	EAFRD	CPR	ESCR	SR	ACHR	MAAE	PVIAC	PVNAC	EAFDAW	TCIDTP	RC	
Observes	P	P	P	P			P	P	P	P	P	

P = Party; S = Signatory; see Appendix for meaning of abbreviations.

Labor Force

Total Labor Force [25]

2.554 million

Labor Force by Occupation [26]

Private services	37.1%
Government services	30.4
Manufacturing and mining	20
Construction	6.3
Agriculture, forestry, and fishing	5.6
Electricity/gas/water	0.6

Date of data: 1991

Unemployment Rate [27]

9.5% (1995)

For sources, notes, and explanations, see Annotated Source Appendix, page 1061.

259

Production Sectors

Commercial Energy Production and Consumption

Data are shown in quadrillion (10^{15}) BTUs and percent for 1995
Values for hydroelectric, nuclear, geothermal, solar, and wind power refer to electrical generation.

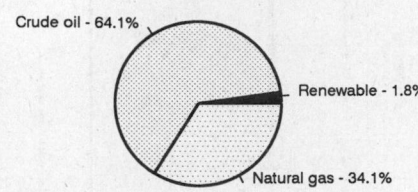

Production [28]

Crude oil - 64.1%
Renewable - 1.8%
Natural gas - 34.1%

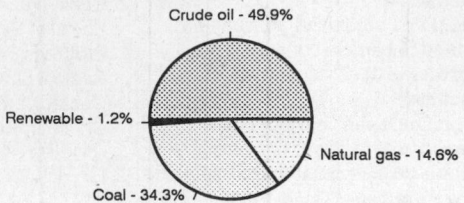

Consumption [29]

Crude oil - 49.9%
Renewable - 1.2%
Natural gas - 14.6%
Coal - 34.3%

Crude oil	0.386		Crude oil	0.471
Dry natural gas	0.205		Dry natural gas	0.138
Geothermal, solar, wind	0.011		Coal	0.323
Total	0.602		Geothermal, solar, wind	0.011
			Total	0.943

Telecommunications [30]

- 4.005 million (1985 est.) telephones; excellent telephone and telegraph services
- Domestic: buried and submarine cables and microwave radio relay form trunk network
- International: 19 submarine coaxial cables; satellite earth stations - 7 Intelsat, NA Eutelsat, and 1 Inmarsat (Atlantic and Indian Ocean Regions); note - Denmark shares the Inmarsat earth station with the other Nordic countries (Finland, Iceland, Norway, and Sweden)
- Radio: Broadcast stations: AM 3, FM 2, shortwave 0
- Television: Broadcast stations: 2 Televisions: 2.04 million (1992 est.)

Top Agricultural Products [32]

Agriculture accounts for 3% of the GDP; produces grain, potatoes, rape, sugar beets; meat, dairy products; fish.

Top Mining Products [33]

Metric tons except as noted	5/1/95[*]
Cement, hydraulic	2,430,000
Chalk	414,000
Steel (mil. cu. meters)	
crude	722,000[76]
semimanufactures	638,000
Lime, hydrated, and quicklime	126,000
Peat	199,000
Salt, all forms	634,000
Limestone, agricultural/industrial	955,000

Transportation [31]

Railways: total: 2,848 km (499 km privately owned and operated); standard gauge: 2,848 km 1.435-m gauge (326 km electrified; 760 km double track) (1995)

Highways: total: 71,042 km; paved: 71,042 km (including 696 km of expressways); unpaved: 0 km (1992 est.)

Merchant marine: total: 334 ships (1,000 GRT or over) totaling 5,013,054 GRT/7,171,871 DWT; ships by type: bulk 13, cargo 114, chemical tanker 25, container 65, liquefied gas tanker 27, livestock carrier 5, oil tanker 31, railcar carrier 1, refrigerated cargo 17, roll-on/roll-off cargo 26, short-sea passenger 9, specialized tanker 1

Airports

Total:	109
With paved runways over 3,047 m:	2
With paved runways 2,438 to 3,047 m:	7
With paved runways 1,524 to 2,437 m:	3
With paved runways 914 to 1,523 m:	13
With paved runways under 914 m:	77

Tourism [34]

	1990	1991	1992	1993	1994
Tourism receipts	3,322	3,475	3,784	3,052	3,174
Tourism expenditures[30]	3,676	3,377	3,779	3,214	3,583
Fare receipts	133	147	146	156	168

Travelers are in thousands, money in million U.S. dollars.

For sources, notes, and explanations, see Annotated Source Appendix, page 1061.

Manufacturing Sector

Manufacturing Summary [35]

	1987		1988		1989		1990		1991	
	$ bil.	%	$ bil.	%	$ bil.	%	$ bil.	%	$ bil.	%
Establishments or enterprises (number)	7,933	0.338	7,823	0.373	7,725	0.412	12,905	0.720	12,565	1.644
Total employment (000)	466	0.344	452	0.330	451	0.366	568	0.513	548	0.789
Production workers (000)	317	0.470	305	0.488	303	0.412	-	-	369	1.120
Output ($ bil.)	45	0.445	49	0.418	48	0.407	59	0.513	59	0.574
Value added ($ bil.)	20	0.473	22	0.454	21	0.438	26	0.524	27	0.781
Capital investment ($ mil.)	2,541	0.583	2,586	0.543	2,233	0.408	2,733	0.492	3,000	0.788
M & E investment ($ mil.)	1,861	0.597	1,837	0.531	1,705	0.437	2,024	0.480	2,268	0.804
Employees per establishment (number)	59	101.580	58	88.566	58	88.812	44	71.271	44	47.973
Production workers per establishment	40	138.897	39	130.825	39	99.838	-	-	29	68.129
Output per establishment ($ mil.)	6	131.729	6	112.300	6	98.687	5	71.279	5	34.948
Capital investment per estab. ($ mil.)	0.320	172.418	0.331	145.701	0.289	98.849	0.212	68.264	0.239	47.924
M & E per establishment ($ mil)	0.235	176.615	0.235	142.424	0.221	105.907	0.157	66.654	0.180	48.921
Payroll per employee ($)	25,913	288.980	27,953	280.578	26,728	239.235	29,999	216.259	30,413	208.296
Wages per production worker ($)	22,036	279.267	24,366	288.362	23,265	231.989	-	-	23,927	217.536
Hours per production worker (hours)	1,597	84.142	1,617	84.149	1,600	87.297	-	-	-	-
Output per employee ($)	97,124	129.681	108,000	126.797	106,496	111.119	103,645	100.011	106,949	72.849
Capital investment per employee ($)	5,451	169.737	5,724	164.510	4,953	111.301	4,809	95.780	5,476	99.898
M & E per employee ($)	3,992	173.868	4,067	160.810	3,783	119.248	3,562	93.522	4,139	101.976

Note: Columns headed % show percent of world total or ratio. Ratios closest to 100 are closest to world average. M & E stands for machinery & equipment.

Output in Manufacturing

	1987		1988		1989		1990		1991	
	$ bil.[f]	%	$ bil.[f]	%	$ bil.[f]	%	$ bil.[f]	%	$ bil.[f]	%
3110 - Food products	12.000	26.67	13.000	26.53	13.000	27.08	16.000	27.12	15.000	25.42
3130 - Beverages	1.050	2.33	1.117	2.28	1.031	2.15	1.259	2.13	1.393	2.36
3140 - Tobacco	0.346	0.77	0.365	0.74	0.353	0.74	0.360	0.61	0.435	0.74
3210 - Textiles	1.234	2.74	1.309	2.67	1.190	2.48	1.317	2.23	1.276	2.16
3211 - Spinning, weaving, etc.	0.398	0.88	0.383	0.78	0.357	0.74	0.355	0.60	0.328	0.56
3220 - Wearing apparel	0.535	1.19	0.492	1.00	0.436	0.91	0.532	0.90	0.522	0.88
3230 - Leather and products	0.058	0.13	0.055	0.11	0.056	0.12	0.032	0.05	0.039	0.07
3240 - Footwear	0.149	0.33	0.137	0.28	0.145	0.30	0.194	0.33	0.200	0.34
3310 - Wood products	0.839	1.86	0.896	1.83	0.877	1.83	1.050	1.78	1.054	1.79
3320 - Furniture, fixtures	0.959	2.13	0.957	1.95	0.985	2.05	1.278	2.17	1.315	2.23
3410 - Paper and products	1.023	2.27	1.133	2.31	1.144	2.38	1.350	2.30	1.407	2.38
3411 - Pulp, paper, etc.	0.259	0.58	0.291	0.59	0.305	0.64	0.343	0.58	0.288	0.49
3420 - Printing, publishing	2.006	4.46	2.141	4.37	1.988	4.14	2.448	4.15	2.391	4.05
3510 - Industrial chemicals	1.675	3.72	1.904	3.89	1.833	3.82	2.172	3.68	2.075	3.52
3513 - Synthetic resins, etc.	0.661	1.47	0.738	1.51	0.721	1.50	0.864	1.46	0.794	1.35
3520 - Chemical products nec	1.874	4.16	2.100	4.29	2.164	4.51	2.611	4.43	2.730	4.63
3522 - Drugs and medicines	1.058	2.35	1.218	2.49	1.272	2.65	1.556	2.64	1.618	2.74
3530 - Petroleum refineries	0.775	1.72	1.132	2.31	1.085	2.26	1.478	2.51	1.773	3.01
3540 - Petroleum, coal products	0.260	0.58	0.355	0.72	0.294	0.61	0.352	0.60	0.360	0.61
3550 - Rubber products	0.171	0.38	0.187	0.38	0.186	0.39	0.213	0.36	0.188	0.32
3560 - Plastic products nec	0.832	1.85	0.925	1.89	0.904	1.88	1.160	1.97	1.151	1.95
3610 - Pottery, china, etc.	0.064	0.14	0.083	0.17	0.063	0.13	0.081	0.14	0.077	0.13
3620 - Glass and products	0.180	0.40	0.169	0.34	0.181	0.38	0.213	0.36	0.206	0.35
3690 - Nonmetal products nec	1.427	3.17	1.422	2.90	1.350	2.81	1.587	2.69	1.510	2.56
3710 - Iron and steel	0.404	0.90	0.472	0.96	0.531	1.11	0.606	1.03	0.519	0.88
3720 - Nonferrous metals	0.155	0.34	0.135	0.28	0.140	0.29	0.183	0.31	0.175	0.30
3810 - Metal products	2.737	6.08	2.980	6.08	2.966	6.18	3.640	6.17	3.724	6.31
3820 - Machinery nec	4.108	9.13	4.581	9.35	4.621	9.63	5.728	9.71	5.661	9.59
3830 - Electrical machinery	2.102	4.67	2.139	4.37	2.070	4.31	2.622	4.44	2.370	4.02
3832 - Radio, television, etc.	1.061	2.36	0.970	1.98	0.964	2.01	1.286	2.18	1.213	2.06
3840 - Transportation equipment	1.845	4.10	1.973	4.03	1.908	3.97	2.606	4.42	2.888	4.89
3841 - Shipbuilding, repair	1.272	2.83	1.322	2.70	1.295	2.70	1.819	3.08	1.976	3.35
3850 - Professional goods	0.773	1.72	0.903	1.84	0.893	1.86	1.042	1.77	1.019	1.73
3900 - Industries nec	0.582	1.29	0.636	1.30	0.661	1.38	0.839	1.42	0.904	1.53

Note: Codes are International Standard Industry codes (ISIC). Percentages are % of total Output. [f]: Factor Prices; [p]: Producer Prices.

For sources, notes, and explanations, see Annotated Source Appendix, page 1061.

261

Finance, Economics, and Trade

Economic Indicators [36]

- **National product**: GDP—purchasing power parity—$112.8 billion (1995 est.)
- **National product real growth rate**: 3.1% (1995 est.)
- **National product per capita**: $21,700 (1995 est.)
- **Inflation rate (consumer prices)**: 2.4% (1995 est.)
- **External debt**: $40.9 billion (1994 est.)

Balance of Payments Summary [37]

Values in millions of dollars.

	1989	1990	1991	1992	1993
Exports of goods (f.o.b.)	28,728	36,072	36,783	40,650	37,070
Imports of goods (f.o.b.)	-26,304	-31,197	-32,035	-33,446	-29,258
Trade balance	2,424	4,875	4,748	7,204	7,812
Services - debits	-17,688	-21,936	-25,019	-32,120	-38,004
Services - credits	14,290	18,841	23,119	30,064	35,411
Private transfers (net)	80	-46	-150	-131	-133
Government transfers (net)	-223	-362	-715	-749	-375
Long-term capital (net)	-3,566	5,985	-1,739	4,904	19,229
Short-term capital (net)	1,207	-1,575	-1,364	-9,042	-21,308
Errors and omissions	-347	-2,407	-2,183	-357	1,220
Overall balance	-3,823	3,375	-3,303	-227	3,852

Exchange Rates [38]

Currency: **Danish krone.**
Symbol: **DKr.**

Data are currency units per $1.

January 1996	5.652
1995	5.602
1994	6.361
1993	6.484
1992	6.036
1991	6.396

Imports and Exports

Top Import Origins [39]

$34 billion (c.i.f., 1994 est.) Data are for 1994.

Origins	%
EU	51
Germany	22
UK	6.5
Sweden	11.6
US	5.2
Norway	5.1
Japan	3.5
FSU	1.7

Top Export Destinations [40]

$39.6 billion (f.o.b., 1994) Data are for 1994.

Destinations	%
EU	49.4
Germany	22.4
UK	8.2
Sweden	10.4
Norway	6.5
US	5.5
Japan	4.1
FSU	1.7

Foreign Aid [41]

Donor: ODA, $1.34 billion (1993).

Import and Export Commodities [42]

Import Commodities	Export Commodities
Petroleum	Meat and meat products
Machinery and equipment	Dairy products
Chemicals	Transport equipment (shipbuilding)
Grain and foodstuffs	Fish
Textiles	Chemicals
Paper	Industrial machinery

For sources, notes, and explanations, see Annotated Source Appendix, page 1061.

Djibouti

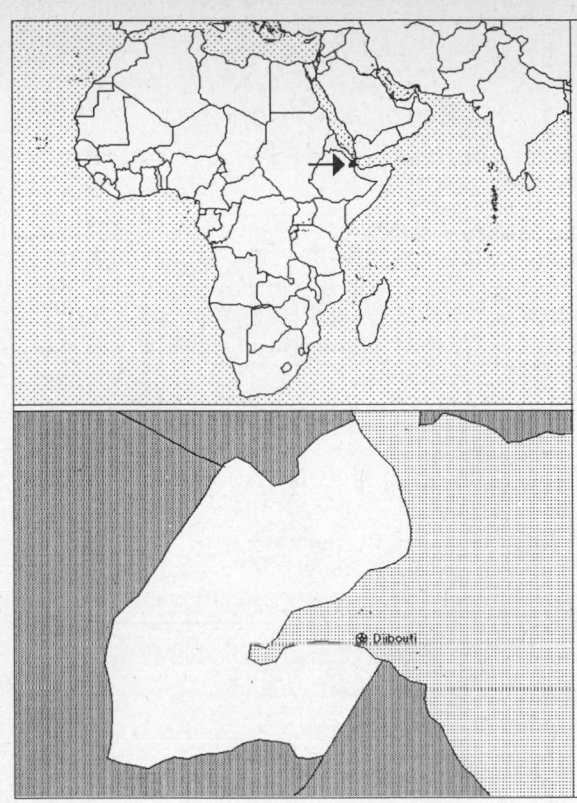

Geography [1]

Total area:
  22,000 sq km   8,494 sq mi
Land area:
  21,980 sq km   8,487 sq mi
Comparative area:
  Slightly larger than Massachusetts
Land boundaries:
  Total 508 km, Eritrea 113 km, Ethiopia 337 km, Somalia 58 km
Coastline:
  314 km
Climate:
  Desert; torrid, dry
Terrain:
  Coastal plain and plateau separated by central mountains
Natural resources:
  Geothermal areas
Land use:
  Arable land: 0%
  Permanent crops: 0%
  Meadows and pastures: 9%
  Forest and woodland: 0%
  Other: 91%

Demographics [2]

	1970	1980	1990	1995[1]	1996	2000	2010	2020	2030
Population	158	279	370	421	428	454	588	751	935
Population density (persons per sq. mi.)	19	33	44	50	50	54	69	88	110
(persons per sq. km.)	7	13	17	19	19	21	27	34	43
Net migration rate (per 1,000 population)	NA	NA	1.9	-12.5	-12.3	-11.6	0.0	0.0	0.0
Births	NA	NA	NA	NA	18	NA	NA	NA	NA
Deaths	NA	NA	NA	NA	7	NA	NA	NA	NA
Life expectancy - males	NA	NA	45.8	47.8	48.2	49.9	54.1	58.3	62.2
Life expectancy - females	NA	NA	49.1	51.6	52.1	54.2	59.4	64.4	69.0
Birth rate (per 1,000)	NA	NA	43.6	42.8	42.5	41.0	37.2	32.6	27.9
Death rate (per 1,000)	NA	NA	16.8	15.5	15.3	14.1	11.6	9.3	7.5
Women of reproductive age (15-49 yrs.)	NA	NA	81	92	93	98	132	183	240
of which are currently married	NA	NA	57	NA	65	68	92	NA	NA
Fertility rate	NA	NA	6.4	6.2	6.1	5.8	5.0	4.1	3.3

Except as noted, values for vital statistics are in thousands; life expectancy is in years.

Health

Health Indicators [3]

Health Expenditures [4]

For sources, notes, and explanations, see Annotated Source Appendix, page 1061.

263

Human Factors

Women and Children [5]	Burden of Disease [6]
	Population per physician (1990) 5,916.67
	Population per nurse (1990) 724.49
	Population per hospital bed NA
	AIDS cases per 100,000 people (1994) 42.7
	Malaria cases per 100,000 people (1992) NA

Ethnic Division [7]

Somali	60%
Afar	35%
French, Arab, Ethiopian, and Italian	5%

Religion [8]

Muslim	94%
Christian	6%

Major Languages [9]

French (official), Arabic (official), Somali, Afar.

Education

Public Education Expenditures [10]

Million (Franc)[21]	1980	1985	1989	1990	1991	1994
Total education expenditure	NA	1,690	2,596	2,614	2,872	NA
as percent of GNP	NA	2.7	NA	3.4	3.8	NA
as percent of total govt. expend.	NA	7.5	1.7	10.5	11.1	NA
Current education expenditure	NA	1,690	2,596	2,614	2,872	NA
as percent of GNP	NA	2.7˙	NA	3.4	3.8	NA
as percent of current govt. expend.	NA	NA	10.9	10.5	11.1	NA
Capital expenditure	NA	-	0	-	-	NA

Educational Attainment [11]

Illiteracy [12]

In thousands and percent[1]	1990	1995	2000
Illiterate population (15+ yrs.)	173	181	187
Illiteracy rate - total pop. (%)	82.4	75.1	71.9
Illiteracy rate - males (%)	57.1	51.2	48.2
Illiteracy rate - females (%)	NA	NA	98.4

Libraries [13]

Daily Newspapers [14]

Culture [15]

Science and Technology

Scientific/Technical Forces [16]

R&D Expenditures [17]

U.S. Patents Issued [18]

For sources, notes, and explanations, see Annotated Source Appendix, page 1061.

Government and Law

Organization of Government [19]

Long-form name:
Republic of Djibouti
Type:
Republic
Independence:
27 June 1977 (from France)
National holiday:
Independence Day, 27 June (1977)
Constitution:
Multiparty constitution approved in referendum 4 September 1992
Legal system:
Based on French civil law system, traditional practices, and Islamic law
Executive branch:
President; Prime Minister; Council of Ministers
Legislative branch:
Unicameral: Chamber of Deputies (Chambre des Deputes)
Judicial branch:
Supreme Court (Cour Supreme)

Elections [20]

National Assembly	% of seats
People's Progress Assembly	100.0

Government Budget [21]

For 1993 est.

	$ mil.
Revenues	164
Expenditures	201
Capital expenditures	16

Military Expenditures and Arms Transfers [22]

	1990	1991	1992	1993	1994
Military expenditures					
Current dollars (mil.)[e]	27	42	40	28	25
1994 constant dollars (mil.)[e]	29	45	42	29	25
Armed forces (000)	4	3	8	8	8
Gross national product (GNP)					
Current dollars (mil.)	373	460	483	468	NA
1994 constant dollars (mil.)	415	493	504	478	NA
Central government expenditures (CGE)					
1994 constant dollars (mil.)[e]	147	233	NA	206	NA
People (mil.)	0.4	0.4	0.4	0.4	0.4
Military expenditure as % of GNP	7.1	9.1	8.3	6.0	NA
Military expenditure as % of CGE	20.1	19.2	NA	13.9	NA
Military expenditure per capita (1994 $)	80	117	107	71	NA
Armed forces per 1,000 people (soldiers)	11.3	8.9	20.5	19.9	19.4
GNP per capita (1994 $)	1,121	1,296	1,289	1,189	NA
Arms imports[6]					
Current dollars (mil.)	5	5	10	0	0
1994 constant dollars (mil.)	6	5	10	0	0
Arms exports[6]					
Current dollars (mil.)	0	0	0	0	0
1994 constant dollars (mil.)	0	0	0	0	0
Total imports[7]					
Current dollars (mil.)	215	214	219	NA	384[e]
1994 constant dollars (mil.)	239	229	228	NA	384[e]
Total exports[7]					
Current dollars (mil.)	25	17	16	NA	NA
1994 constant dollars (mil.)	28	18	17	NA	NA
Arms as percent of total imports[8]	2.3	2.3	4.6	0	0
Arms as percent of total exports[8]	0	0	0	0	0

Crime [23]

	1990
Crime volume	
Cases known to police	2,436
Attempts (percent)	41
Percent cases solved	72
Crimes per 100,000 persons	487.2
Persons responsible for offenses	
Total number offenders	2,285
Percent female	11
Percent juvenile (12-18 yrs.)	13
Percent foreigners	60

Human Rights [24]

	SSTS	FL	FAPRO	PPCG	APROBC	TPW	PCPTW	STPEP	PHRFF	PRW	ASST	AFL
Observes		P	P		P	P	P	P			P	P
	EAFRD	CPR	ESCR	SR	ACHR	MAAE	PVIAC	PVNAC	EAFDAW	TCIDTP		RC
Observes				P			P	P				P

P = Party; S = Signatory; see Appendix for meaning of abbreviations.

Labor Force

Total Labor Force [25]

282,000

Labor Force by Occupation [26]

Agriculture	75%
Industry	11
Services	14

Date of data: 1991 est.

Unemployment Rate [27]

Over 30% (1994 est.)

For sources, notes, and explanations, see Annotated Source Appendix, page 1061.

265

Production Sectors

Energy Resource Summary [28]

Energy resources: Geothermal areas. **Electricity:** Capacity: 90,000 kW. Production: 170 million kWh. Consumption per capita: 398 kWh (1993).

Telecommunications [30]

- 7,200 (1986 est.) telephones; telephone facilities in the city of Djibouti are adequate as are the microwave radio relay connections to outlying areas of the country
- Domestic: microwave radio relay network
- International: submarine cable to Saudi Arabia; satellite earth stations - 1 Intelsat (Indian Ocean) and 1 Arabsat
- Radio: Broadcast stations: AM 2, FM 2, shortwave 0
- Television: Broadcast stations: 1 Televisions: 17,000 (1993 est.)

Top Agricultural Products [32]

Agriculture accounts for 3% of the GDP; produces fruits, vegetables; goats, sheep, camels.

Top Mining Products [33]

Detailed information is not available. A summary of mineral resources follows. **Mineral Resources:** Marine salt, limestone, construction clays, sand, gravel, crushed stone, marble, and granite.

Transportation [31]

Railways: total: 97 km (Djibouti segment of the Addis Ababa-Djibouti railroad); narrow gauge: 97 km 1.000-m gauge

Highways: total: 2,879 km; paved: 363 km; unpaved: 2,516 km (1991 est.)

Merchant marine: total: 1 cargo ship (1,000 GRT or over) totaling 1,369 GRT/3,030 DWT (1995 est.)

Airports

Total: 11
With paved runways over 3,047 m: 1
With paved runways 2,438 to 3,047 m: 1
With paved runways under 914 m: 2
With unpaved runways 1,524 to 2,437 m: 2
With unpaved runways 914 to 1,523 m: 5 (1995 est.)

Tourism [34]

	1990	1991	1992	1993	1994
Tourists	33	33	28	25	NA
Tourism receipts	NA	NA	10	13	NA
Tourism expenditures	NA	NA	7	7	NA
Fare receipts	NA	NA	2	2	NA
Fare expenditures	NA	NA	15	15	NA

Travelers are in thousands, money in million U.S. dollars.

For sources, notes, and explanations, see Annotated Source Appendix, page 1061.

Finance, Economics, and Trade

GDP and Manufacturing Summary [35]

	1980	1985	1990	1991	1992
Gross Domestic Product					
Millions of 1980 dollars	339	357	397	404	415[e]
Growth rate in percent	4.72	0.85	2.00	1.60	2.91[e]
Manufacturing Value Added					
Millions of 1980 dollars	34	36	43	45	47[e]
Growth rate in percent	2.98	0.49	5.10	4.22	3.98[e]
Manufacturing share in percent of current prices	9.7	9.4[e]	NA	NA	NA

Economic Indicators [36]

- **National product**: GDP—purchasing power parity—$500 million (1994 est.)

- **National product real growth rate**: - 3% (1994 est.)

- **National product per capita**: $1,200 (1994 est.)

- **Inflation rate (consumer prices)**: 6% (1993 est.)

- **External debt**: $227 million (1993 est.)

Balance of Payments Summary [37]

Values in millions of dollars.

	1989	1990	1991	1992	1993
Exports of goods (f.o.b.)	NA	NA	NA	192.0	211.2
Imports of goods (f.o.b.)	NA	NA	NA	-383.4	-402.3
Trade balance	NA	NA	NA	-191.4	-191.1
Services - debits	NA	NA	NA	-95.8	-94.8
Services - credits	NA	NA	NA	174.4	183.9
Private transfers (net)	NA	NA	NA	-89.6	-88.7
Government transfers (net)	NA	NA	NA	113.8	102.4
Long-term capital (net)	NA	NA	NA	24.1	25.7
Short-term capital (net)	NA	NA	NA	36.7	-19.2
Errors and omissions	NA	NA	NA	11.4	73.3
Overall balance	NA	NA	NA	-16.4	-8.5

Exchange Rates [38]

Currency: **Djiboutian franc.**
Symbol: **DF.**

Data are currency units per $1.

Fixed rate since 1973   177.721

Imports and Exports

Top Import Origins [39]

$384 million (f.o.b., 1994 est.).

Origins	%
France	NA
UK	NA
Saudi Arabia	NA
Bahrain	NA
South Korea	NA

Top Export Destinations [40]

$184 million (f.o.b., 1994 est.).

Destinations	%
Somalia	48
Yemen	42

Foreign Aid [41]

Recipient: ODA, $NA.

Import and Export Commodities [42]

Import Commodities	**Export Commodities**
Foods	Hides and skins
Beverages	Coffee (in transit)
Transport equipment	
Chemicals	
Petroleum products	

Dominica

Geography [1]

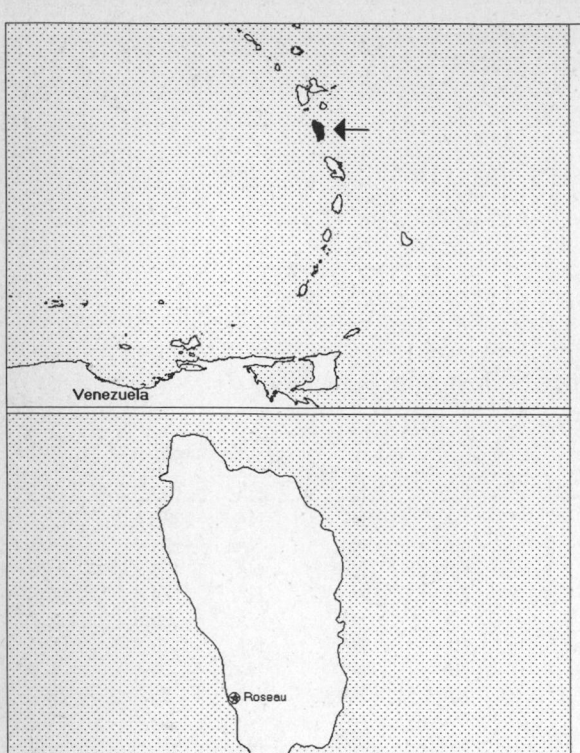

Total area:
 750 sq km   290 sq mi
Land area:
 750 sq km   290 sq mi
Comparative area:
 More than four times the size of Washington, DC
Land boundaries:
 0 km
Coastline:
 148 km
Climate:
 Tropical; moderated by northeast trade winds; heavy rainfall
Terrain:
 Rugged mountains of volcanic origin
Natural resources:
 Timber
Land use:
 Arable land: 9%
 Permanent crops: 13%
 Meadows and pastures: 3%
 Forest and woodland: 41%
 Other: 34%

Demographics [2]

	1970	1980	1990	1995[1]	1996	2000	2010	2020	2030
Population	71	75	81	83	83	84	89	96	102
Population density (persons per sq. mi.)	244	260	279	285	286	290	306	331	351
(persons per sq. km.)	94	100	108	110	110	112	118	128	135
Net migration rate (per 1,000 population)	NA	NA	-9.3	-9.4	-9.3	-9.2	0.0	0.0	0.0
Births	2	2	NA	NA	2	NA	NA	NA	NA
Deaths	0	0	NA	NA	Z	NA	NA	NA	NA
Life expectancy - males	NA	NA	73.2	74.4	74.6	75.4	77.0	78.1	79.0
Life expectancy - females	NA	NA	79.0	80.2	80.4	81.2	82.9	84.1	85.0
Birth rate (per 1,000)	37.9	22.7	20.0	18.6	18.4	16.9	14.0	12.6	11.3
Death rate (per 1,000)	10.0	6.6	5.5	5.3	5.3	5.3	5.2	5.6	6.9
Women of reproductive age (15-49 yrs.)	NA	NA	21	22	22	23	25	24	23
of which are currently married	4	NA	10	NA	12	13	14	NA	NA
Fertility rate	NA	NA	2.1	2.0	1.9	1.9	1.8	1.8	1.8

Except as noted, values for vital statistics are in thousands; life expectancy is in years.

Health

Health Indicators [3]

% of population with access to	
safe water (1990-95)	76
adequate sanitation (1990-95)	78
health services (1985-95)	80
% of 1-year-olds immunized (1990-94) against	
TB (tuberculosis)	64
DPT (diphtheria, pertussis, tetanus)	83
polio	98
measles	87
% of contraceptive prevalence (1980-94)	56
ORT use rate (1990-94)	37

Health Expenditures [4]

For sources, notes, and explanations, see Annotated Source Appendix, page 1061.

Human Factors

Women and Children [5]

% of pregnant women immunized (tetanus 1990-94)	85
% of births attended by trained health personnel (1983-94)	92
Maternal mortality rate (1980-92)	NA
Under-5 mortality rate (1994)	45
% under-5 moderately/severely underweight (1980-1994)	10

Burden of Disease [6]

Population per physician (1984)	2,960.00
Population per nurse	NA
Population per hospital bed (1993)	385.03
AIDS cases per 100,000 people (1994)	5.7
Malaria cases per 100,000 people (1992)	NA

Ethnic Division [7]

Black, Carib. Indians.

Religion [8]

Roman Catholic	77%
Protestant	15%
None	2%
Other	6%

Major Languages [9]

English (official), French patois.

Education

Public Education Expenditures [10]

	1980	1986	1989	1990	1992	1994
Million (E. Carib. Dollar)						
Total education expenditure	NA	17	22	NA	NA	NA
as percent of GNP	NA	5.9	5.8	NA	NA	NA
as percent of total govt. expend.	NA	16.7	10.6	NA	NA	NA
Current education expenditure	NA	16	20	NA	NA	NA
as percent of GNP	NA	5.7	5.2	NA	NA	NA
as percent of current govt. expend.	NA	18.5	19.9	NA	NA	NA
Capital expenditure	NA	1	2	NA	NA	NA

Educational Attainment [11]

Age group (1981)	25+
Total population	27,508
Highest level attained (%)	
No schooling	6.6
First level	
Not completed	80.5
Completed	NA
Entered second level	
S-1	11.1
S-2	NA
Postsecondary	1.7

Illiteracy [12]

In thousands and percent[3]	1970	1980	1990
Illiterate population (15+ yrs.)	2,083	NA	NA
Illiteracy rate - total pop. (%)	5.9	NA	NA
Illiteracy rate - males (%)	6.0	NA	NA
Illiteracy rate - females (%)	5.8	NA	NA

Libraries [13]

	Admin. Units	Svc. Pts.	Vols. (000)	Shelving (meters)	Vols. Added	Reg. Users
National	NA	NA	NA	NA	NA	NA
Nonspecialized	NA	NA	NA	NA	NA	NA
Public (1992)	1	3	29	400	2,727	9,470
Higher ed.	NA	NA	NA	NA	NA	NA
School	NA	NA	NA	NA	NA	NA

Daily Newspapers [14]

Culture [15]

Science and Technology

Scientific/Technical Forces [16]

R&D Expenditures [17]

U.S. Patents Issued [18]

Values show patents issued to citizens of the country by the U.S. Patents Office.

	1990	1991	1992
Number of patents	1	2	0

For sources, notes, and explanations, see Annotated Source Appendix, page 1061.

269

Government and Law

Organization of Government [19]

Long-form name:
Commonwealth of Dominica
Type:
Parliamentary democracy
Independence:
3 November 1978 (from UK)
National holiday:
Independence Day, 3 November. (1978)
Constitution:
3 November 1978
Legal system:
Based on English common law
Executive branch:
President; Prime Minister; Cabinet
Legislative branch:
Unicameral: House of Assembly
Judicial branch:
Eastern Caribbean Supreme Court; Court of Summary Jurisdiction

Elections [20]

House of Assembly	% of seats
United Workers Party (UWP)	52.4
Dominica Freedom Party (DFP)	23.8
Dominica Labor Party (DLP)	23.8

Government Budget [21]

For FY95/96 est.

	$ mil.
Revenues	80
Expenditures	95.8
Capital expenditures	NA

Defense Summary [22]

Branches: Commonwealth of Dominica Police Force (includes Special Service Unit, Coast Guard)

Manpower Availability: Males age 15-49 NA; males fit for military service NA

Defense Expenditures: NA

Crime [23]

Human Rights [24]

	SSTS	FL	FAPRO	PPCG	APROBC	TPW	PCPTW	STPEP	PHRFF	PRW	ASST	AFL
Observes	P	P	P		P	P	P			1	P	P
	EAFRD	CPR	ESCR	SR	ACHR	MAAE	PVIAC	PVNAC	EAFDAW	TCIDTP	RC
Observes		P	P	P		P	P	P	P		P

P = Party; S = Signatory; see Appendix for meaning of abbreviations.

Labor Force

Total Labor Force [25]

25,000

Labor Force by Occupation [26]

Agriculture	40%
Industry and commerce	32
Services	28

Date of data: 1984

Unemployment Rate [27]

15% (1992 est.)

Production Sectors

Energy Resource Summary [28]

Electricity: Capacity: 7,000 kW. Production: 30 million kWh. Consumption per capita: 347 kWh (1993).

Telecommunications [30]

- 14,613 (1993 est.) telephones
- Domestic: fully automatic network
- International: microwave radio relay and SHF radiotelephone links to Martinique and Guadeloupe; VHF and UHF radiotelephone links to Saint Lucia
- Radio: Broadcast stations: AM 3, FM 2, shortwave 0 Radios: 45,000 (1993 est.)
- Television: Broadcast stations: 1 cable Televisions: 5,200 (1993 est.)

Transportation [31]

Railways: 0 km

Highways: total: 800 km; paved: 500 km; unpaved: 300 km

Merchant marine: none

Airports

Total: 2
With paved runways 914 to 1,523 m: 1
With paved runways under 914 m: 1 (1995 est.)

Top Agricultural Products [32]

Agriculture accounts for 26% of the GDP; produces bananas, citrus, mangoes, root crops, coconuts; forestry and fisheries potential not exploited.

Top Mining Products [33]

Detailed information is not available. A summary of mineral resources follows. **Mineral Resources**: Clay, limestone, pumice, volcanic ash, sand and gravel, and crushed stone.

Tourism [34]

	1990	1991	1992	1993	1994
Visitors	59	120	145	147	192
Tourists[12]	45	46	47	52	57
Excursionists	7	9	8	7	9
Cruise passengers	7	65	90	88	126
Tourism receipts	20	24	25	28	31
Tourism expenditures	4	5	6	5	4
Fare expenditures	3	4	6	7	9

Travelers are in thousands, money in million U.S. dollars.

For sources, notes, and explanations, see Annotated Source Appendix, page 1061.

271

Finance, Economics, and Trade

Industrial Summary [35]

Industrial Production: Growth rate - 10% (1994 est.). **Industries:** Soap, coconut oil, tourism, copra, furniture, cement blocks, shoes.

Economic Indicators [36]

- **National product:** GDP—purchasing power parity—$200 million (1995 est.)
- **National product real growth rate:** - 1% (1995 est.)
- **National product per capita:** $2,450 (1995 est.)
- **Inflation rate (consumer prices):** 0.4% (1995)
- **External debt:** $92.8 million (1992)

Balance of Payments Summary [37]

Values in millions of dollars.

	1989	1990	1991	1992	1993
Exports of goods (f.o.b.)	46.3	56.1	55.6	54.6	48.3
Imports of goods (f.o.b.)	-94.4	-104.0	-96.5	-97.5	-98.8
Trade balance	-48.1	-47.9	-40.9	-42.9	-50.6
Services - debits	-34.6	-39.9	-42.5	-42.9	-41.4
Services - credits	27.5	35.0	40.0	44.9	49.3
Private transfers (net)	11.7	12.8	12.7	12.2	13.7
Government transfers (net)	11.0	9.1	9.5	5.8	6.1
Long-term capital (net)	24.2	19.4	22.8	23.8	11.5
Short-term capital (net)	8.4	10.7	3.0	1.7	7.2
Errors and omissions	0.1	6.0	-0.3	0.8	4.8
Overall balance	0.2	5.0	4.2	3.4	0.6

Exchange Rates [38]

Currency: **EC dollar.**
Symbol: **EC$.**

Data are currency units per $1.

Fixed rate since 1976	2.70

Imports and Exports

Top Import Origins [39]

$98.8 million (f.o.b., 1993).

Origins	%
US	25
Caricom	NA
UK	NA
Japan	NA
Canada	NA

Top Export Destinations [40]

$48.3 million (f.o.b., 1993).

Destinations	%
UK	55
Caricom	NA
Italy	NA
US	NA

Foreign Aid [41]

Recipient: ODA, $NA.

Import and Export Commodities [42]

Import Commodities	Export Commodities
Manufactured goods	Bananas
Machinery and equipment	Soap
Food	Bay oil
Chemicals	Vegetables
	Grapefruit
	Oranges

Dominican Republic

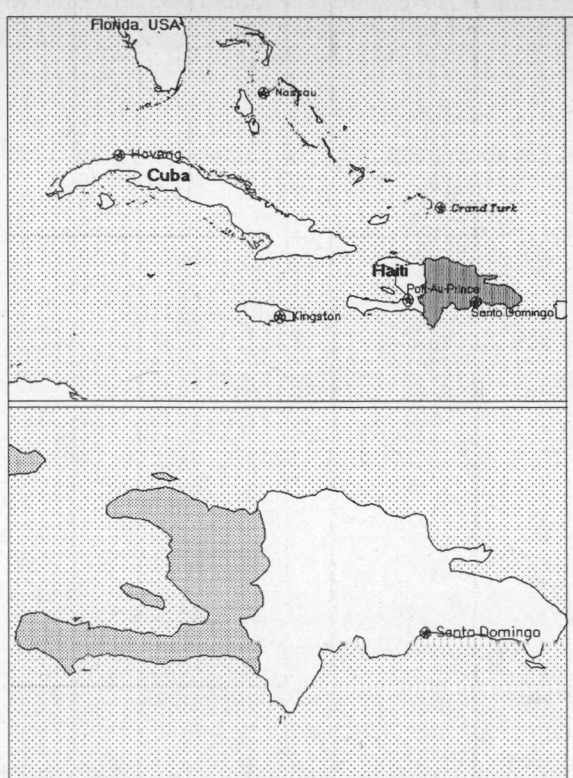

Geography [1]

Total area:
48,730 sq km   18,815 sq mi
Land area:
48,380 sq km   18,680 sq mi
Comparative area:
Slightly more than twice the size of New Hampshire
Land boundaries:
Total 275 km, Haiti 275 km
Coastline:
1,288 km
Climate:
Tropical maritime; little seasonal temperature variation; seasonal variation in rainfall
Terrain:
Rugged highlands and mountains with fertile valleys interspersed
Natural resources:
Nickel, bauxite, gold, silver
Land use:
Arable land: 23%
Permanent crops: 7%
Meadows and pastures: 43%
Forest and woodland: 13%
Other: 14%

Demographics [2]

	1970	1980	1990	1995[1]	1996	2000	2010	2020	2030
Population	4,373	5,697	7,213	7,948	8,089	8,635	9,928	11,152	12,171
Population density (persons per sq. mi.)	234	305	386	425	433	462	531	597	652
(persons per sq. km.)	90	118	149	164	167	178	205	231	252
Net migration rate (per 1,000 population)	1.0	0.5	-0.9	-0.6	-0.5	-0.4	0.0	0.0	0.0
Births	NA	NA	NA	NA	190	NA	NA	NA	NA
Deaths	NA	NA	NA	NA	46	NA	NA	NA	NA
Life expectancy - males	NA	61.2	64.8	66.6	66.9	68.2	71.0	73.3	75.2
Life expectancy - females	NA	65.1	69.1	71.0	71.3	72.8	75.9	78.4	80.5
Birth rate (per 1,000)	43.6	33.5	28.1	24.1	23.5	21.1	18.2	16.0	14.1
Death rate (per 1,000)	13.7	7.7	6.3	5.7	5.7	5.4	5.4	5.8	6.8
Women of reproductive age (15-49 yrs.)	NA	1,340	1,841	2,084	2,133	2,331	2,747	2,920	3,011
of which are currently married	253	NA	1,050	NA	1,252	1,381	1,660	NA	NA
Fertility rate	6.7	4.3	3.2	2.7	2.7	2.4	2.1	2.0	2.0

Except as noted, values for vital statistics are in thousands; life expectancy is in years.

Health

Health Indicators [3]

Health Expenditures [4]

Total health expenditure, 1990 (official exchange rate)
Millions of dollars	263
Dollars per capita	37

Health expenditures as a percentage of GDP
Total	3.7
Public sector	2.1
Private sector	1.6

Development assistance for health
Total aid flows (millions of dollars)[1]	11
Aid flows per capita (dollars)	1.5
Aid flows as a percentage of total health expenditure	4.1

For sources, notes, and explanations, see Annotated Source Appendix, page 1061.

Human Factors

Women and Children [5]	Burden of Disease [6]	
	Population per physician (1993)	937.00
	Population per nurse (1993)	1,330.06
	Population per hospital bed (1993)	500.00
	AIDS cases per 100,000 people (1994)	4.0
	Malaria cases per 100,000 people (1992)	9

Ethnic Division [7]

White	16%
Black	11%
Mixed	73%

Religion [8]

Roman Catholic	95%
Other	5%

Major Languages [9]

Spanish.

Education

Public Education Expenditures [10]

Million (Peso)[22]

	1980	1985	1990	1992	1993	1994
Total education expenditure	139	234	NA	1,502	2,007	2,606
as percent of GNP	2.2	1.8	NA	1.4	1.7	1.9
as percent of total govt. expend.	16.0	14.0	NA	8.9	9.9	12.2
Current education expenditure	104	204	NA	970	1,289	1,621
as percent of GNP	1.6	1.6	NA	0.9	1.1	1.2
as percent of current govt. expend.	NA	NA	NA	NA	NA	NA
Capital expenditure	10	2	NA	24	63	124

Educational Attainment [11]

Illiteracy [12]

In thousands and percent[1]

	1990	1995	2000
Illiterate population (15+ yrs.)	905	908	911
Illiteracy rate - total pop. (%)	19.9	17.5	15.5
Illiteracy rate - males (%)	19.8	17.7	15.9
Illiteracy rate - females (%)	20.0	17.3	15.0

Libraries [13]

Daily Newspapers [14]

	1980	1985	1990	1994
Number of papers	7	7	12	11
Circ. (000)	220	216	230[e]	264

Culture [15]

Science and Technology

Scientific/Technical Forces [16]

R&D Expenditures [17]

U.S. Patents Issued [18]

Values show patents issued to citizens of the country by the U.S. Patents Office.

	1993	1994	1995
Number of patents	1	0	1

Government and Law

Organization of Government [19]

Long-form name:
Dominican Republic
Type:
Republic
Independence:
27 February 1844 (from Haiti)
National holiday:
Independence Day, 27 February (1844)
Constitution:
28 November 1966
Legal system:
Based on French civil codes
Executive branch:
President; Vice President; Cabinet
Legislative branch:
Bicameral National Congress (Congreso Nacional): Senate (Senado) and Chamber of Deputies (Camara de Diputados)
Judicial branch:
Supreme Court (Corte Suprema)

Crime [23]

Elections [20]

Chamber of Deputies	% of seats
Dominican Revolutionary	47.5
Social Christian Reformist	41.7
Dominican Liberation	10.8

Government Expenditures [21]

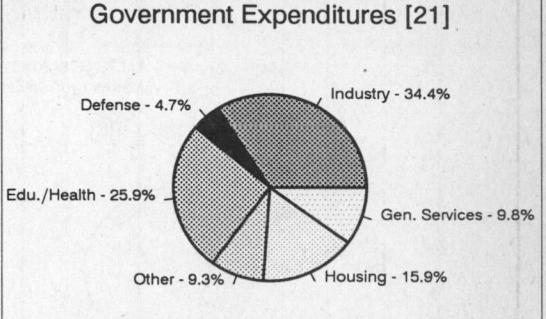

(% distribution). Expend. for CY94: 5,717 (Sucre bil.)

Military Expenditures and Arms Transfers [22]

	1990	1991	1992	1993	1994
Military expenditures					
Current dollars (mil.)	74	57	79	112	115
1994 constant dollars (mil.)	82	61	83	114	115
Armed forces (000)	21	21	22	22	22
Gross national product (GNP)					
Current dollars (mil.)	7,669	8,045	9,089	9,648	10,270
1994 constant dollars (mil.)	8,535	8,623	9,478	9,847	10,270
Central government expenditures (CGE)					
1994 constant dollars (mil.)	970	964	1,314	1,763	1,686[e]
People (mil.)	7.2	7.4	7.5	7.7	7.8
Military expenditure as % of GNP	1.0	0.7	0.9	1.2	1.1
Military expenditure as % of CGE	8.5	6.3	6.3	6.5	6.8
Military expenditure per capita (1994 $)	11	8	11	15	15
Armed forces per 1,000 people (soldiers)	2.9	2.9	2.9	2.9	2.8
GNP per capita (1994 $)	1,183	1,171	1,262	1,285	1,315
Arms imports[6]					
Current dollars (mil.)	5	5	5	5	30
1994 constant dollars (mil.)	6	5	5	5	30
Arms exports[6]					
Current dollars (mil.)	0	0	0	0	0
1994 constant dollars (mil.)	0	0	0	0	0
Total imports[7]					
Current dollars (mil.)	2,062	1,988	2,501	2,430	2,626
1994 constant dollars (mil.)	2,295	2,131	2,608	2,486	2,626
Total exports[7]					
Current dollars (mil.)	735	658	562	511	633
1994 constant dollars (mil.)	818	705	586	522	633
Arms as percent of total imports[8]	0.2	0.3	0.2	0.2	1.1
Arms as percent of total exports[8]	0	0	0	0	0

Human Rights [24]

	SSTS	FL	FAPRO	PPCG	APROBC	TPW	PCPTW	STPEP	PHRFF	PRW	ASST	AFL
Observes		P	P	S	P	P	P			P	P	P
	EAFRD	CPR	ESCR	SR	ACHR	MAAE	PVIAC	PVNAC	EAFDAW	TCIDTP	RC	
Observes	P	P	P	P	P		P	P	P	S	P	

P = Party; S = Signatory; see Appendix for meaning of abbreviations.

Labor Force

Total Labor Force [25]

2.3 million to 2.6 million

Labor Force by Occupation [26]

Agriculture	50%
Services and government	32
Industry	18

Date of data: 1991 est.

Unemployment Rate [27]

30% (1995 est.)

For sources, notes, and explanations, see Annotated Source Appendix, page 1061.

275

Production Sectors

Commercial Energy Production and Consumption

Data are shown in quadrillion (10^{15}) BTUs and percent for 1995
Values for hydroelectric, nuclear, geothermal, solar, and wind power refer to electrical generation.

Production [28]

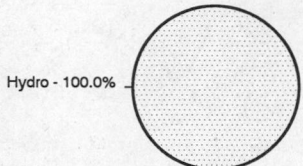

Hydro - 100.0%

Consumption [29]

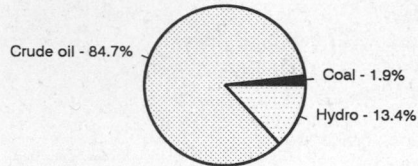

Crude oil - 84.7%
Coal - 1.9%
Hydro - 13.4%

Net hydroelectric power	0.021
Total	0.021

Crude oil	0.133
Coal	0.003
Net hydroelectric power	0.021
Total	0.157

Telecommunications [30]

- 190,000 (1987 est.) telephones
- Domestic: relatively efficient system based on islandwide microwave radio relay network
- International: 1 coaxial submarine cable; satellite earth station - 1 Intelsat (Atlantic Ocean)
- Radio: Broadcast stations: AM 120, FM 0, shortwave 6
- Television: Broadcast stations: 18 (1987 est.) Televisions: 728,000 (1993 est.)

Transportation [31]

Railways: total: 757 km; standard gauge: 375 km 1.435-m gauge (Central Romana Railroad); narrow gauge: 142 km 0.762-m gauge (Dominica Government Railway); 240 km operated by sugar companies in various gauges (0.558-m, 0.762-m, 1.067-m gauges) (1995)

Highways: total: 11,931 km; paved: 5,766 km; unpaved: 6,165 km (1987 est.)

Merchant marine: total: 1 cargo ship (1,000 GRT or over) totaling 1,587 GRT/1,165 DWT (1995 est.)

Airports

Total:	31
With paved runways over 3,047 m:	2
With paved runways 1,524 to 2,437 m:	6
With paved runways 914 to 1,523 m:	3
With paved runways under 914 m:	14
With unpaved runways 2,438 to 3,047 m:	1

Top Agricultural Products [32]

Agriculture accounts for 13% of the GDP; produces sugarcane, coffee, cotton, cocoa, tobacco, rice, beans, potatoes, corn, bananas; cattle, pigs, dairy products, meat, eggs.

Top Mining Products [33]

Metric tons except as noted[e]	3/31/95*
Cement, hydraulic (000 tons)	1,200
Gold (kg.)	1,300
Ferronickel	70,000
Limestone	550,000
Nickel, mine output, Ni content	30,500
Petroleum refinery products (000 42-gal. bls.)	10,500
Salt	10,000[26]
Sand (000 tons)	5,500
Silver (kg.)	10,000

Tourism [34]

	1990	1991	1992	1993	1994
Visitors	1,405	1,281	1,465	1,637	1,767
Tourists[31]	1,305	1,181	1,415	1,609	1,717
Cruise passengers[32]	100	100	50	28	50
Tourism receipts[33]	890	877	1,054	1,234	1,148
Tourism expenditures[34]	101	109	115	118	NA
Fare expenditures	15	20	21	23	NA

Travelers are in thousands, money in million U.S. dollars.

For sources, notes, and explanations, see Annotated Source Appendix, page 1061.

Manufacturing Sector

GDP and Manufacturing Summary [35]

	1980	1985	1989	1990	% change 1980-1990	% change 1989-1990
GDP (million 1980 $)	6,631	7,159	8,358	7,913	19.3	-5.3
GDP per capita (1980 $)	1,164	1,116	1,191	1,102	-5.3	-7.5
Manufacturing as % of GDP (current prices)	15.3	13.6	15.6[e]	13.5	-11.8	-13.5
Gross output (million $)	2,376	1,822[e]	2,887[e]	3,034[e]	27.7	5.1
Value added (million $)	1,013	783[e]	1,252[e]	1,298[e]	28.1	3.7
Value added (million 1980 $)	1,015	986	1,176	1,067	5.1	-9.3
Industrial production index	100	99	117	102	2.0	-12.8
Employment (thousands)	146	131	133[e]	139[e]	-4.8	4.5

Note: GDP stands for Gross Domestic Product. 'e' stands for estimated value.

Profitability and Productivity

	1980	1985	1989	1990	% change 1980-1990	% change 1989-1990
Intermediate input (%)	57	57[c]	57[e]	57[e]	0.0	0.0
Wages, salaries, and supplements (%)	11	7[e]	6[e]	6[e]	-45.5	0.0
Gross operating surplus (%)	31	36[e]	38[e]	37[e]	19.4	-2.6
Gross output per worker ($)	16,284	13,877[e]	21,786[e]	21,898[e]	34.5	0.5
Value added per worker ($)	6,940	5,966[e]	9,451[e]	9,373[e]	35.1	-0.8
Average wage (incl. benefits) ($)	1,867	998	1,230[e]	1,348[e]	-27.8	9.6

Profitability is in percent of gross output. Productivity is in U.S. $. 'e' stands for estimated value.

Profitability - 1990

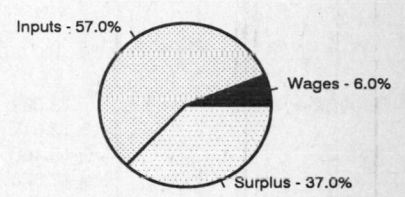

Inputs - 57.0%
Wages - 6.0%
Surplus - 37.0%

The graphic shows percent of gross output.

Value Added in Manufacturing

	1980 $ mil.	1980 %	1985 $ mil.	1985 %	1989 $ mil.	1989 %	1990 $ mil.	1990 %	% change 1980-1990	% change 1989-1990
311 Food products	510	50.3	293[e]	37.4	400[e]	31.9	414[e]	31.9	-18.8	3.5
313 Beverages	103	10.2	110[e]	14.0	167[e]	13.3	179[e]	13.8	73.8	7.2
314 Tobacco products	50	4.9	42[e]	5.4	81[e]	6.5	67[e]	5.2	34.0	-17.3
321 Textiles	29	2.9	26[e]	3.3	53[e]	4.2	45[e]	3.5	55.2	-15.1
322 Wearing apparel	13	1.3	9[e]	1.1	11[e]	0.9	16[e]	1.2	23.1	45.5
323 Leather and fur products	11	1.1	8[e]	1.0	15[e]	1.2	14[e]	1.1	27.3	-6.7
324 Footwear	13	1.3	13[e]	1.7	26[e]	2.1	25[e]	1.9	92.3	-3.8
331 Wood and wood products	2	0.2	3[e]	0.4	3[e]	0.2	2[e]	0.2	0.0	-33.3
332 Furniture and fixtures	11	1.1	11[e]	1.4	23[e]	1.8	19[e]	1.5	72.7	-17.4
341 Paper and paper products	19	1.9	21[e]	2.7	38[e]	3.0	37[e]	2.9	94.7	-2.6
342 Printing and publishing	14	1.4	13[e]	1.7	24[e]	1.9	22[e]	1.7	57.1	-8.3
351 Industrial chemicals	18	1.8	16[e]	2.0	20[e]	1.6	21[e]	1.6	16.7	5.0
352 Other chemical products	41	4.0	27[e]	3.4	49[e]	3.9	44[e]	3.4	7.3	-10.2
353 Petroleum refineries	66	6.5	81[e]	10.3	138[e]	11.0	209[c]	16.1	216.7	51.4
354 Miscellaneous petroleum and coal products	1	0.1	NA	0.0	NA	0.0	1[e]	0.1	0.0	NA
355 Rubber products	6	0.6	6[e]	0.8	11[e]	0.9	10[e]	0.8	66.7	-9.1
356 Plastic products	21	2.1	12[e]	1.5	28[e]	2.2	21[e]	1.6	0.0	-25.0
361 Pottery, china, and earthenware	1	0.1	1[e]	0.1	1[e]	0.1	1[e]	0.1	0.0	0.0
362 Glass and glass products	3	0.3	5[e]	0.6	8[e]	0.6	8[e]	0.6	166.7	0.0
369 Other nonmetal mineral products	32	3.2	29[e]	3.7	47[e]	3.8	46[e]	3.5	43.8	-2.1
371 Iron and steel	10	1.0	15[e]	1.9	25[e]	2.0	24[e]	1.8	140.0	-4.0
372 Nonferrous metals	1	0.1	1[e]	0.1	2[e]	0.2	3[e]	0.2	200.0	50.0
381 Metal products	21	2.1	28[e]	3.6	55[e]	4.4	48[e]	3.7	128.6	-12.7
382 Nonelectrical machinery	5	0.5	3[e]	0.4	8[e]	0.6	6[e]	0.5	20.0	-25.0
383 Electrical machinery	7	0.7	6[e]	0.8	13[e]	1.0	11[e]	0.8	57.1	-15.4
384 Transport equipment	NA	0.0	NA	0.0	NA	0.0	1[e]	0.1	NA	NA
385 Professional and scientific equipment	1	0.1	1[e]	0.1	1[e]	0.1	2[e]	0.2	100.0	100.0
390 Other manufacturing industries	2	0.2	1[e]	0.1	3[e]	0.2	3[e]	0.2	50.0	0.0

Note: The industry codes shown are International Standard Industry codes (ISIC). Percentages are percent of total Value Added. 'e' stands for estimated value

For sources, notes, and explanations, see Annotated Source Appendix, page 1061.

277

Finance, Economics, and Trade

Economic Indicators [36]

- **National product**: GDP—purchasing power parity— $26.8 billion (1995 est.)

- **National product real growth rate**: 3.5% (1995 est.)

- **National product per capita**: $3,400 (1995 est.)

- **Inflation rate (consumer prices)**: 9.5% (1995)

- **External debt**: $4.6 billion (1994)

Balance of Payments Summary [37]

Values in millions of dollars.

	1989	1990	1991	1992	1993
Exports of goods (f.o.b.)	924.4	734.5	658.3	562.5	511.5
Imports of goods (f.o.b.)	-1,963.8	-1,792.8	-1,728.8	-2,174.3	-2,118.4
Trade balance	-1,039.4	-1,058.3	-1,070.5	-1,611.8	-1,609.9
Services - debits	-706.4	-775.4	-759.2	-850.7	-876.0
Services - credits	1,162.8	1,356.9	1,407.5	1,585.5	1,880.3
Private transfers (net)	300.5	314.8	329.5	346.6	361.8
Government transfers (net)	83.9	55.8	57.0	85.2	79.8
Long-term capital (net)	243.1	-93.4	908.5	148.4	125.3
Short-term capital (net)	141.6	558.2	-961.6	18.1	203.5
Errors and omissions	-188.1	-294.1	426.7	305.8	-69.1
Overall balance	-2.0	64.5	337.9	27.1	98.7

Exchange Rates [38]

Currency: **Dominican peso.**
Symbol: **RD$.**

Data are currency units per $1.

December 1995	13.589
1995	13.617
1994	13.160
1993	12.676
1992	12.774
1991	12.692

Imports and Exports

Top Import Origins [39]

$2.867 billion (f.o.b., 1995) Data are for 1993.

Origins	%
US	60

Top Export Destinations [40]

$837.7 million (f.o.b., 1995) Data are for 1994.

Destinations	%
US	47.5
EU	22
Puerto Rico	8.4
Asia	6.7

Foreign Aid [41]

Recipient: ODA, $21 million (1993).

Import and Export Commodities [42]

Import Commodities	Export Commodities
Foodstuffs	Ferronickel
Petroleum	Sugar
Cotton and fabrics	Gold
Chemicals and pharmaceuticals	Coffee
	Cocoa

Ecuador

Geography [1]

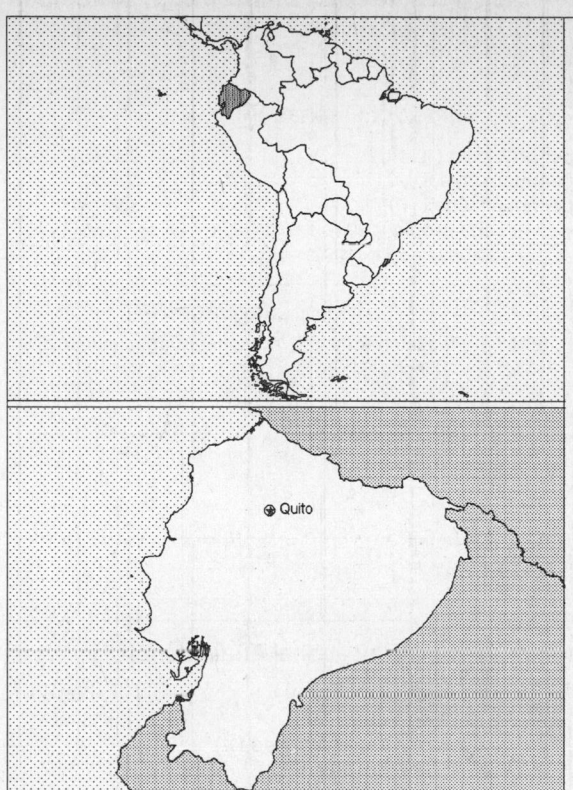

Total area:
283,560 sq km   109,483 sq mi
Land area:
276,840 sq km   106,889 sq mi
Comparative area:
Slightly smaller than Nevada
Note: Includes Galapagos Islands
Land boundaries:
Total 2,010 km, Colombia 590 km, Peru 1,420 km
Coastline:
2,237 km
Climate:
Tropical along coast, becoming cooler inland
Terrain:
Coastal plain (costa), inter-Andean central highlands (sierra), and flat to rolling eastern jungle (oriente)
Natural resources:
Petroleum, fish, timber
Land use:
Arable land: 6%
Permanent crops: 3%
Meadows and pastures: 17%
Forest and woodland: 51%
Other: 23%

Demographics [2]

	1970	1980	1990	1995[1]	1996	2000	2010	2020	2030
Population	6,146	8,315	10,116	11,242	11,466	12,360	14,534	16,546	18,324
Population density (persons per sq. mi.)	57	78	95	105	107	116	136	155	171
(persons per sq. km.)	22	30	37	41	41	45	52	60	66
Net migration rate (per 1,000 population)	0.0	NA	0.0	0.0	0.0	0.0	0.0	0.0	0.0
Births	NA	NA	NA	NA	287	NA	NA	NA	NA
Deaths	NA	NA	50	NA	63	NA	NA	NA	NA
Life expectancy - males	57.4	NA	66.2	68.1	68.5	69.9	72.9	75.1	76.8
Life expectancy - females	60.5	NA	71.5	73.5	73.8	75.3	78.3	80.7	82.5
Birth rate (per 1,000)	41.2	NA	28.6	25.5	25.1	23.0	19.2	16.5	14.7
Death rate (per 1,000)	11.2	NA	6.4	5.6	5.5	5.1	4.7	5.0	5.8
Women of reproductive age (15-49 yrs.)	NA	NA	2,535	2,959	3,044	3,375	4,047	4,465	4,618
of which are currently married	NA	NA	1,537	NA	1,886	2,127	2,650	NA	NA
Fertility rate	6.1	NA	3.5	3.0	2.9	2.6	2.2	2.1	2.0

Except as noted, values for vital statistics are in thousands; life expectancy is in years.

Health

Health Indicators [3]

% of population with access to	
safe water (1990-95)	71
adequate sanitation (1990-95)	48
health services (1985-95)	88
% of 1-year-olds immunized (1990-94) against	
TB (tuberculosis)	100
DPT (diphtheria, pertussis, tetanus)	80
polio	78
measles	100
% of contraceptive prevalence (1980-94)	53
ORT use rate (1990-94)	70

Health Expenditures [4]

Total health expenditure, 1990 (official exchange rate)	
Millions of dollars	441
Dollars per capita	43
Health expenditures as a percentage of GDP	
Total	4.1
Public sector	2.6
Private sector	1.6
Development assistance for health	
Total aid flows (millions of dollars)[1]	31
Aid flows per capita (dollars)	3.0
Aid flows as a percentage of total health expenditure	7.0

For sources, notes, and explanations, see Annotated Source Appendix, page 1061.

279

Human Factors

Women and Children [5]

% of pregnant women immunized (tetanus 1990-94)	NA
% of births attended by trained health personnel (1983-94)	84
Maternal mortality rate (1980-92)	170
Under-5 mortality rate (1994)	57
% under-5 moderately/severely underweight (1980-1994)	17

Burden of Disease [6]

Population per physician (1990)	957.46
Population per nurse (1990)	599.32
Population per hospital bed (1990)	608.24
AIDS cases per 100,000 people (1994)	1.0
Malaria cases per 100,000 people (1992)	383

Ethnic Division [7]

Mestizo	55%
Indian	25%
Black	10%
Spanish	10%

Religion [8]

Roman Catholic	95%
Other	5%

Major Languages [9]

Spanish (official), Indian languages (especially Quechua).

Education

Public Education Expenditures [10]

	1980	1985	1991	1992	1993	1994
Million (Sucre)						
Total education expenditure	15,580	38,009	298,126	492,252	790,964	NA
as percent of GNP	5.6	3.7	2.6	2.7	3.0	NA
as percent of total govt. expend.	33.3	20.6	17.5	19.2	NA	NA
Current education expenditure	14,649	35,611	260,332	416,411	720,441	NA
as percent of GNP	5.3	3.5	2.3	2.3	2.7	NA
as percent of current govt. expend.	36.0	25.8	23.9	24.0	NA	NA
Capital expenditure	931	2,397	37,794	75,841	70,523	NA

Educational Attainment [11]

Age group (1990)	25+
Total population	3,953,452
Highest level attained (%)	
No schooling	1.7
First level	
Not completed	43.7
Completed	NA
Entered second level	
S-1	22.6
S-2	NA
Postsecondary	12.7

Illiteracy [12]

In thousands and percent[1]	1990	1995	2000
Illiterate population (15+ yrs.)	691	719	678
Illiteracy rate - total pop. (%)	11.2	10.0	8.2
Illiteracy rate - males (%)	9.0	8.2	6.7
Illiteracy rate - females (%)	13.3	11.8	9.6

Libraries [13]

	Admin. Units	Svc. Pts.	Vols. (000)	Shelving (meters)	Vols. Added	Reg. Users
National	NA	NA	NA	NA	NA	NA
Nonspecialized	NA	NA	NA	NA	NA	NA
Public (1988)	210	210	142	NA	55,200	NA
Higher ed. (1987)	128	128	531	NA	110,029	224,517
School	NA	NA	NA	NA	NA	NA

Daily Newspapers [14]

	1980	1985	1990	1994
Number of papers	18	26	25	24
Circ. (000)	558	800[e]	820[e]	808

Culture [15]

Cinema (seats per 1,000)	7.2
Annual attendance per person	0.6
Gross box office receipts (mil. Sucre)	4,949
Museums (reporting)	NA
Visitors (000)	NA
Annual receipts (000 Sucre)	NA

Science and Technology

Scientific/Technical Forces [16]

Scientists/engineers	1,732
Number female	NA
Technicians	2,204
Number female	NA
Total	3,936

R&D Expenditures [17]

	Sucre (000) 1990
Total expenditure	8,443,000
Capital expenditure	NA
Current expenditure	NA
Percent current	NA

U.S. Patents Issued [18]

Values show patents issued to citizens of the country by the U.S. Patents Office.

	1993	1994	1995
Number of patents	1	2	0

For sources, notes, and explanations, see Annotated Source Appendix, page 1061.

Government and Law

Organization of Government [19]

Long-form name:
Republic of Ecuador
Type:
Republic
Independence:
24 May 1822 (from Spain)
National holiday:
Independence Day, 10 August (1809)
(independence of Quito)
Constitution:
10 August 1979
Legal system:
Based on civil law system; has not
accepted compulsory ICJ jurisdiction
Executive branch:
President; Vice President; Cabinet
Legislative branch:
Unicameral: National Congress (Congreso
Nacional)
Judicial branch:
Supreme Court (Corte Suprema)

Elections [20]

National Congress	% of seats
Social Christian Party (PSC)	32.9
Roldista Party (PRE)	25.6
Popular Democracy (DP)	12.2
Pachakutik Movement	8.5
Democratic Left (ID)	6.1
Independent and Others	4.9
Radical Liberal Party (PLRE)	3.7
Popular Democratic (MPD)	2.4
Popular Revolutionary Action	2.4
Concentration of Popular Forces	1.2

Government Expenditures [21]

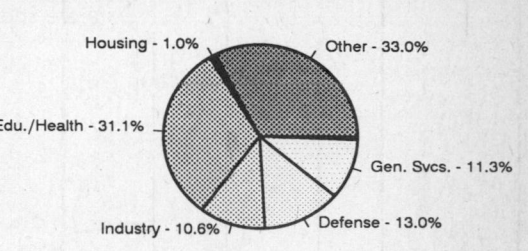

Housing - 1.0%
Other - 33.0%
Edu./Health - 31.1%
Gen. Svcs. - 11.3%
Industry - 10.6%
Defense - 13.0%

(% distribution). Expend. for CY92: 3,145 (Sucre bil.)

Military Expenditures and Arms Transfers [22]

	1990	1991	1992	1993	1994
Military expenditures					
Current dollars (mil.)[e,1]	380	460	500	450	550
1994 constant dollars (mil.)[e,1]	423	493	521	459	550
Armed forces (000)	53	53	57	57	57
Gross national product (GNP)					
Current dollars (mil.)	11,880	13,200	14,220	14,970	15,810
1994 constant dollars (mil.)	13,220	14,150	14,830	15,280	15,810
Central government expenditures (CGE)					
1994 constant dollars (mil.)	2,107	2,001	2,052	2,185	2,519
People (mil.)	9.8	10.0	10.2	10.5	10.7
Military expenditure as % of GNP	3.2	3.5	3.5	3.0	3.5
Military expenditure as % of CGE	20.1	24.6	25.4	21.0	21.8
Military expenditure per capita (1994 $)	43	49	51	44	52
Armed forces per 1,000 people (soldiers)	5.4	5.3	5.6	5.4	5.3
GNP per capita (1994 $)	1,348	1,412	1,448	1,460	1,480
Arms imports[6]					
Current dollars (mil.)	20	20	30	40	30
1994 constant dollars (mil.)	22	21	31	41	30
Arms exports[6]					
Current dollars (mil.)	0	0	0	0	0
1994 constant dollars (mil.)	0	0	0	0	0
Total imports[7]					
Current dollars (mil.)	1,862	2,399	2,501	2,562	3,642
1994 constant dollars (mil.)	2,072	2,571	2,608	2,615	3,642
Total exports[7]					
Current dollars (mil.)	2,714	2,852	3,007	2,904	3,717
1994 constant dollars (mil.)	3,020	3,057	3,136	2,964	3,717
Arms as percent of total imports[8]	1.1	0.8	1.2	1.6	0.8
Arms as percent of total exports[8]	0	0	0	0	0

Crime [23]

	1994
Crime volume	
Cases known to police	53,421
Attempts (percent)	NA
Percent cases solved	71.00
Crimes per 100,000 persons	466.15
Persons responsible for offenses	
Total number offenders	33,165
Percent female	4.80
Percent juvenile (1-18 yrs.)	4.90
Percent foreigners	1.70

Human Rights [24]

	SSTS	FL	FAPRO	PPCG	APROBC	TPW	PCPTW	STPEP	PHRFF	PRW	ASST	AFL
Observes	P	P	P	P	P	P	P	P		P	P	P
	EAFRD	CPR	ESCR	SR	ACHR	MAAE	PVIAC	PVNAC	EAFDAW	TCIDTP	RC	
Observes	P	P	P	P	P	P	P	P	P	P	P	

P = Party; S = Signatory; see Appendix for meaning of abbreviations.

Labor Force

Total Labor Force [25]

2.8 million

Labor Force by Occupation [26]

Agriculture	35%
Manufacturing	21
Commerce	16
Services and other activities	28

Date of data: 1982

Unemployment Rate [27]

7.1% (1994)

For sources, notes, and explanations, see Annotated Source Appendix, page 1061.

281

Production Sectors

Commercial Energy Production and Consumption

Data are shown in quadrillion (10^{15}) BTUs and percent for 1995
Values for hydroelectric, nuclear, geothermal, solar, and wind power refer to electrical generation.

Production [28]	Consumption [29]
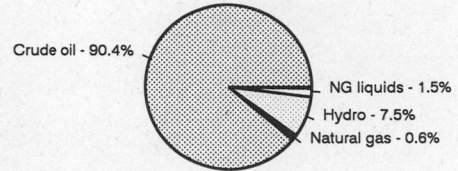	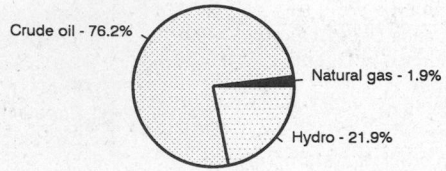

Production [28] pie chart labels:
Crude oil - 90.4%
NG liquids - 1.5%
Hydro - 7.5%
Natural gas - 0.6%

Consumption [29] pie chart labels:
Crude oil - 76.2%
Natural gas - 1.9%
Hydro - 21.9%

Crude oil	0.857	Crude oil	0.247
Natural gas liquids	0.014	Dry natural gas	0.006
Dry natural gas	0.006	Net hydroelectric power	0.071
Net hydroelectric power	0.071	Total	0.324
Total	0.948		

Telecommunications [30]

- 586,300 (1994 est.) telephones
- Domestic: facilities generally inadequate and unreliable
- International: satellite earth station - 1 Intelsat (Atlantic Ocean)
- Radio: Broadcast stations: AM 272, FM 0, shortwave 39
- Television: Broadcast stations: 33 Televisions: 940,000 (1992 est.)

Transportation [31]

Railways: total: 965 km (single track); narrow gauge: 965 km 1.067-m gauge

Highways: total: 43,709 km; paved: 5,245 km; unpaved: 38,464 km (1991 est.)

Merchant marine: total: 19 ships (1,000 GRT or over) totaling 114,701 GRT/171,240 DWT; ships by type: container 2, liquefied gas tanker 1, oil tanker 12, passenger 3, refrigerated cargo 1 (1995 est.)

Airports

Total:	188
With paved runways over 3,047 m:	2
With paved runways 2,438 to 3,047 m:	7
With paved runways 1,524 to 2,437 m:	8
With paved runways 914 to 1,523 m:	13
With paved runways under 914 m:	121

Top Agricultural Products [32]

Agriculture accounts for 13% of the GDP; produces bananas, coffee, cocoa, rice, potatoes, manioc, plantains, sugarcane; cattle, sheep, pigs, beef, pork, dairy products; balsa wood; fish, shrimp.

Top Mining Products [33]

Metric tons except as noted[e]	10/95[*]
Steel	
crude	21,800
semimanufactures	214,000
Sand	
silica (glass sand)	20,000
ferruginous	10,000
Pumice	13,000
Sulfur	14,000
Petroleum, crude (000 42-gal. bls.)	138,000

Tourism [34]

	1990	1991	1992	1993	1994
Visitors[8]	362	365	403	471	482
Tourism receipts	188	189	192	230	252
Tourism expenditures	175	177	178	190	203
Fare receipts	62	63	78	55	57
Fare expenditures	62	64	76	66	91

Travelers are in thousands, money in million U.S. dollars.

For sources, notes, and explanations, see Annotated Source Appendix, page 1061.

Manufacturing Sector

Manufacturing Summary [35]

	1987		1988		1989		1990		1991	
	$ bil.	%	$ bil.	%	$ bil.	%	$ bil.	%	$ bil.	%
Establishments or enterprises (number)	1,746	0.074	1,733	0.083	1,712	0.091	1,702	0.095	-	-
Total employment (000)	126	0.093	128	0.094	129	0.105	131	0.119	-	-
Production workers (000)	92	0.136	92	0.147	93	0.126	94	0.132	-	-
Output ($ bil.)	4	0.041	4	0.037	4	0.034	4	0.039	-	-
Value added ($ bil.)	1	0.028	1	0.024	1	0.021	1	0.027	-	-
Capital investment ($ mil.)	658	0.151	626	0.131	480	0.088	570	0.103	-	-
M & E investment ($ mil.)	-	-	-	-	-	-	-	-	-	-
Employees per establishment (number)	72	125.179	74	113.414	75	114.627	77	124.743	-	-
Production workers per establishment	53	182.848	53	177.689	54	138.067	55	138.888	-	-
Output per establishment ($ mil.)	2	55.140	2	44.679	2	36.711	3	40.990	-	-
Capital investment per estab. ($ mil.)	0.377	202.926	0.361	159.155	0.280	95.861	0.335	107.970	-	-
M & E per establishment ($ mil)	-	-	-	-	-	-	-	-	-	-
Payroll per employee ($)	4,061	45.289	3,646	36.594	3,056	27.355	3,288	23.706	-	-
Wages per production worker ($)	-	-	-	-	-	-	-	-	-	-
Hours per production worker (hours)	-	-	-	-	-	-	-	-	-	-
Output per employee ($)	32,990	44.049	33,555	39.395	30,694	32.027	34,054	32.860	-	-
Capital investment per employee ($)	5,206	162.109	4,883	140.331	3,722	88.629	4,346	86.554	-	-
M & E per employee ($)	-	-	-	-	-	-	-	-	-	-

Note: Columns headed % show percent of world total or ratio. Ratios closest to 100 are closest to world average. M & E stands for machinery & equipment.

Output in Manufacturing

	1987		1988		1989		1990		1991	
	$ bil.	%	$ bil.	%	$ bil.	%	$ bil.	%	$ bil.	%
3110 - Food products	1.173	29.33	1.176	29.40	1.093	27.33	1.235	30.88	-	-
3130 - Beverages	0.130	3.25	0.114	2.85	0.110	2.75	0.138	3.45	-	-
3140 - Tobacco	0.079	1.97	0.069	1.72	0.078	1.95	0.022	0.55	-	-
3210 - Textiles	0.253	6.33	0.267	6.67	0.243	6.07	0.249	6.22	-	-
3211 - Spinning, weaving, etc.	0.202	5.05	0.208	5.20	0.190	4.75	0.194	4.85	-	-
3220 - Wearing apparel	0.028	0.70	0.030	0.75	0.032	0.80	0.034	0.85	-	-
3230 - Leather and products	0.017	0.43	0.014	0.35	0.017	0.43	0.013	0.33	-	-
3240 - Footwear	0.028	0.70	0.017	0.43	0.017	0.43	0.017	0.43	-	-
3310 - Wood products	0.057	1.43	0.050	1.25	0.044	1.10	0.047	1.18	-	-
3320 - Furniture, fixtures	0.037	0.93	0.030	0.75	0.030	0.75	0.029	0.73	-	-
3410 - Paper and products	0.164	4.10	0.173	4.32	0.182	4.55	0.228	5.70	-	-
3411 - Pulp, paper, etc.	0.054	1.35	0.047	1.18	0.057	1.43	0.053	1.33	-	-
3420 - Printing, publishing	0.090	2.25	0.088	2.20	0.078	1.95	0.087	2.17	-	-
3510 - Industrial chemicals	0.091	2.28	0.100	2.50	0.076	1.90	0.065	1.63	-	-
3511 - Basic chemicals, excl fertilizers	0.031	0.78	0.041	1.02	0.029	0.73	0.029	0.73	-	-
3513 - Synthetic resins, etc.	0.031	0.78	0.037	0.93	0.034	0.85	0.029	0.73	-	-
3520 - Chemical products nec	0.216	5.40	0.209	5.22	0.224	5.60	0.237	5.93	-	-
3522 - Drugs and medicines	0.073	1.82	0.071	1.78	0.068	1.70	0.060	1.50	-	-
3530 - Petroleum refineries	0.231	5.78	0.320	8.00	0.249	6.22	0.561	14.03	-	-
3540 - Petroleum, coal products	0.025	0.63	0.031	0.78	0.025	0.63	0.023	0.58	-	-
3550 - Rubber products	0.060	1.50	0.054	1.35	0.063	1.57	0.052	1.30	-	-
3560 - Plastic products nec	0.139	3.47	0.154	3.85	0.131	3.28	0.133	3.33	-	-
3610 - Pottery, china, etc.	0.017	0.43	0.022	0.55	0.017	0.43	0.018	0.45	-	-
3620 - Glass and products	0.026	0.65	0.024	0.60	0.023	0.58	0.022	0.55	-	-
3690 - Nonmetal products nec	0.171	4.28	0.161	4.03	0.135	3.38	0.156	3.90	-	-
3710 - Iron and steel	0.095	2.38	0.122	3.05	0.116	2.90	0.107	2.67	-	-
3720 - Nonferrous metals	0.025	0.63	0.027	0.67	0.023	0.58	0.025	0.63	-	-
3810 - Metal products	0.158	3.95	0.168	4.20	0.141	3.53	0.148	3.70	-	-
3820 - Machinery nec	0.009	0.23	0.009	0.23	0.008	0.20	0.009	0.23	-	-
3830 - Electrical machinery	0.125	3.13	0.131	3.28	0.108	2.70	0.112	2.80	-	-
3832 - Radio, television, etc.	0.023	0.58	0.019	0.48	0.017	0.43	0.025	0.63	-	-
3840 - Transportation equipment	0.126	3.15	0.142	3.55	0.139	3.47	0.145	3.63	-	-
3841 - Shipbuilding, repair	-	-	-	-	0.002	0.05	0.001	0.02	-	-
3843 - Motor vehicles	0.125	3.13	0.140	3.50	0.137	3.42	0.142	3.55	-	-
3850 - Professional goods	0.051	1.27	0.019	0.48	0.013	0.33	0.012	0.30	-	-
3900 - Industries nec	0.010	0.25	0.012	0.30	0.010	0.25	0.010	0.25	-	-

Note: Codes are International Standard Industry codes (ISIC). Percentages are % of total Output. [f]: Factor Prices; [p]: Producer Prices.

For sources, notes, and explanations, see Annotated Source Appendix, page 1061.

283

Finance, Economics, and Trade

Economic Indicators [36]

- **National product**: GDP—purchasing power parity—$44.6 billion (1995 est.)
- **National product real growth rate**: 2.3% (1995 est.)
- **National product per capita**: $4,100 (1995 est.)
- **Inflation rate (consumer prices)**: 25% (1995)
- **External debt**: $12.6 billion (1995 est.)

Balance of Payments Summary [37]

Values in millions of dollars.

	1989	1990	1991	1992	1993
Exports of goods (f.o.b.)	2,354	2,714	2,851	3,008	2,903
Imports of goods (f.o.b.)	-1,693	-1,711	-2,207	-2,048	-2,325
Trade balance	661	1,003	644	960	578
Services - debits	-1,808	-1,839	-1,808	-1,722	-1,748
Services - credits	536	563	587	652	680
Private transfers (net)	NA	NA	NA	NA	NA
Government transfers (net)	97	107	110	120	130
Long-term capital (net)	356	-424	-508	-490	-234
Short-term capital (net)	294	727	918	520	1,023
Errors and omissions	67	71	198	82	132
Overall balance	203	208	141	122	561

Exchange Rates [38]

Currency: **sucre.**
Symbol: **S/.**

Data are currency units per $1.

31 December 1995	2,914.8
1995	2,564.5
1994	2,196.7
1993	1,919.1
1992	1,534.0
1991	1,046.25

Imports and Exports

Top Import Origins [39]

$3.7 billion (c.i.f., 1994).

Origins	%
Latin America	31
US	28
EU	17
Caribbean	NA
Japan	NA

Top Export Destinations [40]

$4 billion (f.o.b., 1994).

Destinations	%
US	42
Latin America	29
EU countries	17
Caribbean	NA

Foreign Aid [41]

	U.S. $	
ODA (1993)	153	million
US (1995)	12.7	million
Other countries (1995)	160	million

Import and Export Commodities [42]

Import Commodities	Export Commodities
Transport equipment	Petroleum 39%
Consumer goods	Bananas 17%
Vehicles	Shrimp 16%
Machinery	Cocoa 3%
Chemicals	Coffee 6%

Egypt

Geography [1]

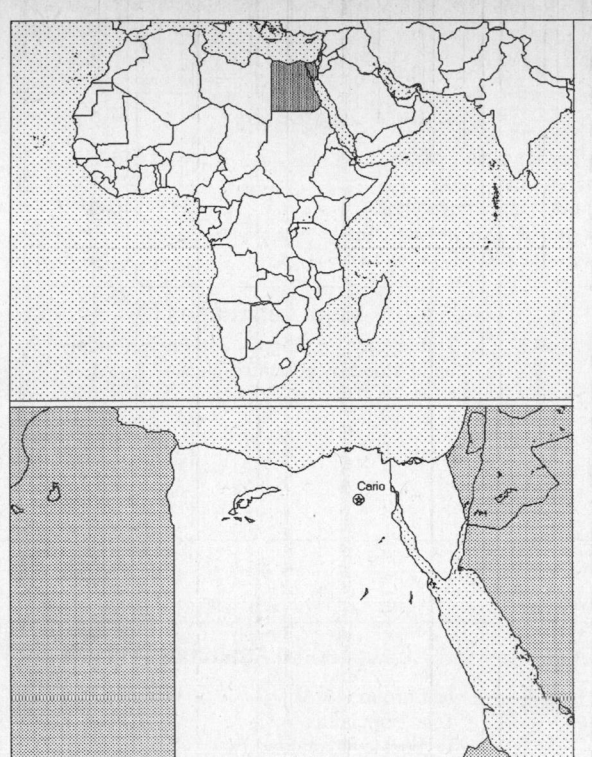

Total area:
 1,001,450 sq km   386,662 sq mi
Land area:
 995,450 sq km   384,345 sq mi
Comparative area:
 Slightly more than three times the size of New Mexico
Land boundaries:
 Total 2,689 km, Gaza Strip 11 km, Israel 255 km, Libya 1,150 km, Sudan 1,273 km
Coastline:
 2,450 km
Climate:
 Desert; hot, dry summers with moderate winters
Terrain:
 Vast desert plateau interrupted by Nile valley and delta
Natural resources:
 Petroleum, natural gas, iron ore, phosphates, manganese, limestone, gypsum, talc, asbestos, lead, zinc
Land use:
 Arable land: 3%
 Permanent crops: 2%
 Meadows and pastures: 0%
 Forest and woodland: 0%
 Other: 95%

Demographics [2]

	1970	1980	1990	1995[1]	1996	2000	2010	2020	2030
Population	33,574	42,441	56,106	62,360	63,575	68,437	80,689	92,350	102,482
Population density (persons per sq. mi.)	87	110	146	162	165	178	210	240	267
(persons per sq. km.)	34	43	56	63	64	69	81	93	103
Net migration rate (per 1,000 population)	NA	4.1	6.1	-0.4	-0.4	-0.4	0.0	0.0	0.0
Births	NA	NA	NA	NA	1,792	NA	NA	NA	NA
Deaths	NA	NA	NA	NA	553	NA	NA	NA	NA
Life expectancy - males	NA	NA	57.7	59.2	59.5	60.7	63.5	66.1	68.4
Life expectancy - females	NA	NA	61.4	63.1	63.5	64.8	68.1	71.0	73.7
Birth rate (per 1,000)	NA	40.2	32.0	28.7	28.2	26.2	22.4	19.0	16.7
Death rate (per 1,000)	NA	12.6	9.9	8.9	8.7	8.1	7.4	7.2	7.6
Women of reproductive age (15-49 yrs.)	NA	NA	13,552	15,554	15,963	17,553	21,471	24,703	26,946
of which are currently married[13]	NA	8	8,915	NA	10,492	11,593	14,548	NA	NA
Fertility rate	NA	5.5	4.2	3.7	3.6	3.2	2.6	2.3	2.1

Except as noted, values for vital statistics are in thousands; life expectancy is in years.

Health

Health Indicators [3]

% of population with access to	
safe water (1990-95)	80
adequate sanitation (1990-95)	50
health services (1985-95)	99
% of 1-year-olds immunized (1990-94) against	
TB (tuberculosis)	95
DPT (diphtheria, pertussis, tetanus)	90
polio	91
measles	90
% of contraceptive prevalence (1980-94)	47
ORT use rate (1990-94)	34

Health Expenditures [4]

Total health expenditure, 1990 (official exchange rate)	
Millions of dollars	921
Dollars per capita	18
Health expenditures as a percentage of GDP	
Total	2.6
Public sector	1.0
Private sector	1.6
Development assistance for health	
Total aid flows (millions of dollars)[1]	111
Aid flows per capita (dollars)	2.1
Aid flows as a percentage of total health expenditure	12.1

For sources, notes, and explanations, see Annotated Source Appendix, page 1061.

285

Human Factors

Women and Children [5]

% of pregnant women immunized (tetanus 1990-94)	64
% of births attended by trained health personnel (1983-94)	41
Maternal mortality rate (1980-92)	270
Under-5 mortality rate (1994)	52
% under-5 moderately/severely underweight (1980-1994)	9

Burden of Disease [6]

Population per physician (1990)	1,316.31
Population per nurse (1990)	488.67
Population per hospital bed (1993)	516.62
AIDS cases per 100,000 people (1994)	*
Malaria cases per 100,000 people (1992)	NA

Ethnic Division [7]

Eastern Hamitic stock (Egyptians, Bedouins, and Berbers)	99%
Greek, Nubian, Armenian, other European (primarily Italian and French)	1%

Religion [8]

Muslim (mostly Sunni)	94%
Coptic Christian and other (official estimate)	6%

Major Languages [9]

Arabic (official), English and French widely understood by educated classes.

Education

Public Education Expenditures [10]

Million (Pound)[23]	1980	1985	1990	1991	1992	1994
Total education expenditure	NA	1,878	3,737	4,557	5,839	NA
as percent of GNP	NA	6.3	4.9	4.7	5.0	NA
as percent of total govt. expend.	NA	NA	NA	9.7	11.0	NA
Current education expenditure	NA	1,775	3,229	3,941	4,683	NA
as percent of GNP	NA	5.9	4.2	4.1	4.0	NA
as percent of current govt. expend.	NA	10.8	NA	11.0	11.0	NA
Capital expenditure	NA	103	508	616	1,156	NA

Educational Attainment [11]

Age group (1986)[7]	25+
Total population	19,441,903
Highest level attained (%)	
No schooling	64.1
First level	
Not completed	16.5
Completed	NA
Entered second level	
S-1	14.8
S-2	NA
Postsecondary	4.6

Illiteracy [12]

In thousands and percent[1]	1990	1995	2000
Illiterate population (15+ yrs.)	17,816	18,954	20,047
Illiteracy rate - total pop. (%)	52.2	48.5	45.2
Illiteracy rate - males (%)	39.4	36.7	34.0
Illiteracy rate - females (%)	65.2	60.4	56.4

Libraries [13]

	Admin. Units	Svc. Pts.	Vols. (000)	Shelving (meters)	Vols. Added	Reg. Users
National (1991)	1	52	2,195	NA	NA	NA
Nonspecialized (1991)	105	105	832	NA	NA	NA
Public (1988)	836	NA	8,523	NA	NA	1,644
Higher ed. (1989)	272	272	35,790	NA	NA	65,900
School	NA	NA	NA	NA	NA	NA

Daily Newspapers [14]

	1980	1985	1990	1994
Number of papers	12	12	14	17
Circ. (000)	1,701	2,383	2,400[e]	3,949

Culture [15]

Cinema (seats per 1,000)	2.2
Annual attendance per person	0.3
Gross box office receipts (mil. Pound)	23
Museums (reporting)	50
Visitors (000)	4,241
Annual receipts (000 Pound)	7,781

Science and Technology

Scientific/Technical Forces [16]

Scientists/engineers	26,415
Number female	NA
Technicians	19,607
Number female	NA
Total[9]	46,022

R&D Expenditures [17]

	Pound (000) 1991
Total expenditure[10]	955,273[e]
Capital expenditure	281,463[e]
Current expenditure	673,810[e]
Percent current	70.5[e]

U.S. Patents Issued [18]

Values show patents issued to citizens of the country by the U.S. Patents Office.

	1993	1994	1995
Number of patents	1	4	3

For sources, notes, and explanations, see Annotated Source Appendix, page 1061.

Government and Law

Organization of Government [19]

Long-form name:
Arab Republic of Egypt
Type:
Republic
Independence:
28 February 1922 (from UK)
National holiday:
Anniversary of the Revolution, 23 July (1952)
Constitution:
11 September 1971
Legal system:
Based on English common law, Islamic law, and Napoleonic codes; judicial review by Supreme Court and Council of State (oversees validity of administrative decisions); accepts compulsory ICJ jurisdiction, with reservations
Executive branch:
President; Prime Minister; Cabinet
Legislative branch:
Bicameral: People's Assembly (Majlis al-Cha'b) and Advisory Council (Majlis al-Shura)
Judicial branch:
Supreme Constitutional Court

Elections [20]

People's Assembly	% of votes
National Democratic Party (NDP)	72.0
Opposition parties	3.0
Independents	25.0

Government Expenditures [21]

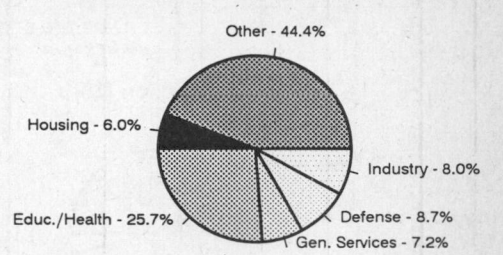

Other - 44.4%
Housing - 6.0%
Industry - 8.0%
Defense - 8.7%
Gen. Services - 7.2%
Educ./Health - 25.7%

(% distribution). Expend. for FY93: 56,143 (Pound mil.)

Crime [23]

	1990
Crime volume	
Cases known to police	1,785,838
Attempts (percent)	0.02
Percent cases solved	95
Crimes per 100,000 persons	3,314.41
Persons responsible for offenses	
Total number offenders	1,509,090
Percent female	1
Percent juvenile (0-18 yrs.)	1.41
Percent foreigners	0.02

Military Expenditures and Arms Transfers [22]

	1990	1991	1992	1993	1994
Military expenditures					
Current dollars (mil.)	1,403	1,526	1,634[e]	1,728[e]	1,742
1994 constant dollars (mil.)	1,561	1,636	1,704[e]	1,764[e]	1,742
Armed forces (000)	434	434	424	424	430
Gross national product (GNP)					
Current dollars (mil.)	34,990	37,660	39,140	40,260	42,120
1994 constant dollars (mil.)	38,940	40,370	40,820	41,090	42,120
Central government expenditures (CGE)					
1994 constant dollars (mil.)	15,450	16,330	20,120	18,900	19,480[e]
People (mil.)	56.1	57.5	58.7	59.9	61.1
Military expenditure as % of GNP	4.0	4.1	4.2	4.3	4.1
Military expenditure as % of CGE	10.1	10.0	8.5	9.3	8.9
Military expenditure per capita (1994 $)	28	28	29	29	28
Armed forces per 1,000 people (soldiers)	7.7	7.5	7.2	7.1	7.0
GNP per capita (1994 $)	694	702	695	686	689
Arms imports[6]					
Current dollars (mil.)	800	900	1,100	1,400	1,500
1994 constant dollars (mil.)	890	965	1,147	1,429	1,500
Arms exports[6]					
Current dollars (mil.)	50	0	40	10	10
1994 constant dollars (mil.)	56	0	42	10	10
Total imports[7]					
Current dollars (mil.)	9,216	7,862	8,245	8,184	10,180
1994 constant dollars (mil.)	10,260	8,427	8,598	8,353	10,180
Total exports[7]					
Current dollars (mil.)	2,585	3,659	3,051	2,244	3,463
1994 constant dollars (mil.)	2,877	3,922	3,182	2,290	3,463
Arms as percent of total imports[8]	8.7	11.4	13.3	17.1	14.7
Arms as percent of total exports[8]	1.9	0	1.3	0.4	0.3

Human Rights [24]

	SSTS	FL	FAPRO	PPCG	APROBC	TPW	PCPTW	STPEP	PHRFF	PRW	ASST	AFL
Observes	P	P	P	P	P	P	P	P		P	P	P
	EAFRD	CPR	ESCR	SR	ACHR	MAAE	PVIAC	PVNAC	EAFDAW	TCIDTP	RC	
Observes	P	P	P	P	P	P	P	P	P	P	P	

P = Party; S = Signatory; see Appendix for meaning of abbreviations.

Labor Force

Total Labor Force [25]

16 million (1994 est.)

Labor Force by Occupation [26]

	%
Government, public sector, and armed forces	36
Agriculture	34
Service and manufacturing	20

Date of data: 1984

Unemployment Rate [27]

20% (1995 est.)

For sources, notes, and explanations, see Annotated Source Appendix, page 1061.

287

Production Sectors

Commercial Energy Production and Consumption

Data are shown in quadrillion (10^{15}) BTUs and percent for 1995
Values for hydroelectric, nuclear, geothermal, solar, and wind power refer to electrical generation.

Production [28]

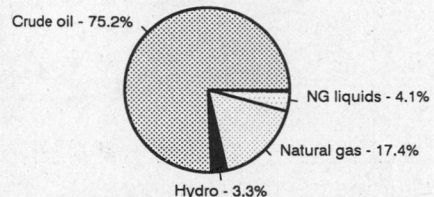

Crude oil - 75.2%
NG liquids - 4.1%
Natural gas - 17.4%
Hydro - 3.3%

Consumption [29]

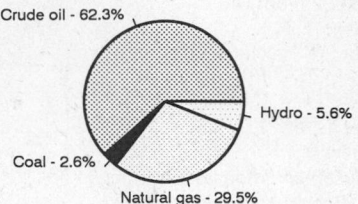

Crude oil - 62.3%
Hydro - 5.6%
Coal - 2.6%
Natural gas - 29.5%

Crude oil	1.989
Natural gas liquids	0.108
Dry natural gas	0.460
Net hydroelectric power	0.088
Total	2.645

Crude oil	0.971
Dry natural gas	0.460
Coal	0.040
Net hydroelectric power	0.088
Total	1.559

Telecommunications [30]

- 2.2 million (1993) telephones; large system by Third World standards but inadequate for present requirements and undergoing extensive upgrading
- Domestic: principal centers at Alexandria, Cairo, Al Mansurah, Ismailia, Suez, and Tanta are connected by coaxial cable and microwave radio relay
- International: satellite earth stations - 2 Intelsat (Atlantic Ocean and Indian Ocean), 1 Arabsat, and 1 Inmarsat; 5 coaxial submarine cables; tropospheric scatter to Sudan; microwave radio relay to Israel; participant in Medarabtel
- Radio: Broadcast stations: AM 39, FM 6, shortwave 0
- Television: Broadcast stations: 41 Televisions: 5 million (1993 est.)

Top Agricultural Products [32]

Produces cotton, rice, corn, wheat, beans, fruits, vegetables; cattle, water buffalo, sheep, goats; annual fish catch about 140,000 metric tons.

Transportation [31]

Railways: total: 4,751 km; standard gauge: 4,751 km 1,435-m gauge (42 km electrified; 951 km double track)

Highways: total: 47,387 km; paved: 34,593 km; unpaved: 12,794 km (1992 est.)

Merchant marine: total: 164 ships (1,000 GRT or over) totaling 1,187,290 GRT/1,833,108 DWT; ships by type: bulk 22, cargo 74, liquefied gas tanker 1, oil tanker 14, passenger 33, refrigerated cargo 1, roll-on/roll-off cargo 15, short-sea passenger 4 (1995 est.)

Airports

Total:	80
With paved runways over 3,047 m:	11
With paved runways 2,438 to 3,047 m:	34
With paved runways 1,524 to 2,437 m:	16
With paved runways 914 to 1,523 m:	2
With paved runways under 914 m:	9

Top Mining Products [33]

Metric tons except as noted[e]	3/95[*]
Aluminum metal	180,000
Ferrosilicon	40,000
Fire clay	420,000
Kaolin	156,000
Feldspar, crude	39,000
Lime	750,000
Soda ash	50,000
Sulfuric acid	100,000
Petroleum, crude (000 42-gal. bls.)	320,000

Tourism [34]

	1990	1991	1992	1993	1994
Visitors[1]	2,600	2,214	3,207	2,508	2,582
Tourists	2,411	2,112	2,944	2,291	2,356
Excursionists	189	102	263	217	226
Tourism receipts	1,994	2,029	2,730	1,332	1,384
Tourism expenditures	129	225	918	1,048	1,067
Fare receipts	430	300	382	346	213
Fare expenditures	46	57	31	23	22

Travelers are in thousands, money in million U.S. dollars.

For sources, notes, and explanations, see Annotated Source Appendix, page 1061.

Manufacturing Sector

Manufacturing Summary [35]

	1987		1988		1989		1990		1991	
	$ bil.	%	$ bil.	%	$ bil.	%	$ bil.	%	$ bil.	%
Establishments or enterprises (number)	7,560	0.322	7,841	0.374	-	-	-	-	-	-
Total employment (000)	1,339	0.987	1,391	1.016	-	-	-	-	-	-
Production workers (000)	1,028	1.524	1,088	1.742	-	-	-	-	-	-
Output ($ bil.)	38	0.371	49	0.424	-	-	-	-	-	-
Value added ($ bil.)	10	0.241	16	0.325	-	-	-	-	-	-
Capital investment ($ mil.)	3,186	0.731	4,174	0.877	-	-	-	-	-	-
M & E investment ($ mil.)	-	-	-	-	-	-	-	-	-	-
Employees per establishment (number)	177	306.211	177	272.035	-	-	-	-	-	-
Production workers per establishment	136	472.821	139	466.398	-	-	-	-	-	-
Output per establishment ($ mil.)	5	115.235	6	113.442	-	-	-	-	-	-
Capital investment per estab. ($ mil.)	0.421	226.838	0.532	234.685	-	-	-	-	-	-
M & E per establishment ($ mil)	-	-	-	-	-	-	-	-	-	-
Payroll per employee ($)	4,464	49.779	5,118	51.376	-	-	-	-	-	-
Wages per production worker ($)	3,210	40.683	3,504	41.469	-	-	-	-	-	-
Hours per production worker (hours)	-	-	-	-	-	-	-	-	-	-
Output per employee ($)	28,185	37.632	35,519	41.701	-	-	-	-	-	-
Capital investment per employee ($)	2,379	74.079	3,002	86.270	-	-	-	-	-	-
M & E per employee ($)	-	-	-	-	-	-	-	-	-	-

Note: Columns headed % show percent of world total or ratio. Ratios closest to 100 are closest to world average. M & E stands for machinery & equipment.

Output in Manufacturing

	1987		1988		1989		1990		1991	
	$ bil.[f]	%	$ bil.[f]	%	$ bil.[f]	%	$ bil.[f]	%	$ bil.[f]	%
3110 - Food products	5.636	14.83	8.517	17.38	-	-	-	-	-	-
3130 - Beverages	0.626	1.65	0.749	1.53	-	-	-	-	-	-
3140 - Tobacco	1.523	4.01	1.684	3.44	-	-	-	-	-	-
3210 - Textiles	4.666	12.28	5.944	12.13	-	-	-	-	-	-
3211 - Spinning, weaving, etc.	3.884	10.22	4.799	9.79	-	-	-	-	-	-
3220 - Wearing apparel	0.264	0.69	0.377	0.77	-	-	-	-	-	-
3230 - Leather and products	0.097	0.26	0.103	0.21	-	-	-	-	-	-
3240 - Footwear	0.177	0.47	0.199	0.41	-	-	-	-	-	-
3310 - Wood products	0.221	0.58	0.209	0.43	-	-	-	-	-	-
3320 - Furniture, fixtures	0.167	0.44	0.363	0.74	-	-	-	-	-	-
3410 - Paper and products	0.617	1.62	0.800	1.63	-	-	-	-	-	-
3411 - Pulp, paper, etc.	0.316	0.83	0.421	0.86	-	-	-	-	-	-
3420 - Printing, publishing	0.470	1.24	0.639	1.30	-	-	-	-	-	-
3510 - Industrial chemicals	1.240	3.26	1.524	3.11	-	-	-	-	-	-
3511 - Basic chemicals, excl fertilizers	0.400	1.05	0.589	1.20	-	-	-	-	-	-
3513 - Synthetic resins, etc.	0.434	1.14	0.374	0.76	-	-	-	-	-	-
3520 - Chemical products nec	2.370	6.24	3.289	6.71	-	-	-	-	-	-
3522 - Drugs and medicines	0.949	2.50	1.260	2.57	-	-	-	-	-	-
3530 - Petroleum refineries	0.327	0.86	0.509	1.04	-	-	-	-	-	-
3540 - Petroleum, coal products	0.470	1.24	0.603	1.23	-	-	-	-	-	-
3550 - Rubber products	0.190	0.50	0.256	0.52	-	-	-	-	-	-
3560 - Plastic products nec	0.780	2.05	0.950	1.94	-	-	-	-	-	-
3610 - Pottery, china, etc.	0.117	0.31	0.300	0.61	-	-	-	-	-	-
3620 - Glass and products	0.276	0.73	0.286	0.58	-	-	-	-	-	-
3690 - Nonmetal products nec	1.700	4.47	2.161	4.41	-	-	-	-	-	-
3710 - Iron and steel	1.290	3.39	2.340	4.78	-	-	-	-	-	-
3720 - Nonferrous metals	1.501	3.95	2.306	4.71	-	-	-	-	-	-
3810 - Metal products	1.177	3.10	1.359	2.77	-	-	-	-	-	-
3820 - Machinery nec	1.184	3.12	1.697	3.46	-	-	-	-	-	-
3825 - Office, computing machinery	-	-	0.087	0.18	-	-	-	-	-	-
3830 - Electrical machinery	1.303	3.43	1.290	2.63	-	-	-	-	-	-
3832 - Radio, television, etc.	0.496	1.31	0.351	0.72	-	-	-	-	-	-
3840 - Transportation equipment	1.419	3.73	1.530	3.12	-	-	-	-	-	-
3841 - Shipbuilding, repair	0.236	0.62	0.223	0.46	-	-	-	-	-	-
3843 - Motor vehicles	0.971	2.56	0.999	2.04	-	-	-	-	-	-
3850 - Professional goods	0.213	0.56	0.216	0.44	-	-	-	-	-	-
3900 - Industries nec	0.040	0.11	0.093	0.19	-	-	-	-	-	-

Note: Codes are International Standard Industry codes (ISIC). Percentages are % of total Output. [f]: Factor Prices; [p]: Producer Prices.

For sources, notes, and explanations, see Annotated Source Appendix, page 1061.

289

Finance, Economics, and Trade

Economic Indicators [36]

- **National product**: GDP—purchasing power parity— $171 billion (1995 est.)
- **National product real growth rate**: 4% (1995 est.)
- **National product per capita**: $2,760 (1995 est.)
- **Inflation rate (consumer prices)**: 9.4% (yearend 1995)
- **External debt**: $33.6 billion (FY93/94 est.)

Balance of Payments Summary [37]

Values in millions of dollars.

	1989	1990	1991	1992	1993
Exports of goods (f.o.b.)	2,907	3,604	3,856	3,400	3,243
Imports of goods (f.o.b.)	-8,841	-10,303	-9,831	-8,901	-9,923
Trade balance	-5,934	-6,699	-5,975	-5,501	-6,680
Services - debits	-4,672	-5,667	-5,507	-7,664	-7,334
Services - credits	5,123	7,148	7,951	8,901	9,307
Private transfers (net)	3,293	4,284	4,054	6,104	5,664
Government transfers (net)	880	13,871	3,218	1,430	2,018
Long-term capital (net)	1,870	-9,344	98	1,248	533
Short-term capital (net)	-1,387	-1,667	-1,793	1,015	820
Errors and omissions	414	630	730	716	-1,519
Overall balance	-413	2,556	2,776	6,249	2,809

Exchange Rates [38]

Currency: **Egyptian pound.**
Symbol: **#E.**

Data are currency units per $1.

January 1996 (market rate)	3.3920
1995 (market rate)	3.3900
November 1994	3.4
1994 (market rate)	3.3910
November 1993	3.369
1993 (market rate)	3.3718
November 1992	3.345
1992 (market rate)	3.3386
1991 (market rate)	3.3322
1990	2.7072

Imports and Exports

Top Import Origins [39]

$15.2 billion (c.i.f., FY94/95 est.).

Origins	%
US	NA
EU	NA
Japan	NA

Top Export Destinations [40]

$5.4 billion (f.o.b., FY94/95 est.).

Destinations	%
EU	NA
US	NA
Japan	NA

Foreign Aid [41]

Recipient: ODA, $1.713 billion (1993).

Import and Export Commodities [42]

Import Commodities	Export Commodities
Machinery and equipment	Crude oil and petroleum products
Foods	Cotton yarn
Fertilizers	Raw cotton
Wood products	Textiles
Durable consumer goods	Metal products
Capital goods	Chemicals

For sources, notes, and explanations, see Annotated Source Appendix, page 1061.

El Salvador

Geography [1]

Total area:
21,040 sq km 8,124 sq mi
Land area:
20,720 sq km 8,000 sq mi
Comparative area:
Slightly smaller than Massachusetts
Land boundaries:
Total 545 km, Guatemala 203 km, Honduras 342 km
Coastline:
307 km
Climate:
Tropical; rainy season (May to October); dry season (November to April)
Terrain:
Mostly mountains with narrow coastal belt and central plateau
Natural resources:
Hydropower, geothermal power, petroleum
Land use:
Arable land: 27%
Permanent crops: 8%
Meadows and pastures: 29%
Forest and woodland: 6%
Other: 30%

Demographics [2]

	1970	1980	1990	1995[1]	1996	2000	2010	2020	2030
Population	3,583	4,602	5,219	5,724	5,829	6,252	7,332	8,473	9,677
Population density (persons per sq. mi.)	448	575	652	715	729	781	917	1,059	1,210
(persons per sq. km.)	173	222	252	276	281	302	354	409	467
Net migration rate (per 1,000 population)	NA	-22.7	-6.7	-4.5	-4.4	-4.1	-2.1	-0.6	0.0
Births	NA	NA	NA	NA	165	NA	NA	NA	NA
Deaths	NA	NA	NA	NA	34	NA	NA	NA	NA
Life expectancy - males	52.6	49.9	62.5	65.0	65.4	67.2	70.8	73.7	75.8
Life expectancy - females	57.0	62.0	69.1	72.1	72.5	74.0	77.1	79.6	81.5
Birth rate (per 1,000)	NA	38.5	31.4	28.6	28.3	26.5	22.1	19.3	17.3
Death rate (per 1,000)	NA	12.1	6.7	5.9	5.8	5.5	5.0	4.8	5.0
Women of reproductive age (15-49 yrs.)	NA	1,021	1,219	1,450	1,491	1,637	2,039	2,330	2,498
of which are currently married	NA	NA	660	NA	812	924	1,191	NA	NA
Fertility rate	NA	5.3	3.8	3.3	3.2	2.9	2.5	2.3	2.3

Except as noted, values for vital statistics are in thousands; life expectancy is in years.

Health

Health Indicators [3]

% of population with access to	
safe water (1990-95)	55
adequate sanitation (1990-95)	81
health services (1985-95)	40
% of 1-year-olds immunized (1990-94) against	
TB (tuberculosis)	83
DPT (diphtheria, pertussis, tetanus)	92
polio	92
measles	81
% of contraceptive prevalence (1980-94)	53
ORT use rate (1990-94)	45

Health Expenditures [4]

Total health expenditure, 1990 (official exchange rate)	
Millions of dollars	317
Dollars per capita	61
Health expenditures as a percentage of GDP	
Total	5.9
Public sector	2.6
Private sector	3.3
Development assistance for health	
Total aid flows (millions of dollars)[1]	44
Aid flows per capita (dollars)	8.5
Aid flows as a percentage of total health expenditure	13.9

For sources, notes, and explanations, see Annotated Source Appendix, page 1061.

291

Human Factors

Women and Children [5]

% of pregnant women immunized (tetanus 1990-94)	79
% of births attended by trained health personnel (1983-94)	66
Maternal mortality rate (1980-92)	160
Under-5 mortality rate (1994)	56
% under-5 moderately/severely underweight (1980-1994)	11

Burden of Disease [6]

Population per physician (1984)	2,816.71
Population per nurse (1984)	930.33
Population per hospital bed (1990)	698.82
AIDS cases per 100,000 people (1994)	6.7
Malaria cases per 100,000 people (1992)	84

Ethnic Division [7]

Mestizo	94%
Indian	5%
White	1%

Religion [8]

There is extensive activity by Protestant groups throughout the country; by the end of 1992, there were an estimated 1 million Protestant evangelicals in El Salvador.

Roman Catholic	75%
Other	25%

Major Languages [9]

Spanish, Nahua (among some Indians).

Education

Public Education Expenditures [10]

Million (Colon)	1980	1984	1990	1991	1992	1994
Total education expenditure	340	336	722	785	884	NA
as percent of GNP	3.9	3.0	1.8	1.7	1.6	NA
as percent of total govt. expend.	17.1	12.5	NA	NA	NA	NA
Current education expenditure	320	293	715	780	883	NA
as percent of GNP	3.7	2.6	1.8	1.7	1.6	NA
as percent of current govt. expend.	22.9	16.3	NA	NA	NA	NA
Capital expenditure	20	43	7	5	2	NA

Educational Attainment [11]

Age group (1980)[8]	10+
Total population	3,132,400
Highest level attained (%)	
No schooling	30.2
First level	
Not completed	60.7
Completed	NA
Entered second level	
S-1	6.9
S-2	NA
Postsecondary	2.3

Illiteracy [12]

In thousands and percent[1]	1990	1995	2000
Illiterate population (15+ yrs.)	914	975	1,015
Illiteracy rate - total pop. (%)	30.8	27.9	25.8
Illiteracy rate - males (%)	27.8	25.5	23.8
Illiteracy rate - females (%)	33.7	30.1	27.7

Libraries [13]

	Admin. Units	Svc. Pts.	Vols. (000)	Shelving (meters)	Vols. Added	Reg. Users
National (1989)	1	3	10	334	12,309	NA
Nonspecialized	NA	NA	NA	NA	NA	NA
Public (1989)	44	83	55	NA	8,145	21,490
Higher ed. (1987)	110	695	220	NA	7,000	23,000
School (1987)	360	360	NA	NA	1,320	3,900

Daily Newspapers [14]

	1980	1985	1990	1994
Number of papers	7	4	5	6
Circ. (000)	291	243	270	284

Culture [15]

Science and Technology

Scientific/Technical Forces [16]

Scientists/engineers	102
Number female	NA
Technicians	1,612
Number female	NA
Total[11]	1,714

R&D Expenditures [17]

	Colon (000) 1992
Total expenditure[11]	1,083,559
Capital expenditure	131,377
Current expenditure	382,603
Percent current	74.4

U.S. Patents Issued [18]

Values show patents issued to citizens of the country by the U.S. Patents Office.

	1993	1994	1995
Number of patents	0	0	1

For sources, notes, and explanations, see Annotated Source Appendix, page 1061.

Government and Law

Organization of Government [19]

Long-form name:
Republic of El Salvador
Type:
Republic
Independence:
15 September 1821 (from Spain)
National holiday:
Independence Day, 15 September (1821)
Constitution:
20 December 1983
Legal system:
Based on civil and Roman law, with traces of common law; judicial review of legislative acts in the Supreme Court; accepts compulsory ICJ jurisdiction, with reservations
Executive branch:
President; Vice President; Council of Ministers
Legislative branch:
Unicameral: Legislative Assembly (Asamblea Legislativa)
Judicial branch:
Supreme Court (Corte Suprema)

Elections [20]

Legislative Assembly	% of votes
National Republican Alliance (ARENA)	46.4
Farabundo Marti National Liberation Front (FMLN)	25.0
Christian Democratic Party (PDC)	21.4
National Conciliation Party (PCN)	9.0
Other	2.4

Government Expenditures [21]

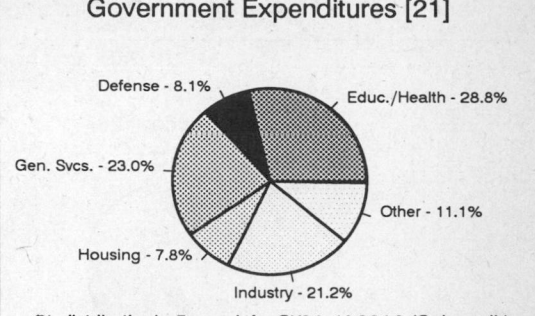

Defense - 8.1%
Educ./Health - 28.8%
Gen. Svcs. - 23.0%
Other - 11.1%
Housing - 7.8%
Industry - 21.2%

(% distribution). Expend. for CY94: 10,264.3 (Colon mil.)

Crime [23]

Military Expenditures and Arms Transfers [22]

	1990	1991	1992	1993	1994
Military expenditures					
Current dollars (mil.)	211[e]	201[e]	135	108	106
1994 constant dollars (mil.)	235[e]	216[e]	141	111	106
Armed forces (000)	55	60	49	49	30
Gross national product (GNP)					
Current dollars (mil.)	6,434	6,915	7,542	8,121	8,815
1994 constant dollars (mil.)	7,160	7,412	7,864	8,288	8,815
Central government expenditures (CGE)					
1994 constant dollars (mil.)	759	907	872	1,068	1,158
People (mil.)	5.3	5.4	5.5	5.6	5.8
Military expenditure as % of GNP	3.3	2.9	1.8	1.3	1.2
Military expenditure as % of CGE	31.0	23.8	16.1	10.4	9.1
Military expenditure per capita (1994 $)	44	40	25	20	18
Armed forces per 1,000 people (soldiers)	10.4	11.1	8.9	8.7	5.2
GNP per capita (1994 $)	1,350	1,370	1,424	1,470	1,532
Arms imports[6]					
Current dollars (mil.)	80	80	50	40	30
1994 constant dollars (mil.)	89	86	52	41	30
Arms exports[6]					
Current dollars (mil.)	0	0	0	0	0
1994 constant dollars (mil.)	0	0	0	0	0
Total imports[7]					
Current dollars (mil.)	1,263	1,406	1,699	1,912	2,249
1994 constant dollars (mil.)	1,406	1,507	1,772	1,951	2,249
Total exports[7]					
Current dollars (mil.)	582	588	598	732	844
1994 constant dollars (mil.)	648	630	624	747	844
Arms as percent of total imports[8]	6.3	5.7	2.9	2.1	1.3
Arms as percent of total exports[8]	0	0	0	0	0

Human Rights [24]

	SSTS	FL	FAPRO	PPCG	APROBC	TPW	PCPTW	STPEP	PHRFF	PRW	ASST	AFL
Observes				P		P	P			S	S	P
	EAFRD	CPR	ESCR	SR	ACHR	MAAE	PVIAC	PVNAC	EAFDAW	TCIDTP	RC	
Observes	P	P	P	P	P		P	P	P		P	

P = Party; S = Signatory; see Appendix for meaning of abbreviations.

Labor Force

Total Labor Force [25]

1.7 million (1982 est.)

Labor Force by Occupation [26]

Agriculture	40%
Commerce	16
Manufacturing	15
Government	13
Financial services	9
Transportation	6
Other	1

Unemployment Rate [27]

6.7% (1993)

For sources, notes, and explanations, see Annotated Source Appendix, page 1061.

293

Production Sectors

Commercial Energy Production and Consumption

Data are shown in quadrillion (10^{15}) BTUs and percent for 1995
Values for hydroelectric, nuclear, geothermal, solar, and wind power refer to electrical generation.

Production [28]

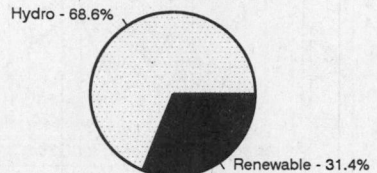

Hydro - 68.6%

Renewable - 31.4%

Consumption [29]

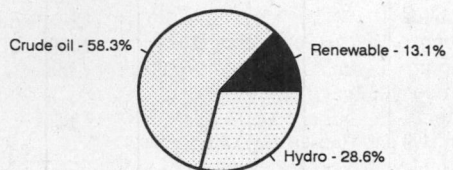

Crude oil - 58.3%

Renewable - 13.1%

Hydro - 28.6%

Net hydroelectric power	0.024
Geothermal, solar, wind	0.011
Total	0.035

Crude oil	0.049
Net hydroelectric power	0.024
Geothermal, solar, wind	0.011
Total	0.084

Telecommunications [30]

- 116,000 (1984 est.) telephones
- Domestic: nationwide microwave radio relay system
- International: satellite earth station - 1 Intelsat (Atlantic Ocean); connected to Central American Microwave System
- Radio: Broadcast stations: AM 77, FM 0, shortwave 2
- Television: Broadcast stations: 5 (1986 est.) Televisions: 500,700 (1993 est.)

Transportation [31]

Railways: total: 602 km (single track; note - some sections abandoned, unusable, or operating at reduced capacity); narrow gauge: 602 km 0.914-m gauge

Highways: total: 12,251 km; paved: 1,740 km (including 107 km of expressways); unpaved: 10,511 km (1992 est.)

Merchant marine: none

Airports

Total:	73
With paved runways over 3,047 m:	1
With paved runways 1,524 to 2,437 m:	1
With paved runways 914 to 1,523 m:	2
With paved runways under 914 m:	48
With unpaved runways 914 to 1,523 m:	21 (1995 est.)

Top Agricultural Products [32]

Produces coffee, sugarcane, corn, rice, beans, oilseed; beef, dairy products; shrimp.

Top Mining Products [33]

Metric tons except as noted	12/94[*]
Aluminum, incl. alloys, semimanufactures	2,400
Cement	924,000
Fertilizer materials	66,000
Gypsum	5,000[e]
Steel, crude	12,000[e]
Limestone (000 tons)	2,600
Petroleum refinery products (000 42-gal. bls.)	6,100
Salt, marine	30,000

Tourism [34]

	1990	1991	1992	1993	1994
Tourists[8]	194	199	314	267	181
Tourism receipts	145	157	128	121	86
Tourism expenditures	61	57	58	61	70
Fare receipts	49	51	79	80	56
Fare expenditures	11	11	12	13	15

Travelers are in thousands, money in million U.S. dollars.

For sources, notes, and explanations, see Annotated Source Appendix, page 1061.

Manufacturing Sector

GDP and Manufacturing Summary [35]

	1980	1985	1989	1990	% change 1980-1990	% change 1989-1990
GDP (million 1980 $)	3,567	3,247	3,375	3,562	-0.1	5.5
GDP per capita (1980 $)	788	681	657	681	-13.6	3.7
Manufacturing as % of GDP (current prices)	15.0	16.4	18.1	18.6	24.0	2.8
Gross output (million $)	1,130	860	1,587[e]	1,274[e]	12.7	-19.7
Value added (million $)	448	393	735[e]	603[e]	34.6	-18.0
Value added (million 1980 $)	536	471	525	541	0.9	3.0
Industrial production index	100	83	93	95	-5.0	2.2
Employment (thousands)	39	25	25[e]	26[e]	-33.3	4.0

Note: GDP stands for Gross Domestic Product. 'e' stands for estimated value.

Profitability and Productivity

	1980	1985	1989	1990	% change 1980-1990	% change 1989-1990
Intermediate input (%)	60	54	54[e]	53[e]	-11.7	-1.9
Wages, salaries, and supplements (%)	15[e]	12[e]	9[e]	12[e]	-20.0	33.3
Gross operating surplus (%)	24[e]	34[e]	38[e]	35[e]	45.8	-7.9
Gross output per worker ($)	28,857	34,129	63,062[e]	47,118[e]	63.3	-25.3
Value added per worker ($)	11,426	15,595	29,222[e]	22,472[e]	96.7	-23.1
Average wage (incl. benefits) ($)	4,383[e]	3,990[e]	5,427[e]	6,024[e]	37.4	11.0

Profitability is in percent of gross output. Productivity is in U.S. $. 'e' stands for estimated value.

Profitability - 1990

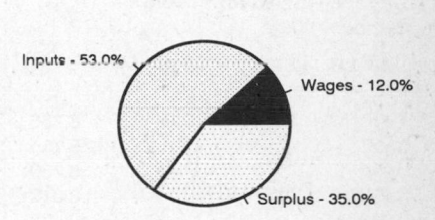

Inputs - 53.0%
Wages - 12.0%
Surplus - 35.0%

The graphic shows percent of gross output.

Value Added in Manufacturing

	1980 $ mil.	1980 %	1985 $ mil.	1985 %	1989 $ mil.	1989 %	1990 $ mil.	1990 %	% change 1980-1990	% change 1989-1990
311 Food products	78	17.4	55	14.0	81[e]	11.0	63[e]	10.4	-19.2	-22.2
313 Beverages	63	14.1	59	15.0	108[e]	14.7	105[e]	17.4	66.7	-2.8
314 Tobacco products	26	5.8	29	7.4	47[e]	6.4	42[e]	7.0	61.5	-10.6
321 Textiles	62	13.8	40	10.2	59[e]	8.0	55[e]	9.1	-11.3	-6.8
322 Wearing apparel	16	3.6	10	2.5	19[e]	2.6	15[e]	2.5	-6.3	-21.1
323 Leather and fur products	5	1.1	5	1.3	8[e]	1.1	7[e]	1.2	40.0	-12.5
324 Footwear	13	2.9	1	0.3	1[e]	0.1	3[e]	0.5	-76.9	200.0
331 Wood and wood products	1	0.2	NA	0.0	NA	0.0	NA	0.0	NA	NA
332 Furniture and fixtures	3	0.7	4	1.0	7[e]	1.0	6[e]	1.0	100.0	-14.3
341 Paper and paper products	40	8.9	24	6.1	51[e]	6.9	39[e]	6.5	-2.5	-23.5
342 Printing and publishing	8	1.8	8	2.0	12[e]	1.6	16[e]	2.7	100.0	33.3
351 Industrial chemicals	4	0.9	7	1.8	11[e]	1.5	11[e]	1.8	175.0	0.0
352 Other chemical products	46	10.3	57	14.5	129[e]	17.6	87[e]	14.4	89.1	-32.6
353 Petroleum refineries	14	3.1	20	5.1	47[e]	6.4	30[e]	5.0	114.3	-36.2
354 Miscellaneous petroleum and coal products	2	0.4	NA	0.0	2[e]	0.3	2[e]	0.3	0.0	0.0
355 Rubber products	4	0.9	3	0.8	4[e]	0.5	4[e]	0.7	0.0	0.0
356 Plastic products	13	2.9	15	3.8	36[e]	4.9	25[e]	4.1	92.3	-30.6
361 Pottery, china, and earthenware	NA	0.0	NA	0.0	NA	0.0	NA	0.0	NA	NA
362 Glass and glass products	NA	0.0	NA	0.0	NA	0.0	NA	0.0	NA	NA
369 Other nonmetal mineral products	11	2.5	13	3.3	23[e]	3.1	22[e]	3.6	100.0	-4.3
371 Iron and steel	9	2.0	7	1.8	14[e]	1.9	11[e]	1.8	22.2	-21.4
372 Nonferrous metals	1	0.2	1	0.3	1[e]	0.1	1[e]	0.2	0.0	0.0
381 Metal products	10	2.2	12	3.1	26[e]	3.5	20[e]	3.3	100.0	-23.1
382 Nonelectrical machinery	6	1.3	7	1.8	15[e]	2.0	11[e]	1.8	83.3	-26.7
383 Electrical machinery	9	2.0	12	3.1	27[e]	3.7	21[e]	3.5	133.3	-22.2
384 Transport equipment	1	0.2	NA	0.0	1[e]	0.1	NA	0.0	NA	NA
385 Professional and scientific equipment	NA	0.0	1	0.3	1[e]	0.1	1[e]	0.2	NA	0.0
390 Other manufacturing industries	4	0.9	2	0.5	5[e]	0.7	4[e]	0.7	0.0	-20.0

Note: The industry codes shown are International Standard Industry codes (ISIC). Percentages are percent of total Value Added. 'e' stands for estimated value

For sources, notes, and explanations, see Annotated Source Appendix, page 1061.

295

Finance, Economics, and Trade

Economic Indicators [36]

- **National product**: GDP—purchasing power parity— $11.4 billion (1995 est.)
- **National product real growth rate**: 6.3% (1995 est.)
- **National product per capita**: $1,950 (1995 est.)
- **Inflation rate (consumer prices)**: 11.4% (1995 est.)
- **External debt**: $2.6 billion (December 1992)

Balance of Payments Summary [37]

Values in millions of dollars.

	1989	1990	1991	1992	1993
Exports of goods (f.o.b.)	497.8	580.2	588.0	597.5	731.7
Imports of goods (f.o.b.)	-1,089.5	-1,180.0	-1,294.1	-1,558.8	-1,766.9
Trade balance	-591.7	-599.8	-706.1	-961.3	-1,035.2
Services - debits	-463.8	-428.8	-474.9	-493.1	-524.6
Services - credits	336.7	323.2	341.9	408.4	437.2
Private transfers (net)	207.8	324.0	469.9	708.8	823.2
Government transfers (net)	337.4	244.6	201.4	228.3	222.1
Long-term capital (net)	193.8	26.9	62.0	103.1	112.3
Short-term capital (net)	-43.3	-6.9	-90.0	31.8	-13.3
Errors and omissions	126.3	270.3	125.9	65.5	90.3
Overall balance	103.2	153.5	-69.9	91.5	112.0

Exchange Rates [38]

Currency: **Salvadoran colon.**
Symbol: **C.**

Data are currency units per $1.

December 1995	8.755
1995	8.755
1994	8.750
1993	8.670
1992	9.170
1991	8.080

Imports and Exports

Top Import Origins [39]

$3.3 billion (c.i.f., 1995 est.).

Origins	%
US	NA
Guatemala	NA
Mexico	NA
Venezuela	NA
Germany	NA

Top Export Destinations [40]

$1.6 billion (f.o.b., 1995 est.).

Destinations	%
US	NA
Guatemala	NA
Costa Rica	NA
Germany	NA

Foreign Aid [41]

	U.S. $	
ODA (1993)	777	million
US commitment (1992-96)	250	million

Import and Export Commodities [42]

Import Commodities	Export Commodities
Raw materials	Coffee
Consumer goods	Sugarcane
Capital goods	Shrimp

For sources, notes, and explanations, see Annotated Source Appendix, page 1061.

Equatorial Guinea

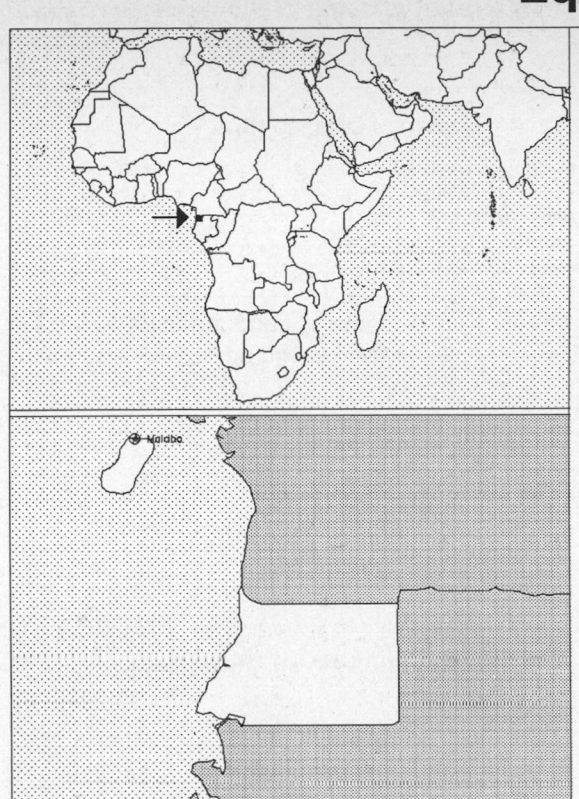

Geography [1]

Total area:
28,050 sq km   10,830 sq mi
Land area:
28,050 sq km   10,830 sq mi
Comparative area:
Slightly larger than Maryland
Land boundaries:
Total 539 km, Cameroon 189 km, Gabon 350 km
Coastline:
296 km
Climate:
Tropical; always hot, humid
Terrain:
Coastal plains rise to interior hills; islands are volcanic
Natural resources:
Timber, petroleum, small unexploited deposits of gold, manganese, uranium
Land use:
Arable land: 8%
Permanent crops: 4%
Meadows and pastures: 4%
Forest and woodland: 51%
Other: 33%

Demographics [2]

	1970	1980	1990	1995[1]	1996	2000	2010	2020	2030
Population	270	256	369	420	431	478	615	783	974
Population density (persons per sq. mi.)	25	24	34	39	40	44	57	72	90
(persons per sq. km.)	10	9	13	15	15	17	22	28	35
Net migration rate (per 1,000 population)	NA	NA	0.0	0.0	0.0	0.0	0.0	0.0	0.0
Births	NA	NA	NA	NA	17	NA	NA	NA	NA
Deaths	NA	NA	NA	NA	6	NA	NA	NA	NA
Life expectancy - males	NA	NA	48.4	50.4	50.8	52.5	56.6	60.5	64.2
Life expectancy - females	NA	NA	52.3	54.8	55.3	57.4	62.4	67.1	71.3
Birth rate (per 1,000)	NA	NA	42.5	40.2	39.8	38.1	34.8	31.0	27.0
Death rate (per 1,000)	NA	NA	16.2	14.4	14.0	12.7	9.9	7.8	6.4
Women of reproductive age (15-49 yrs.)	NA	NA	88	98	101	113	152	197	253
of which are currently married	NA	NA	54	NA	63	70	94	NA	NA
Fertility rate	NA	NA	5.5	5.2	5.2	4.9	4.4	3.8	3.3

Except as noted, values for vital statistics are in thousands; life expectancy is in years.

Health

Health Indicators [3]

Health Expenditures [4]

For sources, notes, and explanations, see Annotated Source Appendix, page 1061.

297

Human Factors

Women and Children [5]

Burden of Disease [6]

Population per physician (1990)	3,555.56
Population per nurse (1990)	469.96
Population per hospital bed	NA
AIDS cases per 100,000 people (1994)	3.3
Malaria cases per 100,000 people (1992)	NA

Ethnic Division [7]

Bioko (primarily Bubi, some Fernandinos), Rio Muni (primarily Fang), European (primarily Spanish).

Religion [8]

Nominally Christian, predominantly Roman Catholic, and pagan practices.

Major Languages [9]

Spanish (official), pidgin English, Fang, Bubi, Ibo.

Education

Public Education Expenditures [10]

	1980	1988	1989	1990	1993	1994
Million (Franc C.F.A.)						
Total education expenditure	NA	620	NA	NA	734	NA
as percent of GNP	NA	1.7	NA	NA	1.8	NA
as percent of total govt. expend.	NA	3.9	NA	NA	5.6	NA
Current education expenditure	NA	527	NA	NA	721	NA
as percent of GNP	NA	1.5	NA	NA	1.7	NA
as percent of current govt. expend.	NA	3.5	NA	NA	5.8	NA
Capital expenditure	NA	93	NA	NA	14	NA

Educational Attainment [11]

Illiteracy [12]

In thousands and percent[1]	1990	1995	2000
Illiterate population (15+ yrs.)	54	49	43
Illiteracy rate - total pop. (%)	25.6	20.5	15.8
Illiteracy rate - males (%)	13.1	9.7	7.7
Illiteracy rate - females (%)	36.6	29.4	23.1

Libraries [13]

	Admin. Units	Svc. Pts.	Vols. (000)	Shelving (meters)	Vols. Added	Reg. Users
National	NA	NA	NA	NA	NA	NA
Nonspecialized (1993)	1	1	5	NA	NA	NA
Public (1992)	3	3	NA	NA	NA	NA
Higher ed.	NA	NA	NA	NA	NA	NA
School (1987)	NA	NA	NA	NA	NA	NA

Daily Newspapers [14]

	1980	1985	1990	1994
Number of papers	2	2	2	1
Circ. (000)	2[e]	2[e]	2[e]	1

Culture [15]

Science and Technology

Scientific/Technical Forces [16]

R&D Expenditures [17]

U.S. Patents Issued [18]

For sources, notes, and explanations, see Annotated Source Appendix, page 1061.

Government and Law

Organization of Government [19]

Long-form name:
Republic of Equatorial Guinea
Type:
Republic in transition to multiparty democracy
Independence:
12 October 1968 (from Spain)
National holiday:
Independence Day, 12 October (1968)
Constitution:
New constitution 17 November 1991
Legal system:
Partly based on Spanish civil law and tribal custom
Executive branch:
President; Prime Minister; Vice Prime Minister; Council of Ministers
Legislative branch:
Unicameral: House of People's Representatives (Camara de Representantes del Pueblo)
Judicial branch:
Supreme Tribunal

Elections [20]

Chamber of People's Reps	% of seats
Democratic Party for Equatorial Guinea (PDGE)	87.8
Opposition parties	12.2

Government Budget [21]

For 1992 est.

	$ mil.
Revenues	32.5
Expenditures	35.9
Capital expenditures	3

Crime [23]

Military Expenditures and Arms Transfers [22]

	1990	1991	1992	1993	1994
Military expenditures					
Current dollars (mil.)	NA	NA	NA	NA	2
1994 constant dollars (mil.)	NA	NA	NA	NA	2
Armed forces (000)	1	1	1	1	1
Gross national product (GNP)					
Current dollars (mil.)	83	85	97	107	112
1994 constant dollars (mil.)	93	91	101	110	112
Central government expenditures (CGE)					
1994 constant dollars (mil.)	21	22[e]	24[e]	NA	NA
People (mil.)	0.4	0.4	0.4	0.4	0.4
Military expenditure as % of GNP	NA	NA	NA	NA	2.2
Military expenditure as % of CGE	NA	NA	NA	NA	NA
Military expenditure per capita (1994 $)	NA	NA	NA	NA	6
Armed forces per 1,000 people (soldiers)	2.7	2.6	2.6	2.5	2.4
GNP per capita (1994 $)	251	241	260	274	274
Arms imports[6]					
Current dollars (mil.)	0	0	0	0	0
1994 constant dollars (mil.)	0	0	0	0	0
Arms exports[6]					
Current dollars (mil.)	0	0	0	0	0
1994 constant dollars (mil.)	0	0	0	0	0
Total imports[7]					
Current dollars (mil.)	61	117	93	60	NA
1994 constant dollars (mil.)	68	125	97	61	NA
Total exports[7]					
Current dollars (mil.)	62	83	42	62	NA
1994 constant dollars (mil.)	69	89	44	63	NA
Arms as percent of total imports[8]	0	0	0	0	0
Arms as percent of total exports[8]	0	0	0	0	0

Human Rights [24]

	SSTS	FL	FAPRO	PPCG	APROBC	TPW	PCPTW	STPEP	PHRFF	PRW	ASST	AFL
Observes						P	P					

	EAFRD	CPR	ESCR	SR	ACHR	MAAE	PVIAC	PVNAC	EAFDAW	TCIDTP	RC
Observes		P	P	P		P	P	P	P		P

P = Party; S = Signatory; see Appendix for meaning of abbreviations.

Labor Force

Total Labor Force [25]

172,000 (1986 est.)

Labor Force by Occupation [26]

Agriculture	66%
Services	23
Industry	11
Date of data: 1980	

Unemployment Rate [27]

For sources, notes, and explanations, see Annotated Source Appendix, page 1061.

299

Production Sectors

Commercial Energy Production and Consumption

Data are shown in quadrillion (10^{15}) BTUs and percent for 1995
Values for hydroelectric, nuclear, geothermal, solar, and wind power refer to electrical generation.

Production [28]	Consumption [29]
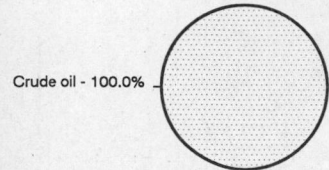	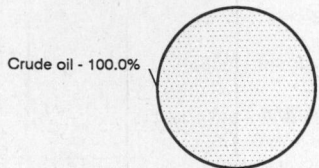
Crude oil - 100.0%	Crude oil - 100.0%

Crude oil	0.014		Crude oil	0.002
Total	0.014		Total	0.002

Telecommunications [30]

- 2,000 (1987 est.) telephones; poor system with adequate government services
- International: international communications from Bata and Malabo to African and European countries; satellite earth station - 1 Intelsat (Indian Ocean)
- Radio: Broadcast stations: AM 2, FM 0, shortwave 0
- Television: Broadcast stations: 1 Televisions: 4,000 (1992 est.)

Transportation [31]

Railways: total: 0 km

Highways: total: 2,744 km; paved: 330 km; unpaved: 2,414 km (1988 est.)

Merchant marine: total: 2 ships (1,000 GRT or over) totaling 6,412 GRT/6,699 DWT; ships by type: cargo 1, passenger-cargo 1 (1995 est.)

Airports

Total: 3
With paved runways 2,438 to 3,047 m: 1
With paved runways 1,524 to 2,437 m: 1
With paved runways under 914 m: 1 (1995 est.)

Top Agricultural Products [32]

Agriculture accounts for 47% of the GDP; produces coffee, cocoa, rice, yams, cassava (tapioca), bananas, palm oil nuts, manioc; livestock; timber.

Top Mining Products [33]

Detailed information is not available. A summary of mineral resources follows. **Mineral Resources:** Petroleum, and exploration of gold, phosphates, uranium, bauxite, copper, ilmenite sand, lead, and zinc deposits.

Tourism [34]

Manufacturing Sector

Manufacturing Summary [35]

	1987		1988		1989		1990		1991	
	$ bil.	%	$ bil.	%	$ bil.	%	$ bil.	%	$ bil.	%
Establishments or enterprises (number)	45	0.002	-	-	-	-	59	0.003	-	-
Total employment (000)	0.183	0.000	-	-	-	-	0.395	0.000	-	-
Production workers (000)	0.156	0.000	-	-	-	-	0.280	0.000	-	-
Output ($ bil.)	0.002	0.000	-	-	-	-	0.002	0.000	-	-
Value added ($ bil.)	0.001	0.000	-	-	-	-	-	-	-	-
Capital investment ($ mil.)	0.001	0.000	-	-	-	-	-	-	-	-
M & E investment ($ mil.)	-	-	-	-	-	-	-	-	-	-
Employees per establishment (number)	4	7.029	-	-	-	-	7	10.835	-	-
Production workers per establishment	3	12.057	-	-	-	-	5	11.978	-	-
Output per establishment ($ mil.)	0.035	0.817	-	-	-	-	0.032	0.493	-	-
Capital investment per estab. ($ mil.)	0.000	0.009	-	-	-	-	-	-	-	-
M & E per establishment ($ mil)	-	-	-	-	-	-	-	-	-	-
Payroll per employee ($)	1,260	14.056	-	-	-	-	16,397	118.204	-	-
Wages per production worker ($)	1,281	16.235	-	-	-	-	4,282	36.057	-	-
Hours per production worker (hours)	380	20.050	-	-	-	-	250	13.338	-	-
Output per employee ($)	8,709	11.629	-	-	-	-	4,714	4.549	-	-
Capital investment per employee ($)	4	0.125	-	-	-	-	-	-	-	-
M & E per employee ($)	-	-	-	-	-	-	-	-	-	-

Note: Columns headed % show percent of world total or ratio. Ratios closest to 100 are closest to world average. M & E stands for machinery & equipment.

Output in Manufacturing

	1987		1988		1989		1990		1991	
	$ bil.	%	$ bil.	%	$ bil.	%	$ bil.	%	$ bil.	%
3110 - Food products	0.001	50.00	-	-	-	-	0.001	50.00	-	-
3130 - Beverages	-	-	-	-	-	-	0.000	0.00	-	-
3220 - Wearing apparel	0.000	0.00	-	-	-	-	0.000	0.00	-	-
3320 - Furniture, fixtures	0.000	0.00	-	-	-	-	0.001	50.00	-	-
3420 - Printing, publishing	0.000	0.00	-	-	-	-	0.000	0.00	-	-
3520 - Chemical products nec	0.000	0.00	-	-	-	-	0.000	0.00	-	-
3690 - Nonmetal products nec	-	-	-	-	-	-	0.000	0.00	-	-
3810 - Metal products	-	-	-	-	-	-	0.000	0.00	-	-

Note: Codes are International Standard Industry codes (ISIC). Percentages are % of total Output. [f]: Factor Prices; [p]: Producer Prices.

For sources, notes, and explanations, see Annotated Source Appendix, page 1061.

301

Finance, Economics, and Trade

Economic Indicators [36]

- **National product**: GDP—purchasing power parity—$325 million (1995 est.)
- **National product real growth rate**: 10% (1995 est.)
- **National product per capita**: $800 (1995 est.)
- **Inflation rate (consumer prices)**: 41% (1994 est.)
- **External debt**: $268 million (1993 est.)

Balance of Payments Summary [37]

Values in millions of dollars.

	1987	1988	1989	1990	1991
Exports of goods (f.o.b.)	38.5	44.7	32.7	37.8	35.8
Imports of goods (f.o.b.)	-47.9	-56.5	-43.6	-53.2	-59.6
Trade balance	-9.4	-11.9	-10.9	-15.4	-23.8
Services - debits	-47.2	-56.6	-39.8	-46.0	-52.4
Services - credits	6.1	5.9	5.8	4.5	6.2
Private transfers (net)	-2.5	-3.8	-13.0	-17.0	-16.5
Government transfers (net)	28.8	49.9	39.3	59.0	64.4
Long-term capital (net)	13.1	13.8	16.6	21.5	49.2
Short-term capital (net)	7.8	8.1	7.0	-1.1	5.2
Errors and omissions	0.8	-1.7	-4.5	-2.4	-30.7
Overall balance	-2.5	3.7	0.6	3.2	1.5

Exchange Rates [38]

Currency: **Communaute Financiere Africaine franc.**
Symbol: **CFAF.**

Data are currency units per $1.

January 1996	500.56
1995	499.15
1994	555.20
1993	283.16
1992	264.69
1991	282.11

Imports and Exports

Top Import Origins [39]

$60 million (f.o.b., 1993).

Origins	%
Cameroon	NA
Spain	NA
France	NA
US	NA
Italy	NA
Netherlands	NA

Top Export Destinations [40]

$62 million (f.o.b., 1993).

Destinations	%
Spain	NA
Nigeria	NA
Cameroon	NA
Japan	NA
Portugal	NA

Foreign Aid [41]

Recipient: ODA, $NA.

Import and Export Commodities [42]

Import Commodities	Export Commodities
Petroleum	Coffee
Food	Cocoa beans
Beverages	Timber
Clothing	Petroleum
Machinery	

For sources, notes, and explanations, see Annotated Source Appendix, page 1061.

Eritrea

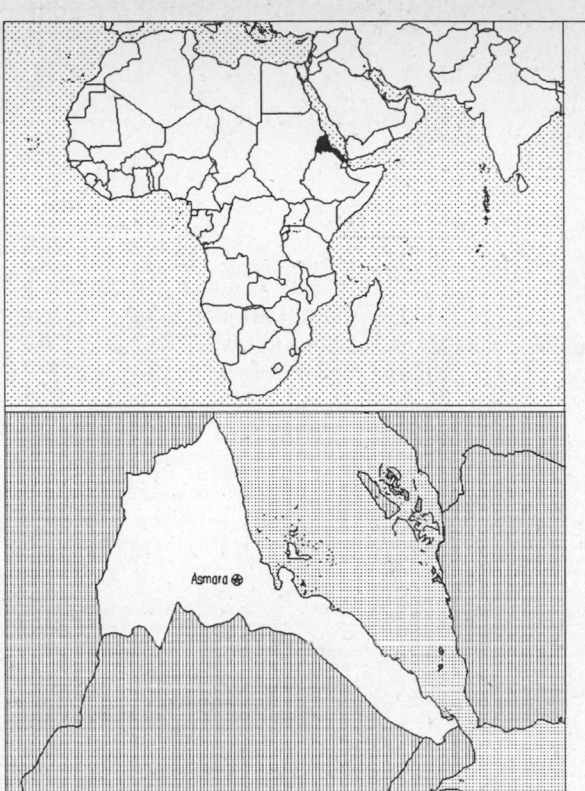

Geography [1]

Total area:
121,320 sq km   46,842 sq mi
Land area:
121,320 sq km   46,842 sq mi
Comparative area:
Slightly larger than Pennsylvania
Land boundaries:
Total 1,630 km, Djibouti 113 km, Ethiopia 912 km, Sudan 605 km
Coastline:
1,151 km (land and island coastline is 2,234 km)
Climate:
Hot, dry desert strip along Red Sea coast; cooler and wetter in the central highlands (up to 61 cm of rainfall annually); semiarid in western hills and lowlands; rainfall heaviest during June-September except on coastal desert
Terrain:
Dominated by extension of Ethiopian north-south trending highlands, descending on the East to a coastal desert plain, on the Northwest to hilly terrain and on the Southwest to flat-to-rolling plains
Natural resources:
Gold, potash, zinc, copper, salt, probably oil (petroleum geologists are prospecting for it), fish
Land use:
Arable land: 3%
Permanent crops: 2% (coffee)
Meadows and pastures: 40%
Forest and woodland: 5%
Other: 50%

Demographics [2]

	1970	1980	1990	1995[1]	1996	2000	2010	2020	2030
Population	2,153	2,555	2,896	3,579	3,910	4,537	6,018	7,674	9,551
Population density (persons per sq. mi.)	46	55	62	76	83	97	128	164	204
(persons per sq. km.)	18	21	24	29	32	37	50	63	79
Net migration rate (per 1,000 population)	NA	NA	-15.9	61.8	56.5	0.0	0.0	0.0	0.0
Births	NA	NA	NA	NA	178	NA	NA	NA	NA
Deaths	NA	NA	NA	NA	61	NA	NA	NA	NA
Life expectancy - males	NA	NA	46.9	48.3	48.6	49.7	52.7	55.6	58.5
Life expectancy - females	NA	NA	50.2	51.8	52.1	53.4	56.7	59.9	63.1
Birth rate (per 1,000)	NA	NA	41.9	44.3	45.6	45.2	38.0	33.3	29.2
Death rate (per 1,000)	NA	NA	17.1	15.7	15.6	14.9	12.1	10.2	8.9
Women of reproductive age (15-49 yrs.)	NA	NA	635	839	936	1,067	1,372	1,895	2,433
of which are currently married	NA	NA	438	NA	647	753	978	NA	NA
Fertility rate	NA	NA	6.7	6.5	6.5	6.4	5.5	4.5	3.6

Except as noted, values for vital statistics are in thousands; life expectancy is in years.

Health

Health Indicators [3]

% of population with access to	
safe water (1990-95)	NA
adequate sanitation (1990-95)	NA
health services (1985-95)	NA
% of 1-year-olds immunized (1990-94) against	
TB (tuberculosis)	46
DPT (diphtheria, pertussis, tetanus)	36
polio	36
measles	27
% of contraceptive prevalence (1980-94)	NA
ORT use rate (1990-94)	68

Health Expenditures [4]

For sources, notes, and explanations, see Annotated Source Appendix, page 1061.

303

Human Factors

Women and Children [5]

% of pregnant women immunized (tetanus 1990-94)	21
% of births attended by trained health personnel (1983-94)	NA
Maternal mortality rate (1980-92)	NA
Under-5 mortality rate (1994)	200
% under-5 moderately/severely underweight (1980-1994)	NA

Burden of Disease [6]

Ethnic Division [7]

Ethnic Tigrinya	50%
Tigre and Kunama	40%
Afar	4%
Saho	3%

Religion [8]

Muslim, Coptic Christian, Roman Catholic, Protestant.

Major Languages [9]

Afar, Amharic, Arabic, Italian, Tigre and Kunama, Tigrinya, minor tribal languages.

Education

Public Education Expenditures [10]

	1980	1985	1987	1990	1991	1994
Million (Birr)[24]						
Total education expenditure	NA	NA	NA	NA	NA	181
as percent of GNP	NA	NA	NA	NA	NA	NA
as percent of total govt. expend.	NA	NA	NA	NA	NA	NA
Current education expenditure	NA	NA	NA	NA	NA	160
as percent of GNP	NA	NA	NA	NA	NA	NA
as percent of current govt. expend.	NA	NA	NA	NA	NA	NA
Capital expenditure	NA	NA	NA	NA	NA	21

Educational Attainment [11]

Illiteracy [12]

Libraries [13]

Daily Newspapers [14]

Culture [15]

Cinema (seats per 1,000)	NA
Annual attendance per person	NA
Gross box office receipts (mil. Birr)	NA
Museums (reporting)	4
Visitors (000)	300
Annual receipts (000 Birr)	NA

Science and Technology

Scientific/Technical Forces [16]

R&D Expenditures [17]

U.S. Patents Issued [18]

For sources, notes, and explanations, see Annotated Source Appendix, page 1061.

Government and Law

Organization of Government [19]

Long-form name:
State of Eritrea
Type:
Transitional government
Independence:
27 May 1993 (from Ethiopia; formerly the Eritrea Autonomous Region)
National holiday:
National Day (independence from Ethiopia), 24 May (1993)
Constitution:
Transitional "constitution" decreed 19 May 1993; the promulgation of a draft constitution is expected in 1996
Legal system:
NA
Executive branch:
President (who is head of the State Council and National Assembly); State Council (the collective executive authority)
Legislative branch:
Unicameral: National Assembly
Judicial branch:
Judiciary

Elections [20]

National Assembly. 75 members of the PFJD (previously the EPLF) Central Committee and 75 directly elected members serve as the country's legislative body until country-wide elections are held in 1997. The only political party recognized is the People's Front for Democracy and Justice (PFJD).

Government Budget [21]

For 1993.

	$ mil.
Revenues	NA
Expenditures	60.8
Capital expenditures	NA

Defense Summary [22]

Branches: Army, Navy, Air Force

Manpower Availability: Males age 15-49 NA; males fit for military service NA

Defense Expenditures: represents 36% of total government expenditure (1993)

Crime [23]

Human Rights [24]

	SSTS	FL	FAPRO	PPCG	APROBC	TPW	PCPTW	STPEP	PHRFF	PRW	ASST	AFL
Observes												
	EAFRD	CPR	ESCR	SR	ACHR	MAAE	PVIAC	PVNAC	EAFDAW	TCIDTP	RC	
Observes										P		P

P = Party; S = Signatory; see Appendix for meaning of abbreviations.

Labor Force

Total Labor Force [25]

Labor Force by Occupation [26]

Agriculture	80%
Industry and commerce	20

Date of data: 1995

Unemployment Rate [27]

For sources, notes, and explanations, see Annotated Source Appendix, page 1061.

305

Production Sectors

Energy Resource Summary [28]

Energy resources: Probably oil (petroleum geologists are prospecting for it).

Telecommunications [30]

- Domestic: very inadequate; about 4 telephones per 100 families, most of which are in Asmara; government is seeking international tenders to improve the system
- Radio: Broadcast stations: AM NA, FM NA, shortwave 0
- Television: Broadcast stations: 1 (government controlled)

Transportation [31]

Railways: total: 307 km; note - nonoperational since 1978 except for about 5 km that was reopened in Massawa in 1994; rehabilitation of the remainder and of the rolling stock is under way; links Ak'ordat and Asmara (formerly Asmera) with the port of Massawa (formerly Mits'iwa); narrow gauge: 307 km 0.950-m gauge (1995 est.)

Highways: total: 3,845 km; paved: 807 km; unpaved: 3,038 km (1993 est.)

Merchant marine: total: 1 cargo ship (1,000 GRT or over) totaling 11,573 GRT/13,593 DWT (1995 est.)

Airports

Total:	14
With paved runways over 3,047 m:	1
With paved runways 2,438 to 3,047 m:	1
With paved runways under 914 m:	2
With unpaved runways over 3,047 m:	1
With unpaved runways 2,438 to 3,047 m:	1

Top Agricultural Products [32]

Produces sorghum, lentils, vegetables, maize, cotton, tobacco, coffee, sisal (for making rope); livestock (including goats); fish.

Top Mining Products [33]

Metric tons unless otherwise specified[*]

Cement	45,000
Gold (kg.)	300
Petroleum refinery products (000 42-gal. bls.)	5,000,000

Tourism [34]

Finance, Economics, and Trade

Industrial Summary [35]

Industrial Production: Growth rate not available. **Industries**: Food processing, beverages, clothing and textiles.

Economic Indicators [36]

- **National product**: GDP—purchasing power parity—$2 billion (1995 est.)

- **National product real growth rate**: 10% (1995 est.)

- **National product per capita**: $570 (1995 est.)

- **Inflation rate (consumer prices)**: 10% (1995 est.)

- **External debt**: $NA

Balance of Payments Summary [37]

Exchange Rates [38]

Currency: **birr.**
Symbol: **Br.**

Data are currency units per $1.
Following independence from Ethiopia, Eritrea continued to use Ethiopian currency, the official rate of which was pegged to the $1 at 5.000 birr.

1995 est.	6.2
September 1994	5.600
Fixed rate 1992-93	5.000

Imports and Exports

Top Import Origins [39]

$360 million (1994 est.).

Origins	%
Saudi Arabia	NA
Ethiopia	NA
Italy	NA
UAE	NA

Top Export Destinations [40]

$33 million (1995 est.).

Destinations	%
Ethiopia	NA
Italy	NA
Saudi Arabia	NA
UK	NA
US	NA
Yemen	NA

Foreign Aid [41]

	U.S. $
ODA	NA
US assistance (FY93)	6 million
USAID programs (FY95)	16 million

Import and Export Commodities [42]

Import Commodities	Export Commodities
Processed goods	Livestock
Machinery	Sorghum
Petroleum products	Textiles

Estonia

Geography [1]

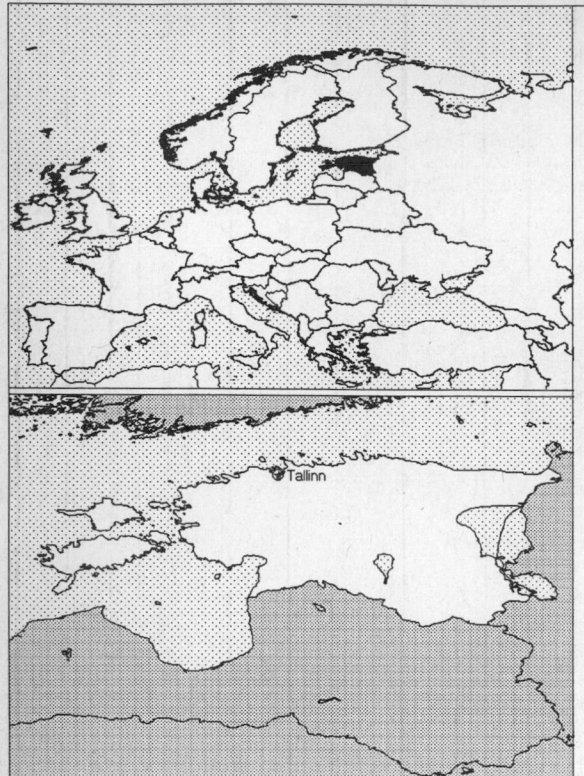

Total area:
  45,100 sq km   17,413 sq mi
Land area:
  43,200 sq km   16,680 sq mi
Comparative area:
  Slightly larger than New Hampshire and Vermont combined
  Note: Includes 1,520 islands in the Baltic Sea
Land boundaries:
  Total 557 km, Latvia 267 km, Russia 290 km
Coastline:
  1,393 km
Climate:
  Maritime, wet, moderate winters, cool summers
Terrain:
  Marshy, lowlands
Natural resources:
  Shale oil, peat, phosphorite, amber
Land use:
  Arable land: 22%
  Permanent crops: 0%
  Meadows and pastures: 11%
  Forest and woodland: 31%
  Other: 36%

Demographics [2]

	1970	1980	1990	1995[1]	1996	2000	2010	2020	2030
Population	1,363	1,482	1,573	1,478	1,459	1,422	1,401	1,370	1,355
Population density (persons per sq. mi.)	78	85	90	85	84	82	80	79	78
(persons per sq. km.)	30	33	35	33	32	32	31	30	30
Net migration rate (per 1,000 population)	NA	NA	-2.6	-9.5	-8.0	-1.4	-0.8	-0.3	0.0
Births	NA	NA	NA	NA	16	NA	NA	NA	NA
Deaths	NA	NA	NA	NA	21	NA	NA	NA	NA
Life expectancy - males	NA	NA	64.7	62.2	62.5	63.7	66.4	70.1	73.2
Life expectancy - females	NA	NA	74.9	74.0	74.1	74.3	75.0	78.0	80.4
Birth rate (per 1,000)	NA	NA	14.1	9.9	10.7	14.3	12.9	10.8	11.1
Death rate (per 1,000)	NA	NA	12.4	14.1	14.1	14.3	14.1	12.5	11.9
Women of reproductive age (15-49 yrs.)	NA	NA	381	365	362	354	331	321	308
of which are currently married	NA	NA	240	NA	227	219	213	NA	NA
Fertility rate	NA	NA	2.0	1.4	1.6	2.0	1.8	1.8	1.7

Except as noted, values for vital statistics are in thousands; life expectancy is in years.

Health

Health Indicators [3]

% of population with access to	
safe water (1990-95)	NA
adequate sanitation (1990-95)	NA
health services (1985-95)	NA
% of 1-year-olds immunized (1990-94) against	
TB (tuberculosis)	99
DPT (diphtheria, pertussis, tetanus)	79
polio	87
measles	76
% of contraceptive prevalence (1980-94)	NA
ORT use rate (1990-94)	NA

Health Expenditures [4]

For sources, notes, and explanations, see Annotated Source Appendix, page 1061.

Human Factors

Women and Children [5]

% of pregnant women immunized (tetanus 1990-94)	NA
% of births attended by trained health personnel (1983-94)	NA
Maternal mortality rate (1980-92)	NA
Under-5 mortality rate (1994)	23
% under-5 moderately/severely underweight (1980-1994)	NA

Burden of Disease [6]

Population per physician (1992)	253.11
Population per nurse (1992)	126.56
Population per hospital bed (1992)	104.32
AIDS cases per 100,000 people (1994)	0.1
Heart disease cases per 1,000 people (1990-93)	437.5

Ethnic Division [7]

Estonian	61.5%
Russian	30.3%
Ukrainian	3.2%
Byelorussian	1.8%
Finn	1.1%
Other (1989)	2.1%

Religion [8]

Lutheran, Orthodox Christian.

Major Languages [9]

Estonian (official), Latvian, Lithuanian, Russian, other.

Education

Public Education Expenditures [10]

Million (Kroon)	1980	1988	1990	1992	1993	1994
Total education expenditure	NA	NA	NA	795	1,540	1,988
as percent of GNP	NA	NA	NA	6.2	7.1	5.8
as percent of total govt. expend.	NA	NA	NA	31.3	20.1	23.8
Current education expenditure	NA	NA	NA	729	1,344	1,748
as percent of GNP	NA	NA	NA	5.7	6.2	5.1
as percent of current govt. expend.	NA	NA	NA	NA	20.9	25.7
Capital expenditure	NA	NA	NA	66	196	240

Educational Attainment [11]

Age group (1989)	25+
Total population	1,001,198
Highest level attained (%)	
No schooling	2.2
First level	
Not completed	NA
Completed	39.0
Entered second level	
S-1	45.1
S-2	NA
Postsecondary	13.7

Illiteracy [12]

	1985	1989	1995
Illiterate population (15+ yrs.)[2]	NA	3,329	3,000
Illiteracy rate - total pop. (%)	NA	0.3	0.2
Illiteracy rate - males (%)	NA	0.1	0.2
Illiteracy rate - females (%)	NA	0.4	0.2

Libraries [13]

	Admin. Units	Svc. Pts.	Vols. (000)	Shelving (meters)	Vols. Added	Reg. Users
National (1992)	2	2	2,688	NA	72,364	1,690
Nonspecialized (1992)	1	1	2,790	NA	21,200	18,316
Public (1994)	605	NA	10,148	NA	NA	344,900
Higher ed. (1993)[12]	13	27	4,815	95,004	130,579	67,863
School	NA	NA	NA	NA	NA	NA

Daily Newspapers [14]

	1980	1985	1990	1994
Number of papers	NA	NA	NA	4
Circ. (000)	NA	NA	NA	373

Culture [15]

Cinema (seats per 1,000)	44.2
Annual attendance per person	2.2
Gross box office receipts (mil. Kroon)	3.2
Museums (reporting)	88
Visitors (000)	793
Annual receipts (000 Kroon)	26,261

Science and Technology

Scientific/Technical Forces [16]

Scientists/engineers	5,079
Number female	2,098
Technicians	848
Number female	588
Total[1]	5,927

R&D Expenditures [17]

	Kroon (000) 1994
Total expenditure[12]	216,798
Capital expenditure	14,767
Current expenditure	201,693
Percent current	93.2

U.S. Patents Issued [18]

Values show patents issued to citizens of the country by the U.S. Patents Office.

	1993	1994	1995
Number of patents	0	1	2

For sources, notes, and explanations, see Annotated Source Appendix, page 1061.

Government and Law

Organization of Government [19]

Long-form name:
Republic of Estonia
Type:
Republic
Independence:
6 September 1991 (from Soviet Union)
National holiday:
Independence Day, 24 February (1918)
Constitution:
Adopted 28 June 1992
Legal system:
Based on civil law system; no judicial review of legislative acts
Executive branch:
President; Prime Minister; Council of Ministers
Legislative branch:
Unicameral: Parliament (Riigikogu)
Judicial branch:
National Court

Elections [20]

Parliament	% of votes
Coalition Party (KMU)	32.2
Reform-Party Liberals	16.2
Center Party (K)	14.2
Pro Patria, National Indep.	7.9
Moderates (M)	6.0
Our Home is Estonia	5.0

Government Expenditures [21]

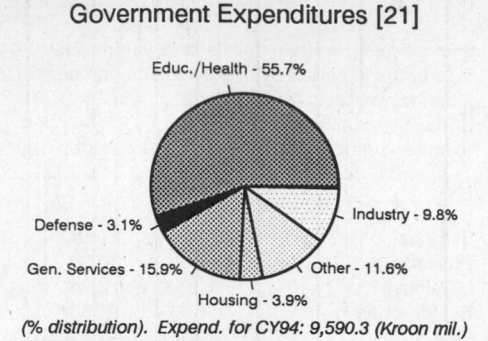

Educ./Health - 55.7%
Industry - 9.8%
Defense - 3.1%
Other - 11.6%
Gen. Services - 15.9%
Housing - 3.9%

(% distribution). Expend. for CY94: 9,590.3 (Kroon mil.)

Crime [23]

	1994
Crime volume	
Cases known to police	35,739
Attempts (percent)	2.50
Percent cases solved	26.70
Crimes per 100,000 persons	2,382.60
Persons responsible for offenses	
Total number offenders	9,316
Percent female	7.30
Percent juvenile (13-18 yrs.)	17.60
Percent foreigners	1.40

Military Expenditures and Arms Transfers [22]

	1990	1991	1992[14]	1993	1994
Military expenditures					
Current dollars (mil.)[e]	NA	NA	48	77	96
1994 constant dollars (mil.)[e]	NA	NA	50	79	96
Armed forces (000)	NA	NA	3	5	10
Gross national product (GNP)					
Current dollars (mil.)[e]	NA	NA	10,900	10,610	10,400
1994 constant dollars (mil.)[e]	NA	NA	11,360	10,820	10,400
Central government expenditures (CGE)					
1994 constant dollars (mil.)[e]	NA	NA	3,694	3,182	3,286
People (mil.)	NA	NA	1.6	1.6	1.6
Military expenditure as % of GNP	NA	NA	0.4	0.7	0.9
Military expenditure as % of CGE	NA	NA	1.4	2.5	2.9
Military expenditure per capita (1994 $)	NA	NA	31	49	60
Armed forces per 1,000 people (soldiers)	NA	NA	1.6	3.1	6.2
GNP per capita (1994 $)	NA	NA	7,103	6,730	6,432
Arms imports[6]					
Current dollars (mil.)	NA	NA	0	20	30
1994 constant dollars (mil.)	NA	NA	0	20	30
Arms exports[6]					
Current dollars (mil.)	NA	NA	0	0	5
1994 constant dollars (mil.)	NA	NA	0	0	5
Total imports[7]					
Current dollars (mil.)	NA	NA	432	896	1,660
1994 constant dollars (mil.)	NA	NA	451	914	1,660
Total exports[7]					
Current dollars (mil.)	NA	NA	463	805	1,299
1994 constant dollars (mil.)	NA	NA	483	822	1,299
Arms as percent of total imports[8]	NA	NA	0	2.2	1.8
Arms as percent of total exports[8]	NA	NA	0	0	0.4

Human Rights [24]

	SSTS	FL	FAPRO	PPCG	APROBC	TPW	PCPTW	STPEP	PHRFF	PRW	ASST	AFL
Observes	2			P					P			
	EAFRD	CPR	ESCR	SR	ACHR	MAAE	PVIAC	PVNAC	EAFDAW	TCIDTP	RC	
Observes	P	P	P					P	P	P	P	P

P = Party; S = Signatory; see Appendix for meaning of abbreviations.

Labor Force

Total Labor Force [25]

750,000 (1992)

Labor Force by Occupation [26]

Industry and construction	42%
Agriculture and forestry	20
Other	38

Date of data: 1990

Unemployment Rate [27]

8% (1994 est.)

Production Sectors

Commercial Energy Production and Consumption

Data are shown in quadrillion (10^{15}) BTUs and percent for 1995
Values for hydroelectric, nuclear, geothermal, solar, and wind power refer to electrical generation.

Production [28]

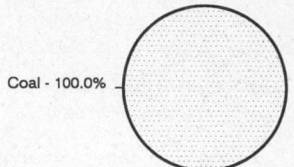

Coal - 100.0%

Consumption [29]

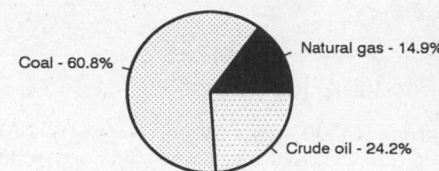

Coal - 60.8%   Natural gas - 14.9%

Crude oil - 24.2%

Coal	0.104
Total	0.104

Crude oil	0.047
Dry natural gas	0.029
Coal	0.118
Total	0.194

Telecommunications [30]

- 400,000 telephones; antiquated system; improvements being made piecemeal, with emphasis on business needs and international connections; about 150,000 unfulfilled requests for subscriber service
- Domestic: substantial investment has been made in cellular systems which are operational throughout Estonia
- International: international traffic is carried to the other former Soviet republics by landline or microwave radio relay and to other countries partly by leased connection to the Moscow international gateway switch and partly by a new Tallinn-Helsinki fiber-optic submarine cable which gives Estonia access to international circuits everywhere; access to the international packet-switched digital network via Helsinki
- Radio: Broadcast stations: AM NA, FM NA, shortwave 0 Radios: 710,000 (1992 est.)
- Television: Broadcast stations: 3 Televisions: 600,000 (1993 est.)

Top Agricultural Products [32]

Agriculture accounts for 10% of the GDP; produces potatoes, fruits, vegetables; livestock and dairy products; fish.

Top Mining Products [33]

Metric tons except as noted	5/29/95*
Ammonia, nitrogen content	100,000
Cement	500,000
Clay	
for bricks (cu. meters)	90,000
for cement	60,000
Oil shale	19,000,000
Peat	600,000
Sand and gravel	14,000,000
Silica sand, industrial	25,000

Transportation [31]

Railways: total: 1,018 km common carrier lines only; does not include dedicated industrial lines; broad gauge: 1,018 km 1.520-m gauge (132 km electrified) (1995)

Highways: total: 14,771 km; paved: 8,124 km (including 62 km of expressways); unpaved: 6,647 km (1993)

Merchant marine: total: 52 ships (1,000 GRT or over) totaling 353,140 GRT/467,086 DWT; ships by type: bulk 6, cargo 33, oil tanker 3, roll-on/roll-off cargo 6, short-sea passenger 4 (1995 est.)

Airports
Total:	22
With paved runways 2,438 to 3,047 m:	7
With paved runways 914 to 1,523 m:	3
With unpaved runways 2,438 to 3,047 m:	1
With unpaved runways 1,524 to 2,437 m:	2
With unpaved runways 914 to 1,523 m:	4

Tourism [34]

	1990	1991	1992	1993	1994
Visitors[1]	NA	NA	NA	1,600	1,900
Tourists	530	388	372	470	550
Excursionists	NA	NA	NA	765	865
Cruise passengers	NA	NA	NA	367	485
Tourism receipts	NA	NA	27	50	92
Tourism expenditures	NA	NA	19	25	48
Fare receipts	NA	NA	19	59	77
Fare expenditures	NA	NA	1	13	19

Travelers are in thousands, money in million U.S. dollars.

For sources, notes, and explanations, see Annotated Source Appendix, page 1061.

311

Finance, Economics, and Trade

Industrial Summary [35]

Industrial Production: Growth rate not available; accounts for 37% of the GDP. **Industries**: Oil shale, shipbuilding, phosphates, electric motors, excavators, cement, furniture, clothing, textiles, paper, shoes, apparel.

Economic Indicators [36]

- **National product**: GDP—purchasing power parity—$12.3 billion (1995 estimate as extrapolated from World Bank estimate for 1994)
- **National product real growth rate**: 6% (1995 est.)
- **National product per capita**: $7,600 (1995 est.)
- **Inflation rate (consumer prices)**: 29% (1995 est.)
- **External debt**: $270 million (January 1996)

Balance of Payments Summary [37]

Values in millions of dollars.

	1989	1990	1991	1992[1]	1993
Exports of goods (f.o.b.)	NA	NA	NA	459	803
Imports of goods (f.o.b.)	NA	NA	NA	-448	-900
Trade balance	NA	NA	NA	10	-97
Services - debits	NA	NA	NA	-185	-340
Services - credits	NA	NA	NA	202	370
Private transfers (net)	NA	NA	NA	NA	NA
Government transfers (net)	NA	NA	NA	97	106
Long-term capital (net)	NA	NA	NA	152	247
Short-term capital (net)	NA	NA	NA	-91	-20
Errors and omissions	NA	NA	NA	-128	-66
Overall balance	NA	NA	NA	58	200

Exchange Rates [38]

Currency: **Estonian kroon.**
Symbol: **EEK.**

Data are currency units per $1. Krooni are tied to the German deutsche mark at a fixed rate of 8 to 1.

December 1995	11.523
1995	11.465
1994	12.991
1993	13.223

Imports and Exports

Top Import Origins [39]

$2.5 billion (c.i.f., 1995).

Origins	%
Finland	NA
Russia	NA
Germany	NA
Sweden	NA

Top Export Destinations [40]

$1.8 billion (f.o.b., 1995).

Destinations	%
Russia	NA
Finland	NA
Sweden	NA
Germany	NA

Foreign Aid [41]

	U.S. $	
ODA (1993)	147	million
Western commitments, including international financial institutions	285	million

Import and Export Commodities [42]

Import Commodities
Machinery 18%
Fuels 15%
Vehicles 14%
Textiles 10%

Export Commodities
Textiles 14%
Food products 11%
Vehicles 11%
Metals 11%

For sources, notes, and explanations, see Annotated Source Appendix, page 1061.

Ethiopia

Geography [1]

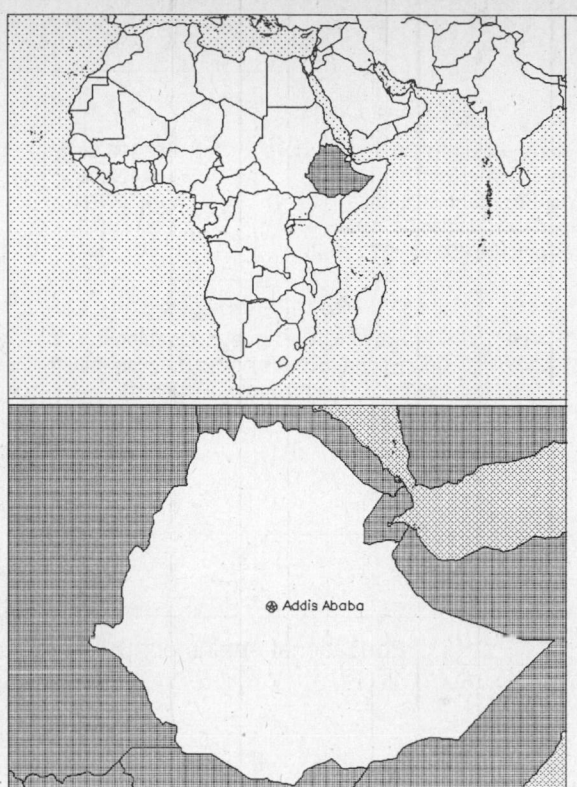

Total area:
 1,127,127 sq km 435,186 sq mi
Land area:
 1,119,683 sq km 432,312 sq mi
Comparative area:
 Slightly less than twice the size of Texas
Land boundaries:
 Total 5,311 km, Djibouti 337 km, Eritrea 912 km, Kenya 830 km,
 Somalia 1,626 km, Sudan 1,606 km
Coastline:
 0 km (landlocked)
Climate:
 Tropical monsoon with wide topographic-induced variation
Terrain:
 High plateau with central mountain range divided by Great Rift Valley
Natural resources:
 Small reserves of gold, platinum, copper, potash
Land use:
 Arable land: 12%
 Permanent crops: 1%
 Meadows and pastures: 41%
 Forest and woodland: 24%
 Other: 22%

Demographics [2]

	1970	1980	1990	1995[1]	1996	2000	2010	2020	2030
Population	29,673	36,413	48,242	55,588	57,172	63,514	81,169	100,813	124,250
Population density (persons per sq. mi.)	69	84	112	129	132	147	188	233	287
(persons per sq. km.)	27	33	43	50	51	57	72	90	111
Net migration rate (per 1,000 population)	NA	NA	10.2	0.0	-1.4	-1.2	0.0	0.0	0.0
Births	NA	NA	NA	NA	2,633	NA	NA	NA	NA
Deaths	NA	NA	NA	NA	1,002	NA	NA	NA	NA
Life expectancy - males	NA	NA	45.9	46.0	45.7	44.8	42.7	45.7	53.3
Life expectancy - females	NA	NA	49.4	48.2	48.0	47.2	45.0	48.3	57.0
Birth rate (per 1,000)	NA	NA	46.9	46.5	46.1	44.3	40.6	36.8	31.7
Death rate (per 1,000)	NA	NA	17.4	17.5	17.5	17.6	18.1	15.7	11.0
Women of reproductive age (15-49 yrs.)	NA	NA	10,799	12,378	12,730	14,108	18,553	24,317	31,821
of which are currently married	NA	NA	7,566	NA	8,880	9,789	12,754	NA	NA
Fertility rate	NA	NA	7.1	7.1	7.0	6.8	5.9	4.9	3.9

Except as noted, values for vital statistics are in thousands; life expectancy is in years.

Health

Health Indicators [3]

% of population with access to	
safe water (1990-95)	25
adequate sanitation (1990-95)	19
health services (1985-95)	46
% of 1-year-olds immunized (1990-94) against	
TB (tuberculosis)	50
DPT (diphtheria, pertussis, tetanus)	37
polio	36
measles	29
% of contraceptive prevalence (1980-94)	2
ORT use rate (1990-94)	68

Health Expenditures [4]

Total health expenditure, 1990 (official exchange rate)	
Millions of dollars	229
Dollars per capita	4
Health expenditures as a percentage of GDP	
Total	3.8
Public sector	2.3
Private sector	1.5
Development assistance for health	
Total aid flows (millions of dollars)[1]	43
Aid flows per capita (dollars)	0.8
Aid flows as a percentage of total health expenditure	18.8

For sources, notes, and explanations, see Annotated Source Appendix, page 1061.

Human Factors

Women and Children [5]

% of pregnant women immunized (tetanus 1990-94)	16
% of births attended by trained health personnel (1983-94)	14
Maternal mortality rate (1980-92)	560
Under-5 mortality rate (1994)	200
% under-5 moderately/severely underweight (1980-1994)[1]	48

Burden of Disease [6]

Population per physician (1988)[1]	32,498.64
Population per nurse (1984)[1]	5,402.02
Population per hospital bed (1990)[1]	4,140.78
AIDS cases per 100,000 people (1994)	10.7
Malaria cases per 100,000 people (1992)	NA

Ethnic Division [7]

Oromo	40%
Amhara and Tigrean	32%
Sidamo	9%
Shankella	6%
Somali	6%
Afar	4%
Gurage	2%
Other	1%

Religion [8]

Muslim	45%-48%
Ethiopian Orthodox	35%-38%
Animist	12%
Other	5%

Major Languages [9]

Amharic (official), Tigrinya, Orominga, Guaraginga, Somali, Arabic, English (major foreign language taught in schools).

Education

Public Education Expenditures [10]

Million (Birr)	1980	1985	1990	1992	1993	1994
Total education expenditure	279	420	600	704	1,107	NA
as percent of GNP	NA	NA	NA	NA	NA	NA
as percent of total govt. expend.	10.4	9.5	9.4	11.9	13.1	NA
Current education expenditure	222	354	494	564	790	NA
as percent of GNP	NA	NA	NA	NA	NA	NA
as percent of current govt. expend.	12.7	14.3	11.1	17.8	NA	NA
Capital expenditure	57	66	105	140	317	NA

Educational Attainment [11]

Illiteracy [12]

In thousands and percent[1]	1990	1995	2000
Illiterate population (15+ yrs.)	17,815	19,052	20,406
Illiteracy rate - total pop. (%)	67.4	63.2	59.6
Illiteracy rate - males (%)	57.6	53.6	50.1
Illiteracy rate - females (%)	77.3	72.8	69.1

Libraries [13]

	Admin. Units	Svc. Pts.	Vols. (000)	Shelving (meters)	Vols. Added	Reg. Users
National (1994)	1	9	100	NA	NA	420
Nonspecialized	NA	NA	NA	NA	NA	NA
Public (1986)	4	17	124	NA	12,197	11,680
Higher ed. (1986)	5	13	530	10,100	25,000	9,400
School	NA	NA	NA	NA	NA	NA

Daily Newspapers [14]

	1980	1985	1990	1994
Number of papers	3	3	6	4
Circ. (000)	40	41	100[e]	81[e]

Culture [15]

Cinema (seats per 1,000)	NA
Annual attendance per person	NA
Gross box office receipts (mil. Birr)	NA
Museums (reporting)	1
Visitors (000)	10
Annual receipts (000 Birr)	37

Science and Technology

Scientific/Technical Forces [16]

Scientists/engineers	47,113
Number female	12,476
Technicians	NA
Number female	NA
Total[36]	NA

R&D Expenditures [17]

U.S. Patents Issued [18]

Government and Law

Organization of Government [19]

Long-form name:
Federal Democratic Republic of Ethiopia
Type:
Federal republic
Independence:
Oldest independent country in Africa and
one of the oldest in the world - at least
2,000 years
National holiday:
National Day, 28 May (1991) (defeat of
Mengistu regime)
Constitution:
New constitution promulgated in
December 1994
Legal system:
NA
Executive branch:
President; Prime Minister; Council of
Ministers
Legislative branch:
Bicameral legislature: Federal Council
and Council of People's Representatives
Judicial branch:
Supreme Court

Elections [20]

Constituent Assembly. Election results
- government parties swept almost all
the seats; in December 1994 the
Constituent Assembly ratified the new
constitution with few changes; the new
constitution prescribes two chambers
for the New Assembly - one which is
elected by popular vote and one which
represents the ethnic interests of the
regional governments.

Government Expenditures [21]

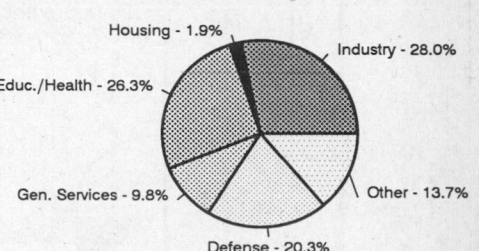

Housing - 1.9%
Industry - 28.0%
Educ./Health - 26.3%
Other - 13.7%
Gen. Services - 9.8%
Defense - 20.3%

(% distribution). Expend. for FY92: 3,708.1 (Birr mil.)

Crime [23]

	1994
Crime volume	
Cases known to police	140,294
Attempts (percent)	2.79
Percent cases solved	5.88
Crimes per 100,000 persons	231.57
Persons responsible for offenses	
Total number offenders	219,967
Percent female	9.52
Percent juvenile (9-17 yrs.)	1.49
Percent foreigners	NA

Military Expenditures and Arms Transfers [22]

	1990	1991	1992	1993	1994
Military expenditures					
Current dollars (mil.)	485	382	NA	142[e]	128
1994 constant dollars (mil.)	540	409	NA	145[e]	128
Armed forces (000)	250	120	120	120	120
Gross national product (GNP)					
Current dollars (mil.)	4,382	4,245	4,220	4,824	4,947
1994 constant dollars (mil.)	4,877	4,551	4,401	4,923	4,947
Central government expenditures (CGE)					
1994 constant dollars (mil.)	1,354	1,034	785	975	1,402
People (mil.)	48.3	49.8	51.1	52.6	54.3
Military expenditure as % of GNP	11.1	9.0	NA	2.9	2.6
Military expenditure as % of CGE	39.8	39.6	NA	14.8	9.1
Military expenditure per capita (1994 $)	11	8	NA	3	2
Armed forces per 1,000 people (soldiers)	5.2	2.4	2.3	2.3	2.2
GNP per capita (1994 $)	101	91	86	94	91
Arms imports[6]					
Current dollars (mil.)	410	80	0	5	0
1994 constant dollars (mil.)	456	86	0	5	0
Arms exports[6]					
Current dollars (mil.)	10	0	0	0	0
1994 constant dollars (mil.)	11	0	0	0	0
Total imports[7]					
Current dollars (mil.)	1,081	472	799	787	1,033
1994 constant dollars (mil.)	1,203	506	833	803	1,033
Total exports[7]					
Current dollars (mil.)	298	189	169	199	372
1994 constant dollars (mil.)	332	203	176	203	372
Arms as percent of total imports[8]	37.9	16.9	0	0.6	0
Arms as percent of total exports[8]	3.4	0	0	0	0

Human Rights [24]

	SSTS	FL	FAPRO	PPCG	APROBC	TPW	PCPTW	STPEP	PHRFF	PRW	ASST	AFL
Observes	P		P	P	P	P	P	P		P	P	
		EAFRD	CPR	ESCR	SR	ACHR	MAAE	PVIAC	PVNAC	EAFDAW	TCIDTP	RC
Observes		P	P	P	P			P	P	P	P	P

P=Party; S=Signatory; see Appendix for meaning of abbreviations.

Labor Force

Total Labor Force [25]

18 million

Labor Force by Occupation [26]

Agriculture and animal husbandry	80%
Government and services	12
Industry and construction	8

Date of data: 1985

Unemployment Rate [27]

For sources, notes, and explanations, see Annotated Source Appendix, page 1061.

315

Production Sectors

Commercial Energy Production and Consumption

Data are shown in quadrillion (10^{15}) BTUs and percent for 1995
Values for hydroelectric, nuclear, geothermal, solar, and wind power refer to electrical generation.

Production [28]

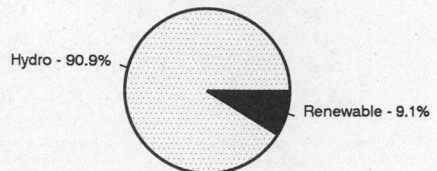

Hydro - 90.9%
Renewable - 9.1%

Consumption [29]

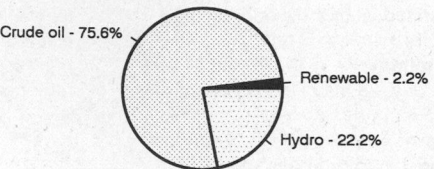

Crude oil - 75.6%
Renewable - 2.2%
Hydro - 22.2%

Net hydroelectric power	0.010
Geothermal, solar, wind	0.001
Total	0.011

Crude oil	0.034
Net hydroelectric power	0.010
Geothermal, solar, wind	0.001
Total	0.045

Telecommunications [30]

- 100,000 (1983 est.) telephones; open wire and microwave radio relay system adequate for government use
- Domestic: open wire and microwave radio relay
- International: open wire to Sudan and Djibouti; microwave radio relay to Kenya and Djibouti; satellite earth stations - 3 Intelsat (1 Atlantic Ocean and 2 Pacific Ocean)
- Radio: Broadcast stations: AM 4, FM 0, shortwave 0 Radios: 9.9 million (1992 est.)
- Television: Broadcast stations: 1 Televisions: 100,000 (1993 est.)

Top Agricultural Products [32]

Agriculture accounts for 48% of the GDP; produces cereals, pulses, coffee, oilseed, sugarcane, potatoes, other vegetables; hides, cattle, sheep, goats.

Top Mining Products [33]

Metric tons except as noted[e]	10/1/95*
Cement, hydraulic	300,000
Clay, brick	10,000
Gold, mine output, Au content (kg.)	2,370[r,27]
Gypsum and anhydrite, crude	30,700
Pumice	113,000
Salt, rock	5,000
Scoria	7,000
Construction stone, crushed (000 tons)	3,720
Dimension stone	3,000
Sand (000 tons)	3,770[28]

Transportation [31]

Railways: total: 681 km (Ethiopian segment of the Addis Ababa-Djibouti railroad); narrow gauge: 681 km 1.000-m gauge

Highways: total: 24,127 km; paved: 3,289 km; unpaved: 20,838 km (1993 est.)

Merchant marine: total: 12 ships (1,000 GRT or over) totaling 62,627 GRT/88,908 DWT; ships by type: cargo 8, oil tanker 2, roll-on/roll-off cargo 2 (1995 est.)

Airports

Total:	58
With paved runways over 3,047 m:	2
With paved runways 2,438 to 3,047 m:	3
With paved runways 1,524 to 2,437 m:	1
With paved runways 914 to 1,523 m:	1
With paved runways under 914 m:	6

Tourism [34]

	1990	1991	1992	1993	1994
Tourists[35]	79	82	83	93	98
Tourism receipts[36]	26	20	23	20	23
Tourism expenditures	11	7	10	11	NA
Fare expenditures	NA	8	4	4	NA

Travelers are in thousands, money in million U.S. dollars.

Manufacturing Sector

Manufacturing Summary [35]

	1987		1988		1989		1990		1991	
	$ bil.	%	$ bil.	%	$ bil.	%	$ bil.	%	$ bil.	%
Establishments or enterprises (number)	433	0.018	436	0.021	-	-	-	-	-	-
Total employment (000)	126	0.092	130	0.095	-	-	-	-	-	-
Production workers (000)	91	0.135	96	0.154	-	-	-	-	-	-
Output ($ bil.)	2	0.017	2	0.016	-	-	-	-	-	-
Value added ($ bil.)	0.856	0.020	0.854	0.018	-	-	-	-	-	-
Capital investment ($ mil.)	32	0.007	41	0.009	-	-	-	-	-	-
M & E investment ($ mil.)	-	-	-	-	-	-	-	-	-	-
Employees per establishment (number)	290	501.101	299	458.417	-	-	-	-	-	-
Production workers per establishment	211	733.562	220	739.378	-	-	-	-	-	-
Output per establishment ($ mil.)	4	94.651	4	76.381	-	-	-	-	-	-
Capital investment per estab. ($ mil.)	0.075	40.192	0.095	41.907	-	-	-	-	-	-
M & E per establishment ($ mil)	-	-	-	-	-	-	-	-	-	-
Payroll per employee ($)	1,936	21.593	2,006	20.132	-	-	-	-	-	-
Wages per production worker ($)	1,526	19.339	1,492	17.658	-	-	-	-	-	-
Hours per production worker (hours)	-	-	-	-	-	-	-	-	-	-
Output per employee ($)	14,147	18.889	14,192	16.662	-	-	-	-	-	-
Capital investment per employee ($)	258	8.021	318	9.142	-	-	-	-	-	-
M & E per employee ($)	-	-	-	-	-	-	-	-	-	-

Note: Columns headed % show percent of world total or ratio. Ratios closest to 100 are closest to world average. M & E stands for machinery & equipment.

Output in Manufacturing

	1987		1988		1989		1990		1991	
	$ bil.	%	$ bil.	%	$ bil.	%	$ bil.	%	$ bil.	%
3110 - Food products	0.290	14.50	0.306	15.30	-	-	-	-	-	-
3130 - Beverages	0.225	11.25	0.249	12.45	-	-	-	-	-	-
3140 - Tobacco	0.079	3.95	0.081	4.05	-	-	-	-	-	-
3210 - Textiles	0.223	11.15	0.227	11.35	-	-	-	-	-	-
3211 - Spinning, weaving, etc.	0.191	9.55	0.197	9.85	-	-	-	-	-	-
3220 - Wearing apparel	0.030	1.50	0.037	1.85	-	-	-	-	-	-
3230 - Leather and products	0.060	3.00	0.081	4.05	-	-	-	-	-	-
3240 - Footwear	0.027	1.35	0.032	1.60	-	-	-	-	-	-
3310 - Wood products	0.012	0.60	0.014	0.70	-	-	-	-	-	-
3320 - Furniture, fixtures	0.008	0.40	0.008	0.40	-	-	-	-	-	-
3410 - Paper and products	0.025	1.25	0.021	1.05	-	-	-	-	-	-
3420 - Printing, publishing	0.032	1.60	0.030	1.50	-	-	-	-	-	-
3510 - Industrial chemicals	0.003	0.15	0.003	0.15	-	-	-	-	-	-
3520 - Chemical products nec	0.061	3.05	0.052	2.60	-	-	-	-	-	-
3530 - Petroleum refineries	0.278	13.90	0.279	13.95	-	-	-	-	-	-
3550 - Rubber products	0.029	1.45	0.030	1.50	-	-	-	-	-	-
3560 - Plastic products nec	0.025	1.25	0.025	1.25	-	-	-	-	-	-
3620 - Glass and products	0.003	0.15	0.004	0.20	-	-	-	-	-	-
3690 - Nonmetal products nec	0.037	1.85	0.043	2.15	-	-	-	-	-	-
3710 - Iron and steel	0.035	1.75	0.039	1.95	-	-	-	-	-	-
3810 - Metal products	0.029	1.45	0.028	1.40	-	-	-	-	-	-
3830 - Electrical machinery	0.001	0.05	0.001	0.05	-	-	-	-	-	-
3840 - Transportation equipment	0.034	1.70	0.030	1.50	-	-	-	-	-	-
3843 - Motor vehicles	0.034	1.70	0.030	1.50	-	-	-	-	-	-

Note: Codes are International Standard Industry codes (ISIC). Percentages are % of total Output. [f]: Factor Prices; [p]: Producer Prices.

For sources, notes, and explanations, see Annotated Source Appendix, page 1061.

317

Finance, Economics, and Trade

Economic Indicators [36]

- **National product**: GDP—purchasing power parity—$24.2 billion (1995 est.)
- **National product real growth rate**: 2.7% (1995 est.)
- **National product per capita**: $400 (1995 est.)
- **Inflation rate (consumer prices)**: 10% (FY93/94)
- **External debt**: $3.7 billion (1993 est.)

Balance of Payments Summary [37]

Values in millions of dollars.

	1989	1990	1991	1992	1993
Exports of goods (f.o.b.)	443.8	292.0	167.6	169.9	198.8
Imports of goods (f.o.b.)	-817.9	-912.1	-470.8	-992.7	-706.0
Trade balance	-374.1	-620.1	-303.2	-822.8	-507.2
Services - debits	-408.4	-436.6	-381.1	-472.4	-377.8
Services - credits	301.8	313.8	282.7	290.1	299.6
Private transfers (net)	145.7	229.1	222.4	341.5	251.8
Government transfers (net)	190.5	220.0	353.0	543.9	279.6
Long-term capital (net)	170.8	132.0	6.1	-12.3	138.9
Short-term capital (net)	51.2	286.9	118.6	287.0	97.1
Errors and omissions	-32.0	-134.6	-254.9	-81.1	66.3
Overall balance	45.5	-9.5	43.6	73.9	248.3

Exchange Rates [38]

Currency: **birr.**
Symbol: **Br.**

Data are currency units per $1. Fixed at 2.070 before 1992; the official rate is pegged to the US$.

December 1995	6.3200
1995	6.3200
1994	5.9500
Fixed rate 1992-93	5.0000

Imports and Exports

Top Import Origins [39]

$972 million (c.i.f., 1994 est.).

Origins	%
US	NA
Germany	NA
Italy	NA
Saudi Arabia	NA
Japan	NA

Top Export Destinations [40]

$296 million (f.o.b., 1994 est.).

Destinations	%
Germany	NA
Japan	NA
Saudi Arabia	NA
France	NA
Italy	NA

Foreign Aid [41]

Recipient: ODA, $1.036 billion (1993).

Import and Export Commodities [42]

Import Commodities	Export Commodities
Capital goods	Coffee
Consumer goods	Leather products
Fuel	Gold

For sources, notes, and explanations, see Annotated Source Appendix, page 1061.

Fiji

Geography [1]

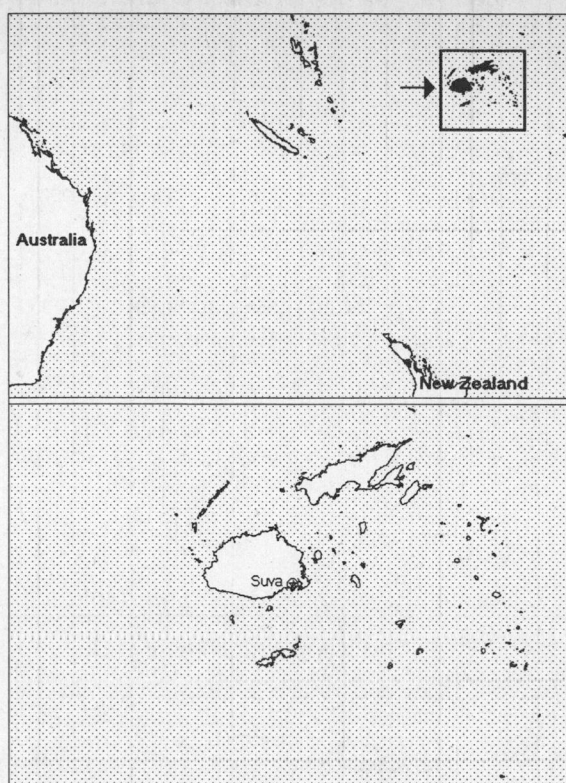

Total area:
  18,270 sq km   7,054 sq mi
Land area:
  18,270 sq km   7,054 sq mi
Comparative area:
  Slightly smaller than New Jersey
Land boundaries:
  0 km
Coastline:
  1,129 km
Climate:
  Tropical marine; only slight seasonal temperature variation
Terrain:
  Mostly mountains of volcanic origin
Natural resources:
  Timber, fish, gold, copper, offshore oil potential
Land use:
  Arable land: 8%
  Permanent crops: 5%
  Meadows and pastures: 3%
  Forest and woodland: 65%
  Other: 19%

Demographics [2]

	1970	1980	1990	1995[1]	1996	2000	2010	2020	2030
Population	521	635	738	773	782	823	933	1,037	1,133
Population density (persons per sq. mi.)	74	90	105	110	111	117	132	147	161
(persons per sq. km.)	29	35	40	42	43	45	51	57	62
Net migration rate (per 1,000 population)	NA	NA	-11.8	-5.7	-4.2	-3.6	-2.4	-1.1	0.0
Births	NA	NA	NA	NA	18	NA	NA	NA	NA
Deaths	NA	NA	NA	NA	5	NA	NA	NA	NA
Life expectancy - males	NA	NA	61.9	63.1	63.4	64.5	66.7	68.8	70.6
Life expectancy - females	NA	NA	66.3	67.8	68.1	69.4	72.1	74.5	76.6
Birth rate (per 1,000)	NA	NA	26.7	23.7	23.4	22.6	20.5	17.3	16.0
Death rate (per 1,000)	NA	NA	7.0	6.4	6.4	6.2	6.3	6.8	7.6
Women of reproductive age (15-49 yrs.)	NA	NA	188	201	205	221	251	271	293
of which are currently married	NA	NA	124	NA	134	144	169	NA	NA
Fertility rate	NA	NA	3.1	2.9	2.8	2.7	2.4	2.2	2.1

Except as noted, values for vital statistics are in thousands; life expectancy is in years.

Health

Health Indicators [3]

Health Expenditures [4]

For sources, notes, and explanations, see Annotated Source Appendix, page 1061.

319

Human Factors

Women and Children [5]	Burden of Disease [6]	
	Population per physician (1989)	2,792.31
	Population per nurse	NA
	Population per hospital bed	NA
	AIDS cases per 100,000 people (1994)	0.3
	Malaria cases per 100,000 people (1992)	NA

Ethnic Division [7]

Other includes European, other Pacific
Islanders, overseas Chinese, and others.

Fijian	49%
Indian	46%
Other	5%

Religion [8]

Christian	52%
Methodist	37%
Roman Catholic	9%
Other	6%
Hindu	38%
Muslim	8%
Other	2%
(1986)	

Major Languages [9]

English (official), Fijian, Hindustani.

Education

Public Education Expenditures [10]

Million (Dollar)[25]	1980	1986	1990	1991	1992	1994
Total education expenditure	NA	85	94	105	128	NA
as percent of GNP	NA	6.0	4.7	4.8	5.4	NA
as percent of total govt. expend.	NA	NA	NA	NA	18.6	NA
Current education expenditure	NA	83	93	98	124	NA
as percent of GNP	NA	5.9	4.7	4.5	5.2	NA
as percent of current govt. expend.	NA	NA	NA	NA	NA	NA
Capital expenditure	NA	2	1	7	4	NA

Educational Attainment [11]

Age group (1986)	25+
Total population	287,175
Highest level attained (%)	
No schooling	10.9
First level	
Not completed	35.9
Completed	23.9
Entered second level	
S-1	24.9
S-2	NA
Postsecondary	4.5

Illiteracy [12]

In thousands and percent[1]	1990	1995	2000
Illiterate population (15+ yrs.)	49	43	37
Illiteracy rate - total pop. (%)	10.7	8.7	6.7
Illiteracy rate - males (%)	7.9	6.5	5.1
Illiteracy rate - females (%)	13.6	10.9	8.7

Libraries [13]

	Admin. Units	Svc. Pts.	Vols. (000)	Shelving (meters)	Vols. Added	Reg. Users
National	NA	NA	NA	NA	NA	NA
Nonspecialized	NA	NA	NA	NA	NA	NA
Public (1986)	1	4	71	NA	3,374	24,677
Higher ed. (1989)	7	7	65	1,905	2,250	800
School (1987)	1	2	18	1,005	4,500	17,000

Daily Newspapers [14]

	1980	1985	1990	1994
Number of papers	3	3	1	1
Circ. (000)	64	68	27	35

Culture [15]

Cinema (seats per 1,000)	NA
Annual attendance per person	NA
Gross box office receipts (mil. Dollar)	NA
Museums (reporting)	1
Visitors (000)	12
Annual receipts (000 Dollar)	348

Science and Technology

Scientific/Technical Forces [16]

Scientists/engineers	36
Number female	4
Technicians	90
Number female	10
Total[1,12]	126

R&D Expenditures [17]

	Dollar (000) 1986
Total expenditure[13]	3,800
Capital expenditure	800
Current expenditure	3,000
Percent current	78.9

U.S. Patents Issued [18]

For sources, notes, and explanations, see Annotated Source Appendix, page 1061.

Government and Law

Organization of Government [19]

Long-form name:
Republic of Fiji

Type:
Republic. Military coup leader Maj. Gen. Sitiveni Rabuka formally declared Fiji a republic on 6 October 1987

Independence:
10 October 1970 (from UK)

National holiday:
Independence Day, 10 October (1970)

Constitution:
10 October 1970 (suspended 1 October 1987); a new Constitution was proposed on 23 September 1988 and promulgated on 25 July 1990; the 1990 Constitution is under review; the review is scheduled to be complete by 1997

Legal system:
Based on British system

Executive branch:
President; First Vice President; Second Vice President; Prime Minister; Deputy Prime Minister; Presidential Council; Great Council of Chiefs; Cabinet

Legislative branch:
Bicameral Parliament: Senate and House of Representatives

Judicial branch:
Supreme Court

Elections [20]

	% of seats
Fijian Political Party (SVT)	44.3
Fijian Nationalist Party (NFP)	28.6
Fijian Labor Party (FLP)	10.0
Fijian Association (FAP)	7.1
General Voters Party (GVP)	5.7
All National Congress (ANC)	1.4
Independents	2.9

Government Expenditures [21]

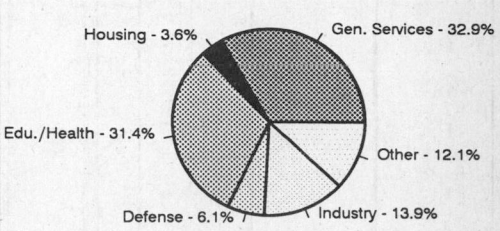

Housing - 3.6%
Gen. Services - 32.9%
Edu./Health - 31.4%
Other - 12.1%
Defense - 6.1%
Industry - 13.9%

(% distribution). Expend. for CY95: 801.14 (Dollar mil.)

Crime [23]

	1994
Crime volume	
Cases known to police	19,730
Attempts (percent)	NA
Percent cases solved	41.36
Crimes per 100,000 persons	2,518.37
Persons responsible for offenses	
Total number offenders	6,659
Percent female	1.23
Percent juvenile (0-17 yrs.)	2.63
Percent foreigners	0.14

Military Expenditures and Arms Transfers [22]

	1990	1991	1992	1993	1994
Military expenditures					
Current dollars (mil.)	34	35	32	27	34
1994 constant dollars (mil.)	38	37	34	28	34
Armed forces (000)	5	5	5	5	4
Gross national product (GNP)					
Current dollars (mil.)	1,502	1,576	1,662	1,727	1,819
1994 constant dollars (mil.)	1,671	1,689	1,733	1,763	1,819
Central government expenditures (CGE)					
1994 constant dollars (mil.)	491	436	444	534	608
People (mil.)	0.7	0.7	0.7	0.8	0.8
Military expenditure as % of GNP	2.3	2.2	1.9	1.6	1.9
Military expenditure as % of CGE	7.7	8.5	7.6	5.2	5.5
Military expenditure per capita (1994 $)	51	50	45	37	44
Armed forces per 1,000 people (soldiers)	6.8	6.7	6.7	6.6	5.2
GNP per capita (1994 $)	2,264	2,270	2,311	2,330	2,379
Arms imports[6]					
Current dollars (mil.)	0	0	0	0	0
1994 constant dollars (mil.)	0	0	0	0	0
Arms exports[6]					
Current dollars (mil.)	0	0	0	0	0
1994 constant dollars (mil.)	0	0	0	0	0
Total imports[7]					
Current dollars (mil.)	743	643	614	634	724
1994 constant dollars (mil.)	827	689	640	647	724
Total exports[7]					
Current dollars (mil.)	615	462	407	405	544
1994 constant dollars (mil.)	684	495	424	413	544
Arms as percent of total imports[8]	0	0	0	0	0
Arms as percent of total exports[8]	0	0	0	0	0

Human Rights [24]

	SSTS	FL	FAPRO	PPCG	APROBC	TPW	PCPTW	STPEP	PHRFF	PRW	ASST	AFL
Observes	P	P		P	P	P	P			P	P	P
		EAFRD	CPR	ESCR	SR	ACHR	MAAE	PVIAC	PVNAC	EAFDAW	TCIDTP	RC
Observes		P		P						P		P

P = Party; S = Signatory; see Appendix for meaning of abbreviations.

Labor Force

Total Labor Force [25]

235,000

Labor Force by Occupation [26]

Subsistence agriculture	67%
Wage earners	18
Salary earners	15

Date of data: 1987

Unemployment Rate [27]

5.4% (1992)

For sources, notes, and explanations, see Annotated Source Appendix, page 1061.

321

Production Sectors

Commercial Energy Production and Consumption

Data are shown in quadrillion (10^{15}) BTUs and percent for 1995
Values for hydroelectric, nuclear, geothermal, solar, and wind power refer to electrical generation.

Production [28]

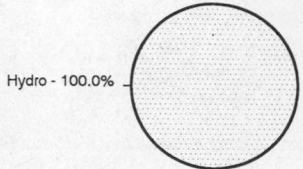

Hydro - 100.0%

Consumption [29]

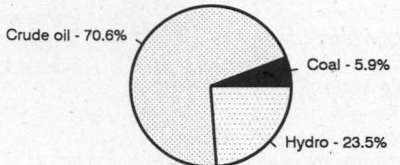

Crude oil - 70.6%
Coal - 5.9%
Hydro - 23.5%

Net hydroelectric power	0.004
Total	0.004

Crude oil	0.012
Coal	0.001
Net hydroelectric power	0.004
Total	0.017

Telecommunications [30]

- 60,017 (1987 est.) telephones; modern local, interisland, and international (wire/radio integrated) public and special-purpose telephone, telegraph, and teleprinter facilities; regional radio communications center
- International: access to important cable link between US and Canada and NZ and Australia; satellite earth station - 1 Intelsat (Pacific Ocean)
- Radio: Broadcast stations: AM 7, FM 1, shortwave 0
- Television: Broadcast stations: 0 Televisions: 12,000 (1992 est.)

Transportation [31]

Railways: total: 597 km; note - belongs to the government-owned Fiji Sugar Corporation; narrow gauge: 597 km 0.610-m gauge (1995)

Highways: total: 4,800 km; paved: NA km; unpaved: NA km

Merchant marine: total: 5 ships (1,000 GRT or over) totaling 16,267 GRT/17,884 DWT; ships by type: chemical tanker 2, oil tanker 1, roll-on/roll-off cargo 2 (1995 est.)

Airports

Total:	21
With paved runways over 3,047 m:	1
With paved runways 1,524 to 2,437 m:	1
With paved runways 914 to 1,523 m:	1
With paved runways under 914 m:	15
With unpaved runways 914 to 1,523 m:	3 (1995 est.)

Top Agricultural Products [32]

Agriculture accounts for 22% of the GDP; produces sugarcane, coconuts, cassava (tapioca), rice, sweet potatoes, bananas; cattle, pigs, horses, goats; fish catch nearly 33,000 tons (1989).

Top Mining Products [33]

Metric tons except as noted	7/7/95[*]
Cement, hydraulic	94,000
Gold, mine output, Au content (kg.)	3,440
Silver, mine output, Ag content (kg.)	1,390
Coral sand for cement manufacture	66,900
River sand for cement manufacture	15,000[e]
Quarried stone (cu. meters)	63,200

Tourism [34]

	1990	1991	1992	1993	1994
Visitors	307	287	308	295	336
Tourists[37]	279	259	279	287	319
Cruise passengers	28	27	30	8	17
Tourism receipts	199	194	218	236	298
Tourism expenditures	31	36	35	39	55
Fare receipts	72	84	85	84	76
Fare expenditures	21	15	15	14	15

Travelers are in thousands, money in million U.S. dollars.

For sources, notes, and explanations, see Annotated Source Appendix, page 1061.

Manufacturing Sector

Manufacturing Summary [35]

	1987		1988		1989		1990		1991	
	$ bil.	%	$ bil.	%	$ bil.	%	$ bil.	%	$ bil.	%
Establishments or enterprises (number)	614	0.026	489	0.023	551	0.029	525	0.029	-	-
Total employment (000)	9	0.007	10	0.008	12	0.010	12	0.011	-	-
Production workers (000)	8	0.012	9	0.015	11	0.015	11	0.016	-	-
Output ($ bil.)	0.396	0.004	0.366	0.003	0.480	0.004	0.501	0.004	-	-
Value added ($ bil.)	0.103	0.002	0.071	0.001	0.099	0.002	0.106	0.002	-	-
Capital investment ($ mil.)	29	0.007	12	0.003	15	0.003	34	0.006	-	-
M & E investment ($ mil.)	-	-	-	-	-	-	-	-	-	-
Employees per establishment (number)	15	25.089	21	32.698	22	33.377	24	38.334	-	-
Production workers per establishment	14	46.955	19	64.183	20	50.342	22	55.141	-	-
Output per establishment ($ mil.)	0.645	14.885	0.749	13.480	0.871	13.832	0.954	14.900	-	-
Capital investment per estab. ($ mil.)	0.047	25.341	0.025	11.226	0.027	9.148	0.064	20.724	-	-
M & E per establishment ($ mil)	-	-	-	-	-	-	-	-	-	-
Payroll per employee ($)	4,937	55.061	3,598	36.118	3,362	30.096	3,489	25.148	-	-
Wages per production worker ($)	4,259	53.973	2,701	31.969	2,712	27.043	2,929	24.661	-	-
Hours per production worker (hours)	-	-	-	-	-	-	-	-	-	-
Output per employee ($)	44,434	59.329	35,114	41.225	39,717	41.442	40,281	38.869	-	-
Capital investment per employee ($)	3,244	101.004	1,195	34.333	1,220	27.409	2,715	54.062	-	-
M & E per employee ($)	-	-	-	-	-	-	-	-	-	-

Note: Columns headed % show percent of world total or ratio. Ratios closest to 100 are closest to world average. M & E stands for machinery & equipment.

Output in Manufacturing

	1987		1988		1989		1990		1991	
	$ bil.	%	$ bil.	%	$ bil.	%	$ bil.	%	$ bil.	%
3110 - Food products	0.301	76.01	0.259	70.77	0.342	71.25	0.332	66.27	-	-
3310 - Wood products	0.012	3.03	0.022	6.01	0.032	6.67	0.036	7.19	-	-
3320 - Furniture, fixtures	0.004	1.01	0.005	1.37	0.008	1.67	0.011	2.20	-	-
3410 - Paper and products	0.010	2.53	0.010	2.73	0.014	2.92	0.016	3.19	-	-
3420 - Printing, publishing	0.010	2.53	0.009	2.46	0.014	2.92	0.018	3.59	-	-
3520 - Chemical products nec	0.020	5.05	0.020	5.46	0.020	4.17	0.029	5.79	-	-
3550 - Rubber products	0.003	0.76	0.003	0.82	0.003	0.63	0.004	0.80	-	-
3560 - Plastic products nec	0.007	1.77	0.007	1.91	0.008	1.67	0.010	2.00	-	-
3810 - Metal products	0.016	4.04	0.021	5.74	0.023	4.79	0.026	5.19	-	-
3840 - Transportation equipment	0.004	1.01	0.003	0.82	0.004	0.83	0.005	1.00	-	-
3841 - Shipbuilding, repair	0.003	0.76	0.002	0.55	0.003	0.63	0.004	0.80	-	-
3843 - Motor vehicles	0.001	0.25	0.001	0.27	0.001	0.21	0.002	0.40	-	-
3900 - Industries nec	0.003	0.76	0.004	1.09	0.007	1.46	0.009	1.80	-	-

Note: Codes are International Standard Industry codes (ISIC). Percentages are % of total Output. [f]: Factor Prices; [p]: Producer Prices.

For sources, notes, and explanations, see Annotated Source Appendix, page 1061.

323

Finance, Economics, and Trade

Economic Indicators [36]

- **National product**: GDP—purchasing power parity—$4.7 billion (1995 est.)
- **National product real growth rate**: 2.2% (1995 est.)
- **National product per capita**: $6,100 (1995 est.)
- **Inflation rate (consumer prices)**: 2% (1995)
- **External debt**: $670 million (1994 est.)

Balance of Payments Summary [37]

Values in millions of dollars.

	1989	1990	1991	1992	1993
Exports of goods (f.o.b.)	399.4	469.6	427.4	417.1	422.5
Imports of goods (f.o.b.)	-495.2	-644.6	-549.5	-539.5	-653.5
Trade balance	-95.8	-175.0	-122.1	-122.4	-231.0
Services - debits	-274.5	-320.7	-354.5	-371.5	-376.4
Services - credits	384.1	467.3	493.1	516.1	578.3
Private transfers (net)	-13.1	-22.2	-24.6	-16.3	-11.1
Government transfers (net)	28.4	16.7	30.0	25.9	27.1
Long-term capital (net)	6.0	39.8	-10.4	24.3	-4.7
Short-term capital (net)	0.4	27.1	16.2	18.4	-27.1
Errors and omissions	-44.7	21.1	-14.6	-15.2	-15.2
Overall balance	-9.2	54.1	13.1	59.3	-60.1

Exchange Rates [38]

Currency: **Fijian dollar.**
Symbol: **F$.**

Data are currency units per $1.

January 1996	1.4347
1995	1.4063
1994	1.4641
1993	1.5418
1992	1.5030
1991	1.4756

Imports and Exports

Top Import Origins [39]

$864.3 million (c.i.f., 1995).

Origins	%
Australia	30
New Zealand	17
Japan	13
EU	6
US	6

Top Export Destinations [40]

$571.8 million (f.o.b., 1995).

Destinations	%
EU	26
Australia	15
Pacific Islands	11
Japan	6

Foreign Aid [41]

Recipient: ODA, $NA.

Import and Export Commodities [42]

Import Commodities	Export Commodities
Machinery and transport equipment	Sugar 40%
Petroleum products	Clothing
Food	Gold
Consumer goods	Processed fish
Chemicals	Lumber

Finland

Geography [1]

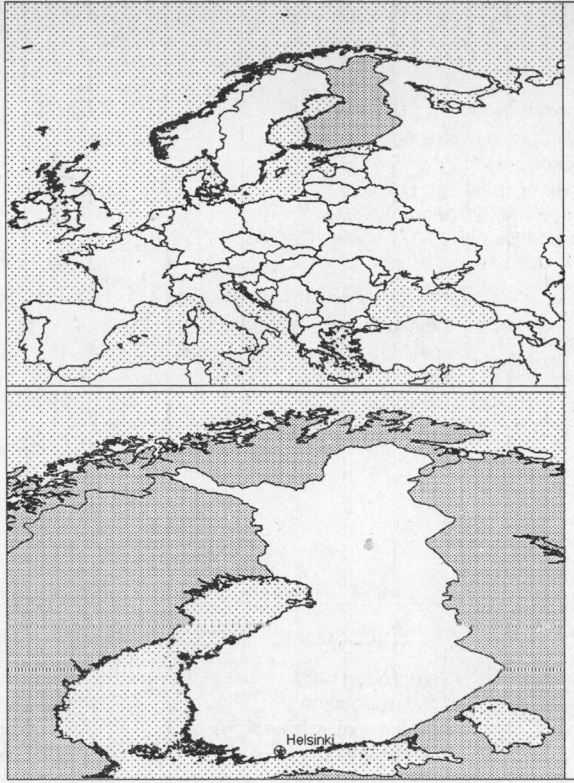

Total area:
  337,030 sq km   130,128 sq mi
Land area:
  305,470 sq km   117,943 sq mi
Comparative area:
  Slightly smaller than Montana
Land boundaries:
  Total 2,628 km, Norway 729 km, Sweden 586 km, Russia 1,313 km
Coastline:
  1,126 km (excludes islands and coastal indentations)
Climate:
  Cold temperate; potentially subarctic, but comparatively mild because of moderating influence of the North Atlantic Current, Baltic Sea, and more than 60,000 lakes
Terrain:
  Mostly low, flat to rolling plains interspersed with lakes and low hills
Natural resources:
  Timber, copper, zinc, iron ore, silver
Land use:
  Arable land: 8%
  Permanent crops: 0%
  Meadows and pastures: 0%
  Forest and woodland: 76%
  Other: 16%

Demographics [2]

	1970	1980	1990	1995[1]	1996	2000	2010	2020	2030
Population	4,606	4,780	4,986	5,099	5,105	5,115	5,109	5,075	4,940
Population density (persons per sq. mi.)	39	41	42	43	43	43	43	43	42
(persons per sq. km.)	15	16	16	17	17	17	17	17	16
Net migration rate (per 1,000 population)	NA	-0.3	1.7	0.6	0.6	0.6	0.6	0.2	0.0
Births	65	63	NA	NA	61	NA	NA	NA	NA
Deaths[14]	44	44	50	NA	50	NA	NA	NA	NA
Life expectancy - males	NA	69.2	70.9	73.7	73.8	74.5	75.9	78.0	79.5
Life expectancy - females	NA	77.6	78.9	77.0	77.2	77.8	79.3	82.8	85.0
Birth rate (per 1,000)	14.0	13.2	13.2	11.9	11.3	10.3	9.9	9.5	8.3
Death rate (per 1,000)	9.6	9.3	10.1	11.0	10.9	10.6	11.0	10.9	12.5
Women of reproductive age (15-49 yrs.)	NA	NA	1,258	1,269	1,259	1,216	1,143	1,051	986
of which are currently married[15]	690	685	675	NA	682	649	600	NA	NA
Fertility rate	NA	1.6	1.8	1.7	1.7	1.6	1.6	1.6	1.5

Except as noted, values for vital statistics are in thousands; life expectancy is in years.

Health

Health Indicators [3]

% of population with access to	
safe water (1990-95)	NA
adequate sanitation (1990-95)	NA
health services (1985-95)	NA
% of 1-year-olds immunized (1990-94) against	
TB (tuberculosis)	99
DPT (diphtheria, pertussis, tetanus)	99
polio	100
measles	99
% of contraceptive prevalence (1980-94)[1]	80
ORT use rate (1990-94)	NA

Health Expenditures [4]

Total health expenditure, 1990 (official exchange rate)	
Millions of dollars	10,200
Dollars per capita	2,046
Health expenditures as a percentage of GDP	
Total	7.4
Public sector	6.2
Private sector	1.2
Development assistance for health	
Total aid flows (millions of dollars)[1]	NA
Aid flows per capita (dollars)	NA
Aid flows as a percentage of total health expenditure	NA

For sources, notes, and explanations, see Annotated Source Appendix, page 1061.

325

Human Factors

Women and Children [5]

% of pregnant women immunized (tetanus 1990-94)	NA
% of births attended by trained health personnel (1983-94)[1]	100
Maternal mortality rate (1980-92)	11
Under-5 mortality rate (1994)	5
% under-5 moderately/severely underweight (1980-1994)	NA

Burden of Disease [6]

Population per physician (1989)	405.59
Population per nurse (1985)	59.10
Population per hospital bed (1990)	93.02
AIDS cases per 100,000 people (1994)	0.9
Heart disease cases per 1,000 people (1990-93)	358.5

Ethnic Division [7]

Finn, Swede, Lapp, Gypsy, Tatar.

Religion [8]

Evangelical Lutheran	89%
Greek Orthodox	1%
None	9%
Other	1%

Major Languages [9]

Official figures. Small Lapp- and Russian speaking minorities also present.

Finnish	93.5%
Swedish	6.3%

Education

Public Education Expenditures [10]

Million (Markka)	1980	1990	1991	1992	1993	1994
Total education expenditure	10,036	28,770	32,412	33,086	38,197	NA
as percent of GNP	5.3	5.7	6.8	7.2	8.4	NA
as percent of total govt. expend.	NA	11.9	11.9	11.6	12.8	NA
Current education expenditure	9,182	26,757	30,670	31,791	35,933	NA
as percent of GNP	4.9	5.3	6.5	7.0	7.9	NA
as percent of current govt. expend.	12.0	12.1	12.2	11.9	NA	NA
Capital expenditure	855	2,013	1,742	1,295	2,264	NA

Educational Attainment [11]

Age group (1990)	25+
Total population	3,387,384
Highest level attained (%)	
No schooling	NA
First level	
Not completed	49.4
Completed	NA
Entered second level	
S-1	35.3
S-2	NA
Postsecondary	15.4

Illiteracy [12]

Libraries [13]

	Admin. Units	Svc. Pts.	Vols. (000)	Shelving (meters)	Vols. Added	Reg. Users
National (1993)	1	1	2,952	NA	NA	NA
Nonspecialized (1993)	31	427	130,363	362,370	339,271	NA
Public (1993)	444	1,339	36,300	NA	2 mil	2 mil
Higher ed. (1993)	30	NA	12,279	NA	NA	NA
School (1990)	5,349	5,349	7,428	NA	194,000	697,329

Daily Newspapers [14]

	1980	1985	1990	1994
Number of papers	58	65	66	56
Circ. (000)	2,414	2,661	2,780	2,405

Culture [15]

Cinema (seats per 1,000)	12.0
Annual attendance per person	1.1
Gross box office receipts (mil. Markka)	196
Museums (reporting)	249
Visitors (000)	3,677
Annual receipts (000 Markka)	317,089

Science and Technology

Scientific/Technical Forces [16]

Scientists/engineers	18,588
Number female	NA
Technicians[2]	11,939
Number female	NA
Total[2]	30,527

R&D Expenditures [17]

	Markka (000) 1993
Total expenditure	10,677,100
Capital expenditure	943,100
Current expenditure	9,734,000
Percent current	91.2

U.S. Patents Issued [18]

Values show patents issued to citizens of the country by the U.S. Patents Office.

	1993	1994	1995
Number of patents	311	341	387

Government and Law

Organization of Government [19]

Long-form name:
Republic of Finland
Type:
Republic
Independence:
6 December 1917 (from Soviet Union)
National holiday:
Independence Day, 6 December (1917)
Constitution:
17 July 1919
Legal system:
Civil law system based on Swedish law;
Supreme Court may request legislation
interpreting or modifying laws; accepts
compulsory ICJ jurisdiction, with
reservations
Executive branch:
President; Prime Minister; Deputy Prime
Minister; Council of State (Valtioneuvosto)
Legislative branch:
Unicameral: Parliament (Eduskunta)
Judicial branch:
Supreme Court (Korkein Oikeus)

Elections [20]

Parliament	% of votes
Social Democratic Party	28.3
Center Party	19.9
National Coalition Party	17.9
Leftist Alliance (Communist)	11.2
Green League	6.5
Swedish People's Party	5.1
Finnish Christian League	3.0
Rural	1.3
Ecology Party	0.3
Liberal People's Party	0.6

Government Expenditures [21]

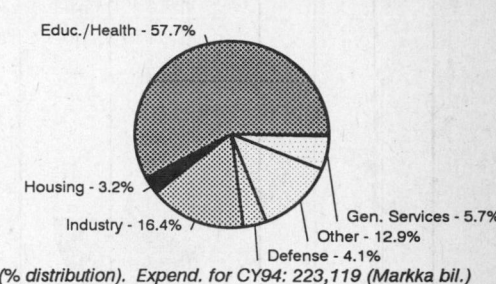

Educ./Health - 57.7%
Housing - 3.2%
Industry - 16.4%
Gen. Services - 5.7%
Other - 12.9%
Defense - 4.1%

(% distribution). Expend. for CY94: 223,119 (Markka bil.)

Crime [23]

	1994
Crime volume	
Cases known to police	754,542
Attempts (percent)	NA
Percent cases solved	66.20
Crimes per 100,000 persons	14,798.56
Persons responsible for offenses	
Total number offenders	572,215
Percent female	13.60
Percent juvenile (0-20 yrs.)	21.10
Percent foreigners	NA

Military Expenditures and Arms Transfers [22]

	1990	1991	1992	1993	1994
Military expenditures					
Current dollars (mil.)	1,521	1,873	1,943	1,878	1,966
1994 constant dollars (mil.)	1,692	2,008	2,026	1,917	1,966
Armed forces (000)	31	32	33	31	35
Gross national product (GNP)					
Current dollars (mil.)	93,410	89,520	88,040	87,920	93,230
1994 constant dollars (mil.)	104,000	95,960	91,810	89,730	93,230
Central government expenditures (CGE)					
1994 constant dollars (mil.)	33,820	38,930	46,680	45,210	NA
People (mil.)	5.0	5.0	5.0	5.1	5.1
Military expenditure as % of GNP	1.6	2.1	2.2	2.1	2.1
Military expenditure as % of CGE	5.0	5.2	4.3	4.2	NA
Military expenditure per capita (1994 $)	339	401	403	379	388
Armed forces per 1,000 people (soldiers)	6.2	6.4	6.5	6.1	6.9
GNP per capita (1994 $)	20,850	19,150	18,250	17,760	18,390
Arms imports[6]					
Current dollars (mil.)	60	50	450	290	60
1994 constant dollars (mil.)	67	54	469	296	60
Arms exports[6]					
Current dollars (mil.)	5	10	0	10	10
1994 constant dollars (mil.)	6	11	0	10	10
Total imports[7]					
Current dollars (mil.)	27,000	21,810	21,210	18,030	23,210
1994 constant dollars (mil.)	30,050	23,380	22,120	18,400	23,210
Total exports[7]					
Current dollars (mil.)	26,570	23,080	23,980	23,450	29,660
1994 constant dollars (mil.)	29,570	24,740	25,010	23,930	29,660
Arms as percent of total imports[8]	0.2	0.2	2.1	1.6	0.3
Arms as percent of total exports[8]	0	0	0	0	0

Human Rights [24]

	SSTS	FL	FAPRO	PPCG	APROBC	TPW	PCPTW	STPEP	PHRFF	PRW	ASST	AFL
Observes	P	P	P	P	P	P	P	P	P	P	P	P

	EAFRD	CPR	ESCR	SR	ACHR	MAAE	PVIAC	PVNAC	EAFDAW	TCIDTP	RC
Observes	P	P	P	P		P	P	P	P	P	P

P = Party; S = Signatory; see Appendix for meaning of abbreviations.

Labor Force

Total Labor Force [25]

2.533 million

Labor Force by Occupation [26]

Public services	30.4%
Industry	20.9
Commerce	15.0
Finance, insurance, and business services	10.2
Agriculture and forestry	8.6
Transport and communications	7.7
Construction	7.2

Unemployment Rate [27]

17% (1995)

For sources, notes, and explanations, see Annotated Source Appendix, page 1061.

327

Production Sectors

Commercial Energy Production and Consumption

Data are shown in quadrillion (10^{15}) BTUs and percent for 1995
Values for hydroelectric, nuclear, geothermal, solar, and wind power refer to electrical generation.

Production [28]

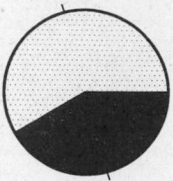

Nuclear - 58.5%

Hydro - 41.5%

Net hydroelectric power	0.132
Net nuclear power	0.186
Total	0.318

Consumption [29]

Crude oil - 42.3%

Hydro - 13.0%

Coal - 13.7%

Natural gas - 12.8%

Nuclear - 18.3%

Crude oil	0.430
Dry natural gas	0.130
Coal	0.139
Net hydroelectric power	0.132
Net nuclear power	0.186
Total	1.017

Telecommunications [30]

- 2.78 million (1986 est.) telephones; good service from cable and microwave radio relay network
- Domestic: cable and microwave radio relay
- International: 1 submarine cable; satellite earth stations - access to Intelsat transmission service via a Swedish satellite earth station, 1 Inmarsat (Atlantic and Indian Ocean Regions); note - Finland shares the Inmarsat earth station with the other Nordic countries (Denmark, Iceland, Norway, and Sweden)
- Radio: Broadcast stations: AM 6, FM 105, shortwave 0 Radios: 4.98 million (1991 est.)
- Television: Broadcast stations: 235 Televisions: 2.1 million (1983 est.)

Top Agricultural Products [32]

Agriculture accounts for 4.6% of the GDP; produces cereals, sugar beets, potatoes; dairy cattle; annual fish catch about 160,000 metric tons.

Top Mining Products [33]

Metric tons except as noted	5/95[*]
Aluminum metal, secondary	45,400
Copper metal	
smelter	98,200
refined	69,200
Nickel, concentrate, gross weight	107,000
Selenium metal (kg.)	29,000
Silver metal (kg.)	26,000
Zinc metal	173,000
Feldspar	41,400
Petroleum refinery products	73,000[e]

Transportation [31]

Railways: total: 5,895 km; broad gauge: 5,895 km 1.524-m gauge (1,993 km electrified; 480 km double- or more-track) (1995)

Highways: total: 76,755 km; paved: 47,588 km (including 318 km of expressways); unpaved: 29,167 km (1992 est.)

Merchant marine: total: 92 ships (1,000 GRT or over) totaling 1,051,231 GRT/1,075,397 DWT; ships by type: bulk 8, cargo 20, chemical tanker 5, oil tanker 12, passenger 2, refrigerated cargo 1, roll-on/roll-off cargo 31, short-sea passenger 12, vehicle carrier 1 (1995 est.)

Airports

Total:	157
With paved runways over 3,047 m:	3
With paved runways 2,438 to 3,047 m:	23
With paved runways 1,524 to 2,437 m:	13
With paved runways 914 to 1,523 m:	21
With paved runways under 914 m:	92

Tourism [34]

	1990	1991	1992	1993	1994
Tourism receipts	1,170	1,247	1,360	1,239	1,436
Tourism expenditures	2,740	2,742	2,449	1,617	1,727
Fare receipts	563	532	570	535	680
Fare expenditures	632	587	412	383	483

Travelers are in thousands, money in million U.S. dollars.

For sources, notes, and explanations, see Annotated Source Appendix, page 1061.

Manufacturing Sector

Manufacturing Summary [35]

	1987		1988		1989		1990		1991	
	$ bil.	%	$ bil.	%	$ bil.	%	$ bil.	%	$ bil.	%
Establishments or enterprises (number)	7,225	0.308	6,917	0.330	6,821	0.364	6,683	0.373	7,100	0.929
Total employment (000)	562	0.414	546	0.399	543	0.441	525	0.474	487	0.701
Production workers (000)	399	0.591	383	0.613	379	0.515	361	0.509	333	1.011
Output ($ bil.)	69	0.676	78	0.672	84	0.714	94	0.822	79	0.777
Value added ($ bil.)	25	0.581	29	0.604	31	0.643	34	0.681	27	0.778
Capital investment ($ mil.)	5,501	1.262	5,785	1.215	7,043	1.286	7,803	1.404	5,342	1.403
M & E investment ($ mil.)	4,255	1.365	4,925	1.423	5,439	1.393	5,904	1.400	3,879	1.376
Employees per establishment (number)	78	134.405	79	121.123	80	121.110	78	127.044	69	75.445
Production workers per establishment	55	191.935	55	186.093	56	141.477	54	136.485	47	108.833
Output per establishment ($ mil.)	10	219.337	11	204.053	12	196.209	14	220.290	11	83.627
Capital investment per estab. ($ mil.)	0.761	409.789	0.836	368.665	1	353.103	1	376.312	0.752	151.030
M & E per establishment ($ mil)	0.589	443.326	0.712	431.815	0.797	382.563	0.883	375.376	0.546	148.108
Payroll per employee ($)	19,419	216.556	22,045	221.275	23,449	209.885	29,071	209.570	28,118	192.575
Wages per production worker ($)	16,854	213.598	19,040	225.331	20,117	200.597	25,012	210.616	24,126	219.352
Hours per production worker (hours)	1,684	88.717	1,669	86.830	1,647	89.840	1,623	86.760	1,556	87.058
Output per employee ($)	122,222	163.191	143,493	168.468	155,268	162.009	179,698	173.397	162,728	110.844
Capital investment per employee ($)	9,791	304.890	10,591	304.372	12,975	291.555	14,874	296.207	10,974	200.185
M & E per employee ($)	7,573	329.843	9,016	356.509	10,020	315.880	11,254	295.470	7,969	196.313

Note: Columns headed % show percent of world total or ratio. Ratios closest to 100 are closest to world average. M & E stands for machinery & equipment.

Output in Manufacturing

	1987		1988		1989		1990		1991	
	$ bil.[f]	%	$ bil.[f]	%	$ bil.[f]	%	$ bil.[f]	%	$ bil.[f]	%
3110 - Food products	9.091	13.18	10.000	12.82	10.000	11.90	12.000	12.77	11.000	13.92
3130 - Beverages	0.824	1.19	0.942	1.21	1.028	1.22	1.253	1.33	1.271	1.61
3140 - Tobacco	0.193	0.28	0.232	0.30	0.219	0.26	0.293	0.31	0.272	0.34
3210 - Textiles	0.971	1.41	0.952	1.22	0.890	1.06	0.913	0.97	0.727	0.92
3211 - Spinning, weaving, etc.	0.380	0.55	0.356	0.46	0.326	0.39	0.303	0.32	0.225	0.28
3220 - Wearing apparel	1.110	1.61	0.997	1.28	0.920	1.10	0.900	0.96	0.645	0.82
3230 - Leather and products	0.152	0.22	0.141	0.18	0.114	0.14	0.120	0.13	0.096	0.12
3240 - Footwear	0.309	0.45	0.280	0.36	0.212	0.25	0.228	0.24	0.168	0.21
3310 - Wood products	3.049	4.42	3.479	4.46	4.055	4.83	4.564	4.86	3.294	4.17
3320 - Furniture, fixtures	0.764	1.11	0.870	1.12	0.927	1.10	1.025	1.09	0.814	1.03
3410 - Paper and products	8.688	12.59	10.000	12.82	11.000	13.10	12.000	12.77	10.000	12.66
3411 - Pulp, paper, etc.	7.442	10.79	8.901	11.41	10.000	11.90	10.000	10.64	8.858	11.21
3420 - Printing, publishing	3.308	4.79	3.820	4.90	4.080	4.86	4.760	5.06	4.273	5.41
3510 - Industrial chemicals	2.518	3.65	2.910	3.73	3.223	3.84	3.533	3.76	2.935	3.72
3511 - Basic chemicals, excl fertilizers	1.040	1.51	1.191	1.53	1.319	1.57	1.428	1.52	1.204	1.52
3513 - Synthetic resins, etc.	0.965	1.40	1.169	1.50	1.335	1.59	1.452	1.54	1.214	1.54
3520 - Chemical products nec	1.074	1.56	1.253	1.61	1.354	1.61	1.554	1.65	1.518	1.92
3522 - Drugs and medicines	0.353	0.51	0.416	0.53	0.454	0.54	0.549	0.58	0.579	0.73
3530 - Petroleum refineries	2.007	2.91	1.707	2.19	1.867	2.22	2.903	3.09	2.517	3.19
3540 - Petroleum, coal products	0.207	0.30	0.213	0.27	0.235	0.28	0.282	0.30	0.314	0.40
3550 - Rubber products	0.209	0.30	0.275	0.35	0.254	0.30	0.277	0.29	0.220	0.28
3560 - Plastic products nec	0.566	0.82	0.669	0.86	0.739	0.88	0.910	0.97	0.712	0.90
3610 - Pottery, china, etc.	0.080	0.12	0.096	0.12	0.107	0.13	0.112	0.12	0.096	0.12
3620 - Glass and products	0.248	0.36	0.304	0.39	0.317	0.38	0.348	0.37	0.321	0.41
3690 - Nonmetal products nec	1.483	2.15	1.700	2.18	1.934	2.30	2.228	2.37	1.696	2.15
3710 - Iron and steel	2.311	3.35	2.950	3.78	3.202	3.81	3.238	3.44	2.831	3.58
3720 - Nonferrous metals	1.149	1.67	1.717	2.20	2.039	2.43	1.954	2.08	1.637	2.07
3810 - Metal products	2.539	3.68	2.955	3.79	3.456	4.11	3.834	4.08	3.175	4.02
3820 - Machinery nec	4.798	6.95	5.602	7.18	6.022	7.17	7.253	7.72	5.381	6.81
3825 - Office, computing machinery	0.512	0.74	0.605	0.78	0.545	0.65	0.745	0.79	0.522	0.66
3830 - Electrical machinery	2.580	3.74	2.900	3.72	3.295	3.92	3.839	4.08	2.930	3.71
3832 - Radio, television, etc.	1.231	1.78	1.222	1.57	1.470	1.75	1.802	1.92	1.259	1.59
3840 - Transportation equipment	3.144	4.56	3.359	4.31	3.365	4.01	3.866	4.11	3.046	3.86
3841 - Shipbuilding, repair	1.461	2.12	1.525	1.96	1.473	1.75	1.734	1.84	1.461	1.85
3843 - Motor vehicles	1.222	1.77	1.310	1.68	1.363	1.62	1.446	1.54	1.014	1.28
3850 - Professional goods	0.421	0.61	0.526	0.67	0.555	0.66	0.654	0.70	0.638	0.81
3900 - Industries nec	0.266	0.39	0.299	0.38	0.301	0.36	0.324	0.34	0.314	0.40

Note: Codes are International Standard Industry codes (ISIC). Percentages are % of total Output. [f]: Factor Prices; [p]: Producer Prices.

For sources, notes, and explanations, see Annotated Source Appendix, page 1061.

Finance, Economics, and Trade

Economic Indicators [36]

- **National product**: GDP—purchasing power parity—$92.4 billion (1995 est.)
- **National product real growth rate**: 5% (1995 est.)
- **National product per capita**: $18,200 (1995 est.)
- **Inflation rate (consumer prices)**: 2% (1995 est.)
- **External debt**: $30 billion (December 1993)

Balance of Payments Summary [37]

Values in millions of dollars.

	1989	1990	1991	1992	1993
Exports of goods (f.o.b.)	22,882	26,101	22,516	23,571	23,135
Imports of goods (f.o.b.)	-23,101	-25,376	-20,195	-19,619	-16,743
Trade balance	-219	725	2,321	3,952	6,392
Services - debits	-11,425	-14,905	-14,929	-14,455	-12,812
Services - credits	6,611	8,193	6,919	6,374	5,991
Private transfers (net)	-252	-341	-308	-286	-99
Government transfers (net)	-510	-634	-768	-532	-452
Long-term capital (net)	1,521	8,541	11,635	8,392	7,157
Short-term capital (net)	1,940	4,721	-6,431	-4,304	-6,828
Errors and omissions	1,276	-2,366	-328	-1,306	927
Overall balance	-1,058	3,934	-1,889	-2,165	276

Exchange Rates [38]

Currency: **markka.**
Symbol: **FMk.**

Data are currency units per $1.

January 1996	4.4425
1995	4.3667
1994	5.2235
1993	5.7123
1992	4.4794
1991	4.0440

Imports and Exports

Top Import Origins [39]

$23.2 billion (c.i.f., 1994) Data are for 1994.

Origins	%
EU	44
Germany	15
UK	8.3
Sweden	10.4
FSU	10.3
US	7.6
Japan	6.5

Top Export Destinations [40]

$29.7 billion (f.o.b., 1994) Data are for 1994.

Destinations	%
EU	46.5
Germany	13.4
UK	10.3
Sweden	11
FSU	8.6
US	7.2
Japan	2.1

Foreign Aid [41]

Donor: ODA, $355 million (1993).

Import and Export Commodities [42]

Import Commodities	Export Commodities
Foodstuffs	Paper and pulp
Petroleum and petroleum products	Machinery
Chemicals	Chemicals
Transport equipment	Metals
Iron and steel	Timber
Machinery	
Textile yarn and fabrics	
Fodder grains	

France

Geography [1]

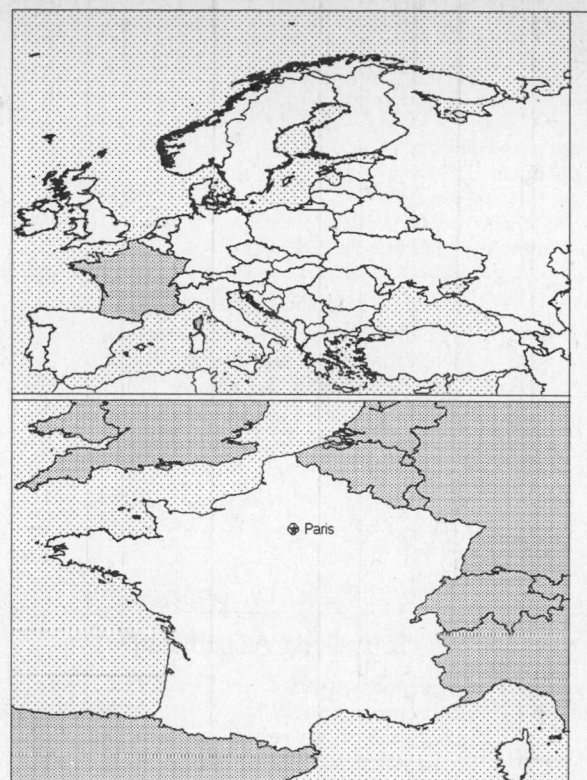

Total area:
547,030 sq km   211,209 sq mi
Land area:
545,630 sq km   210,669 sq mi
Comparative area:
Slightly more than twice the size of Colorado
Note: Includes only metropolitan France (which includes Corsica), but excludes the overseas administrative divisions
Land boundaries:
Total 2,892.4 km, Andorra 60 km, Belgium 620 km, Germany 451 km, Italy 488 km, Luxembourg 73 km, Monaco 4.4 km, Spain 623 km, Switzerland 573 km
Coastline:
3,427 km (mainland 2,783 km, Corsica 644 km)
Climate:
Generally cool winters and mild summers, but mild winters and hot summers along the Mediterranean
Terrain:
Mostly flat plains or gently rolling hills in North and West; remainder is mountainous, especially Pyrenees in South, Alps in East
Natural resources:
Coal, iron ore, bauxite, fish, timber, zinc, potash
Land use:
Arable land: 32%
Permanent crops: 2%
Meadows and pastures: 23%
Forest and woodland: 27%
Other: 16%
Note: Includes Corsica

Demographics [2]

	1970	1980	1990	1995[1]	1996	2000	2010	2020	2030
Population	50,507	53,618	56,484	57,862	58,041	58,816	60,562	61,087	60,114
Population density (persons per sq. mi.)	240	255	268	275	276	279	287	290	285
(persons per sq. km.)	93	98	104	106	106	108	111	112	110
Net migration rate (per 1,000 population)	NA	0.8	1.4	1.3	1.2	0.8	0.8	0.4	0.0
Births	850	800	NA	NA	631	NA	NA	NA	NA
Deaths	540	547	526	NA	541	NA	NA	NA	NA
Life expectancy - males	NA	70.2	72.9	74.2	74.5	75.5	77.4	78.8	79.8
Life expectancy - females	NA	78.4	81.3	82.3	82.5	83.1	84.4	85.4	86.1
Birth rate (per 1,000)	16.7	14.9	13.4	10.9	10.9	12.2	10.6	9.6	8.9
Death rate (per 1,000)	10.7	10.2	9.3	9.1	9.0	9.0	9.6	10.3	11.6
Women of reproductive age (15-49 yrs.)	NA	NA	14,116	14,579	14,622	14,438	13,842	12,914	12,057
of which are currently married	NA	NA	8,908	NA	9,520	9,458	9,120	NA	NA
Fertility rate	NA	1.9	1.8	1.5	1.5	1.7	1.7	1.6	1.6

Except as noted, values for vital statistics are in thousands; life expectancy is in years.

Health

Health Indicators [3]

% of population with access to	
safe water (1990-95)	NA
adequate sanitation (1990-95)	NA
health services (1985-95)	NA
% of 1-year-olds immunized (1990-94) against	
TB (tuberculosis)	78
DPT (diphtheria, pertussis, tetanus)	89
polio	92
measles	76
% of contraceptive prevalence (1980-94)	80
ORT use rate (1990-94)	NA

Health Expenditures [4]

Total health expenditure, 1990 (official exchange rate)	
Millions of dollars	105,467
Dollars per capita	1,869
Health expenditures as a percentage of GDP	
Total	8.9
Public sector	6.6
Private sector	2.3
Development assistance for health	
Total aid flows (millions of dollars)[1]	NA
Aid flows per capita (dollars)	NA
Aid flows as a percentage of total health expenditure	NA

For sources, notes, and explanations, see Annotated Source Appendix, page 1061.

Human Factors

Women and Children [5]

% of pregnant women immunized (tetanus 1990-94)	NA
% of births attended by trained health personnel (1983-94)[1]	94
Maternal mortality rate (1980-92)	9
Under-5 mortality rate (1994)	9
% under-5 moderately/severely underweight (1980-1994)	NA

Burden of Disease [6]

Population per physician (1987)	346.44
Population per nurse	NA
Population per hospital bed (1990)	109.05
AIDS cases per 100,000 people (1994)	9.3
Heart disease cases per 1,000 people (1990-93)	232

Ethnic Division [7]

Celtic and Latin with Teutonic, Slavic, North African, Indochinese, Basque minorities.

Religion [8]

Roman Catholic	90%
Protestant	2%
Jewish	1%
Muslim (North African workers)	1%
Unaffiliated	6%

Major Languages [9]

Rapidly declining regional dialects and languages (Provencal, Breton, Alsatian, Corsican, Catalan, Basque, Flemish).

French	100%

Education

Public Education Expenditures [10]

Million (Franc)[26]	1980	1990	1991	1992	1993	1994
Total education expenditure	142,099	351,867	388,819	393,004	406,772	NA
as percent of GNP	5.0	5.4	5.8	5.7	5.8	NA
as percent of total govt. expend.	NA	NA	NA	NA	10.4	NA
Current education expenditure	131,441	327,427	356,336	362,695	369,289	NA
as percent of GNP	4.7	5.1	5.3	5.2	5.2	NA
as percent of current govt. expend.	NA	NA	NA	NA	NA	NA
Capital expenditure	10,658	24,440	32,483	30,309	37,483	NA

Educational Attainment [11]

Age group (1990)	25+
Total population	37,354,255
Highest level attained (%)	
No schooling	NA
First level	
Not completed	51.1
Completed	NA
Entered second level	
S-1	36.9
S-2	NA
Postsecondary	11.4

Illiteracy [12]

Libraries [13]

	Admin. Units	Svc. Pts.	Vols. (000)	Shelving (meters)	Vols. Added	Reg. Users[15]
National (1992)[13]	1	NA	12,000	NA	NA	NA
Nonspecialized (1991)[14]	1	1	400	15,500	19,500	4 mil
Public (1992)	1,325	NA	NA	NA	NA	17 mil
Higher ed. (1991)	73	232	NA	NA	538,000	858,000
School	NA	NA	NA	NA	NA	NA

Daily Newspapers [14]

	1980	1985	1990	1994
Number of papers	90	92	79	118
Circ. (000)	10,332	10,670	11,792	13,685

Culture [15]

Cinema (seats per 1,000)[5]	16.7
Annual attendance per person	2.3
Gross box office receipts (mil. Franc)	4,519
Museums (reporting)[6]	35
Visitors (000)	14,056
Annual receipts (000 Franc)	199,000

Science and Technology

Scientific/Technical Forces [16]

Scientists/engineers	145,898
Number female	NA
Technicians[2]	168,272
Number female	NA
Total[2]	314,170

R&D Expenditures [17]

	Franc (000) 1993
Total expenditure	173,721,000
Capital expenditure	14,701,000
Current expenditure	159,020,000
Percent current	91.5

U.S. Patents Issued [18]

Values show patents issued to citizens of the country by the U.S. Patents Office.

	1993	1994	1995
Number of patents	3,154	2,985	3,011

For sources, notes, and explanations, see Annotated Source Appendix, page 1061.

Government and Law

Organization of Government [19]

Long-form name:
French Republic
Type:
Republic
Independence:
486 (unified by Clovis)
National holiday:
National Day, Taking of the Bastille, 14 July (1789)
Constitution:
28 September 1958, amended concerning election of president in 1962, amended to comply with provisions of EC Maastricht Treaty in 1992; amended to tighten immigration laws 1993
Legal system:
Civil law system with indigenous concepts; review of administrative but not legislative acts
Executive branch:
President; Prime Minister; Council of Ministers
Legislative branch:
Bicameral Parliament (Parlement): Senate (Senat) and National Assembly (Assemblee Nationale)
Judicial branch:
Supreme Court of Appeals (Cour de Cassation)

Elections [20]

National Assembly	% of seats
Rally for the Republic (RPR)	42.8
Union for French Democracy (UDF)	36.9
Socialist Party (PS)	11.6
Communist Party (PCF)	4.2
Independents	4.5

Government Budget [21]

For 1993.

	$ bil.
Revenues	220.5
Expenditures	249.1
Capital expenditures	47

Crime [23]

	1994
Crime volume	
Cases known to police	3,919,008
Attempts (percent)	NA
Percent cases solved	34.87
Crimes per 100,000 persons	6,782.72
Persons responsible for offenses	
Total number offenders	775,701
Percent female	14.18
Percent juvenile (13-18 yrs.)	14.10
Percent foreigners	19.55

Military Expenditures and Arms Transfers [22]

	1990	1991	1992	1993	1994
Military expenditures					
Current dollars (mil.)	41,020	42,840	42,700	43,060	44,390
1994 constant dollars (mil.)	45,650	45,920	44,520	43,940	44,390
Armed forces (000)	550	542	522	506	506
Gross national product (GNP)					
Current dollars (bil.)	1,146	1,196	1,240	1,256	1,316
1994 constant dollars (bil.)	1,275	1,282	1,293	1,282	1,316
Central government expenditures (CGE)					
1994 constant dollars (mil.)	543,600	571,600	588,000	604,900	617,600
People (mil.)	56.7	57.0	57.3	57.6	57.8
Military expenditure as % of GNP	3.6	3.6	3.4	3.4	3.4
Military expenditure as % of CGE	8.4	8.0	7.6	7.3	7.2
Military expenditure per capita (1994 $)	805	805	777	763	767
Armed forces per 1,000 people (soldiers)	9.7	9.5	9.1	8.8	8.7
GNP per capita (1994 $)	22,480	22,490	22,570	22,260	22,760
Arms imports[6]					
Current dollars (mil.)	310	280	140	80	170
1994 constant dollars (mil.)	345	300	146	82	170
Arms exports[6]					
Current dollars (mil.)	5,000	1,800	1,500	825	800
1994 constant dollars (mil.)	5,565	1,929	1,564	842	800
Total imports[7]					
Current dollars (mil.)	234,400	231,800	239,600	201,800	229,300
1994 constant dollars (mil.)	260,900	248,400	249,900	206,000	229,300
Total exports[7]					
Current dollars (mil.)	216,600	217,100	235,900	209,300	235,500
1994 constant dollars (mil.)	241,000	232,700	246,000	213,700	235,500
Arms as percent of total imports[8]	0.1	0.1	0.1	0	0.1
Arms as percent of total exports[8]	2.3	0.8	0.6	0.4	0.3

Human Rights [24]

	SSTS	FL	FAPRO	PPCG	APROBC	TPW	PCPTW	STPEP	PHRFF	PRW	ASST	AFL
Observes	P	P	P	P	P	P	P	P	P	P	P	P

	EAFRD	CPR	ESCR	SR	ACHR	MAAE	PVIAC	PVNAC	EAFDAW	TCIDTP	RC
Observes	P	P	P	P	P	P	P	P	P	P	P

P=Party; S=Signatory; see Appendix for meaning of abbreviations.

Labor Force

Total Labor Force [25]

24.17 million

Labor Force by Occupation [26]

Services	61.5%
Industry	31.3
Agriculture	7.2

Date of data: 1987

Unemployment Rate [27]

11.7% (yearend 1995)

For sources, notes, and explanations, see Annotated Source Appendix, page 1061.

333

Production Sectors

Commercial Energy Production and Consumption

Data are shown in quadrillion (10^{15}) BTUs and percent for 1995
Values for hydroelectric, nuclear, geothermal, solar, and wind power refer to electrical generation.

Production [28]
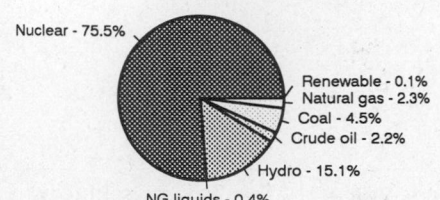

Nuclear - 75.5%
Renewable - 0.1%
Natural gas - 2.3%
Coal - 4.5%
Crude oil - 2.2%
Hydro - 15.1%
NG liquids - 0.4%

Consumption [29]
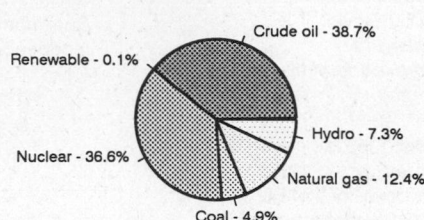

Crude oil - 38.7%
Renewable - 0.1%
Hydro - 7.3%
Nuclear - 36.6%
Natural gas - 12.4%
Coal - 4.9%

Crude oil	0.107
Natural gas liquids	0.018
Dry natural gas	0.111
Coal	0.222
Net hydroelectric power	0.742
Net nuclear power	3.713
Geothermal, solar, wind	0.006
Total	4.919

Crude oil	3.926
Dry natural gas	1.260
Coal	0.493
Net hydroelectric power	0.742
Net nuclear power	3.713
Geothermal, solar, wind	0.006
Total	10.14

Telecommunications [30]

- 35 million (1987 est.) telephones; highly developed
- Domestic: extensive cable and microwave radio relay; extensive introduction of fiber-optic cable; domestic satellite system
- International: satellite earth stations - 2 Intelsat (with total of 5 antennas - 2 for Indian Ocean and 3 for Atlantic Ocean), NA Eutelsat, 1 Inmarsat (Atlantic Ocean Region); HF radiotelephone communications with more than 20 countries
- Radio: Broadcast stations: AM 41, FM 800 (mostly repeaters), shortwave 0 Radios: 49 million (1993 est.)
- Television: Broadcast stations: 846 (mostly repeaters) Televisions: 29.3 million (1993 est.)

Transportation [31]

Railways: total: 33,891 km; standard gauge: 33,524 km 1.435-m gauge; 32,275 km are operated by French National Railways (SNCF); 13,741 km of SNCF routes are electrified and 12,132 km are double- or multiple-tracked; narrow gauge: 367 km 1.000-m gauge

Highways: total: 1,511,200 km; paved: 811,200 km (including 7,700 km of expressways); unpaved: 700,000 km (1992 est.)

Merchant marine: total: 55 ships (1,000 GRT or over) totaling 1,203,086 GRT/1,779,263 DWT; ships by type: bulk 6, cargo 5, chemical tanker 5, container 7, liquefied gas tanker 3, oil tanker 16, passenger 1, roll-on/roll-off cargo 6, short-sea passenger 5, specialized tanker 1

Top Agricultural Products [32]

Agriculture accounts for 2.4% of the GDP; produces wheat, cereals, sugar beets, potatoes, wine grapes; beef, dairy products; fish catch of 850,000 metric tons ranks among world's top 20 countries and is all used domestically.

Airports

Total:	460
With paved runways over 3,047 m:	13
With paved runways 2,438 to 3,047 m:	26
With paved runways 1,524 to 2,437 m:	91
With paved runways 914 to 1,523 m:	73
With paved runways under 914 m:	179

Top Mining Products [33]

Metric tons except as noted[e]	1/95[*]
Lead	
smelter, primary	115,000
refined, primary	165,137
Barite	70,000
Zinc, slab	310,000
Talc	
crude	275,000
powder	200,000
Carbon black	200,000
Petroleum refinery products (000 42-gal. bls.)	562,595[29]

Tourism [34]

	1990	1991	1992	1993	1994
Visitors[38]	93,992	118,584	NA	NA	131,907
Tourists	52,497	55,041	59,740	60,565	61,312
Excursionists	41,495	63,543	NA	NA	70,595
Tourism receipts	20,185	21,375	25,051	23,565	25,629
Tourism expenditures	12,424	12,321	13,914	12,836	13,875

Travelers are in thousands, money in million U.S. dollars.

Manufacturing Sector

Manufacturing Summary [35]

	1987		1988		1989		1990		1991	
	$ bil.	%	$ bil.	%	$ bil.	%	$ bil.	%	$ bil.	%
Establishments or enterprises (number)	-	-	-	-	-	-	-	-	-	-
Total employment (000)	4,237	3.122	4,166	3.044	4,185	3.400	4,126	3.728	4,160	5.989
Production workers (000)	-	-	-	-	-	-	-	-	-	-
Output ($ bil.)	556	5.471	609	5.224	622	5.276	756	6.592	731	7.167
Value added ($ bil.)	208	4.824	229	4.765	229	4.688	281	5.625	272	7.935
Capital investment ($ mil.)	21,129	4.848	23,687	4.974	24,827	4.532	31,807	5.723	30,556	8.024
M & E investment ($ mil.)	16,304	5.232	18,265	5.277	18,981	4.862	26,022	6.172	21,942	7.782
Employees per establishment (number)	-	-	-	-	-	-	-	-	-	-
Production workers per establishment	-	-	-	-	-	-	-	-	-	-
Output per establishment ($ mil.)	-	-	-	-	-	-	-	-	-	-
Capital investment per estab. ($ mil.)	-	-	-	-	-	-	-	-	-	-
M & E per establishment ($ mil)	-	-	-	-	-	-	-	-	-	-
Payroll per employee ($)	29,624	330.359	31,378	314.957	-	-	-	-	-	-
Wages per production worker ($)	-	-	-	-	-	-	-	-	-	-
Hours per production worker (hours)	-	-	-	-	-	-	-	-	-	-
Output per employee ($)	131,243	175.236	146,165	171.605	148,719	155.175	183,257	176.832	175,692	119.675
Capital investment per employee ($)	4,987	155.283	5,686	163.399	5,932	133.301	7,709	153.522	7,345	133.993
M & E per employee ($)	3,848	167.595	4,384	173.354	4,535	142.981	6,307	165.582	5,275	129.944

Note: Columns headed % show percent of world total or ratio. Ratios closest to 100 are closest to world average. M & E stands for machinery & equipment.

Output in Manufacturing

	1987		1988		1989		1990		1991	
	$ bil.	%	$ bil.	%	$ bil.	%	$ bil.	%	$ bil.	%
3110 - Food products	76.000	13.67	80.000	13.14	80.000	12.86	95.000	12.57	94.000	12.86
3130 - Beverages	8.818	1.59	9.653	1.59	10.000	1.61	14.000	1.85	15.000	2.05
3140 - Tobacco	1.880	0.34	2.014	0.33	1.975	0.32	2.406	0.32	2.286	0.31
3210 - Textiles	18.000	3.24	19.000	3.12	18.000	2.89	20.000	2.65	19.000	2.60
3220 - Wearing apparel	12.000	2.16	12.000	1.97	11.000	1.77	13.000	1.72	13.000	1.78
3230 - Leather and products	2.046	0.37	2.149	0.35	2.116	0.34	2.571	0.34	2.233	0.31
3240 - Footwear	2.595	0.47	2.518	0.41	2.414	0.39	2.957	0.39	2.854	0.39
3310 - Wood products	7.354	1.32	8.091	1.33	8.119	1.31	9.917	1.31	9.730	1.33
3320 - Furniture, fixtures	6.688	1.20	7.235	1.19	7.288	1.17	8.999	1.19	8.773	1.20
3410 - Paper and products	14.000	2.52	16.000	2.63	16.000	2.57	19.000	2.51	18.000	2.46
3420 - Printing, publishing	22.000	3.96	24.000	3.94	25.000	4.02	31.000	4.10	31.000	4.24
3510 - Industrial chemicals	23.000	4.14	26.000	4.27	25.000	4.02	28.000	3.70	27.000	3.69
3520 - Chemical products nec	25.000	4.50	27.000	4.43	28.000	4.50	35.000	4.63	36.000	4.92
3522 - Drugs and medicines	10.000	1.80	12.000	1.97	12.000	1.93	15.000	1.98	15.000	2.05
3530 - Petroleum refineries	28.000	5.04	28.000	4.60	28.000	4.50	36.000	4.76	35.000	4.79
3550 - Rubber products	5.357	0.96	5.876	0.96	5.862	0.94	6.721	0.89	6.753	0.92
3560 - Plastic products nec	12.000	2.16	14.000	2.30	14.000	2.25	18.000	2.38	17.000	2.33
3620 - Glass and products	4.209	0.76	4.549	0.75	4.545	0.73	5.748	0.76	5.618	0.77
3690 - Nonmetal products nec	12.000	2.16	14.000	2.30	14.000	2.25	17.000	2.25	17.000	2.33
3710 - Iron and steel	21.000	3.78	24.000	3.94	25.000	4.02	29.000	3.84	26.000	3.56
3720 - Nonferrous metals	12.000	2.16	14.000	2.30	14.000	2.25	15.000	1.98	13.000	1.78
3810 - Metal products	28.000	5.04	31.000	5.09	33.000	5.31	41.000	5.42	39.000	5.34
3820 - Machinery nec	44.000	7.91	48.000	7.88	50.000	8.04	63.000	8.33	59.000	8.07
3830 - Electrical machinery	40.000	7.19	44.000	7.22	44.000	7.07	56.000	7.41	54.000	7.39
3840 - Transportation equipment	62.000	11.15	71.000	11.66	74.000	11.90	91.000	12.04	88.000	12.04
3841 - Shipbuilding, repair	3.411	0.61	3.509	0.58	3.307	0.53	4.444	0.59	4.466	0.61
3843 - Motor vehicles	43.000	7.73	50.000	8.21	52.000	8.36	63.000	8.33	59.000	8.07
3850 - Professional goods	5.374	0.97	5.859	0.96	6.113	0.98	7.713	1.02	7.728	1.06
3900 - Industries nec	6.738	1.21	7.151	1.17	7.179	1.15	8.778	1.16	8.277	1.13

Note: Codes are International Standard Industry codes (ISIC). Percentages are % of total Output. [f]: Factor Prices; [p]: Producer Prices.

For sources, notes, and explanations, see Annotated Source Appendix, page 1061.

335

Finance, Economics, and Trade

Economic Indicators [36]

- **National product**: GDP—purchasing power parity—$1.173 trillion (1995 est.)
- **National product real growth rate**: 2.4% (1995 est.)
- **National product per capita**: $20,200 (1995 est.)
- **Inflation rate (consumer prices)**: 1.7% (1995)
- **External debt**: $300 billion (1993 est.)

Balance of Payments Summary [37]

Values in millions of dollars.

	1989	1990	1991	1992	1993
Exports of goods (f.o.b.)	170,761	206,670	207,129	225,318	195,114
Imports of goods (f.o.b.)	-181,412	-220,341	-217,305	-223,563	-188,117
Trade balance	-10,651	-13,671	-10,176	1,755	6,997
Services - debits	-89,158	-122,502	-140,840	-170,426	-178,197
Services - credits	102,721	134,410	151,966	181,508	187,542
Private transfers (net)	-2,668	-4,011	-3,020	-2,816	-729
Government transfers (net)	-5,863	-9,461	-4,961	-5,684	-5,413
Long-term capital (net)	6,845	12,785	2,196	14,544	-3,539
Short-term capital (net)	-1,889	12,850	-5,879	-34,084	-7,147
Errors and omissions	58,118	1,425	4,897	2,109	2,686
Overall balance	-2,351	11,825	-5,817	-13,094	2,200

Exchange Rates [38]

Currency: **French franc.**
Symbol: **F.**

Data are currency units per $1.

January 1996	5.0056
1995	4.9915
1994	5.5520
1993	5.6632
1992	5.2938
1991	5.6421

Imports and Exports

Top Import Origins [39]

$229.3 billion (c.i.f., 1994).

Origins	%
Germany	17.8
Italy	10.1
Belgium-Luxembourg	9.1
Spain	8.8
US	8.5
UK	7.9
Netherlands	4.9
Japan	3.7
Russia	1.2

Top Export Destinations [40]

$235.5 billion (f.o.b., 1994).

Destinations	%
Germany	17.1
UK	9.9
Italy	9.3
Belgium-Luxembourg	8.7
Spain	7.1
US	7.0
Netherlands	4.6
Japan	2.0
Russia	0.5

Foreign Aid [41]

Donor: ODA, $7.915 billion (1993).

Import and Export Commodities [42]

Import Commodities

Crude oil
Machinery and equipment
Agricultural products
Chemicals
Iron and steel products

Export Commodities

Machinery and transportation equipment
Chemicals
Foodstuffs
Agricultural products
Iron and steel products
Textiles and clothing

For sources, notes, and explanations, see Annotated Source Appendix, page 1061.

Gabon

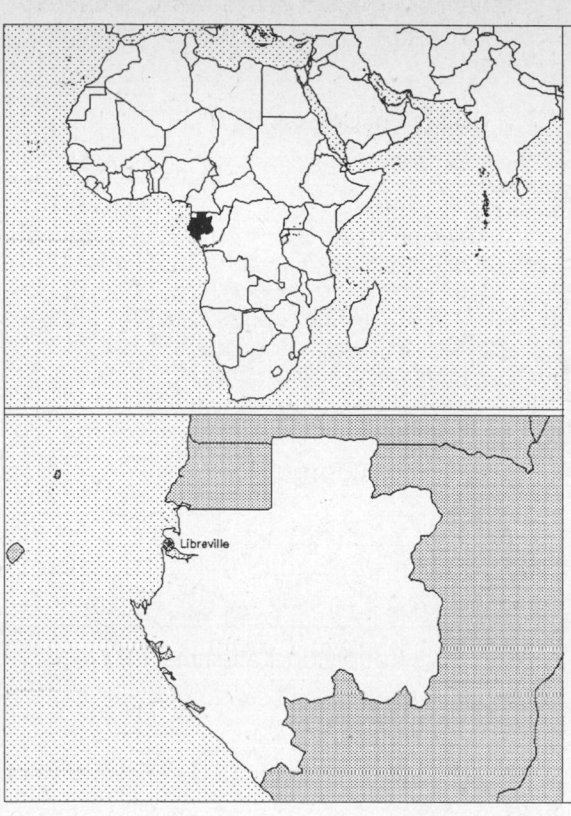

Geography [1]

Total area:
267,670 sq km   103,348 sq mi
Land area:
257,670 sq km   99,487 sq mi
Comparative area:
Slightly smaller than Colorado
Land boundaries:
Total 2,551 km, Cameroon 298 km, Congo 1,903 km, Equatorial Guinea 350 km
Coastline:
885 km
Climate:
Tropical; always hot, humid
Terrain:
Narrow coastal plain; hilly interior; savanna in East and South
Natural resources:
Petroleum, manganese, uranium, gold, timber, iron ore
Land use:
Arable land: 1%
Permanent crops: 1%
Meadows and pastures: 18%
Forest and woodland: 78%
Other: 2%

Demographics [2]

	1970	1980	1990	1995[1]	1996	2000	2010	2020	2030
Population	514	808	1,078	1,156	1,173	1,244	1,445	1,675	1,933
Population density (persons per sq. mi.)	5	8	11	12	12	13	15	17	19
(persons per sq. km.)	2	3	4	4	5	5	6	6	8
Net migration rate (per 1,000 population)	10.1	37.9	-5.6	0.0	0.0	0.0	0.0	0.0	0.0
Births	NA	NA	NA	NA	33	NA	NA	NA	NA
Deaths	NA	NA	NA	NA	16	NA	NA	NA	NA
Life expectancy - males	42.4	46.2	50.2	52.3	52.7	54.4	58.5	62.4	65.9
Life expectancy - females	45.5	50.4	55.5	58.1	58.6	60.6	65.5	69.9	73.7
Birth rate (per 1,000)	31.8	30.7	29.3	28.3	28.2	27.8	26.2	24.2	22.3
Death rate (per 1,000)	21.5	17.4	14.7	13.7	13.6	12.9	11.3	9.7	8.1
Women of reproductive age (15-49 yrs.)	128	202	259	271	274	288	351	428	500
of which are currently married	NA	NA	214	NA	226	238	292	NA	NA
Fertility rate	4.1	4.1	4.1	3.9	3.9	3.7	3.4	3.0	2.8

Except as noted, values for vital statistics are in thousands; life expectancy is in years.

Health

Health Indicators [3]

% of population with access to	
safe water (1990-95)[1]	68
adequate sanitation (1990-95)	NA
health services (1985-95)[1]	90
% of 1-year-olds immunized (1990-94) against	
TB (tuberculosis)	97
DPT (diphtheria, pertussis, tetanus)	66
polio	66
measles	65
% of contraceptive prevalence (1980-94)	NA
ORT use rate (1990-94)	25

Health Expenditures [4]

For sources, notes, and explanations, see Annotated Source Appendix, page 1061.

Human Factors

Women and Children [5]

% of pregnant women immunized (tetanus 1990-94)	NA
% of births attended by trained health personnel (1983-94)	80
Maternal mortality rate (1980-92)	190
Under-5 mortality rate (1994)	151
% under-5 moderately/severely underweight (1980-1994)	NA

Burden of Disease [6]

Population per physician (1988)	4,925.93
Population per nurse	NA
Population per hospital bed (1985)	790.37
AIDS cases per 100,000 people (1994)	15.3
Malaria cases per 100,000 people (1992)	NA

Ethnic Division [7]

Bantu tribes including four major tribal groupings (Fang, Eshira, Bapounou, Bateke), other Africans and Europeans 100,000, including 27,000 French.

Religion [8]

Other is primarily animist.

Christian	55%-75%
Muslim	< 1%
Other	25%-45%

Major Languages [9]

French (official), Fang, Myene, Bateke, Bapounou/Eschira, Bandjabi.

Education

Public Education Expenditures [10]

Million (Franc C.F.A.)[27]	1980	1985	1987	1990	1992	1994
Total education expenditure	22,204	69,500	53,372	NA	41,529	NA
as percent of GNP	2.7	4.5	5.7	NA	3.2	NA
as percent of total govt. expend.	NA	9.4	NA	NA	NA	NA
Current education expenditure	16,055	47,500	47,774	NA	34,407	NA
as percent of GNP	2.0	3.1	5.1	NA	2.7	NA
as percent of current govt. expend.	NA	21.7	NA	NA	NA	NA
Capital expenditure	6,149	22,000	5,598	NA	7,122	NA

Educational Attainment [11]

Illiteracy [12]

In thousands and percent[1]	1990	1995	2000
Illiterate population (15+ yrs.)	320	295	263
Illiteracy rate - total pop. (%)	45.0	38.5	31.8
Illiteracy rate - males (%)	31.4	26.5	21.4
Illiteracy rate - females (%)	59.2	50.8	42.2

Libraries [13]

	Admin. Units	Svc. Pts.	Vols. (000)	Shelving (meters)	Vols. Added	Reg. Users
National (1992)	1	1	NA	NA	454	2,382
Nonspecialized	NA	NA	NA	NA	NA	NA
Public	NA	NA	NA	NA	NA	NA
Higher ed.	NA	NA	NA	NA	NA	NA
School	NA	NA	NA	NA	NA	NA

Daily Newspapers [14]

	1980	1985	1990	1994
Number of papers	1	1	1	1
Circ. (000)	15	20	20[e]	20[e]

Culture [15]

Cinema (seats per 1,000)	2.5
Annual attendance per person	1.6
Gross box office receipts (mil. Franc C.F.A.)	104
Museums (reporting)	NA
Visitors (000)	NA
Annual receipts (000 Franc C.F.A.)	NA

Science and Technology

Scientific/Technical Forces [16]

Scientists/engineers	199
Number female	NA
Technicians	18
Number female	NA
Total[1,13]	217

R&D Expenditures [17]

	(000) 1986
Total expenditure[1]	380,000
Capital expenditure	130,000
Current expenditure	250,000
Percent current	65.8

U.S. Patents Issued [18]

For sources, notes, and explanations, see Annotated Source Appendix, page 1061.

Government and Law

Organization of Government [19]

Long-form name:
Gabonese Republic
Type:
Republic; multiparty presidential regime
(opposition parties legalized 1990)
Independence:
17 August 1960 (from France)
National holiday:
Renovation Day, 12 March (1968)
(Gabonese Democratic Party established)
Constitution:
Adopted 14 March 1991
Legal system:
Based on French civil law system and
customary law; judicial review of
legislative acts in Constitutional Chamber
of the Supreme Court; compulsory ICJ
jurisdiction not accepted
Executive branch:
President; Prime Minister; Council of
Ministers
Legislative branch:
Unicameral: National Assembly
(Assemblee Nationale) Note: The
provision of the constitution for the
establishment of a senate has not been
implemented
Judicial branch:
Supreme Court (Cour Supreme)

Elections [20]

National Assembly	% of seats
Gabonese Democratic Party (PDG)	53.3
Morena-Bucherons/RNB	14.2
Independents	12.5
Gabonese Party for Progress (PGP)	10.0
African Forum for Reconstruction	3.3
People's Unity Party (PUP)	3.3
National Recovery	1.7
Circle of Liberal Reformers (CLR)	0.8
Gabonese Peoples Union (UPG)	0.8
Circle for Renewal	0.8

Government Budget [21]

For 1993 est.

	$ bil.
Revenues	1.3
Expenditures	1.6
Capital expenditures	0.311

Crime [23]

	1990
Crime volume	
Cases known to police	3,231
Attempts (percent)	176
Percent cases solved	1,759
Crimes per 100,000 persons	323.1
Persons responsible for offenses	
Total number offenders	2,180
Percent female	143
Percent juvenile (16-18 yrs.)	49
Percent foreigners	1,011

Military Expenditures and Arms Transfers [22]

	1990	1991	1992	1993	1994
Military expenditures					
Current dollars (mil.)[e]	122	NA	122	132	93
1994 constant dollars (mil.)[e]	136	NA	127	135	93
Armed forces (000)	9	10	7	7	6
Gross national product (GNP)					
Current dollars (mil.)	3,621	4,072	3,895	4,106	3,239
1994 constant dollars (mil.)	4,030	4,365	4,062	4,191	3,239
Central government expenditures (CGE)					
1994 constant dollars (mil.)	998	1,499	1,254[e]	1,403[e]	NA
People (mil.)	1.1	1.1	1.1	1.1	1.1
Military expenditure as % of GNP	3.4	NA	3.1	3.2	2.9
Military expenditure as % of CGE	13.7	NA	10.1	9.6	NA
Military expenditure per capita (1994 $)	126	NA	115	120	82
Armed forces per 1,000 people (soldiers)	8.4	9.2	6.3	6.2	5.3
GNP per capita (1994 $)	3,740	4,003	3,671	3,733	2,844
Arms imports[6]					
Current dollars (mil.)	0	0	0	10	10
1994 constant dollars (mil.)	0	0	0	10	10
Arms exports[6]					
Current dollars (mil.)	0	0	0	0	0
1994 constant dollars (mil.)	0	0	0	0	0
Total imports[7]					
Current dollars (mil.)	772	884	886	835	912[e]
1994 constant dollars (mil.)	859	948	924	852	912[e]
Total exports[7]					
Current dollars (mil.)	2,464	2,273	2,295	2,177	NA
1994 constant dollars (mil.)	2,742	2,436	2,393	2,222	NA
Arms as percent of total Imports[8]	0	0	0	1.2	1.1
Arms as percent of total exports[8]	0	0	0	0	0

Human Rights [24]

	SSTS	FL	FAPRO	PPCG	APROBC	TPW	PCPTW	STPEP	PHRFF	PRW	ASST	AFL
Observes		P	P	P	P	P	P			P		P
	EAFRD	CPR	ESCR	SR	ACHR	MAAE	PVIAC	PVNAC	EAFDAW	TCIDTP	RC	
Observes	P	P	P	P			P	P	P	S	S	

P = Party; S = Signatory; see Appendix for meaning of abbreviations.

Labor Force

Total Labor Force [25]

120,000 salaried

Labor Force by Occupation [26]

Agriculture	65.0%
Industry and commerce	30.0
Services	2.5
Government	2.5

Unemployment Rate [27]

10%-14% (1993 est.)

For sources, notes, and explanations, see Annotated Source Appendix, page 1061.

Production Sectors

Commercial Energy Production and Consumption

Data are shown in quadrillion (10^{15}) BTUs and percent for 1995
Values for hydroelectric, nuclear, geothermal, solar, and wind power refer to electrical generation.

Production [28]

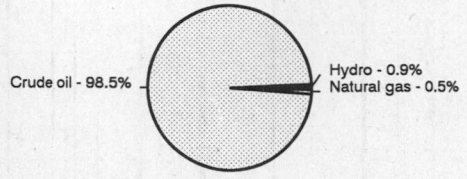

Crude oil - 98.5%
Hydro - 0.9%
Natural gas - 0.5%

Consumption [29]

Crude oil - 78.8%
Natural gas - 7.7%
Hydro - 13.5%

Crude oil	0.741
Dry natural gas	0.004
Net hydroelectric power	0.007
Total	0.752

Crude oil	0.041
Dry natural gas	0.004
Net hydroelectric power	0.007
Total	0.052

Telecommunications [30]

- 22,000 (1991 est.) telephones
- Domestic: adequate system of cable, microwave radio relay, tropospheric scatter, radiotelephone communication stations, and a domestic satellite system with 12 earth stations
- International: satellite earth stations - 3 Intelsat (Atlantic Ocean)
- Radio: Broadcast stations: AM 6, FM 6, shortwave 0 Radios: 250,000 (1993 est.)
- Television: Broadcast stations: 3 (repeaters 5) Televisions: 40,000 (1993 est.)

Top Agricultural Products [32]

Agriculture accounts for 8.2% of the GDP; produces cocoa, coffee, sugar, palm oil; rubber; okoume (a tropical softwood); cattle; small fishing operations (provide a catch of about 20,000 metric tons).

Transportation [31]

Railways: total: 649 km Gabon State Railways (OCTRA); standard gauge: 649 km 1.435-m gauge; single track (1994)

Highways: total: 7,456 km; paved: 560 km; unpaved: 6,896 km (1988 est.)

Merchant marine: total: 3 bulk (1,000 GRT or over) totaling 36,976 GRT/60,319 DWT (1995 est.)

Airports
Total:	54
With paved runways over 3,047 m:	1
With paved runways 2,438 to 3,047 m:	1
With paved runways 1,524 to 2,437 m:	7
With paved runways 914 to 1,523 m:	1
With paved runways under 914 m:	21

Top Mining Products [33]

Metric tons except as noted[e]	6/12/94*
Cement	
clinker	147,000[r]
hydraulic	126,000[30]
Diamond (carats)	500
Gas, natural, gross (mil. cu. meters)	3,000[31]
Gold, mine output, Au content (kg.)	72[r,32]
Manganese	1,540,000
Petroleum, crude (000 42-gal. bls.)	120,000
Uranium oxide (U3O8), content of concentrate	650[r]

Tourism [34]

	1990	1991	1992	1993	1994
Tourists	108	128	133	115	NA
Tourism receipts	3	4	5	4	5
Tourism expenditures	137	112	143	154	143
Fare receipts	30	97	96	79	70
Fare expenditures	42	99	97	74	68

Travelers are in thousands, money in million U.S. dollars.

For sources, notes, and explanations, see Annotated Source Appendix, page 1061.

Manufacturing Sector

GDP and Manufacturing Summary [35]

	1980	1985	1989	1990	% change 1980-1990	% change 1989-1990
GDP (million 1980 $)	4,281	4,459	3,989	3,076	-28.1	-22.9
GDP per capita (1980 $)	5,305	4,522	3,521	2,622	-50.6	-25.5
Manufacturing as % of GDP (current prices)	5.1	5.6	8.6[e]	5.1	0.0	-40.7
Gross output (million $)	690	615[e]	978[e]	843[e]	22.2	-13.8
Value added (million $)	224	182[e]	275[e]	268[e]	19.6	-2.5
Value added (million 1980 $)	239	245	NA	152	-36.4	NA
Industrial production index	100	102	94	90	-10.0	-4.3
Employment (thousands)	18[e]	18[e]	17[e]	15[e]	-16.7	-11.8

Note: GDP stands for Gross Domestic Product. 'e' stands for estimated value.

Profitability and Productivity

	1980	1985	1989	1990	% change 1980-1990	% change 1989-1990
Intermediate input (%)	68[e]	70[e]	72[e]	68[e]	0.0	-5.6
Wages, salaries, and supplements (%)	16[e]	17[e]	17[e]	19[e]	18.8	11.8
Gross operating surplus (%)	16[e]	13[e]	12[e]	13[e]	-18.8	8.3
Gross output per worker ($)	38,481[e]	34,305[e]	57,509[e]	53,905[e]	40.1	-6.3
Value added per worker ($)	12,470[e]	10,360[e]	16,189[e]	17,264[e]	38.4	6.6
Average wage (incl. benefits) ($)	6,283[e]	5,783[e]	9,520[e]	10,333[e]	64.5	8.5

Profitability is in percent of gross output. Productivity is in U.S. $. 'e' stands for estimated value.

Profitability - 1990

Inputs - 68.0%
Surplus - 13.0%
Wages - 19.0%

The graphic shows percent of gross output.

Value Added in Manufacturing

	1980 $ mil.	1980 %	1985 $ mil.	1985 %	1989 $ mil.	1989 %	1990 $ mil.	1990 %	% change 1980-1990	% change 1989-1990
311 Food products	18[e]	8.0	17[e]	9.3	31[e]	11.3	26[e]	9.7	44.4	-16.1
313 Beverages	19	8.5	13[e]	7.1	14[e]	5.1	20[e]	7.5	5.3	42.9
314 Tobacco products	17	7.6	12[e]	6.6	12[e]	4.4	17[e]	6.3	0.0	41.7
321 Textiles	3	1.3	2[e]	1.1	1[e]	0.4	3[e]	1.1	0.0	200.0
322 Wearing apparel	5	2.2	3[e]	1.6	3[e]	1.1	5[e]	1.9	0.0	66.7
323 Leather and fur products	1	0.4	NA	0.0	NA	0.0	1[e]	0.4	0.0	NA
324 Footwear	1	0.4	NA	0.0	NA	0.0	1[e]	0.4	0.0	NA
331 Wood and wood products	64	28.6	36[e]	19.8	27[e]	9.8	53[e]	19.8	-17.2	96.3
332 Furniture and fixtures	9	4.0	5[e]	2.7	4[e]	1.5	7[e]	2.6	-22.2	75.0
341 Paper and paper products	2	0.9	1[e]	0.5	3[e]	1.1	2[e]	0.7	0.0	-33.3
342 Printing and publishing	3	1.3	3[e]	1.6	3[e]	1.1	4[e]	1.5	33.3	33.3
351 Industrial chemicals	6	2.7	6[e]	3.3	6[e]	2.2	7[e]	2.6	16.7	16.7
352 Other chemical products	3	1.3	2[e]	1.1	6[e]	2.2	3[e]	1.1	0.0	-50.0
353 Petroleum refineries	18	8.0	15[e]	8.2	40[e]	14.5	31[e]	11.6	72.2	-22.5
354 Miscellaneous petroleum and coal products	NA	0.0	NA	0.0	NA	0.0	NA	0.0	NA	NA
355 Rubber products	NA	0.0	NA	0.0	NA	0.0	NA	0.0	NA	NA
356 Plastic products	NA	0.0	NA	0.0	NA	0.0	NA	0.0	NA	NA
361 Pottery, china, and earthenware	NA	0.0	NA	0.0	NA	0.0	NA	0.0	NA	NA
362 Glass and glass products	1	0.4	2[e]	1.1	2[e]	0.7	3[e]	1.1	200.0	50.0
369 Other nonmetal mineral products	8	3.6	14[e]	7.7	30[e]	10.9	17[e]	6.3	112.5	-43.3
371 Iron and steel	3	1.3	3[e]	1.6	5[e]	1.8	4[e]	1.5	33.3	-20.0
372 Nonferrous metals	3	1.3	3[e]	1.6	5[e]	1.8	4[e]	1.5	33.3	-20.0
381 Metal products	13	5.8	15[e]	8.2	25[e]	9.1	20[e]	7.5	53.8	-20.0
382 Nonelectrical machinery	2	0.9	2[e]	1.1	4[e]	1.5	3[e]	1.1	50.0	-25.0
383 Electrical machinery	8	3.6	9[e]	4.9	21[e]	7.6	12[e]	4.5	50.0	-42.9
384 Transport equipment	11	4.9	12[e]	6.6	20[e]	7.3	17[e]	6.3	54.5	-15.0
385 Professional and scientific equipment	1	0.4	1[e]	0.5	2[e]	0.7	1[e]	0.4	0.0	-50.0
390 Other manufacturing industries	5	2.2	5[e]	2.7	9[e]	3.3	7[e]	2.6	40.0	-22.2

Note: The industry codes shown are International Standard Industry codes (ISIC). Percentages are percent of total Value Added. 'e' stands for estimated value

For sources, notes, and explanations, see Annotated Source Appendix, page 1061.

341

Finance, Economics, and Trade

Economic Indicators [36]

- **National product**: GDP—purchasing power parity—$6 billion (1995 est.)
- **National product real growth rate**: 2% (1995 est.)
- **National product per capita**: $5,200 (1995 est.)
- **Inflation rate (consumer prices)**: 15% (1995 est.)
- **External debt**: $3.8 billion (1993)

Balance of Payments Summary [37]

Values in millions of dollars.

	1989	1990	1991	1992	1993
Exports of goods (f.o.b.)	1,626.0	2,488.8	2,227.9	2,259.2	2,149.7
Imports of goods (f.o.b.)	-751.7	-805.1	-861.0	-886.3	-845.1
Trade balance	874.3	1,683.7	1,366.9	1,372.9	1,304.6
Services - debits	-1,248.6	-1,643.3	-1,524.3	-1,793.8	-1,765.8
Services - credits	308.0	261.7	352.0	394.8	318.2
Private transfers (net)	-135.2	-158.5	-125.5	-154.1	-140.9
Government transfers (net)	9.3	24.2	5.7	12.1	15.2
Long-term capital (net)	258.4	-260.2	-305.9	-218.0	-149.7
Short-term capital (net)	-145.2	189.5	61.7	37.8	387.4
Errors and omissions	35.0	-38.0	8.6	-55.1	-5.8
Overall balance	-44.0	59.1	-160.8	-403.4	-36.8

Exchange Rates [38]

Currency: **Communaute Financi- ere Africaine franc.**
Symbol: **CFAF.**

Data are currency units per $1.

January 1996	500.56
1995	499.15
1994	555.20
1993	283.16
1992	264.69
1991	282.11

Imports and Exports

Top Import Origins [39]

$800 million (f.o.b., 1994 est.).

Origins	%
France	35
African countries	NA
US	NA
Japan	NA
Netherlands 1994	NA

Top Export Destinations [40]

$2.1 billion (f.o.b., 1994 est.) Data are for 1994 est.

Destinations	%
US	50
France	16
Japan	8
Spain	6
Germany	NA

Foreign Aid [41]

Recipient: ODA, $75 million (1993).

Import and Export Commodities [42]

Import Commodities	Export Commodities
Foodstuffs	Crude oil 80%
Chemical products	Timber 14%
Petroleum products	Manganese 6%
Construction materials	Uranium
Manufactures	
Machinery	

The Gambia

Geography [1]

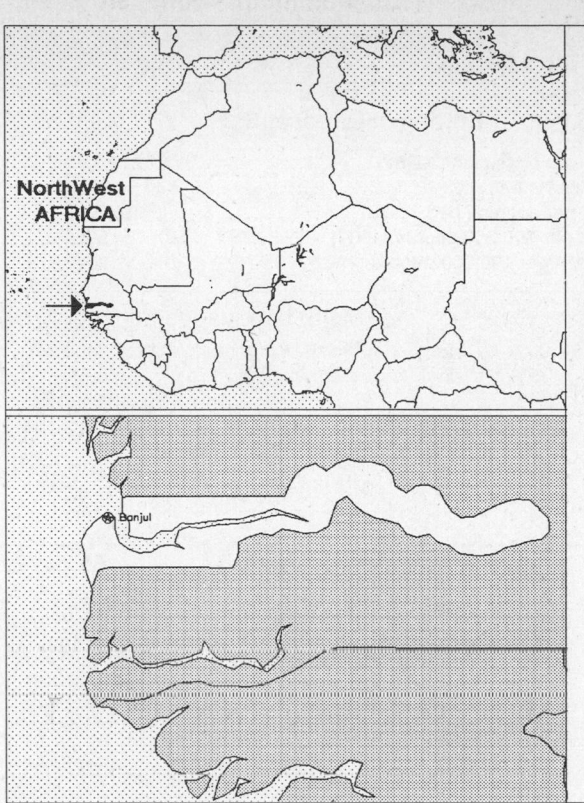

Total area:
 11,300 sq km   4,363 sq mi
Land area:
 10,000 sq km   3,861 sq mi
Comparative area:
 Slightly more than twice the size of Delaware
Land boundaries:
 Total 740 km, Senegal 740 km
Coastline:
 80 km
Climate:
 Tropical; hot, rainy season (June to November); cooler, dry season
 (November to May)
Terrain:
 Flood plain of the Gambia River flanked by some low hills
Natural resources:
 Fish
Land use:
 Arable land: 16%
 Permanent crops: 0%
 Meadows and pastures: 9%
 Forest and woodland: 20%
 Other: 55%

Demographics [2]

	1970	1980	1990	1995[1]	1996	2000	2010	2020	2030
Population	502	676	964	1,163	1,205	1,381	1,864	2,399	2,958
Population density (persons per sq. mi.)	130	175	250	301	312	358	483	621	766
(persons per sq. km.)	50	68	96	116	120	138	186	240	296
Net migration rate (per 1,000 population)	NA	NA	7.2	5.2	4.7	2.9	0.0	0.0	0.0
Births	NA	NA	NA	NA	46	NA	NA	NA	NA
Deaths	NA	NA	NA	NA	15	NA	NA	NA	NA
Life expectancy - males	NA	NA	48.2	50.3	50.7	52.5	56.8	60.9	64.7
Life expectancy - females	NA	NA	52.1	54.7	55.2	57.4	62.6	67.4	71.7
Birth rate (per 1,000)	NA	NA	47.4	45.0	44.4	42.2	36.5	30.5	25.0
Death rate (per 1,000)	NA	NA	16.2	14.0	13.7	12.2	9.3	7.4	6.2
Women of reproductive age (15-49 yrs.)	NA	NA	221	266	275	315	439	600	788
of which are currently married	NA	NA	158	NA	189	215	298	NA	NA
Fertility rate	NA	NA	6.4	6.1	6.1	5.8	4.8	3.7	2.9

Except as noted, values for vital statistics are in thousands; life expectancy is in years.

Health

Health Indicators [3]

% of population with access to	
safe water (1990-95)	48
adequate sanitation (1990-95)	38
health services (1985-95)	93
% of 1-year-olds immunized (1990-94) against	
TB (tuberculosis)	98
DPT (diphtheria, pertussis, tetanus)	90
polio	92
measles	87
% of contraceptive prevalence (1980-94)	NA
ORT use rate (1990-94)	51

Health Expenditures [4]

For sources, notes, and explanations, see Annotated Source Appendix, page 1061.

343

Human Factors

Women and Children [5]

% of pregnant women immunized (tetanus 1990-94)	93
% of births attended by trained health personnel (1983-94)	80
Maternal mortality rate (1980-92)	1,050
Under-5 mortality rate (1994)	213
% under-5 moderately/severely underweight (1980-1994)	NA

Burden of Disease [6]

Population per physician (1985)	12,213.12
Population per nurse	NA
Population per hospital bed (1985)	597.59
AIDS cases per 100,000 people (1994)	5.5
Malaria cases per 100,000 people (1992)	NA

Ethnic Division [7]

African	99%
Mandinka	42%
Fula	18%
Wolof	16%
Jola	10%
Serahuli	9%,
Other	4%
Non-Gambian	1%

Religion [8]

Muslim	90%
Christian	9%
Indigenous beliefs	1%

Major Languages [9]

English (official), Mandinka, Wolof, Fula, other indigenous vernaculars.

Education

Public Education Expenditures [10]

Million (Dalasi)	1980	1985	1989	1990	1991	1994
Total education expenditure	13	31	NA	95	78	NA
as percent of GNP	3.3	3.2	NA	3.8	2.7	NA
as percent of total govt. expend.	NA	NA	NA	11.0	12.9	NA
Current education expenditure	11	25	61	73	75	NA
as percent of GNP	2.9	2.6	2.8	2.9	2.6	NA
as percent of current govt. expend.	NA	16.4	NA	11.6	13.2	NA
Capital expenditure	2	6	NA	22	3	NA

Educational Attainment [11]

Illiteracy [12]

In thousands and percent[1]	1990	1995	2000
Illiterate population (15+ yrs.)	355	403	429
Illiteracy rate - total pop. (%)	68.1	64.1	57.1
Illiteracy rate - males (%)	52.9	48.6	42.4
Illiteracy rate - females (%)	82.8	79.4	72.0

Libraries [13]

	Admin. Units	Svc. Pts.	Vols. (000)	Shelving (meters)	Vols. Added	Reg. Users
National (1992)	1	5	2	220	300	1,350
Nonspecialized	NA	NA	NA	NA	NA	NA
Public (1992)	2	6	94	NA	4,900	NA
Higher ed.	NA	NA	NA	NA	NA	NA
School	NA	NA	NA	NA	NA	NA

Daily Newspapers [14]

	1980	1985	1990	1994
Number of papers	-	6	2	2
Circ. (000)	-	4	2[e]	2[e]

Culture [15]

Science and Technology

Scientific/Technical Forces [16]

R&D Expenditures [17]

U.S. Patents Issued [18]

Government and Law

Organization of Government [19]

Long-form name:
Republic of The Gambia
Type:
Republic under multiparty democratic rule
Note: The Gambia has had a military government since 22 July 1994;
Independence:
18 February 1965 (from UK)
National holiday:
Independence Day, 18 February (1965)
Constitution:
24 April 1970; suspended July 1994
Legal system:
Based on a composite of English common law, Koranic law, and customary law; accepts compulsory ICJ jurisdiction, with reservations
Executive branch:
Chairman of the Armed Forces Provisional Ruling Council; Vice Chairman of the Armed Forces Provisional Ruling Council; Cabinet
Legislative branch:
Unicameral: House of Representatives
Note: Following the military coup on 22 July 1994, all elective offices were dissolved.
Judicial branch:
Supreme Court

Elections [20]

House of Representatives	% of votes
People's Progressive Party (PPP)	58.1
Other	41.9

Government Expenditures [21]

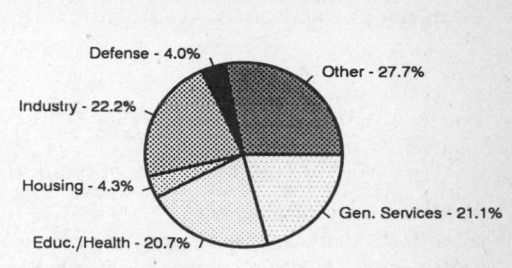

(% distribution). Expend. for 1990: 589.6 (Dalasi mil.)

Defense - 4.0%, Other - 27.7%, Industry - 22.2%, Housing - 4.3%, Educ./Health - 20.7%, Gen. Services - 21.1%

Crime [23]

	1994
Crime volume	
Cases known to police	960
Attempts (percent)	NA
Percent cases solved	NA
Crimes per 100,000 persons	88.76
Persons responsible for offenses	
Total number offenders	1,381
Percent female	NA
Percent juvenile (14-17 yrs.)	NA
Percent foreigners	NA

Military Expenditures and Arms Transfers [22]

	1990	1991	1992	1993	1994
Military expenditures					
Current dollars (mil.)	3	NA	11	13	14
1994 constant dollars (mil.)	3	NA	12	13	14
Armed forces (000)	2	2	1	1	1
Gross national product (GNP)					
Current dollars (mil.)	298	338	338	363	369
1994 constant dollars (mil.)	332	362	353	370	369
Central government expenditures (CGE)					
1994 constant dollars (mil.)	69	84	60[e]	75[e]	NA
People (mil.)	0.8	0.9	0.9	0.9	1.0
Military expenditure as % of GNP	0.9	NA	3.4	3.5	3.7
Military expenditure as % of CGE	4.5	NA	19.7	17.4	NA
Military expenditure per capita (1994 $)	4	NA	13	14	14
Armed forces per 1,000 people (soldiers)	2.4	2.3	1.1	1.1	1.0
GNP per capita (1994 $)	391	414	391	398	385
Arms imports[6]					
Current dollars (mil.)	0	5	5	0	0
1994 constant dollars (mil.)	0	5	5	0	0
Arms exports[6]					
Current dollars (mil.)	0	0	0	0	0
1994 constant dollars (mil.)	0	0	0	0	0
Total imports[7]					
Current dollars (mil.)	200	222	234	NA	209
1994 constant dollars (mil.)	222	238	244	NA	209
Total exports[7]					
Current dollars (mil.)	40	42	63	NA	35
1994 constant dollars (mil.)	45	45	66	NA	35
Arms as percent of total imports[8]	0	2.3	2.1	0	0
Arms as percent of total exports[8]	0	0	0	0	0

Human Rights [24]

	SSTS	FL	FAPRO	PPCG	APROBC	TPW	PCPTW	STPEP	PHRFF	PRW	ASST	AFL
Observes	1			P		P	P				1	
	EAFRD	CPR	ESCR	SR	ACHR	MAAE	PVIAC	PVNAC	EAFDAW	TCIDTP	RC
Observes	P	P	P	P			P	P	P	S	P

P=Party; S=Signatory; see Appendix for meaning of abbreviations.

Labor Force

Total Labor Force [25]

400,000 (1986 est.)

Labor Force by Occupation [26]

Agriculture	75.0%
Industry, commerce, and services	18.9
Government	6.1

Unemployment Rate [27]

Production Sectors

Energy Resource Summary [28]

Electricity: Capacity: 30,000 kW. Production: 70 million kWh. Consumption per capita: 64 kWh (1993).

Telecommunications [30]

- 11,000 (1991 est.) telephones
- Domestic: adequate network of microwave radio relay and open wire
- International: microwave radio relay links to Senegal and Guinea-Bissau; satellite earth station - 1 Intelsat (Atlantic Ocean)
- Radio: Broadcast stations: AM 3, FM 2, shortwave 0 Radios: 180,000 (1993 est.)
- Television:

Transportation [31]

Railways: 0 km

Highways: total: 2,386 km; paved: 764 km; unpaved: 1,622 km (1990 est.)

Merchant marine: none

Airports
Total: 1
With paved runways over 3,047 m: 1 (1995 est.)

Top Agricultural Products [32]

Agriculture accounts for 27% of the GDP; produces peanuts, millet, sorghum, rice, corn, cassava (tapioca), palm kernels; cattle, sheep, goats; forest and fishing resources not fully exploited.

Top Mining Products [33]

Detailed information is not available. A summary of mineral resources follows. **Mineral Resources:** Clays for bricks, laterite, sand and gravel, glass sand, and titaniferous sands.

Tourism [34]

	1990	1991	1992	1993	1994
Visitors	102	NA	NA	NA	NA
Tourists[39]	100	66	64	90	78
Cruise passengers[40]	2	NA	NA	NA	NA
Tourism receipts	26	30	27	26	27
Tourism expenditures	8	15	13	14	14
Fare expenditures	2	2	2	2	2

Travelers are in thousands, money in million U.S. dollars.

Manufacturing Sector

GDP and Manufacturing Summary [35]

	1980	1985	1989	1990	% change 1980-1990	% change 1989-1990
GDP (million 1980 $)	239	320	NA	376	57.3	NA
GDP per capita (1980 $)	374	430	NA	436	16.6	NA
Manufacturing as % of GDP (current prices)	3.6	7.7	NA	5.8	61.1	NA
Gross output (million $)	30	40[e]	NA	52[e]	73.3	NA
Value added (million $)	11	9[e]	NA	13[e]	18.2	NA
Value added (million 1980 $)	16	23	NA	28	75.0	NA
Industrial production index	100	107	NA	125	25.0	NA
Employment (thousands)	2	3[e]	NA	2[e]	0.0	NA

Note: GDP stands for Gross Domestic Product. 'e' stands for estimated value.

Profitability and Productivity

	1980	1985	1989	1990	% change 1980-1990	% change 1989-1990
Intermediate input (%)	62	78[e]	NA	76[e]	22.6	NA
Wages, salaries, and supplements (%)	10	7[e]	NA	8[e]	-20.0	NA
Gross operating surplus (%)	28	14[e]	NA	17[e]	-39.3	NA
Gross output per worker ($)	16,115	13,431[e]	NA	15,916[e]	-1.2	NA
Value added per worker ($)	6,094	3,052[e]	NA	4,230[e]	-30.6	NA
Average wage (incl. benefits) ($)	1,566	1,111[e]	NA	1,628[e]	4.0	NA

Profitability is in percent of gross output. Productivity is in U.S. $. 'e' stands for estimated value.

Profitability - 1990

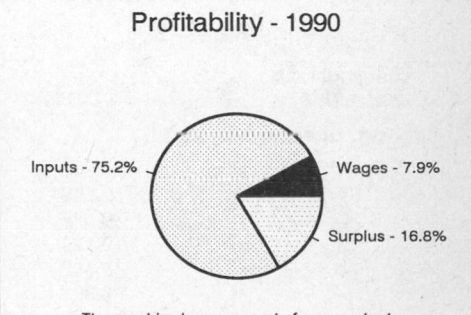

Inputs - 75.2%
Wages - 7.9%
Surplus - 16.8%

The graphic shows percent of gross output.

Value Added in Manufacturing

	1980 $ mil.	1980 %	1985 $ mil.	1985 %	1989 $ mil.	1989 %	1990 $ mil.	1990 %	% change 1980-1990	% change 1989-1990
311 Food products	3	27.3	4[e]	44.4	NA	NA	5[e]	38.5	66.7	NA
313 Beverages	1	9.1	1[e]	11.1	NA	NA	2[e]	15.4	100.0	NA
314 Tobacco products	NA	0.0	NA	0.0	NA	NA	NA	0.0	NA	NA
321 Textiles	NA	0.0	NA	0.0	NA	NA	NA	0.0	NA	NA
322 Wearing apparel	NA	0.0	NA	0.0	NA	NA	NA	0.0	NA	NA
323 Leather and fur products	NA	0.0	NA	0.0	NA	NA	NA	0.0	NA	NA
324 Footwear	NA	0.0	NA	0.0	NA	NA	NA	0.0	NA	NA
331 Wood and wood products	NA	0.0	NA	0.0	NA	NA	NA	0.0	NA	NA
332 Furniture and fixtures	1	9.1	NA	0.0	NA	NA	1[e]	7.7	0.0	NA
341 Paper and paper products	NA	0.0	NA	0.0	NA	NA	NA	0.0	NA	NA
342 Printing and publishing	NA	0.0	NA	0.0	NA	NA	NA	0.0	NA	NA
351 Industrial chemicals	NA	0.0	NA	0.0	NA	NA	NA	0.0	NA	NA
352 Other chemical products	NA	0.0	NA	0.0	NA	NA	NA	0.0	NA	NA
353 Petroleum refineries	NA	0.0	NA	0.0	NA	NA	NA	0.0	NA	NA
354 Miscellaneous petroleum and coal products	NA	0.0	NA	0.0	NA	NA	NA	0.0	NA	NA
355 Rubber products	NA	0.0	NA	0.0	NA	NA	NA	0.0	NA	NA
356 Plastic products	NA	0.0	NA	0.0	NA	NA	NA	0.0	NA	NA
361 Pottery, china, and earthenware	NA	0.0	NA	0.0	NA	NA	NA	0.0	NA	NA
362 Glass and glass products	NA	0.0	NA	0.0	NA	NA	NA	0.0	NA	NA
369 Other nonmetal mineral products	NA	0.0	NA	0.0	NA	NA	NA	0.0	NA	NA
371 Iron and steel	NA	0.0	NA	0.0	NA	NA	NA	0.0	NA	NA
372 Nonferrous metals	NA	0.0	NA	0.0	NA	NA	NA	0.0	NA	NA
381 Metal products	NA	0.0	NA	0.0	NA	NA	NA	0.0	NA	NA
382 Nonelectrical machinery	NA	0.0	NA	0.0	NA	NA	NA	0.0	NA	NA
383 Electrical machinery	NA	0.0	NA	0.0	NA	NA	NA	0.0	NA	NA
384 Transport equipment	NA	0.0	NA	0.0	NA	NA	NA	0.0	NA	NA
385 Professional and scientific equipment	NA	0.0	NA	0.0	NA	NA	NA	0.0	NA	NA
390 Other manufacturing industries	6	54.5	2[e]	22.2	NA	NA	3[e]	23.1	-50.0	NA

Note: The industry codes shown are International Standard Industry codes (ISIC). Percentages are percent of total Value Added. 'e' stands for estimated value

Finance, Economics, and Trade

Economic Indicators [36]

- **National product**: GDP—purchasing power parity—$1.1 billion (1995 est.)
- **National product real growth rate**: 2% (1995 est.)
- **National product per capita**: $1,100 (1995 est.)
- **Inflation rate (consumer prices)**: 1.7% (1994)
- **External debt**: $386 million (1993 est.)

Balance of Payments Summary [37]

Values in millions of dollars.

	1988	1989	1990	1991	1992
Exports of goods (f.o.b.)	83.1	100.2	110.6	142.9	147.0
Imports of goods (f.o.b.)	-105.9	-125.4	-140.5	-185.0	-177.8
Trade balance	-22.9	-25.1	-29.9	-42.1	-30.8
Services - debits	-61.4	-66.4	-64.9	-83.5	-74.2
Services - credits	63.7	67.7	71.4	84.3	85.7
Private transfers (net)	12.8	6.7	14.1	15.1	13.3
Government transfers (net)	41.9	33.9	45.0	43.0	46.1
Long-term capital (net)	11.5	12.9	6.7	28.5	32.7
Short-term capital (net)	-6.6	-16.2	-19.6	-1.3	-1.0
Errors and omissions	-11.3	-20.8	-24.0	-16.7	-36.7
Overall balance	27.7	-7.3	-1.3	27.4	35.1

Exchange Rates [38]

Currency: **dalasi.**
Symbol: **D.**

Data are currency units per $1.

August 1996	9.555
1994	9.576
1993	9.129
1992	8.888
1991	8.803

Imports and Exports

Top Import Origins [39]

$209 million (f.o.b., 1994 est.) Data are for 1989.

Origins	%
Europe	57
Asia	25
USSR and Eastern Europe	9
US	6
Other	3

Top Export Destinations [40]

$35 million (f.o.b., 1994 est.) Data are for 1989.

Destinations	%
Japan	60
Europe	29
Africa	5
US	1
Other	5

Foreign Aid [41]

Recipient: ODA, $NA.

Import and Export Commodities [42]

Import Commodities
Foodstuffs
Manufactures
Raw materials
Fuel
Machinery and transport equipment

Export Commodities
Peanuts and peanut products
Fish
Cotton lint
Palm kernels

For sources, notes, and explanations, see Annotated Source Appendix, page 1061.

Georgia

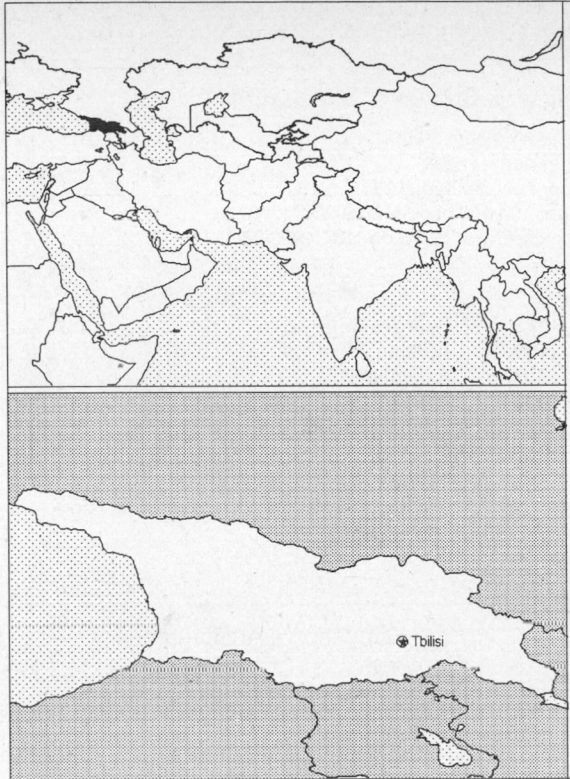

Geography [1]

Total area:
69,700 sq km   26,911 sq mi
Land area:
69,700 sq km   26,911 sq mi
Comparative area:
Slightly larger than South Carolina
Land boundaries:
Total 1,461 km, Armenia 164 km, Azerbaijan 322 km, Russia 723 km, Turkey 252 km
Coastline:
310 km
Climate:
Warm and pleasant; Mediterranean-like on Black Sea coast
Terrain:
Largely mountainous with Great Caucasus Mountains in the North and Lesser Caucasus Mountains in the South; Kolkhida Lowland opens to the Black Sea in the West; Mtkvari River Basin in the East; good soils in river valley flood plains, foothills of Kolkhida Lowland
Natural resources:
Forests, hydropower, manganese deposits, iron ore, copper, minor coal and oil deposits; coastal climate and soils allow for important tea and citrus growth
Land use:
Arable land: 11%
Permanent crops: 4%
Meadows and pastures: 29%
Forest and woodland: 38%
Other: 18%

Demographics [2]

	1970	1980	1990	1995[1]	1996	2000	2010	2020	2030
Population	4,694	5,048	5,457	5,281	5,220	5,132	5,188	5,205	5,301
Population density (persons per sq. mi.)	174	188	203	196	194	191	193	193	197
(persons per sq. km.)	67	72	78	76	75	74	74	75	76
Net migration rate (per 1,000 population)	NA	NA	-2.7	-12.8	-10.8	-2.5	-1.5	-0.5	0.0
Births	NA	NA	NA	NA	67	NA	NA	NA	NA
Deaths	NA	NA	NA	NA	64	NA	NA	NA	NA
Life expectancy - males	NA	NA	68.6	63.3	63.4	64.0	65.5	69.5	72.7
Life expectancy - females	NA	NA	76.1	72.9	73.0	73.4	74.5	77.6	80.1
Birth rate (per 1,000)	NA	NA	17.3	11.7	12.8	16.7	14.5	12.5	12.4
Death rate (per 1,000)	NA	NA	8.6	12.1	12.2	12.5	12.7	11.0	10.3
Women of reproductive age (15-49 yrs.)	NA	NA	1,350	1,350	1,348	1,343	1,288	1,254	1,237
of which are currently married	NA	NA	880	NA	887	881	867	NA	NA
Fertility rate	NA	NA	2.2	1.5	1.7	2.2	2.0	1.9	1.8

Except as noted, values for vital statistics are in thousands; life expectancy is in years.

Health

Health Indicators [3]

% of population with access to	
safe water (1990-95)	NA
adequate sanitation (1990-95)	NA
health services (1985-95)	NA
% of 1-year-olds immunized (1990-94) against	
TB (tuberculosis)	67
DPT (diphtheria, pertussis, tetanus)	58
polio	69
measles	16
% of contraceptive prevalence (1980-94)	NA
ORT use rate (1990-94)	NA

Health Expenditures [4]

For sources, notes, and explanations, see Annotated Source Appendix, page 1061.

Human Factors

Women and Children [5]

% of pregnant women immunized (tetanus 1990-94)	NA
% of births attended by trained health personnel (1983-94)	NA
Maternal mortality rate (1980-92)	NA
Under-5 mortality rate (1994)	27
% under-5 moderately/severely underweight (1980-1994)	NA

Burden of Disease [6]

Population per physician (1993)	181.91
Population per nurse (1990)	85.18
Population per hospital bed (1993)	95.25
AIDS cases per 100,000 people (1994)	*
Heart disease cases per 1,000 people (1990-93)	NA

Ethnic Division [7]

Georgian	70.1%
Armenian	8.1%
Russian	6.3%
Azeri	5.7%
Ossetian	3%
Abkhaz	1.8%
Other	5%

Religion [8]

Christian Orthodox	75%
Georgian Orthodox	65%
Russian Orthodox	10%
Muslim	11%
Armenian Apostolic	8%
Other	6%

Major Languages [9]

Georgian (official)	71%
Russian	9%
Armenian	7%
Azeri	6%
Other	7%

Education

Public Education Expenditures [10]

Million or Trillion (T) (Kupon)	1980	1985	1991	1992	1993	1994
Total education expenditure	NA	NA	NA	NA	NA	6.91 T
as percent of GNP	NA	NA	NA	NA	NA	1.9
as percent of total govt. expend.	NA	NA	NA	NA	NA	6.9
Current education expenditure	NA	NA	NA	NA	NA	5.73 T
as percent of GNP	NA	NA	NA	NA	NA	1.5
as percent of current govt. expend.	NA	NA	NA	NA	NA	7.5
Capital expenditure	NA	NA	NA	NA	NA	1.19 T

Educational Attainment [11]

Illiteracy [12]

	1985	1989	1995
Illiterate population (15+ yrs.)[2]	NA	NA	19,000
Illiteracy rate - total pop. (%)	NA	1.0	0.5
Illiteracy rate - males (%)	NA	0.5	0.3
Illiteracy rate - females (%)	NA	1.5	0.6

Libraries [13]

	Admin. Units	Svc. Pts.	Vols. (000)	Shelving (meters)	Vols. Added	Reg. Users
National (1992)	1	NA	7,524	90,351	37,498	462,285
Nonspecialized	NA	NA	NA	NA	NA	NA
Public (1992)	4,048	NA	32,319	NA	1 mil	3 mil
Higher ed.	NA	NA	NA	NA	NA	NA
School	NA	NA	NA	NA	NA	NA

Daily Newspapers [14]

Culture [15]

Cinema (seats per 1,000)	19.6
Annual attendance per person	5.6
Gross box office receipts (mil. Ruble)	91,093
Museums (reporting)	NA
Visitors (000)	NA
Annual receipts (000 Ruble)	NA

Science and Technology

Scientific/Technical Forces [16]

R&D Expenditures [17]

U.S. Patents Issued [18]

Values show patents issued to citizens of the country by the U.S. Patents Office.

	1993	1994	1995
Number of patents	0	0	1

For sources, notes, and explanations, see Annotated Source Appendix, page 1061.

Government and Law

Organization of Government [19]

Long-form name:
Republic of Georgia
Type:
Republic
Independence:
9 April 1991 (from Soviet Union)
National holiday:
Independence Day, 26 May (1991)
Constitution:
Adopted 17 October 1995
Legal system:
Based on civil law system
Executive branch:
President; Cabinet of Ministers
Legislative branch:
Unicameral: Georgian Parliament
Judicial branch:
Supreme Court

Elections [20]

Georgian Parliament	% of votes
Citizens Union of Georgia(CUG)	24.0
National democratic Party (NDP)	8.0
All Georgia Revival Union	7.0
Others (each)	<5.0

Government Budget [21]

Crime [23]

	1994
Crime volume	
Cases known to police	17,643
Attempts (percent)	NA
Percent cases solved	97.00
Crimes per 100,000 persons	324.95
Persons responsible for offenses	
Total number offenders	7,894
Percent female	NA
Percent juvenile	NA
Percent foreigners	NA

Military Expenditures and Arms Transfers [22]

	1990	1991	1992[14]	1993	1994
Military expenditures					
Current dollars (mil.)[e]	NA	NA	85	94	NA
1994 constant dollars (mil.)[e]	NA	NA	89	96	NA
Armed forces (000)	NA	NA	NA	35	7
Gross national product (GNP)					
Current dollars (mil.)[e]	NA	NA	10,160	7,693	6,426
1994 constant dollars (mil.)[e]	NA	NA	10,600	7,852	6,426
Central government expenditures (CGE)					
1994 constant dollars (mil.)[e]	NA	NA	3,709	NA	NA
People (mil.)	NA	NA	5.6	5.6	5.7
Military expenditure as % of GNP	NA	NA	0.8	1.2	NA
Military expenditure as % of CGE	NA	NA	2.4	NA	NA
Military expenditure per capita (1994 $)	NA	NA	16	17	NA
Armed forces per 1,000 people (soldiers)	NA	NA	NA	6.2	1.2
GNP per capita (1994 $)	NA	NA	1,898	1,394	1,131
Arms imports[6]					
Current dollars (mil.)	NA	NA	0	0	0
1994 constant dollars (mil.)	NA	NA	0	0	0
Arms exports[6]					
Current dollars (mil.)	NA	NA	0	0	0
1994 constant dollars (mil.)	NA	NA	0	0	0
Total imports[7]					
Current dollars (mil.)	NA	NA	NA	63	800
1994 constant dollars (mil.)	NA	NA	NA	64	800
Total exports[7]					
Current dollars (mil.)	NA	NA	NA	39	500
1994 constant dollars (mil.)	NA	NA	NA	40	500
Arms as percent of total imports[8]	NA	NA	0	0	0
Arms as percent of total exports[8]	NA	NA	0	0	0

Human Rights [24]

	SSTS	FL	FAPRO	PPCG	APROBC	TPW	PCPTW	STPEP	PHRFF	PRW	ASST	AFL
Observes				P		P	P					
	EAFRD	CPR	ESCR	SR	ACHR	MAAE	PVIAC	PVNAC	EAFDAW	TCIDTP	RC	
Observes		P	P				P	P	P	P	P	

P=Party; S=Signatory; see Appendix for meaning of abbreviations.

Labor Force

Total Labor Force [25]

2.763 million

Labor Force by Occupation [26]

Industry and construction	31%
Agriculture and forestry	25
Other	44

Date of data: 1990

Unemployment Rate [27]

Officially less than 5% but real unemployment may be more than 20%, with even larger numbers of underemployed workers.

For sources, notes, and explanations, see Annotated Source Appendix, page 1061.

351

Production Sectors

Commercial Energy Production and Consumption

Data are shown in quadrillion (10^{15}) BTUs and percent for 1995
Values for hydroelectric, nuclear, geothermal, solar, and wind power refer to electrical generation.

Production [28]

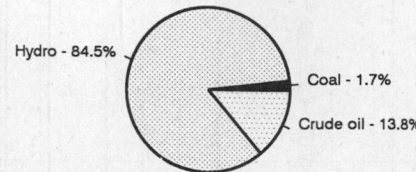

Hydro - 84.5%
Coal - 1.7%
Crude oil - 13.8%

Consumption [29]

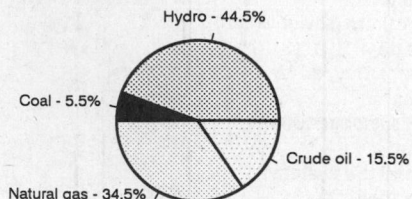

Hydro - 44.5%
Coal - 5.5%
Crude oil - 15.5%
Natural gas - 34.5%

Crude oil	0.008
Coal	0.001
Net hydroelectric power	0.049
Total	0.058

Crude oil	0.017
Dry natural gas	0.038
Coal	0.006
Net hydroelectric power	0.049
Total	0.110

Telecommunications [30]

- 672,000 (1993 est.) telephones; poor service; 339,000 unsatisfied applications for telephones (December 1990 est.)
- International: landline to CIS members and Turkey; satellite earth station - 1 Eutelsat; leased connections with other countries via the Moscow international gateway switch; international electronic mail and telex service available
- Radio: Broadcast stations: AM NA, FM NA, shortwave NA
- Television: Broadcast stations: 3

Transportation [31]

Railways: total: 1,570 km in common carrier service; does not include industrial lines; broad gauge: 1,570 km 1.520-m gauge (1990)

Highways: total: 35,100 km; paved: 31,200 km; unpaved: 3,900 km (1990 est.)

Merchant marine: total: 23 ships (1,000 GRT or over) totaling 307,765 GRT/483,567 DWT; ships by type: bulk 8, cargo 2, oil tanker 12, short-sea passenger 1 (1995 est.)

Airports

Total:	28
With paved runways over 3,047 m:	1
With paved runways 2,438 to 3,047 m:	7
With paved runways 1,524 to 2,437 m:	4
With paved runways 914 to 1,523 m:	1
With paved runways under 914 m:	1

Top Agricultural Products [32]

Agriculture accounts for 70.4% of the GDP; produces citrus, grapes, tea, vegetables, potatoes; small livestock sector.

Top Mining Products [33]

Metric tons except as noted[e]	6/16/95[*]
Barite	20,000
Bentonite	600,000
Cement	400,000
Coal	100,000
Diatomite	30,000
Steel, crude	116,000[r]
Manganese ore, marketable	700,000
Petroleum, crude	80,000

Tourism [34]

Finance, Economics, and Trade

Industrial Summary [35]

Industrial Production: Growth rate - 10% (1995); accounts for 10.2% of the GDP. **Industries**: Steel, aircraft, machine tools, foundry equipment, electric locomotives, tower cranes, electric welding equipment, machinery for food preparation and meat packing, electric motors, process control equipment, trucks, tractors, textiles, shoes, chemicals, wood products, wine.

Economic Indicators [36]

- **National product**: GDP—purchasing power parity—$6.2 billion (1995 estimate as extrapolated from World Bank estimate for 1994)
- **National product real growth rate**: - 11% (1995 est.)
- **National product per capita**: $1,080 (1995 est.)
- **Inflation rate (consumer prices)**: 2.2% monthly average (first half 1995 est.)
- **External debt**: $1.2 billion (of which $135 million to Russia) (1995 est.)

Balance of Payments Summary [37]

Values in millions of dollars.

	1989	1990	1991	1992[1]	1993
Exports of goods (f.o.b.)	NA	NA	NA	266	362
Imports of goods (f.o.b.)	NA	NA	NA	-644	-702
Trade balance	NA	NA	NA	-378	-340
Services - debits	NA	NA	NA	-79	-72
Services - credits	NA	NA	NA	137	116
Private transfers (net)	NA	NA	NA	71	127
Government transfers (net)	NA	NA	NA	NA	NA
Long-term capital (net)	NA	NA	NA	65	159
Short-term capital (net)	NA	NA	NA	NA	NA
Errors and omissions	NA	NA	NA	408	NA
Overall balance	NA	NA	NA	224	-10

Exchange Rates [38]

Currency: **lari.**

Data are currency units per $1. Lari introduced September 1995 replacing the coupon.

End December 1995	1.24

Imports and Exports

Top Import Origins [39]

$250 million (f.o.b., 1995). EU and US send humanitarian food shipments.

Origins	%
Russia	NA
Azerbaijan	NA
Turkey	NA

Top Export Destinations [40]

$140 million (c.i.f., 1995).

Destinations	%
Russia	NA
Turkey	NA
Armenia	NA
Azerbaijan	NA

Foreign Aid [41]

	U.S. $	
ODA (1993)	28	million
Commitments (1992-95)	1,200	million

Import and Export Commodities [42]

Import Commodities
Fuel
Grain and other foods
Machinery and parts
Transport equipment

Export Commodities
Citrus fruits
Tea
Wine
Other agricultural products
Machinery
Ferrous and nonferrous metals
Textiles
Chemicals
Fuel re-exports

For sources, notes, and explanations, see Annotated Source Appendix, page 1061.

353

Germany

Geography [1]

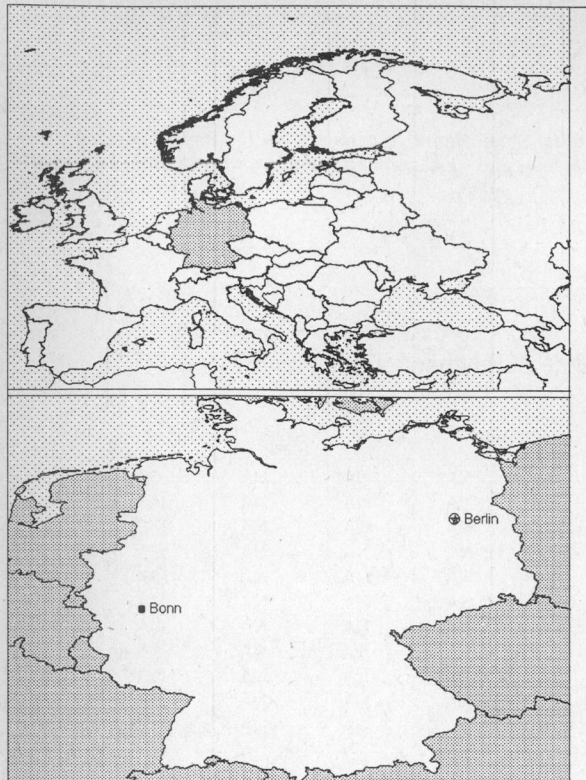

Total area:
356,910 sq km   137,804 sq mi
Land area:
349,520 sq km   134,950 sq mi
Comparative area:
Slightly smaller than Montana
Note: Includes the formerly separate Federal Republic of Germany, the German Democratic Republic, and Berlin, following formal unification on 3 October 1990
Land boundaries:
Total 3,621 km, Austria 784 km, Belgium 167 km, Czech Republic 646 km, Denmark 68 km, France 451 km, Luxembourg 138 km, Netherlands 577 km, Poland 456 km, Switzerland 334 km
Coastline:
2,389 km
Climate:
Temperate and marine; cool, cloudy, wet winters and summers; occasional warm, tropical foehn wind; high relative humidity
Terrain:
Lowlands in North, uplands in center, Bavarian Alps in South
Natural resources:
Iron ore, coal, potash, timber, lignite, uranium, copper, natural gas, salt, nickel
Land use:
Arable land: 34%
Permanent crops: 1%
Meadows and pastures: 16%
Forest and woodland: 30%
Other: 19%

Demographics [2]

	1970	1980	1990	1995[1]	1996	2000	2010	2020	2030
Population	77,783	78,298	79,357	82,948	83,536	85,684	88,975	88,870	85,881
Population density (persons per sq. mi.)	'575	579	587	613	618	634	658	657	635
(persons per sq. km.)	222	224	227	237	238	245	254	254	245
Net migration rate (per 1,000 population)	NA	NA	13.0	8.9	8.3	6.9	3.3	1.1	0.0
Births[16]	811	621	NA	NA	807	NA	NA	NA	NA
Deaths[17]	735	714	713	NA	936	NA	NA	NA	NA
Life expectancy - males	NA	NA	72.0	72.6	72.8	73.6	75.4	77.8	79.4
Life expectancy - females	NA	NA	78.5	79.1	79.3	80.0	81.4	83.8	85.4
Birth rate (per 1,000)	NA	NA	11.4	9.8	9.7	10.5	9.0	8.5	7.9
Death rate (per 1,000)	NA	NA	11.7	11.3	11.2	10.8	11.1	11.3	12.4
Women of reproductive age (15-49 yrs.)	NA	NA	19,399	19,802	20,070	20,460	20,142	17,669	16,252
of which are currently married[18]	8,262	9,532	12,364	NA	13,227	13,497	13,157	NA	NA
Fertility rate	NA	NA	1.5	1.3	1.3	1.6	1.5	1.5	1.5

Except as noted, values for vital statistics are in thousands; life expectancy is in years.

Health

Health Indicators [3]

% of population with access to	
safe water (1990-95)	NA
adequate sanitation (1990-95)	NA
health services (1985-95)	NA
% of 1-year-olds immunized (1990-94) against	
TB (tuberculosis)	NA
DPT (diphtheria, pertussis, tetanus)	70
polio	90
measles	75
% of contraceptive prevalence (1980-94)	75
ORT use rate (1990-94)	NA

Health Expenditures [4]

Total health expenditure, 1990 (official exchange rate)	
Millions of dollars	120,072
Dollars per capita	1,511
Health expenditures as a percentage of GDP	
Total	8.0
Public sector	5.8
Private sector	2.2
Development assistance for health	
Total aid flows (millions of dollars)[1]	NA
Aid flows per capita (dollars)	NA
Aid flows as a percentage of total health expenditure	NA

For sources, notes, and explanations, see Annotated Source Appendix, page 1061.

Human Factors

Women and Children [5]

% of pregnant women immunized (tetanus 1990-94)	NA
% of births attended by trained health personnel (1983-94)	99
Maternal mortality rate (1980-92)	5
Under-5 mortality rate (1994)	7
% under-5 moderately/severely underweight (1980-1994)	NA

Burden of Disease [6]

Population per physician (1988)[2]	366.66
Population per nurse	NA
Population per hospital bed (1990)[2]	117.91
AIDS cases per 100,000 people (1994)	2.6
Heart disease cases per 1,000 people (1990-93)	354.5

Ethnic Division [7]

Other is largely people fleeing the war in the former Yugoslavia.

German	95.1%
Turkish	2.3%
Italians	0.7%
Greeks	0.4%
Poles	0.4%
Other	1.1%

Religion [8]

Protestant	45%
Roman Catholic	37%
Unaffiliated or other	18%

Major Languages [9]

German.

Education

Public Education Expenditures [10]

	1980	1985	1990	1991	1993	1994
Million (Deutsche Mark)[28]						
Total education expenditure	70,099	83,691	98,412	107,497	151,033	NA
as percent of GNP	4.7	4.6	NA	NA	4.8	NA
as percent of total govt. expend.	9.5	9.2	8.6	11.6	9.5	NA
Current education expenditure	60,558	75,566	88,499	97,255	137,345	NA
as percent of GNP	4.1	4.1	NA	NA	4.3	NA
as percent of current govt. expend.	NA	9.4	8.7	12.5	NA	NA
Capital expenditure	9,541	8,125	9,913	10,242	13,688	NA

Educational Attainment [11]

Age group (1981)[9]	25+
Total population	10,714,841
Highest level attained (%)	
No schooling	-
First level	
Not completed	30.1
Completed	NA
Entered second level	
S-1	52.6
S-2	NA
Postsecondary	17.3

Illiteracy [12]

Libraries [13]

	Admin. Units	Svc. Pts.	Vols. (000)	Shelving (meters)	Vols. Added	Reg. Users
National (1992)[16]	8	9	29,438	NA	757,382	150,181
Nonspecialized (1992)	38	45	17,062	NA	439,860	222,102
Public (1992)	14,019	14,019	128,922	NA	NA	9 mil
Higher ed. (1993)[17]	271	NA	126,117	NA	4 mil	2 mil
School	NA	NA	NA	NA	NA	NA

Daily Newspapers [14]

	1980	1985	1990[2]	1994
Number of papers	368[e]	358[e]	414	411
Circ. (000)	29,388[e]	30,428[e]	24,174	25,757

Culture [15]

Cinema (seats per 1,000)	9.2
Annual attendance per person	1.6
Gross box office receipts (mil. Deutsche Mark)	1,170
Museums (reporting)[7]	3,768
Visitors (000)	93,756
Annual receipts (000 Deutsche Mark)	NA

Science and Technology

Scientific/Technical Forces [16]

Scientists/engineers	240,803
Number female	NA
Technicians	128,316
Number female	NA
Total	369,119

R&D Expenditures [17]

	D. Mark (000) 1991
Total expenditure[14]	74,517,000
Capital expenditure	7,704,000
Current expenditure	66,152,000
Percent current	89.6

U.S. Patents Issued [18]

Values show patents issued to citizens of the country by the U.S. Patents Office.

	1993	1994	1995
Number of patents	7,184	6,990	6,875

For sources, notes, and explanations, see Annotated Source Appendix, page 1061.

355

Government and Law

Organization of Government [19]

Long-form name:
Federal Republic of Germany
Type:
Federal republic
Independence:
18 January 1871 (German Empire unification); Federal Republic of Germany and German Democratic Republic created in 1949 from WWII occupation zones; re-unification occurred on 3 October 1990
National holiday:
German Unity Day, 3 October (1990)
Constitution:
23 May 1949, known as Basic Law, became constitution on 3 October 1990
Legal system:
Civil law system with indigenous concepts; judicial review of legislative acts in the Federal Constitutional Court; has not accepted compulsory ICJ jurisdiction
Executive branch:
President; Chancellor; Cabinet
Legislative branch:
Bicameral chamber: Federal Assembly (Bundestag) and Federal Council (Bundesrat)
Judicial branch:
Federal Constitutional Court (Bundesverfassungsgericht)

Elections [20]

Federal Assembly	% of votes
Social Democratic Party (SPD)	36.4
Christian Democratic Union (CDU)	34.2
Alliance 90/Greens	7.3
Christian Social Union (CSU)	7.3
Free Democratic Party (FDP)	6.9
Party of Democratic Socialism (PDS)	4.4
Republicans	1.9

Government Expenditures [21]

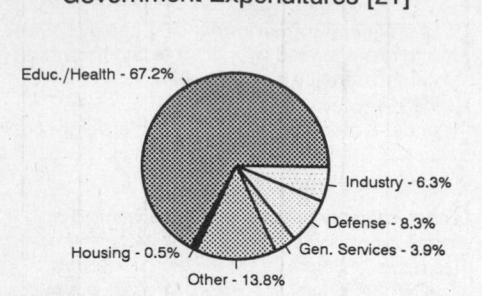

Educ./Health - 67.2%
Industry - 6.3%
Defense - 8.3%
Gen. Services - 3.9%
Housing - 0.5%
Other - 13.8%

(% distribution). Expend. for CY89: 654.91 (Deutsche Mark bil.)

Crime [23]

	1994
Crime volume	
Cases known to police	6,537,748
Attempts (percent)	7.70
Percent cases solved	44.40
Crimes per 100,000 persons	8,037.74
Persons responsible for offenses	
Total number offenders	2,037,729
Percent female	21.60
Percent juvenile (14-17 yrs.)	11.00
Percent foreigners	30.10

Military Expenditures and Arms Transfers [22]

	1990	1991[4]	1992	1993	1994
Military expenditures					
Current dollars (mil.)	38,830	42,180	41,090	37,360	36,310
1994 constant dollars (mil.)	43,220	45,210	42,850	38,130	36,310
Armed forces (000)	545	457	442	398	362
Gross national product (GNP)					
Current dollars (bil.)	1,385	1,854	1,939	1,947	2,045
1994 constant dollars (bil.)	1,542	1,987	2,022	1,987	2,045
Central government expenditures (CGE)					
1994 constant dollars (mil.)	458,800	601,700	677,500	676,800	768,900[e]
People (mil.)	63.1	80.0	80.4	80.8	81.1
Military expenditure as % of GNP	2.8	2.3	2.1	1.9	1.8
Military expenditure as % of CGE	9.4	7.5	6.3	5.6	4.7
Military expenditure per capita (1994 $)	685	565	533	472	448
Armed forces per 1,000 people (soldiers)	8.6	5.7	5.5	4.9	4.5
GNP per capita (1994 $)	24,420	24,850	25,160	24,600	25,220
Arms imports[6]					
Current dollars (mil.)	975	825	975	575	240
1994 constant dollars (mil.)	1,085	884	1,017	587	240
Arms exports[6]					
Current dollars (mil.)	1,700	2,300	1,300	1,700	700
1994 constant dollars (mil.)	1,892	2,465	1,356	1,735	700
Total imports[7]					
Current dollars (mil.)	346,200	389,900	402,400	348,600	373,900
1994 constant dollars (mil.)	385,200	417,900	419,700	355,800	373,900
Total exports[7]					
Current dollars (mil.)	410,100	402,800	422,300	380,200	420,000
1994 constant dollars (mil.)	456,400	431,800	440,300	388,000	420,000
Arms as percent of total imports[8]	0.3	0.2	0.2	0.2	0.1
Arms as percent of total exports[8]	0.4	0.6	0.3	0.4	0.2

Human Rights [24]

	SSTS	FL	FAPRO	PPCG	APROBC	TPW	PCPTW	STPEP	PHRFF	PRW	ASST	AFL
Observes	P	P	P	P	P	P	P		P	P	P	P
	EAFRD	CPR	ESCR	SR	ACHR	MAAE	PVIAC	PVNAC	EAFDAW	TCIDTP	RC	
Observes		P	P	P		P	P	P	P	P	P	

P = Party; S = Signatory; see Appendix for meaning of abbreviations.

Labor Force

Total Labor Force [25]

36.75 million

Labor Force by Occupation [26]

Industry	41%
Agriculture	6
Other	53

Date of data: 1987

Unemployment Rate [27]

For sources, notes, and explanations, see Annotated Source Appendix, page 1061.

Production Sectors

Commercial Energy Production and Consumption

Data are shown in quadrillion (10^{15}) BTUs and percent for 1995
Values for hydroelectric, nuclear, geothermal, solar, and wind power refer to electrical generation.

Production [28]

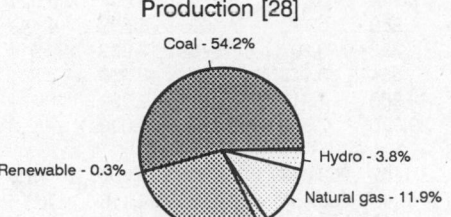

Coal - 54.2%
Renewable - 0.3%
Nuclear - 27.5%
Hydro - 3.8%
Natural gas - 11.9%
Crude oil - 2.3%

Consumption [29]

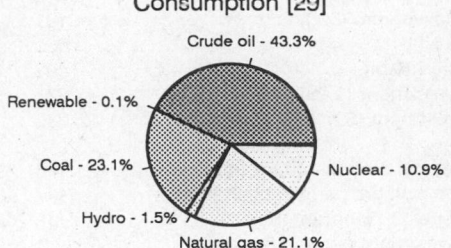

Crude oil - 43.3%
Renewable - 0.1%
Coal - 23.1%
Nuclear - 10.9%
Hydro - 1.5%
Natural gas - 21.1%

Crude oil	0.127
Dry natural gas	0.644
Coal	2.933
Net hydroelectric power	0.206
Net nuclear power	1.491
Geothermal, solar, wind	0.014
Total	5.415

Crude oil	5.908
Dry natural gas	2.880
Coal	3.154
Net hydroelectric power	0.206
Net nuclear power	1.491
Geothermal, solar, wind	0.014
Total	13.653

Telecommunications [30]

- 44 million telephones; has one of the world's most technologically advanced telecommunications systems; the backward system of the eastern area is being rapidly modernized and integrated into the western system
- Domestic: former West German area has an extensive system of automatic exchanges connected by modern networks of fiber-optic cable, coaxial cable, microwave radio relay, and a domestic satellite system; cellular telephone service widely available, including roaming to foreign countries
- International: satellite earth stations - 14 Intelsat, 1 Eutelsat, 1 Inmarsat, 2 Intersputnik; 6 submarine cable connections; 2 HF radiotelephone communication centers; tropospheric scatter links
- Radios: 70 million (1991 est.)
- Television: Broadcast stations: 246 (repeaters 6,000) Televisions: 44.8 million (1992 est.)

Transportation [31]

Railways: total: 43,966 km; standard gauge: 43,531 km 1.435-m; 40,355 km are owned by Deutsche Bahn AG (DB); 17,015 km of the DB system are electrified and 16,941 km are double- or more-tracked; narrow gauge: 389 km 1.000-m gauge (DB operates 146 km of 1.000-m gauge); 7 km 0.900-m gauge; 39 km 0.750-m gauge

Highways: total: 636,282 km; paved: 531,018 km (including 10,955 km of expressways); unpaved: 105,264 km (1991 est.)

Merchant marine: total: 452 ships (1,000 GRT or over) totaling 5,054,327 GRT/6,367,036 DWT; ships by type: bulk 6, cargo 193, chemical tanker 15, combination bulk 4, combination ore/oil 5, container 166, liquefied gas tanker 12, multifunction large-load carrier 6, oil tanker 11, passenger 3, railcar carrier 3, refrigerated cargo 7, roll-on/roll-off cargo 14, short-sea passenger 7 (1995 est.)

Airports

Total:	617
With paved runways over 3,047 m:	13
With paved runways 2,438 to 3,047 m:	65
With paved runways 1,524 to 2,437 m:	67
With paved runways 914 to 1,523 m:	51
With paved runways under 914 m:	351

Top Agricultural Products [32]

Agriculture accounts for 1% of the GDP; produces potatoes, wheat, rye, barley, sugar beets, fruit, cabbage; cattle, pigs, poultry, milk, hides.

Top Mining Products [33]

Thousand metric tons except as noted	7/31/95[*]
Iron and steel metal	29,900
Steel, crude	40,800
Cement, clinker and hydraulic	40,200
Potash, crude, gross weight	34,600
Salt, marketable	12,100
Coal, marketable	207,000
Coke, of anthracite and bituminous coal	11,000
Fuel briquets, of coal	10,900
Lime, quicklime, dead-burned dolomite	8,510
Clays	3,540

Tourism [34]

	1990	1991	1992	1993	1994
Tourists[41]	17,045	15,648	15,913	14,348	14,494
Tourism receipts[42]	11,471	11,666	10,996	10,429	11,091
Tourism expenditures[42]	32,180	34,250	35,702	36,345	43,398
Fare receipts[42]	5,535	5,844	5,618	5,023	5,418
Fare expenditures[42]	5,714	6,223	6,086	5,444	5,920

Travelers are in thousands, money in million U.S. dollars.

For sources, notes, and explanations, see Annotated Source Appendix, page 1061.

357

Manufacturing Sector

Manufacturing Summary [35]

	1987		1988		1989		1990		1991	
	$ bil.	%	$ bil.	%	$ bil.	%	$ bil.	%	$ bil.	%
Establishments or enterprises (number)	35,819	1.527	35,755	1.703	37,345	1.994	38,042	2.124	38,565	5.045
Total employment (000)	6,089	4.487	6,071	4.436	6,229	5.061	6,424	5.804	6,521	9.388
Production workers (000)	4,181	6.199	4,151	6.644	4,260	5.790	4,403	6.203	4,449	13.510
Output ($ bil.)	660	6.497	719	6.170	732	6.207	876	7.633	914	8.967
Value added ($ bil.)	493	11.407	533	11.098	534	10.946	667	13.366	-	-
Capital investment ($ mil.)	45,831	10.516	38,611	8.108	42,269	7.717	51,471	9.261	-	-
M & E investment ($ mil.)	28,385	9.109	26,463	7.646	29,316	7.509	42,287	10.030	-	-
Employees per establishment (number)	170	293.836	170	260.445	167	253.848	169	273.299	169	186.063
Production workers per establishment	117	405.986	116	390.078	114	290.451	116	292.116	115	267.777
Output per establishment ($ mil.)	18	425.513	20	362.220	20	311.337	23	359.447	24	177.725
Capital investment per estab. ($ mil.)	1	688.693	1	476.020	1	387.063	1	436.079	-	-
M & E per establishment ($ mil)	0.792	596.575	0.740	448.885	0.785	376.640	1	472.317	-	-
Payroll per employee ($)	23,308	259.930	24,815	249.082	24,002	214.833	29,106	209.821	30,112	206.232
Wages per production worker ($)	19,786	250.757	21,025	248.827	20,419	203.613	24,664	207.679	25,311	230.120
Hours per production worker (hours)	1,634	86.127	1,642	85.440	1,640	89.471	1,599	85.489	1,577	88.225
Output per employee ($)	108,458	144.813	118,459	139.077	117,544	122.647	136,301	131.522	140,229	95.518
Capital investment per employee ($)	7,527	234.380	6,360	182.772	6,786	152.478	8,012	159.561	-	-
M & E per employee ($)	4,662	203.030	4,359	172.353	4,706	148.372	6,583	172.821	-	-

Note: Columns headed % show percent of world total or ratio. Ratios closest to 100 are closest to world average. M & E stands for machinery & equipment.

Output in Manufacturing

	1987		1988		1989		1990		1991	
	$ bil.	%	$ bil.	%	$ bil.	%	$ bil.	%	$ bil.	%
3110 - Food products	68.000	10.30	72.000	10.01	73.000	9.97	91.000	10.39	96.000	10.50
3130 - Beverages	16.000	2.42	17.000	2.36	16.000	2.19	22.000	2.51	23.000	2.52
3140 - Tobacco	12.000	1.82	12.000	1.67	11.000	1.50	15.000	1.71	17.000	1.86
3210 - Textiles	21.000	3.18	22.000	3.06	22.000	3.01	28.000	3.20	27.000	2.95
3220 - Wearing apparel	11.000	1.67	12.000	1.67	11.000	1.50	14.000	1.60	15.000	1.64
3230 - Leather and products	2.059	0.31	1.993	0.28	1.862	0.25	2.290	0.26	2.169	0.24
3240 - Footwear	3.449	0.52	3.360	0.47	3.191	0.44	3.652	0.42	3.616	0.40
3410 - Paper and products	20.000	3.03	22.000	3.06	23.000	3.14	-	-	-	-
3411 - Pulp, paper, etc.	8.790	1.33	9.851	1.37	10.000	1.37	-	-	-	-
3420 - Printing, publishing	15.000	2.27	16.000	2.23	16.000	2.19	21.000	2.40	22.000	2.41
3550 - Rubber products	9.569	1.45	10.000	1.39	9.628	1.32	11.000	1.26	12.000	1.31
3560 - Plastic products nec	21.000	3.18	25.000	3.48	26.000	3.55	34.000	3.88	37.000	4.05
3610 - Pottery, china, etc.	1.613	0.24	1.708	0.24	1.702	0.23	2.166	0.25	2.109	0.23
3620 - Glass and products	6.120	0.93	6.662	0.93	6.702	0.92	8.294	0.95	8.557	0.94
3690 - Nonmetal products nec	15.000	2.27	16.000	2.23	16.000	2.19	20.000	2.28	22.000	2.41
3710 - Iron and steel	30.000	4.55	35.000	4.87	36.000	4.92	40.000	4.57	37.000	4.05
3720 - Nonferrous metals	14.000	2.12	17.000	2.36	19.000	2.60	20.000	2.28	19.000	2.08
3810 - Metal products	51.000	7.73	56.000	7.79	59.000	8.06	75.000	8.56	80.000	8.75
3820 - Machinery nec	101.000	15.30	111.000	15.44	114.000	15.57	144.000	16.44	147.000	16.08
3825 - Office, computing machinery	9.291	1.41	9.965	1.39	9.202	1.26	12.000	1.37	16.000	1.75
3830 - Electrical machinery	90.000	13.64	97.000	13.49	99.000	13.52	124.000	14.16	127.000	13.89
3840 - Transportation equipment	118.000	17.88	125.000	17.39	128.000	17.49	163.000	18.61	178.000	19.47
3841 - Shipbuilding, repair	3.394	0.51	3.246	0.45	2.872	0.39	5.013	0.57	4.700	0.51
3850 - Professional goods	11.000	1.67	12.000	1.67	12.000	1.64	14.000	1.60	15.000	1.64
3900 - Industries nec	3.672	0.56	3.929	0.55	3.989	0.54	4.951	0.57	5.062	0.55

Note: Codes are International Standard Industry codes (ISIC). Percentages are % of total Output. [f]: Factor Prices; [p]: Producer Prices.

Finance, Economics, and Trade

Economic Indicators [36]

- **National product**: GDP—purchasing power parity— $1.4522 trillion (1995 est.)
- **National product real growth rate**: 1.8% (1995 est.)
- **National product per capita**: $17,900 (1995 est.)
- **Inflation rate (consumer prices)**: 2% (1995 est.)
- **External debt**: $NA

Balance of Payments Summary [37]

Values in millions of dollars.

	1989	1990	1991	1992	1993
Exports of goods (f.o.b.)	324,970	391,290	378,630	406,660	364,150
Imports of goods (f.o.b.)	-247,220	-320,240	-355,410	-373,910	-318,860
Trade balance	77,750	71,050	23,220	32,750	45,290
Services - debits	-102,480	-136,270	-151,780	-182,390	-181,770
Services - credits	100,440	134,150	145,530	160,500	153,230
Private transfers (net)	-5,690	-7,150	-7,020	-8,630	-8,140
Government transfers (net)	-12,270	-15,540	-28,870	-23,410	-23,370
Long-term capital (net)	-12,050	-37,530	-11,930	27,130	98,990
Short-term capital (net)	-30,290	-18,220	25,090	30,230	-86,030
Errors and omissions	4,490	15,000	7,890	6,820	-11,460
Overall balance	-10,620	5,490	2,130	43,000	-13,260

Exchange Rates [38]

Currency: **deutsche mark.**
Symbol: **DM.**

Data are currency units per $1.

January 1996	1.4617
1995	1.4331
1994	1.6228
1993	1.6533
1992	1.5617
1991	1.6595

Imports and Exports

Top Import Origins [39]

$362 billion (f.o.b., 1994) Data are for 1993.

Origins	%
EU	46.4
France	11.3
Netherlands	8.4
Italy	8.1
UK	6.0
Belgium-Luxembourg	5.7
EFTA	14.3
US	7.3
Japan	6.3
Eastern Europe	5.1
OPEC	2.6

Top Export Destinations [40]

$437 billion (f.o.b., 1994) Data are for 1993.

Destinations	%
EU	47.9
France	11.7
Netherlands	7.4
Italy	7.5
UK	7.7
Belgium-Luxembourg	6.6
EFTA	15.5
US	7.7
Eastern Europe	5.2
OPEC	3.0

Foreign Aid [41]

Donor: ODA, $6.954 billion (1993).

Import and Export Commodities [42]

Import Commodities

Manufactures 75.1%
Agricultural products 10.0%
Fuels 8.3%
Raw materials 5.0%

Export Commodities

Manufactures 89.3%
(includes machines,
machine tools, motor
vehicles, iron and
steel products)
Agricultural products 5.5%
Raw materials 2.7%
Fuels 1.3%

Ghana

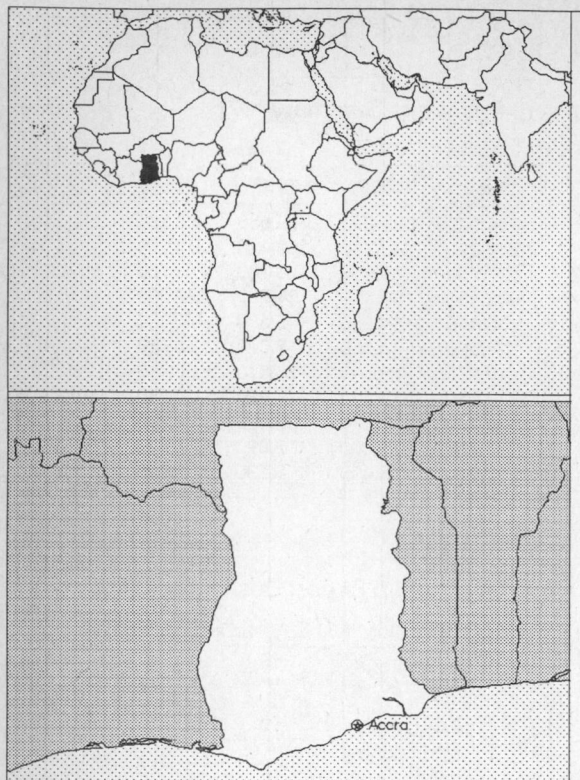

Geography [1]

Total area:
  238,540 sq km   92,101 sq mi
Land area:
  230,020 sq km   88,811 sq mi
Comparative area:
  Slightly smaller than Oregon
Land boundaries:
  Total 2,093 km, Burkina Faso 548 km, Cote d'Ivoire 668 km, Togo 877 km
Coastline:
  539 km
Climate:
  Tropical; warm and comparatively dry along Southeast Coast; hot and humid in Southwest; hot and dry in North
Terrain:
  Mostly low plains with dissected plateau in south-central area
Natural resources:
  Gold, timber, industrial diamonds, bauxite, manganese, fish, rubber
Land use:
  Arable land: 5%
  Permanent crops: 7%
  Meadows and pastures: 15%
  Forest and woodland: 37%
  Other: 36%

Demographics [2]

	1970	1980	1990	1995[1]	1996	2000	2010	2020	2030
Population	8,789	10,880	15,190	17,291	17,698	19,272	22,929	26,516	29,706
Population density (persons per sq. mi.)	99	123	171	195	199	217	258	299	334
(persons per sq. km.)	38	47	66	75	77	84	100	115	129
Net migration rate (per 1,000 population)	-1.1	-16.7	-1.1	-1.0	-0.9	-0.9	-0.4	0.0	0.0
Births	NA	NA	NA	NA	619	NA	NA	NA	NA
Deaths	28	NA	NA	NA	197	NA	NA	NA	NA
Life expectancy - males	46.9	50.9	52.4	53.9	54.2	55.4	58.3	61.2	63.8
Life expectancy - females	50.2	54.4	56.1	57.9	58.2	59.6	63.0	66.3	69.3
Birth rate (per 1,000)	45.5	41.4	41.8	36.1	35.0	30.8	24.6	21.1	17.6
Death rate (per 1,000)	17.0	13.8	12.9	11.4	11.2	10.2	8.7	8.0	7.9
Women of reproductive age (15-49 yrs.)	1,920	2,395	3,565	4,097	4,210	4,751	6,308	7,488	8,331
of which are currently married	NA	NA	2,349	NA	2,848	3,218	4,271	NA	NA
Fertility rate	6.9	6.3	5.7	4.8	4.6	4.0	2.8	2.3	2.1

Except as noted, values for vital statistics are in thousands; life expectancy is in years.

Health

Health Indicators [3]

% of population with access to	
safe water (1990-95)	56
adequate sanitation (1990-95)	42
health services (1985-95)	60
% of 1-year-olds immunized (1990-94) against	
TB (tuberculosis)	61
DPT (diphtheria, pertussis, tetanus)	48
polio	48
measles	49
% of contraceptive prevalence (1980-94)	20
ORT use rate (1990-94)	44

Health Expenditures [4]

Total health expenditure, 1990 (official exchange rate)	
Millions of dollars	204
Dollars per capita	14
Health expenditures as a percentage of GDP	
Total	3.5
Public sector	1.7
Private sector	1.8
Development assistance for health	
Total aid flows (millions of dollars)[1]	29
Aid flows per capita (dollars)	1.9
Aid flows as a percentage of total health expenditure	14.2

For sources, notes, and explanations, see Annotated Source Appendix, page 1061.

Human Factors

Women and Children [5]

% of pregnant women immunized (tetanus 1990-94)	11
% of births attended by trained health personnel (1983-94)	59
Maternal mortality rate (1980-92)	390
Under-5 mortality rate (1994)	131
% under-5 moderately/severely underweight (1980-1994)	27

Burden of Disease [6]

Population per physician (1989)	22,969.75
Population per nurse (1987)	1,669.05
Population per hospital bed (1990)	684.94
AIDS cases per 100,000 people (1994)	13.7
Malaria cases per 100,000 people (1992)	NA

Ethnic Division [7]

Black African	99.8%
Akan	44%
Moshi-Dagomba	16%
Ewe	13%
Ga	8%
European and other	0.2%

Religion [8]

Indigenous beliefs	38%
Muslim	30%
Christian	24%
Other	8%

Major Languages [9]

English (official), African languages (including Akan, Moshi-Dagomba, Ewe, and Ga).

Education

Public Education Expenditures [10]

Million (Cedi)[29]	1980	1985	1989	1990	1991	1994
Total education expenditure	1,319	8,675	47,791	61,900	NA	NA
as percent of GNP	3.1	2.6	3.5	3.1	NA	NA
as percent of total govt. expend.	17.1	19.0	24.3	24.3	NA	NA
Current education expenditure	NA	NA	43,857	53,664	NA	NA
as percent of GNP	NA	NA	3.2	2.7	NA	NA
as percent of current govt. expend.	NA	NA	29.5	27.1	NA	NA
Capital expenditure	NA	NA	3,934	8,236	NA	NA

Educational Attainment [11]

Illiteracy [12]

In thousands and percent[1]	1990	1995	2000
Illiterate population (15+ yrs.)	3,434	3,387	3,300
Illiteracy rate - total pop. (%)	40.2	34.7	29.4
Illiteracy rate - males (%)	28.0	23.6	19.7
Illiteracy rate - females (%)	52.1	45.3	38.7

Libraries [13]

	Admin. Units	Svc. Pts.	Vols. (000)	Shelving (meters)	Vols. Added	Reg. Users
National	NA	NA	NA	NA	NA	NA
Nonspecialized	NA	NA	NA	NA	NA	NA
Public (1989)	13	47	1,576	NA	32,200	56,211
Higher ed.	NA	NA	NA	NA	NA	NA
School	NA	NA	NA	NA	NA	NA

Daily Newspapers [14]

	1980	1985	1990	1994
Number of papers	5	5	2	4
Circ. (000)	500[e]	510[e]	200	310[e]

Culture [15]

Science and Technology

Scientific/Technical Forces [16]

R&D Expenditures [17]

U.S. Patents Issued [18]

Values show patents issued to citizens of the country by the U.S. Patents Office.

	1993	1994	1995
Number of patents	0	1	0

For sources, notes, and explanations, see Annotated Source Appendix, page 1061.

361

Government and Law

Organization of Government [19]

Long-form name:
Republic of Ghana
Type:
Constitutional democracy
Independence:
6 March 1957 (from UK)
National holiday:
Independence Day, 6 March (1957)
Constitution:
New constitution approved 28 April 1992
Legal system:
Based on English common law and
customary law; has not accepted
compulsory ICJ jurisdiction
Executive branch:
President; Vice President; Cabinet
Legislative branch:
Unicameral: Parliament
Judicial branch:
Supreme Court

Elections [20]

Parliament	% of seats
Every Ghanian Living Everywhere (EGLE)	94.5
National Democratic Congress (NDC)	4.0
Independents	1.5

Government Expenditures [21]

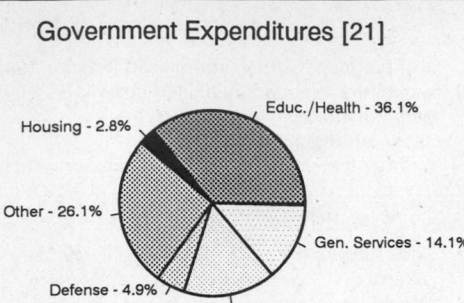

Housing - 2.8%
Educ./Health - 36.1%
Other - 26.1%
Gen. Services - 14.1%
Defense - 4.9%
Industry - 16.0%

(% distribution). Expend. for FY93: 813,526 (New Cedi mil.)

Crime [23]

	1994
Crime volume	
Cases known to police	151,147
Attempts (percent)	NA
Percent cases solved	5.98
Crimes per 100,000 persons	NA
Persons responsible for offenses	
Total number offenders	NA
Percent female	NA
Percent juvenile	NA
Percent foreigners	NA

Military Expenditures and Arms Transfers [22]

	1990	1991	1992	1993	1994
Military expenditures					
Current dollars (mil.)	18	26	37	51	41
1994 constant dollars (mil.)	20	28	38	52	41
Armed forces (000)	9	9	7	7	7
Gross national product (GNP)					
Current dollars (mil.)	4,008	4,384	4,693	5,030	5,319
1994 constant dollars (mil.)	4,461	4,699	4,894	5,134	5,319
Central government expenditures (CGE)					
1994 constant dollars (mil.)	590	653	844	1,086	NA
People (mil.)	15.2	15.7	16.2	16.7	17.2
Military expenditure as % of GNP	0.5	0.6	0.8	1.0	0.8
Military expenditure as % of CGE	3.4	4.3	4.6	4.8	NA
Military expenditure per capita (1994 $)	1	2	2	3	2
Armed forces per 1,000 people (soldiers)	0.6	0.6	0.4	0.4	0.4
GNP per capita (1994 $)	294	300	302	307	309
Arms imports[6]					
Current dollars (mil.)	5	0	0	0	0
1994 constant dollars (mil.)	6	0	0	0	0
Arms exports[6]					
Current dollars (mil.)	0	0	0	0	0
1994 constant dollars (mil.)	0	0	0	0	0
Total imports[7]					
Current dollars (mil.)[e]	1,614	1,611	1,743	1,700	NA
1994 constant dollars (mil.)[e]	1,796	1,727	1,818	1,735	NA
Total exports[7]					
Current dollars (mil.)[e]	925	861	676	623	617
1994 constant dollars (mil.)[e]	1,030	923	705	636	617
Arms as percent of total imports[8]	0.3	0	0	0	0
Arms as percent of total exports[8]	0	0	0	0	0

Human Rights [24]

	SSTS	FL	FAPRO	PPCG	APROBC	TPW	PCPTW	STPEP	PHRFF	PRW	ASST	AFL
Observes	2	P	P	P	P	P	P			P	P	P
	EAFRD	CPR	ESCR	SR	ACHR	MAAE	PVIAC	PVNAC	EAFDAW	TCIDTP	RC	
Observes		P		P				P	P	P		P

P = Party; S = Signatory; see Appendix for meaning of abbreviations.

Labor Force

Total Labor Force [25]

3.7 million

Labor Force by Occupation [26]

Agriculture and fishing	54.7%
Industry	18.7
Sales and clerical	15.2
Professional	3.7
Services, transportation, and communications	7.7

Unemployment Rate [27]

10% (1993 est.)

Production Sectors

Commercial Energy Production and Consumption

Data are shown in quadrillion (10^{15}) BTUs and percent for 1995
Values for hydroelectric, nuclear, geothermal, solar, and wind power refer to electrical generation.

Production [28]

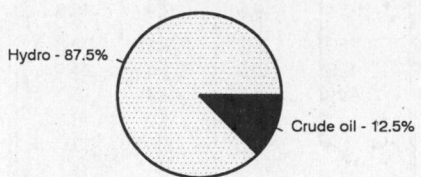

Hydro - 87.5%

Crude oil - 12.5%

Crude oil	0.009
Net hydroelectric power	0.063
Total	0.072

Consumption [29]

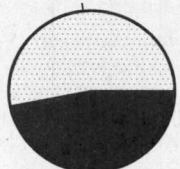

Hydro - 52.9%

Crude oil - 47.1%

Crude oil	0.056
Net hydroelectric power	0.063
Total	0.119

Telecommunications [30]

- 70,000 (1988 est.) telephones; poor to fair system
- Domestic: primarily microwave radio relay
- International: satellite earth station - 1 Intelsat (Atlantic Ocean)
- Radio: Broadcast stations: AM 4, FM 1, shortwave 0
- Television: Broadcast stations: 4 (repeaters 8) Televisions: 250,000 (1993 est.)

Top Agricultural Products [32]

Agriculture accounts for 47% of the GDP; produces cocoa, rice, coffee, cassava (tapioca), peanuts, corn, shea nuts, bananas; timber.

Top Mining Products [33]

Thousand metric tons except as noted[e]	6/23/95[*]
Bauxite, gross weight	426[r]
Arsenic, trioxide (tons)	8,000
Cement, hydraulic	1,350[r,33]
Diamond, gem and industrial (000 carats)	800[34]
Gold (kg.)	44,500[r,35]
Manganese ore, processed	265[r]
Petroleum refinery products (000 42-bls.)	7,500
Silver, content of gold ore (kg.)	2,230

Transportation [31]

Railways: total: 953 km; note - undergoing major renovation (1995 est.); narrow gauge: 953 km 1.067-m gauge; 32 km double track

Highways: total: 38,145 km; paved: 7,476 km (including 21 km of expressways); unpaved: 30,669 km (1990 est.)

Merchant marine: total: 3 ships (1,000 GRT or over) totaling 27,427 GRT/35,894 DWT; ships by type: cargo 2, refrigerated cargo 1 (1995 est.)

Airports

Total:	12
With paved runways 2,438 to 3,047 m:	3
With paved runways 1,524 to 2,437 m:	1
With paved runways 914 to 1,523 m:	2
With paved runways under 914 m:	2
With unpaved runways 1,524 to 2,437 m:	2

Tourism [34]

	1990	1991	1992	1993	1994
Tourists	146	172	213	257	271
Tourism receipts	81	118	167	206	228
Tourism expenditures	13	14	17	20	20
Fare receipts	10	14	14	20	18
Fare expenditures	27	30	39	46	51

Travelers are in thousands, money in million U.S. dollars.

Manufacturing Sector

GDP and Manufacturing Summary [35]

	1980	1985	1989	1990	% change 1980- 1990	% change 1989- 1990
GDP (million 1980 $)	4,788	4,686	5,795	5,893	23.1	1.7
GDP per capita (1980 $)	446	365	398	392	-12.1	-1.5
Manufacturing as % of GDP (current prices)	7.8	11.5	10.4[e]	9.9	26.9	-4.8
Gross output (million $)	505	696	950[e]	1,309[e]	159.2	37.8
Value added (million $)	244	338	498[e]	620[e]	154.1	24.5
Value added (million 1980 $)	374	299	449[e]	409	9.4	-8.9
Industrial production index	100	70	111[e]	104	4.0	-6.3
Employment (thousands)	80	61	68[e]	95[e]	18.8	39.7

Note: GDP stands for Gross Domestic Product. 'e' stands for estimated value.

Profitability and Productivity

	1980	1985	1989	1990	% change 1980- 1990	% change 1989- 1990
Intermediate input (%)	52	51	48[e]	53[e]	1.9	10.4
Wages, salaries, and supplements (%)	10	6	7[e]	7[e]	-30.0	0.0
Gross operating surplus (%)	39	42	46[e]	40[e]	2.6	-13.0
Gross output per worker ($)	6,293	11,306	14,045[e]	13,685[e]	117.5	-2.6
Value added per worker ($)	3,034	5,495	7,360[e]	6,501[e]	114.3	-11.7
Average wage (incl. benefits) ($)	606	711	961[e]	970[e]	60.1	0.9

Profitability is in percent of gross output. Productivity is in U.S. $. 'e' stands for estimated value.

Profitability - 1990

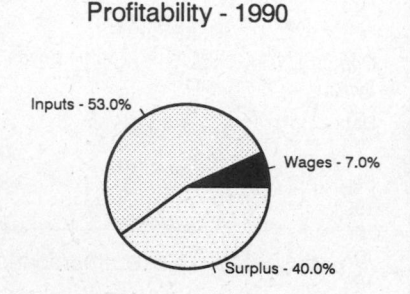

Inputs - 53.0%
Wages - 7.0%
Surplus - 40.0%

The graphic shows percent of gross output.

Value Added in Manufacturing

	1980 $ mil.	1980 %	1985 $ mil.	1985 %	1989 $ mil.	1989 %	1990 $ mil.	1990 %	% change 1980- 1990	% change 1989- 1990
311 Food products	20	8.2	35	10.4	45[e]	9.0	76[e]	12.3	280.0	68.9
313 Beverages	38	15.6	51	15.1	72[e]	14.5	87[e]	14.0	128.9	20.8
314 Tobacco products	32	13.1	68	20.1	77[e]	15.5	70[e]	11.3	118.8	-9.1
321 Textiles	22	9.0	18	5.3	27[e]	5.4	50[e]	8.1	127.3	85.2
322 Wearing apparel	3	1.2	1	0.3	1[e]	0.2	1[e]	0.2	-66.7	0.0
323 Leather and fur products	1	0.4	NA	0.0	1[e]	0.2	1[e]	0.2	0.0	0.0
324 Footwear	1	0.4	NA	0.0	1[e]	0.2	1[e]	0.2	0.0	0.0
331 Wood and wood products	16	6.6	41	12.1	78[e]	15.7	56[e]	9.0	250.0	-28.2
332 Furniture and fixtures	2	0.8	2	0.6	3[e]	0.6	7[e]	1.1	250.0	133.3
341 Paper and paper products	1	0.4	2	0.6	3[e]	0.6	4[e]	0.6	300.0	33.3
342 Printing and publishing	5	2.0	4	1.2	6[e]	1.2	10[e]	1.6	100.0	66.7
351 Industrial chemicals	2	0.8	1	0.3	1[e]	0.2	5[e]	0.8	150.0	400.0
352 Other chemical products	9	3.7	26	7.7	33[e]	6.6	27[e]	4.4	200.0	-18.2
353 Petroleum refineries	37	15.2	34	10.1	63[e]	12.7	45[e]	7.3	21.6	-28.6
354 Miscellaneous petroleum and coal products	NA	0.0	NA	0.0	NA	0.0	NA	0.0	NA	NA
355 Rubber products	5	2.0	2	0.6	2[e]	0.4	5[e]	0.8	0.0	150.0
356 Plastic products	1	0.4	2	0.6	3[e]	0.6	3[e]	0.5	200.0	0.0
361 Pottery, china, and earthenware	1	0.4	NA	0.0	NA	0.0	NA	0.0	NA	NA
362 Glass and glass products	NA	0.0	1	0.3	2[e]	0.4	1[e]	0.2	NA	-50.0
369 Other nonmetal mineral products	6	2.5	16	4.7	16[e]	3.2	23[e]	3.7	283.3	43.8
371 Iron and steel	1	0.4	1	0.3	1[e]	0.2	3[e]	0.5	200.0	200.0
372 Nonferrous metals	29	11.9	16	4.7	45[e]	9.0	121[e]	19.5	317.2	168.9
381 Metal products	7	2.9	8	2.4	11[e]	2.2	14[e]	2.3	100.0	27.3
382 Nonelectrical machinery	NA	0.0	NA	0.0	NA	0.0	2[e]	0.3	NA	NA
383 Electrical machinery	2	0.8	3	0.9	4[e]	0.8	5[e]	0.8	150.0	25.0
384 Transport equipment	3	1.2	2	0.6	2[e]	0.4	4[e]	0.6	33.3	100.0
385 Professional and scientific equipment	1	0.4	1	0.3	1[e]	0.2	1[e]	0.2	0.0	0.0
390 Other manufacturing industries	NA	0.0	NA	0.0	NA	0.0	1[e]	0.2	NA	NA

Note: The industry codes shown are International Standard Industry codes (ISIC). Percentages are percent of total Value Added. 'e' stands for estimated value

Finance, Economics, and Trade

Economic Indicators [36]

- **National product**: GDP—purchasing power parity—$25.1 billion (1995 est.)
- **National product real growth rate**: 5% (1995 est.)
- **National product per capita**: $1,400 (1995 est.)
- **Inflation rate (consumer prices)**: 69% (1995 est.)
- **External debt**: $4.6 billion (December 1993 est.)

Balance of Payments Summary [37]

Values in millions of dollars.

	1988	1989	1990	1991	1992
Exports of goods (f.o.b.)	881.0	807.2	890.6	997.6	986.4
Imports of goods (f.o.b.)	-993.4	-1,002.2	-1,198.9	-1,318.7	-1,456.7
Trade balance	-112.4	-195.0	-308.3	-321.1	-470.3
Services - debits	-399.6	-407.8	-429.0	-462.8	-505.1
Services - credits	77.7	81.9	93.1	110.3	128.9
Private transfers (net)	172.4	202.1	201.9	219.5	254.9
Government transfers (net)	196.1	220.2	213.8	201.4	213.8
Long-term capital (net)	183.9	180.2	334.2	373.5	392.9
Short-term capital (net)	-48.1	-61.0	-43.2	-7.9	-117.6
Errors and omissions	37.9	40.6	8.8	23.8	-0.4
Overall balance	107.9	61.2	71.3	136.7	-102.9

Exchange Rates [38]

Currency: **new cedi.**
Symbol: **C.**

Data are currency units per $1.

September 1995	1,246.11
1994	956.71
1993	649.06
1992	437.09
1991	367.83

Imports and Exports

Top Import Origins [39]

$1.7 billion (f.o.b., 1993 est.) Data are for 1995.

Origins	%
UK	NA
US	NA
Germany	NA
Japan	NA
Netherlands	NA

Top Export Destinations [40]

$1 billion (f.o.b., 1993 est.) Data are for 1995.

Destinations	%
Germany	NA
US	NA
UK	NA
Netherlands	NA
Japan	NA

Foreign Aid [41]

Recipient: ODA, $472 million (1993).

Import and Export Commodities [42]

Import Commodities	Export Commodities
Petroleum	Cocoa 40%
Consumer goods	Gold
Foods	Timber
Intermediate goods	Tuna
Capital equipment	Bauxite
	Aluminum
	Manganese ore
	Diamonds

Greece

Geography [1]

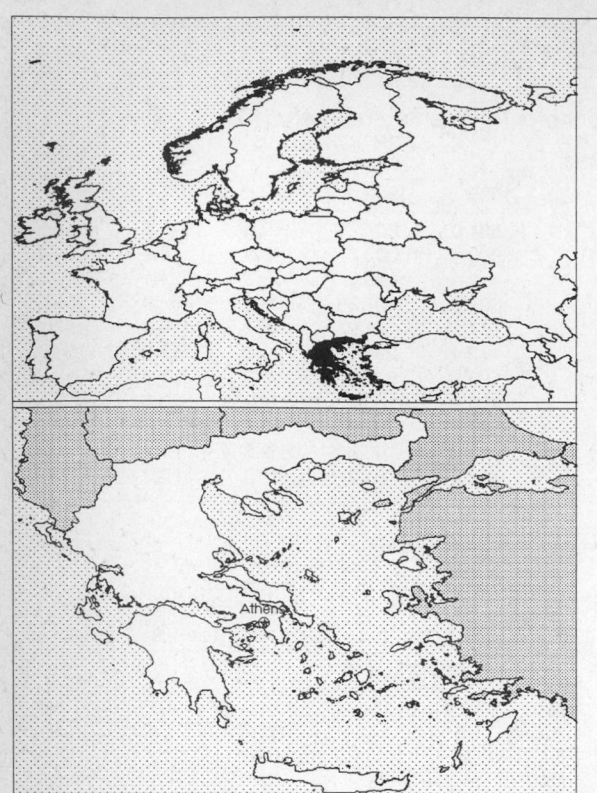

Total area:
131,940 sq km   50,942 sq mi
Land area:
130,800 sq km   50,502 sq mi
Comparative area:
Slightly smaller than Alabama
Land boundaries:
Total 1,210 km, Albania 282 km, Bulgaria 494 km, Turkey 206 km, Macedonia 228 km
Coastline:
13,676 km
Climate:
Temperate; mild, wet winters; hot, dry summers
Terrain:
Mostly mountains with ranges extending into sea as peninsulas or chains of islands
Natural resources:
Bauxite, lignite, magnesite, petroleum, marble
Land use:
Arable land: 23%
Permanent crops: 8%
Meadows and pastures: 40%
Forest and woodland: 20%
Other: 9%

Demographics [2]

	1970	1980	1990	1995[1]	1996	2000	2010	2020	2030
Population	8,793	9,643	10,123	10,494	10,539	10,735	11,135	11,076	10,687
Population density (persons per sq. mi.)	174	191	200	208	209	213	220	219	212
(persons per sq. km.)	67	74	77	80	81	82	85	85	82
Net migration rate (per 1,000 population)	-5.3	5.2	12.4	3.9	4.0	4.1	2.4	0.8	0.0
Births	145	148	NA	NA	114	NA	NA	NA	NA
Deaths	74	87	54	NA	100	NA	NA	NA	NA
Life expectancy - males	70.1	72.2	74.5	75.4	75.6	76.4	78.0	79.1	80.0
Life expectancy - females	73.6	76.4	79.5	80.6	80.8	81.7	83.6	84.9	85.9
Birth rate (per 1,000)	16.5	15.4	10.0	9.8	9.8	10.7	9.4	8.0	7.9
Death rate (per 1,000)	8.4	9.1	9.5	9.5	9.5	9.5	10.1	11.2	12.3
Women of reproductive age (15-49 yrs.)	NA	NA	2,397	2,543	2,554	2,586	2,498	2,295	2,035
of which are currently married	NA	NA	1,677	NA	1,862	1,901	1,839	NA	NA
Fertility rate	2.3	2.2	1.4	1.4	1.4	1.5	1.5	1.5	1.5

Except as noted, values for vital statistics are in thousands; life expectancy is in years.

Health

Health Indicators [3]

% of population with access to	
safe water (1990-95)	NA
adequate sanitation (1990-95)	NA
health services (1985-95)	NA
% of 1-year-olds immunized (1990-94) against	
TB (tuberculosis)	50
DPT (diphtheria, pertussis, tetanus)	78
polio	95
measles	72
% of contraceptive prevalence (1980-94)	NA
ORT use rate (1990-94)	NA

Health Expenditures [4]

Total health expenditure, 1990 (official exchange rate)	
Millions of dollars	3,609
Dollars per capita	358
Health expenditures as a percentage of GDP	
Total	5.5
Public sector	4.2
Private sector	1.3
Development assistance for health	
Total aid flows (millions of dollars)[1]	NA
Aid flows per capita (dollars)	NA
Aid flows as a percentage of total health expenditure	NA

For sources, notes, and explanations, see Annotated Source Appendix, page 1061.

Human Factors

Women and Children [5]

% of pregnant women immunized (tetanus 1990-94)	NA
% of births attended by trained health personnel (1983-94)[1]	97
Maternal mortality rate (1980-92)	5
Under-5 mortality rate (1994)	10
% under-5 moderately/severely underweight (1980-1994)	NA

Burden of Disease [6]

Population per physician (1989)	582.06
Population per nurse (1984)	453.72
Population per hospital bed (1990)	197.42
AIDS cases per 100,000 people (1994)	1.8
Heart disease cases per 1,000 people (1990-93)	302.5

Ethnic Division [7]

Greek	98%
Other	2%

Religion [8]

Greek Orthodox	98%
Muslim	1.3%
Other	0.7%

Major Languages [9]

Greek (official), English, French.

Education

Public Education Expenditures [10]

Million (Drachma)	1980	1985	1989	1990	1991	1994
Total education expenditure	NA	133,091	265,351	325,676	379,864	NA
as percent of GNP	NA	2.9	3.1	3.1	3.0	NA
as percent of total govt. expend.	NA	7.5	NA	NA	NA	NA
Current education expenditure	NA	126,749	248,040	306,303	359,533	NA
as percent of GNP	NA	2.8	2.9	2.9	2.8	NA
as percent of current govt. expend.	NA	8.5	NA	NA	NA	NA
Capital expenditure	NA	6,342	17,311	19,373	20,331	NA

Educational Attainment [11]

Age group (1991)	25+
Total population	6,738,566
Highest level attained (%)	
No schooling	5.7
First level	
Not completed	12.7
Completed	44.2
Entered second level	
S-1	6.7
S-2	22.0
Postsecondary	8.7

Illiteracy [12]

	1989	1991	1995
Illiterate population (15+ yrs.)[2]	NA	389,067	283,000
Illiteracy rate - total pop. (%)	NA	4.8	3.3
Illiteracy rate - males (%)	NA	2.3	1.7
Illiteracy rate - females (%)	NA	7.0	4.7

Libraries [13]

	Admin. Units	Svc. Pts.	Vols. (000)	Shelving (meters)	Vols. Added	Reg. Users
National (1990)	2	2	2,633	36,500	13,043	NA
Nonspecialized (1989)	758	758	7,492	194,430	216,221	1 mil
Public (1990)	680	680	7,400	151,526	227,873	NA
Higher ed. (1990)	39[e]	70[e]	6,482	167,639[e]	NA	128,391[e]
School	NA	NA	NA	NA	NA	NA

Daily Newspapers [14]

	1980	1985	1990	1994
Number of papers	128	140	130	168
Circ. (000)	1,160[e]	1,210[e]	1,250	1,622

Culture [15]

Cinema (seats per 1,000)	38.1
Annual attendance per person	NA
Gross box office receipts (mil. Drachma)	NA
Museums (reporting)[8]	75
Visitors (000)	2,359
Annual receipts (000 Drachma)	9,510

Science and Technology

Scientific/Technical Forces [16]

Scientists/engineers	8,030
Number female	NA
Technicians	3,257
Number female	NA
Total	11,287

R&D Expenditures [17]

	Drachma (000) 1993
Total expenditure	100,460,000
Capital expenditure	NA
Current expenditure	NA
Percent current	NA

U.S. Patents Issued [18]

Values show patents issued to citizens of the country by the U.S. Patents Office.

	1993	1994	1995
Number of patents	8	16	8

For sources, notes, and explanations, see Annotated Source Appendix, page 1061.

Government and Law

Organization of Government [19]

Long-form name:
Hellenic Republic

Type:
Parliamentary republic; monarchy rejected by referendum 8 December 1974

Independence:
1829 (from the Ottoman Empire)

National holiday:
Independence Day, 25 March (1821) (proclamation of the war of independence)

Constitution:
11 June 1975

Legal system:
Based on codified Roman law; judiciary divided into civil, criminal, and administrative courts

Executive branch:
President; Prime Minister; Cabinet

Legislative branch:
Unicameral: Chamber of Deputies (Vouli ton Ellinon)

Judicial branch:
Supreme Judicial Court; Special Supreme Tribunal

Elections [20]

Chamber of Deputies	% of votes
New Democracy (ND)	46.9
Panhellenic Socialist Movement (PASOK)	38.6
Left Alliance	10.3
PASOK/Left Alliance	1.0
Ecologist-Alternative List	0.8
Democratic Renewal (DIANA)	0.7
Muslim independents	0.5

Government Expenditures [21]

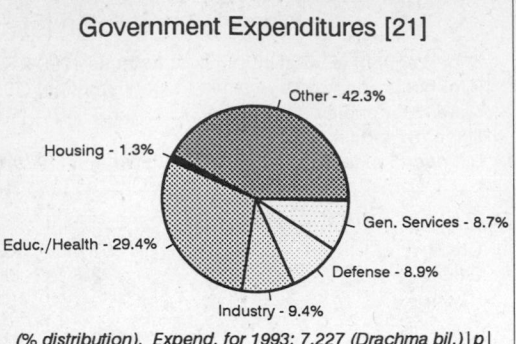

(% distribution). Expend. for 1993: 7,227 (Drachma bil.)|p|

Crime [23]

	1994
Crime volume	
Cases known to police	303,311
Attempts (percent)	0.40
Percent cases solved	82.60
Crimes per 100,000 persons	2,956.28
Persons responsible for offenses	
Total number offenders	237,840
Percent female	12.00
Percent juvenile (0-17 yrs.)	6.00
Percent foreigners	2.40

Military Expenditures and Arms Transfers [22]

	1990	1991	1992	1993	1994
Military expenditures					
Current dollars (mil.)	3,817	3,793	4,113	4,176	4,339
1994 constant dollars (mil.)	4,248	4,065	4,289	4,263	4,339
Armed forces (000)	201	205	208	213	206
Gross national product (GNP)					
Current dollars (mil.)	65,450	70,180	72,760	75,450	78,160
1994 constant dollars (mil.)	72,840	75,220	75,870	77,000	78,160
Central government expenditures (CGE)					
1994 constant dollars (mil.)	48,300	34,150	34,190	35,240	37,600[e]
People (mil.)	10.1	10.2	10.4	10.5	10.6
Military expenditure as % of GNP	5.8	5.4	5.7	5.5	5.6
Military expenditure as % of CGE	8.8	11.9	12.5	12.1	11.5
Military expenditure per capita (1994 $)	420	397	414	407	411
Armed forces per 1,000 people (soldiers)	19.9	20.0	20.1	20.3	19.5
GNP per capita (1994 $)	7,195	7,339	7,320	7,354	7,398
Arms imports[6]					
Current dollars (mil.)	450	280	725	825	270
1994 constant dollars (mil.)	501	300	756	842	270
Arms exports[6]					
Current dollars (mil.)	20	0	20	10	5
1994 constant dollars (mil.)	22	0	21	10	5
Total imports[7]					
Current dollars (mil.)	19,780	21,580	23,220	20,820	NA
1994 constant dollars (mil.)	22,010	23,130	24,210	21,250	NA
Total exports[7]					
Current dollars (mil.)	8,105	8,666	9,509	8,434	NA
1994 constant dollars (mil.)	9,020	9,289	9,916	8,608	NA
Arms as percent of total imports[8]	2.3	1.3	3.1	4.0	NA
Arms as percent of total exports[8]	0.2	0	0.2	0.1	NA

Human Rights [24]

	SSTS	FL	FAPRO	PPCG	APROBC	TPW	PCPTW	STPEP	PHRFF	PRW	ASST	AFL
Observes	P	P	P	P	P	P	P		P	P	P	P
	EAFRD	CPR	ESCR	SR	ACHR	MAAE	PVIAC	PVNAC	EAFDAW	TCIDTP	RC
Observes		P		P	P		P	P	P	P	P	P

P=Party; S=Signatory; see Appendix for meaning of abbreviations.

Labor Force

Total Labor Force [25]

4.077 million

Labor Force by Occupation [26]

Services	52%
Agriculture	23
Industry	25

Date of data: 1994

Unemployment Rate [27]

9.6% (1995 est.)

Production Sectors

Commercial Energy Production and Consumption

Data are shown in quadrillion (10^{15}) BTUs and percent for 1995
Values for hydroelectric, nuclear, geothermal, solar, and wind power refer to electrical generation.

Production [28]

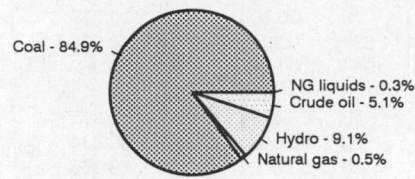

Coal - 84.9%
NG liquids - 0.3%
Crude oil - 5.1%
Hydro - 9.1%
Natural gas - 0.5%

Consumption [29]

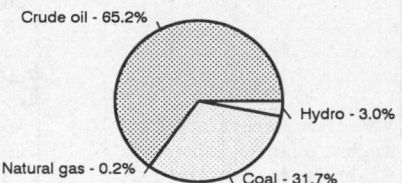

Crude oil - 65.2%
Hydro - 3.0%
Natural gas - 0.2%
Coal - 31.7%

Crude oil	0.019		Crude oil	0.750
Natural gas liquids	0.001		Dry natural gas	0.002
Dry natural gas	0.002		Coal	0.365
Coal	0.316		Net hydroelectric power	0.034
Net hydroelectric power	0.034		Total	1.151
Total	0.372			

Telecommunications [30]

- 5,571,293 (1993 est.) telephones; adequate, modern networks reach all areas; microwave radio relay carries most traffic; extensive open-wire network; submarine cables to off-shore islands
- Domestic: microwave radio relay, open wire, and submarine cable
- International: tropospheric scatter; 8 submarine cables; satellite earth stations - 2 Intelsat (1 Atlantic Ocean and 1 Indian Ocean), 1 Eutelsat, and 1 Inmarsat (Indian Ocean Region)
- Radio: Broadcast stations: AM 29, FM 17 (repeaters 20), shortwave 0
- Television: Broadcast stations: 361 (1987 est.) Televisions: 2.3 million (1993 est.)

Transportation [31]

Railways: total: 2,474 km; standard gauge: 1,565 km 1.435-m gauge (36 km electrified; 100 km double track); narrow gauge: 887 km 1.000-m gauge; 22 km 0.750-m gauge (a rack type railway for steep grades)

Highways: total: 130,000 km; paved: 119,210 km (including 116 km of expressways); unpaved: 10,790 km (1990 est.)

Merchant marine: total: 1,051 ships (1,000 GRT or over) totaling 28,842,200 GRT/52,583,281 DWT; ships by type: bulk 468, cargo 92, chemical tanker 23, combination bulk 22, combination ore/oil 26, container 40, liquefied gas tanker 4, oil tanker 245, passenger 15, passenger-cargo 3, refrigerated cargo 8, roll-on/roll-off cargo 17, short-sea passenger 84, specialized tanker 3, vehicle carrier 1

Top Agricultural Products [32]

Agriculture accounts for 11.8% of the GDP; produces wheat, corn, barley, sugar beets, olives, tomatoes, wine, tobacco, potatoes; meat, dairy products.

Airports

Total:	77
With paved runways over 3,047 m:	5
With paved runways 2,438 to 3,047 m:	15
With paved runways 1,524 to 2,437 m:	16
With paved runways 914 to 1,523 m:	17
With paved runways under 914 m:	21

Top Mining Products [33]

Thousand metric tons except as noted[e]	5/95[*]
Bauxite	2,170
Alumina, A12O3 equivalent	525
Steel, crude	848
Nickel, ore, gross weight	2,000
Bentonite, crude	500
Pozzolan (Santorin earth)	500
Pumice	500
Marble (000 cu. meters)	400
Coal, lignite	48,500
Petroleum, crude (000 metric tons)	500

Tourism [34]

	1990	1991	1992	1993	1994
Visitors[1]	9,310	8,271	9,756	9,913	11,302
Tourists[43]	8,873	8,036	9,331	9,413	10,713
Cruise passengers	437	235	425	500	589
Tourism receipts	2,587	2,567	3,272	3,335	3,905
Tourism expenditures	1,090	1,015	1,186	1,003	1,125
Fare receipts	30	45	92	67	48
Fare expenditures	147	158	164	99	130

Travelers are in thousands, money in million U.S. dollars.

For sources, notes, and explanations, see Annotated Source Appendix, page 1061.

369

Manufacturing Sector

Manufacturing Summary [35]

	1987		1988		1989		1990		1991	
	$ bil.	%	$ bil.	%	$ bil.	%	$ bil.	%	$ bil.	%
Establishments or enterprises (number)	9,063	0.386	9,058	0.431	8,991	0.480	9,020	0.504	-	-
Total employment (000)	406	0.299	407	0.297	407	0.331	400	0.362	-	-
Production workers (000)	267	0.395	263	0.422	258	0.351	250	0.352	-	-
Output ($ bil.)	25	0.247	28	0.237	29	0.245	34	0.298	-	-
Value added ($ bil.)	8	0.181	9	0.183	9	0.186	11	0.219	-	-
Capital investment ($ mil.)	1,236	0.284	1,423	0.299	1,597	0.292	1,801	0.324	-	-
M & E investment ($ mil.)	769	0.247	862	0.249	954	0.244	1,133	0.269	-	-
Employees per establishment (number)	45	77.452	45	68.894	45	68.881	44	71.855	-	-
Production workers per establishment	29	102.290	29	97.698	29	73.045	28	69.894	-	-
Output per establishment ($ mil.)	3	63.910	3	54.967	3	51.001	4	59.106	-	-
Capital investment per estab. ($ mil.)	0.136	73.434	0.157	69.244	0.178	60.733	0.200	64.364	-	-
M & E per establishment ($ mil)	0.085	63.861	0.095	57.697	0.106	50.917	0.126	53.394	-	-
Payroll per employee ($)	8,551	95.362	9,763	98.001	10,458	93.608	13,134	94.680	-	-
Wages per production worker ($)	7,222	91.531	8,165	96.631	8,814	87.891	10,813	91.048	-	-
Hours per production worker (hours)	-		-		-		-		-	-
Output per employee ($)	61,800	82.516	67,957	79.785	70,962	74.042	85,246	82.257	-	-
Capital investment per employee ($)	3,045	94.812	3,497	100.508	3,924	88.170	4,498	89.574	-	-
M & E per employee ($)	1,893	82.453	2,118	83.747	2,345	73.920	2,830	74.307	-	-

Note: Columns headed % show percent of world total or ratio. Ratios closest to 100 are closest to world average. M & E stands for machinery & equipment.

Output in Manufacturing

	1987		1988		1989		1990		1991	
	$ bil.[f]	%	$ bil.[f]	%	$ bil.[f]	%	$ bil.[f]	%	$ bil.[f]	%
3110 - Food products	3.933	15.73	4.232	15.11	4.736	16.33	5.579	16.41	-	-
3130 - Beverages	0.876	3.50	1.054	3.76	1.050	3.62	1.266	3.72	-	-
3140 - Tobacco	0.665	2.66	0.768	2.74	0.755	2.60	1.095	3.22	-	-
3210 - Textiles	3.148	12.59	3.059	10.93	2.827	9.75	3.173	9.33	-	-
3211 - Spinning, weaving, etc.	2.229	8.92	2.169	7.75	1.894	6.53	2.206	6.49	-	-
3220 - Wearing apparel	0.804	3.22	0.895	3.20	0.902	3.11	1.203	3.54	-	-
3230 - Leather and products	0.185	0.74	0.185	0.66	0.215	0.74	0.207	0.61	-	-
3240 - Footwear	0.215	0.86	0.251	0.90	0.219	0.76	0.268	0.79	-	-
3310 - Wood products	0.318	1.27	0.352	1.26	0.393	1.36	0.500	1.47	-	-
3320 - Furniture, fixtures	0.126	0.50	0.138	0.49	0.159	0.55	0.210	0.62	-	-
3410 - Paper and products	0.583	2.33	0.617	2.20	0.627	2.16	0.724	2.13	-	-
3411 - Pulp, paper, etc.	0.263	1.05	0.281	1.00	0.288	0.99	0.319	0.94	-	-
3420 - Printing, publishing	0.368	1.47	0.434	1.55	0.446	1.54	0.558	1.64	-	-
3510 - Industrial chemicals	0.710	2.84	0.756	2.70	0.763	2.63	0.845	2.49	-	-
3511 - Basic chemicals, excl fertilizers	0.181	0.72	0.208	0.74	0.215	0.74	0.242	0.71	-	-
3513 - Synthetic resins, etc.	0.142	0.57	0.152	0.54	0.153	0.53	0.156	0.46	-	-
3520 - Chemical products nec	1.012	4.05	1.201	4.29	1.298	4.48	1.605	4.72	-	-
3522 - Drugs and medicines	0.346	1.38	0.425	1.52	0.454	1.57	0.632	1.86	-	-
3530 - Petroleum refineries	2.172	8.69	2.213	7.90	2.460	8.48	3.156	9.28	-	-
3540 - Petroleum, coal products	0.112	0.45	0.093	0.33	0.094	0.32	0.112	0.33	-	-
3550 - Rubber products	0.187	0.75	0.192	0.69	0.203	0.70	0.223	0.66	-	-
3560 - Plastic products nec	0.471	1.88	0.544	1.94	0.601	2.07	0.746	2.19	-	-
3610 - Pottery, china, etc.	0.098	0.39	0.098	0.35	0.104	0.36	0.126	0.37	-	-
3620 - Glass and products	0.100	0.40	0.101	0.36	0.110	0.38	0.112	0.33	-	-
3690 - Nonmetal products nec	1.055	4.22	1.184	4.23	1.222	4.21	1.527	4.49	-	-
3710 - Iron and steel	0.820	3.28	1.045	3.73	1.261	4.35	1.249	3.67	-	-
3720 - Nonferrous metals	0.642	2.57	1.078	3.85	1.044	3.60	1.192	3.51	-	-
3810 - Metal products	0.941	3.76	1.061	3.79	1.100	3.79	1.238	3.64	-	-
3820 - Machinery nec	0.326	1.30	0.365	1.30	0.411	1.42	0.425	1.25	-	-
3825 - Office, computing machinery	0.003	0.01	0.004	0.01	0.004	0.01	0.004	0.01	-	-
3830 - Electrical machinery	0.808	3.23	0.962	3.44	1.060	3.66	1.211	3.56	-	-
3832 - Radio, television, etc.	0.145	0.58	0.180	0.64	0.207	0.71	0.225	0.66	-	-
3840 - Transportation equipment	0.578	2.31	0.704	2.51	0.845	2.91	0.947	2.79	-	-
3841 - Shipbuilding, repair	0.199	0.80	0.275	0.98	0.329	1.13	0.372	1.09	-	-
3843 - Motor vehicles	0.214	0.86	0.228	0.81	0.285	0.98	0.335	0.99	-	-
3850 - Professional goods	0.022	0.09	0.030	0.11	0.031	0.11	0.034	0.10	-	-
3900 - Industries nec	0.103	0.41	0.115	0.41	0.113	0.39	0.119	0.35	-	-

Note: Codes are International Standard Industry codes (ISIC). Percentages are % of total Output. [f]: Factor Prices; [p]: Producer Prices.

For sources, notes, and explanations, see Annotated Source Appendix, page 1061.

Finance, Economics, and Trade

Economic Indicators [36]

- **National product**: GDP—purchasing power parity—$101.7 billion (1995 est.)

- **National product real growth rate**: 1.7% (1995 est.)

- **National product per capita**: $9,500 (1995 est.)

- **Inflation rate (consumer prices)**: 8.1% (1995 est.)

- **External debt**: $31.2 billion (1995 est.)

Balance of Payments Summary [37]

Values in millions of dollars.

	1989	1990	1991	1992	1993
Exports of goods (f.o.b.)	5,994	6,365	6,797	6,009	5,035
Imports of goods (f.o.b.)	-13,377	-16,543	-16,909	-17,612	-15,592
Trade balance	-7,383	-10,178	-10,112	-11,603	-10,557
Services - debits	-4,352	-5,045	-5,402	-6,331	-5,907
Services - credits	5,191	6,968	7,757	9,319	9,218
Private transfers (net)	1,381	1,817	2,149	2,417	2,414
Government transfers (net)	2,602	2,901	4,034	4,058	4,085
Long-term capital (net)	1,941	2,975	3,587	2,191	5,351
Short-term capital (net)	817	787	-170	614	-954
Errors and omissions	-538	-185	-183	-853	-631
Overall balance	-341	40	1,660	-188	3,019

Exchange Rates [38]

Currency: **drachma.**
Symbol: **Dr.**

Data are currency units per $1.

January 1996	240.21
1995	231.60
1994	242.60
1993	229.26
1992	190.62
1991	182.27

Imports and Exports

Top Import Origins [39]

$21.9 billion (f.o.b., 1994) Data are for 1994.

Origins	%
EU	70
Germany	17
Italy	17
France	8
UK	6
Japan	3

Top Export Destinations [40]

$8.8 billion (f.o.b., 1994) Data are for 1994.

Destinations	%
EU	55
Germany	21
Italy	14
UK	6
France	5.4
US	5

Foreign Aid [41]

Recipient: ODA, $NA.

Import and Export Commodities [42]

Import Commodities	Export Commodities
Manufactured goods 72%	Manufactured goods 53%
Foodstuffs 15%	Foodstuffs 34%
Fuels 10%	Fuels 5%

For sources, notes, and explanations, see Annotated Source Appendix, page 1061.

371

Grenada

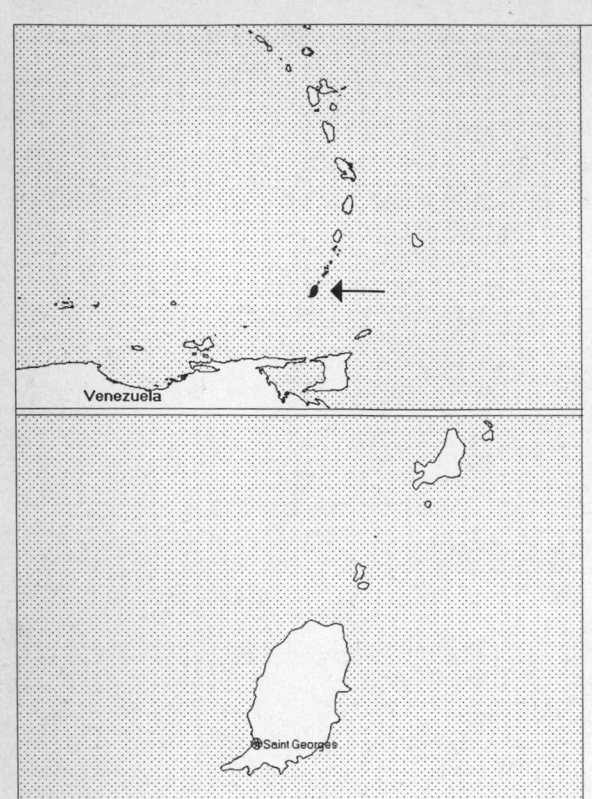

Geography [1]

Total area:
  340 sq km   131 sq mi
Land area:
  340 sq km   131 sq mi
Comparative area:
  Twice the size of Washington, DC
Land boundaries:
  0 km
Coastline:
  121 km
Climate:
  Tropical; tempered by northeast trade winds
Terrain:
  Volcanic in origin with central mountains
Natural resources:
  Timber, tropical fruit, deepwater harbors
Land use:
  Arable land: 15%
  Permanent crops: 26%
  Meadows and pastures: 3%
  Forest and woodland: 9%
  Other: 47%

Demographics [2]

	1970	1980	1990	1995[1]	1996	2000	2010	2020	2030
Population	95	90	94	94	95	98	115	141	167
Population density (persons per sq. mi.)	729	689	715	722	726	748	877	1,080	1,274
(persons per sq. km.)	281	266	276	279	280	289	339	417	492
Net migration rate (per 1,000 population)	-12.3	-35.8	-25.9	-19.2	-17.9	-12.4	0.0	0.0	0.0
Births	2	NA	NA	NA	3	NA	NA	NA	NA
Deaths	0	NA	NA	NA	1	NA	NA	NA	NA
Life expectancy - males	NA	NA	67.2	68.2	68.4	69.2	70.9	72.5	73.8
Life expectancy - females	NA	NA	71.7	73.2	73.4	74.6	77.1	79.1	80.8
Birth rate (per 1,000)	27.8	29.4	32.4	29.7	29.1	27.2	25.7	22.4	18.4
Death rate (per 1,000)	7.8	8.0	7.3	6.0	5.7	5.0	3.9	3.5	3.9
Women of reproductive age (15-49 yrs.)	NA	NA	21	21	21	22	30	38	45
of which are currently married	5	NA	9	NA	10	10	14	NA	NA
Fertility rate	NA	NA	4.2	3.9	3.8	3.5	2.9	2.5	2.3

Except as noted, values for vital statistics are in thousands; life expectancy is in years.

Health

Health Indicators [3]

Health Expenditures [4]

For sources, notes, and explanations, see Annotated Source Appendix, page 1061.

Human Factors

Women and Children [5]

Burden of Disease [6]

Population per physician (1993)	1,769.23
Population per nurse	NA
Population per hospital bed (1993)	125.00
AIDS cases per 100,000 people (1994)	6.7
Malaria cases per 100,000 people (1992)	NA

Ethnic Division [7]
Black African.

Religion [8]
Roman Catholic, Anglican, other Protestant sects.

Major Languages [9]
English (official), French patois.

Education

Public Education Expenditures [10]

Million (E. Carib. Dollar)	1980	1990	1991	1992	1993	1994
Total education expenditure	NA	NA	NA	NA	NA	NA
as percent of GNP	NA	NA	NA	NA	NA	NA
as percent of total govt. expend.	NA	NA	NA	NA	NA	NA
Current education expenditure	NA	30	32	31	31	NA
as percent of GNP	NA	5.9	5.8	5.4	5.3	NA
as percent of current govt. expend.	NA	NA	NA	NA	NA	NA
Capital expenditure	NA	NA	NA	NA	NA	NA

Educational Attainment [11]

Age group (1981)	25+
Total population	33,401
Highest level attained (%)	
No schooling	2.2
First level	
Not completed	87.8
Completed	NA
Entered second level	
S-1	8.5
S-2	NA
Postsecondary	1.5

Illiteracy [12]

	1970	1980	1990
Illiterate population (15+ years)[4]	1,070	NA	NA
Illiteracy rate - total pop. (%)	2.2	NA	NA
Illiteracy rate - males (%)	2.0	NA	NA
Illiteracy rate - females (%)	2.4	NA	NA

Libraries [13]

	Admin. Units	Svc. Pts.	Vols. (000)	Shelving (meters)	Vols. Added	Reg. Users
National	NA	NA	NA	NA	NA	NA
Nonspecialized	NA	NA	NA	NA	NA	NA
Public (1992)	4	8	64	NA	1,540	1,245
Higher ed.	NA	NA	NA	NA	NA	NA
School	NA	NA	NA	NA	NA	NA

Daily Newspapers [14]

	1980	1985	1990	1994
Number of papers	1	-	-	-
Circ. (000)	4[e]	-	-	-

Culture [15]

Science and Technology

Scientific/Technical Forces [16]

R&D Expenditures [17]

U.S. Patents Issued [18]

For sources, notes, and explanations, see Annotated Source Appendix, page 1061.

373

Government and Law

Organization of Government [19]

Long-form name:
None
Type:
Parliamentary democracy
Independence:
7 February 1974 (from UK)
National holiday:
Independence Day, 7 February (1974)
Constitution:
19 December 1973
Legal system:
Based on English common law
Executive branch:
British Monarch (represented by Governor General); Prime Minister; Cabinet
Legislative branch:
Bicameral Parliament: Senate and House of Representatives
Judicial branch:
West Indies Associate States Supreme Court

Elections [20]

House of Representatives	% of seats
New National Party (NNP)	53.3
National Democratic Congress (NDC)	33.3
Grenada United Labor Party (GULP)	13.3

Government Expenditures [21]

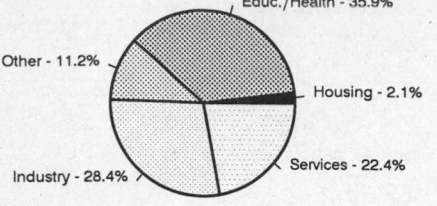

Educ./Health - 35.9%
Other - 11.2%
Housing - 2.1%
Services - 22.4%
Industry - 28.4%

(% distribution). Expend. for CY95: 209.69 (Dollar mil.)

Defense Summary [22]

Branches: Royal Grenada Police Force, Coast Guard

Manpower Availability: Males age 15-49 NA; males fit for military service NA

Defense Expenditures: NA

Crime [23]

	1994
Crime volume	
Cases known to police	7,689
Attempts (percent)	NA
Percent cases solved	82.50
Crimes per 100,000 persons	8,543.33
Persons responsible for offenses	
Total number offenders	NA
Percent female	NA
Percent juvenile (0-20 yrs.)	NA
Percent foreigners	NA

Human Rights [24]

	SSTS	FL	FAPRO	PPCG	APROBC	TPW	PCPTW	STPEP	PHRFF	PRW	ASST	AFL
Observes	1	P		1	P	P	P			1	1	P
	EAFRD	CPR	ESCR	SR	ACHR	MAAE	PVIAC	PVNAC	EAFDAW	TCIDTP	RC
Observes	S	P	P		P				P		P

P = Party; S = Signatory; see Appendix for meaning of abbreviations.

Labor Force

Total Labor Force [25]

36,000

Labor Force by Occupation [26]

Services	31%
Agriculture	24
Construction	8
Manufacturing	5
Other	32

Date of data: 1985

Unemployment Rate [27]

14% (1995 est.)

For sources, notes, and explanations, see Annotated Source Appendix, page 1061.

Production Sectors

Energy Resource Summary [28]

Electricity: Capacity: 12,500 kW. Production: 60 million kWh. Consumption per capita: 639 kWh (1993).

Telecommunications [30]

- 5,650 (1988 est.) telephones; automatic, islandwide telephone system
- Domestic: interisland VHF and UHF radiotelephone links
- International: new SHF radiotelephone links to Trinidad and Tobago and Saint Vincent; VHF and UHF radio links to Trinidad
- Radio: Broadcast stations: AM 1, FM 0, shortwave 0 Radios: 80,000 (1993 est.)
- Television: Broadcast stations: 1 (1988 est.) Televisions: 30,000 (1993 est.)

Top Agricultural Products [32]

Agriculture accounts for 10.2% of the GDP; produces bananas, cocoa, nutmeg, mace, citrus, avocados, root crops, sugarcane, corn, vegetables.

Top Mining Products [33]

Detailed information is not available. A summary of mineral resources follows. **Mineral Resources**: None.

Transportation [31]

Railways: 0 km

Highways: total: 994 km; paved: 597 km; unpaved: 397 km (1988 est.)

Merchant marine: none

Airports

Total:	3
With paved runways 2,438 to 3,047 m:	1
With paved runways 1,524 to 2,437 m:	1
With paved runways under 914 m:	1 (1995 est.)

Tourism [34]

	1990	1991	1992	1993	1994
Visitors	265	288	291	301	318
Tourists[44]	76	85	88	94	109
Excursionists	6	7	7	7	8
Cruise passengers	183	196	196	200	201
Tourism receipts	38	42	38	45	55
Tourism expenditures	5	5	4	4	4
Fare expenditures	6	7	7	6	9

Travelers are in thousands, money in million U.S. dollars.

For sources, notes, and explanations, see Annotated Source Appendix, page 1061.

375

Finance, Economics, and Trade

Industrial Summary [35]

Industrial Production: Growth rate 1.8% (1992 est.); accounts for 40.3% of the GDP. **Industries**: Food and beverages, textiles, light assembly operations, tourism, construction.

Economic Indicators [36]

- **National product**: GDP—purchasing power parity—$284 million (1995 est.)
- **National product real growth rate**: 3% (1995 est.)
- **National product per capita**: $3,000 (1995 est.)
- **Inflation rate (consumer prices)**: 3% (1995 est.)
- **External debt**: $89.1 million (1995 est.)

Balance of Payments Summary [37]

Values in millions of dollars.

	1988	1989	1990	1991	1992
Exports of goods (f.o.b.)	32.8	28.0	26.6	23.2	19.9
Imports of goods (f.o.b.)	-92.2	-99.0	-106.3	-113.6	-103.2
Trade balance	-59.4	-70.8	-79.7	-90.4	-83.3
Services - debits	-38.2	-43.3	-48.6	-47.8	-47.4
Services - credits	56.1	57.4	67.7	76.3	80.0
Private transfers (net)	15.6	16.7	17.0	18.7	17.5
Government transfers (net)	9.4	9.6	15.5	8.4	8.2
Long-term capital (net)	19.0	11.0	23.3	23.9	26.3
Short-term capital (net)	-5.6	23.9	-5.2	3.8	-1.1
Errors and omissions	-2.1	-5.3	12.4	9.6	7.9
Overall balance	-5.2	-0.8	2.5	2.5	8.1

Exchange Rates [38]

Currency: **EC dollar.**
Symbol: **EC$.**

Data are currency units per $1.

Fixed rate since 1976	2.70

Imports and Exports

Top Import Origins [39]

$162.2 million (f.o.b., 1995 est.) Data are for 1991.

Origins	%
US	31.2
Caricom	23.6
UK	13.8
Japan	7.1

Top Export Destinations [40]

$24.2 million (f.o.b., 1995 est.) Data are for 1991.

Destinations	%
Caricom	32.3
UK	20
US	13
Netherlands	8.8

Foreign Aid [41]

Recipient: ODA, $NA.

Import and Export Commodities [42]

Import Commodities	Export Commodities
Food 25%	Bananas
Manufactured goods 22%	Cocoa
Machinery 20%	Nutmeg
Chemicals 10%	Fruit and vegetables
Fuel 6%	Clothing
	Mace

Guatemala

Geography [1]

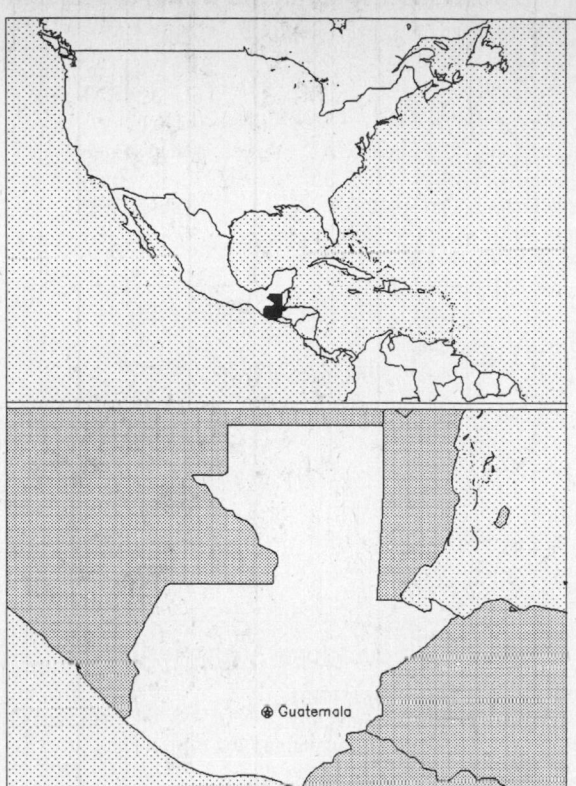

Total area:
108,890 sq km   42,043 sq mi
Land area:
108,430 sq km   41,865 sq mi
Comparative area:
Slightly smaller than Tennessee
Land boundaries:
Total 1,687 km, Belize 266 km, El Salvador 203 km, Honduras 256 km, Mexico 962 km
Coastline:
400 km
Climate:
Tropical; hot, humid in lowlands; cooler in highlands
Terrain:
Mostly mountains with narrow coastal plains and rolling limestone plateau (Peten)
Natural resources:
Petroleum, nickel, rare woods, fish, chicle
Land use:
Arable land: 12%
Permanent crops: 4%
Meadows and pastures: 12%
Forest and woodland: 40%
Other: 32%

Demographics [2]

	1970	1980	1990	1995[1]	1996	2000	2010	2020	2030
Population	5,287	7,232	9,633	10,999	11,278	12,408	15,284	18,131	20,790
Population density (persons per sq. mi.)	126	173	230	263	269	296	365	433	497
(persons per sq. km.)	49	67	89	101	104	114	141	167	192
Net migration rate (per 1,000 population)	-0.4	0.5	-2.4	-2.0	-2.0	-1.8	-1.3	-1.1	-0.9
Births	NA	NA	NA	NA	383	NA	NA	NA	NA
Deaths	NA	NA	NA	NA	81	NA	NA	NA	NA
Life expectancy - males	51.1	55.1	60.3	62.3	62.6	64.2	67.6	70.5	72.9
Life expectancy - females	53.7	59.6	65.3	67.6	68.0	69.7	73.5	76.6	79.1
Birth rate (per 1,000)	45.1	42.6	38.6	34.7	34.0	31.2	25.6	21.4	18.1
Death rate (per 1,000)	14.4	11.5	8.4	7.3	7.2	6.5	5.5	5.0	5.2
Women of reproductive age (15-49 yrs.)	1,173	1,605	2,176	2,561	2,644	2,986	3,949	4,883	5,591
of which are currently married	NA	NA	1,371	NA	1,666	1,893	2,548	NA	NA
Fertility rate	6.5	6.0	5.3	4.6	4.5	4.0	3.0	2.5	2.2

Except as noted, values for vital statistics are in thousands; life expectancy is in years.

Health

Health Indicators [3]

% of population with access to	
safe water (1990-95)	62
adequate sanitation (1990-95)	60
health services (1985-95)	34
% of 1-year-olds immunized (1990-94) against	
TB (tuberculosis)	70
DPT (diphtheria, pertussis, tetanus)	71
polio	73
measles	66
% of contraceptive prevalence (1980-94)	23
ORT use rate (1990-94)	24

Health Expenditures [4]

Total health expenditure, 1990 (official exchange rate)	
Millions of dollars	283
Dollars per capita	31
Health expenditures as a percentage of GDP	
Total	3.7
Public sector	2.1
Private sector	1.6
Development assistance for health	
Total aid flows (millions of dollars)[1]	32
Aid flows per capita (dollars)	3.4
Aid flows as a percentage of total health expenditure	11.1

For sources, notes, and explanations, see Annotated Source Appendix, page 1061.

Human Factors

Women and Children [5]

% of pregnant women immunized (tetanus 1990-94)	11
% of births attended by trained health personnel (1983-94)	51
Maternal mortality rate (1980-92)	200
Under-5 mortality rate (1994)	70
% under-5 moderately/severely underweight (1980-1994)[1]	34

Burden of Disease [6]

Population per physician (1984)	2,183.69
Population per nurse (1984)	851.09
Population per hospital bed (1990)	673.38
AIDS cases per 100,000 people (1994)	1.1
Malaria cases per 100,000 people (1992)	591

Ethnic Division [7]

Mestizo	56%
Amerindian or predominantly Amerindian	44%

Religion [8]

Roman Catholic, Protestant, traditional Mayan.

Major Languages [9]

Spanish	60%
Indian language (23 Indian dialects; primarily Quiche, Cakchiquel, Kekchi)	40%

Education

Public Education Expenditures [10]

	1980	1990	1991	1992	1993	1994
Million (Quetzal)[30]						
Total education expenditure	NA	468	606	774	995	NA
as percent of GNP	NA	1.4	1.3	1.5	1.6	NA
as percent of total govt. expend.	NA	11.8	13.0	NA	12.8	NA
Current education expenditure	12,426	NA	NA	NA	NA	NA
as percent of GNP	4.2	NA	NA	NA	NA	NA
as percent of current govt. expend.	36.8	NA	NA	NA	NA	NA
Capital expenditure	NA	NA	NA	NA	NA	NA

Educational Attainment [11]

Age group (1981)	25+
Total population	2,060,399
Highest level attained (%)	
No schooling	55.0
First level	
Not completed	27.3
Completed	8.6
Entered second level	
S-1	2.9
S-2	4.0
Postsecondary	2.2

Illiteracy [12]

In thousands and percent[1]	1990	1995	2000
Illiterate population (15+ yrs.)	2,346	2,627	2,939
Illiteracy rate - total pop. (%)	44.1	42.0	40.2
Illiteracy rate - males (%)	37.6	35.7	34.1
Illiteracy rate - females (%)	50.5	48.3	46.4

Libraries [13]

	Admin. Units	Svc. Pts.	Vols. (000)	Shelving (meters)	Vols. Added	Reg. Users
National	NA	NA	NA	NA	NA	NA
Nonspecialized	NA	NA	NA	NA	NA	NA
Public (1989)	1	202	1,169	NA	45,271	689,593
Higher ed. (1987)	1	1	133	NA	2,524	54,496
School	NA	NA	NA	NA	NA	NA

Daily Newspapers [14]

	1980	1985	1990	1994
Number of papers	9	9	5	5
Circ. (000)	200[e]	250[e]	190	240

Culture [15]

Cinema (seats per 1,000)	6.8
Annual attendance per person	0.9
Gross box office receipts (mil. Quetzal)	13
Museums (reporting)	NA
Visitors (000)	NA
Annual receipts (000 Quetzal)	NA

Science and Technology

Scientific/Technical Forces [16]

Scientists/engineers	858
Number female	NA
Technicians	925
Number female	NA
Total[14]	1,783

R&D Expenditures [17]

	Quetzal (000) 1988
Total expenditure[15]	31,859
Capital expenditure	NA
Current expenditure	NA
Percent current	NA

U.S. Patents Issued [18]

Values show patents issued to citizens of the country by the U.S. Patents Office.

	1993	1994	1995
Number of patents	1	2	0

378

For sources, notes, and explanations, see Annotated Source Appendix, page 1061.

Government and Law

Organization of Government [19]

Long-form name:
Republic of Guatemala
Type:
Republic
Independence:
15 September 1821 (from Spain)
National holiday:
Independence Day, 15 September (1821)
Constitution:
31 May 1985, effective 14 January 1986
Note: Suspended 25 May 1993 by
President Serrano; reinstated 5 June 1993
following ouster of president
Legal system:
Civil law system; judicial review of
legislative acts; has not accepted
compulsory ICJ jurisdiction
Executive branch:
President; Vice President; Council of
Ministers
Legislative branch:
Unicameral: Congress of the Republic
(Congreso de la Republica)
Judicial branch:
Supreme Court of Justice (Corte Suprema
de Justicia); Court of Constitutionality

Elections [20]

Congress of the Republic	% of votes
National Advancement Party (PAN)	53.8
Guatemalan Republican Front (FRG)	26.3
New Guatemalan Dem. Front (FDNG)	7.5
Christian Democratic Party (DCG)	5.0
National Centrist Union (UCN)	3.8
Democratic Union (UD)	2.5
National Liberation Movement (MLN)	1.3

Government Expenditures [21]

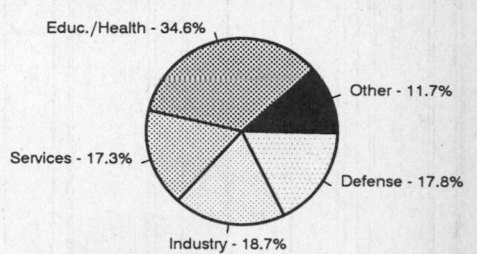

Educ./Health - 34.6%
Other - 11.7%
Services - 17.3%
Defense - 17.8%
Industry - 18.7%

(% distribution). Expend. for CY94: 6,648.98 (Quetzal mil.)

Military Expenditures and Arms Transfers [22]

	1990	1991	1992	1993	1994
Military expenditures					
Current dollars (mil.)	145	150	168	166	175
1994 constant dollars (mil.)	162	160	175	169	175
Armed forces (000)	43	43	44	44	34
Gross national product (GNP)					
Current dollars (mil.)	9,672	10,500	11,360	12,030	12,880
1994 constant dollars (mil.)	10,760	11,250	11,840	12,280	12,880
Central government expenditures (CGE)					
1994 constant dollars (mil.)	1,109	1,035	1,248	1,212	1,156
People (mil.)	9.6	9.9	10.2	10.4	10.7
Military expenditure as % of GNP	1.5	1.4	1.5	1.4	1.4
Military expenditure as % of CGE	14.6	15.5	14.0	14.0	15.2
Military expenditure per capita (1994 $)	17	16	17	16	16
Armed forces per 1,000 people (soldiers)	4.5	4.3	4.3	4.2	3.2
GNP per capita (1994 $)	1,117	1,136	1,164	1,176	1,202
Arms imports[6]					
Current dollars (mil.)	20	0	0	5	0
1994 constant dollars (mil.)	22	0	0	5	0
Arms exports[6]					
Current dollars (mil.)	0	0	0	0	0
1994 constant dollars (mil.)	0	0	0	0	0
Total imports[7]					
Current dollars (mil.)	1,649	1,851	2,532	2,599	2,604
1994 constant dollars (mil.)	1,835	1,984	2,640	2,653	2,604
Total exports[7]					
Current dollars (mil.)	1,163	1,202	1,295	1,340	1,522
1994 constant dollars (mil.)	1,294	1,288	1,350	1,368	1,522
Arms as percent of total imports[8]	1.2	0	0	0.2	0
Arms as percent of total exports[8]	0	0	0	0	0

Crime [23]

	1990
Crime volume	
Cases known to police	46,940
Attempts (percent)	28
Percent cases solved	72
Crimes per 100,000 persons	510.36
Persons responsible for offenses	
Total number offenders	NA
Percent female	NA
Percent juvenile	NA
Percent foreigners	NA

Human Rights [24]

	SSTS	FL	FAPRO	PPCG	APROBC	TPW	PCPTW	STPEP	PHRFF	PRW	ASST	AFL
Observes	P	P	P	P	P	P	P			P	P	P
		EAFRD	CPR	ESCR	SR	ACHR	MAAE	PVIAC	PVNAC	EAFDAW	TCIDTP	RC
Observes		P	P	P	P	P	P	P	P	P	P	P

P = Party; S = Signatory; see Appendix for meaning of abbreviations.

Labor Force

Total Labor Force [25]

3.2 million (1994 est.)

Labor Force by Occupation [26]

Agriculture	60%
Services	13
Manufacturing	12
Commerce	7
Construction	4
Transport	3
Utilities	0.7
Mining	0.3

Date of data: 1985

Unemployment Rate [27]

4.9%; underemployment 30%-40% (1994 est.)

For sources, notes, and explanations, see Annotated Source Appendix, page 1061.

Production Sectors

Commercial Energy Production and Consumption

Data are shown in quadrillion (10^{15}) BTUs and percent for 1995
Values for hydroelectric, nuclear, geothermal, solar, and wind power refer to electrical generation.

Production [28]

Hydro - 51.2%
Crude oil - 48.8%

Consumption [29]

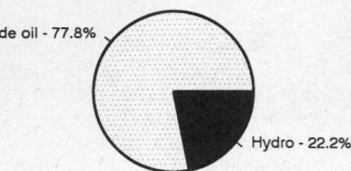

Crude oil - 77.8%
Hydro - 22.2%

Crude oil	0.021
Net hydroelectric power	0.022
Total	0.043

Crude oil	0.077
Net hydroelectric power	0.022
Total	0.099

Telecommunications [30]

- 210,000 (1993 est.) telephones; fairly modern network centered in the city of Guatemala
- International: connected to Central American Microwave System; satellite earth station - 1 Intelsat (Atlantic Ocean)
- Radio: Broadcast stations: AM 91, FM 0, shortwave 15 Radios: 400,000 (1993 est.)
- Television: Broadcast stations: 25 Televisions: 475,000 (1993 est.)

Transportation [31]

Railways: total: 884 km (102 km privately owned); narrow gauge: 884 km 0.914-m gauge (single track)

Highways: total: 12,033 km; paved: 3,117 km (including 125 km of expressways); unpaved: 8,916 km (1992 est.)

Merchant marine: none

Airports

Total: 463
With paved runways over 3,047 m: 1
With paved runways 2,438 to 3,047 m: 1
With paved runways 1,524 to 2,437 m: 2
With paved runways 914 to 1,523 m: 5
With paved runways under 914 m: 320

Top Agricultural Products [32]

Agriculture accounts for 25% of the GDP; produces sugarcane, corn, bananas, coffee, beans, cardamom; cattle, sheep, pigs, chickens.

Top Mining Products [33]

Thousand metric tons except as noted[e]	4/15/95[*]
Cement	1,450
Volcanic sand	100
Salt	100
Limestone	1,000
Sand and gravel	900
Schist	251
Stone, crushed	1,100
Gas, natural, gross (000 cu. meters)	10,000
Petroleum refinery products (000 42-gal. bls.)	5,100

Tourism [34]

	1990	1991	1992	1993	1994
Tourists	509	513	541	562	537
Tourism receipts	185	211	243	265	258
Tourism expenditures	100	67	103	116	161
Fare receipts	NA	NA	15	12	14
Fare expenditures	7	8	6	5	6

Travelers are in thousands, money in million U.S. dollars.

For sources, notes, and explanations, see Annotated Source Appendix, page 1061.

Manufacturing Sector

Manufacturing Summary [35]

	1987		1988		1989		1990		1991	
	$ bil.	%	$ bil.	%	$ bil.	%	$ bil.	%	$ bil.	%
Establishments or enterprises (number)	2,346	0.100	2,514	0.120	2,619	0.140	-	-	-	-
Total employment (000)	74	0.054	105	0.077	-	-	-	-	-	-
Production workers (000)	51	0.075	75	0.120	-	-	-	-	-	-
Output ($ bil.)	1	0.014	3	0.022	-	-	-	-	-	-
Value added ($ bil.)	0.616	0.014	0.960	0.020	-	-	-	-	-	-
Capital investment ($ mil.)	34	0.008	63	0.013	-	-	-	-	-	-
M & E investment ($ mil.)	-		-		-	-	-	-	-	-
Employees per establishment (number)	32	54.497	42	64.327	-	-	-	-	-	-
Production workers per establishment	22	75.135	30	100.024	-	-	-	-	-	-
Output per establishment ($ mil.)	0.591	13.639	1	18.089	-	-	-	-	-	-
Capital investment per estab. ($ mil.)	0.015	7.865	0.025	11.073	-	-	-	-	-	-
M & E per establishment ($ mil)	-		-		-	-	-	-	-	-
Payroll per employee ($)	2,029	22.626	1,936	19.428	-	-	-	-	-	-
Wages per production worker ($)	1,451	18.394	1,424	16.853	-	-	-	-	-	-
Hours per production worker (hours)	2,239	117.981	2,315	120.452	-	-	-	-	-	-
Output per employee ($)	18,744	25.027	23,951	28.120	-	-	-	-	-	-
Capital investment per employee ($)	463	14.431	599	17.214	-	-	-	-	-	-
M & E per employee ($)	-		-		-	-	-	-	-	-

Note: Columns headed % show percent of world total or ratio. Ratios closest to 100 are closest to world average. M & E stands for machinery & equipment.

Output in Manufacturing

	1987		1988		1989		1990		1991	
	$ bil.[p]	%	$ bil.[p]	%	$ bil.[p]	%	$ bil.[p]	%	$ bil.[p]	%
3110 - Food products	0.457	45.70	0.723	24.10	-	-	-	-	-	-
3130 - Beverages	0.102	10.20	0.116	3.87	-	-	-	-	-	-
3140 - Tobacco	0.025	2.50	0.049	1.63	-	-	-	-	-	-
3210 - Textiles	0.064	6.40	0.134	4.47	-	-	-	-	-	-
3211 - Spinning, weaving, etc.	0.051	5.10	0.113	3.77	-	-	-	-	-	-
3220 - Wearing apparel	0.016	1.60	0.063	2.10	-	-	-	-	-	-
3230 - Leather and products	0.007	0.70	0.015	0.50	-	-	-	-	-	-
3240 - Footwear	0.012	1.20	0.024	0.80	-	-	-	-	-	-
3310 - Wood products	0.013	1.30	0.023	0.77	-	-	-	-	-	-
3320 - Furniture, fixtures	0.007	0.70	0.012	0.40	-	-	-	-	-	-
3410 - Paper and products	0.025	2.50	0.042	1.40	-	-	-	-	-	-
3411 - Pulp, paper, etc.	0.010	1.00	0.021	0.70	-	-	-	-	-	-
3420 - Printing, publishing	-	-	0.071	2.37	-	-	-	-	-	-
3510 - Industrial chemicals	0.048	4.80	0.105	3.50	-	-	-	-	-	-
3511 - Basic chemicals, excl fertilizers	0.001	0.10	0.002	0.07	-	-	-	-	-	-
3513 - Synthetic resins, etc.	0.000	0.00	0.002	0.07	-	-	-	-	-	-
3520 - Chemical products nec	0.119	11.90	0.298	9.93	-	-	-	-	-	-
3522 - Drugs and medicines	0.064	6.40	0.107	3.57	-	-	-	-	-	-
3530 - Petroleum refineries	0.068	6.80	0.102	3.40	-	-	-	-	-	-
3540 - Petroleum, coal products	0.000	0.00	0.001	0.03	-	-	-	-	-	-
3550 - Rubber products	0.042	4.20	0.041	1.37	-	-	-	-	-	-
3560 - Plastic products nec	0.024	2.40	0.080	2.67	-	-	-	-	-	-
3610 - Pottery, china, etc.	0.006	0.60	0.010	0.33	-	-	-	-	-	-
3620 - Glass and products	0.030	3.00	0.027	0.90	-	-	-	-	-	-
3690 - Nonmetal products nec	0.065	6.50	0.096	3.20	-	-	-	-	-	-
3710 - Iron and steel	0.048	4.80	0.065	2.17	-	-	-	-	-	-
3720 - Nonferrous metals	0.002	0.20	0.003	0.10	-	-	-	-	-	-
3810 - Metal products	0.029	2.90	0.049	1.63	-	-	-	-	-	-
3820 - Machinery nec	0.007	0.70	0.014	0.47	-	-	-	-	-	-
3825 - Office, computing machinery	-	-	0.001	0.03	-	-	-	-	-	-
3830 - Electrical machinery	0.018	1.80	0.078	2.60	-	-	-	-	-	-
3832 - Radio, television, etc.	0.005	0.50	0.010	0.33	-	-	-	-	-	-
3840 - Transportation equipment	0.009	0.90	0.008	0.27	-	-	-	-	-	-
3841 - Shipbuilding, repair	-	-	0.000	0.00	-	-	-	-	-	-
3843 - Motor vehicles	0.004	0.40	0.005	0.17	-	-	-	-	-	-
3850 - Professional goods	0.002	0.20	0.004	0.13	-	-	-	-	-	-
3900 - Industries nec	0.005	0.50	0.012	0.40	-	-	-	-	-	-

Note: Codes are International Standard Industry codes (ISIC). Percentages are % of total Output. [f]: Factor Prices; [p]: Producer Prices.

For sources, notes, and explanations, see Annotated Source Appendix, page 1061.

Finance, Economics, and Trade

Economic Indicators [36]

- **National product**: GDP—purchasing power parity—$36.7 billion (1995 est.)
- **National product real growth rate**: 4.9% (1995 est.)
- **National product per capita**: $3,300 (1995 est.)
- **Inflation rate (consumer prices)**: 9% (1995 est.)
- **External debt**: $3.1 billion (1995 est.)

Balance of Payments Summary [37]

Values in millions of dollars.

	1989	1990	1991	1992	1993
Exports of goods (f.o.b.)	1,126.1	1,211.4	1,230.0	1,283.7	1,363.2
Imports of goods (f.o.b.)	-1,484.4	-1,428.0	-1,673.0	-2,327.8	-2,384.0
Trade balance	-358.3	-216.6	-443.0	-1,044.1	-1,020.8
Services - debits	-587.3	-600.3	-523.1	-735.4	-765.6
Services - credits	328.7	377.0	522.7	683.1	721.5
Private transfers (net)	178.8	205.3	257.7	338.8	362.0
Government transfers (net)	71.0	21.7	2.0	51.7	1.2
Long-term capital (net)	126.0	33.0	224.0	278.0	314.0
Short-term capital (net)	255.0	114.0	432.0	326.0	427.0
Errors and omissions	54.7	36.2	83.3	81.8	85.2
Overall balance	68.6	-29.7	555.6	-20.1	124.5

Exchange Rates [38]

Currency: **quetzal.**
Symbol: **Q.**

Data are currency units per $1. Free market rates from 1991 forward.

December 1995	5.9346
1995	5.8103
1994	5.7512
1993	5.6354
1992	5.1706
1991	5.0289
May 1989 (BMR)	2.800

Imports and Exports

Top Import Origins [39]

$2.85 billion (c.i.f., 1995 est.).

Origins	%
US	44
Mexico	NA
Venezuela	NA
Japan	NA
Germany	NA

Top Export Destinations [40]

$2.3 billion (f.o.b., 1995 est.).

Destinations	%
US	30
El Salvador	NA
Costa Rica	NA
Germany	NA
Honduras	NA

Foreign Aid [41]

Recipient: ODA, $84 million (1993).

Import and Export Commodities [42]

Import Commodities	Export Commodities
Fuel and petroleum products	Coffee
Machinery	Sugar
Grain	Bananas
Fertilizers	Cardamom
Motor vehicles	Beef

Guinea

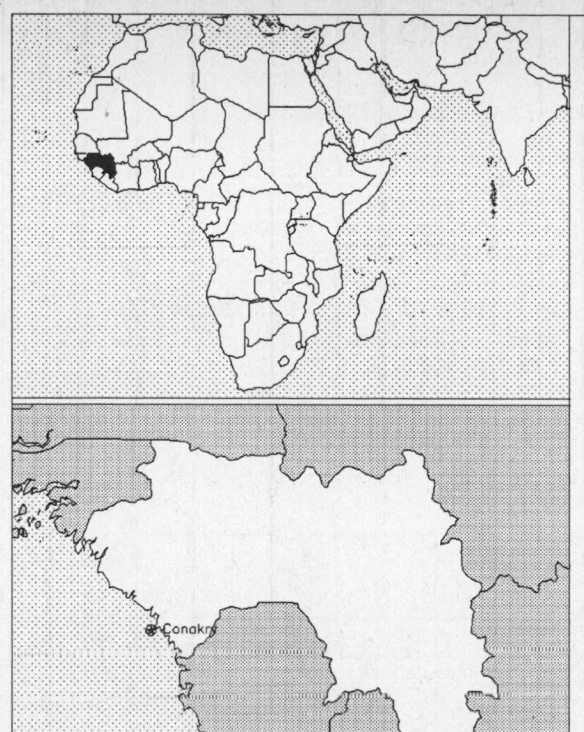

Geography [1]

Total area:
245,860 sq km   94,927 sq mi
Land area:
245,860 sq km   94,927 sq mi
Comparative area:
Slightly smaller than Oregon
Land boundaries:
Total 3,399 km, Guinea-Bissau 386 km, Cote d'Ivoire 610 km, Liberia 563 km, Mali 858 km, Senegal 330 km, Sierra Leone 652 km
Coastline:
320 km
Climate:
Generally hot and humid; monsoonal-type rainy season (June to November) with southwesterly winds; dry season (December to May) with northeasterly harmattan winds
Terrain:
Generally flat coastal plain, hilly to mountainous interior
Natural resources:
Bauxite, iron ore, diamonds, gold, uranium, hydropower, fish
Land use:
Arable land: 6%
Permanent crops: 0%
Meadows and pastures: 12%
Forest and woodland: 42%
Other: 40%

Demographics [2]

	1970	1980	1990	1995[1]	1996	2000	2010	2020	2030
Population	3,587	4,320	5,936	7,194	7,412	7,640	9,450	11,849	14,489
Population density (persons per sq. mi.)	38	46	63	76	78	80	100	125	153
(persons per sq. km.)	15	18	24	29	30	31	38	48	59
Net migration rate (per 1,000 population)	-2.1	-3.2	54.8	17.4	-5.4	-14.2	0.0	0.0	0.0
Births	NA	NA	NA	NA	316	NA	NA	NA	NA
Deaths	NA	NA	NA	NA	139	NA	NA	NA	NA
Life expectancy - males	33.1	36.5	40.3	42.3	42.7	44.5	49.1	53.9	58.7
Life expectancy - females	36.3	40.0	44.4	47.0	47.5	49.6	55.2	60.8	66.1
Birth rate (per 1,000)	48.9	48.1	46.8	43.3	42.6	40.0	36.8	32.0	27.0
Death rate (per 1,000)	29.0	25.3	21.6	19.2	18.7	16.8	13.4	10.5	8.3
Women of reproductive age (15-49 yrs.)	883	1,045	1,388	1,676	1,729	1,808	2,350	3,026	3,840
of which are currently married	NA	NA	1,286	NA	1,602	1,675	2,178	NA	NA
Fertility rate	6.1	6.1	6.1	5.8	5.7	5.5	4.7	3.9	3.2

Except as noted, values for vital statistics are in thousands; life expectancy is in years.

Health

Health Indicators [3]

% of population with access to	
safe water (1990-95)	55
adequate sanitation (1990-95)	21
health services (1985-95)	80
% of 1-year-olds immunized (1990-94) against	.
TB (tuberculosis)	75
DPT (diphtheria, pertussis, tetanus)	70
polio	70
measles	70
% of contraceptive prevalence (1980-94)[1]	1
ORT use rate (1990-94)	82

Health Expenditures [4]

Total health expenditure, 1990 (official exchange rate)	
Millions of dollars	106
Dollars per capita	19
Health expenditures as a percentage of GDP	
Total	3.9
Public sector	2.3
Private sector	1.6
Development assistance for health	
Total aid flows (millions of dollars)[1]	20
Aid flows per capita (dollars)	3.5
Aid flows as a percentage of total health expenditure	23.8

For sources, notes, and explanations, see Annotated Source Appendix, page 1061.

383

Human Factors

Women and Children [5]

% of pregnant women immunized (tetanus 1990-94)	56
% of births attended by trained health personnel (1983-94)	36
Maternal mortality rate (1980-92)	800
Under-5 mortality rate (1994)	223
% under-5 moderately/severely underweight (1980-1994)	NA

Burden of Disease [6]

Population per physician (1990)	6,569.63
Population per nurse (1990)	5,166.07
Population per hospital bed (1990)	1,816.03
AIDS cases per 100,000 people (1994)	7.1
Malaria cases per 100,000 people (1992)	NA

Ethnic Division [7]

Peuhl	40%
Malinke	30%
Soussou	20%
Smaller tribes	10%

Religion [8]

Muslim	85%
Christian	8%
Indigenous beliefs	7%

Major Languages [9]

French (official), each tribe has its own language.

Education

Public Education Expenditures [10]

Million (Syli)	1980	1984	1990	1992	1993	1994
Total education expenditure	NA	1,491	NA	NA	NA	NA
as percent of GNP	NA	NA	NA	NA	NA	NA
as percent of total govt. expend.	NA	15.3	NA	NA	NA	NA
Current education expenditure	NA	1,486	23,483	61,123	65,434	NA
as percent of GNP	NA	NA	1.4	2.4	2.2	NA
as percent of current govt. expend.	NA	17.2	NA	NA	NA	NA
Capital expenditure	NA	5	NA	NA	NA	NA

Educational Attainment [11]

Illiteracy [12]

In thousands and percent[1]	1990	1995	2000
Illiterate population (15+ yrs.)	2,116	2,272	2,432
Illiteracy rate - total pop. (%)	63.9	56.6	56.4
Illiteracy rate - males (%)	52.9	45.5	44.3
Illiteracy rate - females (%)	74.2	67.1	67.7

Libraries [13]

	Admin. Units	Svc. Pts.	Vols. (000)	Shelving (meters)	Vols. Added	Reg. Users
National	NA	NA	NA	NA	NA	NA
Nonspecialized	NA	NA	NA	NA	NA	NA
Public (1987)	1	NA	9	NA	NA	NA
Higher ed. (1988)	6	7	52	1,235	3,315	10,957
School (1987)	6	6	30	585	4,929	7,756

Daily Newspapers [14]

Culture [15]

Cinema (seats per 1,000)	6.9
Annual attendance per person	0.7
Gross box office receipts (mil. Syli)	793
Museums (reporting)	NA
Visitors (000)	NA
Annual receipts (000 Syli)	NA

Science and Technology

Scientific/Technical Forces [16]

Scientists/engineers	1,282
Number female	NA
Technicians	611
Number female	NA
Total	1,893

R&D Expenditures [17]

U.S. Patents Issued [18]

For sources, notes, and explanations, see Annotated Source Appendix, page 1061.

Government and Law

Organization of Government [19]

Long-form name:
Republic of Guinea
Type:
Republic
Independence:
2 October 1958 (from France)
National holiday:
Anniversary of the Second Republic, 3 April (1984)
Constitution:
23 December 1990 (Loi Fundamentale)
Legal system:
Based on French civil law system, customary law, and decree; legal codes currently being revised; has not accepted compulsory ICJ jurisdiction
Executive branch:
President; Council of Ministers
Legislative branch:
Unicameral: People's National Assembly (Assemblee Nationale Populaire)
Judicial branch:
Court of Appeal (Cour d'Appel)

Elections [20]

People's National Assembly	% of seats
Party for Unity and progress PUP)	62.3
Rally for the Guinean People (RGP)	16.7
Party for Renewal and Progress (PRP)	7.9
Union for a New Republic (UNR)	7.9
Union for Progress of Guinea (UPG)	1.8
Others	3.5

Government Budget [21]

For 1990 est.

	$ mil.
Revenues	449
Expenditures	708
Capital expenditures	361

Crime [23]

	1990
Crime volume	
Cases known to police	1,942
Attempts (percent)	1.97
Percent cases solved	19.78
Crimes per 100,000 persons	32.36
Persons responsible for offenses	
Total number offenders	2,811
Percent female	0.91
Percent juvenile (1-15 yrs.)	4.07
Percent foreigners	6.63

Military Expenditures and Arms Transfers [22]

	1990	1991	1992	1993	1994
Military expenditures					
Current dollars (mil.)	31	33	44	NA	50
1994 constant dollars (mil.)	34	36	46	NA	50
Armed forces (000)	15	15	15	15	12
Gross national product (GNP)					
Current dollars (mil.)	2,462	2,656	2,844	3,075	3,277
1994 constant dollars (mil,)	2,740	2,847	2,965	3,139	3,277
Central government expenditures (CGE)					
1994 constant dollars (mil.)	681	654	649	NA	NA
People (mil.)	5.9	6.2	6.2	6.2	6.4
Military expenditure as % of GNP	1.2	1.3	1.5	NA	1.5
Military expenditure as % of CGE	5.0	5.4	7.0	NA	NA
Military expenditure per capita (1994 $)	6	6	7	NA	8
Armed forces per 1,000 people (soldiers)	2.5	2.4	2.4	2.4	1.9
GNP per capita (1994 $)	462	456	475	503	513
Arms imports[6]					
Current dollars (mil.)	20	5	0	0	0
1994 constant dollars (mil.)	22	5	0	0	0
Arms exports[6]					
Current dollars (mil.)	0	0	0	0	0
1994 constant dollars (mil.)	0	0	0	0	0
Total imports[7]					
Current dollars (mil.)[e]	600	732	768	798	NA
1994 constant dollars (mil.)[e]	668	785	801	814	NA
Total exports[7]					
Current dollars (mil.)[e]	605	653	622	NA	NA
1994 constant dollars (mil.)[e]	673	700	649	NA	NA
Arms as percent of total imports[8]	3.3	0.7	0	0	0
Arms as percent of total exports[8]	0	0	0	0	0

Human Rights [24]

	SSTS	FL	FAPRO	PPCG	APROBC	TPW	PCPTW	STPEP	PHRFF	PRW	ASST	AFL
Observes	P	P	P		P	P	P	P		P	P	P
	EAFRD	CPR	ESCR	SR	ACHR	MAAE	PVIAC	PVNAC	EAFDAW	TCIDTP	RC
Observes	P	P	P	P			P	P	P	P	P

P = Party; S = Signatory; see Appendix for meaning of abbreviations.

Labor Force

Total Labor Force [25]

2.4 million (1983)

Labor Force by Occupation [26]

Agriculture	80.0%
Industry and commerce	11.0
Services	5.4
Civil service	3.6

Unemployment Rate [27]

For sources, notes, and explanations, see Annotated Source Appendix, page 1061.

385

Production Sectors

Commercial Energy Production and Consumption

Data are shown in quadrillion (10^{15}) BTUs and percent for 1995
Values for hydroelectric, nuclear, geothermal, solar, and wind power refer to electrical generation.

Production [28]	Consumption [29]
Hydro - 100.0% 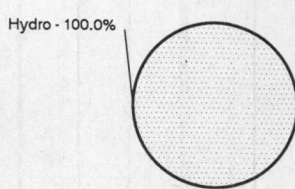	Crude oil - 89.5% 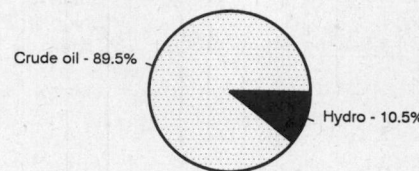 Hydro - 10.5%

Net hydroelectric power	0.002
Total	0.002

Crude oil	0.017
Net hydroelectric power	0.002
Total	0.019

Telecommunications [30]

- 18,000 (1994 est.) telephones; poor to fair system of open-wire lines, small radiotelephone communication stations, and new microwave radio relay system
- Domestic: microwave radio relay and radiotelephone communication
- International: satellite earth station - 1 Intelsat (Atlantic Ocean)
- Radio: Broadcast stations: AM 3, FM 1, shortwave 0 Radios: 257,000 (1992 est.)
- Television: Broadcast stations: 1 Televisions: 65,000 (1993 est.)

Transportation [31]

Railways: total: 1,086 km; standard gauge: 279 km 1.435-m gauge; narrow gauge: 807 km 1.000-m gauge; note - includes 662 km in common carrier service from Kankan to Conakry

Highways: total: 29,750 km; paved: 4,490 km; unpaved: 25,260 km (1991 est.)

Merchant marine: none

Airports

Total:	14
With paved runways over 3,047 m:	1
With paved runways 2,438 to 3,047 m:	2
With paved runways 1,524 to 2,437 m:	1
With paved runways under 914 m:	1
With unpaved runways 1,524 to 2,437 m:	6

Top Agricultural Products [32]

Agriculture accounts for 24% of the GDP; produces rice, coffee, pineapples, palm kernels, cassava (tapioca), bananas, sweet potatoes; cattle, sheep, goats; timber.

Top Mining Products [33]

Thousand metric tons except as noted[36]	5/12/95*
Alumina, hydrated and calcined	1,310
Bauxite, mine production, dry basis	14,100[37]
Diamond carats)[38]	
gem	90
industrial	5
Gold (kg.)	500[39]

Tourism [34]

	1990	1991	1992	1993	1994
Tourists	100	28	33	93	NA
Tourism receipts	30	13	11	6	NA
Tourism expenditures	30	27	17	28	NA
Fare receipts	5	3	3	3	NA
Fare expenditures	19	17	12	21	NA

Travelers are in thousands, money in million U.S. dollars.

For sources, notes, and explanations, see Annotated Source Appendix, page 1061.

Finance, Economics, and Trade

GDP and Manufacturing Summary [35]

	1980	1985	1990	1991	1992
Gross Domestic Product					
Millions of 1980 dollars	1,897	1,807	2,190	2,267	2,314[e]
Growth rate in percent	5.60	3.89	4.00	3.50	2.08[e]
Manufacturing Value Added					
Millions of 1980 dollars	60	76	68	70	72[e]
Growth rate in percent	2.70	33.33	2.90	2.93	2.48[e]
Manufacturing share in percent of current prices	2.9[e]	2.0[e]	4.3[e]	NA	NA

Economic Indicators [36]

- **National product**: GDP—purchasing power parity—$6.5 billion (1995 est.)

- **National product real growth rate**: 4% (1995 est.)

- **National product per capita**: $1,020 (1995 est.)

- **Inflation rate (consumer prices)**: 4.1% (1994 est.)

- **External debt**: $3.02 billion (1994)

Balance of Payments Summary [37]

Values in millions of dollars.

	1989	1990	1991	1992	1993
Exports of goods (f.o.b.)	595.6	671.2	687.1	517.1	561.1
Imports of goods (f.o.b.)	-531.5	-585.7	-694.9	-608.4	-582.6
Trade balance	64.1	85.5	-7.8	-91.3	-21.5
Services - debits	-438.2	-528.7	-529.2	-471.4	-427.3
Services - credits	112.3	170.4	160.2	167.6	317.8
Private transfers (net)	-0.9	-32.5	-26.0	-1.5	64.0
Government transfers (net)	318.6	109.3	114.0	193.1	146.4
Long-term capital (net)	79.67	123.4	171.2	263.3	284.0
Short-term capital (net)	-88.3	38.8	-16.7	-85.5	80.7
Errors and omissions	-44.9	52.1	112.3	18.6	-229.3
Overall balance	2.3	18.3	-22.0	-7.1	53.4

Exchange Rates [38]

Currency: **Guinean franc.**
Symbol: **FG.**

Data are currency units per $1.

August 1995	995.3
1994	976.6
1993	955.5
1992	902.0
1991	753.9

Imports and Exports

Top Import Origins [39]

$688 million (1994 est.).

Origins	%
France	26
Cote d'Ivoire	12
Hong Kong	6
Germany	6

Top Export Destinations [40]

$562 million (1994 est.).

Destinations	%
US	23
Belgium	12
Ireland	12
Spain	12

Foreign Aid [41]

Recipient: ODA, $NA.

Import and Export Commodities [42]

Import Commodities	Export Commodities
Petroleum products	Bauxite
Metals	Alumina
Machinery	Diamonds
Transport equipment	Gold
Textiles	Coffee
Grain and other foodstuffs	Pineapples
	Bananas
	Palm kernels

For sources, notes, and explanations, see Annotated Source Appendix, page 1061.

387

Guinea-Bissau

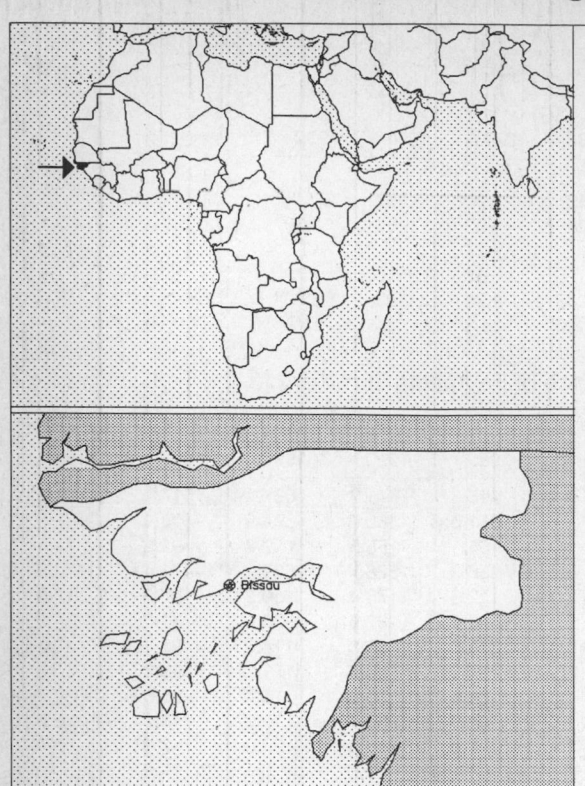

Geography [1]

Total area:
36,120 sq km   13,946 sq mi
Land area:
28,000 sq km   10,811 sq mi
Comparative area:
Slightly less than three times the size of Connecticut
Land boundaries:
Total 724 km, Guinea 386 km, Senegal 338 km
Coastline:
350 km
Climate:
Tropical; generally hot and humid; monsoonal-type rainy season (June to November) with southwesterly winds; dry season (December to May) with northeasterly harmattan winds
Terrain:
Mostly low coastal plain rising to savanna in East
Natural resources:
Phosphates, bauxite, unexploited deposits of petroleum, fish, timber
Land use:
Arable land: 11%
Permanent crops: 1%
Meadows and pastures: 43%
Forest and woodland: 38%
Other: 7%

Demographics [2]

	1970	1980	1990	1995[1]	1996	2000	2010	2020	2030
Population	620	789	998	1,125	1,151	1,263	1,579	1,925	2,280
Population density (persons per sq. mi.)	57	73	92	104	106	117	146	178	211
(persons per sq. km.)	22	28	36	40	41	45	56	69	81
Net migration rate (per 1,000 population)	NA	0.0	0.0	0.0	0.0	0.0	0.0	0.0	0.0
Births	NA	NA	NA	NA	46	NA	NA	NA	NA
Deaths	NA	NA	NA	NA	19	NA	NA	NA	NA
Life expectancy - males	NA	40.4	44.2	46.2	46.6	48.3	52.7	57.1	61.4
Life expectancy - females	NA	43.6	47.5	49.6	50.0	51.7	56.1	60.5	64.7
Birth rate (per 1,000)	NA	47.3	43.2	40.2	39.7	37.8	33.3	28.2	24.1
Death rate (per 1,000)	NA	24.5	18.8	16.6	16.2	14.8	12.0	9.9	8.5
Women of reproductive age (15-49 yrs.)	NA	195	244	276	284	312	393	502	606
of which are currently married	NA	NA	183	NA	212	234	296	NA	NA
Fertility rate	NA	5.9	5.9	5.4	5.3	5.0	4.2	3.4	2.9

Except as noted, values for vital statistics are in thousands; life expectancy is in years.

Health

Health Indicators [3]

% of population with access to	
safe water (1990-95)	53
adequate sanitation (1990-95)	21
health services (1985-95)	40
% of 1-year-olds immunized (1990-94) against	
TB (tuberculosis)	95
DPT (diphtheria, pertussis, tetanus)	74
polio	68
measles	65
% of contraceptive prevalence (1980-94)[1]	1
ORT use rate (1990-94)	26

Health Expenditures [4]

388

Human Factors

Women and Children [5]

% of pregnant women immunized (tetanus 1990-94)	55
% of births attended by trained health personnel (1983-94)	27
Maternal mortality rate (1980-92)	700
Under-5 mortality rate (1994)	231
% under-5 moderately/severely underweight (1980-1994)[1]	23

Burden of Disease [6]

Population per physician (1985)	7,262.29
Population per nurse (1985)	1,128.66
Population per hospital bed	NA
AIDS cases per 100,000 people (1994)	23.5
Malaria cases per 100,000 people (1992)	NA

Ethnic Division [7]

African	99%
Balanta	30%
Fula	20%
Manjaca	14%
Mandinga	13%
Papel	7%
European and mulatto	<1%

Religion [8]

Indigenous beliefs	65%
Muslim	30%
Christian	5%

Major Languages [9]

Portuguese (official), Criolo, African languages.

Education

Public Education Expenditures [10]

	1980	1984	1987	1990	1992	1994
Million (Peso)						
Total education expenditure	NA	NA	2,533	NA	NA	NA
as percent of GNP	NA	NA	2.8	NA	NA	NA
as percent of total govt. expend.	NA	NA	NA	NA	NA	NA
Current education expenditure	208	539	2,473	NA	NA	NA
as percent of GNP	4.0	3.2	2.8	NA	NA	NA
as percent of current govt. expend.	NA	11.2	NA	NA	NA	NA
Capital expenditure	NA	NA	60	NA	NA	NA

Educational Attainment [11]

Age group (1979)	7+
Total population	483,336
Highest level attained (%)	
No schooling	91.1
First level	
Not completed	8.0
Completed	NA
Entered second level	
S-1	0.6
S-2	0.2
Postsecondary	0.1

Illiteracy [12]

In thousands and percent[1]	1990	1995	2000
Illiterate population (15+ yrs.)	287	282	277
Illiteracy rate - total pop. (%)	51.4	44.1	37.7
Illiteracy rate - males (%)	38.8	32.1	26.9
Illiteracy rate - females (%)	62.8	54.9	47.5

Libraries [13]

	Admin. Units	Svc. Pts.	Vols. (000)	Shelving (meters)	Vols. Added	Reg. Users
National	NA	NA	NA	NA	NA	NA
Nonspecialized (1986)	1	1	60	NA	NA	200
Public	NA	NA	NA	NA	NA	NA
Higher ed.	NA	NA	NA	NA	NA	NA
School	NA	NA	NA	NA	NA	NA

Daily Newspapers [14]

	1980	1985	1990	1994
Number of papers	1	1	1	1
Circ. (000)	6	6	6	6

Culture [15]

Science and Technology

Scientific/Technical Forces [16]

R&D Expenditures [17]

U.S. Patents Issued [18]

For sources, notes, and explanations, see Annotated Source Appendix, page 1061.

389

Government and Law

Organization of Government [19]

Long-form name:
Republic of Guinea-Bissau
Type:
Republic, formerly highly centralized, multiparty since mid-1991
Independence:
10 September 1974 (from Portugal)
National holiday:
Independence Day, 10 September (1974)
Constitution:
16 May 1984, amended 4 May 1991 (currently undergoing revision to liberalize popular participation in the government)
Legal system:
NA
Executive branch:
President of the Republic of Guinea-Bissau; Prime Minister; Council of Ministers
Legislative branch:
Unicameral: National People's Assembly
Judicial branch:
None; there is a Ministry of Justice in the Council of Ministers

Elections [20]

National People's Assembly	% of seats
African Party for the Independence of Guinea-Bissau and Cape Verde (PAIGC)	62.0
Guinea-Bissau Resistance (RGB)	19.0
Social Renovation Party (PRS)	12.0
Union for Change Coalition	6.0
Front for the Liberation and Independence of Guinea (FLING)	1.0

Government Expenditures [21]

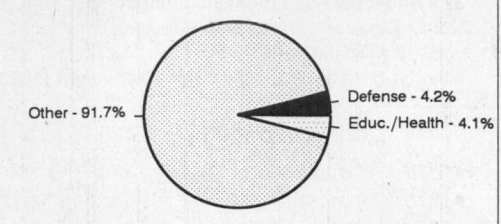

Other - 91.7%
Defense - 4.2%
Educ./Health - 4.1%

(% distribution). Expend. for 1991 est.: 44.8 (Dollars mil.)

Crime [23]

Military Expenditures and Arms Transfers [22]

	1990	1991	1992	1993	1994
Military expenditures					
Current dollars (mil.)	NA	NA	7[e]	8[e]	8
1994 constant dollars (mil.)	NA	NA	7[e]	8[e]	8
Armed forces (000)	12	12	11	11	7
Gross national product (GNP)					
Current dollars (mil.)	197	199	213	220	240
1994 constant dollars (mil.)	220	213	222	224	240
Central government expenditures (CGE)					
1994 constant dollars (mil.)	108	94	90	NA	NA
People (mil.)	1.0	1.0	1.0	1.1	1.1
Military expenditure as % of GNP	NA	NA	3.1	3.6	3.4
Military expenditure as % of CGE	NA	NA	7.6	NA	NA
Military expenditure per capita (1994 $)	NA	NA	7	8	7
Armed forces per 1,000 people (soldiers)	12.0	11.7	10.5	10.3	6.4
GNP per capita (1994 $)	220	208	212	209	218
Arms imports[6]					
Current dollars (mil.)	5	5	0	0	0
1994 constant dollars (mil.)	6	5	0	0	0
Arms exports[6]					
Current dollars (mil.)	0	0	0	0	0
1994 constant dollars (mil.)	0	0	0	0	0
Total imports[7]					
Current dollars (mil.)	68	67	84	62	63
1994 constant dollars (mil.)	76	72	88	63	63
Total exports[7]					
Current dollars (mil.)	19	20	6	16	32
1994 constant dollars (mil.)	21	21	6	16	32
Arms as percent of total imports[8]	7.4	7.5	0	0	0
Arms as percent of total exports[8]	0	0	0	0	0

Human Rights [24]

	SSTS	FL	FAPRO	PPCG	APROBC	TPW	PCPTW	STPEP	PHRFF	PRW	ASST	AFL
Observes		P			P	P	P					P
	EAFRD	CPR	ESCR	SR	ACHR	MAAE	PVIAC	PVNAC	EAFDAW	TCIDTP		RC
Observes			P	P				P	P	P		P

P = Party; S = Signatory; see Appendix for meaning of abbreviations.

Labor Force

Total Labor Force [25]

403,000 (est.)

Labor Force by Occupation [26]

Agriculture	90%
Industry, services, and commerce	5
Government	5

Unemployment Rate [27]

For sources, notes, and explanations, see Annotated Source Appendix, page 1061.

Production Sectors

Energy Resource Summary [28]

Energy resources: Unexploited deposits of petroleum. **Electricity**: Capacity: 22,000 kW. Production: 40 million kWh. Consumption per capita: 37 kWh (1993).

Telecommunications [30]

- 3,000 (1988 est.) telephones; poor system
- Domestic: combination of microwave radio relay, open-wire lines, and radiotelephone communications
- Radio: Broadcast stations: AM 2, FM 3, shortwave 0 Radios: 40,000 (1992 est.)
- Television: Broadcast stations: 1

Transportation [31]

Railways: 0 km

Highways: total: 3,200 km; paved: 416 km; unpaved: 2,784 km (1988 est.)

Merchant marine: none

Airports
Total:	16
With paved runways over 3,047 m:	1
With paved runways 1,524 to 2,437 m:	2
With paved runways 914 to 1,523 m:	1
With paved runways under 914 m:	8
With unpaved runways 914 to 1,523 m:	4 (1995 est.)

Top Agricultural Products [32]

Agriculture accounts for 44% of the GDP; produces rice, corn, beans, cassava (tapioca), cashew nuts, peanuts, palm kernels, cotton; fishing and forest potential not fully exploited.

Top Mining Products [33]

Detailed information is not available. A summary of mineral resources follows. **Mineral Resources**: Low-grade deposits of bauxite and phosphate.

Tourism [34]

For sources, notes, and explanations, see Annotated Source Appendix, page 1061.

391

Finance, Economics, and Trade

GDP and Manufacturing Summary [35]

	1980	1985	1990	1991	1992
Gross Domestic Product					
Millions of 1980 dollars	154	171	218	224	229[e]
Growth rate in percent	-4.19	-2.30	3.04	2.80	2.17[e]
Manufacturing Value Added					
Millions of 1980 dollars	12	11	10	10	10[e]
Growth rate in percent	-5.09	-5.95	1.25	0.14	-0.13[e]
Manufacturing share in percent of current prices	1.6[e]	1.6[e]	NA	NA	NA

Economic Indicators [36]

- **National product**: GDP—purchasing power parity—$1 billion (1994 est.)

- **National product real growth rate**: NA%

- **National product per capita**: $900 (1995 est.)

- **Inflation rate (consumer prices)**: 15% (1994 est.)

- **External debt**: $692 million (December 1993 est.)

Balance of Payments Summary [37]

Values in millions of dollars.

	1989	1990	1991	1992	1993
Exports of goods (f.o.b.)	14.2	19.3	20.4	6.5	16.0
Imports of goods (f.o.b.)	-68.9	-68.1	-67.5	-83.5	-53.8
Trade balance	-54.7	-48.8	-47.0	-77.0	-37.9
Services - debits	-49.0	-35.5	-47.3	-43.8	-40.4
Services - credits	NA	NA	NA	NA	NA
Private transfers (net)	1.2	1.0	-4.1	-0.6	-1.6
Government transfers (net)	63.9	67.0	55.8	53.0	54.5
Long-term capital (net)	29.0	39.6	18.0	12.6	2.5
Short-term capital (net)	11.7	-16.6	32.1	39.0	30.3
Errors and omissions	-9.7	-1.5	-16.3	22.0	-16.0
Overall balance	-7.6	5.2	-8.9	5.1	-8.6

Exchange Rates [38]

Currency: **Guinea-Bissauan peso.**
Symbol: **PG.**

Data are currency units per $1.

December 1995	17,659
1994	12,892
1993	10,082
1992	6,934
1991	3,659
1990	2,185

Imports and Exports

Top Import Origins [39]

$63 million (f.o.b., 1994).

Origins	%
Portugal	NA
Netherlands	NA
China	NA
Germany	NA
Senegal	NA

Top Export Destinations [40]

$32 million (f.o.b., 1994).

Destinations	%
Portugal	NA
Spain	NA
Senegal	NA
India	NA
Nigeria	NA
Cote d'Ivoire	NA

Foreign Aid [41]

Recipient: ODA, $NA.

Import and Export Commodities [42]

Import Commodities	Export Commodities
Foodstuffs	Cashews
Transport equipment	Fish
Petroleum products	Peanuts
Machinery and equipment	Palm kernels

Guyana

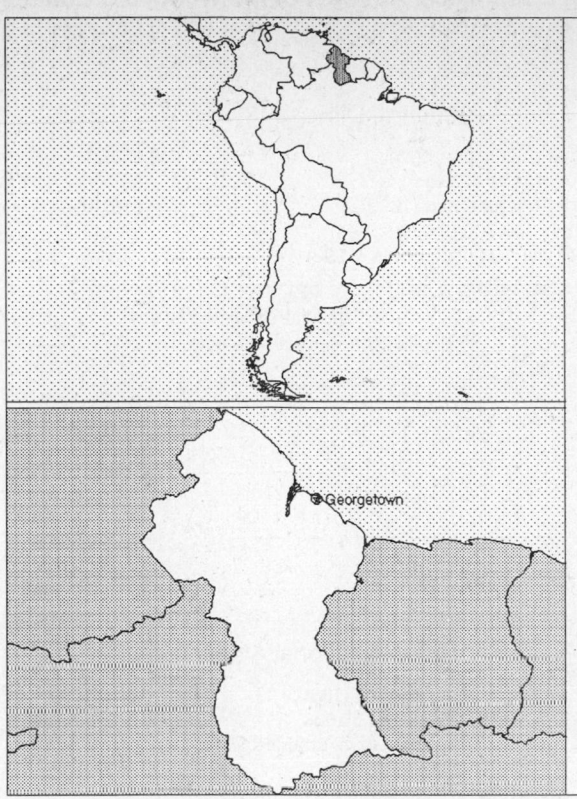

Geography [1]

Total area:
214,970 sq km 83,000 sq mi
Land area:
196,850 sq km 76,004 sq mi
Comparative area:
Slightly smaller than Idaho
Land boundaries:
Total 2,462 km, Brazil 1,119 km, Suriname 600 km, Venezuela 743 km
Coastline:
459 km
Climate:
Tropical; hot, humid, moderated by northeast trade winds; two rainy seasons (May to mid-August, mid-November to mid-January)
Terrain:
Mostly rolling highlands; low coastal plain; savanna in South
Natural resources:
Bauxite, gold, diamonds, hardwood timber, shrimp, fish
Land use:
Arable land: 3%
Permanent crops: 0%
Meadows and pastures: 6%
Forest and woodland: 83%
Other: 8%

Demographics [2]

	1970	1980	1990	1995[1]	1996	2000	2010	2020	2030
Population	715	759	747	719	712	693	695	685	677
Population density (persons per sq. mi.)	9	10	10	9	9	9	9	9	9
(persons per sq. km.)	4	4	4	4	4	4	4	3	3
Net migration rate (per 1,000 population)	-5.7	-19.7	-19.6	-20.3	-18.5	-10.5	0.0	0.0	0.0
Births	NA	NA	NA	NA	14	NA	NA	NA	NA
Deaths	NA	NA	NA	NA	7	NA	NA	NA	NA
Life expectancy - males	NA	58.8	59.7	58.2	57.6	55.1	47.1	50.4	59.0
Life expectancy - females	NA	65.4	66.7	63.9	62.8	58.8	51.2	54.7	64.3
Birth rate (per 1,000)	33.2	29.4	22.0	19.4	19.0	17.9	17.6	15.4	13.9
Death rate (per 1,000)	6.7	8.3	7.9	9.1	9.6	11.5	18.5	17.5	13.6
Women of reproductive age (15-49 yrs.)	NA	183	195	194	193	193	193	184	181
of which are currently married[19]	80	88	99	NA	100	101	105	NA	NA
Fertility rate	5.1	3.7	2.5	2.2	2.2	2.1	1.9	1.8	1.8

Except as noted, values for vital statistics are in thousands; life expectancy is in years.

Health

Health Indicators [3]

Health Expenditures [4]

For sources, notes, and explanations, see Annotated Source Appendix, page 1061.

393

Human Factors

Women and Children [5]

Burden of Disease [6]

Population per physician (1993)	8,956.04
Population per nurse (1993)	893.64
Population per hospital bed (1993)	358.40
AIDS cases per 100,000 people (1994)	9.5
Malaria cases per 100,000 people (1992)	4,914

Ethnic Division [7]

East Indian	51%
Black and mixed	43%
Amerindian	4%
European and Chinese	2%

Religion [8]

Christian	57%
Hindu	33%
Muslim	9%
Other	1%

Major Languages [9]

English, Amerindian dialects.

Education

Public Education Expenditures [10]

	1980	1985	1989	1990	1993	1994
Million (Dollar)						
Total education expenditure	NA	162	478	542	NA	NA
as percent of GNP	NA	9.8	6.3	5.0	NA	NA
as percent of total govt. expend.	NA	10.4	8.9	NA	NA	NA
Current education expenditure	NA	135	342	435	NA	NA
as percent of GNP	NA	8.1	4.5	4.0	NA	NA
as percent of current govt. expend.	NA	13.0	8.4	5.2	NA	NA
Capital expenditure	NA	27	136	107	NA	NA

Educational Attainment [11]

Age group (1980)	25+
Total population	270,849
Highest level attained (%)	
No schooling	8.1
First level	
Not completed	72.9
Completed	NA
Entered second level	
S-1	17.3
S-2	NA
Postsecondary	1.8

Illiteracy [12]

In thousands and percent[1]	1990	1995	2000
Illiterate population (15+ yrs.)	15	11	9
Illiteracy rate - total pop. (%)	3.2	2.3	1.8
Illiteracy rate - males (%)	2.1	1.7	1.2
Illiteracy rate - females (%)	4.2	2.9	2.5

Libraries [13]

	Admin. Units	Svc. Pts.	Vols. (000)	Shelving (meters)	Vols. Added	Reg. Users
National (1986)	1	37	190	NA	6,693	45,233
Nonspecialized	NA	NA	NA	NA	NA	NA
Public	NA	NA	NA	NA	NA	NA
Higher ed. (1987)	1	7	166	6,100	4,237	2,512
School	NA	NA	NA	NA	NA	NA

Daily Newspapers [14]

	1980	1985	1990	1994
Number of papers	1	2	2	2
Circ. (000)	58	78	80[e]	80

Culture [15]

Cinema (seats per 1,000)	NA
Annual attendance per person	NA
Gross box office receipts (mil. Dollar)	NA
Museums (reporting)	2
Visitors (000)	120
Annual receipts (000 Dollar)	NA

Science and Technology

Scientific/Technical Forces [16]

Scientists/engineers	89
Number female	NA
Technicians	178
Number female	NA
Total[15]	267

R&D Expenditures [17]

	Dollar (000) 1982
Total expenditure[16]	2,800
Capital expenditure	NA
Current expenditure	NA
Percent current	NA

U.S. Patents Issued [18]

Government and Law

Organization of Government [19]

Long-form name:
Co-operative Republic of Guyana
Type:
Republic
Independence:
26 May 1966 (from UK)
National holiday:
Republic Day, 23 February (1970)
Constitution:
6 October 1980
Legal system:
Based on English common law with
certain admixtures of Roman-Dutch law;
has not accepted compulsory ICJ
jurisdiction
Executive branch:
Executive President; Prime Minister;
Cabinet of Ministers
Legislative branch:
Unicameral: National Assembly
Judicial branch:
Supreme Court of Judicature

Elections [20]

National Assembly	% of votes
People's Progressive Party (PPP)	53.4
People's National Congress (PNC)	42.3
Working People's Alliance (WPA)	2.0
The United Force (TUF)	1.2

Government Budget [21]

For 1995 est.

	$ mil.
Revenues	209
Expenditures	303
Capital expenditures	109

Crime [23]

	1994
Crime volume	
Cases known to police	70
Attempts (percent)	NA
Percent cases solved	NA
Crimes per 100,000 persons	9.20
Persons responsible for offenses	
Total number offenders	NA
Percent female	NA
Percent juvenile (9-17 yrs.)	NA
Percent foreigners	NA

Military Expenditures and Arms Transfers [22]

	1990	1991	1992	1993	1994
Military expenditures					
Current dollars (mil.)	4	3	6	6	7
1994 constant dollars (mil.)	5	4	6	6	7
Armed forces (000)	4	4	2	2	2
Gross national product (GNP)					
Current dollars (mil.)	250	243	310	388	458
1994 constant dollars (mil.)	279	261	323	396	458
Central government expenditures (CGE)					
1994 constant dollars (mil.)	280	248	209	209	214
People (mil.)	0.7	0.7	0.7	0.7	0.7
Military expenditure as % of GNP	1.7	1.4	1.8	1.6	1.5
Military expenditure as % of CGE	1.7	1.4	2.8	3.1	3.1
Military expenditure per capita (1994 $)	6	5	8	9	9
Armed forces per 1,000 people (soldiers)	4.7	4.7	2.7	2.7	2.7
GNP per capita (1994 $)	373	351	437	540	628
Arms imports[6]					
Current dollars (mil.)	0	0	0	0	0
1994 constant dollars (mil.)	0	0	0	0	0
Arms exports[6]					
Current dollars (mil.)	0	0	0	0	0
1994 constant dollars (mil.)	0	0	0	0	0
Total imports[7]					
Current dollars (mil.)	311	307	443	484	456[e]
1994 constant dollars (mil.)	346	329	462	494	456[e]
Total exports[7]					
Current dollars (mil.)	251	248	302	423	453
1994 constant dollars (mil.)	279	266	315	432	453
Arms as percent of total imports[8]	0	0	0	0	0
Arms as percent of total exports[8]	0	0	0	0	0

Human Rights [24]

	SSTS	FL	FAPRO	PPCG	APROBC	TPW	PCPTW	STPEP	PHRFF	PRW	ASST	AFL
Observes	1	P	P		P	P	P				1	P
	EAFRD	CPR	ESCR	SR	ACHR	MAAE	PVIAC	PVNAC	EAFDAW	TCIDTP	RC	
Observes	P	P	P				P	P	P	P	P	

P = Party; S = Signatory; see Appendix for meaning of abbreviations.

Labor Force

Total Labor Force [25]

268,000

Labor Force by Occupation [26]

Industry and commerce	44.5%
Agriculture	33.8
Services	21.7

Date of data: 1985

Unemployment Rate [27]

12% (1992 est.)

For sources, notes, and explanations, see Annotated Source Appendix, page 1061.

395

Production Sectors

Energy Resource Summary [28]

Electricity: Capacity: 110,000 kW. Production: 230 million kWh. Consumption per capita: 286 kWh (1993).

Telecommunications [30]

- 33,000 (1987 est.) telephones; fair system for long-distance calling
- Domestic: microwave radio relay network for trunk lines
- International: tropospheric scatter to Trinidad; satellite earth station - 1 Intelsat (Atlantic Ocean)
- Radio: Broadcast stations: AM 4, FM 3, shortwave 1 Radios: 398,000 (1992 est.)
- Television: Broadcast stations: 0 (1987 est.) Televisions: 32,000 (1992 est.)

Transportation [31]

Railways: total: 88 km; standard gauge: 40 km 1.435-m gauge (dedicated to ore transport); narrow gauge: 48 km 0.914-m gauge (dedicated to ore transport)

Highways: total: 7,621 km; paved: 547 km; unpaved: 7,074 km (1987 est.)

Merchant marine: total: 1 cargo ship (1,000 GRT or over) totaling 1,317 GRT/2,558 DWT (1995 est.)

Airports

Total:	47
With paved runways 1,524 to 2,437 m:	3
With paved runways 914 to 1,523 m:	1
With paved runways under 914 m:	32
With unpaved runways 1,524 to 2,437 m:	2
With unpaved runways 914 to 1,523 m:	9 (1995 est.)

Top Agricultural Products [32]

Agriculture accounts for 26.5% of the GDP; produces sugar, rice, wheat, vegetable oils; beef, pork, poultry, dairy products; development potential exists for fishing and forestry.

Top Mining Products [33]

Metric tons except as noted[e]	6/1/95[*]
Bauxite, dry equiv., gr. wt. (000 tons)	2,100
Diamond (carats)	34,000
Gold, mine output, Au content (kg.)	11,800[r]
Stone, crushed	136,000

Tourism [34]

	1990	1991	1992	1993	1994
Tourists	64	73	75	107	113
Tourism receipts	27	30	31	45	47

Travelers are in thousands, money in million U.S. dollars.

Finance, Economics, and Trade

GDP and Manufacturing Summary [35]

	1980	1985	1990	1991	1992
Gross Domestic Product					
Millions of 1980 dollars	591	494	430	456	491
Growth rate in percent	1.66	1.02	-6.20	6.06	7.77
Manufacturing Value Added					
Millions of 1980 dollars	64	45	29	28	32[e]
Growth rate in percent	0.76	-3.13	-16.67	-1.34	11.78[e]
Manufacturing share in percent of current prices	12.1	13.9	15.9	9.6	9.6[e]

Economic Indicators [36]

- **National product**: GDP—purchasing power parity—$1.6 billion (1995 est.)

- **National product real growth rate**: 5.1% (1995 est.)

- **National product per capita**: $2,200 (1995 est.)

- **Inflation rate (consumer prices)**: 8.1% (1995)

- **External debt**: $2.2 billion (1994 est.)

Balance of Payments Summary [37]

Values in millions of dollars.

	1970	1973	1975	1980	1985
Exports of goods (f.o.b.)	129.0	135.7	351.4	388.9	214.0
Imports of goods (f.o.b.)	-119.9	-159.4	-305.8	-386.4	-209.1
Trade balance	9.1	-23.7	45.6	2.5	4.9
Services - debits	-48.0	-62.0	-84.5	-151.8	-144.3
Services - credits	17.6	21.9	20.4	21.6	48.0
Private transfers (net)	-0.5	-0.8	-4.4	1.0	-2.0
Government transfers (net)	NA	0.1	-1.7	1.8	-3.2
Long-term capital (net)	17.1	28.7	97.4	79.8	-36.0
Short-term capital (net)	-0.2	4.1	-3.9	5.3	141.5
Errors and omissions	2.6	13.8	-19.1	0.1	-4.3
Overall balance	-2.3	-26.1	49.8	-43.3	4.6

Exchange Rates [38]

Currency: **Guyanese dollar.**
Symbol: **G$.**

Data are currency units per $1.

January 1996	140.3
1995	142.0
1994	138.3
1993	126.7
1992	125.0
1991	111.8

Imports and Exports

Top Import Origins [39]

$456 million (c.i.f., 1994 est.) Data are for 1992.

Origins	%
US	37
Trinidad and Tobago	13
UK	11
Italy	8
Japan	5

Top Export Destinations [40]

$453 million (f.o.b., 1994) Data are for 1992.

Destinations	%
UK	33
US	31
Canada	9
France	5
Japan	3

Foreign Aid [41]

Recipient: ODA, $NA.

Import and Export Commodities [42]

Import Commodities	Export Commodities
Manufactures	Sugar
Machinery	Bauxite/alumina
Petroleum	Rice
Food	Shrimp
	Molasses

For sources, notes, and explanations, see Annotated Source Appendix, page 1061.

397

Haiti

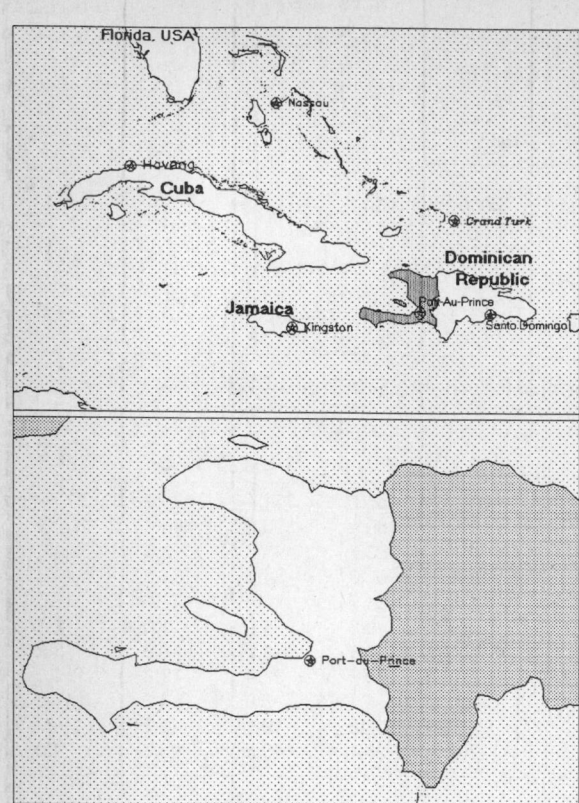

Geography [1]

Total area:
27,750 sq km   10,714 sq mi
Land area:
27,560 sq km   10,641 sq mi
Comparative area:
Slightly larger than Maryland
Land boundaries:
Total 275 km, Dominican Republic 275 km
Coastline:
1,771 km
Climate:
Tropical; semiarid where mountains in East cut off trade winds
Terrain:
Mostly rough and mountainous
Natural resources:
Bauxite
Land use:
Arable land: 20%
Permanent crops: 13%
Meadows and pastures: 18%
Forest and woodland: 4%
Other: 45%

Demographics [2]

	1970	1980	1990	1995[1]	1996	2000	2010	2020	2030
Population	4,605	5,068	6,060	6,613	6,732	7,223	8,681	10,252	11,648
Population density (persons per sq. mi.)	433	476	569	621	633	679	816	963	1,095
(persons per sq. km.)	167	184	220	240	244	262	315	372	423
Net migration rate (per 1,000 population)	NA	NA	-7.0	-4.9	-4.5	-3.0	0.0	0.0	0.0
Births	NA	NA	NA	NA	257	NA	NA	NA	NA
Deaths	NA	NA	NA	NA	107	NA	NA	NA	NA
Life expectancy - males	47.6	NA	46.1	47.1	47.3	48.0	49.9	52.3	57.4
Life expectancy - females	48.3	NA	49.5	51.1	51.4	52.5	55.3	57.9	63.8
Birth rate (per 1,000)	NA	NA	44.4	39.0	38.2	35.5	31.9	26.0	21.3
Death rate (per 1,000)	NA	NA	18.5	16.3	16.0	14.8	13.1	11.8	9.5
Women of reproductive age (15-49 yrs.)	NA	NA	1,365	1,459	1,492	1,669	2,212	2,757	3,315
of which are currently married	NA	NA	833	NA	890	967	1,290	NA	NA
Fertility rate	NA	NA	6.4	5.8	5.7	5.2	3.9	2.9	2.4

Except as noted, values for vital statistics are in thousands; life expectancy is in years.

Health

Health Indicators [3]

% of population with access to	
safe water (1990-95)	28
adequate sanitation (1990-95)	24
health services (1985-95)	50
% of 1-year-olds immunized (1990-94) against	
TB (tuberculosis)	42
DPT (diphtheria, pertussis, tetanus)	41
polio	40
measles	24
% of contraceptive prevalence (1980-94)	18
ORT use rate (1990-94)	20

Health Expenditures [4]

Total health expenditure, 1990 (official exchange rate)	
Millions of dollars	193
Dollars per capita	30
Health expenditures as a percentage of GDP	
Total	7.0
Public sector	3.2
Private sector	3.8
Development assistance for health	
Total aid flows (millions of dollars)[1]	33
Aid flows per capita (dollars)	5.1
Aid flows as a percentage of total health expenditure	17.0

For sources, notes, and explanations, see Annotated Source Appendix, page 1061.

Human Factors

Women and Children [5]

% of pregnant women immunized (tetanus 1990-94)	12
% of births attended by trained health personnel (1983-94)	20
Maternal mortality rate (1980-92)	600
Under-5 mortality rate (1994)	127
% under-5 moderately/severely underweight (1980-1994)	27

Burden of Disease [6]

Population per physician (1993)	10,843.35
Population per nurse (1993)	8,934.81
Population per hospital bed (1993)	1,250.09
AIDS cases per 100,000 people (1994)	*
Malaria cases per 100,000 people (1992)	199

Ethnic Division [7]

Black	95%
Mulatto and European	5%

Religion [8]

Roman Catholic (an overwhelming majority of whom also practice Voodoo)	80%
Protestant	16%
None	1%
Other	3%
(1982)	

Major Languages [9]

French (official) 10%, Creole.

Education

Public Education Expenditures [10]

Million (Gourde)	1980	1985	1989	1990	1991	1994
Total education expenditure	107	118	213	216	NA	NA
as percent of GNP	1.5	1.2	1.8	1.4	NA	NA
as percent of total govt. expend.	14.9	16.5	19.7	20.0	NA	NA
Current education expenditure	86	117	213	216	NA	NA
as percent of GNP	1.2	1.2	1.8	1.4	NA	NA
as percent of current govt. expend.	17.2	16.7	20.0	20.1	NA	NA
Capital expenditure	21	1	NA	-	NA	NA

Educational Attainment [11]

Age group (1986)	25+
Total population	2,229,501
Highest level attained (%)	
No schooling	59.5
First level	
Not completed	30.5
Completed	NA
Entered second level	
S-1	9.3
S-2	NA
Postsecondary	0.7

Illiteracy [12]

In thousands and percent[1]	1990	1995	2000
Illiterate population (15+ yrs.)	2,289	2,360	2,422
Illiteracy rate - total pop. (%)	68.5	66.2	60.1
Illiteracy rate - males (%)	65.0	62.9	57.0
Illiteracy rate - females (%)	71.7	69.3	62.9

Libraries [13]

Daily Newspapers [14]

	1980	1985	1990	1994
Number of papers	4	5	4	4
Circ. (000)	36[e]	50[e]	45	45

Culture [15]

Science and Technology

Scientific/Technical Forces [16]

Scientists/engineers	14,189
Number female	4,530
Technicians	18,020
Number female	8,639
Total[36]	32,209

R&D Expenditures [17]

U.S. Patents Issued [18]

For sources, notes, and explanations, see Annotated Source Appendix, page 1061.

Government and Law

Organization of Government [19]

Long-form name:
Republic of Haiti
Type:
Republic
Independence:
1 January 1804 (from France)
National holiday:
Independence Day, 1 January (1804)
Constitution:
Approved March 1987, suspended June 1988, most articles reinstated March 1989; in October 1991, government claimed to be observing the constitution; return to constitutional rule, October 1994
Legal system:
Based on Roman civil law system; accepts compulsory ICJ jurisdiction
Executive branch:
President; Prime Minister; Cabinet
Legislative branch:
Bicameral National Assembly (Assemblee Nationale): Senate and Chamber of Deputies
Judicial branch:
Court of Appeal (Cour de Cassation)

Crime [23]

Elections [20]

Chamber of Deputies with 83 seats; elections last held 25 June 1995 with reruns on 13 August and runoffs on 17 September; results are not available.

Government Budget [21]

For FY94/95.

	$ mil.
Revenues	242
Expenditures	299.4
Capital expenditures	NA

Military Expenditures and Arms Transfers [22]

	1990	1991	1992	1993	1994
Military expenditures					
Current dollars (mil.)[e]	33	30	28	32	30
1994 constant dollars (mil.)[e]	37	32	29	33	30
Armed forces (000)	8	8	8	8	0
Gross national product (GNP)					
Current dollars (mil.)	1,975	1,991	1,746	1,734	1,581
1994 constant dollars (mil.)	2,199	2,134	1,821	1,770	1,581
Central government expenditures (CGE)					
1994 constant dollars (mil.)	265	NA	NA	NA	122
People (mil.)	6.0	6.1	6.2	6.3	6.4
Military expenditure as % of GNP	17	1.5	1.6	1.8	1.9
Military expenditure as % of CGE	13.8	NA	NA	NA	24.8
Military expenditure per capita (1994 $)	6	5	5	5	5
Armed forces per 1,000 people (soldiers)	1.3	1.3	1.3	1.3	0
GNP per capita (1994 $)	365	349	293	280	246
Arms imports[6]					
Current dollars (mil.)	0	0	0	0	50
1994 constant dollars (mil.)	0	0	0	0	50
Arms exports[6]					
Current dollars (mil.)	0	0	0	0	0
1994 constant dollars (mil.)	0	0	0	0	0
Total imports[7]					
Current dollars (mil.)	279	420	327	265	292
1994 constant dollars (mil.)	311	450	341	270	292
Total exports[7]					
Current dollars (mil.)	158	181	84	89	73
1994 constant dollars (mil.)	176	194	88	91	73
Arms as percent of total imports[8]	0	0	0	0	17.1
Arms as percent of total exports[8]	0	0	0	0	0

Human Rights [24]

	SSTS	FL	FAPRO	PPCG	APROBC	TPW	PCPTW	STPEP	PHRFF	PRW	ASST	AFL
Observes	2	P	P	P	P	P	P	P		P	P	P
		EAFRD	CPR	ESCR	SR	ACHR	MAAE	PVIAC	PVNAC	EAFDAW	TCIDTP	RC
Observes		P	P		P				P			P

P = Party; S = Signatory; see Appendix for meaning of abbreviations.

Labor Force

Total Labor Force [25]

2.3 million

Labor Force by Occupation [26]

Agriculture	66%
Services	25
Industry	9

Date of data: 1982

Unemployment Rate [27]

60% (1995 est.)

Production Sectors

Commercial Energy Production and Consumption

Data are shown in quadrillion (10^{15}) BTUs and percent for 1995
Values for hydroelectric, nuclear, geothermal, solar, and wind power refer to electrical generation.

Production [28]

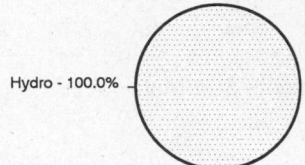

Hydro - 100.0%

Consumption [29]

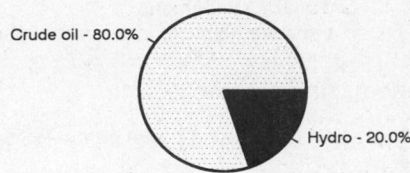

Crude oil - 80.0%

Hydro - 20.0%

Net hydroelectric power	0.002
Total	0.002

Crude oil	0.008
Net hydroelectric power	0.002
Total	0.010

Telecommunications [30]

- 50,000 (1990 est.) telephones; domestic facilities barely adequate, international facilities slightly better
- International: satellite earth station - 1 Intelsat (Atlantic Ocean)
- Radio: Broadcast stations: AM 33, FM 0, shortwave 2 Radios: 320,000 (1992 est.)
- Television: Broadcast stations: 4 (1987 est.) Televisions: 32,000 (1992 est.)

Transportation [31]

Railways: total: 40 km (single track; privately owned industrial line)-closed in early 1990's; narrow gauge: 40 km 0.760-m gauge

Highways: total: 3,978 km; paved: 944 km; unpaved: 3,034 km (1987 est.)

Merchant marine: none

Airports

Total:	11
With paved runways 2,438 to 3,047 m:	2
With paved runways 1,524 to 2,437 m:	1
With paved runways under 914 m:	4
With unpaved runways 914 to 1,523 m:	4 (1995 est.)

Top Agricultural Products [32]

Agriculture accounts for 34.8% of the GDP; produces coffee, mangoes, sugarcane, rice, corn, sorghum; wood.

Top Mining Products [33]

Metric tons except as noted	3/31/95*
Cement, hydraulic	75,000
Clays, for cement	10,000
Gravel (cu. meters)	750,000
Sand (cu. meters)	250,000
Limestone, for cement	75,000
Marble (cu. meters)	100

Tourism [34]

	1990	1991	1992	1993	1994
Visitors	222	203	NA	NA	NA
Tourists[45]	144	119	90	77	70
Cruise passengers	78	84	NA	NA	NA
Tourism receipts	82	71	38	78	46
Tourism expenditures	32	33	25	20	15
Fare receipts	4	4	2	2	1
Fare expenditures	37	39	24	20	18

Travelers are in thousands, money in million U.S. dollars.

For sources, notes, and explanations, see Annotated Source Appendix, page 1061.

401

Finance, Economics, and Trade

GDP and Manufacturing Summary [35]

	1980	1985	1990	1991	1992
Gross Domestic Product					
Millions of 1980 dollars	1,437	1,365	1,389	1,377	1,240
Growth rate in percent	7.34	0.26	-0.70	-0.81	-10.00
Manufacturing Value Added					
Millions of 1980 dollars	274	228	220	215	173[e]
Growth rate in percent	14.69	-2.87	-0.51	-2.37	-19.32[e]
Manufacturing share in percent of current prices	18.3	16.0	NA	NA	NA

Economic Indicators [36]

- **National product**: GDP—purchasing power parity—$6.5 billion (1995 est.)

- **National product real growth rate**: 4.5% (1995 est.)

- **National product per capita**: $1,000 (1995 est.)

- **Inflation rate (consumer prices)**: 14.5% (FY 94/95)

- **External debt**: $827 million (September 1995 est.)

Balance of Payments Summary [37]

Values in millions of dollars.

	1989	1990	1991	1992	1993
Exports of goods (f.o.b.)	148.3	265.8	202.0	75.6	81.6
Imports of goods (f.o.b.)	-259.3	-442.6	-448.6	-214.1	-266.6
Trade balance	-111.0	-176.8	-246.6	-138.5	-185.0
Services - debits	-219.0	-163.5	-170.5	-97.4	-104.1
Services - credits	93.1	59.1	59.6	39.5	37.8
Private transfers (net)	59.3	61.0	69.5	70.0	73.4
Government transfers (net)	114.9	131.9	164.7	85.0	100.0
Long-term capital (net)	30.0	56.7	42.1	-10.9	-15.3
Short-term capital (net)	39.9	-20.5	-4.4	7.9	13.7
Errors and omissions	-5.6	20.1	110.1	55.7	98.6
Overall balance	1.6	-32.0	24.5	11.3	19.1

Exchange Rates [38]

Currency: **gourde.**
Symbol: **G.**

Data are currency units per $1.

January 1996	16.783
1995	16.160
1994	12.947
1993	12.805
1992	10.953
1991	8.240

Imports and Exports

Top Import Origins [39]

$537 million (f.o.b., 1995 est.) Data are for 1993.

Origins	%
US	51
Latin America	18
Europe	16

Top Export Destinations [40]

$161 million (f.o.b., 1995 est.) Data are for 1993.

Destinations	%
US	81
Europe	12

Foreign Aid [41]

Recipient: ODA, $NA.

Import and Export Commodities [42]

Import Commodities	**Export Commodities**
Machines and manufactures 34%	Light manufactures 65%
Food and beverages 22%	Coffee 19%
Petroleum products 14%	Other agriculture 8%
Chemicals 10%	Other 8%
Fats and oils 9%	

For sources, notes, and explanations, see Annotated Source Appendix, page 1061.

Honduras

Geography [1]

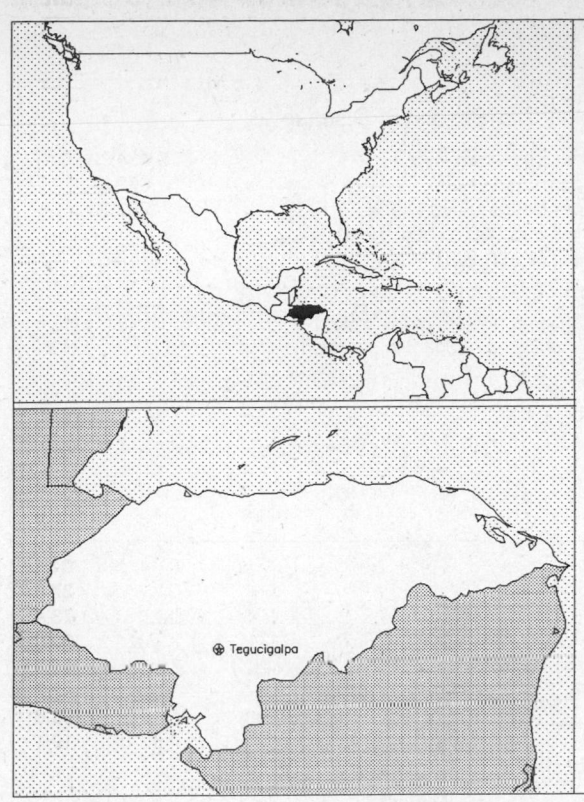

Total area:
112,090 sq km   43,278 sq mi
Land area:
111,890 sq km   43,201 sq mi
Comparative area:
Slightly larger than Tennessee
Land boundaries:
Total 1,520 km, Guatemala 256 km, El Salvador 342 km, Nicaragua 922 km
Coastline:
820 km
Climate:
Subtropical in lowlands, temperate in mountains
Terrain:
Mostly mountains in interior, narrow coastal plains
Natural resources:
Timber, gold, silver, copper, lead, zinc, iron ore, antimony, coal, fish
Land use:
Arable land: 14%
Permanent crops: 2%
Meadows and pastures: 30%
Forest and woodland: 34%
Other: 20%

Demographics [2]

	1970	1980	1990	1995[1]	1996	2000	2010	2020	2030
Population	2,683	3,625	4,741	5,460	5,605	6,192	7,643	9,042	10,345
Population density (persons per sq. mi.)	62	84	110	126	130	143	177	209	239
(persons per sq. km.)	24	32	42	49	50	55	68	81	92
Net migration rate (per 1,000 population)	NA	7.0	-1.2	-1.6	-1.5	-1.4	-1.2	-1.1	-0.8
Births	NA	NA	NA	NA	187	NA	NA	NA	NA
Deaths	NA	NA	NA	NA	33	NA	NA	NA	NA
Life expectancy - males	NA	NA	63.6	65.6	66.0	67.5	70.8	73.4	75.4
Life expectancy - females	NA	NA	68.3	70.6	71.0	72.6	76.2	79.1	81.3
Birth rate (per 1,000)	NA	44.9	37.6	34.1	33.4	30.3	24.2	20.3	17.3
Death rate (per 1,000)	NA	10.2	7.1	6.0	5.8	5.2	4.4	4.2	4.6
Women of reproductive age (15-49 yrs.)	NA	NA	1,077	1,290	1,336	1,526	2,027	2,488	2,792
of which are currently married	NA	NA	650	NA	810	931	1,265	NA	NA
Fertility rate	NA	6.4	5.2	4.6	4.4	3.8	2.8	2.3	2.1

Except as noted, values for vital statistics are in thousands; life expectancy is in years.

Health

Health Indicators [3]

% of population with access to	
safe water (1990-95)	65
adequate sanitation (1990-95)	75
health services (1985-95)	64
% of 1-year-olds immunized (1990-94) against	
TB (tuberculosis)	95
DPT (diphtheria, pertussis, tetanus)	95
polio	95
measles	94
% of contraceptive prevalence (1980-94)	47
ORT use rate (1990-94)	70

Health Expenditures [4]

Total health expenditure, 1990 (official exchange rate)	
Millions of dollars	134
Dollars per capita	26
Health expenditures as a percentage of GDP	
Total	4.5
Public sector	2.9
Private sector	1.6
Development assistance for health	
Total aid flows (millions of dollars)[1]	20
Aid flows per capita (dollars)	4.0
Aid flows as a percentage of total health expenditure	15.1

For sources, notes, and explanations, see Annotated Source Appendix, page 1061.

403

Human Factors

Women and Children [5]

% of pregnant women immunized (tetanus 1990-94)	88
% of births attended by trained health personnel (1983-94)	81
Maternal mortality rate (1980-92)	220
Under-5 mortality rate (1994)	54
% under-5 moderately/severely underweight (1980-1994)	21

Burden of Disease [6]

Population per physician (1991)	2,434.72
Population per nurse (1984)	672.54
Population per hospital bed (1990)	993.38
AIDS cases per 100,000 people (1994)	13.7
Malaria cases per 100,000 people (1992)	1,368

Ethnic Division [7]

Mixed Indian and European	90%
Indian	7%
Black	2%
White	1%

Religion [8]

Roman Catholic	97%
Protestant minority	3%

Major Languages [9]

Spanish, Indian dialects.

Education

Public Education Expenditures [10]

	1980	1985	1989	1990	1991	1994
Million (Lempira)						
Total education expenditure	155	290	416	NA	621	988
as percent of GNP	3.2	4.2	4.2	NA	4.1	4.0
as percent of total govt. expend.	14.2	13.8	15.9	NA	NA	16.0
Current education expenditure	141	286	404	NA	606	971
as percent of GNP	2.9	4.1	4.1	NA	4.0	4.0
as percent of current govt. expend.	19.5	NA	NA	NA	NA	NA
Capital expenditure	14	4	12	NA	14	17

Educational Attainment [11]

Age group (1983)[10]	25+
Total population	NA
Highest level attained (%)	
No schooling	33.5
First level	
Not completed	51.3
Completed	NA
Entered second level	
S-1	4.3
S-2	7.6
Postsecondary	3.3

Illiteracy [12]

In thousands and percent[1]	1990	1995	2000
Illiterate population (15+ yrs.)	817	869	925
Illiteracy rate - total pop. (%)	31.4	28.1	25.2
Illiteracy rate - males (%)	31.3	28.4	26.0
Illiteracy rate - females (%)	31.7	27.7	24.5

Libraries [13]

Daily Newspapers [14]

	1980	1985	1990	1994
Number of papers	6	7	5	5
Circ. (000)	212	293	199	240[e]

Culture [15]

Science and Technology

Scientific/Technical Forces [16]

R&D Expenditures [17]

U.S. Patents Issued [18]

Values show patents issued to citizens of the country by the U.S. Patents Office.

	1993	1994	1995
Number of patents	0	1	1

Government and Law

Organization of Government [19]

Long-form name:
Republic of Honduras
Type:
Republic
Independence:
15 September 1821 (from Spain)
National holiday:
Independence Day, 15 September (1821)
Constitution:
11 January 1982, effective 20 January
1982
Legal system:
Rooted in Roman and Spanish civil law;
some influence of English common law;
accepts ICJ jurisdiction, with reservations
Executive branch:
President: First Vice President; Second
Vice President; Third Vice President;
Cabinet
Legislative branch:
Unicameral: National Congress (Congreso
Nacional)
Judicial branch:
Supreme Court of Justice (Corte Suprema
de Justica)

Elections [20]

National Congress	% of votes
PNH	53.0
Liberal Party (PLH)	41.0
PINU-SD	2.5
Christian Democratic Party (PDCH)	1.0
Other	2.5

Government Budget [21]

For 1993 est.

	$ mil.
Revenues	527
Expenditures	668
Capital expenditures	166

Crime [23]

	1994
Crime volume	
Cases known to police	15,885
Attempts (percent)	NA
Percent cases solved	NA
Crimes per 100,000 persons	353.00
Persons responsible for offenses	
Total number offenders	26,737
Percent female	NA
Percent juvenile (1-18 yrs.)	NA
Percent foreigners	NA

Military Expenditures and Arms Transfers [22]

	1990	1991	1992	1993	1994
Military expenditures					
Current dollars (mil.)	50	41	39	46[e]	45[e]
1994 constant dollars (mil.)	56	44	41	47[e]	45[e]
Armed forces (000)	18	17	17	17	17
Gross national product (GNP)					
Current dollars (mil.)	2,382	2,467	2,686	2,852	2,914
1994 constant dollars (mil.)	2,651	2,644	2,801	2,911	2,914
Central government expenditures (CGE)					
1994 constant dollars (mil.)	510	482	505	558	546
People (mil.)	4.7	4.9	5.0	5.2	5.3
Military expenditure as % of GNP	2.1	1.7	1.5	1.6	1.6
Military expenditure as % of CGE	11.0	9.1	8.1	8.5	8.3
Military expenditure per capita (1994 $)	12	9	8	9	9
Armed forces per 1,000 people (soldiers)	3.8	3.5	3.4	3.3	3.2
GNP per capita (1994 $)	559	542	557	563	548
Arms imports[6]					
Current dollars (mil.)	30	30	20	20	10
1994 constant dollars (mil.)	33	32	21	20	10
Arms exports[6]					
Current dollars (mil.)	0	0	0	0	0
1994 constant dollars (mil.)	0	0	0	0	0
Total imports[7]					
Current dollars (mil.)	935	955	1,037	1,130	1,056
1994 constant dollars (mil.)	1,041	1,024	1,081	1,153	1,056
Total exports[7]					
Current dollars (mil.)	831	792	802	814	843
1994 constant dollars (mil.)	925	849	836	831	843
Arms as percent of total imports[8]	3.2	3.1	1.9	1.8	0.9
Arms as percent of total exports[8]	0	0	0	0	0

Human Rights [24]

	SSTS	FL	FAPRO	PPCG	APROBC	TPW	PCPTW	STPEP	PHRFF	PRW	ASST	AFL
Observes	P	P	P	P	P	P	P					P
	EAFRD	CPR	ESCR	SR	ACHR	MAAE	PVIAC	PVNAC	EAFDAW	TCIDTP	RC
Observes	S	P	P	P	P	P	P	P			P

P = Party; S = Signatory; see Appendix for meaning of abbreviations.

Labor Force

Total Labor Force [25]

1.3 million

Labor Force by Occupation [26]

Agriculture	62%
Services	20
Manufacturing	9
Construction	3
Other	6

Date of data: 1985

Unemployment Rate [27]

10%; underemployed 30%-40% (1992)

For sources, notes, and explanations, see Annotated Source Appendix, page 1061.

Production Sectors

Commercial Energy Production and Consumption

Data are shown in quadrillion (10^{15}) BTUs and percent for 1995
Values for hydroelectric, nuclear, geothermal, solar, and wind power refer to electrical generation.

Production [28]

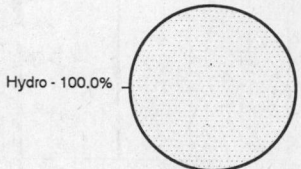

Hydro - 100.0%

Consumption [29]

Crude oil - 59.1%

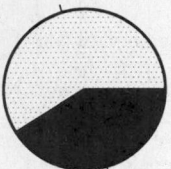

Hydro - 40.9%

Net hydroelectric power	0.027
Total	0.027

Crude oil	0.039
Net hydroelectric power	0.027
Total	0.066

Telecommunications [30]

- 105,000 (1992 est.) telephones; inadequate system
- International: satellite earth stations - 2 Intelsat (Atlantic Ocean); connected to Central American Microwave System
- Radio: Broadcast stations: AM 176, FM 0, shortwave 7 Radios: 2.115 million (1992 est.)
- Television: Broadcast stations: 28 Televisions: 400,000 (1992 est.)

Top Agricultural Products [32]

Agriculture accounts for 30% of the GDP; produces bananas, coffee, citrus; beef; timber; shrimp;.

Top Mining Products [33]

Metric tons except as noted[e]	4/20/95[*]
Cement	645,000
Copper, Cu content of lead and zinc concentrates	1,000
Gypsum	25,500
Lead, mine output, Pb content	2,810[r]
Petroleum refinery products (000 42-gal. bls.)	1,900[40]
Salt	25,000
Silver (kg.)	24,900[r]
Limestone	400,000
Marble	90,000
Zinc, mine output, Zn content	16,700[r]

Transportation [31]

Railways: total: 595 km; narrow gauge: 190 km 1.067-m gauge; 128 km 1.057-m gauge; 277 km 0.914-m gauge

Highways: total: 14,203 km; paved: 2,533 km; unpaved: 11,670 km (1993 est.)

Merchant marine: total: 257 ships (1,000 GRT or over) totaling 769,518 GRT/1,148,423 DWT; ships by type: bulk 29, cargo 165, chemical tanker 2, combination bulk 1, container 7, liquefied gas tanker 1, livestock carrier 3, oil tanker 19, passenger 1, passenger-cargo 3, refrigerated cargo 16, roll-on/roll-off cargo 7, short-sea passenger 2, vehicle carrier 1

Airports

Total:	111
With paved runways 2,438 to 3,047 m:	3
With paved runways 1,524 to 2,437 m:	2
With paved runways 914 to 1,523 m:	5
With paved runways under 914 m:	79
With unpaved runways 2,438 to 3,047 m:	1

Tourism [34]

	1990	1991	1992	1993	1994
Visitors[1]	290	226	244	261	290
Tourists	202	198	230	222	198
Excursionists	87	28	14	39	92
Tourism receipts[46]	29	31	32	32	33
Tourism expenditures[46]	38	37	38	39	39
Fare receipts	15	15	14	11	13
Fare expenditures	7	8	8	8	NA

Travelers are in thousands, money in million U.S. dollars.

For sources, notes, and explanations, see Annotated Source Appendix, page 1061.

Manufacturing Sector

GDP and Manufacturing Summary [35]

	1980	1985	1989	1990	% change 1980-1990	% change 1989-1990
GDP (million 1980 $)	2,544	2,689	3,055	3,101	21.9	1.5
GDP per capita (1980 $)	695	613	613	603	-13.2	-1.6
Manufacturing as % of GDP (current prices)	15.1	15.0	14.0	17.1	13.2	22.1
Gross output (million $)	1,021[e]	1,611	2,468[e]	1,464[e]	43.4	-40.7
Value added (million $)	280[e]	493	715[e]	425[e]	51.8	-40.6
Value added (million 1980 $)	344	363	415	453	31.7	9.2
Industrial production index	100	111	138	134	34.0	-2.9
Employment (thousands)	55[e]	64	69[e]	69[e]	25.5	0.0

Note: GDP stands for Gross Domestic Product. 'e' stands for estimated value.

Profitability and Productivity

	1980	1985	1989	1990	% change 1980-1990	% change 1989-1990
Intermediate input (%)	73[e]	69	71[e]	71[e]	-2.7	0.0
Wages, salaries, and supplements (%)	12[e]	13	11[e]	12[e]	0.0	9.1
Gross operating surplus (%)	16[e]	18	17[e]	17[e]	6.3	0.0
Gross output per worker ($)	18,518[e]	25,167	35,870	20,727[e]	11.9	-42.2
Value added per worker ($)	5,073[e]	7,707	10,386[e]	6,051[e]	19.3	-41.7
Average wage (incl. benefits) ($)	2,147[e]	3,173	4,112[e]	2,443[e]	13.8	-40.6

Profitability is in percent of gross output. Productivity is in U.S. $. 'e' stands for estimated value.

Profitability - 1990

Inputs 71.0%
Wages - 12.0%
Surplus - 17.0%

The graphic shows percent of gross output.

Value Added in Manufacturing

	1980 $ mil.	1980 %	1985 $ mil.	1985 %	1989 $ mil.	1989 %	1990 $ mil.	1990 %	% change 1980-1990	% change 1989-1990
311 Food products	75[e]	26.8	129	26.2	213[e]	29.8	120[e]	28.2	60.0	-43.7
313 Beverages	57[e]	20.4	78	15.8	111[e]	15.5	70[e]	16.5	22.8	-36.9
314 Tobacco products	19[e]	6.8	42	8.5	47[e]	6.6	27[e]	6.4	42.1	-42.6
321 Textiles	12[e]	4.3	13[e]	2.6	22[e]	3.1	16[e]	3.8	33.3	-27.3
322 Wearing apparel	6[e]	2.1	14	2.8	17[e]	2.4	10[e]	2.4	66.7	-41.2
323 Leather and fur products	2[e]	0.7	2	0.4	4[e]	0.6	2[e]	0.5	0.0	-50.0
324 Footwear	1[e]	0.4	2	0.4	4[e]	0.6	3[e]	0.7	200.0	-25.0
331 Wood and wood products	20[e]	7.1	30	6.1	37[e]	5.2	19[e]	4.5	-5.0	-48.6
332 Furniture and fixtures	5[e]	1.8	8	1.6	10[e]	1.4	6[e]	1.4	20.0	-40.0
341 Paper and paper products	4[e]	1.4	9	1.8	18[e]	2.5	11[e]	2.6	175.0	-38.9
342 Printing and publishing	8[e]	2.9	13	2.6	18[e]	2.5	10[e]	2.4	25.0	-44.4
351 Industrial chemicals	1[e]	0.4	2	0.4	3[e]	0.4	2[e]	0.5	100.0	-33.3
352 Other chemical products	11[e]	3.9	20	4.1	31[e]	4.3	19[e]	4.5	72.7	-38.7
353 Petroleum refineries	9[e]	3.2	38	7.7	41[e]	5.7	23[e]	5.4	155.6	-43.9
354 Miscellaneous petroleum and coal products	NA	0.0	NA	0.0	NA	0.0	NA	0.0	NA	NA
355 Rubber products	5[e]	1.8	8	1.6	12[e]	1.7	7[e]	1.6	40.0	-41.7
356 Plastic products	8[e]	2.9	18	3.7	28[e]	3.9	16[e]	3.8	100.0	-42.9
361 Pottery, china, and earthenware	NA	0.0	NA	0.0	NA	0.0	NA	0.0	NA	NA
362 Glass and glass products	NA	0.0	NA	0.0	NA	0.0	NA	0.0	NA	NA
369 Other nonmetal mineral products	16[e]	5.7	24	4.9	40[e]	5.6	26[e]	6.1	62.5	-35.0
371 Iron and steel	NA	0.0	1	0.2	4[e]	0.6	3[e]	0.7	NA	-25.0
372 Nonferrous metals	NA	0.0	1	0.2	1[e]	0.1	1[e]	0.2	NA	0.0
381 Metal products	13[e]	4.6	21	4.3	27[e]	3.8	16[e]	3.8	23.1	-40.7
382 Nonelectrical machinery	1[e]	0.4	3	0.6	5[e]	0.7	4[e]	0.9	300.0	-20.0
383 Electrical machinery	3[e]	1.1	8	1.6	9[e]	1.3	6[e]	1.4	100.0	-33.3
384 Transport equipment	NA	0.0	2	0.4	3[e]	0.4	2[e]	0.5	NA	-33.3
385 Professional and scientific equipment	NA	0.0	1	0.2	1[e]	0.1	1[e]	0.2	NA	0.0
390 Other manufacturing industries	1[e]	0.4	5	1.0	8[e]	1.1	5[e]	1.2	400.0	-37.5

Note: The industry codes shown are International Standard Industry codes (ISIC). Percentages are percent of total Value Added. 'e' stands for estimated value

For sources, notes, and explanations, see Annotated Source Appendix, page 1061.

407

Finance, Economics, and Trade

Economic Indicators [36]

- **National product**: GDP—purchasing power parity—$10.8 billion (1995 est.)

- **National product real growth rate**: 4% (1995 est.)

- **National product per capita**: $1,980 (1995 est.)

- **Inflation rate (consumer prices)**: 30% (1994 est.)

- **External debt**: $3.7 billion (1994)

Balance of Payments Summary [37]

Values in millions of dollars.

	1989	1990	1991	1992	1993
Exports of goods (f.o.b.)	903.2	886.9	834.7	833.1	846.0
Imports of goods (f.o.b.)	-955.7	-907.0	-912.5	-990.2	-943.9
Trade balance	-52.5	-20.1	-77.8	-157.1	-97.9
Services - debits	-492.4	-477.3	-512.6	-586.6	-587.0
Services - credits	182.0	166.3	220.7	269.6	295.2
Private transfers (net)	34.2	48.8	53.0	61.1	61.3
Government transfers (net)	158.4	230.9	144.3	155.0	110.6
Long-term capital (net)	37.9	135.9	-44.6	168.8	194.1
Short-term capital (net)	226.1	59.1	132.1	71.8	-0.5
Errors and omissions	-138.9	-107.4	152.0	29.2	-81.8
Overall balance	-45.2	36.2	67.1	11.8	-108.0

Exchange Rates [38]

Currency: **lempira.**
Symbol: **L.**

Data are currency units per $1. The lempira was allowed to float in 1992.

December 1994	10.3432
1995	10.3432
1994	9.4001
1993	7.2600
1992	5.8300
1991	5.4000

Imports and Exports

Top Import Origins [39]

$1.1 billion (c.i.f. 1994).

Origins	%
US	50
Mexico	8
Guatemala	6

Top Export Destinations [40]

$843 million (f.o.b., 1994).

Destinations	%
US	53
Germany	11
Belgium	8
UK	5

Foreign Aid [41]

Recipient: ODA, $NA.

Import and Export Commodities [42]

Import Commodities	Export Commodities
Machinery and transport equipment	Bananas
Chemical products	Coffee
Manufactured goods	Shrimp
Fuel and oil	Lobster
Foodstuffs	Minerals
	Meat
	Lumber

For sources, notes, and explanations, see Annotated Source Appendix, page 1061.

Hong Kong

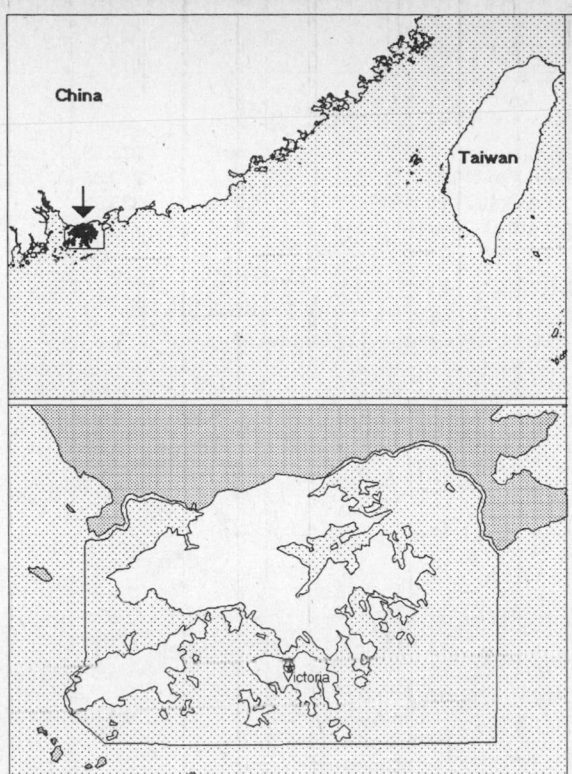

Geography [1]

Total area:
 1,040 sq km   402 sq mi
Land area:
 990 sq km   382 sq mi
Comparative area:
 Six times the size of Washington, DC
Land boundaries:
 Total 30 km, China 30 km
Coastline:
 733 km
Climate:
 Tropical monsoon; cool and humid in winter, hot and rainy from spring through summer, warm and sunny in fall
Terrain:
 Hilly to mountainous with steep slopes; lowlands in North
Natural resources:
 Outstanding deepwater harbor, feldspar
Land use:
 Arable land: 7%
 Permanent crops: 1%
 Meadows and pastures: 1%
 Forest and woodland: 12%
 Other: 79%

Demographics [2]

	1970	1980	1990	1995[1]	1996	2000	2010	2020	2030
Population	3,959	5,063	5,688	6,188	6,305	6,685	7,401	7,967	8,220
Population density (persons per sq. mi.)	10,368	13,259	14,896	16,205	16,513	17,506	19,381	20,864	21,527
(persons per sq. km.)	4,003	5,119	5,751	6,257	6,376	6,759	7,483	8,056	8,312
Net migration rate (per 1,000 population)	8.0	15.7	-1.6	13.9	12.4	7.0	6.4	5.1	3.3
Births	NA	NA	NA	NA	66	NA	NA	NA	NA
Deaths	NA	NA	29	NA	33	NA	NA	NA	NA
Life expectancy - males	NA	NA	76.4	78.7	78.9	79.6	80.5	80.8	80.9
Life expectancy - females	NA	NA	83.4	85.6	85.7	86.3	86.8	87.0	87.0
Birth rate (per 1,000)	20.0	17.0	11.9	11.2	10.5	10.2	9.4	8.6	7.8
Death rate (per 1,000)	5.1	5.2	5.1	5.2	5.2	5.6	6.9	8.2	10.5
Women of reproductive age (15-49 yrs.)	NA	NA	1,534	1,690	1,723	1,802	1,753	1,673	1,636
of which are currently married[20]	NA	NA	881	NA	1,064	1,116	1,079	NA	NA
Fertility rate	NA	2.1	1.3	1.3	1.3	1.4	1.4	1.4	1.4

Except as noted, values for vital statistics are in thousands; life expectancy is in years.

Health

Health Indicators [3]

% of population with access to	
safe water (1990-95)	100
adequate sanitation (1990-95)	88
health services (1985-95)[1]	99
% of 1-year-olds immunized (1990-94) against	
TB (tuberculosis)	99
DPT (diphtheria, pertussis, tetanus)	83
polio	81
measles	77
% of contraceptive prevalence (1980-94)	81
ORT use rate (1990-94)	NA

Health Expenditures [4]

For sources, notes, and explanations, see Annotated Source Appendix, page 1061.

409

Human Factors

Women and Children [5]

% of pregnant women immunized (tetanus 1990-94)	NA
% of births attended by trained health personnel (1983-94)	100
Maternal mortality rate (1980-92)	6
Under-5 mortality rate (1994)	6
% under-5 moderately/severely underweight (1980-1994)	NA

Burden of Disease [6]

Population per physician (1986)	1,073.25
Population per nurse (1986)	240.36
Population per hospital bed (1990)	233.95
AIDS cases per 100,000 people (1994)	0.6
Malaria cases per 100,000 people (1992)	NA

Ethnic Division [7]

Chinese	95%
Other	5%

Religion [8]

Eclectic mixture of local religions	90%
Christian	10%

Major Languages [9]

Chinese (Cantonese), English.

Education

Public Education Expenditures [10]

Million (Dollar)	1980	1990	1991	1992	1993	1994
Total education expenditure	3,446	16,566	19,552	22,193	25,005	NA
as percent of GNP	NA	NA	NA	NA	NA	NA
as percent of total govt. expend.	14.6	17.4	18.1	17.4	17.0	NA
Current education expenditure	3,036	NA	16,915	20,349	NA	NA
as percent of GNP	NA	NA	NA	NA	NA	NA
as percent of current govt. expend.	25.5	NA	NA	NA	NA	NA
Capital expenditure	410	NA	2,637	1,844	NA	NA

Educational Attainment [11]

Age group (1991)	25+
Total population	3,530,524
Highest level attained (%)	
No schooling	15.7
First level	
Not completed	30.3
Completed	NA
Entered second level	
S-1	17.8
S-2	25.5
Postsecondary	10.6

Illiteracy [12]

In thousands and percent[1]	1990	1995	2000
Illiterate population (15+ yrs.)	423	370	311
Illiteracy rate - total pop. (%)	9.4	7.4	5.6
Illiteracy rate - males (%)	4.5	3.8	3.1
Illiteracy rate - females (%)	14.6	11.3	8.3

Libraries [13]

	Admin. Units	Svc. Pts.	Vols. (000)	Shelving (meters)	Vols. Added	Reg. Users
National	NA	NA	NA	NA	NA	NA
Nonspecialized	NA	NA	NA	NA	NA	NA
Public (1992)[6]	2	55	4,189	NA	348,158	3 mil
Higher ed. (1990)	17	33	3,370	NA	225,078	123,125
School (1990)	374[e]	374[e]	3,266	81,637	268,931	417,365

Daily Newspapers [14]

	1980	1985	1990	1994
Number of papers	41	46	38	43
Circ. (000)	3,600[e]	4,100[e]	4,250[e]	4,200[e]

Culture [15]

Cinema (seats per 1,000)	NA
Annual attendance per person	10.3
Gross box office receipts (mil. Dollar)	1,367
Museums (reporting)	NA
Visitors (000)	NA
Annual receipts (000 Dollar)	NA

Science and Technology

Scientific/Technical Forces [16]

Scientists/engineers	206,970
Number female	69,820
Technicians	119,446
Number female	54,112
Total[36]	326,416

R&D Expenditures [17]

U.S. Patents Issued [18]

Values show patents issued to citizens of the country by the U.S. Patents Office.

	1993	1994	1995
Number of patents	182	220	248

Government and Law

Organization of Government [19]

Long-form name:
None
Type:
Dependent territory of the UK scheduled to revert to China on 1 July 1997
Independence:
None (dependent territory of the UK; the UK signed an agreement with China on 19 December 1984 to return Hong Kong to China on 1 July 1997; in the joint declaration, China promises to respect Hong Kong's existing social and economic systems and lifestyle)
National holiday:
Liberation Day, 29 August (1945)
Constitution:
Unwritten; partly statutes, partly common law and practice; Basic Law approved in March 1990 in preparation for 1997
Legal system:
Based on English common law
Executive branch:
British Monarch; Governor and President of the Executive Council; Chief Secretary; Executive Council
Legislative branch:
Unicameral: Legislative Council
Judicial branch:
Supreme Court

Elections [20]

Legislative Council	% of seats
Democratic Party	35.0
Liberal Party	16.7
Democratic Alliance for the Betterment of Hong Kong	10.0
Independents and Others	38.3

Government Budget [21]

For FY94/95.

	$ bil.
Revenues	19
Expenditures	14.1
Capital expenditures	0.289

Defense Summary [22]

Branches: Headquarters of British Forces, Army, Royal Navy, Royal Air Force, Royal Hong Kong Auxiliary Air Force, Royal Hong Kong Police Force

Manpower Availability: Males age 15-49 1,895,535; males fit for military service 1,442,072; males reach military age (18) annually 46,248 (1996 est.)

Defense Expenditures: $207 million, 0.2% of GDP (FY92/93); this represents 65% of the total cost of defending the colony, the remainder being paid by the UK

Note: Defense is the responsibility of the UK until 1 July 1997, when China will assume command

Crime [23]

	1994
Crime volume	
Cases known to police	87,804
Attempts (percent)	NA
Percent cases solved	50.30
Crimes per 100,000 persons	1,448.58
Persons responsible for offenses	
Total number offenders	49,784
Percent female	15.70
Percent juvenile (7-15 yrs.)	14.10
Percent foreigners	6.30

Human Rights [24]

Labor Force

Total Labor Force [25]

2.915 million (1994)

Labor Force by Occupation [26]

Manufacturing	28.5%
Wholesale and retail trade, restaurants and hotels	27.9
Services	17.7
Financing, insurance, and real estate	9.2
Transport and communications	4.5
Construction	2.5
Other	9.7

Date of data: 1989

Unemployment Rate [27]

3.5% (1995 est.)

For sources, notes, and explanations, see Annotated Source Appendix, page 1061.

411

Production Sectors

Energy Resource Summary [28]

Electricity: Capacity: 8,930,000 kW. Production: 33 billion kWh. Consumption per capita: 4,628 kWh (1993).

Telecommunications [30]

- 4.13 million (1995 est.) telephones; modern facilities provide excellent domestic and international services
- Domestic: microwave radio relay links and extensive fiber-optic network
- International: satellite earth stations - 3 Intelsat; coaxial cable to Guangzhou, China; access to 5 submarine cables providing connections to ASEAN member nations, Japan, Taiwan, Australia, Middle East, and Western Europe
- Radio: Broadcast stations: AM 6, FM 6 Radios: 3 million (1992 est.)
- Television: Broadcast stations: 4 (2 repeaters) Televisions: 1.75 million (1992 est.)

Top Agricultural Products [32]

Agriculture accounts for 0.2% of the GDP; produces fresh vegetables; poultry.

Top Mining Products [33]

Detailed information is not available. A summary of natural resources follows. **Mineral Resources**: Feldspar.

Transportation [31]

Railways: total: 35 km; standard gauge: 35 km 1.435-m gauge

Highways: total: 1,661 km; paved: 1,661 km; unpaved: 0 km (1994 est.)

Merchant marine: total: 238 ships (1,000 GRT or over) totaling 8,632,224 GRT/14,820,657 DWT; ships by type: bulk 129, cargo 32, chemical tanker 1, combination bulk 4, combination ore/oil 3, container 39, liquefied gas tanker 3, multifunction large load carrier 1, oil tanker 17, refrigerated cargo 5, short-sea passenger 1, vehicle carrier 3

Airports

Total: 2
With paved runways over 3,047 m: 1
With paved runways under 914 m: 1 (1995 est.)

Tourism [34]

	1990	1991	1992	1993	1994
Visitors	6,581	6,795	8,011	8,938	9,331
Cruise passengers[47]	9	4	4	3	NA
Tourism receipts[48]	5,032	5,078	6,037	7,562	8,317

Travelers are in thousands, money in million U.S. dollars.

Manufacturing Sector

Manufacturing Summary [35]

	1987		1988		1989		1990		1991	
	$ bil.	%	$ bil.	%	$ bil.	%	$ bil.	%	$ bil.	%
Establishments or enterprises (number)	54,415	2.320	55,173	2.628	55,938	2.986	55,445	3.095	-	-
Total employment (000)	1,052	0.775	989	0.722	914	0.742	827	0.747	-	-
Production workers (000)	870	1.290	799	1.279	733	0.997	658	0.927	-	-
Output ($ bil.)	45	0.443	50	0.431	51	0.430	50	0.439	-	-
Value added ($ bil.)	12	0.279	13	0.274	14	0.286	14	0.288	-	-
Capital investment ($ mil.)	1,889	0.433	2,246	0.472	2,023	0.369	1,960	0.353	-	-
M & E investment ($ mil.)	1,065	0.342	1,242	0.359	1,078	0.276	1,001	0.237	-	-
Employees per establishment (number)	19	33.408	18	27.484	16	24.859	15	24.140	-	-
Production workers per establishment	16	55.628	14	48.652	13	33.374	12	29.939	-	-
Output per establishment ($ mil.)	0.828	19.112	0.911	16.409	0.907	14.400	0.907	14.171	-	-
Capital investment per estab. ($ mil.)	0.035	18.687	0.041	17.948	0.036	12.367	0.035	11.395	-	-
M & E per establishment ($ mil)	0.020	14.732	0.023	13.652	0.019	9.246	0.018	7.670	-	-
Payroll per employee ($)	6,784	75.655	8,032	80.622	9,429	84.396	10,898	78.560	-	-
Wages per production worker ($)	-	-	-	-	-	-	-	-	-	-
Hours per production worker (hours)	2,459	129.578	2,486	129.322	2,424	132.245	2,404	128.527	-	-
Output per employee ($)	42,846	57.209	50,853	59.704	55,516	57.926	60,836	58.703	-	-
Capital investment per employee ($)	1,796	55.935	2,272	66.301	2,214	49.749	2,370	47.203	-	-
M & E per employee ($)	1,012	44.096	1,256	49.671	1,180	37.192	1,210	31.775	-	-

Note: Columns headed % show percent of world total or ratio. Ratios closest to 100 are closest to world average. M & E stands for machinery & equipment.

Output in Manufacturing

	1987		1988		1989		1990		1991	
	$ bil.	%	$ bil.	%	$ bil.	%	$ bil.	%	$ bil.	%
3110 - Food products	0.836	1.86	0.991	1.98	1.053	2.06	1.236	2.47	-	-
3130 - Beverages	0.303	0.67	0.383	0.77	0.439	0.86	0.484	0.97	-	-
3140 - Tobacco	0.285	0.63	0.353	0.71	0.444	0.87	0.684	1.37	-	-
3210 - Textiles	5.719	12.71	6.075	12.15	7.039	13.80	6.590	13.18	-	-
3211 - Spinning, weaving, etc.	3.085	6.86	3.318	6.64	3.497	6.86	3.423	6.85	-	-
3220 - Wearing apparel	7.836	17.41	7.653	15.31	7.901	15.49	8.247	16.49	-	-
3230 - Leather and products	0.183	0.41	0.174	0.35	0.180	0.35	0.174	0.35	-	-
3240 - Footwear	0.188	0.42	0.256	0.51	0.254	0.50	0.123	0.25	-	-
3310 - Wood products	0.202	0.45	0.185	0.37	0.164	0.32	0.177	0.35	-	-
3320 - Furniture, fixtures	0.237	0.53	0.191	0.38	0.223	0.44	0.171	0.34	-	-
3410 - Paper and products	0.777	1.73	0.983	1.97	0.924	1.81	1.093	2.19	-	-
3411 - Pulp, paper, etc.	0.133	0.30	0.129	0.26	0.088	0.17	0.148	0.30	-	-
3420 - Printing, publishing	1.093	2.43	1.531	3.06	1.669	3.27	2.079	4.16	-	-
3510 - Industrial chemicals	0.234	0.52	0.356	0.71	-	-	-	-	-	-
3511 - Basic chemicals, excl fertilizers	0.067	0.15	0.081	0.16	0.093	0.18	0.089	0.18	-	-
3513 - Synthetic resins, etc.	0.164	0.36	0.271	0.54	0.346	0.68	0.367	0.73	-	-
3520 - Chemical products nec	0.318	0.71	0.363	0.73	-	-	-	-	-	-
3522 - Drugs and medicines	0.069	0.15	0.081	0.16	0.089	0.17	0.084	0.17	-	-
3540 - Petroleum, coal products	-	-	0.015	0.03	0.032	0.06	0.032	0.06	-	-
3550 - Rubber products	0.051	0.11	0.055	0.11	0.048	0.09	0.047	0.09	-	-
3560 - Plastic products nec	3.064	6.81	3.197	6.39	3.179	6.23	2.465	4.93	-	-
3610 - Pottery, china, etc.	0.016	0.04	0.017	0.03	0.015	0.03	0.015	0.03	-	-
3620 - Glass and products	0.091	0.20	0.070	0.14	0.067	0.13	0.053	0.11	-	-
3690 - Nonmetal products nec	0.437	0.97	0.509	1.02	0.537	1.05	0.548	1.10	-	-
3810 - Metal products	2.221	4.94	2.745	5.49	2.658	5.21	2.294	4.59	-	-
3820 - Machinery nec	1.368	3.04	1.990	3.98	2.507	4.92	3.820	7.64	-	-
3825 - Office, computing machinery	0.584	1.30	1.042	2.08	1.452	2.85	2.598	5.20	-	-
3830 - Electrical machinery	6.422	14.27	6.894	13.79	5.986	11.74	4.836	9.67	-	-
3832 - Radio, television, etc.	4.499	10.00	4.938	9.88	4.256	8.35	3.152	6.30	-	-
3840 - Transportation equipment	0.370	0.82	0.435	0.87	0.510	1.00	0.539	1.08	-	-
3841 - Shipbuilding, repair	0.212	0.47	0.206	0.41	0.270	0.53	0.285	0.57	-	-
3843 - Motor vehicles	0.006	0.01	0.025	0.05	0.042	0.08	0.034	0.07	-	-
3850 - Professional goods	2.393	5.32	2.917	5.83	2.804	5.50	2.794	5.59	-	-
3900 - Industries nec	1.599	3.55	1.844	3.69	1.958	3.84	1.631	3.26	-	-

Note: Codes are International Standard Industry codes (ISIC). Percentages are % of total Output. [f]: Factor Prices; [p]: Producer Prices.

For sources, notes, and explanations, see Annotated Source Appendix, page 1061.

413

Finance, Economics, and Trade

Economic Indicators [36]

- **National product**: GDP—purchasing power parity— $152.4 billion (1995 est.)
- **National product real growth rate**: 5% (1995 est.)
- **National product per capita**: $27,500 (1995 est.)
- **Inflation rate (consumer prices)**: 8.4% (1995)
- **External debt**: None (1995)

Balance of Payments Summary [37]

Exchange Rates [38]

Currency: **Hong Kong dollar.**
Symbol: **HK$.**

Data are currency units per $1. Linked to the US dollar at the rate of about 7.8 HK$ per 1 US$ since 1985.

1995	7.800
1994	7.800
1993	7.800
1992	7.741
1991	7.771

Imports and Exports

Top Import Origins [39]

$195.4 billion (c.i.f., 1995) Data are for 1993.

Origins	%
China	38
Japan	17
Taiwan	9
US	7

Top Export Destinations [40]

$177.1 billion (including re-exports)(f.o.b., 1995 est.) Data are for 1993.

Destinations	%
China	33
US	22
Germany	5
Japan	5
UK	3

Foreign Aid [41]

	U.S. $	
US commitments, including Ex-Im (FY70-87)	152	million
Western (non-US) countries, ODA and OOF bilateral commitments (1970-89)	923	million

Import and Export Commodities [42]

Import Commodities	Export Commodities
Foodstuffs	Clothing
Transport equipment	Textiles
Raw materials	Yarn and fabric
Semimanufactures	Footwear
Petroleum	Electrical appliances
	Watches and clocks
	Toys

For sources, notes, and explanations, see Annotated Source Appendix, page 1061.

Hungary

Geography [1]

Total area:
93,030 sq km   35,919 sq mi
Land area:
92,340 sq km   35,653 sq mi
Comparative area:
Slightly smaller than Indiana
Land boundaries:
Total 2,009 km, Austria 366 km, Croatia 329 km, Romania 443 km, Yugoslavia 151 km (all with Serbia), Slovakia 515 km, Slovenia 102 km, Ukraine 103 km
Coastline:
0 km (landlocked)
Climate:
Temperate; cold, cloudy, humid winters; warm summers
Terrain:
Mostly flat to rolling plains; hills and low mountains on the Slovakian border
Natural resources:
Bauxite, coal, natural gas, fertile soils
Land use:
Arable land: 51%
Permanent crops: 6%
Meadows and pastures: 13%
Forest and woodland: 18%
Other: 12%

Demographics [2]

	1970	1980	1990	1995[1]	1996	2000	2010	2020	2030
Population	10,337	10,711	10,352	10,072	10,003	9,795	9,456	9,103	8,863
Population density (persons per sq. mi.)	290	300	290	283	281	275	265	255	249
(persons per sq. km.)	112	116	112	109	108	106	102	99	96
Net migration rate (per 1,000 population)	-0.2	0.0	-2.4	-2.5	-2.5	-1.3	-0.8	-0.3	0.0
Births	152	149	NA	NA	107	NA	NA	NA	NA
Deaths	120	145	146	NA	151	NA	NA	NA	NA
Life expectancy - males	66.8	66.0	65.2	64.1	64.2	64.8	66.1	70.9	74.6
Life expectancy - females	72.6	73.2	73.8	73.9	74.0	74.7	76.3	79.6	82.0
Birth rate (per 1,000)	14.7	13.9	12.1	10.5	10.7	13.1	10.9	9.7	9.6
Death rate (per 1,000)	11.6	13.6	14.1	15.1	15.1	14.8	14.5	12.6	12.4
Women of reproductive age (15-49 yrs.)	NA	NA	2,535	2,539	2,531	2,439	2,189	2,134	1,891
of which are currently married	1,886	1,881	1,656	NA	1,637	1,616	1,476	NA	NA
Fertility rate	2.0	1.9	1.8	1.5	1.5	1.8	1.7	1.6	1.6

Except as noted, values for vital statistics are in thousands; life expectancy is in years.

Health

Health Indicators [3]

% of population with access to	
safe water (1990-95)	NA
adequate sanitation (1990-95)	NA
health services (1985-95)	NA
% of 1-year-olds immunized (1990-94) against	
TB (tuberculosis)	100
DPT (diphtheria, pertussis, tetanus)	99
polio	99
measles	99
% of contraceptive prevalence (1980-94)	73
ORT use rate (1990-94)	NA

Health Expenditures [4]

Total health expenditure, 1990 (official exchange rate)	
Millions of dollars	1,958
Dollars per capita	185
Health expenditures as a percentage of GDP	
Total	6.0
Public sector	5.0
Private sector	0.9
Development assistance for health	
Total aid flows (millions of dollars)[1]	NA
Aid flows per capita (dollars)	NA
Aid flows as a percentage of total health expenditure	NA

For sources, notes, and explanations, see Annotated Source Appendix, page 1061.

415

Human Factors

Women and Children [5]

% of pregnant women immunized (tetanus 1990-94)	NA
% of births attended by trained health personnel (1983-94)[1]	99
Maternal mortality rate (1980-92)	15
Under-5 mortality rate (1994)	14
% under-5 moderately/severely underweight (1980-1994)	NA

Burden of Disease [6]

Population per physician (1989)	329.71
Population per nurse (1985)	172.23
Population per hospital bed (1990)	97.23
AIDS cases per 100,000 people (1994)	0.2
Heart disease cases per 1,000 people (1990-93)	283

Ethnic Division [7]

Hungarian	89.9%
Gypsy	4%
German	2.6%
Serb	2%
Slovak	0.8%
Romanian	0.7%

Religion [8]

Roman Catholic	67.5%
Calvinist	20%
Lutheran	5%
Atheist and other	7.5%

Major Languages [9]

Hungarian	98.2%
Other	1.8%

Education

Public Education Expenditures [10]

Million (Forint)	1980	1985	1990	1992	1993	1994
Total education expenditure	33,099	54,061	122,120	193,772	229,000	278,322
as percent of GNP	4.7	5.5	6.1	6.9	6.7	6.7
as percent of total govt. expend.	5.2	6.4	7.8	7.7	7.4	6.9
Current education expenditure	27,516	48,125	110,382	178,954	215,195	262,917
as percent of GNP	3.9	4.9	5.5	6.4	6.3	6.3
as percent of current govt. expend.	6.4	7.4	8.6	8.6	8.3	7.9
Capital expenditure	5,583	5,936	11,738	14,818	13,805	15,405

Educational Attainment [11]

Age group (1990)[11]	25+
Total population	6,798,765
Highest level attained (%)	
No schooling	1.3
First level	
Not completed	24.3
Completed	33.6
Entered second level	
S-1	NA
S-2	30.7
Postsecondary	10.1

Illiteracy [12]

	1980	1989	1995
Illiterate population (15+ yrs.)[2]	95,542	NA	69,000
Illiteracy rate - total pop. (%)	1.1	NA	0.8
Illiteracy rate - males (%)	0.7	NA	0.7
Illiteracy rate - females (%)	1.5	NA	1.0

Libraries [13]

	Admin. Units	Svc. Pts.	Vols. (000)	Shelving (meters)	Vols. Added	Reg. Users
National (1992)	1	NA	2,660	NA	NA	25,460
Nonspecialized (1992)	1	NA	1,296	NA	NA	6,856
Public (1993)	3,032	5,264	49,102	NA	NA	2 mil
Higher ed. (1993)[10]	29	NA	12,803	NA	NA	132,751
School (1992)	3,792	NA	31,495	NA	NA	NA

Daily Newspapers [14]

	1980	1985	1990	1994
Number of papers	27	28	34	27
Circ. (000)	2,648	2,717	2,460	2,321

Culture [15]

Cinema seats per 1,000	13.1[e]
Annual attendance per person	1.4
Gross box office receipts (mil. Forint)	1,504
Museums (reporting)	NA
Visitors (000)	7,415
Annual receipts (000 Forint)	NA

Science and Technology

Scientific/Technical Forces [16]

Scientists/engineers	11,818
Number female	NA
Technicians[17]	6,003
Number female	NA
Total[16,17]	17,821

R&D Expenditures [17]

	Forint (000) 1993
Total expenditure[17]	34,686,000
Capital expenditure	3,593,000
Current expenditure	31,093,000
Percent current	89.2

U.S. Patents Issued [18]

Values show patents issued to citizens of the country by the U.S. Patents Office.

	1993	1994	1995
Number of patents	62	48	51

Government and Law

Organization of Government [19]

Long-form name:
Republic of Hungary
Type:
Republic
Independence:
1001 (unification by King Stephen I)
National holiday:
St. Stephen's Day (National Day), 20 August (commemorates the founding of Hungarian state circa 1000 AD)
Constitution:
18 August 1949, effective 20 August 1949, revised 19 April 1972; 18 October 1989 revision ensured legal rights for individuals and constitutional checks on the authority of the prime minister and also established the principle of parliamentary oversight
Legal system:
In process of revision, moving toward rule of law based on Western model
Executive branch:
President; Prime Minister; Council of Ministers
Legislative branch:
Unicameral: National Assembly (Orszaggyules)
Judicial branch:
Constitutional Court

Elections [20]

National Assembly	% of seats
Hungarian Socialist Party	54.1
Alliance of Free Democrats	18.1
Hungarian Democratic Forum	9.6
Independent Smallholders	6.7
Christian Democratic People's Party	5.7
Federation of Young Democrats	5.2
Other	0.5

Government Budget [21]

For 1995.

	$ bil.
Revenues	12.6
Expenditures	13.8
Capital expenditures	NA

Crime [23]

	1994
Crime volume	
Cases known to police	389,451
Attempts (percent)	5.30
Percent cases solved	54.00
Crimes per 100,000 persons	3,789.17
Persons responsible for offenses	
Total number offenders	119,494
Percent female	10.30
Percent juvenile (14-18 yrs.)	12.10
Percent foreigners	4.60

Military Expenditures and Arms Transfers [22]

	1990	1991	1992	1993	1994
Military expenditures					
Current dollars (mil.)	1,277[r]	1,261[r]	1,387[r]	1,261[r]	1,220[e]
1994 constant dollars (mil.)	1,421[r]	1,352[r]	1,446[r]	1,287[r]	1,220[e]
Armed forces (000)	94	87	78	NA	60
Gross national product (GNP)					
Current dollars (mil.)[e]	62,590	59,260	63,620	64,350	64,120
1994 constant dollars (mil.)[e]	69,660	63,520	66,340	65,680	64,120
Central government expenditures (CGE)					
1994 constant dollars (mil.)	34,850	NA	NA	22,400[e]	22,900[e]
People (mil.)	10.4	10.3	10.3	10.3	10.3
Military expenditure as % of GNP	2.0	2.1	2.2	2.0	1.9
Military expenditure as % of CGE	4.1	NA	NA	5.7	5.3
Military expenditure per capita (1994 $)	137	131	140	125	118
Armed forces per 1,000 people (soldiers)	9.1	8.4	7.5	NA	5.8
GNP per capita (1994 $)	6,721	6,139	6,420	6,362	6,213
Arms imports[6]					
Current dollars (mil.)	0	0	0	925	10
1994 constant dollars (mil.)	0	0	0	944	10
Arms exports[6]					
Current dollars (mil.)	110	40	30	10	10
1994 constant dollars (mil.)	122	43	31	10	10
Total imports[7]					
Current dollars (mil.)	8,621	11,450	11,120	12,600	14,320
1994 constant dollars (mil.)	9,595	12,270	11,600	12,860	14,320
Total exports[7]					
Current dollars (mil.)	9,730	10,180	10,680	8,918	10,590
1994 constant dollars (mil.)	10,830	10,910	11,140	9,102	10,590
Arms as percent of total imports[8]	0	0	0	7.3	0.1
Arms as percent of total exports[8]	1.1	0.4	0.3	0.1	0.1

Human Rights [24]

	SSTS	FL	FAPRO	PPCG	APROBC	TPW	PCPTW	STPEP	PHRFF	PRW	ASST	AFL
Observes	P	P	P	P	P	P	P	P	P	P	P	

	EAFRD	CPR	ESCR	SR	ACHR	MAAE	PVIAC	PVNAC	EAFDAW	TCIDTP	RC
Observes	P	P	P	P			P	P	P	P	P

P = Party; S = Signatory; see Appendix for meaning of abbreviations.

Labor Force

Total Labor Force [25]

4.8 million (1995)

Labor Force by Occupation [26]

Services, trade, government, and other	47.2%
Industry	29.7
Agriculture	16.1
Construction	7.0

Date of data: 1991

Unemployment Rate [27]

10.4% (year-end 1995)

For sources, notes, and explanations, see Annotated Source Appendix, page 1061.

417

Production Sectors

Commercial Energy Production and Consumption

Data are shown in quadrillion (10^{15}) BTUs and percent for 1995
Values for hydroelectric, nuclear, geothermal, solar, and wind power refer to electrical generation.

Production [28]	Consumption [29]
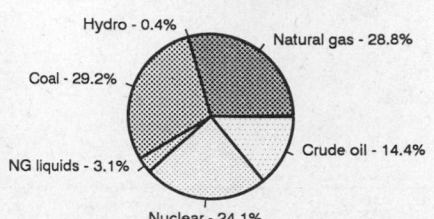	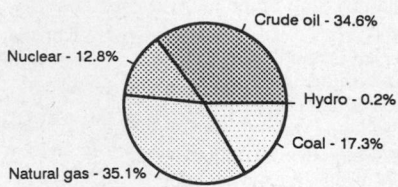

Production [28] pie chart labels: Hydro - 0.4%, Natural gas - 28.8%, Coal - 29.2%, Crude oil - 14.4%, NG liquids - 3.1%, Nuclear - 24.1%

Consumption [29] pie chart labels: Crude oil - 34.6%, Nuclear - 12.8%, Hydro - 0.2%, Coal - 17.3%, Natural gas - 35.1%

Crude oil	0.080
Natural gas liquids	0.017
Dry natural gas	0.160
Coal	0.162
Net hydroelectric power	0.002
Net nuclear power	0.134
Total	0.555

Crude oil	0.361
Dry natural gas	0.366
Coal	0.181
Net hydroelectric power	0.002
Net nuclear power	0.134
Total	1.044

Telecommunications [30]

- 1.52 million (1993 est.) telephones; 14,213 telex lines; automatic telephone network based on microwave radio relay system; 608,000 phones on order, 12-15 year wait; 49% of all telephones are in Budapest (1991 est.); the former state-owned telecommunications firm is now privatized and managed by a US/German consortium - it has plans to upgrade the system, including a contract with Siemens and Ericsson companies to provide 600,000 new phone lines during 1996-98
- Domestic: microwave radio relay
- International: satellite earth stations - 1 Intelsat and 1 Intersputnik (Atlantic Ocean Region)
- Radio: Broadcast stations: AM 32, FM 15, shortwave 0 Radios: 6 million (1993 est.)
- Television: Broadcast stations: 41 (Russian repeaters 8) Televisions: 4.38 million (1993 est.)

Top Agricultural Products [32]

Agriculture accounts for 7.3% of the GDP; produces wheat, corn, sunflower seed, potatoes, sugar beets; pigs, cattle, poultry, dairy products.

Transportation [31]

Railways: total: 7,685 km; broad gauge: 35 km 1.524-m gauge; standard gauge: 7,474 km 1.435-m gauge (2,162 km electrified; 1,236 km double track); narrow gauge: 176 km mostly 0.760-m gauge (1995)

Highways: total: 158,711 km; paved: 69,992 km (including 441 km of expressways); unpaved: 88,719 km (1992 est.)

Merchant marine: total: 10 cargo ships (1,000 GRT or over) totaling 46,121 GRT/61,613 DWT (1995 est.)

Airports

Total:	78
With paved runways over 3,047 m:	2
With paved runways 2,438 to 3,047 m:	7
With paved runways 1,524 to 2,437 m:	4
With paved runways under 914 m:	1
With unpaved runways 2,438 to 3,047 m:	7

Top Mining Products [33]

Thousand metric tons except as noted[e]	3/94[*]
Bauxite, gross weight	830
Pig iron	1,590
Ferroalloys (tons)	8,000
Steel, crude	1,940[r]
Cement, hydraulic	2,810[r]
Stone, dimension, all types	4,000
Limestone	4,000
Coal	13,500
Petroleum refinery products (000 42-gal. bls.)	41,000[41]

Tourism [34]

	1990	1991	1992	1993	1994
Visitors[49]	37,632	33,265	33,491	40,599	39,836
Tourists[50]	20,510	21,860	20,188	22,804	21,425
Excursionists	10,670	7,074	8,155	11,719	13,026
Tourism receipts	824	1,002	1,231	1,181	1,428
Tourism expenditures	477	443	640	741	925

Travelers are in thousands, money in million U.S. dollars.

For sources, notes, and explanations, see Annotated Source Appendix, page 1061.

Manufacturing Sector

Manufacturing Summary [35]

	1987 $ bil.	%	1988 $ bil.	%	1989 $ bil.	%	1990 $ bil.	%	1991 $ bil.	%
Establishments or enterprises (number)	2,552	0.109	2,857	0.136	3,854	0.206	7,646	0.427	15,469	2.024
Total employment (000)	1,479	1.090	1,437	1.050	1,382	1.123	1,301	1.175	1,126	1.621
Production workers (000)	1,143	1.695	1,111	1.778	1,095	1.488	1,003	1.413	873	2.651
Output ($ bil.)	30	0.300	30	0.254	29	0.246	29	0.249	26	0.251
Value added ($ bil.)	8	0.190	8	0.168	9	0.175	-	-	-	-
Capital investment ($ mil.)	1,246	0.286	1,168	0.245	1,268	0.232	1,271	0.229	-	-
M & E investment ($ mil.)	826	0.265	764	0.221	844	0.216	862	0.205	-	-
Employees per establishment (number)	580	1,001.752	503	771.506	359	545.739	170	275.384	73	80.097
Production workers per establishment	448	1,557.795	389	1,306.588	284	723.432	131	331.083	56	130.995
Output per establishment ($ mil.)	12	275.556	10	186.888	8	119.426	4	58.273	2	12.404
Capital investment per estab. ($ mil.)	0.488	262.754	0.409	180.163	0.329	112.556	0.166	53.579	-	-
M & E per establishment ($ mil)	0.324	243.557	0.268	162.253	0.219	105.115	0.113	47.930	-	-
Payroll per employee ($)	1,851	20.647	2,216	22.242	2,259	20.223	2,631	18.965	2,822	19.327
Wages per production worker ($)	1,664	21.088	1,927	22.800	1,898	18.926	2,250	18.942	2,407	21.884
Hours per production worker (hours)	1,684	88.715	1,673	87.063	1,599	87.238	1,637	87.496	1,620	90.634
Output per employee ($)	20,602	27.507	20,633	24.224	20,973	21.883	21,930	21.161	22,735	15.486
Capital investment per employee ($)	842	26.229	813	23.362	918	20.024	977	19.456	-	-
M & E per employee ($)	558	24.313	532	21.031	611	19.261	663	17.405	-	-

Note. Columns headed % show percent of world total or ratio. Ratios closest to 100 are closest to world average. M & E stands for machinery & equipment.

Output in Manufacturing

	1987 $ bil.	%	1988 $ bil.	%	1989 $ bil.	%	1990 $ bil.	%	1991 $ bil.	%
3110 - Food products	4.411	14.70	4.368	14.56	4.466	15.40	5.215	17.98	5.016	19.29
3130 - Beverages	0.566	1.89	0.504	1.68	0.542	1.87	0.666	2.30	0.712	2.74
3140 - Tobacco	0.160	0.53	0.145	0.48	0.151	0.52	0.182	0.63	0.194	0.75
3210 - Textiles	1.303	4.34	1.307	4.36	1.204	4.15	1.111	3.83	0.852	3.28
3211 - Spinning, weaving, etc.	0.960	3.20	0.984	3.28	0.907	3.13	0.816	2.81	0.582	2.24
3220 - Wearing apparel	0.509	1.70	0.502	1.67	0.449	1.55	0.443	1.53	0.452	1.74
3230 - Leather and products	0.224	0.75	0.200	0.67	0.168	0.58	0.157	0.54	0.163	0.63
3240 - Footwear	0.405	1.35	0.351	1.17	0.295	1.02	0.274	0.94	0.241	0.93
3310 - Wood products	0.196	0.65	0.198	0.66	0.222	0.77	0.250	0.86	0.305	1.17
3320 - Furniture, fixtures	0.358	1.19	0.347	1.16	0.337	1.16	0.348	1.20	0.310	1.19
3410 - Paper and products	0.456	1.52	0.452	1.51	0.425	1.47	0.454	1.57	0.306	1.52
3420 - Printing, publishing	0.349	1.16	0.325	1.08	0.366	1.26	0.426	1.47	0.488	1.88
3510 - Industrial chemicals	1.754	5.85	1.831	6.10	1.732	5.97	1.753	6.04	1.535	5.90
3520 - Chemical products nec	0.967	3.22	1.000	3.33	0.967	3.33	0.924	3.19	1.060	4.08
3522 - Drugs and medicines	0.879	2.93	0.948	3.16	0.877	3.02	0.807	2.78	0.935	3.60
3530 - Petroleum refineries	1.571	5.24	1.440	4.80	1.332	4.59	1.685	5.81	1.793	6.90
3550 - Rubber products	0.281	0.94	0.280	0.93	0.284	0.98	0.245	0.84	0.198	0.76
3560 - Plastic products nec	0.345	1.15	0.361	1.20	0.383	1.32	0.416	1.43	0.468	1.80
3610 - Pottery, china, etc.	0.117	0.39	0.115	0.38	0.112	0.39	0.136	0.47	0.140	0.54
3620 - Glass and products	0.226	0.75	0.214	0.71	0.213	0.73	0.220	0.76	0.217	0.83
3690 - Nonmetal products nec	0.613	2.04	0.587	1.96	0.547	1.89	0.628	2.17	0.519	2.00
3710 - Iron and steel	1.671	5.57	1.738	5.79	1.835	6.33	1.821	6.28	1.323	5.09
3720 - Nonferrous metals	0.803	2.68	0.954	3.18	1.212	4.18	1.138	3.92	0.787	3.03
3810 - Metal products	0.747	2.49	0.776	2.59	0.733	2.53	0.742	2.56	0.728	2.80
3820 - Machinery nec	1.833	6.11	1.781	5.94	1.889	6.51	1.819	6.27	1.822	7.01
3830 - Electrical machinery	2.495	8.32	2.323	7.74	2.220	7.66	1.911	6.59	1.547	5.95
3832 - Radio, television, etc.	1.446	4.82	1.353	4.51	1.283	4.42	1.013	3.49	0.733	2.82
3840 - Transportation equipment	2.091	6.97	1.742	5.81	1.569	5.41	1.172	4.04	1.218	4.68
3841 - Shipbuilding, repair	0.126	0.42	0.107	0.36	0.058	0.20	0.044	0.15	-	-
3843 - Motor vehicles	1.412	4.71	1.271	4.24	1.122	3.87	0.769	2.65	-	-
3850 - Professional goods	0.766	2.55	0.766	2.55	0.764	2.63	0.669	2.31	0.624	2.40
3900 - Industries nec	0.432	1.44	0.379	1.26	0.322	1.11	0.277	0.96	0.240	0.92

Note: Codes are International Standard Industry codes (ISIC). Percentages are % of total Output. [f]: Factor Prices; [p]: Producer Prices.

For sources, notes, and explanations, see Annotated Source Appendix, page 1061.

419

Finance, Economics, and Trade

Economic Indicators [36]

- **National product**: GDP—purchasing power parity—$72.5 billion (1995 est.)

- **National product real growth rate**: 1.5% (1995)

- **National product per capita**: $7,000 (1995 est.)

- **Inflation rate (consumer prices)**: 28.3% (1995)

- **External debt**: $32.7 billion (October 1995)

Balance of Payments Summary [37]

Values in millions of dollars.

	1989	1990	1991	1992	1993
Exports of goods (f.o.b.)	10,493	9,151	9,688	10,097	8,119
Imports of goods (f.o.b.)	-9,450	-8,617	-9,330	-10,108	-12,140
Trade balance	1,043	534	358	-11	-4,021
Services - debits	-3,283	-4,107	-3,669	-4,325	-4,275
Services - credits	1,522	3,164	2,847	3,829	3,301
Private transfers (net)	130	794	834	843	711
Government transfers (net)	NA	-7	34	15	21
Long-term capital (net)	1,280	241	2,221	-217	6,391
Short-term capital (net)	-379	-1,042	-747	633	-308
Errors and omissions	-141	10	-82	2	724
Overall balance	172	-413	1,796	769	2,544

Exchange Rates [38]

Currency: **forint.**
Symbol: **Ft.**

Data are currency units per $1.

January 1996	144
1995	125.681
1994	105.160
1993	91.933
1992	78.988
1991	74.735

Imports and Exports

Top Import Origins [39]

$15 billion (f.o.b., 1995 est.) Data are for 1994.

Origins	%
Germany	23.4
Austria	12.0
Russia	12.0
Italy	7.0
UK	4.0

Top Export Destinations [40]

$13 billion (f.o.b., 1995 est.) Data are for 1994.

Destinations	%
Germany	28.2
Austria	10.9
Italy	8.5
Russia	7.5
US	4.0

Foreign Aid [41]

	U.S. $	
ODA (1993)	136	million
OECD countries and international organizations (1990-93)	3,700	million

Import and Export Commodities [42]

Import Commodities

Fuels and energy 11.0%
Raw materials and semifinished goods 36.9%
Capital goods 23.3%
Consumer goods 22.0%
Food and agriculture 6.8%

Export Commodities

Raw materials and semifinished goods 36.4%
Consumer goods 26.7%
Food and agriculture 20.5%
Capital goods 13.1%
Fuels and energy 3.3%

Iceland

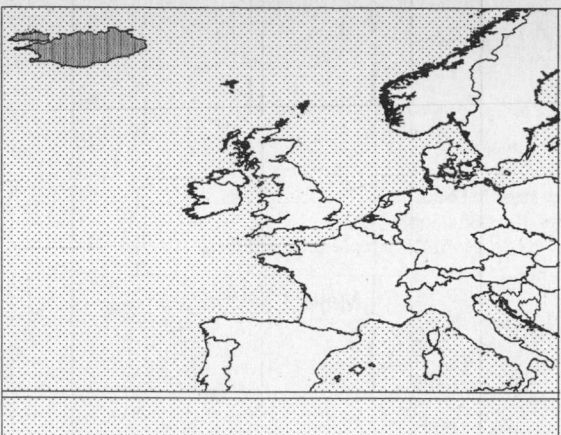

Geography [1]

Total area:
  103,000 sq km   39,769 sq mi
Land area:
  100,250 sq km   38,707 sq mi
Comparative area:
  Slightly smaller than Kentucky
Land boundaries:
  0 km
Coastline:
  4,988 km
Climate:
  Temperate; moderated by North Atlantic Current; mild, windy winters; damp, cool summers
Terrain:
  Mostly plateau interspersed with mountain peaks, icefields; coast deeply indented by bays and fiords
Natural resources:
  Fish, hydropower, geothermal power, diatomite
Land use:
  Arable land: 1%
  Permanent crops: 0%
  Meadows and pastures: 20%
  Forest and woodland: 1%
  Other: 78%

Demographics [2]

	1970	1980	1990	1995[1]	1996	2000	2010	2020	2030
Population	204	228	255	268	270	280	303	325	342
Population density (persons per sq. mi.)	5	6	7	7	7	7	8	8	9
(persons per sq. km.)	2	2	3	3	3	3	3	3	3
Net migration rate (per 1,000 population)	NA	NA	-3.8	-2.8	-2.5	-1.5	-0.3	0.0	0.0
Births	4	5	NA	NA	4	NA	NA	NA	NA
Deaths	1	2	NA	NA	2	NA	NA	NA	NA
Life expectancy - males	NA	NA	76.1	77.4	77.7	79.1	80.5	80.9	81.0
Life expectancy - females	NA	NA	80.5	82.2	82.6	84.5	86.3	86.8	87.0
Birth rate (per 1,000)	NA	NA	18.7	17.2	16.9	15.9	14.2	13.5	12.6
Death rate (per 1,000)	NA	NA	6.8	6.2	6.2	5.9	6.4	7.3	8.7
Women of reproductive age (15-49 yrs.)	NA	NA	65	69	69	70	71	72	73
of which are currently married	NA	NA	35	NA	38	39	40	NA	NA
Fertility rate	NA	NA	2.3	2.2	2.2	2.2	2.1	2.1	2.0

Except as noted, values for vital statistics are in thousands; life expectancy is in years.

Health

Health Indicators [3]

Health Expenditures [4]

For sources, notes, and explanations, see Annotated Source Appendix, page 1061.

421

Human Factors

Women and Children [5]

Burden of Disease [6]

Population per physician	NA
Population per nurse	NA
Population per hospital bed	NA
AIDS cases per 100,000 people (1994)	1.1
Heart disease cases per 1,000 people (1990-93)	NA

Ethnic Division [7]

Homogeneous mixture of descendants of Norwegians and Celts.

Religion [8]

Evangelical Lutheran	96%
Other Protestant and Roman Catholic	3%
None	1%
(1988)	

Major Languages [9]

Icelandic.

Education

Public Education Expenditures [10]

	1980	1985	1988	1990	1993	1994
Million (Krona)						
Total education expenditure	699	5,684	13,357	19,747	21,615	NA
as percent of GNP	4.4	4.9	5.4	5.6	5.4	NA
as percent of total govt. expend.	14.0	13.8	14.1	NA	12.8	NA
Current education expenditure	NA	NA	NA	14,584	19,202	NA
as percent of GNP	NA	NA	NA	4.2	4.8	NA
as percent of current govt. expend.	NA	NA	NA	NA	NA	NA
Capital expenditure	NA	NA	NA	5,163	2,413	NA

Educational Attainment [11]

Illiteracy [12]

Libraries [13]

	Admin. Units	Svc. Pts.	Vols. (000)	Shelving (meters)	Vols. Added	Reg. Users
National (1992)	1	1	431	NA	7,271	10,332
Nonspecialized	NA	NA	NA	NA	NA	NA
Public (1993)	137	NA	1,945	NA	NA	93,263
Higher ed. (1991)	6	6	382	11,472	13,757	NA
School (1991)	85	85	536	NA	NA	NA

Daily Newspapers [14]

	1980	1985	1990	1994
Number of papers	6	6	6	5
Circ. (000)	125	113	130[e]	137[e]

Culture [15]

Cinema (seats per 1,000)	23.2
Annual attendance per person	4.7
Gross box office receipts (mil. Krona)	NA
Museums (reporting)	56
Visitors (000)	486
Annual receipts (000 Krona)	NA

Science and Technology

Scientific/Technical Forces [16]

Scientists/engineers	773
Number female	NA
Technicians[2]	404
Number female	NA
Total[2]	1,177

R&D Expenditures [17]

	Krona (000) 1989
Total expenditure	3,123,000
Capital expenditure	NA
Current expenditure	NA
Percent current	NA

U.S. Patents Issued [18]

Values show patents issued to citizens of the country by the U.S. Patents Office.

	1993	1994	1995
Number of patents	5	4	4

Government and Law

Organization of Government [19]

Long-form name:
Republic of Iceland
Type:
Republic
Independence:
17 June 1944 (from Denmark)
National holiday:
Anniversary of the Establishment of the Republic, 17 June (1944)
Constitution:
16 June 1944, effective 17 June 1944
Legal system:
Civil law system based on Danish law; does not accept compulsory ICJ jurisdiction
Executive branch:
President; Prime Minister; Cabinet
Legislative branch:
Unicameral: Parliament (Althing)
Judicial branch:
Supreme Court (Haestirettur)

Elections [20]

Althing	% of votes
Independence Party	37.1
Progressive Party	23.3
Social Democratic Party	11.4
People's Movement	7.2
Women's Party	4.9

Government Expenditures [21]

Educ./Health - 56.9%
Housing - 1.0%
Other - 14.9%
Gen. Services - 9.4%
Industry - 17.8%

(% distribution). Expend. for CY94: 142,756.2 (Krona mil.)

Crime [23]

Military Expenditures and Arms Transfers [22]

	1990	1991	1992	1993	1994
Military expenditures					
Current dollars (mil.)	0	0	0	0	0
1994 constant dollars (mil.)	0	0	0	0	0
Armed forces (000)	0	0	0	0	0
Gross national product (GNP)					
Current dollars (mil.)	5,255	5,537	5,527	5,698	6,013
1994 constant dollars (mil.)	5,848	5,935	5,763	5,816	6,013
Central government expenditures (CGE)					
1994 constant dollars (mil.)	1,944	2,084	2,018	2,037	2,099[e]
People (mil.)	0.3	0.3	0.3	0.3	0.3
Military expenditure as % of GNP	0	0	0	0	0
Military expenditure as % of CGE	0	0	0	0	0
Military expenditure per capita (1994 $)	0	0	0	0	0
Armed forces per 1,000 people (soldiers)	0	0	0	0	0
GNP per capita (1994 $)	22,960	23,110	22,250	22,260	22,810
Arms imports[6]					
Current dollars (mil.)	0	0	0	0	0
1994 constant dollars (mil.)	0	0	0	0	0
Arms exports[6]					
Current dollars (mil.)	0	0	0	0	0
1994 constant dollars (mil.)	0	0	0	0	0
Total imports[7]					
Current dollars (mil.)	1,680	1,760	1,684	1,349	1,472
1994 constant dollars (mil.)	1,870	1,887	1,756	1,377	1,472
Total exports[7]					
Current dollars (mil.)	1,592	1,550	1,528	1,399	1,623
1994 constant dollars (mil.)	1,772	1,661	1,593	1,428	1,623
Arms as percent of total imports[8]	0	0	0	0	0
Arms as percent of total exports[8]	0	0	0	0	0

Human Rights [24]

	SSTS	FL	FAPRO	PPCG	APROBC	TPW	PCPTW	STPEP	PHRFF	PRW	ASST	AFL
Observes		P	P	P	P	P	P		P	P	P	P
	EAFRD	CPR	ESCR	SR	ACHR	MAAE	PVIAC	PVNAC	EAFDAW	TCIDTP	RC	
Observes		P	P	P			P	P	P	S	P	

P=Party; S=Signatory; see Appendix for meaning of abbreviations.

Labor Force

Total Labor Force [25]

127,900

Labor Force by Occupation [26]

Commerce, transport, & services	60.0%
Manufacturing	12.5
Fishing and fish processing	11.8
Construction	10.8
Agriculture	4.0
Other	0.9

Date of data: 1990

Unemployment Rate [27]

3.9% (December 1995)

For sources, notes, and explanations, see Annotated Source Appendix, page 1061.

423

Production Sectors

Commercial Energy Production and Consumption

Data are shown in quadrillion (10^{15}) BTUs and percent for 1995
Values for hydroelectric, nuclear, geothermal, solar, and wind power refer to electrical generation.

Production [28]

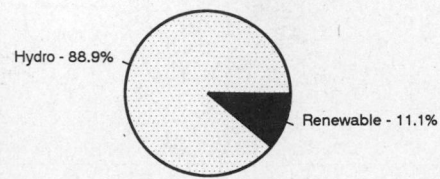

Hydro - 88.9%
Renewable - 11.1%

Consumption [29]

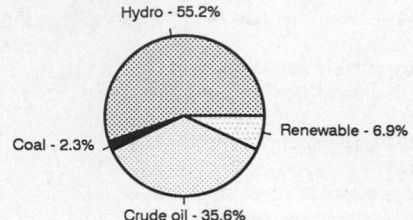

Hydro - 55.2%
Renewable - 6.9%
Coal - 2.3%
Crude oil - 35.6%

Net hydroelectric power	0.048
Geothermal, solar, wind	0.006
Total	0.054

Crude oil	0.031
Coal	0.002
Net hydroelectric power	0.048
Geothermal, solar, wind	0.006
Total	0.087

Telecommunications [30]

- 143,600 (1993 est.) telephones; adequate domestic service
- Domestic: the trunk network consists of coaxial and fiber-optic cables and microwave radio relay links
- International: satellite earth stations - 2 Intelsat (Atlantic Ocean), 1 Inmarsat (Atlantic and Indian Ocean Regions); note - Iceland shares the Inmarsat earth station with the other Nordic countries (Denmark, Finland, Norway, and Sweden)
- Radio: Broadcast stations: AM 5, FM 147 (transmitters and repeaters), shortwave 0 Radios: 91,500 licensed (1993 est.)
- Television: Broadcast stations: 202 (transmitters and repeaters) Televisions: 96,100 licensed (1993 est.)

Top Agricultural Products [32]

Agriculture accounts for 9.6% of the GDP; produces potatoes, turnips; cattle, sheep; fish catch of about 1.1 million metric tons in 1992.

Transportation [31]

Railways: 0 km

Highways: total: 11,373 km; paved: 2,513 km; unpaved: 8,860 km (1992 est.)

Merchant marine: total: 6 ships (1,000 GRT or over) totaling 30,025 GRT/40,410 DWT; ships by type: cargo 1, chemical tanker 1, oil tanker 1, refrigerated cargo 1, roll-on/roll-off cargo 2 (1995 est.)

Airports

Total:	84
With paved runways over 3,047 m:	1
With paved runways 1,524 to 2,437 m:	3
With paved runways 914 to 1,523 m:	5
With paved runways under 914 m:	49
With unpaved runways 1,524 to 2,437 m:	4

Top Mining Products [33]

Metric tons except as noted	3/95*
Aluminum metal, primary	93,300[42]
Ferrosilicon	66,000
Diatomite	25,000[e]
Cement, hydraulic	83,100
Pumice	230,000[e]
Sand, calcareous, shell (cu. meters)	81,500
Silica dust	10,000[e,43]
Stone, crushed	
basaltic	100,000[e]
rhyolite (cu. meters)	16,400

Tourism [34]

	1990	1991	1992	1993	1994
Tourists	142	143	143	157	179
Cruise passengers	NA	13	12	16	18
Tourism receipts	139	136	128	128	137
Tourism expenditures	277	292	287	257	249
Fare receipts	82	92	91	86	103
Fare expenditures[51]	79	84	88	73	76

Travelers are in thousands, money in million U.S. dollars.

For sources, notes, and explanations, see Annotated Source Appendix, page 1061.

Manufacturing Sector

Manufacturing Summary [35]

	1987 $ bil.	1987 %	1988 $ bil.	1988 %	1989 $ bil.	1989 %	1990 $ bil.	1990 %	1991 $ bil.	1991 %
Establishments or enterprises (number)	2,267	0.097	2,141	0.102	2,064	0.110	2,061	0.115	-	-
Total employment (000)	27	0.020	24	0.018	23	0.018	22	0.020	-	-
Production workers (000)	-	-	-	-	-	-	-	-	-	-
Output ($ bil.)	2	0.024	3	0.022	2	0.019	3	0.022	-	-
Value added ($ bil.)	0.811	0.019	0.840	0.017	0.735	0.015	0.787	0.016	-	-
Capital investment ($ mil.)	-	-	-	-	-	-	-	-	-	-
M & E investment ($ mil.)	-	-	-	-	-	-	-	-	-	-
Employees per establishment (number)	12	20.725	11	17.315	11	16.784	11	17.116	-	-
Production workers per establishment	-	-	-	-	-	-	-	-	-	-
Output per establishment ($ mil.)	1	25.194	1	21.440	1	17.663	1	19.466	-	-
Capital investment per estab. ($ mil.)	-	-	-	-	-	-	-	-	-	-
M & E per establishment ($ mil)	-	-	-	-	-	-	-	-	-	-
Payroll per employee ($)	21,650	241.441	22,887	229.728	22,363	200.168	31,444	226.679	-	-
Wages per production worker ($)	-	-	-	-	-	-	-	-	-	-
Hours per production worker (hours)	-	-	-	-	-	-	-	-	-	-
Output per employee ($)	91,047	121.567	105,465	123.821	100,854	105.233	117,861	113.729	-	-
Capital investment per employee ($)	-	-	-	-	-	-	-	-	-	-
M & E per employee ($)	-	-	-	-	-	-	-	-	-	-

Note: Columns headed % show percent of world total or ratio. Ratios closest to 100 are closest to world average. M & E stands for machinery & equipment.

Output in Manufacturing

	1987 $ bil.[f]	1987 %	1988 $ bil.[f]	1988 %	1989 $ bil.[f]	1989 %	1990 $ bil.[f]	1990 %	1991 $ bil.[f]	1991 %
3110 - Food products	1.425	71.25	1.428	47.60	1.298	64.90	1.476	49.20	-	-
3130 - Beverages	0.039	1.95	0.044	1.47	0.041	2.05	0.059	1.97	-	-
3210 - Textiles	0.092	4.60	0.074	2.47	0.061	3.05	0.061	2.03	-	-
3211 - Spinning, weaving, etc.	0.030	1.50	0.023	0.77	0.019	0.95	0.019	0.63	-	-
3220 - Wearing apparel	0.037	1.85	0.036	1.20	0.025	1.25	0.029	0.97	-	-
3230 - Leather and products	0.020	1.00	0.015	0.50	0.014	0.70	0.018	0.60	-	-
3240 - Footwear	0.002	0.10	0.000	0.00	0.001	0.05	0.002	0.07	-	-
3310 - Wood products	0.002	0.10	0.003	0.10	0.003	0.15	0.004	0.13	-	-
3320 - Furniture, fixtures	0.096	4.80	0.101	3.37	0.087	4.35	0.097	3.23	-	-
3410 - Paper and products	0.022	1.10	0.024	0.80	0.021	1.05	0.024	0.80	-	-
3420 - Printing, publishing	0.132	6.60	0.151	5.03	0.141	7.05	0.154	5.13	-	-
3510 - Industrial chemicals	0.035	1.75	0.039	1.30	0.033	1.65	0.041	1.37	-	-
3511 - Basic chemicals, excl fertilizers	0.015	0.75	0.019	0.63	0.015	0.75	0.022	0.73	-	-
3520 - Chemical products nec	0.038	1.90	0.040	1.33	0.037	1.85	0.044	1.47	-	-
3540 - Petroleum, coal products	0.001	0.05	0.001	0.03	0.001	0.05	0.001	0.03	-	-
3560 - Plastic products nec	0.054	2.70	0.058	1.93	0.053	2.65	0.065	2.17	-	-
3610 - Pottery, china, etc.	0.002	0.10	0.002	0.07	0.001	0.05	0.002	0.07	-	-
3620 - Glass and products	0.009	0.45	0.009	0.30	0.008	0.40	0.010	0.33	-	-
3690 - Nonmetal products nec	0.079	3.95	0.087	2.90	0.069	3.45	0.079	2.63	-	-
3710 - Iron and steel	0.035	1.75	0.059	1.97	0.054	2.70	0.040	1.33	-	-
3720 - Nonferrous metals	0.125	6.25	0.151	5.03	0.171	8.55	0.165	5.50	-	-
3840 - Transportation equipment	0.054	2.70	0.052	1.73	0.034	1.70	0.033	1.10	-	-
3841 - Shipbuilding, repair	0.054	2.70	0.052	1.73	0.034	1.70	0.033	1.10	-	-
3900 - Industries nec	0.078	3.90	0.085	2.83	0.074	3.70	0.090	3.00	-	-

Note: Codes are International Standard Industry codes (ISIC). Percentages are % of total Output. [f]: Factor Prices; [p]: Producer Prices.

For sources, notes, and explanations, see Annotated Source Appendix, page 1061.

425

Finance, Economics, and Trade

Economic Indicators [36]

- **National product**: GDP—purchasing power parity—$5 billion (1995 est.)
- **National product real growth rate**: 3.2% (1995 est.)
- **National product per capita**: $18,800 (1995 est.)
- **Inflation rate (consumer prices)**: 2.5% (1995 est.)
- **External debt**: $2.5 billion (1993 est.)

Balance of Payments Summary [37]

Values in millions of dollars.

	1989	1990	1991	1992	1993
Exports of goods (f.o.b.)	1,401	1,589	1,551	1,523	1,399
Imports of goods (f.o.b.)	-1,267	-1,509	-1,599	-1,522	-1,218
Trade balance	134	80	-47	1	181
Services - debits	-764	-853	-899	-888	-866
Services - credits	549	639	642	685	682
Private transfers (net)	1	6	3	2	6
Government transfers (net)	-4	-6	-8	-8	-8
Long-term capital (net)	261	278	292	222	2,191
Short-term capital (net)	-136	-68	8	55	614
Errors and omissions	14	-2	20	2	-853
Overall balance	55	74	10	70	-188

Exchange Rates [38]

Currency: **Icelandic krona.**
Symbol: **IKr.**

Data are currency units per $1.

January 1996	65.970
1995	64.692
1994	69.944
1993	67.603
1992	57.546
1991	58.996

Imports and Exports

Top Import Origins [39]

$1.5 billion (c.i.f., 1994) Data are for 1992.

Origins	%
EU	53
Germany	14
Denmark	10
UK	9
Norway	14
US	9

Top Export Destinations [40]

$1.6 billion (f.o.b., 1994) Data are for 1992.

Destinations	%
EU	68
UK	25
Germany	12
US	11
Japan	8

Foreign Aid [41]

	U.S. $	
US commitments, including Ex-Im (FY70-81)	19.1	million

Import and Export Commodities [42]

Import Commodities

Machinery and transportation equipment
Petroleum products
Foodstuffs
Textiles

Export Commodities

Fish and fish products
Animal products
Aluminum
Ferrosilicon
Diatomite

India

Geography [1]

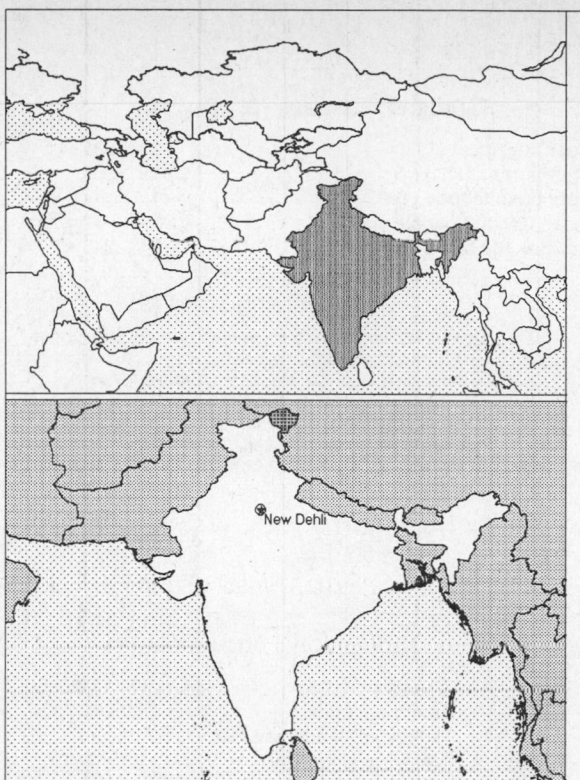

Total area:
3,287,590 sq km   1,269,346 sq mi
Land area:
2,973,190 sq km   1,147,955 sq mi
Comparative area:
Slightly more than one-third the size of the US
Land boundaries:
Total 14,103 km, Bangladesh 4,053 km, Bhutan 605 km, Myanmar 1,463 km, China 3,380 km, Nepal 1,690 km, Pakistan 2,912 km
Coastline:
7,000 km
Climate:
Varies from tropical monsoon in South to temperate in North
Terrain:
Upland plain (Deccan Plateau) in South, flat to rolling plain along the Ganges, deserts in West, Himalayas in North
Natural resources:
Coal (fourth largest reserves in the world), iron ore, manganese, mica, bauxite, titanium ore, chromite, natural gas, diamonds, petroleum, limestone
Land use:
Arable land: 55%
Permanent crops: 1%
Meadows and pastures: 4%
Forest and woodland: 23%
Other: 17%

Demographics [2]

	1970	1980	1990	1995[1]	1996	2000	2010	2020	2030
Population (mil.)	555	692	856	936	952	1,013	1,156	1,289	1,403
Population density (persons per sq. mi.)	484	603	745	816	829	882	1,007	1,123	1,222
(persons per sq. km.)	187	233	288	315	320	341	389	434	472
Net migration rate (per 1,000 population)	NA	0.0	0.3	0.0	0.0	0.0	0.0	0.0	0.0
Births (mil.)	NA	NA	NA	NA	25	NA	NA	NA	NA
Deaths (mil.)	NA	NA	NA	NA	9	NA	NA	NA	NA
Life expectancy - males (yrs.)	NA	52.9	56.9	58.7	59.1	60.7	64.5	67.8	70.6
Life expectancy - females (yrs.)	NA	52.1	57.5	59.8	60.3	62.3	67.0	71.2	74.8
Birth rate (per 1,000)	NA	34.8	30.7	26.5	25.9	23.5	20.0	17.4	15.3
Death rate (per 1,000)	NA	13.6	10.9	9.8	9.6	8.9	8.0	7.7	8.0
Women of reproductive age (15-49 yrs.)	NA	161	209	233	238	259	307	336	355
of which are currently married (mil.)	NA	NA	163	NA	188	204	244	NA	NA
Fertility rate	NA	4.7	3.8	3.3	3.2	2.9	2.4	2.2	2.1

Except as noted, values for vital statistics are in thousands; life expectancy is in years.

Health

Health Indicators [3]

% of population with access to	
safe water (1990-95)	81
adequate sanitation (1990-95)	29
health services (1985-95)	85
% of 1-year-olds immunized (1990-94) against	
TB (tuberculosis)	96
DPT (diphtheria, pertussis, tetanus)	91
polio	91
measles	86
% of contraceptive prevalence (1980-94)	43
ORT use rate (1990-94)	37

Health Expenditures [4]

Total health expenditure, 1990 (official exchange rate)	
Millions of dollars	17,740
Dollars per capita	21
Health expenditures as a percentage of GDP	
Total	6.0
Public sector	1.3
Private sector	4.7
Development assistance for health	
Total aid flows (millions of dollars)[1]	286
Aid flows per capita (dollars)	0.3
Aid flows as a percentage of total health expenditure	1.6

For sources, notes, and explanations, see Annotated Source Appendix, page 1061.

427

Human Factors

Women and Children [5]

% of pregnant women immunized (tetanus 1990-94)	81
% of births attended by trained health personnel (1983-94)	33
Maternal mortality rate (1980-92)	460
Under-5 mortality rate (1994)	119
% under-5 moderately/severely underweight (1980-1994)[1]	69

Burden of Disease [6]

Population per physician (1988)	2,459.34
Population per nurse (1984)	1,700.83
Population per hospital bed (1990)	1,371.44
AIDS cases per 100,000 people (1994)	*
Malaria cases per 100,000 people (1992)	240

Ethnic Division [7]

Indo-Aryan	72%
Dravidian	25%
Mongoloid and other	3%

Religion [8]

Hindu	80%
Muslim	14%
Christian	2.4%
Sikh	2%
Buddhist	0.7%
Jains	0.5%
Other	0.4%

Major Languages [9]

Hindi is the national language and the primary tongue of 30% of the people. Official languages include: Bengali, Telugu, Marathi, Tamil, Urdu, and others, including many unofficial languages. English is the most important language for national, political, and commercial communication.

Education

Public Education Expenditures [10]

Million (Rupee)	1980	1989	1990	1991	1992	1994
Total education expenditure	37,924	181,926	207,897	232,425	260,169	NA
as percent of GNP	2.8	4.1	3.9	3.8	3.8	NA
as percent of total govt. expend.	10.0	11.2	10.9	11.9	11.5	NA
Current education expenditure	37,462	179,501	205,330	228,878	257,377	NA
as percent of GNP	2.7	4.0	3.9	3.8	3.7	NA
as percent of current govt. expend.	NA	14.2	13.7	13.4	13.4	NA
Capital expenditure	463	2,425	2,567	3,547	2,792	NA

Educational Attainment [11]

Age group (1981)	25+
Total population	280,599,720
Highest level attained (%)	
No schooling	72.5
First level	
Not completed	11.3
Completed	NA
Entered second level	
S-1	13.7
S-2	NA
Postsecondary	2.5

Illiteracy [12]

In thousands and percent[1]	1990	1995	2000
Illiterate population (15+ yrs.)	279,610	290,705	300,833
Illiteracy rate - total pop. (%)	51.3	47.6	44.0
Illiteracy rate - males (%)	37.2	34.1	31.2
Illiteracy rate - females (%)	66.5	62.1	57.8

Libraries [13]

	Admin. Units	Svc. Pts.	Vols. (000)	Shelving (meters)	Vols. Added	Reg. Users
National (1986)	8	11	1,893	51,488	32,621	33,220
Nonspecialized	NA	NA	NA	NA	NA	NA
Public	NA	NA	NA	NA	NA	NA
Higher ed.	NA	NA	NA	NA	NA	NA
School	NA	NA	NA	NA	NA	NA

Daily Newspapers [14]

	1980	1985	1990	1994
Number of papers	NA	NA	NA	NA
Circ. (000)	14,531	19,804	NA	NA

Culture [15]

Cinema (seats per 1,000)	7.8[e]
Annual attendance per person	5.0[e]
Gross box office receipts (mil. Rupee)	NA
Museums (reporting)	NA
Visitors (000)	NA
Annual receipts (000 Rupee)	NA

Science and Technology

Scientific/Technical Forces [16]

Scientists/engineers	128,036[e]
Number female	7,710
Technicians	96,737
Number female	6,138
Total[18]	224,773[e]

R&D Expenditures [17]

	Rupee (000) 1990
Total expenditure	41,864,300
Capital expenditure	8,475,300
Current expenditure	33,389,000
Percent current	79.8

U.S. Patents Issued [18]

Values show patents issued to citizens of the country by the U.S. Patents Office.

	1993	1994	1995
Number of patents	30	28	38

For sources, notes, and explanations, see Annotated Source Appendix, page 1061.

Government and Law

Organization of Government [19]

Long-form name:
Republic of India
Type:
Federal republic
Independence:
15 August 1947 (from UK)
National holiday:
Anniversary of the Proclamation of the Republic, 26 January (1950)
Constitution:
26 January 1950
Legal system:
Based on English common law; limited judicial review of legislative acts; accepts compulsory ICJ jurisdiction, with reservations
Executive branch:
President; Vice President; Prime Minister; Council of Ministers
Legislative branch:
Bicameral Parliament (Sansad): Council of States (Rajya Sabha) and People's Assembly (Lok Sabha)
Judicial branch:
Supreme Court

Elections [20]

People's Assembly	% of seats
Congress (I) Party	45.1
Bharatiya Janata Party	21.9
Janata Dal Party	7.2
Janata Dal (Ajit Singh)	3.7
Communist Party of India/Marxist	6.4
Communist Party of India	2.6
Telugu Desam	2.4
All-India Anna Dravida	2.0
Samajwadi Janata Party	0.9
Other (incl. 9 vacant)	7.7

Government Expenditures [21]

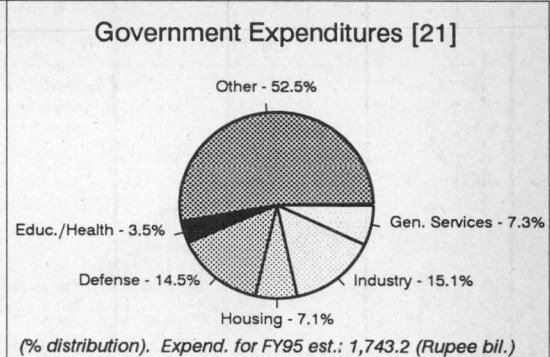

Other - 52.5%
Gen. Services - 7.3%
Industry - 15.1%
Housing - 7.1%
Defense - 14.5%
Educ./Health - 3.5%

(% distribution). Expend. for FY95 est.: 1,743.2 (Rupee bil.)

Crime [23]

	1990
Crime volume	
Cases known to police	654,565
Attempts (percent)	NA
Percent cases solved	483.8
Crimes per 100,000 persons	76.5
Persons responsible for offenses	
Total number offenders	3,167,006
Percent female	NA
Percent juvenile	NA
Percent foreigners	NA

Military Expenditures and Arms Transfers [22]

	1990	1991	1992	1993	1994
Military expenditures					
Current dollars (mil.)	6,546	6,297	6,384	7,430	8,233
1994 constant dollars (mil.)	7,285	6,749	6,657	7,583	8,233
Armed forces (000)	1,262	1,265	1,265	1,265	1,305
Gross national product (GNP)					
Current dollars (mil.)	225,900	235,100	253,100	267,100	287,100
1994 constant dollars (mil.)	251,500	252,000	263,900	272,600	287,100
Central government expenditures (CGE)					
1994 constant dollars (mil.)	55,360	52,330	53,810	58,590	56,600
People (mil.)	852.7	869.5	886.3	903.2	919.9
Military expenditure as % of GNP	2.9	2.7	2.5	2.8	2.9
Military expenditure as % of CGE	13.2	12.9	12.4	12.9	14.5
Military expenditure per capita (1994 $)	9	8	8	8	9
Armed forces per 1,000 people (soldiers)	1.5	1.5	1.4	1.4	1.4
GNP per capita (1994 $)	295	290	298	302	312
Arms imports[6]					
Current dollars (mil.)	1,800	925	650	260	140
1994 constant dollars (mil.)	2,003	991	678	265	140
Arms exports[6]					
Current dollars (mil.)	10	5	0	10	10
1994 constant dollars (mil.)	11	5	0	10	10
Total imports[7]					
Current dollars (mil.)	23,640	20,420	23,580	22,760	26,670
1994 constant dollars (mil.)	26,310	21,890	24,590	23,230	26,670
Total exports[7]					
Current dollars (mil.)	17,970	17,660	19,560	21,550	25,050
1994 constant dollars (mil.)	20,000	18,930	20,400	22,000	25,050
Arms as percent of total imports[8]	7.6	4.5	2.8	1.1	0.5
Arms as percent of total exports[8]	0.1	0	0	0	0

Human Rights [24]

	SSTS	FL	FAPRO	PPCG	APROBC	TPW	PCPTW	STPEP	PHRFF	PRW	ASST	AFL
Observes	P	P		P		P	P	P		P	P	
	EAFRD	CPR	ESCR	SR	ACHR	MAAE	PVIAC	PVNAC	EAFDAW	TCIDTP	RC
Observes		P	P	P					P		P

P = Party; S = Signatory; see Appendix for meaning of abbreviations.

Labor Force

Total Labor Force [25]

314.751 million (1990)

Labor Force by Occupation [26]

Agriculture   65%
Date of data: 1993 est.

Unemployment Rate [27]

For sources, notes, and explanations, see Annotated Source Appendix, page 1061.

429

Production Sectors

Commercial Energy Production and Consumption

Data are shown in quadrillion (10^{15}) BTUs and percent for 1995
Values for hydroelectric, nuclear, geothermal, solar, and wind power refer to electrical generation.

Production [28]

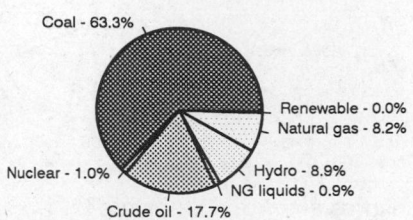

Coal - 63.3%
Renewable - 0.0%
Natural gas - 8.2%
Nuclear - 1.0%
Hydro - 8.9%
NG liquids - 0.9%
Crude oil - 17.7%

Consumption [29]

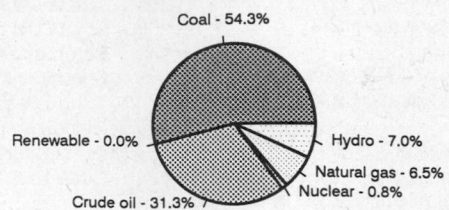

Coal - 54.3%
Renewable - 0.0%
Hydro - 7.0%
Natural gas - 6.5%
Nuclear - 0.8%
Crude oil - 31.3%

Crude oil	1.471		Crude oil	3.282
Natural gas liquids	0.079		Dry natural gas	0.681
Dry natural gas	0.681		Coal	5.698
Coal	5.273		Net hydroelectric power	0.738
Net hydroelectric power	0.738		Net nuclear power	0.087
Net nuclear power	0.087		Geothermal, solar, wind	0.001
Geothermal, solar, wind	0.001		Total	10.487
Total	8.330			

Telecommunications [30]

- 9.8 million (1995) telephones; least adequate telephone system of any industrializing country; 75% of villages have no service, 95% have no long-distance; poor service greatly impedes commercial growth and penalizes India in global markets; slow improvement due to recent admission of private investors; demand is growing rapidly
- Domestic: local service mostly by open wire and obsolete electromechanical and manual switchboard systems; since 1985 a substantial amount of digital switch gear introduced and trunk capacity increased by a fiber-optic cable and satellite system; long-distance is by open wire, coaxial cable, and low-capacity microwave relay
- International: satellite earth stations - 8 Intelsat and 1 Inmarsat; submarine cables to Malaysia and UAE
- Radio: Broadcast stations: AM 96, FM 4 Radios: 70 million (1992 est.)
- Television: Broadcast stations: 274 Televisions: 33 million (1992 est.)

Top Agricultural Products [32]

Produces rice, wheat, oilseed, cotton, jute, tea, sugarcane, potatoes; cattle, water buffalo, sheep, goats, poultry; fish catch of about 3 million metric tons ranks India among the world's top 10 fishing nations.

Top Mining Products [33]

Thousand metric tons except as noted[e]	9/1/95[*]
Iron ore and concentrate, gross weight	57,000
Steel, crude	18,200[r]
Silver, mine and smelter output (kg.)	52,000
Cement, hydraulic	54,000
Diamond, gem and industrial (000 carats)	18
Limestone	80,000
Coal, bituminous and lignite	263,000
Petroleum (000 42-gal. bls.)	
crude	197,000
refinery products	345,000

Transportation [31]

Railways: total: 62,462 km (11,793 km electrified; 12,617 km double track); broad gauge: 37,824 km 1.676-m gauge; narrow gauge: 20,653 km 1.000-m gauge; 3,985 km 0.762-m and 0.610-m gauge (1995 est.)

Highways: total: 2.037 million km; paved: 981,834 km; unpaved: 1,055,166 km (1995 est.)

Merchant marine: total: 310 ships (1,000 GRT or over) totaling 6,787,834 GRT/11,296,222 DWT; ships by type: bulk 133, cargo 65, chemical tanker 10, combination bulk 2, combination ore/oil 3, container 11, liquefied gas tanker 6, oil tanker 73, passenger-cargo 5, roll-on/roll-off cargo 1, short-sea passenger 1 (1995 est.)

Airports

Total:	288
With paved runways over 3,047 m:	11
With paved runways 2,438 to 3,047 m:	48
With paved runways 1,524 to 2,437 m:	59
With paved runways 914 to 1,523 m:	68
With paved runways under 914 m:	62

Tourism [34]

	1990	1991	1992	1993	1994
Visitors	1,721	1,685	1,886	1,778	1,900
Tourists[37]	1,707	1,678	1,868	1,765	1,886
Excursionists	1	1	1	NA	NA
Cruise passengers	13	6	17	13	14
Tourism receipts	1,513	1,757	2,120	2,001	2,265
Tourism expenditures	393	NA	NA	NA	NA

Travelers are in thousands, money in million U.S. dollars.

Manufacturing Sector

Manufacturing Summary [35]

	1987		1988		1989		1990		1991	
	$ bil.	%	$ bil.	%	$ bil.	%	$ bil.	%	$ bil.	%
Establishments or enterprises (number)	115,154	4.909	116,611	5.555	-	-	-	-	-	-
Total employment (000)	8,728	6.431	8,691	6.350	-	-	-	-	-	-
Production workers (000)	6,930	10.275	6,881	11.014	-	-	-	-	-	-
Output ($ bil.)	131	1.293	148	1.270	-	-	-	-	-	-
Value added ($ bil.)	23	0.543	27	0.561	-	-	-	-	-	-
Capital investment ($ mil.)	9,290	2.131	9,310	1.955	-	-	-	-	-	-
M & E investment ($ mil.)	-		-		-	-	-	-	-	-
Employees per establishment (number)	76	131.011	75	114.320	-	-	-	-	-	-
Production workers per establishment	60	209.314	59	198.265	-	-	-	-	-	-
Output per establishment ($ mil.)	1	26.337	1	22.855	-	-	-	-	-	-
Capital investment per estab. ($ mil.)	0.081	43.421	0.080	35.194	-	-	-	-	-	-
M & E per establishment ($ mil)	-		-		-	-	-	-	-	-
Payroll per employee ($)	1,545	17.231	1,544	15.497	-	-	-	-	-	-
Wages per production worker ($)	1,271	16.103	1,330	15.735	-	-	-	-	-	-
Hours per production worker (hours)	-		-		-	-	-	-	-	-
Output per employee ($)	15,056	20.103	17,028	19.992	-	-	-	-	-	-
Capital investment per employee ($)	1,064	33.143	1,071	30.786	-	-	-	-	-	-
M & E per employee ($)	-		-		-	-	-	-	-	-

Note: Columns headed % show percent of world total or ratio. Ratios closest to 100 are closest to world average. M & E stands for machinery & equipment.

Output in Manufacturing

	1987		1988		1989		1990		1991	
	$ bil.[f]	%	$ bil.[f]	%	$ bil.[f]	%	$ bil.[f]	%	$ bil.[f]	%
3110 - Food products	16.000	12.21	18.000	12.16	-	-	-	-	-	-
3130 - Beverages	0.818	0.62	0.956	0.65	-	-	-	-	-	-
3140 - Tobacco	1.381	1.05	1.538	1.04	-	-	-	-	-	-
3210 - Textiles	12.000	9.16	13.000	8.78	-	-	-	-	-	-
3211 - Spinning, weaving, etc.	11.000	8.40	12.000	8.11	-	-	-	-	-	-
3220 - Wearing apparel	0.887	0.68	1.135	0.77	-	-	-	-	-	-
3230 - Leather and products	0.741	0.57	0.790	0.53	-	-	-	-	-	-
3240 - Footwear	0.316	0.24	0.352	0.24	-	-	-	-	-	-
3310 - Wood products	0.440	0.34	0.539	0.36	-	-	-	-	-	-
3320 - Furniture, fixtures	0.039	0.03	0.050	0.03	-	-	-	-	-	-
3410 - Paper and products	1.998	1.53	2.170	1.47	-	-	-	-	-	-
3411 - Pulp, paper, etc.	1.659	1.27	1.796	1.21	-	-	-	-	-	-
3420 - Printing, publishing	1.327	1.01	1.286	0.87	-	-	-	-	-	-
3510 - Industrial chemicals	8.193	6.25	8.874	6.00	-	-	-	-	-	-
3511 - Basic chemicals, excl fertilizers	2.530	1.93	2.903	1.96	-	-	-	-	-	-
3513 - Synthetic resins, etc.	1.605	1.23	1.768	1.19	-	-	-	-	-	-
3520 - Chemical products nec	7.676	5.86	8.278	5.59	-	-	-	-	-	-
3522 - Drugs and medicines	2.708	2.07	3.004	2.03	-	-	-	-	-	-
3530 - Petroleum refineries	8.232	6.28	8.486	5.73	-	-	-	-	-	-
3540 - Petroleum, coal products	1.126	0.86	1.531	1.03	-	-	-	-	-	-
3550 - Rubber products	2.446	1.87	2.917	1.97	-	-	-	-	-	-
3560 - Plastic products nec	1.281	0.98	1.588	1.07	-	-	-	-	-	-
3610 - Pottery, china, etc.	0.193	0.15	0.201	0.14	-	-	-	-	-	-
3620 - Glass and products	0.409	0.31	0.424	0.29	-	-	-	-	-	-
3690 - Nonmetal products nec	3.472	2.65	4.024	2.72	-	-	-	-	-	-
3710 - Iron and steel	12.000	9.16	14.000	9.46	-	-	-	-	-	-
3720 - Nonferrous metals	1.805	1.38	2.616	1.77	-	-	-	-	-	-
3810 - Metal products	2.430	1.85	2.910	1.97	-	-	-	-	-	-
3820 - Machinery nec	6.465	4.94	7.006	4.73	-	-	-	-	-	-
3825 - Office, computing machinery	0.062	0.05	0.079	0.05	-	-	-	-	-	-
3830 - Electrical machinery	7.067	5.39	8.335	5.63	-	-	-	-	-	-
3832 - Radio, television, etc.	1.782	1.36	2.213	1.50	-	-	-	-	-	-
3840 - Transportation equipment	6.588	5.03	7.961	5.38	-	-	-	-	-	-
3841 - Shipbuilding, repair	0.147	0.11	0.101	0.07	-	-	-	-	-	-
3843 - Motor vehicles	3.495	2.67	4.239	2.86	-	-	-	-	-	-
3850 - Professional goods	0.548	0.42	0.553	0.37	-	-	-	-	-	-
3900 - Industries nec	0.424	0.32	0.424	0.29	-	-	-	-	-	-

Note: Codes are International Standard Industry codes (ISIC). Percentages are % of total Output. [f]: Factor Prices; [p]: Producer Prices.

For sources, notes, and explanations, see Annotated Source Appendix, page 1061.

Finance, Economics, and Trade

Economic Indicators [36]

- **National product**: GDP—purchasing power parity—$1.4087 trillion (1995 est.)
- **National product real growth rate**: 5.5% (1995 est.)
- **National product per capita**: $1,500 (1995 est.)
- **Inflation rate (consumer prices)**: 9% (1995)
- **External debt**: $97.9 billion (March 1995)

Balance of Payments Summary [37]

Values in millions of dollars.

	1986	1987	1988	1989	1990
Exports of goods (f.o.b.)	10,248	11,884	13,510	16,144	18,286
Imports of goods (f.o.b.)	-15,686	-17,661	-20,091	-22,254	-23,437
Trade balance	-5,438	-5,777	-6,581	-6,110	-5,151
Services - debits	-5,526	-6,235	-7,537	-8,372	-9,783
Services - credits	3,746	3,813	4,218	4,586	5,061
Private transfers (net)	2,223	2,636	2,295	2,567	2,337
Government transfers (net)	428	391	487	516	500
Long-term capital (net)	4,496	4,511	5,915	6,161	4,760
Short-term capital (net)	-504	1,223	1,328	1,188	911
Errors and omissions	197	-409	-112	-285	-571
Overall balance	-378	153	13	251	-1,936

Exchange Rates [38]

Currency: **Indian rupee.**
Symbol: **Re.**

Data are currency units per $1.

January 1996	35.766
1995	32.427
1994	31.374
1993	30.493
1992	25.918
1991	22.742

Imports and Exports

Top Import Origins [39]

$33.5 billion (c.i.f., 1995).

Origins	%
US	NA
Germany	NA
Saudi Arabia	NA
UK	NA
Belgium	NA
Japan	NA

Top Export Destinations [40]

$29.96 billion (f.o.b., 1995).

Destinations	%
US	NA
Japan	NA
Germany	NA
UK	NA
Hong Kong	NA

Foreign Aid [41]

	U.S. $	
ODA (1993)	1.237	billion
US ODA bilateral commitments	171	million
US Ex-Im bilateral commitments	680	million
Western (non-US) countries, ODA bilateral commitments	2.48	billion
OPEC bilateral aid	200	million
World Bank (IBRD) multilateral commitments	2.8	billion
Asian Development Bank multilateral commitments	760	million
International Finance Corporation multilateral commitments	200	million
Other multilateral commitments (1995-96)	554	million

Import and Export Commodities [42]

Import Commodities	Export Commodities
Crude oil and petroleum products	Clothing
Machinery	Gems and jewelry
Gems	Engineering goods
Fertilizer	Chemicals
Chemicals	Leather manufactures
	Cotton yarn and fabric

For sources, notes, and explanations, see Annotated Source Appendix, page 1061.

Indonesia

Geography [1]

Total area:
1,919,440 sq km   741,100 sq mi
Land area:
1,826,440 sq km   705,192 sq mi
Comparative area:
Slightly less than three times the size of Texas
Land boundaries:
Total 2,602 km, Malaysia 1,782 km, Papua New Guinea 820 km
Coastline:
54,716 km
Climate:
Tropical; hot, humid; more moderate in highlands
Terrain:
Mostly coastal lowlands; larger islands have interior mountains
Natural resources:
Petroleum, tin, natural gas, nickel, timber, bauxite, copper, fertile
soils, coal, gold, silver
Land use:
Arable land: 8%
Permanent crops: 3%
Meadows and pastures: 7%
Forest and woodland: 67%
Other: 15%

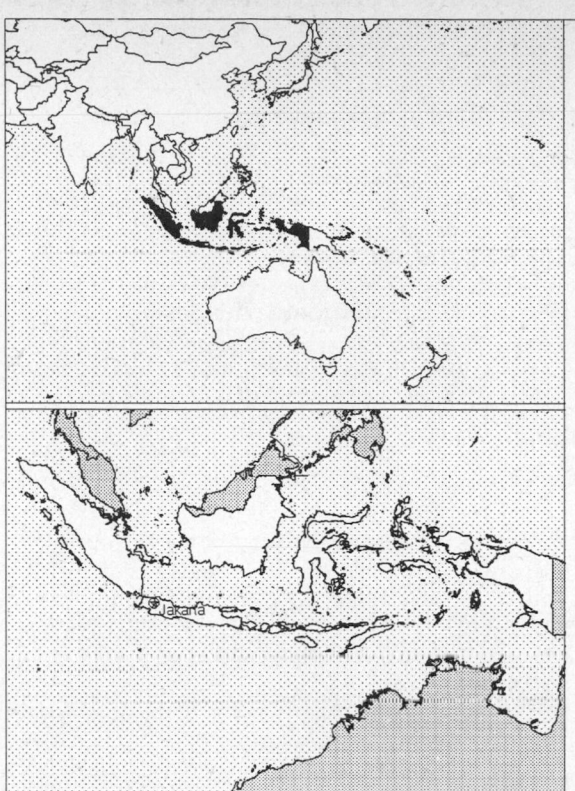

Demographics [2]

	1970	1980	1990	1995[1]	1996	2000	2010	2020	2030
Population	122,889	154,936	187,728	203,459	206,612	219,267	249,679	276,017	299,126
Population density (persons per sq. mi.)	174	220	266	289	293	311	354	391	424
(persons per sq. km.)	67	85	103	111	113	120	137	151	164
Net migration rate (per 1,000 population)	NA	0.0	0.0	0.0	0.0	0.0	0.0	0.0	0.0
Births	NA	NA	NA	NA	4,890	NA	NA	NA	NA
Deaths	NA	NA	NA	NA	1,731	NA	NA	NA	NA
Life expectancy - males	NA	52.0	57.0	59.1	59.5	61.1	64.7	67.9	70.6
Life expectancy - females	NA	55.0	60.8	63.4	63.9	65.8	70.1	73.9	76.9
Birth rate (per 1,000)	NA	34.8	25.9	23.9	23.7	22.4	18.9	16.4	15.1
Death rate (per 1,000)	NA	12.0	9.1	8.5	8.4	8.1	7.5	7.5	8.0
Women of reproductive age (15-49 yrs.)	NA	38,699	48,926	55,364	56,670	61,535	68,602	73,568	74,994
of which are currently married[21]	NA	25,080	33,387	NA	39,002	43,121	49,509	NA	NA
Fertility rate	NA	4.4	3.0	2.7	2.7	2.5	2.3	2.1	2.1

Except as noted, values for vital statistics are in thousands; life expectancy is in years.

Health

Health Indicators [3]

% of population with access to	
safe water (1990-95)	62
adequate sanitation (1990-95)	51
health services (1985-95)	80
% of 1-year-olds immunized (1990-94) against	
TB (tuberculosis)	100
DPT (diphtheria, pertussis, tetanus)	94
polio	93
measles	92
% of contraceptive prevalence (1980-94)	55
ORT use rate (1990-94)	78

Health Expenditures [4]

Total health expenditure, 1990 (official exchange rate)	
Millions of dollars	2,148
Dollars per capita	12
Health expenditures as a percentage of GDP	
Total	2.0
Public sector	0.7
Private sector	1.3
Development assistance for health	
Total aid flows (millions of dollars)[1]	159
Aid flows per capita (dollars)	0.9
Aid flows as a percentage of total health expenditure	7.4

For sources, notes, and explanations, see Annotated Source Appendix, page 1061.

433

Human Factors

Women and Children [5]

% of pregnant women immunized (tetanus 1990-94)	74
% of births attended by trained health personnel (1983-94)	36
Maternal mortality rate (1980-92)	450
Under-5 mortality rate (1994)	111
% under-5 moderately/severely underweight (1980-1994)	40

Burden of Disease [6]

Population per physician (1989)	7,028.10
Population per nurse	NA
Population per hospital bed (1990)	1,503.24
AIDS cases per 100,000 people (1994)	*
Malaria cases per 100,000 people (1992)	7

Ethnic Division [7]

Javanese	45%
Sundanese	14%
Madurese	7.5%
Coastal Malays	7.5%
Other	26%

Religion [8]

Muslim	87%
Protestant	6%
Roman Catholic	3%
Hindu	2%
Buddhist	1%
Other	1%
(1985)	

Major Languages [9]

Bahasa Indonesia (official, modified form of Malay), English, Dutch, local dialects the most widely spoken of which is Javanese.

Education

Public Education Expenditures [10]

	1980	1988	1990	1992	1993	1994
Million or Trillion (T) (Rupiah)[31]						
Total education expenditure	808,087	1.2	2.09 T	5.48 T	3.64 T	4.67 T
as percent of GNP	1.7	0.9	1.1	2.2	1.2	1.3
as percent of total govt. expend.	NA	4.3	NA	NA	NA	NA
Current education expenditure	NA	1.1	1.44 T	3.59 T	2.29 T	3.11 T
as percent of GNP	NA	0.8	0.8	1.5	0.7	0.9
as percent of current govt. expend.	NA	5.5	NA	NA	NA	NA
Capital expenditure	NA	142,822	647,821	1.89 T	1.34 T	1.56 T

Educational Attainment [11]

Age group (1990)	25+
Total population	78,497,680
Highest level attained (%)	
No schooling	54.5
First level	
Not completed	26.4
Completed	NA
Entered second level	
S-1	16.8
S-2	NA
Postsecondary	2.3

Illiteracy [12]

In thousands and percent[1]	1990	1995	2000
Illiterate population (15+ yrs.)	20,899	21,507	18,740
Illiteracy rate - total pop. (%)	17.3	15.6	12.2
Illiteracy rate - males (%)	11.0	10.0	7.6
Illiteracy rate - females (%)	23.5	21.2	16.8

Libraries [13]

	Admin. Units	Svc. Pts.	Vols. (000)	Shelving (meters)	Vols. Added	Reg. Users
National	NA	NA	NA	NA	NA	NA
Nonspecialized	NA	NA	NA	NA	NA	NA
Public	NA	NA	NA	NA	NA	NA
Higher ed. (1989)	45	137	1,735	NA	NA	534,798
School	NA	NA	NA	NA	NA	NA

Daily Newspapers [14]

	1980	1985	1990	1994
Number of papers	84	97	64	56
Circ. (000)	2,281	3,010	5,144	3,800[e]

Culture [15]

Science and Technology

Scientific/Technical Forces [16]

Scientists/engineers	32,038
Number female	NA
Technicians	NA
Number female	NA
Total[1]	NA

R&D Expenditures [17]

	Rupiah (000) 1988
Total expenditure[18]	259,283,000
Capital expenditure	64,645,000
Current expenditure	194,638,000
Percent current	75.1

U.S. Patents Issued [18]

Values show patents issued to citizens of the country by the U.S. Patents Office.

	1993	1994	1995
Number of patents	5	9	7

For sources, notes, and explanations, see Annotated Source Appendix, page 1061.

Government and Law

Organization of Government [19]

Long-form name:
Republic of Indonesia
Type:
Republic
Independence:
17 August 1945 (proclaimed
independence; on 12/27/49 became
legally independent from the Netherlands)
National holiday:
Independence Day, 17 August (1945)
Constitution:
August 1945, abrogated by Federal
Constitution of 1949 and Provisional
Constitution of 1950, restored 5 July 1959
Legal system:
Based on Roman-Dutch law, substantially
modified by indigenous concepts and by
new criminal procedures code; has not
accepted compulsory ICJ jurisdiction
Executive branch:
President; Vice President; Cabinet
Legislative branch:
Unicameral: House of Representatives
(Dewan Perwakilan Rakyat or DPR) Note:
People's Consultative Assembly (DPR
plus 500 indirectly elected members)
meets every five years
Judicial branch:
Supreme Court (Mahkamah Agung)

Elections [20]

House of Representatives	% of votes
GOLKAR	68.0
Development Unity Party	17.0
Indonesia Democracy Party	15.0

Government Expenditures [21]

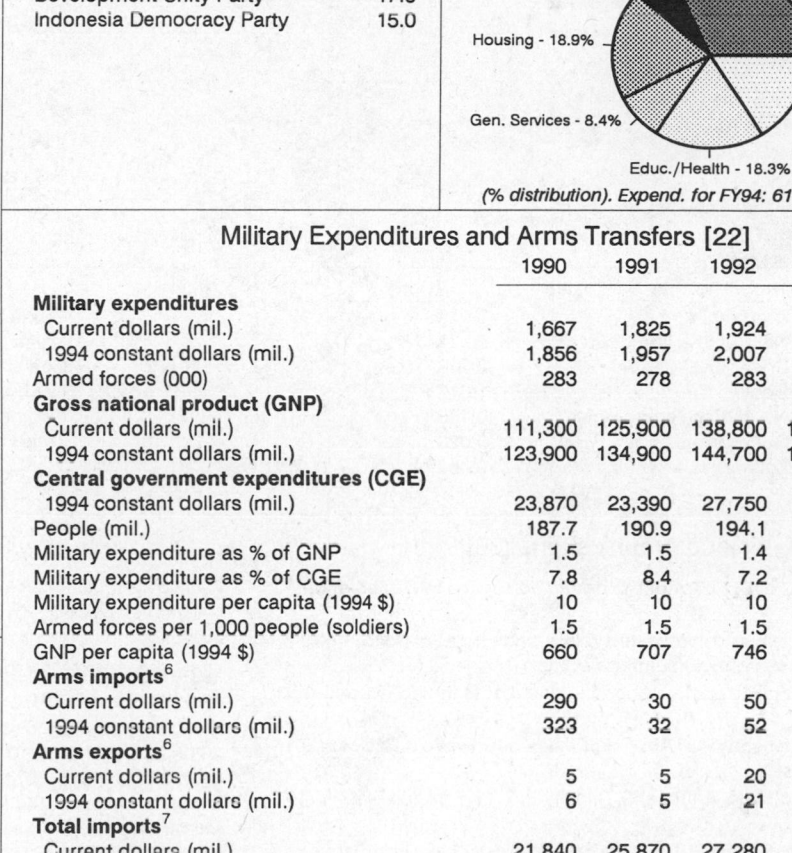

Defense - 6.9%
Industry - 31.3%
Housing - 18.9%
Other - 16.2%
Gen. Services - 8.4%
Educ./Health - 18.3%

(% distribution). Expend. for FY94: 61,866 (Rupiah bil.)

Crime [23]

	1994
Crime volume	
Cases known to police	114,682
Attempts (percent)	NA
Percent cases solved	42.64
Crimes per 100,000 persons	59.66
Persons responsible for offenses	
Total number offenders	NA
Percent female	NA
Percent juvenile (14-17 yrs.)	NA
Percent foreigners	NA

Military Expenditures and Arms Transfers [22]

	1990	1991	1992	1993	1994
Military expenditures					
Current dollars (mil.)	1,667	1,825	1,924	2,043	2,318
1994 constant dollars (mil.)	1,856	1,957	2,007	2,086	2,318
Armed forces (000)	283	278	283	271	280
Gross national product (GNP)					
Current dollars (mil.)	111,300	125,900	138,800	152,900	168,000
1994 constant dollars (mil.)	123,900	134,900	144,700	156,100	168,000
Central government expenditures (CGE)					
1994 constant dollars (mil.)	23,870	23,390	27,750	26,710	32,800[e]
People (mil.)	187.7	190.9	194.1	197.2	200.4
Military expenditure as % of GNP	1.5	1.5	1.4	1.3	1.4
Military expenditure as % of CGE	7.8	8.4	7.2	7.8	7.1
Military expenditure per capita (1994 $)	10	10	10	11	12
Armed forces per 1,000 people (soldiers)	1.5	1.5	1.5	1.4	1.4
GNP per capita (1994 $)	660	707	746	791	838
Arms imports[6]					
Current dollars (mil.)	290	30	50	90	40
1994 constant dollars (mil.)	323	32	52	92	40
Arms exports[6]					
Current dollars (mil.)	5	5	20	20	10
1994 constant dollars (mil.)	6	5	21	20	10
Total imports[7]					
Current dollars (mil.)	21,840	25,870	27,280	28,090	31,980
1994 constant dollars (mil.)	24,300	27,730	28,450	28,660	31,980
Total exports[7]					
Current dollars (mil.)	25,670	29,540	33,860	36,820	40,050
1994 constant dollars (mil.)	28,570	31,670	35,310	37,580	40,050
Arms as percent of total imports[8]	1.3	0.1	0.2	0.3	0.1
Arms as percent of total exports[8]	0	0	0.1	0.1	0

Human Rights [24]

	SSTS	FL	FAPRO	PPCG	APROBC	TPW	PCPTW	STPEP	PHRFF	PRW	ASST	AFL
Observes	P				P	P	P			P		
	EAFRD	CPR	ESCR	SR	ACHR	MAAE	PVIAC	PVNAC	EAFDAW	TCIDTP	RC	
Observes									P	S	P	

P = Party; S = Signatory; see Appendix for meaning of abbreviations.

Labor Force

Total Labor Force [25]

67 million

Labor Force by Occupation [26]

Agriculture	55%
Manufacturing	10
Construction	4
Transport and communications	3

Date of data: 1985 est.

Unemployment Rate [27]

3% official rate; underemployment 40% (1994 est.)

For sources, notes, and explanations, see Annotated Source Appendix, page 1061.

435

Production Sectors

Commercial Energy Production and Consumption

Data are shown in quadrillion (10^{15}) BTUs and percent for 1995
Values for hydroelectric, nuclear, geothermal, solar, and wind power refer to electrical generation.

Production [28]

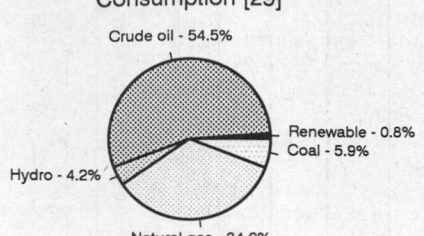

Crude oil - 47.3%
Renewable - 0.3%
NG liquids - 1.7%
Coal - 15.2%
Hydro - 1.9%
Natural gas - 33.5%

Consumption [29]

Crude oil - 54.5%
Renewable - 0.8%
Coal - 5.9%
Hydro - 4.2%
Natural gas - 34.6%

Crude oil	3.148
Natural gas liquids	0.113
Dry natural gas	2.228
Coal	1.012
Net hydroelectric power	0.128
Geothermal, solar, wind	0.023
Total	6.652

Crude oil	1.667
Dry natural gas	1.059
Coal	0.182
Net hydroelectric power	0.128
Geothermal, solar, wind	0.023
Total	3.059

Telecommunications [30]

- 1,276,600 (1993 est.) telephones; domestic service fair, international service good
- Domestic: interisland microwave system and HF radio police net; domestic satellite communications system
- International: satellite earth stations - 2 Intelsat (1 Indian Ocean and 1 Pacific Ocean)
- Radio: Broadcast stations: AM 618, FM 38, shortwave 0 Radios: 28.1 million (1992 est.)
- Television: Broadcast stations: 9 Televisions: 11.5 million (1992 est.)

Top Agricultural Products [32]

Agriculture accounts for 17% of the GDP; produces rice, cassava (tapioca), peanuts, rubber, cocoa, coffee, palm oil, copra, other tropical products; poultry, beef, pork, eggs.

Top Mining Products [33]

Thousand metric tons except as noted[e]	5/18/95[*]
Gold, mine output, Au content (kg.)	45,000[44]
Silver, mine output, Ag content (kg.)	107,000[r]
Cement, hydraulic	19,000
Diamond, gem and industrial (000 carats)	28
Coal	30,900[r]
Limestone	11,700[r]
Gas, natural, gross (mil. cu. ft.)	2,940,000[r]
Petroleum (000 42-gal. bls.)	
crude, incl. field condensate	588,000[r]
refinery products	297,000

Transportation [31]

Railways: total: 6,458 km; narrow gauge: 5,961 km 1.067-m gauge (101 km electrified; 101 km double track); 497 km 0.750-m gauge (1995)

Highways: total: 283,516 km; paved: 125,051 km; unpaved: 158,465 km (1995 est.)

Merchant marine: total: 457 ships (1,000 GRT or over) totaling 2,098,958 GRT/3,056,040 DWT; ships by type: bulk 30, cargo 265, chemical tanker 6, container 11, liquefied gas tanker 5, livestock carrier 1, oil tanker 98, passenger 5, passenger-cargo 12, roll-on/roll-off cargo 7, short-sea passenger 6, specialized tanker 7, vehicle carrier 4 (1995 est.)

Airports

Total:	414
With paved runways over 3,047 m:	4
With paved runways 2,438 to 3,047 m:	9
With paved runways 1,524 to 2,437 m:	35
With paved runways 914 to 1,523 m:	41
With paved runways under 914 m:	299

Tourism [34]

	1990	1991	1992	1993	1994
Tourists	2,178	2,570	3,064	3,403	4,006
Tourism receipts	2,105	2,522	3,278	3,987	4,785
Tourism expenditures	836	969	1,166	1,539	1,900

Travelers are in thousands, money in million U.S. dollars.

For sources, notes, and explanations, see Annotated Source Appendix, page 1061.

Manufacturing Sector

Manufacturing Summary [35]

	1987		1988		1989		1990		1991	
	$ bil.	%	$ bil.	%	$ bil.	%	$ bil.	%	$ bil.	%
Establishments or enterprises (number)	13,422	0.572	15,406	0.734	15,366	0.820	-	-	-	-
Total employment (000)	1,945	1.434	2,230	1.629	2,445	1.986	-	-	-	-
Production workers (000)	1,416	2.099	1,652	2.644	1,823	2.478	-	-	-	-
Output ($ bil.)	22	0.221	28	0.244	37	0.310	-	-	-	-
Value added ($ bil.)	7	0.152	8	0.167	10	0.212	-	-	-	-
Capital investment ($ mil.)	1,264	0.290	2,051	0.431	4,936	0.901	-	-	-	-
M & E investment ($ mil.)	-	-	-	-	-	-	-	-	-	-
Employees per establishment (number)	145	250.546	145	221.978	159	242.123	-	-	-	-
Production workers per establishment	105	366.833	107	360.314	119	302.146	-	-	-	-
Output per establishment ($ mil.)	2	38.626	2	33.227	2	37.763	-	-	-	-
Capital investment per estab. ($ mil.)	0.094	50.670	0.133	58.696	0.321	109.848	-	-	-	-
M & E per establishment ($ mil)	-	-	-	-	-	-	-	-	-	-
Payroll per employee ($)	963	10.743	1,051	10.550	1,073	9.605	-	-	-	-
Wages per production worker ($)	-	-	-	-	-	-	-	-	-	-
Hours per production worker (hours)	-	-	-	-	-	-	-	-	-	-
Output per employee ($)	11,546	15.417	12,749	14.969	14,948	15.597	-	-	-	-
Capital investment per employee ($)	649	20.224	920	20.442	2,019	45.369	-	-	-	-
M & E per employee ($)	-	-	-	-	-	-	-	-	-	-

Note: Columns headed % show percent of world total or ratio. Ratios closest to 100 are closest to world average. M & E stands for machinery & equipment.

Output in Manufacturing

	1987		1988		1989		1990		1991	
	$ bil.[p]	%	$ bil.[p]	%	$ bil.[p]	%	$ bil.[p]	%	$ bil.[p]	%
3110 - Food products	3.301	15.00	4.122	14.72	4.853	13.12	-	-	-	-
3130 - Beverages	0.189	0.86	0.242	0.86	0.257	0.69	-	-	-	-
3140 - Tobacco	2.182	9.92	2.178	7.78	3.260	8.81	-	-	-	-
3210 - Textiles	2.366	10.75	2.948	10.53	4.043	10.93	-	-	-	-
3211 - Spinning, weaving, etc.	0.886	4.03	1.173	4.19	3.353	9.06	-	-	-	-
3220 - Wearing apparel	0.364	1.65	0.518	1.85	0.712	1.92	-	-	-	-
3230 - Leather and products	0.043	0.20	0.071	0.25	0.074	0.20	-	-	-	-
3240 - Footwear	0.057	0.26	0.104	0.37	0.171	0.46	-	-	-	-
3310 - Wood products	2.258	10.26	3.165	11.30	3.613	9.76	-	-	-	-
3320 - Furniture, fixtures	0.052	0.24	0.082	0.29	0.113	0.31	-	-	-	-
3410 - Paper and products	0.544	2.47	0.765	2.73	0.951	2.57	-	-	-	-
3411 - Pulp, paper, etc.	0.405	1.84	0.608	2.17	0.764	2.06	-	-	-	-
3420 - Printing, publishing	0.277	1.26	0.335	1.20	0.404	1.09	-	-	-	-
3510 - Industrial chemicals	1.185	5.39	1.316	4.70	1.446	3.91	-	-	-	-
3511 - Basic chemicals, excl fertilizers	0.285	1.30	0.369	1.32	0.424	1.15	-	-	-	-
3513 - Synthetic resins, etc.	0.008	0.04	0.010	0.04	0.011	0.03	-	-	-	-
3520 - Chemical products nec	0.989	4.50	1.201	4.29	1.345	3.64	-	-	-	-
3522 - Drugs and medicines	0.478	2.17	0.525	1.87	0.558	1.51	-	-	-	-
3550 - Rubber products	1.161	5.28	1.634	5.84	1.581	4.27	-	-	-	-
3560 - Plastic products nec	0.485	2.20	0.658	2.35	0.764	2.06	-	-	-	-
3610 - Pottery, china, etc.	0.054	0.25	0.079	0.28	0.108	0.29	-	-	-	-
3620 - Glass and products	0.170	0.77	0.179	0.64	0.195	0.53	-	-	-	-
3690 - Nonmetal products nec	0.645	2.93	0.669	2.39	0.761	2.06	-	-	-	-
3810 - Metal products	0.923	4.20	1.446	5.16	1.752	4.74	-	-	-	-
3820 - Machinery nec	0.175	0.80	0.243	0.87	0.294	0.79	-	-	-	-
3830 - Electrical machinery	0.697	3.17	0.928	3.31	1.147	3.10	-	-	-	-
3832 - Radio, television, etc.	0.206	0.94	0.234	0.84	0.307	0.83	-	-	-	-
3840 - Transportation equipment	1.075	4.89	1.374	4.91	2.037	5.51	-	-	-	-
3841 - Shipbuilding, repair	0.188	0.85	0.128	0.46	0.122	0.33	-	-	-	-
3843 - Motor vehicles	0.717	3.26	0.994	3.55	0.927	2.51	-	-	-	-
3850 - Professional goods	0.015	0.07	0.024	0.09	0.031	0.08	-	-	-	-
3900 - Industries nec	0.080	0.36	0.103	0.37	0.163	0.44	-	-	-	-

Note: Codes are International Standard Industry codes (ISIC). Percentages are % of total Output. [f]: Factor Prices; [p]: Producer Prices.

For sources, notes, and explanations, see Annotated Source Appendix, page 1061.

437

Finance, Economics, and Trade

Economic Indicators [36]

- **National product**: GDP—purchasing power parity—$710.9 billion (1995 est.)
- **National product real growth rate**: 7.5% (1995 est.)
- **National product per capita**: $3,500 (1995 est.)
- **Inflation rate (consumer prices)**: 8.6% (1995 est.)
- **External debt**: $97.6 billion (1995 est.)

Balance of Payments Summary [37]

Values in millions of dollars.

	1989	1990	1991	1992	1993
Exports of goods (f.o.b.)	22,974	26,807	29,635	33,796	36,607
Imports of goods (f.o.b.)	-16,310	-21,455	-24,834	-26,774	-28,376
Trade balance	6,664	5,352	4,801	7,022	8,231
Services - debits	-10,548	-11,655	-13,062	-14,582	-15,798
Services - credits	2,437	2,897	3,739	4,209	4,923
Private transfers (net)	167	166	130	229	346
Government transfers (net)	172	252	132	342	282
Long-term capital (net)	3,016	4,724	5,483	5,992	5,458
Short-term capital (net)	-98	-229	214	137	223
Errors and omissions	-1,315	744	91	-1,279	-3,078
Overall balance	495	2,251	1,528	2,070	587

Exchange Rates [38]

Currency: **Indonesian rupiah.**
Symbol: **Rp.**

Data are currency units per $1.

January 1996	2,306.3
1995	2,248.6
1994	2,160.8
1993	2,087.1
1992	2,029.9
1991	1,950.3

Imports and Exports

Top Import Origins [39]

$32 billion (f.o.b., 1994).

Origins	%
Japan	24.2
US	11.2
Germany	7.7
South Korea	6.8
Singapore	5.9
Australia	4.8
Taiwan	4.5
China	4.3

Top Export Destinations [40]

$39.9 billion (f.o.b., 1994).

Destinations	%
Japan	27.4
US	14.6
Singapore	10.1
South Korea	6.5
Taiwan	4.1
Netherlands	3.3
China	3.3
Hong Kong	3.3
Germany	3.2

Foreign Aid [41]

Recipient: ODA, $1.542 billion (1993).

Import and Export Commodities [42]

Import Commodities	Export Commodities
Manufactures 75.3%	Manufactures 51.9%
Raw materials 9.0%	Fuels 26.4%
Foodstuffs 7.8%	Foodstuffs 12.7%
Fuels 7.7%	Raw materials 9.0%

Iran

Geography [1]

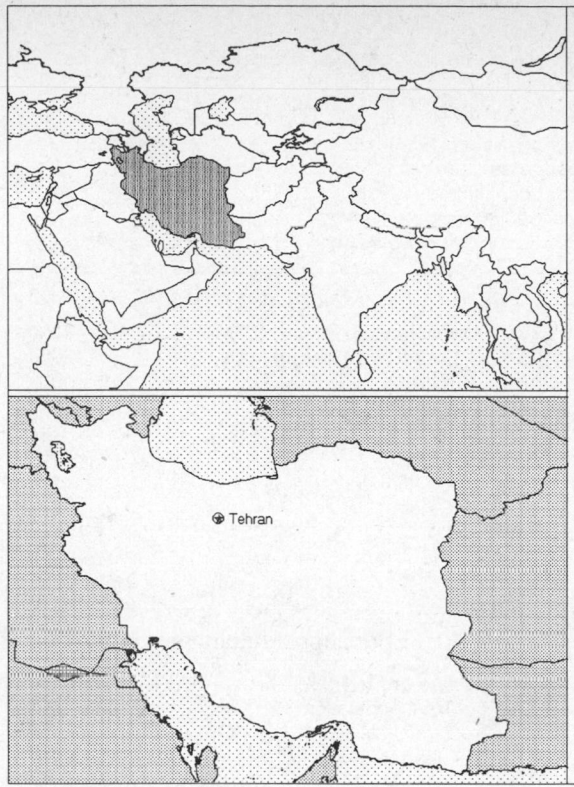

Total area:
  1,648,000 sq km   636,296 sq mi
Land area:
  1,636,000 sq km   631,663 sq mi
Comparative area:
  Slightly larger than Alaska
Land boundaries:
  Total 5,440 km, Afghanistan 936 km, Armenia 35 km, Azerbaijan-proper 432 km, Azerbaijan-Naxcivan exclave 179 km, Iraq 1,458 km, Pakistan 909 km, Turkey 499 km, Turkmenistan 992 km
Coastline:
  2,440 km
  Note: Iran also borders the Caspian Sea (740 km)
Climate:
  Mostly arid or semiarid, subtropical along Caspian coast
Terrain:
  Rugged, mountainous rim; high, central basin with deserts, mountains; small, discontinuous plains along both coasts
Natural resources:
  Petroleum, natural gas, coal, chromium, copper, iron ore, lead, manganese, zinc, sulfur
Land use:
  Arable land: 8%
  Permanent crops: 0%
  Meadows and pastures: 27%
  Forest and woodland: 11%
  Other: 54%

Demographics [2]

	1970	1980	1990	1995[1]	1996	2000	2010	2020	2030
Population	28,933	39,274	56,946	64,625	66,094	71,879	88,231	104,282	119,017
Population density (persons per sq. mi.)	46	62	90	102	105	114	140	165	188
(persons per sq. km.)	18	24	35	40	40	44	54	64	73
Net migration rate (per 1,000 population)	NA	NA	8.2	-5.1	-5.0	0.0	0.0	0.0	0.0
Births	NA	NA	NA	NA	2,225	NA	NA	NA	NA
Deaths	NA	NA	NA	NA	437	NA	NA	NA	NA
Life expectancy - males	NA	NA	62.6	65.8	66.1	67.6	70.3	72.8	74.7
Life expectancy - females	NA	NA	63.8	68.2	68.7	70.8	74.7	78.3	80.8
Birth rate (per 1,000)	NA	NA	41.1	34.9	33.7	29.2	23.0	19.4	16.3
Death rate (per 1,000)	NA	NA	9.0	6.9	6.6	5.8	4.9	4.4	4.7
Women of reproductive age (15-49 yrs.)	NA	NA	12,110	14,183	14,651	16,811	23,441	28,756	32,439
of which are currently married[22]	NA	NA	9,171	NA	11,113	12,674	18,046	NA	NA
Fertility rate	NA	NA	6.0	4.9	4.7	3.9	2.6	2.2	2.0

Except as noted, values for vital statistics are in thousands; life expectancy is in years.

Health

Health Indicators [3]

% of population with access to	
safe water (1990-95)	84
adequate sanitation (1990-95)	67
health services (1985-95)	80
% of 1-year-olds immunized (1990-94) against	
TB (tuberculosis)	100
DPT (diphtheria, pertussis, tetanus)	95
polio	95
measles	97
% of contraceptive prevalence (1980-94)	49
ORT use rate (1990-94)	85

Health Expenditures [4]

Total health expenditure, 1990 (official exchange rate)	
Millions of dollars	3,024
Dollars per capita	54
Health expenditures as a percentage of GDP	
Total	2.6
Public sector	1.5
Private sector	1.1
Development assistance for health	
Total aid flows (millions of dollars)[1]	2
Aid flows per capita (dollars)	NA
Aid flows as a percentage of total health expenditure	NA

For sources, notes, and explanations, see Annotated Source Appendix, page 1061.

439

Human Factors

Women and Children [5]

% of pregnant women immunized (tetanus 1990-94)	51
% of births attended by trained health personnel (1983-94)	70
Maternal mortality rate (1980-92)	120
Under-5 mortality rate (1994)	51
% under-5 moderately/severely underweight (1980-1994)	NA

Burden of Disease [6]

Population per physician (1988)	3,142.35
Population per nurse (1987)	1,154.21
Population per hospital bed (1990)	724.40
AIDS cases per 100,000 people (1994)	*
Malaria cases per 100,000 people (1992)	123

Ethnic Division [7]

Persian	51%
Azerbaijani	24%
Gilaki and Mazandarani	8%
Kurd	7%
Lur and Turkmen	4%
Arab	3%
Baloch	2%
Other	1%

Religion [8]

Shi'a Muslim	89%
Sunni Muslim	10%
Zoroastrian, Jewish, Christian, and Baha'i	1%

Major Languages [9]

Persian and Persian dialects	58%
Turkic and Turkic dialects	26%
Kurdish	9%
Luri	2%
Arabic	1%
Baloch	1%
Turkish	1%
Other	2%

Education

Public Education Expenditures [10]

	1980	1985	1990	1992	1993	1994
Million or Trillion (T) (Rial)						
Total education expenditure	498,268	575,519	1.49 T	3.14 T	4.98 T	5.84 T
as percent of GNP	7.5	3.6	4.1	4.7	5.3	5.9
as percent of total govt. expend.	15.7	17.2	22.4	28.2	22.8	18.1
Current education expenditure	440,298	510,081	1.23 T	2.58 T	4.07 T	4.79 T
as percent of GNP	6.6	3.2	3.4	3.9	4.3	4.9
as percent of current govt. expend.	20.1	20.6	28.8	31.4	26.4	22.7
Capital expenditure	57,970	65,438	261,799	560,393	909,676	1.05 T

Educational Attainment [11]

Age group (1987)	10+
Total population	10,628,447
Highest level attained (%)	
No schooling	52.8
First level	
Not completed	21.6
Completed	NA
Entered second level	
S-1	11.6
S-2	NA
Postsecondary	4.1

Illiteracy [12]

In thousands and percent[2]	1991[6]	1994	1995
Illiterate population (15+ yrs.)	10,652	9,789	NA
Illiteracy rate - total pop. (%)	34.3	27.7	NA
Illiteracy rate - males (%)	25.6	21.6	NA
Illiteracy rate - females (%)	43.6	34.2	NA

Libraries [13]

	Admin. Units	Svc. Pts.	Vols. (000)	Shelving (meters)	Vols. Added	Reg. Users
National (1992)	1	5	392	10,364	17,616	2,902
Nonspecialized	NA	NA	NA	NA	NA	NA
Public (1990)	488	NA	NA	NA	NA	NA
Higher ed. (1994)	113	168	2,323	NA	NA	280,877
School	NA	NA	NA	NA	NA	NA

Daily Newspapers [14]

	1980	1985	1990	1994
Number of papers	45[e]	15	21	12
Circ. (000)	970[e]	1,250[e]	1,500[e]	1,200[e]

Culture [15]

Cinema (seats per 1,000)	2.7
Annual attendance per person	0.5
Gross box office receipts (mil. Rial)	12,616
Museums (reporting)	NA
Visitors (000)	NA
Annual receipts (000 Rial)	NA

Science and Technology

Scientific/Technical Forces [16]

Scientists/engineers	3,194
Number female	NA
Technicians	1,854
Number female	NA
Total	5,048

R&D Expenditures [17]

	Rial (000) 1985
Total expenditure[19]	22,010,713
Capital expenditure	9,464,315
Current expenditure	12,546,398
Percent current	57.0

U.S. Patents Issued [18]

Values show patents issued to citizens of the country by the U.S. Patents Office.

	1993	1994	1995
Number of patents	1	2	2

For sources, notes, and explanations, see Annotated Source Appendix, page 1061.

Government and Law

Organization of Government [19]

Long-form name:
Islamic Republic of Iran
Type:
Theocratic republic
Independence:
1 April 1979 (Islamic Republic of Iran
proclaimed)
National holiday:
Islamic Republic Day, 1 April (1979)
Constitution:
2-3 December 1979; revised 1989 to
expand powers of the presidency and
eliminate the prime ministership
Legal system:
The Constitution codifies Islamic
principles of government
Executive branch:
Leader of the Islamic Revolution;
President; First Vice President; Council of
Ministers
Legislative branch:
Unicameral: Islamic Consultative
Assembly (Majles-e-Shura-ye-Eslami)
Judicial branch:
Supreme Court

Crime [23]

Elections [20]

Islamic Consultative Assembly. There
are at least 76 licensed parties; the three
most important are - Tehran Militant
Clergy Association, Militant Clerics
Association, and Servants of
Reconstruction. Last held 8 March and
19 April 1996; results - percent of vote
by party NA; seats - (270 seats total)
number of seats by party NA.

Government Expenditures [21]

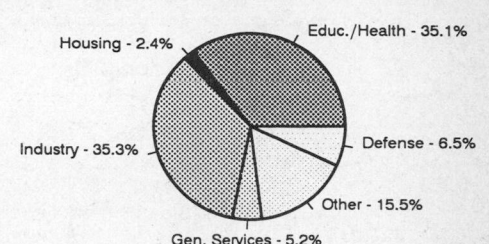

(% distribution). Expend. for FY94: 32,295 (Rial bil.)

Military Expenditures and Arms Transfers [22]

	1990	1991	1992	1993	1994
Military expenditures					
Current dollars (mil.)[r,1]	6,394	6,154	3,964	4,705	3,042
1994 constant dollars (mil.)[r,1]	7,117	6,597	4,133	4,802	3,042
Armed forces (000)	440	465	528	528	528
Gross national product (GNP)					
Current dollars (mil.)	107,000	122,700	133,800	139,800	125,000
1994 constant dollars (mil.)	119,100	131,500	139,600	142,700	125,000
Central government expenditures (CGE)					
1994 constant dollars (mil.)	23,620	24,900	27,830	34,170	41,040
People (mil.)	56.9	59.1	60.9	62.0	63.1
Military expenditure as % of GNP	6.0	5.0	3.0	3.4	2.4
Military expenditure as % of CGE	30.1	26.5	14.9	14.1	7.4
Military expenditure per capita (1994 $)	125	112	68	77	48
Armed forces per 1,000 people (soldiers)	7.7	7.9	8.7	8.5	8.4
GNP per capita (1994 $)	2,092	2,226	2,293	2,301	1,981
Arms imports[6]					
Current dollars (mil.)	1,800	2,100	360	1,000	390
1994 constant dollars (mil.)	2,003	2,251	375	1,021	390
Arms exports[6]					
Current dollars (mil.)	0	40	0	5	90
1994 constant dollars (mil.)	0	43	0	5	90
Total imports[7]					
Current dollars (mil.)	15,720	21,690	23,080	16,040	12,070[e]
1994 constant dollars (mil.)	17,490	23,250	24,070	16,370	12,070[e]
Total exports[7]					
Current dollars (mil.)	15,320	15,920	19,870[e]	18,750[e]	16,700[e]
1994 constant dollars (mil.)	17,050	17,060	20,720[e]	18,950[e]	16,700[e]
Arms as percent of total imports[8]	11.5	9.7	1.6	6.2	3.2
Arms as percent of total exports[8]	0	0.3	0	0	0.5

Human Rights [24]

	SSTS	FL	FAPRO	PPCG	APROBC	TPW	PCPTW	STPEP	PHRFF	PRW	ASST	AFL
Observes	S	P		P		P	P	S			P	P
	EAFRD	CPR	ESCR	SR	ACHR	MAAE	PVIAC	PVNAC	EAFDAW	TCIDTP	RC
Observes	P	P	P	P			S	S			S

P = Party; S = Signatory; see Appendix for meaning of abbreviations.

Labor Force

Total Labor Force [25]

15.4 million

Labor Force by Occupation [26]

Agriculture	33%
Manufacturing	21

Date of data: 1988 est.

Unemployment Rate [27]

Over 30% (1995 est.)

For sources, notes, and explanations, see Annotated Source Appendix, page 1061.

441

Production Sectors

Commercial Energy Production and Consumption

Data are shown in quadrillion (10^{15}) BTUs and percent for 1995
Values for hydroelectric, nuclear, geothermal, solar, and wind power refer to electrical generation.

Production [28]

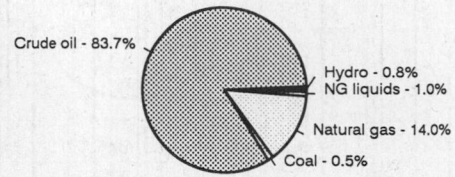

Crude oil - 83.7%
Hydro - 0.8%
NG liquids - 1.0%
Natural gas - 14.0%
Coal - 0.5%

Consumption [29]

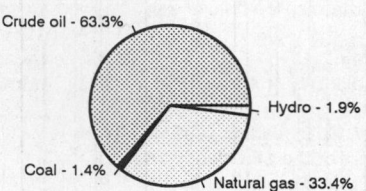

Crude oil - 63.3%
Hydro - 1.9%
Coal - 1.4%
Natural gas - 33.4%

Crude oil	7.830
Natural gas liquids	0.095
Dry natural gas	1.309
Coal	0.043
Net hydroelectric power	0.073
Total	9.350

Crude oil	2.472
Dry natural gas	1.305
Coal	0.055
Net hydroelectric power	0.073
Total	3.905

Telecommunications [30]

- 3.02 million (1992 est.) telephones
- Domestic: microwave radio relay extends throughout country; system centered in Tehran
- International: satellite earth stations - 3 Intelsat (2 Atlantic Ocean and 1 Indian Ocean) and 1 Inmarsat (Indian Ocean Region); HF radio and microwave radio relay to Turkey, Pakistan, Syria, Kuwait, Tajikistan, and Uzbekistan; submarine fiber-optic cable to UAE
- Radio: Broadcast stations: AM 77, FM 3, shortwave 0 Radios: 14.3 million (1992 est.)
- Television: Broadcast stations: 28 Televisions: 3.9 million (1992 est.)

Transportation [31]

Railways: total: 5,093 km; broad gauge: 96 km 1.676-m gauge; standard gauge: 4,997 km 1.432-m gauge (146 km electrified) (1995)

Highways: total: 140,200 km; paved: 42,700 km; unpaved: 97,500 km (1995 est.)

Merchant marine: total: 130 ships (1,000 GRT or over) totaling 2,791,892 GRT/4,891,615 DWT; ships by type: bulk 47, cargo 41, chemical tanker 5, combination bulk 2, liquefied gas tanker 1, multifunction large-load carrier 1, oil tanker 19, refrigerated cargo 3, roll-on/roll-off cargo 9, short-sea passenger 1, specialized tanker 1 (1995 est.)

Airports

Total:	212
With paved runways over 3,047 m:	30
With paved runways 2,438 to 3,047 m:	11
With paved runways 1,524 to 2,437 m:	31
With paved runways 914 to 1,523 m:	17
With paved runways under 914 m:	22

Top Agricultural Products [32]

Agriculture accounts for 21% of the GDP; produces wheat, rice, other grains, sugar beets, fruits, nuts, cotton; dairy products, wool; caviar.

Top Mining Products [33]

Thousand metric tons except as noted[45]	6/30/95[*]
Copper ore, mine output (1%-1.2% Cu) gr. wt.	12,100
Cement, hydraulic	20,000[e]
Iron ore/concentrate, gross weight	8,690
Turquoise (kg.)	5,000[e]
Gypsum	8,430
Limestone	28,000[e]
Gas, natural, gross (mil. cu. meters)	54,000[e]
Natural gas plant liquids (000 42-gal. bls.)	23,500
Petroleum, crude (000 42-gal. bls.)	1,090,000

Tourism [34]

	1990	1991	1992	1993	1994
Visitors	298	368	432	476	478
Tourists[12]	154	212	279	304	362
Tourism receipts	61	88	121	131	153
Tourism expenditures	340	734	1,109	NA	NA
Fare receipts	36	61	40	NA	NA
Fare expenditures	64	127	49	NA	NA

Travelers are in thousands, money in million U.S. dollars.

For sources, notes, and explanations, see Annotated Source Appendix, page 1061.

Manufacturing Sector

Manufacturing Summary [35]

	1987		1988		1989		1990		1991	
	$ bil.	%	$ bil.	%	$ bil.	%	$ bil.	%	$ bil.	%
Establishments or enterprises (number)	12,645	0.539	10,245	0.488	11,554	0.617	-	-	-	-
Total employment (000)	848	0.625	631	0.461	659	0.535	-	-	-	-
Production workers (000)	-		-		-		-	-	-	-
Output ($ bil.)	47	0.463	49	0.422	63	0.531	-	-	-	-
Value added ($ bil.)	25	0.581	24	0.508	28	0.578	-	-	-	-
Capital investment ($ mil.)	1,930	0.443	2,691	0.565	3,394	0.620	-	-	-	-
M & E investment ($ mil.)	-		-		-		-	-	-	-
Employees per establishment (number)	67	115.964	62	94.429	57	86.765	-	-	-	-
Production workers per establishment	-		-		-		-	-	-	-
Output per establishment ($ mil.)	4	85.877	5	86.405	5	86.123	-	-	-	-
Capital investment per estab. ($ mil.)	0.153	82.171	0.263	115.800	0.294	100.465	-	-	-	-
M & E per establishment ($ mil)	-		-		-		-	-	-	-
Payroll per employee ($)	19,383	216.158	17,077	171.411	19,085	170.826	-	-	-	-
Wages per production worker ($)	-		-		-		-	-	-	-
Hours per production worker (hours)	-		-		-		-	-	-	-
Output per employee ($)	55,463	74.054	77,938	91.503	95,130	99.260	-	-	-	-
Capital investment per employee ($)	2,276	70.859	4,267	122.632	5,153	115.790	-	-	-	-
M & E per employee ($)	-		-		-		-	-	-	-

Note: Columns headed % show percent of world total or ratio. Ratios closest to 100 are closest to world average. M & E stands for machinery & equipment.

Output in Manufacturing

	1987		1988		1989		1990		1991	
	$ bil.	%	$ bil.	%	$ bil.	%	$ bil.	%	$ bil.	%
3110 - Food products	6.689	14.23	8.714	17.78	11.000	17.46	-	-	-	-
3130 - Beverages	0.737	1.57	1.016	2.07	1.050	1.67	-	-	-	-
3140 - Tobacco	3.034	6.46	0.735	1.50	0.532	0.84	-	-	-	-
3210 - Textiles	6.272	13.34	9.125	18.62	11.000	17.46	-	-	-	-
3211 - Spinning, weaving, etc.	5.133	10.92	-	-	-	-	-	-	-	-
3220 - Wearing apparel	0.631	1.34	0.748	1.53	1.090	1.73	-	-	-	-
3230 - Leather and products	0.607	1.29	1.005	2.05	0.964	1.53	-	-	-	-
3240 - Footwear	0.693	1.47	0.872	1.78	1.148	1.82	-	-	-	-
3310 - Wood products	0.490	1.04	0.834	1.70	1.105	1.75	-	-	-	-
3320 - Furniture, fixtures	0.157	0.33	0.246	0.50	0.325	0.52	-	-	-	-
3410 - Paper and products	0.557	1.19	0.826	1.69	1.161	1.84	-	-	-	-
3411 - Pulp, paper, etc.	0.084	0.18	-		-		-	-	-	-
3420 - Printing, publishing	0.416	0.89	0.514	1.05	0.978	1.55	-	-	-	-
3510 - Industrial chemicals	0.730	1.55	0.920	1.88	1.603	2.54	-	-	-	-
3511 - Basic chemicals, excl fertilizers	0.174	0.37	-		-		-	-	-	-
3513 - Synthetic resins, etc.	0.211	0.45	-		-		-	-	-	-
3520 - Chemical products nec	2.140	4.55	2.254	4.60	4.159	6.60	-	-	-	-
3522 - Drugs and medicines	0.834	1.77	-		-		-	-	-	-
3530 - Petroleum refineries	0.192	0.41	0.176	0.36	0.196	0.31	-	-	-	-
3540 - Petroleum, coal products	0.209	0.44	0.175	0.36	0.249	0.40	-	-	-	-
3550 - Rubber products	0.750	1.60	0.711	1.45	1.236	1.96	-	-	-	-
3560 - Plastic products nec	0.739	1.57	1.337	2.73	2.143	3.40	-	-	-	-
3610 - Pottery, china, etc.	0.129	0.27	0.191	0.39	0.206	0.33	-	-	-	-
3620 - Glass and products	0.438	0.93	1.067	2.18	0.589	0.93	-	-	-	-
3690 - Nonmetal products nec	4.363	9.28	4.343	8.86	4.759	7.55	-	-	-	-
3710 - Iron and steel	2.019	4.30	3.094	6.31	4.112	6.53	-	-	-	-
3720 - Nonferrous metals	0.722	1.54	1.127	2.30	1.634	2.59	-	-	-	-
3810 - Metal products	1.390	2.96	1.573	3.21	3.365	5.34	-	-	-	-
3820 - Machinery nec	2.583	5.50	3.739	7.63	4.580	7.27	-	-	-	-
3825 - Office, computing machinery	0.031	0.07	-		-		-	-	-	-
3830 - Electrical machinery	0.814	1.73	1.300	2.65	1.346	2.14	-	-	-	-
3832 - Radio, television, etc.	0.507	1.08	-		-		-	-	-	-
3840 - Transportation equipment	1.254	2.67	2.338	4.77	2.173	3.45	-	-	-	-
3841 - Shipbuilding, repair	0.077	0.16	-		-		-	-	-	-
3843 - Motor vehicles	1.017	2.16	-		-		-	-	-	-
3850 - Professional goods	0.084	0.18	0.106	0.22	0.158	0.25	-	-	-	-
3900 - Industries nec	0.146	0.31	0.068	0.14	0.226	0.36	-	-	-	-

Note: Codes are International Standard Industry codes (ISIC). Percentages are % of total Output. [f]: Factor Prices; [p]: Producer Prices.

For sources, notes, and explanations, see Annotated Source Appendix, page 1061.

443

Finance, Economics, and Trade

Economic Indicators [36]

- **National product**: GDP—purchasing power parity—$323.5 billion (1995 est.)

- **National product real growth rate**: - 2% (1995 est.)

- **National product per capita**: $4,700 (1995 est.)

- **Inflation rate (consumer prices)**: 60% (1995 est.)

- **External debt**: $30 billion (1995 est.)

Balance of Payments Summary [37]

Values in millions of dollars.

	1988	1989	1990	1991	1992
Exports of goods (f.o.b.)	10,709	13,081	19,305	18,661	19,868
Imports of goods (f.o.b.)	-10,608	-13,448	-18,330	-25,190	-23,274
Trade balance	101	-367	975	-6,529	-3,406
Services - debits	-2,436	-3,122	-4,040	-5,800	-5,940
Services - credits	467	798	892	881	846
Private transfers (net)	NA	2,500	2,500	2,000	1,996
Government transfers (net)	NA	NA	NA	NA	NA
Long-term capital (net)	-41	-1,036	101	1,351	4,898
Short-term capital (net)	476	2,171	194	4,682	-195
Errors and omissions	421	-682	-946	1,321	1,636
Overall balance	-1,012	2,334	-324	-2,094	-165

Exchange Rates [38]

Currency: **Iranian rials.**
Symbol: **IR.**

Data are currency units per $1. As of May 1995, the "official rate" of 1,750 rials per US$1 is used for imports of essential goods and services and for oil exports, wheras the "official export rate" of 3,000 rials per US$1 is used for non-oil exports and imports not covered by the official rate.

January 1996	1,750
1995	1,747.93
1994	1,748.75
1993	1,267.77
1992	65.55
1991	67.51
December 1995 (BMR)	4,000

Imports and Exports

Top Import Origins [39]

$13 billion (c.i.f., 1994 est.).

Origins	%
Germany	NA
Japan	NA
Italy	NA
UK	NA
UAE	NA

Top Export Destinations [40]

$16 billion (f.o.b., 1994 est.).

Destinations	%
Japan	NA
Italy	NA
France	NA
Netherlands	NA
Belgium/Luxembourg	NA
Spain	NA
Germany	NA

Foreign Aid [41]

Recipient: ODA, $40 million (1993).

Import and Export Commodities [42]

Import Commodities	Export Commodities
Machinery	Petroleum 85%
Military supplies	Carpets
Metal works	Fruits
Foodstuffs	Nuts
Pharmaceuticals	Hides
Technical services	Iron
Refined oil products	Steel

Iraq

Geography [1]

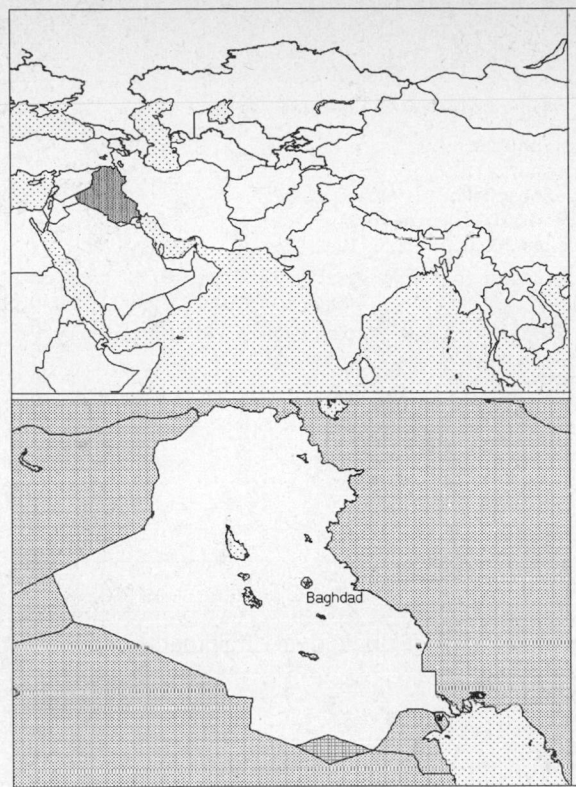

Total area:
437,072 sq km   168,754 sq mi
Land area:
432,162 sq km   166,859 sq mi
Comparative area:
Slightly more than twice the size of Idaho
Land boundaries:
Total 3,631 km, Iran 1,458 km, Jordan 181 km, Kuwait 242 km, Saudi Arabia 814 km, Syria 605 km, Turkey 331 km
Coastline:
58 km
Climate:
Mostly desert; mild to cool winters with dry, hot, cloudless summers; northern mountainous regions along Iranian and Turkish borders experience cold winters with occasionally heavy snows which melt in early spring, sometimes causing extensive flooding in central and southern Iraq
Terrain:
Mostly broad plains; reedy marshes along Iranian border in South; mountains along borders with Iran and Turkey
Natural resources:
Petroleum, natural gas, phosphates, sulfur
Land use:
Arable land: 12%
Permanent crops: 1%
Meadows and pastures: 9%
Forest and woodland: 3%
Other: 75%

Demographics [2]

	1970	1980	1990	1995[1]	1996	2000	2010	2020	2030
Population	9,414	13,233	18,425	20,644	21,422	24,731	34,545	46,260	59,172
Population density (persons per sq. mi.)	56	79	110	123	128	148	206	276	353
(persons per sq. km.)	22	30	42	48	49	57	80	107	136
Net migration rate (per 1,000 population)	NA	NA	-115.8	0.4	0.4	0.0	0.0	0.0	0.0
Births	NA	NA	NA	NA	923	NA	NA	NA	NA
Deaths	NA	NA	NA	NA	141	NA	NA	NA	NA
Life expectancy - males	NA	NA	65.5	65.5	65.9	67.5	69.7	71.6	73.2
Life expectancy - females	NA	NA	67.6	67.6	68.0	70.0	73.1	75.9	78.2
Birth rate (per 1,000)	NA	NA	46.2	43.6	43.1	40.8	36.0	30.6	26.0
Death rate (per 1,000)	NA	NA	7.5	6.8	6.6	5.6	4.5	3.8	3.6
Women of reproductive age (15-49 yrs.)	NA	NA	3,915	4,492	4,688	5,496	8,108	11,392	15,121
of which are currently married[23]	NA	NA	2,665	NA	3,203	3,786	5,636	NA	NA
Fertility rate	NA	NA	7.3	6.6	6.4	5.8	4.8	3.9	3.2

Except as noted, values for vital statistics are in thousands; life expectancy is in years.

Health

Health Indicators [3]

% of population with access to	
safe water (1990-95)	44
adequate sanitation (1990-95)	70
health services (1985-95)	93
% of 1-year-olds immunized (1990-94) against	
TB (tuberculosis)	90
DPT (diphtheria, pertussis, tetanus)	67
polio	67
measles	98
% of contraceptive prevalence (1980-94)	18
ORT use rate (1990-94)	70

Health Expenditures [4]

Total health expenditure, 1990 (official exchange rate)	
Millions of dollars	NA
Dollars per capita	NA
Health expenditures as a percentage of GDP	
Total	NA
Public sector	NA
Private sector	NA
Development assistance for health	
Total aid flows (millions of dollars)[1]	4
Aid flows per capita (dollars)	0.2
Aid flows as a percentage of total health expenditure	NA

For sources, notes, and explanations, see Annotated Source Appendix, page 1061.

445

Human Factors

Women and Children [5]

% of pregnant women immunized (tetanus 1990-94)	60
% of births attended by trained health personnel (1983-94)	50
Maternal mortality rate (1980-92)	120
Under-5 mortality rate (1994)	71
% under-5 moderately/severely underweight (1980-1994)	12

Burden of Disease [6]

Population per physician (1990)	1,658.53
Population per nurse (1990)	98.86
Population per hospital bed (1990)	602.56
AIDS cases per 100,000 people (1994)	0.1
Malaria cases per 100,000 people (1992)	97

Ethnic Division [7]

Arab	75%-80%
Kurdish	15%-20%
Turkoman, Assyrian or other	5%

Religion [8]

Muslim	97%
Shi'a	60%-65%
Sunni	32%-37%
Christian or other	3%

Major Languages [9]

Arabic, Kurdish (official in Kurdish regions), Assyrian, Armenian.

Education

Public Education Expenditures [10]

Million (Dinar)	1980	1985	1990	1991	1992	1994
Total education expenditure	418	551	NA	804	902	NA
as percent of GNP	3.0	4.0	NA	NA	NA	NA
as percent of total govt. expend.	NA	NA	NA	NA	NA	NA
Current education expenditure	NA	NA	NA	711	896	NA
as percent of GNP	NA	NA	NA	NA	NA	NA
as percent of current govt. expend.	NA	NA	NA	NA	NA	NA
Capital expenditure	NA	NA	NA	93	6	NA

Educational Attainment [11]

Illiteracy [12]

In thousands and percent[1]	1990	1995	2000
Illiterate population (15+ yrs.)	4,808	4,848	4,982
Illiteracy rate - total pop. (%)	49.2	45.0	38.0
Illiteracy rate - males (%)	34.5	31.6	26.3
Illiteracy rate - females (%)	64.7	58.4	49.8

Libraries [13]

	Admin. Units	Svc. Pts.	Vols. (000)	Shelving (meters)	Vols. Added	Reg. Users
National (1989)	1	3	NA	NA	NA	NA
Nonspecialized	NA	NA	NA	NA	NA	NA
Public	NA	NA	NA	NA	NA	NA
Higher ed. (1988)	106	106	2,745	NA	NA	807,942
School	NA	NA	NA	NA	NA	NA

Daily Newspapers [14]

	1980	1985	1990	1994
Number of papers	5	6	6	4
Circ. (000)	340[e]	600[e]	650[e]	532

Culture [15]

Cinema (seats per 1,000)	NA
Annual attendance per person	NA
Gross box office receipts (mil. Dinar)	NA
Museums (reporting)	25
Visitors (000)	311
Annual receipts (000 Dinar)	NA

Science and Technology

Scientific/Technical Forces [16]

R&D Expenditures [17]

U.S. Patents Issued [18]

Government and Law

Organization of Government [19]

Long-form name:
Republic of Iraq
Type:
Republic
Independence:
3 October 1932 (from League of Nations mandate under British administration)
National holiday:
Anniversary of the Revolution, 17 July (1968)
Constitution:
22 September 1968, effective 16 July 1970 (provisional Constitution); new constitution drafted in 1990 but not adopted
Legal system:
Based on Islamic law in special religious courts, civil law system elsewhere; has not accepted compulsory ICJ jurisdiction
Executive branch:
President; 2 Vice Presidents; Prime Minister; Deputy Prime Minister; Revolutionary Command Council; Council of Ministers
Legislative branch:
Unicameral: National Assembly (Majlis al-Watani):
Judicial branch:
Court of Cassation

Elections [20]

National Assembly (Majlis al-Watani). Elections last held 24 March 1996; results - percent of vote by party NA; seats - (250 seats total) number of seats by party NA, 30 seats are appointed by Saddam Husayn to represent the three northern provinces of Dahuk, Abril, and As Sulaymaniyah. In norhtern Iraq, A "Kurdish Assembly" was elected in May 1992. It is not recoqnized by the government.

Government Budget [21]

Crime [23]

Military Expenditures and Arms Transfers [22]

	1990	1991	1992	1993	1994
Military expenditures					
Current dollars (mil.)[r,1]	14,210	8,828	NA	NA	NA
1994 constant dollars (mil.)[r,1]	15,820	9,462	NA	NA	NA
Armed forces (000)	1,390	475	407	407	425
Gross national product (GNP)					
Current dollars (mil.)[e]	23,180	11,790	15,660	NA	NA
1994 constant dollars (mil.)[e]	25,800	12,640	16,330	NA	NA
Central government expenditures (CGE)					
1994 constant dollars (mil.)	NA	NA	NA	NA	NA
People (mil.)	18.4	17.9	18.5	19.2	19.9
Military expenditure as % of GNP	61.3	74.9	NA	NA	NA
Military expenditure as % of CGE	NA	NA	NA	NA	NA
Military expenditure per capita (1994 $)	859	528	NA	NA	NA
Armed forces per 1,000 people (soldiers)	75.4	26.5	22.0	21.2	21.4
GNP per capita (1994 $)	1,400	705	885	NA	NA
Arms imports[6]					
Current dollars (mil.)	2,800	0	0	0	0
1994 constant dollars (mil.)	3,116	0	0	0	0
Arms exports[6]					
Current dollars (mil.)	20	0	0	0	0
1994 constant dollars (mil.)	22	0	0	0	0
Total imports[7]					
Current dollars (mil.)	6,526	423	603	520	297
1994 constant dollars (mil.)	7,262	453	628	531	297
Total exports[7]					
Current dollars (mil.)[e]	10,380	468	595	NA	612
1994 constant dollars (mil.)[e]	11,560	502	620	NA	612
Arms as percent of total imports[8]	42.9	0	0	0	0
Arms as percent of total exports[8]	0.2	0	0	0	0

Human Rights [24]

	SSTS	FL	FAPRO	PPCG	APROBC	TPW	PCPTW	STPEP	PHRFF	PRW	ASST	AFL
Observes	P	P		P	P	P	P	P			P	P
	EAFRD	CPR	ESCR	SR	ACHR	MAAE	PVIAC	PVNAC	EAFDAW	TCIDTP	RC	
Observes		P	P	P			P			P		P

P = Party; S = Signatory; see Appendix for meaning of abbreviations.

Labor Force

Total Labor Force [25]

4.4 million (1989)

Labor Force by Occupation [26]

Services	48%
Agriculture	30
Industry	22

Unemployment Rate [27]

Production Sectors

Commercial Energy Production and Consumption

Data are shown in quadrillion (10^{15}) BTUs and percent for 1995
Values for hydroelectric, nuclear, geothermal, solar, and wind power refer to electrical generation.

Production [28]

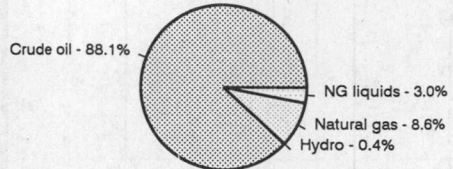

Crude oil - 88.1%
NG liquids - 3.0%
Natural gas - 8.6%
Hydro - 0.4%

Consumption [29]

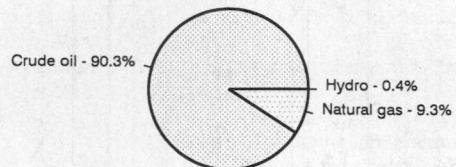

Crude oil - 90.3%
Hydro - 0.4%
Natural gas - 9.3%

Crude oil	1.190
Natural gas liquids	0.040
Dry natural gas	0.116
Net hydroelectric power	0.005
Total	1.351

Crude oil	1.124
Dry natural gas	0.116
Net hydroelectric power	0.005
Total	1.245

Telecommunications [30]

- 632,000 (1987 est.) telephones; reconstitution of damaged telecommunication facilities began after the Gulf war; most damaged facilities have been rebuilt
- Domestic: the network consists of coaxial cables and microwave radio relay links
- International: satellite earth stations - 2 Intelsat (1 Atlantic Ocean and 1 Indian Ocean), 1 Intersputnik (Atlantic Ocean Region) and 1 Arabsat; coaxial cable and microwave radio relay to Jordan, Kuwait, Syria, and Turkey; Kuwait line is probably nonoperational
- Radio: Broadcast stations: AM 16, FM 1, shortwave 0 Radios: 4.02 million (1991 est.)
- Television: Broadcast stations: 13 Televisions: 1 million (1992 est.)

Top Agricultural Products [32]

Produces wheat, barley, rice, vegetables, dates, other fruit, cotton; cattle, sheep.

Top Mining Products [33]

Thousand metric tons except as noted[e]	3/1/95[*]
Cement, hydraulic	12,000
Nitrogen, N content of ammonia	500
Phosphate rock, beneficiated	1,000[46]
Sulfur, elemental	800
Gas, natural, gross (mil. cu. meters)	5,000
Natural gas plant liquids (000 42-gal. bls.)	10,000
Petroleum (000 42-gal. bls.)	
crude (incl. lease condensate)	225,000
refinery products	175,000

Transportation [31]

Railways: total: 2,032 km; standard gauge: 2,032 km 1.435-m gauge

Highways: total: 45,554 km; paved: 38,402 km (including 976 km of expressways); unpaved: 7,152 km (1989 est.)

Merchant marine: total: 36 ships (1,000 GRT or over) totaling 795,346 GRT/1,432,292 DWT; ships by type: cargo 14, oil tanker 16, passenger 1, passenger-cargo 1, refrigerated cargo 1, roll-on/roll-off cargo 3 (1995 est.)

Airports

Total:	102
With paved runways over 3,047 m:	21
With paved runways 2,438 to 3,047 m:	34
With paved runways 1,524 to 2,437 m:	8
With paved runways 914 to 1,523 m:	6
With paved runways under 914 m:	16

Tourism [34]

	1990	1991	1992	1993	1994
Visitors[52]	748	268	504	400	330

Travelers are in thousands, money in million U.S. dollars.

For sources, notes, and explanations, see Annotated Source Appendix, page 1061.

Manufacturing Sector

GDP and Manufacturing Summary [35]

	1980	1985	1989	1990	% change 1980-1990	% change 1989-1990
GDP (million 1980 $)	52,749	39,316	16,272	35,145	-33.4	116.0
GDP per capita (1980 $)	3,969	2,473	890	1,857	-53.2	108.7
Manufacturing as % of GDP (current prices)	4.5	9.2	16.6[e]	NA	NA	-100.0
Gross output (million $)	5,393[e]	7,162	11,771[e]	7,056[e]	30.8	-40.1
Value added (million $)	2,068[e]	3,676	6,119[e]	3,807[e]	84.1	-37.8
Value added (million 1980 $)	2,363	2,520	1,584[e]	1,838	-22.2	16.0
Industrial production index	100	107	126[e]	118[e]	18.0	-6.3
Employment (thousands)	177	174	195[e]	169[e]	-4.5	-13.3

Note: GDP stands for Gross Domestic Product. 'e' stands for estimated value.

Profitability and Productivity

	1980	1985	1989	1990	% change 1980-1990	% change 1989-1990
Intermediate input (%)	62[e]	49	48[e]	46[e]	-25.8	-4.2
Wages, salaries, and supplements (%)	12[e]	13	13[e]	13[e]	8.3	0.0
Gross operating surplus (%)	26[e]	39	39[e]	41[e]	57.7	5.1
Gross output per worker ($)	30,443[e]	41,090	60,482[e]	41,407[e]	36.0	-31.5
Value added per worker ($)	11,673[e]	21,088	31,439[e]	22,386[e]	91.8	-28.8
Average wage (incl. benefits) ($)	3,700	5,242	8,126[e]	5,390[e]	45.7	-33.7

Profitability is in percent of gross output. Productivity is in U.S. $. 'e' stands for estimated value.

Profitability - 1990

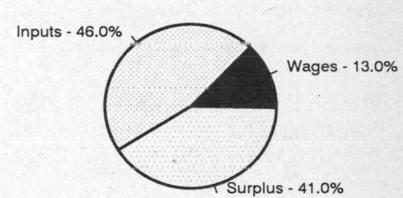

Inputs - 46.0%
Wages - 13.0%
Surplus - 41.0%

The graphic shows percent of gross output.

Value Added in Manufacturing

	1980 $ mil.	1980 %	1985 $ mil.	1985 %	1989 $ mil.	1989 %	1990 $ mil.	1990 %	% change 1980-1990	% change 1989-1990
311 Food products	225[e]	10.9	396	10.8	530[e]	8.7	288[e]	7.6	28.0	-45.7
313 Beverages	74[e]	3.6	125	3.4	162[e]	2.6	112[e]	2.9	51.4	-30.9
314 Tobacco products	105[e]	5.1	140	3.8	226[e]	3.7	124[e]	3.3	18.1	-45.1
321 Textiles	230[e]	11.1	248	6.7	376[e]	6.1	219[e]	5.8	-4.8	-41.8
322 Wearing apparel	30[e]	1.5	53	1.4	74[e]	1.2	45[e]	1.2	50.0	-39.2
323 Leather and fur products	24[e]	1.2	1	0.0	2[e]	0.0	1[e]	0.0	-95.8	-50.0
324 Footwear	19[e]	0.9	81	2.2	88[e]	1.4	49[e]	1.3	157.9	-44.3
331 Wood and wood products	1[e]	0.0	1	0.0	2[e]	0.0	2[e]	0.1	100.0	0.0
332 Furniture and fixtures	10[e]	0.5	13	0.4	30[e]	0.5	14[e]	0.4	40.0	-53.3
341 Paper and paper products	48[e]	2.3	52	1.4	155[e]	2.5	79[e]	2.1	64.6	-49.0
342 Printing and publishing	27[e]	1.3	33	0.9	88[e]	1.4	42[e]	1.1	55.6	-52.3
351 Industrial chemicals	78[e]	3.8	151	4.1	291[e]	4.8	153[e]	4.0	96.2	-47.4
352 Other chemical products	192[e]	9.3	389	10.6	876[e]	14.3	370[e]	9.7	92.7	-57.8
353 Petroleum refineries	392[e]	19.0	868	23.6	1,386[e]	22.7	1,185[e]	31.1	202.3	-14.5
354 Miscellaneous petroleum and coal products	27[e]	1.3	40	1.1	91[e]	1.5	41[e]	1.1	51.9	-54.9
355 Rubber products	6[e]	0.3	10	0.3	21[e]	0.3	11[e]	0.3	83.3	-47.6
356 Plastic products	11[e]	0.5	33	0.9	57[e]	0.9	38[e]	1.0	245.5	-33.3
361 Pottery, china, and earthenware	1[e]	0.0	1	0.0	2[e]	0.0	1[e]	0.0	0.0	-50.0
362 Glass and glass products	21[e]	1.0	35	1.0	51[e]	0.8	31[e]	0.8	47.6	-39.2
369 Other nonmetal mineral products	190[e]	9.2	565	15.4	828[e]	13.5	587[e]	15.4	208.9	-29.1
371 Iron and steel	5[e]	0.2	20[e]	0.5	25[e]	0.4	19[e]	0.5	280.0	-24.0
372 Nonferrous metals	NA	0.0	NA	0.0	NA	0.0	NA	0.0	NA	NA
381 Metal products	55[e]	2.7	47	1.3	105[e]	1.7	75[e]	2.0	36.4	-28.6
382 Nonelectrical machinery	160[e]	7.7	149	4.1	211[e]	3.4	129[e]	3.4	-19.4	-38.9
383 Electrical machinery	122[e]	5.9	185	5.0	340[e]	5.6	150[e]	3.9	23.0	-55.9
384 Transport equipment	12[e]	0.6	40	1.1	103[e]	1.7	42[e]	1.1	250.0	-59.2
385 Professional and scientific equipment	1[e]	0.0	NA	0.0	NA	0.0	NA	0.0	NA	NA
390 Other manufacturing industries	1[e]	0.0	NA	0.0	NA	0.0	NA	0.0	NA	NA

Note: The industry codes shown are International Standard Industry codes (ISIC). Percentages are percent of total Value Added. 'e' stands for estimated value

Finance, Economics, and Trade

Economic Indicators [36]

- **National product**: GDP—purchasing power parity—$41.1 billion (1995 est.)

- **National product real growth rate**: NA%

- **National product per capita**: $2,000 (1995 est.)

- **Inflation rate (consumer prices)**: NA%

- **External debt**: $50 billion (1989 est.), excluding debt of about $35 billion owed to Gulf Arab states

Balance of Payments Summary [37]

Values in millions of dollars.

	1969	1970	1973	1975	1977
Exports of goods (f.o.b.)	NA	1,098	2,204	8,301	10,838
Imports of goods (f.o.b.)	NA	-459	-849	-4,162	-5,867
Trade balance	NA	639	1,355	4,139	4,971
Services - debits	NA	-679	-800	-1,712	-2,707
Services - credits	NA	143	256	543	761
Private transfers (net)	NA	1	1	1	NA
Government transfers (net)	NA	1	-11	-266	-35
Direct investments	NA	18	297	-436	-5
Short-term capital (net)	NA	-2	18	-2,041	-2
Errors and omissions	NA	-127	-254	-726	-510
Overall balance	NA	-6	663	-498	2,473

Exchange Rates [38]

Currency: **Iraqi dinar.**
Symbol: **ID.**

Data are currency units per $1.

Fixed official rate since 1982	3.2169
Semi-official rate	1,000
December 1995 (BMR)	2,900

Imports and Exports

Top Import Origins [39]

$6.6 billion (c.i.f., 1990) Data are for 1990.

Origins	%
Germany	NA
US	NA
Turkey	NA
France	NA
UK	NA

Top Export Destinations [40]

$10.4 billion (f.o.b., 1990) Data are for 1990.

Destinations	%
US	NA
Brazil	NA
Turkey	NA
Japan	NA
Netherlands	NA
Spain	NA

Foreign Aid [41]

Recipient: ODA, $NA.

Import and Export Commodities [42]

Import Commodities	Export Commodities
Manufactures	Crude oil and refined products
Food	Fertilizer
	Sulfur

For sources, notes, and explanations, see Annotated Source Appendix, page 1061.

Ireland

Geography [1]

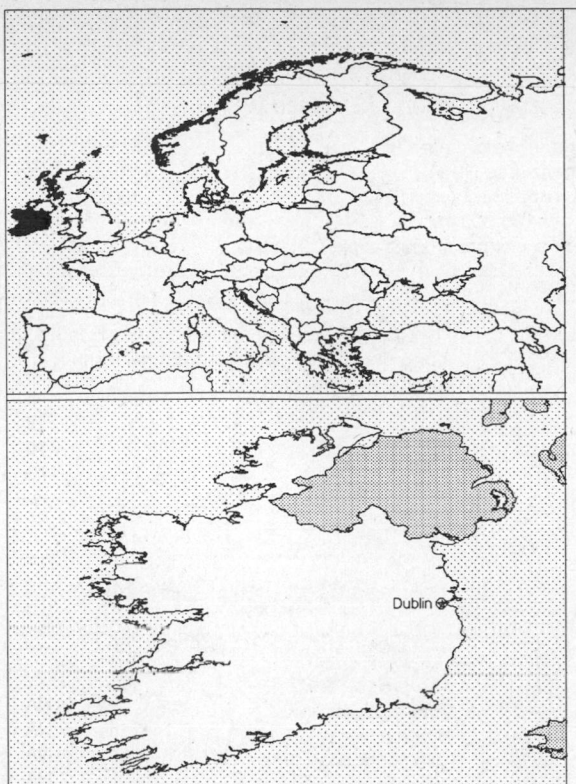

Total area:
70,280 sq km   27,135 sq mi
Land area:
68,890 sq km   26,599 sq mi
Comparative area:
Slightly larger than West Virginia
Land boundaries:
Total 360 km, UK 360 km
Coastline:
1,448 km
Climate:
Temperate maritime; modified by North Atlantic Current; mild winters, cool summers; consistently humid; overcast about half the time
Terrain:
Mostly level to rolling interior plain surrounded by rugged hills and low mountains; sea cliffs on West Coast
Natural resources:
Zinc, lead, natural gas, petroleum, barite, copper, gypsum, limestone, dolomite, peat, silver
Land use:
Arable land: 14%
Permanent crops: 0%
Meadows and pastures: 71%
Forest and woodland: 5%
Other: 10%

Demographics [2]

	1970	1980	1990	1995[1]	1996	2000	2010	2020	2030
Population	2,950	3,401	3,508	3,571	3,567	3,493	3,452	3,570	3,606
Population density (persons per sq. mi.)	111	128	132	134	134	131	130	134	136
(persons per sq. km.)	43	49	51	52	52	51	50	52	52
Net migration rate (per 1,000 population)	NA	NA	-4.5	-5.1	-6.5	-9.8	0.0	0.0	0.0
Births	64	74	NA	NA	49	NA	NA	NA	NA
Deaths	34	33	32	NA	30	NA	NA	NA	NA
Life expectancy - males	NA	NA	71.7	72.7	72.9	73.6	75.1	77.7	79.3
Life expectancy - females	NA	NA	77.5	78.3	78.5	79.1	80.4	83.3	85.2
Birth rate (per 1,000)	NA	NA	15.1	13.9	13.2	12.6	12.6	10.9	9.2
Death rate (per 1,000)	NA	NA	9.1	9.0	8.9	8.9	8.9	8.4	9.5
Women of reproductive age (15-49 yrs.)	NA	NA	851	919	924	907	852	835	783
of which are currently married[24]	NA	NA	471	NA	504	523	575	NA	NA
Fertility rate	NA	NA	2.1	1.9	1.8	1.7	1.7	1.6	1.6

Except as noted, values for vital statistics are in thousands; life expectancy is in years.

Health

Health Indicators [3]

% of population with access to	
safe water (1990-95)	NA
adequate sanitation (1990-95)	NA
health services (1985-95)	NA
% of 1-year-olds immunized (1990-94) against	
TB (tuberculosis)	NA
DPT (diphtheria, pertussis, tetanus)	65
polio	63
measles	78
% of contraceptive prevalence (1980-94)	NA
ORT use rate (1990-94)	NA

Health Expenditures [4]

Total health expenditure, 1990 (official exchange rate)	
Millions of dollars	3,068
Dollars per capita	876
Health expenditures as a percentage of GDP	
Total	7.1
Public sector	5.8
Private sector	1.4
Development assistance for health	
Total aid flows (millions of dollars)[1]	NA
Aid flows per capita (dollars)	NA
Aid flows as a percentage of total health expenditure	NA

For sources, notes, and explanations, see Annotated Source Appendix, page 1061.

Human Factors

Women and Children [5]

% of pregnant women immunized (tetanus 1990-94)	NA
% of births attended by trained health personnel (1983-94)	NA
Maternal mortality rate (1980-92)	2
Under-5 mortality rate (1994)	7
% under-5 moderately/severely underweight (1980-1994)	NA

Burden of Disease [6]

Population per physician (1988)	631.66
Population per nurse (1984)	139.70
Population per hospital bed (1990)	100.61
AIDS cases per 100,000 people (1994)	1.5
Heart disease cases per 1,000 people (1990-93)	336.5

Ethnic Division [7]

Celtic, English.

Religion [8]

Roman Catholic	93%
Anglican	3%
None	1%
Unknown	2%
Other	1%
(1981)	

Major Languages [9]

Irish (Gaelic), spoken mainly in areas located along the western seaboard, English is the language generally used.

Education

Public Education Expenditures [10]

Million (Pound)	1980	1990	1991	1992	1993	1994
Total education expenditure	595	1,372	1,479	1,645	1,816	NA
as percent of GNP	6.3	5.7	5.8	6.2	6.4	NA
as percent of total govt. expend.	NA	10.2	9.7	NA	NA	NA
Current education expenditure	515	1,303	1,416	1,568	1,734	NA
as percent of GNP	5.4	5.4	5.6	5.9	6.1	NA
as percent of current govt. expend.	NA	12.1	12.2	NA	NA	NA
Capital expenditure	80	68	63	77	82	NA

Educational Attainment [11]

Age group (1991)	25+
Total population	1,983,547
Highest level attained (%)	
No schooling	0.0
First level	
Not completed	0.0
Completed	38.5
Entered second level	
S-1	43.7
S-2	NA
Postsecondary	14.6

Illiteracy [12]

Libraries [13]

	Admin. Units	Svc. Pts.	Vols. (000)	Shelving (meters)	Vols. Added	Reg. Users
National (1990)	1	2	750	14,530	10,000	28,300
Nonspecialized (1990)	2	4	383	5,993[e]	344	10,323
Public (1991)	NA	364	11,046	NA	NA	808,042
Higher ed. (1990)	15[e]	33	5,018	177,832	113,000	51,049[e]
School	NA	NA	NA	NA	NA	NA

Daily Newspapers [14]

	1980	1985	1990	1994
Number of papers	7	7	7	8
Circ. (000)	779	685	591	600

Culture [15]

Science and Technology

Scientific/Technical Forces [16]

Scientists/engineers	6,592
Number female	NA
Technicians	1,797
Number female	NA
Total	8,389

R&D Expenditures [17]

	Pound (000) 1993
Total expenditure	400,201
Capital expenditure	NA
Current expenditure	NA
Percent current	NA

U.S. Patents Issued [18]

Values show patents issued to citizens of the country by the U.S. Patents Office.

	1993	1994	1995
Number of patents	59	52	61

Government and Law

Organization of Government [19]

Long-form name:
None
Type:
Republic
Independence:
6 December 1921 (from UK)
National holiday:
Saint Patrick's Day, 17 March
Constitution:
29 December 1937; adopted 1 July 1937
by plebiscite
Legal system:
Based on English common law,
substantially modified by indigenous
concepts; judicial review of legislative acts
in Supreme Court; has not accepted
compulsory ICJ jurisdiction
Executive branch:
President; Prime Minister; Cabinet
Legislative branch:
Bicameral Parliament (Oireachtas):
Senate (Seanad Eireann) and House of
Representatives (Dail Eireann)
Judicial branch:
Supreme Court

Elections [20]

House of Representatives	% of votes
Fianna Fail	39.1
Fine Gael	24.5
Labor Party	19.3
Progressive Democrats	4.7
Democratic Left	2.8
Sinn Fein	1.6
Workers' Party	0.7
Independents	5.9

Government Expenditures [21]

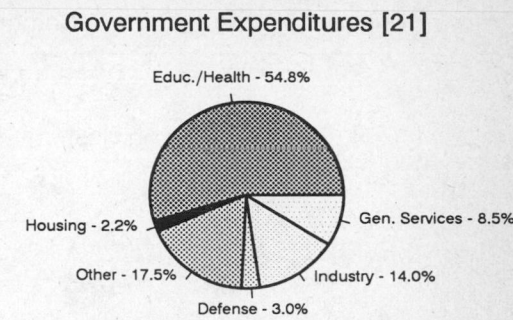

Educ./Health - 54.8%
Gen. Services - 8.5%
Industry - 14.0%
Defense - 3.0%
Other - 17.5%
Housing - 2.2%

(% distribution). Expend. for CY93 est.: 13,677 (Pound mil.)

Military Expenditures and Arms Transfers [22]

	1990	1991	1992	1993	1994
Military expenditures					
Current dollars (mil.)	515	573	595	543	608
1994 constant dollars (mil.)	574	614	621	554	608
Armed forces (000)	13	13	13	13	17
Gross national product (GNP)					
Current dollars (mil.)	34,400	37,400	40,120	42,350	45,810
1994 constant dollars (mil.)	38,280	40,090	41,820	43,220	45,810
Central government expenditures (CGE)					
1994 constant dollars (mil.)	14,470	15,340	16,480	16,850	17,950
People (mil.)	3.5	3.5	3.5	3.5	3.5
Military expenditure as % of GNP	1.5	1.5	1.5	1.3	1.3
Military expenditure as % of CGE	4.0	4.0	3.8	3.3	3.4
Military expenditure per capita (1994 $)	163	175	176	157	172
Armed forces per 1,000 people (soldiers)	3.7	3.7	3.7	3.7	4.8
GNP per capita (1994 $)	10,910	11,410	11,880	12,240	12,940
Arms imports[6]					
Current dollars (mil.)	5	10	5	5	60
1994 constant dollars (mil.)	6	11	5	5	60
Arms exports[6]					
Current dollars (mil.)	0	10	0	0	0
1994 constant dollars (mil.)	0	11	0	0	0
Total imports[7]					
Current dollars (mil.)	20,670	20,770	22,480	21,390	25,480
1994 constant dollars (mil.)	23,000	22,260	23,440	21,830	25,480
Total exports[7]					
Current dollars (mil.)	23,740	24,220	28,330	28,610	34,320
1994 constant dollars (mil.)	26,420	25,970	29,540	29,200	34,320
Arms as percent of total imports[8]	0	0	0	0	0.2
Arms as percent of total exports[8]	0	0	0	0	0

Crime [23]

	1994
Crime volume	
Cases known to police	101,036
Attempts (percent)	NA
Percent cases solved	38.70
Crimes per 100,000 persons	2,867.01
Persons responsible for offenses	
Total number offenders	5,849
Percent female	10.70
Percent juvenile (7-17 yrs.)	11.90
Percent foreigners	NA

Human Rights [24]

	SSTS	FL	FAPRO	PPCG	APROBC	TPW	PCPTW	STPEP	PHRFF	PRW	ASST	AFL
Observes	P	P	P	P	P	P	P		P	P	P	P
	EAFRD	CPR	ESCR	SR	ACHR	MAAE	PVIAC	PVNAC	EAFDAW	TCIDTP	RC	
Observes	S	P	P	P		P	S	S	P	S	P	

P = Party; S = Signatory; see Appendix for meaning of abbreviations.

Labor Force

Total Labor Force [25]

1.37 million

Labor Force by Occupation [26]

Services	57.0%
Manufacturing and construction	28
Agriculture, forestry, and fishing	13.5
Energy and mining	1.5

Date of data: 1992

Unemployment Rate [27]

13.5% (1995 est.)

For sources, notes, and explanations, see Annotated Source Appendix, page 1061.

453

Production Sectors

Commercial Energy Production and Consumption

Data are shown in quadrillion (10^{15}) BTUs and percent for 1995
Values for hydroelectric, nuclear, geothermal, solar, and wind power refer to electrical generation.

Production [28]

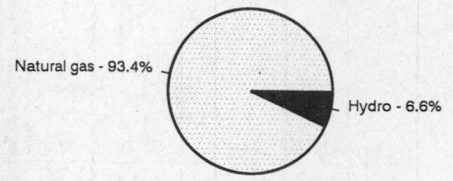

Natural gas - 93.4%
Hydro - 6.6%

Dry natural gas	0.099
Net hydroelectric power	0.007
Total	0.106

Consumption [29]

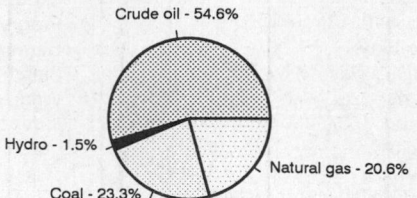

Crude oil - 54.6%
Hydro - 1.5%
Natural gas - 20.6%
Coal - 23.3%

Crude oil	0.262
Dry natural gas	0.099
Coal	0.112
Net hydroelectric power	0.007
Total	0.480

Telecommunications [30]

- 900,000 (1987 est.) telephones; modern digital system using cable and microwave radio relay
- Domestic: microwave radio relay
- International: satellite earth station - 1 Intelsat (Atlantic Ocean)
- Radio: Broadcast stations: AM 9, FM 45, shortwave 0 Radios: 2.2 million (1991 est.)
- Television: Broadcast stations: 86 (1987 est.) Televisions: 1.025 million (1990 est.)

Top Agricultural Products [32]

Agriculture accounts for 6.8% of the GDP; produces turnips, barley, potatoes, sugar beets, wheat; meat and dairy products.

Transportation [31]

Railways: total: 1,944 km; broad gauge: 1,944 km 1.600-m gauge (37 km electrified; 485 km double track) (1995)

Highways: total: 92,327 km; paved: 86,787 km (including 32 km of expressways); unpaved: 5,540 km (1992 est.)

Merchant marine: total: 42 ships (1,000 GRT or over) totaling 129,027 GRT/155,371 DWT; ships by type: bulk 4, cargo 27, chemical tanker 1, container 3, oil tanker 2, short-sea passenger 3, specialized tanker 2 (1995 est.)

Airports

Total:	40
With paved runways over 3,047 m:	1
With paved runways 2,438 to 3,047 m:	1
With paved runways 1,524 to 2,437 m:	3
With paved runways 914 to 1,523 m:	3
With paved runways under 914 m:	29

Top Mining Products [33]

Thousand metric tons except as noted	12/94[*]
Alumina	1,000
Silver, mine output, Ag content (kg.)	15,000
Cement, hydraulic	2,000[e]
Lime	150,000[e]
Sand and gravel	10,000[e,47]
Limestone	10,000
Peat, for fuel use	9,000[e]
Petroleum refinery products (000 42-gal. bls.)	15,500[e,48]

Tourism [34]

	1990	1991	1992	1993	1994
Tourists	3,666	3,571	3,724	3,888	4,309
Tourism receipts[53]	1,453	1,514	1,620	1,600	1,765
Tourism expenditures[54]	1,163	1,128	1,361	1,224	1,575
Fare receipts[55]	436	444	478	407	433
Fare expenditures[56]	277	292	319	272	296

Travelers are in thousands, money in million U.S. dollars.

For sources, notes, and explanations, see Annotated Source Appendix, page 1061.

Manufacturing Sector

Manufacturing Summary [35]

	1987		1988		1989		1990		1991	
	$ bil.	%	$ bil.	%	$ bil.	%	$ bil.	%	$ bil.	%
Establishments or enterprises (number)	4,790	0.204	4,780	0.228	4,751	0.254	-	-	-	-
Total employment (000)	188	0.138	191	0.139	196	0.159	-	-	-	-
Production workers (000)	143	0.213	145	0.233	149	0.202	-	-	-	-
Output ($ bil.)	24	0.239	29	0.245	30	0.256	-	-	-	-
Value added ($ bil.)	10	0.241	12	0.257	13	0.275	-	-	-	-
Capital investment ($ mil.)	666	0.153	901	0.189	1,020	0.186	-	-	-	-
M & E investment ($ mil.)	516	0.166	671	0.194	765	0.196	-	-	-	-
Employees per establishment (number)	39	67.733	40	61.259	41	62.786	-	-	-	-
Production workers per establishment	30	104.198	30	102.275	31	79.640	-	-	-	-
Output per establishment ($ mil.)	5	116.892	6	107.568	6	100.924	-	-	-	-
Capital investment per estab. ($ mil.)	0.139	74.829	0.189	83.134	0.215	73.427	-	-	-	-
M & E per establishment ($ mil)	0.108	81.109	0.140	85.104	0.161	77.267	-	-	-	-
Payroll per employee ($)	16,364	182.487	17,585	176.514	16,684	149.333	-	-	-	-
Wages per production worker ($)	14,860	188.320	15,803	187.020	15,504	154.597	-	-	-	-
Hours per production worker (hours)	-	-	-	-	-	-	-	-	-	-
Output per employee ($)	129,251	172.577	149,564	175.596	154,056	160.744	-	-	-	-
Capital investment per employee ($)	3,548	110.475	4,722	135.708	5,205	116.949	-	-	-	-
M & E per employee ($)	2,749	119.747	3,513	138.925	3,904	123.065	-	-	-	-

Note: Columns headed % show percent of world total or ratio. Ratios closest to 100 are closest to world average. M & E stands for machinery & equipment.

Output in Manufacturing

	1987		1988		1989		1990		1991	
	$ bil.[f]	%	$ bil.[f]	%	$ bil.[f]	%	$ bil.[f]	%	$ bil.[f]	%
3110 - Food products	7.987	33.28	8.801	30.35	8.981	29.94	-	-	-	-
3130 - Beverages	0.844	3.52	0.990	3.41	1.009	3.36	-	-	-	-
3140 - Tobacco	0.186	0.78	0.195	0.67	0.182	0.61	-	-	-	-
3210 - Textiles	0.704	2.93	0.760	2.62	0.751	2.50	-	-	-	-
3220 - Wearing apparel	0.359	1.50	0.354	1.22	0.336	1.12	-	-	-	-
3230 - Leather and products	0.077	0.32	0.084	0.29	0.089	0.30	-	-	-	-
3240 - Footwear	0.040	0.17	0.043	0.15	0.035	0.12	-	-	-	-
3310 - Wood products	0.280	1.17	0.325	1.12	0.333	1.11	-	-	-	-
3320 - Furniture, fixtures	0.134	0.56	0.151	0.52	0.166	0.55	-	-	-	-
3410 - Paper and products	0.318	1.33	0.382	1.32	0.390	1.30	-	-	-	-
3420 - Printing, publishing	0.576	2.40	0.685	2.36	0.714	2.38	-	-	-	-
3550 - Rubber products	0.159	0.66	0.183	0.63	0.193	0.64	-	-	-	-
3560 - Plastic products nec	0.458	1.91	0.583	2.01	0.603	2.01	-	-	-	-
3610 - Pottery, china, etc.	0.040	0.17	0.037	0.13	0.040	0.13	-	-	-	-
3620 - Glass and products	0.217	0.90	0.249	0.86	0.254	0.85	-	-	-	-
3690 - Nonmetal products nec	0.725	3.02	0.865	2.98	0.973	3.24	-	-	-	-
3710 - Iron and steel	0.153	0.64	0.195	0.67	0.227	0.76	-	-	-	-
3720 - Nonferrous metals	0.045	0.19	0.044	0.15	0.044	0.15	-	-	-	-
3810 - Metal products	0.728	3.03	0.829	2.86	0.857	2.86	-	-	-	-
3820 - Machinery nec	3.228	13.45	3.829	13.20	4.114	13.71	-	-	-	-
3825 - Office, computing machinery	2.638	10.99	3.153	10.87	3.397	11.32	-	-	-	-
3830 - Electrical machinery	1.787	7.45	2.488	8.58	2.784	9.28	-	-	-	-
3832 - Radio, television, etc.	1.196	4.98	1.764	6.08	2.031	6.77	-	-	-	-
3840 - Transportation equipment	0.415	1.73	0.476	1.64	0.521	1.74	-	-	-	-
3841 - Shipbuilding, repair	0.028	0.12	0.038	0.13	0.037	0.12	-	-	-	-
3843 - Motor vehicles	0.153	0.64	0.176	0.61	0.219	0.73	-	-	-	-
3850 - Professional goods	0.629	2.62	0.716	2.47	0.742	2.47	-	-	-	-
3900 - Industries nec	0.155	0.65	0.157	0.54	0.173	0.58	-	-	-	-

Note: Codes are International Standard Industry codes (ISIC). Percentages are % of total Output. [f]: Factor Prices; [p]: Producer Prices.

For sources, notes, and explanations, see Annotated Source Appendix, page 1061.

455

Finance, Economics, and Trade

Economic Indicators [36]

- **National product**: GDP—purchasing power parity—$54.6 billion (1995 est.)
- **National product real growth rate**: 7% (1995 est.)
- **National product per capita**: $15,400 (1995 est.)
- **Inflation rate (consumer prices)**: 2.8% (1995 est.)
- **External debt**: $19.5 billion (1994 est.)

Balance of Payments Summary [37]

Values in millions of dollars.

	1989	1990	1991	1992	1993
Exports of goods (f.o.b.)	20,356	23,356	23,660	27,905	28,729
Imports of goods (f.o.b.)	-16,352	-19,387	-19,493	-21,092	-20,557
Trade balance	4,004	3,969	4,167	6,813	8,172
Services - debits	-10,417	-12,499	-12,175	-13,981	-13,217
Services - credits	4,328	5,958	6,265	6,657	5,874
Private transfers (net)	-93	-63	-56	-59	-51
Government transfers (net)	1,670	2,680	3,242	3,022	2,867
Long-term capital (net)	-1,216	-2,999	-2,541	-4,013	879
Short-term capital (net)	-358	1,094	-660	-2,451	-1,269
Errors and omissions	1,145	2,608	2,221	470	659
Overall balance	-937	748	463	-3,542	3,914

Exchange Rates [38]

Currency: **Irish pound.**
Symbol: **#Ir.**

Data are currency units per $1.

January 1996	0.6315
1995	0.6235
1994	0.6676
1993	0.6816
1992	0.5864
1991	0.6190

Imports and Exports

Top Import Origins [39]

$25.3 billion (c.i.f., 1994).

Origins	%
EU	58
UK	36
Germany	7
France	4
US	18

Top Export Destinations [40]

$29.9 billion (f.o.b., 1994).

Destinations	%
EU	73
UK	27
Germany	14
France	9
US	9

Foreign Aid [41]

Donor: ODA, $81 million (1993).

Import and Export Commodities [42]

Import Commodities
Food
Animal feed
Data processing equipment
Petroleum and petroleum products
Machinery
Textiles
Clothing

Export Commodities
Chemicals
Data processing equipment
Industrial machinery
Live animals
Animal products

For sources, notes, and explanations, see Annotated Source Appendix, page 1061.

Israel

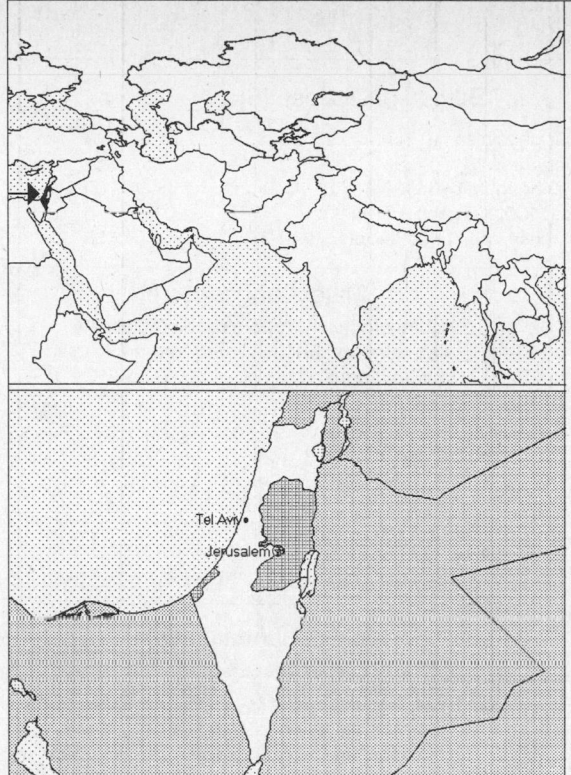

Geography [1]

Total area:
20,770 sq km   8,019 sq mi
Land area:
20,330 sq km   7,849 sq mi
Comparative area:
Slightly larger than New Jersey
Land boundaries:
Total 1,006 km, Egypt 255 km, Gaza Strip 51 km, Jordan 238 km, Lebanon 79 km, Syria 76 km, West Bank 307 km
Coastline:
273 km
Climate:
Temperate; hot and dry in southern and eastern desert areas
Terrain:
Negev desert in the South; low coastal plain; central mountains; Jordan Rift Valley
Natural resources:
Copper, phosphates, bromide, potash, clay, sand, sulfur, asphalt, manganese, small amounts of natural gas and crude oil
Land use:
Arable land: 17%
Permanent crops: 5%
Meadows and pastures: 40%
Forest and woodland: 6%
Other: 32%

Demographics [2]

	1970	1980	1990	1995[1]	1996	2000	2010	2020	2030
Population	2,903	3,737	4,512	5,306	5,422	5,852	6,696	7,439	8,087
Population density (persons per sq. mi.)	370	476	575	676	691	746	853	948	1,030
(persons per sq. km.)	143	184	222	261	267	288	329	366	398
Net migration rate (per 1,000 population)	NA	NA	41.5	8.0	7.0	3.6	0.0	0.0	0.0
Births	NA	NA	NA	NA	106	NA	NA	NA	NA
Deaths	NA	NA	29	NA	33	NA	NA	NA	NA
Life expectancy - males	NA	NA	75.4	76.0	76.2	76.9	78.3	79.2	79.8
Life expectancy - females	NA	NA	79.1	79.8	80.0	80.8	82.6	83.8	84.7
Birth rate (per 1,000)	NA	NA	21.7	20.5	20.3	19.6	17.6	15.7	14.4
Death rate (per 1,000)	NA	NA	6.2	6.3	6.3	6.1	6.0	6.2	7.2
Women of reproductive age (15-49 yrs.)	NA	NA	1,098	1,334	1,364	1,462	1,626	1,795	1,888
of which are currently married[25]	NA	623	696	NA	886	935	1,044	NA	NA
Fertility rate	NA	NA	3.0	2.8	2.8	2.6	2.4	2.2	2.1

Except as noted, values for vital statistics are in thousands; life expectancy is in years.

Health

Health Indicators [3]

% of population with access to	
safe water (1990-95)	NA
adequate sanitation (1990-95)	NA
health services (1985-95)	NA
% of 1-year-olds immunized (1990-94) against	
TB (tuberculosis)	NA
DPT (diphtheria, pertussis, tetanus)	92
polio	93
measles	95
% of contraceptive prevalence (1980-94)	NA
ORT use rate (1990-94)	NA

Health Expenditures [4]

Total health expenditure, 1990 (official exchange rate)	
Millions of dollars	2,301
Dollars per capita	494
Health expenditures as a percentage of GDP	
Total	4.2
Public sector	2.1
Private sector	2.1
Development assistance for health	
Total aid flows (millions of dollars)[1]	3
Aid flows per capita (dollars)	0.6
Aid flows as a percentage of total health expenditure	0.1

For sources, notes, and explanations, see Annotated Source Appendix, page 1061.

457

Human Factors

Women and Children [5]

% of pregnant women immunized (tetanus 1990-94)	NA
% of births attended by trained health personnel (1983-94)	99
Maternal mortality rate (1980-92)	3
Under-5 mortality rate (1994)	9
% under-5 moderately/severely underweight (1980-1994)	NA

Burden of Disease [6]

Population per physician	NA
Population per nurse	NA
Population per hospital bed (1994)	164.00
AIDS cases per 100,000 people (1994)	0.6
Heart disease cases per 1,000 people (1990-93)	332

Ethnic Division [7]

Jewish	82%
Israel-born	50%
Europe/Americas/Oceania-born	20%
Africa-born	7%
Asia-born	5%
Non-Jewish (mostly Arab)	18%

Religion [8]

Judaism	82%
Islam (mostly Sunni Muslim)	14%
Christian	2%
Druze and other	2%

Major Languages [9]

Hebrew (official), Arabic used officially for Arab minority, English most commonly used foreign language.

Education

Public Education Expenditures [10]

	1980	1985	1990	1991	1992	1994
Million (Shekel)						
Total education expenditure	9	1,876	6,239	8,157	10,182	NA
as percent of GNP	7.9	6.4	5.8	5.8	6.0	NA
as percent of total govt. expend.	7.3	8.6	10.5	10.6	11.1	NA
Current education expenditure	8	1,720	5,713	7,420	9,266	NA
as percent of GNP	7.3	5.9	5.3	5.3	5.5	NA
as percent of current govt. expend.	8.9	8.3	10.4	10.9	11.7	NA
Capital expenditure	1	156	526	737	916	NA

Educational Attainment [11]

Age group (1983)	25+
Total population	2,043,720
Highest level attained (%)	
No schooling	10.5
First level	
Not completed	42.4
Completed	NA
Entered second level	
S-1	35.9
S-2	NA
Postsecondary	11.2

Illiteracy [12]

	1989	1992	1995
Illiterate population (15+ yrs.)[2]	NA	183,200	176,000
Illiteracy rate - total pop. (%)	NA	5.1	4.4
Illiteracy rate - males (%)	NA	2.9	2.3
Illiteracy rate - females (%)	NA	7.3	6.4

Libraries [13]

Daily Newspapers [14]

	1980	1985	1990	1994
Number of papers	36	21	30	34
Circ. (000)	1,000[e]	1,100[e]	1,200[e]	1,534

Culture [15]

Cinema (seats per 1,000)	11.5
Annual attendance per person	1.8
Gross box office receipts (mil. Shekel)	NA
Museums (reporting)	NA
Visitors (000)	NA
Annual receipts (000 Shekel)	NA

Science and Technology

Scientific/Technical Forces [16]

Scientists/engineers	20,100
Number female[19]	10,400
Technicians	4,300
Number female	1,400
Total	24,400

R&D Expenditures [17]

	Shekel (000) 1992
Total expenditure[20]	3,663,100
Capital expenditure	200,600
Current expenditure	3,462,500
Percent current	94.5

U.S. Patents Issued [18]

Values show patents issued to citizens of the country by the U.S. Patents Office.

	1993	1994	1995
Number of patents	358	388	432

Government and Law

Organization of Government [19]

Long-form name:
State of Israel

Type:
Republic

Independence:
14 May 1948 (from League of Nations mandate under British administration)

National holiday:
Independence Day, 14 May 1948 (as the Jewish calendar is lunar, the holiday may occur in April or May)

Constitution:
No formal constitution; some of the functions of a constitution are filled by the Declaration of Establishment (1948), the basic laws of the parliament (Knesset), and the Israeli citizenship law

Legal system:
Mixture of English common law, British Mandate regulations, and, in personal matters, Jewish, Christian, and Muslim legal systems; no longer accepts compulsory ICJ jurisdiction

Executive branch:
President; Prime Minister; Cabinet

Legislative branch:
Unicameral: Parliament (Knesset)

Judicial branch:
Supreme Court

Elections [20]

Knesset	% of seats
Labor Party	36.7
Likud bloc	26.7
Meretz	10.0
Tzomet	6.7
National Religious Party	5.0
Shas	5.0
Tzomet	4.2
United Torah Jewry	3.3
Others	9.2

Government Expenditures [21]

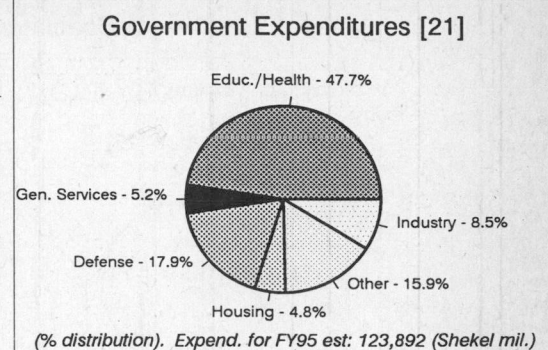

Educ./Health - 47.7%
Gen. Services - 5.2%
Industry - 8.5%
Defense - 17.9%
Other - 15.9%
Housing - 4.8%

(% distribution). Expend. for FY95 est: 123,892 (Shekel mil.)

Crime [23]

	1994
Crime volume	
Cases known to police	282,188
Attempts (percent)	NA
Percent cases solved	30.60
Crimes per 100,000 persons	5,191.00
Persons responsible for offenses	
Total number offenders	75,789
Percent female	9,795
Percent juvenile (12-18 yrs.)	6,477
Percent foreigners	NA

Military Expenditures and Arms Transfers [22]

	1990	1991	1992	1993	1994
Military expenditures					
Current dollars (mil.)	6,721	4,823	6,626	6,382	6,588
1994 constant dollars (mil.)	7,479	5,170	6,909	6,514	6,588
Armed forces (000)	190	190	181	181	185
Gross national product (GNP)					
Current dollars (mil.)	53,280	59,910	65,970	70,390	76,810
1994 constant dollars (mil.)	59,290	64,210	68,800	71,840	76,810
Central government expenditures (CGE)					
1994 constant dollars (mil.)	30,190	25,160	32,390	32,080	34,330
People (mil.)	4.3	4.5	4.7	4.9	5.1
Military expenditure as % of GNP	12.6	8.1	10.0	9.1	8.6
Military expenditure as % of CGE	24.8	20.5	21.3	20.3	19.2
Military expenditure per capita (1994 $)	1,738	1,139	1,455	1,324	1,304
Armed forces per 1,000 people (soldiers)	44.2	41.9	38.1	36.8	36.6
GNP per capita (1994 $)	13,780	14,150	14,490	14,600	15,210
Arms imports[6]					
Current dollars (mil.)	700	625	850	1,100	1,000
1994 constant dollars (mil.)	779	670	886	1,123	1,000
Arms exports[6]					
Current dollars (mil.)	330	430	280	340	470
1994 constant dollars (mil.)	367	461	292	347	470
Total imports[7]					
Current dollars (mil.)	16,790	18,660	20,250	22,620	24,240
1994 constant dollars (mil.)	18,690	20,000	21,120	23,090	24,240
Total exports[7]					
Current dollars (mil.)	11,580	11,920	13,120	14,780	16,440
1994 constant dollars (mil.)	12,880	12,780	13,680	15,080	16,440
Arms as percent of total imports[8]	4.2	3.3	4.2	4.9	4.1
Arms as percent of total exports[8]	2.9	3.6	2.1	2.3	2.9

Human Rights [24]

	SSTS	FL	FAPRO	PPCG	APROBC	TPW	PCPTW	STPEP	PHRFF	PRW	ASST	AFL
Observes	P	P	P	P	P	P	P	P		P	P	P
	EAFRD	CPR	ESCR	SR	ACHR	MAAE	PVIAC	PVNAC	EAFDAW	TCIDTP	RC	
Observes	P	P	P	P		P	P		P	P	P	

P = Party; S = Signatory; see Appendix for meaning of abbreviations.

Labor Force

Total Labor Force [25]

1.9 million (1992)

Labor Force by Occupation [26]

Public services	29.3%
Industry	22.1
Commerce	13.9
Finance and business	10.4
Personal and other services	7.4
Construction	6.5
Transport, storage, communications	6.3
Agriculture, forestry, fishing	3.5

Date of data: 1992

Unemployment Rate [27]

6.3% (1995 est.)

For sources, notes, and explanations, see Annotated Source Appendix, page 1061.

459

Production Sectors

Commercial Energy Production and Consumption

Data are shown in quadrillion (10^{15}) BTUs and percent for 1995
Values for hydroelectric, nuclear, geothermal, solar, and wind power refer to electrical generation.

Production [28]

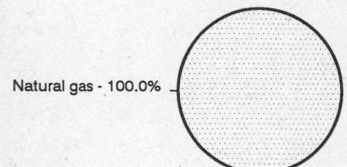

Natural gas - 100.0%

Consumption [29]

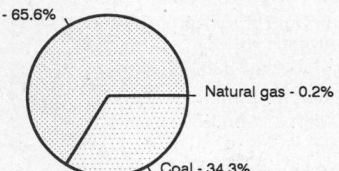

Crude oil - 65.6%

Natural gas - 0.2%

Coal - 34.3%

Dry natural gas	0.001
Total	0.001

Crude oil	0.432
Dry natural gas	0.001
Coal	0.226
Total	0.659

Telecommunications [30]

- 2.425 million (1990 est.) telephones; most highly developed system in the Middle East although not the largest
- Domestic: good system of coaxial cable and microwave radio relay
- International: 3 submarine cables; satellite earth stations - 3 Intelsat (2 Atlantic Ocean and 1 Indian Ocean)
- Radio: Broadcast stations: AM 9, FM 45, shortwave 0 Radios: 2.25 million (1993 est.)
- Television: Broadcast stations: 20 Televisions: 1.5 million (1993 est.)

Transportation [31]

Railways: total: 526 km; standard gauge: 526 km 1.435-m gauge

Highways: total: 13,461 km; paved: 13,461 km (including 56 km of expressways); unpaved: 0 km (1992 est.)

Merchant marine: total: 28 ships (1,000 GRT or over) totaling 577,747 GRT/701,459 DWT; ships by type: cargo 5, container 20, refrigerated cargo 2, roll-on/roll-off cargo 1 (1995 est.)

Airports

Total:	50
With paved runways over 3,047 m:	2
With paved runways 2,438 to 3,047 m:	6
With paved runways 1,524 to 2,437 m:	7
With paved runways 914 to 1,523 m:	8
With paved runways under 914 m:	22

Top Agricultural Products [32]

Agriculture accounts for 3.5% of the GDP; produces citrus and other fruits, vegetables, cotton; beef, poultry, dairy products.

Top Mining Products [33]

Thousand metric tons except as noted[e]	9/14/95[a]
Clays	
flint	40,000
kaolin	40,000
Potash, K20 equivalent (mil. metric tons)	1,300
Glass sand	83,000
Stone, crushed	31,500
Dimension marble	12,000
Gas, natural (000 cu. meters)	32,300
Petroleum refinery products (000 42-gal. bls.)	58,000

Tourism [34]

	1990	1991	1992	1993	1994
Visitors[57]	1,342	1,110	1,805	1,946	2,168
Tourists[37]	1,063	943	1,509	1,656	1,839
Cruise passengers	210	167	296	290	329
Tourism receipts	1,382	1,306	1,842	2,110	2,266
Tourism expenditures	1,442	1,551	1,674	2,313	2,896
Fare receipts	357	369	450	428	397
Fare expenditures	195	235	270	361	467

Travelers are in thousands, money in million U.S. dollars.

Manufacturing Sector

Manufacturing Summary [35]

	1987		1988		1989		1990		1991	
	$ bil.	%	$ bil.	%	$ bil.	%	$ bil.	%	$ bil.	%
Establishments or enterprises (number)	7,736	0.330	7,754	0.369	5,793	0.309	9,276	0.518	9,270	1.213
Total employment (000)	352	0.259	340	0.249	179	0.146	340	0.307	184	0.265
Production workers (000)	-	.	-	.	-	.	-	.	-	.
Output ($ bil.)	20	0.201	23	0.199	13	0.111	27	0.239	16	0.156
Value added ($ bil.)	9	0.208	10	0.217	-	-	12	0.244	-	-
Capital investment ($ mil.)	1,298	0.298	1,215	0.255	-	-	1,286	0.231	-	-
M & E investment ($ mil.)	1,112	0.357	1,028	0.297	-	-	1,060	0.251	-	-
Employees per establishment (number)	45	78.561	44	67.337	31	47.131	37	59.374	20	21.841
Production workers per establishment	-	.	-	.	-	.	-	.	-	.
Output per establishment ($ mil.)	3	60.997	3	53.865	2	35.778	3	46.217	2	12.859
Capital investment per estab. ($ mil.)	0.168	90.319	0.157	69.049	-	-	0.139	44.669	-	-
M & E per establishment ($ mil)	0.144	108.201	0.133	80.425	-	-	0.114	48.551	-	-
Payroll per employee ($)	15,579	173.737	18,798	188.684	14,788	132.367	20,640	148.790	17,969	123.070
Wages per production worker ($)	-	.	-	.	-	.	-	.	-	.
Hours per production worker (hours)	-	.	-	.	-	.	-	.	-	.
Output per employee ($)	58,151	77.643	68,134	79.993	72,754	75.913	80,669	77.840	86,433	58.875
Capital investment per employee ($)	3,692	114.967	3,568	102.542	-	-	3,778	75.234	-	-
M & E per employee ($)	3,162	137.730	3,021	119.436	-	-	3,115	81.772	-	-

Note: Columns headed % show percent of world total or ratio. Ratios closest to 100 are closest to world average. M & E stands for machinery & equipment.

Output in Manufacturing

	1987		1988		1989		1990		1991	
	$ bil.[p]	%	$ bil.[p]	%	$ bil.[p]	%	$ bil.[p]	%	$ bil.[p]	%
3110 - Food products	3.323	16.61	4.197	18.25	-	-	4.232	15.67	-	-
3130 - Beverages	0.271	1.36	0.320	1.39	-	-	0.371	1.37	-	-
3140 - Tobacco	0.066	0.33	0.057	0.25	-	-	0.072	0.27	-	-
3210 - Textiles	0.990	4.95	0.865	3.76	1.081	8.32	1.290	4.78	1.514	9.46
3211 - Spinning, weaving, etc.	0.744	3.72	0.630	2.74	-	-	0.749	2.77	-	-
3220 - Wearing apparel	0.827	4.14	0.877	3.81	1.193	9.18	1.011	3.74	1.372	8.57
3230 - Leather and products	0.053	0.26	0.048	0.21	-	-	0.053	0.20	-	-
3240 - Footwear	0.128	0.64	0.130	0.57	-	-	0.123	0.46	-	-
3310 - Wood products	0.236	1.18	0.268	1.17	-	-	0.303	1.12	-	-
3320 - Furniture, fixtures	0.275	1.38	0.285	1.24	-	-	0.302	1.12	-	-
3410 - Paper and products	0.610	3.05	0.679	2.95	0.675	5.19	0.690	2.56	0.814	5.09
3411 - Pulp, paper, etc.	0.151	0.76	0.198	0.86	-	-	0.219	0.81	-	-
3420 - Printing, publishing	0.681	3.40	0.778	3.38	1.129	8.68	0.917	3.40	1.278	7.99
3520 - Chemical products nec	0.638	3.19	0.791	3.44	-	-	0.865	3.20	-	-
3522 - Drugs and medicines	0.253	1.27	0.274	1.19	-	-	0.368	1.36	-	-
3550 - Rubber products	0.152	0.76	0.142	0.62	-	-	0.169	0.63	-	-
3560 - Plastic products nec	0.788	3.94	0.994	4.32	-	-	1.077	3.99	-	-
3610 - Pottery, china, etc.	0.050	0.25	0.045	0.20	-	-	0.053	0.20	-	-
3620 - Glass and products	0.053	0.26	0.038	0.17	-	-	0.078	0.29	-	-
3690 - Nonmetal products nec	0.479	2.39	0.592	2.57	-	-	0.802	2.97	-	-
3710 - Iron and steel	0.290	1.45	0.299	1.30	-	-	0.422	1.56	-	-
3720 - Nonferrous metals	0.105	0.53	0.156	0.68	-	-	0.174	0.64	-	-
3810 - Metal products	1.940	9.70	2.277	9.90	2.626	20.20	2.507	9.29	3.252	20.33
3820 - Machinery nec	0.487	2.44	0.554	2.41	0.766	5.89	0.669	2.48	0.793	4.96
3825 - Office, computing machinery	0.174	0.87	0.259	1.13	-	-	0.293	1.09	-	-
3830 - Electrical machinery	2.820	14.10	3.171	13.79	4.049	31.15	4.015	14.87	4.967	31.04
3832 - Radio, television, etc.	2.305	11.53	2.604	11.32	-	-	3.312	12.27	-	-
3840 - Transportation equipment	0.877	4.39	0.910	3.96	1.534	11.80	1.403	5.20	1.914	11.96
3841 - Shipbuilding, repair	0.018	0.09	0.023	0.10	-	-	0.036	0.13	-	-
3843 - Motor vehicles	0.122	0.61	0.168	0.73	-	-	0.243	0.90	-	-
3850 - Professional goods	0.235	1.17	0.250	1.09	-	-	0.250	0.93	-	-
3900 - Industries nec	0.305	1.52	0.317	1.38	-	-	0.387	1.43	-	-

Note: Codes are International Standard Industry codes (ISIC). Percentages are % of total Output. [f]: Factor Prices; [p]: Producer Prices.

For sources, notes, and explanations, see Annotated Source Appendix, page 1061.

461

Finance, Economics, and Trade

Economic Indicators [36]

- **National product**: GDP—purchasing power parity—$80.1 billion (1995 est.)

- **National product real growth rate**: 7.1% (1995 est.)

- **National product per capita**: $15,500 (1995 est.)

- **Inflation rate (consumer prices)**: 10.1% (1995)

- **External debt**: $18.5 billion (1995 est.)

Balance of Payments Summary [37]

Values in millions of dollars.

	1989	1990	1991	1992	1993
Exports of goods (f.o.b.)	11,061	12,139	12,029	13,314	14,804
Imports of goods (f.o.b.)	-12,915	-15,120	-16,946	-18,260	-20,411
Trade balance	-1,854	-2,981	-4,917	-4,946	-5,607
Services - debits	-7,703	-8,564	-8,704	-9,186	-9,852
Services - credits	5,688	6,196	6,531	7,466	7,339
Private transfers (net)	1,853	2,123	2,282	2,722	2,853
Government transfers (net)	3,246	3,783	4,392	4,163	3,894
Long-term capital (net)	24	-208	212	-812	1,018
Short-term capital (net)	-1,154	-106	-300	-1,308	983
Errors and omissions	1,298	272	332	445	854
Overall balance	1,398	515	-172	-1,456	1,482

Exchange Rates [38]

Currency: **new shekel.**
Symbol: **NIS.**

Data are currency units per $1.

January 1996	3.1295
1995	3.0113
1994	3.0111
1993	2.8301
1992	2.4591
1991	2.2791

Imports and Exports

Top Import Origins [39]

$40.1 billion (c.i.f., 1995 est.).

Origins	%
EU	NA
US	NA
Japan	NA

Top Export Destinations [40]

$28.4 billion (f.o.b., 1995 est.).

Destinations	%
US	NA
EU	NA
Japan	NA

Foreign Aid [41]

	U.S. $	
Total receipts	12.14	billion
of which US (1990-93)	11.38	billion

Import and Export Commodities [42]

Import Commodities	Export Commodities
Military equipment	Machinery and equipment
Investment goods	Cut diamonds
Rough diamonds	Chemicals
Oil	Textiles and apparel
Other productive inputs	Agricultural products
Consumer goods	Metals

Italy

Geography [1]

Total area:
301,230 sq km   116,306 sq mi
Land area:
294,020 sq km   113,522 sq mi
Comparative area:
Slightly larger than Arizona
Note: Includes Sardinia and Sicily
Land boundaries:
Total 1,935.2 km, Austria 430 km, France 488 km, Holy See (Vatican City) 3.2 km, San Marino 39 km, Slovenia 235 km, Switzerland 740 km
Coastline:
7,600 km
Climate:
Predominantly Mediterranean; Alpine in far North; hot, dry in South
Terrain:
Mostly rugged and mountainous; some plains, coastal lowlands
Natural resources:
Mercury, potash, marble, sulfur, dwindling natural gas and crude oil reserves, fish, coal
Land use:
Arable land: 32%
Permanent crops: 10%
Meadows and pastures: 17%
Forest and woodland: 22%
Other: 19%

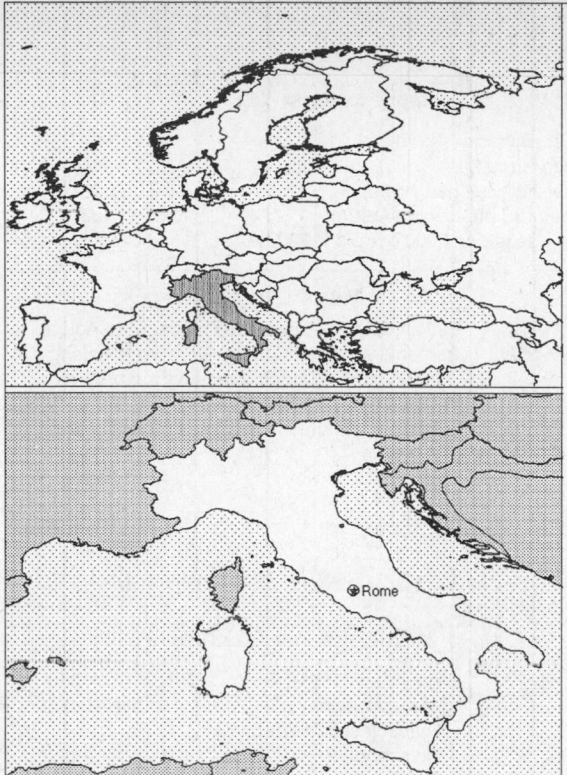

Demographics [2]

	1970	1980	1990	1995[1]	1996	2000	2010	2020	2030
Population	53,661	56,451	57,661	57,384	57,460	57,807	57,660	55,665	53,147
Population density (persons per sq. mi.)	473	497	508	505	506	509	508	490	468
(persons per sq. km.)	183	192	196	195	195	197	196	189	181
Net migration rate (per 1,000 population)	NA	0.8	NA	1.2	1.3	0.5	0.2	0.1	0.0
Births[20]	NA	658	NA	NA	567	NA	NA	NA	NA
Deaths	NA	555	NA	NA	564	NA	NA	NA	NA
Life expectancy - males	NA	NA	NA	74.7	74.9	75.5	76.8	78.5	79.7
Life expectancy - females	NA	NA	NA	81.4	81.5	82.0	83.1	84.7	85.8
Birth rate (per 1,000)	NA	11.7	NA	9.9	9.9	11.4	8.6	7.7	8.1
Death rate (per 1,000)	NA	9.9	NA	9.7	9.8	10.1	11.3	12.0	13.2
Women of reproductive age (15-49 yrs.)	NA	NA	NA	14,420	14,357	13,955	12,775	11,066	9,569
of which are currently married	NA	NA	NA	NA	9,745	9,725	9,032	NA	NA
Fertility rate	NA	1.7	NA	1.3	1.3	1.6	1.5	1.5	1.5

Except as noted, values for vital statistics are in thousands; life expectancy is in years.

Health

Health Indicators [3]

% of population with access to	
safe water (1990-95)	NA
adequate sanitation (1990-95)	NA
health services (1985-95)	NA
% of 1-year-olds immunized (1990-94) against	
TB (tuberculosis)	NA
DPT (diphtheria, pertussis, tetanus)	98
polio	50
measles	50
% of contraceptive prevalence (1980-94)[1]	78
ORT use rate (1990-94)	NA

Health Expenditures [4]

Total health expenditure, 1990 (official exchange rate)	
Millions of dollars	82,214
Dollars per capita	1,426
Health expenditures as a percentage of GDP	
Total	7.5
Public sector	5.8
Private sector	1.7
Development assistance for health	
Total aid flows (millions of dollars)[1]	NA
Aid flows per capita (dollars)	NA
Aid flows as a percentage of total health expenditure	NA

For sources, notes, and explanations, see Annotated Source Appendix, page 1061.

463

Human Factors

Women and Children [5]

% of pregnant women immunized (tetanus 1990-94)	NA
% of births attended by trained health personnel (1983-94)	NA
Maternal mortality rate (1980-92)	4
Under-5 mortality rate (1994)	8
% under-5 moderately/severely underweight (1980-1994)	NA

Burden of Disease [6]

Population per physician (1989)	210.02
Population per nurse	NA
Population per hospital bed (1990)	131.28
AIDS cases per 100,000 people (1994)	9.3
Heart disease cases per 1,000 people (1990-93)	268.5

Ethnic Division [7]

Italian (includes small clusters of German-, French-, and Slovene-Italians in the North and Albanian-Italians and Greek-Italians in the South), Sicilians, Sardinians.

Religion [8]

Roman Catholic	98%
Other	2%

Major Languages [9]

Italian, German (parts of Trentino-Alto Adige region are predominantly German speaking), French (small French-speaking minority in Valle d'Aosta region), Slovene (Slovene-speaking minority in the Trieste-Gorizia area).

Education

Public Education Expenditures [10]

Million or Trillion (T) (Lira)[32]	1980	1990	1991	1992	1993	1994
Total education expenditure	NA	40.85 T	43.14 T	62.01 T	79.31 T	NA
as percent of GNP	NA	3.2	3.1	4.2	5.2	NA
as percent of total govt. expend.	NA	NA	NA	NA	9.0	NA
Current education expenditure[12]	NA	40.40 T	42.66 T	59.09 T	73.77 T	NA
as percent of GNP	NA	3.1	3.0	4.0	4.8	NA
as percent of current govt. expend.	NA	NA	NA	NA	NA	NA
Capital expenditure	NA	446,146	476,737	2.92 T	5.54 T	NA

Educational Attainment [11]

Age group (1981)[8]	25+
Total population	35,596,616
Highest level attained (%)	
No schooling	19.3
First level	
Not completed	47.4
Completed	NA
Entered second level	
S-1	18.0
S-2	11.2
Postsecondary	4.1

Illiteracy [12]

	1981	1989	1995
Illiterate population (15+ yrs.)[2]	1,572,556	NA	932,000
Illiteracy rate - total pop. (%)	3.5	NA	1.9
Illiteracy rate - males (%)	2.5	NA	1.4
Illiteracy rate - females (%)	4.5	NA	2.4

Libraries [13]

	Admin. Units	Svc. Pts.	Vols. (000)	Shelving (meters)	Vols. Added	Reg. Users
National (1992)	2	2	10,960	175,346	98,478	605,417
Nonspecialized (1992)	34	34	7,835	433,448	77,565	752,952
Public (1992)[18]	42	NA	27,518	NA	NA	257,622
Higher ed. (1993)[10]	10	11	6,190	158,140	33,578	841,644
School	NA	NA	NA	NA	NA	NA

Daily Newspapers [14]

	1980	1985	1990	1994
Number of papers	82	72	76	74
Circ. (000)	4,775	5,511	6,000[e]	5,985

Culture [15]

Cinema (seats per 1,000)	NA
Annual attendance per person	1.6
Gross box office receipts (mil. Lira)	758,829
Museums (reporting)[9]	2,497
Visitors (000)	39,882
Annual receipts (mil. Lira)	87,092

Science and Technology

Scientific/Technical Forces [16]

Scientists/engineers	74,434
Number female	NA
Technicians	45,499
Number female	NA
Total	119,933

R&D Expenditures [17]

	Lira (000,000) 1993
Total expenditure	19,518,867
Capital expenditure	2,059,375
Current expenditure	17,459,492
Percent current	89.4

U.S. Patents Issued [18]

Values show patents issued to citizens of the country by the U.S. Patents Office.

	1993	1994	1995
Number of patents	1,454	1,360	1,242

For sources, notes, and explanations, see Annotated Source Appendix, page 1061.

Government and Law

Organization of Government [19]

Long-form name:
Italian Republic
Type:
Republic
Independence:
17 March 1861 (Kingdom of Italy proclaimed)
National holiday:
Anniversary of the Republic, 2 June (1946)
Constitution:
1 January 1948
Legal system:
Based on civil law system, with ecclesiastical law influence; appeals treated as trials de novo; judicial review under certain conditions in Constitutional Court; has not accepted compulsory ICJ jurisdiction
Executive branch:
President; President of the Council of Ministers; Council of Ministers
Legislative branch:
Bicameral Parliament (Parlamento): Senate (Senato della Repubblica) and Chamber of Deputies (Camera dei Deputati)
Judicial branch:
Constitutional Court (Corte Costituzionale)

Elections [20]

Chamber of Deputies	% of seats
Olive Tree	45.1
Freedom Alliance	39.0
Northern League	9.4
Refounded Communist	5.6
Southern Tyrols List	0.5
Autonomous List	0.3
Others	0.2

Government Budget [21]

For 1994 est.

	$ bil.
Revenues	339
Expenditures	431
Capital expenditures	NA

Crime [23]

	1994
Crime volume	
Cases known to police	2,173,477
Attempts (percent)	NA
Percent cases solved	25.51
Crimes per 100,000 persons	3,828.02
Persons responsible for offenses	
Total number offenders	744,866
Percent female	NA
Percent juvenile (14-18 yrs.)	2.99
Percent foreigners	NA

Military Expenditures and Arms Transfers [22]

	1990	1991	1992	1993	1994
Military expenditures					
Current dollars (mil.)	18,930	19,730	19,800	20,440	20,360
1994 constant dollars (mil.)	21,070	21,140	20,650	20,860	20,360
Armed forces (000)	493	473	471	450	436
Gross national product (GNP)					
Current dollars (bil.)	0.8789	0.9213	0.9510	0.9662	1,008
1994 constant dollars (bil.)	0.9781	0.9875	0.9917	0.9861	1,008
Central government expenditures (CGE)					
1994 constant dollars (mil.)	484,300	496,600	530,300	536,000	521,500
People (mil.)	57.7	57.8	57.9	58.0	58.1
Military expenditure as % of GNP	2.2	2.1	2.1	2.1	2.0
Military expenditure as % of CGE	4.4	4.3	3.9	3.9	3.9
Military expenditure per capita (1994 $)	365	366	357	360	350
Armed forces per 1,000 people (soldiers)	8.5	8.2	8.1	7.8	7.5
GNP per capita (1994 $)	16,960	17,090	17,130	17,000	17,330
Arms imports[6]					
Current dollars (mil.)	300	220	110	90	110
1994 constant dollars (mil.)	334	236	115	92	110
Arms exports[6]					
Current dollars (mil.)	200	300	300	350	90
1994 constant dollars (mil.)	223	322	313	357	90
Total imports[7]					
Current dollars (mil.)	182,000	182,700	188,500	148,300	167,700
1994 constant dollars (mil.)	202,500	195,800	196,500	151,300	167,700
Total exports[7]					
Current dollars (mil.)	170,500	169,500	178,200	169,200	189,800
1994 constant dollars (mil.)	189,700	181,700	185,800	172,600	189,800
Arms as percent of total imports[8]	0.2	0.1	0.1	0.1	0.1
Arms as percent of total exports[8]	0.1	0.2	0.2	0.2	0

Human Rights [24]

	SSTS	FL	FAPRO	PPCG	APROBC	TPW	PCPTW	STPEP	PHRFF	PRW	ASST	AFL
Observes	P	P	P	P	P	P	P	P	P	P	P	P
	EAFRD	CPR	ESCR	SR	ACHR	MAAE	PVIAC	PVNAC	EAFDAW	TCIDTP	RC	
Observes	P	P	P	P		P	P	P	P	P	P	

P = Party; S = Signatory; see Appendix for meaning of abbreviations.

Labor Force

Total Labor Force [25]

23.988 million

Labor Force by Occupation [26]

Services	58%
Industry	32.2
Agriculture	9.8

Date of data: 1988

Unemployment Rate [27]

12.2% (January 1995)

For sources, notes, and explanations, see Annotated Source Appendix, page 1061.

465

Production Sectors

Commercial Energy Production and Consumption

Data are shown in quadrillion (10^{15}) BTUs and percent for 1995
Values for hydroelectric, nuclear, geothermal, solar, and wind power refer to electrical generation.

Production [28]

Natural gas - 51.4%
NG liquids - 0.1%
Renewable - 5.2%
Crude oil - 15.1%
Coal - 0.3%
Hydro - 28.0%

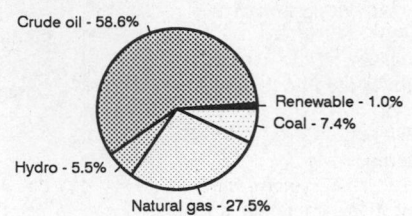

Consumption [29]

Crude oil - 58.6%
Renewable - 1.0%
Coal - 7.4%
Hydro - 5.5%
Natural gas - 27.5%

Crude oil	0.210
Natural gas liquids	0.001
Dry natural gas	0.714
Coal	0.004
Net hydroelectric power	0.389
Geothermal, solar, wind	0.072
Total	1.390

Crude oil	4.112
Dry natural gas	1.929
Coal	0.516
Net hydroelectric power	0.389
Geothermal, solar, wind	0.072
Total	7.018

Telecommunications [30]

- 25.6 million (1987 est.) telephones; modern, well-developed, fast; fully automated telephone, telex, and data services
- Domestic: high-capacity cable and microwave radio relay trunks
- International: satellite earth stations - 3 Intelsat (with a total of 5 antennas - 3 for Atlantic Ocean and 2 for Indian Ocean), 1 Inmarsat (Atlantic Ocean Region), and NA Eutelsat; 21 submarine cables
- Radio: Broadcast stations: AM 135, FM 28 (repeaters 1,840), shortwave 0 Radios: 45.7 million (1992 est.)
- Television: Broadcast stations: 83 (repeaters 1,000) Televisions: 24.35 million (1992 est.)

Top Agricultural Products [32]

Agriculture accounts for 2.9% of the GDP; produces fruits, vegetables, grapes, potatoes, sugar beets, soybeans, grain, olives; meat and dairy products; fish catch of 525,000 metric tons in 1990.

Transportation [31]

Railways: total: 18,961 km; standard gauge: 17,981 km 1.435-m gauge; Italian Railways (FS) operates 16,118 km of the total standard gauge routes (10,560 km electrified); narrow gauge: 113 km 1.000-m gauge (113 km electrified); 867 km 0.950-m gauge (144 km electrified)

Highways: total: 305,388 km (including 45,076 km major roads, 112,111 km secondary roads, 6,301 km motorways); paved: 271,674 km; unpaved: 33,714 km (1991 est.)

Merchant marine: total: 419 ships (1,000 GRT or over) totaling 5,480,320 GRT/7,919,064 DWT; ships by type: bulk 35, cargo 57, chemical tanker 39, combination bulk 1, combination ore/oil 3, container 16, liquefied gas tanker 37, multifunction large-load carrier 1, oil tanker 123, passenger 5, roll-on/roll-off cargo 53, short-sea passenger 31, specialized tanker 11, vehicle carrier 7 (1995 est.)

Airports

Total:	132
With paved runways over 3,047 m:	5
With paved runways 2,438 to 3,047 m:	34
With paved runways 1,524 to 2,437 m:	15
With paved runways 914 to 1,523 m:	24
With paved runways under 914 m:	32

Top Mining Products [33]

Thousand metric tons except as noted[e]	3/95[*]
Pig iron	11,200
Iron and steel semimanufactures	22,800
Silver metal (kg.)	90,000
Cement, hydraulic	42,000
Limestone, crushed	120,000
Coke, metallurgical	5,000
Gas, natural (mil. cu. meters)	20,500
Petroleum (000 42-gal. bls.)	
crude	31,700
refinery products	693,000

Tourism [34]

	1990	1991	1992	1993	1994
Visitors[1]	60,296	51,317	50,089	49,910	51,814
Tourists[37]	26,679	25,878	26,113	26,379	27,480
Excursionists[28]	33,617	25,439	23,976	23,531	24,334
Tourism receipts	20,016	18,421	21,450	20,521	23,927
Tourism expenditures	14,045	11,648	16,530	13,053	12,181
Fare receipts	1,682	1,712	2,026	2,058	2,089
Fare expenditures	1,762	2,114	2,516	2,208	2,633

Travelers are in thousands, money in million U.S. dollars.

For sources, notes, and explanations, see Annotated Source Appendix, page 1061.

Manufacturing Sector

Manufacturing Summary [35]

	1987		1988		1989		1990		1991	
	$ bil.	%	$ bil.	%	$ bil.	%	$ bil.	%	$ bil.	%
Establishments or enterprises (number)	29,471	1.256	30,666	1.461	32,186	1.718	-	-	-	-
Total employment (000)	3,028	2.231	2,770	2.024	2,794	2.270	-	-	-	-
Production workers (000)	2,200	3.262	2,049	3.280	2,056	2.795	-	-	-	-
Output ($ bil.)	328	3.230	389	3.338	404	3.422	-	-	-	-
Value added ($ bil.)	111	2.577	122	2.548	123	2.515	-	-	-	-
Capital investment ($ mil.)	18,124	4.158	21,006	4.411	21,252	3.880	-	-	-	-
M & E investment ($ mil.)	14,608	4.688	17,256	4.986	17,722	4.539	-	-	-	-
Employees per establishment (number)	103	177.596	90	138.553	87	132.114	-	-	-	-
Production workers per establishment	75	259.641	67	224.502	64	162.649	-	-	-	-
Output per establishment ($ mil.)	11	257.139	13	228.471	13	199.158	-	-	-	-
Capital investment per estab. ($ mil.)	0.615	331.014	0.685	301.961	0.660	225.801	-	-	-	-
M & E per establishment ($ mil)	0.496	373.145	0.563	341.280	0.551	264.172	-	-	-	-
Payroll per employee ($)	16,007	178.513	27,103	272.049	28,225	252.634	-	-	-	-
Wages per production worker ($)	14,472	183.403	23,577	279.018	24,416	243.468	-	-	-	-
Hours per production worker (hours)	1,667	87.826	1,669	86.826	1,564	85.338	-	-	-	-
Output per employee ($)	108,439	144.789	140,453	164.898	144,475	150.748	-	-	-	-
Capital investment per employee ($)	5,986	186.385	7,584	217.939	7,606	170.914	-	-	-	-
M & E per employee ($)	4,824	210.109	6,229	246.318	6,343	199.958	-	-	-	-

Note: Columns headed % show percent of world total or ratio. Ratios closest to 100 are closest to world average. M & E stands for machinery & equipment.

Output in Manufacturing

	1987		1988		1989		1990		1991	
	$ bil.	%	$ bil.	%	$ bil.	%	$ bil.	%	$ bil.	%
3110 - Food products	36.000	10.98	38.000	9.77	40.000	9.90	-	-	-	-
3130 - Beverages	5.058	1.54	6.123	1.57	5.672	1.40	-	-	-	-
3140 - Tobacco	2.028	0.62	3.736	0.96	3.822	0.95	-	-	-	-
3210 - Textiles	25.000	7.62	26.000	6.68	27.000	6.68	-	-	-	-
3211 - Spinning, weaving, etc.	11.000	3.35	-	-	-	-	-	-	-	-
3220 - Wearing apparel	12.000	3.66	12.000	3.08	14.000	3.47	-	-	-	-
3230 - Leather and products	4.492	1.37	4.546	1.17	5.016	1.24	-	-	-	-
3240 - Footwear	5.841	1.78	6.079	1.56	6.667	1.65	-	-	-	-
3310 - Wood products	5.119	1.56	4.042	1.04	4.330	1.07	-	-	-	-
3320 - Furniture, fixtures	6.536	1.99	7.056	1.81	7.529	1.86	-	-	-	-
3410 - Paper and products	9.323	2.84	10.000	2.57	11.000	2.72	-	-	-	-
3411 - Pulp, paper, etc.	4.429	1.35	-	-	-	-	-	-	-	-
3420 - Printing, publishing	10.000	3.05	11.000	2.83	12.000	2.97	-	-	-	-
3510 - Industrial chemicals	-	-	20.000	5.14	21.000	5.20	-	-	-	-
3520 - Chemical products nec	-	-	10.000	2.57	9.901	2.45	-	-	-	-
3530 - Petroleum refineries	-	-	22.000	5.66	10.000	2.48	-	-	-	-
3540 - Petroleum, coal products	-	-	1.790	0.46	2.761	0.68	-	-	-	-
3550 - Rubber products	4.645	1.42	4.644	1.19	4.530	1.12	-	-	-	-
3560 - Plastic products nec	11.000	3.35	11.000	2.83	13.000	3.22	-	-	-	-
3610 - Pottery, china, etc.	-	-	4.896	1.26	5.406	1.34	-	-	-	-
3620 - Glass and products	-	-	3.404	0.88	3.601	0.89	-	-	-	-
3690 - Nonmetal products nec	-	-	8.215	2.11	8.597	2.13	-	-	-	-
3710 - Iron and steel	24.000	7.32	26.000	6.68	30.000	7.43	-	-	-	-
3720 - Nonferrous metals	5.864	1.79	7.594	1.95	8.588	2.13	-	-	-	-
3810 - Metal products	14.000	4.27	17.000	4.37	19.000	4.70	-	-	-	-
3820 - Machinery nec	25.000	7.62	47.000	12.08	50.000	12.38	-	-	-	-
3825 - Office, computing machinery	4.500	1.37	-	-	-	-	-	-	-	-
3830 - Electrical machinery	28.000	8.54	29.000	7.46	32.000	7.92	-	-	-	-
3832 - Radio, television, etc.	6.098	1.86	-	-	-	-	-	-	-	-
3840 - Transportation equipment	31.000	9.45	39.000	10.03	41.000	10.15	-	-	-	-
3841 - Shipbuilding, repair	7.810	2.38	-	-	-	-	-	-	-	-
3843 - Motor vehicles	22.000	6.71	-	-	-	-	-	-	-	-
3850 - Professional goods	4.476	1.36	2.774	0.71	3.161	0.78	-	-	-	-
3900 - Industries nec	2.140	0.65	4.095	1.05	4.599	1.14	-	-	-	-

Note: Codes are International Standard Industry codes (ISIC). Percentages are % of total Output. [f]: Factor Prices; [p]: Producer Prices.

For sources, notes, and explanations, see Annotated Source Appendix, page 1061.

467

Finance, Economics, and Trade

Economic Indicators [36]

- **National product**: GDP—purchasing power parity—$1.0886 trillion (1995 est.)
- **National product real growth rate**: 3.2% (1995 est.)
- **National product per capita**: $18,700 (1995 est.)
- **Inflation rate (consumer prices)**: 5.4% (1995)
- **External debt**: $67 billion (1993 est.)

Balance of Payments Summary [37]

Values in millions of dollars.

	1989	1990	1991	1992	1993
Exports of goods (f.o.b.)	140,118	170,304	169,465	178,155	168,456
Imports of goods (f.o.b.)	-142,285	-168,931	-169,911	-175,070	-136,178
Trade balance	-2,167	1,373	-446	3,085	32,278
Services - debits	-53,621	-85,245	-89,046	-115,694	-103,584
Services - credits	47,412	69,370	71,521	90,362	87,864
Private transfers (net)	1,326	1,152	-292	-468	450
Government transfers (net)	-3,829	-3,478	-5,798	-5,192	-5,832
Long-term capital (net)	22,319	36,901	5,060	-8,851	61,675
Short-term capital (net)	2,714	5,739	19,154	20,775	-57,653
Errors and omissions	-2,810	-14,190	-6,871	-8,009	-18,334
Overall balance	11,044	11,622	-6,718	-23,992	-3,136

Exchange Rates [38]

Currency: **Italian lira.**
Symbol: **Lit.**

Data are currency units per $1.

January 1996	1,583.8
1995	1,629.6
1994	1,612.4
1993	1,573.7
1992	1,232.4
1991	1,240.6

Imports and Exports

Top Import Origins [39]

$168.7 billion (c.i.f., 1994) Data are for 1994.

Origins	%
EU	56.3
OPEC	5.3
US	4.6

Top Export Destinations [40]

$190.8 billion (f.o.b., 1994) Data are for 1994.

Destinations	%
EU	53.4
US	7.8
OPEC	3.8

Foreign Aid [41]

Donor: ODA, $3.043 billion (1993).

Import and Export Commodities [42]

Import Commodities
Industrial machinery
Chemicals
Transport equipment
Petroleum
Metals
Food
Agricultural products

Export Commodities
Metals
Textiles and clothing
Production machinery
Motor vehicles
Transportation equipment
Chemicals

For sources, notes, and explanations, see Annotated Source Appendix, page 1061.

Jamaica

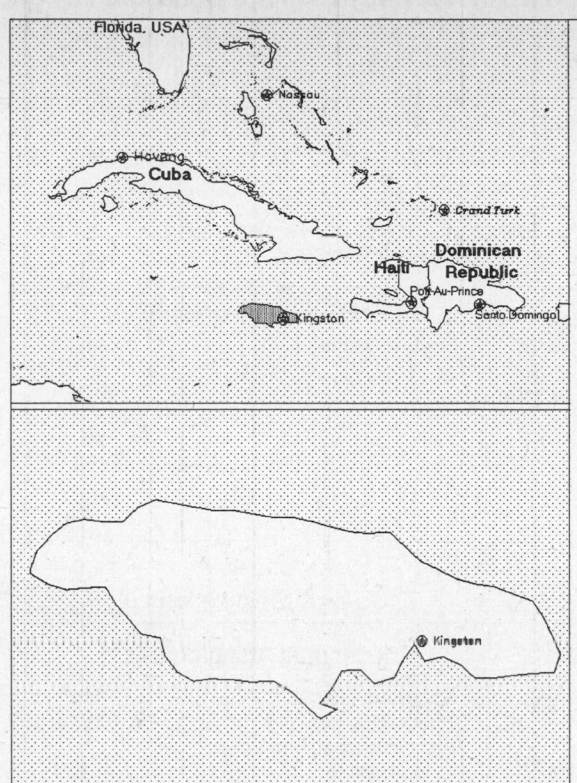

Geography [1]

Total area:
10,990 sq km   4,243 sq mi
Land area:
10,830 sq km   4,181 sq mi
Comparative area:
Slightly smaller than Connecticut
Land boundaries:
0 km
Coastline:
1,022 km
Climate:
Tropical; hot, humid; temperate interior
Terrain:
Mostly mountains with narrow, discontinuous coastal plain
Natural resources:
Bauxite, gypsum, limestone
Land use:
Arable land: 19%
Permanent crops: 6%
Meadows and pastures: 18%
Forest and woodland: 28%
Other: 29%

Demographics [2]

	1970	1980	1990	1995[1]	1996	2000	2010	2020	2030
Population	1,944	2,229	2,466	2,574	2,595	2,669	2,900	3,213	3,475
Population density (persons per sq. mi.)	465	533	590	616	621	638	694	769	831
(persons per sq. km.)	179	206	228	238	240	246	268	297	321
Net migration rate (per 1,000 population)	-7.7	NA	-9.6	-8.7	-8.6	-8.3	0.0	0.0	0.0
Births	65	59	NA	NA	56	NA	NA	NA	NA
Deaths[27]	15	14	NA	NA	14	NA	NA	NA	NA
Life expectancy - males	64.8	NA	71.2	72.4	72.6	73.4	75.2	76.6	77.7
Life expectancy - females	68.3	NA	75.4	77.0	77.3	78.4	80.8	82.5	83.8
Birth rate (per 1,000)	33.5	NA	24.4	22.8	22.2	19.5	16.2	14.6	12.5
Death rate (per 1,000)	7.8	NA	6.1	5.6	5.6	5.3	5.1	5.3	6.2
Women of reproductive age (15-49 yrs.)	NA	NA	634	680	688	725	805	843	844
of which are currently married[28]	102	NA	137	NA	165	184	222	NA	NA
Fertility rate	5.3	NA	2.7	2.5	2.5	2.2	2.0	1.9	1.8

Except as noted, values for vital statistics are in thousands; life expectancy is in years.

Health

Health Indicators [3]

% of population with access to	
safe water (1990-95)	86
adequate sanitation (1990-95)	89
health services (1985-95)	90
% of 1-year-olds immunized (1990-94) against	
TB (tuberculosis)	100
DPT (diphtheria, pertussis, tetanus)	93
polio	93
measles	82
% of contraceptive prevalence (1980-94)	66
ORT use rate (1990-94)	10

Health Expenditures [4]

For sources, notes, and explanations, see Annotated Source Appendix, page 1061.

469

Human Factors

Women and Children [5]

% of pregnant women immunized (tetanus 1990-94)	NA
% of births attended by trained health personnel (1983-94)	82
Maternal mortality rate (1980-92)	120
Under-5 mortality rate (1994)	13
% under-5 moderately/severely underweight (1980-1994)	9

Burden of Disease [6]

Population per physician (1993)	6,420.78
Population per nurse (1993)	489.12
Population per hospital bed (1993)	476.21
AIDS cases per 100,000 people (1994)	13.5
Malaria cases per 100,000 people (1992)	NA

Ethnic Division [7]

African	76.3%
Afro-European	15.1%
White	3.2%
East Indian and Afro-East Indian	3%
Chinese and Afro-Chinese	1.2%
Other	1.2%

Religion [8]

Protestant	55.9%
Roman Catholic	5%
Other (primarily spiritual cults)	39.1%
(1982)	

Major Languages [9]

English, Creole.

Education

Public Education Expenditures [10]

	1980	1985	1990	1992	1993	1994
Million (Jamaican Dollar)[33]						
Total education expenditure	304	550	1,472	3,086	5,516	6,206
as percent of GNP	7.0	5.8	5.4	4.7	6.2	4.7
as percent of total govt. expend.	13.1	12.1	12.8	11.8	12.9	NA
Current education expenditure	303	515	1,276	2,714	5,081	5,551
as percent of GNP	7.0	5.4	4.7	4.1	5.7	4.2
as percent of current govt. expend.	19.4	15.8	17.3	17.9	15.6	NA
Capital expenditure	1	35	196	372	435	655

Educational Attainment [11]

Age group (1982)	25+
Total population	703,714
Highest level attained (%)	
No schooling	3.2
First level	
Not completed	79.8
Completed	NA
Entered second level	
S-1	15.0
S-2	NA
Postsecondary	2.0

Illiteracy [12]

In thousands and percent[1]	1990	1995	2000
Illiterate population (15+ yrs.)	271	254	248
Illiteracy rate - total pop. (%)	16.8	14.7	13.4
Illiteracy rate - males (%)	21.2	19.0	17.5
Illiteracy rate - females (%)	12.5	10.5	9.4

Libraries [13]

	Admin. Units	Svc. Pts.	Vols. (000)	Shelving (meters)	Vols. Added	Reg. Users
National	NA	NA	NA	NA	NA	NA
Nonspecialized	NA	NA	NA	NA	NA	NA
Public (1989)	1	202	1,169	NA	45,271	689,593
Higher ed.	NA	NA	NA	NA	NA	NA
School	NA	NA	NA	NA	NA	NA

Daily Newspapers [14]

	1980	1985	1990	1994
Number of papers	3	4	3	3
Circ. (000)	109	140[e]	160[e]	160[e]

Culture [15]

Science and Technology

Scientific/Technical Forces [16]

Scientists/engineers	18
Number female	10
Technicians	15
Number female	3
Total[20]	33

R&D Expenditures [17]

	Dollar (000) 1986
Total expenditure[21]	4,016
Capital expenditure	130
Current expenditure	3,886
Percent current	96.8

U.S. Patents Issued [18]

Values show patents issued to citizens of the country by the U.S. Patents Office.

	1993	1994	1995
Number of patents	0	0	2

Government and Law

Organization of Government [19]

Long-form name:
None
Type:
Parliamentary democracy
Independence:
6 August 1962 (from UK)
National holiday:
Independence Day (first Monday in August) (1962)
Constitution:
6 August 1962
Legal system:
Based on English common law; has not accepted compulsory ICJ jurisdiction
Executive branch:
British Monarch (represented by Governor General); Prime Minister; Deputy Prime Minister; Cabinet
Legislative branch:
Bicameral Parliament: Senate and House of Representatives
Judicial branch:
Supreme Court

Elections [20]

House of Representatives	% of seats
People's National Party (PNP)	86.7
Jamaica Labor Party (JLP)	13.3

Government Budget [21]

For FY95/96 est.

	$ bil.
Revenues	1.45
Expenditures	2
Capital expenditures	0.732

Crime [23]

	1994
Crime volume	
Cases known to police	43,243
Attempts (percent)	NA
Percent cases solved	64.00
Crimes per 100,000 persons	1,723.10
Persons responsible for offenses	
Total number offenders	27,678
Percent female	NA
Percent juvenile (1-16 yrs.)	NA
Percent foreigners	NA

Military Expenditures and Arms Transfers [22]

	1990	1991	1992	1993	1994
Military expenditures					
Current dollars (mil.)	35	30	25	32	27
1994 constant dollars (mil.)	39	32	27	33	27
Armed forces (000)	3	3	3[e]	3	3
Gross national product (GNP)					
Current dollars (mil.)	3,748	3,925	4,003	4,025	4,021
1994 constant dollars (mil.)	4,172	4,207	4,174	4,108	4,021
Central government expenditures (CGE)					
1994 constant dollars (mil.)	1,205[e]	959[e]	NA	NA	NA
People (mil.)	2.5	2.5	2.5	2.5	2.6
Military expenditure as % of GNP	0.9	0.8	0.6	0.8	0.7
Military expenditure as % of CGE	3.3	3.4	NA	NA	NA
Military expenditure per capita (1994 $)	16	13	11	13	11
Armed forces per 1,000 people (soldiers)	1.2	12	1.2	1.2	1.2
GNP per capita (1994 $)	1,692	1,691	1,663	1,622	1,574
Arms imports[6]					
Current dollars (mil.)	0	10	10	5	5
1994 constant dollars (mil.)	0	11	10	5	5
Arms exports[6]					
Current dollars (mil.)	0	0	0	0	0
1994 constant dollars (mil.)	0	0	0	0	0
Total imports[7]					
Current dollars (mil.)	1,859	1,491	1,668	2,097	2,164
1994 constant dollars (mil.)	2,069	1,598	1,739	2,140	2,164
Total exports[7]					
Current dollars (mil.)	1,135	1,053	1,102	1,069	1,192
1994 constant dollars (mil.)	1,263	1,129	1,149	1,091	1,192
Arms as percent of total imports[8]	0	0.7	0.6	0.2	0.2
Arms as percent of total exports[8]	0	0	0	0	0

Human Rights [24]

	SSTS	FL	FAPRO	PPCG	APROBC	TPW	PCPTW	STPEP	PHRFF	PRW	ASST	AFL
Observes	P	P	P	P	P	P	P			P	P	P
		EAFRD	CPR	ESCR	SR	ACHR	MAAE	PVIAC	PVNAC	EAFDAW	TCIDTP	RC
Observes		P	P	P	P	P		P	P	P		P

P = Party; S = Signatory; see Appendix for meaning of abbreviations.

Labor Force

Total Labor Force [25]

1.062 million

Labor Force by Occupation [26]

Services	41%
Agriculture	22.5
Industry	19

Date of data: 1989

Unemployment Rate [27]

15.4% (1994 est.)

Production Sectors

Commercial Energy Production and Consumption

Data are shown in quadrillion (10^{15}) BTUs and percent for 1995
Values for hydroelectric, nuclear, geothermal, solar, and wind power refer to electrical generation.

Production [28]	Consumption [29]
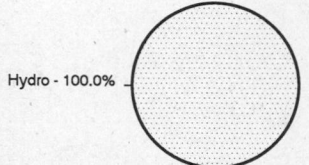	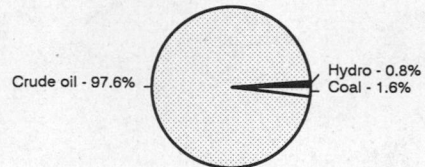
Hydro - 100.0%	Crude oil - 97.6% Hydro - 0.8% Coal - 1.6%

Net hydroelectric power	0.001
Total	0.001

Crude oil	0.123
Coal	0.002
Net hydroelectric power	0.001
Total	0.126

Telecommunications [30]

- 212,257 (1991 est.) telephones; fully automatic domestic telephone network
- International: satellite earth stations - 2 Intelsat (Atlantic Ocean); 3 coaxial submarine cables
- Radio: Broadcast stations: AM 10, FM 17, shortwave 0 Radios: 1.04 million (1992 est.)
- Television: Broadcast stations: 8 Televisions: 330,000 (1992 est.)

Transportation [31]

Railways: total: 272 km; standard gauge: 272 km 1.435-m gauge; note - 207 km belonging to the Jamaica Railway Corporation which were in common carrier service are no longer operational; the remaining track is privately owned and used to transport bauxite

Highways: total: 18,094 km; paved: 12,528 km; unpaved: 5,566 km (1988 est.)

Merchant marine: total: 2 ships (1,000 GRT or over) totaling 3,435 GRT/6,105 DWT; ships by type: oil tanker 1, roll-on/roll-off cargo 1 (1995 est.)

Airports

Total:	27
With paved runways 2,438 to 3,047 m:	2
With paved runways 914 to 1,523 m:	3
With paved runways under 914 m:	21
With unpaved runways 914 to 1,523 m:	1 (1995 est.)

Top Agricultural Products [32]

Agriculture accounts for 7.9% of the GDP; produces sugarcane, bananas, coffee, citrus, potatoes, vegetables; poultry, goats, milk.

Top Mining Products [33]

Thousand metric tons except as noted[e]	3/17/95[*]
Bauxite, dry equivalent, gross weight	11,800[r]
Alumina	3,180[r]
Cement, hydraulic	400
Petroleum refinery products (000 42-gal. bls.)	7,500
Limestone	4,500
Marl and fill	4,000
Sand and gravel	1,600

Tourism [34]

	1990	1991	1992	1993	1994
Visitors	1,226	1,335	1,559	1,609	1,572
Tourists[58]	841	845	909	979	977
Cruise passengers	385	490	650	630	595
Tourism receipts	740	764	858	942	919
Tourism expenditures	114	71	87	82	81
Fare receipts	118	89	96	99	127
Fare expenditures	12	12	13	11	24

Travelers are in thousands, money in million U.S. dollars.

Manufacturing Sector

GDP and Manufacturing Summary [35]

	1980	1985	1989	1990	% change 1980-1990	% change 1989-1990
GDP (million 1980 $)	2,667	2,678	2,887	3,187	19.5	10.4
GDP per capita (1980 $)	1,250	1,158	1,189	1,296	3.7	9.0
Manufacturing as % of GDP (current prices)	16.1	19.3	20.2	18.4	14.3	-8.9
Gross output (million $)	1,661	1,464	4,090^e	2,497^e	50.3	-38.9
Value added (million $)	436	370	554^e	734^e	68.3	32.5
Value added (million 1980 $)	446	469	540	585	31.2	8.3
Industrial production index	100	105	108	122	22.0	13.0
Employment (thousands)	44	46	45^e	64	45.5	42.2

Note: GDP stands for Gross Domestic Product. 'e' stands for estimated value.

Profitability and Productivity

	1980	1985	1989	1990	% change 1980-1990	% change 1989-1990
Intermediate input (%)	74^e	75	87^e	71^e	-4.1	-18.4
Wages, salaries, and supplements (%)	12	10	7^e	11^e	-8.3	57.1
Gross operating surplus (%)	14^e	16	7^e	18^e	28.6	157.1
Gross output per worker ($)	37,512^e	31,521	90,881^e	39,259^e	4.7	-56.8
Value added per worker ($)	9,842^e	7,959	12,279^e	11,543^e	17.3	-6.0
Average wage (incl. benefits) ($)	4,560^e	3,066	6,169^e	4,484^e	-1.7	-27.3

Profitability is in percent of gross output. Productivity is in U.S. $. 'e' stands for estimated value.

Profitability - 1990

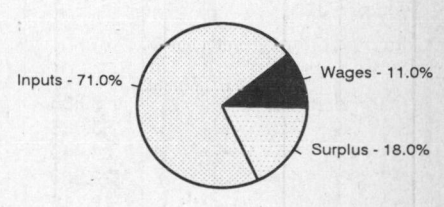

Inputs - 71.0%
Wages - 11.0%
Surplus - 18.0%

The graphic shows percent of gross output.

Value Added in Manufacturing

	1980 $ mil.	1980 %	1985 $ mil.	1985 %	1989 $ mil.	1989 %	1990 $ mil.	1990 %	% change 1980-1990	% change 1989-1990
311 Food products	78	17.9	74	20.0	122^e	22.0	122^e	16.6	56.4	0.0
313 Beverages	63	14.4	51	13.8	77^e	13.9	90^e	12.3	42.9	16.9
314 Tobacco products	61	14.0	48	13.0	71^e	12.8	84^e	11.4	37.7	18.3
321 Textiles	3	0.7	2	0.5	3^e	0.5	5^e	0.7	66.7	66.7
322 Wearing apparel	15	3.4	12	3.2	15^e	2.7	28^e	3.8	86.7	86.7
323 Leather and fur products	2	0.5	3	0.8	4^e	0.7	6^e	0.8	200.0	50.0
324 Footwear	8	1.8	5	1.4	5^e	0.9	11^e	1.5	37.5	120.0
331 Wood and wood products	3	0.7	2	0.5	3^e	0.5	4^e	0.5	33.3	33.3
332 Furniture and fixtures	12	2.8	12	3.2	25^e	4.5	14^e	1.9	16.7	-44.0
341 Paper and paper products	7	1.6	6	1.6	21^e	3.8	10^e	1.4	42.9	-52.4
342 Printing and publishing	15	3.4	14	3.8	13^e	2.3	26^e	3.5	73.3	100.0
351 Industrial chemicals	24^e	5.5	20	5.4	14^e	2.5	56^e	7.6	133.3	300.0
352 Other chemical products	4^e	0.9	4	1.1	50^e	9.0	10^e	1.4	150.0	-80.0
353 Petroleum refineries	55	12.6	28	7.6	46^e	8.3	78^e	10.6	41.8	69.6
354 Miscellaneous petroleum and coal products	2^e	0.5	1	0.3	12^e	2.2	3^e	0.4	50.0	-75.0
355 Rubber products	10^e	2.3	5	1.4	2^e	0.4	15^e	2.0	50.0	650.0
356 Plastic products	11^e	2.5	9	2.4	2^e	0.4	24^e	3.3	118.2	1,100.0
361 Pottery, china, and earthenware	1	0.2	2	0.5	2^e	0.4	5^e	0.7	400.0	150.0
362 Glass and glass products	2	0.5	3	0.8	6^e	1.1	7^e	1.0	250.0	16.7
369 Other nonmetal mineral products	8	1.8	12	3.2	25^e	4.5	27^e	3.7	237.5	8.0
371 Iron and steel	5	1.1	5	1.4	7^e	1.3	9^e	1.2	80.0	28.6
372 Nonferrous metals	NA	0.0	NA	0.0	NA	0.0	NA	0.0	NA	NA
381 Metal products	10	2.3	11	3.0	9^e	1.6	19^e	2.6	90.0	111.1
382 Nonelectrical machinery	6	1.4	7	1.9	2^e	0.4	13^e	1.8	116.7	550.0
383 Electrical machinery	6	1.4	7	1.9	4^e	0.7	14^e	1.9	133.3	250.0
384 Transport equipment	23	5.3	23	6.2	12^e	2.2	44^e	6.0	91.3	266.7
385 Professional and scientific equipment	NA	0.0	NA	0.0	NA	0.0	NA	0.0	NA	NA
390 Other manufacturing industries	4	0.9	4	1.1	5^e	0.9	8^e	1.1	100.0	60.0

Note: The industry codes shown are International Standard Industry codes (ISIC). Percentages are percent of total Value Added. 'e' stands for estimated value

For sources, notes, and explanations, see Annotated Source Appendix, page 1061.

473

Finance, Economics, and Trade

Economic Indicators [36]

- **National product**: GDP—purchasing power parity—$8.2 billion (1995 est.)
- **National product real growth rate**: 0.8% (1995 est.)
- **National product per capita**: $3,200 (1995 est.)
- **Inflation rate (consumer prices)**: 25.5% (1995)
- **External debt**: $3.6 billion (1994 est.)

Balance of Payments Summary [37]

Values in millions of dollars.

	1989	1990	1991	1992	1993
Exports of goods (f.o.b.)	1,000.4	1,157.5	1,150.7	1,053.6	1,044.5
Imports of goods (f.o.b.)	-1,606.4	-1,679.6	-1,575.0	-1,529.2	-1,858.8
Trade balance	-606.0	-522.1	-424.3	-475.6	-814.3
Services - debits	-1,187.2	-1,248.1	-1,182.1	-1,095.2	-1,199.9
Services - credits	1,008.6	1,167.2	1,097.8	1,241.9	1,415.9
Private transfers (net)	299.5	159.0	153.3	248.2	306.4
Government transfers (net)	193.8	116.0	99.5	91.6	65.5
Long-term capital (net)	249.5	285.6	256.5	144.5	131.2
Short-term capital (net)	66.4	107.8	-49.3	173.3	169.6
Errors and omissions	3.9	29.3	-20.4	-59.9	79.2
Overall balance	28.5	94.7	-69.0	268.8	153.6

Exchange Rates [38]

Currency: **Jamaican dollar.**
Symbol: **J$.**

Data are currency units per $1.

December 1995	39.86
1994	33.086
1993	24.949
1992	22.960
1991	12.116

Imports and Exports

Top Import Origins [39]

$2.7 billion (f.o.b., 1995 est.) Data are for 1993.

Origins	%
US	54
Mexico	6
UK	4
Japan	4
Venezuela	3

Top Export Destinations [40]

$2 billion (f.o.b., 1995 est.) Data are for 1993.

Destinations	%
US	47
UK	11
Canada	9
Norway	7
France	4

Foreign Aid [41]

Recipient: ODA, $239 million (1993).

Import and Export Commodities [42]

Import Commodities	Export Commodities
Machinery and transport equipment	Alumina
Construction materials	Bauxite
Fuel	Sugar
Food	Bananas
Chemicals	Rum

Japan

Geography [1]

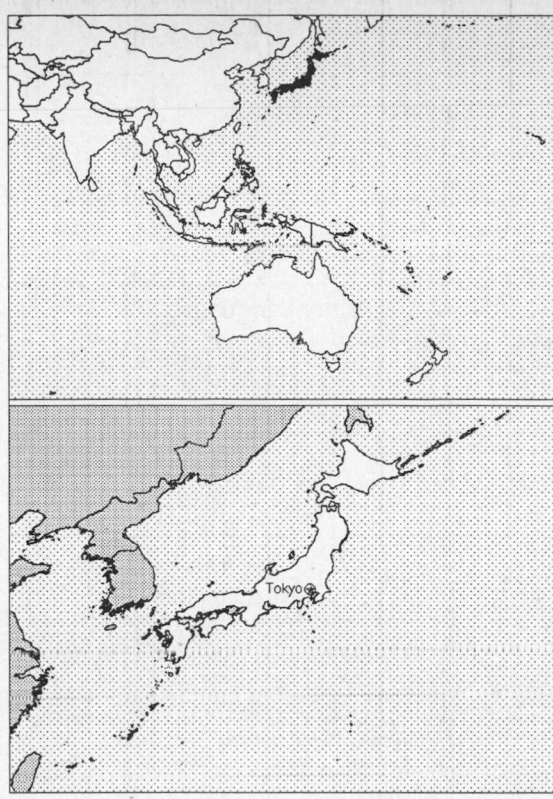

Total area:
377,835 sq km   145,883 sq mi
Land area:
374,744 sq km   144,689 sq mi
Comparative area:
Slightly smaller than California
Note: Includes Bonin Islands (Ogasawara-gunto), Daito-shoto, Minami-jima, Okinotori-shima, Ryukyu Islands (Nansei-shoto), and Volcano Islands (Kazan-retto)
Land boundaries:
0 km
Coastline:
29,751 km
Climate:
Varies from tropical in South to cool temperate in North
Terrain:
Mostly rugged and mountainous
Natural resources:
Negligible mineral resources, fish
Land use:
Arable land: 13%
Permanent crops: 1%
Meadows and pastures: 1%
Forest and woodland: 67%
Other: 18%

Demographics [2]

	1970	1980	1990	1995[1]	1996	2000	2010	2020	2030
Population	104,345	116,807	123,537	125,200	125,450	126,582	127,548	123,620	117,200
Population density (persons per sq. mi.)	685	766	811	821	823	831	837	811	769
(persons per sq. km.)	264	296	313	317	318	321	323	313	297
Net migration rate (per 1,000 population)	NA	NA	0.0	-0.5	-0.4	0.0	0.0	0.0	0.0
Births	1,934	1,577	NA	NA	1,278	NA	NA	NA	NA
Deaths	713	723	820	NA	967	NA	NA	NA	NA
Life expectancy - males	NA	NA	76.0	76.5	76.6	77.0	78.0	78.7	79.3
Life expectancy - females	NA	NA	82.0	82.6	82.7	83.1	84.0	84.7	85.3
Birth rate (per 1,000)	18.7	13.8	10.0	10.0	10.2	10.7	9.2	7.9	8.7
Death rate (per 1,000)	6.9	6.2	6.7	7.5	7.7	8.4	10.4	12.6	14.5
Women of reproductive age (15-49 yrs.)	NA	NA	31,466	31,033	31,038	29,416	27,028	25,081	21,787
of which are currently married	18,409	20,363	18,684	NA	18,529	17,817	17,467	NA	NA
Fertility rate	2.1	1.8	1.5	1.5	1.5	1.5	1.5	1.6	1.6

Except as noted, values for vital statistics are in thousands; life expectancy is in years.

Health

Health Indicators [3]

% of population with access to	
safe water (1990-95)	97
adequate sanitation (1990-95)	NA
health services (1985-95)	NA
% of 1-year-olds immunized (1990-94) against	
TB (tuberculosis)	93
DPT (diphtheria, pertussis, tetanus)	87
polio	94
measles	69
% of contraceptive prevalence (1980-94)	64
ORT use rate (1990-94)	NA

Health Expenditures [4]

Total health expenditure, 1990 (official exchange rate)	
Millions of dollars	189,930
Dollars per capita	1,538
Health expenditures as a percentage of GDP	
Total	6.5
Public sector	4.8
Private sector	1.6
Development assistance for health	
Total aid flows (millions of dollars)[1]	NA
Aid flows per capita (dollars)	NA
Aid flows as a percentage of total health expenditure	NA

For sources, notes, and explanations, see Annotated Source Appendix, page 1061.

Human Factors

Women and Children [5]

% of pregnant women immunized (tetanus 1990-94)	NA
% of births attended by trained health personnel (1983-94)	100
Maternal mortality rate (1980-92)	11
Under-5 mortality rate (1994)	6
% under-5 moderately/severely underweight (1980-1994)	NA

Burden of Disease [6]

Population per physician (1988)	608.02
Population per nurse (1984)	184.17
Population per hospital bed (1990)	63.71
AIDS cases per 100,000 people (1994)	0.2
Heart disease cases per 1,000 people (1990-93)	238.5

Ethnic Division [7]

Japanese	99.4%
Other (mostly Korean)	0.6%

Religion [8]

Observe both Shinto and Buddhist	84%
Other (Christian 0.7%)	16%

Major Languages [9]

Japanese.

Education

Public Education Expenditures [10]

	1980	1985	1989	1990	1991	1994
Million or Trillion (T) (Yen)[34]						
Total education expenditure	13.91 T	16.14 T	18,911	20.26 T	21.30 T	NA
as percent of GNP	5.8	5.0	4.7	4.7	4.7	NA
as percent of total govt. expend.	19.6	17.9	16.5	16.5	16.6	NA
Current education expenditure	9.42 T	15,280	NA	NA	NA	NA
as percent of GNP	3.9	4.8	NA	NA	NA	NA
as percent of current govt. expend.	NA	NA	NA	NA	NA	NA
Capital expenditure	3.92 T	5.14 T	NA	NA	NA	NA

Educational Attainment [11]

Age group (1990)[12]	25+
Total population	81,991,363
Highest level attained (%)	
No schooling	0.3
First level	
Not completed	33.6
Completed	NA
Entered second level	
S-1	43.7
S-2	NA
Postsecondary	20.7

Illiteracy [12]

Libraries [13]

	Admin. Units	Svc. Pts.	Vols. (000)	Shelving (meters)	Vols. Added	Reg. Users
National (1990)[19]	1	3	5,528	NA	NA	452,000
Nonspecialized	NA	NA	NA	NA	NA	NA
Public (1990)	1,475	1,950	161,694	NA	15 mil	16 mil
Higher ed. (1990)[20]	507	704	181,839	NA	NA	1 mil
School (1987)	39,685	39,685	300,827	NA	12 mil	NA

Daily Newspapers [14]

	1980	1985	1990	1994
Number of papers	151	124	125	121
Circ. (000)	66,258	68,296	72,524	71,924

Culture [15]

Cinema (seats per 1,000)	NA
Annual attendance per person	1.1
Gross box office receipts (mil. Yen)	163,700
Museums (reporting)	698
Visitors (000)	73,426
Annual receipts (000 Yen)	NA

Science and Technology

Scientific/Technical Forces [16]

Scientists/engineers	705,346
Number female	NA
Technicians	108,014
Number female	NA
Total[21]	813,360

R&D Expenditures [17]

	Yen (000,000) 1991
Total expenditure[22]	13,771,524
Capital expenditure	2,149,488
Current expenditure	11,622,036
Percent current	84.4

U.S. Patents Issued [18]

Values show patents issued to citizens of the country by the U.S. Patents Office.

	1993	1994	1995
Number of patents	23,410	23,517	22,871

For sources, notes, and explanations, see Annotated Source Appendix, page 1061.

Government and Law

Organization of Government [19]

Long-form name:
  None
Type:
  Constitutional monarchy
Independence:
  660 BC (traditional founding by Emperor Jimmu)
National holiday:
  Birthday of the Emperor, 23 December (1933)
Constitution:
  3 May 1947
Legal system:
  Modeled after European civil law system with English-American influence; judicial review of legislative acts in the Supreme Court; accepts compulsory ICJ jurisdiction, with reservations
Executive branch:
  Emperor; Prime Minister; Deputy Prime Minister; Cabinet
Legislative branch:
  Bicameral Diet (Kokkai): House of Councillors (Sangi-in) and House of Representatives (Shugi-in)
Judicial branch:
  Supreme Court

Elections [20]

House of Representatives	% of seats
Liberal Democratic Party (LDP)	40.5
Shinshinto	33.3
Social Democratic Party (SDP)	12.3
Sakigake	4.3
Others and independents	3.7
Japan Communist Party (JCP)	2.9
Vacant	2.9

Government Expenditures [21]

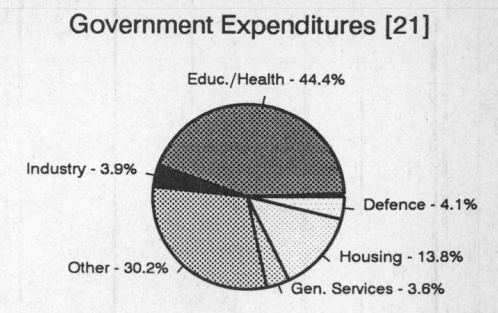

Educ./Health - 44.4%
Industry - 3.9%
Defence - 4.1%
Housing - 13.8%
Gen. Services - 3.6%
Other - 30.2%

(% distribution). Expend. for FY93: 112,655 (Yen bil.)

Crime [23]

	1994
Crime volume	
Cases known to police	1,863,390
Attempts (percent)	3.50
Percent cases solved	45.00
Crimes per 100,000 persons	1,490.31
Persons responsible for offenses	
Total number offenders	376,988
Percent female	18.50
Percent juvenile (14-19 yrs.)	38.10
Percent foreigners	5.30

Military Expenditures and Arms Transfers [22]

	1990	1991	1992	1993	1994
Military expenditures					
Current dollars (mil.)	38,190	41,010	43,000	44,480	45,820
1994 constant dollars (mil.)	42,510	43,960	44,840	45,400	45,820
Armed forces (000)	250	250	242	242	233
Gross national product (GNP)					
Current dollars (bil.)	3,921	4,245	4,421	4,509	4,630
1994 constant dollars (bil.)	4,364	4,550	4,610	4,602	4,630
Central government expenditures (CGE)					
1994 constant dollars (bil.)	0.7061	0.9981	0.9971	1,083	NA
People (mil.)	123.5	123.9	124.3	124.7	125.1
Military expenditure as % of GNP	1.0	1.0	1.0	1.0	1.0
Military expenditure as % of CGE	6.0	4.4	4.5	4.2	NA
Military expenditure per capita (1994 $)	344	355	361	364	366
Armed forces per 1,000 people (soldiers)	2.0	2.0	1.9	1.9	1.9
GNP per capita (1994 $)	35,320	36,720	37,080	36,900	37,000
Arms imports[6]					
Current dollars (mil.)	1,200	1,400	825	525	650
1994 constant dollars (mil.)	1,336	1,501	860	536	650
Arms exports[6]					
Current dollars (mil.)	70	10	10	10	10
1994 constant dollars (mil.)	78	11	10	10	10
Total imports[7]					
Current dollars (mil.)	235,400	237,000	233,200	241,600	275,200
1994 constant dollars (mil.)	261,900	254,000	243,200	246,600	275,200
Total exports[7]					
Current dollars (mil.)	287,600	314,800	339,900	362,200	397,000
1994 constant dollars (mil.)	320,100	337,400	354,400	369,700	397,000
Arms as percent of total imports[8]	0.5	0.6	0.4	0.2	0.2
Arms as percent of total exports[8]	0	0	0	0	0

Human Rights [24]

	SSTS	FL	FAPRO	PPCG	APROBC	TPW	PCPTW	STPEP	PHRFF	PRW	ASST	AFL
Observes		P	P		P	P	P	P		P		
	EAFRD	CPR	ESCR	SR	ACHR	MAAE	PVIAC	PVNAC	EAFDAW	TCIDTP		RC
Observes	P	P	P	P					P			P

P = Party; S = Signatory; see Appendix for meaning of abbreviations.

Labor Force

Total Labor Force [25]

65.87 million (December 1994)

Labor Force by Occupation [26]

Trade and services	54%
Manufacturing, mining, and construction	33
Agriculture, forestry, and fishing	7
Government	3
Other	3

Date of data: 1988

Unemployment Rate [27]

3.1% (1995)

For sources, notes, and explanations, see Annotated Source Appendix, page 1061.

477

Production Sectors

Commercial Energy Production and Consumption

Data are shown in quadrillion (10^{15}) BTUs and percent for 1995
Values for hydroelectric, nuclear, geothermal, solar, and wind power refer to electrical generation.

Production [28]	Consumption [29]
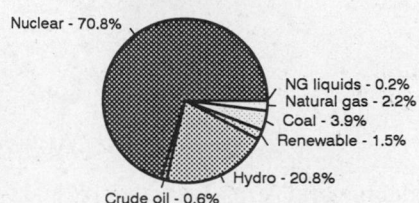	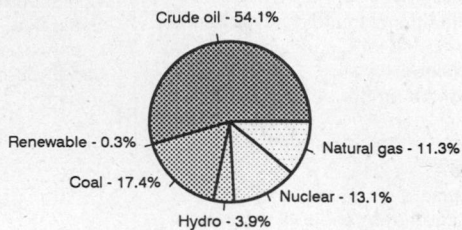

Production [28]:
- Nuclear - 70.8%
- NG liquids - 0.2%
- Natural gas - 2.2%
- Coal - 3.9%
- Renewable - 1.5%
- Hydro - 20.8%
- Crude oil - 0.6%

Consumption [29]:
- Crude oil - 54.1%
- Natural gas - 11.3%
- Nuclear - 13.1%
- Hydro - 3.9%
- Coal - 17.4%
- Renewable - 0.3%

Crude oil	0.024		Crude oil	11.635
Natural gas liquids	0.007		Dry natural gas	2.435
Dry natural gas	0.086		Coal	3.742
Coal	0.156		Net hydroelectric power	0.829
Net hydroelectric power	0.829		Net nuclear power	2.817
Net nuclear power	2.817		Geothermal, solar, wind	0.061
Geothermal, solar, wind	0.061		Total	21.519
Total	3.980			

Telecommunications [30]

- 64 million (1987 est.) telephones; excellent domestic and international service
- International: satellite earth stations - 5 Intelsat (4 Pacific Ocean and 1 Indian Ocean), 1 Intersputnik (Indian Ocean Region), and 1 Inmarsat (Pacific and Indian Ocean Regions); submarine cables to China, Philippines, Russia, and US (via Guam)
- Radio: Broadcast stations: AM 318, FM 58, shortwave 0 Radios: 97 million (1993 est.)
- Television: Broadcast stations: 12,350 (1 kW or greater 196) Televisions: 100 million (1993 est.)

Transportation [31]

Railways: total: 26,506 km; standard gauge: 3,233 km 1.435-m gauge (entirely electrified); narrow gauge: 72 km 1.372-m gauge (72 km electrified); 23,154 km 1.067-m gauge (13,835 km electrified); 47 km 0.762-m gauge (47 km electrified) (1994)

Highways: total: 1,112,844 km; paved: 790,119 km (including 5,054 km of expressways); unpaved: 322,725 km (1992 est.)

Merchant marine: total: 796 ships (1,000 GRT or over) totaling 15,944,137 GRT/23,662,930 DWT; ships by type: bulk 192, cargo 57, chemical tanker 6, combination bulk 2, combination ore/oil 6, container 38, liquefied gas tanker 39, oil tanker 259, passenger 9, passenger-cargo 3, refrigerated cargo 35, roll-on/roll-off cargo 43, short-sea passenger 28, specialized tanker 2, vehicle carrier 77

Top Agricultural Products [32]

Agriculture accounts for 2.1% of the GDP; produces rice, sugar beets, vegetables, fruit; pork, poultry, dairy products, eggs; world's largest fish catch of 10 million metric tons in 1991.

Airports

Total:	164
With paved runways over 3,047 m:	6
With paved runways 2,438 to 3,047 m:	32
With paved runways 1,524 to 2,437 m:	34
With paved runways 914 to 1,523 m:	30
With paved runways under 914 m:	60

Top Mining Products [33]

Thousand metric tons except as noted	8/25/95[*]
Gold metal (kg.)	203,000
Indium metal (kg.)	58,500
Pig iron and blast furnace ferroalloys	73,800
Steel, crude	98,300
Silver (kg.)	
mine output, Ag content	133,000
metal, primary and secondary	2,180,000
Cement, hydraulic	91,500
Limestone	202,000
Petroleum refinery products (000 42-gal. bls.)	1,750,000[e]

Tourism [34]

	1990	1991	1992	1993	1994
Visitors[57]	3,236	3,533	3,582	3,410	3,468
Tourists	1,879	2,104	2,103	1,925	1,915
Excursionists	118	85	90	99	87
Cruise passengers	1	NA	1	1	NA
Tourism receipts	3,578	3,435	3,588	3,557	3,477
Tourism expenditures	24,928	23,983	26,837	26,860	30,715
Fare receipts	1,264	1,260	1,318	1,396	1,520
Fare expenditures	7,253	7,483	8,556	8,905	9,539

Travelers are in thousands, money in million U.S. dollars.

For sources, notes, and explanations, see Annotated Source Appendix, page 1061.

Manufacturing Sector

Manufacturing Summary [35]

	1987		1988		1989		1990		1991	
	$ bil.	%	$ bil.	%	$ bil.	%	$ bil.	%	$ bil.	%
Establishments or enterprises (number)	476,599	20.316	493,915	23.529	476,414	25.433	491,532	27.439	485,352	63.497
Total employment (000)	13,253	9.765	13,421	9.807	13,527	10.990	13,774	12.444	14,035	20.205
Production workers (000)	6,524	9.673	-	-	-	-	6,698	9.437	-	-
Output ($ bil.)	2,335	22.968	2,868	24.604	2,915	24.711	3,006	26.203	3,402	33.364
Value added ($ bil.)	920	21.312	1,143	23.779	1,150	23.583	1,174	23.538	1,331	38.778
Capital investment ($ mil.)	95,098	21.820	121,943	25.607	139,127	25.399	144,913	26.073	189,310	49.715
M & E investment ($ mil.)	71,999	23.106	92,743	26.795	102,493	26.251	105,491	25.022	134,155	47.579
Employees per establishment (number)	28	48.066	27	41.680	28	43.212	28	45.353	29	31.820
Production workers per establishment	14	47.611	-	-	-	-	14	34.392	-	-
Output per establishment ($ mil.)	5	113.051	6	104.570	6	97.161	6	95.495	7	52.544
Capital investment per estab. ($ mil.)	0.200	107.398	0.247	108.833	0.292	99.866	0.295	95.022	0.390	78.295
M & E per establishment ($ mil)	0.151	113.728	0.188	113.884	0.215	103.219	0.215	91.191	0.276	74.931
Payroll per employee ($)	25,954	289.432	30,331	304.453	29,591	264.866	29,639	213.668	32,699	223.953
Wages per production worker ($)	24,546	311.079	-	-	-	-	27,743	233.609	-	-
Hours per production worker (hours)	-	-	-	-	-	-	-	-	-	-
Output per employee ($)	176,154	235.202	213,695	250.889	215,491	224.846	218,212	210.561	242,424	165.130
Capital investment per employee ($)	7,170	223.441	9,080	261.118	10,285	231.107	10,521	209.518	13,488	246.060
M & E per employee ($)	5,433	236.610	6,910	273.237	7,577	238.865	7,659	201.071	9,559	235.485

Note: Columns headed % show percent of world total or ratio. Ratios closest to 100 are closest to world average. M & E stands for machinery & equipment.

Output in Manufacturing

	1987		1988		1989		1990		1991	
	$ bil.	%	$ bil.	%	$ bil.	%	$ bil.	%	$ bil.	%
3110 - Food products	161.000	6.90	187.000	6.52	181.000	6.21	180.000	5.99	203.000	5.97
3130 - Beverages	30.000	1.28	36.000	1.26	33.000	1.13	32.000	1.06	35.000	1.03
3140 - Tobacco	14.000	0.60	15.000	0.52	17.000	0.58	16.000	0.53	18.000	0.53
3210 - Textiles	60.000	2.57	69.000	2.41	66.000	2.26	64.000	2.13	71.000	2.09
3211 - Spinning, weaving, etc.	32.000	1.37	37.000	1.29	34.000	1.17	33.000	1.10	35.000	1.03
3220 - Wearing apparel	20.000	0.86	23.000	0.80	23.000	0.79	23.000	0.77	26.000	0.76
3230 - Leather and products	4.266	0.18	4.947	0.17	4.856	0.17	5.063	0.17	5.605	0.16
3240 - Footwear	3.028	0.13	3.574	0.12	3.559	0.12	3.730	0.12	4.135	0.12
3310 - Wood products	31.000	1.33	37.000	1.29	36.000	1.23	36.000	1.20	39.000	1.15
3320 - Furniture, fixtures	15.000	0.64	19.000	0.66	19.000	0.65	20.000	0.67	22.000	0.65
3410 - Paper and products	48.000	2.06	57.000	1.99	58.000	1.99	57.000	1.90	62.000	1.82
3411 - Pulp, paper, etc.	21.000	0.90	25.000	0.87	25.000	0.86	25.000	0.83	27.000	0.79
3420 - Printing, publishing	71.000	3.04	87.000	3.03	88.000	3.02	91.000	3.03	103.000	3.03
3510 - Industrial chemicals	68.000	2.91	82.000	2.86	83.000	2.85	84.000	2.79	92.000	2.70
3511 - Basic chemicals, excl fertilizers	34.000	1.46	41.000	1.43	40.000	1.37	43.000	1.43	47.000	1.38
3513 - Synthetic resins, etc.	29.000	1.24	36.000	1.26	37.000	1.27	36.000	1.20	40.000	1.18
3520 - Chemical products nec	66.000	2.83	78.000	2.72	79.000	2.71	79.000	2.63	89.000	2.62
3522 - Drugs and medicines	30.000	1.28	36.000	1.26	36.000	1.23	36.000	1.20	40.000	1.18
3530 - Petroleum refineries	43.000	1.84	46.000	1.60	46.000	1.58	53.000	1.76	60.000	1.76
3540 - Petroleum, coal products	4.916	0.21	5.431	0.19	5.364	0.18	5.435	0.18	5.857	0.17
3550 - Rubber products	19.000	0.81	23.000	0.80	22.000	0.75	23.000	0.77	26.000	0.76
3560 - Plastic products nec	59.000	2.53	72.000	2.51	73.000	2.50	74.000	2.46	88.000	2.59
3610 - Pottery, china, etc.	4.660	0.20	5.650	0.20	5.415	0.19	5.477	0.18	5.983	0.18
3620 - Glass and products	12.000	0.51	15.000	0.52	15.000	0.51	15.000	0.50	16.000	0.47
3690 - Nonmetal products nec	44.000	1.88	54.000	1.88	53.000	1.82	54.000	1.80	60.000	1.76
3710 - Iron and steel	95.000	4.07	122.000	4.25	125.000	4.29	126.000	4.19	139.000	4.09
3720 - Nonferrous metals	27.000	1.16	32.000	1.12	38.000	1.30	39.000	1.30	40.000	1.18
3810 - Metal products	101.000	4.33	129.000	4.50	131.000	4.49	138.000	4.59	164.000	4.82
3820 - Machinery nec	197.000	8.44	261.000	9.10	278.000	9.54	300.000	9.98	344.000	10.11
3825 - Office, computing machinery	59.000	2.53	74.000	2.58	79.000	2.71	84.000	2.79	96.000	2.82
3830 - Electrical machinery	260.000	11.13	330.000	11.51	331.000	11.36	337.000	11.21	392.000	11.52
3832 - Radio, television, etc.	152.000	6.51	195.000	6.80	194.000	6.66	196.000	6.52	226.000	6.64
3840 - Transportation equipment	240.000	10.28	290.000	10.11	303.000	10.39	321.000	10.68	361.000	10.61
3841 - Shipbuilding, repair	11.000	0.47	12.000	0.42	13.000	0.45	16.000	0.53	18.000	0.53
3843 - Motor vehicles	219.000	9.38	266.000	9.27	277.000	9.50	293.000	9.75	329.000	9.67
3850 - Professional goods	24.000	1.03	30.000	1.05	30.000	1.03	31.000	1.03	35.000	1.03
3900 - Industries nec	25.000	1.07	31.000	1.08	31.000	1.06	32.000	1.06	37.000	1.09

Note: Codes are International Standard Industry codes (ISIC). Percentages are % of total Output. [f]: Factor Prices; [p]: Producer Prices.

For sources, notes, and explanations, see Annotated Source Appendix, page 1061.

479

Finance, Economics, and Trade

Economic Indicators [36]

- **National product**: GDP—purchasing power parity—$2.6792 trillion (1995 est.)
- **National product real growth rate**: 0.3% (1995 est.)
- **National product per capita**: $21,300 (1995 est.)
- **Inflation rate (consumer prices)**: - 0.1% (1995)
- **External debt**: $NA

Balance of Payments Summary [37]

Values in millions of dollars.

	1989	1990	1991	1992	1993
Exports of goods (f.o.b.)	269,550	280,350	306,580	330,870	351,310
Imports of goods (f.o.b.)	-192,660	-216,770	-203,490	-198,470	-209,740
Trade balance	76,890	63,580	103,090	132,400	141,570
Services - debits	-159,530	-188,150	-206,280	-204,200	-207,620
Services - credits	143,910	165,960	188,590	194,060	203,650
Private transfers (net)	-990	-1,010	-660	-1,310	-2,250
Government transfers (net)	-3,290	-4,510	-11,830	-3,310	-3,840
Long-term capital (net)	-93,760	-53,080	31,390	-30,780	-81,360
Short-term capital (net)	45,830	31,540	-103,240	-75,770	-22,240
Errors and omissions	-21,820	-20,920	-7,680	-10,460	-250
Overall balance	-12,760	-6,590	-6,620	630	27,660

Exchange Rates [38]

Currency: **yen.**
Symbol: **%.**

Data are currency units per $1.

January 1996	105.84
1995	94.06
1994	102.21
1993	111.20
1992	126.65
1991	134.71

Imports and Exports

Top Import Origins [39]

$336.09 billion (c.i.f., 1995).

Origins	%
Southeast Asia	25
US	22
Western Europe	16
China	11

Top Export Destinations [40]

$442.84 billion (f.o.b., 1995).

Destinations	%
Southeast Asia	38
US	27
Western Europe	17
China	5

Foreign Aid [41]

	U.S. $	
ODA (1993)	11.259	billion
ODA and OOF commitments (1970-95)	143	billion

Import and Export Commodities [42]

Import Commodities

Manufactures 52%
Fossil fuels 20%
Foodstuffs and raw materials 28%

Export Commodities

Manufactures 97% (includes machinery 46% motor vehicles 20% consumer electronics 10%)

Jordan

Geography [1]

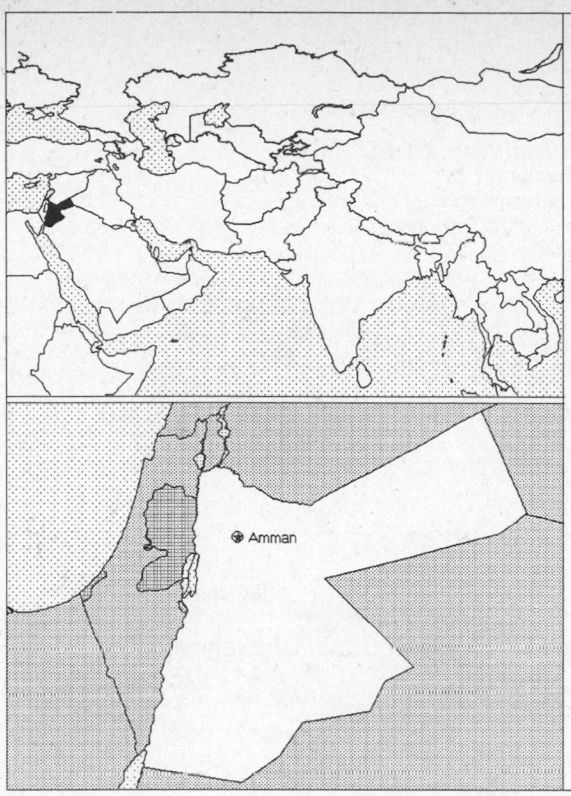

Total area:
89,213 sq km 34,445 sq mi
Land area:
88,884 sq km 34,318 sq mi
Comparative area:
Slightly smaller than Indiana
Land boundaries:
Total 1,619 km, Iraq 181 km, Israel 238 km, Saudi Arabia 728 km,
Syria 375 km, West Bank 97 km
Coastline:
26 km
Climate:
Mostly arid desert; rainy season in West (November to April)
Terrain:
Mostly desert plateau in East, highland area in West; Great Rift Valley
separates East and West Banks of the Jordan River
Natural resources:
Phosphates, potash, shale oil
Land use:
Arable land: 4%
Permanent crops: 0.5%
Meadows and pastures: 1%
Forest and woodland: 0.5%
Other: 94%

Demographics [2]

	1970	1980	1990	1995[1]	1996	2000	2010	2020	2030
Population	1,503	2,168	3,277	4,101	4,212	4,704	6,112	7,529	8,951
Population density (persons per sq. mi.)	43	61	93	116	119	133	173	213	253
(persons per sq. km.)	16	24	36	45	46	51	67	82	98
Net migration rate (per 1,000 population)	NA	5.7	55.4	-6.4	-6.2	0.0	0.0	0.0	0.0
Births	NA	NA	NA	NA	154	NA	NA	NA	NA
Deaths	NA	6	NA	NA	.17	NA	NA	NA	NA
Life expectancy - males	NA	66.3	69.4	70.4	70.6	71.4	73.1	74.5	75.6
Life expectancy - females	NA	69.2	72.9	74.2	74.5	75.4	77.5	79.3	80.7
Birth rate (per 1,000)	NA	43.6	39.6	37.3	36.7	33.4	26.3	22.6	19.1
Death rate (per 1,000)	NA	6.4	4.5	4.0	4.0	3.7	3.4	3.5	3.7
Women of reproductive age (15-49 yrs.)	NA	434	717	927	957	1,093	1,551	1,999	2,357
of which are currently married[29]	NA	NA	399	NA	551	643	947	NA	NA
Fertility rate	NA	7.8	6.1	5.3	5.1	4.5	3.3	2.7	2.3

Except as noted, values for vital statistics are in thousands; life expectancy is in years.

Health

Health Indicators [3]

% of population with access to	
safe water (1990-95)	89
adequate sanitation (1990-95)	95
health services (1985-95)	97
% of 1-year-olds immunized (1990-94) against	
TB (tuberculosis)	NA
DPT (diphtheria, pertussis, tetanus)	96
polio	96
measles	91
% of contraceptive prevalence (1980-94)	35
ORT use rate (1990-94)	53

Health Expenditures [4]

Total health expenditure, 1990 (official exchange rate)	
Millions of dollars	149
Dollars per capita	48
Health expenditures as a percentage of GDP	
Total	3.8
Public sector	1.8
Private sector	2.0
Development assistance for health	
Total aid flows (millions of dollars)[1]	18
Aid flows per capita (dollars)	5.9
Aid flows as a percentage of total health expenditure	12.4

For sources, notes, and explanations, see Annotated Source Appendix, page 1061.

Human Factors

Women and Children [5]

% of pregnant women immunized (tetanus 1990-94)	25
% of births attended by trained health personnel (1983-94)	87
Maternal mortality rate (1980-92)[1]	48
Under-5 mortality rate (1994)	25
% under-5 moderately/severely underweight (1980-1994)	9

Burden of Disease [6]

Population per physician (1991)	741.94
Population per nurse (1991)	480.87
Population per hospital bed (1990)	501.66
AIDS cases per 100,000 people (1994)	0.1
Malaria cases per 100,000 people (1992)	NA

Ethnic Division [7]

Arab	98%
Armenian	1%
Circassian	1%

Religion [8]

Sunni Muslim	92%
Christian	8%

Major Languages [9]

Arabic (official), English widely understood among upper and middle classes.

Education

Public Education Expenditures [10]

Million (Dinar)[35]	1980	1985	1990	1992	1993	1994
Total education expenditure	64	105	101	116	155	155
as percent of GNP	NA	5.5	4.2	3.6	4.2	3.8
as percent of total govt. expend.	11.3	13.0	8.5	9.1	11.6	10.5
Current education expenditure	51	92	92	102	139	NA
as percent of GNP	NA	4.8	3.8	3.1	3.8	NA
as percent of current govt. expend.	15.0	18.9	NA	10.9	13.2	NA
Capital expenditure	13	14	9	14	16	NA

Educational Attainment [11]

Illiteracy [12]

In thousands and percent[1]	1990	1995	2000
Illiterate population (15+ yrs.)	425	414	370
Illiteracy rate - total pop. (%)	24.0	18.1	13.7
Illiteracy rate - males (%)	12.2	8.9	6.6
Illiteracy rate - females (%)	36.4	27.8	21.3

Libraries [13]

	Admin. Units	Svc. Pts.	Vols. (000)	Shelving (meters)	Vols. Added	Reg. Users
National (1992)	1	1	NA	2,800	1,640	NA
Nonspecialized	NA	NA	NA	NA	NA	NA
Public (1986)	5	5	140	NA	14,600	6,865
Higher ed. (1990)	33	44	1,227	NA	36,182	51,369
School	NA	NA	NA	NA	NA	NA

Daily Newspapers [14]

	1980	1985	1990	1994
Number of papers	4	4	4	4
Circ. (000)	66	155	225	250

Culture [15]

Cinema (seats per 1,000)	NA
Annual attendance per person	0.0
Gross box office receipts (mil. Dinar)	NA
Museums (reporting)	11
Visitors (000)	429
Annual receipts (000 Dinar)	546

Science and Technology

Scientific/Technical Forces [16]

Scientists/engineers	418
Number female	54
Technicians	29
Number female	6
Total[9]	447

R&D Expenditures [17]

	Dinar (000,000) 1986
Total expenditure[6]	5,587
Capital expenditure	1,287
Current expenditure	4,300
Percent current	77.0

U.S. Patents Issued [18]

For sources, notes, and explanations, see Annotated Source Appendix, page 1061.

Government and Law

Organization of Government [19]

Long-form name:
Hashemite Kingdom of Jordan
Type:
Constitutional monarchy
Independence:
25 May 1946 (from League of Nations
mandate under British administration)
National holiday:
Independence Day, 25 May (1946)
Constitution:
8 January 1952
Legal system:
Based on Islamic law and French codes;
judicial review of legislative acts in a
specially provided High Tribunal; has not
accepted compulsory ICJ jurisdiction
Executive branch:
King; Prime Minister; Cabinet
Legislative branch:
Bicameral National Assembly (Majlis
al-'Umma): Houoc of Notables (Majlis al-
A'ayan) and House of Representatives
Judicial branch:
Court of Cassation

Elections [20]

House of Representatives	% of seats
Independents	58.8
Islamic Action Front	20.0
Jordanian National Alliance Party	5.0
Awakening Party	2.5
Homeland Party	2.5
Pledge Party	2.5
Jordanian Arab Democratic Party	2.5
Others	6.3

Government Expenditures [21]

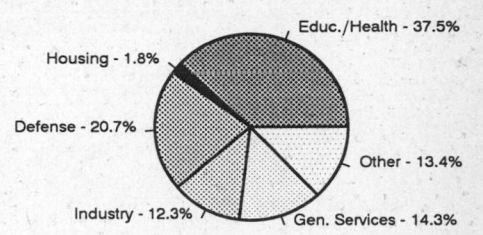

(% distribution). Expend. for CY94: 1,312.8 (Dinar mil.)

Crime [23]

	1990
Crime volume	
Cases known to police	10,628
Attempts (percent)	NA
Percent cases solved	44.7
Crimes per 100,000 persons	324.3
Persons responsible for offenses	
Total number offenders	4,755
Percent female	2.1
Percent juvenile	NA
Percent foreigners	NA

Military Expenditures and Arms Transfers [22]

	1990	1991	1992	1993	1994
Military expenditures					
Current dollars (mil.)	435	447[e]	436[e]	442	434
1994 constant dollars (mil.)	485	479[e]	455[e]	451	434
Armed forces (000)	100	100	100	100	100
Gross national product (GNP)					
Current dollars (mil.)	3,733	4,007	4,893	5,340	5,790
1994 constant dollars (mil.)	4,155	4,295	5,102	5,450	5,790
Central government expenditures (CGE)					
1994 constant dollars (mil.)	1,655	1,761	1,689	1,920	1,978
People (mil.)	3.3	3.6	3.8	3.9	4.0
Military expenditure as % of GNP	11.7	11.2	8.9	8.3	7.5
Military expenditure as % of CGE	29.3	27.2	26.9	23.5	21.9
Military expenditure per capita (1994 $)	148	135	121	116	108
Armed forces per 1,000 people (soldiers)	30.5	28.1	26.6	25.7	25.0
GNP per capita (1994 $)	1,268	1,206	1,356	1,401	1,448
Arms imports[6]					
Current dollars (mil.)	110	50	20	30	40
1994 constant dollars (mil.)	122	54	21	31	40
Arms exports[6]					
Current dollars (mil.)	0	0	0	0	30
1994 constant dollars (mil.)	0	0	0	0	30
Total imports[7]					
Current dollars (mil.)	2,601	2,508	3,255	3,540	3,382
1994 constant dollars (mil.)	2,894	2,688	3,394	3,613	3,382
Total exports[7]					
Current dollars (mil.)	1,064	1,130	1,215	1,232	1,424
1994 constant dollars (mil.)	1,184	1,211	1,267	1,257	1,424
Arms as percent of total imports[8]	4.2	2.0	0.6	0.8	1.2
Arms as percent of total exports[8]	0	0	0	0	2.1

Human Rights [24]

	SSTS	FL	FAPRO	PPCG	APROBC	TPW	PCPTW	STPEP	PHRFF	PRW	ASST	AFL
Observes	P	P		P	P	P	P	P		P	P	P
	EAFRD	CPR	ESCR	SR	ACHR	MAAE	PVIAC	PVNAC	EAFDAW	TCIDTP	RC
Observes		P	P	P				P	P	P	P	P

P=Party; S=Signatory; see Appendix for meaning of abbreviations.

Labor Force

Total Labor Force [25]

600,000 (1992)

Labor Force by Occupation [26]

Industry	11.4%
Commerce, restaurants, and hotels	10.5
Construction	10.0
Transport and communications	8.7
Agriculture	7.4
Other services	52.0

Date of data: 1992

Unemployment Rate [27]

16% (1994 est.)

Production Sectors

Commercial Energy Production and Consumption

Data are shown in quadrillion (10^{15}) BTUs and percent for 1995
Values for hydroelectric, nuclear, geothermal, solar, and wind power refer to electrical generation.

Production [28]	Consumption [29]
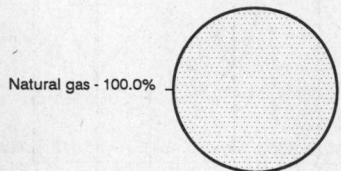	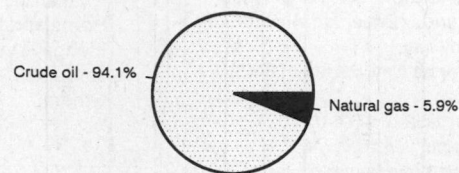

Dry natural gas	0.011	Crude oil	0.177
Total	0.011	Dry natural gas	0.011
		Total	0.188

Telecommunications [30]

- 81,500 (1987 est.) telephones; adequate telephone system
- Domestic: microwave radio relay, cable, and radiotelephone links
- International: satellite earth stations - 2 Intelsat (1 Atlantic Ocean and 1 Indian Ocean) and 1 Arabsat; coaxial cable and microwave radio relay to Iraq, Saudi Arabia, and Syria; microwave radio relay to Lebanon is inactive; participant in Medarabtel
- Radio: Broadcast stations: AM 5, FM 7, shortwave 0 Radios: 1.1 million (1992 est.)
- Television: Broadcast stations: 8 and 1 TV receive-only satellite link Televisions: 350,000 (1992 est.)

Top Agricultural Products [32]

Agriculture accounts for 11% of the GDP; produces wheat, barley, citrus, tomatoes, melons, olives; sheep, goats, poultry.

Top Mining Products [33]

Thousand metric tons except as noted[e]	7/95[*]
Cement, hydraulic	6,600
Petroleum (000 42-gal. bls.)	
crude	50
refinery products	18,700
Phosphate, mine output	
gross weight	4,220
P2O5 content	1,400
Phosphatic fertilizers	491
Potash, K2 equivalent	915

Transportation [31]

Railways: total: 676 km; narrow gauge: 676 km 1.050-m gauge; note - an additional 110 km stretch of the old Hedjaz railroad is out of use

Highways: total: 5,680 km; paved: 5,680 km (including 1,712 km of expressways); unpaved: 0 km (1991 est.)

Merchant marine: total: 3 bulk ships (1,000 GRT or over) totaling 41,960 GRT/67,515 DWT (1995 est.)

Airports

Total:	14
With paved runways over 3,047 m:	10
With paved runways 2,438 to 3,047 m:	3
With paved runways under 914 m:	1 (1995 est.)

Tourism [34]

	1990	1991	1992	1993	1994
Visitors[1]	2,633	2,228	3,243	3,099	3,225
Tourists	572	436	661	765	858
Cruise passengers	4	NA	2	4	5
Tourism receipts	512	317	462	563	582
Tourism expenditures[59]	336	282	350	345	394
Fare receipts	277	187	247	274	263
Fare expenditures	262	127	222	224	228

Travelers are in thousands, money in million U.S. dollars.

For sources, notes, and explanations, see Annotated Source Appendix, page 1061.

Manufacturing Sector

Manufacturing Summary [35]

	1987		1988		1989		1990		1991	
	$ bil.	%	$ bil.	%	$ bil.	%	$ bil.	%	$ bil.	%
Establishments or enterprises (number)	7,081	0.302	10,686	0.509	6,966	0.372	5,593	0.312	7,213	0.944
Total employment (000)	37	0.027	41	0.030	40	0.032	39	0.035	44	0.063
Production workers (000)	-		-		-		-		-	
Output ($ bil.)	2	0.020	2	0.019	2	0.018	2	0.017	2	0.020
Value added ($ bil.)	0.799	0.019	0.826	0.017	0.708	0.015	0.624	0.013	0.649	0.019
Capital investment ($ mil.)	801	0.184	773	0.162	546	0.100	467	0.084	855	0.224
M & E investment ($ mil.)	508	0.163	-		-		200	0.047	-	
Employees per establishment (number)	5	8.911	4	5.887	6	8.735	7	11.335	6	6.658
Production workers per establishment	-		-		-		-		-	
Output per establishment ($ mil.)	0.287	6.619	0.203	3.658	0.304	4.835	0.351	5.486	0.290	2.171
Capital investment per estab. ($ mil.)	0.113	60.894	0.072	31.878	0.078	26.804	0.083	26.901	0.118	23.781
M & E per establishment ($ mil)	0.072	54.003	-		-		0.036	15.195	-	
Payroll per employee ($)	5,392	60.133	5,117	51.365	3,908	34.980	3,225	23.250	3,316	22.708
Wages per production worker ($)	-		-		-		-		-	
Hours per production worker (hours)	-		-		-		-		-	
Output per employee ($)	55,634	74.283	52,919	62.129	53,049	55.352	50,156	48.397	47,874	32.610
Capital investment per employee ($)	21,946	683.368	18,842	541.486	13,656	306.846	11,917	237.318	19,581	357.197
M & E per employee ($)	13,915	606.033	-		-		5,106	134.048	-	

Note: Columns headed % show percent of world total or ratio. Ratios closest to 100 are closest to world average. M & E stands for machinery & equipment.

Output in Manufacturing

	1987		1988		1989		1990		1991	
	$ bil.[p]	%	$ bil.[p]	%	$ bil.[p]	%	$ bil.[p]	%	$ bil.[p]	%
3110 - Food products	0.231	11.55	0.263	13.15	0.221	11.05	0.219	10.95	0.236	11.80
3130 - Beverages	0.070	3.50	0.064	3.20	0.049	2.45	0.049	2.45	0.054	2.70
3140 - Tobacco	0.152	7.60	0.142	7.10	0.090	4.50	0.092	4.60	0.097	4.85
3210 - Textiles	0.040	2.00	0.049	2.45	0.044	2.20	0.046	2.30	0.054	2.70
3211 - Spinning, weaving, etc.	-		0.039	1.95	0.034	1.70	0.035	1.75	0.032	1.60
3220 - Wearing apparel	0.028	1.40	0.031	1.55	0.025	1.25	0.026	1.30	0.033	1.65
3230 - Leather and products	0.006	0.30	0.007	0.35	0.004	0.20	0.008	0.40	0.014	0.70
3240 - Footwear	0.014	0.70	0.015	0.75	0.014	0.70	0.008	0.40	0.009	0.45
3310 - Wood products	-		0.026	1.30	0.025	1.25	-		-	
3320 - Furniture, fixtures	-		0.033	1.65	0.034	1.70	-		-	
3410 - Paper and products	0.063	3.15	0.066	3.30	0.058	2.90	0.066	3.30	0.071	3.55
3411 - Pulp, paper, etc.	-		-		-		-		0.009	0.45
3420 - Printing, publishing	0.035	1.75	0.043	2.15	0.029	1.45	0.032	1.60	0.036	1.80
3510 - Industrial chemicals	0.158	7.90	0.203	10.15	0.212	10.60	0.196	9.80	0.191	9.55
3511 - Basic chemicals, excl fertilizers	0.044	2.20	0.042	2.10	0.036	1.80	0.034	1.70	0.026	1.30
3513 - Synthetic resins, etc.	-		0.008	0.40	0.009	0.45	0.020	1.00	0.018	0.90
3520 - Chemical products nec	0.163	8.15	0.141	7.05	0.164	8.20	0.161	8.05	0.177	8.85
3522 - Drugs and medicines	0.069	3.45	0.069	3.45	0.075	3.75	0.074	3.70	0.074	3.70
3530 - Petroleum refineries	0.703	35.15	0.639	31.95	0.446	22.30	0.417	20.85	0.403	20.15
3550 - Rubber products	0.005	0.25	0.006	0.30	0.007	0.35	0.005	0.25	0.003	0.15
3560 - Plastic products nec	0.064	3.20	0.081	4.05	0.067	3.35	0.064	3.20	0.071	3.55
3610 - Pottery, china, etc.	-		-		0.000	0.00	0.008	0.40	0.010	0.50
3620 - Glass and products	-		-		0.016	0.80	0.012	0.60	0.009	0.45
3690 - Nonmetal products nec	-		-		0.169	8.45	0.158	7.90	0.191	9.55
3710 - Iron and steel	-		-		0.114	5.70	0.088	4.40	0.105	5.25
3720 - Nonferrous metals	-		-		0.026	1.30	0.020	1.00	0.025	1.25
3810 - Metal products	0.107	5.35	0.080	4.00	0.072	3.60	0.059	2.95	0.068	3.40
3820 - Machinery nec	0.016	0.80	0.034	1.70	0.028	1.40	0.022	1.10	0.031	1.55
3830 - Electrical machinery	0.013	0.65	0.018	0.90	0.033	1.65	0.032	1.60	0.029	1.45
3840 - Transportation equipment	0.003	0.15	0.006	0.30	0.005	0.25	0.002	0.10	0.003	0.15
3843 - Motor vehicles	0.003	0.15	0.006	0.30	0.005	0.25	0.002	0.10	0.003	0.15
3850 - Professional goods	-		0.004	0.20	0.005	0.25	0.005	0.25	0.004	0.20
3900 - Industries nec	0.044	2.20	0.055	2.75	0.003	0.15	0.004	0.20	0.002	0.10

Note: Codes are International Standard Industry codes (ISIC). Percentages are % of total Output. [f]: Factor Prices; [p]: Producer Prices.

For sources, notes, and explanations, see Annotated Source Appendix, page 1061.

485

Finance, Economics, and Trade

Economic Indicators [36]

- **National product**: GDP—purchasing power parity—$19.3 billion (1995 est.)
- **National product real growth rate**: 6.5% (1995 est.)
- **National product per capita**: $4,700 (1995 est.)
- **Inflation rate (consumer prices)**: 3% (1995 est.)
- **External debt**: $6.9 billion (1995 est.)

Balance of Payments Summary [37]

Values in millions of dollars.

	1989	1990	1991	1992	1993
Exports of goods (f.o.b.)	1,109.4	1,063.8	1,129.5	1,218.9	1,246.3
Imports of goods (f.o.b.)	-1,882.5	-2,300.7	-2,302.2	-2,998.7	-3,145.2
Trade balance	-773.1	-1,236.9	-1,172.7	-1,779.8	-1,898.9
Services - debits	-1,298.9	-1,549.6	-1,570.2	-1,784.7	-1,756.6
Services - credits	1,278.2	1,514.5	1,465.5	1,561.7	1,672.6
Private transfers (net)	565.4	457.4	408.1	781.3	997.1
Government transfers (net)	613.2	587.6	475.7	386.3	356.6
Long-term capital (net)	184.6	422.3	270.9	60.0	-167.4
Short-term capital (net)	-106.3	188.0	1,814.5	1,143.7	93.1
Errors and omissions	0.3	75.4	321.4	83.1	298.0
Overall balance	463.4	458.7	2,013.2	451.6	-405.5

Exchange Rates [38]

Currency: **Jordanian dinar.**
Symbol: **JD.**

Data are currency units per $1.

January 1996	0.7090
1995	0.7005
1994	0.6987
1993	0.6928
1992	0.6797
1991	0.6808

Imports and Exports

Top Import Origins [39]

$3.8 billion (c.i.f., 1994).

Origins	%
EU	NA
US	NA
Iraq	NA
Japan	NA
Turkey	NA

Top Export Destinations [40]

$1.7 billion (f.o.b., 1994).

Destinations	%
India	NA
Iraq	NA
Saudi Arabia	NA
EU	NA
Indonesia	NA
UAE	NA

Foreign Aid [41]

Recipient: ODA, $238 million (1993).

Import and Export Commodities [42]

Import Commodities	Export Commodities
Crude oil	Phosphates
Machinery	Fertilizers
Transport equipment	Potash
Food	Agricultural products
Live animals	Manufactures
Manufactured goods	

Kazakhstan

Geography [1]

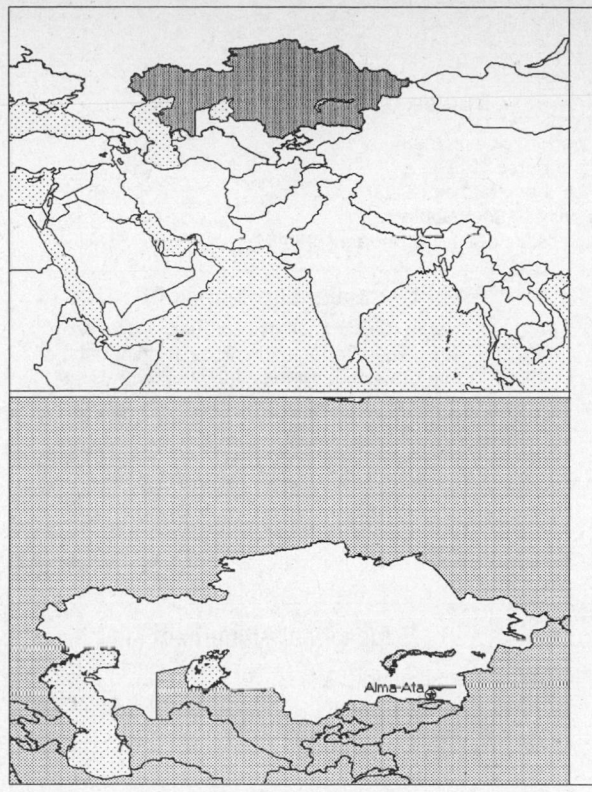

Total area:
2,717,300 sq km   1,049,155 sq mi
Land area:
2,669,800 sq km   1,030,816 sq mi
Comparative area:
Slightly less than four times the size of Texas
Land boundaries:
Total 12,012 km, China 1,533 km, Kyrgyzstan 1,051 km, Russia 6,846 km, Turkmenistan 379 km, Uzbekistan 2,203 km
Coastline:
0 km (landlocked)
Note: Kazakhstan borders the Aral Sea (1,015 km) and the Caspian Sea (1,894 km)
Climate:
Continental, cold winters and hot summers, arid and semiarid
Terrain:
Extends from the Volga to the Altai Mountains and from the plains in Western Siberia to oasis and desert in Central Asia
Natural resources:
Major deposits of petroleum, coal, iron ore, manganese, chrome ore, nickel, cobalt, copper, molybdenum, lead, zinc, bauxite, gold, uranium
Land use:
Arable land: 15%
Permanent crops: Negligible
Meadows and pastures: 57%
Forest and woodland: 4%
Other: 24%

Demographics [2]

	1970	1980	1990	1995[1]	1996	2000	2010	2020	2030
Population	13,106	14,994	16,708	16,950	16,916	16,943	17,564	18,408	19,387
Population density (persons per sq. mi.)	12	14	16	16	16	16	17	18	18
(persons per sq. km.)	5	6	6	6	6	6	6	7	7
Net migration rate (per 1,000 population)	NA	NA	-7.8	-11.9	-10.9	-6.6	-3.8	-1.2	0.0
Births	NA	NA	NA	NA	322	NA	NA	NA	NA
Deaths	NA	NA	NA	NA	163	NA	NA	NA	NA
Life expectancy - males	NA	NA	62.2	58.4	58.6	59.2	60.6	64.7	68.3
Life expectancy - females	NA	NA	71.7	69.7	69.9	70.5	72.0	75.2	78.0
Birth rate (per 1,000)	NA	NA	22.8	19.1	19.0	18.8	18.7	15.4	14.3
Death rate (per 1,000)	NA	NA	8.2	9.6	9.7	9.8	10.3	9.3	9.1
Women of reproductive age (15-49 yrs.)	NA	NA	4,175	4,425	4,464	4,573	4,709	4,787	4,938
of which are currently married	NA	NA	2,724	NA	2,914	2,968	3,133	NA	NA
Fertility rate	NA	NA	2.8	2.4	2.4	2.3	2.2	2.1	2.0

Except as noted, values for vital statistics are in thousands; life expectancy is in years.

Health

Health Indicators [3]

% of population with access to	
safe water (1990-95)	NA
adequate sanitation (1990-95)	NA
health services (1985-95)	NA
% of 1-year-olds immunized (1990-94) against	
TB (tuberculosis)	87
DPT (diphtheria, pertussis, tetanus)	80
polio	75
measles	72
% of contraceptive prevalence (1980-94)	NA
ORT use rate (1990-94)	NA

Health Expenditures [4]

Total health expenditure, 1990 (official exchange rate)	
Millions of dollars	2,572
Dollars per capita	154
Health expenditures as a percentage of GDP	
Total	4.4
Public sector	2.8
Private sector	1.7
Development assistance for health	
Total aid flows (millions of dollars)[1]	NA
Aid flows per capita (dollars)	NA
Aid flows as a percentage of total health expenditure	NA

For sources, notes, and explanations, see Annotated Source Appendix, page 1061.

Human Factors

Women and Children [5]

% of pregnant women immunized (tetanus 1990-94)	NA
% of births attended by trained health personnel (1983-94)	NA
Maternal mortality rate (1980-92)	NA
Under-5 mortality rate (1994)	48
% under-5 moderately/severely underweight (1980-1994)	NA

Burden of Disease [6]

Population per physician (1993)	253.57
Population per nurse (1993)	90.72
Population per hospital bed (1993)	75.40
AIDS cases per 100,000 people (1994)	*
Heart disease cases per 1,000 people (1990-93)	NA

Ethnic Division [7]

Kazak (Qazaq)	41.9%
Russian	37%
Ukrainian	5.2%
German	4.7%
Uzbek	2.1%,
Tatar	2%
Other	7.1%

Religion [8]

Muslim	47%
Russian Orthodox	44%
Protestant	2%
Other	7%

Major Languages [9]

Kazak (Qazaqz) official language spoken by over 40% of population, Russian (language of interethnic communication) spoken by two-thirds of population and used in everyday business.

Education

Public Education Expenditures [10]

Million (Tenge)	1980	1985	1990	1991	1992	1994
Total education expenditure	NA	2,174	3,001	6,252	65,247	NA
as percent of GNP	NA	6.5	6.5	7.7	5.4	NA
as percent of total govt. expend.	NA	18.9	17.6	19.1	25.2	NA
Current education expenditure	NA	NA	NA	NA	60,549	NA
as percent of GNP	NA	NA	NA	NA	5.0	NA
as percent of current govt. expend.	NA	NA	NA	NA	NA	NA
Capital expenditure	NA	NA	NA	NA	4,698	NA

Educational Attainment [11]

Age group (1989)	25+
Total population	8,414,539
Highest level attained (%)	
No schooling	7.7
First level	
Not completed	29.2
Completed	NA
Entered second level	
S-1	50.7
S-2	NA
Postsecondary	12.4

Illiteracy [12]

	1985	1989	1995
Illiterate population (15+ yrs.)[2]	NA	276,835	46,000
Illiteracy rate - total pop. (%)	NA	2.5	0.4
Illiteracy rate - males (%)	NA	0.9	0.3
Illiteracy rate - females (%)	NA	3.9	0.5

Libraries [13]

	Admin. Units	Svc. Pts.	Vols. (000)	Shelving (meters)	Vols. Added	Reg. Users
National (1993)	1	NA	5,209	NA	94,000	46,000
Nonspecialized	NA	NA	NA	NA	NA	NA
Public (1993)	8,770	NA	104,362	NA	5 mil	7 mil
Higher ed.	NA	NA	NA	NA	NA	NA
School	NA	NA	NA	NA	NA	NA

Daily Newspapers [14]

Culture [15]

Cinema (seats per 1,000)	NA
Annual attendance per person	2.3
Gross box office receipts (mil. Tenge)	2.3
Museums (reporting)	180
Visitors (000)	6,536
Annual receipts (000 Tenge)	NA

Science and Technology

Scientific/Technical Forces [16]

R&D Expenditures [17]

U.S. Patents Issued [18]

Values show patents issued to citizens of the country by the U.S. Patents Office.

	1993	1994	1995
Number of patents	0	1	0

Government and Law

Organization of Government [19]

Long-form name:
Republic of Kazakhstan
Type:
Republic
Independence:
16 December 1991 (from the Soviet Union)
National holiday:
Independence Day, 16 December (1991)
Constitution:
Adopted 28 January 1993; has been amended in April 1995 and August 1995
Legal system:
Based on civil law system
Executive branch:
President; Prime Minister; First Deputy Prime Minister; Council of Ministers Note: President has expanded his presidential powers by decree: only he can initiate constitutional amendments, appoint and dismiss the government, dissolve parliament, call referenda at his discretion, and appoint administrative heads of regions and cities
Legislative branch:
Bicameral Parliament: Senate and Majilis
Judicial branch:
Supreme Court

Elections [20]

Majilis Elections last held 9 and 23 December 1995; results - percent of vote by party NA; seats (67 total) by party NA. 172 candidates were put forward by parties and social organizations and 113 candidates were independents.

Government Budget [21]

Crime [23]

	1990
Crime volume	
Cases known to police	106,290
Attempts (percent)	NA
Percent cases solved	44.8
Crimes per 100,000 persons	636.3
Persons responsible for offenses	
Total number offenders	47,666
Percent female	9.5
Percent juvenile	14.4
Percent foreigners	NA

Military Expenditures and Arms Transfers [22]

	1990	1991	1992[14]	1993	1994
Military expenditures					
Current dollars (mil.)[e]	NA	NA	NA	NA	450
1994 constant dollars (mil.)[e]	NA	NA	NA	NA	450
Armed forces (000)	NA	NA	44	44	18
Gross national product (GNP)					
Current dollars (mil.)[e]	NA	NA	71,530	64,730	48,190
1994 constant dollars (mil.)[e]	NA	NA	74,590	66,060	48,190
Central government expenditures (CGE)					
1994 constant dollars (mil.)[e]	NA	NA	20,600	NA	NA
People (mil.)	NA	NA	17.0	17.2	17.3
Military expenditure as % of GNP	NA	NA	NA	NA	0.9
Military expenditure as % of CGE	NA	NA	NA	NA	NA
Military expenditure per capita (1994 $)	NA	NA	NA	NA	26
Armed forces per 1,000 people (soldiers)	NA	NA	2.6	2.6	1.0
GNP per capita (1994 $)	NA	NA	4,378	3,851	2,791
Arms imports[6]					
Current dollars (mil.)	NA	NA	0	80	0
1994 constant dollars (mil.)	NA	NA	0	82	0
Arms exports[6]					
Current dollars (mil.)	NA	NA	0	0	0
1994 constant dollars (mil.)	NA	NA	0	0	0
Total imports[7]					
Current dollars (mil.)	NA	NA	4,080	3,358	3,741
1994 constant dollars (mil.)	NA	NA	4,255	3,427	3,741
Total exports[7]					
Current dollars (mil.)	NA	NA	3,166	3,391	2,950
1994 constant dollars (mil.)	NA	NA	3,302	3,461	2,950
Arms as percent of total imports[8]	NA	NA	0	2.4	0
Arms as percent of total exports[8]	NA	NA	0	0	0

Human Rights [24]

	SSTS	FL	FAPRO	PPCG	APROBC	TPW	PCPTW	STPEP	PHRFF	PRW	ASST	AFL
Observes						P	P					

	EAFRD	CPR	ESCR	SR	ACHR	MAAE	PVIAC	PVNAC	EAFDAW	TCIDTP	RC
Observes					P	P					P

P = Party; S = Signatory; see Appendix for meaning of abbreviations.

Labor Force

Total Labor Force [25]

7.356 million

Labor Force by Occupation [26]

Industry and construction	31%
Agriculture and forestry	26
Other	43

Date of data: 1992

Unemployment Rate [27]

1.4% includes only officially registered unemployed; also large numbers of underemployed workers (September 1995 est.)

For sources, notes, and explanations, see Annotated Source Appendix, page 1061.

489

Production Sectors

Commercial Energy Production and Consumption

Data are shown in quadrillion (10^{15}) BTUs and percent for 1995
Values for hydroelectric, nuclear, geothermal, solar, and wind power refer to electrical generation.

Production [28]	Consumption [29]
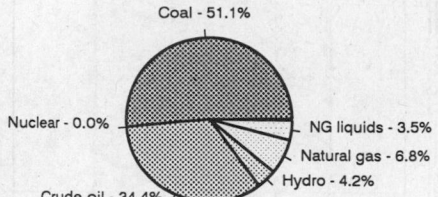	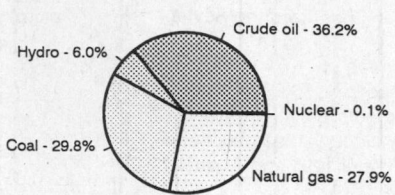

Production [28]:
- Coal - 51.1%
- Nuclear - 0.0%
- NG liquids - 3.5%
- Natural gas - 6.8%
- Hydro - 4.2%
- Crude oil - 34.4%

Consumption [29]:
- Crude oil - 36.2%
- Hydro - 6.0%
- Nuclear - 0.1%
- Coal - 29.8%
- Natural gas - 27.9%

Crude oil	0.777	Crude oil	0.573
Natural gas liquids	0.079	Dry natural gas	0.441
Dry natural gas	0.154	Coal	0.471
Coal	1.156	Net hydroelectric power	0.095
Net hydroelectric power	0.095	Net nuclear power	0.001
Net nuclear power	0.001	Total	1.581
Total	2.262		

Telecommunications [30]

- 2.2 million telephones; service is poor
- Domestic: landline and microwave radio relay
- International: international traffic with other former Soviet republics and China carried by landline and microwave radio relay and with other countries by satellite and through 8 international telecommunications circuits at the Moscow international gateway switch; satellite earth stations - 1 Intelsat and a new satellite earth station established at Almaty of unknown type
- Radio: Broadcast stations: AM NA, FM NA, shortwave NA Radios: 4.088 million (with multiple speakers for program diffusion 6.082 million)
- Television: Broadcast stations: NA; Orbita (TV receive only) earth station Televisions: 4.75 million

Transportation [31]

Railways: total: 13,841 km in common carrier service; does not include industrial lines; broad gauge: 13,841 km 1.520-m gauge (3,299 km electrified) (1992)

Highways: total: 87,873 km public roads; paved: 82,568 km; unpaved: 5,305 km (1994)

Airports

Total:	352
With paved runways over 3,047 m:	7
With paved runways 2,438 to 3,047 m:	23
With paved runways 1,524 to 2,437 m:	11
With paved runways 914 to 1,523 m:	5
With paved runways under 914 m:	9

Top Agricultural Products [32]

Agriculture accounts for 28.5% of the GDP; produces grain, mostly spring wheat, cotton; wool, meat.

Top Mining Products [33]

Thousand metric tons except as noted[e]	8/25/95[*]
Bauxite	2,430[r]
Bismuth metal (000 kg.)	85[r]
Chromite	2,020[r]
Gold, mine output (tons)	26
Steel, crude	2,840[r]
Iron ore, marketable	10,500
Cement	4,000
Coal	104,000[r]
Gas, natural (mil. cu. meters)	4,500[r]
Petroleum, crude	20,300[r]

Tourism [34]

Finance, Economics, and Trade

Industrial Summary [35]

Industrial Production: Growth rate - 8% (1995); accounts for 41.5% of the GDP. **Industries**: Oil, coal, iron ore, manganese, chromite, lead, zinc, copper, titanium, bauxite, gold, silver, phosphates, sulfur, iron and steel, nonferrous metal, tractors and other agricultural machinery, electric motors, construction materials; much of industrial capacity is shut down and/or is in need of repair.

Economic Indicators [36]

- **National product**: GDP—purchasing power parity—$46.9 billion (1995 estimate as extrapolated from World Bank estimate for 1994)

- **National product real growth rate**: - 8.9% (1995 est.)

- **National product per capita**: $2,700 (1995 est.)

- **Inflation rate (consumer prices)**: 60.3% (1995 est.)

- **External debt**: $2.5 billion (of which $1.3 billion to Russia)

Balance of Payments Summary [37]

Values in millions of dollars.

	1989	1990	1991	1992	1993[1]
Exports of goods (f.o.b.)	NA	NA	NA	NA	3,414
Imports of goods (f.o.b.)	NA	NA	NA	NA	-3,236
Trade balance	NA	NA	NA	NA	178
Services - debits	NA	NA	NA	NA	-264
Services - credits	NA	NA	NA	NA	271
Private transfers (net)	NA	NA	NA	NA	115
Government transfers (net)	NA	NA	NA	NA	NA
Long-term capital (net)	NA	NA	NA	NA	1,022
Short-term capital (net)	NA	NA	NA	NA	NA
Errors and omissions	NA	NA	NA	NA	-1,790
Overall balance	NA	NA	NA	NA	-468

Exchange Rates [38]

Currency: **tenge.**

Data are currency units per $1. The tenge was introduced on 15 November 1993.

Year-end 1995	64
Year-end 1994	54

Imports and Exports

Top Import Origins [39]

$3.9 billion (1995).

Origins	%
Russia	NA
Other former Soviet republics	NA
China	NA

Top Export Destinations [40]

$5.1 billion (1995).

Destinations	%
Russia	NA
Ukraine	NA
Uzbekistan	NA

Foreign Aid [41]

	U.S. $	
ODA (1993)	10	million
Commitments (1992-95)	4,780	million

Import and Export Commodities [42]

Import Commodities	Export Commodities
Machinery and parts	Oil
Industrial materials	Ferrous and nonferrous metals
Oil and gas	Chemicals
	Wool
	Meat
	Coal
	Grain

For sources, notes, and explanations, see Annotated Source Appendix, page 1061.

491

Kenya

Geography [1]

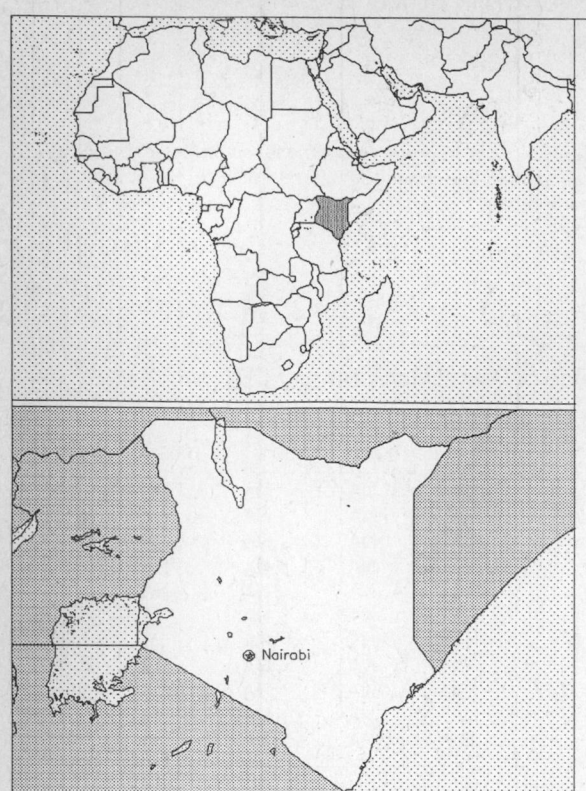

Total area:
582,650 sq km   224,962 sq mi
Land area:
569,250 sq km   219,789 sq mi
Comparative area:
Slightly more than twice the size of Nevada
Land boundaries:
Total 3,446 km, Ethiopia 830 km, Somalia 682 km, Sudan 232 km, Tanzania 769 km, Uganda 933 km
Coastline:
536 km
Climate:
Varies from tropical along coast to arid in interior
Terrain:
Low plains rise to central highlands bisected by Great Rift Valley; fertile plateau in West
Natural resources:
Gold, limestone, soda ash, salt barytes, rubies, fluorspar, garnets, wildlife
Land use:
Arable land: 3%
Permanent crops: 1%
Meadows and pastures: 7%
Forest and woodland: 4%
Other: 85%

Demographics [2]

	1970	1980	1990	1995[1]	1996	2000	2010	2020	2030
Population	11,272	16,685	23,896	27,616	28,177	30,490	33,920	35,236	37,229
Population density (persons per sq. mi.)	51	76	109	126	128	139	154	160	169
(persons per sq. km.)	20	29	42	49	49	54	60	62	65
Net migration rate (per 1,000 population)	NA	0.0	0.0	-7.1	-0.4	-0.3	0.0	0.0	0.0
Births	NA	NA	NA	NA	941	NA	NA	NA	NA
Deaths	NA	NA	NA	NA	290	NA	NA	NA	NA
Life expectancy - males	NA	55.9	59.6	56.9	55.5	50.6	43.7	47.7	59.3
Life expectancy - females	NA	58.9	62.5	56.8	55.7	51.5	42.7	47.2	61.4
Birth rate (per 1,000)	NA	50.0	39.2	34.3	33.4	29.5	24.0	21.1	18.8
Death rate (per 1,000)	NA	11.1	8.3	9.8	10.3	12.6	19.1	17.6	10.9
Women of reproductive age (15-49 yrs.)	NA	3,476	5,252	6,406	6,603	7,479	9,083	9,796	10,511
of which are currently married	NA	NA	2,270	NA	2,852	3,234	4,075	NA	NA
Fertility rate	NA	7.6	5.7	4.6	4.5	3.7	2.6	2.2	2.1

Except as noted, values for vital statistics are in thousands; life expectancy is in years.

Health

Health Indicators [3]

% of population with access to	
safe water (1990-95)	53
adequate sanitation (1990-95)	77
health services (1985-95)	77
% of 1-year-olds immunized (1990-94) against	
TB (tuberculosis)	92
DPT (diphtheria, pertussis, tetanus)	84
polio	84
measles	73
% of contraceptive prevalence (1980-94)	33
ORT use rate (1990-94)	76

Health Expenditures [4]

Total health expenditure, 1990 (official exchange rate)	
Millions of dollars	375
Dollars per capita	16
Health expenditures as a percentage of GDP	
Total	4.3
Public sector	2.7
Private sector	1.6
Development assistance for health	
Total aid flows (millions of dollars)[1]	84
Aid flows per capita (dollars)	3.5
Aid flows as a percentage of total health expenditure	22.3

For sources, notes, and explanations, see Annotated Source Appendix, page 1061.

Human Factors

Women and Children [5]

% of pregnant women immunized (tetanus 1990-94)	72
% of births attended by trained health personnel (1983-94)	54
Maternal mortality rate (1980-92)[1]	170
Under-5 mortality rate (1994)	90
% under-5 moderately/severely underweight (1980-1994)	22

Burden of Disease [6]

Population per physician (1989)	9,851.13
Population per nurse	NA
Population per hospital bed (1990)	601.98
AIDS cases per 100,000 people (1994)	24.8
Malaria cases per 100,000 people (1992)	NA

Ethnic Division [7]

Kikuyu	22%
Luhya	14%
Luo	13%
Kalenjin	12%
Kamba	11%
Kisii and Meru	6%
Asian, European, and Arab	1%
Other	15%

Religion [8]

Protestant (primarily Anglican)	38%
Roman Catholic	28%
Indigenous beliefs	26%
Other	8%

Major Languages [9]

English (official), Swahili (official), numerous indigenous languages.

Education

Public Education Expenditures [10]

	1980	1990	1991	1992	1993	1994
Million (Shilling)[36]						
Total education expenditure	3,526	12,473	11,952	13,203	20,029	NA
as percent of GNP	6.8	6.7	5.7	5.4	6.8	NA
as percent of total govt. expend.	18.1	16.1	NA	NA	NA	NA
Current education expenditure	3,247	11,238	10,989	12,307	18,999	NA
as percent of GNP	6.2	6.1	5.3	5.1	6.4	NA
as percent of current govt. expend.	23.6	19.4	NA	NA	NA	NA
Capital expenditure	279	1,235	963	896	1,030	NA

Educational Attainment [11]

Age group (1979)[13]	25+
Total population	4,818,310
Highest level attained (%)	
No schooling	58.6
First level	
Not completed	32.2
Completed	NA
Entered second level	
S-1	7.9
S-2	1.3
Postsecondary	NA

Illiteracy [12]

In thousands and percent[1]	1990	1995	2000
Illiterate population (15+ yrs.)	3,357	3,237	2,934
Illiteracy rate - total pop. (%)	27.3	21.5	16.5
Illiteracy rate - males (%)	17.5	13.5	10.2
Illiteracy rate - females (%)	36.8	29.3	22.8

Libraries [13]

	Admin. Units	Svc. Pts.	Vols. (000)	Shelving (meters)	Vols. Added	Reg. Users
National (1989)	1	22	603	NA	10,260	178,978
Nonspecialized	NA	NA	NA	NA	NA	NA
Public	NA	NA	NA	NA	NA	NA
Higher ed.	NA	NA	NA	NA	NA	NA
School	NA	NA	NA	NA	NA	NA

Daily Newspapers [14]

	1980	1985	1990	1994
Number of papers	3	4	5	5
Circ. (000)	216	283	330[e]	358[e]

Culture [15]

Cinema (seats per 1,000)	0.3
Annual attendance per person	0.2
Gross box office receipts (mil. Shilling)	NA
Museums (reporting)	NA
Visitors (000)	NA
Annual receipts (000 Shilling)	NA

Science and Technology

Scientific/Technical Forces [16]

Scientists/engineers	16,241
Number female	NA
Technicians	17,741
Number female	NA
Total[36]	33,982

R&D Expenditures [17]

U.S. Patents Issued [18]

Values show patents issued to citizens of the country by the U.S. Patents Office.

	1993	1994	1995
Number of patents	2	1	0

For sources, notes, and explanations, see Annotated Source Appendix, page 1061.

493

Government and Law

Organization of Government [19]

Long-form name:
Republic of Kenya
Type:
Republic
Independence:
12 December 1963 (from UK)
National holiday:
Independence Day, 12 December (1963)
Constitution:
12 December 1963, amended as a
republic 1964; reissued with amendments
1979, 1983, 1986, 1988, 1991, and 1992
Legal system:
Based on English common law, tribal law,
and Islamic law; judicial review in High
Court; accepts compulsory ICJ
jurisdiction, with reservations;
constitutional amendment of 1982 making
Kenya a de jure one-party state repealed
in 1991
Executive branch:
President; Vice President; Cabinet
Legislative branch:
Unicameral: National Assembly (Bunge)
Judicial branch:
Court of Appeal; High Court

Elections [20]

National Assembly	% of seats
Kenya African National Union (KANU)	53.2
Forum for Restoration (FORD-Kenya)	16.5
FORD-Asili	16.5
Democratic Party of Kenya (DP)	12.2
Smaller parties	1.6
Members nominated by President	6.4

Government Expenditures [21]

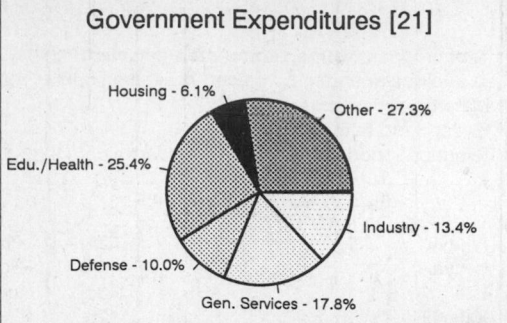

Housing - 6.1%
Other - 27.3%
Edu./Health - 25.4%
Industry - 13.4%
Defense - 10.0%
Gen. Services - 17.8%

(% distribution). Expend. for FY94: 113,721 (Shilling mil.)

Crime [23]

	1990
Crime volume	
Cases known to police	87,400
Attempts (percent)	NA
Percent cases solved	48,175
Crimes per 100,000 persons	364.16
Persons responsible for offenses	
Total number offenders	61,175
Percent female	6,316
Percent juvenile (13-17 yrs.)	3,909
Percent foreigners	NA

Military Expenditures and Arms Transfers [22]

	1990	1991	1992	1993	1994
Military expenditures					
Current dollars (mil.)	166	169	186	162	138
1994 constant dollars (mil.)	185	181	194	166	138
Armed forces (000)	20	20	24	24	22
Gross national product (GNP)					
Current dollars (mil.)	5,707	5,972	6,139	6,093	6,565
1994 constant dollars (mil.)	6,351	6,401	6,402	6,219	6,565
Central government expenditures (CGE)					
1994 constant dollars (mil.)	1,882	2,006	1,689	1,853	2,053
People (mil.)	24.2	25.1	26.2	27.4	28.2
Military expenditure as % of GNP	2.9	2.8	3.0	2.7	2.1
Military expenditure as % of CGE	9.8	9.0	11.5	8.9	6.7
Military expenditure per capita (1994 $)	8	7	7	6	5
Armed forces per 1,000 people (soldiers)	0.8	0.8	0.9	0.9	0.8
GNP per capita (1994 $)	262	255	244	227	232
Arms imports[6]					
Current dollars (mil.)	70	30	0	5	10
1994 constant dollars (mil.)	78	32	0	5	10
Arms exports[6]					
Current dollars (mil.)	0	0	0	0	0
1994 constant dollars (mil.)	0	0	0	0	0
Total imports[7]					
Current dollars (mil.)	2,124	1,798	1,713	1,711	2,156
1994 constant dollars (mil.)	2,364	1,927	1,786	1,746	2,156
Total exports[7]					
Current dollars (mil.)	1,031	1,107	1,339	1,336	1,609
1994 constant dollars (mil.)	1,147	1,187	1,396	1,364	1,609
Arms as percent of total imports[8]	3.3	1.7	0	0.3	0.5
Arms as percent of total imports[8]	0	0	0	0	0

Human Rights [24]

	SSTS	FL	FAPRO	PPCG	APROBC	TPW	PCPTW	STPEP	PHRFF	PRW	ASST	AFL
Observes		P			P	P	P					P
	EAFRD	CPR	ESCR	SR	ACHR	MAAE	PVIAC	PVNAC	EAFDAW	TCIDTP	RC	
Observes		P	P	P		P				P		P

P = Party; S = Signatory; see Appendix for meaning of abbreviations.

Labor Force

Total Labor Force [25]

Labor Force by Occupation [26]

Agriculture	~80%
Nonagriculture	~20

Date of data: 1993 est.

Unemployment Rate [27]

35% urban (1994 est.)

For sources, notes, and explanations, see Annotated Source Appendix, page 1061.

Production Sectors

Commercial Energy Production and Consumption

Data are shown in quadrillion (10^{15}) BTUs and percent for 1995
Values for hydroelectric, nuclear, geothermal, solar, and wind power refer to electrical generation.

Production [28]

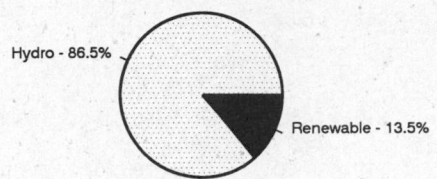

Hydro - 86.5%
Renewable - 13.5%

Consumption [29]

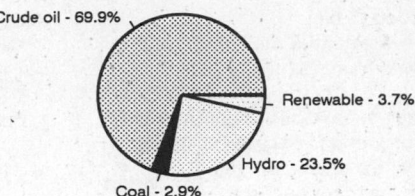

Crude oil - 69.9%
Renewable - 3.7%
Hydro - 23.5%
Coal - 2.9%

Net hydroelectric power	0.032
Geothermal, solar, wind	0.005
Total	0.037

Crude oil	0.095
Coal	0.004
Net hydroelectric power	0.032
Geothermal, solar, wind	0.005
Total	0.136

Telecommunications [30]

- 357,251 (1989 est.) telephones; in top group of African systems
- Domestic: primarily microwave radio relay
- International: satellite earth stations - 2 Intelsat (1 Atlantic Ocean and 1 Indian Ocean)
- Radio: Broadcast stations: AM 16, FM 4, shortwave 0
- Television: Broadcast stations: 6 Televisions: 260,000 (1993 est.)

Transportation [31]

Railways: total: 2,652 km; narrow gauge: 2,652 km 1.000-m gauge

Highways: total: 62,573 km; paved: 8,322 km; unpaved: 54,251 km (1991 est.)

Merchant marine: total: 2 ships (1,000 GRT or over) totaling 4,883 GRT/6,255 DWT; ships by type: oil tanker 1, roll on/roll off 1 (1995 est.)

Airports

Total:	199
With paved runways over 3,047 m:	3
With paved runways 2,438 to 3,047 m:	2
With paved runways 1,524 to 2,437 m:	2
With paved runways 914 to 1,523 m:	22
With paved runways under 914 m:	62

Top Agricultural Products [32]

Agriculture accounts for 27% of the GDP; produces coffee, tea, corn, wheat, sugarcane, fruit, vegetables; dairy products, beef, pork, poultry, eggs.

Top Mining Products [33]

Thousand metric tons except as noted	4/17/95[*]
Feldspar	1,200[e]
Fluorspar (acid grade)	64,000[r]
Amethyst (kg.)	303
Ruby (kg.)	120
Sapphire (kg.)	2,310[r]
Tourmaline (kg.)	229
Petroleum refinery products (000 42-gal. bls.)	16,000
Coral	1,600[e]
Shale	115,000[e]
Vermiculite	1,960

Tourism [34]

	1990	1991	1992	1993	1994
Tourists[60]	814	805	782	826	863
Tourism receipts	466	432	442	413	421
Tourism expenditures	38	24	29	48	115
Fare receipts	113	108	112	124	NA
Fare expenditures	26	46	32	33	NA

Travelers are in thousands, money in million U.S. dollars.

Manufacturing Sector

Manufacturing Summary [35]

	1987		1988		1989		1990		1991	
	$ bil.	%	$ bil.	%	$ bil.	%	$ bil.	%	$ bil.	%
Establishments or enterprises (number)	611	0.026	630	0.030	612	0.033	635	0.035	644	0.084
Total employment (000)	186	0.137	193	0.141	195	0.158	199	0.180	198	0.286
Production workers (000)	-		-		-		-		-	
Output ($ bil.)	7	0.064	7	0.062	7	0.063	8	0.071	8	0.081
Value added ($ bil.)	0.894	0.021	0.967	0.020	0.953	0.020	0.978	0.020	0.961	0.028
Capital investment ($ mil.)	-		-		-		-		-	
M & E investment ($ mil.)	-		-		-		-		-	
Employees per establishment (number)	305	526.656	306	469.522	319	484.937	314	507.484	308	338.993
Production workers per establishment	-		-		-		-		-	
Output per establishment ($ mil.)	11	246.434	12	207.707	12	194.001	13	199.357	13	96.272
Capital investment per estab. ($ mil.)	-		-		-		-		-	
M & E per establishment ($ mil)	-		-		-		-		-	
Payroll per employee ($)	2,027	22.609	2,215	22.232	2,076	18.580	2,030	14.634	1,842	12.618
Wages per production worker ($)	-		-		-		-		-	
Hours per production worker (hours)	-		-		-		-		-	
Output per employee ($)	35,045	46.792	37,680	44.238	38,341	40.005	40,711	39.284	41,693	28.399
Capital investment per employee ($)	-		-		-		-		-	
M & E per employee ($)	-		-		-		-		-	

Note: Columns headed % show percent of world total or ratio. Ratios closest to 100 are closest to world average. M & E stands for machinery & equipment.

Output in Manufacturing

	1987		1988		1989		1990		1991	
	$ bil.[f]	%	$ bil.[f]	%	$ bil.[f]	%	$ bil.[f]	%	$ bil.[f]	%
3110 - Food products	2.341	33.44	2.525	36.07	2.619	37.41	3.139	39.24	3.246	40.58
3130 - Beverages	0.217	3.10	0.229	3.27	0.232	3.31	0.197	2.46	0.193	2.41
3140 - Tobacco	0.142	2.03	0.151	2.16	0.148	2.11	0.158	1.97	0.152	1.90
3210 - Textiles	0.225	3.21	0.222	3.17	0.216	3.09	0.106	1.33	0.120	1.50
3211 - Spinning, weaving, etc.	-		-		0.017	0.24	0.015	0.19	0.012	0.15
3220 - Wearing apparel	0.125	1.79	0.147	2.10	0.120	1.71	0.143	1.79	0.226	2.83
3230 - Leather and products	0.027	0.39	0.025	0.36	0.021	0.30	0.017	0.21	0.035	0.44
3240 - Footwear	0.040	0.57	0.039	0.56	0.038	0.54	0.045	0.56	0.041	0.51
3310 - Wood products	0.069	0.99	0.072	1.03	0.076	1.09	0.081	1.01	0.078	0.98
3320 - Furniture, fixtures	0.028	0.40	0.029	0.41	0.027	0.39	0.033	0.41	0.025	0.31
3410 - Paper and products	0.131	1.87	0.133	1.90	0.136	1.94	0.157	1.96	0.180	2.25
3411 - Pulp, paper, etc.	0.076	1.09	0.073	1.04	0.077	1.10	0.095	1.19	0.144	1.80
3420 - Printing, publishing	0.073	1.04	0.066	0.94	0.063	0.90	0.075	0.94	0.072	0.90
3510 - Industrial chemicals	0.161	2.30	0.151	2.16	0.135	1.93	0.153	1.91	0.133	1.66
3511 - Basic chemicals, excl fertilizers	0.110	1.57	0.107	1.53	0.098	1.40	0.115	1.44	0.104	1.30
3520 - Chemical products nec	0.375	5.36	0.845	12.07	0.930	13.29	0.988	12.35	0.784	9.80
3522 - Drugs and medicines	0.082	1.17	0.100	1.43	0.107	1.53	0.104	1.30	0.134	1.68
3530 - Petroleum refineries	0.724	10.34	0.729	10.41	0.737	10.53	0.804	10.05	0.886	11.08
3550 - Rubber products	0.118	1.69	0.124	1.77	0.117	1.67	0.136	1.70	0.132	1.65
3560 - Plastic products nec	0.083	1.19	0.097	1.39	0.088	1.26	0.086	1.08	0.110	1.38
3610 - Pottery, china, etc.	0.001	0.01	0.001	0.01	0.001	0.01	0.007	0.09	0.001	0.01
3620 - Glass and products	0.007	0.10	0.008	0.11	0.014	0.20	0.011	0.14	0.009	0.11
3690 - Nonmetal products nec	0.169	2.41	0.161	2.30	0.162	2.31	0.183	2.29	0.214	2.67
3810 - Metal products	0.296	4.23	0.296	4.23	0.335	4.79	0.326	4.07	0.365	4.56
3820 - Machinery nec	0.024	0.34	0.025	0.36	0.021	0.30	0.010	0.13	0.010	0.13
3830 - Electrical machinery	0.216	3.09	0.215	3.07	0.254	3.63	0.251	3.14	0.275	3.44
3840 - Transportation equipment	0.380	5.43	0.401	5.73	0.388	5.54	0.354	4.42	0.321	4.01
3841 - Shipbuilding, repair	0.058	0.83	0.063	0.90	0.067	0.96	0.061	0.76	0.051	0.64
3843 - Motor vehicles	0.172	2.46	0.174	2.49	0.149	2.13	0.156	1.95	0.110	1.38
3850 - Professional goods	0.002	0.03	0.002	0.03	0.002	0.03	0.009	0.11	0.008	0.10
3900 - Industries nec	0.052	0.74	0.058	0.83	0.082	1.17	0.091	1.14	0.099	1.24

Note: Codes are International Standard Industry codes (ISIC). Percentages are % of total Output. [f]: Factor Prices; [p]: Producer Prices.

Finance, Economics, and Trade

Economic Indicators [36]

- **National product**: GDP—purchasing power parity—$36.8 billion (1995 est.)

- **National product real growth rate**: 5% (1995 est.)

- **National product per capita**: $1,300 (1995 est.)

- **Inflation rate (consumer prices)**: 1.7% (1995 est.)

- **External debt**: $7 billion (1994 est.)

Balance of Payments Summary [37]

Values in millions of dollars.

	1989	1990	1991	1992	1993
Exports of goods (f.o.b.)	926.1	1,010.5	1,053.8	1,004.0	1,185.6
Imports of goods (f.o.b.)	-1,963.4	-2,005.3	-1,697.3	-1,594.3	-1,492.8
Trade balance	-1,037.3	-994.8	-643.5	-590.3	-307.2
Services - debits	-933.3	-1,122.9	-1,067.4	-937.7	-925.2
Services - credits	1,008.5	1,222.8	1,151.9	1,148.3	1,143.9
Private transfers (net)	101.5	167.8	144.4	68.3	147.1
Government transfers (net)	281.1	206.9	204.5	214.2	93.9
Long-term capital (net)	601.8	223.2	136.8	-162.1	45.0
Short-term capital (net)	32.0	137.6	-40.2	-259.2	53.0
Errors and omissions	67.7	66.9	69.6	110.3	257.5
Overall balance	122.0	-92.5	-43.9	110.2	508.0

Exchange Rates [38]

Currency: **Kenyan shilling.**
Symbol: **KSh.**

Data are currency units per $1.

January 1996	56.715
1995	51.430
1994	56.051
1993	58.001
1992	32.217
1991	27.508

Imports and Exports

Top Import Origins [39]

$2.2 billion (f.o.b., 1994) Data are for 1991.

Origins	%
EU	46
Asia	23
Middle East	20
US	5

Top Export Destinations [40]

$1 6 billion (f.o.b., 1994) Data are for 1991.

Destinations	%
EU	47
Africa	23
Asia	11
US	4
Middle East	3

Foreign Aid [41]

Recipient: ODA, $589 million (1993).

Import and Export Commodities [42]

Import Commodities

Machinery and transportation equipment 29%
Petroleum and petroleum products 15%
Iron and steel 7%
Raw materials
Food and consumer goods

Export Commodities

Tea 25%
Coffee 18%
Petroleum products 11%

Korea, North

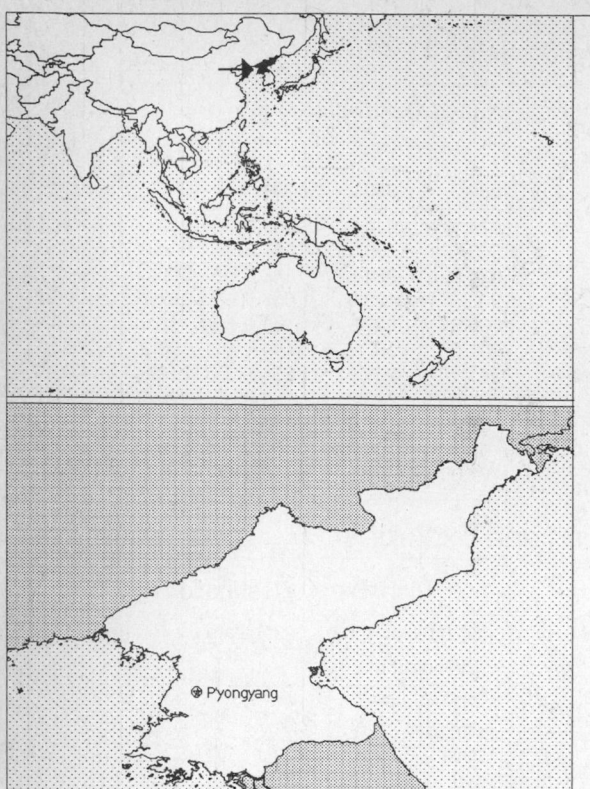

Geography [1]

Total area:
120,540 sq km   46,541 sq mi
Land area:
120,410 sq km   46,491 sq mi
Comparative area:
Slightly smaller than Mississippi
Land boundaries:
Total 1,673 km, China 1,416 km, South Korea 238 km, Russia 19 km
Coastline:
2,495 km
Climate:
Temperate with rainfall concentrated in summer
Terrain:
Mostly hills and mountains separated by deep, narrow valleys; coastal plains wide in West, discontinuous in East
Natural resources:
Coal, lead, tungsten, zinc, graphite, magnesite, iron ore, copper, gold, pyrites, salt, fluorspar, hydropower
Land use:
Arable land: 18%
Permanent crops: 1%
Meadows and pastures: 0%
Forest and woodland: 74%
Other: 7%

Demographics [2]

	1970	1980	1990	1995[1]	1996	2000	2010	2020	2030
Population	14,388	17,999	21,412	23,487	23,904	25,491	28,491	30,969	32,964
Population density (persons per sq. mi.)	309	387	461	505	514	548	613	666	709
(persons per sq. km.)	119	149	178	195	199	212	237	257	274
Net migration rate (per 1,000 population)	0.0	0.0	0.0	0.0	0.0	0.0	0.0	0.0	0.0
Births	NA	NA	NA	NA	546	NA	NA	NA	NA
Deaths	NA	NA	NA	NA	130	NA	NA	NA	NA
Life expectancy - males	56.0	62.7	65.6	67.0	67.2	68.4	70.5	72.4	74.0
Life expectancy - females	62.2	69.0	72.0	73.3	73.6	74.7	76.9	78.8	80.5
Birth rate (per 1,000)	45.2	24.0	24.1	23.3	22.9	19.9	14.8	14.6	12.5
Death rate (per 1,000)	9.7	5.9	5.6	5.5	5.5	5.4	5.9	6.9	8.2
Women of reproductive age (15-49 yrs.)	3,232	4,526	6,213	6,675	6,763	7,092	7,820	7,882	7,740
of which are currently married	NA	NA	3,754	NA	4,690	5,143	5,459	NA	NA
Fertility rate	6.9	3.0	2.5	2.3	2.3	2.2	2.0	1.9	1.8

Except as noted, values for vital statistics are in thousands; life expectancy is in years.

Health

Health Indicators [3]

% of population with access to	
safe water (1990-95)	NA
adequate sanitation (1990-95)	NA
health services (1985-95)	NA
% of 1-year-olds immunized (1990-94) against	
TB (tuberculosis)	100
DPT (diphtheria, pertussis, tetanus)	99
polio	100
measles	99
% of contraceptive prevalence (1980-94)	NA
ORT use rate (1990-94)	85

Health Expenditures [4]

For sources, notes, and explanations, see Annotated Source Appendix, page 1061.

Human Factors

Women and Children [5]

% of pregnant women immunized (tetanus 1990-94)	99
% of births attended by trained health personnel (1983-94)	100
Maternal mortality rate (1980-92)	41
Under-5 mortality rate (1994)	31
% under-5 moderately/severely underweight (1980-1994)	NA

Burden of Disease [6]

Ethnic Division [7]

Racially homogeneous.

Religion [8]

Buddhism and Confucianism, some Christianity and syncretic Chondogyo. Autonomous religious activities now almost nonexistent; government-sponsored religious groups exist.

Major Languages [9]

Korean.

Education

Public Education Expenditures [10]

Educational Attainment [11]

Illiteracy [12]

Libraries [13]

Daily Newspapers [14]

	1980	1985	1990	1994
Number of papers	11	11	11	11
Circ. (000)	4,000[e]	4,500[e]	5,000[e]	5,000[e]

Culture [15]

Science and Technology

Scientific/Technical Forces [16]

R&D Expenditures [17]

	Won (000) 1988
Total expenditure	2,347,000
Capital expenditure	925,000
Current expenditure	1,422,000
Percent current	60.6

U.S. Patents Issued [18]

Values show patents issued to citizens of the country by the U.S. Patents Office.

	1993	1994	1995
Number of patents	0	1	0

For sources, notes, and explanations, see Annotated Source Appendix, page 1061.

499

Government and Law

Organization of Government [19]

Long-form name:
Democratic People's Republic of Korea
Type:
Communist state; Stalinist dictatorship
Independence:
9 September 1948 Note: 15 August 1945, date of independence from the Japanese and celebrated in North Korea as National Liberation Day
National holiday:
DPRK Foundation Day, 9 September (1948)
Constitution:
Adopted 1948, completely revised in 1972, revised again in 1992·
Legal system:
Based on German civil law system with Japanese influences and Communist legal theory; no judicial review of legislative acts; has not accepted compulsory ICJ jurisdiction
Executive branch:
President; Premier; State Administration Council
Legislative branch:
Unicameral: Supreme People's Assembly (Ch'oego Inmin Hoeui)
Judicial branch:
Central Court

Crime [23]

Elections [20]

Supreme People's Assembly. Last held on 7-9 April 1990 (next to be held NA); results - percent of vote by party NA; seats - (687 total); the KWP approves a single list of candidates who are elected without opposition; minor parties hold a few seats.

Government Budget [21]

For 1992 est.

	$ bil.
Revenues	19.3
Expenditures	19.3
Capital expenditures	NA

Military Expenditures and Arms Transfers [22]

	1990	1991	1992	1993	1994
Military expenditures					
Current dollars (mil.)[r]	5,940	4,660	5,500	5,300	5,500
1994 constant dollars (mil.)[r]	6,611	4,995	5,735	5,409	5,500
Armed forces (000)	1,200	1,200	1,200	1,200	1,200
Gross national product (GNP)					
Current dollars (mil.)[e]	29,700	23,300	22,000	20,800	20,900
1994 constant dollars (mil.)[e]	33,050	24,970	22,940	21,230	20,900
Central government expenditures (CGE)					
1994 constant dollars (mil.)	NA	NA	NA	NA	NA
People (mil.)	21.4	21.8	22.2	22.6	23.1
Military expenditure as % of GNP	20.0	20.0	25.0	25.5	26.3
Military expenditure as % of CGE	NA	NA	NA	NA	NA
Military expenditure per capita (1994 $)	309	229	258	239	238
Armed forces per 1,000 people (soldiers)	56.0	55.0	54.0	53.0	52.0
GNP per capita (1994 $)	1,544	1,145	1,032	937	906
Arms imports[6]					
Current dollars (mil.)	200	90	10	5	50
1994 constant dollars (mil.)	223	96	10	5	50
Arms exports[6]					
Current dollars (mil.)	210	180	100	110	30
1994 constant dollars (mil.)	234	193	104	112	30
Total imports[7]					
Current dollars (mil.)	2,620[e]	NA	1,900[e]	1,930	NA
1994 constant dollars (mil.)	2,916[e]	NA	1,981[e]	1,970	NA
Total exports[7]					
Current dollars (mil.)	2,020	1,025[e]	1,300[e]	1,220	NA
1994 constant dollars (mil.)	2,248	1,099[e]	1,356[e]	1,245	NA
Arms as percent of total imports[8]	7.6	NA	0.5	0.3	NA
Arms as percent of total exports[8]	10.4	17.6	7.7	9.0	NA

Human Rights [24]

	SSTS	FL	FAPRO	PPCG	APROBC	TPW	PCPTW	STPEP	PHRFF	PRW	ASST	AFL
Observes				P		P	P					

	EAFRD	CPR	ESCR	SR	ACHR	MAAE	PVIAC	PVNAC	EAFDAW	TCIDTP	RC
Observes		P	P				P				P

P=Party; S=Signatory; see Appendix for meaning of abbreviations.

Labor Force

Total Labor Force [25]

9.615 million

Labor Force by Occupation [26]

Agricultural	36%
Nonagricultural	64

Unemployment Rate [27]

For sources, notes, and explanations, see Annotated Source Appendix, page 1061.

Production Sectors

Commercial Energy Production and Consumption

Data are shown in quadrillion (10^{15}) BTUs and percent for 1995
Values for hydroelectric, nuclear, geothermal, solar, and wind power refer to electrical generation.

Production [28]

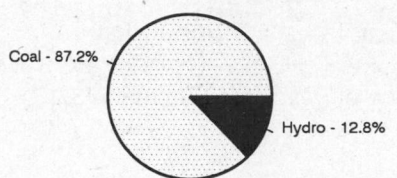

Coal - 87.2%

Hydro - 12.8%

Consumption [29]

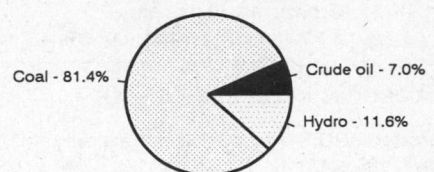

Coal - 81.4%

Crude oil - 7.0%

Hydro - 11.6%

Coal	1.633
Net hydroelectric power	0.239
Total	1.872

Crude oil	0.145
Coal	1.684
Net hydroelectric power	0.239
Total	2.008

Telecommunications [30]

- 30,000 (1990 est.) telephones; system is believed to be available principally for government business
- International: satellite earth stations - 1 Intelsat (Indian Ocean) and 1 Intersputnik (Indian Ocean Region); other international connections through Moscow and Beijing
- Radio: Broadcast stations: AM 18, FM 0, shortwave 0 Radios: 3.5 million
- Television: Broadcast stations: 11 Televisions: 400,000 (1992 est.)

Transportation [31]

Railways: total: 4,915 km; standard gauge: 4,250 km 1.435-m gauge (3,397 km electrified; 159 km double track); narrow gauge: 665 km 0.762-m gauge (1989)

Highways: total: 30,000 km; paved: 4,500 km; unpaved: 25,500 km

Merchant marine: total: 88 ships (1,000 GRT or over) totaling 712,480 GRT/1,140,923 DWT; ships by type: bulk 9, cargo 71, combination bulk 1, oil tanker 3, passenger 2, passenger-cargo 1, short-sea passenger 1

Airports

Total:	49
With paved runways over 3,047 m:	2
With paved runways 2,438 to 3,047 m:	15
With paved runways 1,524 to 2,437 m:	2
With paved runways 914 to 1,523 m:	1
With paved runways under 914 m:	2

Top Agricultural Products [32]

Agriculture accounts for 25% of the GDP; produces rice, corn, potatoes, soybeans, pulses; cattle, pigs, pork, eggs.

Top Mining Products [33]

Thousand metric tons except as noted[e]	6/29/95[*]
Iron ore/concentrate, gr. wt., marketable	11,000
Steel, crude	8,100
Zinc, mine output, Zn content	210
Cement, hydraulic	17,000
Magnesite, crude	1,600
Nitrogen, N content of ammonia	600
Phosphate rock	510
Salt, all types	600
Coal, anthracite and lignite	90,000
Petroleum refinery products (000 42-gal. bls.)	25,000

Tourism [34]

For sources, notes, and explanations, see Annotated Source Appendix, page 1061.

Finance, Economics, and Trade

GDP and Manufacturing Summary [35]

	1980	1985	1990	1991	1992
Gross Domestic Product					
Millions of 1980 dollars	12,730	20,368	26,618	25,234	23,972
Growth rate in percent	9.89	9.59	5.60	-5.20	-5.00
Manufacturing Value Added					
Millions of 1980 dollars	NA	NA	NA	NA	NA
Growth rate in percent	NA	NA	NA	NA	NA
Manufacturing share in percent of current prices	NA	NA	NA	NA	NA

Economic Indicators [36]

- **National product**: GDP—purchasing power parity—$21.5 billion (1995 est.)

- **National product real growth rate**: - 5% (1995 est.)

- **National product per capita**: $920 (1995 est.)

- **Inflation rate (consumer prices)**: NA%

- **External debt**: $8 billion (1992 est.)

Balance of Payments Summary [37]

Exchange Rates [38]

Currency: **North Korean won.**
Symbol: **Wn.**

Data are currency units per $1.

May 1994	2.15
May 1992	2.13
September 1991	2.14
January 1990	2.1
December 1989	2.3

Imports and Exports

Top Import Origins [39]

$1.27 billion (f.o.b., 1994 est.).

Origins	%
China	NA
Japan	NA
Hong Kong	NA
Germany	NA
Russia	NA
Singapore	NA

Top Export Destinations [40]

$840 million (f.o.b., 1994 est.).

Destinations	%
China	NA
Japan	NA
South Korea	NA
Germany	NA
Hong Kong	NA
Russia	NA

Foreign Aid [41]

	U.S. $
ODA	NA
Small amounts of grant aid from Japan and other countries	NA

Import and Export Commodities [42]

Import Commodities	Export Commodities
Petroleum	Minerals
Grain	Metallurgical products
Coking coal	Agricultural and fishery products
Machinery and equipment	Manufactures (including armaments)
Consumer goods	

For sources, notes, and explanations, see Annotated Source Appendix, page 1061.

Korea, South

Geography [1]

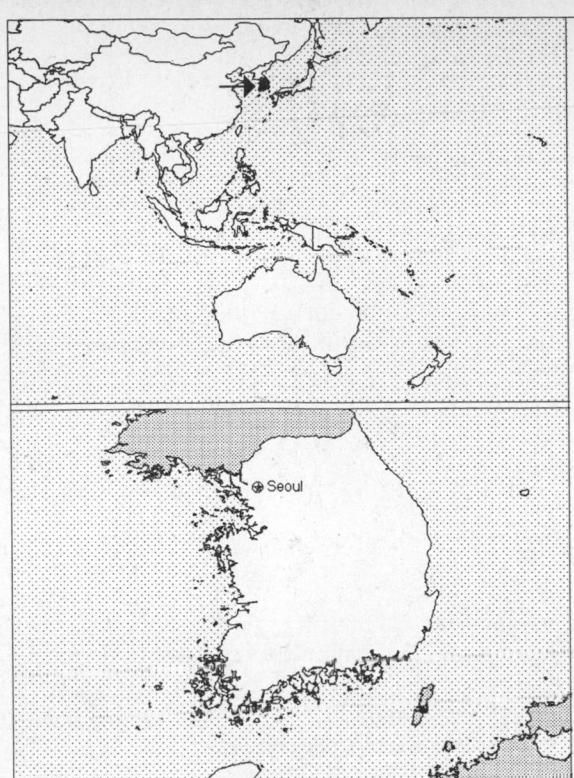

Total area:
  98,480 sq km   38,023 sq mi
Land area:
  98,190 sq km   37,911 sq mi
Comparative area:
  Slightly larger than Indiana
Land boundaries:
  Total 238 km, North Korea 238 km
Coastline:
  2,413 km
Climate:
  Temperate, with rainfall heavier in summer than winter
Terrain:
  Mostly hills and mountains; wide coastal plains in West and South
Natural resources:
  Coal, tungsten, graphite, molybdenum, lead, hydropower
Land use:
  Arable land: 21%
  Permanent crops: 1%
  Meadows and pastures: 1%
  Forest and woodland: 67%
  Other: 10%

Demographics [2]

	1970	1980	1990	1995[1]	1996	2000	2010	2020	2030
Population	32,241	38,124	42,869	45,018	45,482	47,351	51,235	53,451	54,826
Population density (persons per sq. mi.)	850	1,006	1,131	1,187	1,200	1,249	1,351	1,410	1,446
(persons per sq. km.)	328	388	437	458	463	482	522	544	558
Net migration rate (per 1,000 population)	NA	NA	-0.7	-0.4	-0.4	-0.3	-0.1	0.0	0.0
Births	NA	NA	NA	NA	739	NA	NA	NA	NA
Deaths	NA	NA	NA	NA	257	NA	NA	NA	NA
Life expectancy - males	NA	NA	67.1	69.3	69.7	71.1	74.1	76.3	77.8
Life expectancy - females	NA	NA	75.2	77.1	77.4	78.7	81.2	83.0	84.2
Birth rate (per 1,000)	NA	NA	15.0	16.3	16.2	15.8	12.4	10.9	10.7
Death rate (per 1,000)	NA	NA	5.9	5.7	5.7	5.7	6.3	7.7	9.3
Women of reproductive age (15-49 yrs.)	NA	NA	12,115	12,854	12,986	13,414	13,040	12,302	11,297
of which are currently married[30]	NA	5,900	7,279	NA	8,298	8,867	9,021	NA	NA
Fertility rate	NA	NA	1.6	1.8	1.8	1.8	1.8	1.8	1.8

Except as noted, values for vital statistics are in thousands; life expectancy is in years.

Health

Health Indicators [3]

% of population with access to	
safe water (1990-95)	93
adequate sanitation (1990-95)	100
health services (1985-95)	100
% of 1-year-olds immunized (1990-94) against	
TB (tuberculosis)	72
DPT (diphtheria, pertussis, tetanus)	74
polio	74
measles	93
% of contraceptive prevalence (1980-94)	79
ORT use rate (1990-94)	NA

Health Expenditures [4]

Total health expenditure, 1990 (official exchange rate)	
Millions of dollars	16,130
Dollars per capita	377
Health expenditures as a percentage of GDP	
Total	6.6
Public sector	2.7
Private sector	3.9
Development assistance for health	
Total aid flows (millions of dollars)[1]	32
Aid flows per capita (dollars)	NA
Aid flows as a percentage of total health expenditure	0.2

For sources, notes, and explanations, see Annotated Source Appendix, page 1061.

503

Human Factors

Women and Children [5]

% of pregnant women immunized (tetanus 1990-94)	NA
% of births attended by trained health personnel (1983-94)	89
Maternal mortality rate (1980-92)	26
Under-5 mortality rate (1994)	9
% under-5 moderately/severely underweight (1980-1994)	NA

Burden of Disease [6]

Population per physician (1991)	951.03
Population per nurse (1991)	453.85
Population per hospital bed (1991)	300.42
AIDS cases per 100,000 people (1994)	*
Malaria cases per 100,000 people (1992)	NA

Ethnic Division [7]

Homogeneous (except for about 20,000 Chinese).

Religion [8]

Christianity	48.6%
Buddhism	47.4%
Confucianism	3%
Pervasive folk religion (shamanism), Chondogyo (Religion of the Heavenly Way)	0.2%

Major Languages [9]

Korean, English widely taught in high school.

Education

Public Education Expenditures [10]

Million or Trillion (T) (Won)	1980	1990	1991	1992	1993	1994
Total education expenditure	1.37 T	6.15 T	8.54 T	10.02 T	11.76 T	NA
as percent of GNP	3.7	3.5	4.0	4.2	4.5	NA
as percent of total govt. expend.	NA	NA	NA	14.8	16.0	NA
Current education expenditure	1.16 T	5.50 T	6.73 T	8.00 T	9.34 T	NA
as percent of GNP	3.1	3.1	3.1	3.4	3.5	NA
as percent of current govt. expend.	NA	NA	NA	15.3	17.7	NA
Capital expenditure	215,769	663,869	1.81 T	2.02 T	2.41 T	NA

Educational Attainment [11]

Age group (1990)	25+
Total population	23,408,288
Highest level attained (%)	
No schooling	11.0
First level	
Not completed	0.8
Completed	20.9
Entered second level	
S-1	18.9
S-2	35.0
Postsecondary	13.4

Illiteracy [12]

In thousands and percent[1]	1990	1995	2000
Illiterate population (15+ yrs.)	937	697	493
Illiteracy rate - total pop. (%)	2.9	2.0	1.3
Illiteracy rate - males (%)	1.0	0.7	0.5
Illiteracy rate - females (%)	4.8	3.3	2.1

Libraries [13]

	Admin. Units	Svc. Pts.	Vols. (000)	Shelving (meters)	Vols. Added	Reg. Users
National (1994)	1	2	2,343	NA	NA	2 mil
Nonspecialized (1989)	1	1	686	NA	48,709	108,049
Public (1994)	277	277	8,442	NA	NA	30 mil
Higher ed. (1990)	340	NA	35,758	NA	NA	61 mil
School (1990)	6,468	NA	27,675	NA	2 mil	27 mil

Daily Newspapers [14]

	1980	1985	1990	1994
Number of papers	30	35	39	62
Circ. (000)	8,000	10,000[e]	12,000[e]	18,000[e]

Culture [15]

Cinema (seats per 1,000)	4.8
Annual attendance per person	NA
Gross box office receipts (mil. Won)	NA
Museums (reporting)	NA
Visitors (000)	NA
Annual receipts (000 Won)	NA

Science and Technology

Scientific/Technical Forces [16]

Scientists/engineers	117,486
Number female	9,502
Technicians	14,141
Number female	NA
Total[8,9]	131,587

R&D Expenditures [17]

	Won (000,000) 1994
Total expenditure[23]	7,894,746
Capital expenditure	2,332,835
Current expenditure	5,561,911
Percent current	70.5

U.S. Patents Issued [18]

Values show patents issued to citizens of the country by the U.S. Patents Office.

	1993	1994	1995
Number of patents	830	1,008	1,240

Government and Law

Organization of Government [19]

Long-form name:
Republic of Korea
Type:
Republic
Independence:
15 August 1948
National holiday:
Independence Day, 15 August (1948)
Constitution:
25 February 1988
Legal system:
Combines elements of continental European civil law systems, Anglo-American law, and Chinese classical thought
Executive branch:
Chief of state: President; Prime Minister; 2 Deputy Prime Ministers; State Council
Legislative branch:
Unicameral: National Assembly (Kukhoe)
Judicial branch:
Supreme Court

Elections [20]

National Assembly	% of seats
New Korea Party (NKP)	46.5
National Congress for New Politics	26.4
United Liberal Democratic	16.7
Democratic Party	5.0
Independents	5.4

Government Expenditures [21]

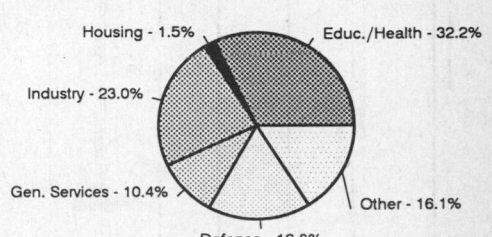

Housing - 1.5%
Educ./Health - 32.2%
Industry - 23.0%
Gen. Services - 10.4%
Other - 16.1%
Defense - 16.8%

(% distribution). Expend. for CY96: 73,582 (Won bil.)

Military Expenditures and Arms Transfers [22]

	1990	1991	1992	1993	1994
Military expenditures					
Current dollars (mil.)	10,190	10,200	11,150	11,680	13,030
1994 constant dollars (mil.)	11,340	10,930	11,620	11,920	13,030
Armed forces (000)	650	750	750	750	750
Gross national product (GNP)					
Current dollars (mil.)	244,100	276,500	298,500	322,000	356,000
1994 constant dollars (mil.)	271,700	296,400	311,300	328,700	356,000
Central government expenditures (CGE)					
1994 constant dollars (mil.)	50,830	55,830	58,600	61,060	74,830
People (mil.)	43.2	43.7	44.1	44.6	45.1
Military expenditure as % of GNP	4.2	3.7	3.7	3.6	3.7
Military expenditure as % of CGE	22.3	19.6	19.8	19.5	17.4
Military expenditure per capita (1994 $)	262	250	263	267	289
Armed forces per 1,000 people (soldiers)	15.0	17.2	17.0	16.8	16.6
GNP per capita (1994 $)	6,284	6,784	7,051	7,367	7,896
Arms imports[6]					
Current dollars (mil.)	1,100	675	700	1,300	1,000
1994 constant dollars (mil.)	1,224	724	730	1,327	1,000
Arms exports[6]					
Current dollars (mil.)	150	40	30	50	40
1994 constant dollars (mil.)	167	43	31	51	40
Total imports[7]					
Current dollars (mil.)	69,840	81,520	81,770	83,800	102,300
1994 constant dollars (mil.)	77,730	87,380	85,280	85,530	102,300
Total exports[7]					
Current dollars (mil.)	65,020	71,870	76,630	82,240	96,010
1994 constant dollars (mil.)	72,360	77,040	79,910	83,930	96,010
Arms as percent of total imports[8]	1.6	0.8	0.9	1.6	1.0
Arms as percent of total exports[8]	0.2	0.1	0	0.1	0

Crime [23]

	1994
Crime volume	
Cases known to police	467,901
Attempts (percent)	NA
Percent cases solved	86
Crimes per 100,000 persons	1,029.26
Persons responsible for offenses	
Total number offenders	NA
Percent female	NA
Percent juvenile (14-18 yrs.)	NA
Percent foreigners	NA

Human Rights [24]

	SSTS	FL	FAPRO	PPCG	APROBC	TPW	PCPTW	STPEP	PHRFF	PRW	ASST	AFL
Observes				P		P	P	P		P		

	EAFRD	CPR	ESCR	SR	ACHR	MAAE	PVIAC	PVNAC	EAFDAW	TCIDTP	RC
Observes	P	P	P	P			P	P	P	P	P

P = Party; S = Signatory; see Appendix for meaning of abbreviations.

Labor Force

Total Labor Force [25]

20 million

Labor Force by Occupation [26]

Services and other	52%
Mining and manufacturing	27
Agriculture, fishing, forestry	21

Date of data: 1991

Unemployment Rate [27]

2% (1995 est.)

For sources, notes, and explanations, see Annotated Source Appendix, page 1061.

505

Production Sectors

Commercial Energy Production and Consumption

Data are shown in quadrillion (10^{15}) BTUs and percent for 1995
Values for hydroelectric, nuclear, geothermal, solar, and wind power refer to electrical generation.

Production [28]	Consumption [29]
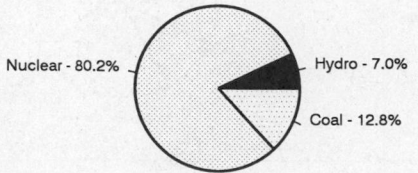	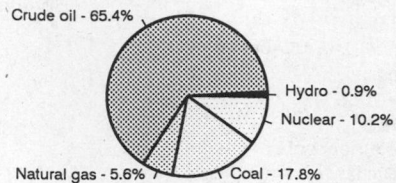

Coal	0.102	Crude oil	4.094
Net hydroelectric power	0.056	Dry natural gas	0.352
Net nuclear power	0.639	Coal	1.115
Total	0.797	Net hydroelectric power	0.056
		Net nuclear power	0.639
		Total	6.256

Telecommunications [30]

- 16.6 million (1993) telephones; excellent domestic and international services
- International: fiber-optic submarine cable to China; satellite earth stations - 3 Intelsat (2 Pacific Ocean and 1 Indian Ocean) and 1 Inmarsat (Pacific Ocean Region)
- Radio: Broadcast stations: AM 79, FM 46, shortwave 0 Radios: 42 million (1993 est.)
- Television: Broadcast stations: 256 (57 of which are 1 kW or greater) (1987 est.) Televisions: 9.3 million (1992 est.)

Transportation [31]

Railways: total: 3,101 km; standard gauge: 3,081 km 1.435-m gauge (560 km electrified); narrow gauge: 20 km 0.762-m gauge

Highways: total: 61,296 km; paved: 51,918 km (including 1,550 km of expressways); unpaved: 9,378 km (1993)

Merchant marine: total: 428 ships (1,000 GRT or over) totaling 6,076,981 GRT/9,822,089 DWT; ships by type: bulk 124, cargo 122, chemical tanker 21, combination bulk 3, combination ore/oil 1, container 59, liquefied gas tanker 12, multifunction large-load carrier 1, oil tanker 61, refrigerated cargo 13, short-sea passenger 1, vehicle carrier 10

Airports

Total:	105
With paved runways over 3,047 m:	1
With paved runways 2,438 to 3,047 m:	20
With paved runways 1,524 to 2,437 m:	13
With paved runways 914 to 1,523 m:	14
With paved runways under 914 m:	54

Top Agricultural Products [32]

Agriculture accounts for 8% of the GDP; produces rice, root crops, barley, vegetables, fruit; cattle, pigs, chickens, milk, eggs; fish catch of 2.9 million metric tons, seventh largest in world.

Top Mining Products [33]

Metric tons except as noted[e]	4/27/95[*]
Gold metal (kg.)	25,000
Pig iron	21,200[r]
Steel, crude	33,700[r]
Silver metal (kg.)	258,000[r]
Cement, hydraulic	52,100[r]
Limestone	82,800[r]
Quartzite	2,400
Coal, anthracite	7,140
Coke	5,700
Petroleum refinery products (000 42-gal. bls.)	453,000

Tourism [34]

	1990	1991	1992	1993	1994
Visitors[61]	2,959	3,196	3,231	3,331	3,580
Tourism receipts[62]	3,559	3,426	3,272	3,475	3,806
Tourism expenditures[62]	3,166	3,784	3,794	3,259	4,088
Fare receipts	753	902	988	1,149	1,317
Fare expenditures	352	490	473	389	452

Travelers are in thousands, money in million U.S. dollars.

For sources, notes, and explanations, see Annotated Source Appendix, page 1061.

Manufacturing Sector

Manufacturing Summary [35]

	1987		1988		1989		1990		1991	
	$ bil.	%	$ bil.	%	$ bil.	%	$ bil.	%	$ bil.	%
Establishments or enterprises (number)	63,527	2.708	70,024	3.336	76,494	4.084	79,766	4.453	-	-
Total employment (000)	3,767	2.775	3,910	2.857	3,858	3.135	3,760	3.397	-	-
Production workers (000)	3,009	4.462	3,073	4.918	2,980	4.051	2,864	4.036	-	-
Output ($ bil.)	183	1.800	242	2.077	289	2.447	330	2.878	-	-
Value added ($ bil.)	65	1.502	86	1.785	105	2.154	129	2.593	-	-
Capital investment ($ mil.)	18,762	4.305	21,904	4.600	31,040	5.667	40,921	7.363	-	-
M & E investment ($ mil.)	13,156	4.222	15,077	4.356	19,962	5.113	25,993	6.166	-	-
Employees per establishment (number)	59	102.486	56	85.642	50	76.760	47	76.288	-	-
Production workers per establishment	47	164.760	44	147.442	39	99.207	36	90.633	-	-
Output per establishment ($ mil.)	3	66.457	3	62.272	4	59.925	4	64.643	-	-
Capital investment per estab. ($ mil.)	0.295	158.961	0.313	137.888	0.406	138.766	0.513	165.346	-	-
M & E per establishment ($ mil)	0.207	155.904	0.215	130.592	0.261	125.209	0.326	138.463	-	-
Payroll per employee ($)	5,295	59.054	7,007	70.333	9,441	84.501	10,586	76.313	-	-
Wages per production worker ($)	4,849	61.448	6,604	78.152	8,835	88.100	10,118	85.200	-	-
Hours per production worker (hours)	2,115	111.438	2,022	105.184	1,903	103.816	1,891	101.077	-	-
Output per employee ($)	48,565	64.845	61,932	72.712	74,820	78.068	87,815	84.736	-	-
Capital investment per employee ($)	4,981	155.106	5,602	161.005	8,045	180.779	10,883	216.740	-	-
M & E per employee ($)	3,493	152.122	3,856	152.485	5,174	163.117	6,913	181.501	-	-

Note: Columns headed % show percent of world total or ratio. Ratios closest to 100 are closest to world average. M & E stands for machinery & equipment.

Output in Manufacturing

	1987		1988		1989		1990		1991	
	$ bil.	%	$ bil.	%	$ bil.	%	$ bil.	%	$ bil.	%
3110 - Food products	9.971	5.45	13.000	5.37	16.000	5.54	17.000	5.15	-	-
3130 - Beverages	2.197	1.20	2.901	1.20	3.477	1.20	3.888	1.18	-	-
3140 - Tobacco	2.289	1.25	2.925	1.21	3.571	1.24	3.647	1.11	-	-
3210 - Textiles	14.000	7.65	17.000	7.02	18.000	6.23	18.000	5.45	-	-
3211 - Spinning, weaving, etc.	10.000	5.46	13.000	5.37	14.000	4.84	13.000	3.94	-	-
3220 - Wearing apparel	5.121	2.80	6.572	2.72	7.603	2.63	7.837	2.37	-	-
3230 - Leather and products	2.492	1.36	2.929	1.21	3.705	1.28	3.900	1.18	-	-
3240 - Footwear	0.818	0.45	1.097	0.45	1.232	0.43	1.338	0.41	-	-
3310 - Wood products	1.264	0.69	1.821	0.75	2.066	0.71	2.458	0.74	-	-
3320 - Furniture, fixtures	0.726	0.40	1.196	0.49	1.716	0.59	2.032	0.62	-	-
3410 - Paper and products	3.814	2.08	5.004	2.07	5.950	2.06	6.255	1.90	-	-
3411 - Pulp, paper, etc.	1.984	1.08	2.630	1.09	3.190	1.10	3.289	1.00	-	-
3420 - Printing, publishing	1.954	1.07	2.604	1.08	3.579	1.24	4.234	1.28	-	-
3510 - Industrial chemicals	6.488	3.55	8.984	3.71	10.000	3.46	13.000	3.94	-	-
3511 - Basic chemicals, excl fertilizers	3.671	2.01	5.441	2.25	5.682	1.97	7.196	2.18	-	-
3513 - Synthetic resins, etc.	3.481	1.90	5.047	2.09	5.533	1.91	6.682	2.02	-	-
3520 - Chemical products nec	4.928	2.69	6.455	2.67	8.709	3.01	9.424	2.86	-	-
3522 - Drugs and medicines	1.820	0.99	2.507	1.04	2.870	0.99	3.627	1.10	-	-
3530 - Petroleum refineries	6.925	3.78	7.560	3.12	8.543	2.96	10.000	3.03	-	-
3540 - Petroleum, coal products	1.736	0.95	1.980	0.82	2.128	0.74	2.153	0.65	-	-
3550 - Rubber products	4.254	2.32	5.698	2.35	6.172	2.14	7.271	2.20	-	-
3560 - Plastic products nec	3.561	1.95	5.169	2.14	6.029	2.09	7.057	2.14	-	-
3610 - Pottery, china, etc.	0.269	0.15	0.323	0.13	0.395	0.14	0.413	0.13	-	-
3620 - Glass and products	0.870	0.48	1.148	0.47	1.340	0.46	1.707	0.52	-	-
3690 - Nonmetal products nec	4.019	2.20	5.516	2.28	7.058	2.44	8.174	2.48	-	-
3710 - Iron and steel	9.523	5.20	13.000	5.37	16.000	5.54	17.000	5.15	-	-
3720 - Nonferrous metals	2.245	1.23	3.522	1.46	4.203	1.45	4.636	1.40	-	-
3810 - Metal products	5.474	2.99	8.168	3.38	11.000	3.81	12.000	3.64	-	-
3820 - Machinery nec	6.847	3.74	10.000	4.13	13.000	4.50	16.000	4.85	-	-
3825 - Office, computing machinery	0.839	0.46	1.368	0.57	1.592	0.55	1.731	0.52	-	-
3830 - Electrical machinery	20.000	10.93	28.000	11.57	33.000	11.42	37.000	11.21	-	-
3832 - Radio, television, etc.	10.000	5.46	13.000	5.37	15.000	5.19	17.000	5.15	-	-
3840 - Transportation equipment	12.000	6.56	16.000	6.61	21.000	7.27	28.000	8.48	-	-
3841 - Shipbuilding, repair	3.202	1.75	2.961	1.22	3.856	1.33	5.195	1.57	-	-
3843 - Motor vehicles	8.628	4.71	12.000	4.96	16.000	5.54	22.000	6.67	-	-
3850 - Professional goods	1.419	0.78	2.097	0.87	2.496	0.86	2.604	0.79	-	-
3900 - Industries nec	3.004	1.64	3.651	1.51	3.892	1.35	3.765	1.14	-	-

Note: Codes are International Standard Industry codes (ISIC). Percentages are % of total Output. [f]: Factor Prices; [p]: Producer Prices.

For sources, notes, and explanations, see Annotated Source Appendix, page 1061.

Finance, Economics, and Trade

Economic Indicators [36]

- **National product**: GDP—purchasing power parity—$590.7 billion (1995 est.)
- **National product real growth rate**: 9% (1995)
- **National product per capita**: $13,000 (1995 est.)
- **Inflation rate (consumer prices)**: 4.3% (1995 est.)
- **External debt**: $77 billion (1995 est.)

Balance of Payments Summary [37]

Values in millions of dollars.

	1989	1990	1991	1992	1993
Exports of goods (f.o.b.)	61,408	63,123	69,581	75,169	80,950
Imports of goods (f.o.b.)	-56,811	-65,127	-76,561	-77,315	-79,090
Trade balance	4,597	-2,004	-6,980	-2,146	1,860
Services - debits	-12,432	-14,712	-17,124	-18,625	-20,220
Services - credits	12,643	14,269	15,531	16,010	18,253
Private transfers (net)	200	266	20	257	633
Government transfers (net)	48	9	-173	-25	-142
Long-term capital (net)	-3,904	-659	6,069	7,522	8,415
Short-term capital (net)	1,278	3,628	756	-368	-5,075
Errors and omissions	690	-2,005	753	1,099	-715
Overall balance	3,120	-1,208	-1,148	3,724	3,009

Exchange Rates [38]

Currency: **South Korean won.**
Symbol: **W.**

Data are currency units per $1.

January 1996	787.27
1995	771.27
1994	803.45
1993	802.67
1992	780.65
1991	733.35

Imports and Exports

Top Import Origins [39]

$135.1 billion (c.i.f., 1995).

Origins	%
Japan	24
US	22
EU	13

Top Export Destinations [40]

$125.4 billion (f.o.b., 1995).

Destinations	%
US	19
Japan	14
EU	13

Foreign Aid [41]

	U.S. $	
US commitments, including Ex-Im (FY70-89)	3.9	billion
Non-US countries (1970-89)	3	billion

Import and Export Commodities [42]

Import Commodities	Export Commodities
Machinery	Electronic and electrical equipment
Electronics and electronic equipment	Machinery
Oil	Steel
Steel	Automobiles
Transport equipment	Ships
Textiles	Textiles
Organic chemicals	Clothing
Grains	Footwear
	Fish

Kuwait

Geography [1]

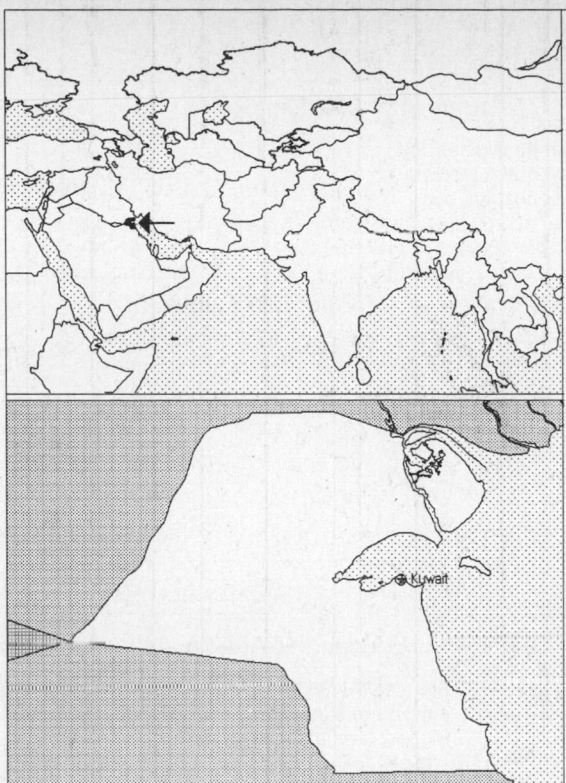

Total area:
17,820 sq km   6,880 sq mi
Land area:
17,820 sq km   6,880 sq mi
Comparative area:
Slightly smaller than New Jersey
Land boundaries:
Total 464 km, Iraq 242 km, Saudi Arabia 222 km
Coastline:
499 km
Climate:
Dry desert; intensely hot summers; short, cool winters
Terrain:
Flat to slightly undulating desert plain
Natural resources:
Petroleum, fish, shrimp, natural gas
Land use:
Arable land: 0%
Permanent crops: 0%
Meadows and pastures: 8%
Forest and woodland: 0%
Other: 92%

Demographics [2]

	1970	1980	1990	1995[1]	1996	2000	2010	2020	2030
Population	748	1,370	2,128	1,817	1,950	2,420	3,160	3,560	3,851
Population density (persons per sq. mi.)	109	199	309	264	283	352	459	517	560
(persons per sq. km.)	42	77	119	102	109	136	177	200	216
Net migration rate (per 1,000 population)	NA	NA	-559.6	55.7	48.5	27.9	0.0	0.0	0.0
Births	34	50	NA	NA	40	NA	NA	NA	NA
Deaths	NA	5	NA	NA	4	NA	NA	NA	NA
Life expectancy - males	NA	NA	72.1	73.3	73.6	74.6	76.5	77.8	78.8
Life expectancy - females	NA	NA	76.6	78.1	78.4	79.6	81.9	83.5	84.7
Birth rate (per 1,000)	45.3	37.3	21.7	21.1	20.3	17.8	16.4	14.8	12.8
Death rate (per 1,000)	5.0	3.6	2.1	2.2	2.2	2.3	3.1	4.7	7.0
Women of reproductive age (15-49 yrs.)	NA	NA	480	420	454	582	786	845	890
of which are currently married	NA	191	321	NA	302	306	527	NA	NA
Fertility rate	6.9	5.4	3.0	2.9	2.8	2.4	2.1	2.0	2.0

Except as noted, values for vital statistics are in thousands; life expectancy is in years.

Health

Health Indicators [3]

% of population with access to	
safe water (1990-95)	NA
adequate sanitation (1990-95)	NA
health services (1985-95)	100
% of 1-year-olds immunized (1990-94) against	
TB (tuberculosis)	NA
DPT (diphtheria, pertussis, tetanus)	98
polio	98
measles	96
% of contraceptive prevalence (1980-94)	35
ORT use rate (1990-94)	10

Health Expenditures [4]

For sources, notes, and explanations, see Annotated Source Appendix, page 1061.

509

Human Factors

Women and Children [5]

% of pregnant women immunized (tetanus 1990-94)	44
% of births attended by trained health personnel (1983-94)	99
Maternal mortality rate (1980-92)	6
Under-5 mortality rate (1994)	14
% under-5 moderately/severely underweight (1980-1994)	6

Burden of Disease [6]

Population per physician (1986)	638.73
Population per nurse (1986)	202.81
Population per hospital bed	NA
AIDS cases per 100,000 people (1994)	0.2
Malaria cases per 100,000 people (1992)	NA

Ethnic Division [7]

Kuwaiti	45%
Other Arab	35%
South Asian	9%
Iranian	4%
Other	7%

Religion [8]

Muslim	85%
Shi'a	30%
Sunni	45%
other	10%
Christian, Hindu, Parsi, and other	15%

Major Languages [9]

Arabic (official), English widely spoken.

Education

Public Education Expenditures [10]

Million (Dinar)	1980	1989	1990	1992	1993	1994
Total education expenditure	219	432	NA	450	466	NA
as percent of GNP	2.4	4.6	NA	6.1	5.6	NA
as percent of total govt. expend.	8.1	14.0	NA	11.4	11.0	NA
Current education expenditure	204	NA	NA	NA	NA	NA
as percent of GNP	2.3	NA	NA	NA	NA	NA
as percent of current govt. expend.	11.2	NA	NA	NA	NA	NA
Capital expenditure	15	NA	NA	NA	NA	NA

Educational Attainment [11]

Age group (1988)	10+
Total population	1,409,065
Highest level attained (%)	
No schooling	17.6
First level	
Not completed	18.4
Completed	NA
Entered second level	
S-1	22.7
S-2	14.6
Postsecondary	11.1

Illiteracy [12]

In thousands and percent[1]	1990	1995	2000
Illiterate population (15+ yrs.)	330	200	222
Illiteracy rate - total pop. (%)	23.5	16.7	13.1
Illiteracy rate - males (%)	20.3	11.6	9.8
Illiteracy rate - females (%)	28.5	24.3	18.0

Libraries [13]

	Admin. Units	Svc. Pts.	Vols. (000)	Shelving (meters)	Vols. Added	Reg. Users
National (1986)	1	4	93	NA	43,508	29,662
Nonspecialized (1992)	3	11	135	NA	18,000	12,500
Public (1992)	1	18	272	4,530	43,440	NA
Higher ed. (1988)	1	14	453	NA	18,181	NA
School (1990)	587	NA	3,117	NA	155,875	820,105

Daily Newspapers [14]

	1980	1985	1990	1994
Number of papers	8	8	9	9
Circ. (000)	305	380	450	655

Culture [15]

Cinema (seats per 1,000)	8.1
Annual attendance per person	0.1
Gross box office receipts (mil. Dinar)	NA
Museums (reporting)	6
Visitors (000)	36
Annual receipts (000 Dinar)	NA

Science and Technology

Scientific/Technical Forces [16]

Scientists/engineers	1,511
Number female	334
Technicians	561
Number female	113
Total[1,22]	2,072

R&D Expenditures [17]

	Dinar (000) 1984
Total expenditure	71,163
Capital expenditure	8,147
Current expenditure	63,016
Percent current	88.6

U.S. Patents Issued [18]

Values show patents issued to citizens of the country by the U.S. Patents Office.

	1993	1994	1995
Number of patents	2	1	1

Government and Law

Organization of Government [19]

Long-form name:
State of Kuwait
Type:
Nominal constitutional monarchy
Independence:
19 June 1961 (from UK)
National holiday:
National Day, 25 February (1950)
Constitution:
Approved and promulgated 11 November 1962
Legal system:
Civil law system with Islamic law significant in personal matters; has not accepted compulsory ICJ jurisdiction
Executive branch:
Emir; Prime Minister; First Deputy Prime Minister; Second Deputy Prime Minister; Council of Ministers
Legislative branch:
Unicameral National Assembly (Majlis al-umma)
Judicial branch:
High Court of Appeal

Elections [20]

National Assembly. No political parties; assembly dissolved 3 July 1986; new elections were held on 5 October 1992 with a second election in the 14th and 16th constituencies held 15 February 1993.

Government Expenditures [21]

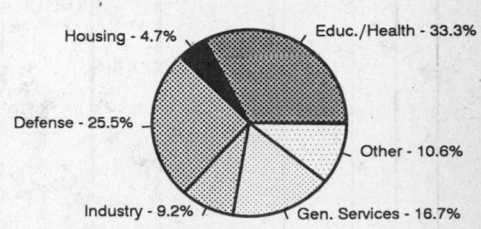

(% distribution). Expend. for FY95: 4,089 (Dinar mil.)

Crime [23]

	1994
Crime volume	
Cases known to police	18,969
Attempts (percent)	NA
Percent cases solved	NA
Crimes per 100,000 persons	1,170.86
Persons responsible for offenses	
Total number offenders	14,812
Percent female	12.00
Percent juvenile	NA
Percent foreigners	53.00

Military Expenditures and Arms Transfers [22]

	1990	1991	1992	1993	1994
Military expenditures					
Current dollars (mil.)	13,240[e]	16,030[e]	19,090[e]	3,604[e]	3,086
1994 constant dollars (mil.)	14,730[e]	17,190[e]	19,910[e]	3,679[e]	3,086
Armed forces (000)	7	10	12	12	15
Gross national product (GNP)					
Current dollars (mil.)[e]	24,950	15,740	24,810	27,900	27,860
1994 constant dollars (mil.)[e]	27,770	16,880	25,870	28,480	27,860
Central government expenditures (CGE)					
1994 constant dollars (mil.)	11,000	23,970	20,670	15,050	13,910
People (mil.)	2.1	0.9	1.4	1.5	1.7
Military expenditure as % of GNP	53.1	101.8	77.0	12.9	11.1
Military expenditure as % of CGE	134.0	71.7	96.3	24.4	22.2
Military expenditure per capita (1994 $)	6,923	18,150	14,260	2,393	1,838
Armed forces per 1,000 people (soldiers)	3.3	10.6	8.6	7.8	8.9
GNP per capita (1994 $)	13,050	17,820	18,530	18,530	16,600
Arms imports[6]					
Current dollars (mil.)	280	480	1,000	750	250
1994 constant dollars (mil.)	312	515	1,043	765	250
Arms exports[6]					
Current dollars (mil.)	0	0	10	0	0
1994 constant dollars (mil.)	0	0	10	0	0
Total imports[7]					
Current dollars (mil.)	3,972	4,761	7,261	7,036	21,720
1994 constant dollars (mil.)	4,421	5,103	7,572	7,181	21,720
Total exports[7]					
Current dollars (mil.)	7,042	1,088	6,660	10,250	11,940
1994 constant dollars (mil.)	7,837	1,166	6,945	10,460	11,940
Arms as percent of total imports[8]	7.0	10.1	13.8	10.7	1.2
Arms as percent of total exports[8]	0	0	0.2	0	0

Human Rights [24]

	SSTS	FL	FAPRO	PPCG	APROBC	TPW	PCPTW	STPEP	PHRFF	PRW	ASST	AFL
Observes	P	P	P	P		P	P	P			P	P
	EAFRD	CPR	ESCR	SR	ACHR	MAAE	PVIAC	PVNAC	EAFDAW	TCIDTP		RC
Observes		P	P	P			P	P	P			P

P = Party; S = Signatory; see Appendix for meaning of abbreviations.

Labor Force

Total Labor Force [25]

1 million (1994 est.)

Labor Force by Occupation [26]

Industry and agriculture	25.0%
Services	25.0
Government and social services	50.0

Date of data: 1994 est.

Unemployment Rate [27]

Negligible (1992 est.)

For sources, notes, and explanations, see Annotated Source Appendix, page 1061.

511

Production Sectors

Commercial Energy Production and Consumption

Data are shown in quadrillion (10^{15}) BTUs and percent for 1995
Values for hydroelectric, nuclear, geothermal, solar, and wind power refer to electrical generation.

Production [28]

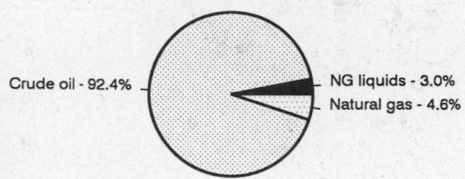

Crude oil - 92.4%
NG liquids - 3.0%
Natural gas - 4.6%

Consumption [29]

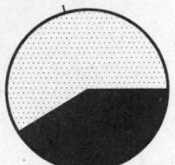

Crude oil - 59.4%
Natural gas - 40.6%

Crude oil	4.446
Natural gas liquids	0.145
Dry natural gas	0.221
Total	4.812

Crude oil	0.324
Dry natural gas	0.221
Total	0.545

Telecommunications [30]

- 548,000 (1991 est.) telephones; civil network suffered some damage during the Gulf war, but most telephone exchanges were intact; by the end of 1994, domestic and international telecommunications had been restored to normal operation; the quality of service is excellent
- Domestic: new telephone exchanges provide a large capacity for new subscribers; trunk traffic is carried by microwave radio relay, coaxial cable, open wire and fiber-optic cable; a cellular telephone system operates throughout Kuwait and the country is well supplied with pay telephones
- International: coaxial cable and microwave radio relay to Saudi Arabia; satellite earth stations - 3 Intelsat, 1 Inmarsat, and 1 Arabsat
- Radio: Broadcast stations: AM 3, FM 0, shortwave 0 Radios: 720,000 (1992 est.)
- Television: Broadcast stations: 3 (1986 est.) Televisions: 800,000 (1993 est.)

Transportation [31]

Railways: 0 km

Highways: total: 4,273 km; paved: NA (including 280 km of expressways) (1989 est.); unpaved: NA

Merchant marine: total: 46 ships (1,000 GRT or over) totaling 2,053,667 GRT/3,242,305 DWT; ships by type: cargo 10, container 3, liquefied gas tanker 7, livestock carrier 4, oil tanker 21, vehicle carrier 1 (1995 est.)

Airports
Total: 4
With paved runways over 3,047 m: 3
With paved runways 2,438 to 3,047 m: 1 (1995 est.)

Top Agricultural Products [32]

Agriculture accounts for 0% of the GDP; produces practically no crops; extensive fishing in territorial waters.

Top Mining Products [33]

Metric tons except as noted	8/1/95*
Cement	500,000
Sand lime bricks (cu. meters)	100,000
Gas, natural (mil. cu. meters)	30,000[49]
Natural gas liquids (000 42-gal. bls.)	40,000
Nitrogen, N content of ammonia	317,000
Petroleum (000 42-gal. bls.)	
crude	742,000[49]
refinery products	173,000[e]
Sulfur, elemental, petroleum byproduct	175,000
Urea	300,000

Tourism [34]

	1990	1991	1992	1993	1994
Visitors[1]	NA	NA	1,568	1,128	NA
Tourists[63]	15	4	65	73	73
Tourism receipts	132	253	273	83	101
Tourism expenditures	1,837	2,012	1,797	1,819	2,146
Fare receipts	340	186	273	149	340
Fare expenditures	173	141	205	209	222

Travelers are in thousands, money in million U.S. dollars.

Manufacturing Sector

Manufacturing Summary [35]

	1987		1988		1989		1990		1991	
	$ bil.	%	$ bil.	%	$ bil.	%	$ bil.	%	$ bil.	%
Establishments or enterprises (number)	4,150	0.177	4,137	0.197	4,139	0.221	-	-	-	-
Total employment (000)	56	0.041	60	0.044	51	0.041	-	-	-	-
Production workers (000)	45	0.066	48	0.076	41	0.055	-	-	-	-
Output ($ bil.)	7	0.072	7	0.062	2	0.015	-	-	-	-
Value added ($ bil.)	3	0.067	3	0.059	0.796	0.016	-	-	-	-
Capital investment ($ mil.)	1,084	0.249	381	0.080	69	0.013	-	-	-	-
M & E investment ($ mil.)	384	0.123	995	0.287	36	0.009	-	-	-	-
Employees per establishment (number)	13	23.235	15	22.350	12	18.707	-	-	-	-
Production workers per establishment	11	37.377	11	38.628	10	25.057	-	-	-	-
Output per establishment ($ mil.)	2	40.546	2	31.242	0.435	6.907	-	-	-	-
Capital investment per estab. ($ mil.)	0.261	140.628	0.092	40.626	0.017	5.727	-	-	-	-
M & E per establishment ($ mil)	0.093	69.702	0.240	145.844	0.009	4.162	-	-	-	-
Payroll per employee ($)	11,082	123.581	11,474	115.175	8,146	72.912	-	-	-	-
Wages per production worker ($)	-	-	-	-	-	-	-	-	-	-
Hours per production worker (hours)	47	2.483	48	2.486	48	2.614	-	-	-	-
Output per employee ($)	130,693	174.502	119,063	139.786	35,383	36.919	-	-	-	-
Capital investment per employee ($)	19,436	605.232	6,325	181.769	1,363	30.615	-	-	-	-
M & E per employee ($)	6,888	299.984	16,503	652.543	706	22.248	-	-	-	-

Note: Columns headed % show percent of world total or ratio. Ratios closest to 100 are closest to world average. M & E stands for machinery & equipment.

Output in Manufacturing

	1987		1988		1989		1990		1991	
	$ bil.	%	$ bil.	%	$ bil.	%	$ bil.	%	$ bil.	%
3110 - Food products	0.301	4.30	0.315	4.50	0.293	14.65	-	-	-	-
3130 - Beverages	0.080	1.14	0.082	1.17	0.088	4.40	-	-	-	-
3210 - Textiles	0.032	0.46	0.036	0.51	0.032	1.60	-	-	-	-
3220 - Wearing apparel	0.163	2.33	0.143	2.04	0.150	7.50	-	-	-	-
3230 - Leather and products	-	-	0.010	0.14	-	-	-	-	-	-
3310 - Wood products	0.030	0.43	0.030	0.43	0.030	1.50	-	-	-	-
3320 - Furniture, fixtures	0.079	1.13	0.073	1.04	0.070	3.50	-	-	-	-
3410 - Paper and products	0.060	0.86	0.072	1.03	0.062	3.10	-	-	-	-
3420 - Printing, publishing	0.115	1.64	0.122	1.74	0.127	6.35	-	-	-	-
3510 - Industrial chemicals	0.133	1.90	0.182	2.60	-	-	-	-	-	-
3511 - Basic chemicals, excl fertilizers	0.024	0.34	0.012	0.17	0.014	0.70	-	-	-	-
3513 - Synthetic resins, etc.	-	-	0.016	0.23	-	-	-	-	-	-
3520 - Chemical products nec	0.057	0.81	0.067	0.96	0.059	2.95	-	-	-	-
3522 - Drugs and medicines	-	-	0.007	0.10	0.009	0.45	-	-	-	-
3530 - Petroleum refineries	5.383	76.90	5.023	71.76	-	-	-	-	-	-
3540 - Petroleum, coal products	0.006	0.09	0.010	0.14	0.009	0.45	-	-	-	-
3550 - Rubber products	0.013	0.19	0.015	0.21	0.014	0.70	-	-	-	-
3560 - Plastic products nec	0.071	1.01	0.089	1.27	0.076	3.80	-	-	-	-
3620 - Glass and products	0.020	0.29	0.022	0.31	0.018	0.90	-	-	-	-
3690 - Nonmetal products nec	0.296	4.23	0.326	4.66	0.293	14.65	-	-	-	-
3710 - Iron and steel	0.031	0.44	0.046	0.66	0.037	1.85	-	-	-	-
3810 - Metal products	0.205	2.93	0.237	3.39	0.211	10.55	-	-	-	-
3820 - Machinery nec	0.048	0.69	0.057	0.81	0.050	2.50	-	-	-	-
3825 - Office, computing machinery	-	-	0.004	0.06	-	-	-	-	-	-
3830 - Electrical machinery	0.066	0.94	0.085	1.21	0.077	3.85	-	-	-	-
3840 - Transportation equipment	0.030	0.43	0.037	0.53	0.032	1.60	-	-	-	-
3841 - Shipbuilding, repair	0.014	0.20	0.020	0.29	0.017	0.85	-	-	-	-
3843 - Motor vehicles	0.016	0.23	0.017	0.24	0.015	0.75	-	-	-	-
3850 - Professional goods	-	-	0.001	0.01	-	-	-	-	-	-
3900 - Industries nec	0.018	0.26	0.022	0.31	0.019	0.95	-	-	-	-

Note: Codes are International Standard Industry codes (ISIC). Percentages are % of total Output. [f]: Factor Prices; [p]: Producer Prices.

For sources, notes, and explanations, see Annotated Source Appendix, page 1061.

513

Finance, Economics, and Trade

Economic Indicators [36]

- **National product**: GDP—purchasing power parity—$30.8 billion (1995 est.)
- **National product real growth rate**: 3% (1995 est.)
- **National product per capita**: $17,000 (1995 est.)
- **Inflation rate (consumer prices)**: 5% (1994 est.)
- **External debt**: $NA

Balance of Payments Summary [37]

Values in millions of dollars.

	1989	1990	1991	1992	1993
Exports of goods (f.o.b.)	11,396	6,989	869	6,548	10,413
Imports of goods (f.o.b.)	-5,525	-3,411	-4,053	-6,292	-6,040
Trade balance	5,871	3,578	-3,184	256	4,373
Services - debits	-4,874	-4,129	-5,745	-5,416	-2,932
Services - credits	10,185	9,665	7,078	7,404	6,262
Private transfers (net)	-1,283	-770	-426	-829	-1,229
Government transfers (net)	-211	-4,181	-23,372	-1,098	-129
Long-term capital (net)	-1,134	-1,428	46	-573	-1,375
Short-term capital (net)	-6,631	2,073	38,727	11,831	-1,776
Errors and omissions	-665	-5,706	-11,849	-9,723	-4,679
Overall balance	1,258	-898	1,275	1,852	-1,485

Exchange Rates [38]

Currency: **Kuwaiti dinar.**
Symbol: **KD.**

Data are currency units per $1.

January 1996	0.2993
1995	0.2984
1994	0.2976
1993	0.3017
1992	0.2934
1991	0.2843

Imports and Exports

Top Import Origins [39]

$6.7 billion (f.o.b., 1994) Data are for 1994 est.

Origins	%
US	14
Japan	12
Germany	8
UK	7
France	6

Top Export Destinations [40]

$11.9 billion (f.o.b., 1994).

Destinations	%
US	23
Japan	13
Germany	10
UK	9
France	8

Foreign Aid [41]

Donor: Pledged bilateral aid to less developed countries (1979-89), $18.3 billion.

Import and Export Commodities [42]

Import Commodities	Export Commodities
Food	Oil
Construction materials	
Vehicles and parts	
Clothing	

Kyrgyzstan

Geography [1]

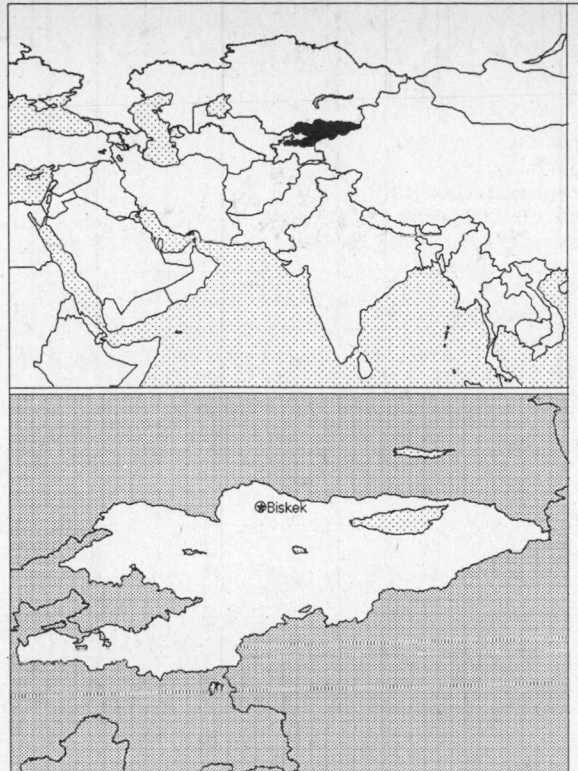

Total area:
  198,500 sq km   76,641 sq mi
Land area:
  191,300 sq km   73,861 sq mi
Comparative area:
  Slightly smaller than South Dakota
Land boundaries:
  Total 3,878 km, China 858 km, Kazakhstan 1,051 km,
  Tajikistan 870 km, Uzbekistan 1,099 km
Coastline:
  0 km (landlocked)
Climate:
  Dry continental to polar in high Tien Shan; subtropical in Southwest
  (Fergana Valley); temperate in northern foothill zone
Terrain:
  Peaks of Tien Shan and associated valleys and basins encompass entire
  nation
Natural resources:
  Abundant hydroelectric potential; significant deposits of gold and
  rare earth metals; locally exploitable coal, oil, and natural gas;
  other deposits of nepheline, mercury, bismuth, lead, and zinc
Land use:
  Arable land: 7%
  Permanent crops: Negligible
  Meadows and pastures: 42%
  Forest and woodland: 0%
  Other: 51%

Demographics [2]

	1970	1980	1990	1995[1]	1996	2000	2010	2020	2030
Population	2,964	3,623	4,390	4,534	4,530	4,664	5,403	6,257	7,138
Population density (persons per sq. mi.)	39	47	57	59	59	61	71	82	93
(persons per sq. km.)	15	18	22	23	23	23	27	32	36
Net migration rate (per 1,000 population)	NA	NA	-4.9	-19.9	-16.5	-2.8	-1.5	-0.4	0.0
Births	NA	NA	NA	NA	118	NA	NA	NA	NA
Deaths	NA	NA	NA	NA	40	NA	NA	NA	NA
Life expectancy - males	NA	NA	62.2	59.0	59.2	59.8	61.2	65.2	68.6
Life expectancy - females	NA	NA	70.9	68.6	68.8	69.4	70.8	74.4	77.4
Birth rate (per 1,000)	NA	NA	30.5	26.3	26.0	25.6	25.2	21.2	19.1
Death rate (per 1,000)	NA	NA	8.1	8.9	8.8	8.7	8.6	7.0	6.5
Women of reproductive age (15-49 yrs.)	NA	NA	1,028	1,108	1,120	1,199	1,454	1,660	1,885
of which are currently married	NA	NA	675	NA	740	785	974	NA	NA
Fertility rate	NA	NA	3.8	3.3	3.2	3.1	2.8	2.6	2.4

Except as noted, values for vital statistics are in thousands; life expectancy is in years.

Health

Health Indicators [3]

% of population with access to	
safe water (1990-95)	NA
adequate sanitation (1990-95)	NA
health services (1985-95)	NA
% of 1-year-olds immunized (1990-94) against	
TB (tuberculosis)	97
DPT (diphtheria, pertussis, tetanus)	82
polio	84
measles	88
% of contraceptive prevalence (1980-94)	NA
ORT use rate (1990-94)	NA

Health Expenditures [4]

Total health expenditure, 1990 (official exchange rate)	
Millions of dollars	517
Dollars per capita	118
Health expenditures as a percentage of GDP	
Total	5.0
Public sector	3.3
Private sector	1.6
Development assistance for health	
Total aid flows (millions of dollars)[1]	NA
Aid flows per capita (dollars)	NA
Aid flows as a percentage of total health expenditure	NA

For sources, notes, and explanations, see Annotated Source Appendix, page 1061.

Human Factors

Women and Children [5]

% of pregnant women immunized (tetanus 1990-94)	81
% of births attended by trained health personnel (1983-94)	NA
Maternal mortality rate (1980-92)	NA
Under-5 mortality rate (1994)	56
% under-5 moderately/severely underweight (1980-1994)	NA

Burden of Disease [6]

Population per physician (1993)	302.91
Population per nurse (1993)	105.23
Population per hospital bed (1993)	91.68
AIDS cases per 100,000 people (1994)	*
Heart disease cases per 1,000 people (1990-93)	384

Ethnic Division [7]

Kirghiz	52.4%
Russian	21.5%
Uzbek	12.9%
Ukrainian	2.5%
German	2.4%
Other	8.3%

Religion [8]

Muslim, Russian Orthodox.

Major Languages [9]

Kirghiz (Kyrgyz) - official language, Russian - official language. In March 1996, the Kyrgyz legislature amended the constitution to make Russian an official language, along with Kyrgyz, in territories and work places where Russian-speaking citizens predominate.

Education

Public Education Expenditures [10]

Million (Som)	1980	1985	1990	1991	1993	1994
Total education expenditure	2	2	4	6	NA	731
as percent of GNP	7.2	7.9	8.5	6.4	NA	6.8
as percent of total govt. expend.	22.2	22.4	22.5	22.7	NA	25.6
Current education expenditure	NA	2	3	5	215	697
as percent of GNP	NA	7.3	7.5	6.2	3.8	6.5
as percent of current govt. expend.	NA	NA	NA	NA	21.0	25.2
Capital expenditure	NA	0.0	1	1	NA	34

Educational Attainment [11]

Illiteracy [12]

	1985	1989	1995
Illiterate population (15+ yrs.)[2]	NA	NA	11,000
Illiteracy rate - total pop. (%)	NA	3.0	0.4
Illiteracy rate - males (%)	NA	1.4	0.3
Illiteracy rate - females (%)	NA	4.5	0.5

Libraries [13]

	Admin. Units	Svc. Pts.	Vols. (000)	Shelving (meters)	Vols. Added	Reg. Users
National (1992)	1	NA	3,076	NA	NA	33,417
Nonspecialized (1992)	1,249	2,015	14,609	NA	265,800	1 mil
Public (1992)	57	3,264	NA	NA	NA	NA
Higher ed.	NA	NA	NA	NA	NA	NA
School	NA	NA	NA	NA	NA	NA

Daily Newspapers [14]

	1980	1985	1990	1994
Number of papers	NA	NA	NA	3
Circ. (000)	NA	NA	NA	53

Culture [15]

Science and Technology

Scientific/Technical Forces [16]

R&D Expenditures [17]

U.S. Patents Issued [18]

For sources, notes, and explanations, see Annotated Source Appendix, page 1061.

Government and Law

Organization of Government [19]

Long-form name:
Kyrgyz Republic
Type:
Republic
Independence:
31 August 1991 (from Soviet Union)
National holiday:
National Day, 2 December; Independence Day, 31 August (1991)
Constitution:
Adopted 5 May 1993 Note: Amendment passed in a national referendum on 10 February 1996 significantly expanded the powers of the president at the expense of the legislature
Legal system:
Based on civil law system
Executive branch:
President; Prime Minister; Cabinet of Ministers
Legislative branch:
Bicameral Supreme Council (Zhogorku Kenesh): Assembly of People's Representatives and Legislative Assembly Note: The legislature became bicameral for the 5 February 1995 elections
Judicial branch:
Supreme Court; Constitutional Court; Higher Court of Arbitration

Elections [20]

Legislative Assembly. Elections last held 5 February 1995 (next to be held NA 2000); results - percent of vote by party NA; seats (35 total) by party NA; not all seats filled by February 1995 elections; therefore runoff elections were scheduled for 19 April 1995. The legislature became bicameral for the 5 February 1995 elections.

Government Budget [21]

Crime [23]

	1990
Crime volume	
Cases known to police	19,279
Attempts (percent)	NA
Percent cases solved	44.6
Crimes per 100,000 persons	439.3
Persons responsible for offenses	
Total number offenders	8,604
Percent female	NA
Percent juvenile	NA
Percent foreigners	NA

Military Expenditures and Arms Transfers [22]

	1990	1991	1992[14]	1993	1994
Military expenditures					
Current dollars (mil.)[e]	NA	NA	47	51	57
1994 constant dollars (mil.)[e]	NA	NA	49	52	57
Armed forces (000)	NA	NA	12	12	9
Gross national product (GNP)					
Current dollars (mil.)[e]	NA	NA	12,120	10,920	7,980
1994 constant dollars (mil.)[e]	NA	NA	12,640	11,140	7,980
Central government expenditures (CGE)					
1994 constant dollars (mil.)[e]	NA	NA	4,286	NA	NA
People (mil.)	NA	NA	4.6	4.6	4.7
Military expenditure as % of GNP	NA	NA	0.4	0.5	0.7
Military expenditure as % of CGE	NA	NA	1.1	NA	NA
Military expenditure per capita (1994 $)	NA	NA	11	11	12
Armed forces per 1,000 people (soldiers)	NA	NA	2.6	2.6	1.9
GNP per capita (1994 $)	NA	NA	2,777	2,409	1,699
Arms imports[6]					
Current dollars (mil.)	NA	NA	0	0	0
1994 constant dollars (mil.)	NA	NA	0	0	0
Arms exports[6]					
Current dollars (mil.)	NA	NA	0	10	0
1994 constant dollars (mil.)	NA	NA	0	10	0
Total imports[7]					
Current dollars (mil.)	NA	NA	419	418	347
1994 constant dollars (mil.)	NA	NA	437	427	347
Total exports[7]					
Current dollars (mil.)	NA	NA	316	268	339
1994 constant dollars (mil.)	NA	NA	330	274	339
Arms as percent of total imports[8]	NA	NA	0	0	0
Arms as percent of total exports[8]	NA	NA	0	3.7	0

Human Rights [24]

	SSTS	FL	FAPRO	PPCG	APROBC	TPW	PCPTW	STPEP	PHRFF	PRW	ASST	AFL
Observes		P	P		P	P	P					
	EAFRD	CPR	ESCR	SR	ACHR	MAAE	PVIAC	PVNAC	EAFDAW	TCIDTP		RC
Observes		P	P	P		P	P	P				P

P = Party; S = Signatory; see Appendix for meaning of abbreviations.

Labor Force

Total Labor Force [25]

1.836 million

Labor Force by Occupation [26]

Agriculture and forestry	38%
Industry and construction	21
Other	41

Date of data: 1990

Unemployment Rate [27]

4.8% includes officially registered unemployed; also large numbers of unregistered, unemployed, and underemployed workers (December 1995)

For sources, notes, and explanations, see Annotated Source Appendix, page 1061.

517

Production Sectors

Commercial Energy Production and Consumption

Data are shown in quadrillion (10^{15}) BTUs and percent for 1995
Values for hydroelectric, nuclear, geothermal, solar, and wind power refer to electrical generation.

Production [28]

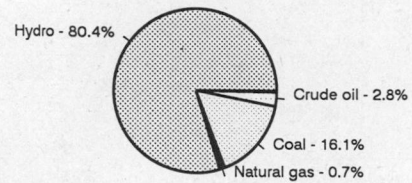

Hydro - 80.4%
Crude oil - 2.8%
Coal - 16.1%
Natural gas - 0.7%

Crude oil	0.004
Dry natural gas	0.001
Coal	0.023
Net hydroelectric power	0.115
Total	0.143

Consumption [29]

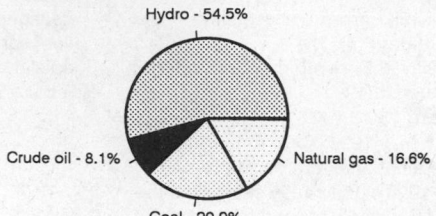

Hydro - 54.5%
Crude oil - 8.1%
Natural gas - 16.6%
Coal - 20.9%

Crude oil	0.017
Dry natural gas	0.035
Coal	0.044
Net hydroelectric power	0.115
Total	0.211

Telecommunications [30]

- 342,000 (1991 est.) telephones; poorly developed; about 100,000 unsatisfied applications for household telephones
- Domestic: principally microwave radio relay
- International: connections with other CIS countries by landline or microwave radio relay and with other countries by leased connections with Moscow international gateway switch and by satellite; satellite earth stations - 1 Intersputnik and 1 Intelsat
- Radio: Broadcast stations: AM NA, FM NA, shortwave NA; note - 1 state-run radio broadcast station Radios: 825,000 (radio receiver systems with multiple speakers for program diffusion 748,000)
- Television: Broadcast stations: 1 Televisions: 875,000

Top Agricultural Products [32]

Produces wool, tobacco, cotton, potatoes, vegetables, grapes, fruits and berries; sheep, goats, cattle.

Transportation [31]

Railways: total: 370 km in common carrier service; does not include industrial lines; broad gauge: 370 km 1.520-m gauge (1990)

Highways: total: 28,400 km; paved: 22,400 km; unpaved: 6,000 km (1990)

Airports

Total:	54
With paved runways over 3,047 m:	1
With paved runways 2,438 to 3,047 m:	3
With paved runways 1,524 to 2,437 m:	9
With paved runways under 914 m:	1
With unpaved runways 1,524 to 2,437 m:	4

Top Mining Products [33]

Metric tons except as noted[e]	7/10/95[*]
Antimony	
mine output, Sb content	1,400
metal	9,000
Cement	600,000
Gold (kg.)	2,000
Mercury	
mine output, Hg content	150
metal	250
Gas, natural (mil. cu. meters)	30
Petroleum, crude	70,000

Tourism [34]

For sources, notes, and explanations, see Annotated Source Appendix, page 1061.

Finance, Economics, and Trade

Industrial Summary [35]

Industrial Production: Growth rate - 12.5% (1995). **Industries**: Small machinery, textiles, food processing, cement, shoes, sawn logs, refrigerators, furniture, electric motors, gold, rare earth metals.

Economic Indicators [36]

- **National product**: GDP—purchasing power parity—$5.4 billion (1995 estimate as extrapolated from World Bank estimate for 1994)
- **National product real growth rate**: - 6% (1995 est.)
- **National product per capita**: $1,140 (1995 est.)
- **Inflation rate (consumer prices)**: 32% (1995 est.)
- **External debt**: $480 million (of which $115 million to Russia) (1995 est.)

Balance of Payments Summary [37]

Values in millions of dollars.

	1989	1990	1991	1992[1]	1993
Exports of goods (f.o.b.)	NA	NA	NA	80	100
Imports of goods (f.o.b.)	NA	NA	NA	70	100
Trade balance	NA	NA	NA	10	NA
Services - debits	NA	NA	NA	NA	NA
Services - credits	NA	NA	NA	NA	NA
Private transfers (net)	NA	NA	NA	15	28
Government transfers (net)	NA	NA	NA	NA	NA
Long-term capital (net)	NA	NA	NA	NA	10
Short-term capital (net)	NA	NA	NA	NA	NA
Errors and omissions	NA	NA	NA	-46	-13
Overall balance	NA	NA	NA	-146	-158

Exchange Rates [38]

Currency: **som.**

Data are currency units per $1. The som was introduced on 10 May 1993.

Year-end 1995	11.2
Year-end 1994	10.6

Imports and Exports

Top Import Origins [39]

$439 million (1995).

Origins	%
Russia	NA
Uzbekistan	NA
Kazakhstan	NA
China	NA
UK	NA

Top Export Destinations [40]

$380 million (1995).

Destinations	%
Russia	NA
Ukraine	NA
Uzbekistan	NA
Kazakhstan	NA
Turkey	NA
Cuba	NA
Germany	NA

Foreign Aid [41]

	U.S. $	
ODA (1993)	56	million
Commitments (1992-95)	1,695	million

Import and Export Commodities [42]

Import Commodities	Export Commodities
Grain	Cotton
Lumber	Wool
Industrial products	Meat
Ferrous metals	Tobacco
Fuel	Mercury
Machinery	Uranium
Textiles	Gold
Footwear	Hydropower
	Machinery
	Shoes

For sources, notes, and explanations, see Annotated Source Appendix, page 1061.

519

Laos

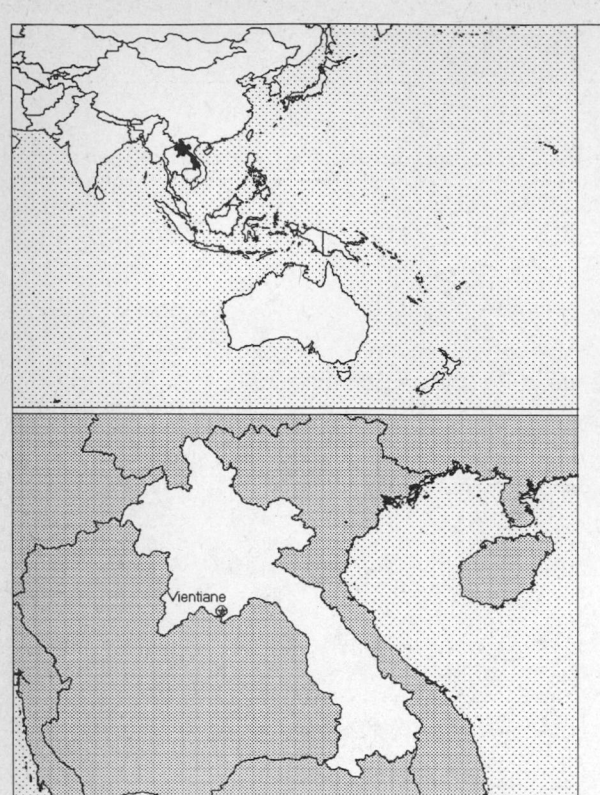

Geography [1]

Total area:
  236,800 sq km   91,429 sq mi
Land area:
  230,800 sq km   89,112 sq mi
Comparative area:
  Slightly larger than Utah
Land boundaries:
  Total 5,083 km, Myanmar 235 km, Cambodia 541 km, China 423 km, Thailand 1,754 km, Vietnam 2,130 km
Coastline:
  0 km (landlocked)
Climate:
  Tropical monsoon; rainy season (May to November); dry season (December to April)
Terrain:
  Mostly rugged mountains; some plains and plateaus
Natural resources:
  Timber, hydropower, gypsum, tin, gold, gemstones
Land use:
  Arable land: 4%
  Permanent crops: 0%
  Meadows and pastures: 3%
  Forest and woodland: 58%
  Other: 35%

Demographics [2]

	1970	1980	1990	1995[1]	1996	2000	2010	2020	2030
Population	2,845	3,293	4,191	4,837	4,976	5,557	7,168	8,923	10,669
Population density (persons per sq. mi.)	32	37	47	54	56	62	80	100	120
(persons per sq. km.)	12	14	18	21	22	24	31	39	46
Net migration rate (per 1,000 population)	NA	NA	0.0	0.0	0.0	0.0	0.0	0.0	0.0
Births	NA	NA	NA	NA	209	NA	NA	NA	NA
Deaths	NA	NA	NA	NA	69	NA	NA	NA	NA
Life expectancy - males	NA	NA	48.2	50.7	51.1	53.1	58.1	62.7	66.8
Life expectancy - females	NA	NA	51.3	53.8	54.3	56.4	61.5	66.4	70.7
Birth rate (per 1,000)	NA	NA	45.7	42.6	41.9	39.3	32.7	26.8	21.9
Death rate (per 1,000)	NA	NA	16.7	14.3	13.8	12.2	9.0	7.0	6.0
Women of reproductive age (15-49 yrs.)	NA	NA	956	1,108	1,142	1,289	1,769	2,329	2,897
of which are currently married	NA	NA	572	NA	683	772	1,056	NA	NA
Fertility rate	NA	NA	6.4	6.0	5.9	5.4	4.2	3.2	2.5

Except as noted, values for vital statistics are in thousands; life expectancy is in years.

Health

Health Indicators [3]

% of population with access to	
safe water (1990-95)	45
adequate sanitation (1990-95)	27
health services (1985-95)	67
% of 1-year-olds immunized (1990-94) against	
TB (tuberculosis)	69
DPT (diphtheria, pertussis, tetanus)	48
polio	57
measles	73
% of contraceptive prevalence (1980-94)	NA
ORT use rate (1990-94)	55

Health Expenditures [4]

Total health expenditure, 1990 (official exchange rate)	
Millions of dollars	22
Dollars per capita	5
Health expenditures as a percentage of GDP	
Total	2.5
Public sector	1.0
Private sector	1.5
Development assistance for health	
Total aid flows (millions of dollars)[1]	5
Aid flows per capita (dollars)	1.2
Aid flows as a percentage of total health expenditure	22.7

For sources, notes, and explanations, see Annotated Source Appendix, page 1061.

Human Factors

Women and Children [5]

% of pregnant women immunized (tetanus 1990-94)	34
% of births attended by trained health personnel (1983-94)	NA
Maternal mortality rate (1980-92)	300
Under-5 mortality rate (1994)	138
% under-5 moderately/severely underweight (1980-1994)	37

Burden of Disease [6]

Population per physician (1990)	4,445.50
Population per nurse (1990)	493.36
Population per hospital bed (1990)	405.35
AIDS cases per 100,000 people (1994)	0.1
Malaria cases per 100,000 people (1992)	935

Ethnic Division [7]

Lao Soung includes the Hmong ("Meo") and the Yao (Mien).

Lao Loum (lowland)	68%
Lao Theung (upland)	22%
Lao Soung (highland)	9%
Ethnic Vietnamese/Chinese	1%

Religion [8]

Buddhist	60%
Animist and other	40%

Major Languages [9]

Lao (official), French, English, and various ethnic languages.

Education

Public Education Expenditures [10]

Million (Kip)	1980	1985	1988	1990	1992	1994
Total education expenditure	24	463	2,782	NA	19,922	NA
as percent of GNP	NA	NA	1.2	NA	2.3	NA
as percent of total govt. expend.	NA	NA	NA	NA	NA	NA
Current education expenditure	NA	NA	2,097	NA	15,094	NA
as percent of GNP	NA	NA	0.9	NA	1.8	NA
as percent of current govt. expend.	NA	NA	NA	NA	NA	NA
Capital expenditure	NA	NA	685	NA	4,828	NA

Educational Attainment [11]

Illiteracy [12]

In thousands and percent[1]	1990	1995	2000
Illiterate population (15+ yrs.)	1,149	1,170	1,190
Illiteracy rate - total pop. (%)	50.1	44.3	38.9
Illiteracy rate - males (%)	36.8	31.6	27.0
Illiteracy rate - females (%)	62.3	56.2	50.1

Libraries [13]

	Admin. Units	Svc. Pts.	Vols. (000)	Shelving (meters)	Vols. Added	Reg. Users
National	NA	NA	NA	NA	NA	NA
Nonspecialized (1986)	1	2	145	1,328	1,500	800
Public	NA	NA	NA	NA	NA	NA
Higher ed. (1987)	5	NA	121	3,870	765	8,370
School (1987)	22	NA	33	880	800	11,000

Daily Newspapers [14]

	1980	1985	1990	1994
Number of papers	3	3	3	3
Circ. (000)	14[e]	13[e]	14[e]	14[c]

Culture [15]

Cinema (seats per 1,000)	1.5
Annual attendance per person	0.2
Gross box office receipts (mil. Kip)	245
Museums (reporting)	NA
Visitors (000)	NA
Annual receipts (000 Kip)	NA

Science and Technology

Scientific/Technical Forces [16]

R&D Expenditures [17]

U.S. Patents Issued [18]

For sources, notes, and explanations, see Annotated Source Appendix, page 1061.

521

Government and Law

Organization of Government [19]

Long-form name:
Lao People's Democratic Republic
Type:
Communist state
Independence:
19 July 1949 (from France)
National holiday:
National Day, 2 December (1975)
(proclamation of the Lao People's
Democratic Republic)
Constitution:
Promulgated 14 August 1991
Legal system:
Based on traditional customs, French
legal norms and procedures, and Socialist
practice
Executive branch:
President; Prime Minister; Deputy Prime
Minister; Council of Ministers
Legislative branch:
Unicameral: National Assembly
Judicial branch:
People's Supreme Court

Elections [20]

Third National Assembly. Last held on
20 December 1992 (next to be held NA);
results - percent of vote by party NA;
seats - (85 total) number of seats by
party NA.

Government Budget [21]

For 1994.

	$ mil.
Revenues	198
Expenditures	351
Capital expenditures	NA

Crime [23]

Military Expenditures and Arms Transfers [22]

	1990	1991	1992	1993	1994
Military expenditures					
Current dollars (mil.)	NA	NA	110	110[r]	114
1994 constant dollars (mil.)	NA	NA	115	112[r]	114
Armed forces (000)	55	53	37	37	45
Gross national product (GNP)					
Current dollars (mil.)	1,087	1,169	1,287	1,392	1,539
1994 constant dollars (mil.)	1,212	1,253	1,342	1,420	1,539
Central government expenditures (CGE)					
1994 constant dollars (mil.)	NA	NA	NA	NA	NA
People (mil.)	4.2	4.3	4.4	4.6	4.7
Military expenditure as % of GNP	NA	NA	8.5	7.9	7.4
Military expenditure as % of CGE	NA	NA	NA	NA	NA
Military expenditure per capita (1994 $)	NA	NA	26	25	24
Armed forces per 1,000 people (soldiers)	13.1	12.3	8.3	8.1	9.6
GNP per capita (1994 $)	289	290	302	311	327
Arms imports[6]					
Current dollars (mil.)	40	10	10	30	90
1994 constant dollars (mil.)	45	11	10	31	90
Arms exports[6]					
Current dollars (mil.)	0	0	0	0	0
1994 constant dollars (mil.)	0	0	0	0	0
Total imports[7]					
Current dollars (mil.)[e]	238	NA	266	338	528
1994 constant dollars (mil.)[e]	265	NA	277	345	528
Total exports[7]					
Current dollars (mil.)[e]	61	79	97	136	277
1994 constant dollars (mil.)[e]	68	85	101	139	277
Arms as percent of total imports[8]	16.8	NA	3.8	8.9	17.0
Arms as percent of total exports[8]	0	0	0	0	0

Human Rights [24]

	SSTS	FL	FAPRO	PPCG	APROBC	TPW	PCPTW	STPEP	PHRFF	PRW	ASST	AFL
Observes		P		P		P	P	P		P	P	
	EAFRD	CPR	ESCR	SR	ACHR	MAAE	PVIAC	PVNAC	EAFDAW	TCIDTP	RC	
Observes	P						P	P	P		P	

P = Party; S = Signatory; see Appendix for meaning of abbreviations.

Labor Force

Total Labor Force [25]

1 million-1.5 million

Labor Force by Occupation [26]

Agriculture   80%
Date of data: 1992 est.

Unemployment Rate [27]

21% (1992 est.)

For sources, notes, and explanations, see Annotated Source Appendix, page 1061.

Production Sectors

Commercial Energy Production and Consumption

Data are shown in quadrillion (10^{15}) BTUs and percent for 1995
Values for hydroelectric, nuclear, geothermal, solar, and wind power refer to electrical generation.

Production [28]

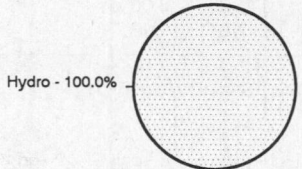

Hydro - 100.0%

Consumption [29]

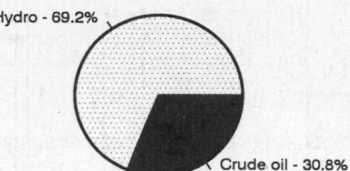

Hydro - 69.2%

Crude oil - 30.8%

Net hydroelectric power	0.009
Total	0.009

Crude oil	0.004
Net hydroelectric power	0.009
Total	0.013

Telecommunications [30]

- 6,600 (1991 est.) telephones; service to general public very poor; radiotelephone communications network provides generally erratic service to government users
- Domestic: radiotelephone communications
- International: satellite earth station - 1 Intersputnik (Indian Ocean Region)
- Radio: Broadcast stations: AM 10, FM 0, shortwave 0 Radios: 560,000 (1992 est.)
- Television: Broadcast stations: 2 Televisions: 32,000 (1993 est.)

Transportation [31]

Railways: 0 km

Highways: total: 14,130 km; paved: 2,261 km; unpaved: 11,869 km (1992 est.)

Merchant marine: total: 1 cargo ship (1,000 GRT or over) totaling 2,370 GRT/3,000 DWT (1995 est.)

Airports

Total:	39
With paved runways over 3,047 m:	1
With paved runways 1,524 to 2,437 m:	5
With paved runways 914 to 1,523 m:	3
With paved runways under 914 m:	16
With unpaved runways 1,524 to 2,437 m:	1

Top Agricultural Products [32]

Agriculture accounts for 50% of the GDP; produces sweet potatoes, vegetables, corn, coffee, sugarcane, cotton; water buffalo, pigs, cattle, poultry.

Top Mining Products [33]

Metric tons except as noted[e]	6/6/95[*]
Coal, all grades	10,000
Cement (from imported clinker)	10,000
Sapphire (carats)	40,000
Gypsum	85,000
Salt, rock	8,000
Tin, mine output, Sn content	200

Tourism [34]

	1990	1991	1992	1993	1994
Visitors	NA	111	92	207	388
Tourists[12]	14	38	88	103	146
Tourism receipts	3	8	18	34	43
Tourism expenditures	1	6	10	11	18
Fare receipts	NA	NA	NA	3	2
Fare expenditures	2	3	3	4	5

Travelers are in thousands, money in million U.S. dollars.

For sources, notes, and explanations, see Annotated Source Appendix, page 1061.

523

Finance, Economics, and Trade

GDP and Manufacturing Summary [35]

	1980	1985	1990	1991	1992
Gross Domestic Product					
Millions of 1980 dollars	462	661	903	942	1,011[e]
Growth rate in percent	1.70	9.83	9.10	4.30	7.30[e]
Manufacturing Value Added					
Millions of 1980 dollars	23	29	34	38	41[e]
Growth rate in percent	7.94	1.99	10.10	11.92	8.00[e]
Manufacturing share in percent of current prices	NA	NA	NA	NA	NA

Economic Indicators [36]

- **National product**: GDP—purchasing power parity—$5.2 billion (1995 est.)
- **National product real growth rate**: 8% (1995 est.)
- **National product per capita**: $1,100 (1995 est.)
- **Inflation rate (consumer prices)**: 20% (1995 est.)
- **External debt**: $2 billion (1995 est.)

Balance of Payments Summary [37]

Values in millions of dollars.

	1989	1990	1991	1992	1993
Exports of goods (f.o.b.)	63.3	78.7	96.6	132.6	231.8
Imports of goods (f.o.b.)	-193.8	-185.5	-197.9	-232.8	-379.5
Trade balance	-130.5	-106.8	-101.3	-100.2	-147.7
Services - debits	-35.7	-33.5	-63.4	-84.1	-95.6
Services - credits	23.3	25.9	41.1	67.0	91.8
Private transfers (net)	8.3	10.9	10.4	8.6	9.5
Government transfers (net)	18.6	55.6	61.0	60.7	98.8
Long-term capital (net)	52.5	51.1	44.6	72.1	78.4
Sort term capital (net)	22.2	34.3	51.6	-10.3	-49.6
Errors and omissions	36.9	-36.3	-28.8	-9.4	30.2
Overall balance	-4.4	1.2	15.2	4.4	15.8

Exchange Rates [38]

Currency: **new kip.**
Symbol: **NK.**

Data are currency units per $1.

1995	920
1994 est.	717
July 1993	720
May 1992	710
December 1991	710
September 1990	700
1989	576

Imports and Exports

Top Import Origins [39]

$486 million (c.i.f., 1994).

Origins	%
Thailand	NA
China	NA
Japan	NA
France	NA
US	NA

Top Export Destinations [40]

$278 million (f.o.b., 1994).

Destinations	%
Thailand	NA
Japan	NA
France	NA
Germany	NA
Netherlands	NA

Foreign Aid [41]

Recipient: ODA, $NA.

Import and Export Commodities [42]

Import Commodities	Export Commodities
Food	Electricity
Fuel oil	Wood products
Consumer goods	Coffee
Manufactures	Tin
	Garments

Latvia

Geography [1]

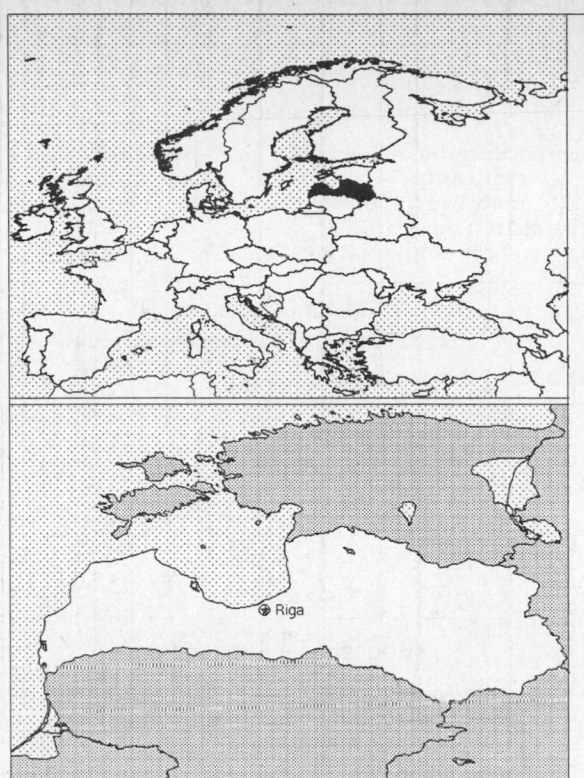

Total area:
64,100 sq km   24,749 sq mi
Land area:
64,100 sq km   24,749 sq mi
Comparative area:
Slightly larger than West Virginia
Land boundaries:
Total 1,078 km, Belarus 141 km, Estonia 267 km, Lithuania 453 km, Russia 217 km
Coastline:
531 km
Climate:
Maritime; wet, moderate winters
Terrain:
Low plain
Natural resources:
Minimal; amber, peat, limestone, dolomite
Land use:
Arable land: 27%
Permanent crops: 0%
Meadows and pastures: 13%
Forest and woodland: 39%
Other: 21%

Demographics [2]

	1970	1980	1990	1995[1]	1996	2000	2010	2020	2030
Population	2,361	2,525	2,672	2,507	2,469	2,380	2,293	2,212	2,167
Population density (persons per sq. mi.)	95	101	107	101	99	96	92	89	87
(persons per sq. km.)	37	39	41	39	38	37	36	34	34
Net migration rate (per 1,000 population)	NA	NA	-3.2	-11.2	-9.7	-2.9	-1.8	-0.6	0.0
Births	NA	NA	NA	NA	27	NA	NA	NA	NA
Deaths	NA	NA	NA	NA	38	NA	NA	NA	NA
Life expectancy - males	NA	NA	64.0	60.4	60.8	62.4	66.1	69.9	73.0
Life expectancy - females	NA	NA	74.2	73.2	73.3	73.7	74.8	77.8	80.3
Birth rate (per 1,000)	NA	NA	14.1	10.3	10.9	13.5	12.8	10.8	11.0
Death rate (per 1,000)	NA	NA	13.1	15.3	15.2	14.9	14.6	13.2	12.4
Women of reproductive age (15-49 yrs.)	NA	NA	648	606	600	584	539	508	495
of which are currently married	NA	NA	410	NA	380	364	346	NA	NA
Fertility rate	NA	NA	2.0	1.5	1.6	2.0	1.8	1.8	1.7

Except as noted, values for vital statistics are in thousands; life expectancy is in years.

Health

Health Indicators [3]

% of population with access to	
safe water (1990-95)	NA
adequate sanitation (1990-95)	NA
health services (1985-95)	NA
% of 1-year-olds immunized (1990-94) against	
TB (tuberculosis)	89
DPT (diphtheria, pertussis, tetanus)	70
polio	72
measles	81
% of contraceptive prevalence (1980-94)	NA
ORT use rate (1990-94)	NA

Health Expenditures [4]

For sources, notes, and explanations, see Annotated Source Appendix, page 1061.

Human Factors

Women and Children [5]

% of pregnant women immunized (tetanus 1990-94)	NA
% of births attended by trained health personnel (1983-94)	NA
Maternal mortality rate (1980-92)	NA
Under-5 mortality rate (1994)	26
% under-5 moderately/severely underweight (1980-1994)	NA

Burden of Disease [6]

Population per physician (1993)	278.06
Population per nurse (1993)	117.55
Population per hospital bed (1993)	82.36
AIDS cases per 100,000 people (1994)	0.1
Heart disease cases per 1,000 people (1990-93)	395.5

Ethnic Division [7]

Latvian	51.8%
Russian	33.8%
Byelorussian	4.5%
Ukrainian	3.4%
Polish	2.3%
Other	4.2%

Religion [8]

Lutheran, Roman Catholic, Russian Orthodox.

Major Languages [9]

Lettish (official), Lithuanian, Russian, other.

Education

Public Education Expenditures [10]

Million (Lat)	1980	1985	1990	1992	1993	1994
Total education expenditure	1	2	2	46	89	125
as percent of GNP	3.3	3.4	3.8	4.5	6.0	6.5
as percent of total govt. expend.	15.3	12.4	10.8	22.1	16.5	16.1
Current education expenditure	NA	NA	NA	NA	87	124
as percent of GNP	NA	NA	NA	NA	5.8	6.4
as percent of current govt. expend.	NA	NA	NA	NA	16.5	16.2
Capital expenditure	NA	NA	NA	NA	2	1

Educational Attainment [11]

Age group (1989)	25+
Total population	1,725,639
Highest level attained (%)	
No schooling	0.6
First level	
Not completed	18.5
Completed	21.2
Entered second level	
S-1	46.3
S-2	NA
Postsecondary	13.4

Illiteracy [12]

	1985	1989	1995
Illiterate population (15+ yrs.)[2]	NA	11,476	5,000
Illiteracy rate - total pop. (%)	NA	0.5	0.3
Illiteracy rate - males (%)	NA	0.2	0.2
Illiteracy rate - females (%)	NA	0.8	0.3

Libraries [13]

	Admin. Units	Svc. Pts.	Vols. (000)	Shelving (meters)	Vols. Added	Reg. Users
National (1993)	1	NA	4,335	NA	518,100	24,800
Nonspecialized	NA	NA	NA	NA	NA	NA
Public (1994)	1,046	NA	21,800	NA	NA	NA
Higher ed. (1993)[21]	15	NA	7,100	NA	221,122	53,814
School	NA	NA	NA	NA	NA	NA

Daily Newspapers [14]

	1980	1985	1990	1994
Number of papers	NA	NA	NA	22
Circ. (000)	NA	NA	NA	589

Culture [15]

Cinema (seats per 1,000)	NA
Annual attendance per person	0.7
Gross box office receipts (mil. Lat)	NA
Museums (reporting)	97
Visitors (000)	1,248
Annual receipts (000 Lat)	2,157

Science and Technology

Scientific/Technical Forces [16]

Scientists/engineers	3,010
Number female	NA
Technicians	944
Number female	NA
Total	3,954

R&D Expenditures [17]

	Lat (000) 1992
Total expenditure[19]	2,380
Capital expenditure	NA
Current expenditure	NA
Percent current	NA

U.S. Patents Issued [18]

For sources, notes, and explanations, see Annotated Source Appendix, page 1061.

Government and Law

Organization of Government [19]

Long-form name:
Republic of Latvia
Type:
Republic
Independence:
6 September 1991 (from Soviet Union)
National holiday:
Independence Day, 18 November (1918)
Constitution:
Newly elected Parliament in 1993 restored the 1933 constitution
Legal system:
Based on civil law system
Executive branch:
President; Prime Minister; Council of Ministers
Legislative branch:
Unicameral: Parliament (Saeima)
Judicial branch:
Supreme Court

Elections [20]

Parliament (Saeima)	% of votes
Saimnieks	18
Latvia's Way (LC)	17
For Latvia	16
Fatherland and Freedom (TB)	14
Latvian Nat Conservative	8
Unity	*
Green/Christian Dem. Union	7
Harmony	6
Socialist	6

Government Expenditures [21]

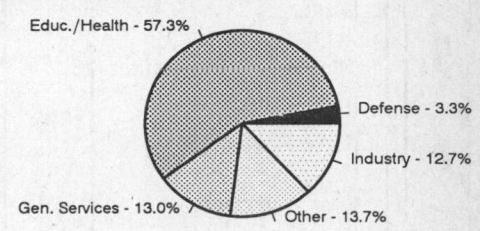

Educ./Health - 57.3%
Defense - 3.3%
Industry - 12.7%
Gen. Services - 13.0%
Other - 13.7%

(% distribution). Expend. for CY94: 564.17 (Lat mil.)

Military Expenditures and Arms Transfers [22]

	1990	1991	1992[14]	1993	1994
Military expenditures					
Current dollars (mil.)[e]	NA	NA	117	122	180
1994 constant dollars (mil.)[e]	NA	NA	122	125	180
Armed forces (000)	NA	NA	5	5	2
Gross national product (GNP)					
Current dollars (mil.)[e]	NA	NA	13,750	13,380	13,350
1994 constant dollars (mil.)[e]	NA	NA	14,340	13,660	13,350
Central government expenditures (CGE)					
1994 constant dollars (mil.)[e]	NA	NA	4,574	NA	4,220
People (mil.)	NA	NA	2.7	2.7	2.7
Military expenditure as % of GNP	NA	NA	0.9	0.9	1.3
Military expenditure as % of CGE	NA	NA	2.7	NA	4.3
Military expenditure per capita (1994 $)	NA	NA	45	46	65
Armed forces per 1,000 people (soldiers)	NA	NA	1.8	1.8	0.7
GNP per capita (1994 $)	NA	NA	5,268	4,992	4,857
Arms imports[6]					
Current dollars (mil.)	NA	NA	0	0	0
1994 constant dollars (mil.)	NA	NA	0	0	0
Arms exports[6]					
Current dollars (mil.)	NA	NA	0	0	0
1994 constant dollars (mil.)	NA	NA	0	0	0
Total imports[7]					
Current dollars (mil.)	NA	NA	944	415	NA
1994 constant dollars (mil.)	NA	NA	984	424	NA
Total exports[7]					
Current dollars (mil.)	NA	NA	825	458	NA
1994 constant dollars (mil.)	NA	NA	860	467	NA
Arms as percent of total imports[8]	NA	NA	0	0	0
Arms as percent of total exports[8]	NA	NA	0	0	0

Crime [23]

	1994
Crime volume	
Cases known to police	40,983
Attempts (percent)	1.70
Percent cases solved	31.00
Crimes per 100,000 persons	1,597.25
Persons responsible for offenses	
Total number offenders	13,350
Percent female	9.00
Percent juvenile (14-17 yrs.)	12.90
Percent foreigners	2.90

Human Rights [24]

	SSTS	FL	FAPRO	PPCG	APROBC	TPW	PCPTW	STPEP	PHRFF	PRW	ASST	AFL
Observes	2		P	P	P	P	P	P	S	P	P	P
	EAFRD	CPR	ESCR	SR	ACHR	MAAE	PVIAC	PVNAC	EAFDAW	TCIDTP	RC	
Observes	P	P	P				P		P	P	P	

P = Party; S = Signatory; see Appendix for meaning of abbreviations.

Labor Force

Total Labor Force [25]

1.407 million

Labor Force by Occupation [26]

Industry and construction	41%
Agriculture and forestry	16
Other	43
Date of data: 1990	

Unemployment Rate [27]

6.5% (1995 est.)

For sources, notes, and explanations, see Annotated Source Appendix, page 1061.

527

Production Sectors

Commercial Energy Production and Consumption

Data are shown in quadrillion (10^{15}) BTUs and percent for 1995
Values for hydroelectric, nuclear, geothermal, solar, and wind power refer to electrical generation.

Production [28]

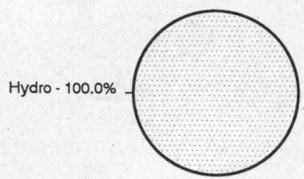

Hydro - 100.0%

Consumption [29]

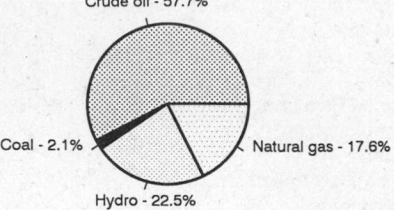

Crude oil - 57.7%
Coal - 2.1%
Natural gas - 17.6%
Hydro - 22.5%

Net hydroelectric power	0.032
Total	0.032

Crude oil	0.082
Dry natural gas	0.025
Coal	0.003
Net hydroelectric power	0.032
Total	0.142

Telecommunications [30]

- 660,000 (1993 est.) telephones; service is better than in most of the other former Soviet republics
- Domestic: an NMT-450 analog cellular telephone network covers 75% of Latvia's population
- International: international traffic carried by leased connection to the Moscow international gateway switch, through the new Ericsson digital telephone exchange in Riga, and through the Finnish cellular net; Sprint data network carries electronic mail
- Radio: Broadcast stations: AM NA, FM NA, shortwave NA; note - there are 25 radio broadcast stations of unknown type Radios: 1.4 million (1993 est.)
- Television: Broadcast stations: 30 Televisions: 1.1 million (1993 est.)

Top Agricultural Products [32]

Agriculture accounts for 9% of the GDP; produces grain, sugar beets, potatoes, vegetables; meat, milk, eggs; fish.

Transportation [31]

Railways: total: 2,412 km; broad gauge: 2,379 km 1.520-m gauge (271 km electrified) (1992); narrow gauge: 33 km 0.750-m gauge (1994)

Highways: total: 66,718 km; paved: 12,076 km; unpaved: 54,642 km (1992 est.)

Merchant marine: total: 56 ships (1,000 GRT or over) totaling 519,859 GRT/678,987 DWT; ships by type: cargo 7, oil tanker 24, refrigerated cargo 18, roll-on/roll-off cargo 7 (1995 est.)

Airports
Total:	50
With paved runways 2,438 to 3,047 m:	6
With paved runways 1,524 to 2,437 m:	2
With paved runways 914 to 1,523 m:	1
With paved runways under 914 m:	27
With unpaved runways 2,438 to 3,047 m:	2

Top Mining Products [33]

Metric tons except as noted[e]	5/25/95[*]
Cement	300,000
Clays (cu. meters)	500,000
Gypsum	300,000
Limestone	700,000
Peat, for fuel use	300,000
Sand and gravel (cu. meters)	1,000,000
Silica sand, industrial (cu. meters)	65,000

Tourism [34]

	1990	1991	1992	1993	1994
Visitors	NA	NA	NA	2,446	1,944
Tourism receipts	NA	NA	7	15	18
Tourism expenditures	NA	NA	13	29	31
Fare receipts	NA	NA	1	3	8
Fare expenditures	NA	NA	1	29	34

Travelers are in thousands, money in million U.S. dollars.

Finance, Economics, and Trade

Industrial Summary [35]

Industrial Production: Growth rate - 9.5% (1994 est.); accounts for 31% of the GDP. **Industries**: Buses, vans, street and railroad cars, synthetic fibers, agricultural machinery, fertilizers, washing machines, radios, electronics, pharmaceuticals, processed foods, textiles; dependent on imports for energy, raw materials, and intermediate products.

Economic Indicators [36]

- **National product**: GDP—purchasing power parity—$14.7 billion (1995 est.)

- **National product real growth rate**: - 1.5% (1995 est.)

- **National product per capita**: $5,300 (1995 estimate as extrapolated from World Bank estimate for 1994)

- **Inflation rate (consumer prices)**: 20% (1995 est.)

- **External debt**: $NA

Balance of Payments Summary [37]

Values in millions of dollars.

	1988	1989	1990	1991	1992[1]
Exports of goods (f.o.b.)	NA	NA	NA	NA	869
Imports of goods (f.o.b.)	NA	NA	NA	NA	914
Trade balance	NA	NA	NA	NA	-45
Services - debits	NA	NA	NA	NA	-177
Services - credits	NA	NA	NA	NA	325
Private transfers (net)	NA	NA	NA	NA	109
Government transfers (net)	NA	NA	NA	NA	NA
Long-term capital (net)	NA	NA	NA	NA	-47
Short-term capital (net)	NA	NA	NA	NA	NA
Errors and omissions	NA	NA	NA	NA	-95
Overall balance	NA	NA	NA	NA	70

Exchange Rates [38]

Currency: **lat.**

Data are currency units per $1. The lat was introduced in March 1993.

January 1996	0.544
1995	0.528
1994	0.560
1993	0.675

Imports and Exports

Top Import Origins [39]

$1.7 billion (c.i.f., 1995 est.).

Origins	%
Russia	NA
Germany	NA
Sweden	NA
Ukraine	NA
UK	NA
Lithuania	NA
Finland	NA

Top Export Destinations [40]

$1.3 billion (f.o.b., 1995 est.).

Destinations	%
Russia	NA
Germany	NA
Sweden	NA
UK	NA
Lithuania	NA

Foreign Aid [41]

	U.S. $	
ODA (1993)	122	million
Commitments from the West and international institutions (1992-95)	525	million

Import and Export Commodities [42]

Import Commodities	**Export Commodities**
Fuels	Timber
Cars	Textiles
Chemicals	Dairy products

Lebanon

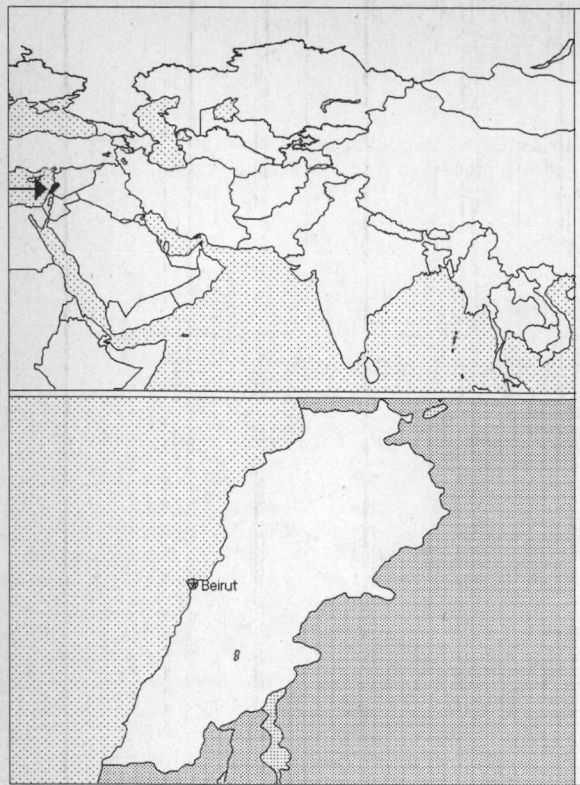

Geography [1]

Total area:
 10,400 sq km   4,015 sq mi
Land area:
 10,230 sq km   3,950 sq mi
Comparative area:
 About 0.8 times the size of Connecticut
Land boundaries:
 Total 454 km, Israel 79 km, Syria 375 km
Coastline:
 225 km
Climate:
 Mediterranean; mild to cool, wet winters with hot, dry summers; Lebanon mountains experience heavy winter snows
Terrain:
 Narrow coastal plain; Al Biqa' (Bekaa Valley) separates Lebanon and Anti-Lebanon Mountains
Natural resources:
 Limestone, iron ore, salt, water-surplus state in a water-deficit region
Land use:
 Arable land: 21%
 Permanent crops: 9%
 Meadows and pastures: 1%
 Forest and woodland: 8%
 Other: 61%

Demographics [2]

	1970	1980	1990	1995[1]	1996	2000	2010	2020	2030
Population	2,383	3,137	3,367	3,696	3,776	4,115	4,973	5,748	6,518
Population density (persons per sq. mi.)	603	794	852	936	956	1,042	1,259	1,455	1,650
(persons per sq. km.)	233	307	329	361	369	402	486	562	637
Net migration rate (per 1,000 population)	30.2	-25.5	6.7	0.0	0.0	0.0	0.0	0.0	0.0
Births	NA	NA	NA	NA	105	NA	NA	NA	NA
Deaths	NA	NA	NA	NA	24	NA	NA	NA	NA
Life expectancy - males	64.2	64.2	65.8	67.2	67.5	68.6	71.1	73.1	74.8
Life expectancy - females	68.4	68.4	70.4	72.3	72.6	74.0	77.0	79.4	81.3
Birth rate (per 1,000)	34.9	31.7	27.9	27.9	27.9	27.1	21.5	18.2	16.6
Death rate (per 1,000)	8.0	7.7	7.0	6.4	6.4	6.0	5.2	4.9	5.1
Women of reproductive age (15-49 yrs.)	518	703	817	971	1,001	1,112	1,356	1,593	1,669
of which are currently married	281	NA	466	NA	583	676	878	NA	NA
Fertility rate	5.5	4.6	3.7	3.3	3.2	3.0	2.5	2.3	2.1

Except as noted, values for vital statistics are in thousands; life expectancy is in years.

Health

Health Indicators [3]

% of population with access to	
safe water (1990-95)	94
adequate sanitation (1990-95)	63
health services (1985-95)	95
% of 1-year-olds immunized (1990-94) against	
TB (tuberculosis)	NA
DPT (diphtheria, pertussis, tetanus)	NA
polio	95
measles	73
% of contraceptive prevalence (1980-94)[1]	55
ORT use rate (1990-94)	45

Health Expenditures [4]

For sources, notes, and explanations, see Annotated Source Appendix, page 1061.

Human Factors

Women and Children [5]

% of pregnant women immunized (tetanus 1990-94)	NA
% of births attended by trained health personnel (1983-94)[1]	45
Maternal mortality rate (1980-92)	NA
Under-5 mortality rate (1994)	40
% under-5 moderately/severely underweight (1980-1994)	NA

Burden of Disease [6]

Population per physician (1991)	754.27
Population per nurse	NA
Population per hospital bed	NA
AIDS cases per 100,000 people (1994)	0.4
Malaria cases per 100,000 people (1992)	NA

Ethnic Division [7]

Arab	95%
Armenian	4%
Other	1%

Religion [8]

A small Jewish population is also present. No data are available.

Islam (5 legally recognized Islamic groups - Alawite or Nusayri, Druze, Isma'ilite, Shi'a, Sunni)	70%
Christian (11 legally recognized Christian groups - 4 Orthodox, 6 Catholic, 1 Protestant)	30%

Major Languages [9]

Arabic (official), French (official), Armenian, English.

Education

Public Education Expenditures [10]

Million (Pound)[37]	1980	1985	1990	1992	1993	1994
Total education expenditure	511	1,639	NA	206,603	251,874	316,678
as percent of GNP	NA	NA	NA	2.1	1.9	2.0
as percent of total govt. expend.	13.2	16.8	NA	12.5	NA	NA
Current education expenditure	NA	NA	NA	204,123	227,049	240,091
as percent of GNP	NA	NA	NA	2.1	1.7	1.5
as percent of current govt. expend.	NA	NA	NA	NA	NA	NA
Capital expenditure	NA	NA	NA	2,480	24,825	96,587

Educational Attainment [11]

Illiteracy [12]

In thousands and percent[1]	1990	1995	2000
Illiterate population (15+ yrs.)	154	151	138
Illiteracy rate - total pop. (%)	7.6	6.4	5.1
Illiteracy rate - males (%)	5.3	4.5	3.7
Illiteracy rate - females (%)	9.7	8.1	6.4

Libraries [13]

	Admin. Units	Svc. Pts.	Vols. (000)	Shelving (meters)	Vols. Added	Reg. Users
National	NA	NA	NA	NA	NA	NA
Nonspecialized	NA	NA	NA	NA	NA	NA
Public	NA	NA	NA	NA	NA	NA
Higher ed. (1993)	72	-	2,075[e]	NA	NA	NA
School	NA	NA	NA	NA	NA	NA

Daily Newspapers [14]

	1980	1985	1990	1994
Number of papers	14	13	14[e]	16
Circ. (000)	290[e]	300[e]	320[e]	500[e]

Culture [15]

Cinema (seats per 1,000)	9.5
Annual attendance per person	35.3
Gross box office receipts (mil. Pound)	691,733
Museums (reporting)	NA
Visitors (000)	NA
Annual receipts (000 Pound)	NA

Science and Technology

Scientific/Technical Forces [16]

Scientists/engineers	180
Number female	NA
Technicians	6
Number female	NA
Total[23]	186

R&D Expenditures [17]

	Pound (000) 1980
Total expenditure[24]	22,000
Capital expenditure	NA
Current expenditure	NA
Percent current	NA

U.S. Patents Issued [18]

Values show patents issued to citizens of the country by the U.S. Patents Office.

	1993	1994	1995
Number of patents	1	1	1

Government and Law

Organization of Government [19]

Long-form name:
Republic of Lebanon
Type:
Republic
Independence:
22 November 1943 (from League of Nations mandate under French administration)
National holiday:
Independence Day, 22 November (1943)
Constitution:
23 May 1926, amended a number of times
Legal system:
Mixture of Ottoman law, canon law, Napoleonic code, and civil law; no judicial review of legislative acts; has not accepted compulsory ICJ jurisdiction
Executive branch:
President; Prime Minister; Deputy Prime Minister; Cabinet
Legislative branch:
Unicameral: National Assembly (Arabic - Majlis Alnuwab French - Assembl)
Judicial branch:
Four Courts of Cassation (three courts for civil and commercial cases and one court for criminal cases)

Elections [20]

National Assembly. Lebanon's first legislative election in 20 years was held in the summer of 1992; the National Assembly is composed of 128 deputies, one-half Christian and one-half Muslim; its mandate expired in 1996. Political party activity is organized along largely sectarian lines; numerous political groupings exist, consisting of individual political figures and followers motivated by religious, clan, and economic considerations.

Government Budget [21]

For 1994 est.

	$ bil.
Revenues	1.4
Expenditures	3.2
Capital expenditures	NA

Crime [23]

	1994
Crime volume	
Cases known to police	32,837
Attempts (percent)	NA
Percent cases solved	24.64
Crimes per 100,000 persons	656.74
Persons responsible for offenses	
Total number offenders	44,828
Percent female	6.07
Percent juvenile (14-18 yrs.)	2.33
Percent foreigners	14.27

Military Expenditures and Arms Transfers [22]

	1990	1991	1992	1993	1994
Military expenditures					
Current dollars (mil.)	246[e]	271[e]	325	307	310[e]
1994 constant dollars (mil.)	274[e]	291[e]	339	313	310[e]
Armed forces (000)	36	36	37	37	50
Gross national product (GNP)					
Current dollars (mil.)	6,201	7,955	8,079	8,659	9,540
1994 constant dollars (mil.)	6,902	8,526	8,425	8,837	9,540
Central government expenditures (CGE)					
1994 constant dollars (mil.)	1,954[e]	2,248[e]	1,835	2,035	3,200[e]
People (mil.)	3.4	3.4	3.5	3.6	3.6
Military expenditure as % of GNP	4.0	3.4	4.0	3.5	3.2
Military expenditure as % of CGE	14.0	12.9	18.5	15.4	9.7
Military expenditure per capita (1994 $)	81	85	97	88	85
Armed forces per 1,000 people (soldiers)	10.7	10.5	10.6	10.4	13.8
GNP per capita (1994 $)	2,050	2,480	2,413	2,488	2,635
Arms imports[6]					
Current dollars (mil.)	5	5	5	10	40
1994 constant dollars (mil.)	6	5	5	10	40
Arms exports[6]					
Current dollars (mil.)	0	0	0	0	0
1994 constant dollars (mil.)	0	0	0	0	0
Total imports[7]					
Current dollars (mil.)	2,521	3,729	4,075	4,371	6,000
1994 constant dollars (mil.)	2,805	3,997	4,250	4,461	6,000
Total exports[7]					
Current dollars (mil.)	489	547	450	517	500
1994 constant dollars (mil.)	544	587	469	528	500
Arms as percent of total imports[8]	0.2	0.1	0.1	0.2	0.7
Arms as percent of total exports[8]	0	0	0	0	0

Human Rights [24]

	SSTS	FL	FAPRO	PPCG	APROBC	TPW	PCPTW	STPEP	PHRFF	PRW	ASST	AFL
Observes	P			P	P	P	P			P		P
	EAFRD	CPR	ESCR	SR	ACHR	MAAE	PVIAC	PVNAC	EAFDAW	TCIDTP		RC
Observes	P	P	P									P

P = Party; S = Signatory; see Appendix for meaning of abbreviations.

Labor Force

Total Labor Force [25]

650,000

Labor Force by Occupation [26]

Services	60%
Industry	28
Agriculture	12
Date of data: 1990 est.

Unemployment Rate [27]

30% (1995 est.)

For sources, notes, and explanations, see Annotated Source Appendix, page 1061.

Production Sectors

Commercial Energy Production and Consumption

Data are shown in quadrillion (10^{15}) BTUs and percent for 1995
Values for hydroelectric, nuclear, geothermal, solar, and wind power refer to electrical generation.

Production [28]

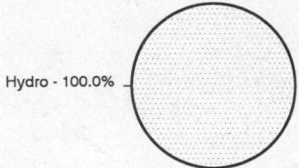

Hydro - 100.0%

Consumption [29]

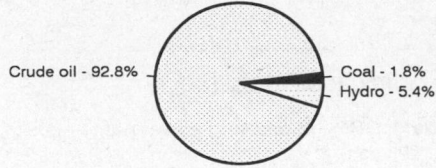

Crude oil - 92.8%
Coal - 1.8%
Hydro - 5.4%

Net hydroelectric power	0.009
Total	0.009

Crude oil	0.155
Coal	0.003
Net hydroelectric power	0.009
Total	0.167

Telecommunications [30]

- 150,000 (1990 est.) telephones; telecommunications system severely damaged by civil war; rebuilding still underway
- Domestic: primarily microwave radio relay and cable
- International: satellite earth stations - 2 Intelsat (1 Indian Ocean and 1 Atlantic Ocean) (erratic operations); coaxial cable to Syria; microwave radio relay to Syria but inoperable beyond Syria to Jordan; 3 submarine coaxial cables
- Radio: Broadcast stations: AM 5, FM 3, shortwave 1 Radios: 2.37 million (1992 est.)
- Television: Broadcast stations: 13 Televisions: 1.1 million (1993 est.)

Top Agricultural Products [32]

Agriculture accounts for 13% of the GDP; produces citrus, vegetables, potatoes, olives, tobacco, hemp (hashish); sheep, goats.

Top Mining Products [33]

Metric tons except as noted[e]	6/15/95[*]
Cement, hydraulic (000 metric tons)	2,800
Gypsum	2,000
Iron and steel, metal manufactures	80,000
Lime	15,000
Salt	3,000

Transportation [31]

Railways: total: 222 km; standard gauge: 222 km 1.435-m (from Beirut to the Syrian border)

Highways: total: 7,370 km; paved: 6,265 km; unpaved: 1,105 km (1990 est.)

Merchant marine: total: 58 ships (1,000 GRT or over) totaling 192,075 GRT/296,256 DWT; ships by type: bulk 4, cargo 39, chemical tanker 1, combination bulk 1, combination ore/oil 1, container 2, livestock carrier 4, refrigerated cargo 1, roll-on/roll-off cargo 2, specialized tanker 1, vehicle carrier 2 (1995 est.)

Airports
Total: 7
With paved runways over 3,047 m: 1
With paved runways 2,438 to 3,047 m: 2
With paved runways 1,524 to 2,437 m: 1
With paved runways under 914 m: 2
With unpaved runways 914 to 1,523 m: 1 (1995 est.)

Tourism [34]

	1990	1991	1992	1993	1994
Tourists[64]	NA	NA	178	266	335
Tourism receipts	NA	NA	NA	600	672

Travelers are in thousands, money in million U.S. dollars.

For sources, notes, and explanations, see Annotated Source Appendix, page 1061.

533

Finance, Economics, and Trade

Industrial Summary [35]

Industrial Production: Growth rate not available; accounts for 28% of the GDP. **Industries**: Banking, food processing, textiles, cement, oil refining, chemicals, jewelry, some metal fabricating.

Economic Indicators [36]

- **National product**: GDP—purchasing power parity— $18.3 billion (1995 est.)
- **National product real growth rate**: 6.5% (1995 est.)
- **National product per capita**: $4,900 (1995 est.)
- **Inflation rate (consumer prices)**: 9% (1995 est.)
- **External debt**: $1.2 billion (July 1995)

Balance of Payments Summary [37]

Exchange Rates [38]

Currency: **Lebanese pound.**
Symbol: **#L.**

Data are currency units per $1.

March 1996	1,584.0
1995	1,621.4
1994	1,680.1
1993	1,741.4
1992	1,712.8
1991	928.2

Imports and Exports

Top Import Origins [39]

$7.3 billion (c.i.f., 1995 est.).

Origins	%
Italy	14
France	9
US	8
Turkey	5
Saudi Arabia	3

Top Export Destinations [40]

$1 billion (f.o.b., 1995 est.).

Destinations	%
Saudi Arabia	13
Switzerland	12
UAE	11
Syria	9
US	5

Foreign Aid [41]

Recipient: ODA, $NA.

Import and Export Commodities [42]

Import Commodities	**Export Commodities**
Consumer goods	Agricultural products
Machinery and transport equipment	Chemicals
Petroleum products	Textiles
	Precious and semiprecious metals and jewelry
	Metals and metal products

Lesotho

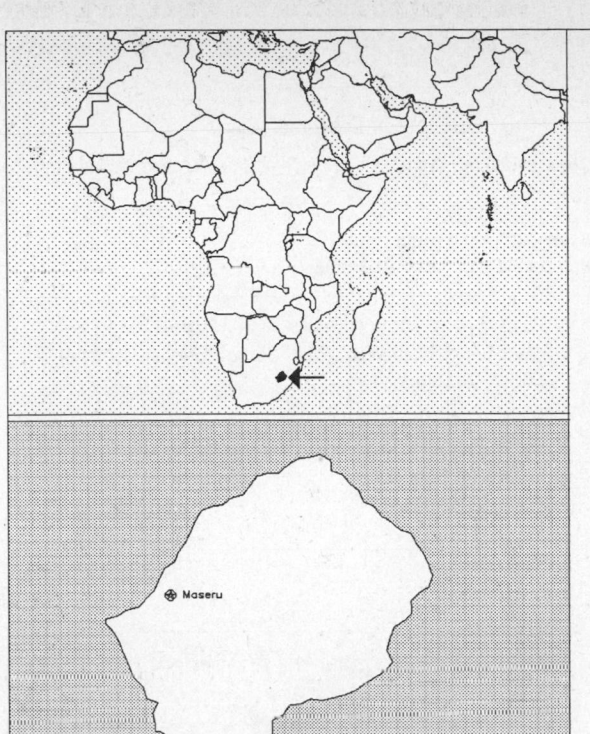

Geography [1]

Total area:
30,350 sq km   11,718 sq mi
Land area:
30,350 sq km   11,718 sq mi
Comparative area:
Slightly larger than Maryland
Land boundaries:
Total 909 km, South Africa 909 km
Coastline:
0 km (landlocked)
Climate:
Temperate; cool to cold, dry winters; hot, wet summers
Terrain:
Mostly highland with plateaus, hills, and mountains
Natural resources:
Water, agricultural and grazing land, some diamonds and other minerals
Land use:
Arable land: 10%
Permanent crops: 0%
Meadows and pastures: 66%
Forest and woodland: 0%
Other: 24%

Demographics [2]

	1970	1980	1990	1995[1]	1996	2000	2010	2020	2030
Population	1,067	1,346	1,735	1,933	1,971	2,114	2,428	2,693	2,974
Population density (persons per sq. mi.)	91	115	148	165	168	180	207	230	254
(persons per sq. km.)	35	44	57	64	65	70	80	89	98
Net migration rate (per 1,000 population)	NA	-1.9	0.0	0.0	0.0	0.0	0.0	0.0	0.0
Births	NA	NA	NA	NA	64	NA	NA	NA	NA
Deaths	NA	NA	NA	NA	27	NA	NA	NA	NA
Life expectancy - males	NA	53.2	52.7	50.7	50.1	47.8	47.0	50.4	59.2
Life expectancy - females	NA	58.0	56.3	54.4	54.1	53.3	51.8	55.5	65.9
Birth rate (per 1,000)	NA	40.9	36.6	33.2	32.7	30.8	27.0	23.3	20.5
Death rate (per 1,000)	NA	14.0	13.1	13.6	13.7	14.5	15.5	13.9	9.8
Women of reproductive age (15-49 yrs.)	NA	312	407	465	477	525	632	727	829
of which are currently married	NA	NA	285	NA	331	363	442	NA	NA
Fertility rate	NA	5.7	4.9	4.4	4.3	3.9	3.1	2.6	2.3

Except as noted, values for vital statistics are in thousands; life expectancy is in years.

Health

Health Indicators [3]

% of population with access to	
safe water (1990-95)	52
adequate sanitation (1990-95)	28
health services (1985-95)	80
% of 1-year-olds immunized (1990-94) against	
TB (tuberculosis)	59
DPT (diphtheria, pertussis, tetanus)	58
polio	59
measles	74
% of contraceptive prevalence (1980-94)	23
ORT use rate (1990-94)	78

Health Expenditures [4]

For sources, notes, and explanations, see Annotated Source Appendix, page 1061.

535

Human Factors

Women and Children [5]

% of pregnant women immunized (tetanus 1990-94)	12
% of births attended by trained health personnel (1983-94)	40
Maternal mortality rate (1980-92)	NA
Under-5 mortality rate (1994)	156
% under-5 moderately/severely underweight (1980-1994)	16

Burden of Disease [6]

Population per physician (1985)	14,305.56
Population per nurse	NA
Population per hospital bed	NA
AIDS cases per 100,000 people (1994)	1.8
Malaria cases per 100,000 people (1992)	NA

Ethnic Division [7]

Sotho	99.7%
Europeans	1,600
Asians	800

Religion [8]

Christian	80%
Indigenous beliefs	20%

Major Languages [9]

Sesotho (southern Sotho), English (official), Zulu, Xhosa.

Education

Public Education Expenditures [10]

Million (Maloti)	1980	1990	1991	1992	1993	1994
Total education expenditure	25	99	168	204	194	NA
as percent of GNP	5.1	3.7	5.6	5.9	4.8	NA
as percent of total govt. expend.	14.8	12.2	17.6	NA	NA	NA
Current education expenditure	20	81	145	160	192	NA
as percent of GNP	4.1	3.0	4.8	4.6	4.8	NA
as percent of current govt. expend.	19.0	17.4	25.3	NA	NA	NA
Capital expenditure	5	18	23	45	2	NA

Educational Attainment [11]

Illiteracy [12]

In thousands and percent[1]	1990	1995	2000
Illiterate population (15+ yrs.)	341	340	338
Illiteracy rate - total pop. (%)	34.2	30.0	26.4
Illiteracy rate - males (%)	23.1	19.9	17.4
Illiteracy rate - females (%)	44.2	39.0	34.6

Libraries [13]

	Admin. Units	Svc. Pts.	Vols. (000)	Shelving (meters)	Vols. Added	Reg. Users
National	NA	NA	NA	NA	NA	NA
Nonspecialized	NA	NA	NA	NA	NA	NA
Public (1989)	1	3	24	672	NA	607
Higher ed.	NA	NA	NA	NA	NA	NA
School	NA	NA	NA	NA	NA	NA

Daily Newspapers [14]

	1980	1985	1990	1994
Number of papers	3	4	4	2
Circ. (000)	44	47	20	14

Culture [15]

Science and Technology

Scientific/Technical Forces [16]

R&D Expenditures [17]

U.S. Patents Issued [18]

Government and Law

Organization of Government [19]

Long-form name:
Kingdom of Lesotho
Type:
Modified constitutional monarchy
Independence:
4 October 1966 (from UK)
National holiday:
Independence Day, 4 October (1966)
Constitution:
2 April 1993
Legal system:
Based on English common law and
Roman-Dutch law; judicial review of
legislative acts in High Court and Court of
Appeal; has not accepted compulsory ICJ
jurisdiction
Executive branch:
King; Prime Minister; Cabinet
Legislative branch:
Bicameral Parliament: Senate and
Assembly
Judicial branch:
High Court; Court of Appeal; Magistrate's
Court; customary or traditional court

Elections [20]

National Assembly. Elections were last
held in March 1993 (first since 1971); all
65 seats were won by the Basotho
Congressl Party (BCP).

Government Expenditures [21]

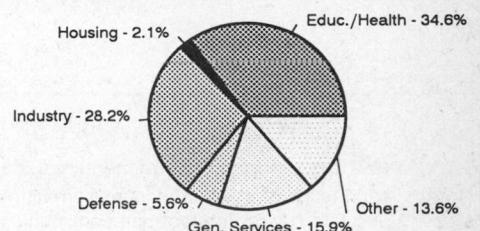

(% distribution). Expend. for FY92: 1,050 (Loti mil.)

Military Expenditures and Arms Transfers [22]

	1990	1991	1992	1993	1994
Military expenditures					
Current dollars (mil.)	56[e]	NA	36	31[e]	26[e]
1994 constant dollars (mil.)	63[e]	NA	38	32[e]	26[e]
Armed forces (000)	2	2	2	2	2
Gross national product (GNP)					
Current dollars (mil.)	1,071	1,107	1,128	1,227	1,351
1994 constant dollars (mil.)	1,192	1,187	1,177	1,252	1,351
Central government expenditures (CGE)					
1994 constant dollars (mil.)	368	384	NA	354	405[e]
People (mil.)	1.8	1.8	1.8	1.9	1.9
Military expenditure as % of GNP	5.3	NA	3.2	2.6	1.9
Military expenditure as % of CGE	17.1	NA	NA	9.1	6.5
Military expenditure per capita (1994 $)	36	NA	20	17	13
Armed forces per 1,000 people (soldiers)	1.1	1.1	1.1	1.1	1.0
GNP per capita (1994 $)	679	659	636	660	695
Arms imports[6]					
Current dollars (mil.)	0	0	0	0	0
1994 constant dollars (mil.)	0	0	0	0	0
Arms exports[6]					
Current dollars (mil.)	0	0	0	0	0
1994 constant dollars (mil.)	0	0	0	0	0
Total imports[7]					
Current dollars (mil.)	640	820	1,058	NA	NA
1994 constant dollars (mil.)	712	879	1,103	NA	NA
Total exports[7]					
Current dollars (mil.)	59	68	65[e]	93[e]	NA
1994 constant dollars (mil.)	66	72	68[e]	95[e]	NA
Arms as percent of total imports[8]	0	0	0	0	0
Arms as percent of total exports[8]	0	0	0	0	0

Crime [23]

	1990
Crime volume	
Cases known to police	NA
Attempts (percent)	NA
Percent cases solved	NA
Crimes per 100,000 persons	NA
Persons responsible for offenses	
Total number offenders	2,229
Percent female	7.9
Percent juvenile	15.6
Percent foreigners	NA

Human Rights [24]

	SSTS	FL	FAPRO	PPCG	APROBC	TPW	PCPTW	STPEP	PHRFF	PRW	ASST	AFL
Observes	P	P	P	P	P	P	P			P	P	
	EAFRD	CPR	ESCR	SR	ACHR	MAAE		PVIAC	PVNAC	EAFDAW	TCIDTP	RC
Observes		P	P	P				P	P	P		P

P = Party; S = Signatory; see Appendix for meaning of abbreviations.

Labor Force

Total Labor Force [25]

689,000 economically active

Labor Force by Occupation [26]

Subsistence agriculture   86.2%

Unemployment Rate [27]

Substantial unemployment and
underemployment.

For sources, notes, and explanations, see Annotated Source Appendix, page 1061.

537

Production Sectors

Energy Resource Summary [28]

Electricity: power supplied by South Africa.

Telecommunications [30]

- 12,000 (1991 est.) telephones; rudimentary system
- Domestic: consists of a few landlines, a small microwave radio relay system, and a minor radiotelephone communication system
- International: satellite earth station - 1 Intelsat (Atlantic Ocean)
- Radio: Broadcast stations: AM 3, FM 4, shortwave 0 Radios: 66,000
- Television: Broadcast stations: 1 Televisions: 11,000 (1992 est.)

Top Agricultural Products [32]

Agriculture accounts for 10.4% of the GDP; produces corn, wheat, pulses, sorghum, barley; livestock.

Top Mining Products [33]

Detailed information is not available. A summary of mineral resources follows. **Mineral Resources**: Some gem-quality diamonds, clay, sand, and gravel, and unexploited coal, limestone, peat, and uranium deposits.

Transportation [31]

Railways: total: 2.6 km; note - owned by, operated by, and included in the statistics of South Africa; narrow gauge: 2.6 km 1.067-m gauge

Highways: total: 5,324 km; paved: 799 km; unpaved: 4,525 km (1993 est.)

Airports

Total:	29
With paved runways over 3,047 m:	1
With paved runways 914 to 1,523 m:	1
With paved runways under 914 m:	23
With unpaved runways 914 to 1,523 m:	4 (1995 est.)

Tourism [34]

	1990	1991	1992	1993	1994
Visitors[1]	242	357	417	349	253
Tourists	171	NA	155	130	97
Excursionists	71	NA	262	219	156
Tourism receipts	17	18	19	17	17
Tourism expenditures	12	11	11	7	7
Fare receipts	4	3	3	2	3
Fare expenditures	8	7	6	6	4

Travelers are in thousands, money in million U.S. dollars.

For sources, notes, and explanations, see Annotated Source Appendix, page 1061.

Finance, Economics, and Trade

GDP and Manufacturing Summary [35]

	1980	1985	1990	1991	1992
Gross Domestic Product					
Millions of 1980 dollars	368	395	552	581	601[e]
Growth rate in percent	8.35	3.49	3.96	5.28	3.40[e]
Manufacturing Value Added					
Millions of 1980 dollars	21	37	65	NA	NA
Growth rate in percent	16.00	4.36	-3.95	NA	NA
Manufacturing share in percent of current prices	6.3	10.4	13.0	15.5	NA

Economic Indicators [36]

- **National product**: GDP—purchasing power parity—$2.8 billion (1994 est.)

- **National product real growth rate**: 13.5% (1994 est.)

- **National product per capita**: $1,430 (1994 est.)

- **Inflation rate (consumer prices)**: 9.5% (January 1995)

- **External debt**: $512 million (1993)

Balance of Payments Summary [37]

Values in millions of dollars.

	1989	1990	1991	1992	1993
Exports of goods (f.o.b.)	66.4	59.5	67.2	109.2	134.0
Imports of goods (f.o.b.)	-592.6	-672.6	-803.5	-932.6	-911.6
Trade balance	-526.2	-613.1	-736.3	-823.4	-777.6
Services - debits	-91.2	-103.3	-104.4	-115.3	-95.1
Services - credits	412.9	495.6	517.7	537.6	493.7
Private transfers (net)	4.2	4.8	2.7	3.9	2.6
Government transfers (net)	210.7	281.1	403.5	434.9	398.0
Long-term capital (net)	45.8	61.4	42.8	43.5	54.4
Short-term capital (net)	-66.0	-106.3	-103.6	-106.0	10.4
Errors and omissions	1.9	-2.8	20.1	74.8	16.1
Overall balance	-7.9	17.4	42.5	50.0	102.5

Exchange Rates [38]

Currency: **loti.**
Symbol: **L.**

Data are currency units per $1. The Basotho loti is at par with the South African rand.

January 1996	3.6417
1995	3.6266
1994	3.5490
1993	3.2636
1992	2.8197
1991	2.7563

Imports and Exports

Top Import Origins [39]

$1 billion (c.i.f., 1994 est.) Data are for 1993.

Origins	%
South Africa	83
Asia	12
EU	3

Top Export Destinations [40]

$142 million (f.o.b., 1994 est.) Data are for 1993.

Destinations	%
South Africa	39
North and South America	33
EU	22

Foreign Aid [41]

Recipient: ODA, $NA.

Import and Export Commodities [42]

Import Commodities	Export Commodities
Mainly corn	Clothing
Building materials	Furniture
Clothing	Footwear
Vehicles	Wool
Machinery	
Medicines	
Petroleum products	

For sources, notes, and explanations, see Annotated Source Appendix, page 1061.

539

Liberia

Geography [1]

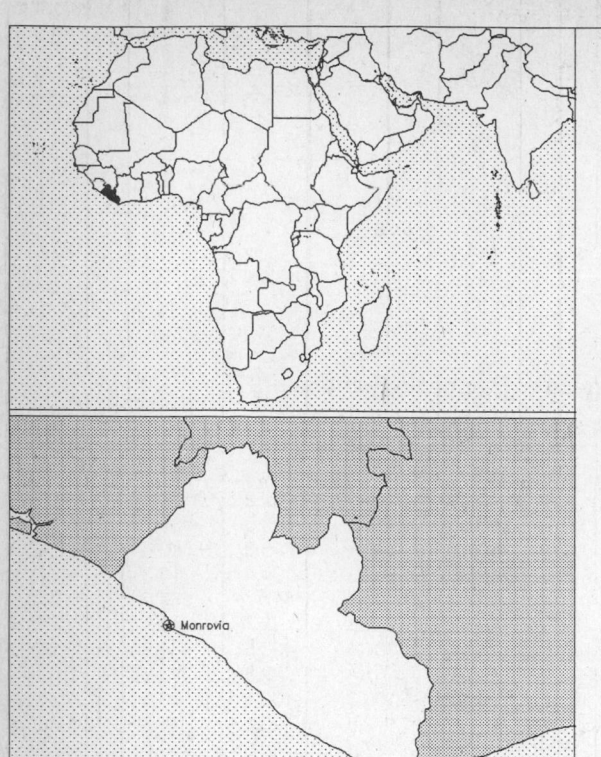

Total area:
  111,370 sq km   43,000 sq mi
Land area:
  96,320 sq km   37,189 sq mi
Comparative area:
  Slightly larger than Tennessee
Land boundaries:
  Total 1,585 km, Guinea 563 km, Cote d'Ivoire 716 km, Sierra Leone 306 km
Coastline:
  579 km
Climate:
  Tropical; hot, humid; dry winters with hot days and cool to cold nights;
  wet, cloudy summers with frequent heavy showers
Terrain:
  Mostly flat to rolling coastal plains rising to rolling plateau and
  low mountains in Northeast
Natural resources:
  Iron ore, timber, diamonds, gold
Land use:
  Arable land: 1%
  Permanent crops: 3%
  Meadows and pastures: 2%
  Forest and woodland: 39%
  Other: 55%

Demographics [2]

	1970	1980	1990	1995[1]	1996	2000	2010	2020	2030
Population	1,397	1,900	2,265	2,182	2,110	3,048	4,540	5,991	7,670
Population density (persons per sq. mi.)	38	51	61	59	57	82	122	161	206
(persons per sq. km.)	14	20	24	23	22	32	47	62	80
Net migration rate (per 1,000 population)	NA	2.2	-323.0	-117.7	-9.5	65.6	0.3	0.0	0.0
Births	NA	NA	NA	NA	90	NA	NA	NA	NA
Deaths	NA	NA	NA	NA	25	NA	NA	NA	NA
Life expectancy - males	45.6	49.7	53.7	55.7	56.1	57.6	61.3	64.7	67.7
Life expectancy - females	48.6	53.2	58.3	60.8	61.2	63.1	67.6	71.6	75.0
Birth rate (per 1,000)	NA	45.5	45.4	43.3	42.7	41.2	37.4	33.3	29.0
Death rate (per 1,000)	NA	15.9	13.2	12.1	12.0	10.8	8.6	7.0	6.0
Women of reproductive age (15-49 yrs.)	NA	421	498	480	463	673	1,040	1,434	1,921
of which are currently married	NA	NA	334	NA	311	451	695	NA	NA
Fertility rate	6.4	6.5	6.6	6.3	6.2	6.0	5.2	4.4	3.6

Except as noted, values for vital statistics are in thousands; life expectancy is in years.

Health

Health Indicators [3]

% of population with access to	
safe water (1990-95)	46
adequate sanitation (1990-95)	30
health services (1985-95)	39
% of 1-year-olds immunized (1990-94) against	
TB (tuberculosis)	84
DPT (diphtheria, pertussis, tetanus)	43
polio	45
measles	44
% of contraceptive prevalence (1980-94)	6
ORT use rate (1990-94)	15

Health Expenditures [4]

For sources, notes, and explanations, see Annotated Source Appendix, page 1061.

Human Factors

Women and Children [5]

% of pregnant women immunized (tetanus 1990-94)	35
% of births attended by trained health personnel (1983-94)	58
Maternal mortality rate (1980-92)	NA
Under-5 mortality rate (1994)	217
% under-5 moderately/severely underweight (1980-1994)[1]	20

Burden of Disease [6]

Ethnic Division [7]

Indigenous (including Kpelle, Bassa, Gio, Kru, Grebo, Mano, Krahn, Gola, Gbandi, Loma, Kissi, Vai, and Bella).

Indigenous African tribes	95%
Americo-Liberians (descendants of former slaves)	5%

Religion [8]

Traditional	70%
Muslim	20%
Christian	10%

Major Languages [9]

English 20% (official), Niger-Congo language group about 20 local languages come from this group.

Education

Public Education Expenditures [10]

	1980	1985	1990	1991	1993	1994
Million (Dollar)						
Total education expenditure	62	NA	NA	NA	NA	NA
as percent of GNP	5.7	NA	NA	NA	NA	NA
as percent of total govt. expend.	24.3	NA	NA	NA	NA	NA
Current education expenditure	53	NA	NA	NA	NA	NA
as percent of GNP	4.9	NA	NA	NA	NA	NA
as percent of current govt. expend.	27.0	NA	NA	NA	NA	NA
Capital expenditure	9	NA	NA	NA	NA	NA

Educational Attainment [11]

Illiteracy [12]

In thousands and percent[1]	1990	1995	2000
Illiterate population (15+ yrs.)	937	1,014	1,094
Illiteracy rate - total pop. (%)	73.8	83.9	64.7
Illiteracy rate - males (%)	55.6	61.7	46.3
Illiteracy rate - females (%)	93.5	NA	83.9

Libraries [13]

Daily Newspapers [14]

	1980	1985	1990	1994
Number of papers	3	5	8	8
Circ. (000)	11	28[e]	35[e]	35[e]

Culture [15]

Science and Technology

Scientific/Technical Forces [16]

R&D Expenditures [17]

U.S. Patents Issued [18]

For sources, notes, and explanations, see Annotated Source Appendix, page 1061.

541

Government and Law

Organization of Government [19]

Long-form name:
Republic of Liberia
Type:
Republic
Independence:
26 July 1847
National holiday:
Independence Day, 26 July (1847)
Constitution:
6 January 1986, Note: Constitutional government ended in September 1990 during civil war; peace accord signed in August 1995; a transitional government was formed in September 1995
Legal system:
Dual system of statutory law based on Anglo-American common law for the modern sector and customary law based on unwritten tribal practices for indigenous sector
Executive branch:
Chairman of the Council of State; Cabinet
Legislative branch:
Unicameral Transitional Legislative Assembly, the members of which are appointed by the leaders of the major factions in the civil war
Judicial branch:
Supreme Court

Crime [23]

Elections [20]

Transitional Legislative Assembly. Members are appointed by the leaders of the major factions in the civil war. The former bicameral legislature no longer exists and it is unlikely to be reconstituted soon.

Government Expenditures [21]

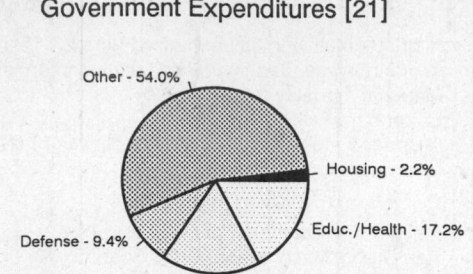

- Other - 54.0%
- Housing - 2.2%
- Educ./Health - 17.2%
- Industry - 17.2%
- Defense - 9.4%

(% distribution). Expend. for 1989: 435.4 (Dollars mil.)

Military Expenditures and Arms Transfers [22]

	1990[19]	1991	1992	1993	1994
Military expenditures					
Current dollars (mil.)	NA	NA	NA	36[e]	30[e]
1994 constant dollars (mil.)	NA	NA	NA	37[e]	30[e]
Armed forces (000)	8	5	2	NA	NA
Gross national product (GNP)					
Current dollars (mil.)	NA	NA	NA	NA	NA
1994 constant dollars (mil.)	NA	NA	NA	NA	NA
Central government expenditures (CGE)					
1994 constant dollars (mil.)	NA	NA	NA	NA	NA
People (mil.)	2.3	2.1	2.5	2.9	3.0
Military expenditure as % of GNP	NA	NA	NA	NA	NA
Military expenditure as % of CGE	NA	NA	NA	NA	NA
Military expenditure per capita (1994 $)	NA	NA	NA	NA	NA
Armed forces per 1,000 people (soldiers)	3.5	2.4	0.8	NA	NA
GNP per capita (1994 $)	NA	NA	NA	NA	NA
Arms imports[6]					
Current dollars (mil.)	10	0	0	0	0
1994 constant dollars (mil.)	11	0	0	0	0
Arms exports[6]					
Current dollars (mil.)	0	0	0	0	0
1994 constant dollars (mil.)	0	0	0	0	0
Total imports[7]					
Current dollars (mil.)	NA	NA	NA	NA	NA
1994 constant dollars (mil.)	NA	NA	NA	NA	NA
Total exports[7]					
Current dollars (mil.)[e]	1,941	482	771	615	NA
1994 constant dollars (mil.)[e]	2,160	517	8,040	628	NA
Arms as percent of total imports[8]	NA	0	0	0	0
Arms as percent of total exports[8]	0	0	0	0	0

Human Rights [24]

	SSTS	FL	FAPRO	PPCG	APROBC	TPW	PCPTW	STPEP	PHRFF	PRW	ASST	AFL
Observes	P	P	P	P	P	P	P	S		S	S	P
	EAFRD	CPR	ESCR	SR	ACHR	MAAE	PVIAC	PVNAC	EAFDAW	TCIDTP	RC
Observes	P	S	S	P			P	P	P		P

P = Party; S = Signatory; see Appendix for meaning of abbreviations.

Labor Force

Total Labor Force [25]

510,000 including 220,000 in the monetary economy

Labor Force by Occupation [26]

Agriculture	70.5%
Services	10.8
Industry and commerce	4.5
Other	14.2

Unemployment Rate [27]

For sources, notes, and explanations, see Annotated Source Appendix, page 1061.

Production Sectors

Commercial Energy Production and Consumption

Data are shown in quadrillion (10^{15}) BTUs and percent for 1995
Values for hydroelectric, nuclear, geothermal, solar, and wind power refer to electrical generation.

Production [28]

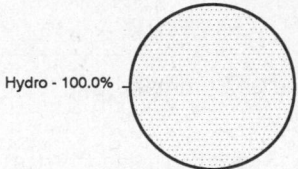

Hydro - 100.0%

Consumption [29]

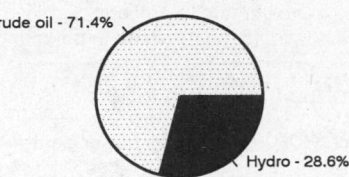

Crude oil - 71.4%

Hydro - 28.6%

Net hydroelectric power	0.002
Total	0.002

Crude oil	0.005
Net hydroelectric power	0.002
Total	0.007

Telecommunications [30]

- Less than 25,000 (1991 est.) telephones; telephone and telegraph service via microwave radio relay network; main center is Monrovia; most telecommunications services inoperable due to insurgency movement
- International: satellite earth station - 1 Intelsat (Atlantic Ocean)
- Radio: Broadcast stations: AM 3, FM 4, shortwave 0 Radios: 622,000 (1992 est.)
- Television: Broadcast stations: 5 (1987 est.) Televisions: 51,000 (1992 est.)

Top Agricultural Products [32]

Produces rubber, coffee, cocoa, rice, cassava (tapioca), palm oil, sugarcane, bananas; sheep, goats; timber.

Top Mining Products [33]

Metric tons except as noted[e]	2/17/95[*]
Diamond (carats)[50]	
gem	40,000
industrial	60,000
Gold (kg.)	700[50]

Transportation [31]

Railways: total: 490 km (single track); note - three rail systems owned and operated by foreign steel and financial interests in conjunction with Liberian Government; one of these, the Lamco Railroad, closed in 1989 after iron ore production ceased; the other two have been shut down by the civil war; standard gauge: 345 km 1.435-m gauge; narrow gauge: 145 km 1.067-m gauge

Highways: total: 10,029 km; paved: 600 km; unpaved: 9,429 km (1987 est.)

Merchant marine: total: 1,601 ships (1,000 GRT or over) totaling 59,449,296 GRT/98,819,081 DWT; ships by type: barge carrier 2, bulk 411, cargo 121, chemical tanker 108, combination bulk 28, combination ore/oil 56, container 143, liquefied gas tanker 77, multifunction large-load carrier 1, oil tanker 463, passenger 42, passenger-cargo 1, refrigerated cargo 64, roll-on/roll-off cargo 23, short-sea passenger 4, specialized tanker 9, vehicle carrier 48

Airports

Total:	39
With paved runways over 3,047 m:	1
With paved runways 1,524 to 2,437 m:	1
With paved runways under 914 m:	29
With unpaved runways 1,524 to 2,437 m:	2
With unpaved runways 914 to 1,523 m:	6 (1995 est.)

Tourism [34]

For sources, notes, and explanations, see Annotated Source Appendix, page 1061.

543

Finance, Economics, and Trade

GDP and Manufacturing Summary [35]

	1980	1985	1990	1991	1992
Gross Domestic Product					
Millions of 1980 dollars	917	843	872	887	876[e]
Growth rate in percent	-6.29	-2.02	-1.99	1.70	-1.24[e]
Manufacturing Value Added					
Millions of 1980 dollars	77	75	80	83	83[e]
Growth rate in percent	-21.21	-1.61	-2.98	3.40	0.29[e]
Manufacturing share in percent of current prices	9.5	6.6	6.9[e]	NA	NA

Economic Indicators [36]

- **National product**: GDP—purchasing power parity—$2.3 billion (1994 est.)
- **National product real growth rate**: 0% (1994 est.)
- **National product per capita**: $770 (1994 est.)
- **Inflation rate (consumer prices)**: 50% (1994 est.)
- **External debt**: $1.9 billion (1993 est.)

Balance of Payments Summary [37]

Values in millions of dollars.

	1975	1980	1985	1986	1987
Exports of goods (f.o.b.)	394.4	600.4	430.4	407.9	374.9
Imports of goods (f.o.b.)	-290.4	-478.0	-263.8	-258.8	-311.7
Trade balance	104.0	122.4	166.6	149.1	63.2
Services - debits	-45.7	-96.7	-212.0	-263.8	-262.5
Services - credits	9.3	13.1	38.3	NA	57.7
Private transfers (net)	-21.7	-29.0	-28.0	-25.4	-21.4
Government transfers (net)	27.8	36.2	91.6	96.4	45.4
Long-term capital (net)	88.4	70.5	-106.5	-207.5	-188.9
Short-term capital (net)	-2.7	8.2	134.1	263.3	275.7
Errors and omissions	-158.9	-175.0	-108.7	-73.8	30.3
Overall balance	0.5	-50.3	-24.6	-2.7	-0.5

Exchange Rates [38]

Currency: **Liberian dollar.**
Symbol: **L$.**

Data are currency units per $1. The market rate floats against the US dollar.

officially fixed rate since 1940	1.0000
October 1995 (market rate)	50
January 1992 (market rate)	7

Imports and Exports

Top Import Origins [39]

NA (c.i.f., 1994 est.).

Origins	%
US	NA
EU	NA
Japan	NA
China	NA
Netherlands	NA
ECOWAS	NA
South Korea	NA

Top Export Destinations [40]

$530 million (f.o.b., 1994 est.).

Destinations	%
US	NA
EU	NA
Netherlands	NA
Singapore	NA

Foreign Aid [41]

Recipient: ODA, $NA.

Import and Export Commodities [42]

Import Commodities	Export Commodities
Mineral fuels	Iron ore 61%
Chemicals	Rubber 20%
Machinery	Timber 11%
Transportation equipment	Coffee
Manufactured goods	
Rice and other foodstuffs	

Libya

Geography [1]

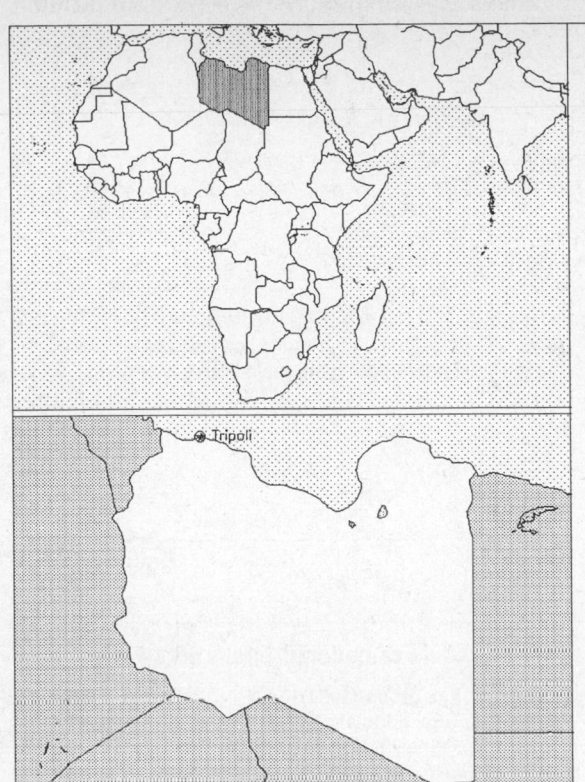

Total area:
  1,759,540 sq km   679,362 sq mi
Land area:
  1,759,540 sq km   679,362 sq mi
Comparative area:
  Slightly larger than Alaska
Land boundaries:
  Total 4,383 km, Algeria 982 km, Chad 1,055 km, Egypt 1,150 km,
  Niger 354 km, Sudan 383 km, Tunisia 459 km
Coastline:
  1,770 km
Climate:
  Mediterranean along coast; dry, extreme desert interior
Terrain:
  Mostly barren, flat to undulating plains, plateaus, depressions
Natural resources:
  Petroleum, natural gas, gypsum
Land use:
  Arable land: 2%
  Permanent crops: 0%
  Meadows and pastures: 8%
  Forest and woodland: 0%
  Other: 90%

Demographics [2]

	1970	1980	1990	1995[1]	1996	2000	2010	2020	2030
Population	2,056	3,119	4,355	5,248	5,445	6,294	8,913	12,391	16,669
Population density (persons per sq. mi.)	3	5	6	8	8	9	13	18	25
(persons per sq. km.)	1	2	2	3	3	4	5	7	9
Net migration rate (per 1,000 population)	NA	5.4	0.0	0.0	0.0	0.0	0.0	0.0	0.0
Births	NA	NA	NA	NA	242	NA	NA	NA	NA
Deaths	NA	NA	NA	NA	42	NA	NA	NA	NA
Life expectancy - males	NA	56.2	60.2	62.1	62.5	64.0	67.3	70.1	72.5
Life expectancy - females	NA	59.9	64.4	66.6	67.0	68.6	72.3	75.5	78.1
Birth rate (per 1,000)	NA	44.5	46.6	44.9	44.4	42.6	39.4	35.7	31.4
Death rate (per 1,000)	NA	11.5	9.1	7.9	7.7	6.9	5.4	4.2	3.6
Women of reproductive age (15-49 yrs.)	NA	630	911	1,098	1,139	1,330	2,007	2,874	4,023
of which are currently married	NA	NA	733	NA	922	1,077	1,622	NA	NA
Fertility rate	NA	6.9	6.6	6.3	6.3	6.0	5.3	4.6	4.0

Except as noted, values for vital statistics are in thousands; life expectancy is in years.

Health

Health Indicators [3]

% of population with access to	
safe water (1990-95)[1]	97
adequate sanitation (1990-95)[1]	98
health services (1985-95)	NA
% of 1-year-olds immunized (1990-94) against	
TB (tuberculosis)	99
DPT (diphtheria, pertussis, tetanus)	91
polio	91
measles	89
% of contraceptive prevalence (1980-94)	NA
ORT use rate (1990-94)	80

Health Expenditures [4]

For sources, notes, and explanations, see Annotated Source Appendix, page 1061.

545

Human Factors

Women and Children [5]

% of pregnant women immunized (tetanus 1990-94)	45
% of births attended by trained health personnel (1983-94)	76
Maternal mortality rate (1980-92)	70
Under-5 mortality rate (1994)	95
% under-5 moderately/severely underweight (1980-1994)	NA

Burden of Disease [6]

Population per physician (1990)	957.04
Population per nurse (1990)	221.26
Population per hospital bed (1990)	245.64
AIDS cases per 100,000 people (1994)	0.1
Malaria cases per 100,000 people (1992)	NA

Ethnic Division [7]

Other includes Greeks, Maltese, Italians, Egyptians, Pakistanis, Turks, Indians, and Tunisians.

Berber and Arab	97%
Other	3%

Religion [8]

Sunni Muslim	97%
Other	3%

Major Languages [9]

Arabic, Italian, English, all are widely understood in the major cities.

Education

Public Education Expenditures [10]

Million (Dinar)

	1980	1985	1990	1992	1993	1994
Total education expenditure	356	575	NA	NA	NA	NA
as percent of GNP	3.4	7.1	NA	NA	NA	NA
as percent of total govt. expend.	NA	19.8	NA	NA	NA	NA
Current education expenditure	224	457	NA	NA	NA	NA
as percent of GNP	2.1	5.7	NA	NA	NA	NA
as percent of current govt. expend.	NA	38.1	NA	NA	NA	NA
Capital expenditure	132	117	NA	NA	NA	NA

Educational Attainment [11]

Age group (1984)[14]	25+
Total population	996,774
Highest level attained (%)	
No schooling	59.7
First level	
Not completed	15.4
Completed	8.5
Entered second level	
S-1	5.2
S-2	8.5
Postsecondary	2.7

Illiteracy [12]

In thousands and percent[1]

	1990	1995	2000
Illiterate population (15+ yrs.)	741	702	649
Illiteracy rate - total pop. (%)	32.0	25.5	19.7
Illiteracy rate - males (%)	17.7	13.4	9.7
Illiteracy rate - females (%)	47.5	38.4	30.1

Libraries [13]

Daily Newspapers [14]

	1980	1985	1990	1994
Number of papers	3	3	3	4
Circ. (000)	55[e]	65[e]	70[e]	70[e]

Culture [15]

Science and Technology

Scientific/Technical Forces [16]

Scientists/engineers	1,100
Number female	NA
Technicians	1,500
Number female	NA
Total	2,600

R&D Expenditures [17]

	Dinar (000) 1980
Total expenditure	22,875
Capital expenditure	NA
Current expenditure	NA
Percent current	NA

U.S. Patents Issued [18]

For sources, notes, and explanations, see Annotated Source Appendix, page 1061.

Government and Law

Organization of Government [19]

Long-form name:
Socialist People's Libyan Arab Jamahiriya

Type:
Jamahiriya (a state of the masses) in theory, governed by the populace through local councils; in fact, a military dictatorship

Independence:
24 December 1951 (from Italy)

National holiday:
Revolution Day, 1 September (1969)

Constitution:
11 December 1969, amended 2 March 1977

Legal system:
Based on Italian civil law system and Islamic law; separate religious courts; no constitutional provision for judicial review of legislative acts; has not accepted compulsory ICJ jurisdiction

Executive branch:
Revolutionary Leader Col. Muammar Abu Minyar al-Qadhafi; Secretary of the General People's Committee (Premier); General People's Committee

Legislative branch:
Unicameral: General People's Congress

Judicial branch:
Supreme Court

Elections [20]

General Peoples Congress No political parties; national elections are indirect through a hierarchy of peoples' committees.

Government Budget [21]

For 1989 est.

	$ bil.
Revenues	8.1
Expenditures	9.8
Capital expenditures	3.1

Crime [23]

	1994
Crime volume	
Cases known to police	45,166
Attempts (percent)	NA
Percent cases solved	NA
Crimes per 100,000 persons	951.26
Persons responsible for offenses	
Total number offenders	52,374
Percent female	4.60
Percent juvenile (14-18 yrs.)	2.60
Percent foreigners	5.30

Military Expenditures and Arms Transfers [22]

	1990	1991	1992	1993	1994
Military expenditures					
Current dollars (mil.)[e,3]	NA	2,257	2,157	1,521	1,399
1994 constant dollars (mil.)[e,3]	NA	2,419	2,249	1,552	1,399
Armed forces (000)	86	86	85	85	80
Gross national product (GNP)					
Current dollars (mil.)[e]	26,570	30,440	31,280	29,580	32,900
1994 constant dollars (mil.)[e]	29,570	32,620	32,620	30,190	32,900
Central government expenditures (CGE)					
1994 constant dollars (mil.)	NA	NA	NA	NA	NA
People (mil.)	4.4	4.5	4.7	4.9	5.1
Military expenditure as % of GNP	NA	7.4	6.9	5.1	4.3
Military expenditure as % of CGE	NA	NA	NA	NA	NA
Military expenditure per capita (1994 $)	NA	535	479	319	277
Armed forces per 1,000 people (soldiers)	19.7	19.0	18.1	17.4	15.8
GNP per capita (1994 $)	6,789	7,215	6,950	6,197	6,506
Arms imports[6]					
Current dollars (mil.)	370	410	80	0	0
1994 constant dollars (mil.)	412	439	83	0	0
Arms exports[6]					
Current dollars (mil.)	60	20	10	0	0
1994 constant dollars (mil.)	67	21	10	0	0
Total imports[7]					
Current dollars (mil.)	5,336	5,361	5,020[e]	5,135	3,916[e]
1994 constant dollars (mil.)	5,939	5,746	5,235[e]	5,240	3,916[e]
Total exports[7]					
Current dollars (mil.)	13,220	11,230	9,948[e]	8,047[e]	7,200[e]
1994 constant dollars (mil.)	14,720	12,040	10,370[e]	8,213[e]	7,200[e]
Arms as percent of total imports[8]	6.9	7.6	1.6	0	0
Arms as percent of total exports[8]	0.5	0.2	0.1	0	0

Human Rights [24]

	SSTS	FL	FAPRO	PPCG	APROBC	TPW	PCPTW	STPEP	PHRFF	PRW	ASST	AFL
Observes	P	P		P	P	P	P	P		P	P	P
		EAFRD	CPR	ESCR	SR	ACHR	MAAE	PVIAC	PVNAC	EAFDAW	TCIDTP	RC
Observes		P	P	P		P	P	P	P	P	P	P

P = Party; S = Signatory; see Appendix for meaning of abbreviations.

Labor Force

Total Labor Force [25]

1 million (includes about 280,000 resident foreigners)

Labor Force by Occupation [26]

Industry	31%
Services	27
Government	24
Agriculture	18

Unemployment Rate [27]

Production Sectors

Commercial Energy Production and Consumption

Data are shown in quadrillion (10^{15}) BTUs and percent for 1995
Values for hydroelectric, nuclear, geothermal, solar, and wind power refer to electrical generation.

Production [28]

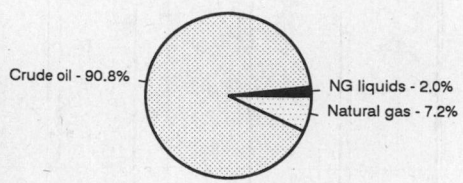

Crude oil - 90.8%
NG liquids - 2.0%
Natural gas - 7.2%

Consumption [29]

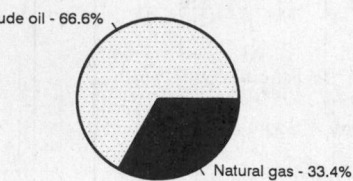

Crude oil - 66.6%
Natural gas - 33.4%

Crude oil	2.930
Natural gas liquids	0.064
Dry natural gas	0.234
Total	3.228

Crude oil	0.357
Dry natural gas	0.179
Total	0.536

Telecommunications [30]

- 370,000 telephones; modern telecommunications system
- Domestic: microwave radio relay, coaxial cable, tropospheric scatter, domestic satellite system with 14 earth stations
- International: satellite earth stations - 2 Intelsat; planned Arabsat and Intersputnik satellite earth stations; submarine cables to France and Italy; microwave radio relay to Tunisia and Egypt; tropospheric scatter to Greece; participant in Medarabtel
- Radio: Broadcast stations: AM 17, FM 3 Radios: 1 million (1993 est.)
- Television: Broadcast stations: 12 (1987 est.) Televisions: 500,000 (1993 est.)

Top Agricultural Products [32]

Produces wheat, barley, olives, dates, citrus, vegetables, peanuts; meat, eggs.

Top Mining Products [33]

Thousand metric tons except as noted[e]	3/94[*]
Cement, hydraulic	2,300
Gas, natural, gross (mil. cu. meters)	14,000
Gypsum	180
Iron, direct-reduced	852
Steel, crude	874
Lime	260
Nitrogen, N content of ammonia	350
Petroleum (000 42-gal. bls.)	
crude	504,000
refinery products	109,000

Transportation [31]

Railways: Note: Libya has had no railroad in operation since 1965, all previous systems having been dismantled; current plans are to construct a 1.435-m standard gauge line from the Tunisian frontier to Tripoli and Misratah, then inland to Sabha, center of a mineral-rich area, but there has been no progress; other plans made jointly with Egypt would establish a rail line from As Sallum, Egypt, to Tobruk with completion set for mid-1994; no progress has been reported

Highways: total: 19,189 km; paved: 10,738 km; unpaved: 8,451 km (1987 est.)

Merchant marine: total: 30 ships (1,000 GRT or over) totaling 686,834 GRT/1,209,263 DWT; ships by type: cargo 10, chemical tanker 1, liquefied gas tanker 2, oil tanker 10, roll-on/roll-off cargo 3, short-sea passenger 4

Airports

Total:	130
With paved runways over 3,047 m:	24
With paved runways 2,438 to 3,047 m:	5
With paved runways 1,524 to 2,437 m:	22
With paved runways 914 to 1,523 m:	6
With paved runways under 914 m:	13

Tourism [34]

	1990	1991	1992	1993	1994
Visitors[65]	789	396	252	659	1,494
Tourists[66]	96	90	89	63	54
Tourism receipts[67]	6	5	6	7	7
Tourism expenditures	424	877	154	NA	NA
Fare receipts	8	NA	NA	NA	NA
Fare expenditures	19	NA	NA	NA	NA

Travelers are in thousands, money in million U.S. dollars.

Manufacturing Sector

GDP and Manufacturing Summary [35]

	1980	1985	1989	1990	% change 1980-1990	% change 1989-1990
GDP (million 1980 $)	35,727	29,777	25,127	31,908	-10.7	27.0
GDP per capita (1980 $)	11,737	7,865	5,730	7,014	-40.2	22.4
Manufacturing as % of GDP (current prices)	1.9	4.5	8.5^e	8.4^e	342.1	-1.2
Gross output (million $)	1,177	1,953^e	3,312^e	3,830^e	225.4	15.6
Value added (million ?)	358	638^e	930^e	1,211^e	238.3	30.2
Value added (mi?)	649	1,262	1,486^e	1,617	149.2	8.8
Industrial product?	100	140	135^e	178	78.0	31.9
Employment (thousands)	18	22^e	24^e	28^e	55.6	16.7

Note: GDP stands for Gross Domestic ?. 'e' stands for estimated value.

Profitability and Productivity

	1980	1985	1989	1990	% change 1980-1990	% change 1989-1990
Intermediate input (%)	70	67^e	72^e	68^e	-2.9	-5.6
Wages, salaries, and supplements (%)	13	12^e	13^e	12^e	-7.7	-7.7
Gross operating surplus (%)	17	20^e	15^e	20^e	17.6	33.3
Gross output per worker ($)	63,982^e	84,676^e	138,218^e	133,325^e	108.4	-3.5
Value added per worker ($)	19,492^e	28,750^e	38,817^e	45,851^e	135.2	18.1
Average wage (incl. benefits) ($)	8,326^e	10,746^e	17,776^e	16,121^e	93.6	-9.3

Profitability is in percent of gross output. Productivity is in U.S. $. 'e' stands for estimated value.

Profitability - 1990

Inputs - 68.0%
Wages - 12.0%
Surplus - 20.0%

The graphic shows percent of gross output.

Value Added in Manufacturing

	1980 $ mil.	1980 %	1985 $ mil.	1985 %	1989 $ mil.	1989 %	1990 $ mil.	1990 %	% change 1980-1990	% change 1989-1990
311 Food products	35	9.8	42^e	6.6	69^e	7.4	67^e	5.5	91.4	-2.9
313 Beverages	17	4.7	20^e	3.1	30^e	3.2	34^e	2.8	100.0	13.3
314 Tobacco products	55	15.4	83^e	13.0	75^e	8.1	131^e	10.8	138.2	74.7
321 Textiles	14	3.9	22^e	3.4	30^e	3.2	33^e	2.7	135.7	10.0
322 Wearing apparel	5^e	1.4	6^e	0.9	5^e	0.5	9^e	0.7	80.0	80.0
323 Leather and fur products	7	2.0	16^e	2.5	26^e	2.8	33^e	2.7	371.4	26.9
324 Footwear	14	3.9	28^e	4.4	40^e	4.3	53^e	4.4	278.6	32.5
331 Wood and wood products	3^e	0.8	6^e	0.9	4^e	0.4	11^e	0.9	266.7	175.0
332 Furniture and fixtures	2^e	0.6	4^e	0.6	5^e	0.5	9^e	0.7	350.0	80.0
341 Paper and paper products	3	0.8	3^e	0.5	6^e	0.6	5^e	0.4	66.7	-16.7
342 Printing and publishing	NA	0.0	1^e	0.2	6^e	0.6	3^e	0.2	NA	-50.0
351 Industrial chemicals	35	9.8	46^e	7.2	102^e	11.0	87^e	7.2	148.6	-14.7
352 Other chemical products	21	5.9	38^e	6.0	19^e	2.0	70^e	5.8	233.3	268.4
353 Petroleum refineries	81	22.6	179^e	28.1	288^e	31.0	374^e	30.9	361.7	29.9
354 Miscellaneous petroleum and coal products	NA	0.0	NA	0.0	NA	0.0	NA	0.0	NA	NA
355 Rubber products	NA	0.0	1^e	0.2	NA	0.0	1^e	0.1	NA	NA
356 Plastic products	2	0.6	5^e	0.8	5^e	0.5	9^e	0.7	350.0	80.0
361 Pottery, china, and earthenware	1	0.3	1^e	0.2	1^e	0.1	1^e	0.1	0.0	0.0
362 Glass and glass products	NA	0.0	NA	0.0	NA	0.0	NA	0.0	NA	NA
369 Other nonmetal mineral products	51	14.2	110^e	17.2	202^e	21.7	222^e	18.3	335.3	9.9
371 Iron and steel	NA	0.0	NA	0.0	NA	0.0	NA	0.0	NA	NA
372 Nonferrous metals	NA	0.0	NA	0.0	NA	0.0	NA	0.0	NA	NA
381 Metal products	3	0.8	8^e	1.3	3^e	0.3	21^e	1.7	600.0	600.0
382 Nonelectrical machinery	NA	0.0	NA	0.0	NA	0.0	NA	0.0	NA	NA
383 Electrical machinery	NA	0.0	NA	0.0	NA	0.0	NA	0.0	NA	NA
384 Transport equipment	NA	0.0	NA	0.0	NA	0.0	NA	0.0	NA	NA
385 Professional and scientific equipment	NA	0.0	NA	0.0	NA	0.0	NA	0.0	NA	NA
390 Other manufacturing industries	9	2.5	19^e	3.0	15^e	1.6	39^e	3.2	333.3	160.0

Note: The industry codes shown are International Standard Industry codes (ISIC). Percentages are percent of total Value Added. 'e' stands for estimated value

For sources, notes, and explanations, see Annotated Source Appendix, page 1061.

Finance, Economics, and Trade

Economic Indicators [36]

- **National product**: GDP—purchasing power parity— $32.9 billion (1994 est.)

- **National product real growth rate**: - 0.9% (1994 est.)

- **National product per capita**: $6,510 (1994 est.)

- **Inflation rate (consumer prices)**: 25% (1993 est.)

- **External debt**: $3.5 billion excluding military debt (1991 est.)

Balance of Payments Summary [37]

Values in millions of dollars.

	1985	1987	1988	1989	1990
Exports of goods (f.o.b.)	10,353	6,292	5,653	7,274	11,352
Imports of goods (f.o.b.)	-5,754	-5,820	-5,762	-6,509	-7,575
Trade balance	4,599	472	-109	765	3,777
Services - debits	-2,314	-1,944	-2,074	-1,869	-1,878
Services - credits	526	912	891	565	783
Private transfers (net)	-859	-508	-497	-472	-446
Government transfers (net)	-45	-60	-37	-16	-35
Long-term capital (net)	91	-3,006	-431	88	-506
Short-term capital (net)	693	2,868	594	1,100	-500
Errors and omissions	-328	184	271	130	-37
Overall balance	2,363	-1,082	-1,392	291	1,158

Exchange Rates [38]

Currency: **Libyan dinar.**
Symbol: **LD.**

Data are currency units per $1.

January 1996	0.3617
1995	0.3532
1994	0.3596
1993	0.3250
1992	0.3013
1991	0.2684

Imports and Exports

Top Import Origins [39]

$6.9 billion (f.o.b., 1994 est.).

Origins	%
Italy	NA
Germany	NA
UK	NA
France	NA
Spain	NA
Turkey	NA
Tunisia	NA
Eastern Europe	NA

Top Export Destinations [40]

$7.2 billion (f.o.b., 1994 est.).

Destinations	%
Italy	NA
Germany	NA
Spain	NA
France	NA
UK	NA
Turkey	NA
Greece	NA
Egypt	NA

Foreign Aid [41]

	U.S. $	
Western (non-US) countries, ODA and OOF bilateral commitments (1970-87)	242	million

Import and Export Commodities [42]

Import Commodities

Machinery
Transport equipment
Food
Manufactured goods

Export Commodities

Crude oil
Refined petroleum products
Natural gas

Liechtenstein

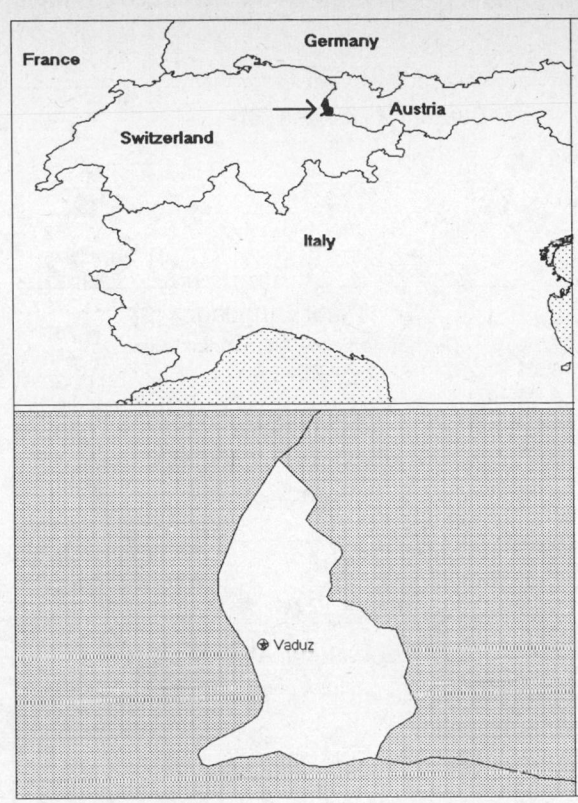

Geography [1]

Total area:
  160 sq km   62 sq mi
Land area:
  160 sq km   62 sq mi
Comparative area:
  About 0.9 times the size of Washington, DC
Land boundaries:
  Total 78 km, Austria 37 km, Switzerland 41 km
Coastline:
  0 km (landlocked)
Climate:
  Continental; cold, cloudy winters with frequent snow or rain; cool to moderately warm, cloudy, humid summers
Terrain:
  Mostly mountainous (Alps) with Rhine Valley in western third
Natural resources:
  Hydroelectric potential
Land use:
  Arable land: 25%
  Permanent crops: 0%
  Meadows and pastures: 38%
  Forest and woodland: 19%
  Other: 18%

Demographics [2]

	1970	1980	1990	1995[1]	1996	2000	2010	2020	2030
Population	21	25	29	31	31	33	35	36	36
Population density (persons per sq. mi.)	339	403	461	495	501	523	568	587	574
(persons per sq. km.)	131	156	178	191	193	202	219	227	221
Net migration rate (per 1,000 population)	12.2	3.6	8.4	5.8	6.1	7.3	4.0	1.3	0.0
Births	0	0	NA	NA	Z	NA	NA	NA	NA
Deaths	0	0	NA	NA	Z	NA	NA	NA	NA
Life expectancy - males	NA	NA	73.3	75.8	75.9	76.5	77.6	78.5	79.2
Life expectancy - females	NA	NA	80.5	82.1	82.2	82.7	83.8	84.6	85.2
Birth rate (per 1,000)	20.0	15.7	13.2	11.6	11.5	10.7	9.2	8.5	7.4
Death rate (per 1,000)	7.7	7.0	6.8	6.8	6.8	6.9	7.6	9.3	12.0
Women of reproductive age (15-49 yrs.)	NA	NA	8	9	9	9	9	8	7
of which are currently married	NA	NA	5	NA	5	5	5	NA	NA
Fertility rate	2.5	1.8	1.4	1.4	1.4	1.4	1.4	1.4	1.5

Except as noted, values for vital statistics are in thousands; life expectancy is in years.

Health

Health Indicators [3]

Health Expenditures [4]

For sources, notes, and explanations, see Annotated Source Appendix, page 1061.

551

Human Factors

Women and Children [5]	Burden of Disease [6]

Ethnic Division [7]

Alemannic	95%
Italian and other	5%

Religion [8]

Roman Catholic	87.3%
Protestant	8.3%
Other	4.4%
(1988)	

Major Languages [9]

German (official), Alemannic dialect.

Education

Public Education Expenditures [10]	Educational Attainment [11]

Illiteracy [12]

In thousands and percent[3]	1985	1991	2000
Illiterate population (15+ yrs.)	883	890	848
Illiteracy rate - total pop. (%)	43.5	36.2	24.0
Illiteracy rate - males (%)	29.9	24.6	16.0
Illiteracy rate - females (%)	59.7	49.6	32.9

Libraries [13]

	Admin. Units	Svc. Pts.	Vols. (000)	Shelving (meters)	Vols. Added	Reg. Users
National (1992)	1	1	160	NA	4,500	15,750
Nonspecialized	NA	NA	NA	NA	NA	NA
Public (1992)	3	3	26	NA	1,145	1,800
Higher ed. (1990)	1	1	30[e]	783[e]	NA	262[e]
School (1990)	8	NA	35	NA	1,454	2,509

Daily Newspapers [14]

	1980	1985	1990	1994
Number of papers	2	2	2[e]	2
Circ. (000)	14	14	18[e]	18

Culture [15]

Cinema (seats per 1,000)	NA
Annual attendance per person	NA
Gross box office receipts (mil. Franc)	NA
Museums (reporting)	4
Visitors (000)	38
Annual receipts (000 Franc)	717

Science and Technology

Scientific/Technical Forces [16]	R&D Expenditures [17]	U.S. Patents Issued [18]
		Values show patents issued to citizens of the country by the U.S. Patents Office.

U.S. Patents Issued [18]

	1993	1994	1995
Number of patents	11	16	18

For sources, notes, and explanations, see Annotated Source Appendix, page 1061.

Government and Law

Organization of Government [19]

Long-form name:
Principality of Liechtenstein
Type:
Hereditary constitutional monarchy
Independence:
23 January 1719 (Imperial Principality of Liechtenstein established)
National holiday:
Assumption Day, 15 August
Constitution:
5 October 1921
Legal system:
Local civil and penal codes; accepts compulsory ICJ jurisdiction, with reservations
Executive branch:
Prince; Heir Apparent Prince; Head of government; Deputy Head of Government; Cabinet
Legislative branch:
Unicameral: Diet (Landtag)
Judicial branch:
Supreme Court (Oberster Gerichtshof) for criminal cases; Superior Court (Obergericht) for civil cases

Elections [20]

Diet	% of votes
Fatherland Union (VU)	50.1
Progressive Citizens' Party (FBP)	41.3
Free Electoral List (FL)	8.5

Government Budget [21]

For 1995 est.

	$ mil.
Revenues	455
Expenditures	442
Capital expenditures	NA

Defense Summary [22]

Note: Defense is the responsibility of Switzerland

Crime [23]

Human Rights [24]

	SSTS	FL	FAPRO	PPCG	APROBC	TPW	PCPTW	STPEP	PHRFF	PRW	ASST	AFL
Observes				P		P	P		P			
	EAFRD	CPR	ESCR	SR	ACHR	MAAE	PVIAC	PVNAC	EAFDAW	TCIDTP	RC
Observes			P				P	P	P	P	P

P = Party; S = Signatory; see Appendix for meaning of abbreviations.

Labor Force

Total Labor Force [25]

20,000 of which 12,000 are foreigners; 6,885 commute from Austria and Switzerland to work each day

Labor Force by Occupation [26]

Industry, trade, and building	48.1%
Services	50.2
Agriculture, fishing, forestry, and horticulture	1.7

Date of data: 1993

Unemployment Rate [27]

0.9% (1995)

For sources, notes, and explanations, see Annotated Source Appendix, page 1061.

553

Production Sectors

Energy Resource Summary [28]

Energy resources: Hydroelectric potential. **Electricity**: Capacity: 23,000 kW. Production: 150 million kWh. Consumption per capita: 5,230 kWh (1992).

Telecommunications [30]

- 18,916 (1993 est.) telephones; limited, but sufficient automatic telephone system
- International: linked to Swiss networks by cable and microwave radio relay
- Radio: Broadcast stations: AM NA, FM NA, shortwave NA Radios: 11,000 (1993 est.)
- Television: Televisions: 10,620 (1993 est.)

Transportation [31]

Railways: total: 18.5 km; note - owned, operated, and included in statistics of Austrian Federal Railways; standard gauge: 18.5 km 1.435-m gauge (electrified)

Highways: total: 238 km; paved: 238 km; unpaved: 0 km (1986 est.)

Airports

Total: none

Top Agricultural Products [32]

Produces vegetables, corn, wheat, potatoes, grapes; livestock.

Top Mining Products [33]

Mineral Resources: None.

Tourism [34]

	1990	1991	1992	1993	1994
Tourists[16]	78	71	72	65	62

Travelers are in thousands, money in million U.S. dollars.

Finance, Economics, and Trade

Industrial Summary [35]

Industrial Production: Growth rate not available. **Industries**: Electronics, metal manufacturing, textiles, ceramics, pharmaceuticals, food products, precision instruments, tourism.

Economic Indicators [36]

- **National product**: GDP—purchasing power parity—$630 million (1990 est.)
- **National product real growth rate**: NA%
- **National product per capita**: $22,300 (1990 est.)
- **Inflation rate (consumer prices)**: 5.4% (1990)
- **External debt**: $NA

Balance of Payments Summary [37]

Exchange Rates [38]

Currency: **Swiss franc, franken, or franco.**
Symbol: **SwF.**

Data are currency units per $1.

January 1996	1.1810
1995	1.1825
1994	1.3677
1993	1.4776
1992	1.4062
1991	1.4340

Imports and Exports

Top Import Origins [39]

Top Export Destinations [40]

$1.636 billion (1993) Data are for 1990.

Destinations	%
EU	42.7
EFTA	20.9
Switzerland	15.4
Other	36.4

Foreign Aid [41]

None.

Import and Export Commodities [42]

Import Commodities
Machinery
Metal goods
Textiles
Foodstuffs
Motor vehicles

Export Commodities
Small specialty machinery
Dental products
Stamps
Hardware
Pottery

Lithuania

Geography [1]

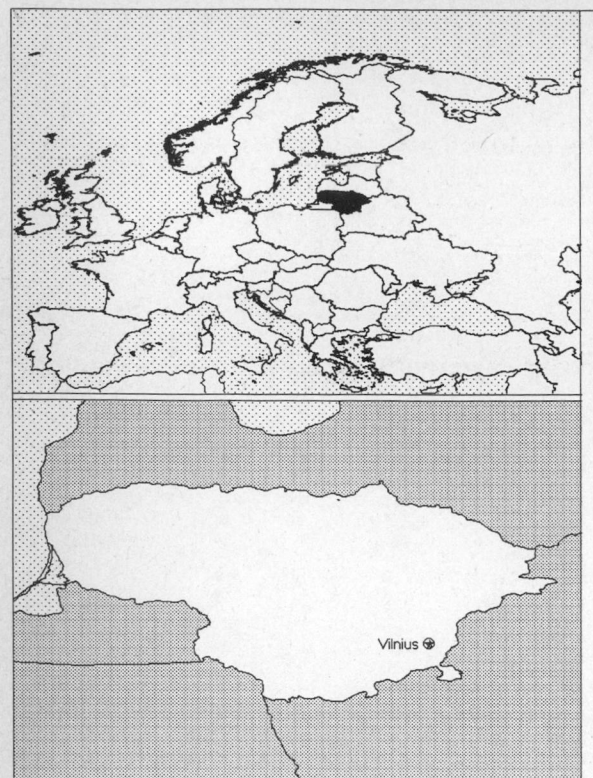

Total area:
  65,200 sq km 25,174 sq mi
Land area:
  65,200 sq km 25,174 sq mi
Comparative area:
  Slightly larger than West Virginia
Land boundaries:
  Total 1,273 km, Belarus 502 km, Latvia 453 km, Poland 91 km,
  Russia (Kaliningrad) 227 km
Coastline:
  108 km
Climate:
  Maritime; wet, moderate winters and summers
Terrain:
  Lowland, many scattered small lakes, fertile soil
Natural resources:
  Peat
Land use:
  Arable land: 49%
  Permanent crops: 0%
  Meadows and pastures: 22%
  Forest and woodland: 16%
  Other: 13%

Demographics [2]

	1970	1980	1990	1995[1]	1996	2000	2010	2020	2030
Population	3,138	3,436	3,702	3,661	3,646	3,629	3,650	3,646	3,642
Population density (persons per sq. mi.)	125	136	147	145	145	144	145	145	145
(persons per sq. km.)	48	53	57	56	56	56	56	56	56
Net migration rate (per 1,000 population)	NA	NA	-2.4	-3.6	-3.1	-1.0	-0.6	-0.2	0.0
Births	NA	NA	NA	NA	47	NA	NA	NA	NA
Deaths	NA	NA	NA	NA	49	NA	NA	NA	NA
Life expectancy - males	NA	NA	66.4	61.7	62.2	63.9	67.7	71.1	73.8
Life expectancy - females	NA	NA	76.2	74.1	74.2	74.8	76.3	78.9	81.0
Birth rate (per 1,000)	NA	NA	15.3	12.2	12.9	14.5	13.0	11.5	11.0
Death rate (per 1,000)	NA	NA	10.8	13.5	13.3	12.7	12.2	11.4	11.2
Women of reproductive age (15-49 yrs.)	NA	NA	923	913	912	916	891	856	830
of which are currently married	NA	NA	600	NA	598	598	589	NA	NA
Fertility rate	NA	NA	2.0	1.7	1.8	2.0	1.8	1.8	1.7

Except as noted, values for vital statistics are in thousands; life expectancy is in years.

Health

Health Indicators [3]

% of population with access to	
safe water (1990-95)	NA
adequate sanitation (1990-95)	NA
health services (1985-95)	NA
% of 1-year-olds immunized (1990-94) against	
TB (tuberculosis)	96
DPT (diphtheria, pertussis, tetanus)	83
polio	88
measles	93
% of contraceptive prevalence (1980-94)	NA
ORT use rate (1990-94)	NA

Health Expenditures [4]

Total health expenditure, 1990 (official exchange rate)	
Millions of dollars	594
Dollars per capita	159
Health expenditures as a percentage of GDP	
Total	3.6
Public sector	2.6
Private sector	1.0
Development assistance for health	
Total aid flows (millions of dollars)[1]	NA
Aid flows per capita (dollars)	NA
Aid flows as a percentage of total health expenditure	NA

For sources, notes, and explanations, see Annotated Source Appendix, page 1061.

Human Factors

Women and Children [5]

% of pregnant women immunized (tetanus 1990-94)	NA
% of births attended by trained health personnel (1983-94)	NA
Maternal mortality rate (1980-92)	NA
Under-5 mortality rate (1994)	20
% under-5 moderately/severely underweight (1980-1994)	NA

Burden of Disease [6]

Population per physician (1992)	235.35
Population per nurse (1992)	92.17
Population per hospital bed (1992)	84.09
AIDS cases per 100,000 people (1994)	*
Heart disease cases per 1,000 people (1990-93)	NA

Ethnic Division [7]

Lithuanian	80.1%
Russian	8.6%
Polish	7.7%
Byelorussian	1.5%
Other	2.1%

Religion [8]

Roman Catholic, Lutheran, other.

Major Languages [9]

Lithuanian (official), Polish, Russian.

Education

Public Education Expenditures [10]

	1980	1985	1990	1992	1993	1994
Million (Lita)						
Total education expenditure	4	5	6	179	531	947
as percent of GNP	5.5	5.3	4.8	5.5	4.4	4.5
as percent of total govt. expend.	15.4	12.9	13.8	22.1	20.1	21.8
Current education expenditure	4	4	6	171	499	886
as percent of GNP	5.1	4.8	4.5	5.2	4.1	4.2
as percent of current govt. expend.	NA	NA	NA	23.5	32.2	23.9
Capital expenditure	0.0	1	0.0	8	32	62

Educational Attainment [11]

Age group (1989)	25+
Total population	2,282,191
Highest level attained (%)	
No schooling	9.1
First level	
Not completed	21.3
Completed	NA
Entered second level	
S-1	57.0
S-2	NA
Postsecondary	12.6

Illiteracy [12]

	1985	1989	1995
Illiterate population (15+ yrs.)[2]	NA	44,308	16,000
Illiteracy rate - total pop. (%)	NA	1.6	0.5
Illiteracy rate - males (%)	NA	0.8	0.4
Illiteracy rate - females (%)	NA	2.2	0.7

Libraries [13]

	Admin. Units	Svc. Pts.	Vols. (000)	Shelving (meters)	Vols. Added	Reg. Users
National	NA	NA	NA	NA	NA	NA
Nonspecialized	NA	NA	NA	NA	NA	NA
Public (1993)	1,623	NA	NA	NA	NA	NA
Higher ed. (1993)	27	NA	12,053	NA	NA	NA
School	NA	NA	NA	NA	NA	NA

Daily Newspapers [14]

	1980	1985	1990	1994
Number of papers	NA	NA	NA	18
Circ. (000)	NA	NA	NA	506

Culture [15]

Cinema (seats per 1,000)	14.4
Annual attendance per person	0.6
Gross box office receipts (mil. Lita)	1.15
Museums (reporting)	57
Visitors (000)	1,246
Annual receipts (000 Lita)	NA

Science and Technology

Scientific/Technical Forces [16]

Scientists/engineers	4,750
Number female	1,518
Technicians	NA
Number female	NA
Total	NA

R&D Expenditures [17]

U.S. Patents Issued [18]

Values show patents issued to citizens of the country by the U.S. Patents Office.

	1993	1994	1995
Number of patents	0	0	1

For sources, notes, and explanations, see Annotated Source Appendix, page 1061.

Government and Law

Organization of Government [19]

Long-form name:
  Republic of Lithuania
Type:
  Republic
Independence:
  6 September 1991 (from Soviet Union)
National holiday:
  Independence Day, 16 February (1918)
Constitution:
  Adopted 25 October 1992
Legal system:
  Based on civil law system; no judicial review of legislative acts
Executive branch:
  President; Premier; Council of Ministers
Legislative branch:
  Unicameral: Seimas (parliament)
Judicial branch:
  Supreme Court; Court of Appeal

Elections [20]

Seimas (parliament)	% of votes
Democratic Labor Party	51.0
Other	49.0

Government Expenditures [21]

Educ./Health - 47.9%
Defense - 1.9%
Gen. Services - 15.1%
Other - 17.6%
Industry - 17.5%

(% distribution). Expend. for CY95: 6,079.1 (Litas mil.)

Crime [23]

	1990
Crime volume	
Cases known to police	31,698
Attempts (percent)	NA
Percent cases solved	24.8
Crimes per 100,000 persons	856.1
Persons responsible for offenses	
Total number offenders	7,870
Percent female	10.0
Percent juvenile	13.5
Percent foreigners	NA

Military Expenditures and Arms Transfers [22]

	1990	1991	1992[14]	1993	1994
Military expenditures					
Current dollars (mil.)	NA	NA	NA	92	71
1994 constant dollars (mil.)	NA	NA	NA	94	71
Armed forces (000)	NA	NA	10	10	10
Gross national product (GNP)					
Current dollars (mil.)	NA	NA	12,750	11,840	12,010
1994 constant dollars (mil.)	NA	NA	13,290	12,090	12,010
Central government expenditures (CGE)					
1994 constant dollars (mil.)	NA	NA	4,473	2,465	2,269
People (mil.)	NA	NA	3.8	3.8	3.8
Military expenditure as % of GNP	NA	NA	NA	0.8	0.6
Military expenditure as % of CGE	NA	NA	NA	3.8	3.1
Military expenditure per capita (1994 $)	NA	NA	NA	25	18
Armed forces per 1,000 people (soldiers)	NA	NA	2.6	2.6	2.6
GNP per capita (1994 $)	NA	NA	3,507	3,164	3,120
Arms imports[6]					
Current dollars (mil.)	NA	NA	0	10	0
1994 constant dollars (mil.)	NA	NA	0	10	0
Arms exports[6]					
Current dollars (mil.)	NA	NA	0	0	0
1994 constant dollars (mil.)	NA	NA	0	0	0
Total imports[7]					
Current dollars (mil.)	NA	NA	1,084	580	2,700[e]
1994 constant dollars (mil.)	NA	NA	1,130	592	2,700[e]
Total exports[7]					
Current dollars (mil.)	NA	NA	1,145	567	2,200[e]
1994 constant dollars (mil.)	NA	NA	1,194	579	2,200[e]
Arms as percent of total imports[8]	NA	NA	0	1.7	0
Arms as percent of total exports[8]	NA	NA	0	0	0

Human Rights [24]

	SSTS	FL	FAPRO	PPCG	APROBC	TPW	PCPTW	STPEP	PHRFF	PRW	ASST	AFL
Observes				P					P			

	EAFRD	CPR	ESCR	SR	ACHR	MAAE	PVIAC	PVNAC	EAFDAW	TCIDTP	RC
Observes		P	P						P	P	P

P = Party; S = Signatory; see Appendix for meaning of abbreviations.

Labor Force

Total Labor Force [25]

1.836 million

Labor Force by Occupation [26]

Industry and construction	42%
Agriculture and forestry	18
Other	40

Date of data: 1990

Unemployment Rate [27]

6.1% (January 1996)

For sources, notes, and explanations, see Annotated Source Appendix, page 1061.

Production Sectors

Commercial Energy Production and Consumption

Data are shown in quadrillion (10^{15}) BTUs and percent for 1995
Values for hydroelectric, nuclear, geothermal, solar, and wind power refer to electrical generation.

Production [28]

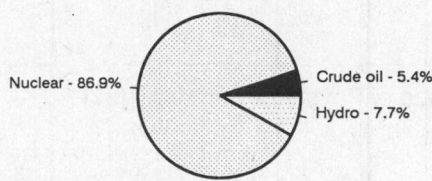

Nuclear - 86.9%

Crude oil - 5.4%

Hydro - 7.7%

Consumption [29]

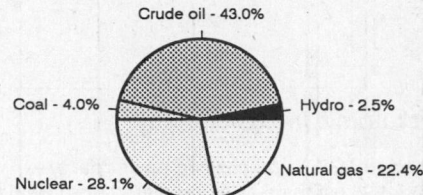

Crude oil - 43.0%

Coal - 4.0%

Hydro - 2.5%

Nuclear - 28.1%

Natural gas - 22.4%

Crude oil	0.007
Net hydroelectric power	0.010
Net nuclear power	0.113
Total	0.130

Crude oil	0.173
Dry natural gas	0.090
Coal	0.016
Net hydroelectric power	0.010
Net nuclear power	0.113
Total	0.402

Telecommunications [30]

- 900,000 telephones; telecommunications system among the most modern of the former Soviet republics
- Domestic: NMT-450 analog cellular telephone network operates in Vilnius and other cities; landlines and microwave radio relay connect switching centers
- International: international connections no longer use the Moscow international gateway switch, but are by satellite through Oslo from Vilnius and through Copenhagen from Kaunas; satellite earth stations - 1 Eutelsat and 1 Intelsat (Atlantic Ocean); cellular network linked internationally through Copenhagen by Eutelsat; international electronic mail available; landlines or microwave radio relay to former Soviet republics
- Radio: Broadcast stations: AM 13, FM 26, shortwave 1, longwave 1 Radios: 1.42 million (1993 est.)
- Television: Broadcast stations: 3 Televisions: 1.77 million (1993 est.)

Transportation [31]

Railways: total: 2,002 km; broad gauge: 2,002 km 1.524-m gauge (122 km electrified) (1994)

Highways: total: 55,603 km; paved: 42,209 km (including 382 km of expressways); unpaved: 13,394 km (1994)

Merchant marine: total: 43 ships (1,000 GRT or over) totaling 264,639 GRT/303,649 DWT; ships by type: cargo 26, combination bulk 11, oil tanker 2, railcar carrier 1, roll-on/roll-off cargo 1, short-sea passenger 2 (1995 est.)

Airports

Total:	96
With paved runways over 3,047 m:	3
With paved runways 2,438 to 3,047 m:	2
With paved runways 1,524 to 2,437 m:	4
With paved runways 914 to 1,523 m:	2
With paved runways under 914 m:	14

Top Agricultural Products [32]

Agriculture accounts for 20% of the GDP; produces grain, potatoes, sugar beets, vegetables; meat, milk, eggs; fish.

Top Mining Products [33]

Metric tons except as noted[e]	5/26/95*
Ammonia, nitrogen content	250,000
Cement	150,000
Clays, brick and concrete agg. (000 cu. meters)	700,000
Limestone	4,000,000
Peat	100,000
Sand and gravel (mil. cu. meters)	10
Sand, for glass	60,000

Tourism [34]

	1990	1991	1992	1993	1994
Visitors[68]	NA	NA	NA	2,440	2,369
Tourists[11]	NA	NA	NA	324	222
Tourism receipts[69]	NA	NA	NA	NA	70
Tourism expenditures[70]	NA	NA	NA	NA	50
Fare receipts	NA	NA	NA	NA	14
Fare expenditures	NA	NA	NA	NA	1

Travelers are in thousands, money in million U.S. dollars.

For sources, notes, and explanations, see Annotated Source Appendix, page 1061.

559

Finance, Economics, and Trade

Industrial Summary [35]

Industrial Production: Growth rate not available; accounts for 42% of the GDP. **Industries:** Metal-cutting machine tools, electric motors, television sets, refrigerators and freezers, petroleum refining, shipbuilding (small ships), furniture making, textiles, food processing, fertilizers, agricultural machinery, optical equipment, electronic components, computers, amber.

Economic Indicators [36]

- **National product:** GDP—purchasing power parity—$13.3 billion (1995 estimate as extrapolated from World Bank estimate for 1994)

- **National product real growth rate:** 1% (1995 est.)

- **National product per capita:** $3,400 (1995 est.)

- **Inflation rate (consumer prices):** 35% (1995 est.)

- **External debt:** $895 million

Balance of Payments Summary [37]

Values in millions of dollars.

	1989	1990	1991	1992[1]	1993
Exports of goods (f.o.b.)	NA	NA	NA	745	1,243
Imports of goods (f.o.b.)	NA	NA	NA	593	1,313
Trade balance	NA	NA	NA	152	-70
Services - debits	NA	NA	NA	-80	-197
Services - credits	NA	NA	NA	208	209
Private transfers (net)	NA	NA	NA	41	115
Government transfers (net)	NA	NA	NA	NA	NA
Long-term capital (net)	NA	NA	NA	9	285
Short-term capital (net)	NA	NA	NA	200	45
Errors and omissions	NA	NA	NA	132	50
Overall balance	NA	NA	NA	662	437

Exchange Rates [38]

Currency: **litas.**

Data are currency units per $1. The convertible litas was introduced in June 1993. Fixed rate since 1 May 1994.

January 1996	4.000
1995	4.000
1994	3.978
1993	4.344
1992	1.773

Imports and Exports

Top Import Origins [39]

$2.7 billion (1994).

Origins	%
Russia	NA
Germany	NA
Belarus	NA

Top Export Destinations [40]

$2.2 billion (1994).

Destinations	%
Russia	NA
Ukraine	NA
Germany	NA

Foreign Aid [41]

	U.S. $	
ODA (1993)	144	million
Commitments from the West and international institutions (1992-95)	765	million

Import and Export Commodities [42]

Import Commodities	Export Commodities
Oil 24%	Electronics 18%
Machinery 14%	Food 10%
Chemicals 8%	Chemicals 6%
Grain NA%	Petroleum products 5%

Luxembourg

Geography [1]

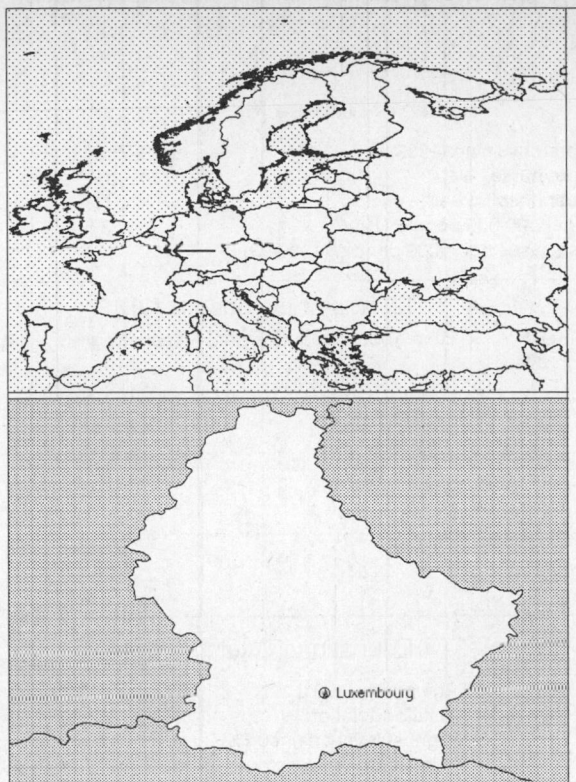

Total area:
  2,586 sq km   998 sq mi
Land area:
  2,586 sq km   998 sq mi
Comparative area:
  Slightly smaller than Rhode Island
Land boundaries:
  Total 359 km, Belgium 148 km, France 73 km, Germany 138 km
Coastline:
  0 km (landlocked)
Climate:
  Modified continental with mild winters, cool summers
Terrain:
  Mostly gently rolling uplands with broad, shallow valleys; uplands to slightly mountainous in the North; steep slope down to Moselle floodplain in the Southeast
Natural resources:
  Iron ore (no longer exploited)
Land use:
  Arable land: 24%
  Permanent crops: 1%
  Meadows and pastures: 20%
  Forest and woodland: 21%
  Other: 34%

Demographics [2]

	1970	1980	1990	1995[1]	1996	2000	2010	2020	2030
Population	339	364	382	409	416	442	495	523	520
Population density (persons per sq. mi.)	340	365	383	410	417	443	496	524	521
(persons per sq. km.)	131	141	148	158	161	171	192	202	201
Net migration rate (per 1,000 population)	3.1	3.7	14.2	10.9	10.9	11.2	6.2	2.0	0.0
Births[31]	4	4	NA	NA	5	NA	NA	NA	NA
Deaths	4	4	2	NA	4	NA	NA	NA	NA
Life expectancy - males	NA	NA	71.9	74.7	75.2	77.7	80.1	80.8	81.0
Life expectancy - females	NA	NA	79.7	81.1	81.6	83.9	86.2	86.8	87.0
Birth rate (per 1,000)	13.0	11.4	12.9	13.4	13.1	11.0	9.7	9.6	8.4
Death rate (per 1,000)	12.3	11.3	9.9	8.6	8.3	7.4	7.7	9.1	11.0
Women of reproductive age (15-49 yrs.)	NA	NA	97	103	105	109	114	109	101
of which are currently married	55	NA	59	NA	65	64	59	NA	NA
Fertility rate	2.0	1.5	1.6	1.8	1.8	1.6	1.6	1.6	1.6

Except as noted, values for vital statistics are in thousands; life expectancy is in years.

Health

Health Indicators [3]

Health Expenditures [4]

For sources, notes, and explanations, see Annotated Source Appendix, page 1061.

561

Human Factors

Women and Children [5]

Burden of Disease [6]

Population per physician (1989)	508.06
Population per nurse	NA
Population per hospital bed	NA
AIDS cases per 100,000 people (1994)	3.5
Heart disease cases per 1,000 people (1990-93)	475

Ethnic Division [7]

Celtic base (with French and German blend), Portuguese, Italian, and European (guest and worker residents).

Religion [8]

Roman Catholic	97%
Protestant and Jewish	3%

Major Languages [9]

Luxembourgisch, German, French, English.

Education

Public Education Expenditures [10]

	1980	1989	1990	1992	1993	1994
Million (Franc)[38]						
Total education expenditure	9,792	16,363	NA	14,027	15,677	16,586
as percent of GNP	5.7	4.1	NA	2.8	3.1	3.1
as percent of total govt. expend.	14.9	16.0	NA	NA	NA	NA
Current education expenditure	9,305	13,440	NA	NA	NA	NA
as percent of GNP	5.4	3.4	NA	NA	NA	NA
as percent of current govt. expend.	19.8	NA	NA	NA	NA	NA
Capital expenditure	486	2,923	NA	NA	NA	NA

Educational Attainment [11]

Age group (1991)	25+
Total population	262,628
Highest level attained (%)	
No schooling	NA
First level	
Not completed	39.7
Completed	NA
Entered second level	
S-1	40.3
S-2	NA
Postsecondary	10.8

Illiteracy [12]

Libraries [13]

	Admin. Units	Svc. Pts.	Vols. (000)	Shelving (meters)	Vols. Added	Reg. Users
National (1990)[5]	1	1	675	26,400	NA	24,650
Nonspecialized	NA	NA	NA	NA	NA	NA
Public (1990)	2	5	613[e]	17,015[e]	NA	43,505[e]
Higher ed. (1990)	1	1	269	7,134[e]	NA	5,321[e]
School	NA	NA	NA	NA	NA	NA

Daily Newspapers [14]

	1980	1985	1990	1994
Number of papers	5	4	5	5
Circ. (000)	135	140	143	154

Culture [15]

Cinema (seats per 1,000)	7.8
Annual attendance per person	1.8
Gross box office receipts (mil. Franc)[10]	134
Museums (reporting)	NA
Visitors (000)	NA
Annual receipts (000 Franc)	NA

Science and Technology

Scientific/Technical Forces [16]

R&D Expenditures [17]

U.S. Patents Issued [18]

Values show patents issued to citizens of the country by the U.S. Patents Office.

	1993	1994	1995
Number of patents	41	44	34

For sources, notes, and explanations, see Annotated Source Appendix, page 1061.

Government and Law

Organization of Government [19]

Long-form name:
Grand Duchy of Luxembourg
Type:
Constitutional monarchy
Independence:
1,839
National holiday:
National Day, 23 June (1921) (public celebration of the Grand Duke's birthday)
Constitution:
17 October 1868, occasional revisions
Legal system:
Based on civil law system; accepts compulsory ICJ jurisdiction
Executive branch:
Grand Duke; Heir Apparent Prince; Prime Minister; Vice Prime Minister; Council of Ministers
Legislative branch:
Unicameral: Chamber of Deputies (Chambre des Deputes) Note: The Council of State (Conseil d'Etat) is an advisory body whose views are considered by the Chamber of Deputies
Judicial branch:
Superior Court of Justice (Cour Superieure de Justice)

Elections [20]

Chamber of Deputies	% of seats
Christian Social Party (CSV)	35.0
Socialist Workers Party (LSAP)	28.3
Liberal (DP)	20.0
Action for Democracy and Pension Rights	8.3
Greens	8.3

Government Expenditures [21]

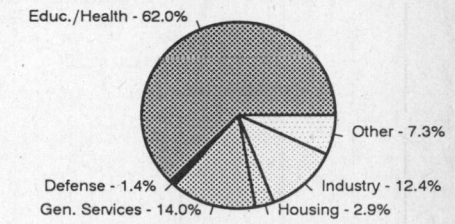

Educ./Health - 62.0%
Other - 7.3%
Defense - 1.4%
Industry - 12.4%
Gen. Services - 14.0%
Housing - 2.9%

(% distribution). Expend. for CY94: 208,455 (Franc mil.)

Crime [23]

	1994
Crime volume	
Cases known to police	29,166
Attempts (percent)	NA
Percent cases solved	34.30
Crimes per 100,000 persons	7,383.80
Persons responsible for offenses	
Total number offenders	12,625
Percent female	14.47
Percent juvenile (0-18 yrs.)	6.18
Percent foreigners	60.12

Military Expenditures and Arms Transfers [22]

	1990	1991	1992	1993	1994
Military expenditures					
Current dollars (mil.)	105	119	126	114	126
1994 constant dollars (mil.)	117	128	131	116	126
Armed forces (000)	1	1	1	1	1
Gross national product (GNP)					
Current dollars (mil.)	14,280	15,310	15,710	15,390	16,130
1994 constant dollars (mil.)	15,900	16,410	16,390	15,710	16,130
Central government expenditures (CGE)					
1994 constant dollars (mil.)	5,123	5,625	5,858	5,920	4,050[e]
People (mil.)	0.4	0.4	0.4	0.4	0.4
Military expenditure as % of GNP	0.7	0.8	0.8	0.7	0.8
Military expenditure as % of CGE	2.3	2.3	2.2	2.0	3.1
Military expenditure per capita (1994 $)	305	329	333	292	313
Armed forces per 1,000 people (soldiers)	2.6	2.6	2.5	2.5	2.5
GNP per capita (1994 $)	41,610	42,270	41,620	39,440	40,140
Arms imports[6]					
Current dollars (mil.)	20	10	10	5	0
1994 constant dollars (mil.)	22	11	10	5	0
Arms exports[6]					
Current dollars (mil.)	0	0	0	0	0
1994 constant dollars (mil.)	0	0	0	0	0
Total imports[7]					
Current dollars (mil.)	7,596	8,044	8,248	NA	NA
1994 constant dollars (mil.)	8,453	8,622	8,601	NA	NA
Total exports[7]					
Current dollars (mil.)	7,041	7,313	6,691	NA	NA
1994 constant dollars (mil.)	7,836	7,838	6,978	NA	NA
Arms as percent of total imports[8]	0.3	0.1	0.1	NA	0
Arms as percent of total exports[8]	0	0	0	0	0

Human Rights [24]

	SSTS	FL	FAPRO	PPCG	APROBC	TPW	PCPTW	STPEP	PHRFF	PRW	ASST	AFL
Observes		P	P	P	P	P	P	P	P	P	P	P
	EAFRD	CPR	ESCR	SR	ACHR	MAAE	PVIAC	PVNAC	EAFDAW	TCIDTP	RC	
Observes		P	P	P		P	P	P	P	P	P	

P = Party; S = Signatory; see Appendix for meaning of abbreviations.

Labor Force

Total Labor Force [25]

200,400 (1992); one-third of labor force are foreign workers, mostly from Portugal, Italy, France, Belgium, and Germany.

Labor Force by Occupation [26]

Trade, restaurants, hotels	20%
Mining, quarrying, manufacturing	18
Other market services	17
Community, social, personal services	14
Construction	11
Finance, insurance, realty, bus. svcs	9
Transport, storage, communications	7
Agriculture, hunting, forestry, fishing	3
Electricity, gas, water	1

Unemployment Rate [27]

2.5% (1995)

For sources, notes, and explanations, see Annotated Source Appendix, page 1061.

Production Sectors

Commercial Energy Production and Consumption

Data are shown in quadrillion (10^{15}) BTUs and percent for 1995
Values for hydroelectric, nuclear, geothermal, solar, and wind power refer to electrical generation.

Production [28]

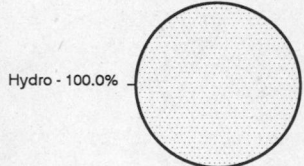

Hydro - 100.0%

Consumption [29]

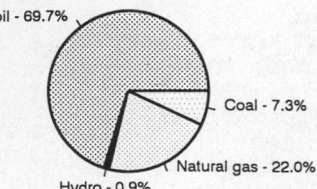

Crude oil - 69.7%

Coal - 7.3%

Natural gas - 22.0%

Hydro - 0.9%

Net hydroelectric power	0.001
Total	0.001

Crude oil	0.076
Dry natural gas	0.024
Coal	0.008
Net hydroelectric power	0.001
Total	0.109

Telecommunications [30]

- 214,821 (1993 est.) telephones; highly developed, completely automated and efficient system, mainly buried cables
- Domestic: nationwide cellular telephone system; buried cable
- International: 3 channels leased on TAT-6 coaxial submarine cable (Europe to North America)
- Radio: Broadcast stations: AM 2, FM 3, shortwave 0 Radios: 230,000 (1993 est.)
- Television: Broadcast stations: 3 (1987 est.) and 1 direct-broadcast satellite link Televisions: 100,500 (1993 est.)

Transportation [31]

Railways: total: 275 km; standard gauge: 275 km 1.435-m gauge (262 km electrified; 178 km double track) (1995)

Highways: total: 5,134 km; paved: 5,088 km (including 121 km of expressways); unpaved: 46 km (1995 est.)

Merchant marine: total: 36 ships (1,000 GRT or over) totaling 825,496 GRT/1,238,354 DWT; ships by type: bulk 3, chemical tanker 4, combination bulk 6, container 2, liquefied gas tanker 6, oil tanker 5, passenger 2, refrigerated cargo 6, roll-on/roll-off cargo 2 (1995 est.)

Airports
Total: 2
With paved runways over 3,047 m: 1
With paved runways under 914 m: 1 (1995 est.)

Top Agricultural Products [32]

Agriculture accounts for 1.4% of the GDP; produces barley, oats, potatoes, wheat, fruits, wine grapes; livestock products.

Top Mining Products [33]

Metric tons except as noted	6/95*
Cement, hydraulic	620,000[e]
Gypsum and anhydrite, crude	400[e]
Pig iron	1,927,000
Steel, crude	3,092,000
Phosphates, Thomas slag	
gross weight	500,000[e]
P2O5 content	75,000[e]

Tourism [34]

	1990	1991	1992	1993	1994
Tourists[71]	820	861	796	831	762

Travelers are in thousands, money in million U.S. dollars.

For sources, notes, and explanations, see Annotated Source Appendix, page 1061.

Manufacturing Sector

Manufacturing Summary [35]

	1987		1988		1989		1990		1991	
	$ bil.	%	$ bil.	%	$ bil.	%	$ bil.	%	$ bil.	%
Establishments or enterprises (number)	466	0.020	459	0.022	448	0.024	453	0.025	-	-
Total employment (000)	24	0.018	23	0.017	22	0.018	22	0.019	-	-
Production workers (000)	19	0.028	17	0.028	17	0.023	16	0.023	-	-
Output ($ bil.)	3	0.027	3	0.028	3	0.029	4	0.033	-	-
Value added ($ bil.)	0.909	0.021	1	0.022	1	0.024	1	0.026	-	-
Capital investment ($ mil.)	139	0.032	133	0.028	123	0.022	264	0.047	-	-
M & E investment ($ mil.)	120	0.039	104	0.030	78	0.020	147	0.035	-	-
Employees per establishment (number)	52	89.857	50	76.454	49	75.022	48	77.078	-	-
Production workers per establishment	40	139.453	38	128.023	38	95.813	36	91.618	-	-
Output per establishment ($ mil.)	6	135.894	7	125.918	8	120.760	8	132.200	-	-
Capital investment per estab. ($ mil.)	0.298	160.381	0.289	127.335	0.275	94.031	0.582	187.508	-	-
M & E per establishment ($ mil)	0.258	194.289	0.227	137.822	0.173	83.004	0.325	138.235	-	-
Payroll per employee ($)	20,458	228.145	21,850	219.322	21,788	195.016	27,462	197.970	-	-
Wages per production worker ($)	16,825	213.226	17,833	211.047	17,602	175.523	21,798	183.549	-	-
Hours per production worker (hours)	1,402	73.869	1,469	76.408	-	-	-	-	-	-
Output per employee ($)	113,266	151.234	140,282	164.698	154,268	160.966	177,748	171.515	-	-
Capital investment per employee ($)	5,732	178.485	5,795	166.552	5,578	125.338	12,216	243.271	-	-
M & E per employee ($)	4,965	216.221	4,559	180.269	3,510	110.639	6,831	179.345	-	-

Note: Columns headed % show percent of world total or ratio. Ratios closest to 100 are closest to world average. M & E stands for machinery & equipment.

Output in Manufacturing

	1987		1988		1989		1990		1991	
	$ bil.[f]	%	$ bil.[f]	%	$ bil.[f]	%	$ bil.[f]	%	$ bil.[f]	%
3110 - Food products	0.263	8.77	0.271	9.03	0.260	8.67	0.313	7.83	-	-
3520 - Chemical products nec	0.054	1.80	0.063	2.10	0.053	1.77	0.079	1.97	-	-
3710 - Iron and steel	1.587	52.90	1.945	64.83	2.106	70.20	2.299	57.47	-	-
3720 - Nonferrous metals	0.220	7.33	0.273	9.10	0.306	10.20	0.290	7.25	-	-
3810 - Metal products	0.364	12.13	0.405	13.50	0.431	14.37	0.518	12.95	-	-
3820 - Machinery nec	0.227	7.57	0.218	7.27	0.213	7.10	0.288	7.20	-	-
3840 - Transportation equipment	0.029	0.97	0.034	1.13	0.038	1.27	0.048	1.20	-	-

Note: Codes are International Standard Industry codes (ISIC). Percentages are % of total Output. [f]: Factor Prices; [p]: Producer Prices.

Finance, Economics, and Trade

Economic Indicators [36]

- **National product**: GDP—purchasing power parity—$10 billion (1995 est.)
- **National product real growth rate**: 2.6% (1995 est.)
- **National product per capita**: $24,800 (1995 est.)
- **Inflation rate (consumer prices)**: 3.6% (1992)
- **External debt**: $800 million (1994 est.)

Balance of Payments Summary [37]

Values in millions of dollars.

	1989	1990	1991	1992	1993
Exports of goods (f.o.b.)	89,988	107,654	106,019	113,638	103,873
Imports of goods (f.o.b.)	-89,020	-107,064	-106,085	-112,307	-99,905
Trade balance	968	590	-66	1,331	3,932
Services - debits	-67,902	-90,646	-101,703	-117,520	-109,283
Services - credits	71,850	96,971	108,251	125,144	120,527
Private transfers (net)	47	-597	-280	-503	-602
Government transfers (net)	-1,765	-1,369	-1,470	-1,984	-1,986
Long-term capital (net)	-2,437	-4	4,049	-7,710	9,503
Short-term capital (net)	-2,767	-1,650	-7,824	60	-23,234
Errors and omissions	-86	-2,844	-992	1,847	-938
Overall balance	-2,092	451	505	665	-2,081

Exchange Rates [38]

Currency: **Luxembourg franc.**
Symbol: **LuxF.**

Data are currency units per $1. The Luxembourg franc is at par with the Belgian franc, which circulates freely in Luxembourg.

January 1996	30.036
1995	29.480
1994	33.456
1993	34.597
1992	32.150
1991	34.148

Imports and Exports

Top Import Origins [39]

$7.5 million (c.i.f., 1993 est.).

Origins	%
Belgium	38
Germany	25
France	11
Netherlands	4

Top Export Destinations [40]

$5.9 million (f.o.b., 1993 est.).

Destinations	%
Germany	28
France	18
Belgium	15
UK	7
Netherlands	5

Foreign Aid [41]

Donor: ODA, $50 million (1993).

Import and Export Commodities [42]

Import Commodities	Export Commodities
Minerals	Finished steel products
Metals	Chemicals
Foodstuffs	Rubber products
Quality consumer goods	Glass
	Aluminum
	Other industrial products

Macedonia

Geography [1]

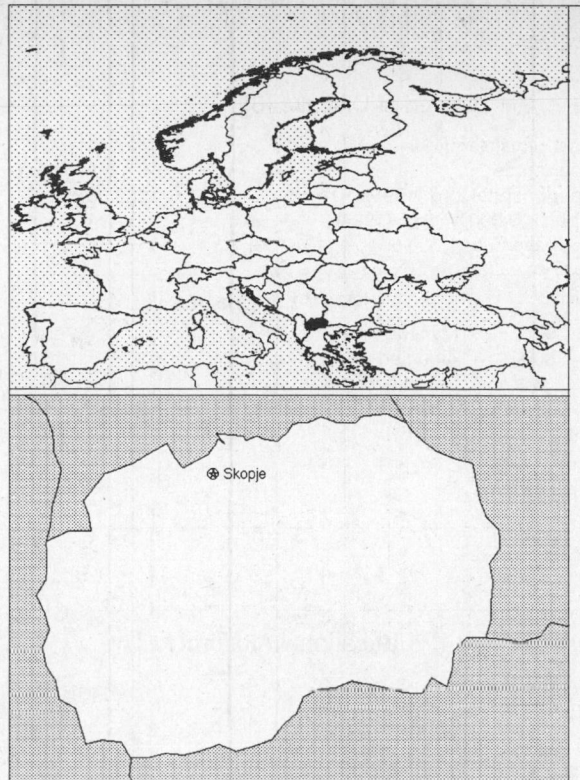

Total area:
25,333 sq km   9,781 sq mi
Land area:
24,856 sq km   9,597 sq mi
Comparative area:
Slightly larger than Vermont
Land boundaries:
Total 748 km, Albania 151 km, Bulgaria 148 km, Greece 228 km, Yugoslavia 221 km (all with Serbia)
Coastline:
0 km (landlocked)
Climate:
Hot, dry summers and autumns and relatively cold winters with heavy snowfall
Terrain:
Mountainous territory covered with deep basins and valleys; there are three large lakes, each divided by a frontier line; country bisected by the Vardar River
Natural resources:
Chromium, lead, zinc, manganese, tungsten, nickel, low-grade iron ore, asbestos, sulfur, timber
Land use:
Arable land: 5%
Permanent crops: 5%
Meadows and pastures: 20%
Forest and woodland: 30%
Other: 40%

Demographics [2]

	1970	1980	1990	1995[1]	1996	2000	2010	2020	2030
Population	1,629	1,893	2,031	2,094	2,104	2,152	2,261	2,296	2,295
Population density (persons per sq. mi.)	164	191	205	211	212	217	228	231	231
(persons per sq. km.)	63	74	79	81	82	84	88	89	89
Net migration rate (per 1,000 population)	NA	NA	NA	-0.3	-0.2	-0.1	0.0	0.0	0.0
Births	NA	NA	NA	NA	28	NA	NA	NA	NA
Deaths	NA	NA	NA	NA	18	NA	NA	NA	NA
Life expectancy - males	NA	NA	NA	69.8	69.9	70.3	71.2	74.1	76.3
Life expectancy - females	NA	NA	NA	74.0	74.2	74.9	76.4	79.6	82.1
Birth rate (per 1,000)	NA	NA	NA	13.3	13.3	15.9	12.7	10.5	9.9
Death rate (per 1,000)	NA	NA	NA	8.4	8.5	8.8	9.9	10.0	10.7
Women of reproductive age (15-49 yrs.)	NA	NA	NA	534	539	547	535	531	509
of which are currently married	NA	NA	NA	NA	374	378	377	NA	NA
Fertility rate	NA	NA	NA	1.8	1.8	2.2	1.8	1.7	1.6

Except as noted, values for vital statistics are in thousands; life expectancy is in years.

Health

Health Indicators [3]

% of population with access to	
safe water (1990-95)	NA
adequate sanitation (1990-95)	NA
health services (1985-95)	NA
% of 1-year-olds immunized (1990-94) against	
TB (tuberculosis)	96
DPT (diphtheria, pertussis, tetanus)	88
polio	91
measles	86
% of contraceptive prevalence (1980-94)	NA
ORT use rate (1990-94)	NA

Health Expenditures [4]

For sources, notes, and explanations, see Annotated Source Appendix, page 1061.

567

Human Factors

Women and Children [5]

% of pregnant women immunized (tetanus 1990-94)	91
% of births attended by trained health personnel (1983-94)	NA
Maternal mortality rate (1980-92)	NA
Under-5 mortality rate (1994)	32
% under-5 moderately/severely underweight (1980-1994)	NA

Burden of Disease [6]

Population per physician (1994)	427.44
Population per nurse	NA
Population per hospital bed (1994)	189.33
AIDS cases per 100,000 people (1994)	NA
Heart disease cases per 1,000 people (1990-93)	NA

Ethnic Division [7]

Macedonian	65%
Albanian	22%
Turkish	4%
Gypsies	3%
Serb	2%
Other	4%

Religion [8]

Eastern Orthodox	67%
Muslim	30%
Other	3%

Major Languages [9]

Macedonian	70%
Albanian	21%
Serbo-Croatian	3%
Turkish	3%
Other	3%

Education

Public Education Expenditures [10]

	1980	1990	1991	1992	1993	1994
Million (Dinar)						
Total education expenditure	NA	NA	NA	605	2,939	6,970
as percent of GNP	NA	NA	NA	5.4	5.0	5.6
as percent of total govt. expend.	NA	NA	NA	NA	21.5	18.3
Current education expenditure	NA	NA	NA	591	2,863	6,807
as percent of GNP	NA	NA	NA	5.2	4.9	5.5
as percent of current govt. expend.	NA	NA	NA	NA	21.3	18.5
Capital expenditure	NA	NA	NA	14	75	164

Educational Attainment [11]

Illiteracy [12]

In thousands and percent[3]	1985	1991	2000
Illiterate population (15+ years)	1,614	1,342	942
Illiteracy rate - total pop. (%)	9.2	7.3	4.7
Illiteracy rate - males (%)	3.5	2.6	1.3
Illiteracy rate - females (%)	14.6	11.9	7.9

Libraries [13]

	Admin. Units	Svc. Pts.	Vols. (000)	Shelving (meters)	Vols. Added	Reg. Users
National (1992)	1	1	2,169	61,655	10,656	152,800
Nonspecialized (1992)	1	1	142	2,582	872	2,428
Public (1992)	62	122	2,729	30,289	68,454	987,483
Higher ed. (1992)	26	43	1,135	26,323	7,833	496,906
School (1989)	7,784	7,784	38,430	NA	1,680	NA

Daily Newspapers [14]

	1980	1985	1990	1994
Number of papers	NA	NA	2	3
Circ. (000)	NA	NA	55[e]	44

Culture [15]

Cinema (seats per 1,000)	6.7
Annual attendance per person[11]	0.2
Gross box office receipts (mil. Dinar)	0.7
Museums (reporting)	20
Visitors (000)	208
Annual receipts (000 Dinar)	47,836

Science and Technology

Scientific/Technical Forces [16]

Scientists/engineers	2,605
Number female	1,008
Technicians	691
Number female	386
Total	3,296

R&D Expenditures [17]

	Dinar (000) 1989
Total expenditure	2,152,032
Capital expenditure	815,082
Current expenditure	1,336,950
Percent current	62.1

U.S. Patents Issued [18]

For sources, notes, and explanations, see Annotated Source Appendix, page 1061.

Government and Law

Organization of Government [19]

Long-form name:
The Former Yugoslav Republic of Macedonia
Type:
Emerging democracy
Independence:
17 September 1991 (from Yugoslavia)
National holiday:
8 September
Constitution:
Adopted 17 November 1991, effective 20 November 1991
Legal system:
Based on civil law system; judicial review of legislative acts
Executive branch:
President; Prime Minister; Council of Ministers
Legislative branch:
Unicameral: Assembly (Sobranje)
Judicial branch:
Constitutional Court; Judicial Court of the Republic

Elections [20]

Assembly (Sobranje)	% of seats
SDSM (former Communist)	48.3
Liberal Party (LP)	24.2
Party for Democratic Prosperity	8.3
Socialist Party	6.7
National Democratic Party	3.3
Independents	5.8
Other	3.3

Government Budget [21]

Crime [23]

	1994
Crime volume	
Cases known to police	18,279
Attempts (percent)	NA
Percent cases solved	NA
Crimes per 100,000 persons	943.74
Persons responsible for offenses	
Total number offenders	18,428
Percent female	NA
Percent juvenile (14-18 yrs.)	NA
Percent foreigners	NA

Military Expenditures and Arms Transfers [22]

	1990	1991	1992[15]	1993	1994
Military expenditures					
Current dollars (mil.)	NA	NA	NA	NA	NA
1994 constant dollars (mil.)	NA	NA	NA	NA	NA
Armed forces (000)	NA	NA	10	10	10
Gross national product (GNP)					
Current dollars (mil.)[e]	NA	NA	3,000	2,240	1,900
1994 constant dollars (mil.)[e]	NA	NA	3,128	2,286	1,900
Central government expenditures (CGE)					
1994 constant dollars (mil.)	NA	NA	NA	NA	NA
People (mil.)	NA	NA	2.1	2.1	2.1
Military expenditure as % of GNP	NA	NA	NA	NA	NA
Military expenditure as % of CGE	NA	NA	NA	NA	NA
Military expenditure per capita (1994 $)	NA	NA	NA	NA	NA
Armed forces per 1,000 people (soldiers)	NA	NA	5.0	4.7	4.7
GNP per capita (1994 $)	NA	NA	1,507	1,082	888
Arms imports[6]					
Current dollars (mil.)	NA	NA	0	0	10
1994 constant dollars (mil.)	NA	NA	0	0	10
Arms exports[6]					
Current dollars (mil.)	NA	NA	0	0	0
1994 constant dollars (mil.)	NA	NA	0	0	0
Total imports[7]					
Current dollars (mil.)	NA	NA	NA	950	1,482
1994 constant dollars (mil.)	NA	NA	NA	970	1,482
Total exports[7]					
Current dollars (mil.)	NA	NA	NA	1,060[e]	NA
1994 constant dollars (mil.)	NA	NA	NA	1,082[e]	NA
Arms as percent of total imports[8]	NA	NA	0	0	0.7
Arms as percent of total exports[8]	NA	NA	0	0	0

Human Rights [24]

	SSTS	FL	FAPRO	PPCG	APROBC	TPW	PCPTW	STPEP	PHRFF	PRW	ASST	AFL
Observes	2			P		P	P		S	P	P	

	EAFRD	CPR	ESCR	SR	ACHR	MAAE	PVIAC	PVNAC	EAFDAW	TCIDTP	RC
Observes	P	P	P	P			P	P	P	P	P

P = Party; S = Signatory; see Appendix for meaning of abbreviations.

Labor Force

Total Labor Force [25]

591,773 (June 1994)

Labor Force by Occupation [26]

Manufacturing and mining   40%
Date of data: 1992

Unemployment Rate [27]

37% (1995 est.)

For sources, notes, and explanations, see Annotated Source Appendix, page 1061.

569

Production Sectors

Commercial Energy Production and Consumption

Data are shown in quadrillion (10^{15}) BTUs and percent for 1995
Values for hydroelectric, nuclear, geothermal, solar, and wind power refer to electrical generation.

Production [28]

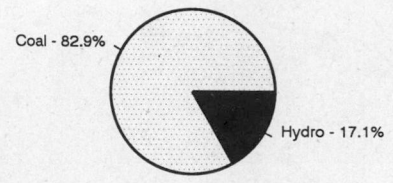

Coal - 82.9%

Hydro - 17.1%

Consumption [29]

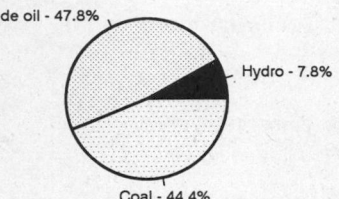

Crude oil - 47.8%

Hydro - 7.8%

Coal - 44.4%

Coal	0.034
Net hydroelectric power	0.007
Total	0.041

Crude oil	0.043
Coal	0.040
Net hydroelectric power	0.007
Total	0.090

Telecommunications [30]

- 125,000 telephones
- Radio: Broadcast stations: AM 6, FM 2, shortwave 0 Radios: 369,000 (1992 est.)
- Television: Broadcast stations: 5 (relays 2) Televisions: 327,011 (1992 est.)

Transportation [31]

Railways: total: 699 km; standard gauge: 699 km 1.435-m gauge (232 km electrified) (1995)

Highways: total: 10,591 km; paved: 5,091 km; unpaved: 5,500 km (1991 est.)

Airports

Total:	16
With paved runways 2,438 to 3,047 m:	2
With paved runways under 914 m:	12
With unpaved runways 914 to 1,523 m:	2 (1995 est.)

Top Agricultural Products [32]

Agriculture accounts for 24% of the GDP; produces rice, tobacco, wheat, corn, millet, cotton, sesame, mulberry leaves, citrus, vegetables; beef, pork, poultry, mutton.

Top Mining Products [33]

Metric tons except as noted[e]	3/95[*]
Copper ore, mine output, gr. wt. (000 tons)	2,000
Iron ore, gross weight (000 tons)	20,000
Steel, crude	90,000
Lead ore, gross weight (Pb, Zn ore)	400,000
Silver (kg.)	10,000
Cement	500,000
Zinc metal, Zn, smelter, primary	30,000
Volcanic tuff	75,000
Dim. stone, ornamental, crude (sq. meters)	200,000
Crushed and brown stone, n.e.s. (cu. meters)	400,000

Tourism [34]

	1990	1991	1992	1993	1994
Tourists	562	294	219	208	185
Tourism receipts	45	9	11	13	29

Travelers are in thousands, money in million U.S. dollars.

For sources, notes, and explanations, see Annotated Source Appendix, page 1061.

Manufacturing Sector

GDP and Manufacturing Summary [35]

	1980	1985	1989	1990	% change 1980-1990	% change 1989-1990
GDP (million 1980 $)	69,958	71,058	72,234	66,371	-5.1	-8.1
GDP per capita (1980 $)	3,136	3,073	3,050	2,786	-11.2	-8.7
Manufacturing as % of GDP (current prices)	30.6	37.2	39.5	42.0	37.3	6.3
Gross output (million $)	72,629	57,020	65,078	62,136[e]	-14.4	-4.5
Value added (million $)	21,750	17,171	30,245	27,660[e]	27.2	-8.5
Value added (million 1980 $)	19,526	22,283	24,021	21,703	11.1	-9.6
Industrial production index	100	116	120	108	8.0	-10.0
Employment (thousands)	2,106	2,467	2,658	2,537[e]	20.5	-4.6

Note: GDP stands for Gross Domestic Product. 'e' stands for estimated value.

Profitability and Productivity

	1980	1985	1989	1990	% change 1980-1990	% change 1989-1990
Intermediate input (%)	70	70	54	55[e]	-21.4	1.9
Wages, salaries, and supplements (%)	14	12	12[e]	18[e]	28.6	50.0
Gross operating surplus (%)	15	18	34[e]	26[e]	73.3	-23.5
Gross output per worker ($)	34,487	23,113	24,484	24,248[e]	-29.7	-1.0
Value added per worker ($)	10,328	6,960	11,379	10,796[e]	4.5	-5.1
Average wage (incl. benefits) ($)	4,991	2,703	2,986[e]	4,488[e]	-10.1	50.3

Profitability is in percent of gross output. Productivity is in U.S. $. 'e' stands for estimated value.

Profitability - 1990

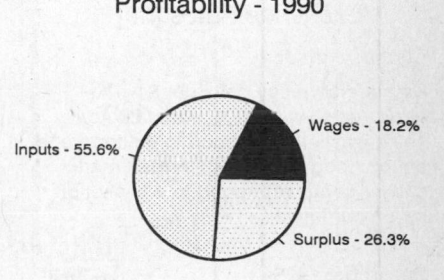

Inputs - 55.6%
Wages - 18.2%
Surplus - 26.3%

The graphic shows percent of gross output.

Value Added in Manufacturing

	1980 $ mil.	1980 %	1985 $ mil.	1985 %	1989 $ mil.	1989 %	1990 $ mil.	1990 %	% change 1980-1990	% change 1989-1990
311 Food products	1,897	8.7	1,458	8.5	3,916	12.9	3,484[e]	12.6	83.7	-11.0
313 Beverages	459	2.1	353	2.1	663	2.2	589[e]	2.1	28.3	-11.2
314 Tobacco products	184	0.8	221	1.3	344	1.1	308[e]	1.1	67.4	-10.5
321 Textiles	1,759	8.1	1,428	8.3	2,881	9.5	2,663[e]	9.6	51.4	-7.6
322 Wearing apparel	903	4.2	718	4.2	1,593	5.3	1,427[e]	5.2	58.0	-10.4
323 Leather and fur products	226	1.0	231	1.3	383	1.3	340[e]	1.2	50.4	-11.2
324 Footwear	482	2.2	503	2.9	1,022	3.4	899[e]	3.3	86.5	-12.0
331 Wood and wood products	977	4.5	530	3.1	794	2.6	706[e]	2.6	-27.7	-11.1
332 Furniture and fixtures	730	3.4	438	2.6	1,030	3.4	1,065[e]	3.9	45.9	3.4
341 Paper and paper products	529	2.4	394	2.3	759	2.5	674[e]	2.4	27.4	-11.2
342 Printing and publishing	876	4.0	462	2.7	761	2.5	678[e]	2.5	-22.6	-10.9
351 Industrial chemicals	694	3.2	631	3.7	1,107	3.7	992[e]	3.6	42.9	-10.4
352 Other chemical products	681	3.1	525	3.1	1,419	4.7	1,315[e]	4.8	93.1	-7.3
353 Petroleum refineries	454	2.1	415	2.4	260	0.9	233[e]	0.8	-48.7	-10.4
354 Miscellaneous petroleum and coal products	101	0.5	101	0.6	104	0.3	91[e]	0.3	-9.9	-12.5
355 Rubber products	276	1.3	269	1.6	479	1.6	456[e]	1.6	65.2	-4.8
356 Plastic products	413	1.9	258	1.5	397	1.3	350[e]	1.3	-15.3	-11.8
361 Pottery, china, and earthenware	128	0.6	72	0.4	162	0.5	144[e]	0.5	12.5	-11.1
362 Glass and glass products	163	0.7	113	0.7	224	0.7	204[e]	0.7	25.2	-8.9
369 Other nonmetal mineral products	906	4.2	513	3.0	683	2.3	604[e]	2.2	-33.3	-11.6
371 Iron and steel	1,221	5.6	1,000	5.8	1,343	4.4	1,171[e]	4.2	-4.1	-12.8
372 Nonferrous metals	480	2.2	509	3.0	944	3.1	927[e]	3.4	93.1	-1.8
381 Metal products	2,105	9.7	1,577	9.2	1,293	4.3	1,130[e]	4.1	-46.3	-12.6
382 Nonelectrical machinery	1,828	8.4	1,463	8.5	2,372	7.8	2,378[e]	8.6	30.1	0.3
383 Electrical machinery	1,600	7.4	1,544	9.0	2,640	8.7	2,334[e]	8.4	45.9	-11.6
384 Transport equipment	1,441	6.6	1,263	7.4	2,389	7.9	2,241[e]	8.1	55.5	-6.2
385 Professional and scientific equipment	101	0.5	93	0.5	154	0.5	146[e]	0.5	44.6	-5.2
390 Other manufacturing industries	134	0.6	88	0.5	128	0.4	114[e]	0.4	-14.9	-10.9

Note: The industry codes shown are International Standard Industry codes (ISIC). Percentages are percent of total Value Added. 'e' stands for estimated value

For sources, notes, and explanations, see Annotated Source Appendix, page 1061.

571

Finance, Economics, and Trade

Economic Indicators [36]

- **National product**: GDP—purchasing power parity—$1.9 billion (1995 est.)
- **National product real growth rate**: 4%
- **National product per capita**: $880 (1995 est.)
- **Inflation rate (consumer prices)**: 14.8% (1995 est.)
- **External debt**: $737.1 million (1994)

Balance of Payments Summary [37]

Exchange Rates [38]

Currency: **dinar.**

Data are currency units per $1. The dinar, which was adopted on 26 April 1992, was initially issued in the form of a coupon pegged to the German mark; subsequently repegged to a basket of seven currencies.

December 1995	38.8
November 1994	39
October 1992	865

Imports and Exports

Top Import Origins [39]

$199 million (1995).

Origins	%
Other former Yugoslav republics	NA
Greece	NA
Albania	NA
Germany	NA
Bulgaria	NA

Top Export Destinations [40]

$916.2 million (1995).

Destinations	%
Yugoslavia	NA
Former Yugoslav republics	NA
Germany	NA
Greece	NA
Albania	NA

Foreign Aid [41]

In December 1995, the EU agreed to provide a credit line of ECU 21.7 million for investment projects.

	U.S. $	
ODA	NA	
US (for humanitarian and technical assistance)	10	million

Import and Export Commodities [42]

Import Commodities

Fuels and lubricants 19%
Manufactured goods 18%
Machinery and transport equipment 15%
Food and live animals 14%
Chemicals 11.4%
Raw materials 10%
Miscellaneous manufactured articles 8.0%
Beverages and tobacco 3.5%

Export Commodities

Manufactured goods 40%
Machinery and transport equipment 14%
Miscellaneous manufactured articles 23%
Raw materials 7.6%
Food (rice) and live animals 5.7%
Chemicals 4.7%
Beverages and tobacco 4.5%
Chemicals 4.7%

Madagascar

Geography [1]

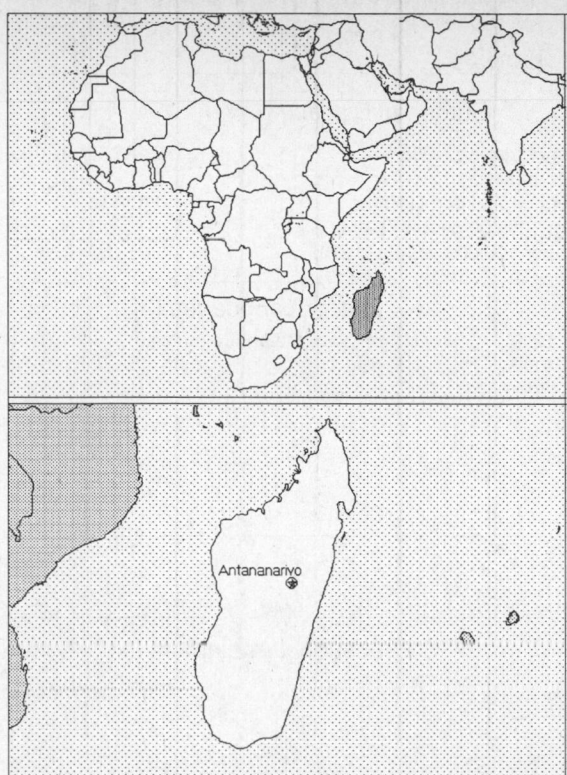

Total area:
587,040 sq km   226,657 sq mi
Land area:
581,540 sq km   224,534 sq mi
Comparative area:
Slightly less than twice the size of Arizona
Land boundaries:
0 km
Coastline:
4,828 km
Climate:
Tropical along coast, temperate inland, arid in South
Terrain:
Narrow coastal plain, high plateau and mountains in center
Natural resources:
Graphite, chromite, coal, bauxite, salt, quartz, tar sands, semiprecious stones, mica, fish
Land use:
Arable land: 4%
Permanent crops: 1%
Meadows and pastures: 58%
Forest and woodland: 26%
Other: 11%

Demographics [2]

	1970	1980	1990	1995[1]	1996	2000	2010	2020	2030
Population	6,766	8,678	11,525	13,289	13,671	15,295	20,096	25,988	32,828
Population density (persons per sq. mi.)	30	39	51	59	61	68	89	116	146
(persons per sq. km.)	12	15	20	23	24	26	35	45	56
Net migration rate (per 1,000 population)	NA	0.0	0.0	0.0	0.0	0.0	0.0	0.0	0.0
Births	NA	NA	NA	NA	583	NA	NA	NA	NA
Deaths	NA	NA	NA	NA	197	NA	NA	NA	NA
Life expectancy - males	NA	46.5	49.3	50.8	51.1	52.3	55.3	58.3	61.1
Life expectancy - females	NA	49.0	50.9	52.9	53.3	54.9	59.0	62.9	66.6
Birth rate (per 1,000)	NA	45.6	44.5	43.0	42.6	41.2	37.5	33.7	29.8
Death rate (per 1,000)	NA	18.0	16.1	14.7	14.4	13.3	10.9	9.1	7.9
Women of reproductive age (15-49 yrs.)	NA	1,926	2,593	3,025	3,119	3,521	4,762	6,333	8,252
of which are currently married	NA	NA	1,546	NA	1,862	2,101	2,850	NA	NA
Fertility rate	NA	6.7	6.2	6.0	5.9	5.6	5.0	4.3	3.7

Except as noted, values for vital statistics are in thousands; life expectancy is in years.

Health

Health Indicators [3]

% of population with access to	
safe water (1990-95)	29
adequate sanitation (1990-95)	3
health services (1985-95)	65
% of 1-year-olds immunized (1990-94) against	
TB (tuberculosis)	81
DPT (diphtheria, pertussis, tetanus)	66
polio	64
measles	54
% of contraceptive prevalence (1980-94)	17
ORT use rate (1990-94)	29

Health Expenditures [4]

Total health expenditure, 1990 (official exchange rate)	
Millions of dollars	79
Dollars per capita	7
Health expenditures as a percentage of GDP	
Total	2.6
Public sector	1.3
Private sector	1.3
Development assistance for health	
Total aid flows (millions of dollars)[1]	17
Aid flows per capita (dollars)	1.5
Aid flows as a percentage of total health expenditure	21.5

For sources, notes, and explanations, see Annotated Source Appendix, page 1061.

Human Factors

Women and Children [5]

% of pregnant women immunized (tetanus 1990-94)	15
% of births attended by trained health personnel (1983-94)	56
Maternal mortality rate (1980-92)	660
Under-5 mortality rate (1994)	164
% under-5 moderately/severely underweight (1980-1994)	39

Burden of Disease [6]

Population per physician (1989)	8,123.56
Population per nurse	NA
Population per hospital bed (1990)	1,139.95
AIDS cases per 100,000 people (1994)	0.1
Malaria cases per 100,000 people (1992)	NA

Ethnic Division [7]

Malayo-Indonesian (Merina and related Betsileo), Cotiers (mixed African, Malayo-Indonesian, and Arab ancestry - Betsimisaraka, Tsimihety, Antaisaka, Sakalava), French, Indian, Creole, Comoran.

Religion [8]

Indigenous beliefs	52%
Christian	41%
Muslim	7%

Major Languages [9]

French (official), Malagasy (official).

Education

Public Education Expenditures [10]

Million (Franc)[39]	1980	1985	1989	1990	1993	1994
Total education expenditure	36,896	52,182	62,111	67,038	116,638	NA
as percent of GNP	4.4	2.9	1.7	1.5	1.9	NA
as percent of total govt. expend.	NA	NA	NA	NA	13.6	NA
Current education expenditure	31,548	49,806	61,071	64,964	104,064	NA
as percent of GNP	3.7	2.8	1.6	1.5	1.7	NA
as percent of current govt. expend.	NA	NA	NA	NA	NA	NA
Capital expenditure	5,348	2,377	1,040	2,074	12,574	NA

Educational Attainment [11]

Illiteracy [12]

	1985	1989	1995
Illiterate population (15+ yrs.)[2]	NA	NA	4,324
Illiteracy rate - total pop. (%)	NA	NA	54.3
Illiteracy rate - males (%)	NA	NA	40.2
Illiteracy rate - females (%)	NA	NA	68.0

Libraries [13]

Daily Newspapers [14]

	1980	1985	1990	1994
Number of papers	6	7	5	7
Circ. (000)	55	67	50	60

Culture [15]

Cinema (seats per 1,000)	NA
Annual attendance per person	0.0
Gross box office receipts (mil. Franc)	209
Museums (reporting)[12]	5
Visitors (000)	134
Annual receipts (000 Franc)	NA

Science and Technology

Scientific/Technical Forces [16]

Scientists/engineers	269
Number female	84
Technicians	956
Number female	NA
Total[1,24]	1,225

R&D Expenditures [17]

	Franc (000) 1988
Total expenditure	14,371,515
Capital expenditure	13,378,000
Current expenditure	993,515
Percent current	6.9

U.S. Patents Issued [18]

For sources, notes, and explanations, see Annotated Source Appendix, page 1061.

Government and Law

Organization of Government [19]

Long-form name:
Republic of Madagascar
Type:
Republic
Independence:
26 June 1960 (from France)
National holiday:
Independence Day, 26 June (1960)
Constitution:
19 August 1992 by national referendum
Legal system:
Based on French civil law system and traditional Malagasy law; has not accepted compulsory ICJ jurisdiction
Executive branch:
President; Prime Minister; Council of Ministers
Legislative branch:
Bicameral Parliament: Senate (Senat) and National Assembly (Assemblee Nationale)
Judicial branch:
Supreme Court (Cour Supreme); High Constitutional Court (Haute Cour Constitutionnelle)

Elections [20]

Popular National Assembly	% of seats
Committee of Living Forces (CFV)	55.1
Military Party for the Development of Madagascar (PMDM/MFM)	11.6
Confederation of Civil Societies for Development	7.8
Famima	7.2
Rally for Social Democracy (RPSD)	5.1
Various pro-Ratsiraka groups	7.2
Other	5.8

Government Expenditures [21]

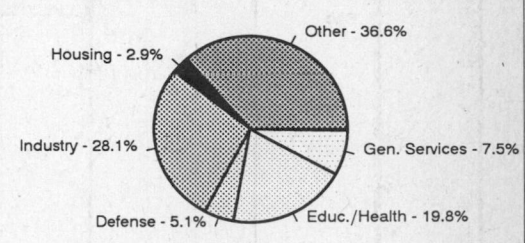

Other - 36.6%
Housing - 2.9%
Industry - 28.1%
Gen. Services - 7.5%
Defense - 5.1%
Educ./Health - 19.8%

(% distribution). Expend. for FY95: 2,344.2 (Franc bil.)

Military Expenditures and Arms Transfers [22]

	1990	1991	1992	1993	1994
Military expenditures					
Current dollars (mil.)	30	29	28[e]	29[e]	24[e]
1994 constant dollars (mil.)	33	31	29[e]	30[e]	24[e]
Armed forces (000)	21	21	21	21	21
Gross national product (GNP)					
Current dollars (mil.)	2,448	2,355	2,492	2,618	2,692
1994 constant dollars (mil.)	2,725	2,525	2,598	2,672	2,692
Central government expenditures (CGE)					
1994 constant dollars (mil.)	405	442	549	572	NA
People (mil.)	11.8	12.2	12.6	13.0	13.4
Military expenditure as % of GNP	1.2	1.2	1.1	1.1	0.9
Military expenditure as % of CGE	8.2	6.9	5.2	5.2	NA
Military expenditure per capita (1994 $)	3	3	2	2	2
Armed forces per 1,000 people (soldiers)	1.8	1.7	1.7	1.6	1.6
GNP per capita (1994 $)	231	207	206	205	200
Arms imports[6]					
Current dollars (mil.)	10	0	0	0	0
1994 constant dollars (mil.)	11	0	0	0	0
Arms exports[6]					
Current dollars (mil.)	0	0	0	0	0
1994 constant dollars (mil.)	0	0	0	0	0
Total imports[7]					
Current dollars (mil.)	571	450	448	468	434
1994 constant dollars (mil.)	635	482	467	478	434
Total exports[7]					
Current dollars (mil.)	319	305	277	261	277
1994 constant dollars (mil.)	355	327	289	266	277
Arms as percent of total imports[8]	1.8	0	0	0	0
Arms as percent of total exports[8]	0	0	0	0	0

Crime [23]

	1994
Crime volume	
Cases known to police	13,382
Attempts (percent)	24.74
Percent cases solved	75.25
Crimes per 100,000 persons	111.52
Persons responsible for offenses	
Total number offenders	7,369
Percent female	11.16
Percent juvenile (0-18 yrs.)	2.44
Percent foreigners	NA

Human Rights [24]

	SSTS	FL	FAPRO	PPCG	APROBC	TPW	PCPTW	STPEP	PHRFF	PRW	ASST	AFL
Observes	P	P	P			P	P			P	P	
	EAFRD	CPR	ESCR	SR	ACHR	MAAE	PVIAC	PVNAC	EAFDAW	TCIDTP	RC
Observes	P	P	P					P	P	P		P

P = Party; S = Signatory; see Appendix for meaning of abbreviations.

Labor Force

Total Labor Force [25]

Labor Force by Occupation [26]

Unemployment Rate [27]

Production Sectors

Commercial Energy Production and Consumption

Data are shown in quadrillion (10^{15}) BTUs and percent for 1995
Values for hydroelectric, nuclear, geothermal, solar, and wind power refer to electrical generation.

Production [28]

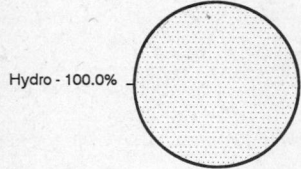

Hydro - 100.0%

Consumption [29]

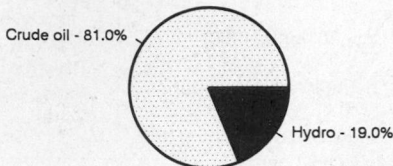

Crude oil - 81.0%

Hydro - 19.0%

Net hydroelectric power	0.004
Total	0.004

Crude oil	0.017
Net hydroelectric power	0.004
Total	0.021

Telecommunications [30]

- 96,000 (1988 est.) telephones; system is above average for Africa
- Domestic: open-wire lines, coaxial cables, microwave radio relay, and tropospheric scatter links
- International: submarine cable to Bahrain; satellite earth stations - 1 Intelsat (Indian Ocean) and 1 Intersputnik (Atlantic Ocean Region)
- Radio: Broadcast stations: AM 17, FM 3, shortwave 0 Radios: 2.565 million (1992 est.)
- Television: Broadcast stations: 1 (repeaters 36) Televisions: 260,000 (1992 est.)

Transportation [31]

Railways: total: 883 km; narrow gauge: 883 km 1.000-m gauge (1994)

Highways: total: 34,750 km; paved: 5,352 km; unpaved: 29,398 km (1991 est.)

Merchant marine: total: 11 ships (1,000 GRT or over) totaling 22,132 GRT/31,261 DWT; ships by type: cargo 5, chemical tanker 1, liquefied gas tanker 1, oil tanker 2, roll-on/roll-off cargo 2 (1995 est.)

Airports

Total:	105
With paved runways over 3,047 m:	1
With paved runways 2,438 to 3,047 m:	2
With paved runways 1,524 to 2,437 m:	3
With paved runways 914 to 1,523 m:	21
With paved runways under 914 m:	31

Top Agricultural Products [32]

Agriculture accounts for 35% of the GDP; produces coffee, vanilla, sugarcane, cloves, cocoa, rice, cassava (tapioca), beans, bananas, peanuts; livestock products.

Top Mining Products [33]

Metric tons except as noted[e]	8/95[*]
Chromite	91,000
Cement, hydraulic	60,000
Clay, kaolin	700
Graphite, all grades	8,000
Mica, phlogopite	774
Salt, marine	30,000
Calcite, industrial	2,000
Dimension stone	3,000
Petroleum refinery products (000 42-gal. bls.)	2,110

Tourism [34]

	1990	1991	1992	1993	1994
Visitors	82	62	89	87	NA
Tourists[12]	53	35	54	55	66
Tourism receipts	40	27	39	41	54
Tourism expenditures	40	32	21	25	23
Fare receipts	38	32	34	33	39
Fare expenditures	20	17	22	18	17

Travelers are in thousands, money in million U.S. dollars.

For sources, notes, and explanations, see Annotated Source Appendix, page 1061.

Manufacturing Sector

GDP and Manufacturing Summary [35]

	1980	1985	1989	1990	% change 1980-1990	% change 1989-1990
GDP (million 1980 $)	3,265	3,086	3,364	3,514	7.6	4.5
GDP per capita (1980 $)	372	301	289	293	-21.2	1.4
Manufacturing as % of GDP (current prices)	11.9[e]	10.2[e]	8.1[e]	12.4	4.2	53.1
Gross output (million $)	569	328	276[e]	353[e]	-38.0	27.9
Value added (million $)	221	132	103[e]	147[e]	-33.5	42.7
Value added (million 1980 $)	365	244	480[e]	282	-22.7	-41.3
Industrial production index	100	81	98[e]	101	1.0	3.1
Employment (thousands)	41	47	45[e]	46[e]	12.2	2.2

Note: GDP stands for Gross Domestic Product. 'e' stands for estimated value.

Profitability and Productivity

	1980	1985	1989	1990	% change 1980-1990	% change 1989-1990
Intermediate input (%)	61	60	63[e]	58[e]	-4.9	-7.9
Wages, salaries, and supplements (%)	15	16	13[e]	13[e]	-13.3	0.0
Gross operating surplus (%)	24	25	25[e]	28[e]	16.7	12.0
Gross output per worker ($)	14,005	6,891	6,170[e]	7,672[e]	-45.2	24.3
Value added per worker ($)	5,439	2,798	2,310[e]	3,200[e]	-41.2	38.5
Average wage (incl. benefits) ($)	2,083	1,099	786[e]	1,010[e]	-51.5	28.5

Profitability is in percent of gross output. Productivity is in U.S. $. 'e' stands for estimated value.

Profitability - 1990

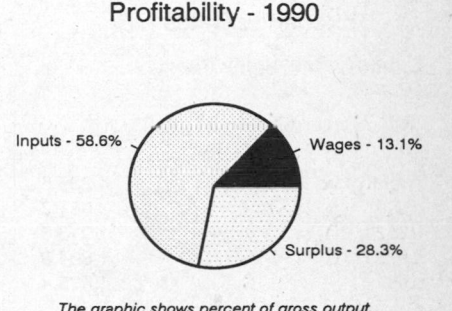

Inputs - 58.6%
Wages - 13.1%
Surplus - 28.3%

The graphic shows percent of gross output.

Value Added in Manufacturing

	1980 $ mil.	1980 %	1985 $ mil.	1985 %	1989 $ mil.	1989 %	1990 $ mil.	1990 %	% change 1980-1990	% change 1989-1990
311 Food products	23	10.4	45	34.1	18[e]	17.5	22[e]	15.0	-4.3	22.2
313 Beverages	34	15.4	16	12.1	16[e]	15.5	16[e]	10.9	-52.9	0.0
314 Tobacco products	3	1.4	3	2.3	1[e]	1.0	1[e]	0.7	-66.7	0.0
321 Textiles	67	30.3	16	12.1	33[e]	32.0	60[e]	40.8	-10.4	81.8
322 Wearing apparel	19	8.6	6	4.5	2[e]	1.9	4[e]	2.7	-78.9	100.0
323 Leather and fur products	3	1.4	1	0.8	1[e]	1.0	1[e]	0.7	-66.7	0.0
324 Footwear	8	3.6	5	3.8	3[e]	2.9	3[e]	2.0	-62.5	0.0
331 Wood and wood products	2	0.9	1	0.8	NA	0.0	NA	0.0	NA	NA
332 Furniture and fixtures	2	0.9	1	0.8	NA	0.0	NA	0.0	NA	NA
341 Paper and paper products	4	1.8	3	2.3	4[e]	3.9	5[e]	3.4	25.0	25.0
342 Printing and publishing	6	2.7	2	1.5	1[e]	1.0	1[e]	0.7	-83.3	0.0
351 Industrial chemicals	1	0.5	1	0.8	NA	0.0	NA	0.0	NA	NA
352 Other chemical products	10	4.5	11	8.3	8[e]	7.8	9[e]	6.1	-10.0	12.5
353 Petroleum refineries	11[e]	5.0	7[e]	5.3	6[e]	5.8	9[e]	6.1	-18.2	50.0
354 Miscellaneous petroleum and coal products	NA	0.0	NA	0.0	NA	0.0	NA	0.0	NA	NA
355 Rubber products	1	0.5	1	0.8	NA	0.0	1[e]	0.7	0.0	NA
356 Plastic products	3	1.4	2	1.5	1[e]	1.0	1[e]	0.7	-66.7	0.0
361 Pottery, china, and earthenware	NA	0.0	NA	0.0	NA	0.0	NA	0.0	NA	NA
362 Glass and glass products	2	0.9	NA	0.0	NA	0.0	1[e]	0.7	-50.0	NA
369 Other nonmetal mineral products	2[e]	0.9	1	0.8	2[e]	1.9	3[e]	2.0	50.0	50.0
371 Iron and steel	NA	0.0	NA	0.0	NA	0.0	NA	0.0	NA	NA
372 Nonferrous metals	NA	0.0	NA	0.0	NA	0.0	NA	0.0	NA	NA
381 Metal products	9	4.1	5	3.8	3[e]	2.9	4[e]	2.7	-55.6	33.3
382 Nonelectrical machinery	NA	0.0	NA	0.0	NA	0.0	NA	0.0	NA	NA
383 Electrical machinery	3	1.4	3	2.3	2[e]	1.9	2[e]	1.4	-33.3	0.0
384 Transport equipment	7	3.2	2[e]	1.5	1[e]	1.0	1[e]	0.7	-85.7	0.0
385 Professional and scientific equipment	NA	0.0	NA	0.0	NA	0.0	NA	0.0	NA	NA
390 Other manufacturing industries	2	0.9	1	0.8	NA	0.0	NA	0.0	NA	NA

Note: The industry codes shown are International Standard Industry codes (ISIC). Percentages are percent of total Value Added. 'e' stands for estimated value

For sources, notes, and explanations, see Annotated Source Appendix, page 1061.

577

Finance, Economics, and Trade

Economic Indicators [36]

- **National product**: GDP—purchasing power parity—$11.4 billion (1995 est.)
- **National product real growth rate**: 2.7% (1995 est.)
- **National product per capita**: $820 (1995 est.)
- **Inflation rate (consumer prices)**: 35% (1994 est.)
- **External debt**: $4.3 billion (1993 est.)

Balance of Payments Summary [37]

Values in millions of dollars.

	1988	1989	1990	1991	1992
Exports of goods (f.o.b.)	284.0	321.0	319.0	338.0	328.0
Imports of goods (f.o.b.)	-319.0	-320.0	-566.0	-440.0	-466.0
Trade balance	-35.0	1.0	-247.0	-102.0	-138.0
Services - debits	-443.0	-437.0	-450.0	-416.0	-411.0
Services - credits	132.0	161.0	209.0	151.0	177.0
Private transfers (net)	38.0	42.0	49.0	52.0	88.0
Government transfers (net)	158.0	161.0	188.0	127.0	148.0
Long-term capital (net)	209.0	215.0	141.0	47.0	-68.0
Short-term capital (net)	-36.0	-64.0	-15.0	155.0	274.0
Errors and omissions	53.0	-46.0	-8.0	2.0	-53.0
Overall balance	76.0	33.0	-133.0	16.0	17.0

Exchange Rates [38]

Currency: **Malagasy franc.**
Symbol: **FMG.**

Data are currency units per $1.

November 1995	4,239.5
1994	3,067.3
1993	1,913.8
1992	1,864.0
1991	1,835.4

Imports and Exports

Top Import Origins [39]

$510 million (f.o.b., 1993 est.).

Origins	%
France	NA
Germany	NA
Japan	NA
UK	NA
Italy	NA
Netherlands	NA

Top Export Destinations [40]

$240 million (f.o.b., 1993 est.).

Destinations	%
France	NA
US	NA
Germany	NA
Japan	NA
Russia	NA

Foreign Aid [41]

Recipient: ODA, $318 million (1993).

Import and Export Commodities [42]

Import Commodities	Export Commodities
Intermediate manufactures 30%	Coffee 45%
Capital goods 28%	Vanilla 20%
Petroleum 15%	Cloves 11%
Consumer goods 14%	Shellfish
Food 13%	Sugar
	Petroleum products

For sources, notes, and explanations, see Annotated Source Appendix, page 1061.

Malawi

Geography [1]

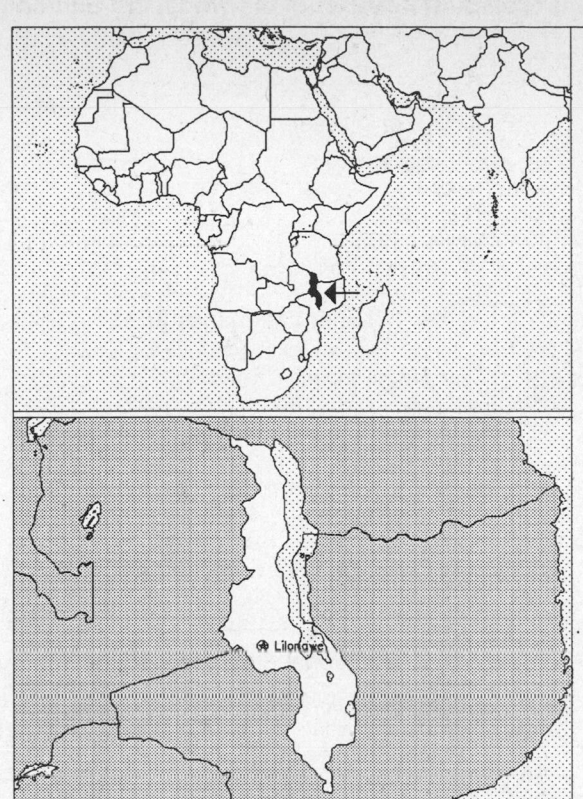

Total area:
118,480 sq km   45,745 sq mi
Land area:
94,080 sq km   36,324 sq mi
Comparative area:
Slightly larger than Pennsylvania
Land boundaries:
Total 2,881 km, Mozambique 1,569 km, Tanzania 475 km, Zambia 837 km
Coastline:
0 km (landlocked)
Climate:
Tropical; rainy season (November to May); dry season (May to November)
Terrain:
Narrow elongated plateau with rolling plains, rounded hills,
some mountains
Natural resources:
Limestone, unexploited deposits of uranium, coal, and bauxite
Land use:
Arable land: 25%
Permanent crops: 0%
Meadows and pastures: 20%
Forest and woodland: 50%
Other: 5%

Demographics [2]

	1970	1980	1990	1995[1]	1996	2000	2010	2020	2030
Population	4,489	6,129	9,136	9,446	9,453	10,011	10,662	10,719	11,370
Population density (persons per sq. mi.)	124	169	252	260	260	276	294	295	313
(persons per sq. km.)	48	65	97	100	100	106	113	114	121
Net migration rate (per 1,000 population)	NA	0.0	11.5	-34.2	0.0	0.0	0.0	0.0	0.0
Births	NA	NA	NA	NA	393	NA	NA	NA	NA
Deaths	NA	NA	NA	NA	231	NA	NA	NA	NA
Life expectancy - males	NA	42.3	41.7	37.0	35.9	32.4	30.8	35.4	50.0
Life expectancy - females	NA	44.0	42.6	37.3	36.5	33.6	28.1	33.4	52.2
Birth rate (per 1,000)	NA	53.1	48.3	42.4	41.6	38.7	32.6	27.7	24.2
Death rate (per 1,000)	NA	24.2	21.6	23.9	24.5	27.1	31.8	26.3	13.4
Women of reproductive age (15-49 yrs.)	NA	1,379	2,078	2,132	2,136	2,303	2,601	2,782	3,211
of which are currently married[32]	745	NA	1,479	NA	1,508	1,611	1,812	NA	NA
Fertility rate	NA	7.7	6.9	6.1	5.9	5.3	3.9	3.0	2.4

Except as noted, values for vital statistics are in thousands; life expectancy is in years.

Health

Health Indicators [3]

% of population with access to	
safe water (1990-95)[1]	47
adequate sanitation (1990-95)	53
health services (1985-95)	80
% of 1-year-olds immunized (1990-94) against	
TB (tuberculosis)	99
DPT (diphtheria, pertussis, tetanus)	98
polio	98
measles	98
% of contraceptive prevalence (1980-94)	13
ORT use rate (1990-94)	50

Health Expenditures [4]

Total health expenditure, 1990 (official exchange rate)	
Millions of dollars	93
Dollars per capita	11
Health expenditures as a percentage of GDP	
Total	5.0
Public sector	2.9
Private sector	2.1
Development assistance for health	
Total aid flows (millions of dollars)[1]	22
Aid flows per capita (dollars)	2.5
Aid flows as a percentage of total health expenditure	23.3

For sources, notes, and explanations, see Annotated Source Appendix, page 1061.

Human Factors

Women and Children [5]

% of pregnant women immunized (tetanus 1990-94)	76
% of births attended by trained health personnel (1983-94)	55
Maternal mortality rate (1980-92)	620
Under-5 mortality rate (1994)	221
% under-5 moderately/severely underweight (1980-1994)	27

Burden of Disease [6]

Population per physician (1990)	45,736.56
Population per nurse (1990)	1,800.04
Population per hospital bed (1990)	644.86
AIDS cases per 100,000 people (1994)	49.2
Malaria cases per 100,000 people (1992)	NA

Ethnic Division [7]

Chewa, Nyanja, Tumbuko, Yao, Lomwe, Sena, Tonga, Ngoni, Ngonde, Asian, European.

Religion [8]

Protestant	55%
Roman Catholic	20%
Muslim	20%
Traditional indigenous beliefs	5%

Major Languages [9]

English (official), Chichewa (official), other languages important regionally.

Education

Public Education Expenditures [10]

	1980	1985	1987	1990	1992	1994
Million (Kwacha)						
Total education expenditure	31	64	85	165	NA	NA
as percent of GNP	3.4	3.5	3.4	3.3	NA	NA
as percent of total govt. expend.	8.4	9.6	9.0	10.3	NA	NA
Current education expenditure	24	46	70	117	227	NA
as percent of GNP	2.6	2.5	2.8	2.4	3.5	NA
as percent of current govt. expend.	NA	NA	9.6	9.8	NA	NA
Capital expenditure	8	18	15	48	NA	NA

Educational Attainment [11]

Age group (1987)	25+
Total population	2,859,826
Highest level attained (%)	
No schooling	55.0
First level	
Not completed	31.8
Completed	8.0
Entered second level	
S-1	2.7
S-2	2.1
Postsecondary	0.4

Illiteracy [12]

In thousands and percent[1]	1990	1995	2000
Illiterate population (15+ yrs.)	2,366	2,587	2,587
Illiteracy rate - total pop. (%)	48.8	51.0	46.8
Illiteracy rate - males (%)	31.6	32.7	29.9
Illiteracy rate - females (%)	64.4	68.3	63.1

Libraries [13]

	Admin. Units	Svc. Pts.	Vols. (000)	Shelving (meters)	Vols. Added	Reg. Users
National	NA	NA	NA	NA	NA	NA
Nonspecialized	NA	NA	NA	NA	NA	NA
Public (1992)	1	7	237	3,000	18,223	36,976
Higher ed. (1988)	5	5	310	NA	7,362	4,669
School (1987)	1	75	2	NA	500	283

Daily Newspapers [14]

	1980	1985	1990	1994
Number of papers	2[e]	1	1	1
Circ. (000)	20[e]	15	25	25

Culture [15]

Science and Technology

Scientific/Technical Forces [16]

R&D Expenditures [17]

U.S. Patents Issued [18]

For sources, notes, and explanations, see Annotated Source Appendix, page 1061.

Government and Law

Organization of Government [19]

Long-form name:
Republic of Malawi
Type:
Multiparty democracy
Independence:
6 July 1964 (from UK)
National holiday:
Independence Day 6 July (1964);
Republic Day 6 July (1966)
Constitution:
18 May 1995; most recent revision
Legal system:
Based on English common law and
customary law; judicial review of
legislative acts in the Supreme Court of
Appeal; has not accepted compulsory ICJ
jurisdiction
Executive branch:
President; Cabinet
Legislative branch:
Unicameral: National Assembly Note: The
constitution of 18 May 1995 provided for a
bicameral legislature; by 1999, in addition
to the existing National Assembly, a
Senate of 80 seats is to be elected
Judicial branch:
High Court; Supreme Court of Appeal

Elections [20]

National Assembly	% of seats
United Democratic Front (UDF)	49.1
Malawi Congress Party (MCP)	31.1
Alliance for Democracy (AFORD)	19.8

Government Expenditures [21]

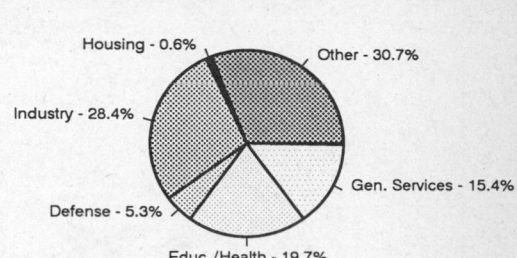

Housing - 0.6%
Other - 30.7%
Industry - 28.4%
Gen. Services - 15.4%
Defense - 5.3%
Educ./Health - 19.7%

(% distribution). Expend. for FY91 est.: 510 (Dollar mil.)

Military Expenditures and Arms Transfers [22]

	1990	1991	1992	1993	1994
Military expenditures					
Current dollars (mil.)[e]	15	15	13	16	13
1994 constant dollars (mil.)[e]	17	16	14	17	13
Armed forces (000)	7	8	10	10	10
Gross national product (GNP)					
Current dollars (mil.)	1,150	1,302	1,241	1,391	1,252
1994 constant dollars (mil.)	1,280	1,396	1,294	1,419	1,252
Central government expenditures (CGE)					
1994 constant dollars (mil.)	348	361[e]	354[e]	NA	NA
People (mil.)	9.3	9.6	9.8	9.8	9.7
Military expenditure as % of GNP	1.3	1.1	1.1	1.2	1.1
Military expenditure as % of CGE	4.8	4.4	3.9	NA	NA
Military expenditure per capita (1994 $)	2	2	1	2	1
Armed forces per 1,000 people (soldiers)	0.8	0.8	1.0	1.0	1.0
GNP per capita (1994 $)	138	145	132	144	129
Arms imports[6]					
Current dollars (mil.)	0	0	0	20	0
1994 constant dollars (mil.)	0	0	0	20	0
Arms exports[6]					
Current dollars (mil.)	0	0	0	0	0
1994 constant dollars (mil.)	0	0	0	0	0
Total imports[7]					
Current dollars (mil.)	581	703	718	546	NA
1994 constant dollars (mil.)	647	754	749	557	NA
Total exports[7]					
Current dollars (mil.)	417	472	383	320	NA
1994 constant dollars (mil.)	464	506	399	327	NA
Arms as percent of total imports[8]	0	0	0	3.7	0
Arms as percent of total exports[8]	0	0	0	0	0

Crime [23]

	1992
Crime volume	
Cases known to police	85,025
Attempts (percent)	NA
Percent cases solved	29.79
Crimes per 100,000 persons	850.25
Persons responsible for offenses	
Total number offenders	25,150
Percent female	3.77
Percent juvenile (7-17 yrs.)	0.96
Percent foreigners	NA

Human Rights [24]

	SSTS	FL	FAPRO	PPCG	APROBC	TPW	PCPTW	STPEP	PHRFF	PRW	ASST	AFL
Observes	P		S		P	P	P	P		P	P	
	EAFRD	CPR	ESCR	SR	ACHR	MAAE	PVIAC	PVNAC	EAFDAW	TCIDTP	RC	
Observes		P	P	P			P	P	P		P	

P = Party; S = Signatory; see Appendix for meaning of abbreviations.

Labor Force

Total Labor Force [25]

428,000 wage earners

Labor Force by Occupation [26]

Agriculture	43%
Manufacturing	16
Personal services	15
Commerce	9
Construction	7
Miscellaneous services	4
Other permanently employed	6

Date of data: 1986

Unemployment Rate [27]

For sources, notes, and explanations, see Annotated Source Appendix, page 1061.

581

Production Sectors

Commercial Energy Production and Consumption

Data are shown in quadrillion (10^{15}) BTUs and percent for 1995
Values for hydroelectric, nuclear, geothermal, solar, and wind power refer to electrical generation.

Production [28]

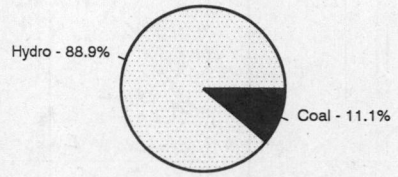

Hydro - 88.9%
Coal - 11.1%

Consumption [29]

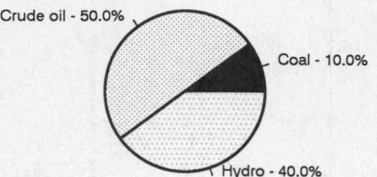

Crude oil - 50.0%
Coal - 10.0%
Hydro - 40.0%

Coal	0.001
Net hydroelectric power	0.008
Total	0.009

Crude oil	0.010
Coal	0.002
Net hydroelectric power	0.008
Total	0.020

Telecommunications [30]

- 43,000 (1985 est.) telephones
- Domestic: fair system of open-wire lines, microwave radio relay links, and radiotelephone communications stations
- International: satellite earth stations - 2 Intelsat (1 Indian Ocean and 1 Atlantic Ocean)
- Radio: Broadcast stations: AM 10, FM 17, shortwave 0 Radios: 1.011 million (1995)
- Television: Broadcast stations: 0 (1987 est.)

Transportation [31]

Railways: total: 789 km; narrow gauge: 789 km 1.067-m gauge

Highways: total: 27,294 km (1990 est.); paved: NA km; unpaved: NA km

Airports

Total:	41
With paved runways over 3,047 m:	1
With paved runways 1,524 to 2,437 m:	1
With paved runways 914 to 1,523 m:	4
With paved runways under 914 m:	20
With unpaved runways 1,524 to 2,437 m:	1

Top Agricultural Products [32]

Agriculture accounts for 31% of the GDP; produces tobacco, sugarcane, cotton, tea, corn, potatoes, cassava (tapioca), sorghum, pulses; cattle, goats.

Top Mining Products [33]

Metric tons except as noted[e]	3/30/95[*]
Cement, hydraulic	130,000
Coal	55,000
Dolomite	2,000
Lime	2,600
Limestone for cement	140,000
Rubiy and sapphire (kg.)	125
Stone, crushed for aggregate	380,000

Tourism [34]

	1990	1991	1992	1993	1994
Tourists[72]	130	127	150	153	NA
Tourism receipts	16	12	8	7	5
Tourism expenditures	16	27	24	11	15
Fare receipts	11	8	4	4	6
Fare expenditures	8	12	13	21	13

Travelers are in thousands, money in million U.S. dollars.

Manufacturing Sector

GDP and Manufacturing Summary [35]

	1980	1985	1989	1990	% change 1980-1990	% change 1989-1990
GDP (million 1980 $)	1,245	1,403	1,496	1,627	30.7	8.8
GDP per capita (1980 $)	201	191	177	186	-7.5	5.1
Manufacturing as % of GDP (current prices)	12.6	12.6	15.6	13.5	7.1	-13.5
Gross output (million $)	340	330	458[e]	586[e]	72.4	27.9
Value added (million $)	123	90	110[e]	133[e]	8.1	20.9
Value added (million 1980 $)	149	155	204	196	31.5	-3.9
Industrial production index	100	116	140	146	46.0	4.3
Employment (thousands)	39	31	41[e]	46[e]	17.9	12.2

Note: GDP stands for Gross Domestic Product. 'e' stands for estimated value.

Profitability and Productivity

	1980	1985	1989	1990	% change 1980-1990	% change 1989-1990
Intermediate input (%)	64	73	76[e]	77[e]	20.3	1.3
Wages, salaries, and supplements (%)	12	10	10[e]	10[e]	-16.7	0.0
Gross operating surplus (%)	24	18	14[e]	13[e]	-45.8	-7.1
Gross output per worker ($)	8,783	10,745	11,097[e]	12,767[e]	45.4	15.0
Value added per worker ($)	3,174	2,923	2,671[e]	3,041[e]	-4.2	13.9
Average wage (incl. benefits) ($)	1,046	1,035	1,116[e]	1,244[e]	18.9	11.5

Profitability is in percent of gross output. Productivity is in U.S. $. 'e' stands for estimated value.

Profitability - 1990

Inputs - 77.0%
Wages - 10.0%
Surplus - 13.0%

The graphic shows percent of gross output.

Value Added in Manufacturing

	1980 $ mil.	1980 %	1985 $ mil.	1985 %	1989 $ mil.	1989 %	1990 $ mil.	1990 %	% change 1980-1990	% change 1989-1990
311 Food products	54	43.9	14	15.6	18[e]	16.4	26[e]	19.5	-51.9	44.4
313 Beverages	8	6.5	7	7.8	9[e]	8.2	12[e]	9.0	50.0	33.3
314 Tobacco products	9	7.3	5	5.6	7[e]	6.4	8[e]	6.0	-11.1	14.3
321 Textiles	12	9.8	14	15.6	15[e]	13.6	18[e]	13.5	50.0	20.0
322 Wearing apparel	2	1.6	1	1.1	1[e]	0.9	1[e]	0.8	-50.0	0.0
323 Leather and fur products	NA	0.0	NA	0.0	NA	0.0	NA	0.0	NA	NA
324 Footwear	1[e]	0.8	3	3.3	4[e]	3.6	4[e]	3.0	300.0	0.0
331 Wood and wood products	2	1.6	2	2.2	1[e]	0.9	2[e]	1.5	0.0	100.0
332 Furniture and fixtures	1	0.8	1	1.1	1[e]	0.9	1[e]	0.8	0.0	0.0
341 Paper and paper products	2	1.6	2	2.2	NA	0.0	1[e]	0.8	-50.0	NA
342 Printing and publishing	8	6.5	6	6.7	7[e]	6.4	9[e]	6.8	12.5	28.6
351 Industrial chemicals	2	1.6	8	8.9	6[e]	5.5	5[e]	3.8	150.0	-16.7
352 Other chemical products	5	4.1	14	15.6	24[e]	21.8	23[e]	17.3	360.0	-4.2
353 Petroleum refineries	NA	0.0	NA	0.0	NA	0.0	NA	0.0	NA	NA
354 Miscellaneous petroleum and coal products	NA	0.0	NA	0.0	NA	0.0	NA	0.0	NA	NA
355 Rubber products	1	0.8	1	1.1	NA	0.0	NA	0.0	NA	NA
356 Plastic products	2	1.6	2	2.2	4[e]	3.6	5[e]	3.8	150.0	25.0
361 Pottery, china, and earthenware	NA	0.0	NA	0.0	NA	0.0	NA	0.0	NA	NA
362 Glass and glass products	NA	0.0	NA	0.0	NA	0.0	NA	0.0	NA	NA
369 Other nonmetal mineral products	3	2.4	1	1.1	7[e]	6.4	8[e]	6.0	166.7	14.3
371 Iron and steel	NA	0.0	NA	0.0	NA	0.0	NA	0.0	NA	NA
372 Nonferrous metals	NA	0.0	NA	0.0	NA	0.0	NA	0.0	NA	NA
381 Metal products	6	4.9	6	6.7	3[e]	2.7	5[e]	3.8	-16.7	66.7
382 Nonelectrical machinery	NA	0.0	1	1.1	2[e]	1.8	3[e]	2.3	NA	50.0
383 Electrical machinery	5	4.1	1	1.1	1[e]	0.9	1[e]	0.8	-80.0	0.0
384 Transport equipment	1[e]	0.8	1	1.1	1[e]	0.9	1[e]	0.8	0.0	0.0
385 Professional and scientific equipment	NA	0.0	NA	0.0	NA	0.0	NA	0.0	NA	NA
390 Other manufacturing industries	NA	0.0	NA	0.0	NA	0.0	NA	0.0	NA	NA

Note: The industry codes shown are International Standard Industry codes (ISIC). Percentages are percent of total Value Added. 'e' stands for estimated value

For sources, notes, and explanations, see Annotated Source Appendix, page 1061.

583

Finance, Economics, and Trade

Economic Indicators [36]

- **National product**: GDP—purchasing power parity—$6.9 billion (1995 est.)
- **National product real growth rate**: 9.9% (1995 est.)
- **National product per capita**: $700 (1995 est.)
- **Inflation rate (consumer prices)**: 83.3% (1995 est.)
- **External debt**: $1.95 billion (December 1994 est.)

Balance of Payments Summary [37]

Values in millions of dollars.

	1975	1980	1985	1987	1988
Exports of goods (f.o.b.)	138.7	280.8	245.5	278.5	297.0
Imports of goods (f.o.b.)	-225.0	-308.0	-176.7	-177.6	-253.0
Trade balance	-86.3	-27.2	68.8	100.9	44.0
Services - debits	-76.6	-329.9	266.8	-243.8	-230.3
Services - credits	73.9	34.0	37.7	43.7	37.8
Private transfers (net)	2.5	13.3	11.1	13.8	15.1
Government transfers (net)	6.8	50.2	25.2	30.3	80.5
Long-term capital (net)	57.2	181.6	5.8	92.5	171.9
Short-term capital (net)	33.6	-29.7	-2.6	6.6	6.7
Errors and omissions	-27.3	86.0	102.8	24.2	-18.0
Overall balance	-16.2	-21.7	-18.0	68.2	107.7

Exchange Rates [38]

Currency: **Malawian kwacha.**
Symbol: **MK.**

Data are currency units per $1.

November 1995	16.3516
1994	8.7364
1993	4.4028
1992	3.6033
1991	2.8033

Imports and Exports

Top Import Origins [39]

$240 million (c.i.f., 1994).

Origins	%
South Africa	NA
Japan	NA
US	NA
UK	NA
Zimbabwe	NA

Top Export Destinations [40]

$365 million (f.o.b., 1994) Data are for 1994.

Destinations	%
US	NA
South Africa	NA
Germany	NA
Japan	NA

Foreign Aid [41]

Recipient: In December 1995, donors pledged $332 million for 1996.

Import and Export Commodities [42]

Import Commodities	Export Commodities
Food	Tobacco
Petroleum products	Tea
Semimanufactures	Sugar
Consumer goods	Coffee
Transportation equipment	Peanuts
	Wood products

For sources, notes, and explanations, see Annotated Source Appendix, page 1061.

Malaysia

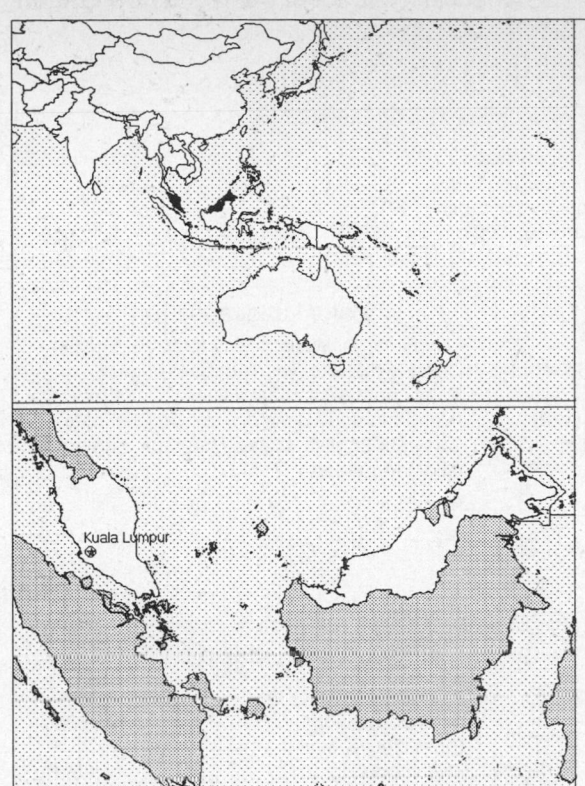

Geography [1]

Total area:
329,750 sq km   127,317 sq mi
Land area:
328,550 sq km   126,854 sq mi
Comparative area:
Slightly larger than New Mexico
Land boundaries:
Total 2,669 km, Brunei 381 km, Indonesia 1,782 km, Thailand 506 km
Coastline:
4,675 km (Peninsular Malaysia 2,068 km, East Malaysia 2,607 km)
Climate:
Tropical; annual Southwest (April to October) and Northeast (October to February) monsoons
Terrain:
Coastal plains rising to hills and mountains
Natural resources:
Tin, petroleum, timber, copper, iron ore, natural gas, bauxite
Land use:
Arable land: 3%
Permanent crops: 10%
Meadows and pastures: 0%
Forest and woodland: 63%
Other: 24%

Demographics [2]

	1970	1980	1990	1995[1]	1996	2000	2010	2020	2030
Population	10,910	13,764	17,507	19,550	19,963	21,610	25,691	29,830	33,685
Population density (persons per sq. mi.)	86	109	138	154	157	170	203	235	266
(persons per sq. km.)	33	42	53	60	61	66	78	91	103
Net migration rate (per 1,000 population)	0.0	0.0	0.0	0.0	0.0	0.0	0.0	0.0	0.0
Births	NA	NA	NA	NA	523	NA	NA	NA	NA
Deaths	NA	NA	83	NA	110	NA	NA	NA	NA
Life expectancy - males	58.6	62.0	65.1	66.6	66.8	67.9	70.3	72.4	74.1
Life expectancy - females	62.1	66.7	70.7	72.6	72.9	74.3	77.2	79.5	81.4
Birth rate (per 1,000)	32.4	31.1	29.1	26.7	26.2	24.1	21.1	19.1	16.7
Death rate (per 1,000)	9.0	7.5	6.1	5.6	5.5	5.2	5.1	5.4	6.1
Women of reproductive age (15-49 yrs.)	NA	3,437	4,518	5,039	5,155	5,629	6,724	7,584	8,413
of which are currently married[33]	1,460	1,931	2,845	NA	3,350	3,637	4,342	NA	NA
Fertility rate	5.3	4.0	3.5	3.3	3.3	3.1	2.7	2.5	2.3

Except as noted, values for vital statistics are in thousands; life expectancy is in years.

Health

Health Indicators [3]

% of population with access to	
safe water (1990-95)	78
adequate sanitation (1990-95)	94
health services (1985-95)	NA
% of 1-year-olds immunized (1990-94) against	
TB (tuberculosis)	97
DPT (diphtheria, pertussis, tetanus)	90
polio	90
measles	81
% of contraceptive prevalence (1980-94)	48
ORT use rate (1990-94)	47

Health Expenditures [4]

Total health expenditure, 1990 (official exchange rate)	
Millions of dollars	1,259
Dollars per capita	67
Health expenditures as a percentage of GDP	
Total	3.0
Public sector	1.3
Private sector	1.7
Development assistance for health	
Total aid flows (millions of dollars)[1]	3
Aid flows per capita (dollars)	0.1
Aid flows as a percentage of total health expenditure	0.2

For sources, notes, and explanations, see Annotated Source Appendix, page 1061.

585

Human Factors

Women and Children [5]

% of pregnant women immunized (tetanus 1990-94)	NA
% of births attended by trained health personnel (1983-94)	87
Maternal mortality rate (1980-92)	59
Under-5 mortality rate (1994)	15
% under-5 moderately/severely underweight (1980-1994)	23

Burden of Disease [6]

Population per physician (1991)	2,441.04
Population per nurse (1990)	480.00
Population per hospital bed (1991)	437.00
AIDS cases per 100,000 people (1994)	0.4
Malaria cases per 100,000 people (1992)	196

Ethnic Division [7]

Malay and other indigenous	59%
Chinese	32%
Indian	9%

Religion [8]

Peninsular Malaysia: Muslim (Malays), Buddhist (Chinese), Hindu (Indians). Sabah: Muslim 38%, Christian 17%, other 45%. Sarawak: Tribal religion 35%, Buddhist and Confucianist 24%, Muslim 20%, Christian 16%, other 5%.

Major Languages [9]

Peninsular Malaysia: Malay (official), English, Chinese dialects, Tamil. Sabah: English, Malay, numerous tribal dialects, Chinese (Mandarin and Hakka dialects predominate). Sarawak: English, Malay, Mandarin, numerous tribal languages.

Education

Public Education Expenditures [10]

Million (Ringgit)[40]	1980	1985	1990	1992	1993	1994
Total education expenditure	3,104	4,754	6,033	7,702	8,074	9,363
as percent of GNP	6.0	6.6	5.4	5.5	5.2	5.3
as percent of total govt. expend.	14.7	16.3	18.3	16.9	15.7	15.5
Current education expenditure	2,575	4,062	4,664	6,656	7,085	7,821
as percent of GNP	5.0	5.6	4.2	4.7	4.6	4.4
as percent of current govt. expend.	18.4	NA	19.3	19.6	17.2	15.9
Capital expenditure	529	692	1,369	1,046	989	1,542

Educational Attainment [11]

Age group (1980)	25+
Total population	5,146,888
Highest level attained (%)	
No schooling	36.6
First level	
Not completed	21.1
Completed	21.1
Entered second level	
S-1	18.1
S-2	1.3
Postsecondary	1.9

Illiteracy [12]

In thousands and percent[1]	1990	1995	2000
Illiterate population (15+ yrs.)	2,190	2,057	1,891
Illiteracy rate - total pop. (%)	19.7	16.5	13.3
Illiteracy rate - males (%)	13.1	11.0	9.1
Illiteracy rate - females (%)	26.2	21.8	17.5

Libraries [13]

	Admin. Units	Svc. Pts.	Vols. (000)	Shelving (meters)	Vols. Added	Reg. Users
National (1992)	1	90	858	NA	81,033	134,956
Nonspecialized	NA	NA	NA	NA	NA	NA
Public (1992)	13	350	8,144	NA	1 mil	1 mil
Higher ed. (1990)	48	73	3,412	NA	383,873	NA
School	NA	NA	NA	NA	NA	NA

Daily Newspapers [14]

	1980	1985	1990	1994
Number of papers	40	32	45	44
Circ. (000)	810[e]	1,500[e]	2,500[e]	2,800[e]

Culture [15]

Cinema (seats per 1,000)	8.2
Annual attendance per person	2.0
Gross box office receipts (mil. Ringgit)	NA
Museums (reporting)	1
Visitors (000)	759
Annual receipts (000 Ringgit)	6,300

Science and Technology

Scientific/Technical Forces [16]

Scientists/engineers	1,633
Number female	501
Technicians	1,655
Number female	507
Total	3,288

R&D Expenditures [17]

	Ringgit (000) 1992
Total expenditure	550,699
Capital expenditure	212,520
Current expenditure	338,179
Percent current	61.4

U.S. Patents Issued [18]

Values show patents issued to citizens of the country by the U.S. Patents Office.

	1993	1994	1995
Number of patents	19	16	8

For sources, notes, and explanations, see Annotated Source Appendix, page 1061.

Government and Law

Organization of Government [19]

Long-form name:
None
Type:
Constitutional monarchy Note: 1963 Federation headed by king and bicameral Parliament; Peninsular states, except Melaka, have hereditary rulers; Sabah and Sarawak self governing, significant powers delegated to federal government
Independence:
31 August 1957 (from UK)
National holiday:
National Day, 31 August (1957)
Constitution:
31 August 1957, amended 16 September 1963
Legal system:
Based on English common law; judicial review of legislative acts in the Supreme Court at request of supreme head
Executive branch:
Paramount Ruler; Deputy Paramount Ruler; Prime Minister; Deputy Prime Minister; Cabinet
Legislative branch:
Bicameral Parliament: Senate and House of Representatives
Judicial branch:
Supreme Court

Crime [23]

	1994
Crime volume	
Cases known to police	90,102
Attempts (percent)	NA
Percent cases solved	29.25
Crimes per 100,000 persons	151.11
Persons responsible for offenses	
Total number offenders	31,775
Percent female	1.77
Percent juvenile (10-18 yrs.)	4.37
Percent foreigners	2.56

Elections [20]

House of Representatives	% of votes
National Front	63
Other	37

Government Expenditures [21]

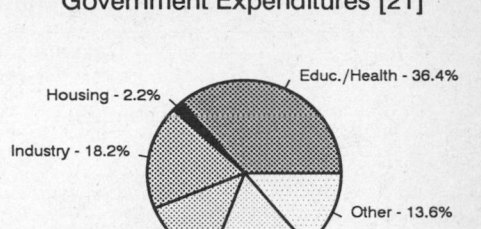

Educ./Health - 36.4%
Housing - 2.2%
Industry - 18.2%
Other - 13.6%
Defense - 12.7%
Gen. Services - 16.9%

(% distribution). Expend. for CY96 est.: 51,296 (Ringgit mil.)

Military Expenditures and Arms Transfers [22]

	1990	1991	1992	1993	1994
Military expenditures					
Current dollars (mil.)	1,208	1,727	1,745	1,934	2,121[e]
1994 constant dollars (mil.)	1,344	1,851	1,820	1,974	2,121[e]
Armed forces (000)	130	128	128	115	115
Gross national product (GNP)					
Current dollars (mil.)	43,950	49,350	54,400	60,460	67,170
1994 constant dollars (mil.)	48,910	52,900	56,730	61,710	67,170
Central government expenditures (CGE)					
1994 constant dollars (mil.)	15,470	16,990	18,400	16,480	17,600
People (mil.)	17.6	18.0	18.4	18.8	19.3
Military expenditure as % of GNP	2.7	3.5	3.2	3.2	3.2
Military expenditure as % of CGE	8.7	10.9	9.9	12.0	12.0
Military expenditure per capita (1994 $)	77	103	99	105	110
Armed forces per 1,000 people (soldiers)	7.4	7.1	7.0	6.1	6.0
GNP per capita (1994 $)	2,786	2,942	3,081	3,274	3,483
Arms imports[6]					
Current dollars (mil.)	40	100	90	110	330
1994 constant dollars (mil.)	45	107	94	112	330
Arms exports[6]					
Current dollars (mil.)	0	0	0	0	50
1994 constant dollars (mil.)	0	0	0	0	50
Total imports[7]					
Current dollars (mil.)	29,260	36,650	39,960	45,660	59,580
1994 constant dollars (mil.)	32,560	39,280	41,670	46,600	59,580
Total exports[7]					
Current dollars (mil.)	29,420	34,350	40,710	47,120	58,760
1994 constant dollars (mil.)	32,740	36,820	42,460	48,090	58,760
Arms as percent of total imports[8]	0.1	0.3	0.2	0.2	0.6
Arms as percent of total exports[8]	0	0	0	0	0.1

Human Rights [24]

	SSTS	FL	FAPRO	PPCG	APROBC	TPW	PCPTW	STPEP	PHRFF	PRW	ASST	AFL
Observes		P		P	P	P	P				P	P
	EAFRD	CPR	ESCR	SR	ACHR	MAAE	PVIAC	PVNAC	EAFDAW	TCIDTP	RC
Observes									P		P

P = Party; S = Signatory; see Appendix for meaning of abbreviations.

Labor Force

Total Labor Force [25]

7.627 million (1993)

Labor Force by Occupation [26]

Unemployment Rate [27]

2.8% (1995 est.)

For sources, notes, and explanations, see Annotated Source Appendix, page 1061.

587

Production Sectors

Commercial Energy Production and Consumption

Data are shown in quadrillion (10^{15}) BTUs and percent for 1995
Values for hydroelectric, nuclear, geothermal, solar, and wind power refer to electrical generation.

Production [28]

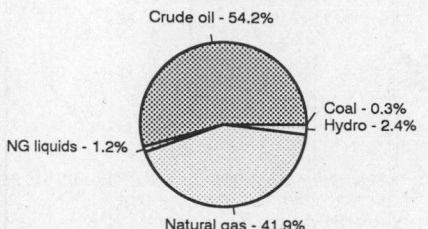

Crude oil - 54.2%
Coal - 0.3%
Hydro - 2.4%
NG liquids - 1.2%
Natural gas - 41.9%

Consumption [29]

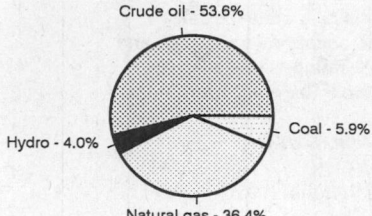

Crude oil - 53.6%
Coal - 5.9%
Hydro - 4.0%
Natural gas - 36.4%

Crude oil	1.419
Natural gas liquids	0.032
Dry natural gas	1.097
Coal	0.007
Net hydroelectric power	0.062
Total	2.617

Crude oil	0.832
Dry natural gas	0.565
Coal	0.092
Net hydroelectric power	0.062
Total	1.551

Telecommunications [30]

- 2,550,957 (1992 est.) telephones; international service good
- Domestic: good intercity service provided on Peninsular Malaysia mainly by microwave radio relay; adequate intercity microwave radio relay network between Sabah and Sarawak via Brunei; domestic satellite system with 2 earth stations
- International: submarine cables to India, Hong Kong and Singapore; satellite earth stations - 2 Intelsat (1 Indian Ocean and 1 Pacific Ocean)
- Radio: Broadcast stations: AM 28, FM 3, shortwave 0 Radios: 8.08 million (1992 est.)
- Television: Broadcast stations: 33 Televisions: 2 million (1993 est.)

Top Agricultural Products [32]

Agriculture accounts for 8% of the GDP. Peninsular Malaysia produces natural rubber, palm oil and rice. The island state of Sabah produces subsistence crops, rubber, timber, coconut, and rice. The island state of Sarawak produces rubber, pepper, and timber.

Top Mining Products [33]

Thousand metric tons except as noted[p]	4/28/95[*]
Bauxite, gross weight	162
Gold, mine output, Au content (kg.)	4,080
Steel, crude	1,850[e]
Silver, mine output, Ag content (kg.)	13,300
Cement, hydraulic	9,970[e]
Limestone	24,000[e]
Gas, natural, gross (mil. cu. meters)	31,500[e]
Petroleum (000 42-gal. bls.)[51]	
crude	238,000
refinery products	87,900[e]

Transportation [31]

Railways: total: 1,806 km (Peninsular Malaysia 1,672 km; Sabah 134 km; Sarawak 0 km); narrow gauge: 1,806 km 1.000-m gauge (Peninsular Malaysia 1,672 km; Sabah 134 km)

Highways: total: 92,545 km; paved: 69,409 km (including 574 km of expressways); unpaved: 23,136 km (1992 est.)

Merchant marine: total: 248 ships (1,000 GRT or over) totaling 3,035,684 GRT/4,494,476 DWT; ships by type: bulk 43, cargo 83, chemical tanker 13, container 31, liquefied gas tanker 12, livestock carrier 1, oil tanker 55, roll-on/roll-off cargo 5, short-sea passenger 1, vehicle carrier 4 (1995 est.)

Airports
Total:	105
With paved runways over 3,047 m:	3
With paved runways 2,438 to 3,047 m:	5
With paved runways 1,524 to 2,437 m:	11
With paved runways 914 to 1,523 m:	6
With paved runways under 914 m:	74

Tourism [34]

	1990	1991	1992	1993	1994
Visitors	15,957	12,042	NA	NA	NA
Tourists[73]	7,446	5,847	6,016	6,504	7,197
Tourism receipts	1,667	1,530	1,768	1,876	3,189
Tourism expenditures	1,450	1,584	1,770	1,838	1,737
Fare receipts	447	591	678	682	835
Fare expenditures	243	292	318	350	365

Travelers are in thousands, money in million U.S. dollars.

                For sources, notes, and explanations, see Annotated Source Appendix, page 1061.

Manufacturing Sector

Manufacturing Summary [35]

	1987 $ bil.	%	1988 $ bil.	%	1989 $ bil.	%	1990 $ bil.	%	1991 $ bil.	%
Establishments or enterprises (number)	6,219	0.265	6,274	0.299	6,651	0.355	7,400	0.413	-	-
Total employment (000)	644	0.475	753	0.550	880	0.715	1,073	0.969	-	-
Production workers (000)	587	0.871	658	1.054	824	1.120	1,004	1.415	-	-
Output ($ bil.)	26	0.252	32	0.273	39	0.330	47	0.411	-	-
Value added ($ bil.)	7	0.160	8	0.169	10	0.205	12	0.242	-	-
Capital investment ($ mil.)	2,062	0.473	1,846	0.388	2,606	0.476	4,504	0.810	-	-
M & E investment ($ mil.)	981	0.315	1,255	0.363	1,772	0.454	3,183	0.755	-	-
Employees per establishment (number)	104	179.105	120	184.047	132	201.457	145	234.564	-	-
Production workers per establishment	94	328.461	105	352.545	124	315.530	136	342.532	-	-
Output per establishment ($ mil.)	4	95.234	5	91.345	6	92.905	6	99.430	-	-
Capital investment per estab. ($ mil.)	0.332	178.448	0.294	129.692	0.392	133.989	0.609	196.174	-	-
M & E per establishment ($ mil)	0.158	118.717	0.200	121.299	0.266	127.801	0.430	182.773	-	-
Payroll per employee ($)	3,947	44.012	3,806	38.201	3,713	33.232	3,871	27.903	-	-
Wages per production worker ($)	3,111	39.427	3,296	39.002	2,966	29.576	3,109	26.179	-	-
Hours per production worker (hours)	-	-	-	-	-	-	-	-	-	-
Output per employee ($)	39,823	53.172	42,274	49.632	44,198	46.117	43,930	42.389	-	-
Capital investment per employee ($)	3,200	99.633	2,452	70.467	2,960	66.510	4,200	83.633	-	-
M & E per employee ($)	1,522	66.283	1,667	65.907	2,012	63.438	2,968	77.920	-	-

Note: Columns headed % show percent of world total or ratio. Ratios closest to 100 are closest to world average. M & E stands for machinery & equipment.

Output in Manufacturing

	1987 $ bil.[f]	%	1988 $ bil.[f]	%	1989 $ bil.[f]	%	1990 $ bil.[f]	%	1991 $ bil.[f]	%
3110 - Food products	4.538	17.45	5.853	18.29	6.260	16.05	5.920	12.60	-	-
3130 - Beverages	0.237	0.91	0.249	0.78	0.272	0.70	0.374	0.80	-	-
3140 - Tobacco	0.444	1.71	0.295	0.92	0.347	0.89	0.323	0.69	-	-
3210 - Textiles	0.680	2.62	0.754	2.36	0.852	2.18	1.054	2.24	-	-
3211 - Spinning, weaving, etc.	0.515	1.98	0.527	1.65	0.575	1.47	0.659	1.40	-	-
3220 - Wearing apparel	0.470	1.81	0.551	1.72	0.703	1.80	0.880	1.87	-	-
3230 - Leather and products	0.009	0.03	0.011	0.03	0.012	0.03	0.026	0.06	-	-
3240 - Footwear	0.010	0.04	0.008	0.02	0.014	0.04	0.012	0.03	-	-
3310 - Wood products	1.024	3.94	1.191	3.72	1.536	3.94	1.951	4.15	-	-
3320 - Furniture, fixtures	0.096	0.37	0.123	0.38	0.145	0.37	0.199	0.42	-	-
3410 - Paper and products	0.227	0.87	0.312	0.98	0.425	1.09	0.484	1.03	-	-
3411 - Pulp, paper, etc.	0.031	0.12	0.076	0.24	0.123	0.32	0.117	0.25	-	-
3420 - Printing, publishing	0.362	1.39	0.398	1.24	0.458	1.17	0.554	1.18	-	-
3510 - Industrial chemicals	1.366	5.25	1.553	4.85	1.636	4.19	1.796	3.82	-	-
3511 - Basic chemicals, excl fertilizers	1.028	3.95	1.100	3.44	1.143	2.93	1.234	2.63	-	-
3513 - Synthetic resins, etc.	0.064	0.25	0.139	0.43	0.133	0.34	0.227	0.48	-	-
3520 - Chemical products nec	0.414	1.59	0.491	1.53	0.569	1.46	0.664	1.41	-	-
3522 - Drugs and medicines	0.061	0.23	0.065	0.20	0.077	0.20	0.080	0.17	-	-
3530 - Petroleum refineries	1.071	4.12	0.977	3.05	1.289	3.31	1.727	3.67	-	-
3540 - Petroleum, coal products	0.044	0.17	0.057	0.18	0.062	0.16	0.082	0.17	-	-
3550 - Rubber products	1.731	6.66	2.398	7.49	2.111	5.41	1.953	4.16	-	-
3560 - Plastic products nec	0.327	1.26	0.431	1.35	0.564	1.45	0.753	1.60	-	-
3610 - Pottery, china, etc.	0.035	0.13	0.046	0.14	0.068	0.17	0.070	0.15	-	-
3620 - Glass and products	0.075	0.29	0.085	0.27	0.111	0.28	0.148	0.31	-	-
3690 - Nonmetal products nec	0.529	2.03	0.607	1.90	0.755	1.94	0.942	2.00	-	-
3710 - Iron and steel	0.673	2.59	0.873	2.73	1.178	3.02	1.466	3.12	-	-
3720 - Nonferrous metals	0.261	1.00	0.319	1.00	0.378	0.97	0.460	0.98	-	-
3810 - Metal products	0.482	1.85	0.692	2.16	0.913	2.34	1.191	2.53	-	-
3820 - Machinery nec	0.457	1.76	0.617	1.93	0.783	2.01	1.090	2.32	-	-
3825 - Office, computing machinery	0.012	0.05	0.006	0.02	0.010	0.03	0.118	0.25	-	-
3830 - Electrical machinery	3.775	14.52	4.927	15.40	6.759	17.33	8.986	19.12	-	-
3832 - Radio, television, etc.	3.412	13.12	4.394	13.73	6.031	15.46	7.882	16.77	-	-
3840 - Transportation equipment	0.531	2.04	0.770	2.41	1.200	3.08	1.701	3.62	-	-
3841 - Shipbuilding, repair	0.079	0.30	0.045	0.14	0.080	0.21	0.100	0.21	-	-
3843 - Motor vehicles	0.339	1.30	0.575	1.80	0.910	2.33	1.274	2.71	-	-
3850 - Professional goods	0.127	0.49	0.163	0.51	0.229	0.59	0.335	0.71	-	-
3900 - Industries nec	0.127	0.49	0.144	0.45	0.201	0.52	0.282	0.60	-	-

Note: Codes are International Standard Industry codes (ISIC). Percentages are % of total Output. [f]: Factor Prices; [p]: Producer Prices.

For sources, notes, and explanations, see Annotated Source Appendix, page 1061.

Finance, Economics, and Trade

Economic Indicators [36]

- **National product**: GDP—purchasing power parity—$193.6 billion (1995 est.)
- **National product real growth rate**: 9.5% (1995)
- **National product per capita**: $9,800 (1995 est.)
- **Inflation rate (consumer prices)**: 5.3% (1995)
- **External debt**: $27.4 billion (1995 est.)

Balance of Payments Summary [37]

Values in millions of dollars.

	1989	1990	1991	1992	1993
Exports of goods (f.o.b.)	24,633	28,636	33,534	39,613	45,984
Imports of goods (f.o.b.)	-20,251	-26,014	-33,007	-36,238	-42,801
Trade balance	4,382	2,622	527	3,375	3,183
Services - debits	-8,390	-9,472	-10,776	-12,385	-13,402
Services - credits	4,185	5,878	5,978	7,042	7,594
Private transfers (net)	-17	3	29	65	84
Government transfers (net)	97	51	8	67	75
Long-term capital (net)	754	1,283	3,754	4,090	5,595
Short-term capital (net)	576	503	1,868	4,694	5,412
Errors and omissions	-358	1,085	-151	-292	2,802
Overall balance	1,229	1,953	1,237	6,656	11,343

Exchange Rates [38]

Currency: **ringgit.**
Symbol: **M$.**

Data are currency units per $1.

January 1996	2.5567
1995	2.5044
1994	2.6243
1993	2.5741
1992	2.5474
1991	2.7501

Imports and Exports

Top Import Origins [39]

$72.2 billion (1995) Data are for 1993.

Origins	%
Japan	26
US	17
Singapore	14
Taiwan	5
Germany	4
UK	3
South Korea	3

Top Export Destinations [40]

$72 billion (1995) Data are for 1994.

Destinations	%
Singapore	21
US	20
Japan	12
UK	4
Thailand	4
Germany	3

Foreign Aid [41]

Recipient: ODA, $45 million (1993).

Import and Export Commodities [42]

Import Commodities
Machinery and equipment
Chemicals
Food
Petroleum products

Export Commodities
Electronic equipment
Petroleum and petroleum products
Palm oil
Wood and wood products
Rubber
Textiles

Maldives

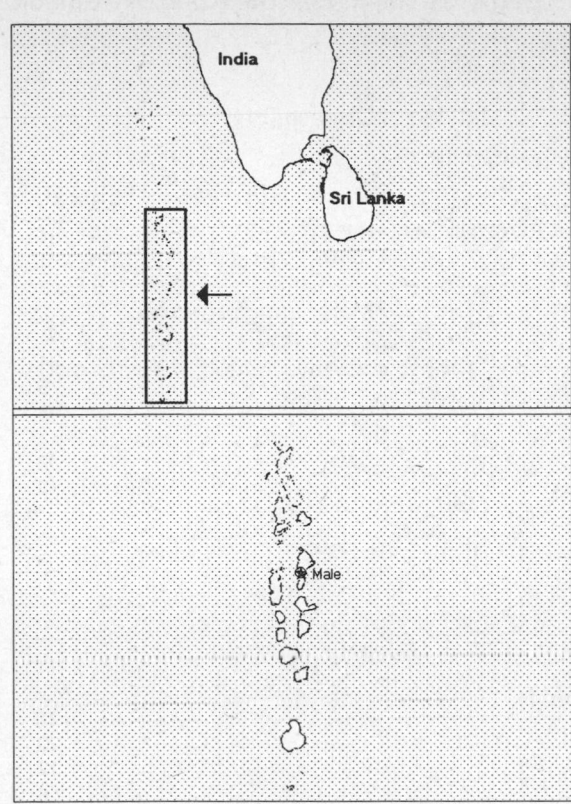

Geography [1]

Total area:
  300 sq km   116 sq mi
Land area:
  300 sq km   116 sq mi
Comparative area:
  Nearly twice the size of Washington, DC
Land boundaries:
  0 km
Coastline:
  644 km
Climate:
  Tropical; hot, humid; dry, northeast monsoon (November to March);
  rainy, southwest monsoon (June to August)
Terrain:
  Flat
Natural resources:
  Fish
Land use:
  Arable land: 10%
  Permanent crops: 0%
  Meadows and pastures: 3%
  Forest and woodland: 3%
  Other: 84%

Demographics [2]

	1970	1980	1990	1995[1]	1996	2000	2010	2020	2030
Population	115	154	218	261	271	310	423	554	692
Population density (persons per sq. mi.)	989	1,329	1,880	2,256	2,338	2,680	3,651	4,785	5,971
(persons per sq. km.)	382	513	726	871	903	1,035	1,410	1,847	2,305
Net migration rate (per 1,000 population)	NA	NA	0.0	0.0	0.0	0.0	0.0	0.0	0.0
Births	NA	NA	NA	NA	11	NA	NA	NA	NA
Deaths	NA	NA	NA	NA	2	NA	NA	NA	NA
Life expectancy - males	NA	NA	60.4	64.0	64.6	67.2	72.3	75.8	77.9
Life expectancy - females	NA	NA	62.7	67.1	67.8	71.0	77.1	81.2	83.7
Birth rate (per 1,000)	NA	NA	46.2	42.8	41.9	38.5	32.6	27.6	22.4
Death rate (per 1,000)	NA	NA	9.3	7.0	6.6	5.3	3.5	2.9	2.8
Women of reproductive age (15-49 yrs.)	NA	NA	47	55	57	67	100	137	179
of which are currently married	NA	NA	35	NA	43	50	74	NA	NA
Fertility rate	NA	NA	6.6	6.2	6.1	5.6	4.4	3.4	2.7

Except as noted, values for vital statistics are in thousands; life expectancy is in years.

Health

Health Indicators [3]

Health Expenditures [4]

For sources, notes, and explanations, see Annotated Source Appendix, page 1061.

591

Human Factors

Women and Children [5]	Burden of Disease [6]	
	Population per physician (1990)	14,400.00
	Population per nurse (1990)	158.47
	Population per hospital bed	NA
	AIDS cases per 100,000 people (1994)	0.5
	Malaria cases per 100,000 people (1992)	NA

Ethnic Division [7]	Religion [8]	Major Languages [9]
Sinhalese, Dravidian, Arab, African.	Sunni Muslim.	Maldivian Divehi (dialect of Sinhala, script derived from Arabic), English spoken by most government officials.

Education

Public Education Expenditures [10]

Million (Rufiyaa)	1980	1990	1991	1992	1993	1994
Total education expenditure	NA	79	176	220	183	NA
as percent of GNP	NA	6.3	11.6	12.0	8.1	NA
as percent of total govt. expend.	NA	10.0	16.0	16.0	13.6	NA
Current education expenditure	NA	NA	NA	NA	132	NA
as percent of GNP	NA	NA	NA	NA	5.8	NA
as percent of current govt. expend.	NA	NA	NA	NA	18.0	NA
Capital expenditure	NA	NA	NA	NA	51	NA

Educational Attainment [11]

Illiteracy [12]

In thousands and percent[1]	1990	1995	2000
Illiterate population (15+ yrs.)	9	9	10
Illiteracy rate - total pop. (%)	7.6	6.5	6.1
Illiteracy rate - males (%)	8.1	7.0	6.0
Illiteracy rate - females (%)	7.0	7.5	6.2

Libraries [13]

	Admin. Units	Svc. Pts.	Vols. (000)	Shelving (meters)	Vols. Added	Reg. Users
National (1986)	1	2	NA	337	720	703
Nonspecialized	NA	NA	NA	NA	NA	NA
Public	NA	NA	NA	NA	NA	NA
Higher ed.	NA	NA	NA	NA	NA	NA
School	NA	NA	NA	NA	NA	NA

Daily Newspapers [14]

	1980	1985	1990	1994
Number of papers	2	2	2[e]	2
Circ. (000)	1	2	3[e]	3

Culture [15]

Science and Technology

Scientific/Technical Forces [16]	R&D Expenditures [17]	U.S. Patents Issued [18]

For sources, notes, and explanations, see Annotated Source Appendix, page 1061.

Government and Law

Organization of Government [19]

Long-form name:
Republic of Maldives
Type:
Republic
Independence:
26 July 1965 (from UK)
National holiday:
Independence Day, 26 July (1965)
Constitution:
4 June 1968
Legal system:
Based on Islamic law with admixtures of English common law primarily in commercial matters; has not accepted compulsory ICJ jurisdiction
Executive branch:
President; Ministry of Atolls
Legislative branch:
Unicameral: Citizens' Council (Majlis)
Judicial branch:
High Court

Elections [20]

Citizens' Council. Elections last held 2 December 1994 (next to be held NA December 1994); results - percent of vote NA; seats - (48 total, 40 elected, 8 appointed by the president) independents 40. Although political parties are not banned, none exist; country governed by the Didi clan for the past eight centuries.

Government Expenditures [21]

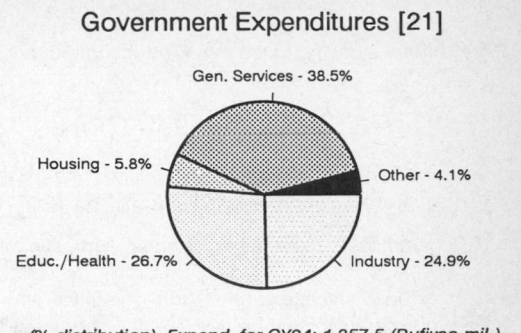

(% distribution). Expend. for CY94: 1,357.5 (Rufiyaa mil.)

Defense Summary [22]

Branches: National Security Service (paramilitary police force)

Manpower Availability: Males age 15-49 59,179; males fit for military service 33,016 (1996 est.)

Defense Expenditures: NA

Crime [23]

	1990
Crime volume	
Cases known to police	5,018
Attempts (percent)	NA
Percent cases solved	99.38
Crimes per 100,000 persons	2,353.49
Persons responsible for offenses	
Total number offenders	1,885
Percent female	11.56
Percent juvenile (0-16 yrs.)	12.41
Percent foreigners	93

Human Rights [24]

	SSTS	FL	FAPRO	PPCG	APROBC	TPW	PCPTW	STPEP	PHRFF	PRW	ASST	AFL
Observes				P		P	P					
	EAFRD	CPR	ESCR	SR	ACHR	MAAE	PVIAC	PVNAC	EAFDAW	TCIDTP	RC	
Observes	P						P	P	P			P

P=Party; S=Signatory; see Appendix for meaning of abbreviations.

Labor Force

Total Labor Force [25]

66,000 (est.)

Labor Force by Occupation [26]

Fishing industry   25%

Unemployment Rate [27]

Negligible

For sources, notes, and explanations, see Annotated Source Appendix, page 1061.

593

Production Sectors

Energy Resource Summary [28]

Electricity: Capacity: 5,000 kW. Production: 30 million kWh. Consumption per capita: 123 kWh (1993).

Telecommunications [30]

- 8,523 (1992 est.) telephones; minimal domestic and international facilities
- International: satellite earth station - 1 Intelsat (Indian Ocean)
- Radio: Broadcast stations: AM 2, FM 1, shortwave 0 Radios: 28,284 (1992 est.)
- Television: Broadcast stations: 1 Televisions: 7,309 (1992 est.)

Top Agricultural Products [32]

Agriculture accounts for 21.5% of the GDP; produces coconuts, corn, sweet potatoes; fishing.

Top Mining Products [33]

Mineral Resources: None.

Transportation [31]

Railways: 0 km

Highways: total: NA km; paved: NA km; unpaved: NA km; note - Male has 9.6 km of coral highways within the city (1988 est.)

Merchant marine: total: 20 ships (1,000 GRT or over) totaling 73,284 GRT/113,669 DWT; ships by type: cargo 17, container 2, oil tanker 1 (1995 est.)

Airports

Total: 2
With paved runways over 3,047 m: 1
With paved runways 2,438 to 3,047 m: 1 (1995 est.)

Tourism [34]

	1990	1991	1992	1993	1994
Tourists[20]	195	196	236	241	280
Tourism receipts	89	95	138	146	181
Tourism expenditures	15	19	22	29	32
Fare expenditures	6	8	8	8	NA

Travelers are in thousands, money in million U.S. dollars.

For sources, notes, and explanations, see Annotated Source Appendix, page 1061.

Finance, Economics, and Trade

Industrial Summary [35]

Industrial Production: Growth rate 6.3% (1994 est.); accounts for 15.3% of the GDP. **Industries**: Fish processing, tourism, shipping, boat building, coconut processing, garments, woven mats, rope, handicrafts, coral and sand mining.

Economic Indicators [36]

- **National product**: GDP—purchasing power parity— $390 million (1994 est.)
- **National product real growth rate**: 6.6% (1994 est.)
- **National product per capita**: $1,560 (1994 est.)
- **Inflation rate (consumer prices)**: 16.5% (1994 est.)
- **External debt**: $137.5 million (1994 est.)

Balance of Payments Summary [37]

Values in millions of dollars.

	1989	1990	1991	1992	1993
Exports of goods (f.o.b.)	51.3	58.1	59.2	51.1	38.5
Imports of goods (f.o.b.)	-111.3	-121.2	-141.8	-167.9	-177.8
Trade balance	-60.0	-63.1	-82.6	-116.8	-139.3
Services - debits	-45.2	-56.5	-60.8	-69.4	-77.8
Services - credits	102.6	124.4	128.9	171.1	181.2
Private transfers (net)	-5.1	-7.4	-16.6	-18.9	-20.0
Government transfers (net)	18.3	11.2	22.1	14.3	8.3
Long-term capital (net)	5.5	10.7	17.7	23.0	9.1
Short-term capital (net)	6.4	-2.6	-12.3	2.5	15.8
Errors and omissions	-19.8	-17.0	2.6	-1.0	22.6
Overall balance	2.7	-0.3	-1.0	4.8	-0.1

Exchange Rates [38]

Currency: **rufiyaa.**
Symbol: **Rf.**

Data are currency units per $1.

January 1996	11.770
1995	11.770
1994	11.586
1993	10.957
1992	10.569
1991	10.253

Imports and Exports

Top Import Origins [39]

$195.1 million (f.o.b., 1994 est.).

Origins	%
Singapore	NA
India	NA
Sri Lanka	NA
Hong Kong	NA
Japan	NA
Thailand	NA

Top Export Destinations [40]

$75.3 million (f.o.b., 1994 est.).

Destinations	%
Sri Lanka	NA
US	NA
Germany	NA
Singapore	NA
UK	NA

Foreign Aid [41]

Recipient: ODA, $NA.

Import and Export Commodities [42]

Import Commodities	Export Commodities
Consumer goods	Fish
Intermediate and capital goods	Clothing
Petroleum products	

Mali

Geography [1]

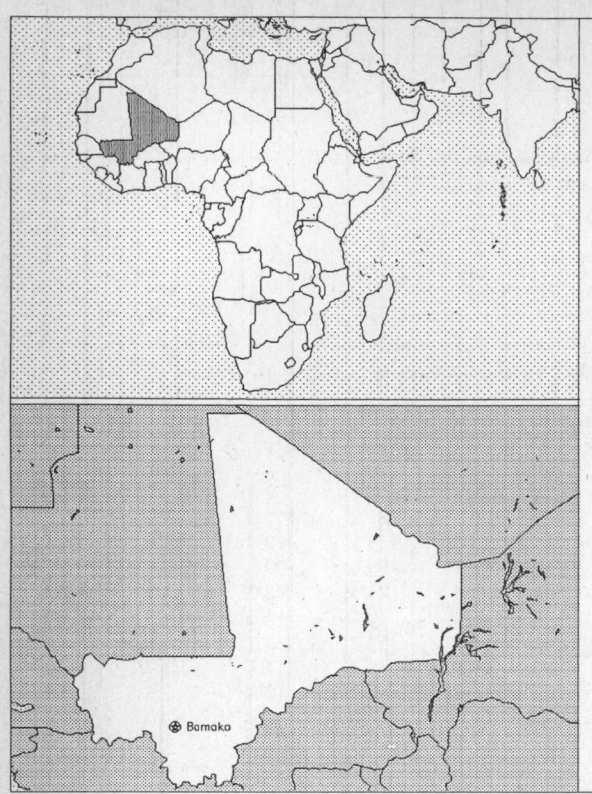

Total area:
1,248,574 sq km   482,077 sq mi
Land area:
1,220,000 sq km   471,045 sq mi
Comparative area:
Slightly less than twice the size of Texas
Land boundaries:
Total 7,243 km, Algeria 1,376 km, Burkina Faso 1,000 km, Guinea 858 km, Cote d'Ivoire 532 km, Mauritania 2,237 km, Niger 821 km, Senegal 419 km
Coastline:
0 km (landlocked)
Climate:
Subtropical to arid; hot and dry February to June; rainy, humid, and mild June to November; cool and dry November to February
Terrain:
Mostly flat to rolling northern plains covered by sand; savanna in South, rugged hills in Northeast
Natural resources:
Gold, phosphates, kaolin, salt, limestone, uranium, bauxite, iron ore, manganese, tin, and copper deposits are known but not exploited
Land use:
Arable land: 2%
Permanent crops: 0%
Meadows and pastures: 25%
Forest and woodland: 7%
Other: 66%

Demographics [2]

	1970	1980	1990	1995[1]	1996	2000	2010	2020	2030
Population	5,525	6,728	8,234	9,375	9,653	10,911	14,966	20,427	27,301
Population density (persons per sq. mi.)	12	14	17	20	20	23	32	43	58
(persons per sq. km.)	5	6	7	8	8	9	12	17	22
Net migration rate (per 1,000 population)	NA	NA	-6.6	-3.0	-2.4	-0.2	0.0	0.0	0.0
Births	NA	NA	NA	NA	496	NA	NA	NA	NA
Deaths	NA	NA	NA	NA	188	NA	NA	NA	NA
Life expectancy - males	NA	NA	42.7	44.7	45.1	46.8	51.3	55.8	60.2
Life expectancy - females	NA	NA	45.6	48.1	48.6	50.7	56.2	61.7	66.8
Birth rate (per 1,000)	NA	NA	51.8	51.9	51.4	49.3	44.8	40.2	34.8
Death rate (per 1,000)	NA	NA	22.2	19.9	19.5	17.5	13.2	9.8	7.3
Women of reproductive age (15-49 yrs.)	NA	NA	1,826	2,081	2,144	2,420	3,385	4,831	6,717
of which are currently married	NA	NA	1,433	NA	1,666	1,880	2,633	NA	NA
Fertility rate	NA	NA	7.3	7.3	7.2	6.9	6.1	5.2	4.4

Except as noted, values for vital statistics are in thousands; life expectancy is in years.

Health

Health Indicators [3]

% of population with access to	
safe water (1990-95)	37
adequate sanitation (1990-95)	31
health services (1985-95)	30
% of 1-year-olds immunized (1990-94) against	
TB (tuberculosis)	67
DPT (diphtheria, pertussis, tetanus)	39
polio	39
measles	46
% of contraceptive prevalence (1980-94)	5
ORT use rate (1990-94)	10

Health Expenditures [4]

Total health expenditure, 1990 (official exchange rate)	
Millions of dollars	130
Dollars per capita	15
Health expenditures as a percentage of GDP	
Total	5.2
Public sector	2.8
Private sector	2.4
Development assistance for health	
Total aid flows (millions of dollars)[1]	36
Aid flows per capita (dollars)	4.3
Aid flows as a percentage of total health expenditure	27.7

For sources, notes, and explanations, see Annotated Source Appendix, page 1061.

Human Factors

Women and Children [5]

% of pregnant women immunized (tetanus 1990-94)	6
% of births attended by trained health personnel (1983-94)	32
Maternal mortality rate (1980-92)	2,000
Under-5 mortality rate (1994)	214
% under-5 moderately/severely underweight (1980-1994)[1]	31

Burden of Disease [6]

Population per physician (1990)	19,448.28
Population per nurse (1990)	1,885.03
Population per hospital bed	NA
AIDS cases per 100,000 people (1994)	5.8
Malaria cases per 100,000 people (1992)	NA

Ethnic Division [7]

Mande (Bambara, Malinke, Sarakole)	50%
Peul	17%
Voltaic	12%
Tuareg and Moor	10%
Songhai	6%
Other	5%

Religion [8]

Muslim	90%
Indigenous beliefs	9%
Christian	1%

Major Languages [9]

Bambara 80%, French (official), numerous African languages.

Education

Public Education Expenditures [10]

Million (Franc C.F.A.)[41]	1980	1985	1987	1990	1993	1994
Total education expenditure	12,903	17,184	18,693	NA	15,369	NA
as percent of GNP	3.8	3.7	3.2	NA	2.1	NA
as percent of total govt. expend.	30.8	NA	17.3	NA	13.2	NA
Current education expenditure	12,752	17,048	18,285	NA	14,994	NA
as percent of GNP	3.7	3.7	3.1	NA	2.0	NA
as percent of current govt. expend.	32.1	NA	18.5	NA	NA	NA
Capital expenditure	151	136	408	NA	375	NA

Educational Attainment [11]

Illiteracy [12]

In thousands and percent[1]	1990	1995	2000
Illiterate population (15+ yrs.)	3,686	3,917	4,111
Illiteracy rate - total pop. (%)	85.1	80.1	72.5
Illiteracy rate - males (%)	77.8	72.1	64.1
Illiteracy rate - females (%)	91.8	87.2	80.1

Libraries [13]

Daily Newspapers [14]

	1980	1985	1990	1994
Number of papers	2	2	2	2
Circ. (000)	4[e]	10[e]	10[e]	40[e]

Culture [15]

Cinema (seats per 1,000)	NA
Annual attendance per person	NA
Gross box office receipts (mil. Franc C.F.A.)	NA
Museums (reporting)	1
Visitors (000)	10
Annual receipts (000 Franc C.F.A.)	NA

Science and Technology

Scientific/Technical Forces [16]

R&D Expenditures [17]

U.S. Patents Issued [18]

For sources, notes, and explanations, see Annotated Source Appendix, page 1061.

Government and Law

Organization of Government [19]

Long-form name:
  Republic of Mali
Type:
  Republic
Independence:
  22 September 1960 (from France)
National holiday:
  Anniversary of the Proclamation of the
  Republic, 22 September (1960)
Constitution:
  Adopted 12 January 1992
Legal system:
  Based on French civil law system and
  customary law; judicial review of
  legislative acts in Constitutional Court
  (which was formally established on 9
  March 1994); has not accepted
  compulsory ICJ jurisdiction
Executive branch:
  President; Prime Minister; Council of
  Ministers
Legislative branch:
  Unicameral: National Assembly
Judicial branch:
  Supreme Court (Cour Supreme)

Crime [23]

Elections [20]

National Assembly	% of seats
Alliance for Democracy (Adema)	65.5
Nat. Com. Dem. Initiative (CNID)	7.8
Sudanese Union (US/RAD)	6.9
Popular Movement for Dev.	5.2
Rally for Democracy (RDP)	3.4
Union for Democracy (UDD)	3.4
Rally for Dem. & Labor (RDT)	2.6
Union Democratic Forces (UFDP)	2.6
Other	2.6

Government Expenditures [21]

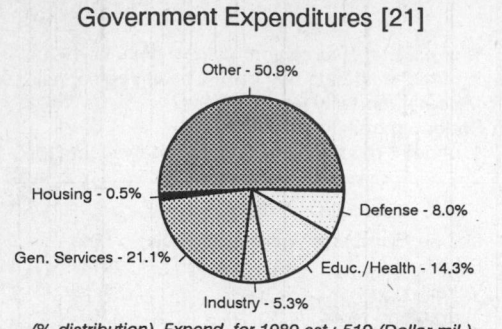

Other - 50.9%
Housing - 0.5%
Defense - 8.0%
Gen. Services - 21.1%
Educ./Health - 14.3%
Industry - 5.3%

(% distribution). Expend. for 1989 est.: 519 (Dollar mil.)

Military Expenditures and Arms Transfers [22]

	1990	1991	1992	1993	1994
Military expenditures					
Current dollars (mil.)	NA	NA	42[e]	39[e]	34
1994 constant dollars (mil.)	NA	NA	43[e]	40[e]	34
Armed forces (000)	13	13	12	12	8
Gross national product (GNP)					
Current dollars (mil.)	1,550	1,575	1,745	1,772	1,838
1994 constant dollars (mil.)	1,725	1,688	1,820	1,808	1,838
Central government expenditures (CGE)					
1994 constant dollars (mil.)	NA	NA	460[e]	NA	NA
People (mil.)	8.2	8.4	8.6	8.9	9.1
Military expenditure as % of GNP	NA	NA	2.4	2.2	1.9
Military expenditure as % of CGE	NA	NA	9.4	NA	NA
Military expenditure per capita (1994 $)	NA	NA	5	4	4
Armed forces per 1,000 people (soldiers)	1.6	1.5	1.4	1.4	0.9
GNP per capita (1994 $)	210	200	211	204	202
Arms imports[6]					
Current dollars (mil.)	10	10	0	0	0
1994 constant dollars (mil.)	11	11	0	0	0
Arms exports[6]					
Current dollars (mil.)	0	0	0	0	0
1994 constant dollars (mil.)	0	0	0	0	0
Total imports[7]					
Current dollars (mil.)	602	756[e]	848[e]	842[e]	NA
1994 constant dollars (mil.)	670	810[e]	884[e]	859[e]	NA
Total exports[7]					
Current dollars (mil.)	359	354	NA	415[e]	NA
1994 constant dollars (mil.)	400	379	NA	424[e]	NA
Arms as percent of total imports[8]	1.7	1.3	0	0	0
Arms as percent of total exports[8]	0	0	0	0	0

Human Rights [24]

	SSTS	FL	FAPRO	PPCG	APROBC	TPW	PCPTW	STPEP	PHRFF	PRW	ASST	AFL
Observes	P	P	P	P	P	P	P	P		P	P	P
		EAFRD	CPR	ESCR	SR	ACHR	MAAE	PVIAC	PVNAC	EAFDAW	TCIDTP	RC
Observes		P	P	P				P	P	P		P

P = Party; S = Signatory; see Appendix for meaning of abbreviations.

Labor Force

Total Labor Force [25]

2.666 million (1986 est.)

Labor Force by Occupation [26]

Agriculture	80%
Services	19
Industry and commerce	1

Date of data: 1981

Unemployment Rate [27]

For sources, notes, and explanations, see Annotated Source Appendix, page 1061.

Production Sectors

Commercial Energy Production and Consumption

Data are shown in quadrillion (10^{15}) BTUs and percent for 1995
Values for hydroelectric, nuclear, geothermal, solar, and wind power refer to electrical generation.

Production [28]

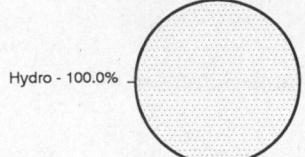

Hydro - 100.0%

Consumption [29]

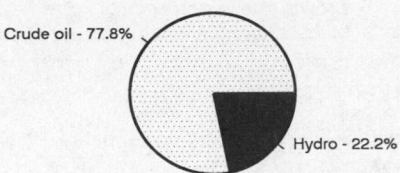

Crude oil - 77.8%

Hydro - 22.2%

Net hydroelectric power	0.002
Total	0.002

Crude oil	0.007
Net hydroelectric power	0.002
Total	0.009

Telecommunications [30]

- 11,000 (1982 est.) telephones; domestic system poor but improving; provides only minimal service
- Domestic: network consists of microwave radio relay, open wire, and radiotelephone communications stations; expansion of microwave radio relay in progress
- International: satellite earth stations - 2 Intelsat (1 Atlantic Ocean and 1 Indian Ocean)
- Radio: Broadcast stations: AM 2, FM 2, shortwave 0 Radios: 430,000 (1992 est.)
- Television: Broadcast stations: 2 (1987 est.) Televisions: 11,000 (1992 est.)

Transportation [31]

Railways: total: 641 km; note - linked to Senegal's rail system through Kayes; narrow gauge: 641 km 1.000-m gauge (1995)

Highways: total: 15,610 km; paved: 1,661 km; unpaved: 13,949 km (1987 est.)

Airports

Total:	24
With paved runways 2,438 to 3,047 m:	4
With paved runways 914 to 1,523 m:	2
With paved runways under 914 m:	7
With unpaved runways 1,524 to 2,437 m:	3
With unpaved runways 914 to 1,523 m:	8 (1995 est.)

Top Agricultural Products [32]

Agriculture accounts for 42.4% of the GDP; produces cotton, millet, rice, corn, vegetables, peanuts; cattle, sheep, goats.

Top Mining Products [33]

Thousand metric tons except as noted[e]	5/31/95[*]
Cement, hydraulic	20
Gold, mine output, Au content (kg.)	5,500[52]
Gypsum (tons)	700
Salt	5
Silver (kg.)	190[53]

Tourism [34]

	1990	1991	1992	1993	1994
Tourists[16]	44	38	38	31	28
Tourism receipts	47	11	11	13	18
Tourism expenditures	62	60	65	60	54
Fare receipts	25	25	27	25	20

Travelers are in thousands, money in million U.S. dollars.

Finance, Economics, and Trade

GDP and Manufacturing Summary [35]

	1980	1985	1990	1991	1992
Gross Domestic Product					
Millions of 1980 dollars	1,629	1,666	2,252	2,248	2,255[e]
Growth rate in percent	4.01	-0.11	2.45	-0.15	0.30[e]
Manufacturing Value Added					
Millions of 1980 dollars	71	105	121	121	125[e]
Growth rate in percent	1.58	-0.47	5.15	0.27	3.34[e]
Manufacturing share in percent of current prices	4.3	7.3	12.2	NA	NA

Economic Indicators [36]

- **National product**: GDP—purchasing power parity—$5.4 billion (1994 est.)

- **National product real growth rate**: 2.4% (1994 est.)

- **National product per capita**: $600 (1994 est.)

- **Inflation rate (consumer prices)**: 8% (1995 est.)

- **External debt**: $2.8 billion (1995 est.)

Balance of Payments Summary [37]

Values in millions of dollars.

	1989	1990	1991	1992	1993
Exports of goods (f.o.b.)	269.3	337.9	354.5	339.3	343.6
Imports of goods (f.o.b.)	-338.8	-432.4	-447.1	-484.0	-463.5
Trade balance	-69.5	-94.5	-92.6	-144.7	-119.9
Services - debits	-353.7	-456.0	-437.6	-461.3	-441.4
Services - credits	75.5	99.5	106.3	99.0	102.8
Private transfers (net)	51.4	66.9	70.0	88.0	84.8
Government transfers (net)	221.6	250.8	312.9	312.8	271.2
Long-term capital (net)	254.2	177.0	153.1	116.4	90.1
Short-term capital (net)	-79.3	-14.2	-17.4	0.3	26.1
Errors and omissions	-1.5	1.1	29.0	-22.7	22.7
Overall balance	98.7	30.6	123.7	-12.2	36.4

Exchange Rates [38]

Currency: **Communaute Financiere Africaine franc.**
Symbol: **CFAF.**

Data are currency units per $1.

January 1996	500.56
1995	499.15
1994	555.20
1993	283.16
1992	264.69
1991	282.11

Imports and Exports

Top Import Origins [39]

$842 million (f.o.b., 1993).

Origins	%
Franc zone	NA
Western Europe	NA

Top Export Destinations [40]

$415 million (f.o.b., 1993).

Destinations	%
Franc zone	NA
Western Europe	NA

Foreign Aid [41]

Recipient: ODA, $NA.

Import and Export Commodities [42]

Import Commodities	Export Commodities
Machinery and equipment	Cotton
Foodstuffs	Livestock
Construction materials	Gold
Petroleum	
Textiles	

For sources, notes, and explanations, see Annotated Source Appendix, page 1061.

Malta

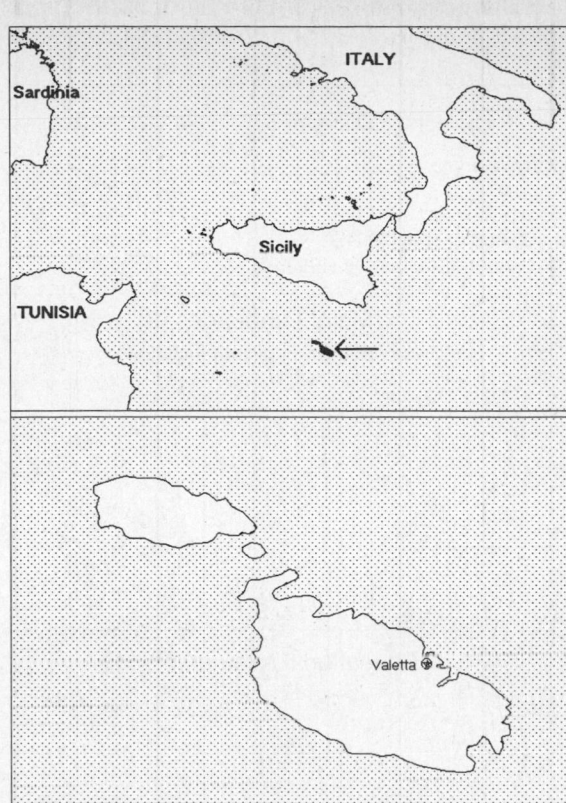

Geography [1]

Total area:
 320 sq km   124 sq mi
Land area:
 320 sq km   124 sq mi
Comparative area:
 Less than twice the size of Washington, DC
Land boundaries:
 0 km
Coastline:
 140 km
Climate:
 Mediterranean with mild, rainy winters and hot, dry summers
Terrain:
 Mostly low, rocky, flat to dissected plains; many coastal cliffs
Natural resources:
 Limestone, salt
Land use:
 Arable land: 38%
 Permanent crops: 3%
 Meadows and pastures: 0%
 Forest and woodland: 0%
 Other: 59%

Demographics [2]

	1970	1980	1990	1995[1]	1996	2000	2010	2020	2030
Population	326	364	354	372	376	391	425	450	462
Population density (persons per sq. mi.)	2,627	2,937	2,857	3,000	3,030	3,152	3,433	3,632	3,730
(persons per sq. km.)	1,014	1,134	1,103	1,158	1,170	1,217	1,325	1,402	1,440
Net migration rate (per 1,000 population)	-7.4	-1.2	2.4	2.3	2.1	1.4	0.8	0.2	0.0
Births[34]	5	6	NA	NA	5	NA	NA	NA	NA
Deaths	3	3	NA	NA	3	NA	NA	NA	NA
Life expectancy - males	NA	68.5	73.7	75.3	75.8	78.0	80.2	80.8	81.0
Life expectancy - females	NA	72.7	78.1	80.0	80.6	83.4	86.1	86.8	86.9
Birth rate (per 1,000)	16.3	16.0	15.2	14.9	14.8	14.2	13.4	12.0	11.5
Death rate (per 1,000)	9.4	8.8	7.8	7.0	6.8	6.1	6.8	8.3	10.1
Women of reproductive age (15-49 yrs.)	NA	NA	91	95	95	95	94	96	96
of which are currently married	NA	NA	57	NA	58	57	57	NA	NA
Fertility rate	NA	2.0	2.0	2.2	2.2	2.1	2.1	2.0	2.0

Except as noted, values for vital statistics are in thousands; life expectancy is in years.

Health

Health Indicators [3]

Health Expenditures [4]

For sources, notes, and explanations, see Annotated Source Appendix, page 1061.

601

Human Factors

Human Factors

Women and Children [5]	Burden of Disease [6]	
	Population per physician	NA
	Population per nurse	NA
	Population per hospital bed	NA
	AIDS cases per 100,000 people (1994)	1.4
	Heart disease cases per 1,000 people (1990-93)	NA

Ethnic Division [7]	Religion [8]		Major Languages [9]
Arab, Sicilian, Norman, Spanish, Italian, English.	Roman Catholic	98%	Maltese (official), English (official).
	Other	2%	

Education

Public Education Expenditures [10]

	1980	1990	1991	1992	1993	1994
Million (Lira)						
Total education expenditure	13	32	35	42	50	NA
as percent of GNP	3.0	4.0	4.1	4.6	5.1	NA
as percent of total govt. expend.	7.8	8.3	8.5	10.9	11.7	NA
Current education expenditure	12	30	33	41	48	NA
as percent of GNP	2.9	3.8	3.9	4.5	4.9	NA
as percent of current govt. expend.	9.7	10.9	11.1	12.5	13.0	NA
Capital expenditure	1	2	2	1	3	NA

Educational Attainment [11]

Illiteracy [12]

	1985	1989	1995
Illiterate population (20+ yrs.)[2]	33,740	NA	25,000
Illiteracy rate - total pop. (%)	14.3	NA	8.7
Illiteracy rate - males (%)	14.8	NA	9.4
Illiteracy rate - females (%)	13.9	NA	8.1

Libraries [13]

	Admin. Units	Svc. Pts.	Vols. (000)	Shelving (meters)	Vols. Added	Reg. Users
National (1993)	1	1	2,950	NA	25,100	16,800
Nonspecialized (1993)	28	28	605	NA	24,100	34,600
Public (1993)	1,598	3,143	18,874	NA	1 mil	1 mil
Higher ed. (1993)	15	NA	5,844	NA	134,444	NA
School (1987)	51	51	127	3,900	2,196	22,648

Daily Newspapers [14]

	1980	1985	1990	1994
Number of papers	5	4	3	3
Circ. (000)	60[e]	56[e]	54[e]	64

Culture [15]

Cinema (seats per 1,000)	19.5
Annual attendance per person	0.8
Gross box office receipts (mil. Lira)	NA
Museums (reporting)	15
Visitors (000)	716
Annual receipts (000 Lira)	NA

Science and Technology

Scientific/Technical Forces [16]

Scientists/engineers	34
Number female	NA
Technicians	5
Number female	NA
Total[25]	39

R&D Expenditures [17]

	Lira (000) 1988
Total expenditure[25]	10
Capital expenditure	1
Current expenditure	9
Percent current	90.0

U.S. Patents Issued [18]

Values show patents issued to citizens of the country by the U.S. Patents Office.

	1993	1994	1995
Number of patents	1	1	1

                                                                 For sources, notes, and explanations, see Annotated Source Appendix, page 1061.

Government and Law

Organization of Government [19]

Long-form name:
  Republic of Malta
Type:
  Parliamentary democracy
Independence:
  21 September 1964 (from UK)
National holiday:
  Independence Day, 21 September (1964)
Constitution:
  1964 constitution substantially amended
  on 13 December 1974
Legal system:
  Based on English common law and
  Roman civil law; has accepted
  compulsory ICJ jurisdiction, with
  reservations
Executive branch:
  President; Prime Minister; Deputy Prime
  Minister; Foreign Minister; Cabinet
Legislative branch:
  Unicameral: House of Representatives
Judicial branch:
  Constitutional Court; Court of Appeal

Elections [20]

House of Representatives	% of votes
Nationalist Party (NP)	51.8
Malta Labor Party (MLP)	46.5

Government Expenditures [21]

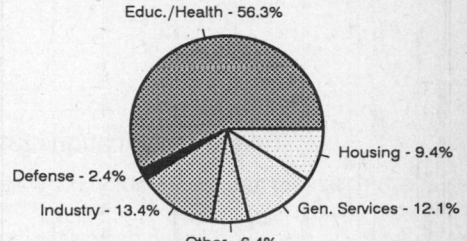

Educ./Health - 56.3%
Housing - 9.4%
Defense - 2.4%
Gen. Services - 12.1%
Industry - 13.4%
Other - 6.4%

(% distribution). Expend. for CY94: 402.47 (Lira mil.)

Crime [23]

	1994
Crime volume	
Cases known to police	6,801
Attempts (percent)	5.81
Percent cases solved	14.32
Crimes per 100,000 persons	1,840.84
Persons responsible for offenses	
Total number offenders	NA
Percent female	NA
Percent juvenile (10-18 yrs.)	NA
Percent foreigners	NA

Military Expenditures and Arms Transfers [22]

	1990	1991	1992	1993	1994
Military expenditures					
Current dollars (mil.)	17	19[e]	23	23	30
1994 constant dollars (mil.)	19	20[e]	24	23	30
Armed forces (000)	2	2	2	2	1
Gross national product (GNP)					
Current dollars (mil.)	2,084	2,275	2,418	2,554	NA
1994 constant dollars (mil.)	2,320	2,439	2,522	2,607	NA
Central government expenditures (CGE)					
1994 constant dollars (mil.)	958	1,025	927	969	1,482[e]
People (mil.)	0.4	0.4	0.4	0.4	0.4
Military expenditure as % of GNP	0.8	0.8	0.9	0.9	NA
Military expenditure as % of CGE	2.0	2.0	2.6	2.4	2.0
Military expenditure per capita (1994 $)	54	56	66	63	82
Armed forces per 1,000 people (soldiers)	5.6	5.6	5.5	5.5	2.7
GNP per capita (1994 $)	6,552	6,823	6,991	7,166	NA
Arms imports[6]					
Current dollars (mil.)	0	0	0	0	0
1994 constant dollars (mil.)	0	0	0	0	0
Arms exports[6]					
Current dollars (mil.)	0	0	0	0	0
1994 constant dollars (mil.)	0	0	0	0	0
Total imports[7]					
Current dollars (mil.)	1,964	2,130	2,331	2,174	2,461
1994 constant dollars (mil.)	2,186	2,283	2,431	2,219	2,461
Total exports[7]					
Current dollars (mil.)	1,133	1,234	1,540	1,355	1,530
1994 constant dollars (mil.)	1,261	1,323	1,606	1,383	1,530
Arms as percent of total imports[8]	0	0	0	0	0
Arms as percent of total exports[8]	0	0	0	0	0

Human Rights [24]

	SSTS	FL	FAPRO	PPCG	APROBC	TPW	PCPTW	STPEP	PHRFF	PRW	ASST	AFL
Observes	P	P	P		P	P	P		P	P	P	P
	EAFRD	CPR	ESCR	SR	ACHR	MAAE	PVIAC	PVNAC	EAFDAW	TCIDTP	RC	
Observes		P	P	P		P	P	P	P	P	P	

P = Party; S = Signatory; see Appendix for meaning of abbreviations.

Labor Force

Total Labor Force [25]

139,600 (1994)

Labor Force by Occupation [26]

Government (excluding job corps)	37%
Services	26
Manufacturing	22
Training programs	9
Construction	4
Agriculture	2

Date of data: 1990

Unemployment Rate [27]

3.4% (third quarter 1995)

For sources, notes, and explanations, see Annotated Source Appendix, page 1061.

603

Production Sectors

Energy Resource Summary [28]

Electricity: Capacity: 250,000 kW. Production: 1.1 billion kWh. Consumption per capita: 2,749 kWh (1993).

Telecommunications [30]

- 191,876 (1992 est.) telephones; automatic system satisfies normal requirements
- Domestic: submarine cable and microwave radio relay between islands
- International: 1 submarine cable; satellite earth station - 1 Intelsat (Atlantic Ocean)
- Radio: Broadcast stations: AM 8, FM 4, shortwave 0 Radios: 189,000 (1992 est.)
- Television: Broadcast stations: 2 (1987 est.) Televisions: 267,000 (1992 est.)

Top Agricultural Products [32]

Produces potatoes, cauliflower, grapes, wheat, barley, tomatoes, citrus, cut flowers, green peppers; pork, milk, poultry, eggs.

Top Mining Products [33]

Metric tons except as noted[e]	3/95[*]
Limestone (000 tons)	2,200
Salt	30

Transportation [31]

Railways: 0 km

Highways: total: 1,582 km; paved: 1,471 km; unpaved: 111 km (1993 est.)

Merchant marine: total: 1,045 ships (1,000 GRT or over) totaling 17,082,925 GRT/28,829,144 DWT; ships by type: bulk 285, cargo 331, chemical tanker 28, combination bulk 28, combination ore/oil 16, container 39, liquefied gas tanker 4, livestock carrier 1, multifunction large-load carrier 6, oil tanker 208, passenger 7, passenger-cargo 3, railcar carrier 1, refrigerated cargo 20, roll-on/roll-off cargo 29, short-sea passenger 23, specialized tanker 4, vehicle carrier 12

Airports

Total: 1

With paved runways over 3,047 m: 1 (1995 est.)

Tourism [34]

	1990	1991	1992	1993	1994
Visitors[72]	917	935	1,054	1,130	1,239
Tourists[72]	872	895	1,002	1,063	1,176
Cruise passengers[72]	45	40	52	67	63
Tourism receipts	495	542	566	607	639
Tourism expenditures	137	132	138	154	176
Fare receipts	68	80	89	84	96
Fare expenditures	11	6	9	24	23

Travelers are in thousands, money in million U.S. dollars.

604

For sources, notes, and explanations, see Annotated Source Appendix, page 1061.

Manufacturing Sector

Manufacturing Summary [35]

	1987		1988		1989		1990		1991	
	$ bil.	%	$ bil.	%	$ bil.	%	$ bil.	%	$ bil.	%
Establishments or enterprises (number)	1,495	0.064	1,544	0.074	-	-	-	-	-	-
Total employment (000)	31	0.023	31	0.023	-	-	-	-	-	-
Production workers (000)	26	0.038	25	0.040	-	-	-	-	-	-
Output ($ bil.)	1	0.011	1	0.012	-	-	-	-	-	-
Value added ($ bil.)	0.439	0.010	0.525	0.011	-	-	-	-	-	-
Capital investment ($ mil.)	42	0.010	77	0.016	-	-	-	-	-	-
M & E investment ($ mil.)	28	0.009	68	0.020	-	-	-	-	-	-
Employees per establishment (number)	21	36.224	20	30.879	-	-	-	-	-	-
Production workers per establishment	17	59.684	16	54.867	-	-	-	-	-	-
Output per establishment ($ mil.)	0.762	17.577	0.905	16.301	-	-	-	-	-	-
Capital investment per estab. ($ mil.)	0.028	15.275	0.050	21.998	-	-	-	-	-	-
M & E per establishment ($ mil)	0.019	14.280	0.044	26.746	-	-	-	-	-	-
Payroll per employee ($)	7,169	79.943	7,780	78.087	-	-	-	-	-	-
Wages per production worker ($)	6,279	79.573	6,648	78.670	-	-	-	-	-	-
Hours per production worker (hours)	-	-	-	-	-	-	-	-	-	-
Output per employee ($)	36,342	48.523	44,964	52.790	-	-	-	-	-	-
Capital investment per employee ($)	1,354	42.169	2,479	71.238	-	-	-	-	-	-
M & E per employee ($)	905	39.423	2,190	86.614	-	-	-	-	-	-

Note: Columns headed % show percent of world total or ratio. Ratios closest to 100 are closest to world average. M & E stands for machinery & equipment.

Output in Manufacturing

	1987		1988		1989		1990		1991	
	$ bil.	%	$ bil.	%	$ bil.	%	$ bil.	%	$ bil.	%
3110 - Food products	0.153	15.30	0.165	16.50	-	-	-	-	-	-
3130 - Beverages	0.058	5.80	0.070	7.00	-	-	-	-	-	-
3140 - Tobacco	0.043	4.30	0.041	4.10	-	-	-	-	-	-
3210 - Textiles	0.021	2.10	0.023	2.30	-	-	-	-	-	-
3211 - Spinning, weaving, etc.	0.009	0.90	0.009	0.90	-	-	-	-	-	-
3220 - Wearing apparel	0.187	18.70	0.182	18.20	-	-	-	-	-	-
3230 - Leather and products	0.001	0.10	0.001	0.10	-	-	-	-	-	-
3240 - Footwear	0.030	3.00	0.032	3.20	-	-	-	-	-	-
3310 - Wood products	0.004	0.40	0.004	0.40	-	-	-	-	-	-
3320 - Furniture, fixtures	0.027	2.70	0.038	3.80	-	-	-	-	-	-
3410 - Paper and products	0.009	0.90	0.011	1.10	-	-	-	-	-	-
3420 - Printing, publishing	0.044	4.40	0.049	4.90	-	-	-	-	-	-
3510 - Industrial chemicals	0.005	0.50	0.005	0.50	-	-	-	-	-	-
3550 - Rubber products	0.022	2.20	0.025	2.50	-	-	-	-	-	-
3560 - Plastic products nec	0.014	1.40	0.017	1.70	-	-	-	-	-	-
3610 - Pottery, china, etc.	0.001	0.10	0.001	0.10	-	-	-	-	-	-
3620 - Glass and products	0.004	0.40	0.003	0.30	-	-	-	-	-	-
3690 - Nonmetal products nec	0.030	3.00	0.033	3.30	-	-	-	-	-	-
3810 - Metal products	0.043	4.30	0.052	5.20	-	-	-	-	-	-
3820 - Machinery nec	0.023	2.30	0.019	1.90	-	-	-	-	-	-
3830 - Electrical machinery	0.145	14.50	0.243	24.30	-	-	-	-	-	-
3832 - Radio, television, etc.	0.119	11.90	0.212	21.20	-	-	-	-	-	-
3840 - Transportation equipment	0.033	3.30	0.044	4.40	-	-	-	-	-	-
3841 - Shipbuilding, repair	0.031	3.10	0.041	4.10	-	-	-	-	-	-
3850 - Professional goods	0.042	4.20	0.044	4.40	-	-	-	-	-	-
3900 - Industries nec	0.041	4.10	0.037	3.70	-	-	-	-	-	-

Note: Codes are International Standard Industry codes (ISIC). Percentages are % of total Output. [f]: Factor Prices; [p]: Producer Prices.

For sources, notes, and explanations, see Annotated Source Appendix, page 1061.

605

Finance, Economics, and Trade

Economic Indicators [36]

- **National product**: GDP—purchasing power parity—$4.4 billion (1995 est.)

- **National product real growth rate**: 5% (1995 est.)

- **National product per capita**: $12,000 (1995 est.)

- **Inflation rate (consumer prices)**: 5% (1995 est.)

- **External debt**: $603 million (1992)

Balance of Payments Summary [37]

Values in millions of dollars.

	1989	1990	1991	1992	1993
Exports of goods (f.o.b.)	866.3	1,154.2	1,283.6	1,557.1	1,351.1
Imports of goods (f.o.b.)	-1,327.7	-1,753.0	-1,897.3	-2,104.0	-1,953.2
Trade balance	-461.4	-598.8	-613.7	-546.9	-602.1
Services - debits	-506.2	-609.5	-638.5	-724.0	-717.8
Services - credits	860.0	1,065.2	1,129.2	1,207.5	1,191.5
Private transfers (net)	54.8	32.4	32.5	9.4	7.3
Government transfers (net)	49.6	55.1	83.2	84.3	53.6
Long-term capital (net)	-20.1	32.1	-189.3	-90.9	NA
Short-term capital (net)	-14.1	-59.8	213.7	121.4	NA
Errors and omissions	64.7	24.2	-93.9	-62.8	NA
Overall balance	27.3	-59.1	-76.8	-2.0	-67.5

Exchange Rates [38]

Currency: **Maltese lira.**
Symbol: **LM.**

Data are currency units per $1.

January 1996	0.3570
1995	0.3529
1994	0.3776
1993	0.3821
1992	0.3178
1991	0.3226

Imports and Exports

Top Import Origins [39]

$2.5 billion (c.i.f., 1994).

Origins	%
Italy	27
Germany	14
UK	13
US	9

Top Export Destinations [40]

$1.5 billion (f.o.b., 1994).

Destinations	%
Italy	32
Germany	16
UK	8

Foreign Aid [41]

Recipient: ODA, $NA.

Import and Export Commodities [42]

Import Commodities	Export Commodities
Food	Machinery and transport equipment
Petroleum	Clothing and footware
Machinery and semimanufactured goods	Printed matter

For sources, notes, and explanations, see Annotated Source Appendix, page 1061.

Marshall Islands

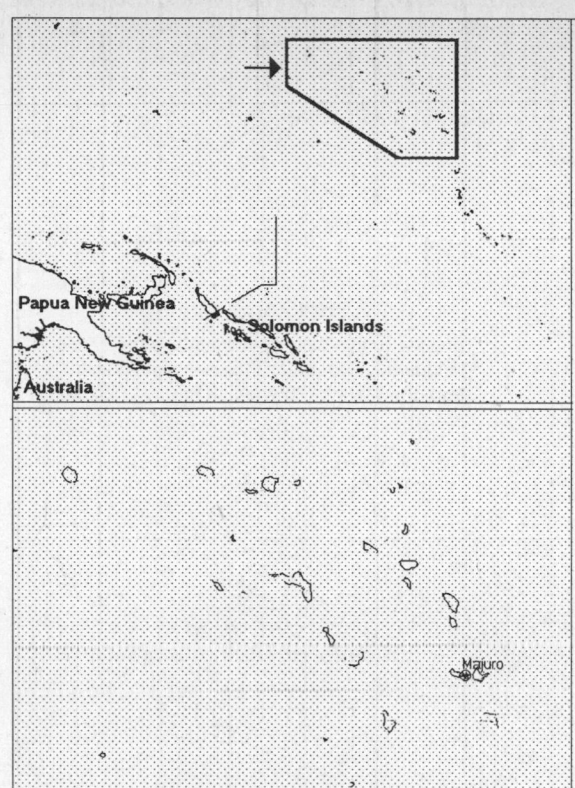

Geography [1]

Total area:
  181.3 sq km   70 sq mi
Land area:
  181.3 sq km   70 sq mi
Comparative area:
  About the size of Washington, DC
  Note: Includes the atolls of Bikini, Enewetak, and Kwajalein
Land boundaries:
  0 km
Coastline:
  370.4 km
Climate:
  Wet season May to November; hot and humid; islands border typhoon belt
Terrain:
  Low coral limestone and sand islands
Natural resources:
  Phosphate deposits, marine products, deep seabed minerals
Land use:
  Arable land: 0%
  Permanent crops: 60%
  Meadows and pastures: 0%
  Forest and woodland: 0%
  Other: 40%

Demographics [2]

	1970	1980	1990	1995[1]	1996	2000	2010	2020	2030
Population	22	31	46	56	58	68	100	144	201
Population density (persons per sq. mi.)	311	439	662	804	835	974	1,435	2,064	2,870
(persons per sq. km.)	120	170	255	310	322	376	554	797	1,108
Net migration rate (per 1,000 population)	NA	NA	0.0	0.0	0.0	0.0	0.0	0.0	0.0
Births	NA	NA	NA	NA	3	NA	NA	NA	NA
Deaths	NA	NA	NA	NA	Z	NA	NA	NA	NA
Life expectancy - males	NA	NA	60.3	61.9	62.3	63.5	66.5	69.1	71.3
Life expectancy - females	NA	NA	63.3	65.1	65.5	66.9	70.1	73.0	75.5
Birth rate (per 1,000)	NA	NA	48.0	46.0	45.8	45.3	43.1	38.8	35.0
Death rate (per 1,000)	NA	NA	8.7	7.5	7.3	6.6	5.1	4.1	3.7
Women of reproductive age (15-49 yrs.)	NA	NA	9	12	12	15	22	32	47
of which are currently married	NA	2	6	NA	8	10	15	NA	NA
Fertility rate	NA	NA	7.1	6.9	6.8	6.6	6.0	5.3	4.6

Except as noted, values for vital statistics are in thousands; life expectancy is in years.

Health

Health Indicators [3]

Health Expenditures [4]

For sources, notes, and explanations, see Annotated Source Appendix, page 1061.

607

Human Factors

Women and Children [5]

Burden of Disease [6]

Ethnic Division [7]
Micronesian.

Religion [8]
Christian (mostly Protestant).

Major Languages [9]
English (universally spoken and is the official language), two major Marshallese dialects from the Malayo-Polynesian family, Japanese.

Education

Public Education Expenditures [10]

Educational Attainment [11]

Illiteracy [12]

Libraries [13]

Daily Newspapers [14]

Culture [15]

Science and Technology

Scientific/Technical Forces [16]

R&D Expenditures [17]

U.S. Patents Issued [18]

Government and Law

Organization of Government [19]

Long-form name:
Republic of the Marshall Islands
Type:
Constitutional government in free association with the US; the Compact of Free Association entered into force 21 October 1986
Independence:
21 October 1986 (from the US-administered UN trusteeship)
National holiday:
Proclamation of the Republic of the Marshall Islands, 1 May (1979)
Constitution:
1 May 1979
Legal system:
Based on adapted Trust Territory laws, acts of the legislature, municipal, common, and customary laws
Executive branch:
President; Cabinet
Legislative branch:
Unicameral: Parliament (Nitijela)
Judicial branch:
Supreme Court; High Court

Elections [20]

Parliament. Elections last held 20 November 1995 (next to be held November 1999); results - percent of vote NA; seats - (33 total) seats by party NA.

Government Budget [21]

For FY94/95 est.

	$ mil.
Revenues	67.2
Expenditures	79.6
Capital expenditures	NA

Defense Summary [22]

Branches: No regular military forces (a coast guard may be established); Police Force

Note: Defense is the responsibility of the US

Crime [23]

	1990
Crime volume	
Cases known to police	NA
Attempts (percent)	NA
Percent cases solved	NA
Crimes per 100,000 persons	NA
Persons responsible for offenses	
Total number offenders	60
Percent female	NA
Percent juvenile	33.3
Percent foreigners	NA

Human Rights [24]

	SSTS	FL	FAPRO	PPCG	APROBC	TPW	PCPTW	STPEP	PHRFF	PRW	ASST	AFL
Observes												
	EAFRD	CPR	ESCR	SR	ACHR	MAAE	PVIAC	PVNAC	EAFDAW	TCIDTP		RC
Observes												P

P = Party; S = Signatory; see Appendix for meaning of abbreviations.

Labor Force

Total Labor Force [25]

4,800 (1986)

Labor Force by Occupation [26]

Unemployment Rate [27]

16% (1991 est.)

For sources, notes, and explanations, see Annotated Source Appendix, page 1061.

609

Production Sectors

Energy Resource Summary [28]

Electricity: Capacity: 42,000 kW. Production: 80 million kWh. Consumption per capita: 1,840 kWh (1990).

Telecommunications [30]

- 800 (1988 est.) telephones; telex services
- Domestic: islands interconnected by shortwave radiotelephone (used mostly for government purposes)
- International: satellite earth stations - 2 Intelsat (Pacific Ocean); US Government satellite communications system on Kwajalein
- Radio: Broadcast stations: AM 1, FM 2, shortwave 1
- Television: Broadcast stations: 1

Transportation [31]

Railways: 0 km

Highways: total: NA km; paved: NA km; unpaved: NA km

Merchant marine: total: 78 ships (1,000 GRT or over) totaling 3,068,782 GRT/5,073,125 DWT; ships by type: bulk carrier 43, cargo 4, combination ore/oil 1, container 17, oil tanker 11, refrigerated cargo 1, vehicle carrier 1 (1995 est.)

Airports

Total:	16
With paved runways 1,524 to 2,437 m:	3
With paved runways 914 to 1,523 m:	1
With paved runways under 914 m:	5
With unpaved runways 914 to 1,523 m:	7 (1995 est.)

Top Agricultural Products [32]

Produces coconuts, cacao, taro, breadfruit, fruits; pigs, chickens.

Top Mining Products [33]

Detailed information is not available. A summary of natural resources follows. **Mineral Resources:** Phosphate deposits and deep seabed minerals.

Tourism [34]

	1990	1991	1992	1993	1994
Tourists	5	7	8	5	5
Tourism receipts	NA	3	3	3	2

Travelers are in thousands, money in million U.S. dollars.

Finance, Economics, and Trade

Industrial Summary [35]

Industrial Production: Growth rate not available. **Industries**: Copra, fish, tourism, craft items from shell, wood, and pearls, offshore banking (embryonic).

Economic Indicators [36]

- **National product**: GDP—purchasing power parity—$94 million (1995 est.)
- **National product real growth rate**: 1.5% (1995 est.)
- **National product per capita**: $1,680 (1995 est.)
- **Inflation rate (consumer prices)**: 4% (1995 est.)
- **External debt**: $170 million (1994)

Balance of Payments Summary [37]

Exchange Rates [38]

Currency: **United States dollar.**
Symbol: **US$.**

US currency is used.

Imports and Exports

Top Import Origins [39]

$69.9 million (c.i.f., 1995 est.).

Origins	%
US	NA
Japan	NA
Australia	NA

Top Export Destinations [40]

$21.3 million (f.o.b., 1995 est.).

Destinations	%
US	NA
Japan	NA
Australia	NA

Foreign Aid [41]

Recipient: Under the terms of the Compact of Free Association, the US is to provide approximately $40 million in aid annually.

Import and Export Commodities [42]

Import Commodities

Foodstuffs
Machinery and equipment
Beverages and tobacco
Fuels

Export Commodities

Coconut oil
Fish
Live animals
Trochus shells

For sources, notes, and explanations, see Annotated Source Appendix, page 1061.

611

Mauritania

Geography [1]

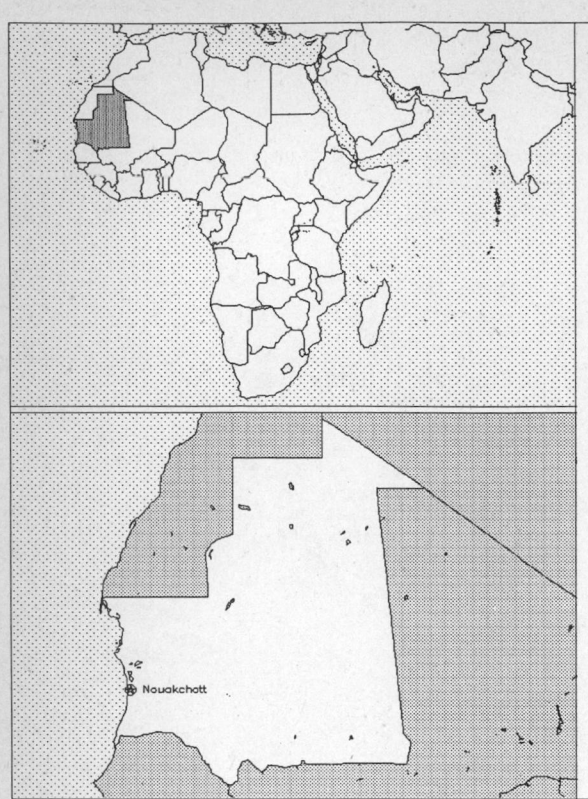

Total area:
1,030,700 sq km   397,955 sq mi
Land area:
1,030,400 sq km   397,840 sq mi
Comparative area:
Slightly larger than three times the size of New Mexico
Land boundaries:
Total 5,074 km, Algeria 463 km, Mali 2,237 km, Senegal 813 km,
Western Sahara 1,561 km
Coastline:
754 km
Climate:
Desert; constantly hot, dry, dusty
Terrain:
Mostly barren, flat plains of the Sahara; some central hills
Natural resources:
Iron ore, gypsum, fish, copper, phosphate
Land use:
Arable land: 1%
Permanent crops: 0%
Meadows and pastures: 38%
Forest and woodland: 5%
Other: 56%

Demographics [2]

	1970	1980	1990	1995[1]	1996	2000	2010	2020	2030
Population	1,227	1,456	1,935	2,263	2,336	2,653	3,630	4,859	6,270
Population density (persons per sq. mi.)	3	4	5	6	6	7	9	12	16
(persons per sq. km.)	1	1	2	2	2	3	4	5	6
Net migration rate (per 1,000 population)	NA	0.0	0.0	0.0	0.0	0.0	0.0	0.0	0.0
Births	NA	NA	NA	NA	110	NA	NA	NA	NA
Deaths	NA	NA	NA	NA	36	NA	NA	NA	NA
Life expectancy - males	NA	39.7	43.6	45.7	46.1	47.8	52.4	56.9	61.3
Life expectancy - females	NA	44.1	49.0	51.5	52.1	54.2	59.6	64.7	69.4
Birth rate (per 1,000)	NA	48.7	49.1	47.3	46.9	45.5	40.8	35.1	29.5
Death rate (per 1,000)	NA	23.5	17.9	15.7	15.2	13.6	10.2	7.6	6.0
Women of reproductive age (15-49 yrs.)	NA	321	427	501	518	594	850	1,199	1,633
of which are currently married	NA	NA	268	NA	325	372	533	NA	NA
Fertility rate	NA	7.3	7.3	6.9	6.8	6.5	5.6	4.5	3.5

Except as noted, values for vital statistics are in thousands; life expectancy is in years.

Health

Health Indicators [3]

% of population with access to	
safe water (1990-95)[1]	66
adequate sanitation (1990-95)	NA
health services (1985-95)	63
% of 1-year-olds immunized (1990-94) against	
TB (tuberculosis)	93
DPT (diphtheria, pertussis, tetanus)	50
polio	50
measles	53
% of contraceptive prevalence (1980-94)	4
ORT use rate (1990-94)	54

Health Expenditures [4]

Human Factors

Women and Children [5]

% of pregnant women immunized (tetanus 1990-94)	28
% of births attended by trained health personnel (1983-94)	40
Maternal mortality rate (1980-92)	NA
Under-5 mortality rate (1994)	199
% under-5 moderately/severely underweight (1980-1994)	48

Burden of Disease [6]

Population per physician (1985)	8,830.00
Population per nurse (1984)	1,203.50
Population per hospital bed (1984)	1,301.81
AIDS cases per 100,000 people (1994)	1.1
Malaria cases per 100,000 people (1992)	NA

Ethnic Division [7]

Mixed Maur/black	40%
Maur	30%
Black	30%

Religion [8]

Muslim	100%

Major Languages [9]

Hasaniya Arabic (official), Pular, Soninke, Wolof (official).

Education

Public Education Expenditures [10]

Million (Ouguiya)[42]	1980	1988	1990	1992	1993	1994
Total education expenditure	NA	NA	NA	NA	NA	NA
as percent of GNP	NA	NA	NA	NA	NA	NA
as percent of total govt. expend.	NA	NA	NA	NA	NA	NA
Current education expenditure	1,546	3,188	3,512	3,869	4,192	4,753
as percent of GNP	5.0	4.8	4.5	4.0	3.9	4.0
as percent of current govt. expend.	NA	22.0	NA	NA	NA	NA
Capital expenditure	NA	NA	NA	NA	NA	NA

Educational Attainment [11]

Age group (1988)	25+
Total population	679,667
Highest level attained (%)	
No schooling	60.8
First level	
Not completed	34.1
Completed	NA
Entered second level	
S-1	3.8
S-2	NA
Postsecondary	1.3

Illiteracy [12]

In thousands and percent[1]	1990	1995	2000
Illiterate population (15+ yrs.)	723	806	904
Illiteracy rate - total pop. (%)	72.0	69.1	65.9
Illiteracy rate - males (%)	59.4	56.6	53.7
Illiteracy rate - females (%)	83.8	80.8	77.3

Libraries [13]

Daily Newspapers [14]

	1980	1985	1990	1994
Number of papers	-	-	1	1
Circ. (000)	-	-	1[e]	1[e]

Culture [15]

Science and Technology

Scientific/Technical Forces [16]

R&D Expenditures [17]

U.S. Patents Issued [18]

For sources, notes, and explanations, see Annotated Source Appendix, page 1061.

Government and Law

Organization of Government [19]

Long-form name:
Islamic Republic of Mauritania
Type:
Republic
Independence:
28 November 1960 (from France)
National holiday:
Independence Day, 28 November (1960)
Constitution:
12 July 1991
Legal system:
Three-tier system: Islamic (Shari'a) courts, special courts, and state security courts (in the process of being eliminated)
Executive branch:
President; Prime Minister; Council of Ministers
Legislative branch:
Bicameral legislature: Senate (Majlis al-Shuyukh) and National Assembly (Majlis al-Watani)
Judicial branch:
Supreme Court (Cour Supreme)

Elections [20]

	% of seats
Union of Democratic Forces New Era (UFD/NE)	84.8
Maur. Party for Renewal (PDR)	1.3
Assembly for Democracy and Unity (RDU)	1.3
Independents	12.7

Government Budget [21]

For 1994 est.

	$ mil.
Revenues	254
Expenditures	280
Capital expenditures	94

Crime [23]

Military Expenditures and Arms Transfers [22]

	1990	1991	1992	1993	1994
Military expenditures					
Current dollars (mil.)[e]	32	30	37	36	36
1994 constant dollars (mil.)[e]	36	32	39	37	36
Armed forces (000)	17	17	16	16	10
Gross national product (GNP)					
Current dollars (mil.)	762	815	851	896	969
1994 constant dollars (mil.)	848	874	887	914	969
Central government expenditures (CGE)					
1994 constant dollars (mil.)	NA	NA	NA	NA	NA
People (mil.)	1.9	2.0	2.1	2.1	2.2
Military expenditure as % of GNP	4.2	3.7	4.4	4.1	3.8
Military expenditure as % of CGE	NA	NA	NA	NA	NA
Military expenditure per capita (1994 $)	18	16	19	17	17
Armed forces per 1,000 people (soldiers)	8.8	8.5	7.8	7.5	4.6
GNP per capita (1994 $)	439	438	431	430	442
Arms imports[6]					
Current dollars (mil.)	0	0	0	0	0
1994 constant dollars (mil.)	0	0	0	0	0
Arms exports[6]					
Current dollars (mil.)	0	0	0	0	0
1994 constant dollars (mil.)	0	0	0	0	0
Total imports[7]					
Current dollars (mil.)	639	502[e]	NA	378[e]	NA
1994 constant dollars (mil.)	711	538[e]	NA	386[e]	NA
Total exports[7]					
Current dollars (mil.)	469	518[e]	444[e]	425[e]	NA
1994 constant dollars (mil.)	522	555[e]	463[e]	434[e]	NA
Arms as percent of total imports[8]	0	0	0	0	0
Arms as percent of total exports[8]	0	0	0	0	0

Human Rights [24]

	SSTS	FL	FAPRO	PPCG	APROBC	TPW	PCPTW	STPEP	PHRFF	PRW	ASST	AFL
Observes	P	P	P			P	P	P		P	P	
	EAFRD	CPR	ESCR	SR	ACHR	MAAE	PVIAC	PVNAC	EAFDAW	TCIDTP	RC
Observes		P			P			P	P			P

P = Party; S = Signatory; see Appendix for meaning of abbreviations.

Labor Force

Total Labor Force [25]

465,000 (1981 est.); 45,000 wage earners (1980)

Labor Force by Occupation [26]

Agriculture	47%
Services	29
Industry and commerce	14
Government	10

Unemployment Rate [27]

20% (1991 est.)

For sources, notes, and explanations, see Annotated Source Appendix, page 1061.

Production Sectors

Energy Resource Summary [28]

Electricity: Capacity: 110,000 kW. Production: 135 million kWh. Consumption per capita: 61 kWh (1993).

Telecommunications [30]

- 17,000 (1991 est.) telephones; poor system of cable and open-wire lines, minor microwave radio relay links, and radiotelephone communications stations (improvements being made)
- Domestic: mostly cable and open-wire lines
- International: satellite earth stations - 1 Intelsat (Atlantic Ocean) and 2 Arabsat
- Radio: Broadcast stations: AM 2, FM 0, shortwave 0 Radios: 300,000 (1993 est.)
- Television: Broadcast stations: 1 (1987 est.) Televisions: 50,000 (1992 est.)

Top Agricultural Products [32]

Agriculture accounts for 27.1% of the GDP; produces dates, millet, sorghum, root crops; cattle, sheep; fish products.

Top Mining Products [33]

Metric tons except as noted[e]	3/15/95[*]
Cement	100,000
Gold (kg.)	1,740
Iron ore, gross weight (000 tons)	11,400[r]
Petroleum refinery products (000 42-gal. bls.)	2,000
Salt	5,500

Transportation [31]

Railways: total: 704 km (single track); note - owned and operated by government mining company; standard gauge: 704 km 1.435-m gauge (1995)

Highways: total: 7,496 km; paved: 1,342 km; unpaved: 6,154 km (1987 est.)

Merchant marine: none

Airports

Total:	24
With paved runways 2,438 to 3,047 m:	3
With paved runways 1,524 to 2,437 m:	4
With paved runways 914 to 1,523 m:	1
With paved runways under 914 m:	2
With unpaved runways 2,438 to 3,047 m:	1

Tourism [34]

Finance, Economics, and Trade

GDP and Manufacturing Summary [35]

	1980	1985	1990	1991	1992
Gross Domestic Product					
Millions of 1980 dollars	829	874	1,023	1,018	1,032[e]
Growth rate in percent	3.93	3.11	4.00	-0.50	1.39[e]
Manufacturing Value Added					
Millions of 1980 dollars	43	66	89	95	101[e]
Growth rate in percent	-1.39	7.80	5.30	6.76	6.28[e]
Manufacturing share in percent of current prices	5.6	12.8	12.9	12.0	NA

Economic Indicators [36]

- **National product**: GDP—purchasing power parity—$2.8 billion (1995 est.)

- **National product real growth rate**: 4% (1995 est.)

- **National product per capita**: $1,200 (1995 est.)

- **Inflation rate (consumer prices)**: 3.5% (1995 est.)

- **External debt**: $1.9 billion (1992 est.)

Balance of Payments Summary [37]

Values in millions of dollars.

	1988	1989	1990	1991	1992
Exports of goods (f.o.b.)	437.6	447.9	443.9	435.8	406.8
Imports of goods (f.o.b.)	-348.9	-349.3	-382.9	-399.1	-461.3
Trade balance	88.7	98.6	61.0	36.7	-54.5
Services - debits	-306.7	-251.7	-187.1	-185.8	-208.9
Services - credits	39.4	39.2	30.6	33.2	21.2
Private transfers (net)	-22.1	-25.0	-15.6	-17.2	24.5
Government transfers (net)	104.8	120.3	101.5	103.3	100.6
Long-term capital (net)	76.9	42.9	36.2	35.3	79.5
Short-term capital (net)	23.0	-2.4	-0.3	-8.6	-4.1
Errors and omissions	-16.0	-3.6	-62.3	19.5	58.8
Overall balance	-12.0	18.3	-36.0	16.4	17.1

Exchange Rates [38]

Currency: **ouguiya.**
Symbol: **UM.**

Data are currency units per $1.

January 1996	135.690
1995	129.768
1994	123.575
1993	120.806
1992	87.027
1991	81.946

Imports and Exports

Top Import Origins [39]

$355 million (c.i.f., 1994 est.).

Origins	%
Algeria	15
China	6
US	3
France	NA
Germany	NA
Spain	NA
Italy	NA

Top Export Destinations [40]

$390 million (f.o.b., 1994 est.).

Destinations	%
Japan	27
Italy	NA
Belgium	NA
Luxembourg	NA

Foreign Aid [41]

Recipient: ODA, $NA.

Import and Export Commodities [42]

Import Commodities	Export Commodities
Foodstuffs	Iron ore
Consumer goods	Fish and fish products
Petroleum products	
Capital goods	

Mauritius

Geography [1]

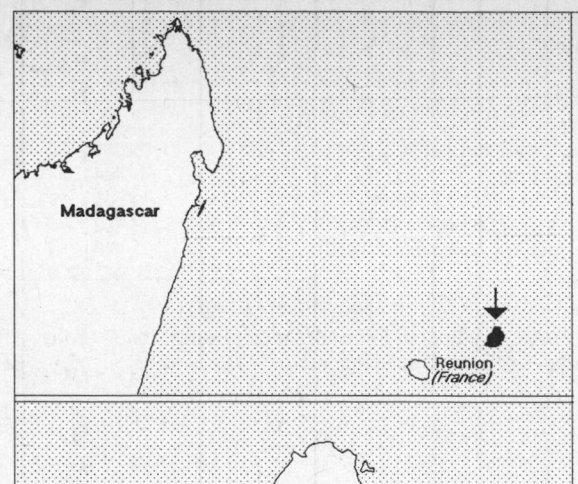

Total area:
1,860 sq km   718 sq mi
Land area:
1,850 sq km   714 sq mi
Comparative area:
Almost 11 times the size of Washington, DC
Note: Includes Agalega Islands, Cargados Carajos Shoals (Saint Brandon), and Rodrigues
Land boundaries:
0 km
Coastline:
177 km
Climate:
Tropical, modified by Southeast trade winds; warm, dry winter (May to November); hot, wet, humid summer (November to May)
Terrain:
Small coastal plain rising to discontinuous mountains encircling central plateau
Natural resources:
Arable land, fish
Land use:
Arable land: 54%
Permanent crops: 4%
Meadows and pastures: 4%
Forest and woodland: 31%
Other: 7%

Demographics [2]

	1970	1980	1990	1995[1]	1996	2000	2010	2020	2030
Population	830	964	1,074	1,128	1,140	1,196	1,328	1,440	1,529
Population density (persons per sq. mi.)	1,162	1,350	1,504	1,580	1,597	1,676	1,860	2,017	2,142
(persons per sq. km.)	449	521	581	610	617	647	718	779	827
Net migration rate (per 1,000 population)	-3.9	-3.9	-3.8	-3.6	0.0	0.0	0.0	0.0	0.0
Births[35]	21	24	NA	NA	21	NA	NA	NA	NA
Deaths[36]	7	7	NA	NA	7	NA	NA	NA	NA
Life expectancy - males	NA	NA	65.5	66.6	66.7	67.4	69.0	70.5	71.9
Life expectancy - females	NA	NA	73.5	74.1	74.3	75.2	76.9	78.4	79.8
Birth rate (per 1,000)	27.5	27.2	21.1	19.2	19.0	18.4	16.1	14.9	14.0
Death rate (per 1,000)	7.9	7.2	6.6	6.6	6.7	6.7	7.0	7.8	9.3
Women of reproductive age (15-49 yrs.)	NA	NA	296	319	324	337	350	357	364
of which are currently married	NA	NA	179	NA	197	208	219	NA	NA
Fertility rate	NA	2.9	2.3	2.2	2.2	2.2	2.1	2.1	2.0

Except as noted, values for vital statistics are in thousands; life expectancy is in years.

Health

Health Indicators [3]

% of population with access to	
safe water (1990-95)	99
adequate sanitation (1990-95)	99
health services (1985-95)	100
% of 1-year-olds immunized (1990-94) against	
TB (tuberculosis)	87
DPT (diphtheria, pertussis, tetanus)	89
polio	89
measles	85
% of contraceptive prevalence (1980-94)	75
ORT use rate (1990-94)	NA

Health Expenditures [4]

For sources, notes, and explanations, see Annotated Source Appendix, page 1061.

617

Human Factors

Women and Children [5]

% of pregnant women immunized (tetanus 1990-94)	78
% of births attended by trained health personnel (1983-94)	85
Maternal mortality rate (1980-92)	99
Under-5 mortality rate (1994)	23
% under-5 moderately/severely underweight (1980-1994)	24

Burden of Disease [6]

Population per physician (1989)	1,166.67
Population per nurse	NA
Population per hospital bed (1985)	300.00
AIDS cases per 100,000 people (1994)	0.8
Malaria cases per 100,000 people (1992)	NA

Ethnic Division [7]

Indo-Mauritian	68%
Creole	27%
Sino-Mauritian	3%
Franco-Mauritian	2%

Religion [8]

Hindu	52%
Christian	28.3%
Roman Catholic	26%
Protestant	2.3%
Muslim	16.6%
Other	3.1%

Major Languages [9]

English (official), Creole, French, Hindi, Urdu, Hakka, Bojpoori.

Education

Public Education Expenditures [10]

Million (Rupee)	1980	1990	1991	1992	1993	1994
Total education expenditure	454	1,384	NA	NA	NA	NA
as percent of GNP	5.3	3.7	NA	NA	NA	NA
as percent of total govt. expend.	11.6	11.8	NA	NA	NA	NA
Current education expenditure	408	1,287	1,451	1,585	1,963	NA
as percent of GNP	4.7	3.4	3.4	3.3	3.6	NA
as percent of current govt. expend.	15.5	14.0	NA	NA	NA	NA
Capital expenditure	46	97	NA	NA	NA	NA

Educational Attainment [11]

Age group (1990)	25+
Total population	540,244
Highest level attained (%)	
No schooling	18.3
First level	
Not completed	42.6
Completed	6.1
Entered second level	
S-1	7.2
S-2	23.9
Postsecondary	1.9

Illiteracy [12]

In thousands and percent[1]	1990	1995	2000
Illiterate population (15+ yrs.)	149	138	126
Illiteracy rate - total pop. (%)	19.8	16.9	14.3
Illiteracy rate - males (%)	14.9	13.0	11.3
Illiteracy rate - females (%)	24.9	20.7	17.3

Libraries [13]

	Admin. Units	Svc. Pts.	Vols. (000)	Shelving (meters)	Vols. Added	Reg. Users
National (1992)	1	1	38	467	2,603	NA
Nonspecialized	NA	NA	NA	NA	NA	NA
Public (1986)	2	2	17	260	1,953	8,147
Higher ed.	NA	NA	NA	NA	NA	NA
School (1990)	26	26	161	NA	7,555	14,345

Daily Newspapers [14]

	1980	1985	1990	1994
Number of papers	10	7	7	6
Circ. (000)	80	70[e]	80	75

Culture [15]

Cinema (seats per 1,000)	12.8
Annual attendance per person	0.7
Gross box office receipts (mil. Rupee)	22[e]
Museums (reporting)	5
Visitors (000)	350
Annual receipts (000 Rupee)	NA

Science and Technology

Scientific/Technical Forces [16]

Scientists/engineers	389
Number female	NA
Technicians	170
Number female	NA
Total[3]	559

R&D Expenditures [17]

	Rupee (000) 1992
Total expenditure[1]	177,000
Capital expenditure	48,000
Current expenditure	129,000
Percent current	72.9

U.S. Patents Issued [18]

Values show patents issued to citizens of the country by the U.S. Patents Office.

	1993	1994	1995
Number of patents	1	0	1

For sources, notes, and explanations, see Annotated Source Appendix, page 1061.

Government and Law

Organization of Government [19]

Long-form name:
Republic of Mauritius
Type:
Parliamentary democracy
Independence:
12 March 1968 (from UK)
National holiday:
Independence Day, 12 March (1968)
Constitution:
12 March 1968; amended 12 March 1992
Legal system:
Based on French civil law system with elements of English common law in certain areas
Executive branch:
President; Vice President; Prime Minister; Deputy Prime Minister; Council of Ministers
Legislative branch:
Unicameral: Legislative Assembly
Judicial branch:
Supreme Court

Elections [20]

Legislative Assembly	% of votes
MMM/MLP	65
MSM/RMM	20
Other	15

Government Expenditures [21]

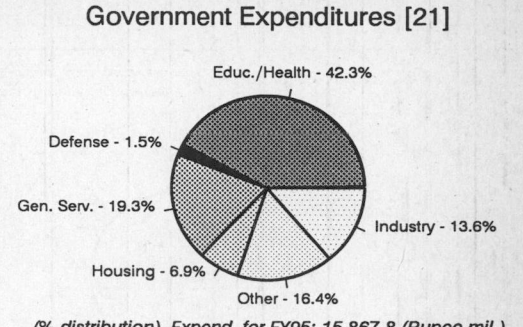

Educ./Health - 42.3%
Defense - 1.5%
Gen. Serv. - 19.3%
Industry - 13.6%
Housing - 6.9%
Other - 16.4%

(% distribution). Expend. for FY95: 15,867.8 (Rupee mil.)

Crime [23]

	1994
Crime volume	
Cases known to police	38,346
Attempts (percent)	0.10
Percent cases solved	40.10
Crimes per 100,000 persons	3,430.40
Persons responsible for offenses	
Total number offenders	23,336
Percent female	3.20
Percent juvenile (8-17 yrs.)	8.50
Percent foreigners	0.30

Military Expenditures and Arms Transfers [22]

	1990	1991	1992	1993	1994
Military expenditures					
Current dollars (mil.)	8	10	11	11	11
1994 constant dollars (mil.)	9	11	11	11	11
Armed forces (000)	1	1	1	1	1
Gross national product (GNP)					
Current dollars (mil.)	2,519	2,753	3,010	3,220	3,454
1994 constant dollars (mil.)	2,803	2,951	3,139	3,286	3,454
Central government expenditures (CGE)					
1994 constant dollars (mil.)	681	699	768	745	776
People (mil.)	1.1	1.1	1.1	1.1	1.1
Military expenditure as % of GNP	0.3	0.4	0.4	0.3	0.3
Military expenditure as % of CGE	1.3	1.5	1.5	1.4	1.4
Military expenditure per capita (1994 $)	8	10	10	10	10
Armed forces per 1,000 people (soldiers)	0.9	0.9	0.9	0.9	0.9
GNP per capita (1994 $)	2,611	2,720	2,864	2,970	3,092
Arms imports[6]					
Current dollars (mil.)	5	0	5	5	0
1993 constant dollars (mil.)	6	0	5	5	0
Arms exports[6]					
Current dollars (mil.)	0	0	0	0	0
1993 constant dollars (mil.)	0	0	0	0	0
Total imports[7]					
Current dollars (mil.)	1,618	1,575	1,630	1,715	1,926
1993 constant dollars (mil.)	1,801	1,688	1,700	1,750	1,926
Total exports[7]					
Current dollars (mil.)	1,194	1,195	1,297	1,299	1,347
1993 constant dollars (mil.)	1,329	1,281	1,353	1,326	1,347
Arms as percent of total imports[8]	0.3	0	0.3	0.3	0
Arms as percent of total exports[8]	0	0	0	0	0

Human Rights [24]

	SSTS	FL	FAPRO	PPCG	APROBC	TPW	PCPTW	STPEP	PHRFF	PRW	ASST	AFL
Observes	P	P			P	P	P	P		P	P	P
	EAFRD	CPR	ESCR	SR	ACHR	MAAE	PVIAC	PVNAC	EAFDAW	TCIDTP	RC	
Observes	P	P	P			P	P	P	P	P	P	

P = Party; S = Signatory; see Appendix for meaning of abbreviations.

Labor Force

Total Labor Force [25]

335,000

Labor Force by Occupation [26]

Government services	29%
Agriculture and fishing	27
Manufacturing	22
Other	22

Unemployment Rate [27]

2.4% (1991 est.)

For sources, notes, and explanations, see Annotated Source Appendix, page 1061.

619

Production Sectors

Commercial Energy Production and Consumption

Data are shown in quadrillion (10^{15}) BTUs and percent for 1995
Values for hydroelectric, nuclear, geothermal, solar, and wind power refer to electrical generation.

Production [28]	Consumption [29]
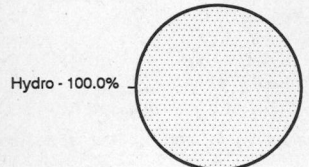	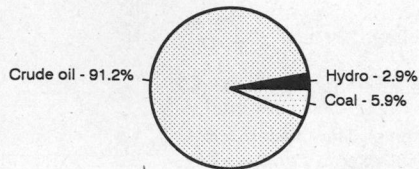
Hydro - 100.0%	Crude oil - 91.2% Hydro - 2.9% Coal - 5.9%

Net hydroelectric power	0.001
Total	0.001

Crude oil	0.031
Coal	0.002
Net hydroelectric power	0.001
Total	0.034

Telecommunications [30]

- 65,000 (1985 est.) telephones; small system with good service
- Domestic: primarily microwave radio relay
- International: satellite earth station - 1 Intelsat (Indian Ocean); new microwave link to Reunion; HF radiotelephone links to several countries
- Radio: Broadcast stations: AM 2, FM 0, shortwave 0 Radios: 395,000 (1992 est.)
- Television: Broadcast stations: 4 (1987 est.) Televisions: 151,096 (1991 est.)

Transportation [31]

Railways: 0 km

Highways: total: 1,831 km; paved: 1,703 km (including 29 km of expressways); unpaved: 128 km (1991 est.)

Merchant marine: total: 17 ships (1,000 GRT or over) totaling 221,446 GRT/308,478 DWT; ships by type: bulk 1, cargo 9, container 4, liquefied gas tanker 1, oil tanker 1, passenger-cargo 1 (1995 est.)

Airports

Total:	4
With paved runways 2,438 to 3,047 m:	1
With paved runways 914 to 1,523 m:	1
With paved runways under 914 m:	2 (1995 est.)

Top Agricultural Products [32]

Produces sugarcane, tea, corn, potatoes, bananas, pulses; cattle, goats; fish.

Top Mining Products [33]

Metric tons except as noted	1994[*]
Lime, from coral	7,000
Salt, marine	6,000
Sand, coral	300,000
Basalt, for construction	1,000,000

Tourism [34]

	1990	1991	1992	1993	1994
Visitors[1]	310	316	346	385	410
Tourists[12]	292	301	335	375	401
Cruise passengers	18	15	11	10	9
Tourism receipts	244	252	299	301	356
Tourism expenditures	94	110	142	128	140
Fare receipts	141	156	172	144	145
Fare expenditures	17	18	22	23	24

Travelers are in thousands, money in million U.S. dollars.

For sources, notes, and explanations, see Annotated Source Appendix, page 1061.

Manufacturing Sector

Manufacturing Summary [35]

	1987		1988		1989		1990		1991	
	$ bil.	%	$ bil.	%	$ bil.	%	$ bil.	%	$ bil.	%
Establishments or enterprises (number)	891	0.038	1,004	0.048	1,007	0.054	1,026	0.057	1,016	0.133
Total employment (000)	106	0.078	112	0.082	114	0.092	113	0.103	116	0.167
Production workers (000)	-	-	-	-	-	-	-	-	-	-
Output ($ bil.)	1	0.012	1	0.012	1	0.011	2	0.014	2	0.016
Value added ($ bil.)	0.336	0.008	0.372	0.008	0.368	0.008	0.443	0.009	0.491	0.014
Capital investment ($ mil.)	-	-	-	-	-	-	-	-	-	-
M & E investment ($ mil.)	-	-	-	-	-	-	-	-	-	-
Employees per establishment (number)	119	205.652	111	170.567	113	171.927	111	179.028	115	125.995
Production workers per establishment	-	-	-	-	-	-	-	-	-	-
Output per establishment ($ mil.)	1	32.786	1	24.144	1	20.967	2	24.027	2	11.980
Capital investment per estab. ($ mil.)	-	-	-	-	-	-	-	-	-	-
M & E per establishment ($ mil)	-	-	-	-	-	-	-	-	-	-
Payroll per employee ($)	2,211	24.654	2,321	23.295	2,308	20.655	2,651	19.113	3,108	21.286
Wages per production worker ($)	-	-	-	-	-	-	-	-	-	-
Hours per production worker (hours)	-	-	-	-	-	-	-	-	-	-
Output per employee ($)	11,940	15.942	12,056	14.155	11,688	12.195	13,909	13.421	13,959	9.509
Capital investment per employee ($)	-	-	-	-	-	-	-	-	-	-
M & E per employee ($)	-	-	-	-	-	-	-	-	-	-

Note: Columns headed % show percent of world total or ratio. Ratios closest to 100 are closest to world average. M & E stands for machinery & equipment.

Output in Manufacturing

	1987		1988		1989		1990		1991	
	$ bil.	%	$ bil.	%	$ bil.	%	$ bil.	%	$ bil.	%
3110 - Food products	0.460	46.00	0.458	45.80	0.438	43.80	0.518	25.90	0.515	25.75
3130 - Beverages	0.058	5.80	0.046	4.60	0.045	4.50	0.054	2.70	0.061	3.05
3140 - Tobacco	0.021	2.10	0.015	1.50	0.014	1.40	0.016	0.80	0.014	0.70
3210 - Textiles	0.051	5.10	0.058	5.80	0.063	6.30	0.076	3.80	0.085	4.25
3220 - Wearing apparel	0.496	49.60	0.546	54.60	0.546	54.60	0.645	32.25	0.669	33.45
3230 - Leather and products	0.008	0.80	0.010	1.00	0.012	1.20	0.014	0.70	0.013	0.65
3240 - Footwear	0.006	0.60	0.006	0.60	0.006	0.60	0.007	0.35	0.007	0.35
3310 - Wood products	0.006	0.60	0.006	0.60	0.006	0.60	0.010	0.50	0.007	0.35
3320 - Furniture, fixtures	0.009	0.90	0.009	0.90	0.011	1.10	0.017	0.85	0.017	0.85
3410 - Paper and products	0.006	0.60	0.007	0.70	0.007	0.70	0.009	0.45	0.010	0.50
3420 - Printing, publishing	0.018	1.80	0.020	2.00	0.019	1.90	0.026	1.30	0.027	1.35
3550 - Rubber products	0.003	0.30	0.004	0.40	0.004	0.40	0.005	0.25	0.006	0.30
3560 - Plastic products nec	0.013	1.30	0.016	1.60	0.016	1.60	0.019	0.95	0.022	1.10
3610 - Pottery, china, etc.	0.000	0.00	0.000	0.00	0.000	0.00	0.000	0.00	0.001	0.05
3620 - Glass and products	0.000	0.00	0.000	0.00	0.000	0.00	0.000	0.00	0.001	0.05
3690 - Nonmetal products nec	0.016	1.60	0.019	1.90	0.023	2.30	0.028	1.40	0.032	1.60
3820 - Machinery nec	0.004	0.40	0.004	0.40	0.006	0.60	0.007	0.35	0.007	0.35
3830 - Electrical machinery	0.016	1.60	0.020	2.00	0.018	1.80	0.020	1.00	0.027	1.35
3840 - Transportation equipment	0.005	0.50	0.006	0.60	0.007	0.70	0.008	0.40	0.010	0.50
3841 - Shipbuilding, repair	0.002	0.20	0.003	0.30	0.004	0.40	0.003	0.15	0.004	0.20
3843 - Motor vehicles	0.003	0.30	0.003	0.30	0.003	0.30	0.003	0.15	0.003	0.15
3850 - Professional goods	0.033	3.30	0.046	4.60	0.041	4.10	0.042	2.10	0.035	1.75
3900 - Industries nec	0.029	2.90	0.042	4.20	0.040	4.00	0.049	2.45	0.053	2.65

Note: Codes are International Standard Industry codes (ISIC). Percentages are % of total Output. [f]: Factor Prices; [p]: Producer Prices.

Finance, Economics, and Trade

Economic Indicators [36]

- **National product**: GDP—purchasing power parity—$10.9 billion (1995 est.)
- **National product real growth rate**: 2.7% (1995 est.)
- **National product per capita**: $9,600 (1995 est.)
- **Inflation rate (consumer prices)**: 9.4% (1993 est.)
- **External debt**: $996.8 million (1993 est.)

Balance of Payments Summary [37]

Values in millions of dollars.

	1989	1990	1991	1992	1993
Exports of goods (f.o.b.)	986.5	1,205.2	1,215.1	1,302.6	1,304.4
Imports of goods (f.o.b.)	-1,196.0	-1,474.8	-1,419.1	-1,473.5	-1,558.6
Trade balance	-209.5	-269.6	-204.0	-170.8	-254.2
Services - debits	-420.0	-519.7	-544.7	-623.3	-605.5
Services - credits	451.0	572.6	649.2	700.6	666.3
Private transfers (net)	67.6	82.3	79.5	86.6	91.7
Government transfers (net)	7.2	14.5	1.9	5.4	8.3
Long-term capital (net)	106.6	144.2	78.3	-23.5	20.2
Short-term capital (net)	-56.8	-5.6	-36.4	8.7	-0.9
Errors and omissions	198.5	213.2	167.2	59.7	81.2
Overall balance	144.6	231.9	191.0	43.4	7.1

Exchange Rates [38]

Currency: **Mauritian rupee.**
Symbol: **MauR.**

Data are currency units per $1.

January 1996	17.842
1995	17.386
1994	17.960
1993	17.648
1992	15.563
1991	15.652

Imports and Exports

Top Import Origins [39]

$1.9 billion (f.o.b., 1994).

Origins	%
EU	NA
US	NA
South Africa	NA
Japan	NA

Top Export Destinations [40]

$1.3 billion (f.o.b., 1994).

Destinations	%
EU	77
US	15

Foreign Aid [41]

Recipient: ODA, $NA.

Import and Export Commodities [42]

Import Commodities

Manufactured goods 50%
Capital equipment 17%
Foodstuffs 13%
Petroleum products 8%
Chemicals 7%

Export Commodities

Textiles 44%
Sugar 40%
Light manufactures 10%

Mexico

Geography [1]

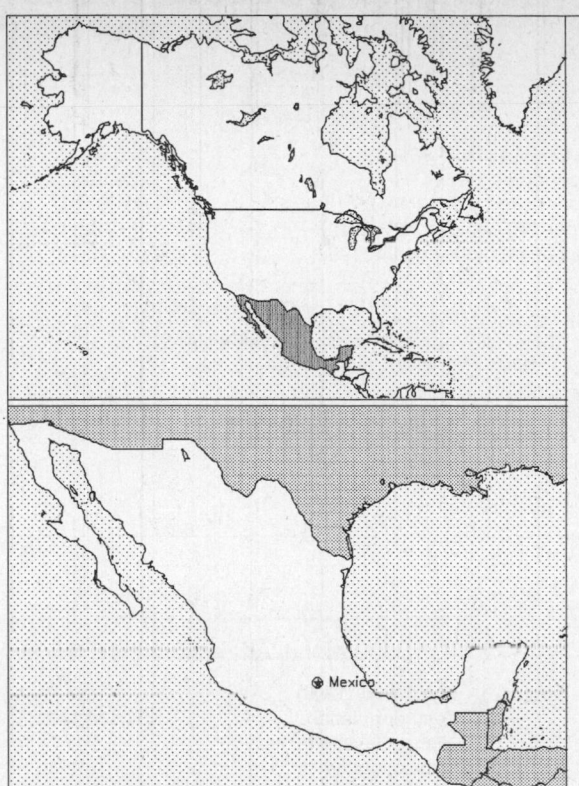

Total area:
1,972,550 sq km   761,606 sq mi
Land area:
1,923,040 sq km   742,490 sq mi
Comparative area:
Slightly less than three times the size of Texas
Land boundaries:
Total 4,538 km, Belize 250 km, Guatemala 962 km, US 3,326 km
Coastline:
9,330 km
Climate:
Varies from tropical to desert
Terrain:
High, rugged mountains, low coastal plains, high plateaus, and desert
Natural resources:
Petroleum, silver, copper, gold, lead, zinc, natural gas, timber
Land use:
Arable land: 12%
Permanent crops: 1%
Meadows and pastures: 39%
Forest and woodland: 24%
Other: 24%

Demographics [2]

	1970	1980	1990	1995[1]	1996	2000	2010	2020	2030
Population	52,236	68,686	85,121	93,986	95,772	102,912	120,115	136,096	150,357
Population density (persons per sq. mi.)	70	93	115	127	129	139	162	183	203
(persons per sq. km.)	27	36	44	49	50	54	62	71	78
Net migration rate (per 1,000 population)	-0.9	-4.4	-3.3	-3.0	-3.0	-2.8	-2.2	-1.8	-1.4
Births	NA	NA	NA	NA	2,513	NA	NA	NA	NA
Deaths	555	NA	423	NA	439	NA	NA	NA	NA
Life expectancy - males	58.6	63.6	67.9	69.7	70.1	71.4	74.1	76.1	77.6
Life expectancy - females	62.7	70.3	75.1	77.1	77.5	78.8	81.5	83.4	84.6
Birth rate (per 1,000)	44.9	34.9	29.0	26.6	26.2	24.3	20.3	17.6	15.5
Death rate (per 1,000)	10.8	6.8	5.1	4.6	4.6	4.4	4.3	4.6	5.4
Women of reproductive age (15-49 yrs.)	NA	15,890	21,559	24,569	25,173	27,450	32,690	36,236	37,851
of which are currently married	5,605	NA	12,857	NA	15,427	17,141	21,002	NA	NA
Fertility rate	NA	4.6	3.5	3.1	3.0	2.8	2.4	2.2	2.1

Except as noted, values for vital statistics are in thousands; life expectancy is in years.

Health

Health Indicators [3]

% of population with access to	
safe water (1990-95)	83
adequate sanitation (1990-95)	50
health services (1985-95)	78
% of 1-year-olds immunized (1990-94) against	
TB (tuberculosis)	98
DPT (diphtheria, pertussis, tetanus)	91
polio	92
measles	94
% of contraceptive prevalence (1980-94)	53
ORT use rate (1990-94)	81

Health Expenditures [4]

Total health expenditure, 1990 (official exchange rate)	
Millions of dollars	7,648
Dollars per capita	89
Health expenditures as a percentage of GDP	
Total	3.2
Public sector	1.6
Private sector	1.6
Development assistance for health	
Total aid flows (millions of dollars)[1]	65
Aid flows per capita (dollars)	0.8
Aid flows as a percentage of total health expenditure	0.9

For sources, notes, and explanations, see Annotated Source Appendix, page 1061.

623

Human Factors

Women and Children [5]

% of pregnant women immunized (tetanus 1990-94)	NA
% of births attended by trained health personnel (1983-94)	77
Maternal mortality rate (1980-92)	110
Under-5 mortality rate (1994)	32
% under-5 moderately/severely underweight (1980-1994)	14

Burden of Disease [6]

Population per physician (1986)	1,770.63
Population per nurse (1984)	838.79
Population per hospital bed (1990)	801.37
AIDS cases per 100,000 people (1994)	4.2
Malaria cases per 100,000 people (1992)	18

Ethnic Division [7]

Mestizo	60%
Amerindian or predominantly Amerindian	30%
Caucasian or predominantly Caucasian	9%
Other	1%

Religion [8]

Nominally Roman Catholic	89%
Protestant	6%
Other	5%

Major Languages [9]

Spanish, various Mayan dialects.

Education

Public Education Expenditures [10]

	1980	1985	1990	1992	1993	1994
Million or Trillion (T) (Peso)[43]						
Total education expenditure	204,326	1.77 T	26.64 T	48.14 T	6.24 T	71.19 T
as percent of GNP	4.7	3.9	4.0	4.8	5.6	5.8
as percent of total govt. expend.	NA	NA	NA	NA	NA	NA
Current education expenditure	129,797	1.23 T	16,617	32,780	43,107	NA
as percent of GNP	3.0	2.7	2.5	3.2	3.9	NA
as percent of current govt. expend.	NA	NA	NA	NA	NA	NA
Capital expenditure	10,174	99,868	1.05 T	2.22 T	3.13 T	NA

Educational Attainment [11]

Age group (1990)	25+
Total population	31,188,180
Highest level attained (%)	
No schooling	18.8
First level	
Not completed	28.6
Completed	19.9
Entered second level	
S-1	12.7
S-2	10.7
Postsecondary	9.2

Illiteracy [12]

In thousands and percent[1]	1990	1995	2000
Illiterate population (15+ yrs.)	6,162	6,246	6,015
Illiteracy rate - total pop. (%)	11.9	10.5	8.9
Illiteracy rate - males (%)	9.2	8.4	7.2
Illiteracy rate - females (%)	14.5	12.5	10.6

Libraries [13]

	Admin. Units	Svc. Pts.	Vols. (000)	Shelving (meters)	Vols. Added	Reg. Users
National (1989)	1	NA	1,500	NA	17,169	110,313
Nonspecialized	NA	NA	NA	NA	NA	NA
Public (1993)	4,894	4,894	19,875	NA	NA	73 mil
Higher ed. (1993)	1,139	1,139	12,775	NA	1 mil	1 mil
School (1990)	3,546	3,546	9,844	NA	533,898	4 mil

Daily Newspapers [14]

	1980	1985	1990	1994
Number of papers	317	332	285[e]	309
Circ. (000)	8,322	9,964	11,237[e]	10,420

Culture [15]

Cinema (seats per 1,000)	NA
Annual attendance per person	3.0
Gross box office receipts (mil. Peso)	NA
Museums (reporting)[13]	1
Visitors (000)	9,883
Annual receipts (000 Peso)	NA

Science and Technology

Scientific/Technical Forces [16]

Scientists/engineers	8,595
Number female	NA
Technicians	2,477
Number female	NA
Total[26]	11,072

R&D Expenditures [17]

	Peso (000,000) 1993
Total expenditure	3,566,158
Capital expenditure	830,260
Current expenditure	2,735,898
Percent current	76.7

U.S. Patents Issued [18]

Values show patents issued to citizens of the country by the U.S. Patents Office.

	1993	1994	1995
Number of patents	50	52	45

For sources, notes, and explanations, see Annotated Source Appendix, page 1061.

Government and Law

Organization of Government [19]

Long-form name:
United Mexican States

Type:
Federal republic operating under a centralized government

Independence:
16 September 1810 (from Spain)

National holiday:
Independence Day, 16 September (1810)

Constitution:
5 February 1917

Legal system:
Mixture of US constitutional theory and civil law system; judicial review of legislative acts; accepts compulsory ICJ jurisdiction, with reservations

Executive branch:
President; Cabinet

Legislative branch:
Bicameral National Congress (Congreso de la Union): Senate (Camara de Senadores) and Chamber of Deputies (Camara de Diputados)

Judicial branch:
Supreme Court of Justice (Corte Suprema de Justicia)

Crime [23]

Elections [20]

Chamber of Deputies	% of seats
Institutional Revolutionary Party (PRI)	60.0
National Action Party (PAN)	23.8
Democratic Revol Party (PRD)	14.2
Workers Party (PT)	2.0

Government Expenditures [21]

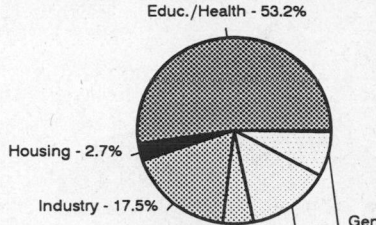

Educ./Health - 53.2%
Housing - 2.7%
Industry - 17.5%
Defense - 4.5%
Other - 14.2%
Gen. Services - 7.9%

(% distribution). Expend. for CY94: 214,161 (New Peso bil.)

Military Expenditures and Arms Transfers [22]

	1990	1991	1992	1993	1994
Military expenditures					
Current dollars (mil.)	1,357[e]	1,390[e]	1,656[e]	1,605	2,246
1994 constant dollars (mil.)	1,510[e]	1,490[e]	1,727[e]	1,638	2,246
Armed forces (000)	175	175	175	175	175
Gross national product (GNP)					
Current dollars (mil.)	293,500	317,600	335,600	344,100	363,300
1994 constant dollars (mil.)	326,600	340,400	349,900	351,200	363,300
Central government expenditures (CGE)					
1994 constant dollars (mil.)	66,800	59,720	57,230	60,540	66,190[e]
People (mil.)	85.1	86.9	88.6	90.4	92.2
Military expenditure as % of GNP	0.5	0.4	0.5	0.5	0.6
Military expenditure as % of CGE	2.3	2.5	3.0	2.7	3.4
Military expenditure per capita (1994 $)	18	17	19	18	24
Armed forces per 1,000 people (soldiers)	2.1	2.0	2.0	1.9	1.9
GNP per capita (1994 $)	3,837	3,918	3,947	3,884	3,940
Arms imports[6]					
Current dollars (mil.)	110	60	110	20	80
1994 constant dollars (mil.)	122	64	115	20	80
Arms exports[6]					
Current dollars (mil.)	20	20	20	20	20
1994 constant dollars (mil.)	22	21	21	20	20
Total imports[7]					
Current dollars (mil.)	29,970	38,120	48,160	50,150	60,980
1994 constant dollars (mil.)	33,350	40,860	50,220	51,180	60,980
Total exports[7]					
Current dollars (mil.)	27,130	27,320	27,720	30,240	34,510
1994 constant dollars (mil.)	30,190	29,280	28,910	30,860	34,510
Arms as percent of total imports[8]	0.4	0.2	0.2	0	0.1
Arms as percent of total exports[8]	0.1	0.1	0.1	0.1	0.1

Human Rights [24]

	SSTS	FL	FAPRO	PPCG	APROBC	TPW	PCPTW	STPEP	PHRFF	PRW	ASST	AFL
Observes	P	P	P	P		P	P	P		P	P	P
	EAFRD	CPR	ESCR	SR	ACHR	MAAE	PVIAC	PVNAC	EAFDAW	TCIDTP	RC	
Observes	P	P	P		P		P		P	P	P	

P = Party; S = Signatory; see Appendix for meaning of abbreviations.

Labor Force

Total Labor Force [25]

33.6 million (1994)

Labor Force by Occupation [26]

Services	31.7%
Agriculture, forestry, hunting, and fishing	28
Commerce	14.6
Manufacturing	11.1
Construction	8.4
Transportation	4.7
Mining and quarrying	1.5

Unemployment Rate [27]

10% (1995 est.) plus considerable underemployment

For sources, notes, and explanations, see Annotated Source Appendix, page 1061.

625

Production Sectors

Commercial Energy Production and Consumption

Data are shown in quadrillion (10^{15}) BTUs and percent for 1995
Values for hydroelectric, nuclear, geothermal, solar, and wind power refer to electrical generation.

Production [28]	Consumption [29]
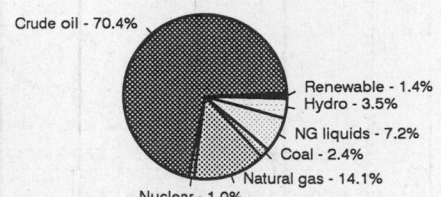	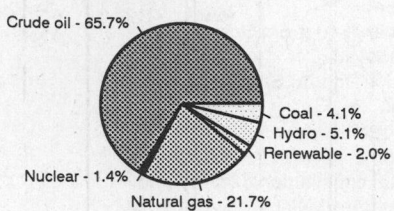

Production [28]:
- Crude oil - 70.4%
- Renewable - 1.4%
- Hydro - 3.5%
- NG liquids - 7.2%
- Coal - 2.4%
- Natural gas - 14.1%
- Nuclear - 1.0%

Consumption [29]:
- Crude oil - 65.7%
- Coal - 4.1%
- Hydro - 5.1%
- Renewable - 2.0%
- Nuclear - 1.4%
- Natural gas - 21.7%

Crude oil	5.742		Crude oil	3.673
Natural gas liquids	0.590		Dry natural gas	1.213
Dry natural gas	1.146		Coal	0.229
Coal	0.198		Net hydroelectric power	0.283
Net hydroelectric power	0.283		Net nuclear power	0.081
Net nuclear power	0.081		Geothermal, solar, wind	0.113
Geothermal, solar, wind	0.113		Total	5.592
Total	8.153			

Telecommunications [30]

- 11,890,868 (1993 est.) telephones; highly developed system with extensive microwave radio relay links; privatized December 1990
- Domestic: adequate telephone service for business and government, but population is poorly served; domestic satellite system with 120 earth stations; extensive microwave radio relay network
- International: satellite earth stations - 5 Intelsat (4 Atlantic Ocean, 1 Pacific Ocean); launched Solidaridad I and II satellites 1993 and 1994 respectively, giving improved access to South America, Central America and much of US as well as enhancing domestic communications; linked to Central American Microwave System of trunk connections
- Radio: Broadcast stations: AM 679, FM 0, shortwave 22 Radios: 22.5 million (1992 est.)
- Television: Broadcast stations: 238 Televisions: 13.1 million (1992 est.)

Transportation [31]

Railways: total: 20,567 km; standard gauge: 20,477 km 1.435-m gauge (246 km electrified); narrow gauge: 90 km 0.914-m gauge (1994)

Highways: total: 245,433 km; paved: 88,601 km (including 4,286 km of expressways); unpaved: 156,832 km (1993 est.)

Merchant marine: total: 51 ships (1,000 GRT or over) totaling 875,314 GRT/1,245,932 DWT; ships by type: cargo 1, chemical tanker 4, container 4, liquefied gas tanker 7, oil tanker 29, refrigerated cargo 1, roll-on/roll-off cargo 2, short-sea passenger 3 (1995 est.)

Airports

Total:	1,411
With paved runways over 3,047 m:	9
With paved runways 2,438 to 3,047 m:	25
With paved runways 1,524 to 2,437 m:	88
With paved runways 914 to 1,523 m:	66
With paved runways under 914 m:	815

Top Agricultural Products [32]

Agriculture accounts for 8.5% of the GDP; produces corn, wheat, soybeans, rice, beans, cotton, coffee, fruit, tomatoes; beef, poultry, dairy products; wood products.

Top Mining Products [33]

Metric tons except as noted	8/31/95[*]
Gold, mine output, Au content (kg.)	13,900
Iron ore, mine output, gr. wt. (000 tons)	15,000[e]
Silver metal, refined, primary (kg.)	1,700,000
Cement, hydraulic (000 tons)	29,700
Sand (000 cu. meters)	48,900
Gravel (000 cu. meters)	44,900
Gas, natural, gross (mil. cu. meters)	37,500
Natural gas liquids (000 42-gal. bls.)	170,000[e]
Petroleum, crude and lease (field) condensate (000 42-gal. bls.)	982,000

Tourism [34]

	1990	1991	1992	1993	1994
Visitors[19]	82,104	80,356	84,187	83,108	83,120
Tourists[19]	17,176	16,281	17,273	16,534	17,113
Excursionists[74]	64,038	62,880	65,511	65,088	64,437
Cruise passengers	890	1,195	1,403	1,486	1,570
Tourism receipts[75,76]	5,467	5,881	6,085	6,167	6,318
Tourism expenditures[76]	5,519	5,814	6,107	5,562	5,363
Fare receipts	448	446	499	455	479
Fare expenditures	473	514	525	489	499

Travelers are in thousands, money in million U.S. dollars.

For sources, notes, and explanations, see Annotated Source Appendix, page 1061.

Manufacturing Sector

Manufacturing Summary [35]

	1987		1988		1989		1990		1991	
	$ bil.	%	$ bil.	%	$ bil.	%	$ bil.	%	$ bil.	%
Establishments or enterprises (number)	3,860	0.165	3,815	0.182	3,789	0.202	3,771	0.211	3,682	0.482
Total employment (000)	1,237	0.912	1,236	0.903	1,270	1.032	1,273	1.150	1,244	1.791
Production workers (000)	868	1.287	868	1.389	891	1.211	902	1.271	789	2.396
Output ($ bil.)	51	0.500	65	0.557	74	0.627	81	0.705	92	0.905
Value added ($ bil.)	-		-		-		-		-	
Capital investment ($ mil.)	-		-		-		-		-	
M & E investment ($ mil.)	-		-		-		-		-	
Employees per establishment (number)	321	553.997	324	497.007	335	510.082	338	546.265	338	371.804
Production workers per establishment	225	782.406	228	764.540	235	598.472	239	603.845	214	497.390
Output per establishment ($ mil.)	13	303.682	17	306.242	20	309.944	21	334.668	25	187.919
Capital investment per estab. ($ mil.)	-		-		-		-		-	
M & E per establishment ($ mil)	-		-		-		-		-	
Payroll per employee ($)	3,028	33.771	3,822	38.367	4,519	40.453	5,162	37.210	6,202	42.480
Wages per production worker ($)	2,288	29.001	2,773	32.821	3,174	31.647	3,495	29.434	4,069	36.992
Hours per production worker (hours)	2,293	120.855	2,316	120.520	2,330	127.113	2,335	124.826	2,344	131.149
Output per employee ($)	41,055	54.817	52,483	61.617	58,235	60.763	63,491	61.265	74,200	50.542
Capital investment per employee ($)	-		-		-		-		-	
M & E per employee ($)	-		-		-		-		-	

Note: Columns headed % show percent of world total or ratio. Ratios closest to 100 are closest to world average. M & E stands for machinery & equipment.

Output in Manufacturing

	1987		1988		1989		1990		1991	
	$ bil.[p]	%	$ bil.[p]	%	$ bil.[p]	%	$ bil.[p]	%	$ bil.[p]	%
3110 - Food products	4.408	8.64	5.513	8.48	6.503	8.79	7.249	8.95	8.303	9.02
3130 - Beverages	2.503	4.91	3.153	4.85	3.683	4.98	4.204	5.19	5.358	5.82
3140 - Tobacco	0.689	1.35	0.712	1.10	0.826	1.12	0.868	1.07	1.211	1.32
3210 - Textiles	1.246	2.44	1.482	2.28	1.617	2.19	1.631	2.01	1.674	1.82
3211 - Spinning, weaving, etc.	0.932	1.83	1.084	1.67	1.196	1.62	1.185	1.46	1.125	1.22
3220 - Wearing apparel	0.301	0.59	0.393	0.60	0.405	0.55	0.416	0.51	0.458	0.50
3240 - Footwear	0.259	0.51	0.336	0.52	0.364	0.49	0.375	0.46	0.420	0.46
3310 - Wood products	0.093	0.18	0.108	0.17	0.123	0.17	0.130	0.16	0.118	0.13
3320 - Furniture, fixtures	0.104	0.20	0.136	0.21	0.160	0.22	0.176	0.22	0.197	0.21
3410 - Paper and products	1.754	3.44	2.208	3.40	2.359	3.19	2.262	2.79	2.345	2.55
3411 - Pulp, paper, etc.	1.254	2.46	1.570	2.42	1.667	2.25	1.599	1.97	1.629	1.77
3420 - Printing, publishing	0.255	0.50	0.346	0.53	0.380	0.51	0.386	0.48	0.426	0.46
3510 - Industrial chemicals	4.151	8.14	5.020	7.72	5.344	7.22	5.413	6.68	5.809	6.31
3511 - Basic chemicals, excl fertilizers	1.910	3.75	2.361	3.63	2.607	3.52	2.615	3.23	2.732	2.97
3513 - Synthetic resins, etc.	1.676	3.29	1.973	3.04	1.927	2.60	1.911	2.36	2.043	2.22
3520 - Chemical products nec	2.632	5.16	3.222	4.96	3.918	5.29	4.331	5.35	4.959	5.39
3522 - Drugs and medicines	0.971	1.90	1.160	1.78	1.430	1.93	1.592	1.97	1.783	1.94
3540 - Petroleum, coal products	0.376	0.74	0.437	0.67	0.486	0.66	0.549	0.68	0.596	0.65
3550 - Rubber products	0.731	1.43	0.910	1.40	0.946	1.28	0.990	1.22	1.037	1.13
3560 - Plastic products nec	0.533	1.05	0.682	1.05	0.841	1.14	0.852	1.05	0.907	0.99
3610 - Pottery, china, etc.	0.147	0.29	0.200	0.31	0.210	0.28	0.226	0.28	0.250	0.27
3620 - Glass and products	0.708	1.39	0.846	1.30	0.956	1.29	1.057	1.30	1.248	1.36
3690 - Nonmetal products nec	1.363	2.67	1.600	2.46	1.585	2.14	1.863	2.30	2.401	2.61
3710 - Iron and steel	3.558	6.98	5.001	7.69	5.182	7.00	5.531	6.83	5.419	5.89
3720 - Nonferrous metals	1.865	3.66	2.157	3.32	2.345	3.17	2.224	2.75	1.853	2.01
3810 - Metal products	1.330	2.61	1.759	2.71	1.896	2.56	2.139	2.64	2.354	2.56
3820 - Machinery nec	0.921	1.81	1.332	2.05	1.486	2.01	1.587	1.96	1.711	1.86
3825 - Office, computing machinery	0.337	0.66	0.466	0.72	0.552	0.75	0.635	0.78	0.700	0.76
3830 - Electrical machinery	1.910	3.75	2.403	3.70	2.668	3.61	2.848	3.52	2.969	3.23
3832 - Radio, television, etc.	0.715	1.40	0.848	1.30	0.947	1.28	1.078	1.33	0.961	1.04
3840 - Transportation equipment	5.533	10.85	7.667	11.80	9.577	12.94	11.000	13.58	15.000	16.30
3841 - Shipbuilding, repair	0.010	0.02	0.009	0.01	0.009	0.01	0.015	0.02	0.016	0.02
3843 - Motor vehicles	5.416	10.62	7.528	11.58	9.443	12.76	11.000	13.58	14.000	15.22
3850 - Professional goods	0.089	0.17	0.111	0.17	0.146	0.20	0.182	0.22	0.241	0.26
3900 - Industries nec	0.113	0.22	0.143	0.22	0.169	0.23	0.175	0.22	0.208	0.23

Note: Codes are International Standard Industry codes (ISIC). Percentages are % of total Output. [f]: Factor Prices; [p]: Producer Prices.

For sources, notes, and explanations, see Annotated Source Appendix, page 1061.

627

Finance, Economics, and Trade

Economic Indicators [36]

- **National product**: GDP—purchasing power parity—$721.4 billion (1995 est.)

- **National product real growth rate**: - 6.9% (1995 est.)

- **National product per capita**: $7,700 (1995 est.)

- **Inflation rate (consumer prices)**: 52% (1995 est.)

- **External debt**: $155 billion (1995 est.)

Balance of Payments Summary [37]

Values in millions of dollars.

	1989	1990	1991	1992	1993
Exports of goods (f.o.b.)	22,765	26,838	26,855	27,516	30,033
Imports of goods (f.o.b.)	-23,410	-31,271	-38,184	-48,193	-48,924
Trade balance	-645	-4,433	-11,329	-20,677	-18,891
Services - debits	-18,592	-21,912	-22,747	-23,957	-24,656
Services - credits	13,204	14,919	16,442	16,807	17,469
Private transfers (net)	1,922	2,679	2,639	2,908	2,591
Government transfers (net)	153	1,296	107	113	96
Long-term capital (net)	2,298	5,679	21,816	20,160	31,450
Short-term capital (net)	-936	2,847	3,343	6,848	609
Errors and omissions	2,775	1,228	-2,278	-457	-1,436
Overall balance	179	2,303	7,993	1,745	7,232

Exchange Rates [38]

Currency: **New Mexican peso.**
Symbol: **Mex$.**

Data are currency units per $1.

December 1995	7.6647
1995	6.4194
1994	3.3751
1993	3.1156
1992	3,094.9
1991	3,018.4

Imports and Exports

Top Import Origins [39]

$72 billion (f.o.b., 1995 est.), includes in-bond industries. Data are for 1994 est.

Origins	%
US	69
EU	12
Japan	6

Top Export Destinations [40]

$80 billion (f.o.b., 1995 est.), includes in-bond industries. Data are for 1994 est.

Destinations	%
US	85
EU	4.6
Japan	1.6

Foreign Aid [41]

	U.S. $	
ODA (1993)	85	million
US commitments, (Emergency Stabilization Fund)	13.5	billion
IMF (1995-96)	13	billion

Import and Export Commodities [42]

Import Commodities	Export Commodities
Metalworking machines	Crude oil
Steel mill products	Oil products
Agricultural machinery	Coffee
Electrical equipment	Silver
Car parts for assembly	Engines
Repair parts for motor vehicles	Motor vehicles
Aircraft and	Cotton
aircraft parts	Consumer electronics

Micronesia

Geography [1]

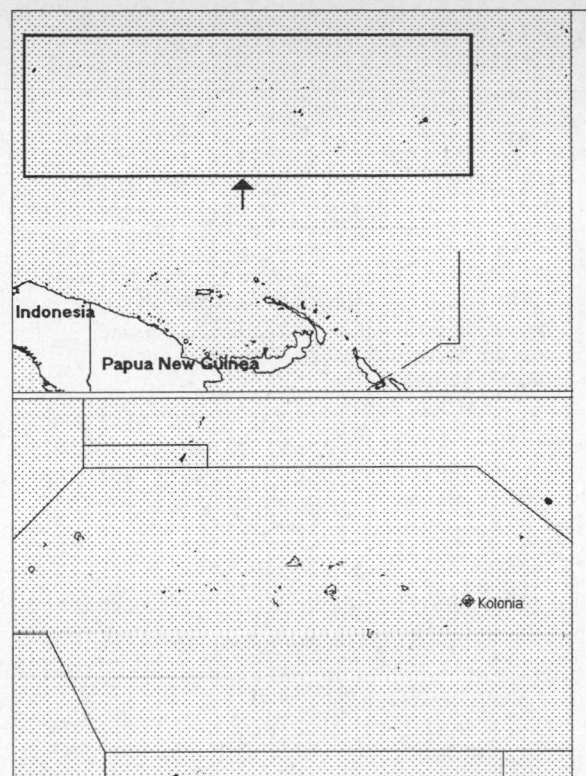

Total area:
  702 sq km   271 sq mi
Land area:
  702 sq km   271 sq mi
Comparative area:
  Four times the size of Washington, DC
  Note: Includes Pohnpei (Ponape), Truk (Chuuk) Islands, Yap Islands,
  and Kosrae
Land boundaries:
  0 km
Coastline:
  6,112 km
Climate:
  Tropical; heavy year-round rainfall, especially in the eastern islands;
  located on southern edge of the typhoon belt with occasional severe damage
Terrain:
  Islands vary geologically from high mountainous islands to low, coral
  atolls; volcanic outcroppings on Pohnpei, Kosrae, and Truk
Natural resources:
  Forests, marine products, deep-seabed minerals
Land use:
  Arable land: NA
  Permanent crops: NA
  Meadows and pastures: NA
  Forest and woodland: NA
  Other: NA

Demographics [2]

	1970	1980	1990	1995[1]	1996	2000	2010	2020	2030
Population	57	77	109	123	125	133	141	143	143
Population density (persons per sq. mi.)	210	284	401	454	463	491	521	527	527
(persons per sq. km.)	81	110	155	175	179	190	201	203	204
Net migration rate (per 1,000 population)	NA	NA	11.7	11.7	11.7	11.7	NA	NA	NA
Births	NA	NA	NA	NA	4	NA	NA	NA	NA
Deaths	NA	NA	NA	NA	1	NA	NA	NA	NA
Life expectancy - males	NA	NA	64.8	65.8	66.0	66.7	NA	NA	NA
Life expectancy - females	NA	NA	68.7	69.8	70.0	70.6	NA	NA	NA
Birth rate (per 1,000)	NA	NA	29.5	28.1	27.9	27.1	NA	NA	NA
Death rate (per 1,000)	NA	NA	6.8	6.3	6.2	6.0	NA	NA	NA
Women of reproductive age (15-49 yrs.)	NA	NA	NA	NA	NA	NA	NA	NA	NA
of which are currently married	NA	6	NA	NA	NA	NA	NA	NA	NA
Fertility rate	NA	NA	4.2	4.0	4.0	3.8	NA	NA	NA

Except as noted, values for vital statistics are in thousands; life expectancy is in years.

Health

Health Indicators [3]

Health Expenditures [4]

For sources, notes, and explanations, see Annotated Source Appendix, page 1061.

629

Human Factors

Women and Children [5]	Burden of Disease [6]

Ethnic Division [7]

Nine ethnic Micronesian and Polynesian groups.

Religion [8]

Roman Catholic	50%
Protestant	47%
Other	3%

Major Languages [9]

English (official and common language), Trukese, Pohnpeian, Yapese, Kosrean.

Education

Public Education Expenditures [10]	Educational Attainment [11]
Illiteracy [12]	Libraries [13]
Daily Newspapers [14]	Culture [15]

Science and Technology

Scientific/Technical Forces [16]	R&D Expenditures [17]	U.S. Patents Issued [18]

For sources, notes, and explanations, see Annotated Source Appendix, page 1061.

Government and Law

Organization of Government [19]

Long-form name:
Federated States of Micronesia

Type:
Constitutional government in free association with the US; the Compact of Free Association entered into force 3 November 1986

Independence:
3 November 1986 (from the US-administered UN Trusteeship)

National holiday:
Proclamation of the Federated States of Micronesia, 10 May (1979)

Constitution:
10 May 1979

Legal system:
Based on adapted Trust Territory laws, acts of the legislature, municipal, common, and customary laws

Executive branch:
President; Vice President; Cabinet

Legislative branch:
Unicameral: Congress

Judicial branch:
Supreme Court

Elections [20]

Congress. No formal parties; last held on 7 March 1995 (next to be held March 1999); results—percent of vote NA; seats (14 total)—independents 14.

Government Budget [21]

For FY94/95 est.

	$ mil.
Revenues	45
Expenditures	31
Capital expenditures	NA

Defense Summary [22]

Note: Defense is the responsibility of the US

Crime [23]

Human Rights [24]

	SSTS	FL	FAPRO	PPCG	APROBC	TPW	PCPTW	STPEP	PHRFF	PRW	ASST	AFL
Observes												
	EAFRD	CPR	ESCR	SR	ACHR	MAAE	PVIAC	PVNAC	EAFDAW	TCIDTP	RC	
Observes							P	P			P	

P = Party; S = Signatory; see Appendix for meaning of abbreviations.

Labor Force

Total Labor Force [25]

Labor Force by Occupation [26]

Two-thirds are government employees

Unemployment Rate [27]

27% (1989)

For sources, notes, and explanations, see Annotated Source Appendix, page 1061.

631

Production Sectors

Energy Resource Summary [28]

Electricity: Capacity: 18,000 kW. Production: 40 million kWh. Consumption per capita: 380 kWh (1990).

Telecommunications [30]

- 960 telephones
- Domestic: islands interconnected by shortwave radiotelephone (used mostly for government purposes)
- International: satellite earth stations - 4 Intelsat (Pacific Ocean)
- Radio: Broadcast stations: AM 5, FM 1, shortwave 1 Radios: 17,000 (1993 est.)
- Television: Broadcast stations: 6 Televisions: 1,290 (1993 est.)

Top Agricultural Products [32]

Produces black pepper, tropical fruits and vegetables, coconuts, cassava (tapioca), sweet potatoes; pigs, chickens.

Top Mining Products [33]

Detailed information is not available. A summary of natural resources follows. **Mineral Resources**: Deep-seabed minerals.

Transportation [31]

Railways: 0 km

Highways: total: 226 km; paved: 39 km; unpaved: 187 km

Merchant marine: none

Airports

Total:	5
With paved runways 1,524 to 2,437 m:	4
With paved runways under 914 m:	1 (1995 est.)

Tourism [34]

For sources, notes, and explanations, see Annotated Source Appendix, page 1061.

Finance, Economics, and Trade

Industrial Summary [35]

Industrial Production: Growth rate not available. **Industries**: Tourism, construction, fish processing, craft items from shell, wood, and pearls.

Economic Indicators [36]

- **National product**: GDP—purchasing power parity— $205 million (1994 est.)
- **National product real growth rate**: 1.4% (1994 est.)
- **National product per capita**: $1,700 (1994 est.)
- **Inflation rate (consumer prices)**: 4% (1994 est.)
- **External debt**: $129 million

Balance of Payments Summary [37]

Exchange Rates [38]

Currency: **United States dollar.**
Symbol: **US$.**

US currency is used.

Imports and Exports

Top Import Origins [39]

$141.1 million (c.i.f., 1994 est.).

Origins	%
US	NA
Japan	NA
Australia	NA

Top Export Destinations [40]

$29.1 million (f.o.b., 1994 est.).

Destinations	%
Japan	NA
US	NA
Guam	NA

Foreign Aid [41]

Recipient: Under terms of the Compact of Free Association, the US will provide $1.3 billion in grant aid during the period 1986-2001.

Import and Export Commodities [42]

Import Commodities	Export Commodities
Food	Fish
Manufactured goods	Garments
Machinery and equipment	Bananas
Beverages	Black pepper

Moldova

Geography [1]

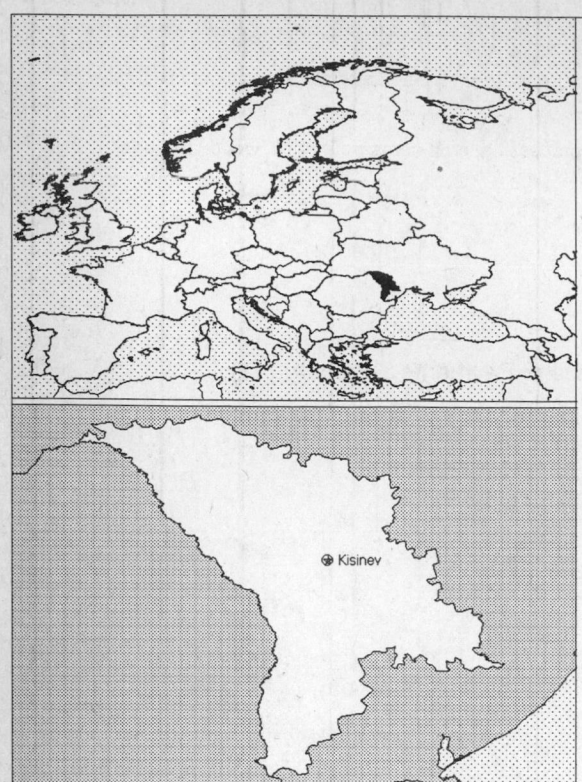

Total area:
  33,700 sq km   13,012 sq mi
Land area:
  33,700 sq km   13,012 sq mi
Comparative area:
  Slightly more than twice the size of Hawaii
Land boundaries:
  Total 1,389 km, Romania 450 km, Ukraine 939 km
Coastline:
  0 km (landlocked)
Climate:
  Moderate winters, warm summers
Terrain:
  Rolling steppe, gradual slope south to Black Sea
Natural resources:
  Lignite, phosphorites, gypsum
Land use:
  Arable land: 50%
  Permanent crops: 13%
  Meadows and pastures: 9%
  Forest and woodland: 0%
  Other: 28%

Demographics [2]

	1970	1980	1990	1995[1]	1996	2000	2010	2020	2030
Population	3,595	3,996	4,398	4,459	4,464	4,543	4,818	5,000	5,184
Population density (persons per sq. mi.)	276	307	338	343	343	349	370	384	398
(persons per sq. km.)	107	119	130	132	132	135	143	148	154
Net migration rate (per 1,000 population)	NA	NA	-0.2	-3.4	-2.8	-0.3	-0.2	-0.1	0.0
Births	NA	NA	NA	NA	73	NA	NA	NA	NA
Deaths	NA	NA	NA	NA	52	NA	NA	NA	NA
Life expectancy - males	NA	NA	63.9	60.6	60.8	61.5	63.3	68.0	71.7
Life expectancy - females	NA	NA	71.0	69.6	69.7	70.3	71.8	75.8	79.0
Birth rate (per 1,000)	NA	NA	18.3	15.3	16.3	18.8	16.3	13.5	12.8
Death rate (per 1,000)	NA	NA	10.1	11.8	11.8	11.8	11.5	9.9	9.5
Women of reproductive age (15-49 yrs.)	NA	NA	1,107	1,160	1,175	1,211	1,214	1,277	1,299
of which are currently married	NA	NA	775	NA	814	828	856	NA	NA
Fertility rate	NA	NA	2.4	2.1	2.2	2.4	2.0	1.9	1.8

Except as noted, values for vital statistics are in thousands; life expectancy is in years.

Health

Health Indicators [3]

% of population with access to	
safe water (1990-95)	NA
adequate sanitation (1990-95)	NA
health services (1985-95)	NA
% of 1-year-olds immunized (1990-94) against	
TB (tuberculosis)	97
DPT (diphtheria, pertussis, tetanus)	86
polio	98
measles	95
% of contraceptive prevalence (1980-94)	NA
ORT use rate (1990-94)	NA

Health Expenditures [4]

Total health expenditure, 1990 (official exchange rate)	
Millions of dollars	623
Dollars per capita	143
Health expenditures as a percentage of GDP	
Total	3.9
Public sector	2.9
Private sector	1.0
Development assistance for health	
Total aid flows (millions of dollars)[1]	NA
Aid flows per capita (dollars)	NA
Aid flows as a percentage of total health expenditure	NA

For sources, notes, and explanations, see Annotated Source Appendix, page 1061.

Human Factors

Women and Children [5]

% of pregnant women immunized (tetanus 1990-94)	NA
% of births attended by trained health personnel (1983-94)	NA
Maternal mortality rate (1980-92)	NA
Under-5 mortality rate (1994)	36
% under-5 moderately/severely underweight (1980-1994)	NA

Burden of Disease [6]

Population per physician (1993)	250.00
Population per nurse (1993)	89.88
Population per hospital bed (1993)	80.11
AIDS cases per 100,000 people (1994)	*
Heart disease cases per 1,000 people (1990-93)	NA

Ethnic Division [7]

Moldavian/Romanian	64.5%
Ukrainian	13.8%
Russian	13%
Gagauz	3.5%
Bulgarian	2%
Jewish	1.5%
Other (1989)	1.7%

Religion [8]

The large majority of churchgoers are ethnic Moldavian. About 1,000 Baptist also present.

Eastern Orthodox	98.5%
Jewish	1.5%
(1991)	

Major Languages [9]

Moldovan (official, virtually the same as the Romanian language), Russian, Gagauz (a Turkish dialect).

Education

Public Education Expenditures [10]

	1980	1985	1990	1992	1993	1994
Million (Ruble)						
Total education expenditure	0.376	0.471	0.706	14	127	426
as percent of GNP	NA	NA	7.1	8.6	6.0	5.5
as percent of total govt. expend.	NA	NA	17.2	26.4	NA	28.9
Current education expenditure	0.315	0.397	0.557	12	118	403
as percent of GNP	NA	NA	5.6	7.0	5.6	5.2
as percent of current govt. expend.	NA	NA	NA	25.4	NA	25.1
Capital expenditure	0.61	0.74	0.149	3	9	23

Educational Attainment [11]

Age group (1989)	25+
Total population	2,499,613
Highest level attained (%)	
No schooling	12.7
First level	
Not completed	17.1
Completed	NA
Entered second level	
S-1	58.9
S-2	NA
Postsecondary	11.3

Illiteracy [12]

	1985	1989	1995
Illiterate population (15+ yrs.)[2]	NA	113,193	34,000
Illiteracy rate - total pop. (%)	NA	3.6	1.1
Illiteracy rate - males (%)	NA	1.4	1.6
Illiteracy rate - females (%)	NA	5.6	0.5

Libraries [13]

Daily Newspapers [14]

	1980	1985	1990	1994
Number of papers	NA	NA	NA	4
Circ. (000)	NA	NA	NA	106

Culture [15]

Cinema (seats per 1,000)	6.8
Annual attendance per person	1.1
Gross box office receipts (mil. Ruble)	0.438
Museums (reporting)	62
Visitors (000)	657
Annual receipts (000 Ruble)	630

Science and Technology

Scientific/Technical Forces [16]

R&D Expenditures [17]

U.S. Patents Issued [18]

For sources, notes, and explanations, see Annotated Source Appendix, page 1061.

Government and Law

Organization of Government [19]

Long-form name:
Republic of Moldova
Type:
Republic
Independence:
27 August 1991 (from Soviet Union)
National holiday:
Independence Day, 27 August 1991
Constitution:
New constitution adopted 28 July 1994;
replaces old Soviet constitution of 1979
Legal system:
Based on civil law system; Constitutional
Court reviews legality of legislative acts
and governmental decisions of resolution;
it is unclear if Moldova accepts
compulsory ICJ jurisdiction but accepts
many UN and OSCE documents
Executive branch:
President; Prime Minister; First Deputy
Prime Minister; Council of Ministers
Legislative branch:
Unicameral: Parliament
Judicial branch:
Supreme Court

Elections [20]

Parliament	% of seats
Agrarian Democratic Party	53.8
Socialist/Yedinstvo Bloc	26.9
Peasants and Intellectual Bloc	10.6
Christian Democratic Popular Front	8.7

Government Budget [21]

Military Expenditures and Arms Transfers [22]

	1990	1991	1992[14]	1993	1994
Military expenditures					
Current dollars (mil.)[e]	NA	NA	NA	NA	92
1994 constant dollars (mil.)[e]	NA	NA	NA	NA	92
Armed forces (000)	NA	NA	9	13	11
Gross national product (GNP)					
Current dollars (mil.)[e]	NA	NA	14,230	13,980	11,900
1994 constant dollars (mil.)[e]	NA	NA	14,840	14,270	11,900
Central government expenditures (CGE)					
1994 constant dollars (mil.)[e]	NA	NA	9,736	NA	NA
People (mil.)	NA	NA	4.4	4.5	4.5
Military expenditure as % of GNP	NA	NA	NA	NA	0.8
Military expenditure as % of CGE	NA	NA	NA	NA	NA
Military expenditure per capita (1994 $)	NA	NA	NA	NA	21
Armed forces per 1,000 people (soldiers)	NA	NA	2.1	2.9	2.5
GNP per capita (1994 $)	NA	NA	3,345	3,203	2,660
Arms imports[6]					
Current dollars (mil.)	NA	NA	5	0	0
1994 constant dollars (mil.)	NA	NA	5	0	0
Arms exports[6]					
Current dollars (mil.)	NA	NA	0	0	80
1994 constant dollars (mil.)	NA	NA	0	0	80
Total imports[7]					
Current dollars (mil.)	NA	NA	576	912	1,251
1994 constant dollars (mil.)	NA	NA	601	931	1,251
Total exports[7]					
Current dollars (mil.)	NA	NA	366	332	398
1994 constant dollars (mil.)	NA	NA	381	339	398
Arms as percent of total imports[8]	NA	NA	0.9	0	0
Arms as percent of total exports[8]	NA	NA	0	0	20.1

Crime [23]

	1990
Crime volume	
Cases known to police	30,150
Attempts (percent)	NA
Percent cases solved	34.6
Crimes per 100,000 persons	687.1
Persons responsible for offenses	
Total number offenders	10,429
Percent female	9.6
Percent juvenile	15.3
Percent foreigners	NA

Human Rights [24]

	SSTS	FL	FAPRO	PPCG	APROBC	TPW	PCPTW	STPEP	PHRFF	PRW	ASST	AFL
Observes				P					S	P		P
	EAFRD	CPR	ESCR	SR	ACHR	MAAE	PVIAC	PVNAC	EAFDAW	TCIDTP	RC	
Observes	P	P	P				P	P	P	P	P	

P = Party; S = Signatory; see Appendix for meaning of abbreviations.

Labor Force

Total Labor Force [25]

2.03 million (January 1994)

Labor Force by Occupation [26]

Agriculture	34.4%
Industry	20.1
Other	45.5

Date of data: 1985 figures

Unemployment Rate [27]

1.2% (includes only officially registered
unemployed; large numbers of
underemployed workers) (December 1995)

Production Sectors

Commercial Energy Production and Consumption

Data are shown in quadrillion (10^{15}) BTUs and percent for 1995
Values for hydroelectric, nuclear, geothermal, solar, and wind power refer to electrical generation.

Production [28]

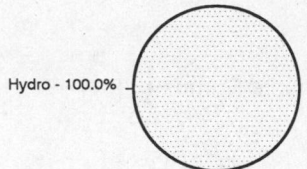

Hydro - 100.0%

Consumption [29]

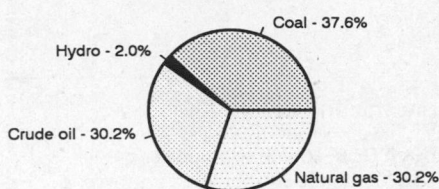

Coal - 37.6%
Hydro - 2.0%
Crude oil - 30.2%
Natural gas - 30.2%

Net hydroelectric power	0.003
Total	0.003

Crude oil	0.045
Dry natural gas	0.045
Coal	0.056
Net hydroelectric power	0.003
Total	0.149

Telecommunications [30]

- 577,000 (1991 est.) telephones; telecommunication system not well developed; 215,000 unsatisfied requests for telephone service (1991 est.)
- International: international connections to other former Soviet republics by landline and microwave radio relay through Ukraine and to other countries by leased connections to the Moscow international gateway switch; satellite earth stations - 1 Eutelsat and 1 Intelsat
- Radio: Broadcast stations: AM 9, FM 5, shortwave NA (1994)
- Television: Broadcast stations: 2 (one national and one private) (1995)

Transportation [31]

Railways: total: 1,328 km; broad gauge: 1,328 km 1.520-m gauge (1992)

Highways: total: 14,508 km; paved: 12,346 km; unpaved: 2,162 km (1992 est.)

Airports

Total:	26
With paved runways over 3,047 m:	1
With paved runways 2,438 to 3,047 m:	2
With paved runways 1,524 to 2,437 m:	2
With paved runways under 914 m:	3
With unpaved runways 2,438 to 3,047 m:	3

Top Agricultural Products [32]

Agriculture accounts for 33% of the GDP; produces vegetables, fruits, wine, grain, sugar beets, sunflower seed, tobacco; meat, milk.

Top Mining Products [33]

Thousand metric tons except as noted[e]	6/9/95*
Cement	1,000
Clays	250
Limestone	1,000
Gypsum	150
Sand and gravel (000 cu. meters)	3,000
Steel, crude	632[r]

Tourism [34]

	1990	1991	1992	1993	1994
Visitors[1]	261	357	293	108	26
Tourists	190	257	226	56	21
Excursionists	70	100	67	52	5
Tourism receipts[77]	NA	NA	4	2	2

Travelers are in thousands, money in million U.S. dollars.

Finance, Economics, and Trade

Industrial Summary [35]

Industrial Production: Growth rate - 6% (1995 est.); accounts for 36% of the GDP. **Industries:** Food processing, agricultural machinery, foundry equipment, refrigerators and freezers, washing machines, hosiery, sugar, vegetable oil, shoes, textiles.

Economic Indicators [36]

- **National product:** GDP—purchasing power parity—$10.4 billion (1995 estimate extrapolated from World Bank estimate for 1994)
- **National product real growth rate:** - 3% (1995 est.)
- **National product per capita:** $2,310 (1995 est.)
- **Inflation rate (consumer prices):** 24% (1995 est.)
- **External debt:** $550 million (of which $250 million to Russia)

Balance of Payments Summary [37]

Values in millions of dollars.

	1989	1990	1991	1992[1]	1993
Exports of goods (f.o.b.)	NA	NA	NA	368	449
Imports of goods (f.o.b.)	NA	NA	NA	506	547
Trade balance	NA	NA	NA	-138	-98
Services - debits	NA	NA	NA	-30	-51
Services - credits	NA	NA	NA	13	24
Private transfers (net)	NA	NA	NA	1	3
Government transfers (net)	NA	NA	NA	NA	19
Long-term capital (net)	NA	NA	NA	33	151
Short-term capital (net)	NA	NA	NA	1	NA
Errors and omissions	NA	NA	NA	NA	NA
Overall balance	NA	NA	NA	-120	48

Exchange Rates [38]

Currency: **leu.**

Data are currency units per $1. The leu was introduced in late 1993.

January 1996	4.5460
1995	4.4990
1994	4.2700
1993	3.6400
1992	0.4145
1991	0.0017

Imports and Exports

Top Import Origins [39]

$822 million (1995).

Origins	%
Russia	NA
Ukraine	NA
Uzbekistan	NA
Romania	NA
Germany	NA

Top Export Destinations [40]

$720 million (1995).

Destinations	%
Russia	NA
Kazakhstan	NA
Ukraine	NA
Romania	NA
Germany	NA

Foreign Aid [41]

	U.S. $	
ODA (1993)	46	million
Commitments (1992-95)	1,335	million

Import and Export Commodities [42]

Import Commodities	Export Commodities
Oil	Foodstuffs
Gas	Wine
Coal	Tobacco
Steel	Textiles and footwear
Machinery	Machinery
Foodstuffs	Chemicals
Automobiles	
Consumer durables	

Monaco

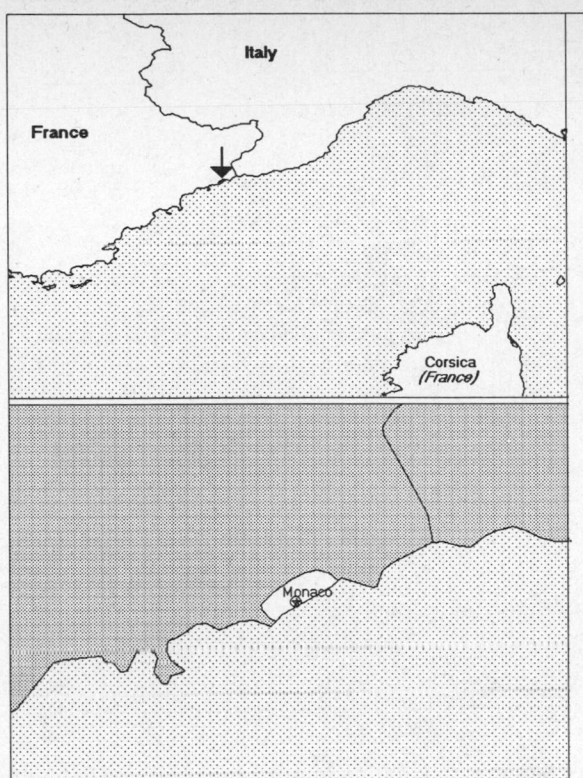

Geography [1]

Total area:
 1.9 sq km   0.7 sq mi
Land area:
 1.9 sq km   0.7 sq mi
Comparative area:
 About three times the size of The Mall in Washington, DC
Land boundaries:
 Total 4.4 km, France 4.4 km
Coastline:
 4.1 km
Climate:
 Mediterranean with mild, wet winters and hot, dry summers
Terrain:
 Hilly, rugged, rocky
Natural resources:
 None
Land use:
 Arable land: 0%
 Permanent crops: 0%
 Meadows and pastures: 0%
 Forest and woodland: 0%
 Other: 100%

Demographics [2]

	1970	1980	1990	1995[1]	1996	2000	2010	2020	2030
Population	24	27	30	32	32	32	33	34	34
Population density (persons per sq. mi.)	30,759	34,400	38,867	40,812	41,076	41,739	42,753	43,569	43,938
(persons per sq. km.)	11,876	13,282	15,007	15,758	15,860	16,116	16,507	16,822	16,965
Net migration rate (per 1,000 population)	NA	14.2	14.2	8.4	7.4	3.1	3.0	3.0	3.0
Births[37]	0	NA	NA	NA	Z	NA	NA	NA	NA
Deaths[38]	0	NA	NA	NA	Z	NA	NA	NA	NA
Life expectancy - males	NA	70.0	73.0	74.2	74.4	75.2	76.9	78.1	79.0
Life expectancy - females	NA	78.0	81.0	81.8	81.9	82.5	83.6	84.5	85.2
Birth rate (per 1,000)	9.0	12.3	10.9	10.7	10.7	10.6	10.8	10.0	9.8
Death rate (per 1,000)	11.1	15.3	12.6	12.1	12.1	11.7	11.5	11.5	12.4
Women of reproductive age (15-49 yrs.)	NA	6	7	7	7	7	7	7	7
of which are currently married	NA	NA	5	NA	5	5	5	NA	NA
Fertility rate	NA	1.8	1.7	1.7	1.7	1.7	1.7	1.7	1.7

Except as noted, values for vital statistics are in thousands; life expectancy is in years.

Health

Health Indicators [3]

Health Expenditures [4]

For sources, notes, and explanations, see Annotated Source Appendix, page 1061.

639

Human Factors

Women and Children [5]

Burden of Disease [6]

Ethnic Division [7]

French	47%
Monegasque	16%
Italian	16%
Other	21%

Religion [8]

Roman Catholic	95%
Other	5%

Major Languages [9]

French (official), English, Italian, Monegasque.

Education

Public Education Expenditures [10]

	1980	1988	1989	1990	1992	1994
Million (French Franc)						
Total education expenditure	NA	106	124	NA	153	NA
as percent of GNP	NA	NA	NA	NA	NA	NA
as percent of total govt. expend.	NA	4.7	5.3	NA	5.6	NA
Current education expenditure	43	96	113	NA	140	NA
as percent of GNP	NA	NA	NA	NA	NA	NA
as percent of current govt. expend.	NA	7.1	7.8	NA	7.7	NA
Capital expenditure	NA	10	11	NA	13	NA

Educational Attainment [11]

Illiteracy [12]

Libraries [13]

	Admin. Units	Svc. Pts.	Vols. (000)	Shelving (meters)	Vols. Added	Reg. Users
National (1992)[6]	1	2	285	6,350	5,700	10,000
Nonspecialized	NA	NA	NA	NA	NA	NA
Public	NA	NA	NA	NA	NA	NA
Higher ed.	NA	NA	NA	NA	NA	NA
School	NA	NA	NA	NA	NA	NA

Daily Newspapers [14]

	1980	1985	1990	1994
Number of papers	2	2	1	1
Circ. (000)	10	10[e]	8	8[e]

Culture [15]

Cinema (seats per 1,000)	53.3
Annual attendance per person	3.7
Gross box office receipts (mil. Franc)	3.9
Museums (reporting)	4
Visitors (000)	1,137
Annual receipts (000 Franc)	44,416

Science and Technology

Scientific/Technical Forces [16]

Scientists/engineers	3,125
Number female	3,125
Technicians	1,374
Number female	NA
Total[36]	NA

R&D Expenditures [17]

U.S. Patents Issued [18]

Values show patents issued to citizens of the country by the U.S. Patents Office.

	1993	1994	1995
Number of patents	5	7	4

For sources, notes, and explanations, see Annotated Source Appendix, page 1061.

Government and Law

Organization of Government [19]

Long-form name:
Principality of Monaco
Type:
Constitutional monarchy
Independence:
1419 (rule by the House of Grimaldi)
National holiday:
National Day, 19 November
Constitution:
17 December 1962
Legal system:
Based on French law; has not accepted compulsory ICJ jurisdiction
Executive branch:
Prince; Heir Apparent Prince; Minister of State; Council of Government
Legislative branch:
Unicameral: National Council (Conseil National)
Judicial branch:
Supreme Tribunal (Tribunal Supreme)

Elections [20]

National Council	% of seats
Campora List	83.3
Medecin List	11.1
Independent	5.6

Government Budget [21]

For 1993 est.

	$ mil.
Revenues	660
Expenditures	586
Capital expenditures	NA

Defense Summary [22]

Note: Defense is the responsibility of France

Crime [23]

	1994
Crime volume	
Cases known to police	1,442
Attempts (percent)	114
Percent cases solved	729
Crimes per 100,000 persons	4,806.67
Persons responsible for offenses	
Total number offenders	991
Percent female	207
Percent juvenile (0-18 yrs.)	81
Percent foreigners	949

Human Rights [24]

	SSTS	FL	FAPRO	PPCG	APROBC	TPW	PCPTW	STPEP	PHRFF	PRW	ASST	AFL
Observes	P			P		P	P					
	EAFRD	CPR	ESCR	SR	ACHR	MAAE	PVIAC	PVNAC	EAFDAW	TCIDTP	RC	
Observes		P								P	P	

P = Party; S = Signatory; see Appendix for meaning of abbreviations.

Labor Force

Total Labor Force [25]

Labor Force by Occupation [26]

Unemployment Rate [27]

3.1% (1994)

For sources, notes, and explanations, see Annotated Source Appendix, page 1061.

641

Production Sectors

Energy Resource Summary [28]

Electricity: Capacity: 10,000 kW standby; power imported from France. Production: NA kWh. Consumption per capita: NA kWh (1993).

Telecommunications [30]

- 34,600 (1987 est.) telephones; automatic telephone system
- International: no satellite earth stations; connected by cable into the French communications system
- Radio: Broadcast stations: AM 3, FM 4, shortwave 0 Radios: 30,000 (1992 est.)
- Television: Broadcast stations: 5 (1987 est.) Televisions: 22,000 (1992 est.)

Transportation [31]

Railways: total: 1.7 km; standard gauge: 1.7 km 1.435-m gauge

Highways: total: 50 km; paved: 50 km; unpaved: 0 km (1994)

Merchant marine: none

Airports

Linked to airport in Nice, France by helicopter service

Top Agricultural Products [32]

Produces none.

Top Mining Products [33]

Mineral Resources: None.

Tourism [34]

	1990	1991	1992	1993	1994
Tourists[16]	245	239	246	208	217

Travelers are in thousands, money in million U.S. dollars.

For sources, notes, and explanations, see Annotated Source Appendix, page 1061.

Finance, Economics, and Trade

Manufacturing [35]

Economic Indicators [36]

- **National product**: GDP—purchasing power parity—$788 million (1994 est.)
- **National product real growth rate**: NA%
- **National product per capita**: $25,000 (1994 est.)
- **Inflation rate (consumer prices)**: NA%
- **External debt**: $NA

Balance of Payments Summary [37]

Exchange Rates [38]

Currency: **French franc.**
Symbol: **F.**

Data are currency units per $1.

January 1996	5.0056
1995	4.9915
1994	5.520
1993	5.6632
1992	5.2938
1991	5.6421

Imports and Exports

Top Import Origins [39]

Full customs integration with France, which collects and rebates Monacan trade duties; also participates in EU market system through customs union with France.

Origins	%
No details available.	NA

Top Export Destinations [40]

Full customs integration with France, which collects and rebates Monacan trade duties; also participates in EU market system through customs union with France.

Destinations	%
No details available.	NA

Foreign Aid [41]

Import and Export Commodities [42]

Mongolia

Geography [1]

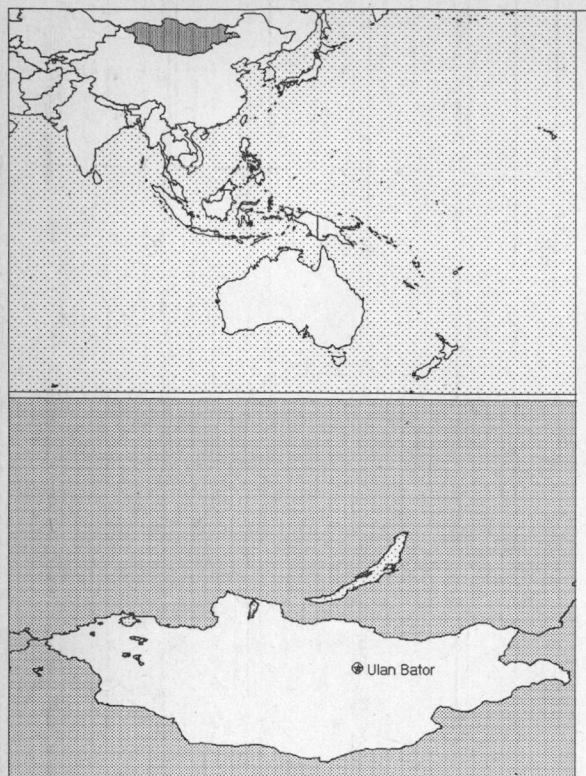

Total area:
  1,565,000 sq km   604,250 sq mi
Land area:
  1,565,000 sq km   604,250 sq mi
Comparative area:
  Slightly larger than Alaska
Land boundaries:
  Total 8,114 km, China 4,673 km, Russia 3,441 km
Coastline:
  0 km (landlocked)
Climate:
  Desert; continental (large daily and seasonal temperature ranges)
Terrain:
  Vast semidesert and desert plains; mountains in West and Southwest;
  Gobi Desert in Southeast
Natural resources:
  Oil, coal, copper, molybdenum, tungsten, phosphates, tin, nickel,
  zinc, wolfram, fluorspar, gold
Land use:
  Arable land: 1%
  Permanent crops: 0%
  Meadows and pastures: 79%
  Forest and woodland: 10%
  Other: 10%

Demographics [2]

	1970	1980	1990	1995[1]	1996	2000	2010	2020	2030
Population	1,248	1,662	2,216	2,454	2,497	2,655	3,018	3,393	3,694
Population density (persons per sq. mi.)	2	3	4	4	4	4	5	6	6
(persons per sq. km.)	1	1	1	2	2	2	2	2	2
Net migration rate (per 1,000 population)	NA	NA	0.0	0.0	0.0	0.0	0.0	0.0	0.0
Births	NA	63	NA	NA	64	NA	NA	NA	NA
Deaths	NA	17	NA	NA	22	NA	NA	NA	NA
Life expectancy - males	NA	NA	58.5	58.5	58.8	60.0	62.9	65.6	68.0
Life expectancy - females	NA	NA	62.4	62.4	62.8	64.4	68.3	71.7	74.7
Birth rate (per 1,000)	40.2	37.9	35.4	26.5	25.6	21.5	19.4	17.2	14.4
Death rate (per 1,000)	12.3	10.4	9.8	8.9	8.7	7.8	6.9	6.9	7.5
Women of reproductive age (15-49 yrs.)	NA	NA	524	618	637	722	907	961	976
of which are currently married	NA	NA	354	NA	440	505	658	NA	NA
Fertility rate	NA	NA	4.5	3.2	3.0	2.5	2.1	2.0	2.0

Except as noted, values for vital statistics are in thousands; life expectancy is in years.

Health

Health Indicators [3]

% of population with access to	
safe water (1990-95)	80
adequate sanitation (1990-95)	74
health services (1985-95)	95
% of 1-year-olds immunized (1990-94) against	
TB (tuberculosis)	90
DPT (diphtheria, pertussis, tetanus)	78
polio	77
measles	80
% of contraceptive prevalence (1980-94)	NA
ORT use rate (1990-94)	65

Health Expenditures [4]

For sources, notes, and explanations, see Annotated Source Appendix, page 1061.

Human Factors

Women and Children [5]

% of pregnant women immunized (tetanus 1990-94)	NA
% of births attended by trained health personnel (1983-94)	99
Maternal mortality rate (1980-92)	240
Under-5 mortality rate (1994)	76
% under-5 moderately/severely underweight (1980-1994)[1]	12

Burden of Disease [6]

Population per physician (1991)	365.90
Population per nurse	NA
Population per hospital bed (1991)	85.85
AIDS cases per 100,000 people (1994)	*
Malaria cases per 100,000 people (1992)	NA

Ethnic Division [7]

Mongol	90%
Kazak	4%
Chinese	2%
Russian	2%
Other	2%

Religion [8]

Predominantly Tibetan Buddhist, Muslim 4%. Note: Previously limited religious activity because of communist regime.

Major Languages [9]

Khalkha Mongol	90%
Turkic, Russian, Chinese	10%

Education

Public Education Expenditures [10]

	1980	1985	1989	1990	1991	1994
Million (Tughrik)						
Total education expenditure	NA	716	859	882	1,598	15,510
as percent of GNP	NA	7.8	8.1	8.6	8.5	5.2
as percent of total govt. expend.	NA	NA	NA	NA	NA	NA
Current education expenditure	NA	NA	NA	NA	NA	15,384
as percent of GNP	NA	NA	NA	NA	NA	5.2
as percent of current govt. expend.	NA	NA	NA	NA	NA	NA
Capital expenditure	NA	NA	NA	NA	NA	126

Educational Attainment [11]

Illiteracy [12]

In thousands and percent[1]	1990	1995	2000
Illiterate population (15+ yrs.)	250	256	250
Illiteracy rate - total pop. (%)	20.0	17.1	14.5
Illiteracy rate - males (%)	13.5	11.5	9.8
Illiteracy rate - females (%)	26.5	22.5	19.1

Libraries [13]

	Admin. Units	Svc. Pts.	Vols. (000)	Shelving (meters)	Vols. Added	Reg. Users
National	NA	NA	NA	NA	NA	NA
Nonspecialized	NA	NA	NA	NA	NA	NA
Public	NA	NA	NA	NA	NA	NA
Higher ed. (1990)[10]	9	NA	1,581	NA	1,100	20,100
School (1990)	530	NA	3,220	NA	NA	450,600

Daily Newspapers [14]

	1980	1985	1990	1994
Number of papers	2	2	1	1
Circ. (000)	177	177	162	207

Culture [15]

Cinema (seats per 1,000)	NA
Annual attendance per person	9.5
Gross box office receipts (mil. Tughrik)	NA
Museums (reporting)	NA
Visitors (000)	NA
Annual receipts (000 Tughrik)	NA

Science and Technology

Scientific/Technical Forces [16]

R&D Expenditures [17]

U.S. Patents Issued [18]

For sources, notes, and explanations, see Annotated Source Appendix, page 1061.

Government and Law

Organization of Government [19]

Long-form name:
None
Type:
Republic
Independence:
13 March 1921 (from China)
National holiday:
National Day, 11 July (1921)
Constitution:
Adopted 13 January 1992
Legal system:
Blend of Russian, Chinese, and Turkish systems of law; no constitutional provision for judicial review of legislative acts; has not accepted compulsory ICJ jurisdiction
Executive branch:
President; Prime Minister; 2 Deputy Prime Ministers; Cabinet
Legislative branch:
Unicameral: State Great Hural
Judicial branch:
Supreme Court

Elections [20]

State Great Hural	% of seats
Mongolian People's Revolutionary Party (MPRP)	93.4
United Party	5.3
Mongolian Social Democratic Party (MSDP)	1.3

Government Expenditures [21]

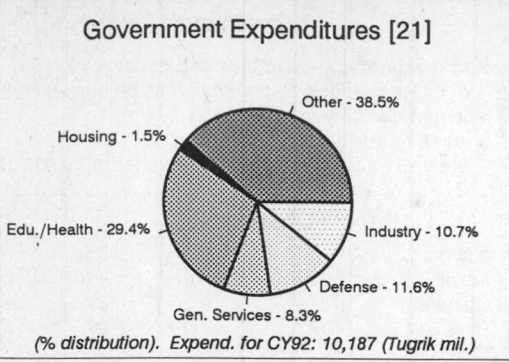

Other - 38.5%
Housing - 1.5%
Edu./Health - 29.4%
Industry - 10.7%
Defense - 11.6%
Gen. Services - 8.3%

(% distribution). Expend. for CY92: 10,187 (Tugrik mil.)

Crime [23]

	1994
Crime volume	
Cases known to police	18,563
Attempts (percent)	NA
Percent cases solved	NA
Crimes per 100,000 persons	823.34
Persons responsible for offenses	
Total number offenders	16,272
Percent female	NA
Percent juvenile (0-17 yrs.)	NA
Percent foreigners	NA

Military Expenditures and Arms Transfers [22]

	1990	1991	1992	1993	1994
Military expenditures					
Current dollars (mil.)	63[e]	32[e]	20	19	17
1994 constant dollars (mil.)	70[e]	35[e]	21	20	17
Armed forces (000)	32	31	21	18	21
Gross national product (GNP)					
Current dollars (mil.)	803	758	705	708	722
1994 constant dollars (mil.)	894	813	735	723	722
Central government expenditures (CGE)					
1994 constant dollars (mil.)	593	388	224	212	174
People (mil.)	2.2	2.2	2.3	2.4	2.4
Military expenditure as % of GNP	7.8	4.3	2.8	2.7	2.4
Military expenditure as % of CGE	11.7	9.0	9.3	9.2	9.8
Military expenditure per capita (1994 $)	32	15	9	8	7
Armed forces per 1,000 people (soldiers)	14.6	13.8	9.2	7.6	8.6
GNP per capita (1994 $)	409	362	319	305	297
Arms imports[6]					
Current dollars (mil.)	0	0	0	0	0
1994 constant dollars (mil.)	0	0	0	0	0
Arms exports[6]					
Current dollars (mil.)	0	0	0	0	0
1994 constant dollars (mil.)	0	0	0	0	0
Total imports[7]					
Current dollars (mil.)	924	361	418	362	223
1994 constant dollars (mil.)	1,028	387	436	369	223
Total exports[7]					
Current dollars (mil.)	661	348	389	381	324
1994 constant dollars (mil.)	736	373	406	389	324
Arms as percent of total imports[8]	0	0	0	0	0
Arms as percent of total exports[8]	0	0	0	0	0

Human Rights [24]

	SSTS	FL	FAPRO	PPCG	APROBC	TPW	PCPTW	STPEP	PHRFF	PRW	ASST	AFL
Observes	P		P	P	P	P	P			P	P	
		EAFRD	CPR	ESCR	SR	ACHR	MAAE	PVIAC	PVNAC	EAFDAW	TCIDTP	RC
Observes		P	P	P				P	P	P		P

P = Party; S = Signatory; see Appendix for meaning of abbreviations.

Labor Force

Total Labor Force [25]

1.115 million (mid-1993 est.)

Labor Force by Occupation [26]

Primarily herding/agricultural

Unemployment Rate [27]

15% (1991 est.)

For sources, notes, and explanations, see Annotated Source Appendix, page 1061.

Production Sectors

Commercial Energy Production and Consumption

Data are shown in quadrillion (10^{15}) BTUs and percent for 1995
Values for hydroelectric, nuclear, geothermal, solar, and wind power refer to electrical generation.

Production [28]

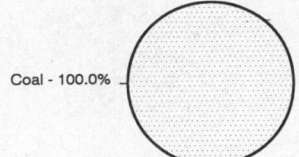

Coal - 100.0%

Consumption [29]

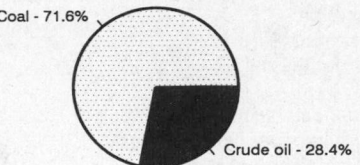

Coal - 71.6%

Crude oil - 28.4%

Coal	0.078
Total	0.078

Crude oil	0.029
Coal	0.073
Total	0.102

Telecommunications [30]

- 89,000 (1995 est.) telephones
- International: satellite earth station - 1 Intersputnik (Indian Ocean Region)
- Radio: Broadcast stations: AM 12, FM 1, shortwave 0 Radios: 220,000
- Television: Broadcast stations: 1 (provincial repeaters 18) Televisions: 120,000 (1993 est.)

Top Agricultural Products [32]

Agriculture accounts for 28% of the GDP; produces wheat, barley, potatoes, forage crops; sheep, goats, cattle, camels, horses.

Transportation [31]

Railways: total: 1,928 km; broad gauge: 1,928 km 1.524-m gauge (1994)

Highways: total: 46,700 km; paved: 1,000 km; unpaved: 45,700 km (1988 est.)

Airports

Total:	34
With paved runways 2,438 to 3,047 m:	7
With paved runways under 914 m:	1
With unpaved runways over 3,047 m:	3
With unpaved runways 2,438 to 3,047 m:	5
With unpaved runways 1,524 to 2,437 m:	10

Top Mining Products [33]

Thousand metric tons except as noted[e]	6/30/95*
Cement, hydraulic	86[r]
Coal	5,000[r]
Copper, mine output, Cu content	99,600[r]
Fluorspar	170[r]
Gold, mine output, Au content (kg.)	2,000[54]
Gypsum	25
Lime, hydrated and quicklime	66[r]
Silver, mine output, Ag content (kg.)	20,000[55]

Tourism [34]

	1987	1988	1989	1990	1991
Tourists	186	240	237	147	NA

Travelers are in thousands, money in million U.S. dollars.

Manufacturing Sector

Manufacturing Summary [35]

	1987		1988		1989		1990		1991	
	$ bil.	%	$ bil.	%	$ bil.	%	$ bil.	%	$ bil.	%
Establishments or enterprises (number)	68	0.003	70	0.003	70	0.004	-	-	-	-
Total employment (000)	23	0.017	20	0.015	21	0.017	-	-	-	-
Production workers (000)	18	0.027	16	0.025	17	0.023	-	-	-	-
Output ($ bil.)	-	-	-	-	0.467	0.004	-	-	-	-
Value added ($ bil.)	-	-	-	-	-	-	-	-	-	-
Capital investment ($ mil.)	-	-	-	-	544	0.099	-	-	-	-
M & E investment ($ mil.)	-	-	-	-	-	-	-	-	-	-
Employees per establishment (number)	340	587.186	284	436.061	300	456.572	-	-	-	-
Production workers per establishment	263	915.564	227	763.192	240	611.092	-	-	-	-
Output per establishment ($ mil.)	-	-	-	-	7	106.032	-	-	-	-
Capital investment per estab. ($ bil.)	-	-	-	-	8	2,655	-	-	-	-
M & E per establishment ($ mil.)	-	-	-	-	-	-	-	-	-	-
Payroll per employee ($)	-	-	-	-	-	-	-	-	-	-
Wages per production worker ($)	-	-	-	-	-	-	-	-	-	-
Hours per production worker (hours)	-	-	-	-	-	-	-	-	-	-
Output per employee ($)	-	-	-	-	22,257	23.223	-	-	-	-
Capital investment per employee ($)	-	-	-	-	25,881	581.543	-	-	-	-
M & E per employee ($)	-	-	-	-	-	-	-	-	-	-

Note: Columns headed % show percent of world total or ratio. Ratios closest to 100 are closest to world average. M & E stands for machinery & equipment.

Output in Manufacturing

	1987		1988		1989		1990		1991	
	$ bil.	%	$ bil.	%	$ bil.	%	$ bil.	%	$ bil.	%
3210 - Textiles	-	-	-	-	0.316	67.67	-	-	-	-
3220 - Wearing apparel	-	-	-	-	0.128	27.41	-	-	-	-
3420 - Printing, publishing	-	-	-	-	0.023	4.93	-	-	-	-

Note: Codes are International Standard Industry codes (ISIC). Percentages are % of total Output. [f]: Factor Prices; [p]: Producer Prices.

Finance, Economics, and Trade

Economic Indicators [36]

- **National product**: GDP—purchasing power parity—$4.9 billion (1995 est.)
- **National product real growth rate**: 6% (1995 est.)
- **National product per capita**: $1,970 (1995 est.)
- **Inflation rate (consumer prices)**: 53% (1995 est.)
- **External debt**: $473.7 million (1994)

Balance of Payments Summary [37]

Values in millions of dollars.

	1970	1975	1980	1985	1990
Exports of goods (f.o.b.)	NA	NA	NA	566.9	468.1
Imports of goods (f.o.b.)	NA	NA	NA	-1,365.7	-1,051.0
Trade balance	NA	NA	NA	-798.8	-582.9
Services - debits	NA	NA	NA	-86.2	-121.3
Services - credits	NA	NA	NA	71.8	53.2
Private transfers (net)	NA	NA	NA	-0.1	NA
Government transfers (net)	NA	NA	NA	NA	7.4
Long-term capital (net)	NA	NA	NA	753.9	522.5
Short-term capital (net)	NA	NA	NA	0.6	24.3
Errors and omissions	NA	NA	NA	83.4	-4.8
Overall balance	NA	NA	NA	24.6	-101.6

Exchange Rates [38]

Currency: tughrik.
Symbol: **Tug.**

Data are currency units per $1.

October 1995	4,465.39
1994	412.72
1992	42.56
1991	9.52
1990	5.63

Imports and Exports

Top Import Origins [39]

$223 million (f.o.b., 1994) Data are for 1991.

Origins	%
USSR	75
Austria	5
China	5

Top Export Destinations [40]

$100 million (f.o.b., 1995 est.) Data are for 1992.

Destinations	%
Former CMEA countries	62
China	17
EU	8

Foreign Aid [41]

Recipient: ODA, $NA.

Import and Export Commodities [42]

Import Commodities	Export Commodities
Machinery and equipment	Copper
Fuels	Livestock
Food products	Animal products
Industrial consumer goods	Cashmere
Chemicals	Wool
Building materials	Hides
Sugar	Fluorspar
Tea	Other nonferrous metals

For sources, notes, and explanations, see Annotated Source Appendix, page 1061.

649

Morocco

Geography [1]

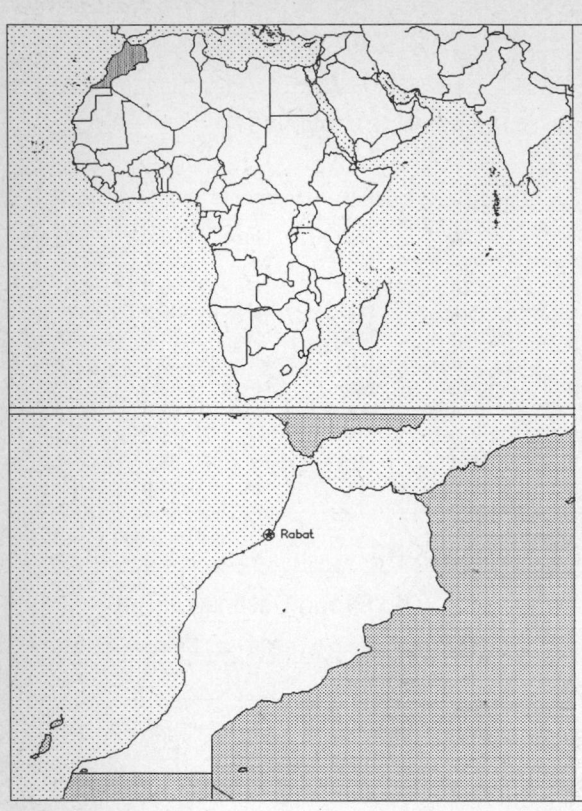

Total area:
446,550 sq km   172,414 sq mi
Land area:
446,300 sq km   172,317 sq mi
Comparative area:
Slightly larger than California
Land boundaries:
Total 2,002 km, Algeria 1,559 km, Western Sahara 443 km
Note: Excludes the length of the boundary between the places of sovereignty and Morocco
Coastline:
1,835 km
Climate:
Mediterranean, becoming more extreme in the interior
Terrain:
Northern coast and interior are mountainous with large areas of bordering plateaus, intermontane valleys, and rich coastal plains
Natural resources:
Phosphates, iron ore, manganese, lead, zinc, fish, salt
Land use:
Arable land: 18%
Permanent crops: 1%
Meadows and pastures: 28%
Forest and woodland: 12%
Other: 41%

Demographics [2]

	1970	1980	1990	1995[1]	1996	2000	2010	2020	2030
Population	15,909	20,457	26,164	29,169	29,779	32,229	38,442	44,519	50,056
Population density (persons per sq. mi.)	92	119	152	169	173	187	223	258	290
(persons per sq. km.)	36	46	59	65	67	72	86	100	112
Net migration rate (per 1,000 population)	NA	NA	-1.0	-1.1	-1.1	-1.0	-0.5	-0.5	-0.4
Births	NA	NA	NA	NA	816	NA	NA	NA	NA
Deaths	NA	NA	NA	NA	172	NA	NA	NA	NA
Life expectancy - males	NA	NA	63.8	67.0	67.5	69.6	73.7	76.4	78.2
Life expectancy - females	NA	NA	67.1	71.0	71.6	74.1	78.9	82.1	84.1
Birth rate (per 1,000)	NA	NA	31.3	27.9	27.4	25.1	21.0	17.8	15.6
Death rate (per 1,000)	NA	NA	7.6	6.0	5.8	5.1	4.3	4.2	4.8
Women of reproductive age (15-49 yrs.)	NA	NA	6,270	7,274	7,499	8,444	10,422	11,937	12,885
of which are currently married	NA	NA	3,444	NA	4,210	4,801	6,189	NA	NA
Fertility rate	NA	5.9	4.4	3.7	3.6	3.1	2.5	2.2	2.1

Except as noted, values for vital statistics are in thousands; life expectancy is in years.

Health

Health Indicators [3]

% of population with access to	
safe water (1990-95)	55
adequate sanitation (1990-95)	41
health services (1985-95)	70
% of 1-year-olds immunized (1990-94) against	
TB (tuberculosis)	93
DPT (diphtheria, pertussis, tetanus)	87
polio	87
measles	NA
% of contraceptive prevalence (1980-94)	42
ORT use rate (1990-94)	48

Health Expenditures [4]

Total health expenditure, 1990 (official exchange rate)	
Millions of dollars	661
Dollars per capita	26
Health expenditures as a percentage of GDP	
Total	2.6
Public sector	0.9
Private sector	1.6
Development assistance for health	
Total aid flows (millions of dollars)[1]	20
Aid flows per capita (dollars)	0.8
Aid flows as a percentage of total health expenditure	3.0

For sources, notes, and explanations, see Annotated Source Appendix, page 1061.

Human Factors

Women and Children [5]

% of pregnant women immunized (tetanus 1990-94)	NA
% of births attended by trained health personnel (1983-94)	31
Maternal mortality rate (1980-92)	330
Under-5 mortality rate (1994)	56
% under-5 moderately/severely underweight (1980-1994)	9

Burden of Disease [6]

Population per physician (1988)	4,718.16
Population per nurse (1987)	1,028.14
Population per hospital bed (1990)	784.59
AIDS cases per 100,000 people (1994)	0.3
Malaria cases per 100,000 people (1992)	2

Ethnic Division [7]

Arab-Berber	99.1%
Jewish	0.2%
Other	0.7%

Religion [8]

Muslim	98.7%
Christian	1.1%
Jewish	0.2%

Major Languages [9]

Arabic (official), Berber dialects, French often the language of business, government, and diplomacy.

Education

Public Education Expenditures [10]

Million (Dirham)[44]	1980	1985	1990	1992	1993	1994
Total education expenditure	4,367	7,697	11,220	13,564	14,589	14,950
as percent of GNP	6.1	6.3	5.5	5.8	6.2	5.4
as percent of total govt. expend.	18.5	22.9	26.1	26.7	25.6	22.6
Current education expenditure	3,529	6,079	10,187	11,944	12,939	13,367
as percent of GNP	4.9	5.0	5.0	5.1	5.5	4.8
as percent of current govt. expend.	23.3	28.6	33.6	32.0	30.9	28.4
Capital expenditure	838	1,618	1,033	1,620	1,650	1,583

Educational Attainment [11]

Illiteracy [12]

In thousands and percent[1]	1990	1995	2000
Illiterate population (15+ yrs.)	9,124	9,730	10,153
Illiteracy rate - total pop. (%)	58.7	54.0	48.7
Illiteracy rate - males (%)	46.0	41.6	37.2
Illiteracy rate - females (%)	71.1	66.1	60.0

Libraries [13]

Daily Newspapers [14]

	1980	1985	1990	1994
Number of papers	11	14	13	13
Circ. (000)	270[e]	320[e]	320[e]	344[e]

Culture [15]

Cinema (seats per 1,000)	5.3
Annual attendance per person	0.8
Gross box office receipts (mil. Dirham)	NA
Museums (reporting)	14
Visitors (000)	222
Annual receipts (000 Dirham)	1,940

Science and Technology

Scientific/Technical Forces [16]

R&D Expenditures [17]

U.S. Patents Issued [18]

Values show patents issued to citizens of the country by the U.S. Patents Office.

	1993	1994	1995
Number of patents	1	0	2

For sources, notes, and explanations, see Annotated Source Appendix, page 1061.

Government and Law

Organization of Government [19]

Long-form name:
Kingdom of Morocco
Type:
Constitutional monarchy
Independence:
2 March 1956 (from France)
National holiday:
National Day, 3 March (1961) (anniversary of King Hassan II's accession to the throne)
Constitution:
10 March 1972, revised 4 September 1992
Legal system:
Based on Islamic law and French and Spanish civil law system; judicial review of legislative acts in Constitutional Chamber of Supreme Court
Executive branch:
Chief of state: King; Prime Minister; Council of Ministers
Legislative branch:
Unicameral: Chamber of Representatives (Majlis Nawab)
Judicial branch:
Supreme Court

Crime [23]

Elections [20]

Chamber of Representatives	% of seats
Constitutional Union (UC)	16.2
Socialist Union (USFP)	15.6
Popular Movement (MP)	15.3
Istiqlal Party (IP)	15.0
National Assembly (RNI)	12.3
Nat Popular Movement (MNP)	7.5
National Democratic (PND)	7.2
Other	10.9

Government Expenditures [21]

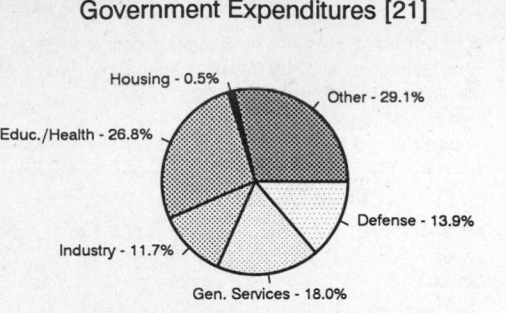

Housing - 0.5%, Other - 29.1%, Educ./Health - 26.8%, Defense - 13.9%, Industry - 11.7%, Gen. Services - 18.0%

(% distribution). Expend. for CY92: 73,008 (Dirham mil.)

Military Expenditures and Arms Transfers [22]

	1990	1991	1992	1993	1994
Military expenditures					
Current dollars (mil.)	1,306[e]	1,157[e]	1,577[e]	1,246	1,228[e]
1994 constant dollars (mil.)	1,454[e]	1,240[e]	1,645[e]	1,272	1,228[e]
Armed forces (000)	195	195	195	195	195
Gross national product (GNP)					
Current dollars (mil.)	24,360	26,840	26,420	26,420	30,200
1994 constant dollars (mil.)	27,110	28,770	27,550	26,970	30,200
Central government expenditures (CGE)					
1994 constant dollars (mil.)	8,129	8,383	8,688	NA	8,897[e]
People (mil.)	26.2	26.8	27.4	28.0	28.6
Military expenditure as % of GNP	5.4	4.3	6.0	4.7	4.1
Military expenditure as % of CGE	17.9	14.8	18.9	NA	13.8
Military expenditure per capita (1994 $)	56	46	60	45	43
Armed forces per 1,000 people (soldiers)	7.5	7.3	7.1	7.0	6.8
GNP per capita (1994 $)	1,036	1,075	1,007	965	1,057
Arms imports[6]					
Current dollars (mil.)	230	60	90	70	130
1994 constant dollars (mil.)	256	64	94	71	130
Arms exports[6]					
Current dollars (mil.)	0	0	0	0	0
1994 constant dollars (mil.)	0	0	0	0	0
Total imports[7]					
Current dollars (mil.)	6,800	6,873	7,348	6,760	7,188
1994 constant dollars (mil.)	7,568	7,367	7,663	6,899	7,188
Total exports[7]					
Current dollars (mil.)	4,265	4,313	3,984	3,991	3,967
1994 constant dollars (mil.)	4,747	4,623	4,155	4,073	3,967
Arms as percent of total imports[8]	3.4	0.9	1.2	1.0	1.8
Arms as percent of total exports[8]	0	0	0	0	0

Human Rights [24]

	SSTS	FL	FAPRO	PPCG	APROBC	TPW	PCPTW	STPEP	PHRFF	PRW	ASST	AFL
Observes	P	P		P	P	P	P	P		P	P	P
	EAFRD	CPR	ESCR	SR	ACHR	MAAE	PVIAC	PVNAC	EAFDAW	TCIDTP	RC
Observes	P	P	P	P			S	S	P	P	P

P=Party; S=Signatory; see Appendix for meaning of abbreviations.

Labor Force

Total Labor Force [25]

7.4 million

Labor Force by Occupation [26]

Agriculture	50%
Services	26
Industry	15
Other	9

Date of data: 1985

Unemployment Rate [27]

16% (1994 est.)

Production Sectors

Commercial Energy Production and Consumption

Data are shown in quadrillion (10^{15}) BTUs and percent for 1995
Values for hydroelectric, nuclear, geothermal, solar, and wind power refer to electrical generation.

Production [28]

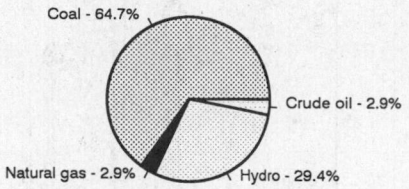

Coal - 64.7%
Crude oil - 2.9%
Natural gas - 2.9%
Hydro - 29.4%

Consumption [29]

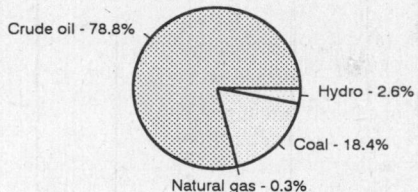

Crude oil - 78.8%
Hydro - 2.6%
Coal - 18.4%
Natural gas - 0.3%

Crude oil	0.001
Dry natural gas	0.001
Coal	0.022
Net hydroelectric power	0.010
Total	0.034

Crude oil	0.309
Dry natural gas	0.001
Coal	0.072
Net hydroelectric power	0.010
Total	0.392

Telecommunications [30]

- 270,100 (1987 est.) telephones
- Domestic: good system of open-wire lines, cables, and microwave radio relay links; principal centers: Casablanca, Rabat; secondary centers: Fes, Marrakech, Oujda, Tangier, Tetouan
- International: 5 submarine cables; satellite earth stations - 2 Intelsat (Atlantic Ocean) and 1 Arabsat; microwave radio relay to Gibraltar, Spain, and Western Sahara; coaxial cable and microwave radio relay to Algeria; participant in Medarabtel
- Radio: Broadcast stations: AM 20, FM 7, shortwave 0 Radios: 5.527 million (1992 est.)
- Television: Broadcast stations: 26 (repeaters 26) Televisions: 1.21 million (1993 est.)

Top Agricultural Products [32]

Agriculture accounts for 14.3% of the GDP; produces barley, wheat, citrus, wine, vegetables, olives; livestock.

Transportation [31]

Railways: total: 1,907 km; standard gauge: 1,907 km 1.435-m gauge (1003 km electrified; 246 km double track) (1994)

Highways: total: 59,474 km; paved: 29,440 km (including 73 km of expressways); unpaved: 30,034 km (1991 est.)

Merchant marine: total: 37 ships (1,000 GRT or over) totaling 175,962 GRT/257,449 DWT; ships by type: cargo 8, chemical tanker 7, container 2, oil tanker 4, refrigerated cargo 9, roll-on/roll-off cargo 6, short-sea passenger 1 (1995 est.)

Airports

Total:	63
With paved runways over 3,047 m:	11
With paved runways 2,438 to 3,047 m:	4
With paved runways 1,524 to 2,437 m:	7
With paved runways 914 to 1,523 m:	2
With paved runways under 914 m:	12

Top Mining Products [33]

Metric tons except as noted[e]	5/95[*]
Mercury, byproduct (kg.)	20,000
Silver (kg.)	326,000
Barite	264,000
Cement, hydraulic (000 tons)	6,300
Gypsum	450,000
Phosphate rock, incl. W. Sahara (000 tons)	20,400
Salt, rock	177,000
Coal, anthracite	650,000
Gas, natural, gross (mil. cu. meters)	24
Petroleum refinery products (000 42-gal. bls.)	46,300

Tourism [34]

	1990	1991	1992	1993	1994
Visitors	4,138	4,211	4,505	4,150	3,599
Tourists[19]	4,024	4,162	4,390	4,027	3,465
Cruise passengers	114	49	115	123	134
Tourism receipts	1,259	1,052	1,360	1,243	1,265
Tourism expenditures	184	190	242	245	302
Fare receipts	38	74	99	131	142
Fare expenditures	31	33	42	57	44

Travelers are in thousands, money in million U.S. dollars.

For sources, notes, and explanations, see Annotated Source Appendix, page 1061.

Manufacturing Sector

Manufacturing Summary [35]

	1987		1988		1989		1990		1991	
	$ bil.	%	$ bil.	%	$ bil.	%	$ bil.	%	$ bil.	%
Establishments or enterprises (number)	4,019	0.171	4,341	0.207	3,618	0.193	3,835	0.214	3,178	0.416
Total employment (000)	162	0.119	191	0.139	179	0.145	194	0.176	233	0.335
Production workers (000)	-	-	-	-	-	-	-	-	-	-
Output ($ bil.)	3	0.034	4	0.033	4	0.035	5	0.043	6	0.062
Value added ($ bil.)	1	0.031	2	0.035	2	0.034	2	0.039	2	0.064
Capital investment ($ mil.)	199	0.046	302	0.063	378	0.069	426	0.077	-	-
M & E investment ($ mil.)	-	-	-	-	-	-	-	-	-	-
Employees per establishment (number)	40	69.731	44	67.353	49	75.114	51	82.061	73	80.545
Production workers per establishment	-	-	-	-	-	-	-	-	-	-
Output per establishment ($ mil.)	0.865	19.962	0.889	16.002	1	18.247	1	20.318	2	14.874
Capital investment per estab. ($ mil.)	0.049	26.612	0.070	30.678	0.104	35.734	0.111	35.791	-	-
M & E per establishment ($ mil)	-	-	-	-	-	-	-	-	-	-
Payroll per employee ($)	3,130	34.909	3,217	32.291	3,703	33.143	4,194	30.235	-	-
Wages per production worker ($)	-	-	-	-	-	-	-	-	-	-
Hours per production worker (hours)	-	-	-	-	-	-	-	-	-	-
Output per employee ($)	21,440	28.627	20,236	23.758	23,281	24.292	25,659	24.760	27,111	18.467
Capital investment per employee ($)	1,226	38.164	1,585	45.548	2,117	47.573	2,190	43.616	-	-
M & E per employee ($)	-	-	-	-	-	-	-	-	-	-

Note: Columns headed % show percent of world total or ratio. Ratios closest to 100 are closest to world average. M & E stands for machinery & equipment.

Output in Manufacturing

	1987		1988		1989		1990		1991	
	$ bil.	%	$ bil.	%	$ bil.	%	$ bil.	%	$ bil.	%
3110 - Food products	1.165	38.83	1.224	30.60	1.234	30.85	1.452	29.04	1.563	26.05
3210 - Textiles	0.877	29.23	0.974	24.35	1.048	26.20	1.153	23.06	1.231	20.52
3220 - Wearing apparel	0.374	12.47	0.452	11.30	0.466	11.65	0.678	13.56	0.778	12.97
3410 - Paper and products	-	-	-	-	-	-	-	-	0.535	8.92
3411 - Pulp, paper, etc.	-	-	-	-	-	-	-	-	0.182	3.03
3420 - Printing, publishing	-	-	-	-	-	-	-	-	0.140	2.33
3810 - Metal products	0.403	13.43	0.524	13.10	0.561	14.03	0.647	12.94	0.736	12.27
3820 - Machinery nec	0.126	4.20	-	-	-	-	-	-	-	-
3830 - Electrical machinery	0.241	8.03	0.295	7.38	0.401	10.02	0.511	10.22	0.471	7.85
3840 - Transportation equipment	0.272	9.07	0.363	9.07	0.417	10.42	0.513	10.26	0.643	10.72
3850 - Professional goods	0.014	0.47	0.017	0.43	0.020	0.50	0.024	0.48	0.029	0.48
3900 - Industries nec	0.004	0.13	0.007	0.18	0.010	0.25	0.010	0.20	-	-

Note: Codes are International Standard Industry codes (ISIC). Percentages are % of total Output. [f]: Factor Prices; [p]: Producer Prices.

For sources, notes, and explanations, see Annotated Source Appendix, page 1061.

Finance, Economics, and Trade

Economic Indicators [36]

- **National product**: GDP—purchasing power parity—$87.4 billion (1995 est.)

- **National product real growth rate**: - 6.5% (1995 est.)

- **National product per capita**: $3,000 (1995 est.)

- **Inflation rate (consumer prices)**: 5.4% (1994)

- **External debt**: $20.5 billion (1994 est.)

Balance of Payments Summary [37]

Values in millions of dollars.

	1989	1990	1991	1992	1993
Exports of goods (f.o.b.)	3,312	4,210	4,277	3,956	3,682
Imports of goods (f.o.b.)	-4,992	-6,282	-6,253	-6,692	-6,062
Trade balance	-1,680	-2,072	-1,976	-2,736	-2,380
Services - debits	-2,431	-2,572	-2,688	-2,873	-3,045
Services - credits	1,700	2,111	1,975	2,643	2,609
Private transfers (net)	1,356	2,012	2,013	2,179	2,138
Government transfers (net)	265	326	282	357	150
Long-term capital (net)	600	1,713	1,345	853	815
Short-term capital (net)	211	175	130	518	151
Errors and omissions	-11	3	86	6	-367
Overall balance	10	1,696	1,167	947	71

Exchange Rates [38]

Currency: **Moroccan dirham.**
Symbol: **DH.**

Data are currency units per $1.

January 1996	8.607
1995	8.540
1994	9.203
1993	9.299
1992	8.538
1991	8.707

Imports and Exports

Top Import Origins [39]

$7.2 billion (c.i.f., 1994) Data are for 1993.

Origins	%
EU	59
US	8
Saudi Arabia	5
UAE	3
Russia	2

Top Export Destinations [40]

$4 billion (f.o.b., 1994) Data are for 1993.

Destinations	%
EU	70
Japan	5
US	4
Libya	3
India	2

Foreign Aid [41]

	U.S. $	
ODA (1993)	297	million
Saudi Arabia debt cancellation (1991)	2.8	billion

Import and Export Commodities [42]

Import Commodities

Capital goods 24%
Semiprocessed goods 22%
Raw materials 16%
Fuel and lubricants 16%
Food and beverages 13%
Consumer goods 9%

Export Commodities

Food and beverages 30%
Semiprocessed goods 23%
Consumer goods 21%
Phosphates 17%

Mozambique

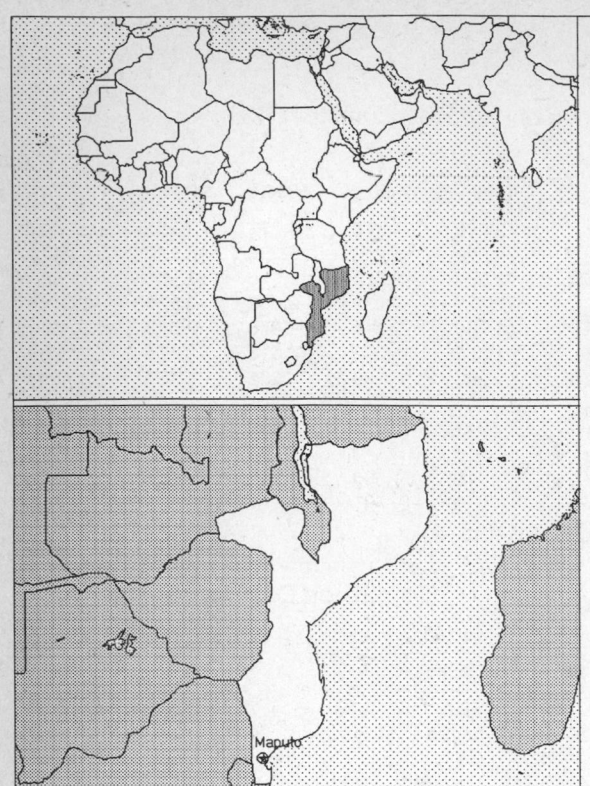

Geography [1]

Total area:
801,590 sq km   309,496 sq mi
Land area:
784,090 sq km   302,739 sq mi
Comparative area:
Slightly less than twice the size of California
Land boundaries:
Total 4,571 km, Malawi 1,569 km, South Africa 491 km, Swaziland 105 km, Tanzania 756 km, Zambia 419 km, Zimbabwe 1,231 km
Coastline:
2,470 km
Climate:
Tropical to subtropical
Terrain:
Mostly coastal lowlands, uplands in center, high plateaus in Northwest, mountains in West
Natural resources:
Coal, titanium, natural gas
Land use:
Arable land: 4%
Permanent crops: 0%
Meadows and pastures: 56%
Forest and woodland: 20%
Other: 20%

Demographics [2]

	1970	1980	1990	1995[1]	1996	2000	2010	2020	2030
Population	9,304	12,103	14,056	17,097	17,878	19,829	25,116	30,810	36,800
Population density (persons per sq. mi.)	31	40	46	56	59	66	83	102	122
(persons per sq. km.)	12	15	18	22	23	25	32	39	47
Net migration rate (per 1,000 population)	NA	0.2	-8.9	36.8	0.0	0.0	0.0	0.0	0.0
Births	NA	NA	NA	NA	814	NA	NA	NA	NA
Deaths	NA	NA	NA	NA	339	NA	NA	NA	NA
Life expectancy - males	NA	41.9	40.8	42.7	43.2	45.3	50.7	56.4	61.8
Life expectancy - females	NA	44.7	43.1	45.0	45.5	47.6	53.3	59.1	64.7
Birth rate (per 1,000)	49.1	49.2	44.7	46.3	45.5	42.3	34.6	28.7	24.0
Death rate (per 1,000)	NA	20.6	20.8	19.5	19.0	17.0	12.7	9.7	7.7
Women of reproductive age (15-49 yrs.)	NA	2,851	3,320	4,047	4,237	4,725	6,214	8,132	10,054
of which are currently married	1,389	2,003	2,493	NA	3,187	3,544	4,670	NA	NA
Fertility rate	6.8	6.8	6.2	6.4	6.2	5.8	4.5	3.4	2.7

Except as noted, values for vital statistics are in thousands; life expectancy is in years.

Health

Health Indicators [3]

% of population with access to	
safe water (1990-95)	33
adequate sanitation (1990-95)[1]	20
health services (1985-95)	39
% of 1-year-olds immunized (1990-94) against	
TB (tuberculosis)	78
DPT (diphtheria, pertussis, tetanus)	55
polio	55
measles	65
% of contraceptive prevalence (1980-94)	4
ORT use rate (1990-94)	60

Health Expenditures [4]

Total health expenditure, 1990 (official exchange rate)	
Millions of dollars	85
Dollars per capita	5
Health expenditures as a percentage of GDP	
Total	5.9
Public sector	4.4
Private sector	1.5
Development assistance for health	
Total aid flows (millions of dollars)[1]	45
Aid flows per capita (dollars)	2.9
Aid flows as a percentage of total health expenditure	52.9

For sources, notes, and explanations, see Annotated Source Appendix, page 1061.

Human Factors

Women and Children [5]

% of pregnant women immunized (tetanus 1990-94)	37
% of births attended by trained health personnel (1983-94)	25
Maternal mortality rate (1980-92)	300
Under-5 mortality rate (1994)	277
% under-5 moderately/severely underweight (1980-1994)	NA

Burden of Disease [6]

Population per physician (1985)	48,534.05
Population per nurse	NA
Population per hospital bed (1990)	1,156.23
AIDS cases per 100,000 people (1994)	3.1
Malaria cases per 100,000 people (1992)	NA

Ethnic Division [7]

Indigenous includes Shangaan, Chokwe, Manyika, Sena, Makua, and others.

Indigenous	99.66%
Euro-Africans	0.2%
Indians	0.08%
Europeans	0.06%

Religion [8]

Indigenous beliefs	50%
Christian	30%
Muslim	20%

Major Languages [9]

Portuguese (official), indigenous dialects.

Education

Public Education Expenditures [10]

Million (Meticai)[45]	1980	1985	1989	1990	1993	1994
Total education expenditure	2,900	4,400	44,571	72,264	NA	NA
as percent of GNP	4.4	4.2	5.5	6.3	NA	NA
as percent of total govt. expend.	12.1	10.6		12.0	NA	NA
Current education expenditure	2,500	4,100	30,371	46,064	NA	NA
as percent of GNP	3.8	3.9	3.7	4.0	NA	NA
as percent of current govt. expend.	17.7	12.3	12.3	17.5	NA	NA
Capital expenditure	400	300	14,200	26,200	NA	NA

Educational Attainment [11]

Age group (1980)[8]	25+
Total population	4,242,819
Highest level attained (%)	
No schooling	81.0
First level	
Not completed	18.1
Completed	NA
Entered second level	
S-1	0.8
S-2	NA
Postsecondary	0.1

Illiteracy [12]

In thousands and percent[1]	1990	1995	2000
Illiterate population (15+ yrs.)	5,158	5,298	5,700
Illiteracy rate - total pop. (%)	67.6	57.8	52.8
Illiteracy rate - males (%)	51.3	42.6	37.0
Illiteracy rate - females (%)	82.3	71.0	67.0

Libraries [13]

Daily Newspapers [14]

	1980	1985	1990	1994
Number of papers	2	2	2	2
Circ. (000)	54	81	81	81

Culture [15]

Science and Technology

Scientific/Technical Forces [16]

R&D Expenditures [17]

U.S. Patents Issued [18]

For sources, notes, and explanations, see Annotated Source Appendix, page 1061.

657

Government and Law

Organization of Government [19]

Long-form name:
Republic of Mozambique
Type:
Republic
Independence:
25 June 1975 (from Portugal)
National holiday:
Independence Day, 25 June (1975)
Constitution:
30 November 1990
Legal system:
Based on Portuguese civil law system and customary law
Executive branch:
President; Prime Minister; Cabinet
Legislative branch:
Unicameral: Assembly of the Republic (Assembleia da Republica)
Judicial branch:
Supreme Court

Crime [23]

Elections [20]

Assembly of the Republic. Election last held 27-29 October 1994 (next to be held October 1999); results—percent vote by party NA; seats (250 total) by party Frelimo won a slim majority.

Government Budget [21]

For 1992 est.

	$ mil.
Revenues	252
Expenditures	607
Capital expenditures	NA

Military Expenditures and Arms Transfers [22]

	1990	1991	1992	1993	1994
Military expenditures					
Current dollars (mil.)	74	104	90[e]	97	104
1994 constant dollars (mil.)	82	111	94[e]	99	104
Armed forces (000)	65	65	50	50	11
Gross national product (GNP)					
Current dollars (mil.)	801	883	888	1,113	1,195
1994 constant dollars (mil.)	891	947	926	1,136	1,195
Central government expenditures (CGE)					
1994 constant dollars (mil.)	NA	NA	565[e]	NA	NA
People (mil.)	14.4	14.8	15.5	16.3	17.3
Military expenditure as % of GNP	9.2	11.7	10.1	8.7	8.7
Military expenditure as % of CGE	NA	NA	16.6	NA	NA
Military expenditure per capita (1994 $)	6	7	6	6	6
Armed forces per 1,000 people (soldiers)	4.5	4.4	3.2	3.1	0.6
GNP per capita (1994 $)	62	64	60	70	69
Arms imports[6]					
Current dollars (mil.)	140	50	5	0	0
1994 constant dollars (mil.)	156	54	5	0	0
Arms exports[6]					
Current dollars (mil.)	0	0	0	0	0
1994 constant dollars (mil.)	0	0	0	0	0
Total imports[7]					
Current dollars (mil.)	878	899	855	955	1,140[e]
1994 constant dollars (mil.)	977	964	892	975	1,140[e]
Total exports[7]					
Current dollars (mil.)	126	162	139	132	150[e]
1994 constant dollars (mil.)	140	174	145	135	150[e]
Arms as percent of total imports[8]	15.9	5.6	0.6	0	0
Arms as percent of total exports[8]	0	0	0	0	0

Human Rights [24]

	SSTS	FL	FAPRO	PPCG	APROBC	TPW	PCPTW	STPEP	PHRFF	PRW	ASST	AFL
Observes				P		P	P					P
	EAFRD	CPR	ESCR	SR	ACHR	MAAE	PVIAC	PVNAC	EAFDAW	TCIDTP	RC
Observes	P	P		P			P				P

P = Party; S = Signatory; see Appendix for meaning of abbreviations.

Labor Force

Total Labor Force [25]

Labor Force by Occupation [26]

Agriculture   90%

Unemployment Rate [27]

50% (1989 est.)

Production Sectors

Commercial Energy Production and Consumption

Data are shown in quadrillion (10^{15}) BTUs and percent for 1995
Values for hydroelectric, nuclear, geothermal, solar, and wind power refer to electrical generation.

Production [28]

Coal - 50.0%

Hydro - 50.0%

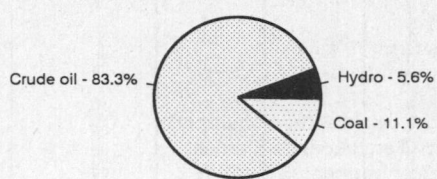

Consumption [29]

Crude oil - 83.3%

Hydro - 5.6%

Coal - 11.1%

Coal	0.001
Net hydroelectric power	0.001
Total	0.002

Crude oil	0.015
Coal	0.002
Net hydroelectric power	0.001
Total	0.018

Telecommunications [30]

- 59,000 (1983 est.) telephones; fair system of tropospheric scatter, open-wire lines, and microwave radio relay
- Domestic: microwave radio relay and tropospheric scatter
- International: satellite earth stations - 5 Intelsat (2 Atlantic Ocean and 3 Indian Ocean)
- Radio: Broadcast stations: AM 29, FM 4, shortwave 0 Radios: 700,000 (1992 est.)
- Television: Broadcast stations: 1 Televisions: 44,000 (1992 est.)

Transportation [31]

Railways: total: 3,131 km; narrow gauge: 2,988 km 1.067-m gauge; 143 km 0.762-m gauge (1994)

Highways: total: 27,287 km; paved: 4,693 km; unpaved: 22,594 km (1991 est.)

Merchant marine: total: 4 cargo ships (1,000 GRT or over) totaling 5,694 GRT/9,724 DWT (1995 est.)

Airports

Total:	131
With paved runways over 3,047 m:	1
With paved runways 2,438 to 3,047 m:	4
With paved runways 1,524 to 2,437 m:	10
With paved runways 914 to 1,523 m:	5
With paved runways under 914 m:	67

Top Agricultural Products [32]

Agriculture accounts for 33% of the GDP; produces cotton, cashew nuts, sugarcane, tea, cassava (tapioca), corn, rice, tropical fruits; beef, poultry.

Top Mining Products [33]

Metric tons except as noted	6/9/95[*]
Cement, hydraulic	20,000[e]
Clay, bentonite	3,350
Aquamarine (grams)	38,500
Emerald (grams)	11,400
Garnet, facet-grade (kg.)	1,170[56]
Tourmaline (kg.)	5,270
Marble	
block (cu. meters)	1,500
slab (sq. meters)	52,300
Salt, marine	40,000[e]

Tourism [34]

Manufacturing Sector

Manufacturing Summary [35]

	1987		1988		1989		1990		1991	
	$ bil.	%	$ bil.	%	$ bil.	%	$ bil.	%	$ bil.	%
Establishments or enterprises (number)	-	-	422	0.020	295	0.016	-	-	-	-
Total employment (000)	114	0.084	121	0.088	83	0.068	-	-	-	-
Production workers (000)	66	0.097	63	0.101	-	-	-	-	-	-
Output ($ bil.)	0.212	0.002	-	-	-	-	-	-	-	-
Value added ($ bil.)	-	-	-	-	-	-	-	-	-	-
Capital investment ($ mil.)	-	-	-	-	-	-	-	-	-	-
M & E investment ($ mil.)	-	-	-	-	-	-	-	-	-	-
Employees per establishment (number)	-	-	287	439.810	282	428.714	-	-	-	-
Production workers per establishment	-	-	149	500.699	-	-	-	-	-	-
Output per establishment ($ mil.)	-	-	-	-	-	-	-	-	-	-
Capital investment per estab. ($ mil.)	-	-	-	-	-	-	-	-	-	-
M & E per establishment ($ mil)	-	-	-	-	-	-	-	-	-	-
Payroll per employee ($)	-	-	-	-	-	-	-	-	-	-
Wages per production worker ($)	-	-	-	-	-	-	-	-	-	-
Hours per production worker (hours)	-	-	-	-	-	-	-	-	-	-
Output per employee ($)	1,849	2.469	-	-	-	-	-	-	-	-
Capital investment per employee ($)	-	-	-	-	-	-	-	-	-	-
M & E per employee ($)	-	-	-	-	-	-	-	-	-	-

Note: Columns headed % show percent of world total or ratio. Ratios closest to 100 are closest to world average. M & E stands for machinery & equipment.

Output in Manufacturing

	1987		1988		1989		1990		1991	
	$ bil.	%	$ bil.	%	$ bil.	%	$ bil.	%	$ bil.	%
3110 - Food products	0.030	14.15	-	-	-	-	-	-	-	-
3130 - Beverages	0.011	5.19	-	-	-	-	-	-	-	-
3140 - Tobacco	0.026	12.26	-	-	-	-	-	-	-	-
3210 - Textiles	0.033	15.57	-	-	-	-	-	-	-	-
3211 - Spinning, weaving, etc.	0.023	10.85	-	-	-	-	-	-	-	-
3220 - Wearing apparel	0.013	6.13	-	-	-	-	-	-	-	-
3230 - Leather and products	0.001	0.47	-	-	-	-	-	-	-	-
3240 - Footwear	0.004	1.89	-	-	-	-	-	-	-	-
3310 - Wood products	0.005	2.36	-	-	-	-	-	-	-	-
3320 - Furniture, fixtures	0.002	0.94	-	-	-	-	-	-	-	-
3410 - Paper and products	0.004	1.89	-	-	-	-	-	-	-	-
3420 - Printing, publishing	0.005	2.36	-	-	-	-	-	-	-	-
3510 - Industrial chemicals	0.001	0.47	-	-	-	-	-	-	-	-
3511 - Basic chemicals, excl fertilizers	0.001	0.47	-	-	-	-	-	-	-	-
3520 - Chemical products nec	0.007	3.30	-	-	-	-	-	-	-	-
3522 - Drugs and medicines	0.000	0.00	-	-	-	-	-	-	-	-
3530 - Petroleum refineries	0.002	0.94	-	-	-	-	-	-	-	-
3540 - Petroleum, coal products	-	-	-	-	-	-	-	-	-	-
3550 - Rubber products	0.004	1.89	-	-	-	-	-	-	-	-
3560 - Plastic products nec	0.003	1.42	-	-	-	-	-	-	-	-
3610 - Pottery, china, etc.	0.000	0.00	-	-	-	-	-	-	-	-
3620 - Glass and products	0.002	0.94	-	-	-	-	-	-	-	-
3690 - Nonmetal products nec	0.006	2.83	-	-	-	-	-	-	-	-
3710 - Iron and steel	0.004	1.89	-	-	-	-	-	-	-	-
3720 - Nonferrous metals	0.001	0.47	-	-	-	-	-	-	-	-
3810 - Metal products	0.007	3.30	-	-	-	-	-	-	-	-
3820 - Machinery nec	0.002	0.94	-	-	-	-	-	-	-	-
3825 - Office, computing machinery	0.000	0.00	-	-	-	-	-	-	-	-
3830 - Electrical machinery	0.001	0.47	-	-	-	-	-	-	-	-
3832 - Radio, television, etc.	0.002	0.94	-	-	-	-	-	-	-	-
3840 - Transportation equipment	0.006	2.83	-	-	-	-	-	-	-	-
3841 - Shipbuilding, repair	0.003	1.42	-	-	-	-	-	-	-	-
3843 - Motor vehicles	0.002	0.94	-	-	-	-	-	-	-	-
3850 - Professional goods	0.000	0.00	-	-	-	-	-	-	-	-
3900 - Industries nec	0.000	0.00	-	-	-	-	-	-	-	-

Note: Codes are International Standard Industry codes (ISIC). Percentages are % of total Output. [f]: Factor Prices; [p]: Producer Prices.

Finance, Economics, and Trade

Economic Indicators [36]

- **National product**: GDP—purchasing power parity—$12.2 billion (1995 est.)

- **National product real growth rate**: - 2.5% (1995 est.)

- **National product per capita**: $700 (1995 est.)

- **Inflation rate (consumer prices)**: 50% (1994 est.)

- **External debt**: $5 billion (1992 est.)

Balance of Payments Summary [37]

Values in millions of dollars.

	1988	1989	1990	1991	1992
Exports of goods (f.o.b.)	103.0	104.8	126.4	162.3	139.3
Imports of goods (f.o.b.)	-662.0	-726.9	-789.7	-808.8	-798.5
Trade balance	-559.0	-622.1	-663.3	-646.5	-659.5
Services - debits	-332.9	-392.3	-373.8	-402.3	-444.1
Services - credits	156.6	166.7	173.4	202.8	222.6
Private transfers (net)	52.7	57.5	72.1	78.0	NA
Government transfers (net)	378.4	408.6	470.7	538.3	499.4
Long-term capital (net)	272.1	-33.9	269.8	197.4	545.7
Short-term capital (net)	NA	397.5	50.8	85.7	-222.2
Errors and omissions	84.7	56.7	66.3	-3.9	32.5
Overall balance	52.6	38.7	66.0	49.5	-25.3

Exchange Rates [38]

Currency: **metical.**
Symbol: **Mt.**

Data are currency units per $1.

December 1995	10,908.0
1995	9,024.3
1994	6,038.6
1993	3,874.2
1992	2,516.5
1991	1,434.5

Imports and Exports

Top Import Origins [39]

$1.14 billion (c.i.f., 1994 est.).

Origins	%
South Africa	NA
UK	NA
France	NA
Japan	NA
Portugal	NA

Top Export Destinations [40]

$170 million (f.o.b., 1995 est.).

Destinations	%
Spain	NA
South Africa	NA
US	NA
Portugal	NA
Japan	NA

Foreign Aid [41]

Recipient: ODA, $NA.

Import and Export Commodities [42]

Import Commodities

Food
Clothing
Farm equipment
Petroleum

Export Commodities

Shrimp 40%
Cashews
Cotton
Sugar
Copra
Citrus

Myanmar

Geography [1]

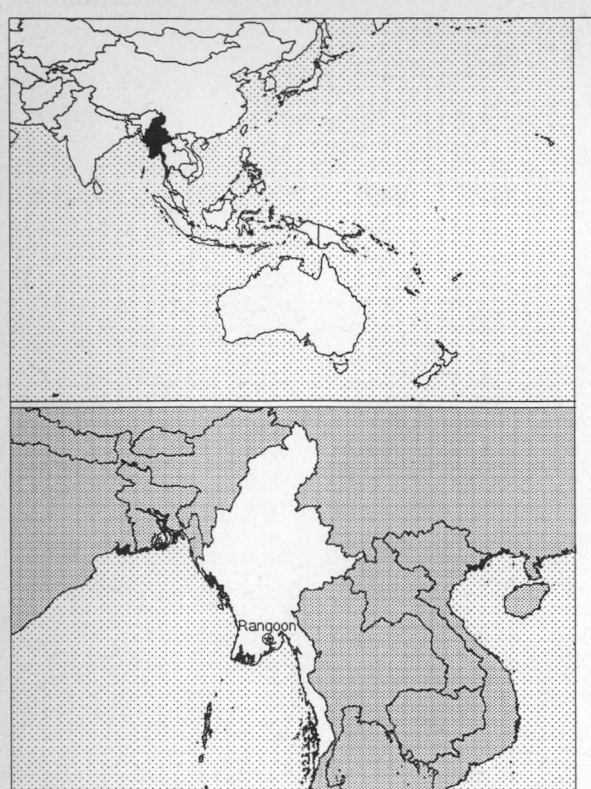

Total area:
678,500 sq km   261,970 sq mi
Land area:
657,740 sq km   253,955 sq mi
Comparative area:
Slightly smaller than Texas
Land boundaries:
Total 5,876 km, Bangladesh 193 km, China 2,185 km, India 1,463 km,
Laos 235 km, Thailand 1,800 km
Coastline:
1,930 km
Climate:
Tropical monsoon; cloudy, rainy, hot, humid summers (southwest monsoon,
June to September); less cloudy, scant rainfall, mild temperatures,
lower humidity during winter (northeast monsoon, December to April)
Terrain:
Central lowlands ringed by steep, rugged highlands
Natural resources:
Petroleum, timber, tin, antimony, zinc, copper, tungsten, lead, coal,
some marble, limestone, precious stones, natural gas
Land use:
Arable land: 15%
Permanent crops: 1%
Meadows and pastures: 1%
Forest and woodland: 49%
Other: 34%

Demographics [2]

	1970	1980	1990	1995[1]	1996	2000	2010	2020	2030
Population	27,386	33,766	41,078	45,135	45,976	49,388	58,236	67,501	76,708
Population density (persons per sq. mi.)	108	133	162	178	181	194	229	266	302
(persons per sq. km.)	42	51	62	69	70	75	89	103	117
Net migration rate (per 1,000 population)	NA	0.0	0.0	0.0	0.0	0.0	0.0	0.0	0.0
Births	NA	NA	NA	NA	1,380	NA	NA	NA	NA
Deaths	NA	NA	NA	NA	536	NA	NA	NA	NA
Life expectancy - males	NA	50.5	51.5	54.0	54.5	56.2	60.5	64.4	67.8
Life expectancy - females	NA	53.1	54.3	57.4	57.9	60.1	65.2	69.8	73.8
Birth rate (per 1,000)	NA	34.7	32.6	30.5	30.0	28.1	24.4	21.5	19.0
Death rate (per 1,000)	NA	14.3	13.6	11.9	11.7	10.7	8.8	7.7	7.3
Women of reproductive age (15-49 yrs.)	NA	8,066	10,157	11,335	11,588	12,648	15,254	17,595	19,829
of which are currently married	NA	NA	5,523	NA	6,418	7,064	8,601	NA	NA
Fertility rate	NA	4.9	4.2	3.9	3.8	3.6	3.1	2.7	2.4

Except as noted, values for vital statistics are in thousands; life expectancy is in years.

Health

Health Indicators [3]

% of population with access to	
safe water (1990-95)	38
adequate sanitation (1990-95)	36
health services (1985-95)	60
% of 1-year-olds immunized (1990-94) against	
TB (tuberculosis)	83
DPT (diphtheria, pertussis, tetanus)	77
polio	77
measles	77
% of contraceptive prevalence (1980-94)	13
ORT use rate (1990-94)	37

Health Expenditures [4]

For sources, notes, and explanations, see Annotated Source Appendix, page 1061.

Human Factors

Women and Children [5]

% of pregnant women immunized (tetanus 1990-94)	68
% of births attended by trained health personnel (1983-94)	57
Maternal mortality rate (1980-92)	460
Under-5 mortality rate (1994)	109
% under-5 moderately/severely underweight (1980-1994)[1]	32

Burden of Disease [6]

Population per physician (1990)	12,897.29
Population per nurse (1990)	1,240.96
Population per hospital bed (1990)	1,590.21
AIDS cases per 100,000 people (1994)	0.5
Malaria cases per 100,000 people (1992)	288

Ethnic Division [7]

Myanmarn	68%
Shan	9%
Karen	7%
Rakhine	4%
Chinese	3%
Indian	2%
Mon	2%
Other	5%

Religion [8]

Buddhist	89%
Christian	4%
Baptist	3%
Roman Catholic	1%
Muslim	4%
Animist beliefs	1%
Other	2%

Major Languages [9]

Burmese, minority ethnic groups have their own languages.

Education

Public Education Expenditures [10]

Million (Kyat)[46]	1980	1985	1989	1990	1991	1994
Total education expenditure	660	1,084	2,948	NA	NA	5,685
as percent of GNP	1.7	2.0	2.4	NA	NA	NA
as percent of total govt. expend.	NA	NA	NA	NA	NA	14.4
Current education expenditure	584	841	2,699	NA	NA	4,436
as percent of GNP	1.5	1.5	2.2	NA	NA	NA
as percent of current govt. expend.	NA	NA	NA	NA	NA	19.0
Capital expenditure	76	243	249	NA	NA	1,249

Educational Attainment [11]

Age group (1983)	25+
Total population	13,948,584
Highest level attained (%)	
No schooling	55.8
First level	
Not completed	27.7
Completed	NA
Entered second level	
S-1	14.5
S-2	NA
Postsecondary	2.0

Illiteracy [12]

In thousands and percent[1]	1990	1995	2000
Illiterate population (15+ yrs.)	4,861	4,913	4,958
Illiteracy rate - total pop. (%)	19.1	17.3	15.7
Illiteracy rate - males (%)	12.4	11.5	10.6
Illiteracy rate - females (%)	25.7	23.1	20.7

Libraries [13]

	Admin. Units	Svc. Pts.	Vols. (000)	Shelving (meters)	Vols. Added	Reg. Users
National (1993)	1	NA	4.2	NA	NA	14,081
Nonspecialized	NA	NA	NA	NA	NA	NA
Public	NA	NA	NA	NA	NA	NA
Higher ed.	NA	NA	NA	NA	NA	NA
School	NA	NA	NA	NA	NA	NA

Daily Newspapers [14]

	1980	1985	1990	1994
Number of papers	7	7	2	5
Circ. (000)	350[e]	511	700[e]	1,032

Culture [15]

Science and Technology

Scientific/Technical Forces [16]

R&D Expenditures [17]

U.S. Patents Issued [18]

For sources, notes, and explanations, see Annotated Source Appendix, page 1061.

663

Government and Law

Organization of Government [19]

Long-form name:
Union of Myanmar
Type:
Military regime
Independence:
4 January 1948 (from UK)
National holiday:
Independence Day, 4 January (1948)
Constitution:
3 January 1974 (suspended since 18
September 1988); national convention
started on 9 January 1993 to draft a new
constitution
Legal system:
Does not accept compulsory ICJ
jurisdiction
Executive branch:
Prime Minister and Chairman of the State
Law and Order Restoration Council; State
Law and Order Restoration Council
Legislative branch:
People's Assembly (Pyithu Hluttaw) Note:
Assembly never convened
Judicial branch:
Limited remnants of the British-era legal
system in place, no guarantee of a fair
public trial; the judiciary is not
independent of the executive

Crime [23]

	1990
Crime volume	
Cases known to police	52,994
Attempts (percent)	NA
Percent cases solved	279.0
Crimes per 100,000 persons	129.0
Persons responsible for offenses	
Total number offenders	147,834
Percent female	15.5
Percent juvenile	1.8
Percent foreigners	NA

Elections [20]

People's Assembly	% of votes
Nat. League for Democr. (NLD)	81.6
National Union Party (NUP)	2.1
Others	16.2

Government Expenditures [21]

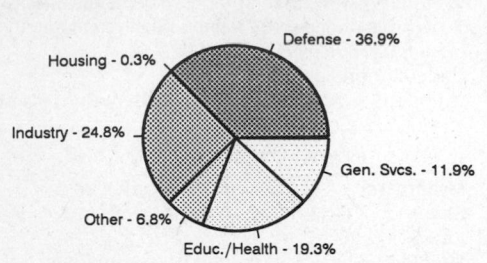

Defense - 36.9%
Housing - 0.3%
Industry - 24.8%
Gen. Svcs. - 11.9%
Other - 6.8%
Educ./Health - 19.3%

(% distribution). Expend. for FY94: 48,021 (Kyat mil.)

Military Expenditures and Arms Transfers [22]

	1990	1991	1992	1993	1994
Military expenditures					
Current dollars (mil.)	1,527	1,972[e]	1,941	2,675	2,618[e]
1994 constant dollars (mil.)	1,699	2,114[e]	2,024	2,730	2,618[e]
Armed forces (000)	230	286	286	286	370
Gross national product (GNP)					
Current dollars (mil.)	42,680	43,710	49,080	NA	NA
1994 constant dollars (mil.)	47,500	46,850	51,180	NA	NA
Central government expenditures (CGE)					
1994 constant dollars (mil.)	7,610	7,183	6,148	NA	NA
People (mil.)	41.0	41.8	42.6	43.5	44.3
Military expenditure as % of GNP	3.6	4.5	4.0	NA	NA
Military expenditure as % of CGE	22.3	29.4	32.9	NA	NA
Military expenditure per capita (1994 $)	41	51	47	63	59
Armed forces per 1,000 people (soldiers)	5.6	6.8	6.7	6.6	8.4
GNP per capita (1994 $)	1,157	1,120	1,200	NA	NA
Arms imports[6]					
Current dollars (mil.)	110	390	150	130	90
1994 constant dollars (mil.)	122	418	156	133	90
Arms exports[6]					
Current dollars (mil.)	0	0	0	0	0
1994 constant dollars (mil.)	0	0	0	0	0
Total imports[7]					
Current dollars (mil.)	270	646	651	814	886
1994 constant dollars (mil.)	300	692	679	831	886
Total exports[7]					
Current dollars (mil.)	325	419	537	583	771
1994 constant dollars (mil.)	362	449	560	595	771
Arms as percent of total imports[8]	40.7	60.4	23.0	16.0	10.2
Arms as percent of total exports[8]	0	0	0	0	0

Human Rights [24]

	SSTS	FL	FAPRO	PPCG	APROBC	TPW	PCPTW	STPEP	PHRFF	PRW	ASST	AFL
Observes	P	P	P	P		P	P	S		S		

	EAFRD	CPR	ESCR	SR	ACHR	MAAE	PVIAC	PVNAC	EAFDAW	TCIDTP	RC
Observes											P

P = Party; S = Signatory; see Appendix for meaning of abbreviations.

Labor Force

Total Labor Force [25]

16.007 million (1992)

Labor Force by Occupation [26]

Agriculture	65.2%
Industry	14.3
Trade	10.1
Government	6.3
Other	4.1

(FY88/89 est.)

Unemployment Rate [27]

Production Sectors

Commercial Energy Production and Consumption

Data are shown in quadrillion (10^{15}) BTUs and percent for 1995
Values for hydroelectric, nuclear, geothermal, solar, and wind power refer to electrical generation.

Production [28]

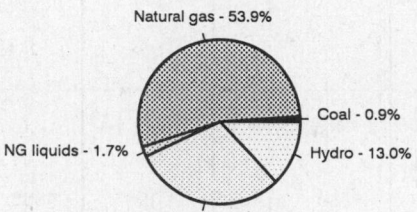

Natural gas - 53.9%
Coal - 0.9%
NG liquids - 1.7%
Hydro - 13.0%
Crude oil - 30.4%

Consumption [29]

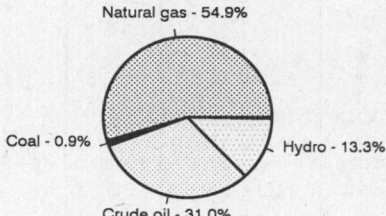

Natural gas - 54.9%
Coal - 0.9%
Hydro - 13.3%
Crude oil - 31.0%

Production		Consumption	
Crude oil	0.035	Crude oil	0.035
Natural gas liquids	0.002	Dry natural gas	0.062
Dry natural gas	0.062	Coal	0.001
Coal	0.001	Net hydroelectric power	0.015
Net hydroelectric power	0.015	Total	0.113
Total	0.115		

Telecommunications [30]

- 122,195 (1993 est.) telephones; meets minimum requirements for local and intercity service for business and government; international service is good
- International: satellite earth station - 1 Intelsat (Indian Ocean)
- Radio: Broadcast stations: AM 2, FM 1, shortwave 0 (1985 est.)
- Television: Broadcast stations: 1 (1988 est.) Televisions: 88,000 (1992 est.)

Top Agricultural Products [32]

Agriculture accounts for 60% of the GDP; produces paddy rice, corn, oilseed, sugarcane, pulses; hardwood.

Transportation [31]

Railways: total: 3,569 km; narrow gauge: 3,569 km 1.000-m gauge (1995)

Highways: total: 26,861 km; paved: 3,181 km; unpaved: 23,680 km (1988 est.)

Merchant marine: total: 40 ships (1,000 GRT or over) totaling 444,957 GRT/610,420 DWT; ships by type: bulk 11, cargo 15, chemical tanker 5, container 1, oil tanker 3, passenger-cargo 3, vehicle carrier 2 (1995 est.)

Airports

Total:	74
With paved runways over 3,047 m:	2
With paved runways 2,438 to 3,047 m:	2
With paved runways 1,524 to 2,437 m:	13
With paved runways 914 to 1,523 m:	10
With paved runways under 914 m:	28

Top Mining Products [33]

Metric tons except as noted[e]	6/2/95[*]
Copper, mine output, Cu content	5,800
Steel, crude	25,000
Barite	16,000
Cement, hydraulic	453,000
Gypsum	32,000
Nitrogen, N content of fertilizer	130,000
Jade (kg.)	250,000
Ruby, sapphire, spinel (carats)	260,000
Coal, lignite	32,000
Petroleum, crude (000 42-gal. bls.)	5,800

Tourism [34]

	1990	1991	1992	1993	1994
Visitors	NA	NA	137	147	167
Tourists[78]	21	22	27	48	80
Excursionists	NA	NA	109	99	82
Cruise passengers	NA	NA	1	NA	2
Tourism receipts	9	13	16	19	24
Tourism expenditures	16	24	NA	NA	NA
Fare receipts	1	3	NA	NA	NA

Travelers are in thousands, money in million U.S. dollars.

For sources, notes, and explanations, see Annotated Source Appendix, page 1061.

665

Finance, Economics, and Trade

Industrial Summary [35]

Industrial Production: Growth rate 4.9% (FY92/93 est.); accounts for 10% of the GDP. **Industries**: Agricultural processing; textiles and footwear; wood and wood products; petroleum refining; copper, tin, tungsten, iron; construction materials; pharmaceuticals; fertilizer.

Economic Indicators [36]

- **National product**: GDP—purchasing power parity—$47 billion (1995 est.)
- **National product real growth rate**: 6.8% (1995 est.)
- **National product per capita**: $1,000 (1995 est.)
- **Inflation rate (consumer prices)**: 38% (1994 est.)
- **External debt**: $5.5 billion (FY94/95 est.)

Balance of Payments Summary [37]

Values in millions of dollars.

	1980	1985	1987	1988	1990
Exports of goods (f.o.b.)	428.6	310.8	219.6	165.7	222.6
Imports of goods (f.o.b.)	-788.28	-512.9	-452.7	-370.2	-524.3
Trade balance	-359.6	-202.1	-233.1	-204.5	-301.7
Services - debits	-134.3	-152.9	-129.3	-113.5	-264.4
Services - credits	66.2	69.0	75.4	49.6	95.9
Private transfers (net)	7.4	5.8	8.4	8.6	10.2
Government transfers (net)	73.3	74.7	98.7	83.9	28.8
Long-term capital (net)	371.1	132.4	188.6	151.1	424.1
Short-term capital (net)	-0.3	16.4	14.3	-11.5	-5.4
Errors and omissions	39.8	41.7	14.7	116.7	21.3
Overall balance	63.6	-15.0	37.7	80.4	8.8

Exchange Rates [38]

Currency: **kyat.**
Symbol: **K.**

Data are currency units per $1.

January 1996	5.8475
1995	5.9170
1994	5.9749
1993	6.1570
1992	6.1045
1991	6.2837
Unofficial rate	120

Imports and Exports

Top Import Origins [39]

$1.5 billion (FY94/95 est.).

Origins	%
Japan	NA
China	NA
Thailand	NA
Singapore	NA
Malaysia	NA

Top Export Destinations [40]

$879 million (FY94/95 est.).

Destinations	%
Singapore	NA
China	NA
Thailand	NA
India	NA
Hong Kong	NA

Foreign Aid [41]

Recipient: ODA, $61 million (1993).

Import and Export Commodities [42]

Import Commodities	Export Commodities
Machinery	Pulses and beans
Transport equipment	Teak
Construction materials	Rice
Food products	Hardwood
Consumer goods	

Namibia

Geography [1]

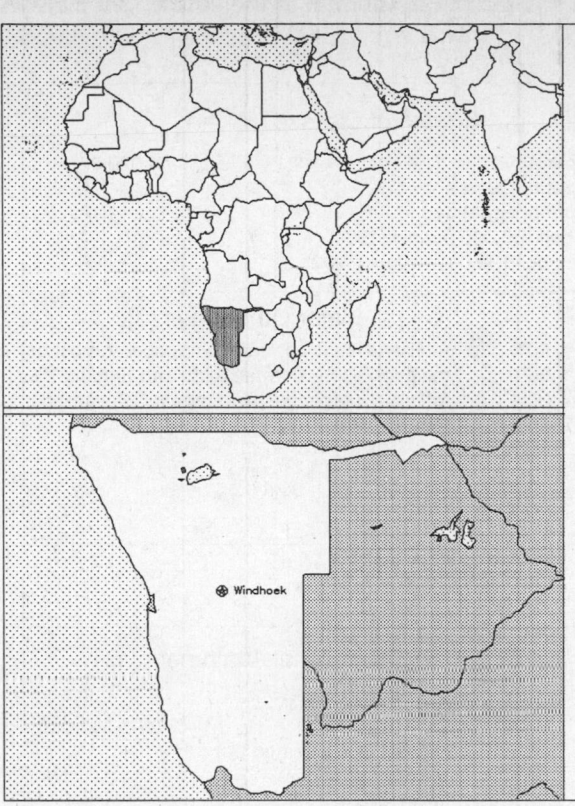

Total area:
825,418 sq km 318,696 sq mi
Land area:
825,418 sq km 318,696 sq mi
Comparative area:
Slightly more than half the size of Alaska
Land boundaries:
Total 3,824 km, Angola 1,376 km, Botswana 1,360 km, South Africa 855 km, Zambia 233 km
Coastline:
1,572 km
Climate:
Desert; hot, dry; rainfall sparse and erratic
Terrain:
Mostly high plateau; Namib Desert along coast; Kalahari Desert in east
Natural resources:
Diamonds, copper, uranium, gold, lead, tin, lithium, cadmium, zinc, salt, vanadium, natural gas, fish; suspected deposits of oil, natural gas, coal, iron ore
Land use:
Arable land: 1%
Permanent crops: 0%
Meadows and pastures: 64%
Forest and woodland: 22%
Other: 13%

Demographics [2]

	1970	1980	1990	1995[1]	1996	2000	2010	2020	2030
Population	765	975	1,409	1,629	1,677	1,886	2,513	3,267	4,124
Population density (persons per sq. mi.)	2	3	4	5	5	6	8	10	13
(persons per sq. km.)	1	1	2	2	2	2	3	4	5
Net migration rate (per 1,000 population)	NA	NA	2.0	0.0	0.0	0.0	0.0	0.0	0.0
Births	NA	NA	NA	NA	63	NA	NA	NA	NA
Deaths	NA	NA	NA	NA	13	NA	NA	NA	NA
Life expectancy - males	NA	NA	60.5	62.5	62.9	64.4	67.8	70.7	73.1
Life expectancy - females	NA	NA	63.2	65.7	66.2	68.1	72.4	75.9	78.7
Birth rate (per 1,000)	NA	NA	38.6	37.5	37.3	36.3	32.8	29.1	25.9
Death rate (per 1,000)	NA	NA	9.9	8.3	8.0	7.0	5.2	4.3	4.0
Women of reproductive age (15-49 yrs.)	NA	NA	321	383	397	456	618	816	1,044
of which are currently married	NA	NA	133	NA	167	194	271	NA	NA
Fertility rate	NA	NA	5.5	5.2	5.1	4.9	4.3	3.8	3.3

Except as noted, values for vital statistics are in thousands; life expectancy is in years.

Health

Health Indicators [3]

% of population with access to	
safe water (1990-95)	57
adequate sanitation (1990-95)	34
health services (1985-95)	62
% of 1-year-olds immunized (1990-94) against	
TB (tuberculosis)	100
DPT (diphtheria, pertussis, tetanus)	79
polio	79
measles	68
% of contraceptive prevalence (1980-94)	29
ORT use rate (1990-94)	75

Health Expenditures [4]

For sources, notes, and explanations, see Annotated Source Appendix, page 1061.

667

Human Factors

Women and Children [5]

% of pregnant women immunized (tetanus 1990-94)	57
% of births attended by trained health personnel (1983-94)	68
Maternal mortality rate (1980-92)	230
Under-5 mortality rate (1994)	78
% under-5 moderately/severely underweight (1980-1994)	26

Burden of Disease [6]

Population per physician (1991)	4,320.87
Population per nurse	NA
Population per hospital bed	NA
AIDS cases per 100,000 people (1994)	*
Malaria cases per 100,000 people (1992)	NA

Ethnic Division [7]

Other includes Baster, Bushmen, Tswana.

Ovambo	50%
Kavangos	9%
Damara	7%
Herero	7%
Nama	5%
Caprivian	4%
Other	18%

Religion [8]

Christian	80%-90%
Lutheran	> 50%
Other denominations	< 30%
Native religions	10%-20%

Major Languages [9]

Afrikaans is the common language of most of the population, German 32%, and English 7% (official). Indigenous languages: Oshivambo, Herero, Nama.

Education

Public Education Expenditures [10]

	1980	1981	1990	1991	1993	1994
Million (Rand)						
Total education expenditure	NA	25	480	NA	NA	909
as percent of GNP	NA	1.5	7.9	NA	NA	8.7
as percent of total govt. expend.	NA	NA	NA	NA	NA	NA
Current education expenditure	NA	NA	NA	NA	NA	NA
as percent of GNP	NA	NA	NA	NA	NA	NA
as percent of current govt. expend.	NA	NA	NA	NA	NA	NA
Capital expenditure	NA	NA	NA	NA	NA	NA

Educational Attainment [11]

Age group (1991)[15]	25+
Total population	340,552
Highest level attained (%)	
No schooling	NA
First level	
Not completed	49.1
Completed	NA
Entered second level	
S-1	43.8
S-2	NA
Postsecondary	4.0

Illiteracy [12]

	1989	1991	1995
Illiterate population (15+ yrs.)[2,7]	NA	198,460	NA
Illiteracy rate - total pop. (%)	NA	24.2	NA
Illiteracy rate - males (%)	NA	22.2	NA
Illiteracy rate - females (%)	NA	26.0	NA

Libraries [13]

Daily Newspapers [14]

	1980	1985	1990	1994
Number of papers	4	3	6	4
Circ. (000)	27	21	220	153

Culture [15]

Science and Technology

Scientific/Technical Forces [16]

R&D Expenditures [17]

U.S. Patents Issued [18]

Government and Law

Organization of Government [19]

Long-form name:
Republic of Namibia
Type:
Republic
Independence:
21 March 1990 (from South African mandate)
National holiday:
Independence Day, 21 March (1990)
Constitution:
Ratified 9 February 1990; effective 12 March 1990
Legal system:
Based on Roman-Dutch law and 1990 constitution
Executive branch:
President; Cabinet
Legislative branch:
Bicameral legislature: National Council and National Assembly
Judicial branch:
Supreme Court

Elections [20]

National Assembly	% of seats
SW Africa People's Org. (SWAPO)	73.6
DTA of Namibia (DTA)	20.8
United Democratic (UDF)	2.8
Monitor Action Group (MAG)	1.4
Democratic Coalition of Namibia (DCN)	1.4

Government Expenditures [21]

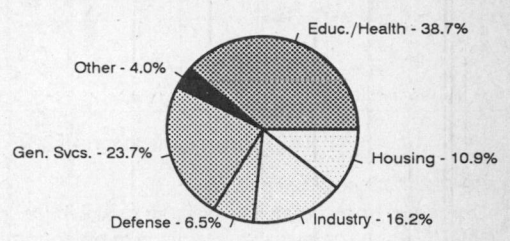

Educ./Health - 38.7%
Other - 4.0%
Gen. Svcs. - 23.7%
Housing - 10.9%
Defense - 6.5%
Industry - 16.2%

(% distribution). Expend. for FY92: 3,311.9 (Dollar mil.)

Crime [23]

	1994
Crime volume	
Cases known to police	40,305
Attempts (percent)	NA
Percent cases solved	NA
Crimes per 100,000 persons	3,358.75
Persons responsible for offenses	
Total number offenders	NA
Percent female	NA
Percent juvenile	NA
Percent foreigners	NA

Military Expenditures and Arms Transfers [22]

	1990	1991	1992	1993	1994
Military expenditures					
Current dollars (mil.)	42[e]	67[e]	64	60	56
1994 constant dollars (mil.)	47[e]	72[e]	67	61	56
Armed forces (000)	NA	8	8	8	8
Gross national product (GNP)					
Current dollars (mil.)	2,237	2,523	2,707	2,745	2,929
1994 constant dollars (mil.)	2,489	2,704	2,823	2,802	2,929
Central government expenditures (CGE)					
1994 constant dollars (mil.)	860	1,093	1,184	1,122	NA
People (mil.)	1.4	1.4	1.5	1.5	1.6
Military expenditure as % of GNP	1.9	2.7	2.4	2.2	1.9
Military expenditure as % of CGE	5.4	6.6	5.6	5.5	NA
Military expenditure per capita (1994 $)	34	50	45	40	35
Armed forces per 1,000 people (soldiers)	NA	5.6	5.4	5.2	5.0
GNP per capita (1994 $)	1,795	1,881	1,896	1,818	1,836
Arms imports[6]					
Current dollars (mil.)	0	0	0	0	10
1994 constant dollars (mil.)	0	0	0	0	10
Arms exports[6]					
Current dollars (mil.)	0	0	0	0	0
1994 constant dollars (mil.)	0	0	0	0	0
Total imports[7]					
Current dollars (mil.)	1,163	1,149	1,283	1,188	1,196
1994 constant dollars (mil.)	1,294	1,232	1,338	1,212	1,196
Total exports[7]					
Current dollars (mil.)	1,084	1,232	1,342	1,290	1,321
1994 constant dollars (mil.)	1,207	1,321	1,399	1,317	1,321
Arms as percent of total imports[8]	0	0	0	0	0.8
Arms as percent of total exports[8]	0	0	0	0	0

Human Rights [24]

	SSTS	FL	FAPRO	PPCG	APROBC	TPW	PCPTW	STPEP	PHRFF	PRW	ASST	AFL
Observes				P		P	P					
	EAFRD	CPR	ESCR	SR	ACHR	MAAE	PVIAC	PVNAC	EAFDAW	TCIDTP	RC
Observes	P	P					P	P	P	P	P

P=Party; S=Signatory; see Appendix for meaning of abbreviations.

Labor Force

Total Labor Force [25]

500,000

Labor Force by Occupation [26]

Agriculture	60%
Industry and commerce	19
Services	8
Government	7
Mining	6

Date of data: 1981 est.

Unemployment Rate [27]

35% in urban areas (1993 est.)

For sources, notes, and explanations, see Annotated Source Appendix, page 1061.

669

Production Sectors

Energy Resource Summary [28]

Energy resources: Natural gas, suspected deposits of oil, coal. **Electricity**: Capacity: 406,000 kW. Production: 1.29 billion kWh. Consumption per capita: 658 kWh (1991).

Telecommunications [30]

- 89,722 (1992 est.) telephones
- Domestic: good urban services; fair rural service; microwave radio relay links major towns; connections to other populated places are by open wire
- Radio: Broadcast stations: AM 4, FM 40, shortwave 0 Radios: 195,000 (1992 est.)
- Television: Broadcast stations: 3 Televisions: 27,000 (1993 est.)

Top Agricultural Products [32]

Produces millet, sorghum, peanuts; livestock; fish catch potential of over 1 million metric tons not being fulfilled.

Top Mining Products [33]

Metric tons except as noted	5/31/95[*]
Copper, mine output (29-30% Cu) gr. wt.	97,900
Lead, mine output (30-32% Pb) gr. wt.	43,800
Silver, mine output, Ag content (kg.)	64,000
Zinc, mine output (49-53% Zn)	64,600
Diamond, gem and industrial (000 carats)	1,310[e]
Fluorspar, concentrate (98% CaF2)	50,600
Salt	400,000[e,57]
Chrysocolla (kg.)	6,500[e]
Marble	15,000[e]
Sulfur, pyrite concentrate, gr. wt.	122,000

Transportation [31]

Railways: total: 2,382 km (1995); narrow gauge: 2,382 km 1.067-m gauge; single track

Highways: total: 54,186 km; paved: 4,056 km; unpaved: 50,130 km (1987 est.)

Merchant marine: none

Airports

Total:	108
With paved runways over 3,047 m:	2
With paved runways 2,438 to 3,047 m:	2
With paved runways 1,524 to 2,437 m:	14
With paved runways 914 to 1,523 m:	3
With paved runways under 914 m:	10

Tourism [34]

	1990	1991	1992	1993	1994
Visitors	NA	213	NA	288	NA
Tourists	NA	NA	NA	255	NA
Excursionists	NA	NA	NA	33	NA
Tourism receipts	69	87	106	156	184
Tourism expenditures	65	71	74	73	75

Travelers are in thousands, money in million U.S. dollars.

Finance, Economics, and Trade

GDP and Manufacturing Summary [35]

	1980	1985	1990	1991	1992
Gross Domestic Product					
Millions of 1980 dollars	2,007	1,871	2,156	2,219	2,241[e]
Growth rate in percent	0.18	0.00	6.00	2.90	1.00[e]
Manufacturing Value Added					
Millions of 1980 dollars	79	83	92	NA	NA
Growth rate in percent	-14.65	-3.54	5.91	NA	NA
Manufacturing share in percent of current prices	4.0	4.3[e]	4.3[e]	NA	NA

Economic Indicators [36]

- **National product**: GDP—purchasing power parity—$5.8 billion (1994 est.)

- **National product real growth rate**: 6.6% (1994 est.)

- **National product per capita**: $3,600 (1994 est.)

- **Inflation rate (consumer prices)**: 11% (1994)

- **External debt**: about $385 million (1994 est.)

Balance of Payments Summary [37]

Values in millions of dollars.

	1988	1989	1990	1991	1992
Exports of goods (f.o.b.)	NA	NA	1,101	1,252	1,288
Imports of goods (f.o.b.)	NA	NA	-1,117	-1,108	-1,177
Trade balance	NA	NA	-16	144	111
Services - debits	NA	NA	-547	-605	-619
Services - credits	NA	NA	308	378	346
Private transfers (net)	NA	NA	25	24	24
Government transfers (net)	NA	NA	192	311	280
Long-term capital (net)	NA	NA	-135	-189	-197
Short-term capital (net)	NA	NA	-62	-23	81
Errors and omissions	NA	NA	275	-54	-33
Overall balance	NA	NA	40	-14	-7

Exchange Rates [38]

Currency: **South African rand.**
Symbol: **R.**

Data are currency units per $1.

January 1996	3.6417
1995	3.6266
1994	3.5490
1993	3.2636
1992	2.8497
1991	2.7653

Imports and Exports

Top Import Origins [39]

$1.2 billion (f.o.b., 1993).

Origins	%
South Africa	NA
Germany	NA
US	NA
Japan	NA

Top Export Destinations [40]

$1.3 billion (f.o.b., 1993).

Destinations	%
Switzerland	NA
South Africa	NA
Germany	NA
UK	NA

Foreign Aid [41]

Recipient: ODA, $NA.

Import and Export Commodities [42]

Import Commodities	Export Commodities
Foodstuffs	Diamonds
Petroleum products and fuel	Copper
Machinery and equipment	Gold
	Zinc
	Lead
	Uranium
	Cattle
	Processed fish
	Karakul skins

For sources, notes, and explanations, see Annotated Source Appendix, page 1061.

671

Nepal

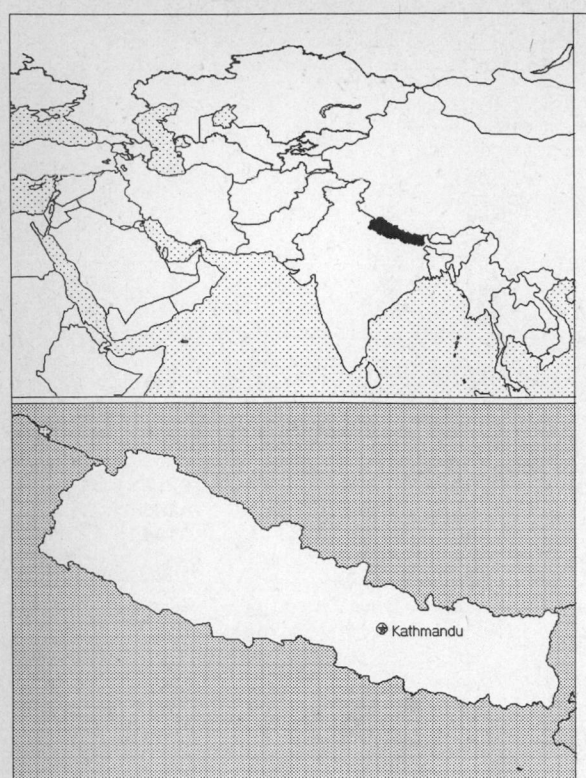

Geography [1]

Total area:
140,800 sq km   54,363 sq mi
Land area:
136,800 sq km   52,819 sq mi
Comparative area:
Slightly larger than Arkansas
Land boundaries:
Total 2,926 km, China 1,236 km, India 1,690 km
Coastline:
0 km (landlocked)
Climate:
Varies from cool summers and severe winters in North to subtropical summers and mild winters in South
Terrain:
Terai or flat river plain of the Ganges in South, central hill region, rugged Himalayas in North
Natural resources:
Quartz, water, timber, hydropower potential, scenic beauty, small deposits of lignite, copper, cobalt, iron ore
Land use:
Arable land: 17%
Permanent crops: 0%
Meadows and pastures: 13%
Forest and woodland: 33%
Other: 37%

Demographics [2]

	1970	1980	1990	1995[1]	1996	2000	2010	2020	2030
Population	11,919	15,001	19,104	21,561	22,094	24,364	30,783	37,767	45,039
Population density (persons per sq. mi.)	226	284	362	408	418	461	583	715	853
(persons per sq. km.)	87	110	140	158	162	178	225	276	329
Net migration rate (per 1,000 population)	NA	0.0	-0.5	0.0	0.0	0.0	0.0	0.0	0.0
Births	NA	349	NA	NA	817	NA	NA	NA	NA
Deaths	NA	NA	NA	NA	278	NA	NA	NA	NA
Life expectancy - males	NA	46.4	50.4	52.9	53.4	55.4	60.2	64.5	68.3
Life expectancy - females	NA	44.6	50.4	53.3	53.9	56.4	62.3	67.7	72.4
Birth rate (per 1,000)	NA	44.3	39.3	37.3	37.0	35.6	30.5	25.7	22.2
Death rate (per 1,000)	NA	19.5	14.9	12.9	12.6	11.2	8.5	6.8	5.9
Women of reproductive age (15-49 yrs.)	NA	3,416	4,293	4,962	5,112	5,750	7,559	9,736	11,888
of which are currently married[39]	NA	NA	3,332	NA	3,929	4,434	5,921	NA	NA
Fertility rate	NA	6.4	5.6	5.2	5.1	4.7	3.8	3.1	2.7

Except as noted, values for vital statistics are in thousands; life expectancy is in years.

Health

Health Indicators [3]

% of population with access to	
safe water (1990-95)	46
adequate sanitation (1990-95)	21
health services (1985-95)	NA
% of 1-year-olds immunized (1990-94) against	
TB (tuberculosis)	61
DPT (diphtheria, pertussis, tetanus)	63
polio	62
measles	57
% of contraceptive prevalence (1980-94)	23
ORT use rate (1990-94)	49

Health Expenditures [4]

Total health expenditure, 1990 (official exchange rate)	
Millions of dollars	141
Dollars per capita	7
Health expenditures as a percentage of GDP	
Total	4.5
Public sector	2.2
Private sector	2.3
Development assistance for health	
Total aid flows (millions of dollars)[1]	33
Aid flows per capita (dollars)	1.8
Aid flows as a percentage of total health expenditure	23.6

For sources, notes, and explanations, see Annotated Source Appendix, page 1061.

Human Factors

Women and Children [5]

% of pregnant women immunized (tetanus 1990-94)	11
% of births attended by trained health personnel (1983-94)	6
Maternal mortality rate (1980-92)	520
Under-5 mortality rate (1994)	118
% under-5 moderately/severely underweight (1980-1994)[1]	70

Burden of Disease [6]

Population per physician (1993)	13,616.57
Population per nurse (1992)	2,256.01
Population per hospital bed (1994)	4,307.96
AIDS cases per 100,000 people (1994)	*
Malaria cases per 100,000 people (1992)	115

Ethnic Division [7]

Newars, Indians, Tibetans, Gurungs, Magars, Tamangs, Bhotias, Rais, Limbus, Sherpas.

Religion [8]

Note: Only official Hindu state in the world, although no sharp distinction between many Hindu and Buddhist groups.

Hindu	90%
Buddhist	5%
Muslim	3%
Other	2%
(1981)	

Major Languages [9]

Nepali (official), 20 other languages divided into numerous dialects.

Education

Public Education Expenditures [10]

	1980	1989	1990	1991	1992	1994
Million (Rupee)[47]						
Total education expenditure	430	2,086	2,079	3,268	4,428	NA
as percent of GNP	1.8	2.3	2.0	2.7	2.9	NA
as percent of total govt. expend.	10.5	10.6	8.5	12.3	13.2	NA
Current education expenditure	NA	NA	NA	NA	NA	NA
as percent of GNP	NA	NA	NA	NA	NA	NA
as percent of current govt. expend.	NA	NA	NA	NA	NA	NA
Capital expenditure	NA	NA	NA	NA	NA	NA

Educational Attainment [11]

Age group (1991)[8]	6+
Total population	15,145,071
Highest level attained (%)	
No schooling	69.6
First level	
Not completed	16.2
Completed	NA
Entered second level	
S-1	8.9
S-2	2.0
Postsecondary	1.5

Illiteracy [12]

In thousands and percent[1]	1990	1995	2000
Illiterate population (15+ yrs.)	8,308	9,149	10,088
Illiteracy rate - total pop. (%)	77.3	73.9	70.5
Illiteracy rate - males (%)	63.4	59.7	56.2
Illiteracy rate - females (%)	91.5	88.5	85.3

Libraries [13]

	Admin. Units	Svc. Pts.	Vols. (000)	Shelving (meters)	Vols. Added	Reg. Users
National (1987)	1	1	71	NA	1,234	NA
Nonspecialized	NA	NA	NA	NA	NA	NA
Public	NA	NA	NA	NA	NA	NA
Higher ed.	NA	NA	NA	NA	NA	NA
School	NA	NA	NA	NA	NA	NA

Daily Newspapers [14]

	1980	1985	1990	1994
Number of papers	28	28	28	28
Circ. (000)	120[e]	130[e]	150[e]	162[e]

Culture [15]

Science and Technology

Scientific/Technical Forces [16]

Scientists/engineers	334
Number female	NA
Technicians	75
Number female	NA
Total[27]	409

R&D Expenditures [17]

U.S. Patents Issued [18]

For sources, notes, and explanations, see Annotated Source Appendix, page 1061.

Government and Law

Organization of Government [19]

Long-form name:
Kingdom of Nepal
Type:
Parliamentary democracy as of 12 May
1991
Independence:
1768 (unified by Prithvi Narayan Shah)
National holiday:
Birthday of His Majesty the King, 28
December (1945)
Constitution:
9 November 1990
Legal system:
Based on Hindu legal concepts and
English common law; has not accepted
compulsory ICJ jurisdiction
Executive branch:
King; Heir Apparent Crown Prince; Prime
Minister; Cabinet
Legislative branch:
Bicameral Parliament: National Council
and House of Representatives
Judicial branch:
Supreme Court (Sarbochha Adalat)

Elections [20]

House of Representatives	% of votes
Nepali Congress Party (NCP)	33.0
Communist Party of Nepal/United Marxist and Leninist (CPN/UML)	31.0
National Democratic Party	18.0
Terai Rights Sadbhavana	3.0
Nepal Workers and Peasants Party (NWPP)	1.0

Government Expenditures [21]

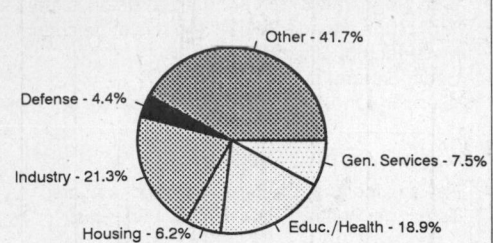

Other - 41.7%
Defense - 4.4%
Gen. Services - 7.5%
Industry - 21.3%
Educ./Health - 18.9%
Housing - 6.2%

(% distribution). Expend. for FY96 est.: 49,485 (Rupee mil.)

Military Expenditures and Arms Transfers [22]

	1990	1991	1992	1993	1994
Military expenditures					
Current dollars (mil.)	30	30	33	42[e]	43
1994 constant dollars (mil.)	34	32	34	42[e]	43
Armed forces (000)	35	35	35	35	35
Gross national product (GNP)					
Current dollars (mil.)	2,970	3,277	3,508	3,698	4,034
1994 constant dollars (mil.)	3,306	3,513	3,658	3,774	4,034
Central government expenditures (CGE)					
1994 constant dollars (mil.)	559	544	688[e]	768[e]	854[e]
People (mil.)	19.1	19.6	20.0	20.5	21.0
Military expenditure as % of GNP	1.0	0.9	0.9	1.1	1.1
Military expenditure as % of CGE	6.0	5.9	5.0	5.5	5.0
Military expenditure per capita (1994 $)	2	2	2	2	2
Armed forces per 1,000 people (soldiers)	1.8	1.8	1.7	1.7	1.7
GNP per capita (1994 $)	173	180	183	184	192
Arms imports[6]					
Current dollars (mil.)	10	0	0	0	10
1994 constant dollars (mil.)	11	0	0	0	10
Arms exports[6]					
Current dollars (mil.)	0	0	0	0	0
1994 constant dollars (mil.)	0	0	0	0	0
Total imports[7]					
Current dollars (mil.)	686	758	792	880	NA
1994 constant dollars (mil.)	763	812	826	898	NA
Total exports[7]					
Current dollars (mil.)	210	264	374	390	NA
1994 constant dollars (mil.)	234	283	390	398	NA
Arms as percent of total imports[8]	1.5	0	0	0	NA
Arms as percent of total exports[8]	0	0	0	0	0

Crime [23]

	1994
Crime volume	
Cases known to police	8,704
Attempts (percent)	NA
Percent cases solved	55.06
Crimes per 100,000 persons	44.23
Persons responsible for offenses	
Total number offenders	15,834
Percent female	6.81
Percent juvenile (8-15 yrs.)	0.42
Percent foreigners	1.67

Human Rights [24]

	SSTS	FL	FAPRO	PPCG	APROBC	TPW	PCPTW	STPEP	PHRFF	PRW	ASST	AFL
Observes	P			P		P	P			P	P	
	EAFRD	CPR	ESCR	SR	ACHR	MAAE	PVIAC	PVNAC	EAFDAW	TCIDTP	RC
Observes		P	P	P					P	P	P

P = Party; S = Signatory; see Appendix for meaning of abbreviations.

Labor Force

Total Labor Force [25]

8.5 million (1991 est.)

Labor Force by Occupation [26]

Agriculture	93%
Services	5
Industry	2

Unemployment Rate [27]

NA; substantial underemployment (1995)

For sources, notes, and explanations, see Annotated Source Appendix, page 1061.

Production Sectors

Commercial Energy Production and Consumption

Data are shown in quadrillion (10^{15}) BTUs and percent for 1995
Values for hydroelectric, nuclear, geothermal, solar, and wind power refer to electrical generation.

Production [28]

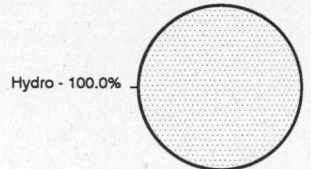

Hydro - 100.0%

Consumption [29]

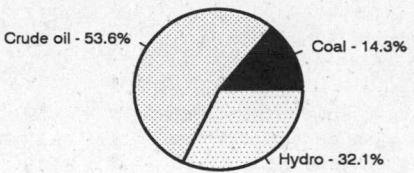

Crude oil - 53.6%
Coal - 14.3%
Hydro - 32.1%

Net hydroelectric power	0.009
Total	0.009

Crude oil	0.015
Coal	0.004
Net hydroelectric power	0.009
Total	0.028

Telecommunications [30]

- 82,774 (1995 est.) telephones; poor telephone and telegraph service; fair radiotelephone communication service
- International: radiotelephone communications; satellite earth station - 1 Intelsat (Indian Ocean)
- Radio: Broadcast stations: AM 88, FM 0, shortwave 0 Radios: 690,000 (1992 est.)
- Television: Broadcast stations: 1 (1988 est.) Televisions: 45,000 (1992 est.)

Transportation [31]

Railways: total: 101 km; note - all in Terai close to Indian border; narrow gauge: 101 km 0.762-m gauge

Highways: total: 9,933 km; paved: 3,421 km; unpaved: 6,512 km (1995 est.)

Airports

Total:	43
With paved runways over 3,047 m:	1
With paved runways 1,524 to 2,437 m:	3
With paved runways 914 to 1,523 m:	1
With paved runways under 914 m:	27
With unpaved runways 1,524 to 2,437 m:	1

Top Agricultural Products [32]

Agriculture accounts for 49.3% of the GDP; produces rice, corn, wheat, sugarcane, root crops; milk, water buffalo meat.

Top Mining Products [33]

Metric tons except as noted[e]	3/31/95[*]
Cement	190,000
Clay, red	8,000
Coal, bituminous and lignite	5,200
Gemstones (kg.)	5,000
Lime, agricultural	25,000
Salt	7,000
Limestone	300,000
Marble, slab (sq. meters)	28,000
Quartzite (sq. meters)	2,500
Talc	1,500

Tourism [34]

	1990	1991	1992	1993	1994
Tourists[79]	255	293	334	294	327
Tourism receipts	109	126	110	157	172
Tourism expenditures	45	38	52	93	112
Fare receipts	6	21	39	45	46
Fare expenditures	19	31	31	15	23

Travelers are in thousands, money in million U.S. dollars.

For sources, notes, and explanations, see Annotated Source Appendix, page 1061.

675

Manufacturing Sector

Manufacturing Summary [35]

	1987		1988		1989		1990		1991	
	$ bil.	%	$ bil.	%	$ bil.	%	$ bil.	%	$ bil.	%
Establishments or enterprises (number)	2,565	0.109	2,586	0.123	-	-	-	-	-	-
Total employment (000)	151	0.111	151	0.110	-	-	-	-	-	-
Production workers (000)	133	0.197	132	0.211	-	-	-	-	-	-
Output ($ bil.)	0.563	0.006	0.593	0.005	-	-	-	-	-	-
Value added ($ bil.)	0.217	0.005	0.230	0.005	-	-	-	-	-	-
Capital investment ($ mil.)	66	0.015	21	0.004	-	-	-	-	-	-
M & E investment ($ mil.)	-		-		-	-	-	-	-	-
Employees per establishment (number)	59	101.833	58	89.417	-	-	-	-	-	-
Production workers per establishment	52	179.719	51	170.962	-	-	-	-	-	-
Output per establishment ($ mil.)	0.219	5.064	0.229	4.126	-	-	-	-	-	-
Capital investment per estab. ($ mil.)	0.026	13.926	0.008	3.587	-	-	-	-	-	-
M & E per establishment ($ mil)	-	-	-		-	-	-	-	-	-
Payroll per employee ($)	503	5.615	509	5.109	-	-	-	-	-	-
Wages per production worker ($)	384	4.861	403	4.771	-	-	-	-	-	-
Hours per production worker (hours)	-		-		-	-	-	-	-	-
Output per employee ($)	3,724	4.972	3,930	4.615	-	-	-	-	-	-
Capital investment per employee ($)	439	13.675	140	4.011	-	-	-	-	-	-
M & E per employee ($)	-	-	-	-	-	-	-	-	-	-

Note: Columns headed % show percent of world total or ratio. Ratios closest to 100 are closest to world average. M & E stands for machinery & equipment.

Output in Manufacturing

	1987		1988		1989		1990		1991	
	$ bil.	%	$ bil.	%	$ bil.	%	$ bil.	%	$ bil.	%
3110 - Food products	0.113	20.07	0.113	19.06	-	-	-	-	-	-
3130 - Beverages	0.030	5.33	0.021	3.54	-	-	-	-	-	-
3140 - Tobacco	0.058	10.30	0.050	8.43	-	-	-	-	-	-
3210 - Textiles	0.089	15.81	0.100	16.86	-	-	-	-	-	-
3211 - Spinning, weaving, etc.	0.040	7.10	0.039	6.58	-	-	-	-	-	-
3220 - Wearing apparel	0.032	5.68	0.033	5.56	-	-	-	-	-	-
3230 - Leather and products	0.007	1.24	0.012	2.02	-	-	-	-	-	-
3240 - Footwear	0.002	0.36	0.002	0.34	-	-	-	-	-	-
3310 - Wood products	0.010	1.78	0.006	1.01	-	-	-	-	-	-
3320 - Furniture, fixtures	0.005	0.89	0.004	0.67	-	-	-	-	-	-
3410 - Paper and products	0.007	1.24	0.007	1.18	-	-	-	-	-	-
3411 - Pulp, paper, etc.	0.007	1.24	0.007	1.18	-	-	-	-	-	-
3420 - Printing, publishing	0.006	1.07	0.006	1.01	-	-	-	-	-	-
3520 - Chemical products nec	0.036	6.39	0.041	6.91	-	-	-	-	-	-
3522 - Drugs and medicines	0.004	0.71	0.003	0.51	-	-	-	-	-	-
3550 - Rubber products	0.005	0.89	0.007	1.18	-	-	-	-	-	-
3560 - Plastic products nec	0.008	1.42	0.012	2.02	-	-	-	-	-	-
3690 - Nonmetal products nec	0.048	8.53	0.057	9.61	-	-	-	-	-	-
3710 - Iron and steel	0.015	2.66	0.034	5.73	-	-	-	-	-	-
3810 - Metal products	0.021	3.73	0.018	3.04	-	-	-	-	-	-
3830 - Electrical machinery	0.016	2.84	0.015	2.53	-	-	-	-	-	-
3850 - Professional goods	-		-		-	-	-	-	-	-
3900 - Industries nec	0.004	0.71	0.004	0.67	-	-	-	-	-	-

Note: Codes are International Standard Industry codes (ISIC). Percentages are % of total Output. [f]: Factor Prices; [p]: Producer Prices.

Finance, Economics, and Trade

Economic Indicators [36]

- **National product**: GDP—purchasing power parity—$25.2 billion (1995 est.)

- **National product real growth rate**: 2.3% (1995 est.)

- **National product per capita**: $1,200 (1995 est.)

- **Inflation rate (consumer prices)**: 6.7% (FY94/95)

- **External debt**: $2.3 billion (FY94/95 est.)

Balance of Payments Summary [37]

Values in millions of dollars.

	1989	1990	1991	1992	1993
Exports of goods (f.o.b.)	156.2	217.9	274.5	376.3	397.0
Imports of goods (f.o.b.)	-571.4	-666.6	-756.9	-752.1	-858.6
Trade balance	-415.2	-448.7	-482.4	-375.8	-461.6
Services - debits	-154.6	-178.6	-200.2	-241.8	-275.5
Services - credits	226.5	229.5	266.9	307.3	362.1
Private transfers (net)	52.0	60.4	53.7	45.7	74.3
Government transfers (net)	48.0	48.2	57.5	83.2	78.2
Long-term capital (net)	213.3	178.7	223.6	130.1	125.7
South term capital (net)	-18.9	125.8	233.5	205.8	157.8
Errors and omissions	4.8	4.9	10.7	0.8	4.6
Overall balance	-44.1	20.2	163.3	155.3	65.6

Exchange Rates [38]

Currency: **Nepalese rupee.**
Symbol: **NR.**

Data are currency units per $1.

January 1996	56.636
1995	51.890
1994	49.398
1993	48.607
1992	42.718
1991	37.255

Imports and Exports

Top Import Origins [39]

$1.4 billion (c.i.f., 1995 est.).

Origins	%
India	NA
Singapore	NA
Japan	NA
Germany	NA

Top Export Destinations [40]

$430 million (f.o.b., 1995 est.) Does not include unrecorded border trade with India.

Destinations	%
India	NA
US	NA
Germany	NA
UK	NA

Foreign Aid [41]

	U.S. $	
ODA (1993)	310	million
Western and Japanese bilateral aid	215	million
Multilateral aid (1994-95)	43	million

Import and Export Commodities [42]

Import Commodities

Petroleum products 20%
Fertilizer 11%
Machinery 10%

Export Commodities

Carpets
Clothing
Leather goods
Jute goods
Grain

For sources, notes, and explanations, see Annotated Source Appendix, page 1061.

677

Netherlands

Geography [1]

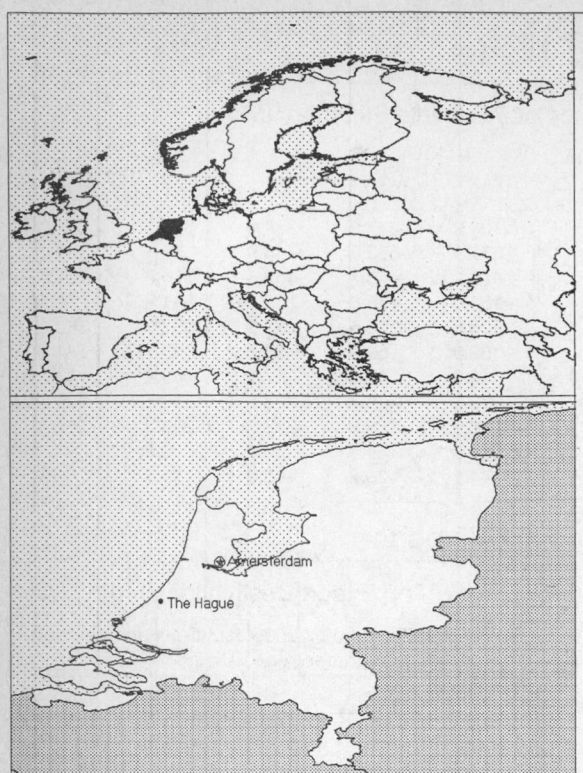

Total area:
37,330 sq km   14,413 sq mi
Land area:
33,920 sq km   13,097 sq mi
Comparative area:
Slightly less than twice the size of New Jersey
Land boundaries:
Total 1,027 km, Belgium 450 km, Germany 577 km
Coastline:
451 km
Climate:
Temperate; marine; cool summers and mild winters
Terrain:
Mostly coastal lowland and reclaimed land (polders); some hills in Southeast
Natural resources:
Natural gas, petroleum, fertile soil
Land use:
Arable land: 26%
Permanent crops: 1%
Meadows and pastures: 32%
Forest and woodland: 9%
Other: 32%

Demographics [2]

	1970	1980	1990	1995[1]	1996	2000	2010	2020	2030
Population	13,032	14,144	14,952	15,478	15,568	15,893	16,382	16,490	16,197
Population density (persons per sq. mi.)	995	1,079	1,141	1,181	1,188	1,213	1,250	1,258	1,236
(persons per sq. km.)	384	417	441	456	459	468	483	486	477
Net migration rate (per 1,000 population)	NA	3.6	3.3	2.3	2.3	2.0	1.9	0.6	0.0
Births	NA	NA	NA	NA	189	NA	NA	NA	NA
Deaths	NA	NA	75	NA	132	NA	NA	NA	NA
Life expectancy - males	NA	NA	73.9	74.8	74.9	75.5	76.7	78.4	79.7
Life expectancy - females	NA	NA	80.2	80.6	80.7	81.1	82.1	84.2	85.5
Birth rate (per 1,000)	NA	12.8	13.2	12.3	12.1	11.5	9.1	9.1	8.5
Death rate (per 1,000)	NA	8.1	8.6	8.7	8.7	8.7	9.4	10.0	11.8
Women of reproductive age (15-49 yrs.)	NA	NA	3,967	4,031	4,021	3,935	3,807	3,451	3,137
of which are currently married[40]	NA	NA	2,215	NA	2,365	2,335	2,178	NA	NA
Fertility rate	NA	NA	1.6	1.5	1.5	1.5	1.5	1.5	1.5

Except as noted, values for vital statistics are in thousands; life expectancy is in years.

Health

Health Indicators [3]

% of population with access to	
safe water (1990-95)	NA
adequate sanitation (1990-95)	NA
health services (1985-95)	NA
% of 1-year-olds immunized (1990-94) against	
TB (tuberculosis)	NA
DPT (diphtheria, pertussis, tetanus)	97
polio	97
measles	95
% of contraceptive prevalence (1980-94)	76
ORT use rate (1990-94)	NA

Health Expenditures [4]

Total health expenditure, 1990 (official exchange rate)	
Millions of dollars	22,423
Dollars per capita	1,500
Health expenditures as a percentage of GDP	
Total	7.9
Public sector	5.7
Private sector	2.2
Development assistance for health	
Total aid flows (millions of dollars)[1]	NA
Aid flows per capita (dollars)	NA
Aid flows as a percentage of total health expenditure	NA

For sources, notes, and explanations, see Annotated Source Appendix, page 1061.

Human Factors

Women and Children [5]

% of pregnant women immunized (tetanus 1990-94)	NA
% of births attended by trained health personnel (1983-94)[1]	100
Maternal mortality rate (1980-92)	10
Under-5 mortality rate (1994)	8
% under-5 moderately/severely underweight (1980-1994)	NA

Burden of Disease [6]

Population per physician (1989)	412.00
Population per nurse	NA
Population per hospital bed (1990)	169.80
AIDS cases per 100,000 people (1994)	3.0
Heart disease cases per 1,000 people (1990-93)	282

Ethnic Division [7]

Dutch	96%
Moroccans, Turks, and other	4%

Religion [8]

Roman Catholic	34%
Protestant	25%
Muslim	3%
Unaffiliated	36%
Other	2%
(1991)	

Major Languages [9]

Dutch.

Education

Public Education Expenditures [10]

	1980	1985	1990	1991	1993	1994
Million (Guilder)						
Total education expenditure	26,016	27,403	30,697	31,709	31,318	NA
as percent of GNP	7.6	6.4	6.0	5.9	5.5	NA
as percent of total govt. expend.	NA	NA	NA	NA	9.2	NA
Current education expenditure	23,947	25,729	29,200	30,312	30,102	NA
as percent of GNP	7.0	6.0	5.7	5.6	5.2	NA
as percent of current govt. expend.	NA	NA	NA	NA	NA	NA
Capital expenditure	2,069	1,674	1,497	1,397	1,216	NA

Educational Attainment [11]

Illiteracy [12]

Libraries [13]

	Admin. Units	Svc. Pts.	Vols. (000)	Shelving (meters)	Vols. Added	Reg. Users
National (1990)[5]	1	4	2,482	54,000[e]	NA	96,770
Nonspecialized (1990)	3[e]	5[e]	876[e]	51,100[e]	NA	80,518[e]
Public (1992)	603	1,192	41,781	NA	NA	5 mil
Higher ed. (1993)	369	NA	25,266	721,000	NA	NA
School	NA	NA	NA	NA	NA	NA

Daily Newspapers [14]

	1980	1985	1990	1994
Number of papers	84	88	45	46
Circ. (000)	4,612	4,496	4,500[e]	5,138

Culture [15]

Cinema (seats per 1,000)	6.6
Annual attendance per person	1.0
Gross box office receipts (mil. Guilder)	182
Museums (reporting)	732
Visitors (000)	22,994
Annual receipts (000 Guilder)	570,980

Science and Technology

Scientific/Technical Forces [16]

Scientists/engineers	40,000
Number female	NA
Technicians[2]	26,710
Number female	NA
Total[2]	66,710

R&D Expenditures [17]

	Guilder (000) 1991
Total expenditure[26]	10,381,000
Capital expenditure	1,059,000
Current expenditure	9,322,000
Percent current	89.8

U.S. Patents Issued [18]

Values show patents issued to citizens of the country by the U.S. Patents Office.

	1993	1994	1995
Number of patents	945	999	894

For sources, notes, and explanations, see Annotated Source Appendix, page 1061.

Government and Law

Organization of Government [19]

Long-form name:
Kingdom of the Netherlands
Type:
Constitutional monarchy
Independence:
1579 (from Spain)
National holiday:
Queen's Day, 30 April (1938)
Constitution:
17 February 1983
Legal system:
Civil law system incorporating French penal theory; judicial review in the Supreme Court of legislation of lower order rather than Acts of the States General; accepts compulsory ICJ jurisdiction, with reservations
Executive branch:
Queen; Heir Apparent Prince; Prime Minister; 2 Vice Prime Ministers; Cabinet
Legislative branch:
Bicameral legislature (Staten Generaal): First Chamber (Eerste Kamer) and Second Chamber (Tweede Kamer)
Judicial branch:
Supreme Court (De Hoge Raad)

Elections [20]

Second Chamber	% of votes
Labor (PvdA)	24.3
Christian Democratic (CDA)	22.3
Liberal (VVD)	20.4
Democrats '66 (D'66)	16.5
Other	16.5

Government Expenditures [21]

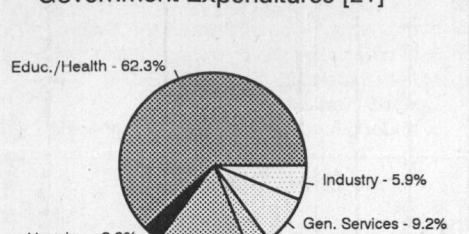

Educ./Health - 62.3%
Industry - 5.9%
Gen. Services - 9.2%
Defense - 3.9%
Housing - 2.9%
Other - 15.8%

Crime [23]

	1990
Crime volume	
Cases known to police	948,053
Attempts (percent)	NA
Percent cases solved	8.0
Crimes per 100,000 persons	6,341.0
Persons responsible for offenses	
Total number offenders	76,266
Percent female	9.3
Percent juvenile (12-17 yrs.)	7.3
Percent foreigners	NA

Military Expenditures and Arms Transfers [22]

	1990	1991	1992	1993	1994
Military expenditures					
Current dollars (mil.)	7,281	7,380	7,589	7,191	7,137
1994 constant dollars (mil.)	8,103	7,910	7,914	7,339	7,137
Armed forces (000)	104	104	90	86	77
Gross national product (GNP)					
Current dollars (mil.)	277,700	294,800	306,200	314,900	331,000
1994 constant dollars (mil.)	309,000	316,000	319,300	321,400	331,000
Central government expenditures (CGE)					
1994 constant dollars (mil.)	159,600	166,000	170,200	167,100	164,000
People (mil.)	15.0	15.1	15.2	15.3	15.4
Military expenditure as % of GNP	2.6	2.5	2.5	2.3	2.2
Military expenditure as % of CGE	5.1	4.8	4.6	4.4	4.4
Military expenditure per capita (1994 $)	542	525	522	480	464
Armed forces per 1,000 people (soldiers)	7.0	6.9	5.9	5.6	5.0
GNP per capita (1994 $)	20,670	20,980	21,040	21,040	21,540
Arms imports[6]					
Current dollars (mil.)	575	430	320	160	220
1994 constant dollars (mil.)	640	461	334	163	220
Arms exports[6]					
Current dollars (mil.)	200	100	180	240	110
1994 constant dollars (mil.)	223	107	188	245	110
Total imports[7]					
Current dollars (mil.)	126,100	125,900	134,600	124,400	143,600
1994 constant dollars (mil.)	140,300	134,900	140,400	127,000	143,600
Total exports[7]					
Current dollars (mil.)	131,800	133,600	140,300	139,100	156,600
1994 constant dollars (mil.)	146,700	143,200	146,300	142,000	156,600
Arms as percent of total imports[8]	0.5	0.3	0.2	0.1	0.2
Arms as percent of total exports[8]	0.2	0.1	0.1	0.2	0.1

Human Rights [24]

	SSTS	FL	FAPRO	PPCG	APROBC	TPW	PCPTW	STPEP	PHRFF	PRW	ASST	AFL
Observes	P	P	P	P	P	P	P		P	P	P	P
	EAFRD	CPR	ESCR	SR	ACHR	MAAE	PVIAC	PVNAC	EAFDAW	TCIDTP	RC	
Observes		P	P	P		P	P	P	P	P	P	

P = Party; S = Signatory; see Appendix for meaning of abbreviations.

Labor Force

Total Labor Force [25]

6.4 million (1993)

Labor Force by Occupation [26]

Services	73%
Manufacturing and construction	23
Agriculture	4

Date of data: 1994

Unemployment Rate [27]

7.1% (fourth quarter 1995)

Production Sectors

Commercial Energy Production and Consumption

Data are shown in quadrillion (10^{15}) BTUs and percent for 1995
Values for hydroelectric, nuclear, geothermal, solar, and wind power refer to electrical generation.

Production [28]

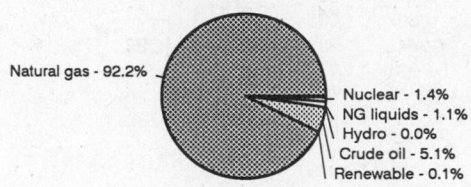

Natural gas - 92.2%
Nuclear - 1.4%
NG liquids - 1.1%
Hydro - 0.0%
Crude oil - 5.1%
Renewable - 0.1%

Consumption [29]

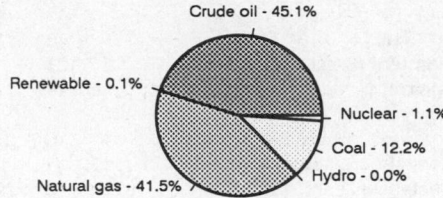

Crude oil - 45.1%
Renewable - 0.1%
Nuclear - 1.1%
Coal - 12.2%
Hydro - 0.0%
Natural gas - 41.5%

Crude oil	0.148	Crude oil	1.650
Natural gas liquids	0.033	Dry natural gas	1.521
Dry natural gas	2.665	Coal	0.446
Net hydroelectric power	0.001	Net hydroelectric power	0.001
Net nuclear power	0.040	Net nuclear power	0.040
Geothermal, solar, wind	0.003	Geothermal, solar, wind	0.003
Total	2.890	Total	3.661

Telecommunications [30]

- 8.272 million (1983 est.) telephones; highly developed and well maintained; extensive redundant system of multiconductor cables, supplemented by microwave radio relay
- Domestic: nationwide cellular telephone system; microwave radio relay
- International: 5 submarine cables; satellite earth stations - 3 Intelsat (1 Indian Ocean and 2 Atlantic Ocean), 1 Eutelsat, and 1 Inmarsat (Atlantic and Indian Ocean Regions)
- Radio: Broadcast stations: AM 3 (relays 3), FM 12 (repeaters 39), shortwave 0 Radios: 13.755 million (1992 est.)
- Television: Broadcast stations: 8 (repeaters 7) Televisions: 7.4 million (1992 est.)

Top Agricultural Products [32]

Agriculture accounts for 3.4% of the GDP; produces grains, potatoes, sugar beets, fruits, vegetables; livestock.

Transportation [31]

Railways: total: 2,891 km; standard gauge: 2,891 km 1.435-m gauge; 2857 km are in common carrier service (1,991 km electrified) and 34 km serve tourists

Highways: total: 104,831 km; paved: 92,251 km (including 2,118 km of expressways); unpaved: 12,580 km (1992 est.)

Merchant marine: total: 352 ships (1,000 GRT or over) totaling 2,681,133 GRT/3,379,762 DWT; ships by type: bulk 1, cargo 206, chemical tanker 21, combination bulk 3, container 34, liquefied gas tanker 13, livestock carrier 1, multifunction large-load carrier 2, oil tanker 38, railcar carrier 1, refrigerated cargo 16, roll-on/roll-off cargo 11, short-sea passenger 3, specialized tanker 2

Airports

Total:	28
With paved runways over 3,047 m:	1
With paved runways 2,438 to 3,047 m:	8
With paved runways 1,524 to 2,437 m:	6
With paved runways 914 to 1,523 m:	3
With paved runways under 914 m:	7

Top Mining Products [33]

Metric tons except as noted	7/31/95[*]
Iron ore, sintered (imported)	4,580,000[e]
Pig iron, incl. ferroalloys	5,440,000
Steel, crude	6,170,000
Cement, hydraulic	3,400,000[e]
Nitrogen, N content of ammonia	2,500,000[e]
Salt, all types	3,500,000[e]
Sand, industrial	20,000,000[e]
Sulfuric acid, 100% H2SO4	1,250,000[e]
Coke, metallurgical	2,750,000[e]
Petroleum refinery products (000 42-gal. bls.)	524,000[e]

Tourism [34]

	1990	1991	1992	1993	1994
Tourists[80]	5,795	5,842	6,083	5,757	6,178
Tourism receipts	3,636	4,246	5,237	4,690	5,612
Tourism expenditures	7,376	8,149	9,649	8,974	10,983
Fare receipts	1,982	2,292	3,652	3,720	3,953
Fare expenditures	1,284	1,451	1,627	1,588	1,578

Travelers are in thousands, money in million U.S. dollars.

For sources, notes, and explanations, see Annotated Source Appendix, page 1061.

681

Manufacturing Sector

Manufacturing Summary [35]

	1987		1988		1989		1990		1991	
	$ bil.	%	$ bil.	%	$ bil.	%	$ bil.	%	$ bil.	%
Establishments or enterprises (number)	49,181	2.096	50,281	2.395	51,854	2.768	52,948	2.956	-	-
Total employment (000)	764	0.563	766	0.560	777	0.631	706	0.638	-	-
Production workers (000)	-	-	-	-	-	-	-	-	-	-
Output ($ bil.)	149	1.463	162	1.392	164	1.394	-	-	-	-
Value added ($ bil.)	46	1.066	52	1.072	50	1.031	-	-	-	-
Capital investment ($ mil.)	7,101	1.629	6,898	1.449	7,074	1.291	8,500	1.529	-	-
M & E investment ($ mil.)	5,744	1.843	5,593	1.616	5,767	1.477	7,006	1.662	-	-
Employees per establishment (number)	16	26.852	15	23.368	15	22.805	13	21.580	-	-
Production workers per establishment	-	-	-	-	-	-	-	-	-	-
Output per establishment ($ mil.)	3	69.785	3	58.103	3	50.343	-	-	-	-
Capital investment per estab. ($ mil.)	0.144	77.711	0.137	60.477	0.136	46.650	0.161	51.743	-	-
M & E per establishment ($ mil)	0.117	87.920	0.111	67.464	0.111	53.360	0.132	56.226	-	-
Payroll per employee ($)	23,417	261.143	24,462	245.537	22,927	205.216	-	-	-	-
Wages per production worker ($)	-	-	-	-	-	-	-	-	-	-
Hours per production worker (hours)	-	-	-	-	-	-	-	-	-	-
Output per employee ($)	194,645	259.891	211,786	248.648	211,569	220.754	-	-	-	-
Capital investment per employee ($)	9,294	289.412	9,005	258.803	9,104	204.560	12,040	239.771	-	-
M & E per employee ($)	7,518	327.429	7,301	288.705	7,422	233.983	9,924	260.547	-	-

Note: Columns headed % show percent of world total or ratio. Ratios closest to 100 are closest to world average. M & E stands for machinery & equipment.

Output in Manufacturing

	1987		1988		1989		1990		1991	
	$ bil.[f]	%	$ bil.[f]	%	$ bil.[f]	%	$ bil.[f]	%	$ bil.[f]	%
3110 - Food products	31.000	20.81	33.000	20.37	32.000	19.51	-	-	-	-
3130 - Beverages	1.989	1.33	2.064	1.27	2.080	1.27	-	-	-	-
3140 - Tobacco	1.471	0.99	1.528	0.94	1.650	1.01	-	-	-	-
3210 - Textiles	2.666	1.79	2.742	1.69	2.641	1.61	-	-	-	-
3211 - Spinning, weaving, etc.	0.168	0.11	0.142	0.09	0.132	0.08	-	-	-	-
3220 - Wearing apparel	0.834	0.56	0.916	0.57	0.957	0.58	-	-	-	-
3230 - Leather and products	0.247	0.17	0.238	0.15	0.217	0.13	-	-	-	-
3240 - Footwear	0.321	0.22	0.324	0.20	0.340	0.21	-	-	-	-
3310 - Wood products	1.654	1.11	1.847	1.14	1.830	1.12	-	-	-	-
3320 - Furniture, fixtures	1.269	0.85	1.376	0.85	1.330	0.81	-	-	-	-
3410 - Paper and products	3.663	2.46	4.078	2.52	4.135	2.52	-	-	-	-
3411 - Pulp, paper, etc.	1.471	0.99	1.750	1.08	1.782	1.09	-	-	-	-
3420 - Printing, publishing	7.736	5.19	8.520	5.26	8.488	5.18	-	-	-	-
3510 - Industrial chemicals	13.000	8.72	16.000	9.88	15.000	9.15	-	-	-	-
3511 - Basic chemicals, excl fertilizers	12.000	8.05	14.000	8.64	14.000	8.54	-	-	-	-
3520 - Chemical products nec	6.472	4.34	7.113	4.39	7.111	4.34	-	-	-	-
3522 - Drugs and medicines	1.866	1.25	1.998	1.23	2.004	1.22	-	-	-	-
3530 - Petroleum refineries	8.165	5.48	7.533	4.65	9.134	5.57	-	-	-	-
3540 - Petroleum, coal products	0.385	0.26	0.430	0.27	0.396	0.24	-	-	-	-
3550 - Rubber products	0.553	0.37	0.612	0.38	0.599	0.37	-	-	-	-
3560 - Plastic products nec	2.932	1.97	3.405	2.10	3.617	2.21	-	-	-	-
3610 - Pottery, china, etc.	0.296	0.20	0.319	0.20	0.307	0.19	-	-	-	-
3620 - Glass and products	0.652	0.44	0.749	0.46	0.773	0.47	-	-	-	-
3690 - Nonmetal products nec	2.537	1.70	2.894	1.79	2.867	1.75	-	-	-	-
3710 - Iron and steel	2.453	1.65	2.919	1.80	3.183	1.94	-	-	-	-
3720 - Nonferrous metals	1.516	1.02	1.872	1.16	2.065	1.26	-	-	-	-
3810 - Metal products	7.711	5.18	8.697	5.37	8.865	5.41	-	-	-	-
3820 - Machinery nec	7.889	5.29	8.413	5.19	8.587	5.24	-	-	-	-
3825 - Office, computing machinery	0.795	0.53	0.870	0.54	0.839	0.51	-	-	-	-
3830 - Electrical machinery	10.000	6.71	10.000	6.17	10.000	6.10	-	-	-	-
3840 - Transportation equipment	6.946	4.66	7.609	4.70	8.271	5.04	-	-	-	-
3841 - Shipbuilding, repair	1.886	1.27	1.816	1.12	1.919	1.17	-	-	-	-
3843 - Motor vehicles	3.732	2.50	4.103	2.53	4.470	2.73	-	-	-	-
3850 - Professional goods	0.617	0.41	0.668	0.41	0.721	0.44	-	-	-	-
3900 - Industries nec	1.175	0.79	1.280	0.79	1.349	0.82	-	-	-	-

Note: Codes are International Standard Industry codes (ISIC). Percentages are % of total Output. [f]: Factor Prices; [p]: Producer Prices.

Finance, Economics, and Trade

Economic Indicators [36]

- **National product**: GDP—purchasing power parity—$301.9 billion (1995 est.)

- **National product real growth rate**: 2.5% (1995 est.)

- **National product per capita**: $19,500 (1995 est.)

- **Inflation rate (consumer prices)**: 2.25% (1995)

- **External debt**: 0

Balance of Payments Summary [37]

Values in millions of dollars.

	1989	1990	1991	1992	1993
Exports of goods (f.o.b.)	101,317	122,071	122,625	129,195	120,303
Imports of goods (f.o.b.)	-93,162	-111,741	-111,885	-117,855	-107,338
Trade balance	8,155	10,330	10,740	11,340	12,915
Services - debits	-45,328	-55,772	-59,945	-64,960	-63,135
Services - credits	49,211	57,614	61,144	65,021	64,805
Private transfers (net)	-944	-1,159	-1,222	-1,619	-1,662
Government transfers (net)	-1,309	-2,084	-3,189	-3,354	-3,553
Long-term capital (net)	4,700	-5,663	-4,862	-17,939	-3,569
Short-term capital (net)	-10,181	1,011	-1,469	10,651	-7,607
Errors and omissions	-3,854	-4,001	-1,121	7,286	8,348
Overall balance	450	276	76	6,426	6,542

Exchange Rates [38]

Currency: **Netherlands guilder, gulden, or florins**
Symbol: **f.**

Data are currency units per $1.

January 1996	1.6365
1995	1.6057
1994	1.8200
1993	1.8573
1992	1.7585
1991	1.8697

Imports and Exports

Top Import Origins [39]

$133 billion (c.i.f., 1995) Data are for 1994.

Origins	%
EU	56
Germany	21
Belgium-Luxembourg	11
UK	8.5
US	8.6

Top Export Destinations [40]

$146 billion (f.o.b., 1995) Data are for 1994.

Destinations	%
EU	73
Germany	28
Belgium-Luxembourg	13
UK	9
Central and Eastern Europe	2
US	5

Foreign Aid [41]

Donor: ODA, $2.525 billion (1993).

Import and Export Commodities [42]

Import Commodities

Raw materials and semifinished products
Consumer goods
Transportation equipment
Crude oil
Food products

Export Commodities

Metal products
Chemicals
Processed food and tobacco
Agricultural products

New Zealand

Geography [1]

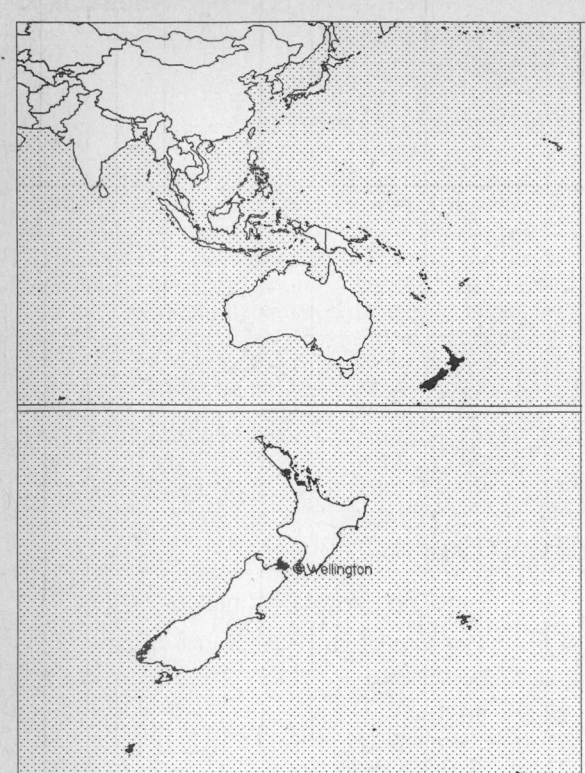

Total area:
268,680 sq km   103,738 sq mi
Land area:
268,670 sq km   103,734 sq mi
Comparative area:
About the size of Colorado
Note: Includes Antipodes Islands, Auckland Islands, Bounty Islands, Campbell Island, Chatham Islands, and Kermadec Islands
Land boundaries:
0 km
Coastline:
15,134 km
Climate:
Temperate with sharp regional contrasts
Terrain:
Predominately mountainous with some large coastal plains
Natural resources:
Natural gas, iron ore, sand, coal, timber, hydropower, gold, limestone
Land use:
Arable land: 2%
Permanent crops: 0%
Meadows and pastures: 53%
Forest and woodland: 38%
Other: 7%

Demographics [2]

	1970	1980	1990	1995[1]	1996	2000	2010	2020	2030
Population	2,811	3,113	3,299	3,508	3,548	3,698	4,029	4,326	4,530
Population density (persons per sq. mi.)	27	30	32	34	34	36	39	42	44
(persons per sq. km.)	10	12	12	13	13	14	15	16	17
Net migration rate (per 1,000 population)	3.8	-11.0	-1.2	3.2	3.2	3.0	2.5	1.9	1.2
Births[41]	NA	51	NA	NA	56	NA	NA	NA	NA
Deaths	NA	NA	27	NA	27	NA	NA	NA	NA
Life expectancy - males	NA	70.0	72.5	73.8	74.0	74.8	76.6	77.9	78.9
Life expectancy - females	NA	75.6	78.6	79.9	80.2	81.6	83.1	84.2	85.0
Birth rate (per 1,000)	22.1	16.2	18.2	16.2	15.8	14.0	12.8	12.3	10.9
Death rate (per 1,000)	8.8	8.6	8.0	7.8	7.7	7.5	7.6	8.0	9.2
Women of reproductive age (15-49 yrs.)	NA	NA	860	910	918	936	1,007	1,035	1,008
of which are currently married	NA	NA	417	NA	461	480	510	NA	NA
Fertility rate	NA	2.0	2.3	2.1	2.0	1.8	1.8	1.8	1.7

Except as noted, values for vital statistics are in thousands; life expectancy is in years.

Health

Health Indicators [3]

% of population with access to	
safe water (1990-95)	97
adequate sanitation (1990-95)	NA
health services (1985-95)	NA
% of 1-year-olds immunized (1990-94) against	
TB (tuberculosis)	20
DPT (diphtheria, pertussis, tetanus)	81
polio	68
measles	82
% of contraceptive prevalence (1980-94)[1]	70
ORT use rate (1990-94)	NA

Health Expenditures [4]

Total health expenditure, 1990 (official exchange rate)	
Millions of dollars	3,150
Dollars per capita	925
Health expenditures as a percentage of GDP	
Total	7.2
Public sector	5.9
Private sector	1.3
Development assistance for health	
Total aid flows (millions of dollars)[1]	NA
Aid flows per capita (dollars)	NA
Aid flows as a percentage of total health expenditure	NA

For sources, notes, and explanations, see Annotated Source Appendix, page 1061.

Human Factors

Women and Children [5]

% of pregnant women immunized (tetanus 1990-94)	NA
% of births attended by trained health personnel (1983-94)	99
Maternal mortality rate (1980-92)	13
Under-5 mortality rate (1994)	9
% under-5 moderately/severely underweight (1980-1994)	NA

Burden of Disease [6]

Population per physician (1986)	570.21
Population per nurse (1986)	80.02
Population per hospital bed (1990)	148.67
AIDS cases per 100,000 people (1994)	1.2
Heart disease cases per 1,000 people (1990-93)	342

Ethnic Division [7]

European	88%
Maori	8.9%
Pacific Islander	2.9%
Other	0.2%

Religion [8]

Anglican	24%
Presbyterian	18%
Roman Catholic	15%
Methodist	5%
Baptist	2%
Other Protestant	3%
Unspecified or none (1986)	33%

Major Languages [9]

English (official), Maori.

Education

Public Education Expenditures [10]

	1980	1985	1990	1991	1992	1994
Million (Dollar)						
Total education expenditure	1,302	2,028	4,451	5,013	5,308	NA
as percent of GNP	5.8	4.7	6.5	7.3	7.3	NA
as percent of total govt. expend.	23.1	18.4	NA	NA	NA	NA
Current education expenditure	1,171	1,849	4,252	4,825	5,107	NA
as percent of GNP	5.2	4.3	6.2	7.0	7.0	NA
as percent of current govt. expend.	27.9	25.5	NA	NA	NA	NA
Capital expenditure	131	179	199	188	201	NA

Educational Attainment [11]

Age group (1991)	25+
Total population	1,992,354
Highest level attained (%)	
No schooling	0.0
First level	
Not completed	36.8
Completed	NA
Entered second level	
S-1	16.3
S-2	7.8
Postsecondary	39.1

Illiteracy [12]

Libraries [13]

	Admin. Units	Svc. Pts.	Vols. (000)	Shelving (meters)	Vols. Added	Reg. Users
National	NA	NA	NA	NA	NA	NA
Nonspecialized	NA	NA	NA	NA	NA	NA
Public	NA	NA	NA	NA	NA	NA
Higher ed. (1990)	7	33	5,910	NA	181,848	NA
School	NA	NA	NA	NA	NA	NA

Daily Newspapers [14]

	1980	1985	1990	1994
Number of papers	32	33	35	31
Circ. (000)	1,059	1,075	1,100[e]	1,050[e]

Culture [15]

Science and Technology

Scientific/Technical Forces [16]

Scientists/engineers	6,198
Number female	NA
Technicians	2,866
Number female	NA
Total	9,064

R&D Expenditures [17]

	Dollar (000) 1993
Total expenditure	825,200
Capital expenditure	96,500
Current expenditure	728,700
Percent current	88.3

U.S. Patents Issued [18]

Values show patents issued to citizens of the country by the U.S. Patents Office.

	1993	1994	1995
Number of patents	51	54	60

For sources, notes, and explanations, see Annotated Source Appendix, page 1061.

Government and Law

Organization of Government [19]

Long-form name:
None
Type:
Parliamentary democracy
Independence:
26 September 1907 (from UK)
National holiday:
Waitangi Day, 6 February (1840) (Treaty of Waitangi established British sovereignty)
Constitution:
No formal, written constitution; consists of various documents, including certain acts of the UK and New Zealand Parliaments; Constitution Act 1986 was to come into force January 1987, but not yet enacted
Legal system:
Based on English law, with special land legislation and land courts for Maoris; accepts compulsory ICJ jurisdiction, with reservations
Executive branch:
British Monarch (represented by Governor General); Prime Minister; Deputy Prime Minister; Executive Council
Legislative branch:
Unicameral: House of Representatives
Judicial branch:
High Court; Court of Appeal

Elections [20]

House of Representatives	% of votes
National Party (NP)	35.2
New Zealand Labor Party (NZLP)	34.7
Alliance	18.3
New Zealand First	8.3

Government Expenditures [21]

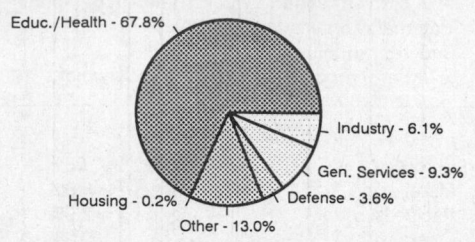

Educ./Health - 67.8%
Industry - 6.1%
Gen. Services - 9.3%
Defense - 3.6%
Housing - 0.2%
Other - 13.0%

(% distribution). Expend. for FY95: 31,428 (Dollar mil.)

Military Expenditures and Arms Transfers [22]

	1990	1991	1992	1993	1994
Military expenditures					
Current dollars (mil.)	793	560	659	713	613
1994 constant dollars (mil.)	883	601	687	728	613
Armed forces (000)	11	11	11	11	10
Gross national product (GNP)					
Current dollars (mil.)	38,760	39,590	42,810	46,020	48,750
1994 constant dollars (mil.)	43,130	42,440	44,640	46,970	48,750
Central government expenditures (CGE)					
1994 constant dollars (mil.)	17,520	16,760	17,290	16,130	17,530
People (mil.)	3.3	3.3	3.3	3.4	3.4
Military expenditure as % of GNP	2.0	1.4	1.5	1.5	1.3
Military expenditure as % of CGE	5.0	3.6	4.0	4.5	3.5
Military expenditure per capita (1994 $)	268	181	205	216	181
Armed forces per 1,000 people (soldiers)	3.3	3.3	3.3	3.3	3.0
GNP per capita (1994 $)	13,070	12,770	13,340	13,940	14,390
Arms imports[6]					
Current dollars (mil.)	40	80	80	60	10
1994 constant dollars (mil.)	45	86	83	61	10
Arms exports[6]					
Current dollars (mil.)	0	0	0	0	0
1994 constant dollars (mil.)	0	0	0	0	0
Total imports[7]					
Current dollars (mil.)	9,501	8,381	9,202	9,636	11,910
1994 constant dollars (mil.)	10,570	8,983	9,596	9,835	11,910
Total exports[7]					
Current dollars (mil.)	9,488	9,598	9,824	10,540	12,180
1994 constant dollars (mil.)	10,560	10,290	10,240	10,750	12,180
Arms as percent of total imports[8]	0.4	1.0	0.9	0.6	0.1
Arms as percent of total exports[8]	0	0	0	0	0

Crime [23]

	1994
Crime volume	
Cases known to police	488,533
Attempts (percent)	NA
Percent cases solved	43.49
Crimes per 100,000 persons	13,853.59
Persons responsible for offenses	
Total number offenders	188,751
Percent female	18.98
Percent juvenile (0-16 yrs.)	22.33
Percent foreigners	NA

Human Rights [24]

	SSTS	FL	FAPRO	PPCG	APROBC	TPW	PCPTW	STPEP	PHRFF	PRW	ASST	AFL
Observes	P	P		P		P	P			P	P	P
	EAFRD	CPR	ESCR	SR	ACHR	MAAE	PVIAC	PVNAC	EAFDAW	TCIDTP	RC	
Observes		P	P	P	P			P	P	P	P	

P=Party; S=Signatory; see Appendix for meaning of abbreviations.

Labor Force

Total Labor Force [25]

1.635 million (September 1995)

Labor Force by Occupation [26]

Services	64.6%
Industry	25.0
Agriculture	10.4

Date of data: 1994

Unemployment Rate [27]

6.1% (October 1995)

Production Sectors

Commercial Energy Production and Consumption

Data are shown in quadrillion (10^{15}) BTUs and percent for 1995
Values for hydroelectric, nuclear, geothermal, solar, and wind power refer to electrical generation.

Production [28]	Consumption [29]
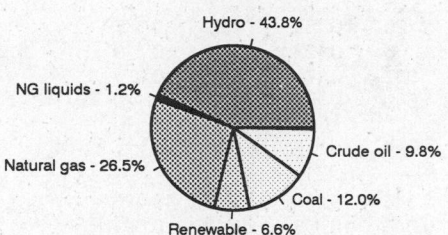	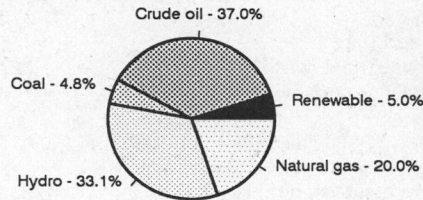

Production [28]

Hydro - 43.8%
NG liquids - 1.2%
Crude oil - 9.8%
Natural gas - 26.5%
Coal - 12.0%
Renewable - 6.6%

Consumption [29]

Crude oil - 37.0%
Coal - 4.8%
Renewable - 5.0%
Natural gas - 20.0%
Hydro - 33.1%

Crude oil	0.063		Crude oil	0.314
Natural gas liquids	0.008		Dry natural gas	0.170
Dry natural gas	0.170		Coal	0.041
Coal	0.077		Net hydroelectric power	0.281
Net hydroelectric power	0.281		Geothermal, solar, wind	0.042
Geothermal, solar, wind	0.042		Total	0.848
Total	0.641			

Telecommunications [30]

- 1.7 million (1986 est.) telephones; excellent international and domestic systems
- International: submarine cables to Australia and Fiji; satellite earth stations - 2 Intelsat (Pacific Ocean)
- Radio: Broadcast stations: AM 64, FM 2, shortwave 0 Radios: 3.215 million (1992 est.)
- Television: Broadcast stations: 14 (1986 est.) Televisions: 1.53 million (1992 est.)

Transportation [31]

Railways: total: 3,973 km; narrow gauge: 3,973 km 1.067-m gauge (504 km electrified)

Highways: total: 93,348 km; paved: 54,142 km (including 141 km of expressways); unpaved: 39,206 km (1992 est.)

Merchant marine: total: 17 ships (1,000 GRT or over) totaling 162,220 GRT/213,749 DWT; ships by type: bulk 6, cargo 1, liquefied gas tanker 1, oil tanker 3, railcar carrier 1, roll-on/roll-off cargo 5 (1995 est.)

Airports

Total:	113
With paved runways over 3,047 m:	2
With paved runways 1,524 to 2,437 m:	8
With paved runways 914 to 1,523 m:	31
With paved runways under 914 m:	50
With unpaved runways 1,524 to 2,437 m:	1

Top Agricultural Products [32]

Agriculture accounts for 7.3% of the GDP; produces wheat, barley, potatoes, pulses, fruits, vegetables; wool, meat, dairy products; fish catch reached a record 503,000 metric tons in 1988.

Top Mining Products [33]

Thousand metric tons except as noted[e]	7/7/95[*]
Gold, mine output, Au content (kg.)	12,000
Iron sand (titaniferous magnetite) gr. wt.	1,100
Silver, mine output, Ag content (kg.)	30,000
Sand and gravel for roads, ballast, bldg. agg.	19,000
Limestone and marl	4,000
Serpentine	23,000
Coal	3,100
Gas, manufactured (000 cu. meters)	11,400
Peat (cu. meters)	110,000
Petroleum, crude (000 42-gal. bls.)	16,400[r]

Tourism [34]

	1990	1991	1992	1993	1994
Tourists[4]	976	963	1,056	1,157	1,323
Tourism receipts[81]	1,019	1,021	1,032	1,165	1,357
Tourism expenditures	958	987	977	1,003	1,101
Fare receipts[82]	469	488	554	573	785
Fare expenditures	371	365	332	309	352

Travelers are in thousands, money in million U.S. dollars.

For sources, notes, and explanations, see Annotated Source Appendix, page 1061.

687

Manufacturing Sector

Manufacturing Summary [35]

	1987		1988		1989		1990		1991	
	$ bil.	%	$ bil.	%	$ bil.	%	$ bil.	%	$ bil.	%
Establishments or enterprises (number)	21,656	0.923	21,267	1.013	20,870	1.114	21,143	1.180	21,579	2.823
Total employment (000)	285	0.210	259	0.189	261	0.212	245	0.222	233	0.335
Production workers (000)	-	-	-	-	-	-	-	-	-	-
Output ($ bil.)	18	0.178	20	0.174	20	0.168	20	0.177	-	-
Value added ($ bil.)	5	0.126	6	0.127	6	0.122	6	0.119	-	-
Capital investment ($ mil.)	1,027	0.236	784	0.165	937	0.171	396	0.071	-	-
M & E investment ($ mil.)	710	0.228	515	0.149	649	0.166	545	0.129	-	-
Employees per establishment (number)	13	22.767	12	18.694	13	19.042	12	18.779	11	11.863
Production workers per establishment	-	-	-	-	-	-	-	-	-	-
Output per establishment ($ mil.)	0.834	19.242	0.953	17.154	0.948	15.061	0.962	15.028	-	-
Capital investment per estab. ($ mil.)	0.047	25.521	0.037	16.246	0.045	15.358	0.019	6.041	-	-
M & E per establishment ($ mil)	0.033	24.693	0.024	14.693	0.031	14.929	0.026	10.956	-	-
Payroll per employee ($)	16,520	184.225	19,121	191.927	19,261	172.406	19,806	142.783	-	-
Wages per production worker ($)	-	-	-	-	-	-	-	-	-	-
Hours per production worker (hours)	-	-	-	-	-	-	-	-	-	-
Output per employee ($)	63,298	84.515	78,156	91.759	75,805	79.096	82,936	80.028	-	-
Capital investment per employee ($)	3,600	112.095	3,024	86.905	3,589	80.652	1,615	32.170	-	-
M & E per employee ($)	2,490	108.462	1,988	78.600	2,487	78.399	2,222	58.345	-	-

Note: Columns headed % show percent of world total or ratio. Ratios closest to 100 are closest to world average. M & E stands for machinery & equipment.

Output in Manufacturing

	1987		1988		1989		1990		1991	
	$ bil.	%	$ bil.	%	$ bil.	%	$ bil.	%	$ bil.	%
3110 - Food products	5.761	32.01	6.805	34.03	6.753	33.77	7.516	37.58	-	-
3210 - Textiles	0.940	5.22	1.042	5.21	0.879	4.39	0.733	3.67	-	-
3211 - Spinning, weaving, etc.	-	-	-	-	-	-	-	-	-	-
3220 - Wearing apparel	-	-	-	-	-	-	-	-	-	-
3230 - Leather and products	-	-	-	-	-	-	-	-	-	-
3240 - Footwear	-	-	-	-	-	-	-	-	-	-
3310 - Wood products	-	-	-	-	-	-	-	-	-	-
3320 - Furniture, fixtures	-	-	-	-	-	-	-	-	-	-
3410 - Paper and products	1.424	7.91	1.547	7.74	1.505	7.53	1.632	8.16	-	-
3411 - Pulp, paper, etc.	0.840	4.67	0.954	4.77	0.951	4.75	1.054	5.27	-	-
3420 - Printing, publishing	1.143	6.35	1.227	6.14	1.192	5.96	1.192	5.96	-	-
3510 - Industrial chemicals	0.658	3.66	0.772	3.86	0.797	3.99	0.789	3.94	-	-
3511 - Basic chemicals, excl fertilizers	0.150	0.83	0.188	0.94	0.155	0.78	0.186	0.93	-	-
3513 - Synthetic resins, etc.	0.272	1.51	0.289	1.45	0.323	1.61	0.303	1.51	-	-
3520 - Chemical products nec	0.721	4.01	0.849	4.25	0.751	3.76	0.795	3.98	-	-
3522 - Drugs and medicines	-	-	-	-	-	-	-	-	-	-
3530 - Petroleum refineries	0.349	1.94	0.371	1.86	0.402	2.01	0.428	2.14	-	-
3540 - Petroleum, coal products	0.053	0.29	0.054	0.27	0.044	0.22	0.037	0.19	-	-
3550 - Rubber products	0.213	1.18	0.235	1.17	0.201	1.01	0.206	1.03	-	-
3560 - Plastic products nec	0.547	3.04	0.616	3.08	0.667	3.34	0.626	3.13	-	-
3610 - Pottery, china, etc.	-	-	-	-	-	-	-	-	-	-
3620 - Glass and products	-	-	-	-	-	-	-	-	-	-
3690 - Nonmetal products nec	-	-	-	-	-	-	-	-	-	-
3710 - Iron and steel	-	-	-	-	-	-	-	-	-	-
3720 - Nonferrous metals	-	-	-	-	-	-	-	-	-	-
3810 - Metal products	1.476	8.20	1.681	8.40	1.560	7.80	1.468	7.34	-	-
3820 - Machinery nec	0.953	5.29	1.151	5.75	0.991	4.96	0.949	4.74	-	-
3825 - Office, computing machinery	-	-	-	-	-	-	-	-	-	-
3830 - Electrical machinery	0.946	5.26	0.843	4.21	0.833	4.17	0.845	4.23	-	-
3832 - Radio, television, etc.	-	-	-	-	-	-	-	-	-	-
3840 - Transportation equipment	1.303	7.24	1.288	6.44	1.481	7.40	1.260	6.30	-	-
3841 - Shipbuilding, repair	-	-	-	-	-	-	-	-	-	-
3843 - Motor vehicles	-	-	-	-	-	-	-	-	-	-
3850 - Professional goods	0.075	0.42	0.084	0.42	0.077	0.39	0.077	0.39	-	-
3900 - Industries nec	0.230	1.28	0.262	1.31	0.236	1.18	0.250	1.25	-	-

Note: Codes are International Standard Industry codes (ISIC). Percentages are % of total Output. [f]: Factor Prices; [p]: Producer Prices.

Finance, Economics, and Trade

Economic Indicators [36]

- **National product**: GDP—purchasing power parity— $62.3 billion (1995 est.)

- **National product real growth rate**: 5.5% (1995 est.)

- **National product per capita**: $18,300 (1995 est.)

- **Inflation rate (consumer prices)**: 2% (FY95/96)

- **External debt**: $38.5 billion (September 1994)

Balance of Payments Summary [37]

Values in millions of dollars.

	1989	1990	1991	1992	1993
Exports of goods (f.o.b.)	8,846	9,191	9,555	9,781	10,463
Imports of goods (f.o.b.)	-7,873	-8,294	-7,483	-8,108	-8,749
Trade balance	973	897	2,072	1,673	1,714
Services - debits	-5,599	-5,590	-5,979	-5,690	-6,067
Services - credits	2,885	3,421	2,564	2,463	2,892
Private transfers (net)	491	650	728	740	575
Government transfers (net)	-27	-46	-43	-56	-46
Long-term capital (net)	-588	-157	-984	-748	-1,046
Short-term capital (net)	1,123	518	-63	5	-29
Errors and omissions	981	1,321	386	1,743	1,933
Overall balance	239	1,014	-1,319	130	-74

Exchange Rates [38]

Currency: **New Zealand dollar.**
Symbol: **NZ$.**

Data are currency units per $1.

January 1996	1.5138
1995	1.5235
1994	1.6844
1993	1.8495
1992	1.8584
1991	1.7265

Imports and Exports

Top Import Origins [39]

$13.62 billion (1995).

Origins	%
Australia	21
US	18
Japan	16
UK	6

Top Export Destinations [40]

$13.41 billion (1995).

Destinations	%
Australia	20
Japan	15
US	12
UK	6

Foreign Aid [41]

Donor: ODA, $98 million (1993).

Import and Export Commodities [42]

Import Commodities	Export Commodities
Machinery and equipment	Wool
Vehicles and aircraft	Lamb
Petroleum	Mutton
Consumer goods	Beef
	Fish
	Cheese
	Chemicals
	Forestry products
	Fruits and vegetables
	Manufactures

For sources, notes, and explanations, see Annotated Source Appendix, page 1061.

689

Nicaragua

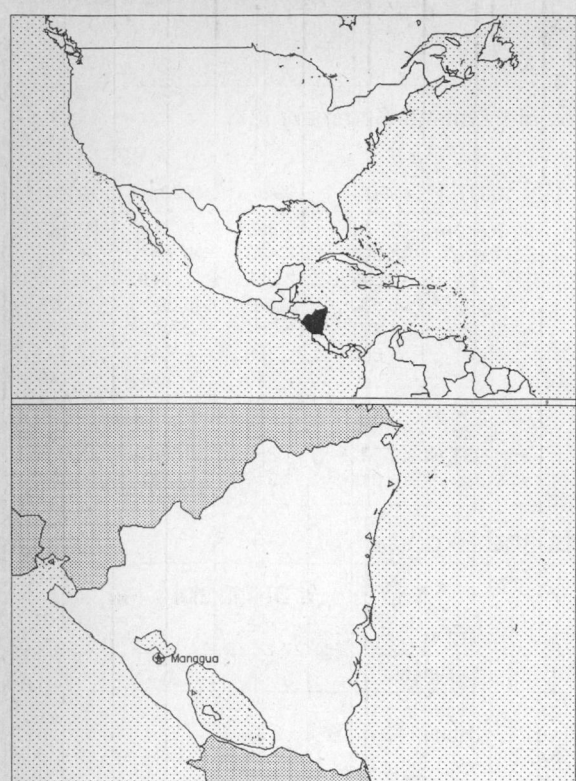

Geography [1]

Total area:
129,494 sq km   49,998 sq mi
Land area:
120,254 sq km   46,430 sq mi
Comparative area:
Slightly larger than New York State
Land boundaries:
Total 1,231 km, Costa Rica 309 km, Honduras 922 km
Coastline:
910 km
Climate:
Tropical in lowlands, cooler in highlands
Terrain:
Extensive Atlantic coastal plains rising to central interior mountains;
narrow Pacific coastal plain interrupted by volcanoes
Natural resources:
Gold, silver, copper, tungsten, lead, zinc, timber, fish
Land use:
Arable land: 9%
Permanent crops: 1%
Meadows and pastures: 43%
Forest and woodland: 35%
Other: 12%

Demographics [2]

	1970	1980	1990	1995[1]	1996	2000	2010	2020	2030
Population	2,053	2,776	3,591	4,159	4,272	4,729	5,863	6,973	8,017
Population density (persons per sq. mi.)	44	60	77	90	92	102	126	150	173
(persons per sq. km.)	17	23	30	35	36	39	49	58	67
Net migration rate (per 1,000 population)	NA	0.8	2.8	-1.2	-1.2	-1.1	-0.8	-0.5	-0.4
Births	NA	NA	NA	NA	145	NA	NA	NA	NA
Deaths	NA	NA	NA	NA	26	NA	NA	NA	NA
Life expectancy - males	NA	53.3	60.7	63.0	63.4	65.1	68.9	72.0	74.5
Life expectancy - females	NA	58.7	65.1	67.7	68.1	70.1	74.2	77.6	80.2
Birth rate (per 1,000)	NA	44.8	39.1	34.7	33.8	30.4	24.3	20.2	17.2
Death rate (per 1,000)	NA	11.2	7.3	6.2	6.0	5.3	4.4	4.1	4.5
Women of reproductive age (15-49 yrs.)	NA	617	823	995	1,034	1,197	1,604	1,964	2,206
of which are currently married	NA	NA	503	NA	634	737	1,014	NA	NA
Fertility rate	NA	6.1	4.9	4.2	4.0	3.5	2.6	2.2	2.1

Except as noted, values for vital statistics are in thousands; life expectancy is in years.

Health

Health Indicators [3]

% of population with access to	
safe water (1990-95)	58
adequate sanitation (1990-95)	60
health services (1985-95)	83
% of 1-year-olds immunized (1990-94) against	
TB (tuberculosis)	89
DPT (diphtheria, pertussis, tetanus)	74
polio	84
measles	74
% of contraceptive prevalence (1980-94)	49
ORT use rate (1990-94)	40

Health Expenditures [4]

Total health expenditure, 1990 (official exchange rate)	
Millions of dollars	133
Dollars per capita	35
Health expenditures as a percentage of GDP	
Total	8.6
Public sector	6.7
Private sector	1.9
Development assistance for health	
Total aid flows (millions of dollars)[1]	27
Aid flows per capita (dollars)	6.6
Aid flows as a percentage of total health expenditure	20.0

For sources, notes, and explanations, see Annotated Source Appendix, page 1061.

Human Factors

Women and Children [5]

% of pregnant women immunized (tetanus 1990-94)	NA
% of births attended by trained health personnel (1983-94)	73
Maternal mortality rate (1980-92)	NA
Under-5 mortality rate (1994)	68
% under-5 moderately/severely underweight (1980-1994)	12

Burden of Disease [6]

Population per physician (1990)	1,491.88
Population per nurse (1984)	531.18
Population per hospital bed (1990)	538.21
AIDS cases per 100,000 people (1994)	0.8
Malaria cases per 100,000 people (1992)	679

Ethnic Division [7]

Mestizo	69%
White	17%
Black	9%
Indian	5%

Religion [8]

Roman Catholic	95%
Protestant	5%

Major Languages [9]

Spanish (official), English- and Indian-speaking minorities on Atlantic coast.

Education

Public Education Expenditures [10]

Million (Cordoba)[48]	1980	1990	1991	1992	1993	1994
Total education expenditure	0.662	NA	242	275	324	340
as percent of GNP	3.4	NA	4.4	4.1	3.9	3.8
as percent of total govt. expend.	10.4	NA	12.7	NA	12.8	12.2
Current education expenditure	0.580	NA	NA	NA	NA	327
as percent of GNP	3.0	NA	NA	NA	NA	3.7
as percent of current govt. expend.	NA	NA	NA	NA	NA	NA
Capital expenditure	0.082	NA	NA	NA	NA	13

Educational Attainment [11]

Illiteracy [12]

In thousands and percent[1]	1990	1995	2000
Illiterate population (15+ yrs.)	689	822	956
Illiteracy rate - total pop. (%)	36.2	35.8	34.7
Illiteracy rate - males (%)	34.8	35.8	35.4
Illiteracy rate - females (%)	37.5	35.8	34.0

Libraries [13]

	Admin. Units	Svc. Pts.	Vols. (000)	Shelving (meters)	Vols. Added	Reg. Users
National	NA	NA	NA	NA	NA	NA
Nonspecialized	NA	NA	NA	NA	NA	NA
Public	NA	NA	NA	NA	NA	NA
Higher ed. (1990)	13	18	187	NA	8,544	27,633
School (1987)	412	412	595	6,834	NA	71,948

Daily Newspapers [14]

	1980	1985	1990	1994
Number of papers	3	3	6	4
Circ. (000)	136	160[e]	180[e]	130

Culture [15]

Science and Technology

Scientific/Technical Forces [16]

Scientists/engineers	725
Number female	NA
Technicians	302
Number female	NA
Total	1,027

R&D Expenditures [17]

	Cordoba (000) 1987
Total expenditure[6]	NA
Capital expenditure	NA
Current expenditure	988,970
Percent current	NA

U.S. Patents Issued [18]

Values show patents issued to citizens of the country by the U.S. Patents Office.

	1993	1994	1995
Number of patents	1	0	0

For sources, notes, and explanations, see Annotated Source Appendix, page 1061.

691

Government and Law

Organization of Government [19]

Long-form name:
Republic of Nicaragua
Type:
Republic
Independence:
15 September 1821 (from Spain)
National holiday:
Independence Day, 15 September (1821)
Constitution:
9 January 1987
Legal system:
Civil law system; Supreme Court may
review administrative acts
Executive branch:
President; Vice President; Cabinet
Legislative branch:
Unicameral: National Assembly
(Asamblea Nacional)
Judicial branch:
Supreme Court (Corte Suprema)

Elections [20]

National Assembly	% of votes
National Opposition Union	53.9
Sandinista National Liberation	40.8
Social Christian	1.6
Revolutionary Unity	1.0

Government Expenditures [21]

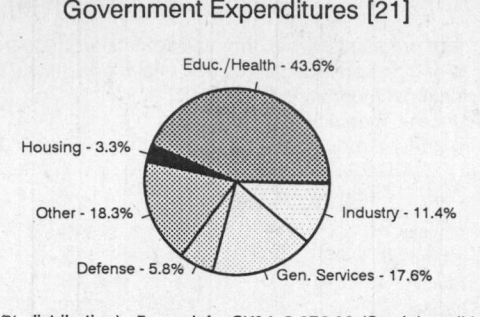

(% distribution). Expend. for CY94: 3,976.23 (Cordoba mil.)

Crime [23]

	1994
Crime volume	
Cases known to police	47,173
Attempts (percent)	2.70
Percent cases solved	70.30
Crimes per 100,000 persons	1,069.17
Persons responsible for offenses	
Total number offenders	26,529
Percent female	9.20
Percent juvenile (0-15 yrs.)	1.10
Percent foreigners	NA

Military Expenditures and Arms Transfers [22]

	1990	1991	1992	1993	1994
Military expenditures					
Current dollars (mil.)	207	46	39	35	34
1994 constant dollars (mil.)	230	50	41	36	34
Armed forces (000)	28	20	15	15	14
Gross national product (GNP)					
Current dollars (mil.)	1,317	1,209	1,254	1,313	1,329
1994 constant dollars (mil.)	1,466	1,295	1,307	1,340	1,329
Central government expenditures (CGE)					
1994 constant dollars (mil.)	807	490	541	535	609
People (mil.)	3.6	3.8	3.9	4.0	4.1
Military expenditure as % of GNP	15.7	3.8	3.1	2.7	2.6
Military expenditure as % of CGE	28.6	10.1	7.6	6.7	5.7
Military expenditure per capita (1994 $)	64	13	11	9	8
Armed forces per 1,000 people (soldiers)	7.7	5.3	3.9	3.8	3.4
GNP per capita (1994 $)	405	344	337	336	324
Arms imports[6]					
Current dollars (mil.)	70	80	5	5	0
1994 constant dollars (mil.)	78	86	5	5	0
Arms exports[6]					
Current dollars (mil.)	0	0	30	0	0
1994 constant dollars (mil.)	0	0	31	0	0
Total imports[7]					
Current dollars (mil.)	638	751	892	746	786[e]
1994 constant dollars (mil.)	710	805	930	761	786[e]
Total exports[7]					
Current dollars (mil.)	331	275	218	267	329[e]
1994 constant dollars (mil.)	368	295	227	273	329[e]
Arms as percent of total imports[8]	11.0	10.7	0.6	0.7	0
Arms as percent of total exports[8]	0	0	13.8	0	0

Human Rights [24]

	SSTS	FL	FAPRO	PPCG	APROBC	TPW	PCPTW	STPEP	PHRFF	PRW	ASST	AFL
Observes	P	P	P	P	P	P	P			P	P	P
	EAFRD	CPR	ESCR	SR	ACHR	MAAE	PVIAC	PVNAC	EAFDAW	TCIDTP	RC	
Observes		P	P	P	P	P	P	S	S	P	S	P

P=Party; S=Signatory; see Appendix for meaning of abbreviations.

Labor Force

Total Labor Force [25]

1.086 million

Labor Force by Occupation [26]

Services	43%
Agriculture	44
Industry	13

Date of data: 1986

Unemployment Rate [27]

20%; substantial underemployment (1995 est.)

For sources, notes, and explanations, see Annotated Source Appendix, page 1061.

Production Sectors

Commercial Energy Production and Consumption

Data are shown in quadrillion (10^{15}) BTUs and percent for 1995
Values for hydroelectric, nuclear, geothermal, solar, and wind power refer to electrical generation.

Production [28]	Consumption [29]
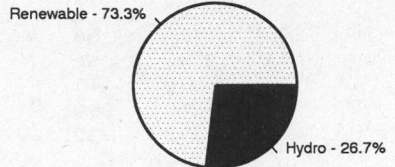	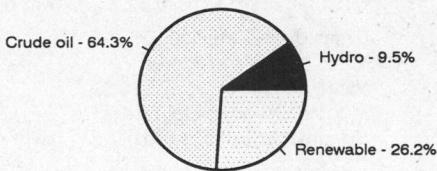
Renewable - 73.3%	Crude oil - 64.3%
Hydro - 26.7%	Hydro - 9.5%
	Renewable - 26.2%

Net hydroelectric power	0.004	Crude oil	0.027
Geothermal, solar, wind	0.011	Net hydroelectric power	0.004
Total	0.015	Geothermal, solar, wind	0.011
		Total	0.042

Telecommunications [30]

- 66,810 (1993 est.) telephones; low-capacity microwave radio relay and wire system being expanded; connected to Central American Microwave System
- Domestic: wire and microwave radio relay
- International: satellite earth stations - 1 Intersputnik (Atlantic Ocean Region) and 1 Intelsat (Atlantic Ocean)
- Radio: Broadcast stations: AM 45, FM 0, shortwave 3 Radios: 1.037 million (1992 est.)
- Television: Broadcast stations: 7 (1994 est.) Televisions: 260,000 (1992 est.)

Transportation [31]

Railways: total: 0 km; narrow gauge: 0 km 1.067-m gauge; note - part of the previous 376 km system was closed and dismantled in 1993 and, in 1994, the remainder was closed, the track and rolling stock being sold for scrap

Highways: total: 26,000 km; paved: 4,000 km; unpaved: 22,000 km (1993 est.)

Merchant marine: none

Airports

Total:	148
With paved runways over 3,047 m:	1
With paved runways 2,438 to 3,047 m:	1
With paved runways 1,524 to 2,437 m:	3
With paved runways 914 to 1,523 m:	3
With paved runways under 914 m:	107

Top Agricultural Products [32]

Produces coffee, bananas, sugarcane, cotton, rice, corn, cassava (tapioca), citrus, beans; beef, veal, pork, poultry, dairy products.

Top Mining Products [33]

Metric tons except as noted[e]	7/7/95[*]
Bentonite	2,200
Cement	570,000
Gold, mine output, Au content (kg.)	1,070[r]
Gypsum and anhydrite, crude	11,000
Lime	3,500
Petroleum refinery products (000 42-gal. bls.)	4,700
Salt, marine	15,000
Sand and gravel (000 tons)	1,200
Silver, mine output, Ag content (kg.)	2,460[r]

Tourism [34]

	1990	1991	1992	1993	1994
Visitors	140	177	195	236	299
Tourists[12]	106	146	167	198	238
Excursionists	22	27	27	37	59
Tourism receipts	12	16	21	30	40
Tourism expenditures	15	28	30	34	38
Fare expenditures	NA	NA	16	15	16

Travelers are in thousands, money in million U.S. dollars.

For sources, notes, and explanations, see Annotated Source Appendix, page 1061.

693

Manufacturing Sector

GDP and Manufacturing Summary [35]

	1980	1985	1989	1990	% change 1980-1990	% change 1989-1990
GDP (million 1980 $)	1,489	1,537	1,389	1,289	-13.4	-7.2
GDP per capita (1980 $)	537	470	371	333	-38.0	-10.2
Manufacturing as % of GDP (current prices)	24.4	27.6	17.0[e]	12.7	-48.0	-25.3
Gross output (million $)	612	1,587	NA	2,733[e]	346.6	NA
Value added (million $)	242	982	NA	1,781[e]	636.0	NA
Value added (million 1980 $)	351	366	270	232	-33.9	-14.1
Industrial production index	100	116	96	121	21.0	26.0
Employment (thousands)	34	39	45[e]	46[e]	35.3	2.2

Note: GDP stands for Gross Domestic Product. 'e' stands for estimated value.

Profitability and Productivity

	1980	1985	1989	1990	% change 1980-1990	% change 1989-1990
Intermediate input (%)	60	38	NA	35[e]	-41.7	NA
Wages, salaries, and supplements (%)	12	10	NA	11[e]	-8.3	NA
Gross operating surplus (%)	28	52	NA	54[e]	92.9	NA
Gross output per worker ($)	18,017	38,009	NA	55,171[e]	206.2	NA
Value added per worker ($)	7,131	23,515	NA	35,959[e]	404.3	NA
Average wage (incl. benefits) ($)	2,078	4,152	NA	6,439[e]	209.9	NA

Profitability is in percent of gross output. Productivity is in U.S. $. 'e' stands for estimated value.

Profitability - 1990

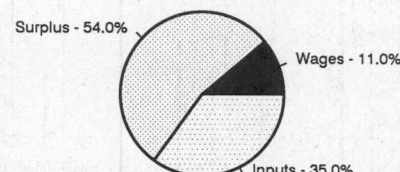

Surplus - 54.0%
Wages - 11.0%
Inputs - 35.0%

The graphic shows percent of gross output.

Value Added in Manufacturing

	1980 $ mil.	1980 %	1985 $ mil.	1985 %	1989 $ mil.	1989 %	1990 $ mil.	1990 %	% change 1980-1990	% change 1989-1990
311 Food products	52	21.5	268	27.3	NA	NA	451[e]	25.3	767.3	NA
313 Beverages	48	19.8	227	23.1	NA	NA	474[e]	26.6	887.5	NA
314 Tobacco products	28	11.6	64	6.5	NA	NA	134[e]	7.5	378.6	NA
321 Textiles	9	3.7	70	7.1	NA	NA	117[e]	6.6	1,200.0	NA
322 Wearing apparel	4	1.7	23	2.3	NA	NA	40[e]	2.2	900.0	NA
323 Leather and fur products	2	0.8	6	0.6	NA	NA	14[e]	0.8	600.0	NA
324 Footwear	4	1.7	27	2.7	NA	NA	48[e]	2.7	1,100.0	NA
331 Wood and wood products	3	1.2	10	1.0	NA	NA	16[e]	0.9	433.3	NA
332 Furniture and fixtures	1	0.4	4	0.4	NA	NA	5[e]	0.3	400.0	NA
341 Paper and paper products	1	0.4	3	0.3	NA	NA	3[e]	0.2	200.0	NA
342 Printing and publishing	4	1.7	22	2.2	NA	NA	41[e]	2.3	925.0	NA
351 Industrial chemicals	11	4.5	23	2.3	NA	NA	36[e]	2.0	227.3	NA
352 Other chemical products	14	5.8	56	5.7	NA	NA	122[e]	6.9	771.4	NA
353 Petroleum refineries	35	14.5	78	7.9	NA	NA	116[e]	6.5	231.4	NA
354 Miscellaneous petroleum and coal products	NA	0.0	1	0.1	NA	NA	2[e]	0.1	NA	NA
355 Rubber products	1	0.4	6	0.6	NA	NA	12[e]	0.7	1,100.0	NA
356 Plastic products	4	1.7	20	2.0	NA	NA	31[e]	1.7	675.0	NA
361 Pottery, china, and earthenware	NA	0.0	2	0.2	NA	NA	NA	0.0	NA	NA
362 Glass and glass products	NA	0.0	1	0.1	NA	NA	2[e]	0.1	NA	NA
369 Other nonmetal mineral products	7	2.9	17	1.7	NA	NA	27[e]	1.5	285.7	NA
371 Iron and steel	NA	0.0	1	0.1	NA	NA	2[e]	0.1	NA	NA
372 Nonferrous metals	NA	0.0	NA	0.0	NA	NA	NA	0.0	NA	NA
381 Metal products	9	3.7	40	4.1	NA	NA	70[e]	3.9	677.8	NA
382 Nonelectrical machinery	NA	0.0	3	0.3	NA	NA	4[e]	0.2	NA	NA
383 Electrical machinery	1	0.4	5	0.5	NA	NA	9[e]	0.5	800.0	NA
384 Transport equipment	1	0.4	3	0.3	NA	NA	5[e]	0.3	400.0	NA
385 Professional and scientific equipment	1	0.4	NA	0.0	NA	NA	NA	0.0	NA	NA
390 Other manufacturing industries	NA	0.0	2	0.2	NA	NA	3[e]	0.2	NA	NA

Note: The industry codes shown are International Standard Industry codes (ISIC). Percentages are percent of total Value Added. 'e' stands for estimated value

For sources, notes, and explanations, see Annotated Source Appendix, page 1061.

Finance, Economics, and Trade

Economic Indicators [36]

- **National product**: GDP—purchasing power parity—$7.1 billion (1995 est.)

- **National product real growth rate**: 4.2% (1995 est.)

- **National product per capita**: $1,700 (1995 est.)

- **Inflation rate (consumer prices)**: 11.4% (1995 est.)

- **External debt**: $11.7 billion (1994)

Balance of Payments Summary [37]

Values in millions of dollars.

	1989	1990	1991	1992	1993
Exports of goods (f.o.b.)	318.7	332.4	268.1	223.1	267.0
Imports of goods (f.o.b.)	-547.3	-569.7	-688.0	-770.8	-659.4
Trade balance	-228.6	-237.3	-419.9	-547.7	-392.4
Services - debits	-330.8	-341.1	-509.2	-650.6	-591.1
Services - credits	28.8	71.6	79.9	93.7	105.6
Private transfers (net)	NA	NA	NA	10.0	25.0
Government transfers (net)	168.9	201.6	844.4	378.6	396.3
Long-term capital (net)	-98.6	-167.9	189.7	-457.1	-390.3
Short-term capital (net)	593.6	615.0	-183.5	1,113.3	639.4
Errors and omissions	-69.2	-181.2	84.7	60.2	128.1
Overall balance	64.1	-39.3	86.1	0.4	-79.4

Exchange Rates [38]

Currency: **gold cordoba.**
Symbol: **C$.**

Data are currency units per $1.

December 1995	7.98
1994	6.72
1993	5.62
1992	5.00
1991	4.27

Imports and Exports

Top Import Origins [39]

$870 million (c.i.f., 1995 est.)

Origins	%
Central America	NA
US	NA
Venezuela	NA
Japan	NA

Top Export Destinations [40]

$525.5 million (f.o.b., 1995 est.).

Destinations	%
US	NA
Central America	NA
Canada	NA
Germany	NA

Foreign Aid [41]

Recipient: ODA, $NA.

Import and Export Commodities [42]

Import Commodities	Export Commodities
Consumer goods	Meat
Machinery and equipment	Coffee
Petroleum products	Cotton
	Sugar
	Seafood
	Gold
	Bananas

Niger

Geography [1]

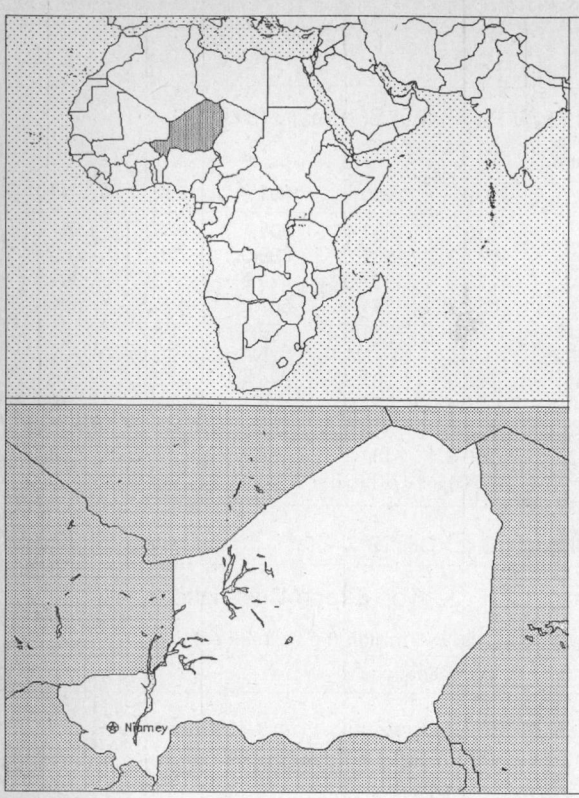

Total area:
1,267,000 sq km   489,191 sq mi
Land area:
1,266,700 sq km   489,076 sq mi
Comparative area:
Slightly less than twice the size of Texas
Land boundaries:
Total 5,697 km, Algeria 956 km, Benin 266 km, Burkina Faso 628 km,
Chad 1,175 km, Libya 354 km, Mali 821 km, Nigeria 1,497 km
Coastline:
0 km (landlocked)
Climate:
Desert; mostly hot, dry, dusty; tropical in extreme South
Terrain:
Predominately desert plains and sand dunes; flat to rolling plains
in South; hills in North
Natural resources:
Uranium, coal, iron ore, tin, phosphates
Land use:
Arable land: 3%
Permanent crops: 0%
Meadows and pastures: 7%
Forest and woodland: 2%
Other: 88%

Demographics [2]

	1970	1980	1990	1995[1]	1996	2000	2010	2020	2030
Population	4,182	5,629	7,644	8,844	9,113	10,260	13,678	17,983	22,993
Population density (persons per sq. mi.)	9	12	16	18	19	21	28	37	47
(persons per sq. km.)	3	4	6	7	7	8	11	14	18
Net migration rate (per 1,000 population)	NA	NA	0.0	0.0	0.0	0.0	0.0	0.0	0.0
Births	NA	NA	NA	NA	496	NA	NA	NA	NA
Deaths	NA	NA	NA	NA	224	NA	NA	NA	NA
Life expectancy - males	NA	NA	38.9	40.7	41.1	42.6	46.9	51.5	56.1
Life expectancy - females	NA	NA	37.6	39.8	40.3	42.2	47.7	53.8	60.1
Birth rate (per 1,000)	NA	NA	55.9	55.2	54.5	51.6	45.3	39.2	32.5
Death rate (per 1,000)	NA	NA	28.0	25.1	24.6	22.2	17.1	13.0	9.7
Women of reproductive age (15-49 yrs.)	NA	NA	1,751	2,013	2,066	2,296	3,079	4,208	5,659
of which are currently married	NA	NA	1,478	NA	1,753	1,951	2,598	NA	NA
Fertility rate	NA	NA	7.5	7.5	7.4	7.2	6.3	5.2	4.1

Except as noted, values for vital statistics are in thousands; life expectancy is in years.

Health

Health Indicators [3]

% of population with access to	
safe water (1990-95)	54
adequate sanitation (1990-95)	15
health services (1985-95)	32
% of 1-year-olds immunized (1990-94) against	
TB (tuberculosis)	32
DPT (diphtheria, pertussis, tetanus)	20
polio	20
measles	19
% of contraceptive prevalence (1980-94)	4
ORT use rate (1990-94)	17

Health Expenditures [4]

Total health expenditure, 1990 (official exchange rate)	
Millions of dollars	126
Dollars per capita	16
Health expenditures as a percentage of GDP	
Total	5.0
Public sector	3.4
Private sector	1.6
Development assistance for health	
Total aid flows (millions of dollars)[1]	43
Aid flows per capita (dollars)	5.6
Aid flows as a percentage of total health expenditure	34.0

For sources, notes, and explanations, see Annotated Source Appendix, page 1061.

Human Factors

Women and Children [5]

% of pregnant women immunized (tetanus 1990-94)	44
% of births attended by trained health personnel (1983-94)	15
Maternal mortality rate (1980-92)	590
Under-5 mortality rate (1994)	320
% under-5 moderately/severely underweight (1980-1994)	36

Burden of Disease [6]

Population per physician (1990)	34,845.45
Population per nurse (1990)	653.82
Population per hospital bed	NA
AIDS cases per 100,000 people (1994)	5.8
Malaria cases per 100,000 people (1992)	NA

Ethnic Division [7]

There are also about 4,000 French expatriates.

Hausa	56%
Djerma	22%
Fula	8.5%
Tuareg	8%
Beri (Kanouri)	4.3%
Arab, Toubou, and Gourmantche	1.2%

Religion [8]

Muslim	80%
Indigenous beliefs and Christians	20%

Major Languages [9]

French (official), Hausa, Djerma.

Education

Public Education Expenditures [10]

Million (Franc C.F.A.)[49]	1980	1989[2]	1990	1991	1993	1994
Total education expenditure	16,533	19,873	NA	20,143	NA	NA
as percent of GNP	3.1	2.9	NA	3.1	NA	NA
as percent of total govt. expend.	22.9	9.0	NA	10.8	NA	NA
Current education expenditure	7,763	15,545	NA	19,493	NA	NA
as percent of GNP	1.5	2.3	NA	3.0	NA	NA
as percent of current govt. expend.	16.8	13.6	NA	18.4	NA	NA
Capital expenditure	8,770	4,328	NA	650	NA	NA

Educational Attainment [11]

Illiteracy [12]

In thousands and percent[1]	1990	1995	2000
Illiterate population (15+ yrs.)	3,576	4,081	4,672
Illiteracy rate - total pop. (%)	88.8	88.5	87.7
Illiteracy rate - males (%)	82.2	81.0	78.9
Illiteracy rate - females (%)	95.0	95.6	96.3

Libraries [13]

	Admin. Units	Svc. Pts.	Vols. (000)	Shelving (meters)	Vols. Added	Reg. Users
National	NA	NA	NA	NA	NA	NA
Nonspecialized	NA	NA	NA	NA	NA	NA
Public	NA	NA	NA	NA	NA	NA
Higher ed. (1987)	1	1	21	NA	638	1,392
School	NA	NA	NA	NA	NA	NA

Daily Newspapers [14]

	1980	1985	1990	1994
Number of papers	1	1	1	4
Circ. (000)	3	4	5	11

Culture [15]

Cinema (seats per 1,000)	NA
Annual attendance per person	NA
Gross box office receipts (mil. Franc C.F.A.)	NA
Museums (reporting)	1
Visitors (000)	450
Annual receipts (000 Franc C.F.A.)	30,000

Science and Technology

Scientific/Technical Forces [16]

R&D Expenditures [17]

U.S. Patents Issued [18]

For sources, notes, and explanations, see Annotated Source Appendix, page 1061.

697

Government and Law

Organization of Government [19]

Long-form name:
Republic of Niger
Type:
Republic
Independence:
3 August 1960 (from France)
National holiday:
Republic Day, 18 December (1958)
Constitution:
The constitution of January 1993 was revised by national referendum on 12 May 1996
Legal system:
Based on French civil law system and customary law; has not accepted compulsory ICJ jurisdiction
Executive branch:
President; Prime Minister; National Salvation Council
Legislative branch:
Unicameral: National Assembly
Judicial branch:
State Court (Cour d'Etat); Court of Appeal (Cour d'Apel)

Elections [20]

	% of seats
National Movement of the Development Society	34.9
Democ. & Social Convention	28.9
Nigerian Party for Democracy & Socialism	14.4
Nigerian Alliance for Democracy & Progress	10.8
Other	11.0

Government Budget [21]

For 1993 est.

	$ mil.
Revenues	188
Expenditures	400
Capital expenditures	125

Crime [23]

Military Expenditures and Arms Transfers [22]

	1990	1991	1992	1993	1994
Military expenditures					
Current dollars (mil.)	NA	19	18[e]	14	14
1994 constant dollars (mil.)	NA	20	19[e]	15	14
Armed forces (000)	5	5	5	5	7
Gross national product (GNP)					
Current dollars (mil.)	1,348	1,428	1,372	1,422	1,507
1994 constant dollars (mil.)	1,500	1,530	1,431	1,451	1,507
Central government expenditures (CGE)					
1994 constant dollars (mil.)	301	237[e]	NA	265[e]	NA
People (mil.)	7.9	8.1	8.4	8.7	9.0
Military expenditure as % of GNP	NA	1.3	1.3	1.0	0.9
Military expenditure as % of CGE	NA	8.4	NA	5.5	NA
Military expenditure per capita (1994 $)	NA	2	2	2	2
Armed forces per 1,000 people (soldiers)	0.6	0.6	0.6	0.6	0.8
GNP per capita (1994 $)	191	188	170	167	168
Arms imports[6]					
Current dollars (mil.)	5	0	0	0	0
1994 constant dollars (mil.)	6	0	0	0	0
Arms exports[6]					
Current dollars (mil.)	0	0	0	0	0
1994 constant dollars (mil.)	0	0	0	0	0
Total imports[7]					
Current dollars (mil.)	389	355	477[e]	503[e]	NA
1994 constant dollars (mil.)	433	381	497[e]	513[e]	NA
Total exports[7]					
Current dollars (mil.)	283	312	185[e]	240[e]	NA
1994 constant dollars (mil.)	315	334	193[e]	245[e]	NA
Arms as percent of total imports[8]	1.3	0	0	0	0
Arms as percent of total exports[8]	0	0	0	0	0

Human Rights [24]

	SSTS	FL	FAPRO	PPCG	APROBC	TPW	PCPTW	STPEP	PHRFF	PRW	ASST	AFL
Observes	P	P	P		P	P	P	P		P	P	P
	EAFRD	CPR	ESCR	SR	ACHR	MAAE	PVIAC	PVNAC	EAFDAW	TCIDTP		RC
Observes		P	P	P	P		P	P	P			P

P = Party; S = Signatory; see Appendix for meaning of abbreviations.

Labor Force

Total Labor Force [25]

2.5 million wage earners (1982)

Labor Force by Occupation [26]

Agriculture	90%
Industry and commerce	6
Government	4

Unemployment Rate [27]

For sources, notes, and explanations, see Annotated Source Appendix, page 1061.

Production Sectors

Commercial Energy Production and Consumption

Data are shown in quadrillion (10^{15}) BTUs and percent for 1995
Values for hydroelectric, nuclear, geothermal, solar, and wind power refer to electrical generation.

Production [28]

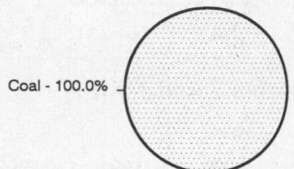

Coal - 100.0%

Consumption [29]

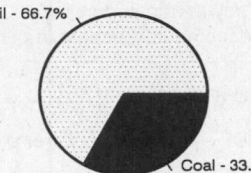

Crude oil - 66.7%

Coal - 33.3%

Coal	0.005
Total	0.005

Crude oil	0.010
Coal	0.005
Total	0.015

Telecommunications [30]

- 14,000 (1991 est.) telephones; small system of wire, radiotelephone communications, and microwave radio relay links concentrated in southwestern area
- Domestic: wire, radiotelephone communications, and microwave radio relay; domestic satellite system with 3 earth stations and 1 planned
- International: satellite earth stations - 2 Intelsat (1 Atlantic Ocean and 1 Indian Ocean)
- Radio: Broadcast stations: AM 15, FM 5, shortwave 0 Radios: 500,000 (1992 est.)
- Television: Broadcast stations: 18 Televisions: 38,000 (1992 est.)

Top Agricultural Products [32]

Agriculture accounts for 38.5% of the GDP; produces cowpeas, cotton, peanuts, millet, sorghum, cassava (tapioca), rice; cattle, sheep, goats.

Top Mining Products [33]

Metric tons except as noted[e]	3/24/95[*]
Cement, hydraulic	29,200
Coal, bituminous	133,500
Gypsum	1,700
Molybdenum concentrate, Mo content	10
Salt	3,000
Tin, mine output, Sn content	20
Uranium, U308 content of concentrate	2,900

Transportation [31]

Railways: 0 km

Highways: total: 11,258 km; paved: 3,265 km; unpaved: 7,993 km (1990 est.)

Airports

Total:	23
With paved runways 2,438 to 3,047 m:	2
With paved runways 1,524 to 2,437 m:	6
With paved runways 914 to 1,523 m:	1
With paved runways under 914 m:	2
With unpaved runways 1,524 to 2,437 m:	1

Tourism [34]

	1990	1991	1992	1993	1994
Tourists[83]	21	16	13	11	11
Tourism receipts	17	16	17	16	16
Tourism expenditures	44	40	30	29	21
Fare receipts	18	18	10	10	5
Fare expenditures	40	43	45	43	23

Travelers are in thousands, money in million U.S. dollars.

For sources, notes, and explanations, see Annotated Source Appendix, page 1061.

699

Finance, Economics, and Trade

GDP and Manufacturing Summary [35]

	1980	1985	1990	1991	1992
Gross Domestic Product					
Millions of 1980 dollars	2,538	2,473	2,639	2,675	2,718[e]
Growth rate in percent	4.90	5.70	2.80	1.35	1.60[e]
Manufacturing Value Added					
Millions of 1980 dollars	94	100	98	99	101[e]
Growth rate in percent	4.68	8.25	1.70	1.73	1.27[e]
Manufacturing share in percent of current prices	3.8	7.4	8.9[e]	NA	NA

Economic Indicators [36]

- **National product**: GDP—purchasing power parity—$5.5 billion (1995 est.)
- **National product real growth rate**: 6.7% (1995 est.)
- **National product per capita**: $600 (1995 est.)
- **Inflation rate (consumer prices)**: 35.6% (1994 est.)
- **External debt**: $1.41 billion (1995 est.)

Balance of Payments Summary [37]

Values in millions of dollars.

	1989	1990	1991	1992	1993
Exports of goods (f.o.b.)	311.0	303.4	283.9	265.6	238.4
Imports of goods (f.o.b.)	-368.6	-337.5	-273.3	-266.3	-244.0
Trade balance	-57.6	-34.1	10.6	-0.7	-5.6
Services - debits	-202.5	-256.4	-195.7	-185.5	-175.5
Services - credits	57.7	71.3	58.5	52.1	49.8
Private transfers (net)	-40.4	-48.8	-37.9	-39.7	-33.9
Government transfers (net)	132.0	184.4	160.6	129.2	136.3
Long-term capital (net)	95.9	105.8	5.3	32.4	-2.5
Short-term capital (net)	NA	-6.6	34.4	41.4	30.8
Errors and omissions	-4.1	-25.2	-40.4	15.6	-9.4
Overall balance	-19.0	-9.6	-4.6	44.8	-10.0

Exchange Rates [38]

Currency: **Communaute Financiere Africaine franc.**
Symbol: **CFAF.**

Data are currency units per $1.

January 1996	500.56
1995	499.15
1994	555.20
1993	283.16
1992	264.69
1991	282.11

Imports and Exports

Top Import Origins [39]

$234 million (c.i.f., 1994 est.).

Origins	%
France	23
Cote d'Ivoire	NA
Germany	NA
Italy	NA
Japan	NA

Top Export Destinations [40]

$232 million (f.o.b., 1994 est.).

Destinations	%
France	77
Nigeria	8
Cote d'Ivoire	NA
Italy	NA

Foreign Aid [41]

Recipient: ODA, $NA.

Import and Export Commodities [42]

Import Commodities	Export Commodities
Consumer goods	Uranium ore 67%
Primary materials	Livestock products 20%
Machinery	Cowpeas
Vehicles and parts	Onions
Petroleum	
Cereals	

Nigeria

Geography [1]

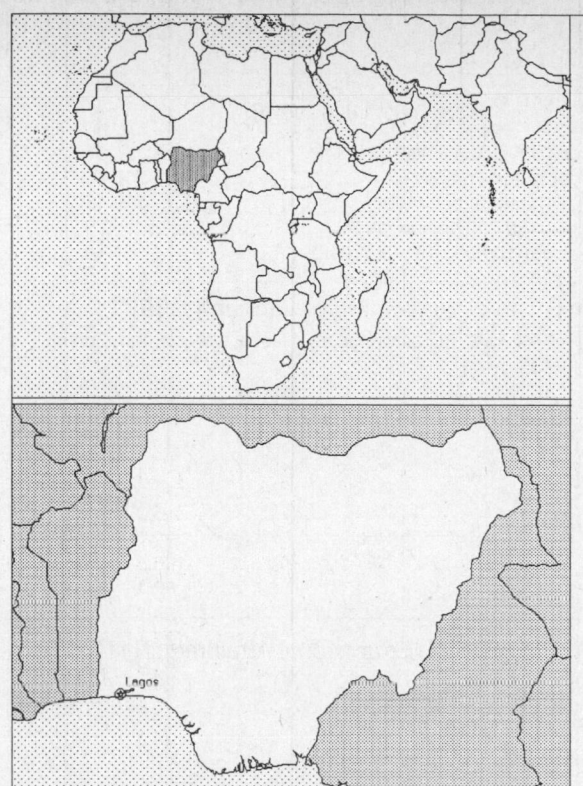

Total area:
923,770 sq km   356,670 sq mi
Land area:
910,770 sq km   351,650 sq mi
Comparative area:
Slightly more than twice the size of California
Land boundaries:
Total 4,047 km, Benin 773 km, Cameroon 1,690 km, Chad 87 km, Niger 1,497 km
Coastline:
853 km
Climate:
Varies; equatorial in South, tropical in center, arid in North
Terrain:
Southern lowlands merge into central hills and plateaus; mountains in Southeast, plains in North
Natural resources:
Petroleum, tin, columbite, iron ore, coal, limestone, lead, zinc, natural gas
Land use:
Arable land: 31%
Permanent crops: 3%
Meadows and pastures: 23%
Forest and woodland: 15%
Other: 28%

Demographics [2]

	1970	1980	1990	1995[1]	1996	2000	2010	2020	2030
Population	49,309	65,699	86,488	100,785	103,912	117,328	157,375	205,160	260,037
Population density (persons per sq. mi.)	140	187	246	287	295	334	448	583	739
(persons per sq. km.)	54	72	95	111	114	129	173	225	286
Net migration rate (per 1,000 population)	0.8	5.3	0.8	0.4	0.3	0.3	0.2	0.0	0.0
Births	NA	NA	NA	NA	4,457	NA	NA	NA	NA
Deaths	NA	NA	NA	NA	1,321	NA	NA	NA	NA
Life expectancy - males	41.4	46.4	50.9	52.8	53.1	54.1	56.9	59.8	66.5
Life expectancy - females	41.3	47.2	53.0	55.3	55.7	57.2	62.5	65.4	72.5
Birth rate (per 1,000)	47.2	46.8	44.8	43.2	42.9	41.5	37.3	32.7	28.1
Death rate (per 1,000)	22.8	18.4	14.6	13.0	12.7	11.7	9.2	7.9	5.4
Women of reproductive age (15-49 yrs.)	11,335	14,914	18,939	22,142	22,866	25,939	36,142	50,005	67,042
of which are currently married	NA	NA	14,375	NA	17,236	19,517	27,263	NA	NA
Fertility rate	6.6	6.6	6.6	6.3	6.2	6.0	5.1	4.2	3.4

Except as noted, values for vital statistics are in thousands; life expectancy is in years.

Health

Health Indicators [3]

% of population with access to	
safe water (1990-95)	40
adequate sanitation (1990-95)	35
health services (1985-95)	66
% of 1-year-olds immunized (1990-94) against	
TB (tuberculosis)	46
DPT (diphtheria, pertussis, tetanus)	41
polio	35
measles	41
% of contraceptive prevalence (1980-94)	6
ORT use rate (1990-94)	35

Health Expenditures [4]

Total health expenditure, 1990 (official exchange rate)	
Millions of dollars	906
Dollars per capita	9
Health expenditures as a percentage of GDP	
Total	2.7
Public sector	1.2
Private sector	1.6
Development assistance for health	
Total aid flows (millions of dollars)[1]	58
Aid flows per capita (dollars)	0.6
Aid flows as a percentage of total health expenditure	6.4

For sources, notes, and explanations, see Annotated Source Appendix, page 1061.

701

Human Factors

Women and Children [5]

% of pregnant women immunized (tetanus 1990-94)	38
% of births attended by trained health personnel (1983-94)	37
Maternal mortality rate (1980-92)	800
Under-5 mortality rate (1994)	191
% under-5 moderately/severely underweight (1980-1994)	36

Burden of Disease [6]

Population per physician (1986)	5,356.37
Population per nurse (1986)	750.86
Population per hospital bed (1990)	599.46
AIDS cases per 100,000 people (1994)	0.5
Malaria cases per 100,000 people (1992)	NA

Ethnic Division [7]

Other includes other Africans and some 27,000 non-Africans.

Hausa, Fulani, Yoruba, and Ibos	65%
Other	35%

Religion [8]

Muslim	50%
Christian	40%
Indigenous beliefs	10%

Major Languages [9]

English (official), Hausa, Yoruba, Ibo, Fulani.

Education

Public Education Expenditures [10]

Million (Naira)[50]	1980	1990	1991	1992	1993	1994
Total education expenditure	NA	2,121	1,558	2,405	7,999	NA
as percent of GNP	NA	0.9	0.5	0.5	1.3	NA
as percent of total govt. expend.	NA	5.3	4.1	6.3	7.3	NA
Current education expenditure	NA	NA	NA	NA	NA	NA
as percent of GNP	NA	NA	NA	NA	NA	NA
as percent of current govt. expend.	NA	NA	NA	NA	NA	NA
Capital expenditure	NA	NA	NA	NA	NA	NA

Educational Attainment [11]

Illiteracy [12]

In thousands and percent[1]	1990	1995	2000
Illiterate population (15+ yrs.)	26,562	26,075	25,171
Illiteracy rate - total pop. (%)	55.7	47.0	38.9
Illiteracy rate - males (%)	41.3	34.4	28.2
Illiteracy rate - females (%)	71.0	60.2	49.9

Libraries [13]

	Admin. Units	Svc. Pts.	Vols. (000)	Shelving (meters)	Vols. Added	Reg. Users
National (1992)	1	12	865	NA	16,384	34,373
Nonspecialized	NA	NA	NA	NA	NA	NA
Public (1992)[7]	12	76	611	NA	24,895	14,927
Higher ed. (1987)	63	144	3,842	NA	105,057	139,938
School (1987)	213	16,714	1,088	-	16,100	13,763

Daily Newspapers [14]

	1980	1985	1990	1994
Number of papers	16	19	31	27
Circ. (000)	1,100[e]	1,400[e]	1,700[e]	1,950[e]

Culture [15]

Cinema (seats per 1,000)	NA
Annual attendance per person	NA
Gross box office receipts (mil. Naira)	NA
Museums (reporting)	12
Visitors (000)	267
Annual receipts (000 Naira)	19

Science and Technology

Scientific/Technical Forces [16]

Scientists/engineers	1,338
Number female	NA
Technicians	6,042
Number female	NA
Total[28]	7,380

R&D Expenditures [17]

	Naira (000) 1987
Total expenditure[27]	86,270
Capital expenditure	16,655
Current expenditure	69,615
Percent current	80.7

U.S. Patents Issued [18]

Values show patents issued to citizens of the country by the U.S. Patents Office.

	1993	1994	1995
Number of patents	0	1	0

For sources, notes, and explanations, see Annotated Source Appendix, page 1061.

Government and Law

Organization of Government [19]

Long-form name:
Federal Republic of Nigeria

Type:
Military government; Nigeria has been ruled by one military regime after another since 31 December 1983

Independence:
1 October 1960 (from UK)

National holiday:
Independence Day, 1 October (1960)

Constitution:
1979 constitution still in force; plan for 1989 constitution to take effect in 1993 was not implemented

Legal system:
Based on English common law, Islamic law, and tribal law

Executive branch:
Chairman of the Provisional Ruling Council and Commander in Chief of Armed Forces and Defense Minister; Vice Chairman of the Provisional Ruling Council; Federal Executive Council

Legislative branch:
Bicameral National Assembly: Senate and House of Representatives: Note: Suspended after military coup of 1993

Judicial branch:
Supreme Court; Federal Court of Appeal

Crime [23]

Elections [20]

House of Representatives Suspended after coup of 17 November 1993; political parties also suspended.

Government Budget [21]

For 1994 est.

	$ bil.
Revenues	2.7
Expenditures	6.4
Capital expenditures	1.8

Military Expenditures and Arms Transfers [22]

	1990	1991	1992	1993	1994
Military expenditures					
Current dollars (mil.)[3]	300[e]	294	228	288	324
1994 constant dollars (mil.)[3]	334[e]	315	238	294	324
Armed forces (000)	94	94	76	76	80
Gross national product (GNP)					
Current dollars (mil.)	32,080	35,880	38,020	39,360	41,270
1994 constant dollars (mil.)	35,700	38,460	39,650	40,170	41,270
Central government expenditures (CGE)					
1994 constant dollars (mil.)	9,320	8,816	8,519	11,080	6,438
People (mil.)	86.6	89.3	92.1	95.1	98.1
Military expenditure as % of GNP	0.9	0.8	0.6	0.7	0.8
Military expenditure as % of CGE	3.6	3.6	2.8	2.7	5.0
Military expenditure per capita (1994 $)	4	4	3	3	3
Armed forces per 1,000 people (soldiers)	1.1	1.1	0.8	0.8	0.8
GNP per capita (1994 $)	412	431	430	423	421
Arms imports[6]					
Current dollars (mil.)	20	130	120	50	0
1994 constant dollars (mil.)	22	139	125	51	0
Arms exports[6]					
Current dollars (mil.)	0	0	0	0	0
1994 constant dollars (mil.)	0	0	0	0	0
Total imports[7]					
Current dollars (mil.)	5,688	9,031	8,119	7,508	5,435[e]
1994 constant dollars (mil.)	6,330	9,680	8,467	7,663	5,435[e]
Total exports[7]					
Current dollars (mil.)	13,670	12,260	11,890	9,914	NA
1994 constant dollars (mil.)	15,210	13,150	12,390	10,120	NA
Arms as percent of total imports[8]	0.4	1.4	1.5	0.7	0
Arms as percent of total exports[8]	0	0	0	0	0

Human Rights [24]

	SSTS	FL	FAPRO	PPCG	APROBC	TPW	PCPTW	STPEP	PHRFF	PRW	ASST	AFL
Observes	P	P	P		P	P	P			P	P	P
	EAFRD	CPR	ESCR	SR	ACHR	MAAE	PVIAC	PVNAC	EAFDAW	TCIDTP	RC	
Observes	P	P	P	P			P	P	P	S	P	

P = Party; S = Signatory; see Appendix for meaning of abbreviations.

Labor Force

Total Labor Force [25]

42.844 million

Labor Force by Occupation [26]

Agriculture	54%
Industry, commerce, and services	19
Government	15

Unemployment Rate [27]

28% (1992 est.)

For sources, notes, and explanations, see Annotated Source Appendix, page 1061.

Production Sectors

Commercial Energy Production and Consumption

Data are shown in quadrillion (10^{15}) BTUs and percent for 1995
Values for hydroelectric, nuclear, geothermal, solar, and wind power refer to electrical generation.

Production [28]

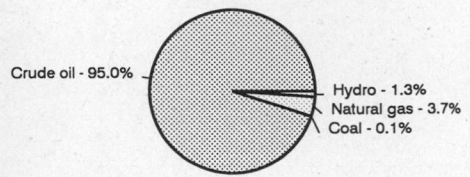

Crude oil - 95.0%
Hydro - 1.3%
Natural gas - 3.7%
Coal - 0.1%

Consumption [29]

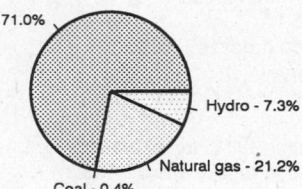

Crude oil - 71.0%
Hydro - 7.3%
Natural gas - 21.2%
Coal - 0.4%

Crude oil	4.277
Dry natural gas	0.165
Coal	0.003
Net hydroelectric power	0.057
Total	4.502

Crude oil	0.552
Dry natural gas	0.165
Coal	0.003
Net hydroelectric power	0.057
Total	0.777

Telecommunications [30]

- 492,204 (1990 est.) telephones; average system limited by poor maintenance; major expansion in progress
- Domestic: microwave radio relay, coaxial cable, and 20 domestic satellite earth stations carry intercity traffic
- International: satellite earth stations - 3 Intelsat (2 Atlantic Ocean and 1 Indian Ocean); 1 coaxial submarine cable
- Radio: Broadcast stations: AM 35, FM 17, shortwave 0 Radios: 20 million (1992 est.)
- Television: Broadcast stations: 28 Televisions: 3.8 million (1992 est.)

Transportation [31]

Railways: total: 3,557 km (1995); narrow gauge: 3,505 km 1.067-m gauge; standard gauge: 52 km 1.435-m gauge

Highways: total: 112,140 km; paved: 31,500 km; unpaved: 80,640 km (1991 est.)

Merchant marine: total: 33 ships (1,000 GRT or over) totaling 387,552 GRT/636,578 DWT; ships by type: bulk 1, cargo 16, chemical tanker 3, oil tanker 12, roll-on/roll-off cargo 1 (1995 est.)

Airports

Total:	66
With paved runways over 3,047 m:	6
With paved runways 2,438 to 3,047 m:	10
With paved runways 1,524 to 2,437 m:	10
With paved runways 914 to 1,523 m:	8
With paved runways under 914 m:	18

Top Agricultural Products [32]

Agriculture accounts for 38% of the GDP; produces cocoa, peanuts, palm oil, rubber, corn, rice, sorghum, millet, cassava (tapioca), yams; cattle, sheep, goats, pigs; fishing and forest resources extensively exploited.

Top Mining Products [33]

Thousand metric tons except as noted[e]	7/14/95[*]
Iron ore, gross weight	400
Cement, hydraulic	2,600[r]
Nitrogen	
N content of ammonia	350
N content of urea	400
Limestone	2,700[r]
Marble	7,300[r]
Coal, bituminous	140
Gas, natural, gross (mil. cu. meters)	31,300
Petroleum, crude (000 42-gal. bls.)	715,000

Tourism [34]

	1990	1991	1992	1993	1994
Visitors[1]	303	322	342	294	327
Tourists	190	214	237	192	NA
Tourism receipts	25	39	29	31	34
Tourism expenditures	576	839	348	234	144
Fare receipts	5	19	22	46	15
Fare expenditures	34	44	70	56	24

Travelers are in thousands, money in million U.S. dollars.

Manufacturing Sector

GDP and Manufacturing Summary [35]

	1980	1985	1989	1990	% change 1980-1990	% change 1989-1990
GDP (million 1980 $)	23,795	21,341	24,009	27,976	17.6	16.5
GDP per capita (1980 $)	303	232	229	325[e]	7.3	41.9
Manufacturing as % of GDP (current prices)	4.7	6.1[e]	7.6[e]	NA	NA	-100.0
Gross output (million $)	4,740	3,454	4,294[e]	5,797[e]	22.3	35.0
Value added (million $)	2,422	1,667	2,283[e]	3,606[e]	48.9	58.0
Value added (million 1980 $)	1,161	1,078	903	1,344	15.8	48.8
Industrial production index	100	85	83	90[e]	-10.0	8.4
Employment (thousands)	432	330	363[e]	418[e]	-3.2	15.2

Note: GDP stands for Gross Domestic Product. 'e' stands for estimated value.

Profitability and Productivity

	1980	1985	1989	1990	% change 1980-1990	% change 1989-1990
Intermediate input (%)	49	52	47[e]	38[e]	-22.4	-19.1
Wages, salaries, and supplements (%)	11[e]	10[e]	10[e]	10[e]	-9.1	0.0
Gross operating surplus (%)	40[e]	38[e]	43[e]	52[e]	30.0	20.9
Gross output per worker ($)	10,273	10,005	11,819[e]	13,472[e]	31.1	14.0
Value added per worker ($)	5,211	4,844	6,283[e]	8,657[e]	66.1	37.8
Average wage (incl. benefits) ($)	1,226[e]	1,037[e]	1,202[e]	1,422[e]	16.0	18.3

Profitability is in percent of gross output. Productivity is in U.S. $. 'e' stands for estimated value.

Profitability - 1990

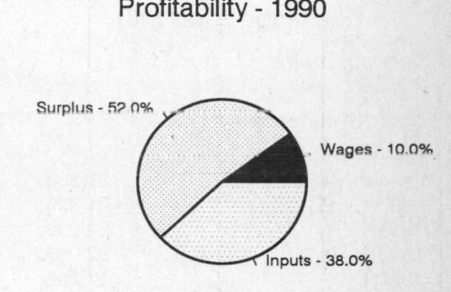

Surplus - 52.0%
Wages - 10.0%
Inputs - 38.0%

The graphic shows percent of gross output.

Value Added in Manufacturing

	1980 $ mil.	1980 %	1985 $ mil.	1985 %	1989 $ mil.	1989 %	1990 $ mil.	1990 %	% change 1980-1990	% change 1989-1990
311 Food products	149	6.2	251	15.1	288[e]	12.6	505[e]	14.0	238.9	75.3
313 Beverages	267	11.0	173[e]	10.4	453[e]	19.8	432[e]	12.0	61.8	-4.6
314 Tobacco products	96	4.0	32[e]	1.9	38[e]	1.7	71[e]	2.0	-26.0	86.8
321 Textiles	231	9.5	233	14.0	370[e]	16.2	449[e]	12.5	94.4	21.4
322 Wearing apparel	3	0.1	1	0.1	5[e]	0.2	2[e]	0.1	-33.3	-60.0
323 Leather and fur products	12	0.5	23	1.4	31[e]	1.4	47[e]	1.3	291.7	51.6
324 Footwear	12	0.5	28	1.7	22[e]	1.0	61[e]	1.7	408.3	177.3
331 Wood and wood products	88	3.6	14	0.8	6[c]	0.3	26[e]	0.7	-70.5	333.3
332 Furniture and fixtures	56	2.3	14	0.8	20[e]	0.9	33[e]	0.9	-41.1	65.0
341 Paper and paper products	38	1.6	51	3.1	62[e]	2.7	115[e]	3.2	202.6	85.5
342 Printing and publishing	75	3.1	45	2.7	60[e]	2.6	107[e]	3.0	42.7	78.3
351 Industrial chemicals	30	1.2	9	0.5	14[e]	0.6	19[e]	0.5	-36.7	35.7
352 Other chemical products	265	10.9	213	12.8	179[e]	7.8	461[e]	12.8	74.0	157.5
353 Petroleum refineries	75[e]	3.1	-7[e]	-0.4	70[e]	3.1	40[e]	1.1	-46.7	-42.9
354 Miscellaneous petroleum and coal products	3[e]	0.1	NA	0.0	7[e]	0.3	2[e]	0.1	-33.3	-71.4
355 Rubber products	26	1.1	31	1.9	35[e]	1.5	67[e]	1.9	157.7	91.4
356 Plastic products	98	4.0	49	2.9	30[e]	1.3	107[e]	3.0	9.2	256.7
361 Pottery, china, and earthenware	NA	0.0	2	0.1	NA	0.0	2[e]	0.1	NA	NA
362 Glass and glass products	24	1.0	7	0.4	14[e]	0.6	18[e]	0.5	-25.0	28.6
369 Other nonmetal mineral products	87	3.6	106	6.4	136[e]	6.0	228[e]	6.3	162.1	67.6
371 Iron and steel	3	0.1	17	1.0	38[e]	1.7	25[e]	0.7	733.3	-34.2
372 Nonferrous metals	33	1.4	27	1.6	57[e]	2.5	73[e]	2.0	121.2	28.1
381 Metal products	140	5.8	92	5.5	176[e]	7.7	201[e]	5.6	43.6	14.2
382 Nonelectrical machinery	23	0.9	19	1.1	28[e]	1.2	42[e]	1.2	82.6	50.0
383 Electrical machinery	46	1.9	36	2.2	52[e]	2.3	78[e]	2.2	69.6	50.0
384 Transport equipment	526	21.7	193	11.6	82[e]	3.6	386[e]	10.7	-26.6	370.7
385 Professional and scientific equipment	NA	0.0	NA	0.0	NA	0.0	1[e]	0.0	NA	NA
390 Other manufacturing industries	13	0.5	6	0.4	10[e]	0.4	10[e]	0.3	-23.1	0.0

Note: The industry codes shown are International Standard Industry codes (ISIC). Percentages are percent of total Value Added. 'e' stands for estimated value

For sources, notes, and explanations, see Annotated Source Appendix, page 1061.

705

Finance, Economics, and Trade

Economic Indicators [36]

- **National product**: GDP—purchasing power parity—$135.9 billion (1995 est.)
- **National product real growth rate**: 2.6% (1995 est.)
- **National product per capita**: $1,300 (1995 est.)
- **Inflation rate (consumer prices)**: 57% (1994 est.)
- **External debt**: $32.5 billion (1993)

Balance of Payments Summary [37]

Values in millions of dollars.

	1988	1989	1990	1991	1992
Exports of goods (f.o.b.)	6,897	7,870	13,585	12,254	11,791
Imports of goods (f.o.b.)	-4,271	-3,692	-4,932	-7,813	-7,181
Trade balance	2,626	4,178	8,653	4,441	4,610
Services - debits	-3,212	-3,918	-4,926	-5,080	-4,304
Services - credits	404	704	1,176	1,097	1,208
Private transfers (net)	-33	-19	1	12	22
Government transfers (net)	21	145	84	732	731
Long-term capital (net)	-2,426	-1,170	-1,248	-2,437	-2,399
Short-term capital (net)	2,398	1,374	-1,497	1,967	-3,474
Errors and omissions	-215	-110	235	-92	-122
Overall balance	-437	1,184	2,478	640	-3,728

Exchange Rates [38]

Currency: **naira.**
Symbol: **N.**

Data are currency units per $1.

January 1996	21.886
1995	21.895
1994	21.996
1993	22.065
1992	17.298
1991	9.909

Imports and Exports

Top Import Origins [39]

$7.5 billion (c.i.f., 1993).

Origins	%
EU	50
US	13
Japan	7

Top Export Destinations [40]

$9.9 billion (f.o.b., 1993).

Destinations	%
US	52
EU	34

Foreign Aid [41]

Recipient: ODA, $NA.

Import and Export Commodities [42]

Import Commodities	Export Commodities
Machinery	Oil 98%
Transportation equipment	Cocoa
Manufactured goods	Rubber
Chemicals	
Food and animals	

Norway

Geography [1]

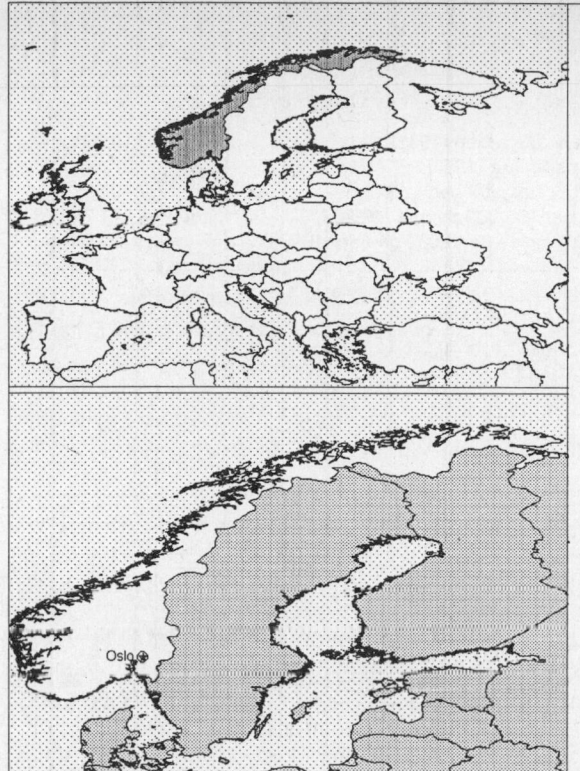

Total area:
324,220 sq km   125,182 sq mi
Land area:
307,860 sq km   118,865 sq mi
Comparative area:
Slightly larger than New Mexico
Land boundaries:
Total 2,515 km, Finland 729 km, Sweden 1,619 km, Russia 167 km
Coastline:
21,925 km (includes mainland 3,419 km, large islands 2,413 km, long fjords, numerous small islands, and minor indentations 16,093 km)
Climate:
Temperate along coast, modified by North Atlantic Current; colder interior; rainy year-round on West Coast
Terrain:
Glaciated; mostly high plateaus and rugged mountains broken by fertile valleys; small, scattered plains; coastline deeply indented by fjords; arctic tundra in North
Natural resources:
Petroleum, copper, natural gas, pyrites, nickel, iron ore, zinc, lead, fish, timber, hydropower
Land use:
Arable land: 3%
Permanent crops: 0%
Meadows and pastures: 0%
Forest and woodland: 27%
Other: 70%

Demographics [2]

	1970	1980	1990	1995[1]	1996	2000	2010	2020	2030
Population	3,877	4,086	4,242	4,362	4,384	4,461	4,577	4,632	4,582
Population density (persons per sq. mi.)	33	34	36	37	37	38	39	39	39
(persons per sq. km.)	13	13	14	14	14	14	15	15	15
Net migration rate (per 1,000 population)	2.5	0.9	0.4	3.4	3.6	3.2	1.6	0.5	0.0
Births	65	51	NA	NA	54	NA	NA	NA	NA
Deaths	39	41	NA	NA	45	NA	NA	NA	NA
Life expectancy - males	NA	NA	73.1	74.5	74.6	75.2	76.5	78.3	79.6
Life expectancy - females	NA	NA	80.1	80.5	80.6	81.1	82.1	84.1	85.5
Birth rate (per 1,000)	16.7	12.5	15.0	12.5	12.0	11.0	9.8	9.6	8.4
Death rate (per 1,000)	10.0	10.1	10.9	10.8	10.7	10.4	9.9	9.6	11.2
Women of reproductive age (15-49 yrs.)	NA	NA	1,056	1,076	1,074	1,072	1,083	1,010	922
of which are currently married	564	NA	530	NA	559	561	543	NA	NA
Fertility rate	NA	NA	2.0	1.7	1.6	1.6	1.5	1.5	1.5

Except as noted, values for vital statistics are in thousands; life expectancy is in years.

Health

Health Indicators [3]

% of population with access to	
safe water (1990-95)	NA
adequate sanitation (1990-95)	NA
health services (1985-95)	NA
% of 1-year-olds immunized (1990-94) against	
TB (tuberculosis)	NA
DPT (diphtheria, pertussis, tetanus)	92
polio	92
measles	93
% of contraceptive prevalence (1980-94)	76
ORT use rate (1990-94)	NA

Health Expenditures [4]

Total health expenditure, 1990 (official exchange rate)	
Millions of dollars	7,782
Dollars per capita	1,835
Health expenditures as a percentage of GDP	
Total	7.4
Public sector	7.0
Private sector	0.3
Development assistance for health	
Total aid flows (millions of dollars)[1]	NA
Aid flows per capita (dollars)	NA
Aid flows as a percentage of total health expenditure	NA

For sources, notes, and explanations, see Annotated Source Appendix, page 1061.

707

Human Factors

Women and Children [5]

% of pregnant women immunized (tetanus 1990-94)	NA
% of births attended by trained health personnel (1983-94)	NA
Maternal mortality rate (1980-92)	3
Under-5 mortality rate (1994)	8
% under-5 moderately/severely underweight (1980-1994)	NA

Burden of Disease [6]

Population per physician (1985)	410.78
Population per nurse (1984)	57.14
Population per hospital bed (1990)	210.52
AIDS cases per 100,000 people (1994)	1.7
Heart disease cases per 1,000 people (1990-93)	327

Ethnic Division [7]

Germanic (Nordic, Alpine, Baltic), Lapps (Sami) 20,000.

Religion [8]

Evangelical Lutheran (state church)	87.8%
Other Protestant and Roman Catholic	3.8%
None	3.2%
Uknown	5.2%
(1980)	

Major Languages [9]

Norwegian (official), Small Lapp- and Finnish-speaking minorities.

Education

Public Education Expenditures [10]

Million (Krone)	1980	1990	1991	1992	1993	1994
Total education expenditure	19,731	51,119	54,709	59,201	65,666	NA
as percent of GNP	7.2	7.9	8.2	8.7	9.2	NA
as percent of total govt. expend.	13.7	14.6	14.6	14.1	NA	NA
Current education expenditure	16,448	44,109	47,807	51,132	61,627	NA
as percent of GNP	6.0	6.9	7.1	7.5	8.7	NA
as percent of current govt. expend.	14.4	14.5	15.7	15.8	NA	NA
Capital expenditure	3,283	7,010	6,902	8,069	4,039	NA

Educational Attainment [11]

Age group (1990)	25+
Total population	2,803,030
Highest level attained (%)	
No schooling	0.1
First level	
Not completed	0.1
Completed	NA
Entered second level	
S-1	32.9
S-2	46.4
Postsecondary	17.9

Illiteracy [12]

Libraries [13]

	Admin. Units	Svc. Pts.	Vols. (000)	Shelving (meters)	Vols. Added	Reg. Users
National (1992)	2	3	2,186	66,377	97,802	NA
Nonspecialized (1992)	20	NA	938	NA	11,042	NA
Public (1992)	439	1,214	19,893	NA	973,399	NA
Higher ed. (1993)[12]	95	209	9,412	267,841	336,708	236,166
School (1990)	3,383	3,383	6,858	NA	307,861	464,557

Daily Newspapers [14]

	1980	1985	1990	1994
Number of papers	85	82	85	83
Circ. (000)	1,892	2,120	2,588	2,623

Culture [15]

Cinema (seats per 1,000)	22.3
Annual attendance per person	2.5
Gross box office receipts (mil. Krone)	410
Museums (reporting)	431
Visitors (000)	8,219
Annual receipts (000 Krone)	979,953

Science and Technology

Scientific/Technical Forces [16]

Scientists/engineers	14,763
Number female	NA
Technicians[2]	7,328
Number female	NA
Total[2]	22,091

R&D Expenditures [17]

	Krone (000) 1993
Total expenditure	14,262,600
Capital expenditure	1,637,800
Current expenditure	12,624,900
Percent current	88.5

U.S. Patents Issued [18]

Values show patents issued to citizens of the country by the U.S. Patents Office.

	1993	1994	1995
Number of patents	127	130	138

Government and Law

Organization of Government [19]

Long-form name:
Kingdom of Norway
Type:
Constitutional monarchy
Independence:
26 October 1905 (from Sweden)
National holiday:
Constitution Day, 17 May (1814)
Constitution:
17 May 1814, modified in 1884
Legal system:
Mixture of customary law, civil law system, and common law traditions; Supreme Court renders advisory opinions to legislature when asked; accepts compulsory ICJ jurisdiction, with reservations
Executive branch:
King; Heir Apparent Crown Prince; Prime Minister; State Council
Legislative branch:
Modified unicameral Parliament: Storting Note: For certain purposes, the Storting divides itself into two chambers and elects one-fourth of its membership to an upper house or Lagting
Judicial branch:
Supreme Court (Hoyesterett)

Crime [23]

	1990
Crime volume	
Cases known to police	299,431
Attempts (percent)	NA
Percent cases solved	5.3
Crimes per 100,000 persons	7,058.7
Persons responsible for offenses	
Total number offenders	15,868
Percent female	9.3
Percent juvenile	24.9
Percent foreigners	NA

Elections [20]

Storting	% of votes
Labor	37.1
Center Party	18.5
Conservative	15.6
Christian People's	8.4
Socialist Left	7.9
Progress	6.0
Left Party	3.6
Red Electoral Alliance	1.2

Government Expenditures [21]

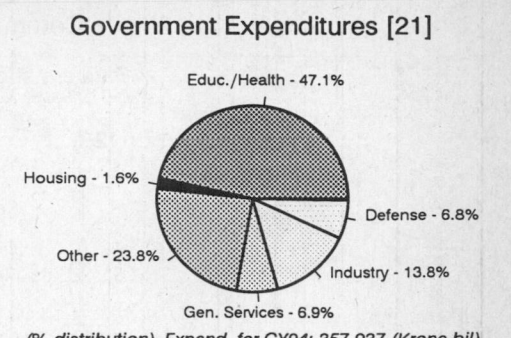

Educ./Health - 47.1%
Housing - 1.6%
Defense - 6.8%
Other - 23.8%
Industry - 13.8%
Gen. Services - 6.9%

(% distribution). Expend. for CY94: 357,937 (Krone bil)

Military Expenditures and Arms Transfers [22]

	1990	1991	1992	1993	1994
Military expenditures					
Current dollars (mil.)	2,819	2,869	3,306	3,247	3,382
1994 constant dollars (mil.)	3,137	3,075	3,448	3,314	3,382
Armed forces (000)	51	41	42	12	12
Gross national product (GNP)					
Current dollars (mil.)	85,350	90,080	95,240	100,700	108,000
1994 constant dollars (mil.)	94,990	96,560	99,320	102,800	108,000
Central government expenditures (CGE)					
1994 constant dollars (mil.)	45,550	49,750	54,180	53,120	55,050[e]
People (mil.)	4.2	4.3	4.3	4.3	4.3
Military expenditure as % of GNP	3.3	3.2	3.5	3.2	3.1
Military expenditure as % of CGE	6.9	6.2	6.4	6.2	6.1
Military expenditure per capita (1994 $)	740	722	806	771	784
Armed forces per 1,000 people (soldiers)	12.0	9.6	9.8	2.8	2.8
GNP per capita (1994 $)	22,390	22,660	23,210	23,910	25,030
Arms imports[6]					
Current dollars (mil.)	470	300	310	140	90
1994 constant dollars (mil.)	523	322	323	143	90
Arms exports[6]					
Current dollars (mil.)	20	80	40	50	50
1994 constant dollars (mil.)	22	86	42	51	50
Total imports[7]					
Current dollars (mil.)	27,230	25,570	25,900	23,900	27,310
1994 constant dollars (mil.)	30,310	27,410	27,010	24,450	27,310
Total exports[7]					
Current dollars (mil.)	34,050	34,110	35,180	31,850	34,690
1994 constant dollars (mil.)	37,890	36,560	36,680	32,510	34,690
Arms as percent of total imports[8]	1.7	1.2	1.2	0.6	0.3
Arms as percent of total exports[8]	0.1	0.2	0.1	0.2	0.1

Human Rights [24]

	SSTS	FL	FAPRO	PPCG	APROBC	TPW	PCPTW	STPEP	PHRFF	PRW	ASST	AFL
Observes	P	P	P	P	P	P	P	P	P	P	P	P
		EAFRD	CPR	ESCR	SR	ACHR	MAAE	PVIAC	PVNAC	EAFDAW	TCIDTP	RC
Observes		P	P	P	P	P	P	P	P	P	P	P

P = Party; S = Signatory; see Appendix for meaning of abbreviations.

Labor Force

Total Labor Force [25]

2.13 million

Labor Force by Occupation [26]

Services	71%
Industry	23
Agriculture, forestry, and fishing	6

Date of data: 1993

Unemployment Rate [27]

8%; including people in job-training programs (November 1995)

For sources, notes, and explanations, see Annotated Source Appendix, page 1061.

709

Production Sectors

Commercial Energy Production and Consumption

Data are shown in quadrillion (10^{15}) BTUs and percent for 1995
Values for hydroelectric, nuclear, geothermal, solar, and wind power refer to electrical generation.

Production [28]

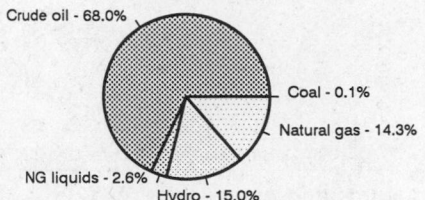

Crude oil - 68.0%
Coal - 0.1%
Natural gas - 14.3%
NG liquids - 2.6%
Hydro - 15.0%

Consumption [29]

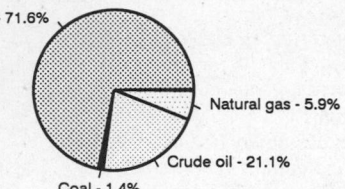

Hydro - 71.6%
Natural gas - 5.9%
Crude oil - 21.1%
Coal - 1.4%

Crude oil	5.677
Natural gas liquids	0.217
Dry natural gas	1.196
Coal	0.008
Net hydroelectric power	1.255
Total	8.353

Crude oil	0.370
Dry natural gas	0.103
Coal	0.024
Net hydroelectric power	1.255
Total	1.752

Telecommunications [30]

- 2.39 million (1986 est.) telephones; high-quality domestic and international telephone, telegraph, and telex services
- Domestic: NA domestic satellite earth stations
- International: 2 buried coaxial cable systems; 4 coaxial submarine cables; satellite earth stations - NA Eutelsat, NA Intelsat (Atlantic Ocean), and 1 Inmarsat (Atlantic and Indian Ocean Regions); note - Norway shares the Inmarsat earth station with the other Nordic countries (Denmark, Finland, Iceland, and Sweden)
- Radio: Broadcast stations: AM 46, FM 493 (350 private and 143 government), shortwave 0 Radios: 3.3 million (1993 est.)
- Television: Broadcast stations: 54 (repeaters 2,100) Televisions: 1.5 million (1993 est.)

Top Agricultural Products [32]

Agriculture accounts for 2.9% of the GDP; produces oats, other grains; beef, milk; livestock output exceeds value of crops; among world's top 10 fishing nations; fish catch of 1.76 million metric tons in 1989.

Transportation [31]

Railways: total: 4,027 km; standard gauge: 4,027 km 1.435-m gauge (2422 km electrified; 96 km double track) (1995)

Highways: total: 88,922 km; paved: 61,356 km (including 75 km of expressways); unpaved: 27,566 km (1990 est.)

Merchant marine: total: 712 ships (1,000 GRT or over) totaling 19,278,205 GRT/32,209,679 DWT; ships by type: bulk 114, cargo 98, chemical tanker 83, combination bulk 10, combination ore/oil 31, container 15, liquefied gas tanker 87, oil tanker 148, passenger 10, passenger-cargo 2, railcar carrier 1, refrigerated cargo 13, roll-on/roll-off cargo 49, short-sea passenger 21, vehicle carrier 30

Airports

Total:	102
With paved runways over 3,047 m:	1
With paved runways 2,438 to 3,047 m:	12
With paved runways 1,524 to 2,437 m:	13
With paved runways 914 to 1,523 m:	11
With paved runways under 914 m:	60

Top Mining Products [33]

Thousand metric tons except as noted	1994[*]
Aluminum, primary	858[r]
Iron ore and concentrate, gross weight	2,364[r]
Metal, ferroalloy	1,090
Platinumgroup metals (kg.)	1,500[e,58]
Cement, hydraulic	1,444[r]
Olivine sand	3,109[r]
Limestone	4,357[r]
Quartz and quartzite	891[r]
Gas, natural, gross (mil. cu. meters)	30,833[r]
Petroleum, crude (000 42-gal. bls.)	913,632[r,59]

Tourism [34]

	1990	1991	1992	1993	1994
Tourists[84]	1,955	2,114	2,375	2,556	2,830
Tourism receipts	1,570	1,646	1,975	1,849	2,157
Tourism expenditures	3,679	3,413	3,870	3,565	3,930
Fare receipts	597	495	516	487	540

Travelers are in thousands, money in million U.S. dollars.

For sources, notes, and explanations, see Annotated Source Appendix, page 1061.

Manufacturing Sector

Manufacturing Summary [35]

	1987		1988		1989		1990		1991	
	$ bil.	%	$ bil.	%	$ bil.	%	$ bil.	%	$ bil.	%
Establishments or enterprises (number)	7,626	0.325	7,268	0.346	7,135	0.381	7,111	0.397	6,889	0.901
Total employment (000)	364	0.269	341	0.249	319	0.259	310	0.280	302	0.434
Production workers (000)	-	-	-	-	-	-	-	-	-	-
Output ($ bil.)	48	0.471	52	0.449	50	0.427	57	0.500	56	0.547
Value added ($ bil.)	14	0.322	15	0.317	14	0.280	15	0.301	14	0.422
Capital investment ($ mil.)	3,061	0.702	2,897	0.608	1,960	0.358	2,448	0.440	2,265	0.595
M & E investment ($ mil.)	1,734	0.556	1,735	0.501	1,397	0.358	1,407	0.334	1,524	0.541
Employees per establishment (number)	48	82.595	47	71.903	45	68.086	44	70.600	44	48.174
Production workers per establishment	-	-	-	-	-	-	-	-	-	-
Output per establishment ($ mil.)	6	144.983	7	129.543	7	112.040	8	125.999	8	60.641
Capital investment per estab. ($ mil.)	0.401	216.040	0.399	175.681	0.275	93.928	0.344	110.965	0.329	66.012
M & E per establishment ($ mil)	0.227	171.179	0.239	144.746	0.196	93.954	0.198	84.088	0.221	59.983
Payroll per employee ($)	23,013	256.633	25,515	256.109	25,567	228.845	30,213	217.807	40,436	276.947
Wages per production worker ($)	-	-	-	-	-	-	-	-	-	-
Hours per production worker (hours)	-	-	-	-	-	-	-	-	-	-
Output per employee ($)	131,467	175.535	153,454	180.163	157,710	164.557	184,953	178.468	184,800	125.879
Capital investment per employee ($)	8,400	261.565	8,502	244.329	6,140	137.955	7,892	157.173	7,512	137.029
M & E per employee ($)	4,759	207.251	5,091	201.306	4,377	137.993	4,537	119.104	5,054	124.513

Note: Columns headed % show percent of world total or ratio. Ratios closest to 100 are closest to world average. M & E stands for machinery & equipment.

Output in Manufacturing

	1987		1988		1989		1990		1991	
	$ bil.	%	$ bil.	%	$ bil.	%	$ bil.	%	$ bil.	%
3110 - Food products	8.233	17.15	8.938	17.19	8.907	17.81	10.000	17.54	11.000	19.64
3130 - Beverages	0.782	1.63	0.859	1.65	-	-	-	-	-	-
3140 - Tobacco	0.453	0.94	0.491	0.94	-	-	-	-	-	-
3210 - Textiles	0.496	1.03	0.488	0.94	0.433	0.87	0.507	0.89	0.499	0.89
3211 - Spinning, weaving, etc.	0.159	0.33	0.149	0.29	0.130	0.26	0.159	0.28	0.158	0.28
3220 - Wearing apparel	0.193	0.40	0.164	0.32	0.123	0.25	0.153	0.27	0.152	0.27
3230 - Leather and products	0.046	0.10	0.041	0.08	0.039	0.08	0.054	0.09	0.052	0.09
3240 - Footwear	0.028	0.06	0.028	0.05	0.022	0.04	0.026	0.05	0.029	0.05
3310 - Wood products	1.980	4.13	1.950	3.75	1.777	3.55	2.036	3.57	1.730	3.09
3320 - Furniture, fixtures	0.692	1.44	0.644	1.24	0.595	1.19	0.682	1.20	0.675	1.21
3410 - Paper and products	2.249	4.69	2.540	4.88	2.721	5.44	2.966	5.20	2.731	4.88
3411 - Pulp, paper, etc.	1.805	3.76	2.067	3.97	2.256	4.51	2.431	4.26	2.194	3.92
3420 - Printing, publishing	2.716	5.66	2.903	5.58	2.789	5.58	3.188	5.59	3.175	5.67
3510 - Industrial chemicals	2.016	4.20	2.176	4.18	2.204	4.41	2.530	4.44	2.421	4.32
3511 - Basic chemicals, excl fertilizers	0.911	1.90	1.010	1.94	1.088	2.18	1.175	2.06	1.113	1.99
3513 - Synthetic resins, etc.	0.494	1.03	0.583	1.12	0.527	1.05	0.626	1.10	0.643	1.15
3520 - Chemical products nec	0.855	1.78	0.899	1.73	0.902	1.80	1.027	1.80	1.001	1.79
3522 - Drugs and medicines	0.295	0.61	0.330	0.63	0.382	0.76	0.447	0.78	0.456	0.81
3530 - Petroleum refineries	1.412	2.94	1.231	2.37	1.654	3.31	2.753	4.83	2.311	4.13
3540 - Petroleum, coal products	0.288	0.60	0.255	0.49	0.239	0.48	0.240	0.42	0.217	0.39
3550 - Rubber products	0.122	0.25	0.138	0.27	0.120	0.24	0.137	0.24	0.119	0.21
3560 - Plastic products nec	0.662	1.38	0.729	1.40	0.672	1.34	0.778	1.36	0.754	1.35
3610 - Pottery, china, etc.	0.043	0.09	0.049	0.09	0.043	0.09	0.048	0.08	0.045	0.08
3620 - Glass and products	0.184	0.38	0.187	0.36	0.162	0.32	0.182	0.32	0.207	0.37
3690 - Nonmetal products nec	0.991	2.06	1.022	1.97	0.860	1.72	0.988	1.73	0.815	1.46
3710 - Iron and steel	1.294	2.70	1.628	3.13	1.477	2.95	1.444	2.53	1.298	2.32
3720 - Nonferrous metals	2.505	5.22	3.657	7.03	4.060	8.12	3.604	6.32	3.231	5.77
3810 - Metal products	1.842	3.84	1.889	3.63	1.692	3.38	2.102	3.69	1.997	3.57
3820 - Machinery nec	5.745	11.97	6.853	13.18	6.771	13.54	6.932	12.16	6.928	12.37
3825 - Office, computing machinery	0.429	0.89	0.439	0.84	0.387	0.77	0.378	0.66	0.187	0.33
3830 - Electrical machinery	2.152	4.48	2.102	4.04	1.854	3.71	2.182	3.83	2.048	3.66
3832 - Radio, television, etc.	0.968	2.02	0.916	1.76	0.708	1.42	0.834	1.46	0.834	1.49
3840 - Transportation equipment	2.470	5.15	2.490	4.79	2.433	4.87	3.336	5.85	3.536	6.31
3841 - Shipbuilding, repair	1.682	3.50	1.685	3.24	1.596	3.19	2.303	4.04	2.630	4.70
3843 - Motor vehicles	0.404	0.84	0.416	0.80	0.390	0.78	0.448	0.79	0.370	0.66
3850 - Professional goods	0.144	0.30	0.175	0.34	0.172	0.34	0.200	0.35	0.266	0.48
3900 - Industries nec	0.168	0.35	0.161	0.31	0.152	0.30	0.200	0.35	0.229	0.41

Note: Codes are International Standard Industry codes (ISIC). Percentages are % of total Output. [f]: Factor Prices; [p]: Producer Prices.

For sources, notes, and explanations, see Annotated Source Appendix, page 1061.

711

Finance, Economics, and Trade

Economic Indicators [36]

- **National product**: GDP—purchasing power parity— $106.2 billion (1995 est.)
- **National product real growth rate**: 4.5% (1995 est.)
- **National product per capita**: $24,500 (1995 est.)
- **Inflation rate (consumer prices)**: 2.5% (1995 est.)
- **External debt**: $NA

Balance of Payments Summary [37]

Values in millions of dollars.

	1989	1990	1991	1992	1993
Exports of goods (f.o.b.)	27,171	34,313	34,212	35,162	31,989
Imports of goods (f.o.b.)	-23,401	-26,552	-25,516	-25,860	-23,974
Trade balance	3,770	7,761	8,696	9,202	8,015
Services - debits	-16,619	-18,954	-18,994	-21,246	-19,426
Services - credits	14,195	16,661	16,871	16,683	15,257
Private transfers (net)	-222	-273	-338	-490	-313
Government transfers (net)	-910	-1,208	-1,185	-1,288	-1,082
Long-term capital (net)	3,021	-1,265	-2,947	2,446	1,539
Short-term capital (net)	-965	504	-4,634	-2,821	5,518
Errors and omissions	-1,305	-2,848	-219	-3,442	-1,659
Overall balance	965	414	-2,750	-856	7,849

Exchange Rates [38]

Currency: **Norwegian krone.**
Symbol: **NKr.**

Data are currency units per $1.

January 1996	6.4160
1995	6.3352
1994	7.0576
1993	7.0941
1992	6.2145
1991	6.4829

Imports and Exports

Top Import Origins [39]

$27.3 billion (c.i.f., 1994) Data are for 1994.

Origins	%
EU	68.9
Germany	13.9
UK	10.4
Denmark	7.4
Sweden	15
US	7.4
Japan	6.0

Top Export Destinations [40]

$34.7 billion (f.o.b., 1994) Data are for 1994.

Destinations	%
EU	77.8
UK	20.8
Germany	12.4
France	8.12
Sweden	9.4
US	6.7
Japan	1.9

Foreign Aid [41]

Donor: ODA, $1.014 billion (1993).

Import and Export Commodities [42]

Import Commodities

Machinery, equipment, and manufactured consumer goods 54%
Chemicals and other industrial inputs 39%
Foodstuffs 6%

Export Commodities

Petroleum and petroleum products 43%
Metals and products 11%
Foodstuffs (mostly fish) 9%
Chemicals and raw materials 25%
Natural gas 6.0%
Ships 5.4%

Oman

Geography [1]

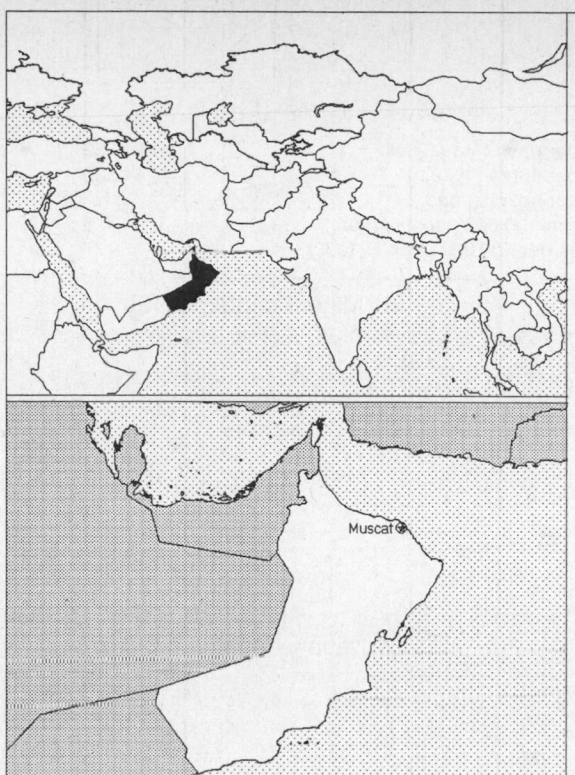

Total area:
212,460 sq km   82,031 sq mi
Land area:
212,460 sq km   82,031 sq mi
Comparative area:
Slightly smaller than Kansas
Land boundaries:
Total 1,374 km, Saudi Arabia 676 km, UAE 410 km, Yemen 288 km
Coastline:
2,092 km
Climate:
Dry desert; hot, humid along coast; hot, dry interior; strong southwest summer monsoon (May to September) in far South
Terrain:
Vast central desert plain, rugged mountains in North and South
Natural resources:
Petroleum, copper, asbestos, some marble, limestone, chromium, gypsum, natural gas
Land use:
Arable land: 2%
Permanent crops: 0%
Meadows and pastures: 5%
Forest and woodland: 0%
Other: 93%

Demographics [2]

	1970	1980	1990	1995[1]	1996	2000	2010	2020	2030
Population	774	1,164	1,751	2,110	2,187	2,512	3,516	4,731	6,029
Population density (persons per sq. mi.)	9	14	21	26	27	31	43	58	73
(persons per sq. km.)	4	5	8	10	10	12	17	22	28
Net migration rate (per 1,000 population)	NA	NA	5.0	2.3	1.8	0.7	0.0	0.0	0.0
Births	NA	NA	NA	NA	83	NA	NA	NA	NA
Deaths	NA	NA	NA	NA	10	NA	NA	NA	NA
Life expectancy - males	NA	NA	66.9	68.3	68.6	69.6	71.9	73.8	75.3
Life expectancy - females	NA	NA	70.8	72.3	72.6	73.7	76.2	78.3	80.1
Birth rate (per 1,000)	NA	NA	39.0	37.9	37.9	37.9	36.0	30.4	25.4
Death rate (per 1,000)	NA	NA	5.1	4.5	4.4	4.2	3.8	3.5	3.6
Women of reproductive age (15-49 yrs.)	NA	NA	342	427	446	529	785	1,116	1,523
of which are currently married	NA	NA	287	NA	372	443	663	NA	NA
Fertility rate	NA	NA	6.5	6.2	6.1	5.8	5.0	4.1	3.2

Except as noted, values for vital statistics are in thousands; life expectancy is in years.

Health

Health Indicators [3]

% of population with access to	
safe water (1990-95)	63
adequate sanitation (1990-95)	78
health services (1985-95)	96
% of 1-year-olds immunized (1990-94) against	
TB (tuberculosis)	96
DPT (diphtheria, pertussis, tetanus)	97
polio	97
measles	97
% of contraceptive prevalence (1980-94)	9
ORT use rate (1990-94)	72

Health Expenditures [4]

For sources, notes, and explanations, see Annotated Source Appendix, page 1061.

713

Human Factors

Women and Children [5]

% of pregnant women immunized (tetanus 1990-94)	99
% of births attended by trained health personnel (1983-94)	60
Maternal mortality rate (1980-92)	NA
Under-5 mortality rate (1994)	27
% under-5 moderately/severely underweight (1980-1994)	NA

Burden of Disease [6]

Population per physician (1989)	1,202.59
Population per nurse (1987)	442.20
Population per hospital bed	NA
AIDS cases per 100,000 people (1994)	0.8
Malaria cases per 100,000 people (1992)	777

Ethnic Division [7]

Arab, Baluchi, South Asian (Indian, Pakistani, Sri Lankan, Bangladeshi), African.

Religion [8]

Ibadhi Muslim	75%
Sunni Muslim, Shi'a Muslim, Hindu	25%

Major Languages [9]

Arabic (official), English, Baluchi, Urdu, Indian dialects.

Education

Public Education Expenditures [10]

Million (Rial)[51]	1980	1985	1990	1992	1993	1994
Total education expenditure	38	123	128	151	170	173
as percent of GNP	2.1	4.0	3.5	3.9	4.5	4.5
as percent of total govt. expend.	4.1	NA	11.1	16.2	15.2	15.5
Current education expenditure	31	77	117	135	151	153
as percent of GNP	1.7	2.5	3.2	3.5	4.0	4.0
as percent of current govt. expend.	4.6	NA	18.8	19.3	19.1	20.0
Capital expenditure	7	46	10	15	19	19

Educational Attainment [11]

Illiteracy [12]

Libraries [13]

	Admin. Units	Svc. Pts.	Vols. (000)	Shelving (meters)	Vols. Added	Reg. Users
National (1992)	1	2	3.6	155	8	168
Nonspecialized	NA	NA	NA	NA	NA	NA
Public	NA	NA	NA	NA	NA	NA
Higher ed. (1993)[10]	1	4	78	1,945	8,852[e]	6,000[e]
School (1987)	130	130	132	NA	-	243,840

Daily Newspapers [14]

	1980	1985	1990	1994
Number of papers	-	3	4	4
Circ. (000)	-	51	62	63

Culture [15]

Cinema (seats per 1,000)	NA
Annual attendance per person	NA
Gross box office receipts (mil. Rial)	NA
Museums (reporting)	5
Visitors (000)	57
Annual receipts (000 Rial)	NA

Science and Technology

Scientific/Technical Forces [16]

R&D Expenditures [17]

U.S. Patents Issued [18]

Values show patents issued to citizens of the country by the U.S. Patents Office.

	1993	1994	1995
Number of patents	1	0	0

For sources, notes, and explanations, see Annotated Source Appendix, page 1061.

Government and Law

Organization of Government [19]

Long-form name:
Sultanate of Oman
Type:
Monarchy
Independence:
1650 (expulsion of the Portuguese)
National holiday:
National Day, 18 November (1940)
Constitution:
None
Legal system:
Based on English common law and Islamic law; ultimate appeal to the sultan; has not accepted compulsory ICJ jurisdiction
Executive branch:
Sultan and Prime Minister; Cabinet
Legislative branch:
Unicameral: Consultative Council (Majlis ash Shura) A 60-member body with advisory powers only
Judicial branch:
None; traditional Islamic judges and a nascent civil court system, administered by region

Elections [20]

No political parties.

Government Expenditures [21]

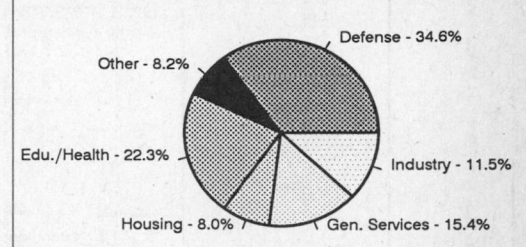

(% distribution). Expend. for CY95: 1,971.2 (Rial mil.)

Crime [23]

	1994
Crime volume	
Cases known to police	3,985
Attempts (percent)	NA
Percent cases solved	NA
Crimes per 100,000 persons	197.51
Persons responsible for offenses	
Total number offenders	6,071
Percent female	3.60
Percent juvenile (9-17 yrs.)	10.30
Percent foreigners	30.00

Military Expenditures and Arms Transfers [22]

	1990	1991	1992	1993	1994
Military expenditures					
Current dollars (mil.)	1,707	1,450	1,767	1,691	1,818
1994 constant dollars (mil.)	1,899	1,554	1,843	1,726	1,818
Armed forces (000)	32	29	35	35	31
Gross national product (GNP)					
Current dollars (mil.)[e]	9,445	9,088	9,996	9,871	10,060
1994 constant dollars (mil.)[e]	10,510	9,741	10,420	10,070	10,060
Central government expenditures (CGE)					
1994 constant dollars (mil.)	4,688	4,373	5,217	5,011	5,001
People (mil.)	1.8	1.8	1.9	2.0	2.0
Military expenditure as % of GNP	18.1	16.0	17.7	17.1	18.1
Military expenditure as % of CGE	40.5	35.5	35.3	34.4	36.4
Military expenditure per capita (1994 $)	1,084	852	972	876	888
Armed forces per 1,000 people (soldiers)	18.3	15.9	18.5	17.8	15.1
GNP per capita (1994 $)	6,002	5,344	5,498	5,112	4,915
Arms imports[6]					
Current dollars (mil.)	10	50	10	120	50
1994 constant dollars (mil.)	11	54	10	122	50
Arms exports[6]					
Current dollars (mil.)	0	0	0	0	0
1994 constant dollars (mil.)	0	0	0	0	0
Total imports[7]					
Current dollars (mil.)	2,681	3,194	3,769	4,114	3,915
1994 constant dollars (mil.)	2,984	3,423	3,930	4,199	3,915
Total exports[7]					
Current dollars (mil.)	5,508	4,871	7,462[e]	7,251[e]	4,600
1994 constant dollars (mil.)	6,130	5,221	7,781[e]	7,400[e]	4,600
Arms as percent of total imports[8]	0.4	1.6	0.3	2.9	1.3
Arms as percent of total exports[8]	0	0	0	0	0

Human Rights [24]

	SSTS	FL	FAPRO	PPCG	APROBC	TPW	PCPTW	STPEP	PHRFF	PRW	ASST	AFL
Observes						P	P					
	EAFRD	CPR	ESCR	SR	ACHR	MAAE	PVIAC	PVNAC	EAFDAW	TCIDTP	RC	
Observes							P	P				

P = Party; S = Signatory; see Appendix for meaning of abbreviations.

Labor Force

Total Labor Force [25]

454,000

Labor Force by Occupation [26]

Agriculture   37%
Date of data: 1993 est.

Unemployment Rate [27]

For sources, notes, and explanations, see Annotated Source Appendix, page 1061.

715

Production Sectors

Commercial Energy Production and Consumption

Data are shown in quadrillion (10^{15}) BTUs and percent for 1995
Values for hydroelectric, nuclear, geothermal, solar, and wind power refer to electrical generation.

Production [28]

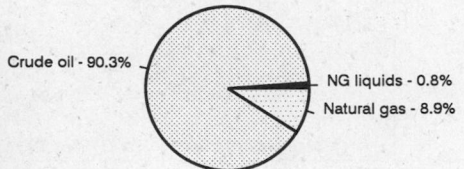

Crude oil - 90.3%
NG liquids - 0.8%
Natural gas - 8.9%

Consumption [29]

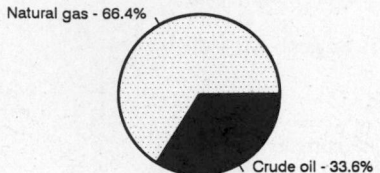

Natural gas - 66.4%
Crude oil - 33.6%

Crude oil	1.824
Natural gas liquids	0.017
Dry natural gas	0.179
Total	2.020

Crude oil	0.086
Dry natural gas	0.170
Total	0.256

Telecommunications [30]

- 150,000 (1994 est.) telephones; modern system consisting of open wire, microwave, and radiotelephone communication stations; limited coaxial cable
- Domestic: open wire, microwave, radiotelephone communications, and a domestic satellite system with 8 earth stations
- International: satellite earth stations - 2 Intelsat (Indian Ocean) and 1 Arabsat
- Radio: Broadcast stations: AM 2, FM 4, shortwave 1 Radios: 1.043 million (1992 est.)
- Television: Broadcast stations: 9 Televisions: 1.195 million (1992 est.)

Transportation [31]

Railways: 0 km

Highways: total: 25,948 km; paved: 4,930 km (including 413 km of expressways); unpaved: 21,018 km (1992 est.)

Merchant marine: total: 3 ships (1,000 GRT or over) totaling 16,306 GRT/8,210 DWT; ships by type: cargo 1, passenger 1, passenger-cargo 1 (1995 est.)

Airports

Total:	129
With paved runways over 3,047 m:	4
With paved runways 1,524 to 2,437 m:	1
With paved runways 914 to 1,523 m:	1
With paved runways under 914 m:	34
With unpaved runways over 3,047 m:	3

Top Agricultural Products [32]

Agriculture accounts for 3% of the GDP; produces dates, limes, bananas, alfalfa, vegetables; camels, cattle; annual fish catch averages 100,000 metric tons.

Top Mining Products [33]

Thousand metric tons except as noted[e]	6/1/95[*]
Cement, hydraulic	1,000
Gas, natural, gross (mil. cu. meters)	5,500
Marble	70
Petroleum, crude (000 42-gal. bls.)	294,000
Sand and gravel	6,500
Silver (kg.)	3,300

Tourism [34]

	1990	1991	1992	1993	1994
Tourists[16]	149	161	192	344	358
Tourism receipts[85]	69	63	85	86	88
Tourism expenditures	47	47	47	47	47

Travelers are in thousands, money in million U.S. dollars.

For sources, notes, and explanations, see Annotated Source Appendix, page 1061.

Finance, Economics, and Trade

GDP and Manufacturing Summary [35]

	1980	1985	1990	1991	1992
Gross Domestic Product					
Millions of 1980 dollars	5,896	11,850	14,054	14,757	15,790[e]
Growth rate in percent	6.03	13.76	9.08	5.00	7.00[e]
Manufacturing Value Added					
Millions of 1980 dollars	45	240	339	NA	NA
Growth rate in percent	19.05	20.39	8.34	NA	NA
Manufacturing share in percent of current prices	0.8	2.4	3.7	4.2	NA

Economic Indicators [36]

- **National product**: GDP—purchasing power parity—$19.1 billion (1995 est.)
- **National product real growth rate**: 3.5% (1995 est.)
- **National product per capita**: $10,800 (1995 est.)
- **Inflation rate (consumer prices)**: - 0.7% (1994 est.)
- **External debt**: $3 billion (1993)

Balance of Payments Summary [37]

Values in millions of dollars.

	1989	1990	1991	1992	1993
Exports of goods (f.o.b.)	4,047	5,508	4,871	5,555	5,365
Imports of goods (f.o.b.)	-2,130	-2,623	-3,112	-3,627	-4,030
Trade balance	1,917	2,885	1,759	1,928	1,335
Services - debits	-1,151	-1,347	-1,547	-1,569	-1,528
Services - credits	381	443	418	341	434
Private transfers (net)	-791	-817	-871	-1,181	-1,329
Government transfers (net)	16	-57	3	-10	18
Long-term capital (net)	162	-241	373	181	148
Short-term capital (net)	-139	-257	148	133	-99
Errors and omissions	-112	-472	253	462	-39
Overall balance	283	137	530	279	-1,060

Exchange Rates [38]

Currency: **Omani rial.**
Symbol: **RO.**

Data are currency units per $1.

Fixed rate since 1986   0.3845

Imports and Exports

Top Import Origins [39]

$4 billion (c.i.f., 1994 est.) Data are for 1993.

Origins	%
UAE (largely re-exports)	27
Japan	20
UK	15
US	5
Germany	4

Top Export Destinations [40]

$4.8 billion (f.o.b., 1994 est.) Data are for 1994.

Destinations	%
Japan	35
South Korea	15.8
US	9
China	8
Thailand	5

Foreign Aid [41]

Recipient: ODA, $82 million (1993).

Import and Export Commodities [42]

Import Commodities	Export Commodities
Machinery	Petroleum 87%
Transportation equipment	Re-exports
Manufactured goods	Fish
Food	Processed copper
Livestock	Textiles
Lubricants	

For sources, notes, and explanations, see Annotated Source Appendix, page 1061.

717

Pakistan

Geography [1]

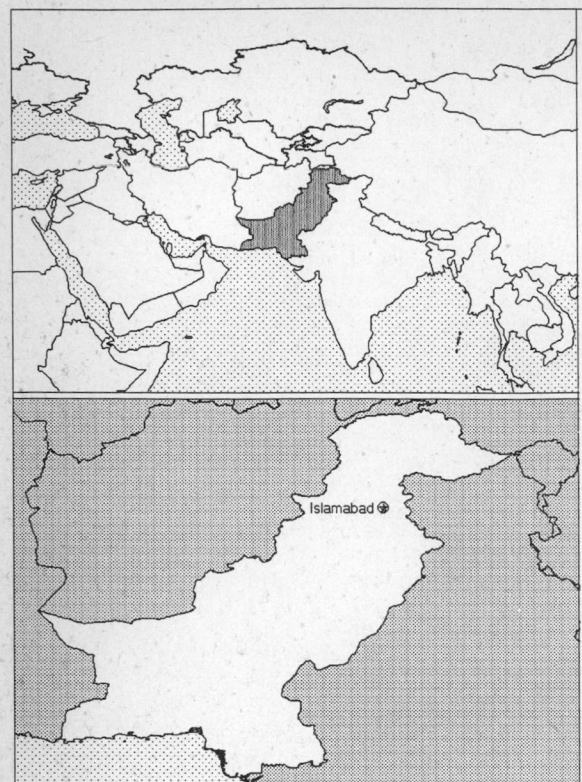

Total area:
803,940 sq km   310,403 sq mi
Land area:
778,720 sq km   300,665 sq mi
Comparative area:
Slightly less than twice the size of California
Land boundaries:
Total 6,774 km, Afghanistan 2,430 km, China 523 km, India 2,912 km, Iran 909 km
Coastline:
1,046 km
Climate:
Mostly hot, dry desert; temperate in Northwest; arctic in North
Terrain:
Flat Indus plain in East; mountains in North and Northwest; Balochistan plateau in West
Natural resources:
Land, extensive natural gas reserves, limited petroleum, poor quality coal, iron ore, copper, salt, limestone
Land use:
Arable land: 23%
Permanent crops: 0%
Meadows and pastures: 6%
Forest and woodland: 4%
Other: 67% (1993)

Islamabad ⊕

Demographics [2]

	1970	1980	1990	1995[1]	1996	2000	2010	2020	2030
Population	65,706	85,219	113,914	126,404	129,276	141,145	170,750	198,722	223,670
Population density (persons per sq. mi.)	219	283	379	420	430	469	568	661	744
(persons per sq. km.)	84	109	146	162	166	181	219	255	287
Net migration rate (per 1,000 population)	NA	NA	-1.5	-3.1	-2.6	-0.9	-0.5	-0.2	0.0
Births	NA	NA	NA	NA	4,675	NA	NA	NA	NA
Deaths	NA	NA	NA	NA	1,450	NA	NA	NA	NA
Life expectancy - males	NA	NA	55.8	57.4	57.7	58.8	61.2	63.5	65.6
Life expectancy - females	NA	NA	56.7	58.9	59.3	60.6	63.9	66.9	69.7
Birth rate (per 1,000)	NA	NA	41.8	37.1	36.2	32.6	25.7	21.2	17.7
Death rate (per 1,000)	NA	NA	13.0	11.5	11.2	10.2	8.5	7.6	7.5
Women of reproductive age (15-49 yrs.)	NA	NA	24,861	28,112	28,862	32,653	43,676	53,498	60,763
of which are currently married	NA	NA	17,032	NA	19,814	22,425	30,327	NA	NA
Fertility rate	NA	NA	6.2	5.4	5.2	4.6	3.2	2.4	2.1

Except as noted, values for vital statistics are in thousands; life expectancy is in years.

Health

Health Indicators [3]

% of population with access to	
safe water (1990-95)	79
adequate sanitation (1990-95)	33
health services (1985-95)	55
% of 1-year-olds immunized (1990-94) against	
TB (tuberculosis)	78
DPT (diphtheria, pertussis, tetanus)	66
polio	66
measles	65
% of contraceptive prevalence (1980-94)	12
ORT use rate (1990-94)	59

Health Expenditures [4]

Total health expenditure, 1990 (official exchange rate)	
Millions of dollars	1,394
Dollars per capita	12
Health expenditures as a percentage of GDP	
Total	3.4
Public sector	1.8
Private sector	1.6
Development assistance for health	
Total aid flows (millions of dollars)[1]	76
Aid flows per capita (dollars)	0.7
Aid flows as a percentage of total health expenditure	5.4

For sources, notes, and explanations, see Annotated Source Appendix, page 1061.

Human Factors

Women and Children [5]

% of pregnant women immunized (tetanus 1990-94)	30
% of births attended by trained health personnel (1983-94)	35
Maternal mortality rate (1980-92)	500
Under-5 mortality rate (1994)	137
% under-5 moderately/severely underweight (1980-1994)	40

Burden of Disease [6]

Population per physician (1987)	2,936.13
Population per nurse (1987)	5,041.83
Population per hospital bed (1990)	1,768.78
AIDS cases per 100,000 people (1994)	*
Malaria cases per 100,000 people (1992)	80

Ethnic Division [7]

Punjabi, Sindhi, Pashtun (Pathan), Baloch, Muhajir (immigrants from India and their descendants).

Religion [8]

Muslim	97%
Sunni	77%
Shi'a	20%
Christian, Hindu, and other	3%

Major Languages [9]

Punjabi	48%
Sindhi	12%
Siraiki (a Punjabi variant)	10%
Pashtu	8%
Urdu (official)	8%
Balochi	3%
Brahui, Burushaski, English (official and lingua franca of government), Hindko	11%

Education

Public Education Expenditures [10]

	1980	1985	1989	1990	1991	1994
Million (Rupee)[52]						
Total education expenditure	4,619	12,645	20,890	23,570	27,790	NA
as percent of GNP	2.0	2.5	2.6	2.6	2.7	NA
as percent of total govt. expend.	5.0	NA	NA	NA	NA	NA
Current education expenditure	3,379	9,390	16,240	19,500	24,090	NA
as percent of GNP	1.5	1.8	2.1	2.2	2.3	NA
as percent of current govt. expend.	5.2	NA	NA	NA	NA	NA
Capital expenditure	1,240	3,255	4,650	4,070	3,700	NA

Educational Attainment [11]

Age group (1990)[16]	25+
Total population	NA
Highest level attained (%)	
No schooling	73.8
First level	
Not completed	9.7
Completed	NA
Entered second level	
S-1	5.8
S-2	8.2
Postsecondary	2.5

Illiteracy [12]

In thousands and percent[1]	1990	1995	2000
Illiterate population (15+ yrs.)	44,805	48,693	53,690
Illiteracy rate - total pop. (%)	69.4	67.1	64.4
Illiteracy rate - males (%)	57.6	55.1	51.8
Illiteracy rate - females (%)	81.8	79.8	77.5

Libraries [13]

	Admin. Units	Svc. Pts.	Vols. (000)	Shelving (meters)	Vols. Added	Reg. Users
National (1992)	1	1	78	621	3,486	NA
Nonspecialized	NA	NA	NA	NA	NA	NA
Public (1992)	4	10	543	6,468	7,339	62,325
Higher ed. (1993)	31	113	3,955	135,367	29,163	49,268
School	NA	NA	NA	NA	NA	NA

Daily Newspapers [14]

	1980	1985	1990	1994
Number of papers	106	118	398	273
Circ. (000)	1,032[e]	1,149	1,826	2,840

Culture [15]

Cinema (seats per 1,000)	NA
Annual attendance per person	NA
Gross box office receipts (mil. Rupee)	NA
Museums (reporting)	12
Visitors (000)	1,305
Annual receipts (000 Rupee)	830

Science and Technology

Scientific/Technical Forces [16]

Scientists/engineers	6,626
Number female	464
Technicians	9,314
Number female	NA
Total[29]	15,940

R&D Expenditures [17]

	Rupee (000) 1987
Total expenditure[28]	5,582,081
Capital expenditure	1,926,257
Current expenditure	3,655,824
Percent current	65.5

U.S. Patents Issued [18]

Values show patents issued to citizens of the country by the U.S. Patents Office.

	1993	1994	1995
Number of patents	1	1	1

For sources, notes, and explanations, see Annotated Source Appendix, page 1061.

Government and Law

Organization of Government [19]

Long-form name:
Islamic Republic of Pakistan
Type:
Republic
Independence:
14 August 1947 (from UK)
National holiday:
Pakistan Day, 23 March (1956)
(proclamation of the republic)
Constitution:
10 April 1973, suspended 5 July 1977,
restored with amendments 30 December
1985
Legal system:
Based on English common law with
provisions to accommodate Pakistan's
stature as an Islamic state; accepts
compulsory ICJ jurisdiction, with
reservations
Executive branch:
President; Prime Minister; Cabinet
Legislative branch:
Bicameral Parliament (Majlis-e-Shoora):
Senate and National Assembly
Judicial branch:
Supreme Court; Federal Islamic (Shari'at)
Court

Elections [20]

National Assembly	% of seats
Pakistan People's Party (PPP)	42.4
Pakistan Muslim League, Nawaz Sharif faction (PML/N)	34.5
Pakistan Muslim League, Junejo faction (PML/J)	2.8
Other	20.3

Government Budget [21]

For FY94/95.

	$ bil.
Revenues	11.9
Expenditures	12.4
Capital expenditures	NA

Crime [23]

Military Expenditures and Arms Transfers [22]

	1990	1991	1992	1993	1994
Military expenditures					
Current dollars (mil.)	2,765	2,614	2,918	3,168	3,068
1994 constant dollars (mil.)	3,077	2,802	3,043	3,233	3,068
Armed forces (000)	550	565	580	580	540
Gross national product (GNP)					
Current dollars (mil.)	39,880	42,940	47,140	48,970	51,410
1994 constant dollars (mil.)	44,390	46,020	49,160	49,980	51,410
Central government expenditures (CGE)					
1994 constant dollars (mil.)	10,800	11,380	12,830	13,120	13,330
People (mil.)	114.8	118.2	121.7	125.2	128.9
Military expenditure as % of GNP	6.9	6.1	6.2	6.5	6.0
Military expenditure as % of CGE	28.5	24.6	23.7	24.7	23.0
Military expenditure per capita (1994 $)	27	24	25	26	24
Armed forces per 1,000 people (soldiers)	4.8	4.8	4.8	4.6	4.2
GNP per capita (1994 $)	387	389	404	399	399
Arms imports[6]					
Current dollars (mil.)	800	220	450	525	260
1994 constant dollars (mil.)	890	236	469	536	260
Arms exports[6]					
Current dollars (mil.)	50	50	30	5	5
1994 constant dollars (mil.)	56	54	31	5	5
Total imports[7]					
Current dollars (mil.)	7,376	8,439	9,379	9,500	8,889
1994 constant dollars (mil.)	8,209	9,046	9,781	9,696	8,889
Total exports[7]					
Current dollars (mil.)	5,589	6,528	7,317	6,688	7,365
1994 constant dollars (mil.)	6,220	6,997	7,630	6,826	7,365
Arms as percent of total imports[8]	10.8	2.6	4.8	5.5	2.9
Arms as percent of total exports[8]	0.9	0.8	0.4	0.1	0.1

Human Rights [24]

	SSTS	FL	FAPRO	PPCG	APROBC	TPW	PCPTW	STPEP	PHRFF	PRW	ASST	AFL
Observes	P	P	P	P	P	P	P	P		P	P	P
	EAFRD	CPR	ESCR	SR	ACHR	MAAE	PVIAC	PVNAC	EAFDAW	TCIDTP	RC
Observes	P						S	S	P		P

P = Party; S = Signatory; see Appendix for meaning of abbreviations.

Labor Force

Total Labor Force [25]

36 million

Labor Force by Occupation [26]

Agriculture	46%
Mining and manufacturing	18
Services	17
Other	19

Extensive export of labor

Unemployment Rate [27]

720

For sources, notes, and explanations, see Annotated Source Appendix, page 1061.

Production Sectors

Commercial Energy Production and Consumption

.Data are shown in quadrillion (10¹⁵) BTUs and percent for 1995

Values for hydroelectric, nuclear, geothermal, solar, and wind power refer to electrical generation.

Production [28]

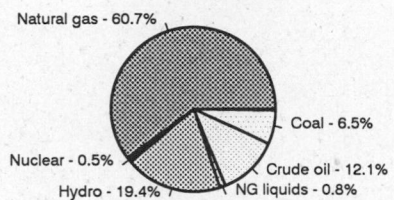

Natural gas - 60.7%
Coal - 6.5%
Nuclear - 0.5%
Crude oil - 12.1%
Hydro - 19.4%
NG liquids - 0.8%

Consumption [29]

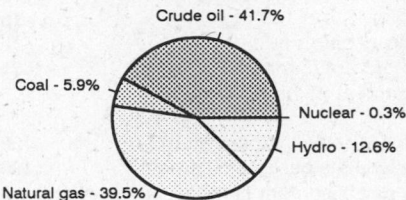

Crude oil - 41.7%
Coal - 5.9%
Nuclear - 0.3%
Hydro - 12.6%
Natural gas - 39.5%

Crude oil	0.120
Natural gas liquids	0.008
Dry natural gas	0.601
Coal	0.064
Net hydroelectric power	0.192
Net nuclear power	0.005
Total	0.990

Crude oil	0.635
Dry natural gas	0.601
Coal	0.090
Net hydroelectric power	0.192
Net nuclear power	0.005
Total	1.523

Telecommunications [30]

- 1.572 million (1993 est.) telephones; the domestic system is mediocre, but adequate for government and business use, in part because major businesses have established their own private systems; since 1988, the government has promoted investment in the national telecommunications system on a priority basis; despite major improvements in trunk and urban systems, telecommunication services are still not readily available to the major portion of the population
- Domestic: microwave radio relay
- International: satellite earth stations - 3 Intelsat (1 Atlantic Ocean and 2 Indian Ocean); microwave radio relay to neighboring countries
- Radio: Broadcast stations: AM 26, FM 8, shortwave 11 Radios: 11.3 million (1992 est.)
- Television: Broadcast stations: 29 Televisions: 2.08 million (1993 est.)

Top Agricultural Products [32]

Agriculture accounts for 24% of the GDP; produces cotton, wheat, rice, sugarcane, fruits, vegetables; milk, beef, mutton, eggs.

Transportation [31]

Railways: total: 8,163 km; broad gauge: 7,718 km 1.676-m gauge (293 km electrified; 1,037 km double track); narrow gauge: 445 km 1.000-m gauge; 661 km less than 1.000-m gauge (1995 est.)

Highways: total: 205,304 km; paved: 104,735 km; unpaved: 100,569 km (1995 est.)

Merchant marine: total: 24 ships (1,000 GRT or over) totaling 345,606 GRT/560,641 DWT; ships by type: bulk 3, cargo 19, oil tanker 1, passenger-cargo 1 (1995 est.)

Airports

Total:	100
With paved runways over 3,047 m:	12
With paved runways 2,438 to 3,047 m:	19
With paved runways 1,524 to 2,437 m:	25
With paved runways 914 to 1,523 m:	11
With paved runways under 914 m:	18

Top Mining Products [33]

Thousand metric tons except as noted^e 6/8/95[*]

Pig iron	1,200
Steel, crude	1,100
Cement, hydraulic	8,300
Clays	1,700
Nitrogen, N content of ammonia	1,500
Salt, rock and marine	913
Limestone	9,000
Coal, all grades	3,000
Gas, natural, gross (mil. cu. ft.)	590,000
Petroleum, crude (000 42-gal. bls.)	22,000

Tourism [34]

	1990	1991	1992	1993	1994
Tourists	424	438	352	379	454
Tourism receipts	156	163	120	111	117
Tourism expenditures	440	555	679	633	398
Fare receipts	468	478	526	421	NA
Fare expenditures	116	169	232	192	NA

Travelers are in thousands, money in million U.S. dollars.

For sources, notes, and explanations, see Annotated Source Appendix, page 1061.

721

Manufacturing Sector

Manufacturing Summary [35]

	1987		1988		1989		1990		1991	
	$ bil.	%	$ bil.	%	$ bil.	%	$ bil.	%	$ bil.	%
Establishments or enterprises (number)	5,938	0.253	6,179	0.294	-	-	-	-	-	-
Total employment (000)	763	0.562	731	0.534	-	-	-	-	-	-
Production workers (000)	595	0.882	574	0.918	-	-	-	-	-	-
Output ($ bil.)	15	0.147	17	0.143	-	-	-	-	-	-
Value added ($ bil.)	5	0.112	5	0.109	-	-	-	-	-	-
Capital investment ($ mil.)	801	0.184	830	0.174	-	-	-	-	-	-
M & E investment ($ mil.)	-	-	-	-	-	-	-	-	-	-
Employees per establishment (number)	129	222.192	118	181.415	-	-	-	-	-	-
Production workers per establishment	100	348.456	93	311.853	-	-	-	-	-	-
Output per establishment ($ mil.)	3	58.138	3	48.669	-	-	-	-	-	-
Capital investment per estab. ($ mil.)	0.135	72.561	0.134	59.195	-	-	-	-	-	-
M & E per establishment ($ mil)	-	-	-	-	-	-	-	-	-	-
Payroll per employee ($)	1,756	19.580	1,956	19.637	-	-	-	-	-	-
Wages per production worker ($)	1,583	20.067	1,728	20.447	-	-	-	-	-	-
Hours per production worker (hours)	-	-	-	-	-	-	-	-	-	-
Output per employee ($)	19,597	26.166	22,850	26.827	-	-	-	-	-	-
Capital investment per employee ($)	1,049	32.657	1,135	32.630	-	-	-	-	-	-
M & E per employee ($)	-	-	-	-	-	-	-	-	-	-

Note: Columns headed % show percent of world total or ratio. Ratios closest to 100 are closest to world average. M & E stands for machinery & equipment.

Output in Manufacturing

	1987		1988		1989		1990		1991	
	$ bil.	%	$ bil.	%	$ bil.	%	$ bil.	%	$ bil.	%
3110 - Food products	2.108	14.05	2.432	14.31	-	-	-	-	-	-
3130 - Beverages	0.111	0.74	0.132	0.78	-	-	-	-	-	-
3140 - Tobacco	0.659	4.39	0.504	2.96	-	-	-	-	-	-
3210 - Textiles	2.585	17.23	2.914	17.14	-	-	-	-	-	-
3211 - Spinning, weaving, etc.	2.382	15.88	2.629	15.46	-	-	-	-	-	-
3220 - Wearing apparel	0.163	1.09	0.231	1.36	-	-	-	-	-	-
3230 - Leather and products	0.283	1.89	0.335	1.97	-	-	-	-	-	-
3240 - Footwear	0.073	0.49	0.085	0.50	-	-	-	-	-	-
3310 - Wood products	0.027	0.18	0.037	0.22	-	-	-	-	-	-
3320 - Furniture, fixtures	0.015	0.10	0.017	0.10	-	-	-	-	-	-
3410 - Paper and products	0.102	0.68	0.164	0.96	-	-	-	-	-	-
3411 - Pulp, paper, etc.	0.027	0.18	0.063	0.37	-	-	-	-	-	-
3420 - Printing, publishing	0.105	0.70	0.108	0.64	-	-	-	-	-	-
3510 - Industrial chemicals	0.651	4.34	0.638	3.75	-	-	-	-	-	-
3511 - Basic chemicals, excl fertilizers	0.080	0.53	0.095	0.56	-	-	-	-	-	-
3513 - Synthetic resins, etc.	0.121	0.81	0.117	0.69	-	-	-	-	-	-
3520 - Chemical products nec	0.687	4.58	0.745	4.38	-	-	-	-	-	-
3522 - Drugs and medicines	0.340	2.27	0.418	2.46	-	-	-	-	-	-
3530 - Petroleum refineries	1.008	6.72	1.103	6.49	-	-	-	-	-	-
3540 - Petroleum, coal products	0.075	0.50	0.079	0.46	-	-	-	-	-	-
3550 - Rubber products	0.108	0.72	0.113	0.66	-	-	-	-	-	-
3560 - Plastic products nec	0.075	0.50	0.074	0.44	-	-	-	-	-	-
3610 - Pottery, china, etc.	0.019	0.13	0.026	0.15	-	-	-	-	-	-
3620 - Glass and products	0.052	0.35	0.062	0.36	-	-	-	-	-	-
3690 - Nonmetal products nec	0.505	3.37	0.563	3.31	-	-	-	-	-	-
3710 - Iron and steel	0.785	5.23	0.895	5.26	-	-	-	-	-	-
3720 - Nonferrous metals	0.003	0.02	0.004	0.02	-	-	-	-	-	-
3810 - Metal products	0.099	0.66	0.117	0.69	-	-	-	-	-	-
3820 - Machinery nec	0.318	2.12	0.309	1.82	-	-	-	-	-	-
3830 - Electrical machinery	0.405	2.70	0.481	2.83	-	-	-	-	-	-
3832 - Radio, television, etc.	0.135	0.90	0.138	0.81	-	-	-	-	-	-
3840 - Transportation equipment	0.438	2.92	0.566	3.33	-	-	-	-	-	-
3841 - Shipbuilding, repair	0.021	0.14	0.025	0.15	-	-	-	-	-	-
3843 - Motor vehicles	0.326	2.17	0.389	2.29	-	-	-	-	-	-
3850 - Professional goods	0.028	0.19	0.039	0.23	-	-	-	-	-	-
3900 - Industries nec	0.038	0.25	0.051	0.30	-	-	-	-	-	-

Note: Codes are International Standard Industry codes (ISIC). Percentages are % of total Output. [f]: Factor Prices; [p]: Producer Prices.

722

For sources, notes, and explanations, see Annotated Source Appendix, page 1061.

Finance, Economics, and Trade

Economic Indicators [36]

- **National product**: GDP—purchasing power parity—$274.2 billion (1995 est.)

- **National product real growth rate**: 4.7% (1995 est.)

- **National product per capita**: $2,100 (1995 est.)

- **Inflation rate (consumer prices)**: 13% (1995 est.)

- **External debt**: $26 billion (1995 est.)

Balance of Payments Summary [37]

Values in millions of dollars.

	1989	1990	1991	1992	1993
Exports of goods (f.o.b.)	4,796	5,380	6,381	6,880	6,760
Imports of goods (f.o.b.)	-7,366	-8,094	-8,642	-9,671	-9,312
Trade balance	-2,570	-2,714	-2,261	-2,791	-2,552
Services - debits	-2,808	-3,238	-3,559	-4,148	-4,237
Services - credits	1,323	1,518	1,596	1,625	1,577
Private transfers (net)	2,207	2,276	2,344	3,068	1,942
Government transfers (net)	524	513	627	382	334
Long-term capital (net)	1,488	1,384	1,478	2,163	2,312
Short-term capital (net)	-106	62	-158	-25	1,118
Errors and omissions	-210	-103	-78	121	-91
Overall balance	-152	-302	-11	395	403

Exchange Rates [38]

Currency: **Pakistani rupee.**
Symbol: **PRe.**

Data are currency units per $1.

January 1996	34.339
1995	31.643
1994	30.567
1993	28.107
1992	25.083
1991	23.801

Imports and Exports

Top Import Origins [39]

$10.7 billion (1995 est.).

Origins	%
Japan	NA
US	NA
Germany	NA
UK	NA
Saudi Arabia	NA
Malaysia	NA
South Korea	NA

Top Export Destinations [40]

$8.7 billion (1995 est.).

Destinations	%
US	NA
Japan	NA
Hong Kong	NA
Germany	NA
UK	NA
UAE	NA
France	NA

Foreign Aid [41]

	U.S. $	
ODA (1993)	697	million
Bilateral/multilateral aid, no US commitments (FY93/94)	2.5	billion
Bilateral/multilateral aid, no US commitments (FY94/95)	3	billion

Import and Export Commodities [42]

Import Commodities	**Export Commodities**
Petroleum	Cotton
Petroleum products	Textiles
Machinery	Clothing
Transportation equipment	Rice
Vegetable oils	Leather
Animal fats	Carpets
Chemicals	

Palau

Geography [1]

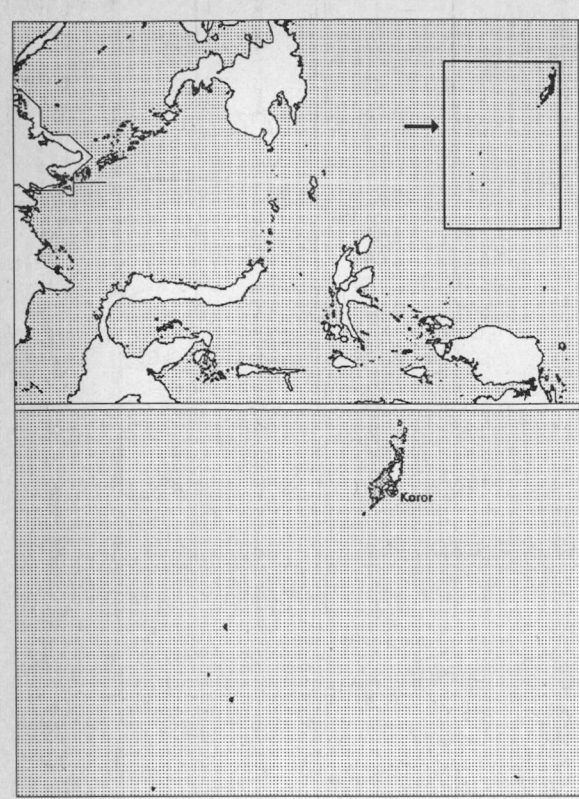

Total area:
  458 sq km   177 sq mi
Land area:
  458 sq km   177 sq mi
Comparative area:
  Slightly more than 2.5 times the size of Washington, DC
Land boundaries:
  0 km
Coastline:
  1,519 km
Climate:
  Wet season May to November; hot and humid
Terrain:
  Varying geologically from the high, mountainous main island of Babelthuap to low, coral islands usually fringed by large barrier reefs
Natural resources:
  Forests, minerals (especially gold), marine products, deep-seabed minerals
Land use:
  Arable land: NA
  Permanent crops: NA
  Meadows and pastures: NA
  Forest and woodland: NA
  Other: NA

Demographics [2]

	1970	1980	1990	1995[1]	1996	2000	2010	2020	2030
Population	12	13	15	17	17	18	20	21	22
Population density (persons per sq. mi.)	68	75	86	94	96	102	111	117	123
(persons per sq. km.)	26	29	33	36	37	39	43	45	48
Net migration rate (per 1,000 population)	NA	-8.2	2.1	2.1	2.1	2.1	NA	NA	NA
Births	NA	NA	NA	NA	Z	NA	NA	NA	NA
Deaths	NA	NA	NA	NA	Z	NA	NA	NA	NA
Life expectancy - males	NA	NA	69.1	69.1	69.1	69.1	NA	NA	NA
Life expectancy - females	NA	NA	73.0	73.0	73.0	73.0	NA	NA	NA
Birth rate (per 1,000)	NA	22.2	23.7	22.1	21.6	19.0	NA	NA	NA
Death rate (per 1,000)	NA	7.1	6.6	6.6	6.6	6.6	NA	NA	NA
Women of reproductive age (15-49 yrs.)	NA	NA	NA	NA	NA	NA	NA	NA	NA
of which are currently married	NA	0.7	NA	NA	NA	NA	NA	NA	NA
Fertility rate	NA	NA	3.1	2.9	2.8	2.4	NA	NA	NA

Except as noted, values for vital statistics are in thousands; life expectancy is in years.

Health

Health Indicators [3]

Health Expenditures [4]

For sources, notes, and explanations, see Annotated Source Appendix, page 1061.

Human Factors

Women and Children [5]

Burden of Disease [6]

Ethnic Division [7]
Palauans are a composite of Polynesian, Malayan, and Melanesian races.

Religion [8]
Christian (Catholics, Seventh-Day Adventists, Jehovah's Witnesses, the Assembly of God, the Liebenzell Mission, and Latter-Day Saints), Modekngei religion (one-third of the population observes this religion which is indigenous to Palau).

Major Languages [9]
English (official in all of Palau's 16 states), Sonsorolese (official in the state of Sonsoral), Angaur and Japanese (in the state of Anguar), Tobi (in the state of Tobi), Palauan (in the other 13 states).

Education

Public Education Expenditures [10]

Educational Attainment [11]

Illiteracy [12]

Libraries [13]

Daily Newspapers [14]

Culture [15]

Science and Technology

Scientific/Technical Forces [16]

R&D Expenditures [17]

U.S. Patents Issued [18]

For sources, notes, and explanations, see Annotated Source Appendix, page 1061.

725

Government and Law

Organization of Government [19]

Long-form name:
Republic of Palau
Type:
Constitutional government in free association with the US; the Compact of Free Association entered into force 1 October 1994
Independence:
1 October 1994 (from the US-administered UN Trusteeship)
National holiday:
Constitution Day, 9 July (1979)
Constitution:
1 January 1981
Legal system:
Based on Trust Territory laws, acts of the legislature, municipal, common, and customary laws
Executive branch:
President; Vice President; Cabinet
Legislative branch:
Bicameral Parliament (Olbiil Era Kelulau or OEK): Senate and House of Delegates
Judicial branch:
Supreme Court; National Court; Court of Common Pleas

Elections [20]

House of Delegates Elections last held 4 November 1992 (next to be held November 1996); results—percent of vot eby party NA; seats—(16 total) number eats by party NA. One party, Palau Nationalist Party.

Government Budget [21]

For 1995 est.

	$ mil.
Revenues	17
Expenditures	57
Capital expenditures	NA

Defense Summary [22]

Note: Defense is the responsibility of the US

Crime [23]

Human Rights [24]

	SSTS	FL	FAPRO	PPCG	APROBC	TPW	PCPTW	STPEP	PHRFF	PRW	ASST	AFL
Observes						P	P					

	EAFRD	CPR	ESCR	SR	ACHR	MAAE	PVIAC	PVNAC	EAFDAW	TCIDTP	RC
Observes							P	P			P

P = Party; S = Signatory; see Appendix for meaning of abbreviations.

Labor Force

Total Labor Force [25]

Labor Force by Occupation [26]

Government	60%
Tourism	

Date of data: 1995 est.

Unemployment Rate [27]

20% (1986)

For sources, notes, and explanations, see Annotated Source Appendix, page 1061.

Production Sectors

Energy Resource Summary [28]

Electricity: Capacity: 16,000 kW. Production: 22 million kWh. Consumption per capita: 1,540 kWh (1990).

Telecommunications [30]

- 1,500 (1988 est.) telephones
- International: satellite earth station - 1 Intelsat (Pacific Ocean)
- Radio: Broadcast stations: AM 1, FM 1, shortwave 0 Radios: 9,000 (1993 est.)
- Television: Broadcast stations: 2 Televisions: 1,600 (1993 est.)

Transportation [31]

Railways: 0 km

Highways: total: 61 km; paved: 36 km; unpaved: 25 km

Merchant marine: none

Airports
Total: 3
With paved runways 1,524 to 2,437 m: 1
With unpaved runways 1,524 to 2,437 m: 2 (1995 est.)

Top Agricultural Products [32]

Produces coconuts, copra, cassava (tapioca), sweet potatoes.

Top Mining Products [33]

Detailed information is not available. A summary of natural resources follows. **Mineral Resources**: gold, deep-seabed minerals.

Tourism [34]

For sources, notes, and explanations, see Annotated Source Appendix, page 1061.

727

Finance, Economics, and Trade

Industrial Summary [35]

Industrial Production: Growth rate not available. **Industries:** Tourism, craft items (from shell, wood, pearls), some commercial fishing and agriculture.

Economic Indicators [36]

- **National product:** GDP—purchasing power parity—$81.8 million (1994 est.)
- **National product real growth rate:** NA%
- **National product per capita:** $5,000 (1994 est.)
- **Inflation rate (consumer prices):** NA%
- **External debt:** about $100 million (1989)

Balance of Payments Summary [37]

Exchange Rates [38]

Currency: **United States dollar.**
Symbol: **US$.**

US currency is used.

Imports and Exports

Top Import Origins [39]

$24.6 million (c.i.f., 1989).

Origins	%
US	NA

Top Export Destinations [40]

$600,000 (f.o.b., 1989).

Destinations	%
US	NA
Japan	NA

Foreign Aid [41]

	U.S. $	
ODA	NA	
US aid (1994-2009)	500	million

Import and Export Commodities [42]

Import Commodities

No details available.

Export Commodities

Trochus (type of shellfish)
Tuna
Copra
Handicrafts

Panama

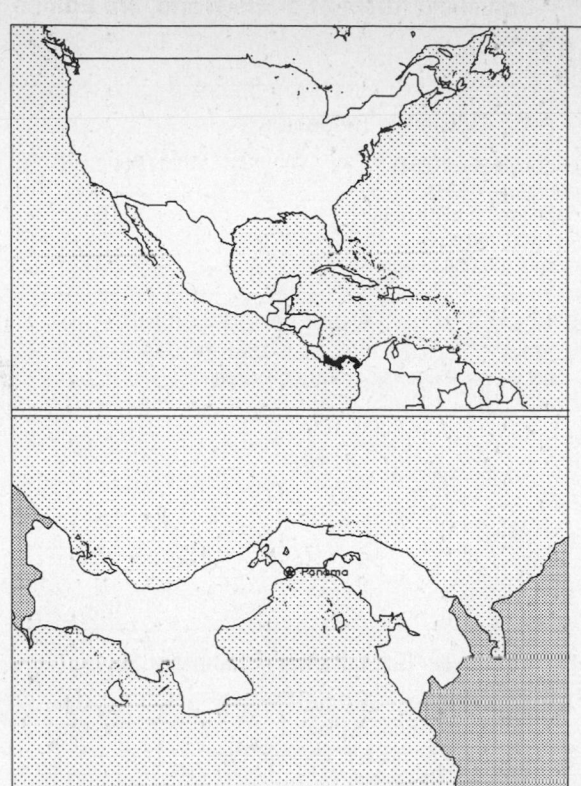

Geography [1]

Total area:
78,200 sq km   30,193 sq mi
Land area:
75,990 sq km   29,340 sq mi
Comparative area:
Slightly smaller than South Carolina
Land boundaries:
Total 555 km, Colombia 225 km, Costa Rica 330 km
Coastline:
2,490 km
Climate:
Tropical; hot, humid, cloudy; prolonged rainy season (May to January), short dry season (January to May)
Terrain:
Interior mostly steep, rugged mountains and dissected, upland plains; coastal areas largely plains and rolling hills
Natural resources:
Copper, mahogany forests, shrimp
Land use:
Arable land: 6%
Permanent crops: 2%
Meadows and pastures: 15%
Forest and woodland: 54%
Other: 23%

Demographics [2]

	1970	1980	1990	1995[1]	1996	2000	2010	2020	2030
Population	1,531	1,956	2,387	2,612	2,655	2,828	3,238	3,625	3,967
Population density (persons per sq. mi.)	52	67	81	89	90	96	110	124	135
(persons per sq. km.)	20	26	31	34	35	37	43	48	52
Net migration rate (per 1,000 population)	NA	-3.0	-1.9	-1.5	-1.4	-1.2	-0.6	-0.2	0.0
Births	NA	NA	NA	NA	62	NA	NA	NA	NA
Deaths	NA	NA	10	NA	14	NA	NA	NA	NA
Life expectancy - males	NA	68.5	71.0	71.0	71.2	71.9	73.4	74.8	75.9
Life expectancy - females	NA	74.8	76.9	76.6	76.8	77.6	79.4	80.8	82.1
Birth rate (per 1,000)	NA	28.4	26.5	23.6	23.2	21.6	18.4	16.5	14.7
Death rate (per 1,000)	NA	5.5	5.2	5.4	5.4	5.4	5.6	6.1	7.0
Women of reproductive age (15-49 yrs.)	NA	459	604	673	685	735	858	930	969
of which are currently married[42]	84	114	341	NA	396	431	511	NA	NA
Fertility rate	NA	3.6	3.1	2.8	2.7	2.6	2.3	2.1	2.1

Except as noted, values for vital statistics are in thousands; life expectancy is in years.

Health

Health Indicators [3]

% of population with access to	
safe water (1990-95)	83
adequate sanitation (1990-95)	88
health services (1985-95)[1]	80
% of 1-year-olds immunized (1990-94) against	
TB (tuberculosis)	95
DPT (diphtheria, pertussis, tetanus)	83
polio	83
measles	84
% of contraceptive prevalence (1980-94)	58
ORT use rate (1990-94)	70

Health Expenditures [4]

For sources, notes, and explanations, see Annotated Source Appendix, page 1061.

729

Human Factors

Women and Children [5]

% of pregnant women immunized (tetanus 1990-94)	28
% of births attended by trained health personnel (1983-94)	96
Maternal mortality rate (1980-92)	75
Under-5 mortality rate (1994)	20
% under-5 moderately/severely underweight (1980-1994)	7

Burden of Disease [6]

Population per physician (1988)	841.00
Population per nurse (1984)	389.77
Population per hospital bed (1985)	300.00
AIDS cases per 100,000 people (1994)	7.2
Malaria cases per 100,000 people (1992)	29

Ethnic Division [7]

Mestizo	70%
West Indian	14%
White	10%
Indian	6%

Religion [8]

Roman Catholic	85%
Protestant	15%

Major Languages [9]

Spanish (official), English 14%. Many Panamanians are bilingual.

Education

Public Education Expenditures [10]

	1980	1985	1990	1991	1992	1994
Million (Balboa)						
Total education expenditure	166	237	248	268	329	346
as percent of GNP	4.8	4.8	5.2	5.1	5.6	5.2
as percent of total govt. expend.	19.0	18.7	20.9	18.8	18.9	20.9
Current education expenditure	156	231	241	262	307	334
as percent of GNP	4.5	4.7	5.1	5.0	5.2	5.0
as percent of current govt. expend.	19.8	19.9	22.2	21.1	21.9	23.4
Capital expenditure	10	6	7	7	22	12

Educational Attainment [11]

Age group (1990)[17]	25+
Total population	1,035,339
Highest level attained (%)	
No schooling	11.7
First level	
Not completed	20.2
Completed	21.8
Entered second level	
S-1	12.6
S-2	16.4
Postsecondary	13.2

Illiteracy [12]

In thousands and percent[1]	1990	1995	2000
Illiterate population (15+ yrs.)	169	161	155
Illiteracy rate - total pop. (%)	10.9	9.2	8.0
Illiteracy rate - males (%)	10.4	8.6	7.3
Illiteracy rate - females (%)	11.5	9.8	8.6

Libraries [13]

Daily Newspapers [14]

	1980	1985	1990	1994
Number of papers	5	7	8	7
Circ. (000)	110[e]	245	234[e]	160

Culture [15]

Cinema (seats per 1,000)	NA
Annual attendance per person	NA
Gross box office receipts (mil. Balboa)	NA
Museums (reporting)	11
Visitors (000)	51
Annual receipts (000 Balboa)	3,809

Science and Technology

Scientific/Technical Forces [16]

R&D Expenditures [17]

	Balboa (000) 1986
Total expenditure[19]	173
Capital expenditure	-
Current expenditure	173
Percent current	100.0

U.S. Patents Issued [18]

Values show patents issued to citizens of the country by the U.S. Patents Office.

	1993	1994	1995
Number of patents	1	0	1

For sources, notes, and explanations, see Annotated Source Appendix, page 1061.

Government and Law

Organization of Government [19]

Long-form name:
Republic of Panama
Type:
Constitutional republic
Independence:
3 November 1903 (from Colombia;
became independent from Spain 28
November 1821)
National holiday:
Independence Day, 3 November (1903)
Constitution:
11 October 1972; major reforms adopted
April 1983
Legal system:
Based on civil law system; judicial review
of legislative acts in the Supreme Court of
Justice; accepts compulsory ICJ
jurisdiction, with reservations
Executive branch:
President; First Vice President; Second
Vice President; Cabinet
Legislative branch:
Unicameral: Legislative Assembly
(Asamblea Legislativa)
Judicial branch:
Supreme Court of Justice (Corte Suprema
de Justicia); five superior courts; three
courts of appeal

Elections [20]

Legislative Assembly	% of seats
Democratic Revolutionary	44.4
Arnulfista Party	19.4
Papa Egoro Movement	8.3
Nationalist Republican	5.6
Solidarity Party	5.6
Others	16.7

Government Expenditures [21]

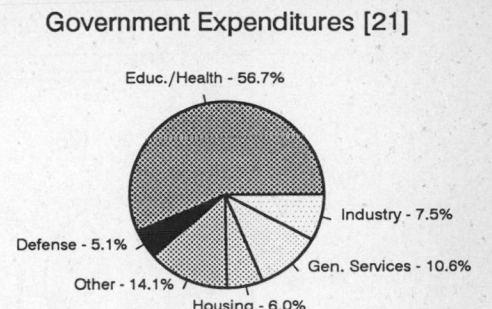

Educ./Health - 56.7%
Industry - 7.5%
Gen. Services - 10.6%
Housing - 6.0%
Other - 14.1%
Defense - 5.1%

(% distribution). Expend. for CY94: 1,969.7 (Balboa mil.)

Military Expenditures and Arms Transfers [22]

	1990	1991	1992	1993	1994
Military expenditures					
Current dollars (mil.)	68	76	0	0	0
1994 constant dollars (mil.)	75	81	0	0	0
Armed forces (000)	11	12	11	11	11
Gross national product (GNP)					
Current dollars (mil.)	4,413	5,052	5,728	6,300	6,688
1994 constant dollars (mil.)	4,912	5,415	5,973	6,429	6,688
Central government expenditures (CGE)					
1994 constant dollars (mil.)	1,286	1,461	1,432	1,683	NA
People (mil.)	2.4	2.5	2.5	2.6	2.6
Military expenditure as % of GNP	1.5	1.5	0	0	0
Military expenditure as % of CGE	5.9	5.5	0	0	0
Military expenditure per capita (1994 $)	31	33	0	0	0
Armed forces per 1,000 people (soldiers)	4.5	4.8	4.4	4.3	4.2
GNP per capita (1994 $)	2,024	2,186	2,363	2,493	2,543
Arms imports[6]					
Current dollars (mil.)	5	5	0	0	0
1994 constant dollars (mil.)	6	5	0	0	0
Arms exports[6]					
Current dollars (mil.)	0	0	10	5	0
1994 constant dollars (mil.)	0	0	10	5	0
Total imports[7]					
Current dollars (mil.)	1,539	1,695	2,024	2,108	2,404
1994 constant dollars (mil.)	1,713	1,817	2,111	2,233	2,404
Total exports[7]					
Current dollars (mil.)	340	358	502	553	583
1994 constant dollars (mil.)	378	384	523	564	583
Arms as percent of total imports[8]	0.3	0.3	0	0	0
Arms as percent of total exports[8]	0	0	2.0	0.9	0

Crime [23]

	1994
Crime volume	
Cases known to police	8,854
Attempts (percent)	NA
Percent cases solved	4,576
Crimes per 100,000 persons	380.11
Persons responsible for offenses	
Total number offenders	2,629
Percent female	240
Percent juvenile	262
Percent foreigners	163

Human Rights [24]

	SSTS	FL	FAPRO	PPCG	APROBC	TPW	PCPTW	STPEP	PHRFF	PRW	ASST	AFL
Observes		P	P	P	P	P	P					P
	EAFRD	CPR	ESCR	SR	ACHR	MAAE	PVIAC	PVNAC	EAFDAW	TCIDTP	RC	
Observes	P	P	P	P	P	P	P	P	P	P	P	

P = Party; S = Signatory; see Appendix for meaning of abbreviations.

Labor Force

Total Labor Force [25]

979,000 (1994 est.)

Labor Force by Occupation [26]

Government and community services	31.8%
Agriculture, hunting, and fishing	26.8
Commerce, restaurants, and hotels	16.4
Manufacturing and mining	9.4
Construction	3.2
Transportation and communications	6.2
Finance, insurance, and real estate	4.3

Unemployment Rate [27]

13.8% (1995)

For sources, notes, and explanations, see Annotated Source Appendix, page 1061.

731

Production Sectors

Commercial Energy Production and Consumption

Data are shown in quadrillion (10^{15}) BTUs and percent for 1995
Values for hydroelectric, nuclear, geothermal, solar, and wind power refer to electrical generation.

Production [28]

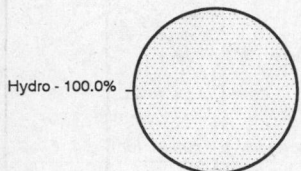

Hydro - 100.0%

Consumption [29]

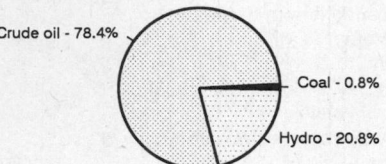

Crude oil - 78.4%
Coal - 0.8%
Hydro - 20.8%

Net hydroelectric power	0.026
Total	0.026

Crude oil	0.098
Coal	0.001
Net hydroelectric power	0.026
Total	0.125

Telecommunications [30]

- 273,000 (1991 est.) telephones; domestic and international facilities well developed
- International: 1 coaxial submarine cable; satellite earth stations - 2 Intelsat (Atlantic Ocean); connected to the Central American Microwave System
- Radio: Broadcast stations: AM 91, FM 0, shortwave 0 Radios: 564,000 (1992 est.)
- Television: Broadcast stations: 23 Televisions: 420,000 (1992 est.)

Transportation [31]

Railways: total: 355 km; broad gauge: 76 km 1.524-m gauge; narrow gauge: 279 km 0.914-m gauge

Highways: total: 10,103 km; paved: 3,233 km; unpaved: 6,870 km (1992 est.)

Merchant marine: total: 3,758 ships (1,000 GRT or over) totaling 69,960,500 GRT/107,632,713 DWT; ships by type: bulk 902, cargo 1,050, chemical tanker 168, combination bulk 40, combination ore/oil 19, container 307, liquefied gas tanker 155, livestock carrier 7, multifunction large-load carrier 3, oil tanker 488, passenger 31, passenger-cargo 5, refrigerated cargo 295, roll-on/roll-off cargo 93, short-sea passenger 34, specialized tanker 11, vehicle carrier 150

Airports

Total:	99
With paved runways over 3,047 m:	1
With paved runways 2,438 to 3,047 m:	1
With paved runways 1,524 to 2,437 m:	5
With paved runways 914 to 1,523 m:	14
With paved runways under 914 m:	60

Top Agricultural Products [32]

Agriculture accounts for 10% of the GDP; produces bananas, rice, corn, coffee, sugarcane, vegetables; livestock; fishing (shrimp).

Top Mining Products [33]

Metric tons except as noted[e]	4/27/95[*]
Cement	350,000
Clays, for cement and products	473,000
Gold (kg.)	275
Petroleum refinery products (000 42-gal. bls.)	10,000
Limestone, for cement and other uses	757,000
Sand and gravel (000 tons)	2,700

Tourism [34]

	1990	1991	1992	1993	1994
Visitors[1]	278	349	361	365	394
Tourists	217	277	291	300	324
Excursionists	63	71	69	65	70
Cruise passengers	3	6	5	7	11
Tourism receipts	172	202	222	228	244
Tourism expenditures	99	109	120	129	131
Fare receipts	7	9	13	13	16
Fare expenditures	36	42	54	49	49

Travelers are in thousands, money in million U.S. dollars.

For sources, notes, and explanations, see Annotated Source Appendix, page 1061.

Manufacturing Sector

Manufacturing Summary [35]

	1987		1988		1989		1990		1991	
	$ bil.	%	$ bil.	%	$ bil.	%	$ bil.	%	$ bil.	%
Establishments or enterprises (number)	924	0.039	917	0.044	822	0.044	-	-	-	-
Total employment (000)	36	0.026	31	0.023	31	0.025	-	-	-	-
Production workers (000)	-	-	-	-	24	0.032	-	-	-	-
Output ($ bil.)	1	0.014	1	0.010	1	0.011	-	-	-	-
Value added ($ bil.)	0.621	0.014	0.463	0.010	0.462	0.009	-	-	-	-
Capital investment ($ mil.)	45	0.010	38	0.008	27	0.005	-	-	-	-
M & E investment ($ mil.)	-	-	-	-	-	-	-	-	-	-
Employees per establishment (number)	39	67.124	34	52.644	38	57.336	-	-	-	-
Production workers per establishment	-	-	-	-	29	73.435	-	-	-	-
Output per establishment ($ mil.)	2	35.735	1	22.618	2	24.965	-	-	-	-
Capital investment per estab. ($ mil.)	0.048	26.098	0.041	18.076	0.033	11.199	-	-	-	-
M & E per establishment ($ mil)	-	-	-	-	-	-	-	-	-	-
Payroll per employee ($)	6,399	71.361	6,172	61.947	6,071	54.344	-	-	-	-
Wages per production worker ($)	-	-	-	-	4,591	45.775	-	-	-	-
Hours per production worker (hours)	-	-	-	-	-	-	-	-	-	-
Output per employee ($)	39,872	53.238	36,594	42.964	41,730	43.542	-	-	-	-
Capital investment per employee ($)	1,249	38.881	1,195	34.336	869	19.533	-	-	-	-
M & E per employee ($)	-	-	-	-	-	-	-	-	-	-

Note: Columns headed % show percent of world total or ratio. Ratios closest to 100 are closest to world average. M & E stands for machinery & equipment.

Output in Manufacturing

	1987		1988		1989		1990		1991	
	$ bil.	%	$ bil.	%	$ bil.	%	$ bil.	%	$ bil.	%
3110 - Food products	0.581	58.10	0.517	51.70	0.591	59.10	-	-	-	-
3130 - Beverages	0.141	14.10	0.117	11.70	0.127	12.70	-	-	-	-
3140 - Tobacco	0.037	3.70	0.030	3.00	0.028	2.80	-	-	-	-
3210 - Textiles	0.011	1.10	0.010	1.00	0.011	1.10	-	-	-	-
3220 - Wearing apparel	0.048	4.80	0.038	3.80	0.048	4.80	-	-	-	-
3230 - Leather and products	0.009	0.90	0.008	0.80	0.007	0.70	-	-	-	-
3240 - Footwear	0.019	1.90	0.012	1.20	0.011	1.10	-	-	-	-
3310 - Wood products	0.014	1.40	0.008	0.80	0.011	1.10	-	-	-	-
3320 - Furniture, fixtures	0.019	1.90	0.009	0.90	0.013	1.30	-	-	-	-
3410 - Paper and products	0.087	8.70	0.083	8.30	0.098	9.80	-	-	-	-
3420 - Printing, publishing	0.051	5.10	0.032	3.20	0.035	3.50	-	-	-	-
3510 - Industrial chemicals	0.030	3.00	0.028	2.80	0.029	2.90	-	-	-	-
3511 - Basic chemicals, excl fertilizers	0.020	2.00	0.019	1.90	0.019	1.90	-	-	-	-
3520 - Chemical products nec	0.091	9.10	0.073	7.30	0.080	8.00	-	-	-	-
3522 - Drugs and medicines	0.019	1.90	0.017	1.70	0.014	1.40	-	-	-	-
3540 - Petroleum, coal products	-	-	-	-	0.007	0.70	-	-	-	-
3550 - Rubber products	0.004	0.40	0.005	0.50	0.004	0.40	-	-	-	-
3560 - Plastic products nec	0.050	5.00	0.040	4.00	0.048	4.80	-	-	-	-
3620 - Glass and products	0.014	1.40	0.008	0.80	0.006	0.60	-	-	-	-
3690 - Nonmetal products nec	0.082	8.20	0.035	3.50	0.031	3.10	-	-	-	-
3810 - Metal products	0.057	5.70	0.029	2.90	0.038	3.80	-	-	-	-
3820 - Machinery nec	0.002	0.20	0.001	0.10	0.001	0.10	-	-	-	-
3830 - Electrical machinery	0.010	1.00	0.006	0.60	0.006	0.60	-	-	-	-
3840 - Transportation equipment	0.015	1.50	0.012	1.20	0.013	1.30	-	-	-	-
3841 - Shipbuilding, repair	0.012	1.20	0.010	1.00	0.011	1.10	-	-	-	-
3843 - Motor vehicles	0.003	0.30	0.002	0.20	0.002	0.20	-	-	-	-
3850 - Professional goods	0.004	0.40	0.004	0.40	0.005	0.50	-	-	-	-

Note: Codes are International Standard Industry codes (ISIC). Percentages are % of total Output. [f]: Factor Prices; [p]: Producer Prices.

Finance, Economics, and Trade

Economic Indicators [36]

- **National product**: GDP—purchasing power parity—$13.6 billion (1995 est.)

- **National product real growth rate**: 2.8% (1995 est.)

- **National product per capita**: $5,100 (1995 est.)

- **Inflation rate (consumer prices)**: 1.1% (1995)

- **External debt**: $6.7 billion (year-end 1993 est.)

Balance of Payments Summary [37]

Values in millions of dollars.

	1989	1990	1991	1992	1993
Exports of goods (f.o.b.)	2,680.8	3,316.3	4,145.7	5,011.7	5,299.4
Imports of goods (f.o.b.)	-3,084.2	-3,804.5	-4,960.4	-5,891.5	-6,152.2
Trade balance	-403.4	-488.2	-814.8	-879.8	-852.8
Services - debits	-1,613.0	-1,764.6	-1,699.3	-1,736.4	-1,604.0
Services - credits	21.1	2,203.5	2,247.3	2,413.8	2,347.3
Private transfers (net)	-35.8	-21.8	-16.1	-27.2	-26.7
Government transfers (net)	105.9	214.1	214.2	330.2	206.5
Long-term capital (net)	-475.8	-386.5	-39.7	-14.7	-225.6
Short-term capital (net)	795.2	182.8	-267.8	-360.4	58.0
Errors and omissions	-420.6	377.1	564.9	443.7	196.3
Overall balance	53.6	316.4	188.7	169.2	99.0

Exchange Rates [38]

Currency: **balboa.**
Symbol: **B.**

Data are currency units per $1.

Fixed rate	1.000

Imports and Exports

Top Import Origins [39]

$2.45 billion (c.i.f., 1995).

Origins	%
US	40
EU	NA
Central America and Caribbean	NA
Japan	NA

Top Export Destinations [40]

$548 million (f.o.b., 1995).

Destinations	%
US	39
EU	NA
Central America and Caribbean	NA

Foreign Aid [41]

Recipient: ODA, $58 million (1993).

Import and Export Commodities [42]

Import Commodities	Export Commodities
Capital goods 21%	Bananas 43%
Crude oil 11%	Shrimp 11%
Foodstuffs 9%	Sugar 4%
Consumer goods	Clothing 5%
Chemicals	Coffee 2%

Papua New Guinea

Geography [1]

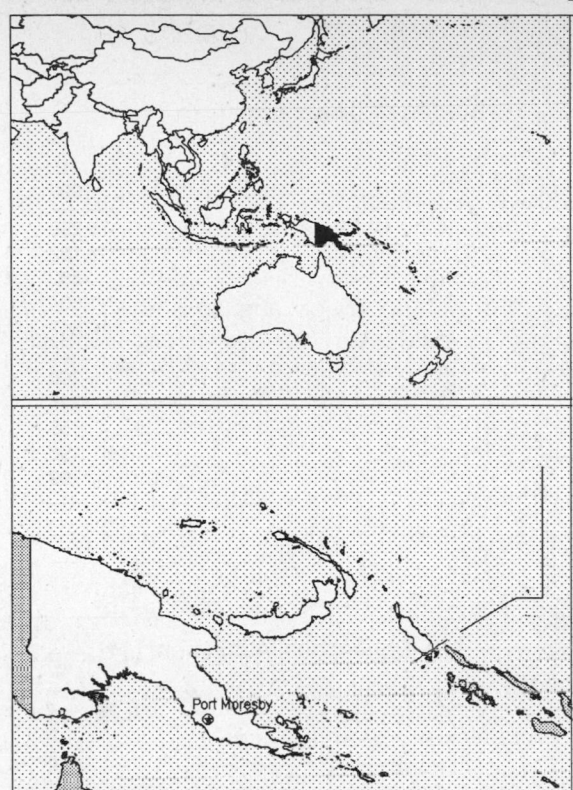

Total area:
  461,690 sq km   178,260 sq mi
Land area:
  451,710 sq km   174,406 sq mi
Comparative area:
  Slightly larger than California
Land boundaries:
  Total 820 km, Indonesia 820 km
Coastline:
  5,152 km
Climate:
  Tropical; northwest monsoon (December to March), southeast monsoon
  (May to October); slight seasonal temperature variation
Terrain:
  Mostly mountains with coastal lowlands and rolling foothills
Natural resources:
  Gold, copper, silver, natural gas, timber, oil potential
Land use:
  Arable land: 0%
  Permanent crops: 1%
  Meadows and pastures: 0%
  Forest and woodland: 71%
  Other: 28%

Demographics [2]

	1970	1980	1990	1995[1]	1996	2000	2010	2020	2030
Population	2,288	2,991	3,823	4,295	4,395	4,812	5,925	7,044	8,140
Population density (persons per sq. mi.)	13	17	22	25	25	28	34	40	47
(persons per sq. km.)	5	7	8	10	10	11	13	16	18
Net migration rate (per 1,000 population)	NA	0.0	0.0	0.0	0.0	0.0	0.0	0.0	0.0
Births	NA	NA	NA	NA	145	NA	NA	NA	NA
Deaths	NA	NA	NA	NA	44	NA	NA	NA	NA
Life expectancy - males	NA	50.9	54.0	56.0	56.4	58.0	61.8	65.2	68.3
Life expectancy - females	NA	51.5	55.6	57.7	58.2	59.8	63.9	67.7	71.1
Birth rate (per 1,000)	NA	39.5	34.7	33.2	32.9	31.6	26.7	22.7	19.8
Death rate (per 1,000)	NA	13.5	11.2	10.2	10.0	9.3	7.8	6.9	6.6
Women of reproductive age (15-49 yrs.)	NA	654	856	1,004	1,033	1,155	1,495	1,865	2,152
of which are currently married	NA	NA	566	NA	679	770	1,023	NA	NA
Fertility rate	NA	5.9	5.1	4.6	4.5	4.1	3.3	2.7	2.4

Except as noted, values for vital statistics are in thousands; life expectancy is in years.

Health

Health Indicators [3]

% of population with access to	
safe water (1990-95)	28
adequate sanitation (1990-95)	22
health services (1985-95)	96
% of 1-year-olds immunized (1990-94) against	
TB (tuberculosis)	91
DPT (diphtheria, pertussis, tetanus)	66
polio	66
measles	39
% of contraceptive prevalence (1980-94)	4
ORT use rate (1990-94)	51

Health Expenditures [4]

Total health expenditure, 1990 (official exchange rate)	
Millions of dollars	142
Dollars per capita	36
Health expenditures as a percentage of GDP	
Total	4.4
Public sector	2.8
Private sector	1.6
Development assistance for health	
Total aid flows (millions of dollars)[1]	7
Aid flows per capita (dollars)	1.8
Aid flows as a percentage of total health expenditure	4.9

For sources, notes, and explanations, see Annotated Source Appendix, page 1061.

735

Human Factors

Women and Children [5]

% of pregnant women immunized (tetanus 1990-94)	13
% of births attended by trained health personnel (1983-94)	20
Maternal mortality rate (1980-92)	900
Under-5 mortality rate (1994)	95
% under-5 moderately/severely underweight (1980-1994)	35

Burden of Disease [6]

Population per physician (1990)	12,754.15
Population per nurse (1990)	1,164.39
Population per hospital bed (1990)	296.63
AIDS cases per 100,000 people (1994)	0.6
Malaria cases per 100,000 people (1992)	NA

Ethnic Division [7]

Melanesian, Papuan, Negrito, Micronesian, Polynesian.

Religion [8]

Indigenous beliefs	34%
Roman Catholic	22%
Lutheran	16%
Presbyterian/Methodist/ London Missionary Society	8%
Anglican	5%
Evangelical Alliance	4%
Other Protestant sects	11%

Major Languages [9]

Motu spoken in Papua region, English spoken by 1%- 2%, pidgin English widespread, 715 indigenous languages.

Education

Public Education Expenditures [10]

Educational Attainment [11]

Age group (1980)	25+
Total population	1,135,783
Highest level attained (%)	
No schooling	82.6
First level	
Not completed	8.2
Completed	5.0
Entered second level	
S-1	3.9
S-2	0.3
Postsecondary	NA

Illiteracy [12]

In thousands and percent[1]	1990	1995	2000
Illiterate population (15+ yrs.)	731	724	710
Illiteracy rate - total pop. (%)	33.3	28.3	24.2
Illiteracy rate - males (%)	23.1	19.4	16.4
Illiteracy rate - females (%)	44.5	38.0	32.7

Libraries [13]

	Admin. Units	Svc. Pts.	Vols. (000)	Shelving (meters)	Vols. Added	Reg. Users
National (1989)	1	4	60	4,000	2,408	3,600
Nonspecialized	NA	NA	NA	NA	NA	NA
Public (1989)	19	NA	151	NA	6,000	46,095
Higher ed.	NA	NA	NA	NA	NA	NA
School	NA	NA	NA	NA	NA	NA

Daily Newspapers [14]

	1980	1985	1990	1994
Number of papers	1	2	2	2
Circ. (000)	27	45	49	65

Culture [15]

Science and Technology

Scientific/Technical Forces [16]

R&D Expenditures [17]

U.S. Patents Issued [18]

Government and Law

Organization of Government [19]

Long-form name:
Independent State of Papua New Guinea
Type:
Parliamentary democracy
Independence:
16 September 1975 (from the Australian-administered UN trusteeship)
National holiday:
Independence Day, 16 September (1975)
Constitution:
16 September 1975
Legal system:
Based on English common law
Executive branch:
British Monarch (represented by Governor General); Prime Minister; Deputy Prime Minister; National Executive Council
Legislative branch:
Unicameral: National Parliament (sometimes referred to as the House of Assembly)
Judicial branch:
Supreme Court

Elections [20]

National Parliament	% of seats
Pangu Party	22.0
People's Democratic Movement (PDM)	15.6
People's Progress Party (PPP)	9.2
People's Action Party (PAP)	9.2
Independents	27.5
Others	16.5

Government Expenditures [21]

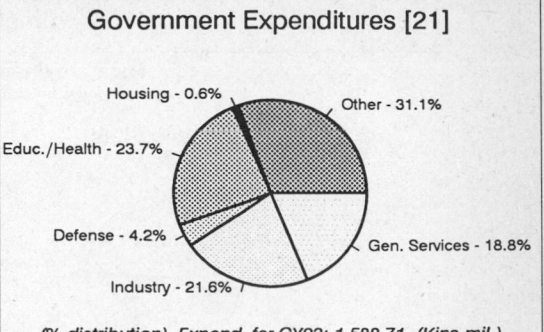

Housing - 0.6%
Other - 31.1%
Educ./Health - 23.7%
Defense - 4.2%
Gen. Services - 18.8%
Industry - 21.6%

(% distribution). Expend. for CY93: 1,588.71 (Kina mil.)

Crime [23]

	1990
Crime volume	
Cases known to police	27,556
Attempts (percent)	NA
Percent cases solved	34.84
Crimes per 100,000 persons	750.60
Persons responsible for offenses	
Total number offenders	11,313
Percent female	5.12
Percent juvenile (8-16 yrs.)	12.89
Percent foreigners	0.15

Military Expenditures and Arms Transfers [22]

	1990	1991	1992	1993	1994
Military expenditures					
Current dollars (mil.)	69	51	58	55	54
1994 constant dollars (mil.)	77	55	61	56	54
Armed forces (000)	4	4	4	NA	4
Gross national product (GNP)					
Current dollars (mil.)	3,127	3,575	3,887	4,690	4,978
1994 constant dollars (mil.)	3,480	3,832	4,053	4,787	4,978
Central government expenditures (CGE)					
1994 constant dollars (mil.)	1,302	1,408	1,440	1,645	1,608
People (mil.)	3.8	3.9	4.0	4.1	4.2
Military expenditure as % of GNP	2.2	1.4	1.5	1.2	1.1
Military expenditure as % of CGE	5.9	3.9	4.2	3.4	3.3
Military expenditure per capita (1994 $)	20	14	15	14	13
Armed forces per 1,000 people (soldiers)	1.0	1.0	0.9	NA	1.0
GNP per capita (1994 $)	910	979	1,012	1,167	1,186
Arms imports[6]					
Current dollars (mil.)	10	10	60	0	0
1994 constant dollars (mil.)	11	11	63	0	0
Arms exports[6]					
Current dollars (mil.)	0	0	0	0	0
1994 constant dollars (mil.)	0	0	0	0	0
Total imports[7]					
Current dollars (mil.)	1,194	1,614	1,485	1,299	1,521
1994 constant dollars (mil.)	1,329	1,730	1,549	1,326	1,521
Total exports[7]					
Current dollars (mil.)	1,144	1,338	1,810	2,491	2,640
1994 constant dollars (mil.)	1,273	1,434	1,887	2,542	2,640
Arms as percent of total imports[8]	0.8	0.6	4.0	0	0
Arms as percent of total exports[8]	0	0	0	0	0

Human Rights [24]

	SSTS	FL	FAPRO	PPCG	APROBC	TPW	PCPTW	STPEP	PHRFF	PRW	ASST	AFL
Observes	P	P		P	P	P	P			P		P
	EAFRD	CPR	ESCR	SR	ACHR	MAAE	PVIAC	PVNAC	EAFDAW	TCIDTP	RC	
Observes	P		P						P		P	

P = Party; S = Signatory; see Appendix for meaning of abbreviations.

Labor Force

Total Labor Force [25]

1.941 million

Labor Force by Occupation [26]

Agriculture   64%
Date of data: 1993 est.

Unemployment Rate [27]

For sources, notes, and explanations, see Annotated Source Appendix, page 1061.

737

Production Sectors

Commercial Energy Production and Consumption

Data are shown in quadrillion (10^{15}) BTUs and percent for 1995
Values for hydroelectric, nuclear, geothermal, solar, and wind power refer to electrical generation.

Production [28]

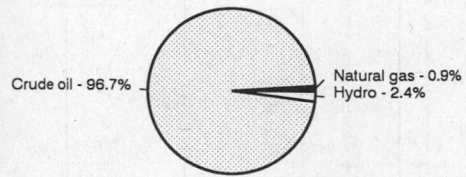

Crude oil - 96.7%   Natural gas - 0.9%   Hydro - 2.4%

Consumption [29]

Crude oil - 82.9%   Natural gas - 4.9%   Hydro - 12.2%

Production	
Crude oil	0.205
Dry natural gas	0.002
Net hydroelectric power	0.005
Total	0.212

Consumption	
Crude oil	0.034
Dry natural gas	0.002
Net hydroelectric power	0.005
Total	0.041

Telecommunications [30]

- 63,212 (1986 est.) telephones; services are adequate and being improved; facilities provide radiotelephone and telegraph, coastal radio, aeronautical radio, and international radio communication services
- Domestic: mostly radiotelephone
- International: submarine cables to Australia and Guam; satellite earth station - 1 Intelsat (Pacific Ocean); international radio communication service
- Radio: Broadcast stations: AM 31, FM 2, shortwave 0 Radios: 298,000 (1992 est.)
- Television: Broadcast stations: 2 (1987 est.) Televisions: 10,000 (1992 est.)

Top Agricultural Products [32]

Produces coffee, cocoa, coconuts, palm kernels, tea, rubber, sweet potatoes, fruit, vegetables; poultry, pork.

Transportation [31]

Railways: 0 km

Highways: total: 19,088 km; paved: 640 km; unpaved: 18,448 km (1988 est.)

Merchant marine: total: 12 ships (1,000 GRT or over) totaling 22,565 GRT/27,114 DWT; ships by type: bulk 2, cargo 3, combination ore/oil 5, container 1, roll-on/roll-off 1 (1995 est.)

Airports

Total:	451
With paved runways 2,438 to 3,047 m:	1
With paved runways 1,524 to 2,437 m:	12
With paved runways 914 to 1,523 m:	5
With paved runways under 914 m:	371
With unpaved runways 1,524 to 2,437 m:	11

Top Mining Products [33]

Metric tons except as noted[e]	5/5/95[*]
Copper, mine output, Cu content	250,000
Gold, mine output, Au content (kg.)	71,200
Petroleum, crude (000 42-gal. bls.)	50,000
Silver, mine output, Ag content (kg.)	125,000

Tourism [34]

	1990	1991	1992	1993	1994
Tourists	41	37	43	34	39
Tourism receipts	41	41	49	45	55
Tourism expenditures	50	57	57	69	70

Travelers are in thousands, money in million U.S. dollars.

Finance, Economics, and Trade

GDP and Manufacturing Summary [35]

	1980	1985	1990	1991	1992
Gross Domestic Product					
Millions of 1980 dollars	2,549	2,739	2,841	3,112	3,391
Growth rate in percent	-1.91	4.31	-3.71	9.55	8.96
Manufacturing Value Added					
Millions of 1980 dollars	242	268	239	243	279[e]
Growth rate in percent	-3.02	2.80	-14.54	1.50	15.06[e]
Manufacturing share in percent of current prices	10.5	11.0	12.2	9.7	9.7[e]

Economic Indicators [36]

- **National product**: GDP—purchasing power parity—$10.2 billion (1995 est.)

- **National product real growth rate**: - 3% (1995 est.)

- **National product per capita**: $2,400 (1995 est.)

- **Inflation rate (consumer prices)**: 15% (1995)

- **External debt**: $3.2 billion (1995)

Balance of Payments Summary [37]

Values in millions of dollars.

	1989	1990	1991	1992	1993
Exports of goods (f.o.b.)	1,318.5	1,173.8	1,482.6	1,950.9	2,504.7
Imports of goods (f.o.b.)	-1,341.3	-1,106.8	-1,403.8	-1,321.7	-1,134.8
Trade balance	-22.8	67.0	78.8	629.2	1,369.9
Services - debits	-674.0	-612.5	-863.0	-1,113.7	-1,204.9
Services - credits	254.9	311.9	374.1	389.0	338.4
Private transfers (net)	-130.7	-107.2	-64.1	-62.6	-129.8
Government transfers (net)	217.3	225.5	323.6	255.0	172.7
Long-term capital (net)	252.0	239.8	62.8	-116.5	-546.9
Short-term capital (net)	13.0	-18.2	-64.1	-27.9	-93.0
Errors and omissions	31.6	-78.6	2.2	-17.5	29.1
Overall balance	-58.7	27.7	-149.7	-65.0	-64.5

Exchange Rates [38]

Currency: **kina.**
Symbol: **K.**

Data are currency units per $1. The government floated the kina on 10 October 1994.

October 1995	0.7552
1994	0.9950
1993	1.0221
1992	1.0367
1991	1.0504

Imports and Exports

Top Import Origins [39]

$1.4 billion (c.i.f., 1995 est.).

Origins	%
Australia	NA
Japan	NA
UK	NA
New Zealand	NA
Netherlands	NA

Top Export Destinations [40]

$2.4 billion (f.o.b., 1995 est.).

Destinations	%
Australia	NA
Japan	NA
US	NA
Singapore	NA
New Zealand	NA

Foreign Aid [41]

Recipient: ODA, $291 million (1993).

Import and Export Commodities [42]

Import Commodities	Export Commodities
Machinery and transport equipment	Gold
Manufactured goods	Copper ore
Food	Oil
Fuels	Logs
Chemicals	Palm oil
	Coffee
	Cocoa
	Lobster

For sources, notes, and explanations, see Annotated Source Appendix, page 1061.

739

Paraguay

Geography [1]

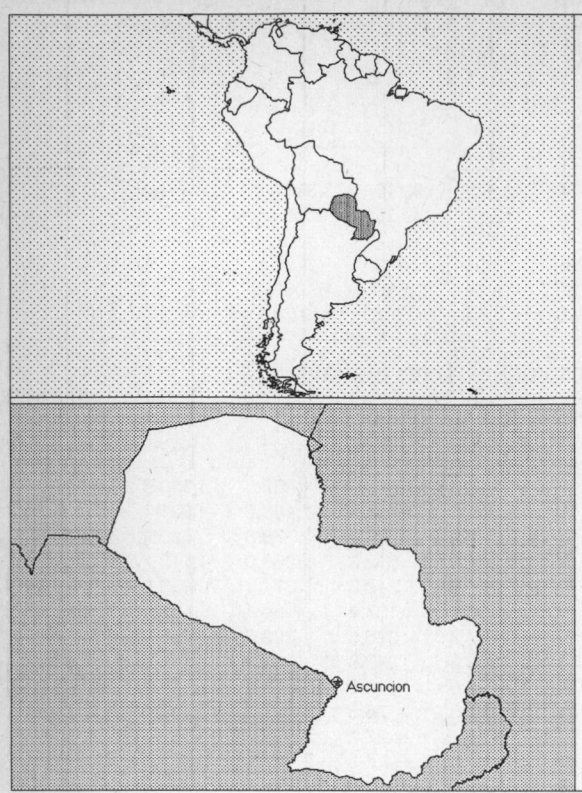

Total area:
406,750 sq km   157,047 sq mi
Land area:
397,300 sq km   153,398 sq mi
Comparative area:
Slightly smaller than California
Land boundaries:
Total 3,920 km, Argentina 1,880 km, Bolivia 750 km, Brazil 1,290 km
Coastline:
0 km (landlocked)
Climate:
Subtropical; substantial rainfall in the eastern portions, becoming
semiarid in the far West
Terrain:
Grassy plains and wooded hills east of Rio Paraguay; Gran Chaco region
west of Rio Paraguay mostly low, marshy plain near the river, and
dry forest and thorny scrub elsewhere
Natural resources:
Hydropower, timber, iron ore, manganese, limestone
Land use:
Arable land: 20%
Permanent crops: 1%
Meadows and pastures: 39%
Forest and woodland: 35%
Other: 5%

Demographics [2]

	1970	1980	1990	1995[1]	1996	2000	2010	2020	2030
Population	2,477	3,379	4,651	5,358	5,504	6,104	7,730	9,474	11,174
Population density (persons per sq. mi.)	16	22	30	35	36	40	50	62	73
(persons per sq. km.)	6	9	12	13	14	15	19	24	28
Net migration rate (per 1,000 population)	NA	2.2	0.1	0.0	0.0	0.0	0.0	0.0	0.0
Births	NA	NA	NA	NA	170	NA	NA	NA	NA
Deaths	NA	NA	NA	NA	24	NA	NA	NA	NA
Life expectancy - males	NA	65.6	70.5	72.1	72.3	73.4	75.7	77.3	78.4
Life expectancy - females	NA	70.0	73.8	75.2	75.4	76.5	78.7	80.5	81.9
Birth rate (per 1,000)	NA	38.5	34.4	31.5	31.0	29.1	26.0	22.2	19.1
Death rate (per 1,000)	NA	6.9	4.9	4.4	4.3	4.1	3.8	3.8	4.4
Women of reproductive age (15-49 yrs.)	NA	788	1,095	1,277	1,318	1,498	1,940	2,397	2,858
of which are currently married	NA	NA	672	NA	813	918	1,204	NA	NA
Fertility rate	NA	5.2	4.6	4.2	4.1	3.9	3.2	2.8	2.5

Except as noted, values for vital statistics are in thousands; life expectancy is in years.

Health

Health Indicators [3]

% of population with access to	
safe water (1990-95)	35
adequate sanitation (1990-95)	62
health services (1985-95)	63
% of 1-year-olds immunized (1990-94) against	
TB (tuberculosis)	97
DPT (diphtheria, pertussis, tetanus)	84
polio	83
measles	79
% of contraceptive prevalence (1980-94)	48
ORT use rate (1990-94)	52

Health Expenditures [4]

Total health expenditure, 1990 (official exchange rate)	
Millions of dollars	160
Dollars per capita	37
Health expenditures as a percentage of GDP	
Total	2.8
Public sector	1.2
Private sector	1.6
Development assistance for health	
Total aid flows (millions of dollars)[1]	10
Aid flows per capita (dollars)	2.4
Aid flows as a percentage of total health expenditure	6.4

For sources, notes, and explanations, see Annotated Source Appendix, page 1061.

Human Factors

Women and Children [5]

% of pregnant women immunized (tetanus 1990-94)	43
% of births attended by trained health personnel (1983-94)	66
Maternal mortality rate (1980-92)	300
Under-5 mortality rate (1994)	34
% under-5 moderately/severely underweight (1980-1994)	4

Burden of Disease [6]

Population per physician (1991)	1,249.86
Population per nurse (1985)	882.02
Population per hospital bed (1990)	1,087.46
AIDS cases per 100,000 people (1994)	0.5
Malaria cases per 100,000 people (1992)	28

Ethnic Division [7]

Mestizo	95%
Whites plus Amerindians	5%

Religion [8]

Roman Catholic	90%
Protestant (primarily Mennonite)	10%

Major Languages [9]

Spanish (official), Guarani.

Education

Public Education Expenditures [10]

Million (Guarani)[53]	1980	1985	1990	1992	1993	1994
Total education expenditure	8,793	20,662	74,387	249,750	338,107	432,812
as percent of GNP	1.5	1.5	1.2	2.6	2.8	2.9
as percent of total govt. expend.	16.4	16.7	9.1	11.9	16.9	NA
Current education expenditure	NA	16,822	72,472	223,139	312,757	401,408
as percent of GNP	NA	1.2	1.1	2.3	2.6	2.7
as percent of current govt. expend.	NA	18.8	NA	NA	17.1	NA
Capital expenditure	NA	3,840	1,915	26,611	25,350	31,404

Educational Attainment [11]

Age group (1992)	15+
Total population	2,427,485
Highest level attained (%)	
No schooling	7.0
First level	
Not completed	38.4
Completed	22.8
Entered second level	
S-1	12.8
S-2	12.2
Postsecondary	6.6

Illiteracy [12]

In thousands and percent[1]	1990	1995	2000
Illiterate population (15+ yrs.)	239	235	229
Illiteracy rate - total pop. (%)	8.8	7.4	6.1
Illiteracy rate - males (%)	7.2	6.2	5.2
Illiteracy rate - females (%)	10.4	8.7	7.1

Libraries [13]

	Admin. Units	Svc. Pts.	Vols. (000)	Shelving (meters)	Vols. Added	Reg. Users
National (1990)	1	NA	NA	NA	NA	NA
Nonspecialized	NA	NA	NA	NA	NA	NA
Public (1990)	28	NA	NA	NA	NA	NA
Higher ed. (1990)	26	NA	NA	NA	NA	NA
School (1990)	145	NA	NA	NA	NA	NA

Daily Newspapers [14]

	1980	1985	1990	1994
Number of papers	5	6	5	5
Circ. (000)	160[e]	170[e]	165[e]	203

Culture [15]

Science and Technology

Scientific/Technical Forces [16]

Scientists/engineers	NA
Number female	NA
Technicians	NA
Number female	NA
Total[1]	807

R&D Expenditures [17]

U.S. Patents Issued [18]

For sources, notes, and explanations, see Annotated Source Appendix, page 1061.

741

Government and Law

Organization of Government [19]

Long-form name:
Republic of Paraguay
Type:
Republic
Independence:
14 May 1811 (from Spain)
National holiday:
Independence Days, 14-15 May (1811)
Constitution:
Promulgated 20 June 1992
Legal system:
Based on Argentine codes, Roman law, and French codes; judicial review of legislative acts in Supreme Court of Justice; does not accept compulsory ICJ jurisdiction
Executive branch:
President; Vice President; Council of Ministers
Legislative branch:
Bicameral Congress (Congreso): Chamber of Senators (Camara de Senadores) and Chamber of Deputies (Camara de Diputados)
Judicial branch:
Supreme Court of Justice (Corte Suprema de Justicia)

Crime [23]

	1994
Crime volume	
Cases known to police	14,717
Attempts (percent)	32.50
Percent cases solved	9.30
Crimes per 100,000 persons	312.55
Persons responsible for offenses	
Total number offenders	3,304
Percent female	19.50
Percent juvenile (0-18 yrs.)	12.60
Percent foreigners	11.30

Elections [20]

Chamber of Deputies	% of seats
Colorado Party	47.5
Authentic Radical Liberal Party	41.3
National Encounter	11.2

Government Expenditures [21]

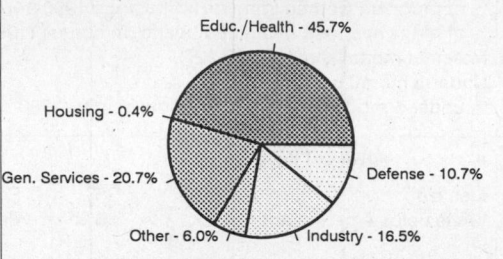

Educ./Health - 45.7%
Housing - 0.4%
Gen. Services - 20.7%
Defense - 10.7%
Other - 6.0%
Industry - 16.5%

(% distribution). Expend. for CY93: 42,001 (Guarani mil.)

Military Expenditures and Arms Transfers [22]

	1990	1991	1992	1993	1994
Military expenditures					
Current dollars (mil.)	80	113	124	102	94
1994 constant dollars (mil.)	90	121	129	104	94
Armed forces (000)	16	16	16	16	15
Gross national product (GNP)					
Current dollars (mil.)	6,131	6,560	6,850	7,330	7,786
1994 constant dollars (mil.)	6,823	7,032	7,144	7,481	7,786
Central government expenditures (CGE)					
1994 constant dollars (mil.)	646	847	978	971	NA
People (mil.)	4.7	4.8	4.9	5.1	5.2
Military expenditure as % of GNP	1.3	1.7	1.8	1.4	1.2
Military expenditure as % of CGE	13.9	14.3	13.2	10.7	NA
Military expenditure per capita (1994 $)	19	25	26	21	18
Armed forces per 1,000 people (soldiers)	3.4	3.3	3.2	3.2	2.9
GNP per capita (1994 $)	1,467	1,468	1,449	1,475	1,493
Arms imports[6]					
Current dollars (mil.)	5	0	10	10	10
1994 constant dollars (mil.)	6	0	10	10	10
Arms exports[6]					
Current dollars (mil.)	0	0	0	0	0
1994 constant dollars (mil.)	0	0	0	0	0
Total imports[7]					
Current dollars (mil.)	1,352	1,460	1,422	1,689	NA
1994 constant dollars (mil.)	1,505	1,565	1,483	1,724	NA
Total exports[7]					
Current dollars (mil.)	959	737	657	725	NA
1994 constant dollars (mil.)	1,067	790	685	740	NA
Arms as percent of total imports[8]	0.4	0	0.7	0.6	NA
Arms as percent of total exports[8]	0	0	0	0	0

Human Rights [24]

	SSTS	FL	FAPRO	PPCG	APROBC	TPW	PCPTW	STPEP	PHRFF	PRW	ASST	AFL
Observes		P	P	S	P	P	P			P		P
	EAFRD	CPR	ESCR	SR	ACHR	MAAE	PVIAC	PVNAC	EAFDAW	TCIDTP	RC	
Observes		P	P	P	P		P	P	P	P	P	

P = Party; S = Signatory; see Appendix for meaning of abbreviations.

Labor Force

Total Labor Force [25]

1.692 million (1993 est.)

Labor Force by Occupation [26]

Agriculture   45%

Unemployment Rate [27]

12% (1995)

For sources, notes, and explanations, see Annotated Source Appendix, page 1061.

Production Sectors

Commercial Energy Production and Consumption

Data are shown in quadrillion (10^{15}) BTUs and percent for 1995

Values for hydroelectric, nuclear, geothermal, solar, and wind power refer to electrical generation.

Production [28]

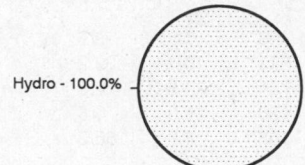

Hydro - 100.0%

Consumption [29]

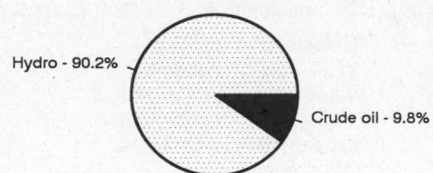

Hydro - 90.2%

Crude oil - 9.8%

Net hydroelectric power	0.416
Total	0.416

Crude oil	0.045
Net hydroelectric power	0.416
Total	0.461

Telecommunications [30]

- 88,730 (1985 est.) telephones; meager telephone service; principal switching center is Asuncion
- Domestic: fair microwave radio relay network
- International: satellite earth station - 1 Intelsat (Atlantic Ocean)
- Radio: Broadcast stations: AM 40, FM 0, shortwave 7 Radios: 775,000 (1992 est.)
- Television: Broadcast stations: 5 Televisions: 370,000 (1992 est.)

Top Agricultural Products [32]

Agriculture accounts for 25.7% of the GDP; produces cotton, sugarcane, soybeans, corn, wheat, tobacco, cassava (tapioca), fruits, vegetables; beef, pork, eggs, milk; timber.

Transportation [31]

Railways: total: 971 km; standard gauge: 441 km 1.435-m gauge; narrow gauge: 60 km 1.000-m gauge

Highways: total: 21,834 km; paved: 1,778 km; unpaved: 20,056 km (1987 est.)

Merchant marine: total: 16 ships (1,000 GRT or over) totaling 21,323 GRT/23,907 DWT; ships by type: cargo 13, oil tanker 2, roll-on/roll-off 1 (1995 est.)

Airports

Total:	739
With paved runways over 3,047 m:	3
With paved runways 1,524 to 2,437 m:	2
With paved runways 914 to 1,523 m:	4
With paved runways under 914 m:	438
With unpaved runways over 3,047 m:	1

Top Mining Products [33]

Metric tons except as noted[e]	12/94*
Clay, kaolin	75,000
Gypsum	5,000
Steel, crude	100,000
Lime	100,000
Petroleum refinery products (000 42-gal. bls.)	2,150
Pigments, mineral: ocher	350
Sand, incl. glass sand (000 tons)	2,000
Limestone, cement and lime	600,000
Marble	750
Talc, soapstone, pyrophyllite	200

Tourism [34]

	1990	1991	1992	1993	1994
Tourists[86]	280	361	334	404	406
Tourism receipts	112	165	154	204	197
Tourism expenditures	58	118	135	138	176
Fare receipts	22	24	29	34	40
Fare expenditures	21	22	26	30	19

Travelers are in thousands, money in million U.S. dollars.

For sources, notes, and explanations, see Annotated Source Appendix, page 1061.

743

Manufacturing Sector

GDP and Manufacturing Summary [35]

	1980	1985	1989	1990	% change 1980-1990	% change 1989-1990
GDP (million 1980 $)	3,844	4,302	5,052	5,227	36.0	3.5
GDP per capita (1980 $)	1,222	1,164	1,215	1,220	-0.2	0.4
Manufacturing as % of GDP (current prices)	16.5	15.2	17.0	17.3	4.8	1.8
Gross output (million $)	1,312	1,395	1,534[e]	1,408	7.3	-8.2
Value added (million $)	575	622[e]	490[e]	633[e]	10.1	29.2
Value added (million 1980 $)	633	669	765	784	23.9	2.5
Industrial production index	100	113	124	125	25.0	0.8
Employment (thousands)	146[e]	128[e]	92[e]	153[e]	4.8	66.3

Note: GDP stands for Gross Domestic Product. 'e' stands for estimated value.

Profitability and Productivity

	1980	1985	1989	1990	% change 1980-1990	% change 1989-1990
Intermediate input (%)	NA	NA	NA	NA	NA	NA
Wages, salaries, and supplements (%)	NA	NA	NA	NA	NA	NA
Gross operating surplus (%)	NA	NA	NA	NA	NA	NA
Gross output per worker ($)	8,962[e]	10,740[e]	16,719[e]	9,102[e]	1.6	-45.6
Value added per worker ($)	4,061[e]	4,824[e]	5,340[e]	4,140[e]	1.9	-22.5
Average wage (incl. benefits) ($)	NA	NA	NA	NA	NA	NA

Profitability is in percent of gross output. Productivity is in U.S. $. 'e' stands for estimated value.

Profitability - 1990

Value Added in Manufacturing

	1980 $ mil.	1980 %	1985 $ mil.	1985 %	1989 $ mil.	1989 %	1990 $ mil.	1990 %	% change 1980-1990	% change 1989-1990
311 Food products	170	29.6	232	37.3	132[e]	26.9	224	35.4	31.8	69.7
313 Beverages	43	7.5	58	9.3	40[e]	8.2	55	8.7	27.9	37.5
314 Tobacco products	6	1.0	9	1.4	7[e]	1.4	9	1.4	50.0	28.6
321 Textiles	44	7.7	54	8.7	29[e]	5.9	41	6.5	-6.8	41.4
322 Wearing apparel	2	0.3	3	0.5	2[e]	0.4	3	0.5	50.0	50.0
323 Leather and fur products	7	1.2	11	1.8	19[e]	3.9	14	2.2	100.0	-26.3
324 Footwear	18	3.1	18	2.9	23[e]	4.7	18	2.8	0.0	-21.7
331 Wood and wood products	95	16.5	87	14.0	75[e]	15.3	92	14.5	-3.2	22.7
332 Furniture and fixtures	6	1.0	10	1.6	9[e]	1.8	9	1.4	50.0	0.0
341 Paper and paper products	NA	0.0	1[e]	0.2	2[e]	0.4	1[e]	0.2	NA	-50.0
342 Printing and publishing	24	4.2	27[e]	4.3	23[e]	4.7	32[e]	5.1	33.3	39.1
351 Industrial chemicals	4	0.7	4[e]	0.6	7[e]	1.4	2[e]	0.3	-50.0	-71.4
352 Other chemical products	10	1.7	8[e]	1.3	5[e]	1.0	6[e]	0.9	-40.0	20.0
353 Petroleum refineries	94	16.3	45	7.2	62[e]	12.7	62	9.8	-34.0	0.0
354 Miscellaneous petroleum and coal products	NA	0.0	NA	0.0	NA	0.0	NA	0.0	NA	NA
355 Rubber products	NA	0.0	NA	0.0	NA	0.0	NA	0.0	NA	NA
356 Plastic products	6	1.0	10[e]	1.6	9[e]	1.8	12[e]	1.9	100.0	33.3
361 Pottery, china, and earthenware	NA	0.0	NA	0.0	NA	0.0	NA	0.0	NA	NA
362 Glass and glass products	1	0.2	2[e]	0.3	3[e]	0.6	2[e]	0.3	100.0	-33.3
369 Other nonmetal mineral products	26	4.5	18	2.9	21[e]	4.3	26	4.1	0.0	23.8
371 Iron and steel	NA	0.0	NA	0.0	NA	0.0	NA	0.0	NA	NA
372 Nonferrous metals	1	0.2	2[e]	0.3	2[e]	0.4	2[e]	0.3	100.0	0.0
381 Metal products	9	1.6	12[e]	1.9	9[e]	1.8	12[e]	1.9	33.3	33.3
382 Nonelectrical machinery	1	0.2	1[e]	0.2	1[e]	0.2	1[e]	0.2	0.0	0.0
383 Electrical machinery	NA	0.0	NA	0.0	NA	0.0	NA	0.0	NA	NA
384 Transport equipment	5	0.9	6[e]	1.0	6[e]	1.2	6[e]	0.9	20.0	0.0
385 Professional and scientific equipment	1	0.2	1[e]	0.2	1[e]	0.2	NA	0.0	NA	NA
390 Other manufacturing industries	2	0.3	3[e]	0.5	2[e]	0.4	3[e]	0.5	50.0	50.0

Note: The industry codes shown are International Standard Industry codes (ISIC). Percentages are percent of total Value Added. 'e' stands for estimated value

Finance, Economics, and Trade

Economic Indicators [36]

- **National product**: GDP—purchasing power parity—$17 billion (1995 est.)
- **National product real growth rate**: 4.2% (1995 est.)
- **National product per capita**: $3,200 (1995 est.)
- **Inflation rate (consumer prices)**: 10.5% (1995)
- **External debt**: $1.38 billion (year-end 1995)

Balance of Payments Summary [37]

Values in millions of dollars.

	1989	1990	1991	1992	1993
Exports of goods (f.o.b.)	1,180.0	1,382.3	1,120.8	1,081.5	1,653.0
Imports of goods (f.o.b.)	-1,015.9	-1,635.8	-1,867.6	-1,950.6	-2,671.6
Trade balance	164.1	-253.5	-746.8	-869.1	-1,018.6
Services - debits	-415.6	-578.7	-661.8	-720.1	-735.6
Services - credits	483.6	604.3	1,012.2	955.2	1,109.2
Private transfers (net)	1.6	7.2	6.7	2.7	4.6
Government transfers (net)	22.3	48.4	65.6	31.2	37.4
Long-term capital (net)	30.5	-66.2	49.0	18.7	160.5
Short-term capital (net)	-50.7	125.4	117.8	-206.9	91.7
Errors and omissions	-90.6	362.4	472.0	457.7	483.2
Overall balance	145.2	249.3	314.7	-330.6	132.4

Exchange Rates [38]

Currency: **guarani.**
Symbol: **G.**

Data are currency units per $1.

January 1996	2,003.8
1995	1,970.4
1994	1,911.5
1993	1,744.3
1992	1,500.3
1991	1,325.2

Imports and Exports

Top Import Origins [39]

$2.871 billion (o.i.f., 1995).

Origins	%
Brazil	30
EU	20
US	18
Argentina	8
Japan	7

Top Export Destinations [40]

$819.5 million (f.o.b., 1995).

Destinations	%
EU	37
Brazil	25
Argentina	10
Chile	6
US	6

Foreign Aid [41]

Recipient: ODA, $38 million (1993).

Import and Export Commodities [42]

Import Commodities	Export Commodities
Capital goods	Cotton
Foodstuffs	Soybeans
Consumer goods	Timber
Raw materials	Vegetable oils
Fuels	Meat products
	Coffee
	Tung oil

Peru

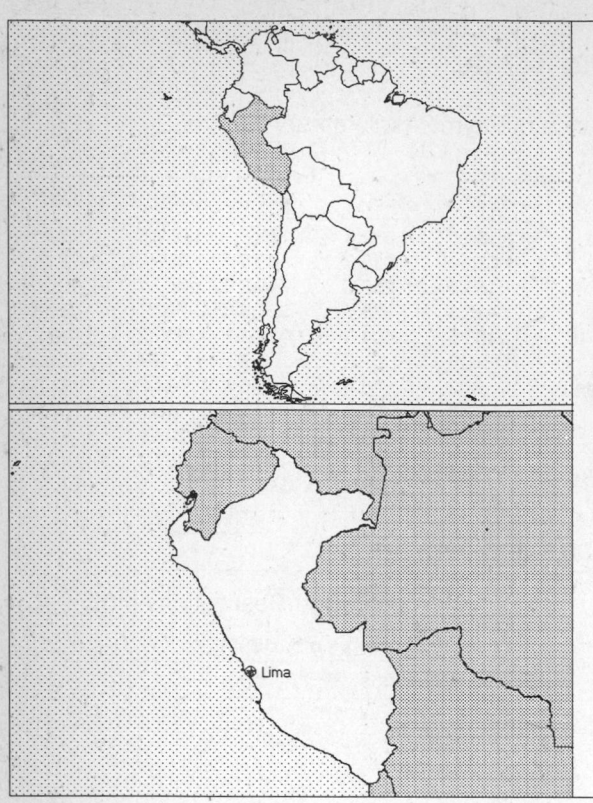

Geography [1]

Total area:
  1,285,220 sq km   496,226 sq mi
Land area:
  1,280,000 sq km   494,211 sq mi
Comparative area:
  Slightly smaller than Alaska
Land boundaries:
  Total 6,940 km, Bolivia 900 km, Brazil 1,560 km, Chile 160 km, Colombia 2,900 km, Ecuador 1,420 km
Coastline:
  2,414 km
Climate:
  Varies from tropical in East to dry desert in West
Terrain:
  Western coastal plain (costa), high and rugged Andes in center (sierra), eastern lowland jungle of Amazon Basin (selva)
Natural resources:
  Copper, silver, gold, petroleum, timber, fish, iron ore, coal, phosphate, potash
Land use:
  Arable land: 3%
  Permanent crops: 0%
  Meadows and pastures: 21%
  Forest and woodland: 55%
  Other: 21%

Demographics [2]

	1970	1980	1990	1995[1]	1996	2000	2010	2020	2030
Population	13,193	17,295	21,841	24,094	24,523	26,198	29,988	33,226	35,752
Population density (persons per sq. mi.)	27	35	44	49	50	53	61	67	72
(persons per sq. km.)	10	14	17	19	19	20	23	26	28
Net migration rate (per 1,000 population)	0.0	NA	-1.2	-0.8	-0.8	-0.6	-0.4	-0.2	-0.1
Births	NA	NA	NA	NA	597	NA	NA	NA	NA
Deaths	NA	NA	NA	NA	150	NA	NA	NA	NA
Life expectancy - males	53.9	NA	63.7	66.6	67.0	68.6	72.0	74.7	76.6
Life expectancy - females	57.3	NA	67.9	71.0	71.4	73.1	76.8	79.7	81.8
Birth rate (per 1,000)	40.5	NA	28.9	24.9	24.3	21.8	17.6	14.8	12.7
Death rate (per 1,000)	12.8	NA	7.1	6.2	6.1	5.8	5.5	5.8	6.7
Women of reproductive age (15-49 yrs.)	NA	NA	5,378	6,157	6,313	6,962	8,300	8,992	9,074
of which are currently married	NA	NA	2,968	NA	3,549	3,967	4,951	NA	NA
Fertility rate	6.0	NA	3.8	3.1	3.0	2.7	2.1	1.9	1.8

Except as noted, values for vital statistics are in thousands; life expectancy is in years.

Health

Health Indicators [3]

% of population with access to	
safe water (1990-95)	71
adequate sanitation (1990-95)	57
health services (1985-95)[1]	75
% of 1-year-olds immunized (1990-94) against	
TB (tuberculosis)	91
DPT (diphtheria, pertussis, tetanus)	87
polio	87
measles	75
% of contraceptive prevalence (1980-94)	59
ORT use rate (1990-94)	31

Health Expenditures [4]

Total health expenditure, 1990 (official exchange rate)	
Millions of dollars	1,065
Dollars per capita	49
Health expenditures as a percentage of GDP	
Total	3.2
Public sector	1.9
Private sector	1.3
Development assistance for health	
Total aid flows (millions of dollars)[1]	29
Aid flows per capita (dollars)	1.4
Aid flows as a percentage of total health expenditure	2.7

Human Factors

Women and Children [5]

% of pregnant women immunized (tetanus 1990-94)	44
% of births attended by trained health personnel (1983-94)	52
Maternal mortality rate (1980-92)	200
Under-5 mortality rate (1994)	58
% under-5 moderately/severely underweight (1980-1994)	11

Burden of Disease [6]

Population per physician (1990)	939.39
Population per nurse	NA
Population per hospital bed (1990)	707.61
AIDS cases per 100,000 people (1994)	1.7
Malaria cases per 100,000 people (1992)	245

Ethnic Division [7]

Indian	45%
Mestizo	37%
White	15%
Black, Japanese, Chinese, and other	3%

Religion [8]

Roman Catholic.

Major Languages [9]

Spanish (official), Quechua (official), Aymara.

Education

Public Education Expenditures [10]

Million (Inti)[54]	1980	1985	1990	1991	1993	1994
Total education expenditure	176	5,042	NA	NA	NA	NA
as percent of GNP	3.1	2.9	NA	NA	NA	NA
as percent of total govt. expend.	15.2	15.7	NA	NA	NA	NA
Current education expenditure	166	4,855	NA	NA	NA	NA
as percent of GNP	2.9	2.8	NA	NA	NA	NA
as percent of current govt. expend.	18.5	17.9	NA	NA	NA	NA
Capital expenditure	10	188	NA	NA	NA	NA

Educational Attainment [11]

Age group (1993)	20+
Total population	9,916,161
Highest level attained (%)	
No schooling	0.4
First level	
Not completed	30.0
Completed	7.6
Entered second level	
S-1	11.8
S-2	22.9
Postsecondary	27.3

Illiteracy [12]

In thousands and percent[1]	1990	1995	2000
Illiterate population (15+ yrs.)	1,847	1,736	1,627
Illiteracy rate - total pop. (%)	13.7	11.2	9.2
Illiteracy rate - males (%)	7.0	5.4	4.2
Illiteracy rate - females (%)	20.4	16.9	14.1

Libraries [13]

	Admin. Units	Svc. Pts.	Vols. (000)	Shelving (meters)	Vols. Added	Reg. Users
National (1992)	1	NA	3,890	NA	163,165	8,696
Nonspecialized	NA	NA	NA	NA	NA	NA
Public (1992)	609	NA	NA	NA	NA	NA
Higher ed.	NA	NA	NA	NA	NA	NA
School (1987)	143	143	18	NA	7,272	NA

Daily Newspapers [14]

	1980	1985	1990	1994
Number of papers	66	70	66	48
Circ. (000)	1,400[e]	1,600[e]	1,700[e]	2,000[e]

Culture [15]

Science and Technology

Scientific/Technical Forces [16]

Scientists/engineers	4,858
Number female	NA
Technicians	NA
Number female	NA
Total[30]	NA

R&D Expenditures [17]

	Inti (000) 1984
Total expenditure[29]	159,024,000
Capital expenditure	NA
Current expenditure	NA
Percent current	NA

U.S. Patents Issued [18]

Values show patents issued to citizens of the country by the U.S. Patents Office.

	1993	1994	1995
Number of patents	2	2	3

For sources, notes, and explanations, see Annotated Source Appendix, page 1061.

Government and Law

Organization of Government [19]

Long-form name:
Republic of Peru
Type:
Republic
Independence:
28 July 1821 (from Spain)
National holiday:
Independence Day, 28 July (1821)
Constitution:
31 December 1993
Legal system:
Based on civil law system; has not accepted compulsory ICJ jurisdiction
Executive branch:
President; Council of Ministers Note: Prime Minister since 3 April 1996 does not exercise executive power; this power is in the hands of the president
Legislative branch:
Unicameral: Congress
Judicial branch:
Supreme Court of Justice (Corte Suprema de Justicia)

Elections [20]

Democratic Constituent Congress	% of votes
Change 90/New Majority	52.1
Union for Peru (UPP)	14.0
11 other parties	33.9

Government Budget [21]

For 1996 est.

	$ bil.
Revenues	8.5
Expenditures	9.3
Capital expenditures	NA

Military Expenditures and Arms Transfers [22]

	1990	1991	1992	1993	1994
Military expenditures					
Current dollars (mil.)	702[e]	475[e]	715	737	797[e]
1994 constant dollars (mil.)	782[e]	509[e]	746	753	797[e]
Armed forces (000)	125	123	112	112	112
Gross national product (GNP)					
Current dollars (mil.)	35,460	38,440	38,600	41,800	48,980
1994 constant dollars (mil.)	39,470	41,200	40,250	42,660	48,980
Central government expenditures (CGE)					
1994 constant dollars (mil.)	5,969	5,013	6,183	6,450	6,008
People (mil.)	21.9	22.3	22.8	23.2	23.7
Military expenditure as % of GNP	2.0	1.2	1.9	1.8	1.6
Military expenditure as % of CGE	13.1	10.1	12.1	11.7	13.3
Military expenditure per capita (1994 $)	36	23	33	32	34
Armed forces per 1,000 people (soldiers)	5.7	5.5	4.9	4.8	4.7
GNP per capita (1994 $)	1,804	1,846	1,768	1,838	2,071
Arms imports[6]					
Current dollars (mil.)	50	50	60	10	20
1994 constant dollars (mil.)	56	54	63	10	20
Arms exports[6]					
Current dollars (mil.)	0	0	0	0	0
1994 constant dollars (mil.)	0	0	0	0	0
Total imports[7]					
Current dollars (mil.)	3,470	4,195	4,860	4,901	6,752
1994 constant dollars (mil.)	3,862	4,497	5,068	5,002	6,752
Total exports[7]					
Current dollars (mil.)	3,231	3,329	3,484	3,463	4,507
1994 constant dollars (mil.)	3,596	3,568	3,633	3,534	4,507
Arms as percent of total imports[8]	1.4	1.2	1.2	0.2	0.3
Arms as percent of total exports[8]	0	0	0	0	0

Crime [23]

	1990
Crime volume	
Cases known to police	102,210
Attempts (percent)	NA
Percent cases solved	40
Crimes per 100,000 persons	474.28
Persons responsible for offenses	
Total number offenders	88,922
Percent female	6
Percent juvenile (1-17 yrs.)	2
Percent foreigners	1

Human Rights [24]

	SSTS	FL	FAPRO	PPCG	APROBC	TPW	PCPTW	STPEP	PHRFF	PRW	ASST	AFL
Observes	P	P	P	P	P	P	P			P	S	P
	EAFRD	CPR	ESCR	SR	ACHR	MAAE	PVIAC	PVNAC	EAFDAW	TCIDTP	RC	
Observes	P	P	P	P	P		P	P	P	P	P	

P = Party; S = Signatory; see Appendix for meaning of abbreviations.

Labor Force

Total Labor Force [25]

8 million (1992)

Labor Force by Occupation [26]

Agriculture
Mining and quarrying
Manufacturing
Construction
Transport
Services

Unemployment Rate [27]

15%; extensive underemployment (1992 est.)

       For sources, notes, and explanations, see Annotated Source Appendix, page 1061.

Production Sectors

Commercial Energy Production and Consumption

Data are shown in quadrillion (10^{15}) BTUs and percent for 1995
Values for hydroelectric, nuclear, geothermal, solar, and wind power refer to electrical generation.

Production [28]

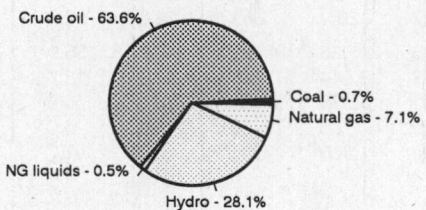

Crude oil - 63.6%
Coal - 0.7%
Natural gas - 7.1%
NG liquids - 0.5%
Hydro - 28.1%

Consumption [29]

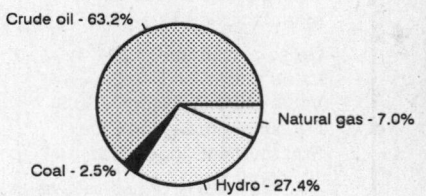

Crude oil - 63.2%
Natural gas - 7.0%
Coal - 2.5%
Hydro - 27.4%

Crude oil	0.276
Natural gas liquids	0.002
Dry natural gas	0.031
Coal	0.003
Net hydroelectric power	0.122
Total	0.434

Crude oil	0.282
Dry natural gas	0.031
Coal	0.011
Net hydroelectric power	0.122
Total	0.446

Telecommunications [30]

- 779,306 (1990 est.) telephones; adequate for most requirements
- Domestic: nationwide microwave radio relay system and a domestic satellite system with 12 earth stations
- International: satellite earth stations - 2 Intelsat (Atlantic Ocean)
- Radio: Broadcast stations: AM 273, FM 0, shortwave 144 Radios: 5.7 million (1992 est.)
- Television: Broadcast stations: 140 Televisions: 2 million (1993 est.)

Transportation [31]

Railways: total: 2,041 km; standard gauge: 1,726 km 1.435-m gauge; narrow gauge: 315 km 0.914-m gauge (1994)

Highways: total: 69,942 km; paved: 13,538 km; unpaved: 56,404 km (1987 est.)

Merchant marine: total: 9 ships (1,000 GRT or over) totaling 77,584 GRT/144,030 DWT; ships by type: bulk 2, cargo 7 (1995 est.)

Airports

Total:	230
With paved runways over 3,047 m:	5
With paved runways 2,438 to 3,047 m:	15
With paved runways 1,524 to 2,437 m:	12
With paved runways 914 to 1,523 m:	6
With paved runways under 914 m:	96

Top Agricultural Products [32]

Produces coffee, cotton, sugarcane, rice, wheat, potatoes, plantains, coca; poultry, red meats, dairy products, wool; fish catch of 6.9 million metric tons (1990).

Top Mining Products [33]

Thousand metric tons except as noted[e]	6/95[*]
Lime	14,000
Salt, all types	230
Flagstone	300
Limestone	1,600
Sand and gravel	1,000
Sulfuric acid, gross weight	220
Gas, natural, gross (mil. cu. meters)	1,200
Petroleum (000 42-gal. bls.)	
crude	46,500[r]
refinery products	55,500

Tourism [34]

	1990	1991	1992	1993	1994
Tourists	317	232	217	272	386
Tourism receipts	259	268	188	265	402
Tourism expenditures	296	263	255	268	323
Fare receipts	45	43	39	49	79
Fare expenditures	116	153	141	115	111

Travelers are in thousands, money in million U.S. dollars.

For sources, notes, and explanations, see Annotated Source Appendix, page 1061.

Manufacturing Sector

GDP and Manufacturing Summary [35]

	1980	1985	1989	1990	% change 1980-1990	% change 1989-1990
GDP (million 1980 $)	20,579	20,167	19,500	18,418	-10.5	-5.5
GDP per capita (1980 $)	1,190	1,039	924	854	-28.2	-7.6
Manufacturing as % of GDP (current prices)	20.4	24.1	20.9	27.1	32.8	29.7
Gross output (million $)	12,977	9,573	22,625[e]	14,267[e]	9.9	-36.9
Value added (million $)	4,984	3,918	8,614[e]	7,265[e]	45.8	-15.7
Value added (million 1980 $)	4,159	3,741	3,635	3,386	-18.6	-6.9
Industrial production index	100	86	80	76	-24.0	-5.0
Employment (thousands)	273	263	300[e]	295[e]	8.1	-1.7

Note: GDP stands for Gross Domestic Product. 'e' stands for estimated value.

Profitability and Productivity

	1980	1985	1989	1990	% change 1980-1990	% change 1989-1990
Intermediate input (%)	62	59	62[e]	49[e]	-21.0	-21.0
Wages, salaries, and supplements (%)	7[e]	6	7[e]	10[e]	42.9	42.9
Gross operating surplus (%)	32[e]	35	31[e]	41[e]	28.1	32.3
Gross output per worker ($)	47,484	36,350	75,312[e]	48,193[e]	1.5	-36.0
Value added per worker ($)	18,238	14,877	28,673[e]	24,575[e]	34.7	-14.3
Average wage (incl. benefits) ($)	3,176[e]	2,154	5,281[e]	4,619[e]	45.4	-12.5

Profitability is in percent of gross output. Productivity is in U.S. $. 'e' stands for estimated value.

Profitability - 1990

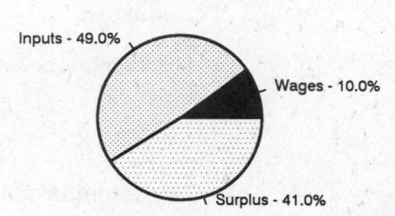

Inputs - 49.0%
Wages - 10.0%
Surplus - 41.0%

The graphic shows percent of gross output.

Value Added in Manufacturing

	1980 $ mil.	1980 %	1985 $ mil.	1985 %	1989 $ mil.	1989 %	1990 $ mil.	1990 %	% change 1980-1990	% change 1989-1990
311 Food products	767	15.4	402	10.3	1,027[e]	11.9	127[e]	1.7	-83.4	-87.6
313 Beverages	379	7.6	303	7.7	1,061[e]	12.3	605[e]	8.3	59.6	-43.0
314 Tobacco products	84	1.7	61	1.6	154[e]	1.8	95[e]	1.3	13.1	-38.3
321 Textiles	466	9.3	352	9.0	820[e]	9.5	704[e]	9.7	51.1	-14.1
322 Wearing apparel	65	1.3	52	1.3	195[e]	2.3	181[e]	2.5	178.5	-7.2
323 Leather and fur products	56	1.1	20	0.5	51[e]	0.6	37[e]	0.5	-33.9	-27.5
324 Footwear	41	0.8	20	0.5	68[e]	0.8	41[e]	0.6	0.0	-39.7
331 Wood and wood products	81	1.6	32	0.8	91[e]	1.1	59[e]	0.8	-27.2	-35.2
332 Furniture and fixtures	40	0.8	19	0.5	65[e]	0.8	49[e]	0.7	22.5	-24.6
341 Paper and paper products	156	3.1	77	2.0	210[e]	2.4	209[e]	2.9	34.0	-0.5
342 Printing and publishing	100	2.0	80	2.0	265[e]	3.1	228[e]	3.1	128.0	-14.0
351 Industrial chemicals	215	4.3	158	4.0	248[e]	2.9	313[e]	4.3	45.6	26.2
352 Other chemical products	289	5.8	193	4.9	471[e]	5.5	440[e]	6.1	52.2	-6.6
353 Petroleum refineries	192	3.9	1,154	29.5	1,479[e]	17.2	1,032[e]	14.2	437.5	-30.2
354 Miscellaneous petroleum and coal products	6	0.1	1	0.0	2[e]	0.0	1[e]	0.0	-83.3	-50.0
355 Rubber products	62	1.2	52	1.3	109[e]	1.3	142[e]	2.0	129.0	30.3
356 Plastic products	89	1.8	90	2.3	311[e]	3.6	233[e]	3.2	161.8	-25.1
361 Pottery, china, and earthenware	15	0.3	8	0.2	21[e]	0.2	7[e]	0.1	-53.3	-66.7
362 Glass and glass products	47	0.9	15	0.4	69[e]	0.8	55[e]	0.8	17.0	-20.3
369 Other nonmetal mineral products	129	2.6	113	2.9	274[e]	3.2	197[e]	2.7	52.7	-28.1
371 Iron and steel	192	3.9	123	3.1	111[e]	1.3	285[e]	3.9	48.4	156.8
372 Nonferrous metals	604	12.1	172	4.4	238[e]	2.8	210[e]	2.9	-65.2	-11.8
381 Metal products	188	3.8	113	2.9	268[e]	3.1	262[e]	3.6	39.4	-2.2
382 Nonelectrical machinery	156	3.1	58	1.5	178[e]	2.1	210[e]	2.9	34.6	18.0
383 Electrical machinery	211	4.2	111	2.8	421[e]	4.9	258[e]	3.6	22.3	-38.7
384 Transport equipment	278	5.6	106	2.7	267[e]	3.1	278[e]	3.8	0.0	4.1
385 Professional and scientific equipment	14	0.3	10	0.3	38[e]	0.4	30[e]	0.4	114.3	-21.1
390 Other manufacturing industries	58	1.2	25	0.6	104[e]	1.2	74[e]	1.0	27.6	-28.8

Note: The industry codes shown are International Standard Industry codes (ISIC). Percentages are percent of total Value Added. 'e' stands for estimated value

For sources, notes, and explanations, see Annotated Source Appendix, page 1061.

Finance, Economics, and Trade

Economic Indicators [36]

- **National product**: GDP—purchasing power parity—$87 billion (1995 est.)
- **National product real growth rate**: 6.8% (1995 est.)
- **National product per capita**: $3,600 (1995 est.)
- **Inflation rate (consumer prices)**: 10.2% (1995 est.)
- **External debt**: $22.4 billion (1994 est.)

Balance of Payments Summary [37]

Values in millions of dollars.

	1989	1990	1991	1992	1993
Exports of goods (f.o.b.)	3,488	3,231	3,330	3,485	3,463
Imports of goods (f.o.b.)	-2,291	-2,892	-3,495	-4,050	-4,043
Trade balance	1,197	339	-165	-565	-580
Services - debits	-2,309	-2,432	-2,478	-2,561	-2,682
Services - credits	971	917	994	983	1,045
Private transfers (net)	0	0	0	0	0
Government transfers (net)	235	275	429	457	449
Long-term capital (net)	-218	-886	5,980	323	2,450
Short-term capital (net)	933	1,761	-5,491	583	-1,515
Errors and omissions	-214	312	1,618	1,348	1,278
Overall balance	595	286	887	568	445

Exchange Rates [38]

Currency: **nuevo sol.**
Symbol: **S/.**

Data are currency units per $1.

January 1996	2.350
1995	2.253
1994	2.195
1993	1.988
1992	1.246
1991	0.773

Imports and Exports

Top Import Origins [39]

$7.4 billion (f.o.b., 1995 est.).

Origins	%
US	21
Colombia	NA
Argentina	NA
Japan	NA
Germany	NA
Brazil	NA

Top Export Destinations [40]

$5.6 billion (f.o.b., 1995 est.).

Destinations	%
US	19
Japan	9
Italy	NA
Germany	NA

Foreign Aid [41]

Recipient: ODA, $363 million (1993).

Import and Export Commodities [42]

Import Commodities	**Export Commodities**
Machinery	Copper
Transport equipment	Zinc
Foodstuffs	Fishmeal
Petroleum	Crude petroleum and
Iron and steel	byproducts
Chemicals	Refined silver
Pharmaceuticals	Lead
	Coffee
	Cotton

For sources, notes, and explanations, see Annotated Source Appendix, page 1061.

751

Philippines

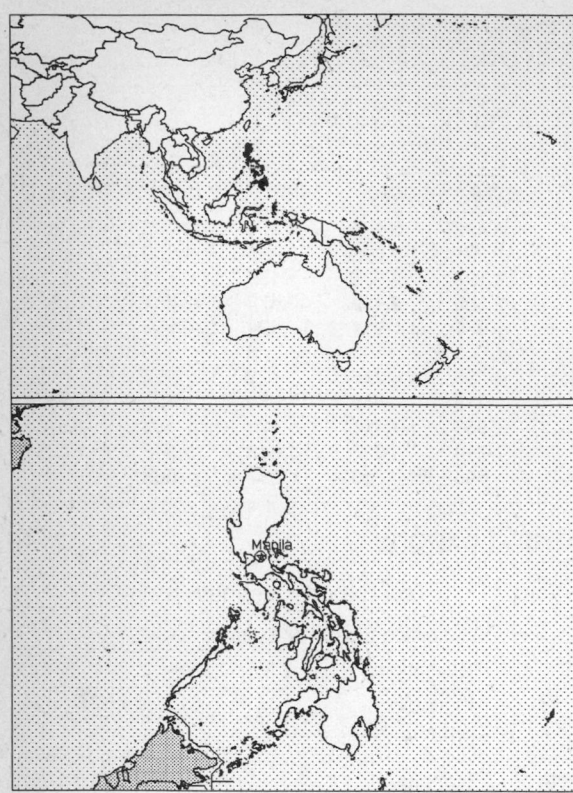

Geography [1]

Total area:
300,000 sq km   115,831 sq mi
Land area:
298,170 sq km   115,124 sq mi
Comparative area:
Slightly larger than Arizona
Land boundaries:
0 km
Coastline:
36,289 km
Climate:
Tropical marine; northeast monsoon (November to April); southwest monsoon (May to October)
Terrain:
Mostly mountains with narrow to extensive coastal lowlands
Natural resources:
Timber, petroleum, nickel, cobalt, silver, gold, salt, copper
Land use:
Arable land: 26%
Permanent crops: 11%
Meadows and pastures: 4%
Forest and woodland: 40%
Other: 19%

Demographics [2]

	1970	1980	1990	1995[1]	1996	2000	2010	2020	2030
Population	38,680	51,092	65,037	72,860	74,481	80,961	97,119	112,963	127,599
Population density (persons per sq. mi.)	336	444	565	633	647	703	844	981	1,108
(persons per sq. km.)	130	171	218	244	250	272	326	379	428
Net migration rate (per 1,000 population)	-0.9	-1.9	-1.7	-1.1	-1.1	-1.0	-0.9	-0.8	-0.7
Births	1,533	NA	NA	NA	2,198	NA	NA	NA	NA
Deaths	388	NA	NA	NA	496	NA	NA	NA	NA
Life expectancy - males	56.0	60.0	61.3	62.9	63.1	64.0	66.1	68.0	69.7
Life expectancy - females	62.4	64.7	67.5	68.6	68.8	69.7	71.8	73.8	75.5
Birth rate (per 1,000)	39.6	36.2	32.3	30.0	29.5	27.3	23.5	20.7	18.0
Death rate (per 1,000)	10.1	8.3	7.3	6.7	6.7	6.4	6.1	6.2	6.7
Women of reproductive age (15-49 yrs.)	NA	12,177	16,241	18,566	19,068	20,988	25,670	29,994	33,269
of which are currently married	4,910	6,997	9,514	NA	11,328	12,628	15,692	NA	NA
Fertility rate	5.8	5.0	4.1	3.8	3.7	3.4	2.9	2.5	2.3

Except as noted, values for vital statistics are in thousands; life expectancy is in years.

Health

Health Indicators [3]

% of population with access to	
safe water (1990-95)	85
adequate sanitation (1990-95)	69
health services (1985-95)	76
% of 1-year-olds immunized (1990-94) against	
TB (tuberculosis)	89
DPT (diphtheria, pertussis, tetanus)	86
polio	88
measles	87
% of contraceptive prevalence (1980-94)	40
ORT use rate (1990-94)	63

Health Expenditures [4]

Total health expenditure, 1990 (official exchange rate)	
Millions of dollars	883
Dollars per capita	14
Health expenditures as a percentage of GDP	
Total	2.0
Public sector	1.0
Private sector	1.0
Development assistance for health	
Total aid flows (millions of dollars)[1]	69
Aid flows per capita (dollars)	1.1
Aid flows as a percentage of total health expenditure	7.8

For sources, notes, and explanations, see Annotated Source Appendix, page 1061.

Human Factors

Women and Children [5]

% of pregnant women immunized (tetanus 1990-94)	69
% of births attended by trained health personnel (1983-94)	53
Maternal mortality rate (1980-92)	210
Under-5 mortality rate (1994)	57
% under-5 moderately/severely underweight (1980-1994)	34

Burden of Disease [6]

Population per physician (1989)	8,116.83
Population per nurse (1984)	2,683.65
Population per hospital bed (1991)	655.19
AIDS cases per 100,000 people (1994)	0.1
Malaria cases per 100,000 people (1992)	174

Ethnic Division [7]

Christian Malay	91.5%
Muslim Malay	4%
Chinese	1.5%
Other	3%

Religion [8]

Roman Catholic	83%
Protestant	9%
Muslim	5%
Buddhist and other	3%

Major Languages [9]

Filipino (official, based on Tagalog), English (official).

Education

Public Education Expenditures [10]

	1980	1990	1991	1992	1993	1994
Million (Peso)[55]						
Total education expenditure	4,191	31,067	37,033	31,687	36,320	NA
as percent of GNP	1.7	2.9	2.9	2.3	2.4	NA
as percent of total govt. expend.	9.1	10.1	10.5	NA	NA	NA
Current education expenditure	4,023	28,713	32,872	NA	NA	NA
as percent of GNP	1.7	2.7	2.6	NA	NA	NA
as percent of current govt. expend.	13.0	11.1	11.1	NA	NA	NA
Capital expenditure	168	2,354	4,161	NA	NA	NA

Educational Attainment [11]

Age group (1990)	25+
Total population	24,156,427
Highest level attained (%)	
No schooling	6.7
First level	
Not completed	46.9
Completed	NA
Entered second level	
S-1	27.2
S-2	NA
Postsecondary	18.7

Illiteracy [12]

In thousands and percent[1]	1990	1995	2000
Illiterate population (15+ yrs.)	2,350	2,234	2,024
Illiteracy rate - total pop. (%)	6.0	5.0	4.0
Illiteracy rate - males (%)	5.7	4.8	3.8
Illiteracy rate - females (%)	6.3	5.2	4.1

Libraries [13]

	Admin. Units	Svc. Pts.	Vols. (000)	Shelving (meters)	Vols. Added	Reg. Users
National (1993)	1	4	902	NA	71,932	189,798
Nonspecialized (1986)	1	1	NA	31	NA	232
Public (1989)	1	517	5,756	NA	NA	2 mil
Higher ed.	NA	NA	NA	NA	NA	NA
School	NA	NA	NA	NA	NA	NA

Daily Newspapers [14]

	1980	1985	1990	1994
Number of papers	22	15	47	42
Circ. (000)	2,000	2,170	3,400[e]	4,286

Culture [15]

Science and Technology

Scientific/Technical Forces [16]

Scientists/engineers	4,830
Number female	2,319
Technicians	1,855
Number female	NA
Total[1]	6,685

R&D Expenditures [17]

	Peso (000) 1984
Total expenditure	613,410
Capital expenditure	98,610
Current expenditure	514,800
Percent current	83.9

U.S. Patents Issued [18]

Values show patents issued to citizens of the country by the U.S. Patents Office.

	1993	1994	1995
Number of patents	5	1	4

For sources, notes, and explanations, see Annotated Source Appendix, page 1061.

753

Government and Law

Organization of Government [19]

Long-form name:
Republic of the Philippines
Type:
Republic
Independence:
4 July 1946 (from US)
National holiday:
Independence Day, 12 June (1898) (from Spain)
Constitution:
2 February 1987, effective 11 February 1987
Legal system:
Based on Spanish and Anglo-American law; accepts compulsory ICJ jurisdiction, with reservations
Executive branch:
President; Vice President; Executive Secretary
Legislative branch:
Bicameral Congress (Kongreso): Senate (Senado) and House of Representatives (Kapulungan Ng Mga Kinatawan)
Judicial branch:
Supreme Court

Elections [20]

House of Representatives	% of seats
Lakas/People Power (NUCD)	63.2
Democratic Filipino Struggle (LDP)	14.2
National People's Coalition (NPC)	12.3
Liberal	2.9
Filipino Democratic Party	1.5
New Society Movement (KBL)	0.5
Results pending	5.4

Government Expenditures [21]

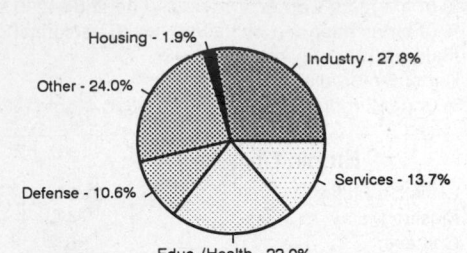

Housing - 1.9%
Industry - 27.8%
Other - 24.0%
Defense - 10.6%
Services - 13.7%
Educ./Health - 22.0%

(% distribution). Expend. for CY93: 282,296 (Peso mil.)

Military Expenditures and Arms Transfers [22]

	1990	1991	1992	1993	1994
Military expenditures					
Current dollars (mil.)	1,165	1,162	1,120	1,324	1,272
1994 constant dollars (mil.)	1,297	1,246	1,168	1,351	1,272
Armed forces (000)	109	107	107	107	109
Gross national product (GNP)					
Current dollars (mil.)	54,100	56,560	58,960	61,820	66,300
1994 constant dollars (mil.)	60,200	60,630	61,490	63,090	66,300
Central government expenditures (CGE)					
1994 constant dollars (mil.)	12,130	11,830	11,480	11,720	12,000
People (mil.)	65.2	66.7	68.4	70.0	71.6
Military expenditure as % of GNP	2.2	2.1	1.9	2.1	1.9
Military expenditure as % of CGE	10.7	10.5	10.2	11.5	10.6
Military expenditure per capita (1994 $)	20	19	17	19	18
Armed forces per 1,000 people (soldiers)	1.7	1.6	1.6	1.5	1.5
GNP per capita (1994 $)	924	908	899	901	926
Arms imports[6]					
Current dollars (mil.)	120	140	130	50	100
1994 constant dollars (mil.)	134	150	136	51	100
Arms exports[6]					
Current dollars (mil.)	0	0	0	0	0
1994 constant dollars (mil.)	0	0	0	0	0
Total imports[7]					
Current dollars (mil.)	13,040	12,790	15,450	18,750	22,550
1994 constant dollars (mil.)	14,510	13,710	16,110	19,140	22,550
Total exports[7]					
Current dollars (mil.)	8,068	8,767	9,752	11,090	13,340
1994 constant dollars (mil.)	8,979	9,397	10,170	11,320	13,340
Arms as percent of total imports[8]	0.9	1.1	0.8	0.3	0.4
Arms as percent of total exports[8]	0	0	0	0	0

Crime [23]

	1990
Crime volume	
Cases known to police	79,711
Attempts (percent)	NA
Percent cases solved	NA
Crimes per 100,000 persons	122.7
Persons responsible for offenses	
Total number offenders	NA
Percent female	NA
Percent juvenile	NA
Percent foreigners	NA

Human Rights [24]

	SSTS	FL	FAPRO	PPCG	APROBC	TPW	PCPTW	STPEP	PHRFF	PRW	ASST	AFL
Observes	P		P	P	P	P	P	P		P	P	P
		EAFRD	CPR	ESCR	SR	ACHR	MAAE	PVIAC	PVNAC	EAFDAW	TCIDTP	RC
Observes		P	P	P	P			S	P	P	P	P

P = Party; S = Signatory; see Appendix for meaning of abbreviations.

Labor Force

Total Labor Force [25]

24.12 million

Labor Force by Occupation [26]

Agriculture	46%
Industry and commerce	16
Services	18.5
Government	10
Other	9.5

Date of data: 1989

Unemployment Rate [27]

9.5% (1995 est.)

Production Sectors

Commercial Energy Production and Consumption

Data are shown in quadrillion (10^{15}) BTUs and percent for 1995
Values for hydroelectric, nuclear, geothermal, solar, and wind power refer to electrical generation.

Production [28]

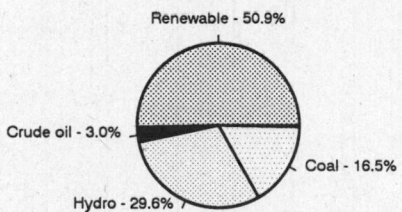

Renewable - 50.9%
Crude oil - 3.0%
Hydro - 29.6%
Coal - 16.5%

Consumption [29]

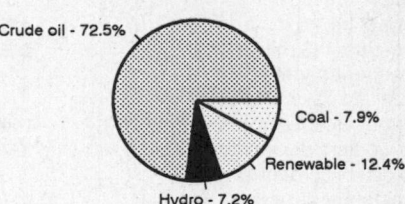

Crude oil - 72.5%
Coal - 7.9%
Renewable - 12.4%
Hydro - 7.2%

Crude oil	0.007	
Coal	0.038	
Net hydroelectric power	0.068	
Geothermal, solar, wind	0.117	
Total	0.230	

Crude oil	0.683
Coal	0.074
Net hydroelectric power	0.068
Geothermal, solar, wind	0.117
Total	0.942

Telecommunications [30]

- 887,229 (1993 est.) telephones; good international radiotelephone and submarine cable services; domestic and interisland service adequate
- Domestic: domestic satellite system with 11 earth stations
- International: submarine cables to Hong Kong, Guam, Singapore, Taiwan, and Japan; satellite earth stations - 3 Intelsat (1 Indian Ocean and 2 Pacific Ocean)
- Radio: Broadcast stations: AM 261, FM 55, shortwave 0 Radios: 9.03 million (1992 est.)
- Television: Broadcast stations: 29 Televisions: 7 million (1993 est.)

Transportation [31]

Railways: total: 499 km; narrow gauge: 499 km 1.067-m gauge (1993)

Highways: total: 160,633 km; paved: 22,489 km; unpaved: 138,144 km (1992 est.)

Merchant marine: total: 535 ships (1,000 GRT or over) totaling 8,033,849 GRT/13,101,188 DWT; ships by type: bulk 230, cargo 126, chemical tanker 3, combination bulk 11, container 12, liquefied gas tanker 9, livestock carrier 12, oil tanker 44, passenger 2, passenger-cargo 12, refrigerated cargo 19, roll-on/roll-off cargo 12, short-sea passenger 18, vehicle carrier 25

Airports

Total:	235
With paved runways over 3,047 m:	2
With paved runways 2,438 to 3,047 m:	7
With paved runways 1,524 to 2,437 m:	25
With paved runways 914 to 1,523 m:	31
With paved runways under 914 m:	104

Top Agricultural Products [32]

Agriculture accounts for 22% of the GDP; produces rice, coconuts, corn, sugarcane, bananas, pineapples, mangoes; pork, eggs, beef; fish catch of 2 million metric tons annually.

Top Mining Products [33]

Thousand metric tons except as noted[e]	4/7/95[*]
Gold, mine output, Au content (kg.)	35,000
Steel, crude	700
Silver, mine output, Ag content (kg.)	52,500
Cement, hydraulic	10,000
Clays, incl. bentonite, red, white, other	958
Pyrite and pyrrhotite, gross weight	430
Salt, marine	786
Limestone	5,400[60]
Coal, all grades	1,500
Petroleum refinery products (000 42-gal. bls.)	102,000

Tourism [34]

	1990	1991	1992	1993	1994
Visitors[87]	1,025	951	1,153	1,372	1,574
Tourists	893	849	1,043	1,246	1,414
Cruise passengers	23	5	4	2	4
Tourism receipts[88]	1,306	1,281	1,674	2,122	2,282
Tourism expenditures	111	61	102	130	196
Fare receipts	40	37	128	71	54
Fare expenditures	149	236	283	278	181

Travelers are in thousands, money in million U.S. dollars.

For sources, notes, and explanations, see Annotated Source Appendix, page 1061.

755

Manufacturing Sector

Manufacturing Summary [35]

	1987		1988		1989		1990		1991	
	$ bil.	%	$ bil.	%	$ bil.	%	$ bil.	%	$ bil.	%
Establishments or enterprises (number)	5,369	0.229	79,168	3.771	-	-	-	-	-	-
Total employment (000)	775	0.571	1,103	0.806	-	-	-	-	-	-
Production workers (000)	574	0.852	690	1.104	-	-	-	-	-	-
Output ($ bil.)	17	0.164	22	0.189	-	-	-	-	-	-
Value added ($ bil.)	5	0.107	8	0.159	-	-	-	-	-	-
Capital investment ($ mil.)	628	0.144	993	0.208	-	-	-	-	-	-
M & E investment ($ mil.)	-		-		-	-	-	-	-	-
Employees per establishment (number)	144	249.538	14	21.363	-	-	-	-	-	-
Production workers per establishment	107	372.105	9	29.284	-	-	-	-	-	-
Output per establishment ($ mil.)	3	71.836	0.278	5.006	-	-	-	-	-	-
Capital investment per estab. ($ mil.)	0.117	62.924	0.013	5.526	-	-	-	-	-	-
M & E per establishment ($ mil)	-		-		-	-	-	-	-	-
Payroll per employee ($)	2,019	22.515	1,983	19.901	-	-	-	-	-	-
Wages per production worker ($)	1,390	17.611	1,285	15.212	-	-	-	-	-	-
Hours per production worker (hours)	2,604	137.235	2,706	140.809	-	-	-	-	-	-
Output per employee ($)	21,560	28.788	19,959	23.432	-	-	-	-	-	-
Capital investment per employee ($)	810	25.216	900	25.869	-	-	-	-	-	-
M & E per employee ($)	-		-		-	-	-	-	-	-

Note: Columns headed % show percent of world total or ratio. Ratios closest to 100 are closest to world average. M & E stands for machinery & equipment.

Output in Manufacturing

	1987		1988		1989		1990		1991	
	$ bil.	%	$ bil.	%	$ bil.	%	$ bil.	%	$ bil.	%
3110 - Food products	3.384	19.91	4.210	19.14	-	-	-	-	-	-
3130 - Beverages	0.953	5.61	1.196	5.44	-	-	-	-	-	-
3140 - Tobacco	0.757	4.45	0.768	3.49	-	-	-	-	-	-
3210 - Textiles	0.745	4.38	0.932	4.24	-	-	-	-	-	-
3211 - Spinning, weaving, etc.	0.531	3.12	0.671	3.05	-	-	-	-	-	-
3220 - Wearing apparel	0.456	2.68	0.763	3.47	-	-	-	-	-	-
3230 - Leather and products	0.013	0.08	0.031	0.14	-	-	-	-	-	-
3240 - Footwear	0.023	0.14	0.048	0.22	-	-	-	-	-	-
3310 - Wood products	0.378	2.22	0.537	2.44	-	-	-	-	-	-
3320 - Furniture, fixtures	0.101	0.59	0.210	0.95	-	-	-	-	-	-
3410 - Paper and products	0.353	2.08	0.488	2.22	-	-	-	-	-	-
3411 - Pulp, paper, etc.	0.209	1.23	0.298	1.35	-	-	-	-	-	-
3420 - Printing, publishing	0.173	1.02	0.266	1.21	-	-	-	-	-	-
3510 - Industrial chemicals	0.573	3.37	0.687	3.12	-	-	-	-	-	-
3511 - Basic chemicals, excl fertilizers	0.204	1.20	0.249	1.13	-	-	-	-	-	-
3513 - Synthetic resins, etc.	0.077	0.45	0.136	0.62	-	-	-	-	-	-
3520 - Chemical products nec	1.004	5.91	1.451	6.60	-	-	-	-	-	-
3522 - Drugs and medicines	0.444	2.61	0.625	2.84	-	-	-	-	-	-
3530 - Petroleum refineries	2.154	12.67	2.109	9.59	-	-	-	-	-	-
3540 - Petroleum, coal products	0.014	0.08	0.024	0.11	-	-	-	-	-	-
3550 - Rubber products	0.210	1.24	0.402	1.83	-	-	-	-	-	-
3560 - Plastic products nec	0.192	1.13	0.319	1.45	-	-	-	-	-	-
3610 - Pottery, china, etc.	0.023	0.14	0.039	0.18	-	-	-	-	-	-
3620 - Glass and products	0.161	0.95	0.143	0.65	-	-	-	-	-	-
3690 - Nonmetal products nec	0.063	0.37	0.372	1.69	-	-	-	-	-	-
3710 - Iron and steel	0.707	4.16	0.938	4.26	-	-	-	-	-	-
3720 - Nonferrous metals	0.392	2.31	0.566	2.57	-	-	-	-	-	-
3810 - Metal products	0.212	1.25	0.316	1.44	-	-	-	-	-	-
3820 - Machinery nec	0.121	0.71	0.178	0.81	-	-	-	-	-	-
3825 - Office, computing machinery	0.020	0.12	0.025	0.11	-	-	-	-	-	-
3830 - Electrical machinery	0.921	5.42	1.216	5.53	-	-	-	-	-	-
3832 - Radio, television, etc.	0.691	4.06	0.879	4.00	-	-	-	-	-	-
3840 - Transportation equipment	0.212	1.25	0.424	1.93	-	-	-	-	-	-
3841 - Shipbuilding, repair	0.031	0.18	0.039	0.18	-	-	-	-	-	-
3843 - Motor vehicles	0.126	0.74	0.297	1.35	-	-	-	-	-	-
3850 - Professional goods	0.016	0.09	0.018	0.08	-	-	-	-	-	-
3900 - Industries nec	0.068	0.40	0.136	0.62	-	-	-	-	-	-

Note: Codes are International Standard Industry codes (ISIC). Percentages are % of total Output. [f]: Factor Prices; [p]: Producer Prices.

 For sources, notes, and explanations, see Annotated Source Appendix, page 1061.

Finance, Economics, and Trade

Economic Indicators [36]

- **National product**: GDP—purchasing power parity—$179.7 billion (1995 est.)
- **National product real growth rate**: 4.8% (1995)
- **National product per capita**: $2,530 (1995 est.)
- **Inflation rate (consumer prices)**: 8.1% (1995)
- **External debt**: $41 billion (1995 est.)

Balance of Payments Summary [37]

Values in millions of dollars.

	1989	1990	1991	1992	1993
Exports of goods (f.o.b.)	7,821	8,186	8,840	9,824	11,375
Imports of goods (f.o.b.)	-10,419	-12,206	-12,051	-14,519	-17,597
Trade balance	-2,598	-4,020	-3,211	-4,695	-6,222
Services - debits	-4,274	-4,231	-4,273	-4,618	-5,294
Services - credits	4,586	4,842	5,623	7,497	7,528
Private transfers (net)	473	357	473	473	398
Government transfers (net)	357	357	354	344	301
Long-term capital (net)	1,424	1,707	1,888	1,630	2,801
Short-term capital (net)	-70	350	1,039	1,578	485
Errors and omissions	402	593	-138	-520	292
Overall balance	300	-45	1,755	1,689	289

Exchange Rates [38]

Currency: **Philippine peso.**
Symbol: **P.**

Data are currency units per $1.

December 1995	26.206
1995	25.714
1994	26.417
1993	27.120
1992	25.512
1991	27.479

Imports and Exports

Top Import Origins [39]

$26.5 billion (f.o.b., 1995) Data are for 1994.

Origins	%
Japan	24
US	18
Singapore	7
Taiwan	6
South Korea	5

Top Export Destinations [40]

$17.4 billion (f.o.b., 1995) Data are for 1994.

Destinations	%
US	39
Japan	15
Germany	5
Hong Kong	5
UK	5

Foreign Aid [41]

Recipient: ODA, $934 million (1993).

Import and Export Commodities [42]

Import Commodities

Raw materials 40%
Capital goods 25%
Petroleum products 10%

Export Commodities

Electronics
Textiles
Coconut products
Copper
Fish

Poland

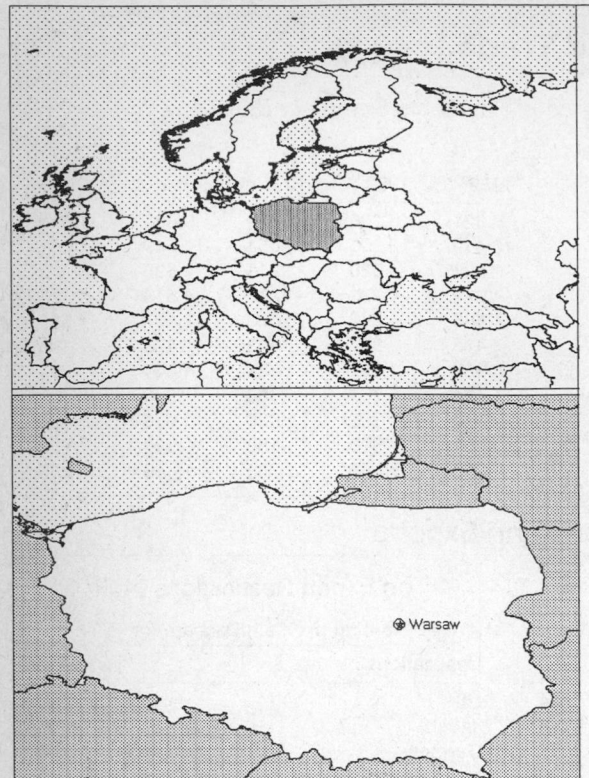

Geography [1]

Total area:
312,683 sq km   120,728 sq mi
Land area:
304,510 sq km   117,572 sq mi
Comparative area:
Slightly smaller than New Mexico
Land boundaries:
Total 2,888 km, Belarus 605 km, Czech Republic 658 km, Germany 456 km, Lithuania 91 km, Russia (Kaliningrad Oblast) 206 km, Slovakia 444 km, Ukraine 428 km
Coastline:
491 km
Climate:
Temperate with cold, cloudy, moderately severe winters with frequent precipitation; mild summers with frequent showers and thundershowers
Terrain:
Mostly flat plain; mountains along southern border
Natural resources:
Coal, sulfur, copper, natural gas, silver, lead, salt
Land use:
Arable land: 48%
Permanent crops: 0%
Meadows and pastures: 13%
Forest and woodland: 29%
Other: 10% (1992)

Demographics [2]

	1970	1980	1990	1995[1]	1996	2000	2010	2020	2030
Population	32,526	35,578	38,109	38,589	38,643	39,010	40,342	40,833	40,827
Population density (persons per sq. mi.)	277	303	324	328	329	332	343	347	347
(persons per sq. km.)	107	117	125	127	127	128	132	134	134
Net migration rate (per 1,000 population)	-0.7	-0.6	-0.4	-0.4	-0.4	-0.3	-0.2	-0.1	0.0
Births	546	693	NA	NA	461	NA	NA	NA	NA
Deaths	267	350	388	NA	390	NA	NA	NA	NA
Life expectancy - males	NA	NA	66.5	67.9	68.0	68.5	69.6	73.1	75.7
Life expectancy - females	NA	NA	75.5	76.3	76.4	77.0	78.3	80.8	82.7
Birth rate (per 1,000)	16.8	19.5	14.3	11.8	11.9	14.2	13.1	10.4	10.1
Death rate (per 1,000)	8.2	9.8	10.2	10.1	10.1	10.2	10.7	10.0	10.6
Women of reproductive age (15-49 yrs.)	NA	NA	9,388	9,989	10,084	10,250	9,626	9,624	9,212
of which are currently married[43]	5,429	6,144	6,485	NA	6,839	6,878	6,757	NA	NA
Fertility rate	2.2	2.3	2.0	1.7	1.7	1.9	1.8	1.7	1.6

Except as noted, values for vital statistics are in thousands; life expectancy is in years.

Health

Health Indicators [3]

% of population with access to	
safe water (1990-95)	NA
adequate sanitation (1990-95)	NA
health services (1985-95)	NA
% of 1-year-olds immunized (1990-94) against	
TB (tuberculosis)	95
DPT (diphtheria, pertussis, tetanus)	95
polio	95
measles	95
% of contraceptive prevalence (1980-94)[1]	75
ORT use rate (1990-94)	NA

Health Expenditures [4]

Total health expenditure, 1990 (official exchange rate)	
Millions of dollars	3,157
Dollars per capita	83
Health expenditures as a percentage of GDP	
Total	5.1
Public sector	4.1
Private sector	1.0
Development assistance for health	
Total aid flows (millions of dollars)[1]	NA
Aid flows per capita (dollars)	NA
Aid flows as a percentage of total health expenditure	NA

For sources, notes, and explanations, see Annotated Source Appendix, page 1061.

Human Factors

Women and Children [5]

% of pregnant women immunized (tetanus 1990-94)	NA
% of births attended by trained health personnel (1983-94)[1]	100
Maternal mortality rate (1980-92)	11
Under-5 mortality rate (1994)	16
% under-5 moderately/severely underweight (1980-1994)	NA

Burden of Disease [6]

Population per physician (1994)	439.44
Population per nurse (1994)	184.80
Population per hospital bed (1993)	179.73
AIDS cases per 100,000 people (1994)	0.2
Heart disease cases per 1,000 people (1990-93)	220.5

Ethnic Division [7]

Polish	97.6%
German	1.3%
Ukrainian	0.6%
Byelorussian	0.5%
(1990 est.)	

Religion [8]

About 75% of Roman Catholics are practicing.

Roman Catholic	95%
Eastern Orthodox, Protestant, and other	5%

Major Languages [9]

Polish.

Education

Public Education Expenditures [10]

	1980	1990	1991	1992	1993	1994
Million or Trillion (T) (Zloty)						
Total education expenditure	NA	NA	41.96 T	62.44 T	83.70 T	NA
as percent of GNP	NA	NA	5.4	5.5	5.5	NA
as percent of total govt. expend.	NA	NA	14.6	14.0	14.0	NA
Current education expenditure	79,984	28,249	38,969	58,468	77,947	NA
as percent of GNP	3.3	5.3	5.0	5.2	5.1	NA
as percent of current govt. expend.	7.0	16.4	15.0	14.2	15.5	NA
Capital expenditure	NA	NA	2.99 T	3.97 T	5.75 T	NA

Educational Attainment [11]

Age group (1988)	25+
Total population	22,986,018
Highest level attained (%)	
No schooling	1.5
First level	
Not completed	5.6
Completed	37.2
Entered second level	
S-1	47.8
S-2	NA
Postsecondary	7.9

Illiteracy [12]

	1970	1978	1980
Illiterate population 15+ years[4]	NA	334,586	NA
Illiteracy rate - total pop. (%)	NA	1.2	NA
Illiteracy rate - males (%)	NA	0.7	NA
Illiteracy rate - females (%)	NA	1.7	NA

Libraries [13]

	Admin. Units	Svo. Pts.	Vols. (000)	Shelving (meters)	Vols. Added	Reg. Users
National (1991)	1	NA	2,209	NA	45,291	6,293
Nonspecialized (1991)	261	NA	10,629	NA	174,839	99,023
Public (1993)	9,605	NA	135,928	NA	5 mil	7 mil
Higher ed. (1993)	126	1,027	43,182	NA	1 mil	988,681
School (1991)	21,538	NA	157,901	NA	5 mil	8 mil

Daily Newspapers [14]

	1980	1985	1990	1994
Number of papers	43	45	67	66
Circ. (000)	8,407	7,714	4,889	5,404

Culture [15]

Cinema (seats per 1,000)	5.5
Annual attendance per person	0.4
Gross box office receipts (mil. Zloty)	346,466
Museums (reporting)	552
Visitors (000)	15,629
Annual receipts (mil. Zloty)	1,011,000

Science and Technology

Scientific/Technical Forces [16]

Scientists/engineers	41,440
Number female	NA
Technicians	52,810
Number female	NA
Total[9]	94,250

R&D Expenditures [17]

	Zloty (000,000) 1992
Total expenditure[6]	9,557,064
Capital expenditure	743,185
Current expenditure	8,813,879
Percent current	92.2

U.S. Patents Issued [18]

Values show patents issued to citizens of the country by the U.S. Patents Office.

	1993	1994	1995
Number of patents	8	9	8

For sources, notes, and explanations, see Annotated Source Appendix, page 1061.

759

Government and Law

Organization of Government [19]

Long-form name:
Republic of Poland
Type:
Democratic state
Independence:
11 November 1918 (independent republic proclaimed)
National holiday:
Constitution Day, 3 May (1791)
Constitution:
Interim "small constitution" in force in December 1992 replacing the communist-imposed constitution of 1952; new democratic constitution being drafted
Legal system:
Mix of Continental (Napoleonic) civil law and holdover communist legal theory; gradual changes being introduced as part of broader democratization; limited judicial review of legislative acts; has not accepted compulsory ICJ jurisdiction
Executive branch:
President; Prime Minister; 3 Deputy Prime Ministers; Council of Ministers
Legislative branch:
Bicameral National Assembly: Sejm and Senate
Judicial branch:
Supreme Court

Elections [20]

Sejm	% of seats
Democratic Left Alliance (SLD)	37.1
Polish Peasant Party (PSL)	28.7
Freedom Union (UW)	16.1
Union of Labour	8.9
Confederation for an Indpendent Poland (KPN)	4.8
Nonparty Bloc for the Support of Reforms (BBWR)	3.5
Ethnic German parties	0.9

Government Budget [21]

For 1995 est.

	$ bil.
Revenues	34.5
Expenditures	37.8
Capital expenditures	NA

Crime [23]

	1994
Crime volume	
Cases known to police	906,157
Attempts (percent)	3.20
Percent cases solved	46.90
Crimes per 100,000 persons	2,351.00
Persons responsible for offenses	
Total number offenders	388,855
Percent female	9.10
Percent juvenile (13-16 yrs.)	15.70
Percent foreigners	1.00

Military Expenditures and Arms Transfers [22]

	1990	1991	1992	1993	1994
Military expenditures					
Current dollars (mil.)	8,752[r]	7,362[r]	4,152[r]	4,334[r]	4,945[e]
1994 constant dollars (mil.)	9,740[r]	7,891[r]	4,329[r]	4,423[r]	4,945[e]
Armed forces (000)	313	305	270	180	255
Gross national product (GNP)					
Current dollars (mil.)[e]	163,300	160,100	181,300	192,600	206,300
1994 constant dollars (mil.)[e]	181,700	171,600	189,000	196,600	206,300
Central government expenditures (CGE)					
1994 constant dollars (mil.)	NA	24,010[e]	50,660[e]	61,920[e]	88,900
People (mil.)	38.1	38.3	38.4	38.5	38.7
Military expenditure as % of GNP	5.4	4.6	2.3	2.3	2.4
Military expenditure as % of CGE	NA	32.9	8.5	7.1	5.6
Military expenditure per capita (1994 $)	256	206	113	115	128
Armed forces per 1,000 people (soldiers)	8.2	8.0	7.0	4.7	6.6
GNP per capita (1994 $)	4,769	4,486	4,924	5,103	5,336
Arms imports[6]					
Current dollars (mil.)	250	0	0	0	5
1994 constant dollars (mil.)	278	0	0	0	5
Arms exports[6]					
Current dollars (mil.)	60	110	20	10	30
1994 constant dollars (mil.)	67	118	21	10	30
Total imports[7]					
Current dollars (mil.)	8,413	15,760	15,700	18,830	21,380
1994 constant dollars (mil.)	9,363	16,890	16,370	19,220	21,380
Total exports[7]					
Current dollars (mil.)	13,630	14,900	13,320	14,140	17,040
1994 constant dollars (mil.)	15,170	15,970	13,890	14,430	17,040
Arms as percent of total imports[8]	3.0	0	0	0	0
Arms as percent of total exports[8]	0.4	0.7	0.2	0.1	0.2

Human Rights [24]

	SSTS	FL	FAPRO	PPCG	APROBC	TPW	PCPTW	STPEP	PHRFF	PRW	ASST	AFL
Observes	2	P	P	P	P	P	P	P	P	P	P	P

	EAFRD	CPR	ESCR	SR	ACHR	MAAE	PVIAC	PVNAC	EAFDAW	TCIDTP	RC
Observes	P	P	P	P		P	P	P	P	P	P

P = Party; S = Signatory; see Appendix for meaning of abbreviations.

Labor Force

Total Labor Force [25]

17.743 million (1994 annual average)

Labor Force by Occupation [26]

Industry and construction	32.0%
Agriculture	27.6
Trade, transport, and communications	14.7
Government and other	25.7

Date of data: 1992

Unemployment Rate [27]

14.9% (December 1995)

For sources, notes, and explanations, see Annotated Source Appendix, page 1061.

Production Sectors

Commercial Energy Production and Consumption

Data are shown in quadrillion (10^{15}) BTUs and percent for 1995
Values for hydroelectric, nuclear, geothermal, solar, and wind power refer to electrical generation.

Production [28]

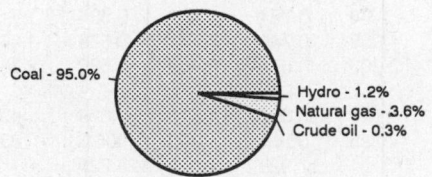

Coal - 95.0%
Hydro - 1.2%
Natural gas - 3.6%
Crude oil - 0.3%

Consumption [29]

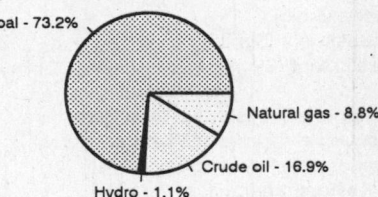

Coal - 73.2%
Natural gas - 8.8%
Crude oil - 16.9%
Hydro - 1.1%

Crude oil	0.010
Dry natural gas	0.135
Coal	3.551
Net hydroelectric power	0.043
Total	3.739

Crude oil	0.648
Dry natural gas	0.337
Coal	2.810
Net hydroelectric power	0.043
Total	3.838

Telecommunications [30]

- 5 million (1994) telephones; underdeveloped and outmoded system; government aims to have 10 million phones in service by the year 2000
- Domestic: cable, open wire, and microwave radio relay
- International: satellite earth stations - NA Intelsat, NA Eutelsat, 1 Inmarsat (Atlantic and Indian Ocean Regions), and 1 Intersputnik (Atlantic Ocean Region)
- Radio: Broadcast stations: AM 27, FM 27, shortwave 0 Radios: 10.9 million (1993 est.)
- Television: Broadcast stations: 40 (Russian repeaters 5) Televisions: 9.6 million

Top Agricultural Products [32]

Agriculture accounts for 7% of the GDP; produces potatoes, milk, fruits, vegetables, wheat; poultry and eggs; pork, beef.

Transportation [31]

Railways: total: 25,166 km; broad gauge: 656 km 1.520-m gauge; standard gauge: 22,655 km 1.435-m gauge (11,496 km electrified; 8,978 km double track); narrow gauge: 1,855 km various gauges including 1.000-m, 0.785-m, 0.750-m, and 0.600-m (1995)

Highways: total: 367,000 km (excluding farm, factory, and forest roads); paved: 235,247 km (including 257 km of expressways); unpaved: 131,753 km (1992 est.)

Merchant marine: total: 131 ships (1,000 GRT or over) totaling 2,093,491 GRT/3,167,660 DWT; ships by type: bulk 73, cargo 36, chemical tanker 4, container 7, oil tanker 1, passenger 1, roll-on/roll-off cargo 4, short-sea passenger 5

Airports

Total:	134
With paved runways over 3,047 m:	2
With paved runways 2,438 to 3,047 m:	30
With paved runways 1,524 to 2,437 m:	27
With paved runways 914 to 1,523 m:	3
With paved runways under 914 m:	7

Top Mining Products [33]

Thousand metric tons except as noted[e]	4/94*
Copper ore, gross weight	26,100
Gold, mine output, Au content (kg.)	30,000
Steel, crude	11,100
Silver, mine output, Ag content (000 kg.)	1,060
Cement, hydraulic	13,900
Magnesite ore, crude	34,000
Limestone	11,000
Coal, bituminous, lignite, and brown	201,000
Coke	10,300
Petroleum refinery products (000 42-gal. bls.)	98,000

Tourism [34]

	1990	1991	1992	1993	1994
Visitors[1]	18,211	36,846	49,015	60,951	74,253
Tourists	NA	11,350	16,200	17,000	18,800
Excursionists	NA	25,496	32,815	43,951	55,453
Tourism receipts[89]	358	2,800	4,100	4,500	6,150
Tourism expenditures	423	143	132	181	316
Fare receipts	334	207	173	270	282
Fare expenditures	177	103	85	107	109

Travelers are in thousands, money in million U.S. dollars.

For sources, notes, and explanations, see Annotated Source Appendix, page 1061.

761

Manufacturing Sector

Manufacturing Summary [35]

	1987 $ bil.	1987 %	1988 $ bil.	1988 %	1989 $ bil.	1989 %	1990 $ bil.	1990 %	1991 $ bil.	1991 %
Establishments or enterprises (number)	5,799	0.247	6,006	0.286	6,955	0.371	6,973	0.389	7,345	0.961
Total employment (000)	4,185	3.084	4,103	2.998	3,947	3.207	3,579	3.234	3,204	4.612
Production workers (000)	2,977	4.414	2,895	4.634	2,744	3.730	2,782	3.920	2,458	7.464
Output ($ bil.)	41	0.403	27	0.228	8	0.067	56	0.487	43	0.422
Value added ($ bil.)	27	0.614	31	0.639	38	0.781	27	0.534	-	-
Capital investment ($ mil.)	3,561	0.817	3,995	0.839	4,059	0.741	3,771	0.678	4,645	1.220
M & E investment ($ mil.)	1,977	0.634	2,328	0.673	2,303	0.590	2,469	0.586	2,331	0.827
Employees per establishment (number)	722	1,247.4	683	1,047.9	568	863.691	513	830.689	436	479.999
Production workers per establishment	513	1,785.5	482	1,619.6	395	1,004.6	399	1,006.9	335	776.772
Output per establishment ($ mil.)	7	162.883	4	79.858	1	17.928	8	125.225	6	43.968
Capital investment per estab. ($ mil.)	0.614	330.563	0.665	293.224	0.584	199.585	0.541	174.294	0.632	126.937
M & E per establishment ($ mil)	0.341	256.611	0.388	235.076	0.331	158.841	0.354	150.467	0.317	86.023
Payroll per employee ($)	0.003	0.000	0.003	0.000	0.004	0.000	0.003	0.000	0.005	0.000
Wages per production worker ($)	0.004	0.000	0.004	0.000	0.005	0.000	0.004	0.000	0.006	0.000
Hours per production worker (hours)	1,813	95.526	1,821	94.754	1,801	98.269	1,514	80.908	1,508	84.340
Output per employee ($)	9,779	13.058	6,491	7.621	1,989	2.076	15,623	15.075	13,448	9.160
Capital investment per employee ($)	851	26.500	974	27.983	1,028	23.108	1,054	20.982	1,450	26.445
M & E per employee ($)	472	20.571	567	22.434	583	18.391	690	18.114	727	17.921

Note: Columns headed % show percent of world total or ratio. Ratios closest to 100 are closest to world average. M & E stands for machinery & equipment.

Output in Manufacturing

	1987 $ bil.	1987 %	1988 $ bil.	1988 %	1989 $ bil.	1989 %	1990 $ bil.	1990 %	1991 $ bil.	1991 %
3110 - Food products	5.996	14.62	3.775	13.98	1.024	12.80	7.924	14.15	6.911	16.07
3130 - Beverages	2.176	5.31	1.355	5.02	0.396	4.95	2.225	3.97	2.158	5.02
3140 - Tobacco	0.269	0.66	0.148	0.55	0.050	0.63	0.556	0.99	0.473	1.10
3210 - Textiles	2.343	5.71	1.559	5.77	0.486	6.07	2.179	3.89	1.583	3.68
3211 - Spinning, weaving, etc.	1.460	3.56	0.968	3.59	0.295	3.69	1.257	2.24	0.836	1.94
3220 - Wearing apparel	1.042	2.54	0.708	2.62	0.209	2.61	1.021	1.82	0.813	1.89
3230 - Leather and products	0.278	0.68	0.186	0.69	0.058	0.73	0.353	0.63	0.229	0.53
3240 - Footwear	0.495	1.21	0.327	1.21	0.092	1.15	0.551	0.98	0.405	0.94
3310 - Wood products	0.594	1.45	0.381	1.41	0.120	1.50	1.095	1.96	1.617	3.76
3320 - Furniture, fixtures	0.579	1.41	0.401	1.49	0.136	1.70	0.591	1.06	0.583	1.36
3410 - Paper and products	0.431	1.05	0.281	1.04	0.082	1.02	0.807	1.44	0.704	1.64
3411 - Pulp, paper, etc.	0.335	0.82	0.216	0.80	0.063	0.79	0.620	1.11	0.497	1.16
3420 - Printing, publishing	0.233	0.57	0.158	0.59	0.049	0.61	0.323	0.58	0.252	0.59
3510 - Industrial chemicals	1.400	3.41	0.919	3.40	0.275	3.44	2.552	4.56	1.747	4.06
3511 - Basic chemicals, excl fertilizers	0.556	1.36	0.371	1.37	0.108	1.35	0.933	1.67	0.764	1.78
3513 - Synthetic resins, etc.	0.382	0.93	0.241	0.89	0.075	0.94	0.587	1.05	0.397	0.92
3520 - Chemical products nec	0.960	2.34	0.622	2.30	0.194	2.42	1.245	2.22	1.234	2.87
3522 - Drugs and medicines	0.378	0.92	0.241	0.89	0.065	0.81	0.475	0.85	0.490	1.14
3530 - Petroleum refineries	1.608	3.92	1.021	3.78	0.280	3.50	2.709	4.84	2.099	4.88
3540 - Petroleum, coal products	0.273	0.67	0.160	0.59	0.045	0.56	0.689	1.23	0.451	1.05
3550 - Rubber products	0.345	0.84	0.223	0.83	0.067	0.84	0.382	0.68	0.296	0.69
3560 - Plastic products nec	0.396	0.97	0.271	1.00	0.085	1.06	0.541	0.97	0.609	1.42
3610 - Pottery, china, etc.	0.114	0.28	0.077	0.29	0.023	0.29	0.164	0.29	0.127	0.30
3620 - Glass and products	0.281	0.69	0.181	0.67	0.058	0.73	0.429	0.77	0.365	0.85
3690 - Nonmetal products nec	0.884	2.16	0.575	2.13	0.173	2.16	1.338	2.39	1.091	2.54
3710 - Iron and steel	2.182	5.32	1.334	4.94	0.370	4.63	4.727	8.44	3.233	7.52
3720 - Nonferrous metals	1.079	2.63	0.710	2.63	0.213	2.66	2.332	4.16	1.725	4.01
3810 - Metal products	1.490	3.63	0.961	3.56	0.303	3.79	2.422	4.32	1.866	4.34
3820 - Machinery nec	3.844	9.38	2.561	9.49	0.768	9.60	4.453	7.95	3.048	7.09
3825 - Office, computing machinery	0.168	0.41	0.125	0.46	0.041	0.51	0.131	0.23	0.065	0.15
3830 - Electrical machinery	2.200	5.37	1.510	5.59	0.466	5.82	2.839	5.07	2.013	4.68
3832 - Radio, television, etc.	1.053	2.57	0.733	2.71	0.229	2.86	1.011	1.81	0.643	1.50
3840 - Transportation equipment	2.746	6.70	1.791	6.63	0.505	6.31	3.667	6.55	2.049	4.77
3841 - Shipbuilding, repair	0.358	0.87	0.223	0.83	0.054	0.67	0.347	0.62	0.279	0.65
3843 - Motor vehicles	1.294	3.16	0.845	3.13	0.243	3.04	1.713	3.06	0.942	2.19
3850 - Professional goods	0.248	0.60	0.167	0.62	0.051	0.64	0.246	0.44	0.180	0.42
3900 - Industries nec	0.455	1.11	0.309	1.14	0.101	1.26	0.480	0.86	0.313	0.73

Note: Codes are International Standard Industry codes (ISIC). Percentages are % of total Output. [f]: Factor Prices; [p]: Producer Prices.

Finance, Economics, and Trade

Economic Indicators [36]

- **National product**: GDP—purchasing power parity—$226.7 billion (1995 est.)

- **National product real growth rate**: 6.5% (1995 est.)

- **National product per capita**: $5,800 (1995 est.)

- **Inflation rate (consumer prices)**: 21.6% (December 1995)

- **External debt**: $42.1 billion (year-end 1995 est.)

Balance of Payments Summary [37]

Values in millions of dollars.

	1989	1990	1991	1992	1993
Exports of goods (f.o.b.)	12,869	15,837	14,393	13,929	13,582
Imports of goods (f.o.b.)	-12,822	-12,248	-15,104	-14,060	-17,087
Trade balance	47	3,589	-711	-131	-3,505
Services - debits	-6,676	-6,836	-6,463	-8,940	-7,823
Services - credits	3,611	3,803	4,260	5,501	4,780
Private transfers (net)	1,521	2,206	723	213	621
Government transfers (net)	512	305	909	2,699	2,229
Long-term capital (net)	-1,426	6,470	-2,318	-270	1,683
Short-term capital (net)	2,780	-7,761	3,192	1,725	2,034
Errors and omissions	-110	133	-767	-148	-106
Overall balance	259	1,909	-1,175	649	-87

Exchange Rates [38]

Currency: **zloty.**
Symbol: **Zl.**

Data are currency units per $1. A currency reform on 1 January 1995 replaced 10,000 old zlotys with 1 new zloty.

January 1996	2.55
1995	2.4250
1994	22,723
1993	18,115
1992	13,626
1991	10,576

Imports and Exports

Top Import Origins [39]

$23.4 billion (f.o.b., 1995 est.) Data are for 1994.

Origins	%
Germany	27.5
Italy	8.4
Russia	6.8
UK	5.3

Top Export Destinations [40]

$22.2 billion (f.o.b., 1995 est.) Data are for 1994.

Destinations	%
Germany	35.7
Netherlands	5.9
Russia	5.4
Italy	4.9

Foreign Aid [41]

Recipient: Western governments and institutions pledged $22 billion in grants and loans during 1990-94, but much of the money has not been disbursed.

Import and Export Commodities [42]

Import Commodities

Machinery and transport equipment 28.9%
Intermediate goods 20.2%
Chemicals 14.7%
Fuels 10.4%
Misc. manufactures 9.9%

Export Commodities

Intermediate goods 27.5%
Machinery and transport equipment 19.8%
Misc. manufactures 20.5%
Foodstuffs 11.6%
Fuels 9.1%

For sources, notes, and explanations, see Annotated Source Appendix, page 1061.

763

Portugal

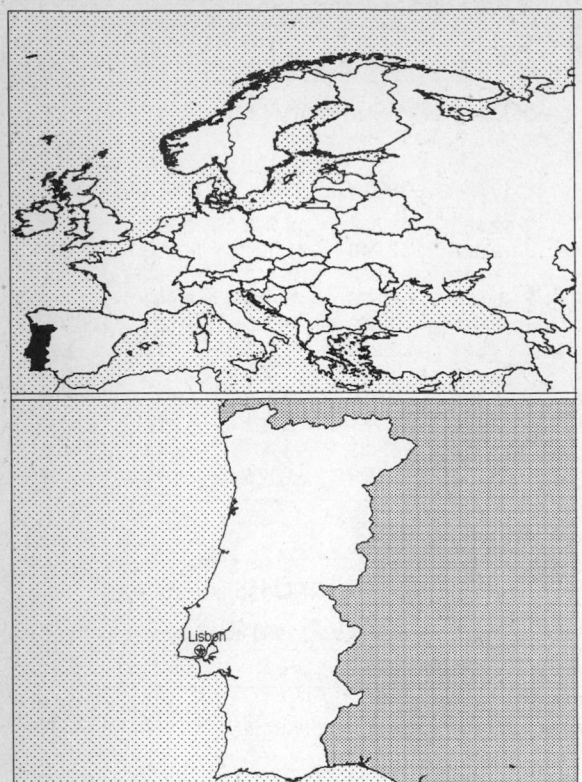

Geography [1]

Total area:
92,080 sq km 35,552 sq mi
Land area:
91,640 sq km 35,382 sq mi
Comparative area:
Slightly smaller than Indiana
Note: Includes Azores and Madeira Islands
Land boundaries:
Total 1,214 km, Spain 1,214 km
Coastline:
1,793 km
Climate:
Maritime temperate; cool and rainy in North, warmer and drier in South
Terrain:
Mountainous north of the Tagus, rolling plains in South
Natural resources:
Fish, forests (cork), tungsten, iron ore, uranium ore, marble
Land use:
Arable land: 32%
Permanent crops: 6%
Meadows and pastures: 6%
Forest and woodland: 40%
Other: 16%

Demographics [2]

	1970	1980	1990	1995[1]	1996	2000	2010	2020	2030
Population	9,044	9,778	9,871	9,865	9,865	9,906	10,080	10,005	9,769
Population density (persons per sq. mi.)	256	276	279	279	279	280	285	283	276
(persons per sq. km.)	99	107	108	108	108	108	110	109	107
Net migration rate (per 1,000 population)	-15.6	2.0	NA	-0.4	-0.2	0.5	1.0	0.3	0.0
Births	173	158	NA	NA	104	NA	NA	NA	NA
Deaths	93	95	103	NA	101	NA	NA	NA	NA
Life expectancy - males	64.2	NA	NA	71.3	71.5	72.5	74.4	77.3	79.2
Life expectancy - females	70.8	NA	NA	79.2	79.3	80.0	81.5	83.8	85.4
Birth rate (per 1,000)	19.1	16.2	NA	10.5	10.5	11.9	10.1	8.5	8.3
Death rate (per 1,000)	10.3	9.7	NA	10.2	10.2	10.3	10.7	10.5	11.4
Women of reproductive age (15-49 yrs.)	NA	NA	NA	2,544	2,550	2,533	2,417	2,238	1,922
of which are currently married	1,297	NA	NA	NA	1,762	1,798	1,776	NA	NA
Fertility rate	NA	2.2	NA	1.4	1.4	1.5	1.5	1.5	1.5

Except as noted, values for vital statistics are in thousands; life expectancy is in years.

Health

Health Indicators [3]

% of population with access to	
safe water (1990-95)	NA
adequate sanitation (1990-95)	NA
health services (1985-95)	NA
% of 1-year-olds immunized (1990-94) against	
TB (tuberculosis)	92
DPT (diphtheria, pertussis, tetanus)	92
polio	92
measles	94
% of contraceptive prevalence (1980-94)[1]	66
ORT use rate (1990-94)	NA

Health Expenditures [4]

Total health expenditure, 1990 (official exchange rate)	
Millions of dollars	3,970
Dollars per capita	383
Health expenditures as a percentage of GDP	
Total	7.0
Public sector	4.3
Private sector	2.7
Development assistance for health	
Total aid flows (millions of dollars)[1]	NA
Aid flows per capita (dollars)	NA
Aid flows as a percentage of total health expenditure	NA

For sources, notes, and explanations, see Annotated Source Appendix, page 1061.

Human Factors

Women and Children [5]

% of pregnant women immunized (tetanus 1990-94)	NA
% of births attended by trained health personnel (1983-94)[1]	90
Maternal mortality rate (1980-92)	10
Under-5 mortality rate (1994)	11
% under-5 moderately/severely underweight (1980-1994)	NA

Burden of Disease [6]

Population per physician (1987)	480.46
Population per nurse	NA
Population per hospital bed (1990)	226.67
AIDS cases per 100,000 people (1994)	5.1
Heart disease cases per 1,000 people (1990-93)	184.5

Ethnic Division [7]

Homogeneous Mediterranean stock in mainland, Azores, Madeira Islands; citizens of black African descent who immigrated to mainland during decolonization number less than 100,000.

Religion [8]

Roman Catholic	97%
Protestant denominations	1%
Other	2%

Major Languages [9]

Portuguese.

Education

Public Education Expenditures [10]

	1980	1990	1991	1992	1993	1994
Million (Escudo)[56]						
Total education expenditure	53,234	412,481	540,503	651,768	739,755	NA
as percent of GNP	3.8	4.3	4.8	5.0	5.4	NA
as percent of total govt. expend.	NA	NA	NA	NA	NA	NA
Current education expenditure	45,443	378,252	508,664	610,807	688,511	NA
as percent of GNP	3.3	4.0	4.6	4.7	5.1	NA
as percent of current govt. expend.	NA	NA	NA	NA	NA	NA
Capital expenditure	7,791	34,229	31,839	40,961	51,244	NA

Educational Attainment [11]

Age group (1991)	25+
Total population	6,280,792
Highest level attained (%)	
No schooling	16.1
First level	
Not completed	61.5
Completed	NA
Entered second level	
S-1	14.8
S-2	NA
Postsecondary	7.7

Illiteracy [12]

	1981	1989	1995
Illiterate population (15+ yrs.)[2]	1,506,206	NA	827,000
Illiteracy rate - total pop. (%)	20.6	NA	10.4
Illiteracy rate - males (%)	15.2	NA	7.5
Illiteracy rate - females (%)	25.4	NA	13.0

Libraries [13]

	Admin. Units	Svc. Pts.	Vols. (000)	Shelving (meters)	Vols. Added	Reg. Users
National (1993)	1	1	2,530[e]	NA	28,983	64,484
Nonspecialized (1993)	8	NA	2,150[e]	NA	12,004	43,433
Public (1993)	161	NA	3,910[e]	NA	294,024	633,077
Higher ed. (1993)[10]	242	343	6,279	NA	200,559	349,111
School (1990)	675	707	2,896	65,563	156,413	175,970

Daily Newspapers [14]

	1980	1985	1990	1994
Number of papers	28	25	24	23
Circ. (000)	480[e]	413	446	404

Culture [15]

Cinema (seats per 1,000)	6.7
Annual attendance per person	0.8
Gross box office receipts (mil. Escudo)	3,122
Museums (reporting)	276
Visitors (000)	5,088
Annual receipts (000 Escudo)	2,762

Science and Technology

Scientific/Technical Forces [16]

Scientists/engineers	5,908
Number female	NA
Technicians	3,755
Number female	NA
Total	9,663

R&D Expenditures [17]

	Escudo (000) 1990
Total expenditure	52,032,200
Capital expenditure	10,483,400
Current expenditure	41,548,800
Percent current	79.9

U.S. Patents Issued [18]

Values show patents issued to citizens of the country by the U.S. Patents Office.

	1993	1994	1995
Number of patents	3	7	3

For sources, notes, and explanations, see Annotated Source Appendix, page 1061.

Government and Law

Organization of Government [19]

Long-form name:
Portuguese Republic
Type:
Republic
Independence:
1140 (independent republic proclaimed 5 October 1910)
National holiday:
Day of Portugal, 10 June (1580)
Constitution:
25 April 1976, revised 30 October 1982 and 1 June 1989
Legal system:
Civil law system; the Constitutional Tribunal reviews the constitutionality of legislation; accepts compulsory ICJ jurisdiction, with reservations
Executive branch:
President; Prime Minister; Council of State; Council of Ministers
Legislative branch:
Unicameral: Assembly of the Republic (Assembleia da Republica)
Judicial branch:
Supreme Court (Supremo Tribunal de Justica)

Elections [20]

Assembly of the Republic	% of votes
Portuguese Socialist Party (PS)	43.8
Social Democratic Party (PSD)	34.0
Democratic Coalition (CDU)	8.6
Other	13.6

Government Expenditures [21]

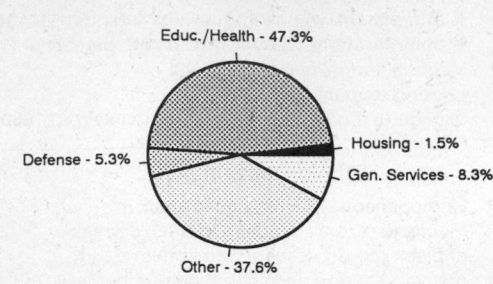

Educ./Health - 47.3%
Housing - 1.5%
Defense - 5.3%
Gen. Services - 8.3%
Other - 37.6%

(% distribution). Expend. for CY90: 3,828.9 (Escudo bil.)

Military Expenditures and Arms Transfers [22]

	1990	1991	1992	1993	1994
Military expenditures					
Current dollars (mil.)	2,117	2,200	2,231	2,186	2,174
1994 constant dollars (mil.)	2,356	2,358	2,327	2,231	2,174
Armed forces (000)	87	86	80	68	122
Gross national product (GNP)					
Current dollars (mil.)	75,670	80,420	84,240	84,370	87,140
1994 constant dollars (mil.)	84,210	86,200	87,840	86,110	87,140
Central government expenditures (CGE)					
1994 constant dollars (mil.)	33,650	37,000	36,170	NA	41,000[e]
People (mil.)	10.4	10.4	10.4	10.5	10.5
Military expenditure as % of GNP	2.8	2.7	2.6	2.6	2.5
Military expenditure as % of CGE	7.0	6.4	6.4	NA	5.3
Military expenditure per capita (1994 $)	227	226	223	213	207
Armed forces per 1,000 people (soldiers)	8.4	8.3	7.7	6.5	11.6
GNP per capita (1994 $)	8,125	8,279	8,407	8,212	8,280
Arms imports[6]					
Current dollars (mil.)	360	600	100	140	320
1994 constant dollars (mil.)	401	643	104	143	320
Arms exports[6]					
Current dollars (mil.)	10	5	20	5	100
1994 constant dollars (mil.)	11	5	21	5	100
Total imports[7]					
Current dollars (mil.)	25,260	26,110	29,580	24,340	26,550
1994 constant dollars (mil.)	28,120	27,990	30,850	24,840	26,550
Total exports[7]					
Current dollars (mil.)	16,420	16,280	18,350	15,430	17,360
1994 constant dollars (mil.)	18,270	17,450	19,140	15,750	17,360
Arms as percent of total imports[8]	1.4	2.3	0.3	0.6	1.2
Arms as percent of total exports[8]	0.1	0	0.1	0	0.6

Crime [23]

	1994
Crime volume	
Cases known to police	97,773
Attempts (percent)	NA
Percent cases solved	34.20
Crimes per 100,000 persons	988.85
Persons responsible for offenses	
Total number offenders	45,163
Percent female	23.00
Percent juvenile	NA
Percent foreigners	NA

Human Rights [24]

	SSTS	FL	FAPRO	PPCG	APROBC	TPW	PCPTW	STPEP	PHRFF	PRW	ASST	AFL
Observes	2	P	P		P	P	P	P	P		P	P
	EAFRD	CPR	ESCR	SR	ACHR	MAAE	PVIAC	PVNAC	EAFDAW	TCIDTP	RC	
Observes	P	P	P	P			P	P	P	P	P	

P = Party; S = Signatory; see Appendix for meaning of abbreviations.

Labor Force

Total Labor Force [25]

4.24 million (1994 est.)

Labor Force by Occupation [26]

Services	54.5%
Manufacturing	24.4
Agriculture, forestry, fisheries	11.2
Construction	8.3
Utilities	1.0
Mining	0.5

Date of data: 1992

Unemployment Rate [27]

7.1% (1995 est.)

For sources, notes, and explanations, see Annotated Source Appendix, page 1061.

Production Sectors

Energy Resource Summary [28]

Energy resources: Uranium ore. **Electricity:** Capacity: 8,220,000 kW. Production: 29.5 billion kWh. Consumption per capita: 2,642 kWh (1993).
Pipelines: crude oil 22 km; petroleum products 58 km.

Telecommunications [30]

- 2,236,411 (1993 est.) telephones
- Domestic: generally adequate integrated network of coaxial cables, open wire, microwave radio relay, and domestic satellite earth stations
- International: 6 submarine cables; satellite earth stations - 3 Intelsat (2 Atlantic Ocean and 1 Indian Ocean), NA Eutelsat; tropospheric scatter to Azores; note - an earth station for Inmarsat (Atlantic Ocean Region) is planned
- Radio: Broadcast stations: AM 57, FM 66 (repeaters 22), shortwave 0 Radios: 2.2 million (1993 est.)
- Television: Broadcast stations: 66 (repeaters 23) Televisions: 2,970,892 (1993 est.)

Top Agricultural Products [32]

Agriculture accounts for 6% of the GDP; produces grain, potatoes, olives, grapes; sheep, cattle, goats, poultry, meat, dairy products.

Top Mining Products [33]

Thousand metric tons except as noted[e]	1/95[*]
Copper, concentrate, gross weight	1,300
Steel, crude	1,000
Silver, mine output, Ag content	40,000
Cement, hydraulic	9,000
Salt, rock and marine	750
Diorite	1,500
Gabbro	3,000
Granite	12,000
Gas, manufactured (mil. cu. meters)	150
Petroleum refinery products (000 42-gal. bls.)	90,300

Transportation [31]

Railways: total: 3,068 km; broad gauge: 2,761 km 1.668-m gauge (464 km electrified; 426 km double track); narrow gauge: 307 km 1.000-m gauge

Highways: total: 70,176 km (statistics for continental Portugal only); paved: 60,351 km (including 519 km of expressways); unpaved: 9,825 km (1992 est.)

Merchant marine: total: 72 ships (1,000 GRT or over) totaling 795,725 GRT/1,418,538 DWT; ships by type: bulk 7, cargo 35, chemical tanker 5, container 5, liquefied gas tanker 4, oil tanker 12, passenger-cargo 1, refrigerated cargo 1, roll-on/roll-off cargo 1, short-sea passenger 1

Airports

Total:	67
With paved runways over 3,047 m:	5
With paved runways 2,438 to 3,047 m:	8
With paved runways 1,524 to 2,437 m:	3
With paved runways 914 to 1,523 m:	18
With paved runways under 914 m:	30

Tourism [34]

	1990	1991	1992	1993	1994
Visitors[90]	18,422	19,641	20,742	20,579	21,728
Tourists[12]	8,020	8,657	8,884	8,434	9,132
Excursionists	10,179	10,755	11,635	11,929	12,384
Cruise passengers	222	230	223	217	211
Tourism receipts	3,555	3,710	3,721	4,102	4,087
Tourism expenditures	867	1,024	1,165	2,058	1,705
Fare receipts	96	156	170	640	482
Fare expenditures	55	36	26	295	341

Travelers are in thousands, money in million U.S. dollars.

For sources, notes, and explanations, see Annotated Source Appendix, page 1061.

767

Manufacturing Sector

Manufacturing Summary [35]

	1987		1988		1989		1990		1991	
	$ bil.	%	$ bil.	%	$ bil.	%	$ bil.	%	$ bil.	%
Establishments or enterprises (number)	13,187	0.562	-	-	-	-	-	-	-	-
Total employment (000)	788	0.580	-	-	-	-	-	-	-	-
Production workers (000)	670	0.993	-	-	-	-	-	-	-	-
Output ($ bil.)	30	0.298	-	-	-	-	-	-	-	-
Value added ($ bil.)	10	0.228	-	-	-	-	-	-	-	-
Capital investment ($ mil.)	1,676	0.384	-	-	-	-	-	-	-	-
M & E investment ($ mil.)	1,338	0.429	-	-	-	-	-	-	-	-
Employees per establishment (number)	60	103.249	-	-	-	-	-	-	-	-
Production workers per establishment	51	176.662	-	-	-	-	-	-	-	-
Output per establishment ($ mil.)	2	53.010	-	-	-	-	-	-	-	-
Capital investment per estab. ($ mil.)	0.127	68.399	-	-	-	-	-	-	-	-
M & E per establishment ($ mil)	0.101	76.357	-	-	-	-	-	-	-	-
Payroll per employee ($)	5,375	59.940	-	-	-	-	-	-	-	-
Wages per production worker ($)	4,775	60.510	-	-	-	-	-	-	-	-
Hours per production worker (hours)	2,037	107.352	-	-	-	-	-	-	-	-
Output per employee ($)	38,452	51.342	-	-	-	-	-	-	-	-
Capital investment per employee ($)	2,127	66.247	-	-	-	-	-	-	-	-
M & E per employee ($)	1,698	73.954	-	-	-	-	-	-	-	-

Note: Columns headed % show percent of world total or ratio. Ratios closest to 100 are closest to world average. M & E stands for machinery & equipment.

Output in Manufacturing

	1987		1988		1989		1990		1991	
	$ bil.	%	$ bil.	%	$ bil.	%	$ bil.	%	$ bil.	%
3110 - Food products	4.206	14.02	-	-	-	-	-	-	-	-
3130 - Beverages	0.424	1.41	-	-	-	-	-	-	-	-
3140 - Tobacco	0.221	0.74	-	-	-	-	-	-	-	-
3210 - Textiles	3.125	10.42	-	-	-	-	-	-	-	-
3211 - Spinning, weaving, etc.	2.088	6.96	-	-	-	-	-	-	-	-
3220 - Wearing apparel	0.932	3.11	-	-	-	-	-	-	-	-
3230 - Leather and products	0.304	1.01	-	-	-	-	-	-	-	-
3240 - Footwear	0.527	1.76	-	-	-	-	-	-	-	-
3310 - Wood products	0.835	2.78	-	-	-	-	-	-	-	-
3320 - Furniture, fixtures	0.163	0.54	-	-	-	-	-	-	-	-
3410 - Paper and products	1.365	4.55	-	-	-	-	-	-	-	-
3411 - Pulp, paper, etc.	1.081	3.60	-	-	-	-	-	-	-	-
3420 - Printing, publishing	0.546	1.82	-	-	-	-	-	-	-	-
3510 - Industrial chemicals	1.701	5.67	-	-	-	-	-	-	-	-
3511 - Basic chemicals, excl fertilizers	0.843	2.81	-	-	-	-	-	-	-	-
3513 - Synthetic resins, etc.	0.576	1.92	-	-	-	-	-	-	-	-
3520 - Chemical products nec	1.250	4.17	-	-	-	-	-	-	-	-
3522 - Drugs and medicines	0.362	1.21	-	-	-	-	-	-	-	-
3530 - Petroleum refineries	1.633	5.44	-	-	-	-	-	-	-	-
3550 - Rubber products	0.200	0.67	-	-	-	-	-	-	-	-
3560 - Plastic products nec	0.441	1.47	-	-	-	-	-	-	-	-
3610 - Pottery, china, etc.	0.256	0.85	-	-	-	-	-	-	-	-
3620 - Glass and products	0.233	0.78	-	-	-	-	-	-	-	-
3690 - Nonmetal products nec	0.805	2.68	-	-	-	-	-	-	-	-
3710 - Iron and steel	0.551	1.84	-	-	-	-	-	-	-	-
3720 - Nonferrous metals	0.128	0.43	-	-	-	-	-	-	-	-
3810 - Metal products	0.847	2.82	-	-	-	-	-	-	-	-
3820 - Machinery nec	0.485	1.62	-	-	-	-	-	-	-	-
3825 - Office, computing machinery	0.045	0.15	-	-	-	-	-	-	-	-
3830 - Electrical machinery	1.129	3.76	-	-	-	-	-	-	-	-
3832 - Radio, television, etc.	0.528	1.76	-	-	-	-	-	-	-	-
3840 - Transportation equipment	1.222	4.07	-	-	-	-	-	-	-	-
3841 - Shipbuilding, repair	0.228	0.76	-	-	-	-	-	-	-	-
3843 - Motor vehicles	0.917	3.06	-	-	-	-	-	-	-	-
3850 - Professional goods	0.057	0.19	-	-	-	-	-	-	-	-
3900 - Industries nec	0.038	0.13	-	-	-	-	-	-	-	-

Note: Codes are International Standard Industry codes (ISIC). Percentages are % of total Output. [f]: Factor Prices; [p]: Producer Prices.

For sources, notes, and explanations, see Annotated Source Appendix, page 1061.

Finance, Economics, and Trade

Economic Indicators [36]

- **National product**: GDP—purchasing power parity—$116.2 billion (1995 est.)
- **National product real growth rate**: 2.8% (1995 est.)
- **National product per capita**: $11,000 (1995 est.)
- **Inflation rate (consumer prices)**: 4.6% (1995 est.)
- **External debt**: $11.8 billion (1995 est.)

Balance of Payments Summary [37]

Values in millions of dollars.

	1989	1990	1991	1992	1993
Exports of goods (f.o.b.)	12,720	16,311	16,231	18,195	15,444
Imports of goods (f.o.b.)	-17,585	-23,141	-24,079	-27,735	-22,330
Trade balance	-4,865	-6,830	-7,848	-9,540	-6,886
Services - debits	-4,152	-5,461	-5,784	-6,188	-8,024
Services - credits	4,630	6,603	6,941	7,718	9,142
Private transfers (net)	3,726	4,509	4,593	4,794	3,842
Government transfers (net)	814	998	1,381	3,032	2,874
Long-term capital (net)	2,796	3,574	3,998	-863	912
Short-term capital (net)	1,208	-1,010	539	-87	-4,536
Errors and omissions	497	1,160	1,893	978	-171
Overall balance	4,654	3,543	5,713	-156	-2,847

Exchange Rates [38]

Currency: **Portuguese escudo.**
Symbol: **Esc.**

Data are currency units per $1.

January 1996	151.61
1995	149.97
1994	165.99
1993	160.80
1992	135.00
1991	144.48

Imports and Exports

Top Import Origins [39]

$24.1 billion (c.i.f., 1995) Data are for 1995.

Origins	%
EU	71
Other developed countries	10.9
US	2.5
Less developed countries	12.9

Top Export Destinations [40]

$18.9 billion (f.o.b., 1995) Data are for 1995.

Destinations	%
EU	75.1
Other developed countries	12.4
US	5.2

Foreign Aid [41]

	U.S. $	
Donor:		
ODA (1993)	248	million
Recipient:		
ODA (1993)	70	million

Import and Export Commodities [42]

Import Commodities	Export Commodities
Machinery and transport equipment	Clothing and footwear
Agricultural products	Machinery
Chemicals	Cork and paper products
Petroleum	Hides
Textiles	

For sources, notes, and explanations, see Annotated Source Appendix, page 1061.

769

Qatar

Geography [1]

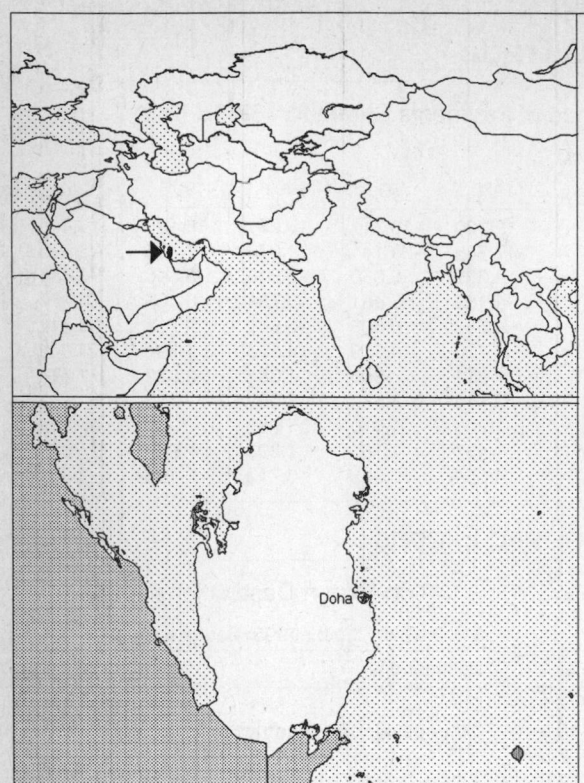

Total area:
  11,000 sq km   4,247 sq mi
Land area:
  11,000 sq km   4,247 sq mi
Comparative area:
  Slightly smaller than Connecticut
Land boundaries:
  Total 60 km, Saudi Arabia 60 km
Coastline:
  563 km
Climate:
  Desert; hot, dry; humid and sultry in summer
Terrain:
  Mostly flat and barren desert covered with loose sand and gravel
Natural resources:
  Petroleum, natural gas, fish
Land use:
  Arable land: 0%
  Permanent crops: 0%
  Meadows and pastures: 5%
  Forest and woodland: 0%
  Other: 95%

Demographics [2]

	1970	1980	1990	1995[1]	1996	2000	2010	2020	2030
Population	113	231	452	534	548	587	660	735	782
Population density (persons per sq. mi.)	27	54	106	126	129	138	155	173	184
(persons per sq. km.)	10	21	41	49	50	53	60	67	71
Net migration rate (per 1,000 population)	NA	NA	19.5	8.3	6.4	0.0	0.0	0.0	0.0
Births	NA	NA	NA	NA	12	NA	NA	NA	NA
Deaths	NA	NA	NA	NA	2	NA	NA	NA	NA
Life expectancy - males	NA	NA	68.7	70.5	70.8	72.0	74.6	76.5	77.9
Life expectancy - females	NA	NA	73.5	75.5	75.8	77.3	80.1	82.2	83.7
Birth rate (per 1,000)	NA	31.3	24.4	22.7	21.0	15.2	16.7	15.8	13.5
Death rate (per 1,000)	NA	NA	3.7	3.6	3.6	3.7	5.1	7.0	9.4
Women of reproductive age (15-49 yrs.)	NA	NA	78	93	97	108	139	164	187
of which are currently married	NA	NA	54	NA	65	70	88	NA	NA
Fertility rate	NA	NA	4.6	4.6	4.3	2.9	2.4	2.1	2.0

Except as noted, values for vital statistics are in thousands; life expectancy is in years.

Health

Health Indicators [3]

Health Expenditures [4]

For sources, notes, and explanations, see Annotated Source Appendix, page 1061.

Human Factors

Women and Children [5]	Burden of Disease [6]	
	Population per physician (1990)	666.67
	Population per nurse (1990)	186.14
	Population per hospital bed	NA
	AIDS cases per 100,000 people (1994)	1.4
	Malaria cases per 100,000 people (1992)	NA

Ethnic Division [7]

Arab	40%
Indian	18%
Pakistani	18%
Iranian	10%
Other	14%

Religion [8]

Muslim	95%
Other	5%

Major Languages [9]

Arabic (official), English commonly used as a second language.

Education

Public Education Expenditures [10]

	1980	1985	1990	1992	1993	1994
Million (Riyal)						
Total education expenditure	792	1,012	929	957	976	889
as percent of GNP	2.6	4.1	3.4	3.4	3.5	NA
as percent of total govt. expend.	7.2	NA	NA	NA	NA	NA
Current education expenditure	598	767	904	896	891	816
as percent of GNP	2.0	3.1	3.3	3.2	3.2	NA
as percent of current govt. expend.	7.8	NA	NA	NA	NA	NA
Capital expenditure	194	246	25	62	85	73

Educational Attainment [11]

Age group (1986)	25+
Total population	211,485
Highest level attained (%)	
No schooling	53.5
First level	
Not completed	9.8
Completed	NA
Entered second level	
S-1	10.1
S-2	13.3
Postsecondary	13.3

Illiteracy [12]

In thousands and percent[1]	1990	1995	2000
Illiterate population (15+ yrs.)	80	82	81
Illiteracy rate - total pop. (%)	25.2	22.0	19.3
Illiteracy rate - males (%)	25.2	22.7	20.8
Illiteracy rate - females (%)	25.3	20.4	16.8

Libraries [13]

	Admin. Units	Svo. Pts.	Vols. (000)	Shelving (meters)	Vols. Added	Reg. Users
National (1992)	1	1	185	NA	4,898	6,292
Nonspecialized (1986)	5	6	NA	NA	NA	NA
Public (1992)	6	6	169	NA	5,088	4,565
Higher ed. (1993)	2	7	329	NA	5,000	6,000
School (1987)	156	NA	471	NA	30,572	NA

Daily Newspapers [14]

	1980	1985	1990	1994
Number of papers	3	4	5	4
Circ. (000)	30[e]	60	80[e]	80

Culture [15]

Cinema (seats per 1,000)	8.6
Annual attendance per person	0.6
Gross box office receipts (mil. Riyal)	NA
Museums (reporting)	6
Visitors (000)	78
Annual receipts (000 Riyal)	NA

Science and Technology

Scientific/Technical Forces [16]

Scientists/engineers[31]	229
Number female	58
Technicians[31]	61
Number female	2
Total[1]	290

R&D Expenditures [17]

	Riyal (000) 1986
Total expenditure	6,650
Capital expenditure	-
Current expenditure	6,650
Percent current	100.0

U.S. Patents Issued [18]

For sources, notes, and explanations, see Annotated Source Appendix, page 1061.

771

Government and Law

Organization of Government [19]

Long-form name:
State of Qatar
Type:
Traditional monarchy
Independence:
3 September 1971 (from UK)
National holiday:
Independence Day, 3 September (1971)
Constitution:
Provisional constitution enacted 2 April 1970
Legal system:
Discretionary system of law controlled by the amir, although civil codes are being implemented; Islamic law is significant in personal matters
Executive branch:
Amir and Prime Minister; Deputy Prime Minister; Council of Ministers
Legislative branch:
Unicameral: Advisory Council (Majlis al-Shura) The constitution calls for elections for part of this consultative body, but no elections have been held since 1970, when there were partial elections to the body; Council members have had their terms extended every four years since
Judicial branch:
Court of Appeal

Elections [20]

Advisory Council. No political parties; constitution calls for elections for part of this consultative body, but no elections have been held since 1970; seats—30 total.

Government Budget [21]

For FY95/96.

	$ bil.
Revenues	2.5
Expenditures	3.5
Capital expenditures	NA

Crime [23]

	1994
Crime volume	
Cases known to police	4,597
Attempts (percent)	1.95
Percent cases solved	84.64
Crimes per 100,000 persons	775.20
Persons responsible for offenses	
Total number offenders	5,106
Percent female	4.29
Percent juvenile (7-16 yrs.)	6.85
Percent foreigners	52.66

Military Expenditures and Arms Transfers [22]

	1990	1991	1992	1993	1994
Military expenditures					
Current dollars (mil.)[e]	NA	934	357	330	302
1994 constant dollars (mil.)[e]	NA	1,001	372	336	302
Armed forces (000)	11	11	8	8	10
Gross national product (GNP)					
Current dollars (mil.)[e]	7,582	7,084	7,644	7,583	7,820
1994 constant dollars (mil.)[e]	8,438	7,593	7,971	7,739	7,820
Central government expenditures (CGE)					
1994 constant dollars (mil.)[e]	NA	3,430	3,568	3,667	NA
People (mil.)	0.5	0.5	0.5	0.5	0.5
Military expenditure as % of GNP	NA	13.2	4.7	4.3	3.9
Military expenditure as % of CGE	NA	29.2	10.4	9.2	NA
Military expenditure per capita (1994 $)	NA	2,131	765	668	582
Armed forces per 1,000 people (soldiers)	24.3	23.4	16.4	15.9	19.3
GNP per capita (1994 $)	18,660	16,160	16,370	15,380	15,070
Arms imports[6]					
Current dollars (mil.)	100	20	40	0	0
1994 constant dollars (mil.)	111	21	42	0	0
Arms exports[6]					
Current dollars (mil.)	0	0	0	10	130
1994 constant dollars (mil.)	0	0	0	10	130
Total imports[7]					
Current dollars (mil.)	1,695	1,720	2,015	2,000	NA
1994 constant dollars (mil.)	1,886	1,844	2,101	2,041	NA
Total exports[7]					
Current dollars (mil.)	3,291[e]	3,180[e]	3,624[e]	3,100	NA
1994 constant dollars (mil.)	3,663[e]	3,409[e]	3,779[e]	3,164	NA
Arms as percent of total imports[8]	5.9	1.2	2.0	0	0
Arms as percent of total exports[8]	0	0	0	0.3	NA

Human Rights [24]

	SSTS	FL	FAPRO	PPCG	APROBC	TPW	PCPTW	STPEP	PHRFF	PRW	ASST	AFL
Observes						P	P					

	EAFRD	CPR	ESCR	SR	ACHR	MAAE	PVIAC	PVNAC	EAFDAW	TCIDTP	RC
Observes	P						P				P

P = Party; S = Signatory; see Appendix for meaning of abbreviations.

Labor Force

Total Labor Force [25]

233,000 (1993 est.)

Labor Force by Occupation [26]

Unemployment Rate [27]

For sources, notes, and explanations, see Annotated Source Appendix, page 1061.

Production Sectors

Commercial Energy Production and Consumption

Data are shown in quadrillion (10^{15}) BTUs and percent for 1995
Values for hydroelectric, nuclear, geothermal, solar, and wind power refer to electrical generation.

Production [28]

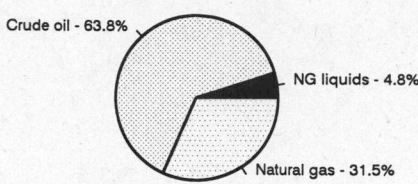

Crude oil - 63.8%

NG liquids - 4.8%

Natural gas - 31.5%

Consumption [29]

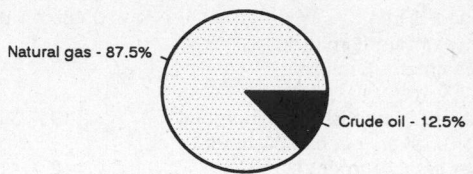

Natural gas - 87.5%

Crude oil - 12.5%

Crude oil	1.019
Natural gas liquids	0.076
Dry natural gas	0.503
Total	1.598

Crude oil	0.072
Dry natural gas	0.503
Total	0.575

Telecommunications [30]

- 160,717 (1992 est.) telephones; modern system centered in Doha
- International: tropospheric scatter to Bahrain; microwave radio relay to Saudi Arabia and UAE; submarine cable to Bahrain and UAE; satellite earth stations - 2 Intelsat (1 Atlantic Ocean and 1 Indian Ocean) and 1 Arabsat
- Radio: Broadcast stations: AM 2, FM 3, shortwave 0 Radios: 201,000 (1992 est.)
- Television: Broadcast stations: 3 (1988 est.) Televisions: 205,000 (1992 est.)

Transportation [31]

Railways: 0 km

Highways: total: 1,191 km; paved: 1,028 km; unpaved: 163 km (1988 est.)

Merchant marine: total: 19 ships (1,000 GRT or over) totaling 467,447 GRT/771,483 DWT; ships by type: combination ore/oil 2, container 3, cargo 11, oil tanker 3 (1995 est.)

Airports
Total: 3
With paved runways over 3,047 m: 1
With paved runways under 914 m: 1
With unpaved runways 914 to 1,523 m: 1 (1995 est.)

Top Agricultural Products [32]

Agriculture accounts for 1% of the GDP; produces fruits, vegetables; poultry, dairy products, beef; fish (all on small scale).

Top Mining Products [33]

Metric tons except as noted	8/95[*]
Cement, hydraulic	544,000
Gas, natural, gross (mil. cu. meters)	18,000
Steel, crude	620,000
Natural gas liquids (000 42-gal. bls.)	18,200
Nitrogen, N content of ammonia	621,000
Petroleum, crude (000 42-gal. bls.)	153,000
Limestone	900,000[e]

Tourism [34]

	1990	1991	1992	1993	1994
Tourists[2]	136	143	141	160	241

Travelers are in thousands, money in million U.S. dollars.

Manufacturing Sector

Manufacturing Summary [35]

	1987		1988		1989		1990		1991	
	$ bil.	%	$ bil.	%	$ bil.	%	$ bil.	%	$ bil.	%
Establishments or enterprises (number)	500	0.021	492	0.023	-	-	507	0.028	511	0.067
Total employment (000)	9	0.007	9	0.006	-	-	9	0.008	9	0.013
Production workers (000)	-	-	-	-	-	-	-	-	-	-
Output ($ bil.)	0.909	0.009	1	0.011	-	-	1	0.011	1	0.014
Value added ($ bil.)	0.483	0.011	0.781	0.016	-	-	0.877	0.018	0.844	0.025
Capital investment ($ mil.)	-	-	-	-	-	-	-	-	-	-
M & E investment ($ mil.)	-	-	-	-	-	-	-	-	-	-
Employees per establishment (number)	19	32.569	18	27.541	-	-	18	29.266	18	20.179
Production workers per establishment	-	-	-	-	-	-	-	-	-	-
Output per establishment ($ mil.)	2	41.948	3	47.121	-	-	3	39.375	3	20.342
Capital investment per estab. ($ mil.)	-	-	-	-	-	-	-	-	-	-
M & E per establishment ($ mil)	-	-	-	-	-	-	-	-	-	-
Payroll per employee ($)	8,440	94.122	9,390	94.251	-	-	8,932	64.390	8,326	57.022
Wages per production worker ($)	-	-	-	-	-	-	-	-	-	-
Hours per production worker (hours)	-	-	-	-	-	-	-	-	-	-
Output per employee ($)	96,464	128.800	145,728	171.092	-	-	139,430	134.541	147,990	100.805
Capital investment per employee ($)	-	-	-	-	-	-	-	-	-	-
M & E per employee ($)	-	-	-	-	-	-	-	-	-	-

Note: Columns headed % show percent of world total or ratio. Ratios closest to 100 are closest to world average. M & E stands for machinery & equipment.

Output in Manufacturing

	1987		1988		1989		1990		1991	
	$ bil.[p]	%	$ bil.[p]	%	$ bil.[p]	%	$ bil.[p]	%	$ bil.[p]	%
3110 - Food products	0.067	7.37	0.064	6.40	-	-	0.057	5.70	0.055	5.50
3130 - Beverages	0.013	1.43	0.015	1.50	-	-	0.016	1.60	0.016	1.60
3210 - Textiles	0.001	0.11	0.001	0.10	-	-	0.000	0.00	0.000	0.00
3230 - Leather and products	0.001	0.11	0.001	0.10	-	-	0.003	0.30	0.002	0.20
3310 - Wood products	0.032	3.52	0.032	3.20	-	-	0.030	3.00	0.024	2.40
3320 - Furniture, fixtures	0.007	0.77	0.004	0.40	-	-	0.002	0.20	0.004	0.40
3410 - Paper and products	0.000	0.00	0.000	0.00	-	-	0.001	0.10	0.001	0.10
3420 - Printing, publishing	0.033	3.63	0.036	3.60	-	-	0.033	3.30	0.024	2.40
3510 - Industrial chemicals	0.249	27.39	0.358	35.80	-	-	0.343	34.30	0.424	42.40
3511 - Basic chemicals, excl fertilizers	0.167	18.37	0.234	23.40	-	-	0.110	11.00	0.193	19.30
3520 - Chemical products nec	0.005	0.55	0.008	0.80	-	-	0.010	1.00	0.015	1.50
3530 - Petroleum refineries	0.201	22.11	0.347	34.70	-	-	0.460	46.00	0.430	43.00
3540 - Petroleum, coal products	0.005	0.55	0.005	0.50	-	-	0.005	0.50	0.009	0.90
3550 - Rubber products	0.001	0.11	0.001	0.10	-	-	0.001	0.10	0.001	0.10
3560 - Plastic products nec	0.011	1.21	0.012	1.20	-	-	0.010	1.00	0.013	1.30
3710 - Iron and steel	0.116	12.76	0.167	16.70	-	-	0.196	19.60	0.173	17.30
3900 - Industries nec	0.002	0.22	0.001	0.10	-	-	0.002	0.20	0.001	0.10

Note: Codes are International Standard Industry codes (ISIC). Percentages are % of total Output. [f]: Factor Prices; [p]: Producer Prices.

Finance, Economics, and Trade

Economic Indicators [36]

- **National product**: GDP—purchasing power parity— $10.7 billion (1994 est.)
- **National product real growth rate**: - 1% (1994 est.)
- **National product per capita**: $20,820 (1994 est.)
- **Inflation rate (consumer prices)**: 3% (1993 est.)
- **External debt**: $1.5 billion (1993 est.)

Balance of Payments Summary [37]

Exchange Rates [38]

Currency: **Qatari riyal.**
Symbol: **QR.**

Data are currency units per $1.

Fixed rate   3.6400 riyals

Imports and Exports

Top Import Origins [39]

$2 billion (c.i.f., 1994 est.) Data are for 1994.

Origins	%
Germany	14
Japan	12
UK	11
US	9
Italy	5

Top Export Destinations [40]

$2.9 billion (f.o.b., 1994 est.) Data are for 1994.

Destinations	%
Japan	61
Australia	5
UAE	4
Singapore	4

Foreign Aid [41]

Donor: Pledged in ODA to less developed countries (1979-88), $2.7 billion.

Import and Export Commodities [42]

Import Commodities

Machinery and equipment
Consumer goods
Food
Chemicals

Export Commodities

Petroleum products 75%
Steel
Fertilizers

Romania

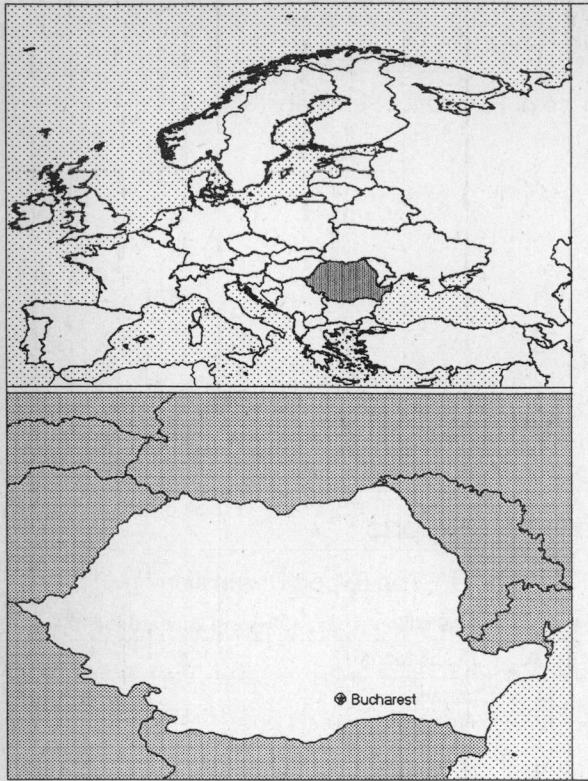

Geography [1]

Total area:
237,500 sq km   91,699 sq mi
Land area:
230,340 sq km   88,935 sq mi
Comparative area:
Slightly smaller than Oregon
Land boundaries:
Total 2,508 km, Bulgaria 608 km, Hungary 443 km, Moldova 450 km, Yugoslavia 476 km (all with Serbia), Ukraine (North) 362 km, Ukraine (South) 169 km
Coastline:
225 km
Climate:
Temperate; cold, cloudy winters with frequent snow and fog; sunny summers with frequent showers and thunderstorms
Terrain:
Central Transylvanian Basin is separated from the Plain of Moldavia in the East by the Carpathian Mountains and separated from the Walachian Plain in the South by the Transylvanian Alps
Natural resources:
Petroleum (reserves declining), timber, natural gas, coal, iron ore, salt
Land use:
Arable land: 43%
Permanent crops: 3%
Meadows and pastures: 19%
Forest and woodland: 28%
Other: 7%

Demographics [2]

	1970	1980	1990	1995[1]	1996	2000	2010	2020	2030
Population	20,253	22,109	22,775	21,924	21,657	20,996	20,741	20,135	19,708
Population density (persons per sq. mi.)	228	249	256	247	244	236	233	226	222
(persons per sq. km.)	88	96	99	95	94	91	90	87	86
Net migration rate (per 1,000 population)	-0.7	-0.8	NA	-9.8	-9.6	-2.4	-0.9	-0.3	0.0
Births	427	399	NA	NA	212	NA	NA	NA	NA
Deaths	193	232	247	NA	266	NA	NA	NA	NA
Life expectancy - males	NA	NA	NA	65.4	65.5	66.0	67.3	71.7	75.0
Life expectancy - females	NA	NA	NA	73.4	73.6	74.3	75.9	79.3	81.9
Birth rate (per 1,000)	21.1	18.0	NA	9.6	9.8	14.3	11.8	9.2	9.4
Death rate (per 1,000)	9.5	10.4	NA	12.2	12.3	12.7	13.3	11.8	11.5
Women of reproductive age (15-49 yrs.)	NA	NA	NA	5,519	5,507	5,352	5,024	4,821	4,333
of which are currently married	NA	NA	NA	NA	3,926	3,889	3,818	NA	NA
Fertility rate	2.9	2.5	NA	1.3	1.3	1.8	1.7	1.6	1.5

Except as noted, values for vital statistics are in thousands; life expectancy is in years.

Health

Health Indicators [3]

% of population with access to	
safe water (1990-95)	NA
adequate sanitation (1990-95)	NA
health services (1985-95)	NA
% of 1-year-olds immunized (1990-94) against	
TB (tuberculosis)	100
DPT (diphtheria, pertussis, tetanus)	98
polio	94
measles	91
% of contraceptive prevalence (1980-94)	57
ORT use rate (1990-94)	NA

Health Expenditures [4]

Total health expenditure, 1990 (official exchange rate)	
Millions of dollars	1,455
Dollars per capita	63
Health expenditures as a percentage of GDP	
Total	3.9
Public sector	2.4
Private sector	1.5
Development assistance for health	
Total aid flows (millions of dollars)[1]	NA
Aid flows per capita (dollars)	NA
Aid flows as a percentage of total health expenditure	NA

For sources, notes, and explanations, see Annotated Source Appendix, page 1061.

Human Factors

Women and Children [5]

% of pregnant women immunized (tetanus 1990-94)	NA
% of births attended by trained health personnel (1983-94)[1]	100
Maternal mortality rate (1980-92)	72
Under-5 mortality rate (1994)	29
% under-5 moderately/severely underweight (1980-1994)	NA

Burden of Disease [6]

Population per physician (1992)	537.69
Population per nurse	NA
Population per hospital bed (1992)	127.40
AIDS cases per 100,000 people (1994)	2.0
Heart disease cases per 1,000 people (1990-93)	NA

Ethnic Division [7]

Romanian	89.1%
Hungarian	8.9%
Ukrainian, Serb, Croat, Russian, Turk, and Gypsy	1.6%
German	0.4%

Religion [8]

3% of the Roman Catholic population is Uniate.

Romanian Orthodox	70%
Roman Catholic	6%
Protestant	6%
Unaffiliated	18%

Major Languages [9]

Romanian, Hungarian, German.

Education

Public Education Expenditures [10]

	1980	1985	1990	1992	1993	1994
Million (Leu)[57]						
Total education expenditure (bil.)	19.9	17.9	24.3	216.5	637.0	14,908.0
as percent of GNP	3.3	2.2	2.8	3.6	3.2	3.1
as percent of total govt. expend.	6.7	NA	7.3	14.2	9.1	13.6
Current education expenditure	17,691	17,345	23,881	208,881	606,683	NA
as percent of GNP	2.9	2.1	2.8	3.5	3.1	NA
as percent of current govt. expend.	NA	NA	9.0	15.5	9.6	NA
Capital expenditure	2,239	596	389	7,584	30,269	NA

Educational Attainment [11]

Age group (1992)	25+
Total population	13,602,159
Highest level attained (%)	
No schooling	5.4
First level	
Not completed	24.4
Completed	NA
Entered second level	
S-1	63.2
S-2	NA
Postsecondary	6.9

Illiteracy [12]

	1989	1992	1995
Illiterate population (15+ yrs.)[2]	NA	577,376	387,000
Illiteracy rate - total pop. (%)	NA	3.3	2.1
Illiteracy rate - males (%)	NA	1.5	1.1
Illiteracy rate - females (%)	NA	5.0	3.1

Libraries [13]

	Admin. Units	Svc. Pts.	Vols. (000)	Shelving (meters)	Vols. Added	Reg. Users
National (1993)	6	NA	20,040	NA	NA	115,000
Nonspecialized	NA	NA	NA	NA	NA	NA
Public (1993)	2,917	NA	46,406	NA	NA	2 mil
Higher ed. (1993)	64	348	20,919	309,300	493,736	614,000
School (1991)	10,246	NA	59,856	NA	2 mil	2 mil

Daily Newspapers [14]

	1980	1985	1990	1994
Number of papers	35	36	65	69
Circ. (000)	4,024[e]	3,601[e]	6,300[e]	6,800[e]

Culture [15]

Cinema (seats per 1,000)[14]	7.3
Annual attendance per person	1.5
Gross box office receipts (mil. Leu)	NA
Museums (reporting)	NA
Visitors (000)	6,518
Annual receipts (000 Leu)	NA

Science and Technology

Scientific/Technical Forces [16]

Scientists/engineers	31,672
Number female	14,048
Technicians	13,272
Number female	7,991
Total	44,944

R&D Expenditures [17]

	Leu (000) 1994
Total expenditure	NA
Capital expenditure	NA
Current expenditure	33,737,900
Percent current	NA

U.S. Patents Issued [18]

Values show patents issued to citizens of the country by the U.S. Patents Office.

	1993	1994	1995
Number of patents	2	1	3

For sources, notes, and explanations, see Annotated Source Appendix, page 1061.

777

Government and Law

Organization of Government [19]

Long-form name:
None
Type:
Republic
Independence:
1881 (from Turkey; republic proclaimed 30 December 1947)
National holiday:
National Day of Romania, 1 December (1990)
Constitution:
8 December 1991
Legal system:
Former mixture of civil law system and communist legal theory; is now based on the Constitution of France's Fifth Republic
Executive branch:
President; Prime Minister; Council of Ministers
Legislative branch:
Bicameral Parliament: Senate (Senat) and House of Deputies (Adunarea Deputatilor)
Judicial branch:
Supreme Court of Justice

Elections [20]

House of Deputies	% of votes
Social Democrats (PSDR)	34.0
The Democratic Convention (CDR)	18.2
Democratic Party (DP-FSN)	12.3
Others	37.3

Government Expenditures [21]

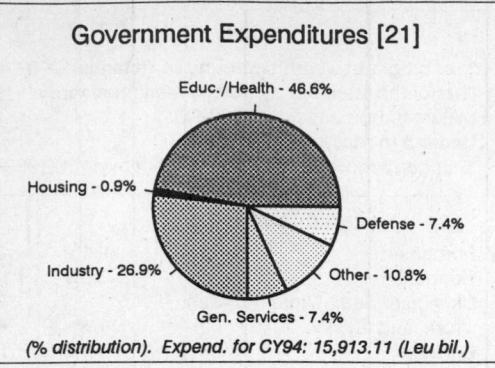

Educ./Health - 46.6%
Housing - 0.9%
Defense - 7.4%
Industry - 26.9%
Other - 10.8%
Gen. Services - 7.4%

(% distribution). Expend. for CY94: 15,913.11 (Leu bil.)

Military Expenditures and Arms Transfers [22]

	1990	1991	1992	1993	1994
Military expenditures					
Current dollars (mil.)	3,869[r]	3,747[r]	2,824[r]	1,676[r]	1,743[e]
1994 constant dollars (mil.)	4,306[r]	4,016[r]	2,945[r]	1,711[r]	1,743[e]
Armed forces (000)	126	201	172	167	200
Gross national product (GNP)					
Current dollars (mil.)	104,900	94,190	64,180	66,230	66,390
1994 constant dollars (mil.)	116,800	101,000	66,930	67,600	66,390
Central government expenditures (CGE)					
1994 constant dollars (mil.)[e]	39,470	35,310	27,910	14,940	5,114
People (mil.)	23.2	23.2	23.2	23.2	23.2
Military expenditure as % of GNP	3.7	4.0	4.4	2.5	2.6
Military expenditure as % of CGE	10.9	11.4	10.6	11.4	34.1
Military expenditure per capita (1994 $)	186	173	127	74	75
Armed forces per 1,000 people (soldiers)	5.4	8.7	7.4	7.2	8.6
GNP per capita (1994 $)	5,035	4,356	2,889	2,917	2,864
Arms imports[6]					
Current dollars (mil.)	825	170	30	0	0
1994 constant dollars (mil.)	918	182	31	0	0
Arms exports[6]					
Current dollars (mil.)	0	0	20	10	40
1994 constant dollars (mil.)	0	0	21	10	40
Total imports[7]					
Current dollars (mil.)	9,843	5,793	6,260	6,522	7,109
1994 constant dollars (mil.)	10,950	6,209	6,528	6,656	7,109
Total exports[7]					
Current dollars (mil.)	5,870	4,266	4,363	4,892	6,151
1994 constant dollars (mil.)	6,533	4,573	4,550	4,993	6,151
Arms as percent of total imports[8]	8.4	2.9	0.5	0	0
Arms as percent of total exports[8]	0	0	0.5	0.2	0.7

Crime [23]

	1994
Crime volume	
Cases known to police	237,004
Attempts (percent)	NA
Percent cases solved	NA
Crimes per 100,000 persons	1,039.04
Persons responsible for offenses	
Total number offenders	174,765
Percent female	18,399
Percent juvenile (14-18 yrs.)	18,612
Percent foreigners	2,107

Human Rights [24]

	SSTS	FL	FAPRO	PPCG	APROBC	TPW	PCPTW	STPEP	PHRFF	PRW	ASST	AFL
Observes	P	P	P	P	P	P	P	P	P	P	P	

	EAFRD	CPR	ESCR	SR	ACHR	MAAE	PVIAC	PVNAC	EAFDAW	TCIDTP	RC
Observes	P	P	P	P		P	S	S	P	P	P

P=Party; S=Signatory; see Appendix for meaning of abbreviations.

Labor Force

Total Labor Force [25]

11.3 million (1992)

Labor Force by Occupation [26]

Industry	38%
Agriculture	28
Other	34
Date of data: 1989	

Unemployment Rate [27]

8.9% (December 1995)

For sources, notes, and explanations, see Annotated Source Appendix, page 1061.

Production Sectors

Commercial Energy Production and Consumption

Data are shown in quadrillion (10^{15}) BTUs and percent for 1995
Values for hydroelectric, nuclear, geothermal, solar, and wind power refer to electrical generation.

Production [28]

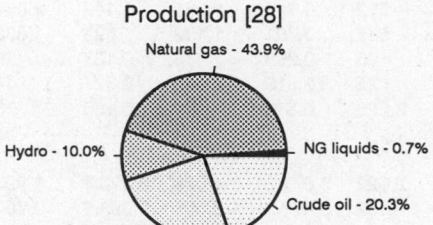

Natural gas - 43.9%
NG liquids - 0.7%
Crude oil - 20.3%
Coal - 25.1%
Hydro - 10.0%

Consumption [29]

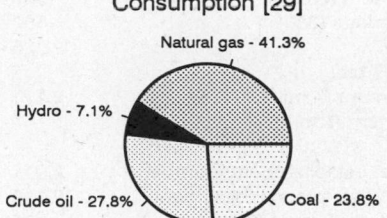

Natural gas - 41.3%
Hydro - 7.1%
Coal - 23.8%
Crude oil - 27.8%

Crude oil	0.285
Natural gas liquids	0.010
Dry natural gas	0.616
Coal	0.352
Net hydroelectric power	0.140
Total	1.403

Crude oil	0.547
Dry natural gas	0.812
Coal	0.469
Net hydroelectric power	0.140
Total	1.968

Telecommunications [30]

- 2.3 million (1990 est.) telephones
- Domestic: poor service; 89% of telephone network is automatic; trunk network is microwave radio relay; roughly 3,300 villages with no service (February 1990 est.)
- International: satellite earth station - 1 Intelsat; new digital international direct-dial exchanges are in Bucharest (1993 est.)
- Radio: Broadcast stations: AM 12, FM 5, shortwave 0 Radios: 4.64 million (1992 est.)
- Television: Broadcast stations: 13 (1990 est.) Televisions: 4.58 million (1992 est.)

Transportation [31]

Railways: total: 11,374 km; broad gauge: 60 km 1.524-m gauge; standard gauge: 10,887 km 1.435-m gauge (3,866 km electrified; 3,060 km double track); narrow gauge: 427 km 0.760-m gauge (1994)

Highways: total: 153,014 km; paved: 78,037 km (including 113 km of expressways); unpaved: 74,977 km (1992 est.)

Merchant marine: total: 233 ships (1,000 GRT or over) totaling 2,425,729 GRT/3,641,741 DWT; ships by type: bulk 39, cargo 166, container 2, oil tanker 13, passenger 1, passenger-cargo 1, railcar carrier 2, roll-on/roll-off cargo 9

Airports

Total:	156
With paved runways over 3,047 m:	4
With paved runways 2,438 to 3,047 m:	9
With paved runways 1,524 to 2,437 m:	14
With unpaved runways 2,438 to 3,047 m:	3
With unpaved runways 1,524 to 2,437 m:	1

Top Agricultural Products [32]

Agriculture accounts for 19.6% of the GDP; produces wheat, corn, sugar beets, sunflower seed, potatoes, grapes; milk, eggs, meat.

Top Mining Products [33]

Thousand metric tons except as noted[e]	3/95*
Gold, mine output, Au content (kg.)	4,000[r]
Pig iron	3,500[r]
Steel, crude	5,570[r]
Cement, hydraulic	6,680[r]
Coal	
run of mine	42,000
washed (produced from above)	37,000
Coke	2,660
Gas, natural, gross (mil. cu. meters)	22,000
Petroleum refinery products (000 42-gal. bls.)	100,000

Tourism [34]

	1990	1991	1992	1993	1994
Visitors[1]	6,533	5,360	6,280	5,786	5,898
Tourists	3,099	3,000	3,798	2,911	2,796
Excursionists	3,416	2,343	2,468	2,861	3,087
Cruise passengers	18	17	14	14	15
Tourism receipts	106	145	262	197	414
Tourism expenditures	103	143	260	195	449
Fare receipts	NA	NA	81	100	211
Fare expenditures	NA	NA	72	78	180

Travelers are in thousands, money in million U.S. dollars.

For sources, notes, and explanations, see Annotated Source Appendix, page 1061.

779

Manufacturing Sector

Manufacturing Summary [35]

	1987		1988		1989		1990		1991	
	$ bil.	%	$ bil.	%	$ bil.	%	$ bil.	%	$ bil.	%
Establishments or enterprises (number)	277	0.012	274	0.013	273	0.015	314	0.018	1,028	0.134
Total employment (000)	557	0.410	554	0.405	552	0.448	1,265	1.143	2,054	2.957
Production workers (000)	587	0.870	583	0.933	582	0.791	1,132	1.595	1,809	5.493
Output ($ bil.)	25	0.243	26	0.222	25	0.208	15	0.127	16	0.161
Value added ($ bil.)	-	-	29	0.598	25	0.515	16	0.327	14	0.409
Capital investment ($ mil.)	2,271	0.521	2,174	0.457	2,026	0.370	999	0.180	773	0.203
M & E investment ($ mil.)	-	-	-	-	-	-	-	-	-	-
Employees per establishment (number)	2,011	3,475.7	2,022	3,101.4	2,022	3,077.3	4,029	6,520.1	1,998	2,198.6
Production workers per establishment	2,119	7,370.6	2,128	7,149.1	2,132	5,428.2	3,605	9,098.9	1,760	4,084.6
Output per establishment ($ mil.)	89	2,054.1	94	1,697.2	90	1,428.3	46	722.6	16	119.4
Capital investment per estab. ($ mil.)	8	4,413.2	8	3,497.8	7	2,538.2	3	1,025.2	0.752	151.0
M & E per establishment ($ mil)	-	-	-	-	-	-	-	-	-	-
Payroll per employee ($)	-	-	-	-	-	-	1,861	13.418	1,207	8.263
Wages per production worker ($)	-	-	-	-	-	-	1,812	15.258	1,171	10.648
Hours per production worker (hours)	2,060	108.565	2,087	108.605	2,037	111.142	1,752	93.653	1,633	91.372
Output per employee ($)	44,262	59.099	46,611	54.724	44,482	46.414	11,486	11.083	7,976	5.433
Capital investment per employee ($)	4,078	126.972	3,924	112.783	3,671	82.482	790	15.724	376	6.866
M & E per employee ($)	-	-	-	-	-	-	-	-	-	-

Note: Columns headed % show percent of world total or ratio. Ratios closest to 100 are closest to world average. M & E stands for machinery & equipment.

Output in Manufacturing

	1987		1988		1989		1990		1991	
	$ bil.	%	$ bil.	%	$ bil.	%	$ bil.	%	$ bil.	%
3140 - Tobacco	-	-	-	-	-	-	-	-	0.187	1.17
3210 - Textiles	5.661	22.64	5.917	22.76	5.619	22.48	3.576	23.84	2.407	15.04
3220 - Wearing apparel	3.150	12.60	3.412	13.12	3.365	13.46	1.890	12.60	0.897	5.61
3310 - Wood products	-	-	-	-	-	-	-	-	0.644	4.03
3320 - Furniture, fixtures	-	-	-	-	-	-	-	-	0.847	5.29
3410 - Paper and products	1.012	4.05	1.068	4.11	1.047	4.19	0.691	4.61	0.534	3.34
3411 - Pulp, paper, etc.	1.012	4.05	1.068	4.11	1.047	4.19	0.691	4.61	-	-
3420 - Printing, publishing	0.197	0.79	0.202	0.78	0.190	0.76	0.152	1.01	0.115	0.72
3510 - Industrial chemicals	-	-	-	-	-	-	-	-	2.832	17.70
3522 - Drugs and medicines	0.475	1.90	0.513	1.97	0.495	1.98	0.294	1.96	-	-
3530 - Petroleum refineries	5.466	21.86	5.908	22.72	5.584	22.34	3.500	23.33	-	-
3690 - Nonmetal products nec	0.281	1.12	0.295	1.13	0.289	1.16	0.152	1.01	-	-
3710 - Iron and steel	1.300	5.20	1.339	5.15	1.328	5.31	0.691	4.61	-	-
3720 - Nonferrous metals	2.314	9.26	2.024	7.78	1.718	6.87	0.950	6.33	-	-
3810 - Metal products	3.786	15.14	4.077	15.68	3.872	15.49	1.944	12.96	1.183	7.39
3820 - Machinery nec	-	-	-	-	-	-	-	-	2.887	18.04
3825 - Office, computing machinery	-	-	-	-	-	-	-	-	0.048	0.30
3830 - Electrical machinery	-	-	-	-	-	-	-	-	1.142	7.14
3832 - Radio, television, etc.	-	-	-	-	-	-	-	-	0.257	1.61
3840 - Transportation equipment	-	-	-	-	-	-	-	-	2.047	12.79
3850 - Professional goods	-	-	-	-	-	-	-	-	0.323	2.02
3900 - Industries nec	-	-	-	-	-	-	-	-	0.031	0.19

Note: Codes are International Standard Industry codes (ISIC). Percentages are % of total Output. [f]: Factor Prices; [p]: Producer Prices.

Finance, Economics, and Trade

Economic Indicators [36]

- **National product**: GDP—purchasing power parity—$105.7 billion (1995 est.)
- **National product real growth rate**: 5.4% (1995 est.)
- **National product per capita**: $4,600 (1995 est.)
- **Inflation rate (consumer prices)**: 25% (1995)
- **External debt**: $4.7 billion (1995)

Balance of Payments Summary [37]

Values in millions of dollars.

	1989	1990	1991	1992	1993
Exports of goods (f.o.b.)	10,487	5,770	4,266	4,364	4,892
Imports of goods (f.o.b.)	-8,437	-9,114	-5,372	-5,558	-6,020
Trade balance	2,050	-3,344	-1,106	-1,194	-1,128
Services - debits	-551	-801	-908	-1,090	-1,118
Services - credits	1,015	785	784	713	862
Private transfers (net)	NA	NA	20	19	103
Government transfers (net)	NA	106	198	46	119
Long term captial (net)	-1,707	40	282	1,207	839
Short-term capital (net)	162	NA	NA	NA	NA
Errors and omissions	114	147	15	-12	140
Overall balance	921	-3,067	-715	-311	-183

Exchange Rates [38]

Currency: **leu.**
Symbol: **L.**

Data are currency units per $1.

January 1996	2,599.24
1995	2,033.28
1994	1,655.09
1993	760.05
1992	307.95
1991	76.39

Imports and Exports

Top Import Origins [39]

$7.1 billion (c.i.f., 1994) Data are for 1994.

Origins	%
OECD	60.0
EU	44.5
US	6.5
Developing countries	16.6
Russia	13.8
East and Central Europe	6.1
Other	3.5

Top Export Destinations [40]

$6.2 billion (f.o.b., 1994) Data are for 1994.

Destinations	%
OECD	57.9
EU	50
US	3.1
Developing countries	30.3
East and Central Europe	8.4
Russia	3.4

Foreign Aid [41]

Recipient: ODA, $81 million (1993).

Import and Export Commodities [42]

Import Commodities

Fuels and minerals 26.8%
Machinery and transport
  equipment 25.1%
Textiles and footwear 12.3%
Food and agricultural
  goods 9.3%
Chemicals 7.9%
Other 18.6%

Export Commodities

Textiles and footwear 23.8%
Metals and metal
  products 17.3%
Fuels and mineral
  products 11.6%
Machinery and transport
  equipment 14.8%
Chemicals 7.9%
Food and agricultural
  goods 6.5%
Other 18.1%

For sources, notes, and explanations, see Annotated Source Appendix, page 1061.

781

Russia

Geography [1]

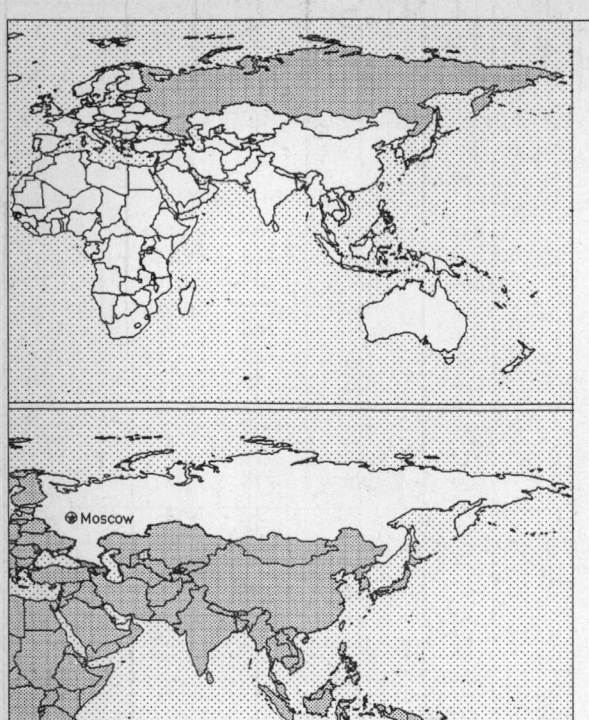

Total area:
 17,075,200 sq km   6,592,772 sq mi
Land area:
 16,995,800 sq km   6,562,115 sq mi
Comparative area:
 Slightly more than 1.8 times the size of the US
Land boundaries:
 Total 19,913 km, Azerbaijan 284 km, Belarus 959 km, China (Southeast) 3,605 km, China (South) 40 km, Estonia 290 km, Finland 1,313 km, Georgia 723 km, Latvia 217 km, Kazakhstan 6,846 km, North Korea 19 km, Lithuania 227 km, Mongolia 3,441 km, Norway 167 km, Poland 206 km, Ukraine 1,576 km
Coastline:
 37,653 km
Climate:
 Ranges from steppes in South to humid continental in much of European Russia; subarctic in Siberia to tundra climate in polar North; winters: cool on Black Sea Coast, frigid in Siberia; summers: warm in steppes, cool on Arctic Coast
Terrain:
 Broad plain with low hills west of Urals; vast coniferous forest and tundra in Siberia; uplands and mountains along southern border regions
Natural resources:
 Includes major deposits of oil, natural gas, coal, strategic minerals, and timber
 Note: Major obstacles of climate, terrain, and distance hinder their exploitation
Land use:
 Arable land: 8%
 Permanent crops: Negligible
 Meadows and pastures: 5%
 Forest and woodland: 45%
 Other: 42%

Demographics [2]

	1970	1980	1990	1995[1]	1996	2000	2010	2020	2030
Population	130,245	139,045	148,081	148,291	148,178	147,938	149,978	149,632	149,111
Population density (persons per sq. mi.)	20	21	22	22	22	22	23	23	23
(persons per sq. km.)	8	8	9	9	9	9	9	9	9
Net migration rate (per 1,000 population)	NA	NA	1.2	5.5	5.5	1.4	0.8	0.3	0.0
Births	NA	NA	NA	NA	1,504	NA	NA	NA	NA
Deaths	NA	NA	NA	NA	2,421	NA	NA	NA	NA
Life expectancy - males	NA	NA	63.4	56.5	56.5	59.4	64.6	68.9	72.3
Life expectancy - females	NA	NA	73.9	70.3	70.3	71.7	74.1	77.4	80.0
Birth rate (per 1,000)	NA	NA	13.9	10.0	10.2	14.5	13.3	10.8	11.1
Death rate (per 1,000)	NA	NA	11.3	16.3	16.3	14.8	13.1	11.7	11.4
Women of reproductive age (15-49 yrs.)	NA	NA	36,024	38,391	38,917	39,733	37,309	35,985	34,508
of which are currently married	NA	NA	24,366	NA	26,100	26,345	25,654	NA	NA
Fertility rate	NA	NA	1.9	1.4	1.4	1.9	1.8	1.7	1.7

Except as noted, values for vital statistics are in thousands; life expectancy is in years.

Health

Health Indicators [3]

% of population with access to	
safe water (1990-95)	NA
adequate sanitation (1990-95)	NA
health services (1985-95)	NA
% of 1-year-olds immunized (1990-94) against	
TB (tuberculosis)	87
DPT (diphtheria, pertussis, tetanus)	65
polio	82
measles	88
% of contraceptive prevalence (1980-94)	NA
ORT use rate (1990-94)	NA

Health Expenditures [4]

Total health expenditure, 1990 (official exchange rate)[4]	
Millions of dollars	23,527
Dollars per capita	157
Health expenditures as a percentage of GDP	
Total	3.0
Public sector	2.0
Private sector	1.0
Development assistance for health	
Total aid flows (millions of dollars)[1]	NA
Aid flows per capita (dollars)	NA
Aid flows as a percentage of total health expenditure	NA

For sources, notes, and explanations, see Annotated Source Appendix, page 1061.

Human Factors

Women and Children [5]

% of pregnant women immunized (tetanus 1990-94)	NA
% of births attended by trained health personnel (1983-94)	NA
Maternal mortality rate (1980-92)	NA
Under-5 mortality rate (1994)	31
% under-5 moderately/severely underweight (1980-1994)	NA

Burden of Disease [6]

Population per physician (1994)	221.70
Population per nurse (1994)	93.00
Population per hospital bed (1994)	78.50
AIDS cases per 100,000 people (1994)	*
Heart disease cases per 1,000 people (1990-93)	362

Ethnic Division [7]

Russian	81.5%
Tatar	3.8%
Ukrainian	3%
Chuvash	1.2%
Bashkir	0.9%
Byelorussian	0.8%
Moldavian	0.7%
Other	8.1%

Religion [8]

Russian Orthodox, Muslim, other.

Major Languages [9]

Russian, other.

Education

Public Education Expenditures [10]

	1980	1990	1991	1992	1993	1994
Million or Trillion (T) (Ruble)						
Total education expenditure	12,689	22,237	49,996	679,434	6.92 T	NA
as percent of GNP	3.5	3.5	3.9	4.0	4.4	NA
as percent of total govt. expend.	NA	NA	NA	NA	9.6	NA
Current education expenditure	NA	NA	NA	NA	6.61 T	NA
as percent of GNP	NA	NA	NA	NA	4.2	NA
as percent of current govt. expend.	NA	NA	NA	NA	10.0	NA
Capital expenditure	NA	NA	NA	NA	309,233	NA

Educational Attainment [11]

Age group (1989)	25+
Total population	86,016,990
Highest level attained (%)	
No schooling	NA
First level	
Not completed	36.9
Completed	NA
Entered second level	
S-1	49.0
S-2	NA
Postsecondary	14.1

Illiteracy [12]

In thousands and percent[2]	1985	1989	1995
Illiterate population (15+ yrs.)	NA	2,275	543
Illiteracy rate - total pop. (%)	NA	2.0	0.5
Illiteracy rate - males (%)	NA	0.5	0.3
Illiteracy rate - females (%)	NA	3.2	0.6

Libraries [13]

	Admin. Units	Svc. Pts.	Vols. (000)	Shelving (meters)	Vols. Added	Reg. Users
National (1993)	2	2	68,271	NA	1 mil	1 mil
Nonspecialized (1993)	7	7	10,266	NA	151,091	153,077
Public (1993)	51,111	NA	884,754	NA	53 mil	62 mil
Higher ed. (1993)	519	NA	324,696	NA	12 mil	4 mil
School (1990)	64,263	NA	739,822	NA	97 mil	NA

Daily Newspapers [14]

	1980	1985	1990	1994
Number of papers	NA	NA	NA	17
Circ. (000)	NA	NA	NA	39,301

Culture [15]

Cinema (seats per 1,000)	1.9
Annual attendance per person	2.6
Gross box office receipts (mil. Ruble)	190,000[e]
Museums (reporting)	1,375
Visitors (000)	79,831
Annual receipts (mil. Ruble)	76,291

Science and Technology

Scientific/Technical Forces [16]

Scientists/engineers	644,900
Number female	377,300
Technicians	133,900
Number female	42,400
Total[1]	778,800

R&D Expenditures [17]

	Ruble (000,000) 1993
Total expenditure	1,313,557[e]
Capital expenditure	102,579[e]
Current expenditure	1,210,978[e]
Percent current	92.2[e]

U.S. Patents Issued [18]

Values show patents issued to citizens of the country by the U.S. Patents Office.

	1993	1994	1995
Number of patents	3	38	99

For sources, notes, and explanations, see Annotated Source Appendix, page 1061.

Government and Law

Organization of Government [19]

Long-form name:
Russian Federation
Type:
Federation
Independence:
24 August 1991 (from Soviet Union)
National holiday:
Independence Day, June 12 (1990)
Constitution:
Adopted 12 December 1993
Legal system:
Based on civil law system; judicial review of legislative acts
Executive branch:
President; Premier and Chairman of the Russian Federation Government; 2 First Deputy Premiers and First Deputy Chairmen of the Government; Security Council; Presidential Administration; Ministries of the Government; Group of Assistants; Council of Heads of Republics; Council of Heads of Administrations; Presidential Council
Legislative branch:
Bicameral Federal Assembly: Federation Council and State Duma
Judicial branch:
Constitutional Court; Supreme Court; Superior Court of Arbitration

Elections [20]

State Duma	% of seats
Comm. Party of the Russian Fed.	34.9
Independents	17.3
Our Home Is Russia	12.2
Liberal Democratic Party	11.3
Yabloko Bloc	10.0
Agrarian Party	4.4
Power To the People	2.0
Russia's Democratic Choice	2.0
Congress of Russian Communities	1.1
Other	4.7

Government Expenditures [21]

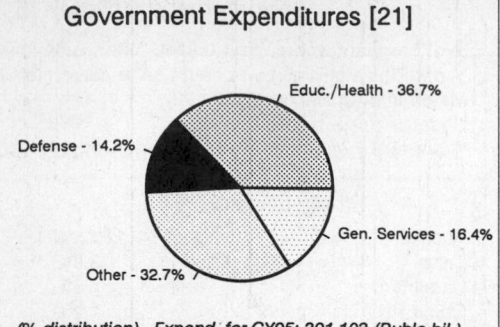

Educ./Health - 36.7%
Defense - 14.2%
Gen. Services - 16.4%
Other - 32.7%

(% distribution). Expend. for CY95: 391,103 (Ruble bil.)

Military Expenditures and Arms Transfers [22]

	1990[5]	1991[5]	1992	1993	1994
Military expenditures					
Current dollars (mil.)	292,000[e]	260,000[e]	139,200[r]	111,500[r]	96,800[r]
1994 constant dollars (mil.)	325,000[e]	278,700[e]	145,200[r]	113,800[r]	96,800[r]
Armed forces (000)	3,400	3,000	2,030	2,250	1,395
Gross national product (GNP)					
Current dollars (bil.)	2,660	2,531[e]	852[e]	777[e]	780[e]
1994 constant dollars (bil.)	2,960	2,713[e]	889[e]	793[e]	780[e]
Central government expenditures (CGE)					
1994 constant dollars (mil.)	754,800[e]	NA	527,800	NA	244,400
People (mil.)	290.9	293.0	149.0	149.3	149.6
Military expenditure as % of GNP[2]	11.0	10.3	16.3	14.3	12.4
Military expenditure as % of CGE[2]	43.1[9]	NA	27.5	NA	39.6
Military expenditure per capita (1994 $)	1,117	951	974	762	647
Armed forces per 1,000 people (soldiers)	11.7	10.2	13.6	15.1	9.3
GNP per capita (1994 $)	10,180	9,259	5,965	5,314	5,216
Arms imports[6]					
Current dollars (mil.)	100	0	0	0	0
1994 constant dollars (mil.)	111	0	0	0	0
Arms exports[6]					
Current dollars (mil.)	16,100	7,400	2,300	3,100	1,300
1994 constant dollars (mil.)	17,920	7,932	2,398	3,164	1,300
Total imports[7]					
Current dollars (mil.)	139,000	78,000	36,900	43,700	51,600
1994 constant dollars (mil.)	154,700	83,610	38,480	44,600	51,600
Total exports[7]					
Current dollars (mil.)	101,000	68,000	42,400	58,200	66,600
1994 constant dollars (mil.)	112,400	72,890	44,220	59,400	66,600
Arms as percent of total imports[8]	0.1	0	0	0	0
Arms as percent of total exports[8]	15.9	10.9	5.4	5.3	2.0

Crime [23]

	1994
Crime volume	
Cases known to police	2,632,708
Attempts (percent)	NA
Percent cases solved	50.40
Crimes per 100,000 persons	1,778.89
Persons responsible for offenses	
Total number offenders	1,441,568
Number female	13.00
Number juvenile (14-18 yrs.)	13.90
Number foreigners	25,656

Human Rights [24]

	SSTS	FL	FAPRO	PPCG	APROBC	TPW	PCPTW	STPEP	PHRFF	PRW	ASST	AFL
Observes	P	P	P	P	P	P	P	P	S	P	P	P
	EAFRD	CPR	ESCR	SR	ACHR	MAAE	PVIAC	PVNAC	EAFDAW	TCIDTP	RC
Observes	P	P	P	P		P	P	P	P	P	P

P = Party; S = Signatory; see Appendix for meaning of abbreviations.

Labor Force

Total Labor Force [25]

85 million (1993)

Labor Force by Occupation [26]

Production and economic services	83.9%
Government	16.1

Unemployment Rate [27]

8.2% (December 1995) with considerable additional underemployment

Production Sectors

Commercial Energy Production and Consumption

Data are shown in quadrillion (10^{15}) BTUs and percent for 1995
Values for hydroelectric, nuclear, geothermal, solar, and wind power refer to electrical generation.

Production [28]

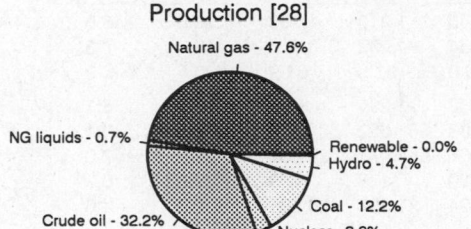

Natural gas - 47.6%
NG liquids - 0.7%
Renewable - 0.0%
Hydro - 4.7%
Coal - 12.2%
Nuclear - 2.6%
Crude oil - 32.2%

Consumption [29]

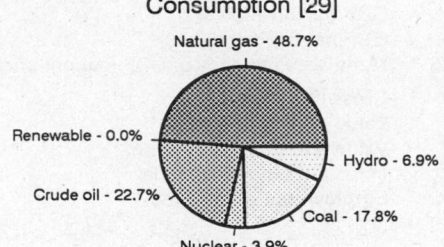

Natural gas - 48.7%
Renewable - 0.0%
Hydro - 6.9%
Coal - 17.8%
Nuclear - 3.9%
Crude oil - 22.7%

Crude oil	12.866
Natural gas liquids	0.273
Dry natural gas	19.01
Coal	4.853
Net hydroelectric power	1.864
Net nuclear power	1.041
Geothermal, solar, wind	0.001
Total	39.908

Crude oil	6.126
Dry natural gas	13.129
Coal	4.803
Net hydroelectric power	1.864
Net nuclear power	1.041
Geothermal, solar, wind	0.001
Total	26.964

Telecommunications [30]

- 25.4 million (1993 est.) telephones; long distance pay phones 34,100; enlisting foreign joint ventures to modernize; 11 mil. unfilled phone requests by 1992; international e-mail service via Sprint; inadequate system a severe economic handicap, especially internationally
- Domestic: NMT-450 analog cellular telephone networks in Moscow and St. Petersburg; limited intercity fiber-optic cable
- International: poorly handled by satellites, landlines, microwave relay, and submarine cables; most use Moscow's gateway switch, which also carries this traffic for others in CIS; a new satellite is to link with Rome for relay to Europe and overseas; satellite earth stations - Intelsat, 4 Intersputnik, Eutelsat, 1 Inmarsat, and Orbita
- Radio: Broadcast stations: ~1,050 stations Radios: 50 million (1993 est.) (with multiple speaker systems for program diffusion 74,300,000)
- Television: Broadcast stations: 7,183 Televisions: 54.85 million (1992 est.)

Top Agricultural Products [32]

Agriculture accounts for 6% of the GDP; produces grain, sugar beets, sunflower seed, vegetables, fruits (because of its northern location does not grow citrus, cotton, tea, and other warm climate products); meat, milk.

Top Mining Products [33]

Thousand metric tons except as noted[e]	8/31/95[*]
Gold, mine output, Au content (kg.)	147,000
Iron ore, 55-63% Fe	73,300[r]
Steel, crude	48,800[r]
Cement, hydraulic	37,200[r]
Diamond, gem and industrial (000 carats)	17,200
Talc	100,000
Coal, bituminous, lignite, and brown	271,000[r,60]
Coke, 6% moisture content	25,400[r]
Gas, natural, marketed (mil. cu. meters)	607,000[r]
Petroleum, crude, converted (000 42-gal. bls.)	2,300,000

Transportation [31]

Railways: total: 154,000 km; note - 87,000 km in common carrier service (38,000 km electrified); 67,000 km serve specific industries and are not available for common carrier use; broad gauge: 154,000 km 1.520-m gauge (1 January 1994)

Highways: total: 934,000 km (including 445,000 km which serve specific industries or farms and are not available for common carrier use); paved: NA km; unpaved: NA km (1994 est.)

Merchant marine: total: 745 ships (1,000 GRT or over) totaling 6,730,178 GRT/9,385,565 DWT; ships by type: barge carrier 2, bulk 25, cargo 406, chemical tanker 6, combination bulk 21, combination ore/oil 17, container 31, multifunction large-load carrier 3, oil tanker 134, passenger 4, passenger-cargo 5, refrigerated cargo 19, roll-on/roll-off cargo 54, short-sea passenger 16, specialized tanker 2

Airports

Total:	2,517
With paved runways over 3,047 m:	54
With paved runways 2,438 to 3,047 m:	202
With paved runways 1,524 to 2,437 m:	108
With paved runways 914 to 1,523 m:	115
With paved runways under 914 m:	151

Tourism [34]

	1990	1991	1992	1993	1994
Visitors	NA	NA	3,009	5,896	4,643

Travelers are in thousands, money in million U.S. dollars.

For sources, notes, and explanations, see Annotated Source Appendix, page 1061.

785

Manufacturing Sector

GDP and Manufacturing Summary [35]

	1980	1985	1989	1990	% change 1980- 1990	% change 1989- 1990
GDP (billion 1980 $)	893	1,045	1,212	1,133	26.9	-6.6
GDP per capita (1980 $)	3,362	3,764	4,232	3,923	16.7	-7.3
Manufacturing as % of GDP (current prices)	45.7	40.0	37.5[e]	34.9	-23.6	-6.9
Gross output (billion $)	834	868	1,295	568[e]	-31.9	-56.1
Value added (million $)	362,425	436,103	501,622	502,119	38.5	0.1
Value added (million 1980 $)	404,805	464,867	564,471	524,952	29.7	-7.0
Industrial production index	100	120	138	139	39.0	0.7
Employment (thousands)	31,464	32,794	31,207	30,596[e]	-2.8	-2.0

Note: GDP stands for Gross Domestic Product. 'e' stands for estimated value.

Profitability and Productivity

	1980	1985	1989	1990	% change 1980- 1990	% change 1989- 1990
Intermediate input (%)	NA	NA	NA	NA	NA	NA
Wages, salaries, and supplements (%)	12	11	NA	12[e]	0.0	NA
Gross operating surplus (%)	NA	NA	NA	NA	NA	NA
Gross output per worker ($)	26,509	26,456	41,500	18,515[e]	-30.2	-55.4
Value added per worker ($)	11,519	13,298	16,074	16,512[e]	43.3	2.7
Average wage (incl. benefits) ($)	3,247	3,002	4,836	2,159[e]	-33.5	-55.4

Profitability is in percent of gross output. Productivity is in U.S. $. 'e' stands for estimated value.

Profitability - 1990

Value Added in Manufacturing

	1980 $ mil.	1980 %	1985 $ mil.	1985 %	1989 $ mil.	1989 %	1990 $ mil.	1990 %	% change 1980- 1990	% change 1989- 1990
311 Food products	66,053	18.2	75,960	17.4	85,208	17.0	85,868	17.1	30.0	0.8
313 Beverages	10,336	2.9	9,303	2.1	8,889	1.8	8,786	1.7	-15.0	-1.2
314 Tobacco products	2,032	0.6	2,866	0.7	2,398	0.5	2,398	0.5	18.0	0.0
321 Textiles	32,553	9.0	34,506	7.9	38,086	7.6	37,435	7.5	15.0	-1.7
322 Wearing apparel	19,633	5.4	21,792	5.0	23,559	4.7	24,345	4.8	24.0	3.3
323 Leather and fur products	2,443	0.7	2,345	0.5	2,345	0.5	2,345	0.5	-4.0	0.0
324 Footwear	3,892	1.1	4,593	1.1	5,371	1.1	5,488	1.1	41.0	2.2
331 Wood and wood products	4,932	1.4	5,771	1.3	6,560	1.3	6,412	1.3	30.0	-2.3
332 Furniture and fixtures	3,457	1.0	4,459	1.0	5,427	1.1	5,669	1.1	64.0	4.5
341 Paper and paper products	2,784	0.8	3,424	0.8	3,981	0.8	4,065	0.8	46.0	2.1
342 Printing and publishing	2,613	0.7	3,214[e]	0.7	3,736	0.7	3,815[e]	0.8	46.0	2.1
351 Industrial chemicals	14,704	4.1	17,939	4.1	20,144	4.0	19,703	3.9	34.0	-2.2
352 Other chemical products	7,584	2.1	8,419	1.9	9,632	1.9	9,632	1.9	27.0	0.0
353 Petroleum refineries	5,490	1.5	6,093	1.4	6,972	1.4	6,972	1.4	27.0	0.0
354 Miscellaneous petroleum and coal products	11,003	3.0	12,214	2.8	13,974	2.8	13,974	2.8	27.0	0.0
355 Rubber products	4,154	1.1	4,861	1.1	5,401	1.1	5,276	1.1	NA	-2.3
356 Plastic products	1,546	0.4	2,273	0.5	2,969	0.6	2,969	0.6	92.0	0.0
361 Pottery, china, and earthenware	2,014	0.6	2,457	0.6	2,860	0.6	3,001	0.6	49.0	4.9
362 Glass and glass products	1,204	0.3	1,517	0.3	1,878	0.4	1,914	0.4	59.0	1.9
369 Other nonmetal mineral products	13,769	3.8	15,696	3.6	18,037	3.6	17,761	3.5	29.0	-1.5
371 Iron and steel	14,418	4.0	15,139	3.5	15,860	3.2	15,139	3.0	5.0	-4.5
372 Nonferrous metals	7,716	2.1	8,333	1.9	8,256	1.6	7,793	1.6	1.0	-5.6
381 Metal products	7,130	2.0	9,625	2.2	11,693	2.3	11,764	2.3	65.0	0.6
382 Nonelectrical machinery	79,367	21.9	107,146	24.6	130,162	25.9	130,956	26.1	65.0	0.6
383 Electrical machinery	9,105	2.5	12,292	2.8	14,932	3.0	15,023	3.0	65.0	0.6
384 Transport equipment	11,574	3.2	15,625	3.6	18,982	3.8	19,097	3.8	65.0	0.6
385 Professional and scientific equipment	9,711	2.7	13,110	3.0	15,927	3.2	16,024	3.2	65.0	0.6
390 Other manufacturing industries	11,210	3.1	15,133[e]	3.5	18,384	3.7	18,496[e]	3.7	65.0	0.6

Note: The industry codes shown are International Standard Industry codes (ISIC). Percentages are percent of total Value Added. 'e' stands for estimated value

Finance, Economics, and Trade

Economic Indicators [36]

- **National product**: GDP—purchasing power parity—$796 billion (1995 estimate as extrapolated from World Bank estimate for 1994)

- **National product real growth rate**: - 4% (1995 est.)

- **National product per capita**: $5,300 (1995 est.)

- **Inflation rate (consumer prices)**: 7% monthly average (1995 est.)

- **External debt**: $130 billion (year-end 1995)

Balance of Payments Summary [37]

Values in millions of dollars.

	1989	1990	1991	1992	1993[1]
Exports of goods (f.o.b.)	NA	NA	NA	NA	43,711
Imports of goods (f.o.b.)	NA	NA	NA	NA	34,163
Trade balance	NA	NA	NA	NA	9,548
Services - debits	NA	NA	NA	NA	-14,580
Services - credits	NA	NA	NA	NA	8,939
Private transfers (net)	NA	NA	NA	NA	2,324
Government transfers (net)	NA	NA	NA	NA	NA
Long-term capital (net)	NA	NA	NA	NA	-883
Short-term capital (net)	NA	NA	NA	NA	-7,323
Errors and omissions	NA	NA	NA	NA	4,850
Overall balance	NA	NA	NA	NA	2,875

Exchange Rates [38]

Currency: **ruble.**
Symbol: **R.**

Data are currency units per $1

29 December 1995	4,640
29 December 1994	3,550
27 December 1993	1,247

Imports and Exports

Top Import Origins [39]

$57.9 billion (c.i.f., 1995)

Origins	%
Europe	NA
North America	NA
Japan	NA
Third World countries	NA
Cuba	NA

Top Export Destinations [40]

$77.8 billion (f.o.b., 1995).

Destinations	%
Europe	NA
North America	NA
Japan	NA
Third World countries	NA
Cuba	NA

Foreign Aid [41]

	U.S. $	
ODA (1993)	2.8	billion
US commitments, including Ex-Im (1990-95)	14	billion
Other countries, ODA and OOF bilateral commitments (1990-95)	125	billion

Import and Export Commodities [42]

Import Commodities

Machinery and equipment
Consumer goods
Medicines
Meat
Grain
Sugar
Semifinished metal products

Export Commodities

Petroleum and petroleum products
Natural gas
Wood and wood products
Metals
Chemicals
Variety of civilian and military manufactures

For sources, notes, and explanations, see Annotated Source Appendix, page 1061.

787

Rwanda

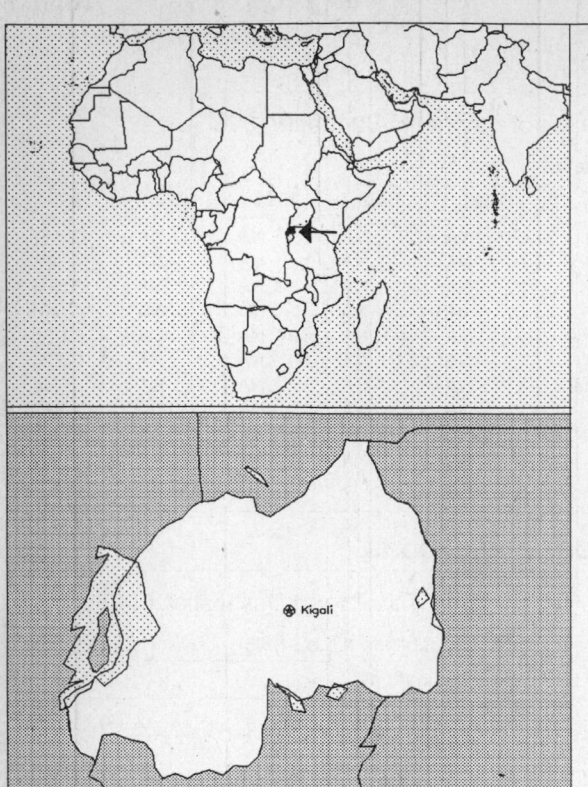

Geography [1]

Total area:
26,340 sq km   10,170 sq mi
Land area:
24,950 sq km   9,633 sq mi
Comparative area:
Slightly smaller than Maryland
Land boundaries:
Total 893 km, Burundi 290 km, Tanzania 217 km, Uganda 169 km, Zaire 217 km
Coastline:
0 km (landlocked)
Climate:
Temperate; two rainy seasons (February to April, November to January);
mild in mountains with frost and snow possible
Terrain:
Mostly grassy uplands and hills; relief is mountainous with altitude
declining from West to East
Natural resources:
Gold, cassiterite (tin ore), wolframite (tungsten ore), natural gas,
hydropower
Land use:
Arable land: 29%
Permanent crops: 11%
Meadows and pastures: 18%
Forest and woodland: 10%
Other: 32%

Demographics [2]

	1970	1980	1990	1995[1]	1996	2000	2010	2020	2030
Population	3,813	5,170	7,145	6,018	6,853	8,900	10,080	11,040	12,725
Population density (persons per sq. mi.)	396	537	742	625	711	924	1,046	1,146	1,321
(persons per sq. km.)	153	207	286	241	275	357	404	443	510
Net migration rate (per 1,000 population)	1.4	0.6	-0.7	74.2	146.4	0.0	0.0	0.0	0.0
Births	NA	NA	NA	NA	266	NA	NA	NA	NA
Deaths	NA	NA	NA	NA	139	NA	NA	NA	NA
Life expectancy - males	NA	45.9	46.0	35.9	39.7	35.9	33.9	38.4	52.8
Life expectancy - females	NA	48.7	47.7	36.4	40.5	37.1	31.4	36.5	54.2
Birth rate (per 1,000)	49.8	54.4	42.9	39.2	38.8	38.5	37.5	32.7	30.3
Death rate (per 1,000)	21.8	19.1	16.8	23.4	20.3	23.5	28.2	22.7	11.6
Women of reproductive age (15-49 yrs.)	NA	1,155	1,539	1,368	1,568	2,113	2,408	2,665	3,333
of which are currently married[44]	565	NA	881	NA	853	1,122	1,309	NA	NA
Fertility rate	7.5	8.5	6.7	6.1	6.0	5.7	5.0	4.2	3.5

Except as noted, values for vital statistics are in thousands; life expectancy is in years.

Health

Health Indicators [3]

% of population with access to	
safe water (1990-95)	66
adequate sanitation (1990-95)	58
health services (1985-95)	80
% of 1-year-olds immunized (1990-94) against	
TB (tuberculosis)	32
DPT (diphtheria, pertussis, tetanus)	23
polio	23
measles	25
% of contraceptive prevalence (1980-94)	21
ORT use rate (1990-94)	47

Health Expenditures [4]

Total health expenditure, 1990 (official exchange rate)	
Millions of dollars	74
Dollars per capita	10
Health expenditures as a percentage of GDP	
Total	3.5
Public sector	1.9
Private sector	1.6
Development assistance for health	
Total aid flows (millions of dollars)[1]	29
Aid flows per capita (dollars)	4.1
Aid flows as a percentage of total health expenditure	39.5

For sources, notes, and explanations, see Annotated Source Appendix, page 1061.

Human Factors

Women and Children [5]

% of pregnant women immunized (tetanus 1990-94)	NA
% of births attended by trained health personnel (1983-94)	26
Maternal mortality rate (1980-92)	210
Under-5 mortality rate (1994)	139
% under-5 moderately/severely underweight (1980-1994)	29

Burden of Disease [6]

Population per physician (1989)	73,902.17
Population per nurse (1989)	4,241.42
Population per hospital bed (1990)	608.11
AIDS cases per 100,000 people (1994)	*
Malaria cases per 100,000 people (1992)	NA

Ethnic Division [7]

Hutu	80%
Tutsi	19%
Twa (Pygmoid)	1%

Religion [8]

Roman Catholic	65%
Protestant	9%
Muslim	1%
Indigenous beliefs and other	25%

Major Languages [9]

Kinyarwanda (official), French (official), Kiswahili (Swahili) used in commercial centers.

Education

Public Education Expenditures [10]

	1980	1984	1989	1990	1993	1994
Million (Franc)						
Total education expenditure	2,880	4,997	7,222	NA	NA	NA
as percent of GNP	2.7	3.1	3.7	NA	NA	NA
as percent of total govt. expend.	21.6	25.1	25.4	NA	NA	NA
Current education expenditure	2,439	4,887	6,793	NA	NA	NA
as percent of GNP	2.3	3.1	3.5	NA	NA	NA
as percent of current govt. expend.	21.5	28.1	29.1	NA	NA	NA
Capital expenditure	442	109	429	NA	NA	NA

Educational Attainment [11]

Illiteracy [12]

In thousands and percent[1]	1990	1995	2000
Illiterate population (15+ yrs.)	1,677	1,695	1,683
Illiteracy rate - total pop. (%)	45.9	52.5	33.9
Illiteracy rate - males (%)	35.4	40.1	26.0
Illiteracy rate - females (%)	55.8	64.4	41.5

Libraries [13]

	Admin. Units	Svc. Pts.	Vols. (000)	Shelving (meters)	Vols. Added	Reg. Users
National	NA	NA	NA	NA	NA	NA
Nonspecialized	NA	NA	NA	NA	NA	NA
Public	NA	NA	NA	NA	NA	NA
Higher ed. (1987)	3	4	163	NA	6,416	1,901
School	NA	NA	NA	NA	NA	NA

Daily Newspapers [14]

	1980	1985	1990	1994
Number of papers	1	1	1	1
Circ. (000)	0.3	0.3	0.5	0.5

Culture [15]

Science and Technology

Scientific/Technical Forces [16]

Scientists/engineers	71
Number female	NA
Technicians	67[e]
Number female	NA
Total	138[e]

R&D Expenditures [17]

	Franc (000) 1985
Total expenditure	918,560
Capital expenditure	819,280
Current expenditure	99,280
Percent current	10.8

U.S. Patents Issued [18]

For sources, notes, and explanations, see Annotated Source Appendix, page 1061.

Government and Law

Organization of Government [19]

Long-form name:
Republic of Rwanda
Type:
Republic; presidential system
Independence:
1 July 1962 (from Belgium-administered UN trusteeship)
National holiday:
Independence Day, 1 July (1962)
Constitution:
18 June 1991
Legal system:
Based on German and Belgian civil law systems and customary law; judicial review of legislative acts in the Supreme Court; has not accepted compulsory ICJ jurisdiction
Executive branch:
President; Vice President; Prime Minister; Council of Ministers
Legislative branch:
Unicameral: National Assembly (Assemblee Nationale)
Judicial branch:
Constitutional Court, consists of the Court of Cassation and the Council of State in joint session

Elections [20]

National Development Council	% of seats
Rwandan Patriotic Front (RPF)	27.1
Democratic Republican Movement	18.6
Democratic and Socialist (PSD)	18.6
Liberal Party	18.6
Christian Democratic	8.6
Rwandan Socialist (PSR)	2.9
Islamic Democratic (PDI)	2.9
Other	2.9

Government Budget [21]

Military Expenditures and Arms Transfers [22]

	1990	1991	1992	1993	1994
Military expenditures					
Current dollars (mil.)	68	120[e]	112[e]	121[e]	114[e]
1994 constant dollars (mil.)	76	128[e]	117[e]	123[e]	114[e]
Armed forces (000)	6	30	30	30	40
Gross national product (GNP)					
Current dollars (mil.)	1,633	1,721	1,781	1,636	NA
1994 constant dollars (mil.)	1,818	1,844	1,857	1,670	NA
Central government expenditures (CGE)					
1994 constant dollars (mil.)	385	413	464	530	NA
People (mil.)	7.4	7.7	7.9	8.1	8.4
Military expenditure as % of GNP	4.2	7.0	6.3	7.4	7.6
Military expenditure as % of CGE	19.8	31.1	25.1	23.3	NA
Military expenditure per capita (1994 $)	10	17	15	15	14
Armed forces per 1,000 people (soldiers)	0.8	3.9	3.8	3.7	4.8
GNP per capita (1994 $)	245	241	235	205	NA
Arms imports[6]					
Current dollars (mil.)	0	0	0	10	80
1994 constant dollars (mil.)	0	0	0	10	80
Arms exports[6]					
Current dollars (mil.)	0	0	0	0	0
1994 constant dollars (mil.)	0	0	0	0	0
Total imports[7]					
Current dollars (mil.)	288	306	287	250[e]	NA
1994 constant dollars (mil.)	321	328	299	255[e]	NA
Total exports[7]					
Current dollars (mil.)	110	93	67	44[e]	NA
1994 constant dollars (mil.)	122	99	70	45[e]	NA
Arms as percent of total imports[8]	0	0	0	4.0	NA
Arms as percent of total exports[8]	0	0	0	0	0

Crime [23]

	1994
Crime volume	
Cases known to police	1,164,000
Attempts (percent)	NA
Percent cases solved	NA
Crimes per 100,000 persons	14,550.00
Persons responsible for offenses	
Total number offenders	NA
Percent female	NA
Percent juvenile (0-14 yrs.)	NA
Percent foreigners	NA

Human Rights [24]

	SSTS	FL	FAPRO	PPCG	APROBC	TPW	PCPTW	STPEP	PHRFF	PRW	ASST	AFL
Observes			P	P	P	P	P					P
	EAFRD	CPR	ESCR	SR	ACHR	MAAE	PVIAC	PVNAC	EAFDAW	TCIDTP		RC
Observes	P	P	P	P		P	P	P	P			P

P = Party; S = Signatory; see Appendix for meaning of abbreviations.

Labor Force

Total Labor Force [25]

3.6 million

Labor Force by Occupation [26]

Agriculture	93%
Government and services	5
Industry and commerce	2

Unemployment Rate [27]

For sources, notes, and explanations, see Annotated Source Appendix, page 1061.

Production Sectors

Commercial Energy Production and Consumption

Data are shown in quadrillion (10^{15}) BTUs and percent for 1995
Values for hydroelectric, nuclear, geothermal, solar, and wind power refer to electrical generation.

Production [28]

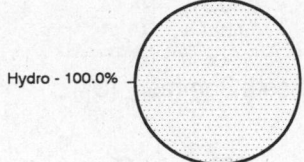

Hydro - 100.0%

Consumption [29]

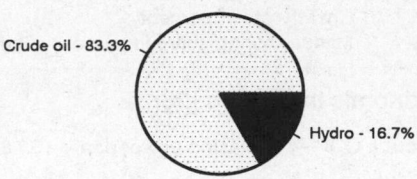

Crude oil - 83.3%

Hydro - 16.7%

Net hydroelectric power	0.002
Total	0.002

Crude oil	0.010
Net hydroelectric power	0.002
Total	0.012

Telecommunications [30]

- 6,400 (1983 est.) telephones; telephone system does not provide service to the general public but is intended for business and government use
- Domestic: the capital, Kigali, is connected to the centers of the prefectures by microwave radio relay; the remainder of the network depends on wire and HF radiotelephone
- International: international connections employ microwave radio relay to neighboring countries and satellite communications to more distant countries; satellite earth stations - 1 Intelsat (Indian Ocean) in Kigali (includes telex and telefax service)
- Radio: Broadcast stations: AM 1, FM 1, shortwave 0 Radios: 630,000 (1993 est.)
- Television: Broadcast stations: 1

Transportation [31]

Railways: 0 km

Highways: total: 13,173 km; paved: 1,186 km; unpaved: 11,987 km (1990 est.)

Airports

Total:	7
With paved runways over 3,047 m:	1
With paved runways 914 to 1,523 m:	2
With paved runways under 914 m:	3
With unpaved runways 914 to 1,523 m:	1 (1995 est.)

Top Agricultural Products [32]

Agriculture accounts for 52% of the GDP; produces coffee, tea, pyrethrum (insecticide made from chrysanthemums), bananas, beans, sorghum, potatoes; livestock.

Top Mining Products [33]

Metric tons except as noted[e]	8/1/95[*]
Cement	10,000
Columbite-tantalite, ore and concentrate gross weight (kg.)	10,000
Gold, mine output, Au content (kg.)	100
Natural gas, gross (000 cu. meters)	100
Tin, mine output, Sn content	50
Tungsten, mine output, W content	30

Tourism [34]

	1990	1991	1992	1993	1994
Tourists[91]	16	3	5	2	NA
Tourism receipts	10	4	4	2	NA
Tourism expenditures	23	17	17	NA	NA
Fare expenditures	11	8	NA	NA	NA

Travelers are in thousands, money in million U.S. dollars.

Finance, Economics, and Trade

GDP and Manufacturing Summary [35]

	1980	1985	1990	1991	1992
Gross Domestic Product					
Millions of 1980 dollars	1,163	1,347	1,365	1,324	1,073
Growth rate in percent	6.01	4.41	-1.65	-2.96	-19.00
Manufacturing Value Added					
Millions of 1980 dollars	178	202	233	215	NA
Growth rate in percent	12.30	6.96	-4.00	-8.00	NA
Manufacturing share in percent of current prices	15.8	14.2	15.2	14.8[e]	NA

Economic Indicators [36]

- **National product**: GDP—purchasing power parity—$3.8 billion (1995 est.)

- **National product real growth rate**: - 2.7% (1995 est.)

- **National product per capita**: $400 (1995 est.)

- **Inflation rate (consumer prices)**: 64% (1994 est.)

- **External debt**: $873 million (1993 est.)

Balance of Payments Summary [37]

Values in millions of dollars.

	1988	1989	1990	1991	1992
Exports of goods (f.o.b.)	117.9	104.7	102.6	95.6	68.5
Imports of goods (f.o.b.)	-278.6	-254.1	-227.7	-228.1	-240.4
Trade balance	-160.7	-149.4	-125.1	-132.5	-171.9
Services - debits	-165.0	-142.0	-152.0	-128.8	-132.0
Services - credits	56.9	52.3	46.6	46.5	36.1
Private transfers (net)	10.5	7.9	5.8	20.9	22.1
Government transfers (net)	139.3	129.2	138.2	159.8	161.1
Long-term capital (net)	87.6	59.6	47.0	79.8	36.8
Short-term capital (net)	6.1	-5.7	8.7	19.3	25.6
Errors and omissions	0.4	1.9	30.3	0.2	18.2
Overall balance	-24.9	-46.2	-0.5	65.2	-4.0

Exchange Rates [38]

Currency: **Rwandan franc.**
Symbol: **RF.**

Data are currency units per $1.

2nd quarter 1994	401.27
1993	168.20
1992	133.35
1991	125.14

Imports and Exports

Top Import Origins [39]

$37 million (1994 est.).

Origins	%
US	NA
Belgium	NA
Germany	NA
Kenya	NA
Japan	NA

Top Export Destinations [40]

$52 million (f.o.b., 1994 est.).

Destinations	%
Germany	NA
Belgium	NA
Italy	NA
Uganda	NA
UK	NA
France	NA
US	NA

Foreign Aid [41]

	U.S. $	
ODA	NA	
Structural Adjustment Program with the IMF		
EC (1991-93)	46	million
US (1991-93)	25	million

Import and Export Commodities [42]

Import Commodities	Export Commodities
Textiles	Coffee 63%
Foodstuffs	Tea
Machines and equipment	Cassiterite
Capital goods	Wolframite
Steel	Pyrethrum
Petroleum products	
Cement and construction material	

Saint Kitts and Nevis

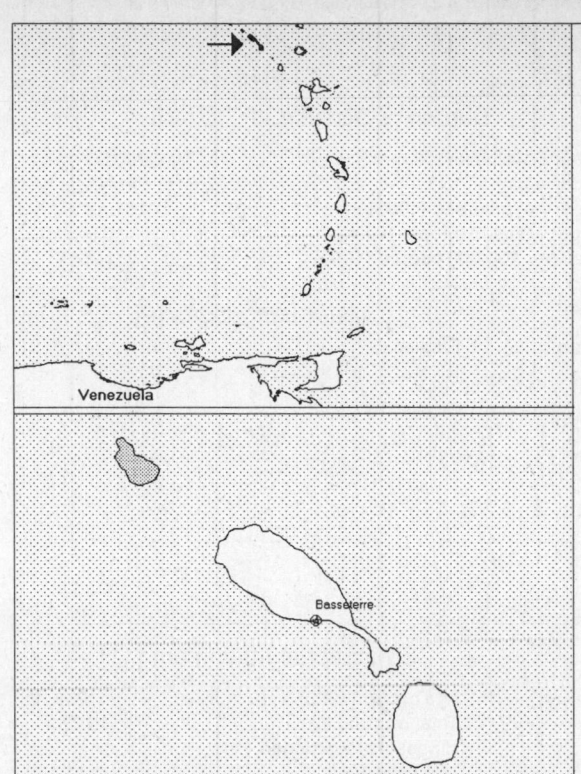

Geography [1]

Total area:
269 sq km   104 sq mi
Land area:
269 sq km   104 sq mi
Comparative area:
Twice the size of Washington, DC
Land boundaries:
0 km
Coastline:
135 km
Climate:
Subtropical tempered by constant sea breezes; little seasonal temperature variation; rainy season (May to November)
Terrain:
Volcanic with mountainous interiors
Natural resources:
Negligible
Land use:
Arable land: 22%
Permanent crops: 17%
Meadows and pastures: 3%
Forest and woodland: 17%
Other: 41%

Demographics [2]

	1970	1980	1990	1995[1]	1996	2000	2010	2020	2030
Population	46	44	40	41	41	43	50	57	62
Population density (persons per sq. mi.)	332	318	287	295	298	313	361	409	450
(persons per sq. km.)	128	123	111	114	115	121	139	158	174
Net migration rate (per 1,000 population)	-20.2	-21.2	-11.1	-5.4	-4.3	0.0	0.0	0.0	0.0
Births	1	1	NA	NA	1	NA	NA	NA	NA
Deaths	0	0	NA	NA	Z	NA	NA	NA	NA
Life expectancy - males	NA	61.4	61.7	63.5	63.8	65.2	68.3	71.0	73.1
Life expectancy - females	NA	67.3	67.6	69.7	70.1	71.6	75.0	77.8	80.0
Birth rate (per 1,000)	25.1	26.7	24.6	23.5	23.3	22.4	19.3	15.9	14.1
Death rate (per 1,000)	13.7	11.5	11.7	9.6	9.2	7.8	5.5	5.1	5.8
Women of reproductive age (15-49 yrs.)	NA	10	9	11	11	12	15	16	16
of which are currently married[45]	2	2	2	NA	3	3	4	NA	NA
Fertility rate	NA	3.3	2.8	2.6	2.5	2.4	2.1	2.0	1.9

Except as noted, values for vital statistics are in thousands; life expectancy is in years.

Health

Health Indicators [3]

Health Expenditures [4]

For sources, notes, and explanations, see Annotated Source Appendix, page 1061.

793

Human Factors

Women and Children [5]	Burden of Disease [6]	
	Population per physician (1984)	2,200.00
	Population per nurse	NA
	Population per hospital bed (1993)	161.42
	AIDS cases per 100,000 people (1994)	10.9
	Malaria cases per 100,000 people (1992)	NA

Ethnic Division [7]	Religion [8]	Major Languages [9]
Black African.	Anglican, other Protestant sects, Roman Catholic.	English.

Education

Public Education Expenditures [10]

Million (E. Carib. Dollar)[58]	1980	1985	1990	1991	1992	1994
Total education expenditure	7	12	11	12	15	NA
as percent of GNP	5.2	5.8	2.7	2.8	3.3	NA
as percent of total govt. expend.	9.4	18.5	NA	11.6	13.5	NA
Current education expenditure	7	12	11	12	14	NA
as percent of GNP	5.2	5.7	2.7	2.7	3.1	NA
as percent of current govt. expend.	13.6	19.1	NA	11.7	NA	NA
Capital expenditure	0.0	0.0	-	0.0	1	NA

Educational Attainment [11]

Age group (1980)	25+
Total population	16,771
Highest level attained (%)	
No schooling	1.1
First leveL	
Not completed	29.0
Completed	NA
Entered second level	
S-1	66.6
S-2	NA
Postsecondary	2.3

Illiteracy [12]

	1980	1989	1995
Illiterate population (15+ yrs.)[2]	674	NA	NA
Illiteracy rate - total pop. (%)	2.7	NA	NA
Illiteracy rate - males (%)	2.9	NA	NA
Illiteracy rate - females (%)	2.5	NA	NA

Libraries [13]

Daily Newspapers [14]

Culture [15]

Science and Technology

Scientific/Technical Forces [16]	R&D Expenditures [17]	U.S. Patents Issued [18]

Government and Law

Organization of Government [19]

Long-form name:
Federation of Saint Kitts and Nevis
Type:
Constitutional monarchy
Independence:
19 September 1983 (from UK)
National holiday:
Independence Day, 19 September (1983)
Constitution:
19 September 1983
Legal system:
Based on English common law
Executive branch:
British Monarch (represented by Governor General); Prime Minister; Cabinet
Legislative branch:
Unicameral: House of Assembly
Judicial branch:
Eastern Caribbean Supreme Court (based on Saint Lucia)

Elections [20]

House of Assembly	% of votes
Labor Party (SKNLP)	58
People's Action Movement (PAM)	41
Other	1

Government Budget [21]

For 1996 est.

	$ mil.
Revenues	100.2
Expenditures	100.1
Capital expenditures	41.4

Defense Summary [22]

Branches: Royal Saint Kitts and Nevis Police Force, Coast Guard

Manpower Availability: Males age 15-49 NA; males fit for military service NA

Defense Expenditures: NA

Crime [23]

	1990
Crime volume	
Cases known to police	1,298
Attempts (percent)	NA
Percent cases solved	17.9
Crimes per 100,000 persons	3,250.1
Persons responsible for offenses	
Total number offenders	232
Percent female	29.3
Percent juvenile	9.5
Percent foreigners	NA

Human Rights [24]

	SSTS	FL	FAPRO	PPCG	APROBC	TPW	PCPTW	STPEP	PHRFF	PRW	ASST	AFL
Observes	1					P	P			1	1	
	EAFRD	CPR	ESCR	SR	ACHR	MAAE	PVIAC	PVNAC	EAFDAW	TCIDTP	RC	
Observes						P	P	P			P	

P = Party; S = Signatory; see Appendix for meaning of abbreviations.

Labor Force

Total Labor Force [25]

18,172 (June 1995)

Labor Force by Occupation [26]

Services	69%
Manufacturing	31

Unemployment Rate [27]

4.3% (May 1995)

For sources, notes, and explanations, see Annotated Source Appendix, page 1061.

795

Production Sectors

Energy Resource Summary [28]

Electricity: Capacity: 15,800 kW. Production: 45 million kWh. Consumption per capita: 990 kWh (1993).

Telecommunications [30]

- 3,800 (1986 est.) telephones; good interisland VHF/UHF/SHF radiotelephone connections and international link via Antigua and Barbuda and Saint Martin (Guadeloupe and Netherlands Antilles)
- Domestic: interisland links are handled by VHF/UHF/SHF radiotelephone
- International: international calls are carried by radiotelephone to Antigua and Barbuda and from there switched to submarine cable or to Intelsat, or carried to Saint Martin (Guadeloupe and Netherlands Antilles) by radiotelephone and switched to Intelsat
- Radio: Broadcast stations: AM 2, FM 0, shortwave 0 Radios: 25,000 (1993 est.)
- Television: Broadcast stations: 4 Televisions: 9,500 (1993 est.)

Top Agricultural Products [32]

Agriculture accounts for 6.2% of the GDP; produces sugarcane, rice, yams, vegetables, bananas; fishing potential not fully exploited.

Top Mining Products [33]

Detailed information is not available. A summary of mineral resources follows. **Mineral Resources**: Salt.

Transportation [31]

Railways: total: 58 km; narrow gauge: 58 km 0.762-m gauge on Saint Kitts to serve sugarcane plantations (1995)

Highways: total: 300 km; paved: 125 km; unpaved: 175 km

Merchant marine: none

Airports

Total: 2
With paved runways 2,438 to 3,047 m: 1
With paved runways under 914 m: 1 (1995 est.)

Tourism [34]

	1990	1991	1992	1993	1994
Visitors[92]	110	137	164	170	209
Tourists[92]	73	83	88	84	94
Excursionists	3	1	2	3	2
Cruise passengers[93]	34	53	74	83	113
Tourism receipts	63	74	67	69	75
Tourism expenditures	4	5	5	5	6
Fare expenditures	6	6	6	6	5

Travelers are in thousands, money in million U.S. dollars.

Finance, Economics, and Trade

Industrial Summary [35]

Industrial Production: Growth rate 5.9% (1992 est.); accounts for 35.7% of the GDP. **Industries**: Sugar processing, tourism, cotton, salt, copra, clothing, footwear, beverages.

Economic Indicators [36]

- **National product**: GDP—purchasing power parity—$220 million (1995 est.)
- **National product real growth rate**: 3% (1995 est.)
- **National product per capita**: $5,380 (1995 est.)
- **Inflation rate (consumer prices)**: - 0.9% (1995)
- **External debt**: $45.3 million (1994 est.)

Balance of Payments Summary [37]

Values in millions of dollars.

	1988	1989	1990	1991	1992
Exports of goods (f.o.b.)	27.4	28.6	27.6	29.3	31.9
Imports of goods (f.o.b.)	-82.0	-86.9	-97.4	-97.1	-92.2
Trade balance	-54.4	-58.2	-69.8	-67.8	-60.3
Services - debits	-34.3	-43.4	-44.7	-46.9	-56.1
Services - credits	51.8	54.3	58.0	71.8	82.2
Private transfers (net)	10.5	14.2	12.8	12.4	13.1
Government transfers (net)	5.0	2.1	-0.5	0.8	0.0
Long-term capital (net)	19.4	46.8	49.2	22.3	15.0
Short-term capital (net)	-1.8	9.6	-1.8	7.3	12.1
Errors and omissions	3.9	-19.1	-3.1	0.7	3.3
Overall balance	NA	6.3	0.1	0.7	9.8

Exchange Rates [38]

Currency: **EC dollar.**
Symbol: **EC$.**

Data are currency units per $1.

Fixed rate since 1976	2.7000

Imports and Exports

Top Import Origins [39]

$112.4 million (f.o.b., 1994 est.) Data are for 1994.

Origins	%
US	45
Caricom	18.8
UK	12.5
Canada	4.2
Japan	4.2

Top Export Destinations [40]

$35.4 million (f.o.b., 1994 est.) Data are for 1994.

Destinations	%
US	46.6
UK	26.4
Caricom	9.8

Foreign Aid [41]

USAID assistance diminished in 1996 when the regional office in Barbados closed.

	U.S. $
ODA	$NA
USAID (1994)	$NA

Import and Export Commodities [42]

Import Commodities	Export Commodities
Machinery	Machinery
Manufactures	Food
Food	Electronics
Fuels	Beverages and tobacco

For sources, notes, and explanations, see Annotated Source Appendix, page 1061.

797

Saint Lucia

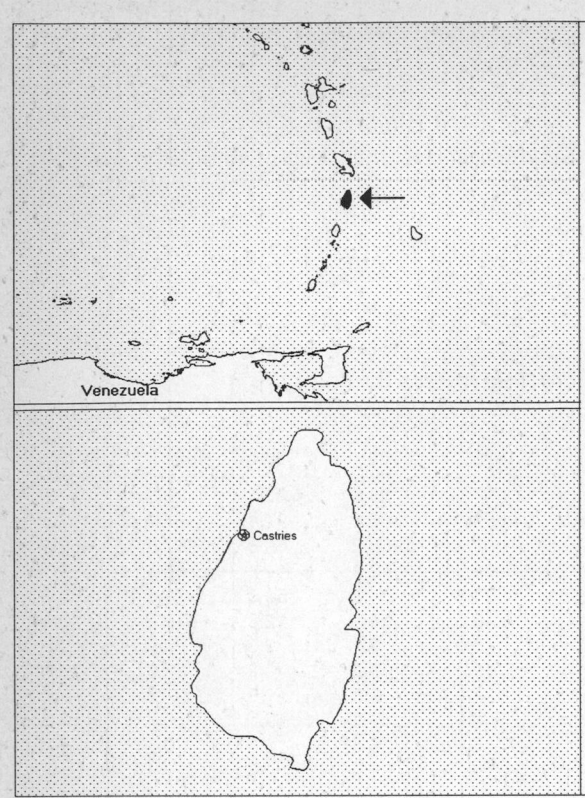

Geography [1]

Total area:
620 sq km   239 sq mi
Land area:
610 sq km   236 sq mi
Comparative area:
3.5 times the size of Washington, DC
Land boundaries:
0 km
Coastline:
158 km
Climate:
Tropical, moderated by northeast trade winds; dry season from January to April, rainy season from May to August
Terrain:
Volcanic and mountainous with some broad, fertile valleys
Natural resources:
Forests, sandy beaches, minerals (pumice), mineral springs, geothermal potential
Land use:
Arable land: 8%
Permanent crops: 20%
Meadows and pastures: 5%
Forest and woodland: 13%
Other: 54%

Demographics [2]

	1970	1980	1990	1995[1]	1996	2000	2010	2020	2030
Population	103	122	146	156	158	165	183	202	217
Population density (persons per sq. mi.)	435	519	619	661	669	698	775	857	919
(persons per sq. km.)	168	200	239	255	258	270	299	331	355
Net migration rate (per 1,000 population)	-23.1	-10.4	-4.6	-4.7	-4.6	-4.4	0.0	0.0	0.0
Births	5	3	NA	NA	3	NA	NA	NA	NA
Deaths	0	0	NA	NA	1	NA	NA	NA	NA
Life expectancy - males	NA	67.2	67.0	66.3	66.5	67.3	69.1	70.8	72.2
Life expectancy - females	NA	70.5	74.3	73.7	73.9	74.9	77.0	78.9	80.4
Birth rate (per 1,000)	48.1	32.2	24.0	22.5	22.0	19.9	17.0	14.3	12.5
Death rate (per 1,000)	8.3	6.9	5.8	6.1	6.0	5.8	5.6	5.9	7.0
Women of reproductive age (15-49 yrs.)	NA	26	37	42	43	47	54	56	55
of which are currently married[45]	6	6	19	NA	23	26	32	NA	NA
Fertility rate	NA	4.5	2.7	2.4	2.3	2.1	1.9	1.8	1.8

Except as noted, values for vital statistics are in thousands; life expectancy is in years.

Health

Health Indicators [3]

Health Expenditures [4]

For sources, notes, and explanations, see Annotated Source Appendix, page 1061.

Human Factors

Women and Children [5]

Burden of Disease [6]

Population per physician (1993)	1,903.61
Population per nurse	NA
Population per hospital bed (1993)	237.95
AIDS cases per 100,000 people (1994)	9.1
Malaria cases per 100,000 people (1992)	NA

Ethnic Division [7]

African descent	90.3%
Mixed	5.5%
East Indian	3.2%
White	0.8%

Religion [8]

Roman Catholic	90%
Protestant	7%
Anglican	3%

Major Languages [9]

English (official), French patois.

Education

Public Education Expenditures [10]

	1980	1986	1990	1991	1992	1994
Million (E. Carib. Dollar)						
Total education expenditure	NA	38	NA	NA	NA	NA
as percent of GNP	NA	5.5	NA	NA	NA	NA
as percent of total govt. expend.	NA	NA	NA	NA	NA	NA
Current education expenditure	19	36	54	55	63	NA
as percent of GNP	6.9	5.2	5.3	5.1	5.2	NA
as percent of current govt. expend.	NA	NA	NA	NA	NA	NA
Capital expenditure	NA	2	NA	NA	NA	NA

Educational Attainment [11]

Age group (1980)	25+
Total population	39,599
Highest level attained (%)	
No schooling	17.5
First level	
Not completed	74.5
Completed	NA
Entered second level	
S-1	6.8
S-2	NA
Postsecondary	1.3

Illiteracy [12]

	1970	1980	1990
Illiterate population (15+ years)[4]	9,195	NA	NA
Illiteracy rate - total pop. (%)	18.3	NA	NA
Illiteracy rate - males (%)	19.2	NA	NA
Illiteracy rate - females (%)	17.6	NA	NA

Libraries [13]

Daily Newspapers [14]

Culture [15]

Cinema (seats per 1,000)	NA
Annual attendance per person	NA
Gross box office receipts (mil. Dollar)	NA
Museums (reporting)	1
Visitors (000)	53
Annual receipts (000 Dollar)	445

Science and Technology

Scientific/Technical Forces [16]

Scientists/engineers	53
Number female	NA
Technicians	86
Number female	NA
Total	139

R&D Expenditures [17]

	E.C. $ (000) 1992
Total expenditure	NA
Capital expenditure	NA
Current expenditure	449[e]
Percent current	NA

U.S. Patents Issued [18]

For sources, notes, and explanations, see Annotated Source Appendix, page 1061.

Government and Law

Organization of Government [19]

Long-form name:
None
Type:
Parliamentary democracy
Independence:
22 February 1979 (from UK)
National holiday:
Independence Day, 22 February (1979)
Constitution:
22 February 1979
Legal system:
Based on English common law
Executive branch:
British Monarch (represented by Governor
General); Prime Minister; Cabinet
Legislative branch:
Bicameral Parliament: Senate and House
of Assembly
Judicial branch:
Eastern Caribbean Supreme Court

Elections [20]

House of Assembly	% of seats
United Workers' Party (UWP)	64.7
Saint Lucia Labor Party (SLP)	35.3

Government Budget [21]

For 1992 est.

	$ mil.
Revenues	121
Expenditures	127
Capital expenditures	104

Defense Summary [22]

Branches: Royal Saint Lucia Police Force, Coast Guard

Manpower Availability: Males age 15-49 NA; males fit for military service NA

Defense Expenditures: $5.0 million, 2.0% of GDP (1991); note - for police forces

Crime [23]

Human Rights [24]

	SSTS	FL	FAPRO	PPCG	APROBC	TPW	PCPTW	STPEP	PHRFF	PRW	ASST	AFL
Observes	P	P	P	1	P	P	P			1		P
		EAFRD	CPR	ESCR	SR	ACHR	MAAE	PVIAC	PVNAC	EAFDAW	TCIDTP	RC
Observes		P	1		1			P	P	P		P

P = Party; S = Signatory; see Appendix for meaning of abbreviations.

Labor Force

Total Labor Force [25]

43,800

Labor Force by Occupation [26]

Agriculture	43.4%
Services	38.9
Industry and commerce	17.7

Date of data: 1983 est.

Unemployment Rate [27]

25% (1993 est.)

Production Sectors

Energy Resource Summary [28]

Energy resources: Geothermal potential. **Electricity**: Capacity: 20,000 kW. Production: 112 million kWh. Consumption per capita: 693 kWh (1993).

Telecommunications [30]

- 26,000 (1992 est.) telephones
- Domestic: system is automatically switched
- International: direct microwave radio relay link with Martinique and Saint Vincent and the Grenadines; tropospheric scatter to Barbados; international calls beyond these countries are carried by Intelsat from Martinique
- Radio: Broadcast stations: AM 4, FM 1, shortwave 0 Radios: 104,000 (1992 est.)
- Television: Broadcast stations: 1 cable Televisions: 26,000 (1992 est.)

Top Agricultural Products [32]

Agriculture accounts for 13.8% of the GDP; produces bananas, coconuts, vegetables, citrus, root crops, cocoa.

Top Mining Products [33]

Detailed information is not available. A summary of natural resources follows. **Mineral Resources**: Pumice.

Transportation [31]

Railways: 0 km

Highways: total: 760 km; paved: 500 km; unpaved: 260 km

Merchant marine: none

Airports

Total: 3
With paved runways 2,438 to 3,047 m: 1
With paved runways 1,524 to 2,437 m: 1
With paved runways under 914 m: 1 (1995 est.)

Tourism [34]

	1990	1991	1992	1993	1994
Visitors	251	312	347	355	396
Tourists[37]	141	152	176	194	219
Excursionists	8	7	6	7	5
Cruise passengers	102	153	165	154	172
Tourism receipts	154	173	208	221	224
Tourism expenditures	17	18	21	20	23
Fare expenditures	12	11	13	12	11

Travelers are in thousands, money in million U.S. dollars.

Finance, Economics, and Trade

Industrial Summary [35]

Industrial Production: Growth rate 3.5% (1990 est.); accounts for 17.4% of the GDP. **Industries**: Clothing, assembly of electronic components, beverages, corrugated cardboard boxes, tourism, lime processing, coconut processing.

Economic Indicators [36]

- **National product**: GDP—purchasing power parity—$640 million (1995 est.)
- **National product real growth rate**: 2% (1995 est.)
- **National product per capita**: $4,080 (1995 est.)
- **Inflation rate (consumer prices)**: 0.8% (1993)
- **External debt**: $222.7 million (1995 est.)

Balance of Payments Summary [37]

Values in millions of dollars.

	1988	1989	1990	1991	1992
Exports of goods (f.o.b.)	119.1	112.0	127.3	110.3	122.8
Imports of goods (f.o.b.)	-194.5	-240.9	-238.7	-261.4	-275.4
Trade balance	-75.4	-128.9	-111.4	-151.1	-152.6
Services - debits	-83.4	-92.8	-121.0	-124.5	-129.9
Services - credits	126.7	146.4	160.2	186.0	204.8
Private transfers (net)	14.2	13.7	15.2	17.3	16.1
Government transfers (net)	5.4	5.3	0.3	4.5	6.4
Long-term capital (net)	25.0	35.3	49.8	62.0	67.3
Short-term capital (net)	-15.0	18.6	5.1	-0.2	2.3
Errors and omissions	5.0	8.2	8.3	13.7	-7.8
Overall balance	2.6	5.7	6.5	7.7	6.5

Exchange Rates [38]

Currency: **EC dollar.**
Symbol: **EC$.**

Data are currency units per $1.

Fixed rate since 1976   2.7000

Imports and Exports

Top Import Origins [39]

$276 million (f.o.b., 1992) Data are for 1991.

Origins	%
US	34
Caricom	17
UK	14
Japan	7
Canada	4

Top Export Destinations [40]

$122.8 million (f.o.b., 1992) Data are for 1991.

Destinations	%
UK	56
US	22
Caricom	19

Foreign Aid [41]

Recipient: ODA, $NA.

Import and Export Commodities [42]

Import Commodities

Manufactured goods 21%
Machinery and transportation
  equipment 21%
Food and live animals
Chemicals
Fuels

Export Commodities

Bananas 60%
Clothing
Cocoa
Vegetables
Fruits
Coconut oil

For sources, notes, and explanations, see Annotated Source Appendix, page 1061.

Saint Vincent and the Grenadines

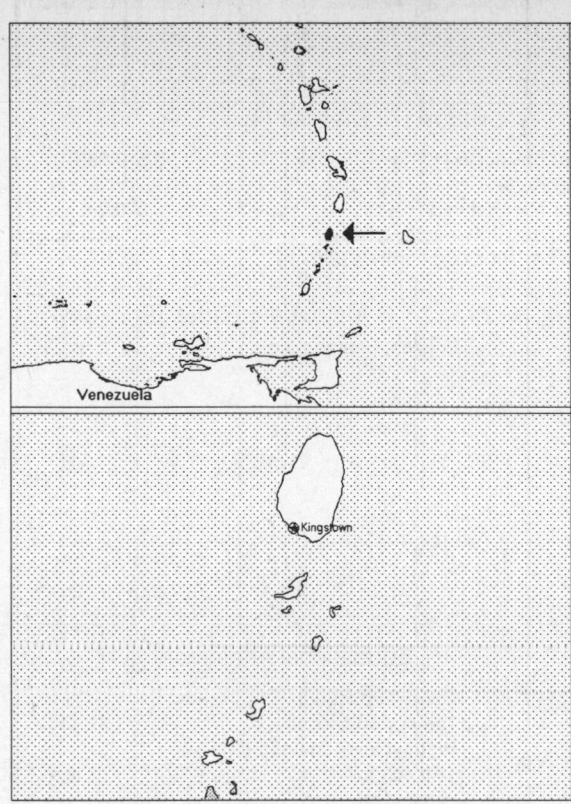

Geography [1]

Total area:
  340 sq km   131 sq mi
Land area:
  340 sq km   131 sq mi
Comparative area:
  Twice the size of Washington, DC
Land boundaries:
  0 km
Coastline:
  84 km
Climate:
  Tropical; little seasonal temperature variation; rainy season (May to November)
Terrain:
  Volcanic, mountainous
Natural resources:
  Negligible
Land use:
  Arable land: 38%
  Permanent crops: 12%
  Meadows and pastures: 6%
  Forest and woodland: 41%
  Other: 3%

Demographics [2]

	1970	1980	1990	1995[1]	1996	2000	2010	2020	2030
Population	88	98	113	118	118	121	132	146	156
Population density (persons per sq. mi.)	669	750	862	898	904	926	1,008	1,112	1,191
(persons per sq. km.)	258	289	333	347	349	357	389	429	460
Net migration rate (per 1,000 population)	-43.6	-9.8	-6.6	-7.6	-7.6	-7.4	0.0	0.0	0.0
Births	3	NA	NA	NA	2	NA	NA	NA	NA
Deaths	0	NA	NA	NA	1	NA	NA	NA	NA
Life expectancy - males	NA	66.4	69.1	71.2	71.4	72.6	74.9	76.7	78.0
Life expectancy - females	NA	69.8	72.3	74.2	74.5	75.7	78.2	80.2	81.8
Birth rate (per 1,000)	38.0	33.5	24.0	19.6	19.4	17.9	16.2	13.4	11.8
Death rate (per 1,000)	8 4	7.5	6.1	5.5	5.4	5.2	4.8	5.1	6.3
Women of reproductive age (15-49 yrs.)	NA	22	28	31	32	35	38	38	38
of which are currently married[45]	4	5	14	NA	16	18	21	NA	NA
Fertility rate	NA	4.2	2.7	2.1	2.0	1.9	1.8	1.8	1.8

Except as noted, values for vital statistics are in thousands; life expectancy is in years.

Health

Health Indicators [3]

Health Expenditures [4]

For sources, notes, and explanations, see Annotated Source Appendix, page 1061.

803

Human Factors

Women and Children [5]	Burden of Disease [6]
	Population per physician (1993) — 2,619.05
	Population per nurse — NA
	Population per hospital bed (1993) — 200.00
	AIDS cases per 100,000 people (1994) — 7.1
	Malaria cases per 100,000 people (1992) — NA

Ethnic Division [7]	Religion [8]	Major Languages [9]
African descent, white, East Indian, Carib Indian.	Anglican, Methodist, Roman Catholic, Seventh-Day Adventist.	English, French patois.

Education

Public Education Expenditures [10]

	1980	1986	1989	1990	1993	1994
Million (E. Carib. Dollar)						
Total education expenditure	NA	19	26	34	NA	NA
as percent of GNP	NA	5.8	5.9	6.7	NA	NA
as percent of total govt. expend.	NA	11.6	10.5	13.8	NA	NA
Current education expenditure	NA	18	23	26	NA	NA
as percent of GNP	NA	5.4	5.2	5.0	NA	NA
as percent of current govt. expend.	NA	16.8	17.1	17.2	NA	NA
Capital expenditure	NA	1	3	8	NA	NA

Educational Attainment [11]

Age group (1980)	25+
Total population	32,444
Highest level attained (%)	
No schooling	2.4
First level	
Not completed	88.0
Completed	NA
Entered second level	
S-1	8.2
S-2	NA
Postsecondary	1.4

Illiteracy [12]

	1970	1980	1990
Illiterate population (15+ years)[4]	1,839	NA	NA
Illiteracy rate - total pop. (%)	4.4	NA	NA
Illiteracy rate - males (%)	4.2	NA	NA
Illiteracy rate - females (%)	4.5	NA	NA

Libraries [13]

	Admin. Units	Svc. Pts.	Vols. (000)	Shelving (meters)	Vols. Added	Reg. Users
National	NA	NA	NA	NA	NA	NA
Nonspecialized	NA	NA	NA	NA	NA	NA
Public (1989)	1	16	100	NA	5,000	NA
Higher ed. (1990)	2	2	12	NA	550	1,100
School (1990)	2	3	NA	NA	NA	1,400

Daily Newspapers [14]

Culture [15]

Science and Technology

Scientific/Technical Forces [16]

R&D Expenditures [17]

U.S. Patents Issued [18]

Values show patents issued to citizens of the country by the U.S. Patents Office.

	1993	1994	1995
Number of patents	0	1	0

Government and Law

Organization of Government [19]

Long-form name:
None
Type:
Constitutional monarchy
Independence:
27 October 1979 (from UK)
National holiday:
Independence Day, 27 October (1979)
Constitution:
27 October 1979
Legal system:
Based on English common law
Executive branch:
British Monarch (represented by Governor General); Prime Minister; Deputy Prime Minister; Cabinet
Legislative branch:
Unicameral: House of Assembly
Judicial branch:
Eastern Caribbean Supreme Court (based on Saint Lucia)

Elections [20]

House of Assembly	% of seats
New Democratic Party (NDP)	80.0
Unity Labor Party (ULP)	20.0

Government Expenditures [21]

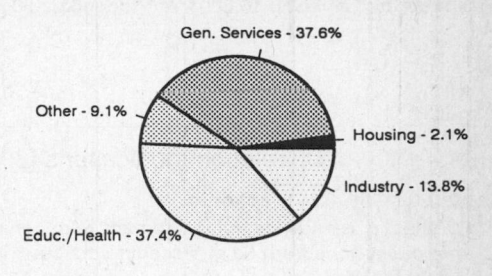

Gen. Services - 37.6%
Other - 9.1%
Housing - 2.1%
Industry - 13.8%
Educ./Health - 37.4%

(% distribution). Expend. for FY95: 222.5 (Dollar mil.)

Defense Summary [22]

Branches: Royal Saint Vincent and the Grenadines Police Force, Coast Guard

Manpower Availability: Males age 15-49 NA; males fit for military service NA

Defense Expenditures: NA

Crime [23]

	1990
Crime volume	
Cases known to police	4,255
Attempts (percent)	NA
Percent cases solved	436.56
Crimes per 100,000 persons	3,976.63
Persons responsible for offenses	
Total number offenders	3,306
Percent female	71.75
Percent juvenile (8-15 yrs.)	29.36
Percent foreigners	17

Human Rights [24]

	SSTS	FL	FAPRO	PPCG	APROBC	TPW	PCPTW	STPEP	PHRFF	PRW	ASST	AFL
Observes	P			P		P	P			1	P	
	EAFRD	CPR	ESCR	SR	ACHR	MAAE	PVIAC	PVNAC	EAFDAW	TCIDTP	RC	
Observes		P	P	P	P		P	P	P		P	

P = Party; S = Signatory; see Appendix for meaning of abbreviations.

Labor Force

Total Labor Force [25]

67,000 (1984 est.)

Labor Force by Occupation [26]

Agriculture
Industry
Services
Date of data: 1980 est.

Unemployment Rate [27]

35%-40% (1994 est.)

For sources, notes, and explanations, see Annotated Source Appendix, page 1061.

Production Sectors

Energy Resource Summary [28]

Electricity: Capacity: 16,600 kW. Production: 50 million kWh. Consumption per capita: 436 kWh (1993).

Telecommunications [30]

- 6,189 (1983 est.) telephones
- Domestic: islandwide, fully automatic telephone system; VHF/UHF radiotelephone from Saint Vincent to the other islands of the Grenadines
- International: VHF/UHF radiotelephone from Saint Vincent to Barbados; new SHF radiotelephone to Grenada and to Saint Lucia; access to Intelsat earth station in Martinique through Saint Lucia
- Radio: Broadcast stations: AM 2, FM 0, shortwave 0 Radios: 76,000 (1992 est.)
- Television: Broadcast stations: 1 cable Televisions: 20,600 (1992 est.)

Top Agricultural Products [32]

Agriculture accounts for 24% of the GDP; produces bananas, coconuts, sweet potatoes, spices; small numbers of cattle, sheep, pigs, goats; small fish catch used locally.

Top Mining Products [33]

Mineral Resources: Negligible.

Transportation [31]

Railways: 0 km

Highways: total: 1,100 km; paved: 330 km; unpaved: 770 km

Merchant marine: total: 611 ships (1,000 GRT or over) totaling 5,690,104 GRT/9,367,014 DWT; ships by type: bulk 106, cargo 305, chemical tanker 20, combination bulk 9, combination ore/oil 4, container 33, liquefied gas tanker 4, livestock carrier 5, oil tanker 58, passenger 1, passenger-cargo 1, refrigerated cargo 35, roll-on/roll-off cargo 25, short-sea passenger 2, specialized tanker 1, vehicle carrier 2

Airports

Total: 6

With paved runways 914 to 1,523 m: 2

With paved runways under 914 m: 4 (1995 est.)

Tourism [34]

	1990	1991	1992	1993	1994
Visitors	158	173	155	163	165
Tourists[12]	54	52	53	57	55
Excursionists	21	29	33	30	31
Cruise passengers	83	92	69	76	79
Tourism receipts	56	53	53	52	51
Tourism expenditures	4	4	4	3	3
Fare receipts	1	1	2	2	2
Fare expenditures	4	4	4	5	6

Travelers are in thousands, money in million U.S. dollars.

For sources, notes, and explanations, see Annotated Source Appendix, page 1061.

Finance, Economics, and Trade

Industrial Summary [35]

Industrial Production: Growth rate not available; accounts for 33.1% of the GDP. **Industries**: Food processing, cement, furniture, clothing, starch.

Economic Indicators [36]

- **National product**: GDP—purchasing power parity— $240 million (1995 est.)

- **National product real growth rate**: 0.4% (1995 est.)

- **National product per capita**: $2,060 (1995 est.)

- **Inflation rate (consumer prices)**: - 0.2% (1995)

- **External debt**: $74.9 million (1993)

Balance of Payments Summary [37]

Values in millions of dollars.

	1988	1989	1990	1991	1992
Exports of goods (f.o.b.)	85.3	74.7	82.7	67.3	77.5
Imports of goods (f.o.b.)	-110.0	-112.2	-119.6	-119.7	-118.6
Trade balance	-24.7	-37.5	-36.9	-52.5	-41.1
Services - debits	-35.2	-47.8	-49.6	-52.9	-51.5
Services - credits	34.4	45.4	52.0	50.5	52.0
Private transfers (net)	20.0	12.6	13.1	12.8	12.9
Government transfers (net)	6.3	8.9	16.1	18.3	11.1
Long-term capital (net)	15.7	15.5	13.7	15.4	22.2
Short-term capital (net)	-12.0	2.7	-11.1	6.4	4.6
Errors and omissions	-2.2	1.6	6.8	-1.8	1.2
Overall balance	2.3	1.4	4.0	-3.8	11.4

Exchange Rates [38]

Currency: **EC dollar.**
Symbol: **EC$.**

Data are currency units per $1.

Fixed rate since 1976	2.7000

Imports and Exports

Top Import Origins [39]

$134.6 million (f.o.b., 1993).

Origins	%
US	36
Caricom	21
UK	18
Trinidad and Tobago	13

Top Export Destinations [40]

$57.1 million (f.o.b., 1993).

Destinations	%
UK	54
Caricom	34
US	10

Foreign Aid [41]

USAID assistance diminished in 1996 when the regional office in Barbados closed.

	U.S. $
ODA	$NA
USAID (1994)	$NA

Import and Export Commodities [42]

Import Commodities	Export Commodities
Foodstuffs	Bananas
Machinery and equipment	Eddoes and dasheen (taro)
Chemicals and fertilizers	Arrowroot starch
Minerals and fuels	Tennis racquets

For sources, notes, and explanations, see Annotated Source Appendix, page 1061.

San Marino

Geography [1]

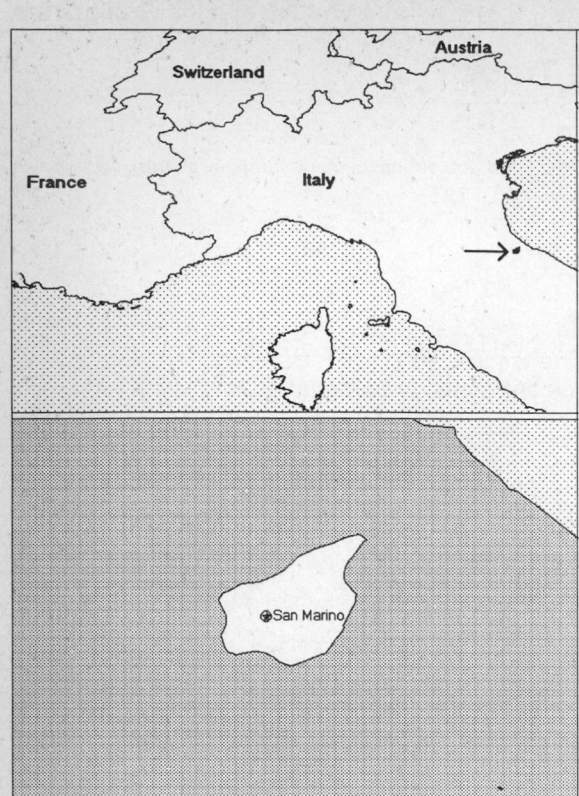

Total area:
  60 sq km   23 sq mi
Land area:
  60 sq km   23 sq mi
Comparative area:
  About 0.3 times the size of Washington, DC
Land boundaries:
  Total 39 km, Italy 39 km
Coastline:
  0 km (landlocked)
Climate:
  Mediterranean; mild to cool winters; warm, sunny summers
Terrain:
  Rugged mountains
Natural resources:
  Building stone
Land use:
  Arable land: 17%
  Permanent crops: 0%
  Meadows and pastures: 0%
  Forest and woodland: 0%
  Other: 83%

Demographics [2]

	1970	1980	1990	1995[1]	1996	2000	2010	2020	2030
Population	19	21	23	24	25	25	26	27	27
Population density (persons per sq. mi.)	828	922	996	1,050	1,058	1,088	1,136	1,159	1,171
(persons per sq. km.)	320	356	385	405	409	420	439	448	452
Net migration rate (per 1,000 population)	NA	6.6	7.2	5.5	5.1	4.0	3.8	3.7	3.7
Births	0	0	NA	NA	Z	NA	NA	NA	NA
Deaths	0	0	NA	NA	Z	NA	NA	NA	NA
Life expectancy - males	NA	73.8	76.8	77.3	77.3	77.7	78.4	78.9	79.4
Life expectancy - females	NA	78.7	85.2	85.3	85.3	85.4	85.5	85.7	85.8
Birth rate (per 1,000)	15.0	11.2	11.5	11.0	10.8	10.3	8.6	9.0	8.9
Death rate (per 1,000)	6.9	7.8	6.7	7.6	7.8	8.4	9.8	11.0	12.4
Women of reproductive age (15-49 yrs.)	NA	5	6	6	6	6	6	6	5
of which are currently married	NA	NA	4	NA	4	4	4	NA	NA
Fertility rate	NA	1.5	1.5	1.5	1.5	1.5	1.5	1.6	1.6

Except as noted, values for vital statistics are in thousands; life expectancy is in years.

Health

Health Indicators [3]

Health Expenditures [4]

Human Factors

Women and Children [5]	Burden of Disease [6]

Ethnic Division [7]	Religion [8]	Major Languages [9]
Sammarinese, Italian.	Roman Catholic.	Italian.

Education

Public Education Expenditures [10]

Million (Lira)	1980	1984	1990	1992	1993	1994
Total education expenditure	7,252	11,470	22,048	30,915	34,112	35,990
as percent of GNP	NA	NA	NA	NA	NA	NA
as percent of total govt. expend.	7.5	10.7	NA	NA	NA	NA
Current education expenditure	6,263	10,474	21,723	29,221	32,981	35,475
as percent of GNP	NA	NA	NA	NA	NA	NA
as percent of current govt. expend.	9.5	10.3	NA	NA	NA	NA
Capital expenditure	989	996	325	1,694	1,131	515

Educational Attainment [11]

Illiteracy [12]

	1976	1980	1990
Illiterate population (15+ years)[4]	640	NA	NA
Illiteracy rate - total pop. (%)	3.9	NA	NA
Illiteracy rate - males (%)	3.2	NA	NA
Illiteracy rate - females (%)	4.7	NA	NA

Libraries [13]

	Admin. Units	Svc. Pts.	Vols. (000)	Shelving (meters)	Vols. Added	Reg. Users
National	NA	NA	NA	NA	NA	NA
Nonspecialized	NA	NA	NA	NA	NA	NA
Public	NA	NA	NA	NA	NA	NA
Higher ed.	NA	NA	NA	NA	NA	NA
School (1987)	5	17	NA	NA	NA	NA

Daily Newspapers [14]

	1980	1985	1990	1994
Number of papers	3	-	-	-
Circ. (000)	1	-	-	-

Culture [15]

Cinema (seats per 1,000)	83.3
Annual attendance per person	1.7
Gross box office receipts (mil. Lira)	191
Museums (reporting)	-
Visitors (000)	NA
Annual receipts (000 Lira)	NA

Science and Technology

Scientific/Technical Forces [16]

Scientists/engineers	566
Number female	213
Technicians	2,004
Number female	883
Total[36]	2,570

R&D Expenditures [17]

U.S. Patents Issued [18]

Values show patents issued to citizens of the country by the U.S. Patents Office.

	1993	1994	1995
Number of patents	0	1	0

For sources, notes, and explanations, see Annotated Source Appendix, page 1061.

Government and Law

Organization of Government [19]

Long-form name:
Republic of San Marino
Type:
Republic
Independence:
301 AD (by tradition)
National holiday:
Anniversary of the Foundation of the
Republic, 3 September
Constitution:
8 October 1600; electoral law of 1926
serves some of the functions of a
constitution
Legal system:
Based on civil law system with Italian law
influences; has not accepted compulsory
ICJ jurisdiction
Executive branch:
2 Captains Regent (Co-Chiefs of S)tate;
Secretary of State for Foreign and Political
Affairs; Congress of State
Legislative branch:
Unicameral: Great and General Council
Judicial branch:
Council of Twelve (Consiglio dei XII)

Elections [20]

Great and General Council	% of votes
Christian Democratic Party (PDCS)	41.4
San Marino Socialist Party (PSS)	23.7
Democratic Progressive Party (PDP)	18.6
Popular Democratic Alliance (ADP)	7.7
Democratic Movement (MD)	5.3
Communist Refoundation (RC)	3.3

Government Budget [21]

For 1995 est.

	$ mil.
Revenues	320
Expenditures	320
Capital expenditures	NA

Defense Summary [22]

Branches: Voluntary Military Force, Police Force

Manpower Availability: Males age 15-49 NA; males fit for military service NA

Defense Expenditures: $3.7 million (1% of GDP) (1992 est.)

Crime [23]

Human Rights [24]

	SSTS	FL	FAPRO	PPCG	APROBC	TPW	PCPTW	STPEP	PHRFF	PRW	ASST	AFL
Observes			P		P	P	P		P		P	
	EAFRD	CPR	ESCR	SR	ACHR	MAAE	PVIAC	PVNAC	EAFDAW	TCIDTP	RC	
Observes		P	P				P	P			P	

P = Party; S = Signatory; see Appendix for meaning of abbreviations.

Labor Force

Total Labor Force [25]

14,874 (1993 est.)

Labor Force by Occupation [26]

Industry	40%
Agriculture	2

Unemployment Rate [27]

4.9% (December 1993)

For sources, notes, and explanations, see Annotated Source Appendix, page 1061.

Production Sectors

Energy Resource Summary [28]

Electricity: supplied by Italy.

Telecommunications [30]

- 22,300 (1992 est.) telephones
- Domestic: automatic telephone system completely integrated into Italian system
- International: microwave radio relay and cable connections to Italian network; no satellite earth stations
- Radio: Broadcast stations: AM NA, FM NA, shortwave NA (1 private radio broadcast station) Radios: 12,535 (1991 est.)
- Television: Broadcast stations: 1 (1991 est.) Televisions: 7,500 (1992 est.)

Transportation [31]

Railways: 0 km; note - there is a 1.5 km cable railway connecting the city of San Marino to Borgo Maggiore

Highways: total: 220 km; paved: NA km; unpaved: NA km

Airports

Total: none

Top Agricultural Products [32]

Produces wheat, grapes, maize, olives; cattle, pigs, horses, meat, cheese, hides.

Top Mining Products [33]

Detailed information is not available. A summary of natural resources follows. **Mineral Resources**: Building stone.

Tourism [34]

	1990	1991	1992	1993	1994
Visitors[104]	2,912	3,113	3,208	3,072	3,104
Tourists[105]	582	NA	NA	NA	533

Travelers are in thousands, money in million U.S. dollars.

Finance, Economics, and Trade

Industrial Summary [35]

Industrial Production: Growth rate not available. **Industries:** Tourism, textiles, electronics, ceramics, cement, wine.

Economic Indicators [36]

- **National product:** GDP—purchasing power parity—$380 million (1993 est.)
- **National product real growth rate:** 2.4% (1993 est.)
- **National product per capita:** $15,800 (1993 est.)
- **Inflation rate (consumer prices):** 5.5% (1993)
- **External debt:** $NA

Balance of Payments Summary [37]

Exchange Rates [38]

Currency: **Italian lire.**
Symbol: **Lit.**

Data are currency units per $1.

January 1996	1,583.8
1995	1,629.2
1994	1,612.4
1993	1,573.7
1992	1,232.4
1991	1,240.6

Imports and Exports

Top Import Origins [39]

Trade data are included with the statistics for Italy.

Origins	%
No details available.	NA

Top Export Destinations [40]

Trade data are included with the statistics for Italy.

Destinations	%
No details available.	NA

Foreign Aid [41]

Recipient: ODA, $NA.

Import and Export Commodities [42]

Import Commodities	Export Commodities
Consumer manufactures	Lime
Food	Wood
	Chestnuts
	Wheat
	Wine
	Baked goods
	Hides
	Ceramics
	Building stone

Sao Tome and Principe

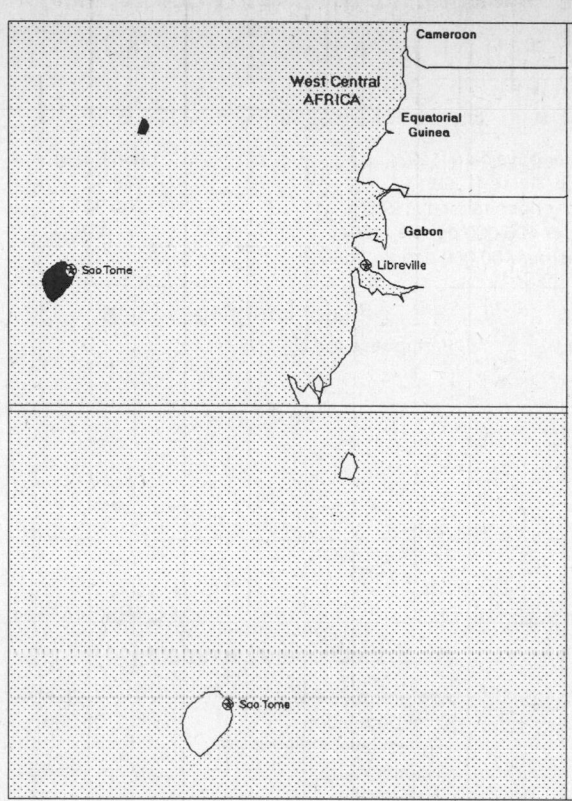

Geography [1]

Total area:
   960 sq km   371 sq mi
Land area:
   960 sq km   371 sq mi
Comparative area:
   More than five times the size of Washington, DC
Land boundaries:
   0 km
Coastline:
   209 km
Climate:
   Tropical; hot, humid; one rainy season (October to May)
Terrain:
   Volcanic, mountainous
Natural resources:
   Fish
Land use:
   Arable land: 1%
   Permanent crops: 20%
   Meadows and pastures: 1%
   Forest and woodland: 75%
   Other: 3%

Demographics [2]

	1970	1980	1990	1995[1]	1996	2000	2010	2020	2030
Population	74	94	123	140	144	159	196	232	266
Population density (persons per sq. mi.)	199	254	332	379	389	429	529	626	718
(persons per sq. km.)	77	98	128	146	150	166	204	242	277
Net migration rate (per 1,000 population)	-17.5	NA	0.0	0.0	0.0	0.0	0.0	0.0	0.0
Births	NA	NA	NA	NA	5	NA	NA	NA	NA
Deaths	NA	NA	NA	NA	1	NA	NA	NA	NA
Life expectancy - males	NA	NA	60.4	61.8	62.0	62.7	64.6	66.4	68.0
Life expectancy - females	NA	NA	63.9	65.6	65.8	66.8	69.2	71.3	73.3
Birth rate (per 1,000)	44.5	NA	35.6	34.9	34.4	31.7	25.2	21.6	18.6
Death rate (per 1,000)	12.5	NA	9.7	8.7	8.6	8.0	6.9	6.3	6.4
Women of reproductive age (15-49 yrs.)	NA	NA	28	34	35	39	52	63	71
of which are currently married	NA	NA	15	NA	19	22	30	NA	NA
Fertility rate	NA	NA	4.9	4.4	4.3	3.9	3.0	2.5	2.2

Except as noted, values for vital statistics are in thousands; life expectancy is in years.

Health

Health Indicators [3]

Health Expenditures [4]

For sources, notes, and explanations, see Annotated Source Appendix, page 1061.

813

Human Factors

Women and Children [5]

Burden of Disease [6]

Population per physician (1990)	1,868.85
Population per nurse (1990)	188.12
Population per hospital bed (1991)	209.30
AIDS cases per 100,000 people (1994)	0.9
Malaria cases per 100,000 people (1992)	NA

Ethnic Division [7]

Mestico, Angolares (descendants of Angolan slaves), Forros (descendants of freed slaves), Servicais (contract laborers from Angola, Mozambique, and Cape Verde), Tongas (children of servicais born on the islands), Europeans (primarily Portuguese).

Religion [8]

Roman Catholic, Evangelical Protestant, Seventh-Day Adventist.

Major Languages [9]

Portuguese (official).

Education

Public Education Expenditures [10]

Million (Dobra)	1980	1981	1986	1990	1993	1994
Total education expenditure	NA	91	100	NA	NA	NA
as percent of GNP	NA	8.0	3.8	NA	NA	NA
as percent of total govt. expend.	NA	NA	18.8	NA	NA	NA
Current education expenditure	NA	NA	NA	NA	NA	NA
as percent of GNP	NA	NA	NA	NA	NA	NA
as percent of current govt. expend.	NA	NA	NA	NA	NA	NA
Capital expenditure	NA	NA	NA	NA	NA	NA

Educational Attainment [11]

Age group (1981)[8]	25+
Total population	33,308
Highest level attained (%)	
No schooling	56.6
First level	
Not completed	18.0
Completed	19.3
Entered second level	
S-1	4.6
S-2	1.3
Postsecondary	0.3

Illiteracy [12]

	1981	1989	1995
Illiterate population (15+ yrs.)[2]	22,080	NA	NA
Illiteracy rate - total pop. (%)	42.6	NA	NA
Illiteracy rate - males (%)	26.8	NA	NA
Illiteracy rate - females (%)	57.6	NA	NA

Libraries [13]

Daily Newspapers [14]

Culture [15]

Science and Technology

Scientific/Technical Forces [16]

R&D Expenditures [17]

U.S. Patents Issued [18]

For sources, notes, and explanations, see Annotated Source Appendix, page 1061.

Government and Law

Organization of Government [19]

Long-form name:
Democratic Republic of Sao Tome and Principe
Type:
Republic
Independence:
12 July 1975 (from Portugal)
National holiday:
Independence Day, 12 July (1975)
Constitution:
Approved March 1990; effective 10 September 1990
Legal system:
Based on Portuguese law system and customary law; has not accepted compulsory ICJ jurisdiction
Executive branch:
President; Prime Minister; Council of Ministers
Legislative branch:
Unicameral: National People's Assembly
Judicial branch:
Supreme Court

Elections [20]

National People's Assembly	% of votes
Movement for the Liberation (MLSTP)	27.0
Dem. Convergence-Reflection Group (PCD-GR)	25.5
Independent Dem. Action (ADI)	25.5

Government Budget [21]

For 1993 est.

	$ mil.
Revenues	58
Expenditures	114
Capital expenditures	54

Crime [23]

Military Expenditures and Arms Transfers [22]

	1990	1991	1992	1993	1994
Military expenditures					
Current dollars (mil.)	NA	NA	NA	NA	NA
1994 constant dollars (mil.)	NA	NA	NA	NA	NA
Armed forces (000)	1	1	1	1	3
Gross national product (GNP)					
Current dollars (mil.)	21	22	23	23	23
1994 constant dollars (mil.)	23	23	24	24	23
Central government expenditures (CGE)					
1994 constant dollars (mil.)	NA	NA	NA	NA	NA
People (mil.)	0.1	0.1	0.1	0.1	0.1
Military expenditure as % of GNP	NA	NA	NA	NA	NA
Military expenditure as % of CGE	NA	NA	NA	NA	NA
Military expenditure per capita (1994 $)	NA	NA	NA	NA	NA
Armed forces per 1,000 people (soldiers)	8.1	7.9	7.7	7.5	21.9
GNP per capita (1994 $)	188	183	183	177	171
Arms imports[6]					
Current dollars (mil.)	5	0	0	0	0
1994 constant dollars (mil.)	6	0	0	0	0
Arms exports[6]					
Current dollars (mil.)	0	0	0	0	0
1994 constant dollars (mil.)	0	0	0	0	0
Total imports[7]					
Current dollars (mil.)	21	25	25	22	NA
1994 constant dollars (mil.)	23	27	26	22	NA
Total exports[7]					
Current dollars (mil.)	4	6	5	5	NA
1994 constant dollars (mil.)	4	6	5	5	NA
Arms as percent of total imports[8]	23.8	0	0	0	0
Arms as percent of total exports[8]	0	0	0	0	0

Human Rights [24]

	SSTS	FL	FAPRO	PPCG	APROBC	TPW	PCPTW	STPEP	PHRFF	PRW	ASST	AFL
Observes			P		P	P	P					

	EAFRD	CPR	ESCR	SR	ACHR	MAAE	PVIAC	PVNAC	EAFDAW	TCIDTP	RC
Observes		S	S	P				P	P	S		P

P = Party; S = Signatory; see Appendix for meaning of abbreviations.

Labor Force

Total Labor Force [25]

Labor Force by Occupation [26]

Subsistence agriculture
Fishing
Plantation labor

Unemployment Rate [27]

For sources, notes, and explanations, see Annotated Source Appendix, page 1061.

815

Production Sectors

Energy Resource Summary [28]

Electricity: Capacity: 5,000 kW. Production: 17 million kWh. Consumption per capita: 105 kWh (1993).

Telecommunications [30]

- 2,200 (1986 est.) telephones
- Domestic: minimal system
- International: satellite earth station - 1 Intelsat (Atlantic Ocean)
- Radio: Broadcast stations: AM 1, FM 2, shortwave 0 Radios: 33,000 (1992 est.)
- Television: Broadcast stations: 1 (1992 est.)

Transportation [31]

Railways: 0 km

Highways: total: 298 km; paved: 198 km; unpaved: 100 km (1987 est.)

Merchant marine: total: 1 cargo ship (1,000 GRT or over) totaling 1,096 GRT/1,105 DWT (1995 est.)

Airports

Total: 2
With paved runways 1,524 to 2,437 m: 1
With paved runways 914 to 1,523 m: 1 (1995 est.)

Top Agricultural Products [32]

Agriculture accounts for 28% of the GDP; produces cocoa, coconuts, palm kernels, copra, cinnamon, pepper, coffee, bananas, papaya, beans; poultry; fish.

Top Mining Products [33]

Detailed information is not available. A summary of mineral resources follows. **Mineral Resources**: Some clay and stone for construction.

Tourism [34]

	1990	1991	1992	1993	1994
Visitors	6.3	6.0	NA	6.0	6.2
Tourists	3.9	3.2	NA	NA	5.2
Cruise passengers	NA	0.1	NA	NA	NA
Tourism receipts	2	NA	NA	NA	NA
Tourism expenditures	2	NA	NA	NA	NA
Fare expenditures	3	NA	NA	NA	NA

Travelers are in thousands, money in million U.S. dollars.

Finance, Economics, and Trade

GDP and Manufacturing Summary [35]

	1980	1985	1990	1991	1992
Gross Domestic Product					
Millions of 1980 dollars	47	33	39	40	40[e]
Growth rate in percent	2.59	-5.01	3.81	1.51	0.80[e]
Manufacturing Value Added					
Millions of 1980 dollars	3	3	3	3	3[e]
Growth rate in percent	0.00	-8.74	3.28	1.91	1.33[e]
Manufacturing share in percent of current prices	7.3	7.2[e]	NA	NA	NA

Economic Indicators [36]

- **National product**: GDP—purchasing power parity—$138 million (1994 est.)

- **National product real growth rate**: 1.5% (1994 est.)

- **National product per capita**: $1,000 (1994 est.)

- **Inflation rate (consumer prices)**: 38% (1994 est.)

- **External debt**: $250 million (1995 est.)

Balance of Payments Summary [37]

Values in millions of dollars.

	1986	1987	1988	1989	1990
Exports of goods (f.o.b.)	9.9	6.5	9.5	4.9	4.2
Imports of goods (f.o.b.)	-23.2	-13.6	-14.1	-13.3	-13.0
Trade balance	-13.3	-7.0	-4.6	-8.4	-8.8
Services - debits	-14.2	-9.3	-9.0	-8.8	-9.3
Services - credits	3.7	1.7	2.0	4.6	3.9
Private transfers (net)	-0.8	-0.3	NA	-0.2	0.1
Government transfers (net)	5.8	1.9	0.9	1.5	2.1
Long-term capital (net)	5.4	3.2	2.0	3.9	11.2
Short term capial (net)	9.8	5.8	4.4	2.8	3.0
Errors and omissions	2.2	NA	NA	-1.0	-2.7
Overall balance	-1.4	-4.0	-4.3	-5.7	-0.5

Exchange Rates [38]

Currency: **dobra.**
Symbol: **Db.**

Data are currency units per $1.

May 1995	1,610
1 July 1993	129.59
1992	230
November 1991	260.0
December 1988	122.48
1987	72.827

Imports and Exports

Top Import Origins [39]

$23.8 million (c.i.f., 1994 est.).

Origins	%
France	NA
Belgium	NA
Japan	NA
Angola	NA
Italy	NA
US	NA

Top Export Destinations [40]

$7.1 million (f.o.b., 1994 est.).

Destinations	%
Netherlands	NA
Germany	NA
China	NA
Portugal	NA

Foreign Aid [41]

Recipient: ODA, $NA.

Import and Export Commodities [42]

Import Commodities	Export Commodities
Machinery and electrical equipment	Cocoa 85%-90%
Food products	Copra
Petroleum	Coffee
	Palm oil

For sources, notes, and explanations, see Annotated Source Appendix, page 1061.

817

Saudi Arabia

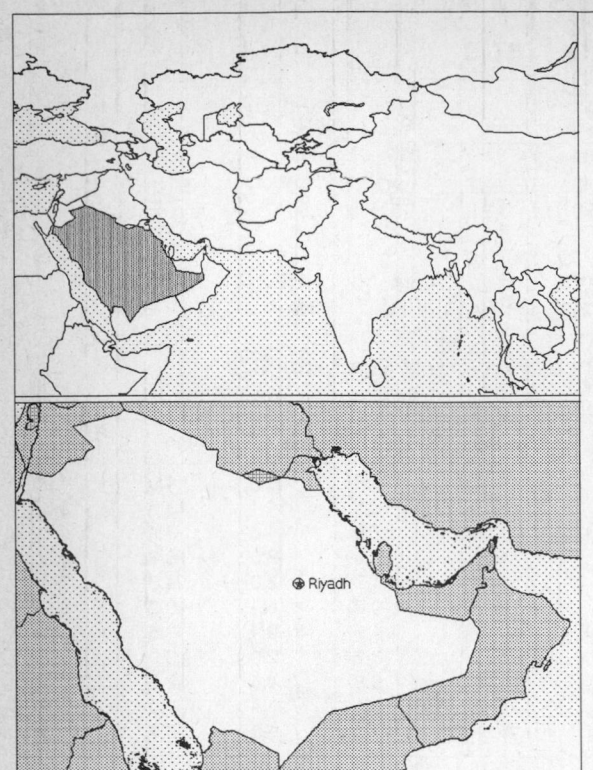

Geography [1]

Total area:
1,960,582 sq km   756,985 sq mi
Land area:
1,960,582 sq km   756,985 sq mi
Comparative area:
Slightly less than one-fourth the size of the US
Land boundaries:
Total 4,415 km, Iraq 814 km, Jordan 728 km, Kuwait 222 km, Oman 676 km, Qatar 60 km, UAE 457 km, Yemen 1,458 km
Coastline:
2,640 km
Climate:
Harsh, dry desert with great extremes of temperature
Terrain:
Mostly uninhabited, sandy desert
Natural resources:
Petroleum, natural gas, iron ore, gold, copper
Land use:
Arable land: 1%
Permanent crops: 0%
Meadows and pastures: 39%
Forest and woodland: 1%
Other: 59%

Demographics [2]

	1970	1980	1990	1995[1]	1996	2000	2010	2020	2030
Population	6,109	9,949	15,871	18,730	19,409	22,246	31,198	43,255	58,250
Population density (persons per sq. mi.)	7	12	19	23	23	27	38	52	70
(persons per sq. km.)	3	5	7	9	9	10	15	20	27
Net migration rate (per 1,000 population)	NA	30.8	-43.1	3.6	1.6	1.4	1.0	0.0	0.0
Births	NA	NA	NA	NA	744	NA	NA	NA	NA
Deaths	NA	NA	NA	NA	104	NA	NA	NA	NA
Life expectancy - males	NA	55.5	64.1	66.8	67.3	69.2	73.0	75.7	77.6
Life expectancy - females	NA	56.9	67.2	70.3	70.8	73.1	77.7	80.9	83.1
Birth rate (per 1,000)	NA	44.0	40.8	38.8	38.3	37.2	36.8	34.8	31.9
Death rate (per 1,000)	NA	11.5	6.7	5.5	5.4	4.7	3.9	3.6	3.5
Women of reproductive age (15-49 yrs.)	NA	1,951	3,027	3,591	3,732	4,372	6,433	9,235	13,275
of which are currently married	NA	NA	1,991	NA	2,433	2,781	4,027	NA	NA
Fertility rate	NA	7.1	6.6	6.5	6.4	6.3	5.7	5.1	4.5

Except as noted, values for vital statistics are in thousands; life expectancy is in years.

Health

Health Indicators [3]

% of population with access to	
safe water (1990-95)[1]	95
adequate sanitation (1990-95)[1]	86
health services (1985-95)	97
% of 1-year-olds immunized (1990-94) against	
TB (tuberculosis)	94
DPT (diphtheria, pertussis, tetanus)	93
polio	94
measles	92
% of contraceptive prevalence (1980-94)	NA
ORT use rate (1990-94)	90

Health Expenditures [4]

Total health expenditure, 1990 (official exchange rate)	
Millions of dollars	4,784
Dollars per capita	322
Health expenditures as a percentage of GDP	
Total	4.8
Public sector	3.1
Private sector	1.7
Development assistance for health	
Total aid flows (millions of dollars)[1]	1
Aid flows per capita (dollars)	0.1
Aid flows as a percentage of total health expenditure	NA

For sources, notes, and explanations, see Annotated Source Appendix, page 1061.

Human Factors

Women and Children [5]

% of pregnant women immunized (tetanus 1990-94)	63
% of births attended by trained health personnel (1983-94)	90
Maternal mortality rate (1980-92)	41
Under-5 mortality rate (1994)	36
% under-5 moderately/severely underweight (1980-1994)	NA

Burden of Disease [6]

Population per physician (1990)	697.64
Population per nurse (1990)	451.15
Population per hospital bed (1990)	400.57
AIDS cases per 100,000 people (1994)	0.2
Malaria cases per 100,000 people (1992)	117

Ethnic Division [7]

Arab	90%
Afro-Asian	10%

Religion [8]

Muslim	100%

Major Languages [9]

Arabic.

Education

Public Education Expenditures [10]

	1980	1985	1990	1992	1993	1994
Million (Riyal)						
Total education expenditure	21,294	23,540	25,460	30,800	31,590	28,817
as percent of GNP	4.1	6.7	6.0	6.3	NA	NA
as percent of total govt. expend.	8.7	12.0	17.8	17.0	NA	NA
Current education expenditure	13,526	19,283	24,033	29,428	29,671	27,386
as percent of GNP	2.6	5.5	5.7	6.0	NA	NA
as percent of current govt. expend.	NA	NA	NA	NA	NA	NA
Capital expenditure	7,768	4,257	1,427	1,372	1,919	1,431

Educational Attainment [11]

Illiteracy [12]

In thousands and percent[1]	1990	1995	2000
Illiterate population (15+ yrs.)	3,821	3,871	4,146
Illiteracy rate - total pop. (%)	40.7	36.1	32.7
Illiteracy rate - males (%)	31.0	27.1	24.5
Illiteracy rate - females (%)	56.5	49.8	44.2

Libraries [13]

	Admin. Units	Svc. Pts.	Vols. (000)	Shelving (meters)	Vols. Added	Reg. Users
National	NA	NA	NA	NA	NA	NA
Nonspecialized	NA	NA	NA	NA	NA	NA
Public (1986)	1	50	630	24,000	NA	NA
Higher ed. (1993)	21	65	4,844	113,849	98,664	483,611
School	NA	NA	NA	NA	NA	NA

Daily Newspapers [14]

	1980	1985	1990	1994
Number of papers	11	13	12	19
Circ. (000)	350[e]	450[e]	600[e]	950[e]

Culture [15]

Cinema (seats per 1,000)	NA
Annual attendance per person	NA
Gross box office receipts (mil. Riyal)	NA
Museums (reporting)	12
Visitors (000)	44,147
Annual receipts (000 Riyal)	NA

Science and Technology

Scientific/Technical Forces [16]

R&D Expenditures [17]

U.S. Patents Issued [18]

Values show patents issued to citizens of the country by the U.S. Patents Office.

	1993	1994	1995
Number of patents	4	11	10

For sources, notes, and explanations, see Annotated Source Appendix, page 1061.

819

Government and Law

Organization of Government [19]

Long-form name:
Kingdom of Saudi Arabia
Type:
Monarchy
Independence:
23 September 1932 (unification)
National holiday:
Unification of the Kingdom, 23 September (1932)
Constitution:
None; governed according to Shari'a (Islamic law)
Legal system:
Based on Islamic law, several secular codes have been introduced; commercial disputes handled by special committees; has not accepted compulsory ICJ jurisdiction
Executive branch:
King and Prime Minister; Crown Prince and First Deputy Prime Minister; Council of Ministers
Legislative branch:
A consultative council composed of 60 members and a chairman who are appointed by the king for a term of four years
Judicial branch:
Supreme Council of Justice

Elections [20]

No political parties allowed; no elections.

Government Budget [21]

For 1996 est.

	$ bil.
Revenues	35.1
Expenditures	40
Capital expenditures	NA

Crime [23]

	1994
Crime volume	
Cases known to police	22,172
Attempts (percent)	NA
Percent cases solved	NA
Crimes per 100,000 persons	130.97
Persons responsible for offenses	
Total number offenders	22,026
Percent female	NA
Percent juvenile (0-15 yrs.)	NA
Percent foreigners	NA

Military Expenditures and Arms Transfers [22]

	1990	1991	1992	1993	1994
Military expenditures					
Current dollars (mil.)	23,160[e]	35,510[e]	35,010[e]	20,480[e]	17,200
1994 constant dollars (mil.)	25,780[e]	38,060[e]	36,510[e]	20,900[e]	17,200
Armed forces (000)	146	191	172	172	164
Gross national product (GNP)					
Current dollars (mil.)[e]	112,600	124,800	128,600	122,700	121,400
1994 constant dollars (mil.)[e]	125,400	133,800	134,100	125,200	121,400
Central government expenditures (CGE)					
1994 constant dollars (mil.)	42,510[e]	NA	50,370[e]	50,420[e]	NA
People (mil.)	15.9	16.1	16.7	17.4	18.0
Military expenditure as % of GNP	20.6	28.5	27.2	16.7	14.2
Military expenditure as % of CGE	60.6	NA	72.5	41.5	NA
Military expenditure per capita (1994 $)	1,624	2,363	2,181	1,202	953
Armed forces per 1,000 people (soldiers)	9.2	11.9	10.3	9.9	9.1
GNP per capita (1994 $)	7,900	8,304	8,014	7,202	6,725
Arms imports[6]					
Current dollars (mil.)[10]	7,800	7,500	8,500	6,800	5,200
1994 constant dollars (mil.)[10]	8,681	8,039	8,864	6,940	5,200
Arms exports[6]					
Current dollars (mil.)	0	0	5	0	10
1994 constant dollars (mil.)	0	0	5	0	10
Total imports[7]					
Current dollars (mil.)	24,070	29,080	33,240	28,220	23,340
1994 constant dollars (mil.)	26,790	31,170	34,670	28,810	23,340
Total exports[7]					
Current dollars (mil.)	44,420	47,800	50,280	42,390	42,000
1994 constant dollars (mil.)	49,430	51,230	52,430	43,270	42,000
Arms as percent of total imports[8]	32.4	25.8	25.6	24.1	22.3
Arms as percent of total exports[8]	0	0	0	0	0

Human Rights [24]

	SSTS	FL	FAPRO	PPCG	APROBC	TPW	PCPTW	STPEP	PHRFF	PRW	ASST	AFL
Observes	P	P		P		P	P				P	P
		EAFRD	CPR	ESCR	SR	ACHR	MAAE	PVIAC	PVNAC	EAFDAW	TCIDTP	RC
Observes								P				P

P = Party; S = Signatory; see Appendix for meaning of abbreviations.

Labor Force

Total Labor Force [25]

6 million-7 million

Labor Force by Occupation [26]

Government	40%
Industry, construction, and oil	25
Services	30
Agriculture	5

Unemployment Rate [27]

6.5% (1992 est.)

For sources, notes, and explanations, see Annotated Source Appendix, page 1061.

Production Sectors

Commercial Energy Production and Consumption

Data are shown in quadrillion (10^{15}) BTUs and percent for 1995
Values for hydroelectric, nuclear, geothermal, solar, and wind power refer to electrical generation.

Production [28]

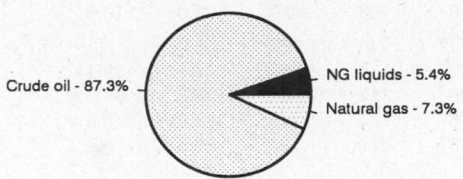

Crude oil - 87.3%
NG liquids - 5.4%
Natural gas - 7.3%

Consumption [29]

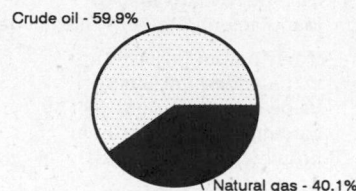

Crude oil - 59.9%
Natural gas - 40.1%

Crude oil	17.756
Natural gas liquids	1.092
Dry natural gas	1.492
Total	20.34

Crude oil	2.228
Dry natural gas	1.492
Total	3.720

Telecommunications [30]

- 1.46 million (1993) telephones; modern system
- Domestic: extensive microwave radio relay and coaxial and fiber-optic cable systems
- International: microwave radio relay to Bahrain, Jordan, Kuwait, Qatar, UAE, Yemen, and Sudan; coaxial cable to Kuwait and Jordan; submarine cable to Djibouti, Egypt and Bahrain; satellite earth stations - 5 Intelsat (3 Atlantic Ocean and 2 Indian Ocean), 1 Arabsat, and 1 Inmarsat (Indian Ocean region)
- Radio: Broadcast stations: AM 43, FM 13, shortwave 0 Radios: 5 million (1993 est.)
- Television: Broadcast stations: 80 Televisions: 4.5 million (1993 est.)

Top Agricultural Products [32]

Agriculture accounts for 9% of the GDP; produces wheat, barley, tomatoes, melons, dates, citrus; mutton, chickens, eggs, milk.

Top Mining Products [33]

Metric tons except as noted[e]	8/15/95[*]
Cement, hydraulic	15,000
Gold	
ore, mine output, gross weight	900
bullion, crude, gross weight (kg.)	9,300
Gas, natural, gross (mil. cu. meters)	67,200[61]
Natural gas liquids (000 42-gal. bls.)	265,000
Petroleum (000 42-gal. bls.)	
crude	2,970,000[r,62]
refinery products	585,000
Silver, Ag content of conc./bullion (kg.)	16,900[61]

Transportation [31]

Railways: total: 1,390 km; standard gauge: 1,390 km 1.435-m gauge (448 km double track) (1992)

Highways: total: 151,532 km; paved: 60,613 km; unpaved: 90,919 km (1992 est.)

Merchant marine: total: 76 ships (1,000 GRT or over) totaling 944,946 GRT/1,322,167 DWT; ships by type: bulk 1, cargo 13, chemical tanker 5, container 3, liquefied gas tanker 1, livestock carrier 4, oil tanker 22, passenger 1, refrigerated cargo 4, roll-on/roll-off cargo 13, short-sea passenger 9 (1995 est.)

Airports

Total:	175
With paved runways over 3,047 m:	30
With paved runways 2,438 to 3,047 m:	11
With paved runways 1,524 to 2,437 m:	22
With paved runways 914 to 1,523 m:	4
With paved runways under 914 m:	13

Tourism [34]

	1990	1991	1992	1993	1994
Visitors[94]	2,811	3,011	3,440	3,733	4,208
Tourists[95]	827	720	750	993	996

Travelers are in thousands, money in million U.S. dollars.

For sources, notes, and explanations, see Annotated Source Appendix, page 1061.

821

Manufacturing Sector

GDP and Manufacturing Summary [35]

	1980	1985	1989	1990	% change 1980-1990	% change 1989-1990
GDP (million 1980 $)	115,962	95,862	84,548	115,756	-0.2	36.9
GDP per capita (1980 $)	12,372	8,267	6,219	8,183	-33.9	31.6
Manufacturing as % of GDP (current prices)	5.0	7.8	8.8	10.0	100.0	13.6
Gross output (million $)	10,798[e]	12,741[e]	NA	17,995[e]	66.7	NA
Value added (million $)	2,594[e]	3,286[e]	13,185[e]	5,387[e]	107.7	-59.1
Value added (million 1980 $)	5,800	9,805	9,252[e]	12,401	113.8	34.0
Industrial production index	100	175	227[e]	245	145.0	7.9
Employment (thousands)	79[e]	138[e]	NA	127[e]	60.8	NA

Note: GDP stands for Gross Domestic Product. 'e' stands for estimated value.

Profitability and Productivity

	1980	1985	1989	1990	% change 1980-1990	% change 1989-1990
Intermediate input (%)	NA	NA	NA	NA	NA	NA
Wages, salaries, and supplements (%)	NA	NA	NA	NA	NA	NA
Gross operating surplus (%)	NA	NA	NA	NA	NA	NA
Gross output per worker ($)	137,239[e]	92,275[e]	NA	141,222[e]	2.9	NA
Value added per worker ($)	NA	NA	NA	NA	NA	NA
Average wage (incl. benefits) ($)	NA	NA	NA	NA	NA	NA

Profitability is in percent of gross output. Productivity is in U.S. $. 'e' stands for estimated value.

Profitability - 1990

Value Added in Manufacturing

	1980 $ mil.	1980 %	1985 $ mil.	1985 %	1989 $ mil.	1989 %	1990 $ mil.	1990 %	% change 1980-1990	% change 1989-1990
311 Food products	306[e]	11.8	241[e]	7.3	834[e]	6.3	317[e]	5.9	3.6	-62.0
313 Beverages	15[e]	0.6	25[e]	0.8	104[e]	0.8	42[e]	0.8	180.0	-59.6
314 Tobacco products	29[e]	1.1	23[e]	0.7	175[e]	1.3	31[e]	0.6	6.9	-82.3
321 Textiles	23[e]	0.9	17[e]	0.5	170[e]	1.3	21[e]	0.4	-8.7	-87.6
322 Wearing apparel	3[e]	0.1	4[e]	0.1	201[e]	1.5	5[e]	0.1	66.7	-97.5
323 Leather and fur products	5[e]	0.2	4[e]	0.1	52[e]	0.4	5[e]	0.1	0.0	-90.4
324 Footwear	1[e]	0.0	1[e]	0.0	80[e]	0.6	1[e]	0.0	0.0	-98.8
331 Wood and wood products	11[e]	0.4	8[e]	0.2	122[e]	0.9	10[e]	0.2	-9.1	-91.8
332 Furniture and fixtures	13[e]	0.5	21[e]	0.6	72[e]	0.5	36[e]	0.7	176.9	-50.0
341 Paper and paper products	51[e]	2.0	75[e]	2.3	119[e]	0.9	110[e]	2.0	115.7	-7.6
342 Printing and publishing	58[e]	2.2	45[e]	1.4	115[e]	0.9	57[e]	1.1	-1.7	-50.4
351 Industrial chemicals	802[e]	30.9	963[e]	29.3	NA	0.0	1,908[e]	35.4	137.9	NA
352 Other chemical products	63[e]	2.4	99[e]	3.0	NA	0.0	167[e]	3.1	165.1	NA
353 Petroleum refineries	432[e]	16.7	582[e]	17.7	8,983	68.1	804	14.9	86.1	-91.0
354 Miscellaneous petroleum and coal products	25[e]	1.0	46[e]	1.4	NA	0.0	119[e]	2.2	376.0	NA
355 Rubber products	3[e]	0.1	4[e]	0.1	NA	0.0	8[e]	0.1	166.7	NA
356 Plastic products	61[e]	2.4	92[e]	2.8	809[e]	6.1	148[e]	2.7	142.6	-81.7
361 Pottery, china, and earthenware	10[e]	0.4	14[e]	0.4	NA	0.0	25[e]	0.5	150.0	NA
362 Glass and glass products	12[e]	0.5	16[e]	0.5	NA	0.0	23[e]	0.4	91.7	NA
369 Other nonmetal mineral products	238[e]	9.2	395[e]	12.0	692[e]	5.2	677[e]	12.6	184.5	-2.2
371 Iron and steel	175[e]	6.7	241[e]	7.3	26[e]	0.2	344[e]	6.4	96.6	1,223.1
372 Nonferrous metals	8[e]	0.3	12[e]	0.4	9[e]	0.1	17[e]	0.3	112.5	88.9
381 Metal products	150[e]	5.8	204[e]	6.2	173[e]	1.3	271[e]	5.0	80.7	56.6
382 Nonelectrical machinery	39[e]	1.5	53[e]	1.6	115[e]	0.9	61[e]	1.1	56.4	-47.0
383 Electrical machinery	35[e]	1.3	60[e]	1.8	125[e]	0.9	106[e]	2.0	202.9	-15.2
384 Transport equipment	11[e]	0.4	18[e]	0.5	148[e]	1.1	32[e]	0.6	190.9	-78.4
385 Professional and scientific equipment	1[e]	0.0	1[e]	0.0	NA	0.0	3[e]	0.1	200.0	NA
390 Other manufacturing industries	16[e]	0.6	24[e]	0.7	61[e]	0.5	38[e]	0.7	137.5	-37.7

Note: The industry codes shown are International Standard Industry codes (ISIC). Percentages are percent of total Value Added. 'e' stands for estimated value

For sources, notes, and explanations, see Annotated Source Appendix, page 1061.

Finance, Economics, and Trade

Economic Indicators [36]

- **National product**: GDP—purchasing power parity— $189.3 billion (1995 est.)

- **National product real growth rate**: 0% (1995 est.)

- **National product per capita**: $10,100 (1995 est.)

- **Inflation rate (consumer prices)**: 5% (1995 est.)

- **External debt**: $18.9 billion (December 1989 est., includes short-term trade credits)

Balance of Payments Summary [37]

Values in millions of dollars.

	1989	1990	1991	1992	1993
Exports of goods (f.o.b.)	28,312	44,246	47,623	47,049	44,918
Imports of goods (f.o.b.)	-19,231	-21,490	-25,968	-30,248	-25,897
Trade balance	9,081	22,756	21,655	16,801	19,021
Services - debits	-20,892	-23,634	-40,737	-33,726	-26,262
Services - credits	13,015	12,398	11,728	10,855	9,680
Private transfers (net)	-8,542	-11,236	-13,746	-13,397	-15,717
Government transfers (net)	-2,200	-4,401	-6,489	-1,501	-940
Long-term capital (net)	-2,441	-1,477	630	-3,725	8,369
Short-term capital (net)	8,470	218	27,008	19,028	7,345
Errors and omissions					
Overall balance	-3,509	-5,376	49	-5,665	1,496

Exchange Rates [38]

Currency: **Saudi riyal.**
Symbol: **SR.**

Data are currency units per $1.

Fixed rate since late 1986   3.7450

Imports and Exports

Top Import Origins [39]

$21.3 billion (f.o.b., 1994 est.) Data are for 1994.

Origins	%
US	21
Japan	12
UK	8
Germany	8
Italy	5

Top Export Destinations [40]

$41.7 billion (f.o.b., 1994 est.) Data are for 1994.

Destinations	%
US	17
Japan	17
South Korea	8
Singapore	7
France	5

Foreign Aid [41]

Donor: Pledged $100 million in 1993 to fund reconstruction of Lebanon.

Import and Export Commodities [42]

Import Commodities

Machinery and equipment
Chemicals
Foodstuffs
Motor vehicles
Textiles

Export Commodities

Petroleum and petroleum products 90%

For sources, notes, and explanations, see Annotated Source Appendix, page 1061.

823

Senegal

Geography [1]

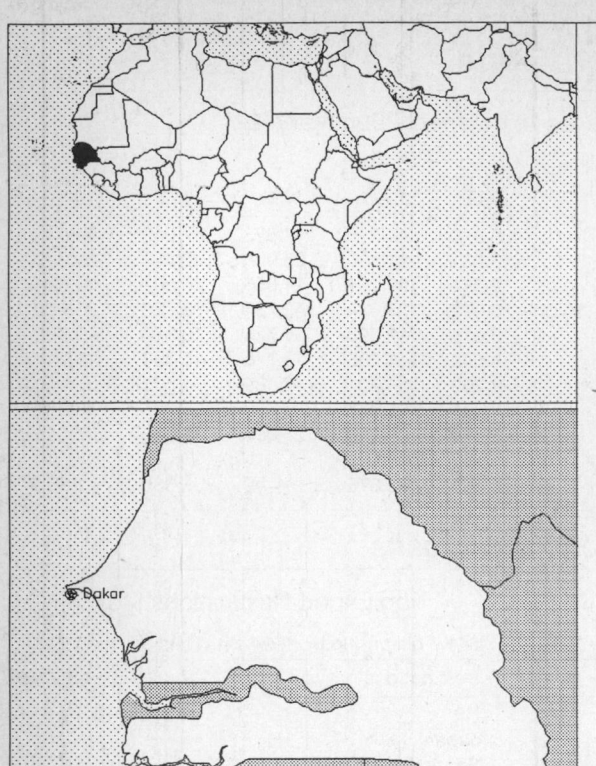

Total area:
196,190 sq km   75,749 sq mi
Land area:
192,000 sq km   74,132 sq mi
Comparative area:
Slightly smaller than South Dakota
Land boundaries:
Total 2,640 km, The Gambia 740 km, Guinea 330 km, Guinea-Bissau 338 km, Mali 419 km, Mauritania 813 km
Coastline:
531 km
Climate:
Tropical; hot, humid; rainy season (December to April) has strong southeast winds; dry season (May to November) dominated by hot, dry, harmattan wind
Terrain:
Generally low, rolling, plains rising to foothills in Southeast
Natural resources:
Fish, phosphates, iron ore
Land use:
Arable land: 27%
Permanent crops: 0%
Meadows and pastures: 30%
Forest and woodland: 31%
Other: 12%

Demographics [2]

	1970	1980	1990	1995[1]	1996	2000	2010	2020	2030
Population	4,318	5,640	7,408	8,790	9,093	10,390	14,362	19,497	25,621
Population density (persons per sq. mi.)	58	76	100	119	123	140	194	263	346
(persons per sq. km.)	22	29	39	46	47	54	75	102	133
Net migration rate (per 1,000 population)	1.4	NA	0.0	0.0	0.0	0.0	0.0	0.0	0.0
Births	NA	NA	NA	NA	413	NA	NA	NA	NA
Deaths	NA	NA	NA	NA	107	NA	NA	NA	NA
Life expectancy - males	43.0	NA	51.4	53.4	53.8	55.4	59.3	62.9	66.3
Life expectancy - females	45.1	NA	56.3	58.8	59.3	61.3	66.0	70.2	73.9
Birth rate (per 1,000)	52.0	NA	48.3	46.0	45.5	43.4	39.5	35.2	30.4
Death rate (per 1,000)	22.0	NA	14.1	12.1	11.8	10.4	7.8	6.1	5.0
Women of reproductive age (15-49 yrs.)	NA	NA	1,686	1,999	2,069	2,383	3,409	4,745	6,493
of which are currently married	772	NA	1,180	NA	1,456	1,672	2,374	NA	NA
Fertility rate	NA	NA	6.6	6.4	6.3	6.0	5.3	4.5	3.8

Except as noted, values for vital statistics are in thousands; life expectancy is in years.

Health

Health Indicators [3]

% of population with access to	
safe water (1990-95)	52
adequate sanitation (1990-95)	58
health services (1985-95)	40
% of 1-year-olds immunized (1990-94) against	
TB (tuberculosis)	71
DPT (diphtheria, pertussis, tetanus)	55
polio	55
measles	49
% of contraceptive prevalence (1980-94)	7
ORT use rate (1990-94)	18

Health Expenditures [4]

Total health expenditure, 1990 (official exchange rate)	
Millions of dollars	214
Dollars per capita	29
Health expenditures as a percentage of GDP	
Total	3.7
Public sector	2.3
Private sector	1.4
Development assistance for health	
Total aid flows (millions of dollars)[1]	36
Aid flows per capita (dollars)	4.9
Aid flows as a percentage of total health expenditure	16.9

For sources, notes, and explanations, see Annotated Source Appendix, page 1061.

Human Factors

Women and Children [5]

% of pregnant women immunized (tetanus 1990-94)	32
% of births attended by trained health personnel (1983-94)	46
Maternal mortality rate (1980-92)	560
Under-5 mortality rate (1994)	115
% under-5 moderately/severely underweight (1980-1994)	20

Burden of Disease [6]

Population per physician (1994)	31,903.47
Population per nurse (1994)	4,246.15
Population per hospital bed (1994)	13,305.96
AIDS cases per 100,000 people (1994)	6.5
Malaria cases per 100,000 people (1992)	NA

Ethnic Division [7]

Wolof	36%
Fulani	17%
Serer	17%
Diola	9%
Mandingo	9%
Toucouleur	9%
European and Lebanese	1%
Other	2%

Religion [8]

Muslim	92%
Indigenous beliefs	6%
Christian (primarily Roman Catholic)	2%

Major Languages [9]

French (official), Wolof, Pulaar, Diola, Mandingo.

Education

Public Education Expenditures [10]

Million (Franc C.F.A.)	1980	1990	1991	1992	1993	1994
Total education expenditure	NA	NA	NA	NA	NA	NA
as percent of GNP	NA	NA	NA	NA	NA	NA
as percent of total govt. expend.	NA	NA	NA	NA	NA	NA
Current education expenditure	26,818	60,467	61,686	67,100	67,008	NA
as percent of GNP	4.4	4.1	4.1	4.3	4.4	NA
as percent of current govt. expend.	23.2	26.8	27.4	32.8	32.6	NA
Capital expenditure	NA	NA	NA	NA	NA	NA

Educational Attainment [11]

Illiteracy [12]

In thousands and percent[1]	1990	1995	2000
Illiterate population (15+ yrs.)	2,844	3,084	3,346
Illiteracy rate - total pop. (%)	73.5	67.8	61.7
Illiteracy rate - males (%)	66.1	60.3	54.2
Illiteracy rate - females (%)	80.1	74.6	68.5

Libraries [13]

	Admin. Units	Svc. Pts.	Vols. (000)	Shelving (meters)	Vols. Added	Reg. Users
National	NA	NA	NA	NA	NA	NA
Nonspecialized	NA	NA	NA	NA	NA	NA
Public (1987)	10	11	15	382	1,721	959
Higher ed. (1994)	7	NA	380	13,894	7,432	12,906
School	NA	NA	NA	NA	NA	NA

Daily Newspapers [14]

	1980	1985	1990	1994
Number of papers	1	3	1	3
Circ. (000)	35	53	50	48

Culture [15]

Science and Technology

Scientific/Technical Forces [16]

Scientists/engineers	1,948
Number female	NA
Technicians	2,662
Number female	NA
Total	4,610

R&D Expenditures [17]

U.S. Patents Issued [18]

For sources, notes, and explanations, see Annotated Source Appendix, page 1061.

Government and Law

Organization of Government [19]

Long-form name:
Republic of Senegal
Type:
Republic under multiparty democratic rule
Independence:
20 August 1960 (from France)
National holiday:
Independence Day, 4 April (1960)
Constitution:
3 March 1963, revised 1991
Legal system:
Based on French civil law system; judicial review of legislative acts in Supreme Court, which also audits the government's accounting office; has not accepted compulsory ICJ jurisdiction
Executive branch:
President; Prime Minister; Council of Ministers
Legislative branch:
Unicameral: National Assembly (Assemblee Nationale)
Judicial branch:
Supreme Court (Cour Supreme)

Elections [20]

National Assembly	% of votes
Socialist Party (PS)	70.0
Senegalese Democratic Party (PDS)	23.0
Other	7.0

Government Budget [21]

For 1996 est.

	$ bil.
Revenues	0.876
Expenditures	197.7
Capital expenditures	NA

Crime [23]

	1994
Crime volume	
Cases known to police	15,355
Attempts (percent)	NA
Percent cases solved	13,626
Crimes per 100,000 persons	NA
Persons responsible for offenses	
Total number offenders	11,209
Percent female	48.35
Percent juvenile (0-18 yrs.)	11.62
Percent foreigners	NA

Military Expenditures and Arms Transfers [22]

	1990	1991	1992	1993	1994
Military expenditures					
Current dollars (mil.)	67	71	101	89	60
1994 constant dollars (mil.)	75	76	105	91	60
Armed forces (000)	18	18	18	18	14
Gross national product (GNP)					
Current dollars (mil.)	3,273	3,503	3,600	3,597	3,711
1994 constant dollars (mil.)	3,643	3,754	3,754	3,672	3,711
Central government expenditures (CGE)					
1994 constant dollars (mil.)	NA	NA	764[e]	NA	NA
People (mil.)	7.7	8.0	8.2	8.5	8.7
Military expenditure as % of GNP	2.1	2.0	2.8	2.5	1.6
Military expenditure as % of CGE	NA	NA	13.7	NA	NA
Military expenditure per capita (1994 $)	10	10	13	11	7
Armed forces per 1,000 people (soldiers)	2.3	2.3	2.2	2.1	1.6
GNP per capita (1994 $)	472	472	457	434	425
Arms imports[6]					
Current dollars (mil.)	0	10	10	10	5
1994 constant dollars (mil.)	0	11	10	10	5
Arms exports[6]					
Current dollars (mil.)	0	0	0	0	0
1994 constant dollars (mil.)	0	0	0	0	0
Total imports[7]					
Current dollars (mil.)	1,314	1,097	1,170	1,228	1,168[e]
1994 constant dollars (mil.)	1,462	1,176	1,220	1,253	1,168[e]
Total exports[7]					
Current dollars (mil.)	762	652	682	573[e]	635[e]
1994 constant dollars (mil.)	848	699	711	584[e]	635[e]
Arms as percent of total imports[8]	0	0.9	0.9	0.8	0.4
Arms as percent of total exports[8]	0	0	0	0	0

Human Rights [24]

	SSTS	FL	FAPRO	PPCG	APROBC	TPW	PCPTW	STPEP	PHRFF	PRW	ASST	AFL
Observes	2	P	P	P	P	P	P	P		P	P	P
		EAFRD	CPR	ESCR	SR	ACHR	MAAE	PVIAC	PVNAC	EAFDAW	TCIDTP	RC
Observes		P	P	P	P			P	P	P	P	P

P=Party; S=Signatory; see Appendix for meaning of abbreviations.

Labor Force

Total Labor Force [25]

2.509 million; 175,000 wage earners

Labor Force by Occupation [26]

Private sector	40%
Government and parapublic	60
Subsistence farming	

Unemployment Rate [27]

For sources, notes, and explanations, see Annotated Source Appendix, page 1061.

Production Sectors

Commercial Energy Production and Consumption

Data are shown in quadrillion (10^{15}) BTUs and percent for 1995
Values for hydroelectric, nuclear, geothermal, solar, and wind power refer to electrical generation.

Production [28]

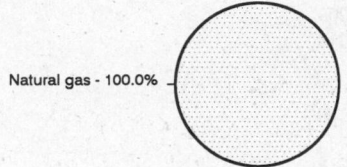

Natural gas - 100.0%

Consumption [29]

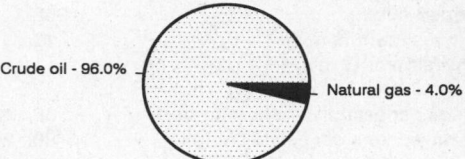

Crude oil - 96.0%   Natural gas - 4.0%

Dry natural gas	0.002
Total	0.002

Crude oil	0.048
Dry natural gas	0.002
Total	0.050

Telecommunications [30]

- 55,000 (1993 est.) telephones
- Domestic: above-average urban system; microwave radio relay and cable trunk system
- International: 3 submarine cables; satellite earth station - 1 Intelsat (Atlantic Ocean)
- Radio: Broadcast stations: AM 8, FM 0, shortwave 0 Radios: 850,000 (1993 est.)
- Television: Broadcast stations: 1 Televisions: 61,000 (1993 est.)

Top Agricultural Products [32]

Produces peanuts, millet, corn, sorghum, rice, cotton, tomatoes, green vegetables; cattle, poultry, pigs; fish catch of 409,000 metric tons in 1992.

Transportation [31]

Railways: total: 904 km; narrow gauge: 904 km 1.000-meter gauge (70 km double track) (1995)

Highways: total: 13,850 km; paved: 3,900 km; unpaved: 9,950 km (1990 est.)

Merchant marine: total: 1 bulk ship (1,000 GRT or over) totaling 1,995 GRT/3,775 DWT (1995 est.)

Airports

Total:	17
With paved runways over 3,047 m:	1
With paved runways 1,524 to 2,437 m:	8
With paved runways 914 to 1,523 m:	1
With paved runways under 914 m:	1
With unpaved runways 1,524 to 2,437 m:	4

Top Mining Products [33]

Thousand metric tons except as noted[e]	8/95[*]
Cement, hydraulic	590
Clays: Fuller's earth (attapulgite)	119
Petroleum refinery products (000 42-gal. bls.)	5,750
Aluminum phosphate	29
Calcium phosphate	1,670
Phosphoric acid	274
Calcium phosphate-based fertilizers	160
Salt	117

Tourism [34]

	1990	1991	1992	1993	1994
Tourists[16]	246	234	246	168	240
Cruise passengers	6	5	3	3	5
Tourism receipts	167	171	182	173	115
Tourism expenditures	105	105	112	106	70
Fare expenditures	83	82	88	82	86

Travelers are in thousands, money in million U.S. dollars.

For sources, notes, and explanations, see Annotated Source Appendix, page 1061.

827

Manufacturing Sector

Manufacturing Summary [35]

	1987		1988		1989		1990		1991	
	$ bil.	%	$ bil.	%	$ bil.	%	$ bil.	%	$ bil.	%
Establishments or enterprises (number)	121	0.005	117	0.006	312	0.017	-	-	-	-
Total employment (000)	8	0.006	7	0.005	31	0.025	-	-	-	-
Production workers (000)	7	0.011	7	0.011	-	-	-	-	-	-
Output ($ bil.)	0.639	0.006	0.678	0.006	1	0.010	-	-	-	-
Value added ($ bil.)	0.262	0.006	0.289	0.006	0.308	0.006	-	-	-	-
Capital investment ($ mil.)	19	0.004	21	0.004	65	0.012	-	-	-	-
M & E investment ($ mil.)	-	-	-	-	-	-	-	-	-	-
Employees per establishment (number)	66	114.282	60	91.771	100	152.669	-	-	-	-
Production workers per establishment	60	208.659	60	200.133	-	-	-	-	-	-
Output per establishment ($ mil.)	5	121.967	6	104.281	4	62.272	-	-	-	-
Capital investment per estab. ($ mil.)	0.161	86.571	0.177	77.845	0.207	70.940	-	-	-	-
M & E per establishment ($ mil)	-	-	-	-	-	-	-	-	-	-
Payroll per employee ($)	6,888	76.819	8,507	85.391	5,668	50.737	-	-	-	-
Wages per production worker ($)	4,397	55.721	4,587	54.283	-	-	-	-	-	-
Hours per production worker (hours)	-	-	-	-	-	-	-	-	-	-
Output per employee ($)	79,931	106.724	96,786	113.632	39,092	40.789	-	-	-	-
Capital investment per employee ($)	2,433	75.752	2,952	84.825	2,068	46.466	-	-	-	-
M & E per employee ($)	-	-	-	-	-	-	-	-	-	-

Note: Columns headed % show percent of world total or ratio. Ratios closest to 100 are closest to world average. M & E stands for machinery & equipment.

Output in Manufacturing

	1987		1988		1989		1990		1991	
	$ bil.	%	$ bil.	%	$ bil.	%	$ bil.	%	$ bil.	%
3110 - Food products	-	-	-	-	0.654	65.40	-	-	-	-
3130 - Beverages	-	-	-	-	0.027	2.70	-	-	-	-
3140 - Tobacco	-	-	-	-	0.030	3.00	-	-	-	-
3310 - Wood products	-	-	-	-	0.002	0.20	-	-	-	-
3320 - Furniture, fixtures	-	-	-	-	0.001	0.10	-	-	-	-
3410 - Paper and products	-	-	-	-	0.014	1.40	-	-	-	-
3420 - Printing, publishing	-	-	-	-	0.027	2.70	-	-	-	-
3510 - Industrial chemicals	-	-	-	-	0.120	12.00	-	-	-	-
3520 - Chemical products nec	-	-	-	-	0.069	6.90	-	-	-	-
3530 - Petroleum refineries	-	-	-	-	0.101	10.10	-	-	-	-
3550 - Rubber products	-	-	-	-	0.000	0.00	-	-	-	-
3560 - Plastic products nec	-	-	-	-	0.026	2.60	-	-	-	-
3690 - Nonmetal products nec	-	-	-	-	0.048	4.80	-	-	-	-
3810 - Metal products	-	-	-	-	0.031	3.10	-	-	-	-
3820 - Machinery nec	-	-	-	-	0.005	0.50	-	-	-	-
3830 - Electrical machinery	-	-	-	-	0.012	1.20	-	-	-	-
3840 - Transportation equipment	-	-	-	-	0.027	2.70	-	-	-	-
3900 - Industries nec	0.639	100.00	0.678	100.00	0.030	3.00	-	-	-	-

Note: Codes are International Standard Industry codes (ISIC). Percentages are % of total Output. [f]: Factor Prices; [p]: Producer Prices.

Finance, Economics, and Trade

Economic Indicators [36]

- **National product**: GDP—purchasing power parity— $14.5 billion (1995 est.)
- **National product real growth rate**: 4.5% (1995 est.)
- **National product per capita**: $1,600 (1995 est.)
- **Inflation rate (consumer prices)**: 6.1% (1995)
- **External debt**: $3.8 billion (1993)

Balance of Payments Summary [37]

Values in millions of dollars.

	1989	1990	1991	1992	1993
Exports of goods (f.o.b.)	758.6	911.6	824.2	831.9	722.6
Imports of goods (f.o.b.)	-998.4	-1,176.1	-1,114.1	-1,200.3	-1,105.4
Trade balance	-239.8	-264.5	-283.9	-368.4	-382.8
Services - debits	-722.5	-831.5	-793.3	-834.9	-790.4
Services - credits	498.0	585.8	584.5	618.5	588.4
Private transfers (net)	6.3	29.4	28.4	36.6	40.3
Government transfers (net)	259.9	355.9	265.1	279.9	239.8
Long-term capital (net)	184.2	275.8	210.2	187.4	117.2
Short-term capital (net)	23.8	-116.3	12.8	42.4	91.6
Errors and omissions	1.6	-16.7	-26.2	82.6	114.8
Overall balance	11.5	17.9	-8.4	44.1	18.9

Exchange Rates [38]

Currency: **Communaute Financi- ere Africaine franc.**

Symbol: **CFAF.**

Data are currency units per $1.

January 1996	500.56
1995	499.15
1994	555.20
1993	283.16
1992	264.69
1991	282.11

Imports and Exports

Top Import Origins [39]

$1.1 billion (c.i.f., 1994 est.).

Origins	%
France	NA
Other EU countries	NA
Nigeria	NA
Cote d'Ivoire	NA
Algeria	NA
China	NA
Japan	NA

Top Export Destinations [40]

$940 million (f.o.b., 1994 est.).

Destinations	%
France	NA
Other EU countries	NA
Cote d'Ivoire	NA
Mali	NA

Foreign Aid [41]

Recipient: ODA, $439 million (1993).

Import and Export Commodities [42]

Import Commodities	Export Commodities
Foods and beverages	Fish
Consumer goods	Ground nuts (peanuts)
Capital goods	Petroleum products
Petroleum	Phosphates
	Cotton

Seychelles

Geography [1]

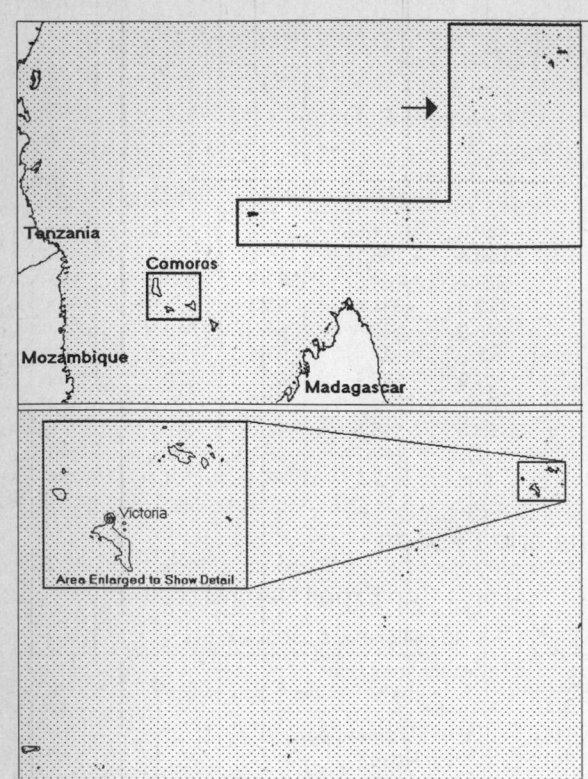

Total area:
  455 sq km   176 sq mi
Land area:
  455 sq km   176 sq mi
Comparative area:
  2.5 times the size of Washington, DC
Land boundaries:
  0 km
Coastline:
  491 km
Climate:
  Tropical marine; humid; cooler season during southeast monsoon (late May to September); warmer season during northwest monsoon (March to May)
Terrain:
  Mahe Group is granitic, narrow coastal strip, rocky, hilly; others are coral, flat, elevated reefs
Natural resources:
  Fish, copra, cinnamon trees
Land use:
  Arable land: 4%
  Permanent crops: 18%
  Meadows and pastures: 0%
  Forest and woodland: 18%
  Other: 60%

Demographics [2]

	1970	1980	1990	1995[1]	1996	2000	2010	2020	2030
Population	54	66	73	77	78	80	84	89	93
Population density (persons per sq. mi.)	307	378	417	437	441	453	479	505	528
(persons per sq. km.)	119	146	161	169	170	175	185	195	204
Net migration rate (per 1,000 population)	4.3	0.4	-4.7	-5.9	-6.1	-6.3	-4.8	-3.4	-2.2
Births	NA	NA	NA	NA	2	NA	NA	NA	NA
Deaths	NA	0	NA	NA	1	NA	NA	NA	NA
Life expectancy - males	NA	66.3	64.0	63.7	64.2	66.6	70.8	73.4	75.4
Life expectancy - females	NA	74.6	74.7	74.3	74.4	74.7	77.1	79.6	81.5
Birth rate (per 1,000)	31.2	28.1	22.0	21.4	21.0	18.9	15.8	14.2	12.3
Death rate (per 1,000)	8.2	6.7	7.4	7.5	7.3	6.6	5.7	5.7	6.5
Women of reproductive age (15-49 yrs.)	NA	14	19	21	22	23	25	24	23
of which are currently married[46]	NA	NA	7	NA	9	10	11	NA	NA
Fertility rate	NA	4.0	2.3	2.1	2.1	2.0	1.8	1.8	1.8

Except as noted, values for vital statistics are in thousands; life expectancy is in years.

Health

Health Indicators [3]

Health Expenditures [4]

For sources, notes, and explanations, see Annotated Source Appendix, page 1061.

Human Factors

Women and Children [5]	Burden of Disease [6]	
	Population per physician (1985)	2,166.67
	Population per nurse	NA
	Population per hospital bed (1985)	200.00
	AIDS cases per 100,000 people (1994)	7.6
	Malaria cases per 100,000 people (1992)	NA

Ethnic Division [7]	Religion [8]		Major Languages [9]
Seychellois (mixture of Asians, Africans, Europeans).	Roman Catholic	90%	English (official), French (official), Creole.
	Anglican	8%	
	Other	2%	

Education

Public Education Expenditures [10]

Million (Rupee)[59]	1980	1985	1990	1992	1993	1994
Total education expenditure	52	125	153	183	148	168
as percent of GNP	5.8	10.7	8.1	8.5	6.5	7.4
as percent of total govt. expend.	14.4	21.3	14.8	12.9	NA	.NA
Current education expenditure	50	120	153	156	NA	NA
as percent of GNP	5.5	10.3	8.1	7.3	NA	NA
as percent of current govt. expend.	14.0	21.9	18.5	12.5	NA	NA
Capital expenditure	2	4	-	26	NA	NA

Educational Attainment [11]

Age group (1987)	25+
Total population	30,912
Highest level attained (%)	
No schooling	12.1
First level	
Not completed	44.9
Completed	NA
Entered second level	
S-1	35.7
S-2	NA
Postsecondary	4.6

Illiteracy [12]

	1987	1989	1995
Illiterate population (15+ yrs.)[2]	7,106	NA	NA
Illiteracy rate - total pop. (%)	15.6	NA	NA
Illiteracy rate - males (%)	16.9	NA	NA
Illiteracy rate - females (%)	14.4	NA	NA

Libraries [13]

	Admin. Units	Svc. Pts.	Vols. (000)	Shelving (meters)	Vols. Added	Reg. Users
National	NA	NA	NA	NA	NA	NA
Nonspecialized	NA	NA	NA	NA	NA	NA
Public (1989)[8]	1	4	42	NA	1,285	16,664
Higher ed. (1990)	1	5	26	510	2,270	1,610[e]
School (1990)	24	24	NA	1,083	8,106	7,940

Daily Newspapers [14]

	1980	1985	1990	1994
Number of papers	1	1	1	1
Circ. (000)	3	3	3	3

Culture [15]

Cinema (seats per 1,000)	NA
Annual attendance per person	NA
Gross box office receipts (mil. Rupee)	NA
Museums (reporting)	2
Visitors (000)	10
Annual receipts (000 Rupee)	NA

Science and Technology

Scientific/Technical Forces [16]

Scientists/engineers	18
Number female	NA
Technicians	6
Number female	NA
Total	24

R&D Expenditures [17]

	Rupee (000) 1983
Total expenditure[6]	12,854
Capital expenditure	6,771
Current expenditure	6,083
Percent current	47.3

U.S. Patents Issued [18]

For sources, notes, and explanations, see Annotated Source Appendix, page 1061.

831

Government and Law

Organization of Government [19]

Long-form name:
Republic of Seychelles
Type:
Republic
Independence:
29 June 1976 (from UK)
National holiday:
National Day, 18 June (1993) (adoption of new constitution)
Constitution:
18 June 1993
Legal system:
Based on English common law, French civil law, and customary law
Executive branch:
President; Council of Ministers
Legislative branch:
Unicameral: People's Assembly (Assemblee du Peuple)
Judicial branch:
Court of Appeal; Supreme Court

Elections [20]

People's Assembly	% of votes
Seychelles People's Progressive Front (SPPF)	82.0
Democratic Party (DP)	15.0
United Opposition (UO)	3.0

Government Expenditures [21]

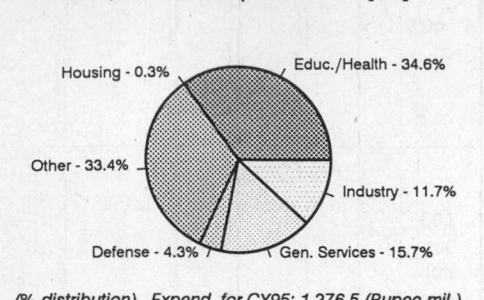

Housing - 0.3% Educ./Health - 34.6% Other - 33.4% Industry - 11.7% Defense - 4.3% Gen. Services - 15.7%

(% distribution). Expend. for CY95: 1,276.5 (Rupee mil.)

Defense Summary [22]

Branches: Army, Coast Guard, Marines, National Guard, Presidential Protection Unit, Police Force

Manpower Availability: Males age 15-49 21,547; males fit for military service 10,883 (1996 est.)

Defense Expenditures: NA

Crime [23]

	1994
Crime volume	
Cases known to police	3,336
Attempts (percent)	NA
Percent cases solved	NA
Crimes per 100,000 persons	4,517.26
Persons responsible for offenses	
Total number offenders	2,783
Percent female	NA
Percent juvenile (7-18 yrs.)	NA
Percent foreigners	NA

Human Rights [24]

	SSTS	FL	FAPRO	PPCG	APROBC	TPW	PCPTW	STPEP	PHRFF	PRW	ASST	AFL
Observes	2	P	P	P		P	P	P		1	P	P
	EAFRD	CPR	ESCR	SR	ACHR	MAAE	PVIAC	PVNAC	EAFDAW	TCIDTP	RC
Observes	P	P	P	P			P	P	P	P	P

P=Party; S=Signatory; see Appendix for meaning of abbreviations.

Labor Force

Total Labor Force [25]

27,700 (1985)

Labor Force by Occupation [26]

Industry and commerce	31%
Services	21
Government	20
Agriculture, forestry, and fishing	12
Other	16

Date of data: 1985

Unemployment Rate [27]

9% (1987)

For sources, notes, and explanations, see Annotated Source Appendix, page 1061.

Production Sectors

Energy Resource Summary [28]

Electricity: Capacity: 30,000 kW. Production: 110 million kWh. Consumption per capita: 1,399 kWh (1993).

Telecommunications [30]

- 8,300 (1982 est.) telephones
- Domestic: radiotelephone communications between islands in the archipelago
- International: direct radiotelephone communications with adjacent island countries and African coastal countries; satellite earth station - 1 Intelsat (Indian Ocean)
- Radio: Broadcast stations: AM 2, FM 0, shortwave 0 Radios: 34,000 (1992 est.)
- Television: Broadcast stations: 2 Televisions: 8,200 (1991 est.)

Top Agricultural Products [32]

Produces coconuts, cinnamon, vanilla, sweet potatoes, cassava (tapioca), bananas; broiler chickens; expansion of tuna fishing under way.

Top Mining Products [33]

Detailed information is not available. A summary of mineral resources follows. **Mineral Resources**: Construction clay, coral, stone, and sand. Exploration of granite, lime, cement, and petroleum deposits.

Transportation [31]

Railways: 0 km

Highways: total: 269 km; paved: 187 km; unpaved: 82 km (1988 est.)

Merchant marine: none

Airports

Total:	14
With paved runways 2,438 to 3,047 m:	1
With paved runways 914 to 1,523 m:	5
With paved runways under 914 m:	6
With unpaved runways 914 to 1,523 m:	2 (1995 est.)

Tourism [34]

	1990	1991	1992	1993	1994
Visitors[12]	112	98	108	126	120
Tourists[12]	104	90	99	116	110
Cruise passengers	8	8	9	10	10
Tourism receipts	120	99	117	116	NA
Tourism expenditures	20	12	16	NA	NA
Fare receipts	19	33	37	NA	NA
Fare expenditures	6	6	7	NA	NA

Travelers are in thousands, money in million U.S. dollars.

Finance, Economics, and Trade

GDP and Manufacturing Summary [35]

	1980	1985	1990	1991	1992
Gross Domestic Product					
Millions of 1980 dollars	147	158	197	201	205[e]
Growth rate in percent	-2.55	10.26	6.62	2.50	1.91[e]
Manufacturing Value Added					
Millions of 1980 dollars	11	11	17	19	20[e]
Growth rate in percent	18.21	8.44	11.04	8.21	7.89[e]
Manufacturing share in percent of current prices	8.0	10.6	9.8	NA	NA

Economic Indicators [36]

- **National product**: GDP—purchasing power parity— $430 million (1993 est.)

- **National product real growth rate**: - 2% (1993 est.)

- **National product per capita**: $6,000 (1993 est.)

- **Inflation rate (consumer prices)**: 3.9% (1993 est.)

- **External debt**: $181 million (1993 est.)

Balance of Payments Summary [37]

Values in millions of dollars.

	1988	1989	1990	1991	1992
Exports of goods (f.o.b.)	17.3	14.5	28.1	18.8	19.6
Imports of goods (f.o.b.)	-135.0	-139.6	-158.4	-146.3	-162.9
Trade balance	-117.7	-125.1	-130.3	-127.5	-143.3
Services - debits	-102.1	-114.1	-130.1	-131.3	-131.2
Services - credits	167.5	189.9	233.2	239.7	251.8
Private transfers (net)	-4.9	-5.2	-2.5	-2.5	-2.8
Government transfers (net)	28.9	31.7	29.3	25.3	23.8
Long-term capital (net)	20.9	28.3	17.1	32.4	11.2
Short-term capital (net)	-0.9	4.6	-2.7	-4.2	1.3
Errors and omissions	4.2	-6.5	-10.0	-21.5	-6.9
Overall balance	-4.2	3.6	4.0	10.5	3.9

Exchange Rates [38]

Currency: **Seychelles rupee.**
Symbol: **SRe.**

Data are currency units per $1.

January 1996	4.9257
1995	4.7620
1994	5.0559
1993	5.1815
1992	5.1220
1991	5.2893

Imports and Exports

Top Import Origins [39]

$261 million (f.o.b., 1993 est.) Data are for 1992.

Origins	%
Singapore	16
Bahrain	16
South Africa	14
UK	13

Top Export Destinations [40]

$50 million (f.o.b., 1993 est.) Data are for 1992.

Destinations	%
France	43
UK	22
Reunion	11

Foreign Aid [41]

Recipient: ODA, $NA.

Import and Export Commodities [42]

Import Commodities	Export Commodities
Manufactured goods	Fish
Food	Cinnamon bark
Petroleum products	Copra
Tobacco	Petroleum products
Beverages	(re-exports)
Machinery and transportation equipment	

Sierra Leone

Geography [1]

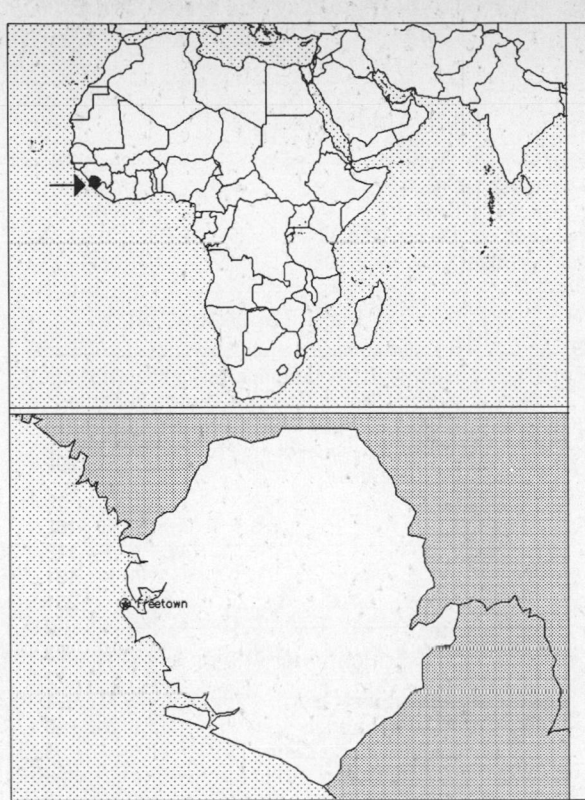

Total area:
  71,740 sq km   27,699 sq mi
Land area:
  71,620 sq km   27,653 sq mi
Comparative area:
  Slightly smaller than South Carolina
Land boundaries:
  Total 958 km, Guinea 652 km, Liberia 306 km
Coastline:
  402 km
Climate:
  Tropical; hot, humid; summer rainy season (May to December); winter dry season (December to April)
Terrain:
  Coastal belt of mangrove swamps, wooded hill country, upland plateau, mountains in East
Natural resources:
  Diamonds, titanium ore, bauxite, iron ore, gold, chromite
Land use:
  Arable land: 25%
  Permanent crops: 2%
  Meadows and pastures: 31%
  Forest and woodland: 29%
  Other: 13%

Demographics [2]

	1970	1980	1990	1995[1]	1996	2000	2010	2020	2030
Population	2,789	3,333	4,283	4,612	4,793	5,580	7,399	9,716	12,440
Population density (persons per sq. mi.)	101	121	155	167	173	202	268	351	450
(persons per sq. km.)	39	47	60	64	67	78	103	136	174
Net migration rate (per 1,000 population)	NA	0.0	29.2	6.5	12.5	-0.2	0.0	0.0	0.0
Births	NA	NA	NA	NA	226	NA	NA	NA	NA
Deaths	NA	NA	NA	NA	87	NA	NA	NA	NA
Life expectancy - males	NA	37.5	41.7	44.1	44.6	46.6	51.9	57.4	62.5
Life expectancy - females	NA	41.8	47.0	49.9	50.5	52.9	59.0	64.9	70.2
Birth rate (per 1,000)	NA	45.9	47.9	47.6	47.1	45.1	39.9	35.2	29.6
Death rate (per 1,000)	NA	25.5	20.9	18.7	18.2	16.3	12.1	8.9	6.6
Women of reproductive age (15-49 yrs.)	NA	765	1,008	1,062	1,099	1,265	1,757	2,418	3,227
of which are currently married	NA	NA	749	NA	820	945	1,305	NA	NA
Fertility rate	NA	6.5	6.5	6.4	6.4	6.1	5.3	4.4	3.5

Except as noted, values for vital statistics are in thousands; life expectancy is in years.

Health

Health Indicators [3]

% of population with access to	
safe water (1990-95)	34
adequate sanitation (1990-95)	11
health services (1985-95)	38
% of 1-year-olds immunized (1990-94) against	
TB (tuberculosis)	60
DPT (diphtheria, pertussis, tetanus)	43
polio	43
measles	46
% of contraceptive prevalence (1980-94)	4
ORT use rate (1990-94)	60

Health Expenditures [4]

Total health expenditure, 1990 (official exchange rate)	
Millions of dollars	22
Dollars per capita	5
Health expenditures as a percentage of GDP	
Total	2.4
Public sector	1.7
Private sector	0.8
Development assistance for health	
Total aid flows (millions of dollars)[1]	7
Aid flows per capita (dollars)	1.7
Aid flows as a percentage of total health expenditure	33.0

For sources, notes, and explanations, see Annotated Source Appendix, page 1061.

835

Human Factors

Women and Children [5]

% of pregnant women immunized (tetanus 1990-94)	61
% of births attended by trained health personnel (1983-94)	25
Maternal mortality rate (1980-92)	450
Under-5 mortality rate (1994)	284
% under-5 moderately/severely underweight (1980-1994)	29

Burden of Disease [6]

Population per physician (1984)	13,389.31
Population per nurse (1984)	1,070.82
Population per hospital bed	NA
AIDS cases per 100,000 people (1994)	0.5
Malaria cases per 100,000 people (1992)	NA

Ethnic Division [7]

13 native African tribes	99%
Temne	30%
Mende	30%
Other	39%
Creole, European, Lebanese, and Asian	1%

Religion [8]

Muslim	60%
Indigenous beliefs	30%
Christian	10%

Major Languages [9]

English (official, regular use limited to literate minority), Mende (principal vernacular in the south), Temne (principal vernacular in the north), Krio (the language of the re-settled ex-slave population of the Freetown area and is lingua franca).

Education

Public Education Expenditures [10]

	1980	1985	1989	1990	1993	1994
Million (Leone)						
Total education expenditure	43	112	604	NA	NA	NA
as percent of GNP	3.8	2.4	1.4	NA	NA	NA
as percent of total govt. expend.	11.8	12.4	NA	NA	NA	NA
Current education expenditure	41	106	577	NA	NA	NA
as percent of GNP	3.6	2.3	1.3	NA	NA	NA
as percent of current govt. expend.	14.5	15.5	NA	NA	NA	NA
Capital expenditure	2	6	27	NA	NA	NA

Educational Attainment [11]

Age group (1985)	5+
Total population	1,315,897
Highest level attained (%)	
No schooling	64.5
First level	
Not completed	18.7
Completed	1.8
Entered second level	
S-1	9.7
S-2	3.8
Postsecondary	1.5

Illiteracy [12]

In thousands and percent[1]	1990	1995	2000
Illiterate population (15+ yrs.)	1,649	1,727	1,805
Illiteracy rate - total pop. (%)	68.2	67.7	59.3
Illiteracy rate - males (%)	55.8	54.3	46.5
Illiteracy rate - females (%)	80.0	80.1	71.1

Libraries [13]

Daily Newspapers [14]

	1980	1985	1990	1994
Number of papers	1	1	1	1
Circ. (000)	10	10	10	10

Culture [15]

Science and Technology

Scientific/Technical Forces [16]

R&D Expenditures [17]

U.S. Patents Issued [18]

Government and Law

Organization of Government [19]

Long-form name:
Republic of Sierra Leone
Type:
Constitutional democracy
Independence:
27 April 1961 (from UK)
National holiday:
Republic Day, 27 April (1961)
Constitution:
1 October 1991; subsequently amended
several times
Legal system:
Based on English law and customary laws
indigenous to local tribes; has not
accepted compulsory ICJ jurisdiction
Executive branch:
President; Ministers of State
Legislative branch:
Unicameral: House of Representatives
Judicial branch:
Supreme Court

Elections [20]

House of Representatives	% of seats
Sierra Leone Peoples (SLPP)	39.7
United National Peoples (UNPP)	25.0
Peoples Democratic (PDP)	17.6
All Peoples Congress (APC)	7.4
National Unity Party (NUP)	5.9
Democratic Center Party (DCP)	4.4

Government Budget [21]

For FY94/95 est.

	$ mil.
Revenues	75
Expenditures	128
Capital expenditures	NA

Crime [23]

Military Expenditures and Arms Transfers [22]

	1990	1991	1992	1993	1994
Military expenditures					
Current dollars (mil.)	7	18[e]	17[e]	16[e]	36[e]
1994 constant dollars (mil.)	8	20[e]	18[e]	16[e]	36[e]
Armed forces (000)	5	5	6	6	13
Gross national product (GNP)					
Current dollars (mil.)	665	678	666	690	728
1994 constant dollars (mil.)	740	726	695	704	728
Central government expenditures (CGE)					
1994 constant dollars (mil.)	82	173	155	162	178
People (mil.)	4.2	4.4	4.5	4.5	4.6
Military expenditure as % of GNP	1.1	2.7	2.5	2.2	4.9
Military expenditure as % of CGE	9.8	11.3	11.4	9.8	20.1
Military expenditure per capita (1994 $)	2	4	4	4	8
Armed forces per 1,000 people (soldiers)	1.2	1.1	1.3	1.3	2.8
GNP per capita (1994 $)	175	165	156	156	157
Arms imports[6]					
Current dollars (mil.)	0	0	10	0	0
1994 constant dollars (mil.)	0	0	10	0	0
Arms exports[6]					
Current dollars (mil.)	0	0	0	0	0
1994 constant dollars (mil.)	0	0	0	0	0
Total imports[7]					
Current dollars (mil.)	149	163	146	147	150
1994 constant dollars (mil.)	166	175	152	150	150
Total exports[7]					
Current dollars (mil.)	138	145	149	118	115
1994 constant dollars (mil.)	154	155	155	120	115
Arms as percent of total imports[8]	0	0	6.8	0	0
Arms as percent of total exports[8]	0	0	0	0	0

Human Rights [24]

	SSTS	FL	FAPRO	PPCG	APROBC	TPW	PCPTW	STPEP	PHRFF	PRW	ASST	AFL
Observes	P	P	P		P	P	P			P	P	P
	EAFRD	CPR	ESCR	SR	ACHR	MAAE		PVIAC	PVNAC	EAFDAW	TCIDTP	RC
Observes	P	P	P	P				P	P	P	S	P

P = Party; S = Signatory; see Appendix for meaning of abbreviations.

Labor Force

Total Labor Force [25]

1.369 million (1981 est.); only about
65,000 wage earners (1985)

Labor Force by Occupation [26]

Agriculture	65%
Industry	19
Services	16
Date of data: 1981 est.

Unemployment Rate [27]

For sources, notes, and explanations, see Annotated Source Appendix, page 1061.

837

Production Sectors

Energy Resource Summary [28]

Electricity: Capacity: 130,000 kW. Production: 220 million kWh. Consumption per capita: 44 kWh (1993).

Telecommunications [30]

- 17,526 (1991 est.) telephones; marginal telephone and telegraph service
- Domestic: national microwave radio relay system made unserviceable by military activities
- International: satellite earth station - 1 Intelsat (Atlantic Ocean)
- Radio: Broadcast stations: AM 1, FM 1, shortwave 0 Radios: 980,000 (1992 est.)
- Television: Broadcast stations: 1 Televisions: 45,000 (1992 est.)

Transportation [31]

Railways: total: 84 km used on a limited basis because the mine at Marampa is closed; narrow gauge: 84 km 1.067-m gauge

Highways: total: 11,674 km; paved: 1,284 km; unpaved: 10,390 km (1992 est.)

Merchant marine: none

Airports
Total: 5
With paved runways over 3,047 m: 1
With paved runways 914 to 1,523 m: 2
With unpaved runways 914 to 1,523 m: 2 (1995 est.)

Top Agricultural Products [32]

Agriculture accounts for 40% of the GDP; produces rice, coffee, cocoa, palm kernels, palm oil, peanuts; poultry, cattle, sheep, pigs; fish catch was 65,000 metric tons in 1994.

Top Mining Products [33]

Thousand metric tons except as noted[e]	3/15/95[*]
Bauxite	735[r]
Diamond, gem and industrial (000 carats)	255[63]
Gold (kg.)	123[63]
Salt	200
Titanium, gross weight	
rutile ore and conc. (96% TiO2)	137[r]
ilmenite ore and conc. (60% TiO2)	47[r]

Tourism [34]

	1990	1991	1992	1993	1994
Tourists	98	96	89	91	72
Tourism receipts	19	18	17	18	10
Tourism expenditures	4	4	3	4	4
Fare expenditures	4	1	1	1	NA

Travelers are in thousands, money in million U.S. dollars.

Finance, Economics, and Trade

GDP and Manufacturing Summary [35]

	1980	1985	1990	1991	1992
Gross Domestic Product					
Millions of 1980 dollars	758	828	888	913	946[e]
Growth rate in percent	3.00	8.53	3.01	2.82	3.54[e]
Manufacturing Value Added					
Millions of 1980 dollars	55	54	44	44	44[e]
Growth rate in percent	-5.57	-13.93	-3.96	0.00	0.81[e]
Manufacturing share in percent of current prices	7.5	4.8	9.7	NA	NA

Economic Indicators [36]

- **National product**: GDP—purchasing power parity—$4.4 billion (1994 est.)

- **National product real growth rate**: - 4% (1994 est.)

- **National product per capita**: $960 (1994 est.)

- **Inflation rate (consumer prices)**: 24% (1994 est.)

- **External debt**: $1.4 billion (yearend 1993)

Balance of Payments Summary [37]

Values in millions of dollars.

	1987	1988	1989	1990	1991
Exports of goods (f.o.b.)	138.9	104.5	139.5	139.8	149.5
Imports of goods (f.o.b.)	-114.8	-138.2	-160.4	-140.3	-138.3
Trade balance	24.1	-33.7	-20.9	-0.6	11.2
Services - debits	-105.2	-29.9	-84.6	-146.2	-78.7
Services - credits	44.1	52.1	38.5	70.5	68.4
Private transfers (net)	NA	0.3	0.1	0.1	2.7
Government transfers (net)	6.8	8.5	7.2	0.8	7.1
Long-term capital (net)	44.0	-63.5	-37.4	-16.1	-25.7
Short-term capital (net)	10.4	135.5	60.7	45.5	21.7
Errors and omissions	-21.9	-62.5	29.2	49.2	11.1
Overall balance	2.3	6.8	-7.2	9.2	17.8

Exchange Rates [38]

Currency: **leone.**
Symbol: **Le.**

Data are currency units per $1.

January 1996	951.63
1995	755.22
1994	586.74
1993	567.46
1992	499.44
1991	295.34

Imports and Exports

Top Import Origins [39]

$150 million (c.i.f., 1994).

Origins	%
US	NA
EU	NA
Japan	NA
China	NA
Nigeria	NA

Top Export Destinations [40]

$115 million (f.o.b., 1994).

Destinations	%
US	NA
UK	NA
Belgium	NA
Germany	NA
Other Western Europe	NA

Foreign Aid [41]

Recipient: ODA, $NA.

Import and Export Commodities [42]

Import Commodities	Export Commodities
Foodstuffs 38%	Rutile 51%
Machinery and equipment 44%	Bauxite 20%
Fuels and lubricants 18%	Diamonds 16%
	Coffee 6%
	Cocoa 7%
	Fish

Singapore

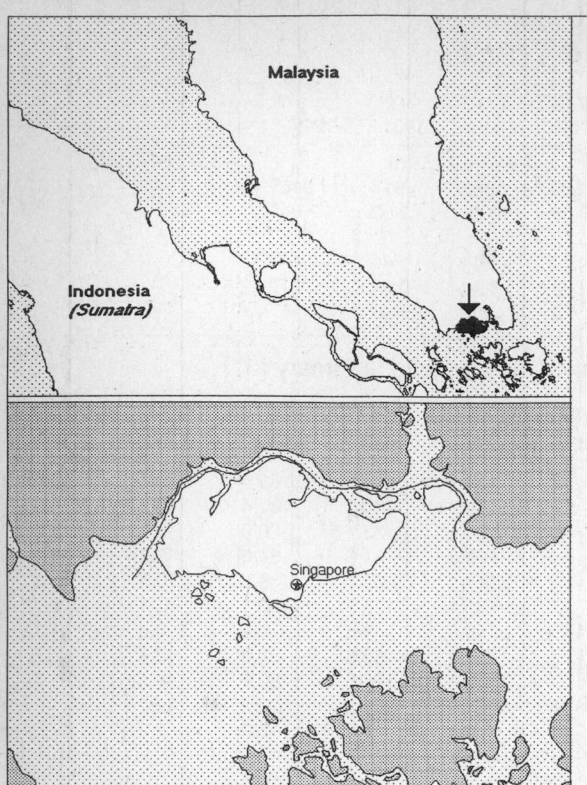

Geography [1]

Total area:
632.6 sq km   244 sq mi
Land area:
622.6 sq km   240 sq mi
Comparative area:
Slightly more than three times the size of Washington, DC
Land boundaries:
0 km
Coastline:
193 km
Climate:
Tropical; hot, humid, rainy; no pronounced rainy or dry seasons; thunderstorms occur on 40% of all days (67% of days in April)
Terrain:
Lowland; gently undulating central plateau contains water catchment area and nature preserve
Natural resources:
Fish, deepwater ports
Land use:
Arable land: 4%
Permanent crops: 7%
Meadows and pastures: 0%
Forest and woodland: 5%
Other: 84%

Demographics [2]

	1970	1980	1990	1995[1]	1996	2000	2010	2020	2030
Population	2,075	2,414	3,039	3,333	3,397	3,620	4,026	4,330	4,567
Population density (persons per sq. mi.)	8,610	10,019	12,612	13,834	14,099	15,027	16,710	17,974	18,956
(persons per sq. km.)	3,325	3,869	4,870	5,341	5,444	5,802	6,452	6,940	7,319
Net migration rate (per 1,000 population)	0.0	NA	6.7	7.4	7.3	2.8	2.5	2.3	2.2
Births	NA	NA	NA	NA	55	NA	NA	NA	NA
Deaths[47]	NA	13	14	NA	15	NA	NA	NA	NA
Life expectancy - males	65.9	69.0	73.6	74.8	75.1	76.1	78.0	79.2	79.9
Life expectancy - females	72.3	74.4	78.8	81.1	81.4	82.5	84.4	85.6	86.2
Birth rate (per 1,000)	22.1	17.1	16.8	16.2	16.3	15.6	11.1	11.2	10.7
Death rate (per 1,000)	5.2	5.2	4.6	4.6	4.6	4.7	5.4	6.9	9.2
Women of reproductive age (15-49 yrs.)	NA	NA	934	1,011	1,025	1,049	1,020	951	936
of which are currently married[48]	271	362	525	NA	630	668	622	NA	NA
Fertility rate	NA	1.7	1.6	1.6	1.7	1.8	1.8	1.8	1.7

Except as noted, values for vital statistics are in thousands; life expectancy is in years.

Health

Health Indicators [3]

% of population with access to	
safe water (1990-95)[1]	100
adequate sanitation (1990-95)[1]	99
health services (1985-95)	100
% of 1-year-olds immunized (1990-94) against	
TB (tuberculosis)	98
DPT (diphtheria, pertussis, tetanus)	92
polio	92
measles	87
% of contraceptive prevalence (1980-94)	74
ORT use rate (1990-94)	NA

Health Expenditures [4]

Total health expenditure, 1990 (official exchange rate)	
Millions of dollars	658
Dollars per capita	219
Health expenditures as a percentage of GDP	
Total	1.9
Public sector	1.1
Private sector	0.8
Development assistance for health	
Total aid flows (millions of dollars)[1]	1
Aid flows per capita (dollars)	0.2
Aid flows as a percentage of total health expenditure	0.1

For sources, notes, and explanations, see Annotated Source Appendix, page 1061.

Human Factors

Women and Children [5]

% of pregnant women immunized (tetanus 1990-94)	NA
% of births attended by trained health personnel (1983-94)	100
Maternal mortality rate (1980-92)	10
Under-5 mortality rate (1994)	6
% under-5 moderately/severely underweight (1980-1994)[1]	14

Burden of Disease [6]

Population per physician (1994)	666.67
Population per nurse	NA
Population per hospital bed (1990)	275.21
AIDS cases per 100,000 people (1994)	1.7
Malaria cases per 100,000 people (1992)	NA

Ethnic Division [7]

Chinese	76.4%
Malay	14.9%
Indian	6.4%
Other	2.3%

Religion [8]

Buddhist (Chinese), Muslim (Malays), Christian, Hindu, Sikh, Taoist, Confucianist.

Major Languages [9]

Chinese (official), Malay (official and national), Tamil (official), English (official).

Education

Public Education Expenditures [10]

	1980	1985	1990	1992	1993	1994
Million (Dollar)						
Total education expenditure	686	1,776	2,055	2,598	2,884	3,416
as percent of GNP	2.8	4.4	3.1	3.2	3.1	3.3
as percent of total govt. expend.	7.3	NA	18.2	21.2	23.0	24.2
Current education expenditure	587	1,388	1,795	2,043	2,192	2,486
as percent of GNP	2.4	3.4	2.7	2.5	2.4	2.4
as percent of current govt. expend.	10.3	NA	25.4	24.1	24.0	25.6
Capital expenditure	99	387	261	555	693	931

Educational Attainment [11]

Age group (1990)[11]	25+
Total population	1,596,600
Highest level attained (%)	
No schooling	64.0
First level	
Not completed	NA
Completed	NA
Entered second level	
S-1	23.2
S-2	8.1
Postsecondary	4.7

Illiteracy [12]

In thousands and percent[1]	1990	1995	2000
Illiterate population (15+ yrs.)	227	196	165
Illiteracy rate - total pop. (%)	9.6	7.5	5.9
Illiteracy rate - males (%)	4.3	3.5	2.8
Illiteracy rate - females (%)	14.8	11.4	8.8

Libraries [13]

	Admin. Units	Svc. Pts.	Vols. (000)	Shelving (meters)	Vols. Added	Reg. Users
National (1992)[6]	1	8	2,819	NA	164,520	875,056
Nonspecialized	NA	NA	NA	NA	NA	NA
Public	NA	NA	NA	NA	NA	NA
Higher ed. (1990)[22]	5	12	2,354	14,865	137,181	77,934
School (1990)	366	NA	4,640	NA	NA	440,042

Daily Newspapers [14]

	1980	1985	1990	1994
Number of papers	12	10	8	8
Circ. (000)	690	706[e]	763	1,027

Culture [15]

Science and Technology

Scientific/Technical Forces [16]

Scientists/engineers	7,086
Number female	NA
Technicians[2]	4,298
Number female	NA
Total[2]	11,384

R&D Expenditures [17]

	Dollar (000) 1994
Total expenditure	1,170,000
Capital expenditure	NA
Current expenditure	NA
Percent current	NA

U.S. Patents Issued [18]

Values show patents issued to citizens of the country by the U.S. Patents Office.

	1993	1994	1995
Number of patents	44	59	61

For sources, notes, and explanations, see Annotated Source Appendix, page 1061.

Government and Law

Organization of Government [19]

Long-form name:
Republic of Singapore
Type:
Republic within Commonwealth
Independence:
9 August 1965 (from Malaysia)
National holiday:
National Day, 9 August (1965)
Constitution:
3 June 1959, amended 1965 (based on preindependence State of Singapore Constitution)
Legal system:
Based on English common law; has not accepted compulsory ICJ jurisdiction
Executive branch:
President; Prime Minister; 2 Deputy Prime Ministers; Cabinet
Legislative branch:
Unicameral: Parliament
Judicial branch:
Supreme Court

Elections [20]

Parliament	% of seats
People's Action Party	95.1
Singapore Democratic Party	3.7
Workers' Party	1.2

Government Expenditures [21]

Housing - 6.6%
Educ./Health - 32.8%
Defense - 28.9%
Industry - 8.9%
Services - 14.9%
Other - 7.9%

(% distribution). Expend. for FY94: 15,059 (Dollar mil)

Crime [23]

	1994
Crime volume	
Cases known to police	40,065
Attempts (percent)	0.50
Percent cases solved	25.00
Crimes per 100,000 persons	1,367.31
Persons responsible for offenses	
Total number offenders	13,873
Percent female	NA
Percent juvenile (7-16 yrs.)	15.10
Percent foreigners	13.50

Military Expenditures and Arms Transfers [22]

	1990	1991	1992	1993	1994
Military expenditures					
Current dollars (mil.)	1,959	2,488	2,775	2,900	3,064
1994 constant dollars (mil.)	2,180	2,667	2,894	2,960	3,064
Armed forces (000)	56	56	56	56	56
Gross national product (GNP)					
Current dollars (mil.)	45,420	50,540	55,400	61,880	68,660
1994 constant dollars (mil.)	50,550	54,180	57,770	63,150	68,660
Central government expenditures (CGE)					
1994 constant dollars (mil.)	10,950	12,290	11,610	12,860	13,100[e]
People (mil.)	2.7	2.8	2.8	2.8	2.9
Military expenditure as % of GNP	4.3	4.9	5.0	4.7	4.5
Military expenditure as % of CGE	19.9	21.7	24.9	23.0	23.4
Military expenditure per capita (1994 $)	802	967	1,036	1,047	1,072
Armed forces per 1,000 people (soldiers)	20.6	20.3	20.1	19.8	19.6
GNP per capita (1994 $)	18,590	19,650	20,690	22,340	24,020
Arms imports[6]					
Current dollars (mil.)	230	360	180	180	270
1994 constant dollars (mil.)	256	386	188	184	270
Arms exports[6]					
Current dollars (mil.)	30	50	30	20	20
1994 constant dollars (mil.)	33	54	31	20	20
Total imports[7]					
Current dollars (mil.)	60,900	66,290	72,180	85,230	102,700
1994 constant dollars (mil.)	67,780	71,060	75,270	86,990	102,700
Total exports[7]					
Current dollars (mil.)	52,750	59,020	63,480	74,010	96,830
1994 constant dollars (mil.)	58,710	63,270	66,200	75,540	96,830
Arms as percent of total imports[8]	0.4	0.5	0.2	0.2	0.3
Arms as percent of total exports[8]	0.1	0.1	0	0	0

Human Rights [24]

	SSTS	FL	FAPRO	PPCG	APROBC	TPW	PCPTW	STPEP	PHRFF	PRW	ASST	AFL
Observes		P		P	P	P	P	P			P	P
	EAFRD	CPR	ESCR	SR	ACHR	MAAE	PVIAC	PVNAC	EAFDAW	TCIDTP	RC	
Observes								P	P		P	

P = Party; S = Signatory; see Appendix for meaning of abbreviations.

Labor Force

Total Labor Force [25]

1.649 million (1994)

Labor Force by Occupation [26]

Financial, business, and other services	33.5%
Manufacturing	25.6
Commerce	22.9
Construction	6.6
Other	11.4

Date of data: 1994

Unemployment Rate [27]

2.6% (1995 est.)

For sources, notes, and explanations, see Annotated Source Appendix, page 1061.

Production Sectors

Energy Resource Summary [28]

Electricity: Capacity: 4,510,000 kW. Production: 17 billion kWh. Consumption per capita: 5,590 kWh (1993).

Telecommunications [30]

- 1.23 million (1993 est.) telephones; good domestic facilities; good international service
- International: submarine cables to Malaysia (Sabah and Peninsular Malaysia), Indonesia, and the Philippines; satellite earth stations - 2 Intelsat (1 Indian Ocean and 1 Pacific Ocean), and 1 Inmarsat (Pacific Ocean region)
- Radio: Broadcast stations: AM 13, FM 4, shortwave 0
- Television: Broadcast stations: 2 (1987 est.) Televisions: 1.05 million (1992 est.)

Top Agricultural Products [32]

Agriculture accounts for a negligible percentage of the GDP; produces rubber, copra, fruit, vegetables; poultry.

Top Mining Products [33]

Detailed information is not available. A summary of mineral resources follows. **Mineral Resources**: Refineries produced more than 1 million barrels of petroleum products per day.

Transportation [31]

Railways: total: 38.6 km; narrow gauge: 38.6 km 1.000-m gauge

Highways: total: 2,989 km; paved: 2,905 km (including 111.6 km of expressways); unpaved: 84 km (1994 est.)

Merchant marine: total: 646 ships (1,000 GRT or over) totaling 12,915,788 GRT/20,292,580 DWT; ships by type: bulk 110, cargo 118, chemical tanker 18, combination bulk 3, combination ore/oil 8, container 92, liquefied gas tanker 13, multifunction large-load carrier 4, oil tanker 234, refrigerated cargo 5, roll-on/roll-off cargo 13, short-sea passenger 1, specialized tanker 3, vehicle carrier 24

Airports

Total:	8
With paved runways over 3,047 m:	2
With paved runways 2,438 to 3,047 m:	1
With paved runways 1,524 to 2,437 m:	3
With paved runways 914 to 1,523 m:	1
With paved runways under 914 m:	1 (1995 est.)

Tourism [34]

	1990	1991	1992	1993	1994
Visitors[96]	5,323	5,415	5,990	6,426	6,899
Tourists	4,842	4,913	5,446	5,804	6,268
Excursionists[28]	481	502	544	622	631
Cruise passengers	8	17	24	10	NA
Tourism receipts	4,596	4,560	5,580	6,289	7,067
Tourism expenditures	1,804	1,944	2,417	3,022	3,665

Travelers are in thousands, money in million U.S. dollars.

Manufacturing Sector

Manufacturing Summary [35]

	1987		1988		1989		1990		1991	
	$ bil.	%	$ bil.	%	$ bil.	%	$ bil.	%	$ bil.	%
Establishments or enterprises (number)	3,851	0.164	4,014	0.191	4,052	0.216	4,083	0.228	4,179	0.547
Total employment (000)	349	0.257	414	0.302	431	0.350	450	0.407	462	0.665
Production workers (000)	-	-	-	-	-	-	-	-	-	-
Output ($ bil.)	24	0.235	32	0.276	37	0.316	44	0.380	48	0.472
Value added ($ bil.)	8	0.187	11	0.219	12	0.247	14	0.279	16	0.457
Capital investment ($ mil.)	1,584	0.364	2,070	0.435	2,590	0.473	2,734	0.492	2,717	0.713
M & E investment ($ mil.)	-	-	-	-	-	-	-	-	-	-
Employees per establishment (number)	91	156.742	103	158.104	106	161.979	110	178.433	111	121.636
Production workers per establishment	-	-	-	-	-	-	-	-	-	-
Output per establishment ($ mil.)	6	143.122	8	144.182	9	145.994	11	166.677	12	86.338
Capital investment per estab. ($ mil.)	0.411	221.457	0.516	227.300	0.639	218.608	0.670	215.785	0.650	130.486
M & E per establishment ($ mil)	-	-	-	-	-	-	-	-	-	-
Payroll per employee ($)	7,714	86.029	8,668	87.009	10,115	90.537	11,697	84.321	13,287	90.998
Wages per production worker ($)	-	-	-	-	-	-	-	-	-	-
Hours per production worker (hours)	-	-	-	-	-	-	-	-	-	-
Output per employee ($)	68,387	91.310	77,675	91.194	86,381	90.131	96,806	93.412	104,206	70.981
Capital investment per employee ($)	4,537	141.287	5,003	143.766	6,006	134.961	6,073	120.934	5,881	107.275
M & E per employee ($)	-	-	-	-	-	-	-	-	-	-

Note: Columns headed % show percent of world total or ratio. Ratios closest to 100 are closest to world average. M & E stands for machinery & equipment.

Output in Manufacturing

	1987		1988		1989		1990		1991	
	$ bil.[f]	%	$ bil.[f]	%	$ bil.[f]	%	$ bil.[f]	%	$ bil.[f]	%
3110 - Food products	0.889	3.70	1.049	3.28	1.143	3.09	1.128	2.56	1.217	2.54
3130 - Beverages	0.194	0.81	0.231	0.72	0.276	0.75	0.282	0.64	0.304	0.63
3140 - Tobacco	0.082	0.34	0.093	0.29	0.113	0.31	0.159	0.36	0.192	0.40
3210 - Textiles	0.151	0.63	0.179	0.56	0.203	0.55	0.218	0.50	0.233	0.49
3211 - Spinning, weaving, etc.	0.085	0.35	0.099	0.31	0.108	0.29	0.116	0.26	0.123	0.26
3220 - Wearing apparel	0.742	3.09	0.842	2.63	0.918	2.48	0.954	2.17	1.007	2.10
3230 - Leather and products	0.018	0.08	0.029	0.09	0.032	0.09	0.030	0.07	0.036	0.08
3240 - Footwear	0.020	0.08	0.022	0.07	0.025	0.07	0.028	0.06	0.031	0.06
3310 - Wood products	0.152	0.63	0.194	0.61	0.204	0.55	0.193	0.44	0.172	0.36
3320 - Furniture, fixtures	0.178	0.74	0.231	0.72	0.250	0.68	0.285	0.65	0.299	0.62
3410 - Paper and products	0.281	1.17	0.348	1.09	0.390	1.05	0.450	1.02	0.504	1.05
3420 - Printing, publishing	0.531	2.21	0.663	2.07	0.796	2.15	0.962	2.19	1.117	2.33
3510 - Industrial chemicals	1.133	4.72	1.619	5.06	1.573	4.25	1.738	3.95	1.790	3.73
3520 - Chemical products nec	0.606	2.53	0.710	2.22	0.817	2.21	0.978	2.22	1.304	2.72
3550 - Rubber products	0.128	0.53	0.190	0.59	0.188	0.51	0.150	0.34	0.135	0.28
3560 - Plastic products nec	0.400	1.67	0.613	1.92	0.692	1.87	0.788	1.79	0.931	1.94
3690 - Nonmetal products nec	0.303	1.26	0.316	0.99	0.370	1.00	0.476	1.08	0.655	1.36
3710 - Iron and steel	0.182	0.76	0.213	0.67	0.270	0.73	0.321	0.73	0.310	0.65
3720 - Nonferrous metals	0.079	0.33	0.102	0.32	0.095	0.26	0.119	0.27	0.115	0.24
3810 - Metal products	1.131	4.71	1.546	4.83	1.819	4.92	2.099	4.77	2.390	4.98
3820 - Machinery nec	0.931	3.88	1.233	3.85	1.567	4.24	1.865	4.24	2.155	4.49
3830 - Electrical machinery	8.667	36.11	12.000	37.50	14.000	37.84	17.000	38.64	18.000	37.50
3832 - Radio, television, etc.	4.755	19.81	6.747	21.08	7.745	20.93	8.852	20.12	9.724	20.26
3840 - Transportation equipment	0.994	4.14	1.351	4.22	1.623	4.39	2.092	4.75	2.323	4.84
3841 - Shipbuilding, repair	0.608	2.53	0.857	2.68	1.148	3.10	1.434	3.26	1.620	3.38
3843 - Motor vehicles	0.040	0.17	0.068	0.21	0.074	0.20	0.100	0.23	0.083	0.17
3850 - Professional goods	0.217	0.90	0.286	0.89	0.345	0.93	0.432	0.98	0.498	1.04
3900 - Industries nec	0.384	1.60	0.491	1.53	0.521	1.41	0.605	1.37	0.582	1.21

Note: Codes are International Standard Industry codes (ISIC). Percentages are % of total Output. [f]: Factor Prices; [p]: Producer Prices.

Finance, Economics, and Trade

Economic Indicators [36]

- **National product**: GDP—purchasing power parity— $66.1 billion (1995 est.)

- **National product real growth rate**: 8.9% (1995)

- **National product per capita**: $22,900 (1995 est.)

- **Inflation rate (consumer prices)**: 1.7% (1995)

- **External debt**: $3.2 million (1994)

Balance of Payments Summary [37]

Values in millions of dollars.

	1989	1990	1991	1992	1993
Exports of goods (f.o.b.)	43,239	51,095	57,156	62,068	71,959
Imports of goods (f.o.b.)	-45,687	-55,812	-60,948	-67,850	-80,025
Trade balance	-2,448	-4,717	-3,792	-5,782	-8,066
Services - debits	-10,970	-14,652	-16,003	-17,075	-19,169
Services - credits	16,334	21,907	24,285	27,200	29,970
Private transfers (net)	-254	-274	-340	-405	-482
Government transfers (net)	-125	-169	-159	-190	-215
Long-term capital (net)	1,660	1,709	2,121	4,918	4,227
Short term captal (net)	-1,318	2,949	-1,187	641	5,225
Errors and omissions	-90	-1,322	-728	-3,208	-3,913
Overall balance	2,737	5,431	4,197	6,099	7,577

Exchange Rates [38]

Currency: **Singapore dollar.**
Symbol: **S$.**

Data are currency units per $1.

January 1996	1.4214
1995	1.4174
1994	1.5274
1993	1.6158
1992	1.6290
1991	1.7276

Imports and Exports

Top Import Origins [39]

$125.9 billion (1995) Data are for 1994.

Origins	%
Japan	22
Malaysia	16
US	15
Taiwan	4
Saudi Arabia	4

Top Export Destinations [40]

$119.6 billion (1995) Data are for 1994

Destinations	%
Malaysia	20
US	19
Hong Kong	9
Japan	7
Thailand	6

Foreign Aid [41]

	U.S. $	
US commitments, including Ex-Im (FY70-83)	590	million
Western (non-US) countries, ODA and OOF bilateral commitments (1970-89)	1	billion

Import and Export Commodities [42]

Import Commodities	**Export Commodities**
Aircraft	Computer equipment
Petroleum	Rubber and rubber products
Chemicals	Petroleum products
Foodstuffs	Telecommunications equipment

Slovakia

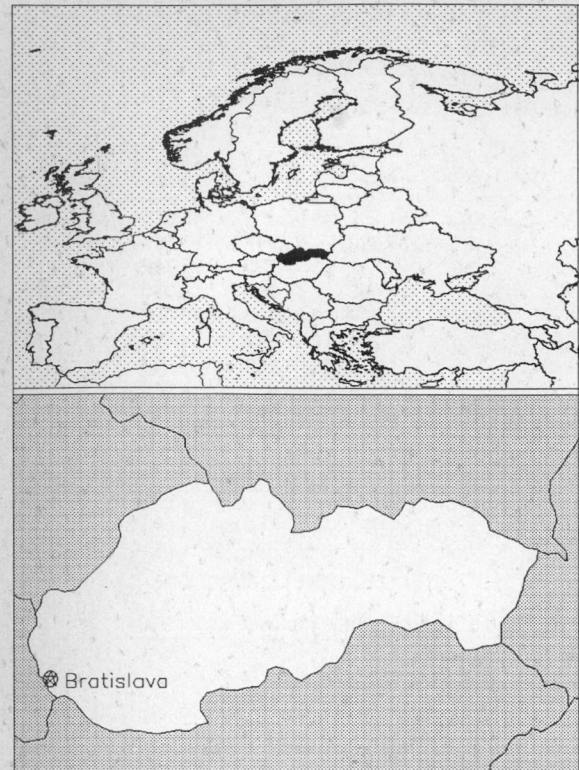

Geography [1]

Total area:
  48,845 sq km   18,859 sq mi
Land area:
  48,800 sq km   18,842 sq mi
Comparative area:
  About twice the size of New Hampshire
Land boundaries:
  Total 1,355 km, Austria 91 km, Czech Republic 215 km, Hungary 515 km, Poland 444 km, Ukraine 90 km
Coastline:
  0 km (landlocked)
Climate:
  Temperate; cool summers; cold, cloudy, humid winters
Terrain:
  Rugged mountains in the central and northern part and lowlands in the South
Natural resources:
  Brown coal and lignite; small amounts of iron ore, copper and manganese ore; salt
Land use:
  Arable land: NA
  Permanent crops: NA
  Meadows and pastures: NA
  Forest and woodland: NA
  Other: NA

Demographics [2]

	1970	1980	1990	1995[1]	1996	2000	2010	2020	2030
Population	4,524	4,966	5,263	5,357	5,374	5,472	5,735	5,837	5,881
Population density (persons per sq. mi.)	239	262	278	283	284	289	303	308	311
(persons per sq. km.)	92	101	107	109	110	112	117	119	120
Net migration rate (per 1,000 population)	NA	NA	NA	0.1	0.1	0.1	0.0	0.0	0.0
Births	NA	NA	NA	NA	68	NA	NA	NA	NA
Deaths	NA	NA	NA	NA	50	NA	NA	NA	NA
Life expectancy - males	NA	NA	NA	68.9	69.0	69.5	70.5	73.6	76.1
Life expectancy - females	NA	NA	NA	77.1	77.2	77.7	79.0	81.2	82.9
Birth rate (per 1,000)	NA	NA	NA	12.5	12.6	15.5	12.8	10.6	10.3
Death rate (per 1,000)	NA	NA	NA	9.3	9.4	9.5	10.0	9.5	10.3
Women of reproductive age (15-49 yrs.)	NA	NA	NA	1,403	1,417	1,444	1,399	1,402	1,315
of which are currently married	NA	NA	NA	NA	939	967	971	NA	NA
Fertility rate	NA	NA	NA	1.7	1.7	2.0	1.8	1.7	1.6

Except as noted, values for vital statistics are in thousands; life expectancy is in years.

Health

Health Indicators [3]

% of population with access to	
safe water (1990-95)	NA
adequate sanitation (1990-95)	NA
health services (1985-95)	NA
% of 1-year-olds immunized (1990-94) against	
TB (tuberculosis)	91
DPT (diphtheria, pertussis, tetanus)	98
polio	98
measles	97
% of contraceptive prevalence (1980-94)	74
ORT use rate (1990-94)	NA

Health Expenditures [4]

Total health expenditure, 1990 (official exchange rate)[2]	
Millions of dollars	2,711
Dollars per capita	173
Health expenditures as a percentage of GDP	
Total	5.9
Public sector	5.0
Private sector	0.9
Development assistance for health	
Total aid flows (millions of dollars)[1]	NA
Aid flows per capita (dollars)	NA
Aid flows as a percentage of total health expenditure	NA

For sources, notes, and explanations, see Annotated Source Appendix, page 1061.

Human Factors

Women and Children [5]

% of pregnant women immunized (tetanus 1990-94)	NA
% of births attended by trained health personnel (1983-94)	NA
Maternal mortality rate (1980-92)	NA
Under-5 mortality rate (1994)	15
% under-5 moderately/severely underweight (1980-1994)	NA

Burden of Disease [6]

Population per physician (1993)	286.64
Population per nurse (1993)	105.24
Population per hospital bed (1993)	11.02
AIDS cases per 100,000 people (1994)	NA
Heart disease cases per 1,000 people (1990-93)	0.1

Ethnic Division [7]

Other includes German, Polish, Ruthenian, Ukranian, and others.

Slovak	85.7%
Hungarian	10.7%
Gypsy	1.5%
Czech	1%
Other	1.1%

Religion [8]

Roman Catholic	60.3%
Atheist	9.7%
Protestant	8.4%
Orthodox	4.1%
Other	17.5%

Major Languages [9]

Slovak (official), Hungarian.

Education

Public Education Expenditures [10]

	1980	1988	1990	1992	1993	1994
Million (Koruna)[20]						
Total education expenditure	NA	32,254	NA	19,829	19,400	19,495
as percent of GNP	NA	4.4	NA	6.5	5.7	4.9
as percent of total govt. expend.	NA	8.0	NA	NA	NA	NA
Current education expenditure	NA	31,005	NA	17,575	16,966	17,152
as percent of GNP	NA	4.2	NA	5.7	5.0	4.3
as percent of current govt. expend.	NA	8.7	NA	NA	NA	NA
Capital expenditure	NA	1,249	NA	2,254	2,434	2,343

Educational Attainment [11]

Age group (1991)	25+
Total population	3,144,143
Highest level attained (%)	
No schooling	0.7
First level	
Not completed	37.9
Completed	NA
Entered second level	
S-1	50.9
S-2	NA
Postsecondary	9.5

Illiteracy [12]

Libraries [13]

	Admin. Units	Svc. Pts.	Vols. (000)	Shelving (meters)	Vols. Added	Reg. Users
National (1992)	1	1	3,580	NA	32,420	4,806
Nonspecialized	NA	NA	NA	NA	NA	NA
Public (1992)	2,682	3,012	19,757	NA	588,012	715,002
Higher ed. (1993)[10]	33	546	4,738	NA	104,183	108,334
School	NA	NA	NA	NA	NA	NA

Daily Newspapers [14]

	1980	1985	1990[2]	1994
Number of papers	NA	NA	18	21
Circ. (000)	NA	NA	1,410	1,363

Culture [15]

Cinema (seats per 1,000)	28.2
Annual attendance per person	4.7
Gross box office receipts (mil. Koruna)	NA
Museums (reporting)	63
Visitors (000)	2,825
Annual receipts (000 Koruna)	13,300

Science and Technology

Scientific/Technical Forces [16]

Scientists/engineers	10,249
Number female	NA
Technicians	4,244
Number female	NA
Total[9]	14,493

R&D Expenditures [17]

	Koruna (000) 1994
Total expenditure[6]	4,473,412
Capital expenditure	385,684
Current expenditure	4,087,728
Percent current	91.4

U.S. Patents Issued [18]

For sources, notes, and explanations, see Annotated Source Appendix, page 1061.

847

Government and Law

Organization of Government [19]

Long-form name:
Slovak Republic
Type:
Parliamentary democracy
Independence:
1 January 1993 (from Czechoslovakia)
National holiday:
Slovak Constitution Day, 1 September
(1992)
Constitution:
Ratified 1 September 1992, fully effective
1 January 1993
Legal system:
Civil law system based on Austro-
Hungarian codes; has not accepted
compulsory ICJ jurisdiction; legal code
modified to comply with the obligations of
Organization on Security and Cooperation
in Europe (OSCE) and to expunge
Marxist-Leninist legal theory
Executive branch:
President; Prime Minister; Cabinet
Legislative branch:
Unicameral: National Parliament (Narodni
Rada)
Judicial branch:
Supreme Court

Elections [20]

National Council	% of votes
Movement for a Dem. Slovakia	35.0
Common Choice	10.4
Hungarian Coalition	10.2
Christian Democratic Movement	10.1
Democratic Union	8.6
Association of Slovak Workers	7.3
Slovak National Party	5.4

Government Budget [21]

For 1995 est.

	$ bil.
Revenues	6.1
Expenditures	6.4
Capital expenditures	NA

Military Expenditures and Arms Transfers [22]

	1990	1991	1992	1993[16]	1994
Military expenditures					
Current dollars (mil.)[e]	NA	NA	NA	843	866
1994 constant dollars (mil.)[e]	NA	NA	NA	860	866
Armed forces (000)	NA	NA	NA	33	47
Gross national product (GNP)					
Current dollars (mil.)[e]	NA	NA	NA	34,470	35,520
1994 constant dollars (mil.)[e]	NA	NA	NA	35,190	35,520
Central government expenditures (CGE)					
1994 constant dollars (mil.)[e]	NA	NA	NA	16,540	13,830
People (mil.)	NA	NA	NA	5.4	5.4
Military expenditure as % of GNP	NA	NA	NA	2.4	2.4
Military expenditure as % of CGE	NA	NA	NA	5.2	6.3
Military expenditure per capita (1994 $)	NA	NA	NA	160	160
Armed forces per 1,000 people (soldiers)	NA	NA	NA	6.1	8.7
GNP per capita (1994 $)	NA	NA	NA	6,546	6,573
Arms imports[6]					
Current dollars (mil.)	NA	NA	NA	230	0
1994 constant dollars (mil.)	NA	NA	NA	235	0
Arms exports[6]					
Current dollars (mil.)	NA	NA	NA	50	10
1994 constant dollars (mil.)	NA	NA	NA	51	10
Total imports[7]					
Current dollars (mil.)	NA	NA	NA	6,655	6,823
1994 constant dollars (mil.)	NA	NA	NA	6,792	6,823
Total exports[7]					
Current dollars (mil.)	NA	NA	NA	5,451	6,587
1994 constant dollars (mil.)	NA	NA	NA	5,563	6,587
Arms as percent of total imports[8]	NA	NA	NA	3.5	0
Arms as percent of total exports[8]	NA	NA	NA	0.9	0.2

Crime [23]

	1994
Crime volume	
Cases known to police	137,712
Attempts (percent)	2.30
Percent cases solved	39.07
Crimes per 100,000 persons	2,571.07
Persons responsible for offenses	
Total number offenders	48,784
Percent female	7.20
Percent juvenile (15-18 yrs.)	11.10
Percent foreigners	1.70

Human Rights [24]

	SSTS	FL	FAPRO	PPCG	APROBC	TPW	PCPTW	STPEP	PHRFF	PRW	ASST	AFL
Observes	2	P	P	P	P	P	P	P	P	P	P	

	EAFRD	CPR	ESCR	SR	ACHR	MAAE	PVIAC	PVNAC	EAFDAW	TCIDTP	RC
Observes	P	P	P	P			P	P		P	P

P = Party; S = Signatory; see Appendix for meaning of abbreviations.

Labor Force

Total Labor Force [25]

2.484 million

Labor Force by Occupation [26]

Industry	33.2%
Agriculture	12.2
Construction	10.3
Communication and other	44.3
Date of data: 1990	

Unemployment Rate [27]

13% (1995 est.)

Production Sectors

Commercial Energy Production and Consumption

Data are shown in quadrillion (10^{15}) BTUs and percent for 1995
Values for hydroelectric, nuclear, geothermal, solar, and wind power refer to electrical generation.

Production [28]	Consumption [29]
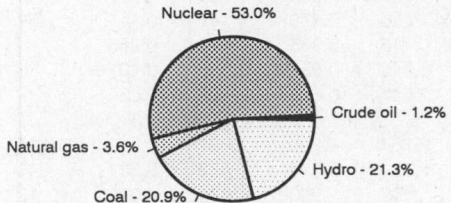	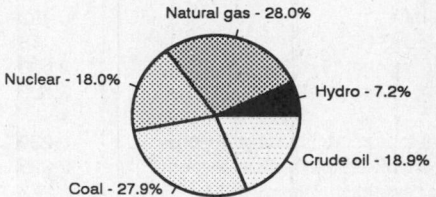

Production [28]
- Nuclear - 53.0%
- Crude oil - 1.2%
- Natural gas - 3.6%
- Hydro - 21.3%
- Coal - 20.9%

Consumption [29]
- Natural gas - 28.0%
- Nuclear - 18.0%
- Hydro - 7.2%
- Crude oil - 18.9%
- Coal - 27.9%

Crude oil	0.003		Crude oil	0.141
Dry natural gas	0.009		Dry natural gas	0.209
Coal	0.053		Coal	0.208
Net hydroelectric power	0.054		Net hydroelectric power	0.054
Net nuclear power	0.134		Net nuclear power	0.134
Total	0.253		Total	0.746

Telecommunications [30]

- 1,362,178 (1992 est.) telephones
- Radio: Broadcast stations: AM NA, FM NA, shortwave NA; note - there is 1 station of NA type Radios: 1.1 million (1992 est.)
- Television: Broadcast stations: 1 Televisions: 1.6 million (1994 est.)

Transportation [31]

Railways: total: 3,660 km; broad gauge: 102 km 1.520-m gauge; standard gauge: 3,507 km 1.435-m gauge (1378 km electrified); narrow gauge: 51 km (46 km 1,000-m gauge; 5 km 0.750-m gauge) (1995)

Highways: total: 17,737 km; paved: NA km; unpaved: NA km (1993 est.)

Merchant marine: total: 4 cargo ships (1,000 GRT or over) totaling 17,010 GRT/22,039 DWT (1995 est.)

Airports

Total:	37
With paved runways over 3,047 m:	1
With paved runways 2,438 to 3,047 m:	3
With paved runways 1,524 to 2,437 m:	2
With paved runways 914 to 1,523 m:	2
With paved runways under 914 m:	4

Top Agricultural Products [32]

Agriculture accounts for 6.7% of the GDP; produces grains, potatoes, sugar beets, hops, fruit; hogs, cattle, poultry; forest products.

Top Mining Products [33]

Thousand metric tons except as noted[e]	7/95*
Gallium (kg.)	600
Pig iron	3,330[r]
Steel, crude	3,790[r]
Diamond, synthetic (carats)	5,000
Dolomite	4,000
Magnesite, crude	1,200
Limestone and other calcareous stones	4,500
Quarry stone (000 cu. meters)	5,000
Gas, manufactured, coke oven (mil. cu. meters)	900
Petroleum refinery products (000 42-gal. bls.)	40,500

Tourism [34]

	1990	1991	1992	1993	1994
Visitors[97]	NA	13,938	15,700	12,900	21,900
Tourists[98]	822	635	566	653	902
Tourism receipts[52]	70	135	213	390	568
Tourism expenditures[52]	NA	NA	155	262	284
Fare receipts	NA	NA	NA	NA	7
Fare expenditures	NA	NA	NA	NA	12

Travelers are in thousands, money in million U.S. dollars.

For sources, notes, and explanations, see Annotated Source Appendix, page 1061.

849

Manufacturing Sector

Manufacturing Summary [35]

	1987		1988		1989		1990		1991	
	$ bil.	%	$ bil.	%	$ bil.	%	$ bil.	%	$ bil.	%
Establishments or enterprises (number)	853	0.036	944	0.045	990	0.053	1,399	0.078	1,743	0.228
Total employment (000)	2,763	2.036	2,755	2.013	2,736	2.223	2,603	2.352	2,081	2.996
Production workers (000)	1,983	2.940	1,969	3.152	1,954	2.656	1,863	2.625	1,465	4.449
Output ($ bil.)	62	0.612	60	0.518	56	0.475	47	0.411	21	0.203
Value added ($ bil.)	18	0.421	18	0.377	16	0.336	13	0.255	-	-
Capital investment ($ mil.)	3,500	0.803	3,673	0.771	3,443	0.629	3,538	0.637	-	-
M & E investment ($ mil.)	2,569	0.824	2,731	0.789	2,547	0.652	2,665	0.632	-	-
Employees per establishment (number)	3,239	5,599	2,918	4,477	2,764	4,206	1,861	3,011	1,194	1,314
Production workers per establishment	2,325	8,086	2,086	7,008	1,974	5,026	1,332	3,361	841	1,951
Output per establishment ($ mil.)	73	1,684	64	1,151	57	900	34	526	12	89
Capital investment per estab. ($ mil.)	4	2,208	4	1,715	3	1,189	3	815	-	-
M & E per establishment ($ mil)	3	2,267	3	1,755	3	1,234	2	809	-	-
Payroll per employee ($)	2,584	28.815	2,536	25.455	2,490	22.285	2,164	15.600	1,587	10.867
Wages per production worker ($)	2,516	31.883	2,458	29.094	2,412	24.053	2,101	17.692	1,508	13.707
Hours per production worker (hours)	1,890	99.616	1,874	97.494	1,843	100.552	1,810	96.761	1,668	93.318
Output per employee ($)	22,525	30.076	21,897	25.708	20,497	21.387	18,093	17.459	9,964	6.787
Capital investment per employee ($)	1,267	39.441	1,333	38.311	1,258	28.278	1,359	27.069	-	-
M & E per employee ($)	930	40.496	991	39.199	931	29.346	1,024	26.875	-	-

Note: Columns headed % show percent of world total or ratio. Ratios closest to 100 are closest to world average. M & E stands for machinery & equipment.

Output in Manufacturing

	1987		1988		1989		1990		1991	
	$ bil.	%	$ bil.	%	$ bil.	%	$ bil.	%	$ bil.	%
3110 - Food products	9.375	15.12	8.918	14.86	10.000	17.86	6.633	14.11	3.818	18.18
3130 - Beverages	0.955	1.54	0.915	1.52	1.017	1.82	0.783	1.67	0.628	2.99
3140 - Tobacco	0.155	0.25	0.150	0.25	0.160	0.29	0.129	0.27	0.088	0.42
3210 - Textiles	2.923	4.71	2.843	4.74	2.740	4.89	2.283	4.86	0.967	4.60
3211 - Spinning, weaving, etc.	2.317	3.74	2.245	3.74	2.178	3.89	1.802	3.83	0.773	3.68
3220 - Wearing apparel	0.795	1.28	0.719	1.20	0.698	1.25	0.575	1.22	0.199	0.95
3230 - Leather and products	0.390	0.63	0.378	0.63	0.403	0.72	0.320	0.68	0.162	0.77
3240 - Footwear	0.841	1.36	0.822	1.37	0.876	1.56	0.640	1.36	0.249	1.19
3310 - Wood products	1.087	1.75	1.058	1.76	1.060	1.89	0.697	1.48	0.381	1.81
3320 - Furniture, fixtures	0.576	0.93	0.566	0.94	0.581	1.04	0.413	0.88	0.222	1.06
3410 - Paper and products	1.329	2.14	1.308	2.18	1.336	2.39	1.040	2.21	0.555	2.64
3420 - Printing, publishing	0.325	0.52	0.335	0.56	0.330	0.59	0.339	0.72	0.166	0.79
3510 - Industrial chemicals	3.687	5.95	3.619	6.03	3.153	5.63	3.090	6.57	0.941	4.48
3520 - Chemical products nec	0.659	1.06	0.646	1.08	0.658	1.18	0.693	1.47	0.255	1.21
3522 - Drugs and medicines	0.456	0.74	0.446	0.74	0.423	0.76	0.379	0.81	0.150	0.71
3530 - Petroleum refineries	4.360	7.03	4.199	7.00	2.829	5.05	2.320	4.94	1.235	5.88
3540 - Petroleum, coal products	0.420	0.68	0.390	0.65	0.407	0.73	0.365	0.78	0.168	0.80
3550 - Rubber products	0.750	1.21	0.724	1.21	0.668	1.19	0.597	1.27	0.215	1.02
3560 - Plastic products nec	0.131	0.21	0.130	0.22	0.109	0.19	0.124	0.26	0.054	0.26
3610 - Pottery, china, etc.	0.079	0.13	0.075	0.13	0.076	0.14	0.074	0.16	0.027	0.13
3620 - Glass and products	0.766	1.24	0.780	1.30	0.817	1.46	0.670	1.43	0.318	1.51
3690 - Nonmetal products nec	1.666	2.69	1.568	2.61	1.443	2.58	1.180	2.51	0.502	2.39
3710 - Iron and steel	6.430	10.37	6.184	10.31	5.385	9.62	5.020	10.68	2.149	10.23
3720 - Nonferrous metals	1.447	2.33	1.405	2.34	1.233	2.20	1.047	2.23	0.384	1.83
3810 - Metal products	2.191	3.53	2.194	3.66	2.055	3.67	1.824	3.88	0.709	3.38
3820 - Machinery nec	8.651	13.95	8.389	13.98	7.073	12.63	6.312	13.43	2.531	12.05
3830 - Electrical machinery	3.115	5.02	3.111	5.18	2.695	4.81	2.971	6.32	0.920	4.38
3840 - Transportation equipment	5.359	8.64	5.234	8.72	4.665	8.33	3.984	8.48	1.595	7.60
3850 - Professional goods	0.262	0.42	0.255	0.43	0.230	0.41	0.192	0.41	0.077	0.37
3900 - Industries nec	0.739	1.19	0.719	1.20	0.744	1.33	0.601	1.28	0.299	1.42

Note: Codes are International Standard Industry codes (ISIC). Percentages are % of total Output. [f]: Factor Prices; [p]: Producer Prices.

Finance, Economics, and Trade

Economic Indicators [36]

- **National product**: GDP—purchasing power parity—$39 billion (1995 est.)
- **National product real growth rate**: 6% (1995 est.)
- **National product per capita**: $7,200 (1995 est.)
- **Inflation rate (consumer prices)**: 7.5% (1995 est.)
- **External debt**: $4.6 billion hard currency indebtedness (1995 est.)

Balance of Payments Summary [37]

Values in millions of dollars.

	1989	1990	1991	1992	1993[2]
Exports of goods (f.o.b.)	NA	NA	NA	NA	43,711
Imports of goods (f.o.b.)	NA	NA	NA	NA	34,163
Trade balance	NA	NA	NA	NA	9,548
Services - debits	NA	NA	NA	NA	-14,580
Services - credits	NA	NA	NA	NA	8,939
Private transfers (net)	NA	NA	NA	NA	2,324
Government transfers (net)	NA	NA	NA	NA	NA
Long-term capital (net)	NA	NA	NA	NA	-883
Short-term capital (net)	NA	NA	NA	NA	-7,323
Errors and omissions	NA	NA	NA	NA	4,850
Overall balance	NA	NA	NA	NA	2,875

Exchange Rates [38]

Currency: **koruna.**
Symbol: **Sk.**

Data are currency units per $1. Values before 1993 reflect Czechoslovak exchange rate.

August 1995	29.587
November 1994	29.447
1994	32.045
1993	30.770
1992	28.26
1991	29.53

Imports and Exports

Top Import Origins [39]

$8.7 billion (f.o.b., January-November 1995)
Data are for January-October 1995.

Origins	%
Czech Republic	28.1
Russia	16.8
Germany	14.3
Austria	5.2
Italy	4.5
Poland	2.9
US	2.3
France	2.3
Hungary	2.2
Netherlands	1.7
Ukraine	1.5

Top Export Destinations [40]

$8.8 billion (f.o.b., January-November 1995)
Data are for January-October 1995.

Destinations	%
Czech Republic	35.4
Germany	18.9
Austria	5.0
Italy	4.7
Hungary	4.6
Poland	4.4
Russia	3.6
Ukraine	2.1
France	2.0
Netherlands	1.7

Foreign Aid [41]

Recipient: ODA, $104 million (1993).

Import and Export Commodities [42]

Import Commodities
Machinery and transport equipment
Fuels and lubricants
Manufactured goods
Raw materials
Chemicals
Agricultural products

Export Commodities
Machinery and transport equipment
Chemicals
Fuels
Minerals and metals
Agricultural products

For sources, notes, and explanations, see Annotated Source Appendix, page 1061.

851

Slovenia

Geography [1]

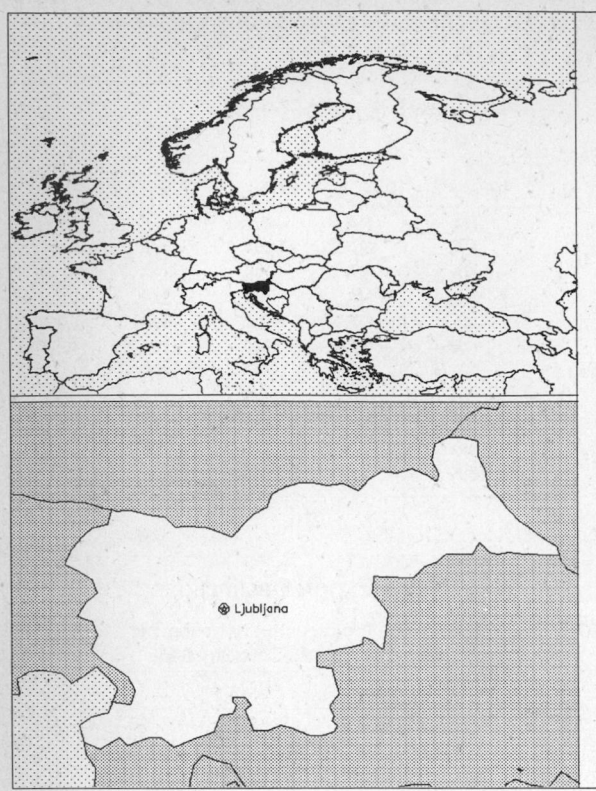

Total area:
20,256 sq km   7,821 sq mi
Land area:
20,256 sq km   7,821 sq mi
Comparative area:
Slightly larger than New Jersey
Land boundaries:
Total 1,207 km, Austria 324 km, Croatia 546 km, Italy 235 km, Hungary 102 km
Coastline:
46.6 km
Climate:
Mediterranean climate on the coast, continental climate with mild
to hot summers and cold winters in the plateaus and valleys to the
east
Terrain:
A short coastal strip on the Adriatic, an alpine mountain region adjacent
to Italy, mixed mountain and valleys with numerous rivers to the East
Natural resources:
Lignite coal, lead, zinc, mercury, uranium, silver
Land use:
Arable land: 10%
Permanent crops: 2%
Meadows and pastures: 20%
Forest and woodland: 45%
Other: 23%

Demographics [2]

	1970	1980	1990	1995[1]	1996	2000	2010	2020	2030
Population	1,718	1,885	1,969	1,957	1,951	1,937	1,926	1,856	1,765
Population density (persons per sq. mi.)	220	241	252	250	250	248	246	237	226
(persons per sq. km.)	85	93	97	97	96	96	95	92	87
Net migration rate (per 1,000 population)	NA	NA	NA	-1.6	-1.6	-0.8	-0.5	-0.2	0.0
Births	NA	NA	NA	NA	16	NA	NA	NA	NA
Deaths	NA	NA	NA	NA	18	NA	NA	NA	NA
Life expectancy - males	NA	NA	NA	71.3	71.4	71.8	72.6	75.0	76.8
Life expectancy - females	NA	NA	NA	78.9	79.0	79.4	80.4	82.1	83.5
Birth rate (per 1,000)	NA	NA	NA	8.3	8.3	10.8	9.9	7.6	8.2
Death rate (per 1,000)	NA	NA	NA	9.3	9.4	10.0	11.7	12.3	13.5
Women of reproductive age (15-49 yrs.)	NA	NA	NA	508	510	503	447	394	346
of which are currently married	NA	NA	NA	NA	359	357	330	NA	NA
Fertility rate	NA	NA	NA	1.1	1.1	1.5	1.6	1.5	1.5

Except as noted, values for vital statistics are in thousands; life expectancy is in years.

Health

Health Indicators [3]

% of population with access to	
safe water (1990-95)	NA
adequate sanitation (1990-95)	NA
health services (1985-95)	NA
% of 1-year-olds immunized (1990-94) against	
TB (tuberculosis)	96
DPT (diphtheria, pertussis, tetanus)	98
polio	98
measles	90
% of contraceptive prevalence (1980-94)	NA
ORT use rate (1990-94)	NA

Health Expenditures [4]

For sources, notes, and explanations, see Annotated Source Appendix, page 1061.

Human Factors

Women and Children [5]

% of pregnant women immunized (tetanus 1990-94)	NA
% of births attended by trained health personnel (1983-94)	NA
Maternal mortality rate (1980-92)	NA
Under-5 mortality rate (1994)	8
% under-5 moderately/severely underweight (1980-1994)	NA

Burden of Disease [6]

Population per physician	NA
Population per nurse	NA
Population per hospital bed (1994)	166.67
AIDS cases per 100,000 people (1994)	NA
Heart disease cases per 1,000 people (1990-93)	NA

Ethnic Division [7]

Slovene	91%
Croat	3%
Serb	2%
Muslim	1%
Other	3%

Religion [8]

Roman Catholic (Uniate 2%)	96%
Muslim	1%
Other	3%

Major Languages [9]

Slovenian	91%
Serbo-Croatian	7%
Other	2%

Education

Public Education Expenditures [10]

Million (Tolar)	1980	1990	1991	1992	1993	1994
Total education expenditure	NA	NA	16,603	55,828	83,183	102,496
as percent of GNP	NA	NA	4.8	5.6	6.2	6.2
as percent of total govt. expend.	NA	NA	16.1	23.2	21.7	12.8
Current education expenditure	NA	NA	15,266	52,595	76,591	92,722
as percent of GNP	NA	NA	4.4	5.3	5.7	5.6
as percent of current govt. expend.	NA	NA	NA	24.6	NA	NA
Capital expenditure	NA	NA	1,337	3,233	6,592	9,774

Educational Attainment [11]

Age group (1991)	25+
Total population	1,272,409
Highest level attained (%)	
No schooling	0.7
First level	
Not completed	45.1
Completed	NA
Entered second level	
S-1	42.4
S-2	NA
Postsecondary	10.4

Illiteracy [12]

	1989	1991	1995
Illiterate population (15+ yrs,)[2]	NA	7,422	NA
Illiteracy rate - total pop. (%)	NA	0.5	NA
Illiteracy rate - males (%)	NA	0.4	NA
Illiteracy rate - females (%)	NA	0.5	NA

Libraries [13]

	Admin. Units	Svc. Pts.	Vols. (000)	Shelving (meters)	Vols. Added	Reg. Users
National (1992)	1	1	1,749	22,300	44,098	9,251
Nonspecialized	NA	NA	NA	NA	NA	NA
Public (1992)	60	833	5,283	NA	233,770	365,888
Higher ed. (1992)	68	69	2,863	79,567	80,866	211,210
School	NA	NA	NA	NA	NA	NA

Daily Newspapers [14]

	1980	1985	1990	1994
Number of papers	3	3	4	6
Circ. (000)	198	216	303	360

Culture [15]

Cinema (seats per 1,000)	14.8
Annual attendance per person	1.2
Gross box office receipts (mil. Tolar)	732
Museums (reporting)	77
Visitors (000)	1,795
Annual receipts (000 Tolar)	2,474,302

Science and Technology

Scientific/Technical Forces [16]

Scientists/engineers	5,789
Number female	1,745
Technicians	4,615
Number female	2,197
Total	10,404

R&D Expenditures [17]

	Tolar (000) 1992
Total expenditure	15,050,524
Capital expenditure	1,365,784
Current expenditure	13,684,740
Percent current	90.9

U.S. Patents Issued [18]

Values show patents issued to citizens of the country by the U.S. Patents Office.

	1993	1994	1995
Number of patents	3	11	5

Government and Law

Organization of Government [19]

Long-form name:
Republic of Slovenia
Type:
Emerging democracy
Independence:
25 June 1991 (from Yugoslavia)
National holiday:
National Statehood Day, 25 June (1991)
Constitution:
Adopted 23 December 1991, effective 23
December 1991
Legal system:
Based on civil law system
Executive branch:
President; Prime Minister; Council of
Ministers
Legislative branch:
Unicameral: National Assembly. Note: The
National Council is an advisory body with
no direct legislative powers; in the
election of 6 December 1992, 40
members were elected to represent local,
professional, and socioeconomic interests
Judicial branch:
Supreme Court; Constitutional Court

Elections [20]

State Assembly	% of seats
Liberal Democratic (LDS)	24.4
Slovene Chris. Democratic (SKD)	16.7
United List	15.6
Slovene National Party	13.3
SN	11.1
Democratic Party	6.7
Greens of Slovenia (ZS)	5.6
Social-Democratic Party (SDSS)	4.4
Hungarian minority	1.1
Italian minority	1.1

Government Budget [21]

For 1993.

	$ bil.
Revenues	6.6
Expenditures	6.6
Capital expenditures	NA

Military Expenditures and Arms Transfers [22]

	1990	1991	1992[15]	1993	1994
Military expenditures					
Current dollars (mil.)	NA	NA	341[r]	195[r]	190[e]
1994 constant dollars (mil.)	NA	NA	355[r]	199[r]	190[e]
Armed forces (000)	NA	NA	15	12	17
Gross national product (GNP)					
Current dollars (mil.)[e]	NA	NA	14,680	15,000	16,000
1994 constant dollars (mil.)[e]	NA	NA	15,310	15,310	16,000
Central government expenditures (CGE)					
1994 constant dollars (mil.)	NA	NA	NA	NA	NA
People (mil.)	NA	NA	2.0	2.0	2.0
Military expenditure as % of GNP	NA	NA	2.3	1.3	1.2
Military expenditure as % of CGE	NA	NA	NA	NA	NA
Military expenditure per capita (1994 $)	NA	NA	176	98	93
Armed forces per 1,000 people (soldiers)	NA	NA	7.4	5.9	8.3
GNP per capita (1994 $)	NA	NA	7,595	7,510	7,819
Arms imports[6]					
Current dollars (mil.)	NA	NA	0	0	10
1994 constant dollars (mil.)	NA	NA	0	0	10
Arms exports[6]					
Current dollars (mil.)	NA	NA	0	0	5
1994 constant dollars (mil.)	NA	NA	0	0	5
Total imports[7]					
Current dollars (mil.)	NA	NA	NA	6,498	7,334
1994 constant dollars (mil.)	NA	NA	NA	6,632	7,334
Total exports[7]					
Current dollars (mil.)	NA	NA	NA	6,088	6,825
1994 constant dollars (mil.)	NA	NA	NA	6,213	6,825
Arms as percent of total imports[8]	NA	NA	0	0	0.1
Arms as percent of total exports[8]	NA	NA	0	0	0.1

Crime [23]

	1994
Crime volume	
Cases known to police	43,635
Attempts (percent)	5.21
Percent cases solved	57.46
Crimes per 100,000 persons	2,209.55
Persons responsible for offenses	
Total number offenders	31,787
Percent female	11.46
Percent juvenile (14-18 yrs.)	24.41
Percent foreigners	8.19

Human Rights [24]

	SSTS	FL	FAPRO	PPCG	APROBC	TPW	PCPTW	STPEP	PHRFF	PRW	ASST	AFL
Observes		P	P	P	P	P	P	P	P	P	P	

	EAFRD	CPR	ESCR	SR	ACHR	MAAE	PVIAC	PVNAC	EAFDAW	TCIDTP	RC
Observes	P	P	P	P	P	P	P	P	P	P	P

P = Party; S = Signatory; see Appendix for meaning of abbreviations.

Labor Force

Total Labor Force [25]

786,036

Labor Force by Occupation [26]

Manufacturing and mining	46%
Agriculture	2

Unemployment Rate [27]

8% (December 1995 est.)

Production Sectors

Commercial Energy Production and Consumption

Data are shown in quadrillion (10^{15}) BTUs and percent for 1995
Values for hydroelectric, nuclear, geothermal, solar, and wind power refer to electrical generation.

Production [28]

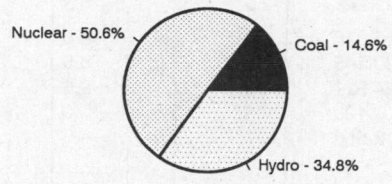

Nuclear - 50.6%
Coal - 14.6%
Hydro - 34.8%

Consumption [29]

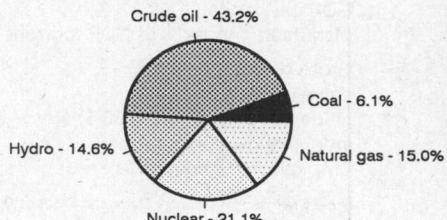

Crude oil - 43.2%
Coal - 6.1%
Natural gas - 15.0%
Hydro - 14.6%
Nuclear - 21.1%

Coal	0.013
Net hydroelectric power	0.031
Net nuclear power	0.045
Total	0.089

Crude oil	0.092
Dry natural gas	0.032
Coal	0.013
Net hydroelectric power	0.031
Net nuclear power	0.045
Total	0.213

Telecommunications [30]

- 527,800 (1993 est.) telephones
- Radio: Broadcast stations: AM 6, FM 5, shortwave 0 Radios: 596,100 (1993 est.)
- Television: Broadcast stations: 7 Televisions: 454,400 (1993 est.)

Transportation [31]

Railways: total: 1,201 km; standard gauge: 1,201 km 1.435-m gauge (electrified 499 km) (1994)

Highways: total: 14,794 km; paved: 13,314 km (including 187 km of expressways); unpaved: 1,480 km (1994 est.)

Merchant marine: total: 14 ships (1,000 GRT or over) totaling 229,727 GRT/290,456 DWT (controlled by Slovenian owners); ships by type: bulk 9, cargo 1, container 4

Airports

Total:	14
With paved runways over 3,047 m:	1
With paved runways 2,438 to 3,047 m:	1
With paved runways 1,524 to 2,437 m:	1
With paved runways 914 to 1,523 m:	2
With paved runways under 914 m:	5

Top Agricultural Products [32]

Agriculture accounts for 5.3% of the GDP; produces potatoes, hops, wheat, sugar beets, corn, grapes; cattle, sheep, poultry.

Top Mining Products [33]

Thousand metric tons except as noted[e]	4/95[*]
Steel, crude, from electric furnaces	300
Cement	1,000
Lime	300
Sand and gravel, excl. glass sand (000 cu. meters)	2,000
Stone (000 cu. meters)	
ornamental	300
crushed and brown	1,000
Coal, brown and lignite	5,100
Gas, natural, gross producing (mil. cu. meters)	20
Petroleum, crude, converted (000 42-gal. bls.)	18,000

Tourism [34]

	1990	1991	1992	1993	1994
Visitors[99]	NA	NA	46,323	59,182	66,781
Tourists[11]	NA	NA	616	624	748
Tourism receipts	NA	NA	671	734	932
Tourism expenditures	NA	NA	282	304	312

Travelers are in thousands, money in million U.S. dollars.

For sources, notes, and explanations, see Annotated Source Appendix, page 1061.

855

Manufacturing Sector

GDP and Manufacturing Summary [35]

	1980	1985	1989	1990	% change 1980-1990	% change 1989-1990
GDP (million 1980 $)	69,958	71,058	72,234	66,371	-5.1	-8.1
GDP per capita (1980 $)	3,136	3,073	3,050	2,786	-11.2	-8.7
Manufacturing as % of GDP (current prices)	30.6	37.2	39.5	42.0	37.3	6.3
Gross output (million $)	72,629	57,020	65,078	62,136[e]	-14.4	-4.5
Value added (million $)	21,750	17,171	30,245	27,660[e]	27.2	-8.5
Value added (million 1980 $)	19,526	22,283	24,021	21,703	11.1	-9.6
Industrial production index	100	116	120	108	8.0	-10.0
Employment (thousands)	2,106	2,467	2,658	2,537[e]	20.5	-4.6

Note: GDP stands for Gross Domestic Product. 'e' stands for estimated value.

Profitability and Productivity

	1980	1985	1989	1990	% change 1980-1990	% change 1989-1990
Intermediate input (%)	70	70	54	55[e]	-21.4	1.9
Wages, salaries, and supplements (%)	14	12	12[e]	18[e]	28.6	50.0
Gross operating surplus (%)	15	18	34[e]	26[e]	73.3	-23.5
Gross output per worker ($)	34,487	23,113	24,484	24,248[e]	-29.7	-1.0
Value added per worker ($)	10,328	6,960	11,379	10,796[e]	4.5	-5.1
Average wage (incl. benefits) ($)	4,991	2,703	2,986[e]	4,488[e]	-10.1	50.3

Profitability is in percent of gross output. Productivity is in U.S. $. 'e' stands for estimated value.

Profitability - 1990

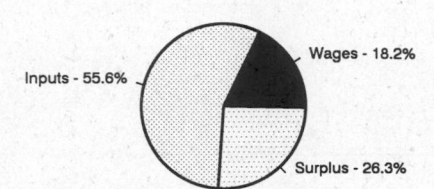

Inputs - 55.6%
Wages - 18.2%
Surplus - 26.3%

The graphic shows percent of gross output.

Value Added in Manufacturing

	1980 $ mil.	1980 %	1985 $ mil.	1985 %	1989 $ mil.	1989 %	1990 $ mil.	1990 %	% change 1980-1990	% change 1989-1990
311 Food products	1,897	8.7	1,458	8.5	3,916	12.9	3,484[e]	12.6	83.7	-11.0
313 Beverages	459	2.1	353	2.1	663	2.2	589[e]	2.1	28.3	-11.2
314 Tobacco products	184	0.8	221	1.3	344	1.1	308[e]	1.1	67.4	-10.5
321 Textiles	1,759	8.1	1,428	8.3	2,881	9.5	2,663[e]	9.6	51.4	-7.6
322 Wearing apparel	903	4.2	718	4.2	1,593	5.3	1,427[e]	5.2	58.0	-10.4
323 Leather and fur products	226	1.0	231	1.3	383	1.3	340[e]	1.2	50.4	-11.2
324 Footwear	482	2.2	503	2.9	1,022	3.4	899[e]	3.3	86.5	-12.0
331 Wood and wood products	977	4.5	530	3.1	794	2.6	706[e]	2.6	-27.7	-11.1
332 Furniture and fixtures	730	3.4	438	2.6	1,030	3.4	1,065[e]	3.9	45.9	3.4
341 Paper and paper products	529	2.4	394	2.3	759	2.5	674[e]	2.4	27.4	-11.2
342 Printing and publishing	876	4.0	462	2.7	761	2.5	678[e]	2.5	-22.6	-10.9
351 Industrial chemicals	694	3.2	631	3.7	1,107	3.7	992[e]	3.6	42.9	-10.4
352 Other chemical products	681	3.1	525	3.1	1,419	4.7	1,315[e]	4.8	93.1	-7.3
353 Petroleum refineries	454	2.1	415	2.4	260	0.9	233[e]	0.8	-48.7	-10.4
354 Miscellaneous petroleum and coal products	101	0.5	101	0.6	104	0.3	91[e]	0.3	-9.9	-12.5
355 Rubber products	276	1.3	269	1.6	479	1.6	456[e]	1.6	65.2	-4.8
356 Plastic products	413	1.9	258	1.5	397	1.3	350[e]	1.3	-15.3	-11.8
361 Pottery, china, and earthenware	128	0.6	72	0.4	162	0.5	144[e]	0.5	12.5	-11.1
362 Glass and glass products	163	0.7	113	0.7	224	0.7	204[e]	0.7	25.2	-8.9
369 Other nonmetal mineral products	906	4.2	513	3.0	683	2.3	604[e]	2.2	-33.3	-11.6
371 Iron and steel	1,221	5.6	1,000	5.8	1,343	4.4	1,171[e]	4.2	-4.1	-12.8
372 Nonferrous metals	480	2.2	509	3.0	944	3.1	927[e]	3.4	93.1	-1.8
381 Metal products	2,105	9.7	1,577	9.2	1,293	4.3	1,130[e]	4.1	-46.3	-12.6
382 Nonelectrical machinery	1,828	8.4	1,463	8.5	2,372	7.8	2,378[e]	8.6	30.1	0.3
383 Electrical machinery	1,600	7.4	1,544	9.0	2,640	8.7	2,334[e]	8.4	45.9	-11.6
384 Transport equipment	1,441	6.6	1,263	7.4	2,389	7.9	2,241[e]	8.1	55.5	-6.2
385 Professional and scientific equipment	101	0.5	93	0.5	154	0.5	146[e]	0.5	44.6	-5.2
390 Other manufacturing industries	134	0.6	88	0.5	128	0.4	114[e]	0.4	-14.9	-10.9

Note: The industry codes shown are International Standard Industry codes (ISIC). Percentages are percent of total Value Added. 'e' stands for estimated value

For sources, notes, and explanations, see Annotated Source Appendix, page 1061.

Finance, Economics, and Trade

Economic Indicators [36]

- **National product**: GDP—purchasing power parity— $22.6 billion (1995 est.)
- **National product real growth rate**: 4.8% (1995 est.)
- **National product per capita**: $11,000 (1995 est.)
- **Inflation rate (consumer prices)**: 8% (December 1995 est.)
- **External debt**: $2.9 billion (1995)

Balance of Payments Summary [37]

Values in millions of dollars.

	1989	1990	1991	1992[3]	1993
Exports of goods (f.o.b.)				6,683	6,083
Imports of goods (f.o.b.)				-5,892	-6,237
Trade balance				791	-154
Services - debits				-1,200	-1,216
Services - credits				1,289	1,518
Private transfers (net)				35	62
Government transfers (net)				10	-17
Long-term capital (net)				90	238
Short-term capital (net)				-106	-298
Errors and omissions				-280	-69
Overall balance				630	65

Exchange Rates [38]

Currency: **tolar.**
Symbol: **SIT.**

Data are currency units per $1.

November 1995	121.27
1995	118.9
1994	128.81
1993	113.24
1992	81.29
1991	27.57

Imports and Exports

Top Import Origins [39]

$9.1 billion (f.o.b., 1995 est.) Data are for January-August 1995.

Origins	%
Germany	23.3
Italy	16.8
Austria	9.7
France	8.5
Former Yugoslavia	7.0

Top Export Destinations [40]

$8.3 billion (f.o.b., 1995 est.) Data are for January-August 1995.

Destinations	%
Germany	30.9
Italy	14.1
Yugoslavia	14.0
France	8.9
Austria	6.4
CEFTA	5

Foreign Aid [41]

Recipient: ODA, $5 million (1993).

Import and Export Commodities [42]

Import Commodities

Machinery and transport equipment 30%
Intermediate manufactured goods 17.6%
Chemicals 11.5%
Raw materials 5.3%
Fuels and lubricants 10.8%
Food 8.4%

Export Commodities

Machinery and transport equipment 27%
Intermediate manufactured goods 26%
Chemicals 9%
Food 4.8%
Raw materials 3%
Consumer goods 26%

For sources, notes, and explanations, see Annotated Source Appendix, page 1061.

857

Solomon Islands

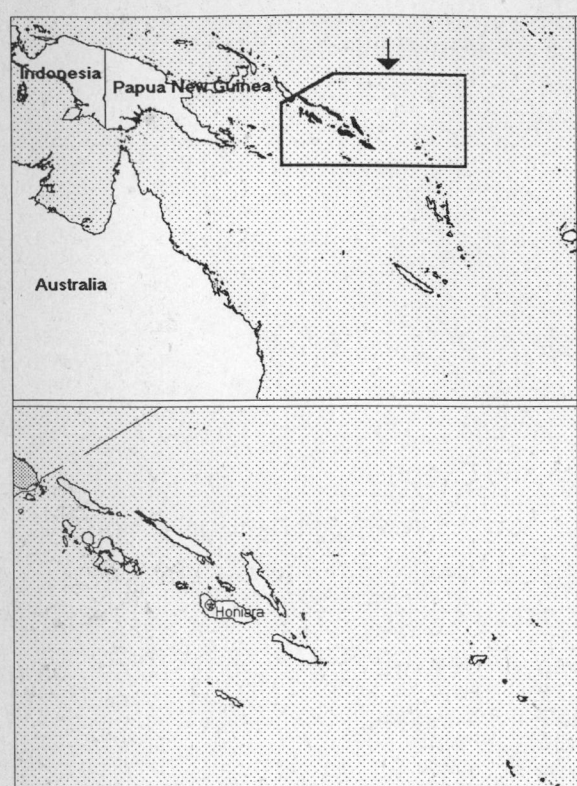

Geography [1]

Total area:
28,450 sq km   10,985 sq mi
Land area:
27,540 sq km   10,633 sq mi
Comparative area:
Slightly larger than Maryland
Land boundaries:
0 km
Coastline:
5,313 km
Climate:
Tropical monsoon; few extremes of temperature and weather
Terrain:
Mostly rugged mountains with some low coral atolls
Natural resources:
Fish, forests, gold, bauxite, phosphates, lead, zinc, nickel
Land use:
Arable land: 1%
Permanent crops: 1%
Meadows and pastures: 1%
Forest and woodland: 93%
Other: 4%

Demographics [2]

	1970	1980	1990	1995[1]	1996	2000	2010	2020	2030
Population	163	233	336	399	413	470	620	767	911
Population density (persons per sq. mi.)	15	22	32	38	39	44	58	72	86
(persons per sq. km.)	6	8	12	14	15	17	23	28	33
Net migration rate (per 1,000 population)	NA	0.0	0.0	0.0	0.0	0.0	0.0	0.0	0.0
Births	NA	NA	NA	NA	16	NA	NA	NA	NA
Deaths	NA	NA	NA	NA	2	NA	NA	NA	NA
Life expectancy - males	NA	63.2	66.8	68.4	68.7	69.9	72.4	74.5	76.1
Life expectancy - females	NA	67.4	71.6	73.4	73.7	75.1	78.0	80.3	82.0
Birth rate (per 1,000)	NA	45.3	40.7	38.5	37.9	35.2	27.5	22.4	19.0
Death rate (per 1,000)	NA	7.3	5.2	4.5	4.4	4.0	3.4	3.4	3.7
Women of reproductive age (15-49 yrs.)	NA	47	71	88	92	107	155	207	248
of which are currently married[49]	22	NA	46	NA	59	70	104	NA	NA
Fertility rate	NA	7.3	6.3	5.6	5.4	4.8	3.4	2.6	2.2

Except as noted, values for vital statistics are in thousands; life expectancy is in years.

Health

Health Indicators [3]

Health Expenditures [4]

For sources, notes, and explanations, see Annotated Source Appendix, page 1061.

Human Factors

Women and Children [5]	Burden of Disease [6]
	Population per physician (1989) 6,595.75
	Population per nurse NA
	Population per hospital bed NA
	AIDS cases per 100,000 people (1994) *
	Malaria cases per 100,000 people (1992) 44,711

Ethnic Division [7]

Melanesian	93%
Polynesian	4%
Micronesian	1.5%
European	0.8%
Chinese	0.3%
Other	0.4%

Religion [8]

Anglican	34%
Roman Catholic	19%
Baptist	17%
United (Methodist/Presbyterian)	11%
Seventh-Day Adventist	10%
Other Protestant	5%
Traditional beliefs	4%

Major Languages [9]

Melanesian pidgin in much of the country is lingua franca, English spoken by 1%- 2% of population. There are 120 indigenous languages.

Education

Public Education Expenditures [10]

	1980	1984	1987	1990	1991	1994
Million (Solomon Isl. Dollar)						
Total education expenditure	5	10	NA	NA	24	NA
as percent of GNP	5.6	4.7	NA	NA	4.2	NA
as percent of total govt. expend.	11.2	12.4	NA	NA	7.9	NA
Current education expenditure	4	4	NA	NA	24	NA
as percent of GNP	4.3	1.9	NA	NA	4.2	NA
as percent of current govt. expend.	14.1	8.0	NA	NA	13.5	NA
Capital expenditure	1	6	NA	NA	-	NA

Educational Attainment [11]

Illiteracy [12]

Libraries [13]

Daily Newspapers [14]

Culture [15]

Science and Technology

Scientific/Technical Forces [16]

R&D Expenditures [17]

U.S. Patents Issued [18]

For sources, notes, and explanations, see Annotated Source Appendix, page 1061.

859

Government and Law

Organization of Government [19]

Long-form name:
None
Type:
Parliamentary democracy
Independence:
7 July 1978 (from UK)
National holiday:
Independence Day, 7 July (1978)
Constitution:
7 July 1978
Legal system:
Common law
Executive branch:
British Monarch (represented by Governor General); Prime Minister; Deputy Prime Minister; Cabinet
Legislative branch:
Unicameral: National Parliament
Judicial branch:
High Court

Elections [20]

National Parliament	% of seats
National Unity and Reconciliation Group (GNUR)	44.7
People's Alliance Party (PAP)	14.9
Independents	12.8
National Action Party (NAPSI)	10.6
Solomon Islands Labor Party	8.5
United Party (UP)	8.5

Government Budget [21]

For 1995 est.

	$ mil.
Revenues	81.3
Expenditures	101.9
Capital expenditures	NA

Defense Summary [22]

Branches: No regular military forces; Solomon Islands National Reconnaissance and Surveillance Force; Royal Solomon Islands Police (RSIP)

Manpower Availability: Males age 15-49 NA; males fit for military service NA

Defense Expenditures: NA

Crime [23]

Human Rights [24]

	SSTS	FL	FAPRO	PPCG	APROBC	TPW	PCPTW	STPEP	PHRFF	PRW	ASST	AFL
Observes	P	P				P	P			P	P	
	EAFRD	CPR	ESCR	SR	ACHR	MAAE	PVIAC	PVNAC	EAFDAW	TCIDTP	RC	
Observes	1		1	P			P	P	S		P	

P = Party; S = Signatory; see Appendix for meaning of abbreviations.

Labor Force

Total Labor Force [25]

26,842

Labor Force by Occupation [26]

Services	41.5%
Agriculture, forestry, and fishing	23.7
Commerce, transport, and finance	21.7
Construction, manufacturing, and mining	13.1

Date of data: 1992 est.

Unemployment Rate [27]

For sources, notes, and explanations, see Annotated Source Appendix, page 1061.

Production Sectors

Energy Resource Summary [28]

Electricity: Capacity: 21,000 kW. Production: 30 million kWh. Consumption per capita: 80 kWh (1993).

Telecommunications [30]

- 5,000 (1991 est.) telephones
- International: satellite earth station - 1 Intelsat (Pacific Ocean)
- Radio: Broadcast stations: AM 4, FM 0, shortwave 0 Radios: 38,000 (1993 est.)
- Television: Broadcast stations: 0 (1987 est.) Televisions: 2,000 (1992 est.)

Transportation [31]

Railways: 0 km

Highways: total: 1,300 km; paved: 30 km; unpaved: 1,270 km

Merchant marine: none

Airports

Total: 30

With paved runways 1,524 to 2,437 m: 1

With paved runways 914 to 1,523 m: 1

With paved runways under 914 m: 18

With unpaved runways 1,524 to 2,437 m: 1

With unpaved runways 914 to 1,523 m: 9 (1995 est.)

Top Agricultural Products [32]

Produces cocoa, beans, coconuts, palm kernels, rice, potatoes, vegetables, fruit; cattle, pigs; timber; fish.

Top Mining Products [33]

Detailed information is not available. A summary of mineral resources follows. **Mineral Resources**: Small amounts of clays, crushed stone, sand and gravel, and alluvial gold.

Tourism [34]

	1990	1991	1992	1993	1994
Visitors	12	38	17	15	NA
Tourists	9	11	12	12	12
Cruise passengers	3	26	5	3	NA
Tourism receipts	4	5	6	6	NA
Tourism expenditures	11	12	11	NA	NA
Fare receipts	1	3	2	NA	NA
Fare expenditures	5	4	3	NA	NA

Travelers are in thousands, money in million U.S. dollars.

Finance, Economics, and Trade

Industrial Summary [35]

Industrial Production: Growth rate - 3.8% (1991 est.). **Industries**: Copra, fishing (tuna).

Economic Indicators [36]

- **National product**: GDP—purchasing power parity—$1 billion (1992 est.)
- **National product real growth rate**: 8% (1992 est.)
- **National product per capita**: $2,590 (1992 est.)
- **Inflation rate (consumer prices)**: 13% (1994)
- **External debt**: $128 million (1988 est.)

Balance of Payments Summary [37]

Values in millions of dollars.

	1988	1989	1990	1991	1992
Exports of goods (f.o.b.)	81.9	74.7	70.1	83.4	101.7
Imports of goods (f.o.b.)	-105.1	-97.6	-77.4	-92.0	-87.4
Trade balance	-23.2	-22.8	-7.2	-8.5	14.3
Services - debits	-74.4	-87.7	-86.3	-98.7	-88.9
Services - credits	29.8	33.9	27.7	33.0	37.0
Private transfers (net)	-1.4	1.2	4.7	1.9	1.8
Government transfers (net)	46.1	37.3	33.3	36.3	36.7
Long-term capital (net)	23.6	20.6	22.8	20.2	21.1
Short-term capital (net)	16.2	0.3	8.5	4.1	1.3
Errors and omissions	-10.3	5.6	-8.6	8.4	-6.2
Overall balance	6.4	-11.6	-5.2	-3.4	17.1

Exchange Rates [38]

Currency: **Solomon Islands dollar.**
Symbol: **SI$.**

Data are currency units per $1.

2d quarter 1995	3.3713
1994	3.2914
1993	3.1877
1992	2.9281
1991	2.7148

Imports and Exports

Top Import Origins [39]

$101 million (c.i.f., 1993).

Origins	%
Australia	34
Japan	16
Singapore	14
New Zealand	9

Top Export Destinations [40]

$94 million (f.o.b., 1993) Data are for 1991.

Destinations	%
Japan	39
UK	23
Thailand	9
Australia	5
US	2

Foreign Aid [41]

Recipient: ODA, $NA.

Import and Export Commodities [42]

Import Commodities	Export Commodities
Plant and machinery	Fish 46%
Manufactured goods	Timber 31%
Food and live animals	Palm oil 5%
Fuel	Cocoa
	Copra

Somalia

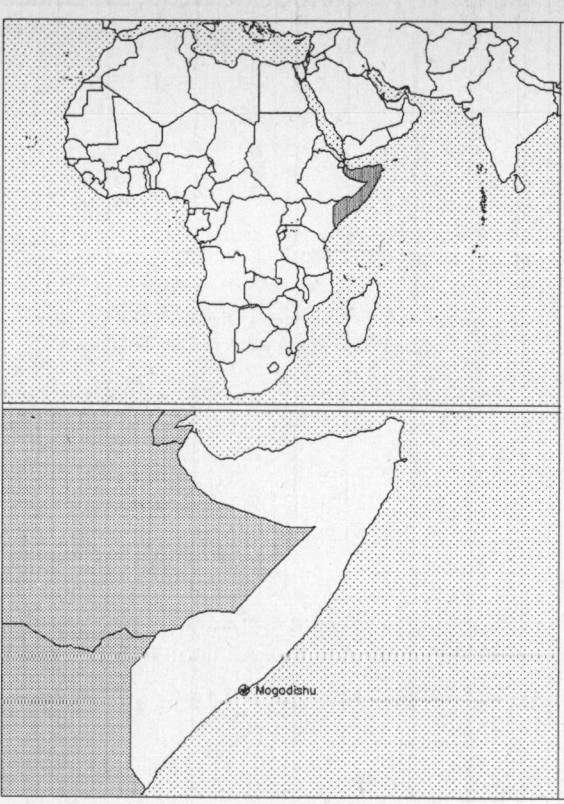

Geography [1]

Total area:
637,660 sq km   246,202 sq mi
Land area:
627,340 sq km   242,217 sq mi
Comparative area:
Slightly smaller than Texas
Land boundaries:
Total 2,366 km, Djibouti 58 km, Ethiopia 1,626 km, Kenya 682 km
Coastline:
3,025 km
Climate:
Principally desert; December to February - northeast monsoon, moderate temperatures in North and very hot in South; May to October - southwest monsoon, torrid in the North and hot in the South, irregular rainfall, hot and humid periods (tangambili) between monsoons
Terrain:
Mostly flat to undulating plateau rising to hills in North
Natural resources:
Uranium and largely unexploited reserves of iron ore, tin, gypsum, bauxite, copper, salt
Land use:
Arable land: 2%
Permanent crops: 0%
Meadows and pastures: 46%
Forest and woodland: 14%
Other: 38%

Demographics [2]

	1970	1980	1990	1995[1]	1996	2000	2010	2020	2030
Population	4,535	6,865	8,334	9,165	9,639	10,880	14,524	18,955	23,633
Population density (persons per sq. mi.)	19	28	34	38	40	45	60	78	98
(persons per sq. km.)	7	11	13	15	15	17	23	30	38
Net migration rate (per 1,000 population)	NA	8.6	0.0	39.9	0.0	0.0	0.0	0.0	0.0
Births	NA	NA	NA	NA	426	NA	NA	NA	NA
Deaths	NA	NA	NA	NA	127	NA	NA	NA	NA
Life expectancy - males	NA	49.2	54.5	54.8	55.2	56.6	60.0	63.2	66.2
Life expectancy - females	NA	49.9	55.0	55.4	55.8	57.4	61.4	65.2	68.7
Birth rate (per 1,000)	NA	44.5	45.4	44.6	44.2	41.7	37.8	32.3	26.3
Death rate (per 1,000)	NA	16.4	14.1	13.6	13.2	12.0	9.6	7.8	6.7
Women of reproductive age (15-49 yrs.)	NA	1,520	1,820	1,977	2,087	2,385	3,318	4,517	6,003
of which are currently married	NA	NA	1,274	NA	1,467	1,668	2,317	NA	NA
Fertility rate	NA	7.3	7.3	7.1	7.0	6.5	5.4	4.2	3.2

Except as noted, values for vital statistics are in thousands; life expectancy is in years.

Health

Health Indicators [3]

% of population with access to	
safe water (1990-95)[1]	37
adequate sanitation (1990-95)[1]	18
health services (1985-95)[1]	27
% of 1-year-olds immunized (1990-94) against	
TB (tuberculosis)	48
DPT (diphtheria, pertussis, tetanus)	23
polio	23
measles	35
% of contraceptive prevalence (1980-94)	1
ORT use rate (1990-94)	78

Health Expenditures [4]

Total health expenditure, 1990 (official exchange rate)	
Millions of dollars	60
Dollars per capita	8
Health expenditures as a percentage of GDP	
Total	1.5
Public sector	0.9
Private sector	0.6
Development assistance for health	
Total aid flows (millions of dollars)[1]	27
Aid flows per capita (dollars)	3.5
Aid flows as a percentage of total health expenditure	45.6

For sources, notes, and explanations, see Annotated Source Appendix, page 1061.

Human Factors

Women and Children [5]

% of pregnant women immunized (tetanus 1990-94)	NA
% of births attended by trained health personnel (1983-94)	2
Maternal mortality rate (1980-92)	1,100
Under-5 mortality rate (1994)	211
% under-5 moderately/severely underweight (1980-1994)	NA

Burden of Disease [6]

Population per physician (1984)	13,450.21
Population per nurse (1984)	1,897.83
Population per hospital bed (1990)	1,332.59
AIDS cases per 100,000 people (1994)	*
Malaria cases per 100,000 people (1992)	NA

Ethnic Division [7]

Somali	85%
Bantu, Arabs	30,000

Religion [8]

Sunni Muslim.

Major Languages [9]

Somali (official), Arabic, Italian, English.

Education

Public Education Expenditures [10]

	1980	1985	1990	1991	1993	1994
Million (Shilling)[60]						
Total education expenditure	169	371	NA	NA	NA	NA
as percent of GNP	1.0	0.5	NA	NA	NA	NA
as percent of total govt. expend.	8.7	4.1	NA	NA	NA	NA
Current education expenditure	154	274	NA	NA	NA	NA
as percent of GNP	0.9	0.3	NA	NA	NA	NA
as percent of current govt. expend.	NA	NA	NA	NA	NA	NA
Capital expenditure	15	97	NA	NA	NA	NA

Educational Attainment [11]

Illiteracy [12]

In thousands and percent[3]	1985	1991	2000
Illiterate population (15+ years)	2,877	3,003	3,235
Illiteracy rate - total pop. (%)	83.1	75.9	61.3
Illiteracy rate - males (%)	73.3	63.9	47.9
Illiteracy rate - females (%)	91.2	86.0	73.5

Libraries [13]

Daily Newspapers [14]

	1980	1985	1990	1994
Number of papers	2	2	1	1
Circ. (000)	5[e]	7[e]	9	9

Culture [15]

Science and Technology

Scientific/Technical Forces [16]

R&D Expenditures [17]

U.S. Patents Issued [18]

For sources, notes, and explanations, see Annotated Source Appendix, page 1061.

Government and Law

Organization of Government [19]

Long-form name:
None
Type:
None
Independence:
1 July 1960 (from a merger of British Somaliland and Italian Somaliland)
National holiday:
NA
Constitution:
25 August 1979, presidential approval 23 September 1979
Legal system:
NA
Executive branch:
Somalia has no functioning government; the United Somali Congress (USC) ousted the regime of Major General Mohamed Siad Barre on 27 January 1991; the present political situation is one of anarchy, marked by interclan fighting and random banditry
Legislative branch:
Unicameral People's Assembly: People's Assembly; Not functioning
Judicial branch:
Supreme Court (not functioning)

Crime [23]

Elections [20]

People's Assembly. Not functioning.

Government Budget [21]

Military Expenditures and Arms Transfers [22]

	1990	1991[18]	1992	1993	1994
Military expenditures					
Current dollars (mil.)	7[e]	NA	NA	NA	NA
1994 constant dollars (mil.)	7[e]	NA	NA	NA	NA
Armed forces (000)	47	NA	NA	NA	NA
Gross national product (GNP)					
Current dollars (mil.)	760	NA	NA	NA	NA
1994 constant dollars (mil.)	846	NA	NA	NA	NA
Central government expenditures (CGE)					
1994 constant dollars (mil.)	NA	NA	NA	NA	NA
People (mil.)	6.8	6.6	6.6	6.5	6.7
Military expenditure as % of GNP	0.9	NA	NA	NA	NA
Military expenditure as % of CGE	NA	NA	NA	NA	NA
Military expenditure per capita (1994 $)	1	NA	NA	NA	NA
Armed forces per 1,000 people (soldiers)	7.0	NA	NA	NA	NA
GNP per capita (1994 $)	125	NA	NA	NA	NA
Arms imports[6]					
Current dollars (mil.)	30	10	20	5	0
1994 constant dollars (mil.)	33	11	21	5	0
Arms exports[6]					
Current dollars (mil.)	0	0	0	0	0
1994 constant dollars (mil.)	0	0	0	0	0
Total imports[7]					
Current dollars (mil.)[e]	249	151	184	205	NA
1994 constant dollars (mil.)[e]	277	162	192	209	NA
Total exports[7]					
Current dollars (mil.)[e]	150	91	118	117	NA
1994 constant dollars (mil.)[e]	167	98	123	119	NA
Arms as percent of total imports[8]	12.0	6.6	10.9	2.4	0
Arms as percent of total exports[8]	0	0	0	0	0

Human Rights [24]

	SSTS	FL	FAPRO	PPCG	APROBC	TPW	PCPTW	STPEP	PHRFF	PRW	ASST	AFL
Observes		P				P	P					P
	EAFRD	CPR	ESCR	SR	ACHR	MAAE	PVIAC	PVNAC	EAFDAW	TCIDTP	RC	
Observes		P	P	P	P						P	

P = Party; S = Signatory; see Appendix for meaning of abbreviations.

Labor Force

Total Labor Force [25]

3.7 million; very few are skilled laborers. (1993 est.)

Labor Force by Occupation [26]

Agriculture (mostly pastoral nomadism)   71%
Industry and services   29

Unemployment Rate [27]

For sources, notes, and explanations, see Annotated Source Appendix, page 1061.

Production Sectors

Energy Resource Summary [28]

Energy resources: Uranium. **Electricity**: Capacity: 75,000 kW prior to the civil war, but now largely shut down due to war damage; some localities operate their own generating plants, providing limited municipal power; note - UN and relief organizations use their own portable power systems. Production: NA kWh. Consumption per capita: NA kWh. **Pipelines**: crude oil 15 km.

Telecommunications [30]

- 9,000 (1991 est.) telephones; the public telecommunications system was completely destroyed or dismantled by the civil war factions; all relief organizations depend on their own private systems
- Domestic: recently, local cellular telephone systems have been established in Mogadishu and in several other population centers
- International: international connections are available from Mogadishu by satellite
- Radio: Broadcast stations: AM NA, FM NA, shortwave NA (there are at least five radio broadcast stations of NA type) Radios: 350,000 (1992 est.)
- Television: Broadcast stations: 0 (Somalia's only TV station was demolished during the civil strife, sometime in 1991) Televisions: 113,000 (1992 est.)

Top Agricultural Products [32]

Produces bananas, sorghum, corn, mangoes, sugarcane; cattle, sheep, goats; fishing potential largely unexploited.

Transportation [31]

Railways: 0 km

Highways: total: 22,500 km; paved: 2,700 km; unpaved: 19,800 km (1992 est.)

Merchant marine: total: 2 ships (1,000 GRT or over) totaling 5,529 GRT/6,892 DWT; ships by type: cargo 1, refrigerated cargo 1 (1995 est.)

Airports

Total:	52
With paved runways over 3,047 m:	3
With paved runways 2,438 to 3,047 m:	1
With paved runways 1,524 to 2,437 m:	2
With paved runways 914 to 1,523 m:	1
With paved runways under 914 m:	6

Top Mining Products [33]

Thousand metric tons except as noted[e]	4/1/95[*]
Cement, hydraulic	25,000
Gypsum	2,000
Limestone	40,000[64]
Salt, marine	1,000
Sepiolite (meerschaum)	5

Tourism [34]

Finance, Economics, and Trade

GDP and Manufacturing Summary [35]

	1980	1985	1990	1991	1992
Gross Domestic Product					
Millions of 1980 dollars	2,755	3,658	4,075	3,260	3,032[e]
Growth rate in percent	-2.25	7.93	-1.60	-20.00	-7.00[e]
Manufacturing Value Added					
Millions of 1980 dollars	123	103	137	NA	NA
Growth rate in percent	9.17	7.55	0.00	NA	NA
Manufacturing share in percent of current prices	4.7	4.9	4.3[e]	NA	NA

Economic Indicators [36]

- **National product**: GDP—purchasing power parity—$3.6 billion (1995 est.)
- **National product real growth rate**: 2% (1995 est.)
- **National product per capita**: $500 (1995 est.)
- **Inflation rate (consumer prices)**: NA
- **External debt**: $1.9 billion (1989)

Balance of Payments Summary [37]

Values in millions of dollars.

	1985	1986	1987	1988	1989
Exports of goods (f.o.b.)	90.6	94.7	94.0	58.4	67.7
Imports of goods (f.o.b.)	-330.7	-342.1	-358.5	-216.0	-346.3
Trade balance	-240.1	-247.4	-264.5	-157.6	-278.6
Services - debits	-123.4	-188.5	-179.7	-164.6	-206.4
Services - credits	37.0	NA	NA	NA	NA
Private transfers (net)	19.4	5.3	-13.1	6.4	-2.9
Government transfers (net)	204.3	304.9	343.3	217.3	331.2
Long-term capital (net)	76.3	33.3	77.2	-33.5	-26.0
Short-term capital (net)	18.2	100.0	11.6	108.0	127.0
Errors and omissions	15.5	19.2	40.7	21.1	35.3
Overall balance	7.2	26.8	15.5	-2.9	-20.4

Exchange Rates [38]

Currency: **Somali shilling.**
Symbol: **So. Sh.**

Data are currency units per $1.

January 1996	7,000
1 January 1995	5,000
1 July 1993	2,616
December 1992	4,200
December 1990	3,800.00
1989	490.7

Imports and Exports

Top Import Origins [39]

$249 million (1990 est.) Data are for 1986.

Origins	%
US	13
Italy	NA
FRG	NA
Kenya	NA
UK	NA
Saudi Arabia	NA

Top Export Destinations [40]

$100 million (1995 est.).

Destinations	%
Saudi Arabia	NA
Other Gulf states	NA
Italy	NA
US	NA

Foreign Aid [41]

Recipient: ODA, $NA.

Import and Export Commodities [42]

Import Commodities
Petroleum products
Foodstuffs
Construction materials

Export Commodities
Bananas
Live animals
Fish
Hides

South Africa

Geography [1]

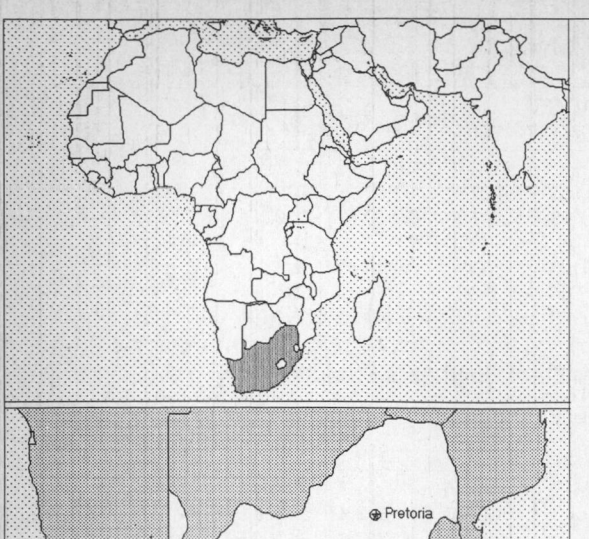

Total area:
1,219,912 sq km   471,011 sq mi
Land area:
1,219,912 sq km   471,011 sq mi
Comparative area:
Slightly less than twice the size of Texas
Note: Includes Prince Edward Islands (Marion Island and Prince Edward Island)
Land boundaries:
Total 4,750 km, Botswana 1,840 km, Lesotho 909 km, Mozambique 491 km, Namibia 855 km, Swaziland 430 km, Zimbabwe 225 km
Coastline:
2,798 km
Climate:
Mostly semiarid; subtropical along East Coast; sunny days, cool nights
Terrain:
Vast interior plateau rimmed by rugged hills and narrow coastal plain
Natural resources:
Gold, chromium, antimony, coal, iron ore, manganese, nickel, phosphates, tin, uranium, gem diamonds, platinum, copper, vanadium, salt, natural gas
Land use:
Arable land: 10%
Permanent crops: 1%
Meadows and pastures: 65%
Forest and woodland: 3%
Other: 21%

Demographics [2]

	1970	1980	1990	1995[1]	1996	2000	2010	2020	2030
Population	22,740	29,252	37,191	40,997	41,743	44,462	49,200	52,264	55,664
Population density (persons per sq. mi.)	48	62	79	87	89	94	104	111	118
(persons per sq. km.)	19	24	30	34	34	36	40	43	46
Net migration rate (per 1,000 population)	1.4	0.6	0.3	0.0	0.0	0.0	0.0	0.0	0.0
Births	NA	NA	NA	NA	1,165	NA	NA	NA	NA
Deaths	NA	NA	NA	NA	431	NA	NA	NA	NA
Life expectancy - males	NA	54.4	58.7	57.8	57.2	55.0	50.5	53.5	61.4
Life expectancy - females	NA	61.5	65.7	62.4	61.8	59.4	53.4	56.8	66.7
Birth rate (per 1,000)	36.6	33.2	30.7	28.5	27.9	25.5	21.9	20.0	18.0
Death rate (per 1,000)	13.7	10.6	8.9	10.0	10.3	11.6	15.2	14.4	10.8
Women of reproductive age (15-49 yrs.)	NA	NA	9,379	10,428	10,638	11,416	12,842	13,820	14,828
of which are currently married[50]	NA	2,634	4,559	NA	5,237	5,636	6,298	NA	NA
Fertility rate	5.4	4.6	3.8	3.5	3.4	3.1	2.6	2.3	2.1

Except as noted, values for vital statistics are in thousands; life expectancy is in years.

Health

Health Indicators [3]

% of population with access to	
safe water (1990-95)	70
adequate sanitation (1990-95)	NA
health services (1985-95)	NA
% of 1-year-olds immunized (1990-94) against	
TB (tuberculosis)	95
DPT (diphtheria, pertussis, tetanus)	73
polio	72
measles	76
% of contraceptive prevalence (1980-94)	50
ORT use rate (1990-94)	NA

Health Expenditures [4]

Total health expenditure, 1990 (official exchange rate)	
Millions of dollars	5,671
Dollars per capita	158
Health expenditures as a percentage of GDP	
Total	5.6
Public sector	3.2
Private sector	2.4
Development assistance for health	
Total aid flows (millions of dollars)[1]	2
Aid flows per capita (dollars)	NA
Aid flows as a percentage of total health expenditure	NA

For sources, notes, and explanations, see Annotated Source Appendix, page 1061.

Human Factors

Women and Children [5]

% of pregnant women immunized (tetanus 1990-94)	26
% of births attended by trained health personnel (1983-94)	NA
Maternal mortality rate (1980-92)	84
Under-5 mortality rate (1994)	68
% under-5 moderately/severely underweight (1980-1994)	NA

Burden of Disease [6]

Population per physician (1987)	1,716.67
Population per nurse	NA
Population per hospital bed	NA
AIDS cases per 100,000 people (1994)	9.1
Malaria cases per 100,000 people (1992)	NA

Ethnic Division [7]

Black	75.2%
White	13.6%
Colored	8.6%
Indian	2.6%

Religion [8]

Christian, Hindu, Muslim 2%.

Major Languages [9]

11 official languages, including Afrikaans, English, Ndebele, Pedi, Sotho, Swazi, Tsonga, Tswana, Venda, Xhosa, Zulu.

Education

Public Education Expenditures [10]

Million (Rand)[61]	1980	1986	1990	1992	1993	1994
Total education expenditure	NA	8,108	17,153	23,221	26,336	NA
as percent of GNP	NA	6.0	6.5	7.0	7.1	NA
as percent of total govt. expend.	NA	NA	NA	22.1	22.9	NA
Current education expenditure	NA	6,844	15,265	21,397	23,849	NA
as percent of GNP	NA	5.0	5.8	6.5	6.4	NA
as percent of current govt. expend.	NA	NA	NA	NA	NA	NA
Capital expenditure	NA	1,264	1,888	1,824	2,487	NA

Educational Attainment [11]

Age group (1985)[18]	25+
Total population	10,388,428
Highest level attained (%)	
No schooling	24.8
First level	
Not completed	41.6
Completed	4.8
Entered second level	
S-1	20.6
S-2	5.9
Postsecondary	2.3

Illiteracy [12]

In thousands and percent[1]	1990	1995	2000
Illiterate population (15+ yrs.)	4,604	4,731	4,847
Illiteracy rate - total pop. (%)	19.7	18.1	16.8
Illiteracy rate - males (%)	19.4	18.1	17.1
Illiteracy rate - females (%)	19.9	18.1	16.5

Libraries [13]

	Admin. Units	Svc. Pts.	Vols. (000)	Shelving (meters)	Vols. Added	Reg. Users
National (1991)	2	5	NA	NA	NA	NA
Nonspecialized	NA	NA	NA	NA	NA	NA
Public (1991)	670	NA	NA	NA	NA	NA
Higher ed. (1989)	84	166	8,120	NA	NA	NA
School	NA	NA	NA	NA	NA	NA

Daily Newspapers [14]

	1980	1985	1990	1994
Number of papers	24[e]	24	22	17
Circ. (000)	1,400[e]	1,440	1,340	1,346

Culture [15]

Science and Technology

Scientific/Technical Forces [16]

Scientists/engineers	12,102
Number female	NA
Technicians	5,006
Number female	NA
Total	17,108

R&D Expenditures [17]

	Rand (000) 1991
Total expenditure	2,786,086
Capital expenditure	240,764
Current expenditure	2,545,322
Percent current	91.4

U.S. Patents Issued [18]

Values show patents issued to citizens of the country by the U.S. Patents Office.

	1993	1994	1995
Number of patents	101	109	127

For sources, notes, and explanations, see Annotated Source Appendix, page 1061.

869

Government and Law

Organization of Government [19]

Long-form name:
Republic of South Africa
Type:
Republic
Independence:
31 May 1910 (from UK)
National holiday:
Freedom Day, 27 April (1994)
Constitution:
27 April 1994 (interim constitution, replacing the constitution of 3 September 1984); note - on 8 May 1996, the Constitutional Assembly voted 421 to two to pass a new constitution which, after certification by the Constitutional Court, will gradually go into effect over a three-year period and come into full force with the next national elections in April 1999
Legal system:
Based on Roman-Dutch law and English common law; accepts compulsory ICJ jurisdiction, with reservations
Executive branch:
President; 2 Deputy Executive Presidents; Cabinet
Legislative branch:
Bicameral: National Assembly and Senate
Judicial branch:
Supreme Court

Elections [20]

National Assembly	% of votes
African National Cong. (ANC)	62.6
National Party (NP)	20.4
Inkatha Feedom Party (IFP)	10.5
Freedom Front (FF)	2.2
Democratic Party (DP)	1.7
Pan Africanist Cong. (PAC)	1.2
African Christian Dem (ACDP)	0.5
Other	0.9

Government Budget [21]

For FY94/95 est.

	$ bil.
Revenues	30.5
Expenditures	38
Capital expenditures	2.6

Crime [23]

	1990
Crime volume	
Cases known to police	NA
Attempts (percent)	NA
Percent cases solved	NA
Crimes per 100,000 persons	NA
Persons responsible for offenses	
Total number offenders	364,518
Percent female	13.2
Percent juvenile	26.2
Percent foreigners	NA

Military Expenditures and Arms Transfers [22]

	1990	1991	1992	1993	1994
Military expenditures					
Current dollars (mil.)	4,444[e]	3,849[e]	3,254	3,482	2,899
1994 constant dollars (mil.)	4,945[e]	4,126[e]	3,393	3,554	2,899
Armed forces (000)	85	80	72	72	102
Gross national product (GNP)					
Current dollars (mil.)	105,000	109,100	109,700	113,900	119,500
1994 constant dollars (mil.)	116,800	117,000	114,400	116,300	119,500
Central government expenditures (CGE)					
1994 constant dollars (mil.)	37,420	35,930	38,590	38,790	39,110
People (mil.)	39.5	40.6	41.7	42.8	43.9
Military expenditure as % of GNP	4.2	3.5	3.0	3.1	2.4
Military expenditure as % of CGE	13.2	11.5	8.8	9.2	7.4
Military expenditure per capita (1994 $)	125	102	81	83	66
Armed forces per 1,000 people (soldiers)	2.1	2.0	1.7	1.7	2.3
GNP per capita (1994 $)	2,954	2,882	2,744	2,717	2,720
Arms imports[6]					
Current dollars (mil.)	0	90	0	0	20
1994 constant dollars (mil.)	0	96	0	0	20
Arms exports[6]					
Current dollars (mil.)	50	10	90	160	50
1994 constant dollars (mil.)	56	11	94	163	50
Total imports[7]					
Current dollars (mil.)	18,400	18,830	19,760	20,020	23,390
1994 constant dollars (mil.)	20,480	20,190	20,610	20,430	23,390
Total exports[7]					
Current dollars (mil.)	23,550	23,310	23,410	24,260	24,990
1994 constant dollars (mil.)	26,210	24,980	24,420	24,760	24,990
Arms as percent of total imports[8]	0	0.5	0	0	0.1
Arms as percent of total exports[8]	0.2	0	0.4	0.7	0.2

Human Rights [24]

	SSTS	FL	FAPRO	PPCG	APROBC	TPW	PCPTW	STPEP	PHRFF	PRW	ASST	AFL
Observes	P					P	P	P		S		
	EAFRD	CPR	ESCR	SR	ACHR	MAAE	PVIAC	PVNAC	EAFDAW	TCIDTP	RC	
Observes	S	S	S	P			P	P	P	S	P	

P = Party; S = Signatory; see Appendix for meaning of abbreviations.

Labor Force

Total Labor Force [25]

14.2 million economically active (1996)

Labor Force by Occupation [26]

Services	35%
Agriculture	30
Industry	20
Mining	9
Other	6

Unemployment Rate [27]

32.6% (1996 est.); an additional 11% underemployment

Production Sectors

Commercial Energy Production and Consumption

Data are shown in quadrillion (10^{15}) BTUs and percent for 1995
Values for hydroelectric, nuclear, geothermal, solar, and wind power refer to electrical generation.

Production [28]

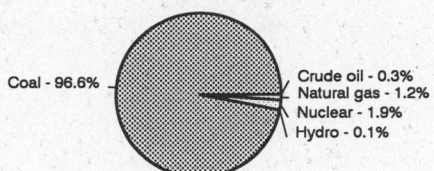

Coal - 96.6%
Crude oil - 0.3%
Natural gas - 1.2%
Nuclear - 1.9%
Hydro - 0.1%

Consumption [29]

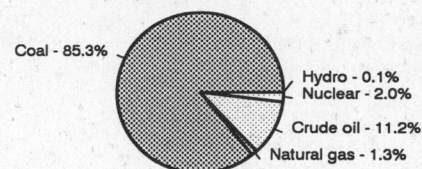

Coal - 85.3%
Hydro - 0.1%
Nuclear - 2.0%
Crude oil - 11.2%
Natural gas - 1.3%

Crude oil	0.018
Dry natural gas	0.072
Coal	5.870
Net hydroelectric power	0.005
Net nuclear power	0.113
Total	6.078

Crude oil	0.620
Dry natural gas	0.072
Coal	4.719
Net hydroelectric power	0.005
Net nuclear power	0.113
Total	5.529

Telecommunications [30]

- 5,206,235 (1993 est.) telephones; the system is the best developed, most modern, and has the highest capacity in Africa
- Domestic: consists of carrier-equipped open-wire lines, coaxial cables, microwave radio relay links, fiber-optic cable, and radiotelephone communication stations; key centers are Bloemfontein, Cape Town, Durban, Johannesburg, Port Elizabeth, and Pretoria
- International: 1 submarine cable; satellite earth stations - 3 Intelsat (1 Indian Ocean and 2 Atlantic Ocean)
- Radio: Broadcast stations: AM 14, FM 286, shortwave 0 Radios: 12.1 million (1992 est.)
- Television: Broadcast stations: 67 (1987 est.) Televisions: 3.45 million (1990 est.)

Top Agricultural Products [32]

Produces corn, wheat, sugarcane, fruits, vegetables; cattle, poultry, sheep, wool, milk, beef.

Transportation [31]

Railways: total: 21,431 km; narrow gauge: 20,995 km 1.067-m gauge (9,087 km electrified); 436 km 0.610-m gauge (1995)

Highways: total: 182,329 km; paved: 55,428 km (including 2,040 km of expressways); unpaved: 126,901 km (1991 est.)

Merchant marine: total: 4 container ships (1,000 GRT or over) totaling 211,276 GRT/198,602 DWT (1995 est.)

Airports

Total:	667
With paved runways over 3,047 m:	10
With paved runways 2,438 to 3,047 m:	4
With paved runways 1,524 to 2,437 m:	44
With paved runways 914 to 1,523 m:	75
With paved runways under 914 m:	221

Top Mining Products [33]

Thousand metric tons except as noted	9/29/95*
Gold, primary (tons)	580
Iron ore and concentrate, gross weight	32,300
Manganese, metallurgical, gross weight	2,800
Uranium oxide (tons)	1,910
Vanadium (tons)	15,700[e]
Diamond, gem and industrial (000 carats)	10,800[e]
Feldspar (tons)	38,400
Tiger's eye (tons)	531
Coal, salable product	196,000
Petroleum refinery products (000 42-gal. bls.)	122,000[e,65]

Tourism [34]

	1990	1991	1992	1993	1994
Tourists[100]	1,029	1,710	2,892	3,358	3,897
Tourism receipts	992	1,131	1,226	1,327	1,424
Tourism expenditures	1,117	1,148	1,544	1,721	1,678
Fare receipts	331	305	362	316	370
Fare expenditures	410	508	494	456	520

Travelers are in thousands, money in million U.S. dollars.

For sources, notes, and explanations, see Annotated Source Appendix, page 1061.

Manufacturing Sector

Manufacturing Summary [35]

	1987		1988		1989		1990		1991	
	$ bil.	%	$ bil.	%	$ bil.	%	$ bil.	%	$ bil.	%
Establishments or enterprises (number)	-	-	-	-	-	-	-	-	-	-
Total employment (000)	1,118	0.824	1,371	1.002	1,384	1.124	1,387	1.253	1,360	1.958
Production workers (000)	-	-	-	-	-	-	-	-	-	-
Output ($ bil.)	53	0.525	59	0.510	62	0.524	67	0.588	67	0.659
Value added ($ bil.)	16	0.382	19	0.390	19	0.396	22	0.437	23	0.661
Capital investment ($ mil.)	-	-	-	-	-	-	-	-	-	-
M & E investment ($ mil.)	-	-	-	-	-	-	-	-	-	-
Employees per establishment (number)	-	-	-	-	-	-	-	-	-	-
Production workers per establishment	-	-	-	-	-	-	-	-	-	-
Output per establishment ($ mil.)	-	-	-	-	-	-	-	-	-	-
Capital investment per estab. ($ mil.)	-	-	-	-	-	-	-	-	-	-
M & E per establishment ($ mil)	-	-	-	-	-	-	-	-	-	-
Payroll per employee ($)	5,998	66.894	6,622	66.469	6,566	58.770	7,666	55.266	7,994	54.750
Wages per production worker ($)	-	-	-	-	-	-	-	-	-	-
Hours per production worker (hours)	-	-	-	-	-	-	-	-	-	-
Output per employee ($)	47,757	63.766	43,363	50.911	44,621	46.558	48,665	46.959	49,429	33.669
Capital investment per employee ($)	-	-	-	-	-	-	-	-	-	-
M & E per employee ($)	-	-	-	-	-	-	-	-	-	-

Note: Columns headed % show percent of world total or ratio. Ratios closest to 100 are closest to world average. M & E stands for machinery & equipment.

Output in Manufacturing

	1987		1988		1989		1990		1991	
	$ bil.	%	$ bil.	%	$ bil.	%	$ bil.	%	$ bil.	%
3110 - Food products	7.774	14.67	7.686	13.03	7.742	12.49	9.043	13.50	9.325	13.92
3130 - Beverages	2.145	4.05	2.382	4.04	2.477	4.00	2.792	4.17	2.958	4.41
3140 - Tobacco	0.550	1.04	0.582	0.99	0.532	0.86	0.662	0.99	0.715	1.07
3210 - Textiles	2.122	4.00	2.233	3.78	2.276	3.67	2.335	3.49	2.293	3.42
3211 - Spinning, weaving, etc.	1.246	2.35	1.281	2.17	1.313	2.12	1.365	2.04	1.363	2.03
3220 - Wearing apparel	1.203	2.27	1.231	2.09	1.301	2.10	1.525	2.28	1.560	2.33
3230 - Leather and products	0.282	0.53	0.296	0.50	0.294	0.47	0.327	0.49	0.308	0.46
3240 - Footwear	0.527	0.99	0.580	0.98	0.573	0.92	0.645	0.96	0.673	1.00
3310 - Wood products	0.843	1.59	0.901	1.53	0.940	1.52	1.085	1.62	1.068	1.59
3320 - Furniture, fixtures	0.599	1.13	0.640	1.08	0.636	1.03	0.716	1.07	0.705	1.05
3410 - Paper and products	2.470	4.66	2.879	4.88	2.919	4.71	3.181	4.75	3.299	4.92
3411 - Pulp, paper, etc.	1.384	2.61	1.654	2.80	1.692	2.73	1.744	2.60	1.730	2.58
3420 - Printing, publishing	1.041	1.96	1.112	1.88	1.165	1.88	1.330	1.99	1.421	2.12
3510 - Industrial chemicals	2.636	4.97	2.930	4.97	3.093	4.99	3.279	4.89	3.360	5.01
3511 - Basic chemicals, excl fertilizers	0.931	1.76	0.971	1.65	1.012	1.63	1.145	1.71	1.151	1.72
3513 - Synthetic resins, etc.	0.984	1.86	1.153	1.95	1.197	1.93	1.232	1.84	1.207	1.80
3522 - Drugs and medicines	0.734	1.38	0.790	1.34	0.816	1.32	1.018	1.52	1.096	1.64
3550 - Rubber products	0.722	1.36	0.829	1.41	0.866	1.40	1.009	1.51	0.906	1.35
3560 - Plastic products nec	0.983	1.85	1.166	1.98	1.233	1.99	1.440	2.15	1.489	2.22
3610 - Pottery, china, etc.	0.070	0.13	0.074	0.13	0.068	0.11	0.073	0.11	0.065	0.10
3620 - Glass and products	0.352	0.66	0.401	0.68	0.395	0.64	0.481	0.72	0.485	0.72
3690 - Nonmetal products nec	1.406	2.65	1.669	2.83	1.590	2.56	1.725	2.57	1.660	2.48
3710 - Iron and steel	4.101	7.74	4.740	8.03	5.038	8.13	5.201	7.76	4.844	7.23
3720 - Nonferrous metals	1.396	2.63	1.639	2.78	1.820	2.94	1.899	2.83	1.734	2.59
3810 - Metal products	3.262	6.15	3.273	5.55	3.346	5.40	3.706	5.53	3.693	5.51
3820 - Machinery nec	2.380	4.49	2.477	4.20	2.556	4.12	2.741	4.09	2.592	3.87
3825 - Office, computing machinery	0.024	0.05	0.026	0.04	0.031	0.05	0.026	0.04	-	-
3830 - Electrical machinery	2.445	4.61	2.715	4.60	2.782	4.49	3.131	4.67	2.968	4.43
3832 - Radio, television, etc.	0.729	1.38	0.737	1.25	0.764	1.23	0.933	1.39	0.830	1.24
3840 - Transportation equipment	4.477	8.45	5.802	9.83	6.240	10.06	6.448	9.62	6.586	9.83
3843 - Motor vehicles	2.812	5.31	3.821	6.48	4.257	6.87	4.290	6.40	4.264	6.36
3850 - Professional goods	0.171	0.32	0.177	0.30	0.185	0.30	0.206	0.31	0.234	0.35
3900 - Industries nec	0.592	1.12	0.604	1.02	0.609	0.98	0.765	1.14	0.641	0.96

Note: Codes are International Standard Industry codes (ISIC). Percentages are % of total Output. [f]: Factor Prices; [p]: Producer Prices.

872

For sources, notes, and explanations, see Annotated Source Appendix, page 1061.

Finance, Economics, and Trade

Economic Indicators [36]

- **National product**: GDP—purchasing power parity—$215 billion (1995 est.)

- **National product real growth rate**: 3.3% (1995 est.)

- **National product per capita**: $4,800 (1995 est.)

- **Inflation rate (consumer prices)**: 8.7% (1995)

- **External debt**: $22 billion (1995 est.)

Balance of Payments Summary [37]

Values in millions of dollars.

	1988	1989	1990	1991	1992
Exports of goods (f.o.b.)	22,432	22,399	23,383	23,715	23,645
Imports of goods (f.o.b.)	-172,100	-16,810	-17,045	-17,449	-18,216
Trade balance	5,222	5,589	6,338	6,266	5,429
Services - debits	-7,424	-7,755	-8,397	-7,871	-8,816
Services - credits	3,244	3,544	4,195	4,200	4,669
Private transfers (net)	90	129	107	101	32
Government transfers (net)	86	72	10	-31	74
Long-term capital (net)	-387	-490	-704	-973	-522
Short-term capital (net)	-640	-13	-789	812	711
Errors and omissions	-965	-575	-416	-1,388	-1,163
Overall balance	-774	501	344	1,116	414

Exchange Rates [38]

Currency: **rand.**
Symbol: **R.**

Data are currency units per $1.

January 1996	3.6417
1995	3.6266
1994	3.5490
1993	3.2636
1992	2.8497
1991	2.7563

Imports and Exports

Top Import Origins [39]

$27 billion (f.o.b., 1995).

Origins	%
Germany	NA
US	NA
Japan	NA
UK	NA
Italy	NA

Top Export Destinations [40]

$27.9 billion (f.o.b., 1995).

Destinations	%
Italy	NA
Japan	NA
Germany	NA
UK	NA
Other EU countries	NA
US	NA
Hong Kong	NA

Foreign Aid [41]

	U.S. $	
ODA	NA	
US pledge (1996)	600	million
UK	150	million
Australia	21	million
Japan (1996)	1.3	biilion
EU	833	million

Import and Export Commodities [42]

Import Commodities

Machinery 32%
Transport equipment 15%
Chemicals 11%
Oil
Textiles
Scientific instruments

Export Commodities

Gold 27%
Other minerals
 and metals 20%
Food 5%
Chemicals 3%

For sources, notes, and explanations, see Annotated Source Appendix, page 1061.

873

Spain

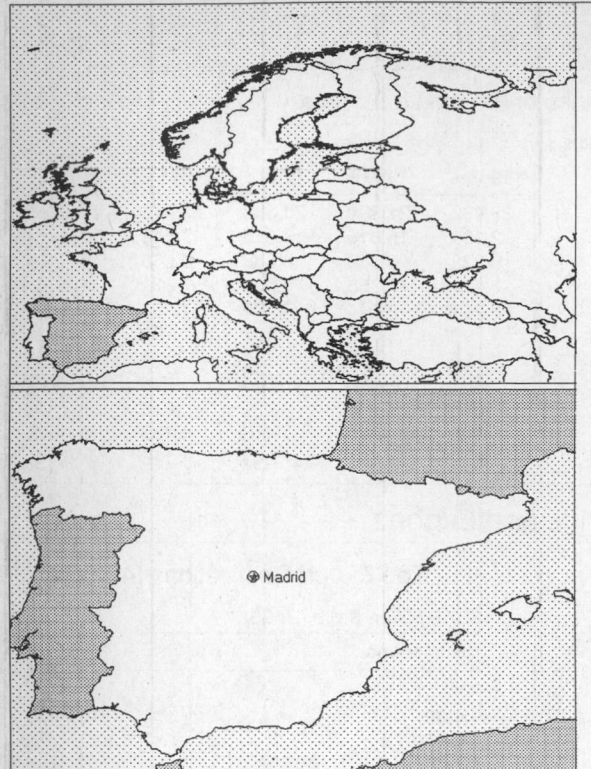

Geography [1]

Total area:
504,750 sq km   194,885 sq mi
Land area:
499,400 sq km   192,819 sq mi
Comparative area:
Slightly more than twice the size of Oregon
Note: Includes Balearic Islands, Canary Islands, and five places of sovereignty (plazas de soberania) on and off the coast of Morocco - Ceuta, Mellila, Islas Chafarinas, Penon de Alhucemas, and Penon de Velez de la Gomera
Land boundaries:
Total 1,903.2 km, Andorra 65 km, France 623 km, Gibraltar 1.2 km, Portugal 1,214 km
Note: Excludes the length of the boundary between the places of sovereignty and Morocco
Coastline:
4,964 km
Climate:
Temperate; clear, hot summers in interior, more moderate and cloudy along coast; cloudy, cold winters in interior, partly cloudy and cool along coast
Terrain:
Large, flat to dissected plateau surrounded by rugged hills; Pyrenees in North
Natural resources:
Coal, lignite, iron ore, uranium, mercury, pyrites, fluorspar, gypsum, zinc, lead, tungsten, copper, kaolin, potash, hydropower
Land use:
Arable land: 31%
Permanent crops: 10%
Meadows and pastures: 21%
Forest and woodland: 31%
Other: 7%

Demographics [2]

	1970	1980	1990	1995[1]	1996	2000	2010	2020	2030
Population	33,876	37,488	38,793	39,118	39,181	39,545	40,398	39,758	38,450
Population density (persons per sq. mi.)	176	194	201	203	203	205	210	206	199
(persons per sq. km.)	68	75	78	78	78	79	81	80	77
Net migration rate (per 1,000 population)	NA	0.3	NA	0.4	0.4	0.5	0.3	0.1	0.0
Births	661	571	NA	NA	314	NA	NA	NA	NA
Deaths	280	289	NA	NA	368	NA	NA	NA	NA
Life expectancy - males	69.6	72.5	NA	74.7	75.0	75.9	77.7	79.0	79.9
Life expectancy - females	75.1	78.6	NA	81.6	81.8	82.6	84.1	85.2	86.0
Birth rate (per 1,000)	19.5	15.2	NA	10.0	10.0	12.0	9.9	7.6	8.2
Death rate (per 1,000)	8.5	7.7	NA	8.8	8.9	9.1	9.8	10.7	11.9
Women of reproductive age (15-49 yrs.)	NA	NA	NA	10,114	10,145	10,131	9,425	8,466	7,198
of which are currently married	5,053	NA	NA	NA	5,926	6,175	6,292	NA	NA
Fertility rate	NA	2.2	NA	1.3	1.3	1.5	1.5	1.5	1.5

Except as noted, values for vital statistics are in thousands; life expectancy is in years.

Health

Health Indicators [3]

% of population with access to	
safe water (1990-95)	NA
adequate sanitation (1990-95)	NA
health services (1985-95)	NA
% of 1-year-olds immunized (1990-94) against	
TB (tuberculosis)	NA
DPT (diphtheria, pertussis, tetanus)	87
polio	88
measles	90
% of contraceptive prevalence (1980-94)	59
ORT use rate (1990-94)	NA

Health Expenditures [4]

Total health expenditure, 1990 (official exchange rate)	
Millions of dollars	32,375
Dollars per capita	831
Health expenditures as a percentage of GDP	
Total	6.6
Public sector	5.2
Private sector	1.4
Development assistance for health	
Total aid flows (millions of dollars)[1]	NA
Aid flows per capita (dollars)	NA
Aid flows as a percentage of total health expenditure	NA

For sources, notes, and explanations, see Annotated Source Appendix, page 1061.

Human Factors

Women and Children [5]

% of pregnant women immunized (tetanus 1990-94)	NA
% of births attended by trained health personnel (1983-94)	96
Maternal mortality rate (1980-92)	5
Under-5 mortality rate (1994)	9
% under-5 moderately/severely underweight (1980-1994)	NA

Burden of Disease [6]

Population per physician (1988)	277.13
Population per nurse (1984)	258.06
Population per hospital bed (1990)	208.74
AIDS cases per 100,000 people (1994)	16.0
Heart disease cases per 1,000 people (1990-93)	256

Ethnic Division [7]

Composite of Mediterranean and Nordic types.

Religion [8]

Roman Catholic	99%
Other sects	1%

Major Languages [9]

Castilian Spanish, Catalan 17%, Galician 7%, Basque 2%.

Education

Public Education Expenditures [10]

	1980	1990	1991	1992	1993	1994
Million or Trillion (T) (Peseta)						
Total education expenditure	NA	2.18 T	2.45 T	2.68 T	2.84 T	NA
as percent of GNP	NA	4.4	4.5	4.6	4.7	NA
as percent of total govt. expend.	NA	9.4	NA	9.3	NA	NA
Current education expenditure	NA	1.94 T	2.17 T	2.42 T	2.59 T	NA
as percent of GNP	NA	3.9	4.0	4.2	4.3	NA
as percent of current govt. expend.	NA	NA	NA	NA	NA	NA
Capital expenditure	NA	245,732	280,328	257,595	252,968	NA

Educational Attainment [11]

Age group (1991)	25+
Total population	24,667,414
Highest level attained (%)	
No schooling	30.4
First level	
Not completed	34.9
Completed	NA
Entered second level	
S-1	25.5
S-2	NA
Postsecondary	8.4

Illiteracy [12]

In thousands and percent[2]	1989	1991	1995
Illiterate population (15+ yrs.)	NA	1,081	957
Illiteracy rate - total pop. (%)	NA	3.5	2.9
Illiteracy rate - males (%)	NA	1.9	1.8
Illiteracy rate - females (%)	NA	4.9	3.9

Libraries [13]

	Admin. Units	Svc. Pts.	Vols. (000)	Shelving (meters)	Vols. Added	Reg. Users[23]
National (1990)	1	2	3,500	71,148[e]	-	55,909
Nonspecialized (1992)	59	NA	NA	NA	NA	NA
Public (1992)	3,993	4,609	29,718	NA	2 mil	4 mil
Higher ed. (1992)	648	1,131	18,618	783,100	959,271	2 mil
School	NA	NA	NA	NA	NA	NA

Daily Newspapers [14]

	1980	1985	1990	1994
Number of papers	111	102	125[e]	148
Circ. (000)	3,487	3,078	3,450	4,100[e]

Culture [15]

Cinema (seats per 1,000)	NA
Annual attendance per person	2.2
Gross box office receipts (mil. Peseta)	40,579
Museums (reporting)	567
Visitors (000)	28,474
Annual receipts (000 Peseta)	NA

Science and Technology

Scientific/Technical Forces [16]

Scientists/engineers	43,367
Number female	11,899
Technicians[32]	13,496
Number female	3,093
Total[32]	56,863

R&D Expenditures [17]

	Peseta (000) 1993
Total expenditure	557,401,895
Capital expenditure	103,605,003
Current expenditure	453,796,892
Percent current	81.4

U.S. Patents Issued [18]

Values show patents issued to citizens of the country by the U.S. Patents Office.

	1993	1994	1995
Number of patents	185	172	168

For sources, notes, and explanations, see Annotated Source Appendix, page 1061.

Government and Law

Organization of Government [19]

Long-form name:
Kingdom of Spain
Type:
Parliamentary monarchy
Independence:
1492 (expulsion of the Moors and unification)
National holiday:
National Day, 12 October
Constitution:
6 December 1978, effective 29 December 1978
Legal system:
Civil law system, with regional applications; does not accept compulsory ICJ jurisdiction
Executive branch:
King; Prime Minister; Deputy Prime Minister; Council of Ministers; Council of State:
Legislative branch:
Bicameral The General Courts or National Assembly (Las Cortes Generales): Senate (Senado) and Congress of Deputies (Congreso de los Diputados)
Judicial branch:
Supreme Court (Tribunal Supremo)

Elections [20]

Congress of Deputies	% of votes
Popular Party	38.9
Spanish Socialist Workers' Party	37.5
United Left	10.7
Convergence and Union	4.6
Other	8.3

Government Expenditures [21]

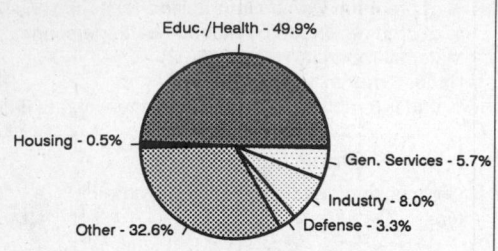

Educ./Health - 49.9%
Housing - 0.5%
Gen. Services - 5.7%
Industry - 8.0%
Defense - 3.3%
Other - 32.6%

(% distribution). Expend. in CY93: 23,998.9 (Peseta bil.)

Crime [23]

	1994
Crime volume	
Cases known to police	901,696
Attempts (percent)	4.20
Percent cases solved	24.84
Crimes per 100,000 persons	2,286.60
Persons responsible for offenses	
Total number offenders	193,083
Percent female	10.27
Percent juvenile (0-16 yrs.)	4.58
Percent foreigners	39.92

Military Expenditures and Arms Transfers [22]

	1990	1991	1992	1993	1994
Military expenditures					
Current dollars (mil.)	7,689	7,651	7,217	8,034	7,425
1994 constant dollars (mil.)	8,557	8,201	7,526	8,199	7,425
Armed forces (000)	263	246	198	204	213
Gross national product (GNP)					
Current dollars (mil.)	414,000	439,000	453,400	458,900	477,700
1994 constant dollars (mil.)	460,800	470,600	472,800	468,300	477,700
Central government expenditures (CGE)					
1994 constant dollars (mil.)	159,300	167,400	124,500	137,700	133,900
People (mil.)	39.0	39.0	39.1	39.2	39.3
Military expenditure as % of GNP	1.9	1.7	1.6	1.8	1.6
Military expenditure as % of CGE	5.4	4.9	6.0	6.0	5.5
Military expenditure per capita (1994 $)	220	210	192	209	189
Armed forces per 1,000 people (soldiers)	6.7	6.3	5.1	5.2	5.4
GNP per capita (1994 $)	11,830	12,050	12,090	11,940	12,150
Arms imports[6]					
Current dollars (mil.)	480	230	270	300	525
1994 constant dollars (mil.)	534	247	282	306	525
Arms exports[6]					
Current dollars (mil.)	350	90	170	160	280
1994 constant dollars (mil.)	390	96	177	163	280
Total imports[7]					
Current dollars (mil.)	87,710	93,310	99,760	78,630	92,510
1994 constant dollars (mil.)	97,620	100,000	10,400	80,250	92,510
Total exports[7]					
Current dollars (mil.)	55,640	60,180	64,330	59,550	73,290
1994 constant dollars (mil.)	61,930	64,500	67,090	60,780	73,290
Arms as percent of total imports[8]	0.5	0.2	0.3	0.4	0.6
Arms as percent of total exports[8]	0.6	0.1	0.3	0.3	0.4

Human Rights [24]

	SSTS	FL	FAPRO	PPCG	APROBC	TPW	PCPTW	STPEP	PHRFF	PRW	ASST	AFL
Observes	P	P	P	P	P	P	P	P	P	P	P	P

	EAFRD	CPR	ESCR	SR	ACHR	MAAE	PVIAC	PVNAC	EAFDAW	TCIDTP	RC
Observes	P	P	P	P		P	P	P	P	P	P

P = Party; S = Signatory; see Appendix for meaning of abbreviations.

Labor Force

Total Labor Force [25]

11.837 million

Labor Force by Occupation [26]

Services	59%
Industry	21
Agriculture	11
Construction	9

Date of data: 1993 est.

Unemployment Rate [27]

22.8% (year-end 1995)

For sources, notes, and explanations, see Annotated Source Appendix, page 1061.

Production Sectors

Commercial Energy Production and Consumption

Data are shown in quadrillion (10^{15}) BTUs and percent for 1995
Values for hydroelectric, nuclear, geothermal, solar, and wind power refer to electrical generation.

Production [28]

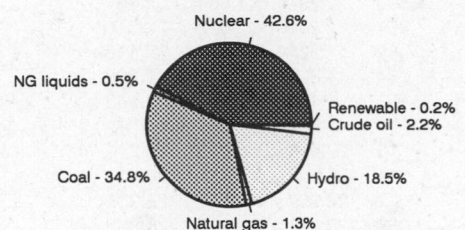

Nuclear - 42.6%
NG liquids - 0.5%
Renewable - 0.2%
Crude oil - 2.2%
Coal - 34.8%
Hydro - 18.5%
Natural gas - 1.3%

Consumption [29]

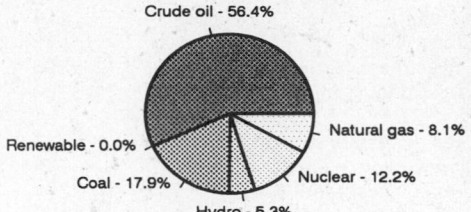

Crude oil - 56.4%
Natural gas - 8.1%
Renewable - 0.0%
Nuclear - 12.2%
Coal - 17.9%
Hydro - 5.3%

Crude oil	0.028
Natural gas liquids	0.006
Dry natural gas	0.016
Coal	0.437
Net hydroelectric power	0.232
Net nuclear power	0.535
Geothermal, solar, wind	0.002
Total	1.256

Crude oil	2.479
Dry natural gas	0.356
Coal	0.788
Net hydroelectric power	0.232
Net nuclear power	0.535
Geothermal, solar, wind	0.002
Total	4.392

Telecommunications [30]

- 12.6 million (1990 est.) telephones; generally adequate, modern facilities
- International: 22 coaxial submarine cables; satellite earth stations - 2 Intelsat (1 Atlantic Ocean and 1 Indian Ocean), NA Eutelsat, NA Inmarsat, and NA Marecs; tropospheric scatter to adjacent countries
- Radio: Broadcast stations: AM 190, FM 406 (repeaters 134), shortwave 0 Radios: 12 million (1992 est.)
- Television: Broadcast stations: 100 (repeaters 1,297) Televisions: 15.7 million (1992 est.)

Transportation [31]

Railways: total: 14,343 km; broad gauge: 12,139 km 1.668-m gauge (6,510 km electrified; 2,295 km double track); standard gauge: 488 km 1.435-m gauge (488 km electrified); narrow gauge: 1,716 km (privately owned: 1,669 km 1.000-m gauge, 489 km electrified; 28 km 0.914-m gauge, 28 km electrified; government owned: 19 km 1.000-m gauge, all electrified)

Highways: total: 331,961 km; paved: 328,641 km (including 2,700 km of expressways); unpaved: 3,320 km (1991 est.)

Merchant marine: total: 147 ships (1,000 GRT or over) totaling 874,688 GRT/1,391,421 DWT; ships by type: bulk 9, cargo 36, chemical tanker 11, combination ore/oil 1, container 8, liquefied gas tanker 4, oil tanker 25, passenger 2, refrigerated cargo 12, roll-on/roll-off cargo 32, short-sea passenger 6, specialized tanker 1 (1995 est.)

Airports

Total:	96
With paved runways over 3,047 m:	15
With paved runways 2,438 to 3,047 m:	11
With paved runways 1,524 to 2,437 m:	15
With paved runways 914 to 1,523 m:	13
With paved runways under 914 m:	28

Top Agricultural Products [32]

Agriculture accounts for 3.6% of the GDP; produces grain, vegetables, olives, wine grapes, sugar beets, citrus; beef, pork, poultry, dairy products; fish catch of 1.4 million metric tons is among top 20 nations.

Top Mining Products [33]

Thousand metric tons except as noted[e]	3/94[*]
Gold, mine output, Au content (kg.)	6,000
Steel, crude	13,500
Mercury metal (kg.)	50,000
Silver, mine output, Ag content (kg.)	160,000
Tantalum, gross weight (kg.)	5,000
Gypsum and anhydrite, crude	7,500
Limestone	150,000
Coal, marketable	34,000
Coke, metallurgical	3,000
Petroleum refinery products (000 42-gal. bls.)	440,290

Tourism [34]

	1990	1991	1992	1993	1994
Visitors[73]	52,044	53,495	55,331	57,263	61,428
Tourists	37,441	38,539	39,638	40,085	43,232
Excursionists	13,758	14,130	15,189	16,425	18,195
Cruise passengers	845	826	504	753	1,952
Tourism receipts	18,593	19,004	22,181	19,425	21,853
Tourism expenditures	4,254	4,530	5,542	4,706	4,188
Fare receipts	1,852	1,596	1,415	1,167	1,488
Fare expenditures	792	752	790	865	951

Travelers are in thousands, money in million U.S. dollars.

For sources, notes, and explanations, see Annotated Source Appendix, page 1061.

877

Manufacturing Sector

Manufacturing Summary [35]

	1987 $ bil.	1987 %	1988 $ bil.	1988 %	1989 $ bil.	1989 %	1990 $ bil.	1990 %	1991 $ bil.	1991 %
Establishments or enterprises (number)	147,818	6.301	146,865	6.996	149,374	7.974	151,428	8.453	-	-
Total employment (000)	2,163	1.594	2,183	1.595	2,249	1.827	2,268	2.049	-	-
Production workers (000)	1,639	2.430	1,661	2.659	1,690	2.297	1,694	2.387	-	-
Output ($ bil.)	204	2.005	242	2.079	267	2.261	323	2.815	-	-
Value added ($ bil.)	72	1.666	82	1.708	89	1.828	107	2.148	-	-
Capital investment ($ mil.)	6,796	1.559	8,029	1.686	10,284	1.877	13,330	2.398	-	-
M & E investment ($ mil.)	5,763	1.849	7,030	2.031	8,457	2.166	10,743	2.548	-	-
Employees per establishment (number)	15	25.293	15	22.800	15	22.914	15	24.240	-	-
Production workers per establishment	11	38.565	11	38.000	11	28.808	11	28.234	-	-
Output per establishment ($ mil.)	1	31.823	2	29.720	2	28.359	2	33.305	-	-
Capital investment per estab. ($ mil.)	0.046	24.747	0.055	24.099	0.069	23.543	0.088	28.372	-	-
M & E per establishment ($ mil)	0.039	29.350	0.048	29.031	0.057	27.163	0.071	30.144	-	-
Payroll per employee ($)	13,123	146.349	15,141	151.982	16,345	146.301	20,864	150.406	-	-
Wages per production worker ($)	11,812	149.699	13,418	158.794	14,651	146.099	18,380	154.770	-	-
Hours per production worker (hours)	1,743	91.866	1,742	90.631	1,749	95.437	1,745	93.253	-	-
Output per employee ($)	94,231	125.818	111,027	130.352	118,613	123.762	142,389	137.396	-	-
Capital investment per employee ($)	3,142	97.840	3,678	105.699	4,573	102.747	5,877	117.045	-	-
M & E per employee ($)	2,664	116.038	3,220	127.330	3,760	118.541	4,737	124.355	-	-

Note: Columns headed % show percent of world total or ratio. Ratios closest to 100 are closest to world average. M & E stands for machinery & equipment.

Output in Manufacturing

	1987 $ bil.[f]	1987 %	1988 $ bil.[f]	1988 %	1989 $ bil.[f]	1989 %	1990 $ bil.[f]	1990 %	1991 $ bil.[f]	1991 %
3110 - Food products	29.000	14.22	34.000	14.05	37.000	13.86	44.000	13.62	-	-
3130 - Beverages	6.989	3.43	7.571	3.13	7.586	2.84	9.556	2.96	-	-
3140 - Tobacco	1.822	0.89	1.897	0.78	1.867	0.70	2.325	0.72	-	-
3210 - Textiles	6.576	3.22	7.108	2.94	7.391	2.77	8.408	2.60	-	-
3211 - Spinning, weaving, etc.	4.041	1.98	4.404	1.82	4.545	1.70	5.121	1.59	-	-
3220 - Wearing apparel	3.264	1.60	3.966	1.64	4.114	1.54	5.484	1.70	-	-
3230 - Leather and products	1.652	0.81	1.811	0.75	1.842	0.69	2.031	0.63	-	-
3240 - Footwear	1.482	0.73	1.846	0.76	1.918	0.72	2.364	0.73	-	-
3310 - Wood products	3.102	1.52	4.155	1.72	4.638	1.74	5.837	1.81	-	-
3320 - Furniture, fixtures	2.243	1.10	2.618	1.08	2.931	1.10	3.748	1.16	-	-
3410 - Paper and products	4.730	2.32	5.382	2.22	5.905	2.21	6.917	2.14	-	-
3411 - Pulp, paper, etc.	2.332	1.14	2.756	1.14	2.957	1.11	3.355	1.04	-	-
3420 - Printing, publishing	4.657	2.28	5.983	2.47	7.535	2.82	9.683	3.00	-	-
3510 - Industrial chemicals	8.528	4.18	9.924	4.10	10.000	3.75	12.000	3.72	-	-
3511 - Basic chemicals, excl fertilizers	4.551	2.23	5.297	2.19	5.575	2.09	6.200	1.92	-	-
3513 - Synthetic resins, etc.	2.762	1.35	3.314	1.37	3.320	1.24	3.816	1.18	-	-
3520 - Chemical products nec	8.309	4.07	10.000	4.13	11.000	4.12	15.000	4.64	-	-
3522 - Drugs and medicines	3.118	1.53	3.829	1.58	4.494	1.68	5.847	1.81	-	-
3530 - Petroleum refineries	7.297	3.58	6.043	2.50	7.273	2.72	9.212	2.85	-	-
3540 - Petroleum, coal products	1.150	0.56	1.253	0.52	1.090	0.41	1.275	0.39	-	-
3550 - Rubber products	2.259	1.11	2.644	1.09	2.636	0.99	2.963	0.92	-	-
3560 - Plastic products nec	3.903	1.91	4.670	1.93	5.305	1.99	7.083	2.19	-	-
3610 - Pottery, china, etc.	0.454	0.22	0.575	0.24	0.608	0.23	0.785	0.24	-	-
3620 - Glass and products	1.417	0.69	1.717	0.71	1.867	0.70	2.296	0.71	-	-
3690 - Nonmetal products nec	5.936	2.91	7.151	2.95	8.312	3.11	11.000	3.41	-	-
3710 - Iron and steel	8.009	3.93	9.683	4.00	11.000	4.12	12.000	3.72	-	-
3720 - Nonferrous metals	3.086	1.51	4.258	1.76	4.528	1.70	4.846	1.50	-	-
3810 - Metal products	7.645	3.75	9.529	3.94	11.000	4.12	13.000	4.02	-	-
3820 - Machinery nec	8.584	4.21	10.000	4.13	12.000	4.49	14.000	4.33	-	-
3825 - Office, computing machinery	0.834	0.41	1.064	0.44	1.132	0.42	1.050	0.33	-	-
3830 - Electrical machinery	8.074	3.96	10.000	4.13	12.000	4.49	15.000	4.64	-	-
3832 - Radio, television, etc.	2.543	1.25	3.374	1.39	4.317	1.62	5.729	1.77	-	-
3840 - Transportation equipment	22.000	10.78	27.000	11.16	30.000	11.24	35.000	10.84	-	-
3841 - Shipbuilding, repair	1.182	0.58	0.996	0.41	2.027	0.76	2.600	0.80	-	-
3843 - Motor vehicles	19.000	9.31	24.000	9.92	25.000	9.36	30.000	9.29	-	-
3850 - Professional goods	0.486	0.24	0.506	0.21	0.524	0.20	0.726	0.22	-	-
3900 - Industries nec	1.109	0.54	1.382	0.57	1.529	0.57	2.168	0.67	-	-

Note: Codes are International Standard Industry codes (ISIC). Percentages are % of total Output. [f]: Factor Prices; [p]: Producer Prices.

For sources, notes, and explanations, see Annotated Source Appendix, page 1061.

Finance, Economics, and Trade

Economic Indicators [36]

- **National product**: GDP—purchasing power parity— $565 billion (1995 est.)

- **National product real growth rate**: 3% (1995 est.)

- **National product per capita**: $14,300 (1995 est.)

- **Inflation rate (consumer prices)**: 4.3% (1995)

- **External debt**: $90 billion (1993 est.)

Balance of Payments Summary [37]

Values in millions of dollars.

	1989	1990	1991	1992	1993
Exports of goods (f.o.b.)	43,301	53,888	58,901	63,921	60,232
Imports of goods (f.o.b.)	-67,797	-83,454	-89,654	-94,954	-75,950
Trade balance	-24,496	-29,566	-30,753	-31,033	-15,718
Services - debits	-19,919	-26,006	-30,201	-38,967	-32,808
Services - credits	28,875	34,496	38,175	45,709	39,321
Private transfers (net)	3,163	3,053	2,201	2,613	1,708
Government transfers (net)	1,444	1,204	3,859	3,197	2,857
Long-term capital (net)	16,879	19,223	34,062	21,526	38,036
Short-term capital (net)	1,463	9,079	-2,086	-15,109	-36,729
Errors and omissions	-2,693	-4,521	-1,117	-5,407	-1,474
Overall balance	4,716	6,962	14,140	-17,472	-4,807

Exchange Rates [38]

Currency: **peseta.**
Symbol: **Pta.**

Data are currency units per $1.

January 1996	123.19
1995	124.69
1994	133.96
1993	127.26
1992	102.38
1991	103.91

Imports and Exports

Top Import Origins [39]

$110 billion (c.i.f., 1995) Data are for 1994.

Origins	%
EU	60.9
Other developed countries	11.5
US	7.3
Middle East	6.2

Top Export Destinations [40]

$85 billion (f.o.b., 1995) Data are for 1994.

Destinations	%
EU	68.7
US	4.9
Other developed countries	7.9

Foreign Aid [41]

Donor: ODA, $1.213 billion (1993).

Import and Export Commodities [42]

Import Commodities	Export Commodities
Machinery	Cars and trucks
Transport equipment	Semifinished manufactured goods
Fuels	Foodstuffs
Semifinished goods	Machinery
Foodstuffs	
Consumer goods	
Chemicals	

Sri Lanka

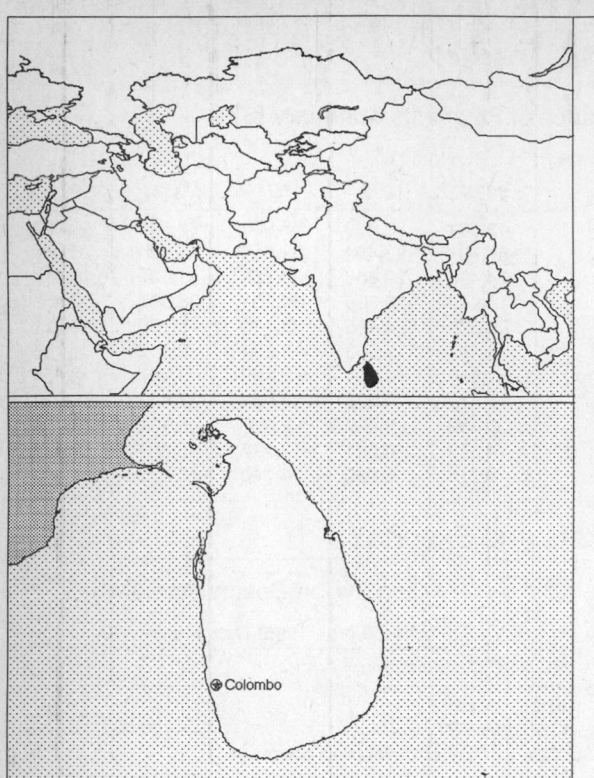

Geography [1]

Total area:
65,610 sq km   25,332 sq mi
Land area:
64,740 sq km   24,996 sq mi
Comparative area:
Slightly larger than West Virginia
Land boundaries:
0 km
Coastline:
1,340 km
Climate:
Tropical monsoon; northeast monsoon (December to March); southwest monsoon (June to October)
Terrain:
Mostly low, flat to rolling plain; mountains in south-central interior
Natural resources:
Limestone, graphite, mineral sands, gems, phosphates, clay
Land use:
Arable land: 16%
Permanent crops: 17%
Meadows and pastures: 7%
Forest and woodland: 37%
Other: 23%

Demographics [2]

	1970	1980	1990	1995[1]	1996	2000	2010	2020	2030
Population	12,532	14,900	17,227	18,343	18,553	19,377	21,331	22,877	23,859
Population density (persons per sq. mi.)	501	596	689	734	742	775	853	915	954
(persons per sq. km.)	194	230	266	283	287	299	329	353	369
Net migration rate (per 1,000 population)	NA	NA	1.2	-0.8	-0.8	-0.5	0.0	0.0	0.0
Births	NA	NA	NA	NA	332	NA	NA	NA	NA
Deaths	NA	NA	NA	NA	108	NA	NA	NA	NA
Life expectancy - males	NA	NA	68.6	69.6	69.8	70.6	72.3	73.8	75.1
Life expectancy - females	NA	NA	73.5	74.8	75.1	76.0	78.1	79.9	81.4
Birth rate (per 1,000)	NA	NA	20.0	18.1	17.9	16.9	15.0	12.7	11.7
Death rate (per 1,000)	NA	NA	5.8	5.8	5.8	5.9	6.4	7.3	8.7
Women of reproductive age (15-49 yrs.)	NA	NA	4,654	5,077	5,171	5,490	5,777	5,769	5,598
of which are currently married	NA	NA	4,282	NA	4,745	5,029	5,276	NA	NA
Fertility rate	NA	NA	2.3	2.1	2.1	2.0	1.8	1.8	1.8

Except as noted, values for vital statistics are in thousands; life expectancy is in years.

Health

Health Indicators [3]

% of population with access to	
safe water (1990-95)	53
adequate sanitation (1990-95)	61
health services (1985-95)[1]	93
% of 1-year-olds immunized (1990-94) against	
TB (tuberculosis)	86
DPT (diphtheria, pertussis, tetanus)	88
polio	88
measles	84
% of contraceptive prevalence (1980-94)	62
ORT use rate (1990-94)	76

Health Expenditures [4]

Total health expenditure, 1990 (official exchange rate)	
Millions of dollars	305
Dollars per capita	18
Health expenditures as a percentage of GDP	
Total	3.7
Public sector	1.8
Private sector	1.9
Development assistance for health	
Total aid flows (millions of dollars)[1]	26
Aid flows per capita (dollars)	1.5
Aid flows as a percentage of total health expenditure	7.4

For sources, notes, and explanations, see Annotated Source Appendix, page 1061.

Human Factors

Women and Children [5]

% of pregnant women immunized (tetanus 1990-94)	79
% of births attended by trained health personnel (1983-94)	94
Maternal mortality rate (1980-92)	80
Under-5 mortality rate (1994)	19
% under-5 moderately/severely underweight (1980-1994)	38

Burden of Disease [6]

Population per physician (1986)	7,269.73
Population per nurse (1985)	1,288.40
Population per hospital bed (1990)	364.50
AIDS cases per 100,000 people (1994)	*
Malaria cases per 100,000 people (1992)	2,260

Ethnic Division [7]

Sinhalese	74%
Tamil	18%
Moor	7%
Burgher, Malay, and Vedda	1%

Religion [8]

Buddhist	69%
Hindu	15%
Christian	8%
Muslim	8%

Major Languages [9]

English is commonly used in government and is spoken by about 10% of the population.

Sinhala (official and national language)	74%
Tamil (national language)	18%

Education

Public Education Expenditures [10]

Million (Rupee)[62]	1980	1985	1990	1992	1993	1994
Total education expenditure	1,799	4,183	8,621	13,883	15,515	18,259
as percent of GNP	2.7	2.6	2.7	3.3	3.1	3.2
as percent of total govt. expend.	7.7	6.9	8.1	8.8	9.0	9.4
Current education expenditure	1,535	3,530	7,024	10,598	12,604	15,314
as percent of GNP	2.3	2.2	2.2	2.5	2.6	2.7
as percent of current govt. expend.	13.5	11.4	10.7	10.6	12.2	13.2
Capital expenditure	264	653	1,597	3,285	2,911	2,945

Educational Attainment [11]

Age group (1981)	25+
Total population	6,490,502
Highest level attained (%)	
No schooling	15.9
First level	
Not completed	48.9
Completed	NA
Entered second level	
S-1	34.1
S-2	NA
Postsecondary	1.1

Illiteracy [12]

In thousands and percent[1]	1990	1995	2000
Illiterate population (15+ yrs.)	1,307	1,241	1,180
Illiteracy rate - total pop. (%)	11.1	9.5	8.2
Illiteracy rate - males (%)	7.3	6.4	5.7
Illiteracy rate - females (%)	14.9	12.6	10.6

Libraries [13]

	Admin. Units	Svc. Pts.	Vols. (000)	Shelving (meters)	Vols. Added	Reg. Users
National (1993)	1	1	157	NA	NA	96
Nonspecialized	NA	NA	NA	NA	NA	NA
Public (1989)	15	154	481	3,030	10,500	98,006
Higher ed. (1990)	11	36	829	NA	26,321	30,727
School	NA	NA	NA	NA	NA	NA

Daily Newspapers [14]

	1980	1985	1990	1994
Number of papers	21	17	18	9
Circ. (000)	450	390	550[e]	450[e]

Culture [15]

Cinema (seats per 1,000)	8.0
Annual attendance per person	1.5
Gross box office receipts (mil. Rupees)	223
Museums (reporting)	NA
Visitors (000)	NA
Annual receipts (000 Rupees)	NA

Science and Technology

Scientific/Technical Forces [16]

Scientists/engineers	2,790
Number female	667
Technicians	693
Number female	188
Total	3,483

R&D Expenditures [17]

	Rupee (000) 1984
Total expenditure	256,799
Capital expenditure	82,464
Current expenditure	174,335
Percent current	67.9

U.S. Patents Issued [18]

Values show patents issued to citizens of the country by the U.S. Patents Office.

	1993	1994	1995
Number of patents	1	0	1

Government and Law

Organization of Government [19]

Long-form name:
Democratic Socialist Republic of Sri
Lanka
Type:
Republic
Independence:
4 February 1948 (from UK)
National holiday:
Independence and National Day, 4
February (1948)
Constitution:
Adopted 16 August 1978
Legal system:
A highly complex mixture of English
common law, Roman-Dutch, Muslim,
Sinhalese, and customary law; has not
accepted compulsory ICJ jurisdiction
Executive branch:
Chief of state and head of government:
President; Prime Minister; Cabinet
Legislative branch:
Unicameral: Parliament
Judicial branch:
Supreme Court

Elections [20]

Parliament	% of votes
People's Alliance	49.0
United National Party	44.0
Sri Lanka Muslim Congress	1.8
Tamil United Liberation Front	1.7
Sri Lanka Freedom Party	1.1
Eelam People's Democratic Party	0.3
Upcountry People's Front	0.3
People's Liberation Organization of Tamil Eelam	0.1
Other	1.7

Government Expenditures [21]

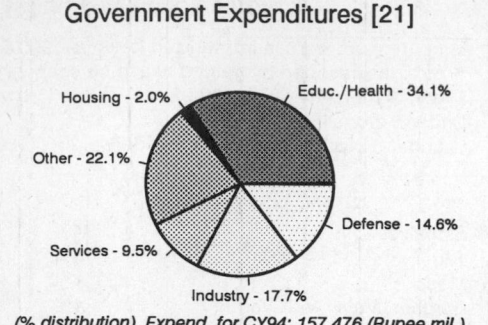

Housing - 2.0%
Educ./Health - 34.1%
Other - 22.1%
Defense - 14.6%
Services - 9.5%
Industry - 17.7%

(% distribution). Expend. for CY94: 157,476 (Rupee mil.)

Military Expenditures and Arms Transfers [22]

	1990	1991	1992	1993	1994
Military expenditures					
Current dollars (mil.)	404[e]	442[e]	369	524[e]	525[e]
1994 constant dollars (mil.)	449[e]	474[e]	385	535[e]	525[e]
Armed forces (000)	22	22	22	22	23
Gross national product (GNP)					
Current dollars (mil.)	8,419	9,142	9,813	10,790	11,550
1994 constant dollars (mil.)	9,370	9,799	10,230	11,020	11,550
Central government expenditures (CGE)					
1994 constant dollars (mil.)	2,928	3,172	2,821	3,091	3,395
People (mil.)	17.2	17.5	17.7	17.9	18.1
Military expenditure as % of GNP	4.8	4.8	3.8	4.9	4.5
Military expenditure as % of CGE	15.3	15.0	13.6	17.3	15.5
Military expenditure per capita (1994 $)	26	27	22	30	29
Armed forces per 1,000 people (soldiers)	1.3	1.3	1.2	1.2	1.3
GNP per capita (1994 $)	544	561	578	615	637
Arms imports[6]					
Current dollars (mil.)	10	60	5	20	100
1994 constant dollars (mil.)	11	64	5	20	100
Arms exports[6]					
Current dollars (mil.)	0	0	0	0	0
1994 constant dollars (mil.)	0	0	0	0	0
Total imports[7]					
Current dollars (mil.)	2,685	3,054	3,445	3,991	4,776
1994 constant dollars (mil.)	2,988	3,274	3,592	4,073	4,776
Total exports[7]					
Current dollars (mil.)	1,983	2,039	2,455	2,859	3,208
1994 constant dollars (mil.)	2,207	2,186	2,560	2,918	3,208
Arms as percent of total imports[8]	0.4	2.0	0.1	0.5	2.1
Arms as percent of total exports[8]	0	0	0	0	0

Crime [23]

	1990
Crime volume	
Cases known to police	56,290
Attempts (percent)	NA
Percent cases solved	NA
Crimes per 100,000 persons	326.9
Persons responsible for offenses	
Total number offenders	NA
Percent female	NA
Percent juvenile (10-16 yrs.)	NA
Percent foreigners	NA

Human Rights [24]

	SSTS	FL	FAPRO	PPCG	APROBC	TPW	PCPTW	STPEP	PHRFF	PRW	ASST	AFL
Observes	P	P		P	P	P	P	P			P	
	EAFRD	CPR	ESCR	SR	ACHR	MAAE	PVIAC	PVNAC	EAFDAW	TCIDTP	RC
Observes	P	P	P						P	P	P

P = Party; S = Signatory; see Appendix for meaning of abbreviations.

Labor Force

Total Labor Force [25]

6.1 million

Labor Force by Occupation [26]

Agriculture	45%
Services	37
Industry	18

Date of data: 1993 est.

Unemployment Rate [27]

13% (1994 est.)

For sources, notes, and explanations, see Annotated Source Appendix, page 1061.

Production Sectors

Commercial Energy Production and Consumption

Data are shown in quadrillion (10^{15}) BTUs and percent for 1995

Values for hydroelectric, nuclear, geothermal, solar, and wind power refer to electrical generation.

Production [28]

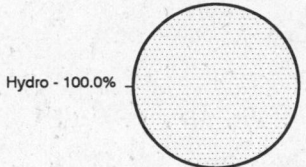

Hydro - 100.0%

Consumption [29]

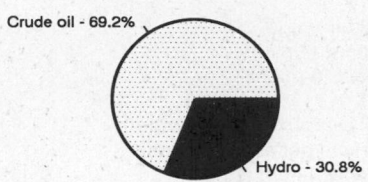

Crude oil - 69.2%

Hydro - 30.8%

Net hydroelectric power	0.044
Total	0.044

Crude oil	0.099
Net hydroelectric power	0.044
Total	0.143

Telecommunications [30]

- 175,000 (1991 est.) telephones; very inadequate domestic service, good international service
- International: submarine cables to Indonesia and Djibouti; satellite earth stations - 2 Intelsat (Indian Ocean)
- Radio: Broadcast stations: AM 12, FM 5, shortwave 0 Radios: 3.525 million (1992 est.)
- Television: Broadcast stations: 5 Televisions: 865,000 (1992 est.)

Top Agricultural Products [32]

Agriculture accounts for 24% of the GDP; produces rice, sugarcane, grains, pulses, oilseed, roots, spices, tea, rubber, coconuts; milk, eggs, hides, meat.

Top Mining Products [33]

Metric tons except as noted[p]	7/20/95*
Cement, hydraulic	925,000
Gemstones, precious and semiprecious	
excl. diamond (value, $000)	60,300[e]
Iron and steel, semimanufactures	55,100
Petroleum refinery products (000 42-gal. bls.)	12,800
Phosphate rock	32,300
Salt	56,200
Limestone	670,000[e]
Ilmenite, gross weight	60,400
Zirconium: zircon concentrate, gross weight	22,300

Transportation [31]

Railways: total: 1,484 km; broad gauge: 1,459 km 1.676-m gauge; narrow gauge: 25 km .762-m gauge (1995)

Highways: total: 94,651 km; paved: 25,749 km; unpaved: 68,902 km (1990)

Merchant marine: total: 26 ships (1,000 GRT or over) totaling 220,508 GRT/329,410 DWT; ships by type: bulk 2, cargo 13, container 1, oil tanker 2, refrigerated cargo 8 (1995 est.)

Airports

Total: 13
With paved runways over 3,047 m: 1
With paved runways 1,524 to 2,437 m: 6
With paved runways 914 to 1,523 m: 6 (1995 est.)

Tourism [34]

	1990	1991	1992	1993	1994
Visitors	302	321	400	398	416
Tourists[37]	298	318	394	392	408
Excursionists	4	3	6	6	8
Tourism receipts	132	157	201	208	224
Tourism expenditures	74	97	111	121	167
Fare receipts	96	110	112	113	123
Fare expenditures	51	51	42	25	61

Travelers are in thousands, money in million U.S. dollars.

Manufacturing Sector

Manufacturing Summary [35]

	1987		1988		1989		1990		1991	
	$ bil.	%	$ bil.	%	$ bil.	%	$ bil.	%	$ bil.	%
Establishments or enterprises (number)	3,372	0.144	3,362	0.160	3,224	0.172	3,246	0.181	-	-
Total employment (000)	240	0.177	249	0.182	277	0.225	308	0.278	-	-
Production workers (000)	-		-		-		-		-	
Output ($ bil.)	2	0.015	2	0.015	2	0.016	2	0.018	-	-
Value added ($ bil.)	0.783	0.018	0.830	0.017	0.845	0.017	0.978	0.020	-	-
Capital investment ($ mil.)	-	-	71	0.015	78	0.014	148	0.027	-	-
M & E investment ($ mil.)	-		-		-		-		-	
Employees per establishment (number)	71	122.820	74	113.814	86	130.580	95	153.618	-	-
Production workers per establishment	-		-		-		-		-	
Output per establishment ($ mil.)	0.465	10.730	0.513	9.233	0.585	9.282	0.652	10.188	-	-
Capital investment per estab. ($ mil.)	-	-	0.021	9.304	0.024	8.263	0.046	14.667	-	-
M & E per establishment ($ mil)	-		-		-		-		-	
Payroll per employee ($)	570	6.354	678	6.808	729	6.521	820	5.908	-	-
Wages per production worker ($)	-		-		-		-		-	
Hours per production worker (hours)	-		-		-		-		-	
Output per employee ($)	6,543	8.736	6,909	8.112	6,813	7.109	6,873	6.632	-	-
Capital investment per employee ($)	-	-	284	8.175	282	6.328	479	9.548	-	-
M & E per employee ($)	-		-		-		-		-	

Note: Columns headed % show percent of world total or ratio. Ratios closest to 100 are closest to world average. M & E stands for machinery & equipment.

Output in Manufacturing

	1987		1988		1989		1990		1991	
	$ bil.[p]	%	$ bil.[p]	%	$ bil.[p]	%	$ bil.[p]	%	$ bil.[p]	%
3110 - Food products	0.439	21.95	0.462	23.10	0.573	28.65	0.451	22.55	-	-
3130 - Beverages	0.112	5.60	0.101	5.05	0.105	5.25	0.052	2.60	-	-
3140 - Tobacco	0.183	9.15	0.182	9.10	0.144	7.20	0.174	8.70	-	-
3210 - Textiles	0.141	7.05	0.136	6.80	0.162	8.10	0.187	9.35	-	-
3211 - Spinning, weaving, etc.	0.087	4.35	0.085	4.25	0.087	4.35	0.115	5.75	-	-
3220 - Wearing apparel	0.132	6.60	0.206	10.30	0.244	12.20	0.355	17.75	-	-
3230 - Leather and products	0.009	0.45	0.003	0.15	0.002	0.10	0.007	0.35	-	-
3240 - Footwear	0.018	0.90	0.020	1.00	0.027	1.35	0.036	1.80	-	-
3310 - Wood products	0.021	1.05	0.013	0.65	0.009	0.45	0.014	0.70	-	-
3320 - Furniture, fixtures	0.003	0.15	0.002	0.10	0.001	0.05	0.002	0.10	-	-
3410 - Paper and products	0.029	1.45	0.037	1.85	0.032	1.60	0.026	1.30	-	-
3411 - Pulp, paper, etc.	0.023	1.15	0.022	1.10	0.016	0.80	0.026	1.30	-	-
3420 - Printing, publishing	0.027	1.35	0.033	1.65	0.034	1.70	0.039	1.95	-	-
3510 - Industrial chemicals	0.009	0.45	0.012	0.60	0.011	0.55	0.023	1.15	-	-
3511 - Basic chemicals, excl fertilizers	0.000	0.00	0.002	0.10	0.004	0.20	0.005	0.25	-	-
3513 - Synthetic resins, etc.	0.003	0.15	-	-	-		0.005	0.25	-	-
3520 - Chemical products nec	0.058	2.90	0.075	3.75	0.069	3.45	0.066	3.30	-	-
3522 - Drugs and medicines	0.008	0.40	0.007	0.35	0.007	0.35	0.008	0.40	-	-
3550 - Rubber products	0.055	2.75	0.064	3.20	0.063	3.15	0.089	4.45	-	-
3560 - Plastic products nec	0.009	0.45	0.022	1.10	0.013	0.65	0.022	1.10	-	-
3610 - Pottery, china, etc.	0.019	0.95	0.031	1.55	0.028	1.40	0.027	1.35	-	-
3620 - Glass and products	0.004	0.20	0.006	0.30	0.005	0.25	0.006	0.30	-	-
3690 - Nonmetal products nec	0.070	3.50	0.074	3.70	0.077	3.85	0.096	4.80	-	-
3710 - Iron and steel	-	-	-		0.014	0.70	0.024	1.20	-	-
3720 - Nonferrous metals	-	-	-		0.006	0.30	0.009	0.45	-	-
3810 - Metal products	0.032	1.60	0.022	1.10	0.032	1.60	0.042	2.10	-	-
3820 - Machinery nec	0.007	0.35	0.006	0.30	0.024	1.20	0.023	1.15	-	-
3825 - Office, computing machinery	-	-	-		0.001	0.05	0.001	0.05	-	-
3830 - Electrical machinery	0.028	1.40	0.016	0.80	0.019	0.95	0.019	0.95	-	-
3832 - Radio, television, etc.	0.001	0.05	0.002	0.10	0.002	0.10	0.005	0.25	-	-
3840 - Transportation equipment	0.006	0.30	0.021	1.05	0.017	0.85	0.042	2.10	-	-
3841 - Shipbuilding, repair	0.003	0.15	0.006	0.30	0.006	0.30	0.034	1.70	-	-
3843 - Motor vehicles	0.001	0.05	0.003	0.15	0.001	0.05	0.007	0.35	-	-
3850 - Professional goods	0.001	0.05	-	-	0.000	0.00	0.001	0.05	-	-
3900 - Industries nec	0.031	1.55	0.054	2.70	0.049	2.45	0.082	4.10	-	-

Note: Codes are International Standard Industry codes (ISIC). Percentages are % of total Output. [f]: Factor Prices; [p]: Producer Prices.

For sources, notes, and explanations, see Annotated Source Appendix, page 1061.

Finance, Economics, and Trade

Economic Indicators [36]

- **National product**: GDP—purchasing power parity— $65.6 billion (1995 est.)

- **National product real growth rate**: 5% (1995 est.)

- **National product per capita**: $3,600 (1995 est.)

- **Inflation rate (consumer prices)**: 8.4% (1994 est.)

- **External debt**: $8.8 billion (1994 est.)

Balance of Payments Summary [37]

Values in millions of dollars.

	1989	1990	1991	1992	1993
Exports of goods (f.o.b.)	1,505.1	1,853.0	2,003.3	2,301.4	2,785.7
Imports of goods (f.o.b.)	-2,055.1	-2,325.6	-2,808.0	-3,016.5	-3,527.8
Trade balance	-550.0	-472.6	-804.7	-715.1	-742.1
Services - debits	-787.1	-898.9	-995.0	-1,069.3	-1,103.6
Services - credits	404.2	532.6	601.1	689.5	745.1
Private transfers (net)	330.7	362.4	401.3	461.7	559.8
Government transfers (net)	188.6	178.1	202.4	182.6	160.2
Long-term capital (net)	184.7	405.7	584.0	406.0	702.1
Short-term capital (net)	392.3	72.4	105.1	95.3	222.3
Errors and omissions	-115.0	-115.1	225.6	173.3	131.1
Overall balance	48.4	64.6	319.8	224.0	674.9

Exchange Rates [38]

Currency: **Sri Lankan rupee.**
Symbol: **SLRe.**

Data are currency units per $1.

January 1996	54.158
1995	51.252
1994	49.415
1993	48.322
1992	43.830
1991	41.372

Imports and Exports

Top Import Origins [39]

$4.8 billion (c.i.f., 1994) Data are for 1994.

Origins	%
Japan	NA
India	NA
UK	NA
Hong Kong	NA
South Korea	NA
Taiwan	NA
Singapore	NA
China	NA

Top Export Destinations [40]

$3.2 billion (f.o.b., 1994) Data are for 1994.

Destinations	%
US	34.7
UK	NA
Germany	NA
Japan	NA
Netherlands	NA
France	NA

Foreign Aid [41]

Recipient: ODA, $423 million (1993).

Import and Export Commodities [42]

Import Commodities	Export Commodities
Textiles and textile materials	Garments and textiles
Machinery and equipment	Teas
Transport equipment	Diamonds
Food	Other gems
Petroleum	Petroleum products
Building materials	Rubber products
	Other agricultural products
	Marine products
	Graphite

For sources, notes, and explanations, see Annotated Source Appendix, page 1061.

885

Sudan

Geography [1]

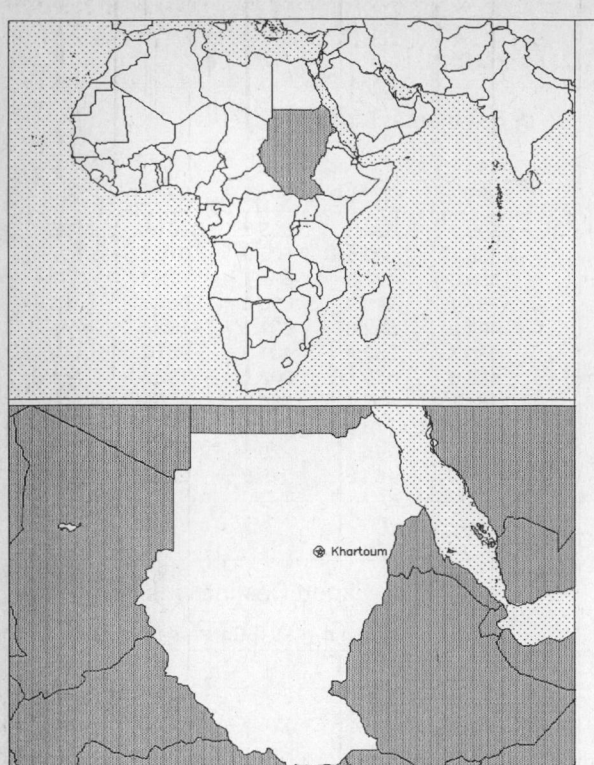

Total area:
  2,505,810 sq km   967,499 sq mi
Land area:
  2,376,000 sq km   917,379 sq mi
Comparative area:
  Slightly more than one-quarter the size of the US
Land boundaries:
  Total 7,687 km, Central African Republic 1,165 km, Chad 1,360 km, Egypt 1,273 km, Eritrea 605 km, Ethiopia 1,606 km, Kenya 232 km, Libya 383 km, Uganda 435 km, Zaire 628 km
Coastline:
  853 km
Climate:
  Tropical in South; arid desert in North; rainy season (April to October)
Terrain:
  Generally flat, featureless plain; mountains in East and West
Natural resources:
  Petroleum; small reserves of iron ore, copper, chromium ore, zinc, tungsten, mica, silver, gold
Land use:
  Arable land: 5%
  Permanent crops: 0%
  Meadows and pastures: 24%
  Forest and woodland: 20%
  Other: 51%

Demographics [2]

	1970	1980	1990	1995[1]	1996	2000	2010	2020	2030
Population	13,788	19,064	26,628	30,312	31,065	35,454	46,512	58,545	70,754
Population density (persons per sq. mi.)	15	21	29	33	34	39	51	64	77
(persons per sq. km.)	6	8	11	13	13	15	20	25	30
Net migration rate (per 1,000 population)	NA	NA	-1.2	-8.2	-1.9	4.6	0.0	0.0	0.0
Births	NA	NA	NA	NA	1,268	NA	NA	NA	NA
Deaths	NA	NA	NA	NA	355	NA	NA	NA	NA
Life expectancy - males	NA	49.6	51.8	53.8	54.2	55.8	59.7	63.3	66.6
Life expectancy - females	NA	49.8	53.4	55.7	56.1	57.9	62.3	66.5	70.3
Birth rate (per 1,000)	NA	NA	45.0	41.5	40.8	38.9	33.3	27.7	23.2
Death rate (per 1,000)	NA	NA	13.5	11.8	11.4	10.3	8.2	6.8	6.1
Women of reproductive age (15-49 yrs.)	NA	NA	6,083	6,858	7,030	8,176	11,281	14,965	18,845
of which are currently married	NA	NA	3,529	NA	4,049	4,687	6,523	NA	NA
Fertility rate	NA	NA	6.5	6.0	5.9	5.5	4.4	3.4	2.8

Except as noted, values for vital statistics are in thousands; life expectancy is in years.

Health

Health Indicators [3]

% of population with access to	
safe water (1990-95)	60
adequate sanitation (1990-95)	22
health services (1985-95)	70
% of 1-year-olds immunized (1990-94) against	
TB (tuberculosis)	78
DPT (diphtheria, pertussis, tetanus)	69
polio	70
measles	76
% of contraceptive prevalence (1980-94)	9
ORT use rate (1990-94)	47

Health Expenditures [4]

Total health expenditure, 1990 (official exchange rate)	
Millions of dollars	300
Dollars per capita	12
Health expenditures as a percentage of GDP	
Total	3.3
Public sector	0.5
Private sector	2.8
Development assistance for health	
Total aid flows (millions of dollars)[1]	39
Aid flows per capita (dollars)	1.5
Aid flows as a percentage of total health expenditure	13.0

For sources, notes, and explanations, see Annotated Source Appendix, page 1061.

Human Factors

Women and Children [5]

% of pregnant women immunized (tetanus 1990-94)	56
% of births attended by trained health personnel (1983-94)	69
Maternal mortality rate (1980-92)	550
Under-5 mortality rate (1994)	122
% under-5 moderately/severely underweight (1980-1994)	20

Burden of Disease [6]

Population per physician (1986)	10685.56
Population per nurse (1984)	1,230.93
Population per hospital bed (1990)	939.08
AIDS cases per 100,000 people (1994)	0.7
Malaria cases per 100,000 people (1992)	NA

Ethnic Division [7]

Black	52%
Arab	39%
Beja	6%
Foreigners	2%
Other	1%

Religion [8]

Sunni Muslim	70%
Indigenous beliefs (primarily in the North)	25%
Christian (primarily in the South and Khartoum)	5%

Major Languages [9]

Arabic (official), Nubian, Ta Bedawie, diverse dialects of Nilotic, Nilo-Hamitic, Sudanic languages, English. Program of Arabization in process.

Education

Public Education Expenditures [10]

Million (Pound)[63]	1980	1985	1990	1991	1993	1994
Total education expenditure	187	NA	NA	NA	NA	NA
as percent of GNP	4.8	NA	NA	NA	NA	NA
as percent of total govt. expend.	9.1	NA	NA	NA	NA	NA
Current education expenditure	172	580	NA	NA	NA	NA
as percent of GNP	4.4	4.0	NA	NA	NA	NA
as percent of current govt. expend.	12.6	15.0	NA	NA	NA	NA
Capital expenditure	15	NA	NA	NA	NA	NA

Educational Attainment [11]

Age group (1983)[13,14]	25+
Total population	6,492,263
Highest level attained (%)	
No schooling	76.7
First level	
Not completed	18.6
Completed	NA
Entered second level	
S-1	1.9
S-2	2.0
Postsecondary	0.8

Illiteracy [12]

In thousands and percent[1]	1990	1995	2000
Illiterate population (15+ yrs.)	8,088	8,507	8,839
Illiteracy rate - total pop. (%)	56.2	52.2	45.1
Illiteracy rate - males (%)	43.7	40.5	34.9
Illiteracy rate - females (%)	69.1	64.1	55.4

Libraries [13]

	Admin. Units	Svc. Pts.	Vols. (000)	Shelving (meters)	Vols. Added	Reg. Users
National	NA	NA	NA	NA	NA	NA
Nonspecialized	NA	NA	NA	NA	NA	NA
Public	NA	NA	NA	NA	NA	NA
Higher ed. (1987)	1	5	500	NA	10,000	20,000
School	NA	NA	NA	NA	NA	NA

Daily Newspapers [14]

	1980	1985	1990	1994
Number of papers	6	5	5	5
Circ. (000)	105	250[e]	610[e]	620[e]

Culture [15]

Science and Technology

Scientific/Technical Forces [16]

R&D Expenditures [17]

U.S. Patents Issued [18]

For sources, notes, and explanations, see Annotated Source Appendix, page 1061.

887

Government and Law

Organization of Government [19]

Long-form name:
Republic of the Sudan
Type:
Transitional - previously ruling military junta; presidential and National Assembly elections held in March 1996
Independence:
1 January 1956 (from Egypt and UK)
National holiday:
Independence Day, 1 January (1956)
Constitution:
12 April 1973, suspended; interim constitution of 10 October 1985, suspended; new constitution to be drafted following elections of March 1996
Legal system:
Based on English common law and Islamic law; some separate religious courts; accepts compulsory ICJ jurisdiction, with reservations
Executive branch:
President; First Vice President; Second Vice President; Cabinet:
Legislative branch:
Unicameral: National Assembly
Judicial branch:
Supreme Court; Special Revolutionary Courts

Elections [20]

National Assembly; Elections last held 6-17 March 1996; results—percent of vote NA; seats—(400 total, 275 directly elected and 125 elected by a supra assembly of interest groups); elections were held on a nonparty basis and parites are banned.

Government Budget [21]

For 1995 est.

	$ bil.
Revenues	0.382
Expenditures	1.06
Capital expenditures	0.91

Crime [23]

Military Expenditures and Arms Transfers [22]

	1990	1991	1992	1993	1994
Military expenditures					
Current dollars (mil.)[e]	456	1,235	2,345	NA	NA
1994 constant dollars (mil.)[e]	507	1,324	2,445	NA	NA
Armed forces (000)	65	65	82	82	82
Gross national product (GNP)					
Current dollars (mil.)	11,290	11,970	13,710	7,912[e]	NA
1994 constant dollars (mil.)	12,560	12,830	14,300	8,075[e]	NA
Central government expenditures (CGE)					
1994 constant dollars (mil.)	NA	NA	NA	NA	1,100[e]
People (mil.)	26.5	27.4	28.1	28.7	29.4
Military expenditure as % of GNP	4.0	10.3	17.1	NA	NA
Military expenditure as % of CGE	NA	NA	NA	NA	NA
Military expenditure per capita (1994 $)	19	48	87	NA	NA
Armed forces per 1,000 people (soldiers)	2.4	2.4	2.9	2.9	2.8
GNP per capita (1994 $)	473	469	509	281	NA
Arms imports[6]					
Current dollars (mil.)	70	90	70	5	0
1994 constant dollars (mil.)	78	96	73	5	0
Arms exports[6]					
Current dollars (mil.)	0	0	0	0	0
1994 constant dollars (mil.)	0	0	0	0	0
Total imports[7]					
Current dollars (mil.)[e]	1,303	1,393	1,193	1,145	1,700
1994 constant dollars (mil.)[e]	1,450	1,493	1,244	1,169	1,700
Total exports[7]					
Current dollars (mil.)[e]	518	368	325	350	411
1994 constant dollars (mil.)[e]	576	394	339	357	411
Arms as percent of total imports[8]	5.4	6.5	5.9	0.4	0
Arms as percent of total exports[8]	0	0	0	0	0

Human Rights [24]

	SSTS	FL	FAPRO	PPCG	APROBC	TPW	PCPTW	STPEP	PHRFF	PRW	ASST	AFL
Observes	P	P			P	P	P				P	P
		EAFRD	CPR	ESCR	SR	ACHR	MAAE	PVIAC	PVNAC	EAFDAW	TCIDTP	RC
Observes		P	P	P	P						S	P

P = Party; S = Signatory; see Appendix for meaning of abbreviations.

Labor Force

Total Labor Force [25]

8.9 million (1993 est.)

Labor Force by Occupation [26]

Agriculture	80%
Industry and commerce	10
Government	6

Date of data: 1983 est.

Unemployment Rate [27]

30% (FY92/93 est.)

For sources, notes, and explanations, see Annotated Source Appendix, page 1061.

Production Sectors

Commercial Energy Production and Consumption

Data are shown in quadrillion (10^{15}) BTUs and percent for 1995
Values for hydroelectric, nuclear, geothermal, solar, and wind power refer to electrical generation.

Production [28]

Hydro - 100.0%

Consumption [29]

Crude oil - 86.5%

Hydro - 13.5%

Net hydroelectric power	0.010
Total	0.010

Crude oil	0.064
Net hydroelectric power	0.010
Total	0.074

Telecommunications [30]

- 77,215 (1983 est.) telephones; large, well-equipped system by African standards, but barely adequate and poorly maintained by modern standards
- Domestic: consists of microwave radio relay, cable, radiotelephone communications, tropospheric scatter, and a domestic satellite system with 14 earth stations
- International: satellite earth stations - 1 Intelsat (Atlantic Ocean) and 1 Arabsat
- Radio: Broadcast stations: AM 11, FM 0, shortwave 0 Radios: 6.67 million (1992 est.)
- Television: Broadcast stations: 3 Televisions: 2.06 million (1992 est.)

Top Agricultural Products [32]

Agriculture accounts for 33% of the GDP; produces cotton, oilseed, sorghum, millet, wheat, gum arabic; sheep.

Transportation [31]

Railways: total: 5,516 km; narrow gauge: 4,800 km 1.067-m gauge; 716 km 1.6096-m gauge plantation line

Highways: total: 19,885 km; paved: 1,989 km; unpaved: 17,896 km (1986 est.)

Merchant marine: total: 5 ships (1,000 GRT or over) totaling 43,024 GRT/57,985 DWT; ships by type: cargo 3, roll-on/roll-off cargo 2 (1995 est.)

Airports

Total:	56
With paved runways 2,438 to 3,047 m:	8
With paved runways 1,524 to 2,437 m:	3
With paved runways under 914 m:	7
With unpaved runways 1,524 to 2,437 m:	13
With unpaved runways 914 to 1,523 m:	25 (1995 est.)

Top Mining Products [33]

Metric tons except as noted[e]	9/1/95[*]
Cement, hydraulic	250,000
Chromite, mine output (48% Cr2O3), gross weight	25,000
Gold, mine output, Au content (kg.)	3,000
Petroleum (000 42-gal. bls.)	
crude, including lease condensate	730
refinery products	7,500
Salt	75,000

Tourism [34]

	1989	1990	1991	1992	1993
Tourists	23	33	16	17	15
Tourism receipts	45	21	8	5	3
Tourism expenditures	144	51	12	33	NA
Fare receipts	36	13	9	10	NA
Fare expenditures	11	7	11	4	NA

Travelers are in thousands, money in million U.S. dollars.

For sources, notes, and explanations, see Annotated Source Appendix, page 1061.

889

Finance, Economics, and Trade

GDP and Manufacturing Summary [35]

	1980	1985	1990	1991	1992
Gross Domestic Product					
Millions of 1980 dollars	7,807	7,688	7,355	7,403	7,551[e]
Growth rate in percent	-3.41	-2.90	-8.04	0.65	2.00[e]
Manufacturing Value Added					
Millions of 1980 dollars	673	744	816	869	878[e]
Growth rate in percent	-7.69	-0.26	4.99	6.45	1.05[e]
Manufacturing share in percent of current prices	8.9	8.8	9.0	9.5	NA

Economic Indicators [36]

- **National product**: GDP—purchasing power parity—$25 billion (1995 est.)
- **National product real growth rate**: 0% (1995 est.)
- **National product per capita**: $800 (1995 est.)
- **Inflation rate (consumer prices)**: 66% (1995 est.)
- **External debt**: $18 billion (year-end 1995 est.)

Balance of Payments Summary [37]

Values in millions of dollars.

	1988	1989	1990	1991	1992
Exports of goods (f.o.b.)	427.0	544.4	326.5	302.5	213.4
Imports of goods (f.o.b.)	-948.5	-1,051.0	-648.8	-1,138.2	-810.2
Trade balance	-521.5	-506.6	-322.3	-835.7	-596.8
Services - debits	-341.3	-495.7	-376.0	-326.4	-297.6
Services - credits	171.6	279.7	184.9	79.7	155.5
Private transfers (net)	216.3	412.4	59.8	45.2	123.7
Government transfers (net)	117.0	161.3	81.4	82.5	109.0
Long-term capital (net)	63.1	114.9	102.7	486.3	268.5
Short-term capital (net)	292.0	197.9	257.1	366.9	177.4
Errors and omissions	7.7	-160.0	9.3	97.8	31.0
Overall balance	4.9	3.9	-3.1	-3.7	-29.3

Exchange Rates [38]

Currency: **Sudanese pound.**
Symbol: **#Sd.**

Data are currency units per $1.

November 1995 (official rate)	750.0
August 1995 (market rate)	571.02
1994 (official rate)	277.8
1994 (market rate)	289.61
1993 (official rate)	153.8
1993 (market rate)	159.31
1992 (official rate)	69.4
1992 (market rate)	97.43
1991 (official rate)	5.4288
1991 (market rate)	6.96

Imports and Exports

Top Import Origins [39]

$1.1 billion (c.i.f., 1995 est.) Data are for 1993.

Origins	%
EU	31
Libya	19
Egypt	5
Saudi Arabia	5
US	5

Top Export Destinations [40]

$535 million (f.o.b., 1995 est.) Data are for 1993.

Destinations	%
EU	39
Saudi Arabia	19
Japan	9
US	3

Foreign Aid [41]

Recipient: ODA, $387 million (1993).

Import and Export Commodities [42]

Import Commodities	Export Commodities
Foodstuffs	Cotton 24%
Petroleum products	Livestock/meat 13%
Manufactured goods	Gum arabic 11%
Machinery and equipment	
Medicines and chemicals	
Textiles	

Suriname

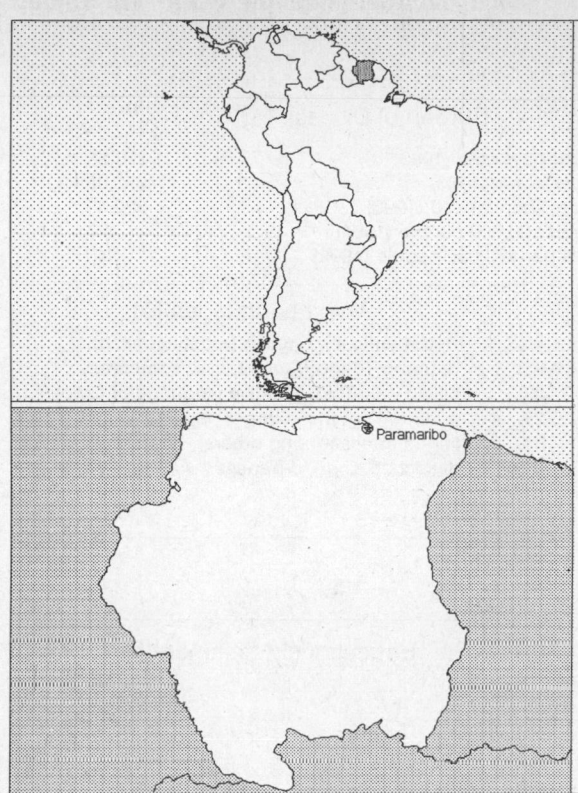

Geography [1]

Total area:
163,270 sq km   63,039 sq mi
Land area:
161,470 sq km   62,344 sq mi
Comparative area:
Slightly larger than Georgia
Land boundaries:
Total 1,707 km, Brazil 597 km, French Guiana 510 km, Guyana 600 km
Coastline:
386 km
Climate:
Tropical; moderated by trade winds
Terrain:
Mostly rolling hills; narrow coastal plain with swamps
Natural resources:
Timber, hydropower potential, fish, shrimp, bauxite, iron ore, and small amounts of nickel, copper, platinum, gold
Land use:
Arable land: Negligible
Permanent crops: 0%
Meadows and pastures: 0%
Forest and woodland: 97%
Other: 3%

Demographics [2]

	1970	1980	1990	1995[1]	1996	2000	2010	2020	2030
Population	373	355	398	430	436	465	534	598	653
Population density (persons per sq. mi.)	6	6	6	7	7	7	9	10	10
(persons per sq. km.)	2	2	2	3	3	3	3	4	4
Net migration rate (per 1,000 population)	NA	-47.1	-6.5	-3.0	-2.4	0.0	0.0	0.0	0.0
Births	NA	NA	NA	NA	NA	11	NA	NA	NA
Deaths	NA	NA	NA	NA	NA	3	NA	NA	NA
Life expectancy - males	NA	62.7	65.8	67.2	67.5	68.6	71.0	73.1	74.7
Life expectancy - females	NA	67.3	70.8	72.4	72.7	73.9	76.5	78.7	80.5
Birth rate (per 1,000)	NA	28.0	27.0	24.7	24.2	21.6	17.8	16.0	13.6
Death rate (per 1,000)	NA	8.0	6.4	5.9	5.8	5.6	5.5	5.8	6.5
Women of reproductive age (15-49 yrs.)	NA	85	103	111	113	123	149	156	162
of which are currently married	NA	32	40	NA	48	53	65	NA	NA
Fertility rate	NA	3.8	3.0	2.7	2.7	2.5	2.2	2.0	1.9

Except as noted, values for vital statistics are in thousands; life expectancy is in years.

Health

Health Indicators [3]

Health Expenditures [4]

For sources, notes, and explanations, see Annotated Source Appendix, page 1061.

891

Human Factors

Women and Children [5]	Burden of Disease [6]	
	Population per physician (1993)	1,273.58
	Population per nurse (1993)	270.00
	Population per hospital bed (1993)	232.49
	AIDS cases per 100,000 people (1994)	4.7
	Malaria cases per 100,000 people (1992)	343

Ethnic Division [7]

Hindustani	37%
Creole	31%
Javanese	15.3%
"Bush Black"	10.3%
Amerindian	2.6%
Chinese	1.7%
Europeans	1%
Other	1.1%

Religion [8]

Hindu	27.4%
Muslim	19.6%
Roman Catholic	22.8%
Protestant (predominantly Moravian)	25.2%
Indigenous beliefs	5%

Major Languages [9]

Dutch (official), English (widely spoken), Sranang Tongo (Surinamese, sometimes called Taki-Taki, is native language of Creoles and much of the younger population and is lingua franca among others), Hindustani (a dialect of Hindi), Javanese.

Education

Public Education Expenditures [10]

	1980	1990	1991	1992	1993	1994
Million (Guilder)						
Total education expenditure	105	250	266	362	384	NA
as percent of GNP	6.7	8.1	7.2	7.2	3.6	NA
as percent of total govt. expend.	22.5	NA	NA	NA	NA	NA
Current education expenditure	105	249	265	360	380	NA
as percent of GNP	6.7	8.1	7.2	7.1	3.6	NA
as percent of current govt. expend.	22.8	NA	NA	NA	NA	NA
Capital expenditure	-	1	1	2	4	NA

Educational Attainment [11]

Illiteracy [12]

In thousands and percent[1]	1990	1995	2000
Illiterate population (15+ yrs.)	22	19	18
Illiteracy rate - total pop. (%)	8.5	6.7	5.7
Illiteracy rate - males (%)	5.4	4.9	3.8
Illiteracy rate - females (%)	10.8	9.2	7.1

Libraries [13]

Daily Newspapers [14]

	1980	1985	1990	1994
Number of papers	4	5	2	3
Circ. (000)	45[e]	55[e]	40	43[e]

Culture [15]

Science and Technology

Scientific/Technical Forces [16]	R&D Expenditures [17]	U.S. Patents Issued [18]

For sources, notes, and explanations, see Annotated Source Appendix, page 1061.

Government and Law

Organization of Government [19]

Long-form name:
Republic of Suriname
Type:
Republic
Independence:
25 November 1975 (from Netherlands)
National holiday:
Independence Day, 25 November (1975)
Constitution:
Ratified 30 September 1987
Legal system:
NA
Executive branch:
President; Vice President; Prime Minister;
Cabinet of Ministers Note: Commander in
Chief of the National Army maintains
significant power
Legislative branch:
Unicameral: National Assembly
(Assemblee Nationale)
Judicial branch:
Supreme Court

Elections [20]

National Assembly	% of seats
The New Front (NF)	58.8
National Democratic Party (NDP)	19.6
DA'91	17.6
Independent	3.9

Government Budget [21]

For 1994 est.

	$ mil.
Revenues	300
Expenditures	700
Capital expenditures	70

Crime [23]

Military Expenditures and Arms Transfers [22]

	1990	1991	1992	1993	1994
Military expenditures					
Current dollars (mil.)	41	41[e]	26[e]	58	47[e]
1994 constant dollars (mil.)	45	43[e]	27[e]	60	47[e]
Armed forces (000)	4	4	2	2	2
Gross national product (GNP)					
Current dollars (mil.)	1,044	1,118	1,196	1,185	1,200
1994 constant dollars (mil.)	1,162	1,198	1,247	1,210	1,200
Central government expenditures (CGE)					
1994 constant dollars (mil.)	NA	NA	NA	NA	NA
People (mil.)	0.4	0.4	0.4	0.4	0.4
Military expenditure as % of GNP	3.9	3.6	2.2	4.9	3.9
Military expenditure as % of CGE	NA	NA	NA	NA	NA
Military expenditure per capita (1994 $)	114	107	66	143	112
Armed forces per 1,000 people (soldiers)	10.0	9.9	4.9	4.8	4.7
GNP per capita (1994 $)	2,918	2,966	3,041	2,906	2,838
Arms imports[6]					
Current dollars (mil.)	5	0	0	0	0
1994 constant dollars (mil.)	6	0	0	0	0
Arms exports[6]					
Current dollars (mil.)	0	0	0	0	0
1994 constant dollars (mil.)	0	0	0	0	0
Total imports[7]					
Current dollars (mil.)	472	470	468[e]	436[e]	NA
1994 constant dollars (mil.)	525	504	488[e]	445[e]	NA
Total exports[7]					
Current dollars (mil.)	472	420	391[e]	375[e]	NA
1994 constant dollars (mil.)	525	450	408[e]	382[e]	NA
Arms as percent of total imports[8]	1.1	0	0	0	0
Arms as percent of total exports[8]	0	0	0	0	0

Human Rights [24]

	SSTS	FL	FAPRO	PPCG	APROBC	TPW	PCPTW	STPEP	PHRFF	PRW	ASST	AFL
Observes	2	P	P			P	P			1	P	P
	EAFRD	CPR	ESCR	SR	ACHR	MAAE	PVIAC	PVNAC	EAFDAW	TCIDTP	RC	
Observes	P	P	P	P	P		P	P	P		P	

P = Party; S = Signatory; see Appendix for meaning of abbreviations.

Labor Force

Total Labor Force [25]

98,240

Labor Force by Occupation [26]

Agriculture
Industry
Services

Unemployment Rate [27]

For sources, notes, and explanations, see Annotated Source Appendix, page 1061.

893

Production Sectors

Commercial Energy Production and Consumption

Data are shown in quadrillion (10^{15}) BTUs and percent for 1995
Values for hydroelectric, nuclear, geothermal, solar, and wind power refer to electrical generation.

Production [28]	Consumption [29]
Crude oil - 53.3%	Crude oil - 60.0%
	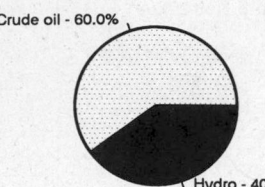
Hydro - 46.7%	Hydro - 40.0%

Crude oil	0.016	Crude oil	0.021
Net hydroelectric power	0.014	Net hydroelectric power	0.014
Total	0.030	Total	0.035

Telecommunications [30]

- 43,522 (1992 est.) telephones; international facilities good
- Domestic: microwave radio relay network
- International: satellite earth stations - 2 Intelsat (Atlantic Ocean)
- Radio: Broadcast stations: AM 5, FM 14, shortwave 1 Radios: 290,256 (1993 est.)
- Television: Broadcast stations: 6 (1987 est.) Televisions: 59,598 (1993 est.)

Transportation [31]

Railways: total: 166 km (single track); standard gauge: 80 km 1.435-m gauge; narrow gauge: 86 km 1.000-m gauge

Highways: total: 4,470 km; paved: 1,162 km; unpaved: 3,308 km (1990)

Merchant marine: total: 2 ships (1,000 GRT or over) totaling 2,421 GRT/2,990 DWT; ships by type: cargo 1, container 1 (1995 est.)

Airports
Total:	38
With paved runways over 3,047 m:	1
With paved runways under 914 m:	31
With unpaved runways 914 to 1,523 m:	6 (1995 est.)

Top Agricultural Products [32]

Agriculture accounts for 21.6% of the GDP; produces paddy rice, bananas, palm kernels, coconuts, plantains, peanuts; beef, chicken; forest products and shrimp of increasing importance.

Top Mining Products [33]

Thousand metric tons except as noted[e]	3/15/95[*]
Bauxite, gross weight	3,300[66]
Alumina	1,600[66]
Aluminum metal, primary	32
Cement, hydraulic	50
Clays, common	20
Gold, mine output, Au content (kg.)	300
Petroleum, crude (000 42-gal. bls.)	1,500
Sand and gravel	195
Stone, crushed and broken	50

Tourism [34]

	1990	1991	1992	1993	1994
Tourists[20]	28	NA	16	18	NA
Tourism receipts	1	1	2	8	11
Tourism expenditures	12	16	11	3	3
Fare receipts	1	2	2	13	25
Fare expenditures	6	7	4	17	22

Travelers are in thousands, money in million U.S. dollars.

For sources, notes, and explanations, see Annotated Source Appendix, page 1061.

Finance, Economics, and Trade

GDP and Manufacturing Summary [35]

	1980	1985	1990	1991	1992
Gross Domestic Product					
Millions of 1980 dollars	891	879	921	895	908[e]
Growth rate in percent	-8.57	2.02	-1.66	-2.81	1.43[e]
Manufacturing Value Added					
Millions of 1980 dollars	140	112	95	88	89[e]
Growth rate in percent	-10.52	6.45	-9.58	-6.95	1.00[e]
Manufacturing share in percent of current prices	17.6	12.5	10.2	9.9[e]	NA

Economic Indicators [36]

- **National product**: GDP—purchasing power parity—$1.3 billion (1995 est.)

- **National product real growth rate**: 0.7% (1995 est.)

- **National product per capita**: $2,950 (1995 est.)

- **Inflation rate (consumer prices)**: 62% (1995)

- **External debt**: $180 million (March 1993 est.)

Balance of Payments Summary [37]

Values in millions of dollars.

	1988	1989	1990	1991	1992
Exports of goods (f.o.b.)	358.4	549.2	465.9	345.9	341.0
Imports of goods (f.o.b.)	-239.4	-330.9	-374.4	-347.1	-272.5
Trade balance	119.0	218.3	91.5	-1.2	68.5
Services - debits	-86.1	-98.0	-106.7	-110.3	-106.8
Services - credits	24.0	24.6	23.0	23.7	23.4
Private transfers (net)	-4.6	-5.6	-7.5	-7.3	-7.3
Government transfers (net)	10.3	23.5	34.4	19.3	33.4
Long-term capital (net)	-90.4	-151.9	-35.7	29.5	-25.4
Short-term capital (net)	-71.5	-188.5	-21.7	13.9	-50.4
Errors and omissions	-1.8	9.9	-9.4	-0.5	25.4
Overall balance	-101.1	-167.7	-32.1	-32.9	-39.2

Exchange Rates [38]

Currency: **Surinamese guilder, gulden, or florin.**
Symbol: **Sf.**

Data are currency units per $1. Beginning in July 1994, the central bank midpoint exchange rate was unified and became market determined.

December 1995	
(central bank rate)	402.32
(parallel rate)	412
1995 (central bank rate)	442.23
December 1994 (parallel rate)	510
1994 (central bank rate)	134.12
January 1994 (parallel rate)	109

Imports and Exports

Top Import Origins [39]

$194.3 million (f.o.b., 1994 est.) Data are for 1992.

Origins	%
US	42
Netherlands	22
Trinidad and Tobago	10
Brazil	5

Top Export Destinations [40]

$293.6 million (f.o.b., 1994 est.) Data are for 1992.

Destinations	%
Norway	33
Netherlands	26
US	13
Japan	6
Brazil	6
UK	3

Foreign Aid [41]

Recipient: ODA, $NA.

Import and Export Commodities [42]

Import Commodities	Export Commodities
Capital equipment	Alumina
Petroleum	Aluminum
Foodstuffs	Shrimp and fish
Cotton	Rice
Consumer goods	Bananas

For sources, notes, and explanations, see Annotated Source Appendix, page 1061.

895

Swaziland

Geography [1]

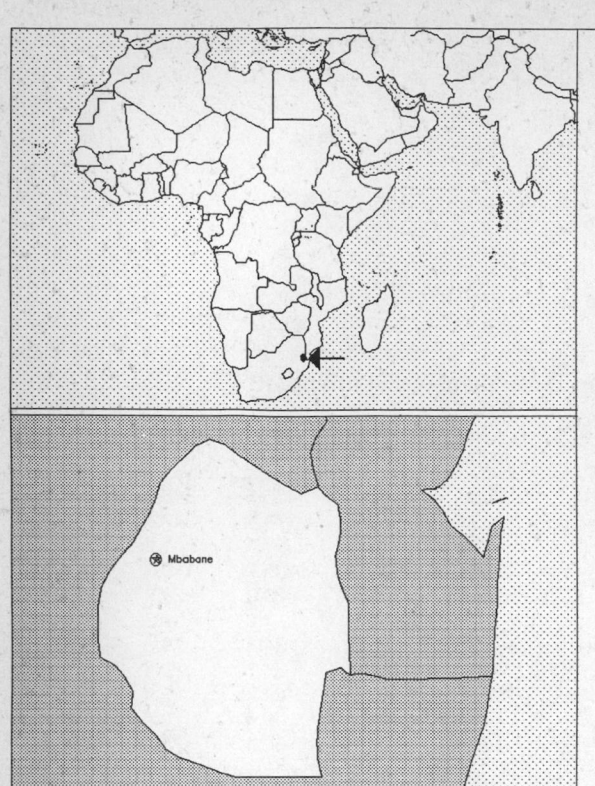

Total area:
17,360 sq km   6,703 sq mi
Land area:
17,200 sq km   6,641 sq mi
Comparative area:
Slightly smaller than New Jersey
Land boundaries:
Total 535 km, Mozambique 105 km, South Africa 430 km
Coastline:
0 km (landlocked)
Climate:
Varies from tropical to near temperate
Terrain:
Mostly mountains and hills; some moderately sloping plains
Natural resources:
Asbestos, coal, clay, cassiterite, hydropower, forests, small gold and diamond deposits, quarry stone, and talc
Land use:
Arable land: 11%
Permanent crops: Negligible
Meadows and pastures: 62%
Forest and woodland: 7%
Other: 20%

Demographics [2]

	1970	1980	1990	1995[1]	1996	2000	2010	2020	2030
Population	455	607	853	967	999	1,137	1,566	2,128	2,844
Population density (persons per sq. mi.)	69	91	128	146	150	171	236	320	428
(persons per sq. km.)	26	35	50	56	58	66	91	124	165
Net migration rate (per 1,000 population)	NA	5.0	8.4	0.0	0.0	0.0	0.0	0.0	0.0
Births	NA	NA	NA	NA	43	NA	NA	NA	NA
Deaths	NA	NA	NA	NA	11	NA	NA	NA	NA
Life expectancy - males	NA	46.6	50.7	52.8	53.3	55.0	59.1	63.0	66.5
Life expectancy - females	NA	54.0	58.7	61.0	61.4	63.2	67.4	71.2	74.5
Birth rate (per 1,000)	NA	48.2	44.3	43.1	42.9	42.0	38.9	36.0	33.1
Death rate (per 1,000)	NA	16.2	12.4	10.8	10.6	9.6	7.5	6.0	5.1
Women of reproductive age (15-49 yrs.)	NA	139	204	231	239	272	376	515	695
of which are currently married	NA	NA	74	NA	86	99	139	NA	NA
Fertility rate	NA	6.8	6.2	6.1	6.1	5.9	5.4	4.9	4.4

Except as noted, values for vital statistics are in thousands; life expectancy is in years.

Health

Health Indicators [3]

Health Expenditures [4]

For sources, notes, and explanations, see Annotated Source Appendix, page 1061.

Human Factors

Women and Children [5]

Burden of Disease [6]

Population per physician (1990)	9,452.38
Population per nurse (1990)	230.95
Population per hospital bed	NA
AIDS cases per 100,000 people (1994)	13.2
Malaria cases per 100,000 people (1992)	NA

Ethnic Division [7]

African	97%
European	3%

Religion [8]

Christian	60%
Indigenous beliefs	40%

Major Languages [9]

English (official, government business conducted in English), Swazi (official).

Education

Public Education Expenditures [10]

Million (Lilangeni)[64]

	1980	1985	1989	1990	1993	1994
Total education expenditure	26	52	101	NA	228	285
as percent of GNP	6.1	5.9	6.0	NA	6.8	NA
as percent of total govt. expend.	NA	20.3	22.5	NA	17.5	NA
Current education expenditure	20	44	88	NA	187	241
as percent of GNP	4.7	5.0	5.2	NA	5.6	NA
as percent of current govt. expend.	23.1	25.9	25.1	NA	21.0	NA
Capital expenditure	6	8	13	NA	41	44

Educational Attainment [11]

Age group (1986)	25+
Total population	221,672
Highest level attained (%)	
No schooling	42.0
First level	
Not completed	24.0
Completed	10.5
Entered second level	
S-1	13.2
S-2	6.3
Postsecondary	3.3

Illiteracy [12]

In thousands and percent[1]

	1990	1995	2000
Illiterate population (15+ yrs.)	114	114	112
Illiteracy rate - total pop. (%)	24.7	21.8	18.0
Illiteracy rate - males (%)	22.6	20.2	16.9
Illiteracy rate - females (%)	26.6	23.3	19.0

Libraries [13]

	Admin. Units	Svc. Pts.	Vols. (000)	Shelving (meters)	Vols. Added	Reg. Users
National (1989)	1	NA	3	32	20	NA
Nonspecialized	NA	NA	NA	NA	NA	NA
Public	NA	NA	NA	NA	NA	NA
Higher ed.	NA	NA	NA	NA	NA	NA
School	NA	NA	NA	NA	NA	NA

Daily Newspapers [14]

	1980	1985	1990	1994
Number of papers	1	2	3	3
Circ. (000)	9	10	11[e]	12[e]

Culture [15]

Science and Technology

Scientific/Technical Forces [16]

R&D Expenditures [17]

U.S. Patents Issued [18]

For sources, notes, and explanations, see Annotated Source Appendix, page 1061.

897

Government and Law

Organization of Government [19]

Long-form name:
Kingdom of Swaziland
Type:
Monarchy; independent member of
Commonwealth
Independence:
6 September 1968 (from UK)
National holiday:
Somhlolo (Independence) Day, 6
September (1968)
Constitution:
None; constitution of 6 September 1968
was suspended 12 April 1973; a new
constitution was promulgated 13 October
1978, but has not been formally presented
to the people
Legal system:
Based on South African Roman-Dutch law
in statutory courts and Swazi traditional
law and custom in traditional courts; has
not accepted compulsory ICJ jurisdiction
Executive branch:
King; Prime Minister; Cabinet
Legislative branch:
Bicameral Parliament (advisory): Senate
and House of Assembly
Judicial branch:
High Court; Court of Appeal

Elections [20]

House of Assembly consists of 65
members (55 elected and 10 appointed
by the king); elections last held NA
October 1993 (next to be held NA);
results—balloting held on a non-party
basis. Political parties are banned by the
Consitution promulgated on 13 October
1978; illegal parties are prohibited from
holding large public gatherings.

Government Expenditures [21]

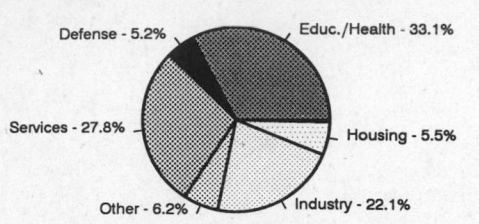

Defense - 5.2%   Educ./Health - 33.1%
Services - 27.8%   Housing - 5.5%
Other - 6.2%   Industry - 22.1%

(% distribution). Expend. for FY94 est.: 410 (Dollar mil.)

Military Expenditures and Arms Transfers [22]

	1990	1991	1992	1993	1994
Military expenditures					
Current dollars (mil.)	14	12	15	20	16
1994 constant dollars (mil.)	15	13	15	20	16
Armed forces (000)	3	3	3	3	3
Gross national product (GNP)					
Current dollars (mil.)	836	899	904	932	967
1994 constant dollars (mil.)	931	963	942	951	967
Central government expenditures (CGE)					
1994 constant dollars (mil.)	233	250	320	328	390
People (mil.)	0.9	0.9	0.9	0.9	0.9
Military expenditure as % of GNP	1.6	1.3	1.6	2.1	1.7
Military expenditure as % of CGE	6.6	5.1	4.8	6.2	4.1
Military expenditure per capita (1994 $)	18	14	17	22	17
Armed forces per 1,000 people (soldiers)	3.5	3.4	3.3	3.3	3.2
GNP per capita (1994 $)	1,091	1,089	1,052	1,049	1,033
Arms imports[6]					
Current dollars (mil.)	0	0	0	0	0
1994 constant dollars (mil.)	0	0	0	0	0
Arms exports[6]					
Current dollars (mil.)	0	0	0	0	0
1994 constant dollars (mil.)	0	0	0	0	0
Total imports[7]					
Current dollars (mil.)	663	718	866	874	937
1994 constant dollars (mil.)	738	770	903	892	937
Total exports[7]					
Current dollars (mil.)	557	596	642	614	658
1994 constant dollars (mil.)	620	639	669	627	658
Arms as percent of total imports[8]	0	0	0	0	0
Arms as percent of total exports[8]	0	0	0	0	0

Crime [23]

	1994
Crime volume	
Cases known to police	44,028
Attempts (percent)	NA
Percent cases solved	40.19
Crimes per 100,000 persons	4,853.12
Persons responsible for offenses	
Total number offenders	22,730
Percent female	11.69
Percent juvenile (10-18 yrs.)	9.39
Percent foreigners	9.56

Human Rights [24]

	SSTS	FL	FAPRO	PPCG	APROBC	TPW	PCPTW	STPEP	PHRFF	PRW	ASST	AFL
Observes	1	P	P		P	P	P			P	1	P
	EAFRD	CPR	ESCR	SR	ACHR	MAAE	PVIAC	PVNAC	EAFDAW	TCIDTP	RC	
Observes		P		P			P	P			P	

P = Party; S = Signatory; see Appendix for meaning of abbreviations.

Labor Force

Total Labor Force [25]

160,355 (1986 est.)

Labor Force by Occupation [26]

Private sector	~65%
Public sector	~35

Unemployment Rate [27]

15% (1992 est.)

Production Sectors

Commercial Energy Production and Consumption

Data are shown in quadrillion (10^{15}) BTUs and percent for 1995

Values for hydroelectric, nuclear, geothermal, solar, and wind power refer to electrical generation.

Production [28]

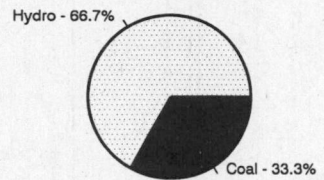

Hydro - 66.7%

Coal - 33.3%

Consumption [29]

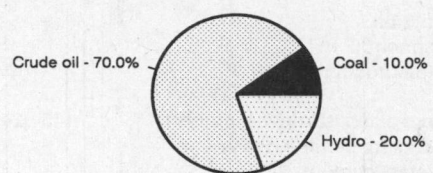

Crude oil - 70.0%

Coal - 10.0%

Hydro - 20.0%

Coal	0.001
Net hydroelectric power	0.002
Total	0.003

Crude oil	0.007
Coal	0.001
Net hydroelectric power	0.002
Total	0.010

Telecommunications [30]

- 30,364 (1993 est.) telephones
- Domestic: system consists of carrier-equipped, open-wire lines and low-capacity, microwave radio relay
- International: satellite earth station - 1 Intelsat (Atlantic Ocean)
- Radio: Broadcast stations: AM 7, FM 6, shortwave 0 Radios: 129,000 (1992 est.)
- Television: Broadcast stations: 10 Televisions: 12,500 (1992 est.)

Transportation [31]

Railways: total: 297 km; note - includes 71 km which are not in use; narrow gauge: 297 km 1.067-m gauge (single track)

Highways: total: 2,960 km; paved: 804 km; unpaved: 2,156 km (1993 est.)

Airports

Total:	17
With paved runways 2,438 to 3,047 m:	1
With paved runways under 914 m:	10
With unpaved runways 914 to 1,523 m:	6 (1995 est.)

Top Agricultural Products [32]

Agriculture accounts for 25% of the GDP; produces sugarcane, cotton, maize, tobacco, rice, citrus, pineapples, corn, sorghum, peanuts; cattle, goats, sheep.

Top Mining Products [33]

Metric tons except as noted	5/19/95*
Asbestos, chrysotile fiber	26,700
Coal, anthracite	228,000
Diamond (carats)	76,100
Stone, quarry products (cu. meters)	292,000

Tourism [34]

	1990	1991	1992	1993	1994
Tourists[101]	294	279	280	288	NA
Tourism receipts	25	26	32	30	29
Tourism expenditures	14	20	17	27	21
Fare receipts	5	6	6	5	5
Fare expenditures	5	5	3	3	3

Travelers are in thousands, money in million U.S. dollars.

For sources, notes, and explanations, see Annotated Source Appendix, page 1061.

Manufacturing Sector

Manufacturing Summary [35]

	1987		1988		1989		1990		1991	
	$ bil.	%	$ bil.	%	$ bil.	%	$ bil.	%	$ bil.	%
Establishments or enterprises (number)	47	0.002	43	0.002	-	-	-	-	-	-
Total employment (000)	7	0.005	7	0.005	-	-	-	-	-	-
Production workers (000)	-	-	-	-	-	-	-	-	-	-
Output ($ bil.)	-	-	0.295	0.003	-	-	-	-	-	-
Value added ($ bil.)	-	-	0.135	0.003	-	-	-	-	-	-
Capital investment ($ mil.)	-	-	17	0.004	-	-	-	-	-	-
M & E investment ($ mil.)	-	-	-	-	-	-	-	-	-	-
Employees per establishment (number)	145	249.899	169	259.440	-	-	-	-	-	-
Production workers per establishment	-	-	-	-	-	-	-	-	-	-
Output per establishment ($ mil.)	-	-	7	123.357	-	-	-	-	-	-
Capital investment per estab. ($ mil.)	-	-	0.390	171.860	-	-	-	-	-	-
M & E per establishment ($ mil)	-	-	-	-	-	-	-	-	-	-
Payroll per employee ($)	-	-	3,693	37.067	-	-	-	-	-	-
Wages per production worker ($)	-	-	-	-	-	-	-	-	-	-
Hours per production worker (hours)	-	-	-	-	-	-	-	-	-	-
Output per employee ($)	-	-	40,499	47.548	-	-	-	-	-	-
Capital investment per employee ($)	-	-	2,305	66.243	-	-	-	-	-	-
M & E per employee ($)	-	-	-	-	-	-	-	-	-	-

Note: Columns headed % show percent of world total or ratio. Ratios closest to 100 are closest to world average. M & E stands for machinery & equipment.

Output in Manufacturing

	1987		1988		1989		1990		1991	
	$ bil.[f]	%	$ bil.[f]	%	$ bil.[f]	%	$ bil.[f]	%	$ bil.[f]	%
3110 - Food products	-	-	0.169	57.29	-	-	-	-	-	-
3130 - Beverages	-	-	0.101	34.24	-	-	-	-	-	-
3810 - Metal products	-	-	0.018	6.10	-	-	-	-	-	-
3900 - Industries nec	-	-	0.006	2.03	-	-	-	-	-	-

Note: Codes are International Standard Industry codes (ISIC). Percentages are % of total Output. [f]: Factor Prices; [p]: Producer Prices.

For sources, notes, and explanations, see Annotated Source Appendix, page 1061.

Finance, Economics, and Trade

Economic Indicators [36]

- **National product**: GDP—purchasing power parity—$3.6 billion (1995 est.)
- **National product real growth rate**: 2.6% (1995 est.)
- **National product per capita**: $3,700 (1995 est.)
- **Inflation rate (consumer prices)**: 14.7% (1995 est.)
- **External debt**: $240 million (1992)

Balance of Payments Summary [37]

Values in millions of dollars.

	1989	1990	1991	1992	1993
Exports of goods (f.o.b.)	493.8	556.6	596.6	637.7	649.8
Imports of goods (f.o.b.)	-515.4	-586.5	-632.3	-764.8	-774.4
Trade balance	-21.6	-29.9	-35.7	-127.1	-124.6
Services - debits	-278.7	-292.7	-295.5	-259.9	-252.3
Services - credits	216.8	273.1	266.9	268.4	225.0
Private transfers (net)	2.4	NA	-1.3	2.2	0.4
Government transfers (net)	85.5	98.3	90.8	121.5	114.4
Long-term capital (net)	66.6	-11.9	39.1	57.8	35.4
Short-term capital (net)	-68.9	-19.7	-14.1	-6.2	-69.4
Errors and omissions	56.1	0.4	-20.0	51.8	26.3
Overall balance	58.2	17.6	30.2	108.5	-44.8

Exchange Rates [38]

Currency: **emalangeni.**
Symbol: **E.**

Data are currency units per $1. The Swazi emalangeni is at par with the South African rand.

January 1996	3.6417
1995	3.6266
1994	3.5490
1993	3.2636
1992	2.8497
1991	2.7563
1990	2.5863

Imports and Exports

Top Import Origins [39]

$827 million (f.o.b., 1994 est.).

Origins	%
South Africa	90
UK	2.6
Switzerland	NA

Top Export Destinations [40]

$798 million (f.o.b., 1994 est.).

Destinations	%
South Africa	50
EU	NA
Canada	NA

Foreign Aid [41]

Recipient: ODA, $NA.

Import and Export Commodities [42]

Import Commodities
Motor vehicles
Machinery
Transport equipment
Petroleum products
Foodstuffs
Chemicals

Export Commodities
Sugar
Edible concentrates
Wood pulp
Cotton yarn
Asbestos

Sweden

Geography [1]

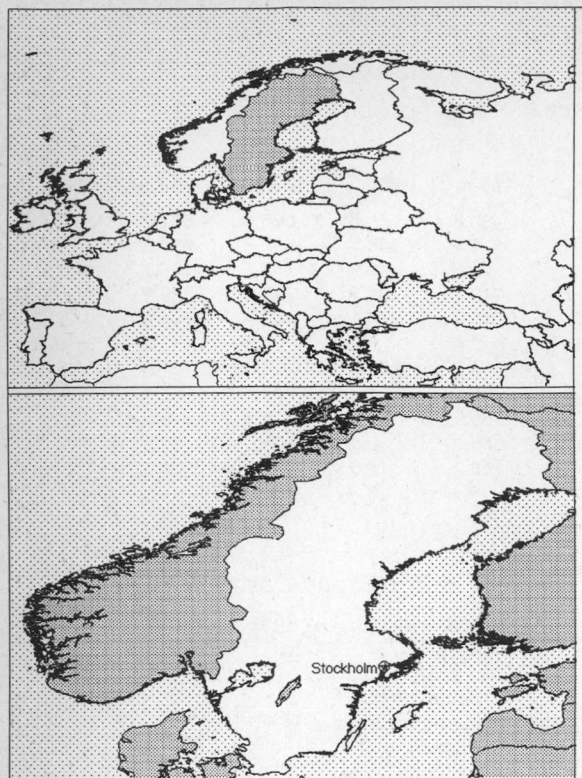

Total area:
449,964 sq km   173,732 sq mi
Land area:
410,928 sq km   158,660 sq mi
Comparative area:
Slightly smaller than California
Land boundaries:
Total 2,205 km, Finland 586 km, Norway 1,619 km
Coastline:
3,218 km
Climate:
Temperate in South with cold, cloudy winters and cool, partly cloudy summers; subarctic in North
Terrain:
Mostly flat or gently rolling lowlands; mountains in West
Natural resources:
Zinc, iron ore, lead, copper, silver, timber, uranium, hydropower potential
Land use:
Arable land: 7%
Permanent crops: 0%
Meadows and pastures: 2%
Forest and woodland: 64%
Other: 27%

Demographics [2]

	1970	1980	1990	1995[1]	1996	2000	2010	2020	2030
Population	8,043	8,310	8,559	8,847	8,901	9,052	9,322	9,515	9,390
Population density (persons per sq. mi.)	51	52	54	56	56	57	59	60	59
(persons per sq. km.)	20	20	21	21	22	22	23	23	23
Net migration rate (per 1,000 population)	NA	1.2	4.1	5.9	5.5	3.9	3.8	1.2	0.0
Births	NA	97	NA	NA	114	NA	NA	NA	NA
Deaths	80	50	95	NA	96	NA	NA	NA	NA
Life expectancy - males	NA	72.8	74.8	75.5	75.6	76.1	77.2	78.7	79.8
Life expectancy - females	NA	78.8	80.5	80.5	80.6	81.1	82.1	84.1	85.5
Birth rate (per 1,000)	13.7	11.7	14.5	12.3	11.6	10.5	9.8	10.0	8.6
Death rate (per 1,000)	10.0	11.1	11.1	11.5	11.4	11.1	10.7	10.4	11.7
Women of reproductive age (15-49 yrs.)	NA	NA	2,048	2,050	2,047	2,036	2,144	2,063	1,925
of which are currently married	1,148	NA	911	NA	922	910	905	NA	NA
Fertility rate	NA	1.7	2.1	1.8	1.7	1.6	1.6	1.6	1.5

Except as noted, values for vital statistics are in thousands; life expectancy is in years.

Health

Health Indicators [3]

% of population with access to	
safe water (1990-95)	NA
adequate sanitation (1990-95)	NA
health services (1985-95)	NA
% of 1-year-olds immunized (1990-94) against	
TB (tuberculosis)	NA
DPT (diphtheria, pertussis, tetanus)	99
polio	99
measles	95
% of contraceptive prevalence (1980-94)	78
ORT use rate (1990-94)	NA

Health Expenditures [4]

Total health expenditure, 1990 (official exchange rate)	
Millions of dollars	20,055
Dollars per capita	2,343
Health expenditures as a percentage of GDP	
Total	8.8
Public sector	7.9
Private sector	0.9
Development assistance for health	
Total aid flows (millions of dollars)[1]	NA
Aid flows per capita (dollars)	NA
Aid flows as a percentage of total health expenditure	NA

For sources, notes, and explanations, see Annotated Source Appendix, page 1061.

Human Factors

Women and Children [5]

% of pregnant women immunized (tetanus 1990-94)	NA
% of births attended by trained health personnel (1983-94)[1]	100
Maternal mortality rate (1980-92)	5
Under-5 mortality rate (1994)	5
% under-5 moderately/severely underweight (1980-1994)	NA

Burden of Disease [6]

Population per physician (1988)	366.78
Population per nurse	NA
Population per hospital bed (1990)	160.77
AIDS cases per 100,000 people (1994)	2.2
Heart disease cases per 1,000 people (1990-93)	372.5

Ethnic Division [7]

White, Lapp (Sami), foreign-born or first-generation immigrants 12% (Finns, Yugoslavs, Danes, Norwegians, Greeks, Turks).

Religion [8]

Evangelical Lutheran	94%
Roman Catholic	1.5%
Pentecostal	1%
Other	3.5%
(1987)	

Major Languages [9]

Swedish, small Lapp- and Finnish-speaking minorities.

Education

Public Education Expenditures [10]

	1980	1985	1990	1991	1992	1994
Million (Krona)						
Total education expenditure	47,322	65,001	101,363	113,123	116,298	NA
as percent of GNP	9.0	7.7	7.7	8.0	8.4	NA
as percent of total govt. expend.	14.1	12.6	13.8	14.0	12.6	NA
Current education expenditure	40,886	57,703	93,083	103,927	105,803	NA
as percent of GNP	7.7	6.8	7.1	7.4	7.6	NA
as percent of current govt. expend.	NA	NA	NA	NA	NA	NA
Capital expenditure	6,437	7,298	8,280	9,196	10,495	NA

Educational Attainment [11]

Age group (1995)	16-74
Total population	6,329,913
Highest level attained (%)	
No schooling	NA
First level	
Not completed	18.2
Completed	NA
Entered second level	
S-1	14.7
S-2	44.1
Postsecondary	21.0

Illiteracy [12]

Libraries [13]

	Admin. Units	Svc. Pts.	Vols. (000)	Shelving (meters)	Vols. Added	Reg. Users
National (1992)	1	4	3,168[e]	79,208	42,880[e]	NA
Nonspecialized (1990)	26	72	18,035[e]	452,390[e]	NA	94,335[e]
Public (1993)	286	1,734	45,147	NA	2 mil	NA
Higher ed. (1991)	25	NA	18,500[e]	460,619	NA	NA
School (1990)	5,240	NA	29,200	NA	NA	NA

Daily Newspapers [14]

	1980	1985	1990	1994
Number of papers	114	115	107	94
Circ. (000)	4,386	4,389	4,499	4,219

Culture [15]

Cinema (seats per 1,000)	NA
Annual attendance per person	1.8
Gross box office receipts (mil. Krone)	860
Museums (reporting)	197
Visitors (000)	18,642
Annual receipts (000 Krone)	1,971

Science and Technology

Scientific/Technical Forces [16]

Scientists/engineers	32,288
Number female	NA
Technicians[2]	27,588
Number female	NA
Total[2]	59,876

R&D Expenditures [17]

	Krona (000) 1993
Total expenditure	48,382,000
Capital expenditure	3,162,000
Current expenditure	45,220,000
Percent current	93.5

U.S. Patents Issued [18]

Values show patents issued to citizens of the country by the U.S. Patents Office.

	1993	1994	1995
Number of patents	741	800	914

For sources, notes, and explanations, see Annotated Source Appendix, page 1061.

903

Government and Law

Organization of Government [19]

Long-form name:
Kingdom of Sweden
Type:
Constitutional monarchy
Independence:
6 June 1523, Gustav Vasa was elected king; 6 June 1809, a constitutional monarchy was established
National holiday:
Day of the Swedish Flag, 6 June
Constitution:
1 January 1975
Legal system:
Civil law system influenced by customary law; accepts compulsory ICJ jurisdiction, with reservations
Executive branch:
King; Heir Apparent Princess; Prime Minister; Cabinet
Legislative branch:
Unicameral: Parliament (Riksdag)
Judicial branch:
Supreme Court (Hogsta Domstolen)

Elections [20]

Riksdag	% of votes
Social Democratic Party	45.4
Moderate Party (Conservatives)	22.3
Center Party	7.7
Liberals	7.2
Left Party	6.2
Greens	5.8
Christian Democrats	4.1
New Democracy	1.2

Government Expenditures [21]

Educ./Health - 57.4%
Housing - 5.1%
Other - 13.1%
Defense - 5.6%
Industry - 10.9%
Gen. Services - 7.9%

(% distribution). Expend. for FY95: 733.87 (Krona bil.)

Military Expenditures and Arms Transfers [22]

	1990	1991	1992	1993	1994
Military expenditures					
Current dollars (mil.)	4,710	5,015	4,775	5,176	5,311
1994 constant dollars (mil.)	5,242	5,375	4,979	5,283	5,311
Armed forces (000)	65	63	45	44	70
Gross national product (GNP)					
Current dollars (mil.)	176,300	181,200	182,100	180,900	188,700
1994 constant dollars (mil.)	196,200	194,300	189,900	184,600	188,700
Central government expenditures (CGE)					
1994 constant dollars (mil.)	87,470	89,390	93,680	103,400	99,480
People (mil.)	8.6	8.6	8.7	8.7	8.8
Military expenditure as % of GNP	2.7	2.8	2.6	2.9	2.8
Military expenditure as % of CGE	6.0	6.0	5.3	5.1	5.3
Military expenditure per capita (1994 $)	612	624	574	605	605
Armed forces per 1,000 people (soldiers)	7.6	7.3	5.2	5.0	8.0
GNP per capita (1994 $)	22,930	22,540	21,890	21,150	21,490
Arms imports[6]					
Current dollars (mil.)	80	160	30	100	30
1994 constant dollars (mil.)	89	172	31	102	30
Arms exports[6]					
Current dollars (mil.)	280	200	240	60	60
1994 constant dollars (mil.)	312	214	250	61	60
Total imports[7]					
Current dollars (mil.)	54,260	49,990	50,020	42,680	51,720
1994 constant dollars (mil.)	60,390	53,580	52,160	43,560	51,720
Total exports[7]					
Current dollars (mil.)	57,540	55,220	56,120	49,860	61,290
1994 constant dollars (mil.)	64,040	59,190	58,520	50,880	61,290
Arms as percent of total imports[8]	0.1	0.3	0.1	0.2	0.1
Arms as percent of total exports[8]	0.5	0.4	0.4	0.1	0.1

Crime [23]

	1994
Crime volume	
Cases known to police	1,112,505
Attempts (percent)	NA
Percent cases solved	30.00
Crimes per 100,000 persons	12,620.33
Persons responsible for offenses	
Total number offenders	101,892
Percent female	17.00
Percent juvenile (15-17 yrs.)	NA
Percent foreigners	NA

Human Rights [24]

	SSTS	FL	FAPRO	PPCG	APROBC	TPW	PCPTW	STPEP	PHRFF	PRW	ASST	AFL
Observes	P	P	P	P	P	P	P		P	P	P	P

	EAFRD	CPR	ESCR	SR	ACHR	MAAE	PVIAC	PVNAC	EAFDAW	TCIDTP	RC
Observes	P	P	P	P		P	P	P	P	P	P

P = Party; S = Signatory; see Appendix for meaning of abbreviations.

Labor Force

Total Labor Force [25]

4.552 million; 84% unionized (1992)

Labor Force by Occupation [26]

	%
Community, social and personal services	38.3
Mining and manufacturing	21.2
Commerce, hotels, and restaurants	14.1
Banking, insurance	9.0
Communications	7.2
Construction	7.0
Agriculture, fishing, and forestry	3.2
Date of data: 1991	

Unemployment Rate [27]

7.8%; plus about 6% in training programs (December 1995)

For sources, notes, and explanations, see Annotated Source Appendix, page 1061.

Production Sectors

Commercial Energy Production and Consumption

Data are shown in quadrillion (10^{15}) BTUs and percent for 1995
Values for hydroelectric, nuclear, geothermal, solar, and wind power refer to electrical generation.

Production [28]

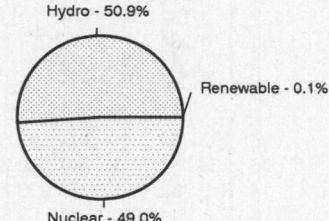

Hydro - 50.9%
Renewable - 0.1%
Nuclear - 49.0%

Consumption [29]

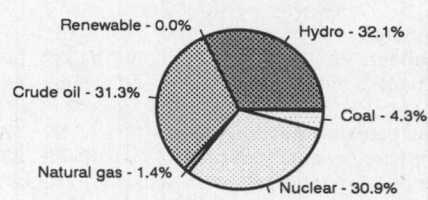

Renewable - 0.0%
Hydro - 32.1%
Crude oil - 31.3%
Coal - 4.3%
Natural gas - 1.4%
Nuclear - 30.9%

Net hydroelectric power	0.698
Net nuclear power	0.671
Geothermal, solar, wind	0.001
Total	1.370

Crude oil	0.681
Dry natural gas	0.030
Coal	0.094
Net hydroelectric power	0.698
Net nuclear power	0.671
Geothermal, solar, wind	0.001
Total	2.175

Telecommunications [30]

- 7.41 million (1986 est.) telephones; excellent domestic and international facilities; automatic system
- Domestic: coaxial and multiconductor cable carry most voice traffic; parallel microwave radio relay network carries some additional telephone channels
- International: 5 submarine coaxial cables; satellite earth stations - 1 Intelsat, 1 Eutelsat, and 1 Inmarsat; shares Inmarsat earth station with the other Nordic countries
- Radio: Broadcast stations: AM 5, FM 360 (mostly repeaters), shortwave 0 Radios: 7.272 million (1993 est.)
- Television: Broadcast stations: 880 (mostly repeaters) Televisions: 3.5 million

Top Agricultural Products [32]

Agriculture accounts for 2% of the GDP; produces grains, sugar beets, potatoes; meat, milk.

Top Mining Products [33]

Thousand metric tons except as noted[e]	6/95[*]
Gold metal, primary (kg.)	8,000[67]
Iron ore concentrate and pellets, gr. wt.	20,000
Pig iron and sponge iron	3,040[r]
Steel, crude	4,950[r]
Silver metal, primary (kg.)	295,000[67]
Granite, crushed	5,000
Limestone, crushed	3,680
Quartzite	1,500
Gas, manufactured (mil. cu. meters)	4,500
Petroleum refinery products (000 42-gal. bls.)	161,000

Transportation [31]

Railways: total: 12,624 km (includes 953 km of privately-owned railways); standard gauge: 11,767 km 1.435-m gauge (7,320 km electrified and 1,152 km double track)

Highways: total: 135,859 km; paved: 97,818 km (including 936 km of expressways); unpaved: 38,041 km (1991 est.)

Merchant marine: total: 169 ships (1,000 GRT or over) totaling 1,993,422 GRT/2,183,215 DWT; ships by type: bulk 10, cargo 35, chemical tanker 24, combination ore/oil 1, liquefied gas tanker 1, oil tanker 32, railcar carrier 2, refrigerated cargo 1, roll-on/roll-off cargo 38, short-sea passenger 7, specialized tanker 4, vehicle carrier 14 (1995 est.)

Airports

Total:	251
With paved runways over 3,047 m:	2
With paved runways 2,438 to 3,047 m:	7
With paved runways 1,524 to 2,437 m:	85
With paved runways 914 to 1,523 m:	26
With paved runways under 914 m:	127

Tourism [34]

	1990	1991	1992	1993	1994
Tourism receipts	2,916	2,704	3,055	2,650	2,826
Tourism expenditures	6,134	6,291	6,969	4,464	4,878
Fare receipts	986	1,001	1,144	812	804
Fare expenditures	969	952	1,028	802	1,067

Travelers are in thousands, money in million U.S. dollars.

For sources, notes, and explanations, see Annotated Source Appendix, page 1061.

Manufacturing Sector

Manufacturing Summary [35]

	1987		1988		1989		1990		1991	
	$ bil.	%	$ bil.	%	$ bil.	%	$ bil.	%	$ bil.	%
Establishments or enterprises (number)	9,848	0.420	9,853	0.469	9,955	0.531	9,828	0.549	-	-
Total employment (000)	975	0.719	963	0.704	953	0.774	908	0.820	-	-
Production workers (000)	671	0.994	669	1.070	661	0.898	627	0.883	-	-
Output ($ bil.)	116	1.143	131	1.120	137	1.162	150	1.310	-	-
Value added ($ bil.)	51	1.191	59	1.219	62	1.264	67	1.342	-	-
Capital investment ($ mil.)	6,339	1.454	-	-	-	-	-	-	-	-
M & E investment ($ mil.)	4,963	1.593	-	-	-	-	-	-	-	-
Employees per establishment (number)	99	171.184	98	149.886	96	145.648	92	149.510	-	-
Production workers per establishment	68	236.843	68	228.067	66	169.066	64	160.889	-	-
Output per establishment ($ mil.)	12	272.358	13	238.593	14	218.733	15	238.759	-	-
Capital investment per estab. ($ mil.)	0.644	346.451	-	-	-	-	-	-	-	-
M & E per establishment ($ mil)	0.504	379.363	-	-	-	-	-	-	-	-
Payroll per employee ($)	17,981	200.519	20,168	202.438	20,900	187.071	24,777	178.618	-	-
Wages per production worker ($)	15,428	195.529	17,354	205.379	18,000	179.489	21,377	180.004	-	-
Hours per production worker (hours)	1,482	78.107	1,490	77.519	1,466	79.999	1,444	77.191	-	-
Output per employee ($)	119,160	159.103	135,585	159.183	143,931	150.180	165,497	159.695	-	-
Capital investment per employee ($)	6,499	202.385	-	-	-	-	-	-	-	-
M & E per employee ($)	5,088	221.611	-	-	-	-	-	-	-	-

Note: Columns headed % show percent of world total or ratio. Ratios closest to 100 are closest to world average. M & E stands for machinery & equipment.

Output in Manufacturing

	1987		1988		1989		1990		1991	
	$ bil.[f]	%	$ bil.[f]	%	$ bil.[f]	%	$ bil.[f]	%	$ bil.[f]	%
3110 - Food products	11.000	9.48	12.000	9.16	12.000	8.76	14.000	9.33	-	-
3130 - Beverages	0.953	0.82	1.076	0.82	1.180	0.86	1.443	0.96	-	-
3140 - Tobacco	0.252	0.22	0.279	0.21	0.293	0.21	0.356	0.24	-	-
3210 - Textiles	1.161	1.00	1.239	0.95	1.168	0.85	1.249	0.83	-	-
3211 - Spinning, weaving, etc.	0.468	0.40	0.501	0.38	0.479	0.35	0.507	0.34	-	-
3220 - Wearing apparel	0.427	0.37	0.436	0.33	0.382	0.28	0.382	0.25	-	-
3230 - Leather and products	0.144	0.12	0.119	0.09	0.121	0.09	0.112	0.07	-	-
3240 - Footwear	0.076	0.07	0.065	0.05	0.056	0.04	0.057	0.04	-	-
3310 - Wood products	5.028	4.33	5.818	4.44	6.428	4.69	7.699	5.13	-	-
3320 - Furniture, fixtures	1.002	0.86	1.105	0.84	1.106	0.81	1.215	0.81	-	-
3410 - Paper and products	9.621	8.29	11.000	8.40	11.000	8.03	12.000	8.00	-	-
3411 - Pulp, paper, etc.	7.982	6.88	9.280	7.08	9.474	6.92	9.635	6.42	-	-
3420 - Printing, publishing	3.908	3.37	4.333	3.31	4.495	3.28	4.994	3.33	-	-
3510 - Industrial chemicals	3.347	2.89	3.907	2.98	3.937	2.87	4.398	2.93	-	-
3511 - Basic chemicals, excl fertilizers	1.768	1.52	2.004	1.53	1.990	1.45	2.186	1.46	-	-
3513 - Synthetic resins, etc.	1.577	1.36	1.903	1.45	1.948	1.42	2.212	1.47	-	-
3520 - Chemical products nec	3.019	2.60	3.600	2.75	3.537	2.58	3.982	2.65	-	-
3522 - Drugs and medicines	1.065	0.92	1.394	1.06	1.436	1.05	1.914	1.28	-	-
3530 - Petroleum refineries	2.913	2.51	2.233	1.70	3.084	2.25	4.168	2.78	-	-
3540 - Petroleum, coal products	0.494	0.43	0.477	0.36	0.487	0.36	0.620	0.41	-	-
3550 - Rubber products	0.686	0.59	0.759	0.58	0.690	0.50	0.735	0.49	-	-
3560 - Plastic products nec	1.161	1.00	1.371	1.05	1.287	0.94	1.458	0.97	-	-
3610 - Pottery, china, etc.	0.151	0.13	0.144	0.11	0.154	0.11	0.174	0.12	-	-
3620 - Glass and products	0.388	0.33	0.436	0.33	0.461	0.34	0.510	0.34	-	-
3690 - Nonmetal products nec	1.472	1.27	1.660	1.27	1.841	1.34	2.141	1.43	-	-
3710 - Iron and steel	4.500	3.88	5.768	4.40	6.375	4.65	6.273	4.18	-	-
3720 - Nonferrous metals	1.815	1.56	2.161	1.65	2.285	1.67	2.240	1.49	-	-
3810 - Metal products	6.072	5.23	7.054	5.38	7.565	5.52	8.402	5.60	-	-
3820 - Machinery nec	9.083	7.83	10.000	7.63	11.000	8.03	12.000	8.00	-	-
3825 - Office, computing machinery	1.158	1.00	1.286	0.98	1.314	0.96	1.478	0.99	-	-
3830 - Electrical machinery	5.780	4.98	6.251	4.77	6.653	4.86	7.407	4.94	-	-
3832 - Radio, television, etc.	2.878	2.48	3.176	2.42	3.535	2.58	3.915	2.61	-	-
3840 - Transportation equipment	13.000	11.21	14.000	10.69	14.000	10.22	15.000	10.00	-	-
3841 - Shipbuilding, repair	0.579	0.50	0.475	0.36	0.445	0.32	0.566	0.38	-	-
3843 - Motor vehicles	10.000	8.62	11.000	8.40	12.000	8.76	12.000	8.00	-	-
3850 - Professional goods	1.031	0.89	1.133	0.86	1.672	1.22	1.909	1.27	-	-
3900 - Industries nec	0.241	0.21	0.289	0.22	0.244	0.18	0.264	0.18	-	-

Note: Codes are International Standard Industry codes (ISIC). Percentages are % of total Output. [f]: Factor Prices; [p]: Producer Prices.

For sources, notes, and explanations, see Annotated Source Appendix, page 1061.

Finance, Economics, and Trade

Economic Indicators [36]

- **National product**: GDP—purchasing power parity—$177.3 billion (1995 est.)
- **National product real growth rate**: 3.5% (1995 est.)
- **National product per capita**: $20,100 (1995 est.)
- **Inflation rate (consumer prices)**: 2.6% (1995)
- **External debt**: $66.5 billion (1994)

Balance of Payments Summary [37]

Values in millions of dollars.

	1989	1990	1991	1992	1993
Exports of goods (f.o.b.)	51,071	56,835	54,543	55,366	49,347
Imports of goods (f.o.b.)	-47,056	-53,433	-48,184	-48,643	-41,679
Trade balance	4,015	3,402	6,359	6,723	7,668
Services - debits	-23,084	-31,218	-33,176	-37,259	-29,661
Services - credits	17,848	23,414	24,171	24,357	19,721
Private transfers (net)	-709	-645	-395	-439	-162
Government transfers (net)	-1,338	-1,645	-1,653	-2,168	-1,624
Long-term capital (net)	-7,681	-17,020	17,919	18,066	18,338
Short-term capital (net)	17,826	36,337	-19,252	-7,596	-6,797
Errors and omissions	-5,654	-5,130	6,159	5,269	-4,955
Overall balance	1,223	7,495	132	6,953	2,528

Exchange Rates [38]

Currency: **Swedish krona.**
Symbol: **SKr.**

Data are currency units per $1.

January 1996	6.7240
1995	7.1333
1994	7.7160
1993	7.7834
1992	5.8238
1991	6.0475

Imports and Exports

Top Import Origins [39]

$51.8 billion (c.i.f., 1994) Data are for 1994.

Origins	%
EU	62.6
Germany	18.4
UK	9.5
Denmark	6.6
France	5.5
Finland	6.3
Norway	6.1
US	8.5

Top Export Destinations [40]

$61.2 billion (f.o.b., 1994) Data are for 1994.

Destinations	%
EU	59.1
Germany	13.2
UK	10.2
Denmark	6.9
France	5.1
Norway	8.1
US	8.0
Finland	4.8

Foreign Aid [41]

Donor: ODA, $1.769 billion (1993).

Import and Export Commodities [42]

Import Commodities
Machinery
Petroleum and petroleum products
Chemicals
Motor vehicles
Foodstuffs
Iron and steel
Clothing

Export Commodities
Machinery
Motor vehicles
Paper products
Pulp and wood
Iron and steel products
Chemicals
Petroleum and petroleum products

Switzerland

Geography [1]

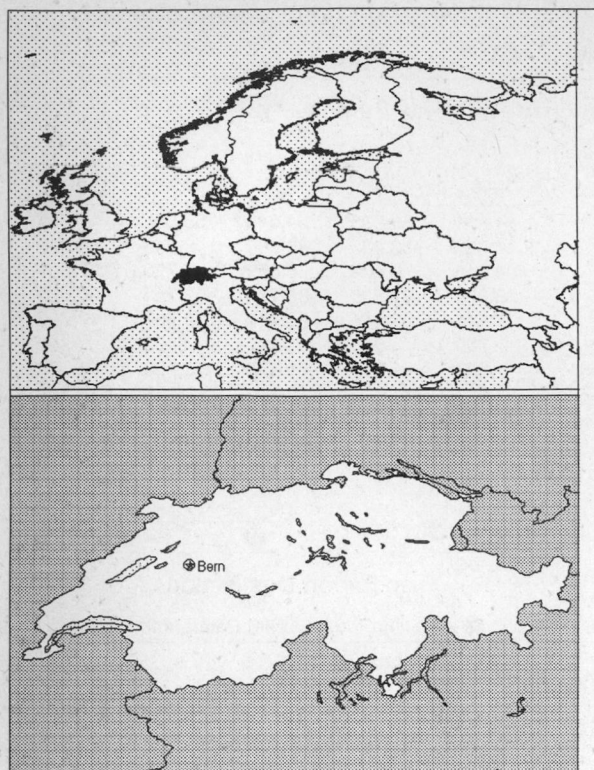

Total area:
41,290 sq km   15,942 sq mi
Land area:
39,770 sq km   15,355 sq mi
Comparative area:
Slightly more than twice the size of New Jersey
Land boundaries:
Total 1,852 km, Austria 164 km, France 573 km, Italy 740 km, Liechtenstein 41 km, Germany 334 km
Coastline:
0 km (landlocked)
Climate:
Temperate, but varies with altitude; cold, cloudy, rainy/snowy winters; cool to warm, cloudy, humid summers with occasional showers
Terrain:
Mostly mountains (Alps in South, Jura in Northwest) with a central plateau of rolling hills, plains, and large lakes
Natural resources:
Hydropower potential, timber, salt
Land use:
Arable land: 10%
Permanent crops: 1%
Meadows and pastures: 40%
Forest and woodland: 26%
Other: 23%

Demographics [2]

	1970	1980	1990	1995[1]	1996	2000	2010	2020	2030
Population	6,267	6,385	6,779	7,164	7,207	7,374	7,674	7,802	7,645
Population density (persons per sq. mi.)	408	416	442	467	469	480	500	508	498
(persons per sq. km.)	158	161	170	180	181	185	193	196	192
Net migration rate (per 1,000 population)	3.7	2.7	8.8	4.3	4.2	3.9	3.8	1.2	0.0
Births	99	74	NA	NA	84	NA	NA	NA	NA
Deaths	57	59	NA	NA	65	NA	NA	NA	NA
Life expectancy - males	NA	72.4	74.0	74.4	74.6	75.2	76.4	78.3	79.6
Life expectancy - females	NA	79.1	81.1	80.7	80.8	81.3	82.2	84.2	85.6
Birth rate (per 1,000)	15.8	11.5	12.4	11.5	11.4	11.2	9.2	9.1	8.3
Death rate (per 1,000)	9.1	9.3	9.4	9.7	9.6	9.5	9.8	10.4	12.1
Women of reproductive age (15-49 yrs.)	NA	1,607	1,734	1,794	1,794	1,788	1,767	1,613	1,485
of which are currently married[51]	963	924	976	NA	1,035	1,032	986	NA	NA
Fertility rate	2.1	1.5	1.6	1.5	1.5	1.6	1.5	1.5	1.5

Except as noted, values for vital statistics are in thousands; life expectancy is in years.

Health

Health Indicators [3]

% of population with access to	
safe water (1990-95)	NA
adequate sanitation (1990-95)	NA
health services (1985-95)	NA
% of 1-year-olds immunized (1990-94) against	
TB (tuberculosis)	NA
DPT (diphtheria, pertussis, tetanus)	89
polio	95
measles	83
% of contraceptive prevalence (1980-94)	71
ORT use rate (1990-94)	NA

Health Expenditures [4]

Total health expenditure, 1990 (official exchange rate)	
Millions of dollars	16,916
Dollars per capita	2,520
Health expenditures as a percentage of GDP	
Total	7.5
Public sector	5.1
Private sector	2.4
Development assistance for health	
Total aid flows (millions of dollars)[1]	NA
Aid flows per capita (dollars)	NA
Aid flows as a percentage of total health expenditure	NA

For sources, notes, and explanations, see Annotated Source Appendix, page 1061.

Human Factors

Women and Children [5]

% of pregnant women immunized (tetanus 1990-94)	NA
% of births attended by trained health personnel (1983-94)[1]	99
Maternal mortality rate (1980-92)	5
Under-5 mortality rate (1994)	7
% under-5 moderately/severely underweight (1980-1994)	NA

Burden of Disease [6]

Population per physician (1988)	627.29
Population per nurse	NA
Population per hospital bed (1990)	93.09
AIDS cases per 100,000 people (1994)	8.0
Heart disease cases per 1,000 people (1990-93)	334.5

Ethnic Division [7]

German	65%
French	18%
Italian	10%
Romansch	1%
Other	6%

Religion [8]

Roman Catholic	47.6%
Protestant	44.3%
Other	8.1%
(1980)	

Major Languages [9]

Figures for Swiss nationals only: German 74%, French 20%, Italian 4%, Romansch 1%, other 1%.

German	65%
French	18%
Italian	12%
Romansch	1%
Other	4%

Education

Public Education Expenditures [10]

	1980	1985	1990	1991	1993	1994
Million (Franc)						
Total education expenditure	8,873	11,696	16,215	18,106	19,933	NA
as percent of GNP	5.0	4.8	5.0	5.2	5.6	NA
as percent of total govt. expend.	18.8	18.6	18.7	18.8	16.1	NA
Current education expenditure	7,937	10,638	14,395	16,061	17,705	NA
as percent of GNP	4.5	4.4	4.4	4.7	5.0	NA
as percent of current govt. expend.	20.0	19.9	NA	19.5	NA	NA
Capital expenditure	936	1,058	1,820	2,045	2,228	NA

Educational Attainment [11]

Age group (1980)	25+
Total population	3,232,206
Highest level attained (%)	
No schooling	NA
First level	
Not completed	75.6
Completed	NA
Entered second level	
S-1	8.9
S-2	NA
Postsecondary	11.5

Illiteracy [12]

Libraries [13]

	Admin. Units	Svc. Pts.	Vols. (000)	Shelving (meters)	Vols. Added	Reg. Users
National (1992)	1	1	2,653	NA	51,825	7,534
Nonspecialized (1992)	34	34	7,697	NA	250,674	382,693
Public (1990)	46[e]	2,555[e]	27,674	768,290[e]	NA	351,444[e]
Higher ed. (1993)[10]	9	NA	14,427	362,308	277,832	200,711
School	NA	NA	NA	NA	NA	NA

Daily Newspapers [14]

	1980	1985	1990	1994
Number of papers	89	97	94	80
Circ. (000)	2,483	3,213	3,063	2,920

Culture [15]

Cinema (seats per 1,000)	14.0
Annual attendance per person	2.3
Gross box office receipts (mil. Franc)	180
Museums (reporting)	520
Visitors (000)	8,792
Annual receipts (000 Franc)	NA

Science and Technology

Scientific/Technical Forces [16]

Scientists/engineers	348,167
Number female	61,729
Technicians	NA
Number female	NA
Total[36]	NA

R&D Expenditures [17]

	Franc (000) 1992
Total expenditure	9,090,000
Capital expenditure	NA
Current expenditure	NA
Percent current	NA

U.S. Patents Issued [18]

Values show patents issued to citizens of the country by the U.S. Patents Office.

	1993	1994	1995
Number of patents	1,197	1,244	1,185

For sources, notes, and explanations, see Annotated Source Appendix, page 1061.

Government and Law

Organization of Government [19]

Long-form name:
Swiss Confederation
Type:
Federal republic
Independence:
1 August 1291
National holiday:
Anniversary of the Founding of the Swiss
Confederation, 1 August (1291)
Constitution:
29 May 1874
Legal system:
Civil law system influenced by customary
law; judicial review of legislative acts,
except with respect to federal decrees of
general obligatory character; accepts
compulsory ICJ jurisdiction, with
reservations
Executive branch:
President; Vice President; Federal Council
(German - Bundesrat, French - Censeil
Federal, Italian - Consiglio Federale)
Legislative branch:
Bicameral Federal Assembly: Council of
States and National Council
Judicial branch:
Federal Supreme Court

Elections [20]

National Council	% of seats
Social Democratic Party (PSS)	27.0
Radical Free Democratic (PRD)	22.5
Democratic People's Party (PDC)	17.0
Swiss People's Party (UDC)	15.0
Green Party (GPS)	4.0
Liberal Party (LPS)	3.5
Freedom Party (FPS)	3.0
Alliance of Independents (LdU)	3.0
Other	5.0

Government Expenditures [21]

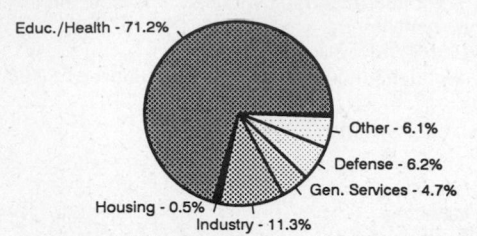

Educ./Health - 71.2%
Other - 6.1%
Defense - 6.2%
Gen. Services - 4.7%
Housing - 0.5%
Industry - 11.3%

(% distribution). Expend. for CY93: 92,918 (Franc mil.)

Military Expenditures and Arms Transfers [22]

	1990	1991	1992	1993	1994
Military expenditures					
Current dollars (mil.)[e]	4,939	4,653	4,463	4,308	4,981
1994 constant dollars (mil.)[e]	5,497	4,988	4,654	4,397	4,981
Armed forces (000)	22	22	31	31	39
Gross national product (GNP)					
Current dollars (mil.)	238,400	247,500	252,700	256,400	267,200
1994 constant dollars (mil.)	265,400	265,300	263,600	261,700	267,200
Central government expenditures (CGE)					
1994 constant dollars (mil.)	NA	NA	NA	NA	NA
People (mil.)	6.8	6.9	6.9	7.0	7.0
Military expenditure as % of GNP	2.1	1.9	1.8	1.7	1.9
Military expenditure as % of CGE	NA	NA	NA	NA	NA
Military expenditure per capita (1994 $)	811	728	672	629	708
Armed forces per 1,000 people (soldiers)	3.2	3.2	4.5	4.4	5.5
GNP per capita (1994 $)	39,140	38,700	38,060	37,460	37,960
Arms imports[6]					
Current dollars (mil.)	675	410	270	230	40
1994 constant dollars (mil.)	751	439	282	235	40
Arms exports[6]					
Current dollars (mil.)	80	150	440	120	110
1994 constant dollars (mil.)	89	161	459	122	110
Total imports[7]					
Current dollars (mil.)	69,680	66,480	61,740	56,720	64,070
1994 constant dollars (mil.)	77,550	71,260	64,380	57,880	64,070
Total exports[7]					
Current dollars (mil.)	63,780	61,520	61,380	58,690	66,230
1994 constant dollars (mil.)	70,990	65,940	64,000	59,900	66,230
Arms as percent of total imports[8]	1.0	0.6	0.4	0.4	0.1
Arms as percent of total exports[8]	0.1	0.2	0.7	0.2	0.2

Crime [23]

	1994
Crime volume	
Cases known to police	362,887
Attempts (percent)	NA
Percent cases solved	NA
Crimes per 100,000 persons	5,168.45
Persons responsible for offenses	
Total number offenders	96,141
Percent female	15.50
Percent juvenile (0-20 yrs.)	15.80
Percent foreigners	40.00

Human Rights [24]

	SSTS	FL	FAPRO	PPCG	APROBC	TPW	PCPTW	STPEP	PHRFF	PRW	ASST	AFL
Observes	P	P	P			P	P		P		P	P
	EAFRD	CPR	ESCR	SR	ACHR	MAAE	PVIAC	PVNAC	EAFDAW	TCIDTP	RC	
Observes	P	P	P	P			P	P	S	P	S	

P = Party; S = Signatory; see Appendix for meaning of abbreviations.

Labor Force

Total Labor Force [25]

3.48 million (900,000 foreign workers;
mostly Italian)

Labor Force by Occupation [26]

Services	50%
Industry and crafts	34
Government	10
Agriculture and forestry	6

Date of data: 1992

Unemployment Rate [27]

3.3% (1995)

For sources, notes, and explanations, see Annotated Source Appendix, page 1061.

Production Sectors

Commercial Energy Production and Consumption

Data are shown in quadrillion (10^{15}) BTUs and percent for 1995
Values for hydroelectric, nuclear, geothermal, solar, and wind power refer to electrical generation.

Production [28]

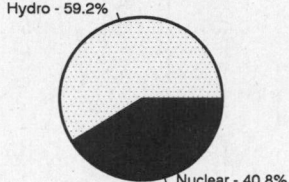

Hydro - 59.2%
Nuclear - 40.8%

Consumption [29]

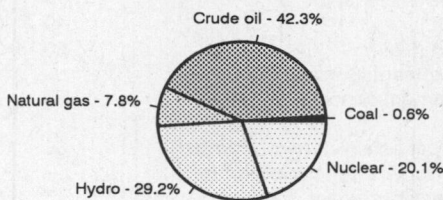

Crude oil - 42.3%
Natural gas - 7.8%
Coal - 0.6%
Nuclear - 20.1%
Hydro - 29.2%

Net hydroelectric power	0.362
Net nuclear power	0.249
Total	0.611

Crude oil	0.525
Dry natural gas	0.097
Coal	0.007
Net hydroelectric power	0.362
Net nuclear power	0.249
Total	1.240

Telecommunications [30]

- 5,622,976 (1986 est.) telephones; excellent domestic and international services
- Domestic: extensive cable and microwave radio relay networks
- International: satellite earth stations - 2 Intelsat (Atlantic Ocean and Indian Ocean)
- Radio: Broadcast stations: AM 7, FM 265, shortwave 0
- Television: Broadcast stations: 18 (repeaters 1,322) Televisions: 2.513 million (1994 est.)

Top Agricultural Products [32]

Agriculture accounts for 3% of the GDP; produces grains, fruits, vegetables; meat, eggs.

Top Mining Products [33]

Thousand metric tons except as noted[e]	5/95[*]
Pig iron	110
Steel, crude	800
Iron/steel semimanufactures, rolled products	700
Cement, hydraulic	4,000
Gypsum	200
Salt	300
Petroleum refinery products (000 42-gal. bls.)	30,500[68]

Transportation [31]

Railways: total: 5,719 km (1,432 km double track); standard gauge: 3,283 km 1.435-m gauge (99% electrified; 310 km nongovernment owned); narrow gauge: 1,255 km 1.000-m gauge (99% electrified; 1,181 km nongovernment owned)

Highways: total: 71,118 km; paved: 71,118 km (including 1,514 km of expressways); unpaved: 0 km (1992 est.)

Merchant marine: total: 23 ships (1,000 GRT or over) totaling 410,581 GRT/727,744 DWT; ships by type: bulk 14, cargo 1, chemical tanker 4, oil tanker 2, roll-on/roll-off cargo 1, specialized tanker 1 (1995 est.)

Airports

Total:	67
With paved runways over 3,047 m:	4
With paved runways 2,438 to 3,047 m:	4
With paved runways 1,524 to 2,437 m:	13
With paved runways 914 to 1,523 m:	5
With paved runways under 914 m:	40

Tourism [34]

	1990	1991	1992	1993	1994
Visitors[52]	129,200	137,000	145,900	130,300	124,800
Tourists[52]	13,200	12,600	12,800	12,400	12,200
Excursionists[102]	116,000	124,400	133,100	117,900	112,500
Tourism receipts	6,789	7,026	7,463	7,011	7,570
Tourism expenditures	5,817	5,682	6,068	5,915	6,325
Fare receipts	1,670	1,744	1,719	1,650	1,734
Fare expenditures	1,129	1,136	1,251	1,356	1,666

Travelers are in thousands, money in million U.S. dollars.

For sources, notes, and explanations, see Annotated Source Appendix, page 1061.

911

Manufacturing Sector

Manufacturing Summary [35]

	1987		1988		1989		1990		1991	
	$ bil.	%	$ bil.	%	$ bil.	%	$ bil.	%	$ bil.	%
Establishments or enterprises (number)	6,719	0.286	6,623	0.315	-	-	-	-	-	-
Total employment (000)	441	0.325	438	0.320	-	-	-	-	-	-
Production workers (000)	-	-	-	-	-	-	-	-	-	-
Output ($ bil.)	49	0.477	52	0.446	48	0.405	60	0.522	59	0.582
Value added ($ bil.)	20	0.471	22	0.450	20	0.406	25	0.495	24	0.708
Capital investment ($ mil.)	-	-	-	-	-	-	-	-	-	-
M & E investment ($ mil.)	-	-	-	-	-	-	-	-	-	-
Employees per establishment (number)	66	113.476	66	101.441	-	-	-	-	-	-
Production workers per establishment	-	-	-	-	-	-	-	-	-	-
Output per establishment ($ mil.)	7	166.706	8	141.485	-	-	-	-	-	-
Capital investment per estab. ($ mil.)	-	-	-	-	-	-	-	-	-	-
M & E per establishment ($ mil)	-	-	-	-	-	-	-	-	-	-
Payroll per employee ($)	-	-	-	-	-	-	-	-	-	-
Wages per production worker ($)	-	-	-	-	-	-	-	-	-	-
Hours per production worker (hours)	-	-	-	-	-	-	-	-	-	-
Output per employee ($)	110,027	146.908	118,799	139.476	-	-	-	-	-	-
Capital investment per employee ($)	-	-	-	-	-	-	-	-	-	-
M & E per employee ($)	-	-	-	-	-	-	-	-	-	-

Note: Columns headed % show percent of world total or ratio. Ratios closest to 100 are closest to world average. M & E stands for machinery & equipment.

Output in Manufacturing

	1987		1988		1989		1990		1991	
	$ bil.	%	$ bil.	%	$ bil.	%	$ bil.	%	$ bil.	%
3210 - Textiles	3.444	7.03	3.543	6.81	3.106	6.47	3.499	5.83	3.338	5.66
3410 - Paper and products	2.656	5.42	2.798	5.38	2.594	5.40	3.135	5.22	2.974	5.04
3420 - Printing, publishing	5.992	12.23	6.400	12.31	5.755	11.99	7.005	11.67	6.681	11.32
3820 - Machinery nec	16.000	32.65	18.000	34.62	16.000	33.33	21.000	35.00	21.000	35.59
3830 - Electrical machinery	15.000	30.61	17.000	32.69	16.000	33.33	20.000	33.33	19.000	32.20
3900 - Industries nec	4.665	9.52	4.812	9.25	4.571	9.52	5.896	9.83	5.833	9.89

Note: Codes are International Standard Industry codes (ISIC). Percentages are % of total Output. [f]: Factor Prices; [p]: Producer Prices.

Finance, Economics, and Trade

Economic Indicators [36]

- **National product**: GDP—purchasing power parity—$158.5 billion (1995 est.)
- **National product real growth rate**: 1.2% (1995 est.)
- **National product per capita**: $22,400 (1995 est.)
- **Inflation rate (consumer prices)**: 1.8% (1995 est.)
- **External debt**: $NA

Balance of Payments Summary [37]

Values in millions of dollars.

	1989	1990	1991	1992	1993
Exports of goods (f.o.b.)	65,366	77,488	73,745	79,353	74,932
Imports of goods (f.o.b.)	-69,690	-83,878	-77,550	-78,863	-72,695
Trade balance	-4,324	-6,390	-3,805	490	2,237
Services - debits	-24,155	-31,563	-30,229	-30,277	-27,670
Services - credits	38,204	47,226	46,980	46,973	44,988
Private transfers (net)	-1,665	-2,183	-2,275	-2,373	-2,227
Government transfers (net)	-18	-146	-346	-624	-633
Long term captial (net)	-11,215	-7,154	-16,435	-11,127	-23,187
Short-term capital (net)	17,943	-4,304	4,758	-4,242	5,744
Errors and omissions	995	5,688	2,322	5,599	1,157
Overall balance	1,419	1,174	970	4,419	409

Exchange Rates [38]

Currency: **Swiss franc, franken, or franco.**
Symbol: **SwF.**

Data are currency units per $1.

January 1996	1.1810
1995	1.1825
1994	1.3677
1993	1.4776
1992	1.4062
1991	1.4340

Imports and Exports

Top Import Origins [39]

$68.2 billion (c.i.f., 1994 est.).

Origins	%
Western Europe	79.2
EU	72.3
Other	6.9
US	6.4

Top Export Destinations [40]

$69.6 billion (f.o.b., 1994 est.).

Destinations	%
Western Europe	63.1
EU countries	56
Other	7.1
US	8.8
Japan	3.4

Foreign Aid [41]

Donor: ODA, $793 million (1993).

Import and Export Commodities [42]

Import Commodities
Agricultural products
Machinery and transportation equipment
Chemicals
Textiles
Construction materials

Export Commodities
Machinery and equipment
Precision instruments
Metal products
Foodstuffs
Textiles and clothing

Syria

Geography [1]

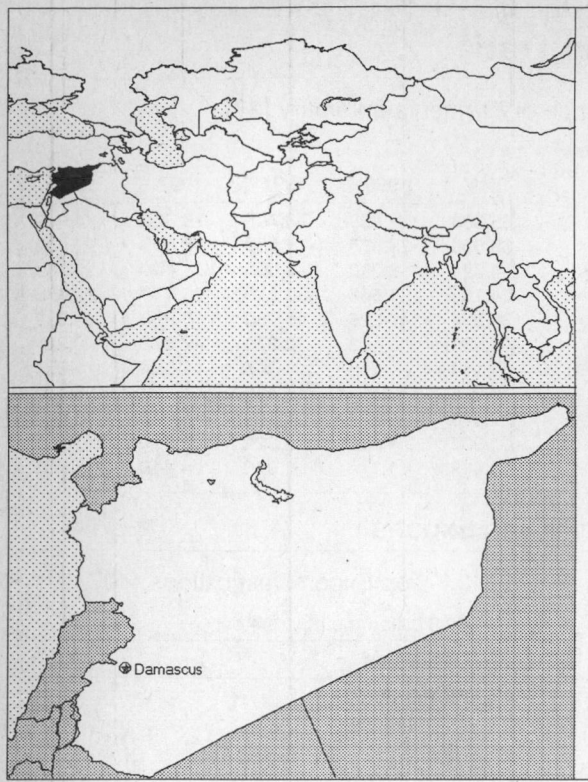

Total area:
185,180 sq km   71,498 sq mi
Land area:
184,050 sq km   71,062 sq mi
Comparative area:
Slightly larger than North Dakota
Note: Includes 1,295 sq km of Israeli-occupied territory
Land boundaries:
Total 2,253 km, Iraq 605 km, Israel 76 km, Jordan 375 km,
Lebanon 375 km, Turkey 822 km
Coastline:
193 km
Climate:
Mostly desert; hot, dry, sunny summers (June to August) and mild,
rainy winters (December to February) along coast; cold weather with
snow or sleet periodically hitting Damascus
Terrain:
Primarily semiarid and desert plateau; narrow coastal plain; mountains
in West
Natural resources:
Petroleum, phosphates, chrome and manganese ores, asphalt, iron ore,
rock salt, marble, gypsum
Land use:
Arable land: 28%
Permanent crops: 3%
Meadows and pastures: 46%
Forest and woodland: 3%
Other: 20%

Demographics [2]

	1970	1980	1990	1995[1]	1996	2000	2010	2020	2030
Population	6,258	8,692	12,620	15,087	15,609	17,759	23,329	28,926	34,352
Population density (persons per sq. mi.)	88	122	178	212	220	250	328	407	483
(persons per sq. km.)	34	47	69	82	85	96	127	157	187
Net migration rate (per 1,000 population)	NA	NA	5.6	0.0	0.0	0.0	0.0	0.0	0.0
Births	NA	NA	NA	NA	617	NA	NA	NA	NA
Deaths	NA	NA	NA	NA	91	NA	NA	NA	NA
Life expectancy - males	NA	NA	64.2	65.7	65.9	67.0	69.5	71.5	73.1
Life expectancy - females	NA	NA	66.1	68.0	68.4	69.9	73.2	76.0	78.2
Birth rate (per 1,000)	45.0	NA	42.8	40.4	39.6	36.1	28.3	23.2	19.3
Death rate (per 1,000)	NA	NA	7.0	6.0	5.9	5.3	4.3	4.0	4.2
Women of reproductive age (15-49 yrs.)	NA	NA	2,660	3,269	3,412	4,014	5,815	7,737	9,371
of which are currently married[52]	856	NA	1,456	NA	1,903	2,260	3,359	NA	NA
Fertility rate	7.6	NA	6.7	6.1	5.9	5.2	3.6	2.7	2.3

Except as noted, values for vital statistics are in thousands; life expectancy is in years.

Health

Health Indicators [3]

% of population with access to	
safe water (1990-95)	85
adequate sanitation (1990-95)	83
health services (1985-95)	90
% of 1-year-olds immunized (1990-94) against	
TB (tuberculosis)	100
DPT (diphtheria, pertussis, tetanus)	89
polio	89
measles	84
% of contraceptive prevalence (1980-94)	52
ORT use rate (1990-94)	95

Health Expenditures [4]

Total health expenditure, 1990 (official exchange rate)	
Millions of dollars	283
Dollars per capita	23
Health expenditures as a percentage of GDP	
Total	2.1
Public sector	0.4
Private sector	1.6
Development assistance for health	
Total aid flows (millions of dollars)[1]	20
Aid flows per capita (dollars)	1.6
Aid flows as a percentage of total health expenditure	7.1

Human Factors

Women and Children [5]

% of pregnant women immunized (tetanus 1990-94)	51
% of births attended by trained health personnel (1983-94)	61
Maternal mortality rate (1980-92)	140
Under-5 mortality rate (1994)	38
% under-5 moderately/severely underweight (1980-1994)	NA

Burden of Disease [6]

Population per physician (1989)	1,158.69
Population per nurse (1987)	874.02
Population per hospital bed (1990)	920.39
AIDS cases per 100,000 people (1994)	*
Malaria cases per 100,000 people (1992)	3

Ethnic Division [7]

Arab	90.3%
Kurds, Armenians, and other	9.7%

Religion [8]

Small Jewish communities present in Damascus, Al Qamishli, and Aleppo. Data are unavailable.

Sunni Muslim	74%
Alawite, Druze, and other Muslim sects	16%
Christian	10%

Major Languages [9]

Arabic (official), Kurdish, Armenian, Aramaic, Circassian, French widely understood.

Education

Public Education Expenditures [10]

Million (Pound)[65]	1980	1985	1990	1991	1992	1994
Total education expenditure	2,347	5,060	10,720	12,025	10,903	17,987
as percent of GNP	4.6	6.1	4.2	4.2	NA	NA
as percent of total govt. expend.	8.1	11.8	17.3	14.2	11.7	12.5
Current education expenditure	1,272	2,799	NA	NA	9,627	15,621
as percent of GNP	2.5	3.4	NA	NA	NA	NA
as percent of current govt. expend.	NA	NA	NA	NA	NA	NA
Capital expenditure	457	846	NA	NA	1,276	2,366

Educational Attainment [11]

Illiteracy [12]

In thousands and percent[1]	1990	1995	2000
Illiterate population (15+ yrs.)	2,197	2,259	2,295
Illiteracy rate - total pop. (%)	33.4	28.3	23.0
Illiteracy rate - males (%)	17.0	13.6	10.8
Illiteracy rate - females (%)	50.5	43.5	37.0

Libraries [13]

	Admin. Units	Svc. Pts.	Vols. (000)	Shelving (meters)	Vols. Added	Reg. Users
National (1992)	1	NA	150	NA	15,000	NA
Nonspecialized (1992)	1	NA	8	NA	20	400
Public	NA	NA	NA	NA	NA	NA
Higher ed. (1993)[24]	1	NA	10	250	25	NA
School	NA	NA	NA	NA	NA	NA

Daily Newspapers [14]

	1980	1985	1990	1994
Number of papers	7	7	10	8
Circ. (000)	114[e]	163[e]	260[e]	261

Culture [15]

Cinema (seats per 1,000)	1.7
Annual attendance per person	0.3
Gross box office receipts (mil. Pound)	12
Museums (reporting)	12
Visitors (000)	1,125
Annual receipts (000 Pound)	3,867

Science and Technology

Scientific/Technical Forces [16]

R&D Expenditures [17]

U.S. Patents Issued [18]

For sources, notes, and explanations, see Annotated Source Appendix, page 1061.

915

Government and Law

Organization of Government [19]

Long-form name:
Syrian Arab Republic
Type:
Republic under military regime since
March 1963
Independence:
17 April 1946 (from League of Nations
mandate under French administration)
National holiday:
National Day, 17 April (1946)
Constitution:
13 March 1973
Legal system:
Based on Islamic law and civil law system;
special religious courts; has not accepted
compulsory ICJ jurisdiction
Executive branch:
President; 3 Vice Presidents; Prime
Minister; 3 Deputy Prime Ministers;
Council of Ministers
Legislative branch:
Unicameral: People's Council (Majlis al-
Chaab)
Judicial branch:
Supreme Constitutional Court; High
Judicial Council; Court of Cassation; State
Security Courts

Elections [20]

People's Council	% of seats
National Progressive Front	66.8
Independents	33.2

Government Expenditures [21]

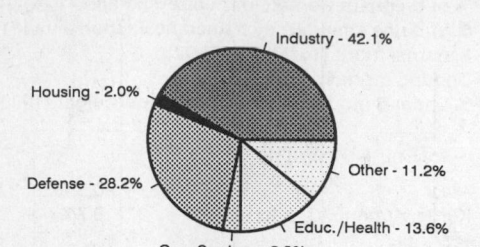

Industry - 42.1%
Housing - 2.0%
Defense - 28.2%
Other - 11.2%
Educ./Health - 13.6%
Gen. Services - 2.9%

(% distribution). Expend. for CY94: 132,016 (Pound mil.)

Military Expenditures and Arms Transfers [22]

	1990	1991	1992	1993	1994
Military expenditures					
Current dollars (mil.)[e,3]	7,526	NA	NA	NA	NA
1994 constant dollars (mil.)[e,3]	8,376	NA	NA	NA	NA
Armed forces (000)	408	408	408	408	320
Gross national product (GNP)					
Current dollars (mil.)	51,390	57,430	NA	NA	NA
1994 constant dollars (mil.)	57,190	61,560	NA	NA	NA
Central government expenditures (CGE)					
1994 constant dollars (mil.)	13,200	16,690	15,390	14,720	NA
People (mil.)	12.8	13.3	13.8	14.3	14.9
Military expenditure as % of GNP	14.6	NA	NA	NA	NA
Military expenditure as % of CGE	63.4	NA	NA	NA	NA
Military expenditure per capita (1994 $)	656	NA	NA	NA	NA
Armed forces per 1,000 people (soldiers)	32.0	30.7	29.5	28.5	21.5
GNP per capita (1994 $)	4,481	4,631	NA	NA	NA
Arms imports[6]					
Current dollars (mil.)	950	825	380	220	10
1994 constant dollars (mil.)	1,057	884	396	225	10
Arms exports[6]					
Current dollars (mil.)	0	0	0	0	0
1994 constant dollars (mil.)	0	0	0	0	0
Total imports[7]					
Current dollars (mil.)	2,400	2,694	3,490	4,140	5,369
1994 constant dollars (mil.)	2,671	2,887	3,640	4,225	5,369
Total exports[7]					
Current dollars (mil.)	4,253	3,618	3,093	3,146	3,547
1994 constant dollars (mil.)	4,733	3,878	3,226	3,211	3,547
Arms as percent of total imports[8]	39.6	30.6	10.9	5.3	0.2
Arms as percent of total exports[8]	0	0	0	0	0

Crime [23]

	1990
Crime volume	
Cases known to police	8,892
Attempts (percent)	100
Percent cases solved	90
Crimes per 100,000 persons	73.40
Persons responsible for offenses	
Total number offenders	3,410
Percent female	1,161
Percent juvenile (7-18 yrs.)	2,249
Percent foreigners	NA

Human Rights [24]

	SSTS	FL	FAPRO	PPCG	APROBC	TPW	PCPTW	STPEP	PHRFF	PRW	ASST	AFL
Observes	P	P	P	P	P	P	P	P			P	P
	EAFRD	CPR	ESCR	SR	ACHR	MAAE	PVIAC	PVNAC	EAFDAW	TCIDTP		RC
Observes		P	P	P				P				P

P = Party; S = Signatory; see Appendix for meaning of abbreviations.

Labor Force

Total Labor Force [25]

4.7 million (1995 est.)

Labor Force by Occupation [26]

Services	42%
Industry	36
Agriculture	22

Date of data: 1990 est.

Unemployment Rate [27]

8% (1994 est.)

For sources, notes, and explanations, see Annotated Source Appendix, page 1061.

Production Sectors

Commercial Energy Production and Consumption

Data are shown in quadrillion (10^{15}) BTUs and percent for 1995
Values for hydroelectric, nuclear, geothermal, solar, and wind power refer to electrical generation.

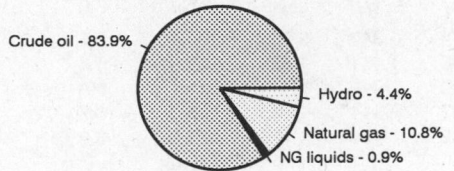

Production [28]

Crude oil - 83.9%
Hydro - 4.4%
Natural gas - 10.8%
NG liquids - 0.9%

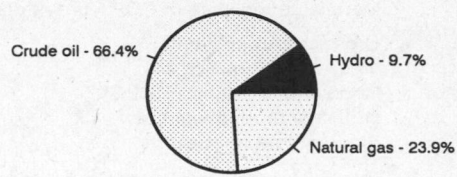

Consumption [29]

Crude oil - 66.4%
Hydro - 9.7%
Natural gas - 23.9%

Crude oil	1.371
Natural gas liquids	0.014
Dry natural gas	0.177
Net hydroelectric power	0.072
Total	1.634

Crude oil	0.493
Dry natural gas	0.177
Net hydroelectric power	0.072
Total	0.742

Telecommunications [30]

- 541,465 (1992 est.) telephones; fair system currently undergoing significant improvement and digital upgrades, including fiber-optic technology
- Domestic: coaxial cable and microwave radio relay network
- International: satellite earth stations - 1 Intelsat (Indian Ocean) and 1 Intersputnik (Atlantic Ocean region); 1 submarine cable; coaxial cable and microwave radio relay to Iraq, Jordan, Lebanon, and Turkey; participant in Medarabtel
- Radio: Broadcast stations: AM 9, FM 1, shortwave 0 Radios: 3.392 million (1992 est.)
- Television: Broadcast stations: 17 Televisions: 700,000 (1993 est.)

Top Agricultural Products [32]

Agriculture accounts for 30% of the GDP; produces wheat, barley, cotton, lentils, chickpeas; beef, lamb, eggs, poultry, milk.

Transportation [31]

Railways: total: 1,998 km; broad gauge: 1,766 km 1.435-m gauge; narrow gauge: 232 km 1.050-m gauge

Highways: total: 31,569 km; paved: 24,308 km (including 712 km of expressways); unpaved: 7,261 km (1991 est.)

Merchant marine: total: 99 ships (1,000 GRT or over) totaling 294,355 GRT/454,990 DWT; ships by type: bulk 12, cargo 85, livestock carrier 1, vehicle carrier 1 (1995 est.)

Airports

Total:	99
With paved runways over 3,047 m:	5
With paved runways 2,438 to 3,047 m:	15
With paved runways 1,524 to 2,437 m:	1
With paved runways 914 to 1,523 m:	1
With paved runways under 914 m:	62

Top Mining Products [33]

Metric tons except as noted[e]	6/1/95[*]
Cement, hydraulic (000 tons)	4,500
Gas, natural, (mil. cu. meters)	7,800
Gypsum	235,000
Natural gas liquids (000 42-gal. bls.)	1,800
Nitrogen	
N content of ammonia	66,700
urea	75,000
Petroleum, crude (000 42-gal. bls.)	211,000
Phosphate rock, gross weight (000 tons)	1,200
Dimension marble (cu. meters)	18,000

Tourism [34]

	1990	1991	1992	1993	1994
Visitors[57]	1,442	1,570	1,740	1,910	2,012
Tourists[16]	562	622	684	703	718
Excursionists	880	948	1,056	1,207	1,294
Tourism receipts[103]	300	410	600	730	800
Tourism expenditures	223	256	260	300	400

Travelers are in thousands, money in million U.S. dollars.

For sources, notes, and explanations, see Annotated Source Appendix, page 1061.

917

Manufacturing Sector

GDP and Manufacturing Summary [35]

	1980	1985	1989	1990	% change 1980-1990	% change 1989-1990
GDP (million 1980 $)	10,593	12,231	11,019	13,598	28.4	23.4
GDP per capita (1980 $)	1,204	1,169	912	1,085	-9.9	19.0
Manufacturing as % of GDP (current prices)	3.6	7.7	6.0[e]	NA	NA	-100.0
Gross output (million $)	3,362	5,914	6,349[e]	9,058	169.4	42.7
Value added (million $)	1,256	1,435	1,461[e]	1,833	45.9	25.5
Value added (million 1980 $)	377	529	NA	449	19.1	NA
Industrial production index	100	147	114[e]	95	-5.0	-16.7
Employment (thousands)	195	182[e]	130	125	-35.9	-3.8

Note: GDP stands for Gross Domestic Product. 'e' stands for estimated value.

Profitability and Productivity

	1980	1985	1989	1990	% change 1980-1990	% change 1989-1990
Intermediate input (%)	63	76	77[e]	80	27.0	3.9
Wages, salaries, and supplements (%)	10[e]	8[e]	6[e]	5	-50.0	-16.7
Gross operating surplus (%)	27[e]	16[e]	17[e]	15	-44.4	-11.8
Gross output per worker ($)	17,278	32,511[e]	48,863[e]	72,252	318.2	47.9
Value added per worker ($)	6,452	7,892[e]	11,243[e]	14,617	126.5	30.0
Average wage (incl. benefits) ($)	1,778[e]	2,738[e]	3,043	3,843	116.1	26.3

Profitability is in percent of gross output. Productivity is in U.S. $. 'e' stands for estimated value.

Profitability - 1990

Inputs - 80.0%
Wages - 5.0%
Surplus - 15.0%

The graphic shows percent of gross output.

Value Added in Manufacturing

	1980 $ mil.	1980 %	1985 $ mil.	1985 %	1989 $ mil.	1989 %	1990 $ mil.	1990 %	% change 1980-1990	% change 1989-1990
311 Food products	214	17.0	235	16.4	235[e]	16.1	325	17.7	51.9	38.3
313 Beverages	37	2.9	42	2.9	40[e]	2.7	58	3.2	56.8	45.0
314 Tobacco products	146	11.6	163	11.4	168[e]	11.5	225	12.3	54.1	33.9
321 Textiles	273	21.7	154	10.7	326[e]	22.3	369	20.1	35.2	13.2
322 Wearing apparel	14	1.1	9	0.6	18[e]	1.2	21	1.1	50.0	16.7
323 Leather and fur products	26	2.1	19	1.3	41[e]	2.8	45	2.5	73.1	9.8
324 Footwear	43	3.4	28	2.0	61[e]	4.2	67	3.7	55.8	9.8
331 Wood and wood products	29	2.3	27	1.9	18[e]	1.2	21	1.1	-27.6	16.7
332 Furniture and fixtures	74	5.9	69	4.8	45[e]	3.1	55	3.0	-25.7	22.2
341 Paper and paper products	6	0.5	8	0.6	4[e]	0.3	8	0.4	33.3	100.0
342 Printing and publishing	14	1.1	16	1.1	9[e]	0.6	18	1.0	28.6	100.0
351 Industrial chemicals	3	0.2	7	0.5	6[e]	0.4	7	0.4	133.3	16.7
352 Other chemical products	31	2.5	73	5.1	64[e]	4.4	75	4.1	141.9	17.2
353 Petroleum refineries	100	8.0	112	7.8	104[e]	7.1	115	6.3	15.0	10.6
354 Miscellaneous petroleum and coal products	4	0.3	4	0.3	4[e]	0.3	4	0.2	0.0	0.0
355 Rubber products	15	1.2	16	1.1	13[e]	0.9	16	0.9	6.7	23.1
356 Plastic products	13	1.0	14	1.0	12[e]	0.8	14	0.8	7.7	16.7
361 Pottery, china, and earthenware	7	0.6	13	0.9	10[e]	0.7	10	0.5	42.9	0.0
362 Glass and glass products	13	1.0	24	1.7	15[e]	1.0	18	1.0	38.5	20.0
369 Other nonmetal mineral products	72	5.7	135	9.4	98[e]	6.7	103	5.6	43.1	5.1
371 Iron and steel	NA	0.0	NA	0.0	NA	0.0	NA	0.0	NA	NA
372 Nonferrous metals	13	1.0	28	2.0	10[e]	0.7	20	1.1	53.8	100.0
381 Metal products	53	4.2	100	7.0	66[e]	4.5	97	5.3	83.0	47.0
382 Nonelectrical machinery	18	1.4	42	2.9	28[e]	1.9	41	2.2	127.8	46.4
383 Electrical machinery	16	1.3	62	4.3	41[e]	2.8	60	3.3	275.0	46.3
384 Transport equipment	3	0.2	11	0.8	8[e]	0.5	11	0.6	266.7	37.5
385 Professional and scientific equipment	NA	0.0	NA	0.0	NA	0.0	NA	0.0	NA	NA
390 Other manufacturing industries	19	1.5	23	1.6	20[e]	1.4	26	1.4	36.8	30.0

Note: The industry codes shown are International Standard Industry codes (ISIC). Percentages are percent of total Value Added. 'e' stands for estimated value

For sources, notes, and explanations, see Annotated Source Appendix, page 1061.

Finance, Economics, and Trade

Economic Indicators [36]

- **National product**: GDP—purchasing power parity—$91.2 billion (1995 est.)
- **National product real growth rate**: 4.4% (1995 est.)
- **National product per capita**: $5,900 (1995 est.)
- **Inflation rate (consumer prices)**: 15.1% (1994 est.)
- **External debt**: $21.2 billion (1995 est.)

Balance of Payments Summary [37]

Values in millions of dollars.

	1989	1990	1991	1992	1993
Exports of goods (f.o.b.)	3,013	4,156	3,438	3,100	3,153
Imports of goods (f.o.b.)	-1,821	-2,062	-2,354	-2,941	-3,475
Trade balance	1,192	2,094	1,084	159	-322
Services - debits	-1,537	-1,723	-2,098	-2,316	-2,503
Services - credits	915	919	1,129	1,349	1,578
Private transfers (net)	430	385	350	550	600
Government transfers (net)	223	88	234	313	40
Long-term capital (net)	-757	-731	-35	173	167
Short-term capital (net)	-951	-1,105	-480	-222	404
Errors and omissions	420	110	-112	70	100
Overall balance	-65	37	72	76	64

Exchange Rates [38]

Currency: **Syrian pound.**
Symbol: **#S.**

Data are currency units per $1.

Official fixed rate	11.225
"Blended rate" used by diplomatic missions	26.6
"Neighboring country rate" for state enterprise imports	42.0
1994 offshore rate	48.0 - 52.0

Imports and Exports

Top Import Origins [39]

$5.4 billion (c.i.f., 1994) Data are for 1993 est.

Origins	%
EU	37
Yugoslavia	17
US and Canada	7
Arab countries	6
Former CEMA countries	NA
China	NA

Top Export Destinations [40]

$3.5 billion (f.o.b., 1994) Data are for 1993 est.

Destinations	%
EU	61
Arab countries	24
Yugoslavia	5
US and Canada	3
Former CEMA countries	NA
China	NA

Foreign Aid [41]

Recipient: ODA, $259 million (1993).

Import and Export Commodities [42]

Import Commodities

Machinery 25%
Metal products 16%
Transport equipment 15%
Foodstuffs 12%
Textiles 10%

Export Commodities

Petroleum 66%
Cotton
Fruits and vegetables 14%
Textiles 9%
Animal products 4%
Industrial products 3%

Taiwan

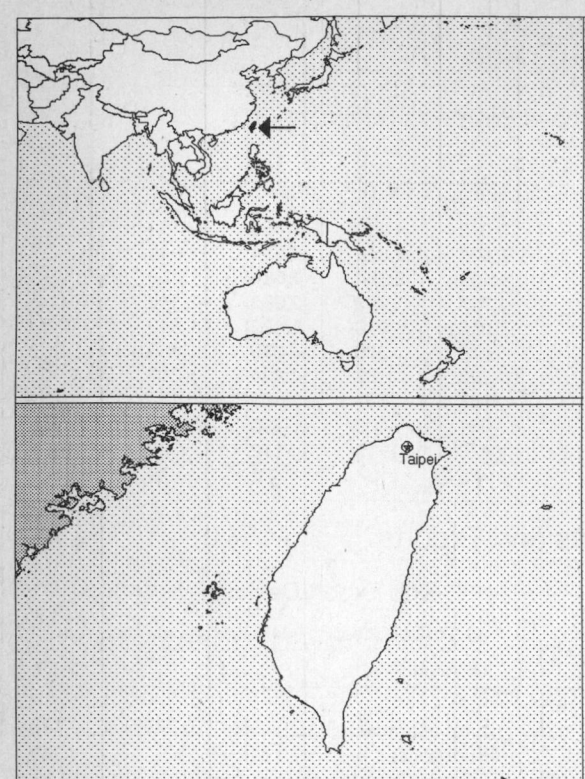

Geography [1]

Total area:
  35,980 sq km   13,892 sq mi
Land area:
  32,260 sq km   12,456 sq mi
Comparative area:
  Slightly smaller than Maryland and Delaware combined
  Note: Includes the Pescadores, Matsu, and Quemoy
Land boundaries:
  0 km
Coastline:
  1,448 km
Climate:
  Tropical; marine; rainy season during southwest monsoon (June to August);
  cloudiness is persistent and extensive all year
Terrain:
  Eastern two-thirds mostly rugged mountains; flat to gently rolling
  plains in West
Natural resources:
  Small deposits of coal, natural gas, limestone, marble, and asbestos
Land use:
  Arable land: 24%
  Permanent crops: 1%
  Meadows and pastures: 5%
  Forest and woodland: 55%
  Other: 15%

Demographics [2]

	1970	1980	1990	1995[1]	1996	2000	2010	2020	2030
Population	14,598	17,848	20,279	21,274	21,466	22,214	23,966	25,155	25,770
Population density (persons per sq. mi.)	1,172	1,433	1,628	1,708	1,723	1,783	1,924	2,020	2,069
(persons per sq. km.)	453	553	629	659	665	689	743	780	799
Net migration rate (per 1,000 population)	NA	NA	0.3	-0.7	-0.6	-0.4	0.0	0.0	0.0
Births	NA	NA	NA	NA	322	NA	NA	NA	NA
Deaths[53]	NA	84	NA	NA	118	NA	NA	NA	NA
Life expectancy - males	NA	NA	71.7	73.1	73.4	74.6	76.9	78.6	79.7
Life expectancy - females	NA	NA	77.1	78.5	78.8	80.1	82.7	84.5	85.7
Birth rate (per 1,000)	NA	NA	16.6	15.2	15.0	14.4	12.8	10.7	10.1
Death rate (per 1,000)	NA	NA	5.2	5.5	5.5	5.7	6.3	7.2	8.9
Women of reproductive age (15-49 yrs.)	NA	NA	5,464	5,952	6,045	6,323	6,136	5,793	5,423
of which are currently married[54]	2,130	2,814	3,685	NA	4,172	4,431	4,488	NA	NA
Fertility rate	NA	NA	1.8	1.8	1.8	1.8	1.7	1.7	1.7

Except as noted, values for vital statistics are in thousands; life expectancy is in years.

Health

Health Indicators [3]

Health Expenditures [4]

For sources, notes, and explanations, see Annotated Source Appendix, page 1061.

Human Factors

Women and Children [5]	Burden of Disease [6]

Ethnic Division [7]

Taiwanese	84%
Mainland Chinese	14%
Aborigine	2%

Religion [8]

Mixture of Buddhist, Confucian, and Taoist	93%
Christian	4.5%
Other	2.5%

Major Languages [9]

Mandarin Chinese (official), Taiwanese (Min), Hakka dialects.

Education

Public Education Expenditures [10]	Educational Attainment [11]

Illiteracy [12]	Libraries [13]

Daily Newspapers [14]	Culture [15]

Science and Technology

Scientific/Technical Forces [16]	R&D Expenditures [17]	U.S. Patents Issued [18]

U.S. Patents Issued [18]

Values show patents issued to citizens of the country by the U.S. Patents Office.

	1993	1994	1995
Number of patents	1,510	1,814	2,087

For sources, notes, and explanations, see Annotated Source Appendix, page 1061.

921

Government and Law

Organization of Government [19]

Long-form name:
None
Type:
Multiparty democratic regime; opposition political parties legalized in March 1989
National holiday:
National Day, 10 October (1911) (Anniversary of the Revolution)
Constitution:
1 January 1947, amended in 1992, presently undergoing revision
Legal system:
Based on civil law system; accepts compulsory ICJ jurisdiction, with reservations
Executive branch:
President; Vice President; Premier; Vice Premier; Executive Yuan
Legislative branch:
Unicameral Legislative Yuan and unicameral National Assembly
Judicial branch:
Judicial Yuan

Crime [23]

Elections [20]

National Assembly	% of votes
Kuomintang (KMT)	55
Democratic Progressive (DPP)	30
Chinese New Party (CNP)	14
Other	1

Government Budget [21]

For 1991 est.

	$ bil.
Revenues	30.3
Expenditures	30.1
Capital expenditures	NA

Military Expenditures and Arms Transfers [22]

	1990	1991	1992	1993	1994
Military expenditures					
Current dollars (mil.)	9,000	9,672	10,390	11,980	11,540
1994 constant dollars (mil.)	10,020	10,370	10,840	12,230	11,540
Armed forces (000)	370	370	360	442	425
Gross national product (GNP)					
Current dollars (mil.)	166,900	185,800	202,500	219,400	238,100
1994 constant dollars (mil.)	185,700	199,100	211,100	223,900	238,100
Central government expenditures (CGE)					
1994 constant dollars (mil.)	NA	32,010	37,260	37,760	NA
People (mil.)	20.4	20.7	20.9	21.1	21.3
Military expenditure as % of GNP	5.4	5.2	5.1	5.5	4.8
Military expenditure as % of CGE	NA	32.4	29.1	32.4	NA
Military expenditure per capita (1994 $)	490	502	519	580	542
Armed forces per 1,000 people (soldiers)	18.1	17.9	17.2	21.0	20.0
GNP per capita (1994 $)	9,087	9,639	10,110	10,620	11,180
Arms imports[6]					
Current dollars (mil.)	650	1,100	850	800	775
1994 constant dollars (mil.)	723	1,179	886	816	775
Arms exports[6]					
Current dollars (mil.)	10	5	10	10	20
1994 constant dollars (mil.)	11	5	10	10	20
Total imports[7]					
Current dollars (mil.)	54,830	63,080	72,180	77,100	85,520
1994 constant dollars (mil.)	61,020	67,610	75,270	78,690	85,520
Total exports[7]					
Current dollars (mil.)	67,140	76,110	81,390	84,680	92,850
1994 constant dollars (mil.)	74,720	81,590	84,880	86,420	92,820
Arms as percent of total imports[8]	1.2	1.7	1.2	1.0	0.9
Arms as percent of total exports[8]	0	0	0	0	0

Human Rights [24]

	SSTS	FL	FAPRO	PPCG	APROBC	TPW	PCPTW	STPEP	PHRFF	PRW	ASST	AFL
Observes	P			P						P	P	
	EAFRD	CPR	ESCR	SR	ACHR	MAAE	PVIAC	PVNAC	EAFDAW	TCIDTP	RC	
Observes		P	S	S	P					P	P	

P = Party; S = Signatory; see Appendix for meaning of abbreviations.

Labor Force

Total Labor Force [25]

8.874 million

Labor Force by Occupation [26]

Services	49%
Industry	39
Agriculture	11

Date of data: 1993 est.

Unemployment Rate [27]

1.6% (1995)

For sources, notes, and explanations, see Annotated Source Appendix, page 1061.

Production Sectors

Commercial Energy Production and Consumption

Data are shown in quadrillion (10^{15}) BTUs and percent for 1995
Values for hydroelectric, nuclear, geothermal, solar, and wind power refer to electrical generation.

Production [28]	Consumption [29]
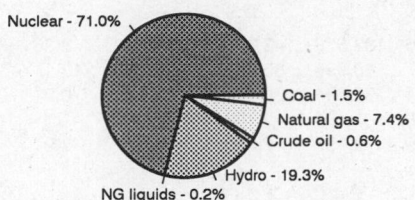	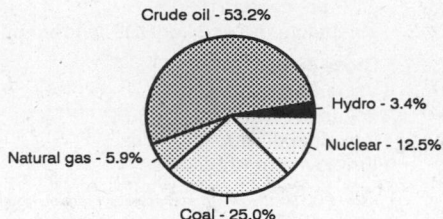

Production [28]

Nuclear - 71.0%
Coal - 1.5%
Natural gas - 7.4%
Crude oil - 0.6%
Hydro - 19.3%
NG liquids - 0.2%

Consumption [29]

Crude oil - 53.2%
Hydro - 3.4%
Nuclear - 12.5%
Natural gas - 5.9%
Coal - 25.0%

Crude oil	0.003	Crude oil	1.428
Natural gas liquids	0.001	Dry natural gas	0.158
Dry natural gas	0.035	Coal	0.671
Coal	0.007	Net hydroelectric power	0.091
Net hydroelectric power	0.091	Net nuclear power	0.335
Net nuclear power	0.335	Total	2.683
Total	0.472		

Telecommunications [30]

- 10,253,773 (1993 est.) telephones; best developed system in Asia outside of Japan
- Domestic: extensive microwave radio relay trunk system on east and west coasts
- International: satellite earth stations - 2 Intelsat (1 Pacific Ocean and 1 Indian Ocean); submarine cables to Japan (Okinawa), Philippines, Guam, Singapore, Hong Kong, Indonesia, Australia, Middle East, and Western Europe
- Radio: Broadcast stations: AM 91, FM 23, shortwave 0 Radios: 8.62 million
- Television: Broadcast stations: 15 (repeaters 13) Televisions: 6.66 million (1993 est.)

Top Agricultural Products [32]

Agriculture accounts for 3.6% of the GDP; produces rice, wheat, corn, soybeans, vegetables, fruit, tea; pigs, poultry, beef, milk; fish catch increasing, reached 1.4 million metric tons in 1988.

Top Mining Products [33]

Thousand metric tons except as noted	5/30/95[*]
Pig iron	5,940
Steel, crude	11,500
Cement, hydraulic	23,700
Lime	650[e]
Dolomite	264
Limestone	13,300
Marble	17,700
Serpentine	475
Gas, natural, gross (mil. cu. meters)	867[e]
Petroleum refinery products (000 42-gal. bls.)	197,000[e]

Transportation [31]

Railways: total: 4,600 km; note - 1,075 km in common carrier service and about 3,525 km is dedicated to industrial use; narrow gauge: 4,600 km 1.067-m

Highways: total: 19,860 km; paved: 17,119 km (including 382 km of expressways); unpaved: 2,741 km (1990 est.)

Merchant marine: total: 198 ships (1,000 GRT or over) totaling 5,812,534 GRT/8,885,092 DWT; ships by type: bulk 50, cargo 29, combination bulk 3, combination ore/oil 1, container 83, oil tanker 19, refrigerated cargo 11, roll-on/roll-off cargo 2 (1995 est.)

Airports

Total:	38
With paved runways over 3,047 m:	8
With paved runways 2,438 to 3,047 m:	12
With paved runways 1,524 to 2,437 m:	4
With paved runways 914 to 1,523 m:	6
With paved runways under 914 m:	7

Tourism [34]

	1990	1991	1992	1993	1994
Tourists	1,934	1,855	1,873	1,850	2,127
Tourism receipts	1,740	2,018	2,449	2,943	3,210
Tourism expenditures	4,984	5,678	7,279	7,585	7,885

Travelers are in thousands, money in million U.S. dollars.

Manufacturing Sector

GDP and Manufacturing Summary [35]

	1980	1985	1989	1990	% change 1980-1990	% change 1989-1990
GDP (million 1980 $)	41,384	57,275	82,933	86,947	110.1	4.8
GDP per capita (1980 $)	2,324	2,974	4,125	4,277	84.0	3.7
Manufacturing as % of GDP (current prices)	36.2	36.9	35.6	32.9	-9.1	-7.6
Gross output (million $)	55,343	69,206	132,678	143,687	159.6	8.3
Value added (million $)	14,907	23,557	48,995	55,424	271.8	13.1
Value added (million 1980 $)	14,907	21,734	30,544	30,484	104.5	-0.2
Industrial production index	100	138	191	188	88.0	-1.6
Employment (thousands)	1,997	2,459	2,453	2,260	13.2	-7.9

Note: GDP stands for Gross Domestic Product. 'e' stands for estimated value.

Profitability and Productivity

	1980	1985	1989	1990	% change 1980-1990	% change 1989-1990
Intermediate input (%)	73	68	63	61	-16.4	-3.2
Wages, salaries, and supplements (%)	10	14	15	16	60.0	6.7
Gross operating surplus (%)	17	20	22	23	35.3	4.5
Gross output per worker ($)	27,719	28,144	54,097	63,575	129.4	17.5
Value added per worker ($)	7,466	9,580	19,977	24,523	228.5	22.8
Average wage (incl. benefits) ($)	2,678	3,862	8,323	10,168	279.7	22.2

Profitability is in percent of gross output. Productivity is in U.S. $. 'e' stands for estimated value.

Profitability - 1990

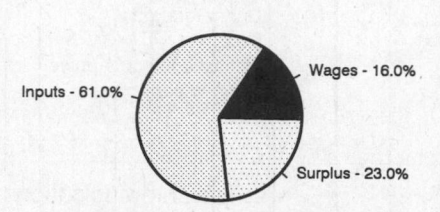

Wages - 16.0%
Inputs - 61.0%
Surplus - 23.0%

The graphic shows percent of gross output.

Value Added in Manufacturing

	1980 $ mil.	1980 %	1985 $ mil.	1985 %	1989 $ mil.	1989 %	1990 $ mil.	1990 %	% change 1980-1990	% change 1989-1990
311 Food products	1,464	9.8	2,444	10.4	4,320	8.8	5,239	9.5	257.9	21.3
313 Beverages	204	1.4	312	1.3	717	1.5	668	1.2	227.5	-6.8
314 Tobacco products	170	1.1	223	0.9	237	0.5	408	0.7	140.0	72.2
321 Textiles	1,885	12.6	2,687	11.4	4,198	8.6	4,680	8.4	148.3	11.5
322 Wearing apparel	337	2.3	720	3.1	1,021	2.1	1,139	2.1	238.0	11.6
323 Leather and fur products	176	1.2	431	1.8	797	1.6	889	1.6	405.1	11.5
324 Footwear	46	0.3	119	0.5	212	0.4	236	0.4	413.0	11.3
331 Wood and wood products	316	2.1	394	1.7	695	1.4	547	1.0	73.1	-21.3
332 Furniture and fixtures	119	0.8	146	0.6	258	0.5	325	0.6	173.1	26.0
341 Paper and paper products	424	2.8	647	2.7	1,662	3.4	1,950	3.5	359.9	17.3
342 Printing and publishing	263	1.8	294	1.2	733	1.5	860	1.6	227.0	17.3
351 Industrial chemicals	718	4.8	1,121	4.8	2,557	5.2	2,435	4.4	239.1	-4.8
352 Other chemical products	502	3.4	890	3.8	2,326	4.7	2,245	4.1	347.2	-3.5
353 Petroleum refineries	834	5.6	1,340	5.7	2,142	4.4	2,768	5.0	231.9	29.2
354 Miscellaneous petroleum and coal products	19	0.1	23	0.1	196	0.4	38	0.1	100.0	-80.6
355 Rubber products	198	1.3	349	1.5	691	1.4	776	1.4	291.9	12.3
356 Plastic products	870	5.8	1,535	6.5	3,605	7.4	3,736	6.7	329.4	3.6
361 Pottery, china, and earthenware	76	0.5	156	0.7	504	1.0	576	1.0	657.9	14.3
362 Glass and glass products	64	0.4	73	0.3	255	0.5	291	0.5	354.7	14.1
369 Other nonmetal mineral products	542	3.6	656	2.8	1,128	2.3	1,290	2.3	138.0	14.4
371 Iron and steel	828	5.6	1,242	5.3	3,074	6.3	3,392	6.1	309.7	10.3
372 Nonferrous metals	139	0.9	146	0.6	349	0.7	385	0.7	177.0	10.3
381 Metal products	584	3.9	1,115	4.7	2,346	4.8	3,052	5.5	422.6	30.1
382 Nonelectrical machinery	524	3.5	827	3.5	1,659	3.4	1,973	3.6	276.5	18.9
383 Electrical machinery	1,890	12.7	2,882	12.2	6,027	12.3	7,247	13.1	283.4	20.2
384 Transport equipment	686	4.6	1,182	5.0	2,411	4.9	2,959	5.3	331.3	22.7
385 Professional and scientific equipment	254	1.7	388	1.6	518	1.1	1,144	2.1	350.4	120.8
390 Other manufacturing industries	774	5.2	1,216	5.2	4,357	8.9	4,176	7.5	439.5	-4.2

Note: The industry codes shown are International Standard Industry codes (ISIC). Percentages are percent of total Value Added. 'e' stands for estimated value

For sources, notes, and explanations, see Annotated Source Appendix, page 1061.

Finance, Economics, and Trade

Economic Indicators [36]

- **National product**: GDP—purchasing power parity—$290.5 billion (1995 est.)

- **National product real growth rate**: 6% (1995 est.)

- **National product per capita**: $13,510 (1995 est.)

- **Inflation rate (consumer prices)**: 4% (1995 est.)

- **External debt**: $620 million (1992 est.)

Balance of Payments Summary [37]

Values in millions of dollars.

	1985	1990	1991	1992	1993
Exports of goods (f.o.b.)	30,469	66,823	75,535	80,723	84,155
Imports of goods (f.o.b.)	-19,296	-51,895	-59,781	-67,956	-72,750
Trade balance	11,173	14,928	15,754	12,767	11,405
Services - debits	-6,681	-17,673	-19,738	-22,478	-24,430
Services - credits	4,955	14,249	16,250	18,072	20,183
Private transfers (net)	-243	-730	-230	-168	-955
Government transfers (net)	-6	-5	-21	-39	-27
Long-term capital (net)	-1,020	-6,601	-2,827	-3,097	-2,560
Short-term capital (net)	-2,143	-8,549	600	-3,450	-2,056
Errors and omissions	491	463	-129	-240	-19
Overall balance	6,526	-3,918	9,659	1,367	1,541

Exchange Rates [38]

Currency: **New Taiwan dollar.**
Symbol: **NT$.**

Data are currency units per $1.

1995	27.4
1994	26.2
1993	26.6
1992	25.4
1991	25.748

Imports and Exports

Top Import Origins [39]

$85.1 billion (c.i.f., 1994) Data are for 1993 est.

Origins	%
Japan	30.1
US	21.7
EU	17.6

Top Export Destinations [40]

$93 billion (f.o.b., 1994) Data are for 1994 est.

Destinations	%
US	27.6
Hong Kong	21.7
EU	15.2
Japan	10.5

Foreign Aid [41]

	U.S. $	
US, including Ex-Im (FY46-82)	4.6	billion
Western (non-US) countries, ODA and OOF bilateral commitments (1970-89)	500	million

Import and Export Commodities [42]

Import Commodities

Machinery and equipment 15.7%
Electronic products 15.6%
Chemicals 9.8%
Iron and steel 8.5%
Crude oil 3.9%
Foodstuffs 2.1%

Export Commodities

Electrical machinery 19.7%
Electronic products 19.6%
Textiles 10.9%
Footwear 3.3%
Foodstuffs 1.0%
Plywood and wood products 0.9%

Tajikistan

Geography [1]

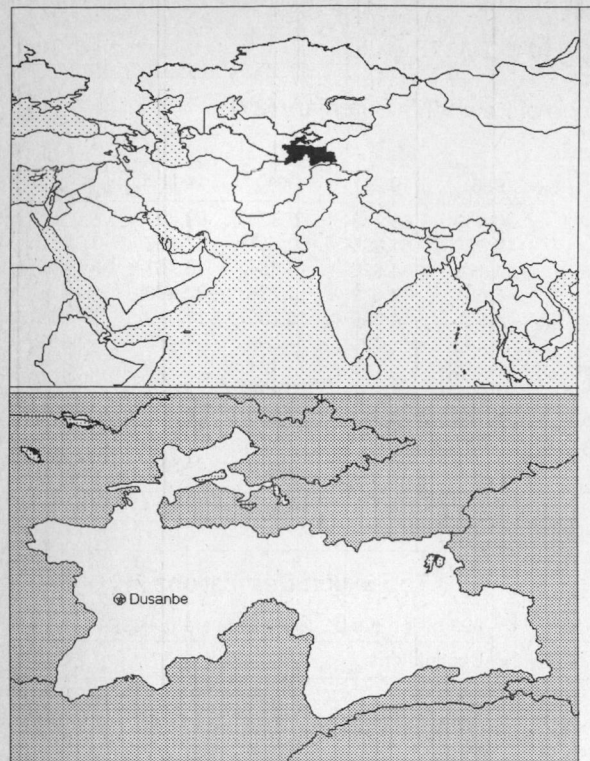

Total area:
143,100 sq km   55,251 sq mi
Land area:
142,700 sq km   55,097 sq mi
Comparative area:
Slightly smaller than Wisconsin
Land boundaries:
Total 3,651 km, Afghanistan 1,206 km, China 414 km, Kyrgyzstan 870 km,
Uzbekistan 1,161 km
Coastline:
0 km (landlocked)
Climate:
Midlatitude continental, hot summers, mild winters; semiarid to polar
in Pamir Mountains
Terrain:
Pamir and Altai Mountains dominate landscape; Western Fergana Valley
in North, Kofarnihon and Vakhsh Valleys in Southwest
Natural resources:
Significant hydropower potential, some petroleum, uranium, mercury,
brown coal, lead, zinc, antimony, tungsten
Land use:
Arable land: 6%
Permanent crops: 0%
Meadows and pastures: 23%
Forest and woodland: 0%
Other: 71%

Demographics [2]

	1970	1980	1990	1995[1]	1996	2000	2010	2020	2030
Population	2,939	3,969	5,332	5,831	5,916	6,384	8,019	10,019	12,232
Population density (persons per sq. mi.)	53	72	97	106	107	116	145	181	221
(persons per sq. km.)	21	28	37	41	41	45	56	70	85
Net migration rate (per 1,000 population)	NA	NA	-7.7	-11.6	-9.9	-3.7	-1.8	-0.5	0.0
Births	NA	NA	NA	NA	200	NA	NA	NA	NA
Deaths	NA	NA	NA	NA	50	NA	NA	NA	NA
Life expectancy - males	NA	NA	64.5	60.7	60.8	61.6	63.2	66.3	69.3
Life expectancy - females	NA	NA	70.1	68.1	68.2	68.9	70.3	74.0	76.8
Birth rate (per 1,000)	NA	NA	40.8	33.6	33.8	34.5	32.6	27.5	24.2
Death rate (per 1,000)	NA	NA	7.7	8.5	8.4	8.3	7.6	6.0	5.4
Women of reproductive age (15-49 yrs.)	NA	NA	1,190	1,344	1,378	1,547	2,064	2,557	3,190
of which are currently married	NA	NA	808	NA	950	1,056	1,438	NA	NA
Fertility rate	NA	NA	5.4	4.4	4.4	4.5	3.9	3.4	3.0

Except as noted, values for vital statistics are in thousands; life expectancy is in years.

Health

Health Indicators [3]

% of population with access to	
safe water (1990-95)	NA
adequate sanitation (1990-95)	NA
health services (1985-95)	NA
% of 1-year-olds immunized (1990-94) against	
TB (tuberculosis)	69
DPT (diphtheria, pertussis, tetanus)	82
polio	74
measles	97
% of contraceptive prevalence (1980-94)	NA
ORT use rate (1990-94)	NA

Health Expenditures [4]

Total health expenditure, 1990 (official exchange rate)	
Millions of dollars	532
Dollars per capita	100
Health expenditures as a percentage of GDP	
Total	6.0
Public sector	4.4
Private sector	1.6
Development assistance for health	
Total aid flows (millions of dollars)[1]	NA
Aid flows per capita (dollars)	NA
Aid flows as a percentage of total health expenditure	NA

For sources, notes, and explanations, see Annotated Source Appendix, page 1061.

Human Factors

Women and Children [5]

% of pregnant women immunized (tetanus 1990-94)	NA
% of births attended by trained health personnel (1983-94)	NA
Maternal mortality rate (1980-92)	NA
Under-5 mortality rate (1994)	81
% under-5 moderately/severely underweight (1980-1994)	NA

Burden of Disease [6]

Population per physician (1993)	423.91
Population per nurse (1993)	139.21
Population per hospital bed (1993)	94.76
AIDS cases per 100,000 people (1994)	*
Heart disease cases per 1,000 people (1990-93)	371

Ethnic Division [7]

Tajik	64.9%
Uzbek	25%
Russian	3.5%
Other	6.6%

Religion [8]

Sunni Muslim	80%
Shi'a Muslim	5%
Other	15%

Major Languages [9]

Tajik (official), Russian widely used in government and business.

Education

Public Education Expenditures [10]

	1980	1990	1991	1992	1993	1994
Million (Ruble)						
Total education expenditure	396	788	1,227	7,207	60,039	NA
as percent of GNP	8.2	10.7	9.2	11.2	9.5	NA
as percent of total govt. expend.	29.2	24.7	24.4	19.2	17.9	NA
Current education expenditure	366	723	1,190	6,976	56,739	NA
as percent of GNP	7.6	9.8	8.9	10.8	9.0	NA
as percent of current govt. expend.	NA	NA	NA	NA	NA	NA
Capital expenditure	30	65	37	231	3,300	NA

Educational Attainment [11]

Age group (1989)	25+
Total population	1,916,494
Highest level attained (%)	
No schooling	9.8
First level	
Not completed	13.0
Completed	NA
Entered second level	
S-1	65.5
S-2	NA
Postsecondary	11.7

Illiteracy [12]

	1985	1989	1995
Illiterate population (15+ yrs.)[2]	NA	66,973	11,000
Illiteracy rate - total pop. (%)	NA	2.3	0.3
Illiteracy rate - males (%)	NA	1.2	0.2
Illiteracy rate - females (%)	NA	3.4	0.4

Libraries [13]

Daily Newspapers [14]

	1980	1985	1990	1994
Number of papers	NA	NA	NA	2
Circ. (000)	NA	NA	NA	80[e]

Culture [15]

Cinema (seats per 1,000)	34.7
Annual attendance per person	2.2
Gross box office receipts (mil. Ruble)	235
Museums (reporting)	NA
Visitors (000)	NA
Annual receipts (000 Ruble)	NA

Science and Technology

Scientific/Technical Forces [16]

Scientists/engineers	3,974
Number female	1,144
Technicians	NA
Number female	NA
Total[30]	NA

R&D Expenditures [17]

U.S. Patents Issued [18]

For sources, notes, and explanations, see Annotated Source Appendix, page 1061.

927

Government and Law

Organization of Government [19]

Long-form name:
Republic of Tajikistan
Type:
Republic
Independence:
9 September 1991 (from Soviet Union)
National holiday:
National Day, 9 September (1991)
Constitution:
New constitution adopted 6 November 1994
Legal system:
Based on civil law system; no judicial review of legislative acts
Executive branch:
President; Prime Minister; Council of Ministers
Legislative branch:
Unicameral: National Assembly (Majlisi Oli)
Judicial branch:
Supreme Court

Elections [20]

National Assembly (Majlisi Oli)	% of seats
Communist Party and affiliates	55.2
People's Party	5.5
Party of People's Unity	3.3
Party of Economic and Political Renewal	0.6
Other	35.4

Government Budget [21]

Crime [23]

	1990
Crime volume	
Cases known to police	13,864
Attempts (percent)	NA
Percent cases solved	37.0
Crimes per 100,000 persons	260.0
Persons responsible for offenses	
Total number offenders	5,124
Percent female	NA
Percent juvenile	NA
Percent foreigners	NA

Military Expenditures and Arms Transfers [22]

	1990	1991	1992[14]	1993	1994
Military expenditures					
Current dollars (mil.)[e]	NA	NA	107	126	68
1994 constant dollars (mil.)[e]	NA	NA	112	129	68
Armed forces (000)	NA	NA	3[e]	3	3
Gross national product (GNP)					
Current dollars (mil.)[e]	NA	NA	9,625	8,128	6,882
1994 constant dollars (mil.)[e]	NA	NA	10,040	8,295	6,882
Central government expenditures (CGE)					
1994 constant dollars (mil.)	NA	NA	NA	NA	NA
People (mil.)	NA	NA	5.7	5.8	6.0
Military expenditure as % of GNP	NA	NA	1.1	1.6	1.0
Military expenditure as % of CGE	NA	NA	NA	NA	NA
Military expenditure per capita (1994 $)	NA	NA	20	22	11
Armed forces per 1,000 people (soldiers)	NA	NA	0.5	0.5	0.5
GNP per capita (1994 $)	NA	NA	1,768	1,421	1,148
Arms imports[6]					
Current dollars (mil.)	NA	NA	0	0	10
1994 constant dollars (mil.)	NA	NA	0	0	10
Arms exports[6]					
Current dollars (mil.)	NA	NA	0	0	0
1994 constant dollars (mil.)	NA	NA	0	0	0
Total imports[7]					
Current dollars (mil.)	NA	NA	100[e]	517	900
1994 constant dollars (mil.)	NA	NA	104[e]	528	900
Total exports[7]					
Current dollars (mil.)	NA	NA	100[e]	314	413
1994 constant dollars (mil.)	NA	NA	104[e]	320	413
Arms as percent of total imports[8]	NA	NA	0	0	1.1
Arms as percent of total exports[8]	NA	NA	0	0	0

Human Rights [24]

	SSTS	FL	FAPRO	PPCG	APROBC	TPW	PCPTW	STPEP	PHRFF	PRW	ASST	AFL
Observes		P	P		P	P	P					
	EAFRD	CPR	ESCR	SR	ACHR	MAAE	PVIAC	PVNAC	EAFDAW	TCIDTP	RC	
Observes	P			P		P	P	P	P	P	P	

P = Party; S = Signatory; see Appendix for meaning of abbreviations.

Labor Force

Total Labor Force [25]

1.95 million (1992)

Labor Force by Occupation [26]

Agriculture and forestry	43%
Government and services	24
Industry	14
Trade and communications	11
Construction	8

Date of data: 1990

Unemployment Rate [27]

3.3% includes only officially registered unemployed; also large numbers of underemployed workers and unregistered unemployed people (December 1995)

Production Sectors

Commercial Energy Production and Consumption

Data are shown in quadrillion (10^{15}) BTUs and percent for 1995
Values for hydroelectric, nuclear, geothermal, solar, and wind power refer to electrical generation.

Production [28]

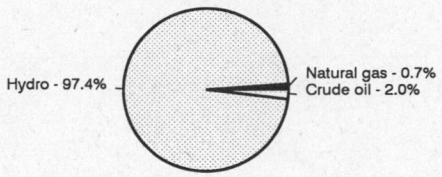

Hydro - 97.4%
Natural gas - 0.7%
Crude oil - 2.0%

Consumption [29]

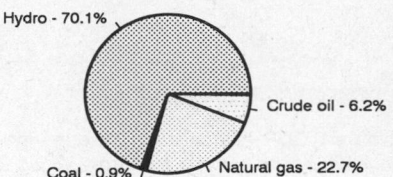

Hydro - 70.1%
Crude oil - 6.2%
Natural gas - 22.7%
Coal - 0.9%

Crude oil	0.003
Dry natural gas	0.001
Net hydroelectric power	0.148
Total	0.152

Crude oil	0.013
Dry natural gas	0.048
Coal	0.002
Net hydroelectric power	0.148
Total	0.211

Telecommunications [30]

- 303,000 (1991 est.) telephones; poorly developed and not well maintained; many towns are not reached by the national network
- Domestic: cable and microwave radio relay
- International: linked by cable and microwave radio relay to other CIS republics, and by leased connections to the Moscow international gateway switch; Dushanbe linked by Intelsat to international gateway switch in Ankara (Turkey); satellite earth stations - 1 Orbita and 2 Intelsat
- Radio: Broadcast stations: AM NA, FM NA, shortwave NA; note - there is one state-owned radio broadcast station
- Television: Broadcast stations: 1

Transportation [31]

Railways: total: 480 km in common carrier service; does not include industrial lines (1990)

Highways: total: 32,752 km; paved: 21,119 km; unpaved: 11,633 km (1992 est.)

Airports

Total:	59
With paved runways over 3,047 m:	1
With paved runways 2,438 to 3,047 m:	5
With paved runways 1,524 to 2,437 m:	7
With paved runways 914 to 1,523 m:	1
With unpaved runways 914 to 1,523 m:	9

Top Agricultural Products [32]

Produces cotton, grain, fruits, grapes, vegetables; cattle, sheep, goats.

Top Mining Products [33]

Metric tons except as noted[e]	6/30/95[*]
Aluminum	250,000
Antimony, metal content of ore	1,000
Cement	200,000
Coal	150,000
Gold (kg.)	1,500
Gypsum	300,000
Lead, metal content of ore	1,200
Mercury, metal content of ore	70
Gas, natural (mil. cu. meters)	40
Petroleum, crude	30,000

Tourism [34]

For sources, notes, and explanations, see Annotated Source Appendix, page 1061.

929

Finance, Economics, and Trade

Industrial Summary [35]

Industrial Production: Growth rate - 5% (1995). **Industries**: Aluminum, zinc, lead, chemicals and fertilizers, cement, vegetable oil, metal-cutting machine tools, refrigerators and freezers.

Economic Indicators [36]

- **National product**: GDP—purchasing power parity—$6.4 billion (1995 estimate as extrapolated from World Bank estimate for 1994)

- **National product real growth rate**: - 12.4% (1995 est.)

- **National product per capita**: $1,040 (1995 est.)

- **Inflation rate (consumer prices)**: 28% monthly average (1995 est.)

- **External debt**: $635 million (of which $250 million to Russia) (1995 est.)

Balance of Payments Summary [37]

Exchange Rates [38]

Currency: **Tajik ruble.**

Data are currency units per $1. Its own currency, the Tajik ruble, was introduced in May 1995.

January 1996   284

Imports and Exports

Top Import Origins [39]

$690 million (1995).

Origins	%
Russia	NA
Uzbekistan	NA
Kazakhstan	NA

Top Export Destinations [40]

$707 million (1995).

Destinations	%
Russia	NA
Kazakhstan	NA
Ukraine	NA
Uzbekistan	NA
Turkmenistan	NA

Foreign Aid [41]

	U.S. $	
ODA (1993)	22	million
Commitments (1992-95)	885	million

Import and Export Commodities [42]

Import Commodities	Export Commodities
Fuel	Cotton
Chemicals	Aluminum
Machinery and transport equipment	Fruits
Textiles	Vegetable oil
Foodstuffs	Textiles

Tanzania

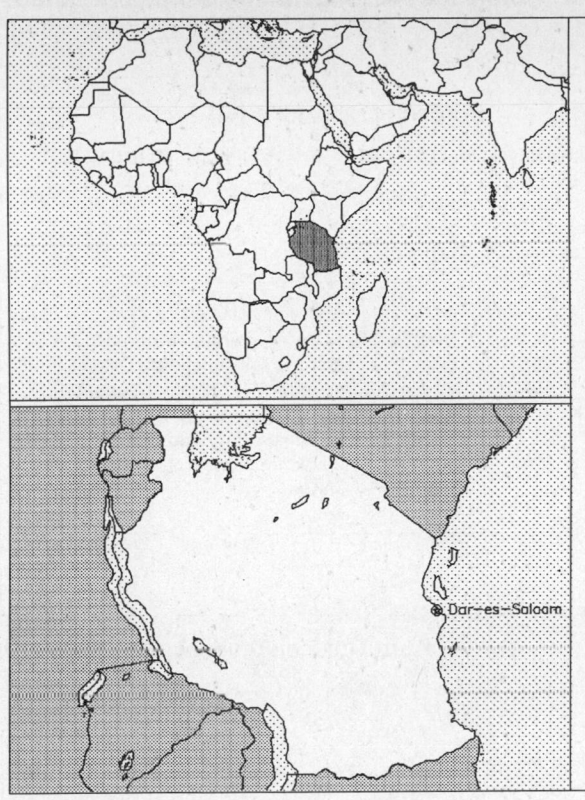

Geography [1]

Total area:
945,090 sq km   364,901 sq mi
Land area:
886,040 sq km   342,102 sq mi
Comparative area:
Slightly larger than twice the size of California
Note: Includes the islands of Mafia, Pemba, and Zanzibar
Land boundaries:
Total 3,402 km, Burundi 451 km, Kenya 769 km, Malawi 475 km, Mozambique 756 km, Rwanda 217 km, Uganda 396 km, Zambia 338 km
Coastline:
1,424 km
Climate:
Varies from tropical along coast to temperate in highlands
Terrain:
Plains along coast; central plateau; highlands in North, South
Natural resources:
Hydropower potential, tin, phosphates, iron ore, coal, diamonds, gemstones, gold, natural gas, nickel
Land use:
Arable land: 5%
Permanent crops: 1%
Meadows and pastures: 40%
Forest and woodland: 47%
Other: 7%

Demographics [2]

	1970	1980	1990	1995[1]	1996	2000	2010	2020	2030
Population	14,038	18,689	24,826	28,569	29,058	31,045	36,076	40,102	45,964
Population density (persons per sq. mi.)	41	55	73	84	85	91	105	117	134
(persons per sq. km.)	16	21	28	32	33	35	41	45	52
Net migration rate (per 1,000 population)	NA	-0.1	0.0	0.0	-10.4	0.0	0.0	0.0	0.0
Births	NA	NA	NA	NA	1,200	NA	NA	NA	NA
Deaths	NA	NA	NA	NA	566	NA	NA	NA	NA
Life expectancy - males	NA	45.1	45.1	41.6	41.0	38.7	36.7	40.9	53.8
Life expectancy - females	NA	49.0	47.9	44.5	43.8	41.4	36.3	41.1	57.5
Birth rate (per 1,000)	NA	49.1	43.5	41.7	41.3	39.8	34.9	30.9	27.4
Death rate (per 1,000)	NA	20.0	17.6	19.1	19.5	20.9	23.8	19.9	11.0
Women of reproductive age (15-49 yrs.)	NA	4,129	5,648	6,621	6,759	7,282	8,607	10,157	12,422
of which are currently married	NA	NA	3,677	NA	4,360	4,692	5,512	NA	NA
Fertility rate	NA	6.9	6.2	5.8	5.7	5.3	4.4	3.6	3.0

Except as noted, values for vital statistics are in thousands; life expectancy is in years.

Health

Health Indicators [3]

% of population with access to	
safe water (1990-95)	50
adequate sanitation (1990-95)	64
health services (1985-95)	80
% of 1-year-olds immunized (1990-94) against	
TB (tuberculosis)	86
DPT (diphtheria, pertussis, tetanus)	79
polio	NA
measles	75
% of contraceptive prevalence (1980-94)	18
ORT use rate (1990-94)	76

Health Expenditures [4]

Total health expenditure, 1990 (official exchange rate)	
Millions of dollars	109
Dollars per capita	4
Health expenditures as a percentage of GDP	
Total	4.7
Public sector	3.2
Private sector	1.5
Development assistance for health	
Total aid flows (millions of dollars)[1]	53
Aid flows per capita (dollars)	2.1
Aid flows as a percentage of total health expenditure	48.3

For sources, notes, and explanations, see Annotated Source Appendix, page 1061.

Human Factors

Women and Children [5]

% of pregnant women immunized (tetanus 1990-94)	23
% of births attended by trained health personnel (1983-94)	53
Maternal mortality rate (1980-92)	340
Under-5 mortality rate (1994)	159
% under-5 moderately/severely underweight (1980-1994)	29

Burden of Disease [6]

Population per physician (1987)[3]	25,870.97
Population per nurse (1987)[3]	5,682.38
Population per hospital bed (1990)[3]	980.96
AIDS cases per 100,000 people (1994)	8.7
Malaria cases per 100,000 people (1992)	NA

Ethnic Division [7]

Mainland:	
Bantu	95%
Other African tribes	4%
Asian, European, and Arab	1%
Zanzibar:	
Arab, mixed Arab and native African, native African	NA

Religion [8]

Mainland:	
Christian	45%
Muslim	35%
Indigenous beliefs	20%
Zanzibar:	
Muslim	> 99%

Major Languages [9]

Swahili (official; widely understood and generally used for communciation between ethnic groups, tought in primary education), English (official, primary language of commerce, administration, and higher education), Arabic (spoken in Zanzibar), and many local languages.

Education

Public Education Expenditures [10]

	1980	1985	1989	1990	1993	1994
Million (Shilling)						
Total education expenditure	1,840	4,234	13,997	23,426	NA	NA
as percent of GNP	4.4	3.6	3.7	5.0	NA	NA
as percent of total govt. expend.	11.2	14.0	14.0	11.4	NA	NA
Current education expenditure	1,522	3,643	13,069	20,599	NA	NA
as percent of GNP	3.6	3.1	3.5	4.4	NA	NA
as percent of current govt. expend.	16.3	15.6	17.0	NA	NA	NA
Capital expenditure	318	591	928	2,827	NA	NA

Educational Attainment [11]

Illiteracy [12]

In thousands and percent[1]	1990	1995	2000
Illiterate population (15+ yrs.)	5,218	5,171	5,011
Illiteracy rate - total pop. (%)	39.2	33.1	29.1
Illiteracy rate - males (%)	25.6	21.3	18.5
Illiteracy rate - females (%)	52.0	44.3	39.1

Libraries [13]

Daily Newspapers [14]

	1980	1985	1990	1994
Number of papers	3	2	3	3
Circ. (000)	208	101	200[e]	220[e]

Culture [15]

Cinema (seats per 1,000)	0.5
Annual attendance per person	0.1[e]
Gross box office receipts (mil. Shilling)	148
Museums (reporting)	4
Visitors (000)	23
Annual receipts (000 Shilling)	87

Science and Technology

Scientific/Technical Forces [16]

R&D Expenditures [17]

U.S. Patents Issued [18]

For sources, notes, and explanations, see Annotated Source Appendix, page 1061.

Government and Law

Organization of Government [19]

Long-form name:
United Republic of Tanzania
Type:
Republic
Independence:
26 April 1964; Tanganyika became independent 9 December 1961 (from UK-administered UN trusteeship); Zanzibar became independent 19 December 1963 (from UK); Tanganyika united with Zanzibar 26 April 1964 to form the United Republic of Tanganyika and Zanzibar; renamed United Republic of Tanzania
National holiday:
Union Day, 26 April (1964)
Constitution:
25 April 1977; major revisions October 1984
Legal system:
Based on English common law; judicial review of legislative acts limited to matters of interpretation; has not accepted compulsory ICJ jurisdiction
Executive branch:
President; Prime Minister; Cabinet
Legislative branch:
Unicameral: National Assembly (Bunge)
Judicial branch:
Court of Appeal; High Court

Elections [20]

National Assembly (Bunge)	% of seats
Revolutionary Party (CCM)	80.2
Opposition parties	19.8

Government Budget [21]

For 1990 est.

	$ mil.
Revenues	495
Expenditures	631
Capital expenditures	118

Crime [23]

	1990
Crime volume	
Cases known to police	289,972
Attempts (percent)	NA
Percent cases solved	32.16
Crimes per 100,000 persons	1,249.88
Persons responsible for offenses	
Total number offenders	27,442
Percent female	NA
Percent juvenile	NA
Percent foreigners	NA

Military Expenditures and Arms Transfers [22]

	1990	1991	1992	1993	1994
Military expenditures					
Current dollars (mil.)	78[e]	NA	82	78	69
1994 constant dollars (mil.)	87[e]	NA	86	80	69
Armed forces (000)	40	40	46	46	50
Gross national product (GNP)					
Current dollars (mil.)	1,656	1,789	1,824	1,907	2,058
1994 constant dollars (mil.)	1,843	1,918	1,902	1,946	2,058
Central government expenditures (CGE)					
1994 constant dollars (mil.)	559	679	699	734	952
People (mil.)	25.2	25.9	26.6	27.3	28.0
Military expenditure as % of GNP	4.7	NA	4.5	4.1	3.3
Military expenditure as % of CGE	15.6	NA	12.3	10.9	7.2
Military expenditure per capita (1994 $)	3	NA	3	3	2
Armed forces per 1,000 people (soldiers)	1.6	1.5	1.7	1.7	1.8
GNP per capita (1994 $)	73	74	72	71	74
Arms imports[6]					
Current dollars (mil.)	30	10	5	5	10
1994 constant dollars (mil.)	33	11	5	5	10
Arms exports[6]					
Current dollars (mil.)	0	0	0	0	0
1994 constant dollars (mil.)	0	0	0	0	0
Total imports[7]					
Current dollars (mil.)	1,027	1,533	1,510	1,497	1,505
1994 constant dollars (mil.)	1,143	1,643	1,575	1,528	1,505
Total exports[7]					
Current dollars (mil.)	415	341	416	450	519
1994 constant dollars (mil.)	462	366	434	459	519
Arms as percent of total imports[8]	2.9	0.7	0.3	0.3	0.7
Arms as percent of total exports[8]	0	0	0	0	0

Human Rights [24]

	SSTS	FL	FAPRO	PPCG	APROBC	TPW	PCPTW	STPEP	PHRFF	PRW	ASST	AFL
Observes	P	P		P	P	P	P			P	P	P
		EAFRD	CPR	ESCR	SR	ACHR	MAAE	PVIAC	PVNAC	EAFDAW	TCIDTP	RC
Observes		P	P	P	P				P	P	P		P

P = Party; S = Signatory; see Appendix for meaning of abbreviations.

Labor Force

Total Labor Force [25]

13.495 million

Labor Force by Occupation [26]

Agriculture	90%
Industry and commerce	10

Date of data: 1986 est.

Unemployment Rate [27]

For sources, notes, and explanations, see Annotated Source Appendix, page 1061.

933

Production Sectors

Commercial Energy Production and Consumption

Data are shown in quadrillion (10^{15}) BTUs and percent for 1995
Values for hydroelectric, nuclear, geothermal, solar, and wind power refer to electrical generation.

Production [28]	Consumption [29]
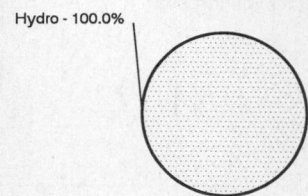	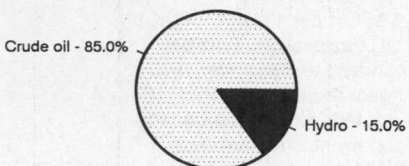
Hydro - 100.0%	Crude oil - 85.0% Hydro - 15.0%

Net hydroelectric power	0.006	Crude oil	0.034
Total	0.006	Net hydroelectric power	0.006
		Total	0.040

Telecommunications [30]

- 137,000 (1989 est.) telephones; fair system operating below capacity
- Domestic: open wire, microwave radio relay, tropospheric scatter
- International: satellite earth stations - 2 Intelsat (1 Indian Ocean and 1 Atlantic Ocean)
- Radio: Broadcast stations: AM 12, FM 4, shortwave 0 Radios: 640,000 (1992 est.)
- Television: Broadcast stations: 2 (1987 est.) Televisions: 45,000 (1992 est.)

Transportation [31]

Railways: total: 3,569 km (1995); narrow gauge: 2,600 km 1.000-m gauge; 969 km 1.067-m gauge

Highways: total: 55,600 km; paved: 20,572 km (including 50 km of expressways); unpaved: 35,028 km (1992 est.)

Merchant marine: total: 8 ships (1,000 GRT or over) totaling 30,371 GRT/41,269 DWT; ships by type: cargo 3, oil tanker 2, passenger-cargo 2, roll-on/roll-off cargo 1 (1995 est.)

Airports

Total:	111
With paved runways over 3,047 m:	2
With paved runways 2,438 to 3,047 m:	2
With paved runways 1,524 to 2,437 m:	6
With paved runways 914 to 1,523 m:	1
With paved runways under 914 m:	28

Top Agricultural Products [32]

Agriculture accounts for 58% of the GDP; produces coffee, sisal, tea, cotton, pyrethrum (insecticide made from chrysanthemums), cashews, tobacco, cloves (Zanzibar), corn, wheat, cassava (tapioca), bananas, fruits, vegetables; cattle, sheep, goats.

Top Mining Products [33]

Metric tons except as noted	6/9/95[*]
Cement, hydraulic	540,000[e]
Coal, bituminous	38,500
Diamond (carats)	15,700[69]
Gemstones, precious and semiprecious excl. diamond (kg.)	33,000[70]
Gold, refined (kg.)	3,370[e]
Gypsum and anhydrite, crude	17,600
Limestone, crushed	991,000
Petroleum refinery products (000 42-gal. bls.)	4,280
Salt, all types	64,000

Tourism [34]

	1990	1991	1992	1993	1994
Visitors	153	187	202	230	262
Tourism receipts	65	95	120	147	192
Tourism expenditures	22	68	82	102	NA

Travelers are in thousands, money in million U.S. dollars.

Manufacturing Sector

GDP and Manufacturing Summary [35]

	1980	1985	1989	1990	% change 1980-1990	% change 1989-1990
GDP (million 1980 $)	5,138	5,327	6,210	6,450	25.5	3.9
GDP per capita (1980 $)	272	234	236	236	-13.2	0.0
Manufacturing as % of GDP (current prices)	10.7	6.1	3.5[e]	3.9	-63.6	11.4
Gross output (million $)	1,266	1,145	464[e]	396[e]	-68.7	-14.7
Value added (million $)	361	278	111[e]	87[e]	-75.9	-21.6
Value added (million 1980 $)	500	387	428[e]	482	-3.6	12.6
Industrial production index	100	81	101[e]	104	4.0	3.0
Employment (thousands)	101	94	110[e]	123[e]	21.8	11.8

Note: GDP stands for Gross Domestic Product. 'e' stands for estimated value.

Profitability and Productivity

	1980	1985	1989	1990	% change 1980-1990	% change 1989-1990
Intermediate input (%)	71	76	76[e]	78[e]	9.9	2.6
Wages, salaries, and supplements (%)	9	9	9[e]	6[e]	-33.3	-33.3
Gross operating surplus (%)	19	16	15[e]	16[e]	-15.8	6.7
Gross output per worker ($)	12,457	12,141	4,206[e]	3,209[e]	-74.2	-23.7
Value added per worker ($)	3,555	2,952	1,008[e]	707[e]	-80.1	-29.9
Average wage (incl. benefits) ($)	1,174	1,042	358[e]	203[e]	-82.7	-43.3

Profitability is in percent of gross output. Productivity is in U.S. $. 'e' stands for estimated value.

Profitability - 1990

Inputs - 78.0%
Wages - 6.0%
Surplus - 16.0%

The graphic shows percent of gross output.

Value Added in Manufacturing

	1980 $ mil.	1980 %	1985 $ mil.	1985 %	1989 $ mil.	1989 %	1990 $ mil.	1990 %	% change 1980-1990	% change 1989-1990
311 Food products	58	16.1	58	20.9	23[e]	20.7	11[e]	12.6	-81.0	-52.2
313 Beverages	14	3.9	21	7.6	9[e]	8.1	5[e]	5.7	-64.3	-44.4
314 Tobacco products	12	3.3	16	5.8	7[e]	6.3	9[e]	10.3	-25.0	28.6
321 Textiles	95	26.3	43	15.5	17[e]	15.3	15[e]	17.2	-84.2	-11.8
322 Wearing apparel	10	2.8	4	1.4	1[e]	0.9	1[e]	1.1	-90.0	0.0
323 Leather and fur products	7	1.9	4	1.4	2[e]	1.8	1[e]	1.1	-85.7	-50.0
324 Footwear	8	2.2	6	2.2	3[e]	2.7	1[e]	1.1	-87.5	-66.7
331 Wood and wood products	7	1.9	6	2.2	2[e]	1.8	2[e]	2.3	-71.4	0.0
332 Furniture and fixtures	6	1.7	3	1.1	1[e]	0.9	1[e]	1.1	-83.3	0.0
341 Paper and paper products	8	2.2	7	2.5	4[e]	3.6	3[e]	3.4	-62.5	-25.0
342 Printing and publishing	14	3.9	12	4.3	5[e]	4.5	2[e]	2.3	-85.7	-60.0
351 Industrial chemicals	11	3.0	9	3.2	4[e]	3.6	12[e]	13.8	9.1	200.0
352 Other chemical products	10	2.8	7	2.5	2[e]	1.8	2[e]	2.3	-80.0	0.0
353 Petroleum refineries	15	4.2	10	3.6	3[e]	2.7	3[e]	3.4	-80.0	0.0
354 Miscellaneous petroleum and coal products	NA	0.0	NA	0.0	NA	0.0	NA	0.0	NA	NA
355 Rubber products	11	3.0	11	4.0	5[e]	4.5	1[e]	1.1	-90.9	-80.0
356 Plastic products	8	2.2	2	0.7	NA	0.0	1[e]	1.1	-87.5	NA
361 Pottery, china, and earthenware	NA	0.0	NA	0.0	NA	0.0	NA	0.0	NA	NA
362 Glass and glass products	NA	0.0	NA	0.0	NA	0.0	NA	0.0	NA	NA
369 Other nonmetal mineral products	11	3.0	4	1.4	1[e]	0.9	4[e]	4.6	-63.6	300.0
371 Iron and steel	2[e]	0.6	6[e]	2.2	3[e]	2.7	2[e]	2.3	0.0	-33.3
372 Nonferrous metals	4[e]	1.1	4[e]	1.4	2[e]	1.8	1[e]	1.1	-75.0	-50.0
381 Metal products	20	5.5	15	5.4	5[e]	4.5	4[e]	4.6	-80.0	-20.0
382 Nonelectrical machinery	3	0.8	4	1.4	1[e]	0.9	1[e]	1.1	-66.7	0.0
383 Electrical machinery	6	1.7	6	2.2	3[e]	2.7	1[e]	1.1	-83.3	-66.7
384 Transport equipment	19	5.3	19	6.8	8[e]	7.2	5[e]	5.7	-73.7	-37.5
385 Professional and scientific equipment	NA	0.0	NA	0.0	NA	0.0	NA	0.0	NA	NA
390 Other manufacturing industries	2	0.6	2	0.7	NA	0.0	NA	0.0	NA	NA

Note: The industry codes shown are International Standard Industry codes (ISIC). Percentages are percent of total Value Added. 'e' stands for estimated value

Finance, Economics, and Trade

Economic Indicators [36]

- **National product**: GDP—purchasing power parity—$23.1 billion (1995 est.)

- **National product real growth rate**: 2.7% (1995 est.)

- **National product per capita**: $800 (1995 est.)

- **Inflation rate (consumer prices)**: 25% (1994 est.)

- **External debt**: $6.7 billion (1993)

Balance of Payments Summary [37]

Values in millions of dollars.

	1989	1990	1991	1992	1993
Exports of goods (f.o.b.)	415.1	407.8	362.2	400.7	462.0
Imports of goods (f.o.b.)	-1,070.1	-1,186.4	-1,284.7	-1,313.6	-1,299.9
Trade balance	-655.0	-778.6	-922.5	-912.9	-837.9
Services - debits	-479.2	-481.1	-502.2	-569.6	-580.4
Services - credits	123.2	141.1	150.0	155.6	290.2
Private transfers (net)	182.4	158.5	269.2	325.0	193.2
Government transfers (net)	469.8	535.1	554.2	580.0	526.5
Long-term capital (net)	411.8	387.8	228.2	162.2	321.9
Short-term capital (net)	47.8	-38.8	253.8	224.2	54.1
Errors and omissions	-114.9	217.0	-20.2	44.6	-18.6
Overall balance	-14.1	141.0	10.5	9.1	-51.0

Exchange Rates [38]

Currency: **Tanzanian shilling.**
Symbol: **TSh.**

Data are currency units per $1.

December 1995	558.18
1995	574.76
1994	509.63
1993	405.27
1992	297.71
1991	219.16

Imports and Exports

Top Import Origins [39]

$1.4 billion (c.i.f., 1994).

Origins	%
Germany	NA
UK	NA
US	NA
Japan	NA
Italy	NA
Denmark	NA

Top Export Destinations [40]

$462 million (f.o.b., 1994).

Destinations	%
Germany	NA
UK	NA
Japan	NA
Netherlands	NA
Kenya	NA
Hong Kong	NA
US	NA

Foreign Aid [41]

Recipient: ODA, $NA.

Import and Export Commodities [42]

Import Commodities	Export Commodities
Manufactured goods	Coffee
Machinery and transportation equipment	Cotton
Cotton piece goods	Sisal
Crude oil	Tobacco
Foodstuffs	Tea
	Cashew nuts

Thailand

Geography [1]

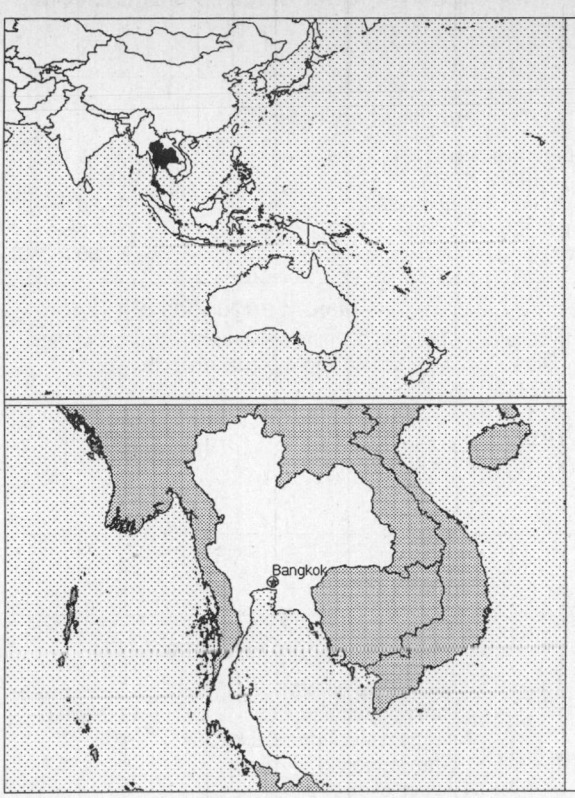

Total area:
514,000 sq km   198,457 sq mi
Land area:
511,770 sq km   197,596 sq mi
Comparative area:
Slightly more than twice the size of Wyoming
Land boundaries:
Total 4,863 km, Myanmar 1,800 km, Cambodia 803 km, Laos 1,754 km, Malaysia 506 km
Coastline:
3,219 km
Climate:
Tropical; rainy, warm, cloudy, southwest monsoon (mid-May to September); dry, cool northeast monsoon (November to mid-March); southern isthmus always hot and humid
Terrain:
Central plain; Khorat Plateau in the East; mountains elsewhere
Natural resources:
Tin, rubber, natural gas, tungsten, tantalum, timber, lead, fish, gypsum, lignite, fluorite
Land use:
Arable land: 34%
Permanent crops: 4%
Meadows and pastures: 1%
Forest and woodland: 30%
Other: 31%

Demographics [2]

	1970	1980	1990	1995[1]	1996	2000	2010	2020	2030
Population	37,091	47,026	55,052	58,241	58,851	61,164	66,092	69,298	70,982
Population density (persons per sq. mi.)	188	238	279	295	298	310	334	351	359
(persons per sq. km.)	72	92	108	114	115	120	129	135	139
Net migration rate (per 1,000 population)	NA	NA	0.0	0.0	0.0	0.0	0.0	0.0	0.0
Births	NA	NA	NA	NA	1,018	NA	NA	NA	NA
Deaths	NA	NA	253	NA	412	NA	NA	NA	NA
Life expectancy - males	NA	NA	65.2	64.7	64.9	65.8	69.4	70.9	73.3
Life expectancy - females	NA	NA	71.1	72.3	72.5	73.2	76.5	78.0	80.6
Birth rate (per 1,000)	NA	NA	18.4	17.5	17.3	16.2	13.7	12.1	11.1
Death rate (per 1,000)	NA	NA	6.6	7.0	7.0	7.2	7.4	8.7	9.7
Women of reproductive age (15-49 yrs.)	NA	NA	15,474	16,912	17,170	17,974	18,021	17,251	16,044
of which are currently married[55]	4,901	6,786	14,456	NA	15,979	16,673	16,606	NA	NA
Fertility rate	NA	NA	2.0	1.9	1.9	1.8	1.8	1.8	1.7

Except as noted, values for vital statistics are in thousands; life expectancy is in years.

Health

Health Indicators [3]

% of population with access to	
safe water (1990-95)[1]	86
adequate sanitation (1990-95)	74
health services (1985-95)	90
% of 1-year-olds immunized (1990-94) against	
TB (tuberculosis)	98
DPT (diphtheria, pertussis, tetanus)	93
polio	93
measles	86
% of contraceptive prevalence (1980-94)	66
ORT use rate (1990-94)	65

Health Expenditures [4]

Total health expenditure, 1990 (official exchange rate)	
Millions of dollars	4,061
Dollars per capita	73
Health expenditures as a percentage of GDP	
Total	5.0
Public sector	1.1
Private sector	3.9
Development assistance for health	
Total aid flows (millions of dollars)[1]	36
Aid flows per capita (dollars)	0.7
Aid flows as a percentage of total health expenditure	0.9

For sources, notes, and explanations, see Annotated Source Appendix, page 1061.

937

Human Factors

Women and Children [5]

% of pregnant women immunized (tetanus 1990-94)	90
% of births attended by trained health personnel (1983-94)	71
Maternal mortality rate (1980-92)	50
Under-5 mortality rate (1994)	32
% under-5 moderately/severely underweight (1980-1994)[1]	26

Burden of Disease [6]

Population per physician (1991)	4,426.99
Population per nurse (1990)	919.75
Population per hospital bed (1990)	614.98
AIDS cases per 100,000 people (1994)	17.5
Malaria cases per 100,000 people (1992)	296

Ethnic Division [7]

Thai	75%
Chinese	14%
Other	11%

Religion [8]

Buddhism	95%
Muslim	3.8%
Christianity	0.5%
Hinduism	0.1%
Other	0.6%
(1991)	

Major Languages [9]

Thai, English the secondary language of the elite, ethnic and regional dialects.

Education

Public Education Expenditures [10]

	1980	1985	1990	1992	1993	1994
Million (Baht)						
Total education expenditure	22,489	39,367	77,420	109,890	128,786	135,358
as percent of GNP	3.4	3.8	3.6	4.0	4.1	3.8
as percent of total govt. expend.	20.6	18.5	20.0	19.6	20.6	18.9
Current education expenditure	15,867	33,830	64,702	NA	104,534	108,485
as percent of GNP	2.4	3.3	3.0	NA	3.3	3.0
as percent of current govt. expend.	19.1	19.2	21.0	NA	22.7	21.1
Capital expenditure	6,622	5,537	12,718	NA	24,252	26,873

Educational Attainment [11]

Age group (1990)	6+
Total population	49,076,100
Highest level attained (%)	
No schooling	10.7
First level	
Not completed	69.6
Completed	NA
Entered second level	
S-1	13.7
S-2	NA
Postsecondary	5.1

Illiteracy [12]

In thousands and percent[1]	1990	1995	2000
Illiterate population (15+ yrs.)	2,572	2,613	2,224
Illiteracy rate - total pop. (%)	6.6	6.1	4.8
Illiteracy rate - males (%)	4.4	3.9	3.2
Illiteracy rate - females (%)	8.8	8.1	6.3

Libraries [13]

	Admin. Units	Svc. Pts.	Vols. (000)	Shelving (meters)	Vols. Added	Reg. Users
National (1992)	1	25	1,528	6,200	138,129	1 mil
Nonspecialized	NA	NA	NA	NA	NA	NA
Public (1992)	589	589	NA	NA	NA	NA
Higher ed.	NA	NA	NA	NA	NA	NA
School	NA	NA	NA	NA	NA	NA

Daily Newspapers [14]

	1980	1985	1990	1994
Number of papers	27	32	34	35
Circ. (000)	2,680	4,350	4,500[e]	2,766

Culture [15]

Cinema (seats per 1,000)	0.1
Annual attendance per person	NA
Gross box office receipts (mil. Baht)	NA
Museums (reporting)	60
Visitors (000)	1,770
Annual receipts (000 Baht)	81,789

Science and Technology

Scientific/Technical Forces [16]

Scientists/engineers	9,752
Number female	NA
Technicians	2,898
Number female	NA
Total[1]	12,650

R&D Expenditures [17]

	Baht (000) 1991
Total expenditure	3,928,100
Capital expenditure	828,600
Current expenditure	3,099,500
Percent current	78.9

U.S. Patents Issued [18]

Values show patents issued to citizens of the country by the U.S. Patents Office.

	1993	1994	1995
Number of patents	17	8	10

For sources, notes, and explanations, see Annotated Source Appendix, page 1061.

Government and Law

Organization of Government [19]

Long-form name:
Kingdom of Thailand
Type:
Constitutional monarchy
Independence:
1238 (traditional founding date; never colonized)
National holiday:
Birthday of His Majesty the King, 5 December (1927)
Constitution:
New constitution approved 7 December 1991; amended 10 June 1992
Legal system:
Based on civil law system, with influences of common law; has not accepted compulsory ICJ jurisdiction; martial law in effect since February 1991 military coup
Executive branch:
King; Heir Apparent Crown Prince; Prime Minister; Council of Ministers; Privy Council
Legislative branch:
Bicameral National Assembly (Rathasapha): Senate (Wuthisapha) and House of Representatives (Sapha Phuthaen Ratsadon)
Judicial branch:
Supreme Court (Sandika)

Elections [20]

House of Representatives	% of seats
Thai Nation Party	23.8
Democrat Party	22.0
NAP	14.3
National Development Party	13.6
Phalang Tham	5.9
Social Action Party	5.9
Thai Leadership Party	4.6
Thai Citizen's Party	4.6
Liberal Democratic Party	2.6
Other	2.8

Government Expenditures [21]

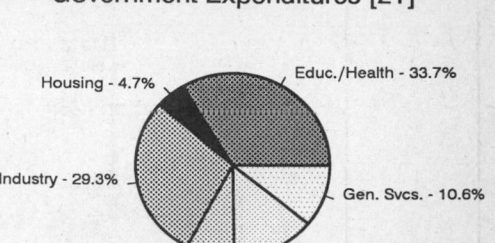

Housing - 4.7%; Educ./Health - 33.7%; Industry - 29.3%; Gen. Svcs. - 10.6%; Other - 7.5%; Defense - 14.2%

(% distribution). Expend. for FY95: 685,546 (Baht mil.)

Military Expenditures and Arms Transfers [22]

	1990	1991	1992	1993	1994
Military expenditures					
Current dollars (mil.)	2,276	2,602	3,020	3,624	3,777
1994 constant dollars (mil.)	2,533	2,789	3,149	3,699	3,777
Armed forces (000)	283	283	283	295	290
Gross national product (GNP)					
Current dollars (mil.)	92,450	103,800	114,800	127,500	141,500
1994 constant dollars (mil.)	102,900	111,300	119,700	130,100	141,500
Central government expenditures (CGE)					
1994 constant dollars (mil.)	14,830	16,440	18,560	21,080	24,060
People (mil.)	56.2	57.1	57.9	58.7	59.5
Military expenditure as % of GNP	2.5	2.5	2.6	2.8	2.7
Military expenditure as % of CGE	17.1	17.0	17.0	17.5	15.7
Military expenditure per capita (1994 $)	45	49	54	63	63
Armed forces per 1,000 people (soldiers)	5.0	5.0	4.9	5.0	4.9
GNP per capita (1994 $)	1,830	1,950	2,067	2,215	2,378
Arms imports[6]					
Current dollars (mil.)	270	575	370	120	360
1994 constant dollars (mil.)	300	616	386	122	360
Arms exports[6]					
Current dollars (mil.)	0	0	0	0	0
1994 constant dollars (mil.)	0	0	0	0	0
Total imports[7]					
Current dollars (mil.)	33,380	37,590	40,690	46,210	54,390
1994 constant dollars (mil.)	37,150	40,290	42,430	47,160	54,390
Total exports[7]					
Current dollars (mil.)	23,070	28,430	32,470	37,170	45,090
1994 constant dollars (mil.)	25,680	30,470	33,860	37,930	45,090
Arms as percent of total imports[8]	0.8	1.5	0.9	0.3	0.7
Arms as percent of total exports[8]	0	0	0	0	0

Crime [23]

	1994
Crime volume	
Cases known to police	204,478
Attempts (percent)	NA
Percent cases solved	NA
Crimes per 100,000 persons	350.52
Persons responsible for offenses	
Total number offenders	NA
Percent female	NA
Percent juvenile	NA
Percent foreigners	NA

Human Rights [24]

	SSTS	FL	FAPRO	PPCG	APROBC	TPW	PCPTW	STPEP	PHRFF	PRW	ASST	AFL
Observes		P				P	P			P		P
	EAFRD	CPR	ESCR	SR	ACHR	MAAE	PVIAC	PVNAC	EAFDAW	TCIDTP		RC
Observes		P							P			P

P = Party; S = Signatory; see Appendix for meaning of abbreviations.

Labor Force

Total Labor Force [25]

32.153 million

Labor Force by Occupation [26]

Agriculture	57%
Industry	17
Commerce	11
Services (including government)	15

Date of data: 1993 est.

Unemployment Rate [27]

2.7% (1995 est.)

For sources, notes, and explanations, see Annotated Source Appendix, page 1061.

939

Production Sectors

Commercial Energy Production and Consumption

Data are shown in quadrillion (10^{15}) BTUs and percent for 1995
Values for hydroelectric, nuclear, geothermal, solar, and wind power refer to electrical generation.

Production [28]	Consumption [29]
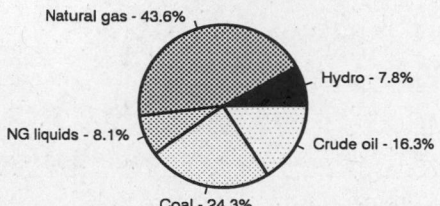	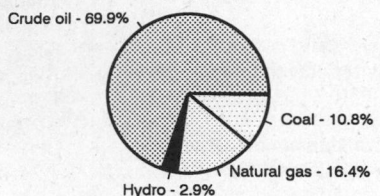

Production [28] pie chart: Natural gas - 43.6%, Hydro - 7.8%, Crude oil - 16.3%, Coal - 24.3%, NG liquids - 8.1%

Consumption [29] pie chart: Crude oil - 69.9%, Coal - 10.8%, Natural gas - 16.4%, Hydro - 2.9%

Crude oil	0.119	Crude oil	1.363
Natural gas liquids	0.059	Dry natural gas	0.319
Dry natural gas	0.319	Coal	0.211
Coal	0.178	Net hydroelectric power	0.057
Net hydroelectric power	0.057	Total	1.950
Total	0.732		

Telecommunications [30]

- 1,553,200 (1994 est.) telephones; service to general public inadequate; bulk of service to government activities provided by multichannel cable and microwave radio relay network
- Domestic: microwave radio relay and multichannel cable; domestic satellite system being developed
- International: satellite earth stations - 2 Intelsat (1 Indian Ocean and 1 Pacific Ocean)
- Radio: Broadcast stations: AM 200 (in government-controlled network), FM 100 (in government-controlled network), shortwave 0 Radios: 10.75 million (1992 est.)
- Television: Broadcast stations: 11 (in government-controlled network) Televisions: 3.3 million (1993 est.)

Top Agricultural Products [32]

Agriculture accounts for 10.2% of the GDP; produces rice, cassava (tapioca), rubber, corn, sugarcane, coconuts, soybeans.

Top Mining Products [33]

Thousand metric tons except as noted[e]	7/28/95*
Steel, crude	1,000
Cement, hydraulic	26,000
Gemstones (000 carats)	4,800
Gypsum	8,100
Dolomite	744
Limestone, cement manufacture only	42,000
Shale, cement manufacture only	3,500
Coal, lignite	17,000
Gas, natural, gr. prod. (mil. cu. meters)	9,700
Petroleum, crude (000 42-gal. bls.)	90,000

Transportation [31]

Railways: total: 4,623 km; narrow gauge: 4,623 km 1.000-m gauge (99 km double track)

Highways: total: 54,388 km; paved: 48,786 km (including 171 km of expressways); unpaved: 5,602 km (1992 est.)

Merchant marine: total: 259 ships (1,000 GRT or over) totaling 1,559,037 GRT/2,498,812 DWT; ships by type: bulk 32, cargo 143, chemical tanker 3, container 11, liquefied gas tanker 12, oil tanker 45, passenger 1, refrigerated cargo 7, roll-on/roll-off cargo 2, short-sea passenger 1, specialized tanker 2 (1995 est.)

Airports

Total:	98
With paved runways over 3,047 m:	6
With paved runways 2,438 to 3,047 m:	9
With paved runways 1,524 to 2,437 m:	12
With paved runways 914 to 1,523 m:	22
With paved runways under 914 m:	36

Tourism [34]

	1990	1991	1992	1993	1994
Tourists[8]	5,299	5,087	5,136	5,761	6,166
Tourism receipts	4,326	3,923	4,829	5,013	5,762
Tourism expenditures	854	1,266	1,590	2,090	2,906
Fare receipts[106]	660	645	958	1,345	822
Fare expenditures[106]	252	251	316	307	384

Travelers are in thousands, money in million U.S. dollars.

For sources, notes, and explanations, see Annotated Source Appendix, page 1061.

Manufacturing Sector

Manufacturing Summary [35]

	1987		1988		1989		1990		1991	
	$ bil.	%	$ bil.	%	$ bil.	%	$ bil.	%	$ bil.	%
Establishments or enterprises (number)	-	-	11,566	0.551	-	-	-	-	-	-
Total employment (000)	-	-	1,264	0.924	-	-	-	-	-	-
Production workers (000)	-	-	1,033	1.653	-	-	-	-	-	-
Output ($ bil.)	-	-	38	0.328	-	-	-	-	-	-
Value added ($ bil.)	-	-	13	0.281	-	-	-	-	-	-
Capital investment ($ mil.)	-	-	-	-	-	-	-	-	-	-
M & E investment ($ mil.)	-	-	-	-	-	-	-	-	-	-
Employees per establishment (number)	-	-	109	167.658	-	-	-	-	-	-
Production workers per establishment	-	-	89	299.974	-	-	-	-	-	-
Output per establishment ($ mil.)	-	-	3	59.539	-	-	-	-	-	-
Capital investment per estab. ($ mil.)	-	-	-	-	-	-	-	-	-	-
M & E per establishment ($ mil)	-	-	-	-	-	-	-	-	-	-
Payroll per employee ($)	-	-	2,618	26.283	-	-	-	-	-	-
Wages per production worker ($)	-	-	-	-	-	-	-	-	-	-
Hours per production worker (hours)	-	-	-	-	-	-	-	-	-	-
Output per employee ($)	-	-	30,247	35.512	-	-	-	-	-	-
Capital investment per employee ($)	-	-	-	-	-	-	-	-	-	-
M & E per employee ($)	-	-	-	-	-	-	-	-	-	-

Note: Columns headed % show percent of world total or ratio. Ratios closest to 100 are closest to world average. M & E stands for machinery & equipment.

Output in Manufacturing

	1987		1988		1989		1990		1991	
	$ bil.	%	$ bil.	%	$ bil.	%	$ bil.	%	$ bil.	%
3110 - Food products	-	-	8.997	23.68	-	-	-	-	-	-
3130 - Beverages	-	-	1.111	2.92	-	-	-	-	-	-
3140 - Tobacco	-	-	0.051	0.13	-	-	-	-	-	-
3210 - Textiles	-	-	3.092	8.14	-	-	-	-	-	-
3211 - Spinning, weaving, etc.	-	-	2.685	7.07	-	-	-	-	-	-
3220 - Wearing apparel	-	-	1.268	3.34	-	-	-	-	-	-
3230 - Leather and products	-	-	0.213	0.56	-	-	-	-	-	-
3240 - Footwear	-	-	0.272	0.72	-	-	-	-	-	-
3310 - Wood products	-	-	0.470	1.24	-	-	-	-	-	-
3320 - Furniture, fixtures	-	-	0.455	1.20	-	-	-	-	-	-
3410 - Paper and products	-	-	0.796	2.09	-	-	-	-	-	-
3411 - Pulp, paper, etc.	-	-	0.095	0.25	-	-	-	-	-	-
3420 - Printing, publishing	-	-	2.074	5.46	-	-	-	-	-	-
3510 - Industrial chemicals	-	-	0.841	2.21	-	-	-	-	-	-
3511 - Basic chemicals, excl fertilizers	-	-	0.163	0.43	-	-	-	-	-	-
3513 - Synthetic resins, etc.	-	-	0.657	1.73	-	-	-	-	-	-
3520 - Chemical products nec	-	-	1.139	3.00	-	-	-	-	-	-
3522 - Drugs and medicines	-	-	0.683	1.80	-	-	-	-	-	-
3530 - Petroleum refineries	-	-	1.897	4.99	-	-	-	-	-	-
3540 - Petroleum, coal products	-	-	-	-	-	-	-	-	-	-
3550 - Rubber products	-	-	1.713	4.51	-	-	-	-	-	-
3560 - Plastic products nec	-	-	0.212	0.56	-	-	-	-	-	-
3610 - Pottery, china, etc.	-	-	0.058	0.15	-	-	-	-	-	-
3620 - Glass and products	-	-	0.468	1.23	-	-	-	-	-	-
3690 - Nonmetal products nec	-	-	0.974	2.56	-	-	-	-	-	-
3710 - Iron and steel	-	-	0.947	2.49	-	-	-	-	-	-
3720 - Nonferrous metals	-	-	0.172	0.45	-	-	-	-	-	-
3810 - Metal products	-	-	1.160	3.05	-	-	-	-	-	-
3820 - Machinery nec	-	-	0.169	0.44	-	-	-	-	-	-
3825 - Office, computing machinery	-	-	0.002	0.01	-	-	-	-	-	-
3830 - Electrical machinery	-	-	1.391	3.66	-	-	-	-	-	-
3832 - Radio, television, etc.	-	-	0.460	1.21	-	-	-	-	-	-
3840 - Transportation equipment	-	-	2.065	5.43	-	-	-	-	-	-
3841 - Shipbuilding, repair	-	-	0.008	0.02	-	-	-	-	-	-
3843 - Motor vehicles	-	-	0.868	2.28	-	-	-	-	-	-
3850 - Professional goods	-	-	0.110	0.29	-	-	-	-	-	-
3900 - Industries nec	-	-	0.503	1.32	-	-	-	-	-	-

Note: Codes are International Standard Industry codes (ISIC). Percentages are % of total Output. [f]: Factor Prices; [p]: Producer Prices.

For sources, notes, and explanations, see Annotated Source Appendix, page 1061.

Finance, Economics, and Trade

Economic Indicators [36]

- **National product**: GDP—purchasing power parity—$416.7 billion (1995 est.)
- **National product real growth rate**: 8.6% (1995 est.)
- **National product per capita**: $6,900 (1995 est.)
- **Inflation rate (consumer prices)**: 5.8% (1995)
- **External debt**: $53.7 billion (1994)

Balance of Payments Summary [37]

Values in millions of dollars.

	1989	1990	1991	1992	1993
Exports of goods (f.o.b.)	19,834	22,811	28,232	32,100	36,410
Imports of goods (f.o.b.)	-22,750	-29,561	-34,222	-36,261	-40,556
Trade balance	-2,916	-6,750	-5,990	-4,161	-4,146
Services - debits	-6,874	-9,222	-11,369	-13,340	-15,449
Services - credits	7,046	8,478	9,526	10,769	12,355
Private transfers (net)	47	26	163	323	281
Government transfers (net)	199	187	98	54	32
Long-term capital (net)	4,271	3,602	5,094	4,410	7,276
Short-term capital (net)	2,380	5,698	7,090	5,823	7,395
Errors and omissions	928	1,419	431	-517	-347
Overall balance	5,081	3,438	5,043	3,361	7,397

Exchange Rates [38]

Currency: **baht.**
Symbol: **B.**

Data are currency units per $1.

January 1996	25.300
1995 est.	25.000
1994	25.150
1993	25.319
1992	25.400
1991	25.517

Imports and Exports

Top Import Origins [39]

$53.9 billion (c.i.f., 1994).

Origins	%
Japan	30.4
US	11.9
Singapore	6.3
Germany	5.8
Taiwan	5.1
Malaysia	4.9
South Korea	3.7
China	2.6

Top Export Destinations [40]

$45.1 billion (f.o.b., 1994).

Destinations	%
US	21.0
Japan	17.1
Singapore	13.6
Hong Kong	5.3
Germany	3.5
UK	3.0
Netherlands	2.8
Malaysia	2.4

Foreign Aid [41]

Recipient: ODA, $624 million (1993).

Import and Export Commodities [42]

Import Commodities

Manufactures 80%
Fuels 6.9%
Raw materials 6.6%
Foodstuffs 4.3%

Export Commodities

Manufactures 73%
Agricultural and fishery products 21%
Raw materials 5%
Fuels 1%

For sources, notes, and explanations, see Annotated Source Appendix, page 1061.

Togo

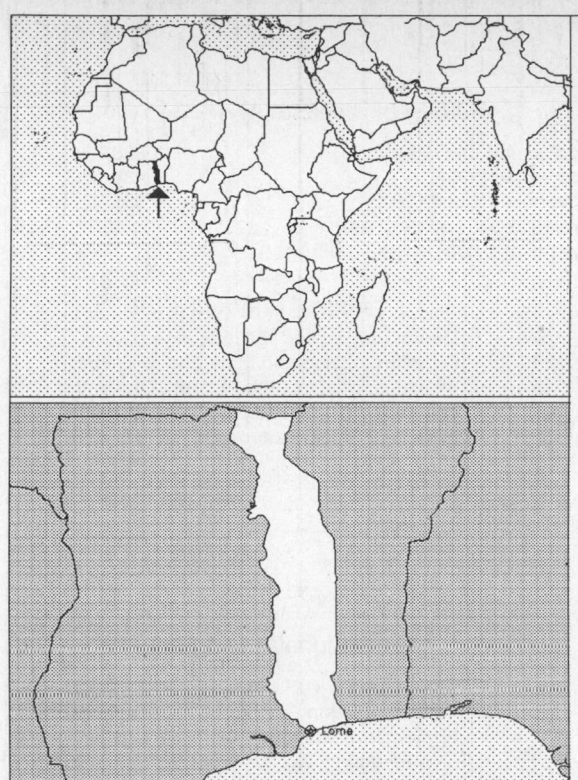

Geography [1]

Total area:
56,790 sq km   21,927 sq mi
Land area:
54,390 sq km   21,000 sq mi
Comparative area:
Slightly smaller than West Virginia
Land boundaries:
Total 1,647 km, Benin 644 km, Burkina Faso 126 km, Ghana 877 km
Coastline:
56 km
Climate:
Tropical; hot, humid in South; semiarid in North
Terrain:
Gently rolling savanna in North; central hills; southern plateau;
low coastal plain with extensive lagoons and marshes
Natural resources:
Phosphates, limestone, marble
Land use:
Arable land: 25%
Permanent crops: 1%
Meadows and pastures: 4%
Forest and woodland: 28%
Other: 42%

Demographics [2]

	1970	1980	1990	1995[1]	1996	2000	2010	2020	2030
Population	1,964	2,596	3,680	4,410	4,571	5,263	7,401	10,146	13,386
Population density (persons per sq. mi.)	94	124	175	210	218	251	352	483	637
(persons per sq. km.)	36	48	68	81	84	97	136	187	246
Net migration rate (per 1,000 population)	NA	NA	0.0	0.0	0.0	0.0	0.0	0.0	0.0
Births	NA	NA	NA	NA	211	NA	NA	NA	NA
Deaths	NA	NA	NA	NA	49	NA	NA	NA	NA
Life expectancy - males	NA	NA	53.2	55.3	55.7	57.4	61.3	64.9	68.1
Life expectancy - females	NA	NA	57.0	59.6	60.1	62.2	67.0	71.3	74.9
Birth rate (per 1,000)	NA	NA	49.8	46.8	46.2	44.4	40.2	35.2	30.2
Death rate (per 1,000)	NA	NA	13.1	11.0	10.7	9.4	7.1	5.5	4.5
Women of reproductive age (15-49 yrs.)	NA	NA	832	984	1,019	1,177	1,708	2,428	3,362
of which are currently married	265	NA	600	NA	731	840	1,218	NA	NA
Fertility rate	NA	NA	7.2	6.8	6.8	6.5	5.6	4.6	3.8

Except as noted, values for vital statistics are in thousands; life expectancy is in years.

Health

Health Indicators [3]

% of population with access to	
safe water (1990-95)	63
adequate sanitation (1990-95)	23
health services (1985-95)	61
% of 1-year-olds immunized (1990-94) against	
TB (tuberculosis)	73
DPT (diphtheria, pertussis, tetanus)	71
polio	71
measles	58
% of contraceptive prevalence (1980-94)	12
ORT use rate (1990-94)	33

Health Expenditures [4]

Total health expenditure, 1990 (official exchange rate)	
Millions of dollars	67
Dollars per capita	18
Health expenditures as a percentage of GDP	
Total	4.1
Public sector	2.5
Private sector	1.6
Development assistance for health	
Total aid flows (millions of dollars)[1]	14
Aid flows per capita (dollars)	3.9
Aid flows as a percentage of total health expenditure	21.0

For sources, notes, and explanations, see Annotated Source Appendix, page 1061.

Human Factors

Women and Children [5]

% of pregnant women immunized (tetanus 1990-94)	72
% of births attended by trained health personnel (1983-94)	54
Maternal mortality rate (1980-92)	420
Under-5 mortality rate (1994)	132
% under-5 moderately/severely underweight (1980-1994)[1]	24

Burden of Disease [6]

Population per physician (1986)	11,866.92
Population per nurse (1984)	1,238.52
Population per hospital bed (1990)	665.35
AIDS cases per 100,000 people (1994)	32.8
Malaria cases per 100,000 people (1992)	NA

Ethnic Division [7]

Native African includes 37 tribes, the largest and most important of which are Ewe, Mina, and Kabye.

Native African	99%
European and Syrian-Lebanese	< 1%

Religion [8]

Indigenous beliefs	70%
Christian	20%
Muslim	10%

Major Languages [9]

French (official and the language of commerce), Ewe and Mina (the two major African languages in the South), Dagomba and Kabye (sometimes spelled Kabiye; the two major African languages in the North).

Education

Public Education Expenditures [10]

Million (Franc C.F.A.)	1980	1985	1989	1990	1992	1994
Total education expenditure	13,049	15,880	22,801	24,420	27,004	NA
as percent of GNP	5.6	5.0	5.5	5.6	6.1	NA
as percent of total govt. expend.	19.4	19.4	24.7	NA	21.6	NA
Current education expenditure	12,575	15,028	21,384	22,720	26,307	NA
as percent of GNP	5.4	4.7	5.1	5.2	6.0	NA
as percent of current govt. expend.	21.0	19.2	NA	NA	29.0	NA
Capital expenditure	474	852	1,417	1,700	697	NA

Educational Attainment [11]

Age group (1981)	25+
Total population	1,084,488
Highest level attained (%)	
No schooling	76.5
First level	
Not completed	13.5
Completed	NA
Entered second level	
S-1	8.7
S-2	NA
Postsecondary	1.3

Illiteracy [12]

In thousands and percent[1]	1990	1995	2000
Illiterate population (15+ yrs.)	1,055	1,085	1,108
Illiteracy rate - total pop. (%)	56.0	48.0	40.5
Illiteracy rate - males (%)	40.6	33.4	27.2
Illiteracy rate - females (%)	70.1	61.6	52.9

Libraries [13]

	Admin. Units	Svc. Pts.	Vols. (000)	Shelving (meters)	Vols. Added	Reg. Users
National (1993)	1	NA	16	NA	600	500
Nonspecialized	NA	NA	NA	NA	NA	NA
Public (1989)	1	26	63	600	2,000	7,706
Higher ed.	NA	NA	NA	NA	NA	NA
School	NA	NA	NA	NA	NA	NA

Daily Newspapers [14]

	1980	1985	1990	1994
Number of papers	3	2	1	1
Circ. (000)	16[e]	11[e]	10	10

Culture [15]

Cinema (seats per 1,000)	NA
Annual attendance per person	NA
Gross box office receipts (mil. Francs C.F.A.)	NA
Museums (reporting)	-
Visitors (000)	NA
Annual receipts (000 Francs C.F.A.)	NA

Science and Technology

Scientific/Technical Forces [16]

R&D Expenditures [17]

U.S. Patents Issued [18]

For sources, notes, and explanations, see Annotated Source Appendix, page 1061.

Government and Law

Organization of Government [19]

Long-form name:
Republic of Togo
Type:
Republic under transition to multiparty democratic rule
Independence:
27 April 1960 (from French-administered UN trusteeship)
National holiday:
Independence Day, 27 April (1960)
Constitution:
Multiparty draft constitution approved by High Council of the Republic 1 July 1992; adopted by public referendum 27 September 1992
Legal system:
French-based court system
Executive branch:
President; Prime Minister; Council of Ministers
Legislative branch:
Unicameral: National Assembly
Judicial branch:
Court of Appeal (Cour d'Appel); Supreme Court (Cour Supreme)

Elections [20]

National Assembly	% of seats
Rally of the Toglese People	43.2
Committee for Renewal	41.9
Togolese Union for Democracy	7.4
Union of Justice and Democracy	2.5
Coordination des Forces Nouvelles	1.2
Unfilled	3.7

Government Budget [21]

For 1995 est.

	$ mil.
Revenues	165
Expenditures	274
Capital expenditures	NA

Crime [23]

Military Expenditures and Arms Transfers [22]

	1990	1991	1992	1993	1994
Military expenditures					
Current dollars (mil.)[e]	29	29	27	32	25
1994 constant dollars (mil.)[e]	33	31	28	32	25
Armed forces (000)	8	8	6	6	6
Gross national product (GNP)					
Current dollars (mil.)	947	974	967	851	929
1994 constant dollars (mil.)	1,054	1,044	1,008	869	929
Central government expenditures (CGE)					
1994 constant dollars (mil.)[e]	238	268	NA	NA	NA
People (mil.)	3.7	3.8	4.0	4.1	4.3
Military expenditure as % of GNP	3.1	3.0	2.8	3.7	2.7
Military expenditure as % of CGE	13.8	11.7	NA	NA	NA
Military expenditure per capita (1994 $)	9	8	7	8	6
Armed forces per 1,000 people (soldiers)	2.2	2.1	1.5	1.5	1.4
GNP per capita (1994 $)	286	274	255	212	218
Arms imports[6]					
Current dollars (mil.)	0	0	0	5	0
1994 constant dollars (mil.)	0	0	0	5	0
Arms exports[6]					
Current dollars (mil.)	0	0	0	0	0
1994 constant dollars (mil.)	0	0	0	0	0
Total imports[7]					
Current dollars (mil.)	581	444	NA	292[e]	NA
1994 constant dollars (mil.)	647	476	NA	298[e]	NA
Total exports[7]					
Current dollars (mil.)	268	253	285	221[e]	NA
1994 constant dollars (mil.)	298	271	297	226[e]	NA
Arms as percent of total imports[8]	0	0	0	1.7	0
Arms as percent of total exports[8]	0	0	0	0	0

Human Rights [24]

	SSTS	FL	FAPRO	PPCG	APROBC	TPW	PCPTW	STPEP	PHRFF	PRW	ASST	AFL
Observes	2	P	P	P	P	P	P	P			P	
	EAFRD	CPR	ESCR	SR	ACHR	MAAE	PVIAC	PVNAC	EAFDAW	TCIDTP	RC	
Observes	P	P	P	P		P	P	P	P	P	P	

P = Party; S = Signatory; see Appendix for meaning of abbreviations.

Labor Force

Total Labor Force [25]

1.538 million (1993 est.)

Labor Force by Occupation [26]

Agriculture	64%
Industry	9
Services	21

Date of data: 1981 est.

Unemployment Rate [27]

6% (1981 est.)

For sources, notes, and explanations, see Annotated Source Appendix, page 1061.

945

Production Sectors

Energy Resource Summary [28]

Electricity: Capacity: 34,000 kW. Production: 41.004 million kWh. Consumption per capita: 9 kWh (1990).

Telecommunications [30]

- 12,000 (1987 est.) telephones; fair system based on network of microwave radio relay routes supplemented by open-wire lines
- Domestic: microwave radio relay and open-wire lines
- International: satellite earth stations - 1 Intelsat (Atlantic Ocean) and 1 Symphonie
- Radio: Broadcast stations: AM 2, FM 0, shortwave 0 Radios: 795,000 (1992 est.)
- Television: Broadcast stations: 3 (relays 2) Televisions: 24,000 (1992 est.)

Transportation [31]

Railways: total: 525 km (1995); narrow gauge: 525 km 1.000-m gauge

Highways: total: 7,545 km; paved: 1,833 km; unpaved: 5,712 km (1993 est.)

Merchant marine: none

Airports

Total:	8
With paved runways 2,438 to 3,047 m:	2
With paved runways under 914 m:	2
With unpaved runways 914 to 1,523 m:	4 (1995 est.)

Top Agricultural Products [32]

Agriculture accounts for 49.2% of the GDP; produces coffee, cocoa, cotton, yams, cassava (tapioca), corn, beans, rice, millet, sorghum; meat; annual fish catch of 10,000-14,000 tons.

Top Mining Products [33]

Metric tons except as noted[e]	7/6/95[*]
Cement	350,000[71]
Iron and steel, semimanufactures	500[72]
Phosphate rock, beneficiated (000 tons)	
gross weight	2,250[r]
P2O5 content	800

Tourism [34]

	1990	1991	1992	1993	1994
Tourists[16]	103	65	49	24	44
Tourism receipts	58	49	39	18	7
Tourism expenditures	43	45	48	30	NA
Fare expenditures	29	22	21	14	NA

Travelers are in thousands, money in million U.S. dollars.

For sources, notes, and explanations, see Annotated Source Appendix, page 1061.

Manufacturing Sector

GDP and Manufacturing Summary [35]

	1980	1985	1989	1990	% change 1980-1990	% change 1989-1990
GDP (million 1980 $)	1,131	1,056	NA	1,317	16.4	NA
GDP per capita (1980 $)	432	349	NA	373	-13.7	NA
Manufacturing as % of GDP (current prices)	7.0	6.6	NA	9.4	34.3	NA
Gross output (million $)	155[e]	104	NA	254	63.9	NA
Value added (million $)	52[e]	43[e]	NA	102[e]	96.2	NA
Value added (million 1980 $)	79	78	NA	91	15.2	NA
Industrial production index	100	92	NA	114	14.0	NA
Employment (thousands)	5[e]	5[e]	NA	6[e]	20.0	NA

Note: GDP stands for Gross Domestic Product. 'e' stands for estimated value.

Profitability and Productivity

	1980	1985	1989	1990	% change 1980-1990	% change 1989-1990
Intermediate input (%)	61[e]	71[e]	NA	79[e]	29.5	NA
Wages, salaries, and supplements (%)	14[e]	12[e]	NA	13[e]	-7.1	NA
Gross operating surplus (%)	25[e]	18[e]	NA	8[e]	-68.0	NA
Gross output per worker ($)	28,454[e]	23,960[e]	NA	53,412[e]	87.7	NA
Value added per worker ($)	9,532[e]	8,800[e]	NA	19,640[e]	106.0	NA
Average wage (incl. benefits) ($)	3,447[e]	2,559[e]	NA	5,207[e]	51.1	NA

Profitability is in percent of gross output. Productivity is in U.S. $. 'e' stands for estimated value.

Profitability - 1990

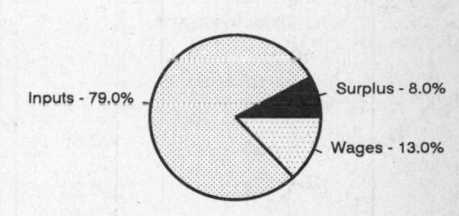

Inputs - 79.0%
Surplus - 8.0%
Wages - 13.0%

The graphic shows percent of gross output.

Value Added in Manufacturing

	1980 $ mil.	1980 %	1985 $ mil.	1985 %	1989 $ mil.	1989 %	1990 $ mil.	1990 %	% change 1980-1990	% change 1989-1990
311 Food products	4	7.7	11[e]	25.6	NA	NA	11[o]	10.8	175.0	NA
313 Beverages	16	30.8	15[e]	34.9	NA	NA	46[e]	45.1	187.5	NA
314 Tobacco products	NA	0.0	NA	0.0	NA	NA	NA	0.0	NA	NA
321 Textiles	8[e]	15.4	5[e]	11.6	NA	NA	12[e]	11.8	50.0	NA
322 Wearing apparel	NA	0.0	-	0.0	NA	NA	-	0.0	NA	NA
323 Leather and fur products	-	0.0	-	0.0	NA	NA	-	0.0	NA	NA
324 Footwear	6	11.5	2[e]	4.7	NA	NA	4[e]	3.9	-33.3	NA
331 Wood and wood products	1	1.9	-	0.0	NA	NA	1[e]	1.0	0.0	NA
332 Furniture and fixtures	NA	0.0	-	0.0	NA	NA	1[e]	1.0	NA	NA
341 Paper and paper products	-	0.0	-	0.0	NA	NA	-	0.0	NA	NA
342 Printing and publishing	3	5.8	1[e]	2.3	NA	NA	3[e]	2.9	0.0	NA
351 Industrial chemicals	3	5.8	1[e]	2.3	NA	NA	4[e]	3.9	33.3	NA
352 Other chemical products	1[e]	1.9	-	0.0	NA	NA	1[e]	1.0	0.0	NA
353 Petroleum refineries	NA	0.0	NA	0.0	NA	NA	NA	0.0	NA	NA
354 Miscellaneous petroleum and coal products	NA	0.0	NA	0.0	NA	NA	NA	0.0	NA	NA
355 Rubber products	NA	0.0	NA	0.0	NA	NA	NA	0.0	NA	NA
356 Plastic products	NA	0.0	NA	0.0	NA	NA	NA	0.0	NA	NA
361 Pottery, china, and earthenware	NA	0.0	3	0.0	NA	NA	9w	0.0	NA	NA
362 Glass and glass products	1[e]	1.9	2[e]	4.7	NA	NA	7[e]	6.9	600.0	NA
369 Other nonmetal mineral products	6	11.5	-	0.0	NA	NA	-	0.0	NA	NA
371 Iron and steel	2	3.8	1	0.0	NA	NA	3	0.0	NA	NA
372 Nonferrous metals	NA	0.0	-	0.0	NA	NA	-	0.0	NA	NA
381 Metal products	1	1.9	-	0.0	NA	NA	-	0.0	NA	NA
382 Nonelectrical machinery	NA	0.0	-	0.0	NA	NA	-	0.0	NA	NA
383 Electrical machinery	NA	0.0	-	0.0	NA	NA	-	0.0	NA	NA
384 Transport equipment	NA	0.0	-	0.0	NA	NA	-	0.0	NA	NA
385 Professional and scientific equipment	NA	0.0	-	0.0	NA	NA	-	0.0	NA	NA
390 Other manufacturing industries	NA	0.0	-	0.0	NA	NA	NA	0.0	NA	NA

Note: The industry codes shown are International Standard Industry codes (ISIC). Percentages are percent of total Value Added. 'e' stands for estimated value

For sources, notes, and explanations, see Annotated Source Appendix, page 1061.

947

Finance, Economics, and Trade

Economic Indicators [36]

- **National product**: GDP—purchasing power parity—$4.1 billion (1995 est.)

- **National product real growth rate**: 6% (1995 est.)

- **National product per capita**: $900 (1995 est.)

- **Inflation rate (consumer prices)**: 8.8% (1995 est.)

- **External debt**: $1.3 billion (1991)

Balance of Payments Summary [37]

Values in millions of dollars.

	1989	1990	1991	1992	1993
Exports of goods (f.o.b.)	411.7	359.2	393.1	322.3	214.7
Imports of goods (f.o.b.)	-407.1	-513.1	-452.3	-417.8	-248.6
Trade balance	-58.4	-117.9	-59.2	-95.5	-33.9
Services - debits	-259.7	-288.0	-273.3	-261.8	-193.2
Services - credits	152.0	180.7	172.3	160.6	88.1
Private transfers (net)	9.2	9.9	14.2	8.7	9.5
Government transfers (net)	126.7	113.9	91.5	81.2	31.1
Long-term capital (net)	58.6	57.4	92.9	61.2	-21.5
Short-term capital (net)	9.0	57.3	17.3	33.6	68.2
Errors and omissions	1.8	10.8	-33.7	-62.1	-42.2
Overall balance	39.2	24.1	22.0	-74.1	-93.9

Exchange Rates [38]

Currency: **Communaute Financi- ere Africaine franc.**

Symbol: **CFAF.**

Data are currency units per $1.

January 1996	500.56
1995	499.15
1994	555.20
1993	283.16
1992	264.69
1991	282.11

Imports and Exports

Top Import Origins [39]

$212 million (c.i.f., 1994) Data are for 1990.

Origins	%
EU	57
Africa	17
US	5
Japan	4

Top Export Destinations [40]

$162.2 (f.o.b., 1994) Data are for 1990.

Destinations	%
EU	40
Africa	16
US	1

Foreign Aid [41]

Recipient: ODA, $NA.

Import and Export Commodities [42]

Import Commodities	Export Commodities
Machinery and equipment	Phosphates
Consumer goods	Cotton
Food	Cocoa
Chemical products	Coffee

For sources, notes, and explanations, see Annotated Source Appendix, page 1061.

Trinidad and Tobago

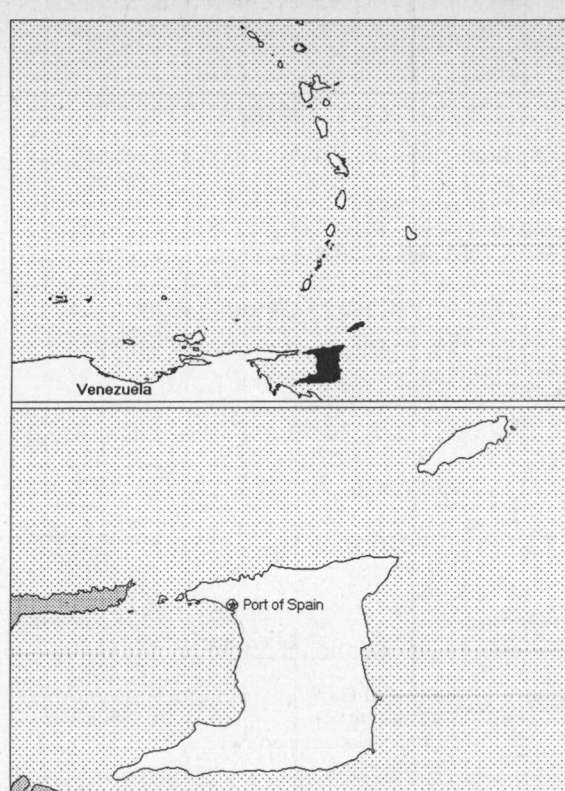

Venezuela

Port of Spain

Geography [1]

Total area:
5,130 sq km   1,981 sq mi
Land area:
5,130 sq km   1,981 sq mi
Comparative area:
Slightly smaller than Delaware
Land boundaries:
0 km
Coastline:
362 km
Climate:
Tropical; rainy season (June to December)
Terrain:
Mostly plains with some hills and low mountains
Natural resources:
Petroleum, natural gas, asphalt
Land use:
Arable land: 14%
Permanent crops: 17%
Meadows and pastures: 2%
Forest and woodland: 44%
Other: 23%

Demographics [2]

	1970	1980	1990	1995[1]	1996	2000	2010	2020	2030
Population	955	1,091	1,256	1,271	1,272	1,273	1,323	1,409	1,449
Population density (persons per sq. mi.)	482	550	634	642	642	643	668	711	731
(persons per sq. km.)	186	213	245	248	248	248	258	275	282
Net migration rate (per 1,000 population)	NA	-5.6	-7.2	-8.6	-8.6	-8.6	0.0	0.0	0.0
Births	25	30	NA	NA	21	NA	NA	NA	NA
Deaths	7	8	NA	NA	9	NA	NA	NA	NA
Life expectancy - males	NA	65.2	67.1	67.8	67.9	68.5	70.0	71.3	72.5
Life expectancy - females	NA	69.7	71.8	72.6	72.8	73.4	75.0	76.5	77.8
Birth rate (per 1,000)	NA	27.4	19.1	16.6	16.3	15.3	15.6	12.8	11.7
Death rate (per 1,000)	NA	7.4	6.7	6.9	6.9	7.1	7.7	8.5	10.0
Women of reproductive age (15-49 yrs.)	NA	277	324	338	340	353	356	351	356
of which are currently married[56]	98	102	180	NA	190	194	208	NA	NA
Fertility rate	NA	3.2	2.2	2.0	2.0	1.9	1.8	1.8	1.8

Except as noted, values for vital statistics are in thousands; life expectancy is in years.

Health

Health Indicators [3]

% of population with access to	
safe water (1990-95)	97
adequate sanitation (1990-95)	79
health services (1985-95)	100
% of 1-year-olds immunized (1990-94) against	
TB (tuberculosis)	NA
DPT (diphtheria, pertussis, tetanus)	85
polio	85
measles	79
% of contraceptive prevalence (1980-94)	53
ORT use rate (1990-94)	75

Health Expenditures [4]

For sources, notes, and explanations, see Annotated Source Appendix, page 1061.

949

Human Factors

Women and Children [5]

% of pregnant women immunized (tetanus 1990-94)	NA
% of births attended by trained health personnel (1983-94)	98
Maternal mortality rate (1980-92)	110
Under-5 mortality rate (1994)	20
% under-5 moderately/severely underweight (1980-1994)[1]	7

Burden of Disease [6]

Population per physician (1993)	1,540.86
Population per nurse (1993)	250.00
Population per hospital bed (1993)	312.53
AIDS cases per 100,000 people (1994)	19.7
Malaria cases per 100,000 people (1992)	NA

Ethnic Division [7]

East Indian is a local term that includes immigrants from northern India and others.

Black	43%
East Indian	40%
Mixed	14%
Chinese	1%
White	1%
Other	1%

Religion [8]

Roman Catholic	32.2%
Hindu	24.3%
Anglican	14.4%
Other Protestant	14%
Muslim	6%
Other	9.1%

Major Languages [9]

English (official), Hindi, French, Spanish.

Education

Public Education Expenditures [10]

	1980	1985	1990	1992	1993	1994
Million (Dollar)						
Total education expenditure	564	1,042	788	811	836	1,170
as percent of GNP	4.0	6.1	4.0	3.8	3.6	4.5
as percent of total govt. expend.	11.5	NA	11.6	10.3	NA	NA
Current education expenditure	431	912	727	772	805	1,061
as percent of GNP	3.0	5.3	3.7	3.6	3.5	4.1
as percent of current govt. expend.	17.6	NA	13.5	10.7	NA	NA
Capital expenditure	133	130	61	39	31	108

Educational Attainment [11]

Age group (1980)	25+
Total population	408,215
Highest level attained (%)	
No schooling	1.3
First level	
Not completed	29.4
Completed	42.6
Entered second level	
S-1	19.7
S-2	4.0
Postsecondary	2.9

Illiteracy [12]

In thousands and percent[1]	1990	1995	2000
Illiterate population (15+ yrs.)	24	19	14
Illiteracy rate - total pop. (%)	2.9	2.2	1.5
Illiteracy rate - males (%)	1.7	1.1	0.8
Illiteracy rate - females (%)	4.1	3.0	2.1

Libraries [13]

	Admin. Units	Svc. Pts.	Vols. (000)	Shelving (meters)	Vols. Added	Reg. Users
National (1986)	1	1	17	NA	NA	NA
Nonspecialized	NA	NA	NA	NA	NA	NA
Public	NA	NA	NA	NA	NA	NA
Higher ed. (1993)[25]	1	10	305	NA	6,864	5,290
School	NA	NA	NA	NA	NA	NA

Daily Newspapers [14]

	1980	1985	1990	1994
Number of papers	4	4	4	4[e]
Circ. (000)	155[e]	173	175	175

Culture [15]

Science and Technology

Scientific/Technical Forces [16]

Scientists/engineers	275
Number female	58
Technicians	254
Number female	44
Total	529

R&D Expenditures [17]

	Dollar (000) 1984
Total expenditure	143,257
Capital expenditure	33,336
Current expenditure	109,921
Percent current	76.7

U.S. Patents Issued [18]

Values show patents issued to citizens of the country by the U.S. Patents Office.

	1993	1994	1995
Number of patents	0	4	0

For sources, notes, and explanations, see Annotated Source Appendix, page 1061.

Government and Law

Organization of Government [19]

Long-form name:
Republic of Trinidad and Tobago
Type:
Parliamentary democracy
Independence:
31 August 1962 (from UK)
National holiday:
Independence Day, 31 August (1962)
Constitution:
1 August 1976
Legal system:
Based on English common law; judicial review of legislative acts in the Supreme Court; has not accepted compulsory ICJ jurisdiction
Executive branch:
President; Prime Minister; Cabinet
Legislative branch:
Bicameral Parliament: Senate and House of Representatives
Judicial branch:
Court of Appeal; Supreme Court

Elections [20]

House of Representatives	% of votes
People's National Movement	52.0
United National Congress	42.2
National Alliance	5.2

Government Budget [21]

For 1996 est.

	$ bil.
Revenues	1.65
Expenditures	1.61
Capital expenditures	NA

Military Expenditures and Arms Transfers [22]

	1990	1991	1992	1993	1994
Military expenditures					
Current dollars (mil.)[e]	NA	NA	60	61	76
1994 constant dollars (mil.)[e]	NA	NA	63	63	76
Armed forces (000)	2	2	3	3	3
Gross national product (GNP)					
Current dollars (mil.)	3,748	4,012	4,073	4,174	4,412
1994 constant dollars (mil.)	4,171	4,301	4,248	4,260	4,412
Central government expenditures (CGE)					
1994 constant dollars (mil.)	NA	1,501	NA	1,567[e]	NA
People (mil.)	1.3	1.3	1.3	1.3	1.3
Military expenditure as % of GNP	NA	NA	1.5	1.5	1.7
Military expenditure as % of CGE	NA	NA	NA	4.0	NA
Military expenditure per capita (1994 $)	NA	NA	50	50	60
Armed forces per 1,000 people (soldiers)	1.6	1.6	2.4	2.4	2.4
GNP per capita (1994 $)	3,322	3,411	3,360	3,363	3,475
Arms imports[6]					
Current dollars (mil.)	0	0	0	0	0
1994 constant dollars (mil.)	0	0	0	0	0
Arms exports[6]					
Current dollars (mil.)	0	0	0	0	0
1994 constant dollars (mil.)	0	0	0	0	0
Total imports[7]					
Current dollars (mil.)	1,121	1,667	1,434	1,448	1,138
1994 constant dollars (mil.)	1,248	1,787	1,495	1,478	1,138
Total exports[7]					
Current dollars (mil.)	1,718	1,985	1,869	1,612	1,867
1994 constant dollars (mil.)	1,912	2,128	1,949	1,645	1,867
Arms as percent of total imports[8]	0	0	0	0	0
Arms as percent of total exports[8]	0	0	0	0	0

Crime [23]

	1994
Crime volume	
Cases known to police	18,614
Attempts (percent)	6.84
Percent cases solved	26.30
Crimes per 100,000 persons	1,382.24
Persons responsible for offenses	
Total number offenders	5,962
Percent female	6.02
Percent juvenile (1-17 yrs.)	9.04
Percent foreigners	NA

Human Rights [24]

	SSTS	FL	FAPRO	PPCG	APROBC	TPW	PCPTW	STPEP	PHRFF	PRW	ASST	AFL
Observes	P	P	P		P	P	P			P	P	P
	EAFRD	CPR	ESCR	SR	ACHR	MAAE	PVIAC	PVNAC	EAFDAW	TCIDTP	RC	
Observes		P	P	P						P		P

P = Party; S = Signatory; see Appendix for meaning of abbreviations.

Labor Force

Total Labor Force [25]

404,500

Labor Force by Occupation [26]

Construction and utilities	13%
Manufacturing, mining, and quarrying	14
Agriculture	11
Services	62

Date of data: 1993 est.

Unemployment Rate [27]

17.8% (December 1995)

For sources, notes, and explanations, see Annotated Source Appendix, page 1061.

951

Production Sectors

Commercial Energy Production and Consumption

Data are shown in quadrillion (10^{15}) BTUs and percent for 1995
Values for hydroelectric, nuclear, geothermal, solar, and wind power refer to electrical generation.

Production [28]

Crude oil - 50.8%

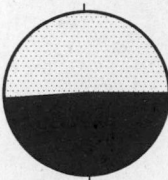

Natural gas - 49.2%

Consumption [29]

Natural gas - 83.4%

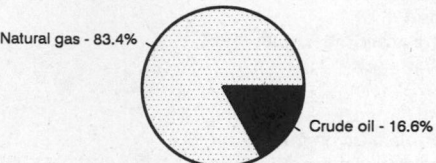

Crude oil - 16.6%

Crude oil	0.289
Dry natural gas	0.280
Total	0.569

Crude oil	0.052
Dry natural gas	0.261
Total	0.313

Telecommunications [30]

- 170,000 (1992 est.) telephones; excellent international service; good local service
- International: satellite earth station - 1 Intelsat (Atlantic Ocean); tropospheric scatter to Barbados and Guyana
- Radio: Broadcast stations: AM 2, FM 4, shortwave 0 Radios: 700,000 (1993 est.)
- Television: Broadcast stations: 5 (1987 est.) Televisions: 400,000 (1992 est.)

Transportation [31]

Railways: minimal agricultural railroad system near San Fernando; railway service was discontinued in 1968

Highways: total: 8,352 km; paved: 3,978 km; unpaved: 4,374 km (1987 est.)

Merchant marine: total: 2 ships (1,000 GRT or over) totaling 2,928 GRT/5,571 DWT; ships by type: cargo 1, oil tanker 1 (1995 est.)

Airports

Total:	5
With paved runways over 3,047 m:	1
With paved runways 2,438 to 3,047 m:	1
With paved runways under 914 m:	2
With unpaved runways 914 to 1,523 m:	1 (1995 est.)

Top Agricultural Products [32]

Agriculture accounts for 4.8% of the GDP; produces cocoa, sugarcane, rice, citrus, coffee, vegetables; poultry.

Top Mining Products [33]

Metric tons except as noted	4/95*
Cement, hydraulic	583,000
Gas, natural gross (mil. cu. meters)	7,700
Iron, sponge	912,000
Steel, crude	631,000
Nitrogen, N content of ammonia (000 tons)	1,560[e]
Petroleum (000 42-gal. bls.)	
crude	43,400
refinery products	37,000
Limestone (000 tons)	1,600[e]

Tourism [34]

	1990	1991	1992	1993	1994
Visitors	238	246	262	283	313
Tourists[20]	195	220	235	250	266
Cruise passengers	43	26	27	33	47
Tourism receipts	95	101	109	82	80
Tourism expenditures	112	113	115	115	90
Fare receipts	127	142	173	169	146
Fare expenditures	26	20	20	16	22

Travelers are in thousands, money in million U.S. dollars.

For sources, notes, and explanations, see Annotated Source Appendix, page 1061.

Manufacturing Sector

Manufacturing Summary [35]

	1987		1988		1989		1990		1991	
	$ bil.	%	$ bil.	%	$ bil.	%	$ bil.	%	$ bil.	%
Establishments or enterprises (number)	687	0.029	685	0.033	-	-	1,277	0.071	-	-
Total employment (000)	22	0.016	22	0.016	-	-	27	0.024	-	-
Production workers (000)	-	-	-	-	-	-	-	-	-	-
Output ($ bil.)	0.883	0.009	0.049	0.000	-	-	-	-	-	-
Value added ($ bil.)	0.356	0.008	0.018	0.000	-	-	-	-	-	-
Capital investment ($ mil.)	49	0.011	-	-	-	-	-	-	-	-
M & E investment ($ mil.)	-	-	-	-	-	-	-	-	-	-
Employees per establishment (number)	32	54.598	32	48.592	-	-	21	33.966	-	-
Production workers per establishment	-	-	-	-	-	-	-	-	-	-
Output per establishment ($ mil.)	1	29.672	0.071	1.276	-	-	-	-	-	-
Capital investment per estab. ($ mil.)	0.072	38.586	-	-	-	-	-	-	-	-
M & E per establishment ($ mil)	-	-	-	-	-	-	-	-	-	-
Payroll per employee ($)	8,091	90.230	2,537	25.467	-	-	-	-	-	-
Wages per production worker ($)	-	-	-	-	-	-	-	-	-	-
Hours per production worker (hours)	-	-	-	-	-	-	-	-	-	-
Output per employee ($)	40,703	54.347	2,236	2.625	-	-	-	-	-	-
Capital investment per employee ($)	2,270	70.673	-	-	-	-	-	-	-	-
M & E per employee ($)	-	-	-	-	-	-	-	-	-	-

Note: Columns headed % show percent of world total or ratio. Ratios closest to 100 are closest to world average. M & E stands for machinery & equipment.

Output in Manufacturing

	1987		1988		1989		1990		1991	
	$ bil.	%	$ bil.	%	$ bil.	%	$ bil.	%	$ bil.	%
3110 - Food products	0.401	45.41	-	-	-	-	-	-	-	-
3130 - Beverages	0.100	11.33	-	-	-	-	-	-	-	-
3140 - Tobacco	0.045	5.10	-	-	-	-	-	-	-	-
3210 - Textiles	0.009	1.02	0.003	6.12	-	-	-	-	-	-
3220 - Wearing apparel	0.030	3.40	0.029	59.18	-	-	-	-	-	-
3240 - Footwear	0.010	1.13	0.006	12.24	-	-	-	-	-	-
3310 - Wood products	0.010	1.13	0.008	16.33	-	-	-	-	-	-
3320 - Furniture, fixtures	0.015	1.70	0.003	6.12	-	-	-	-	-	-
3410 - Paper and products	0.044	4.98	-	-	-	-	-	-	-	-
3420 - Printing, publishing	0.043	4.87	-	-	-	-	-	-	-	-
3522 - Drugs and medicines	0.029	3.28	-	-	-	-	-	-	-	-
3550 - Rubber products	0.021	2.38	-	-	-	-	-	-	-	-
3560 - Plastic products nec	0.022	2.49	-	-	-	-	-	-	-	-
3610 - Pottery, china, etc.	0.000	0.00	-	-	-	-	-	-	-	-
3830 - Electrical machinery	0.037	4.19	-	-	-	-	-	-	-	-
3840 - Transportation equipment	0.035	3.96	-	-	-	-	-	-	-	-
3841 - Shipbuilding, repair	0.001	0.11	-	-	-	-	-	-	-	-
3843 - Motor vehicles	0.034	3.85	-	-	-	-	-	-	-	-

Note: Codes are International Standard Industry codes (ISIC). Percentages are % of total Output. [f]: Factor Prices; [p]: Producer Prices.

Finance, Economics, and Trade

Economic Indicators [36]

- **National product**: GDP—purchasing power parity—$16.2 billion (1995 est.)
- **National product real growth rate**: 3.5% (1995 est.)
- **National product per capita**: $12,100 (1995 est.)
- **Inflation rate (consumer prices)**: 5.4% (1995)
- **External debt**: $2 billion (1994)

Balance of Payments Summary [37]

Values in millions of dollars.

	1989	1990	1991	1992	1993
Exports of goods (f.o.b.)	1,534.6	1,935.2	1,751.3	1,661.9	1,477.2
Imports of goods (f.o.b.)	-1,045.2	-947.6	-1,210.3	-995.6	-952.9
Trade balance	489.4	987.6	541.0	666.3	524.3
Services - debits	-849.9	-915.6	-1,025.1	-1,039.7	-832.4
Services - credits	329.5	393.1	477.1	512.0	416.5
Private transfers (net)	-19.2	-21.0	-15.9	-15.7	-7.0
Government transfers (net)	-5.5	-4.4	2.1	-0.4	0.2
Long-term capital (net)	147.2	-245.3	-113.5	-95.6	184.0
Short-term capital (net)	-62.3	-9.4	6.2	25.2	-62.8
Errors and omissions	45.4	-112.0	-29.0	-72.6	-41.8
Overall balance	74.6	73.0	-157.1	-20.5	181.0

Exchange Rates [38]

Currency: **Trinidad and Tobago dollar.**
Symbol: **TT$.**

Data are currency units per $1. Effective April 13, 1993, the exchange rate is market-determined as opposed to the prior fixed relationship to the US dollar.

January 1996	5.9412
1995	5.9192
1994	5.9249
1993	5.3511
Fixed rate 1989-1992	4.2500

Imports and Exports

Top Import Origins [39]

$996 million (c.i.f., 1994) Data are for 1994.

Origins	%
US	47.7
Venezuela	10
UK	8.3
Other EU	8

Top Export Destinations [40]

$2.2 billion (f.o.b., 1995) Data are for 1994.

Destinations	%
US	48
Caricom	15
Latin America	9
EU	5

Foreign Aid [41]

Recipient: ODA, $10 million (1993).

Import and Export Commodities [42]

Import Commodities	Export Commodities
Machinery	Petroleum and petroleum products
Transportation equipment	Chemicals
Manufactured goods	Steel products
Food	Fertilizer
Live animals	Sugar
	Cocoa
	Coffee
	Citrus
	Flowers

For sources, notes, and explanations, see Annotated Source Appendix, page 1061.

Tunisia

Geography [1]

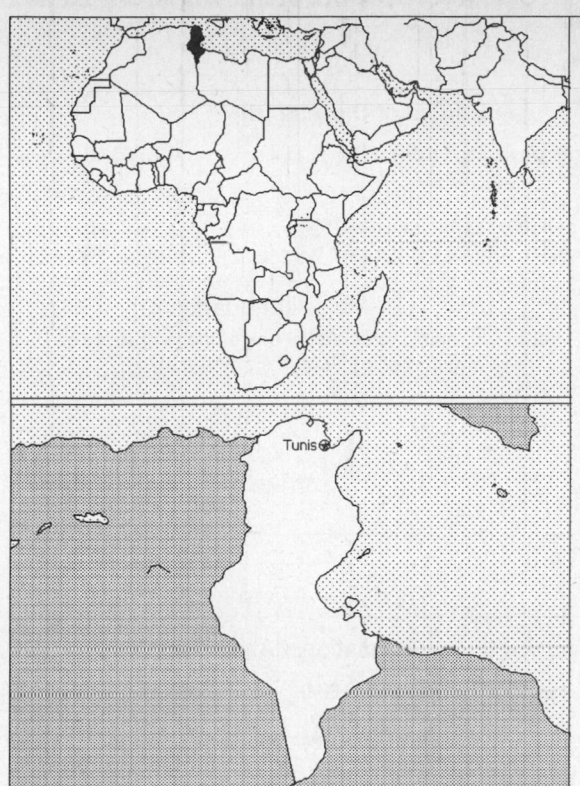

Total area:
163,610 sq km   63,170 sq mi
Land area:
155,360 sq km   59,985 sq mi
Comparative area:
Slightly larger than Georgia
Land boundaries:
Total 1,424 km, Algeria 965 km, Libya 459 km
Coastline:
1,148 km
Climate:
Temperate in North with mild, rainy winters and hot, dry summers; desert in South
Terrain:
Mountains in North; hot, dry central plain; semiarid South merges into the Sahara
Natural resources:
Petroleum, phosphates, iron ore, lead, zinc, salt
Land use:
Arable land: 20%
Permanent crops: 10%
Meadows and pastures: 19%
Forest and woodland: 4%
Other: 47%

Demographics [2]

	1970	1980	1990	1995[1]	1996	2000	2010	2020	2030
Population	5,099	6,443	8,048	8,856	9,020	9,671	11,280	12,751	14,031
Population density (persons per sq. mi.)	85	107	134	148	150	161	188	213	234
(persons per sq. km.)	33	41	52	57	58	62	73	82	90
Net migration rate (per 1,000 population)	-3.8	0.7	-0.6	-0.7	-0.7	-0.7	-0.4	-0.3	-0.3
Births	NA	NA	NA	NA	217	NA	NA	NA	NA
Deaths	NA	38	NA	NA	47	NA	NA	NA	NA
Life expectancy - males	NA	NA	69.8	71.1	71.3	72.2	74.1	75.7	77.0
Life expectancy - females	NA	NA	72.3	73.8	74.0	75.1	77.6	79.6	81.2
Birth rate (per 1,000)	38.1	34.7	25.9	24.4	24.0	22.5	19.3	16.3	14.7
Death rate (per 1,000)	12.5	7.5	5.5	5.2	5.2	5.1	5.1	5.2	6.0
Women of reproductive age (15-49 yrs.)	NA	NA	1,966	2,246	2,308	2,562	3,025	3,311	3,456
of which are currently married	NA	NA	1,102	NA	1,340	1,515	1,898	NA	NA
Fertility rate	NA	NA	3.3	3.0	2.9	2.7	2.3	2.1	2.1

Except as noted, values for vital statistics are in thousands; life expectancy is in years.

Health

Health Indicators [3]

% of population with access to	
safe water (1990-95)	99
adequate sanitation (1990-95)	96
health services (1985-95)[1]	90
% of 1-year-olds immunized (1990-94) against	
TB (tuberculosis)	80
DPT (diphtheria, pertussis, tetanus)	97
polio	97
measles	93
% of contraceptive prevalence (1980-94)	50
ORT use rate (1990-94)	22

Health Expenditures [4]

Total health expenditure, 1990 (official exchange rate)	
Millions of dollars	614
Dollars per capita	76
Health expenditures as a percentage of GDP	
Total	4.9
Public sector	3.3
Private sector	1.6
Development assistance for health	
Total aid flows (millions of dollars)[1]	18
Aid flows per capita (dollars)	2.3
Aid flows as a percentage of total health expenditure	3.0

For sources, notes, and explanations, see Annotated Source Appendix, page 1061.

955

Human Factors

Women and Children [5]

% of pregnant women immunized (tetanus 1990-94)	NA
% of births attended by trained health personnel (1983-94)	69
Maternal mortality rate (1980-92)	70
Under-5 mortality rate (1994)	34
% under-5 moderately/severely underweight (1980-1994)[1]	10

Burden of Disease [6]

Population per physician (1994)	1,756.68
Population per nurse (1994)	389.03
Population per hospital bed (1994)	565.75
AIDS cases per 100,000 people (1994)	0.5
Malaria cases per 100,000 people (1992)	NA

Ethnic Division [7]

Arab-Berber	98%
European	1%
Jewish	< 1%

Religion [8]

Muslim	98%
Christian	1%
Jewish	1%

Major Languages [9]

Arabic (official and one of the languages of commerce), French (commerce).

Education

Public Education Expenditures [10]

	1980	1990	1991	1992	1993	1994
Million (Dinar)[66]						
Total education expenditure	185	648	725	789	895	NA
as percent of GNP	5.4	6.2	6.3	5.9	6.3	NA
as percent of total govt. expend.	16.4	13.5	14.3	14.2	NA	NA
Current education expenditure	162	569	639	701	776	NA
as percent of GNP	4.7	5.5	5.5	5.3	5.5	NA
as percent of current govt. expend.	23.5	16.3	17.4	17.6	NA	NA
Capital expenditure	23	79	86	88	119	NA

Educational Attainment [11]

Age group (1984)	25+
Total population	2,714,100
Highest level attained (%)	
No schooling	66.3
First level	
Not completed	18.9
Completed	NA
Entered second level	
S-1	12.0
S-2	NA
Postsecondary	2.8

Illiteracy [12]

In thousands and percent[1]	1990	1995	2000
Illiterate population (15+ yrs.)	2,005	1,930	1,827
Illiteracy rate - total pop. (%)	39.7	33.3	27.7
Illiteracy rate - males (%)	26.7	21.3	16.8
Illiteracy rate - females (%)	52.9	45.5	38.7

Libraries [13]

	Admin. Units	Svc. Pts.	Vols. (000)	Shelving (meters)	Vols. Added	Reg. Users
National	NA	NA	NA	NA	NA	NA
Nonspecialized	NA	NA	NA	NA	NA	NA
Public (1992)	250	NA	2,493	NA	255,980	NA
Higher ed.	NA	NA	NA	NA	NA	NA
School	NA	NA	NA	NA	NA	NA

Daily Newspapers [14]

	1980	1985	1990	1994
Number of papers	5	6	6	7
Circ. (000)	272	280[e]	345[e]	403[e]

Culture [15]

Science and Technology

Scientific/Technical Forces [16]

Scientists/engineers	3,260[e]
Number female	NA
Technicians	600[e]
Number female	NA
Total[3]	3,860[e]

R&D Expenditures [17]

	Dinar (000) 1992
Total expenditure[1]	45,000,000
Capital expenditure	11,000,000
Current expenditure	34,000,000
Percent current	75.6

U.S. Patents Issued [18]

Values show patents issued to citizens of the country by the U.S. Patents Office.

	1993	1994	1995
Number of patents	1	0	0

Government and Law

Organization of Government [19]

Long-form name:
Republic of Tunisia
Type:
Republic
Independence:
20 March 1956 (from France)
National holiday:
National Day, 20 March (1956)
Constitution:
1 June 1959; amended 12 July 1988
Legal system:
Based on French civil law system and Islamic law; some judicial review of legislative acts in the Supreme Court in joint session
Executive branch:
President; Prime Minister; Council of Ministers
Legislative branch:
Unicameral: Chamber of Deputies (Majlis al-Nuwaab)
Judicial branch:
Court of Cassation (Cour de Cassation)

Crime [23]

Elections [20]

Chamber of Deputies	% of votes
Constitutional Democratic Rally Party	97.7
Movement of Democratic Socialists	1.0
Other	1.3

Government Expenditures [21]

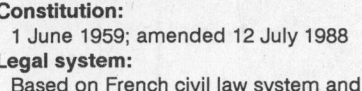

Educ./Health - 37.7%
Other - 3.6%
Defense - 5.6%
Industry - 24.4%
Housing - 4.4%
Services - 24.3%

(% distribution). Expend. for CY91: 4,017.2 (Dinar mil.)

Military Expenditures and Arms Transfers [22]

	1990	1991	1992	1993	1994
Military expenditures					
Current dollars (mil.)	315	338	514	345	543
1994 constant dollars (mil.)	350	363	536	352	543
Armed forces (000)	35	35	35	35	35
Gross national product (GNP)					
Current dollars (mil.)	11,460	12,390	13,820	14,410	15,210
1994 constant dollars (mil.)	12,760	13,280	14,420	14,710	15,210
Central government expenditures (CGE)					
1994 constant dollars (mil.)	4,894	4,618	4,757	5,717[e]	NA
People (mil.)	8.1	8.2	8.4	8.6	8.7
Military expenditure as % of GNP	2.7	2.7	3.7	2.4	3.6
Military expenditure as % of CGE	7.2	7.9	11.3	6.2	NA
Military expenditure per capita (1994 $)	43	44	64	41	62
Armed forces per 1,000 people (soldiers)	4.3	4.2	4.2	4.1	4.0
GNP per capita (1994 $)	1,578	1,610	1,714	1,716	1,742
Arms imports[6]					
Current dollars (mil.)	40	30	20	20	50
1994 constant dollars (mil.)	45	32	21	20	50
Arms exports[6]					
Current dollars (mil.)	0	0	0	0	0
1994 constant dollars (mil.)	0	0	0	0	0
Total imports[7]					
Current dollars (mil.)	5,542	5,190	6,431	6,214	6,581
1994 constant dollars (mil.)	6,168	5,563	6,706	6,342	6,581
Total exports[7]					
Current dollars (mil.)	3,526	3,699	4,019	3,802	4,657
1994 constant dollars (mil.)	3,924	3,965	4,191	3,880	4,657
Arms as percent of total imports[8]	0.7	0.6	0.3	0.3	0.8
Arms as percent of total exports[8]	0	0	0	0	0

Human Rights [24]

	SSTS	FL	FAPRO	PPCG	APROBC	TPW	PCPTW	STPEP	PHRFF	PRW	ASST	AFL
Observes	P	P	P	P	P	P	P			P	P	P
	EAFRD	CPR	ESCR	SR	ACHR	MAAE	PVIAC	PVNAC	EAFDAW	TCIDTP	RC	
Observes		P	P	P	P			P	P	P	P	P

P=Party; S=Signatory; see Appendix for meaning of abbreviations.

Labor Force

Total Labor Force [25]

2.917 million (1993 est.)

Labor Force by Occupation [26]

Services	55%
Industry	23
Agriculture	22

Date of data: 1995 est.

Unemployment Rate [27]

16.2% (1993 est.)

For sources, notes, and explanations, see Annotated Source Appendix, page 1061.

957

Production Sectors

Commercial Energy Production and Consumption

Data are shown in quadrillion (10^{15}) BTUs and percent for 1995
Values for hydroelectric, nuclear, geothermal, solar, and wind power refer to electrical generation.

Production [28]

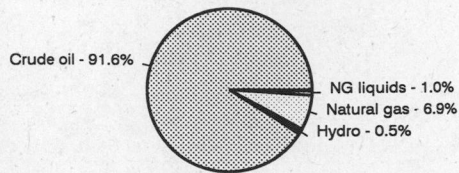

Consumption [29]

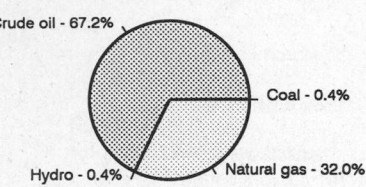

Crude oil	0.185
Natural gas liquids	0.002
Dry natural gas	0.014
Net hydroelectric power	0.001
Total	0.202

Crude oil	0.166
Dry natural gas	0.079
Coal	0.001
Net hydroelectric power	0.001
Total	0.247

Telecommunications [30]

- 233,000 (1987 est.) telephones; the system is above the African average; key centers are Sfax, Sousse, Bizerte, and Tunis
- Domestic: trunk facilities consist of open-wire lines, coaxial cable, and microwave radio relay
- International: 5 submarine cables; satellite earth stations - 1 Intelsat (Atlantic Ocean) and 1 Arabsat with back-up control station; coaxial cable and microwave radio relay to Algeria and Libya; participant in Medarabtel
- Radio: Broadcast stations: AM 7, FM 8, shortwave 0 Radios: 1,693,527 (1991 est.)
- Television: Broadcast stations: 19 Televisions: 670,000 (1992 est.)

Transportation [31]

Railways: total: 2,260 km; standard gauge: 492 km 1.435-m gauge; narrow gauge: 1,758 km 1.000-m gauge

Highways: total: 29,183 km; paved: 17,510 km (including 52 km of expressways); unpaved: 11,673 km (1989 est.)

Merchant marine: total: 19 ships (1,000 GRT or over) totaling 125,840 GRT/164,277 DWT; ships by type: bulk 6, cargo 4, chemical tanker 3, oil tanker 2, roll-on/roll-off cargo 3, short-sea passenger 1 (1995 est.)

Airports

Total:	29
With paved runways over 3,047 m:	3
With paved runways 2,438 to 3,047 m:	6
With paved runways 1,524 to 2,437 m:	3
With paved runways 914 to 1,523 m:	3
With paved runways under 914 m:	6

Top Agricultural Products [32]

Agriculture accounts for 15% of the GDP; produces olives, dates, oranges, almonds, grain, sugar beets, grapes; poultry, beef, dairy products.

Top Mining Products [33]

Thousand metric tons except as noted[e]	1994[*]
Iron ore and concentrate, gross weight	240
Silver metal, primary (kg.)	900
Cement, hydraulic	3,300
Lime	600
Phosphate rock, P2O5 content	1,700
Salt, marine	414
Gas, natural, gross (mil. cu. meters)	200
Petroleum (000 42-gal. bls.)	
crude	33,200
refinery products	11,200

Tourism [34]

	1990	1991	1992	1993	1994
Visitors	3,249	3,268	3,571	3,700	3,931
Tourists[37]	3,204	3,224	3,540	3,656	3,856
Cruise passengers	45	44	31	44	75
Tourism receipts	953	685	1,074	1,114	1,302
Tourism expenditures	179	129	166	203	216
Fare receipts	212	220	328	353	338
Fare expenditures	109	110	139	134	149

Travelers are in thousands, money in million U.S. dollars.

Manufacturing Sector

GDP and Manufacturing Summary [35]

	1980	1985	1989	1990	% change 1980-1990	% change 1989-1990
GDP (million 1980 $)	8,742	10,733	11,699	12,646	44.7	8.1
GDP per capita (1980 $)	1,369	1,478	1,463	1,544	12.8	5.5
Manufacturing as % of GDP (current prices)	13.6	13.5	16.4	17.0	25.0	3.7
Gross output (million $)	3,579	3,449[e]	5,039[e]	5,547[e]	55.0	10.1
Value added (million $)	939	949[e]	1,469[e]	1,612[e]	71.7	9.7
Value added (million 1980 $)	1,030	1,443	1,744[e]	1,982	92.4	13.6
Industrial production index	100	126	128	143	43.0	11.7
Employment (thousands)	125	165[e]	195[e]	213[e]	70.4	9.2

Note: GDP stands for Gross Domestic Product. 'e' stands for estimated value.

Profitability and Productivity

	1980	1985	1989	1990	% change 1980-1990	% change 1989-1990
Intermediate input (%)	74	72[e]	71[e]	71[e]	-4.1	0.0
Wages, salaries, and supplements (%)	12	13[e]	15[e]	15[e]	25.0	0.0
Gross operating surplus (%)	14	14[e]	15[e]	14[e]	0.0	-6.7
Gross output per worker ($)	28,669	20,841[e]	25,899[e]	26,013[e]	-9.3	0.4
Value added per worker ($)	7,525	5,853[e]	7,550[e]	7,905[e]	5.0	4.7
Average wage (incl. benefits) ($)	3,499	2,811[e]	3,784[e]	3,834[e]	9.6	1.3

Profitability is in percent of gross output. Productivity is in U.S. $. 'e' stands for estimated value.

Profitability - 1990

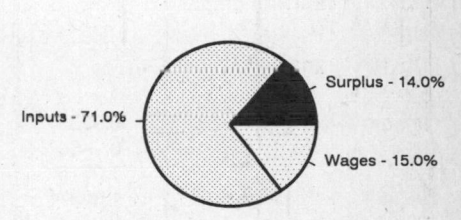

Inputs - 71.0%
Surplus - 14.0%
Wages - 15.0%

The graphic shows percent of gross output.

Value Added in Manufacturing

	1980 $ mil.	1980 %	1985 $ mil.	1985 %	1989 $ mil.	1989 %	1990 $ mil.	1990 %	% change 1980-1990	% change 1989-1990
311 Food products	96	10.2	78[e]	8.2	112[e]	7.6	120[e]	7.4	25.0	7.1
313 Beverages	49	5.2	54[e]	5.7	87[e]	5.9	93[e]	5.8	89.8	6.9
314 Tobacco products	22	2.3	22[e]	2.3	37[e]	2.5	36[e]	2.2	63.6	-2.7
321 Textiles	55	5.9	60[c]	6.3	104[e]	7.1	83[e]	5.1	50.9	-20.2
322 Wearing apparel	92	9.8	87[e]	9.2	171[e]	11.6	122[e]	7.6	32.6	-28.7
323 Leather and fur products	6	0.6	6[e]	0.6	8[e]	0.5	10[e]	0.6	66.7	25.0
324 Footwear	21	2.2	21[e]	2.2	33[e]	2.2	38[e]	2.4	81.0	15.2
331 Wood and wood products	12	1.3	12[e]	1.3	15[e]	1.0	21[e]	1.3	75.0	40.0
332 Furniture and fixtures	13	1.4	12[e]	1.3	20[e]	1.4	16[e]	1.0	23.1	-20.0
341 Paper and paper products	24	2.6	21[e]	2.2	30[e]	2.0	34[c]	2.1	41.7	13.3
342 Printing and publishing	17	1.8	16[e]	1.7	20[e]	1.4	26[e]	1.6	52.9	30.0
351 Industrial chemicals	42[e]	4.5	26[e]	2.7	29[e]	2.0	43[e]	2.7	2.4	48.3
352 Other chemical products	96[e]	10.2	77[e]	8.1	122[e]	8.3	140[e]	8.7	45.8	14.8
353 Petroleum refineries	13	1.4	10[e]	1.1	13[e]	0.9	14[e]	0.9	7.7	7.7
354 Miscellaneous petroleum and coal products	NA	0.0	NA	0.0	NA	0.0	NA	0.0	NA	NA
355 Rubber products	8	0.9	11[e]	1.2	10[e]	0.7	20[e]	1.2	150.0	100.0
356 Plastic products	18	1.9	22[e]	2.3	30[e]	2.0	35[e]	2.2	94.4	16.7
361 Pottery, china, and earthenware	11	1.2	9[e]	0.9	12[e]	0.8	14[e]	0.9	27.3	16.7
362 Glass and glass products	7	0.7	6[e]	0.6	9[e]	0.6	11[e]	0.7	57.1	22.2
369 Other nonmetal mineral products	156	16.6	149[e]	15.7	280[e]	19.1	246[e]	15.3	57.7	-12.1
371 Iron and steel	45	4.8	81[e]	8.5	111[e]	7.6	144[e]	8.9	220.0	29.7
372 Nonferrous metals	8	0.9	7[e]	0.7	6[e]	0.4	12[e]	0.7	50.0	100.0
381 Metal products	53	5.6	77[e]	8.1	120[e]	8.2	151[e]	9.4	184.9	25.8
382 Nonelectrical machinery	2	0.2	2[e]	0.2	2[e]	0.1	4[e]	0.2	100.0	100.0
383 Electrical machinery	35	3.7	38[e]	4.0	52[e]	3.5	85[e]	5.3	142.9	63.5
384 Transport equipment	30	3.2	38[e]	4.0	26[e]	1.8	85[e]	5.3	183.3	226.9
385 Professional and scientific equipment	1	0.1	1[e]	0.1	2[e]	0.1	2[e]	0.1	100.0	0.0
390 Other manufacturing industries	5	0.5	6[e]	0.6	7[e]	0.5	9[e]	0.6	80.0	28.6

Note: The industry codes shown are International Standard Industry codes (ISIC). Percentages are percent of total Value Added. 'e' stands for estimated value

For sources, notes, and explanations, see Annotated Source Appendix, page 1061.

959

Finance, Economics, and Trade

Economic Indicators [36]

- **National product**: GDP—purchasing power parity—$37.1 billion (1994 est.)
- **National product real growth rate**: 4.4% (1994 est.)
- **National product per capita**: $4,250 (1994 est.)
- **Inflation rate (consumer prices)**: 5.5% (1995 est.)
- **External debt**: $7.7 billion (1993 est.)

Balance of Payments Summary [37]

Values in millions of dollars.

	1989	1990	1991	1992	1993
Exports of goods (f.o.b.)	2,931	3,515	3,696	4,014	3,804
Imports of goods (f.o.b.)	-4,139	-5,193	-4,895	-6,077	-5,872
Trade balance	-1,208	-1,678	-1,199	-2,063	-2,068
Services - debits	-1,230	-1,396	-1,448	-1,643	-1,688
Services - credits	1,618	1,782	1,473	2,083	2,138
Private transfers (net)	485	591	574	570	595
Government transfers (net)	213	225	131	88	112
Long-term capital (net)	107	47	353	620	620
Short-term capital (net)	86	335	-15	439	283
Errors and omissions	-5	-28	77	5	16
Overall balance	66	-122	-54	99	8

Exchange Rates [38]

Currency: **Tunisian dinar.**
Symbol: **TD.**

Data are currency units per $1.

January 1996	0.9635
1995	0.9458
1994	1.0116
1993	1.0037
1992	0.8844
1991	0.9246

Imports and Exports

Top Import Origins [39]

$6.6 billion (c.i.f., 1994).

Origins	%
EU	70
US	5
Middle East	2
Japan	2
Switzerland	1
Algeria	1

Top Export Destinations [40]

$4.7 billion (f.o.b., 1994).

Destinations	%
EU	75
Middle East	10
Algeria	2
India	2
US	1

Foreign Aid [41]

Recipient: ODA, $221 million (1993).

Import and Export Commodities [42]

Import Commodities

Industrial goods and equipment 57%
Hydrocarbons 13%
Food 12%
Consumer goods

Export Commodities

Hydrocarbons
Agricultural products
Phosphates and chemicals

For sources, notes, and explanations, see Annotated Source Appendix, page 1061.

Turkey

Geography [1]

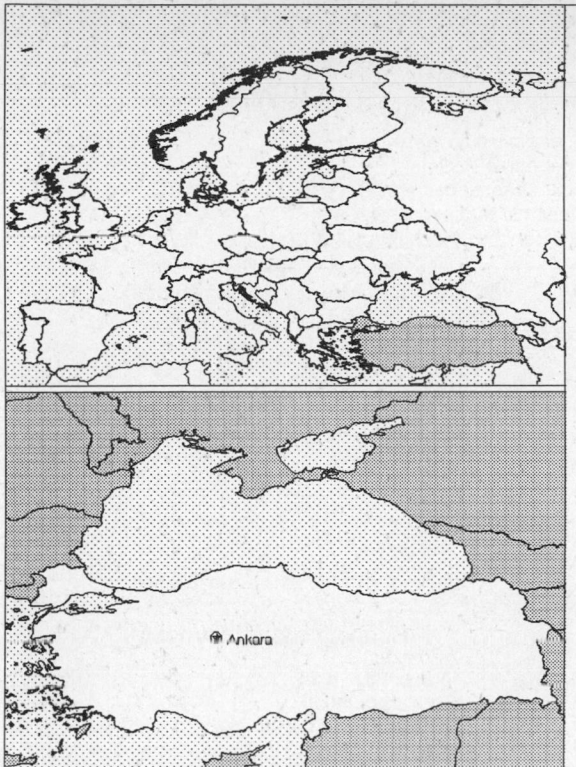

Total area:
780,580 sq km   301,384 sq mi
Land area:
770,760 sq km   297,592 sq mi
Comparative area:
Slightly larger than Texas
Land boundaries:
Total 2,627 km, Armenia 268 km, Azerbaijan 9 km, Bulgaria 240 km, Georgia 252 km, Greece 206 km, Iran 499 km, Iraq 331 km, Syria 822 km
Coastline:
7,200 km
Climate:
Temperate; hot, dry summers with mild, wet winters; harsher in interior
Terrain:
Mostly mountains; narrow coastal plain; high central plateau (Anatolia)
Natural resources:
Antimony, coal, chromium, mercury, copper, borate, sulfur, iron ore
Land use:
Arable land: 30%
Permanent crops: 4%
Meadows and pastures: 12%
Forest and woodland: 26%
Other: 28%

Demographics [2]

	1970	1980	1990	1995[1]	1996	2000	2010	2020	2030
Population	35,758	45,121	56,123	61,437	62,484	66,618	76,570	85,643	93,476
Population density (persons per sq. mi.)	120	152	189	206	210	224	257	288	314
(persons per sq. km.)	46	59	73	80	81	86	99	111	121
Net migration rate (per 1,000 population)	NA	0.0	0.0	0.0	0.0	0.0	0.0	0.0	0.0
Births	NA	NA	NA	NA	1,391	NA	NA	NA	NA
Deaths	NA	NA	NA	NA	345	NA	NA	NA	NA
Life expectancy - males	NA	60.9	66.7	69.1	69.5	71.3	74.6	76.9	78.4
Life expectancy - females	NA	64.5	71.2	74.0	74.4	76.4	80.1	82.7	84.3
Birth rate (per 1,000)	NA	33.7	25.5	22.7	22.3	20.4	17.6	15.3	13.8
Death rate (per 1,000)	NA	9.4	6.4	5.6	5.5	5.2	5.0	5.4	6.2
Women of reproductive age (15-49 yrs.)	NA	10,421	13,795	15,783	16,190	17,753	20,641	22,114	22,549
of which are currently married[57]	5,928	7,576	9,322	NA	11,143	12,413	15,036	NA	NA
Fertility rate	NA	4.6	3.1	2.6	2.6	2.4	2.1	2.0	2.0

Except as noted, values for vital statistics are in thousands; life expectancy is in years.

Health

Health Indicators [3]

% of population with access to	
safe water (1990-95)	80
adequate sanitation (1990-95)	NA
health services (1985-95)	NA
% of 1-year-olds immunized (1990-94) against	
TB (tuberculosis)	72
DPT (diphtheria, pertussis, tetanus)	81
polio	81
measles	76
% of contraceptive prevalence (1980-94)	63
ORT use rate (1990-94)	57

Health Expenditures [4]

Total health expenditure, 1990 (official exchange rate)	
Millions of dollars	4,281
Dollars per capita	76
Health expenditures as a percentage of GDP	
Total	4.0
Public sector	1.5
Private sector	2.5
Development assistance for health	
Total aid flows (millions of dollars)[1]	23
Aid flows per capita (dollars)	0.4
Aid flows as a percentage of total health expenditure	0.5

For sources, notes, and explanations, see Annotated Source Appendix, page 1061.

Human Factors

Women and Children [5]

% of pregnant women immunized (tetanus 1990-94)	29
% of births attended by trained health personnel (1983-94)	76
Maternal mortality rate (1980-92)	150
Under-5 mortality rate (1994)	55
% under-5 moderately/severely underweight (1980-1994)	10

Burden of Disease [6]

Population per physician (1994)	955.09
Population per nurse (1994)	986.13
Population per hospital bed (1994)	406.46
AIDS cases per 100,000 people (1994)	0.1
Malaria cases per 100,000 people (1992)	32

Ethnic Division [7]

Turkish	80%
Kurdish	20%

Religion [8]

Muslim (mostly Sunni)	99.8%
Other (primarily Christian and Jews)	0.2%

Major Languages [9]

Turkish (official), Kurdish, Arabic.

Education

Public Education Expenditures [10]

Million or Trillion (T) (Lira)[67]	1980	1985	1990	1991	1992	1994
Total education expenditure	117,744	627,104	8.51 T	14.94 T	30.36 T	124,763
as percent of GNP	2.8	2.3	2.2	2.4	2.8	3.3
as percent of total govt. expend.	10.5	NA	NA	NA	NA	NA
Current education expenditure	98,593	523,102	7.58 T	13,604	27,895	109,162
as percent of GNP	2.3	1.9	2.0	2.2	2.6	2.9
as percent of current govt. expend.	NA	NA	NA	NA	NA	NA
Capital expenditure	19,152	104,002	925,724	1.34 T	2.46 T	15.60 T

Educational Attainment [11]

Age group (1993)[19]	25+
Total population	NA
Highest level attained (%)	
No schooling	30.6
First level	
Not completed	6.6
Completed	40.6
Entered second level	
S-1	21.9
S-2	NA
Postsecondary	NA

Illiteracy [12]

In thousands and percent[1]	1990	1995	2000
Illiterate population (15+ yrs.)	7,616	7,231	6,611
Illiteracy rate - total pop. (%)	21.0	17.5	14.1
Illiteracy rate - males (%)	10.2	8.3	6.3
Illiteracy rate - females (%)	32.1	26.8	22.2

Libraries [13]

	Admin. Units	Svc. Pts.	Vols. (000)	Shelving (meters)	Vols. Added	Reg. Users
National (1992)	1	1	1,079	45,402	20,505	10,100
Nonspecialized	NA	NA	NA	NA	NA	NA
Public (1992)	NA	910	9,042	NA	545,994	NA
Higher ed. (1993)	212	212	5,700	NA	NA	NA
School	NA	NA	NA	NA	NA	NA

Daily Newspapers [14]

	1980	1985	1990	1994
Number of papers	NA	NA	68[e]	57
Circ. (000)	2,500[e]	3,020[e]	3,499[e]	2,679

Culture [15]

Cinema (seats per 1,000)	3.1
Annual attendance per person	0.3
Gross box office receipts (mil. Lira)	NA
Museums (reporting)	157
Visitors (000)	6,980
Annual receipts (000 Lira)	153,345

Science and Technology

Scientific/Technical Forces [16]

Scientists/engineers	11,948
Number female	NA
Technicians[32]	1,329
Number female	NA
Total[32,33]	13,277

R&D Expenditures [17]

	Lira (000,000) 1991
Total expenditure	3,330,047
Capital expenditure	1,293,048
Current expenditure	2,036,999
Percent current	61.2

U.S. Patents Issued [18]

Values show patents issued to citizens of the country by the U.S. Patents Office.

	1993	1994	1995
Number of patents	0	2	2

For sources, notes, and explanations, see Annotated Source Appendix, page 1061.

Government and Law

Organization of Government [19]

Long-form name:
Republic of Turkey

Type:
Republican parliamentary democracy

Independence:
29 October 1923 (successor state to the Ottoman Empire)

National holiday:
Anniversary of the Declaration of the Republic, 29 October (1923)

Constitution:
7 November 1982

Legal system:
Derived from various continental legal systems; accepts compulsory ICJ jurisdiction, with reservations

Executive branch:
President; Prime Minister; Deputy Prime Minister; National Security Council; Council of Ministers

Legislative branch:
Unicameral: Grand National Assembly of Turkey

Judicial branch:
Constitutional Court; Court of Appeals

Elections [20]

Grand National Assembly	% of votes
Welfare Party (RP)	21.4
Motherland Party (ANAP)	19.7
True Path Party (DYP)	19.2
Democratic Left (DSP)	14.6
Republican People's	10.7
Independent	0.5

Government Expenditures [21]

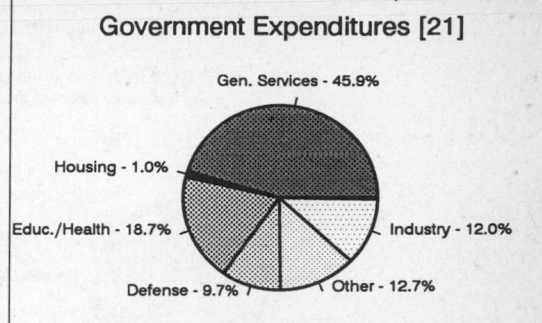

Gen. Services - 45.9%
Housing - 1.0%
Educ./Health - 18.7%
Defense - 9.7%
Other - 12.7%
Industry - 12.0%

(% distribution). Expend. for CY95: 1,723,836 (Lira bil.)

Crime [23]

	1994[2]
Crime volume	
Cases known to police	84,925
Attempts (percent)	NA
Percent cases solved	43.90
Crimes per 100,000 persons	NA
Persons responsible for offenses	
Total number offenders	65,440
Percent female	10.06
Percent juvenile	NA
Percent foreigners	NA

Military Expenditures and Arms Transfers [22]

	1990	1991	1992	1993	1994
Military expenditures					
Current dollars (mil.)	3,815	4,231	4,763	5,321	5,293
1994 constant dollars (mil.)	4,246	4,535	4,967	5,430	5,293
Armed forces (000)	769	804	704	686	811
Gross national product (GNP)					
Current dollars (mil.)	106,800	111,200	121,600	134,200	129,000
1994 constant dollars (mil.)	118,800	119,200	126,800	136,900	129,000
Central government expenditures (CGE)					
1994 constant dollars (mil.)	20,930	25,380	26,450	34,270	30,480
People (mil.)	57.1	58.4	59.6	60.9	62.2
Military expenditure as % of GNP	3.6	3.8	3.9	4.0	4.1
Military expenditure as % of CGE	20.3	17.9	18.8	15.8	17.4
Military expenditure per capita (1994 $)	74	78	83	89	85
Armed forces per 1,000 people (soldiers)	13.5	13.8	11.8	11.3	13.0
GNP per capita (1994 $)	2,080	2,042	2,125	2,249	2,075
Arms imports[6]					
Current dollars (mil.)	1,300	1,300	1,000	1,200	950
1994 constant dollars (mil.)	1,447	1,393	1,043	1,225	950
Arms exports[6]					
Current dollars (mil.)	10	30	20	20	10
1994 constant dollars (mil.)	11	32	21	20	10
Total imports[7]					
Current dollars (mil.)	22,300	21,050	22,870	29,170	23,270
1994 constant dollars (mil.)	24,820	22,560	23,850	29,780	23,270
Total exports[7]					
Current dollars (mil.)	12,960	13,590	14,720	15,340	15,340
1994 constant dollars (mil.)	14,420	14,570	15,350	15,660	15,340
Arms as percent of total imports[8]	5.8	6.2	4.4	4.1	4.1
Arms as percent of total exports[8]	0.1	0.2	0.1	0.1	0.1

Human Rights [24]

	SSTS	FL	FAPRO	PPCG	APROBC	TPW	PCPTW	STPEP	PHRFF	PRW	ASST	AFL
Observes	P	S	P	P	P	P	P		P	P	P	P
	EAFRD	CPR	ESCR	SR	ACHR	MAAE	PVIAC	PVNAC	EAFDAW	TCIDTP	RC	
Observes		S		P					P	P	P	

P=Party; S=Signatory; see Appendix for meaning of abbreviations.

Labor Force

Total Labor Force [25]

20.9 million

Labor Force by Occupation [26]

Agriculture	46%
Services	31
Industry	23

Date of data: 1994

Unemployment Rate [27]

10.2% (1995 est.)

For sources, notes, and explanations, see Annotated Source Appendix, page 1061.

963

Production Sectors

Commercial Energy Production and Consumption

Data are shown in quadrillion (10^{15}) BTUs and percent for 1995
Values for hydroelectric, nuclear, geothermal, solar, and wind power refer to electrical generation.

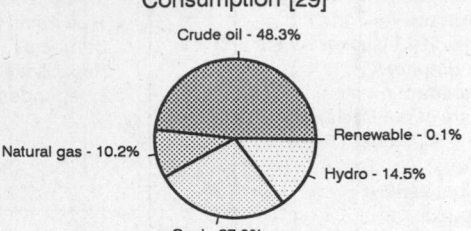

Production [28]
Coal - 49.8%
Natural gas - 0.7%
Renewable - 0.2%
Crude oil - 14.1%
Hydro - 35.2%

Consumption [29]
Crude oil - 48.3%
Natural gas - 10.2%
Renewable - 0.1%
Hydro - 14.5%
Coal - 27.0%

Crude oil	0.147
Dry natural gas	0.007
Coal	0.517
Net hydroelectric power	0.366
Geothermal, solar, wind	0.002
Total	1.039

Crude oil	1.222
Dry natural gas	0.258
Coal	0.683
Net hydroelectric power	0.366
Geothermal, solar, wind	0.002
Total	2.531

Telecommunications [30]

- 6.89 million (1990 est.) telephones; fair domestic and international systems
- Domestic: trunk microwave radio relay network; limited open-wire network
- International: satellite earth stations - 2 Intelsat (Atlantic Ocean), 1 Eutelsat, and 2 Inmarsat (Indian and Atlantic Ocean regions); 1 submarine cable
- Radio: Broadcast stations: AM 15, FM 94, shortwave 0 Radios: 9.4 million (1992 est.)
- Television: Broadcast stations: 357 Televisions: 10.53 million (1993 est.)

Top Agricultural Products [32]

Agriculture accounts for 15.5% of the GDP; produces tobacco, cotton, grain, olives, sugar beets, pulses, citrus; livestock.

Transportation [31]

Railways: total: 10,386 km; standard gauge: 10,386 km 1.435-m gauge (1,088 km electrified)

Highways: total: 386,704 km; paved: 45,683 km (including 862 km of expressways); unpaved: 341,021 km (1992 est.)

Merchant marine: total: 465 ships (1,000 GRT or over) totaling 5,509,741 GRT/9,494,434 DWT; ships by type: bulk 139, cargo 212, chemical tanker 18, combination bulk 7, combination ore/oil 12, container 2, liquefied gas tanker 4, livestock carrier 1, oil tanker 43, passenger-cargo 1, refrigerated cargo 2, roll-on/roll-off cargo 15, short-sea passenger 7, specialized tanker 2

Airports

Total:	104
With paved runways over 3,047 m:	17
With paved runways 2,438 to 3,047 m:	19
With paved runways 1,524 to 2,437 m:	12
With paved runways 914 to 1,523 m:	18
With paved runways under 914 m:	28

Top Mining Products [33]

Thousand metric tons except as noted[e]   8/15/95*

Copper, mine output, gross weight	3,440[r]
Iron ore, gross weight	6,650[r]
Steel, crude, incl. castings	12,100[r]
Silver, mine output, Ag content (kg.)	65,000[73]
Boron minerals, run of mine	2,090
Cement, hydraulic	29,400
Meerschaum (kg.)	3,000[74]
Limestone, other than for cement	11,000
Coal, lignite, run of mine	56,900[r]
Petroleum refinery products (000 42-gal. bls.)	179,910

Tourism [34]

	1990	1991	1992	1993	1994
Visitors[1]	5,389	5,518	7,076	6,500	6,671
Tourists	4,799	5,158	6,549	5,904	6,034
Excursionists	590	360	527	596	637
Tourism receipts	3,225	2,654	3,639	3,959	4,321
Tourism expenditures	520	592	776	934	866
Fare receipts	326	288	NA	NA	NA

Travelers are in thousands, money in million U.S. dollars.

For sources, notes, and explanations, see Annotated Source Appendix, page 1061.

Manufacturing Sector

Manufacturing Summary [35]

	1987		1988		1989		1990		1991	
	$ bil.	%	$ bil.	%	$ bil.	%	$ bil.	%	$ bil.	%
Establishments or enterprises (number)	6,081	0.259	6,274	0.299	6,387	0.341	6,354	0.355	-	-
Total employment (000)	1,174	0.865	1,222	0.893	1,235	1.003	1,244	1.124	-	-
Production workers (000)	935	1.386	941	1.506	969	1.317	986	1.389	-	-
Output ($ bil.)	57	0.563	62	0.528	70	0.591	90	0.788	-	-
Value added ($ bil.)	21	0.477	23	0.489	26	0.535	36	0.723	-	-
Capital investment ($ mil.)	3,121	0.716	3,576	0.751	3,471	0.634	5,532	0.995	-	-
M & E investment ($ mil.)	2,437	0.782	2,954	0.853	2,743	0.702	4,555	1.081	-	-
Employees per establishment (number)	193	333.793	195	298.758	193	294.183	196	316.811	-	-
Production workers per establishment	154	534.845	150	503.834	152	386.298	155	391.611	-	-
Output per establishment ($ mil.)	9	217.136	10	176.528	11	173.231	14	222.179	-	-
Capital investment per estab. ($ mil.)	0.513	276.213	0.570	251.244	0.543	185.858	0.871	280.596	-	-
M & E per establishment ($ mil)	0.401	301.698	0.471	285.531	0.429	206.022	0.717	304.623	-	-
Payroll per employee ($)	3,363	37.498	3,325	33.379	4,770	42.692	7,691	55.444	-	-
Wages per production worker ($)	3,000	38.024	3,200	37.869	4,261	42.492	6,757	56.899	-	-
Hours per production worker (hours)	2,381	125.480	2,590	134.771	2,356	128.524	2,368	126.568	-	-
Output per employee ($)	48,720	65.051	50,328	59.087	56,435	58.886	72,678	70.130	-	-
Capital investment per employee ($)	2,657	82.750	2,926	84.096	2,812	63.178	4,447	88.569	-	-
M & E per employee ($)	2,075	90.385	2,417	95.573	2,221	70.032	3,662	96.153	-	-

Note: Columns headed % show percent of world total or ratio. Ratios closest to 100 are closest to world average. M & E stands for machinery & equipment.

Output in Manufacturing

	1987		1988		1989		1990		1991	
	$ bil.	%	$ bil.	%	$ bil.	%	$ bil.	%	$ bil.	%
3110 - Food products	5.405	9.48	5.739	9.26	7.037	10.05	8.889	9.88	-	-
3130 - Beverages	0.694	1.22	0.802	1.29	1.032	1.47	1.469	1.63	-	-
3140 - Tobacco	1.290	2.26	1.480	2.39	1.615	2.31	2.115	2.35	-	-
3210 - Textiles	5.562	9.76	5.618	9.06	6.449	9.21	7.895	8.77	-	-
3211 - Spinning, weaving, etc.	4.445	7.80	4.562	7.36	5.282	7.55	6.298	7.00	-	-
3220 - Wearing apparel	1.379	2.42	1.804	2.91	2.299	3.28	3.012	3.35	-	-
3230 - Leather and products	0.198	0.35	0.131	0.21	0.197	0.28	0.217	0.24	-	-
3240 - Footwear	0.128	0.22	0.105	0.17	0.134	0.19	0.215	0.24	-	-
3310 - Wood products	0.446	0.78	0.439	0.71	0.436	0.62	0.599	0.67	-	-
3320 - Furniture, fixtures	0.118	0.21	0.128	0.21	0.137	0.20	0.192	0.21	-	-
3410 - Paper and products	0.995	1.75	0.911	1.47	1.107	1.58	1.350	1.50	-	-
3411 - Pulp, paper, etc.	0.609	1.07	0.460	0.74	0.690	0.99	0.738	0.82	-	-
3420 - Printing, publishing	0.457	0.80	0.529	0.85	0.559	0.80	0.856	0.95	-	-
3510 - Industrial chemicals	3.117	5.47	3.796	6.12	3.414	4.88	3.744	4.16	-	-
3511 - Basic chemicals, excl fertilizers	0.514	0.90	0.355	0.57	0.493	0.70	0.444	0.49	-	-
3513 - Synthetic resins, etc.	1.496	2.62	2.268	3.66	1.886	2.69	1.927	2.14	-	-
3520 - Chemical products nec	1.895	3.32	2.031	3.28	2.775	3.96	3.322	3.69	-	-
3522 - Drugs and medicines	0.815	1.43	0.876	1.41	1.224	1.75	1.564	1.74	-	-
3530 - Petroleum refineries	5.129	9.00	5.272	8.50	6.369	9.10	9.076	10.08	-	-
3540 - Petroleum, coal products	1.696	2.98	1.694	2.73	2.282	3.26	2.723	3.03	-	-
3550 - Rubber products	0.652	1.14	0.796	1.28	0.803	1.15	0.909	1.01	-	-
3560 - Plastic products nec	0.563	0.99	0.648	1.05	0.635	0.91	0.973	1.08	-	-
3610 - Pottery, china, etc.	0.413	0.72	0.415	0.67	0.368	0.53	0.703	0.78	-	-
3620 - Glass and products	0.555	0.97	0.594	0.96	0.695	0.99	0.922	1.02	-	-
3690 - Nonmetal products nec	1.641	2.88	1.736	2.80	1.909	2.73	2.464	2.74	-	-
3710 - Iron and steel	4.152	7.28	4.895	7.90	5.749	8.21	5.851	6.50	-	-
3720 - Nonferrous metals	1.044	1.83	1.200	1.94	1.429	2.04	1.557	1.73	-	-
3810 - Metal products	1.353	2.37	1.358	2.19	1.506	2.15	2.037	2.26	-	-
3820 - Machinery nec	2.094	3.67	2.155	3.48	2.040	2.91	3.323	3.69	-	-
3825 - Office, computing machinery	0.093	0.16	0.008	0.01	0.012	0.02	0.028	0.03	-	-
3830 - Electrical machinery	2.266	3.98	2.173	3.50	2.210	3.16	3.554	3.95	-	-
3832 - Radio, television, etc.	1.234	2.16	1.074	1.73	1.019	1.46	1.797	2.00	-	-
3840 - Transportation equipment	2.379	4.17	2.715	4.38	2.918	4.17	4.770	5.30	-	-
3841 - Shipbuilding, repair	0.068	0.12	0.063	0.10	0.099	0.14	0.139	0.15	-	-
3843 - Motor vehicles	2.186	3.84	2.490	4.02	2.625	3.75	4.397	4.89	-	-
3850 - Professional goods	0.034	0.06	0.085	0.14	0.117	0.17	0.169	0.19	-	-
3900 - Industries nec	0.094	0.16	0.096	0.15	0.123	0.18	0.158	0.18	-	-

Note: Codes are International Standard Industry codes (ISIC). Percentages are % of total Output. [f]: Factor Prices; [p]: Producer Prices.

Finance, Economics, and Trade

Economic Indicators [36]

- **National product**: GDP—purchasing power parity— $345.7 billion (1995 est.)
- **National product real growth rate**: 6.8% (1995 est.)
- **National product per capita**: $5,500 (1995 est.)
- **Inflation rate (consumer prices)**: 94% (1995)
- **External debt**: $73.8 billion (1995 est.)

Balance of Payments Summary [37]

Values in millions of dollars.

	1989	1990	1991	1992	1993
Exports of goods (f.o.b.)	11,780	13,026	13,672	14,892	15,610
Imports of goods (f.o.b.)	-15,999	-22,581	-20,998	-23,082	-29,772
Trade balance	-4,219	-9,555	-7,326	-8,190	-14,162
Services - debits	-5,476	-6,496	-6,816	-7,262	-7,829
Services - credits	7,098	8,933	9,315	10,451	11,843
Private transfers (net)	3,135	3,349	2,854	3,147	3,035
Government transfers (net)	423	1,144	2,245	912	733
Long-term capital (net)	1,364	1,037	623	2,252	5,909
Short-term capital (net)	-528	3,000	-3,020	1,396	3,054
Errors and omissions	915	-469	926	-1,222	-2,275
Overall balance	2,712	943	-1,199	1,484	308

Exchange Rates [38]

Currency: **Turkish lira.**
Symbol: **TL.**

Data are currency units per $1.

January 1996	60,502.1
1995	45,845.1
1994	29,608.7
1993	10,984.6
1992	6,872.4
1991	4,171.8

Imports and Exports

Top Import Origins [39]

$32.6 billion (f.o.b., 1995 est.) Data are for 1994.

Origins	%
Germany	16
US	10
Italy	9
Russia	8

Top Export Destinations [40]

$20.7 billion (f.o.b., 1995 est.) Data are for 1994.

Destinations	%
Germany	22
Russia	8
US	8
Italy	6

Foreign Aid [41]

	U.S. $	
ODA (1993)	195	million
Gulf War Allies (1991)	4.1	billion
Pledge for Turkish Defense Fund	2.5	billion

Import and Export Commodities [42]

Import Commodities	Export Commodities
Machinery 25%	Textiles and apparel 37%
Fuels 17%	Steel products 12%
Raw materials 11%	Fruits and vegetables 11%
Foodstuffs 5%	

Turkmenistan

Geography [1]

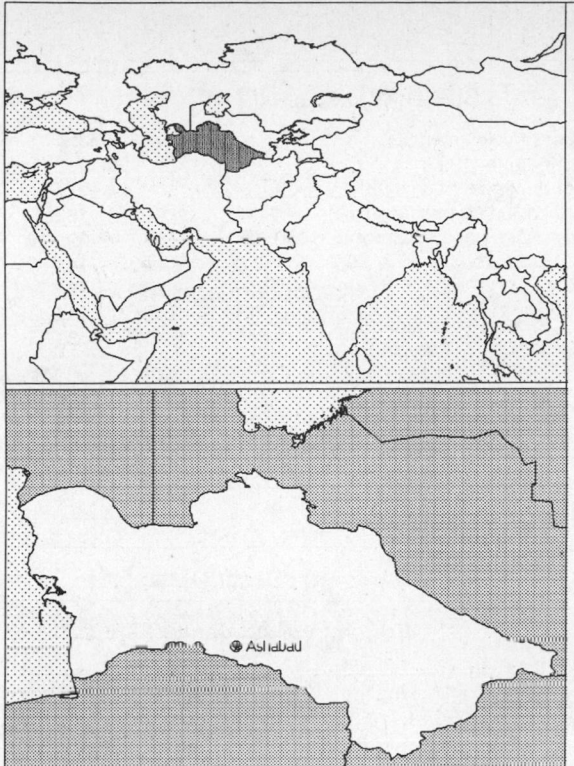

Total area:
  488,100 sq km   188,456 sq mi
Land area:
  488,100 sq km   188,456 sq mi
Comparative area:
  Slightly larger than California
Land boundaries:
  Total 3,736 km, Afghanistan 744 km, Iran 992 km, Kazakhstan 379 km, Uzbekistan 1,621 km
Coastline:
  0 km
  Note: Turkmenistan borders the Caspian Sea (1,768 km)
Climate:
  Subtropical desert
Terrain:
  Flat-to-rolling sandy desert with dunes rising to mountains in the South; low mountains along border with Iran; borders Caspian Sea in West
Natural resources:
  Petroleum, natural gas, coal, sulfur, salt
Land use:
  Arable land: 2%
  Permanent crops: 0%
  Meadows and pastures: 69%
  Forest and woodland: 0%
  Other: 29%

Demographics [2]

	1970	1980	1990	1995[1]	1996	2000	2010	2020	2030
Population	2,181	2,875	3,668	4,075	4,149	4,466	5,362	6,380	7,388
Population density (persons per sq. mi.)	12	15	19	22	22	24	28	34	39
(persons per sq. km.)	4	6	8	8	9	9	11	13	15
Net migration rate (per 1,000 population)	NA	NA	-1.5	-2.3	-2.1	-1.1	-0.6	-0.2	0.0
Births	NA	NA	NA	NA	121	NA	NA	NA	NA
Deaths	NA	NA	NA	NA	37	NA	NA	NA	NA
Life expectancy - males	NA	NA	61.2	56.5	56.7	58.6	60.3	64.6	67.9
Life expectancy - females	NA	NA	68.4	66.4	66.5	67.1	68.6	72.8	75.9
Birth rate (per 1,000)	NA	NA	35.4	29.5	29.1	28.1	26.6	23.0	19.9
Death rate (per 1,000)	NA	NA	8.0	9.0	8.9	8.3	8.0	6.7	6.5
Women of reproductive age (15-49 yrs.)	NA	NA	873	1,015	1,045	1,167	1,457	1,678	1,932
of which are currently married	NA	NA	547	NA	668	746	944	NA	NA
Fertility rate	NA	NA	4.3	3.7	3.6	3.5	3.1	2.8	2.6

Except as noted, values for vital statistics are in thousands; life expectancy is in years.

Health

Health Indicators [3]

% of population with access to	
safe water (1990-95)	NA
adequate sanitation (1990-95)	NA
health services (1985-95)	NA
% of 1-year-olds immunized (1990-94) against	
TB (tuberculosis)	94
DPT (diphtheria, pertussis, tetanus)	71
polio	92
measles	84
% of contraceptive prevalence (1980-94)	NA
ORT use rate (1990-94)	NA

Health Expenditures [4]

Total health expenditure, 1990 (official exchange rate)	
Millions of dollars	459
Dollars per capita	125
Health expenditures as a percentage of GDP	
Total	5.0
Public sector	3.3
Private sector	1.7
Development assistance for health	
Total aid flows (millions of dollars)[1]	2
Aid flows per capita (dollars)	0.5
Aid flows as a percentage of total health expenditure	0.4

For sources, notes, and explanations, see Annotated Source Appendix, page 1061.

Human Factors

Women and Children [5]

% of pregnant women immunized (tetanus 1990-94)	NA
% of births attended by trained health personnel (1983-94)	NA
Maternal mortality rate (1980-92)	NA
Under-5 mortality rate (1994)	87
% under-5 moderately/severely underweight (1980-1994)	NA

Burden of Disease [6]

Population per physician (1993)	305.53
Population per nurse (1993)	100.19
Population per hospital bed (1993)	93.45
AIDS cases per 100,000 people (1994)	*
Heart disease cases per 1,000 people (1990-93)	NA

Ethnic Division [7]

Turkmen	73.3%
Russian	9.8%
Uzbek	9%
Kazak	2%
Other	5.9%

Religion [8]

Muslim	87%
Eastern Orthodox	11%
Unknown	2%

Major Languages [9]

Turkmen	72%
Russian	12%
Uzbek	9%
Other	7%

Education

Public Education Expenditures [10]

	1980	1985	1989	1990	1991	1994
Million (Manat)						
Total education expenditure	NA	431	602	655	1,159	NA
as percent of GNP	NA	7.6	NA	8.6	7.9	NA
as percent of total govt. expend.	NA	28.0	27.1	21.0	19.7	NA
Current education expenditure	NA	NA	NA	NA	NA	NA
as percent of GNP	NA	NA	NA	NA	NA	NA
as percent of current govt. expend.	NA	NA	NA	NA	NA	NA
Capital expenditure	NA	NA	NA	NA	NA	NA

Educational Attainment [11]

Illiteracy [12]

	1985	1989	1995
Illiterate population (15+ yrs.)[2]	NA	NA	8,000
Illiteracy rate - total pop. (%)	NA	2.3	0.3
Illiteracy rate - males (%)	NA	1.2	0.2
Illiteracy rate - females (%)	NA	3.4	0.4

Libraries [13]

Daily Newspapers [14]

Culture [15]

Science and Technology

Scientific/Technical Forces [16]

R&D Expenditures [17]

U.S. Patents Issued [18]

For sources, notes, and explanations, see Annotated Source Appendix, page 1061.

Government and Law

Organization of Government [19]

Long-form name:
None
Type:
Republic
Independence:
27 October 1991 (from the Soviet Union)
National holiday:
Independence Day, 27 October (1991)
Constitution:
Adopted 18 May 1992
Legal system:
Based on civil law system
Executive branch:
President; Prime Minister; 12 Deputy
Prime Ministers; Council of Ministers
Legislative branch:
Under the 1992 constitution, there are two
parliamentary bodies, a unicameral
People's Council (Halk Maslahaty - having
more than 100 members and meeting
infrequently) and a 50-member
unicameral Assembly (Majlis)
Judicial branch:
Supreme Court

Elections [20]

Assembly	% of seats
Democratic Party	90.0
Other	10.0

Government Budget [21]

Crime [23]

Military Expenditures and Arms Transfers [22]

	1990	1991	1992[14]	1993	1994
Military expenditures					
Current dollars (mil.)[e]	NA	NA	144	74	65
1994 constant dollars (mil.)[e]	NA	NA	150	76	65
Armed forces (000)	NA	NA	28	28	15
Gross national product (GNP)					
Current dollars (mil.)[e]	NA	NA	11,800	13,000	13,100
1994 constant dollars (mil.)[e]	NA	NA	12,310	13,270	13,100
Central government expenditures (CGE)					
1994 constant dollars (mil.)[e]	NA	NA	4,098	NA	NA
People (mil.)	NA	NA	3.8	3.9	4.0
Military expenditure as % of GNP	NA	NA	1.2	0.6	0.5
Military expenditure as % of CGE	NA	NA	3.7	NA	NA
Military expenditure per capita (1994 $)	NA	NA	39	19	16
Armed forces per 1,000 people (soldiers)	NA	NA	7.3	7.2	3.8
GNP per capita (1994 $)	NA	NA	3,211	3,389	3,279
Arms imports[6]					
Current dollars (mil.)	NA	NA	0	0	0
1994 constant dollars (mil.)	NA	NA	0	0	0
Arms exports[6]					
Current dollars (mil.)	NA	NA	30	0	0
1994 constant dollars (mil.)	NA	NA	31	0	0
Total imports[7]					
Current dollars (mil.)	NA	NA	1,008	781	1,812
1994 constant dollars (mil.)	NA	NA	1,051	797	1,812
Total exports[7]					
Current dollars (mil.)	NA	NA	2,146	1,286	884
1994 constant dollars (mil.)	NA	NA	2,238	1,313	884
Arms as percent of total imports[8]	NA	NA	0	0	0
Arms as percent of total exports[8]	NA	NA	1.4	0	0

Human Rights [24]

	SSTS	FL	FAPRO	PPCG	APROBC	TPW	PCPTW	STPEP	PHRFF	PRW	ASST	AFL
Observes						P	P					

	EAFRD	CPR	ESCR	SR	ACHR	MAAE	PVIAC	PVNAC	EAFDAW	TCIDTP	RC
Observes	P						P	P			P

P = Party; S = Signatory; see Appendix for meaning of abbreviations.

Labor Force

Total Labor Force [25]

1.642 million (January 1994)

Labor Force by Occupation [26]

Agriculture and forestry	44%
Industry and construction	20
Other	36

Date of data: 1992

Unemployment Rate [27]

For sources, notes, and explanations, see Annotated Source Appendix, page 1061.

969

Production Sectors

Commercial Energy Production and Consumption

Data are shown in quadrillion (10^{15}) BTUs and percent for 1995
Values for hydroelectric, nuclear, geothermal, solar, and wind power refer to electrical generation.

Production [28]

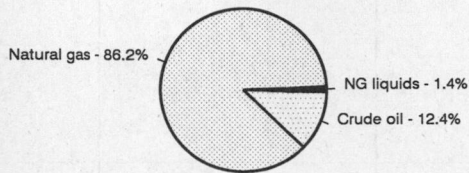

Natural gas - 86.2%
NG liquids - 1.4%
Crude oil - 12.4%

Consumption [29]

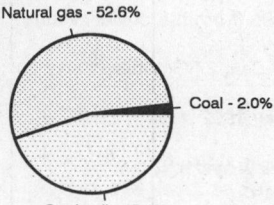

Natural gas - 52.6%
Coal - 2.0%
Crude oil - 45.4%

Crude oil	0.150
Natural gas liquids	0.017
Dry natural gas	1.039
Total	1.206

Crude oil	0.133
Dry natural gas	0.154
Coal	0.006
Total	0.293

Telecommunications [30]

- Poorly developed
- International: linked by cable and microwave radio relay to other CIS republics and to other countries by leased connections to the Moscow international gateway switch; a new telephone link from Ashgabat to Iran has been established; a new exchange in Ashgabat switches international traffic through Turkey via Intelsat; satellite earth stations - 1 Orbita and 1 Intelsat
- Radio: Broadcast stations: AM NA, FM NA, shortwave NA; note - there is at least one state-owned radio broadcast station of NA type
- Television:

Transportation [31]

Railways: total: 2,120 km in common carrier service; does not include industrial lines; broad gauge: 2,120 km 1.520-m gauge (1990)

Highways: total: 23,000 km; paved: NA km; unpaved: NA km (1990 est.)

Airports

Total:	64
With paved runways 2,438 to 3,047 m:	13
With paved runways 1,524 to 2,437 m:	8
With paved runways 914 to 1,523 m:	1
With unpaved runways 914 to 1,523 m:	7
With unpaved runways under 914 m:	35 (1994 est.)

Top Agricultural Products [32]

Agriculture accounts for 32.5% of the GDP; produces cotton, grain; livestock.

Top Mining Products [33]

Metric tons except as noted[e]	6/30/95*
Bentonite	40,000
Cement	800,000
Gypsum	150,000
Iodine	251[r]
Gas, natural (mil. cu. meters)	35,600[r]
Petroleum[r]	
crude	4,100,000
refined	4,300,000
Sodium sulfate	67,500[r]
Sulfur	47,599[r]

Tourism [34]

Finance, Economics, and Trade

Industrial Summary [35]

Industrial Production: Growth rate - 7% (1995); accounts for 33.4% of the GDP. **Industries**: Natural gas, oil, petroleum products, textiles, food processing.

Economic Indicators [36]

- **National product**: GDP—purchasing power parity—$11.5 billion (1995 estimate as extrapolated from World Bank estimate for 1994)

- **National product real growth rate**: - 10% (1995 est.)

- **National product per capita**: $2,820 (1995 est.)

- **Inflation rate (consumer prices)**: 25% monthly average (1994 est.)

- **External debt**: $400 million (of which $275 million to Russia) (1995 est.)

Balance of Payments Summary [37]

Exchange Rates [38]

Currency: **manat.**

Data are currency units per $1. Its own national currency, the manat, was introduced 1 November 1993. The government established an unified rate in mid-January 1996.

January 1996	2,400

Imports and Exports

Top Import Origins [39]

$777 million from States outside the FSU (1995).

Origins	%
Russia	NA
Azerbaijan	NA
Uzbekistan	NA
Kazakhstan	NA
Turkey	NA

Top Export Destinations [40]

$1.9 billion to States outside the FSU (1995).

Destinations	%
Ukraine	NA
Russia	NA
Kazakhstan	NA
Uzbekistan	NA
Georgia	NA
Azerbaijan	NA
Armenia	NA
Eastern Europe	NA
Turkey	NA
Argentina	NA

Foreign Aid [41]

	U.S. $	
ODA (1993)	10	million
Commitments (1992-95)	1,830	million
of which drawn	375	million

Import and Export Commodities [42]

Import Commodities	Export Commodities
Machinery and parts	Natural gas
Grain and food	Cotton
Plastics and rubber	Petroleum products
Consumer durables	Electricity
Textiles	Textiles
	Carpets

Uganda

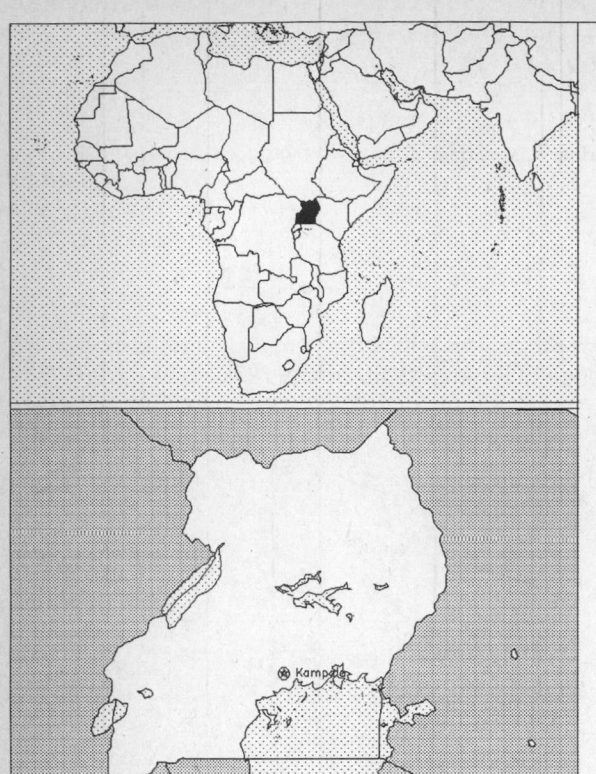

Geography [1]

Total area:
236,040 sq km   91,136 sq mi
Land area:
199,710 sq km   77,108 sq mi
Comparative area:
Slightly smaller than Oregon
Land boundaries:
Total 2,698 km, Kenya 933 km, Rwanda 169 km, Sudan 435 km,
Tanzania 396 km, Zaire 765 km
Coastline:
0 km (landlocked)
Climate:
Tropical; generally rainy with two dry seasons (December to February,
June to August); semiarid in Northeast
Terrain:
Mostly plateau with rim of mountains
Natural resources:
Copper, cobalt, limestone, salt
Land use:
Arable land: 23%
Permanent crops: 9%
Meadows and pastures: 25%
Forest and woodland: 30%
Other: 13%

Demographics [2]

	1970	1980	1990	1995[1]	1996	2000	2010	2020	2030
Population	9,724	12,252	17,040	19,730	20,158	21,891	26,355	30,872	36,704
Population density (persons per sq. mi.)	126	159	221	256	261	284	342	400	476
(persons per sq. km.)	49	61	85	99	101	110	132	155	184
Net migration rate (per 1,000 population)	0.2	-6.9	0.5	-6.0	-2.8	-2.3	0.0	0.0	0.0
Births	NA	NA	NA	NA	926	NA	NA	NA	NA
Deaths	NA	NA	NA	NA	418	NA	NA	NA	NA
Life expectancy - males	45.8	43.2	44.7	40.7	40.0	37.5	35.5	39.8	52.8
Life expectancy - females	47.0	44.7	44.0	41.1	40.6	38.7	35.0	39.8	55.9
Birth rate (per 1,000)	49.1	50.1	50.5	46.9	45.9	43.1	39.9	35.3	29.8
Death rate (per 1,000)	20.6	20.6	19.0	20.5	20.7	21.8	23.7	19.7	10.5
Women of reproductive age (15-49 yrs.)	2,056	2,678	3,738	4,232	4,299	4,675	6,041	7,491	9,775
of which are currently married	NA	NA	2,529	NA	2,903	3,116	3,973	NA	NA
Fertility rate	7.4	7.4	7.1	6.7	6.6	6.2	5.2	4.1	3.3

Except as noted, values for vital statistics are in thousands; life expectancy is in years.

Health

Health Indicators [3]

% of population with access to	
safe water (1990-95)	34
adequate sanitation (1990-95)	57
health services (1985-95)	49
% of 1-year-olds immunized (1990-94) against	
TB (tuberculosis)	100
DPT (diphtheria, pertussis, tetanus)	79
polio	79
measles	77
% of contraceptive prevalence (1980-94)	5
ORT use rate (1990-94)	45

Health Expenditures [4]

Total health expenditure, 1990 (official exchange rate)	
Millions of dollars	95
Dollars per capita	6
Health expenditures as a percentage of GDP	
Total	3.4
Public sector	1.6
Private sector	1.8
Development assistance for health	
Total aid flows (millions of dollars)[1]	46
Aid flows per capita (dollars)	2.8
Aid flows as a percentage of total health expenditure	48.4

For sources, notes, and explanations, see Annotated Source Appendix, page 1061.

Human Factors

Women and Children [5]

% of pregnant women immunized (tetanus 1990-94)	77
% of births attended by trained health personnel (1983-94)	38
Maternal mortality rate (1980-92)	550
Under-5 mortality rate (1994)	185
% under-5 moderately/severely underweight (1980-1994)	23

Burden of Disease [6]

Population per physician (1989)	22,986.88
Population per nurse	NA
Population per hospital bed (1991)	1,091.48
AIDS cases per 100,000 people (1994)	23.2
Malaria cases per 100,000 people (1992)	NA

Ethnic Division [7]

Other includes Acholi, Asian, Arab, Bagisu, Batobo, Bunyoro, European, Lugbara.

Baganda	17%
Karamojong	12%
Basogo	8%
Iteso	8%
Langi and Rwanda	12%
Other	43%

Religion [8]

Protestant	33%
Roman Catholic	33%
Indigenous beliefs	18%
Muslim	16%

Major Languages [9]

English (official), Luganda, Swahili, Bantu languages, Nilotic languages.

Education

Public Education Expenditures [10]

Million (Shilling)[68]	1980	1984	1989	1990	1991	1994
Total education expenditure	15	288	12,971	20,188	35,026	NA
as percent of GNP	1.2	3.0	1.1	1.5	1.9	NA
as percent of total govt. expend.	11.3	NA	10.4	11.5	15.0	NA
Current education expenditure	14	205	12,437	18,527	33,012	NA
as percent of GNP	1.1	2.1	1.1	1.4	1.8	NA
as percent of current govt. expend.	12.8	NA	12.0	15.1	16.5	NA
Capital expenditure	2	83	534	1,661	2,014	NA

Educational Attainment [11]

Age group (1991)	25+
Total population	5,455,582
Highest level attained (%)	
No schooling	46.1
First level	
Not completed	41.4
Completed	NA
Entered second level	
S-1	8.9
S-2	1.3
Postsecondary	0.5

Illiteracy [12]

In thousands and percent[1]	1990	1995	2000
Illiterate population (15+ yrs.)	4,020	4,172	4,142
Illiteracy rate - total pop. (%)	45.7	41.7	37.4
Illiteracy rate - males (%)	31.5	28.3	25.2
Illiteracy rate - females (%)	59.7	55.0	49.7

Libraries [13]

	Admin. Units	Svc. Pts.	Vols. (000)	Shelving (meters)	Vols. Added	Reg. Users
National (1986)	1	2	15	NA	1,100	450
Nonspecialized	NA	NA	NA	NA	NA	NA
Public (1992)	1	17	82	NA	4,184	53,476
Higher ed. (1993)	5	10	1,096	56,300	9,223	34,700
School	NA	NA	NA	NA	NA	NA

Daily Newspapers [14]

	1980	1985	1990	1994
Number of papers	1	1	2	2
Circ. (000)	25	25	30	35[e]

Culture [15]

Science and Technology

Scientific/Technical Forces [16]

R&D Expenditures [17]

U.S. Patents Issued [18]

Values show patents issued to citizens of the country by the U.S. Patents Office.

	1993	1994	1995
Number of patents	2	1	0

For sources, notes, and explanations, see Annotated Source Appendix, page 1061.

973

Government and Law

Organization of Government [19]

Long-form name:
Republic of Uganda
Type:
Republic
Independence:
9 October 1962 (from UK)
National holiday:
Independence Day, 9 October (1962)
Constitution:
8 October 1995; adopted by the interim, 284-member Constituent Assembly, charged with debating 1993 draft constitution
Legal system:
In 1995, the legal system was restored to one based on English common law and customary law and a normal judicial system reinstituted; accepts compulsory ICJ jurisdiction, with reservations
Executive branch:
President; Prime Minister; Cabinet
Legislative branch:
Unicameral: National Assembly Dissolved in 1985 after military coup, succeeded in 1986 by the National Resistance Council; this was dissolved in 1996 for the popular election of a new legislature
Judicial branch:
Court of Appeal; High Court

Elections [20]

National Assembly. Elections last held in 1980. The National Assembly was dissolved following a military coup; the National Resistance Council served as Uganda's acting legislature since 1986; it was dissolved on 15 June 1996 to prepare for popular elections of a new legislature on 27 June 1996 per the provisions of a new constitution. Only officially recognized party until 2001—National Resistance Movement (NRM).

Government Budget [21]

For 1994/95 est.

	$ bil.
Revenues	0.574
Expenditures	1.07
Capital expenditures	0.328

Crime [23]

Military Expenditures and Arms Transfers [22]

	1990	1991	1992	1993	1994
Military expenditures					
Current dollars (mil.)	74	110	87	71	66
1994 constant dollars (mil.)	83	118	91	72	66
Armed forces (000)	60	60	70	70	60
Gross national product (GNP)					
Current dollars (mil.)	3,147	3,466	3,639	4,078	4,414
1994 constant dollars (mil.)	3,502	3,715	3,795	4,163	4,414
Central government expenditures (CGE)					
1994 constant dollars (mil.)	440	743	816	787	869
People (mil.)	16.9	17.4	18.0	18.6	19.1
Military expenditure as % of GNP	2.4	3.2	2.4	1.7	1.5
Military expenditure as % of CGE	18.8	15.9	11.1	9.1	7.6
Military expenditure per capita (1994 $)	5	7	5	4	3
Armed forces per 1,000 people (soldiers)	3.5	3.4	3.9	3.8	3.1
GNP per capita (1994 $)	207	213	211	224	231
Arms imports[6]					
Current dollars (mil.)	10	10	0	0	0
1994 constant dollars (mil.)	11	11	0	0	0
Arms exports[6]					
Current dollars (mil.)	0	0	0	0	0
1994 constant dollars (mil.)	0	0	0	0	0
Total imports[7]					
Current dollars (mil.)	213	196	439	384	870
1994 constant dollars (mil.)	237	210	458	392	870
Total exports[7]					
Current dollars (mil.)	147	200	142	179	421
1994 constant dollars (mil.)	164	214	148	183	421
Arms as percent of total imports[8]	4.7	5.1	0	0	0
Arms as percent of total exports[8]	0	0	0	0	0

Human Rights [24]

	SSTS	FL	FAPRO	PPCG	APROBC	TPW	PCPTW	STPEP	PHRFF	PRW	ASST	AFL
Observes	P	P		P	P	P	P			P	P	P
	EAFRD	CPR	ESCR	SR	ACHR	MAAE	PVIAC	PVNAC	EAFDAW	TCIDTP	RC	
Observes	P	P	P	P			P	P	P	P	P	

P = Party; S = Signatory; see Appendix for meaning of abbreviations.

Labor Force

Total Labor Force [25]

8.361 million (1993 est.)

Labor Force by Occupation [26]

Agriculture	86%
Industry	4
Services	10

Date of data: 1980 est.

Unemployment Rate [27]

For sources, notes, and explanations, see Annotated Source Appendix, page 1061.

Production Sectors

Commercial Energy Production and Consumption

Data are shown in quadrillion (10^{15}) BTUs and percent for 1995
Values for hydroelectric, nuclear, geothermal, solar, and wind power refer to electrical generation.

Production [28]

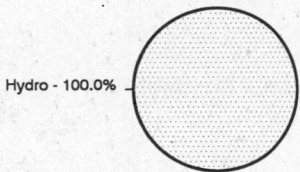

Hydro - 100.0%

Consumption [29]

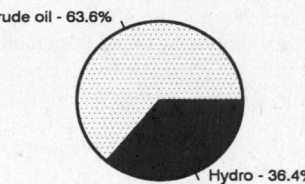

Crude oil - 63.6%

Hydro - 36.4%

Net hydroelectric power	0.008
Total	0.008

Crude oil	0.014
Net hydroelectric power	0.008
Total	0.022

Telecommunications [30]

- 54,900 (1989 est.) telephones; fair system
- Domestic: microwave radio relay and radiotelephone communications stations
- International: satellite earth station - 1 Intelsat (Atlantic Ocean)
- Radio: Broadcast stations: AM 10, FM 0, shortwave 0 Radios: 2.04 million (1992 est.)
- Television: Broadcast stations: 9 (1987 est.) Televisions: 193,000 (1992 est.)

Top Agricultural Products [32]

Agriculture accounts for 55% of the GDP; produces coffee, tea, cotton, tobacco, cassava (tapioca), potatoes, corn, millet, pulses; beef, goat meat, milk, poultry.

Top Mining Products [33]

Metric tons except as noted[e]	3/3/95[*]
Cement, hydraulic	50,000
Gold (kg.)	1,800
Iron ore	130
Lime, hydrated and quick	1,500
Phosphate minerals: Apatite	100
Salt, evaporated	5,000
Tin, mine output, Sn content	30
Tungsten, mine content, W content	60

Transportation [31]

Railways: total: 1,241 km single track; narrow gauge: 1,241 km 1.000-m gauge

Highways: total: 30,320 km; paved: 3,480 km; unpaved: 26,840 km (1987 est.)

Merchant marine: total: 3 roll-on/roll-off cargo ships (1,000 GRT or over) totaling 5,091 GRT/2,743 DWT (1995 est.)

Airports

Total:	21
With paved runways over 3,047 m:	2
With paved runways 1,524 to 2,437 m:	1
With paved runways under 914 m:	7
With unpaved runways 2,438 to 3,047 m:	1
With unpaved runways 1,524 to 2,437 m:	5

Tourism [34]

	1990	1991	1992	1993	1994
Tourists	69	69	76	103	119
Tourism receipts	10	15	38	50	61
Tourism expenditures	8	18	NA	40	78

Travelers are in thousands, money in million U.S. dollars.

For sources, notes, and explanations, see Annotated Source Appendix, page 1061.

975

Finance, Economics, and Trade

GDP and Manufacturing Summary [35]

	1980	1985	1990	1991	1992
Gross Domestic Product					
Millions of 1980 dollars	4,644	5,284	6,846	7,130	7,379
Growth rate in percent	-3.40	2.00	4.30	4.14	3.50
Manufacturing Value Added					
Millions of 1980 dollars	192	194	333	354	374[e]
Growth rate in percent	6.10	-9.80	7.50	6.32	5.74[e]
Manufacturing share in percent of current prices	4.2	2.2	4.1	NA	NA

Economic Indicators [36]

- **National product**: GDP—purchasing power parity—$16.8 billion (1995 est.)

- **National product real growth rate**: 7.1% (1995 est.)

- **National product per capita**: $900 (1995 est.)

- **Inflation rate (consumer prices)**: 6.1% (1995)

- **External debt**: $3.2 billion (1994)

Balance of Payments Summary [37]

Values in millions of dollars.

	1989	1990	1991	1992	1993
Exports of goods (f.o.b.)	277.7	177.8	173.2	151.2	196.7
Imports of goods (f.o.b.)	-588.3	-491.0	-377.1	-421.9	-474.7
Trade balance	-310.6	-313.2	-203.9	-270.7	-278.0
Services - debits	-260.5	-243.1	-318.5	-336.0	-354.3
Services - credits	NA	NA	23.6	38.6	100.0
Private transfers (net)	NA	NA	103.4	207.3	163.3
Government transfers (net)	311.6	293.0	225.6	261.2	261.7
Long-term capital (net)	320.6	251.9	168.8	153.7	183.9
Short-term capital (net)	-4.2	-39.8	-31.2	-38.9	-23.9
Errors and omissions	-38.0	9.5	0.6	9.0	5.5
Overall balance	18.9	-41.7	-31.6	24.2	58.2

Exchange Rates [38]

Currency: **Ugandan shilling.**
Symbol: **USh.**

Data are currency units per $1.

November 1995	1,032.6
1994	979.4
1993	1,195.0
1992	1,133.8
1991	734.0

Imports and Exports

Top Import Origins [39]

$870 million (c.i.f., 1994).

Origins	%
Kenya	25
UK	14
Italy	13

Top Export Destinations [40]

$424 million (f.o.b., 1994).

Destinations	%
US	25
UK	18
France	11
Spain	10

Foreign Aid [41]

Recipient: ODA, $NA.

Import and Export Commodities [42]

Import Commodities	**Export Commodities**
Petroleum products	Coffee 97%
Machinery	Cotton
Cotton piece goods	Tea
Metals	
Transportation equipment	
Food	

Ukraine

Geography [1]

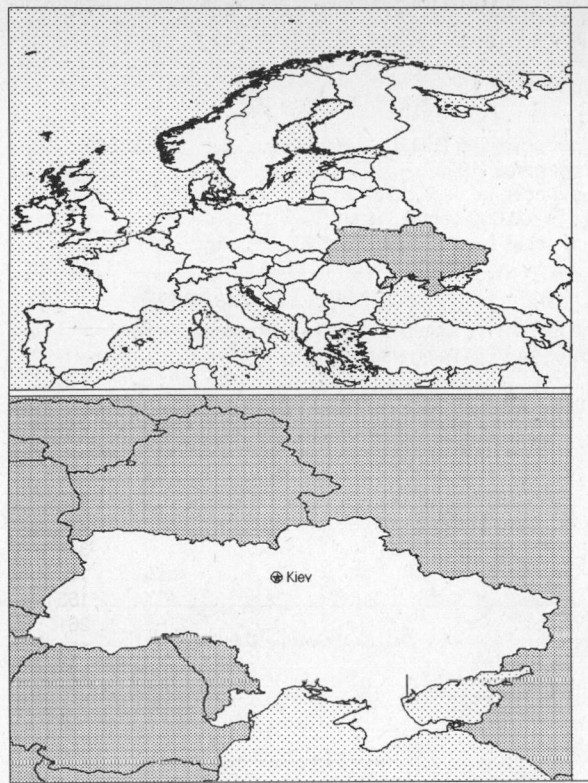

Total area:
603,700 sq km   233,090 sq mi
Land area:
603,700 sq km   233,090 sq mi
Comparative area:
Slightly smaller than Texas
Land boundaries:
Total 4,558 km, Belarus 891 km, Hungary 103 km, Moldova 939 km, Poland 428 km, Romania 531 km, Russia 1,576 km, Slovakia 90 km
Coastline:
2,782 km
Climate:
Temperate continental; Mediterranean only on southern Crimean coast; precipitation disproportionately distributed, highest in West and North, lesser in East and Southeast; winters vary from cool along Black Sea to cold farther inland; summers warm across the greater part of the country, hot in the South
Terrain:
Mostly consists of fertile plains (steppes) and plateaus, mountains being found only in West (the Carpathians), and in Crimean Peninsula in the extreme South
Natural resources:
Iron ore, coal, manganese, natural gas, oil, salt, sulfur, graphite, titanium, magnesium, kaolin, nickel, mercury, timber
Land use:
Arable land: 56%
Permanent crops: 2%
Meadows and pastures: 12%
Forest and woodland: 0%
Other: 30%

Demographics [2]

	1970	1980	1990	1995[1]	1996	2000	2010	2020	2030
Population	47,236	50,047	51,592	51,089	50,864	50,380	49,915	49,038	48,375
Population density (persons per sq. mi.)	203	215	221	219	218	216	214	210	208
(persons per sq. km.)	78	83	85	85	84	83	83	81	80
Net migration rate (per 1,000 population)	NA	NA	0.2	-0.2	0.0	0.8	0.5	0.2	0.0
Births	NA	NA	NA	NA	568	NA	NA	NA	NA
Deaths	NA	NA	NA	NA	771	NA	NA	NA	NA
Life expectancy - males	NA	NA	65.2	61.2	61.5	62.9	66.1	69.9	73.0
Life expectancy - females	NA	NA	74.4	72.2	72.3	72.9	74.3	77.5	80.1
Birth rate (per 1,000)	NA	NA	12.9	10.5	11.2	13.2	12.3	10.7	10.7
Death rate (per 1,000)	NA	NA	12.3	15.2	15.2	14.9	14.3	12.5	11.9
Women of reproductive age (15-49 yrs.)	NA	NA	12,301	12,623	12,705	12,763	12,072	11,583	10,929
of which are currently married	NA	NA	8,504	NA	8,793	8,777	8,537	NA	NA
Fertility rate	NA	NA	1.9	1.5	1.6	1.9	1.8	1.7	1.7

Except as noted, values for vital statistics are in thousands; life expectancy is in years.

Health

Health Indicators [3]

% of population with access to	
safe water (1990-95)	NA
adequate sanitation (1990-95)	NA
health services (1985-95)	NA
% of 1-year-olds immunized (1990-94) against	
TB (tuberculosis)	89
DPT (diphtheria, pertussis, tetanus)	90
polio	91
measles	94
% of contraceptive prevalence (1980-94)	NA
ORT use rate (1990-94)	NA

Health Expenditures [4]

Total health expenditure, 1990 (official exchange rate)	
Millions of dollars	6,803
Dollars per capita	131
Health expenditures as a percentage of GDP	
Total	3.3
Public sector	2.3
Private sector	1.0
Development assistance for health	
Total aid flows (millions of dollars)[1]	NA
Aid flows per capita (dollars)	NA
Aid flows as a percentage of total health expenditure	NA

For sources, notes, and explanations, see Annotated Source Appendix, page 1061.

977

Human Factors

Women and Children [5]

% of pregnant women immunized (tetanus 1990-94)	NA
% of births attended by trained health personnel (1983-94)	NA
Maternal mortality rate (1980-92)	NA
Under-5 mortality rate (1994)	25
% under-5 moderately/severely underweight (1980-1994)	NA

Burden of Disease [6]

Population per physician (1993)	226.87
Population per nurse (1993)	86.96
Population per hospital bed (1993)	76.85
AIDS cases per 100,000 people (1994)	*
Heart disease cases per 1,000 people (1990-93)	300.5

Ethnic Division [7]

Ukrainian	73%
Russian	22%
Jewish	1%
Other	4%

Religion [8]

Ukrainian Orthodox - Moscow Patriarchate, Ukrainian Orthodox - Kiev Patriarchate, Ukrainian Autocephalous Orthodox, Ukrainian Catholic (Uniate), Protestant, Jewish.

Major Languages [9]

Ukrainian, Russian, Romanian, Polish, Hungarian.

Education

Public Education Expenditures [10]

	1980	1985	1990	1992	1993	1994
Million or Trillion (T) (Karbovanets)[69]						
Total education expenditure	5,927	6,721	8,606	327,968	9.01 T	84.71 T
as percent of GNP	5.6	5.2	5.2	7.8	6.1	8.2
as percent of total govt. expend.	24.5	21.2	19.7	17.1	15.7	NA
Current education expenditure	5,114	5,708	6,903	280,690	7.43 T	70,300
as percent of GNP	4.8	4.4	4.2	6.7	5.1	6.8
as percent of current govt. expend.	NA	NA	NA	NA	NA	NA
Capital expenditure	812	1,013	1,704	47,278	1.58 T	14.41 T

Educational Attainment [11]

Illiteracy [12]

	1985	1989	1995
Illiterate population (15+ yrs.)[2]	NA	NA	484,000
Illiteracy rate - total pop. (%)	NA	1.6	1.2
Illiteracy rate - males (%)	NA	0.5	1.8
Illiteracy rate - females (%)	NA	2.6	0.7

Libraries [13]

	Admin. Units	Svc. Pts.	Vols. (000)	Shelving (meters)	Vols. Added	Reg. Users
National	NA	NA	NA	NA	NA	NA
Nonspecialized	NA	NA	NA	NA	NA	NA
Public (1992)	25,300	NA	400,883	NA	21 mil[e]	22 mil
Higher ed.	NA	NA	NA	NA	NA	NA
School (1988)	22,500	NA	299,900	NA	NA	NA

Daily Newspapers [14]

	1980	1985	1990	1994[3]
Number of papers	NA	NA	127	90
Circ. (000)	NA	NA	13,026	6,083

Culture [15]

Cinema (seats per 1,000)	7.3
Annual attendance per person	2.5
Gross box office receipts (mil. Ruble)	27,529
Museums (reporting)	271
Visitors (000)	15,253
Annual receipts (000 Ruble)	77,643

Science and Technology

Scientific/Technical Forces [16]

Scientists/engineers[34]	348,600
Number female	NA
Technicians	NA
Number female	NA
Total	NA

R&D Expenditures [17]

U.S. Patents Issued [18]

Values show patents issued to citizens of the country by the U.S. Patents Office.

	1993	1994	1995
Number of patents	0	9	8

For sources, notes, and explanations, see Annotated Source Appendix, page 1061.

Government and Law

Organization of Government [19]

Long-form name:
None
Type:
Republic
Independence:
1 December 1991 (from Soviet Union)
National holiday:
Independence Day, 24 August (1991)
Constitution:
Adopted 28 June 1996
Legal system:
Based on civil law system; judicial review of legislative acts
Executive branch:
President; Prime Minister; First Deputy Prime Minister; 8 Deputy Prime Ministers; Council of Ministers; National Security Council; Presidential Administration; Council of Regions
Legislative branch:
Unicameral: Supreme Council
Judicial branch:
Supreme Court (highest judicial body); Constitutional Court (exclusive constitutional jurisdiction)

Elections [20]

Supreme Council	% of seats
Independents	52.9
Communists	21.1
Rukh	4.9
Agrarians	4.0
Socialists	3.3
Republicans	2.4
Other parties	5.1
Unfilled seats	6.2

Government Budget [21]

Crime [23]

	1994
Crime volume	
Cases known to police	572,147
Attempts (percent)	NA
Percent cases solved	54.70
Crimes per 100,000 persons	1,096.07
Persons responsible for offenses	
Total number offenders	269,599
Percent female	15.40
Percent juvenile (14-18 yrs.)	12.20
Percent foreigners	NA

Military Expenditures and Arms Transfers [22]

	1990	1991	1992[14]	1993	1994
Military expenditures					
Current dollars (mil.)[e]	NA	NA	NA	824	946
1994 constant dollars (mil.)[e]	NA	NA	NA	841	946
Armed forces (000)	NA	NA	438[e]	510	495
Gross national product (GNP)					
Current dollars (mil.)[e]	NA	NA	238,500	210,100	171,400
1994 constant dollars (mil.)[e]	NA	NA	248,800	214,500	171,400
Central government expenditures (CGE)					
1994 constant dollars (mil.)[e]	NA	NA	180,600	NA	NA
People (mil.)	NA	NA	51.8	51.8	51.8
Military expenditure as % of GNP	NA	NA	NA	0.4	0.6
Military expenditure as % of CGE	NA	NA	NA	NA	NA
Military expenditure per capita (1994 $)	NA	NA	NA	16	18
Armed forces per 1,000 people (soldiers)	NA	NA	8.5	9.8	9.5
GNP per capita (1994 $)	NA	NA	4,804	4,138	3,305
Arms imports[6]					
Current dollars (mil.)	NA	NA	0	0	0
1994 constant dollars (mil.)	NA	NA	0	0	0
Arms exports[6]					
Current dollars (mil.)	NA	NA	0	50	60
1994 constant dollars (mil.)	NA	NA	0	51	60
Total imports[7]					
Current dollars (mil.)	NA	NA	4,218	14,480	10,110
1994 constant dollars (mil.)	NA	NA	4,399	14,780	10,110
Total exports[7]					
Current dollars (mil.)	NA	NA	3,691	8,471	9,572
1994 constant dollars (mil.)	NA	NA	3,849	8,646	9,572
Arms as percent of total imports[8]	NA	NA	0	0	0
Arms as percent of total exports[8]	NA	NA	0	0.6	0.6

Human Rights [24]

	SSTS	FL	FAPRO	PPCG	APROBC	TPW	PCPTW	STPEP	PHRFF	PRW	ASST	AFL
Observes	P	P	P	P	P	P	P	P	S	P	P	P
	EAFRD	CPR	ESCR	SR	ACHR	MAAE	PVIAC	PVNAC	EAFDAW	TCIDTP	RC	
Observes		P	P	P			P	P	P	P	P	

P = Party; S = Signatory; see Appendix for meaning of abbreviations.

Labor Force

Total Labor Force [25]

23.55 million (January 1994)

Labor Force by Occupation [26]

Industry and construction	33%
Agriculture and forestry	21
Health, education, and culture	16
Trade and distribution	7
Transport and communication	7
Other	16

Date of data: 1992

Unemployment Rate [27]

0.7% officially registered; large number of unregistered or underemployed workers (December 1995)

Production Sectors

Commercial Energy Production and Consumption

Data are shown in quadrillion (10^{15}) BTUs and percent for 1995
Values for hydroelectric, nuclear, geothermal, solar, and wind power refer to electrical generation.

Production [28]

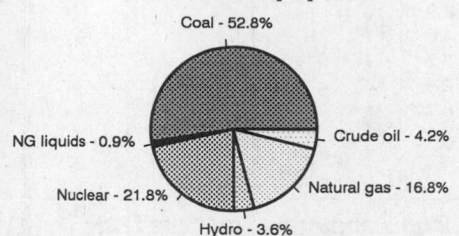

Coal - 52.8%
NG liquids - 0.9%
Crude oil - 4.2%
Nuclear - 21.8%
Natural gas - 16.8%
Hydro - 3.6%

Consumption [29]

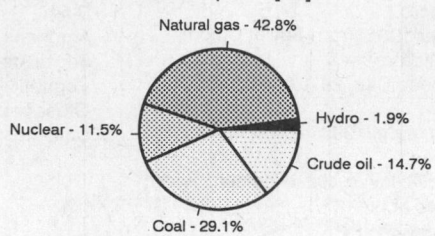

Natural gas - 42.8%
Nuclear - 11.5%
Hydro - 1.9%
Crude oil - 14.7%
Coal - 29.1%

Crude oil	0.140
Natural gas liquids	0.030
Dry natural gas	0.561
Coal	1.765
Net hydroelectric power	0.119
Net nuclear power	0.728
Total	3.343

Crude oil	0.928
Dry natural gas	2.707
Coal	1.842
Net hydroelectric power	0.119
Net nuclear power	0.728
Total	6.324

Telecommunications [30]

- System unsatisfactory both for business and personal use; 3.56 million applications for telephones not satisfied as of January 1991; electronic mail services established in Kiev, Odessa, and Luhans'k by Sprint
- Domestic: an NMT-450 analog cellular telephone network operates in Kiev, allows international direct dialing through Kiev's digital exchange
- International: calls to other CIS countries carried by landline or microwave radio relay; calls to 167 other countries carried by satellite or by 150 leased lines through the Moscow international gateway switch; satellite earth stations - Intelsat, 1 Inmarsat (Atlantic and Indian Ocean Regions), and Intersputnik
- Radio: Broadcast stations: there are at least two radio broadcast stations Radios: 15 million (1990)
- Television: Broadcast stations: at least 2 Televisions: 17.3 million (1992)

Top Agricultural Products [32]

Agriculture accounts for 31% of the GDP; produces grain, sugar beets, vegetables; meat, milk.

Top Mining Products [33]

Thousand metric tons except as noted[e,r]	7/21/95*
Iron ore	51,300
Manganese, marketable ore	2,979
Pig iron	20,120
Steel, crude	23,798
Cement	18,000
Salt	3,940
Coal	95,300
Coke	17,000
Gas, natural (000 cu. meters)	18,300,000
Petroleum, crude	4,200

Transportation [31]

Railways: total: 23,350 km; broad gauge: 23,350 km 1.524-m gauge (8,600 km electrified)

Highways: total: 169,964 km; paved: 168,094 km (including 1,767 km of expressways); unpaved: 1,870 km (1992 est.)

Merchant marine: total: 353 ships (1,000 GRT or over) totaling 3,262,341 GRT/4,356,374 DWT; ships by type: barge carrier 5, bulk 39, cargo 217, chemical tanker 2, combination bulk 1, container 11, multifunction large-load carrier 3, oil tanker 21, passenger 7, passenger-cargo 5, railcar carrier 2, refrigerated cargo 5, roll-on/roll-off cargo 32, short-sea passenger 3 (1995 est.)

Airports

Total:	706
With paved runways over 3,047 m:	14
With paved runways 2,438 to 3,047 m:	55
With paved runways 1,524 to 2,437 m:	34
With paved runways 914 to 1,523 m:	3
With paved runways under 914 m:	57

Tourism [34]

For sources, notes, and explanations, see Annotated Source Appendix, page 1061.

Finance, Economics, and Trade

Industrial Summary [35]

Industrial Production: Growth rate - 11% (1995 est.); accounts for 43% of the GDP. **Industries**: Coal, electric power, ferrous and nonferrous metals, machinery and transport equipment, chemicals, food-processing (especially sugar).

Economic Indicators [36]

- **National product**: GDP—purchasing power parity— $174.6 billion (1995 estimate as extrapolated from World Bank estimate for 1994)

- **National product real growth rate**: - 4% (1995 est.)

- **National product per capita**: $3,370 (1995 est.)

- **Inflation rate (consumer prices)**: 9% monthly average (1995)

- **External debt**: $8.8 billion (including $4.5 billion to Russia) (late 1995 est.)

Balance of Payments Summary [37]

Values in millions of dollars.

	1989	1990	1991	1992	1993[1]
Exports of goods (f.o.b.)	NA	NA	NA	NA	10,862
Imports of goods (f.o.b.)	NA	NA	NA	NA	12,602
Trade balance	NA	NA	NA	NA	-1,740
Services - debits	NA	NA	NA	NA	-1,756
Services - credits	NA	NA	NA	NA	2,623
Private transfers (net)	NA	NA	NA	NA	107
Government transfers (net)	NA	NA	NA	NA	15
Long-term capital (net)	NA	NA	NA	NA	1,639
Short-term capital (net)	NA	NA	NA	NA	NA
Errors and omissions	NA	NA	NA	NA	756
Overall balance	NA	NA	NA	NA	1,644

Exchange Rates [38]

Currency: **hryvnia.**

Data are currency units per $1. On 2 September 1996, Ukraine introduced the hryvnia as its national currency, replacing the karbovanets (in circulation since 12 November 1992) at a rate of 100,000 karbovantsi to 1 hryvnia.

2 September 1996	1.76

Imports and Exports

Top Import Origins [39]

$10.7 billion (1995).

Origins	%
Other FSU countries	NA
Germany	NA
Poland	NA
Czech Republic	NA

Top Export Destinations [40]

$11.3 billion (1995).

Destinations	%
Other FSU countries	NA
China	NA
Italy	NA
Switzerland	NA

Foreign Aid [41]

	U.S. $	
ODA (1993)	220	million
Commitments (1992-95)	4.5	billion
of which drawn	4.1	billion

Import and Export Commodities [42]

Import Commodities

Energy
Machinery and parts
Transportation equipment
Chemicals
Textiles

Export Commodities

Coal
Electric power
Ferrous and nonferrous metals
Chemicals
Machinery and transport equipment
Grain
Meat

For sources, notes, and explanations, see Annotated Source Appendix, page 1061.

981

United Arab Emirates

Geography [1]

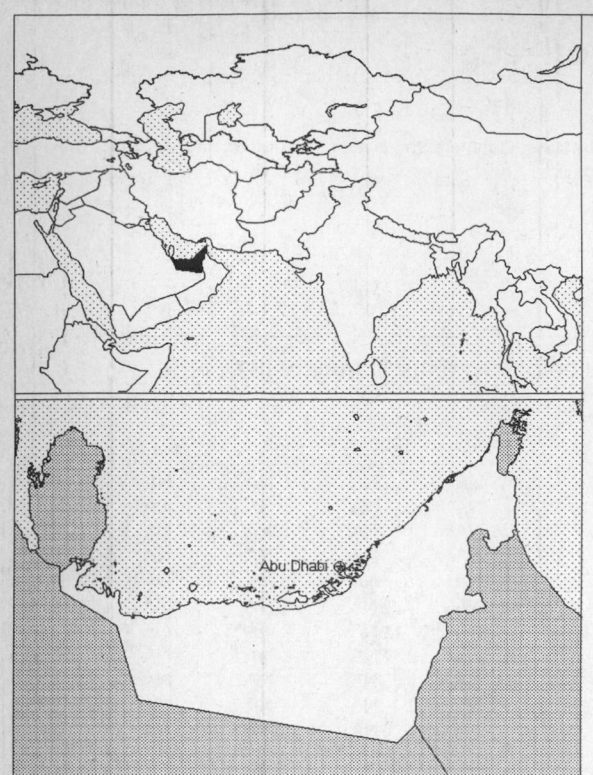

Total area:
75,581 sq km   29,182 sq mi
Land area:
75,581 sq km   29,182 sq mi
Comparative area:
Slightly smaller than Maine
Land boundaries:
Total 867 km, Oman 410 km, Saudi Arabia 457 km
Coastline:
1,318 km
Climate:
Desert; cooler in eastern mountains
Terrain:
Flat, barren coastal plain merging into rolling sand dunes of vast
desert wasteland; mountains in East
Natural resources:
Petroleum, natural gas
Land use:
Arable land: 0%
Permanent crops: 0%
Meadows and pastures: 2%
Forest and woodland: 0%
Other: 98%

Demographics [2]

	1970	1980	1990	1995[1]	1996	2000	2010	2020	2030
Population	249	1,000	2,252	2,925	3,057	3,582	4,873	6,080	7,037
Population density (persons per sq. mi.)	8	31	70	91	95	111	151	188	218
(persons per sq. km.)	3	12	27	35	37	43	58	73	84
Net migration rate (per 1,000 population)	NA	NA	32.6	21.5	19.9	14.7	6.5	1.7	0.0
Births	NA	NA	NA	NA	81	NA	NA	NA	NA
Deaths	NA	NA	NA	NA	9	NA	NA	NA	NA
Life expectancy - males	NA	NA	69.2	70.4	70.6	71.6	73.6	75.2	76.5
Life expectancy - females	NA	NA	73.5	74.7	74.9	75.9	77.9	79.7	81.1
Birth rate (per 1,000)	NA	NA	30.7	27.0	26.4	24.8	23.5	21.2	18.8
Death rate (per 1,000)	NA	NA	3.2	3.0	3.0	3.1	3.8	4.9	6.4
Women of reproductive age (15-49 yrs.)	NA	NA	433	593	626	760	1,053	1,342	1,645
of which are currently married	NA	NA	336	NA	464	549	740	NA	NA
Fertility rate	NA	NA	4.9	4.5	4.5	4.2	3.5	3.0	2.6

Except as noted, values for vital statistics are in thousands; life expectancy is in years.

Health

Health Indicators [3]

% of population with access to	
safe water (1990-95)	95
adequate sanitation (1990-95)	77
health services (1985-95)	99
% of 1-year-olds immunized (1990-94) against	
TB (tuberculosis)	98
DPT (diphtheria, pertussis, tetanus)	90
polio	90
measles	90
% of contraceptive prevalence (1980-94)	NA
ORT use rate (1990-94)	81

Health Expenditures [4]

Human Factors

Human Factors

Women and Children [5]

% of pregnant women immunized (tetanus 1990-94)	NA
% of births attended by trained health personnel (1983-94)	99
Maternal mortality rate (1980-92)	NA
Under-5 mortality rate (1994)	20
% under-5 moderately/severely underweight (1980-1994)	NA

Burden of Disease [6]

Population per physician (1990)	1,095.02
Population per nurse (1990)	576.80
Population per hospital bed	NA
AIDS cases per 100,000 people (1994)	*
Malaria cases per 100,000 people (1992)	204

Ethnic Division [7]

Note: Less than 20% of population are UAE citizens (1982).

South Asian	50%
Other Arab and Iranian	23%
Emiri	19%
Other expatriates (primarily Westerners and East Asians)	8%

Religion [8]

Muslim (Shi'a 16%)	96%
Christian, Hindu, and other	4%

Major Languages [9]

Arabic (official), Persian, English, Hindi, Urdu.

Education

Public Education Expenditures [10]

Million (Dirham)	1980	1985	1990	1992	1993	1994
Total education expenditure	1,460	1,738	2,280	2,637	2,657	2,927
as percent of GNP	1.3	1.7	1.7	2.0	2.0	NA
as percent of total govt. expend.	NA	10.4	14.6	15.2	15.1	16.3
Current education expenditure	1,153	1,637	2,174	2,460	2,457	2,702
as percent of GNP	1.0	1.6	1.7	1.8	1.8	NA
as percent of current govt. expend.	NA	10.6	14.3	15.1	14.9	15.9
Capital expenditure	307	101	106	177	200	225

Educational Attainment [11]

Illiteracy [12]

In thousands and percent[1]	1990	1995	2000
Illiterate population (15+ yrs.)	270	272	269
Illiteracy rate - total pop. (%)	18.3	14.3	11.3
Illiteracy rate - males (%)	18.8	15.4	12.7
Illiteracy rate - females (%)	17.3	12.2	9.0

Libraries [13]

	Admin. Units	Svc. Pts.	Vols. (000)	Shelving (meters)	Vols. Added	Reg. Users
National	NA	NA	NA	NA	NA	NA
Nonspecialized	NA	NA	NA	NA	NA	NA
Public	NA	NA	NA	NA	NA	NA
Higher ed. (1993)	21	65	4,844	113,849	98,664	483,611
School (1990)	290	290	667	17,300	53,000	NA

Daily Newspapers [14]

	1980	1985	1990	1994
Number of papers	9	13	8	8
Circ. (000)	152	290	250	300

Culture [15]

Cinema (seats per 1,000)	18.6
Annual attendance per person	NA
Gross box office receipts (mil. Dirham)	NA
Museums (reporting)	NA
Visitors (000)	NA
Annual receipts (000 Dirham)	NA

Science and Technology

Scientific/Technical Forces [16]

R&D Expenditures [17]

U.S. Patents Issued [18]

Values show patents issued to citizens of the country by the U.S. Patents Office.

	1993	1994	1995
Number of patents	1	1	1

For sources, notes, and explanations, see Annotated Source Appendix, page 1061.

983

Government and Law

Organization of Government [19]

Long-form name:
United Arab Emirates
Type:
Federation with specified powers delegated to the UAE central government and other powers reserved to member emirates
Independence:
2 December 1971 (from UK)
National holiday:
National Day, 2 December (1971)
Constitution:
2 December 1971 (provisional)
Legal system:
Federal court system introduced in 1971; all emirates except Dubayy (Dubai) and Ra's al Khaymah have joined the federal system; all emirates have secular and Islamic law for civil, criminal, and high courts
Executive branch:
President; Vice President; Prime Minister; 2 Deputy Prime Ministers; Supreme Council of Rulers; Council of Ministers
Legislative branch:
Unicameral: Federal National Council - not elected, only reviews legislation
Judicial branch:
Union Supreme Court

Crime [23]

Elections [20]

No political parties; no elections.

Government Expenditures [21]

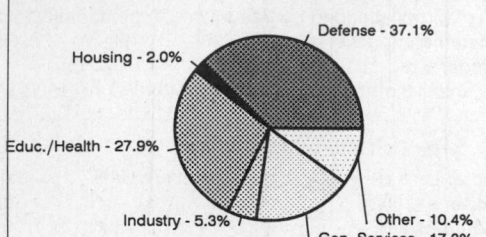

Defense - 37.1%
Housing - 2.0%
Educ./Health - 27.9%
Industry - 5.3%
Gen. Services - 17.3%
Other - 10.4%

(% distribution). Expend. for CY94: 15,693 (Dirham mil.)

Military Expenditures and Arms Transfers [22]

	1990	1991	1992	1993	1994
Military expenditures					
Current dollars (mil.)	2,572[e]	4,867[e]	2,083	2,110	1,907
1994 constant dollars (mil.)	2,863[e]	5,216[e]	2,172	2,154	1,907
Armed forces (000)	66	66	55	55	60
Gross national product (GNP)					
Current dollars (mil.)[e]	35,480	35,740	36,450	36,700	NA
1994 constant dollars (mil.)[e]	39,480	38,310	38,010	37,460	NA
Central government expenditures (CGE)					
1994 constant dollars (mil.)	4,349	4,422	4,393	4,300	NA
People (mil.)	2.3	2.4	2.5	2.7	2.8
Military expenditure as % of GNP	7.3	13.6	5.7	5.7	NA
Military expenditure as % of CGE	65.8	118.0	49.5	50.1	NA
Military expenditure per capita (1994 $)	1,271	2,185	861	811	NA
Armed forces per 1,000 people (soldiers)	29.3	27.6	21.8	20.7	21.5
GNP per capita (1994 $)	17,530	16,050	15,070	14,100	NA
Arms imports[6]					
Current dollars (mil.)	1,400	370	360	430	200
1994 constant dollars (mil.)	1,558	397	375	439	200
Arms exports[6]					
Current dollars (mil.)	0	0	0	0	0
1994 constant dollars (mil.)	0	0	0	0	0
Total imports[7]					
Current dollars (mil.)	11,200	13,750	17,410	19,520	20,000
1994 constant dollars (mil.)	12,460	14,730	18,160	19,920	20,000
Total exports[7]					
Current dollars (mil.)	23,540	24,440	24,760	23,660[e]	24,000[e]
1994 constant dollars (mil.)	26,200	26,190	25,820	24,150[e]	24,000[e]
Arms as percent of total imports[8]	12.5	2.7	2.1	2.2	1.0
Arms as percent of total exports[8]	0	0	0	0	0

Human Rights [24]

	SSTS	FL	FAPRO	PPCG	APROBC	TPW	PCPTW	STPEP	PHRFF	PRW	ASST	AFL
Observes		P				P	P					
	EAFRD	CPR	ESCR	SR	ACHR	MAAE	PVIAC	PVNAC	EAFDAW	TCIDTP	RC	
Observes	P						P	P				

P = Party; S = Signatory; see Appendix for meaning of abbreviations.

Labor Force

Total Labor Force [25]

794,400 (1993 est.)

Labor Force by Occupation [26]

Industry and commerce	56%
Services	38
Agriculture	6

Date of data: 1990 est.

Unemployment Rate [27]

Negligible (1988)

For sources, notes, and explanations, see Annotated Source Appendix, page 1061.

Production Sectors

Commercial Energy Production and Consumption

Data are shown in quadrillion (10^{15}) BTUs and percent for 1995
Values for hydroelectric, nuclear, geothermal, solar, and wind power refer to electrical generation.

Production [28]

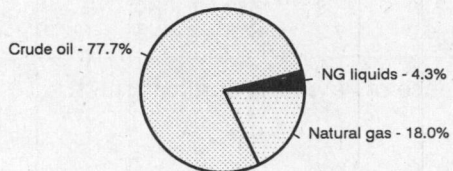

Crude oil - 77.7%
NG liquids - 4.3%
Natural gas - 18.0%

Consumption [29]

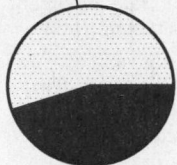

Natural gas - 55.4%
Crude oil - 44.6%

Crude oil	4.816
Natural gas liquids	0.265
Dry natural gas	1.114
Total	6.195

Crude oil	0.704
Dry natural gas	0.873
Total	1.577

Telecommunications [30]

- 677,793 (1993 est.) telephones; modern system consisting of microwave radio relay and coaxial cable; key centers are Abu Dhabi and Dubai
- Domestic: microwave radio relay and coaxial cable
- International: satellite earth stations - 3 Intelsat (1 Atlantic Ocean and 2 Indian Ocean) and 1 Arabsat; submarine cables to Qatar, Bahrain, India, and Pakistan; tropospheric scatter to Bahrain; microwave radio relay to Saudi Arabia
- Radio: Broadcast stations: AM 8, FM 3, shortwave 0 Radios: 545,000 (1992 est.)
- Television: Broadcast stations: 12 Televisions: 170,000 (1993 est.)

Top Agricultural Products [32]

Agriculture accounts for 2% of the GDP; produces dates, vegetables, watermelons; poultry, eggs, dairy products; fish.

Transportation [31]

Railways: 0 km

Highways: total: 3,000 km; paved: 3,000 km; unpaved: 0 km (1993 est.)

Merchant marine: total: 57 ships (1,000 GRT or over) totaling 1,068,980 GRT/1,876,504 DWT; ships by type: bulk 2, cargo 17, chemical tanker 2, container 7, liquefied gas tanker 1, livestock carrier 1, oil tanker 22, refrigerated cargo 2, roll-on/roll-off cargo 3 (1995 est.)

Airports
Total:	36
With paved runways over 3,047 m:	9
With paved runways 2,438 to 3,047 m:	3
With paved runways 1,524 to 2,437 m:	2
With paved runways 914 to 1,523 m:	3
With paved runways under 914 m:	10

Top Mining Products [33]

Thousand metric tons except as noted[e]	6/22/95[*]
Aluminum, metal, primary ingot	247
Cement, hydraulic	3,800
Ammonia, N content	243
Urea, N content	250
Gas, natural, gross (mil. cu. meters)	37,000
Gypsum	100
Natural gas plant liquids (000 42-gal. bls.)	60,000
Petroleum (000 42-gal. bls.)	
crude	805,000[75]
refinery products	77,000

Tourism [34]

	1987	1988	1989	1990	1991
Tourists[107]	543	598	629	633	717

Travelers are in thousands, money in million U.S. dollars.

For sources, notes, and explanations, see Annotated Source Appendix, page 1061.

985

Finance, Economics, and Trade

GDP and Manufacturing Summary [35]

	1980	1985	1990	1991	1992
Gross Domestic Product					
Millions of 1980 dollars	29,629	27,036	30,375	32,654	33,796
Growth rate in percent	26.42	-2.39	17.75	7.50	3.50
Manufacturing Value Added					
Millions of 1980 dollars	1,131	2,547	2,400	2,636	2,910[e]
Growth rate in percent	64.87	-2.20	5.38	9.83	10.40[e]
Manufacturing share in percent of current prices	3.7	9.0	7.2	7.5[e]	NA

Economic Indicators [36]

- **National product**: GDP—purchasing power parity—$70.1 billion (1995 est.)
- **National product real growth rate**: 3.3% (1995 est.)
- **National product per capita**: $24,000 (1995 est.)
- **Inflation rate (consumer prices)**: 4.6% (1994 est.)
- **External debt**: $11.6 billion (1994 est.)

Balance of Payments Summary [37]

Exchange Rates [38]

Currency: **Emirian dirham.**
Symbol: **Dh.**

Data are currency units per $1.

Fixed rate   3.6710

Imports and Exports

Top Import Origins [39]

$21.7 billion (f.o.b., 1994) Data are for 1994.

Origins	%
Japan	11
UK	8
Germany	8
US	8
Italy	7

Top Export Destinations [40]

$25.3 billion (f.o.b., 1994 est.) Data are for 1994.

Destinations	%
Japan	45
India	6
Oman	6
South Korea	5
Iran	5

Foreign Aid [41]

Donor: Pledged in bilateral aid to less developed countries (1979-89) $9.1 billion.

Import and Export Commodities [42]

Import Commodities	Export Commodities
Manufactured goods	Crude oil 66%
Machinery and transport equipment	Natural gas
Food	Re-exports
	Dried fish
	Dates

United Kingdom

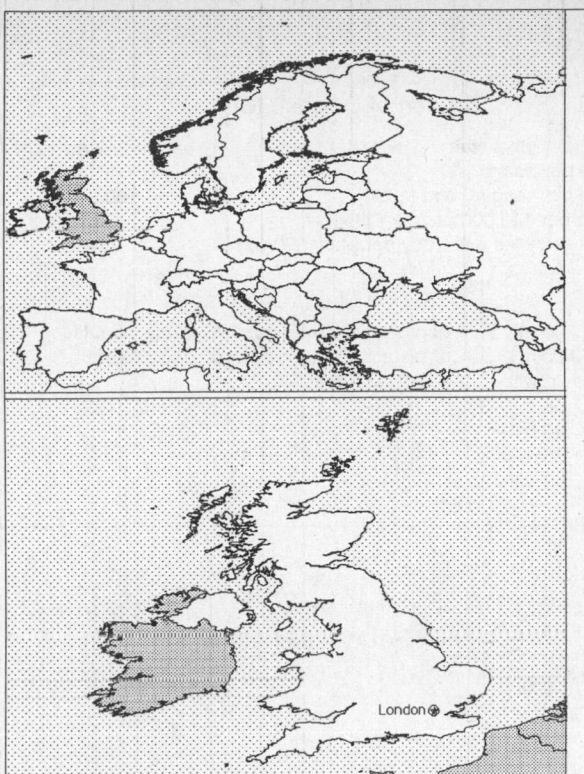

Geography [1]

Total area:
244,820 sq km   94,526 sq mi
Land area:
241,590 sq km   93,278 sq mi
Comparative area:
Slightly smaller than Oregon
Note: Includes Rockall and Shetland Islands
Land boundaries:
Total 360 km, Ireland 360 km
Coastline:
12,429 km
Climate:
Temperate; moderated by prevailing Southwest winds over the North Atlantic Current; more than one-half of the days are overcast
Terrain:
Mostly rugged hills and low mountains; level to rolling plains in East and Southeast
Natural resources:
Coal, petroleum, natural gas, tin, limestone, iron ore, salt, clay, chalk, gypsum, lead, silica
Land use:
Arable land: 29%
Permanent crops: 0%
Meadows and pastures: 48%
Forest and woodland: 9%
Other: 14%

Demographics [2]

	1970	1980	1990	1995[1]	1996	2000	2010	2020	2030
Population	55,632	56,314	57,418	58,361	58,490	58,894	59,159	59,289	58,846
Population density (persons per sq. mi.)	596	604	616	626	627	631	634	636	631
(persons per sq. km.)	230	233	238	242	242	244	245	245	244
Net migration rate (per 1,000 population)	NA	NA	NA	0.2	0.3	0.1	0.0	0.0	0.0
Births	904	754	NA	NA	767	NA	NA	NA	NA
Deaths	652	662	386	NA	657	NA	NA	NA	NA
Life expectancy - males	NA	NA	NA	73.6	73.8	74.5	76.1	78.1	79.5
Life expectancy - females	NA	NA	NA	79.0	79.2	79.9	81.4	83.8	85.4
Birth rate (per 1,000)	NA	NA	NA	13.4	13.1	12.1	10.6	10.4	9.3
Death rate (per 1,000)	NA	NA	NA	11.3	11.2	10.9	10.6	10.3	11.3
Women of reproductive age (15-49 yrs.)	NA	NA	NA	14,160	14,144	13,864	13,653	12,563	11,988
of which are currently married[58]	NA	NA	NA	NA	8,470	8,338	8,052	NA	NA
Fertility rate	NA	NA	NA	1.8	1.8	1.8	1.7	1.7	1.6

Except as noted, values for vital statistics are in thousands; life expectancy is in years.

Health

Health Indicators [3]

% of population with access to	
safe water (1990-95)	NA
adequate sanitation (1990-95)	NA
health services (1985-95)	NA
% of 1-year-olds immunized (1990-94) against	
TB (tuberculosis)	NA
DPT (diphtheria, pertussis, tetanus)	91
polio	93
measles	92
% of contraceptive prevalence (1980-94)	72
ORT use rate (1990-94)	NA

Health Expenditures [4]

Total health expenditure, 1990 (official exchange rate)	
Millions of dollars	59,623
Dollars per capita	1,039
Health expenditures as a percentage of GDP	
Total	6.1
Public sector	5.2
Private sector	0.9
Development assistance for health	
Total aid flows (millions of dollars)[1]	NA
Aid flows per capita (dollars)	NA
Aid flows as a percentage of total health expenditure	NA

For sources, notes, and explanations, see Annotated Source Appendix, page 1061.

987

Human Factors

Women and Children [5]

% of pregnant women immunized (tetanus 1990-94)	NA
% of births attended by trained health personnel (1983-94)[1]	100
Maternal mortality rate (1980-92)	8
Under-5 mortality rate (1994)	7
% under-5 moderately/severely underweight (1980-1994)	NA

Burden of Disease [6]

Population per physician	NA
Population per nurse	NA
Population per hospital bed (1990)	160.91
AIDS cases per 100,000 people (1994)	2.7
Heart disease cases per 1,000 people (1990-93)	NA

Ethnic Division [7]

English	81.5%
Scottish	9.6%
West Indian, Indian, Pakistani, and other	2.8%
Irish	2.4%
Welsh	1.9%
Ulster	1.8%

Religion [8]

Anglican	27.00 mil
Roman Catholic	9.00 mil
Muslim	1.00 mil
Other Protestant	1.56 mil
Sikh	.40 mil
Hindu	.35 mil
Jewish	.30 mil
(1991 est.)	

Major Languages [9]

English, Welsh (about 26% of the population of Wales), Scottish form of Gaelic (about 60,000 in Scotland).

Education

Public Education Expenditures [10]

	1980	1985	1990	1991	1992	1994
Million (Pound Sterling)						
Total education expenditure	12,856	17,501	26,677	29,534	32,162	NA
as percent of GNP	5.6	4.9	4.9	5.2	5.4	NA
as percent of total govt. expend.	13.9	NA	NA	NA	11.2	NA
Current education expenditure	12,094	16,764	25,318	28,045	NA	NA
as percent of GNP	5.2	4.7	4.7	5.0	NA	NA
as percent of current govt. expend.	NA	NA	NA	NA	NA	NA
Capital expenditure	762	737	1,359	1,489	NA	NA

Educational Attainment [11]

Illiteracy [12]

Libraries [13]

	Admin. Units	Svc. Pts.	Vols. (000)	Shelving (meters)	Vols. Added	Reg. Users
National (1990)[26]	3	23	27,500	790,000	542,606	82,945
Nonspecialized (1990)	26[e]	23[e]	8,026[e]	214,295[e]	NA	316,460[e]
Public (1993)	167	5,185	133,134	NA	13 mil	34 mil[e]
Higher ed.	215	860[e]	89,832	2 mil	2 mil	1 mil
School	NA	NA	NA	NA	NA	NA

Daily Newspapers [14]

	1980	1985	1990	1994
Number of papers	113	104	103[e]	103
Circ. (000)	23,472	22,495	22,350[e]	20,372

Culture [15]

Cinema (seats per 1,000)	NA
Annual attendance per person	2.0
Gross box office receipts (mil. Pound Sterling)	343
Museums (reporting)	NA
Visitors (000)	NA
Annual receipts (000 Pound Sterling)	NA

Science and Technology

Scientific/Technical Forces [16]

Scientists/engineers	140,000
Number female	NA
Technicians	59,000
Number female	NA
Total	199,000

R&D Expenditures [17]

	Pound (000) 1993
Total expenditure	13,829,000
Capital expenditure	NA
Current expenditure	NA
Percent current	NA

U.S. Patents Issued [18]

Values show patents issued to citizens of the country by the U.S. Patents Office.

	1993	1994	1995
Number of patents	2,520	2,469	2,678

For sources, notes, and explanations, see Annotated Source Appendix, page 1061.

Government and Law

Organization of Government [19]

Long-form name:
United Kingdom of Great Britain and Northern Ireland
Type:
Constitutional monarchy
Independence:
1 January 1801 (United Kingdom established)
National holiday:
Celebration of the Birthday of the Queen (second Saturday in June)
Constitution:
Unwritten; partly statutes, partly common law and practice
Legal system:
Common law tradition with early Roman and modern continental influences; no judicial review of Acts of Parliament; accepts compulsory ICJ jurisdiction, with reservations
Executive branch:
Queen; Heir Apparent Prince; Prime Minister; Cabinet of Ministers
Legislative branch:
Bicameral Parliament: House of Lords and House of Commons
Judicial branch:
House of Lords

Elections [20]

House of Commons	% of votes
Conservative	41.9
Labor	34.5
Liberal Democratic	17.9
Other	5.7

Government Expenditures [21]

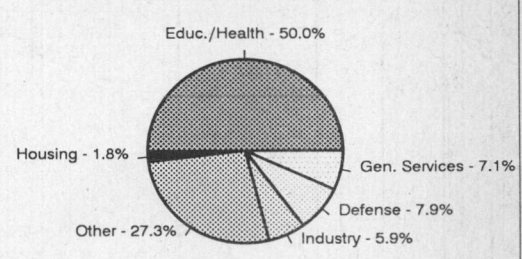

Educ./Health - 50.0%
Housing - 1.8%
Gen. Services - 7.1%
Defense - 7.9%
Industry - 5.9%
Other - 27.3%

(% distribution). Expend. for CY95: 292,652 (Pound mil.)

Crime [23]

	1990
Crime volume	
Cases known to police	5,559,872
Attempts (percent)	NA
Percent cases solved	NA
Crimes per 100,000 persons	31,414.06
Persons responsible for offenses	
Total number offenders	683,310
Percent female	NA
Percent juvenile	NA
Percent foreigners	NA

Military Expenditures and Arms Transfers [22]

	1990	1991	1992	1993	1994
Military expenditures					
Current dollars (mil.)	36,300	38,700	35,750	35,080	34,050
1994 constant dollars (mil.)	40,390	41,490	37,280	35,810	34,050
Armed forces (000)	308	301	293	271	257
Gross national product (GNP)					
Current dollars (bil.)	883.3	898.7	928.7	969.6	1,027.0
1994 constant dollars (bil.)	983.0	963.3	968.5	989.5	1,027.0
Central government expenditures (CGE)					
1994 constant dollars (mil.)	362,000	376,900	401,600	414,300	423,900[e]
People (mil.)	57.4	57.6	57.8	58.0	58.1
Military expenditure as % of GNP	4.1	4.3	3.8	3.6	3.3
Military expenditure as % of CGE	11.2	11.0	9.3	8.6	8.0
Military expenditure per capita (1994 $)	704	720	645	618	586
Armed forces per 1,000 people (soldiers)	5.4	5.2	5.1	4.7	4.4
GNP per capita (1994 $)	17,120	16,720	16,760	17,070	17,670
Arms imports[6]					
Current dollars (mil.)	1,300	650	220	290	200
1994 constant dollars (mil.)	1,447	697	229	296	200
Arms exports[6]					
Current dollars (mil.)	4,600	4,900	4,600	4,400	3,400
1994 constant dollars (mil.)	5,119	5,252	4,797	4,491	3,400
Total imports[7]					
Current dollars (mil.)	223,000	209,900	221,600	205,400	226,800
1994 constant dollars (mil.)	248,200	225,000	231,000	209,600	226,800
Total exports[7]					
Current dollars (mil.)	185,200	185,000	190,000	180,200	204,500
1994 constant dollars (mil.)	206,100	198,300	198,100	183,900	204,500
Arms as percent of total imports[8]	0.6	0.3	0.1	0.1	0.1
Arms as percent of total exports[8]	2.5	2.6	2.4	2.4	1.7

Human Rights [24]

	SSTS	FL	FAPRO	PPCG	APROBC	TPW	PCPTW	STPEP	PHRFF	PRW	ASST	AFL
Observes	P	P	P	P	P	P	P		P	P	P	P
		EAFRD	CPR	ESCR	SR	ACHR	MAAE	PVIAC	PVNAC	EAFDAW	TCIDTP	RC
Observes		P	P	P	P			S	S	P	P	P

P=Party; S=Signatory; see Appendix for meaning of abbreviations.

Labor Force

Total Labor Force [25]

28.048 million

Labor Force by Occupation [26]

Services	62.8%
Manufacturing and construction	25.0
Government	9.1
Energy	1.9
Agriculture	1.2

Date of data: June 1992

Unemployment Rate [27]

8% (December 1995)

Production Sectors

Commercial Energy Production and Consumption

Data are shown in quadrillion (10^{15}) BTUs and percent for 1995
Values for hydroelectric, nuclear, geothermal, solar, and wind power refer to electrical generation.

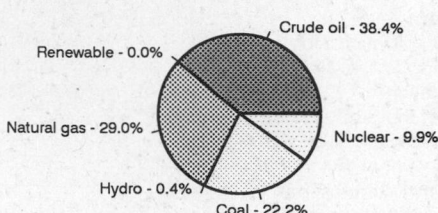

Production [28]
- Crude oil - 49.9%
- Hydro - 0.4%
- Renewable - 0.0%
- Nuclear - 9.0%
- Natural gas - 26.2%
- Coal - 10.3%
- NG liquids - 4.1%

Consumption [29]
- Crude oil - 38.4%
- Renewable - 0.0%
- Nuclear - 9.9%
- Natural gas - 29.0%
- Hydro - 0.4%
- Coal - 22.2%

Production	
Crude oil	5.272
Natural gas liquids	0.437
Dry natural gas	2.775
Coal	1.093
Net hydroelectric power	0.042
Net nuclear power	0.953
Geothermal, solar, wind	0.003
Total	10.575

Consumption	
Crude oil	3.716
Dry natural gas	2.806
Coal	2.147
Net hydroelectric power	0.042
Net nuclear power	0.953
Geothermal, solar, wind	0.003
Total	9.667

Telecommunications [30]

- 29.5 million (1987 est.) telephones; technologically advanced domestic and international system
- Domestic: equal mix of buried cables, microwave radio relay, and fiber-optic systems
- International: 40 coaxial submarine cables; satellite earth stations - 10 Intelsat (7 Atlantic Ocean and 3 Indian Ocean), 1 Inmarsat (Atlantic Ocean region), and 1 Eutelsat; at least 8 large international switching centers
- Radio: Broadcast stations: AM 225, FM 525 (mostly repeaters), shortwave 0 Radios: 70 million
- Television: Broadcast stations: 207 (repeaters 3,210) Televisions: 20 million

Top Agricultural Products [32]

Agriculture accounts for 1.7% of the GDP; produces cereals, oilseed, potatoes, vegetables; cattle, sheep, poultry; fish.

Top Mining Products [33]

Thousand metric tons except as noted[e]	3/95[*]
Steel, crude	17,400[r]
Iron and steel rolled products	14,000
Common sand and gravel	90,000
Dolomite	18,000
Igneous rock	50,000
Limestone	90,000
Coal, anthracite, bituminous, and lignite	57,200[r]
Gas, natural, marketable (mil. cu. meters)	69,700[r,76]
Natrual gas liquids (000 42-gal. bls.)	53,200[r,77]
Petroleum, crude (000 42-gal. bls.)	858,000[78]

Transportation [31]

Railways: total: 17,561 km; broad gauge: 434 km 1.600-m gauge (190 km double track); note - all 1.600-m gauge track, of which 357 km is in common carrier use, is in Northern Ireland; standard gauge: 16,892 km 1.435-m gauge (4,928 km electrified; 12,591 km double or multiple track); note - 16,532 km of 1.435-m routes are in common carrier service; the remaining 360 km are operated by a total of 40 tourist or other private companies; narrow gauge: 235 km 0.260-m, 0.311-m, 0.381-m, 0.600-m, 0.610-m, 0.686-m, 0.760-m, 0.762-m, 0.800-m, 0.825-m, 0.914-m and 1.067-m gauges; note - these short, narrow-gage lines are operated by a total of 25 tourist and other private firms (1995)

Highways: total: 386,243 km (1993 est.); paved: NA km (including 3,237 km of expressways in Great Britain); unpaved: NA km

Merchant marine: total: 151 ships (1,000 GRT or over) totaling 3,191,969 GRT/3,861,239 DWT; ships by type: bulk 10, cargo 21, chemical tanker 2, container 24, liquefied gas tanker 2, oil tanker 56, passenger 8, passenger-cargo 1, roll-on/roll-off cargo 12, short-sea passenger 14, specialized tanker 1 (1995 est.)

Airports

Total:	388
With paved runways over 3,047 m:	9
With paved runways 2,438 to 3,047 m:	29
With paved runways 1,524 to 2,437 m:	103
With paved runways 914 to 1,523 m:	59
With paved runways under 914 m:	166

Tourism [34]

	1990	1991	1992	1993	1994
Visitors[1]	18,013	17,125	18,535	19,398	21,034
Excursionists[1]	990	1,150	1,280	1,350	1,750
Tourism receipts	14,940	13,070	13,932	14,031	15,176
Tourism expenditures	19,063	17,609	19,850	19,058	22,185
Fare receipts	3,866	3,173	3,727	3,573	3,883
Fare expenditures	4,676	4,052	4,606	4,325	5,432

Travelers are in thousands, money in million U.S. dollars.

For sources, notes, and explanations, see Annotated Source Appendix, page 1061.

Manufacturing Sector

Manufacturing Summary [35]

	1987 $ bil.	1987 %	1988 $ bil.	1988 %	1989 $ bil.	1989 %	1990 $ bil.	1990 %	1991 $ bil.	1991 %
Establishments or enterprises (number)	157,815	6.727	160,200	7.631	164,082	8.759	155,303	8.670	-	-
Total employment (000)	5,808	4.280	5,874	4.292	5,880	4.777	5,749	5.194	-	-
Production workers (000)	3,890	5.768	3,928	6.287	3,930	5.342	3,809	5.367	-	-
Output ($ bil.)	528	5.193	636	5.456	642	5.444	723	6.302	-	-
Value added ($ bil.)	230	5.337	279	5.813	280	5.751	314	6.302	-	-
Capital investment ($ mil.)	22,702	5.209	31,025	6.515	33,425	6.102	36,620	6.589	-	-
M & E investment ($ mil.)	19,488	6.254	26,644	7.698	28,255	7.237	30,770	7.299	-	-
Employees per establishment (number)	37	63.614	37	56.242	36	54.539	37	59.911	-	-
Production workers per establishment	25	85.733	25	82.384	24	60.986	25	61.901	-	-
Output per establishment ($ mil.)	3	77.200	4	71.492	4	62.146	5	72.689	-	-
Capital investment per estab. ($ mil.)	0.144	77.427	0.194	85.369	0.204	69.663	0.236	76.000	-	-
M & E per establishment ($ mil)	0.123	92.964	0.166	100.874	0.172	82.620	0.198	84.186	-	-
Payroll per employee ($)	16,796	187.308	19,632	197.054	19,616	175.579	23,433	168.929	-	-
Wages per production worker ($)	15,087	191.200	17,453	206.547	17,401	173.521	20,661	173.974	-	-
Hours per production worker (hours)	-	-	-	-	-	-	-	-	-	-
Output per employee ($)	90,890	121.357	108,270	127.114	109,207	113.948	125,736	121.327	-	-
Capital investment per employee ($)	3,909	121.715	5,282	151.788	5,685	127.732	6,370	126.854	-	-
M & E per employee ($)	3,355	146.138	4,536	179.356	4,805	151.489	5,352	140.517	-	-

Note: Columns headed % show percent of world total or ratio. Ratios closest to 100 are closest to world average. M & E stands for machinery & equipment.

Output in Manufacturing

	1987 $ bil.[f]	1987 %	1988 $ bil.[f]	1988 %	1989 $ bil.[f]	1989 %	1990 $ bil.[f]	1990 %	1991 $ bil.[f]	1991 %
3110 - Food products	55.000	10.42	63.000	9.91	62.000	9.66	72.000	9.96	-	-
3130 - Beverages	11.000	2.08	14.000	2.20	14.000	2.18	15.000	2.07	-	-
3140 - Tobacco	2.131	0.40	3.207	0.50	3.230	0.50	3.641	0.50	-	-
3210 - Textiles	14.000	2.65	16.000	2.52	15.000	2.34	16.000	2.21	-	-
3211 - Spinning, weaving, etc.	6.933	1.31	7.713	1.21	7.002	1.09	7.139	0.99	-	-
3220 - Wearing apparel	7.342	1.39	8.604	1.35	8.133	1.27	9.013	1.25	-	-
3230 - Leather and products	1.623	0.31	1.728	0.27	1.558	0.24	1.606	0.22	-	-
3240 - Footwear	2.163	0.41	2.405	0.38	2.132	0.33	2.516	0.35	-	-
3310 - Wood products	7.621	1.44	8.373	1.32	7.871	1.23	8.335	1.15	-	-
3320 - Furniture, fixtures	6.638	1.26	8.782	1.38	8.477	1.32	9.727	1.35	-	-
3410 - Paper and products	14.000	2.65	17.000	2.67	17.000	2.65	19.000	2.63	-	-
3411 - Pulp, paper, etc.	4.441	0.84	5.130	0.81	5.017	0.78	5.782	0.80	-	-
3420 - Printing, publishing	22.000	4.17	27.000	4.25	28.000	4.36	32.000	4.43	-	-
3510 - Industrial chemicals	29.000	5.49	33.000	5.19	34.000	5.30	36.000	4.98	-	-
3511 - Basic chemicals, excl fertilizers	18.000	3.41	22.000	3.46	22.000	3.43	24.000	3.32	-	-
3513 - Synthetic resins, etc.	6.654	1.26	7.803	1.23	7.903	1.23	7.389	1.02	-	-
3520 - Chemical products nec	19.000	3.60	24.000	3.77	24.000	3.74	28.000	3.87	-	-
3522 - Drugs and medicines	8.227	1.56	10.000	1.57	10.000	1.56	12.000	1.66	-	-
3530 - Petroleum refineries	16.000	3.03	14.000	2.20	16.000	2.49	21.000	2.90	-	-
3540 - Petroleum, coal products	1.606	0.30	2.013	0.32	1.705	0.27	1.856	0.26	-	-
3550 - Rubber products	4.474	0.85	5.148	0.81	5.083	0.79	5.782	0.80	-	-
3560 - Plastic products nec	12.000	2.27	14.000	2.20	15.000	2.34	18.000	2.49	-	-
3610 - Pottery, china, etc.	1.623	0.31	2.013	0.32	1.951	0.30	2.159	0.30	-	-
3620 - Glass and products	2.983	0.56	3.545	0.56	3.493	0.54	3.659	0.51	-	-
3690 - Nonmetal products nec	12.000	2.27	15.000	2.36	16.000	2.49	17.000	2.35	-	-
3710 - Iron and steel	16.000	3.03	20.000	3.14	20.000	3.12	21.000	2.90	-	-
3720 - Nonferrous metals	7.375	1.40	9.887	1.55	10.000	1.56	11.000	1.52	-	-
3810 - Metal products	21.000	3.98	25.000	3.93	26.000	4.05	31.000	4.29	-	-
3820 - Machinery nec	43.000	8.14	54.000	8.49	57.000	8.88	65.000	8.99	-	-
3825 - Office, computing machinery	8.473	1.60	12.000	1.89	13.000	2.02	15.000	2.07	-	-
3830 - Electrical machinery	36.000	6.82	44.000	6.92	43.000	6.70	48.000	6.64	-	-
3832 - Radio, television, etc.	19.000	3.60	23.000	3.62	22.000	3.43	25.000	3.46	-	-
3840 - Transportation equipment	48.000	9.09	59.000	9.28	62.000	9.66	70.000	9.68	-	-
3841 - Shipbuilding, repair	3.360	0.64	3.848	0.61	3.788	0.59	4.408	0.61	-	-
3843 - Motor vehicles	29.000	5.49	37.000	5.82	38.000	5.92	42.000	5.81	-	-
3850 - Professional goods	4.802	0.91	6.057	0.95	6.050	0.94	6.871	0.95	-	-
3900 - Industries nec	4.146	0.79	5.184	0.82	5.296	0.82	5.765	0.80	-	-

Note: Codes are International Standard Industry codes (ISIC). Percentages are % of total Output. [f]: Factor Prices; [p]: Producer Prices.

For sources, notes, and explanations, see Annotated Source Appendix, page 1061.

Finance, Economics, and Trade

Economic Indicators [36]

- **National product**: GDP—purchasing power parity—$1.1384 trillion (1995 est.)
- **National product real growth rate**: 2.7% (1995 est.)
- **National product per capita**: $19,500 (1995 est.)
- **Inflation rate (consumer prices)**: 3.1% (November 1995)
- **External debt**: $16.2 billion (June 1992)

Balance of Payments Summary [37]

Values in millions of dollars.

	1989	1990	1991	1992	1993
Exports of goods (f.o.b.)	150,696	181,729	182,579	188,451	182,084
Imports of goods (f.o.b.)	-191,239	-214,471	-200,853	-211,879	-201,953
Trade balance	-40,542	-32,742	-18,274	-23,428	-19,869
Services - debits	-157,793	-189,082	-184,180	-165,504	-154,294
Services - credits	168,924	197,527	190,274	180,188	-165,167
Private transfers (net)	-491	-535	-531	-485	-405
Government transfers (net)	-6,964	-8,206	-1,931	-8,621	-7,441
Long-term capital (net)	-26,988	15,906	-569	-10,285	-55,353
Short-term capital (net)	43,218	14,250	29,990	8,447	61,652
Errors and omissions	4,346	438	-1,216	12,142	4,306
Overall balance	-16,290	-2,444	13,563	-7,546	-6,237

Exchange Rates [38]

Currency: **British pound.**
Symbol: **£.**

Data are currency units per $1.

January 1996	0.6535
1995	0.6335
1994	0.6529
1993	0.6658
1992	0.5664
1991	0.5652

Imports and Exports

Top Import Origins [39]

$221.9 billion (c.i.f., 1994).

Origins	%
EU	54.9
Germany	14.6
France	10.0
Netherlands	6.7
US	12.2

Top Export Destinations [40]

$200.4 billion (f.o.b., 1994).

Destinations	%
EU	56.4
Germany	12.7
France	9.9
Netherlands	7.0
US	13.1

Foreign Aid [41]

Donor: ODA, $2.908 billion (1993).

Import and Export Commodities [42]

Import Commodities	Export Commodities
Manufactured goods	Manufactured goods
Machinery	Machinery
Semifinished goods	Fuels
Foodstuffs	Chemicals
Consumer goods	Semifinished goods
	Transport equipment

For sources, notes, and explanations, see Annotated Source Appendix, page 1061.

United States

Geography [1]

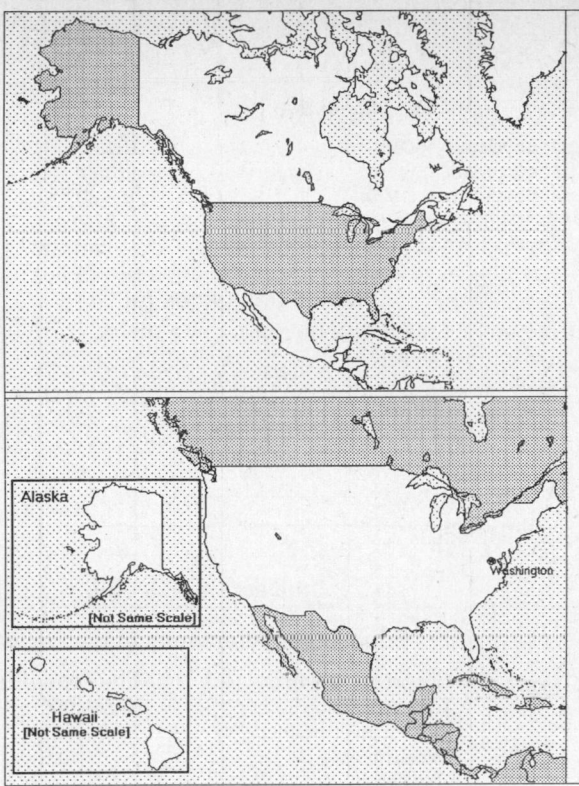

Total area:
9,372,610 sq km   3,618,785 sq mi
Land area:
9,166,600 sq km   3,539,244 sq mi
Comparative area:
About one-half the size of Russia; about one-half the size of South America; slightly smaller than China; about two and one-half times the size of western Europe
Land boundaries:
Total 12,248 km, Canada 8,893 km (including 2,477 km with Alaska), Cuba 29 km (US Naval Base at Guantanamo Bay), Mexico 3,326 km
Coastline:
19,924 km
Climate:
Mostly temperate, but tropical in Hawaii and Florida and arctic in Alaska, semiarid in great plains west of Mississippi River and arid in Great Basin of the Southwest; low Northwest winter temperatures occasionally ameliorated in January and February by warm chinook winds from eastern slopes of Rocky Mountains
Terrain:
Vast central plain, mountains in West, hills and low mountains in East; rugged mountains and broad river valleys in Alaska; rugged volcanic topography in Hawaii
Natural resources:
Coal, copper, lead, molybdenum, phosphates, uranium, bauxite, gold, iron, mercury, nickel, potash, silver, tungsten, zinc, petroleum, natural gas, timber
Land use:
Arable land: 20%
Permanent crops: 0%
Meadows and pastures: 26%
Forest and woodland: 29%
Other: 25%

Demographics [2]

	1970	1980	1990	1995[1]	1996	2000	2010	2020	2030
Population[59]	205,052	227,726	249,913	263,034	265,563	274,943	298,026	323,052	347,209
Population density (persons per sq. mi.)	58	64	71	74	75	78	84	91	98
(persons per sq. km.)	22	25	27	29	29	30	33	35	38
Net migration rate (per 1,000 population)[60]	NA	4.2	2.3	3.1	3.1	3.0	2.8	2.5	2.4
Births	3,731	3,612	NA	NA	3,995	NA	NA	NA	NA
Deaths[61]	1,921	1,990	2,148	NA	2,241	NA	NA	NA	NA
Life expectancy - males[62]	67.1	70.0	71.8	72.6	72.7	73.0	74.2	75.7	77.1
Life expectancy - females[62]	74.7	77.4	78.8	79.3	79.4	79.8	80.7	81.6	82.5
Birth rate (per 1,000)[63]	NA	16.0	16.7	15.0	14.8	14.2	14.3	14.2	13.9
Death rate (per 1,000)[63]	NA	8.6	8.6	8.8	8.8	8.8	8.9	9.0	9.5
Women of reproductive age (15-49 yrs.)[64]	48,956	58,796	65,806	68,365	69,052	69,968	71,029	72,135	77,012
of which are currently married[65]	30,924	34,318	34,516	NA	36,967	37,480	37,100	NA	NA
Fertility rate[66]	2.5	1.8	2.1	2.1	2.1	2.1	2.1	2.1	2.2

Except as noted, values for vital statistics are in thousands; life expectancy is in years.

Health

Health Indicators [3]

% of population with access to	
safe water (1990-95)	NA
adequate sanitation (1990-95)	NA
health services (1985-95)	NA
% of 1-year-olds immunized (1990-94) against	
TB (tuberculosis)	NA
DPT (diphtheria, pertussis, tetanus)	88
polio	79
measles	84
% of contraceptive prevalence (1980-94)	74
ORT use rate (1990-94)	NA

Health Expenditures [4]

Total health expenditure, 1990 (official exchange rate)	
Millions of dollars	690,667
Dollars per capita	2,763
Health expenditures as a percentage of GDP	
Total	12.7
Public sector	5.6
Private sector	7.0
Development assistance for health	
Total aid flows (millions of dollars)[1]	NA
Aid flows per capita (dollars)	NA
Aid flows as a percentage of total health expenditure	NA

For sources, notes, and explanations, see Annotated Source Appendix, page 1061.

993

Human Factors

Women and Children [5]

% of pregnant women immunized (tetanus 1990-94)	NA
% of births attended by trained health personnel (1983-94)	99
Maternal mortality rate (1980-92)	8
Under-5 mortality rate (1994)	10
% under-5 moderately/severely underweight (1980-1994)	NA

Burden of Disease [6]

Population per physician (1988)	420.61
Population per nurse (1984)	73.77
Population per hospital bed (1990)	194.39
AIDS cases per 100,000 people (1994)	22.7
Heart disease cases per 1,000 people (1993)[4]	319.8

Ethnic Division [7]

White	83.4%
Black	12.4%
Asian	3.3%
Native American	0.8%

Religion [8]

Protestant	56%
Roman Catholic	28%
None	10%
Jewish	2%
Other	4%
(1989)	

Major Languages [9]

English, Spanish (spoken by a sizable minority).

Education

Public Education Expenditures [10]

	1980	1985	1989	1990	1992	1994
Million (Dollar)[70]						
Total education expenditure	182,849	199,372	272,498	292,944	328,396	NA
as percent of GNP	6.7	4.9	5.2	5.3	5.5	NA
as percent of total govt. expend.	NA	15.5	12.3	12.3	NA	NA
Current education expenditure	NA	182,875	247,309	265,074	NA	NA
as percent of GNP	NA	4.5	4.7	4.8	NA	NA
as percent of current govt. expend.	NA	16.3	12.4	12.3	NA	NA
Capital expenditure	NA	16,497	25,189	27,870	NA	NA

Educational Attainment [11]

Age group (1994)[20]	25+
Total population	164,511,000
Highest level attained (%)	
No schooling	0.6
First level	
Not completed	8.2
Completed	NA
Entered second level	
S-1	44.6
S-2	NA
Postsecondary	46.5

Illiteracy [12]

	1979	1989	1995
Illiterate population (14+ yrs.)[2,8]	NA	NA	NA
Illiteracy rate - total pop. (%)	0.5	NA	NA
Illiteracy rate - males (%)	NA	NA	NA
Illiteracy rate - females (%)	NA	NA	NA

Libraries [13]

	Admin. Units	Svc. Pts.	Vols. (000)	Shelving (meters)	Vols. Added	Reg. Users
National	NA	NA	NA	NA	NA	NA
Nonspecialized	NA	NA	NA	NA	NA	NA
Public	NA	NA	NA	NA	NA	NA
Higher ed. (1988)	3,438	NA	718,503	NA	22 mil	NA
School (1988)	92,438	NA	738,706	NA	28 mil	NA

Daily Newspapers [14]

	1980	1985	1990	1994
Number of papers	1,745	1,676	1,611	1,548
Circ. (000)	62,200	62,800	62,328	59,305

Culture [15]

Cinema (seats per 1,000)	NA
Annual attendance per person	3.9
Gross box office receipts (mil. Dollar)	4,803
Museums (reporting)[15]	3,800
Visitors (000)	NA
Annual receipts (mil. Dollar)	200

Science and Technology

Scientific/Technical Forces [16]

Scientists/engineers	962,700[e]
Number female	NA
Technicians	NA
Number female	NA
Total[9]	NA

R&D Expenditures [17]

	Dollar (000) 1995
Total expenditure[30]	171,000,000
Capital expenditure	NA
Current expenditure	NA
Percent current	NA

U.S. Patents Issued [18]

Values show patents issued to citizens of the country by the U.S. Patents Office.

	1993	1994	1995
Number of patents	61,225	64,346	64,509

　　　　　　　　For sources, notes, and explanations, see Annotated Source Appendix, page 1061.

Government and Law

Organization of Government [19]

Long-form name:
United States of America
Type:
Federal republic; strong democratic tradition
Independence:
4 July 1776 (from England)
National holiday:
Independence Day, 4 July (1776)
Constitution:
17 September 1787, effective 4 March 1789
Legal system:
Based on English common law; judicial review of legislative acts; accepts compulsory ICJ jurisdiction, with reservations
Executive branch:
President; Vice President; Cabinet
Legislative branch:
Bicameral Congress: Senate and House of Representatives
Judicial branch:
Supreme Court

Elections [20]

House of Representatives	% of seats
Republican Party	53.1
Democratic Party	46.7
Other	0.2

Government Expenditures [21]

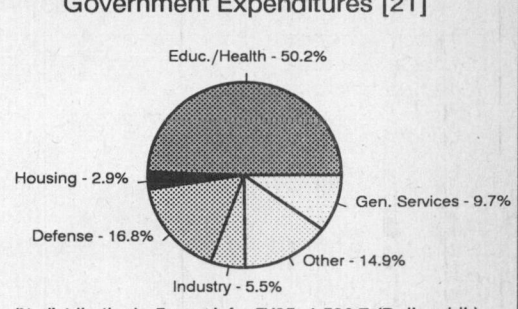

Educ./Health - 50.2%
Gen. Services - 9.7%
Other - 14.9%
Industry - 5.5%
Defense - 16.8%
Housing - 2.9%

(% distribution). Expend. for FY95: 1,590.7 (Dollars bil.)

Crime [23]

	1994
Crime volume	
Cases known to police	13,991,700
Attempts (percent)	NA
Percent cases solved	21.40
Crimes per 100,000 persons	5,374.37
Persons responsible for offenses	
Total number offenders	11,912,411
Percent female	20.00
Percent juvenile (0-18 yrs.)	18.60
Percent foreigners	NA

Military Expenditures and Arms Transfers [22]

	1990	1991	1992	1993	1994
Military expenditures					
Current dollars (mil.)	306,200	280,300	305,100	297,600	288,100
1994 constant dollars (mil.)	340,700	300,400	318,200	303,800	288,100
Armed forces (000)	2,181	2,115	1,919	1,815	1,715
Gross national product (GNP)					
Current dollars (bil.)	5,568	5,741	6,026	6,348	6,727
1994 constant dollars (bil.)	6,197	6,153	6,284	6,479	6,727
Central government expenditures (CGE)					
1994 constant dollars (bil.)	1,451	1,535	1,508	1,523	1,532
People (mil.)	249.9	252.6	255.4	258.1	260.7
Military expenditure as % of GNP	5.5	4.9	5.1	4.7	4.3
Military expenditure as % of CGE	23.5	19.6	21.1	20.0	18.8
Military expenditure per capita (1994 $)	1,363	1,189	1,246	1,177	1,105
Armed forces per 1,000 people (soldiers)	8.7	8.4	7.5	7.0	6.6
GNP per capita (1994 $)	24,790	24,360	24,600	25,100	25,810
Arms imports[6]					
Current dollars (mil.)[11]	1,800	1,900	1,600	1,400	1,100
1994 constant dollars (mil.)[11]	2,003	2,037	1,668	1,429	1,100
Arms exports[6]					
Current dollars (mil.)	16,300	14,000	13,400	13,800	12,400
1994 constant dollars (mil.)	18,140	15,110	14,080	14,080	12,400
Total imports[7]					
Current dollars (mil.)	517,000	508,400	553,900	603,400	689,800
1994 constant dollars (mil.)	575,400	544,900	577,600	615,900	689,800
Total exports[7]					
Current dollars (mil.)	393,600	421,700	448,200	464,800	512,700
1994 constant dollars (mil.)	438,000	452,000	467,300	474,400	512,700
Arms as percent of total imports[8]	0.3	0.4	0.3	0.2	0.2
Arms as percent of total exports[8]	4.1	3.3	3.0	3.0	2.4

Human Rights [24]

	SSTS	FL	FAPRO	PPCG	APROBC	TPW	PCPTW	STPEP	PHRFF	PRW	ASST	AFL
Observes	P			P		P	P			P	P	P
		EAFRD	CPR	ESCR	SR	ACHR	MAAE	PVIAC	PVNAC	EAFDAW	TCIDTP	RC
Observes		P	P	S	P	S		S	S	S	P	S

P = Party; S = Signatory; see Appendix for meaning of abbreviations.

Labor Force

Total Labor Force [25]

132.304 million (includes unemployed) (1995)

Labor Force by Occupation [26]

Managerial and professional	28.3%
Technical, sales, and administrative support	30.0
Services	13.5
Manufacturing, mining, transportation, and crafts	25.3
Farming, forestry, and fishing	2.8

Unemployment Rate [27]

5.6% (December 1995)

Production Sectors

Commercial Energy Production and Consumption

Data are shown in quadrillion (10^{15}) BTUs and percent for 1995
Values for hydroelectric, nuclear, geothermal, solar, and wind power refer to electrical generation.

Production [28]

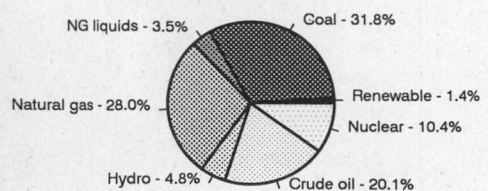

NG liquids - 3.5%
Coal - 31.8%
Natural gas - 28.0%
Renewable - 1.4%
Nuclear - 10.4%
Hydro - 4.8%
Crude oil - 20.1%

Consumption [29]

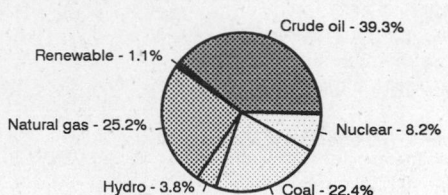

Renewable - 1.1%
Crude oil - 39.3%
Natural gas - 25.2%
Nuclear - 8.2%
Hydro - 3.8%
Coal - 22.4%

Crude oil	13.887
Natural gas liquids	2.442
Dry natural gas	19.331
Coal	21.98
Net hydroelectric power	3.323
Net nuclear power	7.190
Geothermal, solar, wind	0.947
Total	69.1

Crude oil	34.663
Dry natural gas	22.254
Coal	19.761
Net hydroelectric power	3.323
Net nuclear power	7.190
Geothermal, solar, wind	0.947
Total	88.138

Telecommunications [30]

- 182.558 million (1987 est.) telephones
- Domestic: large system of fiber-optic cable, microwave radio relay, coaxial cable, and domestic satellites
- International: 24 ocean cable systems in use; satellite earth stations - 61 Intelsat (45 Atlantic Ocean and 16 Pacific Ocean) (1990 est.), 5 Intersputnik (Atlantic Ocean region), and 4 Inmarsat (Pacific and Atlantic Ocean regions)
- Radio: Broadcast stations: AM 4,987, FM 4,932, shortwave 0 Radios: 540.5 million (1992 est.)
- Television: Broadcast stations: 1,092 (in addition, there are about 9,000 cable TV systems) Televisions: 215 million (1993 est.)

Top Agricultural Products [32]

Agriculture accounts for 2% of the GDP; produces wheat, other grains, corn, fruits, vegetables, cotton; beef, pork, poultry, dairy products; forest products; fish.

Transportation [31]

Railways: total: 240,000 km mainline routes (nongovernment owned); standard gauge: 240,000 km 1.435-m gauge (1989)

Highways: total: 6,284,488 km; paved: 5,574,341 km (in 1991, included 85,267 km of expressways); unpaved: 710,147 km (1993 est.)

Merchant marine: total: 322 ships (1,000 GRT or over) totaling 10,716,000 GRT/15,259,000 DWT; ships by type: bulk 21, cargo 20, chemical tanker 17, intermodal 125, liquefied gas tanker 14, passenger-cargo 2, tanker 110, tanker tug-barge 13

Airports

Total:	13,387
With paved runways over 3,047 m:	179
With paved runways 2,438 to 3,047 m:	201
With paved runways 1,524 to 2,437 m:	1,204
With paved runways 914 to 1,523 m:	2,361
With paved runways under 914 m:	7,720

Top Mining Products [33]

Thousand metric tons except as noted[79]	1994*
Gold (kg.)	331,000[80]
Iron ore (usable)	57,000[81]
Palladium metal (kg.)	6,500
Platinum metal (kg.)	1,800
Silver (metric tons)	1,400[80]
Cement, masonry and portland	79,220
Gas, natural, dry (trillion cu. ft.)	19[82]
Coal, all types (mil. short tons)	1,034[82]
Petroleum, crude (000 barrels/day)	6,662[82]

Tourism [34]

	1990	1991	1992	1993	1994
Tourists[108]	39,539	42,986	47,261	45,779	45,504
Tourism receipts	43,007	48,384	54,742	57,875	60,406
Tourism expenditures	37,349	35,322	38,552	40,713	43,562
Fare receipts	15,298	15,854	16,618	16,611	17,477
Fare expenditures	10,530	10,012	10,556	11,313	12,696

Travelers are in thousands, money in million U.S. dollars.

Manufacturing Sector

Manufacturing Summary [35]

	1987		1988		1989		1990		1991	
	$ bil.	%	$ bil.	%	$ bil.	%	$ bil.	%	$ bil.	%
Establishments or enterprises (number)	374,456	15.962	-	-	-	-	-	-	-	-
Total employment (000)	21,062	15.519	21,276	15.546	21,102	17.144	20,816	18.807	19,890	28.633
Production workers (000)	14,475	21.461	14,444	23.119	14,558	19.788	14,271	20.107	13,534	41.099
Output ($ bil.)	3,100	30.494	3,373	28.935	3,508	29.738	3,585	31.254	3,526	34.578
Value added ($ bil.)	1,451	33.607	1,571	32.689	1,641	33.655	1,656	33.196	1,641	47.830
Capital investment ($ mil.)	105,400	24.183	108,490	22.782	132,220	24.138	140,790	25.331	135,110	35.482
M & E investment ($ mil.)	88,490	28.398	90,380	26.113	110,640	28.338	118,940	28.212	115,530	40.973
Employees per establishment (number)	56	97.224	-	-	-	-	-	-	-	-
Production workers per establishment	39	134.451	-	-	-	-	-	-	-	-
Output per establishment ($ mil.)	8	191.034	-	-	-	-	-	-	-	-
Capital investment per estab. ($ mil.)	0.281	151.502	-	-	-	-	-	-	-	-
M & E per establishment ($ mil)	0.236	177.904	-	-	-	-	-	-	-	-
Payroll per employee ($)	25,437	283.673	26,426	265.248	27,441	245.623	28,422	204.893	29,414	201.453
Wages per production worker ($)	22,311	282.755	23,178	274.303	23,655	235.880	24,495	206.263	25,311	230.125
Hours per production worker (hours)	2,001	105.439	2,021	105.150	2,012	109.788	2,019	107.947	2,022	113.108
Output per employee ($)	147,161	196.490	158,531	186.123	166,237	173.454	172,226	166.187	177,290	120.763
Capital investment per employee ($)	5,004	155.828	5,099	146.542	6,266	140.791	6,764	134.694	6,793	123.917
M & E per employee ($)	4,201	182.984	4,248	167.968	5,243	165.290	5,714	150.012	5,808	143.097

Note: Columns headed % show percent of world total or ratio. Ratios closest to 100 are closest to world average. M & E stands for machinery & equipment.

Output in Manufacturing

	1987		1988		1989		1990		1991	
	$ bil. [f]	%	$ bil. [f]	%	$ bil. [f]	%	$ bil. [f]	%	$ bil. [f]	%
3110 - Food products	287.000	9.26	307.000	9.10	319.000	9.09	337.000	9.40	339.000	9.61
3130 - Beverages	43.000	1.39	45.000	1.33	45.000	1.28	47.000	1.31	49.000	1.39
3140 - Tobacco	21.000	0.68	24.000	0.71	26.000	0.74	30.000	0.84	32.000	0.91
3210 - Textiles	80.000	2.58	82.000	2.43	85.000	2.42	84.000	2.34	84.000	2.38
3211 - Spinning, weaving, etc.	35.000	1.13	37.000	1.10	37.000	1.05	37.000	1.03	37.000	1.05
3220 - Wearing apparel	49.000	1.58	49.000	1.45	47.000	1.34	47.000	1.31	49.000	1.39
3230 - Leather and products	4.500	0.15	4.900	0.15	5.080	0.14	5.090	0.14	4.800	0.14
3240 - Footwear	4.400	0.14	4.600	0.14	4.620	0.13	4.650	0.13	4.200	0.12
3310 - Wood products	51.000	1.65	53.000	1.57	55.000	1.57	54.000	1.51	52.000	1.47
3320 - Furniture, fixtures	30.000	0.97	31.000	0.92	32.000	0.91	33.000	0.92	32.000	0.91
3410 - Paper and products	104.000	3.35	117.000	3.47	126.000	3.59	126.000	3.51	124.000	3.52
3411 - Pulp, paper, etc.	47.000	1.52	55.000	1.63	58.000	1.65	57.000	1.59	54.000	1.53
3420 - Printing, publishing	136.000	4.39	144.000	4.27	150.000	4.28	157.000	4.38	157.000	4.45
3510 - Industrial chemicals	126.000	4.06	147.000	4.36	158.000	4.50	159.000	4.44	156.000	4.42
3511 - Basic chemicals, excl fertilizers	71.000	2.29	82.000	2.43	90.000	2.57	92.000	2.57	91.000	2.58
3513 - Synthetic resins, etc.	41.000	1.32	48.000	1.42	51.000	1.45	48.000	1.34	46.000	1.30
3520 - Chemical products nec	108.000	3.48	118.000	3.50	125.000	3.56	134.000	3.74	142.000	4.03
3522 - Drugs and medicines	39.000	1.26	44.000	1.30	49.000	1.40	54.000	1.51	61.000	1.73
3530 - Petroleum refineries	118.000	3.81	119.000	3.53	131.000	3.73	159.000	4.44	145.000	4.11
3540 - Petroleum, coal products	12.000	0.39	13.000	0.39	13.000	0.37	13.000	0.36	13.000	0.37
3550 - Rubber products	23.000	0.74	24.000	0.71	25.000	0.71	26.000	0.73	25.000	0.71
3560 - Plastic products nec	66.000	2.13	72.000	2.13	76.000	2.17	78.000	2.18	78.000	2.21
3610 - Pottery, china, etc.	2.400	0.08	2.600	0.08	2.610	0.07	2.610	0.07	2.600	0.07
3620 - Glass and products	16.000	0.52	17.000	0.50	17.000	0.48	17.000	0.47	17.000	0.48
3690 - Nonmetal products nec	45.000	1.45	46.000	1.36	46.000	1.31	47.000	1.31	43.000	1.22
3710 - Iron and steel	63.000	2.03	77.000	2.28	78.000	2.22	76.000	2.12	69.000	1.96
3720 - Nonferrous metals	47.000	1.52	60.000	1.78	62.000	1.77	59.000	1.65	53.000	1.50
3810 - Metal products	129.000	4.16	139.000	4.12	143.000	4.08	145.000	4.04	140.000	3.97
3820 - Machinery nec	242.000	7.81	269.000	7.98	278.000	7.92	280.000	7.81	267.000	7.57
3825 - Office, computing machinery	61.000	1.97	68.000	2.02	66.000	1.88	65.000	1.81	59.000	1.67
3830 - Electrical machinery	176.000	5.68	193.000	5.72	201.000	5.73	204.000	5.69	206.000	5.84
3832 - Radio, television, etc.	97.000	3.13	107.000	3.17	111.000	3.16	117.000	3.26	122.000	3.46
3840 - Transportation equipment	350.000	11.29	372.000	11.03	384.000	10.95	385.000	10.74	380.000	10.78
3841 - Shipbuilding, repair	14.000	0.45	15.000	0.44	15.000	0.43	16.000	0.45	15.000	0.43
3843 - Motor vehicles	229.000	7.39	245.000	7.26	249.000	7.10	238.000	6.64	227.000	6.44
3850 - Professional goods	102.000	3.29	109.000	3.23	112.000	3.19	116.000	3.24	118.000	3.35
3900 - Industries nec	30.000	0.97	33.000	0.98	33.000	0.94	35.000	0.98	35.000	0.99

Note: Codes are International Standard Industry codes (ISIC). Percentages are % of total Output. [f]: Factor Prices; [p]: Producer Prices.

For sources, notes, and explanations, see Annotated Source Appendix, page 1061.

997

Finance, Economics, and Trade

Economic Indicators [36]

- **National product**: GDP—purchasing power parity— $7.2477 trillion (1995 est.)
- **National product real growth rate**: 2.1% (1995 est.)
- **National product per capita**: $27,500 (1995 est.)
- **Inflation rate (consumer prices)**: 2.5% (1995)
- **External debt**: $NA

Balance of Payments Summary [37]

Values in millions of dollars.

	1989	1990	1991	1992	1993
Exports of goods (f.o.b.)	361,700	389,310	416,920	440,360	456,870
Imports of goods (f.o.b.)	-477,380	-498,330	-490,980	-536,460	-589,440
Trade balance	-115,680	-109,020	-74,060	-96,100	-132,570
Services - debits	-227,680	-256,570	-239,670	-230,740	-237,840
Services - credits	267,770	307,520	300,160	291,030	298,610
Private transfers (net)	-12,320	-13,040	-13,820	-13,290	-13,720
Government transfers (net)	-13,280	-20,630	20,480	-18,750	-18,400
Long-term capital (net)	85,570	7,510	17,030	-14,040	-49,720
Short-term capital (net)	30,110	14,440	7,830	56,900	63,870
Errors and omissions	2,430	39,980	-39,730	-17,200	21,120
Overall balance	16,920	-29,810	-21,780	-42,190	-68,650

Exchange Rates [38]

Currency: **United States dollar.**
Symbol: **US$.**

Data are currency rates per $1.

Basis for all listed exchange rates   1.00

Imports and Exports

Top Import Origins [39]

$751 billion (c.i.f., 1995 est.) Data are for 1993.

Origins	%
Canada	19.3
Western Europe	18.1
Japan	18.1

Top Export Destinations [40]

$578 billion (f.o.b., 1995 est.) Data are for 1993.

Destinations	%
Western Europe	24.3
Canada	22.1
Japan	10.5

Foreign Aid [41]

Donor: ODA, $9.721 billion (1993).

Import and Export Commodities [42]

Import Commodities
Crude oil and refined petroleum products
Machinery
Automobiles
Consumer goods
Industrial raw materials
Food and beverages

Export Commodities
Capital goods
Automobiles
Industrial supplies and raw materials
Consumer goods
Agricultural products

For sources, notes, and explanations, see Annotated Source Appendix, page 1061.

Uruguay

Geography [1]

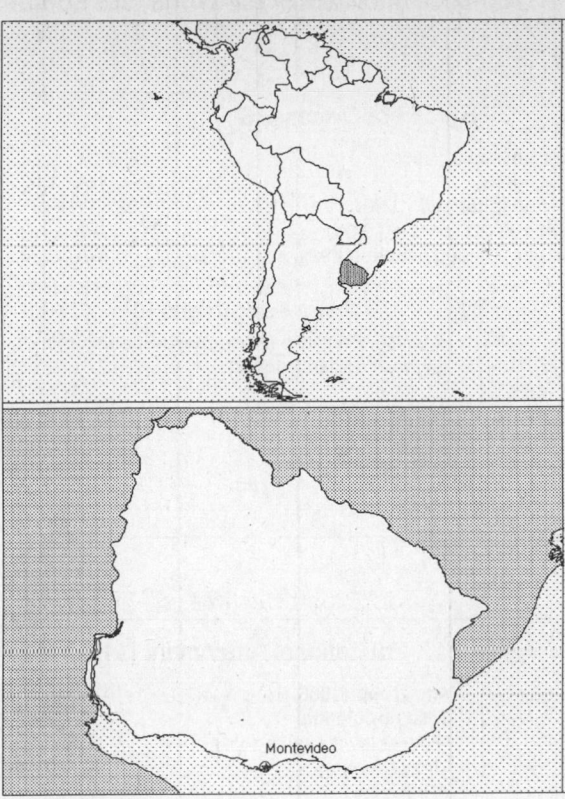

Total area:
176,220 sq km   68,039 sq mi
Land area:
173,620 sq km   67,035 sq mi
Comparative area:
Slightly smaller than Washington State
Land boundaries:
Total 1,564 km, Argentina 579 km, Brazil 985 km
Coastline:
660 km
Climate:
Warm temperate; freezing temperatures almost unknown
Terrain:
Mostly rolling plains and low hills; fertile coastal lowland
Natural resources:
Fertile soil, hydropower potential, minor minerals
Land use:
Arable land: 8%
Permanent crops: 0%
Meadows and pastures: 78%
Forest and woodland: 4%
Other: 10%

Demographics [2]

	1970	1980	1990	1995[1]	1996	2000	2010	2020	2030
Population	2,824	2,920	3,106	3,216	3,239	3,333	3,582	3,811	4,014
Population density (persons per sq. mi.)	42	44	46	48	48	50	53	57	60
(persons per sq. km.)	16	17	18	19	19	19	21	22	23
Net migration rate (per 1,000 population)	-6.2	NA	-1.3	-0.9	-1.0	-0.6	-0.1	0.0	0.0
Births[67]	55	NA	NA	NA	55	NA	NA	NA	NA
Deaths[68]	20	30	NA	NA	29	NA	NA	NA	NA
Life expectancy - males	NA	NA	69.4	71.5	71.8	73.0	75.4	77.1	78.3
Life expectancy - females	NA	NA	76.1	78.0	78.3	79.5	81.8	83.5	84.6
Birth rate (per 1,000)	19.4	18.4	18.2	17.1	17.0	16.7	15.3	14.0	13.1
Death rate (per 1,000)	5.3	10.2	9.9	9.1	9.1	8.7	8.5	8.3	8.6
Women of reproductive age (15-49 yrs.)	NA	NA	740	782	789	811	860	903	909
of which are currently married	NA	NA	446	NA	474	494	533	NA	NA
Fertility rate	NA	NA	2.5	2.3	2.3	2.3	2.2	2.1	2.0

Except as noted, values for vital statistics are in thousands; life expectancy is in years.

Health

Health Indicators [3]

% of population with access to	
safe water (1990-95)[1]	75
adequate sanitation (1990-95)[1]	61
health services (1985-95)	82
% of 1-year-olds immunized (1990-94) against	
TB (tuberculosis)	99
DPT (diphtheria, pertussis, tetanus)	88
polio	88
measles	80
% of contraceptive prevalence (1980-94)	NA
ORT use rate (1990-94)	96

Health Expenditures [4]

Total health expenditure, 1990 (official exchange rate)	
Millions of dollars	383
Dollars per capita	124
Health expenditures as a percentage of GDP	
Total	4.6
Public sector	2.5
Private sector	2.1
Development assistance for health	
Total aid flows (millions of dollars)[1]	5
Aid flows per capita (dollars)	1.7
Aid flows as a percentage of total health expenditure	1.4

For sources, notes, and explanations, see Annotated Source Appendix, page 1061.

999

Human Factors

Women and Children [5]

% of pregnant women immunized (tetanus 1990-94)	13
% of births attended by trained health personnel (1983-94)	96
Maternal mortality rate (1980-92)	36
Under-5 mortality rate (1994)	21
% under-5 moderately/severely underweight (1980-1994)	7

Burden of Disease [6]

Population per physician (1985)	513.31
Population per nurse	NA
Population per hospital bed (1990)	221.11
AIDS cases per 100,000 people (1994)	3.7
Malaria cases per 100,000 people (1992)	NA

Ethnic Division [7]

White	88%
Mestizo	8%
Black	4%

Religion [8]

One-half of the adult Roman Catholic population attends church regularly.

Roman Catholic	66%
Jewish	2%
Protestant	2%
Nonprofessing or other	30%

Major Languages [9]

Spanish, Brazilero (Portuguese-Spanish mix on the Brazilian frontier).

Education

Public Education Expenditures [10]

Million or Trillion (T) (Peso)	1980	1985	1990	1991	1992	1994
Total education expenditure	2,035	12,565	289,354	572,456	959,087	2.05 T
as percent of GNP	2.3	2.8	3.1	2.9	2.8	2.5
as percent of total govt. expend.	10.0	9.3	15.9	16.6	15.4	13.3
Current education expenditure	1,927	12,068	265,660	530,330	873,317	1.98 T
as percent of GNP	2.2	2.7	2.8	2.7	2.5	2.5
as percent of current govt. expend.	NA	9.3	16.7	17.0	15.5	13.3
Capital expenditure	108	497	23,694	42,126	85,770	68,026

Educational Attainment [11]

Age group (1985)	25+
Total population	1,701,705
Highest level attained (%)	
No schooling	4.7
First level	
Not completed	58.0
Completed	NA
Entered second level	
S-1	29.2
S-2	NA
Postsecondary	8.1

Illiteracy [12]

In thousands and percent[1]	1990	1995	2000
Illiterate population (15+ yrs.)	76	65	55
Illiteracy rate - total pop. (%)	3.3	2.7	2.2
Illiteracy rate - males (%)	3.7	3.1	2.5
Illiteracy rate - females (%)	2.9	2.3	1.9

Libraries [13]

	Admin. Units	Svc. Pts.	Vols. (000)	Shelving (meters)	Vols. Added	Reg. Users
National (1986)	1	1	890	25,672	6,658	-
Nonspecialized	NA	NA	NA	NA	NA	NA
Public	NA	NA	NA	NA	NA	NA
Higher ed.	NA	NA	NA	NA	NA	NA
School	NA	NA	NA	NA	NA	NA

Daily Newspapers [14]

	1980	1985	1990	1994
Number of papers	24	25	30	32
Circ. (000)	700[e]	680[e]	720[e]	750[e]

Culture [15]

Science and Technology

Scientific/Technical Forces [16]

Scientists/engineers	2,093
Number female	720
Technicians	NA
Number female	NA
Total[1]	NA

R&D Expenditures [17]

U.S. Patents Issued [18]

Values show patents issued to citizens of the country by the U.S. Patents Office.

	1993	1994	1995
Number of patents	0	0	2

For sources, notes, and explanations, see Annotated Source Appendix, page 1061.

Government and Law

Organization of Government [19]

Long-form name:
  Oriental Republic of Uruguay
Type:
  Republic
Independence:
  25 August 1828 (from Brazil)
National holiday:
  Independence Day, 25 August (1828)
Constitution:
  27 November 1966, effective February
  1967, suspended 27 June 1973, new
  constitution rejected by referendum 30
  November 1980
Legal system:
  Based on Spanish civil law system;
  accepts compulsory ICJ jurisdiction
Executive branch:
  President; Vice President; Council of
  Ministers
Legislative branch:
  Bicameral General Assembly (Asamblea
  General): Chamber of Senators (Camara
  de Senadores) and Chamber of
  Representatives (Camara de
  Representantes)
Judicial branch:
  Supreme Court

Elections [20]

Chamber of Representatives	% of votes
Colorado	32.0
Blanco	31.0
Encuentro Progresista	31.0
New Sector	5.0

Government Expenditures [21]

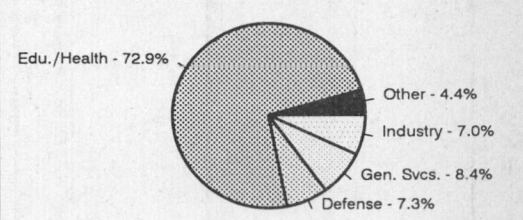

Edu./Health - 72.9%
Other - 4.4%
Industry - 7.0%
Gen. Svcs. - 8.4%
Defense - 7.3%

(% distribution). Expend. for CY94: 28,717 (Peso mil.)

Crime [23]

Military Expenditures and Arms Transfers [22]

	1990	1991	1992	1993	1994
Military expenditures					
Current dollars (mil.)	294	283[e]	312	292[e]	412
1994 constant dollars (mil.)	327	303[e]	326	298[e]	412
Armed forces (000)	25	25	25	25	25
Gross national product (GNP)					
Current dollars (mil.)	11,880	12,840	13,350	14,810	15,990
1994 constant dollars (mil.)	13,220	13,760	13,930	15,110	15,990
Central government expenditures (CGE)					
1994 constant dollars (mil.)	3,601	3,880	4,084	5,328	5,683
People (mil.)	3.1	3.1	3.2	3.2	3.2
Military expenditure as % of GNP	2.5	2.2	2.3	2.0	2.6
Military expenditure as % of CGE	9.1	7.8	8.0	5.6	7.3
Military expenditure per capita (1994 $)	105	97	103	94	129
Armed forces per 1,000 people (soldiers)	8.0	8.0	7.9	7.9	7.8
GNP per capita (1994 $)	4,257	4,398	4,419	4,759	5,000
Arms imports[6]					
Current dollars (mil.)	20	30	10	5	0
1994 constant dollars (mil.)	22	32	10	5	0
Arms exports[6]					
Current dollars (mil.)	0	0	0	0	0
1994 constant dollars (mil.)	0	0	0	0	0
Total imports[7]					
Current dollars (mil.)	1,343	1,637	2,045	2,324	2,773
1994 constant dollars (mil.)	1,495	1,755	2,133	2,372	2,773
Total exports[7]					
Current dollars (mil.)	1,693	1,594	1,703	1,645	1,913
1994 constant dollars (mil.)	1,884	1,709	1,776	1,679	1,913
Arms as percent of total imports[8]	1.5	1.8	0.5	0.2	0
Arms as percent of total exports[8]	0	0	0	0	0

Human Rights [24]

	SSTS	FL	FAPRO	PPCG	APROBC	TPW	PCPTW	STPEP	PHRFF	PRW	ASST	AFL
Observes			P	P	P	P	P			S		P
	EAFRD	CPR	ESCR	SR	ACHR	MAAE	PVIAC	PVNAC	EAFDAW	TCIDTP	RC	
Observes	P	P	P	P	P	P	P	P	P	P	P	

P=Party; S=Signatory; see Appendix for meaning of abbreviations.

Labor Force

Total Labor Force [25]

1.355 million (1991 est.)

Labor Force by Occupation [26]

Government	25%
Manufacturing	19
Agriculture	11
Commerce	12
Utilities, construction, transport, and communications	12
Other services	21

Date of data: 1988 est.

Unemployment Rate [27]

11% (1995)

For sources, notes, and explanations, see Annotated Source Appendix, page 1061.

Production Sectors

Commercial Energy Production and Consumption

Data are shown in quadrillion (10^{15}) BTUs and percent for 1995
Values for hydroelectric, nuclear, geothermal, solar, and wind power refer to electrical generation.

Production [28]

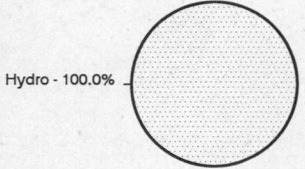

Hydro - 100.0%

Consumption [29]

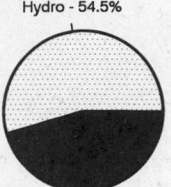

Hydro - 54.5%

Crude oil - 45.5%

Net hydroelectric power	0.078
Total	0.078

Crude oil	0.065
Net hydroelectric power	0.078
Total	0.143

Telecommunications [30]

- 451,000 (1991 est.) telephones; some modern facilities
- Domestic: most modern facilities concentrated in Montevideo; new nationwide microwave radio relay network
- International: satellite earth stations - 2 Intelsat (Atlantic Ocean)
- Radio: Broadcast stations: AM 99, FM 0, shortwave 9 Radios: 1.89 million (1992 est.)
- Television: Broadcast stations: 26 Televisions: 725,000 (1992 est.)

Top Agricultural Products [32]

Agriculture accounts for 10.5% of the GDP; produces wheat, rice, corn, sorghum; livestock; fishing.

Transportation [31]

Railways: total: 2,070 km (461 km closed; additional 460 km only partially operational); standard gauge: 2,070 km 1.435-m gauge

Highways: total: 49,600 km; paved: 6,656 km; unpaved: 42,944 km (1988 est.)

Merchant marine: total: 3 ships (1,000 GRT or over) totaling 71,405 GRT/110,939 DWT; ships by type: cargo 1, container 1, oil tanker 1 (1995 est.)

Airports

Total:	66
With paved runways 2,438 to 3,047 m:	1
With paved runways 1,524 to 2,437 m:	5
With paved runways 914 to 1,523 m:	8
With paved runways under 914 m:	36
With unpaved runways 1,524 to 2,437 m:	2

Top Mining Products [33]

Metric tons except as noted	3/15/95[*]
Cement, hydraulic	600,000
Clays	150,000[e]
Gypsum	145,000
Steel, crude	55,000
Lime	12,000
Petroleum refinery products (000 42-gal. bls.)	11,150
Sand and gravel (000 tons)	2,000
Alum schist	10,000
Dolomite	20,000
Limestone	750,000

Tourism [34]

	1990	1991	1992	1993	1994
Visitors[4]	1,267	1,510	1,802	2,003	2,176
Tourists	NA	NA	NA	NA	1,884
Excursionists	NA	NA	NA	NA	291
Tourism receipts	238	333	381	447	632
Tourism expenditures	111	100	104	129	190
Fare receipts	64	45	78	114	118
Fare expenditures	73	61	91	78	81

Travelers are in thousands, money in million U.S. dollars.

For sources, notes, and explanations, see Annotated Source Appendix, page 1061.

Manufacturing Sector

Manufacturing Summary [35]

	1987		1988		1989		1990		1991	
	$ bil.	%	$ bil.	%	$ bil.	%	$ bil.	%	$ bil.	%
Establishments or enterprises (number)	-	-	-	-	-	-	-	-	-	-
Total employment (000)	152	0.112	149	0.109	-	-	-	-	-	-
Production workers (000)	118	0.175	114	0.183	-	-	-	-	-	-
Output ($ bil.)	2	0.021	6	0.048	-	-	-	-	-	-
Value added ($ bil.)	0.973	0.023	2	0.049	-	-	-	-	-	-
Capital investment ($ mil.)	-	-	-	-	-	-	-	-	-	-
M & E investment ($ mil.)	-	-	-	-	-	-	-	-	-	-
Employees per establishment (number)	-	-	-	-	-	-	-	-	-	-
Production workers per establishment	-	-	-	-	-	-	-	-	-	-
Output per establishment ($ mil.)	-	-	-	-	-	-	-	-	-	-
Capital investment per estab. ($ mil.)	-	-	-	-	-	-	-	-	-	-
M & E per establishment ($ mil)	-	-	-	-	-	-	-	-	-	-
Payroll per employee ($)	1,957	21.826	5,113	51.322	-	-	-	-	-	-
Wages per production worker ($)	1,496	18.964	3,850	45.558	-	-	-	-	-	-
Hours per production worker (hours)	2,115	111.438	2,066	107.498	-	-	-	-	-	-
Output per employee ($)	13,923	18.590	37,300	43.792	-	-	-	-	-	-
Capital investment per employee ($)	-	-	-	-	-	-	-	-	-	-
M & E per employee ($)	-	-	-	-	-	-	-	-	-	-

Note: Columns headed % show percent of world total or ratio. Ratios closest to 100 are closest to world average. M & E stands for machinery & equipment.

Output in Manufacturing

	1987		1988		1989		1990		1991	
	$ bil.	%	$ bil.	%	$ bil.	%	$ bil.	%	$ bil.	%
3110 - Food products	0.467	23.35	1.335	22.25	-	-	-	-	-	-
3130 - Beverages	0.114	5.70	0.280	4.67	-	-	-	-	-	-
3140 - Tobacco	0.047	2.35	0.122	2.03	-	-	-	-	-	-
3210 - Textiles	0.204	10.20	0.536	8.93	-	-	-	-	-	-
3211 - Spinning, weaving, etc.	0.175	8.75	0.470	7.83	-	-	-	-	-	-
3220 - Wearing apparel	0.076	3.80	0.156	2.60	-	-	-	-	-	-
3230 - Leather and products	0.103	5.15	0.254	4.23	-	-	-	-	-	-
3240 - Footwear	0.017	0.85	0.050	0.83	-	-	-	-	-	-
3310 - Wood products	0.010	0.50	0.035	0.58	-	-	-	-	-	-
3320 - Furniture, fixtures	0.004	0.20	0.008	0.13	-	-	-	-	-	-
3410 - Paper and products	0.067	3.35	0.165	2.75	-	-	-	-	-	-
3411 - Pulp, paper, etc.	0.044	2.20	0.105	1.75	-	-	-	-	-	-
3420 - Printing, publishing	0.039	1.95	0.102	1.70	-	-	-	-	-	-
3510 - Industrial chemicals	0.051	2.55	0.154	2.57	-	-	-	-	-	-
3511 - Basic chemicals, excl fertilizers	0.016	0.80	0.047	0.78	-	-	-	-	-	-
3513 - Synthetic resins, etc.	-	-	-	-	-	-	-	-	-	-
3520 - Chemical products nec	0.128	6.40	0.343	5.72	-	-	-	-	-	-
3522 - Drugs and medicines	0.057	2.85	0.161	2.68	-	-	-	-	-	-
3550 - Rubber products	0.043	2.15	0.102	1.70	-	-	-	-	-	-
3560 - Plastic products nec	0.037	1.85	0.090	1.50	-	-	-	-	-	-
3610 - Pottery, china, etc.	0.013	0.65	0.033	0.55	-	-	-	-	-	-
3620 - Glass and products	0.017	0.85	0.025	0.42	-	-	-	-	-	-
3690 - Nonmetal products nec	0.032	1.60	0.083	1.38	-	-	-	-	-	-
3710 - Iron and steel	0.017	0.85	0.043	0.72	-	-	-	-	-	-
3720 - Nonferrous metals	0.004	0.20	0.010	0.17	-	-	-	-	-	-
3810 - Metal products	0.050	2.50	0.128	2.13	-	-	-	-	-	-
3820 - Machinery nec	0.009	0.45	0.019	0.32	-	-	-	-	-	-
3825 - Office, computing machinery	0.001	0.05	0.003	0.05	-	-	-	-	-	-
3830 - Electrical machinery	0.053	2.65	0.152	2.53	-	-	-	-	-	-
3832 - Radio, television, etc.	0.011	0.55	0.019	0.32	-	-	-	-	-	-
3840 - Transportation equipment	0.116	5.80	0.276	4.60	-	-	-	-	-	-
3843 - Motor vehicles	0.092	4.60	0.219	3.65	-	-	-	-	-	-
3850 - Professional goods	0.001	0.05	0.004	0.07	-	-	-	-	-	-
3900 - Industries nec	0.005	0.25	0.014	0.23	-	-	-	-	-	-

Note: Codes are International Standard Industry codes (ISIC). Percentages are % of total Output. [f]: Factor Prices; [p]: Producer Prices.

For sources, notes, and explanations, see Annotated Source Appendix, page 1061.

1003

Finance, Economics, and Trade

Economic Indicators [36]

- **National product**: GDP—purchasing power parity—$24.4 billion (1995 est.)
- **National product real growth rate**: - 2.4% (1995 est.)
- **National product per capita**: $7,600 (1995 est.)
- **Inflation rate (consumer prices)**: 35.4% (1995 est.)
- **External debt**: $4.95 billion (1995)

Balance of Payments Summary [37]

Values in millions of dollars.

	1989	1990	1991	1992	1993
Exports of goods (f.o.b.)	1,599.0	1,692.9	1,604.7	1,801.4	1,731.6
Imports of goods (f.o.b.)	-1,136.2	-1,266.9	-1,543.7	-1,923.2	-2,118.3
Trade balance	462.8	426.0	61.0	-121.8	-386.7
Services - debits	-985.9	-972.1	-889.6	-970.9	-1,031.4
Services - credits	636.2	723.9	830.9	1,055.3	1,166.5
Private transfers (net)	NA	NA	NA	NA	NA
Government transfers (net)	8.0	8.1	40.1	28.6	24.8
Long-term capital (net)	47.2	9.6	-148.2	324.6	345.9
Short-term capital (net)	-18.1	-75.7	-206.0	-365.4	-151.4
Errors and omissions	-28.4	35.7	468.8	238.3	220.8
Overall balance	221.8	155.5	157.0	188.7	188.5

Exchange Rates [38]

Currency: **Uruguayan peso.**
Symbol: **$Ur.**

Data are currency units per $1.

January 1996	7.12
January 1995	5.6
1994	5.0529
1993	3.9484
1992	3.0270
1991	2.0188

Imports and Exports

Top Import Origins [39]

$3.1 billion (c.i.f., 1995 est.).

Origins	%
Brazil	NA
Argentina	NA
US	NA
Nigeria	NA

Top Export Destinations [40]

$2.3 billion (f.o.b., 1995 est.).

Destinations	%
Brazil	NA
Argentina	NA
US	NA
China	NA
Italy	NA

Foreign Aid [41]

Recipient: ODA, $91 million (1993).

Import and Export Commodities [42]

Import Commodities	Export Commodities
Machinery and equipment	Wool and textile manufactures
Vehicles	Beef and other animal products
Chemicals	Leather
Minerals	Rice
Plastics	

1004

For sources, notes, and explanations, see Annotated Source Appendix, page 1061.

Uzbekistan

Geography [1]

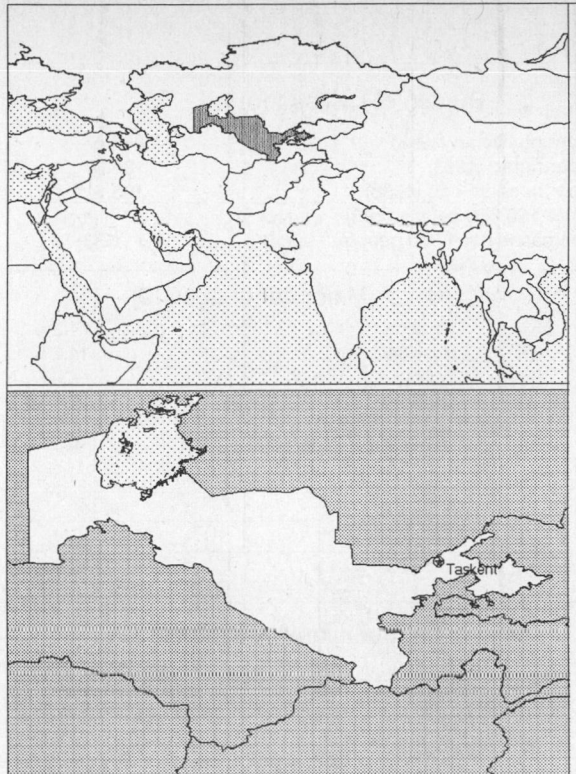

Total area:
447,400 sq km   172,742 sq mi
Land area:
425,400 sq km   164,248 sq mi
Comparative area:
Slightly larger than California
Land boundaries:
Total 6,221 km, Afghanistan 137 km, Kazakhstan 2,203 km, Kyrgyzstan 1,099 km, Tajikistan 1,161 km, Turkmenistan 1,621 km
Coastline:
0 km
Note: Uzbekistan borders the Aral Sea (420 km)
Climate:
Mostly midlatitude desert, long, hot summers, mild winters; semiarid grassland in East
Terrain:
Mostly flat-to-rolling sandy desert with dunes; broad, flat intensely irrigated river valleys along course of Amu Darya and Sirdaryo; Fergana Valley in East surrounded by mountainous Tajikistan and Kyrgyzstan; shrinking Aral Sea in West
Natural resources:
Natural gas, petroleum, coal, gold, uranium, silver, copper, lead and zinc, tungsten, molybdenum
Land use:
Arable land: 10%
Permanent crops: 1%
Meadows and pastures: 47%
Forest and woodland: 0%
Other: 42%

Demographics [2]

	1970	1980	1990	1995[1]	1996	2000	2010	2020	2030
Population	11,940	16,000	20,624	22,984	23,418	25,245	30,536	36,628	42,870
Population density (persons per sq. mi.)	69	93	119	133	136	146	177	212	248
(persons per sq. km.)	27	36	46	51	52	56	68	82	96
Net migration rate (per 1,000 population)	NA	NA	-3.2	-3.4	-3.1	-2.1	-1.1	-0.3	0.0
Births	NA	NA	NA	NA	699	NA	NA	NA	NA
Deaths	NA	NA	NA	NA	188	NA	NA	NA	NA
Life expectancy - males	NA	NA	64.1	60.3	60.4	61.0	62.5	66.1	69.3
Life expectancy - females	NA	NA	70.8	68.8	69.0	69.6	71.0	74.6	77.5
Birth rate (per 1,000)	NA	NA	35.4	30.3	29.9	28.9	27.6	23.5	20.7
Death rate (per 1,000)	NA	NA	7.2	8.1	8.0	7.7	7.4	6.3	5.9
Women of reproductive age (15-49 yrs.)	NA	NA	4,791	5,556	5,725	6,426	8,127	9,532	11,230
of which are currently married	NA	NA	3,228	NA	3,921	4,375	5,618	NA	NA
Fertility rate	NA	NA	4.3	3.7	3.7	3.6	3.2	2.9	2.6

Except as noted, values for vital statistics are in thousands; life expectancy is in years.

Health

Health Indicators [3]

% of population with access to	
safe water (1990-95)	NA
adequate sanitation (1990-95)	NA
health services (1985-95)	NA
% of 1-year-olds immunized (1990-94) against	
TB (tuberculosis)	89
DPT (diphtheria, pertussis, tetanus)	58
polio	51
measles	91
% of contraceptive prevalence (1980-94)	NA
ORT use rate (1990-94)	NA

Health Expenditures [4]

Total health expenditure, 1990 (official exchange rate)	
Millions of dollars	2,388
Dollars per capita	116
Health expenditures as a percentage of GDP	
Total	5.9
Public sector	4.3
Private sector	1.6
Development assistance for health	
Total aid flows (millions of dollars)[1]	NA
Aid flows per capita (dollars)	NA
Aid flows as a percentage of total health expenditure	NA

For sources, notes, and explanations, see Annotated Source Appendix, page 1061.

1005

Human Factors

Women and Children [5]

% of pregnant women immunized (tetanus 1990-94)	NA
% of births attended by trained health personnel (1983-94)	NA
Maternal mortality rate (1980-92)	NA
Under-5 mortality rate (1994)	64
% under-5 moderately/severely underweight (1980-1994)	NA

Burden of Disease [6]

Population per physician (1993)	282.47
Population per nurse (1993)	86.07
Population per hospital bed (1993)	105.52
AIDS cases per 100,000 people (1994)	*
Heart disease cases per 1,000 people (1990-93)	523

Ethnic Division [7]

Uzbek	71.4%
Russian	8.3%
Tajik	4.7%
Kazak	4.1%
Tatar	2.4%
Karakalpak	2.1%
Other	7%

Religion [8]

Muslim (mostly Sunnis)	88%
Eastern Orthodox	9%
Other	3%

Major Languages [9]

Uzbek	74.3%
Russian	14.2%
Tajik	4.4%
Other	7.1%

Education

Public Education Expenditures [10]

	1980	1990	1991	1992	1993	1994
Million (Ruble)						
Total education expenditure	1,457	3,070	5,770	45,216	486,502	NA
as percent of GNP	6.4	9.5	9.4	10.1	11.0	NA
as percent of total govt. expend.	23.0	20.4	17.8	23.3	24.4	NA
Current education expenditure	1,249	2,450	4,893	42,232	475,863	NA
as percent of GNP	5.5	7.6	8.0	9.4	10.7	NA
as percent of current govt. expend.	NA	NA	NA	NA	NA	NA
Capital expenditure	208	620	877	2,984	10,639	NA

Educational Attainment [11]

Illiteracy [12]

	1985	1989	1995
Illiterate population (15+ yrs.)[2]	NA	NA	45,000
Illiteracy rate - total pop. (%)	NA	2.8	0.3
Illiteracy rate - males (%)	NA	1.5	0.2
Illiteracy rate - females (%)	NA	4.0	0.4

Libraries [13]

	Admin. Units	Svc. Pts.	Vols. (000)	Shelving (meters)	Vols. Added	Reg. Users
National	NA	NA	NA	NA	NA	NA
Nonspecialized	NA	NA	NA	NA	NA	NA
Public	NA	NA	NA	NA	NA	NA
Higher ed. (1993)	58	304	2,358	158,738	940,828	263,446
School	NA	NA	NA	NA	NA	NA

Daily Newspapers [14]

	1980	1985	1990	1994
Number of papers	NA	NA	NA	4
Circ. (000)	NA	NA	NA	160

Culture [15]

Cinema (seats per 1,000)	27.9
Annual attendance per person	1.3
Gross box office receipts (mil. Ruble)	1,219
Museums (reporting)	59
Visitors (000)	2,949
Annual receipts (000 Ruble)	19,738

Science and Technology

Scientific/Technical Forces [16]

Scientists/engineers	37,625[e]
Number female	17,005[e]
Technicians	6,687[e]
Number female	NA
Total	44,312[e]

R&D Expenditures [17]

U.S. Patents Issued [18]

Values show patents issued to citizens of the country by the U.S. Patents Office.

	1993	1994	1995
Number of patents	0	1	1

For sources, notes, and explanations, see Annotated Source Appendix, page 1061.

Government and Law

Organization of Government [19]

Long-form name:
Republic of Uzbekistan
Type:
Republic
Independence:
31 August 1991 (from Soviet Union)
National holiday:
Independence Day, 1 September (1991)
Constitution:
New constitution adopted 8 December 1992
Legal system:
Evolution of Soviet civil law; still lacks independent judicial system
Executive branch:
President; Prime Minister; First Deputy Prime Minister; 9 Deputy Prime Ministers; Cabinet of Ministers
Legislative branch:
Unicameral: Supreme Assembly (Oliy Majlis)
Judicial branch:
Supreme Court

Elections [20]

Supreme Assembly (Oliy Majlis)	% of seats
People's Democratic Party	27.6
Social Democratic Party	18.8
Fatherland Progress Party	5.6
Local government	48.0

Government Budget [21]

Crime [23]

	1994
Crime volume	
Cases known to police	73,561
Attempts (percent)	NA
Percent cases solved	NA
Crimes per 100,000 persons	334.37
Persons responsible for offenses	
Total number offenders	59,408
Percent female	11.60
Percent juvenile (14-18 yrs.)	7.40
Percent foreigners	0.30

Military Expenditures and Arms Transfers [22]

	1990	1991	1992[14]	1993	1994
Military expenditures					
Current dollars (mil.)[e]	NA	NA	NA	NA	375
1994 constant dollars (mil.)[e]	NA	NA	NA	NA	375
Armed forces (000)	NA	NA	40	41	20
Gross national product (GNP)					
Current dollars (mil.)[e]	NA	NA	57,410	56,680	53,410
1994 constant dollars (mil.)[e]	NA	NA	59,870	57,850	53,410
Central government expenditures (CGE)					
1994 constant dollars (mil.)[e]	NA	NA	25,620	NA	NA
People (mil.)	NA	NA	21.6	22.1	22.6
Military expenditure as % of GNP	NA	NA	NA	NA	0.7
Military expenditure as % of CGE	NA	NA	NA	NA	NA
Military expenditure per capita (1994)	NA	NA	NA	NA	17
Armed forces per 1,000 people (soldiers)	NA	NA	1.8	1.9	0.9
GNP per capita (1994 $)	NA	NA	2,767	2,614	2,363
Arms imports[6]					
Current dollars (mil.)	NA	NA	0	0	20
1994 constant dollars (mil.)	NA	NA	0	0	20
Arms exports[6]					
Current dollars (mil.)	NA	NA	0	0	20
1994 constant dollars (mil.)	NA	NA	0	0	20
Total imports[7]					
Current dollars (mil.)	NA	NA	1,768	2,245	2,459
1994 constant dollars (mil.)	NA	NA	1,843	2,291	2,459
Total exports[7]					
Current dollars (mil.)	NA	NA	1,506	2,050	3,020
1994 constant dollars (mil.)	NA	NA	1,570	2,092	3,020
Arms as percent of total imports[8]	NA	NA	0	0	0.8
Arms as percent of total exports[8]	NA	NA	0	0	0.7

Human Rights [24]

	SSTS	FL	FAPRO	PPCG	APROBC	TPW	PCPTW	STPEP	PHRFF	PRW	ASST	AFL
Observes						P	P					

	EAFRD	CPR	ESCR	SR	ACHR	MAAE	PVIAC	PVNAC	EAFDAW	TCIDTP	RC
Observes	P		P				P		P	P	

P = Party; S = Signatory; see Appendix for meaning of abbreviations.

Labor Force

Total Labor Force [25]

8.234 million

Labor Force by Occupation [26]

Agriculture and forestry	43%
Industry and construction	22
Other	35

Date of data: 1992

Unemployment Rate [27]

0.4% includes only officially registered unemployed; large numbers of underemployed workers (December 1995)

For sources, notes, and explanations, see Annotated Source Appendix, page 1061.

1007

Production Sectors

Commercial Energy Production and Consumption

Data are shown in quadrillion (10^{15}) BTUs and percent for 1995
Values for hydroelectric, nuclear, geothermal, solar, and wind power refer to electrical generation.

Production [28]	Consumption [29]
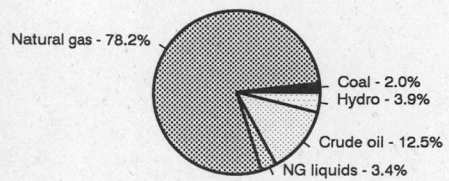	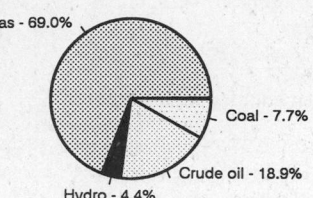
Natural gas - 78.2%	Natural gas - 69.0%
Coal - 2.0%	Coal - 7.7%
Hydro - 3.9%	Crude oil - 18.9%
Crude oil - 12.5%	Hydro - 4.4%
NG liquids - 3.4%	

Crude oil	0.247	Crude oil	0.337
Natural gas liquids	0.068	Dry natural gas	1.230
Dry natural gas	1.546	Coal	0.137
Coal	0.039	Net hydroelectric power	0.078
Net hydroelectric power	0.078	Total	1.782
Total	1.978		

Telecommunications [30]

- 1.458 million (1995 est.) telephones; poorly developed
- Domestic: NMT-450 analog cellular network in Tashkent
- International: linked by landline or microwave radio relay with CIS member states and to other countries via the Moscow international gateway switch; new Intelsat links to Tokyo and Ankara give Uzbekistan international access independent of Russian facilities; satellite earth stations - Orbita and Intelsat
- Radio: Broadcast stations: there is at least one state-owned broadcast station
- Television: Broadcast stations: 2

Transportation [31]

Railways: total: 3,460 km in common carrier service; does not include industrial lines; broad gauge: 3,460 km 1.520-m gauge (1990)

Highways: total: 78,400 km; paved: NA km; unpaved: NA km (1990 est.)

Airports

Total:	261
With paved runways over 3,047 m:	6
With paved runways 2,438 to 3,047 m:	14
With paved runways 1,524 to 2,437 m:	2
With paved runways 914 to 1,523 m:	8
With paved runways under 914 m:	5

Top Agricultural Products [32]

Produces cotton, vegetables, fruits, grain; livestock.

Top Mining Products [33]

Thousand metric tons except as noted[e]	1994[*]
Cement	530
Coal	3,800[83]
Copper, mine output, Cu content	75
Feldspar	70
Fluorspar	90
Kaolin	6,000
Lead, mine output, Pb content	20
Gas, natural (bil. cu. meters)	47,200[83]
Steel, crude	352[83]
Zinc, metal, smelter	50,000

Tourism [34]

For sources, notes, and explanations, see Annotated Source Appendix, page 1061.

Finance, Economics, and Trade

Industrial Summary [35]

Industrial Production: Growth rate 0% (1995 est.). **Industries**: Textiles, food processing, machine building, metallurgy, natural gas.

Economic Indicators [36]

- **National product**: GDP—purchasing power parity—$54.7 billion (1995 estimate as extrapolated from World Bank estimate for 1994)
- **National product real growth rate**: - 1% (1995 est.)
- **National product per capita**: $2,370 (1995 est.)
- **Inflation rate (consumer prices)**: 7.7% monthly average (January-October 1995 est.)
- **External debt**: $1.285 billion (of which $510 million to Russia)

Balance of Payments Summary [37]

Values in millions of dollars.

	1988	1989	1990	1991	1992[1]
Exports of goods (f.o.b.)	NA	NA	NA	NA	869
Imports of goods (f.o.b.)	NA	NA	NA	NA	929
Trade balance	NA	NA	NA	NA	-60
Services - debits	NA	NA	NA	NA	-12
Services - credits	NA	NA	NA	NA	8
Private transfers (net)	NA	NA	NA	NA	2
Government transfers (net)	NA	NA	NA	NA	NA
Long-term capital (net)	NA	NA	NA	NA	129
Short-term capital (net)	NA	NA	NA	NA	NA
Errors and omissions	NA	NA	NA	NA	103
Overall balance	NA	NA	NA	NA	170

Exchange Rates [38]

Currency: **som.**

Data are currency units per $1. Introduced provisional som-coupons 10 November 1993 which circulated parallel to the Russian rubles; became the sole legal currency 31 January 1994; was replaced in July 1994 by the som currency.

End December 1995	35.8
1994 yearend	25

Imports and Exports

Top Import Origins [39]

$2.9 billion (1995).

Origins	%
Other FSU countries	NA
Czech Republic	NA

Top Export Destinations [40]

$3.1 billion (1995).

Destinations	%
Russia	NA
Ukraine	NA
Eastern Europe	NA
US	NA

Foreign Aid [41]

	U.S. $	
ODA (1993)	71	million
Commitments (1992-95)	2.915	billion
Disbursements (1992-95)	135	million

Import and Export Commodities [42]

Import Commodities
Grain
Machinery and parts
Consumer durables
Other foods

Export Commodities
Cotton
Gold
Natural gas
Mineral fertilizers
Ferrous metals
Textiles
Food products

For sources, notes, and explanations, see Annotated Source Appendix, page 1061.

1009

Vanuatu

Geography [1]

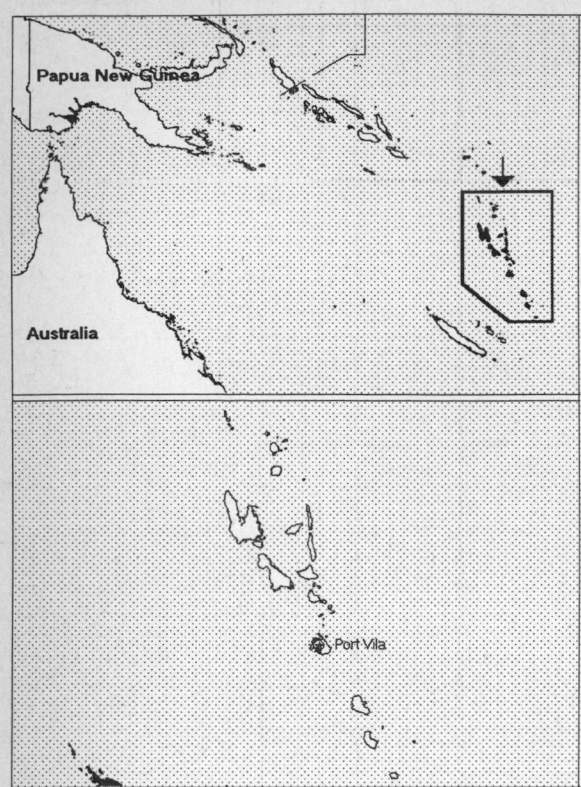

Total area:
14,760 sq km   5,699 sq mi
Land area:
14,760 sq km   5,699 sq mi
Comparative area:
Slightly larger than Connecticut
Note: Includes more than 80 islands
Land boundaries:
0 km
Coastline:
2,528 km
Climate:
Tropical; moderated by southeast trade winds
Terrain:
Mostly mountains of volcanic origin; narrow coastal plains
Natural resources:
Manganese, hardwood forests, fish
Land use:
Arable land: 1%
Permanent crops: 5%
Meadows and pastures: 2%
Forest and woodland: 1%
Other: 91%

Demographics [2]

	1970	1980	1990	1995[1]	1996	2000	2010	2020	2030
Population	85	117	154	174	178	193	230	266	298
Population density (persons per sq. mi.)	15	20	27	30	31	34	40	47	52
(persons per sq. km.)	6	8	10	12	12	13	16	18	20
Net migration rate (per 1,000 population)	NA	0.0	0.0	0.0	0.0	0.0	0.0	0.0	0.0
Births	NA	NA	NA	NA	5	NA	NA	NA	NA
Deaths	NA	NA	NA	NA	2	NA	NA	NA	NA
Life expectancy - males	NA	51.7	56.0	57.9	58.3	59.8	63.4	66.6	69.4
Life expectancy - females	NA	53.6	59.1	61.6	62.1	64.1	68.7	72.7	76.0
Birth rate (per 1,000)	NA	43.3	36.1	31.3	30.6	27.8	22.9	19.2	16.6
Death rate (per 1,000)	NA	14.0	10.4	9.1	8.8	8.1	6.9	6.5	6.7
Women of reproductive age (15-49 yrs.)	NA	26	35	42	43	48	62	73	80
of which are currently married	NA	NA	22	NA	27	31	41	NA	NA
Fertility rate	NA	6.4	5.0	4.1	4.0	3.5	2.6	2.2	2.1

Except as noted, values for vital statistics are in thousands; life expectancy is in years.

Health

Health Indicators [3]

Health Expenditures [4]

For sources, notes, and explanations, see Annotated Source Appendix, page 1061.

Human Factors

Women and Children [5]

Burden of Disease [6]

Population per physician (1989)	7,944.44
Population per nurse (1986)	462.05
Population per hospital bed	NA
AIDS cases per 100,000 people (1994)	*
Malaria cases per 100,000 people (1992)	8,490

Ethnic Division [7]

Indigenous Melanesian	94%
French	4%
Vietnamese, Chinese, Pacific Islanders	2%

Religion [8]

Presbyterian	36.7%
Anglican	15%
Catholic	15%
Indigenous beliefs	7.6%
Seventh-Day Adventist	6.2%
Church of Christ	3.8%
Other	15.7%

Major Languages [9]

English (official), French (official), pidgin (known as Bislama or Bichelama).

Education

Public Education Expenditures [10]

	1980	1989	1990	1991	1992	1994
Million (Vatu)						
Total education expenditure	NA	811	831	929	NA	NA
as percent of GNP	NA	4.8	4.4	4.8	NA	NA
as percent of total govt. expend.	NA	NA	NA	NA	NA	NA
Current education expenditure	NA	811	831	929	NA	NA
as percent of GNP	NA	4.8	4.4	4.8	NA	NA
as percent of current govt. expend.	NA	19.5	19.2	18.8	NA	NA
Capital expenditure	NA	0	-	-	NA	NA

Educational Attainment [11]

Age group (1979)	25 ı
Total population	38,488
Highest level attained (%)	
No schooling	37.2
First level	
Not completed	34.3
Completed	6.5
Entered second level	
S-1	14.7
S-2	7.3
Postsecondary	NA

Illiteracy [12]

	1979	1989	1995
Illiterate population (15+ yrs.)[2]	28,647	NA	NA
Illiteracy rate - total pop. (%)	47.1	NA	NA
Illiteracy rate - males (%)	42.7	NA	NA
Illiteracy rate - females (%)	52.2	NA	NA

Libraries [13]

	Admin. Units	Svc. Pts.	Vols. (000)	Shelving (meters)	Vols. Added	Reg. Users
National (1989)[6]	1	1	50	700	1,000	200
Nonspecialized	NA	NA	NA	NA	NA	NA
Public	NA	NA	NA	NA	NA	NA
Higher ed.	NA	NA	NA	NA	NA	NA
School	NA	NA	NA	NA	NA	NA

Daily Newspapers [14]

Culture [15]

Science and Technology

Scientific/Technical Forces [16]

R&D Expenditures [17]

U.S. Patents Issued [18]

For sources, notes, and explanations, see Annotated Source Appendix, page 1061.

1011

Government and Law

Organization of Government [19]

Long-form name:
Republic of Vanuatu

Type:
Republic

Independence:
30 July 1980 (from France and UK)

National holiday:
Independence Day, 30 July (1980)

Constitution:
30 July 1980

Legal system:
Unified system being created from former dual French and British systems

Executive branch:
President; Prime Minister; Deputy Prime Minister; Council of Ministers to Parliament

Legislative branch:
Unicameral: Parliament Note: The National Council of Chiefs advises on matters of custom and land

Judicial branch:
Supreme Court

Elections [20]

Parliament	% of seats
Union of Moderate Parties	34.0
Vanuatu Party	28.0
National United Party	18.0
Melanesian Progressive Party	10.0
Tan Union Party	4.0
Na-Griamel	2.0
Friend	2.0
Independent	2.0

Government Expenditures [21]

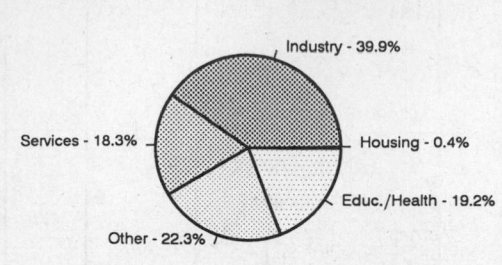

Industry - 39.9%
Housing - 0.4%
Educ./Health - 19.2%
Other - 22.3%
Services - 18.3%

(% distribution). Expend. for 1989 est.: 103 (Dollar mil.)

Defense Summary [22]

Branches: No regular military forces; Vanuatu Police Force (VPF; includes the paramilitary Vanuatu Mobile Force or VMF)

Manpower Availability: Males age 15-49 NA; males fit for military service NA;

Defense Expenditures: NA

Crime [23]

	1990
Crime volume	
Cases known to police	1,836
Attempts (percent)	NA
Percent cases solved	27.6
Crimes per 100,000 persons	1,191.8
Persons responsible for offenses	
Total number offenders	506
Percent female	NA
Percent juvenile	NA
Percent foreigners	NA

Human Rights [24]

	SSTS	FL	FAPRO	PPCG	APROBC	TPW	PCPTW	STPEP	PHRFF	PRW	ASST	AFL
Observes						P	P				1	

	EAFRD	CPR	ESCR	SR	ACHR	MAAE	PVIAC	PVNAC	EAFDAW	TCIDTP	RC
Observes							P	P	P		P

P = Party; S = Signatory; see Appendix for meaning of abbreviations.

Labor Force

Total Labor Force [25]

66,597 (1989 est.)

Labor Force by Occupation [26]

Agriculture	65%
Services	32
Industry	3

Date of data: 1995 est.

Unemployment Rate [27]

For sources, notes, and explanations, see Annotated Source Appendix, page 1061.

Production Sectors

Energy Resource Summary [28]

Electricity: Capacity: 17,000 kW. Production: 30 million kWh. Consumption per capita: 181 kWh (1993).

Telecommunications [30]

- 3,000 (1987 est.) telephones
- International: satellite earth station - 1 Intelsat (Pacific Ocean)
- Radio: Broadcast stations: AM 2, FM 0, shortwave 0
- Television: Broadcast stations: 0 (1987 est.) Televisions: 2,000 (1992 est.)

Top Agricultural Products [32]

Produces coconuts, cocoa, coffee, taro, yams, coconuts, fruits, vegetables; fish.

Top Mining Products [33]

Detailed information is not available. A summary of natural resources follows. **Mineral Resources**: Manganese.

Transportation [31]

Railways: 0 km

Highways: total: 1,021 km; paved: 238 km; unpaved: 783 km (1987 est.)

Merchant marine: total: 112 ships (1,000 GRT or over) totaling 1,587,286 GRT/2,173,970 DWT; ships by type: bulk 38, cargo 29, chemical tanker 3, combination bulk 1, container 3, liquefied gas tanker 5, livestock carrier 1, oil tanker 6, refrigerated cargo 16, vehicle carrier 10

Airports

Total:	31
With paved runways 2,438 to 3,047 m:	1
With paved runways 1,524 to 2,437 m:	1
With paved runways under 914 m:	17
With unpaved runways 1,524 to 2,437 m:	1
With unpaved runways 914 to 1,523 m:	11 (1995 est.)

Tourism [34]

	1990	1991	1992	1993	1994
Tourists	35	40	43	44	42
Cruise passengers	42	37	59	43	41
Tourism receipts	39	47	56	55	55
Tourism expenditures	1	1	1	1	1
Fare expenditures	4	4	4	6	6

Travelers are in thousands, money in million U.S. dollars.

For sources, notes, and explanations, see Annotated Source Appendix, page 1061.

1013

Finance, Economics, and Trade

GDP and Manufacturing Summary [35]

	1980	1985	1990	1991	1992
Gross Domestic Product					
Millions of 1980 dollars	113	171	185	191	191
Growth rate in percent	-11.46	1.11	4.70	3.40	0.00
Manufacturing Value Added					
Millions of 1980 dollars	3	7	12	13	15[e]
Growth rate in percent	13.98	11.23	1.94	13.40	13.69[e]
Manufacturing share in percent of current prices	4.2	3.8	5.9	NA	NA

Economic Indicators [36]

- **National product**: GDP—purchasing power parity—$210 million (1994 est.)
- **National product real growth rate**: 2% (1994 est.)
- **National product per capita**: $1,220 (1994 est.)
- **Inflation rate (consumer prices)**: 7% (1995 est.)
- **External debt**: $38.2 million (year-end 1993)

Balance of Payments Summary [37]

Values in millions of dollars.

	1989	1990	1991	1992	1993
Exports of goods (f.o.b.)	13.7	13.7	14.9	17.8	17.4
Imports of goods (f.o.b.)	-57.9	-79.3	-74.0	-66.8	-64.7
Trade balance	-44.2	-65.6	-59.23	-49.0	-47.3
Services - debits	-48.3	-57.2	-76.1	-74.4	-75.2
Services - credits	63.7	92.1	91.0	87.0	78.6
Private transfers (net)	4.6	11.1	17.7	7.0	10.8
Government transfers (net)	20.7	29.9	35.5	36.3	35.3
Long-term capital (net)	10.1	21.6	38.3	33.8	33.5
Short-term capital (net)	14.2	-7.9	-66.1	-10.1	-17.4
Errors and omissions	-13.0	-19.4	19.3	-26.5	-11.7
Overall balance	7.8	4.7	0.4	4.1	6.7

Exchange Rates [38]

Currency: **vatu**.
Symbol: **VT**.

Data are currency units per $1.

January 1996	114.40
1995	112.11
1994	116.41
1993	121.58
1992	113.39
1991	111.68

Imports and Exports

Top Import Origins [39]

$78.6 million (f.o.b., 1994 est.) Data are for 1992.

Origins	%
Australia	41
France	15
New Zealand	11
Japan	9
Fiji	6

Top Export Destinations [40]

$24.6 million (f.o.b., 1994 est.) Data are for 1993.

Destinations	%
EU	32
Japan	29
Australia	11
New Caledonia	7

Foreign Aid [41]

Recipient: ODA, $NA.

Import and Export Commodities [42]

Import Commodities	Export Commodities
Machines and vehicles	Copra
Food and beverages	Beef
Basic manufactures	Cocoa
Raw materials and fuels	Timber
Chemicals	Coffee

For sources, notes, and explanations, see Annotated Source Appendix, page 1061.

Venezuela

Geography [1]

Total area:
912,050 sq km   352,144 sq mi
Land area:
882,050 sq km   340,561 sq mi
Comparative area:
Slightly more than twice the size of California
Land boundaries:
Total 4,993 km, Brazil 2,200 km, Colombia 2,050 km, Guyana 743 km
Coastline:
2,800 km
Climate:
Tropical; hot, humid; more moderate in highlands
Terrain:
Andes Mountains and Maracaibo Lowlands in Northwest; central plains (llanos); Guiana Highlands in Southeast
Natural resources:
Petroleum, natural gas, iron ore, gold, bauxite, other minerals, hydropower, diamonds
Land use:
Arable land: 3%
Permanent crops: 1%
Meadows and pastures: 20%
Forest and woodland: 39%
Other: 37%

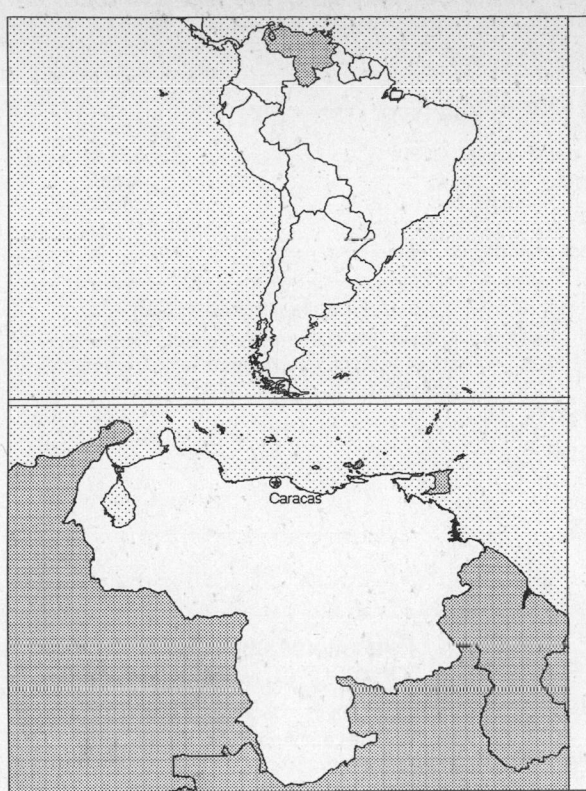

Demographics [2]

	1970	1980	1990	1995[1]	1996	2000	2010	2020	2030
Population	10,758	14,768	19,325	21,564	21,983	23,596	27,345	30,876	33,883
Population density (persons per sq. mi.)	32	43	57	63	65	69	80	91	99
(persons per sq. km.)	12	17	22	24	25	27	31	35	38
Net migration rate (per 1,000 population)	5.8	NA	-0.5	-0.4	-0.4	-0.2	0.0	0.0	0.0
Births	NA	NA	NA	NA	536	NA	NA	NA	NA
Deaths	NA	NA	90	NA	112	NA	NA	NA	NA
Life expectancy - males	63.5	NA	67.9	68.8	69.1	70.3	72.7	74.7	76.3
Life expectancy - females	69.1	NA	74.1	75.0	75.3	76.5	79.0	80.9	82.5
Birth rate (per 1,000)	36.1	NA	29.9	25.1	24.4	21.5	18.1	16.3	14.1
Death rate (per 1,000)	6.5	NA	5.4	5.2	5.1	4.9	4.9	5.4	6.4
Women of reproductive age (15-49 yrs.)	NA	NA	4,935	5,659	5,797	6,310	7,433	8,096	8,356
of which are currently married	NA	NA	2,778	NA	3,306	3,634	4,336	NA	NA
Fertility rate	5.0	NA	3.5	3.0	2.9	2.5	2.2	2.0	2.0

Except as noted, values for vital statistics are in thousands; life expectancy is in years.

Health

Health Indicators [3]

% of population with access to	
safe water (1990-95)	79
adequate sanitation (1990-95)	59
health services (1985-95)	NA
% of 1-year-olds immunized (1990-94) against	
TB (tuberculosis)	95
DPT (diphtheria, pertussis, tetanus)	63
polio	73
measles	94
% of contraceptive prevalence (1980-94)[1]	49
ORT use rate (1990-94)	80

Health Expenditures [4]

Total health expenditure, 1990 (official exchange rate)	
Millions of dollars	1,747
Dollars per capita	89
Health expenditures as a percentage of GDP	
Total	3.6
Public sector	2.0
Private sector	1.6
Development assistance for health	
Total aid flows (millions of dollars)[1]	2.0
Aid flows per capita (dollars)	0.1
Aid flows as a percentage of total health expenditure	0.1

For sources, notes, and explanations, see Annotated Source Appendix, page 1061.

1015

Human Factors

Women and Children [5]

% of pregnant women immunized (tetanus 1990-94)	NA
% of births attended by trained health personnel (1983-94)	69
Maternal mortality rate (1980-92)	NA
Under-5 mortality rate (1994)	24
% under-5 moderately/severely underweight (1980-1994)	6

Burden of Disease [6]

Population per physician (1990)	633.44
Population per nurse (1990)	328.98
Population per hospital bed (1992)	385.00
AIDS cases per 100,000 people (1994)	2.4
Malaria cases per 100,000 people (1992)	105

Ethnic Division [7]

Mestizo	67%
White	21%
Black	10%
Amerindian	2%

Religion [8]

Nominally Roman Catholic	96%
Protestant	2%
Other	2%

Major Languages [9]

Spanish (official), native dialects spoken by about 200,000 Amerindians in the remote interior.

Education

Public Education Expenditures [10]

	1980	1985	1990	1992	1993	1994
Million (Bolivar)						
Total education expenditure	13,162	23,068	69,352	211,659	242,827	434,282
as percent of GNP	4.4	5.1	3.1	5.3	4.6	5.1
as percent of total govt. expend.	14.7	20.3	12.0	23.5	22.0	22.4
Current education expenditure	12,524	NA	NA	NA	NA	419,515
as percent of GNP	4.2	NA	NA	NA	NA	4.9
as percent of current govt. expend.	24.3	NA	NA	NA	NA	31.2
Capital expenditure	639	NA	NA	NA	NA	14,767

Educational Attainment [11]

Age group (1990)[13]	25+
Total population	7,680,427
Highest level attained (%)	
No schooling	21.2
First level	
Not completed	55.0
Completed	NA
Entered second level	
S-1	12.0
S-2	NA
Postsecondary	11.8

Illiteracy [12]

In thousands and percent[1]	1990	1995	2000
Illiterate population (15+ yrs.)	1,131	1,244	1,154
Illiteracy rate - total pop. (%)	9.3	8.9	7.3
Illiteracy rate - males (%)	8.4	8.2	6.9
Illiteracy rate - females (%)	10.2	9.6	7.7

Libraries [13]

	Admin. Units	Svc. Pts.	Vols. (000)	Shelving (meters)	Vols. Added	Reg. Users
National (1993)	1	NA	5,115	NA	NA	99,274
Nonspecialized	NA	NA	NA	NA	NA	NA
Public (1993)	23	672	3,459	NA	NA	13 mil
Higher ed. (1989)	78	168	NA	NA	NA	NA
School (1992)	539	NA	NA	NA	NA	NA

Daily Newspapers [14]

	1980	1985	1990	1994
Number of papers	66	55	54	89
Circ. (000)	2,937	2,700[e]	2,800[e]	4,600[e]

Culture [15]

Cinema (seats per 1,000)	6.3
Annual attendance per person	0.9
Gross box office receipts (mil. Bolivar)	1,896
Museums (reporting)	-
Visitors (000)	NA
Annual receipts (000 Bolivar)	NA

Science and Technology

Scientific/Technical Forces [16]

Scientists/engineers	4,258[e]
Number female	1,490[e]
Technicians	650[e]
Number female	205[e]
Total	4,908[e]

R&D Expenditures [17]

	Bolivar (000) 1992
Total expenditure[7]	19,622,200
Capital expenditure	NA
Current expenditure	NA
Percent current	NA

U.S. Patents Issued [18]

Values show patents issued to citizens of the country by the U.S. Patents Office.

	1993	1994	1995
Number of patents	34	28	31

For sources, notes, and explanations, see Annotated Source Appendix, page 1061.

Government and Law

Organization of Government [19]

Long-form name:
Republic of Venezuela
Type:
Republic
Independence:
5 July 1811 (from Spain)
National holiday:
Independence Day, 5 July (1811)
Constitution:
23 January 1961
Legal system:
Based on Napoleonic code; judicial review of legislative acts in Cassation Court only; has not accepted compulsory ICJ jurisdiction
Executive branch:
President; Council of Ministers
Legislative branch:
Bicameral Congress of the Republic (Congreso de la Republica): Senate (Senado) and Chamber of Deputies (Camara de Diputados)
Judicial branch:
Supreme Court of Justice (Corte Suprema de Justicia)

Elections [20]

Chamber of Deputies	% of votes
Democratic Action	27.9
Social Christian Party	26.9
Movement Toward Socialism	12.4
National Convergence	12.9
Causa R	19.9

Government Budget [21]

For 1995 est.

	$ bil.
Revenues	7.25
Expenditures	9.8
Capital expenditures	NA

Military Expenditures and Arms Transfers [22]

	1990	1991	1992	1993	1994
Military expenditures					
Current dollars (mil.)[e]	940	1,978	1,486	1,029	747
1994 constant dollars (mil.)[e]	1,046	2,120	1,550	1,050	747
Armed forces (000)	75	73	75	75	75
Gross national product (GNP)					
Current dollars (mil.)	45,810	53,090	57,050	58,160	57,710
1994 constant dollars (mil.)	50,980	56,910	59,490	59,360	57,710
Central government expenditures (CGE)					
1994 constant dollars (mil.)	11,670	13,670	13,710	12,690	13,430[e]
People (mil.)	18.8	19.2	19.7	20.1	20.6
Military expenditure as % of GNP	2.1	3.7	2.6	1.8	1.3
Military expenditure as % of CGE	9.0	15.5	11.3	8.3	5.6
Military expenditure per capita (1994 $)	56	110	79	52	36
Armed forces per 1,000 people (soldiers)	4.0	3.8	3.8	3.7	3.6
GNP per capita (1994 $)	2,715	2,960	3,024	2,950	2,807
Arms imports[6]					
Current dollars (mil.)	190	200	80	70	60
1994 constant dollars (mil.)	211	214	83	71	60
Arms exports[6]					
Current dollars (mil.)	5	0	0	0	0
1994 constant dollars (mil.)	6	0	0	0	0
Total imports[7]					
Current dollars (mil.)	7,335	11,150	14,070	12,200	8,879
1994 constant dollars (mil.)	8,163	11,950	14,670	12,450	8,879
Total exports[7]					
Current dollars (mil.)	17,500	15,150	14,180	14,070	15,480
1994 constant dollars (mil.)	19,470	16,240	14,790	14,360	15,480
Arms as percent of total imports[8]	2.6	1.8	0.6	0.6	0.7
Arms as percent of total exports[8]	0	0	0	0	0

Crime [23]

	1994
Crime volume	
Cases known to police	236,341
Attempts (percent)	NA
Percent cases solved	45.16
Crimes per 100,000 persons	1,105.56
Persons responsible for offenses	
Total number offenders	106,751
Percent female	NA
Percent juvenile	NA
Percent foreigners	NA

Human Rights [24]

	SSTS	FL	FAPRO	PPCG	APROBC	TPW	PCPTW	STPEP	PHRFF	PRW	ASST	AFL
Observes		P	P	P	P	P	P	P		P		P
	EAFRD	CPR	ESCR	SR	ACHR	MAAE	PVIAC	PVNAC	EAFDAW	TCIDTP	RC	
Observes	P	P	P	P	P	P			P	P	P	

P = Party; S = Signatory; see Appendix for meaning of abbreviations.

Labor Force

Total Labor Force [25]

7.6 million

Labor Force by Occupation [26]

Services	63%
Industry	25
Agriculture	12

Date of data: 1993

Unemployment Rate [27]

11.7% (1995 est.)

For sources, notes, and explanations, see Annotated Source Appendix, page 1061.

1017

Production Sectors

Commercial Energy Production and Consumption

Data are shown in quadrillion (10^{15}) BTUs and percent for 1995
Values for hydroelectric, nuclear, geothermal, solar, and wind power refer to electrical generation.

Production [28]

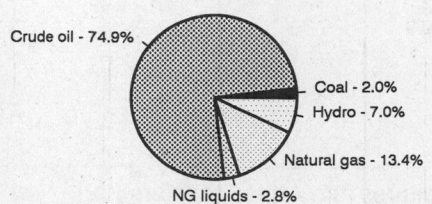

Crude oil - 74.9%
Coal - 2.0%
Hydro - 7.0%
Natural gas - 13.4%
NG liquids - 2.8%

Consumption [29]

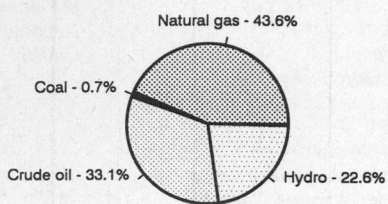

Natural gas - 43.6%
Coal - 0.7%
Crude oil - 33.1%
Hydro - 22.6%

Crude oil	6.158
Natural gas liquids	0.228
Dry natural gas	1.106
Coal	0.161
Net hydroelectric power	0.572
Total	8.225

Crude oil	0.839
Dry natural gas	1.106
Coal	0.019
Net hydroelectric power	0.572
Total	2.536

Telecommunications [30]

- 1.44 million (1987 est.) telephones; modern and expanding
- Domestic: domestic satellite system with 3 earth stations
- International: 3 submarine coaxial cables; satellite earth station - 1 Intelsat (Atlantic Ocean)
- Radio: Broadcast stations: AM 181, FM 0, shortwave 26 Radios: 9.04 million (1992 est.)
- Television: Broadcast stations: 59 Televisions: 3.3 million (1992 est.)

Transportation [31]

Railways: total: 584 km (336 km single track; 248 km privately owned); standard gauge: 584 km 1.435-m gauge

Highways: total: 93,472 km; paved: 29,954 km; unpaved: 63,518 km (1993 est.)

Merchant marine: total: 32 ships (1,000 GRT or over) totaling 612,645 GRT/1,090,707 DWT; ships by type: bulk 4, cargo 9, combination bulk 1, liquefied gas tanker 2, oil tanker 12, passenger-cargo 1, roll-on/roll-off cargo 2, short-sea passenger 1 (1995 est.)

Airports

Total:	377
With paved runways over 3,047 m:	5
With paved runways 2,438 to 3,047 m:	10
With paved runways 1,524 to 2,437 m:	34
With paved runways 914 to 1,523 m:	59
With paved runways under 914 m:	165

Top Agricultural Products [32]

Agriculture accounts for 5% of the GDP; produces corn, sorghum, sugarcane, rice, bananas, vegetables, coffee; beef, pork, milk, eggs; fish.

Top Mining Products [33]

Thousand metric tons except as noted	7/95*
Bauxite	4,790
Gold, mine output, Au content (kg.)	9,990
Iron ore and concentrate	18,300
Cement, hydraulic	6,930
Diamond, gem and industrial (carats)	387,000
Limestone	15,700
Coal, bituminous	4,630
Gas, natural, gross (mil. cu. meters)	44,600
Liquid petroleum gas (000 42-gal. bls.)	42,000[e]
Petroleum, crude (000 42-gal. bls.)	899,000

Tourism [34]

	1990	1991	1992	1993	1994
Visitors	655	724	581	520	506
Tourists[12]	525	598	446	396	429
Excursionists	130	126	135	124	67
Tourism receipts	496	510	437	554	486
Tourism expenditures	1,023	1,227	1,428	2,083	1,429
Fare receipts	153	195	121	110	125
Fare expenditures	98	143	151	110	97

Travelers are in thousands, money in million U.S. dollars.

For sources, notes, and explanations, see Annotated Source Appendix, page 1061.

Manufacturing Sector

Manufacturing Summary [35]

	1987		1988		1989		1990		1991	
	$ bil.	%	$ bil.	%	$ bil.	%	$ bil.	%	$ bil.	%
Establishments or enterprises (number)	10,865	0.463	10,859	0.517	10,574	0.564	10,606	0.592	11,186	1.463
Total employment (000)	542	0.400	572	0.418	536	0.436	531	0.480	567	0.817
Production workers (000)	385	0.571	406	0.649	367	0.499	363	0.512	390	1.186
Output ($ bil.)	29	0.287	38	0.328	25	0.211	27	0.238	30	0.299
Value added ($ bil.)	12	0.286	15	0.310	11	0.230	13	0.269	13	0.382
Capital investment ($ mil.)	1,310	0.301	3,374	0.709	2,315	0.423	1,779	0.320	1,912	0.502
M & E investment ($ mil.)	-		-		-		-		-	
Employees per establishment (number)	50	86.306	53	80.812	51	77.160	50	81.059	51	55.815
Production workers per establishment	35	123.311	37	125.469	35	88.374	34	86.454	35	81.010
Output per establishment ($ mil.)	3	61.921	4	63.351	2	37.335	3	40.216	3	20.406
Capital investment per estab. ($ mil.)	0.121	64.893	0.311	136.968	0.219	74.877	0.168	54.077	0.171	34.312
M & E per establishment ($ mil)	-		-		-		-		-	
Payroll per employee ($)	6,529	72.807	8,026	80.560	4,668	41.780	4,591	33.096	5,310	36.368
Wages per production worker ($)	2,768	35.080	3,238	38.323	2,074	20.678	1,943	16.361	2,302	20.927
Hours per production worker (hours)	-		-		-		-		-	
Output per employee ($)	53,734	71.745	66,771	78.393	46,373	48.386	51,416	49.613	53,674	36.560
Capital investment per employee ($)	2,415	75.109	5,898	109.490	4,319	97.040	3,350	66.713	3,370	61.475
M & E per employee ($)	-		-		-		-		-	

Note: Columns headed % show percent of world total or ratio. Ratios closest to 100 are closest to world average. M & E stands for machinery & equipment.

Output in Manufacturing

	1987		1988		1989		1990		1991	
	$ bil.	%	$ bil.	%	$ bil.	%	$ bil.	%	$ bil.	%
3110 - Food products	4.216	14.54	5.443	14.32	3.679	14.72	3.946	14.61	4.634	15.45
3130 - Beverages	1.105	3.81	1.463	3.85	0.892	3.57	0.951	3.52	1.160	3.87
3140 - Tobacco	0.471	1.62	0.578	1.52	0.385	1.54	0.348	1.29	0.414	1.38
3210 - Textiles	1.007	3.47	1.453	3.82	0.791	3.16	0.711	2.63	0.757	2.52
3211 - Spinning, weaving, etc.	0.884	3.05	1.285	3.38	0.718	2.87	0.636	2.36	0.676	2.25
3220 - Wearing apparel	0.749	2.58	1.014	2.67	0.464	1.86	0.429	1.59	0.466	1.55
3230 - Leather and products	0.185	0.64	0.280	0.74	0.129	0.52	0.127	0.47	0.158	0.53
3240 - Footwear	0.383	1.32	0.519	1.37	0.252	1.01	0.272	1.01	0.313	1.04
3310 - Wood products	0.165	0.57	0.259	0.68	0.122	0.49	0.090	0.33	0.111	0.37
3320 - Furniture, fixtures	0.275	0.95	0.341	0.90	0.180	0.72	0.165	0.61	0.231	0.77
3410 - Paper and products	0.758	2.61	0.963	2.53	0.692	2.77	0.738	2.73	0.957	3.19
3411 - Pulp, paper, etc.	0.512	1.77	0.613	1.61	0.454	1.82	0.513	1.90	0.623	2.08
3420 - Printing, publishing	0.541	1.87	0.776	2.04	0.508	2.03	0.492	1.82	0.576	1.92
3510 - Industrial chemicals	1.069	3.69	1.374	3.62	0.952	3.81	1.041	3.86	1.403	4.68
3511 - Basic chemicals, excl fertilizers	0.383	1.32	0.468	1.23	0.375	1.50	0.389	1.44	0.598	1.99
3513 - Synthetic resins, etc.	0.418	1.44	0.564	1.48	0.367	1.47	0.383	1.42	0.433	1.44
3520 - Chemical products nec	1.444	4.98	1.927	5.07	1.171	4.68	1.281	4.74	1.574	5.25
3522 - Drugs and medicines	0.379	1.31	0.486	1.28	0.330	1.32	0.377	1.40	0.436	1.45
3530 - Petroleum refineries	2.975	10.26	3.275	8.62	4.189	16.76	5.861	21.71	4.254	14.18
3540 - Petroleum, coal products	0.025	0.09	0.037	0.10	0.034	0.14	0.039	0.14	0.053	0.18
3550 - Rubber products	0.323	1.11	0.413	1.09	0.244	0.98	0.290	1.07	0.343	1.14
3560 - Plastic products nec	0.777	2.68	1.095	2.88	0.615	2.46	0.547	2.03	0.708	2.36
3610 - Pottery, china, etc.	0.058	0.20	0.075	0.20	0.039	0.16	0.033	0.12	0.042	0.14
3620 - Glass and products	0.224	0.77	0.283	0.74	0.174	0.70	0.190	0.70	0.232	0.77
3690 - Nonmetal products nec	0.641	2.21	0.905	2.38	0.535	2.14	0.552	2.04	0.651	2.17
3710 - Iron and steel	1.760	6.07	2.045	5.38	1.378	5.51	1.465	5.43	1.581	5.27
3720 - Nonferrous metals	1.116	3.85	1.427	3.76	1.484	5.94	1.619	6.00	1.392	4.64
3810 - Metal products	1.119	3.86	1.509	3.97	0.953	3.81	0.948	3.51	1.105	3.68
3820 - Machinery nec	0.514	1.77	0.838	2.21	0.479	1.92	0.427	1.58	0.646	2.15
3825 - Office, computing machinery	-	-	-	-	0.004	0.02	0.006	0.02	0.007	0.02
3830 - Electrical machinery	0.783	2.70	1.206	3.17	0.693	2.77	0.592	2.19	0.674	2.25
3832 - Radio, television, etc.	0.198	0.68	0.338	0.89	0.154	0.62	0.119	0.44	0.123	0.41
3840 - Transportation equipment	1.743	6.01	2.325	6.12	0.644	2.58	0.772	2.86	1.449	4.83
3841 - Shipbuilding, repair	0.019	0.07	0.037	0.10	0.016	0.06	0.014	0.05	0.018	0.06
3843 - Motor vehicles	1.700	5.86	2.252	5.93	0.598	2.39	0.748	2.77	1.412	4.71
3850 - Professional goods	0.070	0.24	0.107	0.28	0.053	0.21	0.087	0.32	0.123	0.41
3900 - Industries nec	0.163	0.56	0.228	0.60	0.115	0.46	0.114	0.42	0.121	0.40

Note: Codes are International Standard Industry codes (ISIC). Percentages are % of total Output. [f]: Factor Prices; [p]: Producer Prices.

For sources, notes, and explanations, see Annotated Source Appendix, page 1061.

1019

Finance, Economics, and Trade

Economic Indicators [36]

- **National product**: GDP—purchasing power parity— $195.5 billion (1995 est.)
- **National product real growth rate**: 2.2% (1995 est.)
- **National product per capita**: $9,300 (1995 est.)
- **Inflation rate (consumer prices)**: 57% (1995 est.)
- **External debt**: $40.1 billion (1994)

Balance of Payments Summary [37]

Values in millions of dollars.

	1989	1990	1991	1992	1993
Exports of goods (f.o.b.)	12,915	17,444	14,968	13,988	14,019
Imports of goods (f.o.b.)	-7,283	-6,807	-10,131	-12,714	-11,117
Trade balance	5,632	10,637	4,837	1,274	2,902
Services - debits	-5,979	-6,107	-6,357	-7,823	-8,082
Services - credits	2,695	4,032	3,605	3,149	3,274
Private transfers (net)	-171	-259	-316	-347	-310
Government transfers (net)	-16	-24	-33	-6	-7
Long-term capital (net)	-1,318	108	2,490	2,902	2,380
Short-term capital (net)	-2,059	-3,229	-396	-183	-1,347
Errors and omissions	1,418	-1,742	-1,516	-295	407
Overall balance	202	3,416	2,314	-1,329	-783

Exchange Rates [38]

Currency: **bolivar.**
Symbol: **Bs.**

Data are currency units per $1.

January 1996	288.690
1995	176.843
1994	148.503
1993	90.826
1992	68.376
1991	56.816

Imports and Exports

Top Import Origins [39]

$11.6 billion (f.o.b., 1995).

Origins	%
US	40
Germany	NA
Japan	NA
Netherlands	NA
Canada	NA

Top Export Destinations [40]

$18.3 billion (f.o.b., 1995).

Destinations	%
US and Puerto Rico	55
Japan	NA
Netherlands	NA
Italy	NA

Foreign Aid [41]

Recipient: ODA, $46 million (1993).

Import and Export Commodities [42]

Import Commodities	Export Commodities
Raw materials	Petroleum 72%
Machinery and equipment	Bauxite and aluminum
Transport equipment	Steel
Construction materials	Chemicals
	Agricultural products
	Basic manufactures

Vietnam

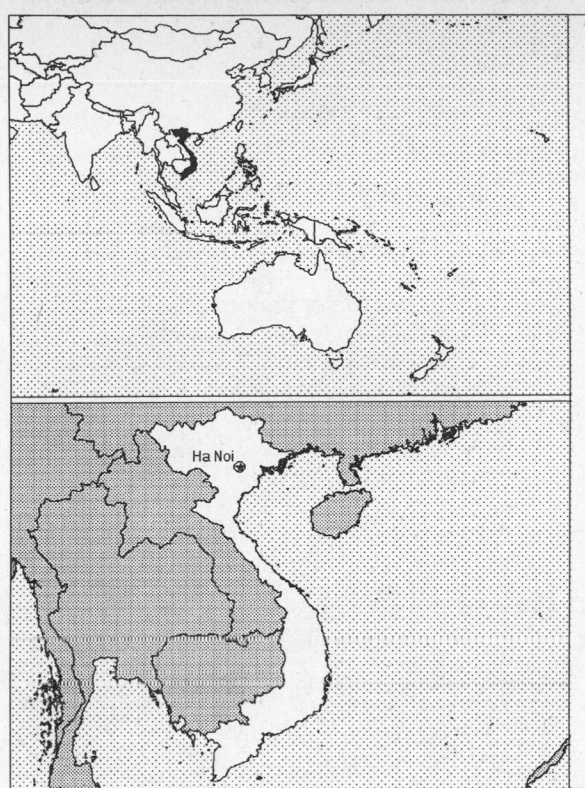

Ha Noi

Geography [1]

Total area:
329,560 sq km   127,244 sq mi
Land area:
325,360 sq km   125,622 sq mi
Comparative area:
Slightly larger than New Mexico
Land boundaries:
Total 3,818 km, Cambodia 982 km, China 1,281 km, Laos 1,555 km
Coastline:
3,444 km (excludes islands)
Climate:
Tropical in South; monsoonal in North with hot, rainy season (mid-May to mid-September) and warm, dry season (mid-October to mid-March)
Terrain:
Low, flat delta in South and North; central highlands; hilly, mountainous in far North and Northwest
Natural resources:
Phosphates, coal, manganese, bauxite, chromate, offshore oil deposits, forests
Land use:
Arable land: 22%
Permanent crops: 2%
Meadows and pastures: 1%
Forest and woodland: 40%
Other: 35%

Demographics [2]

	1970	1980	1990	1995[1]	1996	2000	2010	2020	2030
Population	42,978	54,234	66,314	72,815	73,977	78,350	88,602	99,153	108,117
Population density (persons per sq. mi.)	342	432	528	580	589	624	705	789	861
(persons per sq. km.)	132	167	204	224	227	241	272	305	332
Net migration rate (per 1,000 population)	NA	NA	-1.4	-0.7	-0.4	-0.5	-0.4	0.0	0.0
Births	NA	NA	NA	NA	1,701	NA	NA	NA	NA
Deaths	NA	NA	NA	NA	514	NA	NA	NA	NA
Life expectancy - males	NA	NA	62.6	64.4	64.7	66.1	69.1	71.7	73.8
Life expectancy - females	NA	NA	67.0	69.1	69.5	71.0	74.5	77.3	79.6
Birth rate (per 1,000)	NA	NA	30.5	23.8	23.0	20.0	18.2	16.0	13.8
Death rate (per 1,000)	NA	NA	8.1	7.1	7.0	6.5	6.0	5.9	6.6
Women of reproductive age (15-49 yrs.)	NA	NA	16,519	18,947	19,461	21,739	25,503	26,668	27,194
of which are currently married	NA	NA	10,095	NA	12,204	13,605	16,511	NA	NA
Fertility rate	NA	NA	3.7	2.8	2.7	2.3	2.0	2.0	2.0

Except as noted, values for vital statistics are in thousands; life expectancy is in years.

Health

Health Indicators [3]

% of population with access to	
safe water (1990-95)	36
adequate sanitation (1990-95)	22
health services (1985-95)	90
% of 1-year-olds immunized (1990-94) against	
TB (tuberculosis)	95
DPT (diphtheria, pertussis, tetanus)	94
polio	94
measles	96
% of contraceptive prevalence (1980-94)	53
ORT use rate (1990-94)	52

Health Expenditures [4]

Total health expenditure, 1990 (official exchange rate)	
Millions of dollars	157
Dollars per capita	2
Health expenditures as a percentage of GDP	
Total	2.1
Public sector	1.1
Private sector	1.0
Development assistance for health	
Total aid flows (millions of dollars)[1]	25
Aid flows per capita (dollars)	0.4
Aid flows as a percentage of total health expenditure	15.9

For sources, notes, and explanations, see Annotated Source Appendix, page 1061.

Human Factors

Women and Children [5]

% of pregnant women immunized (tetanus 1990-94)	78
% of births attended by trained health personnel (1983-94)	95
Maternal mortality rate (1980-92)	120
Under-5 mortality rate (1994)	46
% under-5 moderately/severely underweight (1980-1994)	42

Burden of Disease [6]

Population per physician (1992)	2,278.74
Population per nurse (1990)	398.82
Population per hospital bed (1990)	260.90
AIDS cases per 100,000 people (1994)	0.1
Malaria cases per 100,000 people (1992)	304

Ethnic Division [7]

There are also Muong, Thai, Meo, Khmer, Man, Cham.

Vietnamese	85%-90%,
Chinese	3%

Religion [8]

Buddhist, Taoist, Roman Catholic, indigenous beliefs, Islam, Protestant.

Major Languages [9]

Vietnamese (official), French, Chinese, English, Khmer, tribal languages (Mon-Khmer and Malayo-Polynesian).

Education

Public Education Expenditures [10]

Educational Attainment [11]

Age group (1989)	25+
Total population	26,466,214
Highest level attained (%)	
No schooling	16.6
First level	
Not completed	69.8
Completed	NA
Entered second level	
S-1	10.6
S-2	NA
Postsecondary	2.6

Illiteracy [12]

In thousands and percent[1]	1990	1995	2000
Illiterate population (15+ yrs.)	3,797	2,916	2,332
Illiteracy rate - total pop. (%)	9.4	6.3	4.4
Illiteracy rate - males (%)	5.3	3.6	2.5
Illiteracy rate - females (%)	13.0	8.8	6.1

Libraries [13]

	Admin. Units	Svc. Pts.	Vols. (000)	Shelving (meters)	Vols. Added	Reg. Users
National	NA	NA	NA	NA	NA	NA
Nonspecialized	NA	NA	NA	NA	NA	NA
Public (1993)	4	566	12,737	NA	NA	NA
Higher ed.	NA	NA	NA	NA	NA	NA
School	NA	NA	NA	NA	NA	NA

Daily Newspapers [14]

	1980	1985	1990	1994
Number of papers	4	4	5	4
Circ. (000)	520[e]	540[e]	560[e]	570[e]

Culture [15]

Cinema (seats per 1,000)	NA
Annual attendance per person	3.8
Gross box office receipts (mil. Dong)	NA
Museums (reporting)	NA
Visitors (000)	NA
Annual receipts (000 Dong)	NA

Science and Technology

Scientific/Technical Forces [16]

Scientists/engineers	20,000
Number female	NA
Technicians	NA
Number female	NA
Total[35]	NA

R&D Expenditures [17]

	Dong (000) 1985
Total expenditure[19]	498,000
Capital expenditure	NA
Current expenditure	NA
Percent current	NA

U.S. Patents Issued [18]

For sources, notes, and explanations, see Annotated Source Appendix, page 1061.

Government and Law

Organization of Government [19]

Long-form name:
Socialist Republic of Vietnam
Type:
Communist state
Independence:
2 September 1945 (from France)
National holiday:
Independence Day, 2 September (1945)
Constitution:
15 April 1992
Legal system:
Based on communist legal theory and
French civil law system
Executive branch:
President; Prime Minister; First Deputy
Prime Minister; 2 Deputy Prime Ministers;
Cabinet
Legislative branch:
Unicameral: National Assembly (Quoc-
Hoi)
Judicial branch:
Supreme People's Court

Elections [20]

National Assembly. Only party -
Vietnam Communist Party (VCP).
Elections last held 19 July 1992 (next to
be held NA July 1997); results - VCP is
the only party; seats - (395 total) VCP or
VCP-approved 395.

Government Budget [21]

For 1995 est.

	$ bil.
Revenues	4.67
Expenditures	5
Capital expenditures	1.36

Crime [23]

Military Expenditures and Arms Transfers [22]

	1990	1991	1992	1993	1994
Military expenditures					
Current dollars (mil.)[e]	723	720	NA	NA	435
1994 constant dollars (mil.)[e]	804	772	NA	NA	435
Armed forces (000)	1,052	1,041	857	857	857
Gross national product (GNP)					
Current dollars (mil.)[e]	13,900	15,000	16,730	18,490	20,180
1994 constant dollars (mil.)[e]	15,470	16,080	17,450	18,870	20,180
Central government expenditures (CGE)					
1994 constant dollars (mil.)[e]	3,032	1,929	2,086	NA	4,500
People (mil.)	67.7	69.1	70.4	71.8	73.1
Military expenditure as % of GNP	5.2	4.8	NA	NA	2.2
Military expenditure as % of CGE	26.5	40.0	NA	NA	9.7
Military expenditure per capita (1994 $)	12	11	NA	NA	6
Armed forces per 1,000 people (soldiers)	15.5	15.1	12.2	11.9	11.7
GNP per capita (1994 $)	228	233	248	263	276
Arms imports[6]					
Current dollars (mil.)	1,100	200	10	10	80
1994 constant dollars (mil.)	1,224	214	10	10	80
Arms exports[6]					
Current dollars (mil.)	0	0	10	0	0
1994 constant dollars (mil.)	0	0	10	0	0
Total imports[7]					
Current dollars (mil.)	2,752	2,338	2,541	3,415	4,200[e]
1994 constant dollars (mil.)	3,063	2,506	2,650	3,485	4,200[e]
Total exports[7]					
Current dollars (mil.)	2,404	2,087	2,581	2,971	3,600[e]
1994 constant dollars (mil.)	2,675	2,237	2,691	3,032	3,600[e]
Arms as percent of total imports[8]	40.0	8.6	0.4	0.3	1.9
Arms as percent of total exports[8]	0	0	0.4	0	0

Human Rights [24]

	SSTS	FL	FAPRO	PPCG	APROBC	TPW	PCPTW	STPEP	PHRFF	PRW	ASST	AFL
Observes		P		P	P	P	P					
	EAFRD	CPR	ESCR	SR	ACHR	MAAE	PVIAC	PVNAC	EAFDAW	TCIDTP	RC	
Observes	P	P	P				P		P		P	

P = Party; S = Signatory; see Appendix for meaning of abbreviations.

Labor Force

Total Labor Force [25]

32.7 million

Labor Force by Occupation [26]

Agricultural	65%
Industrial and service	35

Date of data: 1990 est.

Unemployment Rate [27]

25% (1995 est.)

For sources, notes, and explanations, see Annotated Source Appendix, page 1061.

1023

Production Sectors

Commercial Energy Production and Consumption

Data are shown in quadrillion (10^{15}) BTUs and percent for 1995
Values for hydroelectric, nuclear, geothermal, solar, and wind power refer to electrical generation.

Production [28]

Crude oil - 51.3%
Renewable - 1.7%
Natural gas - 6.6%
Hydro - 14.0%
Coal - 26.4%

Crude oil	0.381
Dry natural gas	0.049
Coal	0.196
Net hydroelectric power	0.104
Geothermal, solar, wind	0.013
Total	0.743

Consumption [29]

Crude oil - 38.6%
Natural gas - 9.7%
Renewable - 2.6%
Coal - 28.5%
Hydro - 20.6%

Crude oil	0.195
Dry natural gas	0.049
Coal	0.144
Net hydroelectric power	0.104
Geothermal, solar, wind	0.013
Total	0.505

Telecommunications [30]

- 800,000 (1995 est.) telephones; telecommunications lags far behind other countries in Southeast Asia, but considerable progress since 1991 in upgrading the system; 100% of provincial switch boards digitized, fiber-optic and microwave transmission systems extended from Hanoi, Da Nang, and Ho Chi Minh City to all provinces; the density of telephone receivers nationwide doubled from 1993 to 1995, but still far behind others in region; Vietnam's strategy aims to increase telephone density to 30 per 1,000 inhabitants by the year 2000, authorities estimate about $2.7 billion will be spent on upgrades by then
- International: satellite earth stations - 2 Intersputnik (Indian Ocean region)
- Radio: Broadcast stations: FM 228 Radios: 7.215 million (1992 est.)
- Television: Broadcast stations: 36 (repeaters 77) Televisions: 2.9 million (1992 est.)

Top Agricultural Products [32]

Agriculture accounts for 28% of the GDP; produces paddy rice, corn, potatoes, rubber, soybeans, coffee, tea, bananas; poultry, pigs; fish catch of 943,100 metric tons (1989 est.).

Top Mining Products [33]

Thousand metric tons except as noted[e]	6/6/95*
Cement, hydraulic	7,200
Coal, anthracite	6,100[r]
Gold (kg.)	10,000
Gypsum	30
Nitrogen, N content of ammonia	53
Petroleum, crude (000 42-gal. bls.)	51,000[r]
Phosphate rock, P2O5 content	144[r]
Salt	375
Steel, crude	300
Zinc: mine output, Zn content	15

Transportation [31]

Railways: total: 2,835 km (in addition, there are 224 km not restored to service after war damage); standard gauge: 151 km 1.435-m gauge; narrow gauge: 2,454 km 1.000-m gauge

Highways: total: 105,000 km; paved: 10,500 km; unpaved: 94,500 km (1993 est.)

Merchant marine: total: 112 ships (1,000 GRT or over) totaling 569,269 GRT/947,938 DWT; ships by type: bulk 3, cargo 95, oil tanker 10, refrigerated cargo 3, roll-on/roll-off cargo 1

Airports

Total:	48
With paved runways over 3,047 m:	8
With paved runways 2,438 to 3,047 m:	3
With paved runways 1,524 to 2,437 m:	5
With paved runways 914 to 1,523 m:	13
With paved runways under 914 m:	7

Tourism [34]

	1990	1991	1992	1993	1994
Tourists[4]	250	300	440	670	1,018

Travelers are in thousands, money in million U.S. dollars.

Finance, Economics, and Trade

GDP and Manufacturing Summary [35]

	1980	1985	1990	1991	1992
Gross Domestic Product					
Millions of 1980 dollars	5,630	7,791	9,338	9,899	10,700
Growth rate in percent	-4.81	6.20	2.40	6.00	8.10
Manufacturing Value Added					
Millions of 1980 dollars	NA	NA	NA	NA	NA
Growth rate in percent	NA	NA	NA	NA	NA
Manufacturing share in percent of current prices	NA	NA	NA	NA	NA

Economic Indicators [36]

- **National product**: GDP—purchasing power parity—$97 billion (1995 est.)

- **National product real growth rate**: 9.5% (1995 est.)

- **National product per capita**: $1,300 (1995 est.)

- **Inflation rate (consumer prices)**: 14% (1995)

- **External debt**: $7.3 billion Western countries; $4.5 billion CEMA debts primarily to Russia; $9 billion to $18 billion nonconvertible debt (former CEMA, Iraq, Iran)

Balance of Payments Summary [37]

Exchange Rates [38]

Currency: **new dong.**
Symbol: **D.**

Data are currency units per $1.

1995 average	11,193
October 1994	11,000
November 1993	10,800
July 1991	8,100
December 1990	7,280
March 1990	3,996

Imports and Exports

Top Import Origins [39]

$7.5 billion (f.o.b., 1995 est.).

Origins	%
Singapore	NA
South Korea	NA
Japan	NA
France	NA
Hong Kong	NA
Taiwan	NA

Top Export Destinations [40]

$5.3 billion (f.o.b., 1995 est.).

Destinations	%
Japan	NA
Singapore	NA
Taiwan	NA
Hong Kong	NA
France	NA
South Korea	NA

Foreign Aid [41]

	U.S. $	
ODA (1993)	57	million
International credit and grant pledges (1996)	2.31	billion

Import and Export Commodities [42]

Import Commodities	Export Commodities
Petroleum products	Crude oil
Machinery and equipment	Rice
Steel products	Marine products
Fertilizer	Coffee
Raw cotton	Rubber
Grain	Tea
	Garments

For sources, notes, and explanations, see Annotated Source Appendix, page 1061.

1025

Western Samoa

Geography [1]

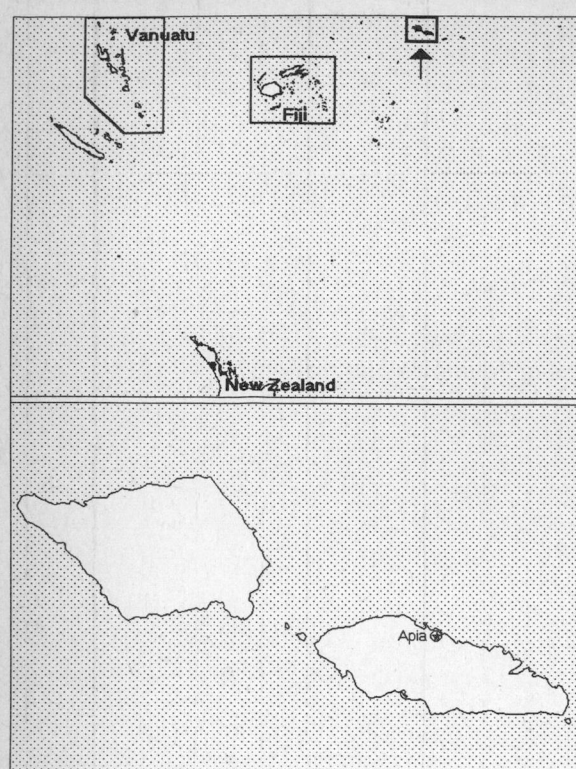

Total area:
 2,860 sq km   1,104 sq mi
Land area:
 2,850 sq km   1,100 sq mi
Comparative area:
 Slightly smaller than Rhode Island
Land boundaries:
 0 km
Coastline:
 403 km
Climate:
 Tropical; rainy season (October to March), dry season (May to October)
Terrain:
 Narrow coastal plain with volcanic, rocky, rugged mountains in interior
Natural resources:
 Hardwood forests, fish
Land use:
 Arable land: 19%
 Permanent crops: 24%
 Meadows and pastures: 0%
 Forest and woodland: 47%
 Other: 10%

Demographics [2]

	1970	1980	1990	1995[1]	1996	2000	2010	2020	2030
Population	142	155	186	209	214	235	288	341	392
Population density (persons per sq. mi.)	129	141	169	190	195	214	262	310	356
(persons per sq. km.)	50	54	65	73	75	83	101	120	137
Net migration rate (per 1,000 population)	NA	NA	-4.8	-2.1	-1.7	0.0	0.0	0.0	0.0
Births	NA	NA	NA	NA	7	NA	NA	NA	NA
Deaths	NA	NA	NA	NA	1	NA	NA	NA	NA
Life expectancy - males	NA	NA	64.0	66.0	66.4	67.8	71.0	73.5	75.5
Life expectancy - females	NA	NA	68.9	70.9	71.2	72.7	75.9	78.5	80.6
Birth rate (per 1,000)	NA	NA	34.2	31.7	31.1	28.0	22.9	19.9	17.0
Death rate (per 1,000)	NA	NA	6.6	5.9	5.8	5.3	4.6	4.4	4.8
Women of reproductive age (15-49 yrs.)	NA	NA	43	48	50	57	76	91	103
of which are currently married	NA	NA	23	NA	29	34	46	NA	NA
Fertility rate	NA	NA	4.7	4.0	3.9	3.5	2.7	2.3	2.2

Except as noted, values for vital statistics are in thousands; life expectancy is in years.

Health

Health Indicators [3]

Health Expenditures [4]

For sources, notes, and explanations, see Annotated Source Appendix, page 1061.

Human Factors

Women and Children [5]

Burden of Disease [6]

Population per physician (1989)	4,818.18
Population per nurse	NA
Population per hospital bed	NA
AIDS cases per 100,000 people (1994)	*
Malaria cases per 100,000 people (1992)	NA

Ethnic Division [7]

Euronesians are persons of European and Polynesian descent.

Samoan	92.6%
Euronesians	7%
Europeans	0.4%

Religion [8]

About one-half of population is associated with the London Missionary Society; includes Congregational, Roman Catholic, Methodist, Latter-Day Saints, Seventh-Day Adventist.

Christian	99.7%
Other	0.3%

Major Languages [9]

Samoan (Polynesian), English.

Education

Public Education Expenditures [10]

Million (Tala)	1980	1985	1990	1991	1993	1994
Total education expenditure	NA	NA	15	NA	NA	NA
as percent of GNP	NA	NA	4.2	NA	NA	NA
as percent of total govt. expend.	NA	NA	10.7	NA	NA	NA
Current education expenditure	NA	NA	14	NA	NA	NA
as percent of GNP	NA	NA	3.9	NA	NA	NA
as percent of current govt. expend.	NA	NA	15.8	NA	NA	NA
Capital expenditure	NA	NA	1	NA	NA	NA

Educational Attainment [11]

Age group (1981)	25+
Total population	48,872
Highest level attained (%)	
No schooling	3.0
First level	
Not completed	53.7
Completed	2.4
Entered second level	
S-1	12.9
S-2	25.4
Postsecondary	2.7

Illiteracy [12]

	1971	1980	1990
Illiterate population (15+ years)[4]	1,581	NA	NA
Illiteracy rate - total pop. (%)	2.2	NA	NA
Illiteracy rate - males (%)	2.2	NA	NA
Illiteracy rate - females (%)	2.1	NA	NA

Libraries [13]

Daily Newspapers [14]

Culture [15]

Science and Technology

Scientific/Technical Forces [16]

R&D Expenditures [17]

U.S. Patents Issued [18]

Government and Law

Organization of Government [19]

Long-form name:
Independent State of Western Samoa
Type:
Constitutional monarchy under native chief
Independence:
1 January 1962 (from New Zealand-administered UN trusteeship)
National holiday:
National Day, 1 June (1962)
Constitution:
1 January 1962
Legal system:
Based on English common law and local customs; judicial review of legislative acts with respect to fundamental rights of the citizen; has not accepted compulsory ICJ jurisdiction
Executive branch:
Chief; Prime Minister; Cabinet
Legislative branch:
Unicameral: Legislative Assembly (Fono)
Note: Only matai (head of family) are able to run for the Legislative Assembly
Judicial branch:
Supreme Court; Court of Appeal

Elections [20]

Legislative Assembly	% of seats
Human Rights Protection Party (HRPP)	59.6
Samoan National Development Party (SNDP)	38.3
Independents	2.1

Government Budget [21]

For 1995 est.

	$ mil.
Revenues	78.6
Expenditures	81.9
Capital expenditures	NA

Defense Summary [22]

Branches: No regular armed services; Western Samoa Police Force

Manpower Availability: Males age 15-49 NA; males fit for military service NA;

Defense Expenditures: NA

Crime [23]

	1994
Crime volume	
Cases known to police	1,503
Attempts (percent)	NA
Percent cases solved	NA
Crimes per 100,000 persons	3,006.00
Persons responsible for offenses	
Total number offenders	169
Percent female	NA
Percent juvenile (13-17 yrs.)	NA
Percent foreigners	NA

Human Rights [24]

	SSTS	FL	FAPRO	PPCG	APROBC	TPW	PCPTW	STPEP	PHRFF	PRW	ASST	AFL
Observes						P	P					

	EAFRD	CPR	ESCR	SR	ACHR	MAAE	PVIAC	PVNAC	EAFDAW	TCIDTP	RC
Observes				P			P	P	P		P

P = Party; S = Signatory; see Appendix for meaning of abbreviations.

Labor Force

Total Labor Force [25]

45,635 (1986 est.)

Labor Force by Occupation [26]

Agriculture	65%
Services	30
Industry	5

Date of data: 1995 est.

Unemployment Rate [27]

For sources, notes, and explanations, see Annotated Source Appendix, page 1061.

Production Sectors

Energy Resource Summary [28]

Electricity: Capacity: 29,000 kW. Production: 50 million kWh. Consumption per capita: 200 kWh (1993).

Telecommunications [30]

- 7,500 (1988 est.) telephones
- International: satellite earth station - 1 Intelsat (Pacific Ocean)
- Radio: Broadcast stations: AM 1, FM 0, shortwave 0 Radios: 76,000 (1992 est.)
- Television: Broadcast stations: 0 (1987 est.) Televisions: 6,000 (1992 est.)

Top Agricultural Products [32]

Agriculture accounts for 50% of the GDP; produces coconuts, bananas, taro, yams.

Top Mining Products [33]

Detailed information is not available. A summary of mineral resources follows. **Mineral Resources**: None.

Transportation [31]

Railways: 0 km

Highways: total: 2,030 km; paved: 373 km; unpaved: 1,657 km (1988 est.)

Merchant marine: total: 1 roll-on/roll-off cargo ship (1,000 GRT or over) totaling 3,838 GRT/5,536 DWT (1995 est.)

Airports

Total: 3
With paved runways 2,438 to 3,047 m: 1
With paved runways under 914 m: 2 (1995 est.)

Tourism [34]

	1990	1991	1992	1993	1994
Tourists	48	35	38	48	50
Tourism receipts	20	18	17	21	NA
Tourism expenditures	2	2	2	2	NA
Fare receipts	2	2	1	1	NA
Fare expenditures	2	3	3	2	NA

Travelers are in thousands, money in million U.S. dollars

For sources, notes, and explanations, see Annotated Source Appendix, page 1061.

1029

Finance, Economics, and Trade

Industrial Summary [35]

Industrial Production: Growth rate not available. **Industries:** Timber, tourism, food processing, fishing.

Economic Indicators [36]

- **National product:** GDP—purchasing power parity—$415 million (1995 est.)
- **National product real growth rate:** 5% (1995 est.)
- **National product per capita:** $1,900 (1995 est.)
- **Inflation rate (consumer prices):** 18% (1994)
- **External debt:** $141 million (June 1993)

Balance of Payments Summary [37]

Values in millions of dollars.

	1989	1990	1991	1992	1993
Exports of goods (f.o.b.)	12.9	8.9	6.5	5.8	6.4
Imports of goods (f.o.b.)	-67.0	-70.0	-77.6	-89.9	-87.4
Trade balance	-54.1	-61.2	-71.1	-84.1	-80.9
Services - debits	-21.2	-26.3	-37.0	-46.0	-42.6
Services - credits	35.4	42.3	38.0	42.8	40.1
Private transfers (net)	38.2	39.7	31.0	34.8	28.4
Government transfers (net)	16.7	14.0	13.7	NA	17.4
Long-term capital (net)	0.9	9.3	18.1	20.2	15.0
Short-term capital (net)	-0.5	0.1	0.5	-0.2	0.6
Errors and omissions	-2.6	-5.7	8.0	19.8	13.8
Overall balance	12.7	12.3	1.0	-12.7	-8.3

Exchange Rates [38]

Currency: **tala.**
Symbol: **WS$.**

Data are currency units per $1.

January 1996	2.5195
1995	2.4722
1994	2.5349
1993	2.5681
1992	2.4655
1991	2.3975

Imports and Exports

Top Import Origins [39]

$11.5 million (c.i.f., 1992 est.).

Origins	%
New Zealand	37
Australia	25
Japan	11
Fiji	9

Top Export Destinations [40]

$6.4 million (f.o.b., 1993).

Destinations	%
New Zealand	34
American Samoa	21
Germany	18
Australia	11

Foreign Aid [41]

Recipient: ODA, $NA.

Import and Export Commodities [42]

Import Commodities	Export Commodities
Intermediate goods 58%	Coconut oil and cream
Food 17%	Taro
Capital goods 12%	Copra
	Cocoa

Yemen

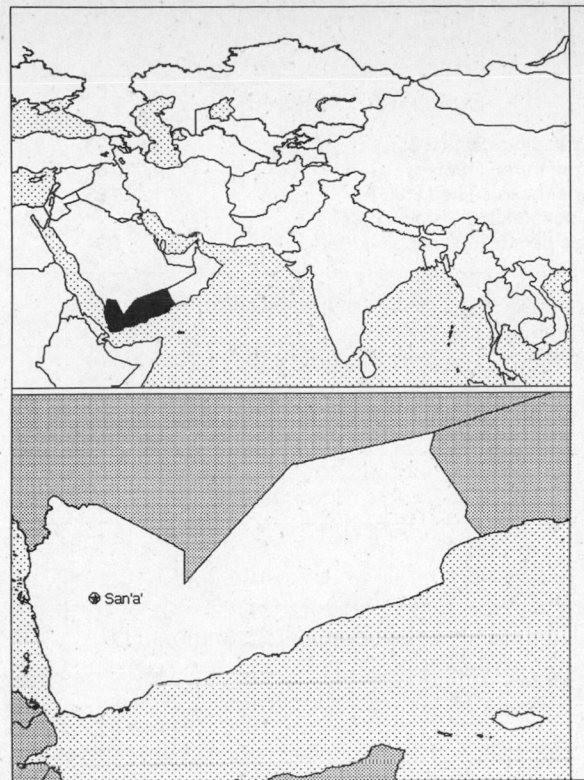

Geography [1]

Total area:
527,970 sq km   203,850 sq mi
Land area:
527,970 sq km   203,850 sq mi
Comparative area:
Slightly larger than twice the size of Wyoming
Note: Includes Perim, Socotra, the former Yemen Arab Republic (YAR or North Yemen), and the former People's Democratic Republic of Yemen (PDRY or South Yemen)
Land boundaries:
Total 1,746 km, Oman 288 km, Saudi Arabia 1,458 km
Coastline:
1,906 km
Climate:
Mostly desert; hot and humid along west coast; temperate in western mountains affected by seasonal monsoon; extraordinarily hot, dry, harsh desert in East
Terrain:
Narrow coastal plain backed by flat-topped hills and rugged mountains; dissected upland desert plains in center slope into desert interior of Arabian Peninsula
Natural resources:
Petroleum, fish, rock salt, marble, small deposits of coal, gold, lead, nickel, and copper, fertile soil in West
Land use:
Arable land: 6%
Permanent crops: 0%
Meadows and pastures: 30%
Forest and woodland: 7%
Other: 57%

Demographics [2]

	1970	1980	1990	1995[1]	1996	2000	2010	2020	2030
Population	5,782	7,439	10,489	13,012	13,483	15,547	21,841	29,469	37,974
Population density (persons per sq. mi.)	28	36	51	64	66	76	107	145	186
(persons per sq. km.)	11	14	20	25	26	29	41	56	72
Net migration rate (per 1,000 population)	NA	-4.9	71.5	0.0	0.0	0.0	0.0	0.0	0.0
Births	NA	NA	NA	NA	610	NA	NA	NA	NA
Deaths	NA	NA	NA	NA	129	NA	NA	NA	NA
Life expectancy - males	NA	47.1	54.0	57.6	58.2	61.0	67.1	71.8	75.1
Life expectancy - females	NA	51.6	56.1	60.2	61.0	64.2	71.3	76.8	80.6
Birth rate (per 1,000)	NA	49.8	47.7	45.5	45.2	43.3	37.2	31.1	25.9
Death rate (per 1,000)	NA	17.2	12.6	10.0	9.6	7.9	5.0	3.5	3.0
Women of reproductive age (15-49 yrs.)	NA	1,633	2,288	2,777	2,876	3,321	4,955	7,099	9,644
of which are currently married	NA	NA	1,582	NA	1,992	2,318	3,488	NA	NA
Fertility rate	NA	7.9	7.7	7.4	7.3	6.9	5.6	4.2	3.2

Except as noted, values for vital statistics are in thousands; life expectancy is in years.

Health

Health Indicators [3]

% of population with access to	
safe water (1990-95)	55
adequate sanitation (1990-95)	65
health services (1985-95)	38
% of 1-year-olds immunized (1990-94) against	
TB (tuberculosis)	61
DPT (diphtheria, pertussis, tetanus)	47
polio	47
measles	45
% of contraceptive prevalence (1980-94)	7
ORT use rate (1990-94)	30

Health Expenditures [4]

Total health expenditure, 1990 (official exchange rate)	
Millions of dollars	217
Dollars per capita	19
Health expenditures as a percentage of GDP	
Total	3.2
Public sector	1.5
Private sector	1.7
Development assistance for health	
Total aid flows (millions of dollars)[1]	25
Aid flows per capita (dollars)	2.2
Aid flows as a percentage of total health expenditure	11.6

For sources, notes, and explanations, see Annotated Source Appendix, page 1061.

Human Factors

Women and Children [5]

% of pregnant women immunized (tetanus 1990-94)	8
% of births attended by trained health personnel (1983-94)	16
Maternal mortality rate (1980-92)	NA
Under-5 mortality rate (1994)	112
% under-5 moderately/severely underweight (1980-1994)	30

Burden of Disease [6]

Population per physician (1985)	5,838.93
Population per nurse (1985)	2,020.85
Population per hospital bed (1990)	1,195.85
AIDS cases per 100,000 people (1994)	*
Malaria cases per 100,000 people (1992)	234

Ethnic Division [7]

Predominantly Arab; Afro-Arab concentrations in western coastal locations; South Asians in southern regions; small European communities in major metropolitan areas.

Religion [8]

Muslim including Sha'fi (Sunni) and Zaydi (Shi'a), small numbers of Jewish, Christian, and Hindu.

Major Languages [9]

Arabic.

Education

Public Education Expenditures [10]

Million (Rial)	1980	1990	1991	1992	1993	1994
Total education expenditure	NA	NA	NA	NA	13,531	16,114
as percent of GNP	NA	NA	NA	NA	NA	NA
as percent of total govt. expend.	NA	NA	NA	NA	20.9	20.8
Current education expenditure	NA	NA	NA	NA	12,939	14,577
as percent of GNP	NA	NA	NA	NA	NA	NA
as percent of current govt. expend.	NA	NA	NA	NA	NA	NA
Capital expenditure	NA	NA	NA	NA	592	1,537

Educational Attainment [11]

Illiteracy [12]

In thousands and percent[3]	1985	1991	2000
Illiterate population (15+ years)	2,423	2,559	2,881
Illiteracy rate - total pop. (%)	67.7	61.5	49.3
Illiteracy rate - males (%)	52.9	46.7	35.5
Illiteracy rate - females (%)	79.5	73.7	61.4

Libraries [13]

Daily Newspapers [14]

	1980	1985	1990	1994
Number of papers	6[e]	4[e]	5[e]	3
Circ. (000)	98[e]	125[e]	135[e]	230

Culture [15]

Science and Technology

Scientific/Technical Forces [16]

R&D Expenditures [17]

U.S. Patents Issued [18]

Values show patents issued to citizens of the country by the U.S. Patents Office.

	1993	1994	1995
Number of patents	0	1	0

Government and Law

Organization of Government [19]

Long-form name:
Republic of Yemen
Type:
Republic
Independence:
22 May 1990 Republic of Yemen
established by merger of the Yemen Arab
Republic (North Yemen) and the People's
Democratic Republic of Yemen (South
Yemen)
National holiday:
Proclamation of the Republic, 22 May
(1990)
Constitution:
16 May 1991; amended 29 September
1994
Legal system:
Based on Islamic law, Turkish law, English
common law, and local tribal customary
law; does not accept compulsory ICJ
jurisdiction
Executive branch:
President; Vice President; Prime Minister;
4 Deputy Prime Ministers; Council of
Ministers
Legislative branch:
Unicameral: House of Representatives
Judicial branch:
Supreme Court

Elections [20]

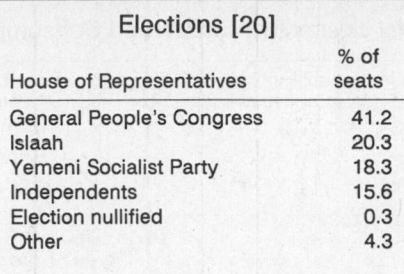

House of Representatives	% of seats
General People's Congress	41.2
Islaah	20.3
Yemeni Socialist Party	18.3
Independents	15.6
Election nullified	0.3
Other	4.3

Government Expenditures [21]

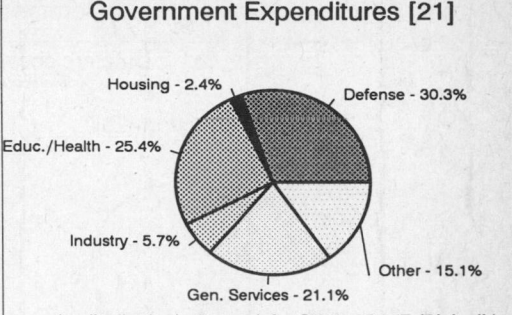

Housing - 2.4% Defense - 30.3%
Educ./Health - 25.4%
Industry - 5.7%
Other - 15.1%
Gen. Services - 21.1%

(% distribution). Expend. for CY93: 65,247 (Rial mil.)

Crime [23]

Military Expenditures and Arms Transfers [22]

	1990[12]	1991[13]	1992	1993	1994
Military expenditures					
Current dollars (mil.)	1,468	1,691[e]	1,288[e]	1,981	2,082
1994 constant dollars (mil.)	1,634	1,812[e]	1,343[e]	2,022	2,082
Armed forces (000)	127	127	64	64	69
Gross national product (GNP)					
Current dollars (mil.)	12,570	12,340	12,100	12,650	14,380
1994 constant dollars (mil.)	13,990	13,220	12,610	12,910	14,380
Central government expenditures (CGE)					
1994 constant dollars (mil.)	5,744	6,220	6,742	6,881	NA
People (mil.)	11.6	12.4	13.0	13.6	14.1
Military expenditure as % of GNP	11.7	13.7	10.6	15.7	14.5
Military expenditure as % of CGE	28.4	29.1	19.9	29.4	NA
Military expenditure per capita (1994 $)	141	146	103	149	147
Armed forces per 1,000 people (soldiers)	11.0	10.2	4.9	4.7	4.9
GNP per capita (1994 $)	1,208	1,063	972	953	1,017
Arms imports[6]					
Current dollars (mil.)	550	40	5	20	230
1994 constant dollars (mil.)	612	43	5	20	230
Arms exports[6]					
Current dollars (mil.)	0	0	0	0	0
1994 constant dollars (mil.)	0	0	0	0	0
Total imports[7]					
Current dollars (mil.)	NA	1,897	2,245	2,156	2,650
1994 constant dollars (mil.)	NA	2,033	2,341	2,200	2,650
Total exports[7]					
Current dollars (mil.)[e]	NA	1,174	999	1,001	NA
1994 constant dollars (mil.)[e]	NA	1,258	1,042	1,022	NA
Arms as percent of total imports[8]	NA	2.1	0.2	0.9	8.7
Arms as percent of total exports[8]	0	0	0	0	0

Human Rights [24]

	SSTS	FL	FAPRO	PPCG	APROBC	TPW	PCPTW	STPEP	PHRFF	PRW	ASST	AFL
Observes	P	P	P	P	P	P	P	P		P		P
	EAFRD	CPR	ESCR	SR	ACHR	MAAE	PVIAC	PVNAC	EAFDAW	TCIDTP	RC
Observes	P	P	P	P			P	P	P	P	P

P = Party; S = Signatory; see Appendix for meaning of abbreviations.

Labor Force

Total Labor Force [25]

No reliable estimates exist.

Labor Force by Occupation [26]

Agriculture
Herding
Expatriate laborers
Services
Construction
Industry
Commerce

Unemployment Rate [27]

30% (1995 est.)

For sources, notes, and explanations, see Annotated Source Appendix, page 1061.

Production Sectors

Commercial Energy Production and Consumption

Data are shown in quadrillion (10^{15}) BTUs and percent for 1995
Values for hydroelectric, nuclear, geothermal, solar, and wind power refer to electrical generation.

Production [28]

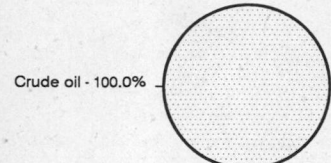

Crude oil - 100.0%

Consumption [29]

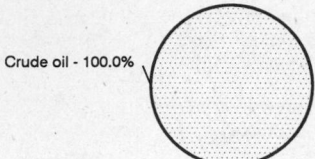

Crude oil - 100.0%

Crude oil	0.721
Total	0.721

Crude oil	0.145
Total	0.145

Telecommunications [30]

- 131,655 (1992 est.) telephones; since unification in 1990, efforts have been made to create a national telecommunications network
- Domestic: the network consists of microwave radio relay, cable, and tropospheric scatter
- International: satellite earth stations - 3 Intelsat (2 Indian Ocean and 1 Atlantic Ocean), 1 Intersputnik (Atlantic Ocean region), and 2 Arabsat; microwave radio relay to Saudi Arabia and Djibouti
- Radio: Broadcast stations: AM 4, FM 1, shortwave 0
- Television: Broadcast stations: 10 Televisions: 350,000 (1992 est.)

Top Agricultural Products [32]

Agriculture accounts for 21% of the GDP; produces grain, fruits, vegetables, qat (mildly narcotic shrub), coffee, cotton; dairy products, poultry, meat; fish.

Top Mining Products [33]

Thousand metric tons except as noted[e]	7/15/95[*]
Cement	500
Gypsum	80
Gas, natural, gross (mil. cu. meters)	100,000
Petroleum (000 42-gal. bls.)	
crude	124,000
refinery products	35,000
Salt	280
Dimension stone (cu. meters)	410,000

Transportation [31]

Railways: 0 km

Highways: total: 51,392 km; paved: 4,831 km; unpaved: 46,561 km (1992 est.)

Merchant marine: total: 3 ships (1,000 GRT or over) totaling 12,059 GRT/18,563 DWT; ships by type: cargo 1, oil tanker 2 (1995 est.)

Airports

Total:	41
With paved runways over 3,047 m:	2
With paved runways 2,438 to 3,047 m:	6
With paved runways 1,524 to 2,437 m:	1
With paved runways under 914 m:	3
With unpaved runways over 3,047 m:	2

Tourism [34]

	1990	1991	1992	1993	1994
Tourists[16]	52	44	72	70	40
Tourism receipts	40	21	47	45	19
Tourism expenditures	64	70	101	80	78
Fare expenditures	2	2	2	2	2

Travelers are in thousands, money in million U.S. dollars.

For sources, notes, and explanations, see Annotated Source Appendix, page 1061.

Manufacturing Sector

Manufacturing Summary [35]

	1987		1988		1989		1990		1991	
	$ bil.	%	$ bil.	%	$ bil.	%	$ bil.	%	$ bil.	%
Establishments or enterprises (number)	112	0.005	118	0.006	-	-	-	-	-	-
Total employment (000)	16	0.012	17	0.012	-	-	-	-	-	-
Production workers (000)	-	-	-	-	-	-	-	-	-	-
Output ($ bil.)	0.744	0.007	0.897	0.008	-	-	-	-	-	-
Value added ($ bil.)	-	-	-	-	-	-	-	-	-	-
Capital investment ($ mil.)	-	-	-	-	-	-	-	-	-	-
M & E investment ($ mil.)	-	-	-	-	-	-	-	-	-	-
Employees per establishment (number)	146	252.224	143	219.657	-	-	-	-	-	-
Production workers per establishment	-	-	-	-	-	-	-	-	-	-
Output per establishment ($ mil.)	7	153.248	8	136.825	-	-	-	-	-	-
Capital investment per estab. ($ mil.)	-	-	-	-	-	-	-	-	-	-
M & E per establishment ($ mil)	-	-	-	-	-	-	-	-	-	-
Payroll per employee ($)	-	-	-	-	-	-	-	-	-	-
Wages per production worker ($)	-	-	-	-	-	-	-	-	-	-
Hours per production worker (hours)	-	-	-	-	-	-	-	-	-	-
Output per employee ($)	45,505	60.758	53,056	62.290	-	-	-	-	-	-
Capital investment per employee ($)	-	-	-	-	-	-	-	-	-	-
M & E per employee ($)	-	-	-	-	-	-	-	-	-	-

Note: Columns headed % show percent of world total or ratio. Ratios closest to 100 are closest to world average. M & E stands for machinery & equipment.

Output in Manufacturing

	1987		1988		1989		1990		1991	
	$ bil.[f]	%	$ bil.[f]	%	$ bil.[f]	%	$ bil.[f]	%	$ bil.[f]	%
3110 - Food products	0.313	42.07	0.352	39.24	-	-	-	-	-	-
3130 - Beverages	0.040	5.38	0.041	4.57	-	-	-	-	-	-
3140 - Tobacco	0.108	14.52	0.151	16.83	-	-	-	-	-	-
3210 - Textiles	0.018	2.42	0.026	2.90	-	-	-	-	-	-
3211 - Spinning, weaving, etc.	0.018	2.42	0.026	2.90	-	-	-	-	-	-
3240 - Footwear	0.002	0.27	0.003	0.33	-	-	-	-	-	-
3310 - Wood products	0.000	0.00	0.001	0.11	-	-	-	-	-	-
3420 - Printing, publishing	0.035	4.70	0.031	3.46	-	-	-	-	-	-
3520 - Chemical products nec	0.080	10.75	0.100	11.15	-	-	-	-	-	-
3522 - Drugs and medicines	0.009	1.21	0.015	1.67	-	-	-	-	-	-
3560 - Plastic products nec	0.045	6.05	0.054	6.02	-	-	-	-	-	-
3690 - Nonmetal products nec	0.063	8.47	0.083	9.25	-	-	-	-	-	-
3810 - Metal products	0.011	1.48	0.015	1.67	-	-	-	-	-	-

Note: Codes are International Standard Industry codes (ISIC). Percentages are % of total Output. [f]: Factor Prices; [p]: Producer Prices.

Finance, Economics, and Trade

Economic Indicators [36]

- **National product**: GDP—purchasing power parity—$37.1 billion (1995 est.)
- **National product real growth rate**: 3.6% (1995 est.)
- **National product per capita**: $2,520 (1995 est.)
- **Inflation rate (consumer prices)**: 71.3% (1994 est.)
- **External debt**: $8 billion (1996)

Balance of Payments Summary [37]

Values in millions of dollars.

	1985	1986	1987	1988	1989
Exports of goods (f.o.b.)	8.2	46.5	119.1	529.2	606.0
Imports of goods (f.o.b.)	-1,078.9	-1,244.5	-1,646.3	-1,905.5	-1,282.7
Trade balance	-1,070.7	-1,198.0	-1,527.2	-1,376.3	-676.7
Services - debits	-259.5	-409.0	-549.0	-766.0	-488.5
Services - credits	190.7	255.9	290.8	350.6	247.6
Private transfers (net)	763.2	819.8	1,010.3	566.5	242.3
Government transfers (net)	89.0	230.1	193.0	126.6	96.2
Long-term capital (net)	111.6	338.0	512.1	764.3	493.8
Short-term capital (net)	123.3	-48.8	51.2	84.5	17.1
Errors and omissions	21.8	41.5	4.1	4.1	57.9
Overall balance	-30.6	29.5	-245.7	-245.7	-10.3

Exchange Rates [38]

Currency: **Yemeni rial.**

Data are currency units per $1.

Official fixed rate	12.010
December 1994 (market rate)	90

Imports and Exports

Top Import Origins [39]

$1.8 billion (c.i.f., 1994 est.) Data are for 1994.

Origins	%
US	11
UK	7
France	7
Germany	5
Japan	5

Top Export Destinations [40]

$1.1 billion (f.o.b., 1994 est.) Data are for 1994.

Destinations	%
US	17
Japan	16
Singapore	15
China	13

Foreign Aid [41]

Recipient: ODA, $148 million (1993).

Import and Export Commodities [42]

Import Commodities	Export Commodities
Textiles and other manufactured consumer goods	Crude oil
Petroleum products	Cotton
Sugar	Coffee
Grain	Hides
Flour	Vegetables
Other foodstuffs	Dried and salted fish
Cement	
Machinery	
Chemicals	

Yugoslavia

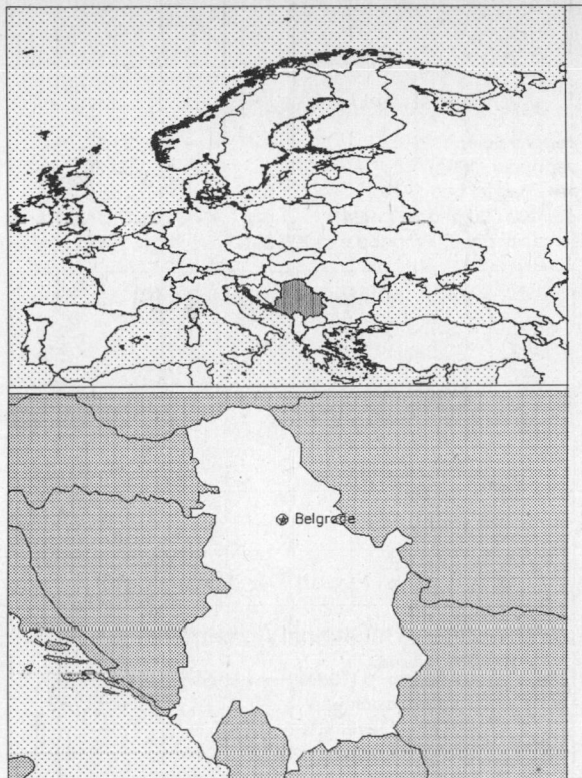

Geography [1]

Total area:
102,350 sq km   39,518 sq mi
Land area:
102,136 sq km   39,435 sq mi
Comparative area:
Slightly larger than Kentucky
Land boundaries:
Total 2,246 km, Albania 287 km, Bosnia and Herzegovina 527 km,
Bulgaria 318 km, Croatia (North) 241 km, Croatia (South) 25 km,
Hungary 151 km, Macedonia 221 km , Romania 476 km
Coastline:
199 km (Montenegro 199 km, Serbia 0 km)
Climate:
North, continental climate (cold winters, hot humid summers with
well distributed rainfall); central portion, continental and
Mediterranean climate; South, Adriatic climate along coast, hot
dry summers and autumns, cold winters with heavy snowfall inland
Terrain:
Varied; North, rich fertile plains; East, limestone ranges and basins; Southeast,
ancient mountains and hills; Southwest, very high shoreline, no islands off coast
Natural resources:
Oil, gas, coal, antimony, copper, lead, zinc, nickel, gold, pyrite, chrome
Land use:
Arable land: 30%
Permanent crops: 5%
Meadows and pastures: 20%
Forest and woodland: 25%
Other: 20%

Demographics [2]

	1970[69]	1980	1990	1995[1]	1996	2000	2010	2020	2030
Population	NA	NA	NA	10,574	10,615	10,787	11,062	11,067	10,997
Population density (persons per sq. mi.)	NA	NA	NA	268	269	273	280	281	279
(persons per sq. km.)	NA	NA	NA	103	104	106	108	108	108
Net migration rate (per 1,000 population)	NA	NA	NA	0.0	0.0	0.0	0.0	0.0	0.0
Births	NA	NA	NA	NA	148	NA	NA	NA	NA
Deaths	NA	NA	NA	NA	107	NA	NA	NA	NA
Life expectancy - males	NA	NA	NA	69.0	69.1	69.5	70.5	73.7	76.1
Life expectancy - females	NA	NA	NA	75.3	75.5	76.1	77.5	80.3	82.4
Birth rate (per 1,000)	NA	NA	NA	12.9	12.9	14.2	11.9	10.1	9.4
Death rate (per 1,000)	NA	NA	NA	8.9	9.0	9.3	10.5	10.5	11.0
Women of reproductive age (15-49 yrs.)	NA	NA	NA	2,587	2,614	2,637	2,578	2,572	2,435
of which are currently married	NA	NA	NA	NA	1,811	1,831	1,824	NA	NA
Fertility rate	NA	NA	NA	1.8	1.8	1.9	1.7	1.6	1.5

Except as noted, values for vital statistics are in thousands; life expectancy is in years.

Health

Health Indicators [3]

% of population with access to	
safe water (1990-95)	NA
adequate sanitation (1990-95)	NA
health services (1985-95)	NA
% of 1-year-olds immunized (1990-94) against	
TB (tuberculosis)	68
DPT (diphtheria, pertussis, tetanus)	85
polio	83
measles	85
% of contraceptive prevalence (1980-94)	NA
ORT use rate (1990-94)	NA

Health Expenditures [4]

Total health expenditure, 1990 (official exchange rate)[3]	
Millions of dollars	4,512
Dollars per capita	205
Health expenditures as a percentage of GDP	
Total	3.0
Public sector	4.0
Private sector	1.0
Development assistance for health	
Total aid flows (millions of dollars)[1]	NA
Aid flows per capita (dollars)	NA
Aid flows as a percentage of total health expenditure	NA

For sources, notes, and explanations, see Annotated Source Appendix, page 1061.

1037

Human Factors

Women and Children [5]

% of pregnant women immunized (tetanus 1990-94)	NA
% of births attended by trained health personnel (1983-94)	NA
Maternal mortality rate (1980-92)	NA
Under-5 mortality rate (1994)	23
% under-5 moderately/severely underweight (1980-1994)	NA

Burden of Disease [6]

Population per physician (1990)	231.52
Population per nurse (1990)	50.28
Population per hospital bed (1990)	73.39
AIDS cases per 100,000 people (1994)	NA
Heart disease cases per 1,000 people (1990-93)	NA

Ethnic Division [7]

Serbs	63%
Albanians	14%
Montenegrins	6%
Hungarians	4%
Other	13%

Religion [8]

Orthodox	65%
Muslim	19%
Roman Catholic	4%
Protestant	1%
Other	11%

Major Languages [9]

Serbo-Croatian	95%
Albanian	5%

Education

Public Education Expenditures [10]

Million (New Dinar)[71]	1980	1990	1991	1992	1993	1994
Total education expenditure	NA	NA	NA	0.292	NA	842
as percent of GNP	NA	NA	NA	NA	NA	NA
as percent of total govt. expend.	NA	NA	NA	NA	NA	NA
Current education expenditure	NA	NA	NA	0.274	NA	750
as percent of GNP	NA	NA	NA	NA	NA	NA
as percent of current govt. expend.	NA	NA	NA	NA	NA	NA
Capital expenditure	NA	NA	NA	0.018	NA	91

Educational Attainment [11]

Age group (1981)[21]	25+
Total population	13,083,762
Highest level attained (%)	
No schooling	15.8
First level	
Not completed	53.9
Completed	NA
Entered second level	
S-1	23.4
S-2	NA
Postsecondary	4.8

Illiteracy [12]

	1989	1991	1995
Illiterate population (15+ yrs.)[2]	NA	463,291	178,000
Illiteracy rate - total pop. (%)	NA	6.7	2.1
Illiteracy rate - males (%)	NA	2.4	1.4
Illiteracy rate - females (%)	NA	10.8	2.7

Libraries [13]

	Admin. Units	Svc. Pts.	Vols. (000)	Shelving (meters)	Vols. Added	Reg. Users
National (1992)	3	3	5,607	180,272	83,829	176,101
Nonspecialized (1992)	10	10	1,171	10,324	15,770	352
Public (1992)[27]	407	924	15,337	NA	267,436	8 mil
Higher ed. (1992)	166	310	7,052	108,242	116,108	2 mil
School	NA	NA	NA	NA	NA	NA

Daily Newspapers [14]

	1980	1985	1990[2]	1994
Number of papers	12	12	11	9
Circ. (000)	537	425	973	966

Culture [15]

Cinema (seats per 1,000)	7.2
Annual attendance per person	0.2
Gross box office receipts (mil. New Dinar)	NA
Museums (reporting)	159
Visitors (000)	1,850
Annual receipts (000 New Dinar)	4,876

Science and Technology

Scientific/Technical Forces [16]

Scientists/engineers	11,246
Number female	NA
Technicians	4,183
Number female	1,858
Total[9]	15,429

R&D Expenditures [17]

	Dina (000) 1992
Total expenditure[6]	335
Capital expenditure	196
Current expenditure	139
Percent current	41.5

U.S. Patents Issued [18]

Values show patents issued to citizens of the country by the U.S. Patents Office.

	1993	1994	1995
Number of patents	25	15	6

Government and Law

Organization of Government [19]

Long-form name:
None
Type:
Republic
Independence:
11 April 1992 (Federal Republic of Yugoslavia formed as self-proclaimed successor to the Socialist Federal Republic of Yugoslavia - SFRY)
National holiday:
St. Vitus Day, 28 June
Constitution:
27 April 1992
Legal system:
Based on civil law system
Executive branch:
Presidents; Prime Minister; 3 Deputy Prime Ministers; Federal Executive Council
Legislative branch:
Bicameral Federal Assembly: Chamber of Republics and Chamber of Citizens
Judicial branch:
Savezni Sud (Federal Court); Constitutional Court

Elections [20]

Chamber of Citizens	% of seats
Serbian Socialist Party (SPS)	34.1
Serbian Radical Party (SRS)	24.6
Democratic Party of Serbia	14.5
Dem. Party of Socialists (DSSCG)	12.3
Democratic Party (DS)	3.6
Socialist Party of Montenegro	3.6
People's Party of Montenegro (NS)	2.9
Dem. Community of Vojvodina Hungarians (DZVM)	2.2
Other	2.2

Government Budget [21]

Crime [23]

	1990[3]
Crime volume	
Cases known to police	98,303
Attempts (percent)	NA
Percent cases solved	45.9
Crimes per 100,000 persons	952.4
Persons responsible for offenses	
Total number offenders	45,083
Percent female	11.9
Percent juvenile	6.7
Percent foreigners	NA

Military Expenditures and Arms Transfers [22]

	1990	1991	1992[17]	1993	1994
Military expenditures					
Current dollars (mil.)	NA	NA	NA	NA	NA
1994 constant dollars (mil.)	NA	NA	NA	NA	NA
Armed forces (000)	NA	NA	137[e]	100	130
Gross national product (GNP)					
Current dollars (mil.)[e]	NA	NA	13,500	10,000	10,000
1994 constant dollars (mil.)[e]	NA	NA	14,080	10,210	10,000
Central government expenditures (CGE)					
1994 constant dollars (mil.)	NA	NA	NA	NA	NA
People (mil.)	NA	NA	10.1	10.3	10.3
Military expenditure as % of GNP	NA	NA	NA	NA	NA
Military expenditure as % of CGE	NA	NA	NA	NA	NA
Military expenditure per capita (1994 $)	NA	NA	NA	NA	NA
Armed forces per 1,000 people (soldiers)	NA	NA	13.6	9.8	12.6
GNP per capita (1994 $)	NA	NA	1,398	995	967
Arms imports[6]					
Current dollars (mil.)	NA	NA	0	0	0
1994 constant dollars (mil.)	NA	NA	0	0	0
Arms exports[6]					
Current dollars (mil.)	NA	NA	100	0	5
1994 constant dollars (mil.)	NA	NA	104	0	5
Total imports[7]					
Current dollars (mil.)	NA	NA	NA	NA	NA
1994 constant dollars (mil.)	NA	NA	NA	NA	NA
Total exports[7]					
Current dollars (mil.)	NA	NA	NA	NA	NA
1994 constant dollars (mil.)	NA	NA	NA	NA	NA
Arms as percent of total imports[8]	NA	NA	0	0	0
Arms as percent of total exports[8]	NA	NA	NA	0	NA

Human Rights [24]

Labor Force

Total Labor Force [25]

2.641 million

Labor Force by Occupation [26]

Industry, mining   40%
Date of data: 1990

Unemployment Rate [27]

More than 40% (1994 est.)

For sources, notes, and explanations, see Annotated Source Appendix, page 1061.

1039

Production Sectors

Commercial Energy Production and Consumption

Data are shown in quadrillion (10^{15}) BTUs and percent for 1995
Values for hydroelectric, nuclear, geothermal, solar, and wind power refer to electrical generation.

Production [28]

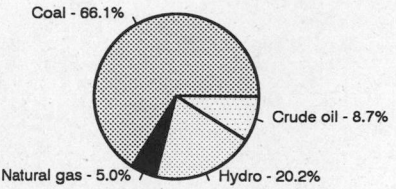

Coal - 66.1%
Crude oil - 8.7%
Natural gas - 5.0%
Hydro - 20.2%

Consumption [29]

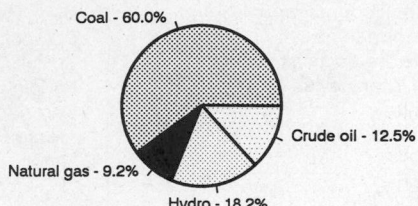

Coal - 60.0%
Crude oil - 12.5%
Natural gas - 9.2%
Hydro - 18.2%

Crude oil	0.047
Dry natural gas	0.027
Coal	0.357
Net hydroelectric power	0.109
Total	0.540

Crude oil	0.075
Dry natural gas	0.055
Coal	0.359
Net hydroelectric power	0.109
Total	0.598

Telecommunications [30]

- 700,000 telephones
- International: satellite earth station - 1 Intelsat (Atlantic Ocean)
- Radio: Broadcast stations: AM 26, FM 9, shortwave 0 Radios: 2.015 million
- Television: Broadcast stations: 18 Televisions: 1 million

Transportation [31]

Railways: total: 3,960 km; standard gauge: 3,960 km 1.435-m gauge (1,341 km electrified) (1992)

Highways: total: 46,019 km; paved: 26,949 km; unpaved: 19,070 km (1990 est.)

Merchant marine: ships by type: bulk 9, cargo 8, container 3, short-sea passenger ferry 1

Airports

Total:	44
With paved runways over 3,047 m:	2
With paved runways 2,438 to 3,047 m:	5
With paved runways 1,524 to 2,437 m:	5
With paved runways 914 to 1,523 m:	2
With paved runways under 914 m:	14

Top Agricultural Products [32]

Produces cereals, fruits, vegetables, tobacco, olives; cattle, sheep, goats.

Top Mining Products [33]

Thousand metric tons except as noted[e]	5/95[*]
Copper ore, mine output, gross weight	17,900[r]
Gold, refined (kg.)	4,000
Palladium (kg.)	70
Platinum (kg.)	10
Selenium (kg.)	25,000
Silver (kg.)	18,300[r]
Sand and gravel (000 cu. meters)	1,810[r]
Dimension stone, crushed/broken (000 cu. meters)	1,800
Coal, bituminous, brown, and lignite	38,300[r]
Petroleum refinery products (000 42-gal. bls.)	13,800

Tourism [34]

	1990	1991	1992	1993	1994
Tourists	1,186	379	156	77	91
Tourism receipts[109,11]	NA	134	88	23	31

Travelers are in thousands, money in million U.S. dollars.

Manufacturing Sector

GDP and Manufacturing Summary [35]

	1980	1985	1989	1990	% change 1980-1990	% change 1989-1990
GDP (million 1980 $)	69,958	71,058	72,234	66,371	-5.1	-8.1
GDP per capita (1980 $)	3,136	3,073	3,050	2,786	-11.2	-8.7
Manufacturing as % of GDP (current prices)	30.6	37.2	39.5	42.0	37.3	6.3
Gross output (million $)	72,629	57,020	65,078	62,136[e]	-14.4	-4.5
Value added (million $)	21,750	17,171	30,245	27,660[e]	27.2	-8.5
Value added (million 1980 $)	19,526	22,283	24,021	21,703	11.1	-9.6
Industrial production index	100	116	120	108	8.0	-10.0
Employment (thousands)	2,106	2,467	2,658	2,537[e]	20.5	-4.6

Note: GDP stands for Gross Domestic Product. 'e' stands for estimated value.

Profitability and Productivity

	1980	1985	1989	1990	% change 1980-1990	% change 1989-1990
Intermediate input (%)	70	70	54	55[e]	-21.4	1.9
Wages, salaries, and supplements (%)	14	12	12[e]	18[e]	28.6	50.0
Gross operating surplus (%)	15	18	34[e]	26[e]	73.3	-23.5
Gross output per worker ($)	34,487	23,113	24,484	24,248[e]	-29.7	-1.0
Value added per worker ($)	10,328	6,960	11,379	10,796[e]	4.5	-5.1
Average wage (incl. benefits) ($)	4,991	2,703	2,986[e]	4,488[e]	-10.1	50.3

Profitability is in percent of gross output. Productivity is in U.S. $. 'e' stands for estimated value.

Profitability - 1990

Wages - 18.2%
Inputs - 55.6%
Surplus - 26.3%

The graphic shows percent of gross output.

Value Added in Manufacturing

	1980 $ mil.	1980 %	1985 $ mil.	1985 %	1989 $ mil.	1989 %	1990 $ mil.	1990 %	% change 1980-1990	% change 1989-1990
311 Food products	1,897	8.7	1,458	8.5	3,916	12.9	3,484[e]	12.6	83.7	-11.0
313 Beverages	459	2.1	353	2.1	663	2.2	589[e]	2.1	28.3	-11.2
314 Tobacco products	184	0.8	221	1.3	344	1.1	308[e]	1.1	67.4	-10.5
321 Textiles	1,759	8.1	1,428	8.3	2,881	9.5	2,663[e]	9.6	51.4	-7.6
322 Wearing apparel	903	4.2	718	4.2	1,593	5.3	1,427[e]	5.2	58.0	-10.4
323 Leather and fur products	226	1.0	231	1.3	383	1.3	340[e]	1.2	50.4	-11.2
324 Footwear	482	2.2	503	2.9	1,022	3.4	899[e]	3.3	86.5	-12.0
331 Wood and wood products	977	4.5	530	3.1	794	2.6	706[e]	2.6	-27.7	-11.1
332 Furniture and fixtures	730	3.4	438	2.6	1,030	3.4	1,065[e]	3.9	45.9	3.4
341 Paper and paper products	529	2.4	394	2.3	759	2.5	674[e]	2.4	27.4	-11.2
342 Printing and publishing	876	4.0	462	2.7	761	2.5	678[e]	2.5	-22.6	-10.9
351 Industrial chemicals	694	3.2	631	3.7	1,107	3.7	992[e]	3.6	42.9	-10.4
352 Other chemical products	681	3.1	525	3.1	1,419	4.7	1,315[e]	4.8	93.1	-7.3
353 Petroleum refineries	454	2.1	415	2.4	260	0.9	233[e]	0.8	-48.7	-10.4
354 Miscellaneous petroleum and coal products	101	0.5	101	0.6	104	0.3	91[e]	0.3	-9.9	-12.5
355 Rubber products	276	1.3	269	1.6	479	1.6	456[e]	1.6	65.2	-4.8
356 Plastic products	413	1.9	258	1.5	397	1.3	350[e]	1.3	-15.3	-11.8
361 Pottery, china, and earthenware	128	0.6	72	0.4	162	0.5	144[e]	0.5	12.5	-11.1
362 Glass and glass products	163	0.7	113	0.7	224	0.7	204[e]	0.7	25.2	-8.9
369 Other nonmetal mineral products	906	4.2	513	3.0	683	2.3	604[e]	2.2	-33.3	-11.6
371 Iron and steel	1,221	5.6	1,000	5.8	1,343	4.4	1,171[e]	4.2	-4.1	-12.8
372 Nonferrous metals	480	2.2	509	3.0	944	3.1	927[e]	3.4	93.1	-1.8
381 Metal products	2,105	9.7	1,577	9.2	1,293	4.3	1,130[e]	4.1	-46.3	-12.6
382 Nonelectrical machinery	1,828	8.4	1,463	8.5	2,372	7.8	2,378[e]	8.6	30.1	0.3
383 Electrical machinery	1,600	7.4	1,544	9.0	2,640	8.7	2,334[e]	8.4	45.9	-11.6
384 Transport equipment	1,441	6.6	1,263	7.4	2,389	7.9	2,241[e]	8.1	55.5	-6.2
385 Professional and scientific equipment	101	0.5	93	0.5	154	0.5	146[e]	0.5	44.6	-5.2
390 Other manufacturing industries	134	0.6	88	0.5	128	0.4	114[e]	0.4	-14.9	-10.9

Note: The industry codes shown are International Standard Industry codes (ISIC). Percentages are percent of total Value Added. 'e' stands for estimated value

Finance, Economics, and Trade

Economic Indicators [36]

- **National product**: GDP—purchasing power parity— $20.6 billion (1995 est.)

- **National product real growth rate**: 4% (1995 est.)

- **National product per capita**: $2,000 (1995 est.)

- **Inflation rate (consumer prices)**: 20% (1994 est.)

- **External debt**: $4.2 billion (1993 est.)

Balance of Payments Summary [37]

Values in millions of dollars.

	1975[3]	1980	1985	1990	1991
Exports of goods (f.o.b.)	4,073	9,093	10,622	14,308	13,799
Imports of goods (f.o.b.)	-7,058	-13,992	-11,210	-16,984	-13,287
Trade balance	-2,984	-4,899	-588	-2,676	512
Services - debits	-1,553	-6,508	-5,316	-16,679	-6,740
Services - credits	2,097	4,740	3,458	7,163	2,935
Private transfers (net)	1,804	4,354	3,281	9,830	2,134
Government transfers (net)	11	-4	-2	-2	-2
Long-term capital (net)	951	1,954	78	-47	-386
Short-term capital (net)	-319	774	-835	3,551	-1,747
Errors and omissions	-314	-723	83	228	497
Overall balance	-308	-312	159	1,368	-2,797

Exchange Rates [38]

Currency: **Yugoslav New Dinar.**
Symbol: **YD.**

Data are currency units per $1.

Early 1995 (official rate)	1.5
Early 1995 (BMR)	2 to 3

Imports and Exports

Top Import Origins [39]

Prior to the imposition of UN sanctions, trade partners were the other former Yugoslav republics, the FSU countries, EU (mainly Italy and Germany), East European countries, US.

Origins	%
No details available.	NA

Top Export Destinations [40]

Prior to the imposition of UN sanctions, trade partners were the other former Yugoslav republics, Italy, Germany, other EU, the FSU countries, East European countries, US.

Destinations	%
No details available.	NA

Foreign Aid [41]

Recipient: ODA, $NA.

Import and Export Commodities [42]

Import Commodities

Machinery and transport equipment
Fuels and lubricants
Manufactured goods
Chemicals
Food and live animals
Coking coal
Raw materials

Export Commodities

Machinery and transport equipment
Manufactured goods
Chemicals
Food and live animals
Raw materials

Zaire

Geography [1]

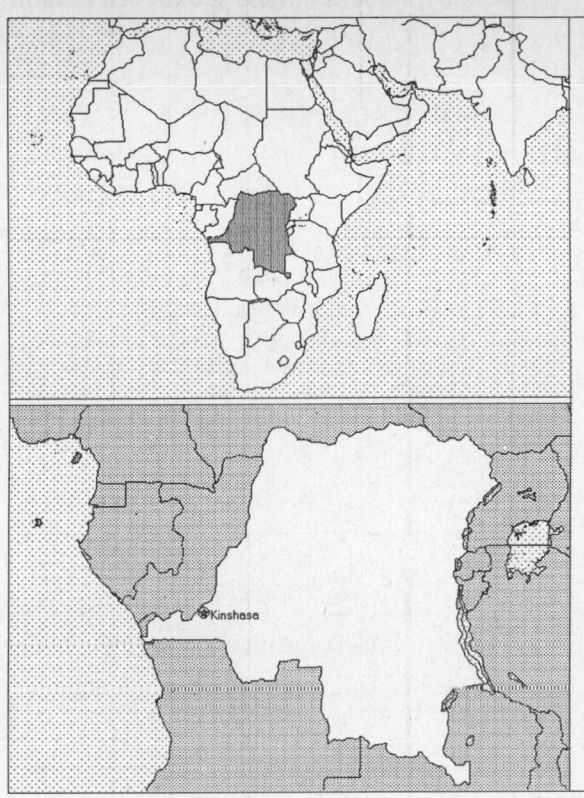

Total area:
  2,345,410 sq km   905,568 sq mi
Land area:
  2,267,600 sq km   875,525 sq mi
Comparative area:
  Slightly more than one-fourth the size of US
Land boundaries:
  Total 10,271 km, Angola 2,511 km, Burundi 233 km, Congo 2,410 km, Uganda 765 km, Central African Republic 1,577 km, Rwanda 217 km, Sudan 628 km, Zambia 1,930 km
Coastline:
  37 km
Climate:
  Tropical; hot and humid in equatorial river basin; cooler and drier in southern highlands; cooler and wetter in eastern highlands; north of equator, wet season April to October, dry season December to February; south of equator, wet season November to March, dry season April to October
Terrain:
  Vast central basin is a low-lying plateau; mountains in East
Natural resources:
  Cobalt, copper, cadmium, petroleum, industrial and gem diamonds, gold, silver, zinc, manganese, tin, germanium, uranium, radium, bauxite, iron ore, coal, hydropower potential
Land use:
  Arable land: 3%
  Permanent crops: 0%
  Meadows and pastures: 4%
  Forest and woodland: 78%
  Other: 15%

Demographics [2]

	1970	1980	1990	1995[1]	1996	2000	2010	2020	2030
Population	20,934	27,954	37,831	45,431	46,499	51,374	69,293	91,548	118,274
Population density (persons per sq. mi.)	24	32	43	52	53	59	79	105	135
(persons per sq. km.)	9	12	17	20	21	23	31	40	52
Net migration rate (per 1,000 population)	NA	NA	2.0	-1.4	-14.6	-1.8	0.0	0.0	0.0
Births	NA	NA	NA	NA	2,237	NA	NA	NA	NA
Deaths	NA	NA	NA	NA	786	NA	NA	NA	NA
Life expectancy - males	NA	NA	44.0	44.8	45.0	45.7	48.6	51.8	59.5
Life expectancy - females	NA	NA	47.4	48.0	48.5	50.4	54.0	57.3	65.9
Birth rate (per 1,000)	NA	NA	48.6	48.5	48.1	46.5	42.2	37.5	31.9
Death rate (per 1,000)	NA	NA	18.1	17.1	16.9	15.6	12.9	11.0	7.3
Women of reproductive age (15-49 yrs.)	NA	NA	8,446	10,095	10,312	11,364	15,728	21,995	30,089
of which are currently married	NA	NA	6,450	NA	7,851	8,660	11,960	NA	NA
Fertility rate	NA	NA	6.7	6.7	6.6	6.4	5.6	4.7	3.8

Except as noted, values for vital statistics are in thousands; life expectancy is in years.

Health

Health Indicators [3]

% of population with access to	
safe water (1990-95)	27
adequate sanitation (1990-95)	23
health services (1985-95)	26
% of 1-year-olds immunized (1990-94) against	
TB (tuberculosis)	43
DPT (diphtheria, pertussis, tetanus)	29
polio	29
measles	33
% of contraceptive prevalence (1980-94)[1]	1
ORT use rate (1990-94)	46

Health Expenditures [4]

Total health expenditure, 1990 (official exchange rate)	
Millions of dollars	179
Dollars per capita	5
Health expenditures as a percentage of GDP	
Total	2.4
Public sector	0.8
Private sector	1.5
Development assistance for health	
Total aid flows (millions of dollars)[1]	48
Aid flows per capita (dollars)	1.3
Aid flows as a percentage of total health expenditure	26.7

For sources, notes, and explanations, see Annotated Source Appendix, page 1061.

Human Factors

Women and Children [5]

% of pregnant women immunized (tetanus 1990-94)	25
% of births attended by trained health personnel (1983-94)	NA
Maternal mortality rate (1980-92)	800
Under-5 mortality rate (1994)	186
% under-5 moderately/severely underweight (1980-1994)[1]	28

Burden of Disease [6]

Population per physician (1988-1991)	14,286
Population per nurse (1988-1991)	1,351
Population per hospital bed (1990)	NA
AIDS cases per 100,000 people (1994)	8.3
Malaria cases per 100,000 people (1992)	NA

Ethnic Division [7]

Over 200 African ethnic groups, the majority are Bantu; the four largest tribes - Mongo, Luba, Kongo (all Bantu), and the Mangbetu-Azande (Hamitic) make up about 45% of the population.

Religion [8]

Roman Catholic	50%
Protestant	20%
Kimbanguist	10%
Muslim	10%
Other syncretic sects and traditional beliefs	10%

Major Languages [9]

French (official), Lingala (a lingua franca trade language), Kingwana (a dialect of Kiswahili or Swahili), Kikongo, Tshiluba.

Education

Public Education Expenditures [10]

	1980	1985	1988	1990	1993	1994
Million (Zaire)						
Total education expenditure	1,015	3,291	15,006	NA	NA	NA
as percent of GNP	2.6	1.0	1.0	NA	NA	NA
as percent of total govt. expend.	24.2	7.3	6.4	NA	NA	NA
Current education expenditure	998	3,239	14,357	NA	NA	NA
as percent of GNP	2.6	1.0	0.9	NA	NA	NA
as percent of current govt. expend.	25.3	7.3	6.4	NA	NA	NA
Capital expenditure	17	52	649	NA	NA	NA

Educational Attainment [11]

Illiteracy [12]

In thousands and percent[1]	1990	1995	2000
Illiterate population (15+ yrs.)	5,524	5,184	4,845
Illiteracy rate - total pop. (%)	27.8	21.9	18.2
Illiteracy rate - males (%)	17.2	13.0	10.6
Illiteracy rate - females (%)	38.8	30.9	26.2

Libraries [13]

Daily Newspapers [14]

	1980	1985	1990	1994
Number of papers	5	4	5	9
Circ. (000)	60[e]	50[e]	75[e]	112[e]

Culture [15]

Science and Technology

Scientific/Technical Forces [16]

R&D Expenditures [17]

U.S. Patents Issued [18]

For sources, notes, and explanations, see Annotated Source Appendix, page 1061.

Government and Law

Organization of Government [19]

Long-form name:
Republic of Zaire
Type:
Republic with a strong presidential
system
Independence:
30 June 1960 (from Belgium)
National holiday:
Anniversary of the Regime (Second
Republic), 24 November (1965)
Constitution:
24 June 1967, amended August 1974,
revised 15 February 1978; amended April
1990; new transitional constitution
promulgated in April 1994
Legal system:
Based on Belgian civil law system and
tribal law; has not accepted compulsory
ICJ jurisdiction
Executive branch:
President; Prime Minister; National
Executive Council
Legislative branch:
Unicameral: Parliament. Note: A single
body consisting of the High Council of the
Republic and the Parliament of the
Transition
Judicial branch:
Supreme Court (Cour Supreme)

Crime [23]

Elections [20]

Parliament. Single body consisting of
the High Council of the Republic and
the Parliament of the Transition with
membership equally divided between
presidential supporters and opponents.
Sole legal party until January 1991 -
Popular Movement of the Revolution
(MPR); other parties include Union for
Democracy and Social Progress,
Democratic Social Christian Party,
Union of Federalists and Independent
Republicans, Unified Lumumbast Party.

Government Expenditures [21]

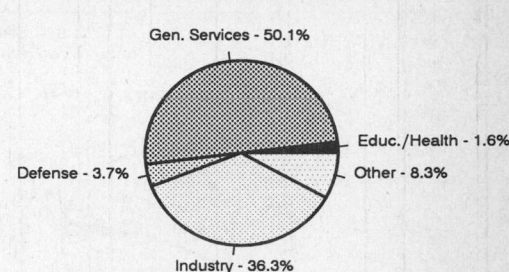

Gen. Services - 50.1%
Educ./Health - 1.6%
Defense - 3.7%
Other - 8.3%
Industry - 36.3%

(% distribution). Expend. for CY95: 3,289 (Zaire bil.)

Military Expenditures and Arms Transfers [22]

	1990	1991	1992	1993	1994
Military expenditures					
Current dollars (mil.)	NA	NA	202	245	117[e]
1994 constant dollars (mil.)	NA	NA	211	250	117[e]
Armed forces (000)	55	60	55	55	53
Gross national product (GNP)					
Current dollars (mil.)	7,832	7,240	6,736	6,195[e]	5,690[e]
1994 constant dollars (mil.)	8,717	7,760	7,025	6,322[e]	5,690[e]
Central government expenditures (CGE)					
1994 constant dollars (mil.)	1,677	1,604	1,313	983	265
People (mil.)	37.9	39.1	40.2	41.3	42.7
Military expenditure as % of GNP	NA	NA	3.0	4.0	2.1
Military expenditure as % of CGE	NA	NA	16.1	25.5	44.2
Military expenditure per capita (1994 $)	NA	NA	5	6	3
Armed forces per 1,000 people (soldiers)	1.5	1.5	1.4	1.3	1.2
GNP per capita (1994 $)	230	198	175	153	133
Arms imports[6]					
Current dollars (mil.)	70	20	0	20	0
1994 constant dollars (mil.)	78	21	0	20	0
Arms exports[6]					
Current dollars (mil.)	0	0	0	0	0
1994 constant dollars (mil.)	0	0	0	0	0
Total imports[7]					
Current dollars (mil.)	886	710	409	377	384
1994 constant dollars (mil.)	986	761	427	385	384
Total exports[7]					
Current dollars (mil.)	999	828	416	317	422
1994 constant dollars (mil.)	1,112	888	434	324	422
Arms as percent of total imports[8]	7.9	2.8	0	5.3	0
Arms as percent of total exports[8]	0	0	0	0	0

Human Rights [24]

	SSTS	FL	FAPRO	PPCG	APROBC	TPW	PCPTW	STPEP	PHRFF	PRW	ASST	AFL
Observes		P		P	P	P	P			P	P	
	EAFRD	CPR	ESCR	SR	ACHR	MAAE	PVIAC	PVNAC	EAFDAW	TCIDTP	RC	
Observes	P	P	P	P			P		P	P	P	

P = Party; S = Signatory; see Appendix for meaning of abbreviations.

Labor Force

Total Labor Force [25]

14.51 million (1993 est.)

Labor Force by Occupation [26]

Agriculture	65%
Industry	16
Services	19

Date of data: 1991 est.

Unemployment Rate [27]

For sources, notes, and explanations, see Annotated Source Appendix, page 1061.

1045

Production Sectors

Commercial Energy Production and Consumption

Data are shown in quadrillion (10^{15}) BTUs and percent for 1995
Values for hydroelectric, nuclear, geothermal, solar, and wind power refer to electrical generation.

Production [28]

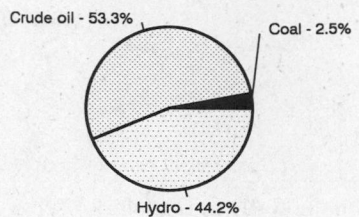

Crude oil - 53.3%
Coal - 2.5%
Hydro - 44.2%

Crude oil	0.064
Coal	0.003
Net hydroelectric power	0.053
Total	0.120

Consumption [29]

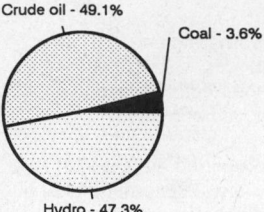

Crude oil - 49.1%
Coal - 3.6%
Hydro - 47.3%

Crude oil	0.055
Coal	0.004
Net hydroelectric power	0.053
Total	0.112

Telecommunications [30]

- 34,000 (1991 est.) telephones
- Domestic: barely adequate wire and microwave radio relay service in and between urban areas; domestic satellite system with 14 earth stations
- International: satellite earth station - 1 Intelsat (Atlantic Ocean)
- Radio: Broadcast stations: AM 10, FM 4, shortwave 0 Radios: 3.87 million (1992 est.)
- Television: Broadcast stations: 18 Televisions: 55,000 (1992 est.)

Top Agricultural Products [32]

Produces coffee, sugar, palm oil, rubber, tea, quinine, cassava (tapioca), palm oil, bananas, root crops, corn, fruits; wood products.

Top Mining Products [33]

Thousand metric tons except as noted[e]	7/21/95[*]
Cobalt, ore milled, gross weight	1,000
Columbite-tantalite concentrate gross weight (kg.)	4,120[r]
Copper, ore mined, gross weight	1,350
Gold (kg.)	6,000
Silver (kg.)	50,000
Cement, hydraulic	150
Diamond, gem and industrial (000 carats)	16,300
Stone, crushed	200
Petroleum, crude (000 42-gal. bls.)	10,600

Transportation [31]

Railways: total: 5,138 km (1995); note - severely reduced trackage in use because of civil strife; narrow gauge: 3,987 km 1.067-m gauge (858 km electrified); 125 km 1.000-m gauge; 1,026 km 0.600-m gauge

Highways: total: 145,000 km; paved: 290 km; unpaved: 144,710 km (1991 est.)

Merchant marine: none

Airports

Total:	217
With paved runways over 3,047 m:	4
With paved runways 2,438 to 3,047 m:	3
With paved runways 1,524 to 2,437 m:	15
With paved runways 914 to 1,523 m:	2
With paved runways under 914 m:	82

Tourism [34]

	1990	1991	1992	1993	1994
Visitors	94	83	72	NA	NA
Tourists[110]	55	33	22	22	18
Excursionists	39	50	50	NA	NA
Tourism receipts	7	NA	7	6	NA
Tourism expenditures	16	NA	16	16	NA

Travelers are in thousands, money in million U.S. dollars.

1046

For sources, notes, and explanations, see Annotated Source Appendix, page 1061.

Manufacturing Sector

GDP and Manufacturing Summary [35]

	1980	1985	1989	1990	% change 1980-1990	% change 1989-1990
GDP (million 1980 $)	6,137	6,653	7,164	6,916	12.7	-3.5
GDP per capita (1980 $)	234	219	208	195	-16.7	-6.3
Manufacturing as % of GDP (current prices)	3.1	1.7	11.4[e]	2.3[e]	-25.8	-79.8
Gross output (million $)	NA	NA	NA	NA	NA	NA
Value added (million $)	170	66[e]	93[e]	96[e]	-43.5	3.2
Value added (million 1980 $)	184	188	169[e]	173	-6.0	2.4
Industrial production index	100	119	119[e]	133	33.0	11.8
Employment (thousands)	50[e]	50[e]	31[e]	50[e]	0.0	61.3

Note: GDP stands for Gross Domestic Product. 'e' stands for estimated value.

Profitability and Productivity

	1980	1985	1989	1990	% change 1980-1990	% change 1989-1990
Intermediate input (%)	NA	NA	NA	NA	NA	NA
Wages, salaries, and supplements (%)	NA	NA	NA	NA	NA	NA
Gross operating surplus (%)	NA	NA	NA	NA	NA	NA
Gross output per worker ($)	NA	NA	NA	NA	NA	NA
Value added per worker ($)	2,929[e]	1,117[e]	2,956[e]	1,616[e]	-44.8	-45.3
Average wage (incl. benefits) ($)	4,535[e]	1,589[e]	3,947[e]	2,092[e]	-53.9	-47.0

Profitability is in percent of gross output. Productivity is in U.S. $. 'e' stands for estimated value.

Profitability - 1990

Value Added in Manufacturing

	1980 $ mil.	1980 %	1985 $ mil.	1985 %	1989 $ mil.	1989 %	1990 $ mil.	1990 %	% change 1980-1990	% change 1989-1990
311 Food products	20	11.8	5[e]	7.6	11[e]	11.8	5[e]	5.2	-75.0	-54.5
313 Beverages	35	20.6	20[e]	30.3	19[e]	20.4	28[e]	29.2	-20.0	47.4
314 Tobacco products	9	5.3	7[e]	10.6	11[e]	11.8	15[e]	15.6	66.7	36.4
321 Textiles	10	5.9	2[e]	3.0	5[e]	5.4	5[e]	5.2	-50.0	0.0
322 Wearing apparel	7	4.1	1[e]	1.5	2[e]	2.2	2[e]	2.1	-71.4	0.0
323 Leather and fur products	NA	0.0	NA	0.0	1[e]	1.1	1[e]	1.0	NA	0.0
324 Footwear	8	4.7	2[e]	3.0	4[e]	4.3	4[e]	4.2	-50.0	0.0
331 Wood and wood products	4	2.4	1[e]	1.5	1[e]	1.1	2[e]	2.1	-50.0	100.0
332 Furniture and fixtures	1	0.6	NA	0.0	1[e]	1.1	NA	0.0	NA	NA
341 Paper and paper products	NA	0.0	NA	0.0	NA	0.0	NA	0.0	NA	NA
342 Printing and publishing	2	1.2	1[e]	1.5	1[e]	1.1	1[e]	1.0	-50.0	0.0
351 Industrial chemicals	12	7.1	6[e]	9.1	12[e]	12.9	8[e]	8.3	-33.3	-33.3
352 Other chemical products	NA	0.0	NA	0.0	NA	0.0	NA	0.0	NA	NA
353 Petroleum refineries	14	8.2	1[e]	1.5	1[e]	1.1	1[e]	1.0	-92.9	0.0
354 Miscellaneous petroleum and coal products	NA	0.0	NA	0.0	NA	0.0	NA	0.0	NA	NA
355 Rubber products	NA	0.0	NA	0.0	NA	0.0	NA	0.0	NA	NA
356 Plastic products	NA	0.0	NA	0.0	NA	0.0	NA	0.0	NA	NA
361 Pottery, china, and earthenware	NA	0.0	NA	0.0	NA	0.0	NA	0.0	NA	NA
362 Glass and glass products	1	0.6	NA	0.0	NA	0.0	NA	0.0	NA	NA
369 Other nonmetal mineral products	4	2.4	1[e]	1.5	1[e]	1.1	2[e]	2.1	-50.0	100.0
371 Iron and steel	4	2.4	1[e]	1.5	NA	0.0	2[e]	2.1	-50.0	NA
372 Nonferrous metals	2	1.2	NA	0.0	NA	0.0	1[e]	1.0	-50.0	NA
381 Metal products	5	2.9	2[e]	3.0	3[e]	3.2	3[e]	3.1	-40.0	0.0
382 Nonelectrical machinery	5	2.9	2[e]	3.0	3[e]	3.2	3[e]	3.1	-40.0	0.0
383 Electrical machinery	3	1.8	1[e]	1.5	1[e]	1.1	2[e]	2.1	-33.3	100.0
384 Transport equipment	5	2.9	3[e]	4.5	5[e]	5.4	3[e]	3.1	-40.0	-40.0
385 Professional and scientific equipment	NA	0.0	NA	0.0	NA	0.0	NA	0.0	NA	NA
390 Other manufacturing industries	15	8.8	7[e]	10.6	9[e]	9.7	9[e]	9.4	-40.0	0.0

Note: The industry codes shown are International Standard Industry codes (ISIC). Percentages are percent of total Value Added. 'e' stands for estimated value

For sources, notes, and explanations, see Annotated Source Appendix, page 1061.

1047

Finance, Economics, and Trade

Economic Indicators [36]

- **National product**: GDP—purchasing power parity—$16.5 billion (1995 est.)

- **National product real growth rate**: - 7.4% (1995 est.)

- **National product per capita**: $400 (1995 est.)

- **Inflation rate (consumer prices)**: 12% monthly average (1995 est.)

- **External debt**: $11.3 billion (December 1993 est.)

Balance of Payments Summary [37]

Values in millions of dollars.

	1986	1987	1988	1989	1990
Exports of goods (f.o.b.)	1,844	1,731	2,178	2,201	2,138
Imports of goods (f.o.b.)	-1,283	-1,376	-1,645	-1,683	-1,539
Trade balance	561	355	533	518	599
Services - debits	-1,286	-1,412	-1,458	-1,461	-1,549
Services - credits	189	262	185	165	171
Private transfers (net)	-62	-70	-67	-109	-81
Government transfers (net)	184	220	226	276	217
Long-term capital (net)	317	639	332	1,150	122
Short-term capital (net)	120	105	420	-442	502
Errors and omissions	-17	13	-134	113	105
Overall balance	6	112	37	210	86

Exchange Rates [38]

Currency: **zaire.**
Symbol: **Z.**

Data are currency units per $1. On 22 October 1993 the new zaire, equal to 3,000,000 old zaires, was introduced.

October 1995	10,618
1994	1,194
1993	3
1992	645,549
1991	15,587

Imports and Exports

Top Import Origins [39]

$382 million (c.i.f., 1994).

Origins	%
South Africa	NA
US	NA
Belgium	NA
France	NA
Germany	NA
Italy	NA
Japan	NA
UK	NA

Top Export Destinations [40]

$419 million (f.o.b., 1994).

Destinations	%
US	NA
Belgium	NA
France	NA
Germany	NA
Italy	NA
UK	NA
Japan	NA
South Africa	NA

Foreign Aid [41]

Recipient: ODA, $NA.

Import and Export Commodities [42]

Import Commodities	Export Commodities
Consumer goods	Copper
Foodstuffs	Coffee
Mining and other machinery	Diamonds
Transport equipment	Cobalt
Fuels	Crude oil

Zambia

Geography [1]

Total area:
752,610 sq km   290,584 sq mi
Land area:
740,720 sq km   285,994 sq mi
Comparative area:
Slightly larger than Texas
Land boundaries:
Total 5,664 km, Angola 1,110 km, Malawi 837 km, Mozambique 419 km,
Namibia 233 km, Tanzania 338 km, Zaire 1,930 km, Zimbabwe 797 km
Coastline:
0 km (landlocked)
Climate:
Tropical; modified by altitude; rainy season (October to April)
Terrain:
Mostly high plateau with some hills and mountains
Natural resources:
Copper, cobalt, zinc, lead, coal, emeralds, gold, silver, uranium,
hydropower potential
Land use:
Arable land: 7%
Permanent crops: 0%
Meadows and pastures: 47%
Forest and woodland: 27%
Other: 19%

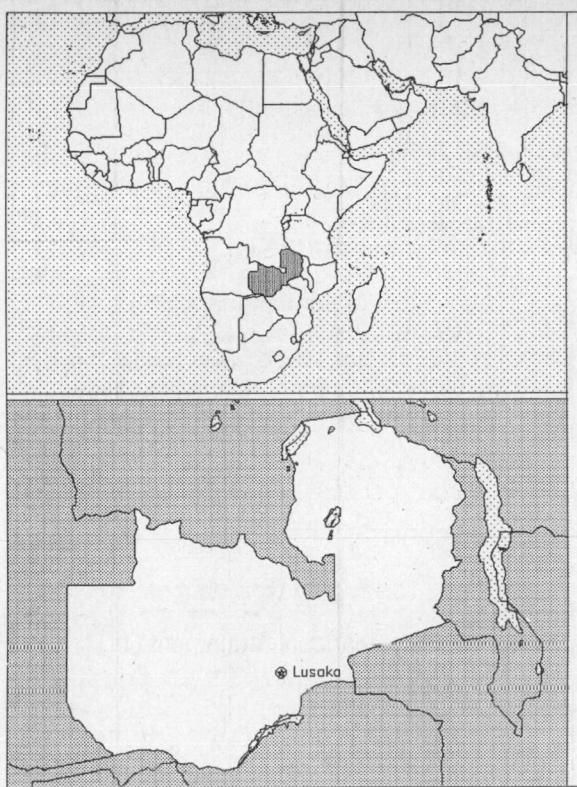

Demographics [2]

	1970	1980	1990	1995[1]	1996	2000	2010	2020	2030
Population	4,247	5,638	8,019	8,969	9,159	9,899	11,471	13,022	15,474
Population density (persons per sq. mi.)	15	20	28	31	32	35	40	46	54
(persons per sq. km.)	6	8	11	12	12	13	15	18	21
Net migration rate (per 1,000 population)	NA	-3.7	-0.2	-1.3	0.0	0.0	0.0	0.0	0.0
Births	NA	NA	NA	NA	410	NA	NA	NA	NA
Deaths	NA	NA	NA	NA	217	NA	NA	NA	NA
Life expectancy - males	NA	50.4	42.5	36.8	36.2	34.0	30.9	35.4	49.6
Life expectancy - females	NA	53.0	42.8	37.4	36.5	33.3	29.7	34.7	51.8
Birth rate (per 1,000)	NA	49.9	48.1	45.2	44.7	43.8	41.2	37.6	33.0
Death rate (per 1,000)	NA	15.0	19.7	23.1	23.7	25.8	29.2	23.8	12.3
Women of reproductive age (15-49 yrs.)	NA	1,282	1,755	1,944	1,983	2,149	2,551	3,075	4,004
of which are currently married	NA	NA	1,160	NA	1,263	1,349	1,575	NA	NA
Fertility rate	NA	7.1	6.9	6.6	6.5	6.3	5.4	4.5	3.7

Except as noted, values for vital statistics are in thousands; life expectancy is in years.

Health

Health Indicators [3]

% of population with access to	
safe water (1990-95)	50
adequate sanitation (1990-95)	37
health services (1985-95)[1]	75
% of 1-year-olds immunized (1990-94) against	
TB (tuberculosis)	100
DPT (diphtheria, pertussis, tetanus)	85
polio	88
measles	88
% of contraceptive prevalence (1980-94)	15
ORT use rate (1990-94)	90

Health Expenditures [4]

Total health expenditure, 1990 (official exchange rate)	
Millions of dollars	117
Dollars per capita	14
Health expenditures as a percentage of GDP	
Total	3.2
Public sector	2.2
Private sector	1.0
Development assistance for health	
Total aid flows (millions of dollars)[1]	6
Aid flows per capita (dollars)	0.7
Aid flows as a percentage of total health expenditure	4.9

For sources, notes, and explanations, see Annotated Source Appendix, page 1061.

1049

Human Factors

Women and Children [5]

% of pregnant women immunized (tetanus 1990-94)	42
% of births attended by trained health personnel (1983-94)	51
Maternal mortality rate (1980-92)	150
Under-5 mortality rate (1994)	203
% under-5 moderately/severely underweight (1980-1994)	25

Burden of Disease [6]

Population per physician (1988-1991)	11,111
Population per nurse (1988-1991)	5,000
Population per hospital bed (1990)	NA
AIDS cases per 100,000 people (1994)	17.3
Malaria cases per 100,000 people (1992)	NA

Ethnic Division [7]

African	98.7%
European	1.1%
Other	0.2%

Religion [8]

Christian	50%-75%
Muslim and Hindu	24%-49%
Indigenous beliefs	1%

Major Languages [9]

English (official), major vernaculars - Bemba, Kaonda, Lozi, Lunda, Luvale, Nyanja, Tonga, and about 70 other indigenous languages.

Education

Public Education Expenditures [10]

	1980	1985	1989	1990	1993	1994
Million (Kwacha)						
Total education expenditure	127	293	1,352	2,737	NA	NA
as percent of GNP	4.5	4.7	2.5	2.6	NA	NA
as percent of total govt. expend.	7.6	13.4	10.9	8.7	NA	NA
Current education expenditure	120	272	1,209	2,382	NA	NA
as percent of GNP	4.2	4.4	2.2	2.3	NA	NA
as percent of current govt. expend.	11.1	14.3	11.6	8.7	NA	NA
Capital expenditure	6	21	143	355	NA	NA

Educational Attainment [11]

Age group (1993)[22]	14+
Total population	NA
Highest level attained (%)	
No schooling	18.6
First level	
Not completed	54.8
Completed	NA
Entered second level	
S-1	12.9
S-2	12.2
Postsecondary	1.5

Illiteracy [12]

In thousands and percent[1]	1990	1995	2000
Illiterate population (15+ yrs.)	1,141	1,082	992
Illiteracy rate - total pop. (%)	28.3	23.8	19.6
Illiteracy rate - males (%)	18.9	15.6	12.9
Illiteracy rate - females (%)	37.0	31.6	26.1

Libraries [13]

	Admin. Units	Svc. Pts.	Vols. (000)	Shelving (meters)	Vols. Added	Reg. Users
National (1992)	1	1	16	NA	98	120
Nonspecialized	NA	NA	NA	NA	NA	NA
Public	NA	NA	NA	NA	NA	NA
Higher ed. (1987)	1	1	10	850	1,737	750
School	NA	NA	NA	NA	NA	NA

Daily Newspapers [14]

	1980	1985	1990	1994
Number of papers	2	2	2	2
Circ. (000)	110	95	99	70

Culture [15]

Cinema (seats per 1,000)	NA
Annual attendance per person	NA
Gross box office receipts (mil. Kwacha)	NA
Museums (reporting)	6
Visitors (000)	185
Annual receipts (000 Kwacha)	NA

Science and Technology

Scientific/Technical Forces [16]

R&D Expenditures [17]

U.S. Patents Issued [18]

For sources, notes, and explanations, see Annotated Source Appendix, page 1061.

Government and Law

Organization of Government [19]

Long-form name:
Republic of Zambia
Type:
Republic
Independence:
24 October 1964 (from UK)
National holiday:
Independence Day, 24 October (1964)
Constitution:
2 August 1991
Legal system:
Based on English common law and customary law; judicial review of legislative acts in an ad hoc constitutional council; has not accepted compulsory ICJ jurisdiction
Executive branch:
President; Vice President; Cabinet
Legislative branch:
Unicameral: National Assembly
Judicial branch:
Supreme Court

Elections [20]

National Assembly	% of seats
Movement for Multiparty Democracy (MMD)	83.3
United National Independence (UNIP)	16.7

Government Expenditures [21]

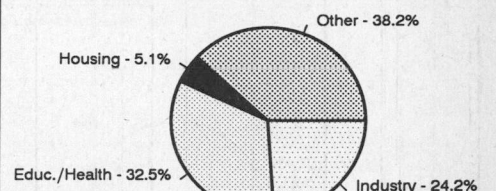

Housing - 5.1%
Other - 38.2%
Educ./Health - 32.5%
Industry - 24.2%

(% distribution). Expend. for CY95: 591,761.3 (Kwacha mil.)

Crime [23]

Military Expenditures and Arms Transfers [22]

	1990	1991	1992	1993	1994
Military expenditures					
Current dollars (mil.)	76[e]	NA	45[e]	56[e]	39
1994 constant dollars (mil.)	85[e]	NA	47[e]	57[e]	39
Armed forces (000)	16	16	16	16	16
Gross national product (GNP)					
Current dollars (mil.)	2,864	2,886	2,964	3,324	3,232
1994 constant dollars (mil.)	3,187	3,093	3,091	3,393	3,232
Central government expenditures (CGE)					
1994 constant dollars (mil.)	874	1,445	956	480	814
People (mil.)	8.2	8.5	8.7	8.9	9.2
Military expenditure as % of GNP	2.7	NA	1.5	1.7	1.2
Military expenditure as % of CGE	9.7	NA	4.9	11.8	4.8
Military expenditure per capita (1994 $)	10	NA	5	6	4
Armed forces per 1,000 people (soldiers)	1.9	1.9	1.8	1.8	1.7
GNP per capita (1994 $)	387	364	354	380	352
Arms imports[6]					
Current dollars (mil.)	5	20	0	0	5
1994 constant dollars (mil.)	6	21	0	0	5
Arms exports[6]					
Current dollars (mil.)	0	0	0	0	0
1994 constant dollars (mil.)	0	0	0	0	0
Total imports[7]					
Current dollars (mil.)	1,220	948	1,107[e]	1,119[e]	NA
1994 constant dollars (mil.)	1,358	1,016	1,154[e]	1,142[e]	NA
Total exports[7]					
Current dollars (mil.)	1,309	745	756	1,043[e]	NA
1994 constant dollars (mil.)	1,457	799	788	1,064[e]	NA
Arms as percent of total imports[8]	0.4	2.1	0	0	NA
Arms as percent of total exports[8]	0	0	0	0	0

Human Rights [24]

	SSTS	FL	FAPRO	PPCG	APROBC	TPW	PCPTW	STPEP	PHRFF	PRW	ASST	AFL
Observes	P	P				P	P			P	P	P
	EAFRD	CPR	ESCR	SR	ACHR	MAAE	PVIAC	PVNAC	EAFDAW	TCIDTP	RC	
Observes		P	P	P		P	P	P	P		P	

P=Party; S=Signatory; see Appendix for meaning of abbreviations.

Labor Force

Total Labor Force [25]

3.4 million

Labor Force by Occupation [26]

Agriculture	85%
Mining, manufacturing, and construction	6
Transport and services	9

Unemployment Rate [27]

22% (1991)

For sources, notes, and explanations, see Annotated Source Appendix, page 1061.

1051

Production Sectors

Commercial Energy Production and Consumption

Data are shown in quadrillion (10^{15}) BTUs and percent for 1995
Values for hydroelectric, nuclear, geothermal, solar, and wind power refer to electrical generation.

Production [28]

Hydro - 87.1%

Coal - 12.9%

Consumption [29]

Hydro - 69.8%

Coal - 10.3%

Crude oil - 19.8%

Coal	0.012
Net hydroelectric power	0.081
Total	0.093

Crude oil	0.023
Coal	0.012
Net hydroelectric power	0.081
Total	0.116

Telecommunications [30]

- 80,900 (1987 est.) telephones; facilities are among the best in Sub-Saharan Africa
- Domestic: high-capacity microwave radio relay connects most larger towns and cities
- International: satellite earth stations - 2 Intelsat (1 Indian Ocean and 1 Atlantic Ocean)
- Radio: Broadcast stations: AM 11, FM 5, shortwave 0 Radios: 1,889,140
- Television: Broadcast stations: 9 Televisions: 215,000 (1995 est.)

Transportation [31]

Railways: total: 2,164 km (1995); narrow gauge: 2,164 km 1.067-m gauge (13 km double track)

Highways: total: 37,359 km; paved: 6,575 km (including 56 km of expressways); unpaved: 30,784 km (1992 est.)

Airports

Total:	104
With paved runways over 3,047 m:	1
With paved runways 2,438 to 3,047 m:	3
With paved runways 1,524 to 2,437 m:	4
With paved runways 914 to 1,523 m:	2
With paved runways under 914 m:	35

Top Agricultural Products [32]

Agriculture accounts for 32% of the GDP; produces corn, sorghum, rice, peanuts, sunflower seed, tobacco, cotton, sugarcane, cassava (tapioca); cattle, goats, beef, eggs.

Top Mining Products [33]

Thousand metric tons except as noted	10/1/95[*]
Cobalt, ore milled, gross weight	5,390[84,85]
Copper, ore milled, gross weight	19,800[84,86]
Selenium, refined, gross weight (kg.)	21,300[84,88]
Silver (kg.)	10,000[84,87]
Cement, hydraulic	280
Amethyst (kg.)	366,000
Emerald (kg.)	160
Limestone (cement and lime)	710[e,89]
Coal, bituminous	163
Petroleum refinery products (000 42-gal. bls.)	5,300[e,84]

Tourism [34]

	1990	1991	1992	1993	1994
Tourists	141	171	159	157	134
Tourism receipts	41	35	51	44	43
Tourism expenditures	54	87	56	NA	NA
Fare receipts	28	21	NA	NA	NA
Fare expenditures	88	77	NA	NA	NA

Travelers are in thousands, money in million U.S. dollars.

Manufacturing Sector

GDP and Manufacturing Summary [35]

	1980	1985	1989	1990	% change 1980-1990	% change 1989-1990
GDP (million 1980 $)	3,883	3,978	3,981	4,307	10.9	8.2
GDP per capita (1980 $)	677	568	489	510	-24.7	4.3
Manufacturing as % of GDP (current prices)	19.0	23.8	31.6[e]	33.2	74.7	5.1
Gross output (million $)	1,671	1,378[e]	2,668	2,610[e]	56.2	-2.2
Value added (million $)	780	575[e]	1,133	1,028[e]	31.8	-9.3
Value added (million 1980 $)	717	789	821[e]	1,098	53.1	33.7
Industrial production index	100	106	113	122	22.0	8.0
Employment (thousands)	59	62[e]	61	61[e]	3.4	0.0

Note: GDP stands for Gross Domestic Product. 'e' stands for estimated value.

Profitability and Productivity

	1980	1985	1989	1990	% change 1980-1990	% change 1989-1990
Intermediate input (%)	53	58[e]	58	61[e]	15.1	5.2
Wages, salaries, and supplements (%)	11	11[e]	11	11[e]	0.0	0.0
Gross operating surplus (%)	35	30[e]	31	29[e]	-17.1	-6.5
Gross output per worker ($)	28,232	22,254[e]	43,950	43,052[e]	52.5	-2.0
Value added per worker ($)	13,184	9,280[e]	18,661	16,966[e]	28.7	-9.1
Average wage (incl. benefits) ($)	3,245	2,542[e]	4,980	4,642[e]	43.1	-6.8

Profitability is in percent of gross output. Productivity is in U.S. $. 'e' stands for estimated value.

Profitability - 1990

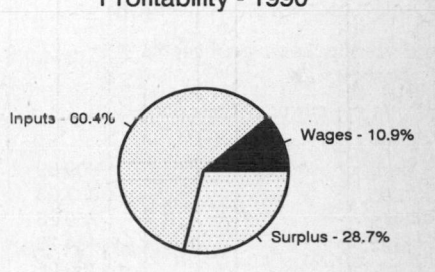

Inputs - 60.4%
Wages - 10.9%
Surplus - 28.7%

The graphic shows percent of gross output.

Value Added in Manufacturing

	1980 $ mil.	1980 %	1985 $ mil.	1985 %	1989 $ mil.	1989 %	1990 $ mil.	1990 %	% change 1980-1990	% change 1989-1990
311 Food products	92	11.8	62[e]	10.8	100	8.8	87[e]	8.5	-5.4	-13.0
313 Beverages	193	24.7	104[e]	18.1	243	21.4	237[e]	23.1	22.8	-2.5
314 Tobacco products	58	7.4	39[e]	6.8	111	9.8	97[e]	9.4	67.2	-12.6
321 Textiles	51	6.5	32[e]	5.6	69	6.1	62[e]	6.0	21.6	-10.1
322 Wearing apparel	34	4.4	23[e]	4.0	47	4.1	46[e]	4.5	35.3	-2.1
323 Leather and fur products	4	0.5	3[e]	0.5	6	0.5	6[e]	0.6	50.0	0.0
324 Footwear	15	1.9	13[e]	2.3	28	2.5	29[e]	2.8	93.3	3.6
331 Wood and wood products	8	1.0	11[e]	1.9	29	2.6	28[e]	2.7	250.0	-3.4
332 Furniture and fixtures	12	1.5	10[e]	1.7	24	2.1	21[e]	2.0	75.0	-12.5
341 Paper and paper products	15	1.9	8[e]	1.4	13	1.1	11[e]	1.1	-26.7	-15.4
342 Printing and publishing	17	2.2	13[e]	2.3	24	2.1	24[e]	2.3	41.2	0.0
351 Industrial chemicals	22	2.8	26[e]	4.5	43	3.8	37[e]	3.6	68.2	-14.0
352 Other chemical products	47	6.0	51[e]	8.9	84	7.4	73[e]	7.1	55.3	-13.1
353 Petroleum refineries	9	1.2	5[e]	0.9	7	0.6	6[e]	0.6	-33.3	-14.3
354 Miscellaneous petroleum and coal products	3	0.4	2[e]	0.3	3	0.3	3[e]	0.3	0.0	0.0
355 Rubber products	20	2.6	16[e]	2.8	27	2.4	23[e]	2.2	15.0	-14.8
356 Plastic products	7	0.9	7[e]	1.2	13	1.1	12[e]	1.2	71.4	-7.7
361 Pottery, china, and earthenware	1	0.1	1[e]	0.2	1	0.1	1[e]	0.1	0.0	0.0
362 Glass and glass products	3	0.4	3[e]	0.5	4	0.4	4[e]	0.4	33.3	0.0
369 Other nonmetal mineral products	33	4.2	45[e]	7.8	60	5.3	55[e]	5.4	66.7	-8.3
371 Iron and steel	10	1.3	5[e]	0.9	8	0.7	7[e]	0.7	-30.0	-12.5
372 Nonferrous metals	2	0.3	1[e]	0.2	1	0.1	1[e]	0.1	-50.0	0.0
381 Metal products	50	6.4	47[e]	8.2	99	8.7	82[e]	8.0	64.0	-17.2
382 Nonelectrical machinery	18	2.3	11[e]	1.9	20	1.8	20[e]	1.9	11.1	0.0
383 Electrical machinery	26	3.3	13[e]	2.3	21	1.9	18[e]	1.8	-30.8	-14.3
384 Transport equipment	28	3.6	24[e]	4.2	45	4.0	39[e]	3.8	39.3	-13.3
385 Professional and scientific equipment	NA	0.0	NA	0.0	NA	0.0	NA	0.0	NA	NA
390 Other manufacturing industries	2	0.3	1[e]	0.2	1	0.1	1[e]	0.1	-50.0	0.0

Note: The industry codes shown are International Standard Industry codes (ISIC). Percentages are percent of total Value Added. 'e' stands for estimated value

Finance, Economics, and Trade

Economic Indicators [36]

- **National product**: GDP—purchasing power parity—$8.9 billion (1995 est.)
- **National product real growth rate**: NA%
- **National product per capita**: $900 (1995 est.)
- **Inflation rate (consumer prices)**: 55% (1994 est.)
- **External debt**: $7 billion (1995 est.)

Balance of Payments Summary [37]

Values in millions of dollars.

	1987	1988	1989	1990	1991
Exports of goods (f.o.b.)	852	1,189	1,340	1,254	1,172
Imports of goods (f.o.b.)	-585	-687	-774	-1,511	-752
Trade balance	267	502	566	-257	420
Services - debits	-553	-893	-953	-825	-1,059
Services - credits	49	61	87	108	93
Private transfers (net)	-20	-25	-30	-18	-22
Government transfers (net)	8	59	109	395	261
Long-term capital (net)	-162	97	160	709	55
Short-term capital (net)	190	209	1,872	-287	187
Errors and omissions	153	40	-1,712	319	126
Overall balance	-68	50	99	144	61

Exchange Rates [38]

Currency: **Zambian kwacha.**
Symbol: **ZK.**

Data are currency units per $1.

December 1995	909.09
1995	833.33
1994	769.23
1993	434.78
1992	156.25
1991	61.7284

Imports and Exports

Top Import Origins [39]

$845 million (f.o.b., 1994 est.).

Origins	%
EU	NA
Japan	NA
Saudi Arabia	NA
South Africa	NA
US	NA

Top Export Destinations [40]

$1.075 billion (f.o.b., 1994 est.).

Destinations	%
EU	NA
Japan	NA
South Africa	NA
US	NA
India	NA
Thailand	NA
Malaysia	NA

Foreign Aid [41]

Recipient: ODA, $734 million (1993).

Import and Export Commodities [42]

Import Commodities	Export Commodities
Machinery	Copper
Transportation equipment	Zinc
Foodstuffs	Cobalt
Fuels	Lead
Manufactures	Tobacco

Zimbabwe

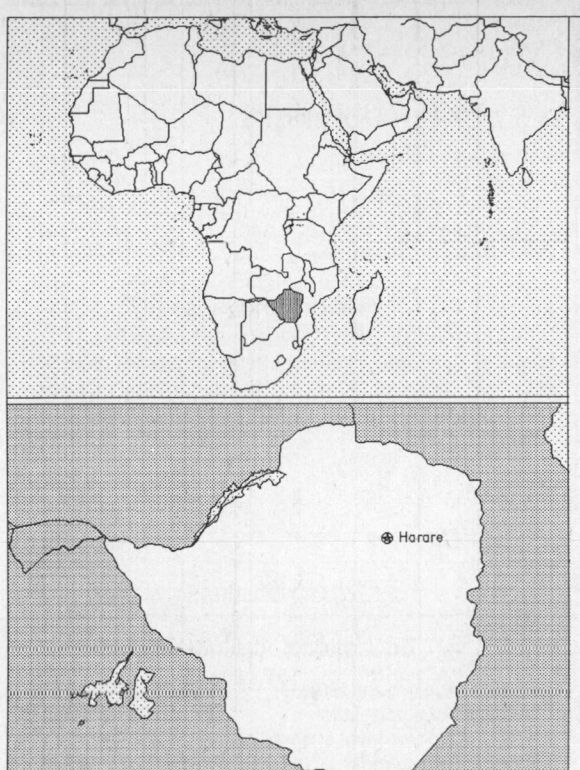

Geography [1]

Total area:
390,580 sq km   150,804 sq mi
Land area:
386,670 sq km   149,294 sq mi
Comparative area:
Slightly larger than Montana
Land boundaries:
Total 3,066 km, Botswana 813 km, Mozambique 1,231 km, South Africa 225 km, Zambia 797 km
Coastline:
0 km (landlocked)
Climate:
Tropical; moderated by altitude; rainy season (November to March)
Terrain:
Mostly high plateau with higher central plateau (high veld); mountains in East
Natural resources:
Coal, chromium ore, asbestos, gold, nickel, copper, iron ore, vanadium, lithium, tin, platinum group metals
Land use:
Arable land: 7%
Permanent crops: Negligible (coffee)
Meadows and pastures: 13%
Forest and woodland: 49%
Other: 31%

Demographics [2]

	1970	1980	1990	1995[1]	1996	2000	2010	2020	2030
Population	5,515	7,298	10,121	11,161	11,271	11,777	11,905	11,344	11,623
Population density (persons per sq. mi.)	37	49	68	75	75	79	80	76	78
(persons per sq. km.)	14	19	26	29	29	30	31	29	30
Net migration rate (per 1,000 population)	NA	NA	0.7	-10.2	0.0	0.0	0.0	0.0	0.0
Births	NA	NA	NA	NA	365	NA	NA	NA	NA
Deaths	NA	NA	NA	NA	205	NA	NA	NA	NA
Life expectancy - males	NA	NA	50.6	43.1	41.9	38.1	34.0	38.9	55.6
Life expectancy - females	NA	NA	49.0	42.8	41.8	38.3	32.0	37.4	57.1
Birth rate (per 1,000)	NA	NA	39.1	33.1	32.3	29.3	24.0	22.4	20.3
Death rate (per 1,000)	NA	NA	14.0	17.4	18.2	21.6	29.4	25.4	12.1
Women of reproductive age (15-49 yrs.)	NA	NA	2,313	2,639	2,692	2,914	3,138	3,159	3,380
of which are currently married	NA	NA	1,441	NA	1,643	1,770	1,924	NA	NA
Fertility rate	NA	NA	5.3	4.2	4.1	3.5	2.4	2.1	2.0

Except as noted, values for vital statistics are in thousands; life expectancy is in years.

Health

Health Indicators [3]

% of population with access to	
safe water (1990-95)	77
adequate sanitation (1990-95)	66
health services (1985-95)	85
% of 1-year-olds immunized (1990-94) against	
TB (tuberculosis)	90
DPT (diphtheria, pertussis, tetanus)	80
polio	80
measles	77
% of contraceptive prevalence (1980-94)	43
ORT use rate (1990-94)	82

Health Expenditures [4]

Total health expenditure, 1990 (official exchange rate)	
Millions of dollars	416
Dollars per capita	42
Health expenditures as a percentage of GDP	
Total	6.2
Public sector	3.2
Private sector	3.0
Development assistance for health	
Total aid flows (millions of dollars)[1]	42
Aid flows per capita (dollars)	4.2
Aid flows as a percentage of total health expenditure	10.0

For sources, notes, and explanations, see Annotated Source Appendix, page 1061.

1055

Human Factors

Women and Children [5]

% of pregnant women immunized (tetanus 1990-94)	NA
% of births attended by trained health personnel (1983-94)	70
Maternal mortality rate (1980-92)	400
Under-5 mortality rate (1994)	81
% under-5 moderately/severely underweight (1980-1994)[1]	12

Burden of Disease [6]

Population per physician (1988-1991)	7,692
Population per nurse (1988-1991)	1,639
Population per hospital bed (1990)	NA
AIDS cases per 100,000 people (1994)	96.7
Malaria cases per 100,000 people (1992)	NA

Ethnic Division [7]

African	98%
Shona	71%
Ndebele	16%
Other	11%
White	1%
Mixed and Asian	1%

Religion [8]

Syncretic (part Christian, part indigenous beliefs)	50%
Christian	25%
Indigenous beliefs	24%
Muslim and other	1%

Major Languages [9]

English (official), Shona, Sindebele (the language of the Ndebele, sometimes called Ndebele), numerous but minor tribal dialects.

Education

Public Education Expenditures [10]

Million (Dollar)[72]	1980	1990	1991	1992	1993	1994
Total education expenditure	224	1,661	1,933	2,444	2,914	NA
as percent of GNP	6.6	10.5	9.4	10.1	8.3	NA
as percent of total govt. expend.	13.7	NA	NA	NA	NA	NA
Current education expenditure	218	1,648	1,913	NA	2,893	NA
as percent of GNP	6.4	10.4	9.3	NA	8.3	NA
as percent of current govt. expend.	14.1	NA	NA	NA	NA	NA
Capital expenditure	6	13	20	NA	21	NA

Educational Attainment [11]

Age group (1992)	25+
Total population	3,445,195
Highest level attained (%)	
No schooling	22.3
First level	
Not completed	53.2
Completed	NA
Entered second level	
S-1	19.4
S-2	NA
Postsecondary	4.9

Illiteracy [12]

In thousands and percent[1]	1990	1995	2000
Illiterate population (15+ yrs.)	972	940	881
Illiteracy rate - total pop. (%)	18.3	15.3	12.9
Illiteracy rate - males (%)	12.3	9.9	8.0
Illiteracy rate - females (%)	23.9	20.5	17.7

Libraries [13]

	Admin. Units	Svc. Pts.	Vols. (000)	Shelving (meters)	Vols. Added	Reg. Users
National (1993)	1	1	96	NA	785	46,605
Nonspecialized	NA	NA	NA	NA	NA	NA
Public (1989)	76	83	1,038	NA	3,195	151,563
Higher ed. (1990)	25	31	764	NA	7,609	30,707
School	NA	NA	NA	NA	NA	NA

Daily Newspapers [14]

	1980	1985	1990	1994
Number of papers	2	3	2	2
Circ. (000)	133	203	206	195

Culture [15]

Cinema (seats per 1,000)	1.2
Annual attendance per person	0.2
Gross box office receipts (mil. Dollar)	7.6
Museums (reporting)	11
Visitors (000)	100
Annual receipts (000 Dollar)	1,000

Science and Technology

Scientific/Technical Forces [16]

R&D Expenditures [17]

U.S. Patents Issued [18]

Values show patents issued to citizens of the country by the U.S. Patents Office.

	1993	1994	1995
Number of patents	1	4	1

For sources, notes, and explanations, see Annotated Source Appendix, page 1061.

Government and Law

Organization of Government [19]

Long-form name:
Republic of Zimbabwe
Type:
Parliamentary democracy
Independence:
18 April 1980 (from UK)
National holiday:
Independence Day, 18 April (1980)
Constitution:
21 December 1979
Legal system:
Mixture of Roman-Dutch and English
common law
Executive branch:
Executive President; 2 Co-Vice
Presidents; Cabinet
Legislative branch:
Unicameral: Parliament
Judicial branch:
Supreme Court

Elections [20]

Parliament	% of seats
African National Union (ZANU-PF)	98.3
African National Union (ZANU-S)	0.7

Government Budget [21]

For FY92/93.

	$ bil.
Revenues	1.7
Expenditures	2.2
Capital expenditures	0.253

Crime [23]

	1994
Crime volume	
Cases known to police	237,653
Attempts (percent)	NA
Percent cases solved	55,952
Crimes per 100,000 persons	2,160.48
Persons responsible for offenses	
Total number offenders	NA
Percent female	NA
Percent juvenile	NA
Percent foreigners	NA

Military Expenditures and Arms Transfers [22]

	1990	1991	1992	1993	1994
Military expenditures					
Current dollars (mil.)	235	262[e]	250[e]	209[e]	188[e]
1994 constant dollars (mil.)	262	281[e]	260[e]	213[e]	188[e]
Armed forces (000)	45	45	48	48	43
Gross national product (GNP)					
Current dollars (mil.)	4,435	4,681	4,547	4,871	5,086
1994 constant dollars (mil.)	4,936	5,018	4,742	4,972	5,086
Central government expenditures (CGE)					
1994 constant dollars (mil.)	2,011	2,010	2,193	1,422[e]	NA
People (mil.)	10.2	10.4	10.7	10.8	11.0
Military expenditure as % of GNP	5.3	5.6	5.5	4.3	3.7
Military expenditure as % of CGE	13.0	14.0	11.9	15.0	NA
Military expenditure per capita (1994 $)	26	27	24	20	17
Armed forces per 1,000 people (soldiers)	4.4	4.3	4.5	4.4	3.9
GNP per capita (1994 $)	485	480	444	459	463
Arms imports[6]					
Current dollars (mil.)	60	50	90	10	5
1994 constant dollars (mil.)	67	54	94	10	5
Arms exports[6]					
Current dollars (mil.)	0	0	5	0	10
1994 constant dollars (mil.)	0	0	5	0	10
Total imports[7]					
Current dollars (mil.)	1,847	2,055	2,220	2,022[e]	NA
1994 constant dollars (mil.)	2,056	2,203	2,315	2,064[e]	NA
Total exports[7]					
Current dollars (mil.)	1,726	1,532	1,445	NA	1,800[e]
1994 constant dollars (mil.)	1,921	1,642	1,507	NA	1,800[e]
Arms as percent of total imports[8]	3.2	2.4	4.1	0.5	NA
Arms as percent of total exports[8]	0	0	0.3	0	0.6

Human Rights [24]

	SSTS	FL	FAPRO	PPCG	APROBC	TPW	PCPTW	STPEP	PHRFF	PRW	ASST	AFL
Observes	1			P		P	P			P		
	EAFRD	CPR	ESCR	SR	ACHR	MAAE	PVIAC	PVNAC	EAFDAW	TCIDTP	RC	
Observes	P	P	P	P			P	P	P		P	

P = Party; S = Signatory; see Appendix for meaning of abbreviations.

Labor Force

Total Labor Force [25]

4.228 million (1993 est.)

Labor Force by Occupation [26]

Agriculture	70%
Transport and services	22
Industry	8

Unemployment Rate [27]

At least 45% (1994 est.)

For sources, notes, and explanations, see Annotated Source Appendix, page 1061.

1057

Production Sectors

Commercial Energy Production and Consumption

Data are shown in quadrillion (10^{15}) BTUs and percent for 1995
Values for hydroelectric, nuclear, geothermal, solar, and wind power refer to electrical generation.

Production [28]

Coal - 83.3%

Hydro - 16.7%

Consumption [29]

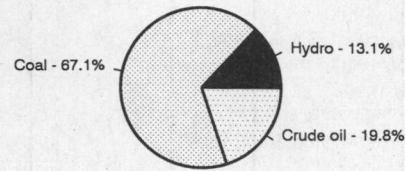

Hydro - 13.1%

Coal - 67.1%

Crude oil - 19.8%

Coal	0.155
Net hydroelectric power	0.031
Total	0.186

Crude oil	0.047
Coal	0.159
Net hydroelectric power	0.031
Total	0.237

Telecommunications [30]

- 301,000 (1990 est.) telephones; system was once one of the best in Africa, but now suffers from poor maintenance
- Domestic: consists of microwave radio relay links, open-wire lines, and radiotelephone communication stations
- International: satellite earth station - 1 Intelsat (Atlantic Ocean)
- Radio: Broadcast stations: AM 8, FM 18, shortwave 0 Radios: 890,000 (1992 est.)
- Television: Broadcast stations: 8 (1986 est.) Televisions: 280,000 (1992 est.)

Transportation [31]

Railways: total: 2,759 km (1995); narrow gauge: 2,759 km 1.067-m gauge (313 km electrified; 42 km double track) (1995 est.)

Highways: total: 91,078 km; paved: 14,572 km; unpaved: 76,506 km (1992 est.)

Airports

Total:	403
With paved runways over 3,047 m:	3
With paved runways 2,438 to 3,047 m:	2
With paved runways 1,524 to 2,437 m:	6
With paved runways 914 to 1,523 m:	8
With paved runways under 914 m:	185

Top Agricultural Products [32]

Agriculture accounts for 18.3% of the GDP; produces corn, cotton, tobacco, wheat, coffee, sugarcane, peanuts; cattle, sheep, goats, pigs.

Top Mining Products [33]

Thousand metric tons except as noted[e]	8/4/95*
Chromite, gross weight	517
Gold (kg.)	20,500
Platinumgroup metals (kg.)	24
Selenium (kg.)	2,010
Silver (kg.)	10,900
Diamond (carats)	174,000
Emerald (kg.)	276[r]
Limestone	1,660
Coal, bituminous	5,520[r]
Coke, metallurgical	550[90]

Tourism [34]

	1990	1991	1992	1993	1994
Visitors[73]	636	697	766	973	1,139
Tourists[12]	606	664	738	943	1,099
Excursionists	30	33	28	30	40
Tourism receipts	64	75	108	138	NA
Tourism expenditures	66	70	55	44	NA
Fare receipts	47	42	41	32	NA
Fare expenditures	38	36	41	34	NA

Travelers are in thousands, money in million U.S. dollars.

For sources, notes, and explanations, see Annotated Source Appendix, page 1061.

Manufacturing Sector

Manufacturing Summary [35]

	1987		1988		1989		1990		1991	
	$ bil.	%	$ bil.	%	$ bil.	%	$ bil.	%	$ bil.	%
Establishments or enterprises (number)	1,341	0.057	1,330	0.063	1,590	0.085	-	-	-	-
Total employment (000)	193	0.142	208	0.152	211	0.172	-	-	-	-
Production workers (000)	-	-	-	-	-	-	-	-	-	-
Output ($ bil.)	4	0.044	5	0.043	5	0.045	-	-	-	-
Value added ($ bil.)	2	0.044	2	0.049	3	0.053	-	-	-	-
Capital investment ($ mil.)	203	0.047	242	0.051	292	0.053	-	-	-	-
M & E investment ($ mil.)	-	-	-	-	-	-	-	-	-	-
Employees per establishment (number)	144	249.030	156	240.001	133	202.251	-	-	-	-
Production workers per establishment	-	-	-	-	-	-	-	-	-	-
Output per establishment ($ mil.)	3	76.317	4	67.250	3	53.493	-	-	-	-
Capital investment per estab. ($ mil.)	0.152	81.670	0.182	80.349	0.183	62.713	-	-	-	-
M & E per establishment ($ mil)	-	-	-	-	-	-	-	-	-	-
Payroll per employee ($)	4,150	46.281	4,266	42.819	4,910	43.946	-	-	-	-
Wages per production worker ($)	-	-	-	-	-	-	-	-	-	-
Hours per production worker (hours)	-	-	-	-	-	-	-	-	-	-
Output per employee ($)	22,952	30.646	23,867	28.021	25,349	26.449	-	-	-	-
Capital investment per employee ($)	1,053	32.795	1,165	33.478	1,380	31.008	-	-	-	-
M & E per employee ($)	-	-	-	-	-	-	-	-	-	-

Note: Columns headed % show percent of world total or ratio. Ratios closest to 100 are closest to world average. M & E stands for machinery & equipment.

Output in Manufacturing

	1987		1988		1989		1990		1991	
	$ bil.[f]	%	$ bil.[f]	%	$ bil.[f]	%	$ bil.[f]	%	$ bil.[f]	%
3110 - Food products	0.952	23.80	0.916	18.32	0.909	18.18	-	-	-	-
3130 - Beverages	0.358	8.95	0.359	7.18	0.357	7.14	-	-	-	-
3140 - Tobacco	0.167	4.17	0.128	2.56	0.113	2.26	-	-	-	-
3210 - Textiles	0.413	10.32	0.454	9.08	0.534	10.68	-	-	-	-
3211 - Spinning, weaving, etc.	0.355	8.88	0.431	8.62	0.444	8.88	-	-	-	-
3220 - Wearing apparel	0.169	4.23	0.202	4.04	0.201	4.02	-	-	-	-
3230 - Leather and products	0.025	0.63	0.026	0.52	0.019	0.38	-	-	-	-
3240 - Footwear	0.095	2.38	0.108	2.16	0.105	2.10	-	-	-	-
3310 - Wood products	0.069	1.72	0.073	1.46	0.073	1.46	-	-	-	-
3320 - Furniture, fixtures	0.049	1.23	0.053	1.06	0.055	1.10	-	-	-	-
3410 - Paper and products	0.107	2.67	0.124	2.48	0.124	2.48	-	-	-	-
3411 - Pulp, paper, etc.	0.032	0.80	0.038	0.76	0.039	0.78	-	-	-	-
3420 - Printing, publishing	0.090	2.25	0.100	2.00	0.101	2.02	-	-	-	-
3511 - Basic chemicals, excl fertilizers	0.027	0.67	0.026	0.52	0.027	0.54	-	-	-	-
3520 - Chemical products nec	0.182	4.55	0.241	4.82	0.205	4.10	-	-	-	-
3522 - Drugs and medicines	0.041	1.02	0.050	1.00	0.045	0.90	-	-	-	-
3550 - Rubber products	0.081	2.03	0.106	2.12	0.124	2.48	-	-	-	-
3560 - Plastic products nec	0.107	2.67	0.125	2.50	0.133	2.66	-	-	-	-
3610 - Pottery, china, etc.	0.005	0.13	0.005	0.10	0.000	0.00	-	-	-	-
3620 - Glass and products	0.014	0.35	0.016	0.32	0.018	0.36	-	-	-	-
3690 - Nonmetal products nec	0.113	2.83	0.115	2.30	0.114	2.28	-	-	-	-
3710 - Iron and steel	0.277	6.93	0.439	8.78	0.623	12.46	-	-	-	-
3720 - Nonferrous metals	0.019	0.48	0.027	0.54	0.032	0.64	-	-	-	-
3810 - Metal products	0.252	6.30	0.267	5.34	0.283	5.66	-	-	-	-
3820 - Machinery nec	0.047	1.18	0.054	1.08	0.050	1.00	-	-	-	-
3830 - Electrical machinery	0.088	2.20	0.117	2.34	0.133	2.66	-	-	-	-
3832 - Radio, television, etc.	0.017	0.43	0.018	0.36	0.018	0.36	-	-	-	-
3840 - Transportation equipment	0.145	3.63	0.186	3.72	0.255	5.10	-	-	-	-
3843 - Motor vehicles	0.114	2.85	0.134	2.68	0.193	3.86	-	-	-	-
3850 - Professional goods	0.003	0.08	0.003	0.06	0.004	0.08	-	-	-	-

Note: Codes are International Standard Industry codes (ISIC). Percentages are % of total Output. [f]: Factor Prices; [p]: Producer Prices.

For sources, notes, and explanations, see Annotated Source Appendix, page 1061.

1059

Finance, Economics, and Trade

Economic Indicators [36]

- **National product**: GDP—purchasing power parity—$18.1 billion (1995 est.)
- **National product real growth rate**: - 2.4% (1995)
- **National product per capita**: $1,620 (1995 est.)
- **Inflation rate (consumer prices)**: 25.8% (1995)
- **External debt**: $4.4 billion (1994)

Balance of Payments Summary [37]

Values in millions of dollars.

	1989	1990	1991	1992	1993
Exports of goods (f.o.b.)	1,693.5	1,748.0	1,694.0	1,528.0	1,609.0
Imports of goods (f.o.b.)	-1,318.3	-1,505.0	-1,646.0	-1,782.0	-1,487.0
Trade balance	375.2	243.0	48.0	-255.0	122.0
Services - debits	-721.3	-782.0	-905.0	-963.0	-851.0
Services - credits	283.7	287.0	300.0	331.0	407.0
Private transfers (net)	-19.2	-3.0	3.0	40.0	27.0
Government transfers (net)	78.5	108.0	95.0	242.0	179.0
Long-term capital (net)	41.7	131.0	282.0	478.0	273.0
Short-term capital (net)	NA	112.0	255.0	-105.0	54.0
Errors and omissions	-91.7	-10.0	-31.0	37.0	15.0
Overall balance	-53.1	-86.0	45.0	-195.0	226.0

Exchange Rates [38]

Currency: **Zimbabwean dollar.**
Symbol: **Z$.**

Data are currency units per $1.

January 1996	0.3633
1995	8.6580
1994	8.1500
1993	6.4725
1992	5.0942
1991	3.4282

Imports and Exports

Top Import Origins [39]

$1.8 billion (c.i.f., 1995 est.) Data are for 1991.

Origins	%
South Africa	25
UK	15
Germany	9
US	6
Japan	5

Top Export Destinations [40]

$2.2 billion (f.o.b., 1995 est.) Data are for 1991.

Destinations	%
UK	14
Germany	11
South Africa	10
Japan	7
US	5

Foreign Aid [41]

Recipient: ODA, $362 million (1993).

Import and Export Commodities [42]

Import Commodities

Machinery and transportation equipment 41%
Other manufactures 23%
Chemicals 16%
Fuels 12%

Export Commodities

Agricultural 35% (tobacco 30% and other 5%)
Manufactures 25%
Gold 12%
Ferrochrome 10%
Textiles 8%

For sources, notes, and explanations, see Annotated Source Appendix, page 1061.

ANNOTATED SOURCE APPENDIX

Table of Contents

Maps

Each national entry has both a regional map and a national map. The regional map indicates the location of the nation within its region; normally a continent or major sub-continental area. The maps of the Regions of the World, provided in the front section, show the location of all nations within a region in one map.

National and regional maps for Africa and North America are Mercator projections produced using the *CIA World Database* and Allison Software's *MAPIT* v1.3 software. The corresponding maps for all other nations are Plate Carée projections using the Defense Mapping Agency's *Digital Chart of the World*. The Regions of the World maps are all Plate Carrée projections.

Each national map is presented at the scale which produces the largest map which fits within the panel. Thus the maps are at different scales. The regional maps of the same region are at identical scales, again determined by the maximum scale which fits the format.

Due to the small size of the maps, no rivers are shown. Only those inland seas and lakes are shown that can be seen easily.

The Mercator projection is probably the most familiar to the public. It projects the earth's sphere onto a cylinder. Lines of latitude and longitude are all parallel straight lines at right angles to each other. Therefore, it is absolutely accurate only at the equator. Distortion of east to west distances increases with the distance from the equator. The Plate Carrée method projects the earth's sphere onto a cone (i.e. a conic projection) with the equator being the standard and absolutely accurate length latitude line. Lines of latitude are concentric circles, and lines of longitude are non-parallel straight lines. The lines of latitude and longitude intersect at right angles. Again, this projection is only absolutely accurate at the equator. However, distortions of distance and shape away from the equator are less than those associated with the Mercator projection. It is useful for maps that have a greater east to west than north to south expanse (as is the case in *SAW*). This projection was used where it was available to provide greater accuracy while providing a view of an area similar to the more familiar Mercator projection.

1 - Geography

Source. U.S. Central Intelligence Agency (CIA) (1996). *The World Factbook 1996* [Online]. Available: http://www.odci.gov//cia/publications/nsolo/factbook/global.html [1997, March 11].

Notes. Following are CIA definitions of terms used in these tables.

Comparative area—based on total area equivalents. Most entities are compared with the entire United States or one of the 50 states. The smaller entities are compared with Wash-

ington, D.C. (178 square km, 69 square miles), or The Mall in Washington, D.C. (0.59 square km, 0.23 square miles, 146 acres).

km—kilometers.

Land area—aggregate of all surfaces delimited by international boundaries and/or coastlines, excluding inland water bodies (lakes, reservoirs, rivers).

Land use—human use of the land surface is categorized as *arable land*—land cultivated for crops that are replanted after each harvest (wheat, maize, rice); *permanent crops*—land cultivated for crops that are not replanted after each harvest (citrus, coffee, rubber); *meadows and pastures*—land permanently used for herbaceous forage crops; *forest and woodland*—land under dense or open stands of trees; and *other*—any land type not specifically mentioned above (urban areas, roads, desert).

mi—miles.

NA—data are not available.

Total area—sum of all land and water areas delimited by international boundaries and/or coastlines.

2 - Demographics

Source. U.S. Bureau of the Census (1996). *International Data Base 1996* (ver. 9605) [Online]. Available: http://www.census.gov/ipc/www/idbnew.html [1996, May].

Source. U.S. Bureau of the Census. Report WP/96. *World Population Profile: 1996*. Special report prepared by Thomas M. McDevitt. Washington, DC: U.S. Government Printing Office, 1996.

Notes. Demographics tables were derived from the *International Data Base* with the exception of data for *Births*, *Deaths*, and *Women of reproductive age . . . currently married*. These figures were derived primarily from *World Population Profile: 1996*.

New estimates and projections of population and vital rates are made for each edition of the sources listed above based on the latest information available. Sometimes the latest information requires making a revision to estimated data for the past as well as new projections for the future. Therefore, the user is cautioned against creating time series of population or vital rates from different issues of the report.

Individual country data are from various sources as referenced in the *International Data Base* and *World Population Profile*.

NA—data are not available.

Z—an indeterminate number of *Births* and *Deaths* of less than 500.

Footnotes

1. The data for 1995 have been included from the above sources for the sole purpose of providing the reader with an additional year of recent demographic data. Therefore, when using these data in a time-series analysis, the 1995 column should be omitted from the set. See **Notes** above for further explanation.

2. The concept of consensual union is not applicable in Afghanistan.

3. 1967 to 1976—Official adjusted registered births.

4. 1967 to 1976—Official adjusted registered deaths.

5. 1970—Data refer to the black population only. Figures were calculated at the U.S. Bureau of the Census based on 1970 census data on the percentages of men and women of known marital status in each age group. No marital status data were reported for the categories "not stated" or "population under age 15". Consensual unions are presumably included in those classified as married.

6. 1966 to 1985—Data are by year of registration.

7. Deaths from 1970-1971 are those registered.

8. Data for married persons include persons legally separated and persons living together by common consent.

9. 1980—Registered deaths adjusted at the U.S. Bureau of the Census for 23.13 percent underregistration.

10. 1970—De jure population. Estimates are based on results of a sample survey of population covering 25,000 people.

11. No distinction was drawn between marriage and consensual union. Any such unions are likely to be included in the "married" category.

12. 1980—Data for "married" include persons who are separated.

13. 1980—Examination of the data indicates an underreporting of young, ever-married women. Data are based on a survey which excludes 136 out of a sample of 27,260 males and 139 out of a sample of 26,466 females who did not report their marital status.

14. Includes nationals temporarily outside the country.

15. De jure population.

16. 1970 and 1980—Data are for the western area of Germany only. 1996—Data are for unified Germany.

17. 1970 and 1980—Data are for the western area of Germany only. 1996—Data are for unified Germany.

18. Figures may not sum to totals due to rounding. 1970 and 1980—Data are for the western area of Germany only. 1996—Data are for unified Germany.

19. 1970 and 1980—Data for consensual unions are included in the "married" category.

20. Persons in consensual unions are included in the "married" category.

21. 1980—Data are from a 5 percent sample census and exclude the population of East Timor (555,350), homeless persons, and persons on board ship (158,475). Persons in consensual unions are included in the "married" category.

22. The concept of consensual union is not applicable in Iran.

23. The concept of consensual union is not applicable in Iraq.

24. Original source of data uses the category "ever married" (excluding widowed).

25. Data include urban non-Jews of East Jerusalem. 1980—Data are official estimates and refer to midyear. "Married" represents "ever married" individuals.

26. Reported births refer to the year of registration for the resident population.

27. Deaths were adjusted for an estimated 7 percent underregistration at the U.S. Bureau of the Census.

28. 1970—The user is cautioned that major discrepancies exist between marital status data published in this source and those data published in other census volumes for the same year. By way of contrast see: U.S. Bureau of the Census, 1977, *Country Demographic Profiles - Jamaica*, Table 10 (Washington, DC).

29. 1990—The only persons responding to the questionnaire were females of childbearing age.

30. Consensual unions are included in the "married" category.

31. The lowest age group is for mothers under 20 years old. Assumed trends in fertility were used to prepare cohort components projections.

32. 1970—African population only.

33. 1970—Excludes 67,677 males and 52,434 females canvassed on self-enumeration forms. "Married" includes those in informal unions.

34. 1973 to 1985—Data include Maltese only.

35. 1955 to 1981—Data refer to the island of Mauritius.

36. Data are tabulated by year of registration rather than year of occurrence. Data refer only to deaths for Mauritius and Rodrigues.

37. Births are based on the de jure population.

38. Deaths are based on the de jure population.

39. The concept of consensual union is not applicable in Nepal.

40. "Married" includes judicial separations. Figures may not sum to totals due to rounding.

41. 1980—Births are tabulated by year of occurrence.

42. Data earlier than 1980 exclude the Canal Zone.

43. 1980—Couples living together by common consent are included in "married" population.

44. Figures may not add to totals due to rounding.

45. 1980—Marital Status—This question was asked only of persons 14 years or over. It emphasizes the presence of legal sanction, rather than union status. The five categories used (plus "not stated") are defined as follows: "never married"—all persons who have never married, either legally or, in the case of East Indians, according to Hindu or Muslim custom; "married"—all persons who are formally married, whether or not they are living with the partners to whom they are legally married; all persons married according to Hindu or Muslim rite, whether or not these have been formally registered; and all persons living apart from, but not legally separated from, their married partner; "widowed"—all persons whose marriage partner has died, where the marriage was either legal or according to customary East Indian rites; "divorced"—all persons whose marriages have been dissolved by legal proceedings; and "legally separated"—all persons who are legally separated, but whose marriages have not yet been dissolved.

46. No distinction was made between "married" and "consensual union".

47. 1979 to 1981—Based on registered deaths.

48. Persons living in consensual unions are included in the "married" category. 1990—Figures may not add up due to rounding.

49. 1970—Consensual unions are included with "married".

50. 1980—Preliminary census results are from a 5 percent sample, which excludes the population of Bophuthatswana, Transkei, and Venda.

51. 1970—Couples living together by common consent are included in "married" population.

52. 1970—Figures refer to Syrian Arabs only; who comprise 96.7 percent of the 1970 total census population.

53. 1980—Figures are registered numbers of deaths tabulated according to date of occurrence.

54. 1970—Population includes Taiwan, Kinmen, and Lienkiang. 1980—"Married" includes consensual unions.

55. 1970—Consensual unions are included in "married". 1980—Total excludes "unknown ever-married" (65,006).

56. 1980—Excludes students still attending school.

57. 1970—Marital status was asked of all persons; however, tabulations included only persons ages 12 years and over. 1980— Individuals were allowed to declare their marital status as single, married, widowed, or divorced.

58. Data have been rounded individually and may not sum to totals.

59. Estimates and projections include Armed Forces overseas. Projections are based on the cohort component method. Data for 1995- 2030 are from middle series projections.

60. Based on official U.S. Bureau of the Census middle series projections. Population base includes Armed Forces overseas estimate.

61. Official data as compiled by the U.S. National Center for Health Statistics and reported in a variety of sources.

62. Interpolated from life tables used for the official U. S. Bureau of the Census middle series projections.

63. Based on official U.S. Bureau of the Census middle series projections. Population base includes Armed Forces overseas estimate.

64. Includes Armed Forces overseas. Projections are based on the cohort component method. 1995-2030—Based on official U.S. Bureau of the Census middle series projections.

65. 1970—"Married" category includes cases where the spouse is absent, with the exception of legal separations. Persons in common-law marriages are classified as married. 1980— "Married" category includes legal separations and cases where the spouse is otherwise absent.

66. Based on official U.S. Bureau of the Census middle series projections. Population base includes Armed Forces overseas estimate.

67. Registered births as reported to the United Nations.

68. Official registered data.

69. Present-day Yugoslavia consists of the republics of Serbia and Montenegro, the result of a formal declaration by those countries on April 17, 1992. At that time, however, Serbia was the main supplier of arms to ethnic Serb fighters against the Croats in Bosnia and Herzegovina. The UN responded by imposing international sanctions

against Serbia and Montenegro. Consequently, there is no data available until after September 23, 1994, when Yugoslavia agreed to cut off support to Bosnian Serbs, and the UN eased sanctions. Data displayed in this table was calculated from figures in individual census reports on Serbia and Montenegro from 1995 forward.

3 - Health Indicators

Source. United Nations Children's Fund (UNICEF) (1996). *The State of the World's Children 1996* [Online]. Available: http://www.unicef.org/sowc96/stat1.html [1996].

Notes. Because the data provided in these tables are derived from so many sources, they will inevitably cover a wide range of data quality. Official government data received by the responsible United Nations agency have been used whenever possible. In the many cases where there are no reliable official figures, estimates made by the responsible United Nations agency have been used.

Where such internationally standardized estimates do not exist, the tables draw on other sources, particularly data received from the appropriate UNICEF field office. Where possible, only comprehensive or representative national data have been used.

Data quality is likely to be adversely affected for countries that have recently suffered from man-made or natural disasters. This is particularly so where basic country infrastructure has been fragmented or major population movements have occurred. Also, because of periodic revisions, some of the data will differ from those found in earlier UNICEF publications.

Data for ORT (oral rehydration therapy) use are undergoing review at WHO and UNICEF, so - with few exceptions - data appearing in table 3 of *The State of the World's Children 1995* have been repeated this year.

Following are UNICEF definitions of acronyms and terms used in these tables.

Access to health services—percentage of the population that can reach appropriate local health services by local means of transport in no more than one hour.

DPT—diphtheria, pertussis (whooping cough), and tetanus.

ORT use rate—percentage of all cases of diarrhea in children under five years of age treated with oral rehydration salts or an appropriate household solution.

Contraceptive prevalence—percentage of married women aged 15-49 years currently using contraception.

NA—data are not available.

4 - Health Expenditures

Source. International Bank for Reconstruction and Development/World Bank. *World Development Report 1993: Investing in Health*. New York: Oxford University Press, 1993. **Reprinted with permission**.

Notes. Health expenditure includes outlays for prevention, promotion, rehabilitation, and care; population activities; nutrition activities; program food aid; and emergency aid specifically for health. It does not include water and sanitation. Per capita expenditures and per capita flows are based on World Bank midyear population estimates.

Total health expenditure is expressed in official exchange rate U.S. dollars. Data on public and private health expenditure for the established market economies and Turkey are from the OECD. For other countries, information on government health expenditures is from national sources, supplemented by *Government Finance Statistics* (published by the International Monetary Fund), World Bank sector studies, and other studies. Data on parastatal expenditures (for health-related social security and social insurance programs) are from the Social Security Division of the International Labour Office (ILO) and the World Bank. Data are drawn from a Murray, Govindaraj, and Chellaraj background paper.

Public sector expenditures include government health expenditures, parastatal expenditures, and foreign aid, making the figures comparable with those for OECD countries. *Private sector* expenditures for countries other than OECD members are based on household surveys carried out by the ILO and other sources, supplemented by information from United Nations National Income Accounts, World Bank studies, and other studies published in scientific literature.

Estimates for countries with incomplete data were calculated in three steps. First, where data on either private or public expenditures were lacking, the missing figures were imputed from data from countries for which information was available. The imputation followed regressions relating public or private expenditure to GDP per capita. Second, for a country with no health expenditure data, it was assumed that the share of GDP spent on health care was the same as the average for the corresponding demographic region. Third, if GDP was also unknown but population was known, it was assumed that per capita health spending was the same as the regional average.

Estimates for **Development assistance for health** are expressed in official exchange rate U.S. dollars. *Total aid flows* represent the sum of all health assistance to each country by bilateral and multilateral agencies and by international nongovernmental organizations (NGOs). Direct bilateral official development assistance (ODA) comes from the OECD countries. Sources of multilateral development assistance include United Nations agencies, development banks (including the World Bank), the European Community, and the Organization of Petroleum Exporting Countries (OPEC). Major international NGOs include the International Committee for the Red Cross (ICRC) and the International Planned Parenthood Federation (IPPF). National NGOs were not included because the available information was not separated by recipient country.

Information on ODA from bilateral and multilateral organizations was completed by data from the OECD's Development Assistance Committee (DAC) and Creditor Reporting System (CRS), and from the Advisory Committee for the Coordination of Information Systems (ACCIS). DAC has compiled annual aggregate ODA statistics, by sector, since 1960. The OECD's CRS, established in 1970, complements the DAC statistics by identifying contributions allocated by sector. The CRS database is the most complete source of information for bilateral ODA, but its completeness varies among OECD countries and from year to year. ACCIS has kept, since 1987, a Register of Development Activities of the United Nations that lists sources of funds and executing agencies for all United Nations projects by sector.

The estimates of development assistance in this table were prepared by the Harvard Center for Population and Development Studies as a background paper for this report.

NA—data are not available.

Footnotes

1. Aid flows are official development assistance and include only a small portion of private flows, that is, nongovernmental organization (NGO) assistance.

2. Refers to former Czechoslovakia because disaggregated data are not yet available.

3. Refers to the former Socialist Federal Republic of Yugoslavia because disaggregated data are not yet available.

4. Refers to the former U.S.S.R. because data for Russia are not yet available.

5 - Women and Children

Source. United Nations Children's Fund (UNICEF) (1996). *The State of the World's Children 1996* [Online]. Available: http://www.unicef.org/sowc96/stat1.html [1996].

Notes. Because the data provided in these tables are derived from so many sources, they will inevitably cover a wide range of data quality. Official government data received by the responsible United Nations agency have been used whenever possible. In the many cases where there are no reliable official figures, estimates made by the responsible United Nations agency have been used.

Where such internationally standardized estimates do not exist, the tables draw on other sources, particularly data received from the appropriate UNICEF field office. Where possible, only comprehensive or representative national data have been used.

Data quality is likely to be adversely affected for countries that have recently suffered from man-made or natural disasters. This is particularly so where basic country infrastructure has been fragmented or major population movements have occurred. Also, because of periodic revisions, some of the data will differ from those found in earlier UNICEF publications.

Following are UNICEF definitions of acronyms and terms used in these tables.

Births attended—percentage of births attended by physicians, nurses, midwives, trained primary health care workers or trained traditional birth attendants.

Maternal mortality rate—annual number of deaths of women from pregnancy-related causes per 100,000 live births.

Under-five mortality rate—probability of dying between birth and exactly five years of age expressed per 1,000 live births.

Underweight, moderate and severe—below minus two standard deviations from median weight for age of reference population.

NA—data are not available.

Footnotes

1. Data refer to years or periods other than those specified in the column heading, differ from the standard definition, or refer to only part of a country.

6 - Burden of Disease

Source. The International Bank for Reconstruction and Development/The World Bank. *Social Indicators of Development 1996*. In *World Bank Data on Diskette 1996* (Stars ver. 3.0) [Diskette]. Prepared by the International Economics Department of the World Bank. Available: International Bank for Reconstruction and Development/The World Bank [1996, February]. **Reprinted with permission**.

Source. United Nations Development Programme (UNDP). *United Nations Development Report 1996*. New York: Oxford University Press, 1996. **Reprinted with permission**.

Source. U.S. Bureau of the Census. *Statistical Abstract of the United States: 1995*. 115th ed. Washington, DC: GPO, 1995.

Notes. *Social Indicators of Development* provided the data for ratios of population to health care personnel.

Human Development Report provided the data for incidence of AIDS, malaria, and heart disease.

The number of reported AIDS cases refers to children and adults.

Statistical Abstract of the United States provided the data for incidence of heart disease in the United States.

Years shown in tables indicate dates of most recent data available.

NA—data are not available.

A dash **(-)**—data are nil or negligible.

An asterisk **(*)**—data equal less than half the unit shown.

Footnotes

1. Data for Ethiopia prior to 1992 include Eritrea.

2. Data for Germany prior to 1990 exclude the former German Democratic Republic (East Germany).

3. Data refer to mainland Tanzania only.

4. Data provided by the *Statistical Abstract of the United States: 1995*. See Sources above.

7 - Ethnic Division

Source. U.S. Central Intelligence Agency (CIA) (1996). *The World Factbook 1996* [Online]. Available: http://www.odci.gov//cia/publications/nsolo/factbook/global.h tml [1997, March 11].

Notes. Tables show the major ethnic divisions of peoples in the given country for the most recent year available. When available, the distribution is shown in percent.

NA—data are not available.

8 - Religion

Source. U.S. Central Intelligence Agency (CIA) (1996). *The World Factbook 1996* [Online]. Available: http://www.odci.gov//cia/publications/nsolo/factbook/global.h tml [1997, March 11].

Notes. Tables show major religious denominations of the peoples of the given country for the most recent year available. When available, the distribution is shown in percent.

NA—data are not available.

9 - Major Languages

Source. U.S. Central Intelligence Agency (CIA) (1996). *The World Factbook 1996* [Online]. Available: http://www.odci.gov//cia/publications/nsolo/factbook/global.h tml [1997, March 11].

Notes. Tables show the major language(s) spoken by inhabitants of the given country for the most recent year available. When available, the distribution is shown in percent.

NA—data are not available.

10 - Public Education Expenditures

Source. United Nations Educational, Scientific, and Cultural Organization (UNESCO). *Statistical Yearbook 1996*. Paris: UNESCO & Bernan, 1997. **Reprinted with permission**.

Notes. Tables present total public expenditure on public education, also distributed between current and capital expenditures, and expressed as a percentage of the gross national product.

Following are definitions of acronyms and terms used in these tables.

Current expenditure—expenditure on administration, emoluments of teaching and other staff, teaching materials, scholarships, and welfare services.

Capital expenditure—expenditure on land, buildings, construction, equipment, etc. This item also includes loan transactions.

Gross national product (GNP)—the sum of all incomes earned by a country's residents, regardless of where those assets are located. For instance, if General Motors has a plant in Mexico, the assets for that business contribute to the United States' GNP. Similarly, the assets of a Mistubishi plant located in the United States would be included in Japan's gross national product.

For almost all countries, data on GNP are supplied by the World Bank. Because these data are revised every year by the World Bank, the percentages of educational expenditure in relation to GNP may sometimes differ from those shown in previous editions of the *UNESCO Statistical Yearbook*.

NA—data are not available.

A dash **(-)**—data are nil or negligible.

T—trillion.

Footnotes

e. Refers to an estimate.

1. From 1990 to 1994, expenditure on third level education is not included.

2. For 1990, data refer to expenditure of the Ministry of Education only.

3. For 1985 and 1990, data refer to expenditure of the Ministry of Education only. From 1980 to 1985, data are expressed in Australes.

4. From 1990 to 1993, expenditure on third level education is not included.

5. Data refer to expenditure of Ministry of Education only.

6. For 1994, data are expressed in B. Ruble.

7. Except for 1993, data refer to expenditure of the Ministry of Education only.

8. For 1990 and 1991, data refer to the expenditure of the Ministry of Education only, and expenditure on universities is not included.

9. From 1990 to 1994, data refer to expenditure of the Ministry of Education only.

10. For 1994, data refer to expenditure of the federal government only and are expressed in Reais (1 Reais ~2,750 thousand Cruzeiros).

11. From 1991 to 1993, data refer to expenditure of the Ministry of Education only, and expenditure on universities is not included.

12. Except for 1992, data refer to expenditure of the Ministry of Education only.

13. Except for 1980, data refer to expenditure of the Ministry of Education only. For 1992 to 1994, expenditure on third level is not included.

14. Data refer to expenditure of the Ministry of Education only.

15. Data refer to expenditure of the Ministry of Education only.

16. From 1992 to 1994, data on current and capital expenditure refer to expenditure of the Ministry of Education only.

17. For 1992, expenditure of the Ministry of Education only. For 1994, expenditure on third level is not included.

18. Expenditure on education is calculated as percentage of global social product.

19. Expenditure of the Office of Greek Education only.

20. On January 1, 1993, Czechoslovakia split into two independent countries: the Czech Republic and Slovakia. Only data previous to 1993 refer to the old Czechoslovakian federation.

21. For 1985, data refer to expenditure of the Ministry of Education only.

22. Data on current and capital expenditure refer to the Ministry of Education only.

23. Expenditure relating to Al-Azhar is not included.

24. Expenditure on third level education is not included.

25. From 1990 to 1992, data refer to expenditure of the Ministry of Education only.

26. Metropolitan France.

27. For 1992, data refer to expenditure of the Ministry of Primary and Secondary Education only.

28. All data previous to 1993 refer to West Germany only. For 1991, data include expenditure for East Berlin.

29. For 1985, data refer to expenditure of the Ministry of Education.

30. From 1990 to 1993, data refer to expenditure of the Ministry of Education only.

31. For 1990, 1993, and 1994, data refer to expenditure of the Ministry of Education only.

32. For 1990 and 1991, data refer to expenditure of the central government only. From 1990 to 1992, data do not include expenditure on third level education.

33. For 1980, 1990, 1993, and 1994, data refer to expenditure of the Ministry of Education only.

34. For 1980, data on current and capital expenditure do not include public subsidies to private education. For 1985, these data refer to total public and private expenditure on education.

35. From 1990 to 1992 and 1994, expenditure on third level education is not included. For 1993, expenditure on universities is not included.

36. From 1990 to 1993, data refer to expenditure of the Ministry of Education only.

37. Data refer to expenditure of the Ministry of Education only.

38. From 1992 to 1994, data refer to expenditure of the Ministry of Education only.

39. For 1990 and 1993, data refer to expenditure of the Ministry of Primary and Secondary Education only.

40. From 1990 to 1994, data refer to expenditure of the Ministry of Education only.

41. For 1993, data refer to expenditure of the Ministry of Education only.

42. From 1990 to 1994, data refer to expenditure of the Ministry of Education only.

43. Data on current and capital expenditure refer to the Ministry of Education only.

44. Data refer to expenditure of the Ministry of Education only.

45. Data include foreign aid received for education.

46. Data refer to expenditure of the Ministry of Education only.

47. Data refer to "regular and development" expenditure.

48. From 1991 to 1994, data refer to expenditure of the Ministry of Primary and Secondary Education only and are expressed in gold Cordobas.

49. For 1989 and 1991, data refer to expenditure of the Ministry of Education only and do not include expenditure on third level education.

50. Except for 1981, data refer to expenditure of the federal government only.

51. For 1994, expenditure on universities is not included.

52. For 1990 and 1991, data do not include expenditure on education by "other ministries" which are not directly related to education.

53. From 1990 to 1994, data refer to expenditure of the Ministry of Education only.

54. Peru's currency changed to the nuevo sol as of 1991.

55. For 1992 and 1993, data refer to expenditure of the Ministry of Education only.

56. For 1990 and 1991, data refer to expenditure of the Ministry of Education only.

57. For 1994, data refer to expenditure of the central government only.

58. For 1990, 1991, and 1993, expenditure on third level education is not included.

59. For 1993 and 1994, data refer to the Ministry of Education only.

60. Expenditure on third level education is not included.

61. For 1990, data do not include expenditure for the following states: Transkei, Bophuthatswana, Venda, and Ciskei (TBVC).

62. For 1980 and 1985, data refer to expenditure of the Ministry of Education only.

63. For 1985, expenditure on third level education is not included.

64. For 1993, data refer to expenditure of the Ministry of Education only.

65. For 1992 and 1994, expenditure on third level education is not included. Data on current and capital expenditure do not include expenditure on third level education.

66. For 1991 and 1992, data refer to expenditure of the Ministry of Education only.

67. From 1990 to 1992, expenditure on third level education is not included.

68. Data refer to expenditure of the Ministry of Education only.

69. Ukraine's new monetary unit, the hryvnya, replaced the karbovanets in 1996.

70. For 1980, data refer to public and private expenditure on education.

71. Data from 1992 forward refer to the Federal Republic of Yugoslavia (Serbia and Montenegro).

11 - Educational Attainment

Source. United Nations Educational, Scientific, and Cultural Organization (UNESCO). *Statistical Yearbook 1996*. Paris: UNESCO & Bernan, 1997. **Reprinted with permission**.

Notes. Data show the distribution of the highest level of educational attainment of the adult population in percent. Data were either provided by the United Nations Statistical Office (national censuses and surveys) or were derived from regional or national publications.

The six levels of educational attainment are based on categories of the Standard Classification of Education (ISCED) and may be defined as follows:

No schooling—applies to those who have completed less than one year of schooling.

First level...not completed—to those who completed at least one year of education at the first level but who did not complete the final grade at this level. The number of years of education included in the first level may vary depending on the country.

First level...completed—to those who completed the final grade of education at the first level (ISCED category 1) but did not go on to second level studies.

Entered second level...S-1—to those who completed no more than the lower stage of education at the second level.

Entered second level...S-2—corresponds to ISCED category 3, and includes persons who moved to the higher stage of the second level of education, but did not proceed to studies at the postsecondary level.

Postsecondary—includes those who undertook third level studies (ISCED categories 5, 6, or 7), whether or not they completed the full course. At the postsecondary education level there is usually a larger number of persons in the 25-34 age group than in the 15-24 age group. This is beacuse many of the persons in the 15-24 age group are too young to have reached entrance age. For this reason the total adult age range is taken as 25+ for the purpose of these data, and not 15+, although some countries vary from this.

NA—data are not available.

A dash (-)—data are nil or negligible.

Unless otherwise indicated, the number of persons whose level of education is not stated has been subtracted from the total population.

Footnotes

1. First stage of second level of education refers to the *Intermedio* level of education. Second stage of second level refers to the *Medio* level, *Tecnica*, and *Normal* education.

2. Not including rural population of the region north of Brazil.

3. The category *First level...Not completed* comprises four first grades of primary education (1 to 4). The category *First level...Completed* comprises the last four grades of primary education (5 to 8).

4. Excluding less than first level and category *No schooling*.

5. Not including expatriate workers.

6. Not including persons whose educational level is unknown.

7. Egyptian population only. Second level also includes third level education not leading to a university degree.

8. Illiteracy data have been used for the category *No schooling*.

9. Data refer to former East Germany only.

10. Based on a sample survey referring to 51,372 persons.

11. Not including persons still enrolled in schools.

12. The distribution by level of education does not take into account those still attending school.

13. Persons who did not state their level of education have been included in the category *No schooling*.

14. Persons who can read and write have been counted with *First level...Not completed*.

15. Excluding the population attending and never having attended school.

16. The category *No schooling* comprises illiterates. Household survey results based on a sample of 6,393 households.

17. Excluding transients and residents of former Canal Zone.

18. Not including Botphuthatswana, Transkei, and Veda.

19. Based on a sample survey referring to 8,619 households (5,563 urban, 3,056 rural).

20. The category *No schooling* refers to those who have completed less than the first grade of the first level.

21. Data refer to the former Yugoslavia.

22. Based on a sample survey referring to 35,502 persons.

12 - Illiteracy

Source. United Nations Educational, Scientific, and Cultural Organization (UNESCO). *Compendium of Statistics on Illiteracy: 1995 Edition.* Paris: UNESCO, 1995. **Reprinted with permission**.

Source. United Nations Educational, Scientific, and Cultural Organization (UNESCO). *Statistical Yearbook 1996.* Paris: UNESCO & Bernan, 1997. **Reprinted with permission**.

Notes. Literacy statistics are concerned with the stock of persons who have successfully acquired the basic reading, writing, and numeracy skills essential for personal growth and cohesion within contemporary societies. Ranging from below 10 percent to almost 100 percent among the countries of the world, the adult literacy rate is a very effective indicator of the general level of development of education and of its achievements.

The 1995 edition of the *Compendium* presents estimates and projections of adult (usually 15+ years) illiteracy rates in 1990, 1995, and 2000, as prepared by UNESCO. The reader should keep in mind the conditional nature of these projections.

Over the past three decades, the Division of Statistics of UNESCO has been systematically monitoring the progress in adult literacy in the world, by collecting on a regular basis and in collaboration with the Statistical Division of the United Nations, data on the number of illiterates by sex and age-group gathered during national population censuses and surveys. Since these censuses take place once every 10 years, estimations and projections are carried out to fill the data gaps for the years in between two censuses, as well as to provide projections showing likely progress in literacy for the future.

The literate adult population in the world has undergone phenomenal expansion during the the period from 1980 to 1995, rising from 2 billions (69.5%) in 1980 to an estimated 3 billions

(77.4%) in 1995. If the current rate of progress continues, the number of adult literates in the world may reach 3.4 billions (80%) in the year 2000.

Despite these signs of positive progress, there remains a large illiterate population in the world of today - numbering some 885 million adults. Almost three-quarters of these illiterate are concentrated in nine countries, each with more than 10 million adult illiterates: India, China, Pakistan, Bangladesh, Nigeria, Indonesia, Ethiopia, Egypt, and Brazil. Among these, India and China account for more than 100 million adult illiterates each, and together they account for more than half of the world's illiterate population.

In addition, most of the world's adult illiterates are women. However, the gender gap narrowed in most countries during the period from 1980 to 1995, reflecting faster progress in the female literacy level as compared to that of the male.

NA—data are not available.

A dash (-)—data are nil or negligible.

Footnotes

e.   Estimated or provisional data.

1.   The UNESCO *Compendium of Statistics on Illiteracy* provided all the data for this table (See Sources listed above).

2.   The UNESCO *Statistical Yearbook 1996* provided all the data for this table (See Sources listed above).

3.   An earlier UNESCO *Compendium* (1990 ed.) provided all the data for this table. These figures represent the most recent data available for this country.

4.   This table represents the most recent data available relating to the illiterate population and percentage illiterate, from the latest census or survey held since 1970. These figures were provided by the United Nations Statistical Office or were derived from regional or national publications.

5.   Based on a sample survey.

6.   In 1991, the total is not the sum of urban and rural areas, as these two areas do not include nomads and unsettled population.

7.   Excluding unemployed population.

8.   According to the CIA *World Factbook 1996*, the literacy rate in the United States in 1995 was 96%.

13 - Libraries

Source. United Nations Educational, Scientific, and Cultural Organization (UNESCO). *Statistical Yearbook 1996*. Paris: UNESCO & Bernan, 1997. **Reprinted with permission.**

Notes. Tables present selected statistics gathered every few years on collections, annual additions, and registered users in several categories of libraries: national and public libraries, libraries of institutions of higher education, and school libraries. Following are UNESCO definititions of acronyms and terms used.

Library—any organized collection of printed books and periodicals or any other graphic or audiovisual materials, and the services of a staff to provide and facilitate the use of such materials necessary to meet the informational, research, educational, or recreational needs of its users.

Libraries thus defined are counted in numbers of *administrative units*, i.e., independent libraries, or groups of libraries, under a single director or a single administration; and *service points*, i.e., libraries which provide in separate quarters a service for users, whether they are independent or part of a larger administrative unit. Libraries are classified as follows:

National—responsible for acquiring and conserving copies of all significant publications produced in the country and functioning as a deposit library, either by law or other arrangement, and normally compiling a national bibliography.

Nonspecialized—those of a learned character which are neither libraries of institutions of higher education nor national libraries, though they may fulfill the functions of a national library for a specified geographical area.

Public—serve the population of a community or region free of charge or for a nominal fee; they may serve the general public or special categories of users such as children, members of the armed forces, hospital patients, prisoners, workers, and employees.

Higher education—primarily serving students and teachers in universities and other institutions of education at the third level.

School—attached to all types of schools below the third level of education and serving primarily the pupils and teachers of such schools, even though they may also be open to the general public.

Annual additions—all materials added to collections during the year whether by purchase, donation, exchange, or any other method. Statistics refer to books and bound periodicals, manuscripts, microforms, audiovisual documents, and other library materials.

Volume—any printed or manuscript work contained in one binding or portfolio.

Registered user—person registered with a library in order to use its materials on or off the premises.

Data refer to latest year available.

Due to space limitations, figures in millions have been rounded to the nearest whole number.

NA—data are not available.

A dash (-)—data are nil or negligible.

Footnotes

e. Estimate.

1. The figure referring to the number of volumes for national libraries includes manuscripts and microforms.

2. Data refer to 19 main or central university libraries, to an unknown number of libraries of institutes or departments, and to two libraries of institutions of higher education which are not part of a university.

3. Data on libraries of institutions of higher education refer only to the library of the University Ahmed Al-Farsi.

4. Data on libraries of institutions of higher education refer only to the library of Erdiston Teacher Training College.

5. The figure referring to the number of registered users for national libraries pertains to the number of readers only.

6. The public libraries also serve as national libraries.

7. Data on public libraries refer only to libraries financed by public authorities.

8. The figure referring to the number of volumes for libraries of institutions of higher education pertains to the number of titles only.

9. Data refer only to the main or central university libraries, and the figure referring to the number of volumes pertains to the number of titles only.

10. Data on libraries of institutions of higher education refer only to main or central university libraries.

11. Data on libraries of institutions of higher education do not include 132 libraries which are not administered by the main library.

12. Data on libraries of institutions of higher education do not include libraries which are not administered by the main library.

13. All data refer to Metropolitan France and overseas departments.

14. Data on nonspecialized libraries refer only to the Bibliotheque publique d'Information (BPI) de Beaubourg.

15. Figures referring to registered users pertain only to the number of visitors.

16. Data on national libraries include central specialized libraries.

17. Data on libraries of institutions of higher education refer to 235 out of 271 libraries.

18. Data relating to public libraries refer only to libraries dependent on the Ministry of Culture and Environment, and the figure referring to the number of volumes includes books and booklets.

19. The figure referring to the number of volumes in the national library includes books only; and to the number of registered users, the number of readers only.

20. Data on libraries of institutions of higher education refer to university libraries only.

21. Data on libraries of institutions of higher education refer to 15 out of 18 libraries.

22. The figure referring to meters of shelving in libraries of institutions of higher education pertains to four libraries only.

23. The figure referring to registered users in public libraries and libraries of institutions of higher education pertains to registered borrowers; and to the figure for national libraries, the number of readers only.

24. Data on libraries of institutions of higher education refer to the University of Damascus only.

25. Data on libraries of institutions of higher education refer only to the main or central library of the University of the West Indies in St. Augustine.

26. The figures referring to registered users in national libraries do not include data for The National Library of Scotland.

27. Data on public libraries refer only to independent libraries and those incorporated in enterprises; the figure referring to the number of registered users pertains to the number of readers only.

14 - Newspapers

Source. United Nations Educational, Scientific, and Cultural Organization (UNESCO). *Statistical Yearbook 1996*. Paris: UNESCO & Bernan, 1997. **Reprinted with permission**.

Notes. It has been UNESCO's policy to publish tables regularly on printed periodic literature which are published in a particular country. Exceptions are publications issued for advertising purposes, those of a transitory character, and those in which the text is not the most important part. Following are UNESCO definitions of terms pertinent to these tables.

Newspapers—periodic publications intended for the general public and mainly designed to be a primary source of written information on current events connected with public affairs, international questions, politics, etc. A newspaper issued at least four times a week is considered to be a *daily newspaper*; those appearing three times a week or less frequently are considered to be *non-daily newspapers*. For 1994, it is known or believed that no daily newspapers were published in these countries: Cape Verde, Comoros, Djibouti, Eritrea, Guinea, Sao Tome and Principe, Antigua and Barbuda, Dominica, Grenada, St. Kitts and Nevis, St. Lucia, St. Vincent and the Grenadines, Bhutan, Cambodia, San Marino, Western Samoa, and the Solomon Islands.

Circulation—figures show the average daily circulation. These figures include the number of copies sold directly, sold by subscription, and mainly distributed free of charge both inside the country and abroad.

When interpreting these data, it should be noted that in some cases, definitions, classifications, and statistical methods applied by certain countries, do not entirely conform to the standards recommended by UNESCO. For example, circulation data refer to the number of copies distributed as defined above. It appears, however, that some countries have reported the number of copies printed, which is usually higher than the distribution figure.

NA—data are not available.

A dash (-)—data are nil or negligible.

Footnotes

e.   Estimate.

1.   Data for 1994 refer to 1992 and only to the territory that is under the control of the government of the Republic of Bosnia and Herzegovina.

2.   Data for 1990 refer to 1991.

3.   Data for 1994 refer to 1992.

---

15 - Culture

Source. United Nations Educational, Scientific, and Cultural Organization (UNESCO). *Statistical Yearbook 1996.* Paris: UNESCO & Bernan, 1997. **Reprinted with permission.**

Source. United Nations Educational, Scientific, and Cultural Organization (UNESCO). *Statistical Yearbook 1995.* Paris: UNESCO & Bernan, 1995. **Reprinted with permission.**

Source. U.S. Bureau of the Census. *Statistical Abstract of the United States: 1995.* 115th ed. Washington, DC: GPO, 1995.

Notes. Following are UNESCO definitions of terms pertinent to these tables.

Cinemas

Statistical Yearbook 1995 (See Sources listed above) provided cinema statistics for these tables.

The statistics shown in the upper half of each table refer to fixed cinemas and mobile units regularly used for commercial exhibition of long films of 16mm and over.

Cinema attendance is calculated from the number of tickets sold for all types of cinemas during a given year. Due to lack of detailed statistical information, it should be noted that for several countries annual attendance per inhabitant does not include attendance at mobile units and/or drive-in cinemas.

As a rule, figures refer only to commercial establishments. Exceptions may occur, however, in countries that include non-commercial units in their report of mobile units. Gross receipts are given in the national currency.

The term *fixed cinema* as used above refers to establishments possessing their own equipment and includes indoor cinemas (those with a permanent fixed roof over most of the seating accommodation), outdoor cinemas, and drive-ins (establishments designed to enable audiences to view films from their automobiles).

Mobile units are defined as projection units equipped and used to serve more than one site.

The capacity for fixed cinemas refers to the number of seats, in the case of cinema houses, and to the number of places for automobiles multiplied by a factor of 4 in the case of drive-ins.

Data refer to latest year available.

Museums

Statistical Yearbook 1996 provided museum data for these tables, supplemented, in the case of the United States, by *Statistical Abstract of the United States: 1995* (See Sources listed above).

The lower half of each table in this category is devoted to museum statistics, i.e., the number of reporting museums, number of visitors, and annual receipts per country. Surveys are conducted at regular intervals of three years in view of the infrequent changes in this field.

The international collection of museum statistics, and thus their comparability, is relatively difficult because of the almost complete absence of generally accepted standards and norms. Consequently, in order to achieve at least a minimum degree of comparability, the categories and terms used in the surveys since 1977 have been based on definitions established by the International Council of Museums (ICOM). For the purpose of these surveys, the term *museum* is defined as a nonprofit, permanent institution in the service of society and of its development, and open to the public, which acquires, conserves, researches, communicates, and exhibits, for purposes of study, education, enjoyment, and material evidence of man and his environment. In addition to museums designated as such, the following entities recognized by ICOM as being of museum nature are also included in the surveys:

a. Conservation institutes and exhibition galleries permanently maintained by libraries and archive centers;

b. Natural, archaeological, and ethnographical monuments and sites, and historical monuments and sites of a museum nature for their acquisition, conservation, and communication activities;

c. Institutions displaying live specimens such as botanical and zoological gardens, aquaria, vivaria, etc.;

d. Nature reserves;

e. Science centers and planetaria.

For statistical purposes, museums and related institutions are counted by number of administrative units rather than by number of collections.

Figures on annual attendance are represented in the *Visitors* category.

Data on total receipts are represented in *Annual receipts* and are given in the national currency.

Museum data (except in the case of the United States) have been obtained through either the 1992 or 1995 UNESCO survey.

Data refer to latest year available.

NA—data are not available.

A dash **(-)**—nil or negligible.

Footnotes

e. Estimated or provisional data.

1. Data refer only to the French community.

2. Receipts do not include taxes.

3. Specialized museums include archaeology, history, ethnology, and anthropology museums.

4. Data on museums include 105 monuments and sites.

5. Data on seating capacity refer to 35mm cinemas only.

6. Data refer only to national and public museums. Of the estimated 1,300 public museums in France, 35 are national museums controlled by the *Direction des musees de France (DMF)*.

7. Specialized museums include ethnology and anthropology museums.

8. Data refer only to national and public museums.

9. Data refer only to government institutions attached to the Ministry of *Beni culturali e ambientali*.

10. Data do not include receipts of one cinema center of five screens.

11. Data on attendance also include attendance at monuments and sites.

12. Data refer to national museums only.

13. Data refer only to the national museum.

14. Data on seating capacity refer to 35mm cinemas only.

15. *Statistical Abstract of the United States: 1995* provided museum data for the United States (See Sources listed above).

16 - Scientific/Technical Forces

Source. United Nations Educational, Scientific, and Cultural Organization (UNESCO). *Statistical Yearbook 1996*. Paris: UNESCO & Bernan, 1997. **Reprinted with permission.**

Notes. Tables present selected results of the data collection effort by UNESCO in the field of science and technology. Most of the data were obtained from replies to the annual statistical questionnaires on manpower and expenditure for research and experimental development (R&D) sent to the Member States of UNESCO during recent years, completed or supplemented by data collected in the earlier surveys and from official reports and publications.

The definitions and concepts used in the R&D questionnaire are based on the *Recommendation Concerning the International Standardization of Statistics on Science and Technology* and can be found in the corresponding *Manual* (doc. UNESCO ST-84/WS/12). Abridged versions of the definitions set out in the above mentioned *Recommendation* are given below.

Scientists/engineers—persons working in those capacities, i.e., as persons with scientific or technological training (usually completion of postsecondary education) in any field of science as defined below, who are engaged in professional work on R&D activities, administrators, and other high-level personnel who direct the execution of R&D activities.

Technicians—persons engaged in that capacity in R&D activities who have received vocational or technical training in any branch of knowledge or technology of a specified standard (usually at least three years after the first stage of second-level education).

Auxiliary personnel—persons whose work is *directly* associated with the performance of R&D activities, i.e., clerical; secretarial and administrative personnel; skilled, semiskilled, and unskilled workers in the various trades; and all other auxiliary personnel. *Excludes* security, janitorial, and maintenance personnel engaged in general housekeeping activities.

It should be noted that, in general, all personnel are considered for inclusion in the appropriate categories regardless of citizenship status or country of origin.

Due to a lack of more specific data, a handful of tables refer to specialists in the national economy, i.e., persons having completed education at the third level (potential scientists and engineers) and secondary specialized education (technicians) (See Footnote 36). These potential scientists, engineers, and technicians are represented in tables for Bahrain, Ethiopia, Haiti, Hong Kong, Kenya, Monaco, San Marino, and Switzerland. Data for these tables were provided by the UNESCO *Statistical Yearbook 1994*.

Data are expressed in full-time equivalent (FTE) except for certain countries where the number of full-time plus part-time R&D personnel have been indicated (See Footnote 1).

NA—data are not available.

A dash **(-)**—data are nil or negligible.

Footnotes

e.   Estimated or provisional data.

1.   Number of full-time plus part-time R&D personnel.

2.   Data include auxiliary personnel.

3.   Not including data for the productive sector.

4.   Data refer to full-time scientists, engineers, and technicians.

5.   Data for scientists and engineers refer to researchers listed in the Directory of Research Groups in Brazil by the *Conselho Nacional de Desenvolvimento Cientifico e Tecnologico (CNPq)*.

6.   Data relate to two research institutes only.

7.   Not including data for the productive sector. 51 of the scientists and engineers are foreigners.

8.   Not including social sciences and humanities in the productive sector (integrated R&D).

9.   Not including military and defense R&D.

10.   206 of the scientists and engineers are foreigners. Not including military and defense R&D.

11.   Data refer to scientists, engineers, and technicians engaged in public enterprise. Not including data for the higher education sector.

12.   Data relate to one research institute only.

13.   Data for scientists and engineers do not include the productive sector. Data for technicians relate only to those in the general service sector.

14.   Data relate to the productive sector (integrated R&D) and the higher education sector only. 88 of the scientists and engineers are foreigners.

15.   Not including military and defense R&D. Data for the general service sector and for medical sciences in the higher education sector are also excluded.

16.   Not including scientists and engineers engaged in the administration of R&D. Of military R&D, only that part carried out in civil establishments is included.

17.   Data include skilled workers.

18.   Data for scientists and engineers include 22,100 (estimate for 1982) personnel in the higher education sector. Data for women scientists and engineers and for technicians in the higher education sector are not included.

19.   The number relating to female scientists and engineers is counted in full-time plus part-time.

20.   Data relate to the Scientific Research Council only.

21.   Data refer to full-time scientists, engineers, and technicians. Not including social sciences and humanities in the productive sector (integrated R&D).

22.   1,027 (F: 179) of the scientists and engineers and 113 (F: 67) of the technicians are foreigners.

23.   Data refer to the Faculty of Science at the University of Lebanon only.

24.   Not including data for the higher education sector.

25.   Data relate to the higher education sector only.

26.   Data relate to the productive sector (integrated R&D) and the higher education sector only.

27.   Data relate to scientific and technological activities (STA) and do not include social sciences and humanities. Data relate to full-time scientists, engineers, and technicians.

28.   Data relate only to 23 out of 26 national research institutes under the Federal Ministry of Science and Technology.

29.   Data relate to R&D activities concentrated mainly in government-financed research establishments only. Not including military and defense R&D.

30.   Data refer to full-time scientists and engineers.

31.   138 of the scientists and engineers and 54 of the technicians are foreigners. Not including social sciences and humanities in the higher education sector.

32.   Data relating to technicians do not include the higher education sector.

33.   Not including social sciences and humanities in the general service sector.

34.   Data refer to all scientific workers, i.e., all persons holding a higher scientific degree or scientific title, regardless of

the nature of their work; persons undertaking research work in scientific establishments; and scientific teaching staff in institutions of higher education. They also include persons undertaking scientific work in industrial enterprises.

35. Not including data for the general service sector.

36. Data refer to specialists in the national economy, i.e., persons having completed education at the third level (potential scientists and engineers) and secondary specialized education (technicians).

17 - R&D Expenditures

Source. United Nations Educational, Scientific, and Cultural Organization (UNESCO). *Statistical Yearbook 1996*. Paris: UNESCO & Bernan, 1997. **Reprinted with permission.**

Notes. In general, *R&D* is defined as any creative systematic activity undertaken in order to increase the stock of knowledge, including knowledge of man, culture, and society; and the use of this knowledge to devise new applications. It includes fundamental research (i.e. experimental or theoretical work undertaken with no immediate practical purpose in mind), applied research in such fields as agriculture, medicine, industrial chemistry, etc. (i.e., research directed primarily towards a special practical aim or objective), and experimental development work leading to new devices, products, or processes.

Total domestic expenditure on R&D activities refers to all expenditure made for this purpose in the course of a reference year in institutions and installations established in the national territory, as well as installations physically situated abroad.

The total *expenditure for R&D* as defined above comprises *current expenditure*, including overheads, and *capital expenditure*.

The following are sources of finance for domestic expenditure of R&D activities pertinent to these tables.

Government funds: include funds provided by the central (federal), state, or local authorities.

Productive enterprise funds and special funds: funds allocated to R&D activities by institutions classified in the productive sector, and all sums received from the "Technical and Economic Progress Fund" and other similar funds.

Foreign funds: funds received from abroad for national R&D activities.

Other funds: funds that cannot be classified under any of the preceding headings.

NA—data are not available.

A dash **(-)**—data are nil or negligible.

Footnotes

e. Estimated or provisional data.

1. Not including data for the productive sector.

2. Data refer to government and productive enterprise funds only.

3. Data refer to two research institutes only.

4. Not including data for the productive sector and labor costs at the Ministry of Public Health.

5. Not including data for the general service sector.

6. Not including data for military and defense R&D.

7. Data relate to government funds only and do not include military and defense R&D.

8. Not including military and defense R&D. Data include depreciation costs.

9. Data for current expenditure do not include those for the productive sector.

10. Data refer to estimated budget for R&D.

11. Data refer to the R&D activities performed in public enterprises. Data include 569,579 thousand colons for which a distribution by sector of performance and by type of expenditure is not known; this amount has been excluded from the percentage calculation.

12. Data include 338,000 kroons for which a distribution by type of expenditure is not available; this figure has been excluded from the percentage calculation.

13. Data relate to one research institute only.

14. Data include 661 million Deutsche marks for which distribution by type of expenditure is not available; this figure has been excluded from the percentage calculation. Not including data for social sciences and humanities in the productive sector.

15. Data refer to the productive sector (integrated R&D) and to the higher education sector only.

16. Not including military and defense R&D. Data for the general service sector and for medical sciences in the higher education sector are also excluded.

17. Of military R&D, only that part carried out in civil establishments is included.

18. Data relate to the general service sector only.

19. Data refer to government expenditure only.

20. Data refer to the civilian sector only.

21. Data relate to the Scientific Research Council only.

22. Not including data for social sciences and humanities in the productive sector (integrated R&D).

23. Excluding military and defense R&D and social sciences and humanities.

24. Data refer to the Faculty of Science at the University of Lebanon only.

25. Data relate to the higher education sector only.

26. Not including data for social sciences and humanities in the productive sector (integrated R&D).

27. Data relate only to 23 out of 26 national research institutes under the Federal Ministry of Science and Technology.

28. Data relate to R&D activities concentrated mainly in government-financed research establishments only. Not including military and defense R&D.

29. Data refer to the budget allotment for science and technology.

30. Total expenditure does not include capital expenditure except that of federal government institutions. Data relating to the general service sector cover only federal government and private nonprofit organizations. R&D expenditure in the productive sector (integrated R&D) includes also depreciation costs. Humanities in the higher education sector is excluded.

18 - U.S. Patents Issued

Source. U.S. Patent and Trademark Office. Office of Electronic Information Products/TAF Program. *Patent Counts by Country/State and Year, All Patents, All Types: January 1, 1977-December 31, 1995.* Washington: PTO, 1995.

Notes. Tables show the number of patents issued by the U.S. Patent and Trademark Office to residents of given countries in featured years 1993 through 1995. Patent data are only displayed when available for these years.

19 - Organization of Government

Source. U.S. Central Intelligence Agency (CIA) (1996). *The World Factbook 1996* [Online]. Available: http://www.odci.gov//cia/publications/nsolo/factbook/global.h tml [1997, March 11].

Notes. ICJ—International Court of Justice.

20 - Elections

Source. U.S. Central Intelligence Agency (CIA) (1996). *The World Factbook 1996* [Online]. Available: http://www.odci.gov//cia/publications/nsolo/factbook/global.h tml [1997, March 11].

Notes. When available, political party representation is shown for the lower house of the legislative branch of government. The lower house was chosen in order to present, in most cases, a picture of the electoral results of voting by the general public. The name of this legislative body is shown in the legend of the given table.

When available, election results are shown as percent distribution of votes in the most recent election. Otherwise, percent distribution of seats by political party is shown. If there are no political parties or there is one-party rule, this information is provided in place of tabular data.

Wherever possible, political party names have been presented in English translation.

NA—data are not available.

21 - Government Budget

Source. U.S. Central Intelligence Agency (CIA) (1996). *The World Factbook 1996* [Online]. Available: http://www.odci.gov//cia/publications/nsolo/factbook/global.h tml [1997, March 11].

Source. Department of State (DOS) (1996). *Background Notes* [Online]. Available: http://www.stat.gov/www/backgroundnotes/index.html [1996].

Source. International Monetary Fund (IMF). *Government Finance Statistics Yearbook 1996.* Washington, DC: IMF, 1996. **Reprinted with permission**.

Notes. IMF data were obtained primarily by means of a detailed questionnaire distributed to government finance statistics correspondents, who are usually located in each country's respective ministry of finance or central bank. Three of the six categories of central government expenditure shown in the IMF tables are comprised of subcategories, whose subtotals have been summed. Below is a list of these subcategories.

Education/Health—also includes *Welfare* and *Social security*.

Industry—includes *Fuel and energy; Agriculture, forestry, fishing, and hunting; Mining, manufacturing, and construction; Transportation and communication*; and *Other economic affairs and services*.

Other—includes *Recreational, cultural, and religious affairs and other expenditures*.

Some of the subcategory data are incomplete for Guatemala, India, and Nepal, and consequently have been calculated as zero (0).

Minor differences between published totals and the sum of components are attributable to rounding.

Following are definitions of acronyms and terms pertinent to these tables.

Central government—all units representing the territorial jurisdiction of the central authority throughout a country.

CY—calendar year: 12-month year beginning January 1 and ending the following December 31.

A dash **(-)**—data are nil or negligible.

est.—estimate.

Expenditure—all nonrepayable payments by government, including both capital and current expenditures and regardless of whether goods or services were received for such expenditures.

FY—fiscal year: presented within the calendar year containing the greatest number of months for that fiscal year. Fiscal years ending June 30 are presented within the same calendar year. For example, the fiscal year July 1, 1995 - June 30, 1996 is shown within the calendar year 1996.

Government—all units that implement public policy by providing nonmarket services and transferring income; these units are financed mainly by compulsory levies on other sectors.

NA—data are not available.

Revenue—all nonrepayable government receipts other than grants.

---

22 - Military Expenditures

Source. U.S. Arms Control and Disarmament Agency (ACDA). *World Military Expenditures and Arms Transfers 1995*. 24th ed. Washington: U.S. Arms Control and Disarmament Agency, 1996.

Source. U.S. Central Intelligence Agency (CIA) (1996). *The World Factbook 1996* [Online]. Available: http://www.odci.gov//cia/publications/nsolo/factbook/global.html [1997, March 11].

Source. U.S. Department of State (DOS) (1996). *Background Notes* [Online]. Available: http://www.stat.gov/www/backgroundnotes/index.html [1996].

Notes. Following are terms pertinent to these tables.

NA—data are not available.

A zero **(0)**—data are nil or negligible.

Notes to ACDA Data

Most of the data are for calendar years. For some countries, however, expenditure data are available only for fiscal years which diverge from calendar years. In such cases, the fiscal year which contains the most months of a given calendar year is assigned to that year; e.g., data for fiscal year April 1993 through March 1994 would be shown under 1993. Data for fiscal years ending on June 30 are normally entered under the calendar year in which they end.

Military Expenditures:

For NATO countries, military expenditures are from NATO publications and are based on the NATO definition. In this definition, (a) civilian expenditures of the defense ministry are excluded and military expenditures of other ministries are included; (b) grant military assistance is included in the expenditures of the donor country; and (c) purchases of military equipment for credit are included at the time the debt is incurred, not at the time of payment.

For non-communist countries, data are generally the expenditures of the ministry of defense. When these are known to include the costs of internal security, an attempt is made to remove these expenditures. A wide variety of data sources is used for these countries, including the publications and data resources of other U.S. government agencies, standardized reporting of countries to the United Nations, and other international sources.

It should be recognized by users of the statistical tables that the military expenditure data are of uneven accuracy and completeness. For example, there are indications that the military expenditures reported by some countries consist mainly or entirely of recurring or operating expenditures and omit all or most capital expenditures, including arms purchases. In the case of several countries (Algeria, Chile, Cuba, Ecuador, Egypt, Iran, Iraq, Libya, Nigeria, and Syria), special note of this possibility is made in the first 12 lines of these tables.

In some of these cases (as noted in subsequent footnotes), it is believed that a better estimate of total military expenditures is obtained by adding to nominal military expenditures the value of arms imports (as shown in lines 13 through 22 of these tables and converted to local currency by current exchange rates). It must be cautioned, however, that this method may over- or underestimate the actual expenditures in a given year due to the fact that payment for arms may not coincide in time with deliveries, which the data in lines 13 through 22 reflect. Also, in some cases arms acquisitions may be financed by, or consist of grants from other countries.

In these tables, the symbol "e" denotes rough estimates such as those described above and others made on the basis of partial or uncertain data. In a few cases of particular interest, *very* rough estimates are also shown marked with the symbol "r". It should be understood that these estimates are based on scant information and are subject to a wide range of error.

For countries that have major clandestine nuclear or other military weapons development programs, such as Iraq, estimation of military expenditures is extremely difficult and especially subject to errors of underestimation. Among the mechanisms commonly used to obscure such expenditures are double-bookkeeping, use of extra-budgetary accounts, highly aggregated budget categories, military assistance, and manipulation of foreign exchange. Further improvements in the quality of the military expenditure data presented for countries throughout the world will be difficult to achieve without better reporting by the countries themselves.

Particular problems arise in estimating the military expenditures of communist countries due to the exceptional scarcity and ambiguity of released information. Data on the military expenditures of the former Soviet Union are based on Central Intelligence Agency (CIA) estimates. Estimates for the most recent year are based on the change in the index of CIA-estimated military expenditures in ruble terms, as reported in the Joint Economic Committee of Congress series, *Allocation of Resources in the Soviet Union and China*.

For former Warsaw Pact countries other than the Soviet Union, the estimates of military expenditures refer only to the officially announced state budget expenditures on national defense. Thus they understate total military expenditures on national defense by excluding possible defense outlays by non-defense agencies of central and local governments, and economic enterprises. Possible subsidization of military procurement may also cause understatement. Dollar estimates were derived by calculating pay and allowances at the current full U.S. average rates for officers and lower ranks. After subtraction of pay and allowances, the remainder of the official defense budgets in national currencies was converted into dollars at overall rates based on comparisons of the various countries' GNPs expressed in dollars and in national currencies. The rates are based on the purchasing power parities estimated by the International Comparison Project of the United Nations, including their latest (Phase V) versions.

Estimates for these countries in 1990 and 1991 are based on total military spending in national currency as reported by the respective governments to the UN (in most cases) or the IMF. These expenditures *in toto* are converted to dollars at the Alton GNP conversion rates for 1989 as adjusted to 1991 by the respective U.S. and national GNP deflators (per World Bank), without estimating personnel compensation separately at U.S. dollar rates as was done for earlier years. The resulting military conversion rates (in national currency per dollar) are substantially higher than the implied rates for previous years, substantially lower than the 1991 market rate, and approximately the same as the implied rate for GNP (see below).

Estimates for the newly independent states of the former Soviet Union, Yugoslavia, and Czechoslovakia and other former Warsaw Pact countries present difficulties due to scarcity of reliable data in national currencies or to problems in converting to dollars. The basic method employed for most of these countries was to establish the ratio of military expenditures to GNP

in national currency and then to multiply this ratio by the World Bank's estimate of GNP in dollars as converted to international dollars by estimated purchasing power parities (PPPs) and reported in the *World Bank Atlas 1996*. This method implicitly converts military spending at the GNP-wide PPP, which, as with conversion by exchange rates, preserves the same ME/GNP ratio in dollars as obtains in national currency.

Data used here for China are based on U.S. government estimates of the yuan costs of Chinese forces, weapons, programs, and activities. Costs in yuan are here converted to dollars using the same estimated coversion rate as used for GNP (see below). Due to the exceptional difficulties in both estimating yuan costs and converting them to dollars, comparisons of Chinese military spending with other data should be treated as having a wide margin of error.

Other published sources used include the *Government Finance Statistics Yearbook*, issued by the International Monetary Fund; *The Military Balance*, issued by the International Institute for Strategic Studies (London); *SIPRI Yearbook: World Armaments and Disarmament*, issued by the Stockholm International Peace Research Institute; and *The World Factbook*, produced annually by the Central Intelligence Agency.

Gross National Product (GNP):

GNP represents the total output of goods and services produced by residents of a country, valued at market prices. The source of GNP data for most non-communist countries is the International Bank for Reconstruction and Development/World Bank.

For a number of countries whose GNP is dominated by oil exports (Bahrain, Kuwait, Libya, Oman, Qatar, Saudi Arabia, and the United Arab Emirates), the World Bank's estimate of deflated (or constant price) GNP in domestic currency tends to underestimate increases in the monetary value of oil exports, and thus of GNP, resulting from oil price increases. These World Bank estimates are designed to measure real (or physical) product. An alternative estimate of constant-price GNP was therefore obtained using the implicit price deflator [the ratio of GNP in current prices to GNP in constant prices] for U.S. GNP (for lack of a better national deflator). This was considered appropriate because a large share of the GNP of these countries is realized in U.S. dollars.

GNP estimates of the Soviet Union for 1990 and of other Warsaw Pact countries for 1990 and 1991 are by the CIA, as published in its *Handbook of Economic Statistics* (1990 and 1992 eds.).

Estimates of GNP in 1992-1994 for successor states to the Soviet Union, Yugoslavia, and Czechoslovakia are based on World Bank estimates of GNP per capita employing PPPs, as published in the *World Bank Atlas 1996*.

GNP data for China are based on World Bank estimates in yuan. These are in line with estimates of GDP in Western accounting terms made by Chinese authorities. Conversion to dollars is highly problematic, however, due to the inappropriateness of the official exchange rate and lack of sufficient yuan price information by which to reliably estimate PPPs. (The ratio of the highest to the lowest estimates by various sources of China's GNP is on the order of 6 or 7 to 1, which would make the world rank of China's GNP vary between about 3rd or 4th and 12th.) The conversion rate used here is based on a PPP estimated for 1981 and moved by respective U.S. and China implicit GNP deflators to 1994.

GNP estimates for a few non-communist countries are from the CIA's *Handbook of Economic Statistics* cited above. Estimates for other communist countries are rough approximations.

Military-Expenditures-to-GNP Ratio:

It should be noted that the meaning of the ratio of military expenditures to GNP, shown in these tables, differs somewhat between most communist or previously communist, and other, countries. For non-communist countries, both military expenditures and GNP are converted from the national currency unit to dollars at the same exchange rate. Consequently, the ratio of military expenditures to GNP is the same in dollars as in the national currency and reflects national relative prices. For communist countries, however, military expenditures and GNP are converted differently. Soviet military expenditures, as already noted, are estimated in a way designed to show the cost of the Soviet armed forces in U.S. prices, as if purchased in this country. On the other hand, the Soviet GNP estimates used here are designed to show average relative size when both U.S. and Soviet GNP are valued and compared at both dollar and ruble prices. The Soviet ratio of military expenditures to GNP in ruble terms, the preferred method of comparison, is estimated to have been 15-18 percent in recent years.

The estimated ratio for Russia derived here in dollars is probably somewhat overstated, since military spending in dollars is related to earlier estimates for the Soviet Union, while GNP estimates (at PPPs) are by the World Bank. Russia's burden ratio in ruble terms is preferably estimated to be in the vicinity of 10 percent or less.

For Eastern European countries, the ratios of military expenditures to GNP in dollars are about twice the ratios that would obtain in domestic currencies. However, since official military budgets in these countries probably substantially understate their actual military expenditures, the larger ratios based on dollar estimates are believed to be the better approximations of the actual ratios.

Central Government Expenditures (CGE):

These expenditures include current and capital (developmental) expenditures plus net lending to government enterprises, by central (or federal) governments. A major source is the International Monetary Fund's *Government Finance Statistics Yearbook*. The category used here is "Total Expenditures and Lending minus Repayment, Consolidated Central Government."

Other sources for these data are the International Monetary Fund monthly, *International Finance Statistics*; OECD, *Economic Surveys*; and CIA, *World Factbook (annual)*. Data for Warsaw Pact countries are from national publications and are supplied by Thad P. Alton and others. For all Warsaw Pact countries and China, conversion to dollars is at the implicit rates used for calculating dollar estimates of GNP.

For all countries, with the same exceptions as noted above for the military-expenditures-to-GNP ratio, military expenditures and central government expenditures are converted to dollars at the same rate; the ratio of the two variables in dollars thus remains the same as in national currency.

It should be noted that for the Soviet Union, China, Iran, Jordan, and possibly others, the ratio of military expenditures to central government expenditures may be overstated, inasmuch as the same estimate for military expenditures is obtained at least in part independently of nominal budget or government expenditure data, and it is possible that not all estimated military expenditures pass through the nominal central government budget.

Population:

Population estimates are for midyear and are made available to ACDA by the U.S. Bureau of the Census.

Armed Forces:

Armed forces refer to active-duty military personnel, including paramilitary forces if those forces resemble regular units in their organization, equipment, training, or mission. Reserve forces are not included unless specifically noted.

Figures for the United States and all other NATO countries are as reported by NATO. Estimates of the number of personnel under arms for other countries are provided by U.S. Government sources. The armed forces series for the Soviet Union includes all special forces judged to have national security missions (e.g., KGB border guards) and excludes uniformed forces primarily performing noncombatant services (construction, railroad, civil defense, and internal security troops).

Arms Transfers:

Arms transfers (arms imports and exports) represent the international transfer (under terms of grant, credit, barter, or cash) of military equipment, usually referred to as "conventional", including weapons of war, parts thereof, ammunition, support equipment, and other commodities designed for military use. Among the items included are tactical guided missiles and rockets, military aircraft, naval vessels, armored and nonarmored military vehicles, communications and electronic equipment, artillery, infantry weapons, small arms, ammunition, other ordnance, parachutes, and uniforms. Dual use equipment, which can have application in both military and civilian sectors, is included when its primary mission is identified as military. The building of defense production facilities and licensing fees paid as royalties for the production of military equipment are included when they are contained in military transfer agreements. There have been no international transfers of purely strategic weaponry. Military services such as training, supply operations, equipment repair, technical assistance, and construction are included where data are available. Excluded are foodstuffs, medical equipment, petroleum products, and other supplies.

Redefinition of U.S. Arms Exports: The scope of U.S. arms exports data has been modified in this edition. These exports include both government-to-government transfers under the Foreign Military Sales (FMS), Military Assistance Program (MAP), and other programs administered by the Department of Defense, and commercial (enterprise-to-government) transfers licensed by the Department of State under International Traffic in Arms Regulations. Under the previous practice, the materiel component (arms, equipment, and "hardware" items) of FMS and MAP sales was included, while the military services component was excluded (although the magnitude and general destination of the omitted services was reported in these Notes). The commercial sales category, covering both materiel and military services, was included in its entirety. Beginning with this edition, both the materiel and the military services components of FMS and other government-to-government sales (such as the International Military Education and Training Program—IMET) are included in total U.S. arms exports as reported here.

The increasing importance of these services and the desire to present a full picture of U.S. arms exports consistent with other sources prompted the change to inclusion. Users should be aware, however, of both the lower true share of services in other countries' arms exports and the tendency to underestimate them. It should also be noted that a portion of the IMET program is devoted to programs that promote improved civil military relations.

The change in scope of U.S. arms exports has increased their overall volume by amounts ranging over the last decade from $2.3 billion (current dollars) to $3.7 billion for deliveries and $2.3 billion to $7.3 billion for agreements.

The statistics contained in the table, lines 13 through 22, are estimates of the value of goods actually delivered during the reference year, in contrast both to payments and the value of programs, agreements, contracts, or orders concluded during the period, which are expected to result in future deliveries. Figures for U.S. arms exports are for fiscal years as reported by the U.S. Departments of Defense and State.

U.S. Arms Imports: Data on U.S. arms imports in this and the previous three editions are revised upward substantially from earlier editions. The present series consists of data obtained from the Department of Commerce, Bureau of Economic Analysis (BEA), including (a) imports of military-type (formerly "special category") goods, as compiled by the Bureau of the Census, and (b) Department of Defense (DOD) direct expenditures abroad for major equipment, as compiled from DOD data by BEA. The goods in (a) include complete military aircraft, all types; engines and turbines for military aircraft; military trucks, armored vehicles, etc.; military (naval) ships and boats; tanks, artillery, missiles, guns, and ammunition; military apparel and footwear; and other military goods, equipment, and parts.—Data on countries other than the United States are estimates by U.S. government sources. Arms transfer data for the Soviet Union and other former communist countries are approximations based on limited information.

It should be noted that the arms transfer estimates for the most recent year, and to a lesser extent for several preceding years, tend to be understated. This applies to both foreign and U.S. arms exports. In the former case, information on transfers comes from a variety of sources and is sometimes acquired and processed with a considerable time lag. In the U.S. case, commercial arms transfer licenses are now valid for three years, causing a delay in the reporting of deliveries made on them to statistical agencies. Data for the most recent two years in the table, lines 13 through 22, therefore, can be expected to undergo some upward revision in succeeding issues.

Close comparisons between the estimated values shown for arms transfers and for GNP and military expenditures are not warranted. Frequently, weapons prices do not reflect true production costs. Furthermore, much of the international arms trade involves offset or barter arrangements, multiyear loans, discounted prices, third-party payments, and partial debt forgiveness. Acquisition of armaments thus may not impose burden on an economy, whether in the same or in other years, that is implied by the estimated equivalent U.S. dollar value of the shipment. Therefore, the value of arms imports should be compared to other categories of data with care.

Total Imports and Exports:

The values for imports and exports cover merchandise transactions and come mainly from *International Financial Statistics*, published by the IMF. The trade figures for communist and formerly communist countries are from the CIA *Handbook of Economic Statistics*, 1995 edition.

Footnotes

e.   Estimate based on partial or uncertain data.

r.   Rough estimate.

1.   Estimated by adding arms imports to data on military expenditures, which are believed to exclude arms purchases. However, it should be noted that the value of arms deliveries in a given year (converted at current exchange rates) may differ significantly from actual expenditures on arms imports in that year.

2.   This ratio is calculated from the dollar values shown in previous lines. In most cases, it also is equal to the ratio that could be calculated from national currency values, since both numerator and denominator are usually converted into dollars by the same exchange rate or other conversion factor. In the case of this country, however, the two variables are converted at differing rates, yielding a different ratio than would obtain in national currency.

The ratio for Russia in rubles terms, for example, is believed to be less than 10 percent in 1994. See Notes, Military Expenditures-to-GNP Ratio for further discussion.

3. This series or entry probably omits a major share of total military expenditures, probably including most expenditures on arms procurement. These tables show estimated annual arms imports. It should be kept in mind, however, that data in these tables represent the estimated value of arms delivered in a given year, not actual expenditures on those arms.

4. Data refer to unified Germany beginning in 1991. 1990 data refer to West Germany only.

5. All 1990 and 1991 data refer to the Soviet Union. All data after 1991 refer to Russia.

6. To avoid the appearance of excessive accuracy, arms transfer data by country are rounded, with greater severity for larger amounts. Thus any country group summation for arms exports and arms imports will consist of rounded country data. Consequently, world totals for arms imports and arms exports will not be equal.

7. Total imports and exports usually are as reported by individual countries and the extent to which arms transfers are included is often uncertain. Imports are reported "cif" (including cost of shipping, insurance, and freight) and exports are reported "fob" (excluding these costs). For these reasons and because of divergent sources, world totals for imports and exports will not be equal.

8. Because some countries exclude arms imports or exports from their trade statistics and their "total" imports and exports are therefore understated and because arms transfers may be estimated independently of trade data, the resulting ratios of arms to total imports or exports may be overstated and may even exceed 100 percent.

9. Some part of estimated total military expenditures may not be included in announced central budget expenditures. The ratio of ME to CGE therefore may be overstated.

10. Includes major equipment purchased by the U.S. Army Corps of Engineers for use in military construction projects in Saudi Arabia, recorded in U.S. accounts as U.S. imports.

11. U.S. arms imports data shown here is revised upward substantially from reports before 1993. (See Notes, Arms Transfers.)

12. Unified Yemen as of 1990 is shown under Yemen (Sanaa); data for Yemen (Aden) end in 1989.

13. Arms purchases made by unified Yemen following May 22, 1990, are shown under Yemen (Sanaa).

14. No data are available until 1992, shortly after the U.S.S.R. had disbanded on December 26, 1991. At that point, this former Soviet republic became an independent state.

15. This country is a former Yugoslav republic. Consequently, no data are available until 1992, shortly after it declared its independence.

16. Czechoslovakia split into two independent states, the Czech Republic and Slovakia, on January 1, 1993. Consequently, no data are available for this new state before 1993.

17. In 1991, the Yugoslav republics of Bosnia and Herzegovina, Macedonia, Croatia, and Slovenia formally declared their independence. Consequently, these data refer only to present-day Yugoslavia, which, as of 1992, is comprised of Serbia and Montenegro.

18. Little data are available because of an ongoing civil war. As of March 3, 1995, the capital Mogadishu still had no functioning government.

19. Little data are available because of an ongoing civil war.

23 - Crime

Source. International Criminal Police Organization (INTERPOL). *International Crime Statistics, 1994.* Lyons: INTERPOL, 1994. **Reprinted with permission**.

Source. United Nations Criminal Justice Information Network (UNCJIN) (1994). *Fourth United Nations Survey of Crime Trends and Operations of Criminal Justice Systems (1986-1990)* [Online]. Available: http://www.ifs.univie.ac.at/~uncjin/wcs.html [1994]. **Reprinted with permission**.

Notes to INTERPOL Data Statistics are based on data collected by the police in ICPO-Interpol member countries and are therefore *police statistics* and not *judicial statistics*.

The form adopted by resolution No AGN/45/RES/6 at the 1976 session of the ICPO-Interpol General Assembly was used to collect information.

The information given is in no way intended for use as a basis for comparisons between different countries.

These statistics cannot take account of the differences that exist between the legal definitions of punishable offenses in various countries, of the different methods of calculation, or of any changes which may have occurred in the countries concerned during the reference period. All these factors obviously have repercussions on the figures supplied.

Police statistics reflect the crimes reported to or detected by the police and therefore cover only part of the total number of offenses actually committed. Moreover, the volume of unreported crimes depends to some extent on action taken by the police, and may therefore vary from one point in time to another and from one country to another.

Consequently, the figures given in these statistics must be interpreted with caution.

Notes to UNCJIN Data

The *Fourth United Nations Survey* has provided crime tables for the following countries: Australia, Belarus, Costa Rica, Czech Republic, India, Jordan, Kazakhstan, Kyrgyzstan, Lithuania, Marshall Islands, Moldova, Myanmar, Netherlands, Norway, Philippines, Saint Kitts and Nevis, South Africa, Sri Lanka, Tajikistan, United Kingdom, Vanuatu, and Yugoslavia.

The findings of the *Fourth United Nations Survey* demonstrate the difficulties of comparing crime internationally. One difficulty is that the vast majority of incidents that become known to the police come from reports by victims, thus credibility becomes a statistical determinant. Another difficulty is that comparison is severely undermined by differences in legal definitions, and by administrative procedures regarding counting, classification, and disclosure. Finally, because this survey is a "work in progress", completeness of data may vary somewhat from country to country. All UNCJIN crime data for the *Statistical Abstract of the World* have been calculated based soley on available source subtotals. The researcher should be aware of these shortcomings when using this data.

NA—data are not available.

Footnotes

1. 1990 data are shown as reported by UNCJIN.

2. INTERPOL data are incomplete for this country. INTERPOL tables normally display several subcategories plus a "Totals" category. For this country, however, data for subcategories "Counterfeit currency offences" and "Drug offences" are omitted (see *International Crime Statistics, 1994*, p. 105). Consequently, figures for this table are summed from available source subcategories only.

3. Figures for this table have been calculated based solely on 1990 UNCJIN source data for Serbia and Montenegro.

24 - Human Rights

Source. U.S. Department of State (DOS) (1997). *Country Reports on Human Rights Practices for 1996 [Online]. Available: http://www.state.gov/www/current/index.html [1997]*.

Notes. The following nations are non-ILO members (ILO stands for International Labor Organization): Albania, Bhutan, Brunei, Gambia, Liechtenstein, Maldives, Marshall Islands, Micronesia, Monaco, North Korea, Oman, South Africa, St. Kitts and Nevis, St. Vincent and Grenadines, Taiwan, Tajikistan, Vanuatu, Vietnam, and Western Samoa.

Human rights conventions, shown in tables as acronyms, are as follows:

ACHR	American Convention on Human Rights of November 22, 1969.
AFL	Convention Concerning the Abolition of Forced Labor of June 25, 1957 (ILO Convention 105).
APROBC	Convention Concerning the Application of the Principles of the Right to Organize and Bargain Collectively of July 1, 1949 (ILO Convention 98).
ASST	Supplementary Convention on the Abolition of Slavery, the Slave Trade, and Institutions and Practices Similar to Slavery of September 7, 1956.
CPR	International Covenant on Civil and Political Rights of December 16, 1966.
EAFDAW	Convention on the Elimination of All Forms of Discrimination Against Women of December 18, 1979.
EAFRD	International Convention on the Elimination of All Forms of Racial Discrimination of December 21, 1965.
ESCR	International Covenant on Economic, Social, and Cultural Rights of December 16, 1966.
FAPRO	Convention Concerning Freedom of Association and Protection of the Right to Organize of July 9, 1948 (ILO Convention 87).
FL	Convention Concerning Forced Labor of June 28, 1930 (ILO Convention 29).
MAAE	Convention Concerning Minimum Age for Admission to Employment of June 26, 1973 (ILO Convention 138).
PCPTW	Geneva Convention Relative to the Protection of Civilian Persons in Time of War of August 12, 1949.
PHRFF	European Convention for the Protection of Human Rights and Fundamental Freedoms of November 4, 1950.
PPCG	Convention on the Prevention and Punishment of the Crime of Genocide of December 9, 1948.
PRW	Convention on the Political Rights of Women of March 31, 1953.

PVIAC   Protocol Additional to the Geneva Conventions of August 12, 1949, and Relating to the Protection of Victims of International Armed Conflicts (Protocol I), of June 8, 1977.

PVNAC   Protocol Additional to the Geneva Conventions of August 12, 1949, and Relating to the Protection of Victims of Non-International Armed Conflicts (Protocol II), of June 8, 1977.

RC   Convention on the Rights of the Child of November 20, 1989.

SR   Protocol Relating to the Status of Refugees of January 31, 1967.

SSTS   Convention to Suppress the Slave Trade and Slavery of September 25, 1926, as amended by the Protocol of December 7, 1953.

STPEP   Convention for the Suppression of the Traffic in Persons and of the Exploitation of the Prostitution of Others of March 21, 1950.

TCIDTP   Convention Against Torture and Other Cruel, Inhuman or Degrading Treatment or Punishment of December 10, 1984.

TPW   Geneva Convention Relative to the Treatment of Prisoners of War of August 12, 1949.

Footnotes

1.   Based on general declaration concerning treaty obligations prior to independence.

2.   Party to 1926 convention only.

25 - Total Labor Force

Source. U.S. Central Intelligence Agency (CIA) (1996). *The World Factbook 1996* [Online]. Available: http://www.odci.gov//cia/publications/nsolo/factbook/global.html [1997, March 11].

Notes. Data show the number of persons in the labor force for the most recent year available.

NA—data are not available.

26 - Labor Force by Occupation

Source. U.S. Central Intelligence Agency (CIA) (1996). *The World Factbook 1996* [Online]. Available: http://www.odci.gov//cia/publications/nsolo/factbook/global.html [1997, March 11].

Source. U.S. Department of State (DOS) (1996). *Background Notes* [Online]. Available: http://www.stat.gov/www/backgroundnotes/index.html [1996].

Notes. Data show distribution of the labor force in percent (when available) by industry.

NA—data are not available.

27 - Unemployment Rate

Source. U.S. Central Intelligence Agency (CIA) (1996). *The World Factbook 1996* [Online]. Available: http://www.odci.gov//cia/publications/nsolo/factbook/global.html [1997, March 11].

Notes. Data show the rate of unemployment in percent for the most recent year available.

NA—data are not available.

28 - Energy Production

Source. U.S. Central Intelligence Agency (CIA) (1996). *The World Factbook 1996* [Online]. Available: http://www.odci.gov//cia/publications/nsolo/factbook/global.html [1997, March 11].

Source. Energy Information Administration (EIA), U.S. Department of Energy (DOE) (1996). *International Energy Annual 1995* [Online]. Available: http://www.eia.doe.gov/emeu/iea/contents.html [1996].

Notes to EIA data EIA attempts to identify and collect the best data available for foreign countries. The most authoritative sources are usually the official national statistical reports of a country. However, when data from official sources are not available, EIA uses data from reputable secondary sources such as the international organizations—the United Nations, the International Energy Agency, the World Bank, and others. In addition, EIA uses industry reports, academic studies, trade publications, and other sources.

Many factors beyond EIA's control affect the reliability and integrity of foreign country data. These include a country's level of economic development, commitment to statistical programs, openness with information, and other considerations.

Btu—British thermal units.

29 - Energy Consumption

Source. Energy Information Administration, U.S. Department of Energy (DOE) (1996). *International Energy Annual 1995* [Online]. Available: http://www.eia.doe.gov/emeu/iea/contents.html [1996].

Notes. EIA attempts to identify and collect the best data available for foreign countries. The most authoritative sources are usually the official national statistical reports of a country. However, when data from official sources are not available, EIA uses data from reputable secondary sources such as the international organizations—the United Nations, the International Energy Agency, the World Bank, and others. In addition, EIA uses industry reports, academic studies, trade publications, and other sources.

Many factors beyond EIA's control affect the reliability and integrity of foreign country data. These include a country's level of economic development, commitment to statistical programs, openness with information, and other considerations.

Btu—British thermal units.

30 - Telecommunications

Source. U.S. Central Intelligence Agency (CIA) (1996). *The World Factbook 1996* [Online]. Available: http://www.odci.gov//cia/publications/nsolo/factbook/global.html [1997, March 11].

Notes. Following are CIA definitions of the acronyms and terms used in these tables.

Arabsat—Arab Satellite Communications Organization (Riyadh, Saudi Arabia).

ASEAN—Association of Southeast Asian Nations.

Central American Microwave System—trunk microwave radio relay system that links the countries of Central America and Mexico to each other.

CIS—Commonwealth of Independent States: established in 1991; refers to former members of the Soviet Union.

Coaxial cable—a multichannel communication cable consisting of a central conducting wire, surrounded by and insulated from a cylindrical conducting shell; a large number of telephone channels can be made available within the insulated space by the use of a large number of carrier frequencies.

Eutelsat—European Telecommunications Satellite Organization (Paris).

Fiber-optic cable—a multichannel communications cable using a thread of optical glass fibers as a transmission medium in which the signal (voice, video, etc.) is in the form of a coded pulse of light.

HF—high-frequency; any radio frequency in the 3,000- to 30,000-kHz range.

Inmarsat—International Mobile Satellite Organization (London); provider of global mobile satellite communications for commercial and distress and safety applications, at sea, in the air, and on land.

Intelsat—International Telecommunications Satellite Organization (Washington, DC).

Intersputnick—International Organization of Space Communications (Moscow); first established in the former Soviet Union and the East European countries, it is now marketing its services worldwide with earth stations in North America, Africa, and East Asia.

Landline—communication wire or cable of any sort that is installed on poles or buried in the ground.

Marecs—Maritime European Communications Satellite used in the Inmarsat system on lease from the European Space Agency.

Medarabtel—Middle East Telecommunications Project of the International Telecommunications Union (ITU), providing a modern telecommunications network, primarily by microwave radio relay, linking Algeria, Djibouti, Egypt, Jordan, Libya, Morocco, Saudi Arabia, Somalia, Sudan, Syria, Tunisia, and Yemen; formerly known as the Middle East Mediterranean Telecommunications Network.

Microwave radio relay—transmission of long-distance telephone calls and television programs by highly directional radio microwaves that are received and sent on from one booster station to another on an optical path.

NA—data are not available.

NMT—Nordic Mobile Telephone; an analog cellular telephone system that was developed jointly by the national telecommunications authorities of the Nordic countries (Denmark, Finland, Iceland, Norway, and Sweden).

NZ—New Zealand.

Orbita—Russian television service; also the trade name of a packet-switched digital telephone network.

Radiotelephone communications—two-way transmission and reception of sounds by broadcast radio authorized frequencies using telephone handsets.

Satellite communication system—two or more earth stations and at least one satellite that provides long-distance transmission of voice, data, and television; the system usually serves as a trunk connection between telephone exchanges; if the earth stations are in the same country, it is a domestic system.

Satellite earth stations—communications facility with a microwave radio transmitting and receiving antenna and required receiving and transmitting equipment for communicating with satellites.

Satellite link—radio connection between a satellite and an earth station permitting communication between them, either one-way (down link from satellite to earth station - television receive-only transmission) or two-way (telephone channels).

SHF—super-high-frequency; any radio frequency in the 3,000- to 30,000-MHz range.

Shortwave—radio frequencies (from 1.605 to 30 MHz) that fall above the commercial broadcast band and are used for communication over long distances.

Solidaridad—geosynchronous satellites in Mexico's system of international telecommunications in the Western Hemisphere.

Submarine cable—cable designed for service under water.

TAT—Trans-Atlantic Telephone; any of a number of high-capacity submarine coaxial telephone cables linking Europe with North America.

Tropospheric scatter—form of microwave radio transmission in which the troposphere is used to scatter and reflect a fraction of the incident radio waves back to earth; powerful, highly directional antennas are used to transmit and receive the microwave signals; reliable over-the-horizon communications are realized for distances of up to 600 miles in a single hop; additional hops can extend the range of this system for very long distances.

Trunk network—network of switching centers, connected by multichannel trunk lines.

UAE—United Arab Emirates.

UHF—ultra-high-frequency; any radio frequency in the 300- to 3,000-MHz range.

VHF—very-high-frequency; any radio frequency in the 30- to 300-MHz range.

31 - Transportation

Source. U.S. Central Intelligence Agency (CIA) (1996). *The World Factbook 1996* [Online]. Available: http://www.odci.gov//cia/publications/nsolo/factbook/global.html [1997, March 11].

Notes. Following are CIA definitions of terms used in these tables.

Airports—only airports with usable runways are included in this listing. Not all airports have facilities for refueling, maintenance, or air traffic control. Paved runways have concrete or asphalt surfaces; unpaved runways have grass, dirt, sand, or gravel surfaces.

DWT—deadweight tons.

GRT—gross register tons.

km—kilometers.

m—meters.

Merchant marine—all ships engaged in the carriage of goods. All commercial vessels (as opposed to all nonmilitary ships), which excludes tugs, fishing vessels, offshore oil rigs, etc. Also, a grouping of merchant ships by nationality or register.

NA—data are not available.

32 - Top Agricultural Products

Source. U.S. Central Intelligence Agency (CIA) (1996). *The World Factbook 1996* [Online]. Available: http://www.odci.gov//cia/publications/nsolo/factbook/global.html [1997, March 11].

Notes. GDP—gross domestic product: the value of all goods and services produced within a nation in a given year.

33 - Top Mining Products

Source. Energy Information Administration, U.S. Department of Energy (DOE). *International Energy Annual 1995*. Washington, DC: DOE, 1996.

Source. U.S. Bureau of Mines, U.S. Department of the Interior (DOI). *Minerals Yearbook, Volume II:—Area Reports: Domestic, 1993- 94*. Washington, DC: DOI, 1996.

Source. U.S. Central Intelligence Agency (CIA) (1996). *The World Factbook 1996* [Online]. Available: http://www.odci.gov//cia/publications/nsolo/factbook/global.html [1997, March 11].

Source. U.S. Geological Survey (USGS). *Minerals Yearbook, Volume III— Area Reports: International, 1994* [Online]. Available: http://www.minerals.er.usgs.gov/minerals/pubs/country/#pubs [1996].

Notes. Top Mining Products indicates that up to 10 mineral commodities were selected from each country's total mining production for this category depending on the importance of the commodities in terms of volume or value for that country.

Many of the tables include data which have been rounded to three significant digits by the U.S. Bureau of Mines.

Footnotes

*. Indicates date through which data are available.

e. stands for estimate based on partial or uncertain data.

p. stands for preliminary data.

r. stands for reported figure.

1. Excludes gas used in reinjection, flaring, venting, transmission losses, and natural gas liquids extraction.

2.　Does not include smuggled production.

3.　Angola has no natural gas distribution system. Most gas is vented, except for a small fraction, from which natural gas liquids are produced. Propane and butane canisters are filled at the well site.

4.　Includes asphalt and natural bitumen.

5.　Data are for year ending November 30 for plants owned by Broken Hill Pty. Co. Ltd.

6.　Excluding stone used by cement and iron and steel industries.

7.　Data are for year ending June 30.

8.　Gross production is not reported; the quantity vented, flared, or reinjected is believed to be negligible.

9.　Includes production of metallic gold.

10.　Includes production of metallic silver.

11.　Ore milled is nickel-copper-cobalt ore shown under "Nickel: mine output, ore milled."

12.　Presumably, principally agate. Reported as sales. Only cut or polished stones could be legally exported after 1989.

13.　Smelter product was granulated nickel-copper-cobalt matte.

14.　From natural soda ash production.

15.　Additional production of sand and gravel from small local operations was periodically reported, but information was inadequate to reliably estimate output.

16.　Direct sales and/or beneficiated (marketable product).

17.　Metal content of concentrates produced.

18.　From all sources, including imports and secondary sources. Excludes intermediate products exported for refining.

19.　Refined sorel slag contained 80% TiO2 in 1990. TiO2 content in 1991-94 is not reported.

20.　Output based entirely on imported clinker.

21.　Does not include artisanal production smuggled out of the country.

22.　Anuario Estadistico de Cuba provides figures of nickel-cobalt content of granular and powder oxide, oxide sinter, and sulfide production. Using an average cobalt content in these products of 0.9% in total granular and powder oxide, 1.1% in total oxide sinter, and 4.5% in total sulfide, the cobalt content of reported Ni-Co production was determined to be 1.16% of granular and powder oxide, 1.21% of oxide sinter, and 7.56% of sulfide. The remain-

der of reported figures would represent the nickel content.

23.　Mineral production data from the northern Turkish-occupied section of the country are not included in this table, as available information is inadequate to make reliable estimates of output levels.

24.　Includes crushed aggregate.

25.　Prodution.

26.　Rock salt only.

27.　Additional artisanal gold reportedly was normally produced (estimated at 1,500 kilograms per year according to a government official in 1994), and there may have been other production, but information is inadequate to reliably estimate output.

28.　May include gravel.

29.　Revised.

30.　Includes cement produced from imported clinker.

31.　Reinjected for repressuring.

32.　Gold production figures do not include production smuggled out of the country. Smuggled production in 1994 was estimated to exceed 400 kilograms.

33.　All from imported clinker.

34.　Production, in thousand carats, includes that of Akwatia Mine (1994: 356), PMMC purchases of artisanal production (1994: 406), and estimates of smuggled artisanal production.

35.　Does not include estimate of smuggled production.

36.　All figures were reported by Bureau de Strategie et de Marketing Minier of Guinea.

37.　Data are for wet-basis ore estimated at 13% water, reduced to dry basis estimated at 3% water.

38.　Figures do not include undocumented artisanal production believed smuggled out of the country.

39.　Figures include undocumented artisanal production. Aurifere de Guinea (AuG) is the only reporting gold mining company, reporting 500 kilograms for 1994.

40.　Includes LPG, aviation and motor gasoline, diesel, kerosene, and distillate fuel oil.

41.　Excludes refinery fuel and losses.

42.　Ingot and rolling billet production.

43.　Byproduct of ferrosilicon.

44. Includes gold content of copper ore and output by government- controlled foreign contractors' operations. Gold output by operators of so-called people's mines and illegal small-scale mines is not available, but may be as much as 18 metric tons per year.

45. Data are for the Iranian year beginning March 21, except data for natural gas, plant liquids, and petroleum, which are for Gregorian calendar years.

46. Estimated to contain 30% phosphorous pentoxide. Last report on crude rock production was for 1988: 3.5 Mmt, estimated to contain 22% phosphorous pentoxide.

47. Excludes output by local authorities and road contractors.

48. From imported crude oil.

49. Includes Kuwait's share of production in the Kuwait-Saudi Arabia Divided Zone.

50. Data for 1994 are estimates of artisanal production, likely smuggled out of Liberia, but which are comparable to that hitherto reported to the government.

51. Includes production from Malaya, Sabah, and Sarawak.

52. Inlcudes estimate of artisanal production and may include some gold smuggled into Mali. The Kalana Mine accounted for nil in 1994. The Syama Mine accounted for 55% of total gold in 1994.

53. Estimated silver content dore bullion.

54. Included gold contained in copper concentrate.

55. Silver contained in copper concentrate.

56. Facet-grade. In addition, there was waste garnet production in 1994, in the amount of 924 kilograms.

57. The increase in 1994 is due to production from Walvis Bay, previously included under South Africa.

58. Data represent exports.

59. Excluding natural gas liquids.

60. Run-of-mine coal.

61. Mahd Adh Dhahab final products include a bulk flotation concentrate containing gold, silver, copper, lead, and zinc, and a crude bullion containing gold, silver, and copper. Ore containing gold and silver from the Sukhaybirat surface mine included since 1991.

62. Includes Saudi Arabian one-half share of production in the Kuwait-Saudi Arabia Divided Zone.

63. Data include only officially reported production.

64. Estimated for cement manufacture only.

65. Excludes refinery fuel and losses, amounting to an estimated 7 to 8 million 42-gallon barrels per year.

66. Estimated capacity on the basis of recent production history.

67. Includes only that recovered from indigenous ores excluding scrap.

68. Total of listed products only.

69. Diamond figures are estimated to represent 70% gem-quality or semigem-quality and 30% industrial-quality stones.

70. Exports

71. Produced from imported clinker.

72. Iron rod production from semifinished metal.

73. Includes estimated content of base metals refinery tankhouse slimes.

74. Data are based on reported units of 50-kilogram boxes.

75. Includes lease condensate.

76. Methane, excluding gas flared or reinjected.

77. Includes ethane, propane, butane, and condensates.

78. Excludes gases and condensates.

79. First six items in column are preliminary data. Production is measured by mine shipments, sales, or marketable production (including consumption by producers).

80. Recoverable content of ores, etc.

81. Placer canvassing discontinued beginning in 1994.

82. Sources: *Minerals Yearbook, Volume II* and *International Energy Annual 1995*. See Source Notes above.

83. Estimates based on information available through July 7, 1995.

84. Data are for year beginning April 1, 1994.

85. Ores from which both a copper concentrate and a cobalt concentrate, or a cobalt concentrate only were produced.

86. From mines operated by Zambia Consolidated Copper Mines Ltd. (ZCCM) only; additional concentrate estimated at about 3,000 metric tons per year is produced by other mines in Zambia. Includes ore and concentrate shown under "Cobalt" entry above, all of which contain copper that was recovered, but separate quantitative data on copper content of cobalt concentrate are not available.

87. From copper and cobalt refinery residue produced by ZCCM only. Additional production probably came from

artisanal operations, but information is inadequate to reliably estimate output. However, total production, presumably from artisanal as well as ZCCM operations, was reported for calendar year 1994, in kilograms: gold—165, silver— 12,200.

88. Presumably recovered from copper and cobalt refinery mud/slimes processed at ZCCM's Ndola Precious Metal plant. A similar quantity may be contained in mud/slimes not processed in-country and possibly sold for treatment elsewhere, but information is inadequate to reliably estimate content.

89. Estimated for cement (about 1.3 tons per ton of cement) and lime (about 1.8 tons per ton of calcined lime) manufacture only.

90. Data represent output by the Wankie Colliery Co. Ltd.; additional output by the Redcliff plant of Zisco Ltd. may total 250,000 metric tons per year of metallurgical coke and coke breeze.

34 - Tourism

Source. World Tourism Organization (WTO). *Compendium of Tourism Statistics 1990-1994, 16th ed..* Madrid: WTO, 1996. **Reprinted with permission**.

Notes. Tourism arrival data refer to the number of arrivals of visitors and not to the number of persons. The same person who makes several trips to a given country during a given period will be counted each time as a new arrival.

For statistical purposes, the term *international visitor* describes any person who travels to a country other than that in which (s)he has his/her usual residence but outside his/her usual environment for a period nor exceeding twelve months and whose main purpose of visit is other than the exercise of an activity remunerated from within the country visited.

International visitors include **Tourists**, visitors who stay at least one night in a collective or private accommodation in the country visited; and **Excursionists**, visitors who do not spend the night in a collective or private accommodation in the country visited.

This includes: 1) **Cruise passengers** who arrive in a country on a cruise ship and return to the ship each night to sleep on board even thought the ship remains in port for several days. Also included in this group are, by extension, owners or passengers of yachts and passengers on a group tour accommodated in a train.

2) **Crew members** who do not spend the night in the country of destination. This group also includes crews of warships on a courtesy visit to a port in the country of destination who spend the night on board ship and not at the destination.

Unless otherwise stated, figures for **Visitors** correspond to the aggregation of the figures for **Tourists** and **Excursionists**. In principle, data for **Cruise passengers** are included in the **Excursionists** category, but in these tables data are provided separately.

Tourism receipts—the expenditure of international inbound visitors including their payments to national carriers for international transport. They include any other prepayments made for goods/services received in the destination country. They also include receipts from excursionists, except in cases when these are important enough to justify a separate classification. The International Monetary Fund (IMF) has advised, that, for the sake of consistency with the Balance of Payments recommendations, international fare receipts be classified separately.

Tourism expenditures—expenditures of outbound visitors in other countries including their payments to foreign carriers for international transport. They include expenditure of residents traveling abroad as excursionists, except in cases when these are important enough to justify a separate classification. As in the case of tourism receipts, the IMF has advised that international expenditure be classified separately.

Fare receipts—any payment made to carriers registered in the survey country of sums owed by nonresident visitors, whether or not traveling to that country. This category corresponds to "Other transportation, passenger services, credits" in the standard reporting form of the IMF.

Fare expenditures—all payments to carriers registered abroad by any resident in the survey country. This category corresponds to "Other transportation, passenger services, debits" in the standard reporting form of the IMF.

NA—data are not available.

Footnotes

1. International visitor arrivals at frontiers.

2. Arrivals in hotels only.

3. Tourist hotels only.

4. Includes nationals of the country residing abroad.

5. 1991 arrivals by air.

6. Air and sea arrivals, excluding nationals of the country residing abroad.

7. Includes cruise ships, windjammer cruises, and yacht arrivals.

8. Excludes nationals of the country residing abroad.

9. Includes the transport of merchandise.

10. Excludes nationals of the country residing abroad and crew members.

11. International tourist arrivals at collective tourism establishments.

12. International tourist arrivals at frontiers.

13. Belgium and Luxembourg.

14. International visitor arrivals at frontiers, including Belizean residents and border permits.

15. In transit and border permits.

16. International tourist arrivals at hotels and similar establishments.

17. Excludes returning residents.

18. Data based on a sample survey conducted by EMBRA-TUR.

19. International tourist arrivals at frontiers, including nationals of the country residing abroad.

20. Arrivals by air.

21. Data based on customs counts and adjusted using questionnaire survey.

22. Excludes crew spending and international fares.

23. Arrivals by air. Figure for 1994 is an estimate.

24. Includes ethnic Chinese arriving from Hong Kong, Macau, Taiwan, and overseas Chinese, of which most excursionists are from Hong Kong and Macau.

25. Excludes ethnic Chinese arriving from Hong Kong, Macau, Taiwan, and overseas Chinese.

26. International tourist arrivals at hotels and similar establishments in Brazzaville, Pointe Noire, Loubomo, Owando, and Sibiti.

27. Air arrivals at the international FHB airport at Port Bouet. Arrivals at land frontiers, Bouake airport, and Air Ivoire airport at Abidjan are not taken into consideration.

28. Includes cruise passengers.

29. International visitor arrivals at frontiers. Up to 1994, the Slovak Republic was not monitored statistically.

30. Includes international fare expenditure.

31. Arrivals by air only, including nationals of the country residing abroad.

32. All arrivals by sea.

33. Revised data from survey on average daily expenditure.

34. Revised data based on surveys of average expenditure of foreigners using air transportation, excluding international fare expenditure.

35. Arrivals to Addis Ababa, Asmara, and Assab airports.

36. Includes revenues from hotel services, tour operators and travel agency services, duty free, gift articles, and souvenir sales.

37. International tourist arrivals at frontiers, excluding nationals of the country residing abroad.

38. Data for 1991-1993 are from border surveys; data for 1994 are provisional.

39. Charter tourists only.

40. Cruise passengers (000).

41. From 1992 data related to the FRG territory after unification.

42. From July 1990, including all transactions of former GDR.

43. Data based on surveys.

44. Hotels, cottages, and guest houses.

45. Port au Prince airport.

46. Registered by the Central Bank.

47. Cruise passengers (included in international visitor arrivals).

48. Receipts from visitors, excluding servicemen, air crew members, and transit passengers.

49. International visitor departures from frontiers.

50. International tourist departures from frontiers, excluding nationals of country residing abroad.

51. Estimated passenger fares paid in Icelandic kronur.

52. Estimate.

53. Includes receipts from Northern Irish tourists and excursionists.

54. Net international tourism/travel expenditure by Irish visitors abroad.

55. Receipts of Irish carriers from visitors.

56. Expenditure of Irish carriers only.

57. International visitor arrivals at frontiers, excluding nationals of country residing abroad.

58. International airline tourist arrivals at frontiers, excluding nationals of country residing abroad.

59. Includes education payments (U.S.$ mil.).

60. Departures, excluding nationals of country residing abroad.

61. Includes nationals of country residing abroad and from June 1988, also crew members.

62. Excludes expenses of students studying abroad.

63. International and domestic tourist arrivals at hotels and similar establishments. 1990 data are from January to June. 1991 data are for December only.

64. Excludes Syrian nationals.

65. Includes all travelers (visitors and other travelers not defined as visitors by WTO).

66. WTO estimates (excludes arrivals from Arab countries).

67. Figures for 1994 are estimated.

68. Excludes visitors by rail.

69. In collective tourism establishments.

70. Data collected at tourism "agentures".

71. International tourist arrivals at collective tourism establishments, youth hostels, tourist private accommodation, and others.

72. Departures.

73. International tourist arrivals at frontiers and foreign tourist departures, including Singapore residents crossing the frontier by road through Johore Causeway.

74. Includes visitors of the U.S. border zone whose duration of stay does not exceed 24 hours.

75. Includes receipts from cruise passengers.

76. Includes receipts/expenditures from frontier visitors (staying less than 24 hours, 24-72 hours, and over).

77. Partial data.

78. Arrivals at Yangon by air. 1994 data includes arrivals from Chinese border to Yangon by overland route (3,000).

79. Includes arrivals from India.

80. International tourist arrivals at collective tourism establishments. 1994 figures are not comparable with previous years due to differences in the survey-population.

81. Includes visits to friends and relatives. Data years 1990-1992 end March 31.

82. Figures for 1990-1994 are provided by the IMF.

83. Air arrivals (Niamey airport).

84. International tourist arrivals at registered hotels and similar establishments.

85. Hotel sales.

86. Arrivals by air and land, excluding nationals of country residing abroad and crew members.

87. International visitor arrivals at frontiers, including nationals of country residing abroad.

88. Visitor receipt figure.

89. 1991-1994 data were estimated through revised methodology.

90. International visitor arrivals at frontiers. Data for 1994 are provisional.

91. 1993 data are estimates. 1993 and 1994 data do not exist due to war in April 1994 (all documents and files were destroyed).

92. International tourist air and sea arrivals at frontiers.

93. Yacht and cruise ship arrivals.

94. Includes travelers with visit visa, Omra visa, and pilgrims.

95. Pilgrims.

96. Excludes arrivals of Malaysian citizens by land, but includes excursionists.

97. Includes the border with Czech Republic.

98. Data for 1990-1994 are only companies registered in Commercial Register, not those registered in Tradesman Register (smaller companies).

99. Includes all categories of travelers irrespective of purpose of visit.

100. Excludes nationals of country residing abroad. Beginning in 1992, contract and border traffic-concession workers are included.

101. Arrivals in hotels, rest camps, and cottage accommodations, caravan parks, and camping sites.

102. Includes persons in transit.

103. Data for 1994 are estimates.

104. Includes Italian visitors.

105. Excludes Italian visitors (1990 = 2,330,290).

106. Preliminary.

107. Arrivals at hotels. Data refer to Dubai only.

108. Includes Mexicans staying one or more nights in the United States.

109. Data refer to new Federal Republic of Yugoslavia made up of Serbia and Montenegro.

110. International tourist arrivals at frontiers, excluding nationals of country residing abroad. 1991 data are for January-June. 1992 data are incomplete. 1994 data are for January-June.

111. Data refer to former Yugoslavia.

35 - Manufacturing

Source. United Nations Industrial Development Organization (UNIDO). *Industry and Development Global Report 1993/94.* Vienna: UNIDO, 1993.

Source. United Nations Statistical Division. *General Industrial Statistics* (series). New York: UN. **Reprinted with permission obtained by Gale Research**.

Source. U.S. Central Intelligence Agency (CIA) (1996). *The World Factbook 1996* [Online]. Available: http://www.odci.gov//cia/publications/nsolo/factbook/global.html [1997, March 11].

Notes. Gross domestic product (GDP)—all economic activity in a given country, including activity engaged in by foreign nationals. For example, assets of a General Motors plant in Mexico would contribute to Mexico's GDP. *Real GDP* measures economic activity in constant prices, that is, after adjustments for inflation.

Value-added manufacturing—the value of output minus the cost of raw materials and other inputs.

A dash **(-)** or **NA**—data are not available.

Because no manufacturing data are available after 1991, manufacturing tables shown for the Czech Republic and Slovakia (formed in 1993) actually refer to the former Czechoslovakia.

Because no manufacturing data are available for Russia since the breakup of the U.S.S.R. in 1991, manufacturing tables shown actually refer to the U.S.S.R.

Because no manufacturing data are available after 1990, manufacturing tables shown for the Federal Republic of Yugoslavia (formed in 1992) actually refer to Serbia and Montenegro.

36 - Economic Indicators

Source. U.S. Central Intelligence Agency (CIA) (1996). *The World Factbook 1996* [Online]. Available: http://www.odci.gov//cia/publications/nsolo/factbook/global.html [1997, March 11].

Notes. Following are CIA definitions of acronyms and terms used in these tables.

est.—estimate.

External debt—the amount of debt owed to foreign entities by the given country.

GDP—gross domestic product: the value of all goods and services produced within a nation in a given year. Methodology: GDP dollar estimates for all countries are derived from purchasing power parity (PPP) calculations rather than from conversions at official currency exchange rates. The PPP method involves the use of international dollar price weights, which are applied to the quantities of goods and services produced in a given economy. The data derived from the PPP method provide a better comparison of economic well-being between countries. The division of a GDP estimate in domestic currency by the corresponding PPP estimate in dollars gives the PPP conversion rate. When priced in PPPs, $1,000 will buy the same market basket of goods in any country. Whereas PPP estimates for OECD countries are quite reliable, PPP estimates for developing countries are often rough approximations. Most of the GDP estimates are based on extrapolation of numbers published by the UN International Comparison Program and by Professors Robert Summers and Alan Heston of the University of Pennsylvania and their colleagues. Note: the numbers for GDP and other economic data can not be chained together from successive volumes of the *Factbook* because of changes in the U.S. dollar measuring rod, revisions of data by statistical agencies, use of new or different sources of information, and changes in national statistical methods and practices.

Inflation rate—an increase in prices unrelated to value.

NA—data are not available.

National product—the total output of goods and services in a given country (See *gross domestic product*).

37 - Balance of Payments

Source. United Nations Conference on Trade and Development (UNCTAD). *Handbook of International Trade and Development Statistics.* Geneva: UNCTAD, 1995.

Notes. Following are UNCTAD definitions of terms pertinent to these tables.

f.o.b.—for free on board, i.e., the value of goods does not include insurance and freight charges.

NA—data are not available or are not separately reported.

A Zero **(0)**—data are nil or negligible.

Balance of payments—the account of all international financial flows for a given country.

Trade balance—the difference in value between merchandise imports and merchandise exports. If the value of imports is greater than the value of exports, the result is a *trade deficit*. If the value of exports is greater, the result is a *trade surplus*.

The balance of trade is measured on an f.o.b./f.o.b. basis and includes transactions in monetary gold.

Services - debits—payments made to another country for services, including interest payments. Direct investment income is another component of the service debit category because it represents earnings (credits) for the foreign investor. Service debit totals include total payments for both factor and non-factor services.

Services - credits—payments received from another country for services. In any relationship between two countries, if one country posts a debit the other country must post the same amount as a credit. Service credit totals Include total receipts for both factor and non-factor services.

Private transfers (net)—consists of migrants' transfers, workers' remittances, gifts, dowries, inheritances, prize monies from non-governmental lotteries, and payment of dues to professional organizations, etc.

Government transfers (net)—inter-official unrequited transfers, i.e., transactions between official sectors of two economies, such as grants, debt cancellations, and reparations; or transactions between official sectors and private non-residents, such as scholarships, licensing fees, and government-sponsored lottery tickets.

Long-term capital (net)—capital with an original contractual maturity of more than one year, or with no stated maturity (e.g., corporate equities). Data rported in this category include repayments on commercial arrears.

Short-term capital (net)—capital with an original contractual maturity of one year or less.

Errors and omissions—balancing item which reflects the fact that, due to errors, omissions, and inconsistencies in reported figures, the sum of credit items does not equal the sum of debit items, as conceptually should be the case. *Errors and omissions* is reported as part of the overall balance, since the remaining items—special drawing rights (SDR) allocation, gold monetization, and changes in reserves and related—are presumed to be relatively free of error.

Overall balance—the balance on goods, services, unrequited transfers and capital, including net errors and omissions but excluding special drawing rights (SDR) allocations and gold monetization.

Footnotes

1. First available data for this country. Former member of the Soviet Union.

2. On January 1, 1993, Czechoslovakia split into two separate states—the Czech Republic and Slovakia. Data prior to 1993 are for Czechoslovakia.

3. On April 17, 1992, the republics of Serbia and Montenegro proclaimed a new *Federal Republic of Yugoslavia*. Data prior to 1992 refer to the former Yugoslav republics of Croatia, Slovenia, Bosnia and Hezegovina, and Macedonia.

---

38 - Exchange Rates

Source. U.S. Central Intelligence Agency (CIA) (1996). *The World Factbook 1996* [Online]. Available: http://www.odci.gov//cia/publications/nsolo/factbook/global.html [1997, March 11].

Notes. Following are CIA definitions of acronyms and terms used in these tables.

Exchange rate—the official value of a nation's monetary unit at a given date or over a given period of time, as expressed in units of local currencey per U.S. dollar and as determined by international market forces or official fiat. These often have little relation to domestic output. In developing countries with weak currencies, the exchange rate estimate in GDP (gross domestic product) in dollars is typically one-fourth to one-half the PPP (purchasing power parity) estimate. Although exchange rates may suddenly go up or down by 10% or more, real output may have remained unchanged. On January 12, 1994, for example, the 14 countries of the African Financial Community (whose currencies are tied to the French franc) devalued their currencies by 50%. This move, of course, did not cut the real output of their countries by half.

BMR—Black Market rate.

NA—data are not available.

---

39 - Top Import Origins

Source. U.S. Central Intelligence Agency (CIA) (1996). *The World Factbook 1996* [Online]. Available: http://www.odci.gov//cia/publications/nsolo/factbook/global.html [1997, March 11].

U.S. Department of State (DOS) (1996). *Background Notes* [Online]. Available: http://www.stat.gov/www/background-notes/index.html [1996].

Notes. Top import origins are distributed in percent when data are available.

Following are CIA definitions of the acronyms and terms used in these tables.

BLEU—Belgium-Luxembourg Economic Union.

Caricom—Caribbean Community and Common Market.

CEMA—Council for Mutual Economic Assistance; also known as *CMEA* or *Comecon*.

c.i.f.—cost, insurance, freight.

CIS—Commonwealth of Independent States.

CMEA—Council for Mutual Economic Assistance; also known as *CEMA* or *Comecon*.

ECOWAS—Economic Community of West African States.

EFTA—European Free Trade Association.

est.—estimate.

EU—European Union.

f.o.b.—free on board

FSU—former Soviet Union.

NA—data are not available.

OECD—Organization for Economic Cooperation and Development.

OECS—Organizaion of Eastern Caribbean States.

OPEC—Organization of Petroleum Exporting Countries.

SACU—South African Customs Union.

UAE—United Arab Emirates.

UK—United Kingdom.

US—United States.

USSR—Union of Soviet Socialist Republics (Soviet Union).

40 - Top Export Destinations

Source. U.S. Central Intelligence Agency (CIA) (1996). *The World Factbook 1996* [Online]. Available: http://www.odci.gov//cia/publications/nsolo/factbook/global.html [1997, March 11].

Notes. Top export destinations are distributed in percent when data are available.

Following are CIA definitions of the acronyms and terms used in these tables.

BLEU—Belgium-Luxembourg Economic Union.

Caricom—Caribbean Community and Common Market.

CEMA—Council for Mutual Economic Assistance; also known as *CMEA* or *Comecon*.

c.i.f.—cost, insurance, freight.

CIS—Commonwealth of Independent States.

CMEA—Council for Mutual Economic Assistance; also known as *CEMA* or *Comecon*.

ECOWAS—Economic Community of West African States.

EFTA—European Free Trade Association.

est.—estimate.

EU—European Union.

f.o.b.—free on board

FSU—former Soviet Union.

NA—data are not available.

OECD—Organization for Economic Cooperation and Development.

OECS—Organizaion of Eastern Caribbean States.

OPEC—Organization of Petroleum Exporting Countries.

SACU—South African Customs Union.

UAE—United Arab Emirates.

UK—United Kingdom.

US—United States.

USSR—Union of Soviet Socialist Republics (Soviet Union).

41 - Foreign Aid

Source. U.S. Central Intelligence Agency (CIA) (1996). *The World Factbook 1996* [Online]. Available: http://www.odci.gov//cia/publications/nsolo/factbook/global.html [1997, March 11].

Source. U.S. Department of State (DOS) (1996). *Background Notes* [Online]. Available: http://www.stat.gov/www/backgroundnotes/index.html [1996].

Notes. Following are CIA definitions of terms used in these tables.

Donor—country that pledges official economic aid to another country.

NA—data are not available.

ODA—official development assistance. ODA refers to financial assistance which is concessional in character, has the main objective of promoting economic development and welfare in less developed countries (LDCs), and contains a grant element of at least 25%.

OOF—other official flows. OOF also refers to official government assistance, but with a main objective other than development and with a grant element less than 25%. Transactions include official export credits (such as Export-Import Bank credits), official equity and portfolio investment, and debt reorganization by the official sector that does not meet concessional terms. Aid is considered to have been committed when the parties involved initial agreements constituting a formal declaration of intent.

Recipient—country that receives official economic aid from another country.

42 - Import and Export Commodities

Source. U.S. Central Intelligence Agency (CIA) (1996). *The World Factbook 1996* [Online]. Available: http://www.odci.gov//cia/publications/nsolo/factbook/global.html [1997, March 11].

Notes. Category 39: *Top Import Origins* and Category 40: *Top Export Destinations* provide corresponding year of commodity imports/exports respectively.

When available, commodities are distributed in percent.

KEYWORD INDEX

The Keyword Index provides access, by page number, to every country in *Statistical Abstract of the World*. Country names are capitalized. Subject references are also provided followed by (1) a listing of countries and page numbers separated by dashes or (2) by page numbers only. Countries are arranged alphabetically within each block. The phrase *pol.* follows index terms that are political parties or entities.

Christian Democratic *pol.* Rwanda 790
Christian Democratic (CDA) *pol.* Netherlands 680
Christian Democratic Movement *pol.* Slovakia 848
Christian Democratic Party (DCG) *pol.* Guatemala 379
Christian Democratic Party (PDC) *pol.* El Salvador 293
Christian Democratic Party (PDCH) *pol.* Honduras 405
Christian Democratic Party (PDCS) *pol.* San Marino 810
Christian Democratic People's Party *pol.* Hungary 417
Christian Democratic Popular Front *pol.* Moldova 636
Christian Democratic Union/ Czech People's (KDU/CSL) *pol.*
  Czech Republic 253
Christian Democratic Union (CDU) *pol.* Germany 356
Christian Democrats *pol.* Sweden 904
Christian Dem. (KDS) *pol.* Czech Republic 253
Christian People's *pol.* Norway 709
Christian People's Party *pol.* Denmark 259
Christian *See* Religion
Christian Social Party (CSV) *pol.* Luxembourg 563
Christian Social Union (CSU) *pol.* Germany 356
Christian Soc. Democrats *pol.* Czech Republic 253
Chromite 490, 576, 889, 1058
Chrysocolla 670
Church of Christ *See* Religion
Church of God *See* Religion
Cinemas *See* Culture
Circle for Renewal *pol.* Gabon 339
Circle of Liberal Reformers (CLR) *pol.* Gabon 339
Citizens Union of Georgia(CUG) *pol.* Georgia 351
Civic Accord Bloc(CAB) *pol.* Belarus 86
Civic Democratic Party (ODS) *pol.* Czech Republic 253
Civic Dem. Alliance (ODA) *pol.* Czech Republic 253
Clay 75, 98, 132, 155, 224, 248, 288, 311, 316, 357, 401,
  460, 528, 559, 576, 637, 659, 675, 721, 732, 743, 755,
  827, 894, 1002
Climate *See* Geography
Coal 4, 126, 143, 195, 201, 254, 352, 357, 369, 418, 430,
  436, 490, 501, 506, 523, 582, 647, 653, 665, 675, 687,
  699, 704, 721, 755, 761, 779, 785, 855, 871, 877, 899,
  929, 934, 940, 964, 980, 990, 996, 1008, 1018, 1024,
  1040, 1052, 1058
Coalition of Parties for Democracy PDC PPD PR *pol.* Chile 194
Coalition Party (KMU) *pol.* Estonia 310
Coastline *See* Geography
Cobalt 242, 1046, 1052
Coke 201, 254, 357, 466, 506, 681, 761, 779, 785, 877,
  980, 1058
COLOMBIA 204
Colon, Costa Rican 226
Colon, Salvadoran 296
Colorado *pol.* Uruguay 1001
Colorado Party *pol.* Paraguay 742
Columbite-tantalite 791, 1046
Committee for Renewal *pol.* Togo 945
Committee of Living Forces (CFV) *pol.* Madagascar 575
Common Choice *pol.* Slovakia 848
Communaute Financiere Africaine franc 105, 151, 168,
  186, 191, 220, 232, 302, 342, 600, 700, 829, 948
Communications *See* Telecommunications
Communications, microwave *See* Telecommunications
Communications, satellite links *See* Telecommunications
Communications, telephone *See* Telecommunications
Communist Party *pol.* Benin 103
Communist Party and affiliates *pol.* Tajikistan 928
Communist Party (KSDM) *pol.* Czech Republic 253
Communist Party of India *pol.* India 429
Communist Party of India/Marxist *pol.* India 429
Communist Party of Nepal/United Marxist and Leninist (CPN/UML) *pol.*
  Nepal 674

Communist Party (PCF) *pol.* France 333
Communist Refoundation (RC) *pol.* San Marino 810
Communists *pol.* Ukraine 979
Comm. Party of the Russian Fed. *pol.* Russia 784
COMOROS 210
Concentration of Popular Forces *pol.* Ecuador 281
Confederation for an Indpendent Poland (KPN) *pol.*
  Poland 760
Confederation of Civil Societies for Development *pol.*
  Madagascar 575
Confucianism *See* Religion
CONGO 215
Congress (I) Party *pol.* India 429
Congress of Russian Communities *pol.* Russia 784
Conscience of the Fatherland (CONDEPA) *pol.* Bolivia 113
Conservative *pol.* Norway 709
Conservative *pol.* United Kingdom 989
Conservative Party *pol.* Denmark 259
Conservatives *pol.* Colombia 206
Constitution
  – Afghanistan 3 – Albania 9 – Algeria 14 – Andorra 20 – Angola
  25 – Antigua and Barbuda 30 – Argentina 35 – Armenia 41 –
  Australia 46 – Austria 52 – Azerbaijan 58 – Bahamas, The 63 –
  Bahrain 69 – Bangladesh 74 – Barbados 80 – Belarus 86 –
  Belgium 91 – Belize 97 – Benin 103 – Bhutan 108 – Bolivia
  113 – Bosnia and Herzegovina 119 – Botswana 125 – Brazil
  131 – Brunei 137 – Bulgaria 142 – Burkina Faso 148 – Burundi
  154 – Cambodia 160 – Cameroon 165 – Canada 171 – Cape
  Verde 177 – Central African Republic 183 – Chad 189 – Chile
  194 – China 200 – Colombia 206 – Comoros 212 – Congo
  217 – Costa Rica 223 – Cote d'Ivoire 229 – Croatia 235 –
  Cuba 241 – Cyprus 247 – Czech Republic 253 – Denmark
  259 – Djibouti 265 – Dominica 270 – Dominican Republic
  275 – Ecuador 281 – Egypt 287 – El Salvador 293 – Equatorial
  Guinea 299 – Eritrea 305 – Estonia 310 – Ethiopia 315 – Fiji
  321 – Finland 327 – France 333 – Gabon 339 – Gambia, The
  345 – Georgia 351 – Germany 356 – Ghana 362 – Greece 368 –
  Grenada 374 – Guatemala 379 – Guinea 385 – Guinea-
  Bissau 390 – Guyana 395 – Haiti 400 – Honduras 405 – Hong
  Kong 411 – Hungary 417 – Iceland 423 – India 429 –
  Indonesia 435 – Iran 441 – Iraq 447 – Ireland 453 – Israel 459 –
  Italy 465 – Jamaica 471 – Japan 477 – Jordan 483 –
  Kazakhstan 489 – Kenya 494 – Korea, North 500 – Korea,
  South 505 – Kuwait 511 – Kyrgyzstan 517 – Laos 522 – Latvia
  527 – Lebanon 532 – Lesotho 537 – Liberia 542 – Libya 547 –
  Liechtenstein 553 – Lithuania 558 – Luxembourg 563 –
  Macedonia 569 – Madagascar 575 – Malawi 581 – Malaysia
  587 – Maldives 593 – Mali 598 – Malta 603 – Marshall Islands
  609 – Mauritania 614 – Mauritius 619 – Mexico 625 –
  Micronesia 631 – Moldova 636 – Monaco 641 – Mongolia
  646 – Morocco 652 – Mozambique 658 – Myanmar 664 –
  Namibia 669 – Nepal 674 – Netherlands 680 – New Zealand
  686 – Nicaragua 692 – Niger 698 – Nigeria 703 – Norway 709 –
  Oman 715 – Pakistan 720 – Palau 726 – Panama 731 – Papua
  New Guinea 737 – Paraguay 742 – Peru 748 – Philippines
  754 – Poland 760 – Portugal 766 – Qatar 772 – Romania 778 –
  Russia 784 – Rwanda 790 – Saint Kitts and Nevis 795 – Saint
  Lucia 800 – Saint Vincent and the Grenadines 805 – San
  Marino 810 – Sao Tome and Principe 815 – Saudi Arabia
  820 – Senegal 826 – Seychelles 832 – Sierra Leone 837 –
  Singapore 842 – Slovakia 848 – Slovenia 854 – Solomon
  Islands 860 – Somalia 865 – South Africa 870 – Spain 876 –
  Sri Lanka 882 – Sudan 888 – Suriname 893 – Swaziland 898 –
  Sweden 904 – Switzerland 910 – Syria 916 – Taiwan 922 –
  Tajikistan 928 – Tanzania 933 – Thailand 939 – Togo 945 –
  Trinidad and Tobago 951 – Tunisia 957 – Turkey 963 –
  Turkmenistan 969 – Uganda 974 – Ukraine 979 – United
  Arab Emirates 984 – United Kingdom 989 – United States

Health, expenditures - *continued*
Uzbekistan 1005 – Venezuela 1015 – Vietnam 1021 – Yemen 1031 – Yugoslavia 1037 – Zaire 1043 – Zambia 1049 – Zimbabwe 1055

Health indicators
– Afghanistan 1 – Albania 7 – Algeria 12 – Angola 23 – Argentina 33 – Armenia 39 – Australia 44 – Austria 50 – Azerbaijan 56 – Bangladesh 72 – Belarus 84 – Belgium 89 – Benin 101 – Bhutan 106 – Bolivia 111 – Bosnia and Herzegovina 117 – Botswana 123 – Brazil 129 – Bulgaria 140 Burkina Faso 146 – Burundi 152 – Cambodia 158 – Cameroon 163 – Canada 169 – Central African Republic 181 – Chad 187 – Chile 192 – China 198 – Colombia 204 – Congo 215 – Costa Rica 221 – Cote d'Ivoire 227 – Croatia 233 – Cuba 239 – Czech Republic 251 – Denmark 257 – Dominica 268 – Ecuador 279 – Egypt 285 – El Salvador 291 – Eritrea 303 – Estonia 308 – Ethiopia 313 – Finland 325 – France 331 – Gabon 337 – Gambia, The 343 – Georgia 349 – Germany 354 – Ghana 360 – Greece 366 – Guatemala 377 – Guinea 383 – Guinea-Bissau 388 – Haiti 398 – Honduras 403 Hong Kong 409 – Hungary 415 – India 427 – Indonesia 433 – Iran 439 – Iraq 445 – Ireland 451 – Israel 457 – Italy 463 – Jamaica 469 – Japan 475 – Jordan 481 – Kazakhstan 487 – Kenya 492 – Korea, North 498 – Korea, South 503 – Kuwait 509 – Kyrgyzstan 515 – Laos 520 – Latvia 525 – Lebanon 530 – Lesotho 535 – Liberia 540 – Libya 545 – Lithuania 556 – Macedonia 567 – Madagascar 573 – Malawi 579 – Malaysia 585 – Mali 596 – Mauritania 612 – Mauritius 617 – Mexico 623 – Moldova 634 – Mongolia 644 – Morocco 650 – Mozambique 656 – Myanmar 662 – Namibia 667 – Nepal 672 – Netherlands 678 – New Zealand 684 – Nicaragua 690 – Niger 696 – Nigeria 701 – Norway 707 – Oman 713 – Pakistan 718 – Panama 729 – Papua New Guinea 735 – Paraguay 740 Peru 746 – Philippines 752 – Poland 758 – Portugal 764 – Romania 776 – Russia 782 – Rwanda 788 – Saudi Arabia 818 Senegal 824 – Sierra Leone 835 – Singapore 840 – Slovakia 846 – Slovenia 852 – Somalia 863 – South Africa 868 – Spain 874 – Sri Lanka 880 – Sudan 886 – Sweden 902 – Switzerland 908 – Syria 914 – Tajikistan 926 – Tanzania 931 – Thailand 937 – Togo 943 – Trinidad and Tobago 949 – Tunisia 955 – Turkey 961 – Turkmenistan 967 – Uganda 972 – Ukraine 977 United Arab Emirates 982 – United Kingdom 987 – United States 993 – Uruguay 999 – Uzbekistan 1005 – Venezuela 1015 – Vietnam 1021 – Yemen 1031 – Yugoslavia 1037 – Zaire 1043 – Zambia 1049 – Zimbabwe 1055

Health services, access to *See* Health indicators
Heart disease *See* Burden of disease
Highways *See* Transportation
Homeland Party *pol.* Jordan 483
HONDURAS 403
HONG KONG 409
Hospital bed, population per *See* Burden of disease
Hryvnia 981
Human rights, agreements on
– Afghanistan 3 – Albania 9 – Algeria 14 – Andorra 20 – Angola 25 – Antigua and Barbuda 30 – Argentina 35 – Armenia 41 – Australia 46 – Austria 52 – Azerbaijan 58 – Bahamas, The 63 – Bahrain 69 – Bangladesh 74 – Barbados 80 – Belarus 86 – Belgium 91 – Belize 97 – Benin 103 – Bhutan 108 – Bolivia 113 – Bosnia and Herzegovina 119 – Botswana 125 – Brazil 131 – Brunei 137 – Bulgaria 142 – Burkina Faso 148 – Burundi 154 – Cambodia 160 – Cameroon 165 – Canada 171 – Cape Verde 177 – Central African Republic 183 – Chad 189 – Chile 194 – China 200 – Colombia 206 – Comoros 212 – Congo 217 – Costa Rica 223 – Cote d'Ivoire 229 – Croatia 235 – Cuba 241 – Cyprus 247 – Czech Republic 253 – Denmark 259 – Djibouti 265 – Dominica 270 – Dominican Republic 275 – Ecuador 281 – Egypt 287 – El Salvador 293 – Equatorial

Human rights, agreements on - *continued*
Guinea 299 – Eritrea 305 – Estonia 310 – Ethiopia 315 – Fiji 321 – Finland 327 – France 333 – Gabon 339 – Gambia, The 345 – Georgia 351 – Germany 356 – Ghana 362 – Greece 368 – Grenada 374 – Guatemala 379 – Guinea 385 – Guinea-Bissau 390 – Guyana 395 – Haiti 400 – Honduras 405 – Hungary 417 – Iceland 423 – India 429 – Indonesia 435 – Iran 441 – Iraq 447 – Ireland 453 – Israel 459 – Italy 465 – Jamaica 471 – Japan 477 – Jordan 483 – Kazakhstan 489 – Kenya 494 – Korea, North 500 – Korea, South 505 – Kuwait 511 – Kyrgyzstan 517 – Laos 522 – Latvia 527 – Lebanon 532 – Lesotho 537 – Liberia 542 – Libya 547 – Liechtenstein 553 – Lithuania 558 – Luxembourg 563 – Macedonia 569 – Madagascar 575 – Malawi 581 – Malaysia 587 – Maldives 593 – Mali 598 – Malta 603 – Marshall Islands 609 – Mauritania 614 – Mauritius 619 – Mexico 625 – Micronesia 631 – Moldova 636 – Monaco 641 – Mongolia 646 – Morocco 652 – Mozambique 658 – Myanmar 664 – Namibia 669 – Nepal 674 – Netherlands 680 – New Zealand 686 – Nicaragua 692 – Niger 698 – Nigeria 703 – Norway 709 – Oman 715 – Pakistan 720 – Palau 726 – Panama 731 – Papua New Guinea 737 – Paraguay 742 – Peru 748 – Philippines 754 – Poland 760 – Portugal 766 – Qatar 772 – Romania 778 – Russia 784 – Rwanda 790 – Saint Kitts and Nevis 795 – Saint Lucia 800 – Saint Vincent and the Grenadines 805 – San Marino 810 – Sao Tome and Principe 815 – Saudi Arabia 820 – Senegal 826 – Seychelles 832 – Sierra Leone 837 – Singapore 842 – Slovakia 848 – Slovenia 854 – Solomon Islands 860 – Somalia 865 – South Africa 870 – Spain 876 – Sri Lanka 882 – Sudan 888 – Suriname 893 – Swaziland 898 – Sweden 904 – Switzerland 910 – Syria 916 – Taiwan 922 – Tajikistan 928 – Tanzania 933 – Thailand 939 – Togo 945 – Trinidad and Tobago 951 – Tunisia 957 – Turkey 963 – Turkmenistan 969 – Uganda 974 – Ukraine 979 – United Arab Emirates 984 – United Kingdom 989 – United States 995 – Uruguay 1001 – Uzbekistan 1007 – Vanuatu 1012 – Venezuela 1017 – Vietnam 1023 – Western Samoa 1028 – Yemen 1033 – Zaire 1045 – Zambia 1051 – Zimbabwe 1057

Human Rights Protection Party (HRPP) *pol.*
Western Samoa 1028
Hungarian Coalition *pol.* Slovakia 848
Hungarian Democratic Forum *pol.* Hungary 417
Hungarian minority *pol.* Slovenia 854
Hungarian Socialist Party *pol.* Hungary 417
HUNGARY 415
ICELAND 421
Igneous rock 990
Illiteracy
– Afghanistan 2 – Algeria 13 – Angola 24 – Argentina 34 – Armenia 40 – Azerbaijan 57 – Bahamas, The 62 – Bahrain 68 Bangladesh 73 – Barbados 79 – Belarus 85 – Belize 96 – Benin 102 – Bhutan 107 – Bolivia 112 – Bosnia and Herzegovina 118 – Botswana 124 – Brazil 130 – Brunei 136 – Bulgaria 141 – Burkina Faso 147 – Burundi 153 – Cambodia 159 – Cameroon 164 – Canada 170 – Cape Verde 176 – Central African Republic 182 – Chad 188 – Chile 193 – China 199 – Colombia 205 – Comoros 211 – Congo 216 – Costa Rica 222 – Cote d'Ivoire 228 – Croatia 234 – Cuba 240 – Cyprus 246 – Djibouti 264 – Dominica 269 – Dominican Republic 274 – Ecuador 280 – Egypt 286 – El Salvador 292 – Equatorial Guinea 298 – Estonia 309 – Ethiopia 314 – Fiji 320 – Gabon 338 – Gambia, The 344 – Georgia 350 – Ghana 361 – Greece 367 – Grenada 373 – Guatemala 378 – Guinea 384 – Guinea-Bissau 389 – Guyana 394 – Haiti 399 – Honduras 404 – Hong Kong 410 – Hungary 416 – India 428 – Indonesia 434 – Iran 440 – Iraq 446 – Israel 458 – Italy 464 – Jamaica 470 – Jordan 482 – Kazakhstan 488 – Kenya 493 – Korea, South 504 – Kuwait 510 – Kyrgyzstan 516 – Laos 521 –

Newspapers - *continued*

Bissau 389 – Guyana 394 – Haiti 399 – Honduras 404 – Hong Kong 410 – Hungary 416 – Iceland 422 – India 428 – Indonesia 434 – Iran 440 – Iraq 446 – Ireland 452 – Israel 458 – Italy 464 – Jamaica 470 – Japan 476 – Jordan 482 – Kenya 493 – Korea, North 499 – Korea, South 504 – Kuwait 510 – Kyrgyzstan 516 – Laos 521 – Latvia 526 – Lebanon 531 – Lesotho 536 – Liberia 541 – Libya 546 – Liechtenstein 552 – Lithuania 557 – Luxembourg 562 – Macedonia 568 – Madagascar 574 – Malawi 580 – Malaysia 586 – Maldives 592 – Mali 597 – Malta 602 – Mauritania 613 – Mauritius 618 – Mexico 624 – Moldova 635 – Monaco 640 – Mongolia 645 – Morocco 651 – Mozambique 657 – Myanmar 663 – Namibia 668 – Nepal 673 – Netherlands 679 – New Zealand 685 – Nicaragua 691 – Niger 697 – Nigeria 702 – Norway 708 – Oman 714 – Pakistan 719 – Panama 730 – Papua New Guinea 736 – Paraguay 741 – Peru 747 – Philippines 753 – Poland 759 – Portugal 765 – Qatar 771 – Romania 777 – Russia 783 – Rwanda 789 – San Marino 809 – Saudi Arabia 819 – Senegal 825 – Seychelles 831 – Sierra Leone 836 – Singapore 841 – Slovakia 847 – Slovenia 853 – Somalia 864 – South Africa 869 – Spain 875 – Sri Lanka 881 – Sudan 887 – Suriname 892 – Swaziland 897 – Sweden 903 – Switzerland 909 – Syria 915 – Tajikistan 927 – Tanzania 932 – Thailand 938 – Togo 944 – Trinidad and Tobago 950 – Tunisia 956 – Turkey 962 – Uganda 973 – Ukraine 978 – United Arab Emirates 983 – United Kingdom 988 – United States 994 – Uruguay 1000 – Uzbekistan 1006 – Venezuela 1016 – Vietnam 1022 – Yemen 1032 – Yugoslavia 1038 – Zaire 1044 – Zambia 1050 – Zimbabwe 1056

Ngultrum 110

NICARAGUA 690

Nickel 126, 207, 242, 276, 328, 369

NIGER 696

NIGERIA 701

Nigerian Alliance for Democracy & Progress *pol.* Niger 698

Nigerian Party for Democracy & Socialism *pol.* Niger 698

Nitrogen 4, 10, 15, 47, 70, 75, 87, 172, 207, 448, 501, 512, 548, 665, 681, 704, 721, 773, 917, 952, 1024

Nonparty Bloc for the Support of Reforms (BBWR) *pol.* Poland 760

Northern League *pol.* Italy 465

NORWAY 707

Nuevo peso 38

Nuevo sol 751

Nurse, population per *See* Burden of disease

Occupations in labor force

– Afghanistan 3 – Albania 9 – Algeria 14 – Andorra 20 – Angola 25 – Antigua and Barbuda 30 – Argentina 35 – Armenia 41 – Australia 46 – Austria 52 – Azerbaijan 58 – Bahamas, The 63 – Bahrain 69 – Bangladesh 74 – Barbados 80 – Belarus 86 – Belgium 91 – Belize 97 – Benin 103 – Bhutan 108 – Bolivia 113 – Bosnia and Herzegovina 119 – Botswana 125 – Brazil 131 – Brunei 137 – Bulgaria 142 – Burkina Faso 148 – Burundi 154 – Cambodia 160 – Cameroon 165 – Canada 171 – Cape Verde 177 – Central African Republic 183 – Chad 189 – Chile 194 – China 200 – Colombia 206 – Comoros 212 – Congo 217 – Costa Rica 223 – Cote d'Ivoire 229 – Croatia 235 – Cuba 241 – Cyprus 247 – Czech Republic 253 – Denmark 259 – Djibouti 265 – Dominica 270 – Dominican Republic 275 – Ecuador 281 – Egypt 287 – El Salvador 293 – Equatorial Guinea 299 – Eritrea 305 – Estonia 310 – Ethiopia 315 – Fiji 321 – Finland 327 – France 333 – Gabon 339 – Gambia, The 345 – Georgia 351 – Germany 356 – Ghana 362 – Greece 368 – Grenada 374 – Guatemala 379 – Guinea 385 – Guinea-Bissau 390 – Guyana 395 – Haiti 400 – Honduras 405 – Hong Kong 411 – Hungary 417 – Iceland 423 – India 429 – Indonesia 435 – Iran 441 – Iraq 447 – Ireland 453 – Israel 459 –

Occupations in labor force - *continued*

Italy 465 – Jamaica 471 – Japan 477 – Jordan 483 – Kazakhstan 489 – Kenya 494 – Korea, North 500 – Korea, South 505 – Kuwait 511 – Kyrgyzstan 517 – Laos 522 – Latvia 527 – Lebanon 532 – Lesotho 537 – Liberia 542 – Libya 547 – Liechtenstein 553 – Lithuania 558 – Luxembourg 563 – Macedonia 569 – Madagascar 575 – Malawi 581 – Malaysia 587 – Maldives 593 – Mali 598 – Malta 603 – Marshall Islands 609 – Mauritania 614 – Mauritius 619 – Mexico 625 – Micronesia 631 – Moldova 636 – Monaco 641 – Mongolia 646 – Morocco 652 – Mozambique 658 – Myanmar 664 – Namibia 669 – Nepal 674 – Netherlands 680 – New Zealand 686 – Nicaragua 692 – Niger 698 – Nigeria 703 – Norway 709 – Oman 715 – Pakistan 720 – Palau 726 – Panama 731 – Papua New Guinea 737 – Paraguay 742 – Peru 748 – Philippines 754 – Poland 760 – Portugal 766 – Qatar 772 – Romania 778 – Russia 784 – Rwanda 790 – Saint Kitts and Nevis 795 – Saint Lucia 800 – Saint Vincent and the Grenadines 805 – San Marino 810 – Sao Tome and Principe 815 – Saudi Arabia 820 – Senegal 826 – Seychelles 832 – Sierra Leone 837 – Singapore 842 – Slovakia 848 – Slovenia 854 – Solomon Islands 860 – Somalia 865 – South Africa 870 – Spain 876 – Sri Lanka 882 – Sudan 888 – Suriname 893 – Swaziland 898 – Sweden 904 – Switzerland 910 – Syria 916 – Taiwan 922 – Tajikistan 928 – Tanzania 933 – Thailand 939 – Togo 945 – Trinidad and Tobago 951 – Tunisia 957 – Turkey 963 – Turkmenistan 969 – Uganda 974 – Ukraine 979 – United Arab Emirates 984 – United Kingdom 989 – United States 995 – Uruguay 1001 – Uzbekistan 1007 – Vanuatu 1012 – Venezuela 1017 – Vietnam 1023 – Western Samoa 1028 – Yemen 1033 – Yugoslavia 1039 – Zaire 1045 – Zambia 1051 – Zimbabwe 1057

Oil *See* Petroleum

Oil shale 311

Olive Tree *pol.* Italy 465

Olivine sand 710

OMAN 713

Oral Rehydration Therapy (ORT) *See* Health indicators

Organization People's Dem. (ODP-MT) *pol.* Burkina Faso 148

Ornamental stone 236

ORT *See* Oral Rehydration Therapy

Ouguiya 616

Our Common Cause (NCC) *pol.* Benin 103

Our Home is Estonia *pol.* Estonia 310

Our Home Is Russia *pol.* Russia 784

Pachakutik Movement *pol.* Ecuador 281

PAKISTAN 718

Pakistan Muslim League, Junejo faction (PML/J) *pol.* Pakistan 720

Pakistan Muslim League, Nawaz Sharif faction (PML/N) *pol.* Pakistan 720

Pakistan People's Party (PPP) *pol.* Pakistan 720

PALAU 724

Palladium 996, 1040

Pan-African Union (UPADS) *pol.* Congo 217

Pan Africanist Cong. (PAC) *pol.* South Africa 870

PANAMA 729

Pangu Party *pol.* Papua New Guinea 737

Panhellenic Socialist Movement (PASOK) *pol.* Greece 368

Papa Egoro Movement *pol.* Panama 731

PAPUA NEW GUINEA, 735

PARAGUAY 740

Parsi *See* Religion

Party for Democratic Convergence (PCD) *pol.* Cape Verde 177

Party for Democratic Prosperity *pol.* Macedonia 569

Party for Renewal and Progress (PRP) *pol.* Guinea 385

Party for Unity and progress PUP) *pol.* Guinea 385

Party of All-Belarusian Unity and Concord (UPNAZ)
Keyword Index

Party of All-Belarusian Unity and Concord (UPNAZ) *pol.*
Belarus 86
Party of Democratic Action (SDA) *pol.*
Bosnia and Herzegovina 119
Party of Democratic Changes *pol.* Bosnia and Herzegovina 119
Party of Democratic Socialism (PDS) *pol.* Germany 356
Party of Economic and Political Renewal *pol.* Tajikistan 928
Party of People's Concord *pol.* Belarus 86
Party of People's Unity *pol.* Tajikistan 928
PASOK/Left Alliance *pol.* Greece 368
Pastures *See* Geography
Patents, U.S. patents issued
– Andorra 19 – Antigua and Barbuda 29 – Argentina 34 –
Armenia 40 – Australia 45 – Austria 51 – Bahamas, The 62 –
Belarus 85 – Belgium 90 – Brazil 130 – Bulgaria 141 – Canada
170 – Chile 193 – China 199 – Colombia 205 – Costa Rica 222 –
Croatia 234 – Cuba 240 – Cyprus 246 – Czech Republic 252 –
Denmark 258 – Dominica 269 – Dominican Republic 274 –
Ecuador 280 – Egypt 286 – El Salvador 292 – Estonia 309 –
Finland 326 – France 332 – Georgia 350 – Germany 355 –
Ghana 361 – Greece 367 – Guatemala 378 – Honduras 404 –
Hong Kong 410 – Hungary 416 – Iceland 422 – India 428 –
Indonesia 434 – Iran 440 – Ireland 452 – Israel 458 – Italy 464 –
Jamaica 470 – Japan 476 – Kazakhstan 488 – Kenya 493 –
Korea, North 499 – Korea, South 504 – Kuwait 510 –
Lebanon 531 – Liechtenstein 552 – Lithuania 557 –
Luxembourg 562 – Malaysia 586 – Malta 602 – Mauritius 618 –
Mexico 624 – Monaco 643 – Morocco 651 – Netherlands 679 –
New Zealand 685 – Nicaragua 691 – Nigeria 702 – Norway
708 – Oman 714 – Pakistan 719 – Panama 730 – Peru 747 –
Philippines 753 – Poland 759 – Portugal 765 – Romania 777 –
Russia 783 – Saint Vincent and the Grenadines 804 – San
Marino 809 – Saudi Arabia 819 – Singapore 841 – Slovenia
853 – South Africa 869 – Spain 875 – Sri Lanka 881 – Sweden
903 – Switzerland 909 – Taiwan 921 – Thailand 938 – Trinidad
and Tobago 950 – Tunisia 956 – Turkey 962 – Uganda 973 –
Ukraine 978 – United Arab Emirates 983 – United Kingdom
988 – United States 994 – Uruguay 1000 – Uzbekistan 1006 –
Venezuela 1016 – Yemen 1032 – Yugoslavia 1038 –
Zimbabwe 1056
Peasants and Intellectual Bloc *pol.* Moldova 636
Peat 87, 155, 172, 260, 311, 454, 528, 559, 687
Pentecostals *See* Religion
People's Action Movement (PAM) *pol.* Saint Kitts and Nevis 795
People's Action Party *pol.* Singapore 842
People's Action Party (PAP) *pol.* Papua New Guinea 737
People's Alliance *pol.* Sri Lanka 882
People's Alliance Party (PAP) *pol.* Solomon Islands 860
People's Democratic Movement *pol.* Cameroon 165
People's Democratic Movement (PDM) *pol.*
Papua New Guinea 737
People's Democratic Party *pol.* Uzbekistan 1007
Peoples Democratic (PDP) *pol.* Sierra Leone 837
People's Front for Democracy and Justice (PFJD) *pol.*
Eritrea 305
People's Liberation Organization of Tamil Eelam *pol.*
Sri Lanka 882
People's Movement *pol.* Iceland 423
People's National Congress (PNC) *pol.* Guyana 395
People's National Movement *pol.* Trinidad and Tobago 951
People's National Party (PNP) *pol.* Jamaica 471
People's Party *pol.* Tajikistan 928
People's Party of Montenegro (NS) *pol.* Yugoslavia 1039
People's Progress Assembly *pol.* Djibouti 265
People's Progress Party (PPP) *pol.* Papua New Guinea 737
People's Progressive Party (PPP) *pol.* Gambia, The 345
People's Progressive Party (PPP) *pol.* Guyana 395
People's Union (PU) *pol.* Bulgaria 142

People's United Party (PUP) *pol.* Belize 97
People's Unity Party (PUP) *pol.* Gabon 339
Perlite 42
PERU 746
Pertussis, immunization against *See* Health indicators
Peseta 879
Peseta, Spanish 22
Peso, Chilean 197
Peso, Colombian 209
Peso, Cuban 244
Peso, Dominican 278
Peso, Guinea-Bissauan 392
Peso, Mexican new 628
Peso, Nuevo 38
Peso, Philippine 757
Peso, Uruguayan 1004
Petroleum 10, 15, 26, 36, 47, 53, 59, 70, 75, 81, 87, 104,
114, 138, 143, 166, 172, 195, 207, 218, 224, 230, 242,
248, 254, 276, 282, 288, 294, 306, 328, 334, 340, 352,
363, 369, 380, 406, 418, 430, 436, 442, 448, 454, 460,
466, 472, 478, 484, 490, 495, 501, 506, 512, 518, 548,
576, 588, 615, 626, 653, 665, 681, 687, 693, 704, 710,
716, 721, 732, 738, 743, 749, 755, 761, 767, 773, 779,
785, 821, 827, 849, 855, 871, 877, 883, 889, 894, 905,
911, 917, 923, 929, 934, 940, 952, 958, 964, 970, 980,
985, 990, 996, 1002, 1018, 1024, 1034, 1040, 1046, 1052
Phalang Tham *pol.* Thailand 939
PHILIPPINES 752
Phosphate 448, 484, 501, 564, 653, 883, 917, 946, 958,
975, 1024
Phosphatic fertilizers 484
Phosphoric acid 827
Physician, population per *See* Burden of disease
Pig iron 10, 92, 120, 418, 466, 478, 506, 564, 681, 721,
779, 849, 905, 911, 923, 980
Pigments 743
PINU-SD *pol.* Honduras 405
Platinum 710, 996, 1040
Platinumgroup metals 1058
PLD *pol.* Central African Republic 183
Pledge Party *pol.* Jordan 483
PNH *pol.* Honduras 405
POLAND 758
Polio, immunization against *See* Health indicators
Polish Peasant Party (PSL) *pol.* Poland 760
Political parties
– Afghanistan 3 – Albania 9 – Algeria 14 – Andorra 20 – Angola
25 – Antigua and Barbuda 30 – Argentina 35 – Armenia 41 –
Australia 46 – Austria 52 – Azerbaijan 58 – Bahamas, The 63 –
Bahrain 69 – Bangladesh 74 – Barbados 80 – Belarus 86 –
Belgium 91 – Belize 97 – Benin 103 – Bhutan 108 – Bolivia
113 – Bosnia and Herzegovina 119 – Botswana 125 – Brazil
131 – Brunei 137 – Bulgaria 142 – Burkina Faso 148 – Burundi
154 – Cambodia 160 – Cameroon 165 – Canada 171 – Cape
Verde 177 – Central African Republic 183 – Chad 189 – Chile
194 – China 200 – Colombia 206 – Comoros 212 – Congo
217 – Costa Rica 223 – Cote d'Ivoire 229 – Croatia 235 –
Cuba 241 – Cyprus 247 – Czech Republic 253 – Denmark
259 – Djibouti 265 – Dominica 270 – Dominican Republic
275 – Ecuador 281 – Egypt 287 – El Salvador 293 – Equatorial
Guinea 299 – Eritrea 305 – Estonia 310 – Ethiopia 315 – Fiji
321 – Finland 327 – France 333 – Gabon 339 – Gambia, The
345 – Georgia 351 – Germany 356 – Ghana 362 – Greece 368 –
Grenada 374 – Guatemala 379 – Guinea 385 – Guinea-
Bissau 390 – Guyana 395 – Haiti 400 – Honduras 405 – Hong
Kong 411 – Hungary 417 – Iceland 423 – India 429 –
Indonesia 435 – Iran 441 – Iraq 447 – Ireland 453 – Israel 459 –
Italy 465 – Jamaica 471 – Japan 477 – Jordan 483 –